Biological and Medical Physics, Biomedical Engineering

BIOLOGICAL AND MEDICAL PHYSICS, BIOMEDICAL ENGINEERING

This series is intended to be comprehensive, covering a broad range of topics important to the study of the physical, chemical and biological sciences. Its goal is to provide scientists and engineers with textbooks, monographs, and reference works to address the growing need for information. The fields of biological and medical physics and biomedical engineering are broad, multidisciplinary and dynamic. They lie at the crossroads of frontier research in physics, biology, chemistry, and medicine.

Books in the series emphasize established and emergent areas of science including molecular, membrane, and mathematical biophysics; photosynthetic energy harvesting and conversion; information processing; physical principles of genetics; sensory communications; automata networks, neural networks, and cellular automata. Equally important is coverage of applied aspects of biological and medical physics and biomedical engineering such as molecular electronic components and devices, biosensors, medicine, imaging, physical principles of renewable energy production, advanced prostheses, and environmental control and engineering.

More information about this series at http://www.springer.com/series/3740

Claudia Tanja Mierke

Cellular Mechanics and Biophysics

Structure and Function of Basic Cellular Components Regulating Cell Mechanics

Springer

Claudia Tanja Mierke
Faculty of Physics and Earth Sciences,
Peter Debye Institute for Soft Matter
Physics, Biological Physics
University of Leipzig
Leipzig, Germany

ISSN 1618-7210 ISSN 2197-5647 (electronic)
Biological and Medical Physics, Biomedical Engineering
ISBN 978-3-030-58531-0 ISBN 978-3-030-58532-7 (eBook)
https://doi.org/10.1007/978-3-030-58532-7

This Springer imprint is published by the registered company Springer Nature Switzerland AG
The registered company address is: Gewerbestrasse 11, 6330 Cham, Switzerland

The Cellular Mechanics and Biophysics book is completed by the great support and encouragement of Thomas M. L. Mierke.

Preface

Dear Reader

My *Cellular Mechanics and Biophysics* book covers the novel and promising field of the analysis of the mechanical properties of cells and cell components and its connection to cell and molecular biological research. Therefore, basic knowledge about cellular structures and processes is recommended to combine them with mechanical features. Beyond that, cell mechanics has become a focus of physical and biological research all over the world. The scope of the Cellular Mechanics and Biophysics book is to present molecular, cellular and biochemical principles and structures and to explain several biophysical approaches used in cellular biophysics. My book presents the basic concepts and processes of biological matter and major findings that contribute significantly to the multidisciplinary field of Cellular Mechanics and Biophysics from a biophysical point of view. The dedicated reader is everyone interested in biophysical research focusing on cells such as readers who study biological physics, physics, biotechnology, life science, biomedicine or biology or work in those fields. The book is suited for upper undergraduate, graduate, doctoral and post-doctoral students, senior scientists, lecturers, principal investigators and professors. I hope that the *Cellular Mechanics and Biophysics* book will serve as a standard work for all those interested in this field of research. To my knowledge, there is no standard work in English, neither as hardcover nor as e-book, which can neither serve as standard work in the field of Cellular Mechanics and Biophysics nor serve to gain deeper knowledge about this highly interesting field of research. Many parts of the book are illustrated by self-designed figures that can be used in future lectures or seminars. In my own lectures, in which I teach physics students, I can use this book as a reference guide and recommend it as further reading, as it covers all aspects and topics. In any case, this book is also suitable for reading in addition to the lectures and internships, which are intended to help students and scientists gain insights into the field of Cellular Mechanics and Biophysics. This book can even serve as a standard work in this field. The main reason for covering all these different topics, which relate to several disciplines, has been to write a large book consisting that help the reader to understand the general mechanisms and principles of biological matter from a biophysical point of view. However, the *Cellular Mechanics and Biophysics* book is for all those who are interested in the field of cellular biomechanics and biological physics. I hope that the book will help to understand why Cellular Mechanics and Biophysics are so important for soft matter and biological research. In summary, all readers, such as students and researchers, should be guided by the book through the broad field of Cellular Mechanics and Biophysics. Each chapter of the 17 chapters is a closed part containing a large number of cited references that can serve as further reading material. Finally, the book will contribute to establishing Cellular Mechanics and Biophysics as an essential part of biological soft matter research.

Sincerely

Leipzig, Germany

Claudia Tanja Mierke

Contents

Part I

Introduction into Biophysics and Basic Biological Processes from a Physical Point-of-view

Part 1 of the book Cellular Mechanics and Biophysics introduces the reader to the biophysics research field and provides insights into fundamental processes of biological research from a biophysical point of view. After the introduction of the field, basic knowledge of classical cell and molecular biology is conveyed, which is necessary for the understanding of the following chapters. This first part discusses the main differences between prokaryotes and eukaryotes and presents a theory for the development of eukaryotes. The most important cell compartments (termed organelles), such as nucleus, mitochondria, endoplasmic reticulum, Golgi apparatus,vacuoles, lysosomes and peroxisomes, are presented, and their individual functions are briefly outlined. More precisely, a special focus lies on the impact of cell organelles on the overall cell mechanical properties.

All chapters of the first part will help the reader, even if they have limited knowledge of biology, to understand the field of biophysics and subsequently to better convey an understanding of the complexity of biological matter. The first part serves also as a basic knowledge reservoir for the other chapters in part 2 and introduces the reader into the complex world of living biological matter.

In any case, this first part explains the most important classical biological components, cellular functions and processes from a biophysical point of view and forms the basis for understanding the following two other volumes of the Cellular Mechanics and Biophysics book. The purely biological view, which is available in many review articles and in the introductory part of research articles, is presented here, if possible, together with the biomechanical aspects. Finally, state-of-the-art biophysical research is included and integrated into the classical field of cellular and molecular biology.

1 The Definition of Biophysics: What Exactly is Biophysics?

Abstract

This chapter describes what kind of research is performed in biophysics and what major breakthroughs biophysics has brought. The big problem for the field of biology and life sciences is that we have too much information about all molecules and their signaling pathways. We are simply lost in all the information and cannot decipher the essential parts. Thus, we need to focus on the framework helping us to organize and categorize all those endless numbers of facts. Biologists and life scientists wonder why a physicist is often seen as a reductionist who wants to leave out all the details that distinguish cells from metals. The reason for the reduction is that the whole unified framework must finally be seen in the context of an overall picture. The reduction to the main characteristics stands in contrast to the diversity and abundance of the mechanisms and interaction pathways that provide a fruitful, but also enormous tension. Thus, future research will at the first glance combine these seemingly contradictory directions and, if necessary, switch back and forth between them in order to understand these living beings. There has been a tremendous development of several physical techniques suitable for the measuring of living biological matter, and there is an approach revealing what happens in the nano- and microenvironment of cells. Moreover, a lot of physical ideas and principles are behind the scientific cartoons and schematic drawings present in cell biology and molecular biology books, which may serve as suitable tests to confirm or reject these hypotheses. In general, the biological question of how living organisms can be highly ordered differs from the physical question of whether the flow of energy can evoke increased order. Basically, both scientists, the biologist and physicist, achieve the same thing, but have very different approaches to getting answers and reaching their goals. In detail, this chapter gives a brief overview of the field of biophysics and also mentions subareas. The main objective of this chapter is to develop an understanding of biological processes and mechanisms and to sensitize the reader to the great diversity observed in the study of living materials in biological processes that can be attained in all fields of biology.

1.1 Introduction to Biophysics

Biophysics (also known as biological physics) is an interdisciplinary science, in which principles of physics and chemistry are employed to understand living matter. In addition, methods of mathematical analysis and computer modeling are used to demonstrate how the mechanisms of biological

C. T. Mierke, *Cellular Mechanics and Biophysics*, Biological and Medical Physics, Biomedical Engineering, https://doi.org/10.1007/978-3-030-58532-7_1

systems work. As the name biophysics implies, biology and physics are involved. In biophysics, both disciplines are included at equal proportions (Fig. 1.1). In general, the studies in biology are about life at all levels and are broadly ranged with a great diversity and maximum complexity. Biology simply examines how living organisms eat, communicate with others, live and interact with their environment and divide or reproduce. Physics usually describes nature through mathematical laws to predict the forces running idealized systems. Biophysics attempts to combine the complexity of life with relatively simple physical laws in order to promote the understanding of basic regulatory mechanisms and to gain insights into biological research topics. In specific, biophysics follows general principles to define patterns in life. These principles are described by physical laws to predict unknown patterns of different biological topics precisely. These hypotheses can be tested by applying physical methods with biophysical devices. Finally, biophysics is the investigation of biological systems and processes based on physical principles, which will be proven and manifested or withdrawn. Biophysics is a molecular science, which aims to explain biological functions through the investigation of the molecular structures and the characterization of the properties of specific molecules. The length scales of these molecular structures vary widely, since the size of these molecules is largely diverse (Fig. 1.2). For example, small fatty acids and sugars are characterized by a contour length of 1 nm, which is the length of a molecule or aggregate at maximum physically possible extension. Moreover, macromolecules such as proteins have a contour length of 5–10 nm, starches possess a contour length of more than 1000 nm, and elongated DNA molecules have a contour length of more than 1 cm and a width of only 20 nm. These biomolecules represent the simple and sole building blocks of all living organisms, since they assemble into cells, aggregates of cells, such as cell clusters and tissues and finally whole organisms. The complex structures formed by biomolecules can be visualized under a conventional light microscope.

What do biophysicists study? Biophysicists study the complexity of life at different levels such as various length scales, which ranges from single atoms or molecules to "real" living cells of

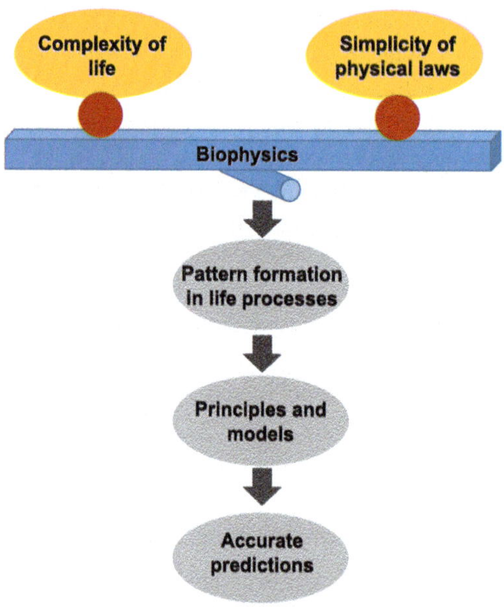

Fig. 1.1 What do we understand by biophysics? Biophysics combines the complexity of life with the simplicity of physical laws in a balanced manner. The study of biophysics leads to principles and models after observation of pattern formation in life processes. Based on these models, accurate predictions are feasible

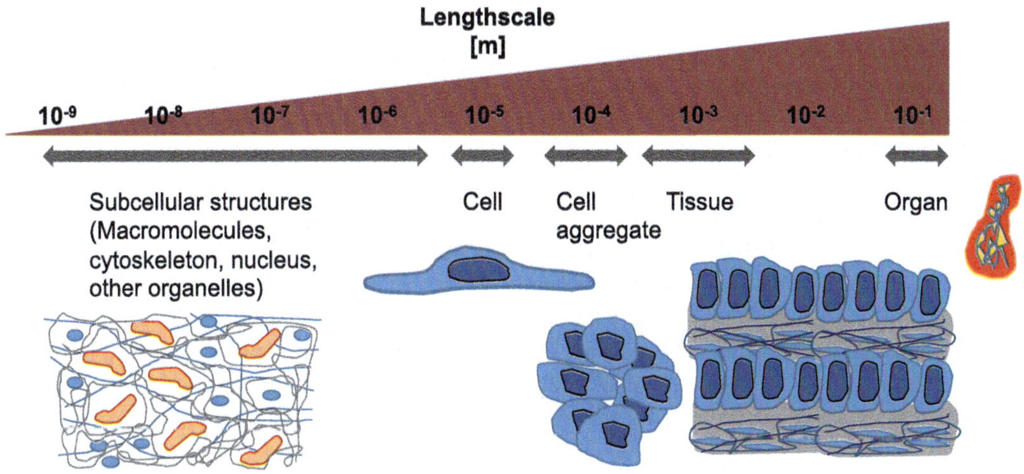

Fig. 1.2 Length scales of biophysical measurements vary largely

prokaryotic or eukaryotic origin, whole tissues or organisms. In physical terms, the length scales range from nanoscopic (molecular processes), mesoscopic (intracellular organization) to microscopic levels (multi-cellular tissues). Innovations, especially those in devices, come from biophysical laboratories, and thus, even new fields of research are opened up. Finally, the scientific work aims to understand how biological systems function. Therefore, biophysicists may ask the following simple questions:

How do cellular machines work to produce proteins? The molecular machines need to be relatively small, but they seem to function based on the same principles of machines used in other disciplines such as several branches of industrial productions.

These molecular machines consume energy to perform work. One example is the kinesin machine that can be seen a machine carrying a load that is transported by using a cellular track such a microtubule filament. This process has been described by biophysical approaches, focusing on how each step is driven forward to transport a cargo.

How do cells communicate among each other? Biophysicists created colored tags being attached to proteins to reveal the location of a certain protein within the cell and determine the protein's turnover dynamics. These colored tags exist in different colors and can be used for co-staining and multiple staining in cells, when the cells are still viable. Thus, an individual protein can be tracked during its function in many pathways.

The overall question is how essential is biophysics for the further development of biological research? Biophysics revealed how atoms are arranged to function in building DNA and proteins. Protein molecules undergo chemical reactions to provide, for example, cellular functions such as the motility of single or collective movement of entire cells through the extracellular matrix of connective tissue. Moreover, proteins provide functions of cells and tissues such as the recognition of transmigrating immune cells through the endothelial lining of blood or lymph vessels during acute inflammation upon tissue injury. In this special case, the proteins help to sense the environment and to facilitate the interaction between immune cells and endothelial cells.

Another major function of proteins is to digest food in order to get energy. In addition, proteins regulate immunological processes such as host defense, cause illness, repair DNA and regulate cellular growth, transmit electrical signals in the brain, read DNA and replicate DNA and pass it on to future generations. Since there are large variations in proteins, their reaction to pharmacological drugs can be very different. The understanding of these differences is useful for the precise diagnosis and in determining differences in drug response. This knowledge can then help to develop and improve drug design, drug vehicles for their efficient and proper delivery, controlled release and storage, which are useful in the treatment of diseases that lead to individualized drug treatments due to the special needs of patients.

What has been achieved by biophysics in the past is still a major breakthrough? In the 1940s, it has been shown that genes are built of simple four DNA bases such as adenine (A), cytosine (C), guanine (G) and thymine (T). How can this simple chemical structure include the information of DNA to be inherited to future generations? What is the mystery of this structure? In 1953, the structure of the DNA as a double helix has been discovered. The discovery of the DNA structure sheds light on how simple it is to get just by small variations in the structure unique individuals and even well-defined species. These major discoveries have provided the path for understanding the molecular machinery of life.

During the 2000s, the human genome was fully sequenced and hence discovered, which was only possible through the further improvement of a biophysical device, namely the technique of sequencing the genome by using a laser-based approach that employs individual, differently colored tags attached to each type of the four DNA bases. Since the entire genome from more than 200 species was sequenced, this knowledge forms the basis for understanding how some genes were conserved in different species. Hence, the relationships between different genes can be analyzed easily based on the knowledge of their homology in the sequence using computational methods. Beyond that, the knowledge gained is regarded as useful for the prevention and cure of diseases and the understanding of cell functions.

Besides these major breakthroughs, biophysics also leads to the development of suitable physical models and pre-supposes the knowledge of biological parameters affecting the system of interest and the implementation of physical or chemical principles. Many books on cellular biology, molecular biology and genetics miss this part and simply describe processes in a very detailed way without pointing out the most important regulatory steps required or the underlying principles. In biophysics, a working model is first created, based on the data obtained from quantitative experiments, and the parameters involved in a particular process are determined. The second step is to decide which characteristics concern a problem and are therefore of central importance or which characteristics are less important and can hence be omitted. Beyond the development of physical models, the facts about an individual system of interest must to be disclosed. There are numerous facts about individual systems, some of them with a high level of certainty, while others are relatively uncertain. Moreover, there are statements of a first category that are easy to understand and confirm, such as the statement that prokaryotes have no nucleus and no organelles. In addition, there are statements of a second category, which can be confirmed by numerous experiments using a variety of experimental methods. An example of such a statement is that genes containing introns need to remove these introns from the coding sequence in a large and complex machinery, which is termed spliceosome. After all, only the exons form the coding sequence. A statement of the third category is a compelling explanation for an observation without a proof by a biological experiment, and it depends exclusively on assumptions and can hence be highly speculative. The endosymbiotic hypothesis falls into this category because it states that the mitochondria of eukaryotic cells are the descendants of a former free-living and independent bacterium which, after entering the eukaryotic host, has developed a symbiotic relationship with a eukaryotic ancestor cell. The hypothesis is based on observations, and its interpretations represent a modeling approach that definitely needs to be improved. These hypotheses must comply with the existing laws of physics and chemistry and hence lead to improved understanding of how biological systems will react. A physical principle is first and foremost that the average energy of a molecule increases with increasing temperature of its microenvironment, and a chemical principle is that oil and water cannot mix.

1.2 Historical Overview of Selected Early Developments in Biophysics with a Focus on Cell Mechanics

The onset of the analysis of protoplasmic physical properties started over 100 years ago, which is the timepoint where we have documentary work available. Nowadays, intricate techniques, such as atomic force microscopy (AFM), cell poking, particle rheology, cell stretching and confocal laser scanning microscopy, have been developed and further improved to access and quantify the mechanical properties of cells, cell clusters, tissues, such as biopsies or surgical resections or extracellular matrix material.

However, historically, the mechanical properties of cells have been solely described qualitatively. Although the techniques used in the past are not suitable for up-to-date the analysis of cell mechanics, the historic view starting with the early roots may help us to understand nanoscientific approaches and nanomedicine techniques employed today or in the future.

Biophysics has often been considered a thriving discipline, as it continues to develop and expand as an interdisciplinary science through the combination of several disciplines. Due to its rapid growth and many directions of research that are implicated in the numerous different research topics, many researchers have a problem to describe what biophysics means to them or what the general meaning of biophysics is. Hence, biophysicists often fail to describe what biophysics represents. In the following, a historical view on biophysics is presented in order to provide an encompassing description of biophysical research topics.

It all started more than 100 years ago, in the seventeenth century, when Robert Hooke and Antony van Leeuwenhoek explored very small living matter using simple optical microscopy, in which fluid and cellular movements seemed to be of the extreme type and depended on different length and time scales (Fig. 1.3). More precisely, the ciliate named Vorticella appeared to move in a bell-shape manner combined with a stirring of water molecules in the middle of the round opening. The motion of particles within this environment was also seen by others, such as the Brownian motion of particles and even organelles within living cells (Heilbrunn 1926, 1927). More precisely, it seems to be suitable to measure the motion quantitatively in order to determine the viscosity. However, techniques

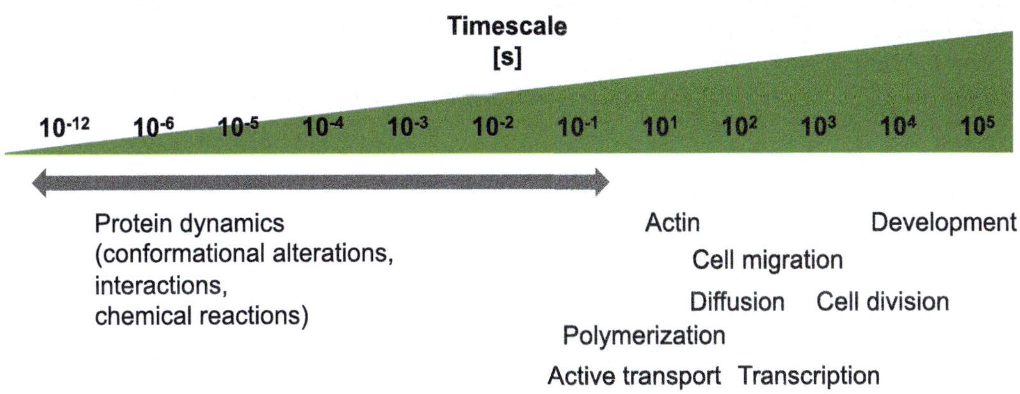

Fig. 1.3 Time scales of biophysical analysis can be performed over a wide range

to measure the microrheological properties or to perform nanoindentation analysis were not invented that time. Based on the material testing techniques, the living matter, such as cells and tissues, can be characterized in detail, and hence, theories can be formulated, such as indentation techniques, the bending of beams and the Hertz model in the nineteenth century (Thompson 1917). More precisely, the material properties for different kinds of wood or alloys have been revealed and have led to the concept of the adaption of tissue, such as the Wolff's law that is applied to the skeleton (Thompson 1917). As the nineteenth century was almost over, the mechanical properties of living matter, such as cells, were investigated with a broad range of biophysical techniques that are related to engineering mechanics on a macroscopic length scale. Although much time has passed, nanoscale probing techniques and theoretical treatment of biological organisms are still based on nineteenth-century techniques and models (Wang and Discher 2007; Boal 2002; Mofrad and Kamm 2006).

However, the living matter, such as a cell, represents a universe itself. This universe seems to be very complex and is subject to forces and dynamics that are somehow connected with its function. In former times, the role of mechanical forces in biology has been under debate (Thompson 1917), whereas now there is agreement about the important function of mechanical forces in cells and tissues. Hence, mechanical laws can be applied to living matter and mechanical properties of cells have been started to be determined. In the next step, the theoretical understanding revealed that a compelling mechanical picture of the cell is required, but under debate that time (Heilbrunn 1924, 1926, 1927; Bingham 1933). In detail, at that time cells were seen as homogeneous gels or sols, and as viscoelastic and plastic fluids. These descriptions are still commonly used to describe living cells, although multiple models have been presented to describe cell mechanics in a different manner, such as a connected discrete mechanical elements, a viscoelastic continuum and a combination of a viscoelastic fluid in a dense meshwork (Wang and Discher 2007; Boal 2002; Mofrad and Kamm 2006; Bao and Suresh 2003; Charras et al. 2005; Dai and Sheetz 1998; Heidemann et al. 2000; Huang et al. 2004; Ingber 2000; Kasza et al. 2007). For all these, different models seem to be an experimental proof or disproof that leads to support or reject of the model, such as the soft glass rheology phenomenon (Trepat et al. 2007; Stamenovic et al. 2007). Since the experiments have been refined in the last century, viscosity elasticity, plasticity and motion have been analyzed more precisely. Hence, there exists no comprehensive theoretical model of cell mechanics, which can account time-dependent alterations and predict changes accurately.

Moreover, there is no agreement as to whether the mechanical phenomena and properties are exclusively by-products of biological functions and processes or whether they are regulated at genetic and physiological level by feedback loops, actuation and/or response signal transduction pathways. In addition, the mechanical properties have been supposed to regulate cellular functions, such as cell adhesion and motility. Previously, some studies have tried to answer this question (Even-Ram et al. 2006; Engler et al. 2006; Paszek et al. 2005; Ingber 2005, 2006). A major focus is AFM studies of cell mechanics, although there are many other biophysical techniques probing cell mechanics (Wang and Discher 2007; Boal 2002; Mofrad and Kamm 2006; Mierke et al. 2008a, b; Mierke 2011). The cell mechanics research is currently still growing and connected to other disciplines, including cell biology, molecular biology, biochemistry, cell (nano-)physiology. This chapter focuses the historic view on the development of cell mechanics using the AFM technique.

1.2.1 Isolated Protoplasmic Mechanics

It has been demonstrated that resected blood vessels and nerves can be probed mechanically in lectures (Stuart 1738). Thereby, concepts of hydrostatics, elasticity and viscoelastic fluids were mentioned, and it has turned out that nerves are inelastic. In living matter, these mechanical oscillations were largely investigated, such as the spontaneous heart contractions outside the organism (Paget 1857). These heart contractions can pertain without a proper nervous system. In line with these results, heart and muscle cells showed the same behavior using AFM (Domke et al. 1999; Shroff et al. 1995; Pelling et al. 2007a, b; Haupt et al. 2006). These oscillations have been found in many other organisms, such as in vacuoles, cell walls of plants or the cilia movement in microorganisms (Paget 1857). All these structures have in common that they possess no muscle structure or a nervous system. There was no explanation why these organisms perform mechanical oscillations. The concept of biological mechanics still not yet fully formulated. These early investigations of mechanical phenotypes were focused on viscosity of the protoplasm due to the unavailability of microscopic techniques. However, the cytoplasmic streaming (circular flow of eukaryotic cytoplasm) was detected early (Heilbrunn 1927; Ewart 1901) and had been employed to determine protoplasmic viscosity in a qualitative manner. The application of particle tracking in cell mechanics started with viscosity measurements (Heilbrunn 1926, 1927) and hence is regarded as the earliest uses of particle tracking in cell mechanics that led to the development of nanoscale particle tracking and microrheological measurements (Weihs et al. 2006; Mason et al. 1997).

There has been discussion about the influence of the particle size, the mesh size of the protoplasm, cell damage and experimental conditions such as temperature on the mechanical properties (Heilbrunn 1926, 1927; Valentine et al. 2004). These points are still under investigation in modern particle tracking approaches and need to be addressed. In addition, there is also an effect that the particles supplied externally and taken up by the cells affect their function, such as their motility in 3D confinements (Mierke 2013).

Another approach has been reported in 1924 using a newly developed early magnetic microscope (Seifritz 1924), which was used to oscillate nickel particles (of approximately 16 μm in diameter) in living cells. This approach is similar to modern particle microrheology (Trepat et al. 2007; Stamenovic et al. 2007; Weihs et al. 2006; Mason et al. 1997) and conceptually similar to magnetic bead-twisting cytometry (Valberg and Albertini 1985; Lele et al. 2007; Massiera et al. 2007). Another experimental approach of magnetic manipulation employed the injection of iron particles into living bacteria that are altered in their position by an external electromagnet (Heilbrunn 1927). Another approach to analyze the viscosity at the time was the centrifugation of the cells. Thereby, the cell granules are displaced to one end of the cell and their migration back to their original position is monitored that revealed an estimated protoplasm viscosity (Heilbrunn 1927). Alterations in the range of two orders of magnitude in the viscosity were observed during the mitosis and fertilization of sea urchin eggs (Heilbrunn 1927). In detail, when the viscosity is kept constant, the mitosis can be impaired (Heilbrunn 1920a, b). More precisely the protoplasmic viscosity can be altered by temperature, distinct chemicals, such as anesthetic agents, salt or organic solvents, radiation and electric currents (Heilbrunn 1926, 1927; Seifriz 1931; Heilbrunn 1920a; Heilbrunn 1925a, b; Heilbrunn and Wilson 1948, 1957, Heilbrunn et al. 1954; Seifriz and Uraguchi 1941; Scarth1924; Packard 1931; Lepeschkin 1932; Jacobs 1922; Forbes and Thacher 1925; Carlson 1946; Bayliss 1920; Angerer 1939; Addoms 1927). Within the last past two decades, the Young's moduli (synonymously termed elasticity or stiffness) have been determined using AFM. The Young's modulus is fundamentally different to the viscosity, but somehow related to it. Thereby, anti-cytoskeletal drugs (Rotsch and Radmacher 2000; Pelling et al. 2007a, b; Northern 1950), chemotherapy reagents (Heilbrunn and Wilson 1957; Lam et al. 2007) and electrical stimulation (Greely 1904; Zhang et al. 2001) and their impact on the Young's modulus has been determined. More precisely, the majority of AFM measurements with living cells rely on the employment of the nano-indentation mode in order to reveal mechanical parameters from the resulting force-distance (or synonymously termed force–displacement) curves (Binning et al. 1986). The major mechanical parameter is the elasticity, but other rheological parameters were also determined in living cells (Alcaraz et al. 2003; Smith et al. 2005,

2007). Indentation approaches have been utilized together with scanning of the cell to obtain full force maps (Rotsch and Radmacher 2000; Rotsch et al. 1997, 1999; Radmacher 2002, 2007). Alternatively, single spots on living cells have been analyzed to investigate time dependence (Pelling et al. 2007a, b; Lam et al. 2007).

A century ago, the initial indentation approaches on living cells deal with glass microneedles that can be slowly inserted into multiple cell types to estimate their viscosity (Seifriz 1918; Kite 1913; Chambers 1915; Chambers and Fell 1931). The microdissection technique and its variants have often been used to study the protoplasm mechanics, although it is exclusively a qualitative analysis. In addition, a modified microneedle experiment has been performed by injecting the microneedle in living cells in order to push and into organelles (Chambers and Fell 1931). In 2005, a similar approach has been reported, in which a modified AFM cantilever tip (termed nanoneedle) was used to push and penetrate the largest organelle in the living cell, the nucleus (Obataya et al. 2005). These two techniques have been employed to measure similar mechanical properties, such as the penetration and deformation of the nucleus. An advantage of the AFM techniques over the "old" microneedle technique was that quantitative measurements of the force can be performed. Moreover, the simultaneous laser scanning confocal during the AFM measurement procedure delivered an accurate three-dimensional information during the mechanical stress application.

Before the AFM was developed in 1986, another technique termed cell poking was presented that deals with calibrated microneedles (McConnaughey and Petersen 1980; Petersen et al. 1982; Daily et al. 1984) (Fig. 1.4). Dissimilar to earlier techniques that pushed the needle through the entire cell, the cell poking needle solely indented the plasma membrane of the cell in order to determine the cell deformation and elasticity. More precisely, the contribution of cytoskeletal components was investigated using anti-cytoskeletal drugs (Petersen et al. 1982). Early examples for bulk cell mechanical properties were plant cells whose elasticity has been revealed (Treitel 1944). In detail, plant tissue was clamped with one end in a stretching device and the tissue was stretched using calibrated weights in order to obtain stress–strain curves of the tissue. This concept is similar to current and future directions of cell and tissue mechanics that analyze cell clusters (termed cell spheroids), cell monolayers and distinct tissues (Huang et al. 2004; Ingber 2005, 2006; Bray 1984). In addition, whole cell mechanics (Waugh and Evans 1979; Evans and Hochmuth 1976; Evans and Yeung 1989;

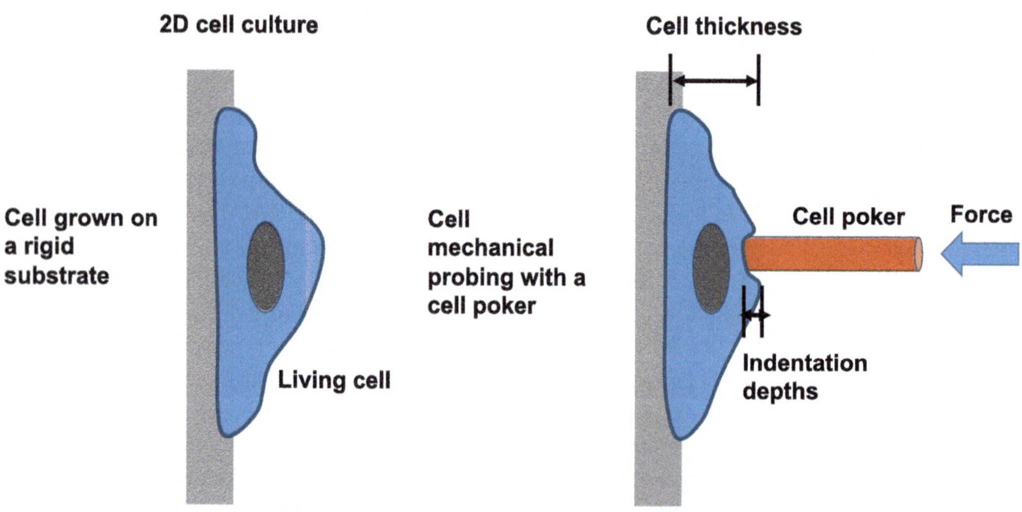

Fig. 1.4 Simple setup of the traditional cell poking technique

Hochmuth 2000) and nuclear mechanics (Lammerding et al. 2004, 2006, 2007; Rowat et al. 2006) can be determined using micropipette aspiration, where the cell is sucked in a micropipette. In addition, microplates (Thoumine et al. 1999; Smith et al. 2000) have been used to analyze cell deformation and elasticity due to force exertion.

The adhesion forces of cells playing an important role in cell spreading and migration have been determined by rupturing cells from a surface using a cantilever (Sagvolden et al. 1999). Since cell migration has been analyzed in order to investigate wound healing processes, it has turned out that this process involves cell mechanics (Herrick 1932). Using AFM , protrusive forces at the edge of motile cells can be determined (Radmacher 2007; Prass et al. 2006). In complement with these measurements, traction forces can be analyzed on these 2D substrates (Pelham and Wang 1999; Jurado et al. 2005) and additionally micropipette and laser trap measurements can be performed (Li et al. 2005). Since cell migration and invasion are key elements in the malignant progression of cancer, such as cancer metastasis, it has been hypothesized that the increased invasive potential of malignant cancer cells is related to altered cell mechanics. Thus, cell mechanics have been analyzed by optically trapping and stretching cells in electromagnetic fields to determine their mechanical properties that are corrected with their metastatic potential (Svoboda and Block 1994; Svoboda et al. 1992; Guck et al. 2001, 2005). These assays have been developed in addition to initial cell deformation assays (Ochalek et al. 1988). Moreover, magnetic traps were employed to examine the rheological properties of cells. Hence, super-paramagnetic beads were coupled to cells in order to probe them mechanically by altering the magnetic field (Stamenovic et al. 2007; Wang and Ingber 1994; Puig-De-Morales et al. 2001). Thereby, the mechanical properties of cells such as the deformation of organelles by external forces and the force transmission processes in cells were revealed using the magnetic (bead-)twisting cytometry (Wang and Ingber 1994; Hu et al. 2004). The concept has similarities to earlier investigations by the Seifriz group (Seifriz 1924). In detail, he reported with the magnetic microscope and the simple tracking of organelles due to force application by indentation of the cells with micropipettes (Puig-De-Morales et al. 2001).

Over more than 100 years ago, a wide variety of techniques have been developed to analyze the mechanical properties of living cells, such as cancer cells. Initially, the mechanical characterization of living cells and organisms was highly conceptual; however, there was an enormous increase in biophysical techniques probing cellular mechanics (Wang and Discher 2007; Boal 2002; Mofrad and Kamm 2006). At that time, an increasing number of laboratories established successfully the measurement of cell mechanics. Many questions that have been asked for years are still the same that needs to be examined today. It is now clear that living soft matter, such as cells and tissues, exhibits distinct mechanical properties that seem to drive physiological processes, such as proliferation and apoptosis and pathological processes in multiple diseases, such as the malignant progression of cancer. A key future development of the field of nanomedicine or nanophysiology seems to be the identification of specific mechanical states and their reliable detection. Since the concepts have been reported for several years, the question can be raised, whether there can be performed another action than just developing and improving accurate biophysical tools to measure cell and tissue mechanics.

1.2.2 Future View on Cell Mechanics

In 1737, Stuart has initiated the discussion whether the heart can be regulated by well-defined stimulation. The accurate measurement of the mechanical parameters is a key feature of biophysical research, and in addition, the controlled regulation of biological pathways seems to be another key feature that needs to be addressed by altering cell mechanics. Mechanical force that is applied by the cantilever tip using AFM has caused multiple chemomechanical responses (Charras and Horton 2002; Charras et al. 2001; Formigli et al. 2005). It has also been reported that the mechanical environment of many cell types, such as cancer cells and stem cells, can be employed to precisely regulate the control of gene expression and differentiation pathways (Even-Ram et al. 2006; Engler et al. 2006; Paszek et al. 2005; Ingber 2005, 2006; Wang and Ingber 1994; Huang and Ingber 2005; Suresh 2007). Moreover, it can also be used to alter gene expression and differentiation pathways (Engler et al. 2006). Since all the tools to analyze the mechanical properties have been available, we can probe cells with well-defined forces and even alter the mechanical properties of the microenvironment of the cells. In addition, we can also monitor the mechanical properties of the microenvironment. It seems to be only a small step from here to the initiation and control of biological pathways in defined cell cultures and additionally in the future in vivo settings. In the future, it seems likely that the emerging field of nanophysiology will become a new branch of the nanomechanical regulation of biological pathways. This ill-conceived area of research, combined with ultrasensitive detection technologies and modern pharmaceutical treatments, can play an important role in the advancement of nanomedicine and the diagnosis and treatment of diseases.

Since the physical principles that regulate cell mechanics are yet clearly understood, the field of nanotechnology can be envisioned to be helpful in medicine, such as cancer research and physiology. However, the entire field is currently under strong debate. In detail, the elasticity concept seems to be rather ill-defined in a living cell. The cell is no static and homogeneous object; instead, the cells are rather heterogeneous and undergo continuously dynamical remodeling of its cytoskeleton and is rather likely highly anisotropic. Hence, the Hertz model, which is usually employed in nanoindentation experiments using AFM, cannot applied in an ideal manner. Moreover, cellular Poisson ratio appears to be rather undefined and is conventionally considered constant, even though this may not be the case. However, there is no evidence presented that the Poisson ratio itself is not affected even during physiological processes and that the variation in the Poisson ratio does not correlate with Young's modulus alterations. Finally, the theoretical descriptions of cell mechanics will still need further improvement. However, there is clearly no doubt that future discussions will take place in this issue. The conclusions in cell mechanics are usually drawn from investigations of an individual cell type, which has been analyzed under limited conditions. Moreover, these conclusions were generalized to a wide range of different cell types or even to all cell types. However, it has also been suggested that cell mechanicals and the biochemical/structural requirement for mechanical properties of cells are cell type-dependent and additionally they even depend on the physiological and mechanical microenvironments of cells (Mierke 2019a, b).

Many different cell types possess the same structural components, such as the cytoplasm, cytoskeleton, nucleus, membranes and other organelles; however, it is not convincingly shown that they all use the same signal transduction pathways during these biological processes. Hence, besides the identification of a unified theory of cell mechanics, it should also be taken into account that there might be different phenotypic mechanical responses including different signal transduction pathways due to the broad heterogeneity of the living cells. The classification of cells may be based on individual mechanical models or on a combination of different models best suited to the mechanical reactions and internal structures required for mechanotransduction or during physiological processes. The trend in nanoscientific research is becoming interdisciplinary and is one of the greatest advances in the field of cell mechanics.

Historically, the research field of biophysics emerged as a distinct discipline and started originally as a group of four including Emil du Bois-Reymond, Ernst von Bruecke, Hermann von Helmholtz and Carl Ludwig, which are all four physicians. Emil du Bois-Reymond, Ernst von Bruecke and Hermann von Helmholtz are former students of the German physiologist Johannes Mueller. In 1847, they all developed a research project which had been rejected with the notion that living animals depend on specific biological laws, and hence, physical laws cannot be applied to them. Moreover, it had been stated that these vital forces of living matter would be different to those forces present in inorganic matter. Thus, the group had to explain why biological processes or phenomena exhibit the same laws that can be used to describe and elucidate physical and chemical phenomena. It has been mentioned by Ludwig and also said by Cranefield (1957) that physiology should be based on a chemico-physical foundation and it should be scientifically ranked equally as physics. Moreover, the term of "Organic Physics" has been created, which had been used by du Bois-Reymond. However, this term did not develop as expected, and hence, in 1982, Karl Pearson created the novel term "Bio-Physics" in "The Grammar of Science" (Pearson 1900) that should describe science connecting physical and biological science. However, he also stated that this novel branch of science has not developed fully at present and improvement is highly needed as it may become important in the future. Indeed, it worked out well for the novel field as Julius Bernstein reported a possible mechanistic basis for the development of transmembrane potential differences, which relies on electrodiffusion work by Nernst and Planck (Bernstein 1902). Several years later, Archibald V. Hill described the Hill equation (Hill 1910). Both studies are prototypical examples of biophysics for quantitative analysis of biological phenomena. The main research branch in biophysics in the twentieth century was neurophysiology and muscle physiology that are well suited for quantitative analysis.

In the second half of the twentieth century, biophysicists are headed in physics, chemistry and mathematics and not only biology and medicine, leading to the improvement and the development of the modern optical and electron microscopes, the generation of fluorescent probes based on small molecules or genetically encoded proteins, the preparation of synthetic oligonucleotides, the development of magnetic resonance and diffraction methods as well as computational methods. All these developments are actual tools for biophysical research. However, it was not yet clear and was therefore intensively discussed how the question "What is biophysics?" can be answered. Hill had stated that the development and implementation of physical instruments in biological research does not mean that the researcher becomes a biophysicist. A biophysicist is a researcher who investigates the biological function, organization and structure by physical and physicochemical ideas (concepts) and methods (Hill 1956). It is the way of thinking that distinguishes a biophysicist from a biologist, for instance when it is all about the importance of a quantitative, theoretically sound analysis of the problem to be investigated. In this context, the focus on theory and quantity is a central part of the methodological developments required for current biophysical research. Biophysics can be seen as a quantitative approach to the study of biological problems.

In fact, biophysics, as envisioned by the four scientists in 1847, is based on the emerging convergence of sophisticated quantitative experimental analyses with computational approaches such as molecular dynamics simulations using classical and statistical mechanics to reveal the function of proteins.

A major breakthrough in the field of biophysics was the development of novel microscopes that enable us to visualize structures with higher spatial resolutions that exceed the limits set by the diffraction barrier. In former times, the diffraction barrier limited the ability of light microscopes to distinguish two points separated by lateral distances of less than half the wavelength of the light used to visualize them. Another major breakthrough was the development of mostly fluorescent probes (also known as tags), which enable us to visualize living cells that can be located deep in tissue sample sections or in the tissue of living animals (Horwitz 2016).

Genetically encoded fluorescent probes such as the green fluorescent protein have evolved as powerful tools for targeting specific cells or intracellular organelles, addressing problems that go far beyond chemical probes. The targeting of probes to specific cell types in a distinct tissue allows us to analyze them in living cells and tissues with high spatial and temporal resolution. To be more specific, focused light impulses can be utilized to manipulate genetically encoded targets that hence allow us to alter cellular functions of the whole organism. The role of the microscope has shifted from a simple tool to observe biological function to an advanced tool for modifying biological functions (Cohen 2016). This field is known as optogenetics. Finally, all these examples offer a limited overview on the importance of biophysical science.

There are two concluding questions: Why do we have physical biology, biophysics and biological physics as synonyms? Or are they still characterizing different research fields? The major agreement about this issue is that biological physics studies physical problems by using biological materials such as living cells (mostly) without considering biological questions. In biophysics, researchers investigate biological problems using "trivial" biophysical approaches, in other words, devices in which the user does not need to be understood physics for answering biological questions. For example, when a scientist examines the structure of proteins, X-ray analysis does not need to be understood in detail because the computer software shows the final structure. In physical biology, however, the researcher still has biological problems, but the physics behind the biophysical device is "not trivial" and must be understood by the researcher and even the device must be improved by the researcher to achieve the right results.

1.3 Major Fields of Biophysics

Biophysics employs physical methods to investigate biological systems. The wide field of biophysics can be divided into three main branches: molecular, cellular and systems biophysics. Molecular biophysicists usually detect and manipulate single molecules in order to analyze the function of molecular motors, the process of protein folding, (bio)polymer physics, the rheology of single molecules and the dynamics of complex cellular processes such as transcription, replication and translation outside of cellular systems. Cellular biophysics unravels the molecular mechanisms inside cells or tissues such as the gating of channels, the structure and assembly of chemical networks within living cells and the imaging of fundamental cellular processes such as cell migration or cell division. Therefore, computational models are required to describe cellular processes such as metabolism and molecular switches and predict an outcome of disturbed cellular processes. The field of systems biophysics aims to describe collective phenomena and analyzes evolutionary and ecological dynamics of populations, the development of biofilms, sensory studies, neural development and signal transmission, the organization and function of synapses, the functional imaging of the brain and the coordination of movements in space and time.

Biophysical research is interdisciplinary and can range from molecules, proteins, polymers, cells, tissues, organs to organisms such as animals, plants or microorganisms. In addition, viruses can also be the field of biological research. Over time, several major disciplines such as molecular and cellular biophysics, systems biophysics and subdisciplines including membrane biophysics, theoretical biophysics, biophysical chemistry and structural biophysics have been developed (Fig. 1.5).

These major disciplines are all roofed under biophysics. Biophysics has developed as physically established approaches and methods applied to biological systems to gain more insights and predict the outcome of an experiment or scientific approach. It has been seen that the potential of biophysics is enormous, and the understanding of the biological subjects under investigation is necessary to further improve biophysical research. Even the field of material sciences dealing with biomimetic materials uses biophysical methods and approaches to get more insights on a special problem involving biological matter. The major breakthroughs and even fundamental improvements are urgently needed such as in the field of medicine where implants and stem cells grown on biomaterials are used.

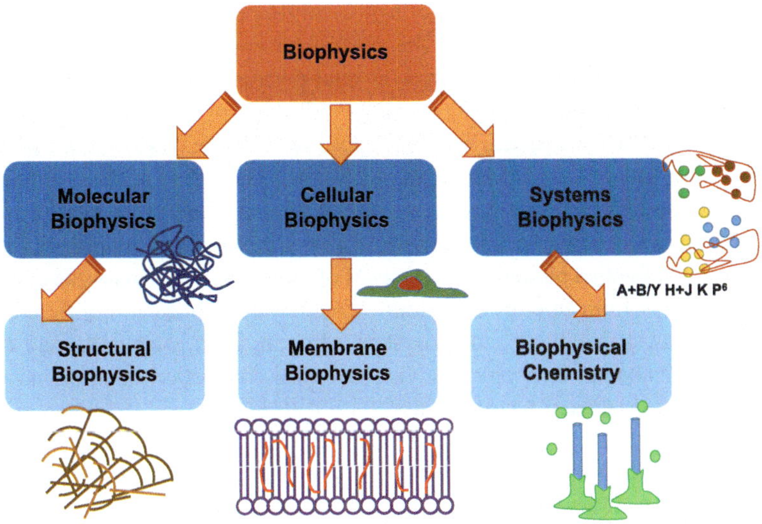

Fig. 1.5 Biophysics or biological physics is divided into at least these three main parts. These subdisciplines may be highly related and partly intertwined

The field of biophysics ranges from nanoscale to macroscale, dealing with widely varying time and spatial scales and even covering all aspects of the biological organization. Biophysical research has significant overlaps with many other disciplines such as biochemistry, nanotechnology, biomaterial science, bioengineering, cell biology and computational systems biology.

1.4 Molecular Biophysics

Molecular biophysics investigates biological questions similar those in biochemistry and molecular biology. The difference is that molecular biophysics tries to get more quantitative results and searches for physical substantiation of the observed biomolecular phenomena. More precisely, molecular biophysicists conduct experiments to elucidate interactions between different systems of a cell involving protein biosynthesis, replication and transcription, and how, for example, interactions between DNA, RNA and proteins are regulated. A variety of biophysical and molecular techniques, mainly derived from molecular biology, are used in the consideration of carrying out this research. In more detail, molecular biophysicists investigate the structure and conformation of biological molecules, the relationship between structure and function, conformational transitions, ligand binding and intermolecular binding, the flow of energy, thermodynamics, statistical mechanics, chemical reactions kinetics and molecular machines. In conclusion, molecular biophysics investigates physical principles that control biomolecular systems. Moreover, it aims to describe biological functions based on molecular structure, dynamics and organization emanating from single molecules and ending with supramolecular structures.

1.4.1 Subdiscipline: Structural Molecular Biophysics

There are different fluorescent imaging techniques, electron microscopy, X-ray crystallography, NMR spectroscopy, atomic force microscopy (AFM) and small-angle scattering (SAS) both with X-rays and neutrons (SAXS/SANS) for the analysis of biological structures. Neutron spin echo spectroscopy can be used to observe protein dynamics. The conformation of biological substances is analyzed using biophysical and physical techniques such as dual polarization interferometry, circular dichroism, SAXS and SANS. In addition, optical tweezers or AFM are employed to investigate the forces in a molecule and to determine distances on the nanoscale. Complex biological events are seen as systems of interacting entities, which are analyzed using statistical mechanics, classical thermodynamics and even chemical kinetics such as enzyme kinetics. The structure of proteins is also modified by genetic alterations induced by specific mutations. By using molecular biological techniques, the interactions between individual proteins or complexes of molecules are also revealed. The models and experimental techniques from physics and biomathematics are applied to highly complex systems such as tissues, organs, populations and ecosystems. In particular, established and confirmed biophysical models are available for people studying electrical conduction in single neurons or neural circuit analysis in both tissue and whole brain.

1.4.2 Subdiscipline: Nanotechnology

The design of synthetic DNA based on physical principles is named DNA origami. The Japanese word "Origami" means folding a plain sheet into any shape with a specific dimension. DNA origami is one of the most recent molecular biological techniques that uses DNA as a building block for the synthesis of designed nanoparticles with specific properties and functions. This novel technique is frequently employed in the field of nanotechnology. In detail, long DNA strands are folded to a complex scaffold that contains staples of DNA strands consisting of 200–300 nucleotides. Finally, a complex structure can be formed that has the desired characteristics defined by its nanoscale dimensions (Tørring et al. 2011). DNA nanostructures are not yet suitable for medical applications or the administration of drugs because they are in the early stages of development. More specifically, biocompatibility and physicochemical properties need to be determined in order to establish them as a suitable delivery tool for pharmacological drugs. In theory, DNA origami seems to have an immense potential to contribute pronouncedly to a wide range of medical fields, such as diagnostic tools for certain diseases, such as cancer, and drug delivery systems (Zhang et al. 2014). Since cancer therapy and diagnosis is such a potential filed, where DNA origami has shown significant efficacy against cancer, it will contribute to tumor oncology immensely. For example, it has been shown in an animal nude mice model system for cancer that doxorubicin-loaded, triangle-shaped DNA origami seems to be an efficient and safe manner for breast cancer treatment (Zhang et al. 2014).

1.5 Systems Biophysics

Biological science performs reductionist analysis of biological matter by using tools of biochemistry, molecular biology and structural biology and thereby reveals increasingly details on individual parts of living systems. Since the amount of molecular data is increasing, bioinformatics is required in order to sort out the relevant and general information that enable biologists to annotate, query, search and integrate these data in an easy and reliable manner. Thus, systems biology comes into play, which aims to identify similarities and reconstruct networks of the molecular signaling pathways that reveal the essential molecular biological, biochemical, biophysical and regulatory functions of cells (Fig. 1.6). In more detail, the unique properties of different cell types are required to build specialized organ systems such as central nervous, musculoskeletal and cardiovascular systems. Based on the increasingly detailed, but still yet complete, molecular network reconstructions, the basis of system

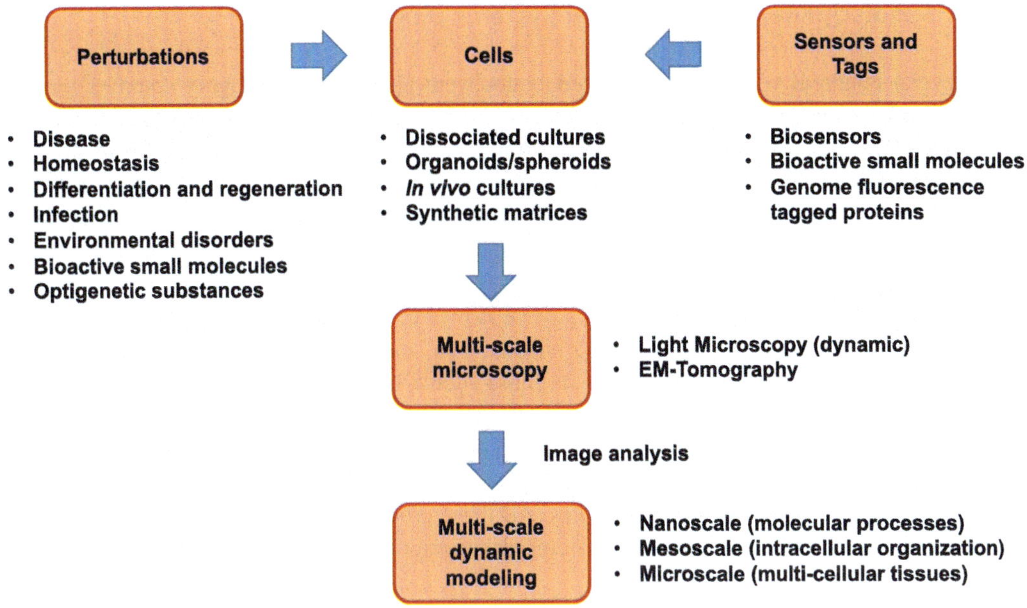

Fig. 1.6 Systems biology approach on the nano- and mesoscales

models of biological functions at all length scales has been created in order to simulate the dynamics of living systems under physiological or pathological conditions, which ranges from cells to organs or organisms that are displayed as large circuit diagrams of functional interactions. Major impacts have the quantitative, computer-driven approaches that enable us to gain a higher level of integrative scientific insight and to identify finally promising novel pharmacological drugs for multiple diseases (McCulloch 2016).

In order to achieve the physiological function of tissues or organs, biological systems are structured in a distinct pattern and the entire architecture depends on their dynamic three-dimensional organization to ensure proper function under normal conditions. However, when the organization is disturbed, a pathological situation occurs causing possibly a diseased state. The advantage of building models is that they can be structurally integrated across the physical length scales of the organization of living matter from molecules and cells to organ systems and organisms. Similar to systems biology, the field of multi-scale modeling is becoming more and more data-intensive, but it is imperative to understand the basic principles. In order to study these different length scales of biological matter, the structural biology, microscopy and medical imaging technologies need to be combined to obtain high-quality and high-resolution datasets on molecular, cellular, tissue and organ structure length scales. However, the multi-scale modeling mainly requires physics to define and isolate molecular and cellular processes leading to tissue formation and even organ-scale physiology.

The requirement for the multi-scale modeling of organ systems seems to be apparent from the viewpoint of physicians. From their viewpoint, the patients have symptoms and possess diseases that are manifested at tissue, organ and entire body scales. As the current medical treatments and drug therapies target solely specific molecules, an overall treatment approach is necessary. Hence, can we ask the question of how we can diagnose and effectively treat diseases without knowledge of the multi-scale relationships between the molecules and the whole body in order achieve the best outcome for the patient?

1.5.1 Subdiscipline: Immunological Systems Biophysics

Over several hundreds of million years, vertebrates were able to build a highly complex immune system in order to defend against pathogens by protective immune responses. The tolerance of this immune system to its own proteins and also to microorganisms, living in symbiosis with the host, must be guaranteed. The immune response is facilitated by the activation of regulatory networks controlled by immune cells involving distinct cytokines and cell surface receptors. All of which enables the vertebrates to initiate the precise timing of the immune response that may finally lead to the clearance of the pathogens from the host and subsequently to long-term immunity.

The knowledge of the controlling of this process is regarded as the key to understand the infection process and subsequently helps to develop efficient new vaccines and immunotherapies for these infectious diseases. Therefore, the immune response should be fully understood. In more detail, we need to understand how the immune response restricts tissue damage of the host, but still works efficiently in killing the pathogens (clearance process). These two goals are archived by keeping a precise balance between effector and regulatory cells, based on immune activating and inhibitory receptors as well as gene regulatory networks. All these processes involve biophysical approaches, which seem to be highly promising in providing deep insights into immunological processes.

1.5.2 Subdiscipline: Developmental and Evolutionary Systems Biophysics

During the animal development and homeostasis of animals and humans, the structure of tissues including muscles, blood or lymph vessels, and connective tissues can adapt to the external mechanical strains within the extracellular matrix. The strains can originate from the differential growth potential of various tissues, or forces caused by the contraction of muscle or by gravity. The active cell traction forces fulfill a crucial role in the alignment of cells caused by static uniaxial stretch such as muscle contraction. Using contact guidance approaches, cells manage to adjust their orientation to the cytoskeletal fibers, which usually align parallel with the strain (Vader et al. 2009; Chaubaroux et al. 2015). When the extracellular matrix is under strain such as by pulling on the matrix, the cells align parallel to the matrix fibers (Klebe et al. 1989). This behavior is termed mechanical cell-fiber feedback that drives the coordination of cellular alignment (Takakuda and Miyairi 1996; Reinhardt et al. 2013; Reinhardt and Gooch 2014) and the formation of strings along a certain strain (Dyson et al. 2016). However, in vitro observations revealed that cell alignment toward uniaxial stretch may not need to be facilitated by sole fiber alignment. Moreover, mesenchymal stem cells are able to align along the cell body orientation toward the direction of strain on a non-fibrous matrix (Liu et al. 2014). In addition, within stretched collagen fiber matrices, fibroblasts can align in parallel to the strain direction even in the absence of fiber alignment (Eastwood et al. 1998; Tondon and Kaunas 2014). In line with this, it has been found that collagen fibers align solely after the cells had aligned in a distinct manner (Lee et al. 2008; Pang et al. 2011). Moreover, fibroblasts manage to orient along the uniaxial stretch direction, although the fibronectin fibers have aligned perpendicular to the stretch direction (Mudera et al. 2000). In conclusion, these results indicate that the cells are able to orient themselves in the direction of the stretch regardless of the orientation of the fibers.

In addition, the mathematical modeling seems to be a useful tool to decipher which biophysical mechanisms are suitable to explain the alignment of cells toward strain. However, the previous mathematical modeling was based on the optimization principles (Bischofs and Schwarz 2003; De et al. 2007). In more detail, it has been suggested that cells try to minimize the amount of work (energy) required for the contraction of the extracellular matrix environment (Bischofs and Schwarz 2003). For dipolar cells, the work (energy) is minimized, when these cells orient their cell axis in parallel with the uniaxial stretch. When the cells are proposed to generate strains in their local microenvironment, the cells build strings that are able to align with an external strain field (Bischofs and Schwarz 2003, 2005, 2006). Based on the finding that cells restructure their focal adhesions and stress fibers to maintain the constant local external stresses, it has been suggested that cells modulate their own contractility and orient to minimize the local stress on them in the extracellular matrix environment (De et al. 2007). It has been found that the local stress gets minimal when a dipolar cell orientates in parallel (its longest cellular axis) toward uniaxial stretch, as in this distinct configuration the cell traction forces counteract the uniaxial stretch maximally. Cellular alignment to strain can be predicted on a mesoscopic scale and hence experimentally tested to identify a cellular mechanism previously revealed by the use of computational analysis (Rens and Merks 2016).

As the study of molecular evolution at the level of protein-coding genes has its limits in terms of large datasets of sequences to reveal evolutionary relationships, hence systems biophysics is required to provide an overall approach. Although a protein's structure and conformational dynamics are crucial for protein function, the biophysical knowledge of the proteins need to be included in those studies. More precisely, biophysical constraints can provide natural selection, such as effects of protein mutations, as the physical basis for marginal stability of natural globular proteins may be required to withstand physical constraints. Moreover, altered kinetic stability may also be required and misfolding as well as misinteractions need to be avoided, as those have large effects on protein evolution and hence must be integrated, when investigating molecular evolution. In the past, the biophysical approaches have underpinned these effects, as they have been addressed by models that deal with an explicit coarse-grained spatial representation of the sole polypeptide chain. However, the sequence–structure mappings based on those models are still powerful conceptual tools to identify the mutational robustness, evolvability, epistasis, the promiscuous function of proteins performed by "hidden" conformational states, the resolution of adaptive conflicts and conformational protein switches that have evolved in the evolution from one protein fold to another. The protein biophysics science seeks to obtain more accurate evolutionary information of sequence data. In more detail, biophysical methods have also been established to get from sequence-based data, the evolutionary information to predict biophysical properties of proteins, which has thereby led to a synergy of protein biophysics and protein evolution (Sikosek and Chan 2014).

1.6 Biophysics-Related Discipline: Biophysical Chemistry

The focus of biophysical chemistry is on the functional analysis, which can be the investigation of electrogenic membrane proteins such as transporters, ion pumps or ion channels. For the analysis, electric, electrophysiological and spectroscopic biophysical methods are needed. In more detail, the proteins of interest are analyzed in their native physiological microenvironment by using methods such as voltage-clamp, patch-clamp and voltage-clamp fluorometry. In some cases, purified proteins reconstituted in artificial membranes are analyzed using various electric biophysical methods such as black lipid membranes and solid supported membranes. The two types of lipid layers build a model lipid bilayer, as they are assembled in vitro. Many modeled lipid bilayers consist of synthetic or natural lipids. A simple model system is formed by a single pure synthetic lipid. More sophisticated and hence more physiological relevant models are those made of mixtures of several synthetic or natural lipids. Different types of model bilayers exist. The first system consisted of a black lipid membrane, termed painted bilayer, which was used to electrical characterization the bilayers. It is composed of a thin layer of hydrophobic material such as Teflon having a small hole with a diameter of a few micrometers. The area around the hole was painted with a solution of lipids, which are in a hydrophobic and highly viscous solvent such as decane. After drying, a salt solution was added from both sides, and thus, a lipid bilayer is spontaneously formed. A detailed view of the transport across the membrane requires the use of membrane-kinetic methods, which allow us to investigate the transport and conformational dynamics of proteins in situ and in vitro. Bacterial transporters have been analyzed in reconstituted systems.

In neuroscience and cell biology, the investigation of the function and structure of the retinal-binding protein channelrhodopsin 2 (ChR2) was of major interest. ChR2 was the first light-controlled channel and has become a brilliant tool in neurobiology, as its insertion into neuronal cells in vitro and in vivo in living animals allows cells to be stimulated simply by light. Thus, ChR2 can be used as a tool for the activation of electrically excitable cells in a non-invasive manner without vehicles, which must be absorbed by the cells and additionally leads to a known temporal and spatial resolution. Finally, this tool is widely used in basic neurobiological research and biomedical applications.

1.7 Cellular Biophysics

Cellular biophysics is a branch of biophysics that investigates cells from the perspective of a physicist. Physical methods are used to analyze cellular structures and functions and to develop models for cells based on physical and physical–chemical principles.

The focus of cellular biophysics is on the investigation of living cells of different types, different organisms of eukaryotic or prokaryotic origin or various tissue origins under healthy or diseased states. Multiple studies explore the effect of gene expression and protein levels on cell mechanics.

What do we currently understand about cellular biophysics? In contemplation of answering this question, we need to focus on principles, methods and certain prospects. As it is part of normal physiological conditions cells generate, transmit and can withstand mechanical forces of their microenvironment. In fact, the cells are active soft matter and are able to sense a mechanical stimulation through the activation of signal transduction pathways after a mechanosensory process activation. More precisely, cells respond to physical cues as they restructure their cytoskeletal architecture and adapt their generation and transmission of forces to the environmental needs. Genetic mutations induced by UV light or pathogens, such as bacteria or viruses, can induce alterations within the cytoskeleton of cells and hence alter the overall cell mechanics such as viscoelasticity, adhesiveness and contractile forces. All these alterations may change the cell's behavior and function pronouncedly. In turn, alterations the cell's microenvironmental mechanical properties can change of the cell's cytoskeletal structure similarly and subsequently alter the entire cellular mechanics. Hence, transformations of cells are a hallmark and a symptom of certain pathologies such as cancer injuries or inflammation.

There exists numerous novel experimental methods and theoretical models that have been adapted from the broad field of soft matter physics and mechanical engineering to biological living material such as cells, cell clusters and tissues. Theoretical physical models are frequently employed to characterize the cellular mechanical properties quantitatively. In addition, these physical models help to predict their alterations of cellular mechanical properties under distinct conditions. The interdisciplinary research in this special field is still growing, which is necessary to study molecular biological aspects from different angles. Together with cell mechanics, molecular and cellular biology, cellular biophysics involves approaches and physical principles in order to characterize cell mechanics and hence provides a fundamental understanding of cellular mechanical properties and processes regulating cellular development, physiology and disease, such as cancer.

More precisely, besides genetic and biochemical alterations, altered cellular morphology, structural and mechanical changes may account for the invasiveness of cancer cells and subsequently lead to malignant progression of cancer such as the formation of metastases. It is now known that together with biochemical signal transduction processes, mechanical confinements or properties regulate the linear propagation of steps within the entire process of cancer metastasis including critical steps such as cancer cell migration initiation, and the spread of cancer cells from the primary tumor mass, their propagation through blood and lymph vessels and their transmigration out of the vessels endothelial cell lining into the targeted tissue for the formation of secondary tumors. Thus, it is highly important to investigate the mechanics of these steps of the metastatic cascade in cell culture systems or in animal models. The genetic and biological details of cancer have been revealed well, whereas the progress on models investigating both the biological and the mechanical parts together is rarely reported. It might be due to the lack of suitable and realistic models and biophysical measurement methods. Indeed, novel in vitro assays for studying cell and matrix mechanics and cell biological parameter simultaneously need to emerge.

1.7.1 Subdiscipline: Membrane Cellular Biophysics

Membrane biophysics focuses on the fundamental features of biological membranes and the individual function of membrane proteins. Cellular membranes cover the entire cellular surface and hold the components of the cell together. Inside the cells, the membranes comprise the compartments of the cell, termed organelles, such as the nucleus, mitochondria, endoplasmic reticulum, Golgi apparatus and lysosomes. The biological membranes are composed of a lipid bilayer and membrane proteins embedded in the lipid bilayer, which acts as a barrier for ions and polar substances. The membrane barrier leads to the formation of ion gradients and electric voltages, termed membrane potentials. Therefore, membrane proteins are inevitable for the specific transport of selected proteins and substances through the membrane. In addition, the cells interact with other cells, sense their microenvironment and recognize signals, which are also performed by the membrane receptors, acting as sensors and receptors with special ligands. These signals can come from outside or inside the cell and activate the membrane receptors, which react by transducing the signal across the membrane. Moreover, membrane proteins play a central role in biological energy conversion such as photosynthesis in chloroplasts and respiration in mitochondria enabling the transport of electrons and protons over the membrane, which is the initial step of energy conversion. The membrane potentials and ion gradients are used to synthesize adenosine-5′-triphosphate (ATP), absorb nutrients, secrete waste and facilitate the flagella motor movement. In addition, even some membrane embedded enzymes can alter their hydrophobic substrates or produce hydrophobic products.

The functions of the membrane proteins are in summary, firstly the passage and transport across the membrane such as passive (through channels, pores or permeases) and primary (ATP-driven) or secondary active (downhill-flow of an ion gradient) transport systems, secondly, signal sensing and transduction such as G-protein-coupled receptors, thirdly, biological energy conversion such as in mitochondria or chloroplasts and fourthly, enzymes with hydrophobic substrates or products. The function of a membrane protein such as a channel is facilitated by different mechanisms, as a channel may work by opening upon ligand binding or the ligand binding may close the channel.

1.7.2 Subdiscipline: Nuclear Cellular Biophysics

Besides the cytoskeleton of the cell, the nucleus has become a focus of biophysical research, termed nuclear biophysics. The cell nucleus encloses a large number of membrane-less bodies that are crucial for the spatiotemporal regulation of gene expression.

The knowledge about the structure and function of the nucleus seems to be essential for deciphering of the computational logic encoded in the genome on top of the sole genetic code. Hence, it is obvious that there seems to be a connection between the 3D structure of the nucleus and the nuclear biological function. The hypothesis of a strongly non-uniform spatial distribution of molecules within the nucleus does not seem to be not appropriate, however, since these variations may have a strong impact on the molecular interactions governing gene expression.

It is rather more likely that the concentration variations within the nucleus are required to adapt reaction rates and regulate the diffusive transport. Diffusion is a stochastic, non-directional process, and hence, the diffusive transport rates are not dependent on the distance (as with directional transport); however, they depend on the distance-squared, which takes much more time to reach a targeted region. In conclusion, the diffusive transport seems to be a rate-limiting step for nuclear processes, especially of those in which a large number of binding partners are involved. The process of "active diffusion," which means driven by imbalance (non-equilibrium) and ATP-dependent dynamics, can partially solve this problem, since the dynamics in the nucleus are accelerated (Weber et al. 2012; Brangwynne et al. 2008). The 3D architecture of the genome is very important for the regulation of the spatial proximity of co-regulated genes and enhancer elements, and subsequently for the timely expression of genes (Nicodemi and Pombo 2014; Branco and Pombo 2006). Finally, the establishment of efficient transport and reaction rates still remains to be identified or increased within the densely crowded nature of the nucleoplasm containing the nucleoskeletal scaffold (Matsuda et al. 2014). A special role is fulfilled by nuclear bodies as they are not membrane bound and hence pattern the concentration of molecules in the nucleoplasm. These structures can locally elevate the concentrations of molecules involved in the remodeling of chromatin, the initiation of gene transcription and the processing of RNA. Moreover, these structures seem to function as biophysical switches, as they act due to local signals to assemble only above a distinct threshold concentration. However, revealing the biophysical principles underlying their assembly and properties may help to enlighten still unknown aspects of gene regulation inside the nucleus.

1.8 Cellular Mechanics

Cellular mechanics (also referred to as cell mechanics) is a subdiscipline of biophysics. It focuses on the mechanical properties and behavior of living cells and how they are associated with cell functions such as adhesion, migration or invasion (Moeendarbary and Harris 2014). Cellular mechanics has some overlaps with cellular biophysics as it encompasses aspects of cell biophysics. In addition, cellular mechanics covers biomechanics, classical rheology, soft matter physics, mechanobiology and cell biology. Scientists working in cellular mechanics are trying to uncover the mechanics and dynamics of assemblies and structures such as plasma membranes, cytoskeleton, organelles and cytoplasm, and how these compartments interact to contribute to the emerging properties of the entire cell (Fletcher and Mullins 2010).

Biomechanics research has become a focus of biophysics because the mechanical behaviors and properties of cells, cell clusters and tissues can be either a direct consequence or a regulating factor of biological function and overall cellular architecture (Fig. 1.7) (Maloney et al. 2010; Mierke 2011; Mierke et al. 2008a, b, 2010, 2017; Sun et al. 2012a). However, even both functions of the cell mechanics can be possible. Therefore, the basic goal of current cell mechanics research is to combine experimental, theoretical and computational approaches (simulations) for a realistic picture of cell mechanics behavior that can open up new perspectives on the role of mechanics in diseases such as cancer and acute or chronic inflammation (Moulding et al. 2012; Mierke et al. 2011a; Park et al. 2010).

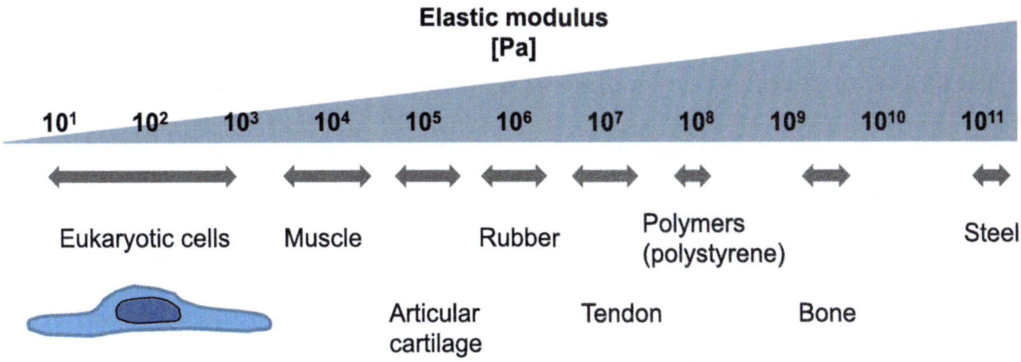

Fig. 1.7 Force scales of biological matter cover a wide range of forces

There is, however, one simple question that can be asked: What do we understand by cellular mechanics? In contemplation of answering this question, we need to focus on principles, methods and certain prospects. As it is part of normal physiological conditions, cells generate, transmit and can withstand mechanical forces of their microenvironment. In detail, the cells are active soft matter which is able to sense a mechanical stimulation through the activation of signal transduction pathways after a mechanosensory process activation. Moreover, cells respond to physical cues as they restructure their cytoskeletal architecture and adapt their generation and transmission of forces to the environmental needs. Genetic mutations and pathogens such as bacteria or viruses can induce alterations within the cytoskeleton of cells and hence alter the overall cellular mechanical, such as elasticity, viscosity and adhesiveness. All these alterations can strongly change cell behavior and function. Alterations in the microenvironmental mechanical properties of cells can also interfere with their cytoskeletal structure and consequently alter the entire cellular mechanics. Hence, transformations of cells are a hallmark and a symptom of certain pathologics such as cancer injuries or inflammation. There exist numerous experimental methods and theoretical models that have been adapted from the broad field of soft matter physics and mechanical engineering to biological living material such as cells, cell clusters and tissues.

Interdisciplinary research in this field continues to grow, and it is still necessary to investigate molecular biological aspects together with cell mechanics in order to finally gain a fundamentally new understanding of cell mechanics and its importance for physiological processes such as development and pathological processes including cancer. Besides genetic and biochemical alterations, altered cellular morphology, structural and mechanical changes may account for the invasiveness of cancer cells and subsequently lead to malignant progression of cancer such as the formation of metastases. It is now known that together with biochemical signal transduction processes, mechanical confinements or properties regulate the linear propagation of steps within the entire process of cancer metastasis including critical steps such as cancer cell migration initiation, and the spread of cancer cells from the primary tumor mass, their propagation through blood and lymph vessels and their transmigration out of the vessels endothelial cell lining into the targeted tissue for the formation of secondary tumors. Thus, it is highly important to investigate the mechanics of these steps of the metastatic cascade in cell culture systems or in animal models. The genetic and biological details of cancer have been revealed well, whereas the progress on models investigating both the biological and the mechanical parts together is rarely reported. It might be due to the lack of suitable and realistic models and biophysical measurement methods. Indeed, novel in vitro assays for studying organelle, cell and matrix mechanics and other cell biological parameter, such as protein expression or intracellular localization, simultaneously need to emerge.

1.9 Biophysical Bioengineering of the Cellular Microenvironment

Physical factors of the cells and the local cellular microenvironment, such as cell shape and geometry, matrix mechanics, external mechanical forces and topographical features of the extracellular matrix on the nanoscale, can all have strong influence on the fate of cell, such as stem cells. These physical factors can influence cells and the microenvironment. In particular, somatic cells can become reprogrammed into pluripotent stem cells by cell shape changes based on environmental cues (Downing et al. 2013). Stem cells sense and react to these intrinsic biophysical signals through cell–matrix receptor-based adhesions, such as integrins, maintain the balance between intracellular cytoskeletal contractile forces and deformation-resistant forces induced by the extracellular matrix. These mechanotransduction processes may be linked to several other growth factor-based signaling pathways for regulating the fate of stem cells. Various bioengineering tools, including micro- and nanoscale devices, have been successfully established to engineer the physical cues of the cellular microenvironment for specific cells, such as stem cells, and these techniques and devices have emerged as powerful tools for the identification of the extrinsic physical factors and their downstream intracellular signaling pathways regulating the functions of stem cells (Sun et al. 2012b).

Central questions still arise about the molecular mechanisms with which stem cells maintain their self-renewability and facilitate their differentiation. These questions must be answered for efficient functional tissue engineering and regenerative medicine. Major efforts have been concentrated on the biochemical components and soluble factors of the microenvironment of stem cells, which are crucial for self-renewal and differentiation of stem cells. However, it has been revealed that stem cells are strongly affected by co-existing insoluble adhesive, mechanical and topological cues of the dynamic stem cell niche. Insoluble biophysical signals, such as cell shape and geometry, external forces and matrix mechanics and nanoscale topography, will induce intracellular programs controlling stem cell fate, possibly by the integrin-based focal adhesion signal and the force equilibrium across the mechanical continuum of the extracellular matrix, integrins and cytoskeleton.

Since the molecular mechanisms of stem cells or similarly specific cancer cells to sense and respond to various biophysical signals are not yet known, it should be kept in mind that they can be cell specific and hence exploit distinct mechanisms that can act together. The predominant effect of specific biophysical signals on the function of stem cells relies on altered experimental settings, so that the fate of stem cells is driven by the intricate interactions and interdependences between soluble factors and insoluble biophysical signals within their local cellular environment. How can these microenvironments be engineered to reveal reproduceable results? How can interdependencies between these factors be separated?

In the future, it needs to borne in mind that the development of tissues from stem cells in vivo is a long-term process in which dynamic alterations of both chemical and physical environment of the cells occur in abundance. It is currently a great challenge how we can create in vitro microenvironments for stem cells to simulate the dynamic character and sophistication of the in vivo stem cell niche. Among these microenvironments are synthetic hydrogels, microcontact printing, microfluidic systems and micro/nanofabrication of 3D scaffolds. These tools extend over widely varying length scales ranging from molecular through cellular to organ levels, all of which have demonstrated extreme performance in detecting extrinsic physical factors and their independent effects on the fate of cells. Hence, dynamics and complex synthetic cellular environments, including molecular, structural, hydrodynamic and mechanical effects, can be envisioned in combination with high-throughput approaches.

Finally, there are at least five major issues that are interesting for further investigation.

1. The fate of stem cells or cancer cells can be strongly affected by physical signals in the nearby cellular environment.
2. A key regulatory factor for the cell fate is the shape of cells that is based mainly on cytoskeletal tension.
3. The behavior of stem cells or cancer cells can be guided by nanotopographical cues that affect the molecular arrangement, organization dynamics and adhesion signal transduction processes.
4. The extracellular mechanical forces and entire matrix mechanics interfere with the behavior of stem cells and cancer cells by disturbing the force balance between the mechanical continuum of the extracellular matrix-integrin-cytoskeleton connection and its regulation by the mechanical signals inside the individual cellular niche.
5. The extracellular mechanical forces and the overall force equilibrium via the connection of the extracellular matrix-integrin-cytoskeleton are mechanically transferred into the cytoplasm and subsequently into the cell nucleus to transmit signal molecules that are essential for the fate determination of cells, including stem cells or cancer cells, as made possible by integrin signaling and mechanosensitive ion channels.

1.10 "Experimental" Biophysics is Guided by Theoretical Biophysics

Life is based on the interactions of proteins, nucleic acids, lipids and other biomolecules. The major aim is to provide detailed and quantitative predictive models of biomolecular processes such as energy conversion, molecular transport, signal transduction and enzymatic catalysis. These models help to understand the overall functions of living organisms at the molecular or cellular. The knowledge of the physical and chemical mechanisms employed in biological systems results in the development of novel technologies and biophysical devices.

In more detail, computational and theoretical methods are used to explore the structure, stability, dynamics and molecular functions of single and complexed biomolecules. The computational and theoretical studies help to get more insights in the interpretation of increasingly complex experimental studies and improve the design and setup of future experiments. Theoretical biophysics combines fundamental physics, chemistry and biology to investigate biomolecular processes over a broad range. It may range from quantum mechanics to chemical kinetics, from atomistic models for physical processes and chemical reactions using molecular dynamics simulations. Finally, it can even include highly coarse-grained models of the non-equilibrium work performed by molecular machines and important for protein interactions.

1.11 Short Introduction into Biophysical Models Quantifying Cell Mechanics

Why Are Biological Cartoons and Models Needed?

Biological cartoons are needed to feature the parts of an observed phenomenon that seem to be essential. Thus, conceptual models are useful to explain the outcome of a yet not known experimental setup. Basically, this approach is not restricted to the field of cell mechanics and biophysics; it is more prominent in physics as theoretical concepts for very different phenomena are described, but can be transferred all other experimental disciplines similarly in contemplation of interpretation of the data. In furtherance of understanding biochemical reactions in cells, we need to use statistical mechanics and thermodynamics. Visual schematic drawings will help to explain the important features in biological and biomechanical processes including the interaction between different processes. This conceptual modeling is essential to reveal the underlying processes.

1.12 Why Are These Quantitative Models Usually Idealized?

Quantitative models are an essential part of all fields of physics and are now transferred to the field of cell mechanics and biophysics. In furtherance of being able to apply appropriate models, we need a tight interplay between theoretic and experimental physics. In the ideal case, the consequences of the model are explored by performing experiments. In turn, these novel experimental results have evoked the formulation and further development of new theories. A prominent example is the famous and fundamental model of a simple spring (synonymously termed harmonic oscillator), which has been used multiple times even in various contexts. The concept of the spring is based on an old physical idea, termed the Hook's law that has been employed initially to describe solid matter mechanics and has been combined with the mathematical idea of a Taylor series.

1.12.1 What is the Conceptual Framework?

All springs have the fundamental mathematical idea in common that the potential energy for nearly every system that undergoes small displacements from its equilibrium state can be appropriately approximated by the employment of a quadratic function for the displacement. Mathematically, it can be expressed as:

$$E\,(\text{energy}) = 1/2\,kx^2. \tag{1.1}$$

The equation states that the potential energy increases with the square of the displacement x, when the system is pushed away from its equilibrium. The stiffness, k, measures what it takes drive the system out of the equilibrium and hence reflects the spring's material mechanics.

An alternative way is to characterize the springs as a restoring force that is proportional to the displacement of the spring from its position at the equilibrium state. Using mathematics, this idea can be presented in an equation:

$$F = -kx. \tag{1.2}$$

where F represents the restoring force, x is the spring's displacement from the spatial equilibrium and k is termed the spring constant representing the stiffness. The restoring force is directed to $x = 0$, and therefore, there is a minus sign in (1.2), which indicates this and denotes the position of the equilibrium. This result is the well-known Hooke's law. When it is stated in this way, the physical model delivers the impression of an abstract example of masses located on frictionless substrates with pulleys and springs. However, there are naturally many different examples of when this precise model can be applied. From a technological point of view, many of the most important single-molecule techniques for investigating macromolecules and their compositions evoke this description. For example, both common biophysical techniques, such as optical tweezers and the atomic force microscopy, can be precisely and unambiguously mapped to spring problems. However, there are many other problems that can be handled as a simple spring problem. In specific detail, simple spring models have been employed to describe the bending of DNA or the membrane fluctuations at the cell surface. All these processes are mechanical processes, and it is quite easy to see how the concept of a spring can work on them.

The real strength of the simple spring model, however, is that it can be applied to non-mechanical problems. For example, when we talk about biochemical reactions in a cell, we will envision how molecules move in an "energy landscape" where they cluster in "potential wells." The mathematics of simple springs can be used to clarify our understanding of the rates that guide these biochemical reactions. Moreover, the ideas of harmonic oscillators and potential wells enlighten the investigation of conformational changes of proteins. Scientific research at the interdisciplinary interface between physics and biology requires that scientists who are educated in one discipline acquire a working knowledge of the other discipline. In the future, interdisciplinary research between physics and biology will contribute to creating a functional basis for selected key physical concepts that are directly relevant to biological problems. In addition, the identification of specific biological phenomena may help to answer important questions in the field of biology by employing classical physical equation representing each an individual specific single class of physical models and specific physical foundations, such as simple harmonic oscillator, random walks, entropy, elastic theory of one-dimensional tracks or two-dimensional sheet-like environments, diffusion, Newtonian fluid model and Navier–Stokes equation.

1.12.2 Unifying Ideas of Biological Processes

Since the field of biology is highly diverse, there are so many different species, cell types, structures, genes and proteins including their mutations that it would be too time consuming to elaborate each of them one by one. Hence, simplifications need to be made in order to reveal the universal mechanism of specific parts in biology.

1.12.3 What Are the Unifying Ideas of Biology?

An interesting aspect of scientific research is the connection of at the first glance unrelated phenomena. Only a few examples from the physical sciences are better able to illustrate this point than the surprising union of the ancient phenomenon of magnetism, known from the iodine stones and the beginnings of compasses for navigation to various manifestations of electricity, such as the lightning that occurs during summer thunderstorms, many and varied phenomena of light including rainbows or the apparent curvature of a spoon in a glass of water. All of these phenomena were combined by the modern electromagnetism theory.

The presented toolbox of basic physical ideas concentrated on the modalities by which distinct key ideas from physics have contributed to the compilation not only of descriptive but also of predictive observations. These unifying concepts are not restricted to physical research. Moreover, biology has even far-reaching ideas. There are, on the one hand, a handful of basic physical models that are usually sufficient to create a rigorous framework for the interpretation quantitative biological datasets of different origins, and, on the other hand, there are also a handful of basic biological ideas that an encompassing basis for everything we think of in biology. In fact, many former authors have opened up various intriguing viewpoints on what might be termed the "great ideas of biology".

Feynman had chosen in this lecture on physics that one statement has the most information for future generations: It is the idea that "all things are made of atoms." An important biological idea is the statement that "all living organisms are related through descent from a common ancestor through the process of evolution by natural selection." Another important and clear statement has been made by Dobzhansky that is still timely and true: "Nothing in biology makes sense except in the light of evolution." In fact, there is a strong argument that evolution is the greatest idea of biology and has left its inerasable stamp on almost everything. In detail, as Darwin and Wallace tried to find an understanding of the evolutionary process, Mendel worked tirelessly to establish the early ideas about the underlying mechanisms of inheritance, leading to discoveries that gave rise to another of the uniquely major ideas of biology, genetics. A major breakthrough in the history of biology was the connection of Mendel's abstract hereditary particles with the molecules of the cell, whereby the successful realization was the realization that DNA is the vehicle of the genetic information.

There is a key concept of physics transferred to the field of biology where all is about cell and their interactions. More precisely, a unifying concept has been applied to cells.

Stating that they are all the same, and subsequently, the cell theory has been formulated. The cell theory states that the insights of all living organisms are composed of single units, such as cells. Hence, individual cells represent the so-called quantum unit of living matter. The biochemical processes, such as common metabolic reactions in cells of several organisms, exhibit a remarkable unity or the molecular machinery that imposes our common genetic heritage is composed of a few common macromolecular compounds, such as polymerases and the ribosome. In this book, the key broad biological foundations, such as the theory of evolution, cell theory, genetics and nature of inheritance and the unity of biochemistry will be addressed.

1.12.4 Why Do We Need Mathematics?

Cells are able to sense concentrations in their microenvironment and can even decide to perform a distinct action based on similar physical measurements. There are questions with many outcomes, such as whether a receptor will be bound by a ligand or not, where the probability theory can be helpful. In detail, four key distributions are mostly employed: firstly, the binomial distribution, secondly the Gaussian distribution, thirdly the Poisson distribution and fourthly the exponential distribution. Each of these distributions will contribute to sharpening the manner in which we think about many diverse biological issues. However, we will require more than simply the powerful uncertainty calculation of modern probability and statistics theory. Commonly, we seek to quantity of some specific parameters in cells, such as the number of microtubules, that varies in time. With a calculus toolkit containing differential equations, we can use one of the preferred tools that deals with the superlative mathematics, i.e., how to identify the largest or smallest values of a function.

The modeling used in physical sciences can be simply transferred to biology. The interesting outcomes of the early years of electron microscopy in biology were simple, such as that, through the observation of the physical region occupied by the DNA molecule from a ruptured bacterial virus (termed bacteriophage), the genomic size can be estimated in a simple manner, which means the length in base pairs of the DNA of bacteriophage.

1.12.5 What Role Play Numbers in Cell Mechanics and Biophysics?

Why is it necessary to count the copies of a molecule regulating the gene expression of a gene of interest in a bacterium such as *E. coli*? What is the value of estimating the rate of the polymerization of actin at the front edge of a moving cell? What is the value of knowing the time it takes a specific protein to diffuse one millimeter? Firstly, simple estimates can help to test the reality of whether our impressions of how a system works are reasonable or not. Secondly, a feel for the numbers in a system can give us an idea of what kinds of physical limitations are out there. If, for instance, we are familiar with the size of a bacterium and its swimming performance, we can determine an indicator (termed Reynolds number) that informs us about the significance of acceleration in the dynamic phenomenon of swimming. In the same way, knowledge of the number of specific DNA-binding proteins (termed transcription factors), which govern the expression of a distinct gene, can provide information on the significance of fluctuations in the process of gene expression. Oscillatory behavior can be found in many biological processes, such as in the migration of cells, the circadian rhythm and the cell cycle process. Using mathematical models of regulatory biology, we will observe that minimal oscillators can be built from genetic networks with only two proteins, an activator and a repressor. Many of the possible parameters we could choose for protein production and degradation rates show no oscillatory behavior at all, but rather stable solutions in which the protein concentrations relax to stationary values after some time. Therefore, even a clear picture of the cabling of a network cannot tell us importantly what role this network has in the everyday life of a cell and requires a quantitative study of how this network interacts with its control parameters.

In conclusion, the focus of this cellular mechanics and biophysics book is on the basic components and processes required to understand fundamental cellular functions, which are subject to biophysical, bioengineering and cellular mechanical-driven research.

References

R.M. Addoms, Toxicity as evidenced by changes in the protoplasmic structure of root hairs of wheat. Am. J. Bot. **14**, 147–165 (1927)

J. Alcaraz, L. Buscemi, M. Grabulosa, X. Trepat, B. Fabry, R. Farre, D. Navajas, Microrheology of human lung epithelial cells measured by atomic force microscopy. Biophys. J. **84**, 2071–2079 (2003)

C.A. Angerer, The effect of electric current on the relative viscosity of sea-urchin egg protoplasm. Bio Bull. **77**, 399–406 (1939)

G. Bao, S. Suresh, Cell and molecular mechanics of biological materials. Nat. Mater. **2**, 715–725 (2003)

W.M. Bayliss, The properties of colloidal systems. IV Reversible gelation in living protoplasm. Proc. R. Soc. London B. **91**, 196–201 (1920)

J. Bernstein, Untersuchungen zur Thermodynamik der bio-elektrischen Ströme. Arch f. Physiologie **92**, 521–562 (1902)

E.C. Bingham, *Fluidity and Plasticity*, 1st edn. (McGraw- Hill, New York, 1933)

G. Binnig, C.F. Quate, C. Gerber, Atomic force microscope. Phys. Rev. Lett. **56**, 930–933 (1986)

I.B. Bischofs, U.S. Schwarz, Cell organization in soft media due to active mechanosensing. Proc. Natl. Acad. Sci. U. S. A. **100**, 9274–9279 (2003)

I.B. Bischofs, U.S. Schwarz, Effect of Poisson ratio on cellular structure formation. Phys. Rev. Lett. **95**, 068102 (2005)

I.B. Bischofs, U.S. Schwarz, collective effects in cellular structure formation mediated by compliant environments: a Monte Carlo study. Acta Biomater. **2**, 253–265 (2006)

D.H. Boal, *Mechanics of the Cell*, 2nd edn. (Cambridge University Press, Cambridge, UK, 2002)

M.R. Branco, A. Pombo, Intermingling of chromosome territories in interphase suggests role in translocations and transcription-dependent associations. PLoS Biol. **4**, e138 (2006)

C.P. Brangwynne, G.H. Koenderink, F.C. MacKintosh, D.A. Weitz, Cytoplasmic diffusion: molecular motors mix it up. J Cell Biol. **183**, 583–587 (2008)

D. Bray, Axonal growth in response to experimentally applied mechanical tension. Dev. Biol. **102**, 379–389 (1984)

J.G. Carlson, Protoplasmic viscosity changes in different regions of the grasshopper neuroblast during mitosis. Bio Bull. **90**, 109–121 (1946)

R. Chambers, Microdissection studies on the germ cell. Science **41**, 290–293 (1915)

R. Chambers, H.B. Fell, Micro-operations on cells in tissue cultures. Proc. R. Soc. London B **109**, 380–403 (1931)

G.T. Charras, M.A. Horton, Single cell mechanotransduction and its modulation analyzed by atomic force microscope indentation. Biophys. J. **82**, 2970–2981 (2002)

G.T. Charras, P.P. Lehenkari, M.A. Horton, Atomic force microscopy can be used to mechanically stimulate osteoblasts and evaluate cellular strain distributions. Ultramicroscopy **86**, 85–95 (2001)

G.T. Charras, J.C. Yarrow, M.A. Horton, L. Mahadevan, T.J. Mitchison, Non-equilibration of hydrostatic pressure in blebbing cells. Nature **435**, 365–369 (2005)

C. Chaubaroux, F. Perrin-Schmitt, B. Senger, L. Vidal, J.C. Voegel, P. Schaaf, Y. Haikel, F. Boulmedais, P. Lavalle, J. Hemmerlé, Cell alignment driven by mechanically induced collagen fiber alignment in collagen/alginate coatings. Tissue Eng. Part C Methods **21**, 881–888 (2015)

A.E. Cohen, Optogenetics: turning the microscope on its head. Biophys. J. **110**, 997–1003 (2016)
P.F. Cranefield, The organic physics of 1847 and the biophysics of today. J. Hist. Med. Allied Sci. **12**, 407–423 (1957)
J. Dai, M.P. Sheetz, Cell membrane mechanics. Methods Cell. Biol. **55**, 157–171 (1998)
B. Daily, E.L. Elson, G.I. Zahalak, Cell poking. Determination. of the elastic area compressibility modulus of the erythrocyte membrane. Biophys. J.**45**, 671–82 (1984)
R. De, A. Zemel, S.A. Safran, Dynamics of cell orientation. Nat. Phys. **3**, 655–659 (2007)
J. Domke, W.J. Parak, M. George, H.E. Gaub, M. Radmacher, Mapping the mechanical pulse of single cardiomyocytes with the atomic force microscope. Eur. J. Biophys. **28**, 179–186 (1999)
T.L. Downing, J. Soto, C. Morez, T. Houssin, A. Fritz, F. Yuan, J. Chu, S. Patel, D.V. Schaffer, S. Li, Biophysical regulation of epigenetic state and cell reprogramming. Nat. Mater. **12**, 1154–1162 (2013)
R.J. Dyson, J.E. Green, J.P. Whiteley, H.M. Byrne, An investigation of the influence of extracellular matrix anisotropy and cell-matrix interactions on tissue architecture. J. Math. Biol. **72**, 1775–1809 (2016)
M. Eastwood, V.C. Mudera, D.A. McGrouther, R.A. Brown, Effect of precise mechanical loading on fibroblast populated collagen lattices: morphological changes. Cell Motil. Cytoskeleton. **40**, 13–21 (1998)
A.J. Engler, S. Sen, H.L. Sweeney, D.E. Discher, Matrix elasticity directs stem cell lineage specification. Cell **126**, 677–689 (2006)
E.A. Evans, R.M. Hochmuth, Membrane viscoelasticity. Biophys. J. **16**, 1–11 (1976)
E. Evans, A. Yeung, Apparent viscosity and cortical tension of blood granulocytes determined by micropipet aspiration. Biophys. J. **56**, 151–160 (1989)
S. Even-Ram, V. Artym, K.M. Yamada, Matrix control of stem cell fate. Cell **126**, 645–647 (2006)
A.J. Ewart, On the physics and physiology of the protoplasmic streaming in plants. Proc. R. Soc. London **69**, 466–470 (1901)
D.A. Fletcher, R.D. Mullins, Cell mechanics and the cytoskeleton. Nature **463**(7280), 485–492 (2010)
A. Forbes, C. Thacher, Changes in the protoplasm of Nereis eggs induced by ß-radiation. Am. J. Physiol. **74**, 567–578 (1925)
L. Formigli, E. Meacci, C. Sassoli, F. Chellini, R. Giannini, F. Quercioli, B. Tiribilli, R. Squecco, P. Bruni, F. Francini, S. Zecchi-Orlandini, Sphingosine 1-phosphate induces cytoskeletal reorganization in C2C12 myoblasts: physiological relevance for stress fibres in the modulation of ion current through stretch-activated channels. J Cell. Sci. **118**, 1161–1171 (2005)
A. Greely, Experiments on the physical structure of the protoplasm of Paramaecium and its relation to the reactions of the organism tothermal, chemical and electrical stimuli. Biol. Bull. **7**, 3–32 (1904)
J. Guck, R. Ananthakrishnan, H. Mahmood, T.J. Moon, C.C. Cunningham, J. Kas, The optical stretcher: a novel laser tool to micromanipulate cells. Biophys. J. **81**, 767–784 (2001)
J. Guck, S. Schinkinger, B. Lincoln, F. Wottawah, S. Ebert, M. Romeyke, D. Lenz, H.M. Erickson, R. Ananthakrishnan, D. Mitchell, J. Kas, S. Ulvick, C. Bilby, Optical deformability as an inherent cell marker for testing malignant transformation and metastatic competence. Biophys. J. **88**, 3689–3698 (2005)
B.J. Haupt, A.E. Pelling, M.A. Horton, Integrated confocal and scanning probe microscopy for biomedical research. Sci. World J. **6**, 1609–1618 (2006)
S.R. Heidemann, P. Lamoureaux, R.E. Buxbaum, Opposing views on tensegrity as a structural framework for understanding cell mechanics. J. Appl. Physiol. **89**, 1670–1678 (2000)
L.V. Heilbrunn, The physical effect of anesthetics upon living protoplasm. Bio Bull. **39**, 307–315 (1920)
L.V. Heilbrunn, An experimental study of cell division. I the physical conditions which determine the appearance of the spindle in sea-urchin eggs. J. Exp. Zool. **30**, 211–237 (1920)
L.V. Heilbrunn, The surface tension theory of membrane elevation. Bio Bull. **46**, 277–280 (1924)
L.V. Heilbrunn, The electrical charges of living cells. Science **1574**, 236–237 (1925)
L.V. Heilbrunn, The action of ether on protoplasm. Bio Bull **49**, 461–476 (1925)
L.V. Heilbrunn, The physical structure of the protoplasm of sea-urchin eggs. Am. Nat. **60**, 143–156 (1926)
L.V. Heilbrunn, The viscosity of the protoplasm. Q. Rev. Biol. **2**, 230–248 (1927)
L.V. Heilbrunn, W.L. Wilson, A rational approach to the problem of cancer chemotherapy. Bio Bull. **113**, 388–396 (1957)
L.V. Heilbrunn, A.B. Chaet, A. Dunn, W.L. Wilson, Antimitotic substances from ovaries. Bio Bull. **106**, 158–168 (1954)
L.V. Heilbrunn, W.L. Wilson, Protoplasmic viscosity changes during mitosis in the egg of chaetopterus. Bio Bull. **95**, 57–68 (1948)
L.V. Heilbrunn, W.L. Wilson, T.R. Tosteson, E. Davidson, R.J. Rutman, The antimitotic and carcinostatic action of ovarian extracts. Bio Bull. **113**, 129–134 (1957)
E.H. Herrick, Mechanism of movement of epidermis, especially its melanophores, in wound healing, and behavior of skin grafts in frog tadpoles. Bio Bull.**63**, 271–286 (1932)
A.V. Hill, The possible effects of the aggregation of the molecules of hemoglobin on its dissociation curves. J. Physiol. **40**, iv–vii (1910)

A.V. Hill, Why biophysics? Science **124**, 1233–1237 (1956)
R.M. Hochmuth, Micropipette aspiration of living cells. J. Biomech. **33**, 15–22 (2000)
R. Horwitz, Cellular biophysics. Biophys. J. **110**, 993–996 (2016)
S. Hu, L. Eberhard, J. Chen, J.C. Love, J.P. Butler, J.J. Fredberg, G.M. Whitesides, N. Wang, Mechanical anisotropy of adherent cells probed by a three-dimensional magnetic twisting device. Am. J. Physiol. Cell Physiol. **287**, C1184–C1191 (2004)
H. Huang, R.D. Kamm, R.T. Lee, Cell mechanics and mechanotransduction: pathways, probes, and physiology. Am. J. Physiol. Cell Physiol. **287**, C1–C11 (2004)
D.E. Ingber, Opposing views on tensegrity as a structural framework for understanding cell mechanics. J. Appl. Physiol. **89**, 1663–1670 (2000)
D.E. Ingber, Mechanical control of tissue growth: function follows form. Proc. Natl. Acad. Sci. U. S. A. **102**, 11571–11572 (2005)
D.E. Ingber, Mechanical control of tissue morphogenesis during embryological development. Int. J. Dev. Biol. **50**, 255–266 (2006)
M.H. Jacobs, The effect of carbon dioxide on the consistency of protoplasm. Bio Bull. **42**, 14–30 (1922)
C. Jurado, J.R. Haserick, J. Lee, Slipping or gripping? Fluorescent speckle microscopy in fish keratocytes reveals two different mechanisms for generating a retrograde flow of actin. Mol. Biol. Cell **16**, 507–518 (2005)
K.E. Kasza, A.C. Rowat, J. Liu, T.E. Angelini, C.P. Brangwynne, G.H. Koenderink, D.A. Weitz, The cell as a material. Curr. Opin. Cell Biol. **19**, 101–107 (2007)
G.L. Kite, Studies on the physical properties of protoplasm. Am. J. Physiol. **32**, 146–164 (1913)
R.J. Klebe, H. Caldwell, S. Milam, Cells transmit spatial information by orienting collagen fibers. Matrix **9**, 451–458 (1989)
W.A. Lam, M.J. Rosenbluth, D.A. Fletcher, Chemotherapy exposure increases leukemia cell stiffness. Blood **109**, 3505–3508 (2007)
J. Lammerding, K.N. Dahl, D.E. Discher, R.D. Kamm, Nuclear mechanics and methods. Methods Cell Biol. **83**, 269–294 (2007)
J. Lammerding, L.G. Fong, J.Y. Ji, K. Reue, C.L. Stewart, S.G. Young, R.T. Lee, Lamins A and C but not lamin B1 regulate nuclear mechanics. J. Biol. Chem.**281**, 25768–25780
J. Lammerding, P.C. Schulze, T. Takahashi, S. Kozlov, T. Sullivan, R.D. Kamm, C.L. Stewart, R.T. Lee, Lamin A/C deficiency causes defective nuclear mechanics and mechanotransduction. J. Clin. Invest. **113**, 370–378 (2004)
E.J. Lee, J.W. Holmes, K.D. Costa, Remodeling of engineered tissue anisotropy in response to altered loading conditions. Ann. Biomed. Eng. **36**, 1322–1334 (2008)
T.P. Lele, J.E. Sero, B.D. Matthews, S. Kumar, S. Xia, M. Montoya- Zavala, T. Polte, D. Overby, N. Wang, D.E. Ingber, Tools to study cell mechanics and mechanotransduction. Methods Cell. Biol. **83**, 443–472 (2007)
W.W. Lepeschkin, The influence of narcotics, mechanical agents, and light upon the permeability of protoplasm. Am. J. Bot.**19**, 568–580 (1932)
S. Li, J.L. Guan, S. Chien, Biochemistry and biomechanics of cell motility. Annu. Rev. Biomed. Eng. **7**, 105–150 (2005)
C. Liu, S. Baek, J. Kim, E. Vasko, R. Pyne, C. Chan, Effect of static pre-stretch induced surface anisotropy on orientation of mesenchymal stem cells. Cell Mol. Bioeng. **7**, 106–121 (2014)
J.M. Maloney, D. Nikova, F. Lautenschläger, E. Clarke, R. Langer, J. Guck, K.J. VanVliet, Mesenchymal stem cell mechanics from the attached to the suspended state. Biophys. J. **99**, 2479–2487 (2010)
T.G. Mason, K. Ganesan, J.H. vanZanten, D. Wirtz, S.C. Kuo, Particle tracking microrheology of complex fluids. Phys. Rev. Lett. **79**, 3282–3285 (1997)
G. Massiera, K.M. Van Citters, P.L. Biancaniello, J. Crocker, Mechanics of single cells: rheology, time dependence and fluctuations. Biophys. J.**93**, 3703–3713 (2007)
H. Matsuda, G.G. Putzel, V. Backman, I. Szleifer, Macromolecular crowding as a regulator of gene transcription. Biophys. J. **106**, 1801–1810 (2014)
W.B. McConnaughey, N.O. Petersen, Cell poker: an apparatus for stress–strain measurements on living cells. Rev. Sci. Instrum.**51**, 575–580 (1980)
A.D. McCulloch, Systems biophysics: multiscale biophysical modeling of organ systems. Biophys. J. **110**(5), 1023–1027 (2016)
C.T. Mierke, Cancer cells regulate biomechanical properties of human microvascular endothelial cells. J. Biol. Chem. **286**, 40025–40037 (2011)
C.T. Mierke, Phagocytized beads reduce the $\alpha 5\beta 1$ integrin facilitated invasiveness of cancer cells by regulating cellular stiffness. Cell Biochem. Biophys. **66**, 599–622 (2013)
C.T. Mierke, The matrix environmental and cell mechanical properties regulate cell migration and contribute to the invasive phenotype of cancer cells. Rep. Prog. Phys. **82**(6), 064602 (2019)
C.T. Mierke, The role of the optical stretcher is crucial in the investigation of cell mechanics regulating cell adhesion and motility. Front. Cell Dev. Biol. **7**, 184 (2019)

C.T. Mierke, N. Bretz, P. Altevogt, Contractile forces contribute to increased GPI-anchored receptor CD24 facilitated cancer cell invasion. J. Biol. Chem. **286**, 34858–34871

C.T. Mierke, T. Fischer, S. Puder, T. Kunschmann, B. Soetje, W.H. Ziegler, Focal adhesion kinase activity is required for actomyosin contractility-based invasion of cells into dense 3D matrices. Sci. Rep. **7**, 42780 (2017)

C.T. Mierke, B. Frey, M. Fellner, M. Herrmann, B. Fabry, Integrin α5β1 facilitates cancer cell invasion through enhanced contractile forces. J. Cell Sci. **124**, 369–83 (2011)

C.T. Mierke, P. Kollmannsberger, D. Paranhos-Zitterbart, G. Diez, T.M. Koch, S. Marg, W.H. Ziegler, W.H. Goldmann, B. Fabry, Vinculin facilitates cell invasion into 3D collagen matrices. J. Biol. Chem. **285**, 13121–13130 (2010)

C.T. Mierke, P. Kollmannsberger, D.P. Zitterbart, J. Smith, B. Fabry, W.H. Goldmann, Mechano-coupling and regulation of contractility by the vinculin tail domain. Biophys. J. **94**(2), 661–670 (2008)

C.T. Mierke, D.P. Zitterbart, P. Kollmannsberger, C. Raupach, U. Schlotzer-Schrehardt, T.W. Goecke, J. Behrens, B. Fabry, Breakdown of the endothelial barrier function in tumor cell transmigration. Biophys. J. **94**, 2832–2846 (2008)

E. Moeendarbary, A.R. Harris, Cell mechanics: principles, practices, and prospects. WIREs Syst. Biol. Med. **6**, 371–388 (2014)

M.R.K. Mofrad, R.D. Kamm, *Cytoskeletal Mechanics: Models and Measurements*, 1st edn. (Cambridge University Press, Cambridge, 2006)

D.A. Moulding, E. Moeendarbary, L. Valon, J. Record, G.T. Charras, A.J. Thrasher, Excess F-actin mechanically impedes mitosis leading to cytokinesis failure in X-linked neutropenia by exceeding Aurora B kinase error correction capacity. Blood **120**, 3803–3811 (2012)

V.C. Mudera, R. Pleass, M. Eastwood, R. Tarnuzzer, G. Schultz, P. Khaw, D.A. McGrouther, R.A. Brown, Molecular responses of human dermal fibroblasts to dual cues: contact guidance and mechanical load. Cell Motil. Cytoskeleton **45**, 1–9 (2000)

M. Nicodemi, A. Pombo, Models of chromosome structure. Curr. Opin. Cell Biol. **28**, 90–95 (2014)

H.T. Northern, Alterations in the structural viscosity of protoplasm by colchicine and their relationship to C-mitosis and C-tumor formation. Am. J. Bot. **37**, 705–711 (1950)

I. Obataya, C. Nakamura, S. Han, N. Nakamura, J. Miyake, Nanoscale operation of a living cell using an atomic force microscope with a nanoneedle. Nano Lett. **5**, 27–30 (2005)

T. Ochalek, F.J. Nordt, K. Tullberg, M.M. Burger, Correlation between cell deformability and metastatic potential in B16-F1 melanoma cell variants. Cancer Res.**48**, 5124–5128 (1988)

C. Packard, The biological effects of short radiations. Q. Rev. Biol. **6**, 253–280 (1931)

J. Paget, Croonian lecture: on the cause of the rhythmic motion of the heart. Proc. R. Soc. London **8**, 473–488 (1857)

Y. Pang, X. Wang, D. Lee, H.P. Greisler, Dynamic quantitative visualization of single cell alignment and migration and matrix remodeling in 3-D collagen hydrogels under mechanical force. Biomaterials **32**, 3776–3783 (2011)

Y. Park, C.A. Best, K. Badizadegan, R.R. Dasari, M.S. Feld, T. Kuriabova, M.L. Henle, A.J. Levine, G. Popescu, Measurement of red blood cell mechanics during morphological changes. Proc. Natl. Acad. Sci. U. S. A. **107**, 6731–6736 (2010)

M.J. Paszek, N. Zahir, K.R. Johnson, J.N. Lakins, G.I. Rozenberg, A. Gefen, C.A. Reinhart-King, S.S. Margulies, M. Dembo, D. Boettiger, D.A. Hammer, V.M. Weaver, Tensional homeostasis and the malignant phenotype. Cancer Cell **8**, 241–254 (2005)

K. Pearson, *The Grammar of Science*, 2nd edn (Adam and Charles Black, London, 1900)

R.J. Pelham Jr, Y. Wang, High resolution detection of mechanical forces exerted by locomoting fibroblasts on the substrate. Mol. Biol. Cell. **10**, 935–945 (1999)

A.E. Pelling, D.W. Dawson, D.M. Carreon, J.J. Christiansen, R.R. Shen, M.A. Teitell, J.K. Gimzewski, Distinct contributions of microtubule subtypes to cell membrane shape and stability. Nanomedicine **3**, 43–52 (2007)

A.E. Pelling, F.S. Veraitch, C. Pui-Kei Chu, B.M. Nicholls, A.L. Hemsley, C. Mason, M.A. Horton, Mapping correlated membrane pulsations and fluctuations in human cells. J. Mol. Recognit. **20**, 467–475 (2007)

N.O. Petersen, W.B. McConnaughey, E.L. Elson, Dependence of locally measured cellular deformability on position on the cell, temperature, and cytochalasin B. Proc. Natl. Acad. Sci. U. S. A. **79**, 5327–5331

M. Prass, K. Jacobson, A. Mogilner, M. Radmacher, Direct measurement of the lamellipodial protrusive force in a migrating cell. J. Cell. Biol. **174**, 767–772 (2006)

M. Puig-De-Morales, M. Grabulosa, J. Alcaraz, J. Mullol, G.N. Maksym, J.J. Fredberg, D. Navajas, Measurement of cell microrheology by magnetic twisting cytometry with frequency domain demodulation. J. Appl. Physiol. **91**, 1152–1159 (2001)

M. Radmacher, Measuring the elastic properties of living cells by the atomic force microscope. Methods Cell. Biol. **68**, 67–90 (2002)

M. Radmacher, Studying the mechanics of cellular processes by atomic force microscopy. Methods Cell. Biol. **83**, 347–372 (2007)

J.W. Reinhardt, K.J. Gooch, Agent-based modeling traction force mediated compaction of cell-populated collagen gels using physically realistic fibril mechanics. J. Biomech. Eng. **136**, 021024 (2014)

J.W. Reinhardt, D.A. Krakauer, K.J. Gooch, Complex matrix remodeling and durotaxis can emerge from simple rules for cell-matrix interaction in agent-based models. J. Biomech. Eng. **135**, 71003 (2013)

E.G. Rens, R.M.H. Merks, Cell contractility facilitates alignment of cells and tissues to static uniaxial stretch. Biophys. J. **112**, 755–766 (2016)

C. Rotsch, M. Radmacher, Drug-induced changes of cytoskeletal structure and mechanics in fibroblasts: an atomic force microscopy study. Biophys. J. **78**, 520–535 (2000)

C. Rotsch, F. Braet, E. Wisse, M. Radmacher, AFM imaging and elasticity measurements on living rat liver macrophages. Cell Biol. Int. **21**, 685–696 (1997)

C. Rotsch, K. Jacobson, M. Radmacher, Dimensional and mechanical dynamics of active and stable edges in motile fibroblasts investigated by using atomic force microscopy. Proc. Natl. Acad. Sci. U. S. A. **96**, 921–926 (1999)

A.C. Rowat, J. Lammerding, J.H. Ipsen, Mechanical properties of the cell nucleus and the effect of emerin deficiency. Biophys. J. **91**, 4649–4664 (2006)

G. Sagvolden, I. Giaever, E.O. Pettersen, J. Feder, Cell adhesion force microscopy. Proc. Natl. Acad. Sci. U. S. A. **96**, 471–476 (1999)

G.W. Scarth, Colloidal changes associated with protoplasmic contraction. Q. J. Exp. Physiol. **14**, 99–113 (1924)

W. Seifriz, Observations on the structure of protoplasm by aid of microdissection. Biol. Bull. **34**, 307–324 (1918)

W. Seifriz, An elastic value of protoplasm, with further observations on the viscosity of protoplasm. J. Exp. Biol. **2**, 1–11 (1924)

W. Seifriz, The structure of protoplasm, Science **1902**, 648–649 (1931)

W. Seifriz, M. Uraguchi, The toxic effects of heavy metals on protoplasm. Am. J. Bot. **28**, 191–197 (1941)

S.G. Shroff, D.R. Saner, R. Lal, Dynamic micromechanical properties of cultured rat atrial myocytes measured by atomic force microscopy. Am. J. Physiol. **269**, C286–C292 (1995)

T. Sikosek, H.S. Chan, Biophysics of protein evolution and evolutionary protein biophysics. J. R. Soc. Interface **11**, 20140419 (2014)

A.E. Smith, Z. Zhang, C.R. Thomas, K.E. Moxham, A.P. Middelberg, The mechanical properties of fs. Proc. Natl. Acad. Sci. U. S. A. **97**, 9871–9874 (2000)

B.A. Smith, H. Roy, P. De Koninck, P. Grutter, Y. De Koninck, Dendritic spine viscoelasticity and soft-glassy nature: balancing dynamic remodeling with structural stability. Biophys. J. **92**, 1419–1430 (2007)

B.A. Smith, B. Tolloczko, J.G. Martin, P. Grutter, Probing the viscoelastic behavior of cultured airway smooth muscle cells with atomic force microscopy: stiffening induced by contractile agonist. Biophys. J. **88**, 2994–3007 (2005)

D. Stamenovic, N. Rosenblatt, M. Montoya-Zavala, B.D. Matthews, S. Hu, B. Suki, N. Wang, D.E. Ingber, Rheological behavior of living cells is timescale-dependent. Biophys. J. **93**, L39–L41 (2007)

A. Stuart, Three lectures on muscular motion. Philos. Trans. (1638–1775) **40**, i–iiv (1738)

Y. Sun, C.S. Chen, J. Fu, Forcing stem cells to behave: a biophysical perspective of the cellular microenvironment. Annu. Rev. Biophys. **41**, 519–542 (2012)

Y. Sun, L.G. Villa-Diaz, R.H. Lam, W. Chen, P.H. Krebsbach, J. Fu, Mechanics regulates fate decisions of human embryonic stem cells. PLoS ONE **7**, e37178 (2012)

S. Suresh, Biomechanics and biophysics of cancer cells. Acta Biomater. **3**, 413–438 (2007)

K. Svoboda, S.M. Block, Biological applications of optical forces. Annu. Rev. Biophys. Biomol. Struct. **23**, 247–285 (1994)

K. Svoboda, C.F. Schmidt, D. Branton, S.M. Block, Conformation and elasticity of the isolated red blood cell membrane skeleton. Biophys. J. **63**, 784–793 (1992)

K. Takakuda, Miyairi, Tensile behaviour of fibroblasts cultured in collagen gel. Biomaterials **17**, 1393–1397 (1996)

D.A.W. Thompson, *On Growth and Form*, 1st edn. (Cambridge University Press, Cambridge, England, 1917)

O. Thoumine, A. Ott, O. Cardoso, J.J. Meister, Microplates: a new tool for manipulation and mechanical perturbation of individual cells. J. Biochem. Biophys. Methods**39**, 47–62 (1999)

A. Tondon, R. Kaunas, The direction of stretch-induced cell and stress fiber orientation depends on collagen matrix stress. PLoS ONE **9**, e89592 (2014)

T. Tørring, N.V. Voigt, J. Nangreave, H. Yan, K.V. Gothelf, DNA origami: a quantum leap for self-assembly of complex structures. Chem. Soc. Rev. **40**(12), 5636–5646 (2011)

O. Treitel, Elasticity of plant tissues. Trans. Kans. Acad. Sci. **47**, 219–239 (1944)

X. Trepat, L. Deng, S.S. An, D. Navajas, D.J. Tschumperlin, W.T. Gerthoffer, J.P. Butler, J.J. Fredberg, Universal physical responses to stretch in the living cell. Nature **447**, 592–595 (2007)

D. Vader, A. Kabla, D. Weitz, L. Mahadevan, Strain-induced alignment in collagen gels. PLoS ONE **4**, 6e5902 (2009)

P.A. Valberg, D.F. Albertini, Cytoplasmic motions, rheology, and structure probed by a novel magnetic particle method. J. Cell Biol.**101**, 130–140 (1985)

M.T. Valentine, Z.E. Perlman, M.L. Gardel, J.H. Shin, P. Matsudaira, T.J. Mitchison, D.A. Weitz, Colloid surface chemistry critically affects multiple particle tracking measurements of biomaterials. Biophys. J. **86**, 4004–4014 (2004)

N. Wang, D.E. Ingber, Control of cytoskeletal mechanics by extracellular matrix, cell shape, and mechanical tension. Biophys. J.**66**, 2181–2189 (1994)

Y. Wang, D.E. Discher, *Cell Mechanics*, 1st edn. (Elsevier Academic, Amsterdam, 2007)

R. Waugh, E.A. Evans, Thermoelasticity of red blood cell membrane. Biophys. J. **26**, 115–131 (1979)

S.C. Weber, A.J. Spakowitz, J.A. Theriot, Nonthermal ATP-dependent fluctuations contribute to the in vivo motion of chromosomal loci. Proc. Natl. Acad. Sci. U. S. A. **109**, 7338–7343 (2012)

D. Weihs, T.G. Mason, M.A. Teitell, Bio-microrheology: a frontier in microrheology. Biophys. J.**91**, 4296–4305 (2006)

F. Zhang, Y. Wen, X. Guo, CRISPR/Cas9 for genome editing: progress, implications and challenges. Hum. Mol. Genet.**23**(R1), R40–R46 (2014)

P.C. Zhang, A.M. Keleshian, F. Sachs, Voltage-induced membrane movement. Nature **413**, 428–432 (2001)

2 Focus on Eukaryotic Cells

Abstract

This Chapter presents what we understand about soft biological matter and living biological matter. It also explores basic tools and simple models that describe biological processes at different levels of certainty. It is discussed in detail why the models for complex processes in living matter need to be simple and why they are useful. The main biological materials, the four classes of macromolecules are introduced, such as proteins that make up living cells and organisms. The general differences between prokaryotes and eukaryotes are mentioned, and the endosymbiotic hypothesis for the development of eukaryotes is presented and discussed. Cell compartments (synonymously termed organelles) are defined, and the structure and function of mitochondria are outlined. In detail, it is presented how mitochondria function by fusion and fission. In summary, the description eukaryotes, including organelles, in this chapter forms the basis for understanding of all other chapters of the Cellular Mechanics and Biophysics book dealing with eukaryotic cells.

2.1 What Is Biological Soft Matter in General?

Biological soft matter can hardly be pressed into a general definition of life and even scientists working with living matter for several years can solely approximate a broad description of life (Fig. 2.1). There exists no unique feature by which we can discriminate living matter from dead matter, whereas commonly all living systems possess certain central and general properties. In more detail, living matter systems grow, consume energy provided by their environment, reproduce to get genetically similar offspring and finally die.

Scientists have tried to understand the nature of life for being able to discriminate between life and dead matter. Since both matter types consist of atoms that obey the same physical laws, however, there exist useful generalizations about the material nature of life. However, there are still differences between life and dead matter. An example for is that many molecules, which assemble living organisms, are large and structurally complex and therefore are termed macromolecules. Another example is that living organisms have large numbers of small and relatively simple molecules, such as water, metal ions and glucose, which are critical for their proper function. Moreover, these small molecules serve as building blocks to assemble macromolecules within cells.

However, there exists two types of macromolecules, the first type is assembled of small molecules that can arise through non-living chemical processes, and the second type is created by characteristic

C. T. Mierke, *Cellular Mechanics and Biophysics*, Biological and Medical Physics, Biomedical Engineering, https://doi.org/10.1007/978-3-030-58532-7_2

Fig. 2.1 What is the difference between living and non-living material? The living material, the crocus, (left) grows, produces and consumes energy, reproduces and dies, whereas the dead material (shrew driver) (right) cannot fulfill these tasks

biological macromolecules that are solely found in living organisms. Although the macromolecules of living matter, such as DNA, RNA and proteins, are largely diverse, the atoms building these macromolecules are mainly composed of carbon (C), hydrogen (H), nitrogen (N) and oxygen (O) and partly of phosphorus (P) and sulfur (S). The connection between the atoms building the macromolecules of living organisms is special, as they are normally not only connected by covalently bonds, and in addition by the hydrogen bonds playing a major role. Moreover, also other numerous molecules are bound to these macromolecules affecting their function and structural stability.

2.2 Four Major Classes of Macromolecules

The organisms are created by four major classes of macromolecules, such as proteins, carbohydrates, nucleic acids and lipids (Fig. 2.2). The four classes of macromolecules fulfill largely different functions within living organisms. In more detail, proteins build the structural elements of cells and catalyze specific chemical reactions in the cells and the entire organism. The membrane is formed by lipids and represents a barrier toward the exterior of a cell. Inside the eukaryotic cell, membrane lipids build their compartments, which are termed the organelles. The carbohydrates serve as energy storage, form specific membrane surface properties outside the cell membranes and built even rigid cell walls in plant cells. The nucleic acids serve as the building blocks of all kinds of macromolecules and hence are critical for the memory and the operating instructions that enable cells to replicate themselves and generate macromolecules. Each of these four classes of macromolecules have a large diversity, since they consist of numerous structurally distinct proteins. The cellular membranes can be formed by hundreds of different lipid species.

The question why these four classes of macromolecules are the stuff of life cannot clearly be answered, as it would still need more knowledge about the specific chemical and physical conditions on the early Earth at the time of the first living cells' origin. However, these molecules are prerequisites of life, as each of these four classes of molecules can be built up from a small number of simpler subunits or precursor molecules within the cell. The combination of these simple subunits to a large assembled complex provides the structural diversity that is needed to store the information about all processes in a living organism. Hence, a cell is able to synthesize with a small limited number of chemical reactions, these few simple subunits directly from the local environmental food.

Proteins and nucleic acids are composed of similar simple subunits that involve the synthesis of polymers. However, different kinds of subunit types are needed to assemble the DNA and its nucleotides. The Cellular Mechanics and Biophysics book presents how special properties, such as the macromolecular behavior of proteins and DNA, depend on physical and chemical interactions. In

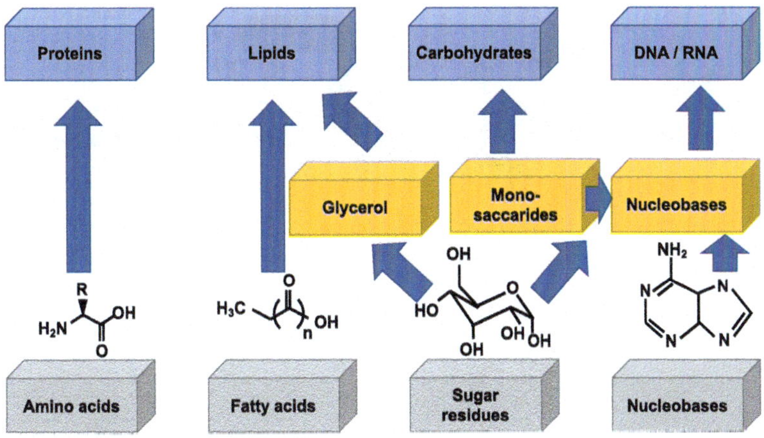

Fig. 2.2 Four major classes of macromolecules

more detail, the sequential order of nucleic acid subunits manifested in the structure of cellular DNA directly encodes directly the sequence of amino acids that assemble the cellular proteins (see Chaps. 12 and 14). Finally, these observation leads to the discovery of the genetic code (see Chap. 11).

The major connection between a DNA sequence and a protein sequence is based on the polymeric nature of these structures. The DNA, nucleic acids and subsequently proteins are built up from a small number of subunits. For example, the nucleic acids are composed of a simple four-letter code in which each element represents a specific nucleotide, such as adenine (A), cytosine (C), guanine (G) and thymine (T) for DNA (Fig. 2.3). In RNA, thymine (T) is replaced by uracil (U). The proteins are built up by 20 distinct building blocks and termed amino acids (Fig. 2.4). It is an extraordinary feature of living organisms that only four nucleotide bases are translated into numerous different proteins assembled by only 20 different amino acids. The important point is that the DNA bases are grouped together to small units of three following bases. All these DNA strands start with the first unique beginning of ATG for the part, which encodes the information for the protein. The following sequential bases are units of exactly three.

Moreover, the three-letter code can be any combination of the four bases, in which each position is occupied by each base except the start codon and the last codon that represents the end of the gene. The so-called stop codon can have three combinations, such as TAG, TAA and TGA and are termed amber, ochre and opal (sometimes named umber), respectively (Fig. 2.5). Moreover, each possible combination of three bases (nucleotides) encodes only one specific amino acid. In turn, one amino acid is not encoded by just only one combination of nucleotides, there are others, for certain amino acids, encoding for the same amino acid, except methionine, which is encoded by solely one combination (ATG). This phenomenon is described by the Wobble hypothesis. The three-letter code builds the fundamental units of the protein's secondary structure such as alpha-helices and beta-strands. In more detail, the three-letter units are grouped together and build the gene encoding for a special protein.

In some cases, the genes can be grouped together in a gene cluster, in which the genes are transcribed together. The genes in a gene cluster encode for proteins that are involved in one pathway or share a generalized similar function. For keeping the nucleic acids and proteins connected, the ribosome serves a stable mechanical linkage by taking messenger RNA (mRNA) as a template and transfer-RNA (t-RNA) as a shuttle serve to deliver the associated (encoded) amino acid for a specific three nucleotide codon, when it has the specific anticodon to bind to the mRNA. The genes and the proteins are connected through the genetic code (Fig. 2.6).

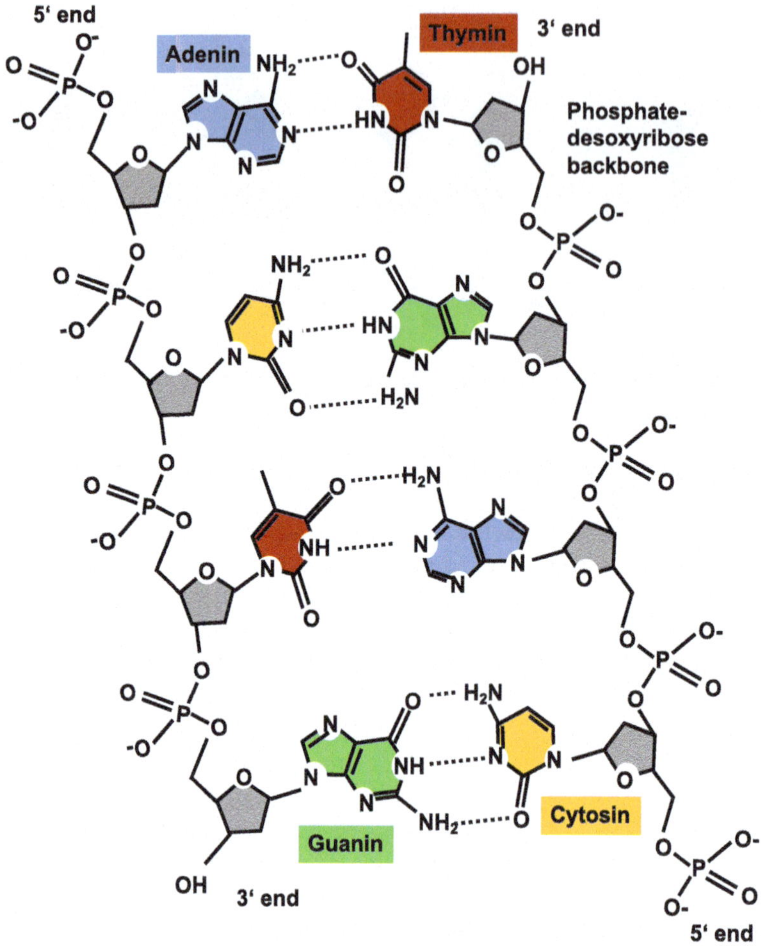

Fig. 2.3 Chemical structure of the nucleotides such as adenine (A), cytosine (C), guanine (G) and thymine (T) of DNA form a double helix. The complementary nucleotides are connected through hydrogen bonds (dotted lines)

2.3 Why Are Physical Models Needed and Kept Simple?

Physical models help to get insights into the cellular mechanics and biophysics, and subsequently, these models are the goals for every biophysical scientist. Insights into cellular mechanics and biophysics are obtained by abstracting and simplifying the highly complex biological soft matter. This procedure helps to develop simple, analytical models that still predict the outcome of future experiments. As it is not possible to develop an individual atomic description for each protein, only the relevant properties of the protein are selected that are important for a specific aspect of the addressed molecule's behavior. However, the model should not be too simple for a highly complex macromolecule, such as DNA or RNA. Hence, to deal with the enormous complexity, simple models seem to be most applicable, since they can serve as projections of the complex natural DNA or RNA molecules.

Models need to be simple in order to represent most of the complex biological material. Moreover, these models are idealized by applying numerous different physical models. When a model is too

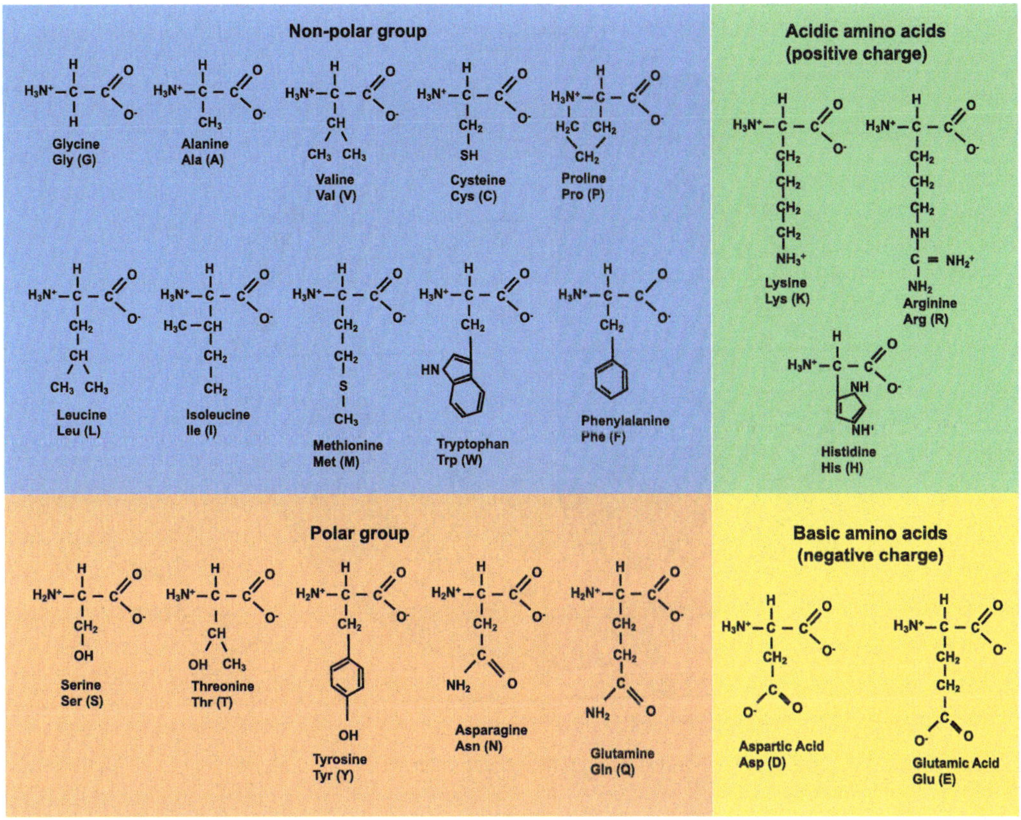

Fig. 2.4 Chemical structures of the 20 amino acids building the proteins and the four major groups of amino acids

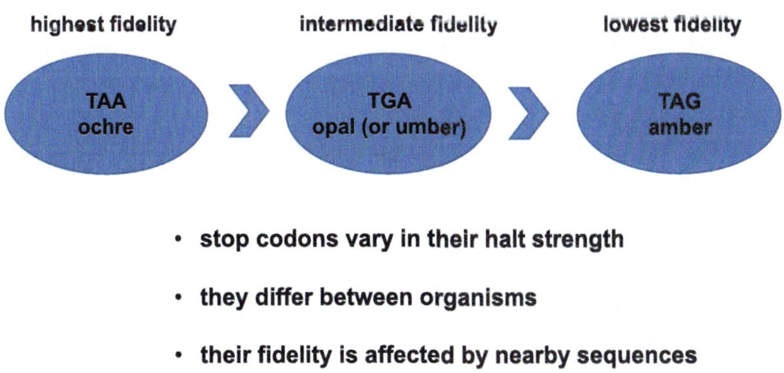

Fig. 2.5 There are three different stop codons that differ in their fidelity

specific, it may cover not many cases. Therefore, it is necessary to identify which are the essential parts that are useful for this model and what can be omitted in order to strengthen the view on the important regulatory factors. All these steps help to create an enlightening and helpful model. The view from different angles on objects or processes helps to understand the overall underlying order or structure and explains complex and diverse problems in simple steps. Finally, many diverse problems may reveal unique steps present that are present in all of them. The degree of simplicity of a model

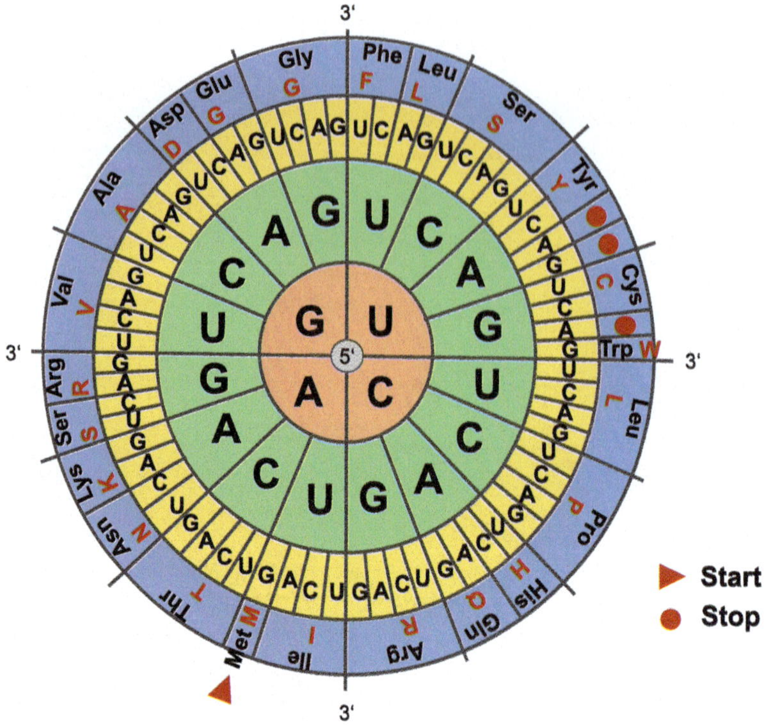

Fig. 2.6 Genes and proteins are connected through the genetic code

depends on the specific question that is being studied and therefore needs to be answered. The different questions lead to various models with diverse complexity on the same issue or problem.

An example of a model-like soft matter material is DNA, which is a highly important molecule, since it encodes the genetic information and has a prominent role in genetic engineering in molecular biology. Each strand of the helical DNA polymer has one polarity, which means that the two ends of the strands are different in structure and chemical composition. Thus, one end of the DNA strand is referred to as 5′ prime end and the other end as a 3′ prime end. Moreover, the entire length of the DNA molecule is heterogeneous, since the chemical identity of the bases differs from that of the neighboring nucleotide. The information given by the choice of nucleotide neighbors and their individual order in the DNA polymer carries the information of the entire genome of the organisms. These organisms can be bacteria up to humans, and still, the information of the genome is given by the order of the nucleotides and not by the molecular details of the DNA polymer itself. Thus, the focus is here to the simple sequence of the nucleotides A's, C's, G's and T's that encode the information.

By knowing the sequence, the history of the mutations in the genome that led to the different species during the evolution can be analyzed. Moreover, the information within a genome can be correlated with the amount of DNA in an organism and correlated with the overall size of the species. Another example of the importance of the sequence knowledge is that variations of sequence parts can be identified as binding sequences with a strong affinity to a special protein-binding partner, such as the RNA polymerase, which copies the sequence of the DNA by building up the mRNA molecule. In addition, in certain diseases, parts of the sequence can be altered and finally be functionally linked to them. These two examples point out to similarities or changes in the chemical structure of the DNA molecule. In these two cases, the DNA is regarded as a physical entity, where the specificity of the DNA sequence is irrelevant. However, the entire mechanical and chemical properties of the molecule are analyzed, such as its total charge distribution, bending elasticity (bending modulus) or its behavior as a long polymer chain that is subject to thermal undulations in solution.

Proteins can be idealized in many different manners. In more detail, since nucleic acids can be regarded as a linear sequence of nucleotides, we can also see a protein as a linear sequence of amino acids that forms the entire protein. Besides the DNA, the other main group of polymers within a cell is the group of proteins regulating all the important functions of the cell, such as the catalysis of certain chemical reactions. The precise atomic structure of the proteins is even more complex than the structure of the DNA because there are 20 instead of four basic subunits that represent proteins. In addition, the chain of amino acids has a three-dimensional (3-D) structure, which is very diverse compared to the relatively simple DNA double helical structure due to charges of the individual amino acids.

From a biophysical point of view, it is solely emphasized on one specific feature of a protein, which is that a protein can be seen as a linear sequence of amino acids by omitting atomic-scale structural details and the complex chemistry including intramolecular interactions. In biology, the focus is often on a very detailed description of a material or on processes and not on a simplified description that leads to the development of models. In the linear sequence of amino acids, the hydrophobic amino acids are grouped and entitled with H. The polar amino acids are grouped and entitled with P (Fig. 2.4). The physical differences of hydrophobic and polar amino acids affect the interactions with solvents, such as water and oil, and regulate how amino acids fold and build up a functional protein. Another simplification of proteins is to present their three-dimensional structure as small cylinders, which are aligned to larger cylinders (entitled alpha-helices) and/or ribbons, which bind together (entitled beta-strands) in a precisely defined manner. Finally, the simplification helps to find out and consider the basic structural elements and their function.

In order to investigate protein folding, a physical point of view on the proteins is suitable by treating them as a class of lattice models of compact polymers. In these lattice models, an amino acid is allowed solely to occupy a regular array of positions in space. The sequence of the amino acids and the folded structure of the protein provides its functions or activity. There are models simply describing the activity of the molecule. Most proteins in cells fulfill specific functions and therefore need to interact with themselves or other binding proteins. The different interactions can be modeled by treating proteins as receptors containing binding sites and others, which bind to these receptors, as binding partners (synonymously termed ligands). However, this description will omit the complex internal protein structure. Models will point on the activity of proteins and treat them as two-state systems, which can be converted into each other. One state is an active state, and the other one is an inactive state. The conversion of the states can be induced by adding ligands or structural modifications such as opening of internally bound structures by ligand binding or phosphorylation of proteins. In addition, an enzyme can catalyze a chemical reaction and can be present in two states, an active and an inactive state, which is regulated by the presence of ligands or structural modifications. These simplifications are artificial and serve to investigate the properties of a protein in isolation from the other properties. At this stage, it is simply not useful in getting solely individual information on one protein property, when the complexity of the proteins is not ignored or the sequence of the amino acids, their folding, protein compaction or conformational changes is not neglected. However, it does not mean that they are always omitted in every model.

From a theoretical point of view, in some cases, it may not be useful to consider only a distinct single level of description at a time, as this leads to estimates that provide intuitive information and produces simplified models that can only serve as a first and fast overview. Therefore, we will extend our primary approach far beyond distinct macromolecules to assemblies of molecules on a larger scale, such as cellular membranes, which consist of large numbers of lipids and proteins. The views on the membrane are different and depend on whether biologists or physicists are the observers. If we have distinct macromolecules, large-scale analysis may be useful by focusing on a single aspect of their physical behavior. In more detail, the membranes can be seen as bendable and spring-like (elastic) sheets, random surfaces or as arrays of electrical elements such as a capacitor or resistor. During the interaction between two cells, membranes represent selective barriers through which distinct molecules can pass, whereas others are hindered in their motion.

The same is true for entire cells migrating through a dense and closed (confluent) cell-layer barrier. The degree of idealization of the biological membrane depends strongly on the individual questions that we have asked us at the beginning of the experimental approach. The idealization can be transferred from plasma membranes to entire living cells. A living cell is highly complex, and we need to extract the essential physical property of the cell to get insights in particular functions by using simple approaches. One example organism is the bacterium *Escherichia coli*, which can be seen as an object with many protein receptors on its cellular surface. A bacterium can swim through a water-based environment, and hence, it can be regarded as an elastohydrodynamic object, which is entirely a mechanical object being bendable and interacting with the flow of the water-like fluid. In more detail, the physical properties of the distinct path the bacterium follows can be modeled by considering the bacterium's large-scale motion representing a biased random walk, which simply ignores all the hydrodynamic details contributing to bacterial motion. As biological research is more focused on alterations in expression of polymers due to environmental conditions, it may be useful to treat a cell as a device processing information such as the cell is regarded as a network that regulates the flow of information through gene expression. This view is not exclusive, and abstractions have been made to improve the clarity. The abstraction concentrates on a single physical property to describe the biological entities, which is also currently used for consideration of other systems such non-living soft matter that include solutions of charged ions. There exist various abstractions of water being the medium of life and a solution representing regular lattice structure. In order to investigate ligands in a solution and their ability to bind to a receptor, the solution is treated as a liquid in which are only discrete positions permitted for the ligands. However, this is an extreme approximation, which is valuable for calculations that involve the chemical equilibrium. Indeed, it turned out that this model is extremely valuable for calculations involving the chemical equilibrium and delivers remarkably accurate quantitative predictions.

Although it is still a huge approximation, it is helpful for calculations about the chemical equilibrium of reactions, as it provides exact quantitative predictions. When living matter interacts with water, the hydrodynamic flow properties are included. When we calculate the chemical reaction rates in water solvents, we need to consider the diffusive fluctuations, since the macromolecules in the cytoplasm of cells, which mostly consists of water, are altered by the specific properties of water. The chemical groups on proteins are often polar or hydrophilic (love water and hence attract water), which can form hydrogen bonds with water, whereas other groups are oil-like and hydrophobic (hate water and hence repel water). In folded proteins, the oil-like groups usually cluster inside of the globular protein; however, charged surface-membrane located molecules can interact with the dielectric property of water molecules, which means that that are neutral in principle, but have a slight charge separation similar to dipoles.

All these simplifications and reductions help to answer many different biological questions and formulate future models. However, none of these simple models will answer all questions and lead to a broad understanding of the diverse behavior of living soft matter such as cells. Each model seems to reveal an insight into certain aspects of living cells, and finally, all these models can help to provide a realistic overall picture of the cells.

2.4 Characteristics of Prokaryotes and Eukaryotes

All cells share four components, such as DNA, plasma membrane, ribosomes and the cytoplasm. The DNA contains the genetic information of the cells, at the ribosomes is the genetic information translated from RNA to proteins, the jelly-like solution of the cytoplasm is termed cytosol and the plasma membrane encompasses the entire cell to separate it from its environment.

There exist two major types of cells that are referred to prokaryotes and eukaryotes.

2.4.1 Prokaryotic Cells

Besides the four similar components, prokaryotes are different from eukaryotic cells in several respects. A prokaryote can be described as a simple, unicellular organism that contains neither an organized nucleus nor any other membrane-bound organelle, which are the major differences to eukaryotes. The DNA of prokaryotes is located in a central part of the cell, which is termed the nucleoid. In general, prokaryotes are surrounded by a peptidoglycan cell wall, and many prokaryotes possess a polysaccharide capsule. In more detail, the cell wall serves as an additional protective layer for the cell's interior that supports the maintenance of the cell's shape and helps to omit dehydration. The cell capsule facilitates the attachment of the cell to surfaces in its local microenvironment. In addition, certain prokaryotes possess flagella, pili or fimbriae on their cell surface. Prokaryotes utilize flagella for movements. The pili can be employed to transfer genetic material from one cell to another cell by direct cell–cell contact during a distinct type of reproduction, which is termed conjugation. Prokaryotic fimbriae can be exploited by bacteria to interact with a host cell.

A commonly investigated simple cell model for the prokaryotic cell is the bacterium *Escherichia coli* (*E. coli*) (Fig. 2.7). *E. coli* lives in the gastrointestinal tract and serves as a model organism for the investigation of bacterial cell motility and the specific function of the flagella from a biophysical point of view. In molecular biology, *E. coli* is used as a laboratory organism helpful for genetic engineering.

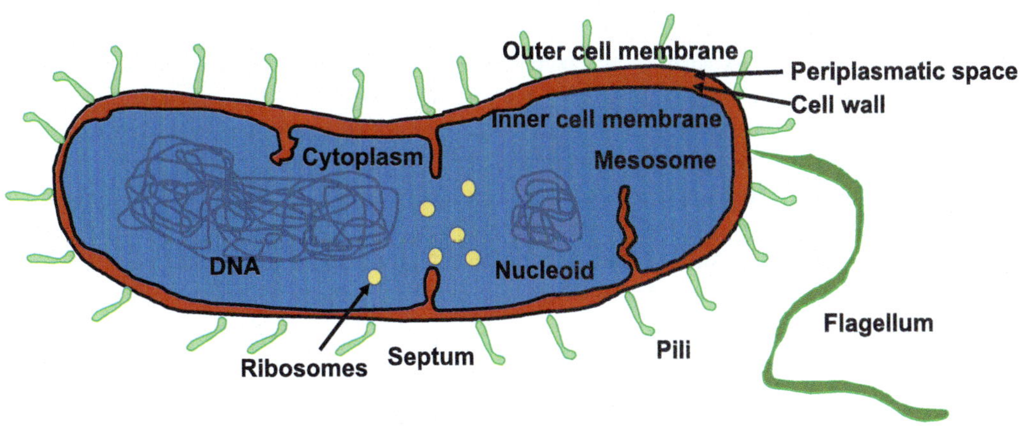

Fig. 2.7 Simple structure of the bacterium *E. coli*

Table 2.1 Differences between prokaryotes and eukaryotes

	Prokaryotes	Eukaryotes
Ribosomes	70S	80S
Nucleus	No	Yes
Chromosomes	No	Yes
Histones	No	Yes
Alternative splicing	No	Yes
Compartments (organelles)	No	Yes
Zygote (spores)	No	Yes

S = Svedberg unit

The main differences between prokaryotes and eukaryotes are listed in Table 2.1.

The Svedberg unit (s) is the sedimentation rate for a particle, such as a ribosome, of an individual size and shape, and shows how fast a particle has moved to the bottom of a tube, when a centrifugal force is applied using a centrifuge. For example, a particle with a sedimentation coefficient of 70S (70×10^{-13} s) has a sedimentation speed of 30 microns per second under the influence of an acceleration of 10^7 m/s^2 (a million gravities). The Svedberg unit depends on the size, shape and cross-sectional area of the particle, such as cell or organelle, and is a measure of time. It is affected by the frictional force on this particle during the sedimentation process.

2.4.2 Eukaryotic Cells

Prokaryotic cells have commonly a diameter in the range of 0.1–5.0 μm, and hence, they are pronouncedly smaller than eukaryotic cells, which cell diameters are in the range of 10–100 μm. Since prokaryotic cells are small in size, ions and organic molecules enter them to rapidly diffuse into all other parts of the cell. Likewise, any waste produces in a prokaryotic cell rapidly diffuse out of the cell into the local environment. In contrast, eukaryotic cells have utilized different structural architectures and mechanisms to perform efficient intracellular transport. However, for all cells encompassing prokaryotic or eukaryotic cells, a small size is envisioned. Since most cells exhibit a spherical shape, they can be approximated as a sphere, with a surface area of $4\pi r^2$ and a volume is $4/3\pi r^3$. As the cell radius increases, its surface area increases as the square of its radius, while its volume increases the cube of its radius (much faster). Hence, as the cell size increases, the ratio of the surface area to volume of the cell decreases. In addition, when a cell grows too strongly, the plasma membrane does not have enough surface area to sustain the diffusion rate demanded for the increased volume. Hence, as a cell grows, it becomes less efficient. In order to overcome this less efficient state, the cell needs to divide. Another option to improve efficiency is to develop organelles that perform distinct tasks. These modifications resulted in the development of more complex cells known as eukaryotic cells. Moreover, these eukaryotic cells are structurally highly complex at a nanoscale and mesoscale level, can grow in clusters and arrange to higher ordered structures such as tissues at a microscale level. These ordered structures are specialized compartments that are termed organelles. Moreover, different organelles fulfill diverse roles within the cell. For example, mitochondria generate energy from molecules that serve as food, such as glucose, lysosomes degrade and recycle organelles and biomacromolecules and the endoplasmic reticulum rebuild membranes and facilitate the protein transport throughout the entire cell. However, there is one central question that needs to be raised here: What are the characteristics that all organelles have share?

2.5 Organelles—What Defines Them?

The cellular interior of eukaryotic cells is rich of organelles. The extraordinary size of eukaryotic cells compared to bacterial cells causes the compartmentation of cells into closed structural units such as the organelles. One major parameter that distinguishes bacteria and archaea from eukaryotes is the membrane confined organelles (Fig. 2.8). Besides the cell nucleus, eukaryotic cells possess several other types of organelles including mitochondria, Golgi apparatus, endoplasmic reticulum (ER), chloroplasts (in plants) and lysosomes.

Organelles are specialized compartments in which the cell carries out specific functions, such the management of the genome (in the nucleus), energy generation (in mitochondria and chloroplasts), protein biosynthesis and the posttranslational modification of proteins (in the ER and Golgi apparatus). The organelles are cellular compartments confined by phospholipid membranes possessing completely different protein and ion compositions compared to cellular membranes. Moreover, the membranes of each of these different membrane systems, which enclose the organelles, possess a specific composition of lipids and proteins. Each individual organelle fulfills a distinct function that is crucial for the survival of the cell. Almost all eukaryotic organelles are surrounded by a membrane that separates the interior of the organelles from the cytoplasm. These organelle membranes are lipid bilayers that are highly similar to the outer plasma membrane of the cell. Hence, a characteristic feature of many organelles is that they are compartmentalized structures, which are separated from the rest of the cell by membranes. Similar to the plasma membrane, the organelle membranes hold the inside "inside" and the outside "outside." This division allows different types of biochemical reactions to take place in different organelles and additionally protects the cytoplasm from dangerous reactions. Although each organelle has its own specific function in the cell, all cell organelles can work together in an integrated way to cover the overall needs of the cell. An example are biochemical reactions in the mitochondria of a cell that transfer energy from fatty acids and pyruvate molecules into a high-energy molecule, such as adenosine triphosphate (ATP). The rest of the organelles of the cell then use this ATP as an energy source for operation.

An advantage of these membrane surrounded compartments is that they can be easily visualized, for example, by high-resolution electron microscopy of a thin cross section or slice of a cell. Thereby,

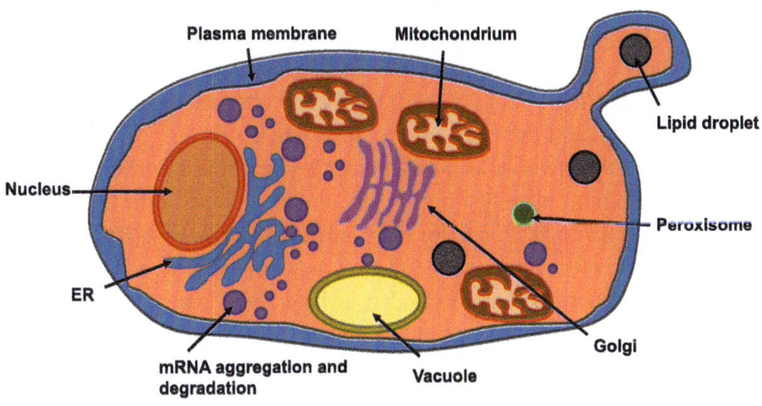

Fig. 2.8 Typical basic structure of a cell containing several membrane confined organelles

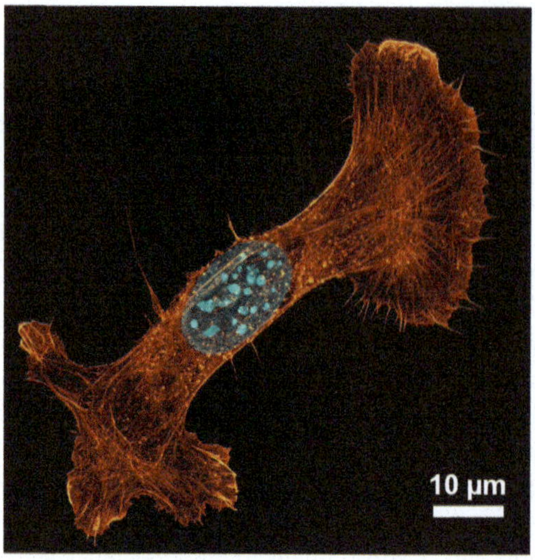

Fig. 2.9 Fluorescent microscopic image of a mouse embryonic fibroblast. The F-actin cytoskeleton is visualized through AlexaFluor548 Phalloidin (red), and the nucleus is stained with Hoechst 33,342. The scale bar is 10 µm

structural details and key characteristics of different organelles, such as the long, thin compartments of the ER or the compacted (condensed) chromatin of the nucleus, can be identified. Since a cell is dynamically remodeled, its organelles are not static and hence undergo constant motion. Sometimes, the organelles move even to a certain location within the cell, merge with other organelles and become larger or smaller.

The nucleus is one of the well-known examples, as it is often easily visible with standard light microscopy. When we use fibroblasts as an example, the cell has dimensions of about 50 µm, while the nucleus has a characteristic linear dimension of about 10 µm (Fig. 2.9). From a functional point of view, the nucleus has a lot more to offer than merely serving as a reservoir for the genetic material. In the nucleus, the chromosomes are highly organized and hence create specific domains, which is discussed in detail (see Chap. 10). Since transcription and various kinds of RNA processing occur in the nucleus, molecules, such as transcription factors, fluctuate in and out the nucleus, while completed RNA molecules are exported out through gateways within the nuclear membrane, known as nuclear pore complexes. The parts of the genome that is involved in synthesis of ribosomal RNA are clustered together in a specific compartment. This compartment is characterized by striking spots, which can be seen in the light microscope, and they are referred to as nucleoli.

When molecules are transported out of the nucleus, they impinge the next membranous organelle, which is referred to the endoplasmic reticulum. In fact, the membrane of the ER is connected with the membrane of the nuclear envelope. In certain cells, such as pancreatic cells, the endoplasmic reticulum occupies most of the cell interior. This complex organelle serves as the site of lipid synthesis, and also, the site of synthesis of proteins extended to be excreted or incorporated into membranes. The endoplasmic reticulum can acquire various geometries in diverse cell types and under varying conditions. There remain still not yet answered questions: How much total membrane area does the ER occupy? How heavily does the special membrane morphology influence the overall size of the organelle?

2.6 Endosymbiotic Hypothesis

Symbiogenesis basically means living together and points out to the decisive role of symbiosis in evolution, which contributes to important innovations. The term is usually earmarked for the major transition from prokaryotes to eukaryotes, such as the photosynthesis of eukaryotic algae and plants through the process of endosymbiosis. In some eukaryotic lineages, however, endosymbionts have been secondary lost, indicating that symbiosis can lead to a major evolutionary innovation even under this rare circumstance. This finding leads to the compelling assumption that symbiosis has played an important role in other major evolutionary innovations, even if not all existing representatives of such groups still have the symbiotic link (Gray 2017).

In 1967, the endosymbiotic theory of organelle origins has been formulated (Sagan 1967) and is now widely accepted (Archibald 2014). The endosymbiotic theory found that three fundamental organelles such as the mitochondria, the photosynthetic plastids and the (9 + 2) basal bodies of flagella were once free-living prokaryotic cells (Sagan 1967). The fact that mitochondria and plastids could endosymbiotically originate from prokaryotic ancestors was not a new idea at that time, which only emerged in various forms at the end of the 19th and beginning of the twentieth centuries before fading out from a biological point of view (Sapp 1994). Therein, it has been stated that the article in 1967 was remarkable in that it presented an all-encompassing view of endosymbiosis as the endpoint of the eukaryotic cell, and hence, this theory seems to be the first coherent theory of eukaryogenesis. More precisely, a third subcellular structural element, the eukaryotic flagellum (formerly termed undulipodium), was proposed to originated from the endocytosis of specific motile prokaryotes, such as spirochaete-like prokaryotes, which may then get connected by symbiosis to their cellular hosts. This overall scenario was later referred to as the serial endosymbiosis theory (Taylor 1974).

In contrast to the proposed endosymbiotic origin of mitochondria and plastids, there seems to be no evidence of the origin of mitosis (Sagan 1967). Since despite efforts to identify one, no genome was linked with the eukaryotic flagellar apparatus (Johnson and Rosenbaum 1991). The genomes of the mitochondrion and the plastid or more precisely the genes they house them and how they are arranged and expressed enable us to know with a high degree of certainty where these organelles come from: the bacterial species α-Proteobacteria and Cyanobacteria (Gray and Doolittle 1982; Gray 1992).

The energetic promotion of the role of symbiosis in eukaryotic cell development (Margulis 1970a, b) triggered a lively debate about autogenous origin (origin from inside) and xenogenous origin (origin from outside) theories of organelle development. At that time, the endosymbiont scenarios for both mitochondria and plastids were controversially discussed (Uzzell and Spolsky 1974), with the concentration in this period being primarily on the mitochondrion (Raff and Mahler 1972). The mitochondrial genetic system was reported to possess a great inter and intraspecies diversity, which leads to the hypothesis that it is a unique system distinct from prokaryotic and eukaryotic cells (Mahler 1981). The comparative analysis of mitochondrial genomes revealed that support for both scenarios endosymbiotic or no endosymbiotic event can be found (Burger et al. 2003). Biochemical, molecular and cell biological findings and the characterization of a group of eukaryotic microbes, the jakobid flagellates, as a gene-rich mitochondrial genome that strongly resembles a reduced bacterial genome (Burger et al. 2013) and thereby delivers an exemplary case for a single,endosymbiotic, α-proteobacterial origin of mitochondria, are all indicators for an endosymbiotic origin (Gray et al. 1999; Gray 2012). However, a compelling case for an endosymbiotic origin is easier to find for the plastid than for the mitochondrion, since the plastid is evolutionarily younger than the mitochondrion. The last eukaryotic common ancestor had still a functional mitochondrion that comes close to its modern counterpart (Koumandou et al. 2013), and several major eukaryotic lineages, containing animals and fungi, are clearly primitively aplastidic and originate from ancestors who never had plastids. Therefore, the similarity between plastic and cyanobacteria structure and biochemistry is much more pronounced in most plastic-containing eukaryotes than in the mitochondria/α-proteobacteria comparison. In addition, plastid genomes usually contain considerably more genes on which such a comparison can be based than do mitochondrial genomes, and hence, the plastid translation system exhibits more bacterial character than its equivalent in most mitochondrial systems (Gray 1992).

The treatment of the plastid has been in that way that eukaryotic plant cells did not develop oxygen-eliminating photosynthesis. However, these cells become photosynthetic active by symbiosis with blue-green algae, such as cyanobacteria (Sagan 1967). More precisely, various photosynthetic eukaryotes (protoplastids) were recorded at different times during the evolution of heterotrophic protozoans that were transformed into obligately symbiotic plastids and persevered their typical photosynthetic pigments and pathways (Sagan 1967; Raven 1970). However, the current consensus for the development of the eukaryotic line, which includes both land plants and green, red and cyanophytic alga, is a single, separate, endosymbiotic origin of mitochondria and plastids, whose primary origin from an endosymbiotic cyanobacterium lies in an ancestor of Archaeplastida. Plastids then penetrated other algae clades via a secondary symbiosis process in which a eukaryotic host takes up a eukaryotic symbiont (a green or red alga) (Archibald and Keeling 2002).

Although there is agreement that mitochondria and chloroplasts are obtained from free-living bacterial ancestors through an endosymbiotic process, however, the mechanism still not clearly revealed. In the case of the mitochondrion, it is still unclear when the triggering event occurred related to the origin of the eukaryotic cell and how long the process of converting bacterial endosymbiont into fully integrated organelles took and by what mechanisms. In fact, in over five decades since the first publication on the theory, a variety of symbiogenetic models have been formulated that arouse various hosts and processes (Martin et al. 2015).

A particularly controversial aspect is the nature of the host. In endosymbiosis, the host is often represented as a primitive and amitochondrial eukaryote, which captures a prokaryotic symbiote through phagotrophy, which is a process of capturing and internalizing other organisms through phagocytosis. Other descriptions show the host as a prokaryote taking another prokaryote by phagocytosis, although the process of phagotrophy is still unknown in free-living prokaryotes. The first step in the development of eukaryotes from prokaryotes was driven by the survival in the new oxygenated atmosphere: An aerobic prokaryotic microbe, such as a protomitochondrion, was incorporated into the cytoplasm of a heterotrophic anaerobic organism. The process of endosymbiosis seems to be mandatory and led to the development of the first aerobic amitoid amoebae. It remained unclear whether the host was a prokaryote or an advanced version of it. Based on ameboid and ingestion, a type of protoeukaryote can be hypothesized, although it still lacks a nucleus and mitotic apparatus, which are both characteristic features if a eukaryote. However, also a prokaryotic host seems to be possible (Margulis 1981). Moreover, protomitochondria may invaded their hosts similar to modern predatory bacteria, such as Bdellovibrio that can even be performed without phagocytosis. In fine detail, Bdellovibrio can effectively destroy its host (the bacterium) during the invading process. However, it remains still an open question of how this route may develop into a stable prokaryote–prokaryote symbiosis.

A central point in various symbiogenesis models of mitochondrial origin and evolution is the nature of the host, which is divided into two major types: These are the mitochondria early type (termed mito-early) and secondly mitochondria late type (termed mito-late), which can be distinguished in timing of the transition from first eukaryotic common ancestor to last eukaryotic common ancestor and with different effects on the entire eukaryotic cell origin (Poole and Gribaldo 2014). Based on comparative genomics, it can be underlined that the last eukaryotic common ancestor A was quite a complex organism with a completely functional mitochondrion (Koumandou et al. 2013) and that all allegedly mitochondrial eukaryotic lineages, except one (Karnkowska et al. 2016), comprise mitochondrion-related organelles and originate from mitochondria-containing ancestors. Thus, the initial acquisition of a bacterial symbiont designated to form the mitochondrion could not have taken place very close to the emergence of the last eukaryotic common ancestor because of the numerous and complex alterations that appeared during the transition from symbiont to organelle, although evidence for a late mitochondrion acquisition has been reported (Pittis and Gabaldón 2016).

Probably, the most famous mitotic early model is the hydrogen hypothesis (Martin and Müller 1998), where the host, an anaerobic, hydrogen-dependent archaeon, picks up an α-proteobacterium that is able to respire but produces molecular hydrogen as a waste product of the anaerobic heterotrophic metabolism. The selection of endosymbiosis is pushed forward here by metabolic syntrophy between the two partners: The waste product (hydrogen) of one partner is an essential metabolic resource for the other. Both mitochondrion and eukaryotic cell origin are simultaneous in this scenario, with the subsequent subcellular characteristics of the latter being directly dependent on a distinctive increase in cell energy.

In contrast, in mito-late models, the underlying mechanism of symbiogenesis is phagotrophy, a hallmark of eukaryotic cells that is widespread within the eukaryotic domain. Phagotrophy is an endocytosis in which the boundary membrane of one organism (host) surrounds another organism (symbiont) and thereby internalizes it in a membrane-bound phagosome (Fig. 2.10). In a number of mito-late models, the host is virtually an amitochondriate eukaryote with the potential for phagocytosis (Cavalier-Smith 1987).

Eukaryotes are closely related to a new group of Archaea, the Asgard superphylum (Zaremba-Niedzwiedzka et al. 2017) that encodes numerous proteins whose homologues have so far only been detected in eukaryotes. Hence, it can be hypothesized that this strain represents an archaeal line, which had originally developed characteristic eukaryotic features, such as the process of phagocytosis, and therefore seems to be the initial host for mitochondrial endosymbiosis. These observations form the basis for the phagocytotic archaeon model of eukaryogenesis, where the mitochondrial endosymbiont was derived by a temporarily complex phagocytotic archaeon (Martijn and Ettema 2013). The transition from symbiont to organelle obviously entailed many steps: The dispersal of the bacterial cell wall, early capture of key metabolite transporters by the symbiont, substantial and

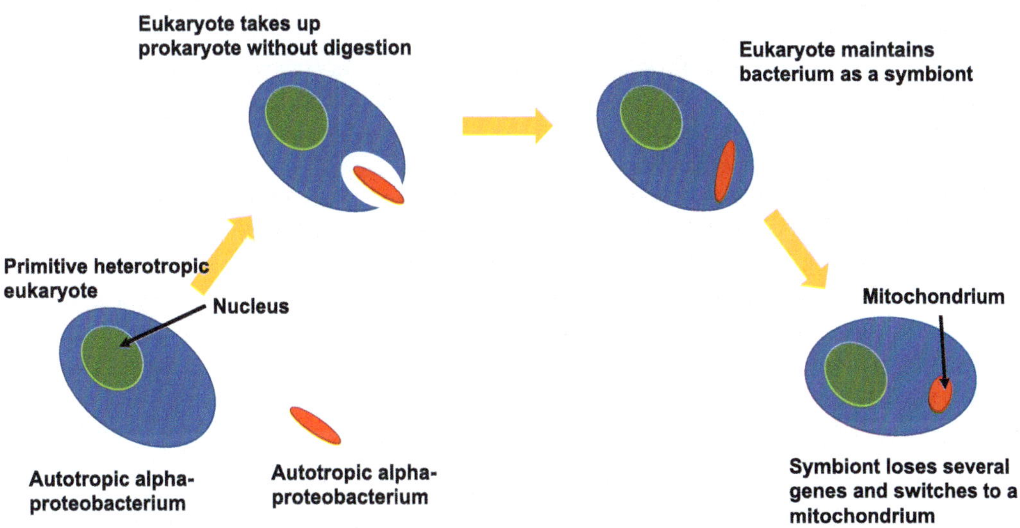

Fig. 2.10 Endosymbiosis

variably reduction of the symbiont genome by total loss or transfer of genes to the nucleus, functional activation of transferred genes in the nucleus and retargeting of their cytoplasmically synthesized proteins back into the developing organelle based on specific organellar targeting sequences or anywhere else in the cell and large-scale recruitment of many supplemental organellar proteins whose origin is obscure. After all, this transformation process seems to be a step-by-step process over a wide period of time.

Although much knowledge has been gained, great efforts are still needed to determine the origin of the mass of the mitochondrial and plastid proteomes, since they seem to have at the first glance neither α-proteobacterial nor cyanobacterial origin. While it is generally accepted that mitochondrion and plastid are the result of bacterial endosymbiosis, however, these organelles consist of mosaic components and functions, which have more than one origin (Cavalier-Smith 1987). In conclusion, symbiosis fulfills a crucial role in organelle origins and the entire eukaryogenesis, but its role is not all-encompassing.

2.7 Single or Serial Endosymbiosis

Does the endosymbosis process occur as a single event or are serial endosymbiosis events possible? Indeed, there is a new concept discussed that predicts serial endosymbiotic events and is hence termed serial endodymbiosis theory (Fig. 2.11) (Archibald 2015). There are many roots of the endosymbiotic theory, and some of tehm are even intertwined. It begun in 1897 with the concept of endosymbisos hypothesized by the Swiss biologist Simon Schwendener, which states that lichens were a composite being composed of a fungus and an alga (Honneger 2002). For systmatists of the nineteenth century, lichens did not fit into the classification schemes (Sapp 1994). In 1879, symbiosis was described as a combined living of two different organisms of different types by the German Anton de Bary (de Bary 1879). This statement was supported by the pioneering work of Polish Franz Kamienski and the German Albert Frank on mycorrhizal fungi, which posses an intimate connection to the roots of plants (Berch et al. 2005; Frank and Trappe 2005). At that time, symbiosis was recognized as a legitimate, yet astounding, biological phenomenon. Thereby, the Russian botanist Constantin Mereschkowsky fulfilled a major role the development of the symbiogenesis concept development, which can be seen of the generation of a new organism from two or more formerly independent organisms that are now interacting to grow and survive together (Mereschkowsky 1920). Mereschkowsky was familiar with subcellular architecture of diatom algae and their chromatophores (synonymously termed plastids) and in lichen biology (Sapp et al. 2002). Hence, in a 1905 publication that has been later translated into English by Martin and Kowallik in 1999 (Martin and Kowallik 1999), Mereschkowsky pointed out that this is the origin for the endosymbiotic plastids.

Mereschkowsky reviewed in this pioneering work the present state of knowledge with respect to symbiosis, including lichens and certain amoebae containing green algae living inside their cytoplasm. Therein has been stated that there are obvious similarities between cellular plastids and free-living cyanobacteria. In addition, it has been pointed out that there exists a continuity of chromatophores in cells by the German Andreas Schimper and the Swiss Carl Wilhelm von Nägeli. Hence, they and others concluded that the plastids are no de novo products of the cell, and they originate rather from other preexisting organelles by the event of organelle division. In line with these results, Mereschkowsky strongly proposed that the plastids had in former times been individual free-living organisms, and similarly, it has been reported by Schimper (1883). Hence, Mereschkowsky is mainly seen as the founder of the famous endosymbiotic theory. However, it has not been noted that mitochondria might be originated by endosymbiosis, although the foundation for the hypothesis of endosymbiosis had been started (Sapp 1994; Archiblad 2014). In the eighteenth century, the concept

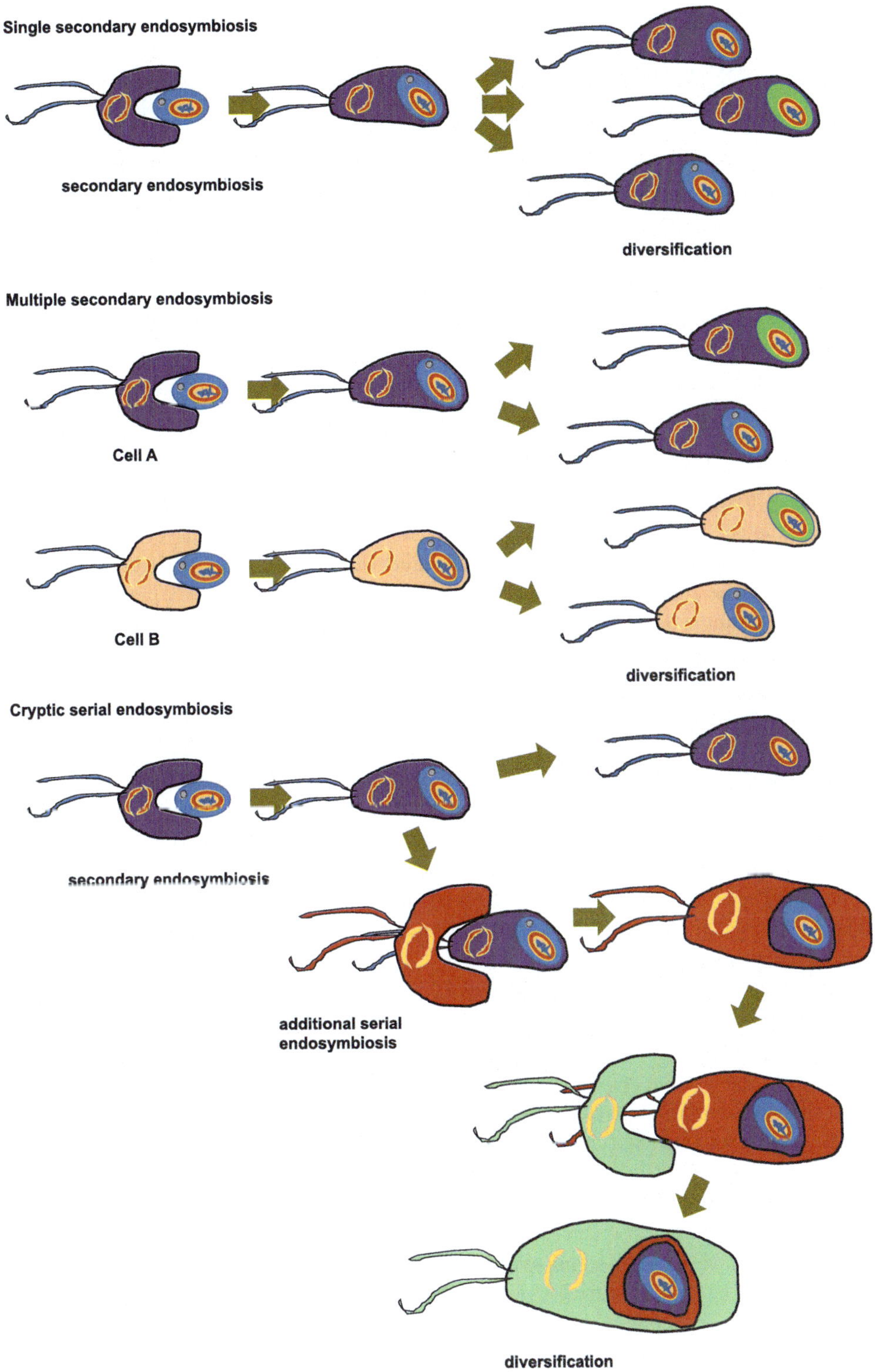

Fig. 2.11 Single or serial endosymbiosis

of symbiosis was forgotten and not mentioned as possible source with great evolutionary capacity. The concept of endosymbiosis was once again put in the spotlight by the American biologist Lynn Margulis (Sapp 1994; Archiblad 2014; Margulis 1998). In the 1950s and 60s, she investigated algae and amoebae and thereby hypothesized that there is properly extranuclear DNA in eukaryotes (Plaut and Sagan 1958).

In 1962, there was the first reported microscopic observation for DNA that is located in plastids (Ris and Plaut 1962) Moreover, the same was identified for mitochondria (Nass and Nass 1963). Since knowledge from different disciplines, such as genetics, bacteriology, cell biology, ecology and even paleontology was gained, Margulis was able to publish a bold, broad-ranging hypothesis for the overall evolution of eukaryotic organisms (Sagan 1967). Therein, the origin of mitosing cells was proposed to be originated from the uptake of free-living cells to create as an ancient endosymbiosis event that then become mitochondria, the 9 + 2 basal bodies of flagella and photosynthetic plastids (Sagan 1967). In 1970, a book entitled "Origin of Eukaryotic Cells by Margulis came out (Margulis 1970a, b) and shed light on symbiosis and especially on endosymbiosis. Thereafter, in the molecular biological age, these ideas became the focus of scientific biological research (Archiblad 2014). Besides Margulis, other researcher reported the importance of endosymbiosis for the evolution of eukaryotic cells (Goksøyr 1967; Raven 1970). Is there an alternative way for how organelles may be developed? Another view on this topic was that mitochondria and plastids originated within the confinement of a photosynthetic eukaryotic cell developed from a cyanobacterial-like prokaryote (Klein and Cronquist 1967; Cavalier-Smith 1975).

In contrast, Margulis did not support this hypothesis (Margulis 1981). In contrast, she explained that oxygen-containing photosynthesis is not an ancestral eukaryotic characteristic, but has only recently developed through endosymbiotic incorporation of a cyanobacterium by an a heterotrophic eukaryote. Other model for the origin in of the oxygen-containing photosynthesis or mitochondria came from report by Lawrence Bogorad (1975), Rudolph Raff and Henry Mahler (1972) as well as Uzzell and Spolsky (1974) that are highly similar to Klein Klein and Cronquist, since they suggest that the intracellular compartments have developed through an mechanism by which finally membrane-bound mitochondria and plastids had formed. All these reports present diverse assumptions about the specific type of prokaryotic population that produced the earliest eukaryotic cells. All these models agreed that endosymbiosis is needlessly radical, since they argue that there is no obvious reason why the initial eukaryotic cell, capable of remarkable evolutionary innovations, should have emerged as a composite of procaryotic cells and cell parts, rather than developing more directly from a partly prokaryotic cell type (Raff and Mahler 1972). In 1974, Max Taylor postulated a so-called Serial Endosymbiosis Theory that provide a framework for all hypothesis, which at the first glance seem to be competing. Thereby, he developed the "autogenous" and "xenogenous" (synonymously termed foreign) models for the eukaryotic evolution, and subsequently, the latter model was the fundament for the "Serial Endosymbiosis Theory" (Archiblad 2014; Taylor 1974).

In the mid-1970s, the first nucleic acid sequences have been identified that support the endosymbiont hypothesis using the laborious RNA cataloguing technique invented by Carl Woese that has been employed to identify the archaea (Woese and Fox 1977). Linda Bonen, Ford Doolittle, Carl Woese and colleagues received extracts of rRNA sequences from algae plastids and cyanobacteria and thus showed a strong evolutionary connection for them (Bonen and Doolittle 1975, 1976; Zablen et al. 1975, see for further reading: Archiblad 2014; Sapp 2009). The rRNA fragments of the mitochondria were confirmed to originate from prokaryotes (Bonen et al. 1977). However, in 1977, these rRNA fragements have not been associated with distinct bacterial lineage, as a phylogenetic linkage between a proteobacteria and mitochondria has been found in 1985 bei Carl Woese and collegues (Yang et al. 1985). The hypothesis of Margulis stating that eukaryotic flagella (synonymously undulipodia) originated through endosymbiosis (Sagan 1967; Margulis 1970a, b, 1981) could not be experimentally examined directly, since DNA was not detected in these organelles providing active cell movement.

Moreover, the evidence of the endosymbiont hypothesis about the of mitochondria and plastids was a conclusion. It has been argued that the molecular sequencing techniques have improved greatly and hence it has been found that mitochondrial and plastid rRNA sequences are more similar to those of specific bacteria (different one for each organelle) than they are the their own nuclear rRNAs. It was the contemplation of a plethora of insights into the biochemistry and molecular biology of organelles accumulated by researchers worldwide that culminated in the death of the autogenic model. In the mid-1980s, it turned out clearly that endosymbiosis seems to be the only explanation for the origin of the organelles (Gray 1992; Gray and Doolittle 1982).

2.8 How Can Eukaryotes Lose Their Mitochondria?

The acquisition of mitochondria enabled eukaryotes to overcome the limits of their structure and hence developed complex organisms (Lane 2017). More precisely, the eukaryotes were able to possess new gene families, and in the last eukaryotic common ancestor, 3000 of these new gene families have been detected (Koonin et al. 2004; Fritz-Laylin et al. 2010). There were even larger proteins (Brocchieri and Karlin 2005), and higher gene expression levels recognized in this ancestor. Moreover, eukaryotes accumulated far more DNA including regulatory proteins (Elliot and Gregory 2015), which at the same time does not indicate that eukaryotes possess an enlarged genome. Instead distinct eukaryotes preformed genomic streamlining, which had never been reported for prokaryotes.

15 Mb is the largest size of the genome in bacteria and archaea, which is 10,000-folds less than the largest eukaryotic genome sizes that can be up to 150,000 Mb (Elliot and Gregory 2015). This large difference in size cannot be explained by a simple continuum of complexity between prokaryotic and eukaryotic cells (Lynch and Marinov 2015, 2017; Lane and Martin 2016). When mitochondria make the difference in complexity, how can eukaryotic cells remain such complex cells if they have lost mitochondria secondarily? The answer refers to amount of pressure for the selection. In bacteria, the advantage is little or no if you are a little bigger, have a little more membrane and more ATP; these cells tend to lose to smaller, more streamlined cells that multiply faster (Vellai et al. 1998). Since bacteria produce enough ATP, more ATP would be no or only a little improvement, which is most prominent in obligate fermenters (Lane 2011, 2015). More precisely, these bacterial fermenters have to be competitive with other cells that can generate more energy from the same substrate and thus grow longer. Moreover, the obligate fermenters among bacteria need to compete with others by growing and multiplying rapidly and are therefore usually among the smallest and most genomically streamlined cells (Makarova and Koonin 2007).

The origin of the eukaryotic cell was not due to gaining more ATP from the endosymbiotic process, since bacteria did not release ATP to the local microenvironment (Martin et al. 2001). It seems to be more likely that endosymbosis was stable for the entire evolution of the eukaryotic cell, except some negligible secondary lists of organelles in a few cell types, due to the effect of the metabolic syntrophy, where the endosymbiont (synonymously the organelle) delivered products for the host cell that are required for cell growth, such as H_2 gas, in the hydrogen hypothesis (Martin and Müller 1998). The hydrogen hypothesis predicts how the first eukaryote was generated. In specific detail, the hypothesis is based on linked biochemistry of energy metabolism. The origin of eukaryotes was based on a symbiosis event between an anaerobic, strictly hydrogen dependent, strictly autotrophic archaebacterium (the host cells) and an eubacterium (the symbiont), which was able to respire, but it still generated molecular hydrogen as an end product serving as waste in an anaerobic heterotrophic metabolism. Hence, the requirement of molecular hydrogen by the host cell that has been generated by the symbionet was the selection principle for the common ancestor of eukaryotic cells. As a result, the more endosymbionts are present, the more substrates are available, and the greater the amount of cellular growth. For the host cells, it was an advantage of being a larger cell and managing to develop transport pathways for organics to bacterial endosymbionts from the early onset, since the selection pressures were broken down due to the fact that the adaptation to the pressure was performed by internal factors. Moreover, from the point of view of the host cell, the larger cell size, the advanced transport systems and the increased complexity of the cytoskeleton from generation to generation seem to be an advantage (Lane 2015).

Ultimately, ATP became an integral factor with the emergence of an ATP/ADP translocator. The eukaryotes then had too much ATP, if any at all, which had to be quickly returned to ADP to effectively control their own mitochondrial membrane potential (Radzvilavicius and Blackstone 2015). The excess of ATP was used for infrastructure programs such as a dynamic remodeling of their cytoskeleton and elevated membrane trafficking that did not impede free flow through metabolic pathways, but rather made it easier to move, endocytose and ultimately phagocytose (Garg and Martin 2016). The process of phagocytosis must have caused the necessary *de novo* evolution and the extremely high expression of hundreds, or even thousands, of new genes, which is a huge energy cost that appears to go far beyond any prokaryote (Lane and Martin 2010, 2016). As soon as all these machines had improved and the advantages were there, the energy needed to sustain a phagocytic lifestyle was certainly transformed. The reason for this behavior is the inherent nature of phagocytosis on its own. While phagocytes reside in an organically rich environment, they are able to remain alive through fermentation only, they do not have to compete in rate of growth, contrary to bacteria, they simply eat the opposition. Phagocytes can therefore continue to survive through fermentation alone and lose their mitochondria, and morphologically, simple phagocytes with slightly enlarged genomes among the archetypes, such as Tritrichomonas Foetus (177 Mb) (Zubacova et al. 2008), all coming from more complex ancestors, are known. A critical difference exists between the origin of phagocytosis, which required mitochondria, supplying both energy and selective propulsion, and the maintenance of phagocytosis, which required only enough food to ferment to allow simpler phagocytes to completely eliminate their mitochondria. This distinction is strict, but has occasionally been forgotten (Booth and Doolittle 2015; Lane and Martin 2016).

References

J.M. Archibald, *One Plus One Equals One: Symbiosis and the Evolution of Complex Life* (Oxford University Press, New York, 2014)

J.M. Archibald, Endosymbiosis and eukaryotic cell evolution. Curr. Biol. **25**, R911–R921 (2015)

J.M. Archibald, P.J. Keeling, Recycled plastids: a 'green movement' in eukaryotic evolution. Trends Genet **18**, 577–584 (2002)

S.M. Berch, H.B. Massicotte, L.E. Tackaberry, Republication of a translation of 'The vegetative organs of Monotropa hypopitys L.' published by F. Kamienski in 1882, with an update on Monotropa mycorrhizas Mycorrhiza, vol. 15, pp. 323–332 (2005)

L. Bogorad, Evolution of organelles and eukaryotic. Genomes Sci. **188**, 891–898 (1975)

L. Bonen, W.F. Doolittle, On the prokaryotic nature of red algal chloroplasts. Proc Natl. Acad. Sci. USA **72**, 2310–2314 (1975)

L. Bonen, W.F. Doolittle, Partial sequences of 16S rRNA and the phylogeny of blue-green algae and chloroplasts. Nature **261**, 669–673 (1976)

L. Bonen, R.S. Cunningham, M.W. Gray, W.F. Doolittle, Wheat embryo mitochondrial 18S ribosomal RNA: evidence for its prokaryotic nature. Nucleic Acids Res. **4**, 663–671 (1977)

A. Booth, W.F. Doolittle, Eukaryogenesis, how special really? Proc. Natl. Acad. Sci. USA **112**, 10278–10285 (2015)

L. Brocchieri, S. Karlin, Protein length in eukaryotic and prokaryotic proteomes. Nucleic Acids Res **33**, 3390–3400 (2005)

G. Burger, M.W. Gray, B.F. Lang, Mitochondrial genomes: anything goes. Trends Genet **19**, 709–716 (2003)

G. Burger, M.W. Gray, L. Forget, B.F. Lang, Strikingly bacteria-like and gene-rich mitochondrial genomes throughout jakobid protists. Genome Biol. Evol. **5**, 418–438 (2013)

T. Cavalier-Smith, The origin of nuclei and of eukaryotic cells. Nature **256**, 463–467 (1975)

T. Cavalier-Smith, The simultaneous symbiotic origin of mitochondria, chloroplasts, and microbodies Ann. NY Acad. Sci. **503**, 55–71 (1987)

A. de Bary, Die Erscheinung der Symbiose (Privately printed in Strasbourg, 1879)

T.A. Elliott, T.R. Gregory, What's in a genome? The C-value enigma and the evolution of eukaryotic genome content . Philos Trans. R. Soc. B **370**, 20140331 (2015)

A.B. Frank, J.M. Trappe, On the nutritional dependence of certain trees on root symbiosis with belowground fungi (an English translation of A.B. Frank's classic paper of 1885). Mycorrhiza **15**, 267–275 (2005)

L.K. Fritz-Laylin, S.E. Prochnik, M.L. Ginger et al., The genome of Naegleria gruberi illuminates early eukaryotic versatility. Cell **140**, 631–642 (2010)

S.G. Garg, W.F. Martin, Mitochondria, the cell cycle, and the origin of sex, via a syncytial eukaryotic common ancestor. Genome Biol. Evol. **8**, 1950–1970 (2016)

J. Goksøyr, Evolution of Eucaryotic Cells. Nature **214**, 1161 (1967)

M.W. Gray, The Endosymbiont Hypothesis Revisited. Int Rev Cytol **141**, 233–357 (1992)

M.W. Gray, Mitochondrial evolution. Cold Spring Harb. Perspect. Biol. **4**, a011403 (2012)

M. W. Gray, Lynn Margulis and the endosymbiont hypothesis: 50 years later. Mol. Bio. Cell **28**, 1285–1287 (2017)

M.W. Gray, W.F. Doolittle, Has the endosymbiont hypothesis been proven? Microbiol. Rev. **46**, 1–42 (1982)

M.W. Gray, G. Burger, B.F. Lang, Mitochondrial evolution. Science **283**, 1476–1481 (1999)

R. Honneger, Simon Schwendener (1829–1919) and the dual hypothesis of lichens. Bryologist **103**, 307–313 (2002)

K.A. Johnson, J.L. Rosenbaum, Basal bodies and DNA. Trends. Cell Biol. **1**, 145–149 (1991)

A. Karnkowska, Vacek V, Zubáčová Z, et al., A eukaryote without a mitochondrial organelle. Curr. Biol. **26**, 1274–1284 (2016)

R. Klein, A. Cronquist, A consideration of the evolutionary and taxonomic significance of some biochemical, micromorphological and physiological characters in the Thallophytes. Quart. Rev. Biol. **42**, 105–296 (1967)

E.V. Koonin, N.D. Fedorova, J.D. Jackson et al., A comprehensive evolutionary classification of proteins encoded in complete eukaryotic genomes. Genome. Biol. **5**, R7 (2004)

V.L. Koumandou, B. Wickstead, M.L. Ginger, M. van der Giezen, J.B. Dacks, M.C. Field, Molecular paleontology and complexity in the last eukaryotic common ancestor. Crit. Rev. Biochem. Mol. Biol. **48**, 373–396 (2013)

N. Lane, Energetics and genetics across the prokaryote-eukaryote divide. Biol. Direct. **6**, 35 (2011)

N. Lane, *The Vital Question: Why is Life the Way it is?* (Profile Books, London, 2015)

N. Lane, Serial endosymbiosis or singular event at the origin of eukaryotes? J. Theor. Biol. **434**, 58–67 (2017)

N. Lane, W. Martin, The energetics of genome complexity. Nature **467**, 929–934 (2010)

N. Lane, W. Martin, Mitochondria, complexity, and evolutionary deficit spending. Proc. Natl. Acad. Sci. USA **113**, E666 (2016)

M. Lynch, G.K. Marinov, The bioenergetic costs of a gene. Proc. Natl. Acad. Sci. USA **112**, 15690–15695 (2015)

M. Lynch, G.K. Marinov, Membranes, energetics, and evolution across the prokaryote-eukaryote divide. eLife **6**, e20437 (2017)

H.R. Mahler, Mitochondrial evolution: organization and regulation of mitochondrial genes. Ann. NY Acad. Sci. **361**, 53–75 (1981)

K.S. Makarova, E.V. Koonin, Evolutionary genomics of lactic acid bacteria. J. Bacteriol. **189**, 1199–1208 (2007)

L. Margulis, *Origin of Eukaryotic Cells* (Yale University Press, 1970)

L. Margulis, *Origin of Eukaryotic Cells* (CT Yale University Press, New Haven, 1970)

L. Margulis, *Symbiosis in Cell Evolution* (W. H. Freeman and Company, 1981)

L. Margulis, *Symbiosis in Cell Evolution* (W. H. Freeman, New York, 1981)

L. Margulis, *Symbiotic Planet* (Basic Books, 1998)

J. Martijn, T.J. Ettema, From archaeon to eukaryote: the evolutionary dark ages of the eukaryotic cell. Biochem. Soc. Trans. **41**, 451–457 (2013)

W. Martin, K.V. Kowallik, Annotated English translation of Mereschkowsky's 1905 paper 'Uber Natur and Ursprung der Chromatophoren im Pflanzenreiche.' Eur. J. Phycol. **34**, 287–295 (1999)

W. Martin, M. Müller, The hydrogen hypothesis for the first eukaryote. Nature **392**, 37–41 (1998)

W.F. Martin, S. Garg, V. Zimorski, Endosymbiotic theories for eukaryote origin. Philos Trans. R. Soc. Lond. B. Biol. Sci. **370**, 20140330 (2015)

W.F. Martin, M. Hoffmeister, C. Rotte, K. Henze, An overview of endosymbiotic models for the origins of eukaryotes, their ATP-producing organelles (mitochondria and hydrogenosomes), and their heterotrophic lifestyle. Biol Chem. **382**, 1521–1539 (2001)

C. Mereschkowsky, La plante considere´ e comme un complexe symbiotique. Societe´ des Sciences Naturelles de l'Ouest de la France. Nante, Bulletin **6**, 17–98 (1920)

M.M.K. Nass, S. Nass, Intramitochondrial fibers with DNA characteristics. I. Fixation and electron staining reactions. J. Cell. Biol. **19**, 593–611 (1963)

A.A. Pittis, T. Gabaldón, Late acquisition of mitochondria by a host with chimaeric prokaryotic ancestry. Nature **531**, 101–104 (2016)

W. Plaut, A.L. Sagan, Incorporation of thymidine in the cytoplasm of Amoeba proteus. J. Biophys. Biochem. Cytol. **4**, 843–846 (1958)

A.M. Poole, S. Gribaldo, Eukaryotic origins: how and when was the mitochondrion acquired? Cold Spring Harb. Perspect. Biol. **6**, a015990 (2014)

A.L. Radzvilavicius, N.W. Blackstone, Conflict and cooperation in eukaryoge- nesis: implications for the timing of endosymbiosis and the evolution of sex. J. R. Soc. Interface **12**, 20150584 (2015)

R.A. Raff, H.R. Mahler, The non symbiotic origin of mitochondria. Science **177**, 575–582 (1972)

P.H. Raven, A multiple origin for plastids and mitochondria. Science **169**, 641–646 (1970)

H. Ris, W. Plaut, Ultrastructure of DNA-containing areas in the chloroplast of Chlamydomonas. J. Cell Biol. **13**, 383–391 (1962)

L. Sagan, On the origin of mitosing cells. J. Theor. Biol. **14**, 225–274 (1967)

J. Sapp, *Evolution by Association A History of Symbiosis* (Oxford University Press, New York, 1994)

J. Sapp, *The New Foundations of Evolution* (Oxford University Press, 2009)

J. Sapp, F. Carrapico, M. Zolotonosov, Symbiogenesis: the hidden face of constantin Mereszhkowsky. Hist. Phil. Life Sci. **24**, 413–440 (2002)

A.F.W. Schimper, Ueber die Entwickelung der Chlorophyllkörner und Farbkörper. Bot. Zeit. **41**, 105–162 (1883)

F.J.R. Taylor, Implications and extensions of the serial endosymbiosis theory of the origin of eukaryotes. Taxon **23**, 229–258 (1974)

T. Uzzell, C. Spolsky, Mitochondria and plastids as endosymbionts: a revival of Special Creation? Am. Sci. **62**, 334–343 (1974)

T. Vellai, K. Takacs, G. Vida, A new aspect to the origin and evolution of eukaryotes. J. Mol. Evol. **46**, 499–507 (1998)

C.R. Woese, G.E. Fox, Phylogenetic structure of the prokaryotic domain. The primary kingdoms. Proc. Natl. Acad. Sci. USA **74**, 5088–5090 (1977)

D. Yang, Y. Oyaizu, H. Oyaizu, G.J. Olsen, C.R. Woese, Mitochondrial origins. Proc. Natl. Acad. Sci. USA **82**, 4443–4447 (1985)

L.B. Zablen, M.S. Kissil, C.R. Woese, D.E. Buetow, Phylogenetic origin of the chloroplast and prokaryotic nature of its ribosomal RNA. Proc. Natl. Acad. Sci. USA **72**, 2418–2422 (1975)

K. Zaremba-Niedzwiedzka, E.F. Caceres, J.H. Saw et al., Asgard archaea illuminate the origin of eukaryotic cellular complexity. Nature **541**, 353–358 (2017)

Z. Zubacova, Z. Cimburek, J. Tachezy, Comparative analysis of trichomonad genome sizes and karyotypes. Mol. Biochem. Parasitol. **161**, 49–54 (2008)

Biomechanical View on the Cytoplasm (and Cytosol) of Cells

3

Abstract

This chapter presents the cytoplasm and the contained organelles from a biomechanical point of view. Thereby the entire cell is treated as gel with associated physical features. General physical principles are applied on cells at a mesoscopic length scale. However, on a submesoscopic scale, the mechanical properties of organelles, such as mitochondria, contribute to the overall mechanical properties of the cell. A cell may even respond to the applied mechanical stress by regulating the fusion of fission of the mitochondria. Hence, mitochondria seem to fulfill a key role in mechanotransduction processes in cells. Finally, it is discussed why mitochondria are predestinated for providing a mechanical response to external mechanical perturbations.

3.1 Biomechanical Picture of the Cytoplasm

From a biomechanical point of view, the cell can be seen as a gel (Pollack 2003). A gel is defined as a cross-linked polymer network that is swollen by the absorption of a liquid. The network maintains the liquid in its spatial position (interstitial space), while the liquid impedes the polymer network from collapsing. The overall properties of a gel are based on its equilibrium state, dynamics and kinetic properties. However, the properties of gels rely on the interaction between the polymer network and the liquid. The swelling of a gel cannot be understood by knowledge of the equilibrium volume in a solvent alone. In addition, thermal fluctuations of local swelling and shrinkage and the time dependency of volume changes due to multiple external perturbations must be taken into account. The relative position of the gel to the boundary of the phase transition affects the characteristics of the gel. The gel state depends on the temperature, osmotic pressure, composition of the solvent and the degree of swelling.

The phase transition of a gel is characterized by a discontinuous volume-phase transition evoked by minute changes in temperature, composition of the solvent, pH, ionic composition or applied electric field. Since these phenomena of gels are universal, there are common physical principles underlying the phase transition of gels.

These gel-like characteristics of the cell's cytoplasm have not been the focus of engineering design, but may possible reveal features for bioengineering or biomaterials. In particular, the knowledge of cell mechanical properties is crucial for coordinated processes including differentiation, morphogenesis and maturations of cell populations in complex tissues (Costantini and Kopan 2010).

C. T. Mierke, *Cellular Mechanics and Biophysics*, Biological and Medical Physics, Biomedical Engineering, https://doi.org/10.1007/978-3-030-58532-7_3

Although systems biological approaches have attempted to understand highly complex developmental process by analyzing the regulatory networks of genes (Davidson et al. 2002; Materna and Davidson 2007, Stathopoulos and Levine 2005), there is not in all cases an advancement over reductionist approaches. The overall characterization of a cell or tissue depends on the processes of morphogenesis, differentiation and homeostasis that rely on the cell's behavior and function, which is controlled by the environmental cues and gene expression. From a biomechanical point of view, when the cell is simply a gel, the cell functions can be understood by understanding the gel function. The knowledge of the principles underlying gel dynamics has increased greatly (Sadati et al. 2013; Pegoraro et al. 2016; Nogucci 2019; Bi et al. 2016; Oswald et al. 2017; Mierke 2018; Mongera et al. 2018). A crucial mechanism in biology is therefore the polymer-gel phase transition, which is a major structural alteration that can be caused by discrete changes of microenvironment. Phase transitions help the cell to perform work required for mechanisms, such as the migration of cells out of a cluster of cells. A polymer-gel-based foundation for cell function can be used to analyze how much this foundation explains how the cell functions and how these principles can be utilized for future engineering design.

What are the Differences Between Cytoplasm and Cytosol?

From the biological point of view, the cytoplasm constitutes a thick liquid, which is a filling in every cell and is surrounded by the cell membrane. The main constituents of cytoplasm are mainly of water, salts and proteins. In eukaryotic cells, all material outside the nucleus and inside the cell compromises the cytoplasm. All of the described organelles of eukaryotic cells, such as endoplasmic reticulum and mitochondria, are located within the cytoplasm. The cell nucleus is surrounded by cytoplasm and contains nucleoplasm. The cytoplasm consists of cytosol, the cell organelles, excluding the nucleus, and several insoluble substances. Besides organelles, the remaining part of the cytoplasm is referred to as the cytosol and represents the liquid part of the cytoplasm.

Although there seems to be no form or structure present in the cytoplasm, it is extremely organized. More precisely, a scaffold of proteins, termed cytoskeleton, structures the cytoplasm and subsequently, the entire cell. In detail, the cytoplasm contains two basic components, which are termed the endoplasm and the ectoplasm. The location of the endoplasm is the central part of the cytoplasm, where the organelles are located too. The outmost region of the cytoplasm is the location of the ectoplasm, which represents a gel-like substance within the cell.

The cytoskeleton is a network of microfilaments that is surrounded by the cytosol. Among internal organelles are mitochondria, Golgi apparatus, lysosomes and vacuoles. However, specific insoluble substances, such glycogen, starch and lipids, are detected additionally in the cytoplasm, which are termed cytoplasmic inclusions. In contrast, salts, sugars, various enzymes, amino acids and fatty acids are dissolved in the cytoplasm.

What are the Functions of Cytoplasm?

Commonly, the cytoplasm surrounds the internal cytoskeleton and the organelles inside the cells and thereby structures the entire cell. Moreover, the cytoplasm contains many proteins that can polymerize to a filamentous scaffold, the cytoskeleton, that helps the cell to maintain its shape and texture. These biopolymeric filaments keep the organelles at place, otherwise in the absence of filaments the organelles would localize to the basal side of the cell. Another function of the cytoskeleton is the regulation of the migration of organelles through the interior of the cell. Since the actin filaments of the cell's ectoplasm are responsible for the migration of the entire cell, the central filaments of the cell promote the movement of organelles and even other cellular structures throughout the cell.

Beyond the storage of multiple nutrients, the cytoplasm is the place for many important cellular functions, such as protein synthesis and anaerobic glycolysis, and activities, such as differentiation, cell division, cell growth, reproduction and responses to their microenvironment. The gel-like substance, the cytoplasm, aids the cell in transport, absorption and the processing of the required nutrients. In detail, the cytosolic enzymes facilitate the cleavage and degradation of large molecules and hence aide the organelles to further process them. An example are mitochondria that are not capable to utilize the cytoplasmatic glucose molecules. Instead, the cytosolic enzymes are required to cleave glucose into pyruvate that can then be further processed by the mitochondria. The activities of the cytoplasm include the exchange of chemicals between organelles and the cytosol. The cytoplasm therefore represents the linking element for the organelles within the cell and synchronizes mitochondrial functions with specific cellular functions. However, there are many more functions of the cytoplasm that are presented in the following chapters.

3.1.1 Cells Serve as an Example for Engineered Designs

The cell-driven engineering design is still developing slowly. One reason for this is the lack of knowledge about how cells perform their tasks, which is due to the less studied physical nature of the cells. The cell represents a network of biopolymers containing basically proteins, sugars and nucleic acids, whose interaction with the solvent, such as water, interaction yields a gel-like structure. This relation is based on a classical view that has been formulated by Frey-Wyssling (1953). Since, the gel-like structure is generally accepted, the so-called gel-sol transition as a crucial mechanism is still under debate (Jones 1999; Berry et al. 2000) similarly as the other consequences of the treatment of the cell's cytoplasm's as a gel-like structure (Janmey et al. 2001; Hochachka 1999). The phenomena and consequence of the gel-sol transition are largely investigated by engineers, polymer and surface scientists, while their findings have not been extensively transferred to biological matter such as cells. An example of such as a transfer is the wetting of tissue phenomenon (Pérez-González et al. 2019; Beaune et al. 2017; Yan et al. 2016). Another example represents the gel-sol transition that can also be employed for cancer cells degrading the extracellular matrix enzymatically in order to migrate into and through it (Berry et al. 2000).

One reason could be that practically all cell biological mechanisms are based on the idea of an aqueous solution or, more precisely, on the free diffusion of dissolved substances (solutes) in an aqueous solution. Moreover, multiple diffusional steps may be required for an action to take place. Among these steps are ions that diffuse into and out of membrane channels or membrane pumps as well as through the cytoplasm and proteins diffusing to other proteins, such as substrates diffusing in the direction of enzymes. This cascade of diffusional steps is influenced by intracellular processes that are largely independent of the gel-like nature of the cytoplasm, where the diffusion process is sufficiently slow and is hence biologically non-relevant. This issue has not been addressed by cell biology, as this discipline is not associated to the gel function of the cytoplasm. Finally, the gel-like nature of the cell's cytoplasm has not been taken into account, while the gel-like nature of the cell's microenvironment has recently been revisited in several reports regarding the engineering of biomaterials (Yan et al. 2016; Echalier et al. 2016, Owensa et al. 2016).

3.1.2 Cells Are Treated as Gels

Gels generally consist of a framework of long-chain polymers, which are usually cross-linked to one another and solvent-containing. The cytoplasm is highly similar to gels. Cellular biopolymers, such as proteins, polysaccharides, and nucleic acids are commonly long-chained molecules, frequently cross-linked to one another to assemble a 3D matrix scaffold. The matrix retains the solvent (water), which is uniform—which is also maintained during the removal of the cell membrane. For instance, "skinned" muscle cells still hold water similar to classical gels. More precisely, the cytoplasm behaves therefore very similar to an ordinary gel.

How the gel cell matrix retains water is of considerable relevance (Rand et al. 2000), and there are minimally two mechanistic choices. The first choice is based on osmosis. More precisely, charged surfaces attract counterions causing the influx of water. The second choice are water–molecule dipoles adsorbing onto charged surfaces, such as the cell membrane, and subsequently these dipoles assemble onto one another to create multilayers. However, the first choice does not seem to be dominant, since firstly, if the gels are in a sufficiently large-sized water bath, the gels may be depleted of counterions. As a consequence, no water can be held and the gels are then be dehydrated. But instead it has been experimentally determined that the gels are still hydrated. The secondly choice is that if the cytoplasm is placed under high-speed centrifugation, it loses ions before it loses water (Ling and Walton 1976). The second hypothesis, that charged surfaces are attracting water dipoles in a multilayered structure, was formulated a long time ago (Ling 1965). More precisely, it is suggested that the water molecules build layer by layer. However, this is still intensively discussed, but is based on important observations. Among these observations, the classic first is that polished quartz surfaces adsorb water films with 600 molecular layers thickness in a humid atmosphere (Pashley and Kitchener 1979). This also means that the water layers are built up one after the other. A second set of observations are that a force is necessary to displace solvents sandwiched between closely spaced parallel mica surfaces (Horn and Israelachvili 1981, Israelachvili and McGuiggan 1988; Israelachvili and Wennerström 1996). The overall behavior can be described by the classical Derjaguin, Landau, Vervey and Overbeek (DLVO) theory. The DLVO theory has been formulated to describe the stability of colloids and has proven to be the key factor in explaining the interactions between colloidal particles and the way they aggregate. More specifically, the DLVO theory can be utilized to rationalize forces, which act between interfaces, and to explain the deposition of particles to planar substrates. In addition, the DLVO theory can interpret the forces between planar substrates, such as in thin liquid films. First of all, the DVLO theory was initially defined for two identical interfaces representing a symmetric system that describes the aggregation of identical particles, the chemical process is known as homoaggregation (Fig. 3.1). In a later stage, this initial concept was refined to two different interfaces, which represents an asymmetric system, and hence the aggregation of diverse particles can be explained for the chemical process termed heteroaggregation. In the borderline situation of large size differences between individual particles, this is similar to particle deposition on a planar substrate (Fig. 3.1).

The well-known DLVO theory is a helpful framework to predict interactions in aqueous colloidal suspensions and their rates of aggregation. More precisely, the theory is based on the assumption that the interaction forces can be estimated well by a superimposing van der Waals and double-layer forces. Within a symmetric system or for homoaggregation, van der Waals forces act attractively and double-layer forces react repulsively. However, when the system is asymmetric and heteroaggregation occurs, it becomes quite more complex. Although the van der Waals forces are still normally attractive, forces of the double-layer forces may be attractive, repulsive or both types. In detail, the charge regulation may have an effect on the overall type of forces. Furthermore, the DLVO theory can help to explain experimental situations in a proper manner. In distinct cases, the DVLO theory can predict the interaction forces and the rate constant for the aggregation rate constants in a quantitatively manner. Even at higher salt concentrations, the deviations can possibly still persist.

A series of regularly spaced peaks and valley can be superimposed to an anticipated monotonic response. Moreover, the distance between the peaks equals the molecular diameter of the fluid that is sandwiched. Hence, it can be concluded that the force oscillations are based on the molecule layering between the surfaces.

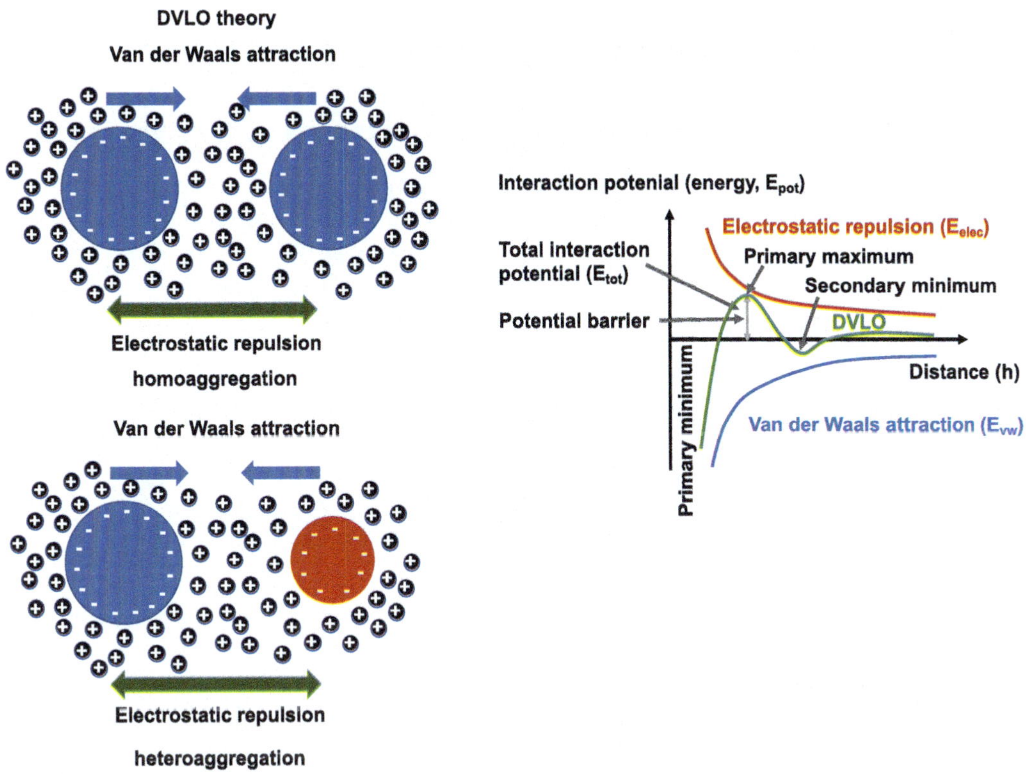

Fig. 3.1 Illustration of the DVLO theory for homoaggregation (top left) and heteroaggregation (bottom left) and an energy distance diagram (right)

The experiments by Israelachvili showed that near charged surfaces, the molecules are layered. However, they do not reveal whether the molecules are interconnected. Other later experiments reported that carbon-nanotube-tipped AFM probes that are approached to flexible monolayer surfaces display a similar molecule layering in water (Jarvis et al. 2000). Hence, these results imply that the ordering of the molecules is not simply based on the constraints of the molecule packing. In addition, the Pashley/Kitchener experiment leads to the suggestion that even multiple layers can arrange. Hence, a distinct kind of layering is diagrammed (Fig. 3.2) and collectively implied by these experiments.

When the two charged polymeric surfaces are located in close proximity, the interfacial layers of water molecules can even bond the surfaces similar to a glue. This phenomenon can be observed in commonly in nature and hence is an everyday experience. As long as the slides are not wet, a separation of the two glass slides that stacked face-to-face is still possible. However, when the slides become wet, a separation is no longer easily obtained, since the sandwiched water molecules hinder the separation. More precisely, they adhere stubbornly to the glass surfaces and to each other and thereby prevent the separation. A similar principle can be found in sand: When walking in dry sand at the beach, your foot will ordinarily sink deeply into the sand, which creates a large imprint. However, when you walk in wet sand instead of dry sand, your foot will not sink deeply in it and hence the imprint is shallow. More precisely, the water adheres to the sand particles and combines them with sufficient strength to carry the full weight.

When we transfer this observation to cells, the cytoplasmic matrix seems to have the appearance of a gel matrix. In both examples, the water molecules are retained, because they have high affinity for

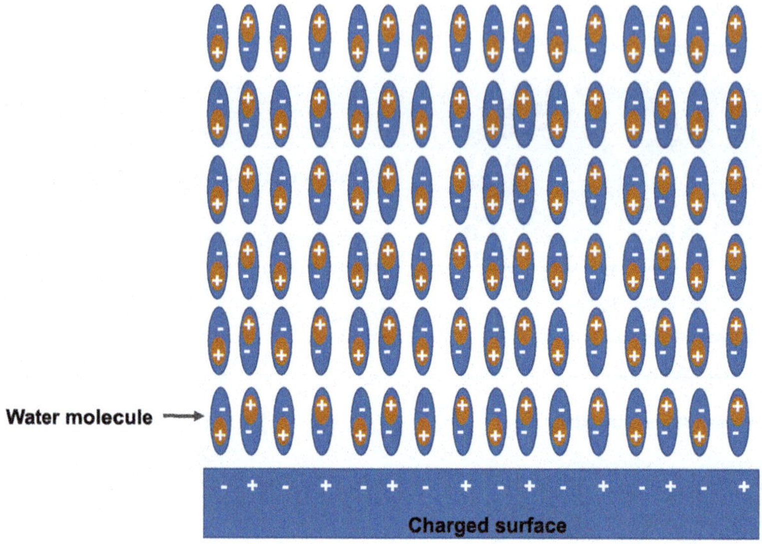

Fig. 3.2 Pashley/Kirchener experiment: The organization of water molecules (force dipoles) in the immediate vicinity of a charged surface is realized in several layers of water molecules

the charged and hence hydrophilic surfaces and their affinity to each another (Fig. 3.3). The gel is mainly composed of the polymer matrix and the adsorbed water, which thereby provides an explanation why cells without a plasma membrane still maintain their overall integrity. This gel-like structure contains several characteristics that provide cell functions. Ion partitioning seems to be an important issue. Why are there ion gradients between the extracellular and intracellular compartments? An obvious explanation can be maintenance of the equilibrium between the passive flow through ion channels and the active transport of ions with ion pumps. Hence, the low intracellular sodium concentration relative to the extracellular seems to be evoked by the sodium pump activities that can transport sodium ions against the concentration gradient from the interior of a cell across the plasma membrane into the exterior environment. However, when we treat the cells as a gel, an alternative explanation seems to be possible. Firstly, there seem to be a difference in the ion solubility of extracellular bulk water and intracellular layered and thereby so-called structured water and secondly, a difference in the affinity of various ions for the charged polymeric surfaces of a living cell (Ling 1992). For example, Na^+ ions possess a larger hydrated diameter than K^+ ions. Thus, Na^+ ions are more strongly excluded from the cytoplasm than K^+ ions, since the hydration layers of Na^+ ions require more energy to be removed compared to hydration layers of K^+ ions. Moreover, K^+ ions possess a higher affinity for the negatively charged polymeric plasma membrane surfaces. Hence, the cytoplasm contains significantly more potassium than sodium ions (Pollack 2001).

In the same manner, the gel treatment of the cell can help to explain the cell potential. Since the cell contains several negatively charged polymers, these polymers attract surrounding cations. The number of cations that penetrate the cell (or gel) is limited by the low solubility of the cations in structured water. The cations that manage to enter a cell must compete with the water dipoles for the fixed anionic charges of a cell. Subsequently, the negative charge of the cell's interior is not completely balanced by cytosolic cations. The residual charge is approximately 0.3 mol/kg (Wiggins 1990). Finally, the cytoplasm has a net negative charge and hence possesses a net negative potential. In fact, depending on specific conditions, cells without a plasma membrane have cell potentials of

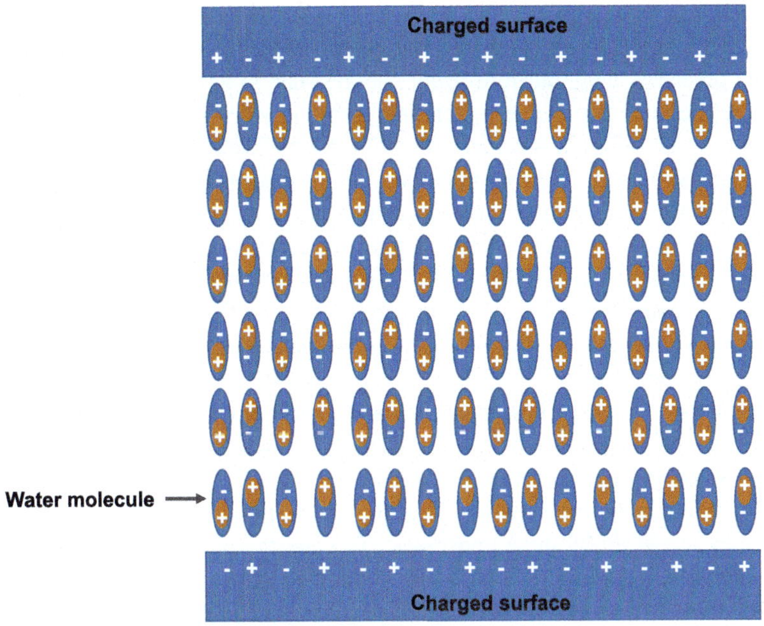

Fig. 3.3 Organization of water molecules between two charged surfaces is arranged in several layers of water molecules

50 mV (Collins and Edwards 1971). In line with this result, gels constituted of negatively charged polymers possess comparable or even larger negative potentials, whereas gels that are constituted of positively charged polymers exhibit equivalent positive potentials.

In conclusion, plasma membranes, ion pumps and ion channels seem to be not important for the generation of these cell potentials. Moreover, the gel paradigm of the cells can go quite far in explaining the cell's most fundamental attributes, such as the distribution of ions and the existence of a cell potential. These two processes represent equilibrium processes and hence no energy for the maintenance is necessary.

3.1.3 Dynamics of Cells

In order understand the function of cells, its dynamical nature needs to be taken into account. Since the cell is obviously not a static structure, it rather represents a machine that is designed for a variety of functions. Among these functions are a broad variety of mechanisms, which do not seem to have a clearly underlying theme. More precisely, each process seems to be driven by different mechanisms. Hence, a question can be raised whether there exists a common underlying theme that controls the tasks of the cell. At the first glance, a cell can be seen as a simple gel and treated like this. However, over time the cell can become specialized and hence differentiated. Thereby the gel structure and even the processes were made more complex. In view of such a line, the potential for a simple, general, underlying, gel-based theme should not necessarily be small. The identification of a common underlying theme is still an ongoing task in other disciplines. In (bio-)physics, the search for a unifying force has never ended when the protégés of Einstein continue. The fact that nature works sparingly and uses variations of a few simple principles to accomplish manifold actions is an attractive concept that, to my knowledge, has not yet been applied to cell functions, although the concept of simplicity is a key principle of technology.

When the cell is a gel, it would be the next step to ask for a common principle that underlies this cell function. Hence, what are the principle that guides the cell's gel function? In fact, gels fulfill a function. More precisely, they transform from one gel state to another gel state. This behavior has been revealed as the famous polymer-gel phase transition, which is similar to the well-known transition from ice to water, where a small alteration of the environment, such as a temperature increase, evokes a pronounced alteration in the structural arrangement of the water molecules (Tanaka et al. 1992; Hofman 1991). This dramatic alteration can lead to work. Similar as the formation of ice possesses the power to even causes the fracture of hardened concrete, the expansion of a gel expansion or its contraction can perform several types of work that can cause the separation of solute from solvent or the generation of forces. Among these highly useful phase transitions are the so-called time-release capsule, where during the gel-sol transition bioactive substances are secreted by the cell. Thereby the gel functions similar as a disposable diaper, when a condensed gel is subject to large hydration and extension to take on the load, such as a variety of man-made muscles. These behaviors are important for advanced research in the Physics of Diseases, since a small alteration of local environment can lead to large structural alterations.

Similar to synthetic gels, the natural gels of the cells similarly perform such useful transitions. However, do they indeed fulfill these transitions? The question may require more adaptation to the situation of natural cell gels, since a cell is not a homogenous gel, but contains various gel-like organelles, each of which has its own purpose. Hence, it is more important to ask whether all or only a selection of these various organelles can perform the phase transition function. The answer is simple yes, and they can undergo a phase transition.

3.1.4 Cells, Gels and Motion

The motion based on phase transitions can be subdivided into two main categories, such as isotropic and linear. Inside isotropic gels, the polymers are distributed in a random manner and sometimes they are even cross-linked. The water is kept largely by its affinity to the biopolymers, such as the proteins of the cell. Since the gel is largely hydrated, it may even possess in an extreme case 99.97% water (Osada and Gong 1993). When the gel is in a transitioned state, the prevailing polymer–water affinity yields to a higher polymer–polymer affinity, triggering subsequent condensation of the entire gel into a concentrated mass which ejects the solvent.

Similar to isotropic gels, linear polymers show a transition from the prolonged to the shortened state. Moreover, the prolonged state represents a stable state, as it increases the polymer–water interactions to a maximal number and thereby increases the energy of the system to a minimum. Water structures itself layer by layer. In the shortened state of the gel, the affinity of the polymer to itself is greater than the affinity of the polymer to water, and consequently the polymer folds. It can fold completely or locally over a partial length. As the gel folds, polymer and water molecules both move accordingly. If a load is located at the end of the shortening filament, the load can also move.

The phase transitions are inevitably cooperative: As soon as they are triggered, they are completed. The reason for this is the razor blade behavior of the transition: As soon as the polymer–polymer affinity (or the polymer–water affinity) prevails, the prevalence of the polymer increases and the transition is completed. An example is presented in Fig. 3.4. There the divalent ion, calcium, can evoke cross-links between neighboring polymer strands (Fig. 3.4). The presence of calcium ions can transform the prevailing affinity of polymer–water to polymer–polymer interactions. When part of the strand is bridged, adjacent segments of the polymer are more closely joined, increasing the tendency

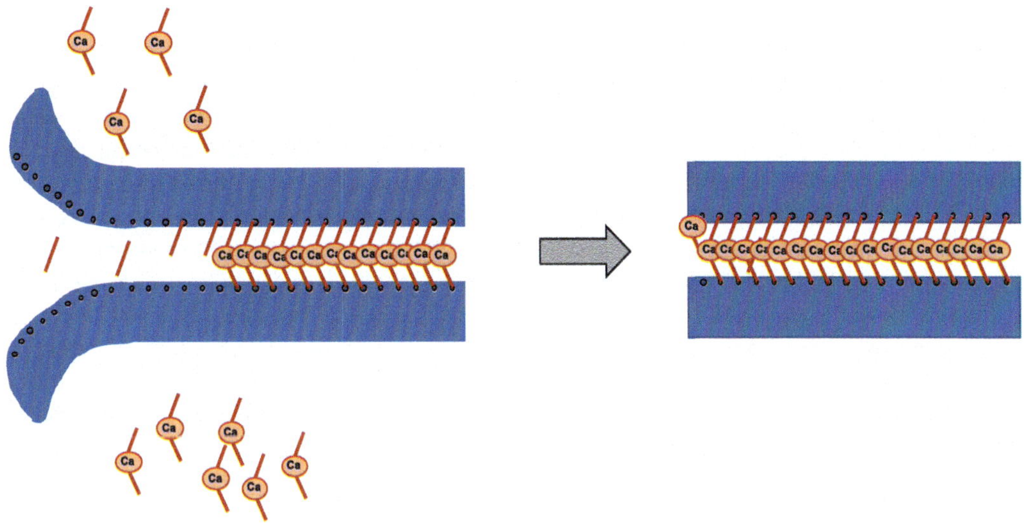

Fig. 3.4 Calcium and divalent cations can bridge the gap between two negatively charged surfaces and be arranged as a zipper-like condensation

for an additional calcium bridge. In this way, the local action increases the propensity to act in a contiguous segment and guarantees that the response is completed. Finally, these transitions tend toward completeness.

Obviously, the polymer-gel phase transition can generate different types of movement. When the cell would use this principle, it could have an easy way to generate a wide variety of movements, depending on the type and configuration of each polymer. To all intents and purposes, a small shift in an environmental parameter such as pH, chemical content and temperature, can result in a cooperative, all-or-nothing reaction that could produce a massive mechanical response. Thus, phase transitions provide more than a theoretical value for cell organelles (Pollack 2001). Although these mechanisms are simpler in its nature than accepted mechanisms, there exists experimental evidence from them. Finally, the phase transition has been hypothesized to be a basic mechanism for cellular motion.

3.2 Anisotropic Mechanics and Dynamics of the Cytoplasm of Living Cells

Cell which under physiological conditions alteration that perturbs the distribution of the cytoskeleton and thereby leads to anisotropic behavior of the entire cell. There is evidence of anisotropy in cells under mechanical stress; however, the role of cytoskeletal remodeling due to alterations in the shape of cells that display mechanical anisotropy and its effects are not clearly understood. The role of cell morphology in the initiation of anisotropy in both features, the intracellular mechanics and dynamics has been addressed (Gupta et al. 2019). When the aspect ratio of the cell is altered by limiting its widths, the cytoskeletal mechanics can be determined with optical tweezers along the major and minor axes of the cell to identify the extent of mechanical anisotropy. In the next step, these active microrheological analyses can be linked with movements of intracellular structures or elements to quantify the overall intracellular force spectrum by force spectrum microscopy (FSM), out of which the grade of anisotropy in force and dynamics can be measured. As suggested, it turned out that unconfined cells displaying an aspect ratio of 1 are isotropic. In turn, when cells lost their symmetry, they become pronouncedly anisotropic in the cytoplasmic mechanics and the dynamic remodeling of the entire cytoplasm (Gupta et al. 2019).

The cytoplasm can be described as an active complex material, in which multiple biochemical reactions take place that are crucial for the proper function of the entire cell. Biological processes including the migration of cells (Lange and Fabry 2013; Recho et al. 2015; Mak et al. 2016; Mierke et al. 2011, 2017; Fischer et al. 2017, Mierke 2019), mechanotransduction pathways (Wang et al. 1993) and the metastatic progression of cancer (Suresh 2007; Heyden and Ortiz 2016; Kim et al. 2018; Mak et al. 2017; Spill et al. 2016; Zaman 2013; Leggett et al. 2017) are associated with mechanical processes and are even crucial for the regulation of cell mechanics. Subsequently, many novel biophysical techniques have been reported that measure distinct mechanical properties of cells (Bao and Suresh 2003; Maloney et al. 2010; Hochmuth 2000; Bausch et al. 1999; Guo et al. 2013; Dao et al. 2003; Gupta and Guo 2017; Hu et al. 2017; Mierke et al. 2017). The majority of studies of the cell mechanics, such as cell stiffness or shear modulus, assume that cells under investigation are isotropic (Bao and Suresh 2003; Maloney et al. 2010; Hochmuth 2000; Bausch et al. 1999; Guo et al. 2013; Dao et al. 2003; Gupta and Guo 2017; Hu et al. 2017; Ehrlicher et al. 2015). Contrary, the cytoplasm of a cell presents a pronouncedly crowded, dynamically remodeling and hence heterogeneous material that is able to transform into a polar and anisotropic material by altering its morphology under specific, but still physiological conditions. In fact, it was suggested that the formation of structural asymmetry in the plasma membrane was the first decisive factor for the direction of cell polarity (Yeaman et al. 1999).

The anisotropy in cells is crucial and seems to be responsible for mediating a various set of cellular functions, including the directional migration of cells (Hidalgo-Carcedo et al. 2011), cell differentiation (Huelsken et al. 2001), the regional membrane growth (Ziman et al. 1993; Wodarz 2000), immune system activation (Hershberg and Mayer 2000) and molecule transport over cell layers (Drubin 1991). In addition, there it has been state that the cell–extracellular matrix and cell–cell interactions facilitate the polarization of cells that pronouncedly causes a restructuring of the intracellular organization and subsequently fulfills a key function in the development of organisms (Thery et al. 2006).

Understanding the directional changes in cells that affect the mechanical properties of the cytoskeleton due to structural disturbances will contribute to a broader knowledge of how cell polarity determines cellular processes. In fact, the restructuring of cytoskeletal elements, such as actin filaments, intermediate filaments and microtubules, is caused by changes in cell morphology that occur during external stimulation by specific substances or physiological processes and can therefore respond to different cellular stresses (Oakes et al. 2014; Zemel et al. 2010; Wang et al. 2002; Tolic-Nørrelykke and Wang 2005) and subsequently manifest the anisotropic mechanical properties of the cell (Thery et al. 2006; Hu et al. 2004; Del Alamo et al. 2008). For example, endothelial cells can be probed with a continuous laminar flow shear stress that causes directional anisotropy within these cells and finally results in temporal and spatial alterations of the creep compliance of the cytoplasm due to the selected direction (Del Alamo et al. 2008). Another example are smooth muscle cells that explore mechanical strain and then adapt their cell mechanics by the structural remodeling of the cytoskeleton, leading subsequently to anisotropy in mechanical properties (Smith et al. 2003; Su et al. 2007). Finally, the mechanical anisotropy of the cell membrane can be evoked by the binding of distinct ligands toward their specific receptors (Irmscher et al. 2012). The majority of the studies analyzed the polarity in the generation of traction forces and the local mechanical alterations of the cell cortex. A systematic investigation of the directional component in the mechanical properties of the cytoplasm is still not yet fully known, but approaches have recently been made (Gupta et al. 2019).

The structural remodeling of the cytoskeleton determines the overall cytoplasmic mechanics, whereas it can additionally impact the dynamical processes inside the cytoskeleton, such as the motor proteins, which act together with the cytoskeletal filaments (Hirokawa 1998; Vale 2003; Howard 1997; Kasza and Zallen 2011; Lau et al. 2003). When the cells function out-of-equilibrium (Gupta and Guo 2017; Guo et al. 2014; Fodor et al. 2015; Fakhri et al. 2014; Wilhelm 2008), the intracellular movement that is generated by coordinated and collective action of the motor proteins is coupled with intracellular forces and of course with mechanics. The intracellular forces facilitate essential functions of the entire cells including cell division (Sharp et al. 2000), migration and invasion of cells (Gardel et al. 2010) and contraction of the cell for cell division or movement (Murrell et al. 2015). Thus, it is necessary to improve the knowledge on how these forces can affect the polarity of cells from a biophysical point of view.

The local restructuring of the cell's cytoplasm caused cell shape alterations, and it has been analyzed in terms of intracellular mechanics and dynamics changes (Gupta et al. 2019). The microcontact printing technique can be employed to limit the cell width in one direction, whereas in the other perpendicular direction, the cells are not confined, which leads to highly polarized cells with increased aspect ratios (greater than 1). In addition, optical tweezers and microscopic particle tracking can be utilized to explore cell mechanics and dynamics within specific regions of the cytoplasm in at least two different directions, such as parallel and perpendicular to the principal axis of the cells. It has been found that the mechanical properties of the cytoplasm and the movement of tracer beads, which have been phagocytized by the cells, can significantly anisotropic due to the cell morphology. Moreover, the spectrum of the intracellular forces can be determined with FSM. As expected, it has been demonstrated that the anisotropic dynamics of the cytoplasm are guided by the alignment of fiber or actomyosin-based forces of the cytoplasm (Gupta and Guo 2017; Guo et al. 2014).

When a cell remodels its actin and microtubules, its entire morphology can change and thereby also become less isotropic. This involves anisotropic features of cytoplasmic mechanics and dynamics. With microcontact printing one-directional confined cells with aspect ratios greater than 1 can be generated. The orientation of actin filaments and microtubules in these highly polarized cells is not randomly distributed, instead they are aligned in their major axis (longitudinal direction) and hence anisotropic distributed (Gupta et al. 2019). Finally, this anisotropy in the intracellular structures causes directionality-dependent mechanics and dynamics of the cell's cytoplasm. Active microrheological measurements with optical tweezers revealed that cells, which possess an aspect ratio of 1, exhibit similar intracellular stiffness values independent of whether the longitudinal and transverse directions are analyzed that suggests isotropic cell behavior. When the aspect ratio of the cell increases, the intracellular mechanics are no longer isotropic, since the stiffness of the cell increases in both directions, the longitudinal and the transverse, but in each at different rates. In particular, the stiffness increases more rapidly in the longitudinal direction compared to the transverse direction that causes the elevation of aspect ratio of the cell (Gupta et al. 2019). An explanation for this behavior may be that the mechanical stress within the cell plane parallel to cytoskeletal filaments leads to a stretching of the filaments. In contrast, mechanical stress in the plane perpendicular to the filaments causes a deformation that results finally in the bending of the filaments. In detail, the bending modulus of these major cytoskeletal components, such as F-actin stress fibers and microtubules, was lower than the elastic modulus of the stretched fibers, and thereby the different stiffness of the two directions can be explained. Consequently, the stiffness in the longitudinal direction (major axis of the cell) correlates with the increasing aspect ratio of the elongated cells (Higuchi et al. 1995; Gittes et al. 1993; Needleman et al. 2005; Howard 2001).

In addition, it needs to be figured out why there is an increase in stiffness in both directions of elongated (increased aspect ratio). Hence, the volume of the cells has been analyzed and correlated with the aspect ratio. In fact, it has turned out that there was a pronouncedly reduction of the volume of the cells, when the aspect ratio is increased (Gupta et al. 2019). More specifically, the relationship between an increase in cell stiffness and a decrease in cell volume is consistent with the relationship between a decrease in cell volume when the outflow of water from the cells is increased, an increase in cytoplasmic stiffness that can be explained by the concentration of intracellular components in the shrinking cells (Guo et al. 2017; Zhou et al. 2009).

By modulation of the cell morphology that in turn causes the remodeling of the cytoskeleton, which then in the case that motor proteins act on cytoskeletal fibers, the intracellular movement can be altered. The motion of markers, such as 500-nm-sized tracer particles can be determined by calculating the mean squared displacement (MSD). When a cell breaks its own symmetry, the intracellular movement is anisotropic. The ball-like cells with aspect ratio of 1 can be treated as an isotropic material, since the MSD values in the longitudinal and transverse directions are very similar. When the aspect ratio of the cell is increased, the MSD is significantly increased in the longitudinal direction, whereas it is even decreased in the transverse direction, which clearly shows that the movements inside the cells switch to anisotropic movement behavior (Gupta et al. 2019). The actin cytoskeleton fulfills a key role, the intracellular fluctuations are random. However, in order to archive anisotropic dynamical movements in the cytoplasm, the actin filament network needs to be destroyed by cytochalasin D. As suggested, the MSDs in both directions, such as longitudinal and transverse cell axis, are broken down, and the dynamical remodeling is abolished (Gupta et al. 2019). Therefore, it seems to be likely that actin filament alignment in longitudinal (highly polarized) cells is a key factor in the regulation of the anisotropy within cytoskeletal dynamics. The hypothesis that the actin cytoskeleton regulates the anisotropic motion of the cytoplasm is not easy to validate, as the active forces of this structure and the scaffold of the entire cytoskeleton are coupled. In fine detail, the molecular motors that are the key producer of intracellular force are interconnected with structure of the cytoskeleton and subsequently, they function together (Schliwa and Woehlke 2003; Spudich 1994). Another point is obvious that the force direction due to the polymerization and depolymerization of the cytoskeletal components, and cytoskeletal filaments also correlate with the overall structure of the cytoskeleton (Desai and Mitchison 1997; Mogilner and Oster 2003).

Consequently, the cell symmetry seems to be disturbed by the alignment of the actin filaments and the associated biased forces dependent on this alignment, leading to anisotropic dynamics in the cells. The anisotropic MSD of cells and its associated increased aspect ratio can be investigated by measurements of the aggregate intracellular force with FSM. In fact, in the longitudinal direction the amount of intracellular force correlates with the increasing aspect ratio. The increased force that facilitates intracellular movement appears to be more pronounced than the increased resistance measured as the stiffness that increases the MSD in the longitudinal direction of the cell due to the cell's increase in aspect ratio. However, the force in the transverse direction of the cells is not altered, when the aspect ratio is elevated. Moreover, an increase in the stiffness of the transverse direction decreases the MSD (Gupta et al. 2019). When the cell is not under structural confinement, the FSM displays that the intracellular forces in the longitudinal and transverse directions are nearly the same magnitude confirming that the cells possess isotropic dynamical properties.

When the cell's aspect ratio increases, the cells are no longer isotropic and instead, they display anisotropy or forces that rely on the direction. This finding indicates that the anisotropy in the fluctuations of intracellular forces seems to be based on alterations of the cell morphology. Another result is that the transition point of the plateau (short-term) to the apparent diffusion behavior (long-term) of the MSD takes place later in the transverse direction. An explanation for this behavior may be the decreased power that is generated by the active elements in the transverse direction (minor axis) of the cell, which has been reported with analysis of the force spectrum (Gupta et al. 2019). Worth highlighting is the recent evidence that cells work far from equilibrium (Guo et al. 2014;

Fodor et al. 2015; Fakhri et al. 2014). For this reason, intracellular motion is characterized by both cytoplasmic mechanics and non-equilibrium forces inside the cell. Therefore, the mere monitoring of intracellular fluctuations does not supply us with an insight into the mechanical properties of the cell (Gupta and Guo 2017; Guo et al. 2014; Fodor et al. 2015; Fakhri et al. 2014); however, an exception applies to high frequencies (Gupta and Guo 2017). An example are combined experiments with active microrheology and particle tracking, which demonstrate that the increase in MSD with increasing aspect ratio of the cell is more due to increased intracellular forces than to a decrease in stiffness of the longitudinal direction of the cell (Gupta et al. 2019).

Finally, these findings demonstrate systematically that the morphology of a cell seem to be a key regulatory element of the anisotropic behavior of the cytoplasm that can be seen in cytoplasmic mechanics, dynamics and forces. The anisotropy in the mechanical properties has been established through the alignment of cytoskeletal and the anisotropy in dynamical properties has been found to be dependent on anisotropic forces and a distorted cytoskeletal structural scaffold (Gupta et al. 2019). In fact, it is known that various cell types including epithelial or endothelial cells and fibroblasts exhibit aligned cytoskeletal fibers, when they are polarized and the aspect ratio is increased (Versaevel et al. 2012; Lee et al. 2016; Ishizaki et al. 2001). The perceived anisotropic mechanics, dynamics and forces are actually a result of the transformation of the cytoskeleton into a better aligned structure when cells crack their symmetry. That is why we assume that these findings should be valid for various cell types in general. In conclusion, these findings highlight the importance of the directionality-based intracellular mechanics, dynamics and forces inside a cell, when the cell exhibits a shape that is distant from isotropicity. This basic insight into how cell polarity affects the mechanics and dynamics of living mammalian cells will provide a deeper knowledge of the physical aspects of biological processes, including cell migration and invasion, cell and tissue differentiation, the intravasation and extravasation of cancer cells, in of which occurs the induction of the polarity of the cells.

3.3 Relation Between the Shape of the Nucleus and Cytoskeletal Mechanics

The connection between the cytoskeleton and the nucleus of cells is crucial for several major cell functions, such as proliferation, differentiation, extracellular induced signal transduction pathways and the formation of tissues. However, the mechanical linkages of the cytoskeleton and the nucleus are not yet clearly revealed. In order to gain more knowledge about this linkage, 2D particle scale tracking microrheology has been performed together with the cell morphological analysis. In particular, distinct human stem cell types in their adhesive state can be identified due to specific cytoplasmic mechanical features under defined cell culture conditions. It has been reported that the shape of the nucleus quantifies the mechanics in the perinuclear cytoskeleton of multiple stem cell types (Lozoya et al. 2016). The perinuclear cytoskeleton region is unique for its specific mechanical properties that cannot be found anywhere else within the entire cytoskeleton, since it contains heterogeneously distributed subregions displaying subdiffusive characteristics and that follow physical relationships, which characterize in a conserved manner multiple stem cell types (Lozoya et al. 2016). In conclusion, different stem cell types seem to be distinguishable by connection of the perinuclear mechanics to the shape of the nucleus.

Indeed, the nucleus and the cytoskeleton are generally physically connected in all eukaryotic cells. These linkages enable cells to physically sense their surrounding microenvironment by employing their cytoskeleton and then transmitting the information to the nucleus, where it induces physiological reactions (Maniotis et al. 1997; Cytrynbaum et al. 2005; Driscoll et al. 2015). How these linkages induce a nuclear reaction depends on the individual cell phenotype and the cytokines it encounters in its specific microenvironment (Pritchard and Guilak 2006; Haudenschild et al. 2009; Godbout et al. 2013). The capacity of the cells to recognize and react to physical impulses is given throughout the entire development process. This property relies not just on the mechanical properties of the various cytoskeletal backbones, but also on their rebuilding capability under stress and on the interplay between the cytoskeleton and the nucleus (Wozniak and Chen 2009; Swift et al. 2013). Cells exploit the mechanical signals that reach this connection to tune their own phenotype during their development and adjust the physical state of the nucleus to their microenvironment (Chowdhury et al. 2010; Ingber et al. 1995).

Based on these physiological interactions, the cells facilitate their organization to coordinate higher-level morphogenic mechanisms such as the collective migration of cell (Angelini et al. 2011; Tambe et al. 2011; Trepat and Fredberg 2011; Anon et al. 2012; Vedula et al. 2014) and the sorting of specific cells (Beysens et al. 2000; Cartwright et al. 2008; Foty and Steinberg 2005; Krieg et al. 2008). During the course of time, these phenomena dictate the morphogenic processes, the differentiation of stem cells and finally regulate the heterogeneity of tissues (Engler et al. 2006; Guilak et al. 2009; Calzolari et al. 2014; Grapin-Botton and Melton 2000; Moore et al. 2005; Lozoya et al. 2011).

Stem cells are capable of reorganizing their cytoskeleton to control intracellular mechanics during maturation and to accommodate altered physical and biochemical conditions. In the course of the remodeling of the cytoskeleton, the cells change their cytoskeletal linkage to the nucleus to maintain their environmental awareness and simultaneously remodel their intracellular structure. In the meantime, the nucleus is able to rearrange its structure to accommodate a reshaping cytoskeleton (Heo et al. 2015). Therefore, the shape of the nucleus can be seen as an architectural signature that creates a delicate balance between the mechanical properties of the nucleus sensing the mechanical signals of the surrounding microenvironment of the cell and the cytoskeleton that is ultimately in charge of transmitting these mechanical signals through the cell, when these cells are physically connected to their immediate environment. This structural connection is based on the correlation between the shape of the nucleus and phenomenon of multipotency, which is generally found in stem cells in vitro (Lozoya et al. 2011; Heo et al. 2015; Chalut et al. 2012, 2010; Kutscheidt et al. 2014; Pagliara et al. 2014).

The structural transformations lead to a path-breaking and heterogeneous physiology in the nucleus and the cytoplasm of stem cells undergoing differentiation (Daniels et al. 2010; Ekpenyong et al. 2012; Li et al. 2014; Young and Engler 2011). Afterward, these specialized characteristics can initially lead sparse pools of progenitors to arrange physiologically demanding tissues with complex multicellular assemblies and spatio-temporally regulated architectures (Foty and Steinberg 2005; Lozoya et al. 2012; Wan et al. 2008; Forgacs and Newman 2005; Forgacs et al. 1998; Foty et al. 1996). In other words, in this light, it seems clear that the cytoskeleton may have different structural characteristics between various stem cell types, but we overlook the way in which these differences are expressed in the intracellular mechanics of stem cells or how they can be manipulated to distinguish stem cell lines by physical evidence. Hence, the mechanics of the cytoplasm are explored in various human stem cells. Thereby it has been hypothesized that stem cells of different multipotency levels or altered stemness display different cytoplasmic mechanical properties. In specific detail, it was investigated whether the cytoplasmic mechanical properties can be utilized as parameters to distinguish the specific phenotypes of stem cells. Therefore, it is determined whether the cytoplasmic mechanical properties: firstly, identify the specific restructuring of the cytoskeleton, when the stem cells are stimulated with distinct soluble cytokines or the polymerization of actin is pharmacologically impaired in vitro, secondly, adjust themselves differently to the stimulated cytoskeletal transformation when the stem cell morphology was committed to a fixed geometry and thirdly, differentiate between stem cells from multiple lines (Lozoya et al. 2016). Following an investigation of the relationships between the cytoplasmic mechanical properties in stem cells and their cytoskeleton, we assessed whether a combination of these mechanical properties is consistent with a predictive relationship shared by stem cells in all experimental regimes, assuming that such relationships, when they exist, may indicate a structural basis present in all stem cells.

In adherent stem cells, cytoplasmatic beads throughout the F-actin cytoskeleton can be localized and tracked dynamically. Thus, the cytoplasmic mechanical properties can be determined in living human stem cells with nanoscale particle tracking microrheology. More precisely, 1 μm-sized spherical beads, which are taken up by the cells through endocytosis, serve as cytoskeletal marker that can be tracked. It has so far been demonstrated that estimates of cytoplasmic mechanics in living cells with beads 1 μm or larger are consistent for microrheology regardless of whether they are trapped inside or outside endosomes (Jonas et al. 2008). Hence, the lipofection-based particle delivery to the cells was adapted for a low-titer of beads minimizing their effect of the viability and growth of the cells. After the transfection of human stem cells with beads, the beads are followed by confocal laser scanning microscopy to determine whether the cytoplasmic beads are entangled by cytoskeletal lattices or separated in vacuoles from the cytoplasm. The live microscopic analysis of actin-GFP expressing cells and the microscopic analysis of phalloidin stained fixated cells revealed that single beads colocalize with dense F-actin stress fibers along the cells' periphery, whereas they cannot be detected within vacuoles (Fig. 3.5). Thus, after the bead endocytosis a subset of beads can be utilized to approximate the cytoskeletal microrheology in living stem cells, when they are clearly entangled with F-actin stress fibers.

In order to perform parameterized intracellular rheology, a nucleus-centered elliptical coordinate system can be employed. The displacements of beads that are not covered by an endosomic membrane inside the cytoplasm displayed a pronounced directional distortion. This biased behavior has been predicted, since the cytoskeleton represents a structural scaffold composed of highly oriented filaments, which possess anisotropic mechanical properties. Moreover, it has been seen that the

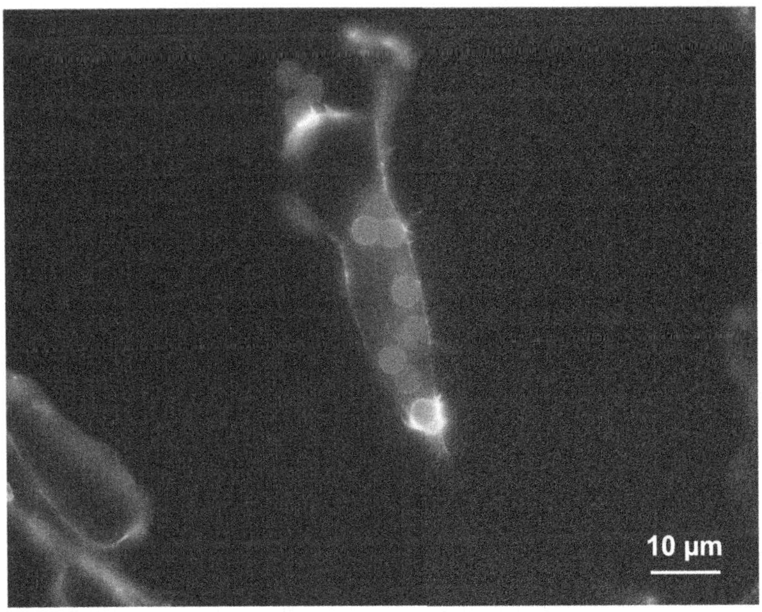

Fig. 3.5 A 4.5-μm fibronectin-coated bead is connected via integrins to the F-actin cytoskeleton of mouse embryonic fibroblasts. Scale is 10 μm as indicated

largest bead displacement directions were roughly parallel to the nuclear surface, which was found to be true for all analyzed cells (Lozoya et al. 2016). These results indicate that the anisotropic bead movement followed an orthotropic foundation, in fact one based on nuclear geometry. This interpretation contains important statistical results for the analysis of bead movement: When the interpretation is sufficient, an orthotropic model can emphasize anisotropic aberrations in the averaged space-dependent properties, which a fixed rectangular coordinate system would instead eliminate. In fact, we verified that a nucleus-centered geometric transformation better reflects the anisotropic behavior of bead displacements than rectangular coordinates. Consequently, we have constructed a curve-going and nucleus-centered elliptical coordinate system for analysis of displacements of individual beads and the estimation of the direction-dependent cytoplasmic mechanical properties.

The generalized Stokes–Einstein relation can predict the behavior of the thermally driven movement of small particles within a complex fluid and the resistance of the surrounding fluid to the movement of the particles. This theory was largely taken over to approximate the cytoplasmic mechanics in living cells, when the displacement of individual particles inside the cells were monitored over time (Daniels et al. 2010; Jonas et al. 2008; Crocker and Hoffman 2006, 2007; Liu et al. 2006; Wirtz 2009; Mason and Weitz 1995; Tseng et al. 2002; Caspi et al. 2000; Fakhri et al. 2014; Weber et al. 2012; Panorchan et al. 2007; Ott et al. 1990; Mierke et al. 2008; Mierke 2011; Raupach et al. 2007; Metzner et al. 2010). In detail, this biophysical technique can be employed to examine anisotropic effects inside the cytoplasmic mechanical properties of living adhesive stem cells. In the first step, the displacements of the individual beads from nucleus-centered elliptical coordinates were broken down into their two orthotropic components. In the second step, the mean squared displacement of beads is characterized that undergo subdiffusive motion in the two orthotropic directions. Ultimately, we evaluated the subdiffuse rheological properties that correspond to the orthotropic displacements of each bead at its precise position inside the cytoplasm.

In order to determine the beads included for the rheological analysis, the mathematical subdiffusive criterion has been employed that is described by the following equation of the mean squared displacement power coefficient:

$$\alpha(t) \equiv \partial[\log(\ll \Delta r^2(t) \gg)]/\partial[\log(t)] \tag{3.1}$$

This equation defines the slope $\alpha(t)$ for the mean squared displacement $\ll\Delta r^2(\mathrm{t})\gg$ in relation to the trace of time intervals t in logarithmic time space.

In specific detail, the displacement profiles of beads, which possess power coefficients of $\alpha(t) \leq 1$ for all t, show subdiffuse movements and were traced back to their distinct cytoplasmic localization inside the cells. Their specific localization is from now on termed subdiffusive loci, which are clearly different to those locations where the beads display a power coefficient of $\alpha(t) > 1$. Under this condition, the generalized Stokes–Einstein relation is mathematically invalid.

While acknowledging that the cytoskeleton is highly dynamic at the molecular level, it is also clear that it must inevitably play a constant structural role at the cellular level in order to improve tissue formation. That is a crucial function that the cytoskeleton performs during early development, where the extracellular matrix is limited and cellular polarization is a critical feature of the cells to control tissue morphogenesis. The argumentation behind the extension of the generalized Stokes–Einstein relation theory to delineate intracellular mechanics in living stem cells does not support predictions of thermally driven cytoplasmic properties—rather, in general the overall non-Brownian character of cytoskeletal mechanics is excellently substantiated (Crocker and Hoffman 2007; Hofman et al. 2006; Liu et al. 2006). Instead, it is intended to detect short-lived patterns of individual bead movements with subdiffuse characteristics at specific sites inside the cytoplasm and especially at the time of measurement. Similarly, this term implicates that the type of movement emanating from each bead, especially when measured, may vary between measurements, test periods and in the course of time. Experience has been consistent with such requirements: Beads with subdiffuse movement at a given range of 30 s did not automatically show a subdiffuse pattern at later points in time. Therefore, we tracked all observed beads and each one and at each test interval, checked whether their motion patterns match the subdiffuse motion during data analysis in every case, and administered the data set based on cases of local subdiffuse characteristics instead of beads themselves. Hence, it can be assumed that the underlying mechanics of the cytoskeleton can be approximated in this manner. When applied to a large number of living cells, this strategy provides other benefits, such as it delivers estimates of the rheological properties of the cytoplasm of individual living cells monitored in space and over time at the physiological level. In addition, there are enough cases of subdiffuse motion to generate statistical distributions of mechanical properties that can be used to distinguish between cell populations. This is faster than searching for cytoplasmic pearls with persistent subdiffuse properties, which is improbable within living cells.

Overall, the combined realization of the generalized Stokes–Einstein relational theory with nucleus-centered elliptic coordinates resulted in this: First, subdiffusive loci are predominantly situated in an area within a nuclear radius of the nuclear perimeter, which we classify as a perinuclear cytoskeleton; second, the perinuclear cytoskeleton of human stem cells maintains a constant anisotropic relationship of radial to angular shear modulus; and third, cytoplasmic rheology of living stem cells pursues power-law characteristics including frequency-invariant viscoelastic damping ranging from 0.2 Hz $< fP <$ 5 Hz. Similar to the nucleus-centered elliptical coordinate system, the power-law rheology is also inherently accessible to non-dimensioning. That is why we have lowered the number of parameters in the analysis by extrapolating the rheology and position of each subdiffuse loci trapped in stem cells to a uniform, dimensionless projection screen with the principal following parameters: first, the nuclear elliptical shape S; second, angle to the bead from the main elliptical axis Ω; third, distance of the bead to the nuclear center of gravity Θ; fourth, the anisotropy K; and fifth, the fluidity N. Beyond that we have also evaluated the damping ratio G''/G' at $fP = 1$ Hz to provide a reference value for viscoelasticity.

In a first experiment, the signaling of cytokines on the cytoskeletal mechanical properties in human adipose-derived mesenchymal stem cells (hASC) has been explored. Concretely, soluble cytokines are promotors of the remodeling processes of the cytoskeleton in adult stem cells, such as hASCs. In order to prove that active modulatory capacity, two different cytokines are explored on hASC cytoplasmic rheology after one hour of incubation of the cells with interleukin 1α (IL1α), which represents an inflammatory cytokine that can trigger the transport of de novo and recycled G-actin monomers of depolymerized F-actin stress fibers toward the cortex of the cell, where these actin fibers polymerize (Pritchard and Guilak 2006), and tumor growth factor β1 (TGFβ1), which is an anabolic growth factor facilitating the spreading of cells, F-actin stress fiber assembly and finally the stabilization of F-actin stress fibers (Jungmann et al. 2012). Both stimulations cause alterations of the cytoplasmic rheology in hASCs compared to untreated control cells, and in all instances, such effects were associated with a decrease in angular slope Ω of subdiffuse locations relative to the principal axis of the nucleus, even without altering the relative distance $R = \Theta - 1$ of the beads from the perimeter of the nucleus, where it is defined: $\Theta = 1$. In particular, the IL1α-stimulation of the cells leads to a pronounced decrease in all G_u, G_v and G^*_{2D} shear moduli including an alteration of the nuclear shape S, enhanced fluidity N and elevated anisotropy K. The losses in the shear moduli after IL1α stimulation are slightly smaller than the losses after TGFβ1 stimulation. The restructuring of the cytoskeleton after TGFβ1 stimulation causes distinct structural characteristics similar to those occurring after IL1α stimulation in hASCs. However, there are differences of both treatments: In more detail, TGFβ1 stimulation does not alter the nuclear shape *S* or the anisotropy *K*, which is dissimilar to IL1α stimulation that causes pronouncedly alterations is the nuclear shape and the anisotropy.

Since there are diverse mechanical features that are stimulated by each of the two cytokines, it can be considered that the collective of non-dimensional perinuclear cytoskeleton parameters for each condition has contributed to typical mechanical markers. In fact, the stimulated hASCs cells exhibit perinuclear cytoskeleton mechanical properties that can be segregated in a good manner form untreated controls using multivariate profiling.

Another second experiment deals with the restructuring of the cytoskeleton in hASCs exhibiting polarized and hence directed cell morphology. In order to analyze the correlation between cell morphology and cytoskeletal mechanics in a quantitative manner, PVA-film microphotopatterning (μPP) techniques (Doyle 2009) were employed to precisely regulate the morphology of hASCs (Lozoya et al. 2016). The hASCs were confined on a flat fibronectin-coated surface of a rectangular size of 15 μm × 70 μm. To alter the structural architecture of the cell cytoskeleton, the pharmacological drug cytochalasin D impairs the polymerization of F-actin stress fibers and thereby alters the entire structure of the actin cytoskeleton of the cells (Cooper 1987; Holzinger 2009).

Cytochalasin D stimulation of the cells causes pronouncedly alterations in the perinuclear cytoskeleton mechanics, since the cell's anisotropy K is decreased, which is facilitated by an increase in G_v. Together with this mechanical result, subdiffusive loci at higher angle inclination Ω with lower R values were detected, which reflects a shift of subdiffuse loci positions accumulating in the proximity of the nucleus. In specific, these effects on morphologically restricted HASCs were not supported by any modification of the nuclear shape S, fluidity N, or radial stiffness G_u. Overall, hASCs treated with cytochalasin D with structured morphologies based on multivariate perinuclear cytoskeletal mechanics were poorly separated from untreated hASCs.

In a third experiment, the evolution of the perinuclear cytoskeleton mechanical properties in hASCs, when in these cells a restructuring of the F-actin stress fibers is stimulated. For the purpose of mimicking the time-dependent activity of perinuclear cytoskeletal mechanics in cells under "continuous" catabolic induction, individual cytoplasmic beads were tracked in single hASCs for 30 s at regular intervals of 15 min over two hours that have been stimulated with IL1α or Cytochalasin D. As expected, both stimulations caused alterations in the cytoskeleton of hASCs involving active restructuring of the cytoplasm for up to two hours. The exchange of growth medium with treatment medium occurred 5–10 min after the single-cell analysis for nanoscale particle tracking. Both treatments affected the fluidity N, radial modulus G_u, and all morphological parameters, such as subdiffusive loci to nucleus distance R, nuclear shape S and subdiffusive loci at higher angle inclination Ω, and showed fast stabilization. In the 5–10 min of initial stimulation with either IL1α or cytochalasin D, evoked a pronounced increase in N (fluid-like rheology), a substantial drop in S (shape of the nucleus) and intermediate reductions in R (subdiffusive loci to nucleus distance). The other parameters, such as G_u, G^*_{2D}, G_v, K and Ω, were distinct between the two substances: The treatment with IL1α leads to increased anisotropy K, due to reduced G_v, while there are losses in Ω, whereas the treatment with Cytochalasin D leads to alterations in G^*_{2D}, K and R. While the findings after IL1α stimulation of spatially confined cells are comparable to the IL1α induction experiment, the findings for cytochalasin D differ due to the spatially confinement. Based on these different results for cytochalasin D due to cell morphological alterations, it can be asked whether also alterations in perinuclear cytoskeleton mechanical properties, as it has been seen after stimulation with cytochalasin D in unconfined hASCs. In specific, multivariate profiling of the influence of cytokines in spatially unconfined hASCs revealed pronounced effects. In order to determine whether multivariate states of perinuclear cytoskeleton mechanical properties across various types of human stem cells were addressed in a fourth experiment. In specific detail, it was investigated whether different types of stem cells with different levels of multipotency have mechanical inequalities. Hence, the mechanical properties of the perinuclear cytoskeleton and the morphologies of hASCs, human bone marrow-derived hMSCs and skin fibroblast-derived human-induced pluripotent stem cells (hiPSCs) were investigated as a monolayer under non-differentiation culture conditions. It was demonstrated that increases in subdiffuse local anisotropy K, fluidity Ň, nuclear shape S, distance from core Ʀ and decreases in tilt angle from the nuclear major axis Ω were associated with distinct types of human stem cells in varying configurations. The various stem cell types did not differ in the order of magnitude of the rheological properties, such as the 2D flow consistency index G*, radial shear modulus and angular shear modulus.

However, does each individual stem cell type possess specific mechanical properties? In fact, the multivariate characterization of the three human stem cells, such as hASCs, hMSCs and hiPSCs and can be performed by discriminant analysis of the perinuclear cytoskeleton mechanical properties. Finally, non-dimensional maps of perinuclear cytoskeleton lead to the identification of diverse multipotent cell types. There are two major characteristics that are obtained by the entire perinuclear cytoskeleton non-dimensional spectrum analyses: firstly, the determination of mechanical and structural features that display pronouncedly nonlinear correlations and secondly, a finite range of bead locations around the nucleus Θ, which becomes a fixed upper limit. These findings indicate that the non-dimensional perinuclear cytoskeleton profiles are treated as configurations of a cytoskeleton that is based on conserved mechanical features in multiple cell types.

It can be hypothesized that the mechanics of the perinuclear cytoskeleton follow a conservative relation in all phenotypes of the cells, and hence, it seems to be likely that there exists a baseline state of perinuclear cytoskeleton mechanics, which is associated with the nuclear shape and the position of the subdiffusive loci inside the cell. In more detail, the baseline configuration of the perinuclear cytoskeleton can be estimated by distributing the perinuclear cytoskeleton conditions after remodeling the cytoskeleton under multiple conditions.

It has turned out that the initial state of the nucleus is that its major axis is 60% longer than its perpendicular minor axis. The nucleus is connected to a perinuclear cytoskeleton that is at most two radii distant from the nuclear perimeter, has subdiffuse properties in perinuclear regions about 35° from the major axis, behaves in a highly anisotropic manner and is radially stiffer and somewhat more solid than liquid in response to stress.

Empirical relationships can be found between perinuclear cytoskeleton conditions and the shape of the nucleus. The experiments with the phenotype of stem cells revealed compensatory reactions between perinuclear cytoskeletal rheology, nuclear shape and the regional heterogeneity of subdiffusive loci. A quantitative relationship between the three parameters can be achieved by fitting an empirical mixture model using recursive Bayesian inferencing. Hence, these findings lead to the following hypotheses: First, the mechanics of the perinuclear cytoskeleton mirror a conserved coherence between rheological and geometric parameters within the subdiffuse regions inside the perinuclear cytoskeleton; and second, the shape of the nucleus and the anisotropy are connected by the positioning of the subdiffusive loci. Moreover, there seems to be a strong relationship between the reference viscoelasticity G''/G' at $fP = 1$ Hz and the fluidity Ň that has been found in materials exhibiting power-law rheology.

When an elliptical coordinate basis is utilized, which was mathematically prescribed by the nuclear shape to characterize the mechanics of the perinuclear cytoskeleton in individual stem cells, it demonstrates that the heterogeneous rheological phenotypes within the perinuclear cytoskeleton are inversely associate with the shape of the nucleus and that the same empirical fit can be produced in a homogeneous, non-dimensional space in all cell types analyzed. These findings are in line with the revealed connection between the contractile history of the entire cytoskeleton, the state of the chromatin and the architecture of the nucleus (Heo et al. 2015; Chalut et al. 2010). Subsequently, this leads to the hypothesis that the structural linkages between the nucleus and the cytoskeleton need to display reciprocal mechanics.

The implication of these results is that the mechanical properties of one, such as the cytoskeleton, can be predicted by the other, such as the nucleus, or vice versa. In view of this and the dimensionless relationships between perinuclear cytoskeletal mechanics and nuclear shape that exist between multiple stem cells with various nuclei, the results emphasize a crucial aspect that we can examine: Are these mathematical interactions between mechanical or spatial cytoplasmic properties of stem cells the same unless they do not seem similar when their nuclei have different forms? In approaching this eventuality, it can be examined whether the localization and rheology of subdiffuse loci in the perinuclear cytoskeleton is predetermined by geometry of the nucleus. Through cumulative frequency analysis, the distributions of Ω, Ɍ and Ň, that is, the position and fluidity of subdiffuse loci in the perinuclear cytoskeleton, were well simulated by fixed lognormal distributions, whose parameters are univariate functions of the nuclear shape S. As the remaining non-dimensional parameters are forecasted by various combinations of Ω, Ɍ and Ň and S, and each one is a function of S itself, the non-dimensional factors can be assumed by simply determining the nuclear shape. In summary, it can be said that perinuclear cytoskeleton mechanics is numerically feasible due to the shape of the nucleus. These are results with a strong impact: It indicates that anisotropy and heterogeneity in the organization of the cytoskeleton are linked mechanical features based on a specific cytoskeletal subdomain that is located around the nucleus, bound to its structure and quantifiable by the nuclear shape.

These results provide a multivariable minimization approach that offers a coherent framework for the assessment of perinuclear mechanics of adherent cells. By this approximation, it is shown that the nuclear shape of stem cells resizes the relationships between orthotropic cytoplasmic mechanics in the vicinity of the nucleus. These relationships include distinct mechanical identifiers for various stem cell types, all numerically derivable by the geometry of nucleus inside the stem cells. To put it briefly, what emerges is that mechanical and morphological profiles, along with genetic and epigenetic, are strong discriminants between phenotypes of stem cells. More specifically, both hASCs and hMSCs, which each exhibit a lower rate of differentiation than hiPSCs, are more elastic and anisotropic, have more extended nuclear cores, and their subdiffusive loci are more distant from the nucleus and closer to the principal axis of the nucleus than hiPSCs. The differences in the distribution of subdiffuse loci and the structural pre-stressing of the nucleus and its environment seem to indicate that cytoplasmic mechanics can differentiate various stem cell types in perinuclear regions. Symmetries in the cytoplasm of stem cells were consistently revealed as a result of analytical methods. The key lies in receiving a single multidimensional mapping that accommodates individuality in the shape of each cell examined and simultaneously gathers heterogeneous cytoplasmic mechanics out of many cells simultaneously.

To this end, we have developed an elliptical coordinate system based on the nuclear shape of each cell and taking over orthotropic mechanics. From a geometric point of view, the strategy permits the coordinate systems of all cells, after geometric normalization in terms of their individual nuclear ellipses, to merge into a nucleus standardized plane, with all perimeters of the nucleus placed one above the other along the unit radius. From a rheological point of view, the implementation of an orthotropic model permits the acquisition of cytoplasmic anisotropic mechanics that are unique for each bead in every cell due to an elliptical and curvilinear coordinate system. The following is considered for the purpose of the investigation: firstly to be characteristic of the original position of each pearl in relation to the nucleus; secondly, to be traced only to the cell under examination if the nuclear shape is recognized; and thirdly, to approximate the cytoskeletal mechanics of beads revealing, measured, a subdiffusive window of movement within the cytoplasm of stem cells, similar to that which otherwise may or may not appear for that specific cytoplasmic bead.

In a broad spectrum, these results suggest that experimental designs with cell-independent coordinates, planar rheology or single-frequency measurements can be especially demanding in identifying the mechanical signs of the differentiation transmitted by cytoskeletal meshes inside stem cells.

An elliptically based method of measuring perinuclear cytoskeletal rheology can be applied experimentally to increase the statistical power of the data to map site-specific variations in properties compared to Cartesian schemes. For instance, shifts in Cartesian x- and y-directions of beads near the nucleus tips are compensated by those in beads on the edges when the cytoplasmic mechanics is orthotropic, thereby increasing the sample size is required to detect statistically significant anisotropic effects. For the exact same beads, this is less pronounced in an elliptical system where orthotropic distortions are statistically additive—reducing the variance, increasing the statistical performance, and reducing the number of samples needed to match the statistical significance found in Cartesian coordinates.

Hence, it can be said in a conclusive manner that the mechanical properties of the perinuclear regions within specific stem cells are altered through biochemical and morphological effects. Finally, the structural evidence after investigating the effects of biochemical stimulations of the remodeling of the cytoskeleton in hASCs under non-confined or confined, e.g., micropatterned, morphologies reveals that: Firstly, these cytokines remodel the localization of the subdiffusive loci within the cell that is commonly associated with alterations in the shape of the nucleus; secondly, the relocation of the subdiffusive loci within multiple cell morphologies leads to modifications in the anisotropy in cell mechanics; and thirdly, perinuclear cytoskeleton network senses the disruption of F-actin stress fibers, nevertheless create a stable mechanical profile within a few minutes, which remains intact despite the ongoing depolymerization of peripheral F-Actin.

During all these alterations within the cell, perinuclear mesh networks are able to transform the shape of the nucleus and dynamically remodel the cell mechanics, which leads to the hypothesis that the perinuclear cytoskeleton works as an endogenous element in various stem cells. To this end, this non-dimensioning approach to perinuclear cytoskeletal mechanics, which quantitatively determines how cells collectively compensate for their mechanical phenotype and the architecture of the nucleus, provides a functional mapping of biophysical states in perinuclear cytoskeletal mechanics. This information provides an opportunity to consistently quantify the structural properties of the perinuclear cytoskeleton and possesses the potential to test cells based exclusively on their perinuclear cytoskeletal mechanics to discriminate between multipotential phenotypes.

For the delivery of beads into cells, an endocytosis-based approach is utilized. An advantage of this approach is that it provides more cytoplasmic beads for testing than traditional bio-ballistic bead delivery. A disadvantage is that the endocytic release also yields more beads with driven or non-thermal movement, as it has been found for the shuttling of endosomes by motor proteins (Wirtz 2009; Tseng et al. 2002; Caspi et al. 2000) or in cellular locations exhibiting enormous energetic fluctuations within cells (Fakhri et al. 2014; Weber et al. 2012; Ott et al. 1990). These movements do not fulfill the generalized Stokes–Einstein relational theory, because the random thermal shifts of a spherical bead are opposed by their own resistance within a passive environment (Mason and Weitz 1995).

Beads with thermally driven motions can be traced over a wide range of different time intervals to show a mean squared displacement at $\ll \Delta r^2(t) \gg$ with a logarithmic slope $\alpha(t) \equiv \partial[\log(\ll \Delta r^2(t) \gg)] / \partial[\log(t)] \leq 1$, where $\alpha(t) = 0$ for a purely elastic medium that exhibits no movements within the solid material, $\alpha(t) = 1$ for a purely viscous medium that displays free diffusion inside the Newtonian fluid, and $0 < \alpha(t) < 1$ for a viscoelastic medium that exhibits both no movements and free diffusion. In contrast, spherical sensors displaying enthalpic or persistent movement with $\alpha(t) > 1$ break down the generalized Stokes–Einstein relational theory and this behavior frequently at large t.

However, since the advantages of bio-ballistic techniques decrease with probe size—for example, rheological assessments are comparable between the two methods when 1 μm beads are employed (Jonas et al. 2008) and bio-ballistic techniques also produce super diffuse beads that have to be excluded according to similar $\alpha(t)$-based criteria, endocytic delivery is still the preferred method (Panorchan et al. 2007).

There is an observation that medium surrounding endocytosed beads within living cells is not passive: It rather consists of bioactive monomers, cytoskeletal fibers, hydrolyzable molecules, shuttle proteins, metallic ions and enzymes with catalytic activity in a mixture that is linked by the cytoskeleton (Jonas et al. 2008; Crocker and Hoffmann 2007; Hoffman et al. 2006; Liu et al. 2006; Wirtz 2009; Mason and Weitz 1995; Tseng et al. 2002; Caspi et al. 2000; Fakhri et al. 2014; Weber et al. 2012; Panorchan et al. 2007; Ott et al. 1990). All this is heterogeneously dispersed in the cytoplasm and leads to locally different cytoskeletal fluctuation rates from a few seconds for lamellipodia to several minutes for stress fibers. These local levels depend on the presence of specific pools and kinetics of cytoskeletal constituents and may vary over time (Baumann et al. 2012; Coue et al. 1987; Theriot 1994). For this reason, how can the generalized Stokes–Einstein relation theory be used for the shifting of beads? The cytoskeletal rearrangements oscillate between Brownian and non-thermal phases of energy transfer and differ spatially. The generalized Stokes–Einstein relation-based cytoplasmic rheology is justified if measurements originate from bead shifts that are recorded at short intervals in relation to local cytoskeletal fluctuation times, such as seconds instead of minutes in the close vicinity of stress fibers (Baumann et al. 2012; Coue et al. 1987; Theriot 1994), and correspond to a Brownian pattern. Therefore, in conducting these analyses, the following factors have been discerned: firstly, individual beads within large vesicles that display a movement dominated by the aqueous environment of each vesicle rather than cytoskeletal deflections; secondly, groups of beads that together do not present a spherical resistance pattern; and thirdly, beads bearing non-thermal movement. Therefore, brightfield or epifluorescence microscopy images need to be recorded from individual cells before measurement. To order to exclude beads in vesicles and grouped beads, they are not both traced, the latter is recognizable by the light scattering of the vesicle membranes, which can be separated from the bead contour enclosed by them. In the last step, beads with non-thermal movement ($\alpha(t) > 1$) were removed from the analysis. Hence, the captured bead patterns can be used to reveal cytoplasmic rheology: The data contain cytoplasmic beads that are closely connected to F-actin, monitored during a subdiffusive motion phase and spread heterogeneously within the cytoplasm, which present the reference group to those beads in subdiffusive loci.

Anisotropic cytoplasmic mechanics anticipated from filamentous polymers including the cytoskeleton of living cells were previously identified based on the Cartesian coordinate systems (Tseng et al. 2002; Fakhri et al. 2014). Anisotropy within the cytoskeleton is spatially different, independent of lag time and characterized by a well-defined power law. These characteristics correspond to a contractile active gel mechanically linked to a deformable but relatively stiffer core (Hoffman et al. 2006). But here a single reference frame for cells with different geometries is provided that includes the radial distance and angle of the nucleus, the nuclear shape and the level of perinuclear anisotropy in the cytoplasm through a compositional empirical relationship, which seems to be highly conserved in different populations of stem cells (Lozoya et al. 2016). These results are consistent with a polymeric cytoskeletal network model complemented by radially interlaced filaments, an interpretation experimentally consistent with perinuclear actin cables entangled around the nucleus and connected by transmembrane actin-associated nuclear (TAN) lines (Kutscheidt et al. 2014). From a structural point of view, these physical characteristics are similar to a hydrostat (Kier and Smith 1985; Wainwright 1976), except in this case microscopically on the stem cell nucleus and with phenotype-specific mechanical conditions. The perinuclear cytoskeleton is a fiber-reinforced mechanical structure that bypasses physical cross-actions between the cytoplasm and the nucleus of human stem cells. In order to gain a deeper understanding of the relevance of the implementation of this model in the analysis of nanoscale particle tracking microscopy obtained from intracellular mechanics experiments, the elliptical mapping of perinuclear cytoskeletal mechanics can be converted into a rectangular coordinate system. With such a perspective, it is feasible to extract an empirical relationship that depicts the behavior of the "nucleocytoskeleton" as a coupled entity in which the displacement of perinuclear subdiffusive sites and nuclear shape changes are mutually reciprocal. This evidence-based relationship can be exploited to derive hypotheses in stem cell mechanics that can, for example, elucidate how physical responses between nucleus and perinuclear cytoskeleton develop under stress and deliver verifiable predictions in cell morphology and cytoplasmic organization. As a matter of fact, when conducting vectorial analyses of infinitesimal shear stress–strain ratios and fitting the empirical connections between subdiffuse loci localization and the shape of the nucleus, perinuclear cytoskeletal deformations tended to be mechanically linked to alterations in the shape of the nucleus. Local strains at subdiffuse sites in the perinuclear cytoskeleton of rounded cells, for example, tend to prolong the entire nucleus under stress, while an enlarged perinuclear cytoskeleton in elongated cells tends to pull inward and round the nucleus; both behaviors were immediately noticed (Driscoll et al. 2015; Heo et al. 2015). To this end, the analysis proposes that the mechanically coupled perinuclear cytoskeleton and the nucleus form a single mechanical ensemble, in which a nucleus "core" functions in combination with an enveloping and deformable perinuclear cytoskeleton, and this shows: a first, mechanical equilibrium in a pre-stressed perinuclear cytoskeleton state; second, uniform pulling through the perinuclear cytoskeleton at the nucleus along stress carrying sites; and third, stress carrying sites, which are at about $30.4 < \theta < 56.6°$ from the major axis when in equilibrium. Beyond that, subdiffuse locations in this mechanical model are suitable stress locations where the mechanical stress reaches the nucleus.

With a view to the future, a practical role for the nuclear form is offered as a quantifiable marker for transient cytoplasmic mechanics in stem cells, especially as an accessible tool for dynamic estimation of perinuclear cytoskeletal properties. Cytokines have been demonstrated to modify the spatial dispersion of subdiffuse loci, usually associated with alterations in the shape of the nucleus and to compensate for the redistribution of subdiffuse loci in different cell morphologies by alterations in mechanical anisotropy. In other respects, perinuclear cytoskeletal networks are sensitive to F-actin disturbances, but within minutes they produce a stable mechanical pattern that remains intact despite prolonged peripheral F-actin degradation, indicating that the nucleus and its immediate environment were better able to withstand the structural effects of induced actin depolymerization than the outer regions of the cytoplasm where cytoskeletal integrity was severely compromised. It is therefore probable that the perinuclear cytoskeleton is an impenetrable structure that reshapes its structure to maintain physical stress, balance a structurally weakened "outer" cytoskeleton, and enclose the nucleus of the stem cell to provide mechanical protection. In this role, the perinuclear subdomain of the cell cytoskeleton actually behaves like a structural link that "buffers" the mechanical signaling on the cells on their path to the nucleus. This does not mean that this perinuclear domain is not embedded in the cytoskeleton, but that its mechanical properties vary from those of the more distant cytoskeleton. While the composition of the perinuclear region may also be relevant, it is assumed that this is due to differences in the way cytoskeletal networks are built close to and far from the nucleus.

The nucleus and cytoskeleton are interactive entities that contain traceable information about the mechanical condition of cells. There are many biological processes in stem cells that affect the cytoskeleton and have strong connections between nuclear shape and cell mechanics, such as the spreading of cells (Driscoll et al. 2015; Heo et al. 2015). Not only do these variables confirm that nuclear form, cytoplasmic rheology and anisotropy are mutual: First, these variables converge to a master mapping of non-dimensional mechanics; second, stem cell types differ mainly in how they balance these properties; and third, non-dimensional cytoplasmic mechanics in stem cells can be assessed from the shape of their nucleus. It has also been noted how perinuclear cytoskeletal mechanics can rapidly achieve stability, even if peripheral cytoskeletal networks continue to become degraded. These observations point to perinuclear networks maintaining the ability to interact mechanically with the nucleus, especially during biological processes that cause large-scale structural alterations inside and outside cells, such as the differentiation of stem cells. Finally, it can be said that the perinuclear cytoskeleton of the stem cell in combination with the nucleus acts as a mechanically independent endogenous unit.

In conclusion, it can be said that there exists a specialized function for a precisely defined perinuclear domain, the perinuclear cytoskeleton, that coordinates the linkage of the nuclear structure with the overall cell mechanical properties. In addition, evidence is presented for intracellular shape–function relationships that quantitatively determine the homeostatic mechanical equilibrium between nuclear shape and perinuclear cytoskeletal structure among a series of exogenously induced phenotypes. Finally, in human stem cells, the shape of the nucleus is adequate to reflect both the spatial dispersion and the rheological scale of abundant rheological properties within subdiffuse domains in the perinuclear cytoskeleton and to reflect different mechanical conditions within a shared parameter map which is representative of multiple cell phenotypes. This work develops a new concept for the investigation of intracellular mechanics in stem cells of different lines with different biophysical characteristics or under different metabolic regulation modes as functions of mechanical perinuclear cytoskeletal conditions, all within the probabilistic patterns as predicted by the shape of the nucleus and as such detectable solely by standard microscopy.

3.4 Cytoplasm Mechanics Are Size- and Velocity-Dependent

Many physiological processes fulfill the active transport within the cytoplasm crucial functions in living cells. However, the mechanical resistance of the intracellular compartments, which is determined by the cytoplasmic characteristics of the material, still cannot be grasped precisely, especially the dependence on size and speed. In this case, an optical tweezer is required to pull a bead into the cytoplasm and directly measure the mechanical resistance with different size a and velocity V. A method combining direct measurement and simple scaling analysis is outlined to identify different origins of size- and velocity-dependent resistance in mammalian cytoplasm of living cells (Hu et al. 2017). The cytoplasm demonstrates a size-independent viscoelasticity when the effective strain rate V/a is kept in a relatively narrow range ($0.1\ s^{-1} < V/a < 2\ s^{-1}$) and a size-dependent poroelasticity when the effective strain rate mode is high ($5\ s^{-1} < V/a < 80\ s^{-1}$) (Hu et al. 2017). In addition, it is shown that the cytoplasmic modulus is positively correlated with merely V/a in the viscoelastic regime, but also rises with the bead size at constant V/a in the poroelastic mode. Based on these measurements, a complete state pattern of the mammalian living cytoplasm is generated, demonstrating that the cytoplasm undergoes a change from a viscous liquid to an elastic solid, and from a compressible material to an incompressible material, with rising values of the two-dimensional parameters. These state diagrams are valuable for gaining an insight into the underlying mechanical properties of the cytoplasm in a wide range of cellular functions over a broad spectrum of velocity and size scales.

The cytoplasm of living mammalian cells presents an overcrowded but dynamic surrounding (Alberts et al. 2008). There exist ongoing intracellular motions that are crucial for cell physiology, such as vesicle transport and transport of other organelles. While biological motors and other enzymatic reactions drive these activities, the mechanical properties of the cytoplasm are decisive in determining the mechanical resistance encountered by cell compartments. In fact, both the active propulsive force and a suitable mechanical surrounding are decisive for the design of the living cell machinery. But while the force produced by molecular motors both individually and collectively has been extensively explored (Hendricks et al. 2012; Guo et al. 2014), the mechanics of the cytoplasmic surroundings are still not yet clearly revealed. In addition, the influence of object size and velocity on the mechanical resistance that active forces must withstand in order to permit transport is still uncertain. The characterization is essential for gaining an insight into the physical landscape and various crucial dynamic functions of living cells.

The cytoplasm possesses a cytoskeletal network and multiple proteins, various organelles and vesicles. Such complex microstructured materials are supposed to exhibit time-dependent or frequency-dependent characteristics (Larson 1999). Consequently, it has been detected in various biophysical approaches that the mechanics of living cells display a pronounced frequency dependency (Hoffman and Crocker 2009; Zhu et al. 2000) and the cells' reaction show a power-law rheology behavior in a broad range of frequencies (Guo et al. 2014; Fabry et al. 2001; Hoffman et al. 2006). Based on these findings, a common agreement of all these biophysical approaches is that cells can be seen as viscoelastic materials (Janmey et al. 2009; Bausch et al. 1999; Guck et al. 2001; Lu et al. 2006; Nawaz et al. 2012; Mierke et al. 2008) and additionally, the revealed mechanical properties rely on the time period over which the deformation lasted throughout the measurement. An essential characteristic of a viscoelastic medium is the fact that its mechanical property differs according to the length scale of the measurement (Hu and Suo 2012). An interesting fact has recently been shown that living cells are able to react similar to a poroelastic gel within a short time frame (Moeendarbary et al. 2013; Rosenbluth et al. 2008; Charras et al. 2005). Moreover, the stress relaxation of cells can be completely controlled by cytosol migration through cytoskeletal grids. In terms of poroelasticity, the measured mechanical property is highly influenced by the probe's size, as it requires more time for the cytosol to move over a greater distance, thus reducing stress relaxation. The size-dependent poroelastic behavior stands in direct contrast to viscoelasticity, whose relaxation is a material property and is independent of the size of the probe, adjusted by the time-dependent reaction of the macromolecular and supramolecular components of the materials. Even more important is the fact that most previous efforts to examine cell mechanical sensor cells from the outside, e.g., with the help of atomic force microscopy or an optical stretcher, and thus the measurement rather depends on the stiff actin-rich cell cortex than on the far softer cytoplasm (Nawaz et al. 2012; Moeendarbary et al. 2013; Guo et al. 2013). It remains uncertain whether viscoelasticity or poroelasticity adequately reflects the rate-dependent resistance of the cytoplasm of living mammalian cells or whether both are necessary.

Therefore, optical tweezers are employed to pull a plastic bead inside the cytoplasm of a living mammalian cell and directly detect the force (designated by F) and displacement (designated by x) that reflect the cytoplasmic mechanical behavior. Taking into account viscoelasticity and poroelasticity, two independent dimensional parameters emerge from the experiments: V/a and V_a, where V and a represent the velocity and diameter of the probe bead. With these two reference values and a combination of experimental measurement and the scaling analysis technique, various origins of cytoplasmic resistance can be identified, varying from viscous, viscoelastic, poroelastic to purely elastic. In a final step, various cytoplasmic mechanical modes are categorized in an overall state diagram that visualizes the diverse origins of the mechanical resistance of a multitude of physiological pathways in cells of dissimilar characteristic size and velocity.

To elucidate the mechanical opposition that intracellular specimens encounter in a living mammalian cytoplasm, intracellular polystyrene beads are delivered into living normal rat kidney epithelial cells by endocytosis. The sizes of these beads are from 0.5 μm to 1.5 μm, and they are randomly dispersed within the cells and their size is pronouncedly larger than the typical mesh size of the cytoskeleton, which is about 50 nm (Luby-Phelps 2000). These beads are capable of examining the cytoplasm as a continuous fluid. An optical tweezer is then employed to catch and pull a single bead unidirectionally at a constant speed to the cell boundary. To prevent any interference with the mechanically separated cell cortex and nucleus, beads larger than 1.5 μm distant from the cell boundary and both from the thin lamellar region and from the nucleus are preferred. The beads are propelled unidirectionally with constant velocity. Therefore, the intracellular resistance force on the bead in the direction of movement equals the traction force measured. Finally, a force versus displacement curve of the cytoplasm can be drawn. In order to compare measurements with varying bead sizes, the force as F/S and the displacement as x/a need to be normalized, where *S* represents the cross-sectional area and a represents bead the diameter, and subsequently, a normalized force–displacement curve can be drawn.

Understanding the mechanical resistance in the cytoplasm, it is estimated that the resistance of the bead is primarily the result of the restoring force of the deformed cytoskeleton and the inhomogeneously distributed nuclear pressure within the cytoplasm. The deformation of the cytoskeleton microstructure itself is typically viscoelastic, resulting from the rearrangement of the cytoskeleton microstructure and the breaking of protein bonds (Kumar et al. 2006). The viscoelasticity kinetics are characterized by intrinsic relaxation time scales or intrinsic time spectra (Balland et al. 2006; Wang et al. 2014), which are unaffected by stress conditions such as bead diameter and velocity. Cytoplasm local distortion leads to inhomogeneous pressure distribution of pores within the porous structure generated by the cytoskeleton as a result of poroelasticity (Moeendarbary et al. 2013, Detournay and Cheng 2014). In order to achieve a rehomogenization of the pore pressure dispersion, the intracellular liquid moves over representative distances according to the bead size, with typical poroelastic relaxation time a^2/D. *D* represents the effective poroelastic diffusivity of the cytosol that scales as $D \sim E\kappa$ (Detournay and Cheng 2014), in which E is the equilibrium elastic modulus of the cytoplasm under equilibrium conditions and κ is the permeability that is defined as the cytosol flux through cytoskeletal scaffold under a uniform gradient of the pressure. In the analysis of the mechanical strength of the cytoplasm, however, viscoelasticity and poroelasticity are taken into account. The viscoelastic relaxation times and the effective poroelastic diffusivity can serve as parameters defining normal rat kidney epithelial cells.

Normally, only one time scale is assigned to the loading condition, namely a/V. On the foundation of a simple scaling calculation, the ratio between the resistance force and the bead displacement, $F = f(E, t_i, D, v, a, x, V)$, may be converted into a dimensionless form, $F/ES = f(x/a, V_a/D, Vt_i/a, v)$, wherein *S* represents the cross-sectional area of the bead, ν constitutes the equilibrium of Poisson's ratio of the cytoplasm and t_i comprises a series of typical viscoelastic time intervals. For a specific cell type, *E*, *D*, t_i and *v* represent all constants. Vt_i/a $(i = 1, 2, 3\ldots)$ have no dimensions here, and the ratio of the typical viscoelastic time intervals to the experimental time interval is designated as Deborah number (Reiner 1964). V_a/D denotes the ratio of the typical poroelastic time interval to the experimental time interval. If the predicted rate-dependent mechanical response found to be due only to viscoelasticity and a constant V/a is sustained, the measured relationships between F/S and x/a break down to a single curve; if the observed rate-dependent response found to be due only to poroelasticity and a constant V_a is sustained, the measured relationships should break down to a single curve.

In order to validate the ability of the experimental method to discern between viscoelastic and poroelastic properties, unidirectional particle stretching is carried out in classical poroelastic and viscoelastic materials with established material properties. To measure the poroelastic behavior, polyacrylamide (PA) (3% acrylamide and 0.05% bisacrylamide cross-linker) is measured, which represents a hydrogel with covalently cross-linked meshes exhibiting linear elastic behavior over a wide frequency range (Lourenço et al. 2016). Nevertheless, if the gel is exposed to a high strain rate deformation, the fluid flow within the gel network influences the material resistivity against the applied strain, and the velocity dependence is well captured by the linear poroelasticity theory (Hu et al. 2010). Beads are pre-mixed in two distinct sizes (0.5 μm and 1 μm diameter), and unidirectional stretching is performed at comparatively high velocities (5, 10, 20 and 40 μm/s). When V_a is a constant, the measured normalized force–displacement curves appear to be close to each other. In addition, higher values of V_a at the same strain yield higher stresses which correspond to the characteristics of poroelastic material. In accordance with the analytical solution of the inclusion problem of a spherical rigid inclusion in an elastic environment (Selvadurai 2016), the modulus of elasticity of the medium E can be calculated as $E = [(5–6\nu)(1 + \nu)/24(1 - \nu)]EA$, where ν gives the Poisson's ratio and EA expresses the apparent modulus, which is measured as the average slope of the normalized force–displacement curve of 0–10% normalized deformation. At a Poisson's ratio of 0.457 (Takigawa et al. 1996), the computed elastic modulus of the PA gel is approximately 80 Pa, in accordance with the bulk rheology measurements performed on the same gel. Ionic cross-linked alginate gel made from 5 mg/ml alginate solution cross-linked with 3 mM calcium sulfate is employed to check whether stress–strain curves with the same V/a collapse in the viscoelastic regime. In fact, the viscoelastic properties of alginate gel are well defined for a wide frequency range equivalent to low strain rates (Mitchell and Blanshard 1976).

The beads are pulled at relatively low speeds (0.4, 0.8, 1, 2 and 4 μm/s) where the viscoelasticity prevails over the stress–strain response. In fact, assuming constant U/a, the measured normalized force–displacement curves are near each other. With the same bead size, a higher pulling velocity leads to a higher resistance force, which also corresponds to the reaction of most viscoelastic materials. At high elongation rates, stresses in the material leave no time to relax. These results show that the method can actually discriminate between poroelasticity and viscoelasticity within a material.

Beyond that, finite element simulations were also carried out to check the scaling analysis. For the simulation, a rigid bead in an elastic mass that is far wider than the bead is embedded. A constant speed on the bead is imposed. Then, the resistant force on the bead is determined. During the simulation, the elastic material is characterized either as a viscoelastic material with a Prony series (Park and Schapery 1999) or as a linear poroelastic material. Unsurprisingly, the numerical simulations once again validate the scaling analysis, and hence, this method has been employed to normal rat kidney cells. In precise detail, the normalized force–displacement curves at different loading speeds and with three different bead sizes, such as 0.5, 1, and 1.5 μm, were determined. In fact, the mechanical response of the cytoplasm strongly relies on the 2D numbers, such as V/a and V_a, in two effective strain rate regimes. At low effective strain rates ($0.1\ s^{-1} < V/a < 2\ s^{-1}$), the cytoplasm behaves viscoelastic, since the normalized force–displacement curves lie close together during measurements with the same V/a, such as 1, 0.5 and 0.2 s^{-1}. If we compare measurements within this regime with the same V_a, such as 0.5 $\mu m^2/s$, they differ considerably from each other. By similar analyses, we observe that the cytoplasm looks poroelastic under high effective strain rate ($5\ s^{-1} < V/a < 80\ s^{-1}$), since the normalized force–displacement curves for measurements with the same V_a, such as 20, 15 and 10 $\mu m^2/s$ are of the same nature. Within this regime, when measurements with the same V/a, such as 20 s^{-1} are compared, the curves can be discriminated from one another. In addition, if the effective strain rate continues to be decreased ($V/a < 0.1\ s^{-1}$), the resistance stays approximately constant as the displacement increases. This is a typical characteristic of a viscous fluid having an effective viscosity, which is approximated as 12 Pa·s, which is similar to previous studies (Bausch et al. 1999; Berret 2016). Similar results can be obtained with HeLa cells. Similarly, it is no longer valid to consider the cytoplasm as a continuum and to use the above scaling analysis to investigate the mechanical interaction between the bead and the cytoplasm when the probe size becomes comparable or even smaller than the mesh size of the cytoskeleton mesh. However, most of the usual cytoplasmic organelles, such as vesicles and mitochondria, have a size considerably larger than the typical cytoplasmic mesh size and are indeed similar to the beads used in this case. For this reason, these measurements offer direct insights into the knowledge of the mechanical resistance that conventional organelles undergo in living cells.

The cytoplasm of mammalian cells consists of a porous structure consisting of cytoskeletal structures and other proteins with cytosol filling (Moeendarbary et al. 2013). The results indicate that the cytoplasm of living mammals displays viscous fluid behavior, viscoelasticity or poroelasticity with varying combinations of bead size and velocity. At sufficiently high effective strain rates, the characteristic time for movement over a bead diameter is similar to the time it takes for cytosol to migrate around the bead; this results in the rate dependence seen to be caused by poroelasticity. However, at low effective stretching rates, the associated slow cytosolic movement is coupled with an essentially immediate rehomogenization of pore pressure; consequently, any remaining rate-dependent resistance to bead movement is attributable to the viscoelastic response of the cytoskeletal system. In extremely low bead velocities, the cytoplasm also behaves like a viscous fluid, since the cytoskeleton has sufficient time to depolymerize and transform over very long periods of time, allowing the fully relaxed cytoplasm to experience a viscous flow during loading. In the experiments, the poroelastic relaxation time of the cytoplasm is approximately 0.02–0.1 s, assumed from the range of effective strain rates where poroelasticity is monitored; it is much less than the apparent viscoelastic relaxation time, $t = 0.79 \pm 0.13$ s, recorded in the normal rat kidney epithelial cytoplasm. t is achieved by one-time exponential adaptation of the relaxation curve obtained from the cytoplasm. While the cytoplasm may exhibit a range of relaxation times (Balland et al. 2006), the exponential fit with a characteristic single time is somewhat in agreement with the experimental data indicating that, although there is a range of relaxation times, there is a dominant region in the time scale sharing the same order of magnitude with the matched apparent relaxation time t. During this apparent relaxation time, the viscoelastic response within the cytoplasm relaxes pronouncedly. For simplification, the apparent relaxation time t is used in the following, instead of the time spectrum t_i ($i = 1, 2, 3\ldots$), when the cytoplasmic viscoelasticity is addressed. The poroelastic diffusion coefficient of the cytosol inside the cells is estimated from the experimental timescale, T ($=a/V$), where the poroelastic regime can be detected with $D = a^2/T \approx 50$ µm²/s, which is in line with reported measurements in mammalian cells (Moeendarbary et al. 2013). These measurements show that cytoplasmic mechanics can possess various origins at diverse timescales: Viscoelastic and viscous behavior may be present upon relaxation, and the restructuring of the cytoplasmic scaffold can dominate cytoplasmic mechanics at larger timescales, while poroelasticity is based on fluid–solid, such as cytosol–cytoskeleton, frictional interactions and the dissipation of energy prevails at relatively short timescales.

In order to reveal the impact of the cytoskeletal scaffold to the mechanical phenotype of the cytoplasm, 5 µg/mL cytochalasin D can be added to living cells to depolymerize F-actin structures, which represents a key element of the cytoskeleton, and thereafter unidirectional stretching is applied. In fact, a clear decrease in the characteristic viscoelastic timescale t can be detected and a pronounced increase of the effective poroelastic diffusivity D of the cytoplasm can be seen, which then reduces the poroelastic relaxation time. The poroelastic diffusion coefficient, representing the effective diffusion of the fluid within a porous environment due to internal pressure gradients, distinguishes itself from the usual diffusion coefficient, which takes into account the collective movement of the solutes in response to concentration gradients in the solvent environment. The depolymerization of F-actin increases the mesh size and porosity of the cytoskeleton in the cytoplasm and thus enhances the poroelastic diffusivity. The decrease in the viscoelastic time scale is probably attributable to the loss of friction between F-actin structures and other cytoplasmic components (Ferry 1980) or various actin-associated interactions. The disruption of the F-actin network thus accelerates both cytoplasmic viscoelastic relaxation and cytosolic diffusion during cell deformation. Consequently, the critical effective strain rate, V/a, which triggers poroelastic activity, rises. In addition, the expansion rate of the transition from an apparently viscous fluid behavior to a viscoelastic material also rises. The biological activity of molecular motors which seems to impact pronouncedly affects the mechanical properties of reconstituted cytoskeletal scaffolds (Humphrey et al. 2002, Koenderink et al. 2009) and elevates the cytoplasmic fluidity inside bacteria (Parry et al. 2014).

To reveal how such the biological activity impacts the mechanics of the cytoplasm of mammalian cells, force-relaxation measurements in HeLa cells were performed in the presence or absence of with myosin II motor inhibitors, such as blebbistatin or depletion of ATP. As expected, the characteristic relaxation time in the cytoplasm significantly is elevated in normal HeLa cells after blebbistatin treatment cells and even more after ATP depletion. The inhibition of myosin by blebbistatin and the depletion of ATP depletion decrease the remodeling of the cytoskeleton, which hence elevates the specific viscoelastic relaxation time t. This finding is in line with the relaxation time of reconstituted cytoskeletal scaffolds and of the cytoplasm in bacteria (Humphrey et al. 2002; Parry et al. 2014).

This particular elevation in t leads to a reduction in the transition of the effective strain rate V/a of a viscous fluid toward a viscoelastic gel. These data show that the cytoskeleton alters together with various active components of the cells the mechanical reaction of the cytoplasm through adaption of the cytoplasmic mechanical factors including the viscoelastic relaxation time and the poroelastic diffusivity. To analyze the mechanical resistance in a quantitative manner, which is sensed by a particle that moves through the cytoplasm, the rate-dependent apparent modulus of the cytoplasm can be determined from the slope of the stress–strain curve. In the next step, the rate-dependent modulus can be divided into the storage modulus and the loss modulus that are derived from specific rheological models and the Boltzmann's superposition principle (Koenderink et al. 2009). The storage modulus and the loss modulus are frequently employed to characterize soft materials and complex fluids (Guo et al. 2014, Shuck and Advani 1972); however, the apparent modulus can be used to determine the cytoplasmic mechanics, as it is highly important for the rate- and size-dependent resistance force that a particle undergoes when moving in the cytoplasm.

In fact, it has been found that all measured normalized force–displacement curves of the cytoplasm are commonly linear, if the normalized displacement is small, which means that the displacement is smaller than the bead radius. Under these conditions, an apparent elastic modulus, E_A, can be determined from the average slope from 0 to 10% strain. The apparent elastic modulus seems to rely on the size of the probe and speed of the loading. To discover whether E_A depends on the probe size, the 3D plot can be projected onto 2D axes, such as E_A versus S. Within the viscoelastic regime, which means, when the effective strain rate ($0.1\ s^{-1} < V/a < 2\ s^{-1}$) is small, E_A is constant independent of the distinct probe size. When the effective strain rate ($V/a > 5\ s^{-1}$) is relatively large, E_A depends strongly on the probe size, as it increases with the probe size, which means that the poroelasticity begins to affect the measurement of the apparent modulus and thereby it will become increasingly difficult to achieve a relaxation of the differential pore pressure as the probe size rises. Various origins of cytoplasmic elasticity can be further demonstrated by recording the apparent cytoplasmic modulus in dependence of V/a. The apparent modulus of all measurements at low effective strain rate regardless of bead size precisely matches the same pattern; E_A in this regime is entirely defined by V/a, which verifies that the resistance is primarily attributable to viscoelasticity. In this regime, the measured dependence of E_A on the effective strain rate is also in accordance with the previously monitored power-law rheology with a rate of 0.2% (Guo et al. 2014; Fabry et al. 2001). But if the effective strain rate is high, there is no evident correlation between the apparent modulus and the number V/a. Instead, if V_a is constant, the same module is retained; E_A stays the same although the value of V/a varies several orders of magnitude. The rate dependence of the resistance in this regime is mainly determined by the poroelasticity in the cytoplasm. The transition time between these two systems is $V/a = 3\ s^{-1}$; both viscoelasticity and poroelasticity account for this transition. The unequal dependence of the apparent modulus on V/a and V_a according to various regimes strengthens the idea that the viscoelasticity prevails over the size-independent (at a fixed effective strain rate, V/a) mechanical behavior at low effective strain rate and predominates if the experimental time scale a/V is similar to the viscoelastic time scale t, that is $Vt/a \sim 1$. In contrast, the poroelasticity prevails the size-dependent at a fixed effective strain rate, V/a, mechanical behavior at a high effective strain rate if the experimental time frame is similar to the poroelastic time frame a^2/D, that is $V_a/D \sim 1$.

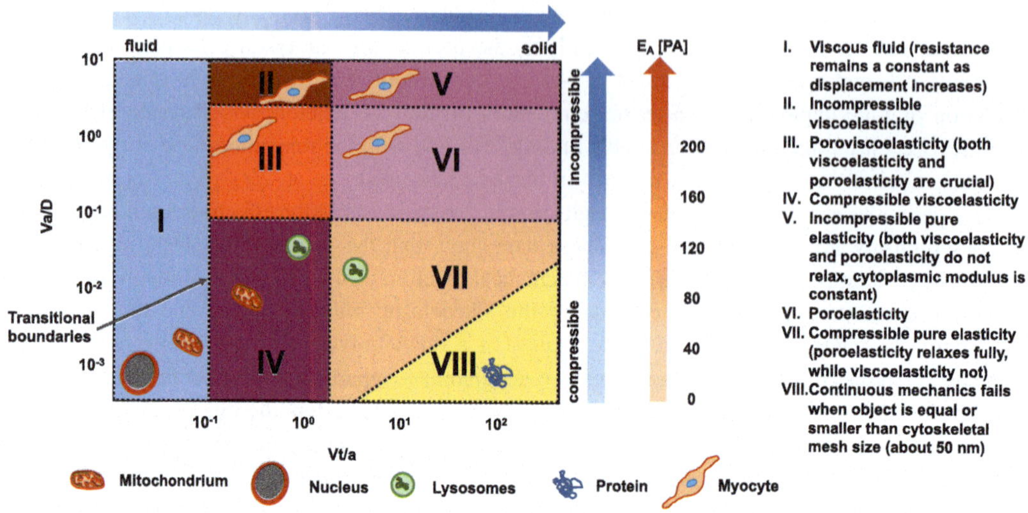

Fig. 3.6 State diagram of living mammalian cytoplasm as a function of the two dimensionless numbers Vt/a and Va/D

In the end, the various cytoplasmic mechanical characteristics within a state diagram can be revealed as a function of the two dimensionless parameters, V_t/a and V_a/D (Fig. 3.6). There are eight different types of mechanical cytoplasm responses operating at various load conditions, each corresponding to one of the eight sections in the state diagram. At $V_t/a << 1$ in Section I, the cytoplasm acts as a viscous fluid (Fig. 3.6). At $V_t/a \sim 1$ (Sections II–IV), the viscoelasticity determines the rate-dependent mechanics of cells. At $V_t/a >> 1$ (Sections V–VII), the cytoplasm acts similar as an unrelaxed elastic material; if the experimental time frame is far less than the viscoelastic relaxation time, the cytoskeleton is not able to rearrange, displace or break, and thus no viscoelastic dissipation is possible. Similarly, the cytoplasm at $V_a/D << 1$ (Sections IV and VII) is extremely compressible, because if the experimental time frame is considerably extended than the poroelastic relaxation time, the infiltrated fluid moves freely around in cells, equivalent to significant volumetric alterations. In this case, the apparent bulk modulus corresponds to the bulk modulus of the solid part in the cytoplasm, considered to be extremely compressible. The poroelasticity adds to the size-dependent mechanical properties of the cytoplasm in the same way as $V_a/D \sim 1$ (Sections III and VI). At $V_a/D >> 1$ (Sections II and V), the experimental requirement demands a high loading size and fast loading speed, impossible to reach with optical tweezers within the cytoplasm. However, based on the experimental results gained in other areas of the state diagram and the theoretical analysis of poroelasticity, the time frame of the loading seems to be too short compared to the poroelastic relaxation time in these two areas and thus the intracellular fluid is confined in the porous cytoskeleton. Under this condition, there are no changes in volumetric strength and can therefore be described as incompressible, with a bulk modulus substantially greater than its shear modulus. An interesting fact is that the cytoplasm exhibits a purely elastic behavior even if the dimensionless parameter V/a is extremely large and V_a is either extremely large (Section V) or extremely small (Section VII). In these two cases, a subsequent alteration of the bead diameter and velocity fails to modify the resulting normalized force–displacement curve, pointing to a size and rate-independent elastic behavior. At the same time, the probe in section VIII is identical or smaller than the characteristic mesh size of the cytoskeleton of approximately 50 nm (Luby-Phelps 2000), and consequently, a continuum mechanics model cannot be used. This state diagram visualizes various features and multiple origins of cytoplasmic mechanics due to different characteristic sizes and speeds.

Based on this diagram, the origin of mechanical resistance can be figured out that a wide range of physiological functions and cellular processes evoke inside the cytoplasm of mammalians. Multiple organelles exist in the cytoplasm that continuously move through it; hence, the mechanical characteristic of this resistance, which one after another commonly senses, relies on their individual sizes and velocities. Using the common size of an organelle and the typical velocity, the precise values of two dimensionless numbers for the motion of the cell nucleus, the mitochondria, the lysosome and the myocyte contraction can be determined. These values specify the distinct region that each process usually occupies within the state diagram. Based on this result, the fundamental roots of the mechanical resistance of different cellular processes across a broad range of velocity and size scales emerge.

The combined analysis of unidirectional force–displacement in the cytoplasm of living mammalian cells can be performed with optical tweezers and a simple scaling analysis. Hence, a technique is presented that can discriminate between various origins of cytoplasmic mechanical resistance due to the intracellular transport. Thereby it can be revealed that mechanical resistance of the cytoplasm can be based on either viscosity, viscoelasticity, poroelasticity or pure elasticity, which relies on the velocity of object and its size. In summary, the mechanical resilience and the apparent cytoplasmic modulus rely both on the velocity and size of the probe and, even bolder, on different modes, while various mechanisms prevail over the apparent cytoplasmic mechanics. Beyond that, in a state diagram with two dimensionless variables, two types of cytoplasmic mechanical behavior are summarized; this state diagram depicts that the living cytoplasm increases from a viscous fluid to an elastic solid as a dimensionless parameter, V_t/a, and from a compressible material to an incompressible material as another dimensionless parameter, V_a/D. Not only the cytoskeleton but also bioactive cell motors are able to control the viscoelastic relaxation time t as well as the poroelastic diffusivity D directly and thus influence the switch between distinct mechanical states. The cytoplasmic state diagram delivers a glimpse into the underlying mechanical nature of a wide range of intracellular transport and other cellular events over a wide variety of velocities and sizes. Finally, an extension of this state diagram to evaluate the mechanical properties of other soft materials of viscoelastic and poroelastic character can be made.

References

B. Alberts, A. Johnson, J. Lewis, M. Raff, K. Roberts, P. Walter, *Molecular Biology of the Cell*, 5th edn. (Garland Science, New York, 2008), pp. 965–1052

T.E. Angelini, E. Hannezo, X. Trepat, M. Marquez, J.J. Fredberg, D.A. Weitz, Glass-like dynamics of collective cell migration. Proc. Natl. Acad. Sci. U.S.A. **108**, 4714–4719 (2011)

E. Anon, X. Serra-Picamal, P. Hersen, N.C. Gauthier, M.P. Sheetz, X. Trepat, B. Ladoux, Cell crawling mediates collective cell migration to close undamaged epithelial gaps. Proc. Natl. Acad. Sci. U.S.A. **109**, 10891–10896 (2012)

M. Balland, N. Desprat, D. Icard, S. Féréol, A. Asnacios, J. Browaeys, S. Hénon, F. Gallet, Power laws in microrheology experiments on living cells: comparative analysis and modeling. Phys. Rev. E: Stat., Nonlin, Soft Matter Phys. **74**, 021911 (2006)

G. Bao, S. Suresh, Cell and molecular mechanics of biological materials. Nat. Mater. **2**(11), 715–725 (2003)

S. Baumann, T. Pohlmann, M. Jungbluth, A. Brachmann, M. Feldbrugge, Kinesin-3 and dynein mediate microtubule-dependent co-transport of mRNPs and endosomes. J. Cell Sci. **125**, 2740–2752 (2012)

A.R. Bausch, W. Möller, E. Sackmann, Measurement of local viscoelasticity and forces in living cells by magnetic tweezers. Biophys. J. **76**(1 Pt 1), 573–579 (1999)

G. Beaune, G. Duclos, N. Khalifat, T.V. Stirbat, D.M. Vignjevic, F. Brochard-Wyart, Reentrant wetting transition in the spreading of cellular aggregates. Soft Matter **13**, 8474–8482 (2017)

J.F. Berret, Local viscoelasticity of living cells measured by rotational magnetic spectroscopy. Nat. Commun. **7**, 10134 (2016)

H. Berry, J. Pelta, D. Lairez, V. Larreta-Garde, Gel–sol transition can describe the proteolysis of extracellular matrix gels. Biochimica et Biophysica Acta General Subjects **1524**(2–3), 110–117 (2000)
D.A. Beysens, G. Forgacs, J.A. Glazier, Cell sorting is analogous to phase ordering in fluids. Proc. Natl. Acad. Sci. U.S.A. **97**, 9467–9471 (2000)
D. Bi, X. Yang, M.C. Marchetti, M.L. Manning, Motility-driven glass and jamming transitions in biological tissues. Phys. Rev. X **6**(2) pii 021011 (2016)
S. Calzolari, J. Terriente, C. Pujades, Cell segregation in the vertebrate hindbrain relies on actomyosin cables located at the interhombomeric boundaries. EMBO J. **33**, 686–701 (2014)
J.H. Cartwright, N. Piro, O. Piro, I. Tuval, Fluid dynamics of nodal flow and left-right patterning in development. Dev. Dyn. **237**, 3477–3490 (2008)
A. Caspi, R. Granek, M. Elbaum, Enhanced diffusion in active intracellular transport. Phys. Rev. Lett. **85**, 5655–5658 (2000)
K.J. Chalut, M. Höpfler, F. Lautenschläger, L. Boyde, C.J. Chan, A. Ekpenyong, A. Martinez-Arias, J. Guck, Chromatin decondensation and nuclear softening accompany Nanog downregulation in embryonic stem cells. Biophys. J. **103**, 2060–2070 (2012)
K.J. Chalut, K. Kulangara, M.G. Giacomelli, A. Wax, K.W. Leong, Deformation of stem cell nuclei by nanotopographical cues. Soft Matter **6**, 1675–1681 (2010)
G.T. Charras, J.C. Yarrow, M.A. Horton, L. Mahadevan, T.J. Mitchison, Non-equilibration of hydrostatic pressure in blebbing cells. Nature **435**, 365–369 (2005)
F. Chowdhury, S. Na, D. Li, Y.C. Poh, T.S. Tanaka, F. Wang, N. Wang, Material properties of the cell dictate stress-induced spreading and differentiation in embryonic stem cells. Nat. Mater. **9**, 82–88 (2010)
E.W. Collins Jr., C. Edwards, Role of Donnan equilibrium in the resting potentials in glycerol-extracted muscle. Am. J. Physiol. **22**(4), 1130–1133 (1971)
J.A. Cooper, Effects of cytochalasin and phalloidin on actin. J. Cell Biol. **105**, 1473–1478 (1987)
F. Costantini, R. Kopan, Patterning a complex organ: Branching morphogenesis and nephron segmentation in kidney development. Dev. Cell **18**, 698–712 (2010)
M. Coue, S.L. Brenner, I. Spector, E.D. Korn, Inhibition of actin polymerization by latrunculin A. FEBS Lett. **213**, 316–318 (1987)
J.C. Crocker, B.D. Hoffman, Multiple-particle tracking and two-point microrheology in cells. Methods Cell Biol. **83**, 141–178 (2007)
E.N. Cytrynbaum, P. Sommi, I. Brust-Mascher, J.M. Scholey, A. Mogilner, Early spindle assembly in Drosophila embryos: role of a force balance involving cytoskeletal dynamics and nuclear mechanics. Mol. Biol. Cell **16**, 4967–4981 (2005)
B.R. Daniels, C.M. Hale, S.B. Khatau, S. Kusuma, T.M. Dobrowsky, S. Gerecht, D. Wirtz, Differences in the microrheology of human embryonic stem cells and human induced pluripotent stem cells. Biophys. J. **99**, 3563–3570 (2010)
M. Dao, C.T. Lim, S. Suresh, Mechanics of the human red blood cell deformed by optical tweezers. J. Mech. Phys. Solids **51**(11), 2259–2280 (2003)
E.H. Davidson, J.P. Rast, P. Oliveri et al., A genomic regulatory network for development. Science **295**, 1669–1678 (2002)
J.C. Del Alamo, G.N. Norwich, Y.S.J. Li, J.C. Lasheras, S. Chien, Anisotropic rheology and directional mechanotransduction in vascular endothelial cells. Proc. Natl. Acad. Sci. U.S.A. **105**(40), 15411–15416 (2008)
A. Desai, T.J. Mitchison, Microtubule polymerization dynamics. Annu. Rev. Cell Dev. Biol. **13**(1), 83–117 (1997)
E. Detournay, AH-D. Cheng, Fundamentals of poroelasticity, in *Comprehensive Rock Engineering: Principles, Practice & Projects*. Comprehensive Rock Engineering, ed Fairhurst C (Pergamon, New York, 2014), pp. 113–172
A.D. Doyle, Generation of micropatterned substrates using micro photopatterning. Curr. Protoc. Cell Biol. Chapter 10: Unit 10.15 (2009)
T.P. Driscoll, B.D. Cosgrove, S.J. Heo, Z.E. Shurden, R.L. Mauck, Cytoskeletal to nuclear strain transfer regulates YAP signaling in mesenchymal stem cells. Biophys. J. **108**, 2783–2793 (2015)
D.G. Drubin, Development of cell polarity in budding yeast. Cell **65**(7), 1093–1096 (1991)
C. Echalier, S. Jebors, G. Laconde et al., Sol–gel synthesis of collagen-inspired peptide hydrogel. Mater. Today **20**(2), 59–66 (2016)
A.J. Ehrlicher, R. Krishnan, M. Guo, C.M. Bidan, D.A. Weitz, M.R. Pollak, Alpha-actinin binding kinetics modulate cellular dynamics and force generation. Proc. Natl. Acad. Sci. U.S.A. **112**(21), 6619–6624 (2015)
A.E. Ekpenyong, G. Whyte, K. Chalut, S. Pagliara, F. Lautenschläger, C. Fiddler, S. Paschke, U.F. Keyser, E.R. Chilvers, J. Guck, Viscoelastic properties of differentiating blood cells are fate- and function-dependent. PLoS ONE **7**, e45237 (2012)
A.J. Engler, S. Sen, H.L. Sweeney, D.E. Discher, Matrix elasticity directs stem cell lineage specification. Cell **126**, 677–689 (2006)

B. Fabry, G.N. Maksym, J.P. Butler, M. Glogauer, D. Navajas, J. Jeffrey, J.J. Fredberg, Scaling the microrheology of living cells. Phys. Rev. Lett. **87** 148102 (2001)

N. Fakhri, A.D. Wessel, C. Willms, M. Pasquali, D.R. Klopfenstein, F.C. MacKintosh, C.F. Schmidt, High-resolution mapping of intracellular fluctuations using carbon nanotubes. Science **344**(6187), 1031–1035 (2014)

J.D. Ferry, *Viscoelastic Properties of Polymers* (Wiley, New York, 1980)

T. Fischer, N. Wilharm, A. Hayn, C.T. Mierke, Matrix and cellular mechanical properties are the driving factors for facilitating human cancer cell motility into 3D engineered matrices. Convergent Sci. Phys. Oncol. **3**(4), 044003 (2017)

E. Fodor, M. Guo, N.S. Gov, P. Visco, D.A. Weitz, F. van Wijland, Activity-driven fluctuations in living cells. EPL **110** (4), 48005 (2015)

G. Forgacs, S.A. Newman, *Biological Physics of the Developing Embryo*. Cambridge University Press (2005)

G. Forgacs, R.A. Foty, Y. Shafrir, M.S. Steinberg, Viscoelastic properties of living embryonic tissues: a quantitative study. Biophys. J. **74**, 2227–2234 (1998)

R.A. Foty, M.S. Steinberg, The differential adhesion hypothesis: a direct evaluation. Dev. Biol. **278**, 255–263 (2005)

R.A. Foty, C.M. Pfleger, G. Forgacs, M.S. Steinberg, Surface tensions of embryonic tissues predict their mutual envelopment behavior. Development **122**, 1611–1620 (1996)

A. Frey-Wyssling, *Submicroscopic Morphology of Protoplasm* (Elsevier, Amsterdam, 1953)

M.L. Gardel, I.C. Schneider, Y. Aratyn-Schaus, C.M. Waterman, Mechanical integration of actin and adhesion dynamics in cell migration. Annu. Rev. Cell Dev. Biol. **26**, 315–333 (2010)

F. Gittes, B. Mickey, J. Nettleton, J. Howard, Flexural rigidity of microtubules and actin filaments measured from thermal fluctuations in shape. J. Cell Biol. **120**, 923 (1993)

C. Godbout, L. Follonier Castella, E.A. Smith, N. Talele, M.L. Chow, A. Garonna, B. Hinz, The mechanical environment modulates intracellular calcium oscillation activities of myofibroblasts. PLoS ONE **8**, e64560 (2013)

A. Grapin-Botton, D.A. Melton, Endoderm development: from patterning to organogenesis. Trends Genet. **16**, 124–130 (2000)

J. Guck, R. Ananthakrishnan, H. Mahmood, T.J. Moon, C.C. Cunningham, J. Käs, The optical stretcher: a novel laser tool to micromanipulate cells. Biophys. J. **81**, 767–784 (2001)

F. Guilak, D.M. Cohen, B.T. Estes, J.M. Gimble, W. Liedtke, C.S. Chen, Control of stem cell fate by physical interactions with the extracellular matrix. Cell Stem Cell **5**, 17–26 (2009)

M. Guo, A.J. Ehrlicher, M.H. Jensen, M. Renz, J.R. Moore, R.D. Goldman, J. Lippincott-Schwartz, F.C. Mackintosh, D.A. Weitz (2014) Probing the stochastic, motor-driven properties of the cytoplasm using force spectrum microscopy. Cell **158**, 822–832 (2014)

M. Guo, A.J. Ehrlicher, S. Mahammad, H. Fabich, M.H. Jensen, J.R. Moore, J.J. Fredberg, R.D. Goldman, D.A. Weitz, The role of Vimentin intermediate filaments in cortical and cytoplasmic mechanics. Biophys. J. **105**(7), 1562–1568 (2013)

M. Guo, A.F. Pegoraro, A. Mao et al., Cell volume change through water efflux impacts cell stiffness and stem cell fate. Proc. Natl. Acad. Sci. U. S. A. **2**(114), E8618–E8627 (2017)

S.K. Gupta, M. Guo, Equilibrium and out-of-equilibrium mechanics of living mammalian cytoplasm. J. Mech. Phys. Solids **107**, 284–293 (2017)

S.K. Gupta, Y. Li, M. Guo, Anisotropic mechanics and dynamics of a living mammalian cytoplasm. Soft Matter **15**, 190–199 (2019)

D.R. Haudenschild, J. Chen, N. Steklov, M.K. Lotz, D.D. D'Lima, Characterization of the chondrocyte actin cytoskeleton in living three-dimensional culture: response to anabolic and catabolic stimuli. Mol. Cell Biomech. **6**, 135–144 (2009)

A.G. Hendricks, E.L.F. Holzbaur, Y.E. Goldman, Force measurements on cargoes in living cells reveal collective dynamics of microtubule motors. Proc. Natl. Acad. Sci. U.S.A. **109**, 18447–18452 (2012)

S.J. Heo, S.D. Thorpe, T.P. Driscoll, R.L. Duncan, D.A. Lee, R.L. Mauck, Biophysical regulation of chromatin architecture instills a mechanical memory in mesenchymal stem cells. Sci. Rep. **5**, 16895 (2015)

R.M. Hershberg, L.F. Mayer (2000) Antigen processing and presentation by intestinal epithelial cells–polarity and complexity. Immunol. Today **21**(3), 123–128 (2000)

S. Heyden, M. Ortiz, Oncotripsy: targeting cancer cells selectively via resonant harmonic excitation. J. Mech. Phys. Solids **92**, 164–175 (2016)

C. Hidalgo-Carcedo, S. Hooper, S.I. Chaudhry, P. Williamson, K. Harrington, B. Leitinger, E. Sahai, Collective cell migration requires suppression of actomyosin at cell-cell contacts mediated by DDR1 and the cell polarity regulators Par3 and Par6. Nat. Cell Biol. **13**(1), 49 (2011)

H. Higuchi, T. Yanagida, Y.E. Goldman, Compliance of thin filaments in skinned fibers of rabbit skeletal muscle. Biophys. J. **69**(3), 1000–1010 (1995)

N. Hirokawa, Kinesin and dynein superfamily proteins and the mechanism of organelle transport. Science **279**(5350), 519–526 (1998)

P.W. Hochachka, The metabolic implications of intracellular circulation. Proc. Natl. Acad. Sci. U. S. A. **96**(22), 12233–12239 (1999)
R.M. Hochmuth, Micropipette aspiration of living cells. J. Biomech. **33**(1), 15–22 (2000)
A.S. Hoffman, Conventionally and environmentally sensitive hydrogels for medical and industrial use: a review paper. Polym. Gels **268**(5), 82–87 (1991)
B.D. Hoffman, J.C. Crocker, Cell mechanics: dissecting the physical responses of cells to force. Annu. Rev. Biomed. Eng. **11**, 259–288 (2009)
B.D. Hoffman, G. Massiera, K.M. Van Citters, J.C. Crocker, The consensus mechanics of cultured mammalian cells. Proc. Natl. Acad. Sci. U. S. A. **103**, 10259–10264 (2006)
A. Holzinger, Jasplakinolide: an actin-specific reagent that promotes actin polymerization. Methods Mol. Biol. **586**, 71–87 (2009)
R.G. Horn, J.N. Israelachvili, Direct measurement of structural forces between two surfaces in a nonpolar liquid. J. Chem. Phys. **75**(3), 1400–1411 (1981)
J. Howard, Molecular motors: structural adaptations to cellular functions. Nature **389**(6651), 561–567 (1997)
J. Howard, *Mechanics of motor proteins and the cytoskeleton Sinauer Associates* (Publishers. xvi, Sunderland, Mass, 2001), p. 367
J. Hu, S. Jafari, Y. Han, A.J. Grodzinsky, S. Cai, M. Guo, Size-and speed-dependent mechanical behavior in living mammalian cytoplasm. Proc. Natl. Acad. Sci. U. S. A. **114**(36), 9529–9534 (2017)
S. Hu, L. Eberhard, J. Chen, J.C. Love, J.P. Butler, J.J. Fredberg, G.M. Whitesides, N. Wang, Mechanical anisotropy of adherent cells probed by a three-dimensional magnetic twisting device. Am. J. Physiol. Cell Physiol. **287**(5), C1184–C1191 (2004)
Y. Hu, Z. Suo, Viscoelasticity and poroelasticity in elastomeric gels. Acta Mech. Solida Sin. **25**, 441–458 (2012)
Y. Hu, X. Zhao, J.J. Vlassak, Z. Suo, Using indentation to characterize the poroelasticity of gels. Appl. Phys. Lett. **96**, 121904 (2010)
J. Huelsken, R. Vogel, B. Erdmann, G. Cotsarelis, W. Birchmeier, b-Catenin controls hair follicle morphogenesis and stem cell differentiation in the skin. Cell **105**(4), 533–545 (2001)
D. Humphrey, C. Duggan, D. Saha, D. Smith, J. Käs, Active fluidization of polymer networks through molecular motors. Nature **416**, 413–416 (2002)
D.E. Ingber, D. Prusty, Z. Sun, H. Betensky, N. Wang, Cell shape, cytoskeletal mechanics, and cell cycle control in angiogenesis. J. Biomech. **28**, 1471–1484 (1995)
M. Irmscher, A.M. de Jong, H. Kress, M.W.J. Prins, Probing the cell membrane by magnetic particle actuation and Euler angle tracking. Biophys. J. **102**(3), 698–708 (2012)
T. Ishizaki, Y. Morishima, M. Okamoto, T. Furuyashiki, T. Kato, S. Narumiya, Coordination of microtubules and the actin cytoskeleton by the Rho effector mDia1. Nat. Cell Biol. **3**(1), 8 (2001)
J.N. Israelachvili, P.M. McGuiggan, Forces between surfaces in liquids. Science **241**, 795–800 (1988)
J.N. Israelachvili, H. Wennerström, Role of hydration and water structure in biological and colloidal interactions. Nature **379**, 219–225 (1996)
P.A. Janmey, J.V. Shah, J.X. Tang, T.P. Stossel, Actin filament networks. Results Probl. Cell Differ. **32**, 181–199 (2001)
P.A. Janmey, J.P. Winer, M.E. Murray, Q. Wen, The hard life of soft cells. Cell Motil. Cytoskeleton **66**, 597–605 (2009)
S.P. Jarvis, T. Uchihashi, T. Ishida, H. Tokumoto, Local solvation shell measurement in water using a carbon nanotube probe. J. Phys. Chem. B **104**, 6091–6097 (2000)
M. Jonas, H. Huang, R.D. Kamm, P.T. So, Fast fluorescence laser tracking microrheometry, II: quantitative studies of cytoskeletal mechanotransduction. Biophys. J. **95**, 895–909 (2008)
D.S. Jones, Dynamic mechanical analysis of polymeric systems of pharmaceutical and biomedical significance. Int. J. Pharm. **179**(2), 167–178 (1999)
P.M. Jungmann, A.T. Mehlhorn, H. Schmal, H. Schillers, H. Oberleithner, N.P. Südkamp, Nanomechanics of human adipose-derived stem cells: small GTPases impact chondrogenic differentiation. Tissue Eng. Part A **18**, 1035–1044 (2012)
K.E. Kasza, J.A. Zallen, Dynamics and regulation of contractile actin–myosin networks in morphogenesis. Curr. Opin. Cell Biol. **23**(1), 30–38 (2011)
W.M. Kier, K.K. Smith, Tongues, tentacles and trunks—the biomechanics of movement in muscular-hydrostats. Zool. J. Linn. Soc. **83**, 307–324 (1985)
J.E. Kim, D.S. Reynolds, M.H. Zaman, M. Mak, Characterization of the mechanical properties of cancer cells in 3D matrices in response to collagen concentration and cytoskeletal inhibitors. Integr. Biol. **10**(4), 232–241 (2018)
G.H. Koenderink, Z. Dogic, F. Nakamura, P.M. Bendix, F.C. MacKintosh, J.H. Hartwig, T.P. Stossel, D.A. Weitz, An active biopolymer network controlled by molecular motors. Proc. Natl. Acad. Sci. U. S. A. **106**, 15192–15197 (2009)

M. Krieg, Y. Arboleda-Estudillo, P.H. Puech, J. Käfer, F. Graner, D.J. Müller, C.P. Heisenberg, Tensile forces govern germ-layer organization in zebrafish. Nat. Cell Biol. **10**, 429–436 (2008)

S. Kumar, I.Z. Maxwell, A. Heisterkamp, T.R. Polte, T.P. Lele, M. Salanga, E. Mazur, D.E. Ingber, Viscoelastic retraction of single living stress fibers and its impact on cell shape, cytoskeletal organization, and extracellular matrix mechanics. Biophys. J. **90**, 3762–3773 (2006)

S. Kutscheidt, R. Zhu, S. Antoku, G.W. Luxton, I. Stagljar, O.T. Fackler, G.G. Gundersen, FHOD1 interaction with nesprin-2G mediates TAN line formation and nuclear movement. Nat. Cell Biol. **16**, 708–715 (2014)

J.R. Lange, B. Fabry, Cell and tissue mechanics in cell migration. Exp. Cell Res. **319**(16), 2418–2423 (2013)

R.G. Larson, *The Structure and Rheology of Complex Fluids* (Oxford Univ Press, New York, 1999)

A.W. Lau, B.D. Hoffman, A. Davies, J.C. Crocker, T.C. Lubensky, Microrheology, stress fluctuations, and active behavior of living cells. Phys. Rev. Lett. **91**(19), 198101 (2003)

K. Lee, E.H. Kim, N. Oh, N.A. Tuan, N.H. Bae, S.J. Lee, K.G. Lee, C.-Y. Eom, E.K. Yim, S. Park, Contribution of actin filaments and microtubules to cell elongation and alignment depends on the grating depth of microgratings. J. Nanobiotechnol. **14**(1), 35 (2016)

S.E. Leggett, A.S. Khoo, I.Y. Wong, Multicellular tumor invasion and plasticity in biomimetic materials. Biomater. Sci. **5**(8), 1460–1479 (2017)

Q. Li, A. Kumar, E. Makhija, G.V. Shivashankar, The regulation of dynamic mechanical coupling between actin cytoskeleton and nucleus by matrix geometry. Biomaterials **35**, 961–969 (2014)

G.N. Ling, The physical state of water in living cell and model systems. Ann. N. Y. Acad. Sci. **125**, 401–417 (1965)

G.N. Ling, A Revolution in the Physiology of the Living Cell. Krieger Publ. Co Malabar FL (1992)

G.N. Ling, C.L. Walton, What retains water in living cells? Science **191**, 293–295 (1976)

J. Liu, M.L. Gardel, K. Kroy, E. Frey, B.D. Hoffman, J.C. Crocker, A.R. Bausch, D.A. Weitz, Microrheology probes length scale dependent rheology. Phys. Rev. Lett. **96**, 118104 (2006)

T. Lourenço, J. Paes de Faria, C.A. Bippes, J. Maia, J.A. Lopes-da-Silva, J.B. Relvas, M. Grãos, Modulation of oligodendrocyte differentiation and maturation by combined biochemical and mechanical cues. Sci. Rep. **6**, 21563 (2016)

O.A. Lozoya, S.R. Lubkin, Mechanical control of spheroid growth: distinct morphogenetic regimes. J. Biomech. **45**, 319–325 (2012)

O.A. Lozoya, C.L. Gilchrist, F. Guilak, Universally conserved relationships between nuclear shape and cytoplasmic mechanical properties in human stem cells. Sci. Rep. **6**, 23047 (2016)

O.A. Lozoya, E. Wauthier, R.A. Turner, C. Barbier, G.D. Prestwich, F. Guilak, R. Superfine, S.R. Lubkin, L.M. Reid, Regulation of hepatic stem/progenitor phenotype by microenvironment stiffness in hydrogel models of the human liver stem cell niche. Biomaterials **32**, 7389–7402 (2011)

Y.-B. Lu, K. Franze, G. Seifert et al., Viscoelastic properties of individual glial cells and neurons in the CNS. Proc. Natl. Acad. Sci. U. S. A. **103**, 17759–17764 (2006)

K. Luby-Phelps, Cytoarchitecture and physical properties of cytoplasm: volume, viscosity, diffusion, intracellular surface area. Int. Rev. Cytol. **192**, 189–221 (2000)

M. Mak, S. Anderson, M.C. McDonough, F. Spill, J.E. Kim, A. Boussommier-Calleja, M.H. Zaman, R.D. Kamm, Integrated analysis of intracellular dynamics of MenaINV cancer cells in a 3D matrix. Biophys. J. **112**(9), 1874–1884 (2017)

M. Mak, F. Spill, R.D. Kamm, M.H. Zaman, Single-cell migration in complex microenvironments: mechanics and signaling dynamics. J. Biomech. Eng. **138**(2), 021004 (2016)

J.M. Maloney, D. Nikova, F. Lautenschläger, E. Clarke, R. Langer, J. Guck, K.J. Van Vliet, Mesenchymal stem cell mechanics from the attached to the suspended state. Biophys. J. **99**(8), 2479–2487 (2010)

A.J. Maniotis, C.S. Chen, D.E. Ingber, Demonstration of mechanical connections between integrins, cytoskeletal filaments, and nucleoplasm that stabilize nuclear structure. Proc. Natl. Acad. Sci. U. S. A. **94**, 849–854 (1997)

T.G. Mason, D.A. Weitz, Optical measurements of frequency-dependent linear viscoelastic moduli of complex fluids. Phys. Rev. Lett. **74**, 1250–1253 (1995)

S.C. Materna, E.H. Davidson, Logic of gene regulatory networks. Curr. Opinion Biotechnol. **18**, 351–354 (2007)

C. Metzner, C. Raupach, C.T. Mierke, B. Fabry, Fluctuations of cytoskeleton-bound microbeads—the effect of bead–receptor binding dynamics. J. Phys. Condensed Matter **22**(19), 194105 (2010)

C.T. Mierke, Cancer cells regulate biomechanical properties of human microvascular endothelial cells. J. Biol. Chem. **286**(46), 40025–40037 (2011)

C.T. Mierke, *Physics of Cancer, Interplay Between Tumor Biology, Inflammation and Cell Mechanics, Chapter 1: Initiation of a Neoplasm or Tumor*, Vol. 1, 2nd edn. (IOP Publishing Ltd., 2018), pp. 1-1 to 1-121

C.T. Mierke, The matrix environmental and cell mechanical properties regulate cell migration and contribute to the invasive phenotype of cancer cells. Rep. Progress Phys. **82**(6), 064602 (2019)

C.T. Mierke, T. Fischer, S. Puder, T.Kunschmann, B. Soetje, W.H. Ziegler, Focal adhesion kinase activity is required for actomyosin contractility-based invasion of cells into dense 3D matrices. Sci. Rep. **7**, 42780 (2017)

C.T. Mierke, B. Frey, M. Fellner, M. Herrmann, B. Fabry, Integrin α5β1 facilitates cancer cell invasion through enhanced contractile forces. J. Cell Sci. **124**(3), 369–383 (2011)

C.T. Mierke, D. Paranhos Zitterbart, P. Kollmannsberger, C. Raupach, U. Schlötzer-Schrehardt, T.W. Goecke, J. Behrens, B. Fabry, Breakdown of the endothelial barrier function in tumor cell transmigration. Biophys. J. **94**(7), 2832–2846 (2008)

J. Mitchell, J. Blanshard, Rheological properties of alginate gels. J. Texture Stud. **7**, 219–234 (1976)

E. Moeendarbary, L. Valon, M. Fritzsche, A.R. Harris, D.A. Moulding, A.J. Thrasher, E. Stride, L. Mahadevan, G.T. Charras, The cytoplasm of living cells behaves as a poroelastic material. Nat. Mater. **12**, 253–261 (2013)

A. Mogilner, G. Oster, Force generation by actin polymerization II: the elastic ratchet and tethered filaments. Biophys. J. **84**(3), 1591–1605 (2003)

A. Mogilner, G. Oster, Polymer motors: pushing out the front and pulling up the back. Curr. Biol. **13**(18), R721–R733 (2003)

A. Mongera, P. Rowghanian, H.J. Gustafson, E. Shelton, D.A. Kealhofer, E.K. Carn, F. Serwane, A.A. Lucio, J. Giammona, O. Campàs, A fluid-to-solid jamming transition underlies vertebrate body axis elongation. Nature **561** (7723), 401–405 (2018)

K.A. Moore, T. Polte, S. Huang, B. Shi, E. Alsberg, M.E. Sunday, D.E. Ingber, Control of basement membrane remodeling and epithelial branching morphogenesis in embryonic lung by Rho and cytoskeletal tension. Dev. Dyn. **232**, 268–281 (2005)

M. Murrell, P.W. Oakes, M. Lenz, M.L. Gardel, Forcing cells into shape: the mechanics of actomyosin contractility. Nat. Rev. Mol. Cell Biol. **16**(8), 486–498 (2015)

S. Nawaz, P. Sánchez, K. Bodensiek, S. Li, M. Simons, I.A.T. Schaap, Cell visco-elasticity measured with AFM and optical trapping at sub-micrometer deformations. PLoS ONE **7**, e45297 (2012)

D.J. Needleman, M.A. Ojeda-Lopez, U. Raviv, K. Ewert, H.P. Miller, L. Wilson, C.R. Safinya, Radial compression of microtubules and the mechanism of action of taxol and associated proteins. Biophys. J. **89**(5), 3410–3423 (2005)

H. Nogucci, Dynamic and static analyses of glass-like properties of three-dimensional tissues. Biophys. Physicobiol. **16**, 9–17 (2019)

P.W. Oakes, S. Banerjee, M.C. Marchetti, M.L. Gardel, Geometry regulates traction stresses in adherent cells. Biophys. J. **107**(4), 825–833 (2014)

Y. Osada, J. Gong, Stimuli-responsive polymer gels and their application to chemomechanical systems. Prog. Polym. Sci. **18**, 187–226 (1993)

L. Oswald, S. Grosser, D.M. Smith, J.A. Käs, Jamming transitions in cancer. J. Phys. D Appl. Phys. **50**(48), 483001 (2017)

A. Ott, J.P. Bouchaud, D. Langevin, W. Urbach, Anomalous diffusion in living polymers—a genuine levy flight. Phys. Rev. Lett. **65**, 2201–2204 (1990)

G.J. Owensa, R.K. Singh, F. Foroutan, M. Alqaysi, C.-M. Han, C. Mahapatra, H.-W. Kim, J.C. Knowles, Sol–gel-based materials for biomedical applications. Prog. Mater. Sci. **77**, 1–79 (2016)

S. Pagliara, K. Franze, C.R. McClain, G. Wylde, C.L. Fisher, R.J.M. Franklin, A.J. Kabla, U.F. Keyser, K.J. Chalut, Auxetic nuclei in embryonic stem cells exiting pluripotency. Nat. Mater. **13**, 638–644 (2014)

P. Panorchan, J.S. Lee, B.R. Daniels, T.P. Kole, Y. Tseng, D. Wirtz, Probing cellular mechanical responses to stimuli using ballistic intracellular nanorheology. Methods Cell Biol. **83**, 115–140 (2007)

S. Park, R. Schapery, Methods of interconversion between linear viscoelastic material functions. Part I—a numerical method based on Prony series. Int. J. Solids Struct. **36**(1), 653–1675 (1999)

B.R. Parry, I.V. Surovtsev, M.T. Cabeen, C.S. O'Hern, E.R. Dufresne, C. Jacobs-Wagner, The bacterial cytoplasm has glass-like properties and is fluidized by metabolic activity. Cell **156**, 183–194 (2014)

R.M. Pashley, J.A. Kitchener, Surface forces in adsorbed multilayers of water on quartz. J. Colloid Interface Sci. **71**, 491–500 (1979)

A.F. Pegoraro, J.J. Fredberg, J.A. Park, Problems in biology with many scales of length: Cell-cell adhesion and cell jamming in collective cellular migration. Exp. Cell Res. **343**(1), 54–59 (2016)

C. Pérez-González, R. Alert, C. Blanch-Mercader, M. Gómez-González, T. Kolodziej, E. Bazellieres, J. Casademunt, X. Trepat, Active wetting of epithelial tissues. Nat. Phys. **15**, 79–88 (2019)

G.H. Pollack, *Cells, Gels and the Engines of Life: A New, Unifying Approach to Cell Function* (Ebner and Sons, Seattle, 2001)

G.H. Pollack, The role of aqueous interfaces in the cell. Adv. Colloid Interface Sci. **103**(2), 173–196 (2003)

S. Pritchard, F. Guilak, Effects of interleukin-1 on calcium signaling and the increase of filamentous actin in isolated and in situ articular chondrocytes. Arthritis Rheum **54**, 2164–2174 (2006)

R.P. Rand, V.A. Parsegian, D.C. Rau, Intracellular osmotic action. Cell. Mol. Life Sci. **57**(7), 1018–1032 (2000)

C. Raupach, D.P. Zitterbart, C.T. Mierke, C. Metzner, F.A. Müller, B. Fabry, Stress fluctuations and motion of cytoskeletal-bound markers. Phys. Rev. E **76**(1), 011918 (2007)

P. Recho, T. Putelat, L. Truskinovsky, Mechanics of motility initiation and motility arrest in crawling cells. J. Mech. Phys. Solids **84**, 469–505 (2015)

M. Reiner, The Deborah number. Phys. Today **17**, 62 (1964)
M.J. Rosenbluth, A. Crow, J.W. Shaevitz, D.A. Fletcher, Slow stress propagation in adherent cells. Biophys. J. **95**, 6052–6059 (2008)
S.M. Sadati, N.T. Qazvini, R. Krishnan, C.Y. Park, J. Jeffrey, J.J. Fredberg, Collective migration and cell jamming. Differentiation **86**(3), 121–125 (2013)
M. Schliwa, G. Woehlke, Molecular motors. Nature **422**(6933), 759–765 (2003)
A.P.S. Selvadurai, Indentation of a spherical cavity in an elastic body by a rigid spherical inclusion: Influence of non-classical interface conditions. Contin. Mech. Thermodyn. **28**, 617–632 (2016)
D.J. Sharp, G.C. Rogers, J.M. Scholey, Microtubule motors in mitosis. Nature **407**(6800), 41–47 (2000)
L. Shuck, S. Advani, Rheologioal response of human brain tissue in shear. J. Basic Eng. **94**, 905–911 (1972)
P.G. Smith, L. Deng, J.J. Fredberg, G.N. Maksym, Mechanical strain increases cell stiffness through cytoskeletal filament reorganization. Am. J. Physiol.: Lung Cell. Mol. Physiol. **285**(2), L456–L463 (2003)
F. Spill, D.S. Reynolds, R.D. Kamm, M.H. Zaman, Impact of the physical microenvironment on tumor progression and metastasis. Curr. Opin. Biotechnol. **40**, 41–48 (2016)
J.A. Spudich, How molecular motors work. Nature **372**(6506), 515–518 (1994)
J. Su, R.R. Brau, X. Jiang, G.M. Whitesides, M.J. Lang, P.T.C. So, Geometric confinement influences cellular mechanical properties II—intracellular variances in polarized. Cell. Mol. Cell. Biomech. **4**, 105–118 (2007)
S. Suresh, Biomechanics and biophysics of cancer cells. Acta Biomater. **3**(4), 413–438 (2007)
J. Swift, I.L. Ivanovska, A. Buxboim et al., Nuclear lamin-A scales with tissue stiffness and enhances matrix-directed differentiation. Science **341**, 1240104 (2013)
T. Takigawa, Y. Morino, K. Urayama, T. Masuda, Poisson's ratio of polyacrylamide (PAAm) gels. Polym. Gels Netw. **4**, 1–5 (1996)
D.T. Tambe, C.C. Hardin, T.E. Angelini et al., Collective cell guidance by cooperative intercellular forces. Nat. Mater. **10**, 469–475 (2011)
T. Tanaka, M. Anaka, F. Ilmain, K. Ishii, E. Kokfuta, A. Suzuki, M. Tokita, *Phase Transitions in Gels: Mechanics of Swelling NATO ASI Series*, vol. H64 (Springer, Berlin, 1992)
J.A. Theriot, Regulation of the actin cytoskeleton in living cells. Semin. Cell Biol. **5**, 193–199 (1994)
M. Thery, V. Racine, M. Piel, A. Pépin, A. Dimitrov, Y. Chen, J.B. Sibarita, M. Bornens, Anisotropy of cell adhesive microenvironment governs cell internal organization and orientation of polarity. Proc. Natl. Acad. Sci. U. S. A. **103** (52), 19771–19776 (2006)
I.M. Tolic-Nørrelykke, N. Wang, Traction in smooth muscle cells varies with cell spreading. J. Biomech. **38**(7), 1405–1412 (2005)
X. Trepat, J.J. Fredberg, Plithotaxis and emergent dynamics in collective cellular migration. Trends Cell Biol. **21**, 638–646 (2011)
Y. Tseng, T.P. Kole, D. Wirtz, Micromechanical mapping of live cells by multiple-particle-tracking microrheology. Biophys. J. **83**, 3162–3176 (2002)
R.D. Vale, The molecular motor toolbox for intracellular transport. Cell **112**(4), 467–480 (2003)
S.R. Vedula, H. Hirata, M.H. Nai, A. Brugués, Y. Toyama, X. Trepat, C.T. Lim, B. Ladoux, Epithelial bridges maintain tissue integrity during collective cell migration. Nat. Mater. **13**, 87–96 (2014)
M. Versaevel, T. Grevesse, S. Gabriele, Spatial coordination between cell and nuclear shape within micropatterned endothelial cells. Nat. Commun. **3**, 671 (2012)
S.A. Wainwright, Mechanical design in organisms. Wiley (1976)
X. Wan, Z. Li, S.R. Lubkin, Mechanics of mesenchymal contribution to clefting force in branching morphogenesis. Biomech. Model. Mechanobiol. **7**, 417–426 (2008)
N. Wang, J.P. Butler, D.E. Ingber, Mechanotransduction across the cell surface and through the cytoskeleton. Science **260**(5111), 1124–1127 (1993)
N. Wang, H. Hirata, M.H. Nai, A. Brugués, Y. Toyama, X. Trepat, C.T. Lim, B. Ladoux, Micropatterning tractional forces in living cells. Cytoskeleton **52**(2), 97–106 (2002)
Q.-M. Wang, A.C. Mohan, M.L. Oyen, X.-H. Zhao, Separating viscoelasticity and poroelasticity of gels with different length and time scales. Acta. Mech. Sin. **30**, 20–27 (2014)
S.C. Weber, A.J. Spakowitz, J.A. Theriot, Nonthermal ATP-dependent fluctuations contribute to the in vivo motion of chromosomal loci. Proc. Natl. Acad. Sci. U. S. A. **109**, 7338–7343 (2012)
P.M. Wiggins, Role of water in some biological processes. Microbiol. Rev. **54**(4), 432–449 (1990)
C. Wilhelm, Out-of-equilibrium microrheology inside living cells. Phys. Rev. Lett. **101**(2), 028101 (2008)
D. Wirtz, Particle-tracking microrheology of living cells: principles and applications. Annu. Rev Biophys. **38**, 301–326 (2009)
A. Wodarz, Tumor suppressors: linking cell polarity and growth control. Curr. Biol. **10**(17), R624–R626 (2000)
M.A. Wozniak, S. Chen, Mechanotransduction in development: a growing role for contractility. Nat. Rev. Mol. Cell Biol. **10**, 34–43 (2009)

J. Yan, F. Chen, B.G. Amsden, Cell sheets prepared via gel-sol transition of calcium RGD-alginate. Acta Biomater. **30**, 277–284 (2016)

C. Yeaman, K.K. Grindstaff, W.J. Nelson, New perspectives on mechanisms involved in generating epithelial cell polarity. Physiol. Rev. **79**(1), 73–98 (1999)

J.L. Young, A.J. Engler, Hydrogels with time-dependent material properties enhance cardiomyocyte differentiation in vitro. Biomaterials **32**, 1002–1009 (2011)

M.H. Zaman, The role of engineering approaches in analysing cancer invasion and metastasis. Nat. Rev. Cancer **13**(8), 596 (2013)

A. Zemel, F. Rehfeldt, A.E. Brown, D.E. Discher, S.A. Safran, Optimal matrix rigidity for stress-fibre polarization in stem cells. Nat. Phys. **6**(6), 468–473 (2010)

E.H. Zhou, X. Trepat, C.Y. Park, G. Lenormand, M.N. Oliver, S.M. Mijailovich, C. Hardin, D.A. Weitz, J.P. Butler, J. J. Fredberg, Universal behavior of the osmotically compressed cell and its analogy to the colloidal glass transition. Proc. Natl. Acad. Sci. U. S. A. **106**(26), 10632–10637 (2009)

C. Zhu, G. Bao, N. Wang, Cell mechanics: mechanical response, cell adhesion, and molecular deformation. Annu. Rev. Biomed. Eng. **2**, 189–226 (2000)

M. Ziman, D. Preuss, J. Mulholland, J.M. O'Brien, D. Botstein, D.I. Johnson, Subcellular localization of Cdc42p, a Saccharomyces cerevisiae GTP-binding protein involved in the control of cell polarity. Mol. Biol. Cell **4**(12), 1307–1316 (1993)

4 Focal Adhesion Proteins Regulate Cell–Matrix and Cell–Cell Adhesion and Act as Force Sensors

Abstract

This chapter presents focal adhesions and the proteins they contain by selecting the most appropriate focal adhesion protein, vinculin, as a representative example. The role of a key focal adhesion protein vinculin is presented for the processes of cell–matrix adhesion and cell–cell adhesion in precise detail. Mechanotransduction signaling is also included and the role of intracellular force sensors is highlighted. Since force sensors can act as mechanical markers, their function is presented. There are different force sensors described, and an alternative myosin-independent force sensor, the lamellipodium, is presented. Finally, it is discussed why a cell needs to mechanically sense their environment and what impact it has on the cell itself.

4.1 Introduction to Cell–Matrix Adhesion

The cell adhesion represents a fundamental feature for the survival and function of various cell types and facilitates the function of basic cellular processes such as cell division, cell survival and cell migration. The dynamical interaction of cells with their surrounding microenvironment is facilitated by cell–matrix adhesions that drives many basic cellular processes including the recruitment of immune cells, wound healing after tissue injury and the malignant progression of cancer. The focus of current biophysical research is on the investigation of focal adhesions with newly developed imaging techniques that have shed light on the precise analysis of the molecular composition and dynamics of the focal adhesions. Several specific types of cell adhesions are formed by adhesive cells upon the contact with the extracellular matrix environment. A classification of these structures can be performed due to various factors including their protein composition, the lifespan of the adhesion and the proteolytic activity of the assembly.

4.1.1 The Major Adhesion Types

The size of the adhesive structures in cells can vary considerably within a single cell and on different time scales (Izzard and Lochner 1980). Subsequently, these adhesive structures can be defined by two different types based on their composition of proteins, the size of these adhesive structures, their lifespan and their ability to degrade proteolytically the extracellular matrix environment (Puklin-

C. T. Mierke, *Cellular Mechanics and Biophysics*, Biological and Medical Physics, Biomedical Engineering, https://doi.org/10.1007/978-3-030-58532-7_4

Faucher and Sheetz 2009; Zaidel-Bar et al. 2004, 2007). Among different types of adhesive structures, the family of integrins plays a key role for the initiation and stabilization of them. The binding of integrins to the extracellular matrix microenvironment is driven by the localization of talin on the cytoplasmic side of the cell membrane to the cytoplasmic tail of the integrins. The binding of talin to integrins initiates the further recruitment of other focal adhesion proteins, such as vinculin, focal adhesion kinase (FAK) and paxillin. Each of these plays a decisive role in the signal transduction pathways and in the entire lifetime of the structure.

4.1.2 Focal Point Contacts

The focal point contacts (synonymously termed dot-like contacts or simply point contacts) represent the shortest-lived adhesive structures of the cell. They have a relatively small size of 0.5–1.0 μm (Bershadsky et al. 1985; Tawil et al. 1993) and are located at the edges of lamellipodia of adhesive cells. Moreover, these structures are also described as focal complexes (Nobes and Hall 1995). Focal point contacts possess a small size, and they are typically directly located after the leading edge of a cell that is just spreading or migrating. Moreover, these focal point contacts can be assembled and disassembled at relatively short time scales in the order of seconds or minutes. Therefore, they are supposed to have "scanned" the local extracellular matrix before either decomposing themselves or converting to more stable structures such as focal adhesions.

4.1.3 Focal Adhesions

The size of focal adhesions is much larger than that of focal contacts (synonymously termed focal contacts, which are in some studies referred to smaller sized structures). These elongated streak-like structures have a length of 3–10 μm and are connected to actin filament and myosin filament bundles that are termed stress fibers (Heath and Dunn 1978; Rottner et al. 1999; Zamir et al. 2000). In addition, focal adhesions possess a higher stability than focal point contacts and are therefore available in the order of 10 min. Focal adhesions encompass a wide variety of proteins to maintain the stability of the cells and generate and exert traction forces toward their surrounding extracellular matrix environment from the cell and vice versa.

4.1.4 Fibrillar Adhesions

Even larger and more stable focal adhesions can be found that are termed fibrillar adhesions. These fibrillar adhesions can be enriched with tensin (Zamir et al. 1999; Zamir et al. 2000; Katz et al. 2000) that is engaged with the fibrillogenesis of the extracellular matrix protein fibronectin (Pankov et al. 2000). These long, stable structures, which are located aligned to bundles of fibronectin in vivo, fibrillar adhesions contain besides tensin, highly enriched in $\alpha5\beta1$ integrins (Green and Yamada 2007). At fibrillar adhesions, extracellular matrix can be repositioned and the fibrillogenesis of fibronectin can be induced. Two other adhesion structure classes can be distinguished by their capacity to interact with the surrounding extracellular matrix by degrading it. One such structure is podosomes that are generally present in monocytic originated cells that have differentiated in the tissue into macrophages. In specific detail, these podosomes contain F-actin stress fibers and several actin-binding proteins that are surrounding by a ring-like structure of integrins. Another type of structure is invadopodia, which can be characterized in their composition as similar to podosomes, but only occur in malignant cells (Linder 2007), where they emerge as more punctual, finger-like projections into the extracellular matrix and potentially endow these cells with differences in mechanical stability (Albiges-Rizo et al. 2009).

Adhesive structure formation is controlled temporally and locally by numerous different proteins, and it is presumed that adhesion formation is initiated by the activation of the integrin family of extracellular matrix receptors. The interaction of integrins and the extracellular matrix environment leads to the controlled assembly of intracellular signaling sites within the regions, where the cytoplasmic integrin domains are clustered. These signaling sites recruit adaptor proteins, kinases, phosphatases and other cell surface receptors. The activation of integrins causes fast alterations of the associated cytoskeleton in an indirect manner that thereby facilitates mechanically the maturation of adhesion sites. In a similar way, the disassembly of these cell adhesions is also closely coordinated in terms of space and time and is essential for ensuring effective cell migration.

Most of what is momentarily understood in the localization and turnover of cell–matrix adhesions has been characterized by the visualization of protein dynamics with a variety of microscopy techniques. However, most studies to date have been performed with cells plated on 2D surfaces. However, there are countless questions that have not yet been answered regarding the characteristics and dynamics of cell–matrix adhesions in intricate 3D matrices. In fact, the actual importance of various adhesion classes specified in cells located on 2D matrices for those encountered in cells in 3D environments is still questionable. This chapter explores how multiple techniques that are frequently utilized to investigate issues of adhesion patterns in living cells are employed and reviews specific results obtained by using these techniques.

Finally, these adhesion sites that couple cells with their adjacent cells or with the surrounding extracellular matrix made of large complexes with multiple proteins, all of which can mechanical couple and thereby sense the matrix mechanical properties of their microenvironment (Bershadsky et al. 2003; Chen et al. 2004; Geiger et al. 2001; Sastry and Burridge 2000). The challenging interaction between the mechanical function of cell adhesions and their “instructive role”, which manifests itself through the activation of various signaling pathways, is conveyed by an array of proteins collectively referred to as “adhesome.” The combined action of the components of the adhesome essentially impacts across all cell functions encompassing cell morphogenesis, proliferation, viability, migration and differentiation (Berrier and Yamada 2007; Streuli 2009; Thiery 2003; Vicente-Manzanares et al. 2009). Cell–matrix adhesions range from small, short-lived focus contacts to punctiform, finger-based projections referred to as invadopodia.

4.2 Focal Adhesions and Focal Adhesion Proteins

Approximately 40 years after vinculins identification as a component of focal adhesions and the adherens junctions of cells, a huge effort of research has been put into the investigation of vinculin, including its activation, regulation and function (Bays and DeMali 2017). In this chapter, the present knowledge of the function of vinculin in cell–cell and cell–matrix adhesions opens up. The focus is on the recruitment, activation and regulation of vinculin. However, another recent insight is how vinculin reacts to integrin- and cadherin-containing adhesion complexes and exerts force on the underlying cytoskeleton. In parallel, topics of discussion include the role of vinculin in the attachment and restructuring of the actin cytoskeleton.

Tissues in intricate multicellular organisms consist only of one or more cell types. Cells with epithelial or endothelial roots connect to adjacent cells over cell–cell contacts and attach to the subjacent basement membrane by cell–matrix coupling. Both kinds of adhesions are crucial for the development of embryos, the rebuilding of tissues, the migration of cells and many other homeostatic functions. The misregulation of adhesion sites results in a wide array of diseases including cancer, diabetes and cardiovascular disease.

The process of cell adhesion undergoes continuously restructuring due to receptor engagement, receptor clustering and the dynamic turnover of adhesion receptors, such as integrins. These integrins are found strongly assembled within cell matrix adhesions, such as focal complexes and focal adhesions. The focal complexes represent small transient adhesions on the cell circumference that can increase to focal adhesions that become then bigger and more stable structures. In opposite, areas where cells adhere to adjacent cells are referred to as adherens junctions and are augmented with transmembrane adhesion receptors, referred to as cadherin.

The pivotal role of adhesion receptors is played by the acquisition of proteins connecting the adhesion receptor to the actin cytoskeleton. Although a large number of proteins are present in adhesive complexes, vinculin, which is a cytoplasmic protein that can bind to actin, is accumulated in cell–cell and cell–matrix adhesions and is one of the most investigated in these areas. Vinculin possesses no enzymatic activity. It governs the adhesion by directly linking to actin, promotes the polymerization of actin and targets actin remodeling proteins. When vinculin is absent, cell–matrix and cell–cell adhesion are severely compromised, suggesting that vinculin is a key factor in human physiology.

4.2.1 Vinculin and Interacting Proteins

Vinculin is a 116 kDa protein that has been identified in the laboratories of Benjamin Geiger and Keith Burridge as enriched in areas in which cells interact with each other and the subjacent medium (Geiger 1979; Burridge and Feramisco 1980). The cloning of the cDNA of vinculin showed that it encodes a protein with 1066 amino acids, and consecutive crystallization studies revealed that vinculin consists of eight anti-parallel α helical bundles divided into five different regions (Bakolitsa et al. 2004). The domains 1–3 (referred to as D1–3) of vinculin are organized in a tri-lobar head of 80 Å diameter (Molony and Burridge 1985) and possess a molecular weight of 95 kDa (Gimona et al. 1988). The vinculin head domain can interact with several proteins, such as α-actinin, talin, β-catenin, α-catenin and IpaA (Carisey and Ballestrem 2011). The head and tail domain of vinculin are linked by a 61 amino acid proline-rich linkage region that encompasses the residues 837–878 (Bakolitsa et al. 2004). These linker regions interact with the vasodilator-stimulated phosphoprotein (VASP) (Brindle et al. 1996), ponsin (Mandai et al. 1999), Arp2/3 (DeMali et al. 2002) and vinexin (Kioka et al. 1999). Finally, the vinculin tail (amino acids 879–1066) consists of a 30 kDa spiral bundle with five helices interconnected with short loops (3–8 residues) (Bakolitsa et al. 2004; Groesch and Otto 1990). The tail has binding sites for the vinculin head domain, actin, paxillin and acidic phospholipids (Fig. 4.1).

4.2.2 Activation of Vinculin

In cells, vinculin can be detected in two vastly different conformations: Firstly, it can be found as an open and hence active protein form, and secondly, it can be a closed and hence auto-inhibited protein state, where the head domain of vinculin interacts stably with its tail domain. In this head-to-tail connection, the access of vinculin-binding proteins within this hidden region seems to be no longer possible. In fact, vinculin FRET probes, which are based on the two different conformational states of vinculin, revealed that vinculin is present in its active and extended conformation in focal adhesion sites, whereas vinculin can be found strongly folded by the interaction of its head and tail region and hence in its inactive form throughout the cytoplasm (Chen 2005).

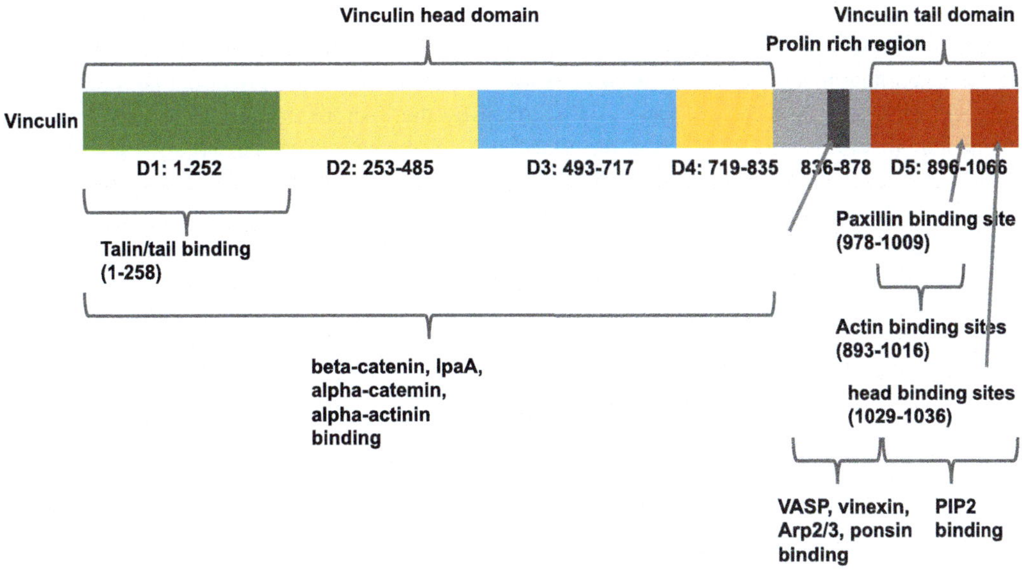

Fig. 4.1 Structure of vinculin, including domains and binding sites for vinculin interacting proteins and the internal head to tail binding regions of vinculin

Several hypotheses have been presented to elucidate how vinculin is triggered within the cell. The close bond between vinculin head and tail is considered too strong to be broken by a single ligand. In fact, the tail establishes two contacts with the head and one with the linker having a total $K_d < 1$ nM (Bakolitsa et al. 2004; Cohen et al. 2005). This close interaction culminated in the suggestion of a combined activation pathway where two or more ligands are necessary to alleviate intramolecular head–tail interactions (Fig. 4.2). In such a model, the linkage of actin to the tail and talin, α-actinin or

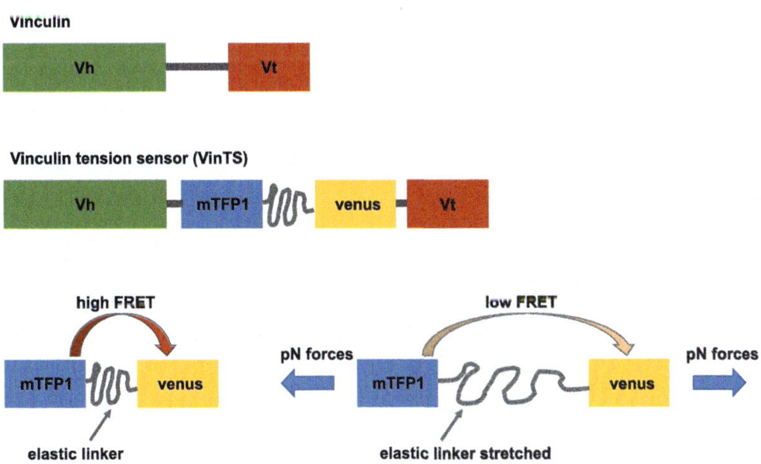

Fig. 4.2 Head-to-tail domain interaction of vinculin can be measured using FRET. The upper image shows the basic structure of vinculin. The intermediate image displays a typical vinculin tension sensor that can be employed for FRET. The lower two images show a high (lower left) and a low FRET (lower right) signal. The low FRET signal of vinculin is observed, when the vinculin molecule is stretched out and both fluorescent probes are increasingly distant from one another

α-catenin to the vinculin head encourages an openness in conformation (Chen et al. 2006; Janssen et al. 2006; Bois et al. 2006; Peng et al. 2012; Izard et al. 2004). The molecular dynamical modeling has revealed how this mechanism can be activated. Hence, it can be proposed that talin can interact with vinculin head through a linkage that is driven by the hydrophobic surface interactions. This interaction makes it possible to free the vinculin head domain from the tail root and supports conformational alterations which enable the talin to fit completely into the center of the vinculin head domain (Golji and Mofrad 2010; Golji et al. 2011).

There is other evidence to the effect that a single ligand is sufficient for vinculin to acquire an openness of conformation. Talin or α-actinin binding has been seen to be able to support alone the induction of conformational alterations that can break the connection of the vinculin head from the tail in vitro by a process that is known as the conversion of the helical bundle (Izard et al. 2004). But this model is mainly derived from experiments with purified vinculin head domain D1 and tail. It is now known that the Vinculin head connects the tail with a 1000 times higher affinity than the D1 domain on its own (Cohen et al. 2005). The activation of vinculin by a solitary ligand in the context of the whole molecule or within the cell therefore probably is not feasible.

However, more recent investigations indicate that factors other than protein binding are able to affect the activation of vinculin. Molecular dynamic simulations, for instance, indicate that the phosphorylation of vinculin at Y100, Y1065, S1033 and S1045 activates the cell and stimulates the attachment of talin and actin (Golji and Mofrad 2010; Golji et al. 2012). There is other evidence to suggest that force encourages the activation of vinculin. To substantiate this claim, the force initiates the activation of conformational transitions in vinculin. By the opposite, a loss of tension results in the fast inactivation of vinculin (Grashoff et al. 2010; Carisey et al. 2013). Lastly, a third option identified is phosphorylation which improves the mechanical activation process and vice versa (Golji et al. 2012). In line with this notion, stretching uncovers tyrosine phosphorylation sites in other proteins, such as p130 Cas (Janostiak et al. 2011, 2014) and the binding sites of vinculin with talin (del Rio et al. 2009). For instance, vinculin activation is probably more ingenious than the combined activation or bundle inversion models suggest.

4.2.3 Role of Vinculin in Cell–Matrix Adhesions

The cell–matrix adhesions or focal adhesions accumulate a wealth of adhesion receptors, including integrins. More than fifty proteins seem specifically to be targeted in the integrin cytoplasmic tail (Zamir and Geiger 2001). A more recent investigation carried out using super-resolution microscopy shows that these proteins are located in 3D nanodomains. The detected areas are: a membrane-applied integrin signaling layer, an actin-binding and force-transmitting intermediate layer and a top actin-regulating layer (Kanchanawong et al. 2010). Vinculin is located in the signal layer, but is quickly transferred to the actin-binding layer when autoinhibitory head–tail interactions are alleviated (Case et al. 2015). At the actin-binding level, vinculin eases the acquisition of supplemental proteins controlling focal adhesion dynamics and permitting efficacious cell migration. Thereby, the function of vinculin in the process of cell migration is presented below. Since migration is strongly susceptible to the reordering of the actin cytoskeleton, a short discussion on the relationship between vinculin and F-actin is provided in this chapter. Finally, the mechanisms for the effective regulation of vinculin in focal adhesion are described.

4.2.3.1 Vinculin Functions in Cell Migration

Vinculin-depleted cells possess reduced migratory capacities (Xu et al. 1998a; Mierke et al. 2010). In specific detail, the migration of cells can be divided into four major essential steps: The first step encompasses the exertion of a protrusion of the leading edge of the cell, the second step involves the adhesion of the cell to its underlying substrate, the third step is the traction force generation that helps the cell to propel itself forward in a specific direction and the fourth and last step is the deadhesion of older adhesions at the rear of the cell in order to move the entire cell body forward. In all these steps, vinculin fulfills a crucial function.

The Protrusion of the Cell at Its Leading Edge

At the cell's leading edge, vinculin is present in first adhesions (Zimerman et al. 2004). These emerging adhesions contain a large amount of vinculin linked to the Arp2/3 complex, which is a powerful nucleator of polymerization of actin (DeMali et al. 2002). The interaction between vinculin and the Arp2/3 complex ensures that the polymerization of actin is initiated at the newly assembled adhesions (DeMali et al. 2002; Chorev et al. 2014) and thereby provides a strong linkage between the integrins on the cell surface and machinery promoting actin polymerization. Thereafter, these newly assembled adhesions induce the enhanced protrusion of the entire membrane of the cell in the direction of migration (DeMali and Burridge 2003). However, it is not yet clearly known in great detail how the linkage of vinculin with the Arp2/3 complex is spatially restricted to the leading edge of the cell. In fact, there exists the hypothesis that phosphorylation processes are performing the regulation of their localization. Vinculin at Y1065 is highly phosphorylated in emerging nascent adhesions, and phosphorylation at Y1065 controls its attachment to the Arp2/3 complex (Moese et al. 2007; Mohl et al. 2009). In addition, mutant deficient of vinculin cannot attract the Arp2/3 complex or facilitate the protrusion of lamellipodia. As a result, the tyrosine phosphorylation-based recruitment of the Arp2/3 complex to vinculin seem to be a possible regulatory mechanism to provide the protrusion of the plasma membrane.

4.2.3.2 Vinculin Functions in Cell–Matrix Adhesion

Directly after the cell's leading edge, vinculin fulfills a key role in the maintenance of the focal adhesions. In fact, early experimental evidence leads to the assumption that cells exhibiting large focal adhesions overexpress vinculin (Rodriguez Fernandez et al. 1993), whereas cells that lack vinculin build up focal adhesions that become smaller and fewer (Mierke et al. 2008; Xu et al. 1998a, b). It has been exhaustingly investigated how vinculin regulates the size and the number of focal adhesions. The enrollment of vinculin to talin causes a stabilization of focal adhesions and drives the clustering of integrins and the overall extension of focal adhesions (Cohen et al. 2006; Humphries et al. 2007). Vinculin is known to directly facilitate the activation of integrins by its interaction with talin (Humphries et al. 2007; Ohmori et al. 2010; Nanda et al. 2014). Studies of how this happens point to talin being diverted from the cytoplasm to the plasma membrane when a cell signposted elevated integrin activation. The Rap1-interacting molecule (RIAM) regulates the relocalization of talin (Han et al. 2006), whereas the binding of vinculin to talin causes the disruption of RIAM that then enables transient RIAM-positive emerging adhesions to convert into vinculin-rich and further change toward mature and strong focal adhesions (Lee et al. 2013; Goult et al. 2013).

Specifically, there is evidence that talin pinpoints vinculin to focal adhesions. While there are many trials encouraging this model, others indicate the involvement of other vinculin-binding partners. For example, talin may need support in situations where recruitment of vinculin is fast and robust, as for instance in cells experiencing force. Paxillin eases the enrollment of vinculin in cells undergoing external tension (Pasapera et al. 2010). In fact, the activity of myosin II enhances the binding of paxillin to vinculin. Based upon this background knowledge, a two-step manual model for the localization of vinculin in focal adhesions has been introduced (Pasapera et al. 2010). Paxillin binds vinculin in this model and delivers it to focal adhesions. In the next step, paxillin hands over vinculin to talin (Pasapera et al. 2010).

In complement to the modulation of talin activities, vinculin acts to stabilize integrins by connecting to the actin cytoskeleton. How this happens is not fully clearly understood. Near the anterior edge of migrating cells, the vinculin tail domain collects backward flowing actin fibers (Thievessen et al. 2013). A result of the attachment to flowing actin filaments may be an elevated tension over the vinculin molecule. Tension enhancements can make vinculin to take on an open conformation and alter actin and focal adhesion dynamics on the anterior edge (Thievessen et al. 2013).

4.3 Transmission and Generation of Cellular Traction Forces

Focal adhesions close to the leading edge of migrating cells can transmit myosin-generated forces of the actin cytoskeleton toward the surrounding extracellular matrix, thus creating traction forces that draw the entire cell body to the front during cell migration (Thievessen et al. 2015). A crucial role for vinculin in the process of traction force generation and transmission has been initially demonstrated in focal adhesions that recruit vinculin and thereby boost its strength in response to force (Galbraith et al. 2002). Moreover, the key role for vinculin in the generation and transmission of traction force is based on investigations of migrating cells deficient of vinculin in 3D collagen fiber matrices (Mierke et al. 2008; Mierke 2009; Mierke et al. 2010; Thievessen et al. 2015). In specific detail, the loss of vinculin pronouncedly reduces the generation and transmission of traction forces and subsequently motility in 3D environments. It is interesting to mention that while a loss of vinculin reduces their motility in 3D matrices, fibroblasts from mouse embryos without vinculin display increased migration rates, when cultured on 2D substrates (Xu et al. 1998a, b). This evidence may imply that vinculin does not play a role in the generation of traction or that the generation of contractile forces is only necessary for the migration through 3D microenvironments, where the migration of cells may be sterically restricted. Nevertheless, direct measurements of the generation of traction force in the vinculin null mouse embryo fibroblasts have shown that vinculin is necessary for the generation of traction forces in 2D (Dumbauld et al. 2013). However, there may still be another migration type involved in the motility on flat 2D substrates. In line with this result, another role for vinculin in providing the generation of traction forces has been revealed. This research showed that focal adhesions have two types of tensile forces: a highly dynamic state in which primarily tugging traction forces are generated and a second state in which stable traction forces are generated and transmitted (Plotnikov et al. 2012). In both states of traction forces, vinculin is essential and even both tugging and stable traction forces are needed for the directional migration of cells. In line with these findings, cells with reduced expression of vinculin display a more random migration type than control cells (Rahman et al. 2016). Finally, it can be concluded that vinculin exhibits a key role not only in the generation of traction forces, but also in the directionality of the cell migration process.

In fact, it is still an active research area how vinculin facilitates the generation and transmission of traction forces. While initial adhesions are developing, talin represents one of the first molecules enrolled for integrin-containing sites. In cell–matrix adhesions, talin is exposed to a tension that releases binding sites for vinculin (del Rio et al. 2009). The elongation of talin induces conformational changes in vinculin that strengthen F-actin anchoring and thus enable the formation of additional connections between integrins and the actin cytoskeleton (Ciobanasu et al. 2014). The final consequence of this is improved integration of clustering of integrins and the maturation of focal adhesions (Humphries et al. 2007; Hirata et al. 2014). Consequently, vinculin facilitates the generation of traction force through the stabilization of the linkage of integrins with the actin cytoskeleton.

4.3.1 Breakdown of Cell Adhesions at the Rear End

Considering the critical roles of vinculin in controlling the activation of integrins and initiating the focal adhesion assembly, in turn, the disassembly of focus adhesions may entail a loss of vinculin or a reduction of its activation. How vinculin undergoes inactivation inside focal adhesions is still not yet discovered. There is a number of indications that phosphatidylinositol 4,5-bisphosphate (PIP2) and calpain have an important role to play. PIP2 binding triggers a conformational shift which diminishes the vinculin tail region of the binding to actin (Chandrasekar et al. 2005; Saunders et al. 2006). To continue fostering this hypothesis, the increase in PIP2 content in cells induces the removal of focus adhesions (Chandrasekar et al. 2005). Another mechanism to encourage focal adhesion degradation occurs in the interruption of vinculin-talin interactions. Calpain, which is a calcium-dependent protease, splits talin and thus helps to reduce focus adhesions. If the calpain cleavage site in talin is mutated, the protease becomes inactive. The vinculin sticks longer in the focus adhesions, and disassembly is prevented by this (Franco et al. 2004). Therefore, a loss of vinculin, vinculin that binds to actin, or vinculin that binds to talin, increases the turnaround of focal adhesions.

4.3.2 Interaction of Vinculin and Actin in Focal Adhesions

The binding of vinculin to F-actin is crucial when it comes to cell–matrix adhesion. Disturbances in the interaction of vinculin and F-actin influence cell morphology, cell adhesion, cell motility and cell stiffness (Ezzell et al. 1997). The underlying reason for such failures is an inability to sustain, transduce and generate forces (Mierke et al. 2008, 2010; Thievessen et al. 2013; Thompson et al. 2014). It is explored in the following how vinculin facilitates the binding to F-actin and how this specific binding event can induce crucial processes.

Among the initial investigations of vinculin is the notion that vinculin can bind to F-actin (Burridge and Feramisco 1980). Admittedly, the vinculin–actin boundary was strongly discussed. At first, the vinculin binding to F-actin was rebuilt using negative dye electron microscopy and diffraction data (Janssen et al. 2006). In specific detail, the F-actin has been hypothesized to interact with two distinct regions within the tail domain of vinculin, such as in the upper (amino acids 925–952) and lower vinculin monomer site (amino acids 1050–1056) (Janssen et al. 2006). These predicted residues of vinculin to interact with F-actin have been confirmed by others as vinculin-F-actin interaction regions (Cohen et al. 2005; Janssen et al. 2006; Shen et al. 2011; Jannie et al. 2015). Nevertheless, other investigations turned out, when residues outside the upper and lower monomer are mutated, and the vinculin binding to F-actin is significantly perturbed (Thievessen et al. 2013; Thompson et al. 2014). In conclusion, these different findings enlightened that the binding of vinculin binding to F-actin is far more complicated and seems to involve multiple sites within vinculin being essential for the functional interaction.

Due to the binding of vinculin, diverse impacts on actin have been detected. Based on this interaction, it seems to be widely accepted that vinculin and the bundled actin filaments interact with one another (Shen et al. 2011; Tolbert et al. 2014; Zhang et al. 2004; Tolbert et al. 2013). In addition, it has been revealed that vinculin is able to alter the actin bundles (Wen et al. 2009) and even induces the assembly of new actin bundles (Wen et al. 2009). Besides this function, vinculin caps actin filaments (Le Clainche et al. 2010) and nucleates the de novo polymerization of new actin filaments

(Le Clainche et al. 2010; Wen and Janmey 2011). At the end, vinculin is able to promote the attraction of actin modifying proteins including the vasodilator-activating phosphoprotein (VASP protein), which acts as an anti-capping protein (Brindle et al. 1996) and the well-known Arp2/3 complex that acts as an actin nucleator for the branching of new actin filaments (DeMali et al. 2002). These functional features of vinculin make it suitable to serve as an ideal candidate for the generation of new actin filament assemblies and the alteration of the existing actin filament structures.

4.3.3 Vinculin Regulation in Cell–Matrix Adhesions

How the regulation of vinculin with in cell–matrix adhesions is performed is still not yet clearly known. It has been hypothesized by a growing number of experimental studies that the tensional state plays a crucial role (Grashoff et al. 2010; Carisey et al. 2013; Pasapera et al. 2010; Plotnikov et al. 2012; Ciobanasu et al. 2014). Other additional factors seem to be the processes of phosphorylation and the oligomerization of the vinculin determining its function.

What Role Plays the Intracellular Tension?

In fact, a new topic is that vinculin is governed by tension. The force over the vinculin encourages the assembly of focal adhesion and promotes their own growth (Grashoff et al. 2010). However, it is still not clearly revealed how tension affects the activity of vinculin. The tension in vinculin has the potential to induce conformational changes that could promote interactions with its binding partners (Golji and Mofrad 2010). Similarly, conformational alterations within the binding partners may also foster vinculin binding. For example, the force exerted on the talin activates the unfolding of its rod domain and shows binding sites for vinculin. Vinculin linkage to talin dams talin in an active conformation constituting focal adhesions (Bachir et al. 2014; Atherton et al. 2015a). For this reason, tension-dependent conformational alterations govern the binding to vinculin.

4.3.4 Phosphorylation of Vinculin

Former studies point out that modifications in the tyrosine phosphorylation of vinculin are associated with a loss of cell–matrix adhesion. Subsequent trials demonstrated that Y100 and Y1065 of vinculin are phosphorylated by Src kinase during development and maturation of focal adhesion (Sefton et al. 1981). The mutation of one of these tyrosine residues does not affect the targeting of vinculins to focal adhesions, but rather impedes cell spreading, migration of cells and the generation of traction forces (Zhang et al. 2004; Auernheimer et al. 2015; Kupper et al. 2010).

How the tyrosine phosphorylation of vinculin affects its function is still a major field of research. Based on the full-length crystal structure of vinculin, the Y100 residue is freely accessible in the solvent, while Y1065 is closed by the linker domain (Borgon et al. 2004). More precisely, it has been hypothesized that both residues, such as Y100 and/or Y1065 of vinculin, represent crucial sites of the interaction of vinculin with phospholipids (Ito et al. 1983; Diez et al. 2009) and enzymes altering actin (DeMali et al. 2002; Moese et al. 2007). It is necessary to note that the phosphorylation of vinculin at these tyrosine residues has no impact the interaction with actin. Nevertheless, the phosphorylation changes the protein binding through the conformational alteration of vinculin. In line with this finding, mutation of the tyrosine residues Y100F and Y1065F abolishes the interaction between the head and tail domain of vinculin (Auernheimer et al. 2015; Diez et al. 2009; Huang et al. 2014), and consequently, the phosphorylation of vinculin at these sites supports the activation of vinculin by adapting an active conformation (Golji et al. 2012; Auernheimer et al. 2015). Simulations of the molecular dynamic of vinculin lead to the hypothesis that the phosphorylation causes the remodeling of charged residues within the head domain of vinculin that are located at the interaction sites the D1 region and the tail domain (Golji et al. 2012). Finally, all these studies support the hypothesis that the phosphorylation at the two tyrosine residues Y100 and Y1065 represents a critical determinant for the function of vinculin.

Similar to the analysis of the tyrosine phosphorylation, the serine phosphorylation of vinculin has been examined. It has been discovered by in vitro phosphopeptide mapping that two sites, such as the serine residues S1033 and S1045 within the tail domain of vinculin are a substrate of a specific kinase, protein kinase C (PKC) (Ziegler et al. 2002). The phosphorylation of the residue S1033 of vinculin is mostly investigated. In specific detail, cells that express the S1033A phosphodeficient mutant of vinculin are more bendable and less able to generate tensile forces (Auernheimer and Goldmann 2014). Nevertheless, it is still highly discussed how the phosphorylation at the residue S1033 of vinculin can alter the structure of the entire vinculin molecule. With computational simulations, it has been detected that the phosphorylation of S1033 of vinculin may somehow weaken the interaction between the head and the tail domains of vinculin. Hence, it can be proposed that the serine phosphorylation of these residues induces a conformational shift of vinculin similarly to the tyrosine phosphorylation (Golji et al. 2012). However, the validation of this hypothesis is still not provided and requires further experiments.

4.3.5 Interaction of Vinculin and PIP2

Similar to the phosphorylation of vinculin, there is some still growing evidence that the phosphatidylinositol 4,5-bisphosphate (PIP2) contributes to the regulation of the function of vinculin. More precisely, the tail domain of vinculin binds PIP2 and hence induces a conformational shift in vinculin that frees the head domain of vinculin to facilitate the binding of talin (Scott et al. 2006; Diez et al. 2008; Wirth et al. 2010). As an alternative way, PIP2 acts together with a ligand, such as talin or α-actinin and/or F-actin, of the head domain of vinculin that is required for the activation of vinculin in vitro (Bakolitsa et al. 2004; Izard et al. 2004; Thompson et al. 2017; Bakolitsa et al. 1999; Weekes et al. 1996). At the end, vinculin capacity to assemble to oligomers seems to be critical. Vinculin's tail domain binds to F-actin and thereby induces the oligomerization of PIP2. In the course of vinculin's binding to actin, it assembles to dimers that drive the dimer formation of PIP2 and even other high-ordered structures, such as trimers and tetramers (Chinthalapudi et al. 2014, 2015). Any physiological implications of the diverse oligomeric states and the findings of PIP2 binding still require additional analyses.

4.4 Vinculin Functions in Cell–Cell Adherence Junctions

The functional role of vinculin in cell–cell adhesions is not yet precisely known than the function in cell–matrix adhesions. Nevertheless, developmental investigations revealed that vinculin is highly required in the proper function of cell–cell adhesions. In fact, mice, in which vinculin has been genetically knocked-out, are embryonic lethal at day 10.5 (Xu et al. 1998a). In specific detail, these mice display multiple developmental abnormalities, such as closure defect of the neural tube, and their hearts show failure in bringing them together properly (Xu et al. 1998a). These two different phenotypes may be based on the impaired cell–cell adhesive interaction. In the same way, in 49% of the mice suffering from a cardiomyocyte-specific deletion of vinculin, there is a spontaneous mortality in the course of the first three months after birth (Zemljic-Harpf et al. 2007). Before the beginning of death, ultrastructural analysis on the hearts of these mice yielded aberrant adherens junctions (Zemljic-Harpf et al. 2007). Based on these finding, vinculin's presence in cell–cell adherence junctions seems to be essential for multiple physiological functions.

It is still largely elusive how vinculin is assembled into cadherin junctions. It has been hypothesized that α-catenin manages the recruitment of vinculin to these sites. The hypothesis comes from the experimental result that cells deficient of α-catenin or in hearts lacking α-catenin display no vinculin in their cell–cell adherence junctions (Sheikh et al. 2006; Watabe-Uchida et al. 1998). But cells expressing vinculin mutants incapable of binding to β-catenin while retaining the ability to bind to α-catenin do not locate on cadherin contacts. Therefore, β-catenin plays an extremely important role (Peng et al. 2010). In continuing to underpin this concept, vinculin cannot be detected in adherence junctions within cells deficient of β-catenin (Ray et al. 2013). These observations indicate evidence that the recruitment of vinculin to cell–cell interactions can be a multi-stage procedure comprising both α-catenin and β-catenin. Two actual hypotheses for the localization imply: Firstly, β-catenin facilitates the recruitment of vinculin and passes it on to α-catenin, or secondly, β-catenin mediates the recruitment of vinculin, whereas α-catenin contributes to the stabilization of this interaction. However, there is still more research effort needed to figure out which hypothesis can be confirmed.

Most of the first trials that investigated the recruitment of vinculin used cell-cultured cells, which were not subjected to force or had a low force effect. Follow-up research revealed that force has a major role in recruiting proteins for the cadherin-containing adhesion complex. Vinculin is known to be located in cadherin-containing sites due to myosin II-driven contractility (Huveneers et al. 2012), myosin VI-driven contractility (Leerberg et al. 2014) and externally exerted tension (Barry et al. 2014; Thomas et al. 2013). A number of various proteins interfere with a force-based recruitment of vinculin. Among these are α-catenin that is subject to a force-dependent conformational alteration and thereby uncovers vinculin-binding sites (Yao et al. 2014a, b; Yonemura et al. 2010) and Eplin, which is a protein that causes the exclusion of vinculin from cell–cell adherence junctions, when it is impaired. Finally, α-actinin has been reported to localize vinculin toward cell–cell adherence junctions due to tension (Kannan and Tang 2015). Decreased expression of α-actinin or its upstream activator synaptopodin impairs the tension-based targeting of vinculin to cell–cell adherence junctions (Kannan and Tang 2015). However, it is uncertain about whether α-actinin targets vinculin directly or indirectly via a α-catenin-based interaction.

Another major factor in the recruitment of vinculin into cadherin-containing cell–cell adherence junctions represents the tyrosine phosphorylation. Vinculin has been shown to be phosphorylated at its tyrosine residue Y822 due to the force on E-cadherin but not on integrins. Vinculin mutant Y822F cannot bind β-catenin and is not targeted to cell–cell adherence junctions, which leads to the hypothesis that the phosphorylation of vinculin at Y822 seems to be impact the conformation of vinculin (Bays et al. 2014). This concept is underpinned by employing super-resolution microscopy to determine the localization of vinculin in cadherin-rich cell–cell contacts and a FRET biosensor approach that monitors the confirmation of vinculin (Bertocchi et al. 2017). A phosphomimetic Y822E passes through an alteration in FRET and repositions itself in an actin-binding layer, pointing to a conformational shift in vinculin (Bertocchi et al. 2017). Finally, the tyrosine phosphorylation of the Y822 residue of vinculin seems to effectively regulate the precisely targeting of vinculin localization through the determination of the conformational phenotype.

There are functions identified for vinculin in cell–cell adherence junctions. The targeted removal of vinculin from cell–cell adherence junctions, without affecting its function in cell–matrix adhesions, reduces cell–cell adhesion and leads to a loss of E-cadherin from the cell surface (Peng et al. 2010). The vinculin binding to β-catenin is decisive for this outcome. On support of this thought, mutant versions of vinculin, which cannot interact with β-catenin, are not able to rescue the membrane expression of E-cadherin (Peng et al. 2010). This role recalls the function of vinculin in cell–matrix adhesions, in which vinculin fixes integrins in the cytoskeleton during the maturation of focal adhesion and prolongs the retention time of integrins in focal adhesions. Consequently, the function of vinculin in the stabilization of adhesion receptors seems to be highly conserved in cell–matrix and cell–cell adhesions.

Besides the facilitation of adhesion, new findings indicate that the transfer of force represents another major function of vinculin in cell–cell connections. E-cadherin represents a mechanosensory, which transfers the external forces toward the actin cytoskeleton, and vinculin fulfills a prominent role therein (Le Duc et al. 2010). The evidence comes from experiments pointing out that mechanosensors are missing in cells without vinculin or expressing a mutant form of vinculin that cannot be phosphorylated in the Y822 residue (Bays et al. 2014). However, vinculin is probably not the only promoter of E-cadherin mechanotransduction, as other work proposes a part for α catenin and/or eplin in E-cadherin mechanotransduction and in the recruitment and retention of vinculin in cell–cell adherence junctions in cells subject to tension (Chervin-Petinot et al. 2012).

It is highly speculative how vinculin performs the precise regulation cell–cell adhesive actions. A probable possibility is that vinculin regulates actin dynamics to link to the actin cytoskeleton. In strengthen this concept, VASP has been shown to interact with vinculin, which is required for vinculin's regulatory function on the junctional actin assembly, when the cells are exposed to force (Leerberg et al. 2014). Alternatively, but not necessarily exclusive, vinculin gives α-catenin stability in a conformational state can bind to F-actin. In other words, α-catenin, not vinculin, is the most relevant link between the cadherin adherence junctions and the actin cytoskeleton.

4.5 Unanswered Questions Regarding Vinculin for Future Investigations

In a larger number of reports, the functional roles of vinculin within cell–cell and cell–matrix adhesions are usually described as separate units. However, a series of investigations indicate that balanced cell–cell adhesion and cell–matrix adhesion are crucial for correct development (McCain et al. 2012). In line with this assumption, there is significant cross talk occurring between the two splices with variations in adhesion/force transfer at one point influencing the variation at the other (DeMali et al. 2014; Weber et al. 2011). In fact, the binding of integrin to the matrix reinforces the adhesion provided by cadherin (Martinez-Rico et al. 2010). In the same way, increases in integrin-mediated traction forces are associated with an augmentation of the myosin-dependent tension at cadherin adherence junctions (Maruthamuthu et al. 2011; de Rooij et al. 2005). The reverse is also happening, and the tension on cadherins subsequently alters integrins. Cells that come into contact with their neighboring cells are capable of generating higher traction forces than single cells (Jasaitis et al. 2012; Mertz et al. 2013). In order to fully reveal how these interactions of these activities are precisely coordinated at these two different types of adhesion sites, more investigations are highly required.

The adhesion/force transfer at cell–matrix adhesions is not based solely on alterations in the cell–cell adhesion/force transfer and the other way around. For instance, increasing the density of vascular endothelial cells deposited on a substrate increases cell–cell contact and reduces cell–matrix adhesion (Nelson et al. 2004). The mechanisms how cells can differentiate between the different functions of the specific adhesion site are not clearly known. The exertion of force toward cadherins, but not toward integrins, induces the activation of the Abl tyrosine kinase and thereby promotes the phosphorylation of the Y822 residue of vinculin (Bays et al. 2014). In addition, the degree of Y822 vinculin phosphorylation defines the amount of force transferred by cadherins. In contrary, vinculin phosphorylation at the residue Y822 has no impact on the transmission of integrin forces. Thus, the phosphorylation can be used to differentiate between the function of vinculin in cell–matrix and cell–

cell adhesions. It is most probable that there are alternative mechanisms to synchronize the vinculin's functions and that of vinculin's-binding partners. Finally, there is still future effort required to figures out the fine detail of these major regulatory signaling pathways.

4.6 Mechanotransduction

The cells possess the capacity to sense and react to the mechanical cues, which seems to be essential for the proper development and the functioning of healthy tissues. There are multiple diseases that are based on either perturbed mechanics of the tissue, or perturbances of the capacity of cells to sense the mechanical environmental cues (Jansen et al. 2017). To a certain degree, the sensing can be seen at integrin-associated complexes assembling attachment sites of the cell toward the extracellular matrix microenvironment. The complex mechanical signal transduction processes of the extracellular matrix environment are presented in the following. It is explored how integrin-associated complexes contribute to the cellular sensing of these mechanical cues by point out to the associated molecular mechanisms of major adhesion molecules. In the end, the cellular mechanotransduction processes including mechanosensing and signaling aspects of key proteins within focal adhesions are highlighted.

In the tissues of our body, the cells are exposed to various mechanical irritations, such as shear stresses from the blood circulation or stretching and compressive forces exerted by a number of tissues connected to muscle activity (Jaalouk and Lammerding 2009). In general, the cells are circumvented by an extracellular matrix environment consisting of multiple proteins, such as fibronectin, laminins and various types of collagens. Cells are able to directly fell the intrinsic mechanical cues of the surrounding extracellular matrix through the exertion of traction forces. The capacity of the cells to react to external force exertions, to explore and evaluate the mechanical features of the extracellular matrix and to generate and redesign the extracellular matrices, the cells need to adapt their functions in many areas of cellular performances (Jansen et al. 2015). For instance, to the stiffening of the extracellular matrix during various states of different diseases including cancer and fibrosis (Jaalouk and Lammerding 2009; Lu et al. 2012) or during the process of aging (Snedeker et al. 2014) can negatively influence cellular processes spanning over cell migration, differentiation and proliferation. As an alternative strategy, abnormal intracellular signals impairing the ability of cells to detect and react to extracellular mechanical stimuli may also cause disease states such as cancer (Jaalouk and Lammerding 2009).

The feature of cells to sense and react to external mechanical stimuli is referred to mechanotransduction. The process of mechanotransduction necessitates the capacity to detect the external environmental forces or biomechanical phenomena and the transfer of this information initiating the onset a distinct intracellular signaling responses. The cytoskeleton plays a critical role in mechanotransduction by linking cellular compartments, such as the cell's cytoskeleton and the nucleus (Cho et al. 2017), to the machinery of force-sensing.

Specific mechanical terms are defined briefly:

Mechanotransduction means the overall process of how the cells sense mechanical cues and convert them to biochemical and intracellular responses.

Mechanoresponse is the special response of cells toward mechanical cues, such as elevated phosphorylation of proteins, alterations in the transcription of genes and alterations in the behavior of cells.

Mechanosensing means the cellular process of sensing a mechanical cue.

Mechanosignaling is an intracellular signaling phenomenon that occurs as a reaction to a mechanical trigger.

Mechanosensitive means that a protein or a specific domain of a protein detects to force, whereby it alters its conformation in a force-dependent manner.

Mechanotransmission is the act of transmitting a force, such as the transmission of intracellular forces to the outside environment to explore the extracellular matrix properties, or vice versa.

A highlight is on the complexity of the mechanical phenotypes of the extracellular matrix that are sensed by the cells. Cells contain multiple force-sensing machines ranging from force-sensitive channels (Kobayashi and Sokabe 2010) to cell–cell adhesions (Huveneers and de Rooij 2013); however, a special focus is on mechanotransduction processes at integrin-associated complexes. These complexes can directly couple the extracellular matrix to the cell's actin cytoskeleton, and thereby, they represent major contributors of the mechanotransduction processes in a cell. The relevance of the integrin-associated complex protein dynamics is discussed and correlated with the cell's functional role in mechanosensing and mechanosignaling issues.

4.6.1 Besides Functioning as a Passive Supporter for Cells, the Extracellular Matrix Can Fulfill Additional Tasks

Besides a passive scaffold representing ligands for cell–matrix adhesion receptors for the adhesion of the cells, the underlying substrate, the extracellular matrix possesses various types of mechanical stimuli and even offers dimensionality for the cells. An emphasis will be placed on the role of the physical properties of the extracellular matrix environment in the light of cellular mechanosensors. There are large differences between the 2D and 3D mechanosensing mechanisms that are also addressed below (Doyle and Yamada 2016; Hogrebe et al. 2017).

4.6.1.1 Mechanics of Extracellular Matrix Scaffolds

2D substrates that have been used for cell culturing experiments for many decades revealed that various cell types can able to react to the nearby mechanical properties such as the elastic properties of the underlying surface medium. The elasticity of these substrates has been shown to impact multiple fundamental cellular functions and processes, such as the motility of a wide range of different cells, the proliferation, differentiation and death of cells (Engler et al. 2006), and even the phenomenon of axon routing of neuronal cells (Koser et al. 2016). Besides the elastic properties of the matrices, such as the purely elastic polyacrylamide gels, that are crucial for the alteration of cell behavior, tissues display not only purely elastic behavior and display instead also a stress relaxation behavior (Chaudhuri et al. 2016). In fact, the spreading behavior of fibroblasts on soft flat substrates is increased when the remodeling capacity of the extracellular matrix is enhanced by imposing stress relaxation on the underneath interface or raising its viscous properties (Chaudhuri et al. 2015). Hydrogels that exhibit more or less stress relaxation behavior govern the fate of stem cells, which is not dependent on other parameters, such as the elasticity or density of ligands (Chaudhuri et al. 2016). These combined features, elasticity and the viscous properties referred to as viscoelasticity, appear to be partially critical for environments closer to in vivo tissue environments, such as collagen fiber frameworks, which exhibit highly viscous properties in cell-relevant time periods (Mohammadi and McCulloch 2014).

It becomes even more complex when the mechanical properties of extracellular matrix networks are taken into account, since these networks have more than one stiffness value because they stiffen under external load (Storm et al. 2005). This strain stiffening of extracellular matrix meshes can also occur under the impact of cell forces, which renders the entire meshwork stiffer (Jansen et al. 2013). Strain-stiffening can imply that cells measure a stiffer environment relative to the initial polymerization conditions in vitro and can generate a favorable positive feedback loop for the increased generation of cellular forces (Hall et al. 2016). The resulting stiffness of the microenvironment can thus be defined in terms of a cumulative set of factors: extracellular matrix framework mechanics, cell density and the capacity of these cells to react and apply traction. In fact, low traction cells, such as neuronal cells, will use this mechanism to stiffen the extracellular matrix considerably less than cells with high traction force potential, such as fibroblasts by employing this strain-stiffening feedback loop (Georges et al. 2005).

The forces also spread more widely in extracellular matrix meshes than in traditional polyacrylamide or PDMS substrates on account of their fibrous character (Ma et al. 2013; Rudnicki et al. 2013). Hence, the mechanical interaction distances between different cells within this scaffold increase pronouncedly (Reinhart-King et al. 2008; Sapir and Tzlil 2017).

4.6.1.2 What Microenvironment Can a Cell Feel?

Besides mechanical properties, the extracellular matrix delivers dimensionality. Under in vivo conditions, the extracellular matrix may exhibit various structural architectures. For instance, eye collagen can polymerize to thin fibrils of about 30 nm in, whereas collagen in tendons is assembled to form relatively large bundles of several micrometers in diameter (Fratzl 2017). Hence, these results lead to the hypothesis how the dimensionality is important for cell functions and their structural arrangement. Embedded in a meshwork of fibers, cells have the option to directedly connect to several fibers of the extracellular matrix environment and moreover to sense the functional and mechanical properties of it. Thick fibers, on the other side, could rather remind of a 2D surface. In diseases involving migration of cells, such as cancer, cell migration can be performed along microtracks that possess both, features from 2D and 3D microenvironments (Rahman et al. 2016). Specifically, experiments with fibroblasts, which migrate on 2D line structures or in 3D collagen networks, revealed that the cells perceive under both conditions a 1D situation (Doyle et al. 2009, 2012).

Since this variation of cell behavior is a function of the type of dimensionality of a cell, the question is whether the molecular mechanisms of mechanotransduction deviate accordingly. In specific detail, it can be raised the following question of whether the mechanisms employed by cells cultured on top of 2D substrates are similarly important for the cells that are embedded within in a 3D extracellular matrix microenvironment?

The morphology of cells is pronouncedly altered when they are experiencing a 3D microenvironment. Adhesive cells cultured on 2D surfaces usually spread to a thin pancake-like structure, whereas cells embedded in 3D extracellular matrix scaffolds display multiple morphologies, including spindle-like or roundish shapes (Hakkinen et al. 2011). Although microtubules have proven to be negligible for cell propagation in 2D, they are necessary for 3D cell migration within collagen grids (Kraning-Rush et al. 2011; Rhee et al. 2007), even when this depends on the tension state of the grids. In 3D, actin stress fibers are lessened and focus adhesions are typically weaker than in 2D microenvironments (Hakkinen et al. 2011). Nonetheless, an uninterrupted remodeling of the extracellular matrix could improve cell spreading by boosting the regional ligand density and thus raising the probability of developing more integrin-associated complexes in 3D settings (Kubow et al. 2013). In reality, the fibrous character of the extracellular matrix can improve cell spreading provided the cells have the capacity to attract adjacent fibers (Baker et al. 2015). Of interest is that the focal adhesion composition is similar to the 2D conditions (Hogrebe et al. 2017), although it may require the accurate experimental parameters as the level of phosphorylation of focal adhesion proteins in 3D tends to be somewhat lower. On the downside, the focal adhesions accumulated in 3D are generally stronger (Doyle et al. 2012). In view of the complex nature of the integrin-associated complexes, it is essential to conduct follow-up experiments in order to clarify the source of this elevated stability. Beyond that, it is uncertain as to whether the binding kinetics of all focal adhesion proteins undergoes a modification in the course of a 3D environment. In 3D environments, a number of focal adhesion proteins apparently try to perform differently than in 2D (Fraley et al. 2010).

4.6.1.3 Integrin-Facilitated Cell Adhesions

Integrin-associated complexes couple extracellular matrix proteins to the intracellular actin cytoskeleton through transmembrane receptors, referred to as integrins. Integrins with an α and a β chain (Hynes 2002) are heterodimeric proteins that have specific sites in the extracellular matrix such as RGD or LDV (Humphries et al. 2006). The extracellular head domain of integrins is the binding domain of the extracellular matrix (Hynes 2002), whereas the intracellular tail of integrin is linked to the actin cytoskeleton through several plaque accumulating proteins. The creation, maturation and dismantling of these plaques are strictly controlled with regard to time and space. Small integrin-associated complexes emerging at the leading edge are referred to as focus complexes, capable of developing larger focus adhesions (Vicente-Manzanares and Horwitz 2011). Focal complexes build up with no need for actomyosin-induced forces. They disintegrate quickly when no force is exerted, but as soon as they connect to the force vehicle, the focal complexes age toward focus adhesions. Specifically, in matrix-secreting fibroblasts, focal adhesions are able to differentiate into fibrillar adhesions assisting extracellular matrix biosynthesis and the structural remodeling (Zamir et al. 2000; Pankov et al. 2000; Georgiadou et al. 2017). The maturing of integrin-associated complexes comprises highly dynamical processes entailing the enrollment and the activation of certain proteins.

At the same time as there is a deep insight into protein interaction systems in integrin-associated complexes, a limited knowledge of how this structure regulates cellular responses and how it is implicated in mechanotransduction mechanisms.

4.6.2 Mechanisms of Mechanotransduction

Integrin-associated complexes provide a connection between the intracellular actomyosin meshwork and the surrounding extracellular matrix environment. Through their capability to govern extracellular matrix production and rearrangement, these cells define the physical and biochemical properties of the tissues on which they react and ultimately induce cell performance.

4.6.2.1 The Composition of the Force-Sensing Integrin-Associated Complexes

Integrins operating with extracellular matrix attract a wide range of proteins that modify the binding to the actin cytoskeleton. A first collection on the basis of the published literature identified 156 components of the "adhesive" with 690 connections that is binding, activation and inhibitory interactions between the different compounds (Zaidel-Bar et al. 2007), even though much more constituents within the adhesome have now been discovered (Schiller et al. 2011; Schiller et al. 2013; Kuo et al. 2011; Ng et al. 2014). A most recently conducted study merged various proteomics datasets into a "consensus adhesome" that pinpointed 60 components depicting the core components of the integrin-associated complexes found on all surfaces of the integrin ligands (Horton et al. 2015).

Proteomic investigations yielded valuable information on the complexity of integrin-associated complexes as signaling nodes. Almost half of the many focal adhesion proteins are phosphorylated in adhesions (Robertson et al. 2015). Among them, two subsets of phospho proteins were identified within integrin-associated complexes: some that are specifically phosphorylated as a reaction to the integrin extracellular matrix interference, and others that are contiguously phosphorylated, which are then recruited to integrin-activated complexes when the integrin extracellular matrix interference occurs. Integrin-associated complexes are naturally intricate entities with a high amount of plasticity in terms of their components. However, the proteins seem to be arranged hierarchically than just being placed randomly.

4.6.2.2 The Structure and Architecture of Focal Adhesions

The super-resolution microscopy investigations revealed how the precise organization of focal adhesion structures on the nanoscale is achieved (Kanchanawong et al. 2010) and even determines molecular protein layers within mature focal adhesions. The nearest protein film to the integrins in the plasma membrane comprises the highly phosphorylated signal proteins focal adhesion kinase (FAK) and paxillin. The next nearby film is referred to as force transduction layer that consists of vinculin and talin that are termed adapter proteins coupling the integrins to the actomyosin cytoskeleton. Among these adapter proteins, talin can span over the entire layer. Thereby, talin's head domain is associated to integrins and its rod binds actin, which can be located even up to 30 nm away from the integrins (Liu et al. 2015). In specific detail, the actin regulatory binding proteins including VASP, Zyxin and α-actinin and actin stress are located in the most internal layer that can even be located 60 nm distant from the integrins. These proteins can also interact with F-actin fibers.

Based on two initial adhesome reports (Zaidel-Bar et al. 2007; Zaidel-Bar and Geiger 2010), the adhesome has been subdivided in specific layers, which help to identify the specific function of adhesome proteins due to their distinct localization (Geiger and Yamada 2011). For instance, kindlin is able to interact with integrins and paxillin (Theodosiou et al. 2016). Paxillin is found to be located exclusively uppermost signaling film nearby the cytoplasmic domain of the integrins, whereas kindlin can be counted to the same signaling layer. The ILK-PINCH-parvin complex represents another candidate that can be added easily. This protein complex is connected to integrins and F-actin fibers (Radovanac et al. 2013), thus seems to be an essential part of the force transduction layer.

In order to reveal the 3D organization of mature focal adhesions, the super-resolution microscopy can be employed. Besides focal adhesion, other integrin-associated complex structures, such as focal complexes or fibrillar adhesions, are not yet analyzed. The molecular assembly of these structures can be altered during the maturation that is evoked by the interaction of actin and myosin, and consequently, they are differently composed and structurally organized. In precise detail, zyxin has be detected to be usually not present in focal complexes, whereas it can be found in focal adhesions and fibrillar adhesions (Zaidel-Bar et al. 2003) and tensin is identified to be accumulated in fibrillar adhesions (Georgiadou et al. 2017).

Another issue is that the differences may even be larger in various cell types, such as neurons can assembly solely small point contacts, which seem to be similar to focal complexes due to their limited size (Myers et al. 2011), but they are still different to them, as these small point contacts can transmit traction forces to their microenvironment similar to larger focal adhesions (Athamneh and Suter 2015). However, whether they contain a similar molecular architecture to focal adhesions is not yet clearly revealed.

4.6.2.3 Transmission of Forces and the Molecular Clutch

In the early stages of cell adhesion, such as in the formation of focal complexes, integrin-activated complexes recruited solely a small number of proteins, which are not ordered precisely. Afterward, to assemble more proteins in maturated focal adhesion is precisely regulated in a hierarchical manner, which facilitated by actomyosin-driven contractility (Zaidel-Bar et al. 2003; Ballestrem 2001; Balaban et al. 2001; Riveline et al. 2001). However, the specific mechanism that regulates the ordered recruitment of the focal adhesion proteins is still subject of ongoing research (Serrels and Frame 2012; Lawson and Schlaepfer 2012) and may even turn out to be cell type specific.

The integrin function combined with the actin-associated force generator is decisive, as it governs the cell movement and is important for mechanosensor functions. This connection is referred to as the "molecular clutch" mechanism and is required for the mechanosensory process. At the periphery of the cell nearby the plasma membrane at the cell's anterior leading edge, the polymerization of actin leads to a filamentous actin (F-actin) structural scaffold. The continuous polymerization of actin generates a reverse (synonymously termed retrograde) flow of actin (Case and Waterman 2015). The engagement of the actin filaments to integrin-associated complexes causes a reduction of this retrograde flow, whereas the still continuing polymerization of actin deforms the plasma membrane in the direction of the external environment, which finally cumulates in the exertion of specific membrane protrusions. Based on the myosin II-driven pulling forces, a cell can move itself toward other locations by the translocation of the entire cell body that leads to a forward movement in targeted directions.

The molecular clutch associated proteins within focal adhesions, such as talin, vinculin and α-actinin provide the functional coupling with the cell's actin cytoskeleton and move in the same direction as the retrograde flow (Hu et al. 2007). Among these proteins, talin and vinculin, which belong to the force transducing film, seem to be major components of the molecular clutch (Elosegui-Artola et al. 2016; Thievessen et al. 2013; Humphries et al. 2007). More experimental findings are in line with this interpretation, such as the cells lacking talin that cannot facilitate the formation of adhesions. Moreover, when the connection with the actin cytoskeletion is abolished, the focal complexes cannot mature (Elosegui-Artola et al. 2016; Atherton et al. 2015a, b; Zhang et al. 2008; Austen et al. 2015). In contrast, cells lacking vinculin are still able to assemble focal adhesions, in which the generation of forces and their transmission is impaired together with the cells' inability to react to forces (Plotnikov et al. 2012; Dumbauld et al. 2013; Carisey et al. 2013; Holle et al. 2013; Rubashkin et al. 2014; Yamashita et al. 2014; Jannie et al. 2015). Since the operation of integrins with the contractile actomyosin apparatus is crucial for the analysis of substrate mechanics, the identical proteins governing the coupling are decisive for the mechanosensing mechanisms.

4.6.2.4 In the Process of Mechanosensing, Talin and Vinculin Connect the Actomyosin Cytoskeleton to the Integrins

The acquisition of the mechanical input is a decisive initial step in the process of cell decision. In this initial step, triggering impulses such as stress or shear forces (Friedland et al. 2009) or cellular forces are transmitted to the matrix (Plotnikov et al. 2012). The two scenarios concern the well-known adapter proteins talin and vinculin. Talin and vinculin pass through both activation mechanisms to ultimately combine integrins with F-actin. Both proteins target focal adhesions when they are activated and both are maintained in an inactive, closed conformation by autoinhibitory linkages in the cytoplasm. Talin possesses an N-terminal integrin-binding four-point-one, ezrin, radixin, moesin (FERM) domain that is connected to a C-terminal rod domain, which consists of 13 helical bundles (known as R1-R13) ending up in a dimerization motif. An interplay between F3 of the FERM domain and R9 of the rod constitutes the autoinhibitory linkage inside talin (Calderwood et al. 2013; Goksoy et al. 2008; Goult et al. 2009). In vinculin, the tail domain (Vt) is retained in a forceps-like manner between the areas of the head (D1-D4) (Izard et al. 2004; Bakolitsa et al. 2004). Nevertheless, in vitro investigations carried out have demonstrated that the strength of this autoinhibition is significantly stronger for vinculin (Bakolitsa et al. 2004) than for talin (Goksoy et al. 2008).

The mechanisms providing the activation of both proteins seem to be closely related. In their inactive state talin and vinculin can assemble a complex inside the cytoplasm (Bachir et al. 2014) that can be targeted toward the regions of the integrin–extracellular matrix linkages to drive the assembly of adhesions. An alternative pathway represents the GTPase Rap1 and its effector RIAM that are able to facilitate the targeting of talin toward the plasma membrane (Lee et al. 2009; Yang et al. 2014). In both scenarios, there is a need for enhanced activation mechanisms to bring both proteins into their fully activated conformations.

For talin, the exertion of force on the molecule initiates the stretching of helical bundles that uncover cryptic-binding sites and thereby facilitate activation (del Rio et al. 2009; Yao et al. 2016). The force for this extension derives from the actomyosin engine, probably by binding to the third of three actin-binding sites (ABS3) at talin's C-terminus. A modification of this site, but not of the pivotal ABS2 to decrease the binding of actin, can significantly impede the formation of adhesions (Atherton et al. 2015a, b). Coexpression of active vinculin that binds to talin at the vinculin-binding sites (VBS) and fosters ABS2 attachment to the actin cytoskeleton can, however, repair this deficiency (Atherton et al. 2015a, b). For this reason, force is possibly not the only factor that enables complete activation of talin. Kank2, a component of adhesions newly identified, binds directly to talin, activates talin and enhances the binding of talin to integrins. Curiously, this interplay decreases talin–actin binding in ABS2 and encourages adhesion (Sun et al. 2016).

Talin's ability to perform stretching operations is crucial for its function as a mechanosensory protein. Talin is capable of altering the accessibility of binding sites for interacting focal adhesion proteins in reaction to shifts in the forces induced by actomyosin interactions. The stretching of talin in a step-driven manner due to the stretching of each individual helical bundle (Yao et al. 2016) evokes the recruitment of vinculin to focal adhesions (Yao et al. 2014a, b). The open-shaped and hence activated vinculin facilitates the further targeting of multiple other focal adhesions proteins (Carisey et al. 2013), such as a large number of signaling proteins which impact the Rho and Rac GTPase signal transduction pathways (Carisey and Ballestrem 2011; Atherton et al. 2015a, b). It has been demonstrated that purely, forces can lead to an activation of vinculin (Carisey et al. 2013; Case et al. 2015) and vinculin are subject of forces in the focal adhesions of living cells (Case et al. 2015; Grashoff et al. 2010). However, vinculin's activation state in focal adhesions seem to be either active or inactive, since no intermediate states have been found so far, which is in contrast to talin that can exhibited rather different stretched states and therefore may exhibit intermediate activation states (Margadant et al. 2011). One way in which vinculin can guide mechanosensors is by adjusting its cyclic turnover to and from the adhesion complex.

In fact, focal adhesions are reported to be vastly dynamic structures. In the same way, the adhesion proteins themselves are subject to a fast transition into and out of the adhesion site (Stutchbury et al. 2017). As part of "molecular clutch," both talin and vinculin experience forces in living cells demonstrably through the employment of FRET-based tension sensors (Austen et al. 2015; Grashoff et al. 2010; Kumar et al. 2016). Through contact with actin, these two proteins are probably triggered by the small forces to retard the retrograde flow (Chen 2005). Photokinetic studies of talin and vinculin show that their conversion is diminished once they are active (Humphries et al. 2007; Atherton et al. 2015a; Kumar et al. 2016). The active talin–vinculin complex keeps integrins in an activated conformational state (Carisey et al. 2013). As a result of the orderly recruitment and liberation of other focal adhesion signal proteins such as FAK and Paxillin, such a mechanosensory core unit seems to initiate downstream signal phenomena.

4.6.2.5 FAK and Paxillin Fulfill Key Functions in Mechanosignaling

An increase in the tyrosine phosphorylation of FAK and paxillin as the first response to several mechanical stimuli on the surface has been identified (Mitra et al. 2005). Both FAK and paxillin exhibit a relatively high tyrosine phosphorylation based on the interaction of FAK with Src and on possibly Abl kinases (Gotoh et al. 1995). However, the tyrosine phosphorylation is regarded to be crucial for the process of mechanosensing (Fig. 4.3). For instance, the inhibition of FAK and Src signaling impairs the reaction of the cells toward a stretch in a cyclic application (Wang et al. 2001a, b, 2005; Sai et al. 1999). Phosphorylation of FAK at Y397 proved to be higher on stiff substrates than on soft ones (Friedland et al. 2009; Bae et al. 2014), and the phosphorylation of paxillin after the activation of myosin II has been shown to facilitate the localization of the adapter protein vinculin

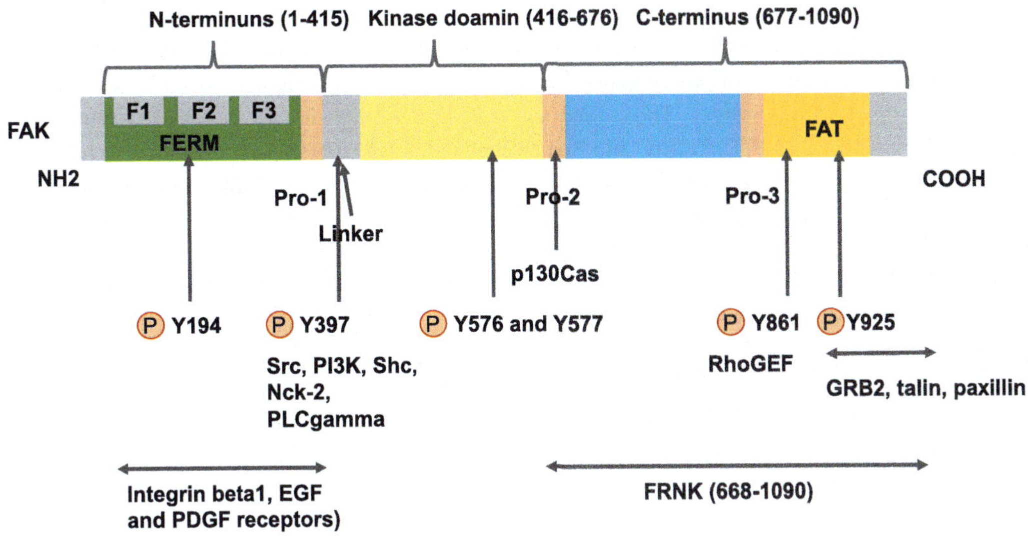

Fig. 4.3 Structure of focal adhesion kinase (FAK) and the FAK interacting molecules

toward focal adhesions (Case et al. 2015; Pasapera et al. 2010). The crucial role of the phosphoprotein FAK has been revealed by analyzing FAK-null cells that were not able to perform durotaxis that is usually the cell migration in the direction to a stiffer substrate (Wang et al. 2001a, b). An interesting aspect is that FAK isoforms can be phosphorylated more rapidly (Toutant et al. 2002). Hence, it can be hypothesized that the activation of FAK in a faster manner can serve as are regulatory cue that cells employ to pre-trigger themselves for the softer stiffness phenotype. This seem to be of high interest in relatively soft environments (≤ 1 kPa) that specific cell types, such as neurons, experience and which are known to contain various FAK isoforms compared to other cells of the connective tissue, such as fibroblasts (Armendariz et al. 2014).

In contrast, the treatment of cells with Src and FAK inhibitors which abolished the tyrosine phosphorylation completely with in focal adhesions revealed that focal adhesions are formed in a normal way and contain the same composition of focal adhesion proteins in control cells displaying normal phosphorylation levels (Horton et al. 2016). Although the impairment of the phosphorylation did not impact the maturation of focal adhesions or the transmission of traction force toward elastic substrates, the migration of cells is fully inhibited (Stutchbury et al. 2017; Horton et al. 2016). To investigate the turnover of vinculin, we used vinculin constructs tagged with photoactivatable-GFP and assessed fluorescence loss after photoactivation (FLAP). These FLAP experiments determine the turnover of vinculin under specific conditions and indicate that the mechanosensing, which is the identification of slower turnover of vinculin on stiffer substrates, was still functional, whereas this mechanical sensing was not translated efficiently to a response by the cell (Stutchbury et al. 2017). The exposure of cells to specific frequency of ultrasound has been employed in clinical trials to support healing processes, and on the protein level, it impairs the turnover of vinculin in focal adhesions, although the signaling of FAK is impaired (Atherton et al. 2017). Downstream of the mechanosensing action of vinculin, FAK fulfills a crucial and necessary function in providing a cellular reaction (Atherton et al. 2017).

After these observations, a model is offered that incorporates the molecular clutching and dynamics of focal adhesion proteins and how this interferes with adhesion signaling. Talin and vinculin deal with the retrograde flow of actin and thereby adapt their turnover rate and activation state. In turn, the recruitment of other focal adhesion-associated signaling proteins is facilitated. Through the regulation of the molecular stoichiometry of adhesions, talin and vinculin therefore operate as major regulators of the adhesive signaling, although they cannot control the generation of the signaling process themselves. The subsequent signaling through focal adhesion proteins such as FAK and paxillin on the GTPases Rho and Rac (Mitra et al. 2005) governs contraction which leads to increased growth of the adhesion site (Ridley and Hall 1992) or the polymerization of actin at the cell's anterior front, which induces the formation of new focal adhesions. Consequently, a complicated feedback loop system is established, in which adhesion reacts to and impact the entire actin cytoskeleton. Within this model, adhesion proteins are able to work together by sensing the mechanical cues (mechanosensing) and by reacting to mechanical alterations through providing signaling processes throughout the cell (mechanosignaling). Finally, it can be concluded that the combination of both elements, the mechanosensing and mechanosignaling events, is necessary for the proper mechanotransduction cascade.

4.6.2.6 Other Mechanotransduction Components

The adhesion proteins described so far are not exclusively involved in the mechanotransduction processes. In fact, a subset of LIM-domain containing proteins, including paxillin, Hic-5 and zyxin, performs essential tasks in the regulation of a cellular reaction upon stimulation with mechanical forces and subsequently, also in the facilitation of the downstream signal transduction processes that may additionally cause alterations in the transcription of distinct genes (Schiller et al. 2011; Smith et al. 2014; Kadrmas and Beckerle 2004).

Moreover, integrin-associated complexes consist of multiple actin-regulating proteins that are able to drive the polymerization and/or the bundling of actin. These proteins have also been revealed to fulfill crucial roles in the process of mechanotransduction. For instance, the α-actinin associates with actin (Youssoufian et al. 1990), integrins (Otey et al. 1990) and vinculin (McGregor et al. 1994) and hence is supposed to facilitate the transmission of forces and the maturation of the integrin-associated complex (Roca-Cusachs et al. 2013), which points out to a critical role in mechanotransduction.

Multiple other proteins that can bind to actin are located inside integrin-associated complexes, which encompass VASP, vinexin, ponsin, Arp2/3 and ArgBP2, and all of them can interact with vinculin's neck region serving as a mechanosensory protein (Carisey and Ballestrem 2011). Besides this interaction, these actin interacting proteins are capable to reorganize actin. Nonetheless, vinexin seems to be crucial for the sensing of the rigidity in mouse embryonic fibroblasts (Yamashita et al. 2014). Apart from vinculin, the functional roles of VASP and Arp2/3 are less well known; however, they contribute locally to the precise regulation of the polymerization of actin at the focal adhesion sites (Chorev et al. 2014). Arp2/3 has been shown to interact and hence colocalizes with FAK (Serrels et al. 2007) and vinculin (DeMali et al. 2002) within integrin-associated complexes at the cell's anterior edge; however, Arp2/3 fulfills a functional role in directing both vinculin and FAK toward actin-rich lamellipodia, where they need to initiate the formation and assembly of new adhesions.

4.7 Integrin-Associated Complex-Facilitated Mechanotransduction Causes Transcription and Differentiation

The extracellular matrix features can affect the motility of cells and the expression of distinct genes in an integrin-associated complex-based manner. Traction force experiments performed on polyacrylamide gels revealed a pronounced cellular response to stiffness alterations in terms of the expression of distinct genes that regulate finally the differentiation of cells. For instance, mesenchymal stem cells (MSCs), which adhered to soft substrates of approximately 1 kPa, start to differentiate toward a neuronal lineage, whereas MSCs adhered to more rigid substrates of approximately 10 kPa differentiate to a myoblast or on even stiffer substrates of approximately 30 kPa to osteoid phenotypes (Engler et al. 2006). In addition, this behavior can be impaired by the treatment of cells with the myosin inhibitor blebbistatin, which suggests that the intracellular actomyosin-dependent tension is required for the mechanosensitivity feature of cells.

The plasticity in the differentiation of cells due to the mechanical characteristics of the polyacrylamide gels, such as various stiffnesses, is evoked by the signal transduction to the basically two transcriptional co-activators, the Yes-associated protein (YAP) and the transcriptional co-activator with PDZ-binding motif (TAZ). These YAP/TAZ co-activators can shuttle between the nucleus and the cytoplasm due to stiffness alterations of the cell's microenvironment. For instance, a stiff substrate induces the localization of YAZ/TAZ in the nucleus, which is associated with higher transcriptional activity (Dupont et al. 2011). Another important finding is that the stiffness-based differentiation of MSCs into either adipocytes, when placed on a soft substrate, or osteoblasts, when placed on a stiff substrate can be fully abolished through the knockdown of vinculin that even causes a perturbed nuclear localization of the YAP/TAZ proteins (Kuroda et al. 2017).

Mechanosensing by talin seems to determine the localization and the activity of YAP/TAZ on 2D microenvironments. Fibroblasts lacking both isoforms of talin such as talin1 and talin2 cannot react the changes in the stiffness of their underlying substrate through the translocation of YAP in the nucleus (Elosegui-Artola et al. 2016). Moreover, there seems to be a force threshold for the unfolding of talin, which thereby enables a mechanoresponse. When four threonine residues within the mechanosensitive R2R3 domain of talin are mutated, the hydrophobic region is narrowed (Goult et al. 2013), the threshold for the force-based unfolding of talin is shifted, and also, the induction of the nuclear targeting of YAP is elevated. In another experimental approach, the expression of a vinculin fragment (referred to as vinD1) which carries only the talin-binding region without any other functional domains similarly impaired the stiffness-induced nuclear translocation of YAP. Hence, this finding leads to the hypothesis that talin and vinculin act together to perform the (Elosegui-Artola et al. 2016).

YAP/TAZ proteins provide a phenomenon that is referred to as a "mechanomemory effect," where the cells maintain to react to the increased stiffness levels, although they have meanwhile been reduced. For instance, MSCs and myofibroblasts exhibited elevated fibrotic activity, which is associated by increased α-SMA expression, when the cells are seeded and cultured on rather stiff surfaces. A special feature is that the cells continue to possess increased higher fibrotic activity even when they are placed on substrates with lower stiffness, which resembles a healthy microenvironment. Thus, the cells maintain their activity after they have experienced the stiff microenvironment for a specific time frame (Li et al. 2017; Balestrini et al. 2012; Yang et al. 2014a, b).

There is a lot of evidence that there exists a deep cross talk between integrin-associated complexes and the nucleus that itself acts a mechanosensory compartment or organelle (Cho et al. 2017). Although a lot of insights have been revealed about the involved molecules facilitating this event, still many questions remain unanswered. Hence, it is still not precisely clear which proteins are triggering the transmission of the mechanosignal from an integrin-associated complex toward the nucleus after the mechanosignal has been sensed the two focal adhesion molecules vinculin and talin. It seems to be possible that there is also a cross talk between other mechanosensitive cellular proteins or structures, such as those at cell–cell adherence junctions. It needs to be still to be resolved how these proteins or structures interconnect to precisely provide the mechanical signaling.

The exact mechanism through which the cells detect and transfer the mechanical signal is a complicated event that associates various proteins at several cellular compartments. Within the cell–extracellular matrix adhesions, the process of mechanotransduction seems to be further divided into specific modules. The mechanosensing unit encompasses talin, vinculin and possible additional other distinct proteins that directly connect integrins to actin. Besides FAK and paxillin, other mechanosignaling proteins, such as the actin polymerization factors including Arp2/3 and VASP or actin-cross-linking proteins including α-actinin, all of which can interact with actin to drive the mechanotransduction processes that regulate vastly distinct physiological functions in living cells.

There are still numerous open questions that are still not yet answered completely and which are not entirely new; instead, they need be gained new information from 3D microenvironments or the in vivo situation in living tissues in order to fully answer them. In a 3D microenvironment, a major difficulty is to develop an experimental setup, in which these complex parameters can be assed and monitored easily and precisely. In part, the difficult is based on the complexity of the extracellular matrix meshworks, such as purified systems including fibrin or collagen fiber networks that seem to be highly heterogenous. Nonetheless, the increasing information about the chemical and physical nature of the extracellular matrix scaffolds, and how these parameters impact the behavior of the cells, will finally contribute to a much clearer picture of how cells can adapt to varying conditions in vivo. A different approach may be to work with synthetic biomaterials, in which the mechanical features of the fibers can be distinguished from the features of the overall network structure in order to decipher the individual and independent contribution of the nature of the fibers and the overall network.

During the last decade, biophysical techniques such as mass spectrometry and super-resolution microscopy have gained information of the precise composition of the protein network during mechanotransduction events. Although many advances have been reported of those networks, there exists that still major blind spot in our understanding of the wiring of the networks is regulated. The combination of the in vitro biochemical structural protein analysis connected with the in cellulo and in vivo employment of biosensors, which rely on the FRET technique (Grashoff et al. 2010; Brenner et al. 2016), and analysis of protein dynamics will heavily contribute to the identification of the activation mechanisms that initiate process of mechanotransduction.

Biotechnological improvements, such as the discovery of the CRISPR technique, have initiated novel in vivo approaches. The usage of conditional knock-outs and knock-ins with mutant proteins in a wide range of different model organisms will help to reveal the impact of specific domains within a distinct protein on the alteration of the mechanical properties of tissues and how these tissue mechanics changes can be detected in healthy and disease conditions. Model organisms including Drosophila have helped to identify partly the in vivo relevance in mechanotransduction processes (Hakonardottir et al. 2015; Klapholz et al. 2015). Before the discovery of the CRISPR technique, the model organisms need to be quite simple, but with this important technique, the model systems can be even mammalian species.

However, there is still much work needed to reveal the overall complex mechanotransduction in 3D microenvironments that seem to be definitely required to precisely investigate this mechanotransduction event. At the same time, the structural analysis of the mecahnotransduction event seems to be similarly required in vitro. In fact, the combination of in vitro analysis of individual proteins or their domains with the cellular analysis in 3D model systems will improve the knowledge of how mechanotransduction is performed in vivo and how they become malfunction in diseases, such as cancer.

4.8 Force Sensors Serve as Mechanical Markers

The precise quantification of traction force or stress transmitted by the cells in native tissues necessitates the knowledge of the tissue's material properties and hence deserves a biophysical technique for tracking the deformations induced by the cells in real time. Hence, the 3D traction force microscopy has been employed to determine the mechanical stresses, including the convergence and expansion process, that are transmitted during the gastrulation step in Xenopus laevis (Zhou et al. 2015). In fact, the traction stresses can be explored at the level of tissues. In order to reveal a precise spatial resolution, oil droplets can be transferred into tissues that have a diameter of a few µm (Campàs et al. 2014; Lucio et al. 2015). The surface of those oil droplets can be covered with specific ligands for cell surface receptors, which enables cells to exert forces on the droplets. The deformation of a droplet leads to shape alterations. The material properties of the oil droplet available, the deformation of the oil droplet, which causes by the nearby cells, can be converted into mechanical stresses. Initial experiments have been performed in embryos of Drosophila melanogaster, but can be easily transferred to other tissues and organisms.

Forces play a role during the physiological development of tissues in embryos, tissue repair upon injury, immunological reactions involving the migration of immune cells or during pathological conditions, such as cancer. In order to measure forces of cells, a stretch sensitive fluorescence resonance energy transfer (FRET) can be used, which offers the inherent capability to perform non-invasive tension measurements in living tissue. A FRET-based actinin tension sensor was inserted into Xenopus laevis embryos, and this sensor detected tension alterations during differentiating ectoderm. The actinin tension sensor, which comprises a spider silk protein that connects mCherry and EGFP, has been tested in human embryonic kidney cells (HEK) and embryos. In fact, this specific tension sensor colocalizes with actin filaments, and the FRET signal is altered in this amount due to the remodeling of actin filaments, the inactivation of myosin or the perturbation by osmotic pressure. A time-lapse FRET analysis revealed that the future neuronal ectoderm carries a higher tension during gastrulation and neurulation than the epidermal ectoderm, and that, the morphogenetic characteristics of the cells are correlated with the tension alterations (Yamashita et al. 2016).

Several biophysical techniques are available to measure physical properties of single cells with atomic force microscopy or atomic force microscopy (cellular stiffness) or micropipette aspiration (membrane tension). These two biophysical techniques are none invasive. A disadvantage of these two is that the direction of forces cannot be determined, and it hence cannot be addressed. In order to investigate the intercellular tension in an embryo, laser ablation, hole punching (Varner et al. 2010) and the calculation from cellular geometry (Ishihara and Sugimura 2012) seem to be suitable (Yamashita et al. 2016). Moreover, the measurement of intracellular forces is much more complicated and requires the development of appropriate force sensors that can be transfected transiently or stably in living cells.

4.8.1 Force Sensor

One option is to construct a molecular force sensor that can detect in real time the traction voltages transmitted by living cells in their vicinity. Such an approach potentially relies on the identification of intracellular proteins that stretch in response to an applied force, the amount of stretch resulting in an increase or decrease in Förster Resonance Energy Transfer (FRET) of fluorescent peptides attached to the protein (Grashoff et al. 2010; Borghi et al. 2012; Cai et al. 2013; Yamashita et al. 2016). The majority of FRET-based cellular force sensors are based on the knowledge of the organization of focal adhesion and adherence junctions and hence utilize focal adhesion proteins, such as vinculin, talin, α-actinin, paxillin and E-cadherin, to quantitative measure the forces that are exerted them within the cytoplasm. Hence, the knowledge on how the actomyosin machinery functions at specific points creating cell–matrix or cell–cell adhesions can be gained via these sensors, and even, the usage of distinct force sensors in living embryos or even in adult animals has been reported (Kelley et al. 2015).

FRET-based tension sensors represent a non-invasive technique to determine the forces of cells in in vitro cell culture systems (Grashoff et al. 2010; Borghi et al. 2012; Cai et al. 2013; Yamashita et al. 2016). In specific detail, the FRET-based tension sensors are compromising a cell adhesion molecule, in which a FRET domain is incorporated that includes two fluorescent domains linked by an elastic element. Among these cell adhesion molecules are vinculin (Grashoff et al. 2010), α-actinin (Borghi et al. 2012), β-spectrin (Krieg et al. 2014) or even the prominent cell–cell adhesion receptor E-cadherin (Cai et al. 2013; Yamashita et al. 2016). When the sensor experiences tension, the elastic part of the sensor is stretched, and thereby, the distance of the fluorescent domains increases, which subsequently diminishes the efficiency of the FRET system. In fact, these tension sensors deliver the forces within a cell at a subcellular resolution; however, they are rarely used in living tissues (Krieg et al. 2014; Kelley et al. 2015) and an adapted tension sensor for in situ force measurements is required.

4.8.2 Besides Vinculin, Talin Acts as a Force Sensor

Integrin-dependent cell adhesions are identified as mechanosensitive elements, where talin facilitates the connection of actin filaments to integrins either directly or indirectly by the recruitment of vinculin. Hence, it seems to be suitable to develop a force sensor based on talin. It has been reported that talin is under tension, when it is located in focal adhesions. Moreover, the tension of talin is increased, when it is present in peripheral than in central focal adhesions of the cells. Vinculin manages to enhance the tension on talin, which relies basically on the actin-binding site 2 (ABS2) that is located in the center of the rod domain, and rather not on the ABS3 the outermost regions of the C-terminus. Dissimilar to vinculin, talin is less tensioned when the cells experience rather soft substrates. The central and peripheral focal adhesions of cells are distinguishable, since the central one needs the ABS3 region of vinculin to interact with actin, whereas the peripheral ones require ABS2. In contrast to vinculin, talin utilizes the ABS region in vinculin for the sensing of stiffness cues and not the ABS3 region (Kumar et al. 2016). These results lead to the hypothesis that the central and the peripheral focal adhesions need to highly organized and regulated in different manners as well as that the two domains of ABS2 and ABS3 can adapt their functions to spatial alterations and the sensing of the stiffness. Finally, these results have elucidated the function of talin and triggered the development of constraining models for the process of cellular mechanosensing.

Integrins couple the extracellular matrix to the actin cytoskeleton by involving a complex assembly of connections in which the focal adhesion protein talin fulfills a crucial task (Ziegler et al. 2008; Calderwood et al. 2013). The head domain (synonymously termed N-terminal FERM domain) of talin is located in its N-terminal part and can directly interact with the cytoplasmic part of the integrin β subunit and is also needed for conformation-based activation of integrins that can then interact with extracellular matrix proteins in a high affinity manner. In specificity, talin has three F-actin-binding sites (ABSs). Among the ABS, the far C-terminal-binding site within the rod domain, the ABS3, seems to be the key sites. The rod domain of talin possesses several vinculin-binding sites that are somehow hidden in four and five α-helical bundles. Under mechanical stimulation, the domains of talin uncover the binding regions for the head domain of vinculin that in turn causes a reinforcement of connections to actin by the ABS region within the tail domain of vinculin. The lack of talin in selected organisms causes distinct phenotypes, which closely resemble the phenotypes caused by deletion or mutation of the integrins on the cell's surface indicating key function of talin (Monkley et al. 2000; Brown et al. 2002; Cram et al. 2003).

Due to mechanical alterations in the microenvironment, tissues can adapt their mechanosensitivity through integrin-based focal adhesions by altering their function and the expression of specific genes

(Orr et al. 2006; Costa et al. 2012). For instance, cells have the capacity to detect the mechanical stiffness of the surrounding environment, such as the extracellular matrix scaffold, and can accordingly adapt their cellular contractility, the mechanosignaling, and the expression of genes, which is commonly known as stiffness-sensing behavior (Humphrey et al. 2014). These effects encompass the alteration of the production of extracellular matrix through changes in the stiffness of the matrix and forces exerted externally. The mechanosensing process facilitated by integrins is essential for the development and multiple diseases, such as fibrosis, cancer and hypertension (Orr et al. 2006; Butcher et al. 2009).

The force transmission is based on highly dynamics couplings between integrins and actin, whereby F-actin can flow over the focal adhesions, when a force is exerted by the polymerization of actin and the filament sliding induced by myosin (Case and Waterman 2015). At focal adhesions close to the cell borders, the actin flows backward over the rather non-moving integrins, whereas talin and vinculin can move also backward with an intermediate velocity. Hence, the connections between the integrins and the F-actin fibers need to be dynamically arranged, which in turn requires that the adaptor proteins vinculin and talin need to be highly dynamic, since they need to quickly be associated or dissociated upon the transmission of forces to drive the transmission of forces that is known as a focal adhesion clutch. There is still a major question to be raised of how this dynamic assembly facilitates the process of mediates mechanotransduction.

The establishment of a biophysical technique to determine the forces over a distinct molecule, such as the FRET-based approach, where two fluorophores are coupled to a calibrated spring region, revealed that vinculin is directly subject to mechanical tension inside focal adhesions (Grashoff et al. 2010). Additionally, a talin tension sensor has been invented that revealed the functional role of talin during the mechanical force exertion at integrin-facilitated focal adhesion and the process of mechanotransduction. FRET-based tension sensor modules can be utilized to determine how specific proteins deal with mechanical forces in living cells. Nevertheless, the forces in the single-pikonewton (pN) regime are still cumbersome to solve, and tools for multiplexed tension measurement are unavailable. A genetically coded, FRET-based biosensor, termed FRET-based tension sensor modules, which is characterized by an almost digital force behavior and an elevated sensitivity at 3–5 pN, is generated and calibrated (Ringer et al. 2017). Moreover, a technique is reported that enables the simultaneous analysis of coexpressed tension sensor modules based on two-color fluorescence lifetime microscopy. Finally, a procedure for calculating the proportion of mechanically linked molecules in cells is presented. The transfer of these techniques to new talin biosensors revealed that there is an intramolecular tension gradient over talin-1 due to integrin-driven cell adhesion (Ringer et al. 2017). In specific detail, the tension gradient is based on the actomyosin machinery and on the linkage through vinculin and can in a sensitive manner detect the rigidity of the extracellular matrix microenvironment.

The cells have the capacity to adhere and sense alterations in the stiffness of tissue that is essential for the proper development of organs and their function. There is still a lot to explore, since the key mechanisms through which adherent cells feel the compliance of the extracellular matrix scaffold are not clearly understood. With two single-molecule calibrated biosensors, which permit the analysis of a hitherto unattainable, but physiologically extremely important force regime in cells, it is demonstrated that the integrin activator talin produces mechanical connections during cell adhesion, which are indispensable for cells to examine tissue stiffness. Talin connections can be subject to various piconewton (pN) forces, but usually, they carry about 7–10 pN during the cell adhesion, which relies on the coupling strength to F-actin and vinculin (Austen et al. 2015). The break of the mechanical coupling of talin leads inevitably to the activation of integrins and the initial cell adhesion, although it avoids the strengthening of focal adhesion and thus the acquisition of extracellular stiffness. The talin mechanics are isoform-dependent, and therefore, the expression of either talin-1 or talin-2 influences the sensing of the extracellular stiffness.

The rigidity of tissues represents an epigenetic factor guiding the patterning of tissues and the proper development of organs (Heisenberg and Bellaiche 2013; Wozniak and Chen 2009; Franze et al. 2013), whereas alterations in tissue mechanics seem to be connected to various disease states, such as development of cancer, cardiovascular disorders or injury of the spinal cord (Ingber 2003; Lu et al. 2011). To discriminate between tissue stiffness variations, the cells continuously evaluate the mechanical properties of their environment by binding and contracting the surrounding extracellular matrix (Discher et al. 2005; Plotnikov et al. 2012; Schoen et al. 2013). This attachment-dependent rigidity analysis is facilitated by focal adhesions, subcellular structures in which extracellular cell–matrix receptors, such as integrins, are linked to the cell's actin cytoskeleton through adapter proteins (Geiger et al. 2009; Hoffman et al. 2011; Elosegui-Artola et al. 2014). Although the essential role of single integrin subunits and various focal adhesion molecules, such as focal adhesion kinase (FAK), vinculin and paxillin, has been recognized (Plotnikov et al. 2012; Elosegui-Artola et al. 2014; Grashoff et al. 2010), the key mechanism linking the cell adhesion to mechanosensing activity is still unclear.

One of the involved regulators of focal adhesion mechanosensors is talins, which are mainly characterized by their fundamental role in integrin activation (Critchley 2009). Talins connect and activate integrin receptors directly with an N-terminal head domain and are supposed to transmit mechanical signals by linking to the actin cytoskeleton at the same time with their C-terminal rod domain (Kanchanawong et al. 2010; del Rio et al. 2009; Liu et al. 2015). However, in the absence of suitable techniques for measuring subcellular talin forces, there was no quantitative proof of the mechanical tension via the talin in the cells. For this reason, biosensors have been developed to study the piconewton (pN) mechanics of talin compounds in living cells.

4.9 The Lamellipodium Acts as an Alternative Force Sensor

The capacity of adherent cells to perceive alterations in the mechanical properties of their extracellular environment is crucial for many aspects of their physiology. There exists evidence that cell adhesion and spreading are both sensitive to the rigidity of the substrate. It has been shown that this behavior is biphasic, since a transition point is approximately at a Young's modulus of 7 kPa. In contrast to established hypotheses, it has been shown that this behavior is independent of the activity of myosin II. Instead, the cell spreading on soft substrate seems to be impaired through the decreased nascent adhesion formation in the absence of myosin II throughout the lamellipodium. Cells cultured on soft substrates exhibit a normal activity of their leading-edge protrusions; however, these protrusions are not further stabilized, since the ongoing assembly of the adhesions is blocked. Increasing the integrin–extracellular matrix affinity by the addition of Mn^{2+} ions can induce the assembly of nascent adhesions and spreading of cells on soft substrates. Employing a computational model for the simulation of the assembly of nascent adhesions, the biophysical properties of the integrin–extracellular matrix have been identified to strengthen the binding interactions over a distinct matrix stiffness threshold, which is in line with experimental results (Oakes et al. 2018). Based on these findings, it can be proposed that myosin II-independent forces inside the lamellipodium facilitate the mechanosensation process that subsequently supports the formation of new adhesions, which then regulate the proper cell spreading. Moreover, this proposed mechanism for myosin II-independent sensing of the substrate stiffness may facilitate a variety of several stiffness-sensitive processes.

The capacity of cells to feel the mechanical forces and switch them into biochemical signals drives a wide variety of physiological functions (Paluch et al. 2015; Iskratsch et al. 2014; Charras and Sahai

2014). In specific detail, the cells react to alteration in the extracellular matrix stiffness through the adaption of several adhesion-dependent behaviors, including the processes of cell spreading (Lo et al. 2000; Yeung et al. 2005; Prager-Khoutorsky et al. 2011; Oakes et al. 2009; Califano and Reinhart-King 2010; Han et al. 2012; Ghibaudo et al. 2008; Tee et al. 2011; Engler et al. 2004), proliferation (Wang et al. 2000), differentiation (Engler et al. 2006; Fu et al. 2010), cell migration (Lo et al. 2000; Breckenridge et al. 2014; Raab et al. 2012), and metastasis (Paszek et al. 2005; Mekhdjian et al. 2017). Matrix mechanosensing process seems to be driven basically by focal adhesions, which are hierarchical organized organelles that include approximately over 150 proteins providing the dynamic and force-sensitive interplay between actin cytoskeleton and the extracellular matrix scaffold (Zaidel-Bar et al. 2007; Schiller et al. 2011; Kanchanawong et al. 2010). These dynamical remodeling organelles facilitate the mechanical sensing of their local environment in a wide range of physiological conditions, and their detailed function is still not clearly revealed.

Many studies addressed myosin II-driven mechanisms for the sensing of the substrate stiffness (Chan and Odde 2008; Macdonald et al. 2008; Bangasser and Odde 2013; Elosegui-Artola et al. 2014, 2016; Plotnikov et al. 2012). The stresses, which are produced by myosin motors and act on the actin cytoskeleton, are transduced through focal adhesion toward the surrounding extracellular matrix environment. In fact, these stresses associated with matrix stiffness influence the deformation and affinity of proteins to bind within focus adhesion (Yan et al. 2015; Thievessen et al. 2013; Sawada et al. 2006; Oakes et al. 2014). Alterations in the composition of focal adhesions and the kinetics of proteins therein seem to regulate the transfer of forces originating from the actin cytoskeleton toward the surrounding extracellular matrix (Hu et al. 2007; Aratyn-Schaus and Gardel 2010; Gardel et al. 2008), which initiates the handling of focal adhesions as molecular clutches. The initial assembly of an adhesion takes place at the cell's leading edge, which is the outermost region of the lamellipodium and represents a myosin-independent process (Bachir et al. 2014; Alexandrova et al. 2008). The emerging structures are termed nascent adhesions and are themselves stimulated by forces, which are produced by the actin filament polymerization. However, the role of nascent adhesions for the sensing of the stiffness of the substrate has not yet been determined precisely.

Ligand-coated substrates represent one of the best-characterized metrics for the microenvironmental sensing through the attachment and spreading of adhesive cells. The density and spatial organization of matrix ligands determines the degree of cell spreading (Cavalcanti-Adam et al. 2007; Reinhart-King et al. 2005; Dubin-Thaler et al. 2004) together with overall the rigidity of the substrate that resembles the ligands on the surface (Lo et al. 2000; Yeung et al. 2005; Prager-Khoutorsky et al. 2011; Oakes et al. 2009; Califano and Reinhart-King 2010; Han et al. 2012; Ghibaudo et al. 2008; Tee et al. 2011; Engler et al. 2004). Moreover, it has been hypothesized that the stress-relaxing features of the matrix promote the spreading of cells (Chaudhuri et al. 2015; Bauer et al. 2017). When the softness of the substrates lies below a Young's modulus of 500 Pa, the spreading of the cells impaired. When the stiffness of the substrate increases, the spreading area is accordingly increased, and finally, it reaches a plateau (Califano and Reinhart-King 2010; Han et al. 2012; Ghibaudo et al. 2008; Tee et al. 2011; Engler et al. 2004). Based on slight variations in the experimental approach, differences in the exact stiffness range controlling this behavior can be observed (Denisin and Pruitt 2016); however, cell spreading still is a robust method to analyze the sensing of substrate stiffness.

In order to reveal the mechanism that controls the substrate stiffness-based spreading of cells, NIH 3T3 fibroblasts were analyzed. Indeed, the spreading of NIH 3T3 fibroblasts is precisely altered by the Young's modulus of the substrate that is elevated from 5 to 8 kPa. In specific detail, cells cannot sufficiently spread out fully on substrates with a weak stiffness of about 5 kPa. When the substrate stiffness increased to values above 5 kPa, the spreading area is increased until a plateau region is reached, when the substrate stiffness is higher than 8 kPa. The spreading area stays constant far beyond this limit. Unexpectedly, it was found that this stiffness-dependent alteration of cell spread was independent of the motor activity of myosin II. In contrast, on soft substrates, the spreading is compromised by the reduced accumulation of nascent, myosin-independent adhesions on the cell periphery. The improvement of the integrin ligand affinity by the addition of Mn^{2+} was necessary to stabilize the nascent adhesions as well as to enlarge the cell spreading area on soft substrates. Subsequently, a computational model was applied to determine how shifts in the catch-bond kinetics of the integrin influence the integrin binding on substrates of various stiffnesses. It was observed that the biophysical properties of integrin matrix catch bonds were improved to detect alterations in substrate stiffness at approximately 6 kPa, in accordance with experimental outcomes. Finally, these findings show that the assembly of nascent adhesions within the lamellipodium functions as a myosin II-independent mechanosensor to regulate the adhesion and spreading of cells.

4.9.1 The Spread Area Results from a Biphasic Response of the Stiffness of the Substrate Which Is Independent of Myosin Activity

To analyze the mechanisms for the sensing of the substrate stiffness, the spreading area measurement of adhesive cells can be employed. Specifically, NIH 3T3 fibroblasts were seeded on a wide range of polyacrylamide gels with covalently bound fibronectin, a ligand for integrins, displaying Young's moduli from 0.6 to 150 kPa. For control, the cells were seeded on glass with absorbed fibronectin. In line with the previous studies (Lo et al. 2000; Prager-Khoutorsky et al. 2011; Califano and Reinhart-King 2010; Han et al. 2012; Ghibaudo et al. 2008; Tee et al. 2011), the spreading area of the cells area correlates with the stiffness of the substrate. In contrast, it has been revealed that there are basically two regimes: Firstly, there was solely poor spreading detected on soft (below approximately 5 kPa) substrates, and secondly, a high degree of cell spreading can be seen on stiff (above approximately 8 kPa) substrates, with a well-defined transition point between these two values and no statistical difference in the spreading behavior between cell populations within each these two regime. The morphology of cells on soft and stiff substrates has been revealed to be vastly variable (Yeung et al. 2005). In specific detail, cells adhered to soft substrates display a more rounded shape, and their cytoskeleton is rather disturbed and purely organized. An opposite behavior can be observed, when cells are positioned on stiff substrates, on which the cells possess a rather polarized cell shape and their cytoskeleton appears to be rather organized by F-actin stress fibers spanning over the entire cytoplasm.

Since myosin II activity is associated with mechanosensing (Plotnikov et al. 2012), it can be proposed that the impairment of myosin II has no effect on the spreading area on a wide variety of substrate stiffness values. Nevertheless, cells treated with 50 μM of the myosin ATPase inhibitor blebbistatin displayed still biphasic response due to varying substrate stiffness. After treatment with blebbistatin, the cells possess an increased spreading area than untreated controls for a wide range of substrate stiffnesses, whereas they displayed the same behavior independently of whether the experiment has been performed on soft and stiff regimes. Myosin II-inhibited cells exhibited on all substrates increased cell protrusions, whereas when placed on stiff substrates, the cells exerted increased spindle-like membrane extensions. When the Rho-kinase inhibitor Y26732 has been employed, the cells showed similar phenotypes and additionally when they were cultured on other proteins of the extracellular matrix scaffold. Finally, alterations in the spreading area of cells have not been observed between soft and stiff regimes and are independent of the myosin II activity.

4.9.2 Dynamic Remodeling of Lamellipodia Is Not Altered by the Material Stiffness

To decipher how the stiffness of the substrate affects the spreading area of cells, we analyzed how the substrate stiffness impacts on the dynamics of protrusions. The formation of lamellipodia can be analyzed by recording time-lapse images of cells that are transiently or stably transfected with a fluorescent membrane marker, such as GFP-stargazin and treated with 20 μM Y27632 on representative soft, such as 2.1 kPa, and stiff, such as 48 kPa, substrates 30 min after cell seeding. Cells located on soft substrates had repeated cycles of protrusion and retraction, which restricted their spreadability. However, cells on rigid substrates produced continuous and uniform protrusions which led to advanced progress of the leading edge. Using cell contours derived from the fluorescence pictures, we pinpointed protrusive regions and measured their morphology and properties. No statistically significant discrepancy was noted among soft and rigid substrates for average protrusion area or average protrusion width area measurements. From these observations, it appears that substrate stiffness influences the stability of leading-edge protrusions, whereas the protrusion dynamics are not concerned. The Arp2/3-induced lamellipodium generation remains necessary for propagation as cells on soft and rigid substrates, which were treated with CK869, an Arp2/3 inhibitor, could not be distinguished from control cells on soft substrates. Finally, these outcomes clarify that it is the stabilization, rather than the formation, of Arp2/3-dependent lamellipodial protrusions hampered on soft substrates.

Soft substrates can inhibit the assembly of nascent adhesions. To enlighten the regulatory mechanism for the substrate stiffness-driven alterations in the stabilization of myosin II-independent cell protrusions, the assembly of these myosin II-independent nascent adhesions is investigated, which are built at the basis of the emerging lamellipodium. After two hours of cell seeding, 20 μM of the Y27632 ROCK inhibitor were added to these cells for 30 min. After fixation, the cells were triple stained for actin, the p34 subunit of the Arp2/3 complex and the adaptor protein paxillin. The p34 protein is observed at the cell periphery on both soft and stiff substrates and hence marks the Arp2/3-based lamellipodium. Paxillin can assemble on stiff substrates small point-like nascent adhesions at the cell's leading edge (Choi et al. 2008). In contrast, paxillin-enriched nascent adhesions were rarely detected on soft substrates and when present they are located far distant from the cell's leading edge. To quantize these discrepancies in protein localization, the average actin, p34 and paxillin intensities in roughly 0.5 μm belts taken radially from the edge of the cell were evaluated. The peak of p34 intensity was located directly at the edge of the cell independently of the substrate. Paxillin was found on stiff substrates at approximately 0.5 μm of the p34 peak. On soft substrates, the accumulation of paxillin was considerably decreased, the peak of which was found approximately 5 μm beneath the leading edge. Finally, it can be summarized that cells on soft substrates display decreased nascent adhesions.

4.9.3 Mn^{2+}-Based Integrin Activation Induces Cell Spreading on Soft Substrates

Since the density of nascent adhesions on soft substrates is pronouncedly decreased, it seems to be necessary to determine whether the extent to which alterations in the integrin–ligand affinity affect the nascent adhesion assembly. The presence of 3 μM Mn^{2+} prolonged the lifespan of integrin–fibronectin bonds (Kong et al. 2009), but had no effect on cell contractility. When cells were deposited on soft substrates in the presence of 3 μM Mn^{2+} ions, they had more than a double increase in the spreading area on soft substrates, similar to their spreading area on rigid substrates either in the presence or absence of Mn^{2+} ions. For a direct evaluation of the effect of Mn^{2+} ions on adhesion assembly on soft substrates, immunofluorescence of paxillin and actin was conducted. Adding Mn^{2+} ions to cells on soft substrates encouraged the generation of paxillin-rich adhesions close to the cell periphery and the formation of lamellar actin bundles. To assess how fast Mn^{2+} initiated alterations in adhesion formation and cell spreading area, real-time cell imaging of EGFP paxillin and mApple-actin was conducted in cells deposited on a soft substrate during the administration of Mn^{2+} to the culture medium. Before the addition of Mn^{2+} ions, the protrusive activity on soft substrates was significant, without any alteration in surface area or cell shape. Upon addition of 3 μM Mn^{2+}, the cell protrusions are stabilized, new focal adhesions can be seen, and the spreading area of the cells is enhanced. After one incubation hour, the medium was exchanged with a Mn^{2+} ions-depleted medium and the cells start directly retracking themselves to their initial spreading area. These results show that the presence of Mn^{2+} ions can induce the spreading on rather soft substrates. This result leads to the hypothesis that the integrin–fibronectin-binding strength enables cells to detect the stiffness of the underlying substrate facilitate their own spreading.

The integrin catch-bond kinetics drive the sensing of the stiffness of the substrate. In order to reveal how the integrin–fibronectin interaction kinetics provides the sensing of the substrate stiffness, a computational model for the nascent adhesion assembly at the anterior edge of the edge is established. Similar to other approaches (Chan and Odde 2008; Macdonald et al. 2008; Bangasser and Odde 2013; Elosegui-Artola et al. 2014, 2016), this computational model included biophysical properties of the cell–matrix adhesions, the retrograde flow of actin and the rigidity of the substrate. Therein, individual integrins can serve as molecular clutches, and intermittently, transmitting force is evoked the retrograde flow of actin toward the substrate. Integrin–fibronectin bonds are demonstrated to act as catch bonds, which means that the lifetime of these catch bonds is extended under load (Kong et al. 2009). In order to examine the special features of these catch-bond kinetics that are required to sense the stiffness of the surrounding substrate in nascent adhesions, a simple model is employed.

In the model, each of which integrins and ligands are depicted as single point objects. At first, a certain number of fibronectin molecules were randomly bound to a substrate to which integrins could bind. The integrins endured successive rounds of diffusion, binding and debonding along a quasi-2D surface, which mimics the ventral membrane of cells above the substrate. Integrins with a connection to actin are subjected to retrograde flow, such as within the lamellipodium (Kong et al. 2009), whereas unbound integrins can diffuse over the membrane surface (Rossier et al. 2012). When an integrin approached a free fibronectin, it generated a harmonic potential interaction imitating the bond with the stiffness dictated by substrate stiffness. The hypothesis of concomitant binding of integrin to substrate ligands and actin was justified by the requirement to generate tension on the integrin–fibronectin bond, which governed the lifetime of the bond. Maintaining a steady flow of actin, the forces on the bonds produced were proportional to the substrate stiffness. All model variables are based on experimental results (Alexandrova et al. 2008; Rossier et al. 2012; Xu et al. 2016; Burnette et al. 2011). In specific, the life-force relationships of integrin–fibronectin bonds from single-molecule experiments of atomic force microscopy were directly included (Kong et al. 2009). In order to quantify the integrin binding amount, the average fraction of bound integrins has been determined during the simulations with each condition ranging between 10 and 300 s.

Initially, by simulating the life-force ratio of the integrin–fibronectin bond as a step function, a catch-bond behavior has been determined to be required to measure substrate stiffness. The longevity of the binding was kept constant below a peak force of 30 pN and below 0 at higher forces. The variation of the amount of integrin fibronectin lifetime had no impact on the proportion of bound integrins in dependence of substrate stiffness. Thus, in the lack of a force-dependent catch-bond mechanism, the integrin-binding kinetics remained unaffected by substrate stiffness. Subsequently, one of the questions was how alterations in the biophysical properties of catch bonds led to a measurement of stiffness. To simulate the life-force ratio as a typical catch bond, and to adjust the maximum life for the 30-pN peak force, the unloaded lifetime requires to be maintained as a constant. Under these conditions, since the lifetime of the integrin–fibronectin bond accelerated with rising force and the force was constrained by the stiffness of the substrate, the actin flow improved the amount of bound integrins on stiff substrates. However, alterations in the maximum bond lifespan showed very little effect on the proportion of bound integrins for each tested substrate stiffness. Unexpectedly, the transition between the different systems of integrin binding occurred in the model around the Young's module of approximately 7 kPa utilizing known biophysical parameters of integrin catch bonds that have been experimentally determined.

When Mn^{2+} ions are available, integrin–fibronectin bonds exhibit both an increased affinity and a longer lifetime of the bond reacting to elevated stress related to wild-type conditions (Kong et al. 2009; Gailit and Ruoslahti 1988; Smith et al. 1994; Mould et al. 1995). In order to imitate these phenomena in the model, both results have been combined to increase both the unloaded lifetime and the peak force lifetime. Through the enhancement of the integrin–fibronectin bond lifetime for forces up to 30 pN, the actin flow elevates the number of integrins that are bound to soft substrates. In this situation, the proportion of bound integrins on soft substrates in the presence of Mn^{2+} ions was quantitatively comparable to the proportion of bound integrins on rigid substrates in the absence of Mn^{2+} ions. Therefore, although an enhancement of unloaded lifetime or peak force lifetime alone is not enough to capture the impact of Mn^{2+}, its cumulative effect is sufficient to eliminate the effects of substrate stiffness on cell spreading. Moreover, this observation did not rely on the ligand number employed in this model.

In summary, these results show that the stiffness measurement in the lamellipodium is determined by the catch-bond kinetics of integrin–fibronectin linkages and that the proportion of bound integrins is susceptible to the unloaded lifetime and the maximum lifespan of the catch-bond curve. The addition of Mn^{2+} ions causes an increased lifespan of the integrins on soft substrates, which in turn elevates the average portion of bound integrins. However, when this behavior of the integrin binding kinetics is altered, the cells can even spread out well on soft substrates.

For many cellular processes, such as the viability, growth, cell migration and differentiation, the capability of cells to feel the stiffness of the surrounding extracellular environment is crucial (Paluch et al. 2015; Iskratsch et al. 2014; Charras and Sahai 2014). Fibroblasts have been seen to display a biphasic response in spreading on matrices with variable stiffnesses. For extracellular matrices, which have a Young's modulus of less than approximately 5 kPa, the cells cannot spread out well and hence display a minimal assembly of adhesion with solely less organized actin filament structures. Above approximately 8 kPa, the fibroblasts possess a maximal spreading area and exhibit a typical adhesion assembly with a highly structured actin cytoskeleton. This stiffness transition seems to in the same range as the physiological stiffness of tissues (Engler et al. 2006; Califano and Reinhart-King 2010; Han et al. 2012; Ghibaudo et al. 2008; Tee et al. 2011; Engler et al. 2004). It has been proposed that the cell spreading area is regulated by the stiffness of the substrate and the dependence seems to be a power-law behavior (Han et al. 2012; Engler et al. 2004), or when the number of substrates is increased, the behavior can be seen as biphasic.

Based on myosin II's key role in the cellular force generations, myosin II has been hypothesized to fulfill a key role in the stiffness sensing of the underlying substrates by adhesive cells (Chan and Odde 2008; Macdonald et al. 2008; Bangasser and Odde 2013; Elosegui-Artola et al. 2014, 2016; Kraning-Rush et al. 2011). It has been shown that a myosin-independent sensing mechanism of stiffness is regulated by the spreading area and is based on forces that are originating from the polymerization of actin in the lamellipodium. Integrins have been demonstrated to function as catch bonds, since they couple the cytoskeleton with the extracellular matrix and facilitate thereby the transmission of stress (Kong et al. 2009; Choquet et al. 1997). Moreover, the lifetime of these catch bonds depends on the amount of the applied physical load. When the load on theses integrin is elevated, the lifetime of the catch bonds is prolonged similarly (Kong et al. 2009). The prolonged lifetime of these catch-bonds can even induce the clustering and the formation of focal adhesions. In contrast, on soft substrates, the decreased stiffness causes decreased catch-bond lifetimes that impair the clustering and subsequently the formation of focal adhesions. Both experimental and simulation studies point to the fact that integrin-force-dependent-binding kinetics is mainly sensitive to substrates with a stiffness ranging between approximately 5 and 8 kPa. The addition of Mn^{2+} ions modifies the kinetics of integrin–extracellular matrix bonds as it prolongs the longevity of unloaded and peak force (Kong et al. 2009; Gailit and Ruoslahti 1988; Smith et al. 1994; Mould et al. 1995). Thereby, the amount of integrins is increased, and the average distance between integrins is reduced. Therefore, these effects induced the assembly of adhesions and facilitate the spreading of cells on soft substrates and impact the cell morphology on stiff substrates.

The concept of this framework is reminiscent of the general motor clutch model (Suter et al. 1998) proposed as a mechanism for explaining mechanosensitivity (Chan and Odde 2008; Macdonald et al. 2008; Elosegui-Artola et al. 2014, 2016). As opposed to forces generated by myosin motors, the force acting through the integrin bonds is generated by actin polymerization within the lamellipodium. These polymerization forces modify the integrin extracellular matrix-binding kinetics and provide a remarkably simple and elegant mechanism for gaining an understanding of the measurement of substrate stiffness. Former work has shown that a minimum distance between the integrins is necessary for adhesion formation (Cavalcanti-Adam et al. 2007). The binding of integrins to their ligands also restricts their membrane diffusion (Rossier et al. 2012) and promotes nanoscale cluster formation (Bachir et al. 2014). When a nanoscale cluster of integrins emerges, the force needed to rupture adhesion, such as the adhesion strength, is over an order of magnitude larger than typical tensions induced in the cytoskeleton (Coyer et al. 2012). Adhesion stabilization is therefore improved by raising the density of the bound integrins, and the cell can spread.

Finally, these results imply that the lamellipodium functions as a myosin-independent mechanosensor and exerts force on bound integrins through an actin polymerization-based retrograde flow. On soft substrates, the enhanced flexibility of the matrix results in less load on the integrin extracellular matrix binding, leading to a reduced lifetime. The shorter lifetime eliminates integrin clustering and prevents adhesion stabilization, resulting in poor spreadability. On stiff substrates, the integrin–extracellular matrix bonds can undergo a higher load and thus possess a longer lifetime, which supports the stabilization of the adhesion and thereby allows the cells to spread. Although these findings do not rule out the feasibility that myosin-generated forces can be a mechanism to examine the stiffness of the substrate, the results indicate that stiffness sensorics are obtained passively through the characteristics of integrin-binding kinetics. Since a simple shift of this kinetics is able to trigger the spreading on soft substrates, in the future it becomes useful to examine whether this concept is adequate to restore other functions compromised by soft substrates such as development, differentiation and disease.

The physiology of cells may be based on the regulation of the mechanical properties of the extracellular microenvironment. The spreading of cells is basically a mechanosensitive process that is driven by weak forces that are provided by the cell's periphery and are not based on the activity of motor proteins. Finally, it has been demonstrated that the sensing of the environmental stiffness relies on the kinetics of the initial formation of adhesion bonds, which are probed by forces based on the polymerization of proteins. On the basis of this analysis, the binding kinetics of adhesion molecules becomes sensitive to a wide variety of forces that permit mechanosensation.

References

C. Albiges-Rizo, O. Destaing, B. Fourcade, E. Planus, M.R. Block, Actin machinery and mechanosensitivity in invadopodia, podosomes and focal adhesions. J. Cell Sci. **122**, 3037–3304 (2009)

A.Y. Alexandrova, K. Arnold, S. Schaub, J.M. Vasiliev, J.-J. Meister, A.D. Bershadsky, A.B. Verkhovsky, Comparative dynamics of retrograde actin flow and focal adhesions: formation of nascent adhesions triggers transition from fast to slow flow. PLoS One **3**, e3234 (2008)

Y. Aratyn-Schaus, M.L. Gardel, Transient frictional slip between integrin and the ECM in focal adhesions under myosin II tension. Curr. Biol. **20**, 1145–1153 (2010)

B.G. Armendariz, M. Masdeu Mdel, E. Soriano, J.M. Ureña, F. Burgaya, The diverse roles and multiple forms of focal adhesion kinase in brain. Eur. J. Neurosci. **40**(11), 3573–3590 (2014)

A.I. Athamneh, D.M. Suter, Quantifying mechanical force in axonal growth and guidance. Front. Cell. Neurosci. **9**, 359 (2015)

P. Atherton, B. Stutchbury, D.Y. Wang et al., Vinculin controls talin engagement with the actomyosin machinery. Nat. Commun. **6**, 10038 (2015a)

P. Atherton, B. Stutchbury, D. Jethwa, C. Ballestrem, Mechanosensitive components of integrin adhesions: role of vinculin. Exp. Cell Res. **343**(1), 21–27 (2015b)

P. Atherton, F. Lausecker, A. Harrison, C. Ballestrem, Low intensity pulsed ultrasound (LIPUS) promotes cell motility through vinculin-controlled rac1 GTPase activity. J. Cell Sci. **30**(14), 2277–2291 (2017)

V. Auernheimer, W.H. Goldmann, Serine phosphorylation on position 1033 of vinculin impacts cellular mechanics. Biochem. Biophys. Res. Commun. **450**(2), 1095–1098 (2014)

V. Auernheimer, L.A. Lautscham, M. Leidenberger, O. Friedrich, B. Kappes, B. Fabry, W.H. Goldmann, Vinculin phosphorylation at residues Y100 and Y1065 is required for cellular force transmission. J. Cell Sci. **128**(18), 3435–3443 (2015)

K. Austen, P. Ringer, A. Mehlich, A. Chrostek-Grashoff, C. Kluger, C. Klingner, B. Sabass, R. Zent, M. Rief, C. Grashoff, Extracellular rigidity sensing by talin isoform-specific mechanical linkages. Nat. Cell Biol. **17**(12), 1597–1606 (2015)

A.I. Bachir, J. Zareno, K. Moissoglu, E.F. Plow, E. Gratton, A.R. Horwitz, Integrin-associated complexes form hierarchically with variable stoichiometry in nascent adhesions. Curr. Biol. **24**(16), 1845–1853 (2014)

Y.H. Bae, K.L. Mui, B.Y. Hsu, S.L. Liu, A. Cretu, Z. Razinia, T. Xu, E. Pure, R.K. Assoian, A FAK-Cas-Rac-lamellipodin signaling module transduces extracellular matrix stiffness into mechanosensitive cell cycling. Sci. Signal. **7**, ra57 (2014)

B.M. Baker, B. Trappmann, W.Y. Wang, M.S. Sakar, I.L. Kim, V.B. Shenoy, J.A. Burdick, C.S. Chen, Cell-mediated fibre recruitment drives extracellular matrix mechanosensing in engineered fibrillar microenvironments. Nat. Mater. **14**(12), 1262–1268 (2015)

C. Bakolitsa, J.M. de Pereda, C.R. Bagshaw, D.R. Critchley, R.C. Liddington, Crystal structure of the vinculin tail suggests a pathway for activation. Cell **99**(6), 603–613 (1999)

C. Bakolitsa, D.M. Cohen, L.A. Bankston, A.A. Bobkov, G.W. Cadwell, L. Jennings, D.R. Critchley, S.W. Craig, R.C. Liddington, Structural basis for vinculin activation at sites of cell adhesion. Nature **430**(6999), 583–586 (2004)

N.Q. Balaban, U.S. Schwarz, D. Riveline et al., Force and focal adhesion assembly: a close relationship studied using elastic micropatterned substrates. Nat. Cell Biol. **3**(5), 466–472 (2001)

J.L. Balestrini, S. Chaudhry, V. Sarrazy, A. Koehler, B. Hinz, The mechanical memory of lung myofibroblasts. Integr. Biol. (Camb.) **4**(4), 410–421 (2012)

C. Ballestrem, Marching at the front and dragging behind: differential alphaVbeta3-integrin turnover regulates focal adhesion behavior. J. Cell Biol. **155**(7), 1319–1332 (2001)

B.L. Bangasser, D.J. Odde, Master equation-based analysis of a motor-clutch model for cell traction force. Cell Mol. Bioeng. **6**, 449–459 (2013)

A.K. Barry, H. Tabdili, I. Muhamed, J. Wu, N. Shashikanth, G.A. Gomez, A.S. Yap, C.J. Gottardi, J. de Rooij, N. Wang, D.E. Leckband, Alpha-catenin cytomechanics—role in cadherin-dependent adhesion and mechanotransduction. J. Cell Sci. **127**(Pt 8), 1779–1791 (2014)

A. Bauer, L. Gu, B. Kwee, W.A. Li, M. Dellacherie, A.D. Celiz, D.J. Mooney, Hydrogel substrate stress-relaxation regulates the spreading and proliferation of mouse myoblasts. Acta Biomater. **62**, 82–90 (2017)

J.L. Bays, K.A. DeMali, Vinculin in cell-cell and cell-matrix adhesions. Cell Mol. Life Sci. **74**(16), 2999–3009 (2017)

J.L. Bays, X. Peng, C.E. Tolbert, C. Guilluy, A.E. Angell, Y. Pan, R. Superfine, K. Burridge, K.A. DeMali, Vinculin phosphorylation differentially regulates mechanotransduction at cell-cell and cell–matrix adhesions. J. Cell Biol. **205**(2), 251–263 (2014)

A.L. Berrier, K.M. Yamada, Cell-matrix adhesion. J. Cell. Physiol. **213**, 565–573 (2007)

A.D. Bershadsky, I.S. Tint, A.A. Neyfakh Jr., J.M. Vasiliev, Focal contacts of normal and RSV-transformed quail cells. Hypothesis of the transformation-induced deficient maturation of focal contacts. Exp. Cell Res. **158**(2), 433–444 (1985)

A.D. Bershadsky, N.Q. Balaban, B. Geiger, Adhesion-dependent cell mechanosensitivity. Annu. Rev. Cell Dev. Biol. **19**, 677–695 (2003)

C. Bertocchi, Y. Wang, A. Ravasio et al., Nanoscale architecture of cadherin-based cell adhesions. Nat. Cell Biol. **19**(1), 28–37 (2017)

P.R. Bois, B.P. O'Hara, D. Nietlispach, J. Kirkpatrick, T. Izard, The vinculin binding sites of talin and alpha-actinin are sufficient to activate vinculin. J. Biol. Chem. **281**(11), 7228–7236 (2006)

N. Borghi, M. Sorokina, O.G. Shcherbakova, W.I. Weis, B.L. Pruitt, W.J. Nelson, A.R. Dunn, E-cadherin is under constitutive actomyosin-generated tension that is increased at cell-cell contacts upon externally applied stretch. Proc. Natl. Acad. Sci. USA **109**, 12568–12573 (2012)

R.A. Borgon, C. Vonrhein, G. Bricogne, P.R. Bois, T. Izard, Crystal structure of human vinculin. Structure **12**(7), 1189–1197 (2004)

M.T. Breckenridge, R.A. Desai, M.T. Yang, J. Fu, C.S. Chen, Substrates with engineered step changes in rigidity induce traction force polarity and durotaxis. Cell Mol. Bioeng. **7**, 26–34 (2014)

M.D. Brenner, R. Zhou, D.E. Conway, L. Lanzano, E. Gratton, M.A. Schwartz, T. Ha, Spider silk peptide is a compact, linear nanospring ideal for intracellular tension sensing. Nano Lett. **16**(3), 2096–2102 (2016)

N.P. Brindle, M.R. Holt, J.E. Davies, C.J. Price, D.R. Critchley, The focal-adhesion vasodilator-stimulated phosphoprotein (VASP) binds to the proline-rich domain in vinculin. Biochem. J. **318**(Pt 3), 753–757 (1996)

N.H. Brown, S.L. Gregory, W.L. Rickoll, L.I. Fessler, M. Prout, R.A. White, J.W. Fristrom, Talin is essential for integrin function in Drosophila. Dev. Cell **3**, 569–579 (2002)

D.T. Burnette, S. Manley, P. Sengupta, R. Sougrat, M.W. Davidson, B. Kachar, J. Lippincott-Schwartz, A role for actin arcs in the leading-edge advance of migrating cells. Nat. Cell Biol. **13**, 371–381 (2011)

K. Burridge, J.R. Feramisco, Microinjection and localization of a 130 K protein in living fibroblasts: a relationship to actin and fibronectin. Cell **19**(3), 587–595 (1980)

D.T. Butcher, T. Alliston, V.M. Weaver, A tense situation: forcing tumour progression. Nat. Rev. Cancer **9**, 108–122 (2009)

D.A. Calderwood, I.D. Campbell, D.R. Critchley, Talins and kindlins: partners in integrin-mediated adhesion. Nat. Rev. Mol. Cell Biol. **14**(8), 503–517 (2013)

J.P. Califano, C.A. Reinhart-King, Substrate stiffness and cell area predict cellular traction stresses in single cells and cells in contact. Cell Mol. Bioeng. **3**, 68–75 (2010)

O. Campàs, T. Mammoto, S. Hasso, R.A. Sperling, D. O'Connell, A.G. Bischof, R. Maas, D.A. Weitz, L. Mahadevan, D.E. Ingber, Quantifying cell-generated mechanical forces within living embryonic tissues. Nat. Methods **11**, 183–189 (2014)

A. Carisey, C. Ballestrem, Vinculin, an adapter protein in control of cell adhesion signalling. Eur. J. Cell Biol. **90**(2–3), 157–163 (2011)

A. Carisey, R. Tsang, A.M. Greiner, N. Nijenhuis, N. Heath, A. Nazgiewicz, R. Kemkemer, B. Derby, J. Spatz, C. Ballestrem, Vinculin regulates the recruitment and release of core focal adhesion proteins in a force-dependent manner. Curr. Biol. **23**(4), 271–281 (2013)

L.B. Case, C.M. Waterman, Integration of actin dynamics and cell adhesion by a three-dimensional, mechanosensitive molecular clutch. Nat. Cell Biol. **17**, 955–963 (2015)

L.B. Case, M.A. Baird, G. Shtengel, S.L. Campbell, H.F. Hess, M.W. Davidson, C.M. Waterman, Molecular mechanism of vinculin activation and nanoscale spatial organization in focal adhesions. Nat. Cell Biol. **17**(7), 880–892 (2015)

E.A. Cavalcanti-Adam, T. Volberg, A. Micoulet, H. Kessler, B. Geiger, J.P. Spatz, Cell spreading and focal adhesion dynamics are regulated by spacing of integrin ligands. Biophys. J. **92**, 2964–2974 (2007)

C.E. Chan, D.J. Odde, Traction dynamics of filopodia on compliant substrates. Science **322**, 1687–1691 (2008)

I. Chandrasekar, T.E. Stradal, M.R. Holt, F. Entschladen, B.M. Jockusch, W.H. Ziegler, Vinculin acts as a sensor in lipid regulation of adhesion-site turnover. J. Cell Sci. **118**(Pt 7), 1461–1472 (2005)

G. Charras, E. Sahai, Physical influences of the extracellular environment on cell migration. Nat. Rev. Mol. Cell Biol. **15**, 813–824 (2014)

O. Chaudhuri, L. Gu, M. Darnell, D. Klumpers, S.A. Bencherif, J.C. Weaver, N. Nathaniel Huebsch, D.J. Mooney, Substrate stress relaxation regulates cell spreading. Nat. Commun. **6**, 6364 (2015)

O. Chaudhuri, L.G.D. Klumpers, M. Darnell et al., Hydrogels with tunable stress relaxation regulate stem cell fate and activity. Nat. Mater. **15**, 326–334 (2016)

C.S. Chen, J. Tan, J. Tien, Mechanotransduction at cell-matrix and cell-cell contacts. Annu. Rev. Biomed. Eng. **6**, 275–302 (2004)

H. Chen, Spatial distribution and functional significance of activated vinculin in living cells. J. Cell Biol. **169**(3), 459–470 (2005)

H. Chen, D.M. Choudhury, S.W. Craig, Coincidence of actin filaments and talin is required to activate vinculin. J. Biol. Chem. **281**(52), 40389–40398 (2006)

A. Chervin-Petinot, M. Courcon, S. Almagro, A. Nicolas, A. Grichine, D. Grunwald, M.H. Prandini, P. Huber, D. Gulino-Debrac, Epithelial protein lost in neoplasm (EPLIN) interacts with alpha-catenin and actin filaments in endothelial cells and stabilizes vascular capillary network in vitro. J. Biol. Chem. **287**(10), 7556–7572 (2012)

K. Chinthalapudi, E.S. Rangarajan, D.N. Patil, E.M. George, D.T. Brown, T. Izard, Lipid binding promotes oligomerization and focal adhesion activity of vinculin. J. Cell Biol. **207**(5), 643–656 (2014)

K. Chinthalapudi, D.N. Patil, E.S. Rangarajan, C. Rader, T. Izard, Lipid-directed vinculin dimerization. BioChemistry **54**(17), 2758–2768 (2015)

S. Cho, J. Irianto, D.E. Discher, Mechanosensing by the nucleus: from pathways to scaling relationships. J. Cell Biol. **216**(2), 305–315 (2017)

C.K. Choi, M. Vicente-Manzanares, J. Zareno, L.A. Whitmore, A. Mogilner, A.R. Horwitz, Actin and alpha-actinin orchestrate the assembly and maturation of nascent adhesions in a myosin II motor-independent manner. Nat. Cell Biol. **10**, 1039–1050 (2008)

D. Choquet, D.P. Felsenfeld, M.P. Sheetz, Extracellular matrix rigidity causes strengthening of integrin-cytoskeleton linkages. Cell **88**, 39–48 (1997)

D.S. Chorev, O. Moscovitz, B. Geiger, M. Sharon, Regulation of focal adhesion formation by a vinculin-Arp2/3 hybrid complex. Nat. Commun. **5**, 3758 (2014)

C. Ciobanasu, B. Faivre, C. Le Clainche, Actomyosin-dependent formation of the mechanosensitive talin-vinculin complex reinforces actin anchoring. Nat. Commun. **5**, 3095 (2014)

D.M. Cohen, H. Chen, R.P. Johnson, B. Choudhury, S.W. Craig, Two distinct head-tail interfaces cooperate to suppress activation of vinculin by talin. J. Biol. Chem. **280**(17), 17109–17117 (2005)

D.M. Cohen, B. Kutscher, H. Chen, D.B. Murphy, S.W. Craig, A conformational switch in vinculin drives formation and dynamics of a talin-vinculin complex at focal adhesions. J. Biol. Chem. **281**(23), 16006–16015 (2006)

P. Costa, F.V. Almeida, J.T. Connelly, Biophysical signals controlling cell fate decisions: how do stem cells really feel? Int. J. Biochem. Cell Biol. **44**, 2233–2237 (2012)

S.R. Coyer, A. Singh, D.W. Dumbauld, D.A. Calderwood, S.W. Craig, E. Delamarche, A.J. García, Nanopatterning reveals an ECM area threshold for focal adhesion assembly and force transmission that is regulated by integrin activation and cytoskeleton tension. J. Cell Sci. **125**, 5110–5123 (2012)

E.J. Cram, S.G. Clark, J.E. Schwarzbauer, Talin loss-of-function uncovers roles in cell contractility and migration in *C. elegans*. J. Cell Sci. **116**, 3871–3878 (2003)

D.R. Critchley, Biochemical and structural properties of the integrin-associated cytoskeletal protein talin. Annu. Rev. Biophys. **38**, 235–254 (2009)

J. de Rooij, A. Kerstens, G. Danuser, M.A. Schwartz, C.M. Waterman-Storer, Integrin-dependent actomyosin contraction regulates epithelial cell scattering. J. Cell Biol. **171**(1), 153–164 (2005)

A. del Rio, R. Perez-Jimenez, R. Liu, P. Roca-Cusachs, J.M. Fernandez, M.P. Sheetz, Stretching single talin rod molecules activates vinculin binding. Science **323**(5914), 638–641 (2009)

K.A. DeMali, K. Burridge, Coupling membrane protrusion and cell adhesion. J. Cell Sci. **116**(Pt 12), 2389–2397 (2003)

K.A. DeMali, C.A. Barlow, K. Burridge, Recruitment of the Arp2/3 complex to vinculin: coupling membrane protrusion to matrix adhesion. J. Cell Biol. **159**(5), 881–891 (2002)

K.A. DeMali, X. Sun, G.A. Bui, Force transmission at cell-cell and cell–matrix adhesions. BioChemistry **53**(49), 7706–7717 (2014)

A.K. Denisin, B.L. Pruitt, Tuning the range of polyacrylamide gel stiffness for mechanobiology applications. ACS Appl. Mater. Interfaces **8**, 21893–21902 (2016)

G. Diez, F. List, J. Smith, W.H. Ziegler, W.H. Goldmann, Direct evidence of vinculin tail-lipid membrane interaction in beta-sheet conformation. Biochem. Biophys. Res. Commun. **373**(1), 69–73 (2008)

G. Diez, P. Kollmannsberger, C.T. Mierke, T.M. Koch, H. Vali, B. Fabry, W.H. Goldmann, Anchorage of vinculin to lipid membranes influences cell mechanical properties. Biophys. J. **97**(12), 3105–3112 (2009)

D.E. Discher, P. Janmey, Y.L. Wang, Tissue cells feel and respond to the stiffness of their substrate. Science **310**, 1139–1143 (2005)

A.D. Doyle, K.M. Yamada, Mechanosensing via cell-matrix adhesions in 3D microenvironments. Exp. Cell Res. **343**(1), 60–66 (2016)

A.D. Doyle, F.W. Wang, K. Matsumoto, K.M. Yamada, One-dimensional topography underlies three-dimensional fibrillar cell migration. J. Cell Biol. **184**(4), 481–490 (2009)

A.D. Doyle, M.L. Kutys, M.A. Conti, K. Matsumoto, R.S. Adelstein, K.M. Yamada, Micro-environmental control of cell migration–myosin IIA is required for efficient migration in fibrillar environments through control of cell adhesion dynamics. J. Cell Sci. **125**(Pt 9), 2244–2256 (2012)

B.J. Dubin-Thaler, G. Giannone, H.-G. Döbereiner, M.P. Sheetz, Nanometer analysis of cell spreading on matrix-coated surfaces reveals two distinct cell states and STEPs. Biophys. J. **86**, 1794–1806 (2004)

D.W. Dumbauld, T.T. Lee, A. Singh, J. Scrimgeour, C.A. Gersbach, E.A. Zamir, J. Fu, C.S. Chen, J.E. Curtis, S.W. Craig, A.J. Garcia, How vinculin regulates force transmission. Proc. Natl. Acad. Sci. USA **110**(24), 9788–9793 (2013)

S. Dupont, L. Morsut, M. Aragona et al., Role of YAP/TAZ in mechanotransduction. Nature **474**(7350), 179–183 (2011)

A. Elosegui-Artola, E. Bazellières, M.D. Allen et al., Rigidity sensing and adaptation through regulation of integrin types. Nat. Mater. **13**, 631–637 (2014)

A. Elosegui-Artola, R. Oria, Y. Chen, A. Kosmalska, C. Pérez-González, N. Castro, C. Zhu, X. Trepat, P. Roca-Cusachs, Mechanical regulation of a molecular clutch defines force transmission and transduction in response to matrix rigidity. Nat. Cell Biol. **18**(5), 540–548 (2016)

A. Engler, L. Bacakova, C. Newman, A. Hategan, M. Griffin, D. Discher, Substrate compliance versus ligand density in cell on gel responses. Biophys. J. **86**, 617–628 (2004)

A.J. Engler, S. Sen, H.L. Sweeney, D.E. Discher, Matrix elasticity directs stem cell lineage specification. Cell **126**(4), 677–689 (2006)

R.M. Ezzell, W.H. Goldmann, N. Wang, N. Parashurama, D.E. Ingber, Vinculin promotes cell spreading by mechanically coupling integrins to the cytoskeleton. Exp. Cell Res. **231**(1), 14–26 (1997)

S.I. Fraley, Y. Feng, R. Krishnamurthy, D.H. Kim, A. Celedon, G.D. Longmore, D.A. Wirtz, distinctive role for focal adhesion proteins in three-dimensional cell motility. Nat. Cell Biol. **12**(6), 598–604 (2010)

S.J. Franco, M.A. Rodgers, B.J. Perrin, J. Han, D.A. Bennin, D.R. Critchley, A. Huttenlocher, Calpain-mediated proteolysis of talin regulates adhesion dynamics. Nat. Cell Biol. **6**(10), 977–983 (2004)

K. Franze, P.A. Janmey, J. Guck, Mechanics in neuronal development and repair. Annu. Rev. Biomed. Eng. **15**, 227–251 (2013)

P. Fratzl, *Collagen Structure and Mechanics* (Springer Science + Business Media, 2017)

J.C. Friedland, M.H. Lee, D. Boettiger, Mechanically activated integrin switch controls alpha5beta1 function. Science **323**(5914), 642–644 (2009)

J. Fu, Y.K. Wang, M.T. Yang, R.A. Desai, X. Yu, Z. Liu, C.S. Chen, Mechanical regulation of cell function with geometrically modulated elastomeric substrates. Nat. Methods **7**, 733–736 (2010)

J. Gailit, E. Ruoslahti, Regulation of the fibronectin receptor affinity by divalent cations. J. Biol. Chem. **263**, 12927–12932 (1988)

C.G. Galbraith, K.M. Yamada, M.P. Sheetz, The relationship between force and focal complex development. J. Cell Biol. **159**(4), 695–705 (2002)

M.L. Gardel, B. Sabass, L. Ji, G. Danuser, U.S. Schwarz, C.M. Waterman, Traction stress in focal adhesions correlates biphasically with actin retrograde flow speed. J. Cell Biol. **183**, 999–1005 (2008)

B.A. Geiger, 130 K protein from chicken gizzard: its localization at the termini of microfilament bundles in cultured chicken cells. Cell **18**(1), 193–205 (1979)

B. Geiger, K.M. Yamada, Molecular architecture and function of matrix adhesions. Cold Spring Harbor Perspect. Biol. **3**(5), a005033 (2011)

B. Geiger, A. Bershadsky, R. Pankov, K.M. Yamada, Transmembrane crosstalk between the extracellular matrix-cytoskeleton. Nat. Rev. Mol. Cell Biol. **2**, 793–805 (2001)

B. Geiger, J.P. Spatz, A.D. Bershadsky, Environmental sensing through focal adhesions. Nat. Rev. Mol. Cell Biol. **10**, 21–33 (2009)

P.C. Georges, M. McCormick, L.A. Flanagan, Y.-E. Ju, E.S. Sawyer, P.A. Janmey, Tuning the elasticity of biopolymer gels for optimal wound healing. Mater. Res. Soc. Symp. Proc. **897E**, 0897-J02-01 (2005)

M. Georgiadou, J. Lilja, G. Jacquemet, AMPK negatively regulates tensin-dependent integrin activity. J. Cell Biol. **216** (4), 1107–1121 (2017)

M. Ghibaudo, A. Saez, L. Trichet, A. Xayaphoummine, J. Browaeys, P. Silberzan, A. Buguin, B. Ladoux, Traction forces and rigidity sensing regulate cell functions. Soft Matter **4**, 1836–1843 (2008)

M. Gimona, J.V. Small, M. Moeremans, J. Van Damme, M. Puype, J. Vandekerckhove, Porcine vinculin and metavinculin differ by a 68-residue insert located close to the carboxy-terminal part of the molecule. EMBO J. **7**(8), 2329–2334 (1988)

E. Goksoy, Y.-Q. Ma, X. Wang, X. Kong, D. Perera, E.F. Plow, J. Qin, Structural basis for the autoinhibition of talin in regulating integrin activation. Mol. Cell **31**(1), 124–133 (2008)

J. Golji, M.R. Mofrad, A molecular dynamics investigation of vinculin activation. Biophys. J. **99**(4), 1073–1081 (2010)

J. Golji, J. Lam, M.R. Mofrad, Vinculin activation is necessary for complete talin binding. Biophys. J. **100**(2), 332–340 (2011)

J. Golji, T. Wendorff, M.R. Mofrad, Phosphorylation primes vinculin for activation. Biophys. J. **102**(9), 2022–2030 (2012)

A. Gotoh, K. Miyazawa, K. Ohyashiki, T. Tauchi, H.S. Boswell, H.E. Broxmeyer, K. Toyama, Tyrosine phosphorylation and activation of focal adhesion kinase (p125FAK) by BCR-ABL oncoprotein. Exp. Hematol. **23**(11), 1153–1159 (1995)

B.T. Goult, N. Bate, N.J. Anthis, K.L. Wegener, A.R. Gingras, B. Patel, I.L. Barsukov, I.D. Campbell, G.C. Roberts, D. R. Critchley, The structure of an interdomain complex that regulates talin activity. J. Biol. Chem. **284**(22), 15097–15106 (2009)

B.T. Goult, T. Zacharchenko, N. Bate et al., RIAM and vinculin binding to talin are mutually exclusive and regulate adhesion assembly and turnover. J. Biol. Chem. **288**(12), 8238–8249 (2013)

C. Grashoff, B.D. Hoffman, M.D. Brenner, R. Zhou, M. Parsons, M.T. Yang, M.A. McLean, S.G. Sligar, C.S. Chen, T. Ha, M.A. Schwartz, Measuring mechanical tension across vinculin reveals regulation of focal adhesion dynamics. Nature **466**(7303), 263–266 (2010)

J.A. Green, K.M. Yamada, Three-dimensional microenvironments modulate fibroblast signaling responses. Adv. Drug Deliv. Rev. **59**(13), 1293–1298 (2007)

M.E. Groesch, J.J. Otto, Purification and characterization of an 85 kDa talin-binding fragment of vinculin. Cell Motil. Cytoskeleton. **15**(1), 41–50 (1990)

K.M. Hakkinen, J.S. Harunaga, A.D. Doyle, K.M. Yamada, Direct comparisons of the morphology, migration, cell adhesions, and actin cytoskeleton of fibroblasts in four different three-dimensional extracellular matrices. Tissue Eng. Part A **17**(5–6), 713–724 (2011)

G.K. Hakonardottir, P. López-Ceballos, A.D. Herrera-Reyes, R. Das, D. Coombs, G. Tanentzapf, In vivo quantitative analysis of Talin turnover in response to force. Mol. Biol. Cell **26**(22), 4149–4162 (2015)

M.S. Hall, F. Alisafaei, E. Ban, Hui C.Y. Feng, V.B. Shenoy, M. Wu, Fibrous nonlinear elasticity enables positive mechanical feedback between cells and ECMs. Proc. Natl. Acad. Sci. USA **113**(49), 14043–14048 (2016)

J. Han, C.J. Lim, N. Watanabe et al., Reconstructing and deconstructing agonist-induced activation of integrin alphaIIbbeta3. Curr. Biol. **16**(18), 1796–1806 (2006)

S.J. Han, K.S. Bielawski, L.H. Ting, M.L. Rodriguez, N.J. Sniadecki, Decoupling substrate stiffness, spread area, and micropost density: a close spatial relationship between traction forces and focal adhesions. Biophys. J. **103**, 640–648 (2012)

J.P. Heath, G.A. Dunn, ell to substratum contacts of chick fibroblasts and their relation to the microfilament system. A correlated interference-reflexion and high-voltage electron-microscope study. J. Cell Sci. **29**, 197–212 (1978)

C.P. Heisenberg, Y. Bellaiche, Forces in tissue morphogenesis and patterning. Cell **153**, 948–962 (2013)

H. Hirata, H. Tatsumi, C.T. Lim, M. Sokabe, Force-dependent vinculin binding to talin in live cells: a crucial step in anchoring the actin cytoskeleton to focal adhesions. Am. J. Physiol. Cell Physiol. **306**(6), C607–C620 (2014)

B.D. Hoffman, C. Grashoff, M.A. Schwartz, Dynamic molecular processes mediate cellular mechanotransduction. Nature **475**, 316–323 (2011)

N.J. Hogrebe, J.W. Reinhardt, K.J. Gooch, Biomaterial microarchitecture: apotent regulator of individual cell behavior and multicellular organization. J. Biomed. Mater. Res. A **105**(2), 640–661 (2017)

A.W. Holle, X. Tang, D. Vijayraghavan, L.G. Vincent, A. Fuhrmann, Y.S. Choi, J.C. del Álamo, A.J. Engler, situ mechanotransduction via vinculin regulates stem cell differentiation. Stem. Cells **31**(11), 2467–2477 (2013)

E.R. Horton, A. Byron, J.A. Askari et al., Definition of a consensus integrin adhesome and its dynamics during adhesion complex assembly and disassembly. Nat. Cell Biol. **17**(12), 1577–1587 (2015)

E.R. Horton, J.D. Humphries, B. Stutchbury, G. Jacquemet, C. Ballestrem, S.T. Barry, M.J. Humphries, Modulation of FAK and Src adhesion signaling occurs independently of adhesion complex composition. J. Cell Biol. **212**(3), 349–364 (2016)

K. Hu, L. Ji, K.T. Applegate, G. Danuser, C.M. Waterman-Storer, Differential transmission of actin motion within focal adhesions. Science **315**(5808), 111–115 (2007)

Y. Huang, R.N. Day, S.J. Gunst, Vinculin phosphorylation at Tyr1065 regulates vinculin conformation and tension development in airway smooth muscle tissues. J. Biol. Chem. **289**(6), 3677–3688 (2014)

J.D. Humphrey, E.R. Dufresne, M.R. Schwartz, Mechanotransduction and extracellular matrix homeostasis. Nat. Rev. Mol. Cell Biol. **15**, 802–812 (2014)

D.J. Humphries, A. Byron, M.J. Humphries, Integrin ligands at a glance. J. Cell Sci. **119**, 3901–3903 (2006)

J.D. Humphries, P. Wang, C. Streuli, B. Geiger, M.J. Humphries, C. Ballestrem, Vinculin controls focal adhesion formation by direct interactions with talin and actin. J. Cell Biol. **179**(5), 1043–1057 (2007)

S. Huveneers, J. de Rooij, Mechanosensitive systems at the cadherin-F-actin interface. J. Cell Sci. **126**(Pt 2), 403–413 (2013)

S. Huveneers, J. Oldenburg, E. Spanjaard, G. van der Krogt, I. Grigoriev, A. Akhmanova, H. Rehmann, J. de Rooij, Vinculin associates with endothelial VE-cadherin junctions to control force-dependent remodeling. J. Cell Biol. **196** (5), 641–652 (2012)

R.O. Hynes, Intergrins: bidirectional, allosteric signaling machines. Cell **110**, 673–687 (2002)

D.E. Ingber, Mechanobiology and diseases of mechanotransduction. Ann. Med. **35**, 564–577 (2003)

S. Ishihara, K. Sugimura, Bayesian inference of force dynamics during morphogenesis. J. Theor. Biol. **313**, 201–211 (2012)

T. Iskratsch, H. Wolfenson, M.P. Sheetz, Appreciating force and shape—the rise of mechanotransduction in cell biology. Nat. Rev. Mol. Cell Biol. **15**, 825–833 (2014)

S. Ito, D.K. Werth, N.D. Richert, I. Pastan, Vinculin phosphorylation by the src kinase. Interaction of vinculin with phospholipid vesicles. J. Biol. Chem. **258**(23), 14626–14631 (1983)

T. Izard, G. Evans, R.A. Borgon, C.L. Rush, G. Bricogne, P.R. Bois, Vinculin activation by talin through helical bundle conversion. Nature **427**(6970), 171–175 (2004)

C.S. Izzard, L.R. Lochner, Formation of cell-to-substrate contacts during fibroblast motility: an interference-reflexion study. J. Cell Sci. **42**, 81–116 (1980)

D.E. Jaalouk, J. Lammerding, Mechanotransduction gone awry. Nat. Rev. Mol. Cell Biol. **10**(1), 63–73 (2009)

K.M. Jannie, S.M. Ellerbroek, D.W. Zhou, S. Chen, D.J. Crompton, A.J. Garcia, K.A. DeMali, Vinculin-dependent actin bundling regulates cell migration and traction forces. Biochem. J. **465**(3), 383–393 (2015)

R. Janostiak, O. Tolde, Z. Bruhova, M. Novotny, S.K. Hanks, D. Rosel, J. Brabek, Tyrosine phosphorylation within the SH3 domain regulates CAS subcellular localization, cell migration, and invasiveness. Mol. Biol. Cell **22**(22), 4256–4267 (2011)

R. Janostiak, J. Brabek, V. Auernheimer et al., CAS directly interacts with vinculin to control mechanosensing and focal adhesion dynamics. Cell. Mol. Life Sci. **71**(4), 727–744 (2014)

K.A. Jansen, R.G. Bacabac, I.K. Piechocka, G.H. Koenderink, Cells actively stiffen fibrin networks by generating contractile stress. Biophys. J. **105**(10), 2240–2251 (2013)

K.A. Jansen, D.M. Donato, H.E. Balcioglu, T. Schmidt, E.H. Danen, G.H. Koenderink, A guide to mechanobiology: where biology and physicsmeet. Biochim. Biophys. Acta **1853**(11 Pt B), 3043–3052 (2015)

K.A. Jansen, P. Atherton, C. Ballestrem, Mechanotransduction at the cell-matrix interface. Semin. Cell Dev. Biol. **71**, 75–83 (2017)

M.E. Janssen, E. Kim, H. Liu, L.M. Fujimoto, A. Bobkov, N. Volkmann, D. Hanein, Three-dimensional structure of vinculin bound to actin filaments. Mol. Cell **21**(2), 271–281 (2006)

A. Jasaitis, M. Estevez, J. Heysch, B. Ladoux, S. Dufour, E-cadherin-dependent stimulation of traction force at focal adhesions via the Src and PI3K signaling pathways. Biophys. J. **103**(2), 175–184 (2012)

J.L. Kadrmas, M.C. Beckerle, The LIM domain: from the cytoskeleton to the nucleus. Nat. Rev. Mol. Cell Biol. **5**(11), 920–931 (2004)

P. Kanchanawong, G. Shtengel, A.M. Pasapera, E.B. Ramko, M.W. Davidson, H.F. Hess, C.M. Waterman, Nanoscale architecture of integrin-based cell adhesions. Nature **468**(7323), 580–584 (2010)

N. Kannan, V.W. Tang, Synaptopodin couples epithelial contractility to alpha-actinin-4-dependent junction maturation. J. Cell Biol. **211**(2), 407–434 (2015)

B.-Z. Katz, M. Zohar, H. Teramoto, K. Matsumoto, J.S. Gutkind, D.C. Lin, S. Lin, K.M. Yamada, Tensin can induce JNK and p38 activation. Biochem. Biophy. Res. Commun. **272**, 717–720 (2000)

M. Kelley, J. Yochem, M. Krieg et al., Fbn-1, a fibrillin-related protein, is required for resistance of the epidermis to mechanical deformation during *C. elegans* embryogenesis. Elife **4**, e06565 (2015)

N. Kioka, S. Sakata, T. Kawauchi, T. Amachi, S.K. Akiyama, K. Okazaki, C. Yaen, K.M. Yamada, S. Aota, Vinexin: a novel vinculin-binding protein with multiple SH3 domains enhances actin cytoskeletal organization. J. Cell Biol. **144**(1), 59–69 (1999)

B. Klapholz, S.L. Herbert, J. Wellmann, R. Johnson, M. Parsons, N.H. Brown, Alternative mechanisms for talin to mediate integrin function. Curr. Biol. **25**(7), 847–857 (2015)

K. Kobayashi, M. Sokabe, Sensing substrate rigidity by mechanosensitive ion channels with stress fibers and focal adhesions. Curr. Opin. Cell Biol. **22**(5), 669–676 (2010)

F. Kong, A.J. García, A.P. Mould, M.J. Humphries, C. Zhu, Demonstration of catch bonds between an integrin and its ligand. J. Cell Biol. **185**, 1275–1284 (2009)

D.E. Koser, A.J. Thompson, S.K. Foster et al., Mechanosensing is critical for axon growth in the developing brain. Nat. Neurosci. **19**(12), 1592–1598 (2016)

C.M. Kraning-Rush, S.P. Carey, J.P. Califano, B.N. Smith, C.A. Reinhart-King, The role of the cytoskeleton in cellular force generation in 2D and 3D environments. Phys. Biol. **8**(1), 015009 (2011)

M. Krieg, A.R. Dunn, M.B. Goodman, Mechanical control of the sense of touch by β-spectrin. Nat. Cell Biol. **16**, 224–233 (2014)

K.E. Kubow, S.K. Conrad, A.R. Horwitz, Matrix microarchitecture and myosin II determine adhesion in 3D matrices. Curr. Biol. **23**(17), 1607–1619 (2013)

A. Kumar, M. Ouyang, K. Van den Dries, E.J. McGhee, K. Tanaka, M.D. Anderson, A. Groisman, B.T. Goult, K.I. Anderson, M.A. Schwartz, Talin tension sensor reveals novel features of focal adhesion force transmission and mechanosensitivity. J. Cell Biol. **213**(3), 371–383 (2016)

J.-C. Kuo, X. Han, C.T. Hsiao, J.R. Yates III, C.M. Waterman, Analysis of the myosin-II-responsive focal adhesion proteome reveals a role for Pix in negative regulation of focal adhesion maturation. Nat. Cell Biol. **13**(4), 383–393 (2011)

K. Kupper, N. Lang, C. Mohl, N. Kirchgessner, S. Born, W.H. Goldmann, R. Merkel, B. Hoffmann, Tyrosine phosphorylation of vinculin at position 1065 modifies focal adhesion dynamics and cell tractions. Biochem. Biophys. Res. Commun. **399**(4), 560–564 (2010)

M. Kuroda, H. Wada, Y. Kimura, K. Ueda, N. Kioka, Vinculin promotes nuclear localization of taz to inhibit ECM stiffness-dependent differentiation into adipocytes. J. Cell Sci. **130**(5), 989–1002 (2017)

C. Lawson, D. Schlaepfer, Integrin adhesions: who's on first? What's on second? Connections between FAK and talin. Cell Adhes. Migr. **6**(4), 302–306 (2012)

C. Le Clainche, S.P. Dwivedi, D. Didry, M.F. Carlier, Vinculin is a dually regulated actin filament barbed end-capping and side-binding protein. J. Biol. Chem. **285**(30), 23420–23432 (2010)

Q. le Duc, Q. Shi, I. Blonk, A. Sonnenberg, N. Wang, D. Leckband, J. de Rooij, Vinculin potentiates E-cadherin mechanosensing and is recruited to actin-anchored sites within adherens junctions in a myosin II-dependent manner. J. Cell Biol. **189**(7), 1107–1115 (2010)

H.S. Lee, C.J. Lim, W. Puzon-McLaughlin, S.J. Shattil, M.H. Ginsberg, RIAM activates integrins by linking talin to ras GTPase membrane-targeting sequences. J. Biol. Chem. **284**(8), 5119–5127 (2009)

H.S. Lee, P. Anekal, C.J. Lim, C.C. Liu, M.H. Ginsberg, Two modes of integrin activation form a binary molecular switch in adhesion maturation. Mol. Biol. Cell **24**(9), 1354–1362 (2013)

J.M. Leerberg, G.A. Gomez, S. Verma, R. Priya, B.D. Hoffman, C. Grashoff, M.A. Schwartz, A.S. Yap, Tension-sensitive actin assembly supports contractility at the epithelial zonula adherens. Curr. Biol. **24**(15), 1689–1699 (2014)

C.X. Li, N.P. Talele, S. Boo, A. Koehler, E. Knee-Walden, J.L. Balestrini, P. Speight, A. Kapus, B. Hinz, MicroRNA-21 preserves the fibrotic mechanical memory of mesenchymal stem cells. Nat. Mater. **16**, 379–389 (2017)

S. Linder, The matrix corroded: podosomes and invadopodia in extracellular matrix degradation. Trends Cell Biol. **17** (3), 107–117 (2007)

J. Liu, Y. Wang, W.I. Goh, H. Goh, M.A. Baird, S. Ruehland, S. Teo, N. Bate, D.R. Critchley, M.W. Davidson, P. Kanchanawong, Talin determines the nanoscale architecture of focal adhesions. Proc. Natl. Acad. Sci. USA **112**, E4864–E4873 (2015)

C.-M. Lo, H.-B. Wang, M. Dembo, Y.-L. Wang, Cell movement is guided by the rigidity of the substrate. Biophys. J. **79**, 144–152 (2000)

P. Lu, K. Takai, V.M. Weaver, Z. Werb, Extracellular matrix degradation and remodeling in development and disease. Cold Spring Harb. Perspect. Biol. **3**(12), a005058 (2011)

P. Lu, V.M. Weaver, Z. Werb, The extracellular matrix: a dynamic niche in cancer progression. J. Cell Biol. **196**(4), 395–406 (2012)

A.A. Lucio, D.E. Ingber, O. Campàs, Generation of biocompatible droplets for in vivo and in vitro measurement of cell-generated mechanical stresses. Methods Cell Biol. **125**, 373–390 (2015)

X. Ma, M.E. Schickel, M.D. Stevenson, A.L. Sarang-Sieminski, K.J. Gooch, S.N. Ghadiali, R.T. Hart, Fibers in the extracellular matrix enable long-range stress transmission between cells. Biophys. J. **104**(7), 1410–1418 (2013)

A. Macdonald, A.R. Horwitz, D.A. Lauffenburger, Kinetic model for lamellipodal actin-integrin 'clutch' dynamics. Cell Adhes. Migr. **2**, 95–105 (2008)

K. Mandai, H. Nakanishi, A. Satoh, K. Takahashi, K. Satoh, H. Nishioka, A. Mizoguchi, Y. Takai, Ponsin/SH3P12: an l-afadin- and vinculin-binding protein localized at cell-cell and cell–matrix adherens junctions. J. Cell Biol. **144**(5), 1001–1017 (1999)

F. Margadant, L.L. Chew, X. Hu, H. Yu, N. Bate, X. Zhang, M. Sheetz, Mechanotransduction in vivo by repeated talin stretch-relaxation events depends upon vinculin. PLoS Biol. **9**(12), e1001223 (2011)

C. Martinez-Rico, F. Pincet, J.P. Thiery, S. Dufour, Integrins stimulate E-cadherin-mediated intercellular adhesion by regulating Src-kinase activation and actomyosin contractility. J. Cell Sci. **123**(Pt 5), 712–722 (2010)

V. Maruthamuthu, B. Sabass, U.S. Schwarz, M.L. Gardel, Cell-ECM traction force modulates endogenous tension at cell-cell contacts. Proc. Natl. Acad. Sci. USA **108**(12), 4708–4713 (2011)

M.L. McCain, H. Lee, Y. Aratyn-Schaus, A.G. Kleber, K.K. Parker, Cooperative coupling of cell–matrix and cell-cell adhesions in cardiac muscle. Proc. Natl. Acad. Sci. USA **109**(25), 9881–9886 (2012)

A. McGregor, A.D. Blanchard, A.J. Rowe, D.R. Critchley, Identification of the vinculin-binding site in the cytoskeletal protein α-actinin. Biochem. J. **301**, 225–233 (1994)

A.H. Mekhdjian, F. Kai, M.G. Rubashkin et al., Integrin-mediated traction force enhances paxillin molecular associations and adhesion dynamics that increase the invasiveness of tumor cells into a three-dimensional extracellular matrix. Mol. Biol. Cell **28**, 1467–1488 (2017)

A.F. Mertz, Y. Che, S. Banerjee, J.M. Goldstein, K.A. Rosowski, S.F. Revilla, C.M. Niessen, M.C. Marchetti, E.R. Dufresne, V. Horsley, Cadherin-based intercellular adhesions organize epithelial cell–matrix traction forces. Proc. Natl. Acad. Sci. USA **110**(3), 842–847 (2013)

C.T. Mierke, The role of vinculin in the regulation of the mechanical properties of cells. Cell Biochem. Biophys. **53**(3), 115–126 (2009)

C.T. Mierke, P. Kollmannsberger, D.P. Zitterbart, J. Smith, B. Fabry, W.H. Goldmann, Mechano-coupling and regulation of contractility by the vinculin tail domain. Biophys. J. **94**(2), 661–670 (2008)

C.T. Mierke, P. Kollmannsberger, D.P. Zitterbart, G. Diez, T.M. Koch, S. Marg, W.H. Ziegler, W.H. Goldmann, B. Fabry, Vinculin facilitates cell invasion into three-dimensional collagen matrices. J. Biol. Chem. **285**(17), 13121–13130 (2010)

S.K. Mitra, D.A. Hanson, D.D. Schlaepfer, Focal adhesion kinase: in command and control of cell motility. Nat. Rev. Mol. Cell Biol. **6**(1), 56–68 (2005)

S. Moese, M. Selbach, V. Brinkmann, A. Karlas, B. Haimovich, S. Backert, T.F. Meyer, The Helicobacter pylori CagA protein disrupts matrix adhesion of gastric epithelial cells by dephosphorylation of vinculin. Cell Microbiol. **9**(5), 1148–1161 (2007)

H. Mohammadi, C.A. McCulloch, Impact of elastic and inelastic substrate behaviors on mechanosensation. Soft Matter **10**(3), 408–420 (2014)

C. Mohl, N. Kirchgessner, C. Schafer, K. Kupper, S. Born, G. Diez, W.H. Goldmann, R. Merkel, B. Hoffmann, Becoming stable and strong: the interplay between vinculin exchange dynamics and adhesion strength during adhesion site maturation. Cell Motil. Cytoskeleton. **66**(6), 350–364 (2009)

L. Molony, K. Burridge, Molecular shape and self-association of vinculin and metavinculin. J. Cell. Biochem. **29**(1), 31–36 (1985)

S.J. Monkley, X.H. Zhou, S.J. Kinston et al., Disruption of the talin gene arrests mouse development at the gastrulation stage. Dev. Dyn. **219**, 560–574 (2000)

A.P. Mould, S.K. Akiyama, M.J. Humphries, Regulation of integrin alpha 5 beta 1-fibronectin interactions by divalent cations. Evidence for distinct classes of binding sites for Mn^{2+}, Mg^{2+}, and Ca^{2+}. J. Biol. Chem. **270**, 26270–26277 (1995)

J.P. Myers, M. Santiago-Medina, T.M. Gomez, Regulation of axonal outgrowth and pathfinding by integrin-ECM interactions. Dev. Neurobiol. **71**(11), 901–923 (2011)

S.Y. Nanda, T. Hoang, P. Patel, H. Zhang, Vinculin regulates assembly of talin: beta3 integrin complexes. J. Cell Biochem. **115**(6), 1206–1216 (2014)

C.M. Nelson, D.M. Pirone, J.L. Tan, C.S. Chen, Vascular endothelial-cadherin regulates cytoskeletal tension, cell spreading, and focal adhesions by stimulating RhoA. Mol. Biol. Cell **15**(6), 2943–2953 (2004)

D.H. Ng, J.D. Humphries, A. Byron, A. Millon-Fremillon, M.J. Humphries, Microtubule-dependent modulation of adhesion complex composition. PLoS ONE **9**, e115213 (2014)

C.D. Nobes, A. Hall, Rho, rac, and cdc42 GTPases regulate the assembly of multimolecular focal complexes associated with actin stress fibers, lamellipodia, and filopodia. Cell **81**(1), 53–62 (1995)

P.W. Oakes, D.C. Patel, N.A. Morin, D.P. Zitterbart, B. Fabry, J.S. Reichner, J.X. Tang, Neutrophil morphology and migration are affected by substrate elasticity. Blood **114**, 1387–1395 (2009)

P.W. Oakes, S. Banerjee, M.C. Marchetti, M.L. Gardel, Geometry regulates traction stresses in adherent cells. Biophys. J. **107**, 825–833 (2014)

P.W. Oakes, T.C. Bidone, Y. Beckham, A.V. Skeeters, G.R. Ramirez-San Juan, S.P. Winter, G.A. Voth, M.L. Gardel, Lamellipodium is a myosin-independent mechanosensor. PNAS **115**(11), 2646–2651 (2018)

T. Ohmori, Y. Kashiwakura, A. Ishiwata, S. Madoiwa, J. Mimuro, S. Honda, T. Miyata, Y. Sakata, Vinculin activates inside-out signaling of integrin alphaIIbbeta3 in Chinese hamster ovary cells. Biochem. Biophys. Res. Commun. **400**(3), 323–328 (2010)

A.W. Orr, B.P. Helmke, B.R. Blackman, M.A. Schwartz, Mechanisms of mechanotransduction. Dev. Cell **10**, 11–20 (2006)

C.A. Otey, F.M. Pavalko, K. Burridge, An interaction between α-actinin and the β1 integrin subunit in vitro. J. Cell Biol. **111**, 721–729 (1990)

E.K. Paluch, C.M. Nelson, N. Biais, B. Fabry, J. Moeller, B.L. Pruitt, C. Wollnik, G. Kudryasheva, F. Rehfeldt, W. Federle, Mechanotransduction: use the force(s). BMC Biol. **13**, 47 (2015)

R. Pankov, E. Cukierman, B.Z. Katz, K. Matsumoto, D.C. Lin, S. Lin, C. Hahn, K.M. Yamada, Integrin dynamics and matrix assembly: tensin-dependent translocation of alpha(5)beta(1) integrins promotes early fibronectin fibrillogenesis. J. Cell Biol. **148**(5), 1075–1090 (2000)

A.M. Pasapera, I.C. Schneider, E. Rericha, D.D. Schlaepfer, C.M. Waterman, Myosin II activity regulates vinculin recruitment to focal adhesions through FAK-mediated paxillin phosphorylation. J. Cell Biol. **188**(6), 877–890 (2010)

M.J. Paszek, N. Zahir, K.R. Johnson, J.N. Lakins, G.I. Rozenberg, A. Gefen, C.A. Reinhart-King, S.S. Margulies, M. Dembo, D. Boettiger, D.A. Hammer, V.M. Weaver, Tensional homeostasis and the malignant phenotype. Cancer Cell **8**, 241–254 (2005)

X. Peng, L.E. Cuff, C.D. Lawton, K.A. DeMali, Vinculin regulates cell-surface E-cadherin expression by binding to beta-catenin. J. Cell Sci. **123**(Pt 4), 567–577 (2010)

X. Peng, J.L. Maiers, D. Choudhury, S.W. Craig, K.A. DeMali, alpha-Catenin uses a novel mechanism to activate vinculin. J. Biol. Chem. **287**(10), 7728–7737 (2012)

S.V. Plotnikov, A.M. Pasapera, B. Sabass, C.M. Waterman, Force fluctuations within focal adhesions mediate ECM-rigidity sensing to guide directed cell migration. Cell **151**(7), 1513–1527 (2012)

M. Prager-Khoutorsky, A. Lichtenstein, R. Krishnan, K. Rajendran, A. Mayo, Z. Kam, B. Geiger, A.D. Bershadsky, Fibroblast polarization is a matrix-rigidity-dependent process controlled by focal adhesion mechanosensing. Nat. Cell Biol. **13**, 1457–1465 (2011)

E. Puklin-Faucher, M.P. Sheetz, The mechanical integrin cycle. J. Cell Sci. **122**(Pt 2), 179–186 (2009)

M. Raab, J. Swift, P.C. Dingal, P. Shah, J.W. Shin, D.E. Discher, Crawling from soft to stiff matrix polarizes the cytoskeleton and phosphoregulates myosin-II heavy chain. J. Cell Biol. **199**, 669–683 (2012)

K. Radovanac, J. Morgner, J.N. Schulz et al., Stabilization of integrin-linked kinase by theHsp90-CHIP axis impacts cellular force generation, migration and the fibrotic response. EMBO J. **32**(10), 1409–1424 (2013)

A. Rahman, S.P. Carey, C.M. Kraning-Rush, Z.E. Goldblatt, F. Bordeleau, M.C. Lampi, D.Y. Lin, A.J. Garcia, C.A. Reinhart-King, Vinculin regulates directionality and cell polarity in 2D, 3D matrix and 3D microtrack migration. Mol. Biol. Cell **27**(9), 1431–1441 (2016)

S. Ray, H.P. Foote, T. Lechler, Beta catenin protects the epidemis from mechanical stresses. J. Cell Biol. **202**(1), 45–52 (2013)

C.A. Reinhart-King, M. Dembo, D.A. Hammer, The dynamics and mechanics of endothelial cell spreading. Biophys. J. **89**, 676–689 (2005)

C.A. Reinhart-King, M. Dembo, D.A. Hammer, Cell-cell mechanical communication through compliant substrates. Biophys. J. **95**(12), 6044–6051 (2008)

S. Rhee, H. Jiang, C.H. Ho, F. Grinnell, Microtubule function in fibroblast spreading is modulated according to the tension state of cell-matrix interactions. Proc. Natl. Acad. Sci. USA **104**(13), 5425–5430 (2007)

A.J. Ridley, A. Hall, The small GTP-binding protein rho regulates the assembly of focal adhesions and actin stress fibers in response to growth factors. Cell **70**(3), 389–399 (1992)

P. Ringer, A. Weißl, A.L. Cost, A. Freikamp, B. Sabass, A. Mehlich, M. Tramier, M. Rief, C. Grashoff, Multiplexing molecular tension sensors reveals piconewton force gradient across talin-1. Nat. Methods **14**(11), 1090–1096 (2017)

D. Riveline, E. Zamir, N.Q. Balaban, U.S. Schwarz, T. Ishizaki, S. Narumiya, Z. Kam, B. Geiger, A.D. Bershadsky, Focal contacts as mechanosensors: externally applied local mechanical force induces growth of focal contacts by an mDia1-dependentand ROCK-independent mechanism. J. Cell Biol. **153**(6), 1175–1186 (2001)

J. Robertson, G. Jacquemet, A. Byron, M.C. Jones, S. Warwood, J.N. Selley, D. Knight, J.D. Humphries, M. J. Humphries, Defining the phospho-adhesome through the phosphoproteomic analysis of integrin signalling. Nat. Commun. **6**, 6265 (2015)

P. Roca-Cusachs, A. del Rio, E. Puklin-Faucher, N.C. Gauthier, N. Biais, M.P. Sheetz, Integrin-dependent force transmission to the extracellular matrix by α-actinin triggers adhesion maturation. Proc. Natl. Acad. Sci. USA **110** (15), E1361–E1370 (2013)

J.L. Rodriguez Fernandez, B. Geiger, D. Salomon, A. Ben-Ze'ev, Suppression of vinculin expression by antisense transfection confers changes in cell morphology, motility, and anchorage-dependent growth of 3T3 cells. J. Cell Biol. **122**(6), 1285–1294 (1993)

O. Rossier, V. Octeau, J.B. Sibarita et al., Integrins β1 and β3 exhibit distinct dynamic nanoscale organizations inside focal adhesions. Nat. Cell Biol. **14**, 1057–1067 (2012)

K. Rottner, A. Hall, J.V. Small, Interplay between Rac and Rho in the control of substrate contact dynamics. Curr. Biol. **9**(12), 640–648 (1999)

M.G. Rubashkin, L. Cassereau, R. Bainer, C.C. DuFort, Y. Yui, G. Ou, M.J. Paszek, M.W. Davidson, Y.Y. Chen, V.M. Weaver, Force engages vinculin and promotes tumor progression by enhancing PI3K activation of phosphatidylinositol(3,4,5)-triphosphate. Cancer Res. **74**(17), 4597–4611 (2014)

M.S. Rudnicki, H.A. Cirka, M. Aghvami, E.A. Sander, Q. Wen, K.L. Billiar, Nonlinear strain stiffening is not sufficient to explain how far cells can feel on fibrous protein gels. Biophys. J. **105**(1), 11–20 (2013)

X. Sai, K. Naruse, M. Sokabe, Activation of pp60(src) is critical for stretch-induced orienting response in fibroblasts. J. Cell Sci. **112**(Pt 9), 1365–1373 (1999)

L. Sapir, S. Tzlil, Talking over the extracellular matrix: how do cells communicate mechanically? Semin. Cell Dev. Biol. **71**, 99–105 (2017)

S.K. Sastry, K. Burridge, Focal adhesions: a nexus for intracellular signaling and cytoskeletal dynamics. Exp. Cell Res. **261**, 25–36 (2000)

R.M. Saunders, M.R. Holt, L. Jennings et al., Role of vinculin in regulating focal adhesion turnover. Eur. J. Cell Biol. **85**(6), 487–500 (2006)

Y. Sawada, M. Tamada, B.J. Dubin-Thaler, O. Cherniavskaya, R. Sakai, S. Tanaka, M.P. Sheetz, Force sensing by mechanical extension of the Src family kinase substrate p130Cas. Cell **127**, 1015–1026 (2006)

H.B. Schiller, C.C. Friedel, C. Boulegue, R. Fässler, Quantitative proteomics of the integrin adhesome show a myosin II-dependent recruitment of LIM domain proteins. EMBO Rep. **12**, 259–266 (2011)

H.B. Schiller, M.R. Hermann, J. Polleux et al., beta1- and alphav-class integrins cooperate to regulate myosin II during rigidity sensing of fibronectin-based microenvironments. Nat. Cell Biol. **15**(6), 625–636 (2013)

I. Schoen, B.L. Pruitt, V. Vogel, The Yin-Yang of rigidity sensing: how forces and mechanical properties regulate the cellular response to materials. Annu. Rev. Mater. Res. **43**, 589–618 (2013)

D.L. Scott, G. Diez, W.H. Goldmann, Protein-lipid interactions: correlation of a predictive algorithm for lipid-binding sites with three-dimensional structural data. Theor. Biol. Med. Model. **3**, 17 (2006)

B.M. Sefton, T. Hunter, E.H. Ball, S.J. Singer, Vinculin: a cytoskeletal target of the transforming protein of Rous sarcoma virus. Cell **24**(1), 165–174 (1981)

B. Serrels, M.C. Frame, FAK and talin: who is taking whom to the integrin engagement party? J. Cell Biol. **196**(2), 185–187 (2012)

B. Serrels, A. Serrels, V.G. Brunton, M. Holt, G.W. McLean, C.H. Gray, G.E. Jones, M.C. Frame, Focal adhesion kinase controls actin assembly via aFERM-mediated interaction with the Arp2/3 complex. Nat. Cell Biol. **9**(9), 1046–1056 (2007)

F. Sheikh, Y. Chen, X. Liang, A. Hirschy, A.E. Stenbit, Y. Gu, N.D. Dalton, T. Yajima, Y. Lu, K.U. Knowlton, K.L. Peterson, J.C. Perriard, J. Chen, Alpha-E-catenin inactivation disrupts the cardiomyocyte adherens junction, resulting in cardiomyopathy and susceptibility to wall rupture. Circulation **114**(10), 1046–1055 (2006)

K. Shen, C.E. Tolbert, C. Guilluy, V.S. Swaminathan, M.E. Berginski, K. Burridge, R. Superfine, S.L. Campbell, The vinculin C-terminal hairpin mediates F-actin bundle formation, focal adhesion, and cell mechanical properties. J. Biol. Chem. **286**(52), 45103–45115 (2011)

J.W. Smith, R.S. Piotrowicz, D. Mathis, A mechanism for divalent cation regulation of beta 3-integrins. J. Biol. Chem. **269**, 960–967 (1994)

M.A. Smith, L.M. Hoffman, M.C. Beckerle, LIM proteins in actin cytoskeleton mechanoresponse. Trends Cell Biol. **24** (10), 575–583 (2014)

J.G. Snedeker, A. Gautieri, The role of collagen crosslinks in ageing and diabetes—the good, the bad, and the ugly. Muscles Ligaments Tendons J. **4**(3), 303–308 (2014)

C. Storm, F.C. MacKintosh, T.C. Lubensky, P.A. Janmey, Nonlinear elasticity in biological gels. Nature **435**(7039), 191–194 (2005)

C.H. Streuli, Integrins and cell-fate determination. J. Cell Sci. **122**, 171–177 (2009)

B. Stutchbury, P. Atherton, R. Tsang, D.-Y. Wang, C. Ballestrem, Distinct focal adhesion protein modules control different aspects of mechanotransduction. J. Cell Sci. **130**, 1612–1624 (2017)

Z. Sun, H.Y. Tseng, S. Tan et al., Kank2 activates talin, reduces force transduction acrossintegrins and induces central adhesion formation. Nat. Cell Biol. **18**(9), 941–953 (2016)

D.M. Suter, L.D. Errante, V. Belotserkovsky, P. Forscher, The Ig superfamily cell adhesion molecule, apCAM, mediates growth cone steering by substrate-cytoskeletal coupling. J. Cell Biol. **141**, 227–240 (1998)

N. Tawil, P. Wilson, S. Carbonetto, Integrins in point contacts mediate cell spreading: factors that regulate integrin accumulation in point contacts vs. focal contacts. J. Cell Biol. **120**(1), 261–271 (1993)

S.-Y. Tee, J. Fu, C.S. Chen, P.A. Janmey, Cell shape and substrate rigidity both regulate cell stiffness. Biophys. J. **100**, L25–L27 (2011)

M. Theodosiou, M. Widmaier, R.T. Bottcher et al., Kindlin-2cooperates with talin to activate integrins and induces cell spreading by directly binding paxillin Fassler. Elife **5**, 10130 (2016)

J.P. Thiery, Cell adhesion in development: a complex signaling network. Curr. Opin. Genet. Dev. **13**, 365–371 (2003)

I. Thievessen, P.M. Thompson, S. Berlemont, K.M. Plevock, S.V. Plotnikov, A. Zemljic-Harpf, R.S. Ross, M.W. Davidson, G. Danuser, S.L. Campbell, C.M. Waterman, Vinculin-actin interaction couples actin retrograde flow to focal adhesions, but is dispensable for focal adhesion growth. J. Cell Biol. **202**(1), 163–177 (2013)

I. Thievessen, N. Fakhri, J. Steinwachs et al., Vinculin is required for cell polarization, migration, and extracellular matrix remodeling in 3D collagen. FASEB J. **29**(11), 4555–4567 (2015)

W.A. Thomas, C. Boscher, Y.S. Chu et al., alpha-Catenin and vinculin cooperate to promote high E-cadherin-based adhesion strength. J. Biol. Chem. **288**(7), 4957–4969 (2013)

P.M. Thompson, C.E. Tolbert, K. Shen et al., Identification of an actin binding surface on vinculin that mediates mechanical cell and focal adhesion properties. Structure **22**(5), 697–706 (2014)

P.M. Thompson, S. Ramachandran, L.B. Case, C.E. Tolbert, A. Tandon, M. Pershad, N.V. Dokholyan, C.M. Waterman, S.L. Campbell, A structural model for vinculin insertion into PIP2-containing membranes and the effect of insertion on vinculin activation and localization. Structure **25**(2), 264–275 (2017)

C.E. Tolbert, K. Burridge, S.L. Campbell, Vinculin regulation of F-actin bundle formation: what does it mean for the cell? Cell Adh. Migr. **7**(2), 219–225 (2013)

C.E. Tolbert, P.M. Thompson, R. Superfine, K. Burridge, S.L. Campbell, Phosphorylation at Y1065 in vinculin mediates actin bundling, cell spreading, and mechanical responses to force. BioChemistry **53**(34), 5526–5536 (2014)

M. Toutant, A. Costa, J.M. Studler, G. Kadaré, M. Carnaud, J.A. Girault, Alternative splicing controls the mechanisms of FAK autophosphorylation. Mol. Cell. Biol. **22**(22), 7731–7743 (2002)

V.D. Varner, D.A. Voronov, L.A. Taber, Mechanics of head fold formation: investigating tissue-level forces during early development. Development **137**, 3801–3811 (2010)

M. Vicente-Manzanares, A.R. Horwitz, Adhesion dynamics at a glance. J. Cell Sci. **124**(Pt 23), 3923–3927 (2011)

M. Vicente-Manzanares, C.K. Choi, A.R. Horwitz, Integrins in cell migration-the actin connection. J. Cell Sci. **122**, 199–206 (2009)

H.B. Wang, M. Dembo, Y.L. Wang, Substrate flexibility regulates growth and apoptosis of normal but not transformed cells. Am. J. Physiol. Cell Physiol. **279**, C1345–C1350 (2000)

H.B. Wang, M. Dembo, S.K. Hanks, Y. Wang, Focal adhesion kinase is involved in mechanosensing during fibroblast migration. Proc. Natl. Acad. Sci. USA **98**(20), 11295–11300 (2001a)

J.G. Wang, M. Miyazu, E. Matsushita, M. Sokabe, K. Naruse, Uniaxial cyclic stretch induces focal adhesion kinase (FAK) tyrosine phosphorylation followed by mitogen-activated protein kinase (MAPK) activation. Biochem. Biophys. Res. Commun. **288**(2), 356–361 (2001b)

J.G. Wang, M. Miyazu, P. Xiang, S.N. Li, M. Sokabe, K. Naruse, Stretch-induced cell proliferation is mediated by FAK-MAPK pathway. Life Sci. **76**(24), 2817–2825 (2005)

M. Watabe-Uchida, N. Uchida, Y. Imamura, A. Nagafuchi, K. Fujimoto, T. Uemura, S. Vermeulen, F. van Roy, E.D. Adamson, M. Takeichi, alpha-Catenin-vinculin interaction functions to organize the apical junctional complex in epithelial cells. J. Cell Biol. **142**(3), 847–857 (1998)

G.F. Weber, M.A. Bjerke, D.W. DeSimone, Integrins and cadherins join forces to form adhesive networks. J. Cell Sci. **124**(Pt 8), 1183–1193 (2011)

J. Weekes, S.T. Barry, D.R. Critchley, Acidic phospholipids inhibit the intramolecular association between the N- and C-terminal regions of vinculin, exposing actin-binding and protein kinase C phosphorylation sites. Biochem. J. **314** (Pt 3), 827–832 (1996)

Q. Wen, P.A. Janmey, Polymer physics of the cytoskeleton. Curr. Opin. Solid State Mater. Sci. **15**(5), 177–182 (2011)

K.K. Wen, P.A. Rubenstein, K.A. DeMali, Vinculin nucleates actin polymerization and modifies actin filament structure. J. Biol. Chem. **284**(44), 30463–30473 (2009)

V.F. Wirth, F. List, G. Diez, W.H. Goldmann, Vinculin's C-terminal region facilitates phospholipid membrane insertion. Biochem. Biophys. Res. Commun. **398**(3), 433–437 (2010)

M.A. Wozniak, C.S. Chen, Mechanotransduction in development: a growing role for contractility. Nat. Rev. Mol. Cell Biol. **10**, 34–43 (2009)

W. Xu, H. Baribault, E.D. Adamson, Vinculin knockout results in heart and brain defects during embryonic development. Development **125**(2), 327–337 (1998a)

W. Xu, J.L. Coll, E.D. Adamson, Rescue of the mutant phenotype by reexpression of full-length vinculin in null F9 cells; effects on cell locomotion by domain deleted vinculin. J. Cell Sci. **111**(Pt 11), 1535–1544 (1998b)

X.-P. Xu, E. Kim, M. Swift, J.W. Smith, N. Volkmann, D. Hanein, Three-dimensional structures of full-length, membrane-embedded human α(IIb)β(3) integrin complexes. Biophys. J. **110**, 798–809 (2016)

H. Yamashita, T. Ichikawa, D. Matsuyama, Y. Kimura, K. Ueda, S.W. Craig, I. Harada, N. Kioka, The role of the interaction of the vinculin proline-rich linker region with vinexin alpha in sensing the stiffness of the extracellular matrix. J. Cell Sci. **127**(Pt 9), 1875–1886 (2014)

S. Yamashita, T. Tsuboi, N. Ishinabe, T. Kitaguchi, T. Michiue, Wide and high resolution tension measurement using FRET in embryo. Sci. Rep. **6**, 28535 (2016)

J. Yan, M. Yao, B.T. Goult, M.P. Sheetz, Talin dependent mechanosensitivity of cell focal adhesions. Cell Mol. Bioeng. **8**, 151–159 (2015)

C. Yang, M.W. Tibbitt, L. Basta, K.S. Anseth, Mechanical memory and dosing influence stem cell fate. Nat. Mater. **13**, 645–652 (2014a)

J. Yang, L. Zhu, H. Zhang et al., Conformational activation of talin by RIAM triggers integrin-mediated cell adhesion. Nat. Commun. **5**, 5880 (2014b)

M. Yao, B.T. Goult, H. Chen, P. Cong, M.P. Sheetz, J. Yan, Mechanical activation of vinculin binding to talin locks talin in an unfolded conformation. Sci. Rep. **4**, 4610 (2014a)

M. Yao, W. Qiu, R. Liu, A.K. Efremov, P. Cong, R. Seddiki, M. Payre, C.T. Lim, B. Ladoux, R.M. Mege, J. Yan, Force-dependent conformational switch of alpha-catenin controls vinculin binding. Nat. Commun. **5**, 4525 (2014b)

M. Yao, B.T. Goult, B. Klapholz, X. Hu, C.P. Toseland, Yingjian Guo, Peiwen Cong, M.P. Sheetz, J. Yan, The mechanical response of talin. Nat. Commun. **7**, 11966 (2016)

T. Yeung, P.C. Georges, L.A. Flanagan et al., Effects of substrate stiffness on cell morphology, cytoskeletal structure, and adhesion. Cell Motil. Cytoskeleton. **60**, 24–34 (2005)

S. Yonemura, Y. Wada, T. Watanabe, A. Nagafuchi, M. Shibata, alpha-Catenin as a tension transducer that induces adherens junction development. Nat. Cell Biol. **12**(6), 533–542 (2010)

H. Youssoufian, M. McAfee, D.J. Kwiatkowski, Cloning and chromosomal localization of the human cytoskeletal α-actinin gene reveals linkage to the β-spectrin gene. Am. J. Hum. Genet. **47**, 62–72 (1990)

R. Zaidel-Bar, B. Geiger, The switchable integrin adhesome. J. Cell Sci. **123**, 1385–1388 (2010)

R. Zaidel-Bar, C. Ballestrem, Z. Kam, B. Geiger, Early molecular events in the assembly of matrix adhesions at the leading edge of migrating cells. J. Cell Sci. **116**(Pt 22), 4605–4613 (2003)

R. Zaidel-Bar, M. Cohen, L. Addadi, B. Geiger, Hierarchical assembly of cell-matrix adhesion complexes. Biochem. Soc. Trans. **32**(Pt3), 416–420 (2004)

R. Zaidel-Bar, S. Itzkovitz, A. Maayan, R. Iyengar, B. Geiger, Functional atlas of the integrin adhesome. Cell Biol. **9**(8), 858–867 (2007)

E. Zamir, B. Geiger, Molecular complexity and dynamics of cell–matrix adhesions. J. Cell Sci. **114**(Pt 20), 3583–3590 (2001)

E. Zamir, B.Z. Katz, S. Aota, K.M. Yamada, B. Geiger, Z. Kam, Molecular diversity of cell-matrix adhesions. J. Cell Sci. **112**(Pt 11), 1655–1669 (1999)

E. Zamir, M. Katz, Y. Posen, N. Erez, K.M. Yamada, B.Z. Katz, S. Lin, D.C. Lin, A. Bershadsky, Z. Kam, B. Geiger, Dynamics and segregation of cell-matrix adhesions in cultured fibroblasts. Nat. Cell Biol. **2**(4), 191–196 (2000)

A.E. Zemljic-Harpf, J.C. Miller, S.A. Henderson, A.T. Wright, A.M. Manso, L. Elsherif, N.D. Dalton, A.K. Thor, G.A. Perkins, A.D. McCulloch, R.S. Ross, Cardiac-myocyte-specific excision of the vinculin gene disrupts cellular junctions, causing sudden death or dilated cardiomyopathy. Mol. Cell Biol. **27**(21), 7522–7537 (2007)

Z. Zhang, G. Izaguirre, S.Y. Lin, H.Y. Lee, E. Schaefer, B. Haimovich, The phosphorylation of vinculin on tyrosine residues 100 and 1065, mediated by SRC kinases, affects cell spreading. Mol. Biol. Cell **15**(9), 4234–4247 (2004)

X. Zhang et al., Talin depletion reveals independence of initial cell spreading from integrin activation and traction. Nat. Cell Biol. **10**(9), 1062–1068 (2008)

J. Zhou, S. Pal, S. Maiti, L.A. Davidson, Force production and mechanical accommodation during convergent extension. Development **142**, 692–701 (2015)

W.H. Ziegler, U. Tigges, A. Zieseniss, B.M. Jockusch, A lipid-regulated docking site on vinculin for protein kinase C. J. Biol. Chem. **277**(9), 7396–7404 (2002)

W.H. Ziegler, A.R. Gingras, D.R. Critchley, J. Emsley, Integrin connections to the cytoskeleton through talin and vinculin. Biochem. Soc. Trans. **36**, 235–239 (2008)

B. Zimerman, T. Volberg, B. Geiger, Early molecular events in the assembly of the focal adhesion-stress fiber complex during fibroblast spreading. Cell Motil Cytoskeleton **58**(3), 143–159 (2004)

Structure and Function of the Mitochondrion

5

Abstract

This chapter introduces the organelle mitochondrion. The structure and function are outlined, described and related to the mechanical properties of the cell. It also explores basic structures, concepts and processes of mitochondria. In detail, it is presented how mitochondria function by fusion and fission. Besides the description of mitochondria in its static state, the dynamic state is included that plays a role in processes undergoing dynamically remodeling, such as cell migration and invasion. In summary, the description of mitochondria in this chapter forms the basis for understanding of Chap. 6 of the Cellular Mechanics and Biophysics book dealing with the mechanical properties of mitochondria.

5.1 Introduction to the Mitochondrion

Mitochondria are organelles with a rod-like morphology and supply the cells with energy by serving as a power generator. They convert oxygen and nutrients into adenosine triphosphate (ATP), which stores the chemical energy of the cell required for cell metabolism. The process is known as aerobic respiration and is the reason why some organisms (complex animals such as humans) breathe oxygen. Higher animals cannot exist without aerobic respiration. Another option of respiration is the anaerobic respiration, which is less efficient than aerobic respiration because aerobic respiration results in 15 times higher ATP levels. This high amount of ATP enables these complex animals to survive because they require large amounts of energy. In particular, the number of mitochondria in a cell is adapted to the special energy and hence metabolic requirements of the cells. A cell can have one mitochondrion or more than thousands of mitochondria. Mitochondria occur in animals, plants, fungi and protists and have been observed under a conventional light microscope. The isolation of mitochondria from cells leads to an understanding of mitochondrial function. The structure of the mitochondrion is important for its function. As for all organelles, the two specialized membranes represent the boundaries of the mitochondrion. The organelles are divided into a narrow intermembrane space and a wider internal matrix, both consisting of highly specialized proteins (Fig. 5.1).

C. T. Mierke, *Cellular Mechanics and Biophysics*, Biological and Medical Physics, Biomedical Engineering, https://doi.org/10.1007/978-3-030-58532-7_5

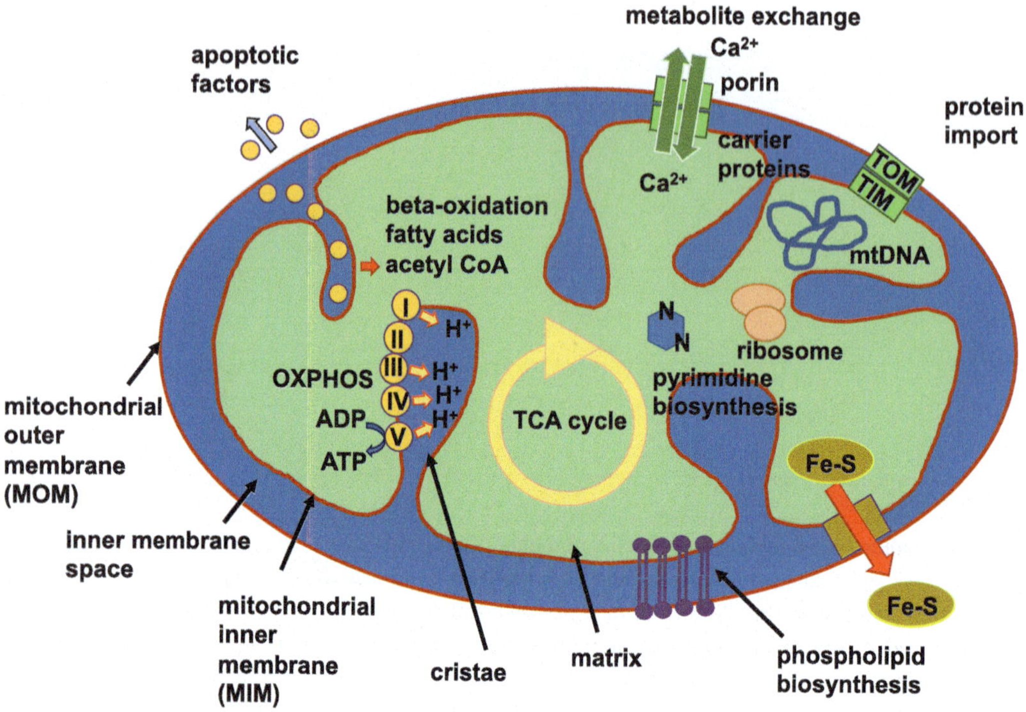

Fig. 5.1 Structure and membrane compartments in the mitochondrion. The outer membrane covers the mitochondrial matrix from the cytoplasm, and the inner membrane contains several cristae, at which the mitochondrial energy is generated

5.2 Mitochondrial Shape and Function in Energy Production

Two billion years ago, mitochondria are developed evolutionary by the engulfment of an α-proteobacterium by an early eukaryotic cell precursor (Lane and Martin 2010). The mitochondria have retained the double membrane of their ancestors and the basic ATP production, while their entire shape and composition have been pronouncedly altered, and they have additionally claimed a myriad of cellular functions. Through the process of gaining new functions during evolution, most of the genomic material of the α-proteobacterium progenitor was wasted quickly or even transferred to the nuclear genome of the former host (Gabaldón and Huynen 2004).

In human cells, a small, circular genome of approximately 16 kilobases is left, which is present in cells in a vast abundance of copies compared to nuclear chromosomes. The human mitochondrial genome stores genetic-coding information for 13 proteins that are central components of the mitochondrial respiratory complexes I–IV incorporated in the inner membrane. The respiratory chain, together with the Krebs' cycle, generates an electrochemical gradient in the matrix through the joint transfer of electrons to oxygen and protons from the matrix via the inner membrane into the intermembrane space. The electrochemical gradient drives the end complex V of the chain, ATP synthase, an old rotary turbine machine that induces the synthesis of most cellular ATP. The electrochemical potential is exploited for key mitochondrial functions, such as buffering the signal ion Ca^{2+} through a uniporter-based uptake in the inner membrane (Baughman et al. 2011; De Stefani et al. 2011).

A reduction in the electrochemical potential of mitochondria in cells has emerged as a measure of the mitochondrial functional status, which transduces signals to activate pathways to repair or eliminate defective mitochondria.

Proteomics, genomics and bioinformatics revealed that mitochondria consist of over 1000 proteins; the composition is plastic and varies within and across species in response to cellular and tissue-specific organismal demands (Pagliarini et al. 2008; Sickmann et al. 2003; Forner et al. 2006). The origin of the mitochondrial proteome is a combination of bacterial (old) and eukaryotic (new) proteins (Gabaldón and Huynen 2004). The replication and translation of machineries for mitochondrial DNA (mtDNA) differ in their evolutionary origins in bacteriophages (Lecrenier et al. 1997; Stumpf et al. 2011; Tiranti et al. 1997), while the mitochondrial translational mechanism exhibits a distinctive evolutionary relationship to bacteria (Christian and Spremulli 2012). In supplement to protein components, the mitochondrial genome codifies 22 transfer RNAs and two mitochondrial ribosome-coding RNAs, essential components of the genome's own translation apparatus. The mitochondrial ribosome arrangement in the mitochondrial matrix is a fairly complex and highly regulated process comprising the processing and maturation of mitochondrial ribosomal encoding RNA and the assembly of mitochondrial ribosomal proteins into small and large subunits (Fung et al. 2013). Nevertheless, there is only a fraction of mitochondrial ribosome proteins that possess identifiable homologs in bacteria (Sharma et al. 2003). The roles of mitochondrial specific ribosomal proteins are not comprehended, but it is assumed that these proteins were designed to coordinate mitochondrial translation with extra-mitochondrial pathways occurring in eukaryotic cells. As with many mitochondrial machines, the ribosome represents a mixture of old and new innovations (Friedman and Nunnari 2014).

The nuclear-coded proteins constituting the major of the mitochondrial proteome are translated on cytosolic ribosomes and actively imported by outer and inner membrane translocase machines and sorted into mitochondrial subdomains depending on the electrochemical potential (Neupert and Herrmann 2007; Schmidt et al. 2010). Transcriptional, posttranscriptional and post-translational regulatory mechanisms exist for nuclear-coded mitochondrial proteins. In humans, transcriptional regulation of mitochondrial biogenesis is achieved through the action of the PGC-1 family of co-activators, which react to alterations in nutrient status, such as NAD+/NADH and AMP/ATP ratios (both recorded by SIRT1 and AMPK) and environmental factors (Jäger et al. 2007; Jeninga et al. 2010). Interactions between PGC-1 co-activators and specific transcription factors, such asNRF1, NRF2 and ERR, balance and determine the key functional pathways in mitochondria. These interactions coordinate the nuclear and mitochondrial genomes through the induction of nuclear genes that directly affect the preservation of mtDNA (Scarpulla et al. 2012). The evidence in yeast indicated that nuclear transcribed messenger RNAs coding for mitochondrial proteins are post-transcribed in a highly regulated spatial and temporal manner at the mitochondrial outer membrane and translated coordinately (Gadir et al. 2011; Garcia et al. 2007). The knowledge of pathways underlying molecular mechanisms of mRNA targeting to mitochondria is suggested to be crucial in polarized cells such as neurons. Cytosolic kinases are capable for post-translational modifications, such as phosphorylation, of import machinery for mitochondrial components and hence can even fine-tune the proteome due to metabolic requirements (Schmidt et al. 2011).

Mutations in mtDNA genes or nuclear genes encoding the mitochondrial proteins essential for aerobic production of ATP evoke a multifaceted and often disastrous array of human mitochondrial diseases that can target any organ in the body at any time in a person's life (Nunnari and Suomalainen 2012). Mitochondrial diseases are characterized by a large degree of clinical heterogeneity. Some of these heterogeneities can be attributed to the fact that human cells may have a variable ratio of mutant and wild mtDNA, a condition known as heteroplasia. This applies to mtDNA mutations in protein-coding regions of the mitochondrial genome, where an increase in mutant workload leads to more severe disease phenotypes. Disease heterogeneity, which cannot be attributed to heteroplasia, also appears when mutations are detected in non-coding mitochondrial tRNAs. In particular, mutations in genes with a common function, such as genes encoding subunits of complex I of the respiratory chain, manifest very different diseases, such as optic nerve atrophy in adults or encephalopathy in infants (Wallace et al. 1988; Morris et al. 1996). There appears to be a correlation between mutations in mtDNA and aging, possibly due to mtDNA-bound defects in somatic stem cells (Ross et al. 2013; Ahlqvist et al. 2012).

The role of mitochondria in diseases has been broadened beyond the respiratory chain, since mitochondrial aberrations in their functions and behaviors have been associated with cancer, metabolic disorders and neurodegenerative diseases, including Alzheimer's, Parkinson's and Huntington's disease (Nunnari and Suomalainen 2012). In any case, generally speaking, our current knowledge on the underlying connection between the mitochondrial phenotype and disease is limited and warrants a better comprehension of mitochondrial organization and of the compounds that mitochondria exhibit with the nuclear genome and extra-mitochondrial pathways in different cell types and at the organismic level. To remedy this deficit, a revival in mitochondrial research has begun, accelerated by recent advances in genetics, systemic approaches and our capacity to image mitochondria in high temporal and spatial resolution.

5.3 Mitochondrial Chromosome

In view of the relevance of mtDNA-encoded genes for mitochondrial function, it is not astonishing that there are special mechanisms controlling the structure and distribution of mitochondria and mtDNA actively; whereas in higher eukaryotes, these mechanisms differ from those of their ancestors (Nunnari and Suomalainen 2012). Dissimilar to bacteria, in the majority of the cell types, individual mitochondria are not detectable, in contrast, they embrace an interconnected network that comprises multiple copies of the mitochondrial chromosome and forms a syncytium (Fig. 5.2). Similar to bacterial and nuclear chromosomes, mtDNA is highly concentrated within the mitochondrial matrix, and therefore mtDNA–protein complexes can be seen as punctate structures in these networks, which are known as nucleoids. The mechanism of mtDNA condensation has been elucidated by researchers who have revealed the crystal structure of the most common mtDNA-associated protein in mammalian cells, the mitochondrial transcription factor A (TFAM). The structure demonstrates that TFAM both binds and bends short sections of mtDNA and forms loops facilitating mtDNA packaging (Ngo et al. 2011; Rubio-Cosials et al. 2011). TFAM also fulfills a crucial role in mtDNA transcription, and its expression guidance of mtDNA copy levels in cells renders it a key regulator in the maintenance and transmission of mtDNA (Ekstrand et al. 2004; Shi et al. 2012).

Additional proteins, which are decisive for the maintenance of mtDNA, are pinpointed on nucleoids. Among these proteins are replication and repair machines, which in humans contain DNA polymerase γ and its secondary proteins, such as the replication helicase twinkle (Bogenhagen 2012). Mutations in genes coding for these and additional factors essential for mtDNA maintenance result in a wide range of human mitochondrial diseases, even though it is at the molecular level not clear how these and the multiple other identified mtDNA-associated proteins are assembled and organized to form nucleoids (Copeland 2012). For many species, proteomic populations of mtDNA-associated proteins have been identified (Bogenhagen et al. 2003; Kaufman et al. 2000; He et al. 2012). Both the higher-order nucleoid organization and the mode of mtDNA replication are varying between biological kingdoms, leading to greater complexity. All these differences are a direct result of the components of the genome and nucleoid. The yeast, for instance, owns an active recombination machine in the same way as Rad52 recombination proteins, and replication is expected to be primarily achieved by recombination (Mbantenkhu et al. 2011). Compared to the human mitochondrial

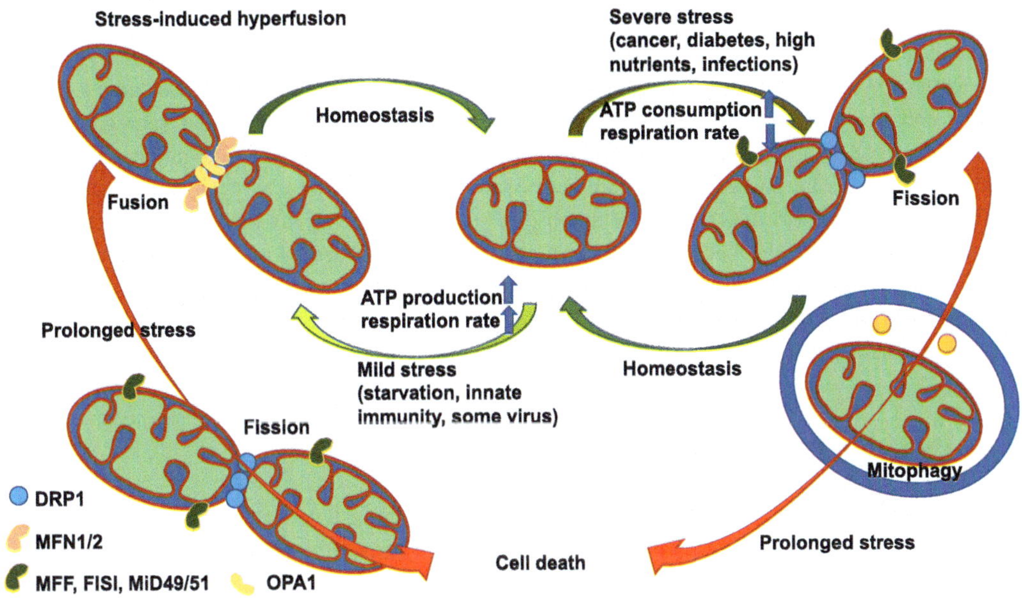

Fig. 5.2 Fission and fusion of mitochondria due to environmental cues, such as severe or mild external stresses

genome, the larger 80 kilobase yeast genome is therefore packaged in several copies within the nucleoid. Conversely, the mechanism of human mtDNA replication is in most tissues independent of recombination and is achieved by strand shifting (Holt and Reyes 2012, Brown et al. 2008). High-resolution imaging repeatedly demonstrates that human nucleoids retain a comparatively small number of mtDNA molecules and are therefore more solitary (Brown et al. 2011; Kukat et al. 2011). These differences in the organization and transmission modes of the mitochondrial chromosome have a major impact on the segregation behavior of mtDNA.

5.4 Segregation Modes of the mtDNA

Multiple copying of mtDNA denotes that the mode of mtDNA transmission is considered stochastic or relaxed in most cell types and differs radically from that of nulcear genes (Birky 2001). At organismic level, including humans, the mtDNA is inherited in a uniparentally on the mother's side and paternal mtDNA undergoes active destruction after fertilization (Al Rawi et al. 2011; Sato and Sato 2011). Mitochondria in oocytes and sperm vary in functional states, morphology and cellular distributions, and these variations are presumably important to increase fitness. In other respects, the mtDNA genotypes separate rapidly from generation to generation to eliminate severely damaged mitochondria and/or mtDNA. In line with this, oocytes have a bottleneck partly due to the way mtDNA is replicated and as a result of mitochondrial organization. The oocyte mitochondria are arranged in a temporary structure, which is known as Balbiani body, consisting of other organelles, such as the endoplasmic reticulum (ER) and Golgi apparatus, however the biogenesis of this structure is barely known (Pepling et al. 2007). When reprogramming fibroblasts to induced pluripotent stem cells, heteroplasmic mtDNA genotypes also separate through a bottleneck and mitochondria are clustered into a Balbiani-like structure (Hämäläinen et al. 2013), indicating that induced pluripotent stem cells have the potential to be a useful instrument to explore the mechanisms of mtDNA genotype segregation during development on cellular and molecular length scales.

Nuclear genes are replicated during a S-phase of the cell cycle and segregated by the combined action of a microtubule-based spindle apparatus and an actin-driven cytokinesis machinery, which lead to the physical partition of the chromosomes into the two daughter cells. Bacterial cells also exhibit cell cycle mechanisms to synchronize cell division with chromosome segregation by a tubulin-like FtsZ cell division mechanism. In contrast, the replication and segregation of mitochondrial chromosomes in most eukaryotes are not necessarily linked to the cell cycle and hence mtDNA replication can be performed for solely a subset of nucleoids at various time points (Meeusen and Nunnari 2003). Since the bacterial cytoskeletal machinery has been lost during the evolution of mitochondria, a question can be raised: What mechanisms are employed to locate sites of division and segregate mtDNA? These mechanisms are likely to be relevant to understanding the cell- and tissue-specific mechanisms underlying diseases associated with specific mtDNA mutations.

5.5 Dynamin-Facilitated Mitochondrial Dynamics

In higher eukaryotes, the segregation of mtDNA on the cellular level relies in part on continuous division and fusion events (Figure 5.3), whose rates tend to respond to the demands of a distinct cell type (Hoppins et al. 2007). A fundamental role of mitochondrial fusion is in facilitating communication between organelles, possibly to provide access to mtDNA expression products (Chen et al. 2007, 2010; Hermann 1998). Mitochondrial fusion is a mechanism to cushion transient defects occurring in mitochondria (Rolland et al. 2013). The division of mitochondria counteracts the fusion-driven network mitochondrial assembly to promote their distribution and transport by motor proteins on cytoskeletal pathways to and from various distinct locations (Verstreken et al. 2005). An equilibrium between division and fusion is crucial for the distribution and conservation of mtDNA. The loss of mitochondrial fusion leads to a fragmentation of a normally connected network into numerous small mitochondria due to the non-consistent division and the loss of mtDNA either in whole or in part in cells with the consequent severe defects in oxidative phosphorylation (Hermann 1998; Chen et al. 2005). The reduction of mitochondrial division induces the elongation of mitochondria and the assembly of highly cross-linked reticular structures, along with defects in oxidative phosphorylation and the loss of mtDNA during cell division (Ishihara et al. 2009; Parone et al. 2008; Wakabayashi et al. 2009; Hanekamp et al. 2002). The connection between mitochondrial dynamics and mtDNA transmission is in line with the elementary role of dynamics in mitochondrial copy control. The more distributive character of mitochondrial division combined with opposite fusion has therefore advanced to substitute the old bacterial cytoskeletal machines.

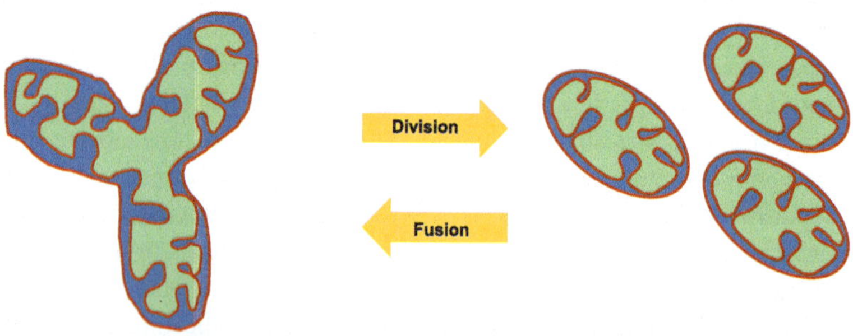

Fig. 5.3 Fission and fusion of mitochondria are highly conserved and contribute to the mitochondrial organization of higher eukaryotes

The mitochondrial division and fusion events are driven by the highly conserved dynamin-related proteins (DRPs), which, through their ability to self-assemble and hydrolyze GTP themselves, assist the remodeling of various intracellular membranes (Faelber et al. 2013). The division of mitochondria is catalyzed by a single DRP, the DRP1 in mammals and Dnm1 in yeast. DRP1 and Dnm1 build up helical structures through stalk domains that are twines around the outer surface of mitochondria at the constriction sites, whose diameters are identical to the diameters of the division helix (Ingerman et al. 2005; Labrousse et al. 1999). Moreover, interactions between the GTPase domains of division DRPs spiral-shaped sprouts promote the hydrolysis of GTP, which seem to be designated for conformational alterations transmitted by the DRP helix to enable coordinate division of the outer and inner membranes (Ford et al. 2011; Fröhlich et al. 2013; Faelber et al. 2011). The fusion of the outer and inner membranes of the mitochondria necessitates two evolutionarily distinct integral membrane proteins, DRPs, which are synonymously termed Mitofusin (MFN)1/MFN2 in mammals or Fzo1 in yeast, and OPA1 in mammals (Mgm1 in yeast) necessary (Meeusen et al. 2004). It is still largely unclear how the fusion of the DRPs is carried out at a mechanistic level, despite the fact that interactions between the GTPase domains on opposite membranes seem to be used for membrane anchoring and that self-assembly through the proposed stalk-like regions within a membrane can be employed for fusion.

The DRP family has its origin in bacteria, implying that the members also operate in membrane-linked processes (Low et al. 2009; Lewis et al. 2016). However, the acquired roles of DRPs as primary engines controlling the mitochondrial copy number represent a radical aberration from the bacterial FtsZ-dependent division machine that operates from the cytosolic face of the plasma membrane to facilitate constriction and cleavage. The cleavage machines of primitive eukaryotic organelles in organisms, such as the red alga cyanidioschyzon merolae, and of endosymbiotic plastids and chloroplasts in most photosynthetic eukaryotic organisms yield insights into this transition to DRP-controlled cleavage. These endosymbiotic organelles contain hybrid division machines with both internal FtsZ and external DRP components (Osteryoung and Nunnari 2003; Nishida et al. 2003). The FtsZ engine operates by narrowing and splitting the inner and outer membranes and the division site; whereas, the DRP engine is engaged on the outer surface of the organelle, in particular, at the narrowing, and operates relatively late in the process to complete the cleavage of the outer membrane (Ji et al. 2017). In bacteria, the FtsZ-driven placement of the division site in these mitochondria and plastids is critical for the transfer of organellar genomes.

5.6 Endoplasmatic Reticulum-Driven Division of the Mitochondrion

The loss of the FtsZ-like mechanism in higher eukaryotes lets to the questions of how and where division sites are positioned in mitochondria and whether the placement of the division site is critical for mtDNA transfer. These questions can be partially answered, since a key interorganellar interaction with the ER is responsible to the positioning of the mitochondrial division site (Friedman et al. 2011). Before DRP1 recruited to the mitochondrial outer membrane, the ER tubules are entwined around mitochondria and thereby determine the sites of mitochondrial division, which is a highly conserved (from yeast to humans) phenomenon that is known as ER-associated mitochondrial division (ERMD). At these points, the mitochondria are narrowed and hence geometric prerequisites are created for the assembly of the division DRP helix. In the next step, the integral outer membrane DRP1 receptor and effector MFF (Otera et al. 2010) are recruited to the contact sites, which delivers a spatial linkage of DRP1 recruitment with its activation and assembly into a division process. However, the underlying mechanism of this ER–mitochondrial microdomain and the ER-driven mitochondrial constriction is not still unclear. The ER may be able to directly change the composition and/or morphology of the mitochondrial membrane to regulate the recruitment of factors localized on the outside and/or inside of mitochondria performing the narrowing and the division.

In mammals, the actin cytoskeleton has been involved in ERMD, possibly through the ER-localized isoform of the formin INF2, which leads to the question of whether the mitochondrial constriction during the division is regulated by actin (Korobova et al. 2013). Additionally, the ERMD must include a linkage or tether between the mitochondrion and the ER. In yeast, the molecular basis of this connection has been revealed and is facilitated by a multi-protein complex, which is known as ER–mitochondrial encounter structure (ERMES). The ERMES provides a discrete and finite number of connections between the ER and mitochondria (Kornmann et al. 2009, Murley et al. 2013). Besides marking the division sites, ERMES is tightly coupled to a subset of nucleoids involved in the replication of mtDNA (Meeusen and Nunnari 2003, Hobbs et al. 2001) and acts as components of a larger structural network connecting multiple membranes. At sites of an ERMES complex, the nucleoids separated by an unknown mechanism and mostly are localized into both ends of divided mitochondria (Murley et al. 2013). The ERMES complex is included in a bridge between mitochondria and the actin cytoskeleton, implying that it can connect and coordinate the nucleoid segregation, mitochondrial narrowing during division and mitochondrial distribution after division (Boldogh et al. 2003). Hence, the process of ERMD and nucleoid segregation in yeast seems to be crucially related to the role of actin in ERMD in mammals.

After ERMD in yeast, the distribution of daughter mitochondria demands the highly conserved Miro GTPase Gem1 (Murley et al. 2013). Gem1 appears to promote with ERMES the cleavage of daughter mitochondria by attracting motility factors to mitochondrial tips after division. Metazoan Gem1 orthologues, MIRO1 and MIRO2 (synonymously referred to as Miro), similarly act in the distribution of mitochondria. Thereby, Miro proteins seem to connect mitochondria to a Milton/TRAK protein family of kinesin-1 adaptors member for their microtubule-based transport (Glater et al. 2006; Fransson et al. 2006). Although the Miro GTPase family is highly conserved, the mitochondrial transport mechanisms are distinct in mammals from yeast, as the mammalian mitochondrial motility is actin-based. Hence, the Gem1 and Miro GTPases may provide the motility in the distribution of mitochondria by directly facilitate molecular tethers to disengage mitochondria from the ER at division sites. However, Gem1-driven distribution of daughter mitochondria after division represents an intracellular mechanism for the coordinated mitochondria and mtDNA distribution.

5.7 What Determines the Division of the Mitochondria?

The position of ERMD is in the division plane neighboring mitochondrial nucleoids to influence their distribution into newly produced daughter mitochondria. In mammalian cells, nucleoids are positioned similarly to ERMD at mitochondrial division sites and tips (Iborra et al. 2004; Garrido et al. 2003), and in the absence of the division, DRP1 nucleoids cluster within hyperfused mitochondria (Ban-Ishihara et al. 2013). This finding leads to the hypothesis that ERMD's role in the nucleoid distribution is highly conserved, although the molecular nature of the tethers between the ER and the mitochondria at division sites in mammals is not yet clear. What regulates the localization of division sites? It is still not clear what prerequisites are necessary for the number and localization of ER–mitochondria interaction sites, which are associated with the nucleoid segregation. These questions refer to whether there exists a mechanism in mitochondria capable of easing mitochondrial division, in an analogous manner to bacterial FtsZ. In yeast, the inner membrane protein Mdm represents a good candidate for the internal membrane cleavage mechanism (Bogenhagen et al. 2003), which comprises matrix-localized coiled-coil regions operating in trans across inner membranes to confer constriction (Messerschmitt et al. 2003).

Based on the endosymbiotic origin of mitochondria, it seems to be likely that a spatial mark inside the organelles exists for the positioning of division plane. There is indeed evidence for an autonomous structure within the mitochondria that is similar to the DNA-replicating replisome acting as such a marker. In the absence of mtDNA in yeast and mammalian cells, nucleoid proteins necessary for mtDNA management stay localized on discrete spotted structures in mitochondrial tubules, implying an intrinsic potential to self-organize in a mtDNA-independent structure (Meeusen and Nunnari 2003; Spelbrink et al. 2001). In yeast, the replisomes dissociate within mitochondria and persist their link with ERMES-marked ER–mitochondria interaction sites even if there are no mitochondrial genomes (Meeusen and Nunnari 2003).

Mitochondrial skeletal structures can additionally act as internal spatial marks. Although bacterial-like cytoskeletal elements seem to have been lost, there exist still several scaffold-like structures in mitochondria, which aid the formation of their intricate external and internal structure. Mitochondrial scaffolds collaborate to establish a higher-level organization of the organelle encoding spatial labels for the placement of nucleoids and/or division sites. These scaffolds incorporate the conserved prohibitin complex, forming ring-like structures in the inner membrane, which, together with mitochondrial lipids, such as cardiolipin and phosphatidylethanolamine, enable the organization of the domains of the inner membrane (Osman et al. 2011). Another primary skeletal element in mitochondria represents the conserved multi-component inner membrane-associated complex MitOS (synonymously termed MICOS and MINOS) (Hoppins et al. 2011a; Harner et al. 2011; von der Malsburg et al. 2011). Evidence suggests that MitOS constitutes an extended heteromorphic structure organizing and shaping the mitochondrial inner membrane, which is subdivided into several regions exhibiting structural, compositional and functional diversities. The region in close proximity to the outer membrane, known as the boundary region, promotes the traffic of lipids, the import of mitochondrial proteins and the assembly of the respiratory complex. Inner membrane cisternal invaginations, known as cristae house, build respiratory complexes and created strongly curved edges, which are fixed by dimerization or multimerization of ATP synthase complexes (Davies et al. 2012). Rather narrow tubules, named crista junctions, associate cristae with the boundary membrane and remove soluble intermembrane space components from the boundary regions. In apoptosis, these junctions undergo restructuring to evoke the release of intermembrane space-localized cell-death inducers into the cytosol during mitochondrial outer membrane permeabilization (MOMP) (Frezza et al. 2006). Super-resolution imaging of mammalian nucleoids highlighted that they are closely coupled with inner membrane cristae (Brown et al. 2011). Thus, the MitOS complex can also fulfill a prominent role in nucleoid positioning and can partly contribute to a spatial label connecting the inside of mitochondria to the outside. In accordance with this possibility, elements of the MitOS complex appear in the yeast adjacent to nucleoids, and the loss of an intact MitOS complex results in the aggregation of the nucleoids (Itoh et al. 2013). This complex also conveys mitochondrial biogenesis through association with components of the import and sorting machineries within the outer mitochondrial membrane (van der Malsburg et al. 2011). Thus, MitOS seem to possess a broader global function in the form of a blueprint of the mitochondrial organization and may serve as a form-to-function integrator in this sense.

5.8 Role of the ERMD Microdomain

Although the basic role of ER-associated division seems to be the facilitation of the mtDNA distribution, there is evidence that ERMD domains are utilized for additional and diverse functions in cells and can therefore act as integrators. In addition to the ERMD, the ERMES complex is functionally involved in the biogenesis of outer membrane proteins and in lipid transport between the ER and mitochondria, which is essential for the synthesis of important mitochondrial lipids, such as phosphatidylethanolamine and cardiolipin (Kornmann et al. 2009; Voss et al. 2012). ERMD domains could therefore also control the cellular status by providing the communication between mitochondrial behaviors and cellular signaling pathways, such as mitochondrial division, fusion and cell death. In line with this is the fact that DRP1 drives the recruitment and activation of the pro-apoptotic Bcl-2 protein BAX to the mitochondrial outer membrane to confer MOMP (Montessuit et al. 2010). In addition, ER-synthesized sphingolipids support in vitro the mitochondrial assembly of BAX and the activation of MOMP (Chipuk et al. 2012).

In contrast, the mitochondrial fusion holds a negative regulatory role in MOMP since cytosolic BAX stimulates mitochondrial fusion in vitro via the DRP MFN2 (Hoppins et al. 2011b), which increases the chance that fusion DRPs also govern apoptosis by ERMD domains. The ERMD domains seem to span both into inner mitochondrial compartments and the ER lumen to accommodate the functional status of both organelles, as proposed for the regulation of MOMP under ER stress. In accordance with this, the apoptotic regulator CDIP1 works together with the ER protein BAP31, when the ER is stressed resulting in a BAX arrangement on mitochondria (Namba et al. 2013). Therefore, it needs to be revealed whether the interaction sites of BAP31 match with sites of interaction with ERMD. The relationship between the ER and mitochondria is evidenced by their role in multiple diseases linked to altered mitochondrial dynamics. Among them are Huntington's disease, optic atrophy and spinocerebellar ataxias (Schon and Przedborski 2011, in which proteins are targeted to ER–mitochondrial interaction sites. In line with this, alteration of ER–mitochondrial interaction has been observed in Alzheimer's disease (Area-Gomez et al. 2012; Hedskog et al. 2013). Finally, the dysfunction of the ERMD domain seems to be a major contributory factor in many diseases.

However, the ERMD domains are solely one special type of ER–mitochondrial interaction. In yeast, for example, the ER is part of two distinct Num1 and Mmr1 bonds that selectively direct the mitochondria to the cortex of mother and daughter cells independently of ERMES and ERMD (Lackner et al. 2013; Swayne et al. 2011). In addition, the fusion DRP MFN2, which is not required to the formation of the ERMD contact, seems to function in mammalian cells as an ER–mitochondrial bound (Friedman et al. 2011; de Brito et al. 2008). However, there is still more effort necessary, especially in mammalian cells, to uncover the molecular basis of ER–mitochondria interactions and whether specific interaction sites and bounds are formed for distinct shared ER–mitochondrial functions, including lipid biosynthesis, Ca^{2+} homeostasis, ERMD and autophagy (Rowland et al. 2012; Hamasaki et al. 2013). This is an exciting research field in mitochondrial biology that is regarded of great importance for the understanding of the etiology underlying mitochondrial dysfunction-related diseases.

5.9 Coordination of Various Mitochondrial Behaviors

Mitochondrial division and fusion represent major regulatory elements of the mitochondrial distribution; however, the mitochondrial network in cells is driven by additional activities, such as tether formation and motility. Neuronal cells represent an example of how the mitochondrial behavioral networks function coordinately to maintain cellular functions. More precisely, neurons exhibit a long cell shape and its interior is highly compartmentalized and for their appropriate function mitochondria are required to be distributed to fulfill the specific spatial and temporal demands of cells. At the ends of axons, increased mitochondrial ATP production and Ca^{2+} buffering are needed. The ends of the axons are subject to dynamic remodeling of their structures, which necessitate the pinpointed localization of mitochondria for the synaptic transmission. Since the mitochondrial biogenesis is found mainly in the neuronal stoma, active mechanisms are needed either for the transport or the immobilization of mitochondria at the distal synaptic ends.

Results on neuron-specific protein syntaphilin, which binds specifically to the mitochondrial outer membrane and interacts with immobilized axonal mitochondria at active terminal ends of synapses, have yielded new insights into how these two diverging processes are coordinated and function together to selectively align mitochondria at active synaptic ends (Chen et al. 2013). Mitochondria targeted for axons are produced by mitochondrial division in the soma and then transported along the microtubules to the synapse. The spatial couplings between division and nucleoids, and nucleoids and cytoskeletal elements, seem to help to direct mtDNA to mitochondria labeled for transport in neurons. Syntaphilin acts as a brake, and therefore it employs at least two distinct mechanisms. Firstly, it combines directly to the microtubule-based kinesin motor KIF5 in vitro and thereby impairs its motor activity by potentially transforming KIF5 into a component of a static microtubule-dependent mitochondrial attachment, such as a tether. However, whether the ER fulfills a role in the syntaphilin–KIF5 tether biogenesis remains an important question. To ease tethering, syntaphilin also rivaled for KIF5 binding with the adaptor Milton/TRAK. Therefore, a comprehensive interaction between the motility and tethering mechanisms exists to regulate mitochondrial distribution in an activity dependent and spatially specific way.

The mitochondrial dynamics are presumably also coordinated at a molecular level with the transport and tethering systems in various cell types. In mammalian cells, the mitochondria are devoid of the fusion DRP MFN2 and hence exhibit severe defects in motility (Baloh et al. 2007), comparable to the motility defects found in cells lacking Miro. This result is not surprising, since MFN2 should physically engage with Miro and Milton/TRAK (Misko et al. 2010). This connection between mitochondrial fusion and motility appears to be necessary for understanding why mutations in MFN2 and OPA1 in humans evoke the tissue-specific neurodegenerative diseases, Charcot Marie Tooth Type 2A (CMT2A) and dominant optic atrophy (DOA), respectively (Nunnari and Suomalainen 2012). With a high frequency, mutation in mtDNA and nuclear genes leads to mitochondrial dysfunction, which can selectively occur in neurons and contribute to various neurodegenerative diseases (Copeland 2012).

5.10 How are Mitochondrial Pathways Controlled?

Mitochondrial behaviors are part of multiple stress or quality control pathways in cells that perceive and respond to mitochondrial and cellular dysfunction. More precisely, within mitochondria, molecular chaperones and quality control proteases work together to facilitate the assembly of protein complexes consisting of mtDNA- and nucleus-encoded proteins, to monitor the unfloding of proteins and subsequently to degrade them (Baker et al. 2011). The imbalance between nuclear and mitochondrial proteomes and/or the accumulation of unfolded mitochondrial proteins induces a transcriptional response in metazoans, which is known as the unfolded protein stress response pathway in mitochondria (UPRmt) (Houtkooper et al. 2013; Zhao et al. 2002; Martinus et al. 1996; Melber et al. 2018). Moreover, UPRmt appears to be associated with several age-based diseases, including Alzheimer's disease, Huntington's disease and Parkinson's disease (Kikis et al. 2010; Jovaisaite et al. 2014). To restore the homeostasis of the organelle, a response is triggered by signals generated at the mitochondrial level regulating the transcriptional activation of nucleus-encoded mitochondrial chaperone genes and others (Melber et al. 2018). The pathway has been molecularly identified in *Caenorhabditis elegans* requiring both the mitochondrial inner membrane peptide transporter HAF1 and the bZip transcription factor ATFS1 for the UPRmt signal transduction process (Haynes et al. 2010). In the mitochondria of healthy cells, ATFS1 is actively imported into the mitochondrial matrix, where it is constitutively degraded (Nargund et al. 2012). When the electron transport chain is interrupted, the membrane potential is diminished and thus the import efficiency of ATFS1 is reduced in a way that depends to some extent on HAF1. ATFS1 localized outside the mitochondria is stabilized and enters the nucleus, in which it induces the transcription of distinct genes, such as mitochondrial chaperones and the import machinery, and leads to a remodeling of the metabolism to be even less dependent on the respiratory system. In *C. elegans*, the UPRmt activation correlates with an extended lifespan, and there is some support for the hypothesis that activation of this pathway in mammals also prolongs their lives, indicating that mitochondria are a key element in aging, which further identifies mitochondria as a crucial factor in aging (Houtkooper et al. 2013). However, it needs to be clarified whether the molecular mechanisms that drive the UPRmt are conserved in mammalian systems.

Perturbations in the function of the electron transport chain and/or a decrease in the membrane potential can be induced by additional mitochondrial stress-induced pathways. In detail, the mitochondrial inner membrane fusion DRP OPA1 functions as a switchover between two pathways. In normal and healthy cells, OPA1 processing is constitutively performed by the i-AAA protease YME1L to produce both long transmembrane connected and short soluble isoforms necessary for membrane fusion (Griparic et al. 2007). The progressive decrease of mitochondrial membrane potential allows on the one hand the metalloprotease OMA1 to transform the long OPA1 isoforms into short isoforms, impairing mitochondrial fusion and subsequent mitochondrial fragmentation (Head et al. 2009; Ehses et al. 2009). On the other hand, another stress-induced response, known as mitochondrial hyperfusion, requires long OPA1 isoforms. More specifically, mitochondrial hyperfusion facilitates the assembly of a highly cross-linked mitochondrial network and designed to mitigate the potentially harmful effects of stress, including UV irradiation and nutrient deficiency. During starvation, hyperfusion seems to prevent mitochondria from autophagic degradation or mitophagy by steric obstruction (Rambold et al. 2011; Gomes et al. 2011). Additionally, mitochondrial hyperfusion also provides a homeostatic function to preserve ATP production, when the complex IV of the electron transport chain is disrupted (Rolland et al. 2013). Nevertheless, the hyperfusion response is transient and therefore cannot compensate long-term defects in the activity of the electron transport chain. A more terminal reaction to mitochondrial dysfunction is mitophagy, a condition that is also caused by a decrease in membrane potential-based protein import. In this process, the kinase PINK1 is transported into healthy mitochondria and undergoes progressive degradation. A mitochondrial dysfunction-based reduction in import leads to PINK1 accumulation on the outer membrane, where it attracts the E3 ligase Parkin (Matsuda et al. 2010; Narendra et al. 2010). Parkin's target molecules for ubiquitination compromise the transport factor Miro and the mitochondrial fusion protein MFN2 (Wang et al. 2011; Chan et al. 2011; Tanaka et al. 2010).

The proteasomal degradation of ubiquitinated mitochondrial outer membrane proteins is based on the AAA-ATPase p97 in a similar way to ER-associated degradation (Tanaka et al. 2010; Xu et al. 2011). Parkin-facilitated degradation of factors associated with mitochondrial motility and fusion increases the selective removal of defective mitochondria by autophagy. In supplement to detailing defective mitochondria for degradation, an in vivo study in Drosophila points out that the PINK1–Parkin pathway may also be selectively targeting respiratory complexes for break-down (Vincow et al. 2013). In furtherance of this assumption, the selective targeting of complex I for dissociation has also been detected in cell culture models (Hämäläinen et al. 2013), whereas the mechanisms driving this phenomenon are still elusive.

Mitochondrial stress pathways seem to fulfill crucial roles in the manifestation of diseases involving mitochondrial dysfunction and in turn may represent interesting therapeutic targets. The PINK1–Parkin mitophagy pathway has mutations in each gene that are associated with genetically inherited Parkinson's disease (Kitada et al. 1998; Valente et al. 2004), meaning that this pathway is of relevance. However, are all these stress-based pathways including those with a reduction of the membrane potential physiologically relevant? In contrast to these multiple stress pathways, mtDNA mutations can concentrate in differentiated post-mitotic cells, at the cost of functional wild-type mtDNA, subsequently causing diseases (Sterky et al. 2011). Therefore, animal models for diseases based on mitochondrial dysfunction will be developed in the future to demonstrate the physiological contributions of these pathways. Further basic biological research is required to investigate how the cell properly senses, activates and coordinates these pathways differentially, both among themselves and with other signal transduction pathways, including cell death-related processes. There is no clarity as to how cells adequately assimilate the UPRmt and mitophagy pathways, both governed by the basic import level. Beyond that, it is not known whether OPA1 can work as a molecular integrator of stresses or purely as a simple modulator of mitochondrial shape or how OPA1-driven stress pathways are synchronized with UPRmt and mitophagy.

The resulting view of mitochondria is that of a highly organized structural area for construction of an organelle, which behaves to respond to cellular needs and to its own dysfunction. The combination of system-based approaches with super-resolution microscopy and new genetic tools will clarify our understanding for the molecular basis of mitochondrial structure. Precisely as mitochondrial super-organization is constructed, the fundamental question will be whether the primary determinants of organization emanate from within the organelle itself and are closely related to the oldest trait, the genome.

5.11 Controversial Discussion of the Drp1/Dnm1-Independent Mitophagy

It is still a major issue and controversially discussed whether or not mitophagy relies on prior mitochondrial fragmentation facilitated by the canonical mitochondrial division process. In detail, it has been reported that mitochondrial fragments can begin to bud and divide from mitochondrial tubules, when they are tightly connected with autophagosomes; whereas, the mitochondrial division factor Drp1/Dnm1 is not required (Yamashita et al. 2016).

The process of macroautophagy is usually referred to simply as autophagy, which constitutes a highly conserved catabolic pathway that is crucial to intracellular homeostasis, quality control and responses to stress. A hallmark of autophagy are double-membraned vesicular structures known as autophagosomes, which assemble *de novo* and convert parts of the cytoplasm into lysosomes or vacuoles for degradation and recycling (Feng et al. 2014). A multi-component core autophagy system appears to coordinate membranes and membrane-remodeling components of diverse origin to propagate nucleation, extension and sealing of a single-membrane structure, as well as the isolation of a membrane, to build the double-membrane layered autophagosome. The process of autophagy can take place without any selection pressure, with autophagosomes randomly enclosing parts of the cytoplasm. However, there are selective autophagy modes that permit the cells to target distinct cargoes in a preferred manner or even explicitly for degradation. Selectivity is obtained by receptor-based physical interactions that couple the autophagy system with the cargo, resulting in the formation of an isolation membrane that closely wraps the bound cargo. The process of selective turnover of mitochondria by autophagy, which is synonymously termed mitophagy, is crucial in the quality control of mitochondria (Graef 2016).

Mitochondria commonly assemble to highly dynamic networks of interconnected tubules that constantly fuse and divide. The structural and functional integrity is obtained by fusion and division of the outer and inner membrane of mitochondria and is performed spatially and temporally coordinated that is facilitated by evolutionarily conserved dynamin-related GTPases such as Mfn1/2 and Fzo1 and Opa1 and Mgm1 for fusion and Drp1 and Dnm1 for division in mammals and yeasts, respectively. During the division of mitochondria, Drp1/Dnm1 is recruited into mitochondria by the effectors/adaptors, such as Mdv1 and Fis1 in yeasts and Fis1, Mff, MiD49 and MiD51 in mammals. In its GTP-bound state, Drp1 assembles into spiral-shaped oligomeric structures entwined around mitochondrial constrictions that are created and labeled by the ER (Lackner 2014). The hydrolysis of GTP causes conformational alterations in Drp1/Dnm1 compounds that finally lead to the helical constriction and division of the two mitochondrial membranes. The mitochondrial architecture is closely determined by the functional state of the organelle and the entire cell. Hence, the mitochondrial fragmentation based on increased Drp1/Dnm1-driven division is generally accompanied by mitochondrial stress and dysfunction (Nunnari and Suomalainen 2012).

In mammals, dysfunctional mitochondria cause activation of the ubiquitin ligase parkin and the ubiquitin kinase PINK1 (the PINK1–parkin pathway), which thereby produce phospho-ubiquitinated mitochondrial outer membrane proteins. Phospho-ubiquitin acts as a signal for autophagy and attracts the core autophagy system via the two mitophagy receptors, NDP52 and optineurin (Narendra et al. 2008; Lazarou et al. 2015). In other respects, the mitochondrial outer membrane proteins Bcl2-L-13 in mammals, and the homolog Atg32 in yeast, also resemble the autophagy system in mitochondrial dysfunction, which leads to the hypothesis whether redundant pathways are available to induce mitophagy (Kanki et al. 2009; Okamoto et al. 2009; Murakawa et al. 2015). Since mitochondria are intrinsically devoured by autophagosomes of an apparently limited size, a pivotal role was assumed for Drp1/Dnm1-driven division in achieving mitochondrial fragments of variable and suitable size for turnover in yeasts and mammals (Kanki et al. 2009; Tanaka et al. 2010; Rambold et al. 2011; Abeliovich et al. 2013; Mao et al. 2013; Kageyama et al. 2014; Ikeda et al. 2015). However, a small number of studies revealed that Drp1/Dnm1-independent mitophagy may be seen in yeasts and mammals (Mendl et al. 2011; Bernhardt et al. 2015; Murakawa et al. 2015). A controversial issue remained whether Drp1/Dnm1-driven mitochondrial cleavage is necessary for mitophagy or not.

Thus, the role of Drp1/Dnm1-driven mitochondrial cleavage for mitophagy has been examined in two yeast systems, *Saccharomyces cerevisiae* and *Pichia pastoris*, and three different cell culture systems, including HeLa cells, SH-SY5Y cells and mouse embryonic fibroblasts, under various mitophagy-inducing conditions (Yamashita et al. 2016). In both yeasts, a slightly delayed but still pronounced turnover of mitochondria anchoring the Dnm1 protein was observed during starvation in the absence of Dnm1 or Fis1 (Yamashita et al. 2016). This finding suggests a minor role for Dnm1-facilitated division during mitophagy, as yeast cells dnm1Δ and fis1Δ in fluorescence or electron microscopy exhibited several small mitochondrial fragments in autophagosomes, implying that these small mitochondrial fragments are produced and picked up by autophagosomes apart from the canonical mitochondrial division mechanism. In the next step, various cell culture systems are analyzed for the effect of loss of Drp1-driven mitochondrial division on the process of mitophagy. Mitophagy was therefore principally triggered by exposure of cells to hypoxic environments and by the treatment of cells with deferiprone, an iron-chelating reagent, or by loss of the mitochondrial membrane potential by treatment of cells with carbonyl cyanide m-chlorophenylhydrazone (Yamashita et al. 2016). Under both experimental conditions, turnover of small mitochondrial fragments was demonstrated by mitophagy in the absence of the Drp1-driven mitochondrial division mechanism. All in all, these data suggest that Drp1/Dnm1-independent mitophagy is evolutionarily preserved and appears under a wide array of mitophagia-inducing circumstances.

However, these findings lead to two main questions: How can the mitochondrial fragments be present in autophagosomes in the absence of Drp1/Dnm1? Can the Drp1/Dnm1-independent mechanism simultaneously function even in the presence of the canonical mitochondrial division mechanism? Therefore, the assembly of autophagosomes and the presence of mitochondrial fragments in wild-type and Drp1-knockout HeLa cells were monitored during hypoxia-facilitated mitophagy using fluorescence-based live-cell imaging (Yamashita et al. 2016). The biogenesis of autophagosomes is initiated by mitochondrial tubules and small mitochondrial fragments that bud and divide from mitochondrial tubules simultaneously with the expansion and closure of the isolation membrane to generate autophagosomes regardless of the presence or absence of Drp1 in these cells. In practice, the use of triple color imaging demonstrated that Drp1, as predicted, creates focal points at canonical division sites on mitochondrial tubules, but was not found at mitochondrial bud sites in conjunction with autophagosome formation (Yamashita et al. 2016). Thus, Drp1/Dnm1 is not only unnecessary for mitophagy, but similarly appears to be lacking from sites of mitochondrial narrowing and division during mitophagy. Hence, these findings are crucial for the establishment of a novel view on the process of mitophagy. However, unlike sequential models (Kalia et al. 2018), which suggest that mitochondrial fragments form in a Drp1/Dnm1-dependent mode before they can later be captured by autophagy, the new model is based on the principles that mitochondrial fragments are formed independently of Drp1/Dnm1 and that autophagosome formation and mitochondrial budding and division are spatially and temporally coordinated processes (Yamashita et al. 2016). Since the production of the mitochondrial fragments depends on the autophagy mechanism's integrity, it can be questioned whether the autophagy mechanism itself facilitates budding and division of mitochondria by nucleation and expansion of the isolation membrane at these sites. However, it is also possible that a mechanism not yet revealed regulates the membrane cleavage in the autophagosome formation during mitophagy.

The novel model of the Drp1/Dnm1-independent mitochondrial division during mitophagy raises new questions: How are the sites of autophagosome formation on mitochondrial tubules identified for mitophagy? How mitochondria interact with the autophagy system? After Parkin's accumulation along mitochondrial tubules with increased levels of reactive oxygen species (ROS), the autophagy system is built up in a point pattern corresponding to the focal ubiquitous sites (Yang and Yang 2013). In detail, a short-ranged ROS signal may induce the formation of autophagosomes by precisely targeting ROS-rich regions on mitochondrial tubules by a not yet known mechanism for identifying damaged mitochondria and then facilitating the assembly of the autophagosome.

Finally, is it possible to combine both models in a unifying one? Even though the Drp1/Dnm1-driven mitochondrial division is not required, it can still promote mitophagy by providing mitochondrial fragments, which are simply targeted for degradation. In addition, the type and strength of the stress on the cells and subsequently on the mitochondria seem to decide whether the Drp1/Dnm1-dependent mitochondrial fragmentation is preferred for the turnover of mitochondria through mitophagy or not. When the stress triggers no or only less efficient Drp1/Dnm1-dependent fragmentation, Drp1/Dnm1-independent fragmentation of mitochondria by the autophagy system appears to be the rate-limiting step in mitochondrial turnover by mitophagy. In future experiments, it must to be determined from which types and quantities of cellular stresses one of these two somewhat redundant mitophagy pathways is selected.

References

H. Abeliovich, M. Zarei, K.T. Rigbolt, R.J. Youle, J. Dengjel, Involvement of mitochondrial dynamics in the segregation of mitochondrial matrix proteins during stationary phase mitophagy. Nat. Commun. **4**, 2789 (2013)

K.J. Ahlqvist, R.H. Hämäläinen, S. Yatsuga et al., Somatic progenitor cell vulnerability to mitochondrial DNA mutagenesis underlies progeroid phenotypes in Polg mutator mice. Cell Metab. **15**, 100–109 (2012)

S. Al Rawi, S. Louvet-Vallée, A. Djeddi, M. Sachse, E. Culetto, C. Hajjar, L. Boyd, R. Legouis, V. Galy, Postfertilization autophagy of sperm organelles prevents paternal mitochondrial DNA transmission. Science **334**, 1144–1147 (2011)

E. Area-Gomez, M. Del Carmen Lara Castillo, M.D. Tambini et al., Upregulated function of mitochondria-associated ER membranes in Alzheimer disease EMBO. J. **31**, 4106–4123 (2012)

M.J. Baker, T. Tatsuta, T. Langer, Quality control of mitochondrial proteostasis. Cold Spring Harb. Perspect. Biol. **3**, a007559 (2011)
R.H. Baloh, R.E. Schmidt, A. Pestronk, J. Milbrandt, Altered axonal mitochondrial transport in the pathogenesis of Charcot-Marie-Tooth disease from mitofusin 2 mutations. J. Neurosci. **27**, 422–430 (2007)
R. Ban-Ishihara, T. Ishihara, N. Sasaki, K. Mihara, N. Ishihara, Dynamics of nucleoid structure regulated by mitochondrial fission contributes to cristae reformation and release of cytochrome c. Proc. Natl Acad. Sci. USA **110**, 11863–11868 (2013)
J.M. Baughman, F. Perocchi, H.S. Girgis et al., Integrative genomics identifies MCU as an essential component of the mitochondrial calcium uniporter. Nature **476**, 341–345 (2011)
D. Bernhardt, M. Müller, A.S. Reichert, H.D. Osiewacz, Simultaneous impairment of mitochondrial fission and fusion reduces mitophagy and shortens replicative lifespan. Sci. Rep. **5**, 7885 (2015)
C.W. Jr Birky, The inheritance of genes in mitochondria and chloroplasts: laws, mechanisms, and models. Annu. Rev. Genet. **35**, 125–148 (2001)
D.F. Bogenhagen, Mitochondrial DNA nucleoid structure. Biochim. Biophys. Acta **1819**, 914–920 (2012)
D.F. Bogenhagen, Y. Wang, E.L. Shen, R. Kobayashi, Protein components of mitochondrial DNA nucleoids in higher eukaryotes. Mol. Cell. Proteomics **2**, 1205–1216 (2003)
I.R. Boldogh, D.W. Nowakowski, H.C. Yang, H. Chung, S. Karmon, P. Royes, L.A. Pon, A protein complex containing Mdm10p, Mdm12p, and Mmm1p links mitochondrial membranes and DNA to the cytoskeleton-based segregation machinery. Mol. Biol. Cell **14**, 4618–4627 (2003)
T.A. Brown, A.N. Tkachuk, D.A. Clayton, Native R-loops persist throughout the mouse mitochondrial DNA genome. J. Biol. Chem. **283**, 36743–36751 (2008)
T.A. Brown, A.N. Tkachuk, G. Shtengel, B.G. Kopek, D.F. Bogenhagen, H.F. Hess, D.A. Clayton, Superresolution fluorescence imaging of mitochondrial nucleoids reveals their spatial range, limits, and membrane interaction. Mol. Cell. Biol. **31**, 4994–5010 (2011)
N.C. Chan, A.M. Salazar, A.H. Pham, M.J. Sweredoski, N.J. Kolawa, R.L. Graham, S. Hess, D.C. Chan, Broad activation of the ubiquitin-proteasome system by Parkin is critical for mitophagy. Hum. Mol. Genet. **20**, 1726–1737 (2011)
H. Chen, A. Chomyn, D.C. Chan, Disruption of fusion results in mitochondrial heterogeneity and dysfunction. J. Biol. Chem. **280**, 26185–26192 (2005)
H. Chen, J.M. McCaffery, D.C. Chan, Mitochondrial fusion protects against neurodegeneration in the cerebellum. Cell **130**, 548–562 (2007)
H. Chen, M. Vermulst, Y.E. Wang, A. Chomyn, T.A. Prolla, J.M. McCaffery, D.C. Chan, Mitochondrial fusion is required for mtDNA stability in skeletal muscle and tolerance of mtDNA mutations. Cell **141**, 280–289 (2010)
Y. Chen, Z.H. Sheng, Kinesin-1-syntaphilin coupling mediates activity-dependent regulation of axonal mitochondrial transport. J. Cell Biol. **202**, 351–364 (2013)
J.E. Chipuk, G.P. McStay, A. Bharti, T. Kuwana, C.J. Clarke, L.J. Siskind, L.M. Obeid, D.R. Green, Sphingolipid metabolism cooperates with BAK and BAX to promote the mitochondrial pathway of apoptosis. Cell **148**, 988–1000 (2012)
B.E. Christian, L.L. Spremulli, Mechanism of protein biosynthesis in mammalian mitochondria. Biochim. Biophys. Acta **1819**, 1035–1054 (2012)
W.C. Copeland, Defects in mitochondrial DNA replication and human disease. Crit. Rev. Biochem. Mol. Biol. **47**, 64–74 (2012)
K.M. Davies, C. Anselmi, I. Wittig, J.D. Faraldo-Gomez, W. Kuhlbrandt, Structure of the yeast F1Fo- ATP synthase dimer and its role in shaping the mitochondrial cristae Proc. Natl Acad. Sci. USA **109**, 13602–13607 (2012)
O. M.. de Brito, L. Scorrano, Mitofusin 2 tethers endoplasmic reticulum to mitochondria. Nature **456**, 605–610 (2008)
D. De Stefani, A. Raffaello, E. Teardo, I. Szabo, R. Rizzuto, A forty-kilodalton protein of the inner membrane is the mitochondrial calcium uniporter. Nature **476**, 336–340 (2011)
S. Ehses, I. Raschke, G. Mancuso, A. Bernacchia, S. Geimer, D. Tondera, J.C. Martinou, B. Westermann, E.I. Rugarli, T. Langer, Regulation of OPA1 processing and mitochondrial fusion by m-AAA protease isoenzymes and OMA1. J. Cell Biol. **187**, 1023–1036 (2009)
M.I. Ekstrand, M. Falkenberg, A. Rantanen, C.B. Park, M. Gaspari, K. Hultenby, P. Rustin, C.M. Gustafsson, N.G. Larsson, Mitochondrial transcription factor A regulates mtDNA copy number in mammals. Hum. Mol. Genet. **13**, 935–944 (2004)
K. Faelber, S. Gao, M. Held, Y. Posor, V. Haucke, F. Noé, O. Daumke, Oligomerization of dynamin superfamily proteins in health and disease. Prog. Mol. Biol. Transl. Sci. **117**, 411–443 (2013)
K. Faelber, Y. Posor, S. Gao, M. Held, Y. Roske, D. Schulze, V. Haucke, F. Noé, O. Daumke, Crystal structure of nucleotide-free dynamin. Nature **477**, 556–560 (2011)
Y. Feng, D. He, Z. Yao, D.J. Klionsky, The machinery of macroautophagy. Cell Res. **24**, 24–41 (2014)
M.G. Ford, S. Jenni, J. Nunnari, The crystal structure of dynamin. Nature **477**, 561–566 (2011)

F. Forner, L.J. Foster, S. Campanaro, G. Valle, M. Mann, Quantitative proteomic comparison of rat mitochondria from muscle, heart, and liver. Mol. Cell. Proteomics **5**, 608–619 (2006)

S. Fransson, A. Ruusala, P. Aspenstrom, The atypical Rho GTPases Miro-1 and Miro-2 have essential roles in mitochondrial trafficking. Biochem. Biophys. Res. Commun. **344**, 500–510 (2006)

C. Frezza, S. Cipolat, O. Martins de Brito, et al. OPA1 controls apoptotic cristae remodeling independently from mitochondrial fusion. Cell **126** 177–189 (2006)

J.R. Friedman, J. Jodi Nunnari, Mitochondrial form and function. Nature **505**(7483), 335–343, (2014)

J.R. Friedman, L.L. Lackner, M. West, J.R. DiBenedetto, J. Nunnari, G.K. Voeltz, ER tubules mark sites of mitochondrial division. Science **334**, 358–362 (2011)

C. Fröhlich, S. Grabiger, D. Schwefel, K. Faelber, E. Rosenbaum, J. Mears, O. Rocks, O. Daumke, Structural insights into oligomerization and mitochondrial remodelling of dynamin 1-like protein. EMBO J. **32** 1280–1292 (2013)

S. Fung, T. Nishimura, F. Sasarman, E.A. Shoubridge, The conserved interaction of C7orf30 with MRPL14 promotes biogenesis of the mitochondrial large ribosomal subunit and mitochondrial translation. Mol. Biol. Cell **24**, 184–193 (2013)

T. Gabaldón, M.A. Huynen, Shaping the mitochondrial proteome. Biochim. Biophys. Acta **1659**, 212–220 (2004)

N. Gadir, L. Haim-Vilmovsky, J. Kraut-Cohen, J.E. Gerst, Localization of mRNAs coding for mitochondrial proteins in the yeast *Saccharomyces cerevisiae*. RNA **17**, 1551–1565 (2011)

M. Garcia, X. Darzacq, T. Delaveau, L. Jourdren, R.H. Singer, C. Jacq, Mitochondria-associated yeast mRNAs and the biogenesis of molecular complexes. Mol. Biol. Cell **18**, 362–368 (2007)

N. Garrido, L. Griparic, E. Jokitalo, J. Wartiovaara, A.M. van der Bliek, J.N. Spelbrink, Composition and dynamics of human mitochondrial nucleoids. Mol. Biol. Cell **14**, 1583–1596 (2003)

E.E. Glater, L.J. Megeath, R.S. Stowers, T.L. Schwarz, Axonal transport of mitochondria requires milton to recruit kinesin heavy chain and is light chain independent. J. Cell Biol. **173**, 545–557 (2006)

L.C. Gomes, G. Di Benedetto, L. Scorrano, During autophagy mitochondria elongate, are spared from degradation and sustain cell viability. Nat. Cell Biol. **13** 589–598 (2011)

M. Graef, A dividing matter: Drp1/Dnm1-independent mitophagy. J. Cell Biol. **215**, 599–601 (2016)

L. Griparic, T. Kanazawa, A.M. van der Bliek, Regulation of the mitochondrial dynamin-like protein Opa1 by proteolytic cleavage. J. Cell Biol. **178**, 757–764 (2007)

R.H. Hämäläinen, T. Manninen, H. Koivumäki, M. Kislin, T. Otonkoski, A. Suomalainen, Tissue- and cell-type-specific manifestations of heteroplasmic mtDNA 3243A>G mutation in human induced pluripotent stem cell-derived disease model. Proc. Natl Acad. Sci. USA **110**, E3622–E3630 (2013)

M. Hamasaki, N. Furuta, A. Matsuda et al., Autophagosomes form at ER-mitochondria contact sites. Nature **495**, 389–393 (2013)

T. Hanekamp, M.K. Horsness, I. Rebbapragada, E.M. Fisher, C. Seebart, M.R. Darland, J.A. Coxbill, D.L. Updike, P.E. Thorsness, Maintenance of mitochondrial morphology is linked to maintenance of the mitochondrial genome in *Saccharomyces cerevisiae*. Genetics **162**, 1147–1156 (2002)

M. Harner, C. Körner, D. Walther, D. Mokranjac, J. Kaesmacher, U. Welsch, J. Griffith, M. Mann, F. Reggiori, W. Neupert, The mitochondrial contact site complex, a determinant of mitochondrial architecture. EMBO J. **30**, 4356–4370 (2011)

C.M. Haynes, Y. Yang, S.P. Blais, T.A. Neubert, D. Ron, The matrix peptide exporter HAF-1 signals a mitochondrial UPR by activating the transcription factor ZC376.7 in *C. elegans*. Mol. Cell **37**, 529–540 (2010)

J. He, H.M. Cooper, A. Reyes, M. Di Re, H. Sembongi, T.R. Litwin, J. Gao, K.C. Neuman, I.M. Fearnley, A. Spinazzola, J.E. Walker, I.J. Holt, Mitochondrial nucleoid interacting proteins support mitochondrial protein synthesis. Nucleic Acids Res. **40**, 6109–6121 (2012)

B. Head, L. Griparic, M. Amiri, S. Gandre-Babbe, A.M. van der Bliek, Inducible proteolytic inactivation of OPA1 mediated by the OMA1 protease in mammalian cells. J. Cell Biol. **187**, 959–966 (2009)

L. Hedskog, C.M. Pinho, R. Filadi, et al., Modulation of the endoplasmic reticulum-mitochondria interface in Alzheimer's disease and related models. Proc. Natl Acad. Sci. USA **110**, 7916–7921 (2013)

G.J. Hermann, J.W. Thatcher, J.P. Mills, K.G. Hales, M.T. Fuller, J. Nunnari, J.M. Shaw, Mitochondrial fusion in yeast requires the transmembrane GTPase Fzo1p. J. Cell Biol. **143**, 359–373 (1998)

A.E. Hobbs, M. Srinivasan, J.M. McCaffery, R.E. Jensen, Mmm1p, a mitochondrial outer membrane protein, is connected to mitochondrial DNA (mtDNA) nucleoids and required for mtDNA stability. J. Cell Biol. **152**, 401–410 (2001)

I.J. Holt, A. Reyes, Human mitochondrial DNA replication. Cold Spring Harb. Perspect. Biol. **4**, a012971 (2012)

S. Hoppins, S.R. Collins, A. Cassidy-Stone et al., A mitochondrial-focused genetic interaction map reveals a scaffold-like complex required for inner membrane organization in mitochondria. J. Cell Biol. **195**, 323–340 (2011a)

S. Hoppins, F. Edlich, M.M. Cleland, S. Banerjee, J.M. McCaffery, R.J. Youle, J. Nunnari, The soluble form of Bax regulates mitochondrial fusion via MFN2 homotypic complexes. Mol. Cell **41**, 150–160 (2011b)

S. Hoppins, L. Lackner, J. Nunnari, The machines that divide and fuse mitochondria Annu. Rev. Biochem. **76**, 751–780 (2007)

R.H. Houtkooper, L. Mouchiroud, D. Ryu, N. Moullan, E. Katsyuba, G. Knott, R.W. Williams, J. Auwerx, Mitonuclear protein imbalance as a conserved longevity mechanism. Nature **497**, 451–457 (2013)
F.J. Iborra, H. Kimura, P.R. Cook, The functional organization of mitochondrial genomes in human cells. BMC Biol. **2**, 9 (2004)
Y. Ikeda, A. Shirakabe, Y. Maejima et al., Endogenous Drp1 mediates mitochondrial autophagy and protects the heart against energy stress. Circ. Res. **116**, 264–278 (2015)
E. Ingerman, E.M. Perkins, M. Marino, J.A. Mears, J.M. McCaffery, J.E. Hinshaw, J. Nunnari, Dnm1 forms spirals that are structurally tailored to fit mitochondria. J. Cell Biol. **170**, 1021–1027 (2005)
N. Ishihara, M. Nomura, A. Jofuku et al. Mitochondrial fission factor Drp1 is essential for embryonic development and synapse formation in mice. Nat. Cell Biol. **11**, 958–966 (2009)
K. Itoh, Y. Tamura, M. Iijima, H. Sesaki, Effects of Fcj1-Mos1 and mitochondrial division on aggregation of mitochondrial DNA nucleoids and organelle morphology. Mol. Biol. Cell **24**, 1842–1851 (2013)
S. Jäger, C. Handschin, J. St-Pierre, B.M. Spiegelman, AMP-activated protein kinase (AMPK) action in skeletal muscle via direct phosphorylation of PGC-1α. Proc. Natl Acad. Sci. USA **104**, 12017–12022 (2007)
E.H. Jeninga, K. Schoonjans, J. Auwerx, Reversible acetylation of PGC-1: connecting energy sensors and effectors to guarantee metabolic flexibility. Oncogene **29**, 4617–4624 (2010)
W.K. Ji, R. Chakrabarti, X. Fan, L. Schoenfeld, S. Strack, H.N. Higgs, Eceptor-mediated Drp1 oligomerization on endoplasmic reticulum. J Cell Biol. **216**(12), 4123–4139 (2017)
V. Jovaisaite, L. Mouchiroud, J. Auwerx, The mitochondrial unfolded protein response, a conserved stress response pathway with implications in health and disease. J. Exp. Biol. **217**(Pt 1), 137–143 (2014)
R. Kalia, R.Y.-R. Wang, A. Yusuf, P.V. Thomas, D.A. Agard, J.M. Shaw, A. Frost, Structural basis of mitochondrial receptor binding and constriction by DRP1. Nature **558**(7710), 401–405 (2018)
T. Kanki, K. Wang, Y. Cao, M. Baba, D.J. Klionsky, Atg32 is a mitochondrial protein that confers selectivity during mitophagy. Dev. Cell **17**, 98–109 (2009)
B.A. Kaufman, S.M. Newman, R.L. Hallberg, C.A. Slaughter, P.S. Perlman, R.A. Butow, In organello formaldehyde crosslinking of proteins to mtDNA: identification of bifunctional proteins. Proc. Natl Acad. Sci. USA **97**, 7772–7777 (2000)
Kageyama, Y., Hoshijima, M., Seo, K., et al. Parkin-independent mitophagy requires Drp1 and maintains the integrity of mammalian heart and brain. EMBO J. **33**, 2798–2813 (2014)
E.A. Kikis, T. Gidalevitz, R.I. Morimoto, Protein homeostasis in models of aging and age-related conformational disease. Adv. Exp. Med. Biol. **694**, 138–159 (2010)
T. Kitada, S. Asakawa, N. Hattori, H. Matsumine, Y. Yamamura, S. Minoshima, M. Yokochi, Y. Mizuno, N. Shimizu, Mutations in the parkin gene cause autosomal recessive juvenile parkinsonism. Nature **392**, 605–608 (1998)
B. Kornmann, E. Currie, S.R. Collins, M. Schuldiner, J. Nunnari, J.S. Weissman, P. Walter, An ER-mitochondria tethering complex revealed by a synthetic biology screen. Science **325**, 477–481 (2009)
F. Korobova, V. Ramabhadran, H.N. Higgs, An actin-dependent step in mitochondrial fission mediated by the ER-associated formin INF2. Science **339**, 464–467 (2013)
C. Kukat, C.A. Wurm, H. Spåhr, M. Falkenberg, N.G. Larsson, S. Jakobs, Super-resolution microscopy reveals that mammalian mitochondrial nucleoids have a uniform size and frequently contain a single copy of mtDNA. Proc. Natl. Acad. Sci. USA **108**, 13534–13539 (2011)
A.M. Labrousse, M.D. Zappaterra, D.A. Rube, A.M. van der Bliek, C. elegans dynamin-related protein DRP-1 controls severing of the mitochondrial outer membrane. Mol. Cell **4**, 815–826 (1999)
L.L. Lackner, Shaping the dynamic mitochondrial network BMC. Biol. **12**, 35 (2014)
L.L. Lackner, H. Ping, M. Graef, A. Murley, J. Nunnari, Endoplasmic reticulum-associated mitochondria-cortex tether functions in the distribution and inheritance of mitochondria. Proc. Natl. Acad. Sci. USA **110**, E458–E467 (2013)
N. Lane, W. Martin, The energetics of genome complexity. Nature **467**, 929–934 (2010)
M. Lazarou, D.A. Sliter, L.A. Kane, S.A. Sarraf, C. Wang, J.L. Burman, D.P. Sideris, A.I. Fogel, R.J. Youle, The ubiquitin kinase PINK1 recruits autophagy receptors to induce mitophagy. Nature **524**, 309–314 (2015)
N. Lecrenier, P. Van Der Bruggen, F. Foury, Mitochondrial DNA polymerases from yeast to man: a new family of polymerases. Gene **185**, 147–152 (1997)
S.C. Lewis, L.F. Uchiyama, J. Nunnari, ER-mitochondria contacts couple mtDNA synthesis with mitochondrial division in human cells. Science **353**, aaf5549 (2016)
H.H. Low, C. Sachse, L.A. Amos, J. Lowe, Structure of a bacterial dynaminlike protein lipid tube provides a mechanism for assembly and membrane curving. Cell **139**, 1342–1352 (2009)
K. Mao, K. Wang, X. Liu, D.J. Klionsky, The scaffold protein Atg11 recruits fission machinery to drive selective mitochondria degradation by autophagy. Dev. Cell **26**, 9–18 (2013)
R.D. Martinus, G.P. Garth, T.L. Webster, P. Cartwright, D.J. Naylor, P.B. Høj, N.J. Hoogenraad, Selective induction of mitochondrial chaperones in response to loss of the mitochondrial genome. Eur. J. Biochem. **240**, 98–103 (1996)
N. Matsuda, S. Sato, K. Shiba et al., PINK1 stabilized by mitochondrial depolarization recruits Parkin to damaged mitochondria and activates latent Parkin for mitophagy. J. Cell Biol. **189**, 211–221 (2010)

M. Mbantenkhu, X. Wang, J.D. Nardozzi, S. Wilkens, E. Hoffman, A. Patel, M.S. Cosgrove, X.J. Chen, Mgm101 is a Rad52-related protein required for mitochondrial DNA recombination. J. Biol. Chem. **286**, 42360–42370 (2011)
S. Meeusen, J. Nunnari, Evidence for a two membrane-spanning autonomous mitochondrial DNA replisome. J. Cell Biol. **163**, 503–510 (2003)
S. Meeusen, J.M. McCaffery, J. Nunnari, Mitochondrial fusion intermediates revealed in vitro. Science **305**, 1747–1752 (2004)
A. Melber, M. Cole, C.M. Haynes, UPRmt regulation and output: a stress response mediated by mitochondrial-nuclear communication. Cell Res. **28**, 281–295 (2018)
N. Mendl, A. Occhipinti, M. Müller, P. Wild, I. Dikic, A.S. Reichert, Mitophagy in yeast is independent of mitochondrial fission and requires the stress response gene WHI2. J. Cell Sci. **124**, 1339–1350 (2011)
M. Messerschmitt, S. Jakobs, F. Vogel, S. Fritz, K.S. Dimmer, W. Neupert, B. Westermann, The inner membrane protein Mdm33 controls mitochondrial morphology in yeast. J. Cell Biol. **160**, 553–564 (2003)
A. Misko, S. Jiang, I. Wegorzewska, J. Milbrandt, R.H. Baloh, Mitofusin 2 is necessary for transport of axonal mitochondria and interacts with the Miro/Milton complex. J. Neurosci. **30**, 4232–4240 (2010)
S. Montessuit, S.P. Somasekharan, O. Terrones, et al. Membrane remodeling induced by the dynamin-related protein Drp1 stimulates Bax oligomerization. Cell **142**, 889–901 (2010)
A.A. Morris, J.V. Leonard, G.K. Brown et al., Deficiency of respiratory chain complex I is a common cause of Leigh disease. Ann. Neurol. **40**, 25–30 (1996)
T. Murakawa, O. Yamaguchi, A. Hashimoto et al., Bcl-2-like protein 13 is a mammalian Atg32 homologue that mediates mitophagy and mitochondrial fragmentation. Nat. Commun. **6**, 7527 (2015)
A. Murley, L.L. Lackner, C. Osman, M. West, G.K. Voeltz, P. Walter, J. Nunnari, ER-associated mitochondrial division links the distribution of mitochondria and mitochondrial DNA in yeast. eLife **2**, e00422 (2013)
T. Namba, F. Tian, K. Chu, S.Y. Hwang, K.W. Yoon, S. Byun, M. Hiraki, A. Mandinova, S.W. Lee CDIP1-BAP31 complex transduces apoptotic signals from endoplasmic reticulum to mitochondria under endoplasmic reticulum stress. Cell Rep. **5**, 331–339 (2013)
D.P. Narendra, S.M. Jin, A. Tanaka, D.F. Suen, C.A. Gautier, J. Shen, M.R. Cookson, R.J. Youle, PINK1 is selectively stabilized on impaired mitochondria to activate Parkin. PLoS Biol. **8**, e1000298 (2010)
D. Narendra, A. Tanaka, D.F. Suen, R.J. Youle, Parkin is recruited selectively to impaired mitochondria and promotes their autophagy. J. Cell Biol. **183**, 795–803 (2008)
A.M. Nargund, M.W. Pellegrino, C.J. Fiorese, B.M. Baker, C.M. Haynes mitochondrial import efficiency of ATFS-1 regulates mitochondrial UPR activation. Science, **337**, 587–590 (2012)
Neupert, W., Herrmann, J.M. Translocation of proteins into mitochondria. Annu. Rev. Biochem. **76**, 723–749 (2007)
H.B. Ngo, J.T. Kaiser, D.C. Chan, The mitochondrial transcription and packaging factor Tfam imposes a U-turn on mitochondrial DNA. Nature Struct. Mol. Biol. **18**, 1290–1296 (2011)
K. Nishida, M. Takahara, S.Y. Miyagishima, H. Kuroiwa, M. Matsuzaki, T. Kuroiwa, Dynamic recruitment of dynamin for final mitochondrial severance in a primitive red alga. Proc. Natl. Acad. Sci. USA **100**, 2146–2151 (2003)
J. Nunnari, A. Suomalainen, Mitochondria: in sickness and in health. Cell **148**, 1145–1159 (2012)
K. Okamoto, N. Kondo-Okamoto, Y. Ohsumi, Mitochondria-anchored receptor Atg32 mediates degradation of mitochondria via selective autophagy. Dev. Cell **17**, 87–97 (2009)
C. Osman, D.R. Voelker, T. Langer, Making heads or tails of phospholipids in mitochondria. J. Cell Biol. **192**, 7–16 (2011)
K.W. Osteryoung, J. Nunnari, The division of endosymbiotic organelles. Science **302**, 1698–1704 (2003)
H. Otera, C. Wang, M.M. Cleland, K. Setoguchi, S. Yokota, R.J. Youle, K. Mihara, Mff is an essential factor for mitochondrial recruitment of Drp1 during mitochondrial fission in mammalian cells. J. Cell Biol. **191**, 1141–1158 (2010)
D.J. Pagliarini, S.E. Calvo, B. Chang et al., A mitochondrial protein compendium elucidates complex I disease biology. Cell **134**, 112–123 (2008)
P.A. Parone, S. Da Cruz, D. Tondera, Y. Mattenberger, D.I. James, P. Maechler, F. Barja, J.C. Martinou, Preventing mitochondrial fission impairs mitochondrial function and leads to loss of mitochondrial DNA. PLoS ONE **3**, e3257 (2008)
M.E. Pepling, J.E. Wilhelm, A.L. O'Hara, G.W. Gephardt, A.C. Spradling, Mouse oocytes within germ cell cysts and primordial follicles contain a Balbiani body. Proc. Natl. Acad. Sci. USA **104**, 187–192 (2007)
A.S. Rambold, B. Kostelecky, N. Elia, J. Lippincott-Schwartz, Tubular network formation protects mitochondria from autophagosomal degradation during nutrient starvation. Proc. Natl. Acad. Sci. USA **108**, 10190–10195 (2011)
S.G. Rolland, E. Motori, N. Memar, J. Hench, S. Frank, K.F. Winklhofer, B. Conradt, Impaired complex IV activity in response to loss of LRPPRC function can be compensated by mitochondrial hyperfusion. Proc. Natl. Acad. Sci. USA **110**, E2967–E2976 (2013)
J.M. Ross, J.B. Stewart, E. Hagström et al., Germline mitochondrial DNA mutations aggravate ageing and can impair brain development. Nature **501**, 412–415 (2013)

A.A. Rowland, G.K. Voeltz, Endoplasmic reticulum-mitochondria contacts: function of the junction. Nature Rev. Mol. Cell Biol. **13**, 607–625 (2012)

A. Rubio-Cosials, J.F. Sidow, N. Jiménez-Menéndez, P. Fernández-Millán, J. Montoya, H.T. Jacobs, M. Coll, P. Bernadó, M. Solà, Human mitochondrial transcription factor A induces a U-turn structure in the light strand promoter. Nat. Struct. Mol. Biol. **18**, 1281–1289 (2011)

M. Sato, K. Sato, Degradation of paternal mitochondria by fertilizationtriggered autophagy in *C. elegans* embryos. Science **334**, 1141–1144 (2011)

R.C. Scarpulla, R.B. Vega, D.P. Kelly, Transcriptional integration of mitochondrial biogenesis. Trends. Endocrinol. Metab. **23**, 459–466 (2012)

O. Schmidt, A.B. Harbauer, S. Rao et al., Regulation of mitochondrial protein import by cytosolic kinases. Cell **144**, 227–239 (2011)

O. Schmidt, N. Pfanner, C. Meisinger, Mitochondrial protein import: from proteomics to functional mechanisms. Nat. Rev. Mol. Cell Biol. **11**, 655–667 (2010)

E.A. Schon, S. Przedborski, Mitochondria: the next (neurode) generation. Neuron, **70**, 1033–1053 (2011)

M.R. Sharma, E.C. Koc, P.P. Datta, T.M. Booth, L.L. Spremulli, R.K. Agrawal, Structure of the mammalian mitochondrial ribosome reveals an expanded functional role for its component proteins. Cell **115**, 97–108 (2003)

Y. Shi, A. Dierckx, P.H. Wanrooij, S. Wanrooij, N.G. Larsson, L.M. Wilhelmsson, M. Falkenberg, C.M. Gustafsson, Mammalian transcription factor A is a core component of the mitochondrial transcription machinery. Proc. Natl. Acad. Sci. USA **109**, 16510–16515 (2012)

A. Sickmann, J. Reinders, Y. Wagner, C. Joppich, R. Zahedi, H.E. Meyer, B. Schönfisch, I. Perschil, A. Chacinska, B. Guiard, P. Rehling, N. Pfanner, C. Meisinger, The proteome of *Saccharomyces cerevisiae* mitochondria. Proc. Natl. Acad. Sci. USA **100**, 13207–13212 (2003)

J.N. Spelbrink, F.Y. Li, V. Tiranti, et al., Human mitochondrial DNA deletions associated with mutations in the gene encoding Twinkle, a phage T7 gene 4-like protein localized in mitochondria. Nat. Genet. **28**, 223–231 (2001)

F.H. Sterky, S. Lee, R. Wibom, L. Olson, N.G. Larsson, Impaired mitochondrial transport and Parkin-independent degeneration of respiratory chain-deficient dopamine neurons in vivo. Proc. Natl. Acad. Sci. USA **108**, 12937–12942 (2011)

J.D. Stumpf, W.C. Copeland, Mitochondrial DNA replication and disease: insights from DNA polymerase γ mutations. Cellular and molecular life sciences. Cell Mol. Life Sci. **68**, 219–233 (2011)

T.C. Swayne, C. Zhou, I.R. Boldogh et al., Role for cER and Mmr1p in anchorage of mitochondria at sites of polarized surface growth in budding yeast. Curr. Biol. **21**, 1994–1999 (2011)

A. Tanaka, M.M. Cleland, S. Xu, D.P. Narendra, D.F. Suen, M. Karbowski, R.J. Youle, Proteasome and p97 mediate mitophagy and degradation of mitofusins induced by Parkin. J. Cell Biol. **191**, 1367–1380 (2010)

V. Tiranti, A. Savoia, F. Forti, M.-F. D'Apolito, M. Centra, M. Rocchi, M. Zeviani, Identification of the gene encoding the human mitochondrial RNA polymerase (h-mtRPOL) by cyberscreening of the expressed sequence tags database. Hum. Mol. Genet. **6**, 615–625 (1997)

E.M. Valente, P.M. Abou-Sleiman, V. Caputo, et al., Hereditary early-onset Parkinson's disease caused by mutations in PINK1. Science **304** 1158–1160 (2004)

P. Verstreken, C.V. Ly, K.J. Venken, T.W. Koh, Y. Zhou, H.J. Bellen, Synaptic mitochondria are critical for mobilization of reserve pool vesicles at Drosophila neuromuscular junctions. Neuron **47**, 365–378 (2005)

E.S. Vincow, G. Merrihew, R.E. Thomas, N.J. Shulman, R.P. Beyer, M.J. MacCoss, L.J. Pallanck, The PINK1-Parkin pathway promotes both mitophagy and selective respiratory chain turnover in vivo. Proc. Natl. Acad. Sci. USA **110**, 6400–6405 (2013)

K. von der Malsburg, J.M. Müller, M. Bohnert et al., Dual role of mitofilin in mitochondrial membrane organization and protein biogenesis. Dev. Cell **21**, 694–707 (2011)

C. Voss, S. Lahiri, B.P. Young, C.J. Loewen, W.A. Prinz, ER-shaping proteins facilitate lipid exchange between the ER and mitochondria in *S. cerevisiae*. J. Cell Sci. **125**, 4791–4799 (2012)

J. Wakabayashi, Z. Zhang, N. Wakabayashi, Y. Tamura, M. Fukaya, T.W. Kensler, M. Iijima, H. Sesaki, The dynamin-related GTPase Drp1 is required for embryonic and brain development in mice. J. Cell Biol. **186**, 805–816 (2009)

D.C. Wallace, G. Singh, M.T. Lott, J.A. Hodge, T.G. Schurr, A.M. Lezza, L.J. Elsas, E.K. Nikoskelainen, Mitochondrial DNA mutation associated with Leber's hereditary optic neuropathy. Science **242**, 1427–1430 (1988)

X. Wang, D. Winter, G. Ashrafi, J. Schlehe, Y.L. Wong, D. Selkoe, S. Rice, J. Steen, M.J., T.L. Schwarz, PINK1 and Parkin target Miro for phosphorylation and degradation to arrest mitochondrial motility. Cell **147**, 893–906 (2011)

S. Xu, G. Peng, Y. Wang, S. Fang, M. Karbowski, The AAA-ATPase p97 is essential for outer mitochondrial membrane protein turnover. Mol. Biol. Cell **22**, 291–300 (2011)

S.-I. Yamashita, X. Jin, K. Furukawa, M. Hamasaki, A. Nezu, H. Otera, T. Saigusa, T. Yoshimori, Y. Sakai, K. Mihara, T. Kanki, Mitochondrial division occurs concurrently with autophagosome formation but independently of Drp1 during mitophagy. J. Cell Biol. **215**(5), 649–665 (2016)

J.Y. Yang, W.Y. Yang, Bit-by-bit autophagic removal of parkin-labelled mitochondria. Nat. Commun. **4**, 2428 (2013)

Q. Zhao, J. Wang, I.V. Levichkin, S. Stasinopoulos, M.T. Ryan, N.J. Hoogenraad, A mitochondrial specific stress response in mammalian cells. EMBO J. **21**, 4411–4419 (2002)

Mechanical View on the Mitochondria 6

Abstract

This chapter presents the cytoplasm and the contained organelles from a biomechanical point of view. Thereby, the entire cell is treated as gel with associated physical features. General physical principles are applied on cells at a mesoscopic length scale. However, on a submesoscopic scale, the mechanical properties of organelles, such as mitochondria, contribute to the overall mechanical properties of the cell. A cell may even respond to the applied mechanical stress by regulating the fusion of fission of the mitochondria. Hence, mitochondria seem to fulfill a key role in mechanotransduction processes in cells. In this chapter, it is firstly described how mechanical factors regulate the structure and function of mitochondria. Secondly, it is presented how the structures of mitochondria impact ATP production and the formation of mitochondrial clusters is described. Thirdly, the properties of such mitochondrial networks are discussed in detail. Fourthly, interaction between the mitochondria and cytoskeletal microenvironment, such as actin filaments, microtubules and intermediate filaments, is highlighted. Fifthly, the mechanobiological aspects of the structure and function of mitochondria are discussed. Finally, it is enlightened how mitochondria are predestinated for providing a mechanical response to external mechanical perturbations.

6.1 Mechanical Factors Regulate the Structure and Function of Mitochondria

The mitochondria serve as a supplier for energy that is provided to the cells as adenosine triphosphate (ATP). Besides this major function, the mitochondria guide the apoptosis, the programmed cell deaths of the cells, support the buffering of calcium and produce the reactive oxygen species (ROS). In order to facilitate all these different functions, mitochondria create an extensive network that contains smaller clusters, which are able to move along microtubules guided by motor proteins, such as dyneins. Mitochondria can interact with the actin cytoskeletal scaffold that is helpful in various cellular responses and for several mechanical factors (Bartolák-Suki et al. 2017).

In this chapter, the mitochondrial structure and function are presented and correlated to the cytoskeleton and several mechanical parameters affecting the functions of cells. The morphological characteristics of mitochondria are presented by specifically pointing out to the fission and fusion processes and how the mitochondrial network features guide the function. In particular, the interaction of mitochondria and the cytoskeletal architecture is highlighted, which focuses also in

C. T. Mierke, *Cellular Mechanics and Biophysics*, Biological and Medical Physics, Biomedical Engineering, https://doi.org/10.1007/978-3-030-58532-7_6

mechanical interactions between these elements. More precisely, it is described how mitochondria stretch and how their dynamic pattern regulates the architectural structure of the mitochondria and subsequently their function. At the end, it is provided how the stiffness of the extracellular matrix environment affects the shape of mitochondria and their generation of ATP. The relatively general role of mitochondria in the process of mechanobiology and how the mechanosensitive properties of mitochondria can induce the onset and further development of specific diseases and cell aging are discussed.

The eukaryotic cells contain an extremely complex internal structure to fulfill specialized functions, including the migration of the cells, cell contraction and the division of cells. Moreover, they need to act multiple chemical and mechanical cues evoked by the microenvironment. The cell functions and activities require energy of the mitochondria that is provided in the form of ATP through the oxidative phosphorylation cycle. Besides this key role, mitochondria facilitate the management of apoptosis, buffer calcium and generate ROS (Kluge et al. 2013).

To facilitate diverse cell functions, the mitochondria built an expanding interrelated network within the cell that contains somewhat smaller clusters of mitochondria. These smaller clusters possess the capacity to move along the cytoskeletal fibers driven by motor proteins (Boldogh and Pon 2007). In precise detail, the mitochondrial architecture itself is undergoing a highly dynamic transformation as it is continuously subject to division and fusion (Bereiter-Hahn and Voth 1994; Palmer et al. 2011). All these fundamental processes are essential for either the cell integrity or the survival of the entire organism. When proteins, which facilitate the fission (Ishihara et al. 2009) or fusion of mitochondria (Chen et al. 2003) in mice, are genetically knocked out, they produce embryos that are dead before their birth. Although there is strong knowledge about the mitochondrial framework or cluster properties and their dynamic remodeling (Palmer et al. 2011; Aon et al. 2004; Bach et al. 2003; Rambold et al. 2011; Santel et al. 2003; Sukhorukov et al. 2012), there is less understanding of the interaction between intracellular structures, such as the cell's cytoskeleton (Anesti and Scorrano 2006) and external mechanical parameters, such as the cell stretching (Ali et al. 2004) and mitochondria, which influences subsequently mitochondrial functions. As the impacts of stretching of the cells are predominantly transduced by the cytoskeletal scaffold to which mitochondria are connected, it is imaginable that the organization of the cytoskeleton and mechanical features commonly interfere with the structure and function of the mitochondrial framework.

The main objective is to give an overview of two relevant regulatory determinants of mitochondria, the intracellular cytoskeleton and the external mechanical parameters. An essential mechanical element is the deformation to which the cell in the body is subjected during its normal tissue function. The in vivo natural dynamic character of the stretch pattern has recently demonstrated to be beneficial in sustaining the general mitochondrial function (Bartolak-Suki et al. 2015). Meanwhile, the mechanical stiffness (roughly defined as the response of the extracellular matrix under stress to a unit alteration in the strain of the specimen) has proven to be an important regulator of multiple functions of the cell (Discher et al. 2005). Thereby, the important question of whether the extracellular matrix mechanics additionally impact the function and structure of mitochondria should be answered.

In the following, the morphological features of the mitochondria are presented briefly and with respect to their impact on mechanical properties and a special emphasis on the distinct function guided by the special configuration of the framework. In specific detail, the interaction of mitochondria and the structural architecture of the cytoskeleton is pointed out with a focus on the gained potential of the interaction of the two structures. It is presented how the stretching and the dynamical remodeling of mitochondrial patterns alters their overall structure and function. In the end, it is provided how the stiffness of the surrounding extracellular matrix governs the shape of the mitochondria and the generation of energy in form of ATP. Finally, it is discussed which general functional role mitochondria fulfill in mechanobiology and how mitochondrial mechanosensitivity leads to the emergence of several diseases and the aging process.

6.2 How Are Mitochondrial Structure and Production of ATP Related?

At first glance, it is assumed that the mitochondria emanate from free-living alpha-proteobacteria that evolved a symbiotic association with the target host cell (Margulis 1975). However, there is compelling phylogenetic evidentiary proof now that this scenario exists (Wang and Wu 2015), substantiated in that mitochondria possess their very own DNA, designated by mtDNA, and that multiple mitochondrial proteins originate from bacteria (Gray 2015). These organelles are bound by an outer membrane and an inner membrane, in a similar way as in bacteria. The outer membrane enables the metabolite transfer between the inner membrane and the cytosol, while also preventing mitochondria from releasing pollutants such as ROS and mtDNA into the cytosol (Zhang et al. 2010; Pernas and Scorrano 2016). Nonetheless, mitochondrial ROS serves in subtoxic quantities as post-release signal molecules in the cytosol (Suzuki et al. 1997).

The inner membrane of mitochondria is composed of specific morphological sites which encompass the boundary region to the membrane boundary, the cristae and the junctions of cristae (Vogel et al. 2006). The cristae represent the invagination of the inner membrane of mitochondria, which pronouncedly enlarges the internal surface of mitochondria. The proteins providing the electron transport chain are targeted to this surface. Mitofilins can contribute to the organization of the morphology of cristae and can be accumulated between inner and outer membranes of mitochondria (John et al. 2005). The luminal side of the membrane harbors the mitochondrial matrix, an which the Krebs cycle is active. In specific detail, the Krebs cycle delivers NADH and FAD into two transmembrane complexes, such as the respiratory complex I and II. Energetic electrons can travel through the respiration complexes of the electron transport chain, whereas protons are transported from the matrix into the intermembrane space, resulting in a formation of charge and proton gradient over the inner membrane that is termed the electromotive force and functions similar to a battery storing electrochemical energy. The final component of the electron transport chain is the ATP synthase that utilizes the electromotive force to add an inorganic phosphate group to ADP which results in ATP. This process is termed the oxidative phosphorylation. Thereby a part of the ATP amount is employed by the mitochondria and the remaining ATP is secreted into the cytoplasm, which can be employed as chemical energy for multiple processes inside the cell (Fig. 6.1).

6.3 Cluster Formation of Mitochondria

Mitochondria seem to be rather isolated organelles that take food from their surrounding environment, such as the cytoplasm. In addition, mitochondria continuously undergo highly dynamically remodeling and perform interactions with other organelles (Pernas and Scorrano 2016). At the luminal side of the mitochondria are the cristae located that are able to pronouncedly restructure themselves due to environmental factors and mechanical stresses or biochemical alterations. For instance, the supply with energy-rich substances alters the entire architecture of the mitochondrial cristae (Patten et al. 2014) that in turn can promote the assembly of the respiratory complex and subsequently trigger the overall growth of the cell dependent on the proper function of the mitochondria (Cogliati et al. 2013). The dimerized form of the ATP synthase can fall into two parts due to cell aging, which is coupled to a loss of the cristae invagination (Daum et al. 2013) that leads to the hypothesis that the ATP synthase activity and the structure of cristae are closely coupled. Specifically, the dynamin-related GTPase

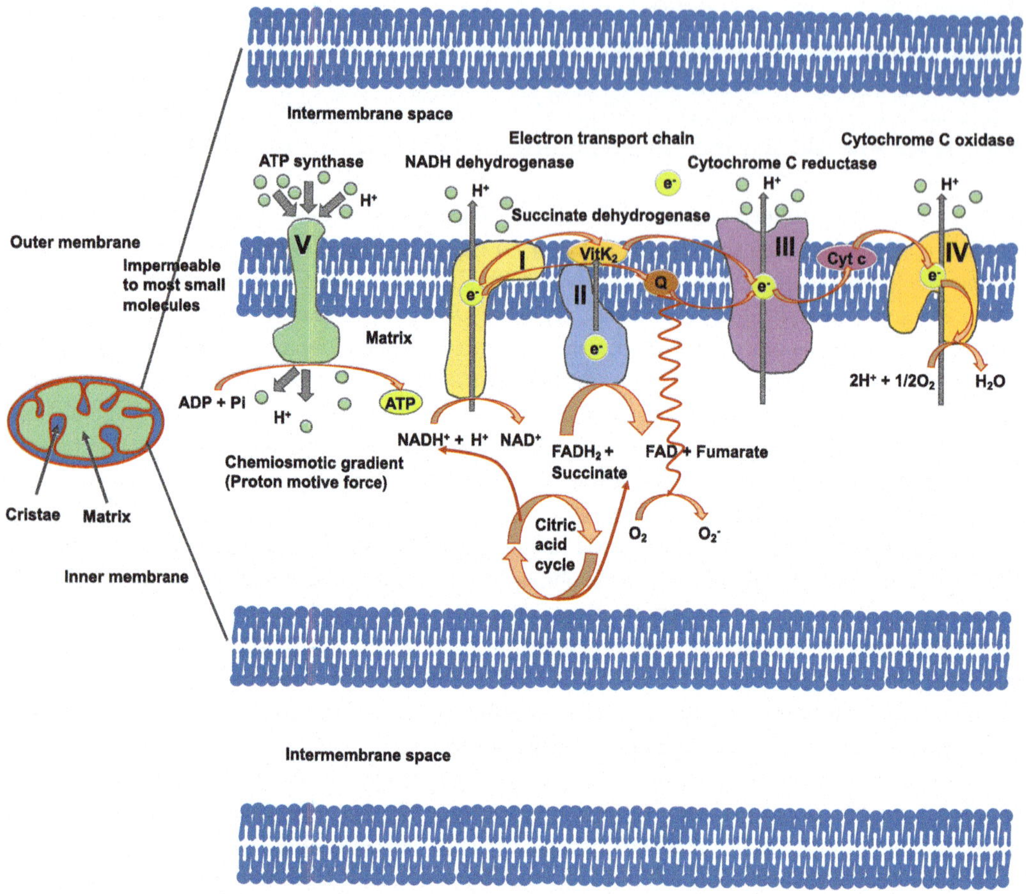

Fig. 6.1 Mitochondria represent the powerhouses of the cell. A sketch of a mitochondrion displays its outer and inner membrane, the cristae and the matrix. Fat and sugar can enter the mitochondria through specific channels of the outer membrane (left). The citric acid cycle can feed the respiratory complex chain and thereby generates an electrical and proton (H^+) gradient, which serves as an electromotive force across the inner membrane. The ATP synthetase employs these electromotive forces to produce ATP from ADP and inorganic phosphate (Pi) (right)

protein Optic Atrophy 1 (OPA1) has been identified as the primary regulator of the maintenance of the cristae. OPA1 seems to fulfill specific tasks in the fusion of mitochondria and in the remodeling process of cristae, which is properly most recognized in the programmed cell death, referred to as apoptosis (Frezza et al. 2006).

Cluster of mitochondria possesses the capacity to fuse with other single clusters of mitochondria to create a large organelle. Subsequently, these mitochondrial clusters assemble a dynamically reticular network containing multiple interconnections that are distributed through the whole cellular volume. The elements of the mitochondrial network elements possess a cylindrical shape that have a diameter of about several hundred nanometers. The formation of the mitochondrial network by fusion requires the combination of the outer and inner membranes of two distinct mitochondrial clusters, and their constituents, together with mtDNA, are thoroughly intermixed over 12 h (Legros et al. 2002). The fusion of the outer membrane of mitochondria is guided by the fusion proteins, such as the mitofusin1 (Mfn1) and mitofusion2 (Mfn2) (Chen et al. 2003). However, the fusion of the inner membrane of mitochondria is guided by the OPA1 protein (Olichon et al. 2003). Specifically, while fusion of outer membranes occurs regardless of oxidative phosphorylation, inner membrane fusion necessitates enzymatic cleavage of OPA1 induced by high membrane potential (Mishra et al. 2014), implying only healthy and active mitochondria are able to fuse properly.

The large mitochondrial clusters are found to split into two or several more parts (Fig. 6.2). After this process of mitochondrial cleavage, the smaller mitochondrial clusters have an almost spherical shape with diameters of several hundred nanometers. The fragmentation of mitochondria is necessary for their orderly distribution of mitochondria during the division of cells and the growth of embryos (Yaffe 1999). However, if cleavage is not managed and compensated by fusion, the network becomes overfragmented, resulting in glucose break-down, mitochondrial internal membrane potential decrease and consequently reduced ATP production (Bach et al. 2003). The process of mitochondrial cleavage appears to be strictly controlled by a combined sequence of events to which cytoplasmic components, cytoskeletal elements and organelles all participate in three steps: firstly, labeling of the cleavage site; secondly, building the cytosolic dynamin-related protein 1 (DRP1) into a superstructure at the cleavage site (Fig. 6.3); and thirdly, narrowing the membranes at the cleavage site to divide the mitochondrial cluster into subsidiary clusters (Pernas and Scorrano 2016; Hoppins et al. 2007).

The cleavage and fusion rates of mitochondria seem to be strongly kept in a balance (Lackner 2014). In specific detail, the association of two heptad-repeat regions of Mfn2 to DRP1 leads to the fusion of mitochondria, while the association of these two regions of Mfn2 impairs the fusion process (Huang et al. 2011). To be more even specific, the long version of OPA1 is able to induce the fusion of mitochondria, whereas the short version of OPA1, which has been enzymatically cleaved, supports the opposite behavior of mitochondria, as it causes the fission of mitochondria (Anand et al. 2014). Moreover, the short version of OPA1 has been detected to be co-localizes with the interaction sites of the mitochondria with the endoplasmic reticulum (ER). Consequently, the cleavage of mitochondria

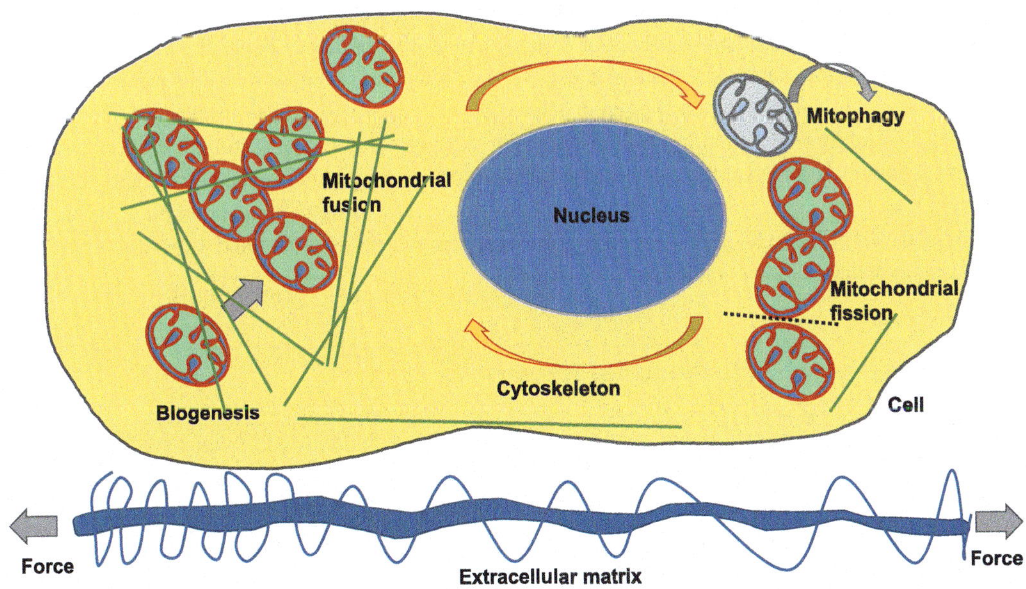

Fig. 6.2 Intracellular and extracellular events contribute to the mitochondrial architecture and function. Intracellular mitochondrial dynamics include processes, such as fusion, fission, mitophagy and biogenesis. The gray colored mitochondrion is damaged and undergoes degradation by mitophagy. The green lines indicate the site of microtubules, on which small mitochondria can migrate along with the help of motor proteins. The dashed line indicates the site of mitochondrial fission. Cells are coupled to the extracellular matrix and are subject of external mechanical forces, which can trigger fusion and fission of mitochondria

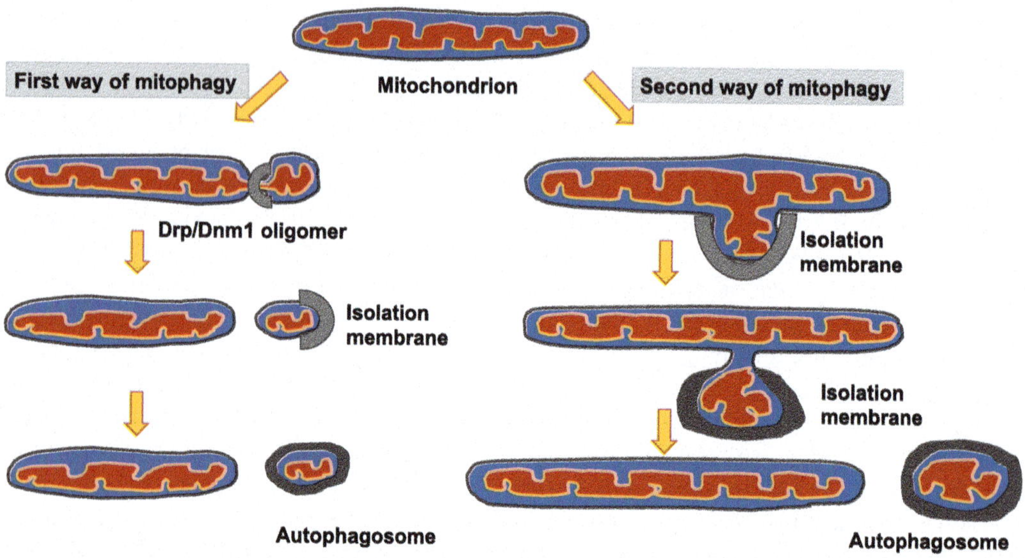

Fig. 6.3 Drp1/Dnm1-dependent (first way) and -independent (second way) mitophagy. The first way facilitates the mitochondrial division through the production of a mitochondrial fragment that is enclosed by an isolation membrane and later on it is engulfed in an autophagosome. The second way involves the autophagosome biogenesis through the initiation of mitochondrial tubules and creates a budding event and finally the full division of a mitochondrial fragment, which is enveloped by an isolation membrane. The autophagy mechanism generates the scission and closure of the isolation membrane to generate the autophagosome

is facilitated by the interactions between mitochondria and relatively narrow ER tubules. Aquaporin-driven water channels provide the swelling of ER tubules, and hence, the constriction of the tubules bears the mechanical force needed for the cleavage of mitochondria (Lee et al. 2013). In fact, the mitochondrial fission can take place directly at the interaction region of mitochondria with ER tubules, where the constriction is triggered before DRP1 is recruited to the cleavage site (Friedman et al. 2011). Additional functions of the tethering between the ER and mitochondria have been reported (Lackner 2014). Among them are at least two processes providing the complete dynamics of mitochondria: firstly, the biogenesis that drives the generation of mitochondrial content through the nucleus and the mitochondria and secondly, the mitophagy that degrades and subsequently phagocytized male-functioned mitochondria.

6.4 Properties of Mitochondrial Networks

Due to the highly organized and complex architecture of mitochondrial clusters, fluorescence microscopy and mathematical analyses have been employed to analyze this structural network. The fission and fusion processes seem to be not fully independent of one another, and consequently, they seem to create a cycle (Fig. 6.4), where the probability that a fusion succeeds the fission and a fission joins the fusion is approximately 0.8 (Cagalinec et al. 2013). In neuronal cells, it was found that cleavage is controlled by the length of the mitochondrial cluster, while fusion is managed by cluster motility to ensure that the total rates of cleavage and fusion are achieved and preserved in a homeostatic state (Cagalinec et al. 2013). The exact rates and probabilities for a fusion or fission event depend on the specific cell type. The energy state of mitochondria is crucially connected to their structure because they dissipate the inner membrane potential, inhibit complex III and complex V

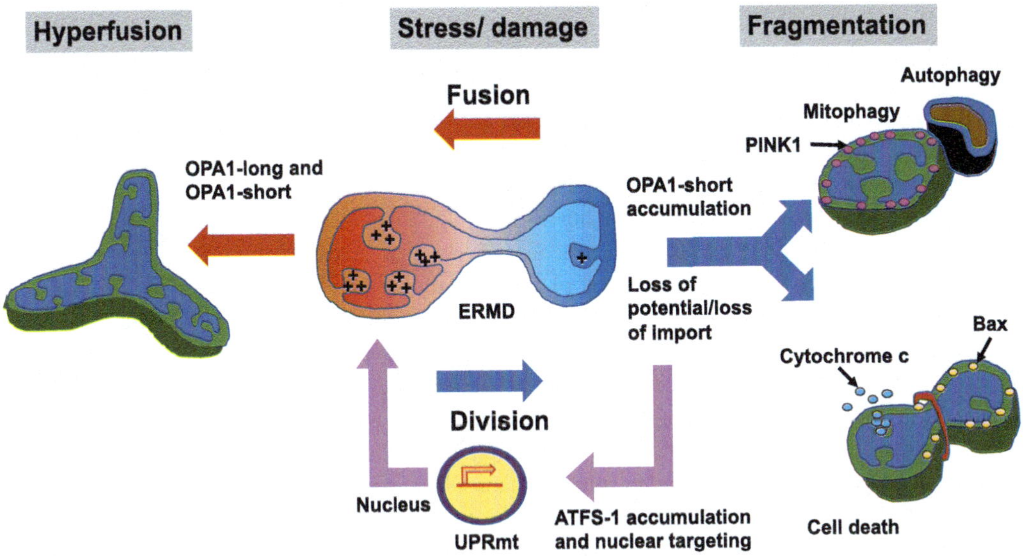

Fig. 6.4 Mitochondrial stress response is coordinated with the shape of mitochondria. Multiple different mitochondrial pathways react to stress or damage (loss of mitochondrial membrane potential and subsequently loss of protein import capabilities) and are connected to mitochondrial dynamics. During unfolded protein stress response (UPRmt) pathway, loss of import leads to accumulation of transcription factor ATFS1 in the nucleus, which triggers the mitochondrial repair mechanisms and the metabolic adaption reaction. Loss of membrane potential induces OMA-1 dependent proteolysis of long isoforms of the membrane fusion DRP OPA1, which leads to mitochondrial fusion and potentially elevates ER-mediated division (ERMD) that results in mitochondrial fragmentation accumulating PINK1 kinase and hence induces mitophagy. In addition, ERMD domain may change to directly induce BAX oligomerization on the outer mitochondrial membrane, which in turn permeabilizes the membrane and releases cytochrome c that causes cell deaths. Mitochondrial dysfunction and stress can initiate the mitochondrial hyperfusion depending on the mitochondrial membrane potential and the presence of long and short OPA1 isoforms. Hyperfused mitochondria circumvent mitophagy and buffer dysfunctions

(ATP synthase) of the electron transport chain, or inhibit cytosolic ATP by limiting glycolysis. All of which causes a decreased rate of fissions and fusions, diminishes their motility, which is determined by the diffusion coefficient analysis, decreases the likelihood of cluster movements bursting and increases the number of fragment of mitochondria (Giedt et al. 2012).

In order to reveal the precise structure of individual clusters, the length of the cluster backbone can be determined (Cagalinec et al. 2013) and the apparent aspect ratio (AR), which is the ratio of major to minor axis of an ellipse fit to the mitochondrial cluster shape, or the form factor (FF), which presents the perimeter squared divided by the product of 4π and the area, can be calculated (Koopman et al. 2005a, b). More precisely, the AR reveals how spherical the shape of the cluster is, while the FF correlates to the degree of branching across the backbone of the cluster. The total mitochondrial complexity of the network is expressed as the fractal dimension D_f, which refers to the network's room filling capacity (Palmer et al. 2011; Bartolak-Suki et al. 2015). When in 2D approaches, the factor D_f almost 1, the network is mainly composed of lines, and when D_f is nearly 2, the network is more of a dense 2D object just like a filled circle. In order to investigate what is responsible for the structural properties of the mitochondrial network, it is appropriate to display some fluorescent labels linked to the mitochondria and to calculate the aforementioned indices before and after the administration of various stimuli or inhibitors. For instance, the mitochondrial average cluster size in vascular smooth muscle cells (VSMCs) in culture free of stretching appears to depend on a number of inhibitors, including blebbistatin, an inhibitor of non-muscular myosin II, dynasore, an inhibitor of DRP1, and paprotrain, an inhibitor of mitotic kinesin-like protein 2 that permits mitochondria to travel down microtubules (Bartolak-Suki et al. 2015). In extension, the chronic suppression of complex I with rotenone in human skin fibroblasts as a model of mitochondrial dysfunction resulted in a marked elevation of FF, which points out to the hyperproduction of ROS leading to mitochondrial protuberances (Koopman et al. 2005a, b).

An overall view of the entire mitochondrial network shows that this network spreads efficiently across the cell, probably an evolutionary outcome of the ideal spatial distribution of energy delivery in the guise of ATP. One of the features of this overall network is its connectivity by means of percolation theory (Stauffer and Aharony 1992). Examine a grid in which each neighboring location is covered with the probability p. A cluster can be defined as a group of connected occupied sites. When p rises from 0 to 1 and specifies a completely empty and separate grid, the size of the clusters gradually increases. There is a stage where a large cluster crosses the grid and offers full interconnectivity from one end to the other. The transition from a separate grid to a grid containing a connected cluster covering the entire system takes place when p exceeds a critical percolation threshold, which is characterized by p_c. The microstructure of the percolation pattern at $p = p_c$ is a self-similar fractal. A percolation transition of this type can therefore be expected when a critical mass density of mitochondria is fused to a global network. In order to test this concept, in isolated ventricular myocytes, it was shown (Aon et al. 2004) that the mitochondria build a network close to the critical point and that the fractal properties of the network match with a percolating mitochondrial network perfectly. The question of how signals are sent over such a network is even more challenging. When mitochondria aggregate levels of ROS that are above a certain threshold, a small additional release of ROS generated locally by electron transport chain leakage initiates a ROS wave propagating through the interconnected cluster, depolarizing initially the inner membrane potential and subsequently switching to oscillations. The biological implications include that a shift to oscillatory patterns may destabilize the heart. Precisely, it means that the potential for action of repolarization in the entire heart, implying that there are microscopic levels of criticality, can be converted into the organism's death on the microscopic length scale (Aon et al. 2004). Moreover, this scenario leads to the hypothesis that mitochondria have the potential to signalize across cells and reach the tissue level to interfere with the fate of organs and the whole organism.

6.5 Interactions Between the Cytoskeleton and Mitochondria

Besides the structure, dynamics and regulation of mitochondria, there exists an interplay between mitochondria and the cytoskeleton. A function of the cytoskeleton, which is a filamentous protein scaffold, is to prevent deformation of the entire cell in order to resist external forces and allow shape alterations during cell motility, the transport of cell cargo, such as mitochondria. Moreover, the cytoskeleton signals in mechanotransduction processes through the conversion of mechanical signals into biochemical signals (Fletcher and Mullins 2010). In turn besides transducing signals, such as physical forces, from the outside to the inside of the cell, the cytoskeleton can transmit forces from organelles, such as the nucleus and the mitochondria to the outward environment or to other neighboring cells. The cytoskeletal network is able to react to external forces by displaying hysteresis and memory, and long-lived cytoskeletal structures are epigenetically passed on to future generations after cell division (Fletcher and Mullins 2010). Due to this fundamental mechanical function of the cytoskeleton in general cell behavior, the focus will be on the mechanical issues of the interface between the cytoskeleton and mitochondrial clusters.

The cytoskeleton consists of three major types of biopolymers: actin filaments synonymously referred to microfilaments, intermediate filaments and microtubules. Based on the connection via cross-linkers, stabilizers and motor proteins, these filaments can assemble interconnected architectures. The shape of cells and their mechanical phenotype depends on the amount of the structural organization and the predominant network type. These dynamical remodeling scaffolds help the cells to adapt their structural resistance to the amount of externally applied forces toward the cell and support the communication of the cell with other adjacent cells or multiple cell organelles by the generation and transmission of cell mechanical forces. The structural reorganization of the cell's cytoskeleton usually requires the polymerization and depolymerization of filaments that is driven specific factors, including nucleation promoting factors supporting the initial growth of filaments, filament capping proteins, impairing the further polymerization process, polymerases, increasing the polymer elongation and depolymerizing factors that disrupt filaments and networks through promoting the disassembly (Fletcher and Mullins 2010). Consequently, all three major cytoskeletal components are linked to several mitochondrial functions.

6.5.1 Interactions Between Mitochondria and the Actin Cytoskeleton

Initially, an early study proposed a spatial co-localization of gamma-actin with mitochondria of skeletal muscle cells (Pardo et al. 1983), while a later trial revealed that mitochondria in sympathetic neurons of chicks can move along the axon in bilateral directions and their movement demands either microtubules or actin, depending on which cytoskeletal network is available (Morris and Hollenbeck 1995). It was nevertheless later observed that in axons and dendrites, the mitochondria exhibited a preferential movement across the microtubules, even though a restricted movement across the actin remained feasible (Ligon and Steward 2000). While the movement of mitochondria necessitates actin in plants and fungi or microtubules in mammalian cells, actin assists in the immobilization of mitochondria in neurons at sites where ATP is demanded (Boldogh and Pon 2006), by the reinforcing its Ca^{2+}-dependent interplay with actin (Kremneva et al. 2013). Actin contributes to the rearrangement of the mitochondrial system during mitosis. The delivery of mitochondria to the daughter cell at the terminal stage of mitosis is facilitated by the cell cycle-dependent enrollment of Cenp-F, a cytoskeleton-associated protein, by a mitochondrial protein known as Miro (Kanfer et al. 2015).

It has been revealed that the cortical actin structure relies on the presence of ATP and non-muscle myosin II, and consequently, the degree of the cross-linking of the actin network by myosin II correlates with the overall stiffness of the cell (Parameswaran et al. 2014). The inhibition of non-muscular myosin II alters cortical actin (Bondzie et al. 2016); however, this inhibition decreases the average mitochondrial cluster size in VSMCs (Bartolak-Suki et al. 2015). In turn, mitochondria are able to affect multiple actin-based cell functions. For instance, as the production of ATP depends mostly on the mitochondria, the entire cell stiffness seems to rely additionally on mitochondrial production of ATP. The same factor seems to account for the cell contraction behavior, since ATP is highly needed for the myosin-based cell contraction. In fact, in VSMCs and in aorta rings, it has been revealed that the impairment of the ATP synthase by oligomycin causes a decrease in the active generation of forces (Bartolak-Suki et al. 2015). Consequently, there appears to be a subtle and double-directional association between cortical actin and mitochondrial architecture leading to general cell-mechanical functions such as stiffness and contractility.

6.5.2 Can Microtubules Impact the Function of Mitochondria?

For over four decades, it is known that mitochondria strongly associate with microtubules in multiple cell types (Heggeness et al. 1978). Both alpha and beta tubulins can be targeted to the lumen of mitochondria, and they are closely connected to mitochondrial voltage-dependent anion channels (Carre et al. 2002). The close storage of beta-tubulin II in the neighborhood of voltage-dependent anion channels enables it to facilitate the transition of the mitochondrial permeability (Kuznetsov et al. 2013), where the pores, which are mainly voltage-dependent anion channels, located on the outer membrane open and subsequently cause their necrosis and apoptosis (Kim et al. 2003).

Microtubule filaments act as train rails on which mitochondrial clusters move inside the cell, using motor proteins such as dyneins and kinesins that move to the minus or plus ends of the microtubules (Vale 2003). The disassembly of microtubules fully removes the mitochondrial motility (Ligon and Steward 2000), which is required for fusion and cleavage and for the maintenance of a healthy mitochondrial microstructure and for cellular bioenergetics. Theoretical analysis and mathematical modeling yield empirical results proving that, in addition to fusion and cleavage rates, the mitochondrial structure is determined by the retrograde and anterograde motions and the equilibrium between these rates is responsible for the heterogeneous dispersion of mitochondria throughout the cell (Sukhorukov and Meyer-Hermann 2015). Microtubules are also responsible for cell shape and stability through their ability to absorb compressive forces (Stamenovic et al. 2002). External forces are capable of changing cell shape, leading to a reorganization of microtubules; effectively, cyclic uniaxial stretching has affected cell orientation and microtubular microstructure (Morioka et al. 2011), which may influence mitochondrial cluster and system properties and consequently ATP generation.

6.5.3 How Can Intermediate Filaments Affect Function and Structure of Mitochondria?

Apart from mitochondria, there are indications that intermediate filaments play a role in mitochondrial architecture and function (Anesti and Scorrano 2006). In this way, for instance, plectin, a cytoskeletal cross-linker, is physically affiliated to desmin, an intermediate filament that is in orderly fashion associated with mitochondria along the length of the sarcomere of striped muscles, pointing to the possibility that these proteins support the branching habit of mitochondria (Reipert et al. 1999). Vimentin represents another intermediate filament linked to mitochondria, as vimentin null cells demonstrated mitochondrial fission and reorganization, possibly by altering the connection between mitochondria and microtubules (Tang et al. 2008). Vimentin is involved in cell mechanics and safeguards the cell against compressive stress (Mendez et al. 2014), indicating the feasibility of external mechanical stress controlling mitochondrial architecture and function.

6.6 Mechanobiological Aspects of the Structure and Function of Mitochondria

Almost all cell types in the human body experience mechanical forces, including external pressure, shear stress, tensile stress and the environmental stiffness of the local extracellular matrix network. These mechanical parameters impact cell functions through the coupling of cell surface receptors including integrins at focal adhesions that represent interaction regions with the surrounding extracellular matrix, which is composed of several fibers, mostly collagen type I with the conserved-binding motif Arg-Gly-Asp (RGD). Cells are constantly exposed to the extracellular matrix, and hence, they can adapt to the matrix itself and react to such specific stimuli. Consequently, a cell tries to continuously reach and maintain a homeostatic balance with the extracellular matrix. In fact, vascular endothelial cells can mechanically sense forces, such as the shear stress in blood or lymph vessels that is evoked by the blood flow (James et al. 1995). In contrast, epithelial kidney cilia cells (Praetorius et al. 2005), bone cells including osteoblasts, osteocytes and osteoclasts (Jacobs et al. 2010), cartilage cells, such as chondrocytes (Shao et al. 2012), and eye (Luo et al. 2014) cells, such as cone or rod cells, can react to pressure in their specific tissues. The contraction of muscles exerts stresses on the muscle cells (Martineau and Gardiner 2001) and on adjacent nerve cells (Tock et al. 2002). The cyclic alteration of the blood pressure causes in the arteries circumferential stresses that are probing the vessel cells, such as VSMCs (Osol 1995). The process of breathing exerts also cyclically stretches toward all lung cells (Waters et al. 2002). With regard to the stiffness of the extracellular matrix, stem cells are a good candidate for highly adaptive cells because, when placed on soft extracellular matrices, which should resemble the neurogenic mechanical properties of the brain, they exhibit a soft phenotype. When the stem cells are placed on stiffer extracellular matrices, which resemble the mechanical phenotype of myogenic cells, they display a stiffer mechanical phenotype. Consequently, when the stem cells are cultured on even very stiff osteogenic extracellular matrices, which represent the mechanical phenotype of the muscle, they increase their stiffness (Engler et al. 2006).

6.7 Impact of Transient and Monotonous Stretching

Due to the fact that the cytoskeleton is the primary supporting element that reacts to all external mechanical impulses, the strong connection between the cytoskeletal and mitochondrial systems implies that mitochondria also need to be mechanosensitive. Despite this, mitochondrial reactions to mechanical irritations were only taken into account in a relatively short time. In an in vitro enhanced stretch model (24 h with 20% surface loading) of an abnormal mechanical environment, cardiomyocytes were subjected to apoptosis as cytochrome c was liberated out of mitochondria (Liao et al. 2004). The mitochondrial membrane potential also decreased, which apparently led to mitochondrial splitting (Liao et al. 2004). Bcl 2 proteins lead to stretch facilitated mitochondrial apoptosis. The breathing muscle vulnerability of the intensive care unit due to compromised membrane contractility was studied with a five-day mechanical ventilation model in farrows (Fredriksson et al. 2005). Whereas the mitochondrial amount did not vary, the electron transport chain complex IV activity dropped by 21%. Mechanical respiration, by contrast, absorbs the inherent flexibility of conventional tidal respiration and therefore indicates strongly that long-lasting abnormal monotonous mechanical impulses lead to distinct molecular alterations in the mitochondria. Periodic mechanical loading enhanced ROS synthesis in endothelial cells in an actin cytoskeleton-dependent way (Sukhorukov et al. 2012). In the same direction, a cyclic extension of lung epithelial cell types increased the expression of ROS in a manner dependent on duration and amplitude, indicating that lung hyperdistension during mechanical respiration can cause mitochondrial ROS-based lung injury (Chapman et al. 2005). Specifically, the trial also provided direct proof based on imaging that a total equibiaxial strain, with a 17% strain on the elastic membrane where the cells adhered to, led to local mitochondrial strain with up to 32% linear strain. When pulmonary fibroblasts were subjected to large transient equibiaxial extensions of up to 30%, mitochondria were torn at distinct sites instantly after extension (Imsirovic et al. 2015). Cells can be stained with the dye tetramethylrhodamine methyl ester (TMRM), whose intensity is regulated by the inner mitochondrial membrane potential and consequently by the production of ATP (Kadenbach et al. 2010) and thus ultimately represents a reliable mitochondrial marker (Ehrenberg et al. 1988). The findings indicate that external mechanical stresses can initiate cleavage directly and instantaneously. The culprit is that the cytoskeleton remains in a pre-stressed condition with tensile forces on actin fibers as a result of myosin motor activity (Ingber 2006), predisposing the cytoskeleton for the fast transfer of mechanical forces over long distances. In actual fact, it reminds of the quick and far-reaching mechanical transmission of force from focal adhesions to the nucleus (Hu et al. 2005). Although it has been confirmed that external mechanical forces, both temporary and long-term monotonic, affect the morphology and function of mitochondria, the internal mechanical microenvironment has also been found to propel mitochondrial cleavage via elastocapillary instability (Gonzalez-Rodriguez et al. 2015).

6.8 Fluctuations Affect Mechanotransduction Behavior

Tissues and cells of the body can react to external mechanical stresses by variations and fluctuations of their cellular constituents and structures. Beat-by-beat blood pressure alterations display significant variability, which is increased with hypertension (Mancia et al. 2001; Schillaci et al. 2012). Breathing produces a large breath-to-breath flexibility of the lateral volume (Dellaca et al. 2015). As the quantities and timing of mechanical stresses both govern the response of the actual cellular signal (Hoffman et al. 2011), fluctuations in mechanical impulses are anticipated to interfere with the modalities of mechanotransduction, a process referred to as fluctuation-driven mechanotransduction (Suki et al. 2016). Mechanotransduction is a mechanism that can be integrated into all mechanosensitive cellular processes over the course of evolutionary time scales. Under standard laboratory conditions, however, mechanotransduction is examined with either static or cyclic monotonic strain. It has been reported in several recent studies that fluctuations in cyclic loading or shear stress, referred to as variable strain or shear stress, profoundly affect cell function, including cytoskeletal organization, bioenergetics and transmission of signals (Bartolak-Suki et al. 2015; Arold et al. 2009; Uzarski et al. 2013).

A recent report has demonstrated that variation-induced mechanotransduction influences mitochondrial morphology and functional properties directly (Bartolák-Suki et al. 2017). In concrete terms, the ATP production rate determined by the quantitative analysis with a tetramethylrhodamine methyl ester labeling in mitochondria, which is a dye, whose intensity is correlated with the inner mitochondrial membrane potential and subsequently with the production of ATP, was twice as high for VSMCs grown on elastic membranes and equibiaxially elongated as for monotonic elongation after four hours of variable elongation (Bartolak-Suki et al. 2015). In addition, variable strain also directly involved constituents and phosphorylation of electron transport chain complexes: ATP synthase and cytochrome c oxidase and their phosphorylated form together with Mfn1 and Mfn2 were elevated, however not DRP1. Remarkably, variable strain has also initiated mitochondrial biogenesis, as the master regulator of biogenesis stimulated by external factors, the peroxisome proliferator-activated γ coactivator (Wu et al. 1999), PGC-1α, has risen relative to monotonic strain (Bartolak-Suki et al. 2015). These biochemical alterations were characterized by various structural shifts, such as an enhanced organization of actin, microtubule and mitochondrial meshes, which are distinguished by their fractal dimension and by their coefficient of variation (Bartolak-Suki et al. 2015). To investigate the potential mechanisms of fluctuation-driven mechanotransduction, inhibitory pharmacological drugs of the polymerization of actin, the depolymerization of microtubules, the ATP synthase, focal adhesion kinase (FAK), or the presence of calcium were employed. Both ATP synthesis and the size of the mitochondrial cluster were reduced, whereas the variable stretch can still keep an increased membrane potential of mitochondria than those stretched monotonously. Inhibitors of the microtubule and vimentin assembly caused the abolishment of the membrane potential of mitochondria and thereby differences between monotonously stretched and variable stretched cells vanished. In the case, when non-muscular myosin II, DRP1, which regulates the cleavage of mitochondria (Otera et al. 2013), or the mitotic kinesin-like protein 2, lowered the membrane potential in variably stretched cells, it enters the membrane potential levels of monotonically stretched cells. Consequently, the functional impact of fluctuation-driven, mechanotransduction-induced ATP generation during myosin light-chain phosphorylation in VSMCs in culture and aorta rings was elevated, resulting in turn in an increased contractile force generation during variable stretch in aortic rings (Bartolak-Suki et al. 2015). Thus, VSMCs are able to exploit fluctuations in their mechanical surroundings, and the extracted excess energy becomes apparent in enhanced chemical energy stored in ATP as a consequence of the reorganization and interaction of cytoskeletal and mitochondrial systems. The structure complexity-function relationship is a result of macroscopic mechanical fluctuations that alter the complexity of the cytoskeletal and mitochondrial framework, control oxidative phosphorylation and improve bioenergetics. The network complexity and the production of ATP of unstretched, monotonous stretched and variable stretched cells need to be compared to reveal similarities or dissimilarities. To shed light on this, unstretched cells generate and require the smallest amount of ATP, whereas variably stretched cells generate and require themselves the largest amount of ATP. Therefore, unstretched cells are nearest, while variable stretched cells are most distant from thermodynamic equilibrium as they are able to use energy from microenvironmental fluctuations to load mitochondria, the battery of life. The far from equilibrium operation is facilitated by a higher complexity of the mitochondrial system architecture. But it remains to wait and see whether the fluctuation-induced mechanotransduction has comparable results in certain other cell types.

6.9 Role of the Stiffness of the Extracellular Matrix on Mitochondria

There is not much knowledge about the regulatory function of the extracellular matrix stiffness on the structure and function of mitochondria. An evidence was supplied by an experiment in cardiac myocytes, basic metabolism is manipulated by extracellular matrix stiffness and, even bolder, the capacity of cells to adjust to metabolic stress is governed by both extracellular matrix stiffness and orientation of fibers (referred to as fiber alignment (Lyra-Leite et al. 2017, 2019). In order to finish these investigations, the hypothesis is that there exists a linkage between the mitochondrial structure and function of VSMCs and the stiffness of the extracellular matrix scaffold, since there has been similar evidence for this linkage in cardiac myocytes.

The influence of the substrate stiffness on the structure and function of mitochondria has been determined employing elastic gel (synonymously termed NuSil® 8100) with tunable stiffnesses due to the alteration of the ratios of the polymer and its cross-linker (Yoshie et al. 2018). Hence, several ratios of the polymer to cross-linker, such as 1:1, 1:2 and 1:5, have been analyzed after they have been incubated at 70 °C for one day. Stiffer gels can be generated by the addition of silicone elastomer Sylgard 184, which is mixed at a ratio of 1:10 to the 1:1 NuSil at 20 and 33% by weight. The stiffness of the gels was revealed by employing a uniaxial stretch (Araujo et al. 2011). The size of the gel was measured before fixing in a stretching device that applied a known displacement and recorded the corresponding force. Stress and strain were obtained using the measured force and physical dimensions of the specimen, and the slope of a linear fit was used as the stiffness. In the next step, new gel layers were polymerized and ligated using collagen type I. VSMC derived from bovine thoracic aortae was placed on top of these gels that possess stiffnesses in a range of 1–75 kPa. Thereafter, the cell's mitochondria were stained by using the membrane potential marker tetramethylrhodamine. In fact, the effects of stiffness on the size of the mitochondrial cluster were major, although the maximal stiffness difference was 18% for cells seeded on 12.5 kPa to all other stiffness values examined, such as 1.0, 12.5, 25 and 75 kPa. The stiffness impact on the tetramethylrhodamine intensity was even detected to be more pronounced, where the maximum intensity has been identified at 12.5 kPa with a

27% difference compared to the other values. Moreover, it should be mentioned that a stiffness of 12.5 kPa is highly related to the in vivo stiffness of the vascular wall, on which these cells are adhered to (Sehgel et al. 2015). Subsequently, the entire structural organization and the proper function of mitochondria within VSMCs seems to be adapted to this optimal stiffness value of the extracellular matrix stiffness creating the local vascular wall.

6.10 How Contributes the Mitochondrial Structure and Function to Diseases and Aging?

A focus lies on the linkage between the structure and function of mitochondria and the cytoskeleton and the impact on mechanical features on the cell body. It seems to be obvious that the fission dynamics and the fusion events fulfill a crucial role in the entire organization of the mitochondrial network structure that in turn controls the intramitochondrial processes including the generation of ATP and even external mitochondrial processes including apoptosis can be regulated. In specific detail, mitochondria can be seen as a network that seems to be critically connected to multiple cytoskeletal filamentous scaffolds, such as actin filaments. As the cell's cytoskeleton is known to be associated with all mechanosensitive functions, it seems to be possible that the entire mitochondrial scaffold as an organelle fulfills the same mechanosensitive tasks as the cytoskeleton, and hence, it should be treated as an integral section of the entire mechanosensing machine of the cell. In addition, the mitochondrial scaffold seems to be associated with the bidirectional mechanotransduction. In the first place, the network reacts to elongation and its time fluctuations applied to mitochondria on the cell scale by modifying biochemical signaling to generate ATP and ROS on a scale considerably lower than that of the network as a whole. But on the opposite, the mitochondrial system is able to control whole cell and possibly tissue and organ processes, including tissue contraction, apoptosis and organ-based male function, by regulating the cytosolic availability of ATP, ROS and cytochrome.

Mitochondria are associated with a wide spectrum of diseases and processes of cell aging (Zhang et al. 2010; Olichon et al. 2003; Kim et al. 2003; Bonnet et al. 2006; Bratic and Larsson 2013; Chistiakov et al. 2014; Gomes et al. 2013; Irwin et al. 2003; Mora et al. 2017; Nunnari and Suomalainen 2012; Reddy 2011; Sutendra et al. 2011; Zhou et al. 2012). In multiple diseases, mechanical features play a key or a prominent role; therefore, it seems possible that this is a reason why the function and the structure of mitochondria are found abnormal in many diseases when the mechanical properties of the extracellular matrix environment also become also abnormal. In fibrosis, the stiffness of the extracellular matrix rises, which in return modifies the cytoskeleton and therefore induces a mitochondrial reaction (Mora et al. 2017). A feature of asthma is the pathologically very strong contractional capability of airway smooth muscle cells. As the contraction of the muscle is known to require ATP and the extracellular matrix stiffness of the airway wall additionally needs to be elevated in asthma based on the redistribution of the tissue, mitochondria seem to be associated with the initiation and the further development of asthma (Reddy 2011). In the same way, the physiological process of aging involves a stiffening of the vascular wall (Roccabianca et al. 2014), and also the stiffness of the vascular wall (Mitchell et al. 2007) and similarly the variability in blood pressure (Mancia et al. 2007) are both elevated hypertension. Each of which factors seem to affect the mitochondrial processes, and thus, mechanism for the alteration of mitochondria seems to be novel in promoting different processes of aging and hypertension. Moreover, it can be hypothesized that almost all other body cells are subject to mechanical perturbation, and hence, the mitochondria of them may be associated with diseases where mechanical features are impacted. At present, the importance of mitochondrial structure and function in mechanobiology is underestimated, while mechanotransduction certainly lacked any influence on medicine. In the future, it requires to be examined that mitochondria and the mechanobiology are closely linked together and insights in this linkage may help to find new pathways for the treatment of several human diseases.

6.11 Mechanical Force Evokes Mitochondrial Fission

It is the present knowledge that eukaryotic cells are densely filled complexes that are assembled by macromolecules and entangled organelles, which are moved continuously through the cytoplasm and undergo ongoing reshaping. An interesting feature of organelles is that they try not to be entangled with each other or smashed against one another due to the constrained space. Apart from staying single, mitochondria can assemble complicated networks that are under continuous restructuring by division (synonymously termed fission) and fusion of mitochondria. It has been demonstrated that the fusion of mitochondria is induced by mechanical forces acting on them. The mechanical probing of mitochondria, which are evoked by migrating intracellular pathogens, can be mimicked by exertion of external pressure using an AFM or the migration of the entire cells over uneven microsurfaces. All of which leads finally to the involvement of the fission mechanism of mitochondria and their subsequent division. It seems plausible that the mitochondrial fission factor, due to its affinity for narrow mitochondria, serves as a membrane-bound force sensor to engage the fission apparatus in mechanically loaded places (Helle et al. 2017). Therefore, mitochondria conform to their environment by perceiving and reacting to biomechanical signals. The results that mechanical events can be connected with biochemical reactions in membrane dynamics allow to elucidate how organelles in the overfilled cytoplasm live in an ordered way together (Helle et al. 2017).

Eukaryotic cells are tightly filled with macromolecular complexes and entangled membranous organelles (Marsh et al. 2001). Some organelles, such as ER and mitochondria, gather to form highly intricate and dynamic interconnections and increase the complexity of the microcellular architecture. In addition, the cells are continuously reconfiguring their cytoplasm. In this way, for example, the vesicular and membrane-like turnover constantly travels large assemblies and organelles over long distances throughout the cytoplasm. In view of the restricted volume, organelles are often exposed to large complexes without any interference or intervention. In the specific situation of mitochondria, crashes and entanglements have catastrophic implications, such as cytochrome c secreting into the cytosol and the triggering of apoptosis. Therefore, at this point, it is assumed that cells are endowed with mechanisms that enable mitochondria to release potential tensions arising from collisions with other molecular structures, and hence, it needs to be analyzed how mitochondria deal with intra-cellular confinement.

6.11.1 Mitochondria Are Affected by the Actin-Driven Movement of *Shigella Flexneri*

The question is how the mitochondria deal with being attacked by an intracellular, rapidly moving specimen (Helle et al. 2017). *Shigella flexneri* microorganisms are pathogenic bacteria from the family Enterobacteriaceae, and an infection in humans results in diarrhea and heavy inflammation in the intestines. When entering the cytoplasm of diseased cells, a specific subpopulation of bacteria captures the actin cytoskeleton and activates its polymerization on the bacterial surface, creating an actin comet tail (Ray et al. 2009), allowing them to drift rapidly through the cytoplasm and reach velocities of up to 0.5 µm/s (Gouin et al. 1999). U2OS or COS7 cells were virally infected with fluorescent labeled *S. flexneri* and mitochondria viewed using BFP labeled mitochondrial matrix (mtBFP) (Kanfer et al. 2015). Time-lapse microscopy enabled us to monitor that bacteria frequently collided with mitochondria and displaced the mitochondrial tubules to the side, upward or downward. Clashes in some instances led to a noticeable decrease in mitochondrial fluorescence, suggesting that the matrix was narrowed. The mitochondria were cleaved within one to five minutes at the narrowed site in 60% of these occasions. In contrast, only 4% of non-stimulated mitochondria were cleaved within five minutes.

Mitochondrial cleavage and fusion are two contradictory phenomena that govern mitochondrial morphology and interconnectivity. Both biochemical processes are strongly regimented and peak in the specific enrollment of dynamin-related GTPases that drive mitochondrial cleavage and fusion (van der Bliek et al. 2013). The cleavage GTPase DRP1 builds up as homogeneous rings around the mitochondria and exploits the energy of GTP hydrolysis to crush and cleave the mitochondria (Francy et al. 2015).

To judge whether the collision-associated events of mitochondrial partitioning concerned the canonical splitting apparatus, bacterial motions were mapped cells lacking DRP1. The depletion of DRP1 can be performed by three different methods: firstly, the treatment of cells with siDRP1, secondly, the lentiviral transduction of shDRP1 and thirdly, the CRISPR-driven mutagenesis of the exon 2 with DRP1CRISPR. All possible experimental conditions led to an effective decrease of DRP1 levels and to an increased presence of mitochondria in both mock-infected and Shigella-infected cells. Mitochondria originating from DRP1-depleted cells were severely concerned by bacterial motility. They were pressed and towed and occasionally conspicuously diluted by narrowing. Contrary to wild type cells, mitochondria recovered from DRP1CRISPR cells in 100% of those cells without cleavage, and despite the severe loss of matrix staining during mechanical activation of mitochondria, mitochondria persisted permanently united. In the same way, DRP1 siRNA treatment abolished mobile Shigella-induced cleavage in wild type cells. These results suggest that DRP1 is required for bacterial-driven cleavage.

To verify that the cleavage occurrences seen in wild type cells were true cleavage phenomena (Friedman et al. 2011), the targeting of the mitochondrial cleavage apparatus to these fission sites needs to be observed. Consistent with other reports, in uninfected cells, the fluorescent protein-tagged DRP1 has been seen mainly to be diffusely localized in the cytosol with brighter spots in the mitochondria that seem to be stably connected to mitochondria, whereas a subset of this marker fusion-protein highlighted cleavage sites.

After the infection with Shigella, the DRP1-driven arrangement of focal points was directed to sites where moving bacteria intersected mitochondrial tubules. In fact, at these specific sites, mitochondria were cleaved. Incidents were also reported in which Shigella encountered mitochondrial areas that were characterized by a low DRP1 level signal that evolved into a more intense puncture on impact and ultimately resulted in cleavage. Together with the DRP1 deletion patterns, these findings point to the fact that mitochondria respond to collisions with bacteria by undergoing active cleavage. The variability in elapsed time between Shigella influence and possible cleavage can represent stochastic variations in DRP1 enrollment and activating kinetics.

6.11.2 The Fission of Mitochondria Can Be Triggered with an Atomic Force Microscope

How did the mitochondrial splitting device sense the presence of the bacterium? A possibility is that the verification is biochemical, due to factors located on the bacterial membrane surface. An alternative hypothesis is based on a mechanical verification that the mechanical forces caused by the collision initiated mitochondrial cleavage.

To identify whether mechanical probing is able to directly impose the cleavage of mitochondria, atomic force microscopy (AFM) was utilized (Binnig et al. 1986). More precisely, using AFM, the forces sensing or transmitting can be analyzed in the range of nanonewtons through a cantilever that carries a tip to probe the cells (Müller and Dufrene 2011). It is known that adhesive cells possess a flat periphery (typically 100–200 nm thick), when they are fully spread (Xu et al. 2012); thus, it seems to be possible that a pressure probing of these peripheral regions may cause a deformation of the cell's membrane surface and thereby transduce forces toward nearby mitochondria.

With round cantilever tip, a specific amount of force, such as 15 nN, can be exerted to cells, which are transfected with an outer membrane marker, such as mCherry-Fis1TM that is fusion protein of the mCherry fluorescent molecule to the transmembrane domain of Fis1 and hence identifies the outer mitochondrial membrane and co-stained with Mitotracker Deep Red identifying the mitochondrial matrix. The force probing with the AFM technique causes a constriction of the mitochondria similarly as it has been observed with the decrease of the mitotracker and the mCh-Fis1TM intensity. It is common for mitochondria to be split within one minute of approaching the tip, rarely between 1 and 5 min. To verify that the mitochondria had truly been cleaved, the AFM tip was withdrawn 50 s after the cleavage event and the cells remained imaged for a further approximately 100 s. The cells were then clearly divided by the AFM tip and a constriction ring of the mitochondrial matrix has been seen at the division site.

As noticed in the Shigella model, the time lag between peak attachment and mitochondrial cleavage tended to vary, probably mirroring the splitting apparatus recruitment. In fact, these results depended on DRP1, as the repetition of these experiments in DRP1-deficient cells caused a reduction of cleavage processes during force exertion. The mitochondria in DRP1-deficient cells were noticeably constrained at and during 5 min of force exertion, whereas they generally relaxed after retraction of the tip from the force impulse. Although the force exertion can be precisely determined at the AFM tip, however, it is impossible to reveal the force fraction that is exerted to underlying mitochondria, since the cortical actin scaffold can additionally be exerted by a force fraction. The forces required to deform the actin cytoskeleton are in the range of several nanonewtons (Hadjiantoniou et al. 2012).

6.11.3 Mitochondrial Cleavage of Cells Placed on Patterned Substrates

External forces exerted by an AFM cantilever tip may be irrelevant for the cleavage of mitochondria. For this reason, an experimental system was needed in which the cells themselves were directly involved in force generation. It can be hypothesized that the morphological plasticity is a physical property of multiple cell types to remodel themselves due to environmental cues. When cells grown inside of dense and complex tissues, they are subject to ongoing deformation that can even impact their mitochondria. These environmental conditions can be rebuilt in vitro by culturing U2OS cells on uneven surfaces with grooves that are approximately 40 μm deep and 80 μm wide and possess flat edge between neighboring grooves, such as vinyl records (Read 1952). Hence, the question can be raised whether cells grown on vinyl records spreading over the edge into the groove can restrict the edge region to peripheral cytoplasmic content without mitochondria. In order to proof this hypothesis, cells have been analyzed that express an ER marker, such as GFP-Sec61β, marking the cytoplasm (Kanfer et al. 2015). In fact, the 3D reconstruction of the spread cell over the edge region revealed that the cytoplasm seems to be confined over the edge region. These constricted sites seem to be regions with elevated mitochondrial cleavage activity. In fact, the cytoplasmic regions of the cells covering the edge of the groove tended to be free of mitochondria. In order to analyze whether these regions display elevated cleavage activity of mitochondria, time-lapse microscopy can be employed, and subsequently, a focus can be set on the subset of cells that still possess intact mitochondria. This subset of cells has just begun to cover the edge region and start immediately to cleave their mitochondria. These results support the hypothesis that cells in constricted areas due to their adhesion to a patterned surface can cleave heir mitochondria.

The mitochondrial matrix at the edge of the groove showed complete division into about 80% of wild type cells, whereas the cleavage rates in DRP1 siRNA-treated and DRP1CRISPR knockout cells are reduced to only 15 and 43% that exhibited a complete cleavage of the mitochondrial matrix, suggesting that the canonical cleavage system is essential for this process. The milder effect of the CRISPR-mediated DRP1 knockout compared to the siRNA-mediated knockdown can be a result of adjustments to long-term DRP1 depletion. Cell adhesion to uneven surfaces can cause mitochondria to undergo DRP-mediated cleavage in narrowed cytoplasmic areas.

6.12 The Inhibition of the ER or the Dynamics of Actin May Not Significantly Alter the Force-Induced Fission

Without any stimulation, the ER tubules can highlight the sites for mitochondrial fission (Friedman et al. 2011). The ER seems to be entangled around mitochondria and thereby can facilitate the constriction of mitochondria before their fission starts that appears to be regulated by the locally restricted polymerization of actin through the activity of inverted formin 2 (INF2) (Korobova et al. 2013) and Spire1C (Manor et al. 2015). However, in AFM activation of U2OS cells, which stably express both a mitochondrial (mtBFP) and an ER (GFP-Sec61β) marker, it was found that the tip-facilitated indentation of cells displaced the ER from the site of force exertion and subsequent causes the mitochondrial cleavage. However, this seems to be contrary to the hypothesis that the ER fulfills an important role in the force-based cleavage of mitochondria. Hence, it has been investigated the effect of perturbed dynamics of the ER on mitochondria function and thereby employed cells that overexpress the dominant-negative Atlastin (ATL) mutant or the CLIMP-63 mutant. More precisely, ATLs represent large GTPases facilitating the fusion of ER tubules through the assembly of *trans*-homooligomers on the cytoplasmic ER. The overexpression of a GTPase-dead mutant, such as ATL1-K80A, or overexpression of the cytoplasmic domain, such as cyto-ATL2, had both a dominant-negative effect on endogenous ATLs and subsequently they impaired the fusion of ER tubules (Goyal and Blackstone 2013; Pawar et al. 2017). In contrast, CLIMP-63 is able to sheets of the ER, and hence, its overexpression fuses ER tubules mainly to ER sheets (Goyal and Blackstone 2013).

Obviously, in cells overexpressing ATL1-K80A or Cyto-ATL2, ER assumed a hair-like, hypoconnected morphology, which raised the number of cytoplasmic regions lacking ER. Inversely, the ER was transformed from tubules to plates in CLIMP-63-overexpressing cells. The mitochondrial matrix in these cells displayed typical morphology. In cells located at the edge of vinyl grooves, mitochondria were cleaved at the edge, similar to wild type cells. Some of these cleavages arose in areas with ER sheets only or without ER. The vast majority of cells overexpressing ATL1-K80A and CLIMP-63 exhibited a full cleavage of their mitochondrial matrix. Subsequently, cells expressing cyto-ATL2 were also exposed to Shigella, and a motile Shigella triggered the mitochondrial cleavage even in regions without an ER signal. Although ER could play a role in force-induced cleavage, this function does not seem absolutely essential.

In addition, the role of actin in mechanically facilitated mitochondrial cleavage has been studied, since actin polymerization appears to be a key downstream step in ER-tubule-based cleavage through the ER-anchored isoform of inverted formin 2 (INF2). The knockdown of the ER-associated isoform of INF2 had no apparent influence on collision-induced cleavage into Shigella-infected cells or cells cultured on patterned substrates, such as vinyl records. In fact, the collisions found in INF2-deficient cells causing a cleavage process are not pronouncedly altered to cells treated with scrambled siRNA. It is interesting to note that actin-based Shigella motility was not impaired by INF2 depletion, probably due to the usage of Shigella N-WASP and Arp2/3 to encourage actin polymerization (Gouin et al. 1999; Ray et al. 2009). In the same way, 83% of cells cultured on gramophone recordings and exhibiting a constricted cytoplasm had isolated mitochondrial populations, similar to what in control cells. These results suggest that the INF2-facilitated route is not required for the mechanically triggered cleavage. To tackle more generally when actin fibers were essential for force-induced cleavage, cells were exposed to cytochalasin D (cytD), a widely applied drug that abolishes the actin polymerization. Since the actin cytoskeleton is crucial for Shigella motility and for the correct distribution of cells on vinyl plates, the actin cytoskeleton has to be disturbed in AFM experiments. In effective concentration, the drug strongly alters the actin cytoskeleton as evaluated by LifeAct-GFP. Therefore, actin polymerization seems not to be a mandatory step in force-induced cleavage of mitochondria.

6.13 Force-Sensing Mechanism and the Recruitment of DRP1

It was observed that mechanical forces induced the recruitment and activation of the cleavage effector DRP1 and how such a mechanical impulse was perceived at the molecular level. DRP1 is present in the cytosol and is enrolled into the mitochondria by employing integral mitochondrial membrane adaptor molecules such as the mitochondrial cleavage factor (Gandre-Babbe and van der Bliek 2008; van der Bliek et al. 2013) and mitochondrial dynamic protein of 49/51 kDa (Mid49/51) (Palmer et al. 2011; Zhao et al. 2011). These two adaptor proteins seem to able to target DRP1 to the predefined cleavage regions (Friedman et al. 2011). The mitochondrial fission factor can be highly concentrated in dot-like structures within mitochondria, which is regulated in an independent way to the availability of DRP1 (Otera et al. 2010). Hence, it can be hypothesized that adaptor proteins function as mechanosensory molecules located on the membrane surface of mitochondria. To verify this hypothesis, the localization of the mitochondrial fission factor can be determined on the mitochondrial surface using immunofluorescence. This analysis was examined in cells lacking DRP1 in order to rule out that DRP1 did not drive the localization of the mitochondrial fission factor. It has been observed that the mitochondrial fission factor is localized in specific foci independent of DRP1, which is in line with a previous finding (Otera et al. 2010). Moreover, the mitochondrial fission factor seems to be concentrated at constriction regions that are built even on non-perturbed tubules of the mitochondria over time. This behavior has been seen in living cells, which were transfected with GFP-tagged mitochondrial fission factor (Friedman et al. 2011). In order to figure out whether mitochondrial fission factor manages to be highly concentrated in force-initiated constrictions, DRP1-depleted cells were placed on patterned surfaces, such as vinyl records. In fact, the mitochondrial fission factor has been accurately detected in constricted regions of the mitochondria adhering to the groove edge, suggesting that the mitochondrial fission factor can sense the mechanical properties at the constricted region in the mitochondria through the use of factors located upstream of DRP1. In the same way, pronouncedly reduced cleavage events have been determined in cells lacking the mitochondrial fission factor than in wild type cells, when the either the Shigella setup or vinyl records have been employed.

To detect this phenomenon in living cells, lentiviruses expressing GFP-tagged mitochondrial fission factor have been transfected to cells in order to monitor the localization upon mechanical probing with bacteria or patterned topologies. In addition, the expression of DRP1 can be reduced by employing siDRP1. In fact, a pronounced mitochondrial constriction can be seen, when the mtBFP signal is reduced. In addition, the GFP-mitochondrial fission factor was targeted to these constriction regions in both mechanical model systems. Hence, the mitochondrial fission factor accumulates referentially to flat mitochondria and may act as a mechanosensor for the cleavage system.

In these experiments, it has been seen that there are mitochondrial regions to which the mitochondrial fission factor is mainly targeted, independently of Shigella- or pattern-based mechanical probing and even in the absence of mechanical probing. At these sites, a decreased mitochondrial matrix-targeted mtBFP signal has been detected, which points out to a decreased diameter of the mitochondria. The outer mitochondrial membrane (OMM)-directed fluorescence marker, termed mCh-Fis1TM, revealed that these mitochondria still possess connectivity, since these thin, GFP-mitochondrial fission factor-positive, matrix-negative tubules are associated to thick regions of mitochondria. These two heterogeneous regions remained interconnected even for several minutes, while they were still clearly and stably isolated topologically into two distinct mitochondrial zones. However, these two distinct structures can solely be detected when GFP-mitochondrial fission factor is highly overexpressed, which may be just can artifact of the overexpression and no actual cell property. It seemed that the mitochondrial fission factor not only functions as a sensor, but also potentially serves as an initiator of mitochondrial constriction.

Since it appeared paradoxical that the mitochondrial cleavage factor operated both as a sensor and as a trigger for mitochondrial constrictions, the question arose whether these two properties might be mutually connected. In order to answer this question, computer-aided Monte Carlo (MC) simulations were used to model a typical protein with specific affinity for narrow regions of the mitochondria. To model the intended affinity, a protein with a curved membrane-binding surface emerged and multiple copies were laid on a membrane tube whose radius exceeded that of the protein's attachment surface. Once the optimal ratio of the protein and tube diameters needed for protein compounding had been determined, a first simulation was carried out in which the protein density was arbitrarily low. Under these specific conditions, the protein can diffuse freely on the mitochondrial membrane tube, and hence, it is still homogenously distributed. The simulation was repeated; however, this time a narrowing in the membrane tube to imitate mechanical stimuli had been chosen and it has been seen that the proteins are targeted to the narrowing region, which clearly indicates that the mitochondrial fission factor was recruited to mechanically probed regions. The next step was to repeat the simulation using a high protein density. Under these conditions, the proteins spontaneously tightened the membrane tube without prior narrowing. The simulation outcomes closely rebuilt the mitochondrial cleavage factor-stabilized, matrix-free thin mitochondrial regions, which were characterized by a remarkably high level of overexpression of the mitochondrial cleavage factor.

Consequently, these findings imply that mitochondria react not only to biochemical but also to mechanical signals. The mechanical force, causing the deformation of mitochondrial membranes, facilitated the cleavage of mitochondria that connects a mechanical event with a biochemical reaction. How is this reaction organized on the molecular scale? A previously unknown property of mitochondrial fission factor is reported to have a preference for smaller diameter mitochondrial tubules. Thus, a simple model for force-based cleavage is that the physical narrowing could decrease the mitochondrial diameter that leads to the concentration of the mitochondrial fission factor and consequently to the assembly of DRP1 and the cleavage of mitochondria.

The mitochondrial fission factor can not only have a mechanosensory effect, but can also additionally trigger the narrowing in the mitochondria. The Monte Carlo simulations demonstrate that a protein with an increased affinity for mitochondria with a narrower diameter can be quickly aggregated at narrow locations. At high concentrations, on the contrary, it will initiate and stabilize narrowing, two behaviors that imitate what is experienced with the GFP mitochondrial cleavage factor in living cells. Therefore, the tendency of a protein to pinpoint narrow sites is inextricably linked to its ability to stabilize narrowed regions. The distinct behaviors, which differ in the initiating and sensing reactions, seem to be regulated by the amount of available protein. In fact, the mitochondrial fission factor performs a stabilization functions of narrowed region in mitochondria, when it is overexpressed, which is similarly to the function of other membrane curvature-sensing proteins (Baumgart et al. 2011; Simunovic et al. 2015; Sorre et al. 2012). It is noteworthy that the protein simulated is not intended to reproduce the form of the mitochondrial fission factor. However, the simulations show that capturing and inducing narrowrings are two sides of the same medal and that the difference between capturing and induction is in the degree of expression of the factor. As a matter of fact, it is unclear whether mitochondrial fission factor monomers or oligomers possess a curved membrane-binding interfacial region, such as the model protein in the simulations. The affinity of the mitochondrial fission factor for narrowed mitochondria may arise from other factors such as the affinity for specific lipid species or the properties of its transmembrane domain. There are several other DRP1 adapters in the cell that point to a large number of factors that influence DRP1 enrollment. The mitochondrial cleavage factor is probably not the only force sensor present on the mitochondrial surface. Similarly, it is improbable that force is the only cause of mitochondrial cleavage.

In undisturbed adhesive cells, mitochondrial cleavage occurs almost exclusively at sites that come into direct physical contact with the ER (Friedman et al. 2011). The force-induced mitochondrial cleavage is a mechanistic justification for this phenomenon. In fact, adherent cells have a very flat circumference (Xu et al. 2012), with only a single ER tubuli layer, which renders them particularly suitable for microscopic investigations. Therefore, the ER-mediated mitochondrial division is mainly examined in peripheral areas of adherent cells. Mitochondrial tubules tend to be wider in diameter of about 500–700 nm (Youle and Karbowski 2005) than the cytoplasm thickness in these peripheral areas. In fact, high-resolution imaging techniques demonstrate that peripheral mitochondria are flattened by the proximity of the "ventral" and "dorsal" actin cortical fibers (Huang et al. 2008, 2016), assume an elliptical—and not a tubular—cross section, and slightly curve away from the cell surface (Wojcik et al. 2015). In high-speed AFM measurements of living cells, it can be detected that mitochondria distort the plasma membrane from the nearby actin cytoskeleton of about 50 nm (Yoshida et al. 2015). It is obvious that peripheral mitochondria trapped between the dorsal and ventral actin cortices are mechanically restricted in these regions. Why would the pressure not divide the peripheral mitochondria? Presumably, the pressure of the actin cortices is uniform over the entire length of the mitochondria and does not cause local contractions in which the mitochondrial fission factor and DRP1 accumulate. In this connection, it is appealing to suggest that crossing-over ER tubules, solidified by an INF2- and Spire1C-mediated actin envelope, could exert localized mechanical forces on the underlying mitochondria, initiating cleavage at the ER–mitochondria contact points. As it is the most extensive intracellular membrane system, the ER network physically collides with the mitochondrial network, which accounts for the frequent occurrence of cleavage at ER contact sites. In furtherance of this concept, the mechanically stimulated cleavage process no longer looks to be dependent on ER tubules. An encouraging hypothesis is that ER is an intrinsic reservoir of mechanical force and therefore seems to be dispensable under the conditions of this experiment because an external mechanical force is available. In turn, this implies that the ER exerts a force rather than providing biochemical signals in ER-mediated cleavage events. Since actin is an essential force generator in the cell, the model could also help elucidate why it is crucial for mitochondrial cleavage (Korobova et al. 2013; Manor et al. 2015), except when cells are mechanically exited by an AFM.

The role of mitochondrial fusion and cleavage has been left unclear. Both processes are suggested to serve in the repair of damaged mitochondria by merging with healthy mitochondria, in the severance of terminal injured mitochondria for waste disposal by mitophagy and in the proper distribution of mitochondria to daughter cells in mitosis (Katajisto et al. 2015; Twig et al. 2008). Mitochondria are cleaved by DRP1 and mitochondrial fission factor-mediated cleavage when mechanically triggered. Cells continuously pass through incidents such as organelle transport, expansion and contraction of the cytoskeleton and the overall reconstruction of the cytoplasm. Organelles which are as extensive as the ER, and the mitochondria must deal with a constantly evolving natural environment and prevent conflicts and enmeshment. By eliminating all the consequential mechanical stresses, cleavage could allow mitochondria to avoid shearing and rupturing, leading to cytochrome c secretion and apoptotic cell death and preventing entanglements. In this property, DRP1 activity can occur in parallel with the activity of topoisomerase-II during genome unbundling. By using serial block surface scanning electron microscopy, interlinkages between mitochondria and ER in mammalian cells are detected. Due to the dynamics of these organelles, it is likely that entanglements lead to a mechanical loading of the mitochondria and thus to their dissolution by cleavage. In contrast to mitochondrial cleavage, ER cleavage is not frequently encountered. Therefore, it is assumed that mitochondrial cleavage is the principal mechanism for loosening these linkages. Mechanically initiated mitochondrial alterations might be of particular relevance for cell migration through narrow passages such as leukocyte extravasation and immune monitoring (Muller 2011; Nourshargh and Alon 2014) as well as for the malignant progression of cancer (Friedl and Wolf 2003).

In conclusion, it has been demonstrated that the mechanical force can be connected to a distinct cellular process that is critical for the shaping and the localization of the intracellular membrane and therefore seem to offer a new view on the dynamics of mitochondria and the interaction of mitochondria with other organelles.

References

M.H. Ali, D.P. Pearlstein, C.E. Mathieu, P.T. Schumacker, Mitochondrial requirement for endothelial responses to cyclic strain: implications for mechanotransduction. Am, J. Physiol. Lung Cell Mol. Physiol. **287**, L486–L496 (2004)

R. Anand, T. Wai, M.J. Baker, N. Kladt, A.C. Schauss, E. Rugarli, T. Langer, The i-AAA protease YME1L and OMA1 cleave OPA1 to balance mitochondrial fusion and fission. J. Cell Biol. **204**, 919–929 (2014)

V. Anesti, L. Scorrano, The relationship between mitochondrial shape and function and the cytoskeleton. Biochim. Biophys. Acta **1757**, 692–699 (2006)

M.A. Aon, S. Cortassa, B. O'Rourke, Percolation and criticality in a mitochondrial network. Proc. Natl. Acad. Sci. USA **101**, 4447–4452 (2004)

A.D. Araujo, A. Majumdar, H. Parameswaran, E. Yi, J.L. Spencer, M.A. Nugent, B. Suki, Dynamics of enzymatic digestion of elastic fibers and networks under tension. Proc. Natl. Acad. Sci. USA **108**, 9414–9419 (2011)

S.P. Arold, E. Bartolak-Suki, B. Suki, Variable stretch pattern enhances surfactant secretion in alveolar type II cells in culture. Am. J. Physiol. Lung Cell Mol. Physiol. **296**, L574–L581 (2009)

D. Bach, S. Pich, F.X. Soriano et al., Mitofusin-2 determines mitochondrial network architecture and mitochondrial metabolism. A novel regulatory mechanism altered in obesity. J. Biol. Chem. **278**, 17190–17197 (2003)

E. Bartolák-Suki, J. Imsirovic, Y. Nishibori, R. Krishnan, B. Suki, Regulation of mitochondrial structure and dynamics by the cytoskeleton and mechanical factors. Int. J. Mol. Sci. **18**(8). pii: E1812 (2017)

E. Bartolak-Suki, J. Imsirovic, H. Parameswaran, T.J. Wellman, N. Martinez, P.G. Allen, U. Frey, B. Suki, Fluctuation-driven mechanotransduction regulates mitochondrial-network structure and function. Nat. Mater. **14**, 1049–1057 (2015)

T. Baumgart, B.R. Capraro, C. Zhu, S.L. Das, Thermodynamics and mechanics of membrane curvature generation and sensing by proteins and lipids. Annu. Rev. Phys. Chem. **62**, 483–506 (2011)

J. Bereiter-Hahn, M. Voth, Dynamics of mitochondria in living cells: shape changes, dislocations, fusion, and fission of mitochondria. Microsc. Res. Tech. **27**, 198–219 (1994)

G. Binnig, C.F. Quate, C. Gerber, Atomic force microscope. Phys. Rev. Lett. **56**, 930–933 (1986)

I.R. Boldogh, L.A. Pon, Mitochondria on the move. Trends Cell Biol. **17**, 502–510 (2007)

I.R. Boldogh, L.A. Pon, Interactions of mitochondria with the actin cytoskeleton. Biochim. Biophys. Acta **1763**, 450–462 (2006)

P.A. Bondzie, H.A. Chen, M.Z. Cao, J.A. Tomolonis, F. He, M.R. Pollak, J.M. Henderson, Non-muscle myosin-IIA is critical for podocyte f-actin organization, contractility, and attenuation of cell motility. Cytoskeleton **73**, 377–395 (2016)

S. Bonnet, E.D. Michelakis, C.J. Porter et al., An abnormal mitochondrial-hypoxia inducible factor-1alpha-Kv channel pathway disrupts oxygen sensing and triggers pulmonary arterial hypertension in fawn hooded rats: similarities to human pulmonary arterial hypertension. Circulation **113**, 2630–2641(2006)

A. Bratic, N.G. Larsson, The role of mitochondria in aging. J. Clin. Investig. **123**, 951–957 (2013)

M. Cagalinec, D. Safiulina, M. Liiv, J. Liiv, V. Choubey, P. Wareski, V. Veksler, A. Kaasik, Principles of the mitochondrial fusion and fission cycle in neurons. J. Cell Sci. **126**, 2187–2197 (2013)

M. Carre, N. Andre, G. Carles, H. Borghi, L. Brichese, C. Briand, D. Braguer, Tubulin is an inherent component of mitochondrial membranes that interacts with the voltage-dependent anion channel. J. Biol. Chem. **277**, 33664–33669 (2002)

K.E. Chapman, S.E. Sinclair, D. Zhuang, A. Hassid, L.P. Desai, C.M. Waters, Cyclic mechanical strain increases reactive oxygen species production in pulmonary epithelial cells. Am. J. Physiol. Lung Cell Mol. Physiol. **289**, L834–L841 (2005)

H. Chen, S.A. Detmer, A.J. Ewald, E.E. Griffin, S.E. Fraser, D.C. Chan, Mitofusins Mfn1 and Mfn2 coordinately regulate mitochondrial fusion and are essential for embryonic development. J. Cell Biol. **160**, 189–200 (2003)

D.A. Chistiakov, I.A. Sobenin, V.V. Revin, A.N. Orekhov, Y.V. Bobryshev, Mitochondrial aging and age-related dysfunction of mitochondria. BioMed. Res. Int. **2014**, 238463 (2014)

S. Cogliati, C. Frezza, M.E. Soriano et al., Mitochondrial cristae shape determines respiratory chain supercomplexes assembly and respiratory efficiency. Cell **155**, 160–171 (2013)

B. Daum, A. Walter, A. Horst, H.D. Osiewacz, W. Kuhlbrandt, Age-dependent dissociation of ATP synthase dimers and loss of inner-membrane cristae in mitochondria. Proc. Natl. Acad. Sci. USA **110**, 15301–15306 (2013)

R.L. Dellaca, A. Aliverti, A.L. Mauro, K.R. Lutchen, A. Pedotti, B. Suki, Correlated variability in the breathing pattern and end-expiratory lung volumes in conscious humans. PLoS ONE **10**, e0116317 (2015)

D.E. Discher, P. Janmey, Y.L. Wang, Tissue cells feel and respond to the stiffness of their substrate. Science **310**, 1139–1143 (2005)

B. Ehrenberg, V. Montana, M.D. Wei, J.P. Wuskell, L.M. Loew, Membrane potential can be determined in individual cells from the nernstian distribution of cationic dyes. Biophys. J. **53**, 785–794 (1988)

A.J. Engler, S. Sen, H.L. Sweeney, D.E. Discher, Matrix elasticity directs stem cell lineage specification. Cell **126**, 677–689 (2006)

D.A. Fletcher, R.D. Mullins, Cell mechanics and the cytoskeleton. Nature **2010**(463), 485–492 (2010)

C.A. Francy, F.J. Alvarez, L. Zhou, R. Ramachandran, J.A. Mears, The mechanoenzymatic core of dynamin-related protein 1 comprises the minimal machinery required for membrane constriction. J. Biol. Chem. **290**, 11692–11703 (2015)

K. Fredriksson, P. Radell, L.I. Eriksson, K. Hultenby, O. Rooyackers, Effect of prolonged mechanical ventilation on diaphragm muscle mitochondria in piglets. Acta Anaesthesiol. Scand. **49**, 1101–1107 (2005)

C. Frezza, S. Cipolat, O. Martins de Brito et al., OPA1 controls apoptotic cristae remodeling independently from mitochondrial fusion. Cell **126**, 177–189 (2006)

P. Friedl, K. Wolf, Tumour-cell invasion and migration: diversity and escape mechanisms. Nat. Rev. Cancer **3**, 362–374 (2003)

J.R. Friedman, L.L. Lackner, M. West, J.R. DiBenedetto, J. Nunnari, G.K. Voeltz, ER tubules mark sites of mitochondrial division. Science **334**, 358–362 (2011)

S. Gandre-Babbe, A.M. van der Bliek, The novel tail-anchored membrane protein Mitochondrial fission factor controls mitochondrial and peroxisomal fission in mammalian cells. Mol. Biol. Cell **19**, 2402–2412 (2008)

R.J. Giedt, D.R. Pfeiffer, A. Matzavinos, C.Y. Kao, B.R. Alevriadou, Mitochondrial dynamics and motility inside living vascular endothelial cells: role of bioenergetics. Ann. Biomed. Eng. **40**, 1903–1916 (2012)

A.P. Gomes, N.L. Price, A.J. Ling et al., Declining NAD(+) induces a pseudohypoxic state disrupting nuclear-mitochondrial communication during aging. Cell **155**, 1624–1638 (2013)

D. Gonzalez-Rodriguez, S. Sart, A. Babataheri, D. Tareste, A.I. Barakat, C. Clanet, J. Husson, Elastocapillary instability in mitochondrial fission. Phys. Rev. Lett. **115**, 088102 (2015)

E. Gouin, H. Gantelet, C. Egile, I. Lasa, H. Ohayon, V. Villiers, P. Gounon, P.J. Sansonetti, P. Cossart, A comparative study of the actin-based motilities of the pathogenic bacteria Listeria monocytogenes, *Shigella flexneri* and *Rickettsia conorii*. J. Cell Sci. **112**(Pt 11), 1697–1708 (1999)

U. Goyal, C. Blackstone, Untangling the web: mechanisms underlying ER network formation. Biochim. Biophys. Acta (BBA)—Mol. Cell Res. **1833**, 2492–2498 (2013)

M.W. Gray, Mosaic nature of the mitochondrial proteome: Implications for the origin and evolution of mitochondria. Proc. Natl. Acad. Sci. USA **112**, 10133–10138 (2015)

S. Hadjiantoniou, L. Guolla, A.E. Pelling, Mechanically induced deformation and strain dynamics in actin stress fibers. Commun. Integr. Biol. **5**, 627–630 (2012)

M.H. Heggeness, M. Simon, S.J. Singer, Association of mitochondria with microtubules in cultured cells. Proc. Natl. Acad. Sci. USA **75**, 3863–3866 (1978)

S.C.J. Helle, Q. Feng, M.J. Aebersold et al., Mechanical force induces mitochondrial fission. eLife **6**, e30292 (2017)

B.D. Hoffman, C. Grashoff, M.A. Schwartz, Dynamic molecular processes mediate cellular mechanotransduction. Nature **475**, 316–323 (2011)

S. Hoppins, L. Lackner, J. Nunnari, The machines that divide and fuse mitochondria. Annu. Rev. Biochem. **76**, 751–780 (2007)

S. Hu, J. Chen, J.P. Butler, N. Wang, Prestress mediates force propagation into the nucleus. Biochem. Biophys. Res. Commun. **329**, 423–428 (2005)

B. Huang, S.A. Jones, B. Brandenburg, X. Zhuang, Whole-cell 3D STORM reveals interactions between cellular structures with nanometer-scale resolution. Nat. Methods **5**, 1047–1052 (2008)

F. Huang, G. Sirinakis, E.S. Allgeyer et al., Ultra-high resolution 3D imaging of whole cells. Cell **166**, 1028–1040 (2016)

P. Huang, C.A. Galloway, Y. Yoon, Control of mitochondrial morphology through differential interactions of mitochondrial fusion and fission proteins. PLoS ONE **6**, e20655 (2011)

J. Imsirovic, T.J. Wellman, J.R. Mondonedo, E. Bartolak-Suki, B. Suki, Design of a novel equi-biaxial stretcher for live cellular and subcellular imaging. PLoS ONE **10**, e0140283 (2015)

D.E. Ingber, Cellular mechanotransduction: putting all the pieces together again. FASEB J. **20**, 811–827 (2006)

W.A. Irwin, N. Bergamin, P. Sabatelli et al., Mitochondrial dysfunction and apoptosis in myopathic mice with collagen VI deficiency. Nat. Genet. **35**, 367–371 (2003)

N. Ishihara, M. Nomura, A. Jofuku et al., Mitochondrial fission factor Drp1 is essential for embryonic development and synapse formation in mice. Nat. Cell Biol. **11**, 958–966 (2009)

C.R. Jacobs, S. Temiyasathit, A.B. Castillo, Osteocyte mechanobiology and pericellular mechanics. Annu. Rev. Biomed. Eng. **12**, 369–400 (2010)

N.L. James, D.G. Harrison, R.M. Nerem, Effects of shear on endothelial cell calcium in the presence and absence of ATP. FASEB. J. **9**, 968–973 (1995)

G.B. John, Y. Shang, L. Li, C. Renken, C.A. Mannella, J.M. Selker, L. Rangell, M.J. Bennett, J. Zha, The mitochondrial inner membrane protein mitofilin controls cristae morphology. Mol. Biol. Cell **16**, 1543–1554 (2005)

B. Kadenbach, R. Ramzan, L. Wen, S. Vogt, New extension of the Mitchell theory for oxidative phosphorylation in mitochondria of living organisms. Biochim. Biophys. Acta **1800**, 205–212 (2010)

G. Kanfer, T. Courtheoux, M. Peterka et al., Mitotic redistribution of the mitochondrial network by Miro and Cenp-F. Nat. Commun. **6**, 8015 (2015)

P. Katajisto, J. Döhla, C.L. Chaffer, N. Pentinmikko, N. Marjanovic, S. Iqbal, R. Zoncu, W. Chen, R.A. Weinberg, D. M. Sabatini, Stem cells. Asymmetric apportioning of aged mitochondria between daughter cells is required for stemness. Science **348**, 340–343 (2015)

J.S. Kim, L. He, J.J. Lemasters, Mitochondrial permeability transition: a common pathway to necrosis and apoptosis. Biochem. Biophys. Res. Commun. **304**, 463–470 (2003)

M.A. Kluge, J.L. Fetterman, J.A. Vita, Mitochondria and endothelial function. Circ. Res. **112**, 1171–1188 (2013)

W.J. Koopman, S. Verkaart, H.J. Visch, F.H. van der Westhuizen, M.P. Murphy, L.W. van den Heuvel, J.A. Smeitink, P.H. Willems, Inhibition of complex I of the electron transport chain causes O_2–mediated mitochondrial outgrowth. Am. J. Physiol. Cell Physiol. **288**, C1440–C1450 (2005a)

W.J. Koopman, H.J. Visch, S. Verkaart, L.W. van den Heuvel, J.A. Smeitink, P.H. Willems, Mitochondrial network complexity and pathological decrease in complex I activity are tightly correlated in isolated human complex I deficiency. Am. J. Physiol. Cell Physiol. **289**, C881–C890 (2005b)

F. Korobova, V. Ramabhadran, H.N. Higgs, An actin-dependent step in mitochondrial fission mediated by the ER-associated formin INF2. Science **339**, 464–467 (2013)

E. Kremneva,M. Kislin, X.L. Kang, Khiroug, Motility of astrocytic mitochondria is arrested by Ca^2+-dependent interaction between mitochondria and actin filaments. Cell Calcium **53**, 85–93 (2013)

A.V. Kuznetsov, S. Javadov, R. Guzun, M. Grimm, V. Saks, Cytoskeleton and regulation of mitochondrial function: the role of beta-tubulin II. Front. Physiol. **4**, 82 (2013)

L.L. Lackner, Shaping the dynamic mitochondrial network. BMC Biol. **12**, 35 (2014)

J.S. Lee, X. Hou, N. Bishop et al., Aquaporin-assisted and ER-mediated mitochondrial fission: a hypothesis. Micron **47**, 50–58 (2013)

F. Legros, A. Lombes, P. Frachon, M. Rojo, Mitochondrial fusion in human cells is efficient, requires the inner membrane potential, and is mediated by mitofusins. Mol. Biol. Cell **13**, 4343–4354 (2002)

X.D. Liao, X.H. Wang, H.J. Jin, L.Y. Chen, Q. Chen, Mechanical stretch induces mitochondria-dependent apoptosis in neonatal rat cardiomyocytes and G2/M accumulation in cardiac fibroblasts. Cell Res. **14**, 16–26 (2004)

L.A. Ligon, O. Steward, Role of microtubules and actin filaments in the movement of mitochondria in the axons and dendrites of cultured hippocampal neurons. J. Comp. Neurol. **427**, 351–361 (2000)

N. Luo, M.D. Conwell, X. Chen et al., Primary cilia signaling mediates intraocular pressure sensation. Proc. Natl. Acad. Sci. USA **111**, 12871–12876 (2014)

D.M. Lyra-Leite, A.M. Andres, A.P. Petersen, N.R. Ariyasinghe, N. Cho, J.A. Lee, R.A. Gottlieb, M.L. McCain, Mitochondrial function in engineered cardiac tissues is co-regulated by extracellular matrix elasticity and tissue alignment. Am. J. Physiol. Heart Circ. Physiol. **313**(4), H757–H767 (2017)

D.M. Lyra-Leite, A.M. Andres, N. Cho, A.P. Petersen, N.R. Ariyasinghe, S.S. Kim, R.A. Gottlieb, M.L. McCain, Matrix-guided control of mitochondrial function in cardiac myocytes. Acta Biomater. **97**, 281–295 (2019)

G. Mancia, G. Parati, M. Hennig et al., Relation between blood pressure variability and carotid artery damage in hypertension: baseline data from the European Lacidipine Study on Atherosclerosis (ELSA). J. Hypertens. **19**, 1981–1989 (2001)

G. Mancia, M. Bombelli, R. Facchetti, F. Madotto, G. Corrao, F.Q. Trevano, G. Grassi, R. Sega, Long-term prognostic value of blood pressure variability in the general population: results of the Pressioni Arteriose Monitorate e Loro Associazioni Study. Hypertension **49**, 1265–1270 (2007)

U. Manor, S. Bartholomew, G. Golani, E. Christenson, M. Kozlov, H. Higgs, J. Spudich, J. Lippincott-Schwartz, A mitochondria-anchored isoform of the actin-nucleating spire protein regulates mitochondrial division. eLife **4**, 1–27 (2015)

L. Margulis, Symbiotic theory of the origin of eukaryotic organelles; criteria for proof. Symp. Soc. Exp. Biol. 21–38 (1975)

B.J. Marsh, D.N. Mastronarde, K.F. Buttle, K.E. Howell, J.R. McIntosh, Organellar relationships in the Golgi region of the pancreatic beta cell line, HIT-T15, visualized by high resolution electron tomography. PNAS **98**, 2399–2406 (2001)

L.C. Martineau, P.F. Gardiner, Insight into skeletal muscle mechanotransduction: MAPK activation is quantitatively related to tension. J. Appl. Physiol. **91**, 693–702 (2001)

M.G. Mendez, D. Restle, P.A. Janmey, Vimentin enhances cell elastic behavior and protects against compressive stress. Biophys. J. **107**, 314–323 (2014)

P. Mishra, V. Carelli, G. Manfredi, D.C. Chan, Proteolytic cleavage of Opa1 stimulates mitochondrial inner membrane fusion and couples fusion to oxidative phosphorylation. Cell Metab. **19**, 630–641 (2014)

G.F. Mitchell, C.Y. Guo, E.J. Benjamin, M.G. Larson, M.J. Keyes, J.A. Vita, R.S. Vasan, D. Levy, Cross-sectional correlates of increased aortic stiffness in the community: the Framingham Heart Study. Circuation **115**, 2628–2636 (2007)

A.L. Mora, M. Bueno, M. Rojas, Mitochondria in the spotlight of aging and idiopathic pulmonary fibrosis. J. Clin. Investig. **127**, 405–414 (2017)

M. Morioka, H. Parameswaran, K. Naruse, M. Kondo, M. Sokabe, Y. Hasegawa, B. Suki, S. Ito, Microtubule dynamics regulate cyclic stretch-induced cell alignment in human airway smooth muscle cells. PLoS ONE **6**, e26384 (2011)

R.L. Morris, P.J. Hollenbeck, Axonal transport of mitochondria along microtubules and F-actin in living vertebrate neurons. J. Cell Biol. **131**, 1315–1326 (1995)

D.J. Müller, Y.F. Dufrene, Atomic force microscopy: a nanoscopic window on the cell surface. Trends Cell Biol. **21**, 461–469 (2011)

W.A. Muller, Mechanisms of leukocyte transendothelial migration. Ann. Rev. Pathol.: Mech. Dis. **6**, 323–344 (2011)

S. Nourshargh, R. Alon, Leukocyte migration into inflamed tissues. Immunity **41**, 694–707 (2014)

J. Nunnari, A. Suomalainen, Mitochondria: in sickness and in health. Cell **148**, 1145–1159 (2012)

A. Olichon, L. Baricault, N. Gas, E. Guillou, A. Valette, P. Belenguer, G. Lenaers, Loss of OPA1 perturbates the mitochondrial inner membrane structure and integrity, leading to cytochrome c release and apoptosis. J. Biol. Chem. **278**, 7743–7746 (2003)

G. Osol, Mechanotransduction by vascular smooth muscle. J. Vasc. Res. **32**, 275–292 (1995)

H. Otera, C. Wang, M.M. Cleland, K. Setoguchi, S. Yokota, R.J. Youle, K. Mihara, Mff is an essential factor for mitochondrial recruitment of Drp1 during mitochondrial fission in mammalian cells. J. Cell Biol. **191**, 1141–1158 (2010)

H. Otera, N. Ishihara, K. Mihara, New insights into the function and regulation of mitochondrial fission. Biochim. Biophys. Acta **1833**, 1256–1268 (2013)

C.S. Palmer, L.D. Osellame, D. Stojanovski, M.T. Ryan, The regulation of mitochondrial morphology: intricate mechanisms and dynamic machinery. Cell Signal. **23**, 1534–1545 (2011)

H. Parameswaran, K.R. Lutchen, B. Suki, A computational model of the response of adherent cells to stretch and changes in substrate stiffness. J. Appl. Physiol. **116**, 825–834 (2014)

J.V. Pardo, M.F. Pittenger, S.W. Craig, Subcellular sorting of isoactins: selective association of gamma actin with skeletal muscle mitochondria. Cell **32**, 1093–1103 (1983)

D.A. Patten, J. Wong, M. Khacho et al., OPA1-dependent cristae modulation is essential for cellular adaptation to metabolic demand. EMBO J. **33**, 2676–2691 (2014)

S. Pawar, R. Ungricht, P. Tiefenboeck, J.C. Leroux, U. Kutay, Efficient protein targeting to the inner nuclear membrane requires Atlastin-dependent maintenance of ER topology. eLife **6**, e28202 (2017)

L. Pernas, L. Scorrano, Mito-morphosis: mitochondrial fusion, fission, and cristae remodeling as key mediators of cellular function. Annu. Rev. Physiol. **78**, 505–531 (2016)

H.A. Praetorius, J. Frokiaer, J. Leipziger, Transepithelial pressure pulses induce nucleotide release in polarized MDCK cells. Am. J. Physiol. Renal. Physiol. **288**, F133–F141 (2005)

A.S. Rambold, B. Kostelecky, N. Elia, J. Lippincott-Schwartz, Tubular network formation protects mitochondria from autophagosomal degradation during nutrient starvation. Proc. Natl. Acad. Sci. USA **108**, 10190–10195 (2011)

K. Ray, B. Marteyn, P.J. Sansonetti, C.M. Tang, Life on the inside: the intracellular lifestyle of cytosolic bacteria. Nat. Rev. Microbiol. **7**, 333–340 (2009)

O. Read, The Recording and Reproduction of Sound HW Sams (1952)

P.H. Reddy, Mitochondrial dysfunction and oxidative stress in asthma: implications for mitochondria-targeted antioxidant therapeutics. Pharmaceutical **4**, 429–456 (2011)

S. Reipert, F. Steinbock, I. Fischer, R.E. Bittner, A. Zeold, G. Wiche, Association of mitochondria with plectin and desmin intermediate filaments in striated muscle. Exp. Cell Res. **252**, 479–491 (1999)

S. Roccabianca, C.A. Figueroa, G. Tellides, J.D. Humphrey, Quantification of regional differences in aortic stiffness in the aging human. J. Mech. Behav. Biomed. Mater. **29**, 618–634 (2014)

A. Santel, S. Frank, B. Gaume, M. Herrler, R.J. Youle, M.T. Fuller, Mitofusin-1 protein is a generally expressed mediator of mitochondrial fusion in mammalian cells. J. Cell Sci. **116**, 2763–2774 (2003)

G. Schillaci, G. Bilo, G. Pucci et al., Relationship between short-term blood pressure variability and large-artery stiffness in human hypertension: findings from 2 large databases. Hypertension **60**, 369–377 (2012)

N.L. Sehgel, Z. Sun, Z. Hong, W.C. Hunter, M.A. Hill, D.E. Vatner, S.F. Vatner, G.A. Meininger, Augmented vascular smooth muscle cell stiffness and adhesion when hypertension is superimposed on aging. Hypert **65**, 370–377 (2015)

Y.Y. Shao, L. Wang, J.F. Welter, R.T. Ballock, Primary cilia modulate Ihh signal transduction in response to hydrostatic loading of growth plate chondrocytes. Bone **50**, 79–84 (2012)

M. Simunovic, G.A. Voth, A. Callan-Jones, P. Bassereau, When physics takes over: BAR proteins and membrane curvature. Trends Cell Biol. **25**, 780–792 (2015)

B. Sorre, A. Callan-Jones, J. Manzi, B. Goud, J. Prost, P. Bassereau, A. Roux, Nature of curvature coupling of amphiphysin with membranes depends on its bound density. PNAS **109**, 173–178 (2012)

D. Stamenovic, S.M. Mijailovich, I.M. Tolic-Norrelykke, J. Chen, N. Wang, Cell prestress. II. Contribution of microtubules. Am. J. Physiol. Cell Physiol. **282**, C617–C624 (2002)

D. Stauffer, A. Aharony, *Introduction to Percolation Theory*, 2nd edn (Taylor & Francis, London, UK, Washington, DC, USA, 1992) 181p

V.M. Sukhorukov, D. Dikov, A.S. Reichert, M. Meyer-Hermann, Emergence of the mitochondrial reticulum from fission and fusion dynamics. PLoS. Comput. Biol. **8**, e1002745 (2012)

V.M. Sukhorukov, M. Meyer-Hermann, Structural heterogeneity of mitochondria induced by the microtubule cytoskeleton. Sci. Rep. **5**, 13924 (2015)

B. Suki, H. Parameswaran, J. Imsirovic, E. Bartolak-Suki, Regulatory roles of fluctuation-driven mechanotransduction in cell function. Physiology **31**, 346–358 (2016)

G. Sutendra, P. Dromparis, P. Wright et al., The role of Nogo and the mitochondria-endoplasmic reticulum unit in pulmonary hypertension. Sci. Transl. Med. **3**, 88ra55 (2011)

Y.J. Suzuki, H.J. Forman, A. Sevanian, Oxidants as stimulators of signal transduction. Free Radic. Biol. Med. **22**, 269–285 (1997)

H.L. Tang, H.L. Lung, K.C. Wu, A.H. Le, H.M. Tang, M.C. Fung, Vimentin supports mitochondrial morphology and organization. Biochem. J. **410**, 141–146 (2008)

Y. Tock, M. Ljubisavljevic, J. Thunberg, U. Windhorst, G.F. Inbar, H. Johansson, Information-theoretic analysis of de-efferented single muscle spindles. Biol. Cybern. **87**, 241–248 (2002)

G. Twig, A. Elorza, A.J. Molina et al., Fission and selective fusion govern mitochondrial segregation and elimination by autophagy. EMBO J. **27**, 433–446 (2008)

J.S. Uzarski, E.W. Scott, P.S. McFetridge, Adaptation of endothelial cells to physiologically-modeled, variable shear stress. PLoS ONE **8**, e57004 (2013)

R.D. Vale, The molecular motor toolbox for intracellular transport. Cell **112**, 467–480 (2003)

A.M. van der Bliek, Q. Shen, S. Kawajiri, Mechanisms of mitochondrial fission and fusion. Cold Spring Harb. Perspect. Biol. **5**, a011072 (2013)

F. Vogel, C. Bornhovd, W. Neupert, A.S. Reichert, Dynamic subcompartmentalization of the mitochondrial inner membrane. J. Cell Biol. **175**, 237–247 (2006)

Z. Wang, M. Wu, An integrated phylogenomic approach toward pinpointing the origin of mitochondria. Sci. Rep. **5**, 7949 (2015)

C.M. Waters, P.H. Sporn, M. Liu, J.J. Fredberg, Cellular biomechanics in the lung. Am. J. Physiol. Lung Cell Mol. Physiol. **283**, L503–L509 (2002)

M. Wojcik, M. Hauser, W. Li, S. Moon, K. Xu, Graphene-enabled electron microscopy and correlated super-resolution microscopy of wet cells. Nat. Commun. **6**, 7384 (2015)

Z. Wu, P. Puigserver, U. Andersson et al., Mechanisms controlling mitochondrial biogenesis and respiration through the thermogenic coactivator PGC-1. Cell **98**, 115–124 (1999)

K. Xu, H.P. Babcock, X. Zhuang, Dual-objective STORM reveals three-dimensional filament organization in the actin cytoskeleton. Nat. Methods **9**, 185–188 (2012)

M.P. Yaffe, The machinery of mitochondrial inheritance and behavior. Science **283**, 1493–1497 (1999)

A. Yoshida, N. Sakai, Y. Uekusa, K. Deguchi, J.L. Gilmore, M. Kumeta, S. Ito, K. Takeyasu, Probing in vivo dynamics of mitochondria and cortical actin networks using high-speed atomic force/fluorescence microscopy. Genes Cells **20**, 85–94 (2015)

H. Yoshie, N. Koushki, R. Kaviani et al., Traction force screening enabled by compliant PDMS elastomers. Biophys. J. **114**(9), 2194–2199 (2018)

R.J. Youle, M. Karbowski, Mitochondrial fission in apoptosis. Nat. Rev. Mol. Cell Biol. **6**, 657–663 (2005)

Q. Zhang, M. Raoof, Y. Chen, Y. Sumi, T. Sursal, W. Junger, K. Brohi, K. Itagaki, C.J. Hauser, Circulating mitochondrial DAMPs cause inflammatory responses to injury. Nature **464**, 104–107 (2010)

J. Zhao, T. Liu, S. Jin, X. Wang, M. Qu, P. Uhlen, N. Tomilin, O. Shupliakov, U. Lendahl, M. Nister, Human MIEF1 recruits Drp1 to mitochondrial outer membranes and promotes mitochondrial fusion rather than fission. EMBO J. **30**, 2762–2778 (2011)

R.H. Zhou, A.E. Vendrov, I. Tchivilev et al., Mitochondrial oxidative stress in aortic stiffening with age: the role of smooth muscle cell function. Arterioscler. Thromb. Vasc. Biol. **32**, 745–755 (2012)

Mechanical View on the Endoplasmatic Reticulum and Golgi

7

Abstract

This chapter characterizes the most important cellular organelles such as the endoplasmatic reticulum and the biological and physical functions based on the Golgi apparatus and relates the properties of these organelles tothe overall mechanical properties of the cell. More specific, the connections between the endoplasmatic reticulum and the Golgi apparatus are presented. All cell organelles consist of highly specialized subcompartments fulfilling distinct tasks, and hence, they provide on this discrete level a higher organization of the overall cell. What we understand about the function of organelles is mainly based on microscopic observations and less on the effects of stress on the structure and function of organelles, which is also addressed and discussed.

7.1 Introduction to the Endoplasmic Reticulum (ER)

A characteristic feature of the endoplasmic reticulum (ER) is the enormous surface area of the organelle. In detail, the ER can be described as an organelle consisting of a series of concentric spheres located around the nucleus of the cells. The ER was originally classified as a lamellar system of relatively flat and thin cavities with uniform and parallel localization (Fawcett 1966) (Fig. 7.1). In addition, the ER represents a large structure that is dynamically remodeled.

The ER fulfills several roles in the cell such as the storage of intracellular calcium, protein synthesis and the metabolism of lipids. Different functions of the ER are spatially separated and hence performed in distinct domains of the ER, such as ER tubules, ER sheets and ER-drived nuclear envelope. However, the overall architecture and dynamics of the ER are based on different proteins, and how the ER changes in shape due to cellular signals, cell type, cell cycle phase and during the development of the organism, remains open (Schwarz and Blower 2016). Hence, the current knowledge of the ER dynamics is presented and discussed. Finally, it is described the responses of the ER are coordinated in the differed layered structures of the organelle.

The ER constitutes the largest organelle of the cell and plays a key role in various cellular processes. Among them are firstly the protein synthesis, folding and transport, secondly, the lipid, steroid and carbohydrate metabolism and thirdly the deposition of calcium (Reid and Nicchitta 2015; Rapoport 2007; Braakman and Hebert 2013; Fagone and Jackowski 2009; Hebert et al. 2005; Clapham 2007; Westrate et al. 2015). Since the ER fulfills multiple different functions, numerous specific proteins are necessary for these different tasks and unique physical structures enable the

C. T. Mierke, *Cellular Mechanics and Biophysics*, Biological and Medical Physics, Biomedical Engineering, https://doi.org/10.1007/978-3-030-58532-7_7

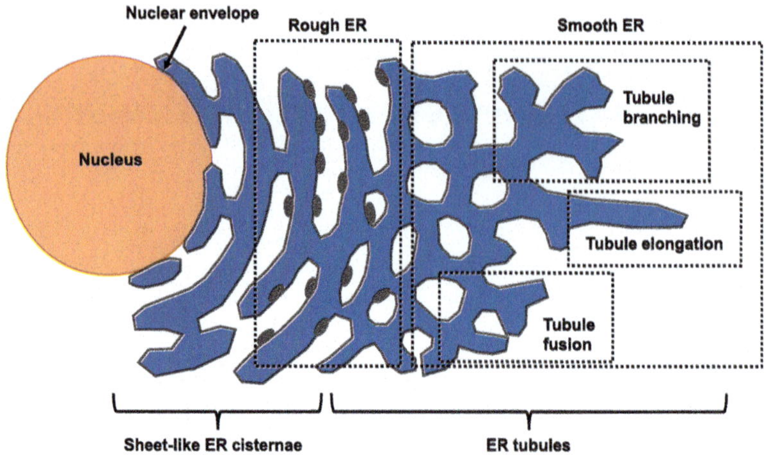

Fig. 7.1 The structure of the ER. The ER can be separated into three major domains, the sheet-like ER, the tubular ER and the nuclear envelope. The sheet-like ER contains numerous ribosomes and is termed the rough ER. The tubular ER possesses only a few ribosomes and is termed the smooth ER. The sheet-like and tubular ERs are highly dynamically remodeled and can switch between the two configurational states. The ER has the capacity to fuse, elongate and branch

coordination of the ER with the intracellular microenvironment and facilitate the response to varying intracellular cues. The ER consists of several different structural areas, each of which is assigned to one or more functions. However, how these subdomains function and how they are organized is still elusive. Hence, it needs to be identified how the function of an individual domain determines its structural arrangement. Are the different functions of the subdmains the determining factors for the structural arrangements of certain subdomains?

7.2 The Structure of the ER

There general structure of the ER structure has been reviewed in great detail (Westrate et al. 2015; Engish and Voeltz 2013; Friedman and Voeltz 2011; English et al. 2009; Shibata et al. 2006; Hu et al. 2011). The structure of the ER has been coupled to alterations of the ER shape due to cellular or environmental signals (Schwarz and Blower 2016). The basic structure of the ER is characterized by two distinct elements, the nuclear envelope and the peripheral ER. More precisely, the peripheral ER consists of smooth tubules and rough sheets. The ER represents an interconnected network that is enclosed by a continuous membrane. Different structures of the ER enable it to fulfill various special tasks within the cell.

In specific detail, the envelope of the nucleus is composed of two lipid bilayers: the inner nuclear membrane (INM) and outer nuclear membrane (ONM). The nuclear envelope exhibited a common lumen with the surrounding peripheral ER. The several hundred pores located within the ONM and spanning to the underlying INM of the nuclear envelope to provide molecular transport that encompasses proteins and RNAs with different degrees of diffusion or controlled transport according to molecule size. The nuclear envelope is linked to sheets or cisterns that comprise part of the peripheral ER. The sheets are thin and made of two lipid bilayers with a lumen between them, with the curved areas positioned exclusively at the membrane edges. The size of peripheral ER leaves is subject to fluctuation, although the luminal spacing is very uniform, typically approximately 50 nm for mammals and 30 nm for yeasts (Bernales et al. 2006) (Fig. 7.2). Sheets are generally surveyed in

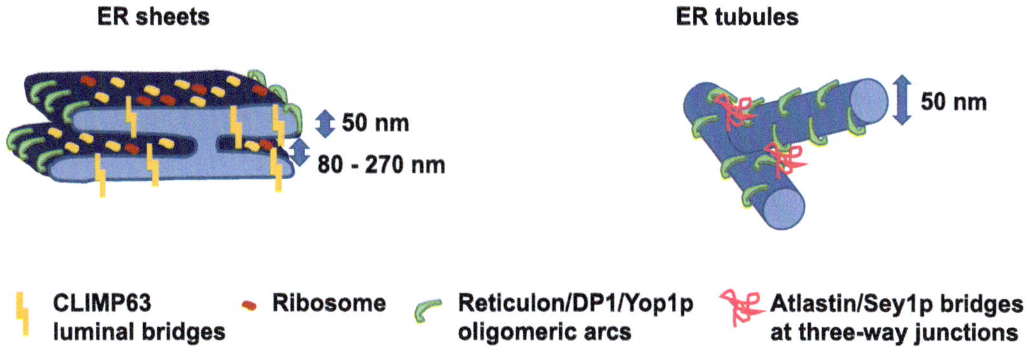

Fig. 7.2 ER sheets are characterized by CLIMP63 luminal bridges and contain several ribosomes on the membrane. ER tubules at three-way junctions are stabilized through Atlastin/Sey1p bridges. Both ER structures display Reticulin/DP1/Yop1p oligomeric arcs

a layered conformation and linked to helical edges via areas of twisted membranes (Terasaki et al. 2013). The rough sheets of the ER are containing a high density of ribosomes on their cytosolic surface (Shibata et al. 2010; West et al. 2011). Moreover, these rough sheets are the major sites of the synthesis of proteins, their precise folding and the post-transcriptional modification that can traged proteins to the membrane, nucleus, ER, Golgi apparatus or even for secretion by the cells. However, less ribosomes are located on the outer membrane surface of ER tubules (West et al. 2011) that exhibits a curved and smooth surface that cannot support the interaction with large polysomal structures. In fact, the tubular scaffold is highly dynamic, since it undergoes continuously rearranging and even rebuilding, and is more specifically arranged in three-way junctions linking the individual tubules (Fig. 7.3). Although tubules and sheets exhibit widely divergent structural characteristics and therefore participate in a variety of cell processes, the luminal distance between tubules and sheets is similar (Bernales et al. 2006; West et al. 2011).

Remarkably, ER tubules and sheets are present in all eukaryotic cells (Staehelin 1997), although the ratio of sheets to tubules differs in various cell types and mirrors the specific features of these cells. The ER framework of dedicated cells that produce high amounts of secreted proteins, such as pancreatic secretory cells and B cells, is largely made up of sheets. Cells engaged in functions such as lipid synthesis, calcium signaling and points of contact for other organelles inherit an ER consisting mainly of tubules. Adrenal, liver and muscle cells exemplify specialized cells with a generally tubular structure and accurately mirror the functionality of these cells (Baumann and Walz 2001).

Another peripherally located ER arrangement is the cortical ER, which is attached to the plasma membrane and has an intermediate phenotype between sheets and tubules with both curved and flat membranes. Calcium signaling takes place at the interface between the plasma membrane and the adjacent cortical ER and is required for contraction of the muscle (Block et al. 1988; Takeshima et al. 2000). For this reason, the structural morphology and intracellular distribution of ER subdomains determine the function of these subdomains and consequently the function of the differentiated cell within they are situated.

Advanced microscopic techniques have made it possible to distinguish between various ER structures and the proportions of these features in highly specialized cell types. When the function of these cells in the organism is examined, the type and quantity of peripheral ER present clearly mirrors their function. It is not yet understood how these proportions are achieved and in what manner cellular signaling pathways interfere with the characterization of the ER type which will be dominant in a specific cell type.

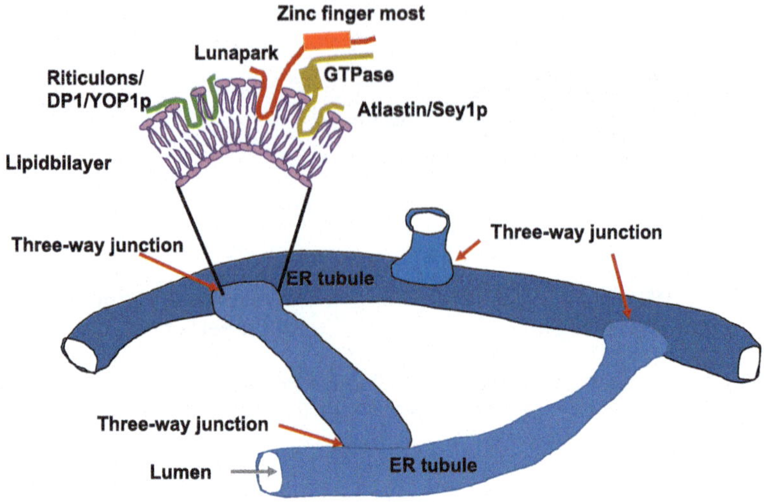

Fig. 7.3 Three-way functions are presented in detail, which are formed through ER tubules

7.3 The Shaping of ER Proteins

7.3.1 ER Tubules

Peripheral ER structures are as different and multifaceted as the amount of proteins that provide their conformation. In precise detail, various proteins have been designated that favor specific ER structures, although probably the brightest group of proteins investigated comprises the reticulone family of proteins targeting tubules and the strongly curved edges of ER sheets (Shibata et al. 2010; Voeltz et al. 2006). By producing a transmembrane hairpin topology serving as a spike, these integral membrane proteins promote membrane bending and push lipids into the outer wings of the bilayers, resulting in membrane bending (Voeltz et al. 2006). These proteins usually arrange to oligomers and are pronouncedly less mobile than other proteins located in the ER (Shibata et al. 2008). The overexpression of some reticulone isoforms results in the development of long ER tubules at the cost of the sheets (Shibata et al. 2008). The degradation of reticulones and thus the capacity to bend membranes subsequently results in a decrease in the amount of ER tubules, which entails an expansion of the peripheral sheets (Voeltz et al. 2006; De Craene et al. 2006; Anderson and Hetzler 2008). For this reason, the content of reticulones inside a cell defines the occurrence and microstructure of ER tubules.

Reticulones are not only effective in the forming of ER tubules. Members of the DP1/Yop1/REEP5/6 and REEP1-4 families, who are ample ER-resident proteins targeting specific tubules and edges of sheets, also serve as stimulating factors for tubules. The two proteins DP1/Yop1 and REEP5/6 (Hu et al. 2009) possess a very similar transmembrane hairpin structure in common with the reticulons that causes the stabilization of the bended membranes of tubules (Voeltz et al. 2006; Shibata et al. 2008; Hu et al. 2008). Specifically, the topology of REEP1-4 proteins is somewhat altered from that of REEP5/6, implying that these proteins may perform slightly differing functions in the structuring of ER from closely associated REEP5/6 proteins (Park et al. 2010). Beyond that, purified reticulones and proteins of the DP1/Yop1 family succeeded in inducing the generation of tubules out of the purified vesicles (Hu et al. 2008), revealing that these proteins have a substantial role in the growth of ER tubules.

Reticulons and DP1/Yop1 encourage tubule development; however, other factors are necessary to foster the establishment of the tubular mesh and typical three-way connections by homotypic fusion. Atlastins belong to the dynamin-like family of GTPases that facilitate the homotypic fusion processes. Depletion by RNAi or the expression of dominant-negative atlastine within cells results in a deficiency of fusion occurrences, causing a plethora of elongated, straight tubules (Hu et al. 2009). When a dominant-negative cytoplasmic fragment of Xenopus comprising the GTPase domain and lacking the transmembrane domain and cytoplasmic tail (Bian et al. 2011) is transferred into Xenopus interphase extracts, the establishment of the ER framework was inhibited (Wang et al. 2013). Similar point mutations, impeding dimerization of the cytoplasmic partition of human atlastin (Byrnes and Sondermann 2011), were carried out in the cytoplasmic part of the atlastin protein of Xenopus, injected into interphase extract and had no influence on the generation of the ER matrix (Wang et al. 2013). In addition, antibodies directed against atlastin prevent the development of ER systems when inserted into extracts of Xenopus oocytes (Hu et al. 2009). In Drosophila, exhaustion of atlastine fragments the ER and purified atlastine is capable of triggering GTP-dependent fusion of proteoliposomes (Bian et al. 2011; Byrnes and Sondermann 2011; Orso et al. 2009). Hence, investigations of a high number of organisms, extracts and purified proteins revealed that atlastin seem to be crucial for the induction of the homotypic fusion of vesicles between the ER membranes that is required for the general function of the entire meshwork.

Several new major constituents of the ER dynamics have been revealed. It has been shown that purified ER vesicles of Xenopus eggs need GTP for the homotypic fusion of ER vesicles, when there are no cytosolic factors available (Voeltz et al. 2006; Dreier and Rapoport 2000). These results are in line with other reports stating that GTPases are necessary for the process of ER fusion (Turner et al. 1997; Audhya et al. 2007), and even more recently, it has been employed a proteomics approach that determined Rab10 as a parameter essential for the assembly of the ER (English and Voeltz 2013). Moreover, a knockdown of Rab10 or the overexpression of a dominant-negative point mutant locking GDP in cultured human cells leads to elevated levels of RAb10 in ER sheets, whereas it is reduced within ER tubules (English and Voeltz 2013). The fusion between ER–ER regions occurs at locations with increased Rab10 levels. Rab10 has been seen to be co-localized with specific lipid-synthesizing enzymes such as phosphoinositol synthase (IS) and choline/ethanolamine phosphotransferase (CEPT1) (English and Voeltz 2013), which can be not yet identified subdomain or subcompartment of the ER. However, it is still not well understood what function the Rab10 fulfills during the fusion of the ER vesicle or even how the homotypic fusion of ER vesicles is connected to the synthesis of lipids.

It has been demonstrated that Rab18 performs a functional role in the process of ER dynamics, since it is positioned in the ER nearby the Rab3 GTPase-activating protein (GAP) complex. When Rab18 is eliminated, the phenotype is close to that of the phenotype seen after Rab10 inhibition (Gerondopoulos et al. 2014). When Rab10 is deficient, Rab18 is able to redisperse into peripheral sheets (Gerondopoulos et al. 2014). Thus, it also seems that the depletion of Rab10 or Rab18 impedes the stability of ER tubule fusion and decreases the density of tubules, with the result of an increase in ER sheets. The depletion of Caenorhabditis elegans RAB-5, previously involved in early endosome function (Stenmark 2009), reproduces the peripheral ER imperfections appearing in RET-1 and YOP-1 (homologs of Rtn4a and DP1) (Audhya et al. 2007). Besides the role played by RAB-5 in the assembly of the peripheral ER, the kinetics of disassembly of the nuclear envelope in these mutants is concerned (Audhya et al. 2007). More precisely, beyond the role of GTPases, which fulfill a direct function in the homotypic vesicular membrane fusion, there seems to be a key role of lipid synthesizing enzymes in the regulation of the ER shape and its overall organization. The impairment of the C-terminal domain (CTD) of the nuclear envelope phosphatase-1 (CNEP-1), which is highly concentrated in the vicinity of nuclear envelope and thereby induces the membrane phospholipid synthesis, elicits the formation of ectopic leaves covering the nuclear envelope and distorting the degradation of the nuclear envelope (Bahmanyar et al. 2014). Consequently, these findings show that the interlinked scaffold of proteins and their functions contribute to the shape of the ER structural architecture.

It is widely known that the ER represents a highly dynamic meshwork, which continuously undergoes the remodeling and reordering (Lee and Chen 1998). The tubules of the ER fuse and branch constantly, leading to the formation of new three-way connections. In a concurrent effort, the sliding interconnects and tubule ring closure mechanisms lead to the disappearance of three-way connections and the inherent polygonal structure (Friedman et al. 2010). Although there is less understood how these complexes regulate the precise function of this process, the Lunapark (Lnp1) protein has been found to be positioned in these complexes and thereby can provide a stabilization of the three-way junctions (Chen et al. 2012, 2015). Lnp1 interacts with reticulons and Yop1, and its close positioning to junctions is controlled by Sey1p, which is the homolog of atlastin in yeast (Chen et al. 2012). The lack of Lnp1 evokes a collapsed and tightly reticulated ER network in yeast and human cells in in vitro cultures (Chen et al. 2012, 2015), although even though only half of the crossings are linked to Lnp1 (Chen et al. 2015), mirroring the fluidity of the ER mesh. When Lnp1 is overexpressed, it becomes targeted to the peripheral ER, where is causes the assembly of a complex polygonal tubular scaffold (Moriya et al. 2013). Moreover, the assembly a network structure is impaired by mutations of Lnp1 inhibiting the N-myristoylation (Moriya et al. 2013) that is an attachment of the myristic acid (consisting of a 14-carbon saturated fatty acid), which reveals that this alteration seems to be crucial for the Lnp1-driven processes on the ER morphology. In terms of the membrane translocation, formation of topology and the targeting of the protein to the ER, the N-myristoylation is not necessary; however, it can fulfill still an important function in providing protein–protein or protein–lipid interactions needed for morphological alterations of the ER, although the precise molecular mechanism of this functions is still elusive (Moriya et al. 2013).

The intrinsic mechanism for the Lnp1-facilitated stabilization of three-way compounds is not understood, although recent research and knowledge of the structure and domains within the protein have revealed how Lnp1 fixes the compounds (Chen et al. 2012, 2015). Firstly, Lnp1 possesses two transmembrane domains and a zinc-finger domain that is positioned at the cytoplasmic site of the ER membrane (Chen et al. 2015). However, the polygons are smaller in cysteine mutations within the zinc-finger domain and the regions without cortical ER appear more visible due to an elevated number of mutated cysteines (Chen et al. 2012). Mutations in the zinc-finger domain can hence impact interactions between individual proteins, affect complex generation or disrupt the dispersion of intracellular lipids on the cytoplasmic side of the membrane and have harmful effects on the stabilization of the junctional linkage. Beyond that, the transmembrane domains function as an upside-down cotter which enhances the local negative bending feature of three-way connections (Chen et al. 2015) and counteracts the positive bending feature of reticulons. It is also conceivable that several Lnp1 proteins can also work together collaboratively to fix the connection, or that Lnp1 intervenes temporarily to fix or alter lipids or other proteins at the connections (Chen et al. 2015).

More and more evidence emerged that, in contrast to proteins controlling membrane morphology and dynamics, a modification of the nucleic acid level of ER can also affect the ER form. Initial experiments revealed that a short incubation of cultured tissue cells with the translation inhibitor puromycin, dissociating mRNA-ribosome complexes, results in the removal of ribosomes from ER and the destruction of ER sheets (Shibata et al. 2010; Puhka et al. 2007). Evidence points to the fact that the availability of mRNA-ribosome complexes can help to stabilize ER sheets. To corroborate this hypothesis, an ER-localized ribonuclease, XendoU (Laneve et al. 2003) has been characterized which alters the RNA content of ER in reaction to alterations in the Ca^{2+} concentration (Schwarz and Blower 2014; Seidel and Peck 1994). These alterations take place at physiological concentrations of approximately 1.5 μM, which resembles the release of Ca^{2+} ions from intracellular and extracellular storage sites at the time of fertilization (Busa and Nuccitelli 1985; Lorca et al. 1991). A subpopulation of XendoU can target the ER, and moreover, it can be co-immunoprecipitated with a large number of proteins that are located in the ER (Schwarz and Blower 2014). The exhaustion of XendoU causes the creation of long, unbranched tubules in Xenopus leavis egg cell extracts, and the recovery of this phenotype demands an intact catalytic activity of the protein, suggesting that proper nuclease activity is crucial for the correct development of an ER framework (Schwarz and Blower 2014). In parallel, the addition of antibodies to purified vesicles induces a barrier in mesh generation, which shows that XendoU affects the surface of ER membranes to adjust the ER architecture (Schwarz and Blower 2014). Curiously, the addition of 5′5′-dibromo BAPTA, which is a powerful calcium chelator, inhibited the fusion of vesicles in this complex (Dreier and Rapoport 2000). The depletion of XendoU results also in a retardation of the replication and sealing of the nuclear envelope (Schwarz and Blower 2014), and BAPTA prevents the production of the nuclear envelope in experiments on the regeneration of Xenopus egg extracts (Sullivan et al. 1993). Collectively, these results point to XendoU working on membranes to decompose RNAs.

Vesicle fusion revealed that RNAs were cleaved and liberated from the membrane surface, signaling XendoU's ability to break down these RNAs and liberate proteins to purify membrane areas that permit a vesicle generation that triggers cross-linking (Schwarz and Blower 2014). Specifically, where purified vesicles with rising concentrations of RNaseA are exposed to the same experiment and handled, an ever more abnormal system of major vesicles that could not merge is created (Schwarz and Blower 2014). Findings from in vitro trials demonstrated that XendoU at membranes becomes activated in conjunction with calcium secretion to decompose local RNAs and vacant membrane sites conducive to controlled fusion to optimize mesh building. Finally, the decay of the human homolog EndoU in cultured human cells, in a similar way to the decay of other proteins involved in tubule generation, causes an enlargement of the sheets (Schwarz and Blower 2014). In addition, the recovery of the phenotype of the expanded sheet was reliant on the integrity of the catalytic activity monitored with recombinant protein in the extract environment. Thus, XendoU exemplifies a protein being activated to respond to cellular signals in order to control proper ER generation, and additional trials may uncover other proteins which are controlled in this fashion to optimize organelle architecture.

7.3.2 ER Sheets

Since the assembly and preservation of tubules are discussed, the role of sheets and other peripheral ER structures need to be pointed out. At the first glance, it needs to be figured out how the sheets are generated. In fact, various mechanisms have been proposed, among them the concept of integral membrane proteins crossing the intraluminal space and building bridges linking the lipid bilayers (Shibata et al. 2009, 2010; Senda and Yoshinaga-Hirabayashi 1998). These proteins can fulfill at least two functions, such as the stabilization of the structural architecture of the lipid bilayers or the precisely set distance of the two lipid bilayers. Moreover, all these proteins or complexes are able to build a backbone supporting sheet stabilization or moving the two lipid membranes closer together (Shibata et al. 2009). Several proteins, such as Climp63, p180 and kinectin have been revealed to fulfill their tasks in the generation of ER sheets, their stabilization and maintenance (Shibata et al. 2010).

Besides considerably elevated membrane proteins and key factors of the translocon, Climp63, which is a coiled coiled protein with a single transmembrane domain, was investigated together with kinectin and p180 in a mass spectrometric assay for the presence of rich integral ER membrane-proteins (Shibata et al. 2010). Employing many different approaches and using vastly varying cell types Climp63 can be confirmed to be a largely abundant protein (Foster et al. 2006; Nagaraj et al. 2011; Beck et al. 2011), which is positioned in the perinuclear ER and cannot be found in the nuclear envelope (Klopfenstein et al. 1998, 2001). Extremely stable oligomers of Climp63 are able to accumulate, which limits the mobility of the protein across the membrane and facilitates its localization to the rough ER (Klopfenstein et al. 2001). Climp63 overexpression causes a huge proliferation of ER sheets, whereas the decrease in expression unexpectedly does not produce a reduction in sheets, but rather a narrowing of the gap between sheets (Shibata et al. 2010). In contrast, these sheets are diffusely scattered in the cytoplasm, recalling the phenotype of cells exposed to the translation inhibitor puromycin (Shibata et al. 2010). However, the key elements of the translocon, the protein channel interfering with ribosomes and leading to the translocation of nascent peptides into the ER or the attachment of transmembrane slices of newly synthesized proteins, have been accumulated on sheets (Nikonov et al. 2007). Accordingly, these results indicate that the role of Climp63 in forming sheets appears plausible that supplemental determinants are associated that serve as an overall sophisticated regulatory framework that governs the balance of sheet and tubule growth.

7.4 The Synthesis and Folding of Proteins in the ER

A key function of the ER is that it can be a major site for the production of proteins that can be either secreted proteins or integral membrane proteins (Walter and Blobel 1981), and additionally a specific subset of cytosolic proteins (Reid and Nicchitta 2015). Protein synthesis necessitates the targeting of ribosomes to the cytosolic surface of the ER, and the canonical pathway governing protein synthesis entails the co-translational binding of the mRNA-ribosome composite to the membrane of the ER. The translation of secretory or integral membrane proteins is induced in the cytosol, in which case ribosomes comprising these mRNAs are enrolled into the ER membrane via a signal sequence within the amino terminus of the resulting polypeptide detected and linked by the signal recognition particle (SRP) (Walter et al. 1981; Noriega et al. 2014). The complex of the mRNA-ribosome–nascent polypeptide SRP is aimed at the ER where it attaches to the SRP receptor (Gilmore et al. 1982; Meyer et al. 1982). The translation proceeds on the ER and the developing polypeptide is able to penetrate the ER co-translationally over the translocon (Rapoport 2007), a channel comprising several Sec proteins stretching across the lipid bilayer (Blobel and Dobberstein 1975).

During the translation time, or in exceptional cases upon the completed translation (Braakman and Hebert 2013), a signal peptidase can cut off the short signal peptide that then enables the free protein to access lumen of the ER (Siekevitz and Palade 1960). When the newly translated protein is targeted to become an integral membrane protein, which is characterized by a stretch of hydrophobic residues or stop-transfer membrane linkage sequence, the translocation will be held on standby (Potter et al. 2001). Then the protein is transferred laterally and become connected to the phospholipid bilayer where it stays (Potter et al. 2001). There is one type of transmembrane proteins compromising a single hydrophobic stretch domain, which are termed classical single-pass transmembrane proteins. Another type of transmembrane proteins consists of proteins possessing multiple regions that transverse the membrane, which are referred to as multispanning transmembrane proteins (Braakman and Hebert 2013).

If the freshly polymerized protein is not a membrane protein and has to be integrated into the secretory pathway or even into the lumen of membrane-bound organelles, the protein is transported directly. After the complete translation process, the signal peptide can be cut off the ribosomes and secreted into the cytoplasm (Seiser and Nicchitta 2000; Potter and Nicchitta 2000). In the case of mRNAs that are translated through stably ER connected ribosomes, the mRNAs are freely secreted into the cytoplasm, while the ribosomes still maintained in an ER connected state and hence can perform even multiple rounds of translation processes (Reid et al. 2014; Palade 1955). For cytoplasmic proteins that are translated through the ER connected ribosomes, it is less understood how the targeting of these mRNAs toward the ER is regulated or what type of specific ribosomes are employed to trigger the translational process. However, it has been reported that the ER inherent protein p180 fulfills a crucial role during the targeting process of mRNAs to the ER, which seems to be separated from the translational process (Hicks et al. 1969).

After the protein synthesis and the translocation of the entire protein into the lumen of the ER, the protein targeted to secretion need to be folded precisely and post-translational modified through the utilization of chaperones or enzymes facilitating the protein folding. Among these modifications are the N-linked glycosylation, the assembly of disulfide bonds and the oligomerization of the proteins (Fig. 7.4) (Braakman and Hebert 2013). At this stage, these proteins are targeted for their secretion. When the protein performs a function within the ER, such as a chaperone, the correct folding begins. When the protein is intended for secretion, it is freed from the chaperones and boxed for passage through the Golgi to a final location, such as the plasma membrane or secreted to the microenvironment, or is delivered to peroxisomes (Mueckler and Pitot 1981). In addition, the cytosolic parts of the transmembrane proteins can interfere with cytosolic proteins or chaperones to correctly bend these regions.

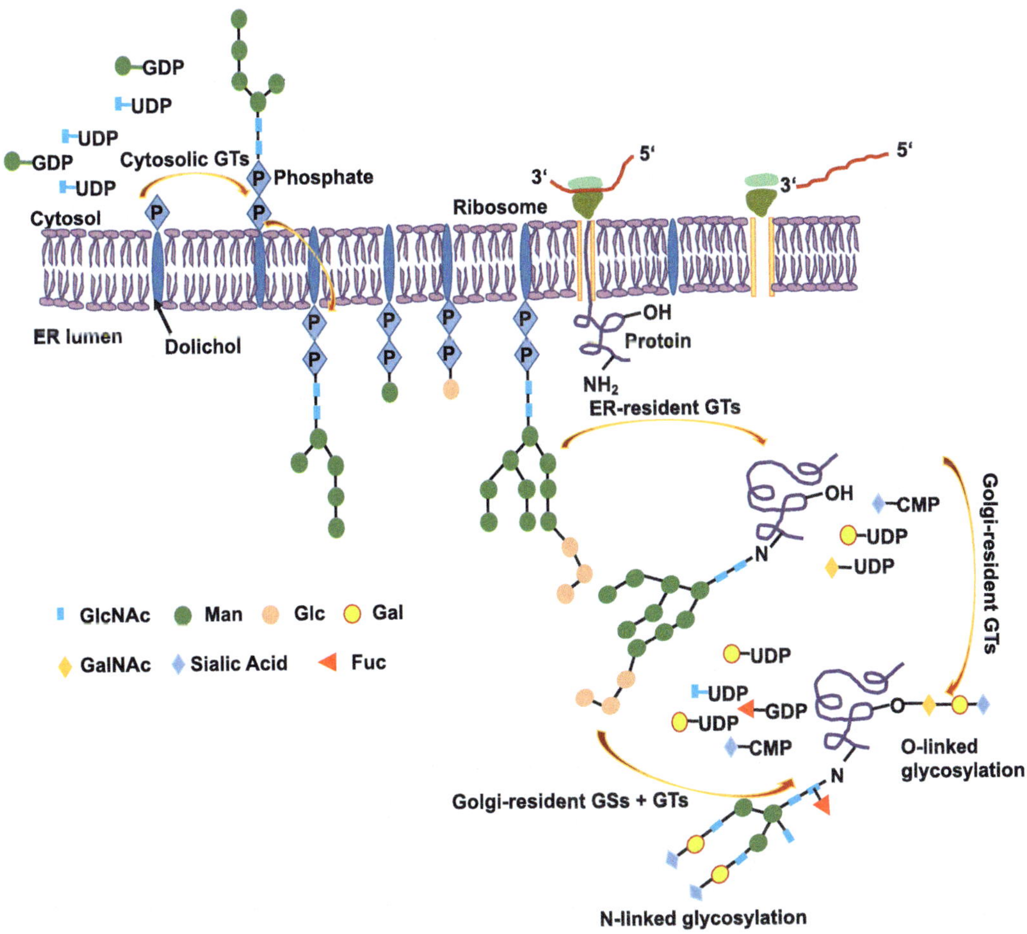

Fig. 7.4 N- and O-glycosylation of proteins in the ER and Golgi

However, although there exists a number of proteins and complexes that aid the precise folding of proteins, still a certain fraction of proteins is not in its native and functional state, as this fraction contains proteins, which are misfolded or even clustered in aggregates (Mechler and Vassalli 1975a). There are at least two options, firstly these proteins still continue to be located within the ER or secondly, they manage to undergo degradation via the ER-associated degradation (ERAD) process, which is facilitated by the proteasome that impairs the entry of aberrant polypeptides into the secretory system (Mechler and Vassalli 1975b). The identification of wrongly folded proteins is associated with their degradation through the ERAD pathway, which requires to be strongly regulated in order not to interfere with the functions of the cell (Mechler and Vassalli 1975b). Intriguingly, there exist various pathways for the initiation of an ER stress reaction and subsequently for the creation of pathological conditions for humans. The activation of ER stress reaction signal transduction pathways supports various neurodegenerative diseases that are based on the misfolding of specific proteins, including as Alzheimer's disease. There is also evidence of activation of the ER stress reaction pathway in diabetes, inflammatory bowel disease and several types of cancer. The role of ER stress reaction pathways in these disorders is an ongoing field of research and several components of stress reaction pathways appear as possible therapeutic objectives (Kopczynski et al. 1998). Under physiological conditions, the protein synthesis in the ER is performed within the restricted ER sheets and it is frequently observed that the structure of the ER is based on the positioning of the RNA and on the stress of the ER; however, it will be discussed after the biogenesis of the lipids.

7.4.1 The Biogenesis of Lipids

Since the ER has been revealed as a major site of the synthesis of proteins, it is employed as a site of bulk membrane lipid biogenesis (Fagone and Jackowski 2009) that is located within the endomembrane compartment connecting the ER and the Golgi apparatus. The major constituents of the membranes are proteins and phospholipids and they get transferred and biochemically altered in the region of the ER, which in close neighborhood to the Golgi apparatus (Diehn et al. 2000). This specific region is termed the ER–Golgi intermediate compartment (ERGIC), which contains multiple tubules and vesicles (Fagone and Jackowski 2009) (Fig. 7.5). When the lipids are directed at the ERGIC, they are dispersed in the cell either via organelle contacts or secretory vesicles (Diehn et al. 2006). The *cis*-Golgi is nearest to the ERGIC compartment and enters the *trans*-Golgi system, which contains vesicles with freshly produced secretory proteins from the ER form and the bud (Fagone and Jackowski 2009). The *trans*-Golgi system has historically been regarded as the primary sorting location in the cell, attracting cytosolic freight adapters to link, transport indirectly or directly, proteins or lipids (Lerner et al. 2003).

7.4.2 The Metabolism of Calcium (Ca^{2+})

Since ER has been discovered as an eminent place for the synthesis and transport of a large number of biomolecules, the ER system is also a key storage site for intracellular Ca^{2+} (Reid and Nicchitta 2012; Stephens et al. 2008). The general physical concentration of Ca^{2+} ions in the cytosol is around 100 nM, whereas the Ca^{2+} ion concentration in the ER lumen of the ER is in between 100 and 800 nM, and even in the extracellular microenvironment, the Ca^{2+} concentration is much higher, such as approximately 2 mM (Clapham 2007; Jagannathan et al. 2011). The ER possesses multiple calcium channels, ryanodine receptors and inositol 1,4,5-trisphosphate receptors (IP3R), which provide the freeing of Ca^{2+} ions from the ER in the cytoplasm at relatively low levels of intracellular Ca^{2+}

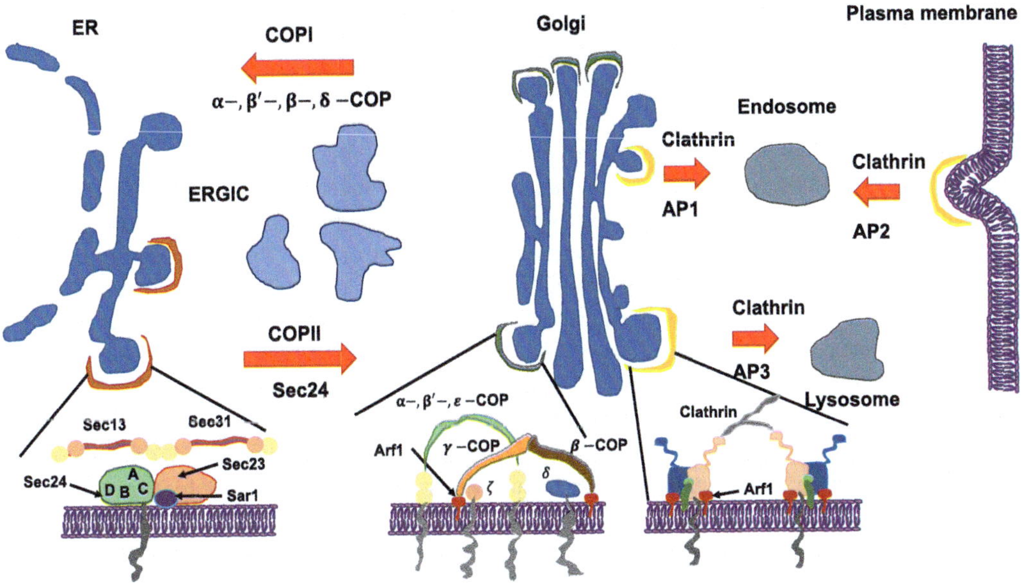

Fig. 7.5 Overview of the intracellular transport routes. Schematic view of the secretory route and presentation of major coat proteins that facilitate the protein sorting within various cellular compartments. Secretory cargoes are trafficked in an anterograde direction from the ER to the Golgi in COPII-coated vesicles. Sec24 represents the cargo adapter with various cargo-binding sites, such as A, B, C and D, to foster a various set of cargo proteins. The COPI coat drives retrograde transport from the Golgi to the ER and between different Golgi compartments. The cargo-binding subunits of COPI vesicles build an arch-like structure contacting the membrane through the N-terminal domains that interfere with Cargo proteins. Clathrin-coated vesicles bud from multiple organelles and transport proteins between the TGN, endosomes and plasma membrane. Different cargo adaptors function at the multiple donor membranes, such as AP1, AP2 and AP3. The general structure of the AP complexes is composed of an accurately folded domain containing the trunk domains of the two large subunits, which interfere with the membrane, cargo proteins and two unstructured sequence motifs, which are able to attach to clathrin and other accessory proteins

ions (Clapham 2007). The release of Ca^{2+} ions is triggered upon the induction of the phospholipase C (PLC) by the activation of the G-protein-coupled receptor (GPCR) (Jagannathan et al. 2014a, b), and subsequently, it causes the cleavage of phosphatidylinositol 4,5 bisphosphate (PIP2) into diacylglycerol (DAG) and IP3 that is then able to attach to the IP3R causing a release of Ca^{2+} ions and thereby a transient elevation in intracellular Ca^{2+} ion amounts (Clapham 2007). Ryanodine receptors (RyRs) function based on a voltage-induced Ca^{2+} ion release (CICR), when these receptors bind Ca^{2+} due to Ca^{2+} amounts (Martin and Ephrussi 2009).

The depolarization of t-tubule membranes may also result in conformational alterations in voltage-dependent Ca^{2+} channels, such as dyhydropyridine receptors (DHPRs), interfering and activating RyRs, which in turn liberate Ca^{2+} ions (Palacios and Johnston 2001). In addition, Ca^{2+} ions can either the cytoplasm through ER leakages and they can be pumped back actively into the ER through sarcoendoplasmic reticular Ca^{2+} ATPases (SERCAs), or they can penetrate from the extracellular surroundings into the cell and contribute to the levels of regulation (Clapham 2007). When ER reservoirs of Ca^{2+} ions are fast exhausted by IP3 receptor (IP3R)-induced release, a mechanism for Ca^{2+} entrance into the cell is initiated, referred to as store-operated Ca^{2+} entry (SOCE) (Clapham 2007; Cui et al. 2013). After luminal Ca^{2+} ion depletion of the ER, STIM1 proteins accumulate in areas of the ER adjacent to the plasma membrane. In these regions, concentrated STIM1 captures plasma membrane-distributing Orai1 subunits (Loya et al. 2008; Nicchitta et al. 2005) and assemblies them into active Ca^{2+} release-activated channels (CRAC), which enable the incorporation of extracellular Ca^{2+} into the ER lumen to reestablish Ca^{2+} levels (Pyhtila et al. 2008; Zhou et al. 2014; Krogh et al. 2001). Curiously, SOCE and the activation of CRAC are not linked to alterations of Ca^{2+} ion levels in the cytoplasm (Clapham 2007), but rather to the recognition and response to alterations of luminal Ca^{2+} ion concentrations.

Calcium is a widely distributed signaling molecule that can affect various processes, such as the targeting, functional activation and interaction of proteins, entire organelles or nucleic acids. The release of Ca^{2+} ions may induce a Ca^{2+} ion front, which is able to propagate across the complete cell (Mueckler and Pitot 1982), and thus, a concentration gradient of Ca^{2+} ions from the origin of the release or a spatially limited front from bundled channels, which are known as a Ca^{2+} spark, appears (Mechler and Rabbitts 1981). At the fertilization process after the entry of the sperm, the eventually mostly investigated Ca^{2+} ion release can be observed (Mueckler and Pitot 1982; Chen et al. 2011), whereas such an release can be detected in a similar manner upon the contraction of muscle, during secretion processes (Clapham 2007) or neuronal processes, such as the release of neurotransmitters for signal transduction (Jagannathan et al. 2014a, b). However, the Ca^{2+} ion release can also fulfill a prominent role during the reshaping the ER upon cellular signal transduction initiation.

7.5 How Is the Shape and the Function of the ER Regulated?

The ER represents a highly complex organelle that regulates the synthesis of proteins and lipids, controls the calcium level and facilitates the interactions with other cell organelles. The physical structural architecture mirrors the large complexity of the ER. The general architecture of the ER is compromised by a continuous membrane network, which contains the nuclear envelope and the peripheral ER that is characterized by branched tubules and flat sheets. A large number of integral membrane proteins and interactions of the ER with other organelles and the cytoskeleton govern the shape and spread of these ER complexes. These interventions are of a dynamic character and mirror alterations within the cell, either by cell cycle or developmental stage, cellular differentiation, internal cellular signals or interactions with proteins. Although it is widely accepted how the basic forms of ER sheets and tubules are characterized, how alterations in shape or the ratio of sheets to tubules arise as a reaction to distinct cellular stimuli is quite ambiguous.

In the following, the assembly of the ER structure, the control of the ER dynamics and the dynamical alterations due to the phase in the cell cycle or specific environmental or cellular cues are discussed. Beyond that, there are instances of how the proteins participating in the design of ER are impacted by these molecular signals, such as calcium release, and how this is mirrored in the dynamics of ER and finally in the function of differentiated cells with varying ratios of sheets to tubules.

7.6 The Effect on Stress on the ER

As previously underlined, the ER is an organelle having many distinct functions that need to be strictly regimented to accomplish the correct functions. The most important function of the ER is doubtless the synthesis of proteins. The accumulation of unfolded or misfolded proteins in the lumen of the ER can still persist despite the presence of multiple chaperones and folding enzymes. When the cell is exposed to this kind of stress, there are a number of issues that must arise to maintain equilibrium and correct function, among them the translational retardation, the break-down of unfolded or incorrectly folded proteins, and an augmentation of chaperone and folding enzyme levels to re-establish proper ER and cell function. When the equilibrium is not regained, cell death or apoptosis can occur (Tabas and Ron 2011), thus attaining normal function is crucial for cell survival.

When a peptide is targeted for secretion and accessed the lumen of the cell, distinct alterations take place, such as the N-linked glycosylation, the creation of disulfide bonds and subsequently the oligomerization (Braakman and Hebert 2013). The N-linked glycosylation can take place co-translationally because the protein is transferred to the ER lumen. The oligosaccharyltransferase (OST) is able to alter the Asparagine residue in the Asn-X-Ser/Thr sequence after approximately 13 amino acids of the protein have entered the ER lumen (Nilsson and von Heijne 1993), which elevates the kinetics and improves the thermodynamics values of protein folding (Jitsuhara et al. 2002; Hanson et al. 2009). The misfolding of proteins in the lumen of the ER often arises as a result of the specific luminal conditions and the high protein concentration of newly produced proteins, mature secretion proteins as well as proteins acting as molecular chaperones and folding enzymes. Because of the elevated protein concentration and encapsulation in the lumen, the folding enzymes must logistically first recognize and identify the correct target protein for folding. Failure to properly alter proteins will cause the ER to perceive the deficiency of glucose residues and proteins such as UDP glucose-glycoprotein-glucosyltransferase (UGGT) to reglycosylate the protein (Kaufman 1999; Kaufman et al. 2002; Lee 1992). When the physiological folding process cannot be regained, the hydrophobic residues are still exposed and can interact with Grp78, which leads to the clustering of these proteins and the unfolded protein response (UPR) is switched on (Ron and Walter 2007; Walter and Ron 2011). The first role of the UPR is to raise the ER loading in order to satisfy the requirements of the cell to properly fold the proteins, leading to an extension of the ER through the production of sheets (Schuck et al. 2009) and to an augmentation of the ER folding devices.

The UPR comprises three parallel branches that are stress-activated and comprise inositol-requiring enzyme 1 (IRE1) by non-conventional splicing, double-stranded RNA-activated protein kinase R (PKR)-like ER kinase (PERK) by translational guidance by phosphorylation of eIF2a and activating transcription factor 6 (ATF6) through controlled proteolysis (Walter and Ron 2011). In summary, the activation of these three pathways causes the generation of the b-ZIP transcription factors promoting the activation of the UPR genes (Walter and Ron 2011). Firstly, an ER-located protein IRE1, which is a transmembrane endoribonuclease, facilitates the post-transcriptional and non-canonical splicing of the XBP1 mRNA that is targeted to the ER (Yanagitani et al. 2009; Yoshida et al. 2001; Lee et al. 2002) and encodes a transcription factor that drives the upregulating of additional stress sensitive and reactive genes. The nuclease activity of IRE1 is also implicated in the decomposition of a fraction of ER-associated RNAs in a pathway referred to as IRE1-dependent decay (RIDD) (Hollien and Weissman 2006; Gaddam et al. 2013). The cell has developed this mechanism to alleviate the translational burden on ER through the removal of mRNAs that otherwise might be translated and may provide a mechanism for the cell to elevate the stress reaction genes required in UPR. While it is understood that ER stress causes large-scale alterations in the protein and RNA levels of ER, it is still questionable whether this will result in instant restructuring to accommodate the new demands of organelles. Beyond all this, it is uncertain whether the activation of stress-reactive signal pathways induces the alteration of the intrinsic structural elements of ER. Specifically, XBP1 splicing during meiosis has been found to be activated in both Xenopus and budding yeast (Cao et al. 2006; Brar et al. 2012), pointing to a potential link between modifications in ER structure during meiosis and the ER stress response. Both of these are intriguing paths for prospective research examining structural abnormalities in ER triggered directly by cellular signaling features.

7.7 The General Structure and Function of Golgi Apparatus

The Golgi apparatus is the place where the transport and modification of proteins including their processing occur. The proteins processed are produced in the ER and directly transported toward the Golgi apparatus. In specific detail, the proteins can get into the Golgi apparatus at the surface facing the ER, termed *cis* region of the Golgi and leave the Golgi on the other end of the stack that is next to the plasma membrane of the cell, termed *trans*-region of the Golgi apparatus (Fig. 7.6). A typical feature of the Golgi apparatus is its highly dynamic nature that is even more prominent in the compartments of the Golgi apparatus, termed cisternae. Another feature of the Golgi apparatus is the wide range of morphologies of the Golgi due to specific cell types in disntinct organisms. However, it is not yet clear whether the different morphologies of the Golgi apparatus affect its function. A question can be raised of how molecules can move through the Golgi apparatus.

In concrete terms, the Golgi apparatus possesses multiple different classes of cisterns, which exhibit altered structures, compositions and functions. There is agreement on the total number and the precise definition of these various classes. A promising way to classify the Golgi cisternae is based on the trafficking pathways that facilitate the cisternae import and export events (Day et al. 2013). With this classification parameter, the Golgi cisternae can be subdivided into three major classes based on their individual function and their maturation state. Firstly, the cisternae compromising the cisternal assembly region become COPII vesicles, which are delivered from the ER and commonly contain constituents for recycling processes that need to be transported back to the ER through COPI vesicles. At this individual stage, even new cisternae are produced. Secondly, cisternae involved in the carbohydrate synthesis system can interchange material with each other by way of COPI vesicles. During this stage, the majority of reactions of glycosylation and polysaccharide synthesis take place. Thirdly, cisternae in the carrier generation step manufacture clathrin-coated vesicles and swapping material with endosomes. At this step, biosynthetic load proteins are wrapped around various transport vehicles and the cisternae are subsequently dismantled. Distinct transitions can take place when the cisterna is maturated from one step to another next step. In each step, the construction and the configuration of a cistern can continue to improve, although the distribution routes stay the same. This model affords a coherent and uniform framework for grasping the characteristics of Golgi in different organisms.

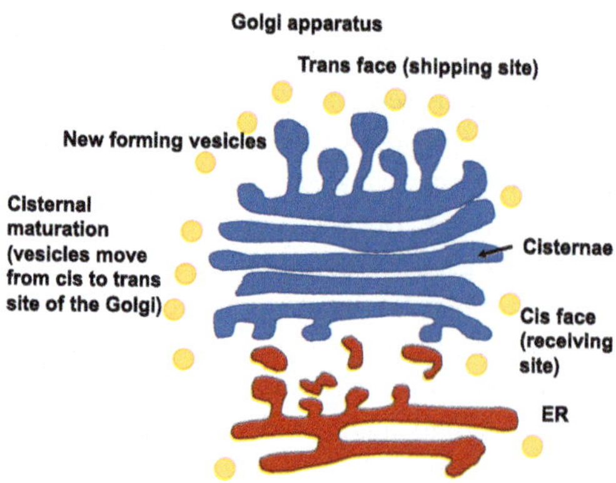

Fig. 7.6 The Golgi apparatus alters and sorts proteins trafficking throughout the cell. The Golgi is closely situated in the vicinity of the ER. Protein cargo goes from the ER to the Golgi, where it is transported to multiple targeting sites, such as plasma membrane or lysosomes

The Golgi apparatus ranks among the most complex organelles inside the cell (Farquhar and Palade 1981). It is made up of flattened membrane bags, the known cisterns, which are usually, but not necessarily, arranged in polarized stacks (Mowbrey and Dacks 2009). According to organism and cell type, a Golgi stack can have up to 3 or up to 20 cisternae (Becker and Melkonian 1996; Mogelsvang et al. 2003; Rambourg and Clermont 1997). Most organisms have single Golgi stacks; however, in many vertebrate cells, the Golgi stacks are connected by lateral links to create a continuous band (Klumperman 2011; Marsh et al. 2001). Even with this varying morphology, the Golgi still works on conserved fundamentals in various eukaryotes (Duden and Schekman 1997; Hawes 2005).

Inside a Golgi stack, the cisternae vary in construction, assembly and functionality. Biosynthetic load proteins arrive in the Golgi at the *cis* end of the staple and exit it at the *trans* end (Farquhar and Palade 1981). During the transit through the Golgi staple, biosynthetic load proteins experience glycan conversion and various other transformations (Ruiz-May et al. 2012; Stanley 2011). The complex polysaccharides are also made inside the Golgi (Dick et al. 2012; Parsons et al. 2012). The *trans*-most cisterns are known as tans-Golgi networks (TGNs) and are in charge of wrapping biosynthetic load proteins and polysaccharides in transport containers for shipment to downstream locations (Anitei and Hoflack 2011; Kang et al. 2011; Mellman and Simons 1992; Viotti et al. 2010). The membranes on the *cis* and *trans* surfaces of the stack are frequently tubulovesicular patterns that may not be tightly connected to the remaining parts of the stack (Rambourg and Clermont 1997; Staehelin and Kang 2008), but for the sake of clarity, all these patterns will be referred to as cisternae.

The functions of the Golgi stack are performed by enzymes and trafficking proteins, which are preferably positioned to a specific group of cisternae. The unstacked Golgi cisternae of the budding yeast *Saccharomyces cerevisiae* equally exhibit variations in their constitution and functional role (Papanikou and Glick 2009; Preuss et al. 1992). These results gave rise to the concept that the Golgi is composed of several classes of cisternae. However, there is still no consent on the precise number of such classes or the molecular entities underlying them.

The easiest route to the classification of the Golgi cisternae is to pay attention to the routes of human trafficking. In this view, cisternae in a certain class share the same mechanisms for importing and exporting their components, and these distribution routes mirror preserved key functions. This approach will enable experimental knowledge of the Golgi organization to be integrated into a wide range of organisms.

7.7.1 Why Are There Various Classes of Golgi Cisternae?

In earlier grading regimes, the main factor distinguishing Golgi cisterns was the polarity of the distribution of processing and biosynthetic enzymes. The N-linked oligosaccharides are altered by a series of glycosylation steps driven by Golgi inherent glycosylation enzymes located at distinctive places within the staple (Ruiz-May et al. 2012; Stanley 2011). It was elucidated by density gradient fracturing and immunolocalization analysis of the mammalian Golgi that early reacting glycosylation enzymes generally are clustered in cisterns on the *cis* side of the staple, whereas late reacting glycosylation enzymes are clustered in cisterns on the *trans*-side (Dunphy and Rothman 1985; Kornfeld and Kornfeld 1985; Rabouille et al. 1995). In mammalian and plant cells, selected glycosylation enzymes are located directly in the middle of the Golgi staple that is referred to as the medial cisternae (Donohoe et al. 2013; Dunphy and Rothman 1985; Rabouille et al. 1995). The highly ordered, and hence, polarized positioning of glycosylation enzymes is suppose to elevate the efficiency due to the fact that a biosynthetic cargo protein undergoes processing in exactly this ordered manner. During another processing type, termed TGN, which is employed by mammalian and fungal cells, the peptidases initiate the activation of prohormones (Rockwell et al. 2002).

Why are Golgi enzymes located in distinct cisternae? This separation allows the cell to fine-tune pH, ion content and substrate concentrations for individual sets of enzymes. In addition, the separation of enzymes can be indispensable to avoid two reactions compelling each other. For instance, lysosomal hydrolases of mammals with a mannose 6-phosphate sorting label are altered, and this adaptation must be made before the lysosomal hydrolases collide with Golgi-localized mannosidases (Goldberg and Kornfeld 1983). In plant cells, the segregation of pectic and xyloglucan polysaccharide backbone synthesis reactions to various cisterns impedes the capture of the resulting molecules prior to methyl esterification of the pectic polysaccharides (Atmodjo et al. 2013; Zhang and Staehelin 1992). However, these demands are doubtless unusual, as most carbohydrate synthesis reactions can take place in efficient manner in a unique mixed compartment (Dunphy and Rothman 1985). The more widespread justification for the Golgi cisternae configuration is that it serves as a delay timer (Glick and Malhotra 1998). Biosynthetic freight proteins require a certain amount of time to cross the Golgi, and this retardation offers the possibility to remove resident ER proteins and completely handle cargo proteins (Becker and Melkonian 1996; Rothman 1981). In cells with large Golgi stacks, for that reason, a certain physical and temporal function can be distributed over several cisternae.

Further indications for various classes of cisterns stem from morphological investigations of Golgi stacks (Donohoe et al. 2013; Klumperman 2011; Rambourg and Clermont 1997; Staehelin and Kang 2008). In plant cells, electron microscopy showed that the *cis* most common cisternae tend to stain considerably weaker than the medial and *trans*-cisternae, and this staining pattern corresponds to the limited localization of glycosylation and biosynthetic enzymes to the medial and *trans*-cisternae (Donohoe et al. 2013). Beyond that, in the case of numerous organisms, the cisternae on the *trans*-surface are partially or entirely isolated from the Golgi stack (Kang et al. 2011; Mogelsvang et al. 2003; Mollenhauer and Morré 1991; Rambourg and Clermont 1997; Viotti et al. 2010). According to electron tomographic findings, Golgi stacks of mammals exhibit a sharp difference between the *trans*-most cisternae, which manufactures vesicles coated with clathrin, and earlier cisternae, which generate COPI vesicles (Ladinsky et al. 1999; Mogelsvang et al. 2004).

The risk exists that advances in analytical methods will lead to ever more fine-grained subdivisions of the Golgi. For instance, a recent report on TGN yeast showed that the GGA and AP-1 clathrin adapters are sequentially enrolled (Daboussi et al. 2012), but a separation of the TGN cistern into two groups after the enrollment of adapters cannot be constructively feasible. In most cases, structural and compository variations among cisterns can masquerade the fundamental functional similarities.

Existent Classes of Golgi Cisternae

The general consensus is that the Golgi cistern can be grouped into four classes: cis, medial, *trans* and TGN (Farquhar and Hauri 1997). Each class of cisternae is regarded as the location for a distinct set of processing enzymes. This model is supported experimentally to some degree. For instance, at each step, a trial was conducted on the yeast pro-α factor mating the pheromone precursor, that tracks serial processing by four Golgi enzymes, and the Sec18/NSF vesicle fusion protein, implying that pro-α factor traverses four categories of Golgi cisternae (Brigance et al. 2000). However, studies with mammalian glycosylation enzymes have not resulted in these unambiguous characteristics. As judged by quantitative immunolabeling, individual glycosylation enzymes are most concentrated in particular cisternae, but they have overlapping distributions across the stack (Nilsson et al. 1993; Rabouille et al. 1995). In addition, a particular glycosylation enzyme may have varying intra-Golgi placements in a variety of cell types (Velasco et al. 1993). The polarized disposition of Golgi processing enzymes tends to mirror gradients of enzyme topicalization rather than accurate partitions (Glick et al. 1997).

Besides the major four possible classes of the Golgi cisternae, mammalian cells possess an additional ER–Golgi intermediate compartment (ERGIC) (Appenzeller-Herzog and Hauri 2006). The ERGIC appears to be stably connected to peripheral ER exit regions; however, it produces mobile membrane carriers, which are transported alongside the microtubules toward the Golgi band (Appenzeller-Herzog and Hauri 2006; Lippincott-Schwartz et al. 2000). These ERGIC-originated membranes can integrate at the *cis* part of the Golgi stack, where they built an addditional layer that is referred to as the *cis*-Golgi network (CGN) (Bannykh and Balch 1997; Ladinsky et al. 1999; Mellman and Simons 1992). However, the term CGN can be confusing, since the ERGIC-originated membranes at the *cis* part of the Golgi stack are not truly a cross-linked scaffold (Ladinsky et al. 1999); hence, these membranes are often termed as the "pre-cis layer." In plants and algae, the membrane structures, which are referred to as "*cis* initiators" at the *cis* region of the Golgi stack, possess morphological and functional characteristics that are similar to those observed in the mammalian ERGIC or synonymously termed in the pre-cis layer (Donohoe et al. 2013).

In order to describe the cisternal structure, composition and function of the Golgi in a complete manner, a precise and detailed overall view of the Golgi organization requires to address the following major questions. How can the Golgi cistern transport biosynthetic secretions when material is transferred to one another and even to other organelles? How can the Golgi residual proteins be pinpointed on specific cistern species? What types of trafficking vehicles function on the Golgi? These questions will be investigated in the light of the classification approach offered for Golgi cisterns.

7.7.2 Organization of the Golgi and Models for Golgi-Based Trafficking

It is a controversial issue that the properties of the Golgi cisternae are connected to the routes of Golgi membrane traffic (Glick and Luini 2011; Rabouille and Klumperman 2005). At present, the predominant model is cistern maturation, where the Golgi cisternae is postulated to be created de novo and then gradually matures into a TGN cisternae. In the simplest type of the maturation model, a TGN cisterna represents solely an older version of a *cis* cisterna, and the Golgi seems to be composed of a set of cisternae that mature in a continuous manner (Glick et al. 1997). A more differentiated view is that the ripening process takes place in a series of separate steps, with the various classes of Golgi cisternae constituting sequential stages of ripening (Glick and Nakano 2009).

The maturation model can take up most experimental data from a wide range of cell types (Glick and Luini 2011; Rizzo et al. 2013; Staehelin and Kang 2008). Under this model, the residing Golgi proteins can be recirculated from older to younger cisternae and thus remain in the organelle during the progression of the biosynthetic cargo. The mechanism for recycling residential Golgi proteins is still in doubt. One way is to recirculate Golgi membrane proteins in COPI vesicles that emerge from the Golgi cisternae (Donohoe et al. 2007; Glick and Malhotra 1998; Rabouille and Klumperman 2005). There is strong conclusive evidence that COPI vesicles carry at least some Golgi membrane proteins, although contradictory data have been received on the occurrence of glycosylation enzymes in COPI vesicles (Cosson et al. 2002; Gilchrist et al. 2006; Kweon et al. 2004; Malsam et al. 2005; Martínez-Menárguez et al. 2001; Orci et al. 1997, 2000a; Sönnichsen et al. 1998). Transient tubular connections between cisternae were proposed in supplement to the COPI vesicles to facilitate the recirculation of Golgi membrane proteins (Glick and Luini 2011; Marsh et al. 2004; Trucco et al. 2004).

Alternative models for Golgi transport consider the cisternae in a completely contrasting light. The fast division model suggests that the Golgi is a continuity structure with glycosylation and export taking place across all layers of the stack (Patterson et al. 2008). This model handles the Golgi as one mixed compartment and eliminates the discrimination among the various cisternae. The progenitor and marginal progression models of the cisternal suggest that the Golgi cisternae is a durable structure and that large portions of the cisternae are cleaved and then fused to transfer biosynthetic load, whereas the residual Golgi enzymes stay in the static sections of the cisternae (Lavieu et al. 2013; Pfeffer 2010).

These different models for the trafficking at the Golgi seem to indicate that Golgi cisternae can be either transient or static, and either sharply separated or intermixed. Hence, the classification of the Golgi cisternae may be impacted by the individual trafficking model. In the following, it is discussed whether the cisternal maturation model seems to be most suitable for interpretating the experimental results.

7.7.3 Organization of the Golgi with the Positioning of Resident Proteins

Specific Golgi cisterns have different habitation proteins; therefore, a classification system ought to consider the mechanisms for identifying the resident Golgi proteins. Localization signs for a series of Golgi glycosylation enzymes have been recognized (Banfield 2011; Machamer 1993). These proteins exhibit a type II morphology including a cytosolic N-terminus and a single transmembrane chain. The transmembrane sequences together with the flanking sequences are frequently essential and adequate for localization in the Golgi region. But the mechanisms of the localization were difficult to grasp. On the basis of the maturation model, glycosylation enzymes are constantly recirculated, probably in a way that is COPI-dependent. The yeast protein Vps74 was involved in the association of specific glycosylation enzymes with COPI (Tu et al. 2008). The size of the transmembrane sequence is relevant for the Golgi regionalization, pointing out that the segmentation into lipid domains is of importance (Sharpe et al. 2010). The length of the transmembrane sequence depending on the intended localization mechanism can encourage or impede the separation into a bended vesicle membrane.

It is much more obvious for the specific class of TGN proteins, such as the mammalian TGN38 and the yeast Kex2 (Machamer 1993). In a more detailed way, these proteins feature a type I topology, and they manage to target the TGN through the use of retrieval mechanisms that include adaptor-facilitated detection of signals in the C-terminal cytosolic tails.

The separation of the Golgi-based membrane proteins into type II glycosylation enzymes and type I TGN proteins is unequivocal. Conversely, the type I targeting signals of various Golgi glycosylation enzymes appear to be similar (Sharpe et al. 2010). The well-known features of the Golgi localization markers reveal an essential discrimination between the TGN and earlier cisternae.

7.7.4 Organized Golgi and the Trafficking System

Trafficking pathways limit the options for the classification of the Golgi cisternae. A trafficking step typically includes: a shell complex that sorts load and culpates vesicles, a regulatory Rab-GTPase, binding proteins and a fusogenic SNARE complex (Bonifacino and Glick 2004). The Golgi system utilizes a restricted amount of these compounds. Arguably, the Golgi-related elements of trafficking can be categorized into three groups, based on their location of operation either at the ER–Golgi intersection, or inside the Golgi, or in transport to and from the TGN.

The conserved envelops of the Golgi vesicles are COPI, COPII and clathrin. COPII vesicles transport proteins from the ER into the *cis*-Golgi and also merge with each other to produce ERGIC units (Barlowe and Miller 2013). COPI vesicles have two separate functions: The vesicles reclaim proteins from the Golgi cistern and transport them between the different Golgi cisternae (Barlowe and Miller 2013; Popoff et al. 2011). In plant and algae cells, Golgi-to-ER and intra-Golgi transport activities are performed by various subtypes of COPI vesicles, the so-called COPIa and COPIb vesicles (Donohoe et al. 2007, 2013). COPIa vesicles are restricted to the ER–Golgi boundary, while COPIb vesicles encompass the medial and *trans*-Golgi cisternae. Mammalian cells include COPI vesicle subtypes, which are similar to COPIa and COPIb (Lanoix et al. 2001; Malsam et al. 2005; Moelleken et al. 2007; Orci et al. 1997). Clathrin-coated vesicles emerge from the TGN to transport proteins to endosomes and lysosomes/vacuoles and can also reclaim proteins from mature to newly produced TGN cisternae (Myers and Payne 2013; Valdivia et al. 2002; Viotti et al. 2010). A rational assumption is that various classes of cisternae are characterized by COPIa, COPIb and clathrin.

The amount of Golgi-associated Rab proteins differs according to the cell type. *S. cerevisiae* represents the most simplest model organism that can be studied since it has a streamlined set of Rab proteins. Ypt1, which is homologous to mammalian Rab1, works together in the ER-to-Golgi transport and additionally in the internal Golgi transport (Barlowe and Miller 2013; Jedd et al. 1995; Suvorova et al. 2002). Ypt6, which is homologous to mammalian Rab6, functions in the transport of membrane carriers from endosomes toward the TGN (Liu and Storrie 2012; Storrie et al. 2012). Subsequently, the Ypt31/Ypt32 pair, which is homologous to mammalian Rab11, fulfills tasks in transport of the TGN toward the surface of the plasma membrane (Jedd et al. 1997).

During ER-to-Golgi transport, the conserved binding proteins comprise the coiled-coil protein referred to as Uso1 or p115 and the multi-subunit TRAPP complex (Kang and Staehelin 2008; Lord et al. 2013). The links of the GRASP family are also effective during ER-to-Golgi transportation in mammalian and fungal cells, although they are not present in plant cell (Behnia et al. 2007; Kinseth et al. 2007; Levi et al. 2010; Marra et al. 2001). The retrograde translocation of COPIa vesicles to the ER contains the Dsl1 binding complex (Lerich et al. 2012; Ren et al. 2009; Spang 2012). Intra-Golgi transport mainly occurs to be reliant on the conserved COG complex (Smith and Lupashin 2008; Ungar et al. 2006), despite the involvement of "Golgin" coiled-coil proteins (Faso et al. 2009; Munro 2011). The supply of means of transport from the endosomes to the TGN is carried out with the GARP compound and extra Golgins (Munro 2011; Pfeffer 2011). In conclusion, Golgi-based tethers facilitate the trafficking either between the ER and Golgi, or inside the Golgi, or between endosomes and the TGN.

SNARE proteins combine in different permutations to advance the fusion of the membrane. Recent findings point to the fact that in mammalian and yeast cells, one SNARE complex conveys ER-to-Golgi anterograde transport, a second SNARE complex transmits Golgi-to-ER retrograde traffic, a third SNARE composite communicates intra-Golgi transport, and one or two SNARE complexes confer retrograde transport to TGN (Malsam and Söllner 2011). The SNARE repository of the Golgi regime is therefore relatively limited (Pelham 1998).

With this overview of the traveling mechanism, the concept of classifying the Golgi into cis, medial, *trans* and TGN classes of cisterns as well as a possibly ERGIC/pre-*cis* class of layers is not supported, as the number of traveling elements is inadequate. Instead, it is proposed that the Golgi cisternae can be classified into three categories (Mellman and Simons 1992), mirroring three structural and functional phases of maturation: cisternae construction, synthesis of carbohydrates and carrier generation.

7.7.4.1 Stage I: The Cisternal Assembly

Membranes in cisternal assembly are equivalent to the ERGIC and pre-cis layers plus one or more *cis* cisternae in mammalian cells (Farquhar and Hauri 1997), or to the *cis* initiators with one or more additional *cis* cisternae in plants and algae (Donohoe et al. 2013). When cisterns are built, the COPII vesicles merge homotypically into larger vesicles, which then merge with one another and with extra COPII vesicles to form a fully fledged *Cis* cisterna (Bentley et al. 2006; Donohoe et al. 2013; Zeuschner et al. 2006). This pattern matches the historical denomination of the Golgi's *cis* surface as a "sculpting" surface (Mollenhauer and Whaley 1963). Electron tomographically oriented models of the *cis* surface of plant Golgi stacks with distinct structures representing successive stages of cisternae arrangement (Donohoe et al. 2013). From these images, it appears that 3–5 COPII vesicles merge to one *cis* initiator, which then expands by merging additional COPII vesicles to a full-fledged *cis* cisterna. The synthesis of few carbohydrates takes place in the assembly phase, but the assembling of cisterns is presumably in charge of adding mannose 6-phosphate to lysosomal hydrolases in mammalian cells (Pelham 1988) and constructing flake protein complexes in algae (Donohoe et al. 2013).

The capacity to absorb COPII vesicles is the key characteristic of membranes during the assembly period. In the assembly phase, COPIa vesicles are also employed to reuse the components of the transport vehicles in the ER. COPIa vesicle production in mammalian cells appears to be limited to ERGIC (Ladinsky et al. 1999), indicating that the recycling to ER precedes the transport of ERGIC-based membranes into the Golgi band.

7.7.4.2 Stage II: The Synthesis of Carbohydrate

Cisternae functioning in the carbohydrate synthesis are generally named medial and *trans*-cisternae. The primary function of these cisternae lies in the glycosylation of proteins and lipids and the synthesis of intricate polysaccharides (Atmodjo et al. 2013; Dick et al. 2012; Stanley 2011). In certain cell types, carbohydrate metabolism is associated with a considerable number of cisternae. Various types of glycosylation enzymes are frequently found concentrated in younger (medial) or older (trans) cisternae, and these distributions often remain a sharp one (Donohoe et al. 2013; Driouich and Staehelin 1997; Dunphy and Rothman 1985).

The intercistern transport at the carbohydrate synthesis level is obviously facilitated by COPIb vesicles (Donohoe et al. 2013; Rothman and Wieland 1996; Staehelin and Kang 2008), although the exact function of COPI in intra-Golgi transport is a permanent enigma (Glick and Luini 2011; Rabouille and Klumperman 2005). A number of variants of the maturation paradigm take Golgi polarity into account by assuming that the COPIb vesicles travel in an only retrograde direction (Glick et al. 1997). But a molecular foundation for the unidirectional transport of COPIb vesicles has not been obtained and is difficult to imagine if cisterns all share the same transport equipment during the carbohydrate synthesis phase. An alternative scenario is that COPIb vesicles could percolate by chance throughout the Golgi (Orci et al. 2000b). However, how can the percolation of vesicles enable the concentration of an enzyme in certain cisterna when COPIb vesicles carry glycosylation enzymes? A potential mechanism includes gradual alterations in lipid content. The production of lipids and their restructuring represents an active process within the Golgi (Bankaitis et al. 2012), thereby older cisternae possess an altered content of lipids than younger cisternae. If a certain glycosylation enzyme ends up in a cisterna in which the lipid environment is beneficial for the transmembrane sequence (Sharpe et al. 2010), the enzyme probably will not divide into buds of COPIb vesicles and would therefore accumulate in this cisterna. The concept seems to be highly hypothetical, while it enlightens that stochastic COPIb-facilitated transport may lead to an asymmetric dispersion of Golgi residential proteins.

Since older cisternae undergo a transition from the carbohydrate production phase to the carrier assembly phase, they cease their ability to absorb COPIb vesicles. When they still continue to produce COPIb vesicles during the transition step, the cisternae will lose content and hence shrink, as it has been seen in plant cells with electron tomography (Kang et al. 2011). Intermediately, when younger cisterns move from cistern construction to carbohydrate production, they gain the capability to absorb COPI vesicles. Finally, this results in a net material transport from older toward younger cisternae.

7.7.4.3 Stage III: The Formation of the Carrier

Cisternae working on the carrier formation stage can even interact with the TGN (Griffiths and Simons 1986; Mellman and Simons 1992). However, the boundaries of the TGN have examined in various ways that usually associate multiple cisternae at the *trans*-surface of the staple (Kang et al. 2011; Mogelsvang et al. 2004). To overcome this uncertainty, it is advisable to redefine the phase of carrier building through the ability of a cisterna to wrap biosynthetic cargo in a variety of transport vehicles (Bard and Malhotra 2006; De Matteis and Luini 2008). A certain processing and biosynthesis can also take place during carrier development. For instance, type I TGN proteins are able to initiate the enzymatic break-down of prohormones (Rockwell et al. 2002), and the nearest *trans*-surface cisternae of the staple contain the mammalian sialyltransferase and specific biosynthetic enzymes of the plant's cell wall (Atmodjo et al. 2013; Rabouille et al. 1995). The assembly of carriers represents a sorting process, which is connected to the condensation of distinct cargo proteins, and to disturbances of the Ca^{2+} ion concentration, the pH and the lipid content (Bard and Malhotra 2006; Chanat and Huttner 1991; Dettmer et al. 2006; Graham and Burd 2011; Kang et al. 2011; Schmidt and Moore 1995; von Blume et al. 2012). Cisternae be ripped off from Golgi staple during the carrier development phase, and hence, they are free-floating TGN within plant cells and the yeast Pichia pastoris (Bevis et al. 2002; Mogelsvang et al. 2003; Staehelin and Kang 2008). Subsequnetly, the final fate of Golgi cisternae in mammalian cells is not clearly understood, whereas transport vehicles are assembled at the *trans*-surface of the Golgi that then move toward the plasma membrane (Bard and Malhotra 2006; Polishchuk et al. 2000; Wakana et al. 2012). In plant cells, mature cisternae of carrier formation break down into vesicles and remnant membrane fragments and thus finish the life cycle of a Golgi cisterna (Kang and Staehelin 2008).

Cisternae, which are at the step of the carrier assembly, generate constitutive and regulated secretory vesicles targeted to the plasma membrane, and they additionally built clathrin-coated vesicles targeted toward endosomes, lysosomes or vacuoles (De Matteis and Luini 2008). In addition, older carrier forming cisternae can exchange specific compounds with younger carrier forming cisternae through a molecular transduction pathway associated with the clathrin adaptor AP-1 (Valdivia et al. 2002). The creation of clathrin-coated vesicles instead of COPI vesicles is a clear difference between the carrier development phase and earlier phases. Besides, in contrast to earlier cisternae, the carrier forming cisternae interchange membrane with the endosomal membrane system (Pfeffer 2011; Viotti et al. 2010). This disparity is obvious from the reaction of mammalian and plant cells to brefeldin A, which induces *cis* to fuse with the ER through *trans*-cisternae, while the TGN merges with endosomes (Klausner et al. 1992; Nebenführ et al. 2002).

7.7.5 The Overall Maturation Versus Micromaturation

The Golgi maturation process is characterized as the transition between each of the three phases: cisternal construction, carbohydrate production and carrier generation. All these transitions are characterized by alterations in their trafficking ways. Maturing incidents as visualized by live-cell imaging of yeast probably mirror the phase transition between the carbohydrate production and carrier generation (Losev et al. 2006; Matsuura-Tokita et al. 2006). It is anticipated that the transitions between the phases of the Golgi will be relatively rapid, although some studies on the transitional intermediates will probably provide insights. An appealing scenario is that a cisterna in the phase transition is of a mixed composition (Pfeffer 2010), whereby a domain of the earlier phase diminishes while a domain of the later phase extends.

Some residential Golgi proteins are singularly present at a specific stage, while others occur at two or even all three of them. For instance, the yeast-guanine nucleotide exchange factor Gea2 has been found on both old (early) and new (late) Golgi cisternae (Chantalat et al. 2004; Spang et al. 2001; Tsai et al. 2013), and the homologous GBF1 protein of the mammals is seemingly connected to Golgi membranes from cisternal assembling to carrier development (Lowery et al. 2013). This coincidence mirrors the interaction between precursor and product in sequential phases between cisternae.

Within each step or synonymously termed stage of the Golgi maturation, alterations of the Golgi can be observed during the aging of the cisterna. Examples are given in the following:

The first step or stage is termed cisternal assembly step, where the membrane structure increases in size from initially a few fused COPII vesicles to a cisterna of full size.
The second step is called the carbohydrate synthesis step, in the course of which the enzymatic and lipidic constitution of the cisternae develops.

During the step of the carrier building, the clathrin adaptors are being serially recruited.

Alterations at each step of the maturation can be seen as a micromaturation substep. These fairly small alterations deviate mechanistically from one another and from the major alterations in migratory routes that emerge during transitions between phases. This concept forms the foundation for the categorization of the numerous processes that transform the Golgi cisternae.

In the end, the three-step model needs to be validated and fine-tuned by examining how the Golgi cisternae ripen. As a possible mechanism for Golgi maturation, a process known as Rab transformation has arisen, which is to some extent rooted in analogy to maturation of endosomes (Glick and Nakano 2009; Mizuno-Yamasaki and Rivera-Molina 2012; Rink et al. 2005). Cascades including the functions of Arf GTPases seem to be crucial for the maturation of the Golgi apparatus (Lowery et al. 2013; Richardson et al. 2012; Stalder and Antonny 2013). The constituents that trigger the transitions between the three steps or stages of the Golgi are supposed to be highly conserved in nearly most of all eukaryotes. Supplementary elements present in specific organisms, such as species-specific Rab proteins, that are linked to the mammal and plant Golgi (Liu and Storrie 2012; Woollard and Moore 2008), are expected to govern the microripening processes and endow the Golgi with its unique characteristics in different cell types.

7.8 Golgi Apparatus and Vesicle-Based Protein Trafficking

Golgi apparatus the first cellular organelle revealed in the year 1898 by an Italian biologist Camillo Golgi. After him, the organelle was named Golgi apparatus. Due to its large size, the Golgi apparatus was investigated easily. Similar to the ER, the Golgi apparatus is associated with the processing, trafficking of proteins that can be either secreted or bound to the membrane. Hence, the ER can be described as the central organelle facilitating the transport of proteins and lipids through the cytoplasm of a eukaryotic cell.

An important prerequisite of this organelle is its highly dynamic structure and its broad variety of morphologies in different cell types. The variety is manifested in the Golgi compartments (termed cisternae) being of different morphology due to the cell type. The Golgi apparatus is typically composed of a pancake-like stack of flattened compartments (Fig. 7.6) with possess specific types of enzymes. Since proteins need to be further processed before secretion, such as through the addition and structural remodeling of attached sugars, they seem to transit in an ordered manner through each part of the Golgi stack. The mitochondria are especially conspicuous organelles with a smooth outer surface, which absorb an extensively folded network of internal membrane structures. Mitochondria are the main place of ATP synthesis for cells growing in the vicinity of oxygen, and their intriguing physiology and microstructure is well understood.

It is not yet understood why there exist these diverse structures of the Golgi apparatus and how the different morphologies impair its function. In order to reveal individual functions of the Golgi apparatus, different Golgi morphologies of different cell types are analyzed in great detail.

7.8.1 How Are Proteins Altered When They Travel Through the Golgi Apparatus?

The Golgi processes proteins produced by the ER before distributing them throughout the cell. Proteins initially penetrade the Golgi on the side facing the ER (synonymously termed *cis* site) and exit the Golgi on the opposite side of the stack, which faces the plasma membrane (synonymously termed *trans*-site). Proteins need to find their path through the stack of intervening cisternae on which they are altered and packaged for the targeted transport throughout the cell (Fig. 7.6). The cisternae of the Golgi apparatus can be different in total number, shape and organization due to the type of the cell. The commonly employed typical diagrammatic division of the Golgi apparatus into three major cisternae, referred to as cis, medial and trans, is a major simplification of the structural organization and the associated functions of the Golgi apparatus. In some cases, also additional parts can be associated to the Golgi, such as the CGN or the TGN. These Golgi associate networks structures exhibit an increased variable structure, which are compromised by several cisterna-like parts and several vesiculated parts.

Individual cisternae or cisternal regions of the Golgi possess a distinct set of protein modifying enzymes. What type of role does these enzymes fulfill? The Golgi enzymes are able to trigger the addition or removement of sugar residues of load proteins (termed glycosylation), the addition of sulfate groups (termed sulfation) and the addition of phosphate groups (termed phosphorylation). The load proteins can be altered by enzymes, which are termed resident enzymes that are typically present in each cisterna. In a serial manner, the enzymes perform appropriately alterations of load proteins. Specific Golgi-facilitated alteration can serve as signals for sorting out proteins directly toward their targeted location inside the cells, such as the lysosome and the plasma membrane. What occurs upon defects in the function of the Golgi apparatus? Defects in several aspects of Golgi function may cause congenital glycosylation disorders, including distinct versions of muscular dystrophy and may promote diabetes, cystic fibrosis and cancer (Ungar 2009).

How can load proteins manage to travel between the different Golgi cisternae? There seem to be at least two possible mechanisms: on the one hand, the vesicular transport mechanism and on the other hand the cisternal maturation mechanism. It is important to note that the two models can both provide the Golgi's balanced steady state and contribute to the processes, although they are largely divers. In 2002, it has been revealed that the membrane system and the vesicles facilitate the secretion mechanisms inside cells. There is strong evidence that general molecules and processes are employed during the fusion and fission of the membrane in eukaryotic cells. Moreover, it has been shown that biochemically reconstituted Golgi membranes of mammalian cells and isolated vesicles are able to transmigrate from one cisterna to the adjacent cisterna.

In a vastly different approach, by employing yeast genetics, crucial proteins in the secretion pathway had been enlightened and precisely characterized in their function during the entire secretion process. These identified molecules associated in the generation and fusion of vesicles formed the basics for the vesicular transport model.

There exist currently two models how protein trafficking through the Golgi apparatus can be performed, the cisternal maturation model and the vesicle vehicle model (Fig. 7.7). The first model deals with the movement of proteins through the Golgi and is referred to as cisternal maturation model. When a new *cis* cisterna is created, it will transmigrate through the entire Golgi stack, alter itself by maturation, when it is clustered medial, and finally, the *trans*-enzymes can be moved by

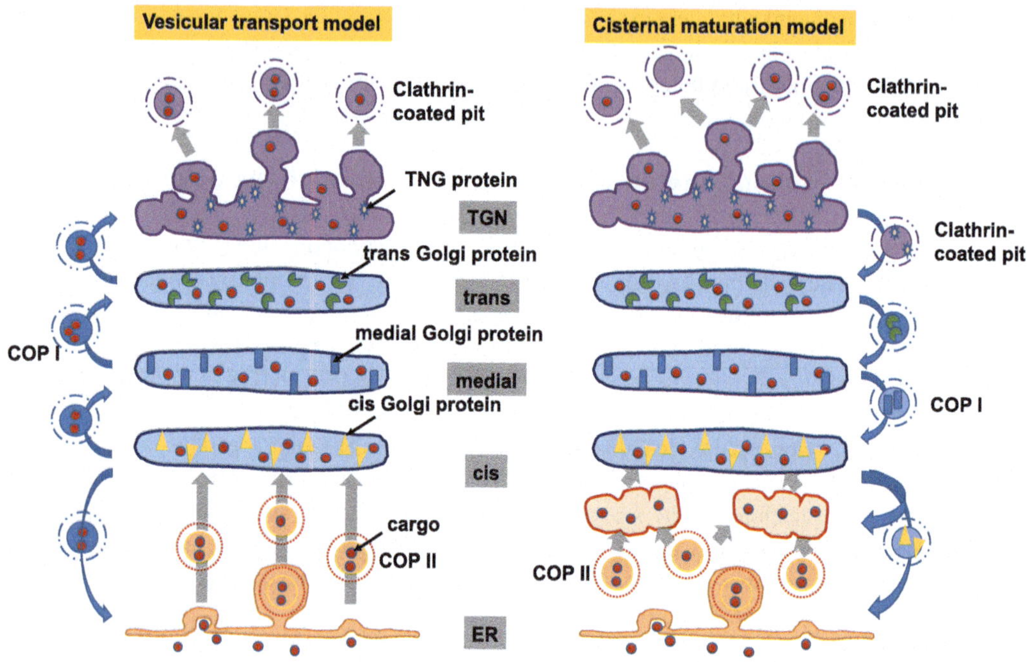

Fig. 7.7 The cisternal maturation model and the vesicle vehicle model

vesicle carriers from later to earlier cisternae that is termed retrograde trafficking. The second model, which is named vesicular transport model, deals with cisterna that stay at their position and contain unaltered enzyme content. Instead the proteins are transported through the Golgi staple via vesicles moving from earlier to later cisternae that is referred to as anterograde trafficking.

What Is the Evidence for a Vesicular Transport Model?

A principal observation was that the vesicles created inside the Golgi can transport load proteins between the cisternae from the *cis* side toward the *trans*-side. These findings clearly point out to the vesicular transport model (Farquhar and Palade 1998). The vesicular transport model states that the Golgi cisternae represent stable regions that contain distinct protein modifying enzymes, which can add or cut off sugars, add sulfate groups and fulfill other types of post-translational modifications. Vesicles reaching at an individual cisterna possess loaded proteins, which can be altered by the resident enzymes concentrated inside the cisterna. The next step represents the budding of new vesicles that possess the load proteins and mange the transport toward the next stable cisterna, in which another set of enzymes can process and alter the cargo protein (Rothman and Wieland 1996).

7.8.2 The Cisternal Maturation Model

Before the enlightening findings that vesicles contain proteins to be transported toward the Golgi cisternae, it has been hypothesized that each Golgi cisterna was rather transient and that the cisternae themselves travelling from the *cis* toward the *trans*-surface of the Golgi can be altered over time. The protein movement as passengers of cisternae through the Golgi staple is termed the cisternal maturation model. This model states that the enzymes located in an individual cisterna can be altered over time, whereas the cargo proteins still remain within the cisterna. Once this model was largely supported, it has been revealed that there are large numbers of small transport vesicles nearby the Golgi, and hence, the vesicular transport model has been proposed that seems to replace the old model by a novel updated model. However, it can even be revisited an old idea that can become important in a new manner.

What Type of Novel Evidence Exists for the Cisternal Maturation Model?
In the 1990s, multiple cell types have been analyzed to widen the knowledge of the Golgi apparatus. Using mammalian cells, it has been shown in vitro how large protein complexes are travelling through the Golgi. With immunoelectron microscopy, it has been revealed that rigid, 300 nm wide, rod-shaped, procollagen trimers travel through the Golgi apparatus of mammalian fibroblasts. Moroever, it has been reported that procollagen can solely been found in Golgi cisternae, whereas it is absent in vesicles that usually possess a much smaller diameter, even less than 100 nm, and hence, these vesicles are smaller in diameter that the diameter of procollagen (Bonfanti et al. 1998). Beyond that, similar results have been found in the Golgi apparatus of algae. Flagellated protists of vastly different types built and export scales, which are in contact with the cell membrane of these organisms. In specific detail, the scales seem to be different, but possess a defined size and morphology. In various species of algae exporting scales of relatively large size (1.5–2 mm) and even moderately sized (about 40 nm), the scales can be continuously detected in the Golgi, while there are not found within transport vesicles (Becker et al. 1995; Becker and Melkonian 1996). Finally, these findings of various cell types point out to the cisternal maturation model for the transport of proteins across the Golgi apparatus.

What roles fulfill all the vesicles that have been identified inside the Golgi? The favored cisternal maturation model states that the vesicles serve as transport machinery for enzymes of the Golgi rather than for other protein load. The so-called retrograde vesicles can move load backward from the Golgi budding site toward the cisterna to deliver enzymes toward younger cisternae. Thereby other vesicles, which came from older cisternae, transfer the enzymes required for the following next steps of the post-translational alteration of proteins (Glick and Malhotra 1998; Pelham 1998).

7.8.3 Which Model Is Best Suited for the Precise Description of the Golgi Functions?

There is a large agreement that the cisternal maturation model is the favorite candidate (Emr et al. 2009). There has been experimental support that strongly underpins the process of the cisternal maturation. Live-cell fluorescence microscopy has been employed to visualize the cisternal maturation in the Golgi apparatus of living *S. cerevisiae* yeast cells (Losev et al. 2006; Matsuura-Tokita et al. 2006; Malhotra and Mayor 2006). The Golgi apparatus of *S. cerevisiae* possesses a specific structural lack, since the Golgi stack has not the typical pita bread appearance and is instead rather less organized. In specific detail, the cisternae are distributed in an individual and irregular manner throughout the entire cell. Based on this unusual structure, it is well suited for light microscopic analysis of alterations in individual cisternae on different time scales.

In contrast, the vesicular transport model states that each individual *cis* cisterna stays a *cis* cisterna, which contains specific *cis* enzymes throughout its whole lifespan. Conversly, the cisternal maturation model states that a newly created *cis* cisterna can possibly maturate toward a medial and at a later time point toward a *trans*-cisterna, before it breaks apart upon packaging of the entire content for the distribution to their final destinations inside the cell. In order to reveal what type of model predicts the experimental results appropriately, green and red fluorescent proteins are coupled to proteins, which are located in different, individual cisternae of *S. cerevisiae*, and these fluorescently labeled proteins are tracked over specific time spans. The experiments are planned to either probe the vesicular transport model or the cisternal maturation model. When the vesicular transport model is applicable, the cisternae will stably possess the same fluorescently labeled Golgi proteins on different time scales. Conversly, when the cisternal maturation model is not applicable, every cisterna will display a different set of Golgi proteins on different time scales. In fact, it has been seen that individual cisternae altered their composition of fluorescently labeled proteins on different time scales. These findings supported pronouncedly the cisternal maturation model.

7.8.4 What Are the Major Questions About the Trafficking of Proteins Through the Golgi Apparatus?

There is substantial agreement that the cisternal maturation model is well suited to describe the experimental results. However, it is not yet fully understood whether all cargo proteins use the same pathway. The dynamics of cellular membrane proteins, such as Golgi membrane proteins, can be analyzed quantitatively and it has been demonstrated that some cargo proteins can move through the Golgi more slowly than the rates with which the cisternae mature (Patterson et al. 2008). Thus, these results lead to the conclusion that the cisternal maturation model cannot accurately describe the experimental results. Since the cisternal maturation model is not applicable in all cases, an additional model is proposed, in which a two-phase system of membranes defines what kind of cargo proteins and Golgi enzymes need to be dispersed during their transport. In order to be more precise, specific cell types possess linkages between different cisternae in a single Golgi stack, such as between between *cis* and medial cisternae. For instance, intercisternal connections have been identified during the propagating waves of protein traffic within mammalian cells (Trucco et al. 2004). Therefore, an advanced and improved version of the cisternal maturation model is required.

Distinct aspects of the protein transport machinery over the Golgi are clearly revealed, while there is still less knowledge about the specifics of the Golgi within various other organisms. There is a question that remains unanswered of whether there are universal characteristics in all Golgi. There are selected important question about the Golgi to be raised for future ongoing research:

Can various types of secretory load employ the same routes through the Golgi staple or are different routes developed?
What type of molecular mechanisms facilitate and control the maturation process of the Golgi cisternae?
Do specific specialized regions within the Golgi cisternae exist? How are these special regions generated and what particular roles do they fulfill during the sorting of cargo and its export?
How are the individual compartments of the Golgi compartments designed and restructured?
Is stacking architecture of the Golgi crucial for the membrane trafficking? If the answer is yes, how can organisms such as *S. cerevisiae* circumvent this demand? (Emr et al. 2009)

In conclusion, the structural architecture of the Golgi apparatus can be different due to the individual cell type. In the yeast *S. cerevisiae*, based on the vastly distributed feature of the Golgi cisternae, individual one can be enrolled. Using fluorescently labeled proteins targeting different cisternae, it has been evidenced that the Golgi cisternae can alter with time, which underpins the famous cisternal maturation model for the protein traveling across the Golgi apparatus. However, in the end, there is much not yet enlightened about the structure and function of the Golgi.

7.9 The Interface Between Endoplasmatic Reticulum and Golgi Serves for Protein Sorting

The trafficking of proteins is crucial for the physiological cell functions under normal healthy conditions. In eukaryotic cells, the spherical transport vesicles can deliver proteins and lipids that are located in one internal membrane-bound compartment toward another compartment associated with the secretory pathway. When an individual protein is transported to a defined target region, which is commonly termed protein sorting, it represents an important function that is inherently connected to the biogenesis of vesicles (Gomez-Navarro and Miller 2016). The principles of the sorting of cargo by the vesicle traffic machinery are presented and it is discussed how these various mechanisms that transport the cargo proteins are selected and how they are subdivided into distinct transport vesicles. A major focus of current research lies on the secretory process and especially on the two initial compartments: the ER and the Golgi apparatus. Emphasis is placed on the great complexity and the diverse function and precise regulation of the cargo adaptors, which are presented in compressed form, with a focus on the mechanistic insight into the protein sorting process in living cells.

Since eukaryotic cells are known to be highly compartmentalized, they contain distinct organelles with specific protein and lipid arrangements. Due to the continuous exchange of membranes and cargo proteins during the highly dynamic protein trafficking, the physically linked compartments of the entire secretory pathway are remodeled. In specific detail, roughly one-fourth to one-third of the cell's proteome seems to be associated with the secretion into extracellular microenvironment or with the interior membrane system. The first step of the secretory signal transduction pathway represents the ER-to-Golgi transport. Inside the ER, the proteins targeted toward the extracellular microenvironment or to other organelles along their transport way they are packaged into specific types of vesicles that deliver them toward the Golgi apparatus. During this stage, all cells may be able to identify them as native and non-native proteins, which subsequently leads exclusively to the transport progression of appropriately folded and assembled cargo proteins. Within the export to the ER, a majority of secretory proteins is actively sorted. Conversely, trafficking of proteins can alternatively be performed without a selection, which is termed bulk flow. In the end, the transport from Golgi to ER contributes to the efficient return of immature charges or even escaped ER-resident proteins to the ER. In the following, it will be presented how cells fulfill the sorting requirements of the largely diverse set of proteins that step over the ER–Golgi interface, which requires a huge cellular effort due to the vastly heterogenous cargo proteins.

7.9.1 What Are the Underlying Principles of Selective Capture by Vesicles?

The guided transport of proteins among organelles of the secretory pathway is performed by spherical membrane-bounded vesicles, which extent as a bud from a donor organelle carrying the transport protein and fuse with an acceptor organelle in different part inside the cell. The fission and fusion transport mechanisms enable secretory proteins to overcome the membrane barriers, which are thus not disturbed in their specific functional segregation by organelles. The specific classes of transport vesicles are generated by conserved sets of cytoplasmic proteins, which are grouped based on their protein coating that facilitate their assembly. There exist basically three major vesicular coating proteins that can be detected across all eukaryotic life, which are clathrin, COPI and COPII proteins derived from evolutionarily closely resembling envelope proteins. In specific detail, COPII-coated vesicles deliver cargo proteins of the ER toward the Golgi. In contrast, COPI-coated vesicles can deliver cargo in the opposite and hence retrograde direction, which means from the *cis*-Golgi site back toward the ER, and also inbetween different Golgi cisternae. Moreover, clathrin-coated vesicles are built by the plasma membrane and the TGN in order to provide the fusion with endosomes or lysosomes (Fig. 7.5). The coatings of the vesicles fulfill two important tasks: firstly, they bend the vesicle membrane into a spherical object and secondly, they fill the vesicle with distinct cargo protein. By the coupling of the cargo selection with the formation of the vesicle, the cells manage to efficiently sort the proteins as an inherent feature of the specific transport route.

7.9.2 Recognition of Sorting Signal by Coat Adaptors

The analysis of the cell surface receptor internalization through clathrin-based endocytosis first revealed the principle that distinct protein-based signals facilitate the targeting of specific cargo to vesicles. The biochemical, structural and genetic discrimination of clathrin and other related vesicle types has identified how these different envelope configurations link the sorting of cargo with the overall assembly of vesicles. A key feature of the appropriate sorting of cargo represents specific coat subunits (synonymously termed cargo adaptors) that contain binding surfaces for the recognition of specific sorting signals, which are located within the cytoplasmic domains of cargo proteins. The interplay between the coating of the vesicle and the signal triggers the engulfment of cargo by the forming vesicles. The majority of the binary cargo–coat interactions determined in vitro have a relatively low affinity, which can be relevant with respect to coat dynamics in trafficking. During the entire vesicle lifetime, the envelope proteins can be detached from its membrane surface in order to uncover the fusion machinery. Hence, the linkages between the enveloping proteins and the components of the vesicle need to be reversible. Nevertheless, cargo adapters frequently also exhibit an affinity to lipids, which in turn leads to both specific recruitment and overall affinity of the adapter to a donor organelle. Consequently, adaptors are able to interact directly with structural scaffolding components of the numerous coatings, which thereby aid in connecting the membrane, cargo proteins and the membrane deformation system. Knowledge of how the local cooperative interplay between cargo proteins, membrane lipids, adaptor proteins and scaffolding proteins can mutually contribute to affect the assembly of vesicles is a key determinant for nearly all cellular transports.

7.9.3 Coupling of Cargo and Coat by Cargo Receptors

A limitation of the signal-facilitated sorting of the coat-linked cargo adaptors is the accessibility of the signal. The numerous soluble secreted and lysosomal proteins cannot therefore interact directly with the subunits of the vesicle envelope, since the barrier formed by the lipid bilayer is in the way. The sorting of these cargo proteins is performed by a receptor-facilitated transport. The cargo receptors represent proteins, which are distributed over the entire membrane and bind simultaneously to cargo proteins and coat adaptor proteins in order to efficiently target soluble proteins toward nascent vesicles (Dancourt and Barlowe 2010; Geva and Schuldiner 2014; Guo et al. 2014; Barlowe and Helenius 2016). A key feature of the function of cargo receptors is their capacity to bind cargo in a particular compartment and release it by fusion with the downstream organelle. Thereby, the cargo receptor can be reused by returning it to the donor compartment to facilitate future trafficking. Cargo receptors are long-lived proteins that facilitate several rounds of transport events. However, it is still difficult to understand how the majority of charge receptors manage the direction of vector movement between the compartments. One of the most characterized examples is the KDEL receptor, whose name is based on the conserved amino acid sequence Lys-Asp-Glu-Leu (KDEL) serving as an ER retrieval signal and that can cycle between the ER and Golgi in order to bring escaped ER resident proteins back. The binding and release of the ligand depends on the pH. In specific detail, in the lower-pH environment of the *cis*-Golgi, the receptor can interact with the retrieval motif, whereas in the neutral pH of the ER lumen, this interaction is impaired. Moreover, for other receptor-cargo interaction pairs, additional factors of the lumenal enivornment, such as ion concentration, can impact the conditional interaction. In addition, there seems to be a cargo-based alteration of the localization of the KDEL receptor (Lewis and Pelham 1992). These results indicate that the KDEL receptor represents not solely an inert platform but can also react to cargo sorting requirements that may be facilitated by allosteric interactions of the cargo binding.

While integral membrane cargo proteins are in principle capable of interacting directly with envelope adapters and therefore are not supposed to require cargo receptors, several examples of such proteins with cargo receptors have been reported in several different routes. These membrane cargo receptors seem to provide a sorting signal that is not present in the cargo, but they may additionally fulfill functions in chaperoning transmembrane domains, such as Erv14/CNIH, and the identification or recovery of unassembled components of oligomeric complexes, such as Rer1. It is still not yet clear how the sensing and clearance of membrane-embedded interactions is regulated and coordinated in each cellular compartment.

7.9.4 Sorting Mechanisms at the ER–Golgi Connection Site

7.9.4.1 How Is the Cargo Recruited into COPII Vesicles? What Initiates the ER Export?

The entry of the nascent proteins in the secretory mechanism is triggered by the decisive incorporation of precisely folded and assembled secretory and membrane proteins into vesicles that are generated by the cytoplasmic COPII coating. The coating consists of five subunits, including Sar1–GTP, dimeric Sec23/Sec24 and tetrameric Sec13/Sec31 (Barlowe et al. 1994). The budding process of COPII vesicles is triggered when Sar1–GTP starts to direct Sec23/24 to the ER in order to create a so-called prebudding complex to which at a later time point Sec13/31 is recruited through the interaction of Sec23/Sar1 with Sec31 (Bi et al. 2007). Sec13/31 facilitates the bending of the membrane and creates a cage-like external coat of the nascent vesicle (Stagg et al. 2006; Fath et al. 2007; Bhattacharya et al. 2012), whereas Sec23/24 drives the specific selection of the cargo proteins. The assembly of the COPII coat proteins takes place at specific membrane regions, which are referred to as ER exit sites (ERESs) (Lee et al. 2004; Sato and Nakano 2007). The cargo proteins are able to accumulate temporarily at specific sites, such as ERESs, before they are embedded in a nascent vesicle or accompany the embedding. There is a certain degree of quality control mechanisms in ERESs, since ER residents and misfolded cargo proteins, such as the thermosensitive vesicular stomatitis virus G-protein, are specifically excluded from ERESs (Mezzacasa and Helenius 2002).

The Sec24 subunit represents a cargo adaptor protein of the COPII coating (Miller et al. 2002). Genetic, biochemical and structural experimental analysis revealed a specific and precisely defined set of cargo–Sec24 interactions that seem to form the basis for the cargo specificity in the assembly of COPII vesicles (Miller et al. 2003; Mossessova et al. 2003; Mancias and Goldberg 2007, 2008; Pagant et al. 2015). Yeast Sec24 possesses at least four specific cargo-binding sites, each of which identifies different sorting signals and thus broadens the range of cargo that can be processed with the same type of vesicle. Mammalian Sec24 has an additional cargo-binding site, which is characterized by its distinct structure and recognizes even another signal (Mancias and Goldberg 2008). More diversity of the cargo is provided by the expression of several isoforms of the Sec24 subunits. For instance, there are two isoforms of Sec24 expressed in yeasts, such as Iss1/Sfb2 and Lst1/Sfb3 and four isoforms of Sec24 expressed in mammals, such as Sec24A, Sec24B, Sec24C and Sec24D. Each of them possesses specific preferences of distinct cargo proteins. Besides an expansion of the cargo repertoire, various Sec24 isoforms can also affect the structural features of the vesicle assembly. The yeast Sec24 paralogue Lst1/Sfb3 is specific for the envelope of the large oligomeric protein Pma1 and thus seems to form larger vesicles than pure Sec24 vesicles (Roberg et al. 1999; Kurihara et al. 2000; Shimoni et al. 2000). In fact, vesicles can be preprogrammed to obtain a specific geometry that is determined by the distinct combinations of coat adaptors and suitable cargo proteins. From the viewpoint of the cargo protein, the recruitment of the appropriate type of coat adaptor, the cargo protein itself can impact the vesicle's geometry.

The sorting signals sensed by Sec24 cover two simple amino acid motifs, such as the DxE or LxxL/M/E motifs in yeast Bet1, Sys1, and Sed5 proteins, and folded epitopes, such as the longin/SNARE domain in the Sec22 protein (Mossessova et al. 2003; Liu et al. 2004; Mancias and Goldberg 2007). However, it is still unknown how many different sorting signals of a specific Sec24 molecule can interact at the same time. The major knowledge of the structural cargo binding by Sec24 has been obtained from cocrystalization studies of coat proteins, which are bound to synthetic peptides that comprise the sorting signal. In addition, the full impact that every cargo protein has on interacting surface with Sec24 is still elusive. In specific detail, sec24 represents a cargo-binding surface, which is not allosterically alterating its conformation upon the binding of cargo proteins. In this aspect, the COPII vesicle transport machinery is dissimilar to the clathrin cargo protein adaptor complexes that alter their conformation due to the binding of the cargo protein and thereby induces the formation of the entire vesicle. In contrast, the initiation of COPII vesicle formation is triggered by accessory proteins, which are not necessarily components of the vesicle coating.

7.9.4.2 How Is the Cargo Enveloped by COPI Vesicles? What Delivers the Cargo from the Golgi to the ER?

COPI vesicles manage the recovery of lost ER resident proteins and transport components that continuously translocate from ER to Golgi and vice versa. The COPI complex, referred to as coatomer, is composed of seven subunits, such as α, β, β′, γ, δ, ε, and ζ-COP, which are targeted toward membranes as an intact assembly (Hara-Kuge et al. 1994). In specific detail, based on biochemical features, the coatomer is subdivided into two major subcomplexes: firstly, as a trimer that is consistant of α, β′, and ε, referred to as the B-subcomplex and secondly, a tetramer that encompasses β, γ, δ, and ζ and refers to the F-subcomplex (Waters et al. 1991; Eugster et al. 2000). Based on bioinformatic and crystallographic analyses, it has been shown that the COPI coat displays structural motifs, which are resembling those in the clathrin and COPII coats, such as the extended α-helical solenoid domains connected with β-propeller structures (Devos et al. 2004; Lee and Goldberg 2010). In contrast, the structural assembly of these various elements differs between the coat systems.

Although the tetrameric F-subcomplex is similar to the cargo-binding clathrin adaptor protein (AP) complexes, the structural analysis revealed clearly identifiable β-propeller domains of α-COPI and β′-COP that serves as interaction region with dilysine motifs, which represent the predominant retrieval sign for nearly all ER resident membrane proteins (Jackson et al. 2012). Moreover, the F-subcomplex seem to fulfills tasks in the interaction with the cargo, in which the β- and δ-subunits are crucial for sorting of Arg-driven retrieval signals (Michelsen et al. 2007) and the aromatic amino acids within the small cytoplasmic tails of the p24 family proteins can interact directly with the F-subcomplex (Fiedler et al. 1996). Additionally, the δ-COP can bind to a WxW motif of the cytosolic accessory protein Dsl1, although this interaction domain is located distal to the surface of the membrane (Suckling et al. 2015). However, there is still the possibility for additional cargo-binding surfaces to be recognized on the F-subcomplex. In an alternative manner, the protein–protein interaction sites on the COPI AP-like complex seem to provide specific tasks in coat regulation that are beyond that of simple cargo binding.

The knowledge about the arrangement of COPI subunits inside the assembled coat was gained by cryo-electron tomography of COPI-coated vesicles and the general fitting of crystal structures into the EM density (Dodonova et al. 2015). The cargo-binding site surfaces of an α- and β′-COP dimer were facing the membrane and lie in close proximity. In specific detail, their α-solenoid domains built an arch-like structure that is orientated away from the membrane and seem to provide the membrane scaffolding, when these structures are connected through the F-subcomplex. A tetramer, which is composed of γ-ζ-COP and β-δ-COP dimmers, can additionally built a hinge-like structure that is analogous to that of clathrin AP complexes, but exhibits a more extended conformational shape. The importance of the different structures of the COPI F-subcomplex and the AP complexes seems to be clearly demonstrated. However, there may another explanation for this difference that is based on the cargo coupling to the vesicle formation. In specific detail, the clathrin adaptors seem to expose a "closed" conformation within the cytosol that can turn into an "open" conformation due to the binding to the membrane, which thereby leads to the exposure of clathrin-binding sites to initiate the assembly of a vesicle (Jackson et al. 2010; Kelly et al. 2014). Subsequnetly, the assembly of vesicles in the clathrin system is mechanistically associated with the binding of the cargo. In fact, when an adaptor complex is connected to its ligand, the assembly of the vesicle is exclusively further progressed. Conversly, the COPI structure, which has been built in the absence of cargo proteins and in the presence of a lipid bilayer, represents a "hyperopen" conformation (Dodonova et al. 2015). However, it is still elusive how these putative structural rearrangements facilitate the binding of cargo, membrane binding and the assembly of the vesicle coat, whereas it seems to be based on allosteric effects that impact the assembly of COPI.

7.10 What Mechanistic Roles Fulfill Cargo Receptors in Sorting Processes?

7.10.1 Cargo Receptors Provide ER Export

There exists a high diversity among the cargo proteins, which can traverse the entire ER, when they are soluble proteins and may also depend on cargo receptors to interfere with the COPII coat. There are many review articles about the ER export receptors that present and discuss the diversity of these trafficking factors (Dancourt and Barlowe 2010; Geva and Schuldiner 2014; Barlowe and Helenius 2016). The receptor-driven ER export enables the transport of a wide variety of cargo proteins, which can be targeted toward vesicles through the utilization of distinct adaptors, receptors and accessory factors (Geva and Schuldiner 2014). Another important feature of the receptor-driven export represents the precise regulation of the transport due to the degree of protein folding. When a cargo receptor solely interacts with its ligand, which therefore needs to be folded in a correct manner, hence only mature secretory proteins will be exclusively transported and depart from the chaperone-rich lumen of the ER. Mechanistic insights into how certain cargo receptors identify their cargo proteins are still increasing, and there is little knowledge as to whether the binding of client proteins triggers allosteric alterations that then induce envelope binding, thereby guaranteeing that only cargo-loaded receptors are transported.

The properly most investigated mammalian ER export receptor is the ERGIC-53/LMAN1. In fact, the ERGIC-53 represents a single-pass transmembrane protein (Schweizer et al. 1988; Schindler et al. 1993), which possesses a large N-terminal lumenal lectin domain for its interaction with glycosylated cargo proteins (Appenzeller et al. 1999). The cathepsin Z–related glycoprotein can interact with the

ERGIC-53 protein in the ER and it can be released at later time points in the ERGIC by utilizing a well-established pH- and calcium ion-facilitated mechanism (Appenzeller-Herzog et al. 2004). ERGIC-53 additionally transports secreted serum coagulation factors and disorders of human blood clotting are lined to specific mutations in the ERGIC-53 (Nichols et al. 1998). ERGL, VIPL and VIP36 represent related lectins with specific intracellular localizations within the secretory pathway, also interacting with different glycoproteins in a Ca^{2+} ion-, pH- and sugar-dependent fashion (Kamiya et al. 2008). Another role of ERGIC-53 in protein assembly has been supposed to be in the process of the oligomerization of IgM molecules. ERGIC-53 fulfills the tasks of an export receptor for fully assembled IgM "monomers," which contain two heavy and two light chains that are connected through disulfide bonds (Anelli et al. 2007). The oligomerization of these subunits to polymeric IgM is mediated by ERGIC-53, which may be facilitated by the oligomeric state of ERGIC-53 itself, building finally a hexamer. In guiding the trafficking of different sets of unrelated glycoproteins, ERGIC-53 can employ of accessory proteins help to improve the specificity for certain clients, such as ERp44 that aids in exporting IgM and MCFD2 that triggers the interaction with coagulation factors (Cortini and Sitia 2010). Through the use of accessory factors, the diversity of customers served by an individual cargo receptor can be increased.

Cargo receptors are able to facilitate the interplay between transmembrane cargo proteins and coat proteins. Under certain conditions, these transmembrane cargos do not have their own ER export motif and therefore require a receptor that binds to the coat. That is the specific case with lipid-anchored proteins facing the ER lumen. After their synthesis and integration into the ER, the glycosylphosphatidylinositol (GPI)-anchored proteins are embedded in the ER membrane by a distinct lipid moiety. Hence, they miss cytosolic domains for the interaction with the COPII coat. Instead, they need the support of a receptor family referred to as the p24 family. Subsequnetly, these abundant oligomeric proteins are able to shuttle between the ER and ERGIC/*cis*-Golgi through their efficient incorporation into both types of vesicles, such as COPI and COPII vesicles (Wada et al. 1991; Stamnes et al. 1995; Belden and Barlowe 1996; Blum et al. 1996; Emery et al. 2000; Gommel et al. 2001; Langhans et al. 2008). The p24 proteins have been conserved across all eukaryotes and all possess a stereotypical organization of their domains: a lumenal N-terminal GOLD (Golgi dynamics) domain, a coiled-coil region, a single transmembrane domain and a short cytoplasmic tail, which can interact and support the nucleation of both COPI and COPII elements (Fiedler et al. 1996; Goldberg 2000; Belden and Barlowe 2001; Anantharaman and Aravind 2002; Strating et al. 2009). While the molecular features of cargo detection are still not known, p24 proteins appear to act as lectin in the detection of the carbohydrate fraction of the GPI anchor (Manzano-Lopez et al. 2015). Since the cargo does not come into contact with the receptor complex until the GPI binding has been rearranged accordingly, the p24 family also appears to act as an indicator for protein maturation.

Certain plasma membrane proteins have a need for a cargo receptor, although they carry their own ER export motifs. An efficient ER export in this situation demands several signals: one signal is found in the cargo protein itself, a second signal originates from a cargo receptor. The best-characterized examples are the yeast Erv14, the cornichon in flies and the CNIH in mammals, which act as receptors for various plasma membrane resident proteins (Powers and Barlowe 1998, 2002; Nakanishi et al. 2007; Herzig et al. 2012; Pagant et al. 2015). Erv14 probably fulfills several functions in its role as cargo receptor. First, it offers affinity to the COPII mantle through a cytoplasmic loop. This function is necessary for all its cargo customers. In multiple instances, this Erv14 signal enhances the interplay between Sec24 and endogenous sorting signals of cargo proteins, thus improving the affinity of the envelope for Erv14-bound cargo (Powers and Barlowe 2002; Pagant et al. 2015). Second, Erv14 can chaperonise long transmembrane domains typical of plasma membrane proteins. In fact, the Erv14 reliance for ER export is associated with the length of the transmembrane domain (Herzig et al. 2012). Ultimately, based on the interaction of Erv14 with a specific subset of cargo through a binding site embedded in the lipid bilayer, it appears to act as a quality controller, since it only binds when the transmembrane domains of the cargo protein are correctly assembled, thereby linking the COPII recruitment with the protein folding (Pagant et al. 2015).

7.10.2 Cargo Receptors Act in the Retrograde COPI Trafficking

The receptor that facilitates the Golgi–ER retrieval of ER-resident proteins represents one of the first initially identified cargo-binding receptors acting in protein trafficking. The yeast Erd2 is able to interact with the tetrapeptide motifs, such as KDEL (Lys-Asp-Glu-Leu) or HDEL (His-Asp-Glu-Leu), which have been identified on ER residents (Semenza et al. 1990). In humans, there are three homologs of the receptor, such as ERD21, ERD22 and ERD23, each of which seem to possess slightly distinct preferences for specific substrates (Raykhel et al. 2007). All of which represent proteins that possess seven transmembrane domains to directly interfere with both cargo proteins and COPI components (Majoul et al. 2001). The association with cargo proteins and their release through KDEL receptors is managed through the differences in the luminal pH of the Golgi and the ER. In specific detail, the lower-pH milieu induces the binding to the cargo, whereas the increased pH milieu supports the release of the cargo proteins (Scheel and Pelham 1996). A different soluble cargo-receptor complex that targets ER-resident proteins deficient in known sorting signals is the Erv41-Erv46 complex. Erv41 and Erv46 are closely related proteins comprising two transmembrane domains, each with a large lumenal domain. The Erv41-Erv46 complex is known to interfere with cargo proteins such as Fpr2 and Gls1 in a pH-dependent fashion comparable to the KDEL receptors (Shibuya et al. 2015).

Similar to the anterograde transport, not all membrane proteins to be retrieved carry their individual sorting signal. The ER retrieval receptor Rer1 facilitates the trafficking between the Golgi and the ER of a vastly variable group of ER membrane proteins lacking both KKXX and K/HDEL signals (Sato et al. 1997). In yeast, these cargos encompass resident ER system proteins, such as Sec12, Sed4, Mns1, Sec71 and Sec63, as well as monomeric subunits of complexes, which are commonly needed for the forward trafficking (Sato et al. 2003, 2004; Kaether et al. 2007; Valkova et al. 2011). Among the latter proteins is Rer1 that can interact with a polar residue of the transmembrane domain, which is generally hidden within the assembled complex (Sato et al. 2003). Similarly, human Rer1 can function as an ER retrieval receptor (Füllekrug et al. 1997).

Consequently, the ER retrieval utilizes accessory proteins to support and/or expand the binding of cargo. The soluble cytoplasmic Vps74 protein functions as a vesicle loading protein of glycosyl transferases, since it can interact with a semi-conserved motif (F/L-L/V/I-X-X-RK) of their cytosolic tail and the coat of COPI vesicles (Tu et al. 2008, 2012). ERp44 fulfills at least two functions, firstly it works as an effector for quality control, and secondly, it serves as a retrieval adapter for thiol-assisted retention of misfolded or unmounted heavy and light IgM chains (Anelli et al. 2003). ERp44 is able to shuttle between the ER and the Golgi, whereby it performs its quality control cycle in a pH-driven fashion (Anelli et al. 2007; Cortini and Sitia 2010; Vavassori et al. 2013). In the ER with a pH of 7.1, ERp44 is located therein in a closed conformational shape and hence cannot interact with the cargos or the KDEL receptor. In the *cis*-Golgi with a pH of 6.7, the ERp44 can perform a conformational alteration that prefers the interaction with cargo proteins containing unpaired cysteines and exposing a non-burried KDEL motif (Cortini and Sitia 2010; Vavassori et al. 2013; Sannino et al. 2014). Linked to the KDEL receptor, the ERp44 protein is transported backwards to the ER, where its cargo protein is released and the chaperone are prepared for additional quality control cycles.

7.10.3 Regulatory Processes of Cargo Sorting and Impact on Trafficking by the Cargo

One of the still unanswered questions in the protein sorting field is how the biogenesis of vesicles is regulated to adapt to the physiological requirement of each individual cell. Are vesicles made up locally solely upon the delivery of the cargo? Can the vesicles be designed in such a way that they are only specific for a certain cargo load? A central key to answering these questions is the comprehension of mechanistic effects of coat composition in relation to different cargo proteins. The clathrin system definitely involves large-scale movements by adapter complexes during cargo binding to initiate framework recruitment. It is yet to be seen whether a comparable localized activation arises in the COPI and COPII systems. Similarities in the structure of the COPI and clathrin-AP complex indicate that allosteric alterations probably govern the assembly of COPI (Dodonova et al. 2015); however, there are additional discoveries about how the various coat subcomplexes interact with one another before answering this question. Conversely, the COPII system seems to suffer from a lack of allosteric alterations in cargo binding, but the charge-related modification of coat construction can be accomplished instead by regulating coat stabilization. In this case, both envelope and accessory proteins can modify the GTP cycle of the envelope and thereby control the residence time of the envelope on the membrane. The peripheral ER protein Sec16 is in competition with Sec31 for its binding to Sec23/Sec24, which delays the enrollment of Sec31 and avoids the complete activation of the GTPase activity of Sar1 (Kung et al. 2012; Yorimitsu and Sato 2012). Mutations in Sec24 that cancel this process lead to the assumption that this process could be affected by the cargo loading on Sec24 (Kung et al. 2012), even though this has yet to be proven. Although much remains to be known about how charge can affect vesicle generation, there is ample evidence that vesicle production may not be fixed and static, but rather sufficiently flexible to meet different cellular demands.

7.10.4 Large Carriers Represent a Specialized ER Export Mechanism

Both yeast and mammalian cells are able to secrete unusually large proteins, which hinder the efficient assembly of small, spherical vesicles for transport issues. Yeast Pma1 is a plentiful oligomeric protein from ~1 MD, which has a specific need for the Sec24 paralogue Lst1/Sfb1, generating slightly bigger vesicles than those produced with Sec24 on its own (Shimoni et al. 2000; Miller et al. 2002). In the same way, GPI-APs depend on Lst1 for their highly efficient ER export; the binding of GPI-APs triggers the p24 complex for the targeted recruitment of Lst1 (Castillon et al. 2009, 2011; Manzano-Lopez et al. 2015). GPI-APs are likely to necessitate custom adjustments due to the nature of their asymmetric spread across the membrane, which probably tends to antagonize vesicle production (Copic et al. 2012). In both examples, the cargo itself, by recruiting a mantle component (Lst1) with the ability to produce larger vesicles, is responsible for entering a vesicle that can hold it.

In mammalian cells, the trafficking of fibrillar procollagens causes a similar physical problem: the cargo protein is too large to be transported by a canonical COPII vesicle. Cargo transport therefore requires transmembrane accompanying factors such as MIA3/TANGO1 and cTAGE5, which themselves are not trapped in vesicles (Malhotra and Erlmann 2011, 2015). Similar to a traditional cargo receptor, TANGO1 can bind to collagen through its SH3 domain, which supports the packing of collagens into vesicles (Saito et al. 2009, 2011). It has recently been implied, however, that SH3 alone is not capable of conducting collagens in COPII vesicles. Rather, Hsp47 appears to serve as an anchor molecule between TANGO1 and several different types of collagens (Ishikawa et al. 2016). cTAGE5 represents a TANGO1-related protein that is able to interact with Sec12. Sec12 is the nucleotide exchange factor that can induce the assembly of the COPII coat through the attaching of

GTP to Sar1. Through the local linkage of Sar1 activation to a certain cargo, TANGO1/cTAGE5 is expected to moderate the acquisition of adequate amounts of the COPII envelope to produce a large "megacarrier" that can carry procollagen. Besides the simple stimulation of COPII enrollment, TANGO1/cTAGE5 can also directly affect the coat geometry. Both proteins possess several interspersed polyproline motifs scattered across their cytoplasmic domains, which are attached to Sec23 (Ma and Goldberg 2016). By concurrently recruiting several Sec23/Sec24 dimers, TANGO1/cTAGE5 could place these mantle elements in a grid-like configuration similar to that encountered in COPII-coated tubular structures (Zanetti et al. 2013). Another special feature of the polyproline bond of Sec23 is that Sec31 also incorporates these interspersed motifs, which are able to battle for the bond with those of TANGO1/cTAGE5 (Ma and Goldberg 2016). The result is a model in which Sec31, when enrolling at the site of beginning vesicle formation, dislodges the "receptor" that provided the cargo acquired during local mantle formation. In this instance, the interactions between the cargo and the coating can determine both the geometry of the coating and the timing of the coating removal, which directly affects the final architecture of the substrate. Other regulatory layers also appear to be involved in the modulation of COPII envelope activity during procollagen export. For instance, the ubiquitination of Sec31 is necessary for collagen transport (Jin et al. 2012) and the alteration of the Sar1 GTPase activity through a TANGO1-associated protein sedlin seems to be additionally crucial (Venditti et al. 2012). How these factors could directly or indirectly impact the emergence of megacarriers is still to be clarified.

7.10.5 Cargo-Driven Signaling into the *cis*-Golgi

A contribution of cargo to the direct encouragement of COPI vesicle production has been proposed by several unrelated pathways. The p24 protein family includes COPI binding cues that can modify the catalytic activity of the envelope, stabilize it possibly during assembly and thereby foster vesicle generation (Goldberg 2000). Analogous responses have been suggested for a family of proteins that circulate cyclically in yeast between ER and Golgi. In specific detail, the overexpression of Mst27/Mst28 represses the lethality linked to COPI mutants, which indicates that vesicle development can be encouraged by increasing COPI binding events (Sandmann et al. 2003). More precisely, the binding of a KDEL-containing ligand to the KDEL receptor triggers the oligomerization, as it has been seen through the fluorescence resonance energy transfer, and thereby triggers the COPI interaction with Golgi membranes (Majoul et al. 2001). How these multiple charge-driven effects are mechanistically handled, is not yet completely clarified. A second facet of cargo-driven traffic manipulation is proposed from recent research on Src-based cytoplasmic signaling events subsequent to the activation of the KDEL receptor (Pulvirenti et al. 2008). It is assumed that the binding of clients within the Golgi activates the kinases of the Src family, which themselves affect membrane transport. How this signal path could directly stimulate the COPI function needs to be studied more thoroughly, while more direct retardation testing would tend to predict a more general impact on Golgi traffic than a specific retrograde inducement (Bard et al. 2003).

7.10.6 How Is the Process of Sorting Dynamically Regulated?

With many new advances in the structural properties of coating assembly and the interaction between cargo and coating, a deeper understanding of the mechanisms of protein sorting is being achieved. However, understanding the cellular control of these processes is not yet complete, and an appreciation of cargo–coat allostery (or other types of regulation) will be the key to how trafficking strategies can be adapted to suit specific cellular conditions. Further progress results from the investigation of non-canonical traffic, such as procollagen secretion, an area where the plasticity of established systems and the operational implications of new intrinsic features are only now starting to be estimated. However, new examples of the different sorting procedures are needed to develop the field continuously in a forward direction.

Actually, an example of "non-canonical" transport involves the reactivation of an old model: the non-selective bulk flow. Several cargoes leave the ER with no concentration in vesicles or interference with the vesicle envelope, albeit probably remaining in a COPII-dependent mode (Wieland et al. 1987). Since the established cargo receptors transport just a fraction of the secretome, the stochastic transport via the bulk flow has to be reassessed as an important mechanism of cargo movement. In fact, a newly performed quantification of bulk flow in mammalian cells using a novel stable inert and quickly folding tracer indicates that there is a substantial flow of fluid and protein through the secretory route in a non-selective mode (Thor et al. 2009). Whether this also applies to any other cell types and organisms has yet to be clarified. Equally, it has not been verified whether the bulk flow of fluid and membrane takes place in retrograde vesicles, and it may be technically challenging to distinguish them from anterograde incidents. In view of bulk flow export, it is unclear how incompletely folded proteins and ER-residents could be impeded from being trapped in vesicles and needs to be further characterized.

Similar to the problem of selectivity, it is difficult to understand how the folding of proteins impacts the ER–Golgi trafficking, in terms of both the signal-driven sorting and the bulk flow. This is an important issue, as emerging cargo proteins that are exported prematurely lose access to ER folding and dismantling machines. In essence, cargo adapters and receptors in both the ER and the Golgi are capable of directly assessing transport reliability. ER receptors can distinguish between correctly folded proteins and those proteins, which require the action of chaperones. Golgi receptors can bind misfolded cargo proteins in order to retrieve them back toward the ER. While in practice this model may be adequate for some misfolded proteins, it is improbable that it can be extended to all proteins. First, some ER export signals represent simple peptides that are probably not altered when a protein folds on the luminal part of the membrane. Probably even more relevant and enigmatic is the finding that a large number of imperfectly folded and resident proteins are in fact capable of being transported (Hong et al. 1996; Jenness et al. 1997; Spear and Ng 2003; Liu et al. 2006; Kincaid and Cooper 2007; Ashok and Hegde 2009; Wang and Ng 2010).

Thereby, it seems to be more accurate to propose that the functional ER export signals can even coexist with ERAD and retention signals and moreover, the forward trafficking is dynamically governed by numerous competing interplays. Exposed to stress, the ER can utilize sorting signals as an ER detoxification mechanism (Kawaguchi et al. 2010; Satpute-Krishnan et al. 2014). However, similar questions still remain to be answered as to how misfolded proteins can be identified within the Golgi, and to what extent retrieval maintains the steady-state ER retention requires to be quantitively measured. Various misfolding evoked diseases seek to reveal the interaction between these pathways that stem from defects in quality control mechanisms during the sorting of proteins at the ER–Golgi interaction side.

7.11 Ribosomes

The ribosomes represent the universal translating machines, which convert each mRNAs into the encoded proteins. The genetic material of eukaryotic cells is enclosed by a nuclear envelope that serves as a barrier for the free travelling of proteins and nucleic acids between the nucleus and its surrounding cytoplasm. The resulting physical separation of transcription and translation enables the cells to govern the level of control of gene expression in eukaryotes, which is not possible in prokaryotes. Moreover, this also needs to the presence of an efficient transport system to selectively shuttle macromolecules between these two compartments. Such a specific system must be able to distinguish between a large number of structurally and functionally different molecules and can also react to different growth and environmental circumstances.

The biosynthesis and transport of ribosomes in a rapidly growing cell, such as the baker's yeast *S. cerevisiae*, represents a fundamental paradigm for the extent of the difficulties of the nucleocytoplasmic transport system. Every individual mRNA coding for a ribosomal protein first has to be transported into the cytoplasm where it is translated. The resultant ribosomal proteins then enter the cell nucleus and subsequently the nucleolus, where they combine with newly synthesized ribosomal RNA (rRNA). The preribosomes created in this way then pass through a series of sophisticated changes before they are re-exported into the cytoplasm to fulfill their function in translation. In a rapidly growing culture of yeast that double their ribosomal content every 1.5 h, each cell must import about 150,000 ribosomal proteins per minute via the nuclear envelope and it needs to simultaneously export about 4000 ribosomal subunits per minute (Warner 1999).

There were tremendous improvements in the methodology that enabled us to follow a sketchy, sequential route throughout the constituents and mechanisms of ribosome biogenesis (Kressler et al. 1999). This route starts with the manufacture of ribosomal proteins, proceeds through their entry into the nucleus and their combination with the nascent rRNA in the nucleolus, and leads back toward the nuclear envelope during the not yet defined steps of prereibosomal subunit maturation. But so far, the road has had a blatant pothole: the proteins that specifically convey the export of the ribosomal subunit have been difficult to grasp. An idea of this process was gathered by examining the process of ribosome biogenesis in yeast (Ho et al. 2000). Such a finding has highlighted actors in this export process and potentially connected the last stages of maturation of the ribosomal subunit with its nuclear export, underlining the continuum of the ribosome's life cycle starting from its birth and maturation in the nucleus to the functional involvement in translation in the cytoplasm.

7.11.1 Ribosomes Are Transported via the Nuclear Transport System

Transportation of proteins and RNA between the nucleus and the cytoplasm is facilitated by soluble transport factors that are able to interact with cargos and drive them through multiple pores within the nuclear envelope. In the pores, there are enormous macromolecular arrangements, referred to as nuclear pore complexes, which function as gate guards of the nucleus (Wente 2000). The nuclear pore complexes permit the free diffusion of small molecules such as water and ions, although they shut out all macromolecules beyond the diffusion limit ($\sim$9 nm), with the exception of those bearing specific nucleocytoplasmic target sequences. Thus, proteins bearing a nuclear localization sequence (NLS) can be intrinsically transported across the nuclear pore complex, whereas macromolecules intended to be transferred out of the nucleus host nuclear export sequences (NES). These signals can be detected through karyopherins (sometimes abbreviated as kaps), which are synonymously referred to as importins, transportins and exportins (Mattaj and Englmeier 1998). Karyopherins are able to connect to the import or export signals of their cargoes, facilitate the connection of these cargoes toward the nuclear pore complex and consequently chaperone the cargoes via the pores into the nucleus. The GTPase Ran is responsible for the transfer of cargo between the nucleus and the cytoplasm. Ran is kept through the aid of cofactors in an GTP-bound state inside the nucleus and maintained in an GDP-bound state throughout the cytoplasm. This distribution establishes an energy gradient over the nuclear pore complex that drives the transport between the nucleus and the cytoplasm and can also be utilized by transport players to identify on what side of the nuclear envelope they are localized. Thus, the establishment of an import compound between a karyopherin and its cargo is stabilized in the cytoplasm by cytoplasmic Ran–GDP, whereas in the nucleoplasm, Ran–GTP stimulates the breakdown of this compound. On the contrary, the creation of an export compound in the nucleus becomes more stable by Ran-GTP, but once this compound enters the cytoplasm, the GTP on Ran is hydrolyzed and the compound decays. At this point, it is obvious that there are various transport routes that make effective usage of different related karyopherins. There are at least 14 structurally closely related karyopherins in yeast that are involved in the import and export of certain classes of molecules, and it is expected that this number will be considerably higher in metazoans. In this way, numerous diverse, but sometimes mutually overlapping and redundant transport pathways are merged in the nuclear pore complex. One of the most widely trafficked routes is the exporting of ribosomal subunits.

7.11.2 What Function Fulfills the Ribosome?

In general, ribosomes are regarded as one of the most fundamental complexes that can be found in nearly all organisms. As offspring of an ancient RNA universe, they consist mainly of a catalytically active rRNA scaffold on which the about 80 proteins are built. It appears that the principal function of most proteins within the ribosome (at least for the large subunit of the bacterium Halarula marisomortui) lies in the three-dimensional structural fixation (Ban et al. 2000). But it is also probable that numerous proteins linked to the ribosome participate in guiding it through the numerous steps in its biogenesis, both in time and space (Aitchison and Rout 2000). Ribosomal proteins, just as other proteins, are generally produced in the cytoplasm. Despite the fact that most of them are small and therefore lie below the nominal cutoff point of the nuclear pore complex, they are still actively introduced by capping-driven processes. In yeast, Kap123p is the primary source of this, but it is obvious that the process is superfluous and other karyopherins can absorb a large part of the sludge in the absence of Kap123p (Rout et al. 1997). When introduced into the cell nucleus, ribosomal proteins combine with newly transcribed rRNA inside the cell nucleus to produce a pre-ribosomal particle.

Due to the difficulty of distinguishing the mechanism of ribosome export from its biogenesis, understanding the complex nature of this process is crucial. In yeast, more than 60 *trans*-acting elements are required for proper biogenesis of ribosomes (Kressler et al. 1999). Most of these proteins and small RNA components are stored inside the nucleolus, although they are equally abundant in the nucleoplasm, the cytoplasm and to some extent even in the nuclear pore complex. However, the precise roles of many of these factors are elusive and some factors may even fulfill other functions within the nucleus, such as the splicing of intron-carrying pre-mRNAs. In line with this, it is uncertain as for when and where all ribosomal proteins attach themselves to the rRNA. Nevertheless, the assembly of the preribosomes is closely synchronized with rRNA maturation and rRNA transport and can be regarded as a succession of consecutive steps, from the sites of nucleolar rRNA transcription through the nucleoplasm to the nuclear pore complex and beyond (Kressler et al. 1999).

Across all eukaryotic cells, three of the four rRNAs are transcribed as a single RNA polymerase I transcript that undergoes comprehensive splicing, trimming and other covalent alterations to generate the 18, 5.8, 25/28S rRNA species. The fourth, the 5S RNA, is also pronouncedly altered, albeit transcribed by RNA polymerase III and recruited individually into the mounting ribosome. First, the primary RNA polymerase I transcript combines with a variety of large and small subunit proteins to yield a pre-90S particle. The transcript is thereafter split to obtain precursors of the 60S and 40S subunits. These appear to be transferred separately into the cytoplasm. The pre-40S particle passes through a further cytoplasmic ripening of its 20S rRNA to obtain the fully mature 18S rRNA, while the rRNA within the 60S particle is expected to ripen shortly before export. Once inside the cytoplasm, the 40S subunit attaches to the 5′ end of the mRNA and scanning to the initiation codon, where this translation-initiation complex will be awaiting the 60S subunit to adhere and initiate translation. The 60S subunit has to complete a series of cytoplasmic maturation cycles prior to drive a translation. Such fine-tuning probably requires the addition of several proteins, including Rpl10p. The inclusion of the 60S subunit in the translation engine is also synchronized with the release of factors, including eIF2 from the initiation engine. In the lack of effective 80S ribosome assembly, eliciting, for example, an inadequate amount of functional 60S subunits, the 40S subunit jams, resulting in clustering of half-meters. These kinetic intermediates of polysomes with a blocked 40S opening compound are typical for 60S biogenesis disorders.

The ribosome represents the factory of cells for the synthesis of proteins. With rates of protein synthesis reaching up to 20 amino acids per second and an accuracy of 99.99%, the exceptional catalytic power of the bacterial translation engine stimulated considerable development, reconstruction and reuse efforts for biochemical trials and new functionalities. Notwithstanding these endeavors, the full exploitation potential of the translation device for the production of bio-based products beyond natural boundaries continues to be untapped, and fundamental limitations of the chemistry that the active center of the RNA-based ribosome can perform are largely unidentified.

Ribosomes constitute macromolecular units, which are the key locations of protein synthesis or translation within every cell. The principal chemical step in protein synthesis at ribosomes is peptidyl transfer, whereby the expanding or developing peptide is passed from one tRNA molecule to the attached amino acid on another tRNA. Amino acids are inserted into the expanding polypeptide on the ribosome following the codon sequence of an mRNA. The ribosome therefore contains binding locations for one mRNA and a minimum of two tRNAs. Ribosomes consist of two different subunits, the large and the small subunit, comprising a handful of rRNA molecules and a diverse number of ribosomal proteins. Several protein triggers trigger various steps of protein synthesis. The accuracy of the genetic code translation is crucial for the manufacture of functional proteins and the viability of the entire cell.

7.12 Ribosome Mechanics Sheds Light on the Mechanism

The fundamental characteristics of the ribosome mechanism are derived from coarse-grained simulations, involving the ratchet movement, the joint movement of critical bases in the decoding hub, and the displacement of the peptide tunnel lining that helps eject the synthesized peptide. Due to its huge size, the coarse grain contributes to simplifying and better comprehension of the mechanism. The findings reported here exploit coarse-grained elastic network modeling to identify the dynamics, and both RNAs and proteins are coarse-grained. To verify the results obtained so far, the well-known ratchet movements and the movements in the peptide tunnel and in the mRNA tunnel are examined. The movements of the casing of the peptide tunnel apparently support the ejection of the increasing peptide chain, and brackets at the ends of the mRNA tunnel consisting of three proteins assure that the mRNA is retained firmly throughout the decoding phase and plays an indispensable part for helicase activity at the entry point (Zimmermann et al. 2016). The input block can also aid in base identification to guarantee the appropriate choice of entering tRNA (Zimmermann et al. 2016). The total accuracy of the ribosome as a machine is noteworthy.

The translation of the genetic code of DNA into protein compounds is the main activity performed by the ribosome. The ribosome is an amazingly complicated ribonucleoprotein construct that engages with a variety of distinct substrates and cofactors along the translation mechanism. Knowledge of the dynamics of the ribosome is crucial to a better grasp of its mechanism. Great efforts have been made to elicit the structural elements of the ribosome in its various shapes, the ribosome structures of different organisms and the binding sites of pharmaceutical drugs. All these investigations result in the conclusion that the ribosome is a very rugged molecular device with numerous mobile components. In fact, the number of different functional ribosomal conformations (states) needed throughout translation is completely undefined and, notwithstanding the significance of the ribosome for cell viability, it is possible that experiments may leave it imperfectly defined. Calculations can play a role by detecting supplementary modes.

In fact, there are several approaches made to reveal more insights into the detailed mechanism of the ribosome's work. Cryo-electron microscopy (cryo-EM) has helped to discover the distinct steps of the protein synthesis pathway (Agrawal et al. 1999; Frank et al. 1999; Frank and Agrawal 2000; Valle et al. 2002; VanLoock et al. 2000; Wriggers et al. 2000). Based on an analysis of 3-D cryo-EM instantaneous images of the functional state of the intact ribosome, in its different functional states, it was reported (Frank and Agrawal 2000) that ratchet-like rotations of the 30S subunit relative to the 50S subunit during translocation are very similar to the movements observed in the dynamics simulations (Fig. 7.8) (Wang et al. 2004). Although many specific details have been revealed, such as the sequential steps during the translocation, there are still many specificities that have not yet been fully uncovered (Agrawal et al. 2000; Noller and Baucom 2002). Other biophysical techniques, such as NMR and FRET, have delivered additional specific details (Blanchard et al. 2004a, b; Johnson 2004; Lynch et al. 2003; Woolhead et al. 2004; Cooperman and Goldman 2006; Vanzi et al. 2003, 2005; Wang et al. 2007).

Experimental structure identification may yield the patterns of several states of ribosomal protein and RNA compounds equivalent to several distinct states with various bound ligands, fixed in several manners, in multiple surroundings, and so on, but overall, identifying all potential structural stages to capture all the intricacies of the ribosome mechanism is a tremendous challenge. Computation can be crucial in complementing the existing structural and functional insights with simulations that give further specifics on the dynamics of the ribosome. A comprehensive knowledge of the mechanism will only be gained by collecting dynamic properties from a wide range of various computer simulations. The movements of the ribosome are effectively modeled by atomic molecular dynamics simulations over extraordinarily long periods of time. However, these simulations sometimes do not give an elementary description of the mechanism. Various facets of the ribosomal dynamics and its components have been examined with a variety of computational approaches including fully atomic molecular dynamics (Sanbonmatsu 2006a, b, 2012, 2014; Sanbonmatsu and Joseph 2003;

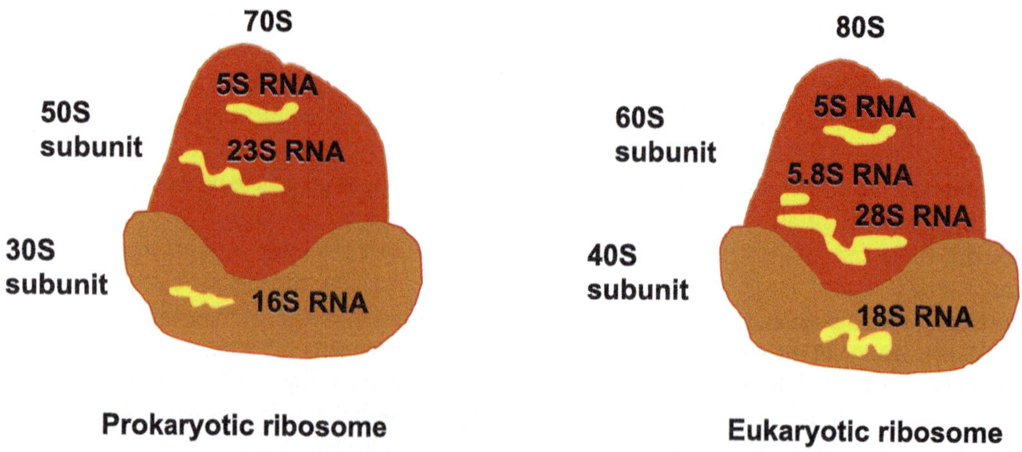

Fig. 7.8 The ribosome 30S and 50S subunits of prokaryotes and 60S and 40S of eukaryotes

Sanbonmatsu et al. 2005; Sanbonmatsu and Tung 2006; Tung and Sanbonmatsu 2004), coarse-grained molecular dynamics (Trylska et al. 2005), normal mode analysis employing a coarse-grained potential (Wang et al. 2004, 2007; Kurkcuoglu et al. 2008, 2009a, b; Wang and Jernigan 2005), elastic framework models (Tama 2003; Tama et al. 2003) and subsequently the combination of experimental data and computational data (Mitra and Frank 2006).

The focus here is on the mechanical properties of the ribosome in a coarse-grained formulation. The application of basic elastic network models is a suitable and an instructive tool for this task, which can deliver straightforward findings to enable the understanding of a number of mechanistic features of translation and the elongation of proteins. More in-depth examinations of a number of models with several included or excluded constituents are essential for the comprehension of ribosomal function. These simulations can supplement and improve the knowledge that can be acquired experimentally.

The research into the dynamics of biological processes is the ideal basis for understanding the functional principles of the ribosome. For the description of harmonic and anharmonic movements of proteins and their assemblies, computer simulation techniques are often employed that make use of fully atomic empirical capabilities, such as molecular dynamics simulations (Levitt et al. 1985; McCammon and Harvey 1987; Tirion 1996; Kitao and Go 1999). The low-frequency movements in which large segments of proteins are involved, typically movements of entire domains, are frequently linked to their biological functionalities (Berendsen and Hayward 2000), such as the crossovers in proteins between the open and closed conformations (Kim et al. 2002a, b, 2003, 2005; Schuyler and Chirikjian 2005; Schuyler et al. 2009; Yang et al. 2005, 2007, 2009; Tama and Sanejouand 2001) and the ratchet-like movement of the ribosome in the course of protein synthesis (Frank et al. 2000).

Nevertheless, with the aid of coarse-grained models, these domain motions are far better understood, particularly in systems considerably smaller than the ribosome. For exceptionally large-scale systems such as the ribosome, complete atomic simulations not only require enormous computational resources, but are also extremely complicated to evaluate and interpret when it comes to pinpoint the major steps of the mechanisms. Besides the atomic MD, difficulties also exist when it comes to ensuring sufficiently long trajectories for a profound investigation of their overall movements, once again due to the high requirements on computing capabilities. The coarse-grained scale models allow easier, more immediate evaluations with a higher degree of confidence that all potentially relevant movements are fully taken into account.

At this point, a review of what has been achieved so far from previous ribosome simulations, but also some new findings that demonstrate how the ribosome pins the incoming mRNA, which is needed for its helicase activity to unroll the mRNA (Takyar et al. 2005) and to establish a binding site that can perceive the bases that are primed for delivery in front of the anticodon of the tRNA. Consequently, there are exists multiple diverse structures of the ribosome and its compounds within the Protein Data Bank (PDB) (Berman et al. 2000). The results described belong are based on the T. thermophilus structure, which has been the first completely identified structure (referred to pdb id::4V42). As with other bacterial ribosomes, it is made up of two large subunits, the 30S and the 50S, which rotate relative to each other in their dominant ratcheting movement at successive stages of synthesis.

7.12.1 Physical Models Can Be Employed for Elastic Networks

Elastic network models, such as the Gaussian network model (GNM) (Bahar et al. 1997), can be employed for scalar descriptions of motions and the anisotropic network model (ANM) (Atilgan et al. 2001) can be utilized to describe the directions of motion. It is assumed that the initial structure has the least energy and the energy is in Gaussian format, which increases with the square of the displacements from the initial structure, as described in (7.1):

$$U(t) = \left(\frac{\gamma}{2}\right) tr\left[\Delta R(t)^T \Gamma \Delta R(t)\right] \langle \Delta R_i \cdot \Delta R_j \rangle = \left(\frac{1}{Z_N}\right) \int (\Delta R_i \cdot \Delta R_j) \exp\left(\frac{-U}{kT}\right) d\Delta R$$
$$= \left(\frac{3kT}{\gamma}\right) \left[\Gamma^{-1}\right]_{ij} \tag{7.1}$$

where U stands for the energy, γ represents the spring constant and ΔR_i indicates the displacement of the point i within the structure.

The input to the movements can be broken down into normal modes, defined by the eigenvalues and the eigenvectors of this matrix Γ^{-1}. This is an excellent way to measure the harmonic movements of a structure around its original structure. It considers the geometric restrictions of the construction and provides a mechanical template. The stunning thing to learn from many implementations of these techniques is simply that it is the type of structure that governs the movements (Lu and Ma 2005; Doruker and Jernigan 2003; Doruker et al. 2002a, b).

Generally, elastic network models are employed on coarse-grained constructions, and the findings demonstrate the mechanical movements of a system by using a uniform potential (all springs are assumed as being identical) for the interfering node couples in the system (Bahar and Rader 2005; Ma 2005; Jernigan et al. 2008). Coarse grain is usually carried out as one point per amino acid for proteins, but models with lower resolution, i.e., with several residues per node, have also yielded acceptable performance for the identification of low-frequency collective modes (Doruker et al. 2002a, b, 2005; Kurkcuoglu et al. 2004), while the general structure of the molecule is maintained. Other alternative tasks for coarse-grained normal mode investigations are the building block approach (Tama et al. 2000) and the minimalist network model (Lu and Ma 2008).

It is easy to calculate the correlations of the variations between any two points in the structure ΔR_i and ΔR_j by inverting the Kirchhoff matrix, which is then defined by inserting a one in the matrix for points in the structure which are very close to one another, using a limiting distance to this end, and 0 elsewhere in the matrix, where the diagonal is assumed to be the negative sum of the other items in a particular line. The Kirchhoff matrix of the contacts Γ is concerned, and normally, a spring is positioned between close points in the framework which are defined by a limiting gap (Wang et al. 2004; Atilgan et al. 2001). For this example, by using Cα atoms for the proteins and two atoms for each nucleotide, the P and O_4* atoms, coarsely engraved, the cutoff spacing is 15 Å. In specific detail, the dimensions of the square matrices which are to be diagonalized are 16,266 for the 30S subunit and 29,238 for the 70S subunit. Alternative approaches to coarse graining have been tested, and these show very small differences. One of the great advantages of these methods is their insensitivity to detail.

A simulation of the complete ribosome provides the calculated correlations between the various parts of the ribosome microstructure and helps to decipher its underlying principles. The eigenvectors define the direction of movement of the individual items in the structure. The available results comprise the collective findings for the vectors of the first 100 movement modes of the ribosome. The correlation function can be defined as a cosine function, and therefore, a value of one indicates a perfect correlation in the direction of motion of the two points, and zero indicates either no correlation or a tangential movement. The prevalence of the ratchet movement is evident in the strong anti-correlation between the 30S and 50S subunits. All minor separate constituents—the tRNAs and the mRNAs—are more strongly positively associated. The tRNAs have almost zero correlation to the 30S and 50S subunits, whereas the mRNA appears to be strongly linked to the 30S and uncorrelated with the 50S subunits.

7.12.2 Ratchet Motion Within a Complex of Three tRNAs and One mRNA

In general, the most dominant movement seen is the ratchet movement between the large and small subunits of the ribosome. In the simulations, the first 10 normal modes, which depict the most collective movements of the ribosome, display the ratchet-like rotation of the two subunits, the head rotation of the small subunit and various types of anticorrelated movements between the large rods. Moreover, the ratchet-like rotation associated with certain countermovements of the L1 and L7/L12 stalks was found in various diverse normal modes. The large 50S substructures are architecturally intertwined with its proteins, and it possesses twice the amount of protein contacts with the rRNA than the proteins in the small subunit. The only exceptions are the L1 and L7/L12 rods, which are elongated without rRNA connections and which were experimentally challenging to image. The small subunit possesses pronounced internal domain flexibility (Liljas 2004), whereas the large subunit seems to be stiffer and more rigid. It has been shown that the global ratchet movement can be reproduced reliable through the employment of elastic network models (Wang et al. 2004) and has been revealed whether the tRNAs and mRNA are detectable (Kurkcuoglu et al. 2008). There is an inherent feature of the ribosomal structure that it facilitates its ratchet movement, and this capacity does not rely on the availability of the tRNAs or mRNA.

7.12.3 Complex Motions at an Early Decoding Stage

The modeling of the ribosome architecture within the complex with three tRNAs and mRNAs demonstrated the computational power of the elastic network models. Beyond the ratchet movement, several other characteristic insights are detected that are discussed in the following. The examination of the movements in the decoding region provides numerous interesting features of the reading apparatus of the ribosome codon (Kurkcuoglu et al. 2008). For instance, there is a larger mobility of the anticodon stem loop (ASL) of the A-tRNA compared to the movements of P-tRNA and mRNA. The A site of mRNA is actually far more mobile than the P site, which is in good agreement with the experimental B factors, and also with the overall capacity of the A site to house a variety of proteins and RNAs, as well as the demand to maintain the P site more rigid in order to guarantee fidelity through strong recognition between codon and anticodon. In the simulation model, a high-resolution atomic region (Kurkcuoglu et al. 2009a, b; Kurkcuoglu et al. 2004) for the codon and anticodon regions of the mRNA and tRNAs at the A and P sites and the decoding site A1492, A1493 of the 16S rRNA in the small subunit have been incorporated. The ratchetic rotation and countermovements of the L1 and L7/L12 rods were monitored concurrently with the movement of the atoms in the decoding center and the movement of the codon/anticodon region. The lacking proteins on the L7/L12 handle in the crystalline structure were incorporated to provide density for this area of the elastic mesh. The elongated protein L9 on the large subunit is erased in the structure to eliminate its massive motility, that otherwise leads to a dominance of the calculated movements with the lowest frequency. Via 3/4 the movements from the entirely coarse-grained model are maintained in this mixed coarse-grained model; thus, it is an approximate but still offers the most essential subtleties. For the accumulative initial 10 modes, the mean squared fluctuations were computed for several levels of mixed coarse grain sizes. The third longest-running normal mode of this mixed coarse-grained ribosome model most closely resembles the ratchet-like movement of the subunits which is seen in cryo-EM. The side chains of A1492, A1493 and the A-tRNA show greater mobility in the slower modes than in the P-tRNA and mRNA. The analyses indicated that G34 of the A-tRNA and U43 of the mRNA exhibit smaller relationships with the decoding center (Kurkcuoglu et al. 2008).

The benefit of incorporating these pieces as atoms is to monitor meaningful features of the movements closest related to its functional operations. For instance, the bases A1492 and A1493 are much more agile than their backbone, something not evident in a coarse-grained model.

7.12.4 The mRNA Helicase Function of the Ribosome

The residues Arg131, Arg132, and Lys135 on S3, and additionally the residues Arg47 and Arg50 on S4 possess all helicase activity (Takyar et al. 2005), they are closely located nearby the mRNA and they coordinately perform their motion with the associated mRNA in the culminating ten slowest movements. S3 is able to perform an individual movement nearby the mRNA during the overall movement, in such a fashion that each of its side chains Glu161, Gln162 and Arg164 and the Arg49 of S4 is sufficiently close to the mRNA to be capable of interfering with it. These residues and any new contacts established in the alternative configurations could play an essential part in the activity of the ribosome helicase (Kurkcuoglu et al. 2008).

7.12.4.1 Protein S5 Orientates the mRNA for the Translation

Variations in the mean squared distance between S5 residues and nucleotide A27 indicate that the entire ribosomal S5 protein stays nearer to the mRNA than the S3 and S4 proteins. These findings point to the fact that the close positional alignment of S5 with the mRNA seems to be functionally essential to guide the strand for the accurate reading of the frames at the A site, a fact confirmed by earlier work (Kurkcuoglu et al. 2008). However, the opening and closing of the other two proteins surrounding the mRNA may be examined since these interactions are less persistent in dynamics simulations.

7.12.4.2 Proteins S3, S4, and S5 Serve as Gate for the mRNA

A new result has revealed how the three proteins S3, S4 and S5 seem to govern the entrance of the mRNA. It has been observed that protein S3 possesses the largest mobility. In a normal mode in which it operates, the movements interact alternately with these three proteins to clip the mRNA where it joins the ribosome, with a related group of three proteins functioning as the clip when the mRNA leaves the ribosome. This movement is aligned with the head swivel movement of the 30S subunit. In the case of any of the tracked entries, 5′ end will be opened and the end of 3′ will be shut down alternately, or vice versa. The mRNA at the entry of the tunnel for the mRNA is then confined by the proteins S3, S4 and S5, that restrict the mRNA in a bracket-like manner. The S3 movements are mainly accountable for this restriction. When the entrance to the mRNA tunnel is shut, the geometry of the opening alters. It gets thinner on one side, indicating the possibility that it possibly even scans the base of the ingoing mRNA. This capture of the mRNA bases may be structurally transferred to align the selection of the tRNA and assure that the anticodon of the tRNA is matched to this codon of the mRNA.

Allostery in the structure facilitates the transfer of mRNA sensing to the incoming tRNA to guarantee the correspondence between codon and anticodon. That is a novel feature and necessitates additional examination of allostery within the ribosome. It is an excellent example of how certain ribosomal movements can result in structures that perceive certain regions and transmit knowledge across long ranges, in which case to assure proofreading of the ingoing tRNA anticodon. At the outlet of the mRNA tube the ribosomal proteins S7, S11 and S18 are responsible for the restriction of the mRNA. This lock may refer to the fact that the mRNA is decoded sideways by the rigid retention of the mRNA itself. The mRNA interfaces with the 3′ end of the 16S RNA and creates the Shine–Dalgarno complex for the initiation step. The 3′ end of the 16S RNA can act as a "hook" to roll up the mRNA and ease its release. (Kurkcuoglu et al. 2008). It is probable that this change between 2 modes —open input—closed output and closed input—open output—all in all has a major impact on the overall performance of the decoding accuracy. The structure of ribosomes is seemingly under strong scrutiny and guides the dynamics of mRNA entering, translocating and exiting.

7.12.4.3 The Dynamics of tRNAs Inside the Ribosome

The ribosome experiences large movements to induce the translocation of tRNAs and mRNAs. The movements experimentally experienced in translocation are intrinsically governed by the ribosomal structure, causing the ratchet movement to propel these interior elements on an almost linear track. Normal mode results also demonstrate that the mobility of A- and P-tRNAs enhances without the presence of E-tRNA. The dynamics of the E-tRNA are also significantly influenced by the lack of the ribosomal protein L1. According to the simulations, the L1 arm deforms and seems to adhere to the E-tRNA, which probably helps to detach the E-tRNA out of the ribosome.

7.12.4.4 The Ribosomal Peptide Tunnel Exhibits Collective Dynamical Properties

The collective dynamic al properties of the polypeptide exit tunnel have been additionally examined. Low-frequency fluctuations have been found in three specific regions in the tunnel, such as the entrance, the neck and the exit. The linings of these three specific regions exhibit different types of motions. In broad terms, it can be said that the lining of the entry region travels in the output direction to aid in the elimination of the peptide, although the neck itself has much more complex movements. The exit area has much larger movements that rotate with the exit vector as an axis of rotation. It is noteworthy that the generally conserved prolongations of the ribosomal proteins L4 and L22, which are located at the narrowest entrance area of the tunnel, usually perform opening and closing movements, which can play an essential role in the emerging polypeptide gate mechanism to guarantee the stiffness of the reaction center in the closed state; this movement largely conforms to the circular peristaltic movements shown at this place (Kurkcuoglu et al. 2009a, b).

7.12.4.5 Intrinsic Movements of the tRNAs Are Closedly Linked to Their Movements Inside the Ribosome

The movements apparent from various tRNA structures are closely similar to those of simulations of the elastic network (Zimmermann and Jernigan 2014). The first three major normal mode movements of tRNA have been revealed (Bahar et al. 1997). For every structure, it is demonstrated how it travels according to its conformational modification route. The first mode features the acceptor arm with large movements and may be necessary to ease amino acid adaptation as the tRNA entries the ribosome to seek its proper binding position. The second mode contains a bending of the structure resembling the movements experienced during the ratchet movement and constituting Noller's hybrid state (Ramrath et al. 2013; Spiegel et al. 2007). The third movement mode entails unrolling and stretching the anticodon stalk loop, and this can obviously be instrumental in achieving the exact matching between codon and anticodon. Moreover, the movements of the tRNA at all three sites A, P and E are almost identical after their rigid body movements were eliminated (Wang and Jernigan 2005). The tRNA elbows interfere with areas of the 50S subunit and suffer large conformational alterations, while the anticodon ribbon loops interact in accordance with the 30S. The bending of the tRNA during the ratchet movement is fundamental and intrinsic to the structure of tRNA. These curved tRNA structural mid-products seem to be identical to the hybrid states of Noller (Ramrath et al. 2013; Spiegel et al. 2007).

The first two movements of the independent tRNAs referred to are probably affected when the tRNA initially joins the ribosome to assure the proper alignment of the acceptor arm and the anticodon of the tRNA. Once the tRNA is completely bound in the ribosome, these pieces cannot move, and actually, these pieces of tRNA are considered the stiffest pieces of the tRNAs when located in the ribosome. Since the charged amino acid ends of A- and P-tRNAs are attached close to the peptidyltransferase center in the center of the 70S complex, these are the least fluctuating regions in the tRNAs. The anticodon is also fixed very rigidly to the codon of the mRNA.

There are many findings from simulations that refer directly to single features of the mechanisms of the ribosome. The ratchet movement is extremely powerful and hard to undermine, and even after removal of all proteins (Kurkcuoglu et al. 2008). The dominant movement guarantees that the active compounds—the tRNAs and the mRNA—pass through the ribosome correctly. Many results obtained during the simulations are in agreement with what is commonly accepted knowledge of the ribosome and its movements, including the hybrid bent state of the tRNAs and the peristaltic movement inside the tunnel of the peptides. Moreover, even additional conformations have been obtained by simulations that may provide novel functional roles, such as the partial shut down of the mRNA tunnel opening in order to detect the bases of the codon and utilize this gained results for choosing the suitable tRNA, which still needs to be experimentally confirmed. In fact, these novel distinct interactions of possible functional impact that have been revealed through simulations need to be verified by employing mutagenesis. In general, the high level of precision in these precise allosteric interactions inside the ribosome delivers a high fidelity of entire translation process.

The ribosome represents a complex machinery, and hence, coarse-grained simulations seem to uncover specific features of its mechanism

The famous ratchet movement occurring between the two large subunits is the principal movement, and this is seen in the simulations where the t-RNAs constitute the hybrid state

The compounds lining the mRNA tunnel can pin the mRNA into its position through their performance of distinct movements

The lining of the peptide tunnel contains regions with significantly varying movements that support the emergence of the resulting peptide

Coarse-grained dynamics models seem to be useful approaches for figuring out the molecular mechanistic insights of the largest molecular structures.

7.13 ER Mechanics

The ER represents a single organelle in eukaryotic cells that covers the entire cell and fulfills a wide variety of functions inside the cell. Employing a set of fixed and living cells (human MRC5 lung cells) in diffraction limited and super-resolved fluorescence microscopy (STORM) experiments, the average persistence length of ER tubules was found to be 3.03 ± 0.24 μm. The removal of the branched network intersections from the analysis resulted in a modest elevation of the average persistence length to 4.71 ± 0.14 μm and imparts a moderate dependence of the tubule persistence length on the length scale. The average tubulus radius has been sown to be 44.1 ± 3.2 nm. The bending rigidity of the membranes of ER tubules has been revealed to be 10.9 ± 1.2 kT and 17.0 ± 1.3 kT without any connecting points. The dynamic nature of ER tubules has been investigated in live cells, and it has turned out that the ER tubules exhibited a behavior of semi-flexible fibere that are tensioned (Georgiades et al. 2017). Most ER tubules undergo equilibrium transverse fluctuations, when they experience tension, whereas a minority of them performed active super-diffusive movements powered by motor proteins. In addition, the cells are able to actively modulate the dynamic properties of the ER in a carefully coordinated manner, which presumably influences its multiple functionalities.

The ER represents a remarkably intricate organelle that occurs in eukaryotic cells and is surrounded by an uninterrupted membrane that traverses the complete cell. Two diverse morphological areas are discernable in the ER; an area that encircles the nucleus and seems flatter and leaf-like, and an area that straddles the peripheral areas in the cell, which comprises a pattern of tubules (Shibata et al. 2006, 2010), with incidental lamellar spots (Shemesh et al. 2014). Accordingly, the ER mesh cannot be regarded as being static over time, it is instead extremely dynamic, constantly experiencing redistributions, channel configurations, enlargements and fluctuations in their movement (Chen et al. 2013; Borgese et al. 2006; Wozniak et al. 2009). Being the biggest organelle in eukaryotic cells, the ER demonstrably participates in numerous functional processes of the cell, such as the synthesis of proteins, the monitoring of quality and the break-down of proteins (Sita and Braakman 2003), exchange of ions and lipids (Eden et al. 2016), the induction of the endosomal translocation (Raiborg et al. 2015) and the regulation of apoptotic events (Csordas et al. 2006).

The progress made in imaging techniques and the consequent availability of higher resolution images has led to new insights into the structure of supposedly flat ER plates (Shibata et al. 2010; Shemesh et al. 2014). Combining high-resolution fluorescence imaging techniques for living and fixed cells, it was possible to detect ER tubules in areas previously considered ER leaflets by identifying gaps in the leaf-shaped peripheral ER (Nixon-Abell et al. 2016). Until now, these have been unexplored because of the limits of spatial and temporal resolution of the imaging procedures, and their presence has been verified by means of electron microscopy. Moreover, studies on the dynamic characteristics of the ER have been conducted by tracing three-way crossings between the tubules, emphasizing the dynamic oscillations of the ER tubules and the crossings along with its extremely flexible shape and morphology (Nixon-Abell et al. 2016; Valm et al. 2017).

The polymer physics of semi-flexible fibers is today a comprehensive and well-established research area (Broedersz and MacKintosh 2014; Granek 1997). The standard calculations employing the equipartition theorem enable the determination of the persistence length (L_p) of semi-flexible fibers to be associated with the elasticity of the fibers. The persistence length (L_p) is defined as the length scale over which angular correlations in the tangent direction over the entire backbone are not corrected with the distance along the contour of the backbone. Hence, the persistence length provides a key property of every polymers and aids in the quantification of their rigidity. In reality, a polymer chain exhibits flexible properties when its length extends beyond its persistence length, while it acts as a rigid bar for lengths far smaller than its persistence length and as a semi-flexible chain on length scales of the same order of magnitude as its persistence length (Gittes et al. 1993, 16, 17). Supportive to the end-to-end distance and cross-sectional diameter, the persistence length represents one of the most important quantitative measurements for the description of the conformational state of a fiber. The majority of the cytoskeletal proteins within cells, such as actin with a L_p of approximately 10 ± 1 μm (Tassieri 2008) and microtubules with a L_p of approximately 5200 μm (Gittes et al. 1993), have been investigated precisely to uncover their impact on the cell mechanics, such as assessment of the energy deposited when bending a microtubular fiber during the deformation-driven reshaping of a cell. In this context, numerous biomacromolecules were investigated to gain an insight into the relationships between their conformations and their specific functional role, for example, the persistence length of DNA is ~0.050 μm, having a slight reliance on the individual distinct sequence (Waigh 2014), and the persistence length can be exploited to comprehend the molecular interaction of DNA and histones. It is anticipated that the persistence length is potentially also a valuable geometric instrument for the characterization of ER tubules, helping to identify their function in various cell types which are characterized by distinct dedicated structures. There has been a little work undertaken on the geometry of the ER in cells derived from tobacco leaves obtained by confocal microscopy, but here polygons have been employed to comprehend the shape of the underlying network, and the flexibility of the tubules in this case has not been assessed (Lin et al. 2014).

Recent advances with semi-flexible polymers broadened the theoretical framework for calculating the effects of hydrodynamics on transverse displacements $\left(\langle \Delta r_{\perp}^{2}(t)\rangle\right)$, the transverse mean squared displacement as a function of time interval t of semi-flexible fiber without the implementation of tension $\left(\langle \Delta r_{\perp}^{2}(t)\rangle \sim t^{3/4}\right)$ and in tension-controlled regimes $\left(\langle \Delta r_{\perp}^{2}(t)\rangle \sim t^{1/2}\right)$ (Granek 1997). These kinds of theories for fiber dynamics coincide neatly with a number of experimental procedures such as fluorescence microscopy, photon correlation spectroscopy, bright field microscopy and a sequence of microrheology techniques (Jahnel et al. 2008; Carrick et al. 2005; Morse 1998). The newer theoretical concepts focus on the active movement of semi-flexible fibers linked to motor proteins (Ghosh and Gov 2014; Eisenstecken et al. 2016; Weber et al. 2015; Isele-Holder et al. 2015). While these theories are at a more early state of evolution than those for the passive fluctuations of fibers, they forecast a propelled super-diffusive movement of semi-flexible fibers on long-term time scales $\left(\langle \Delta r_{\perp}^{2}(t)\rangle \sim t^{\alpha}\,\alpha > 1\right)$ (Ghosh and Gov 2014; Eisenstecken et al. 2016; Weber et al. 2015; Isele-Holder et al. 2015).

A combination of live-cell imaging, conventional diffraction-limited epifluorescence microscopy and stochastic optical reconstruction microscopy (STORM) has been employed to analyze the mechanical properties of the ER tubules and the dynamic nature of single tubules. There is substantial evidence available to corroborate the idea that ER tubules act as hollow, tubular, semi-flexible fibers, with an average outer radius of 44.1 ± 3.2 nm, a persistence length, L_p, of 3.03 ± 0.24 μm (4.71 ± 0.14 μm on small length scales within the branching nodes), which is equivalent to a super soft membrane bending stiffness κ_{mem}, of 10.9 ± 1.2 kT (17.0 ± 1.3 kT on small length scales between the branching nodes), which is a characteristic feature of biological lipids in bilayer membranes (Boal 2012).

In the past, data for the ER persistence length have never been provided, and hence, data have been recently revealed that lie between those of actin filaments, which means that are slightly smaller than actin Lp by a factor of 3 or 4 (Tassieri 2008) and DNA, which seems to be much larger by a factor of 60 (Philips 2013). Additionally, the values of the ER persistence length are usually one order of magnitude smaller than the overall size of the cell, in which they are determined. It is clear that the ER has an extremely soft microstructure and its input into the mechanics of whole cells is probably far less than that of microtubules, actins or intermediate filaments (Mofrad and Kamm 2006). There is ample proof of the active regulation of ER dynamics by motor proteins, and these active movements are anticipated to perform an essential task for ER functions (Nixon-Abell et al. 2016), such as the synthesis rate and the following alteration of proteins in a rocked reaction vessel model.

7.13.1 The Fit of the Fibers and Analysis of Fiber Mechanics

Super-resolved and diffraction-limited pictures of the ER in fixed MRC5 cells as well as single videos of transiently transfected living MRC5 cells expressing EGFP-ER were adjusted using FiberApp, an open-source MATLAB toolbox capable of extracting structural data of fibrillar specimens in atomic force microscopy and fluorescence pictures (Usov and Mezzenga 2015). The contour of the ER tubules was adapted by the embedded algorithms based on active contour models, allowing the x–y position of the tubules to be computed.

To determine the persistence length of the ER tubules, the contour length and the end-to-end distance of the active contour of ER tubules needs to be taken from the fluorescent images employing the FiberApp (Valle et al. 2005; Cox et al. 2017). The mean squared end-to-end distance $\langle R^2\rangle$ can be described with (7.2):

$$\left\langle R^2 \right\rangle = 2L_{\rm p}\left(L_{\rm c} - L_{\rm p}\left(1 - {\rm e}^{-L_{\rm c}/L_{\rm p}}\right)\right) \tag{7.2}$$

where L_p is the persistence length and L_c represents the contour length (Mantelli et al. 2011).

The end-to-end distance and the contour length can be determined using FiberApp, and subsequently, the persistence length can be extracted by fitting (7.2). As expected, the mean persistence length was 3.03 ± 0.24 μm. The mean squared end-to-end distance has been revealed to be a more robust for the calculation of the persistence lengths based on fluorescence microscopy images than the direct analysis of angular correlations, as it is less perturbed by the small length scale noise, such as for a semi-flexible fiber, where $L_p \sim L_c$, and most of the angular deflections is based on the largest wavelength bending modes that also revealed $\langle R^2 \rangle$ (Cox et al. 2017). Super-resolved images of fixated MRC5 cells were employed to analyze the outer radius R of the ER tubules, which is on average 44.1 ± 3.2 nm (Georgiades et al. 2017).

Besides the persistence length and the cross-sectional radius, an additional crucial property of polymers is the bending rigidity, κ_{pol}, that determines how the applied bending moment, M, scales in line with the curvature d^2wdr^2, where r stands for the deflection and x represents the distance along the beam during the deformations (Mofrad and Kamm 2006), such as $M = \kappa_{pol}d^2\, w\, dr^2$. For a semi-flexible fiber, the L_p can be correlated with the polymeric bending rigidity, when the equipartition theorem is used to analyze the bending energy, as given in (7.3):

$$L_{\rm p} = \frac{\kappa_{\rm pol}}{k_B T} \tag{7.3}$$

where k_BT represents the thermal energy (Gittes et al. 1993; Philips 2013).

It is common knowledge that the ER fibers are no uniform rigid (solid) cylinders; hence, they are hollow cylinders, which wall is 3–5 nm thick (Nagle and Tristram-Nagle 2000). When the ER is seen as an elastic hollow cylinder with a radius R of negligible thickness, which represents a thin hollow cylinder approximation, (7.3) can be replaced. Hence, the persistence length (L_p) of (7.4) relies on the anisotropic elasticity of membranes (Derenyi et al. 2002):

$$L_{\rm p} = \frac{2\kappa_{\rm mem}\pi R}{k_B T} \tag{7.4}$$

where κ_{mem} represents the membrane's bending rigidity, which is a two-dimensional entity and is a proportionality constant between the free membrane energy and the averaged mean squared membrane curvature, R is the tubule radius and k_BT represents the thermal energy.

This particular approach has two advantages: Firstly, it provides the determination of the bending rigidity of the membranes that can be related to other experimental in the field of nanomechanics. Secondly, it has been extensively investigated with in vitro experiments, such as lipid nanotubules of variable radii (Yamamoto and Ichikawa 2012; Baroji et al. 2014; Tanaka-Takiguchi et al. 2012). When values for the persistence length and the ER tubules radii are inserted in (7.4), bending rigidity for the ER tubule membranes of 10.9 ± 1.2 kT (44.8 ± 4.9 × 10^{-21} J) can be obtained with the branch points included and without the branch point the value increases to 17.0 ± 1.3 kT (69.9 ± 5.3 × 10^{-21} J) (Georgiades et al. 2017). However, both values lie still inside the range of 3–225 kT, which is usually detected for lipids in biological membranes (Boal 2012), and hence, they are in line with the predicted values.

The membrane bending rigidity (κ_{mem}) can be determined from the fitted ER tubules and the data are log normal distributed. The bending rigidities, such as those with an average 10.9 ± 1.2 kT, imply very soft materials; however, they lie in the range of lipid bilayers. For comparison propose, the bending rigidity of a graphene bilayer is 1380 kT (Lindahl et al. 2012).

7.13.2 How Can ER Tubules Be Monitored?

To characterize the ER tubular network in more detail, live-cell videos were utilized, where the ER motion can be tracked. In specific detail, the *x*–*y* coordinates of the ER tubules along a line perpendicularly to the ER tubule can be employed to finally determine the mean squared displacement (MSD) as a function of the lag time for all given tracks. The MSD is defined by (7.5):

$$\langle \Delta r^2(\tau) \rangle = \left\langle \left(x(t+\tau) - x(t)^2 \right) + (y(t+\tau) - y(t))^2 \right\rangle_t \tag{7.5}$$

where t represents the time and τ is referred to as lag time.

Based on the tiny longitudinal movements of the fibers, the MSD seems to be a suitable approximation to the transverse MSD ($\langle \Delta r_\perp^2(\tau) \rangle$) of the fiber's movements, such as $\langle \Delta r^2(\tau) \rangle = \langle \Delta r_\perp^2(\tau) \rangle + \left\langle \Delta r_\parallel^2(\tau) \right\rangle \approx \langle \Delta r_\perp^2(\tau) \rangle$ (Broedersz and MacKintosh 2014). To characterize the dynamic movement of the ER tubules, a power law of the form $\langle \Delta r^2(\tau) \rangle \sim \tau^\alpha$ can be fitted, which provides the determination of power exponent α values of the MSD curves. There are found two distinct populations of MSDs that are detected at longer time scales, such as those above 2 s: one population displays subdiffusive behavior, where $\alpha < 1$ and the other population displays super-diffusive behavior, where the following applies: $1 < \alpha < 2$. The collective motion of the ER system on longer time scales, such as above 2 s, contributes to the increase of the exponent. The two subpopulations can be found due to their discrepancies in their power-law exponents at long time scales with individually averaged MSDs. A universal power law of $\tau = 0.48 \pm 0.02$ determines the average MSD curve for the subdiffusive MSD subpopulation, whereas a power law of $\tau = 0.58 \pm 0.04$ was characteristic for the super-diffusive subpopulation in the regime of short lag times, such as up to 2 s and a $\tau = 1.53 \pm 0.03$ can be observed for the regime of longer lag times, such as those above 2 s.

The ER represents a critical organelle in eukaryotic cells, since it regulates many basic cellular processes and its malfunction causes a variety of diseases (Roussel et al. 2013). Beyond that, the structural and mechanical characteristics of the ER are still not understood in detail. However, there are advances based on super-resolution microscopy and electron microscopy that enable researchers determine precisely previously unreachable small length scales to discover new insights in the ER's tubular architectural network and the dynamical restructuring processes. An example is the finding that the ER sheets within the cell periphery are not positioned inside sheet-like morphologies, rather tightly packed tubules with triangular connections, so-called ER matrices, and moreover, morphological evidence has been found for the highly dynamic and transferable phenotype of these complex ER networks (Nixon-Abell et al. 2016; Valm et al. 2017).

While the architecture, morphology and allocation of the ER grid system in a variety of cell lines are well defined, the mechanical properties of organelles have so far been poorly understood. The ER is endowed with a high number of proteins, ribosomes and active motor proteins that have the potential to modify its mechanical properties and dynamics (Lin et al. 2012; Voeltz et al. 2006; English et al. 2009). A persistence length of the ER tubules of 3.03 ± 0.24 μm was observed, both in living and fixed cells. The persistence lengths determined in the four unrelated experiments, such as live-cell imaging, diffraction-limited KLC3 antibody labeling and super-resolution imaging with varying labeling strategies, match well, eliminating artifacts from sample preparation. Besides the persistence length of the ER tubules, we also calculated their average membrane bending stiffness, which amounted to 10.9 ± 1.2 kT. To verify the potential influence of ER contacts on the persistence length that is, an L_p dependent on the length scale caused by chemical heterogeneity, such as contact promoting proteins (Shemesh et al. 2014; Lin et al. 2012), the investigation was replicated for the stretches of the tubules between the connection points. A small incremental increase in the persistence length was noted of 4.71 ± 0.14 μm. The respective membrane bending stiffness obtained with (7.3) is also somewhat higher in the case of 17.0 ± 1.3 kT, which has to be anticipated, as the effects of branching proteins have been eliminated. The membrane bending stiffness value is linearly related to the persistence length (see 7.3), and vice versa to the tubule radius. This offers an obvious benefit of super-resolution imaging, as the detected radius of 44.1 ± 3.2 nm is a subdiffraction length scale for a standard optical microscope, but is far above the resolution limit of about 20 nm of the STORM microscope.

The persistence lengths and membrane bending stiffnesses calculated from these experiments displayed a widespread distribution pointing to the wide spectrum of ER tubule rigidities available, which with wide fluctuations in both morphology and the constitution of the ER networks across the cell as a whole seems to be coherent. It has been proposed that the tubules' curvature is guided through the interplay of two sets of proteins, the R-type and the S-type proteins. In specific detail, the R-type proteins, such as reticulons and DP1/Yop1p, support the assembly of tubules, while the S-type proteins, such as atlastins, Sey1p and Climp63, induce junctions within tubules and so-termed sheets (Shemesh et al. 2014; Lin et al. 2012). However, the full set of proteins that govern the morphology of the ER is not yet clearly known. In extension, the ER physically interfaces with a range of organelles, such as endosomes, and the movement of the ER has been associated directly with the microtubular cytoskeleton of the cell (Wozniak et al. 2009), both types of interference feeding into the general structure of the network (English et al. 2009). It is consequently deduced that the monitored distribution of persistence lengths and membrane bending stiffnesses is due to a complicated interference between the abovementioned elements, all of which disturb the ultrasoft tubular structures forming the ER.

Besides the mechanical properties of the ER mesh, its dynamic characteristics were examined with live-cell imaging and cell tracking software. Two specific mean squared displacement (MSD) populations were revealed to be a function of delay time (τ) for longer delay times, such as $\tau > 2$ s: the first one displays subdiffusive motion with $0 < \alpha < 1$ and the second one shows super-diffusive subballistic motion with $1 < \alpha < 2$. The subpopulation of super-diffusive subballistic tracked ER tubules may be impacted by the collective motion of the entire ER grid due to the overall movement of the cell or, more probably, the immediate outcome of an active mechanism fueled by motor proteins or the translocation of the tubule due to a temporary interplay with microtubules or vesicles (Nixon-Abell et al. 2016). By extrapolating tracks with super-diffusive movement in their MSDs at long times (>2 s), a universal power law was found with an exponent α of 0.48 ± 0.02 for any other tracks, which is consistent with the theoretical error forecast for semi-flexible fibers under tension that fixed at both ends, as provided in (7.6) (Granek 1997). Short-lived traces (<5 s) were not taken into account in the analysis. The frame rate was not high enough to track the fastest moving tubules, and therefore, they were not added to the records. These tubules may have been in touch with other structural elements of the cell that experienced sudden translocation. Consequently, a third population of high-speed trajectories carrying super-diffusive subballistic movements is anticipated. To determine whether this third population differs from the other populations of the previously observed super-diffusive subballistic trajectories, additional investigations seem to be necessary.

A part of the ER tubular framework is subjected to transverse fluctuations. Hence, a distinct prediction for the transverse MSD $(\langle \Delta r^2 \perp(\tau) \rangle)$ of a fiber exposed to tension is given by (7.6):

$$\langle \Delta r_{\perp}^{2}(\tau) \rangle = \left(\frac{k_B}{\sigma} \right)^{3/4} \frac{L_p^{1/4} \tau^{1/2}}{2\eta^{1/2}} \tag{7.6}$$

where τ represents the lag time, σ is the stress within the fiber and η is termed the solvent's viscosity (Granek 1997).

Matching the gradient of the MSD, which has a τ 1/2 scale, results in the average stress (σ) in the ER tubules of 58.3 ± 9.1 Pa. This value appears to be rational, as it is equivalent to an average tensile force of about 10 fN exerted longitudinally on every ER tubule and lies comfortably in the margin of what can be achieved by individual motor proteins, fN-pN. It entails a gentle correction of the persistence lengths calculated with (7.2). The effective strain can be estimated by employing an analytic expression for the nonlinear elasticity of a semi-flexible filament (Broedersz and MacKintosh 2014) leading to an effective strain value of 0.03 for tensioned tubules. This strain, on the other hand, suggests a ~3% correction of the persistence length specified by (7.2), which is overestimated by this amount, but lies inside the experimental errors and is therefore negligible. The tension force has been found to be three orders of magnitude smaller than that necessary to form bonds (that is to say, large morphological alterations) in reconstituted ER in in vitro measurements performed with optical tweezers (18.6 ± 2.8 pN) (Upadhyaya and Sheetz 2004). A typical length scale is connected to the tensile stress, which is determined by $l_t \sim \sqrt{\kappa_{\text{pol}}\sigma}$, as a result of the rivalry between bending and tension, and it is equivalent to (3.7 ± 0.6) × 10^{-14} m in the case of tubules (Broedersz and MacKintosh 2014). For length scales above lt, it is anticipated that the dynamics will be controlled by the stress, and therefore, the ER tubules are characterized by (7.6), because the movement of the micrometer sized fibers is far above this threshold. In line with this, the mentioned stresses are largely above the $\sigma \gg k_B T L_p / L^2$ threshold, and hence, (7.6) is valid for the semi-flexible dynamics experiencing passive tension (Broedersz and MacKintosh 2014; Granek 1997) (L represents the contour length of the fiber).

The time dependence in (7.6) seems to be Rouse-like, such as it displays $\langle \Delta r^2(\tau) \rangle \sim \tau\, 1/2$ that represents a scaling result for flexible polymer chains without tension (Rubinstein and Colby 2003), but it applies to semi-flexible polymers (image analysis gives clear evidence that ER tubules are semi-flexible because their persistence length is in the order of magnitude of the contour length) the prefactors within the equation are scaled at a different scale as compared to the outcome of the flexible polymer (Granek 1997). Subtle fluctuations in the persistence length of semi-flexible fibers are reported to exist under stress (Dobrynin et al. 2010); however, (7.4) is considered to provide a good first approximation for the membrane bending stiffness (κmem) of ER tubules and the calculated persistence lengths based on the geometric definition of persistence length and the utilization of direct image analysis and are independent of tension phenomena. There was no meaningful population of ER tubules identified that was not stressed, i.e., a population that did not conform to the predictions of (7.6) did it have super-diffusive movements powered by motor proteins.

On the basis of the analogy with a hexagonal honeycomb foam, a first approximation for the elastic modulus and the shear modulus of the full ER network structure can be given (Gibson and Ashby 1999). The in-plane Young's modulus, E_{in}, of the ER system, for example, in the plane of the 2D honeycomb, scales as in (7.7):

$$E_{\text{in}} \sim \left(\frac{2r}{L} \right)^3 E \tag{7.7}$$

where $2r$ represents the tubule thickness, E refers to the modulus of the tubules and L displays the distance between branches.

This leads to the prediction that the in-plane Young's modulus of the ER scaffold seems to be a million times weaker than that of a single tubule, such as E in of approximately 10^{-6} E. It is also anticipated that the shear modulus in the plane is ultrasoft with a G of approximately 0.25×10^{-6} E. The out-of-plane modulus of elasticity has a $(2r/L)^1$ scale, meaning that E_{out} is approximately 10^{-2} E and the out-of-plane shear moduli (G_{out}) also lie in a similar scale. In summary, the ER network exhibits an ultrasoft microstructure compared to other cytoskeletal materials, such a actin filaments, intermediate filaments and microtubules, and that the in-plane deformations in the ER structure level, which may coincident with the plane of the cell, are likely to be more severe than the off-plane deformations (outside the ER structure level). More complex models are available to characterize the nonlinear mechanical properties of ER tubules in an analogous way to a foam structure, with more precise local geometric characteristics for their interconnected networks, such as bond angles, coordination numbers, tubule thicknesses and so on (Broedersz and MacKintosh 2014). Under stress, nonlinear elasticity is anticipated for the ER tubules in the case of large deformations, while in the case of compression, linear elasticity is anticipated for large deformations, followed by buckling and compression of the tubules (Gibson and Ashby 1999), which subsequently represents a severe collapse of the entire structural scaffold that is associated with a significant reduction in volume. Despite this, however, ER microstructures remain sufficiently soft to make only a very minor physical contribution to the total mechanics of a discrete cell, compared to the input of cytoskeletal filaments, such as actin filaments and microtubules, which carry the bulk of the exerted forces (Boal 2012).

Besides, modeling of how motor proteins induce both passive stresses and active movements of biological fibers has been initiated (Ghosh and Gov 2014; Weber et al. 2015), but additional progress is required to render these trials directly applicable to ER tubules. In this case, a robust and simple experimental measurement of the mechanical features of individual tubules has been made available, and it is expected that these measurements can be further enhanced in future investigations. It is assumed that the very soft dynamic mesh of the ER tubules performs an essential task for the biological functions of the ER, for instance, it seems to provide an effective mechanism of convective transport for the mixing of reagents during the synthesis of biomolecules, such as the synthesis of proteins by ribosomes linked to the tubular membranes and the protein processing inside the lumen of the ER tubules, which means that it represents a shaken reaction vessel model.

The structural architecture, the mechanical properties and the dynamic remodeling of the ER are crucial for the proper understanding of its functional roles. Further exploration is necessary to fully characterize the set of proteins that dominate the shape of the tubular network and dictate the functions of the ER (Chen et al. 2013). In addition, the morphology of ER and its functional roles may differ between cell types; therefore, it is essential to examine its characteristics in a broader spectrum of cells, such as the sarcoplasmic reticulum in heart cells.

In conclusion, the ER constitutes a notably complicated and sophisticated interconnected network of tubules in eukaryotic cells that extends from the nuclear envelope toward the cell periphery. A unique combination of diffraction-limited fluorescence imaging of fixed and living cells, and super-resolution STORM imaging of fixed cells was employed to evaluate the mechanical properties of the ER and to explore the dynamic properties of the tubules. The ER tubule radius was identified to be 44.1 ± 3.2 nm, the average persistence length amounted to 3.03 ± 0.24 μm (4.71 ± 0.14 μm between branching points) and the average membrane bending stiffness seemed to be 10.9 ± 1.2 kT (17.0 ± 1.3 kT without branching points). Beyond that, in combination with MSD analysis, it has been ascertained that there is a population of active ER tubules where motility is either directly or indirectly powered by motor proteins inside the cytoplasm, as well as a larger population of ER tubules receiving only passive tension. These movements can affect the performance of the ER, such as the speed of protein synthesis in a rocked reaction tube model. The transverse fluctuations of the tubes under tension correspond quite well to a theory for the hydrodynamics of semi-flexible fibers. There was no population of ER tubules detected that was not in tension.

7.14 Golgi Mechanics

There have been significant new insights in the mechanical function of actin in the organization of the Golgi organization and trafficking of vesicles. In an elegant trial, optical tweezers and live-cell imaging are employed to prove the effects of mechanical restriction on the dynamics of secretory membrane transport, whereby physical experimental techniques are coupled with cellulo investigations of endomembranes (Egea and Serra-Peinado 2019).

Enormous progress has been made over the last 30 years in the recognition and mechanistic comprehension of the molecular constituents involved in the transport of lipids and proteins through the pathway of secretion. The significance of the membrane trafficking in terms of health and disease was acknowledged with the Nobel Prize for Medicine and Physiology awarded to well-known and acclaimed pioneers in this field: James Rothman, Randy Scheckman and Thomas Südhof. Thus, it could be misguided to assume that the field of membrane trafficking offers no major emerging "surprises" or discoveries and may even enter a state of total lethargy.

Well, that is a misconception. At least there is just one factor that is still widely excluded from the membrane trafficking: the impact and functionality of the physical forces on the morphology, the organization and the proper functioning of endomembranes. It has been demonstrated that there has been considerable progress in tackling the impact of membrane tension and bending on coat-induced budding and molecular sorting (Gauthier et al. 2012; Römer et al. 2010; Boulant et al. 2011) and on membrane cleavage of transport vehicles (Bassereau and Goud 2011). There exists an elegant investigation (Guet et al. 2014) combining physical techniques with confocal microscopy in living cells, which is reported that Golgi membranes are flexible and mechanically linked, actin imparts stiffness to the Golgi apparatus, and a mechanical stress induces a shift from vesicular to tubular transportation, connecting the forces with membrane cleavage.

On one side, considering "forces" in the cell, the most important subcellular contribution is made by the cytoskeleton, which consists mainly of two extremely dynamic (and tightly regulated) polymers—microtubules and actin filaments. On the other side, considering the membrane trafficking in the secretion system, the Golgi apparatus instantly pops up in front of us.

In the majority of organisms, the Golgi consists of one or more layers of tightly packed, flattened membranes referred to as cisterns. In animal cells, these layers are positioned end to end to create the "Golgi band." As is generally recognized, the cytoskeleton is involved in the patterning of the Golgi apparatus: Microtubules are involved in the lateral interconnection of the Golgi band and its polarity, and actin filaments are responsible for maintaining the flattened morphology of the cisternae (Gurel et al. 2014; Egea et al. 2013). Both cytoskeletal elements are supported by the Golgi matrix, the structural framework that forms protein-containing transverse bridges connecting adjacent Golgi cisternae. The members of the Golgi Reassembly and Stacking Protein (GRASP) and Golgin families of proteins are part of the matrix (Xiang and Wang 2011). Together with microtubules, actin filaments and their corresponding motor proteins, these peripheral membrane proteins accumulate the Golgi cisternae.

To obtain an in-depth understanding of cell membranes and their organizational structure, a careful mix of physical and cell biological approaches is necessary. Optical tweezers (Norregaard et al. 2014) permit precise regulation of membrane tension and bending and provide the possibility to examine every stage of membrane transport—from the budding of a vesicle to its movement across the cytoskeleton—and simultaneously quantify the forces implicated in these processes. A number of newer studies have yielded useful insights into the membrane trafficking, but have concentrated primarily on in vitro investigations. In contrary, in the recent investigation (Guet et al. 2014), optical tweezers are combined with confocal fluorescence microscopy to evaluate the mechanical properties of the Golgi apparatus in cellulo. The experimental setup is built on the endocytic internalization of beads of a micrometric size which are pressed against the Golgi membranes that are identified by the presence of a GFP-labeled variant of the Golgi marker Rab6 or by BodipyFL ceramide staining, and the translocation of the beads is monitored by particle tracking. Compared to the cytoplasm, the Golgi is rigid and forces of more than 100 pN are required to induce deformations.

The mechanical properties of the Golgi have been assessed and it has been revealed that the actin cytoskeleton seems to act as a key element for providing the stiffness of the Golgi membranes and the peri-Golgi region. Several pharmacological drugs that modify actin dynamics have been proven to influence this, such as cytochalasin D (a traditional actin-depolymerizing agent) and jasplakinolide (an actin stabilizer in vitro inhibiting actin dynamics in vivo by promoting the unregulated polymerization of actin into amorphous assemblies), and with actomyosin contractility by retarding or enabling the actin-based myosin II motor activity. The findings demonstrated that a disturbance of actin dynamics resulted in a decrease in Golgi stiffness, which then was related to the motor activity of myosin II. In this section, the impact of mechanical restriction of the trafficking processes is revealed. Due to the exertion force on the Golgi apparatus, the Rab6+ vesicle production was pronouncedly diminished. Once the force was withdrawn, there was a gradual recovery in vesicular Rab6 traffic. Parallel to the forcibly induced Rab6 vesicular diminution, $Rab6^+$ tubules were attached to the Golgi. This finding suggests that the employed mechanical force is involved in the splitting of the membrane. To connect mechanical force and deficiencies in membrane cleavage, in order to map actin dynamics in living cells before and after the exertion of force, LifeAct mCherry has been used to identify dynamic transient actin dots depicting critical areas of membrane cleavage (Miserey-Lenkei et al. 2010). The force applied decreased the frequency of actin staining on the Golgi surface, which strongly associated the mechanical limitation with deficiencies in membrane cleavage. Ultimately, a mechanical assessment of the behavior of the actin polymerization machinery and matrix proteins in the Golgi after mechanical forcing was conducted. The number of cortactin patches has been examined, indicative of the involvement of N-WASP-Arp2/3-facilitated actin nucleation.

Simultaneously, it has been analyzed whether the actin-regulating RhoA-mDia1 signaling pathway is also involved in Golgi mechanics by targeting the small molecule inhibitor of the formin homology domain FH2 (SMIFH2). A reduction in the number of cortactin bandages after mechanical force was noted, and exposure to SMIFH2 (a cortactin inhibitor) decreased Golgi stiffness. To investigate the involvement of the Golgi matrix proteins, microrheological measurements were carried out in cells depleted of the matrix protein giantin and it has been detected that the Golgi stiffness dropped to values comparable to those obtained after a disturbance in actin dynamics.

This new study combines the biological and physical approaches to address a biological problem, such as the membrane trafficking (Guet et al. 2014). It is clearly doubtless that it is major step forward in this research field and seems to be supporting similar approaches of possible of other different endomembrane systems. It is unfortunate that the problem with this sort of study, intersecting two fields of science that are so dissimilar in their experimental approaches, concepts and techniques, is that neither side is completely satisfied. From the physical point of view, biological matter, such as molecules, cells or organelles are addressed in the light of a mechanical parameter that refers to a "softness index" and changes due to specific pharmacological stimulations of living cells. The softness index represents a general parameter to distinguish between the rigidities of the Golgi microenvironment due to specific conditions. An advantage of this quantitative biophysical approach lies in its independency of any distinct viscoelastic model, although the softness index is not classically employed in the wide field of rheology. A major concern about this softness index is that it may be a too simple or reductionist approach when rather complex and vastly divers molecularly composed Golgi membranes and their surrounding microenvironment, including various matrix proteins, the cytoskeleton and cytoplasm are measured. In line with this, it needs to be questioned whether the (sub)cellular injuries, which are evoked by the laser beam employed for the bead trapping, can cause perturbations of the mechanical measurements. From the biological point of view, the simple usage of pharmacological agents with variable target specificity may not appropriately mirror the linkage between specific components of the Golgi apparatus and organelle or cell mechanics in complex process of membrane trafficking. Another concern about the optical tweezer approach seems to be the possible interference of membrane that encloses endocytosed beads on the overall membrane system and the luminal transport of cargo proteins or other cell functions (Mierke 2013). Instead, the injection of uncoated beads directly into the cytoplasm or antibody coated beads that target a specific Golgi protein seem to be of advantage, since the force can be applied directly on the Golgi membranes.

However, this problem can be solved easily in future approaches. It is still a sophisticated problem to be solved, when the mechanical properties of distinct organelles need to be addressed. There is the requirement of precise combination of biological and physical approaches in order to cover the entire complexity of these measurements accurately for revealing reliable and substantial results.

In specific detail, the optical tweezer technique seems to be promising for mechanical probing of endomembrane compartments in living cells; however, the heating of the cells due to the focused laser beam needs to be kept in mind. Nonetheless, this problem will be solved and the organization and the dynamic remodeling of endomembranes can be accessed and analyzed in a physiological relevant microenvironment. Initial results have been reported for actin that seems to fulfill a novel role in the secretory process of cells. There are still more proteins in this process to be discovered soon.

References

R.K. Agrawal, A.B. Heagle, P. Penczek, R.A. Grassucci, J. Frank, EF-G-dependent GTP hydrolysis induces translocation accompanied by large conformational changes in the 70S ribosome. Nat. Struct. Biol. **6**, 643–647 (1999)

J.D. Aitchison, M.P. Rout, The road to ribosomes. Filling potholes in the export pathway. J. Cell Biol. **151**(5), 23–26 (2000)

V. Anantharaman, L. Aravind, The GOLD domain, a novel protein module involved in Golgi function and secretion. Genome Biol. **3**, 0023.1–0023.7 (2002)

D.J. Anderson, M.W. Hetzer, Reshaping of the endoplasmic reticulum limits the rate for nuclear envelope formation. J. Cell Biol. **182**(5), 911–924 (2008)

T. Anelli, M. Alessio, A. Bachi, L. Bergamelli, G. Bertoli, S. Camerini, A. Mezghrani, E. Ruffato, T. Simmen, R. Sitia, Thiol-mediated protein retention in the endoplasmic reticulum: the role of ERp44. EMBO J. **22**, 5015–5022 (2003)

T. Anelli, S. Ceppi, L. Bergamelli, M. Cortini, S. Masciarelli, C. Valetti, R. Sitia, Sequential steps and checkpoints in the early exocytic compartment during secretory IgM biogenesis. EMBO J. **26**, 4177–4188 (2007)

M. Anitei, B. Hoflack, Exit from the *trans*-Golgi network: from molecules to mechanisms. Curr. Opin. Cell Biol. **23**, 443–451 (2011)

C. Appenzeller, H. Andersson, F. Kappeler, H.P. Hauri, The lectin ERGIC-53 is a cargo transport receptor for glycoproteins. Nat. Cell Biol. **1**, 330–334 (1999)

C. Appenzeller-Herzog, H.P. Hauri, The ER-Golgi intermediate compartment (ERGIC): in search of its identity and function. J. Cell Sci. **119**, 2173–2183 (2006)

C. Appenzeller-Herzog, A.C. Roche, O. Nufer, H.P. Hauri, pH-induced conversion of the transport lectin ERGIC-53 triggers glycoprotein release. J. Biol. Chem. **279**, 12943–12950 (2004)

A. Ashok, R.S. Hegde, Selective processing and metabolism of disease-causing mutant prion proteins. PLoS Pathog. **5**, e1000479 (2009)

A.R. Atilgan, S.R. Durell, R.L. Jernigan, M.C. Demirel, O. Keskin, I. Bahar, Anisotropy of fluctuation dynamics of proteins with an elastic network model. Biophys. J. **80**, 505–515 (2001)

M.A. Atmodjo, Z. Hao, D. Mohnen, Evolving views of pectin biosynthesis. Annu. Rev. Plant Biol. **64**, 747–779 (2013)

A. Audhya, A. Desai, K. Oegema, A role for Rab5 in structuring the endoplasmic reticulum. J. Cell Biol. **178**(1), 43–56 (2007)

I. Bahar, A.J. Rader, Coarse-grained normal mode analysis in structural biology. Curr. Opin. Struct. Biol. **15**, 586–592 (2005)

I. Bahar, A.R. Atilgan, B. Erman, Direct evaluation of thermal fluctuations in proteins using a single-parameter harmonic potential. Fold Des. **2**, 173–181 (1997)

S. Bahmanyar, R. Biggs, A.L. Schuh, A. Desai, T. Muller-Reichert, A. Audhya, J.E. Dixon, K. Oegema, Spatial control of phospholipid flux restricts endoplasmic reticulum sheet formation to allow nuclear envelope breakdown. Genes Dev. **28**(2), 121–126 (2014)

N. Ban, P. Nissen, J. Hansen, P.B. Moore, T.A. Steitz, The complete atomic structure of the large ribosomal subunit at 2.4 Å resolution. Science **289**, 905–920 (2000)

D.K. Banfield, Mechanisms of protein retention in the Golgi. Cold Spring Harb. Perspect. Biol. **3**, a005264 (2011)

V.A. Bankaitis, R. Garcia-Mata, C.J. Mousley, Golgi membrane dynamics and lipid metabolism. Curr. Biol. **22**, R414–R424 (2012)

S.I. Bannykh, W.E. Balch, Membrane dynamics at the endoplasmic reticulum-Golgi interface. J. Cell Biol. **138**, 1–4 (1997)

F. Bard, V. Malhotra, The formation of TGN-to-plasma membrane transport carriers. Annu. Rev. Cell Dev. Biol. **22**, 439–455 (2006)

F. Bard, L. Mazelin, C. Péchoux-Longin, V. Malhotra, P. Jurdic, Src regulates Golgi structure and KDEL receptor-dependent retrograde transport to the endoplasmic reticulum. J. Biol. Chem. **278**, 46601–46606 (2003)

C. Barlowe, A. Helenius, Cargo capture and bulk flow in the early secretory pathway. Annu. Rev. Cell Dev. Biol. **32**, 197–222 (2016)

C.K. Barlowe, E.A. Miller, Secretory protein biogenesis and traffic in the early secretory pathway. Genetics **193**, 383–410 (2013)

C. Barlowe, L. Orci, T. Yeung, M. Hosobuchi, S. Hamamoto, N. Salama, M.F. Rexach, M. Ravazzola, M. Amherdt, R. Schekman, COPII: a membrane coat formed by Sec proteins that drive vesicle budding from the endoplasmic reticulum. Cell **77**, 895–907 (1994)

Y.F. Baroji, L.B. Oddershede, S.N.S. Reihani, P.M. Bendix, Fluorescent quantification of size and lamellarity of membrane nanotubes Euorpean. Biophys. J. **43**, 595–602 (2014)

P. Bassereau, B. Goud, Physics, biology and the right chemistry F1000. Biol. Rep. **3**, 1–5 (2011)

O. Baumann, B. Walz, Endoplasmic reticulum of animal cells and its organization into structural and functional domains. Int. Rev. Cytol. **205**, 149–214 (2001)

M. Beck, A. Schmidt, J. Malmstroem, M. Claassen, A. Ori, A. Szymborska, F. Herzog, O. Rinner, J. Ellenberg, R. Aebersold, The quantitative proteome of a human cell line. Mol. Syst. Biol. **7**, 549 (2011)

B. Becker, M. Melkonian, The secretory pathway of protists: spatial and functional organization and evolution. Microbiol. Rev. **60**, 697–721 (1996)

B. Becker, B. Bolinger, M. Melkonian, Anterograde transport of algal scales through the Golgi complex is not mediated by vesicles. Trends Cell Biol. **5**, 305–307 (1995)

R. Behnia, F.A. Barr, J.J. Flanagan, C. Barlowe, S. Munro, The yeast orthologue of GRASP65 forms a complex with a coiled-coil protein that contributes to ER to Golgi traffic. J. Cell Biol. **176**, 255–261 (2007)

W.J. Belden, C. Barlowe, Erv25p, a component of COPII-coated vesicles, forms a complex with Emp24p that is required for efficient endoplasmic reticulum to Golgi transport. J. Biol. Chem. **271**, 26939–26946 (1996)

W.J. Belden, C. Barlowe, Deletion of yeast p24 genes activates the unfolded protein response. Mol. Biol. Cell **12**, 957–969 (2001)

M. Bentley, Y. Liang, K. Mullen, D. Xu, E. Sztul, J.C. Hay, SNARE status regulates tether recruitment and function in homotypic COPII vesicle fusion. J. Biol. Chem. **281**, 38825–38833 (2006)

H.J. Berendsen, S. Hayward, Collective protein dynamics in relation to function. Curr. Opin. Struct. Biol. **10**, 165–169 (2000)

H.M. Berman, J. Westbrook, Z. Feng et al., The protein data bank. Nucleic Acids Res. **28**, 235–242 (2000)

S. Bernales, K.L. McDonald, P. Walter, Autophagy counterbalances endoplasmic reticulum expansion during the unfolded protein response. PLoS Biol. **4**(12), e423 (2006)

B.J. Bevis, A.T. Hammond, C.A. Reinke, B.S. Glick, De novo formation of transitional ER sites and Golgi structures in *Pichia pastoris*. Nat. Cell Biol. **4**, 750–756 (2002)

N. Bhattacharya, J.O. Donnell, S.M. Stagg, The structure of the Sec13/31 COPII cage bound to Sec23. J. Mol. Biol. **420**, 324–334 (2012)

X. Bi, J.D. Mancias, J. Goldberg, Insights into COPII coat nucleation from the structure of Sec23. Sar1 complexed with the active fragment of Sec31. Dev. Cell **13**, 635–645 (2007)

X. Bian, R.W. Klemm, T.Y. Liu, M. Zhang, S. Sun, X. Sui, X. Liu, T.A. Rapoport, J. Hu, Structures of the atlastin GTPase provide insight into homotypic fusion of endoplasmic reticulum membranes. Proc. Natl. Acad Sci. USA **108**(10), 3976–3981 (2011)

S.C. Blanchard, R.L. Gonzalez Jr., H.D. Kim, S. Chu, J.D. Puglisi, tRNA selection and kinetic proofreading in translation. Nat. Struct. Biol. **11**, 1008–1014 (2004)

S.C. Blanchard, H.D. Kim, R.L. Gonzalez, J.D. Puglisi, tRNA dynamics on the ribosome during translation. Proc. Natl. Acad. Sci. USA **101**, 12893–12898 (2004)

G. Blobel, B. Dobberstein, Transfer of proteins across membranes. I Presence of proteolytically processed and unprocessed nascent immunoglobulin light chains on membrane-bound ribosomes of murine myeloma. J. Cell Biol. **67**, 835–851 (1975)

B.A. Block, T. Imagawa, K.P. Campbell, C. Franzini-Armstrong, Structural evidence for direct interaction between the molecular components of the transverse tubule/sarcoplasmic reticulum junction in skeletal muscle. J. Cell Biol. **107** (6 Pt 2), 2587–2600 (1988)

R. Blum, P. Feick, M. Puype, J. Vandekerckhove, R. Klengel, W. Nastainczyk, I. Schulz, Tmp21 and p24A, two type I proteins enriched in pancreatic microsomal membranes, are members of a protein family involved in vesicular trafficking. J. Biol. Chem. **271**, 17183–17189 (1996)

D. Boal, *Mechanics of the Cell*, 2nd edn (CUP, 2012)

L. Bonfanti, A.A. Mironov Jr., J.A. Martínez-Menárguez, O. Martella, A. Fusella, M. Baldassarre, R. Buccione, H. J. Geuze, A.A. Mironov, A. Luini, Procollagen traverses the Golgi stack without leaving the lumen of cisternae: evidence for cisternal maturation. Cell **95**, 993–1003 (1998)

J.S. Bonifacino, B.S. Glick, The mechanisms of vesicle budding and fusion. Cell **116**, 153–166 (2004)

N. Borgese, M. Francolini, E. Snapp, Endoplasmic reticulum architecture: structures in flux. Curr. Opin. Cell Biol. **18**, 358–364 (2006)

S. Boulant, C. Kural, J.C. Zeeh, F. Ubelmann, T. Kirchhausen, Actin dynamics counteract membrane tension during clathrin-mediated endocytosis. Nat. Cell Biol. **13**, 1124–1131 (2011)

I. Braakman, D.N. Hebert, Protein folding in the endoplasmic reticulum. Cold Spring Harb. Perspect. Biol. **5**(5), a013201 (2013)

G.A. Brar, M. Yassour, N. Friedman, A. Regev, N.T. Ingolia, J.S. Weissman, High-resolution view of the yeast meiotic program revealed by ribosome profiling. Science **335**(6068), 552–557 (2012)

W.T. Brigance, C. Barlowe, T.R. Graham, Organization of the yeast Golgi complex into at least four functionally distinct compartments. Mol. Biol. Cell **11**, 171–182 (2000)

C.P. Broedersz, F.C. MacKintosh, Modelling semiflexible polymer networks. Rev. Mod. Phys. **86**, 995–1033 (2014)

W.B. Busa, R. Nuccitelli, An elevated free cytosolic Ca2+ wave follows fertilization in eggs of the frog, Xenopus laevis. CellBiol. **100**, 1325–1329 (1985)

L.J. Byrnes, H. Sondermann, Structural basis for the nucleotide-dependent dimerization of the large G protein atlastin-1/SPG3A. Proc. Natl. Acad. Sci. USA **108**(6), 2216–2221 (2011)

Y. Cao, S. Knochel, F. Oswald, C. Donow, H. Zhao, W. Knochel, XBP1 forms a regulatory loop with BMP-4 and suppresses mesodermal and neural differentiation in Xenopus embryos. Mech. Dev. **123**(1), 84–96 (2006)

L. Carrick, M. Tassieri, T.A. Waigh, A. Aggeli, N. Boden, C. Bell, J. Fisher, E. Ingham, R.M.L. Evans, Internal dynamic modes of charged self-assembled peptide fibrils. Langmuir **21**, 3733–3737 (2005)

G.A. Castillon, R. Watanabe, M. Taylor, T.M.E. Schwabe, H. Riezman, Concentration of GPI-anchored proteins upon ER exit in yeast. Traffic **10**, 186–200 (2009)

E. Chanat, W.B. Huttner, Milieu-induced, selective aggregation of regulated secretory proteins in the *trans*-Golgi network. J. Cell Biol. **115**, 1505–1519 (1991)

S. Chantalat, S.K. Park, Z. Hua, K. Liu, R. Gobin, A. Peyroche, A. Rambourg, T.R. Graham, C.L. Jackson, The Arf activator Gea2p and the P-type ATPase Drs2p interact at the Golgi in *Saccharomyces cerevisiae*. J. Cell Sci. **117**, 711–722 (2004)

Q. Chen, S. Jagannathan, D.W. Reid, T. Zheng, C.V. Nicchitta, Hierarchical regulation of mRNA partitioning between the cytoplasm and the endoplasmic reticulum of mammalian cells. Mol. Biol. Cell **22**, 2646–2658 (2011)

S. Chen, T. Desai, J.A. McNew, P. Gerard, P.J. Novick, S. Ferro-Novick, Lunapark stabilizes nascent three-way junctions in the endoplasmic reticulum. Proc. Natl. Acad. Sci. USA **112**(2), 418–423 (2015)

S. Chen, P. Novick, S. Ferro-Novick, ER network formation requires a balance of the dynamin-like GTPase Sey1p and the Lunapark family member Lnp1p. Nat. Cell Biol. **14**(7), 707–716 (2012)

S. Chen, P. Novick, S. Ferro-Novick, ER structure and function. Curr. Opin. Cell Biol. **25**, 428–433 (2013)

D.E. Clapham, Calcium signaling. Cell **131**(6), 1047–1058 (2007)

B.S. Cooperman, Y.E. Goldman, No longer lost in translation. Nat. Biotech. **24**, 526–527 (2006)

A. Copic, C.F. Latham, M.A. Horlbeck, J.D. D'Arcangelo, E.A. Miller, ER cargo properties specify a requirement for COPII coat rigidity mediated by Sec13p. Science **335**, 1359–1362 (2012)

M. Cortini, R. Sitia, ERp44 and ERGIC-53 synergize in coupling efficiency and fidelity of IgM polymerization and secretion. Traffic **11**, 651–659 (2010)

P. Cosson, M. Amherdt, J.E. Rothman, L. Orci, A resident Golgi protein is excluded from peri-Golgi vesicles in NRK cells. Proc. Natl. Acad. Sci. USA **99**, 12831–12834 (2002)
H. Cox, P. Georgiades, H. Xu, T.A. Waigh, J.R. Lu, Self-assembly of mesoscopic peptide surfactant fibrils investigated by STORM super-resolution fluorescence microscopy. Biomacromol **18**(11), 3481–3491 (2017)
G. Csordas, C. Renken, P. Várnai, L. Walter, D. Weaver, K.F. Buttle, T. Balla, C.A. Mannella, G. Hajnóczky, Structural and functional features and significance of the physical linkage between ER and mitochondria. J. Cell. Biol. **174**, 915–921 (2006)
X.A. Cui, Y. Zhang, S.J. Hong, A.F. Palazzo, Identification of a region within the placental alkaline phosphatase mRNA that mediates p180-dependent targeting to the endoplasmic reticulum. J. Biol. Chem. **288**, 29633–29641 (2013)
L. Daboussi, G. Costaguta, G.S. Payne, Phosphoinositide-mediated clathrin adaptor progression at the *trans*-Golgi network. Nat. Cell Biol. **14**, 239–248 (2012)
J. Dancourt, C. Barlowe, Protein sorting receptors in the early secretory pathway. Annu. Rev. Biochem. **79**, 777–802 (2010)
K.J. Day, L.A. Staehelin, B.S Glick, A three-stage model of Golgi structure andfunction. Histochem. Cell Biol. **140**(3), 239–249 (2013)
J.O. De Craene, J. Coleman, P. Estrada de Martin, M. Pypaert, S. Anderson, J.R. Yates 3rd, S. Ferro-Novick, P. Novick, Rtn1p is involved in structuring the cortical endoplasmic reticulum. Mol. Biol. Cell **17**(7), 3009–3020 (2006)
M.A. De Matteis, A. Luini, Exiting the Golgi complex. Nat. Rev. Mol. Cell Biol. **9**, 273–284 (2008)
I. Derenyi, F. Julicher, J. Prost, Formation and interaction of membrane tubes. Phys. Rev. Lett. **88**, 238101 (2002)
J. Dettmer, A. Hong-Hermesdorf, Y.D. Stierhof, K. Schumacher, Vacuolar H+-ATPase activity is required for endocytic and secretory trafficking in *Arabidopsis*. Plant Cell **18**, 715–730 (2006)
D. Devos, S. Dokudovskaya, F. Alber, R. Williams, B.T. Chait, A. Sali, M.P. Rout, Components of coated vesicles and nuclear pore complexes share a common molecular architecture. PLoS Biol. **2**, e380 (2004)
G. Dick, L.K. Akslen-Hoel, F. Grøndahl, I. Kjos, K. Prydz, Proteoglycan synthesis and Golgi organization in polarized epithelial cells. J. Histochem. Cytochem. **60**, 926–935 (2012)
M. Diehn, R. Bhattacharya, D. Botstein, P.O. Brown, Genome-scale identification of membrane-associated human mRNAs. PLoS Genet. **2**, e11 (2006)
M. Diehn, M.B. Eisen, D. Botstein, P.O. Brown, Large-scale identification of secreted and membrane-associated gene products using DNA microarrays. Nature Genet **25**, 58–62 (2000)
A.V. Dobrynin, J.M.Y. Carrillo, M. Rubinstein, Chains are more flexible under tension. Macromolecules **43**, 9181–9190 (2010)
S.O. Dodonova, P. Diestelkoetter-Bachert, A. von Appen, W.J. Hagen, R. Beck, M. Beck, F. Wieland, J.A. Briggs, Vesicular transport. A structure of the COPI coat and the role of coat proteins in membrane vesicle assembly. Science **349**, 195–198 (2015)
B.S. Donohoe, B.H. Kang, L.A. Staehelin, Identification and characterization of COPIa- and COPIb-type vesicle classes associated with plant and algal Golgi. Proc. Natl. Acad. Sci. USA **104**, 163–168 (2007)
B.S. Donohoe, B.H. Kang, M.J. Gerl, Z.R. Gergely, C.M. McMichael, S.Y. Bednarek, L. Staehelin, *cis*-Golgi cisternal assembly and biosynthetic activation occur sequentially in plants and algae. Traffic **14**, 551–567 (2013)
P. Doruker, R.L. Jernigan, Functional motions can be extracted from on-lattice construction of protein structures. Proteins **53**, 174–181 (2003)
P. Doruker, R.L. Jernigan, I. Bahar, Dynamics of large proteins through hierarchical levels of coarse-grained structures. J. Comp. Chem. **23**, 119–127 (2002a)
P. Doruker, R.L. Jernigan, I. Navizet, R. Hernandez, Important fluctuation dynamics of large protein structures are preserved upon coarse-grained renormalization. Int. J. Quant. Chem. **90**, 822–837 (2002b)
L. Dreier, T.A. Rapoport, In vitro formation of the endoplasmic reticulum occurs independently of microtubules by a controlled fusion reaction. J. Cell Biol. **148**(5), 883–898 (2000)
A. Driouich, L.A. Staehelin, in *The Plant Golgi Apparatus: Structural Organization and Functional Properties*, ed. by E.G. Berger, J. Roth (The Golgi Apparatus. Birkhäuser Basel, 1997), pp. 275–301
R. Duden, R. Schekman, *Insights into Golgi Function Through Mutants in Yeast and Animal Cells*, ed. by E.D. Berger, J. Roth (The Golgi Apparatus Birkhäuser Verlag Basel, 1997), pp. 219–246
W.G. Dunphy, J.E. Rothman, Compartmental organization of the Golgi stack. Cell **42**, 13–21 (1985)
E.R. Eden, E. Sanchez-Heras, A. Tsapara, A. Sobota, T.P. Levine, C.E. Futter, Annexin A1 tethers membrane contact sites that mediate ER to endosome cholesterol transport. Dev. Cell **37**(5), 473–483 (2016)
G. Egea, C. Serra-Peinado, Golgi apparatus: finally mechanics comes to play in the secretory pathway. Curr. Biol. **24** (16), R742 (2019)
G. Egea, C. Serra-Peinado, L. Salcedo-Sicilia, E. Gutierrez-Martinez, Actin acting at the Golgi. Histochem. Cell Biol. **140**, 347–360 (2013)
T. Eisenstecken, G. Gompper, R.G. Winkler, Conformational properties of active semiflexible polymers. Polymers **8**, 304 (2016)

G. Emery, M. Rojo, J. Gruenberg, Coupled transport of p24 family members. J. Cell Sci. **113**, 2507–2516 (2000)

S. Emr, B.S. Glick, A.D. Linstedt et al., Journeys through the Golgi—taking stock in a new era. J. Cell Biol. **187**, 449–453 (2009)

A.R. English, G.K. Voeltz, Endoplasmic reticulum structure and interconnections with other organelles. Cold Spring Harb. Perspect. Biol. **5**(4), a013227 (2013)

A.R. English, G.K. Voeltz, Rab10 GTPase regulates ER dynamics and morphology. Nat. Cell Biol. **15**(2), 169–178 (2013)

A.R. English, N. Zurek, G.K. Voeltz, Peripheral ER structure and function. Curr. Opin. Cell Biol. **21**(4), 596–602 (2009)

A. Eugster, G. Frigerio, M. Dale, R. Duden, COP I domains required for coatomer integrity, and novel interactions with ARF and ARF-GAP. EMBO J. **19**, 3905–3917 (2000)

P. Fagone, S. Jackowski, Membrane phospholipid synthesis and endoplasmic reticulum function. J. Lipid Res. **50** (Suppl), S311–S316 (2009)

M.G. Farquhar, H.-P. Hauri, *Protein Sorting and Vesicular Traffic in the Golgi Apparatus*, ed. by E.G. Berger, J. Roth (The Golgi Apparatus Birkhäuser Verlag Basel, 1997), pp. 63–129

M.G. Farquhar, E. Palade, The Golgis apparatus: 100 years of progress and controversy. Trends in Cell Biology **8**, 2–10 (1998)

M.G. Farquhar, G.E. Palade, The Golgi apparatus (complex)—(1954–1981)—from artifact to center stage. J. Cell Biol. **91**, 77s–103s (1981)

C. Faso, A. Boulaflous, F. Brandizzi, The plant Golgi apparatus: last 10 years of answered and open questions. FEBS Lett. **583**, 3752–3757 (2009)

S. Fath, J.D. Mancias, X. Bi, J. Goldberg, Structure and organization of coat proteins in the COPII cage. Cell **129**, 1325–1336 (2007)

D.W. Fawcett, On the occurrence of a fibrous lamina on the inner aspect of the nuclear envelope in certain cells of vertebrates. Dev. Dyn. **119**, 129–145 (1966)

K. Fiedler, M. Veit, M.A. Stamnes, J.E. Rothman, Bimodal interaction of coatomer with the p24 family of putative cargo receptors. Science **273**, 1396–1399 (1996)

L.J. Foster, C.L. de Hoog, Y. Zhang, X. Xie, V.K. Mootha, M. Mann, A mammalian organelle map by protein correlation profiling. Cell **125**(1), 187–199 (2006)

J. Frank, R.K. Agrawal, A ratchet-like inter-subunit reorganization of the ribosome during translocation. Nature **406**, 318–322 (2000)

J. Frank, A.B. Heagle, R.K. Agrawal, Animation of the dynamical events of the elongation cycle based on cryoelectron microscopy of functional complexes of the ribosome. J. Struct. Biol. **128**, 15–18 (1999)

J.R. Friedman, G.K. Voeltz, The ER in 3D: a multifunctional dynamic membrane network Trends Cell Biol **21**(12), 709–717 (2011)

J.R. Friedman, B.M. Webster, D.N. Mastronarde, K.J. Verhey, G.K. Voeltz, ER sliding dynamics and ER-mitochondrial contacts occur on acetylated microtubules. J. Cell Biol. **190**(3), 363–375 (2010)

J. Füllekrug, J. Boehm, S. Röttger, T. Nilsson, G. Mieskes, H.D. Schmitt, Human Rer1 is localized to the Golgi apparatus and complements the deletion of the homologous Rer1 protein of *Saccharomyces cerevisiae* Eur. J. Cell Biol. **74**, 31–40 (1997)

D. Gaddam, N. Stevens, J. Hollien, Comparison of mRNA localization and regulation during endoplasmic reticulum stress in Drosophila cells. Mol. Biol. Cell **24**(1), 14–20 (2013)

N.C. Gauthier, T.A. Masters, M.P. Sheetz, Mechanical feedback between membrane tension and dynamics. Trends Cell Biol. **22**, 527–535 (2012)

P. Georgiades, V.J. Allan, G.D. Wright, P.G. Woodman, P. Udommai, M.A. Chung, T.A. Waigh, The flexibility and dynamics of the tubules in the endoplasmic reticulum. Sci. Rep. **7**, 16474 (2017)

A. Gerondopoulos, R.N. Bastos, S. Yoshimura, R. Anderson, S. Carpanini, I. Aligianis, M.T. Handley, F.A. Barr, Rab18 and a Rab18 GEF complex are required for normal ER structure. J. Cell Biol. **205**(5), 707–720 (2014)

Y. Geva, M. Schuldiner, The back and forth of cargo exit from the endoplasmic reticulum. Curr. Biol. **24**, R130–R136 (2014)

A. Ghosh, N.S. Gov, Dynamics of active semi-flexible polymers. Biophys. J. **107**, 1065–1073 (2014)

L. Gibson, M. Ashby, *Cellular Solids: Structure and Properties*, 2nd edn (CUP, 1999)

A. Gilchrist, C.E. Au, J. Hiding et al., Quantitative proteomics analysis of the secretory pathway. Cell **127**, 1265–1281 (2006)

R. Gilmore, G. Blobel, P. Walter, Protein translocation across the endoplasmic reticulum. I Detection in the microsomal membrane of a receptor for the signal recognition particle. J. Cell Biol. **95**, 461–469 (1982)

F. Gittes, B. Mickey, J. Nettleton, J. Howard, Flexural rigidity of microtubules and actin filaments measured from thermal fluctuations in shape. J. Cell Biol. **120**, 923–934 (1993)

B.S. Glick, A. Luini, Models for Golgi traffic: a critical assessment. Cold Spring Harb. Perspect. Biol. **3**, a005215 (2011)

B.S. Glick, V. Malhotra, The curious status of the Golgi apparatus. Cell **95**, 883–889 (1998)
B.S. Glick, A. Nakano, Membrane traffic within the Golgi stack. Annu. Rev. Cell Dev. Biol. **25**, 113–132 (2009)
B.S. Glick, T. Elston, G. Oster, A cisternal maturation mechanism can explain the asymmetry of the Golgi stack. FEBS Lett. **414**, 177–181 (1997)
D.E. Goldberg, S. Kornfeld, Evidence for extensive subcellular organization of asparagine-linked oligosaccharide processing and lysosomal enzyme phosphorylation. J. Biol. Chem. **258**, 3159–3165 (1983)
J. Goldberg, Decoding of sorting signals by coatomer through a GTPase switch in the COPI coat complex. Cell **100**, 671–679 (2000)
N. Gomez-Navarro, E. Miller, Protein sorting at the ER–Golgi interface. J. Cell Biol. **215**(6), 769–778 (2016)
D.U. Gommel, A.R. Memon, A. Heiss, F. Lottspeich, J. Pfannstiel, J. Lechner, C. Reinhard, J.B. Helms, W. Nickel, F. T. Wieland, Recruitment to Golgi membranes of ADP-ribosylation factor 1 is mediated by the cytoplasmic domain of p23. EMBO J. **20**, 6751–6760 (2001)
T.R. Graham, C.G. Burd, Coordination of Golgi functions by phosphatidylinositol 4-kinases. Trends Cell Biol. **21**, 113–121 (2011)
R. Granek, From semi-flexible polymers to membranes: anomalous diffusion and reptation. J. Phys. II France **7**, 1761–1788 (1997)
G. Griffiths, K. Simons, The *trans* Golgi network: sorting at the exit site of the Golgi complex. Science **234**, 438–443 (1986)
D. Guet, K. Mandal, M. Pinot, J. Hoffmann, Y. Abidine, W. Sigaut, S. Bardin, K. Schauer, B. Goud, J.-B. Manneville, Mechanical role of actin dynamics in the rheology of the Golgi complex and in Golgi-associated trafficking events. Curr. Biol. **24**, 1700–1711 (2014)
Y. Guo, D.W. Sirkis, R. Schekman, Protein sorting at the *trans*-Golgi network. Annu. Rev. Cell Dev. Biol. **30**, 169–206 (2014)
P.S. Gurel, A.L. Hatch, H.N. Higgs, Connecting the cytoskeleton to the endoplasmic reticulum and Golgi. Curr. Biol. **24**, R660–R672 (2014)
S.R. Hanson, E.K. Culyba, T.L. Hsu, C.H. Wong, J.W. Kelly, E.T. Powers, The core trisaccharide of an N-linked glycoprotein intrinsically accelerates folding and enhances stability. Proc. Natl. Acad. Sci. USA **106**(9), 3131–3136 (2009)
S. Hara-Kuge, O. Kuge, L. Orci, M. Amherdt, M. Ravazzola, F.T. Wieland, J.E. Rothman, En bloc incorporation of coatomer subunits during the assembly of COP-coated vesicles. J. Cell Biol. **124**, 883–892 (1994)
C. Hawes, Cell biology of the plant Golgi apparatus. New Phytol. **165**, 29–44 (2005)
D.N. Hebert, S.C. Garman, M. Molinari, The glycan code of the endoplasmic reticulum: asparagine-linked carbohydrates as protein maturation and quality-control tags. Trends Cell Biol. **15**(7), 364–370 (2005)
Y. Herzig, H.J. Sharpe, Y. Elbaz, S. Munro, M.A. Schuldiner, systematic approach to pair secretory cargo receptors with their cargo suggests a mechanism for cargo selection by Erv14. PLoS Biol. **10**, e1001329 (2012)
S.J. Hicks, J.W. Drysdale, H.N. Munro, Preferential synthesis of ferritin and albumin by different populations of liver polysomes. Science **164**, 584–585 (1969)
J. Ho, H.-N. Kallstrom, G. Johnson, Nmd3p is a Crm1p-dependent adapter protein for nuclear export of the large ribosomal subunit. J. Cell Biol. **151**, 1057–1066 (2000)
J. Hollien, J.S. Weissman, Decay of endoplasmic reticu- lum-localized mRNAs during the unfolded protein response. Science **313**(5783), 104–107 (2006)
E. Hong, A.R. Davidson, C.A. Kaiser, A pathway for targeting soluble misfolded proteins to the yeast vacuole. J. Cell Biol. **135**, 623–633 (1996)
J. Hu, W.A. Prinz, T.A. Rapoport, Weaving the web of ER tubules. Cell **147**(6), 1226–1231 (2011)
J. Hu, Y. Shibata, C. Voss, T. Shemesh, Z. Li, M. Coughlin, M.M. Kozlov, T.A. Rapoport, W.A. Prinz, Membrane proteins of the endoplasmic reticulum induce high-curvature tubules. Science **319**(5867), 1247–1250 (2008)
J. Hu, Y. Shibata, P.P. Zhu, C. Voss, N. Rismanchi, W.A. Prinz, T.A. Rapoport, C. Blackstone, A class of dynamin-like GTPases involved in the generation of the tubular ER network. Cell **138**(3), 549–561 (2009)
R.E. Isele-Holder, J. Elgeti, G. Gompper, Self-propelled worm-like filaments: spontaneous spiral formation, structure and dynamics Soft Matter **11**, 7181–7190 (2015)
Y. Ishikawa, S. Ito, K. Nagata, L.Y. Sakai, H.P. Bächinger, Intracellular mechanisms of molecular recognition and sorting for transport of large extracellular matrix molecules. Proc. Natl. Acad. Sci. USA **113**, E6036–E6044 (2016)
L.P. Jackson, M. Lewis, H.M. Kent, M.A. Edeling, P.R. Evans, R. Duden, D.J. Owen, Molecular basis for recognition of dilysine trafficking motifs by COPI. Dev. Cell **23**, 1255–1262 (2012)
L.P. Jackson, B.T. Kelly, A.J. McCoy, T. Gaffry, L.C. James, B.M. Collins, S. Höning, P.R. Evans, D.J. Owen, A large-scale conformational change couples membrane recruitment to cargo binding in the AP2 clathrin adaptor complex. Cell **141**, 1220–1229 (2010)
S. Jagannathan, J.C. Hsu, D.W. Reid, Q. Chen, W.J. Thompson, A.M. Moseley, C.V. Nicchitta, Multifunctional roles for the protein translocation machinery in RNA anchoring to the endoplasmic reticulum. J. Biol Chem. **289**, 25907–25924 (2014a)

S. Jagannathan, C. Nwosu, C.V. Nicchitta, Analyzing mRNA localization to the endoplasmic reticulum via cell fractionation. Methods Mol. Biol. **714**, 301–321 (2011)

S. Jagannathan, D.W. Reid, A.H. Cox, C.V. Nicchitta, De novo translation initiation on membrane-bound ribosomes as a mechanism for localization of cytosolic protein mRNAs to the endoplasmic reticulum. RNA **120**, 489–498 (2014b)

M. Jahnel, T.A. Waigh, J. Lu, Thermal fluctuations of fibrin fibres at short time scales. Soft Matter **4**, 1438–1442 (2008)

G. Jedd, J. Mulholland, N. Segev, Two new Ypt GTPases are required for exit from the yeast *trans*-Golgi compartment. J. Cell Biol. **137**, 563–580 (1997)

G. Jedd, C.J. Richardson, R.J. Litt, N. Segev, The Ypt1 GTPase is essential for the first two steps of the yeast secretory pathway. J. Cell Biol. **131**, 583–590 (1995)

D.D. Jenness, Y. Li, C. Tipper, P. Spatrick, Elimination of defective alpha-factor pheromone receptors. Mol. Cell. Biol. **17**, 6236–6245 (1997)

R.L. Jernigan, L. Yang, G. Song, O. Kurkcuoglu, P. Doruker, Elastic network models of coarse-grained proteins are effective for studying the structural control exerted over their dynamics, ed. by G. Voth, *Coarse-Graining of Condensed Phase and Biomolecular Systems* (Taylor and Francis Group LLC, 2008), pp. 237–254

L. Jin, K.B. Pahuja, K.E. Wickliffe, A. Gorur, C. Baumgärtel, R. Schekman, M. Rape, Ubiquitin-dependent regulation of COPII coat size and function. Nature **482**, 495–500 (2012)

Y. Jitsuhara, T. Toyoda, T. Itai, H. Yamaguchi, Chaperone-like functions of high-mannose type and complex-type N-gly- cans and their molecular basis. J. Biochem. **132**(5), 803–811 (2002)

A.E. Johnson, Functional ramifications of FRET-detected nascent chain folding far inside the membrane-bound ribosome. Biochem. Soc. Trans. **32**, 668–672 (2004)

C. Kaether, J. Scheuermann, M. Fassler, S. Zilow, K. Shirotani, C. Valkova, B. Novak, S. Kacmar, H. Steiner, C. Haass, Endoplasmic reticulum retention of the gamma-secretase complex component Pen2 by Rer1. EMBO Rep. **8**, 743–748 (2007)

Y. Kamiya, D. Kamiya, K. Yamamoto, B. Nyfeler, H.-P. Hauri, K. Kato, Molecular basis of sugar recognition by the human L-type lectins ERGIC-53, VIPL, and VIP36. J. Biol. Chem. **283**, 1857–1861 (2008)

B.H. Kang, L.A. Staehelin, ER-to-Golgi transport by COPII vesicles in Arabidopsis involves a ribosome-excluding scaffold that is transferred with the vesicles to the Golgi matrix. Protoplasma **234**, 51–64 (2008)

B.H. Kang, E. Nielsen, M.L. Preuss, D. Mastronarde, L.A. Staehelin, Electron tomography of RabA4b- and PI-4 Kβ1-labeled *trans* Golgi network compartments in *Arabidopsis*. Traffic **12**, 313–329 (2011)

R.J. Kaufman, Stress signaling from the lumen of the endoplasmic reticulum: coordination of gene transcriptional and translational controls. Genes Dev. **13**(10), 1211–1233 (1999)

R.J. Kaufman, D. Scheuner, M. Schroder, X. Shen, K. Lee, C.Y. Liu, S.M. Arnold, The unfolded protein response in nutrient sensing and differentiation. Nat. Rev. Mol. Cell Biol. **3**(6), 411–421 (2002)

S. Kawaguchi, C.-L. Hsu, D.T.W. Ng, Interplay of substrate retention and export signals in endoplasmic reticulum quality control. PLoS One **5**, e15532 (2010)

B.T. Kelly, S.C. Graham, N. Liska, P.N. Dannhauser, S. Höning, E.J. Ungewickell, D.J. Owen, Clathrin adaptors. AP2 controls clathrin polymerization with a membrane-activated switch. Science **345**, 459–463 (2014)

M.K. Kim, G.S. Chirikjian, R.L. Jernigan, Elastic models of conformational transitions in macromolecules. J. Mol. Graph. Model. **21**, 151–160 (2002a)

M.K. Kim, R.L. Jernigan, G.S. Chirikjian, Efficient generation of feasible pathways for protein conformational transitions. Biophys. J. **83**, 1620–1630 (2002b)

M.K. Kim, R.L. Jernigan, G.S. Chirikjian, An elastic network model of HK97 capsid maturation. J. Struct. Biol. **143**, 107–117 (2003)

M.K. Kim, R.L. Jernigan, G.S. Chirikjian, Rigid-cluster models of conformational transitions in macromolecular machines and assemblies. Biophys. J. **89**, 43–55 (2005)

M.M. Kincaid, A.A. Cooper, Misfolded proteins traffic from the endoplasmic reticulum (ER) due to ER export signals. Mol. Biol. Cell **18**, 455–463 (2007)

M.A. Kinseth, C. Anjard, D. Fuller, G. Guizzunti, W.F. Loomis, V. Malhotra, The Golgi associated protein GRASP is required for unconventional secretion during development. Cell **130**, 524–534 (2007)

A. Kitao, N. Go, Investigating protein dynamics in collective coordinate space. Curr. Opin. Struct. Biol. **9**, 164–169 (1999)

R.D. Klausner, J.G. Donaldson, J. Lippincott-Schwartz, Brefeldin A: insights into the control of membrane traffic and organelle structure. J. Cell Biol. **116**, 1071–1080 (1992)

D.R. Klopfenstein, F. Kappeler, H.P. Hauri, A novel direct interaction of endoplasmic reticulum with microtubules. EMBO J. **17**(21), 6168–6177 (1998)

D.R. Klopfenstein, J. Klumperman, A. Lustig, R.A. Kammerer, V. Oorschot, H.P. Hauri, Subdomain-specific localization of CLIMP-63 (p63) in the endoplasmic reticulum is mediated by its luminal alpha-helical segment. J. Cell Biol. **153**(6), 1287–1300 (2001)

J. Klumperman, Architecture of the mammalian Golgi. Cold Spring Harb. Perspect. Biol. **3**, a005181 (2011)

C.C. Kopczynski, J.N. Noordermeer, T.L. Serano, W.-Y. Chen, J.D. Pendleton, S. Lewis, C.S. Goodman, G.M. Rubin, A high throughput screen to identify secreted and transmembrane proteins involved in Drosophila embryogenesis. Proc. Natl. Acad. Sci. USA **95**, 9973–9978 (1998)

R. Kornfeld, S. Kornfeld, Assembly of asparagine-linked oligosaccharides. Annu. Rev. Biochem. **54**, 631–664 (1985)

D. Kressler, P. Linder, J. de La Cruz, Protein *trans*-acting factors involved in ribosome biogenesis in *Saccharomyces cerevisiae*. Mol. Cell Biol. **19**, 7897–7912 (1999)

A. Krogh, B. Larsson, G. von Heijne, E.L. Sonnhammer, Predicting transmembrane protein topology with a hidden Markov model: application to complete genomes. J. Mol Biol. **305**, 567–580 (2001)

L.F. Kung, S. Pagant, E. Futai, et al., Sec24p and Sec16p cooperate to regulate the GTP cycle of the COPII coat. EMBO J. **31**, 1014–1027 (2012)

T. Kurihara, S. Hamamoto, R.E. Gimeno, C.A. Kaiser, R. Schekman, T. Yoshihisa, Sec24p and Iss1p function interchangeably in transport vesicle formation from the endoplasmic reticulum in *Saccharomyces cerevisiae*. Mol. Biol. Cell **11**, 983–998 (2000)

O. Kurkcuoglu, P. Doruker, T.Z. Sen, A. Kloczkowski, R.L. Jernigan, The ribosome structure controls and directs mRNA entry, translocation and exit dynamics. Phys. Biol. **5**, 046005 (2008)

O. Kurkcuoglu, R.L. Jernigan, P. Doruker, Mixed levels of coarse-graining of large proteins using elastic network model succeeds in extracting the slowest motions. Polymer **45**, 649–657 (2004)

O. Kurkcuoglu, Z. Kurkcuoglu, P. Doruker, R.L. Jernigan, Collective dynamics of the ribosomal tunnel revealed by elastic network modeling. Proteins **75**, 837–845 (2009a)

O. Kurkcuoglu, O.T. Turgut, S. Cansu, R.L. Jernigan, P. Doruker, Focused functional dynamics of supramolecules by use of a mixed-resolution elastic network model. Biophys. J. **97**, 1178–1187 (2009b)

H.S. Kweon, G.V. Beznoussenko, M. Micaroni et al., Golgi enzymes are enriched in perforated zones of Golgi cisternae but are depleted in COPI vesicles. Mol. Biol. Cell **15**, 4710–4724 (2004)

M.S. Ladinsky, D.N. Mastronarde, J.R. McIntosh, K.E. Howell, L.A. Staehelin, Golgi structure in three dimensions: functional insights from the normal rat kidney cell. J. Cell Biol. **144**, 1135–1149 (1999)

P. Laneve, F. Altieri, M.E. Fiori, A. Scaloni, I. Bozzoni, E. Caffarelli, Purification, cloning, and characterization of XendoU, a novel endoribonuclease involved in processing of intron-encoded small nucleolar RNAs in Xenopus laevis. J. Biol. Chem. **278**(15), 13026–13032 (2003)

M. Langhans, M.J. Marcote, P. Pimpl, G. Virgili-López, D.G. Robinson, F. Aniento, In vivo trafficking and localization of p24 proteins in plant cells. Traffic **9**, 770–785 (2008)

J. Lanoix, J. Ouwendijk, A. Stark, E. Szafer, D. Cassel, K. Dejgaard, M. Weiss, T. Nilsson, Sorting of Golgi resident proteins into different subpopulations of COPI vesicles: a role for ArfGAP1. J. Cell Biol. **155**, 1199–1212 (2001)

G. Lavieu, H. Zheng, J.E. Rothman, Stapled Golgi cisternae remain in place as cargo passes through the stack. eLife **2**, e00558 (2013)

A.S. Lee, Mammalian stress response: induction of the glucose-regulated protein family. Curr. Opin. Cell Biol. **4**(2), 267–273 (1992)

C. Lee, L.B. Chen, Dynamic behavior of endoplasmic reticulum in living cells. Cell **54**(1), 37–46 (1998)

C. Lee, J. Goldberg, Structure of coatomer cage proteins and the relationship among COPI, COPII, and clathrin vesicle coats. Cell **142**, 123–132 (2010)

K. Lee, W. Tirasophon, X. Shen, M. Michalak, R. Prywes, T. Okada, H. Yoshida, K. Mori, R.J. Kaufman, IRE1-mediated unconventional mRNA splicing and S2P-mediated ATF6 cleavage merge to regulate XBP1 in signaling the unfolded protein response. Genes Dev. **16**(4), 452–466 (2002)

M.C.S. Lee, E.A. Miller, J. Goldberg, L. Orci, R. Schekman, Bi-directional protein transport between the ER and Golgi. Annu. Rev. Cell Dev. Biol. **20**, 87–123 (2004)

A. Lerich, S. Hillmer, M. Langhans, D. Scheuring, P. van Bentum, D.G. Robinson, ER import sites and their relationship to ER exit sites: a new model for bidirectional ER-Golgi transport in higher plants. Front. Plant Sci. **3**, 143 (2012)

R.S. Lerner, R.M. Seiser, T. Zheng, P.J. Lager, M.C. Reedy, J.D. Keene, C.V. Nicchitta, Partitioning and translation of mRNAs encoding soluble proteins on membrane-bound ribosomes. RNA **9**, 1123–1137 (2003)

S.K. Levi, D. Bhattacharyya, R.L. Strack, J.R.I. Austin, B.S. Glick, The yeast GRASP Grh1 colocalizes with COPII and is dispensable for organizing the secretory pathway. Traffic **11**, 1168–1179 (2010)

M. Levitt, C. Sander, P.S. Stern, Protein normal-mode dynamics: trypsin inhibitor, crambin, ribonuclease and lysozyme. J. Mol. Biol. **181**, 423–447 (1985)

M.J. Lewis, H.R. Pelham, Ligand-induced redistribution of a human KDEL receptor from the Golgi complex to the endoplasmic reticulum. Cell **68**, 353–364 (1992)

A. Liljas, *Structural Aspects of Protein Synthesis* (World Scientific Publishing Co., Pte. Ltd., Singapore, 2004)

C. Lin, Y. Zhang, I. Sparkes, P. Ashwin, Structure and dynamics of ER: minimal networks and biophysical constraints. Biophys. J. **107**, 763–772 (2014)

S. Lin, S. Sun, J. Hu, Molecular basis for sculpting the endoplasmic reticulum membrane. Int. J. Biochem. Cell Biol. **44**, 1436–1443 (2012)

N. Lindahl, D. Midtvedt, J. Svensson, O.A. Nerushev, N. Lindvall, A. Isacsson, E.E.B. Campbell, Determination of the bending rigidity of graphene via electrostatic actuation of buckled membranes. Nanoletters **12**, 3526–3531 (2012)

J. Lippincott-Schwartz, T.H. Roberts, K. Hirschberg, Secretory protein trafficking and organelle dynamics in living cells. Annu. Rev. Cell Dev. Biol. **16**, 557–589 (2000)

S. Liu, B. Storrie, Are Rab proteins the link between Golgi organization and membrane trafficking? Cell Mol. Life Sci. **69**, 4093–4106 (2012)

Y. Liu, J.J. Flanagan, C. Barlowe, Sec22p export from the endoplasmic reticulum is independent of SNARE pairing. J. Biol. Chem. **279**, 27225–27232 (2004)

Y. Liu, S. Sitaraman, A. Chang, Multiple degradation pathways for misfolded mutants of the yeast plasma membrane ATPase, Pma1. J. Biol. Chem. **281**, 31457–31466 (2006)

T. Lorca, S. Galas, D. Fesquet, A. Devault, J.C. Cavadore, M. Doree, Degradation of the proto-oncogene product p39mos is not necessary for cyclin proteolysis and exit from meiotic metaphase: requirement for a Ca(2+)-calmodulin dependent event. EMBO J. **10**(8), 2087–2093 (1991)

C. Lord, S. Ferro-Novick, E.A. Miller, The highly conserved COPII coat complex sorts cargo from the endoplasmic reticulum and targets it to the Golgi. Cold Spring Harb. Perspect. Biol. **5**, a013367 (2013)

E. Losev, C.A. Reinke, J. Jellen, D.E. Strongin, B.J. Bevis, B.S. Glick, Golgi maturation visualized in living yeast. Nature **22**, 1002–1006 (2006)

J. Lowery, T. Szul, M. Styers, Z. Holloway, V. Oorschot, J. Klumperman, E. Sztul, The Sec7 guanine nucleotide exchange factor GBF1 regulates membrane recruitment of BIG1 and BIG2 guanine nucleotide exchange factors to the *trans*-Golgi network (TGN). J. Biol. Chem. **288**, 11532–11545 (2013)

A. Loya, L. Pnueli, Y. Yosefzon, Y. Wexler, M. Ziv-Ukelson, Y. Arava, The 3′-UTR mediates the cellular localization of an mRNA encoding a short plasma membrane protein. RNA **14**, 1352–1365 (2008)

M. Lu, J. Ma, The role of shape in determining molecular motions. Biophys. J. **89**, 2395–2401 (2005)

M. Lu, J. Ma, A minimalist network model for coarse-grained normal mode analysis and its application to biomolecular x-ray crystallography. Proc. Natl. Acad. Sci. USA **105**, 15358–15363 (2008)

S.R. Lynch, R.L. Gonzalez, J.D. Puglisi, Comparison of X-ray crystal structure of the 30S subunit-antibiotic complex with NMR structure of decoding site oligonucleotide-paromomycin complex. Structure **11**, 43–53 (2003)

J. Ma, Usefulness and limitations of normal mode analysis in modeling dynamics of biomolecular complexes. Structure **13**, 373–380 (2005)

W. Ma, J. Goldberg, TANGO1/cTAGE5 receptor as a polyvalent template for assembly of large COPII coats. Proc. Natl. Acad. Sci. USA **113**, 10061–10066 (2016)

C.E. Machamer, Targeting and retention of Golgi proteins. Curr. Opin. Cell Biol. **5**, 606–612 (1993)

I. Majoul, M. Straub, S.W. Hell, R. Duden, H.D. Söling, KDEL-cargo regulates interactions between proteins involved in COPI vesicle traffic: measurements in living cells using FRET. Dev. Cell **1**, 139–153 (2001)

V. Malhotra, P. Erlmann, Protein export at the ER: loading big collagens into COPII carriers. EMBO J. **30**, 3475–3480 (2011)

V. Malhotra, P. Erlmann, The pathway of collagen secretion. Annu. Rev. Cell Dev. Biol. **31**, 109–124 (2015)

V. Malhotra, S. Mayor, Cell biology: the Golgi grows up. Nature **441**, 939–940 (2006)

J. Malsam, T.H. Söllner, Organization of SNAREs within the Golgi stack. Cold Spring Harb. Perspect. Biol. **3**, a005249 (2011)

J. Malsam, A. Satoh, L. Pelletier, G. Warren, Golgin tethers define subpopulations of COPI vesicles. Science **307**, 1095–1098 (2005)

J.D. Mancias, J. Goldberg, The transport signal on Sec22 for packaging into COPII-coated vesicles is a conformational epitope. Mol. Cell **26**, 403–414 (2007)

J.D. Mancias, J. Goldberg, Structural basis of cargo membrane protein discrimination by the human COPII coat machinery. EMBO J. **27**, 2918–2928 (2008)

S. Mantelli, P. Muller, S. Harlepp, M. Maaloum, Conformational analysis and estimation of the persistence length of DNA using atomic force microscopy in solution. Soft Matter **7**, 3412–3416 (2011)

J. Manzano-Lopez, A.M. Perez-Linero, A. Aguilera-Romero et al., COPII coat composition is actively regulated by luminal cargo maturation. Curr. Biol. **25**, 152–162 (2015)

P.Marra, T. Maffucci, T. Daniele, G. Di Tullio, Y. Ikehara, E.K.L. Chan, A. Luini, G. Beznoussenko, A. Mironov, M.A. De Matteis, The GM130 and GRASP65 Golgi proteins cycle through and define a subdomain of the intermediate compartment. Nat. Cell Biol. 3, 1101–1113 (2001)

B.J. Marsh, D.N. Mastronarde, K.F. Buttle, K.E. Howell, J.R. McIntosh, Organellar relationships in the Golgi region of the pancreatic beta cell line, HIT-T15, visualized by high resolution electron tomography. Proc. Natl. Acad. Sci. USA 98, 2399–2406 (2001)

B.J. Marsh, N. Volkmann, J.R. McIntosh, K.E. Howell, Direct continuities between cisternae at different levels of the Golgi complex in glucose-stimulated mouse islet beta cells. Proc. Natl. Acad. Sci. USA **101**, 5565–5570 (2004)

K.C. Martin, A. Ephrussi, mRNA localization: gene expression in the spatial dimension. Cell 136, 719–730 (2009)

J.A. Martínez-Menárguez, R. Prekeris, V.M.J. Oorschot, R. Scheller, J.W. Slot, H.J. Geuze, J. Klumperman, Peri-Golgi vesicles contain retrograde but not anterograde proteins consistent with the cisternal progression model of intra-Golgi transport. J. Cell Biol. **155**, 1213–1224 (2001)

K. Matsuura-Tokita, M. Takeuchi, A. Ichihara, K. Mikuriya, A. Nakano, Live imaging of yeast Golgi cisternal maturation. Nature **22**, 1007–1010 (2006)

I.W. Mattaj, L. Englmeier, Nucleocytoplasmic transportthe soluble phase. Annu. Rev. Biochem. **67**, 265–306 (1998)

A. McCammon, S.C. Harvey, *Dynamics of Proteins and Nucleic Acids* (Cambridge University Press, 1987)

B. Mechler, T.H. Rabbitts, Membrane-bound ribosomes of myeloma cells. IV mRNA complexity of free and membrane-bound polysomes. J. Cell Biol. **88**, 29–36 (1981)

B. Mechler, P. Vassalli, Membrane-bound ribosomes of myeloma cells. III The role of the messenger RNA and the nascent polypeptide chain in the binding of ribosomes to membranes. J. Cell Biol. **67**, 25–37 (1975a)

B. Mechler, P. Vassalli, Membrane-bound ribosomes of myeloma cells. II Kinetic studies on the entry of newly made ribosomal subunits into the free and the membrane-bound ribosomal particles. J. Cell Biol. **67**, 16–24 (1975b)

I. Mellman, K. Simons, The Golgi complex: in vitro veritas? Cell **68**, 829–840 (1992)

D.I. Meyer, E. Krause, B. Dobberstein, Secretory protein translocation across membranes-the role of the 'docking protein'. Nature 297, 647–650 (1982)

A. Mezzacasa, A. Helenius, The transitional ER defines a boundary for quality control in the secretion of tsO45 VSV glycoprotein. Traffic 3, 833–849 (2002)

K. Michelsen, V. Schmid, J. Metz, K. Heusser, U. Liebel, T. Schwede, A. Spang, B. Schwappach, Novel cargo-binding site in the beta and delta subunits of coatomer. J. Cell Biol. **179**, 209–217 (2007)

C.T. Mierke, Phagocytized beads reduce the $\alpha5\beta1$ integrin facilitated invasiveness of cancer cells by regulating cellular stiffness. Cell Biochem. Biophys. 66(3), 599–622 (2013)

E.A. Miller, T.H. Beilharz, P.N. Malkus, M.C.S. Lee, S. Hamamoto, L. Orci, R. Schekman, Multiple cargo binding sites on the COPII subunit Sec24p ensure capture of diverse membrane proteins into transport vesicles. Cell 114, 497–509 (2003)

E. Miller, B. Antonny, S. Hamamoto, R. Schekman, Cargo selection into COPII vesicles is driven by the Sec24p subunit. EMBO J. 21, 6105–6113 (2002)

S. Miserey-Lenkei, G. Chalancon, S. Bardin, E. Formstecher, B. Goud, A. Echard, Rab and actomyosin-dependent fission of transport vesicles at the Golgi complex. Nat. Cell Biol. **12**, 645–654 (2010)

K. Mitra, J. Frank, Ribosome dynamics: insights from atomic structure modeling into cryo-electron microscopy maps. Annu. Rev. Biophys. Biomol. Struct. **35**, 299–317 (2006)

E. Mizuno-Yamasaki, F.P.N. Rivera-Molina, GTPase networks in membrane traffic. Annu. Rev. Biochem. **81**, 637–659 (2012)

J. Moelleken, J. Malsam, M.J. Betts et al., Differential localization of coatomer complex isoforms within the Golgi apparatus. Proc. Natl. Acad. Sci. USA **104**, 4425–4430 (2007)

R.K.M. Mofrad, R.D. Kamm, *Cytoskeletal Mechanics: Models and Measurements* (CUP, 2006)

S. Mogelsvang, N. Gomez-Ospina, J. Soderholm, B.S. Glick, L.A. Staehelin, Tomographic evidence for continuous turnover of Golgi cisternae in Pichia pastoris. Mol. Biol Cell **14**, 2277–2291 (2003)

S. Mogelsvang, B.J. Marsh, M.S. Ladinsky, K.E. Howell, Predicting function from structure: 3D structure studies of the mammalian Golgi complex. Traffic 5, 338–345 (2004)

H.H. Mollenhauer, D.J. Morré, Perspectives on Golgi apparatus form and function. J. Electron. Microsc. Tech. **17**, 2–14 (1991)

H.H. Mollenhauer, W.G. Whaley, An observation on the functioning of the Golgi apparatus. J. Cell Biol. **17**, 222–225 (1963)

K. Moriya, K. Nagatoshi, Y. Noriyasu, T. Okamura, E. Takamitsu, T. Suzuki, T. Utsumi, Protein N-myristoylation plays a critical role in the endoplasmic reticulum morphological change induced by overexpression of protein Lunapark, an integral membrane protein of the endoplasmic reticulum. PLoS One 8(11), e78235 (2013)

D.C. Morse, Viscoelasticity of concentrated isotropic solutions of semiflexible polymers. 2. Linear response. Macromolecules **31**, 7044–7067 (1998)

E. Mossessova, L.C. Bickford, J. Goldberg, SNARE selectivity of the COPII coat. Cell **114**, 483–495 (2003)

K. Mowbrey, J.B. Dacks, Evolution and diversity of the Golgi body. FEBS Lett. **583**, 3738–3745 (2009)

M.M. Mueckler, H.C. Pitot, Structure and function of rat liver polysome populations. I Complexity, frequency distribution, and degree of uniqueness of free and membrane-bound polysomal polyadenylate-containing RNA populations. J. Cell Biol. **90**, 495–506 (1981)

M.M. Mueckler, H.C. Pitot, Structure and function of rat liver polysome populations. II Characterization of polyadenylate-containing mRNA associated with subpopulations of membrane-bound particles. J. Cell Biol. **94**, 297–307 (1982)

S. Munro, The golgin coiled-coil proteins of the Golgi apparatus. Cold Spring Harb. Perspect. Biol. **3**, a005256 (2011)

M.D. Myers, G.S. Payne, Clathrin, adaptors and disease: insights from the yeast *Saccharomyces cerevisiae*. Front. Biosci. 18, 862–891 (2013)

N. Nagaraj, J.R. Wisniewski, T. Geiger, J. Cox, M. Kircher, J. Kelso, S. Paabo, M. Mann, Deep proteome and transcriptome mapping of a human cancer cell line. Mol. Syst. Biol. **7**, 548 (2011)

J.F. Nagle, S. Tristram-Nagle, Structure of lipid bilayers. Biochim. Biophys. Acta-Rev. Biomembr. **1469**, 159–195 (2000)

H. Nakanishi, Y. Suda, A.M. Neiman, Erv14 family cargo receptors are necessary for ER exit during sporulation in *Saccharomyces cerevisiae*. J. Cell Sci. **120**, 908–916 (2007)

A. Nebenführ, C. Ritzenthaler, D.G. Robinson, Brefeldin A: deciphering an enigmatic inhibitor of secretion. Plant Physiol. **130**, 1102–1108 (2002)

C.V. Nicchitta, R.S. Lerner, S.B. Stephens, R.D. Dodd, B. Pyhtila, Pathways for compartmentalizing protein synthesis in eukaryotic cells: the template-partitioning model. Biochem. Cell Biol. **83**, 687–695 (2005)

W.C. Nichols, U. Seligsohn, A. Zivelin et al., Mutations in the ER-Golgi intermediate compartment protein ERGIC-53 cause combined deficiency of coagulation factors V and VIII. Cell 93, 61–70 (1998)

A.V. Nikonov, H.P. Hauri, B. Lauring, G. Kreibich, Climp- 63-mediated binding of microtubules to the ER affects the lateral mobility of translocon complexes. J. Cell Sci. 120(Pt 13), 2248–2258 (2007)

I.M. Nilsson, G. von Heijne, Determination of the distance between the oligosaccharyltransferase active site and the endoplasmic reticulum membrane. J. Biol. Chem. **268**(8), 5798–5801 (1993)

T. Nilsson, M. Pypaert, M.H. Hoe, P. Slusarewicz, E.G. Berger, G. Warren, Overlapping distribution of two glycosyltransferases in the Golgi apparatus of HeLa cells. J. Cell Biol. **120**, 5–13 (1993)

J. Nixon-Abell, C.J. Obara, A.V. Weigel et al., Increased spatiotemporal resolution reveals highly dynamic dense tubular matrices in the peripheral ER. Science 354, aaf3928 (2016)

H.F. Noller, A. Baucom, Structure of the 70 S ribosome: implications for movement. Biochem. Soc. Trans. 30, 1159–1161 (2002)

T.R. Noriega, J. Chen, P. Walter, J.D. Puglisi, Real-time observation of signal recognition particle binding to actively translating ribosomes. Elife **3**, e04418 (2014)

K. Norregaard, L. Jauffred, K. Berg-Sørensen, L.B. Oddershede, Optical manipulation of single molecules in the living cell. Phys. Chem. Chem. Phys. **16**, 12614–12624 (2014)

L. Orci, M. Amherdt, M. Ravazzola, A. Perrelet, J.E. Rothman, Exclusion of Golgi residents from transport vesicles budding from Golgi cisternae in intact cells. J. Cell Biol. **150**, 1263–1270 (2000)

L. Orci, M. Ravazzola, A. Volchuk, T. Engel, M. Gmachl, M. Amherdt, A. Perrelet, T.H. Söllner, J.E. Rothman, Anterograde flow of cargo across the Golgi stack potentially mediated via bidirectional "percolating" COPI vesicles. Proc. Natl. Acad. Sci. USA 97, 10400–10405 (2000b)

L. Orci, M. Stamnes, M. Ravazzola, M. Amherdt, A. Perrelet, T.H. Söllner, J.E. Rothman, Bidirectional transport by distinct populations of COPI-coated vesicles. Cell **90**, 335–349 (1997)

G. Orso, D. Pendin, S. Liu, J. Tosetto, T.J. Moss, J.E. Faust, M. Micaroni, A. Egorova, A. Martinuzzi, J.A. McNew, A. Daga, Homotypic fusion of ER membranes requires the dynamin-like GTPase atlastin. Nature **460**(7258), 978–983 (2009)

S. Pagant, A. Wu, S. Edwards, F. Diehl, E.A. Miller, Sec24 is a coincidence detector that simultaneously binds two signals to drive ER export. Curr. Biol. **25**, 403–412 (2015)

I.M. Palacios, D.S. Johnston, Getting the message across: the intracellular localization of mRNAs in higher eukaryotes. Annu. Rev. Cell Dev. Biol. **17**, 569–614 (2001)

G.E. Palade, A small particulate component of the cytoplasm. J. Biophys. Biochem. Cytol. **1**, 59–68 (1955)

E. Papanikou, B.S. Glick, The yeast Golgi apparatus: insights and mysteries. FEBS Lett. **583**, 3746–3751 (2009)

S.H. Park, P.P. Zhu, R.L. Parker, C. Blackstone, Hereditary spastic paraplegia proteins REEP1, spastin, and atlastin-1 coordinate microtubule interactions with the tubular ER network. J. Clin. Invest. 120(4), 1097–1110 (2010)

H.T. Parsons, K. Christiansen, B. Knierim et al., Isolation and proteomic characterization of the Arabidopsis Golgi defines functional and novel components involved in plant cell wall biosynthesis. Plant Physiol. **159**, 12–26 (2012)

G.H. Patterson, K. Hirschberg, R.S. Polishchuk, D. Gerlich, R.D. Phair, J. Lippincott-Schwartz, Transport through the Golgi apparatus by rapid partitioning within a two-phase membrane system. Cell **133**, 1055–1067 (2008)

H.R.B. Pelham, Evidence that luminal ER proteins are sorted from secreted proteins in a post-ER compartment. EMBO J. **7**, 913–918 (1988)

H.R.B. Pelham, Getting through the Golgi complex. Trends Cell Biol. **8**, 45–49 (1998)

S.R. Pfeffer, Entry at the *trans*-face of the Golgi. Cold Spring Harb. Perspect. Biol. **3**, a005272 (2011)

S.R. Pfeffer, How the Golgi works: a cisternal progenitor model. Proc Natl Acad Sci USA **107**, 19614–19618 (2010)

R. Philips, Physical biology of the cell Garland (2013)

R.S. Polishchuk, E.V. Polishchuk, P. Marra, S. Alberti, R. Buccione, A. Luini, A.A. Mironov, Correlative light-electron microscopy reveals the tubular-saccular ultrastructure of carriers operating between the Golgi apparatus and plasma membrane. J. Cell Biol. **148**, 45–58 (2000)

V. Popoff, F. Adolf, B. Brügger, F. Wieland, COPI budding within the Golgi stack. Cold Spring Harb. Perspect. Biol. **3**, a005231 (2011)

M.D. Potter, C.V. Nicchitta, Regulation of ribosome detachment from the mammalian endoplasmic reticulum membrane. J. Biol. Chem. **275**, 33828–33835 (2000)
M.D. Potter, R.M. Seiser, C.V. Nicchitta, Ribosome exchange revisited: a mechanism for translation-coupled ribosome detachment from the ER membrane. Trends Cell Biol. **11**, 112–115 (2001)
J. Powers, C. Barlowe, Erv14p directs a transmembrane secretory protein into COPII-coated transport vesicles. Mol. Biol. Cell **13**, 880–891 (2002)
J. Powers, C. Barlowe, Transport of axl2p depends on erv14p, an ER-vesicle protein related to the Drosophila cornichon gene product. J. Cell Biol. **142**, 1209–1222 (1998)
D. Preuss, J. Mulholland, A. Franzusoff, N. Segev, D. Botstein, Characterization of the Saccharomyces Golgi complex through the cell cycle by immunoelectron microscopy. Mol. Biol. Cell **3**, 789–803 (1992)
M. Puhka, H. Vihinen, M. Joensuu, E. Jokitalo, Endoplasmic reticulum remains continuous and undergoes sheet-to-tubule transformation during cell division in mammalian cells. J. Cell Biol. **179**(5), 895–909 (2007)
T. Pulvirenti, M. Giannotta, M. Capestrano et al., A traffic-activated Golgi-based signalling circuit coordinates the secretory pathway Nat. Cell Biol. **10**, 912–922 (2008)
B. Pyhtila, T. Zheng, P.J. Lager, J.D. Keene, M.C. Reedy, C.V. Nicchitta, Signal sequence- and translation-independent mRNA localization to the endoplasmic reticulum. RNA **14**, 445–453 (2008)
C. Rabouille, J. Klumperman, Opinion: the maturing role of COPI vesicles in intra-Golgi transport. Nat. Rev. Mol. Cell Biol. 6, 812–817 (2005)
C. Rabouille, N. Hui, F. Hunte, R. Kieckbusch, E.G. Berger, G. Warren, T. Nilsson, Mapping the distribution of Golgi enzymes involved in the construction of complex oligosaccharides. J. Cell Sci. **108**, 1617–1627 (1995)
C. Raiborg, E.M. Wenzel, N.M. Pedersen et al., Repeated ER-endosome contacts promote endosome translocation and neurite outgrowth. Nature **520**, 234–238 (2015)
Rambourg A and Clermont Y, *Three-dimensional Structure of the Golgi Apparatus in Mammalian Cells*, ed. by E.G. Berger, J. Roth (The Golgi Apparatus Birkhäuser Verlag Basel, 1997), pp. 37–61
D.J. Ramrath, L. Lancaster, T. Sprink, T. Mielke, J. Loerke, H.F. Noller, C.M. Spahn, Visualization of two transfer RNAs trapped in transit during elongation factor G-mediated translocation. Proc. Natl. Acad. Sci. U S A **110**, 20964–20969 (2013)
T.A. Rapoport, Protein translocation across the eukaryotic endoplasmic reticulum and bacterial plasma membranes. Nature **450**(7170), 663–669 (2007)
I. Raykhel, H. Alanen, K. Salo, J. Jurvansuu, V.D. Nguyen, M. Latva-Ranta, L. Ruddock, A molecular specificity code for the three mammalian KDEL receptors. J. Cell Biol. **179**, 1193–1204 (2007)
D.W. Reid, C.V. Nicchitta, Primary role for endoplasmic reticulum-bound ribosomes in cellular translation identified by ribosome profiling. J. Biol. Chem. **287**, 5518–5527 (2012)
D.W. Reid, C.V. Nicchitta, Diversity and selectivity in mRNA translation on the endoplasmic reticulum. Nat. Rev. Mol. Cell Biol. **16**(4), 221–231 (2015)
D.W. Reid, Q. Chen, A.S. Tay, S. Shenolikar, C.V. Nicchitta, The unfolded protein response triggers selective mRNA release from the endoplasmic reticulum. Cell **158**, 1362–1374 (2014)
Y. Ren, C.K. Yip, A. Tripathi, D. Huie, P.D. Jeffrey, T. Walz, F.M. Hughson, A structure-based mechanism for vesicle capture by the multisubunit tethering complex Dsl1. Cell 139, 1119–1129 (2009)
B.C. Richardson, C.M. McDonold, J.C. Fromme, The Sec7 Arf-GEF is recruited to the *trans*-Golgi network by positive feedback. Dev. Cell 22, 799–810 (2012)
J. Rink, E. Ghigo, Y. Kalaidzidis, M. Zerial, Rab conversion as a mechanism of progression from early to late endosomes. Cell **122**, 735–749 (2005)
R. Rizzo, S. Parashuraman, P. Mirabelli, C. Puri, J. Lucocq, A. Luini, The dynamics of engineered resident proteins in the mammalian Golgi complex relies on cisternal maturation. J. Cell Biol. **201**, 1027–1036 (2013)
K.J. Roberg, M. Crotwell, P. Espenshade, R. Gimeno, C.A. Kaiser, LST1 is a SEC24 homologue used for selective export of the plasma membrane ATPase from the endoplasmic reticulum. J. Cell Biol. **145**, 659–672 (1999)
N.C. Rockwell, D.J. Krysan, T. Komiyama, R.S. Fuller, Precursor processing by Kex2/furin proteases. Chem. Rev. 102, 4525–4548 (2002)
W. Römer, L.L. Pontani, B. Sorre et al., Actin dynamics drive membrane reorganization and scission in clathrin-independent endocytosis. Cell **140**, 540–553 (2010)
D. Ron, P. Walter, Signal integration in the endoplasmic reticulum unfolded protein response. Nat. Rev. Mol. Cell Biol. **8**(7), 519–529 (2007)
J.E. Rothman, The Golgi apparatus: two organelles in tandem. Science **213**, 1212–1219 (1981)
J.E. Rothman, F.T. Wieland, Protein sorting by transport vesicles. Science **272**, 227–234 (1996)
B.D. Roussel, A.J. Kruppa, E. Miranda, D.C. Crowther, D.A. Lomas, S.J. Marciniak, Endoplasmic reticulum dysfunction in neurological disease. Lancet Neurol. **12**, 105–118 (2013)
M.P. Rout, G. Blobel, J.D. Aitchison, A distinct nuclear import pathway used by ribosomal proteins. Cell **89**, 715–725 (1997)
M. Rubinstein, R.H. Colby, *Polymer Physics* (OUP, 2003)

E. Ruiz-May, S.J. Kim, F. Brandizzi, J.K. Rose, The secreted plant N-glycoproteome and associated secretory pathways. Front. Plant Sci. **3**, 117 (2012)

K. Saito, M. Chen, F. Bard, S. Chen, H. Zhou, D. Woodley, R. Polischuk, R. Schekman, V. Malhotra, TANGO1 facilitates cargo loading at endoplasmic reticulum exit sites. Cell 136, 891–902 (2009)

K. Saito, K. Yamashiro, Y. Ichikawa, P. Erlmann, K. Kontani, V. Malhotra, T. Katada, cTAGE5 mediates collagen secretion through interaction with TANGO1 at endoplasmic reticulum exit sites. Mol. Biol. Cell **22**, 2301–2308 (2011)

K.Y. Sanbonmatsu, Alignment/misalignment hypothesis for tRNA selection by the ribosome. Biochimie **88**, 1075–1089 (2006a)

K.Y. Sanbonmatsu, Energy landscape of the ribosomal decoding center. Biochimie **88**, 1053–1059 (2006b)

K.Y. Sanbonmatsu, Computational studies of molecular machines: the ribosome. Curr. Opin. Struct. Biol. **22**, 168–174 (2012)

K.Y. Sanbonmatsu, Flipping through the genetic code: new developments in discrimination between cognate and near-cognate tRNAs and the effect of antibiotics. J. Mol. Biol. **426**, 3197–3200 (2014)

K.Y. Sanbonmatsu, S. Joseph, Understanding discrimination by the ribosome: stability testing and groove measurement of codon-anticodon pairs. J. Mol. Biol. **328**, 33–47 (2003)

K.Y. Sanbonmatsu, C.S. Tung, Large-scale simulations of the ribosome: a new landmark in computational biology. J. Phys.: Conf. Ser. **46**, 334–342 (2006)

K.Y. Sanbonmatsu, C.S. Tung, High performance computing in biology: multimillion atom simulations of nanoscale systems. J. Struct. Biol. **157**, 470–480 (2007)

K.Y. Sanbonmatsu, S. Joseph, C.S. Tung, Simulating movement of tRNA into the ribosome during decoding. Proc. Natl. Acad. Sci. U S A **102**, 15854–15859 (2005)

T. Sandmann, J.M. Herrmann, J. Dengjel, H. Schwarz, A. Spang, Suppression of coatomer mutants by a new protein family with COPI and COPII binding motifs in *Saccharomyces cerevisiae*. Mol. Biol. Cell **14**, 3097–3113 (2003)

S. Sannino, T. Anelli, M. Cortini, S. Masui, M. Degano, C. Fagioli, K. Inaba, R. Sitia, Progressive quality control of secretory proteins in the early secretory compartment by ERp44. J. Cell Sci. **127**, 4260–4269 (2014)

K. Sato, A. Nakano, Mechanisms of COPII vesicle formation and protein sorting. FEBS Lett. **581**, 2076–2082 (2007)

K. Sato, M. Sato, A. Nakano, Rer1p as common machinery for the endoplasmic reticulum localization of membrane proteins. Proc. Natl. Acad. Sci. USA **94**, 9693–9698 (1997)

K. Sato, M. Sato, A. Nakano, Rer1p, a retrieval receptor for ER membrane proteins, recognizes transmembrane domains in multiple modes. Mol. Biol. Cell **14**, 3605–3616 (2003)

M. Sato, K. Sato, A. Nakano, Endoplasmic reticulum quality control of unassembled iron transporter depends on Rer1p-mediated retrieval from the golgi. Mol. Biol. Cell **15**, 1417–1424 (2004)

P. Satpute-Krishnan, M. Ajinkya, S. Bhat, E. Itakura, R.S. Hegde, J. Lippincott-Schwartz, ER stress-induced clearance of misfolded GPI-anchored proteins via the secretory pathway. Cell **158**, 522–533 (2014)

A.A. Scheel, H.R. Pelham, Purification and characterization of the human KDEL receptor. Biochemistry **35**, 10203–10209 (1996)

R. Schindler, C. Itin, M. Zerial, F. Lottspeich, H.P. Hauri, ERGIC-53, a membrane protein of the ER-Golgi intermediate compartment, carries an ER retention motif. Eur. J. Cell Biol. **61**, 1–9 (1993)

W.K. Schmidt, H.P. Moore, Ionic milieu controls the compartment-specific activation of pro-opiomelanocortin processing in AtT-20 cells Mol Biol Cell 6, 1271–1285 (1995)

S. Schuck, W.A. Prinz, K.S. Thorn, C. Voss, P. Walter, Membrane expansion alleviates endoplasmic reticulum stress independently of the unfolded protein response. J. Cell Biol. **187**(4), 525–536 (2009)

A.D. Schuyler, G.S. Chirikjian, Efficient determination of low-frequency normal modes of large protein structures by cluster-NMA. J. Mol. Gr. **24**, 46–58 (2005)

A.D. Schuyler, R.L. Jernigan, P.K. Qasba, B. Ramakrishnan, G.S. Chirikjian, Iterative cluster-NMA: a tool for generating conformational transitions in proteins. Proteins **74**, 760–776 (2009)

D.S. Schwarz, M.D. Blower, The calcium-dependent ribonuclease XendoU promotes ER network formation through local RNA degradation. J. Cell Biol. **207**(1), 41–57 (2014)

D.S. Schwarz, M.D. Blower, The endoplasmic reticulum: structure, function and response to cellular signaling. Cell Mol. Life Sci. **73**, 79–94 (2016)

A. Schweizer, J.A. Fransen, T. Bächi, L. Ginsel, H.P. Hauri, Identification, by a monoclonal antibody, of a 53-kD protein associated with a tubulo-vesicular compartment at the *cis*-side of the Golgi apparatus. J. Cell Biol. **107**, 1643–1653 (1988)

C.W. Seidel, L.J. Peck, Purification of a calcium dependent ribonuclease from Xenopus laevis. Nucleic Acids Res. **22** (8), 1456–1462 (1994)

R.M. Seiser, C.V. Nicchitta, The fate of membrane-bound ribosomes following the termination of protein synthesis. J. Biol. Chem. **275**, 33820–33827 (2000)

J.C. Semenza, K.G. Hardwick, N. Dean, H.R. Pelham, ERD2, a yeast gene required for the receptor-mediated retrieval of luminal ER proteins from the secretory pathway. Cell 61, 1349–1357 (1990)

T. Senda, T. Yoshinaga-Hirabayashi, Intermembrane bridges within membrane organelles revealed by quick-freeze deep-etch electron microscopy. Anat. Rec. **251**(3), 339–345 (1998)

H.J. Sharpe, T.J. Stevens, S. Munro, A comprehensive comparison of transmembrane domains reveals organelle-specific properties. Cell **142**, 158–169 (2010)

T. Shemesh, R.W. Klemm, F.B. Romano, S. Wang, J. Vaughan, X. Zhuang, H. Tukachinsky, M.M. Kozlov, T.A. Rapoport, A model for generation and interconversion of ER morphologies. PNAS **111**, E5243–E5251 (2014)

Y. Shibata, J. Hu, M.M. Kozlov, T.A. Rapoport, Mechanisms shaping the membranes of cellular organelles. Annu. Rev. Cell Dev. Biol. **25**, 329–354 (2009)

Y. Shibata, T. Shemesh, W.A. Prinz, A.F. Palazzo, M.M. Kozlov, T.A. Rapoport, Mechanisms determining the morphology of the peripheral ER. Cell **143**(5), 774–788 (2010)

Y. Shibata, G.K. Voeltz, T.A. Rapoport, Rough sheets and smooth tubules. Cell **126**(3), 435–439 (2006)

Y. Shibata, C. Voss, J.M. Rist, J. Hu, T.A. Rapoport, W.A. Prinz, G.K. Voeltz, The reticulon and DP1/Yop1p proteins form immobile oligomers in the tubular endoplasmic reticulum. J. Biol. Chem. 283(27), 18892–18904 (2008)

A. Shibuya, N. Margulis, R. Christiano, T.C. Walther, C. Barlowe, The Erv41-Erv46 complex serves as a retrograde receptor to retrieve escaped ER proteins. J. Cell Biol. **208**, 197–209 (2015)

Y. Shimoni, T. Kurihara, M. Ravazzola, M. Amherdt, L. Orci, R. Schekman, Lst1p and Sec24p cooperate in sorting of the plasma membrane ATPase into COPII vesicles in *Saccharomyces cerevisiae*. J. Cell Biol. **151**, 973–984 (2000)

P. Siekevitz, G.E. Palade, A cytochemical study on the pancreas of the guinea pig. 5 In vivo incorporation of leucine-1-C14 into the chymotrypsinogen of various cell fractions. J. Biophys. Biochem. Cytol. 7, 619–630 (1960)

R. Sitia, I. Braakman, Quality control in the endoplasmic reticulum protein factory. Nature **426**, 891–894 (2003)

R.D. Smith, V.V. Lupashin, Role of the conserved oligomeric Golgi (COG) complex in protein glycosylation. Carbohydr. Res. 343, 2024–2031 (2008)

B. Sönnichsen, M. Lowe, T. Levine, E. Jämsä, B. Dirac-Svejstrup, G. Warren, A role for giantin in docking COPI vesicles to Golgi membranes. J. Cell. Biol. **140**, 1013–1021 (1998)

A. Spang, The DSL1 complex: the smallest but not the least CATCHR. Traffic 13, 908–913 (2012)

A. Spang, J.M. Herrmann, S. Hamamoto, R. Schekman, The ADP ribosylation factor-nucleotide exchange factors Gea1p and Gea2p have overlapping, but not redundant functions in retrograde transport from the Golgi to the endoplasmic reticulum. Mol. Biol. Cell 12, 1035–1045 (2001)

E.D. Spear, D.T.W. Ng, Stress tolerance of misfolded carboxypeptidase Y requires maintenance of protein trafficking and degradative pathways. Mol. Biol. Cell **14**, 2756–2767 (2003)

P.C. Spiegel, D.N. Ermolenko, H.F. Noller, Elongation factor G stabilizes the hybrid-state conformation of the 70S ribosome. RNA 13, 1473–1482 (2007)

L.A. Staehelin, B.H. Kang, Nanoscale architecture of endoplasmic reticulum export sites and of Golgi membranes as determined by electron tomography. Plant Physiol. **147**, 1454–1468 (2008)

L.A. Staehelin, The plant ER: a dynamic organelle composed of a large number of discrete functional domains. Plant J. **11**(6), 1151–1165 (1997)

S.M. Stagg, C. Gürkan, D.M. Fowler, P. LaPointe, T.R. Foss, C.S. Potter, B. Carragher, W.E. Balch, Structure of the Sec13/31 COPII coat cage. Nature 439, 234–238 (2006)

D. Stalder, B. Antonny, Arf GTPase regulation through cascade mechanisms and positive feedback loops. FEBS Lett. **587**, 2028–2035 (2013)

M.A. Stamnes, M.W. Craighead, M.H. Hoe, N. Lampen, S. Geromanos, P. Tempst, J.E. Rothman, An integral membrane component of coatomer-coated transport vesicles defines a family of proteins involved in budding. Proc. Natl. Acad. Sci. USA **92**, 8011–8015 (1995)

P. Stanley, Golgi glycosylation. Cold Spring Harb. Perspect. Biol. **3**, a005199 (2011)

H. Stenmark, Rab GTPases as coordinators of vesicle traffic. Nat. Rev. Mol. Cell. Biol. **10**(8), 513–525 (2009)

S.B. Stephens, R.D. Dodd, R.S. Lerner, B.M. Pyhtila, C.V. Nicchitta, Analysis of mRNA partitioning between the cytosol and endoplasmic reticulum compartments of mammalian cells. Methods Mol. Biol. **419**, 197–214 (2008)

B. Storrie, M. Micaroni, G.P. Morgan, N. Jones, J.A. Kamykowski, N. Wilkins, T.H. Pan, B.J. Marsh, Electron tomography reveals Rab6 is essential to the trafficking of *trans*-Golgi clathrin and COPI-coated vesicles and the maintenance of Golgi cisternal number. Traffic 13, 727–744 (2012)

J.R.P.M. Strating, T.G.M. Hafmans, G.J.M. Martens, COP-binding sites in p24δ2 are necessary for proper secretory cargo biosynthesis. Int. J. Biochem. Cell Biol. **41**, 1619–1627 (2009)

R.J. Suckling, P.P. Poon, S.M. Travis, I.V. Majoul, F.M. Hughson, P.R. Evans, R. Duden, D.J. Owen, Structural basis for the binding of tryptophan-based motifs by δ-COP. Proc. Natl. Acad. Sci. USA **112**, 14242–14247 (2015)

K.M. Sullivan, W.B. Busa, K.L. Wilson, Calcium mobilization is required for nuclear vesicle fusion in vitro: implications for membrane traffic and IP3 receptor function. Cell 73(7), 1411–1422 (1993)

E.S. Suvorova, R. Duden, V.V. Lupashin, The Sec34/Sec35p complex, a Ypt1p effector required for retrograde intra-Golgi trafficking, interacts with Golgi SNAREs and COPI vesicle coat proteins. J. Cell Biol. 157, 631–643 (2002)

I. Tabas, D. Ron, Integrating the mechanisms of apoptosis induced by endoplasmic reticulum stress. Nat. Cell Biol. **13** (3), 184–190 (2011)

H. Takeshima, S. Komazaki, M. Nishi, M. Iino, K. Kangawa, Junctophilins: a novel family of junctional membrane complex proteins. Mol. Cell 6(1), 11–22 (2000)

S. Takyar, R.P. Hickerson, H.F. Noller, MRNA helicase activity of the ribosome. Cell **120**, 49–58 (2005)

F. Tama, Normal mode analysis with simplified models to investigate the global dynamcis of biological systems. Protein Pept. Lett. **10**, 119–132 (2003)

F. Tama, Y.H. Sanejouand, Conformational change of proteins arising from normal mode calculations. Protein Eng. **14**, 1–6 (2001)

F. Tama, M. Valle, J. Frank, C.L. Brooks, III Dynamic reorganization of the functionally active ribosome explored by normal mode analysis and cryo-electron microscopy. Proc. Natl. Acad. Sci. USA **100**, 9319–9323 (2003)

Y. Tanaka-Takiguchi, T. Itoh, K. Tsujita, S. Yamada, M. Yanagisawa, K. Fujiwara, A. Yamamoto, M. Ichikawa, K. Takiguchi, Physicochemical analysis from real-time imaging of liposome tubulation reveals the characteristics ofindividual F-BAR doman proteins. Langmuir **29**, 328–336 (2012)

M. Tassieri, Dynamics of semiflexible polymer solutions in the highly entangled regime. Phys. Rev. Lett. **101**, 198301 (2008)

M. Terasaki, T. Shemesh, N. Kasthuri et al., Stacked endoplasmic reticulum sheets are connected by helicoidal membrane motifs. Cell **154**(2), 285–296 (2013)

F. Thor, M. Gautschi, R. Geiger, A. Helenius, Bulk flow revisited: transport of a soluble protein in the secretory pathway. Traffic **10**, 1819–1830 (2009)

M.M. Tirion, Large amplitude elastic motions in proteins from a single-parameter, atomic analysis. Phys. Rev. Lett. **77**, 1905–1908 (1996)

A. Trucco, R.S. Polishchuk, O. Martella et al., Secretory traffic triggers the formation of tubular continuities across Golgi sub-compartments. Nat. Cell Biol. **6**, 1071–1081 (2004)

J. Trylska, V. Tozzini, J.A. McCammon, Exploring global motions and correlations in the ribosome. Biophys. J. **89**, 1455–1463 (2005)

P.C. Tsai, J.W. Hsu, Y.W. Liu, K.Y. Chen, F.J. Lee, Arl1p regulates spatial membrane organization at the *trans*-Golgi network through interaction with Arf-GEF Gea2p and flippase Drs2p. Proc. Natl. Acad. Sci. USA 110, E668–E677 (2013)

L. Tu, L. Chen, D.K. Banfield A conserved N-terminal arginine-motif in GOLPH3-family proteins mediates binding to coatomer. Traffic 13, 1496–1507 (2012)

L. Tu, W.C.S. Tai, L. Chen, D.K. Banfield, Signal-mediated dynamic retention of glycosyltransferases in the Golgi. Science **321**, 404–407 (2008)

M.D. Turner, H. Plutner, W.E. Balch, A Rab GTPase is required for homotypic assembly of the endoplasmic reticulum. J. Biol. Chem. **272**(21), 13479–13483 (1997)

D. Ungar, Golgi linked protein glycosylation and associated diseases. Semin. Cell Dev. Biol. **20**, 762–769 (2009)

D. Ungar, T. Oka, M. Krieger, F.M. Hughson, Retrograde transport on the COG railway. Trends Cell Biol. **16**, 113–120 (2006)

A. Upadhyaya, M.P. Sheetz, Tension in tubolovesicular networks of golgi and endoplasmic reticulum membranes. Biophys. J. **86**, 2923–2928 (2004)

I. Usov, R. Mezzenga, FiberApp: an open-source software for tracking and analyzing polymers, filaments and fibrous objects. Macromolecules 48, 1269–1280 (2015)

R.H. Valdivia, D. Baggott, J.S. Chuang, R. Schekman, The yeast clathrin adaptor protein complex 1 is required for the efficient retention of a subset of late Golgi membrane proteins. Dev. Cell 2, 283–294 (2002)

C. Valkova, M. Albrizio, I.V. Röder, M. Schwake, R. Betto, R. Rudolf, C. Kaether, Sorting receptor Rer1 controls surface expression of muscle acetylcholine receptors by ER retention of unassembled alpha-subunits. Proc. Natl. Acad. Sci. USA. **108**, 621–625 (2011)

F. Valle, M. Favre, Rios P. De Los, A. Rosa, G. Dietler, Scaling exponents and probability distributions of DNA end-to-end distance. Phys. Rev. Lett. **95**, 1–4 (2005)

M. Valle, J. Sengupta, N.K. Swami, R.A. Grassucci, N. Burkhardt, K.H. Nierhaus, R.K. Agrawal, J. Frank, Cryo-EM reveals an active role for aminoacyl-tRNA in the accommodation process. EMBO J. **21**, 3557–3567 (2002)

A.M. Valm, S. Cohen, W.R. Legant, J. Melunis, U. Hershberg, E. Wait, A.R. Cohen, M.W. Davidson, E. Betzig, J. Lippincott-Schwartz, Applying systems-level spectral imaging and analysis to reveal the organell interactome. Nature **546**, 162–167 (2017)

M.S. VanLoock, R.K. Agrawal, I.S. Gabashvili, L. Qi, J. Frank, S.C. Harvey, Movement of the decoding region of the 16 S ribosomal RNA accompanies tRNA translocation. J. Mol. Biol. 304, 507–515 (2000)

F. Vanzi, Y. Takagi, H. Shuman, B.S. Cooperman, Y.E. Goldman, Mechanical studies of single ribosome/mRNA complexes. Biophys. J. **89**, 1909–1919 (2005)

F. Vanzi, S. Vladimirov, C.R. Knudsen, Y.E. Goldman, B.S. Cooperman, Protein synthesis by single ribosomes. RNA **9**, 1174–1179 (2003)

S. Vavassori, M. Cortini, S. Masui et al., A pH-regulated quality control cycle for surveillance of secretory protein assembly. Mol. Cell **50**, 783–792 (2013)

A. Velasco, L. Hendricks, K.W. Moremen, D.R.P. Tulsiani, O. Touster, M.G. Farquhar, Cell type-dependent variations in the subcellular distribution of α-mannosidase I and II. J. Cell Biol. **122**, 39–51 (1993)
R. Venditti, T. Scanu, M. Santoro et al., Sedlin controls the ER export of procollagen by regulating the Sar1 cycle. Science 337, 1668–1672 (2012)
C. Viotti, J. Bubeck, Y.D. Stierhof et al., Endocytic and secretory traffic in Arabidopsis merge in the *trans*-Golgi network/early endosome, an independent and highly dynamic organelle. Plant Cell 22, 1344–1357 (2010)
G.K. Voeltz, W.A. Prinz, M. Shibata, J.M. Rist, T.A. Rapoport, A class of membrane proteins shaping the tubular endoplasmic reticulum. Cell **124**, 573–586 (2006)
J. von Blume, A.M. Alleaume, C. Kienzle, A. Carreras-Sureda, M. Valverde, V. Malhotra, Cab45 is required for Ca^{2+}-dependent secretory cargo sorting at the *trans*-Golgi network. J. Cell Biol. 199, 1057–1066 (2012)
I. Wada, D. Rindress, P.H. Cameron, W.J. Ou, J.J. Doherty, D. Louvard, A.W. Bell, D. Dignard, D.Y. Thomas, J. J. Bergeron, SSR alpha and associated calnexin are major calcium binding proteins of the endoplasmic reticulum membrane. J. Biol. Chem. **266**, 19599–19610 (1991)
T.A. Waigh, *The Physics of Living Processes: A Mesoscopic Approach* (Wiley, 2014)
Y. Wakana, J. van Galen, F. Meissner, M. Scarpa, R.S. Polishchuk, M. Mann, V. Malhotra, A new class of carriers that transport selective cargo from the *trans* Golgi network to the cell surface. EMBO J. **31**, 3976–3990 (2012)
P. Walter, G. Blobel, Translocation of proteins across the endoplasmic reticulum. II. Signal recognition protein (SRP) mediates the selective binding to microsomal membranes of in-vitro-assembled polysomes synthesizing secretory protein. J. Cell Biol. 91, 551–556 (1981)
P. Walter, D. Ron, The unfolded protein response: from stress pathway to homeostatic regulation. Science **334**(6059), 1081–1086 (2011)
P. Walter, I. Ibrahimi, G. Blobel, Translocation of proteins across the endoplasmic reticulum. I Signal recognition protein (SRP) binds to in-vitro-assembled polysomes synthesizing secretory protein. J. Cell Biol. 91, 545–550 (1981)
S. Wang, D.T.W. Ng, Evasion of endoplasmic reticulum surveillance makes Wsc1p an obligate substrate of Golgi quality control. Mol. Biol. Cell **21**, 1153–1165 (2010)
S. Wang, F.B. Romano, C.M. Field, T.J. Mitchison, T.A. Rapoport, Multiple mechanisms determine ER network morphology during the cell cycle in Xenopus egg extracts. J. Cell Biol. **203**(5), 801–814 (2013)
Y. Wang, R.L. Jernigan, Comparison of tRNA motions in the free and ribosomal bound structures. Biophys. J. **89**, 3399–3409 (2005)
Y. Wang, H. Qin, R.D. Kudaravalli, S.V. Kirillov, G.T. Dempsey, D. Pan, B.S. Cooperman, Y.E. Goldman, Single-molecule structural dynamics of EF-G–ribosome interaction during translocation. Biochemistry 46, 10767–10775 (2007)
Y. Wang, A.J. Rader, I. Bahar, R.L. Jernigan, Global ribosome motions revealed with elastic network model. J. Struct. Biol. **147**, 302–314 (2004)
J.R. Warner, The economics of ribosome biosynthesis in yeast. Trends Biochem. Sci. **24**, 437–440 (1999)
M.G. Waters, T. Serafini, J.E. Rothman, 'Coatomer': a cytosolic protein complex containing subunits of non-clathrin-coated Golgi transport vesicles. Nature 349, 248–251 (1991)
C.A. Weber, R. Suzuki, V. Schaller, I.S. Aranson, A.R. Bausch, E. Frey, Random bursts determine the dynamics of active filaments. PNAS **112**, 10703–10707 (2015)
S.R. Wente, Gatekeepers of the nucleus. Science **288**, 1374–1377 (2000)
M. West, N. Zurek, A. Hoenger, G.K. Voeltz, A 3D analysis of yeast ER structure reveals how ER domains are organized by membrane curvature. J. Cell Biol. 193(2), 333–346 (2011)
L.M. Westrate, J.E. Lee, W.A. Prinz, G.K. Voeltz, Form follows function: the importance of endoplasmic reticulum shape. Annu. Rev. Biochem. **84**, 791–811 (2015)
F.T. Wieland, M.L. Gleason, T.A. Serafini, J.E. Rothman, The rate of bulk flow from the endoplasmic reticulum to the cell surface. Cell **50**, 289–300 (1987)
C.A. Woolhead, P.J. McCormick, A.E. Johnson, Nascent membrane and secretory proteins differ in FRET-detected folding far inside the ribosome and in their exposure to ribosomal proteins. Cell **116**, 725–736 (2004)
A.A. Woollard, I. Moore, The functions of Rab GTPases in plant membrane traffic. Curr. Opin. Plant Biol. **11**, 610–619 (2008)
M. Wozniak, B. Bola, K. Brownhill, Y.C. Yang, V. Levakova, V.J. Allan, Role of kinesin-1 and cytoplasmic dynein in endoplasmic reticulum movement in VERO cells. J. Cell Sci. **122**, 1979–1989 (2009)
W. Wriggers, R.K. Agrawal, D.L. Drew, A. McCammon, J. Frank, Domain motions of EF-G bound to the 70S ribosome: insights from a handshaking between multi-resolution structures. Biophys. J. 79, 1670–1678 (2000)
Y. Xiang, Y. Wang, New components of the Golgi matrix. Cell Tissue Res. **344**, 365–379 (2011)
A. Yamamoto, M. Ichikawa, Direct measurement of single soft lipid nanotubes: nanoscale information extracted in a noninvasive manner. Phys. Rev. E **86**, 061905 (2012)
K. Yanagitani, Y. Imagawa, T. Iwawaki, A. Hosoda, M. Saito, Y. Kimata, K. Kohno, Cotranslational targeting of XBP1 protein to the membrane promotes cytoplasmic splicing of its own Mrna. Mol. Cell 34(2), 191–200 (2009)

L.W. Yang, X. Liu, C.J. Jursa, M. Holliman, A. Rader, H.A. Karimi, I. Bahar, iGNM: a database of protein functional motions based on Gaussian network model. Bioinformatics 21, 2978–2987 (2005)

L. Yang, G. Song, R.L. Jernigan, How well can we understand large-scale protein motions using normal modes of elastic network models? Biophys. J. **93**, 920–929 (2007)

L. Yang, G. Song, R.L. Jernigan, Protein elastic network models and the ranges of cooperativity. Proc. Natl. Acad. Sci. U S A **106**, 12347–12352 (2009)

T. Yorimitsu, K. Sato, Insights into structural and regulatory roles of Sec16 in COPII vesicle formation at ER exit sites. Mol. Biol. Cell **23**, 2930–2942 (2012)

H. Yoshida, T. Matsui, A. Yamamoto, T. Okada, K. Mori XBP1 mRNA is induced by ATF6 and spliced by IRE1 in response to ER stress to produce a highly active transcription factor. Cell 107(7), 881–891 (2001)

G. Zanetti, S. Prinz, S. Daum, A. Meister, R. Schekman, K. Bacia, J.A.G. Briggs, The structure of the COPII transport-vesicle coat assembled on membranes. eLife **2**, e00951 (2013)

D. Zeuschner, W.J. Geerts, E. van Donselaar, B.M. Humbel, J.W. Slot, A.J. Koster, J. Klumperman, Immuno-electron tomography of ER exit sites reveals the existence of free COPII-coated transport carriers. Nat. Cell Biol. **8**, 377–383 (2006)

G.F. Zhang, L.A. Staehelin, Functional compartmentation of the Golgi apparatus of plant cells: immunocytochemical analysis of high-pressure frozen and freeze-substituted sycamore maple suspension culture cells. Plant Physiol. **99**, 1070–1083 (1992)

C. Zhou, B.D. Slaughter, J.R. Unruh, F. Guo, Z. Yu, K. Mickey, A. Narkar, R.T. Ross, M. McClain, R. Li, Organelle-based aggregation and retention of damaged proteins in asymmetrically dividing cells. Cell **159**, 530–542 (2014)

M.T. Zimmermann, R.L. Jernigan, Elastic network models capture the motions apparent within ensembles of RNA structures. RNA **20**, 792–804 (2014)

M.T. Zimmermann, K. Jia, R.L. Jernigan, Ribosome mechanics informs about mechanism. J. Mol. Biol. 428(5 Pt A), 802–810 (2016)

Mechanical View on Vacuoles

8

Abstract

This chapter describes the structure of a major cell organelle, the vacuole, that plays a key role in the regulation of the size of the cell. Besides its waste storage function, the vacuole is characterized in structure and function, and its contribution to the mechanical properties of the entire cell is discussed. More specific, the connections between the vacuoles and other cytoskeletal structures, such as actomyosin filaments, are presented. As for all organelles, vacuoles are highly specialized subcompartments of the cells that deliver specific functions, and therefore, they add on this subcellular level a higher cellular organization. In the past, what we learned on the function of organelles depended primarily on microscopic phenomena and not so much on the impact of stress on the structural and functional properties of organelles, something that is pointed out here.

8.1 Introduction to Vacuoles

Vacuoles are membrane-confined organelles and represent enclosed compartments filled with mostly water-containing soluble inorganic and organic molecules such as enzymes or engulfed solids. They are built by the fusion of multiple membrane vesicles. The shape or size of vacuoles is not uniform and can be variably adapted to the content. Within the vacuoles, the materials harmful to the cell can be isolated, and hence, they contain often waste products such as unwanted substance and small molecules. Vacuoles maintain the internal hydrostatic pressure (termed turgor) of the entire cells. Moreover, vacuoles can have an acidic luminal pH. A vacuole can grow fast by influx of water. It plays a role in autophagy, balances the biogenesis (the production) and the degradation (the turnover) of substances and cellular structures. In addition, cellular lysis processes are promoted, and misfolded proteins are misfolded. Vacuoles store food and help to digestive and waste management. The function of vacuoles is in the development of a functional circularly system in vertebrates relying in vivo on the vascular lumenization based on longitudinal vacuoles which are extruded and combined from one cell to the next neighboring cell (Kamei et al. 2006). The intracellular vacuoles are built through integrin-dependent and cdc42/Rac1-dependent pinocytic events induced by integrin–extracellular matrix dependent signal transduction processes.

C. T. Mierke, *Cellular Mechanics and Biophysics*, Biological and Medical Physics, Biomedical Engineering, https://doi.org/10.1007/978-3-030-58532-7_8

8.2 Volume Regulation Mechanism in Vacuoles

Cells regulate the size of their compartments due to the alterations in the cell volume, but there exists also a similar-size regulation inside each individual organelle. Yeast vacuoles neither explode nor crumble into a disheveled morphology, implying that the volume of the organellar envelope is matched to the quantity of content. However, it is not yet clearly revealed how the vacuole regulates the adjustment of it size to the amount of content. The increasing content of yeast vacuoles triggers the fusion with other vacuoles and thus improves the volume-to-surface ratio. The polyphosphate synthesis of a dominant compound stored in vacuoles triggers the binding of the chaperone Sec18/NSF to the vacuolar SNAREs, thereby inducing their activation and fusion. The activation of SNAREs is solely possible by lumenal, not by cytosolic polyphosphate (polyP). In the cytosol, the lumenal polyP is precisely regulated to control the activation of SNARE which involves in an essential manner both cyclin-dependent kinase Pho80–Pho85 and R-SNARE Nyv1. These findings indicate that cells can adjust the volume of vacuoles to their level as they provide feedback of the vacuole lumen toward the SNAREs that are located at the cytosolic membrane surface of the organelle.

8.2.1 The Increasing Vacuole Content Activates SNAREs to Enlarge the Volume of the Organelle Volume by Inducing Their Homotypic Fusion

Preserving the membrane tension of an organelle by avoiding its lysis demands agreement between the amount of its content and the volume of the boundary membrane. The maintenance of this communication may be functionally crucial, as the activity of a large number of membrane proteins, such as transporters and channels, may rely on the tension of the membrane in which they are located (Hamill and Martinac 2001; Sukharev and Sachs 2012), and macromolecular overcrowding heavily impacts the complexes and activities of soluble proteins within the lumen of the organelle (Zhou et al. 2008). In some cases, the concentration of macromolecules in the organelle also interferes with their specific interactions. The choice of the surface-to-volume ratio of an organelle ensures that the membrane tension is ideal and that concentrations of the soluble content of the organelle are adequate. The fusion and cleavage of the multiple copies of a compartment offers a fast and efficient route to accomplish this as they are independent of the new membrane synthesis and easily reversible. These mechanisms seem also to be expected to help in the adaptation of organelle volume to cell size, a ratio that is unexpectedly maintained stable under stationary growth conditions, although it can vary substantially when the growth conditions of a cell are altered (Marshall 2012). It seems possible that other mechanisms may also be involved, such as the net inflow and outflow of the membrane or variations in organelle shape. Alterations in organellar size and/or organellar surface-to-volume ratio for Golgi apparatus, vacuoles, mitochondria and endosomes occurred (Banta et al. 1988; Bevis et al. 2002; Bhave et al. 2014; Chan and Marshall 2014; Gaynor et al. 1998; Seeley et al. 2002).

It has been said that the size of multiple organelles or protein compounds tends to be actively regulated (Katsura 1987; Levy and Heald 2010; Ludington et al. 2012), but it is still little knowledge what size control is accomplished at the molecular level. The size of the membrane-closed organelles scales with the cell size at constant growth conditions (Chan and Marshall 2012). When growth conditions are altered, the macromolecular amount and the volume required by an organelle type are subject to modifications both absolute and relative to the cell surface. Since organelles stay intact and do not rupture during such alterations, cells may be able to control the volume of the organellar boundary membrane in relation to the content of organelles. A way to accomplish this is to vary the shape, since any departure from a spherical shape reduces the volume-to-surface ratio. Non-spherical shapes of membranes can be defined by proteins or by variations in the distribution of lipids (Derganc et al. 2013; Frolov et al. 2011; Heald and Cohen-Fix 2014; Rafelski and Marshall 2008). The manipulation of the equilibrium between membrane inflow and outflow may also raise or lower the volume of the organellar envelope. For instance, a mutation that decreases vesicular traffic from vacuoles to endosomes or the Golgi (atg18Δ) elevates the volume of yeast vacuoles, whereas a lowering of membrane inflow (apl5Δ) reduces them (Chan and Marshall 2014; Dove et al. 2004; Efe et al. 2007; Tamura et al. 2013; Zieger and Mayer 2012). Another option is the fusion or fission of the organelle. In specific detail, the fusion of several organelle of the same type leads to an elevation of the volume-to-surface ratio without the requirement of de novo synthesis of the membrane, and the fission takes place inversely. Until now, however, these mechanisms have not been involved in the adaptation of the volume of the organellar envelope to the volume of the content to be encompassed. The lysosome-like vacuoles of yeast cells have been investigated.

Vacuoles and lysosomes represent highly dynamics organelles (Settembre and Ballabio 2014; Weisman 2003). During the starvation and autophagy, lysosomal hydrolases are extensively induced. The resulting increase in macromolecular content is associated with an expansion in the volume of the lysosomal boundary membrane (Yu et al. 2010). During this rise in volume as well as during its inverse, the boundary membrane neither bursts nor crumbles into a disheveled morphology. It seems to keep a well-inflated appearance, pointing out that the volume-to-surface ratio of the organelle envelope is controllable. Thus, the membrane tension and the concentration of lumenal macromolecules, essential for organelle function, can be sustained (Ellis 2001; García-Pérez et al. 1999; Heald and Cohen-Fix 2014; Upadhyaya and Sheetz 2004).

The hypothesis is that the volume of the boundary membrane of vacuoles adjusts to the quantity of content to be entrapped by membrane fusion or fission. Vacuoles represent simple objects that can be employed for the investigation of this issue. Their size and number of copies are affected by an equilibrium of membrane fusion and fission (Baars et al. 2007; Efe et al. 2007; Zieger and Mayer 2012) and by microenvironmental alterations (Peters et al. 2004; Weisman 2003). A decisive advantage is that the content of the vacuole is controlled by a few compounds, polyphosphate (polyP) and the associated basic amino acids (Saito et al. 2005). Polyphosphate represents a polymer that consists of up to hundreds of inorganic phosphate units connected through phospho-anhydride bonds. In specific detail, it is a source of phosphate that is a compound limiting the growth rates in plants, bacteria and fungi. Polyphosphate has been found additionally in mammalian lysosomes; however, its role there is not yet clear (Pisoni and Lindley 1992). Yeast cells can synthesize polyphosphate via a structure termed the vacuolar transporter chaperone (VTC) complex that is placed within the vacuolar membrane and contains the three subunits, Vtc1, Vtc3 and Vtc4 (Hothorn et al. 2009). Vacuolar polyphosphate can account for up to 15% of the yeast cell's dry weight. Since the vacuole makes up less than a third of the yeast cell's volume, whereas it includes virtually all cellular polyphosphate, the polyphosphate can account for over 50% of the dry weight of the vacuole. An alteration of the polyphosphate synthesis largely alters the content of the vacuole.

Vacuoles react to an augmentation of their osmotic pressure by fusion, and in vitro experiments have characterized the stimulation of Rab-GTPase-driven vacuole docking as one of the major mechanisms that cause this alteration (Brett and Merz 2008). The VTC complex has been revealed to be required for vacuole fusion and the priming of SNARE proteins (Müller et al. 2002, 2003). Since the VTC complex has been interacting with parts of the vacuum fusion machine, such as the R-SNARE Nyv1 and the V0 sector of the V-ATPase, the effects of VTC deletion on the fusion were considered as a result of the absence of these interactions (Desfougeres et al. 2016). Since VTC has been revealed as the major polyphosphate polymerase in yeast (Hothorn et al. 2009), a role of VTC in the vacuole fusion has been suggested and it has been probed whether VTC triggers the fusion of vacuoles by alteration of the level of vacuole constituents. To uncover a possible connection between the vacuolar fusion system and the storage of polyphosphate in the lumen of the vacuoles, the availability of structural knowledge about the catalytic center of the VTC and techniques for reliable analysis of the synthesis and translocation of polyphosphate in yeast vacuoles was investigated (Gerasimaite et al. 2014).

8.2.2 The Fusion of Homotypic Vacuoles Requires Polyphosphate Synthesis and Accumulation

It has been questioned whether the polyphosphate synthesis initiates an enhancement of the vacuolar storage space of this compound. Fusing a variety of vacuoles in a cell to a single larger organelle can increased the volume. The vacuoles in yeast cells can be visualized by a vital dye FM4-64 stain (Vida and Emr 1995). Hence, the vacuoles in yeast cells cannot accumulate polyphosphate, since the VTC complexes remained inactive. When yeast cells are cultured in synthetic Hartwell's complete medium, countless small vacuoles can be detected in cells that possess the catalytically inactive vtc4R264A variant or the deletion mutants vtc4Δ and vtc1Δ. Normal and healthy (wild-type) cells possess generally one or two large vacuoles. These findings are contrary to observations that similar to nyv1Δ cells, vtc4Δ cells, which are cultured in yeast extract peptone dextrose medium, contain a single large vacuole (Müller et al. 2001). In order to solve this contradiction, an imbalance of the osmotic pressure can be assumed that is due to the constituents of the yeast extract peptone dextrose medium consisting of the quadruple of the amino acids Arg, Lys and His as the total medium of Hartwell. The three amino acids are able to concentrate in vacuoles to very high concentrations and for yeast extract peptone dextrose this reaction is much more intense than for synthetic medium (Kitamoto et al. 1988). With no polyphosphate, that usually sequesters and osmotically inactivates these amino acids (Dürr et al. 1979), the excessive amount of vacuolar amino acids causes osmotic pressure, which impairs the fragmentation of the entire vacuole in the yeast extract peptone dextrose cultured vtc mutant strains.

In vtc1Δ and vtc4Δ cells, it has been revealed that they display increased fragmented vacuoles, which seem to be indicator for the decreased fusion activity of vacuoles. It has been investigated whether the vacuolar polyphosphate triggers the fusion of vacuoles through the parallel analysis of the fusion activity of isolated vacuoles due to the polyphosphate accumulation rate (Fig. 8.1). The fusion of purified vacuoles can be induced by ATP (Wickner 2010). In 20 min, polyphosphate can be produced efficiently, and a concentration of 0.4 μg of polyphosphate per μg of vacuolar protein can be achieved. Structure-function assays of the VTC complex pinpointed missense mutations and/or single-residue substitutions that interfere with or promote polyphosphate synthesis without compromising the structural integrity of the complex (Hothorn et al. 2009). The substitutions are found within the catalytic domain of Vtc4, which is positioned to the cytosolic site and transfers ATP to polyphosphate and ADP. However, in the cytosolic SPX, domains that are suppose to possess regulatory functions can additional substitutions be identified and also in the transmembrane domains of Vtc1 that seems to facilitate the transport of the polyphosphate chain through the vacuolic membrane (Gerasimaite et al. 2014; Hothorn et al. 2009; Wild et al. 2016). Among these group of mutations are the catalytic domain substitution vtc4K200A that enables Vtc4 to hydrolyze ATP, whereas no polyphosphate is generated (Hothorn et al. 2009). The fusion activity of vacuoles from these mutants is elevated due to the enhanced polyphosphate accumulation. As single amino acid substitutions in the Vtc proteins control the assembly of the entire VTC complex and its targeting toward vacuoles (Hothorn et al. 2009; Wild et al. 2016), it can be proposed that the vacuolar fusion

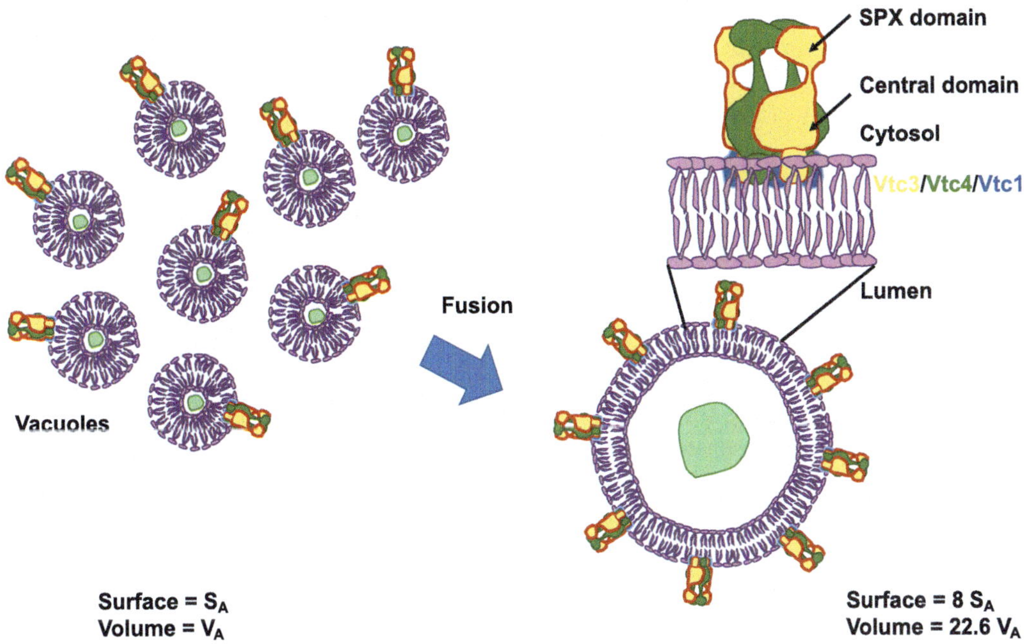

Fig. 8.1 Schematic representation of the vacuole accumulation, where the overall volume of the resulting large vacuole is increased by over 20-fold. The VTC complex contributes to the vacuole fusion process

machinery seem to be regulated through the accumulation of polyphosphate in the lumen of vacuoles independently of the availability of VTC complex.

8.2.3 The Fusion of Vacuoles is not Altered by the Cytosolic Polyphosphate Concentration

The effects of the lumenal fraction of polyphosphate and an artificially generated cytosolic pool of polyphosphate are analyzed in terms of vacuole fusion. VTC has been suggested to function as an obligatorily coupled synthase and translocase (Gerasimaite et al. 2014). In specific detail, VTC's cytosolic domains can synthesize the polyphosphate and deliver it directly to the vacuolar lumen through the transmembrane domains of VTC. Hence, it can be said that VTC generates lumenal polyphosphate rather than the freely diffusible cytosolic polyphosphate (Saito et al. 2005). The luminal pool of polyphosphatase can even be decreased, when a delivery of a polyphosphatase into the lumen of vacuoles has been arranged. As a result, the VTC complex stays the same and still produces polyphosphate, but polyphosphate accumulation in the lumen is decreased. A polyphosphatase Ppx1 fusion with the pre-Pro sequence of vacuolar carboxypeptidase Y (Prc1) was utilized to transfer the fusion protein through ER, Golgi and endosomes into the vacuolar lumen (Gerasimaite et al. 2014). However, this construct has to be applied to strains that are insufficient in vacuolar proteases, such as BJ3505, as it is degraded in protease-capable vacuoles. Vacuolar Ppx1 turns polyphosphate into inorganic phosphate and oligophosphate (Wurst and Kornberg 1994). Inorganic phosphates are transported to the cytosol by Pho91 (Hürlimann et al. 2007).

This construct was GFP tagged and hence showed a pronounced staining of the lumen of the vacuole, a very slight staining of the cell periphery and the nearby the nuclei, which seems to be a staining of the ER and even no signal has been detected in the cytosol (Gerasimaite et al. 2014). Moreover, whole cell protein extracts displayed no pronounced degradation of the fusion protein, which hence is supposed to be targeted to the vacuoles. Ppx1, which is targeted to vacuoles, can decrease the in vivo polyphosphate in BJ3505 cells to 10%. Therefore, it reduces the in vitro polyphosphate storage in the vacuoles by 80%, and also the activity of vacuole fusion is solely lowered be 40%, which leads to the hypothesis that the lumenal amount of polyphosphate is affected by the vacuole's fusion activity. There is an important limitation of this approach, since the polyphosphate accumulation can exclusively be decreased in one of the fusion partners, the protease-deficient BJ3505 vacuoles; however, in the other fusion partner from DKY6281 cells, where is simply destroyed by degradation. This can answer why the fusion activity is only slightly altered than the accumulated storage of polyphosphate. In line with this, the decreased polyphosphate levels seem to be overestimated, since Ppx1 possesses over 100 times higher K_m values for short chains (under 10 residues) than for long chains (over 50 residues), and it is also not functional in degrading short end-products including pyrophosphate (Wurst and Kornberg 1994). These short end-product may stay in the vacuoles of cells expressing the vacuole-targeted Ppx1, whereas they did not alter the DAPI fluorescence, which serves as the fundament for the detection in the polyphosphate assay (Omelon et al. 2016).

When the accumulation of polyphosphate at the cytosolic side induces a signal for the fusion of the vacuoles, the addition of polyphosphate to the buffer needs to support the polyphosphate synthesis by VTC. The vacuoles from vtc4Δ and vtc4R264A cells can be extracted and incubated in fusion reactions that consist of synthetic polyphosphate with an average length of 60 (polyphosphate-60) to 300 phosphate residues (polyphosphate-300), which represent a broad variety of chain lengths generated by the VTC complex (Hothorn et al. 2009; Lonetti et al. 2011), however, they are not transported into the vacuoles (Gerasimaite et al. 2014). A large variety of concentrations of polyphosphate-60 and polyphosphate-300 have been analyzed for their function to trigger the fusion of the vacuoles at the cytosolic side. However, none of which were not able to induce the vacuole fusion. Finally, these findings indicate that exclusively the lumenal polyP induces the promotion of vacuoles fusion.

8.2.4 In the Absence of Polyphosphate Synthesis, the Priming Step Remains the Only Fusion Step that is Interrupted

What step of the fusion reaction is altered by polyphosphate? The fusion of vacuoles is composed of specific subreactions that compromise the activation of SNARE, pairing of the trans-SNARE, the hemifusion and the opening of the fusion pore (Ostrowicz et al. 2008; Wickner 2010). The hydrolysis of ATP by Sec18/NSF facilitates the activation of SNARE. This chaperone can dissociate and hence activate SNAREs, which are present in cis-complexes on the vacuoles (Mayer et al. 1996; Ungermann et al. 1998). The activation of SNARE can be avoided by incorporating the purified recombinant Qc-SNARE subunit Vam7 that enrolls an existing pool of unpaired free SNAREs into trans-SNARE complexes and promotes fusion without Sec18/NSF activity (Thorngren et al. 2004). The vacuole fusion activity of vtc4Δ, vtc4R264A and vtc4K200A can be reused by the addition of Vam7 and reached the level of wild-type cells. This indicated that trans-SNARE pairing, hemifusion and fusion pore opening occur in the absence of polyphosphate and that SNARE activation may be the only step affected. The activation of SNARE can be directed modulated by addition of Sec17/α-SNAP, which is a cofactor of Sec18/NSF. The membrane release of Sec17 is associated with the disruption of cis-SNARE complexes (Mayer et al. 1996; Ungermann et al. 1998). Vacuoles have undergone incubation and sedimentation at fusion regimes and Sec17 has been investigated within the membrane pellets and supernatants. The release of Sec17 in vacuoles from vtc1Δ, vtc4Δ and the recombinant Qc-SNARE subunit Vam7 was decreased by 80–95%, recruiting an existing pool of unpaired, free SNAREs into trans-SNARE complexes and propelling fusion in the absence of Sec18/NSF activity (Thorngren et al. 2004). Vam7 supplement reestablished the fusion activity of the vacuoles vtc4Δ, vtc4R264A and vtc4K200A at wild-type level. This indicated that trans-SNARE pairing, hemifusion and fusion pore opening occur in the absence of polyphosphate and that the activation of SNARE possibly is the only step involved. The SNARE activation procedure was directly evaluated with Sec17/α-SNAP, a co-factor of Sec18/NSF. The release of Sec17 from the membrane coincides with the break-down of cis-SNARE complexes (Mayer et al. 1996; Ungermann et al. 1998). Vacuoles were kept under fusion conditions and sedimented, and then, Sec17 was analyzed in the pellets of the membrane and the supernatants. In fact, the release of Sec17 was pronouncedly reduced by 80–95% in vacuoles from vtc1Δ and vtc4Δ and subsequently, the break-down of the SNARE complexes relies on the polyphosphate synthesis and its transfer in the lumen of the vacuoles.

The Sec18 is localized to SNARE complexes when the synthesis of polyphosphate is performed. Vacuoles of mutants that are insufficient in polyphosphate synthesis bear normal quantities of Sec18/NSF. That is why it has been examined whether their deficit of SNARE activation can be attributed to an abnormal interaction of Sec18/NSF with the SNARE complex. Vacuoles were extracted from cells expressing SNARE Vam3 that carries a HA-tag at its C-terminus (Pieren et al. 2010). The membranes underwent treatment with the cleavable cross-linker DTSSP were dissolved in detergents and immunoprecipitated using anti-HA antibodies. In wild-type vacuoles, the Vam3–HA was shown to be linked with Sec18/NSF (Ostrowicz et al. 2008). This association of VAM3 and SEC18/NSF can exclusively be seen under cross-linking conditions. On vtc4R264A and vtc4Δ vacuoles, the Vam3–HA protein cannot bind to Sec18/NSF; however, it can still functionally interact with Nyv1, Sec17 and Vam7 similarly as seen in wild-type vacuoles. Based on this result, it can be proposed that the accumulation of polyphosphate promotes the action of SNARE through the interaction of Sec18/NSF and the vacuolar SNARE complex. Thus, it has been analyzed whether the fusion between polyphosphate lacking vacuoles is inducible by an excess of Sec18/NSF. The overexpression of SEC18 in a vtc1Δ mutants by at least tenfold above wild-type levels can reconstitute the fragmented vacuole phenotype of the vtc1Δ mutant and even also provide one or two large vacuoles in the cells. The in vitro fusion of vacuoles isolated from vtc4R264A, vtc4K200A and vtc4Δ cells yielded a similar result by incubation with purified recombinant Sec18/NSF in concentrations that exceeded the endogenous values by a factor of 50.

8.2.5 Functional Interaction of the Vacuolar SNARE Nyv1 and the Cyclin Pho80 with the VTC Complex

The VTC complex can physically interplay with Nyv1 (Müller et al. 2003) and genetically associate with Pho80 (Huang et al. 2002), whereas the cyclin interacts with the kinase Pho85. It has been assayed whether these proteins are involved in the polyphosphates dependent priming. When Nyv1 is not present, the SNAREs still present on nyv1Δ vacuoles, the proteins Vam3, Vam7 and Vti1 need to be switched to an active state by Sec18/NSF and Sec17 (Mayer and Wickner 1997; Nichols et al. 1997; Ungermann et al. 1998). Hence, the activation of SNARE activation can be assayed in nyv1Δ vacuoles in the same manner as in wild-type vacuoles. Moreover, even when analyzing vacuoles from a nyv1Δ vtc1Δ double mutant (ΔΔ), a pronounced reconstitution of the release of Sec17 up to about 70% of the wild-type level can be observed. When considering whether the deletion of NYV1 could save the fusion activity of a vtc1Δ vacuole, it was necessary to understand that vacuole fusion necessitated Nyv1 in at least one of the fusion partners (Nichols et al. 1997) and thus had to compare the fusion of vtc1Δ and ΔΔ vacuoles with wild-type organelles. Even though a vtc1Δ × wild-type mix displayed only 40–50% of the fusion signal of a wild type × wild type mix, the fusion in a ΔΔ × wild type mix returned 80–90%.

There was no fusion seen between two identical ΔΔ or two identical vtc1Δ vacuoles. Evidence points to the need for Nyv1 to interfere with the interaction of Sec18 with the SNAREs in the absence of polyphosphate. For *in vitro* vacuole fusion, a lipid mixture assay was conducted to evaluate available mutants that were deficient for Pho85 cyclin genes (Reese et al. 2005). Eight cyclin mutants have been investigated, and exclusively vacuoles from pho80Δ strains are impaired in the mixing of lipids. In the content mixing approaches, the pho80Δ vacuoles displaced 20% of the wild-type vacuole activity. Specifically, the vacuoles from pho80Δ demonstrated normal Sec17 secretion and SNARE complex break-down, implying that Pho80 was not necessary for SNARE activation. It can still negatively control the action of SNARE. In these circumstances, a pho80Δ double knockout can resolve the defects of the vtc1Δ vacuoles. Vacuoles from pho80Δ vtc1Δ cells did actually exhibit a fractional rescue of Sec17 release and fusion to 30–40% of wild-type level. In vivo, a pho80Δ vtc1Δ mutant had fewer fragmented vacuoles than vtc1Δ cells. Based on these findings, the hypothesis is supported that polyphosphate accumulation in vacuoles results in Pho80-Pho85-dependent alterations in the cytosol that impact SNARE activation.

Therefore, it was examined whether the priming inhibition is due to a strong alteration of the SNARE complex by immune adsorption of Vam3-HA-tagged SNARE complexes to protein-G agarose beads. The adsorbed complexes failed to degrade by simple administration of an ATP repair system, implying that the activity of Sec18/NSF of the complex was diminished during purification. To determine whether SNARE complexes from vtc1Δ and wild-type cells are similarly sensitive to activation by this protein, pure Sec18/NSF was administered in increasing concentrations. The activation of SNARE, quantified by the release of Sec17 and Vam7 from the beads, exhibited the exact same concentration dependence for wild-type and vtc1Δ complexes, pointing to the fact that vtc1Δ complexes had not maintained a stable variation that prevented Sec18/NSF from targeting them. SNARE complex disorders are also likely to be triggered by the addition of wild-type cytosol and ATP, regardless of whether the vacuoles originate from vtc1Δ or wild-type cells. Cytosol of vtc1Δ cells was not sufficient to cause SNARE complex disorders, although it included the same amount of Sec18 as wild-type cytosol. Conversely, cytosol of pho80Δ vtc1Δ cells facilitated SNARE complex disorder, which confirmed the fractional recovery of vacuole fusion on pho80Δ vtc1Δ vacuoles. The heating of the cytosol to 95 °C for 5 min eliminated the activity of the cytosol, indicating that the activating function is not attributable to the availability of heat-stable low-molecular mass compounds. Collectively, these findings substantiate the hypothesis that cytosolic factors govern SNARE activation as a function of polyphosphate accumulation in vacuoles and moreover they imply Pho80-Pho85 kinase and Nyv1 in this regulation.

Vacuole fusion raises the volume of the vacuolar system due to lack of energy, accumulation of storage substances such as polyphosphate or under hypotonic shock. The stimulation of the vacuole's volume enlargement depends on the type of stimulus. After starvation, the vacuolar hydrolases are massively elevated, the vacuoles merge and the organelle volumes are enlarged, probably to meet the demand for enhanced degradation by autophagy. The reaction of the vacuole fusions is performed over hours and is driven by the TORC1 kinase. The protein TORC1 is required for the fusion of vacuoles, but not for their division (Michaillat et al. 2012). As the activity of TORC1 reduces after starvation, starvation triggers the balance of fusion-fission to fusion, resulting in a prolonged network fusion of vacuoles. Lysosomes of mammalian species merge and experience a similar increase in volume during the stimulation of autophagy (Sardiello et al. 2009; Yu et al. 2010). In opposition to the slow adjustment to starvation, the hypotonic shock raises fast the tension of the vacuole membrane. Vacuoles are able to fuse within seconds to minutes, possibly to prevent their disruption. The fast fusion is facilitated by the induction of the Rab-GTPase-dependent vacuole docking system (Brett and Merz 2008). The triggering of SNAREs is undisturbed by the membrane tension produced by hypotension (Brett and Merz 2008). The accumulation of a large vacuolar storage compound induces fusion through SNARE activation and directs the interaction of Sec18/NSF and cis-SNARE complexes. In this way, the vacuolar surface-volume ratio obtained by fusion fission equilibrium can be tuned over at least three distinct molecular routes. These routes are not self-exclusive. But it is improbable that the polyphosphate-induced fusion of vacuoles relies on the osmotic water flow into the lumen of the organelle, since the polyphosphate chains produced by the vacuoles are ranging in length from dozens to hundreds of phosphate residues. The enhancement of vacuum osmolarity through such macromolecules appears insignificant. Polyphosphate, a highly charged polyanion, may also elevate the osmotic value of vacuoles promoting an influx of compensating cations including lysine or arginine. While these amino acids conglomerate in vacuoles to molar concentrations, it is not expected that they will raise osmotic pressure accordingly, probably by inerting themselves osmotically through polyphosphate sequestration (Dürr et al. 1979). Another argument opposing the activation of fusion by an osmotic impact of polyphosphate accumulation is the fact that the polyphosphate selectively alters the activation of SNARE and the binding of NSF/Sec18, while the osmotic elevation of the membrane tension cannot impact this process (Brett and Merz 2008). Polyphosphate is also improbable to support vacuole fusion by TORC1 as vacuole fusion needs no TORC1 activity (Michaillat et al. 2012). Therefore, the stimulation of NSF-dependent SNARE activation by synthesis and lumenal concentration of polyphosphate provides a hitherto unknown regulatory route for the size and number of vacuoles, distinguishing it completely from the effects of starvation and osmotic stress.

The fusion defect of vtcΔ vacuoles has been interpreted initially as a physical interplay of VTC with the vacuolar H^+-ATPase (Müller et al. 2002, 2003), which is necessary for the fusion of vacuoles (Bayer et al. 2003; Peters et al. 2001; Sreelatha et al. 2014; Strasser et al. 2011). A function of VTC as a vacuoolar polyphosphate polymerase has been abandoned because the complex is fully cytosolic while polyphosphate is housed within the vacuoles. Advances in our knowledge of VTC structure and polyphosphate metabolism (Gerasimaite et al. 2014; Hothorn et al. 2009) have now led to the realization that the decisive factor is actually the vacuolar polyphosphate. The effect of point mutants in the catalytic core and the synthetic expression of a polyphosphatase in the vacuole lumen reinforce this. The results of our observations underpin a model in which polyphosphate accumulation governs fusion through the interplay of Sec18/NSF with the SNARE complex. Subsequently, it needs to be assumed that polyphosphate accumulation inside the organelle is transferred to the cytosolic side of the membrane where Sec18/NSF and SNAREs fulfill their interplay. An argument for this term results out of the fact that a cytosolic extract from a wild-type strain enhances SNARE dissociation, avtc1Δ strain extract cannot trigger the activation of SNARE, whereas an extract of a vtc1Δ pho80Δ mutant strain can restore the activation of SNARE. Based on these results, it seems to be that a critical factor related to the accumulation of polyphosphate in vacuoles is present in the soluble fraction. How can then the signal be transported? It has been shown that the VTC complex can directly react with Nyv1 and genetically with the Pho80–Pho85 kinase (Huang et al. 2002; Müller et al. 2003). These findings imply that both proteins regulate the activation of SNARE, since the absence of one of them leads to the deletion of one of the genes coding for Nyv1, or Pho80 partially saved the priming failure of the vtcΔ vacuoles.

The role of Pho80 seems to be in the Pho80–Pho85 kinase activated by the presence of cytosolic phosphate (Schneider et al. 1994). Since the polyphosphate synthesis delivers large level of phosphate into vacuoles that is associated with a decrease in the cytosolic phosphate concentration (Thomas and O'Shea 2005), the increased storage of polyphosphate in vacuoles seems to be detectable in an indirect manner through measuring the cytosolic phosphate amount. The decreased cytosolic phosphate concentration based on polyphosphate synthesis appears to diminish the Pho80-Pho85 activity, while both parameters are expected to increase, when the polyphosphate synthesis is discontinued. The deletion of Pho80 leads to an abolishment of the Pho80–Pho85 kinase activity and thus represents a state where polyphosphate is exclusively synthesized through the activation of SNARE. The identification of important Pho80–Pho85-based phosphorylation sites were highly complicated due to the broad complexity of the pattern of phosphorylation. There are over twenty phosphorylation sites solely on the VTC complex, which are pronouncedly deregulated in pho80Δ vacuoles relative to the levels of wild-type organelles (Desfougeres et al. 2016). It has been identified that the phosphorylation of residue Ser197 of Nyv1 is increased in vtc4Δ vacuoles. When the Ser197 is exchanged with Ala or Asp, the morphology of the vacuole, their fusion activity or the inhibition of priming in vtcΔ mutants was not altered. In future experiments, a comprehensive approach should be considered to enlighten the regulation of the vacuolar fusion system through its phosphorylation.

8.3 Mechanics of Vacuoles

The influence of the cellular components and organelles have been suggested to heavily impact the mechanical properties of cells; however, they are still not yet clearly determined. However, there are studies around that have started to investigate the effect of large cellular components, such as vacuoles of cells, on the overall cell mechanical properties. Cellular compartments such as vacuoles influence the overall mechanical properties of cells that have been determined with atomic force microscopy, such as that vacuoles lower the stiffness of the cells (Nikolaev et al. 2013). Although the atomic force microscopy (AFM) is at the first glance a biophysical surface technique, it has been adapted appropriately for the analysis of the elasticity of intracellular vacuoles (Riethmüller et al. 2007), combined with the visualization of actin filaments within cells (Radmacher 2002), the imaging of mitochondria (Yoshida et al. 2015) and specific localization of the nucleus (Parpura et al. 1993). Single organelles have been shown to impact the measurements of cell stiffness suing AFM (Janel et al. 2019). Due to this fact, the stiffness tomography seems to provide a clearer picture of the measurements of cell elasticity, and it will help to compare the elasticity of different cell types. In concrete terms, the stiffer the organelles within a relatively soft cytoplasm, the stronger the contrast, even if soft organelles in a stiff material are theoretically verifiable for minor indentations (Roduit et al. 2009).

This concept has been proven by experimental measurements. First of all, fixed PtK2 cells have been measured through the identification of organelles using photonic and electron microscopy (Janel et al. 2019). After validating the identification of the intracellular components, the second aim was to measure the dynamical alterations of the stiffness of these internal cellular components in living HeLa cancer cells to investigate the mechanical properties of these structures and components locally inside cells (Janel et al. 2019). Since it is common knowledge that the elasticity of cells is altered during the transition of normal healthy cells to cancer cells (Rebelo et al. 2013; Plodinec et al. 2012), whereas it has not yet been analyzed in detail how these overall mechanical alterations of cancer cells can be attributed to alterations of the mechanical properties of single organelles. In conclusion, the identification of the specific localization of distinct cell organelles inside the cells and their mechanical properties will help us to precisely compare the mechanical properties of cancer cells with their normal healthy counterparts.

References

T.L. Baars, S. Petri, C. Peters, A. Mayer, Role of the V-ATPase in regulation of the vacuolar fission-fusion equilibrium. Mol. Biol. Cell **18**, 3873–3882 (2007)

M.J. Bayer, C. Reese, S. Bühler, C. Peters, A. Mayer, Vacuole embrane fusion: V0 functions after trans-SNARE pairing and is coupled to the Ca^{2+}-releasing channel. J. Cell Biol. **162**, 211–222 (2003)

B.J. Bevis, A.T. Hammond, C.A. Reinke, B.S. Glick, De novo formation of transitional ER sites and Golgi structures in Pichia pastoris. Nat. Cell Biol. **4**, 750–756 (2002)

M. Bhave, E. Papanikou, P. Iyer et al., Golgi enlargement in Arf- depleted yeast cells is due to altered dynamics of cisternal maturation. J. Cell Sci. **127**, 250–257 (2014)

C.L. Brett, A.J. Merz, Osmotic regulation of Rab-mediated organelle docking. Curr. Biol. **18**, 1072–1077 (2008)

Y.-H.M. Chan, W.F. Marshall, How cells know the size of their organelles. Science **337**, 1186–1189 (2012)

J. Derganc, B. Antonny, A. Copič, Membrane bending: the power of protein imbalance. Trends Biochem. Sci. **38**, 576–584 (2013)

Y. Desfougeres, H. Neumann, A. Mayer, Organelle size control—increasing vacuole content activates SNAREs to augment organelle volume through homotypic fusion. J. Cell Sci. **129**, 2817–2828 (2016)

S.K. Dove, R.C. Piper, R.K. McEwen et al., Svp1p defines a family of phosphatidylinositol 3,5-bisphosphate effectors. EMBO J. **23,** 1922–1933 (2004)

M. Dürr, K. Urech, T. Boller, A. Wiemken, J. Schwencke, M. Nagy, Sequestration of arginine by polyphosphate in vacuoles of yeast (Saccharomyces cerevisiae). Arch. Microbiol. **121**, 169–175 (1979)

J.A. Efe, R.J. Botelho, S.D. Emr, Atg18 regulates organelle morphology and Fab1 kinase activity independent of its membrane recruitment by phosphatidylinositol 3,5-bisphosphate. Mol. Biol. Cell **18**, 4232–4244 (2007)

R.J. Ellis, Macromolecular crowding: an important but neglected aspect of the intracellular environment. Curr. Opin. Struct. Biol. **11**, 114–119 (2001)

V.A. Frolov, A.V. Shnyrova, J. Zimmerberg, Lipid polymorphisms and membrane shape. Cold Spring Harb. Perspect. Biol. **3**, a004747 (2011)

A.I. García-Pérez, E.A. López-Beltrán, P. Klüner, J. Luque, P. Ballesteros, S. Cerdán, Molecular crowding and viscosity as determinants of translational diffusion of metabolites in subcellular organelles. Arch. Biochem. Biophys. **362**, 329–338 (1999)

E.C. Gaynor, C.Y. Chen, S.D. Emr, T.R. Graham, ARF is required for maintenance of yeast Golgi and endosome structure and function. Mol. Biol. Cell **9**, 653–670 (1998)

R. Gerasimaite, S. Sharma, Y. Desfougeres, A. Schmidt, A. Mayer, Coupled synthesis and translocation restrains polyphosphate to acidocalcisome-like vacuoles and prevents its toxicity. J. Cell Sci. **127**, 5093–5104 (2014)

O.P. Hamill, B. Martinac, Molecular basis of mechanotransduction in living cells. Physiol. Rev. **81**, 685–740 (2001)

R. Heald, O. Cohen-Fix, Morphology and function of membrane-bound organelles. Curr. Opin. Cell Biol. **26**, 79–86 (2014)

M. Hothorn, H. Neumann, E.D. Lenherr et al., Catalytic core of a membrane-associated eukaryotic polyphosphate polymerase. Science **324**, 513–516 (2009)

D. Huang, J. Moffat, B. Andrews, Dissection of a complex phenotype by functional genomics reveals roles for the yeast cyclin-dependent protein kinase Pho85 in stress adaptation and cell integrity. Mol. Cell. Biol. **22**, 5076–5088 (2002)

H.C. Hürlimann, M. Stadler-Waibel, T.P. Werner, F.M. Freimoser, Pho91 is a vacuolar phosphate transporter that regulates phosphate and polyphosphate metabolism in Saccharomyces cerevisiae. Mol. Biol. Cell **18**, 4438–4445 (2007)

S. Janel, M. Popoff, N. Barois, E. Werkmeister, S. Divoux, F. Perez, F. Lafont, Stiffness tomography of eukaryotic intracellular compartments by atomic force microscopy. Nanoscale **11**, 10320–10328 (2019)

M. Kamei, W.B. Saunders, K.J. Bayless, L. Dye, G.E. Davis, B.M. Weinstein, Endothelial tubes assemble from intracellular vacuoles in vivo. Nature **442**(7101), 453–456 (2006)

I. Katsura, Determination of bacteriophage λ tail length by a protein ruler. Nature **327**, 73–75 (1987)

K. Kitamoto, K. Yoshizawa, Y. Ohsumi, Y. Anraku, Dynamic aspects of vacuolar and cytosolic amino acid pools of Saccharomyces cerevisiae. J. Bacteriol. **170**, 2683–2686 (1988)

D.L. Levy, R. Heald, Nuclear size is regulated by importin α and Ntf2 in Xenopus. Cell **143**, 288–298 (2010)

A. Lonetti, Z. Szijgyarto, D. Bosh, O. Loss, C. Azevedo, A. Saiardi, Identification of an evolutionary conserved family of inorganic polyphosphate endopolyphosphatases. J. Biol. Chem. **286**, 31966–31974 (2011)

W.B. Ludington, L.Z. Shi, Q. Zhu, M.W. Berns, W.F. Marshall, Organelle size equalization by a constitutive process. Curr. Biol. **22**, 2173–2179 (2012)

W.F. Marshall, Organelle size control systems: from cell geometry to organelle-directed medicine. BioEssays **34**, 721–724 (2012)

A. Mayer, W. Wickner, Docking of yeast vacuoles is catalyzed by the Ras-like GTPase Ypt7p after symmetric priming by Sec18p (NSF). J. Cell Biol. **136**, 307–317 (1997)

A. Mayer, W. Wickner, A. Haas, Sec18p (NSF)-driven release of Sec17p (alpha-SNAP) can precede docking and fusion of yeast vacuoles. Cell **85**, 83–94 (1996)

L. Michaillat, T.L. Baars, A. Mayer, Cell-free reconstitution of vacuole membrane fragmentation reveals regulation of vacuole size and number by TORC1. Mol. Biol. Cell **23**, 881–895 (2012)

O. Müller, M.J. Bayer, C. Peters, J.S. Andersen, M. Mann, A. Mayer, The Vtc proteins in vacuole fusion: coupling NSF activity to V(0) transcomplex formation. EMBO J. **21**, 259–269 (2002)

O. Müller, D.I. Johnson, A. Mayer, Cdc42p functions at the docking stage of yeast vacuole membrane fusion. EMBO J. **20**, 5657–5665 (2001)

O. Müller, H. Neumann, M.J. Bayer, A. Mayer, Role of the Vtc proteins in V-ATPase stability and membrane trafficking. J. Cell Sci. **116**, 1107–1115 (2003)

B.J. Nichols, C. Ungermann, H.R.B. Pelham, W.T. Wickner, A. Haas, Homotypic vacuolar fusion mediated by t- and v-SNAREs. Nature **387**, 199–202 (1997)

N.I. Nikolaev, T. Müller, D.J. Williams, Y. Liu, Changes in the stiffness of human mesenchymal stem cells with the progress of cell death as measured by atomic force microscopy. J. Biomech. **47**(2014), 625–630 (2013)

S. Omelon, J. Georgiou, W. Habraken, A cautionary (spectral) tail: red-shifted fluorescence by DAPI-DAPI interactions. Biochem. Soc. Trans. **44**, 46–49 (2016)

C.W. Ostrowicz, C.T.A. Meiringer, C. Ungermann, Yeast vacuole fusion: a model system for eukaryotic endomembrane dynamics. Autophagy **4**, 5–19 (2008)

V. Parpura, P.G. Haydon, E. Henderson, Three-dimensional imaging of living neurons and glia with the atomic force microscope. J. Cell Sci. **104**(Pt 2), 427–432 (1993)

C. Peters, T.L. Baars, S. Buhler, A. Mayer, Mutual control of membrane fission and fusion proteins. Cell **119**, 667–678 (2004)

C. Peters, M.J. Bayer, S. Bühler, J.S. Andersen, M. Mann, A. Mayer, Trans-complex formation by proteolipid channels in the terminal phase of membrane fusion. Nature **409**, 581–588 (2001)

M. Pieren, A. Schmidt, A. Mayer, The SM protein Vps33 and the t- SNARE H(abc) domain promote fusion pore opening. Nat. Struct. Mol. Biol. **17**, 710–717 (2010)

R.L. Pisoni, E.R. Lindley, Incorporation of [32P]orthophosphate into long chains of inorganic polyphosphate within lysosomes of human fibroblasts. J. Biol. Chem. **267**, 3626–3631 (1992)

M. Plodinec, M. Loparic, C.A. Monnier, E.C. Obermann, R. Zanetti-Dallenbach, P. Oertle, J.T. Hyotyla, U. Aebi, M. Bentires-Alj, L.R.Y. Hand, C.A. Schoenenberger, The nanomechanical signature of breast cancer. Nat. Nanotechnol. **7**, 757–765 (2012)

M. Radmacher, Measuring the elastic properties of living cells by the atomic force microscope. Methods Cell Biol. **68**, 67–90 (2002)

S.M. Rafelski, W.F. Marshall, Building the cell: design principles of cellular architecture. Nat. Rev. Mol. Cell Biol. **9**, 593–602 (2008)

L.M. Rebelo, J.S. de Sousa, J.M. Filho, M. Radmacher, Comparison of the viscoelastic properties of cells from different kidney cancer phenotypes measured with atomic force microscopy. Nanotechnology **24**, 055102 (2013)

C. Reese, F. Heise, A. Mayer, Trans-SNARE pairing can precedea hemifusion intermediate in intracellular membrane fusion. Nature **436**, 410–414 (2005)

C. Riethmüller, T.E. Schäffer, F. Kienberger, W. Stracke, H. Oberleithner, Vacuolar structures can be identified by AFM elasticity mapping. Ultramicroscopy **107**, 895–901 (2007)

C. Roduit, S. Sekatski, G. Dietler, S. Catsicas, F. Lafont, S. Kasas, Stiffness tomography by atomic force microscopy. Biophys. J. **97**, 674–677 (2009)

K. Saito, R. Ohtomo, Y. Kuga-Uetake, T. Aono, M. Saito, Direct labeling of polyphosphate at the ultrastructural level in Saccharomyces cerevisiae by using the affinity of the polyphosphate binding domain of *Escherichia coli* exopolyphosphatase. Appl. Environ. Microbiol. **71**, 5692–5701 (2005)

M. Sardiello, M. Palmieri, A. di Ronza et al., A gene network regulating lysosomal biogenesis and function. Science **325**, 473–477 (2009)

K.R. Schneider, R.L. Smith, E.K. O'Shea, Phosphate-regulated inactivation of the kinase PHO80-PHO85 by the CDK inhibitor PHO81. Science **266**, 122–126 (1994)

E.S. Seeley, M. Kato, N. Margolis, W. Wickner, G. Eitzen, Genomic analysis of homotypic vacuole fusion. Mol. Biol. Cell **13**, 782–794 (2002)

C. Settembre, A. Ballabio, Lysosomal adaptation: how the lysosome responds to external cues. Cold Spring Harb. Perspect. Biol. **6**(6) (2014)

A. Sreelatha, T.L. Bennett, E.M. Carpinone, K.M. O'Brien, K.D. Jordan, D.L. Burdette, K. Orth, V.J. Starai, Vibrio effector protein VopQ inhibits fusion of V-ATPase-containing membranes. Proc. Natl. Acad. Sci. USA **112**, 100–105 (2014)

B. Strasser, J. Iwaszkiewicz, O. Michielin, A. Mayer, The V-ATPase proteolipid cylinder promotes the lipid-mixing stage of SNARE-dependent fusion of yeast vacuoles. EMBO J. **30**, 4126–4141 (2011)

S. Sukharev, F. Sachs, Molecular force transduction by ion channels: diversity and unifying principles. J. Cell Sci. **125**, 3075–3083 (2012)

N. Tamura, M. Oku, M. Ito, N.N. Noda, F. Inagaki, Y. Sakai, Atg18 phosphoregulation controls organellar dynamics by modulating its phosphoinositide-binding activity. J. Cell Biol. **202**, 685–698 (2013)

M.R. Thomas, E.K. O'Shea, An intracellular phosphate buffer filters transient fluctuations in extracellular phosphate levels. Proc. Natl. Acad. Sci. USA **102**, 9565–9570 (2005)

N. Thorngren, K.M. Collins, R.A. Fratti, W. Wickner, A.J. Merz, A soluble SNARE drives rapid docking, bypassing ATP and Sec17/18p for vacuole fusion. EMBO J. **23**, 2765–2776 (2004)

C. Ungermann, B.J. Nichols, H.R.B. Pelham, W. Wickner, A vacuolar v-t-SNARE complex, the predominant form in vivo and on isolated vacuoles, is disassembled and activated for docking and fusion. J. Cell Biol. **140**, 61–69 (1998)

A. Upadhyaya, M.P. Sheetz, Tension in tubulovesicular networks of Golgi and endoplasmic reticulum membranes. Biophys. J. **86**, 2923–2928 (2004)

T.A. Vida, S.D. Emr, A new vital stain for visualizing vacuolar membrane dynamics and endocytosis in yeast. J. Cell Biol. **128**, 779–792 (1995)

L.S. Weisman, Yeast Vacuole Inheritance and dynamics annu. Rev. Genet **37**, 435–460 (2003)

W. Wickner, Membrane fusion: five lipids, four SNAREs, three chaperones, two nucleotides, and a Rab, all dancing in a ring on yeast vacuoles. Annu. Rev. Cell Dev. Biol. **26**, 115–136 (2010)

R. Wild, R. Gerasimaite, J.-Y. Jung et al., Control of eukaryotic phosphate homeostasis by inositol polyphosphate sensor domains. Science **352**, 986–990 (2016)

H. Wurst, A. Kornberg, A soluble exopolyphosphatase of Saccharomyces cerevisiae purification and characterization. J. Biol. Chem. **269**, 10996–11001 (1994)

A. Yoshida, N. Sakai, Y. Uekusa, K. Deguchi, J.L. Gilmore, M. Kumeta, S. Ito, K. Takeyasu, Probing in vivo dynamics of mitochondria and cortical actin networks using high-speed atomic force/fluorescence microscopy. Genes Cells **20**, 85–94 (2015)

L. Yu, C.K. McPhee, L. Zheng et al., Termination of autophagy and reformation of lysosomes regulated by mTOR. Nature **465**, 942–946 (2010)

H.-X. Zhou, G. Rivas, A.P. Minton, Macromolecular crowding and confinement: biochemical, biophysical, and potential physiological consequences. Annu. Rev. Biophys. **37**, 375–397 (2008)

M. Zieger, A. Mayer, Yeast vacuoles fragment in an asymmetrical two-phase process with distinct protein requirements. Mol. Biol. Cell **23**, 3438–3449 (2012)

Lysosomes and Peroxisomes

9

Abstract

Besides the major organelles of the cell, such as nucleus, mitochondria, ER and Golgi apparatus, the lysosomes and peroxisomes may fulfill prominent functions in the response to mechanical property alterations of their surrounding environment and possible also in providing the overall cell mechanics. However, their biological functions are still not less important and are presented here additionally. In specific detail, the structure and function of lysosomes and peroxisomes are presented. The effect of environmental mechanics on organelles, such as lysosomes and peroxisomes, is discussed.

9.1 Structure and Function of Lysosomes

Lysosomes represent membrane-enclosed cytoplasmic organelles that facilitate sequestration of a large number of biomacromolecules within the endomembrane system of all animal cells (Braulke and Bonifacino 2009; Saftig and Klumperman 2009). Moreover, lysosomes fulfill a major role in the degradation of various components, such as bacterial products of invaders in macrophagic lysosomes. More than 60 luminal hydrolases with specificity for a variety of substrates perform the digestible role of lysosomes. The majority of the hydrolases require an acidic optimum pH that is similar to the pH value of the lysosome lumen ranging between 4.5–5.0. The limiting membrane of lysosomes contains more than 200 integral membrane proteins inclusive of a proton-importing V-type ATPase which preserves the acid pH of the lumen. These include ion transporters and channels that regulate luminal ion composition, as well as several other transporters that carry degradation products to the cytosol. In addition, a set of highly-glycosylated lysosome-associated membrane proteins (LAMPs) is present in the membrane to protect the entire membrane from lysosomal degradation. Beyond that, there are tethering factors and SNARE proteins supporting the interaction and fusion of lysosomes with even other specific cytoplasmic organelles. Consequently, the cytosolic site of the lysosomal membrane functions is an assembly region for multiple proteins and protein complexes that facilitate the nutrient and stress signal transduction and drive the interactions with the cell's cytoskeleton.

Lysosomes decompose the substances delivered to them by endocytosis, biosynthetic transport, autophagy and phagocytosis. Due to their outstanding role in intracellular degradation, lysosomes have gained a solid standing as the cell's "waste disposal tool." However, this recognition of lysosomes as pure degradable organelles has begun to shift in recent years with the recognition that they

C. T. Mierke, *Cellular Mechanics and Biophysics*, Biological and Medical Physics, Biomedical Engineering, https://doi.org/10.1007/978-3-030-58532-7_9

are involved in several other cellular processes. These processes include elimination of intracellular pathogens, cell adhesion and migration, cancer cell invasion and cancer metastasis, presentation of antigens, repair of plasma membranes, release of exosomes, apoptosis, metabolism signaling and gene regulation (Ballabio 2016; Mukhopadhyay 2016) (Fig. 9.1). Lysosomal abnormalities have been found both in rare lysosomal storage disorders (LSDs) and in common neurodegenerative diseases such as Alzheimer's, Parkinson's and Huntington's diseases, and in amyotrophic lateral sclerosis (ALS). (Parenti et al. 2015; Fraldi et al. 2016). These results have reawakened researchers' fascination with lysosomes as decisive factors influencing several cell functions.

Although lysosomes are confined by a double-layered membrane, they are relatively dynamic organelles, which contain macromolecules from several pathways regulating secretory, endocytic, autophagic and phagocytic membrane-trafficking processes to degrade them. These membrane-bound organelles contain mainly acids, proteases, and other degradative enzymes. The shapes of lysosomes are polymorphous with resting lysosomes exhibiting a simple and nearly spherical shape and active lysosomes display a pronouncedly larger and more complex shape. These active lysosomes can degrade distinct components of the cells and even exocytose particles from the external microenvironment. Subsequently, they are regarded as an end point of endocytosis. However, the function of lysosomes is not only the degradation of proteins. They are able to fuse with the cell's plasma membrane upon cell injury, and some cell types possess lysosomes with specialized secretory functions. As dynamic organelles lysosomes can fuse with several targets and are rebuild after these fusion processes. Moreover, they sense nutrients and transmit signals into the cell's nucleus. Lysosomes are additionally secretory vesicles consisting of a specialized machinery facilitating the secretion of proteins in certain cell types.

From a historic point of view, the lysosome has been established as a cell compartment that performs recycling processes, whereas recent research points to a function of lysosomes in the metabolism of cells (Settembre et al. 2013). Lysosomes represent small organelles and possess an acidic interior, where a wide variety of reactions can take place and which therefore harbors several

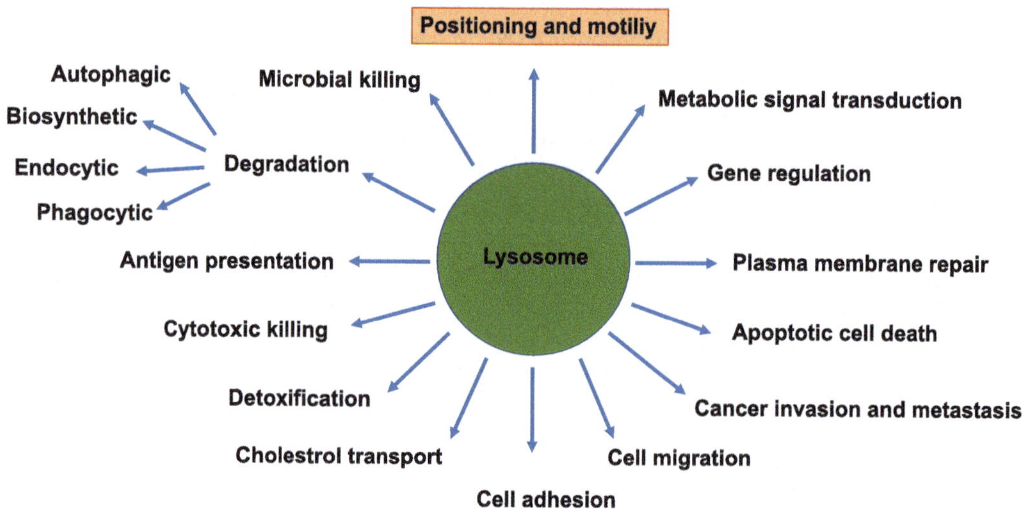

Fig. 9.1 Multiple functions of lysosomes are impacted by their positioning and motility. Besides degradation, lysosomes fulfill multiple other functions, such as endocytic, phagocytic and autophagic degradation, presentation of antigens causing cell death of target cells by cytotoxic T cells, metabolic signal transduction, adhesion and migration of cells, cancer invasion and metastasis, repair mechanisms of the plasma membrane, which relies on the motility of lysosomes within the cytoplasm

hydrolytic enzymes including proteases, nucleases and lipases. The hydrolases are able to facilitate the degradation and the recycling of macromolecules or of entire organelles, where they govern their endocytosis and autophagy (Saftig and Klumperman 2009). The role of lysosomes is crucial for large different processes and includes even a malfunctional role in the development of disease, including cancer (Polishchuk et al. 2019). Intriguingly, cancer cells exhibit elevated lysosomal activity and possess a higher level of autophagy than non-malignant cells, which underlines that lysosomes are key actors in resistance to energetic stress conditions (Piao and Amaravadi 2016). Moreover, the finding that lysosomes fulfill a key function in lipid metabolism leads to the hypothesis that these organelles can be a target for novel cancer treatments.

Nonetheless, cellular lipid and cholesterol metabolism have become novel target for the treatment of cancer, as these metabolic processes are found to be frequently disturbed in cancerous tissue. However, the mechanistic findings on how the manipulation of lipid metabolism restricts the survival of cancer cells are still relatively scarce (Santos and Schulze 2012; Rohrig and Schulze 2016). Whereas numerous descriptive studies demonstrated alterations in the lipid metabolism of cancer cells in comparison with healthy tissue, and several studies even associate the deregulation in lipid metabolism with cancer malignancy; however, the influence of specific alterations on cellular organelles has not been thoroughly investigated (Long et al. 2018; Zhang and Du 2012). Most recently, lysosomes have been shown to play a key role in cholesterol metabolism. It has evidenced that lysosomal V-ATPase, a proton pump required for lysosomal acidification, limits intracellular cholesterol supply and hence suppresses tumor progression and connects lipid metabolism to lysosomes (Bartel et al. 2017). The V-ATPase can be impaired with several natural compounds, including archazolid, bafilomycin, iejimalide and concanamycin and thereby trigger the mitochondrial-based process of programmed cell death, such as apoptosis in various cancer cells and to alter the process of autophagy; however, the precise mechanisms how the mitochondrial apoptosis can be initiated are still not yet clearly understood (von Schwarzenberg et al. 2013; McHenry et al. 2010; Teplova et al. 2007; Aiko et al. 2002).

The function of mitochondria in the process of the regulation of cell death has been largely examined. The crucial role of mitochondria lies in the secretion of proteins from the intermembrane gap into the cytosol and their capacity to fully activate caspases. However, the interplay between mitochondria and other organelles, such as lysosomes, is less clearly understood (Murley and Nunnari 2016; Todkar et al. 2017; Twig et al. 2008). However, there is evidence that lysosomes directly interact with mitochondria, where the cellular lipid metabolism of the cells is crucially involved. A novel mechanistic insight is presented stating that a triangle of lysosomes, mitochondria and the lipid metabolism represents the missing connection between lysosomal triggered and mitochondria-driven death of cells.

There is evidence for a novel function of the interorganelle cross-talk between lysosomes and mitochondria in the cellular lipid metabolism. The initiation of lysosomal<Query ID="Q1" Text="The term 'programmed' has been retained as such throughout the book.`` –> stress, such as the inhibition of the V-ATPase, causes a decrease of the mitochondrial cardiolipin amount, activates the fission, impairs the overall mitochondrial function and subsequently leads to the mitochondria-based induction of programmed cell death, termed apoptosis. Thereby the knowledge of the interplay between lysosomes and mitochondria is increased and especially the mechanism of how the impairment of the V-ATPase causes apoptosis is enlightened. Moreover, this specific knowledge may lead to the development of new anti-cancer therapeutic strategies that aim to disrupt or modify the lysosome-mitochondria interaction (Bartel et al. 2019).

The biogenesis of lysosomes and lysosome-related organelles is presented with a focus on their dynamic nature and various functions. Lysosomes are enclosed by a double-lipid-layered membrane, as every organelle, and they possess over 50 acid hydrolases that fulfill their tasks in the degradation of macromolecules, which are recruited through endocytic, phagocytic and autophagic mechanisms.

The lysosomes had been identified as subcellular compartments by Christian De Duve. More precisely, it was found that the activity of acid hydrolases is significantly increased after positioning subcellular fractions in hypotonic solutions, detergents or other offensives of membrane integrity (Bainton 1981; de Duve 2005). By electron microscopy, it has been revealed that the made up to 5% of the intracellular volume within animal cells. They display heterogeneous sizes and shapes, contain many times electron dense compounds and even carry sometimes multilamellar membrane stirrers (Klumperman and Raposo 2014). Lysosomes can be discriminated from late endosomes by the lack of mannose-6-phosphate receptors (MPRs) (Brown et al. 1986). Lysosomes have become a rediscovered focus of research because there are results that identify them as signal organelles that can sense the availability of nutrients and induce activation of the lysosome-to-nucleus signaling pathway that conveys the hunger response and governs energy metabolism (Settembre et al. 2013). Moreover, the precisely regulated exocytosis of conventional lysosomes has been identified and it seems to be a crucial property of multiple lysosome-related organelles (LROs), which represent a group of functionally very different, cell type-specific organelles that resemble similar to conventional lysosomes.

The transport mechanisms of macromolecules via the endocytic signaling pathway through the lysosomes are well-known (Luzio et al. 2007; Woodman and Futter 2008; Huotari and Helenius 2011). While intercourse through early endosomes to late endosomes or synonymously is referred to as multivesicular bodies (MVBs), administration of lysosomal hydrolases is easiest to consider as a maturation process, since administration of lysosomal hydrolases involves kiss and run events and full fusions between late endosomes and lysosomes. The fusion processes of the endosomes and lysosomes have been investigated by employing electron microscopy (Futter et al. 1996; Bright et al. 1997), the cell-free mixing of the two organelles (Mullock et al. 1994) and the direct investigation of these events in living cells using live-cell microscopic imaging (Bright et al. 2005; Gan et al. 2009). A consequence of the fusion of the two organelles lies in the creation of the "endolysosome", which represents a hybrid organelle that shares intermediate properties composed of the late endosome and the lysosome. For instance, although it is assumed that late endosomes are relatively depleted in mannose-6-phosphate receptors when they prepare for fusion with lysosomes (Hirst et al. 1998), endolysosomes still comprise mannose-6-phosphate receptors and are consequently still prelysosomal compartments as traditionally defined (Griffiths et al. 1988). In the next step, the lysosomes are altered by maturated endolysosomes and thereby reshaped (Fig. 9.2). While in the past lysosomes were considered the terminal degradation component of the endocytic pathway, the majority of degradation is probably in endolysosomes once they are generated and as they undergo the process of newly regenerating lysosomes. In fact, the classical electron dense lysosome may rather resemble a secretory granule that retains hydrolases that are ready to be transferred to the place of macromolecule breakdown as the fusion with late endosomes takes place. The highly dynamic character of the terminal partitions of the endocytic pathway needs to be considered when examining lysosomal biogenesis, as it has been demonstrated that numerous newly synthesized lysosomal proteins are initially transported to the endosomes.

9.1.1 Fusion and Reformation of Lysosomes

Experimental techniques have been engineered to examine the molecular mechanism of cargo delivery to lysosomes from endocytic, phagocytic and autophagous tracks within mammalian cells. Proteins necessary for the fusion of late endosomes with lysosomes have been discovered by various different strategies, including inhibitors in experiments with permeabilized and cell-free cells, overexpression and RNA interference/knockdown trials in cultivated cells. The fusion engine consists of cytosolic factors and trans-SNARE compounds (soluble N-ethylmaleimide-sensitive factor-binding

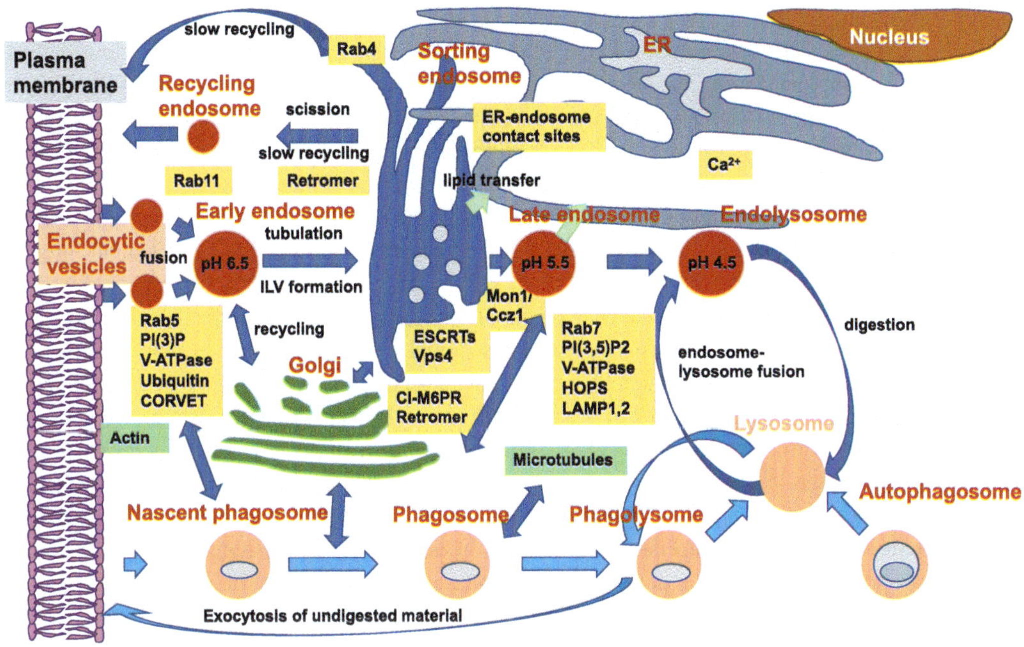

Fig. 9.2 Maturation of endolysosomes and involved molecules

protein receptor); however, for proper fusion also Ca^{2+} ions are required, which are liberated out of the lumen of the fusing organelles (Mullock et al. 1998; Pryor et al. 2000; Ward et al. 2000b). Cytosolic factors comprise NSF (N-ethyl-maleimide-sensitive factor), dissolved NSF-binding proteins, the small GTPase Rab7 (Vanlandingham and Ceresa 2009), and some tethers consisting of the HOPS proteins (homotypic fusion and vacuolar protein sorting) (Balderhaar and Ungermann 2013; Pols et al. 2013a). Antibody blocking experiments in a cell-free system clearly demonstrated that the trans-SNARE complex is responsible for the heterotypic fusion of late endosomes with lysosomes and compromises syntaxin 7, Vti1b, syntaxin 8 and VAMP7 (vesicle-associated membrane protein 7) (Pryor et al. 2004).

This is different from the trans-SNARE complex necessary for homotypic late endosome fusion, where VAMP8 substitutes VAMP7 (Antonin et al. 2000). VAMP7 is at the same time necessary for the fusion of lysosomes with the plasma membrane (Rao et al. 2004) and is involved in the fusion of lysosomes with phagosomes (Braun et al. 2004) and with autophagosomes (Fader et al. 2009). Despite the fact that VAMP7 seems to be relevant for a wide range of lysosomal fusion incidents, the absence of apparent developmental abnormalities or the lysosomal phenotype in VAMP7 knockout mice (Sato et al. 2011; Danglot et al. 2012) and the disability to inhibit the endocytic supply to lysosomes by VAMP7 knockdown in cultured cells (Pols et al. 2013a) strongly recommends at least careful interpretation of its probable significance for lysosome fusion.

The fusion event of late endosomes with lysosomes requires both organelles when no recovery event is present, and hence to rebuilding of the lysosome out of endolysosomes seems to be a needed step; however, even though this step is still not clearly revealed. In specific detail, lysosomes possess a more-condensed lumenal amount of proteins than endolysosomes, and an in vitro investigation has figured out that the degree of condensation necessitates a proton-pumping ATPase and lumenal Ca^{2+} ions (Pryor et al. 2000). Hence, the V-ATPase in the lysosomal membrane seems to be not the only responsible molecule for the well-known function of the establishment of the acidic microenvironment for the hydrolysis of macromolecules through lysosomal hydrolases, but additionally it provides a dense-core within lysosomes (Hirota et al. 2004). The reconstruction of lysosomes from existing endolysosomes also involves the recovery of the membrane through the removal of endosomal membrane proteins and the recycling of SNAREs. In line with this, living cell microscopy has proven that vesicular tubular structures abandon the endolysosomes after intracellular endosome–lysosome fusion (Bright et al. 2005). During the remodeling of lysosomes from autolysosomes formed by autophagosome–lysosome fusion (Jahreiss et al. 2008), protolysosomal tubules expand and ripen into lysosomes, the event is controlled by mTOR (mammalian target of rapamycin) (Yu et al. 2010).

Several additional remarks on the mechanism of lysosome reformation are derived from measurements of cells from patients suffering from lysosomal storage diseases. In many cases, these defects are uncommon, inherited genetic disorders that cause imperfections in specific lysosomal acid hydrolases, whereas in other cases they lead to failures in lysosomal membrane proteins or non-enzymatic, soluble lysosomal proteins. Cells obtained from these patients carry membrane-bound heterogeneous storage lesions, probably abnormal endolysosomes/autolysosomes, which are loaded with varying amounts in a variety of diseases (Platt et al. 2012). The deficiency of proper break-down of macromolecules can itself impede the generation of new lysosomes (Bright et al. 1997; Schmid et al. 1999) and even decrease the performance of fusion through the action of membrane cholesterol (Fraldi et al. 2010). A failure in lysosome reformation from endolysosomes has been suggested as a principal cellular deficiency in Niemann-Pick type C2 deficient cells, as opposed to a failure in endosome–lysosome fusion in Niemann-Pick type C1 deficient cells, whereby the later is presumably a sequel to the modified luminal Ca^{2+} levels (Lloyd-Evans et al. 2008). The mechanism by which Niemann—Pick type C2 can hinder the reformation of lysosomes remains a matter of uncertainty. In Chediak-Higashi syndrome, the absence of the protein Lyst (lysosomal trafficking regulator) similarly has no effect on lysosome fusion, but instead influences cleavage through an unidentified pathway (Durchfort et al. 2012). Ultimately, the lack of the lysosomal cation transporter mucolipin-1 in the cells of patients with mucolipidosis type IV was claimed to be the reason for the collapse of lysosome reformation out of endolysosomes (Treusch et al. 2004).

9.1.2 Newly Synthesized Proteins Are Targeted to Lysosomes

Only recently has it been demonstrated that the biosynthesis of lysosomes requires the controlled transcription of genes exclusively coding for lysosomal proteins. Many genes which encode lysosomal enzymes and membrane proteins contain a palindromic 10-bp (base pair) GTCACGTGAC motif in their promoter region that can bind the transcription factor EB (TFEB) and has been referred to as the coordinated lysosomal expression and regulation element (CLEAR) (Sardiello et al. 2009). Non-active TFEB is heavily phosphorylated and attaches to the cytosolic surface of lysosomes, and under certain conditions, such as hunger or lysosome malfunction, it is dephosphorylated and swiftly relocated into the nucleus (Settembre et al. 2013; Settembre and Ballabio 2014), leading to an up-regulation of the synthesis of acid hydrolases and other proteins detected in lysosomes, as well as it is the case with lysosomal membrane proteins.

The pathway through which the majority of newly synthesized acid hydrolases are secreted into lysosomes in mammalian cells has been established for a number of years (Kornfeld and Mellman 1989; Ghosh et al. 2003; Braulke and Bonifacino 2009; Saftig and Klumperman 2009).

After introduction into the lumen of the endoplasmic reticulum, separation of the signal sequence and core glycosylation, these hydrolases enter the Golgi complex, in which the removal of the mask in the cis-Golgi leads to the display of a lysosomal targeted mannose-6-phosphate sugar. When these hydrolases enter the trans-Golgi network (TGN), they are in the next step targeted by one of two mannose-6-phosphate receptors and subsequently are transported towards the endosomes, where they are able to disconnect from the receptors, since the alteration of the pH in the lumen of these endosomes towards an acidic milieu enables them provide the recycling of the receptors to the TGN. Then, the acid hydrolases are allowed to head on towards the endolysosomes and lysosomes, where they are further processed, which causes the switching of these hydrolases to an enzymatically active state. There has been gained huge effort on the knowledge of the molecular insights of the mannose-6-phosphate receptor trafficking process. In the TGN, the cargo-burdened mannose-6-phosphate receptors are separated into clathrin-coated vesicles for transport purpose to the endosomes, whereby the adaptors AP1 and GGAs (Golgi-localized, g-ear-containing, ADP-ribosylation factor-binding proteins) are employed, which can interfere with specific sequence motifs within the cytosolic tails of the mannose-6-phosphate receptors. The way returning from the endosomes to the TGN for the unoccupied mannose-6-phosphate receptors demands the retromer machinery (Attar and Cullen 2010). Although the pathway across the mannose-6-phosphate receptor is the main pathway for the delivery of soluble luminal proteins into the lysosome, it may not be the one and only pathway. It is obvious that in patients with I-cell disease the lysosomal hydrolases fail to gain mannose-6-phosphate tags owing to a deficit of N-acetylglucosamine phosphotransferase activity, but in certain cells, for instance hepatocytes and lymphocytes, a considerable amount of newly produced acid hydrolases achieves the lysosomes. Beyond that, some MPR-independent ways to target lysosomal enzymes have been outlined (Coutinho et al. 2012). In this way, for instance, beta-glucocerebrosidase is made available to lysosomes, utilizing the lysosomal integral membrane protein LIMP-2 as a trafficking and hence targeting receptor.

The boundary membrane of the lysosome comprises over 100 proteins, inclusive of the V-ATPase, which is needed to guarantee an acidic surrounding for the function of lysosomal hydrolases, and many transporters (Schwake et al. 2013). The most common type 1 transmembrane proteins are LAMP1 and LAMP2, which have been proposed to account for 50% of the lysosomal membrane protein fraction. The supply of freshly synthesized integral membrane proteins to the lysosome dispenses with the need for labeling with mannose-6-phosphate (Andrews 2002; Braulke and Bonifacino 2009; Saftig and Klumperman 2009; Schwake et al. 2013). The delivery pathways of the most widely investigated lysosomal membrane proteins have in the past been subdivided into direct and indirect pathways, according to whether the transport from the TGN to the lysosome passes across the plasma membrane (the indirect pathway), even though this can vary between cell types for the identical protein. The so far best characterized sequence motifs targeting membrane proteins from TGN to lysosomes have been defined as YXXØ or [DE]XXXL[LI] types, which also act as endocytic signals. However, delivery to lysosomes is encouraged by some specific characteristics such as a glycine residue immediately preceding the YXXØ residues, and it is noteworthy that these sequence motifs are generally located near the transmembrane domain and frequently near the tip of the short cytosolic tails. When lysosomal membrane proteins comprising those sequence motifs are indirectly transferred to lysosomes, attachment to the AP2-clathrin adapter on the plasma membrane permits rapid sorting into clathrin-coated vesicles for transfer into the endosomal system. These sequence motifs actually link successfully to the clathrin adapter AP3, located primarily on tubulovesicular early sort/recycling endosomes, which guarantees transfer to endolysosomes/lysosomes. Despite the fact that lysosomal membrane proteins had been identified in AP1/clathrin-positive vesicles of the TGN, recently efforts have highlighted a direct pathway from the TGN to the late endosomes, facilitated by a specific type of uncoated vesicles (Pols et al. 2013b). These lysosome membrane protein vehicles have been proven not to incorporate MPRs or endocytic markers, although the HOPS protein hVps41 and VAMP7 are present. The knockdown of hVps41 or VAMP7 led to an aggregation of the carriers. There is comparatively less knowledge of how lysosomal SNAREs are supplied, even though they are crucial for the establishment of a functional lysosome. As with most SNAREs, VAMP7 include no classic tyrosine- or dileucine-based sorting/targeting motifs. However, its longin domain remains accessible for interaction with the trafficking devices if the VAMP7 is located in a cis-SNARE complex after the fusion of the membrane. The longin domain has been reported to be bound to Hrb, which is a protein that can interplay with clathrin and AP2, and additionally with AP3 (Kent et al. 2012). Moreover, these specific interactions are necessary in retrieving and delivering VAMP7 towards the late endocytic organelles.

9.1.3 Secretory Lysosomes

What turns a lysosome into a secretory organelle? Even though lysosomes are not traditional secretory organelles, there is a lot of evidence available that strongly promotes the idea that lysosomes can be secreted at the plasma membrane in practically all cell types. It has been implicated in the repair of plasma membranes and the protection against parasites, in the release of gliotransmitters and ATP out of astrocytes (Reddy et al. 2001; Andrews 2002; Jaiswal et al. 2002; Zhang et al. 2007; Li et al. 2008; Divangahi et al. 2009; Laulagnier et al. 2011). The release of common lysosomes seems to entail the release of a rather small fraction of lysosomes, which are probably located near the plasma membrane. The release lysosomes is calcium-dependent and has been found to affect synaptotagmin VII and the SNARE proteins VAMP7, SNAP23 (23-kDa synaptosome-associated protein) and syntaxin 4 (Rao et al. 2004). Most recent research has demonstrated that TFEB regulates lysosomal secretion at the plasma membrane and induces its translocation towards the plasma membrane and thereafter its release, despite the fact that the exact mechanisms for this are still not fully known (Medina et al. 2011). In several cellular models of lysosomal storage diseases, it has been reported that overexpression of TFEB results in an increase in lysosomal docking and exocytosis, a reduction in lysosomal size, and subsequently in the degradation of accumulated metabolites (Medina et al. 2011; Spampanato et al. 2013).

However, despite some evidence that conventional lysosomes are able to be integrated in the plasma membrane when they are stimulated through injury or other inflammatory substances, there exist still distinct specialized secretory lysosomes, which possess specific mechanisms for their delivery to the plasma membrane and the process of exocytosis. In this type of cell, the lysosomes carry proteins intended for secretion in supplement to the lysosomal hydrolases necessary for protein degradation inside the cell. These specific lysosomes are referred to as "secretory lysosomes" or "lysosome-related organelles" (LROs).

Secretory Lysosomes and Related Organelles

What turns a lysosome into a specialist for secretion? Cells with secretory lysosomes aim for supplementary secretory proteins to the traditional lysosomes and exhibit a machinery for movement towards the plasma membrane which includes the recruitment of supplementary proteins to their outer membrane for both movement and fusion. In cell types, including CTL, mast cells and osteoclasts, the secretory lysosome biogenesis is strikingly similar to the biogenesis of traditional lysosomes. For other cell types utilizing LROs, such as melanocytes and endothelial cells, there are also traditional lysosomes present, and the biogenesis of LROs differs from that of traditional lysosomes. Proteins intended for secretion are specifically delivered to LROs for the dedicated secretion pathway.

Many cell types inherited from the hematopoietic system, including immune cells and osteoclasts, rely on lysosomal secretion as a mechanism for the specialized secretion, whereby various effector proteins are encapsulated for release. Melanocytes and endothelial cells represent the most well characterized non-hematopoietic cell types that employ a very closely resembling mechanism for the generation and fusion of melanosomes and Weibel–Palade bodies (WPBs), respectively. The former releases melanin, which is important for pigmentation, whereas since the ingredients released from Weibel–Palade bodies play a key function in coagulation of the blood. More recently, other cell types have been added to this index, among them astrocytes, catecholaminergic neurons and pulmonary type II cells (Weaver et al. 2002; Tribl et al. 2006; Zhang et al. 2007; Li et al. 2008).

9.1.4 How Travel Secretory Proteins to Lysosomes?

However, there exist several review articles about the trafficking of proteins to lysosomes in melanocytes (Raposo et al. 2007; Sitaram and Marks 2012; Marks et al. 2013). In the following, the aim is the summarized presentation of mechanisms for sorting and secretion of the secretory lysosomes, which are located in cytotoxic T lymphocytes (CTLs) and natural killer (NK) cells. Perforin, a multi-domain protein crucial for the function of cytotoxic T lymphocytes and NK cell effectors, is one of the core proteins retained in the secretory lysosomes of cytotoxic T lymphocytes and NK cells. Moreover, perforin is similar to the C9 component of the membrane attack complex in that it is able to oligomerize at neutral pH and in the vicinity of Ca^{2+} ions, thereby producing 50–300 A° pores over the lipid bilayers (Henkart et al. 1984, 1986; Podack et al. 1985; Lowrey et al. 1989; Baran et al. 2009; Law et al. 2010). The remaining set of proteins secreted by the specialized lysosomes in the CTL is a set of serine proteases referred to as granzymes which break down caspases just the one time in the target cell and induce rapid apoptosis.

However, initial research on the targeting of these proteins indicated that perforin and granzyme pursued quite distinct routes of trafficking (Burkhardt et al. 1989). Granzymes are altered in the same way as lysosomal hydrolases by a mannose-6-phosphate, and their ordering is disturbed if the phosphotransferase which brings about this modification is not present in mucolipidosis II (I-cell disease) (Griffiths and Isaaz 1993). However, perforin is not altered with mannose-6-phosphate, but receives only complex glycans and is unable to use the MPR pathway (Burkhardt et al. 1989; Uellner et al. 1997). The exact way in which perforin is transported to the secretory lysosome is still a puzzle as the interruption of the protein either by cutting or removing the glycans leads to misfolding and degradation when it is expressed in cells (Uellner et al. 1997; Brennan et al. 2011). It has been implied that the AP1 subunit (loss of adaptability) or the LAMP1 inhibits the perforin transport to the granules by interfering with the movement of the granules instead of functioning as a transport receptor (Krzewski et al. 2013).

The transmembrane protein Fas ligand can be additionally transported and hence sorted towards the secretory lysosomes of cytotoxic T lymphocytes (Bossi and Griffiths 1999; Blott et al. 2001; Kojima et al. 2002). The key sorting motif is a proline-rich domain in the cytoplasmic tail of the Fas ligand, with the ability to attach the kinase Fgr and phosphorylate the Fas ligand. It is essential that the proline-rich domain is accompanied by a dileukin motif which can be ubiquitinated and eases sorting to intraluminal vesicles (Blott et al. 2001; Zuccato et al. 2007). Thus, the Fas ligand utilizes the ubiquitination pathway in MVBs and the lysosome, not to degrade but to retain and secrete. It seems unusual that phosphorylation is not necessary for the mono-ubiquitination of the Fas ligand, although the deletion of either tyrosines or lysines interferes with the internalization into intraluminal vesicles. While intraluminal vesicle localization is so far only demonstrated in the expression of the Fas ligand in mast cells (Zuccato et al. 2007), it appears probable that the sorting motifs would sort the Fas ligand according to the same localization in cytotoxic T lymphocytes and NKs. Screening for proteins that attach to the cytoplasmic tail of the Fas ligand has uncovered many binding partners, some of whom have been proposed to be involved in the direction of the Fas ligand to the lysosome (Baum et al. 2005; Thornhill et al. 2007).

Other cell types use related pathways to traffic proteins to LROs. Among them are the acidic hydrolases, which are deposited in LROs of osteoclasts and transferred by the MPR. Another example is AP1 is connected to tyrosinase. Moreover, TYRP-1 and Rab32/38 work together with AP3 during the trafficking within melanocytes and fulfill a crucial task in the assembly of WPBs in endothelial cells. The BLOC complexes, termed BLOC-1, -2, and -3, are critical for the assembly of melanosomes and their subsequent maturation (Marks et al. 2013). The tetraspanin CD63 serves as a platform for the recruitment of elastases into neutrophils; and other proteins (von Willebrand factor, proteins in azurophilic granules of neutrophils) accumulate and thus take on different structures that can induce the development of an LRO around them (Harrison-Lavoie et al. 2006; Doyle et al. 2011).

9.1.4.1 The Common Mechanism Based on Immunodeficiencies

The information of how lytic granules are created, how are they polarized, and finally secreted at the immunological synapse has brought substantially advanced by the analysis of the molecular mechanisms, which are underlying human immunodeficiencies, including familial haemophagocytic lymphohistiocytosis (FHL), Griscelli syndrome, and Hermansky–Pudlak and Chediak–Higashi syndromes. These autosomal recessive diseases have in common the clinical phenotype of continuous infections and the overactivation of the whole immune system, that frequently lead to the death of the affected person. Griscelli, Hermansky–Pudlak and Chediak–Higashi syndrome also lead to albinism. This delivered valuable insights that connecting the biogenesis of secretory lysosomes of immune cells and LROs of melanocytes and endothelial cells.

Chediak–Higashi syndrome (CHS) is phenotypically characterized by an increase in the size of the lysosomes in all different cell types. Nevertheless, only those cell types with secretory lysosomes or LROs have been found to be functionally impaired, indicating that it is the secretory stage of the disease which is selectively impacted. Mutations in CHS were imaged on Lyst (or beige in the mouse), a 430 kDa protein comprising Heat domains, a WD40 repeat, and belonging to the BEACH (beige and Chediak) protein family. Lytic granules in cytotoxic T lymphocytes originating from CHS patients perform properly in protein break-down, but they are not secreted upon recognition of the target cells (Baetz et al. 1995; Ward et al. 2000a). Lyst is able to interfere with several SNARE proteins, either directly or potentially through the CALM protein (Clathrin-assembly lymphoid myeloid leukemia), which participates in the recycling of SNAREs out of the plasma membrane (Miller et al. 2011), the signaling protein 14–3-3, casein kinase II and Hrs. The evidence that secretory lysosomes in cytotoxic T lymphocytes of CHS patients hold back ER-specific membrane proteins and show autophagous inclusions could point to the finding that lyst deficiency leads to normal vesicle fusion but compromised cleavage of lysosomes (Miller et al. 2011; Durchfort et al. 2012). Why exactly these granules cannot secrete in CTLs and melanocytes remains unclear. A possibility that has not been excluded is that the granules are merely too large to fuse correctly.

The AP3-adapter complex is also needed for melanocytes, endothelial cells and platelets needed and not exclusively for CTLs, despite the fact that the precise mechanisms appear to vary between CTLs and melanocytes. The mutations in the b3A subunit of the AP3 heterotetramer lead to Hermansky–Pudlak syndrome type 2 (HPS2) and are mainly connected to partial albinism and platelet dysfunction, as well as to immunodeficiency (Hermansky and Pudlak 1959; Schmid et al. 1999; Huizing et al. 2002). Even though AP3 is essential for the sorting of tyrosinase and melanin-processing proteins in melanosomes (Theos et al. 2005), it may help to figure out why loss of AP3 leads to albinism and reveal the precise mechanism through which cytotoxic T lymphocytes are involved, which is currently only partly clear. However, cytotoxic T lymphocytes of an HPS2 patient exhibit an elevation of the tubulovesicular endosomes and cannot polarize in the direction of the immunological synapse. However, patients with other mutations in the b3A subunit present in comparison only a slight loss of secretion (Clark et al. 2003; Wenham et al. 2010). Since AP3 is notoriously crucial for cargo selection into late endosomes/lysosomes (Kent et al. 2012), it appears probable that AP3 conveys the trafficking of a microtubule motor protein or its adapter to the membrane surface of secretory lysosomes in cytotoxic T lymphocytes.

Other adaptor proteins also have a decisive role in the formation of LROs. In endothelial cells, AP1 arranges the production of Weibel–Palade bodies (Lui-Roberts et al. 2005); however, apart from a possible connection to perforin transfer to secretory lysosomes, little is understood on the role of AP1 in cytotoxic T lymphocytes (Krzewski et al. 2013). Notably, the AP1, AP2 and the AP1 recruitment factor gadkin, and not AP3, are also key for traditional lysosome secretion (Laulagnier et al. 2011), indicating that the participation of different adaptor protein assemblies signifies the selective trafficking of proteins towards either LROs or lysosomes.

9.1.4.2 The Secretion at the Immunological Synapse is a Model for Lysosome Secretion

Lyst and the adaptor complexes fulfill crucial tasks in lysosome biogenesis and additionally in the biogenesis of secretory lysosomes and LROs. However, the next few proteins identified as crucial for the secretion of cytotoxic T lymphocytes seem to be specialized only in secretory lysosomes and LROs. Figure 9.3 provides a summary of these protein functions in the final steps of the secretion at the immunological synapse that is created between cytotoxic T lymphocytes and the target cells, which are then destroyed by the coordinated release of perforin and granzymes. Cytotoxic T lymphocytes and NK cells are key cells of the immunity system that identify and kill virally infected and

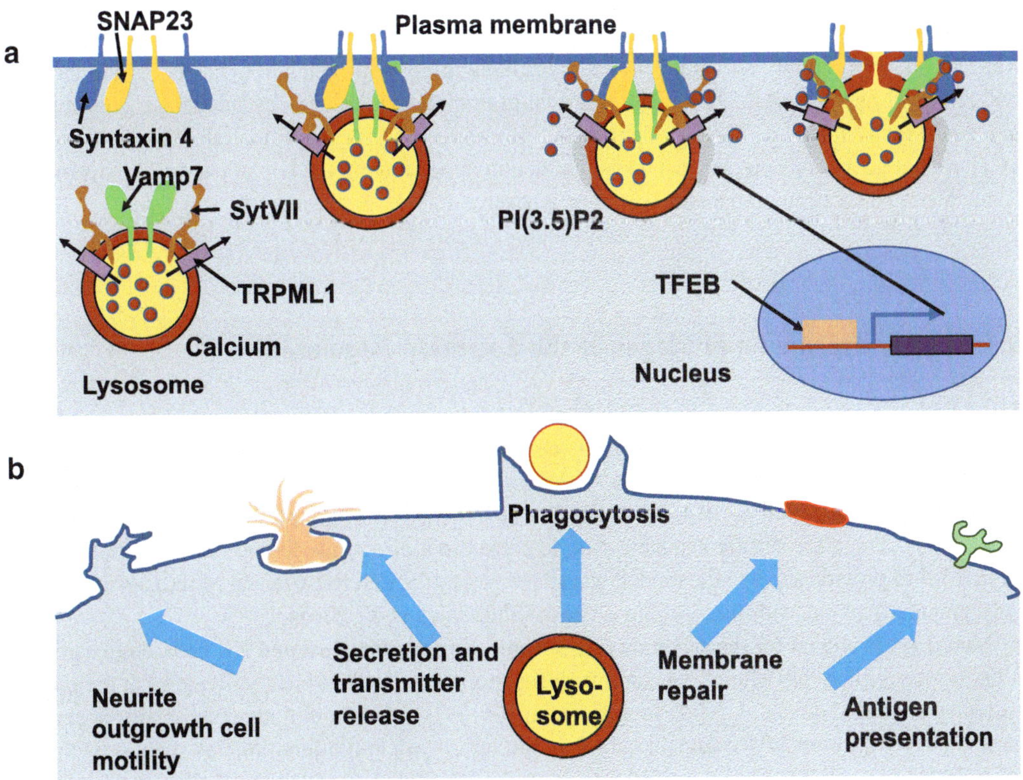

Fig. 9.3 Lysosomal functions. A) Model of lysosome secretion. B) Several functions of lysosomes in close proximity to the plasma membrane

cancer cells. Since their killing is so effective, it is therefore essential that the killing is concentrated on the exact place of recognition between killer and target. This is achieved by polarizing the centrosome to the plasma membrane at the contact point, which is termed the immunological synapse. As soon as cytotoxic T lymphocytes and NKs identify their targets, the secretory lysosomes travel alongside microtubules to the synapse, then attach and release their cargo to the target cell. Immunodeficiencies leading to the four known types of familial hemophagocytic lymphohistiocytosis (FHL) and Griscelli syndrome, which unites FHL with albinism, have led to the identification of four important proteins which are essential for these final steps of secretion.

Two Faces of Rab27A

Rab27A protein is a member of the family of small GTPases that fulfill functions as molecular switches in several diverse membrane transport-based cellular processes (Hutagalung and Novick 2011; Park 2013). Rab27a represents a key molecules in the secretory release of secretory lysosomes from a wide range of different cell types, including CTLs (Kornfeld and Mellman 1989; Haddad et al. 2001; Stinchcombe et al. 2001), endothelial cells (Nightingale et al. 2009; Rojo Pulido et al. 2011), neutrophilic granulocytes (Brzezinska et al. 2008; Herrero-Turrion et al. 2008), platelets (Shirakawa et al. 2004) and melanocytes (Hume et al. 2001; Wu et al. 2001). The absence of functional Rab27A protein leads to the development of the Griscelli syndrome type 2 (GS2), which is associated with a partial albinism and severe immunodeficiency. In specific detail, the lytic granules inside cytotoxic T lymphocytes from GS2 patients can perform a normal polarization towards the immunological synapse, whereas they are still possibly associated with the microtubules, which arranges them similar to pearls on a thread (Stinchcombe et al. 2004). These findings lead to the hypothesis that Rab27a drives the release and the docking of the lysosome. The function of Rab27a in lysosome secretion is connected to additional effectors, which it connects to in other cell types (Fukuda 2013). While active Rab27A attaches directly to Munc13-4 and synaptotagmin-like proteins in cytotoxic T lymphocytes, NK cells and platelets, Rab27a binds melanophilin in melanocytes, thus establishing a connection to myosin V for attachment. Through the binding of various effector proteins, Rab27a can be regarded as a hub that governs its precise mechanism of action depending on the bound partner. In this manner, Rab27a controls various modes of secretory lysosome release in a variety of cell types.

9.1.5 How Are the Final Stages of the Secretion Regulated?

The functional role of Rab27a during the release of lytic granules from cytotoxic T lymphocyte and NK cells is connected to Munc13-4, which seems to drive their marking for the secretion pathway. Mutations within Munc13-4 are known to drive the FHL type 3 establishment that is featured by secretory granules, which are able to attach, but cannot fuse with the plasma membrane (Feldmann et al. 2003). There are results available that demonstrate a direct interference between Rab27A and Munc13-4 to provide support the model, where the two molecules Rab27a and Munc13-4 connect the lytic granule docking with the priming process (Shirakawa et al. 2005).

Munc13-4 is crucial for the secretion of lysosomes from neutrophils and NK cells. In this process, it has been revealed that Munc13-4 can impair the motion of Rab27a-positive vesicles at the place of secretory action (Elstak et al. 2011; Johnson et al. 2011). This function of tethering is almost certainly carried out by the core MUN domain, which is similar to vesicle-binding proteins such as Sec6p (Pei et al. 2009; Li et al. 2011), and is therefore possibly linked directly or through other mechanisms of tethering to membranes (Guan et al. 2008). Alternatively, two C2 domains bordering the core MUN domain might also contribute to the secretory lysosome and LROs tethering at the targeting membrane by coupling phospholipid head groups or t-SNARE complexes existing on the targeting membrane. In addition, several reports suggest an interplay between the MUN domain and syntaxin 11, which could trigger the formation of SNARE complexes (Boswell et al. 2012).

In the end, the final steps of lysosomal secretory pathway are driven by SNARE proteins and additional accessory proteins, which govern the activity of SNARE (Rizo and Sudhof 2012). It has been shown that patients with Syntaxin 11 (Stx11) deficiency, rated as FHL4, have decreased effector functionality of cytotoxic T lymphocytes and NK cells and compromised secretion of platelet-dense granules (zur Stadt et al. 2005; Rudd et al. 2006). The exact mechanism of Stx11 during the lytic granule release from CTLs and NK cells is difficult to comprehend because various studies indicate varying cellular patterns for Stx11 (Advani et al. 1998; Valdez et al. 1999; Prekeris et al. 2000; Arneson et al. 2007; Zhang et al. 2008). Moreover, Stx11 has been demonstrated to assembly a SNARE complex together with two other molecules referred to SNAP23 and VAMP3/8 (Ye et al. 2012), which indicates an active function of the Stx11 molecule in the process of the membrane fusions; however, other investigations pointed out that there exists an inhibitory function of the SNARE complex that breaks down proteins (Grote et al. 2000). Whether this occurs during the fusion of secretory lysosomes at the plasma membrane or possibly sooner during the maturation of secretory lysosomes, as previously proposed for cytotoxic T lymphocytes (Dabrazhynetskaya et al. 2012) has yet to be ascertained.

One other genetic location connected with primary immunodeficiency, that is coding for the syntaxin-binding protein Munc18-2 (FHL5), has been identified (Cote et al. 2009; zur Stadt et al. 2009). The lack of Munc18-2 not only compromises the secretion of cytotoxic T lymphocytes and NK cells, but also of mast cells (Bin et al. 2013), neutrophils (Zhao et al. 2013) and platelets (Al Hawas et al. 2012). It is remarkable that the degranulation of NK cells and cytotoxic T lymphocytes of FHL5 patients can frequently be reconstituted after a cultivation of the cells in the presence of IL2, similarly to FHL4 patients (Cote et al. 2009; zur Stadt et al. 2009; Pagel et al. 2012). This connection is consistent with the insight that Munc18-2 engages Stx11 and possibly serves as its chaperone, which is assessed by a decreased level of Stx11 protein in the cells of FHL5 patients (Cote et al. 2009; zur Stadt et al. 2009; Cetica et al. 2010). Since the polarization and attachment of lytic granules in cytotoxic T lymphocytes of FHL5 patients appears to be a normal process, however, it seems that the function of Munc18-2 occurs late-stage during the secretion, much resembling its neuronal homologue Munc18-1. So far, however, little is generally understood about the interactive relationship between Stx11 and Munc18-2 and about whether it operates in a similar manner to neuronal homologues.

In conclusion, the biosynthesis of lysosomes demands multiple transport ways out of the TGN to supply newly synthesized lysosomal proteins directly or indirectly to the endosomes. The current scientific knowledge is in agreement with the production of lysosomes, which are the consequence of the new formation or budding of endolysosomes/autolysosomes generated by merging preexisting lysosomes with late endosomes/autophagosomes. The new roles for the lysosome underline the fact that it not only represents the end point for the endocytosis. Latest findings indicate that the lysosome has become an organelle that is essential for nutrient uptake, nuclear signaling and multiple types of specialized secretion. These multiple features place the lysosome at the heart of the system, and the way these multiple features are managed will be a fascinating issue for the near and far future.

There are still a lot of remaining issues to be resolved to gain a comprehensive knowledge of lysosome biogenesis. There is a need for improved tools to discriminate endolysosomes from lysosomes and to identify the crucial steps and processes involved in the ripening of endolysosomes to form lysosomes. The great diversity of fusion phenomena in late endocytic and autophagous paths demands a vastly improved comprehension of the SNARE complexes concerned and how they are governed. Moreover, unfortunately there is a lack of familiarity with the splitting mechanism which would permit the production of several lysosomes from one endolysosome and/or the cleavage of these organelles.

While a number of key proteins needed for lysosomal release have been successfully isolated, the sequence of events causing secretion and the exact set of molecular linkages that affect these proteins are still uncertain. There are probably other compounds remaining to be discovered, and from the evidence currently available, it appears probable that these compounds will differ from individual cell type to individual cell type. Rab27a exemplifies this idea perfectly and intervenes with melanophilin in melanocytes and with Munc13-4 in cytotoxic T lymphocytes to ease the release of various lysosome-related organelles. It is obvious that a topic is subject to numerous variations in a number of different cell types, which offer ideal mechanisms for establishing appropriate lysosomal secretion in each of them.

The emergence of new microscopic imaging tools to track lysosomal biogenesis in real time and their capacity to disrupt specific proteins will certainly bring about new findings in this field. As the capability to investigate more specialized cell types improves, a more comprehensive knowledge of lysosomal biogenesis and secretion might arise.

9.2 Lysosomes Can Respond to Environment Mechanics

From the softest tissue, such as the brain (below 1 kPa stiffness) up to the hardest tissues, such as bones (up to 100 kPa), the stiffness of microenvironment of a cell can pronouncedly change within the diverse tissues of the body of mammals. It has been demonstrated that the elasticity of the matrix influences the ability of stem cells to differentiate (Engler et al. 2006), the spread and morphology of cells, and the ability to migrate (Tzvetkova-Chevolleau et al. 2008). Individual cells attach to the substrate and record the stiffness of the material through integrin-facilitated focal adhesions (Chen et al. 2015a, b; Sun et al. 2016). The adhesion of cells is known to guide the control of the tension of the cytoskeleton and the plasma membrane (Sens and Plastino 2015). It has been seen that processes such as exocytosis and endocytosis are tightly governed by the spreading behavior of the cells and the osmotic pressure from the environment (Gauthier et al. 2011). Moreover, the tension of the plasma membrane is driven by the attachment of secretory vesicles through a mechanism associated with Munc18-a (Papadopulos et al. 2015). How can the stiffness of the substrate acquisition guide the process of exocytosis, that in turn can impact the tension of the plasma membrane, is not yet clearly understood. The key drivers of the secretion of mechanisms are SNAREs that drive the fusion of intracellular membranes (Südhof and Rothman 2009). The process of exocytosis is associated with the assembly of the SNARE complex that is composed of a vesicular SNARE (v-SNARE) on the part of the vesicle. The two main v-SNAREs can facilitate specific exocytic mechanisms: clostridial neurotoxin-sensitive VAMP2 (and also the closely related VAMP1 and VAMP3) fulfills a function in the synaptic vesicle and the early endosomal exocytosis, whereas the clostridial neurotoxin-insensitive VAMP7 guides the Golgi-based, late endosomal and lysosomal secretory routes (Proux-Gillardeaux et al. 2005a) that highly characterized by the presence of the tetraspanin CD63 (Chiaruttini et al. 2016; Coco et al. 1999). In the past, the lysosomal exocytosis has been regarded as a universe mechanism, which can be detected in every cell type (Ghosh et al. 2016; Jaiswal et al. 2002; Li et al. 2008; Verderio et al. 2012). However, VAMP2 and VAMP3 have been revealed to drive the recycling process of integrins (Hasan and Hu 2010; Proux-Gillardeaux et al. 2005b; Skalski and Coppolino 2005; Tayeb et al. 2005), and VAMP7 has been identified to fulfill a crucial task in the migration and invasion of cells (Proux-Gillardeaux et al. 2007; Steffen et al. 2008; Williams and Coppolino 2011). Moreover, VAMP7 can additionally aid to facilitate the regulation of the compositional structure of the membrane, which involves the recruitment of sphingolipids and glycosylphosphatidylinositol (GPI)-anchored proteins (Molino et al. 2015) that in turn can alter the dynamics of the integrins within the membrane and the overall adhesion of cells (Eich et al. 2016;

van Zanten et al. 2009). VAMP7 provides the release of ATP (Fader et al. 2012; Verderio et al. 2012) and interleukin-12 (Chiaruttini et al. 2016), and it is necessary for the delivery of vesicles containing the cold-sensing channel TRPM8- and lysosomal-associated membrane protein 1 to sensory neurons (Ghosh et al. 2016). In these neurons, the exocytosis of VAMP2 has been reported to be based on an integrin-facilitated mechanism and the exocytosis of VAMP7 through an integrin-, FAK-, and Src-dependent pathway (Gupton and Gertler 2010).

Atomic force microscopy, micropatterned surfaces and the utilization of pHluorin-based imaging can be employed to determine the effect of substrate sigidity and composition on the process of lysosomal endocytosis due to cellular reaction towards mechanical stress or confinements (Balaji and Ryan 2007). The pHluorin-mediated imaging technique is based on the usage of a GFP variant, which has been generated through structure-directed combinatorial mutagenesis, and can facilitate live imaging of lysosomal exocytosis. In specific detail, the lumen of synaptic vesicles is acidic (a pH of approximately 5.7) represents an environment, which switches the pHluorins off permanently. When the vesicles are fused to the plasma membrane during the process of exocytosis, the pH increases to extracellular levels, which switches the pHluorin on and leads to a fluorescent signal. An advantage of the usage of synapto-pHluorins is that numerous rounds of vesicle release and recycling can be monitored, which consequently permits vesicle recycling also to be visualized. The VAMP7-based lysosomal secretion process can indeed adapt to alterations in rigidity through the action of its partners, such as LRRK1 and VARP, of the peripheral cluster of secretory lysosomes. The stiffness of the cell environment varies considerably between tissues and under pathological conditions. How this characteristic can influence intracellular dynamics of the membrane is widely uncertain. With the help of atomic force microscopy, it can be demonstrated that cells lacking the secretory lysosome v-SNARE VAMP7 are impeded in their adjustment to the substrate stiffness. In contrast, VAMP7-facilitated secretion is encouraged by a stiffer substrate, and this regulation relies on the Longin domain of VAMP7. Furthermore, it has been found that the Longin domain is capable of binding the kinase and the retrograde trafficking adapter LRRK1 and that LRRK1 negatively adjusts VAMP7-driven exocytosis. In the opposite direction, VARP, a VAMP7- and kinesin-1-interacting protein, continues to direct the accessibility for the secretion of peripheral VAMP7 vesicles and the reaction of the cells to mechanical restraints. LRRK1 and VARP operate with VAMP7 in a competition-oriented fashion. A mechanism is suggested by which biomechanical restraints modulate VAMP7-dependent lysosomal secretion through LRRK1 and VARP tension regulation of the peripheral reservoir of secretory lysosomes.

It has been found that VAMP7-based exocytosis of lysosomes is necessary for cells to perceive substrate stiffness and the substrate rigidity redirects VAMP7 to the periphery of the cell in an LD-, VARP- and LRRK1-based fashion. LRRRK1 and VARP seemed to rival for VAMP7 attachment through a conserved region in one of their ankyrin repeats and act through an opposing availability check for the release of peripheral VAMP7 vesicles following mechanical restraints, suggesting a pull-back mechanism.

VAMP7 KO COS7 cells exhibited notably enhanced cell elasticity on soft substrate and reduced elasticity on stiffer substrate in comparison with the control cells. This probably indicates that the absence of VAMP7 could impede the cells from reacting correctly to mechanical restrictions. In contrast, stiffness of the substrate is able to enhance the event of the exocytosis of VAMP7, where the exocytosis of VAMP2 is not affected. This result seems to demonstrate the requirement various membrane types that are delivered towards the surface of the cell due to the mechanical properties of the cellular microenvironment, enclosing especially the rigidity. VAMP7 has been reported to be crucial for the assembly of a phagophore and the secretion of autophagosomes (Fader et al. 2012; Moreau et al. 2011), whereby the rigidity has been seen to elevate the levels of autophagy (Ulbricht et al. 2013). However, no pronounced induction of LC3-II, under the various conditions analyzed, has been observed, and hence, it can be argued that the substrate stiffness can pronouncedly initiate the activation of the autophagy in these approaches.

Cell tension represents a combination of two major factors: firstly, the cortical tension provided by the cell's actomyosin cytoskeleton and secondly, the in-plane tension evoked by the plasma membrane (Sens and Plastino 2015). It was found that the hyperosmotic shocking suppresses the frequency of exocytosis of VAMP7 after a remarkably fast adjustment of the frequency of exocytosis in response to strong alterations in the tension of the membrane. The impact of a hyper-osmotic event on the persistence of the signal at plasma membrane seems to be appropriately provided by reduced fusion pore flattening, since the growth of the fusion pore is facilitated or even pronouncedly elevated by the tension of the membrane (Bretou et al. 2014) and a possibly elevated plasma membrane recovery through the process of endocytosis due to the osmotic shock. Therefore, the data match well with the idea that exocytosis expands the surface area and thus reduces membrane tension and therefore must be switched off to balance the reduced membrane tension after a hyperosmotic shock (Gauthier et al. 2011; Keren 2011; Sens and Plastino 2015). Despite this, however, similar results of a hyperosmotic shock on VAMP7 in which the LD was canceled were detected, while this was not the norm with enhanced substrate stiffness. These findings imply that acute alterations in cell tension, such as osmotic shocks, which are probably acting on membrane tension through a direct impact, and secretory vesicles in the immediate vicinity of the plasma membrane, are triggered by mechanisms that distinguish them from substrate stiffness. Overall, from the two components of cell tension (membrane tension and cortical tension), the first could be related to hyperosmotic shock and not specific for VAMP7, as it was not dependent on Longin, while the second could be the principal regulator activated by substrate stiffness when laminin is present.

It is fascinating that the substrate rigidity and compound composition influenced the exocytosis of VAMP7, meaning the late endosomal and lysosomal CD63+ exocytosis, but not that of VAMP2, meaning the fast endosomal recycling. Probably this suggests that various types of membranes have to be transferred to the surface of the cell, according to the environment of the cell. However, it seems that our results are connected to the earlier evidence that VAMP7 is capable of facilitating the transport of GPI-anchored proteins and lipid microdomains to the plasma membrane (Lafont et al. 1999; Molino et al. 2015; Pocard et al. 2007), thereby in turn facilitating integrin dynamics and adhesion (Eich et al. 2016; van Zanten et al. 2009), the epidermal growth factor (EGFR)-consisting microdomains, and the EGF signal transduction (Danglot et al. 2010). Consequently, an elevated exocytosis of GPI-anchored proteins was observed in the secondary contractile stage in the course of cell spreading (Gauthier et al. 2011). An attractive hypothesis seems to be that VAMP7+/CD63 +secretory late endosomes and lysosomes supply the plasma membrane with lipids which suit best to a stressed membrane, for instance on stiffer substrates and broader transport membrane microdomains, essential for cells to adjust their own mechanical properties and to assess substrate stiffness. Now further investigations are needed to decode the exact signaling pathways ranging from rigidity measurement and cortical tension up to VAMP7 exocytosis.

The mechanism elucidated herein also refers to the inclusion of two members of the VAMP7 subgroup in the mechanosensor-dependent control of transport and exocytosis. In this case, it was observed that LRRK1 strongly reacts with the LD and SNARE domain of VAMP7, whereby the interaction with LD is especially intense in vitro. LRRRK1 and VAMP7 together were transferred to the center of the cell following the uptake of EGF. By shutting down LRRK1, the control of VAMP7 exocytosis through substrate stiffness was eliminated. The overexpression of LRRK1 focused VAMP7 in the center of the cell. This phenomenon prevailed over the substrate stiffness and was further enhanced through kinase activity, as previously shown for EGFR (Ishikawa et al. 2012). VARP arranges the transportation of VAMP7 to the periphery of the cell (Burgo et al. 2009; 2012; Hesketh et al. 2014). At this point, it was observed that VARP efficiently binds DLD VAMP7 and its overexpression reduces the perinuclear reservoir of VAMP7, whereas the peripheral reservoir is increased. Thus, these data provide a useful statement for the elevated exocytosis incidence of DLD VAMP7, as the latter would continue to adhere efficiently to VARP and to a lesser extent to LRRK1. VARP interfaces with the retromer and kinesin-1, and these interactions are essential for the transport of VAMP7 in the direction of the cell periphery on microtubules (Burgo et al. 2009; Burgo et al. 2012; Hesketh et al. 2014; Schäfer et al. 2012). On the contrary, LRRK1 engages with the dynein complex, which is responsible for the retrograde transportation on microtubules in the direction of the center of the cell (Kedashiro et al. 2015). Beyond that, the displacement of the lysosomes in the direction of the peripheral cytoplasm relies on the BORC/BLOC-driven linkage to the plus end of microtubules that is final controlled kinesin motors (Dennis et al. 2016; Guardia et al. 2016; Pu et al. 2015). Besides, it was stated that Rab3 and its effector myosin IIA are important in the transportation to the periphery and in the lysosomal secretion (Encarnacao et al. 2016). In this case, COS7 cells were employed in which a deficiency of myosin IIA has been identified (Even-Ram et al. 2007). Future trials ought therefore to focus on the possible interplay between the VARP-LRKK1 traction mechanism elucidated here, which is probably related to motility on microtubules, and myosin IIA (Encarnacao et al. 2016), and probably the VARP recruitment retromer-associated WASH complex (Hesketh et al. 2014; McGough et al. 2014) and the motion of lysosomes on actin microfilaments. Future studies will surely have to deal with the possible adjustment of these mutual influences by measuring substrate stiffness and signaling.

Overall, the available data prompt us to recommend a molecular pulling mechanism in which LRRK1 and VARP contend for the binding of VAMP7, with LRRK1 targeting the retrograde end and VARP targeting the anterograde end of the VAMP7 trafficking route. Moreover, the stiffness of the substrate would be able to govern the tension between LRRK1 and VARP for peripheral lysosome placement and exocytosis. The effects of LRRK1 and VARP indicate that their intracellular levels are relevant for the spreading of VAMP7 from center towards periphery, which suits the idea of a pulling match well. The signaling cascades capable of controlling the expression of LRRK1 and VARP also need further exploration, since the data indicate that the specific relative quantities of these proteins are crucial for the location of secretory lysosome. A very interesting fact is that VARP is binding Rab40C, and this protein enhances the proteasomal break-down of VARP (Yatsu et al. 2015), and RACK1 hinders the interplay between Varp and Rab40C, thereby antagonizing the adverse impact of Rab40C (Marubashi et al. 2016). RACK1 acts downstream of the β 1 integrin (Lee et al. 2002; Liliental and Chang 1998) and IGFR1 (Kiely et al. 2002; Zhang et al. 2006), which has been revealed to triggers the release of VAMP7 through exocytosis (Burgo et al. 2013). Therefore, it is attractive to assume that the conversion of VARP might be adjusted by integrin-dependent mechanosensors, a hypothesis that now needs to be examined.

Finally, it can be concluded that the VAMP7 lysosomal secretion is governed by the biomechanical restrictions relayed by LRRK1 and VARP, which represents a mechanism that has a broad impact on the dynamics of the plasma membrane under normal physiological (Koseoglu et al. 2015), under infections (Chiaruttini et al. 2016; Larghi et al. 2013), the differentiation of stem cells (Engler et al. 2006), and the initiation and development of cancer (Kostic et al. 2009; Steffen et al. 2008), which are finally all connected to the mechanical constraints of the microenvironment.

9.3 Structure and Function of Peroxisomes

Peroxisomes represent key metabolic organelles that are involved in the lipid metabolism including the β-oxidation of fatty acids and myelin sheath synthesis or the maintenance of the redox balance of the cell (Fig. 9.4). Peroxisomes are often described as ubiquitous, single-membrane-bound

a

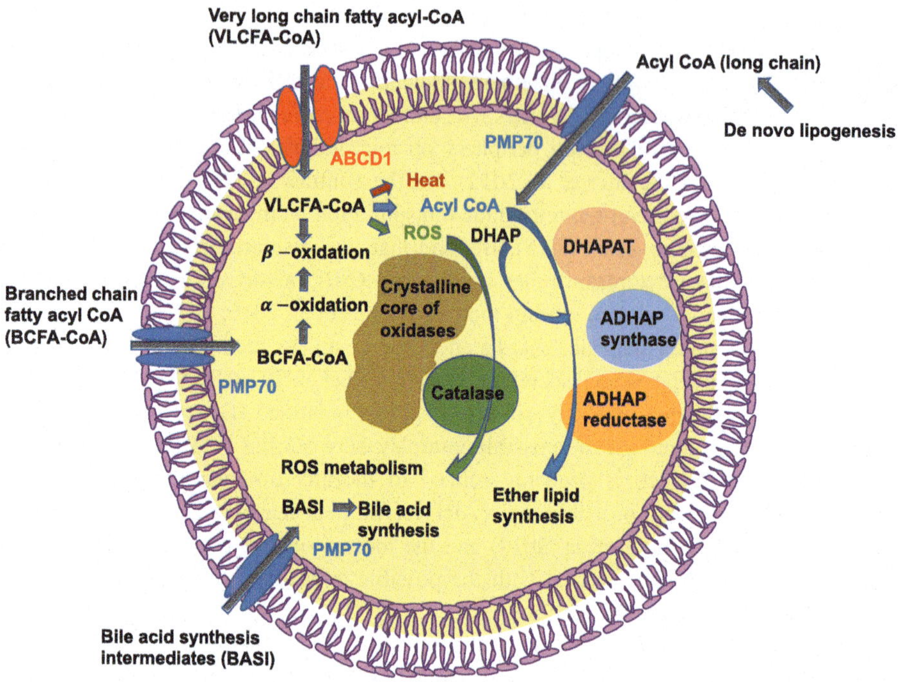

b

$$R-CH_2-CH_2-\overset{O}{\overset{\|}{C}}-S-CoA + O_2 \rightarrow R\ \ CH=CH-\overset{O}{\overset{\|}{C}}-S-CoA + H_2O_2$$

1. $2\,H_2O_2 \xrightarrow{\text{Catalase}} 2\,H_2O + O_2$ **conversion to water**

2. $H_2O_2 + AH_2 \xrightarrow{\text{Catalase}} 2\,H_2O + A$ **oxidation of another organic compound such as AH_2**

Fig. 9.4 **a** Structure and function of peroxisomes. The main metabolic functions of peroxisomes in mammalian cells encompass beta-oxidation of very long-chain fatty acids, alpha-oxidation of branched-chain fatty acids, synthesis of bile acids and ether-linked phospholipids and removal of reactive oxygen species. Many peroxisomes possess a crystalline core of oxidative enzymes. **b** The catalase reaction takes place in the peroxisome

intracellular organelles that possess a fine-structured granular matrix and perform crucial metabolic tasks in the hydrogen peroxide and lipid metabolism, which are in fact highly needed for maintenance of human health and development. The disorders of peroxisomal functions have been connected to severe metabolic malfunctions in humans, whereas peroxisomes are now additionally seen as protective organelles with a broader human health relevance and possible implications for a wide range of broadly relevant human diseases including neurodegeneration, obesity, cancer and age-related disorders. Therefore, research on peroxisomes and their physiological functions has grown considerably in recent years.

In contrast to the mitochondria, peroxisomes are less well-explored over the last five decades. Peroxisomes are considered to be the waste bin of the cell. It has been shown that peroxisomes are dynamic, interrelated organelles which, besides other functions, actively participate in signaling, developmental processes, aging and the protection against pathogens. Peroxisomes exhibit unique properties that have frequently been in conflict with existing concepts in cell biology. It is noteworthy that peroxisomes can completely integrate folded or even oligomeric proteins and under special circumstances produce them de novo directly from the endoplasmic reticulum (ER). The peroxisome organelle has been initially isolated from rat liver and several H_2O_2-producing oxidases have been co-localized within the organelle matrix, including another enzyme, the catalase, which is a H_2O_2-degrading enzyme. Peroxisomes are multifunctional organelles, as their functions vary greatly based on the organism, cell type and stage of development of an individual organism. These organelles participate in numerous metabolic pathways and are essential for the maintenance of cellular homeostasis. Among the common features are the metabolism of hydrogen peroxide and the fatty acid oxidation. However, various specialized features have been developed, such as the biosynthesis of penicillin in fungi, the glycolysis in trypanosomes, the photorespiration and glyoxylate cycle in plants and the synthesis of plasmalogen in mammals (ether lipids, such as plasmalogens, are essential for the neuronal myelin sheaths within the brain). To perform these tasks, peroxisomes interfere, cooperate and intercommunicate with subcellular domains such as the mitochondria, the smooth ER or lipid droplets. While in yeast and plants the β oxidation of fatty acids solely relies on peroxisomes, in animals this function is found both in peroxisomes and mitochondria. Both organelles contain their own batch of β oxidation enzymes, but exhibit substrate sensitivity for selected fatty acids. Extremely long-chain fatty acids and phytanic acid (from dairy products), for instance, can only be broken down in peroxisomes and are poisonous if they become concentrated. In animals, peroxisomes are also implicated in the production of bile acids, inflammatory mediators, for example, leukotrienes and docosahexanoic acid, which is a regulator of neuronal function.

In this chapter, the recent insights on peroxisomes are highlight together with the advancements and trends in peroxisome research. Lastly, an actualization is provided that deal with the unsolved enigmas of these amazing organelles. To summarize new insights into the biological roles of peroxisomes, their biogenesis, development, the membrane dynamics and division as well as links and interactions between peroxisome and organelles. In addition, new peroxisomal proteins and mechanisms at the peroxisomal membrane are subject of discussion. Ultimately, the latest knowledge about the function of peroxisomes in the brain, in neurological disturbances and in the development of cancer has been addressed (Islinger et al. 2018).

Interest in peroxisomes and their (patho-)physiological importance for health and disease is steadily rising, and there is little dispute that peroxisomes are on the upswing in the research field. Since their detection over 60 years ago (Rhodin 1958), major metabolic processes of peroxisomes (for instance in fat metabolism including fatty acid β-oxidation and synthesis of myelin sheath lipids or the metabolism of reactive oxygen species (ROS), especially hydrogen peroxide) have been discovered, showing that peroxisomes are important organelles in metabolism and their malfunction has been associated with severe metabolic disturbances in humans. In addition to the enzymatic activity of catalase, peroxisomes carry several other enzymes that participate in the generation and capture of reactive oxygen species (ROS), such as superoxide dismutases, peroxiredoxins and glutathione peroxidase, and thus may be involved in ROS metabolism, oxidative stress, aging, neurodegeneration and carcinogenesis. In accordance with this, these organelles are involved in the signaling of ROS and reactive nitrogen species. A novel biological feature for peroxisomes has also been detected in antiviral innate immunity and the antiviral signaling by the outer mitochondrial membrane protein MAVS.

How are the proteins of peroxisomes recruited? Peroxisomal matrix proteins are all produced on free ribosomes inside the cytoplasm and subsequently transferred to the organelle (Fig. 9.5). Two peroxisomal markers, such as PTS1 and PTS2, are identified. The signals are identified by soluble receptor peroxins (Pex5p or Pex7p), interfering with a receptor attachment site on the peroxisomal

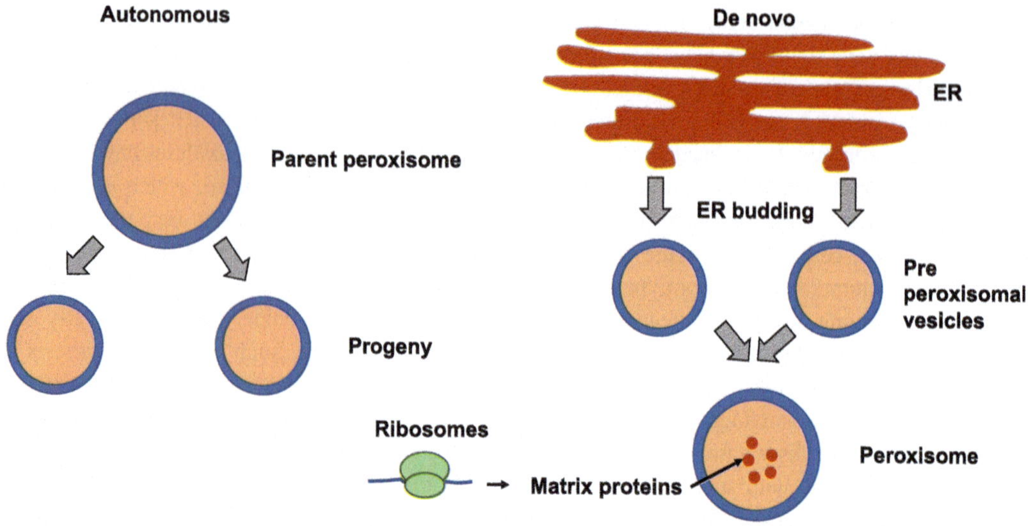

Fig. 9.5 Potential pathways of the biogenesis of peroxisomes. Peroxisomes are formed autonomously through the division of preexisting organelles (left) or through a de novo process covering the budding from the ER subsequently to the import of matrix proteins (right)

membrane. The import of matrix proteins includes a sophisticated protein relocation system that can process folded, cofactor-bound and oligomeric proteins. It is proposed that the receptor Pex5p fits into the peroxisomal membrane to produce a large transient pore for freight relocation. In this case, the recycling of the Pex5p receptor is facilitated by ubiquitous reactions. The majority of the well-known peroxins are implicated in the acquisition of peroxisomal matrix proteins. The targeting of peroxisomal membrane proteins (PMPs) demands a totally separate, poorly defined mechanism, which involves the cytosolic receptor/chaperone Pex19p. When a nascent PMP is attached to Pex19p, the PMPs Pex3p and Pex16p interoperates with Pex19p to convey the membrane integration of the nascent protein. It is presently not clear whether all PMPs are transferred to the cytoplasm or whether the PMPs are first introduced into the ER and then sprouted from the ER into specialized vesicles that are ultimately passed on to peroxisomes (Fig. 9.5).

How can a peroxisome proliferate? Peroxisomes constitute dynamic organelles acting in response to physiological alterations in their cellular environment and adjusting their number, enzymatic concentration and metabolic processes appropriately. An augmentation of the peroxisome number and/or enzymes (peroxisome proliferation) can be encouraged, for instance by fatty acids, cold adjustment or hypolipidemic substances, while hypoxia or hunger lead to their reduction. The classical peroxisome amplifiers in mammals such as fatty acids and hypolipidemic fibrates attach to the nuclear hormone receptor PPARα, which stimulates the transcription of peroxisomal genes participating in β-oxidation and proliferation, such as Pex11. Prolonged treatment with peroxisome proliferators produces a carcinogenic response in rodents resulting in liver tumors. Supplementary mechanisms independent of PPARα have been proposed. Peroxisome subpopulations with different morphology, protein constitution and functionality are probably created following a variety of stimuli and the activation of specific transcription factors.

In the last few years, it has emerged that peroxisomes also perform decisive non-metabolic functions, for instance in cellular stress reactions, the fight against pathogens and antiviral defense, serving as cellular signaling sites and in the process of healthy aging. These results suggest that peroxisomes are also “protective” organelles, which have a broader relevance for human health and have potential effects on a wide range of key universal human diseases including neurodegeneration, obesity, cancer and age-related disorders. However, many aspects of the physiological functions of the peroxisome are still puzzling. The insights, advancements and future directions of the peroxisomal research effort will be highlighted, which will then guide through the current developments in the field of peroxisome biology. Hence, an update and the continuation of the research efforts is presented (Schrader and Fahimi 2008; Islinger et al. 2012a, 2012b). New progress in comprehension of pexophagy, the guided elimination of peroxisomes, has been discussed (Eberhart and Kovacs 2018).

Peroxisomes are small single-membrane-confined organelles and possess at least 50 different enzymes regulating metabolic reactions in furtherance of getting energy (Schrader and Fahimi 2008). The similarity to lysosomes is quite high. The assembly of peroxisomes is similarly to that of mitochondria or chloroplasts: The proteins are synthesized on free ribosomes and transported into the peroxisomes in which they are polymerized to polypeptide chains. Peroxisomes are originally free of nucleic acids and ribosomes. They contain no DNA building an own genome and replicate by cell division. The peroxisomes perform oxidation reactions which lead to the harmful end product hydrogen peroxide. Hence, the peroxisomes contain the enzyme catalase decomposing hydrogen peroxide to water or using it for the conversion of another organic compound. Many substrates such as uric acid, amino acids or fatty acids are degraded by oxidative reactions within peroxisomes providing a major source of energy (Fig. 9.4).

Peroxisomes are involved in the synthesis of lipids. Cholesterol and dolichol are both synthesized in peroxisomes and the endoplasmatic reticulum (ER). Most proteins are carrying the short amino acid sequence Ser-Lys-Leu (SKL) at their carboxy terminus, termed peroxisome targeting signal 1 = PST1. Some proteins carry a N-terminal targeting sequence of nine amino acids, termed PTS2 and not yet known sequences. Both targeting sequences PTS1 and PTS2 are seen by distinct receptors and when recognized they are transferred to a translocation complex facilitating the transport into the peroxisome. Cytosolic Hsp 70 is involved in the protein import into peroxisomes and thus acts as a chaperone. The imported proteins are mostly partially folded rather than extended. Some components of peroxisome import processes are identified to cause serious human diseases by evoking disorders of peroxisomes. In some cases, one single mutated enzyme is the cause of a defect in peroxisome function. One example is the Zellweger syndromes (possessing mutations in at least ten different genes regulating peroxisomal protein import such as the receptor for TS1 receptor) which is lethal in the first ten year after birth.

Peroxisomes perform lipid metabolism and are involved in cellular signal transduction processes (Lodhi and Semenkovich 2014). In more detail, they clean up reactive oxygen chemical debris produced by other organelles. They act far beyond hydrogen peroxide metabolism. Peroxisomes are associated with lipid droplets and mitochondria. Oxidation of fatty acids and synthesis of lipids such as ether lipids belong to their functions. They are critical for cellular signal transduction processes for normal physiological processes. Moreover, they are important for diseases such as cancer, obesity-related diabetes or neurologic diseases due to degeneration. In all eukaryotic cells, peroxisomes are present as multifunctional organelles. Peroxisomes are highly plastic and adapt to cellular or environmental cues by adapting their size, number, morphology and function (Schrader et al. 2013). They are enclosed by a single membrane containing a granular matrix with a crystalline core (Rhodin 1958). Peroxisomes contain oxidases and catalases to perform hydrogen peroxide metabolism (De Duve and Baudhuin 1966). They function in the metabolism of hydrogen peroxide, which seems to be their primary function. Secondly, they have been characterized as microbodies using a clinical staining procedure with the 3,3-diaminobenzadine (DAB) dye revealing their visualization based on peroxidase activity of catalase at alkaline pH (Fahimi 1969, Novikoff and Goldfischer 1969). The shape of peroxisomes can vary largely in shape such as their diameter can have values from 0.1 to 0.5 μm. In adipocytes, the peroxisomes are smaller and are increased in number during the differentiation of adipogenic cells. They are fundamentally important in lipid metabolism (Hajra and Das 1996). In more detail, they perform beta-oxidation of very long-chain acids, which is also performed in mitochondria. The alpha-oxidation of branched fatty acids and the synthesis of ether-linked phospholipids and bile acids are additionally performed in peroxisomes.

The origin of peroxisomes remains controversial (Dimitrov et al. 2013). Peroxisomes are autonomous organelles, and they are derived from growth and division of the preexisting organelles, templated assembly using a mast copy or de novo synthesis (Lazarow and Fujiki 1985). They are visualized as dumbbell-shaped interconnected structures. In yeast cells with a gene mutation impairing peroxisome formation the reintroduction of the wild type of the gene complements the defect and restores peroxisome formation (Baerends et al. 1996; Wiemer et al. 1996). However, peroxisomes are not exclusively autonomous, instead they are semiautonomous as matrix proteins from the cytosol and membrane proteins as well as lipids from the ER are transported into the peroxisomes (Tabak et al. 2013). As numerous interactions such as contacts between the smooth ER and peroxisomes exist, it has been suggested that peroxisomes arise from the dilated region of the ER consisting high levels of peroxisomal proteins such as catalase (Novikoff and Novikoff 1972; Redely and Svoboda 1971). However, preperoxisomal vesicles (Precursors of peroxisomes) are build off the ER, s luminal continuity is needed for the transport of peroxisomal contents (van der Zand et al. 2012).

In more detail, two distinct preperoxisomal vesicles containing half of the peroxisomal translocation complex fuse heterotypically and assemble a functional translocon taking up matrix proteins from the cytosol (van der Zand et al. 2012).

Peroxisomes contain at least 12 peroxins encoded by Pex genes regulating peroxisomal membranes, factors interacting with peroxisomal targeting sequences (PTSs) for protein shuttling to peroxisomes and factors acting as docking receptors for peroxisomal proteins.

Pex 3, Pex 16 and Pex 19 are critical for the assembly of the peroxisomal membrane and needed of the peroxisomal membrane and needed for importing peroxisomal membrane proteins (PMPs). Pex 19 is a soluble chaperone-receptor for newly synthesized PMPs (Jones et al. 2004). Pex 3 buds a preperoxisomal vesicle and docking receptor for Pex 19 (Fang et al. 2004). Pex 16 builds a docking site on the peroxisomal membrane to recruit Pex 3 (Matsuzaki and Fujiki 2008). Peroxisomal matrix proteins are translocated at free ribosomes in the cytoplasm with PTSs ribosomes in the cytoplasm with PTSs at the carboxyl (PTS1) or amino (PTS2) terminus (Gould et al. 1987).

9.3.1 Functions of Peroxisomes in Metabolism

The subcellular architecture is characterized by a single-layer membrane that envelops a granular, homogeneous matrix, detected in the kidney cells of rodents and later on in the liver, and is referred to as "microbody" in accordance with its morphology (Rhodin 1958; Rouiller and Bernhard 1956). Initially, the microbody was seen as a peculiarity of the cell, without an identified role in essential functions and intermediary metabolism. In the following decades, however, there was growing scientific evidence that the obscure "Cinderella" among the famous cell organelles gradually developed into a multifunctional and multi-functional global player with deep and extensive significance for the health and disease of animal and plant organisms.

Initiated by the groundbreaking work on the unambiguous biochemical individualization and characterization of microbodies, which have been renamed to peroxisomes (De Duve 1965; De Duve and Baudhuin 1966; Vamecq et al. 2014). Moreover, the finding Zellweger patients possess no peroxisomes has been established their important role inside of cells (Goldfischer et al. 1973). Therefore, it is now well-established that peroxisomes are essential for eukaryotic cells and are thus practically ubiquitously distributed. They have a unique heterogeneity in shape and morphology and exhibit a striking functional plasticity in both anabolic and catabolic pathways. Their proteome is tailored to the cell type, growth requirements and varying environment. Not least, they are engaged in basic essential functions such as the detoxification of dangerous oxygen/nitrogen species (Fransen et al. 2012), provide signaling stages (Mast et al. 2015), play crucial roles for innate immunity (Dixit et al. 2010) development and differentiation (Titorenko and Rachubinski 2004), and communicate closely with other organelles (Schrader et al. 2013; Shai et al. 2016). Malfunctions or even the absence of peroxisomes have not only been the cause of the common peroxisomal dysfunctions, but also participate in physio- and pathophysiological mechanisms such as aging and related diseases (Deori et al. 2018) or even cancer.

In order to jointly support the wellness of a cell and improve its multiple roles, peroxisomes work together and interact with other cell organelles. Multiple mechanisms have developed that facilitate such communication, including signal transduction pathways, trafficking of vesicles and contact points (Shai et al. 2016). Peroxisome, ER and mitochondrial interactions are most frequent, but neither the fundamental mechanisms nor the functionality have been experimentally confirmed in depth (Shai et al. 2016).

The influence of peroxisomes on lipid metabolism can most effectively be illustrated by the accumulation of very long-chain fatty acids (VLCFA) in plasma and the complete lack of plasmalogen in the tissues of Zellweger patients (Brown et al. 1982; Heymans et al. 1983). The synthesis of ether lipids is performed in peroxisomes and starts with the esterification of dihydroxyacetone phosphate (DHAP) with a long-chain fatty acid by the enzyme DHAP-acyltransferase (DHAPAT) and the consequent substitution of the fatty acid by a fatty alcohol to form alkyl-DHAP by alkyl-glycerone-phosphate synthase (AGPS). Noteworthy, the crucial AGPS enzyme is elevated in aggressive cancer cells and primary human breast tumors, and its genetic ablation severely impairs the aggressiveness of cancer and tumor development (Benjamin et al. 2013). Since it has been established that AGPS knockdown dramatically affects tumor growth in mice and blocking AGPS activity reduces ether lipids and decreases cancer pathogenicity in various types of human cancer cells (Piano et al. 2015), it could be decisive to investigate the development of effective suitable inhibitors in cancer therapy.

Peroxisomes and mitochondria can excessively interplay the metabolism of fatty acids (Wanders 2014) and the ROS metabolism (Fransen et al. 2012; Lismont et al. 2015), and during the detoxification of glyoxylate and phytanic acid (Wanders et al. 2011). Perhaps most significantly, peroxisomes exclusively β-oxidize VLCFA. Elevated levels of VLCFA have been detected in body fluids and tissues of patients with X-ALD and acyl-CoA oxidase 1 (ACOX1) deficiency, with specific implications for the nervous system (Wanders et al. 2010). Suitable cytotoxic characteristics of the VLCFA are inflammatory demyelination and axonopathy, cell death of oligodendrocytes and astrocytes, a deregulation of the intracellular Ca^{2+} homeostasis and a significant reduction of the membrane potential of mitochondria inside oligodendrocytes (Hein et al. 2008).

In mammalian organisms, including humans, the α oxidation of 3-methyl-branched fatty acids such as phytanic acid represents a strictly peroxisomal event. In order to clarify the toxic characteristics of phytanic acid when processed improperly, the hypothesis was first put forward that its integration into membranes disturbs the arrangement of lipids and their interactions with the proteins and thus impairs their integrity. In the alternative, assuming that the chemical structure of phytanic acid has similarities with that of vitamins A, E and K, phytanic acid can function as an antimetabolite in respect of these isoprenoids. Successive in vitro investigations have concentrated primarily on the impact of phytic acid on mitochondria (Schönfeld and Struy 1999), but it is not yet clear whether these in vitro properties also correspond to the in vivo status of Refsum disease.

9.3.2 Heterogeneity of Peroxisomes

The heterogeneity of peroxisomes has been mentioned in early electron microscopic experiments, when these organelles were described simply as "microbodies and related particles" (Hruban and Rechcigl 1969). The detection of the hydrogen peroxide metabolism and its denomination as "peroxisome" underline the similarity and the strong relatedness of this group of organelles in animal and plant cells (De Duve and Baudhuin 1966). But following investigations highlighted the typical characteristics of peroxisomes of various organs, such as the clear distinction between peroxisomes from rat liver and brain (Gaunt and de Duve 1976). Beyond that, the changes in the enzymes of the peroxisomes in the progression of pre- and postnatal stages demonstrated the ability of these organelles to adjust to the varying metabolic demands of the organism (Krahling et al. 1979). The diversity of peroxisomes within the animal kingdom has been reported in a comprehensive review (Islinger et al. 2010a, b). The heterogeneity of the peroxisomes can be determined using the cytochemical technique for D-amino acid oxidase with the use of cerium (Angermüller and Fahimi 1988; Angermüller 1989). A mosaic structure with strongly and weakly reactive peroxisomes is seen in rat hepatocytes, with the total staining occurring more strongly in the peri-portal (oxygen-rich conc.) parts of the liver lobules than in their peri-central (oxygen-poor conc.) parts. Within the kidney, the proximal tubules of the renal cortex are heavily dyed, while the remainder of the nephron is strictly negative. Especially in certain cells, strongly and weakly stained peroxisomes are next to each other in the selfsame cells (Angermüller and Fahimi 1988). The occurrence of heterogeneous subpopulations of peroxisomes in biochemical analyses has also been found in isolated peroxisomes (Lüers et al. 1993; Islinger et al. 2012a, b), in morphological investigations of mammalian cell cultures (Schrader et al. 1994) and in the development of fungi (Takano-Rojas et al. 2016). These variations have been associated with the production and maturation of peroxisomes. It is of interest that de novo generation of peroxisomes from the ER through preperoxisomal vesicles as well as from preexisting organelles through membrane growth and division results in the establishment of membrane sections, ripening through consecutive implantation of matrix proteins (Hoepfner et al. 2005; Delille et al. 2010). The matrix constituents of preexisting peroxisomes are consequently not uniformly spread over new organelles, suggesting that peroxisome production by division appears to be an asymmetric process (Huybrechts et al. 2009; Delille et al. 2010).

Peroxisomes possess an age-dependent heterogeneity due to their capability to take up newly synthesized proteins (Huybrechts et al. 2009) and the segregation of proteins in the course of the cell cycling at the final cell division event (Kumar et al. 2018). This segregation function is also an issue for peroxisomal membrane proteins (PMPs), since they remodel the themselves within the peroxisomal membrane during the growth of the membrane and the process of cell division (Delille et al. 2010; Cepińska et al. 2011). Super-resolution microscopy supported has been supported the finding that the PMPs are compartmentalized (Galiani et al. 2016; Soliman et al. 2018). There can be even a higher degree of heterogeneity and also PMP compartmentalization be reached through the dynamic interaction of membrane tether-based contact sites towards other organelles, such as the ER. Moreover, the tethering interaction affects the motility of the entire peroxisome, which may provide an explanation why only a certain fraction of peroxisomes can perform a long-range movement by using cytoskeletal routes and finally cumulates in heterogeneous movement events (Costello et al. 2017; Castro et al. 2018a, b).

9.3.3 Peroxins, Import of Proteins and Membrane Adaptors

The biogenesis of peroxisome requires the production of the peroxisomal membrane, the subsequent homing and insertion of PMPs into this peroxisomal lipid bilayer, and the entry of enzymes/proteins into the peroxisomal matrix. Contrarily to other organelles, such as mitochondria or the ER, peroxisomes are able to import fully folded, oligomeric or cofactor-bound proteins by the guidance of the dynamic protein translocon (Meinecke et al. 2010; Montilla-Martinez et al. 2015; Dias et al. 2017). More precisely, the import of matrix proteins and almost all PMPs requires the function of largely conserved, but rather diverse import systems each of which with individual properties. Key biogenesis components, referred to as peroxins (Pex proteins), constitute the import mechanisms. During the last years, the number of peroxins has been increased to 36, whereby peroxin 36 is the newest discovered one (Farré et al. 2017). Pex9 represents a new Pex5-like yeast peroxisomal targeting receptor which can bind a specific subset of PTS1 (peroxisomal targeting signal)-containing matrix proteins during the growth phase in oleate (Effelsberg et al. 2016; Yifrach et al. 2016). There are two diverse PTS1 receptors, termed Pex5 and Pex9 (additionally to the PTS2-driven import pathways), that enables yeast cells to coordinate the metabolic capacity of peroxisomes to microenvironmental alterations. The Pex35 molecule has been revealed to be a new peroxisomal membrane protein in yeast and acts as a regulatory protein of peroxisome abundance (Yofe et al. 2017), whereas the new yeast peroxin Pex36 represents a functional homolog of the mammalian Pex16, which fulfills tasks in the trafficking of PMPs from the ER towards the peroxisome (Farré et al. 2017). In the end, a role in the import of matrix proteins is inherently necessary for the β-oxidation required of fatty acid and the synthesis of bile acid has been addressed to the peroxisomal transmembrane protein TMEM135 (PMP52) that is highly homological o the TIM17 family proteins, which drive the translocation of proteins over the membranes of mitochondria (Renquist et al. 2018). Initially, TMEM135 has been detected to be a novel target of liver x receptors (LXRs) that all are assembled to the nuclear receptor superfamily and act as crucial major regulators of the cholesterol and fatty acid metabolism (Renquist et al. 2018).

With relation to the import of matrix proteins, the processes of freight translocation and the recycling of receptors continue to be discussed (Francisco et al. 2017). Advances have been achieved in gaining an understanding of the intrinsic structure and unique molecular function of the peroxisomal AAA-ATPase Pex1/Pex6 complex, which participates in the export and subsequent reuse of the ubiquitinated import receptors Pex5 and Pex7 (Ciniawsky et al. 2015; Blok et al. 2015; Schwerter et al. 2017). It has been previously demonstrated that the AAA-ATPase Pex1/Pex6 deploys substrates by continuous threading (Gardner et al. 2018), and that monoubiquitinated Pex5, which is able to interfere with the AAA-ATPases Pex1 and Pex6, is deployed during its displacement in the cytosol (Pedrosa et al. 2018).

Intriguingly, more proof has now been offered that human Pex5 can act as a redox/stress sensor to trap peroxisomal catalase within the cytosol and thus control the oxidative stress of non-peroxisomal source (Walton et al. 2017). Notably, small molecule inhibitors have been designed for the import of peroxisomal (glycosomal) proteins (against Pex14), which are capable of intercepting the import of glycosomal matrix proteins in trypanosoma parasites. As a consequence, the glycosomal enzymes are incorrectly localized, resulting in a metabolic disaster and the parasite's death (Dawidowski et al. 2017). The import of peroxisomal proteins is associated with redox homeostasis and healthy aging processes, as well as with the control of parasites and the emergence of new therapies to combat trypanosomiases.

The import/entry of PMPs strongly relies on the three membrane biogenesis factors Pex19, Pex16 and Pex3. There exist several reviews on peroxisomal membrane biogenesis, the PMP-dependent targeting and the integration in the lipid bilayer (Erdmann 2016). New insights have been gained into the targeting, the insertion and the regulation of the quality control mechanisms of tail-anchored membrane proteins on peroxisomes. More recent investigations confirm a direct, Pex19-dependent path (Yagita et al. 2013; Costello et al. 2017a, b, c) and a hydrophobic transfer feature for membrane insertion mechanism (Chen et al. 2014). Targeted specific information in peroxisomal TA proteins was also identified and new peroxisomal TA proteins prophezied and predicted (Buentzel et al. 2015; Costello et al. 2017a, b, c). These comprise the peroxisome ER tether ACBD4 and MIRO1, which represents an adaptor protein for the motor proteins. A role for the AAA protein Msp1/ATAD1 in the removal of abundant tail-anchored proteins within the peroxisomal membrane has also been identified (Weir et al. 2017). A large proportion of these TA proteins, which function as membrane adaptors for key, disease-related cellular mechanisms, are exchanged with the mitochondria (Schrader et al. 2015a, b).

9.3.4 Peroxisomal Membrane Proteins

Since many Pex mutants (excluding Pex3, Pex16 and Pex19) are only broken when peroxisomal matrix proteins are inserted and they still carry peroxisomal residues or ghosts bearing PMPs, sorting PMPs demands compounds that differ from those participating in the implantation of peroxisomal matrix proteins. PMPs perform a wide range of purposes, such as acting as components of the peroxisomal translocon or exporting of material, role as membrane transporters for metabolites and ions, act in the quality control or the organelle division system, serves as redox proteins, signaling molecules, organelle membrane linker and so forth. PMPs comprise one or more mPTSs (Jones et al. 2001) which are either Pex19-dependent or Pex19-independent ordered to the peroxisome membrane (Giannopoulou et al. 2016).

Depending on whether Pex19 is necessary for their membrane incorporation step or not, these PMPs are roughly divided into two basic classes:

Class I or direct pathway of PMPs: They utilize the Pex19-driven membrane integration of PMPs, which is a main pathway for the majority of PMPs (Sacksteder et al. 2000; Snyder et al. 2000; Jones et al. 2004).

Class II or indirect pathway of PMPs: This alternative was originally recommended to treat the PEX19-independent membrane incorporation of PEX3 in mammalian cells (Jones et al. 2004; Kim et al. 2006), which leads to peroxisomes through the ER (Fakieh et al. 2013; Mayerhofer et al. 2016). But since these early studies have confirmed that many PMPs enter the peroxisomes through the ER; therefore, it is preferable to refer to this as the indirect route (i.e., through the ER) of PMP trafficking towards the peroxisomes (van der Zand et al. 2010).

Also, it should be considered that the identical PMP can directly or indirectly enter peroxisomes. Consequently, the mammalian PEX3 can additionally be targeted directly towards peroxisomes by employing a PEX16- and PEX19-facilitated way (Matsuzaki and Fujiki 2008). It is possible that most PMP (or even all of them) possess the flexibility to aim at peroxisomes directly or indirectly through the ER (van der Zand et al. 2010), the latter mode of action is the only possible mode in the absence of preexisting peroxisomes.

The tail-anchored (TA) proteins are a subclass of PMPs that is characterized by the integral membrane proteins possessing a short, C-terminal sequence, which positioned adjacent to their transmembrane domain (TMD) (Borgese et al. 2007). The insertion of these TA proteins into the membranes (of either peroxisomal membranes or ER membranes) can be Pex19-dependent or Pex19-independent, either by direct integration into pre-existing peroxisomes or in an indirect fashion through the prior integration into membranes of other subcellular organelles, wherefrom subsequently peroxisomes are generated.

9.3.5 The Assembly of the Peroxisome

The peroxisomes can be generated by growth and division of preexisting peroxisomes (Farré et al. 2019). In addition, the peroxisomes can be made of the former ER. The de novo formation of peroxisomes has been found in mutant cells without any peroxisome after the reintroduction of the gene, such as Pex19, Pex3 or Pex16, that been mutated. The division of peroxisomes is performed by the expansion of the organelle membrane through a mechanism that is facilitated by the action of the conserved PMP Pex11p. In the end, the fission necessitates dynamin-related proteins, which have GTPase activity, and associated adaptor proteins, such as DLP1/Drp1 and the adaptors Fis1 and MFF in mammals. It is interesting to note, these key elements are exchanged with mitochondria, which seems to be generally acquired strategy of mammals, fungi and plants. The generation of peroxisomes either from other organelles, such as the ER, or by the division of enlarged existing increasingly growing peroxisomes seems to be based on a strict a maturation process, which induces the recruitment of new membrane and matrix proteins to these sites.

It is meanwhile acknowledged that peroxisomes can be generated through the traditional path of growth and division of pre-existent organelles or through an alternative path of de novo generation of emerging peroxisomes (Hettema et al. 2014; Agrawal and Subramani 2016). The latter approach is supported by examinations on mutated cells which are lacking peroxisomes as a result of the loss of membrane biogenesis factors Pex3, Pex16 or Pex19. However, the peroxisome counts seem to be controlled primarily through growth and division (Motley and Hettema 2007). The de novo scheme implies that several key PMPs, including Pex3, are directed to the ER, which then sequester into pre-peroxisomal vesicles that are liberated and build import competent peroxisomes which then continue to grow and cleave to proliferate (Hoepfner et al. 2005). There was some discussion about initiating the de novo formation pf peroxisomes at the ER, since pre-peroxisomal vesicles were also detected in yeast cells deficient in Pex3 or Pex19. These vesicles were decomposed by autophagy and had therefore been overseen (Knoops et al. 2014; Wróblewska et al. 2017). There has been seen by recent investigations in yeast that the reticulon-like proteins Pex30 and Pex31 proteins possess a crucial role in the creation of an ER subdomain, at which region pre-peroxisomal vesicles can bud (David et al. 2013; Mast et al. 2016; Joshi et al. 2016). In addition, it has been established that the ESCRT III proteins Vps20 and Snf7 can be involved in the clearance of pre-peroxisomal vesicles out of the ER (Mast et al. 2018), promoting the ER origin path of pre-peroxisomal vesicles. Besides, Pex36, a novel mammalian yeast peroxin and functional homologue of Pex16, has been discovered that works on ER peroxisome transport of membrane-bound peroxisomal proteins (Farré et al. 2017).

The investigations of human patient fibroblasts, which are deficient of Pex3 or Pex16 and hence consequently free of peroxisomes, revealed another interesting idea that finally turns the existing model of de novo biogenesis into another direction (Sugiura et al. 2017). In specific detail, when Pex3 has been reexpressed in these cells, it becomes targeted to mitochondria and is secreted of pre-peroxisomal vesicles. However, when Pex16 has been reexpressed, it gets targeted towards the ER, where it is then secreted by pre-peroxisomal vesicles, which seem to fuse with the mitochondria-based pre-peroxisomes to produce a new, import-potent peroxisome (Sugiura et al. 2017). In summary, the ER and mitochondria are competent to facilitate both to the de novo creation of peroxisomes in mammalian cells. Therefore, the initial organelle-targeting of these PMPs in the absence of peroxisomes seems to be a crucial step in the course of the de novo assembly of peroxisomes (Hettema and Gould 2017; Schrader and Pellegrini 2017; Costello and Schrader 2018). However, the ER-driven biogenic path and the physiological function of the de novo route are presently under strong debate, however, there is agreement of the finding that the ER provides a strong contribution to the biogenesis of the peroxisomes and that the peroxisomes represent semi-autonomous organelles, which rely on other compartments of the cells, such as the ER to be supplied with lipids or even certain proteins (Titorenko and Rachubinski 2014).

9.3.6 Division of Peroxisomes

Peroxisomes are revealed to be dynamic organelles that are able to multiply themselves by membrane growth and division of preexisting organelles (Islinger et al. 2012a, b; Schrader et al. 2016). The generation of new peroxisomes requires the dynamical remodeling and elongation of the peroxisomal membrane. In specific detail, the peroxisomal membrane extrudes the membrane to generate tubular extensions of the membrane. In the next step, these tubular structures undergo constriction and subsequently they are cutoff to generate new peroxisomes. In mammalian cells, this process seems to be asymmetric, since the new peroxisomes are generated through the production of a membrane compartment and subsequently the newly translated matrix proteins are incorporated (Huybrechts et al. 2009; Delille et al. 2010). The membrane peroxin Pex11β is regarded as a crucial factor in controlling the amount of peroxisome in mammals, which is included at each step of the growth and division of peroxisomes. Due to oligomerization and engagement with membrane lipids through N-terminal amphipathic helices, Pex11β functions as a membrane-forming protein that reshapes, distorts and elongates the peroxisomal membrane before it cleaves (Opaliński et al. 2011; Yoshida et al. 2015; Su et al. 2018). Pex11β binds to the membrane adapters Fis1 (fission protein 1) and Mff (mitochondrial cleavage factor) in the peroxisomal membrane, attracting the dynamine-related fission GTPase Drp1 and thus participating in the assembly of the peroxisomal division mechanism (Koch and Brocard 2012; Itoyama et al. 2013; Schrader et al. 2016). Beyond that, Pex11β has been found to act as a GTPase-activating protein (GAP) for Drp1 during peroxisomal cleavage (Williams et al. 2015). How peroxisomal membranes contract before the final membrane cleavage by Drp1 is not yet fully understood, but it is conceivable that Pex11β is participating in the constriction as well. For mitochondria, a functional role of the ER in membrane division was discovered (Friedman et al. 2011; Lewis et al. 2016). Whether this also accounts for peroxisomes is not known at present. Although there has been a significant increase in the knowledge of core proteins involved in peroxisome division and amplification, it will be a huge challenge in the coming years to gain an understanding of their combined interaction and regulation.

An exciting breakthrough in this field has been the discovery that peroxisomes and mitochondria jointly use proteins of their division machinery, for instance Fis1 (Koch et al. 2005), Mff (Gandre-Babbe and van der Bliek 2008), the ganglioside-induced differentiation-associated protein GDAP1 (Huber et al. 2013), and Drp1 (Li and Gould 2003; Koch et al. 2003) in mammalian cells. The co-use of division factors between peroxisomes and mitochondria has also been documented in other organisms, for instance for the plant-specific division factor PMD1 (peroxisomal and mitochondrial division factor 1) (Aung and Hu 2011), and for the adaptors Mdv1 and Caf4 and the dynamine-related GTPase Dnm1 in the yeast S. cerevisiae (Kuravi et al. 2006, Motley et al. 2008). PMD1 has only recently been shown to affect peroxisome proliferation in the case of salt stress in Arabidopsis thaliana (Frick and Strader 2018; Hu 2010; Saraya et al. 2010). The mutual use of dividing elements between peroxisomes and mitochondria is considered to be a commonly adopted, evolutionarily conserved mechanism across mammals, fungi and plants that supports the "peroxisome-mitochondrial junction", impacting on their cooperative roles and involvement in disease, and encouraging healthy lifespans (Waterham et al. 2007; Shamseldin et al. 2012; Schrader et al. 2015a, b; Koch et al. 2016; Weir et al. 2017a, 2017b).

In the meantime, a number of patients with imperfections in the peroxisomal division/dynamic proteins Drp1, Mff and Pex11β have been detected (Costello et al. 2018). Drp1 and Mff defects normally affect both peroxisomal and mitochondrial division, leading to greatly prolonged organelles. Drp1 insufficiency, the first disease ever reported with a defect in both mitochondrial and peroxisomal cleavage (Waterham et al. 2007), combined clinical characteristics of peroxisomal, including dysmyelination and severity, and mitochondrial diseases, such as autosomal dominant optic atrophy and neuropathy. The genetic screening of this first patient, dying a few weeks after birth, yielded a heterozygous dominant-negative missense mutation (Ala395Asp) in the mid-domain of Drp1, which blocks Drp1 oligomerization and consequent its activity in membrane cleavage (Chang et al. 2010). Other Drp1 patients who showed developmental retardation, refractory epilepsy or infantile encephalopathy have recently been identified (Yoon et al. 2016; Chao et al. 2016; Sheffer et al. 2016; Vanstone et al. 2016; Fahrner et al. 2016; Nasca et al. 2016; Zaha et al. 2016). The genetic analysis yielded firstly, missense variants in the central (oligomerization) domain of Drp1 (Gly362Asp, G350R, E379K), which suggests a dominant-negative mechanism, secondly, recessive nonsense mutations resulting in a shortened instable protein (Chao et al. 2016; Sheffer et al. 2016; Vanstone et al. 2016), or thirdly, the first dominantly inherited mutations in Drp1 involving conserved amino acids inside the Drp1 GTPase region (Gerber et al. 2017). The last Drp1 missense mutations were associated with blinding optic atrophy. Although the mitochondria in the patients' fibroblasts were lengthened, the morphology of the peroxisome seemed normal (Gerber et al. 2017). The very first patients with Mff deficiency caused by a functional loss mutation in the Mff gene have been identified (Shamseldin et al. 2012; Koch et al. 2016). The patients had delayed development, peripheral neuropathy, optic atrophy and leight-like encephalopathy. Mitochondria and peroxisomes are severely stretched in patients' fibroblasts as a result of organelle division break-down. It is noteworthy that Mff has also been found to be a key player in energy-sensitive adenosine monophosphate (AMP)-activated protein kinase (AMPK)-facilitated mitochondrial cleavage (Toyama et al. 2016). Contrary to the neurological characteristics seen in Mff patients, Mff-deficient mice are dying as a consequence of severe dilated cardiomyopathy that causes cardiac insufficiency, which is presumably the outcome of mitochondrial disorders (Chen et al. 2015a, b). While mitochondria and peroxisomes were highly extended in Mff-deficient mouse embryonic fibroblasts, their elongation was not significantly modified in Mff-deficient mouse cardiomyocytes. Nevertheless, an enhanced heterogeneity in the shape and frequency of mitochondria was detected (Chen et al. 2015a, b). This may suggest that the peroxisomal (mitochondrial) morphology and division is influenced cell type specifically. Recently, a mathematical modeling effort has been made to account for and predict changes in the morphology and dynamics of peroxisomes in health and disease states (Castro et al. 2018a, 2018b).

Patients suffering from Pex11β show short stature, eye problems (congenital cataract), progressive hearing loss and neurological defects (Ebberink et al. 2012; Taylor et al. 2017). The number of peroxisomes and their morphology in the patients' fibroblasts are changed. Similar to Drp1 and Mff deficiency, the metabolic capacities of peroxisomes are not seriously impaired. This contrasts with the classical peroxisome biogenesis defects, such as the Zellweger syndrome, and can impede the diagnosis by metabolic biomarkers, such as VLCFA. It also indicates that the patients' symptoms are associated with deficiencies in both peroxisome dynamics and plasticity, underlining the relevance of proper monitoring of peroxisome incidence and membrane dynamics for cell performance. Most interestingly, it has lately been stated that an abnormal peroxisome frequency in Pex11β -deficient epidermal cells leads to irregular mitosis and organelle heredity and thus influences decisions about cell destiny (Asare et al. 2017).

9.3.7 Motility and Distribution of Peroxisomes

Advances have also been achieved in terms of a better knowledge of peroxisome motility and the influence of the cytoskeleton on the dynamics and dispersion of peroxisomes. In baker's yeast, peroxisomes travel across the actin filaments by enrolling the myosin V motor Myo2 through the PMP Inp2 (Inheritance protein 2). This is decisive for the transportation of peroxisomes into the bud and thus for heredity in peroxisomes. For a more balanced distribution, Inp1, another hereditary protein, connects peroxisomes to the peripheral ER and thus holds back some peroxisomes in the parent cell (Knoblach and Rachubinski 2015, 2016). Plant cells are also capable of moving peroxisomes through actin filaments and myosin motors (Sparkes and Gao 2014). PMD1, which is needed for NaCl-induced division of peroxisomes, is another actin-binding protein and can be responsible for establishing the peroxisome-cytoskeleton linkage in plants (Frick and Strader 2018). On the contrary, peroxisomes in mammalian cells travel bidirectionally across microtubules, using both kinesin and dynein motors (Schrader et al. 2003; Neuhaus et al. 2016). How microtubule motors are attracted to peroxisomes in mammalian cells has long been poorly understood, however, a function for the mitochondrial Rho-GTPase Miro1 has recently been uncovered (Castro et al. 2018a, 2018b; Okumoto et al. 2018). Miro1, which was originally characterized as a mitochondrial membrane adapter for kinesin, has also been directed towards peroxisomes and accounts for peroxisome dispersion and microtubule-facilitated motility. Similar to Mff, Fis1, GDAP1, MAVS and AtPMD1, Miro1 also acts as a tail-anchored membrane adapter jointly used by mitochondria and peroxisomes (Costello et al. 2017a, 2017b, 2017c). Notably, Miro1-facilitated tensile forces also promote stretching and proliferation of the peroxisome membrane of cellular models for peroxisome disorders (Castro et al. 2018a, 2018b). These findings in conjunction with a mathematical model of peroxisome dynamics now enable us to connect the cytoskeleton of the microtubules and the motor-driven traction forces with peroxisome generation by growth and division in mammalian cells (Castro et al. 2018a, 2018b). Since peroxisome stretching and division can also take place without microtubules, this connection was controversially discussed. It is, however, proposed that autonomous but cooperative mechanisms are in place and that motor forces assist membrane dynamics by ensuring directional dependency. This is in total agreement with findings in yeast, where actomyosin-facilitated pulling forces provide the expansion of peroxisome and their spatial separation in cells with mutated dynamin (Hoepfner et al. 2001; Nagotu et al. 2008). In accordance with these results, it is even conceivable that mechanical forces can split peroxisomes, as it was announced recently for mitochondria (Helle et al. 2017). Under this assumption, it is proposed that Mff functions as a membrane-bound force sensor to stimulate the splitting process at mechanically stressed mitochondrial locations.

In a similar way to mammalian cells, peroxisomes in filamentous fungi travel across microtubules (Knoblach and Rachubinski 2016; Steinberg 2016; Salogiannis and Reck-Peterson 2017). Instead of directly binding motor proteins, however, they interfere with mobile early endosomes (EEs) (Guimaraes et al. 2015). In A. nidulans, this "hitchhiking" on EEs necessitates PxdA, an EE-bound linkage protein that facilitates the peroxisome-EEE cooperation (Salogiannis et al. 2016). In the corn fungus Ustilago maydis, the consistent EE motility enhances also the diffusive movements of the peroxisomes, which is said to affect local mixing and the interactions among organelles (Lin et al. 2016). Mathematical simulation of different facets of intracellular distribution in the filamentous fungus U. maydis has provided new knowledge about the spatial organization of peroxisomes and other organelles (Lin et al. 2016; Lin and Steinberg 2017). The peroxisome mobility and mixing has been found to demand both the active diffusion and targeted transport. These mechanisms regulate the precise distribution of peroxisomes and enable their frequent interplay, which is necessary for the appropriate function of the entire cell.

9.3.8 Tether-Based Interactions of the Peroxisomes

An upcoming topic in cell biology is the cooperation, interaction and communication between various organelles, which are frequently associated with physical contacts through membrane contact interfaces. Peroxisomes are not separate units within the cell, but rather interact and exchange signals, metabolites and proteins with other compartments such as the ER, mitochondria, lipid droplets and lysosomes (Schrader et al. 2015; Wanders et al. 2016; Shai et al. 2016; Yoboue et al. 2018; Castro et al. 2018). The study of interactions between organelles and the identification of proteins that facilitate organelle connections represents an active field of research that demands new approaches and techniques. For instance, the application of multispectral imaging permitted the simultaneous visualization of six organelles (including peroxisomes) and the imaging of their interaction (Valm et al. 2017). The systematic mapping of the contact points in baker's yeast using a proximity detection technique based on cleaved fluorophores resulted in new connections on the one hand between peroxisomes and the plasma membrane and on the other hand between peroxisomes and the vacuole (Shai et al. 2018). In addition, individual tethering mechanisms for the yeast mitofusin Fzo1 and PMP Pex34 in peroxisome–mitochondrial contacts were elicited. Moreover, this study showed a physiological relevance for peroxisome–mitochondrial interactions in the β-oxidation of fatty acids, a process that demands a metabolic interplay between both organelles (Schrader et al. 2015a, 2015b; Wanders et al. 2016; Pascual-Ahuir et al. 2017; Fransen et al. 2017). In this regard, it is of interest that in S. cerevisiae peroxisomes can be found adjacent to a specific mitochondrial cavity in the vicinity of the ER–mitochondrial interface, proximal to the site where the pyruvate dehydrogenase complex is located in the mitochondrial matrix, indicating a three-way organelle connection. Peroxisomal Pex11 and mitochondrial Mdm34, which is one of the proteins that form the ER–mitochondrial ether (ERMES), are thought to be responsible for peroxisome–mitochondrial linkage (Cohen et al. 2014). Similar linked three-way organelle complexes among peroxisomes, chloroplasts and mitochondria were observed in green cotyledon cells of A. thaliana under light irradiation (Hayashi et al. 2001). Pex10 and an unknown counterpart on the chloroplast are suspected to be promoters of the interaction, which apparently function to transport photosynthesis products across the photorespiratory pathway in the course of photorespiration. Even to date, there is no conclusive evidence of direct communication between peroxisomes and mitochondria in mammalian cells. However, hormone-induced, regulated steroid hormone biosynthesis necessitates cross-organ interaction between peroxisomes and mitochondria. By means of immunofluorescence staining and imaging of living cells, it was demonstrated that the di-butyryl-cAMP treatment of Leydig cells of the MA-10 mouse tumor swiftly forces the peroxisomes to align with the mitochondria (Fan et al. 2016). It has been hypothesized that the isoform A of the endogenous acyl-CoA-binding protein ACBD2/ECI2, incorporated from head to tail in peroxisomes and mitochondria, could potentially be involved in facilitating bidirectional interorganell communication for the delivery of cholesterol required for the biosynthesis of steroid hormones.

Efforts to forecast by employing machine learning together with mutation analyses identified new tail-anchored adaptor proteins on peroxisomes and other organelles (Costello et al. 2017a, 2017b, 2017c). Subsequently, this resulted in the molecular analysis of the first peroxisome ER membrane interface in mammalian cells with the peroxisomal acyl-CoA-binding domain proteins ACBD5 and ACBD4, interfering through FFAT-like domains, which display two phenylalanines (FF) in an acidic region, with ER-resident VAPA/B proteins, that constitute tail-anchored membrane adapters (Costello et al. 2017b, 2017c; Hua et al. 2017). Interaction of ACBD5 with VAPA/B fulfills a function in the biosynthesis of plasmalogen, which necessitates the combined action of the ER and the peroxisomes in the metabolism (Hua et al. 2017; Herzog et al. 2017). In specific detail, the peroxisome–ER interface facilitates the movement of peroxisomes, which leads to a novel function for the peroxisome–ER connection in the governing of the peroxisome movement and the dynamical remodeling of membranes in mammalian cells (Costello et al. 2017; Hua et al. 2017; Castro et al. 2018a, 2018b). Moreover, a specific function of the peroxisome–ER connection within lipid transfer for the membrane of peroxisomes and biogenesis has been enlightened. The enlargement and increased growth of the peroxisomal membrane seem to be a crucial initial prerequisite for their division and proliferation. This role necessitates lipids that are suggested to be transported by the ER through an avesicular route (Raychaudhuri and Prinz 2008; Costello et al. 2017). Malfunctions in the division of peroxisomes, such as those that are associated with impaired function of Mff or Drp1, lead to increased elongated peroxisomes, which leads to the hypothesis that there seem to be a continuously delivery of lipids by the ER towards the peroxisomes. Impairment of the interplay between peroxisomes and the ER decreased the aid of these peroxisome ER connections in the delivery of lipids to the biogenesis of peroxisomes. In order to figure out how the lipids are transported seems to be sophisticated project for future investigations in this research field. Since approximately 70–80% of the peroxisomes of mammalian cells, cultured in vitro, can interfere with the ER, the ER-based pre-peroxisomal vesicles do not seem to play a crucial functional role for the delivery of lipids towards peroxisomes.

Based on these findings, patients carrying mutations in the ACBD5 gene need to be detected that display retinal dystrophy and white matter disease (Yagita et al. 2017; Ferdinandusse et al. 2017). In line with this, mutations of VAPB are even associated with amyotrophic lateral sclerosis (Taylor et al. 2016). Hence, this result suggests that there may a connection between the peroxisome interfering regions and the disordered function of the cell (Castro et al. 2018).

A previously unexplored organellar interaction, "collaboration", between peroxisomes and the lysosomal components that facilitate the intracellular pathway of cholesterol has been demonstrated (Du et al. 2015). Cholesterol is a major factor governing membrane fluidity, permeability and the organization within animal cells (Chang et al. 2006). Since the overwhelming majority of the cholesterol is localized in the plasma membrane (Maxfield and Wüstner 2002), it is obtained from the ER through de novo synthesis (Horton et al. 2002; Kovacs et al. 2002), and also from lysosomes through a receptor-based endocytosis pathway of the plasma LDL (Brown and Goldstein 1986). Cholesterol fulfills major tasks in steroidogenesis, bile acid biosynthesis and the transduction of signals through regulatory oxysterols (Yeagle 1988; Lingwood and Simons 2010), which suggests a dynamic intracellular distribution. Hence, a basic question can be raised whether cholesterol can be delivered from one compartment to another compartment. However, peroxisomes seem to fulfill crucial tasks in the transport of the free cholesterol of lysosomes towards the plasma membrane (Chu et al. 2015). A not yet detected interference between lysosomes and peroxisomes has been seen, which has been partly provided by the interaction of the integral lysosomal membrane protein synaptotagmin 7 and the lipid PI(4,5)P2 on the membrane of the peroxisome. In this important to mention that the efficient tether formation is needed for NPC1, which seem to be transiently existent and depend on cholesterol. A malfunction of crucial peroxisomal genes causes the aggregation of cholesterol in lysosomes, similarly as it has been found in the Niemann–Pick disease type C (NPC). NPC represents a fatal and foremost neurodegenerative disease (Carstea et al. 1997), based on mutations in NPC1 and NPC2, which together enable the release of free cholesterol from the lumen towards the boundary membrane of the lysosomes (Sleat et al. 2004). How cholesterol subsequently reaches the plasma membrane is still not known. Anyway, the peroxisomes fulfill a crucial function for the intracellular transport of cholesterol and derivatives thereof.

Dietary intake, the endogenous de novo synthesis, the outflow and conversion of cholesterol into derivatives such as bile acids strictly adjusts the level of cellular cholesterol. A sophisticated feedback system records the actual level and adapts it through both transcriptional and post-transcriptional mechanisms. Of central importance for transcriptional regulation are firstly the sterol regulating element-binding protein (SREBP) family, secondly, the SREBP cleavage-activating protein (SCAP) which acts a sensory protein for sterols and thirdly the Insigs (insulin-induced genes) that govern the SREBPs over a broad spectrum of cholesterol concentrations. The Insig–SCAP–SREBP complex is located in the ER, where the sterol concentrations are relatively low. When a mouse model for Zellweger syndrome (Pex2$^{-/-}$ mice) is utilized, there are low levels of cholesterol in the plasma and the liver of the mice were detected. In addition, the mice failed to sustain normal cholesterol homeostasis notwithstanding the activation of the principal regulator SREBP and elevated activities of biosynthetic cholesterol enzymes. Finally, the SREBP complex is importantly still in an active state even after the normalization of cholesterol levels in the liver due to bile acid feeding (Faust and Kovacs 2014). To support these studies (Kovacs et al. 2009), on the other hand, it is pointed out that peroxisome defect stimulates hepatic ER stress paths causing the endogenous sterol reaction mechanism to become dysregulated.

9.3.9 Signal Transduction in Peroxisomes and Defense Mechanisms for Viruses

It is important that the peroxisomes work together in antiviral signaling and defense using the tail-anchored MAVS proteins (mitochondrial antiviral signaling) (Dixit et al. 2010; Kagan 2012). Thereby, peroxisomal MAVS vastly trigger the expression of a subset of antiviral genes, which that inhibit viral replication, whereas mitochondrial MAVS initiate a persistent antiviral response. Noteably, peroxisomes are also necessary for the devouring of bacteria by Drosophila and mouse macrophages and the clearance of bacterial infections through the modification of canonical innate immunity routes through ROS and RNA signaling (Di Cara et al. 2017). The role of peroxisomes was examined in adult Drosophila flies and the closely related S2 cell line including mice where the key peroxins Pex5 and Pex7 were consistently compromised. A decreased reactivity to microbial pathogens, deficiencies in immune signaling and diminished viability have been reported in flies and S2 cells, and a need for peroxisomes in microbial entrapment by murine macrophages has been identified. When interpreting the results in connection with the disturbed expression of peroxins, an impairment of bacterial phagocytosis, defective reorganization of the cytoskeleton, necessary for the formation of phagosomes and progressive phagocytosis, and a regulation of the ROS/RNS level to trigger an immune response are assumed. According to the available reports (Dixit et al. 2010), it is attractive to assign peroxisomes an important subcellular node in the enhancement of innate immune reactions.

HIV viruses are especially effective in undermining the antiviral responses of the host. Recently, peroxisomes have been shown to be part of these effective interventions. (Xu et al. 2017). In particular, HIV-infected cells are expressing large amounts of microRNAs, a portion of which is aimed at peroxisome biogenesis factors (PEX2, PEX7, PEX11ß, PEX13), leading to a lower number of peroxisomes. However, interestingly, the content of these microRNAs was found to be elevated in both the brain tissue of HIV patients and similarly in HIV-infected macrophages. Therefore, enhancing the expression of microRNAs that depress peroxisomes may be a novel mechanism to intervene with the early antiviral signals emitted by these organelles. The development of neurological disturbances in AIDS patients could be linked to this mechanism.

Reactive oxygen/nitrogen species (ROS/NOS), in combination with other reactive molecules, have become important regulators of numerous physiological and pathological processes in recent decades, contribute to and complement with superior regulatory systems within a living organism. ROS/NOS act as signal transducers conveying a variety of biological reactions, such as gene expression, cell proliferation, angiogenesis, innate immunity, programmed cell death and senescence (Dowling and Simmons 2009; Scherz-Shouval and Elazar 2011). Conversely, elevated concentrations of these short-lived reactive molecules or any interference with ROS/NOS homeostasis due to oxidative stress can have detrimental consequences (Salmon and Pérez 2010).

In recent years, peroxisomes have been identified as pivotal controllers of the entire lipid and ROS/NOS cellular metabolism, allowing them to tightly interoperate both functionally and physically with other cell organelles, especially mitochondria (Fransen et al. 2017). It has been shown that plant peroxisomes harbor the main enzymes implicated in the formation of ROS/NOS, which are related to the resistance to oxidative stress (Corpas et al. 2017). The seed sprouting in the dark demands enzymes that catalyze β oxidation and gluconeogenesis to catalyze the conversion of fatty acids to sugars, as well as the components of the ascorbate–glutathione cycle to prevent the oil bodies from being damaged by oxidative caused by H_2O_2, which is generated when fatty acids are decomposed (Eastmond 2007; Goepfert and Poirier 2007). Within the leaves, the stomatal motion is regulated to a high degree by external stimuli, for instance light, and internal molecules, such as hormones. NO initiates the stomatal occlusion and limits the entrance of pathogenic microorganisms (Neill et al. 2008). Peroxisomal NO and ROS are implicated in the senescence of leaves, which is marked by a reduction in catalase activity and a reduction in NO synthesis (Corpas et al. 2004). An increase in peroxisomal NO content has been observed in A. thaliana seeds cultivated under salinity stress or subjected to cadmium (Corpas and Barroso 2014).

In conclusion, these results underline the significance of peroxisomal NO metabolism during abiotic stress situations in plants. ROS levels in plant cells increase dramatically in the period of senescence and in the presence of biotic and abiotic stress (Zentgraf 2007). H_2O_2 plays a crucial role in this setting. Ascorbate peroxidase is perhaps the main enzyme that captures H_2O_2 generated in chloroplasts, but is also abundant in the cytoplasm, peroxisomes and mitochondria (Narendra et al. 2006). Various catalase isoforms are known for Arabidopsis, whereby the Cat3 concentration fluctuates considerably during the lifetime of the plant and increases especially in the leaves of aging plants (Zimmermann et al. 2006). In line with these variations, in vivo imaging showed that peroxisomal H_2O_2 in the leaves continues to be strongly affected during the life cycle (Costa et al. 2010). Most interesting, there was clear correlation between Cat3 expression intensities and effective H_2O_2 removal as a function of intraperoxisomal Ca^{2+} levels. It seems that the initiation of Cat3, induced by an elevation of Ca^{2+} levels within peroxisomes, is an extremely powerful cellular process to tightly monitor H_2O_2 levels.

The aggregation of senescent cells and concomitant secretions and the resulting failure to renew stem cells is widely acknowledged to be a factor in tissue aging (Collado et al. 2007). The availability of these cells in tissues of aging primates has been verified (Herbig et al. 2006), and their removal in a mouse model actually retarded the occurrence of age-related abnormalities (Baker et al. 2011). Senescent cells are no longer able to replicate, are magnified and start expressing marker proteins called senescence markers. Besides, these substances are completely immune to apoptosis and excrete bioactive molecules such as cytokines (Giordano and Terlecky 2012). Aging is regarded as a natural phenotype in that the cells move into a senescence phase in order to prevent transformation into cancer cells (Campisi and Robert 2014). Several factors contribute to cell aging, among them the truncation of telomeres, changes in protein expression, failures in the DNA repair mechanism and the clustering of cellular ROS—in specific H_2O_2—which is proposed as the "primary driving factor" of in vitro senescence and in vivo aging (Lu and Finkel 2008). In view of their multiple metabolic roles, especially their function as both a generator and a reservoir of ROS in a cell, peroxisomes are a privileged node in the process of cell aging. In fact, a large number of investigations with cell lines of different origin, animals and yeast cultures indicate far-reaching alterations in the biogenesis and proliferation of organelles, in the level of expression and of the localization of peroxisomal enzymes participating in ROS metabolism (catalase), and finally in the interplay with other cell organelles, especially mitochondria (Deori et al. 2018). They unanimously demonstrate that peroxisomes decisively promote aging, longevity and age-related disturbances.

The theory of free radicals of aging postulates oxidative degradation of macromolecules to be the primary life-threatening determinant. In certain instances, however, longevity is increased by the inactivation of the defense mechanisms against oxidative stress or alternatively appears to increase rather than decrease ROS and oxidative injury. With S. cerevisiae, it has been demonstrated that caloric restriction or catalase blockage causes oxidative stress through H_2O_2 and yet fosters survival despite enhanced oxidative injury to macromolecules (Mesquita et al. 2010). An inducement of superoxide dismutase through H_2O_2, which decreases the content of oxygen radicals, has been suggested to explain this unexpected finding, which indicates a hormone producing effect of H_2O_2 in encouraging longevity.

A further regulatory mechanism of peroxisomal ROS with a deep influence on cell growth has been communicated (Zhang et al. 2013). It was discovered that two components (TSC1 and TSC2) of the tuberous sclerosis complex (TSC) are pinpointed on peroxisomes. In reaction to the ROS, both affect mTorC1, which, with diverse imports of insulin, glucose or amino acids, affects the alternation between growth and autophagy. Based on these findings, peroxisomes fulfill a role as a key regulator of cellular growth in mammalian tissues.

Type 2 diabetes represents a complicated disease, which is associated with enhanced levels of non-esterified fatty acids (NEFAs). The latter are thought to interfere with the operation of β cells and trigger the destruction of these cells, a process referred to as lipotoxicty. In a trial with primary rat islet cells and related cell lines, experimental confirmation was obtained that NEFA-induced β cell lipotoxicity is closely related to the peroxisomal metabolism of NEFA (Elsner et al. 2011). As the expression of H2O2-inactivating catalase is practically unavailable in peroxisomes of insulin-secreting β cells (Lenzen et al. 1996), the deactivation of H2O2 generated in peroxisomes by the β oxidation of NEFAs is strongly hampered, which accounts for the extraordinary proneness of pancreatic β cells to lipotoxicity.

When summarizing the insights into the functional plasticity of peroxisomes, the view of a leftover "fossil organelle" becomes crucial for the optimal operation of a cell. Despite major progress in elucidating their manifold contributions to the vitality-dependent abiosis of cells, it is obvious that peroxisomes will remain a key contributor to these fundamental characteristics.

9.4 Interplay Between Peroxisomes and Cancer

Cancer cells are exerted to a highly altered microenvironment compared to normal cells, and hence, they need to manage an adaption of their entire cellular metabolism to the hypoxic and hypo-nutrient environmental conditions within a solid primary tumor (Yoshida 2015). This regulatory element of the metabolic reprogramming of the cancer cells is proposed as one of the famous hallmarks of cancer and due to alterations in glucose and amino acid metabolism, and there are additionally changed for the lipid metabolism revealed (Ghaffari et al. 2015). Nonetheless, malfunctions in the redox homeostasis of the cells can not only be addressed to be pro-tumorigenic, they may also cause a severe failure of the tumor chemotherapy by inducing resistance within cancer cells (Glasauer and Chandel 2014). In specific detail, mitochondria have been regarded as a key organelle that can support these dramatic changes in the cell's metabolism and its redox homeostasis (Valcarcel-Jimenez et al. 2017; Ježek et al. 2018). Moreover, the functional roles of mitochondria and peroxisomes are closely connected (Schrader et al. 2015a, 2015b); hence, the peroxisome physiology seems to be additionally important in the transition of healthy somatic cells into cancer cells that can even undergo further malignant progression.

Peroxisomes, which harbor a wide range of oxidases, function as potent H_2O_2 producers, create an imbalance in H_2O_2 generation and break-down during the PPARα-facilitated initiation of the peroxisomal fatty acid β-oxidation, which has all them associated to the development and growth of liver tumors during the ongoing and hence chronic exposure to pharmacological drugs leading peroxisome proliferation (Yu et al. 2003). Nonetheless, these investigations pinpoint towards a functional role of peroxisomes in spontaneously developing tumor tissue seem to be restricted. In a pronounced number of neoplastic tissues, the peroxisomal function appears to be reduced in comparison with unaffected tissues. A decreased number of peroxisomes have been detected in colon carcinomas employing electron microscopy in order to analyze the number of catalase positive staining using alkaline DAB in various probes (Cablé et al. 1992). These data were confirmed by several scientific studies showing a diminished peroxisomal protein abundance, such as the catalase, ABCD3, ACOX1 and PXMP2, or enzymatic activities, including catalase, D-amino acid oxidase, polyamine oxidase and peroxisomal β oxidation, in colon tumor tissue, indicating a general decrease in peroxisome incidence and functionality in neoplastic tissue (Baur and Wendel 1980; Cablé et al. 1992; Keller et al. 1993; Lauer et al. 1999). A diminution of both peroxisomal enzyme activities and protein levels in breast and hepatocellular carcinomas has similarly been documented (Keller et al. 1993; Litwin et al. 1999). Importantly, the findings of elevated mRNA levels of the correspondent enzymes (Lauer et al. 1999) lead to the hypothesis of an imbalance in the peroxisome remodeling dynamics which may be attributed to inefficient peroxisome biogenesis or elevated levels of peroxisome break-down (pexophagy). However, contradictory results have been presented in a recent report stating that peroxisomes are essential for the longevity of liver cancer cells (Cai et al. 2018). Since previous studies demonstrated an upregulation of Pex2 at the mRNA level in liver carcinomas (Chen et al. 2002; Wurmbach et al. 2007), the silencing of Pex2 using RNAi in hepatocellular carcinoma xenografts caused a pronouncedly decrease in the growth of the tumor due to the treatment (Cai et al. 2018). Moreover, it has been found that the impairment of peroxisome function enhances the levels of ROS due to the mislocalization of catalase into the cytosol. Consequently, the stress of the ER of cancer cells may lead to the suppression of mTORC1 signaling pathway and an elevation of the autophagy pathway causes subsequently the death of the cell (Cai et al. 2018). In contrast to these interpretations, it has been reported that decreased import of the catalase represents a cellular mechanism for the protection of the cells from redox stress (Fujiki et al. 2017; Walton et al. 2017). Newly released data on the possible involvement of ACOX1, the rate-limiting and H_2O_2-producing oxidase in peroxisomal β oxidation, in hepatocellular tumors emphasize the intricacies of gaining an accurate and comprehensive insight into peroxisomal function in hepatocarcinogenesis (Chen et al. 2018). In this trial, it was observed that ACOX1 enzyme activity is reduced by SIRT5-dependent lysine desuccinylation. Following the SIRT5 knockdown, therefore, an increase in intracellular H_2O_2 concentrations was noted. Importantly, when compared to the surrounding liver tissue, ACOX1 protein levels were similar or marginally reduced, and it has been seen that ACOX1 activity was increased and SIRT5 expression dropped in most tumor samples. Based on these findings, elevated ACOX1 succinylation could lead to excessive H2O2 production in cancer cells, thus encouraging the conversion of healthy cells into cancer cells.

Peroxisomes are most frequent in the epithelial cells of the proximal tubule within the healthy kidney (Islinger et al. 2010a, 2010b). In clear cell renal tumors that reconstruct the tissue from the proximal tubules (Prasad et al. 2007), however, peroxisomes were missing after alkaline DAB staining and immunofluorescence detection of catalase (Frederiks et al. 2010). In accordance with this observation, a reduced catalase activity was noted in renal tumors (Pljesa-Ercegovac et al. 2008). The hypoxia-inducible transcription factor Hif2a lately proved to enhance the break-down of peroxisomes by autophagy (Walter et al. 2014). Strikingly, the Hif2a concentration correlates adversely with the frequency of peroxisomes in renal clear cell carcinoma, indicating that its exposure to a hypoxic tumor environment exhausts the peroxisomes through increased pexophagy levels. However, since all current knowledge about peroxisome frequency in renal tumors is based on the detection of catalase, future studies with additional peroxisome markers will have to demonstrate a decrease in peroxisome frequency. In summary, these contributions conclude that a disturbance of the peroxisomal environment seems to be a key general hallmark of cancer biology. The reduction of the peroxisomes could trigger some local changes in the configuration of the membrane lipids and thus modify the integration of the neoplastic cells into the adjacent tissue environment. Furthermore, the lack of peroxisomes possibly disturbs the intracellular ROS equilibrium and results in increased local toxic and partly oxidized VLCFA metabolites. In this context, it should be noted that the latter propose the hypothesis that the ACOX1 knockout mice induce an ER stress response that could eventually induce the generation of liver tumors that are regularly seen in this mouse line (Huang et al. 2011).

A general loss of peroxisome function during the progression of the tumor has been described above, findings from prostate cancer tissue suggest a more intricate involvement of peroxisomes in the perpetuation of tumor growth. The peroxisomal expression of α methyl acyl-CoA racemase (AMACR) was strongly increased in prostate cancer tissue compared to benign prostate tissue (Jiang et al. 2001). In the time since this first publication, multiple studies have validated elevated levels of AMACR as a credible marker for prostate cancer (Lloyd et al. 2008). AMACR acts as an auxiliary enzyme in the α-oxidation pathway for phytic acid and is needed for the transition from 2R-methylacyl-CoA to 2S-methylacyl-CoA (Wanders and Waterham 2006). 2S-Pristanoyl-CoA is broken down by means of the peroxisomal β-oxidation pathway for 2-methyl-branched-chain fatty acids. It has been shown that the expression of other enzymes, which play a role in the peroxisomal branched-chain fatty acid break-down pathway, such as ACOX3, D-PBE and the 3-ketoacyl-CoA thiolase ACAA1, is elevated in prostate tumors (Zha et al. 2005; Valença et al. 2015). In addition, it was demonstrated lately that the monocarboxylate transporter 2 (MCT2) reassigns in peroxisomal membranes of malignant prostate cancer cells (Valença et al. 2015). Monocarboxylate transporters may be involved in the transport of lactate pyruvate for reoxidation of NADH to produce NAD^+ as a cofactor for the peroxisomal β-oxidation process (McClelland et al. 2003). The results could indicate a selective increase in peroxisomal branched-chain fatty acid metabolism within prostate tumors. Increased AMACR expression has also been observed in colon, gastric, breast, renal and hepatocellular carcinomas (Jiang et al. 2003; Witkiewicz et al. 2005; Chen et al. 2005; Went et al. 2006; Jindal et al. 2016), indicating that peroxisomal branched-chain metabolism is probably linked to a wider range of tumors.

In this context, it is tantalizing to guess at the reasons why the break-down of branched-chain fatty acids in the prostate and other tumors may be increased. A correlation between elevated serum phytanic acid levels, which may result in the initiation of AMACR, and the incidence of prostate cancer has been reported (Xu et al. 2005) but is currently being debated (Kataria et al. 2015). In addition, a decrease in AMACR expression through RNAi to reduce the growth rates of the prostate cancer cell line LAPC-4 was shown (Zha et al. 2003). Human AMACR deficiency has variable phenotypes such as childhood cholestasis, late onset peripheral neuropathy, pigmented retinopathy or seizures, although it is not linked to elevated tumor development (Ferdinandusse et al. 2000). These results indicate that metabolites downstream of AMACR could be the active substances that induce cancer development, for instance by triggering receptors of important signaling routes such as PPARs or RXRs. More general modifications of the fat metabolism have been postulated to affect the development of the tumor (Wu et al. 2014). To support the statement of the latter, information on the levels of other enzymes and lipid metabolites involved in peroxisomal, mitochondrial and ER metabolism are needed to enhance our knowledge of the pathology of prostate cancer. Besides the enzymes of peroxisomal branched-chain lipid metabolism, the peroxisomal membrane protein PMP24/PXMP4 has been linked to the formation of prostate cancer. However, in contrast, PMP24 has been shown to be silenced by methylation of a single intronic CpG during the transition of the prostate cancer cell line LNCaP from androgen dependence to androgen independence (Wu and Ho 2004; Zhang et al. 2010). Moreover, PMP24 belongs to the TIM17 family of membrane proteins, whereas its functional role is not yet discovered. Nonetheless, the findings suggest that peroxisomes may be implicated at various levels in the transition from healthy prostate tissue to malignant cancer cells.

Peroxisomes in glioblastomas were examined with regard to the progression of the tumor (Benedetti et al. 2010). In relation to the progressing tumor grade, an enhanced staining for peroxisomes was found using immunocytochemistry. These results are confirmed by the detection of the proteins Pex14, PMP70, ACOX1 and 3-ketothiolase through immunoblotting, which suggest a peroxisome proliferation resulting in an increased organelle population. At the same time, the tumors displayed several lipid droplets and an increased expression of PPARα. In a follow-up investigation, it was examined how hypoxic environments affect peroxisomes in primary cultures of Glioblastoma multiforme (GBM) tumors (Laurenti et al. 2011). Following PMP70 immunofluorescence staining an elevation of peroxisomes and lipid droplets was recorded. In conjunction with this, an inducement of the hypoxia-inducible factor Hif1α and PPARα in response to the hypoxia was monitored. Consequently, the activation of Hif1α in a hypoxic environment, such as those in a tumor, may cause the increased expression of PPARα, which then finally induces the proliferation of peroxisomes. However, it needs to be determined why the induction of PPARα that in turn increases the catabolic lipid metabolism, is correlated with the elevated occurance of lipid droplets within the GBM cells. After monitoring an elevation of HMG-CoA reductase and cholesterol and triglyceride levels in the cell cultures, it has been supposed that peroxisomal β-oxidation in a hypoxic microenvironment can be utilized to generate acetyl-CoA as a substrate for the de novo synthesis of lipids. Another lately identified peroxisomal protein, hydroxysteroid dehydrogenase-like 2 (HSDL2) is apparently also upregulated in glioblastomas (Ruokun et al. 2016) and ovarian cancer (Sun et al. 2018). The knockdown of HSDL2 leads to reduced growth of glioblastoma cell lines, and to impaired proliferation, formation of colonies, motility and tumorigenesis in ovarian cancer cells, which underpins the crucial function of peroxisomes in these specific cancer types. At this stage, more data are needed to decode whether peroxisomes contribute to the metabolic conversion of glial cells into a malignant glioblastoma.

Peroxisomes constitute dynamically changing organelles, capable of adapting their quantity and enzyme level to the special needs of their cellular surroundings (Schrader et al. 2015). Thus, it must be taken into account that pharmaceuticals used in therapeutic cancer treatment could rearrange peroxisomes and consequently alter the physiology of the tumor. The reaction of peroxisomes towards Vorinostat, a HDAC inhibitor used for lymphoma treatment, has been explored, which induces the ROS-driven apoptosis in order to reveal their potential functional task in resistance to tumor intervention (Dahabieh et al. 2017). In fact, the study showed that peroxisomes in lymphoma cells in vitro are increased expressed due to treatment with Vorinostat. In line with this, the knockdown of Pex3 or even more specifically knockdown of the catalase pronouncedly elevated the levels of ROS-based apoptosis in the lymphoma cells in the presence of Vorinostat. Therefore, peroxisomes, which can ROS-degrading organelles, may be hypothesized to fulfill important tasks in the resistance to therapeutic tumor treatment, when pharmacological substance are administered that cause the induction of ROS-facilitated apoptosis.

An elevated expression of the peroxisomal Lon protease LonP2 has been detected in cervical cancer tissue (Wu et al. 2018). Functionally, LonP2 performs the same function as a combined chaperone/protease that refolds or breaks down committed peroxisomal proteins (Bartoszewska et al. 2012). A downregulation of LonP2 in the cancer cell lines HeLa and SiHA decreased oxidative stress and prevented the proliferation and movement of cervical cancer cells (Wu et al. 2018). However, since the study does not include any further analysis of peroxisomal condition in cervical cancer cells or LonP2-depleted cell lines, it is not yet feasible to mechanistically elucidate how alterations in LonP2 expression could alter peroxisomal physiology in terms of carcinogenesis.

Most investigations outlined earlier concentrate on the metabolic role of peroxisomes in the development of tumors. Most interestingly, a recent publication could bring an unanticipated role of peroxisomes in controlling proper cell division (Asare et al. 2017). The study shows that the accurate placement of the peroxisome during mitosis is necessary for the adequate asymmetric cell division in the skin epithelial cells. The RNAi-facilitated knockdown of Pex11β and Pex14 caused a mitotic retardation in the target cells and produced an unbalanced growth and differentiation into basal and suprabasal skin cells, concomitant with a decrease in the amount of terminal differentiation biomarkers in the tissue. Noteworthy, mitotic deregulation was not connected with a disturbance of the peroxisomal metabolic functions, but was instead induced by a misplacement of peroxisomes in spindle formation. In normal circumstances, peroxisomes are placed at the spindle poles, while a divergent localization led to erratic spindle rotation and caused a standstill at a mitotic control point for the separation of organelles (Asare et al. 2017). The authors summarize their findings and conclude that proper peroxisome segregation is involved in regulating the equilibrium between cell growth and differentiation. Disorders in this system result in the formation of basal daughter cells with differentiating markers, although still proliferating characteristics that are commonly linked to cancer (Asare et al. 2017).

In conclusion, the current perception of the role of peroxisomes in cancer cells is still patchy, and while previous studies primarily referred to a reduction in peroxisome activity in a large number of tumors, more recent studies also indicate that certain peroxisome activities are necessary for proper tumor growth. These tentatively contradictory results could be due to the large heterogeneity of the tumors examined. In addition, a single tumor itself is not a single-cell type, which further hampers the evaluation of the results. To corroborate our present knowledge about the status of peroxisomes in tumors, comprehensive comparative investigations are needed to link potential peroxisome malfunction to tumor type and grade and metabolic state of the tissue.

9.5 Intracellular Mechanics and Peroxisomes

The cytoplasm is very complex and highly dynamically organized which provides helps to omit possible negative outcomes of clashes and entanglements between various organelles within the cell. Since there are only a few cytoskeletal "highways", organelles and other loads unavoidably collide and may even sometimes compete. Motor-based transports produce forces on the membrane surface. In extension, many organelles also attach themselves to different motors that move in opposite directions (Hancock 2014), which can cause mechanical collisions. Apart from motor-driven transport, cytoskeletal compounds can also directly transfer forces to organelles as they pass through polymerization and depolymerization events (Gurel et al. 2014; Svitkina 2018). In the same way, organelle contact points can connect two organelles in a stable manner—even if in many cases temporarily—in order to facilitate the replacement of structural or signal molecules (Feng and Kornmann 2018). It is likely that a particular organelle also has a large number of interorganellar connections at different sites in the cytoplasm simultaneously. Such binding behavior adds a further dimension to the mechanical stress on these dynamic structures. Therefore, the organelles are permanently subjected to both tensile and compressive forces and have to withstand the associated mechanical stresses. This can be illustrated using the example of the cell nucleus and mitochondria.

The nucleus is by no means a sphere positioned more or less randomly in the middle of a cell. In contrast, nuclear positioning is closely monitored and involves dynamic displacement of these organelles throughout many processes, such as cell division (in eukaryotes experiencing closed mitosis), polarity establishment, differentiation, fertilization and migration (Gundersen and Worman 2013). The speed of intracellular nuclear motion ranges from 0.1 to 1 μm per minute, but in extreme cases can exceed 10 μm/min (Gundersen and Worman 2013). In addition, the nuclear shape must also adjust to the (local) cell geometry and often necessitates compaction of the organelle. This is illustrated to an extreme level during extravasation of white blood cells and metastasis of cancer cells, in which cells deform significantly to penetrate through close tight junctions of tissue cells and dense extracellular matrix (Barzilai et al. 2017; Karreman et al. 2016). It is not an easy process because the nucleus is quite rigid, largely due to the presence of the nuclear lamina, a nucleoskeleton that lines the inside of the nuclear envelope and consists mainly of lamin proteins. Finally, the nucleus requires to be mechanically regulated (Feng and Kornmann 2018, see Chap. 10).

References

R.J. Advani, H.R. Bae, J.B. Bock, D.S. Chao, Y.C. Doung, R. Prekeris, J.S. Yoo, R.H. Scheller, Seven novel mammalian SNARE proteins localize to distinct membrane compartments. J Biol Chem **273**, 10317–10324 (1998)

G. Agrawal, S. Subramani, De novo peroxisome biogenesis: evolving concepts and conundrums. Biochim. Biophys. Acta Mol. Cell Res. **1863**, 892–901 (2016)

K. Aiko, T. Tsujisawa, T. Koseki, S. Hashimoto, Y. Morimoto, T. Amagasa, T. Nishihara, Involvement of cytochrome c and caspases in apoptotic cell death of human submandibular gland ductal cells induced by concanamycin A. Cell Signal **14**(8), 717–722 (2002)

R. Al Hawas, Q. Ren, S. Ye, Z.A. Karim, A.H. Filipovich, S.W. Whiteheart, Munc18b/STXBP2 is required for platelet secretion. Blood **120**, 2493–2500 (2012)

N.W. Andrews, Lysosomes and the plasma membrane: Trypanosomes reveal a secret relationship. J. Cell Biol. **158**, 389–394 (2002)

S. Angermüller, Peroxisomal oxidases: cytochemical localization and biological relevance. Prog. Histochem. Cytochem. **20**, 1–65 (1989)

S. Angermüller, H.D. Fahimi, Heterogenous staining of D-amino acid oxidase in peroxisomes of rat liver and kidney a light and electron microscopic study. Histochemistry **88**, 277–285 (1988)

W. Antonin, C. Holroyd, D. Fasshauer, S. Pabst, G.F. Von Mollard, R. Jahn, A SNARE complex mediating fusion of late endosomes defines conserved properties of SNARE structure and function. EMBO J. **19**, 6453–6464 (2000)

L.N. Arneson, A. Brickshawana, C.M. Segovis, R.A. Schoon, C.J. Dick, P.J. Leibson, Cutting edge: syntaxin 11 regulates lymphocyte-mediated secretion and cytotoxicity. J. Immunol. **179**, 3397–3401 (2007)

A. Asare, J. Levorse, E. Fuchs, Coupling organelle inheritance with mitosis to balance growth and differentiation. Science **355** eaah4701 (2017)

N. Attar, P.J. Cullen, The retromer complex. Adv. Enzyme Regul. **50**, 216–236 (2010)

K. Aung, J. Hu, The arabidopsis tail-anchored protein peroxisomal and mitochondrial division factor1 is involved in the morphogenesis and proliferation of peroxisomes and mitochondria. The Plant Cell **23**(12), 4446–4461 (2011)

K. Baetz, S. Isaaz, G.M. Griffiths, Loss of cytotoxic T lymphocyte function in Chediak-Higashi syndrome arises from a secretory defect that prevents lytic granule exocytosis. J. Immunol. **154**, 6122–6131 (1995)

D.F. Bainton, The discovery of lysosomes. J Cell Biol **91**, 66s–76s (1981)

D.J. Baker, T. Wijshake, T. Tchkonia, N.K. LeBrasseur, B.G. Childs, B. van de Sluis, J.L. Kirkland, J.M. van Deursen, Clearance of p16Ink4a-positive senescent cells delays ageing-associated disorders. Nature **479**, 232–236 (2011)

J. Balaji, T.A. Ryan, Single-vesicle imaging reveals that synaptic vesicle exocytosis and endocytosis are coupled by a single stochastic mode. Proc. Natl. Acad. Sci. U.S.A. **104**, 20576–20581 (2007)

H.J. Balderhaar, C. Ungermann, CORVET and HOPS tethering complexes—coordinators of endosome and lysosome fusion. J. Cell Sci. **126**, 1307–1316 (2013)

A. Ballabio, The awesome lysosome. EMBO Mol. Med. **8**, 73–76 (2016)

K. Baran, M. Dunstone, J. Chia et al., The molecular basis for perforin oligomerization and transmembrane pore assembly. Immunity **30**, 684–695 (2009)

K. Bartel, H. Pein, B. Popper et al., Connecting lysosomes and mitochondria—a novel role for lipid metabolism in cancer cell death. Cell Commun. Signaling **17**, 87 (2019)

K. Bartel, M. Winzi, M. Ulrich, A. Koeberle, D. Menche, O. Werz, R. Müller, J. Guck, A.M. Vollmar, K. von Schwarzenberg, V-ATPase inhibition increases cancer cell stiffness and blocks membrane related ras signaling—a new option for HCC therapy. Oncotarget. **8**(6), 9476–9487 (2017)

M. Bartoszewska, C. Williams, A. Kikhney, Ł Opaliński, C.W. van Roermund, R. de Boer, M. Veenhuis, I.J. van der Klei, Peroxisomal proteostasis involves a lon family protein that functions as protease and chaperone. J. Biol. Chem. **287**, 27380–27395 (2012)
S. Barzilai, S.K. Yadav, S. Morrell et al., Leukocytes breach endothelial barriers by insertion of nuclear lobes and disassembly of endothelial actin filaments. Cell Rep. **18**, 685–699 (2017)
W. Baum, V. Kirkin, S.B. Fernandez, R. Pick, M. Lettau, O. Janssen, M. Zornig, Binding of the intracellular Fas ligand (FasL) domain to the adaptor protein PSTPIP results in a cytoplasmic localization of FasL. J. Biol. Chem. **280**, 40012–40024 (2005)
G. Baur, A. Wendel, The activity of the peroxide-metabolizing system in human colon carcinoma. J. Cancer Res. Clin. Oncol. **97**, 267–273 (1980)
E. Benedetti, R. Galzio, G. Laurenti et al., Lipid metabolism impairment in human gliomas: expression of peroxisomal proteins in human gliomas at different grades of malignancy. Int. J. Immunopathol. Pharmacol. **23**, 235–246 (2010)
D.I. Benjamin, A. Cozzo, X. Ji, L.S. Roberts, S.M. Louie, M.M. Mulvihill, K. Luo, D.K. Nomura, Ether lipid generating enzyme AGPS alters the balance of structural and signaling lipids to fuel cancer pathogenicity. Proc. Natl. Acad. Sci. U.S.A. **110**, 14912–14917 (2013)
N.R. Bin, C.H. Jung, C. Piggott, S. Sugita, Crucial role of the hydrophobic pocket region of Munc18 protein in mast cell degranulation. Proc. Natl. Acad. Sci. **110**, 4610–4615 (2013)
N.B. Blok, D. Tan, R.Y.-R. Wang, P.A. Penczek, D. Baker, F. DiMaio, T.A. Rapoport, T. Walz, Unique double-ring structure of the peroxisomal Pex1/Pex6 ATPase complex revealed by cryo-electron microscopy. Proc. Natl. Acad. Sci. **112**, E4017–E4025 (2015)
E.J. Blott, G. Bossi, R. Clark, M. Zvelebil, G.M. Griffiths, Fas ligand is targeted to secretory lysosomes via a proline-rich domain in its cytoplasmic tail. J. Cell Sci. **114**, 2405–2416 (2001)
N. Borgese, S. Brambillasca, S. Colombo, How tails guide tail-anchored proteins to their destinations. Curr. Opin. Cell Biol. **19**, 368–375 (2007)
G. Bossi, G.M. Griffiths, Degranulation plays an essen- tial part in regulating cell surface expression of Fas ligand in T cells and natural killer cells. Nat. Med. **5**, 90–96 (1999)
K.L. Boswell, D.J. James, J.M. Esquibel, S. Bruinsma, R. Shirakawa, H. Horiuchi, T.F. Martin, Munc13-4 reconstitutes calcium-dependent SNARE-mediated membrane fusion. J. Cell Biol. **197**, 301–312 (2012)
T. Braulke, J.S. Bonifacino, Sorting of lysosomal proteins. Biochim. Biophys. Acta **1793**, 605–614 (2009)
V. Braun, V. Fraisier, G. Raposo, I. Hurbain, J.B. Sibarita, P. Chavrier, T. Galli, F. Niedergang, TI-VAMP/VAMP7 is required for optimal phagocytosis of opsonised particles in macrophages. EMBO J. **23**, 4166–4176 (2004)
A.J. Brennan, J. Chia, K.A. Browne et al., Protection from endogenous perforin: glycans and the C terminus regulate exocytic trafficking in cytotoxic lymphocytes. Immunity **34**, 879–892 (2011)
M. Bretou, O. Jouannot, I. Fanget et al., Cdc42 controls the dilation of the exocytotic fusion pore by regulating membrane tension. Mol. Biol. Cell **25**, 3195–3209 (2014)
N.A. Bright, M.J. Gratian, J.P. Luzio, Endocyticdeliveryto lysosomes mediated by concurrent fusion and kissing events in living cells. Curr. Biol. **15**, 360–365 (2005)
N.A. Bright, B.J. Reaves, B.M. Mullock, J.P. Luzio, Dense core lysosomes can fuse with late endosomes and are reformed from the resultant hybrid organelles. J. Cell Sci. **110**, 2027–2040 (1997)
F.R. Brown, A.J. McAdams, J.W. Cummins, R. Konkol, I. Singh, A.B. Moser, H.W. Moser, Cerebro-hepato-renal (Zellweger) syndrome and neonatal adrenoleukodystrophy: similarities in phenotype and accumulation of very long chain fatty acids. Johns Hopkins Med. J. **151**, 344–351 (1982)
M.S. Brown, J.L. Goldstein, A receptor-mediated pathway for cholesterol homeostasis. Science **232**, 34–47 (1986)
W.J. Brown, J. Goodhouse, M.G. Farquhar, Mannose- 6-phosphate receptors for lysosomal enzymes cycle between the Golgi complex and endosomes. J. Cell Biol. **103**, 1235–1247 (1986)
A.A. Brzezinska, J.L. Johnson, D.B. Munafo, K. Crozat, B. Beutler, W.B. Kiosses, B.A. Ellis, S.D. Catz, The Rab27a effectors JFC1/Slp1 and Munc13–4 regulate exocytosis of neutrophil granules. Traffic **9**, 2151–2164 (2008)
J. Buentzel, F. Vilardi, A. Lotz-Havla, J. Gärtner, S. Thoms, Conserved targeting information in mammalian and fungal peroxisomal tail-anchored proteins. Sci. Rep. **5**, 17420 (2015)
A. Burgo, V. Proux-Gillardeaux, E. Sotirakis, P. Bun, A. Casano, A. Verraes, R.K.H. Liem, E. Formstecher, M. Coppey-Moisan, T. Galli, A molecular network for the transport of the TI-VAMP/VAMP7 vesicles from cell center to periphery. Dev Cell **23,** 166–180 (2012)
A. Burgo, E. Sotirakis, M.-C. Simmler, A. Verraes, C. Chamot, J.C. Simpson, L. Lanzetti, V. Proux-Gillardeaux, T. Galli, Role of Varp, a Rab21 exchange factor and TI-VAMP/VAMP7 partner, in neurite growth. EMBO Rep. **10,** 1117–1124 (2009s)
A. Burgo, A.M. Casano, A. Kuster, S.T. Arold, G. Wang, S. Nola, A. Verraes, F. Dingli, D. Loew, T. Galli, Increased activity of the vesicular soluble N-ethylmaleimide-sensitive factor attachment protein receptor TI-VAMP/VAMP7 by tyrosine phosphorylation in the Longin domain. J. Biol. Chem. **288**, 11960–11972 (2013)
J.K. Burkhardt, S. Hester, Y. Argon, Two proteins targeted to the same lytic granule compartment undergo very different posttranslational processing. Proc. Natl. Acad. Sci. **86**, 7128–7132 (1989)

S. Cablé, J.M. Keller, S. Colin, K. Haffen, M. Kédinger, R.M. Parache, M. Dauça, Peroxisomes in human colon carcinomas a Cytochemical and Biochemical Study. Virchows Arch. B Cell Pathol. Incl. Mol. Pathol. **62**, 221–226 (1992)

M. Cai, X. Sun, W. Wang, Z. Lian, P. Wu, S. Han, H. Chen, P. Zhang, Disruption of peroxisome function leads to metabolic stress, mTOR inhibition, and lethality in liver cancer cells. Cancer Lett. **421**, 82–93 (2018)

J. Campisi, L. Robert, Cell senescence: role in aging and age-related diseases, in *Robert L.* ed. by T. Fulop (Karger, Interdisciplinary Topics in Gerontology and Geriatrics Basel, 2014), pp. 45–61

E.D. Carstea, J.A. Morris, K.G. Coleman et al., Niemann–Pick C1 disease gene: homology to mediators of cholesterol homeostasis. Science **277**, 228–231 (1997)

I.G. Castro, D.M. Richards, J. Metz, J.L. Costello, J.B. Passmore, T.A. Schrader, A. Gouveia, D. Ribeiro, M. Schrader, A role for Mitochondrial Rho GTPase 1 (MIRO1) in motility and membrane dynamics of peroxisomes. Traffic **19**, 229–242 (2018a)

I.G. Castro, M. Schuldiner, E. Zalckvar, Mind the organelle gap—peroxisome contact sites in disease. Trends Biochem. Sci. **43**, 199–210 (2018b)

M.N. Cepińska, M. Veenhuis, I.J. van der Klei, S. Nagotu, Peroxisome fission is associated with reorganization of specific membrane proteins. Traffic **12**, 925–937 (2011)

V. Cetica, A. Santoro, K.C. Gilmour et al., STXBP2 mutations in children with familial haemophagocytic lymphohistiocytosis type 5. J. Med. Genet **47**, 595–600 (2010)

C.-R. Chang, C.M. Manlandro, D. Arnoult, J. Stadler, A.E. Posey, R.B. Hill, C. Blackstone, A lethal de novo mutation in the middle domain of the dynamin-related GTPase Drp1 impairs higher order assembly and mitochondrial division. J. Biol. Chem. **285**, 32494–32503 (2010)

T.-Y. Chang, C.C.Y. Chang, N. Ohgami, Y. Yamauchi, Cholesterol sensing, trafficking, and esterification. Annu. Rev. Cell Dev. Biol. **22**, 129–157 (2006)

Y.-H. Chao, L.A. Robak, F. Xia, M.K. Koenig, A. Adesina, C.A. Bacino, F. Scaglia, H.J. Bellen, M.F. Wangler, Missense variants in the middle domain of DNM1L in cases of infantile encephalopathy alter peroxisomes and mitochondria when assayed in Drosophila. Hum. Mol. Genet. **25**, 1846–1856 (2016)

B. Chen, B. Ji, H. Gao, Modeling active mechanosensing in cell-matrix interactions. Annu Rev. Biophys. **44**, 1–32 (2015a)

H. Chen, S. Ren, C. Clish, M. Jain, V. Mootha, M.J. McCaffery, D.C. Chan, Titration of mitochondrial fusion rescues Mff-deficient cardiomyopathy. J. Cell Biol. **211**, 795–805 (2015b)

X.-F. Chen, M.-X. Tian, R.-Q. Sun et al., SIRT5 inhibits peroxisomal ACOX1 to prevent oxidative damage and is downregulated in liver cancer. EMBO Rep. **19**, e45124 (2018)

X. Chen, S.T. Cheung, S. So et al., Gene expression patterns in human liver cancers. Mol. Biol. Cell **13**, 1929–1939 (2002)

Y. Chen, L. Pieuchot, R.A. Loh, J. Yang, T.M. Kari, J.Y. Wong, G. Jedd, Hydrophobic handoff for direct delivery of peroxisome tail-anchored proteins. Nat Commun **5**, 5790 (2014)

Z.-M.E. Chen, J.H. Ritter, H.L. Wang, Differential expression of alpha-methylacyl coenzyme a racemase in adenocarcinomas of the small and large intestines. Am. J. Surg. Pathol. **29**, 890–896 (2005)

G. Chiaruttini, G.M. Piperno, M. Jouve et al., The SNARE VAMP7 regulates exocytic trafficking of interleukin-12 in dendritic cells. Cell Rep. **14**, 2624–2636 (2016)

B.-B. Chu, Y.-C. Liao, W. Qi, C. Xie, X. Du, J. Wang, H. Yang, H.H. Miao, B.L. Li, B.L. Song, Cholesterol transport through lysosome-peroxisome membrane contacts. Cell **161**, 291–306 (2015)

S. Ciniawsky, I. Grimm, D. Saffian, W. Girzalsky, R. Erdmann, P. Wendler, Molecular snapshots of the Pex1/6 AAA + complex in action. Nat. Commun.**6**, 7331 (2015)

R.H. Clark, J.C. Stinchcombe, A. Day, E. Blott, S. Booth, G. Bossi, T. Hamblin, E.G. Davies, G.M. Griffiths, Adaptor protein 3-dependent microtubule-mediated movement of lytic granules to the immunological synapse. Nat. Immunol. **4** 1111–1120 (2003)

S. Coco, G. Raposo, S. Martinez, J.J. Fontaine, S. Takamori, A. Zahraoui, R. Jahn, M. Matteoli, D. Louvard, T. Galli, Subcellular localization of tetanus neurotoxin-insensitive vesicle-associated membrane protein (VAMP)/ VAMP7 in neuronal cells: evidence for a novel membrane compartment. J. Neurosci. **19**, 9803–9812 (1999)

Y. Cohen, Y.A. Klug, L. Dimitrov et al., Peroxisomes are juxtaposed to strategic sites on mitochondria. Mol. Biosyst. **10**, 1742–1748 (2014)

M. Collado, M.A. Blasco, M. Serrano, Cellular senescence in cancer and aging. Cell **130**, 223–233 (2007)

F.J. Corpas, J.B. Barroso, Peroxynitrite (ONOO-) is endogenously produced in arabidopsis peroxisomes and is overproduced under cadmium stress. Ann. Bot. **113**, 87–96 (2014)

F.J. Corpas, J.B. Barroso, A. Carreras et al., Cellular and subcellular localization of endogenous nitric oxide in young and senescent pea plants. Plant Physiol. **136**, 2722–2733 (2004)

F.J. Corpas, J.B. Barroso, J.M. Palma, M. Rodriguez-Ruiz, Plant peroxisomes: a nitro-oxidative cocktail. Redox Biol. **11**, 535–542 (2017)

A. Costa, I. Drago, S. Behera, M. Zottini, P. Pizzo, J.I. Schroeder, T. Pozzan, F. Lo Schiavo, H_2O_2 in plant peroxisomes: an in vivo analysis uncovers a Ca(2+)-dependent scavenging system. Plant J. **62**, 760–772 (2010)
J.L. Costello, M. Schrader, Unloosing the Gordian knot of peroxisome formation. Curr. Opin. Cell Biol. **50**, 50–56 (2018)
J.L. Costello, I.G. Castro, F. Camões et al., Predicting the targeting of tail-anchored proteins to subcellular compartments in mammalian cells. J. Cell Sci. **130**, 1675–1687 (2017)
J.L. Costello, I.G. Castro, C. Hacker et al., ACBD5 and VAPB mediate membrane associations between peroxisomes and the ER. J. Cell Biol. **216**, 331–342 (2017b)
J.L. Costello, I.G. Castro, T.A. Schrader, M. Islinger, M. Schrader, Peroxisomal ACBD4 interacts with VAPB and promotes ER-peroxisome associations. Cell Cycle **16**, 1039–1045 (2017c)
J.L. Costello, J.B. Passmore, M. Islinger, M. Schrader, Multi-localized proteins: the peroxisome–mitochondria connection, in Subcellular biochemistry, ed. By L.A. del Rio, M. Schrader (Springer, Dordrecht, 2018b)
M. Cote, M.M. Menager, A. Burgess et al., Munc18-2 deficiency causes familial hemophagocytic lymphohistiocytosis type 5 and impairs cytotoxic granule exocytosis in patient NK cells. J. Clin. Invest **119**, 3765–3773 (2009)
M.F. Coutinho, M.J. Prata, S. Alves, A shortcut to the lysosome: the mannose-6-phosphate-independent pathway. Mol. Genet Metab. **107,** 257–266 (2012)
A. Dabrazhynetskaya, J. Ma, A.O. Guerreiro-Cacais, Z. Arany, E. Rudd, J.I. Henter, K. Karre, J. Levitskaya, V. Levitsky, Syntaxin 11 marks a distinct intracellular compartment recruited to the immunological synapse of NK cells to colocalize with cytotoxic granules. J. Cell Mol. Med. **16**, 129–141 (2012)
M.S. Dahabieh, Z. Ha, E. Di Pietro et al., Peroxisomes protect lymphoma cells from HDAC inhibitor-mediated apoptosis. Cell Death Differ. **24**, 1912–1924 (2017)
L. Danglot, M. Chaineau, M. Dahan, M.-C. Gendron, N. Boggetto, F. Perez, T. Galli, Role of TI-VAMP and CD82 in EGFR cell-surface dynamics and signaling. J. Cell Sci. **123**, 723–735 (2010)
L. Danglot, K. Zylbersztejn, M. Petkovic et al., Absence of TI-VAMP/Vamp7 leads to increased anxiety in mice. J. Neurosci. **32**, 1962–1968 (2012)
C. David, J. Koch, S. Oeljeklaus, A. Laernsack, S. Melchior, S. Wiese, A. Schummer, R. Erdmann, B. Warscheid, C. Brocard, A combined approach of quantitative interaction proteomics and live-cell imaging reveals a regulatory role for endoplasmic reticulum (ER) reticulon homology proteins in peroxisome biogenesis. Mol. Cell Proteom. **12**, 2408–2425 (2013)
M. Dawidowski, L. Emmanouilidis, V.C. Kalel et al., Inhibitors of PEX14 disrupt protein import into glycosomes and kill Trypanosoma parasites. Science **355**, 1416–1420 (2017)
C. De Duve, Functions of microbodies (peroxisomes). J. Cell Biol. **27,** 25A–26A (1965)
C. De Duve, The lysosome turns fifty. Nat. Cell Biol. **7**, 847–849 (2005)
C. De Duve, P. Baudhuin, Peroxisomes (microbodies and related particles). Physiol. Rev. **46**, 323–357 (1966)
H.K. Delille, B. Agricola, S.C. Guimaraes, H. Borta, G.H. Lüers, M. Fransen, M. Schrader, Pex11pbeta-mediated growth and division of mammalian peroxisomes follows a maturation pathway. J. Cell Sci. **123**, 2750–2762 (2010)
M.K. Dennis, C. Delevoye, A. Acosta-Ruiz et al., BLOC-1 and BLOC-3 regulate VAMP7 cycling to and from melanosomes via distinct tubular transport carriers. J. Cell Biol. **214**, 293–308 (2016)
N.M. Deori, A. Kale, P.K. Maurya, S. Nagotu, Peroxisomes: role in cellular ageing and age related disorders. Biogerontology **9**(5), 303–324 (2018)
F. Di Cara, A. Sheshachalam, N.E. Braverman, R.A. Rachubinski, A.J. Simmonds, Peroxisome-mediated metabolism is required for immune response to microbial infection. Immunity **47**, 93-106.e7 (2017)
A.F. Dias, T.A. Rodrigues, A.G. Pedrosa, A. Barros-Barbosa, T. Francisco, J.E. Azevedo, The peroxisomal matrix protein translocon is a large cavity-forming protein assembly into which PEX5 protein enters to release its cargo. J. Biol. Chem. **292**, 15287–15300 (2017)
M. Divangahi, M. Chen, H. Gan, D. Desjardins, T.T. Hickman, Mycobacterium tuberculosis evades macrophage defenses by inhibiting plasma membrane repair. Nat. Immunol. **10**, 899–906 (2009)
E. Dixit, S. Boulant, Y. Zhang et al., Peroxisomes are signaling platforms for antiviral innate immunity. Cell **141**, 668–681 (2010)
D.K. Dowling, L.W. Simmons, Reactive oxygen species as universal constraints in life-history evolution. Proc. Biol. Sci. **276**, 17371745 (2009)
E.L. Doyle, V. Ridger, F. Ferraro, M. Turmaine, P. Saftig, D.F. Cutler, CD63 is an essential cofactor to leukocyte recruitment by endothelial P-selectin. Blood **118**, 4265–4273 (2011)
X. Du, A.J. Brown, H. Yang, Novel mechanisms of intracellular cholesterol transport: oxysterol-binding proteins and membrane contact sites. Curr. Opin. Cell Biol. **35**, 37–42 (2015)
N. Durchfort, S. Verhoef, M.B. Vaughn, R. Shrestha, D. Adam, J. Kaplan, D.M. Ward, The enlarged lysosomes in beigej cells result from decreased lysosome fission and not increased lysosome fusion. Traffic **13**, 108–119 (2012)
P.J. Eastmond, Monodehydroascorbate reductase 4 is required for seed storage, oil hydrolysis and postgerminative growth in Arabidopsis. Plant Cell **19**, 1376–1387 (2007)

M.S. Ebberink, J. Koster, G. Visser, F. Spronsen, I. Stolte-Dijkstra, G.P. Smit, J.M. Fock, S. Kemp, R.J. Wanders, H.R. Waterham, A novel defect of peroxisome division due to a homozygous non-sense mutation in the PEX11β gene. J Med Genet. **49**, 307–313 (2012)
T. Eberhart, W. Kovacs, Pexophagy in yeast and mammals: an update on mysteries. Histochem. Cell Biol. **150**(5), 473–488 (2018)
D. Effelsberg, L.D. Cruz-Zaragoza, W. Schliebs, R. Erdmann, Pex9p is a new yeast peroxisomal import receptor for PTS1-containing proteins. J. Cell Sci. **129**, 4057–4066 (2016)
C. Eich, C. Manzo, S. de Keijzer, G.-J. Bakker, I. Reinieren-Beeren, M.F. Garcia-Parajo, A. Cambi, Changes in membrane sphingolipid composition modulate dynamics and adhesion of integrin nanoclusters. Sci Rep. **6**, 20693 (2016)
M. Elsner, W. Gehrmann, S. Lenzen, Peroxisome-generated hydrogen peroxide as important mediator of lipotoxicity in insulin-producing cells. Diabetes **60**, 200–208 (2011)
E.D. Elstak, M. Neeft, N.T. Nehme et al., The munc13-4-rab27 complex is specifically required for tethering secretory lysosomes at the plasma membrane. Blood **118**, 1570–1578 (2011)
M. Encarnacao, L. Espada, C. Escrevente et al., A Rab3a-dependent complex essential for lysosome positioning and plasma membrane repair. J. Cell Biol. **213**, 631–640 (2016)
A.J. Engler, S. Sen, H.L. Sweeney, D.E. Discher, Matrix elasticity directs stem cell lineage specification. Cell **126**, 677–689 (2006)
R. Erdmann, Assembly, maintenance and dynamics of peroxisomes. Biochim. Biophys. Acta Mol. Cell Res. **1863**, 787–789 (2016)
S. Even-Ram, A.D. Doyle, M.A. Conti, K. Matsumoto, R.S. Adelstein, K.M. Yamada, Myosin IIA regulates cell motility and actomyosin-microtubule crosstalk. Nat. Cell Biol. **9**, 299–309 (2007)
C.M. Fader, M.O. Aguilera, M.I. Colombo, ATP is released from autophagic vesicles to the extracellular space in a VAMP7-dependent manner. Autophagy **8**, 1741–1756 (2012)
C.M. Fader, D.G. Sanchez, M.B. Mestre, M.I. Colombo, TI-VAMP/VAMP7 and VAMP3/cellubrevin: two v- SNARE proteins involved in specific steps of the autoph- agy/multivesicular body pathways. Biochim. Biophys. Acta **1793**, 1901–1916 (2009)
J.A. Fahrner, R. Liu, M.S. Perry et al., A novel de novo dominant negative mutation in DNM1L impairs mitochondrial fission and presents as childhood epileptic encephalopathy. Am. J. Med. Genet A **170**, 2002–2011 (2016)
M.H. Fakieh, P.J. Drake, J. Lacey, J.M. Munck, A.M. Motley, E.H. Hettema, Intra-ER sorting of the peroxisomal membrane protein Pex3 relies on its luminal domain. Biol. Open **2**, 829–837 (2013)
J. Fan, X. Li, L. Issop, M. Culty, V. Papadopoulos, ACBD2/ECI2-mediated peroxisome–mitochondria interactions in leydig cell steroid biosynthesis. Mol. Endocrinol. **30**, 763–782 (2016)
J.-C. Farré, K. Carolino, O.V. Stasyk et al., A new yeast peroxin, Pex36, a functional homolog of mammalian PEX16, functions in the ER-to-peroxisome traffic of peroxisomal membrane proteins. J. Mol. Biol. **429**, 3743–3762 (2017)
J.-C. Farré, S. Mahalingam, M. Proietto, S. Subramani, Peroxisome biogenesis, membrane contact sites, and quality control. EMBO Rep. **20**, e46864 (2019)
P.L. Faust, W.J. Kovacs, Cholesterol biosynthesis and ER stress in peroxisome deficiency. Biochimie **98**, 75–85 (2014)
J. Feldmann, I. Callebaut, G. Raposo et al., Munc13-4 is essential for cyto-lytic granules fusion and is mutated in a form of familial hemophagocytic lymphohistiocytosis (FHL3). Cell **115**, 461–473 (2003)
Q. Feng, B. Kornmann, Mechanical forces on cellular organelles. J. Cell Sci. **131**, jcs218479 (2018)
S. Ferdinandusse, S. Denis, P.T. Clayton et al., Mutations in the gene encoding peroxisomal α-methylacyl-CoA racemase cause adult-onset sensory motor neuropathy. Nat. Genet **24**, 188–191 (2000)
S. Ferdinandusse, K.D. Falkenberg, J. Koster et al., ACBD5 deficiency causes a defect in peroxisomal very long-chain fatty acid metabolism. J. Med. Genet **54**, 330–337 (2017)
A. Fraldi, F. Annunziata, A. Lombardi et al., Lysosomal fusion and SNARE function are impaired by cholesterol accumulation in lysosomal storage disorders. EMBO J. **29**, 3607–3620 (2010)
A. Fraldi, A.D. Klein, D.L. Medina, C. Settembre, Brain disorders due to lysosomal dysfunction. Annu. Rev. Neurosci. **39**, 277–295 (2016)
T. Francisco, T.A. Rodrigues, A.F. Dias, A. Barros-Barbosa, D. Bicho, J.E. Azevedo, Protein transport into peroxisomes: knowns and unknowns. BioEssays **39**, 1700047 (2017)
M. Fransen, C. Lismont, P. Walton, The peroxisome–mitochondria connection: how and why? Int. J. Mol. Sci. **18**, 1126–1151 (2017)
M. Fransen, M. Nordgren, B. Wang, O. Apanasets, Role of peroxisomes in ROS/RNS-metabolism: implications for human disease. Biochim. Biophys. Acta **1822**, 1363–1373 (2012)
W.M. Frederiks, K.S. Bosch, K.A. Hoeben, J. van Marle, S. Langbein, Renal cell carcinoma and oxidative stress: the lack of peroxisomes. Acta Histochem. **112**, 364–371 (2010)
E.M. Frick, L.C. Strader, Kinase MPK17 and the peroxisome division factor PMD1 influence salt-induced peroxisome proliferation. Plant Physiol. **176**, 340–351 (2018)

J.R. Friedman, L.L. Lackner, M. West, J.R. DiBenedetto, J. Nunnari, G.K. Voeltz, ER tubules mark sites of mitochondrial division. Science **334**, 358–362 (2011)

Y. Fujiki, N. Miyata, S. Mukai, K. Okumoto, E.H. Cheng, BAK regulates catalase release from peroxisomes. Mol. Cell Oncol. **4**, e1306610 (2017)

M. Fukuda, Rab27 effectors, pleiotropic regulators in secretory pathways. Traffic **14**, 949–963 (2013)

C.E. Futter, A. Pearse, L.J. Hewlett, C.R. Hopkins, Multi-vesicular endosomes containing internalized EGF–EGF receptor complexes mature and then fuse directly with lysosomes. J. Cell Biol. **132,** 1011–1023 (1996)

S. Galiani, D. Waithe, K. Reglinski, L.D. Cruz-Zaragoza, E. Garcia, M.P. Clausen, W. Schliebs, R. Erdmann, C. Eggeling, Super-resolution microscopy reveals compartmentalization of peroxisomal membrane proteins. J. Biol. Chem. **291**16948–16962 (2016)

Z. Gan, S. Ram, C. Vaccaro, R.J. Ober, E.S. Ward, Analyses of the recycling receptor, FcRn, in live cells reveal novel pathways for lysosomal delivery. Traffic **10**, 600–614 (2009)

S. Gandre-Babbe, A.M. van der Bliek, The novel tail-anchored membrane protein Mff controls mitochondrial and peroxisomal fission in mammalian cells. Mol. Biol. Cell **19**, 2402–2412 (2008)

B.M. Gardner, D.T. Castanzo, S. Chowdhury, G. Stjepanovic, M.S. Stefely, J.H. Hurley, G.C. Lander, A. Martin, The peroxisomal AAA-ATPase Pex1/Pex6 unfolds substrates by processive threading. Nat. Commun. **9**, 135 (2018)

G.L. Gaunt, C. de Duve, Subcellular distribution of D-amino acid oxidase and catalase in rat brain. J. Neurochem. **26**, 749–759 (1976)

N.C. Gauthier, M.A. Fardin, P. Roca-Cusachs, M.P. Sheetz, Temporary increase in plasma membrane tension coordinates the activation of exocytosis and contraction during cell spreading. Proc. Natl. Acad. Sci. USA **108**, 14467–14472 (2011)

S. Gerber, M. Charif, A. Chevrollier et al., Mutations in DNM1L, as in OPA1, result in dominant optic atrophy despite opposite effects on mitochondrial fusion and fission. Brain**140**, 2586–2596 (2017)

P. Ghaffari, A. Mardinoglu, J. Nielsen, Cancer metabolism: a modeling perspective. Front Physiol. **6**, 382 (2015)

D. Ghosh, S. Pinto, L. Danglot et al., VAMP7 regulates constitutive membrane incorporation of the cold-activated channel TRPM8. Nat. Commun. **7**, 10489 (2016)

P. Ghosh, N.M. Dahms, S. Kornfeld, Mannose 6-phosphate receptors: New twists in the tale. Nat. Rev. Mol. Cell Biol. **4** 202–212 (2003)

E.A. Giannopoulou, L. Emmanouilidis, M. Sattler, G. Dodt, M. Wilmanns, Towards the molecular mechanism of the integration of peroxisomal membrane proteins. Biochim. Biophys. Acta **1863**, 863–869 (2016)

C.R. Giordano, S.R. Terlecky, Peroxisomes, cell senescence, and rates of aging. BBA Mol. Basis Dis. **1822**, 1358–1362 (2012)

A. Glasauer, N.S. Chandel, Targeting antioxidants for cancer therapy. Biochem. Pharmacol. **92**, 90–101 (2014)

S. Goepfert, Y. Poirier, Beta-oxidation in fatty acid degradation and beyond. Curr. Opin. Plant Biol. **10**, 245–251 (2007)

S. Goldfischer, C.L. Moore, A.B. Johnson, A.J. Spiro, M.P. Valsamis, H.K. Wisniewski, R.H. Ritch, W.T. Norton, I. Rapin, L.M. Gartner, Peroxisomal and mitochondrial defects in the cerebro-hepato-renal syndrome. Science **182**, 62–64 (1973)

G.M. Griffiths, S. Isaaz, Granzymes A and B are targeted to the lytic granules of lymphocytes by the mannose- 6-phosphate receptor. J. Cell Biol. **120**, 885–896 (1993)

G. Griffiths, B. Hoflack, K. Simons, I. Mellman, S. Kornfeld, The mannose 6-phosphate receptor and the bio-genesis of lysosomes. Cell **52**, 329–341 (1988)

E. Grote, M. Baba, Y. Ohsumi, P.J. Novick, Geranylger-anylated SNAREs are dominant inhibitors of membrane fusion. J. Cell Biol. **151**, 453–466 (2000)

R. Guan, H. Dai, J. Rizo, Binding of the Munc13–1 MUN domain to membrane-anchored SNARE complexes. Biochemistry **47**, 1474–1481 (2008)

C.M. Guardia, G.G. Farias, R. Jia, J. Pu, J.S. Bonifacino, BORC functions upstream of kinesins 1 and 3 to coordinate regional movement of lysosomes along different microtubule tracks. Cell Rep. **17**, 1950–1961 (2016)

S.C. Guimaraes, M. Schuster, E. Bielska, G. Dagdas, S. Kilaru, B.R. Meadows, M. Schrader, G. Steinberg, Peroxisomes, lipid droplets, and endoplasmic reticulum "hitchhike" on motile early endosomes. J. Cell Biol. **211**, 945–954 (2015)

G.G. Gundersen, H.J. Worman, Nuclear positioning. Cell **152**, 1376–1389 (2013)

S.L. Gupton, F.B. Gertler, Integrin signaling switches the cytoskeletal and exocytic machinery that drives neuritogenesis. Dev. Cell **18**, 725–736 (2010)

P.S. Gurel, A.L. Hatch, H.N. Higgs, Connecting the cytoskeleton to the endoplasmic reticulum and Golgi. Curr. Biol. **24**, 660–672 (2014)

E.K. Haddad, X. Wu, J.A. Hammer 3rd, P.A. Henkart, Defective granule exocytosis in Rab27a-deficient lymphocytes from Ashen mice. J. Cell Biol. **152,** 835–842 (2001)

W.O. Hancock, Bidirectional cargo transport: moving beyond tug of war. Nat. Rev. Mol. Cell Biol. **15**, 615–628 (2014)

K.J. Harrison-Lavoie, G. Michaux, L. Hewlett, J. Kaur, J. Hannah, W. Lui-Roberts, E. Norman, D.F. Cutler, P-selectin and CD63 use different mechanisms for delivery to Weibel–Palade bodies. Traffic **7**, 647–662 (2006)

N. Hasan, C. Hu, Vesicle-associated membrane protein 2 mediates trafficking of alpha5beta1 integrin to the plasma membrane. Exp. Cell Res. **316**, 12–23 (2010)

Y. Hayashi, M. Hayashi, H. Hayashi, I. Hara-Nishimura, M. Nishimura, Direct interaction between glyoxysomes and lipid bodies in cotyledons of the Arabidopsis thaliana ped1 mutant. Protoplasma **218**, 83–94 (2001)

S. Hein, P. Schönfeld, S. Kahlert, G. Reiser, Toxic effects of X-linked adrenoleukodystrophy-associated, very long chain fatty acids on glial cells and neurons from rat hippocampus in culture. Hum. Mol. Genet **17**, 1750–1761 (2008)

S.C.J. Helle, Q. Feng, M.J. Aebersold et al., Mechanical force induces mitochondrial fission. Elife **6**, e30292 (2017)

P.A. Henkart, P.J. Millard, C.W. Reynolds, M.P. Henkart, Cytolytic activity of purified cytoplasmic granules from cytotoxic rat large granular lymphocyte tumors. J. Exp. Med. **160**, 75–93 (1984)

P.A. Henkart, C.C. Yue, J. Yang, S.A. Rosenberg, Cytolytic and biochemical properties of cytoplasmic granules of murine lymphokine-activated killer cells. J. Immunol. **137**, 2611–2617 (1986)

U. Herbig, M. Ferreira, L. Condel, D. Carey, J.M. Sedivy, Cellular senescence in aging primates. Science **311**, 1257 (2006)

F. Hermansky, P. Pudlak, Albinism associated with hemorrhagic diathesis and unusual pigmented reticular cells in the bone marrow: report of two cases with histochemical studies. Blood **14**, 162–169 (1959)

M.J. Herrero-Turrion, J. Calafat, H. Janssen, M. Fukuda, F. Mollinedo, Rab27a regulates exocytosis of tertiary and specific granules in human neutrophils. J. Immunol. **181**, 3793–3803 (2008)

G.G. Hesketh, I. Perez-Dorado, L.P. Jackson et al., VARP is recruited on to endosomes by direct interaction with retromer, where together they function in export to the cell surface. Dev. Cell **29**, 591–606 (2014)

E.H. Hettema, R. Erdmann, I.J. van der Klei, M. Veenhuis, Evolving models for peroxisome biogenesis. Curr. Opin. Cell Biol. **29**, 25–30 (2014)

H.S. Heymans, R.B. Schutgens, R. Tan et al., Severe plasmalogen deficiency in tissues of infants without peroxisomes (Zellweger syndrome). Nature **306**, 69–70 (1983)

Y. Hirota, N. Masuyama, T. Kuronita, H. Fujita, M. Himeno, Y. Tanaka, Analysis of post-lysosomal compartments. Biochem. Biophys. Res. Commun. **314**, 306–312 (2004)

J. Hirst, C.E. Futter, C.R. Hopkins, The kinetics of man- nose 6-phosphate receptor trafficking in the endocytic pathway in HEp-2 cells: the receptor enters and rapidly leaves multivesicular endosomes without accumulating in a prelysosomal compartment. Mol. Biol. Cell **9**, 809–816 (1998)

D. Hoepfner, D. Schildknegt, I. Braakman, P. Philippsen, H.F. Tabak, Contribution of the endoplasmic reticulum to peroxisome formation. Cell **122**, 85–95 (2005)

D. Hoepfner, M. van den Berg, P. Philippsen, H.F. Tabak, E.H. Hettema, A role for Vps1p, actin, and the Myo2p motor in peroxisome abundance and inheritance in Saccharomyces cerevisiae. J. Cell Biol. **155**, 979–990 (2001)

J.D. Horton, J.L. Goldstein, M.S. Brown, SREBPs: activators of the complete program of cholesterol and fatty acid synthesis in the liver. J. Clin. Invest **109**, 1125–1131 (2002)

Z. Hruban, M. Rechcigl, Microbodies and related particles. Morphology, biochemistry, and physiology. Int. Rev. Cytol. Suppl. **1**, 1–296 (1969)

R. Hua, D. Cheng, É. Coyaud et al., VAPs and ACBD5 tether peroxisomes to the ER for peroxisome maintenance and lipid homeostasis. J. Cell Biol. **216**, 367–377 (2017)

J. Huang, N. Viswakarma, S. Yu et al., Progressive endoplasmic reticulum stress contributes to hepatocarcinogenesis in fatty acyl-CoA oxidase 1-deficient mice. Am. J. Pathol. **179**, 703–713 (2011)

N. Huber, S. Guimaraes, M. Schrader, U. Suter, A. Niemann, Charcot-marie-tooth disease-associated mutants of GDAP1 dissociate its roles in peroxisomal and mitochondrial fission. EMBO Rep. **14**, 545–552 (2013)

M. Huizing, C.D. Scher, E. Strovel, D.L. Fitzpatrick, L.M. Hartnell, Y. Anikster, W.A. Gahl, Nonsense mutations in ADTB3A cause complete deficiency of the b3A subunit of adaptor complex-3 and severe Hermansky–Pudlak syndrome type 2. Pediatr Res. **51**, 150–158 (2002)

A.N. Hume, L.M. Collinson, A. Rapak, A.Q. Gomes, C.R. Hopkins, M.C. Seabra, Rab27a regulates the peripheral distribution of melanosomes in melanocytes. J. Cell Biol. **152**, 795–808 (2001)

J. Huotari, A. Helenius, Endosome maturation. EMBO J. **30**, 3481–3500 (2011)

A.H. Hutagalung, P.J. Novick, Role of Rab GTPases in membrane traffic and cell physiology. Phys. Rev. **91**, 119–149 (2011)

S.J. Huybrechts, P.P. Van Veldhoven, C. Brees, G.P. Mannaerts, G.V. Los, M. Fransen, Peroxisome dynamics in cultured mammalian cells. Traffic **10**, 1722–1733 (2009)

K. Ishikawa, A. Nara, K. Matsumoto, H. Hanafusa, EGFR-dependent phosphorylation of leucine-rich repeat kinase LRRK1 is important for proper endosomal trafficking of EGFR. Mol. Biol. Cell **23**, 1294–1306 (2012)

M. Islinger, M.J.R. Cardoso, M. Schrader, Be different—the diversity of peroxisomes in the animal kingdom. Biochim. Biophys. Acta**1803** 881–897 (2010a)

M. Islinger, K.W. Li, M. Loos, S. Liebler, S. Angermüller, C. Eckerskorn, G. Weber, A. Abdolzade, A. Völkl, Peroxisomes from the heavy mitochondrial fraction: isolation by zonal free flow electrophoresis and quantitative mass spectrometrical characterization. J. Proteome Res. **9**, 113–124 (2010b)

M. Islinger, A. Abdolzade-bavil, S. Liebler, M. Schrader, Assessing heterogeneity of peroxisomes: isolation of two subpopulations from rat liver, in *Liver proteomics: methods and protocols*. ed. by D. Josic, D.C. Hixson et al. (Humana, Totowa, 2012a), pp. 83–96

M. Islinger, S. Grille, H.D. Fahimi, M. Schrader, The peroxisome: an update on mysteries. Histochem. Cell Biol. **137**, 547–574 (2012b)

M. Islinger, A. Voelkl, H.D. Fahimi, M. Schrader, The peroxisome: an update on mysteries 2.0. Histochem. Cell Biol. **150**(5), 443–471 (2018)

A. Itoyama, S. Michiyuki, M. Honsho, T. Yamamoto, A. Moser, Y. Yoshida, Y. Fujiki, Mff functions with Pex11p and DLP1 in peroxisomal fission. Biol. Open **2**, 998–1006 (2013)

L. Jahreiss, F.M. Menzies, D.C. Rubinsztein, The itinerary of autophagosomes: From peripheral formation to kiss-and-run fusion with lysosomes. Traffic **9**, 574–587 (2008)

J.K. Jaiswal, N.W. Andrews, S.M. Simon, Membrane proximal lysosomes are the major vesicles responsible for calcium-dependent exocytosis in nonsecretory cells. J. Cell Biol. **159**, 625–635 (2002)

J. Ježek, K. Cooper, R. Strich, Reactive oxygen species and mitochondrial dynamics: the Yin and Yang of mitochondrial dysfunction and cancer progression. Antioxidants **7**, 13 (2018)

Z. Jiang, B.A. Woda, K.L. Rocks et al., P504S: a new molecular marker for the detection of prostate carcinoma. Am. J. Surg. Pathol. **25**, 139ssss7–1404 (2001)

Y. Jindal, A. Singh, R. Kumar, K. Varma, V. Misra, S.P. Misra, M. Dwivedi, Expression of alpha methylacyl CoA racemase (AMACR) in gastric adenocarcinoma and its correlation with helicobacter pylori infection. J. Clin. Diagn. Res. **10**, EC10–EC12 (2016)

J.L. Johnson, H. Hong, J. Monfregola, W.B. Kiosses, S.D. Catz, Munc13-4 restricts motility of Rab27a-expressing vesicles to facilitate lipopolysaccharide-induced priming of exocytosis in neutrophils. J. Biol. Chem. **286**, 5647–5656 (2011)

J.M. Jones, J.C. Morrell, S.J. Gould, Multiple distinct targeting signals in integral peroxisomal membrane proteins. J. Cell Biol. **153**, 1141–1150 (2001)

J.M. Jones, J.C. Morrell, S.J. Gould, PEX19 is a predominantly cytosolic chaperone and import receptor for class 1 peroxisomal membrane proteins. J. Cell Biol. **164**, 57–67 (2004)

J.C. Kagan, Signaling organelles of the innate immune system. Cell **151**, 1168–1178 (2012)

M.A. Karreman, L. Mercier, N.L. Schieber, G. Solecki, G. Allio, F. Winkler, B. Ruthensteiner, J.G. Goetz, Y. Schwab, Fast and precise targeting of single tumor cells in vivo by multimodal correlative microscopy. J. Cell Sci. **129**, 444–456 (2016)

Y. Kataria, M. Wright, R.J. Deaton, E.E. Rueter, B.A. Rybicki, A.B. Moser, V. Ananthanrayanan, P.H. Gann, Dietary influences on tissue concentrations of phytanic acid and AMACR expression in the benign human prostate. Prostate **75**, 200–210 (2015)

S. Kedashiro, S.I. Pastuhov, T. Nishioka, T. Watanabe, K. Kaibuchi, K. Matsumoto, H. Hanafusa, LRRK1-phosphorylated CLIP-170 regulates EGFR trafficking by recruiting p150Glued to microtubule plus ends. J. Cell Sci. **128**, 385–396 (2015)

J.M. Keller, S. Cablé, F. et Bouhtoury, S. Heusser, C. Scotto, L. Armbruster, E. Ciolek, S. Colin, J. Schilt, M. Dauça, Peroxisome through cell differentiation and neoplasia. Biol. Cell **77**, 77–88 (1993)

H.M. Kent, P.R. Evans, I.B. Schafer, S.R. Gray, C.M. Sanderson, J.P. Luzio, A.A. Peden, D.J. Owen, Structural basis of the intracellular sorting of the SNARE VAMP7 by the AP3 adaptor complex. Dev. Cell **22**, 979–988 (2012)

K. Keren, Membrane tension leads the way. Proc. Natl. Acad. Sci. USA **108**, 14379–14380 (2011)

P.A. Kiely, A. Sant, R. O'Connor, RACK1 is an insulin-like growth factor 1 (IGF-1) receptor-interacting protein that can regulate IGF-1-mediated Akt activation and protection from cell death. J. Biol. Chem. **277**, 22581–22589 (2002)

P.K. Kim, R.T. Mullen, U. Schumann, J. Lippincott-Schwartz, The origin and maintenance of mammalian peroxisomes involves a de novo PEX16-dependent pathway from the ER. J. Cell Biol. **173**, 521–532 (2006)

J. Klumperman, G. Raposo, The complex ultrastructure of the endolysosomal system. Cold Spring Harb. Perspect. Biol. **6**(10), a016857 (2014)

B. Knoblach, R. Rachubinski, Sharing the cell's bounty—organelle inheritance in yeast. J. Cell Sci. **128**, 621–630 (2015)

B. Knoblach, R.A. Rachubinski, How peroxisomes partition between cells a story of yeast, mammals and filamentous fungi. Curr. Opin. Cell Biol. **41**, 73–80 (2016)

A. Koch, M. Thiemann, M. Grabenbauer, Y. Yoon, M.A. McNiven, M. Schrader, Dynamin-like protein 1 is involved in peroxisomal fission. J. Biol. Chem. **278**, 8597–8605 (2003)

A. Koch, Y. Yoon, N.A. Bonekamp, M.A. McNiven, M. Schrader, A role for Fis1 in both mitochondrial and peroxisomal fission in mammalian cells. Mol. Biol. Cell **16**, 5077–5086 (2005)

J. Koch, C. Brocard, PEX11 proteins attract Mff and hFis1 to coordinate peroxisomal fission. J. Cell Sci. **125**, 3813–3826 (2012)

J. Koch, R.G. Feichtinger, P. Freisinger, M. Pies, F. Schrödl, A. Iuso, W. Sperl, J.A. Mayr, H. Prokisch, T.B. Haack, Disturbed mitochondrial and peroxisomal dynamics due to loss of MFF causes Leigh-like encephalopathy, optic atrophy and peripheral neuropathy. J. Med. Genet **53**, 270–278 (2016)

Y. Kojima, A. Kawasaki-Koyanagi, N. Sueyoshi, A. Kanai, H. Yagita K. Okumura, Localization of Fas ligand in cytoplasmic granules of CD8þ cytotoxic T lymphocytes and natural killer cells: participation of Fas ligand in granule exocytosis model of cytotoxicity. Biochem. Biophys. Res. Commun. **296**, 328–336 (2002)

S. Kornfeld, I. Mellman, The biogenesis of lysosomes. Annu. Rev. Cell Biol. **5**, 483–525 (1989)

S. Koseoglu, C.G. Peters, J.L. Fitch-Tewfik, O. Aisiku, L. Danglot, T. Galli, R. Flaumenhaft, VAMP-7 links granule exocytosis to actin reorganization during platelet activation. Blood **126**, 651–660 (2015)

A. Kostic, C.D. Lynch, M.P. Sheetz, Differential matrix rigidity response in breast cancer cell lines correlates with the tissue tropism. PLoS ONE **4**, e6361 (2009)

W.J. Kovacs, L.M. Olivier, S.K. Krisans, Central role of peroxisomes in isoprenoid biosynthesis. Prog. Lipid Res. **41**, 369–391 (2002)

W.J. Kovacs, K.N. Tape, J.E. Shackelford, T.M. Wikander, M.J. Richards, S.J. Fliesler, S.K. Krisans, P.L. Faust, Peroxisome deficiency causes a complex phenotype because of hepatic SREBP/Insig dysregulation associated with endoplasmic reticulum stress. J. Biol. Chem. **284**, 7232–7245 (2009)

J.B. Krahling, R. Gee, J.A. Gauger, N.E. Tolbert, Postnatal development of peroxisomal and mitochondrial enzymes in rat liver. J. Cell. Physiol. **101**, 375–390 (1979)

K. Krzewski, A. Gil-Krzewska, V. Nguyen, G. Peruzzi, J.E. Coligan, LAMP1/CD107a is required for efficient perforin delivery to lytic granules and NK-cell cytotoxicity. Blood **121**, 4672–4683 (2013)

S. Kumar, R. de Boer, I.J. van der Klei, Yeast cells contain a heterogeneous population of peroxisomes that segregate asymmetrically during cell division. J. Cell Sci. **131**, jcs207522 (2018)

K. Kuravi, S. Nagotu, A.M. Krikken, K. Sjollema, M. Deckers, R. Erdmann, M. Veenhuis, I.J. van der Klei, Dynamin-related proteins Vps1p and Dnm1p control peroxisome abundance in Saccharomyces cerevisiae. J. Cell Sci. **119**, 3994–4001 (2006)

F. Lafont, P. Verkade, T. Galli, C. Wimmer, D. Louvard, K. Simons, Raft association of SNAP receptors acting in apical trafficking in Madin-Darby canine kidney cells. Proc. Natl. Acad. Sci. USA **96**, 3734–3738 (1999)

C. Lauer, A. Völkl, S. Riedl, H.D. Fahimi, K. Beier, Impairment of peroxisomal biogenesis in human colon carcinoma. Carcinogenesis **20**, 985–989 (1999)

K. Laulagnier, N.L. Schieber, T. Maritzen, V. Haucke, R.G. Parton, J. Gruenberg, Role of AP1 and Gadkin in the traffic of secretory endo-lysosomes. Mol. Biol. Cell **22**, 2068–2082 (2011)

G. Laurenti, E. Benedetti, B. D'Angelo et al., Hypoxia induces peroxisome proliferator-activated receptor α (PPARα) and lipid metabolism peroxisomal enzymes in human glioblastoma cells. J. Cell Biochem. **112**, 3891–3901 (2011)

R.H. Law, N. Lukoyanova, I. Voskoboinik et al., The structural basis for membrane binding and pore formation by lymphocyte perforin. Nature **468**, 447–451 (2010)

H.S. Lee, S.J. Millward-Sadler, M.O. Wright, G. Nuki, R. Al-Jamal, D.M. Salter, Activation of Integrin-RACK1/PKC alpha signalling in human articular chondrocyte mechanotransduction. Osteoarthritis Cartilage **10**, 890–897 (2002)

S. Lenzen, J. Drinkgern, M. Tiedge, Low antioxidant enzyme gene expression in pancreatic islets compared with various other mouse tissues. Free Radic. Biol. Med. **20**, 463–466 (1996)

S.C. Lewis, L.F. Uchiyama, J. Nunnari, ER–mitochondria contacts couple mtDNA synthesis with mitochondrial division in human cells. Science **353**, aaf5549 (2016)

D. Li, N. Ropert, A. Koulakoff, C. Giaume, M. Oheim, Lysosomes are the major vesicular compartment undergoing Ca^{2+}-regulated exocytosis from cortical astrocytes. J. Neurosci. **28**, 7648–7658 (2008)

W. Li, C. Ma, R. Guan, Y. Xu, D.R. Tomchick, J. Rizo, The crystal structure of a Munc13 C-terminal module exhibits a remarkable similarity to vesicle tethering factors. Structure **19**, 1443–1455 (2011)

X. Li, S.J. Gould, The dynamin-like GTPase DLP1 is essential for peroxisome division and is recruited to peroxisomes in part by PEX11. J. Biol. Chem. **278**, 17012–17020 (2003)

J. Liliental, D.D. Chang, Rack1, a receptor for activated protein kinase C, interacts with integrin beta subunit. J. Biol. Chem. **273**, 2379–2383 (1998)

C. Lin, G. Steinberg, Spatial organization of organelles in fungi: insights from mathematical modelling. Fungal Genet Biol. **103**, 55–59 (2017)

C. Lin, M. Schuster, S.C. Guimaraes, P. Ashwin, M. Schrader, J. Metz, C. Hacker, S.J. Gurr, G. Steinberg, Active diffusion and microtubule-based transport oppose myosin forces to position organelles in cells. Nat. Commun. **7**, 11814 (2016)

D. Lingwood, K. Simons, Lipid rafts as a membrane-organizing principle. Science **327**, 46–50 (2010)

C. Lismont, M. Nordgren, P.P. Van Veldhoven, M. Fransen, Redox interplay between mitochondria and peroxisomes. Front. Cell Dev. Biol. **3**, 35 (2015)

J.A. Litwin, K. Beier, A. Völkl, W.J. Hofmann, H.D. Fahimi, Immunocytochemical investigation of catalase and peroxisomal lipid beta-oxidation enzymes in human hepatocellular tumors and liver cirrhosis. Virchows Arch. **435**, 486–495 (1999)

M.D. Lloyd, D.J. Darley, A.S. Wierzbicki, M.D. Threadgill, Alpha-methylacyl-CoA racemase—an "obscure" metabolic enzyme takes centre stage. FEBS J. **275**, 1089–1102 (2008)

E. Lloyd-Evans, A.J. Morgan, X. He, D.A. Smith, E. Elliot-Smith, D.J. Sillence, G.C. Churchill, E.H. Schuchman, A. Galione, F.M. Platt, Niemann–Pick disease type C1 is a sphingosine storage disease that causes deregulation of lysosomal calcium. Nat. Med. **14**, 1247–1255 (2008)

J. Long, C.-J. Zhang, N. Zhu, K. Du, Y.-F. Yin, X. Tan, K. Du, Y.F. Yin, X. Tan, D.F. Liao, L. Qin, Lipid metabolism and carcinogenesis, cancer development. Am. J. Cancer Res. **8**(5), 778–791 (2018)

D.M. Lowrey, T. Aebischer, K. Olsen, M. Lichtenheld, F. Rupp, H. Hengartner, E.R. Podack, Cloning, analysis, and expression of murine perforin 1 cDNA, a component of cytolytic T-cell granules with homology to complement component C9. Proc. Natl. Acad. Sci. **86**, 247–251 (1989)

T. Lu, T. Finkel, Free radicals and senescence. Exp. Cell Res. **314**, 1918–1922 (2008)

G. Lüers, T. Hashimoto, H.D. Fahimi, A. Völkl, Biogenesis of peroxisomes: isolation and characterization of two distinct peroxisomal populations from normal and regenerating rat liver. J. Cell Biol. **121**, 1271–1280 (1993)

W.W. Lui-Roberts, L.M. Collinson, L.J. Hewlett, G. Michaux, D.F. Cutler, An AP1/clathrin coat plays a novel and essential role in forming the Weibel–Palade bodies of endothelial cells. J. Cell Biol. **170**, 627–636 (2005)

J.P. Luzio, P.R. Pryor, N.A. Bright, Lysosomes: fusion and function. Nat. Rev. Mol. Cell Biol. **8**, 622–632 (2007)

M.S. Marks, H.F. Heijnen, G. Raposo, Lysosome-related organelles: unusual compartments become mainstream. Curr. Opin. Cell Biol. **25**, 495–505 (2013)

S. Marubashi, N. Ohbayashi, M. Fukuda, A varp-binding protein, RACK1, regulates dendrite outgrowth through stabilization of varp protein in mouse melanocytes. J. Invest. Dermatol. **136**, 1672–1680 (2016)

F.D. Mast, T. Herricks, K.M. Strehler, L.R. Miller, R.A. Saleem, R.A. Rachubinski, J.D. Aitchison, ESCRT-III is required for scissioning new peroxisomes from the endoplasmic reticulum. J. Cell Biol. **217**, 2087–2102 (2018)

F.D. Mast, A. Jamakhandi, R.A. Saleem, D.J. Dilworth, R.S. Rogers, R.A. Rachubinski, J.D. Aitchison, Peroxins Pex30 and Pex29 dynamically associate with reticulons to regulate peroxisome biogenesis from the endoplasmic reticulum. J. Biol. Chem. **291**, 15408–15427 (2016)

F.D. Mast, R.A. Rachubinski, J.D. Aitchison, Signaling dynamics and peroxisomes. Curr. Opin. Cell Biol. **35**, 131–136 (2015)

T. Matsuzaki, Y. Fujiki, The peroxisomal membrane protein import receptor Pex3p is directly transported to peroxisomes by a novel Pex19p- and Pex16p-dependent pathway. J. Cell Biol. **183**, 1275–1286 (2008)

F.R. Maxfield, D. Wüstner, Intracellular cholesterol transport. J. Clin. Invest. **110**(7), 891–898 (2002)

P.U. Mayerhofer, M. Bano-Polo, I. Mingarro, A.E. Johnson, Human peroxin PEX3 Is co-translationally integrated into the ER and exits the ER in budding vesicles. Traffic **17**, 117–130 (2016)

G.B. McClelland, S. Khanna, G.F. González, C.E. Butz, G.A. Brooks, Peroxisomal membrane monocarboxylate transporters: evidence for a redox shuttle system? Biochem. Biophys. Res. Commun. **304**, 130–135 (2003)

I.J. McGough, F. Steinberg, M. Gallon, A. Yatsu, N. Ohbayashi, K.J. Heesom, M. Fukuda, P.J. Cullen, Identification of molecular heterogeneity in SNX27-retromer-mediated endosome-to-plasma-membrane recycling. J. Cell Sci. **127**, 4940–4953 (2014)

P. McHenry, W.L. Wang, E. Devitt, N. Kluesner, V.J. Davisson, E. McKee, D. Schweitzer, P. Helquist, M. Tenniswood, Iejimalides a and B inhibit lysosomal vacuolar H+-ATPase (V-ATPase) activity and induce S-phase arrest and apoptosis in MCF-7 cells. J. Cell Biochem. **109**(4), 634–642 (2010)

D.L. Medina, A. Fraldi, V. Bouche et al., Transcriptional activation of lysosomal exocytosis promotes cellular clearance. Dev. Cell **21**, 421–430 (2011)

M. Meinecke, C. Cizmowski, W. Schliebs, V. Krüger, S. Beck, R. Wagner, R. Erdmann, The peroxisomal importomer constitutes a large and highly dynamic pore. Nat. Cell Biol. **12**, 273–277 (2010)

A. Mesquita, M. Weinberger, A. Silva, B. Sampaio-Marques, B. Almeida, C. Leão, V. Costa, F. Rodrigues, W.C. Burhans, P. Ludovico, Caloric restriction or catalase inactivation extends yeast chronological lifespan by inducing H_2O_2 and superoxide dismutase activity. Proc. Natl. Acad. Sci. USA **107**, 15123–15128 (2010)

S.E. Miller, D.A. Sahlender, S.C. Graham, S. Honing, M.S. Robinson, A.A. Peden, D.J. Owen, The molecular basis for the endocytosis of small R-SNAREs by the clathrin adaptor CALM. Cell **147**, 1118–1131 (2011)

D. Molino, S. Nola, S.M. Lam et al., Role of tetanus neurotoxin insensitive vesicle-associated membrane protein in membrane domains transport and homeostasis. Cell Logist. **5**, e1025182 (2015)

M. Montilla-Martinez, S. Beck, J. Klümper, M. Meinecke, W. Schliebs, R. Wagner, R. Erdmann, Distinct pores for peroxisomal import of PTS1 and PTS2 proteins. Cell Rep. **13**, 2126–2134 (2015)

K. Moreau, B. Ravikumar, M. Renna, C. Puri, D.C. Rubinsztein, Autophagosome precursor maturation requires homotypic fusion. Cell **146**, 303–317 (2011)

A.M. Motley, E.H. Hettema, Yeast peroxisomes multiply by growth and division. J. Cell Biol. **178**, 399–410 (2007)

R. Mukhopadhyay, Lively Lysosomes. ASBMB Today **15**, 7 (2016)

B.M. Mullock, N.A. Bright, C.W. Fearon, S.R. Gray, J.P. Luzio, Fusion of lysosomes with late endosomes produces a hybrid organelle of intermediate density and is NSF dependent. J. Cell Biol. **140**, 591–601 (1998)

A. Murley, J. Nunnari, The emerging network of mitochondria-organelle contacts. Mol. Cell **61**(5), 648–653 (2016)

S. Nagotu, R. Saraya, M. Otzen, M. Veenhuis, I.J. van der Klei, Peroxisome proliferation in Hansenula polymorpha requires Dnm1p which mediates fission but not de novo formation. Biochim. Biophys. Acta **1783**, 760–769 (2008)

S. Narendra, S. Venkataramani, G. Shen, J. Wang, V. Pasapula, Y. Lin, D. Kornyeyev, A.S. Holaday, H. Zhang, The Arabidopsis ascorbate peroxidase 3 is a peroxisomal membrane-bound antioxidant enzyme and is dispensable for Arabidopsis growth and development. J. Exp. Bot. **57**, 3033–3042 (2006)

A. Nasca, A. Legati, E. Baruffini, C. Nolli, I. Moroni, A. Ardissone, P. Goffrini, D. Ghezzi, Biallelic mutations in DNM1L are associated with a slowly progressive infantile encephalopathy. Hum. Mutat. **37**, 898–903 (2016)

S. Neill, R. Barros, J. Bright, R. Desikan, J. Hancock, J. Harrison, P. Morris, D. Ribeiro, I. Wilson, Nitric oxide, stomatal closure, and abiotic stress. J. Exp. Bot. **59**, 165–176 (2008)

A. Neuhaus, C. Eggeling, R. Erdmann, W. Schliebs, Why do peroxisomes associate with the cytoskeleton? Biochim. Biophys. Acta Mol. Cell Res. **1863**, 1019–1026 (2016)

T.D. Nightingale, K. Pattni, A.N. Hume, M.C. Seabra, D.F. Cutler, Rab27a and MyRIP regulate the amount and multimeric state of VWF released from endothelial cells. Blood **113**, 5010–5018 (2009)

K. Okumoto, T. Ono, R. Toyama, A. Shimomura, A. Nagata, Y. Fujiki, New splicing variants of mitochondrial Rho GTPase-1 (Miro1) transport peroxisomes. J. Cell Biol. **217**, 619–633 (2018)

L. Opaliński, J.A.K.W. Kiel, C. Williams, M. Veenhuis, I.J. van der Kleia, Membrane curvature during peroxisome fission requires Pex11. EMBO J. **30**, 5–16 (2011)

J. Pagel, K. Beutel, K. Lehmberg et al., Distinct mutations in STXBP2 are associated with variable clinical presentations in patients with familial hemophagocytic lymphohistiocytosis type 5 (FHL5). Blood **119**, 6016–6024 (2012)

A. Papadopulos, G.A. Gomez, S. Martin, J. Jackson, R.S. Gormal, D.J. Keating, A.S. Yap, F.A. Meunier, Activity-driven relaxation of the cortical actomyosin II network synchronizes Munc18-1-dependent neurosecretory vesicle docking. Nat. Commun. **6**, 6297 (2015)

G. Parenti, G. Andria, A. Ballabio, Lysosomal storage diseases: from pathophysiology to therapy. Annu. Rev. Med. **66**, 471–486 (2015)

H.H. Park, Structural basis of membrane trafficking by Rab family small G protein. Int. J. Mol. Sci. **14**, 8912–8923 (2013)

A. Pascual-Ahuir, S. Manzanares-Estreder, M. Proft, Pro- and antioxidant functions of the peroxisome–mitochondria connection and its impact on aging and disease. Oxid. Med. Cell Longev. **2017**, 1–17 (2017)

A.G. Pedrosa, T. Francisco, D. Bicho, A.F. Dias, A. Barros-Barbosa, V. Hagmann, G. Dodt, T.A. Rodrigues, J.E. Azevedo, Peroxisomal monoubiquitinated PEX5 interacts with the AAA ATPases PEX1 and PEX6 and is unfolded during its dislocation into the cytosol. J. Biol. Cell **293**, 11553–11563 (2018)

J. Pei, C. Ma, J. Rizo, N.V. Grishin, Remote homology between Munc13 MUN domain and vesicle tethering complexes. J. Mol. Biol. **391**, 509–517 (2009)

V. Piano, D.I. Benjamin, S. Valente, S. Nenci, B. Marrocco, A. Mai, A. Aliverti, D.K. Nomura, A. Mattevi, Discovery of inhibitors for the ether lipid-generating enzyme AGPS as anti-cancer agents. ACS Chem. Biol. **10**, 2589–2597 (2015)

S. Piao, R.K. Amaravadi, Targeting the lysosome in cancer. Ann. N. Y. Acad. Sci. **1371**(1), 45–54 (2016)

F.M. Platt, B. Boland, A.C. van der Spoel, The cell biology of disease: lysosomal storage disorders: the cellular impact of lysosomal dysfunction. J. Cell Biol. **199**, 723–734 (2012)

M. Pljesa-Ercegovac, J. Mimic-Oka, D. Dragicevic, A. Savic-Radojevic, M. Opacic, S. Pljesa, R. Radosavljevic, T. Simic, Altered antioxidant capacity in human renal cell carcinoma: role of glutathione associated enzymes. Urol. Oncol. **26**, 175–181 (2008)

T. Pocard, A. Le Bivic, T. Galli, C. Zurzolo, Distinct v-SNAREs regulate direct and indirect apical delivery in polarized epithelial cells. J. Cell Sci. **120**, 3309–3320 (2007)

E.R. Podack, J.D. Young, Z.A. Cohn, Isolation and biochemical and functional characterization of perforin 1 from cytolytic T-cell granules. Proc. Natl. Acad. Sci. **82**, 8629–8633 (1985)

E.V. Polishchuk, A. Merolla, J. Lichtmannegger et al., Activation of autophagy, observed in liver tissues from patients with Wilson disease and from ATP7B-deficient animals protects hepatocytes from copper-induced apoptosis. Gastroenterology **156**(4), 1173–1189 (2019)

M.S. Pols, E. van Meel, V. Oorschot, C. ten Brink, M. Fukuda, M.G. Swetha, S. Mayor, J. Klumperman, hVps41 and VAMP7 function in direct TGN to late endosome transport of lysosomal membrane proteins. Nat. Commun. **4**, 1361 (2013b)

M.S. Pols, C. ten Brink, P. Gosavi, V. Oorschot, J. Klumperman, The HOPS proteins hVps41 and hVps39 are required for homotypic and heterotypic late endosome fusion. Traffic **14**, 219–232 (2013a)

S.R. Prasad, V.R. Narra, R. Shah, P.A. Humphrey, J. Jagirdar, J.R. Catena, N.C. Dalrymple, C.L. Siegel, Segmental disorders of the nephron: histopathological and imaging perspective. Br. J. Radiol. **80**, 593–602 (2007)

R. Prekeris, J. Klumperman, R.H. Scheller, Syntaxin 11 is an atypical SNARE abundant in the immune system. Eur. J. Cell Biol. **79**, 771–780 (2000)

V. Proux-Gillardeaux, J. Gavard, T. Irinopoulou, R.-M. Mege, T. Galli, Tetanus neurotoxin-mediated cleavage of cellubrevin impairs epithelial cell migration and integrin-dependent cell adhesion. Proc. Natl. Acad. Sci. USA **102**, 6362–6367 (2005)

V. Proux-Gillardeaux, G. Raposo, T. Irinopoulou, T. Galli, Expression of the Longin domain of TI-VAMP impairs lysosomal secretion and epithelial cell migration. Biol. Cell **99**, 261–271 (2007)

V. Proux-Gillardeaux, R. Rudge, T. Galli, The tetanus neurotoxin-sensitive and insensitive routes to and from the plasma membrane: fast and slow pathways? Traffic **6**, 366–373 (2005)

P.R. Pryor, B.M. Mullock, N.A. Bright, S.R. Gray, J.P. Luzio, The role of intraorganellar Ca2þ in late endosome—lysosome heterotypic fusion and in the reformation of lysosomes from hybrid organelles. J. Cell Biol. **149**, 1053–1062 (2000)

P.R. Pryor, B.M. Mullock, N.A. Bright, M.R. Lindsay, S.R. Gray, S.C. Richardson, A. Stewart, D.E. James, R.C. Piper, J.P. Luzio, Combinatorial SNARE complexes with VAMP7 or VAMP8 define different late endocytic fusion events. EMBO Rep. **5**, 590–595 (2004)

J. Pu, C. Schindler, R. Jia, M. Jarnik, P. Backlund, J.S. Bonifacino, BORC, a multisubunit complex that regulates lysosome positioning. Dev. Cell **33**, 176–188 (2015)

S.K. Rao, C. Huynh, V. Proux-Gillardeaux, T. Galli, N.W. Andrews, Identification of SNAREs involved in synaptotagmin VII–regulated lysosomal exocytosis. J. Biol. Chem. **279**, 20471–20479 (2004)

G. Raposo, M.S. Marks, D.F. Cutler, Lysosome-related organelles: driving post-Golgi compartments into specialisation. Curr. Opin. Cell Biol. **19**, 394–401 (2007)

S. Raychaudhuri, W.A. Prinz, Nonvesicular phospholipid transfer between peroxisomes and the endoplasmic reticulum. Proc. Natl. Acad. Sci. USA **105**, 15785–15790 (2008)

A. Reddy, E.V. Caler, N.W. Andrews, Plasma membrane repair is mediated by Ca2þ-regulated exocytosis of lysosomes. Cell **106**, 157–169 (2001)

B.J. Renquist, T.W. Madanayake, S. Ghimire, C.E. Geisler, Y. Xu, R.L. Bogan, Transmembrane protein 135 (TMEM135) is a liver X receptor target gene that mediates an auxiliary peroxisome matrix protein import pathway. bioRxiv (2018). https://doi.org/10.1101/334979

J. Rhodin, Electron microscopy of the kidney. Am. J. Med. **24**(5), 661–675 (1958)

J. Rizo, T.C. Sudhof, The membrane fusion enigma: SNAREs, Sec1/Munc18 proteins, and their accomplices—guilty as charged? Ann. Rev. Cell Dev. Biol. **28**, 279–308 (2012)

F. Rohrig, A. Schulze, The multifaceted roles of fatty acid synthesis in cancer. Nat. Rev. Cancer **16**(11), 732–749 (2016)

I. Rojo Pulido, T.D. Nightingale, F. Darchen, M.C. Seabra, D.F. Cutler, V. Gerke, Myosin Va acts in concert with Rab27a and MyRIP to regulate acute von-Willebrand factor release from endothelial cells. Traffic **12**, 1371–1382 (2011)

C. Rouiller, W. Bernhard, Microbodies and the problem of mitochondrial regeneration in liver cells. J. Biophys. Biochem. Cytol. **2**, 355–360 (1956)

E. Rudd, K. Goransdotter Ericson, C. Zheng, Z. Uysal, A. Ozkan, A. Gurgey, B. Fadeel, M. Nordenskjold, J.I. Henter, Spectrum and clinical implications of syntaxin 11 gene mutations in familial haemophagocytic lymphohistiocytosis: association with disease-free remissions and haematopoietic malignancies. J. Med. Genet **43**, e14 (2006)

C. Ruokun, X. Yake, Y. Fengdong, W.-L. Sun, X. Ma, P. Zhang, B.-Q. Wu, P.-Y. Cui, Lentivirus-mediated silencing of HSDL2 suppresses cell proliferation in human gliomas. Tumour Biol. **37**, 15065–15077 (2016)

K.A. Sacksteder, J.M. Jones, S.T. South, X. Li, Y. Liu, S.J. Gould, PEX19 binds multiple peroxisomal membrane proteins, is predominantly cytoplasmic, and is required for peroxisome membrane synthesis. J. Cell Biol. **148**, 931–944 (2000)

P. Saftig, J. Klumperman, Lysosome biogenesis and lysosomal membrane proteins: trafficking meets function. Nat. Rev. Mol. Cell Biol. **10**, 623–635 (2009)

A.B. Salmon, V.I.R.A. Pérez, Update on the oxidative stress theory of aging: does oxidative stress play a role in aging or healthy aging? Free Radic. Biol. Med. **48**, 642–655 (2010)

J. Salogiannis, S.L. Reck-Peterson, Hitchhiking: a non-canonical mode of microtubule-based transport. Trends Cell Biol. **27**, 141–150 (2017)

J. Salogiannis, M.J. Egan, S.L. Reck-Peterson, Peroxisomes move by hitchhiking on early endosomes using the novel linker protein PxdA. J. Cell Biol. **212**, 289–296 (2016)

C.R. Santos, A. Schulze, Lipid metabolism in cancer. FEBS J. **279**(15), 2610–2623 (2012)

R. Saraya, M. Veenhuis, I.J. van der Klei, Peroxisomes as dynamic organelles: peroxisome abundance in yeast. FEBS J. **277**, 3279–3288 (2010)

M. Sardiello, M. Palmieri, A. di Ronza et al., A gene network regulating lysosomal biogenesis and function. Science **325**, 473–477 (2009)

M. Sato, S. Yoshimura, R. Hirai, et al, The role of VAMP7/TI-VAMP in cell polarity and lysosomal exocytosis in vivo. Traffic **12**, 1383–1393 (2011)

I.B. Schäfer, G.G. Hesketh, N.A. Bright, S.R. Gray, P.R. Pryor, P.R. Evans, J.P. Luzio, D.J. Owen, The binding of Varp to VAMP7 traps VAMP7 in a closed, fusogenically inactive conformation. Nat. Struct. Mol. Biol. **19**, 1300–1309 (2012)

R. Scherz-Shouval, Z. Elazar, Regulation of autophagy by ROS: physiology and pathology. Trends Biochem. Sci. **36**, 30–38 (2011)

J.A. Schmid, L. Mach, E. Paschke, J. Glossl, Accumulation of sialic acid in endocytic compartments interferes with the formation of mature lysosomes. Impaired proteolytic processing of Cathepsin B in fibroblasts of patients with lysosomal sialic acid storage disease. J. Biol. Chem. **274**, 19063–19071 (1999)

P. Schönfeld, H. Struy, Refsum disease diagnostic marker phytanic acid alters the physical state of membrane proteins of liver mitochondria. FEBS Lett. **457**, 179–183 (1999)

M. Schrader, H.D. Fahimi, The peroxisome: still a mysterious organelle. Histochem. Cell Biol. **129**, 421–440 (2008)

M. Schrader, L. Pellegrini, The making of a mammalian peroxisome, version 2.0: mitochondria get into the mix. Cell Death Differ. **24**, 1148–1152 (2017)

M. Schrader, E. Baumgart, A. Volkl, H.D. Fahimi, Heterogeneity of peroxisomes in human hepatoblastoma cell line HepG2 evidence of distinct subpopulations. Eur. J. Cell Biol. **64**, 281–294 (1994)

M. Schrader, J.L. Costello, L.F. Godinho, A.S. Azadi, M. Islinger, Proliferation and fission of peroxisomes—an update. Biochim. Biophys. Acta Mol. Cell Res. **1863**, 971–983 (2016)

M. Schrader, J. Costello, L.F. Godinho, M. Islinger, Peroxisome–mitochondria interplay and disease. J. Inherit. Metab. Dis. **38**, 681–702 (2015a)

M. Schrader, L.F. Godinho, J.L. Costello, M. Islinger, The different facets of organelle interplay—an overview of organelle interactions. Front. Cell Dev. Biol. **3**, 56 (2015b)

M. Schrader, S. Grille, H.D. Fahimi, M. Islinger, Peroxisome interactions and cross-talk with other subcellular compartments in animal cells. Subcell Biochem. **69**, 1–22 (2013)

M. Schrader, M. Thiemann, H.D. Fahimi, Peroxisomal motility and interaction with microtubules. Microsc. Res. Tech. **61**, 171–178 (2003)

M. Schwake, B. Schroder, P. Saftig, Lysosomal membrane proteins and their central role in physiology. Traffic **14**, 739–748 (2013)

D.P. Schwerter, I. Grimm, H.W. Platta, R. Erdmann, ATP-driven processes of peroxisomal matrix protein import. Biol. Chem. **398**, 607–624 (2017)

P. Sens, J. Plastino, Membrane tension and cytoskeleton organization in cell motility. J. Phys. Condens Matter **27**, 273103 (2015)

C. Settembre, A. Ballabio, Lysosomal adaptation: how the lysosome responds to external cues. Cold Spring Harb. Perspect. Biol. **6**(6) (2014)

C. Settembre, A. Fraldi, D.L. Medina, A. Ballabio, Signals from the lysosome: a control Centre for cellular clearance and energy metabolism. Nat. Rev. Mol. Cell Biol. **14**(5), 283–296 (2013)

N. Shai, M. Schuldiner, E. Zalckvar, No peroxisome is an island—peroxisome contact sites. Biochim. Biophys. Acta Mol. Cell Res. **1863**, 1061–1069 (2016)

N. Shai, E. Yifrach, C.W.T. van Roermund et al, Systematic mapping of contact sites reveals tethers and a function for the peroxisome–mitochondria contact. Nat. Commun. **9**, 1761 (2018)

H.E. Shamseldin, M. Alshammari, T. Al-Sheddi, M.A. Salih, H. Alkhalidi, A. Kentab, G.M. Repetto, M. Hashem, F.S. Alkuraya, Genomic analysis of mitochondrial diseases in a consanguineous population reveals novel candidate disease genes. J. Med. Genet **49**, 234–241 (2012)

R. Sheffer, L. Douiev, S. Edvardson, A. Shaag, K. Tamimi, D. Soiferman, V. Meiner, A. Saada, Postnatal microcephaly and pain insensitivity due to a de novo heterozygous DNM1L mutation causing impaired mitochondrial fission and function. Am. J. Med. Genet A **170**, 1603–1607 (2016)

R. Shirakawa, T. Higashi, H. Kondo, A. Yoshioka, T. Kita, H. Horiuchi, Purification and functional analysis of a Rab27 effector Munc 13–4 using a semi-intact platelet dense-granule secretion assay. Methods Enzymol. **403**, 778–788 (2005)

R. Shirakawa, T. Higashi, A. Tabuchi, A. Yoshioka, H. Nishioka, M. Fukuda, T. Kita, H. Horiuchi, Munc13–4 is a GTP-Rab27-binding protein regulating dense core granule secretion in platelets. J. Biol. Chem. **279**, 10730–10737 (2004)

A. Sitaram, M.S. Marks, Mechanisms of protein delivery to melanosomes in pigment cells. Physiology **27**, 85–99 (2012)

M. Skalski, M.G. Coppolino, SNARE-mediated trafficking of alpha5beta1 integrin is required for spreading in CHO cells *Biochem.* Biophys. Res. Commun. **335**, 1199–1210 (2005)

D. E. Sleat, J.A. Wiseman, M. El-Banna, S.M. Price, L. Verot, M.M. Shen, G.S. Tint, M.T. Vanier, S.U. Walkley, P. Lobel, Genetic evidence for nonredundant functional cooperativity between NPC1 and NPC2 in lipid transport. Proc. Natl. Acad. Sci. USA **101**, 5886–5891 (2004)

W.B. Snyder, A. Koller, A.J. Choy, S. Subramani, The peroxin Pex19p interacts with multiple, integral membrane proteins at the peroxisomal membrane. J. Cell Biol. **149**, 1171–1178 (2000)

K. Soliman, F. Göttfert, H. Rosewich, S. Thoms, J. Gärtner, Super-resolution imaging reveals the sub-diffraction phenotype of Zellweger syndrome ghosts and wild-type peroxisomes. Sci. Rep. **8**, 7809 (2018)

Spampanato C, Feeney E, Li L, Cardone M et al. 2013 Transcription factor EB (TFEB) is a new therapeutic target for Pompe disease *EMBO Mol Med* **5** 691–706.

I. Sparkes, H. Gao, Molecular machines involved in peroxisome biogenesis and maintenance. Berlin: *Springer Plant peroxisome dynamics: movement, positioning and connections*, pp. 461–477 (2014)

A. Steffen, G. Le Dez, R. Poincloux, C. Recchi, P. Nassoy, K. Rottner, T. Galli, P. Chavrier, MT1-MMP-dependent invasion is regulated by TI-VAMP/VAMP7. Curr. Biol. **18**, 926–931 (2008)

G. Steinberg, The mechanism of peroxisome motility in filamentous fungi. Fungal Genet Biol. **97**, 33–35 (2016)

J.C. Stinchcombe, D.C. Barral, E.H. Mules, S. Booth, A.N. Hume, L.M. Machesky, M.C. Seabra, G.M. Griffiths, Rab27a is required for regulated secretion in cytotoxic T lymphocytes. J. Cell Biol. **152**, 825–834 (2001)

J. Stinchcombe, G. Bossi, G.M. Griffiths, Linking albinism and immunity: the secrets of secretory lysosomes. Science **305**, 55–59 (2004)

J. Su, A.S. Thomas, T. Grabietz, C. Landgraf, R. Volkmer, S.J. Marrink, C. Williams, M.N. Melo, The N-terminal amphipathic helix of Pex11p self-interacts to induce membrane remodelling during peroxisome fission. Biochim. Biophys. Acta Biomembr. **1860**(6), 1292–1300 (2018)

T.C. Südhof, J.E. Rothman, Membrane fusion: grappling with SNARE and SM proteins. Science **323**, 474–477 (2009)

A. Sugiura, S. Mattie, J. Prudent, H.M. McBride, Newly born peroxisomes are a hybrid of mitochondrial and ER-derived pre-peroxisomes. Nature **542**, 251–254 (2017)

Q. Sun, Y. Zhang, J. Su, T. Li, Y. Jiang, Role of hydroxysteroid dehydrogenase-like 2 (HSDL2) in human ovarian cancer. Med. Sci. Monit. **24**, 3997–4008 (2018)

Z. Sun, S.S. Guo, R. Fässler, Integrin-mediated mechanotransduction. J. Cell Biol. **215**, 445–456 (2016)

T. Svitkina, The actin cytoskeleton and actin-based motility. Cold Spring Harb. Perspect. Biol. **10**, a018267 (2018)

H. Takano-Rojas, D. Zickler, L. Peraza-Reyes, Peroxisome dynamics during development of the fungus Podospora anserina. Mycologia **108**, 590–602 (2016)

M.A. Tayeb, M. Skalski, M.C. Cha, M.J. Kean, M. Scaife, M.G. Coppolino, Inhibition of SNARE-mediated membrane traffic impairs cell migration. Exp. Cell Res. **305**(1), 63–73 (2005)

J.P. Taylor, R.H. Brown, D.W. Cleveland, Decoding ALS: from genes to mechanism. Nature **539**, 197–206 (2016)

R.L. Taylor, M.T. Handley, S. Waller et al., Novel PEX11B mutations extend the peroxisome biogenesis disorder 14B phenotypic spectrum and underscore congenital cataract as an early feature. Invest Ophthalmol. Vis. Sci. **58**, 594–603 (2017)

V.V. Teplova, A.A. Tonshin, P.A. Grigoriev, N.E. Saris, M.S. Salkinoja-Salonen, Bafilomycin A1 is a potassium ionophore that impairs mitochondrial functions. J. Bioenerg. Biomembr. **39** (4), 321–329 (2007)

A.C. Theos,D. Tenza, J.A. Martina et al., Functions of adaptor protein (AP)- 3 and AP1 in tyrosinase sorting from endosomes to melanosomes. Mol. Biol. Cell **16**, 5356–5372 (2005)

P.B. Thornhill, J.B. Cohn, G. Drury, W.L. Stanford, A. Bernstein, J. Desbarats, A proteomic screen reveals novel Fas ligand interacting proteins within nervous system Schwann cells. FEBS Lett. **581**, 4455–4462 (2007)

V.I. Titorenko, R.A. Rachubinski, The peroxisome: orchestrating important developmental decisions from inside the cell. J. Cell Biol. **164**, 641–645 (2004)

V.I. Titorenko, R.A. Rachubinski, Origin and spatiotemporal dynamics of the peroxisomal endomembrane system. Front. Physiol. **5**, 493 (2014)

K. Todkar, H.S. Ilamathi, M. Germain, Mitochondria and lysosomes: discovering bonds. Front. Cell Dev. Biol. **5**(106), 106 (2017)

E.Q. Toyama, S. Herzig, J. Courchet et al., Metabolism AMP-activated protein kinase mediates mitochondrial fission in response to energy stress. Science **351**, 275–281 (2016)

S. Treusch,S. Knuth, S.A. Slaugenhaupt, E. Goldin, B.D. Grant, H. Fares, Caenorhabditis elegans functional orthologue of human protein h-mucolipin-1 is required for lysosome biogenesis. Proc. Natl. Acad. Sci. **101** 4483–4488 (2004)

F. Tribl, K. Marcus, H.E. Meyer, G. Bringmann, M. Gerlach, P. Riederer, Subcellular proteomics reveals neuromelanin granules to be a lysosome-related organelle. J. Neural Trans. **113**, 741–749 (2006)

G. Twig, B. Hyde, O.S. Shirihai Mitochondrial fusion, fission and autophagy as a quality control axis: the bioenergetic view *Biochim Biophys Acta Bioenerg* **1777**(9), 1092–1097 (2008)

T. Tzvetkova-Chevolleau, A. Stephanou, D. Fuard, J. Ohayon, P. Schiavone, P. Tracqui, The motility of normal and cancer cells in response to the combined influence of the substrate rigidity and anisotropic microstructure. Biomaterials **29**, 1541–1551 (2008)

R. Uellner, M.J. Zvelebil, Hopkins J, Jones J, MacDougall L K, Morgan B P, Podack E, Waterfield M D and Griffiths G M, Perforin is activated by a proteolytic cleavage during biosynthesis which reveals a phospholipid-binding C2 domain. EMBO J. **16**, 7287–7296 (1997)

A. Ulbricht, F.J. Eppler, V.E. Tapia et al., Cellular mechanotransduction relies on tension-induced and chaperone-assisted autophagy *Curr.* Biol. **23**, 430–435 (2013)

L. Valcarcel-Jimenez, E. Gaude, V. Torrano, C. Frezza, A. Carracedo, Mitochondrial metabolism: Yin and Yang for tumor progression. Trends Endocrinol Metab **28**, 748–757 (2017)

A.C. Valdez, J.P. Cabaniols, M.J. Brown, P.A. Roche, Syntaxin 11 is associated with SNAP23 on late endosomes and the trans-Golgi network. J. Cell Sci. **112**, 845–854 (1999)

I. Valença, Pértega-Gomes N, Vizcaino J R, Henrique R M, Lopes C, Baltazar F and Ribeiro D, Localization of MCT2 at peroxisomes is associated with malignant transformation in prostate cancer. J. Cell Mol. Med. **19**(4), 723–723 (2015)

A.M. Valm, S. Cohen, W.R. Legant, J. Melunis, U. Hershberg, E. Wait, A.R. Cohen, M.W. Davidson, E. Betzig, J. Lippincott-Schwartz, Applying systems-level spectral imaging and analysis to reveal the organelle interactome. Nature **546**, 162–167 (2017)

J. Vamecq, P. Andreoletti, N. Latruffe, The human peroxisome in health and disease: the story of an oddity becoming a vital organelle. Biochimie **98**, 4–15 (2014)

A. van der Zand, I. Braakman, H.F. Tabak, Peroxisomal membrane proteins insert into the endoplasmic reticulum. Mol Biol Cell **21**, 2057–2065 (2010)

T.S. van Zanten, A. Cambi, M. Koopman, B. Joosten, C.G. Figdor, M.F. Garcia-Parajo, Hotspots of GPI-anchored proteins and integrin nanoclusters function as nucleation sites for cell adhesion *Proc.* Natl. Acad. Sci. USA **106**, 18557–18562 (2009)

P.A. Vanlandingham, B.P. Ceresa Rab7 regulates late endocytic trafficking downstream of multivesicular body biogenesis and cargo sequestration. J. Biol. Chem. **284**, 12110–12124 (2009)

J.R. Vanstone, A..M. Smith, S. McBride et al., DNM1L-related mitochondrial fission defect presenting as refractory epilepsy. Eur. J. Hum. Genet. **24**, 1084–1088 (2016)

C. Verderio, C. Cagnoli, M. Bergami et al., TI-VAMP/VAMP7 is the SNARE of secretory lysosomes contributing to ATP secretion from astrocytes *Biol.* Cell **104**, 213–228 (2012)

K. von Schwarzenberg, Wiedmann R M, Oak P, Schulz S, Zischka H, Wanner G, Efferth T, Trauner D and Vollmar A M, Mode of cell death induction by pharmacological vacuolar H+-ATPase (V-ATPase) inhibition. J. Biol. Chem. **288**(2), 1385–1396 (2013)

K.M. Walter, M.J. Schönenberger, M. Trötzmüller et al., Hif-2α promotes degradation of mammalian peroxisomes by selective autophagy. Cell Metab. **20**, 882–897 (2014)

P.A. Walton, Brees C, Lismont C, Apanasets O and Fransen M, The peroxisomal import receptor PEX5 functions as a stress sensor, retaining catalase in the cytosol in times of oxidative stress. Biochim. Biophys. Acta **1864**, 1833–1843 (2017)

R.J.A. Wanders, Metabolic functions of peroxisomes in health and disease. Biochimie **98**, 36–44 (2014)

R.J.A. Wanders, H.R. Waterham, Biochemistry of mammalian peroxisomes revisited. Annu. Rev. Biochem. **75**, 295–332 (2006)

R.J.A. Wanders, Ferdinandusse S, Brites P and Kemp S, Peroxisomes, lipid metabolism and lipotoxicity. Biochim. Biophys. Acta **1801**, 272–280 (2010)

R.J.A. Wanders, J. Komen, S. Ferdinandusse, Phytanic acid metabolism in health and disease. Biochim Biophys Acta **1811**, 498–507 (2011)

R.J.A. Wanders, H.R. Waterham, S. Ferdinandusse, Metabolic interplay between peroxisomes and other subcellular organelles including mitochondria and the endoplasmic reticulum *Front Cell Dev Biol Front Cell.* Dev Biol **3**, 833383–833389 (2016)

D.M. Ward, Griffiths G M, Stinchcombe J C, Kaplan J, Analysis of the lysosomal storage disease Che´ diak – Higashi syndrome. Traffic **1**, 816–822 (2000a)

D.M. Ward, Pevsner J, Scullion M A, Vaughn M and Kaplan J, Syntaxin 7 and VAMP-7 are soluble N-ethylmaleimide-sensitive factor attachment protein receptors required for late endosome – lysosome and homotypic lysosome fusion in alveolar macrophages. Mol. Biol. Cell **11**, 2327–2333 (2000b)

H.R. Waterham, J. Koster, C.W.T. van Roermund, P.A. Mooyer, R.J. Wanders, J.V. Leonard, A lethal defect of mitochondrial and peroxisomal fission. N Engl J Med **356**, 1736–1741 (2007)

T. Weaver, C. Na, M. Stahlman, Biogenesis of lamellar bodies, lysosome-related organelles involved in storage and secretion of pulmonary surfactant. Cell Dev. Biol. **13**, 263–270 (2002)

H.J. Weir, P. Yao, F.K. Huynh, C.C. Escoubas, R.L. Goncalves, K. Burkewitz, R. Laboy, M.D. Hirschey, W.B. Mair, Dietary restriction and AMPK increase lifespan via mitochondrial network and peroxisome remodeling. Cell Metab **26**, 884–896 (2017)

N.R. Weir, Kamber R A, Martenson J S and Denic V, The AAA protein Msp1 mediates clearance of excess tail-anchored proteins from the peroxisomal membrane. Elife **6**, e28507 (2017)

M. Wenham, Grieve S, Cummins M, Jones M L, Booth S, Kilner R, Ancliff P J, Griffiths G M and Mumford A D, Two patients with Hermansky Pudlak syndrome type 2 and novel mutations in AP3B1. Haematologica **95**, 333– 337 (2010)

P.T. Went, G. Sauter, M. Oberholzer, L. Bubendorf, Abundant expression of AMACR in many distinct tumour types. Pathology **38**, 426–432 (2006)

C. Williams, Opalinski L, Landgraf C, Costello J, Schrader M, Krikken A M, Knoops K, Kram A M, Volkmer R and van der Klei I J, The membrane remodeling protein Pex11p activates the GTPase Dnm1p during peroxisomal fission. Proc. Natl. Acad. Sci. USA **112** 6377–6382 (2015)

K.C. Williams, M.G. Coppolino, Phosphorylation of membrane type 1-matrix metalloproteinase (MT1-MMP) and its vesicle-associated membrane protein 7 (VAMP7)-dependent trafficking facilitate cell invasion and migration. J. Biol. Chem. **286**, 43405–43416 (2011)

A.K. Witkiewicz, S. Varambally, R. Shen, R. Mehra, M.S. Sabel, D. Ghosh, A.M. Chinnaiyan, M.A. Rubin, C.G. Kleer, Alpha-Methylacyl-CoA racemase protein expression is associated with the degree of differentiation in breast cancer using quantitative image analysis. Cancer Epidemiol. Biomark Prev. **14**, 1418–1423 (2005)

P.G. Woodman, C.E. Futter, Multivesicular bodies: Co-ordinated progression to maturity. Curr. Opin. Cell Biol. **20**, 408–414 (2008)

J.P. Wróblewska, Cruz-Zaragoza L D, Yuan W et al., Saccharomyces cerevisiae cells lacking Pex3 contain membrane vesicles that harbor a subset of peroxisomal membrane proteins. Biochim. Biophys. Acta **1864**, 1656–1667 (2017)

M. Wu, S.-M. Ho, PMP24, a gene identified by MSRF, undergoes DNA hypermethylation-associated gene silencing during cancer progression in an LNCaP model. Oncogene **23**, 250–259 (2004)

W. Wu, Liu F, Wu K, Chen Y, Wu H, Dai G and Zhang W, Lon peptidase 2, peroxisomal (LONP2) contributes to cervical carcinogenesis via oxidative stress. Med. Sci. Monit **24**, 1310–1320 (2018)

X. Wu, G. Daniels, P. Lee, M.E. Monaco, Lipid metabolism in prostate cancer. Am. J. Clin. Exp. Urol. **2**, 111–120 (2014)

X. Wu, Rao K, Bowers M B, Copeland N G, Jenkins N A and Hammer J A 3rd, Rab27a enables myosin Va dependent melanosome capture by recruiting the myosin to the organelle. J. Cell Sci. **114** 1091–1100 (2001)

E. Wurmbach, Y. Chen, G. Khitrov et al., Genome-wide molecular profiles of HCV-induced dysplasia and hepatocellular carcinoma. Hepatology **45**, 938–947 (2007)

J. Xu, T. Thornburg, A.R. Turner et al., Serum levels of phytanic acid are associated with prostate cancer risk. Prostate **63**, 209–214 (2005)

Z. Xu, Asahchop E L, Branton W G, Gelman B B, Power C and Hobman T C, MicroRNAs upregulated during HIV infection target peroxisome biogenesis factors: implications for virus biology, disease mechanisms and neuropathology. PLOS Pathog. **13**, e1006360 (2017)

Y. Yagita, T. Hiromasa, Y. Fujiki Y, Tail-anchored PEX26 targets peroxisomes via a PEX19-dependent and TRC40-independent class I pathway. J. Cell Biol. **200**, 651–666 (2013)

Y. Yagita, Shinohara K, Abe Y, Nakagawa K, Al-Owain M, Alkurya F S and Fujiki Y, Deficiency of a retinal dystrophy protein, Acyl-CoA binding domain-containing 5 (ACBD5), impairs peroxisomal β-oxidation of very-long-chain fatty acids. J. Biol. Chem. **292**, 691–705 (2017)

A. Yatsu, H. Shimada, N. Ohbayashi, M. Fukuda, Rab40C is a novel Varp- binding protein that promotes proteasomal degradation of Varp in melanocytes. Biol. Open **4**, 267–275 (2015)

S. Ye, Karim Z A, Al Hawas R, Pessin J E, Filipovich A H and Whiteheart S W, Syntaxin-11, but not syntaxin-2 or syntaxin-4, is required for platelet secretion. Blood **120**, 2484–2492 (2012)

P. Yeagle, *Biology of Cholesterol* (CRC, Boca Ratón, 1988)

E. Yifrach, G. Chuartzman, N. Dahan, S. Maskit, L. Zada, U. Weill, I. Yofe, T. Olender, M. Schuldiner, E. Zalckvar, Characterization of proteome dynamics during growth in oleate reveals a new peroxisome-targeting receptor. J. Cell Sci. **129**, 4067–4075 (2016)

E.D. Yoboue, Sitia R and Simmen T, Redox crosstalk at endoplasmic reticulum (ER) membrane contact sites (MCS) uses toxic waste to deliver messages. Cell Death Dis. **9**, 331 (2018)

I. Yofe, Soliman K, Chuartzman S G et al., Pex35 is a regulator of peroxisome abundance. J. Cell Sci. **130**, 791–804 (2017)

G. Yoon, Malam Z, Paton T et al., Lethal disorder of mitochondrial fission caused by mutations in DNM1L. J. Pediatr. **171**, 313–316 (2016)

G.J. Yoshida, Metabolic reprogramming: the emerging concept and associated therapeutic strategies. J. Exp. Clin. Cancer Res. **34**, 111 (2015)

Y. Yoshida, Niwa H, Honsho M, Itoyama A and Fujiki Y, Pex11mediates peroxisomal proliferation by promoting deformation of the lipid membrane. Biol. Open **4**, 710–721 (2015)

L. Yu, C.K. McPhee, L. Zheng et al., Termination of autophagy and reformation of lysosomes regulated by mTOR. Nature **465**, 942–946 (2010)

S. Yu, Matsusue K, Kashireddy P, Cao W Q, Yeldandi V, Yeldandi A V, Rao M S, Gonzalez F J and Reddy J K, Adipocyte-specific gene expression and adipogenic steatosis in the mouse liver due to peroxisome proliferator-activated receptor γ1 (PPARγ1) overexpression. J. Biol. Chem. **278** 498–505 (2003)

K. Zaha, Matsumoto H, Itoh M et al., DNM1L-related encephalopathy in infancy with Leigh syndrome-like phenotype and suppression-burst. Clin. Genet **90**, 472–474 (2016)

U. Zentgraf, *Senescence Processes in Plants* (Blackwell Oxidative stress and leaf senescence, Oxford, 2007), pp. 69–86

S. Zha, Ferdinandusse S, Denis S, Wanders R J, Ewing C M, Luo J, De Marzo A M and Isaacs W B, Alpha-methylacyl-CoA racemase as an androgen-independent growth modifier in prostate cancer. Cancer Res. **63**, 7365–7376 (2003)

S. Zha, Ferdinandusse S, Hicks J L, Denis S, Dunn T A, Wanders R J, Luo J, De Marzo A M and Isaacs W B, Peroxisomal branched chain fatty acid?—oxidation pathway is upregulated in prostate cancer. Prostate **63**, 316–323 (2005)

F. Zhang, G. Du, Dysregulated lipid metabolism in cancer. World J. Biol. Chem. **3**(8), 167–174 (2012)

J. Zhang, Kim J, Alexander A et al., A tuberous sclerosis complex signalling node at the peroxisome regulates mTORC1 and autophagy in response to ROS. Nat. Cell Biol. **15**, 1186–1196 (2013)

S. Zhang, Ma D, Wang X, Celkan T, Nordenskjold M, Henter I, Fadeel B and Zheng C, Syntaxin-11 is expressed in primary human monocytes/macrophages and acts as a negative regulator of macrophage engulfment of apoptotic cells and IgG-opsonized target cells. Br. J. Haematol. **142**, 469–479 (2008)

W. Zhang, C.S. Zong, U. Hermanto, P. Lopez-Bergami, Z. Ronai, L.-H. Wang, RACK1 recruits STAT3 specifically to insulin and insulin-like growth factor 1 receptors for activation, which is important for regulating anchorage-independent growth. Mol. Cell. Biol. **26**, 413–424 (2006)

X. Zhang, M. Wu, H. Xiao, M.T. Lee, L. Levin, Y.K. Leung, S.M. Ho, Methylation of a single intronic CpG mediates expression silencing of the PMP24 gene in prostate cancer. Prostate **70** 765–776 (2010)

Z. Zhang, G. Chen, W. Zhou, A. Song, T. Xu, Q. Luo, W. Wang, X.S. Gu, S. Duan, Regulated ATP release from astrocytes through lysosome exocytosis. Nat. Cell Biol. **9**, 945–953 (2007)

X.W. Zhao, Gazendam R P, Drewniak A, et al., Defects in neutrophil granule mobilization and bactericidal activity in familial hemophagocytic lymphohistiocytosis type 5 (FHL-5) syndrome caused by STXBP2/Munc18-2 mutations. Blood **122**, 109–111 (122)

P. Zimmermann, Heinlein C, Orendi G and Zentgraf U, Senescence-specific regulation of catalases in Arabidopsis thaliana (L.) Heynh. Plant Cell Environ. **29**, 1049–1060 (2006)

E. Zuccato, E.J. Blott, O. Holt, S. Sigismund, M. Shaw, G. M. Bossi Ga nd Griffiths, Sorting of Fas ligand to secretory lysosomes is regulated by mono-ubiquitylation and phosphorylation. J. Cell Sci. **120**, 191–199 (2007)

U. zur Stadt, J. Rohr, W. Seifert et al., Familial hemophagocytic lymphohistiocytosis type 5 (FHL-5) is caused by mutations in Munc18-s2 and impaired binding to syntaxin 11. Am. J. Hum. Genet **85** 482 – 492 (2009)

U. zur Stadt, S. Schmidt, B. Kasper et al., Linkage of familial hemophagocytic lymphohistiocytosis (FHL) type-4 to chromosome 6q24 and identification of mutations in syntaxin 11 *Hum Mol Genet* **14** 827– 834 (2005)

The Cell Nucleus and Its Compartments

10

Abstract

This chapter presents the basic knowledge about the function of cell division and the cell cycle process. In fine detail, the architecture and function of the largest compartment of the eukaryotic cell, the nucleus and its components, such as the nuclear envelope, nuclear pore complex, chromosomes and nucleoskeleton, are emphasized. Centrioles play an important role in cell division and are being discussed, including their replication. It is discussed how forces are transmitted from the cellular microenvironment, which are sensed by cell–matrix adhesion receptors, coupled via focal adhesion proteins to the cell's cytoskeletal network, and subsequently they have an impact on the mechanical phenotype of the innermost organelle of the cell, the cell nucleus. The biophysical aspects of these main functions of cells under normal physiological conditions and under pathological conditions such as cancer and acute or chronic inflammation are at the center of attention. Finally, the effect of mechanical force on gene expression is addressed, how the cells deal with it.

10.1 Cell Nucleus: Architecture and Function

The nucleus is the largest and stiffest organelle of the cells which is enveloped with a doubled lipid-by-layer membrane. The nuclear membrane represents a porous membrane that enables the transport of substances in and out the nucleus. The nucleus consists of a karyoplasm containing chromatin and a nucleoskeleton. However, there is still a question to be raised: Why represents the cell nucleus such a key organelle for the eukaryotic cell? Among all eukaryotic organelles, the nucleus seems to be the most critical and prominent one. The simple presence of a cell nucleus is a core characteristic element of a eukaryotic cell. In specific detail, the nuclear structure is crucial, since is the location where the DNA of the cell is stored safely, and it subsequently regulates how many times the DNA is processed into RNA (termed transcription process). The gene transcription efficiency is precisely controlled by the specific location of a distinct gene on the entire chromosome and by external stresses including forces of the surrounding microenvironment that act on the cells.

The DNA encodes the specific information that is necessary to polymerize the distinct proteins of the cell. In eukaryotic cells, a double membrane encloses the entire nucleus, which is synonymously termed nuclear envelope. The nuclear envelope ensures that the DNA is separated from the protein synthesis system of the cell, which is positioned inside the cytoplasm. The small pores of the nuclear

C. T. Mierke, *Cellular Mechanics and Biophysics*, Biological and Medical Physics, Biomedical Engineering, https://doi.org/10.1007/978-3-030-58532-7_10

envelope (referred to as nuclear pores) help to decide what type of macromolecule can get into or out of the nucleus. Among these macromolecules are RNA molecules that possess the transcribed information of the cellular DNA for the generation of proteins inside the cytoplasm. The spatial separation of the DNA location and the location of the protein synthesis complex enable the eukaryotic cells to precisely govern the protein production and the generation of the RNA intermediates including pre-mRNAs and mRNAs.

In contrast, the prokaryotic DNA is loosely arranged and less ordered inside the cytoplasm and functions together and in parallel with synthesis of proteins. The close connection between the DNA and the synthesis of proteins in prokaryotic cells provide a fast sense response mechanism to quickly altered microenvironmental cues. In specific detail, the proteins types and their amounts to be produced can be adapted. The eukaryotic cells seem to be developed from simpler prokaryotic that have been chosen to undergo a symbiotic relationship. In fact, one set of prokaryotic DNA may have been enveloped by the nuclear membrane and finally developed to the nucleus. Over long periods of time, the DNA segments of other prokaryotes that are still in the cytoplasm may either be incorporated into the nuclear DNA or not and thus get lost during evolution. Finally, the nucleus contains the chromatin and hence the genetic information of the whole organism.

Major functions of the nucleus have been largely investigated such as the splicing of pre-mRNA to "mature" mRNA and the assembly of ribosomes using biochemical, genetic and molecular analytical approaches. Thus, the molecular mechanisms are largely studied. However, little is known about the integration of these molecular mechanisms into the structural architecture of the cell nucleus and how they are temporally and spatially controlled. There is an important question not yet answered: How it affects the nuclear architecture gene expression? In furtherance of answering this question, we need to reveal how the nuclear architecture imports gene expression and more general the organization of the genome. In more detail, the principles of the nuclear architecture and function should be known. The combination of molecular, biochemical and cell biological approaches seems to be well suited to determine how genes are expressed in vivo, and subsequently, how the nucleus performs its functions.

The nucleus is subdivided into distinct and self-organized regions (termed compartments); however, many nuclear components, such as chromatin, mRNA and nuclear proteins, are highly motile and therefore dynamically organized. The nucleus has been discovered as first intracellular compartment by Franz Bauer in 1802 and thereafter described in more detail by Robert Brown (Harris 1999). The nucleus is an organelle that is of great interest for cell division, as it shows the exact control of the process directly through its own regulation. Moreover, the nucleus is essential for fertilization and inheritance. The process of gene expression has become the focus of nuclear studies, where the nuclear structure was only partly addressed. The development of the fluorescent proteins enabled us to reveal subnuclear structural details, based on their distinct morphologies. However, the function of these structures is still not well understood. In summary, the nucleus consists of distinct subdivisions and is a highly dynamic organelle. These subdivisions are not separated by a membrane; however, they are still nuclear compartments, as they possess specific proteins and can be identified by their specific morphology using light or electron microscopy. In addition, the nuclear compartments can be isolated biochemically (Busch et al. 1967; Mintz et al. 1999). The cytoplasmic counterparts of these nuclear compartments are enclosed by a membrane. Although distinct compartments are present in the nucleus and multitude of nuclear processes require to be precisely organized within nuclear space, it has been suggested that the nuclear matrix arranges the overall nuclear architecture. This hypothesis is still highly controversial discussed (Pederson 2000; Nickerson 2001). However, we still need to know whether proteins are attached to a nuclear matrix and what components build the nuclear matrix. Since indirect experimental results indicate the existence of a nuclear matrix, no matrix elements or components have been identified and even no nuclear compartment components have been found to anchor the nuclear matrix (Dundr and Mistelli 2001).

Nuclear compartments are the nucleolus, small dot-like nuclear speckles, which are termed splicing factor compartments (SFCs), the Cajal bodies and the promyelocytic leukemia oncoprotein (PML) bodies (Fig. 10.1). More precisely, SFCs are perichromatin fibrils and interchromatin granule clusters. The perichromatin fibrils have a variable diameter and are nascent transcripts (Fakan 1994). Interchromatin granules clusters are dense granules of 20–25 nm diameter, which are connected by fibrils (Monneron and Bernhard 1969). SFCs represent no primary sites for pre-mRNA splicing (Misteli 2000); however, most active genes are at their periphery (Xing et al. 1993, 1995). SFCs seem to be the storage room of spliceosomal components. The distribution of splicing factors occurs during expression of intron-containing genes or after viral infection. Thus, specific factors are transported to a newly transcribed gene (Jimenez-Garcia and Spector 1993; Misteli et al. 1997); however, it is still possible that the pre-mRNA moves to splicing factors (Xing et al. 1993; Huang and Spector 1996; Melcak et al. 2000). The recruitment of splicing factors from SFCs to transcription sites activates a cycle of phosphorylation and dephosphorylation events (Misteli 2000).

10.1.1 Nuclear Shape and the Mechanical Stability of the Nucleus

The cell nucleus can be deformed by external cues, such as confinement or external forces that are exerted via cell–matrix adhesion receptors to the cell's cytoskeleton. The cytoskeleton itself is coupled via actin filaments to the nucleus and its nucleoskeleton (Fig. 10.2). More precisely, extensive deformations of the nucleus elevate the rupture event of nuclear envelope and induce DNA damage. The maintenance of the nuclear shape relies on its mechanical strength, which is characterized by decreased nuclear deformation and reduced strain upon the mechanical deformation. In order to reveal the nature of the nuclear deformation and recovery properties of the nucleus, the nuclear viscoelastic properties need to be identified. Hence, the individual contributions to the viscoelastic properties are determined separately for the nucleus, the plasma membrane and the cytoskeleton. As hypothesized, the cytoskeleton seems to impact the mechanical properties of the nucleus in a manner that the cytoskeleton increases the mechanical stability of the nucleus through a

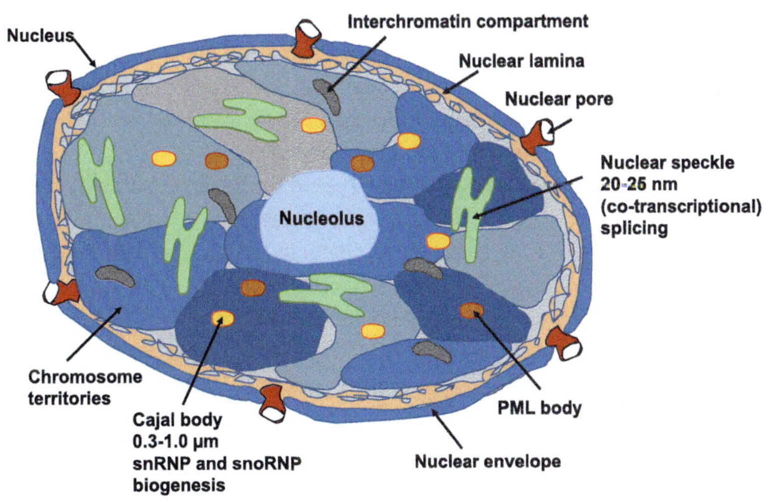

Fig. 10.1 Schematic view of the nuclear compartments

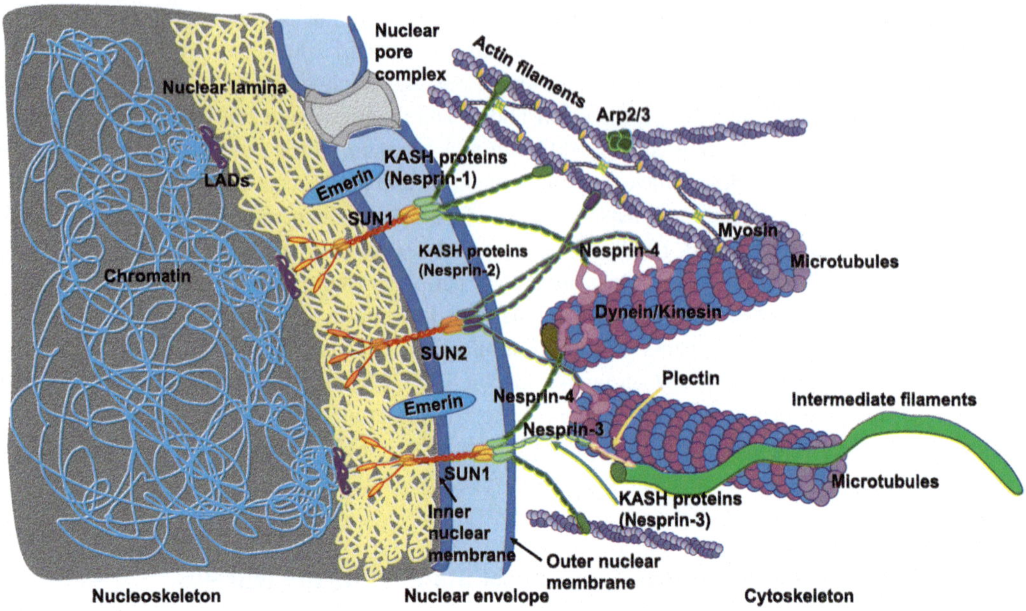

Fig. 10.2 Coupling between the cytoskeleton, the cell nucleus and the nuclear envelope. The connection between the nucleoskeleton is facilitated through the SUN-Nesprins interactions connecting to cytoskeletal filaments, such as F-actin fibers, microtubules and intermediate filaments

decrease of the effective deformability of the nucleus; whereas, there is still the nuclear capacity to sense mechanical stimulation. Moreover, the cytoskeleton reduces the release rate of the strain of the nucleus and hence may help to prevent shape alteration-based damage toward the overall structural arrangement of chromatin.

It is known that extracellular forces are able to mechanically deform the cell nucleus through the plasma membrane originated forces transmitted through the cytoskeleton that act on the nuclear envelope (Haase et al. 2016). Pronouncedly large deformations of the nucleus are able to evoke a local reduction of the integrity of the nuclear envelope, which causes the unregulated alterations of the nucleocytoplasmic constituents, induce damage to DNA and subsequently supports cell death (Denais et al. 2016). The capacity of the nucleus to prevent pronounced deformation and extraordinary release strain energy amounts is crucial of the overall mechanical stability of the nucleus (Rowat et al. 2006). In specific detail, quantitative experimental assessments of the deformation of the nucleus and their recovery seem to be required for gaining knowledge of how the cell nucleus acts to external forces and keeps the stability of the nucleus.

There are a multiple toolset of biophysical techniques and approaches for the investigation of the nuclear mechanics, such as the micropipette aspiration technique (Pajerowski et al. 2007), atomic force microscopy (AFM) based on mechanical indentation with a cantilever (Ivanovska et al. 2017), the employment of magnetic tweezers (Guilluy et al. 2014) or optical tweezers (Schreiner et al. 2015), the analysis of substrate strain approaches (Lombardi et al. 2011) and microfluidic-based techniques (Hanson et al. 2015). Despite these various biophysical techniques, the majority of nuclear measurements has been performed on isolated nuclei or indirectly by the induction of large deformations on a cell to explore the mechanical properties of the nucleus. A sharp AFM cantilever has been used to penetrate the plasma membrane in order to directly determine the elastic modulus of the nuclei in living cells (Liu et al. 2014). However, there is agreement that the elasticity as a single parameter cannot sufficiently account for the deformation behavior of the entire nucleus. In specific detail, the viscoelastic characteristics of the cytoskeleton and the nucleus largely effect the deformation behavior of the nucleus and the recovery of the strain through the dissipation of strain energy that has been stored due to the mechanical deformation of the organelle (Corbin et al. 2016).

To determine the viscoelastic characteristics of intact nuclei, the nucleus needs to be loaded directly in a mechanical manner, for instance by employing an AFM cantilever-based deformation, which can be performed at a broad variety of speeds. The AFM cantilever tip applies at the first step a deformation to the plasma membrane and penetrates the plasma membrane, and in a second step the cell nucleus is loaded until the crossing of the nuclear envelope. The exact position of the cantilever probe can be obtained by recording the entire AFM measurement and performing at the same time confocal z-stack scanning. The force–displacement analysis can be performed at a broad range of different loading velocities in order to reveal the viscoelastic properties of the cell membrane, cytoskeleton and nucleus in a quantitative manner, when the data are fitted to viscoelastic models. In fact, it has shown that the cytoskeleton is able to increase the stiffness of the nucleus, which is facilitated through the connection of the cytoskeleton and the nucleus (Wang et al. 2018; Liu et al. 2014; Corbin et al. 2016). The cell nucleus in turn has the ability to vastly release the strain energy retained under deformation, and the cytoskeleton decelerates this high rate of strain energy liberation to preserve the structures of chromatin.

10.1.2 Quantification of Elasticity and Viscosity of the Plasma Membrane, Cytoskeleton and Nucleus

When environmental-based extracellular forces act on the plasma membrane and subsequently deform it, these forces are directly transferred through the connection of cell–matrix membrane receptors to the cells' cytoskeleton via focal adhesion proteins, such as vinculin, talin, paxillin or focal adhesion kinase (FAK), toward the nucleus (Fig. 10.3). The mechanical model that is presented in the following links plasma membrane, the cytoskeleton and the cell nucleus in series and each of which is either presented as a spring and a damper due to the proposed classical rheological model, which is referred to as the Kevin–Voigt model (K-V model) (Fig. 10.3). There are other physical and molecular models that were utilized in cell mechanical investigations, such as the matrix stiffness regulated lamin A/C-based nuclear tension model or rapid mechanical force transmission through adhesions toward the nucleus (Swift et al. 2013; Guilluy et al. 2014; Mierke 2017).

Another key mechanism of mechanotransduction that affects the cell and must be taken into account is the nuclear envelope, since the force-facilitated deformation of this membrane can cause the entire cell to alter its shape and phenotype (Wang et al. 2009). The nuclear envelope consists of the nuclear lamina proteins and other components that manage to tether the lamina to the nuclear membrane and thereby couple it via the perinuclear space to the cell's cytoskeleton (Gruenbaum et al. 2005). Therefore, in the search for the internal assemblies, structures and dynamics of the nucleus that govern its mechanical performance, it has been identified that the specific properties and intricate linkages between microtubules, lamin-A/C, Lamin-B, myosin II and heterochromatin generate nuclear stresses and stiffness by virtue of their stoichiometric levels within the cell and their inherent or collective mechanical characteristics (Kim et al. 2017; Kirby and Lammerding 2018). As a matter of fact, the importance of these mechanosensitive compounds is so large that a number of severe diseases such as cancer, heart disease, muscular dystrophy and progeria have been attributed to imperfections or mutations of specific lamins (Ferrera et al. 2014). In principle, lamin A/C rises with the apparent microstiffness of the tissue or the stiffness of the extracellular matrix in vitro; whereas, the lamin-B values stay at a low level (Hadden et al. 2017; Swift et al. 2013; Swift and Discher 2014).

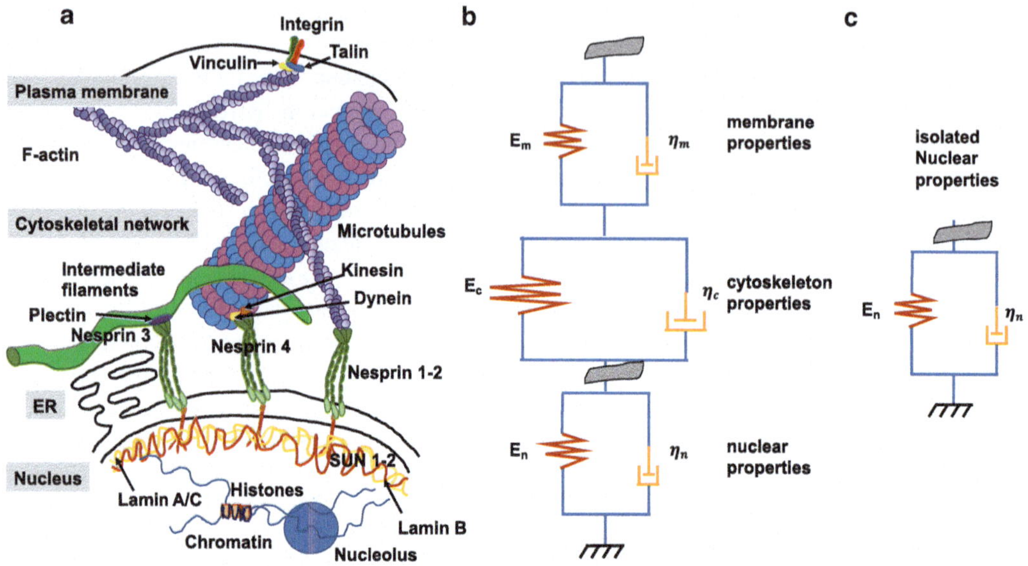

Fig. 10.3 Connections, such as the mechanical coupling, between the plasma membrane, the cytoskeleton and the nucleus (**a**) can be modeled with a Kelvin–Voigt model (**b**) and compared to the mechanical properties of isolated nuclei (**c**)

It was nearly at the same time discovered that the heterochromatin in the cell nucleus and lamin-A levels are mainly in charge of the speedy transmission of mechanical input signals from the cell's external microenvironment, thus contributing to its deformation and keeping the cell nucleus resilient to exogenous stresses (Haase et al. 2016). Curiously, it was later ascertained that these proteins are accountable for two completely distinct types of mechanotransduction reactions: On the one hand, chromatin, which is connected to the nuclear envelope via lamin B receptors, only controls the reaction to small deformations; on the other hand, lamin A/C governs the reaction to larger deformations (Stephens et al. 2017). This dual-regime activity can be witnessed further in the cross-correlation of the lamin A/C protein and lamin B receptors (Solovei et al. 2013). When the overwhelming majority of investigations concentrates on a periodic or reversible mechanotransduction mechanism of the cytoskeleton and seeks for relationships between mechanical evidence and the signaling paths that are in some cases accountable for cellular responsiveness. Experimental evidence has highlighted irreversible modes of behavior that are inextricably intertwined with nuclear deformation phenomena (Tocco et al. 2018). Apart from the involvement of these known nuclear envelope proteins, another crucial aspect of the regulation of the forces exerted on the nucleus has been identified in the rheology of nuclear chromatin, since its squeezable structure has been shown to be plastic and not separated from the organization of the laminar A/C mesh for transferring stress on the nucleus (Miroshnikova et al. 2017; Poh et al. 2012).

In fact, although it is a known fact that chromatin carries up to 30% of the nuclear load, it is both dissipation from the nuclear envelope and chromatin that sustains the load and deformation; in fact, there is a definite relationship experimentally between the stoichiometric values of viscous lamin A/C, elastic lamin B, the microstiffness of the extracellular matrix and the viscoelastic properties of the nucleus (Pérez-Calixto et al. 2019).

However, as explained in great detail below, the K-V model has been selected for describing AFM indentation measurements on viscoelastic solids.

10.1.3 Viscoelastic Model Based on the K-V Model

The viscoelastic model employs elements, such as springs and dashpots, to describe the elastic and viscous features of materials, such as living cells, cell clusters or tissues. The spring dashpot model yields more insights than previous investigations of the mechanical properties of the nucleus, the vast majority of which are limited to the elastic property. The model is often employed to characterize the viscoelastic behavior of cellular structures, since it captures both time-variant (the dashpot) and time-invariant (the spring) stress–strain relationships (Swift et al. 2013; Guilluy et al. 2014).

Among the spring dashpot models, the K-V model, the Maxwell model and the standard linear solid model (SLS model) represent the most frequently encountered models (López-Guerra and Solares 2014). In specific detail, the K-V model, in which a damper and a spring are linked in parallel, is most frequently utilized to characterize the force–displacement–velocity characteristics of viscoelastic solids. Moreover, the parallel combination of the two components, the spring and the damper enables us to distinguish between two parts of the force–displacement curve: One part is the speed-invariant portion (F_{elastic}), and the other part is the speed-dependent portion (F_{viscous}), and hence, it can formulate the following: $F = F_{\text{elastic}} + F_{\text{viscous}}$. By employing various indentation velocities of the AFM cantilever probe, the velocity-invariant part (F_{elastic}) and the velocity-dependent part (F_{viscous}) can be separately determined and revealed with linear regression (see (10.7) and (10.9)) in order to quantitative calculate the elastic modulus E and viscosity η.

The Maxwell model, in which a damper and a spring are linked in series, is utilized to characterize the relaxation behavior of a material; whereas, it cannot provide a description for the creep behavior under the indentation with a cantilever (López-Guerra and Solares 2014). The AFM biophysical method can be employed is primarily a mechanical probing device for the creep behavior (indentation test). Applying the Maxwell model to specify the creep behavior under mechanical compression (indentation) of the cell would lead to an excessively elevated viscosity level and an excessively depressed modulus of elasticity, since the Maxwell model presupposes that the material to be examined actually flows and will not attain the steady state during a creep experiment.

The SLS model can be utilized for characterizing both types of material: the viscoelastic solids and the viscoelastic fluids. However, in comparison with the K-V model (two components) and the Maxwell model (two components), the SLS model comprises three components; whereas, AFM force–displacement velocity measurement data cannot provide the stress rate information required to adjust all three variables in the SLS model. Directly fitted force–displacement velocity data from the AFM measurements with the SLS model yielded a correlation coefficient of solely 0.4 ± 0.2, which relies on 30 force–displacement velocity curves acquired in AFM indentation experiments (Wang et al. 2018). Hence, for these curves, the SLS model has not been selected for further analysis.

Since the nucleus, cytoskeleton and plasma membrane display a behavior more similar to viscoelastic solids, the K-V model was employed to characterize the force–displacement–velocity behavior of the materials. The time constant of the strain recovery rate quantitatively defines how quickly the strain recovery proceeds in the K-V model. Comparatively, findings with only the single nucleus and those with the nucleus connected to the cytoskeleton highlight the contribution of the cytoskeleton in the stretching (strain) recovery process.

For a viscoelastic material, the stress–strain dependence in the K-V model is related as given in (10.1):

$$\sigma = E\varepsilon + \eta \frac{d\varepsilon}{t} \tag{10.1}$$

where σ represent the stress, ε stands for the strain, $(d\varepsilon/t)$ provides the strain rate, E is referred to the elastic modulus, and η provides the sample's viscosity.

When referring to the forces, the K-V model contains the portions of the elastic and viscous parts as given in (10.2):

$$F = F_{\text{elastic}} + F_{\text{viscous}} \tag{10.2}$$

The Hertz model can be used to calculate for F_{elastic} a cylindrical tip, which is the shape of the most fabricated AFM cantilever tips, as it has been described in (Wallace 2012) and given in the following (10.3):

$$F_{\text{elastic}} = \frac{4}{3}\frac{E_s}{1 - v_s^2} R^{\frac{1}{2}} d^{\frac{3}{2}} \tag{10.3}$$

where E_s stands for the experimentally revealed elastic modulus of the sample, v_s is the sample's Poisson's ratio, R gives the radius of the tip, and d represents the tip displacement.

Since the mechanical strain is provided by the calculation of ε as $\varepsilon = ((\Delta l)/l) = ((vt)/l)$, then the following (10.4) can be formulated:

$$F_{\text{viscous}} = S\sigma_{\text{viscous}} = \pi Rd \cdot \eta \frac{d\varepsilon(t)}{t} = \eta \frac{v}{l} \pi Rd \tag{10.4}$$

Alteration of spliceosomalwhere the area of contact is given by $S = \pi RD$ due to the Hertz model.

When the elastic and the viscous parts are connected in the K-V model, the relation between the force, the displacement, and the velocity is given in (10.5):

$$F = F_{\text{elastic}} + F_{\text{viscous}} = \frac{4}{3} ER^{\frac{1}{2}} d^{\frac{3}{2}} + \eta \frac{v}{l} \pi Rd \tag{10.5}$$

10.1.3.1 How Are the Reduced Elastic Modulus and Reduced Viscosity, *E**, *η** and *E*** and *η*** Revealed?

The cell model presented here is at the foundation of the K-V model for viscoelastic materials. The plasma membrane, the cytoskeleton and the cell nucleus are individually depicted by a spring and a damper wired in parallel fashion. The plasma membrane is sustained downward by the cytoskeleton. When determining the mechanical characteristics of the plasma membrane by analyzing the force–displacement velocity measurements, the results mirror the linked mechanical characteristics of the plasma membrane and the cytoskeleton. As outlined in Fig. 10.3, Young's modulus and viscosity extracted from "region A" of Fig. 10.3, the reduced modulus E^* and η^*, which characterize the mixed material properties of the cell membrane and the cytoskeleton, were evaluated. In a similar fashion is the nucleus connected to the cell's cytoskeleton. The obtained mechanical properties from the force–displacement velocity measurement data mirror the combined mechanical features of both the nucleus and the cytoskeleton. The elastic modulus and viscosity can be obtained from "region section B" of Fig. 10.3 were reduced modulus E^{**} and η^{**}, which both represent the linked material properties of the nucleus and the cytoskeleton. Since the tip is sharpened and therefore capable of penetrating the plasma membrane, a clear cutoff between the membrane-pressing process and the nucleus-pressing phase can be seen. It was therefore estimated that no effect emanates from the nucleus when penetrating (indenting) the plasma membrane and that no effect emanates from the plasma membrane when penetrating the nucleus.

In order to obtain the reduced elastic modulus and reduced viscosity, the raw data measured with the AFM can be imported into MATLAB. The baseline value can be subtracted and region of interest (ROI) can be set. The ROI for the quantification of the reduced elastic modulus of the plasma membrane and nucleus is chosen similarly as depicted as the region A and region B in Fig. 10.3. The high-frequency noise was filtered and interpolated, and the resulting data were differentiated for further regression.

The differentiation of (10.5) with respect to d gives the following result in (10.6):

$$\dot{F} = 2E^{*}R^{\frac{1}{2}}d^{\frac{1}{2}} + \eta\frac{v}{l}\pi R \tag{10.6}$$

where E^* is linearly related to $d^{\frac{1}{2}}$ and $\dot{F}$. Based on (10.6), the regression provides (10.7):

$$\dot{F} = k_1 d^{\frac{1}{2}} + b_1 \tag{10.7}$$

with k_1 and b_1 representing linear regression parameters. Subsequently, the reduced elasticity modulus of the cell membrane linked to the cytoskeleton has been calculated by the regression as given in (10.8):

$$E^{*} = \frac{k_1}{2R^{\frac{1}{2}}} \tag{10.8}$$

For the calculation of the reduced viscosity, ΔF needs to be defined as the force difference between two indentation velocities, such as $\Delta F = F_{v_1} - F_{v_2}$. In line with (10.5), the following equation is

$$\Delta F = \eta^{*}\frac{\Delta v}{l}\pi R d \tag{10.9}$$

where Δv is equal to $v_1 - v_2$ and the reduced viscosity η^* correlates linearly to d and ΔF.

In the next step, the regression function is formulated as follows:

$$\Delta F = k_2 d + b_2$$

where k_2 and b_2 represent parameters of the linear regression.

The reduced viscosity of the plasma membrane is linked to the cell's cytoskeleton and can be described by (10.10):

$$\eta^{*} = \frac{k_2}{((\Delta v/l)\,\pi R)} \tag{10.10}$$

The analysis for the calculation of the reduced elastic modulus E^{**} and reduced viscosity η^{**} for the cell's nucleus that is linked to the cytoskeleton is the same as the presented before.

10.1.3.2 Separation of the Modulus of Elasticity and Viscosity from the Reduced Modulus and Viscosity

The relationship of the reduced elastic modulus and reduced viscosity is provided in the following (10.11) and (10.12), respectively:

$$\frac{1}{E^*} = \frac{1}{E_m} + \frac{1}{E_c} \tag{10.11}$$

$$\frac{1}{E^{**}} = \frac{-1}{E_c} + \frac{1}{E_n} \tag{10.12}$$

where E^* and E^{**} stand for the reduced elastic modulus of the plasma membrane and the cell nucleus, respectively. E_m, E_c, and E_n refer to the elastic modulus of the plasma membrane, cytoskeleton, and cell nucleus, respectively.

The reduced viscosity of the plasma membrane and the cell nucleus are given by (10.13) and (10.14):

$$\frac{1}{\eta^*} = \frac{1}{\eta_m} + \frac{1}{\eta_c} \tag{10.13}$$

$$\frac{1}{\eta^{**}} = \frac{-1}{\eta_c} + \frac{1}{\eta_n} \tag{10.14}$$

where η^* and η^{**} refer to the reduced viscosity of the plasma membrane and cell nucleus, respectively. The η_m, η_c, and η_n represent the viscosity of the plasma membrane, cytoskeleton, and cell nucleus, respectively.

The reduced elastic moduli E^* and E^{**} as well as the reduced viscosities η^* and η^{**} were determined due to the aforementioned analysis of the results. The elastic modulus and viscosity of cell nucleus, such as E_n and η_n, respectively, were examined due to the experimental measurement analysis of isolated cell nuclei.

10.1.4 The Propagation of the Error Propagation in This Model

Based on the variations among cell in a cell population and due to measurement errors, the mechanical parameters, which have been determined from experiments possess specific uncertainties. By the analysis of the experimental data, the standard error of reduced modulus E^*, E^{**}, E_n, η^*, η^{**}, η_n, can be quantified employing the one-way ANOVA statistical analysis. When the modulus was separated from the reduced modulus, the error of the reduced modulus is able to propagate to the decoupled modulus.

In line with this previous report (Ku 1966), the elastic modulus E_m and E_c are expressed by (10.15) and (10.16):

$$E_c = \frac{E_n E^{**}}{E^{**} - E_n} \tag{10.15}$$

$$E_m = \frac{E_c E^*}{E_c - E^*} \tag{10.16}$$

In line with these equations, the viscosities η_m and η_c are expressed by the expression of (10.17) and (10.18):

$$\eta_c = \frac{\eta_n \eta^{**}}{\eta^{**} - \eta_n} \tag{10.17}$$

$$\eta_m = \frac{\eta_c \eta^*}{\eta_c - \eta^*} \tag{10.18}$$

The uncertainty, such as the standard error of the mean, which is referred to as Se, has been determined from the equations above and described in (10.19), (10.20), (10.21) and (10.22):

$$\mathrm{Se}(E_c) = \sqrt{\left(\frac{\mathrm{d}E_c}{\mathrm{d}E_n}\mathrm{Se}(E_n)\right)^2 + \left(\frac{\mathrm{d}E_c}{\mathrm{d}E^{**}}\mathrm{Se}(E^{**})\right)^2} \tag{10.19}$$

$$\mathrm{Se}(E_m) = \sqrt{\left(\frac{\mathrm{d}E_m}{\mathrm{d}E_c}\mathrm{Se}\ (E_c)\right)^2 + \left(\frac{\mathrm{d}E_m}{\mathrm{d}E^*}\mathrm{Se}(E^*)\right)^2} \tag{10.20}$$

$$\mathrm{Se}(\eta_c) = \sqrt{\left(\frac{\mathrm{d}\eta_c}{\mathrm{d}\eta_n}\mathrm{Se}(\eta_3)\right)^2 + \left(\frac{\mathrm{d}\eta_c}{\mathrm{d}\eta^{**}}\mathrm{Se}(\eta^{**})\right)^2} \tag{10.21}$$

$$\mathrm{Se}(\eta_m) = \sqrt{\left(\frac{\mathrm{d}\eta_m}{\mathrm{d}\eta_c}\mathrm{Se}(\eta_2)\right)^2 + \left(\frac{\mathrm{d}\eta_m}{\mathrm{d}\eta^*}\mathrm{Se}(\eta^*)\right)^2} \tag{10.22}$$

The proposed model specifies the deformation as a function of both the quantity and the velocity of the force application. The force perturbation rate has been diversified in the AFM indentation, from which the elastic part (spring, rate-independent) and the viscous part (damper, rate-dependent) have been quantitatively determined in the model (10.5).

In this model, the cytoskeleton mechanically sustains the plasma membrane. When the plasma membrane is distorted by the AFM probe, the cytoskeleton adds to the mechanical response measured at the plasma membrane. But instead of treating the acquired findings only as plasma membrane properties, the experimental data mirror the joint impact of the plasma membrane and the cytoskeleton (reduced modulus of elasticity and reduced viscosity of the plasma membrane E^* and η^*); whereas, no nuclear influences are detected. Since the cytoskeleton is linked to the nucleus by the LINC complex, the acquired results reflect the cumulative properties (reduced modulus of elasticity and reduced viscosity of the nucleus E^{**} and η^{**}) of the nucleus and the cytoskeleton when the nucleus is deformed by the probe; however, they do not reflect any membrane properties.

In order to isolate the properties of the plasma membrane, the cytoskeleton and the nucleus on the basis of the acquired cumulative effects (E^* and E^{**}, η^* and η^{**}), on isolated nuclei which only reflect the properties of the nucleus (E_n and η_n), mechanical measurements were also performed. To obtain a quantification of Young's modulus and cytoskeletal viscosity, the reduced modulus computed from the coupled effects of nucleus and cytoskeleton (E^{**} and η^{**}) was replaced by Young's modulus and viscosity of the nucleus in the reduced modulus equations (10.11 to 10.14). In a similar way, the elastic modulus and viscosity of the plasma membrane were separated from the cumulative impact of the plasma membrane and cytoskeleton (E^* and η^*) by inserting the elastic modulus and viscosity of the cytoskeleton into the reduced modulus equations. Elastic modulus values have been obtained to be 2.98 ± 0.04 kPa, 4.28 ± 0.33 kPa and 2.01 ± 0.10 kPa for the plasma membrane, cytoskeleton and nucleus, respectively (Wang et al. 2018). The viscosity amounts were 4.10 ± 0.06 kPa, 5.43 ± 0.55 kPa s and 0.97 ± 0.08 kPa s for the plasma membrane, the cytoskeleton and the nucleus, respectively (values presented as mean ± SEM). These results are in line with other previously measured values (Guilak et al. 2000; Dahl et al. 2005). For the process of disconnecting the cytoskeleton and the nucleus, the cytoskeleton has been treated as a homogeneous material. However, in non-idealized cells, the cortex of the in close proximity to the plasma membrane is enriched with actin and, hence, may help to provide an explanation for the increased elastic modulus and the increased viscosity of the plasma membrane.

A separate experiment with a tipless AFM probe has been conducted, and the results have been comparatively evaluated with those of the finite element simulation. In specific detail, AFM experiments with an indentation depth of 1 μm, 2 μm and 3 μm and with rates of 1 μm/s, 5 μm/s and 10 μm/s were carried out without penetration of the cells (for each condition, ten cells were measured). In order to translate the force–displacement curves into strain–stress curves, the interaction area between the probe and the cell contact area was determined. Therefore, a 3D computational model with multiple layers has been generated (Karcher et al. 2003; Lau et al. 2015). The material properties to be employed for the three layers were taken from experimental data, and the geometries of the cell and nucleus were deduced from confocal z-stack reconstruction. Moreover, the same mechanical stress as in the experiments was used as values for the force exertion on the cell surface in the computational model, and subsequently the strain–stress curves were simulated and validated with experimentally obtained curves. In fact, it turned out that the results of the modelled and calculated data are in good agreement, which provides a proper validation of the measured and subsequently decoupled factors. In conclusion, it can be said that the cytoskeleton stiffens the cell nucleus to protect the nucleus from excessive deformation and its chromosomal content form being damaged severely.

From the plasma membrane, the cytoskeleton and the cell nucleus, the cell nucleus possessed the lowest elastic modulus (2.01 ± 0.10 kPa), which means that the nucleus can be deformed in an easy manner, when it is isolated and not protected by a surrounding cytoskeleton (Wang et al. 2018). The low elastic modulus of isolated nuclei is nearly doubled (3.81 ± 0.21 kPa), when the nucleus has been connected to the cytoskeleton, which indicates that the cytoskeleton seems to stiffen the nucleus to prevent the excessive deformation of the nucleus under exertion of extracellular forces (Wang et al. 2018). The cytoskeleton displays a high elastic modulus and is itself coupled to the cell nucleus, which is referred to as a nucleocytoskeletal coupling. The cytoskeleton can increase the mechanical stability of the cell nucleus through the stiffening of the entire nucleus to prevent excessive deformation of the nucleus. However, when the actin filaments and microtubules within the cytoskeleton were treated with cytochalasin D (impairs actin filaments) and nocodazole (impairs microtubules), respectively, the cytoskeleton undergoes a pronounced decrease of its elastic modulus, and hence consequently, the elastic modulus of the nucleus is reduced. As a proof-of-principle experiment, anti-cytoskeletal drug treatments of isolated nuclei did not lead to a pronounced alteration of the elastic modulus and viscosity of the nucleus, which indicates that the alterations of the cytoskeleton cannot significantly impact the decoupled mechanical features of isolated nuclei.

External environmental forces evoked by the extracellular matrix scaffold are transduced through the cytoskeleton into the cell toward the nucleus and hence lead to the nuclear deformation. In specific detail, the higher elastic modulus of the cytoskeleton relative to the elastic modulus of the nucleus drives the nuclear sensitivity toward the forces transduced by the cytoskeleton for the subsequent deformation of the entire nucleus. The nuclear deformation may initiate the gene expression of tenascin C (Chiquet et al. 2009), which proves subsequently a protein, which can alter the adhesion to distinct extracellular matrix proteins, such as fibronectin. In this case, the adhesion strength to fibronectin is lowed and finally impaired and thereby the force transduced from the surrounding extracellular matrix is reduced. In line with this finding, for isolated nuclei has been shown that the forces applied on nesprin-1 in the nucleus are able to trigger the phosphorylation of emerin and the reinforcement of the lamin A/C network of the nucleus (Guilluy et al. 2014). The overall strengthening of the nuclear lamin network causes a stiffening of isolated nuclei and, hence, provides a protection for chromatin being exerted to massive mechanical stress.

10.1.5 The Cytoskeleton Decreases the Process of the Strain Recovery in the Nucleus

After removing an imposed stress, the deformation of the nucleus progressively decays, the stress on the nuclear structures is dissipated, and the strain energy retained through the deformation of the spring is successively released by the damper. The stress–strain relation for the strain recovery process is be described in (10.23):

$$\varepsilon(t) = ((\sigma_0)/E)\left(1 - e^{-(t/(\eta/E))}\right) \tag{10.23}$$

where $\varepsilon(t)$ represents the strain, σ_0 stands for the maximum stress prior the strain recovery, η denotes the viscosity, and E indicates the elastic modulus. The ratio between viscosity and modulus of elasticity (time constant), η/E, defines the rate of exponential relaxation of the strain. The time constant further specifies the rate of relaxation of strain energy during strain recovery, since the strain energy decays linearly with the power of the square of the strain, as given in (10.24):

$$U = \left(\frac{1}{2}\right) VE\varepsilon(t)^2 \tag{10.24}$$

where U denotes the strain energy and V provides the volume of the nucleus. A higher time constant implies a prolonged period of strain relaxation and a reduced strain energy rate of dissipation (Vincent 2012).

With an isolated nucleus, the strain recovery is fast, since the nucleus usually responds elastically with a small time constant (0.48 ± 0.05 s), which has been noted earlier in micropipette aspiration and AFM indentation measurements (Dahl et al. 2005). Due to the high viscosity, the cytoskeleton (time constant 1.27 ± 0.13 s) takes longer to recover than the nucleus. Therefore, when the nucleus is linked to the cytoskeleton, the cytoskeleton considerably delays the strain relaxation of the nucleus (intact versus isolated nucleus from 0.76 ± 0.02 s versus 0.48 ± 0.05 s, respectively), resulting in a lesser rate of release of strain energy (Wang et al. 2018). A large relaxation rate of the strain energy (in other words, a small time constant) points to an increased probability for damage to the structure. When strain relaxation occurs, the strain energy accumulated in the elastic component is decayed as a result of shear and fracture damage to the structures (Koop and Lewis 2003), possibly rupturing chromatin and chromatin protein linkages and leading to epigenetic alterations (Miroshnikova et al. 2017). In terms of structure, the cytoskeleton attaches to the nuclear lamina, which is associated with the chromatin. The elevated viscosity of the cytoskeleton substantially decelerates the relaxation process of the nuclear load and drops the rate of release of the load energy, possibly creating pillow mechanisms to preserve the structure of the chromatin.

Treatment of cells with cytoskeletal destabilizing drugs uncovered both the superfluous and prominent role of actin filaments and microtubules in nuclear stiffening and retarding strain recovery. From one side, the perturbation of both actin filaments and microtubules yielded considerably softer nuclei than the perturbation of either actin filaments only or microtubules only, implying that both actin filaments and microtubules contribute to the avoidance of severe nuclear distortions. On the other side, the disruption of both the actin filaments and the microtubules led to a very similar viscosity value to the disruption of only one of the two. Therefore, these data demonstrate the need to have both actin filaments and microtubules available to ensure high viscosity of the nucleus and retard the elongation recovery process.

To analyze the functional involvement of the LINC complex (LInker of Nucleoskeleton and Cytoskeleton) in nuclear mechanics, the SUN domain proteins, the principal constituent of the LINC complex, were also knocked down by both a siRNA pool and single siRNA, validated by real-time quantitative PCR (qRT-PCR), and the characteristics of the membrane, cytoskeleton, nucleus and isolated nucleus were assessed. After the knockdown of the SUN proteins, Young's modulus and viscosity were significantly reduced compared to the control, indicating the need for cytoskeleton nucleus linkage for the cytoskeleton to rigidify the nucleus and, thereby, prevent extreme reshaping of the nucleus and retard the strain relaxation. It was found that the exhaustion of SUN1 or SUN2 by itself has a dramatic effect on levels of nuclear stiffness, suggesting that they may play fundamentally divergent parts in nuclear mechanics.

10.1.6 Cancer Cells from a Later Disease Stage Are Characterized by a Lower Elastic Modulus, Decreased Viscosity and Smaller Time Constant

The instability of genes represents a major hallmark of late-stage cancer cells (Simi et al. 2015). Thus, as a next step, the intact nuclei of human bladder cancer cells in an early (RT4, stage I) and a late stage (T24, stage III) of the disease were examined. The intact RT4 and T24 bladder cancer cells were analyzed, and the nuclei of T24 cancer cells displayed a reduced elastic modulus, reduced viscosity and lower time constant than the nuclei of RT4 cancer cells. Both the actin and tubulin dye staining affirmed that T24 cells exhibit a markedly decreased cytoskeletal density, leading to a reduced modulus of elasticity and cytoskeletal viscosity (Kim et al. 2014). After pharmacological drug treatment of the cytoskeleton (cytochalasin D and nocodazole), the decreased elastic modulus of the nuclei dropped from 4.33 ± 0.36 kPa to 3.45 ± 0.20 kPa, the reduced viscosity dropped from 4.47 ± 0.35 kPa s to 2.54 ± 0.21 kPa s, and the time constant fell from 1.35 ± 0.25 s to 0.81 ± 0.40 s. After treatment with anti-cytoskeletal drugs, the decreased Young's modulus and decreased viscosity of RT4 cells did not lead to a marked disparity in RT4 nuclei comparable with those of T24 nuclei, indicating that the cytoskeletal dissimilarity is accountable for the discrepancy in mechanical characteristics between RT4 and T24 nuclei. However, there is another difference between the two cell types: The isolated nuclei of RT4 were pronouncedly stiffer than the T24 nuclei, which cannot be attributed to alterations in their cytoskeletal structures. Instead, the higher stiffness of isolated nuclei of RT4 cancer cells compared to those of T24 cancer cells can be explained by the elevated density of lamin A/C in RT4 nuclei, which represents the major structural protein of the nuclear envelope (Swift et al. 2013).

These findings suggest that the cell nucleus is a rather soft compartment compared to the cytoskeleton and therefore a tight connection between the stiff cytoskeleton protects the cell nucleus for extraordinary deformation resulting in shape changes and subsequently in damage of the chromosomes. In specific detail, the enormous nuclear deformations may exert higher stress toward the nuclear envelope, the lamins, chromatin and other structures within the cell nucleus. The increased stress applied to the nuclear envelope can then trigger the local rupture and enable unregulated exchanges of substances between the cell nucleus and the cytoplasmic environment, and consequently to the damage of DNA (Denais et al. 2016; Irianto et al. 2017a, b). The elevated stress on the cell nucleus possesses specifically the capacity to modify the conformation of the chromatin, the association of transcription factors with the DNA and the alteration of histones (Mattout et al. 2015). A tethering connection between the cytoskeleton and cell nucleus seems to provide lower the probability for the genetic instability due to the external exertion of a force toward the cell.

In previous studies, a large number of late-stage cancer cell lines exhibited anomalies in actin and microtubules (Sun et al. 2015; Sakthivel and Sehgal 2016), which is in agreement with the discrepancies between RT4 and T24 cell lines. The structural variations in their cytoskeleton indicate that there are commonly distinctive mechanical properties in late-stage cancer cells, although further investigations are needed to identify other types of cancer cells to exhibit these specific late-stage mechanical properties. During metastasis, cancer cells have to pass through narrow spaces (Denais et al. 2016), which leads to stretching of the chromatin and injury to the DNA (Irianto et al. 2017a, b). The low modulus of elasticity (in other words the large deformation) and the small time constant (in other words the high strain energy release rate) of cancer cells may be crucial in triggering gene mutations, adjusting cells to new microenvironments and in promoting the process of metastasis (Burrell et al. 2013).

10.2 The Sites of Transcription and Splicing Factor Compartments (SFCs)

The fundamental cell biological question regarding the architecture of the nucleus is where are genes transcribed. The transcription of the RNA polymerase II (RNA pol II) can co-localize with thousands of discrete dots in the cell nucleus, at which its transcription is performed (Jackson et al. 1993; Wansink et al. 1993; Cook 1999). The RNA pol II is also localized outside of these dots at special storage places (Grande et al. 1997). By fluorescent in situ hybridization, several specific pre-mRNAs are visualized in enlarged spots or elongated tracks (Xing et al. 1993; Dirks et al. 1995). At the end of these tracks, the corresponding active genes are located (Xing et al. 1993). Since splicing is often co-transcriptional, these transcription sites serve as additional sites of pre-mRNA splicing (Beyer and Osheim 1988; Bauren and Wieslander 1994; Xing et al. 1995). Nevertheless, the majority of pre-mRNA splicing factors is not concentrated at active transcription sites, but in different separate domains, known as speckles or SFCs.

After the expression of intron-containing genes or a viral infection, splicing factors from SFCs are redirected to the new sites of transcription (Jimenez-Garcia and Spector 1993; Huang and Spector 1996; Misteli et al. 1997). As these results lead to the hypothesis that splicing factors generally move toward a gene, it is most probable that pre-mRNA also travels toward SFCs (Xing et al. 1993; Huang and Spector 1996; Melcak et al. 2000). The recruitment of splicing factors from SFCs to transcription sites is assumed to compromise a cyclic series of phosphorylation and dephosphorylation events (Misteli 2000). In the conveying, the accumulation of splicing factors at sites of transcription is rendered to the largest subunit, the C-terminal domain, of RNA pol II (Du and Warren 1997; Kim et al. 1997; Misteli and Spector 1999). Hence, these observations support the hypothesis that all RNA-processing mechanisms are physically coupled with the transcription process (Bentley 1999; Hirose and Manley 2000).

10.3 The Nucleolus

The nucleolus is the most prominent nuclear compartment (Fig. 10.4), since it encompasses clusters of tandemly repeated and RNA pol I transcribed ribosomal genes (rDNA genes). The human rDNA genes exist as approximately 180 copies of a 47 kb repeat region on chromosomes 13, 14, 15, 21 and

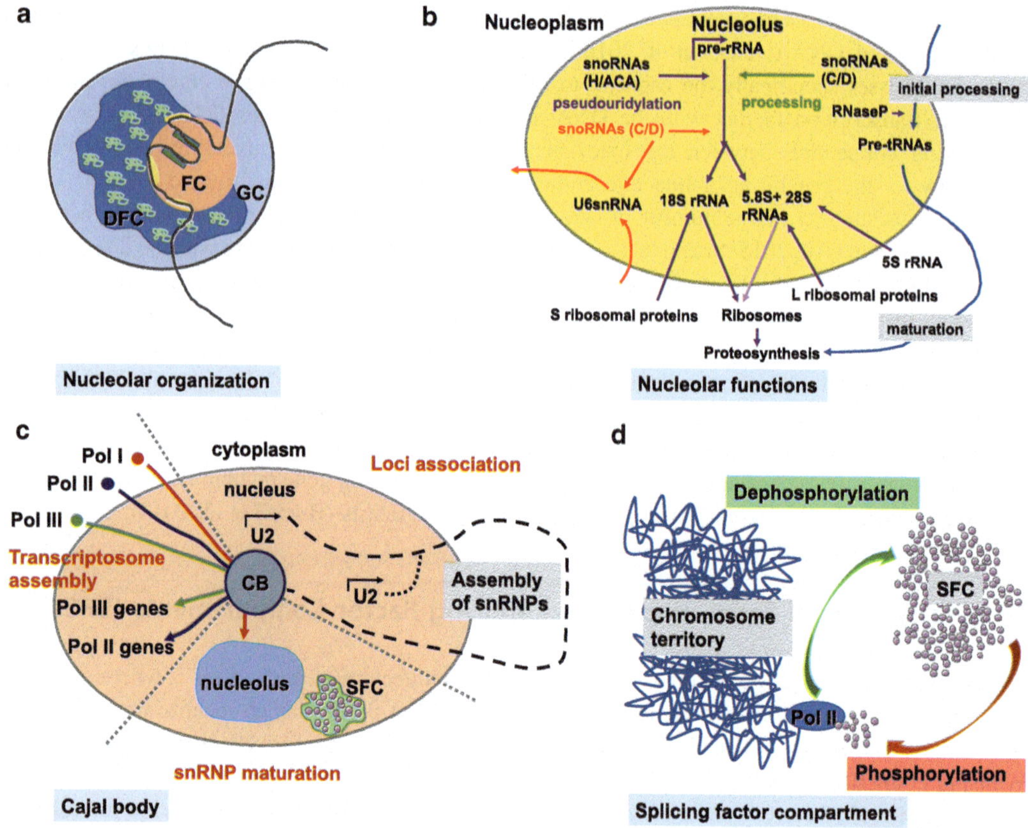

Fig. 10.4 Structure and function of nuclear compartments. **a** The nucleolus is organized from chromatin-containing rDNA genes that are connected to the fibrillar center (FC). Transcriptionally active rDNA genes are positioned at the close proximity of FCs. Nascent pre-rRNA transcripts are able to enter DFCs, where they are subjected to a cascade of processing events. Later processing and assembly events take place in the granular component (GC). **b** There are three distinct functional routes in the nucleolus. Pre-rRNAS are co-transcriptionally pseudouridinylated, 2-0-methylated and split through a sequence of processing steps facilitated by distinct snoRNAs until they are fully processed, and, with association of ribosomal proteins and 5 S rRNA, assembled into large and small pre-ribosomal subunits. In addition, they are transferred to the cytoplasm in order to create a ribosome (black). The nucleolus plays a role in the initial step of the pre-tRNA maturation route. The 5′-end of pre-tRNAs are altered by RNase P prior to further processing steps within the nucleoplasm (blue). Alteration of spliceosomal U6 snRNA takes place in the nucleolus, facilitated by specific snoRNAs (red). **c** Three possible models for CB functions. SnRNA genes are usually linked to CBs (top right), and the organelle fulfills a function in their transcriptional activity. The CB controls the maturation of snRNPs. After transcription, they are transported into the cytoplasm, where they are assembled to snRNP particles. Thereafter, they are transiently retransferred to the CB, from where they pass through the nucleolus and enter the SFCs (top right and bottom). Preassembly of RNA pol I, pol II and pol III transcriptosomes from newly synthesized subunits may be present in CBs before transcriptosomes are associated in specific gene transcriptions (left). **d** Recruitment of splicing factors from SFCs. A cycle of phosphorylation and dephosphorylation regulates the linkage of splicing factors with SFCs. Liberation of factors from the SFC is driven by phosphorylation. Accumulation of factors at active genes is mediated through interaction with the RNA pol I C-terminal domain

22. The copies are located as repeats on these chromosomes and form the basis for the structured organization of the nucleolus. These areas of tandem rDNA gene arrays comprise the nucleolar-organizing regions and are the foundation of the nucleolus' structural organization, which is then the hub of all processing and assembly components necessary for the biosynthesis of the ribosome (Scheer and Hock 1999; Carmo-Fonseca et al. 2000).

The nucleolus contains three, morphologically different main compartment: Fibrillary centers surrounded by dense fibrillary components (DFCs) and granular components radically originating from DFCs, which mirrors the vectorial process of the ribosome biogenesis. For example, in HeLa cells, only about 120–150 rDNA genes of a total of number of about 540 rDNA genes are active (Jackson et al. 1998). A typical nucleolus contains 30 fibrillar structural centers comprising four genes. An active gene is associated with 100–200 engaged RNA pol I molecules. The entire cell possesses 15,000 engaged RNA pol I molecules that synthesize each a primary transcript at 2.5 kb per min. A normal transcription of an average 13.3 kb gene requires five minutes for the entire rDNA (Jackson et al. 1998, 2000). The regions of the nucleolus with tandem arrays of rDNA genes are the nuclear-organizing regions building the basis for structural organization and ribosome biosynthesis (Scheer and Hock 1999; Carmo-Fonseca et al. 2000).

In high-resolution in situ hybridization and Br-UTP-incorporation studies, active rDNA genes are found at the periphery of the fibrillar centers, with the exception of their interior (Dundr and Raska 1993; Hozak et al. 1994; Puvion-Dutilleul et al. 1997; Cmarko et al. 1999). Since the rDNA is not present in DFCs (Puvion-Dutilleul et al. 1992), these nucleolar subdomains seem to serve as transient accumulation sites for both elongating and complete primary transcripts liberated from the template. In fine detail, a series of a minimum of ten separate pre-rRNA cleavage events is started in DFCs and completed in the granular components. In fact, about 115 methylations and 95 conversions of uridines into pseudouridines of pre-rRNA probably are rapidly encountered on emerging transcripts in DFCs (Maden and Hughes 1997). Since these modifications arise, the newly synthesized pre-rRNAs exist in the transcription sites in a radial flow (Thiry et al. 2000).

The processing and posttranscriptional modification of the primary 47 S pre-rRNA transcript are aided by about 200 small nuclear ribonucleoprotein particles (snRNPs), including metabolically stable RNAs (Sollner-Webb et al. 1996; Tollervey and Kiss 1997). Two major classes of small nucleolar RNAs (snoRNAs) have been revealed based on the conservation of structural elements, and they are termed box C/D and box H/ACA (Balakin et al. 1996; Ganot et al. 1999). Additionally, some snoRNAs, such as U3, U8, U14 and U22, are necessary for the cleavage of pre-rRNAs (Sollner-Webb et al. 1996). The proteins associated with snoRNAs or the enzymes that catalyze these modifications are not well understood.

The mechanisms for export of pre-ribosomal particles out of the nucleolus are closely linked to their biogenesis (Aitchison and Rout 2000). In the yeast, more than 60 transacting factors are required for the effective assembly of ribosomes (Kressler et al. 1999). Although the majority of these factors is localized in the nucleolus, some can be found in the nucleoplasm and in close proximity to the nuclear pore complex. The structured arrangement of ribosomal proteins initiates on the primary pre-rRNA transcript and culminates in the formation of a 90 S pre-ribosomal particle. The 90 S particle is subsequently modified by a series of processing steps and terminates in the separation in a pre-60 S large subunit, containing 28 S and 5.8 S rRNAs, and a 43 S small subunit precursor, containing 20 S rRNA (Kressler et al. 1999). In the next step, these two subunits are exported separately to the cytoplasm and further processed to mature ribosomal subunits. More precisely, the export factor Crm1p serves as the export receptor for the 60 S subunit, which is assisted by the adapter protein Nmd3p (Ho et al. 2000). Hence, the export of pre-ribosomal subunits is driven by specific receptors and specific members of the karyopherin family (Aitchison and Rout 2000).

In complement to ribosome biogenesis, the nucleolus appears to perform additional functions such as the construction and the interaction between multiple ribonucleoprotein machines, many of which are engaged in protein biosynthesis (Pederson and Politz 2000). The signal recognition particle (SRP) is also part in this process. Three of the six SRP proteins (SRP19, SRP68 and SRP72) are positioned in the nucleolus with SRP-RNA and in the cytoplasm (Politz et al. 2000). Therefore, on the basis of these results, it can be argued that the SRP pre-assembly takes place during its passage through the nucleolus. In addition, posttranscriptional modifications of spliceosomal U6 snRNA appear in the nucleolus. More precisely, the 2 h-O-ribose methylation and pseudouridylation by three nucleolar box C/D snoRNA are performed in the nucleolus, and hence it can be proposed that U6 snRNA is maturated before leaving the nucleolus and reentering the nucleoplasm (Lange and Gerbi 2000). In line with this, the U5 snRNA also enters the nucleolus in order to be altered (Jady and Kiss 2001).

In yeast, the nucleolus also apparently contributes to the maturation of certain tRNAs (Wolin and Matera 1999). The pre-tRNA 5 h-leader is removed by the RNA-containing endonuclease RNase P, which is mainly located in the nucleolus (Jarrous et al. 1999). In specific, the C-terminal domain of the human La protein, which interacts with the 5 h-ppp ends of nascent RNAs, keeps the pre-tRNA in an unprocessed state by obstructing the 5 h-processing site recognized by RNase P (Intine et al. 2000). In turn, phosphorylation of the human La protein on serine-366 induces the processing and thereby facilitates the tRNA maturation. A close coupling between the syntheses of rRNA and tRNA is observed in yeast (Briand et al. 2001). In temperature-sensitive RNA pol III mutants, the synthesis of RNA pol III transcripts is inhibited, while the synthesis of the mRNA of ribosomal protein genes, which is performed by RNA pol II is not altered. In contrast, the generation of the 25 S, 18 S and 5.8 S rRNAs from the 35 S primary transcript synthesized by RNA pol I is switched off (Briand et al. 2001). Hence, the mutants of RNA pol III impact the synthesis rate of rRNAs. Subsequently, these results lead to the hypothesis that yeast cells modulate the pre-rRNA processing through the activity of the RNA pol III (Briand et al. 2001).

The nucleolus is engaged in the regulation of the tumor suppressor protein, p53 (Prives and Hall 1999). The interaction with murine double minute clone 2 oncoprotein (MDM2) is crucial for the activity regulation of p53, as MDM2 causes the degradation of p53 in the cytoplasm through the ubiquitination pathway. MDM2 functions as a shuttle protein between the nucleus and the cytoplasm. The activity of MDM2 is controlled by the p19ARF protein, which hinders the export of MDM2 from the nucleus to the cytoplasm and thereby sequesters MDM2 inside the nucleolus (Weber et al. 1999). When co-expressed with p19ARF, MDM2 is increasingly positioned in nucleoli. However, the retention of MDM2 in the nucleolus avoids its export to the cytoplasm and subsequently its p53 facilitated degradation. Finally, the nuclear import of p53 is activated and leads to its accumulation in the nucleus (Tao and Levine 1999; Zhang and Xiong 1999).

10.4 The Cajal Bodies and PML Bodies

The Cajal bodies and PML bodies are two organelles of the highly structured nucleus. Cajal bodies are found in certain cells and tissues, such as neuronal and cancer cells that are actively engaged in transcription (Spector et al. 1992; Gall 2000). PML bodies are usually displayed as 10–20 nuclear foci and contain the promyelocytic leukemia (PML) protein together with numerous factors involved in various cellular functions including the regulation of the transcription (Matera 1999; Maul et al. 2000; Strudwick and Borden 2002). Specific classes of PML bodies can move ATP dependent through the nucleus (Muratani et al. 2002; Wiesmeijer et al. 2002) and locate in close proximity to transcriptionally active gene loci (Shiels et al. 2001; Wang et al. 2004).

10.4.1 The Cajal Bodies

The Cajal body (initially named nucleolar accessory bodies) has a characteristic appearance as a tangle of coiled fibrillary strands 40–60 nm in diameter (Monneron and Bernhard 1969). These small spherical structures (1–5 copies per nucleus) are 0.1–1.0 μm in size. Cajal bodies contain components such as spliceosomal snRNPs, U3, U7, U8 and U14 snRNAs, basal transcription factors TFIIF and TFIIH, cleavage stimulation factors and cleavage and polyadenylation specificity factor, nuclear components fibrillarin and Nopp 140 as well as B23 protein (Gall 2000). Despite the knowledge about the proteins stored in Cajal bodies, their function is still difficult to grasp. Cajal bodies are located in the nucleoplasm of normal cells, but they are found in the nucleolus of human breast carcinoma cells (Malatesta et al. 1994; Ochs et al. 1994). Since Cajal bodies are highly motile, they can move toward and out of the nucleolar periphery as well as within the nucleolus (Boudonck et al. 1999; Platani et al. 2000). There exist at least two sizes of Cajal bodies in living cells, the larger are termed Cajal bodies and the smaller are termed mini Cajal bodies. A specific marker of Cajal bodies serves the nucleocytoplasmic-shuttling phosphoprotein, p80-coilin (Raska et al. 1991). A single point mutation at residue 202 (exchange of serine with aspartate) in transiently expressed p80-coillin causes the assembly of Cajal bodies in the nucleolus (Bohmann et al. 1995). Moreover, Nopp140, which interacts with p80-coilin in vitro (Isaac et al. 1998), is located in both Cajal bodies and the nucleolus, and the transient expression of non-functional mutants of Nopp140 alters both nuclear structures (Isaac et al. 1998; Yang et al. 2000). The treatment of human HeLa cervical cancer cells with a specific serine/threonine protein phosphatase inhibitor, such as okadaic acid, resulted in a redirection of p80-coilin into the nucleolus (Lyon et al. 1997). Subsequently, Cajal bodies and the nucleolus seem to be functionally interconnected.

Cajal bodies seem to facilitate the transport and maturation of snRNPs and snoRNPs. When GFP-tagged core Sm proteins associated with U-snRNA are reimported into the nucleus, they first locate on Cajal bodies before passaging through the nucleolus to achieve their final destination in SFCs (Sleeman and Lamond 1999). Newly transcribed U snRNA are impaired to be exported from the nucleoplasm into the cytoplasm, and the reimport of newly assembled snRNPs in the nucleus by leptomycin B causes a progressive decrease of U snRNPs in Cajal bodies (Carvalho et al. 1999). In accordance with this, when p80-coilin was immunodepleted from Xenopus egg extracts prior to assembly in pronuclei, Cajal bodies were still generated in the pronuclei without coilin and Sm proteins (Bauer and Gall 1997). Therefore, there seems to be a separate role of Cajal bodies in the maturation path of snRNPs, which compromises several steps in their modification and assembly process to RNP complexes (Narayanan et al. 1999). More precisely, fluorescently labeled U3, U8 and U14 snoRNAs were microinjected into Xenopus oocytes, and they transiently concentrated firstly in Cajal bodies and secondly in nucleoli, which leads to the hypothesis that snoRNAs can flow from the Cajal bodies toward nucleoli (Narayanan et al. 1999).

Cajal bodies seem to be in close proximity to distinct loci on the chromosomes. They are preferentially interacting with tandemly repeated histone genes on amphibian Lampbrush chromosomes, and also with tandemly repeated genes for U1, U2, U3, U4, U11 and U12 sn(o)RNAs in interphase nuclei of mammalian cells (Callab and Gall 1991; Gao et al. 1997; Jacobs et al. 1999). When artificial tandemly repeated U2 snRNA genes are transcriptionally active, they interact with Cajal bodies (Frey et al. 1999). In addition, when the expression levels of U2 through elevated U2 copy numbers, their interaction with Cajal bodies was also increased (Frey et al. 1999). Hence, it is suggested that targeting of Cajal bodies to a distinct chromosomal site is facilitated by the interaction with nascent snRNA transcripts. Moreover, through the interaction with Cajal bodies, the gene loci seemed to be more transcriptionally active (Frey et al. 1999; Schul et al. 1998). In more detail, Cajal bodies possess the 45 kDa γ-subunit of the proximal element sequence-binding transcription factor (PTF), which is, together with TBP and a subset of RNA pol II subunits, specific for snRNA genes (Schul et al. 1998). Therefore, Cajal bodies act as distribution centers attracting specific transcription and processing factors to snRNA transcription sites (Frey et al. 1999). Cajal bodies recruit RNA 3 h-processing factors to neighboring genes from specific domains, which are linked with Cajal bodies and known as cleavage bodies (Schul et al. 1996, 1999). Besides their association with snRNA genes, Cajal bodies seem to preferably co-localize with histone loci (Schul et al. 1999; Frey and Matera 1995; Shopland et al. 2001). However, for gene expression, a sustained interaction of the HIST2 locus with Cajal bodies is not necessary (Schopland et al. 2001). A reason for this interaction may be that Cajal bodies deliver factors inducing the expression of histone genes. In late G4 and S-phase of the cell cycle, the increased number of Cajal bodies seems to mirror the elevated expression level of histone genes. This is underlined by the discovery that the CDK2–cyclin E complex is locally distributed in Cajal bodies depending on the cell cycle (Liu et al. 2000).

An alternative model for the function of Cajal bodies, which is not mutually exclusive, proposes that Cajal bodies represent the assembly sites for the nuclear transcription machinery (Gall 2000; Morgan et al. 2000). Since transcription, capping, splicing, polyadenylation and cleavage factors of pre-mRNAs are located in Cajal bodies of the oocyte (Gall 2000), it can be suggested that the RNA pol II machinery is pre-assembled in Cajal bodies together with other elements of the processing machinery. Subsequently, transcriptionally competent large multi-subunit complexes are generated, which are known as polymerase II transcriptosomes. The existence of coupled transcription and processing machineries seems to be likely, as the coupling of all RNA processing events to transcription was observed (Bentley 1999; Hirose and Manley 2000). Moreover, gemins, which function in assembly and regeneration of snRNPs, associate with the C-terminal domain of RNA pol II via RNA helicase A (Pellizzoni et al. 2001). The expression of a dominant-negative mutant of survival motor neuron reorganizes snRNPs into large domains, which contain markers of Cajal bodies. Associated with this reorganization is the inhibition of RNA pol I and pol II transcription and accumulation of hypo-phosphorylated RNA pol II within these domains (Pellizzoni et al. 2001). The localization of various pol I and pol III components to Cajal bodies suggests that Cajal bodies serve as a platform for pol I and pol III transcriptosomes assemblies (Gall 2000).

10.4.2 PML Bodies

PML bodies are small spherical domains and are synonymously termed nuclear domain 10, Kremer bodies and PML oncogenic domains. PML bodies are identified to be scattered throughout the nucleoplasm. Cells commonly possess 10–30 PML bodies with a size in the range of 0.2–1.0 μm in their nucleus, and they undergo fluctuations in number and size during the cell cycle (Koken et al. 1995; Terris et al. 1995). PML bodies are often observed to in close neighborhood to both Cajal bodies and cleavage bodies (Schul et al. 1996), which form under distinct circumstances triplets. Ultrastructurally, this nuclear domain appears as a dense ring, which is largely stained by anti-PML antibody (Sterndorf et al. 1997). Besides PML, the bodies possess many other components, including Sp100, retinoblastoma protein, Daxx and the Bloom syndrome protein (Ruggero et al. 2000). PML bodies seem to play a role in terminal differentiation, the regulation of transcription (Ruggero et al. 2000), nuclear storage (Maul 1998), the regulation of growth (Ruggero et al. 2000) and apoptosis (Quignon et al. 1998).

PML bodies are of particular interest in acute promyelocytic leukemia (APL), because they are torn in cell lines of APL patients by the formation of a fusion protein between the PML protein and the retinoic acid receptor, as a result of a reciprocal chromosomal translocation at t(15;17) q(22;21) (de Thé et al. 1990). The disorder of PML bodies is correlated with APL and hence seems to involve of PML bodies in promyelocytes differentiation (Dyck et al. 1994; Koken et al. 1994). Hence, the retinoic acid addition leads to reappearance of the PML bodies and a remission of cancer (Dyck et al. 1994; Koken et al. 1994). Although the PML protein appears to be required for proper formation

PML bodies and their structural integrity, the exact function of PML protein or the entire PML body remains unknown.

PML bodies seem to function in the transcriptional regulation (Ruggero et al. 2000). In addition, a stably expressed reporter construct with the lac operator sequence associates with PML bodies, when the lac repressor protein is expressed, while it is undetectable without the lac repressor expression (Tsukamoto et al. 2000). Therefore, the association seems to be correlated with the local concentration of the expressed lac repressor protein at the gene locus, and not to the locus itself. Hence, PML bodies could recognize highly localized concentrations of exogenously introduced specific proteins and pronouncedly regulate their transcription (Tsukamoto et al. 2000).

10.5 Nuclear Envelope

The nuclear envelope and the nuclear pore complex (NPC) are both highly dynamic processes that control the traveling of substances and hence the content of the nuclear composition. Besides this, the shape and structural architecture is provided by the nuclear envelop and its connection to the nuclear matrix, such as the karyoplasm and the nucleoskeleton (Hetzler 2010). However, there is still a question that needs to be answered: Does a nuclear matrix exist? In 1974, the formerly hypothesized nuclear residue has been termed the nuclear matrix (Berezney and Coffey 1974). More precisely, the hypothesis has been formulated that the chromatin and nascent RNA interact with the nuclear matrix. However, the nuclear matrix is still seen as a concept that needs to be proven, but a network of short- or long-ranged filaments with possibly structural motifs throughout the nucleoplasm can turn out to be true and also reveal to be existent.

The nuclear envelope is a highly restrictive membrane barrier engulfing the nucleus and hence separating it from the cytoplasm in eukaryotic cells (Fig. 10.5). The nuclear envelope is built by a

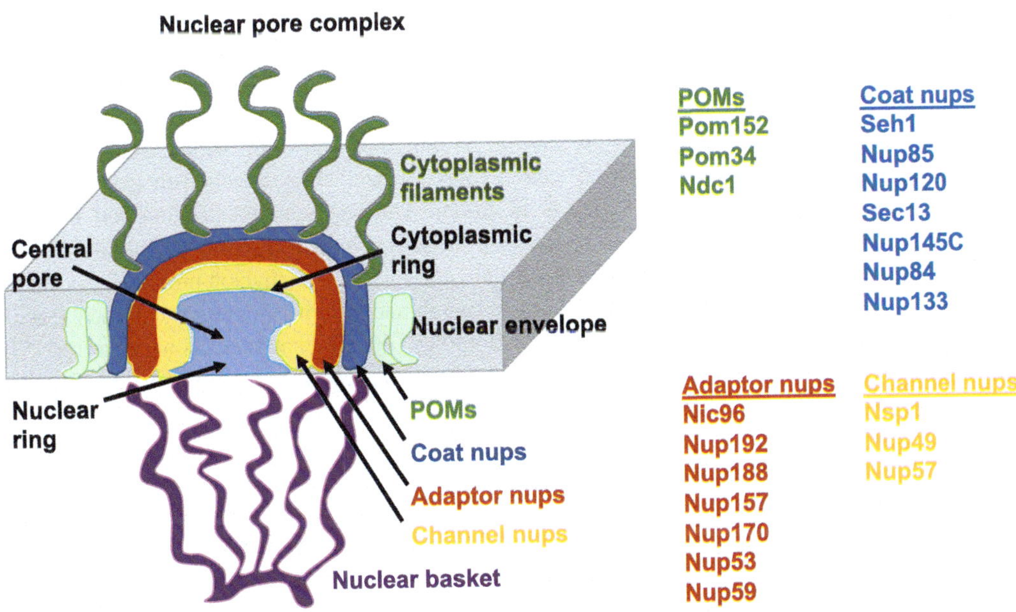

Fig. 10.5 Organization of Nups and nuclear pore complex

large number of different proteins, which contribute to the organization of chromatin organization and regulation of genes. The nuclear membrane facilitates complex levels of gene expression, and it is a challenge for cell division. To provide access for the mitotic spindle to chromatin, the nucleus of metazoans must completely disintegrate during mitosis, which entails the need to restore the nuclear compartment at the completion of each cell division. Hence, the nuclear envelop is dynamically remodeled during the cell cycle.

The nuclear envelope, representing a major hallmark of eukaryotic cells, is composed of a highly organized double-layered membrane containing the nuclear genome (Kite 1913). Electron microscopy demonstrated that the inner (INM) and outer nuclear membranes (ONM) of the nuclear envelope merge in the endoplasmic reticulum (ER) (Watson 1955). Regardless of the lipid continuity between the nuclear envelop and the ER, both ONM and INM consist of different protein groups, which are typically not enriched in the ER (Hetzer et al. 2005). The first group consists of about 30 different polypeptides, which are known as nucleoporins or Nups and assemble approximately 40–70 MD NPCs (Tran and Wente 2006; D'Angelo and Hetzer 2008). NPCs represent aqueous channels with eightfold rotational symmetry, an outer diameter of about 100 nm and a central transport channel with a diameter of 40 nm, which is suitable for bidirectional exchange of proteins, RNA and ribonucleoprotein complexes between nucleoplasm and cytoplasm (Beck et al. 2004, 2007; Terry et al. 2007). A subset of Nups is stably present in the nuclear envelope and thereby provides a scaffold structure or the NPC core (Rabut et al. 2004; D'Angelo et al. 2009), which stabilizes strongly the highly curved and energetically unfavorable membrane of the pore (Alber et al. 2007; Boehmer et al. 2008). This core contains the Nup107/160 complex (termed Nup84 complex in yeast) and the Nup205 complex (termed Nup170 in yeast), which both constitute roughly 50% of the entire NPC (Brohawn et al. 2009).

Firstly, peripheral Nups are attached to this scaffold, many of which are rich in phenylalanine-glycine (FG) rich repeats that build up a permeability barrier and also confer active, receptor-dependent transport across the nuclear envelope (Peters 2009). Secondly, another group of nuclear envelop proteins is located specifically to the INM (Schirmer and Gerace 2005). Although most of these more than 60 integral membrane proteins, synonymously referred to as nuclear envelop transmembrane proteins (NETs) (Schirmer et al. 2003), are still largely uncharacterized, their interaction with lamins and chromatin has been observed, such as lamin B receptor, lamina-associated polypeptide (LAP) 1, LAP2, emerin and MAN1 (Akhtar and Gasser 2007; Dorner et al. 2007; Schirmer and Foisner 2007). INM proteins seem to be important and fulfill various roles in nuclear functions, such as chromatin organization, gene expression and DNA metabolism (Mattout et al. 2006; Heessen and Fornerod 2007; Reddy et al. 2008). The unprecise localization and improper function of INM proteins have been associated with numerous human diseases (Vlcek and Foisner 2007; Worman and Bonne 2007; Neilan 2009). A third class of nuclear envelop proteins is located in the ONM. These highly various groups of integral membrane proteins possess a common feature, such as the small KASH (Klarsicht, ANC-1, Syne Homolgy) domain that interacts with Sad1p/UNC-84 (SUN)-domain proteins of the INM within the periplasmic space of the nuclear envelop (Starr and Han 2003; Wilhelmsen et al. 2006). Two related ONM proteins, the nuclear envelope spectrin repeat (nesprin)-1 and -2, interact directly with the actin cytoskeleton through their amino-terminal actin-binding domain (ABD) (Wilhelmsen et al. 2005). Moreover, these ONM proteins are involved in the positioning of the nucleus that is an essential feature for major cellular processes including cell polarization, pronuclear migration and the organization of syncitia (Fridkin et al. 2009). More precisely, the ONM and INM proteins build bridges across the perinuclear space, which can contribute to the separation of the two nuclear envelope membrane leaflets at a uniform distance of roughly 50 nm

(Voeltz and Prinz 2007). These lumenal protein-containing bridges may generate physical linkages between the cytoskeleton and chromatin that can be crucial for transcription, replication and DNA repair mechanisms (Tzur et al. 2006; Stewart et al. 2007). The final group of nuclear envelope proteins comprises the lamina, which is a meshwork of intermediate filaments including A- and B-type lamins (Gruenbaum et al. 2000). Since the lamina is crucial for the nuclear stability in tissues such as muscle fiber, undergoing stimulation by mechanical forces such as muscle fibers (Cohen et al. 2008), lamins turned out to regulate chromatin function and gene expression (Gruenbaum et al. 2005; Dechat et al. 2008; Reddy et al. 2008). Similar to INM proteins, mutations in lamins are associated with a variety of human diseases (Mounkes et al. 2003; Muchir and Worman 2004; Mattout et al. 2006) and with aging (Gruenbaum et al. 2005; Scaffidi and Misteli 2006), underlining the critical role of the network of nuclear envelope proteins for providing normal cell functions. In resume, it can be said that the nuclear envelope plays a decisive role in protecting the genome from cytoplasmic components, but at the same time serves as a highly specialized membrane that offers anchorages for chromatin and the cytoskeleton (D'Angelo and Hetzer 2006).

10.6 Function of the Nuclear Pore Complex

The nuclear pore complex functions as a highly flexible and dynamic gateway and transport system between the nucleoplasm and the cytoplasm of a cell in a bidirectional manner. Nuclear pore complexes (NPCs) represent holes in the nuclear envelope and therefore carry out the primary transport to exchange molecules between the nucleus and the cytoplasm. The NPC is a megadalton complex and contains elements, which is essential for the overall organization. The NPC can be divided into scaffolding components and the disordered elements attached to them, which form a selective barrier between compartments. These structural elements have a high degree of flexibility that can help to understand the NPC assembly and function. How does structural and compositional heterogeneity of NPCs affect its function?

In general, eukaryotic cells are characterized by membrane-enclosed organelles that form distinct reaction compartments, which are embedded by the cytoplasm. For communication purpose, the different compartments require to exchange signaling molecules across membrane in an efficient manner. While organelles, such as the endoplasmic reticulum, mitochondria and the Golgi apparatus, interact via various mechanisms, among them channels, transporters and vesicle trafficking, the transport of signaling molecules between nucleus and cytoplasm is performed commonly through the nuclear envelope through NPCs (Fig. 10.5).

NPCs are huge accumulations of proteins arranged in circles across the nuclear envelope, where both INM and ONM are fused (Bilokapic and Schwartz 2012a; Grossman et al. 2012; Hurt and Beck 2015). NPCs appear to encase these highly curved fusion membranes and produce a stable protein scaffold integrated in the membrane. The stable framework is equipped with eight extended filaments on the nucleoplasmic side, which are joined together in a distal ring, and subsequently form the nuclear basket. Another set of eight flexible filaments radiate into the cytoplasm. Finally, fiber-like extensions also extend from the scaffold into a central opening and thereby build the main permeability barrier. Since these fiber-like extensions yield a peculiar microenvironment with distinct biochemical properties, the central opening can serve as a gateway for precisely controlled import and export of biomacromolecules (Hülsmann et al. 2012; Mohr et al. 2009). The NPC cannot choose specific ions or metabolites that diffuse freely between the nucleus and cytoplasm not can it select solely small macromolecules, which are smaller than roughly 5 nm or roughly 40 kDa in size (Knockenbauer and Schwartz 2016).

10.6.1 The Ultrastructure of the Nuclear Pore Complex Reveals a Three Stacked Ring Structure

Using electron microscopy, the assembled nuclear pore complexes in the nuclear envelope can be seen due to their large size. Cryo-electron tomographic (cryo-ET) reconstructions of human and Xenopus laevis NPCs revealed that dimensions of the NPCs are nearly identical in both vertebrates (Bui et al. 2013; Eibauer et al. 2015). In more detail, the reconstructed NPC core possesses a diameter of about 110 nm and a height of about 70 nm. Due to scanning electron microscopic analysis, an eightfold rotational symmetry seems to be assumed. Based on this assumption, the resolution of the electron microscopic reconstructions can be improved by subtomogram averaging. The core structure with a resolution of about 2–3 nm appears as three porous ring densities, such as cytoplasmic, inner and nucleoplasmic, ring densities (CR, IR and NR, respectively).

Based on the electron microscopic technique, additional details can be determined. More precisely, on the nucleoplasmic side, the eight rod-shaped extensions are bound to a distal ring and thereby build a nuclear basket (Allen et al. 2000). On the cytoplasmic side, eight flexible extensions originate from a cytoplasmic ring structure. This basic ring structure seems to be highly conserved among all metazoans (Field et al. 2014; Fiserova et al. 2009; Neumann et al. 2010). How similar is the NPC of unicellular organisms in size compared to vertebrates? Proteomic analysis has been demonstrated that the roughly 50 MDa mass of yeast NPC (Alber et al. 2007) is less than half that of the roughly 112 MDa estimate for vertebrate NPC (Reichelt et al. 1990). The entire structure of the yeast NPC is not clearly revealed, and depending on the technique, the height estimations range from approximately 35 nm (Alber et al. 2007; Yang et al. 1998) to approximately 70 nm (Rout et al. 2000). The proteomic analysis of highly enriched NPC preparations yielded that the inventory of NPC components, referred to as nucleoporins or Nups, is highly preserved in all eukaryotic clades (Cronshaw et al. 2002; Degrasse and Devos 2010; Rout et al. 2000; Tamura et al. 2010). This finding leads to the hypothesis that NPCs were present in the last eukaryotic common ancestor about 1.6–2.1 billion years ago or even much earlier (Devos et al. 2014; Koumandou et al. 2013; Mans et al. 2004).

10.6.2 The Module-Based Assembly of the Nuclear Pore Complex

The approximately 30 nucleoporins are organized in a few biochemically well-defined subcomplexes. More precisely, the supramolecular complex is usually less stable than the individual subcomplexes (Alber et al. 2007; Amlacher et al. 2011; Belgareh et al. 2001; Grandi et al. 1993; Siniossoglou et al. 2000). The Y-shaped complex of 0.5–0.75 MDa represents the largest and most accurately characterized NPC subcomplex. The Y-complex consists of six universally conserved Nups (Kelley et al. 2015), whereas I can compromise up to ten members according to the species (Franz et al. 2007; Loïodice et al. 2004; Rasala et al. 2006). In the NPC, the Y-complex represents an essential scaffolding element, since its degradation eliminates NPC formation (Harel et al. 2003; Walther et al. 2003). Cryo-electron tomography suggests that the Y-complex covers the majority of the nuclear and cytoplasmic ring structures (von Appen et al. 2015; Bui et al. 2013; Eibauer et al. 2015). The Nic96-complex (yeast Nup nomenclature, if not further defined) or hsNup93-complex (Homo sapiens) includes a further five to seven Nups and, together with the Y-complex, performs an architectural function probably localized to the inner ring structure (Vollmer and Antonin 2014).

The NPC scaffold is connected to the nuclear envelope by a small number of transmembrane Nups building a functional entity (Eisenhardt et al. 2014; Onischenko et al. 2009; Stavru et al. 2006). These proteins are clustered by topology and are assumed to function as NPC anchors. They are not necessarily as stable as the other subcomplexes. However, the transmembrane Nups between the species are not strongly conserved (Field et al. 2014; Neumann et al. 2010). Proteins constituting the nuclear basket on the nucleoplasmic side and, conversely, those forming the cytoplasmic filaments make up the architecture of the NPC.

This framework provides the NPC with its special shape, while the permeability barrier itself consists of distinct, disordered elements. About ten nucleoporins comprise fibrous extensions of hundreds of residues characterized by up to roughly 50 phenylalanine-glycine (FG) repeat interruptions (Allen et al. 2001; Terry and Wente 2009). These extensions are intended to build a permeable, but selective hydrogel in the middle of the pore, which serves as the primary transport barrier (Bestembayeva 2014; Labokha et al. 2013; Patel et al. 2007; Schmidt and Görlich 2015).

10.6.3 Dynamics and Stable Elements Built the Nuclear Pore Complex

The question arises with such a complex and large-scale molecular network as the NPC, how rigorously it can be specified in terms of its components. Is it possible that there are also fluctuations in its conformational state? All these issues together are matters relating to the flexibility of the NPC complex. The residence time (synonymously referred to dwell time) of the majority of nucleoporins within the supra-assembly was investigated through photobleaching experiments with fluorescence-labeled nucleoporins (Morchoisne-Bolhy et al. 2015; Rabut et al. 2004). The results indicate that the scaffold nucleoporins are generally very stably linked to the NPC, with residence times of hours or days; whereas, the FG- and other nucleoporins are partially mobilized with residence times in the minute scale. The extraordinary stability of the scaffolding nucleoporins is also reflected in the long life cycle of these proteins. Among the longest-lived proteins are eight scaffold nucleoporins within the cell (D'Angelo et al. 2009; Savas et al. 2012; Toyama et al. 2013), which possess extremely slow turnover rates in resting and hence post-mitotic cells measured over several months or even years. Whether this slowly renewing process is actually the cause of the leaky pores found in aging cells, possibly initiated by the aggregation of damaged proteins, is an intriguing concept that deserves more attention. The compositional stability of the NPC scaffolding points to a framework that can be rather uniform in nature.

On the other hand, microscopic imaging tools are not sufficiently good to provide information about the feasibility of compositionally dissimilar NPCs inside a cell. This state of knowledge will necessitate a special technology to identify a single NPC from the ~2000–3000 specimens detected in an average human cell (Kubitscheck et al. 1996; Wild et al. 2009). The modular construction of the NPC creates the conditions for possibly diversely composed NPCs (Schwartz 2005). In view of the wide variety of transport modes, it is an appealing concept that certain NPCs can be customized to transport certain goods more efficiently than others. Cell type-specific variants of the NPCs have actually been identified and bear witness to the fact that the NPCs are not identical (Ori et al. 2013). In a series of investigations, nucleoporins display cell type-specific expression profiles (D'Angelo et al. 2012; Gomez-Cavazos and Hetzer 2015; Lupu et al. 2008), and interestingly, there is also indication of a compounding discrepancy between NPCs in the same cell (Lowe et al. 2015). These findings can be grouped as the compositional variability of NPCs, and it is only just at the beginning of their comprehension in the perspective of maintaining cellular and organismic homeostasis. Conversely, the NPC offers conformational flexibility, which can be more specifically addressed. Over the last ten years, substantial architectural understanding of the main architectural components of the NPC, the Y-complex, the Nic96-complex and lately the Nsp1-complex has been achieved. The nucleoporins implemented in these three architectural arrangements provide several degrees of constraint flexibility that are probably crucial to the functioning of the NPCs, which is discussed in the following. However, this degree of flexibility of the scaffolding blocks does not need to be mixed up with the inherent randomness of the barrier-forming FG elements, which are handled independently.

10.6.4 A Few Architectural Elements Form the Nuclear Pore Complex

The conserved core within the Y-complex consists of six proteins, which are in most fungi referred to Sec13, Nup84, Nup85, Nup120, Nup133 and Nup145C (Kelley et al. 2015). In specific detail, the branched Y shape is formed by Nup84, Nup85, Nup120, Nup133 and Nup145C, whose helical domains are elongated and serve as major building blocks (Bui et al. 2013; Kampmann and Blobel 2009; Kelley et al. 2015; Stuwe et al. 2015b). In addition, β-Propellers are surrounding the branched, helical core element (Berke et al. 2004; Kelley et al. 2015; Kim et al. 2014b; Stuwe et al. 2015b). There are maximally four species-specific nucleoporins that are able to bind to the core, such as Y-Seh1 that binds Nup85 (Brohawn et al. 2008; Debler et al. 2008), whereas Nup37 interacts with Nup120 (referred to Nup160 in metazoa) (Bilokapic and Schwartz 2012b; Liu et al. 2012). Nup43 is associated with Nup85 in metazoa (Kim et al. 2014a; Xu et al. 2015). The interaction of the approximately 250 kDa ELYS is not clearly revealed; however, it seems to be possible that Nup160 plays a role (von Appen et al. 2015; Bilokapic and Schwartz 2012b).

The special character of the elongated helical elements within the Y-complex justifies a more precise investigation. A frequent geometry of these framework proteins is that the helices are piled up in repetitions of pairs or triplets to construct a superhelical, solenoid-like architecture (Kobe and Kajava 2000). This configuration commonly yields consistently flexible proteins. However, within the Y-complex, the flexibility is limited to Nup133 and the C-terminal approximately 40 kDa range of Nup120 (Bilokapic and Schwartz 2012b; Kelley et al. 2015; Whittle and Schwartz 2009). The other three spiral units, Nup84, Nup85 and Nup145C, have in common a special retractable construction in which the N-terminal spirals twist and then snap into a central stapled spiral unit (Brohawn et al. 2008, 2009). This topology is known as the ancestral Coatomer Element 1 (ACE1), because it is present in Sec16 and Sec31, two components of the COPII vesicle mantle structure (Miller and Schekman 2013). The refolding topology probably stiffens the helical protein and successfully removes the permanent flexibility that is typical of helical repetitions. On the downside, ACE1 proteins include prominent hinging areas that permit the formation of spatially defined, but perhaps also distinctive, flexural motions. In a coincidentally tapered tilted shape of the composite Y-complex, these joint displacements are clearly recorded and correspond to regions that are mapped in a newer compound crystal structure (Kampmann and Blobel 2009; Kelley et al. 2015).

The β propellers inside the Y-complex are more or less stiff components and probably provide stability for different 3D configurations. Most remarkable in this context is the Sec13 β propeller, generally preserved and placed in the center of the Y-complex node, where it seems to lock the comparative position of the bases of the Nup85 short arm and the Nup145C-Nup84-Nup133 long stalk (Kelley et al. 2015; Stuwe et al. 2015b). The Nup133 β propeller seems to be the the last one that can be flexibly attached to the Y-complex.

The second, large structural framework complex of the NPC represents the Nic96 complex that consists of Nup53/59, Nic96, Nup157/170, Nup188 and Nup192 (Vollmer and Antonin 2014). Although structural insights into the individual constituents from measurements of large protein segments are available, compositional patterns of Nic96 constituents are not available at present. Nup53/59 is extensively deployed, with the exception of a central 10 kDa dimerization area (Handa et al. 2006). Nic96 includes an ACE1 folding (Jeudy and Schwartz 2007; Schrader et al. 2008) which is predominately followed by short helical segments that give a bond to Nup188, Nup192 and the

Nsp1 complex (Amlacher et al. 2011; Chug et al. 2015; Stuwe et al. 2015b). The Nup157/170 includes an embedded β propeller-α-helical staple region that is far from closely resembling Nup133 of the Y-complex (Seo et al. 2013; Whittle and Schwartz 2009). Nup188 and Nup192 have a similar structure and consist of a mixture of HEAT and arm repeats that produce long-stretched, superhelical and constantly elastic proteins (Andersen et al. 2013; Flemming et al. 2012; Sampathkumar et al. 2013; Stuwe et al. 2014).

While Nic96 and Nup157/170 are definitely closely related to members of the Y-complex, the question remains open whether the Nic96 complex also exhibits a special 3D configuration. The newest biochemical classification of the Nic96 complex indicates that the Nic96 complex is much physically more flexible than the Y-complex. Actually, various proteins seem to be linked by flexible connectors instead of rigid ones (Fischer et al. 2015; Stuwe et al. 2015b).

With this group of architectural nucleoporins, which are arranged in the Y- and Nic96-complexes, the principal skeleton of the NPCs is constructed. How stiff is it or how adaptable must it become? Latest cryo-ET reconstructions of human and frog NPCs provide a few indications of this (Bui et al. 2013; Eibauer et al. 2015). Both reconstructed frameworks, as noted above, exhibit a three-ring design throughout, indicating that these rings comprise the most well-ordered components of the scaffolding. Sixteen specimens of a cropped Y-complex, which lacked the flexible section of the long handle, were fairly well matched for the CR, which prompted a proposal for the reticular two-ring model (Bui et al. 2013). In this model, two eight-headed rings of Y-complexes are arranged with a gentle offset. The two rings are only slightly divergent in diameter, and the Y-complexes are placed in a head–tail formation. This hypothesis was corroborated by the recent docking of a about 400 kDa Y-complex crystal fragment from yeast, which further assumes that the main ring configuration is preserved evolutionarily (Stuwe et al. 2015b). In the recent 2–3 nm cryo-ET structure of the human NPC, the mesh-like two-ring model has been further refined to encompass the long handle of the Y-complex. In addition, the current cryo-ET mapping indicates that NR and CR are both very similarly arranged (von Appen et al. 2015). That is contrary to the structure of the frog NPC, which especially in the NR region differs considerably (Eibauer et al. 2015). However, whether and to what impact these variations rely on species, tissues or on the experimental setup has to be figured out clearly. Although the fact that the supplementation of the transcriptional blocker actinomycinD can substantially modify the NR region of the Xenopus NPC is possibly the strongest evidence that the elasticity of the NPC appears at the scaffold level (Eibauer et al. 2015).

The infrared spectroscopy (IR) also displays a permeable, lattice-like density similar to that of NR and CR, but an alignment of the available crystal structures of the Nic96 complex is not yet feasible. A reduced infrared region resolution can be attributed to an increased inherent flexibility of the nucleoporin compound (von Appen et al. 2015; Eibauer et al. 2015), which implies that these entities either have a considerable amount of flexibility or that they do not conform to an eightfold symmetry.

10.6.5 Evolutionary Conserved Nuclear Pore Complex Structure

The phylogenetic analysis indicates the occurrence of a primal NPC during the period of transition from the first to the last common eukaryotic ancestors around 1.5–2.5 billion years ago (Devos et al. 2014; Koumandou et al. 2013). Nucleoporins are considered to be one of the most diverse proteins inside the cell, but the structural maintenance over the total eukaryotic species range is noteworthy (Brohawn et al. 2008; Devos et al. 2004). Various types contain widely overlapping amounts of nucleoporins (Cronshaw et al. 2002; Degrasse et al. 2009; Rout et al. 2000; Tamura et al. 2010), but numerous variations also exist, which are by no means fully comprehended. The duplication of the entire genome, for example, has produced the scaffold Nup paralog pairs Nup157/170 and Nup53/59 in *Saccharomyces cerevisiae*. A series of organisms is missing Nup37 and/or Nup43, with as yet unidentified consequences in structural or functional terms (Neumann et al. 2010). For instance, the key FG nucleoporin Nup145N possesses the paralogs Nup100 and Nup116 in *S. cerevisiae*; whereas, they are missing in the thermophile fungus *Chaetomium thermophilum* (Amlacher et al. 2011). Moreover, *Schizosaccharomyces pombe* harbors paralogs for Nup133 and Nic96 (Asakawa et al. 2014). *Caenorhabditis elegans* have no Nup188 homolog (Galy et al. 2003).

An opportunity to contemplate these variations is to specify that the basic structure and function of the NPCs are preserved and that these specific characteristics can modify the NPCs in various manners that are influenced or imposed by the particular way of life of the individual species. The fact that it is a module-based construction increases the complexity even more, as type-specific Nups may only work in a limited subset of NPCs.

Overall, the NPC is definitely not a solid monolithic structure. Although it is not a new approach, the flexibilities within the NPCs can be more precisely refined and studied. Heterogeneity becomes apparent in the flexibility of both composition and conformation. It is essential now to take into account the temporal scales and spatial characteristics of these alterations in order to arrive at sensible functional deductions.

10.6.6 How Can the Architecture Determine the Function?

Once the key elements of the NPC structure have been delineated, the next questions that need to be dealt with in order to gain an insight into how this preserved, broadly stable architecture sustains the dynamic nucleus–cytoplasmic interaction.

10.6.6.1 Flexibility of the NPC Scaffold

In cryo-ET investigations (Bui et al. 2013; Eibauer et al. 2015), the unstructured, circular central opening of the NPC has a diameter of approximately 40 nm, which is in line with earlier measurements of the maximum load size (Pante and Kann 2002). A typical load is considerably smaller compared to this. During the import of protein, a compound is built between an NTR and the imported cargo prior to its passage through the main channel. For the export of proteins, NTR, cargo and RanGTP constitute a ternary protein transport complex. One typical export complex is approximately 12 nm in diameter. Therefore, the main channel is sufficiently broad to receive several cargo NTR complexes traveling through simultaneously. The physical results of this analysis are in agreement with transport rate measurements of NPCs, which indicate that several transport incidents take place concomitantly (Ribbeck and Görlich 2001). However, when large charges, such as pre-40S and pre-60S (around 25 nm diameter) ribosomal subunits on their way to the cytoplasm, are involved, the main channel is broad enough to avoid the need to expand the main scaffold. For the biggest known load, the virus capsids with a diameter of up to approximately 40 nm, the transport can be carried out in goose march. The entry of viruses into the nucleus is not well understood at the molecular scale, and various viruses probably interfere with the NPC in a variety of manners (Fay and Pante 2015; Matreyek and Engelman 2013). It is not clear whether the flexibility of the NPC framework could be relevant for the entry of viruses, because this specific research branch will require more scientific effort in the future.

10.6.6.2 What About the High Specific Cargo Transport by NPC Combined with Rapid Transport Speeds?

Another scenario is possible for proteins targeted at the inner nuclear membrane (INM). Most of these membrane proteins are incorporated co-translationally into the endoplasmic reticulum and therefore have to pass through the NPC some way to enter the INM. Ideally, this may be achieved either by a lateral aperture of the NPC, which allows access to the center channel, or through separate, peripheral channels adjacent to the pore membrane. In fact, peripheral channels of adequate size are found in tomographic NPC networks (Maimon et al. 2012). Two main arguments for not including the central channel in the targeting process to the IMN are as follows: First of all, a lateral opening of the NPC scaffolding cannot be considered a simple operation. Such a scenario seems rather unlikely if one takes into account that different scaffold nucleoporins need to be used and coordinated. In second place, the nucleoplasmic domain of the majority of INM proteins imposes a size limitation of about 60 kDa or 5 nm in diameter, which is equivalent to crossing narrowed peripheral channels (Soullam and Worman 1995). A model is being developed to assist recent studies on INM transport, whereby membrane proteins of suitable size can diffuse between ER and the adjacent INM, but not stay at the INM except when they carry a retention signal of a certain type (diffusion retention model). That process is completely independent and distinct from the classic NLS-driven transport (Boni et al. 2015; Ungricht and Kutay 2015; Ungricht et al. 2015). In a reconstruction experiment, it was revealed that Nup188 exerts a direct impact on the transit of INM proteins through the NPC (Theerthagiri et al. 2010).

Newer cryo-ET investigations all agree with narrowed peripheral channels, although their biochemical properties are widely not well understood. The flexibility of the NPC scaffold seems to be a key element in the regulation of proteins dedicated to the INM, for instance by taking specimens of specific biophysical characteristics, but this is a very active field of research. While there is increasing scientific substantiation of the diffusion retention model, it is worth noting that there is also empirical evidence that some INM proteins are carried in a strictly NTR-facilitated fashion, thought to be transported through the central channel (King et al. 2006; Kralt et al. 2015; Meinema et al. 2011). Since a lateral opening of the NPC, as mentioned above, is improbable, the alignment of these INM proteins can be most simply accounted for by their post-translational insertion into the INM.

The flexibility of the NPC scaffold is surely the most crucial factor in the assembling of the NPC. Although this is still in its infancy at the molecular level, the task of tightening up to one hundred subcomplexes in a membrane pore is a huge task (Hetzer and Wente 2009; Rothballer and Kutay 2013a, b). In the more basic membrane layer systems that are concerned with the transport of vesicles, such flexibility is particularly evident in the generation of a protein layer film over a membrane that quickly switches from flat to spherical. The lattice-like mantle, which consists of edges and vertices, must be conformationally flexible to accommodate the emerging membrane geometry (Kirchhausen et al. 2014; Zanetti et al. 2012). When NPCs are installed, it is not yet known whether scaffold Nups are participating in the shaping of the membrane or whether they will be enlisted at a later point in time just to strengthen the strongly bent pore membrane. Pores can be joint-together post-mitotic and in the interphase. It is not clear at this stage if there are distinctions between the two (Doucet and Hetzer 2010; Schooley et al. 2012). Irrespective of this, the merging of such a large number of proteins into a pre-defined framework is a logistical hurdle, even though a chronological sequence nevertheless is available and has been examined (Dultz and Ellenberg 2010). Due to the net-like character of the NPC, as demonstrated in cryo-ET rebuilding, it is fairly straightforward to imagine how the bending and hinging of scaffold Nups can be crucial to the construction of a completely functional NPC.

Although the manner in which flexibility supports the entire channel scaffolding framework architecture is being elaborated, it is recognized that it is of critical concern for cargo transport. The central channel is packed with random and highly flexible fibrous FG-repeats with extensions of about 10 nucleoporins (Cronshaw et al. 2002; Denning et al. 2003; Patel et al. 2007; Rout et al. 2000). These flexible dilatations constitute the permeability barrier, which prevents the free interchange of the nucleus content with the cytoplasm. The barrier provides the transport mechanism with specific features that enable a selective transport. Firstly, it enables the diffusion-driven transport of molecules with a size of up to 40 kDa, which corresponds to approximately 5 nm. Secondly, it is recognized that larger loads pass through the NPC in a nuclear transport receptor (NTR)-facilitated fashion. NTRs identify cargo by nuclear localization signals (NLSs) or nuclear export sequences (NESs) and exhibit an innate affinity for FG repeats (Cook and Conti 2010). NTRs can therefore "merge" with the FG network to tap into the central channel (Hülsmann et al. 2012; Schmidt and Görlich 2015). Finally, it is essential that the mutual interactions between NTR and FG are both low and transient, so that the cargo travels quickly and does not get trapped within the pore. This characteristic has recently been proven in two completely separate approaches (Hough et al. 2015; Milles et al. 2015).

Several models have been suggested for the physical foundation of how the FG repeats fulfill these demands. A common conception of the NBC barrier consists in the fact that the FG repeats, which include dilatations, create an interlinked hydrogel where the FG elements establish hydrophobic interfaces in a normally hydrophilic milieu (Ribbeck and Görlich 2001). Polymer "brushes" (Lim et al. 2007) and entropic boundaries (Rout et al. 2000; Hough et al. 2015) were also proposed as concepts to explain the selective filtering. With the progress of biophysical technologies, these models can be checked with new information to verify the disparities among them.

A key issue is the question: Whether there is a clear organization of the FG network despite its intrinsic flexibility? FG repeating areas are not equivalent; however, they can be identified by several variables, emanating from the surrounding sequence, cohesion, O-GlcNac modification, anchoring (grafting) distance, charge and so on (Ader et al. 2010; Denning et al. 2003; Labokha et al. 2013; Peleg and Lim 2010). Moreover, it can be seen that specific FG regions are more essential than others for certain characteristics, both at the operational and physiological scale, such as transportation or survivability (Hülsmann et al. 2012; Strawn et al. 2004). Based on the results of studies on the crucial Nsp1 complex, it is now clear that there is a separate organization within the FG network. In the solution, the three central FG Nups, Nsp1, Nup57 and Nup49, form a 1:1:1 complex (Ulrich et al. 2014) and are connected by a characteristic trimeric coiled coil element, which anchors this complex specifically to the NPC scaffold via Nic96. The specific configuration is preserved, a fact proven experimentally by the discovery of closely resembling topologies in the fungal and metazoic complex crystal structures (Chug et al. 2015; Stuwe et al. 2015a). In conclusion, these data radically contradict the concept of a pore opening mechanism eased by putative promoting agents of the Nsp1 complex, as originally recommended (Melcák et al. 2007) and subsequently elaborated (Koh and Blobel 2015; Solmaz et al. 2013). In addition, the most recent data on the Nsp1 cluster also indicate that the FG network justifies a more in-depth investigation that may reveal novel transport functions.

10.6.7 Is There a Functional Overlap Between Nups and NTRs?

Whereas up to now, the flexibility of the NPC scaffold and the FG hydrogel have been discussed as distinct subjects and with disconnected functional implications, and the real situation is perhaps more sophisticated. However, as mentioned above, FG repeats mainly attach to each other and also to NTRs. Several framework Nups have also been documented that link FG repeats, particularly Nup188 and Nup192, which are similar in structure to NTRs (Andersen et al. 2013; Sampathkumar et al. 2013; Schrader et al. 2008; Stuwe et al. 2014). Hence, the concept was introduced that specific scaffold Nups and NTRs might perform redundant functions (Devos et al. 2006). FG repeats for this reason may potentially be implicated in the construction of the NPC framework. If that were the case, it seems probable that this will provide flexible features within the NPC. Conversely, NTRs could possibly work alongside as framework modules. To promote this principle, several NTRs join forces with the NPC, even if they are not involved in cargo transport. The transport mechanisms are regulated by these "empty" NTRs, although mechanistically there is no proper knowledge of this (Lowe et al. 2015). It might be a straightforward occlusal phenomenon, or it might be able to modify the FG binding, or even affect the NVK framework on its own.

As the primary channel for information sharing between the cell nucleus and the cytoplasm, the development of a deep insight into the structure and functioning of the NPC is of utmost priority. In recent decades, remarkable advances have been achieved in gaining a better grasp of the structure and function of the NPC. The original idea of a monolithic construction has been questioned, and variations within the pore arrangement and components have been characterized in increasing depth over the years. It is obvious at this stage that the NPC constitutes a dynamic conglomeration which displays a distinct conformational flexibility, with tempting signs of compositorial flexibility also emerging. The flexibility of the central FG hydrogel is mandatory, but why the scaffold nucleoporins must be flexible still depends on a thorough and compelling molecular comprehension. Finding answers to these outstanding questions demands a combined effort by scientists with different scientific backgrounds, strategies and approaches.

10.7 Transport System of the Nucleus

The transport of macromolecules into and out of the nucleus is crucial for the exchange of macromolecules between the cytoplasm and the nucleus and for the overall function of eukaryotic cells. Nuclear pore complexes serve as transport channels for large macromolecules and therefore are located throughout the nuclear envelope to facilitate their bidirectional transport between the nucleus and cytoplasm. The effect of macromolecular trafficking goes beyond the nuclear pore complex toward transcription and RNA processing within the nucleus and signaling pathways reaching into and beyond the cytoplasm (Tran et al. 2014). For understanding the nucleocytoplasmic trafficking, the following questions need to be answered: How work transport receptors and cargoes together facilitate the transport of cargos through the nuclear pore complex? What type of signaling pathways affect the transport pathways? How is the interaction between the nuclear envelope, nuclear pore complexes and transport pathways coordinated? What type of linkages exist between transport pathways and human diseases? What will be the focus of future research in nucleocytoplasmic trafficking?

The characteristic element of all eukaryotic cells is the existence of a nucleus, which is limited by a double membrane and envelops the genetic material in order to physically constrain the process of transcription from the translational systems, which functions in the cytoplasm. The spatial compartmentalization of these cells enables them to selectively and efficiently transport through the nuclear pore complex (NPC), which represents a large macromolecular structural composite of proteins that acts as a selective doorway for the transport through the nuclear envelope (NE). Even though the key determinants necessary for the recognition and transport of macromolecules to and from the nucleus by NPCs are well over a decade old, there are still significant deficiencies in the mechanisms underpinning this targeted transit.

What are the new advances in this field of nuclear research with a focus on transport mechanisms between the cytoplasm and the nucleus? There is still research needed that addresses the transport of substances through the NPC, about the precise assembly of the complicated structure of the NPC and the connection between the nucleus and the cytoskeleton or even with other basic cellular processes. The combination of different disciplines, including cell biology, biophysics, genetics, structural biology and molecular biology, seems to be fruitful for future discoveries int his field. There is some agreement on the transport mechanisms for cargos in and out of the nucleus via NPCs; however, there are still additional stuff, such as functional pathways that involve the identification of NPC interacting proteins and the analysis of the entire NPC. The high complexity of the entire NPC leads to the suggestion that the transport process may be the key element for the regulatory pathways at different levels, which seem to be of primary interest in enlightening the regulatory network around the NPCs selective transport mechanisms in and out of the nucleus.

10.7.1 Nuclear Pore Complex (NPC)

The NE of a eukaryotic cell consists of outer nuclear membrane (ONM) and inner nuclear membrane (INM), which are both surrounding the so-termed luminal space. Since the NPCs are spread all over the NE, they act as connection between the OMN and the IMN by directly fusing them together (Raices and D'Angelo 2012; Grossman et al. 2012). In specific detail, the NPCs are built by a few different protein types (about 30), arranged in an eightfold rotationally symmetric repeated pattern, with an additional double axis of symmetry in the plane of the NE for numerous compounds (Grossman et al. 2012; Wente and Rout 2010). Investigations into the mechanism by which cargoes pass through the NPC are based on compiling a comprehensive inventory of the fragments of the conglomerate that make up the NPC and a picture of how these fragments are put together to complete this extraordinary macromolecular construct.

10.7.2 The Structural Assembly of the Nuclear Pore Complex

The NPC consists of a central core, conveying the passage via the NE, and peripheral elements, consisting of cytoplasmic filaments and a nuclear cage, conveying the docking of transport receptors and cargo. The proteins composing the NPCs, collectively referred to as nucleoporins or Nups, embed the NPCs in the NE and offer transport domains for interaction processes. Since many researchers have focused on the investigation of single molecules of the NPC, there have been some comprehensive studies (Fernandez-Martinez and Rout 2012; Rout et al. 2000; Alber et al. 2007). The overall constituents of the NPC have been revealed (Alber et al. 2007). Moving on, therefore, to refine the way in which specific Nups are organized within the larger structure of the NPC (Fig. 10.5). A detail analysis of mutations in budding yeast in distinct subcomponents of the NPC will shed light on their functions. Combining electron microscopy, protein domain mapping data and yeast genetics, it was shown that mutations in genes encoding components of the Y-shaped complex composed of Nup84, Nup85, Nup120, Nup133, Nup145C, Sec13 and Seh1 in budding yeast (Fernandez-Martinez et al. 2012; Siniossoglou et al. 1996), have varying impacts on mRNA export and distribution of NPC. Such functional investigations remain to identify the roles of different subcomplexes within the NPC, thereby improving and enhancing the knowledge of the assembly and function of the NPC.

One other way to characterize the entire architecture of the NPC is to map the high-resolution three-dimensional structures of individual domains of Nups and subsequently remap the derived structures to lower-resolution models, predominantly from electron microscopy. The specific knowledge of structural insights would help us to reveal the function and the dynamical properties of the larger complex (Brohawn et al. 2009). In fact, there have been structural analyses performed on Nups may lead to insights into the entire organization and the overall function of the NPC. The mammalian FG Nups, Nup62, Nup54 and Nup58, have been investigated by employing X-ray crystallography. It has been demonstrated that the alpha-helical architecture of such proteins, which may suggest that the specific Nups have the ability to bend and slide, thereby easing large diameter alterations along the central transport channel of the NPC (Solmaz et al. 2013). Moreover, the structure of several Nups of the thermophilic fungi, Chaetomium thermophilum, has been revealed.

Structural investigations of the N-terminal domain of Nup192 indicate two rigid half-molecules linked by a flexible hinging bridge (Stuwe et al. 2014). In line with this, it has been found that the two regions consist of HEAT and ARM repeat motifs (Stuwe et al. 2014; Sampathkumar et al. 2013; Andersen et al. 2013), exhibiting structural resemblance to transport receptors of the import/karyopherin family (Conti et al. 1998, 2006; Vetter et al. 1999; Cingolani et al. 1999). Complementary biochemical experiments helped to determine interaction regimes with two other compounds of the NPC, Nup53 and Nic96 (Stuwe et al. 2014). In this way, structural analyses provide insights into how separate compounds of the NPC can be joined together to build this tremendous macromolecular channel.

The efficient assembly of the high-resolution crystal structures of single Nup domains demands progressively more refined pictures of Nup subcomplexes and intact NPCs, especially by electron microscopy (Adams and Wente 2013). The implementation of new technologies in order to rebuild the entire structural architecture of the human NPC seems to be promising. In fact, the entire structure of the human Y-shaped assembly, which is referred to as the Nup107 subcomplex and the Saccharomyces cerevisiae analog structure represents the Nup84 complex, has been revealed by employing single-particle electron tomography (Bui et al. 2013). In addition, the model of the human NPC framework has been upgraded in its resolution, which is now 3.2 nm, when cryo-electron tomography is coupled with advanced image analysis using eightfold symmetry of the NPC (Beck et al. 2007), so that they can position the Y-shaped complex inside the NPC framework utilizing a template matching strategy (Bui et al. 2013). This model was additionally evaluated by cross-linking mass spectrometry (Bui et al. 2013). Through the combination of these efficient techniques, the results of this study conclude that the stoichiometry of the Y-shaped complex inside the NPCs in human cells can vary compared to the Nup84 complex of budding yeast, which offers a reasonable interpretation of the cryptic variation in the mass of NPCs in eukaryotic cells even though the Nup components are quite similar (Bui et al. 2013; Alber et al. 2007). Altogether, these studies show a strikingly similar overall organization of NPCs in the different spheres of life (Fiserova and Goldberg 2010) as the results of the comparison of microscopic field emission images of plant and fungal NPCs show.

Since all these multi-layered efforts are constantly advancing and converging, the high-resolution structure of the NPC is slowly moving to the center of attention (Adams and Wente 2013; Beck and Glavy 2014). Researchers in this area are still shifting the frontiers of technology to identify the complex macromolecular architecture that permits the NPC to act both as an actual physical barrier and as an efficient transit channel that can enable the fast and targeted transport of numerous macromolecular loads.

10.7.3 Transport Through the Nuclear Pore Complex

The movement through the NPC comprises a sequence of operations including docking at the perimeter of the NPC, transit through the channel and ultimately the discharge from the NPC to the target compartment (Wente and Rout 2010). Except for small molecules, metabolites and proteins smaller as about 40 kDa, transport across the NPC depends on both receptors and energy. Nuclear transport receptors, commonly referred to as karyopherins or import/export compounds, promote bidirectional transportation for the overwhelming majority of nucleocytoplasmic transport occurrences due to their capacity to interfere with three different protein counterparts: transport cargo, FG Nups and the small GTPase, Ran (Xu et al. 2010). Ran ensures the alignment of the transport process by managing the interface between the transport receptors and the specific cargo.

Cargo proteins comprise specific transport signal sequences, which are identified by distinct transport receptors (Xu et al. 2010; Guttler and Gorlich 2011). The best known and consequently best characterized transport patterns are the classical nuclear localization signal (cNLS), which is composed of a series of basic moieties (Lange et al. 2007) and the classical nuclear export signal (cNES), which is made of separated leucine and other hydrophobic moieties (Kutay and Guttinger 2005). Cargoes displaying a cNLS motif are captured by an importin/karyopherin α adapter protein to build a complex with the import karyopherin receptor, referred to as importin/karyopherin β. The assembly of the trimeric import cargo complex inside the cytoplasm occurs in the absence of RanGTP and is broken down when RanGTP interacts with importin/karyopherin β within the nucleus (Lee et al. 2005). The cNLS import route is nevertheless a specialized import mechanism insofar as it employs an adapter, importin α. In fact, most transport products are probably directly identified by a specific transport receptor from the importin/cytopherin β family. Export complexes, for instance the cNES-dependent path utilizing the importin/caryopherin β export receptor termed CRM1 (Ossareh-Nazari et al. 1997; Stade et al. 1997; Fornerod et al. 1997), constitute obligatory trimeric complexes composed of the transport receptor, cargo and RanGTP. This export complex is broken down in the cytoplasm at the moment of Ran-driven hydrolysis of GTP to GDP. Consistent with this phenomenon, RanGTP levels are raised in the cell nucleus, when export complexes assemble and import complexes disassemble, because of the nuclear localization of the RanGTP exchange factor (GEF). The nuclear localization of GEF facilitates the generation of RanGTP from RanGDP, and subsequently the cytoplasmic localization of the Ran GTPase activation protein (GAP), promoting the GTP hydrolysis activity of Ran (Quimby and Dasso 2003). Thus, the GTP-bound status of Ran determines the functional requirements in a specific compartment for the loading and delivery of the cargo.

The classical import and export routes provide a gateway for the transit of cargo through the NPCs. Nevertheless, more than 20 transport receptors exist in mammals and 14 in budding yeast (Marelli et al. 2001). The lack of a "classical" transport signature does not exclude the transport receptor-switched passage of a cargo into or out of the nucleus. Several of the signals recognized by the other members of the transport receptor family have been identified by structural investigations of receptor-cargo complexes (Xu et al. 2010; Guttler and Gorlich 2011). A transport signal resulting from such an attempt is referred to as proline-tyrosine (PY)-NLS, which can be detected in a series of RNA-binding proteins (Lee et al. 2006). Of note, PY-NLS is not so closely linked to the primary amino acid sequence, but rather to general characteristics of the structure, rendering it challenging to detect this motif by basic sequence analysis (Suel et al. 2008). A continuing aim for this area is to further specify multiple cargo proteins interfering with specific transport receptors and to gain insights into detection mechanisms at the atomic level. These approaches may result in the design of tools to define transport pathways/receptors for certain cargoes, currently only available for the classical nuclear import and export routes.

10.7.4 The Permeability Barrier of the Nuclear Pore Complex

An important outstanding question in the field of nuclear transport concerns the biochemical foundation of the selective transport wall inside the NPC. The NPC facilitates two basic functions: Firstly, it prevents molecules with a diameter of more than approximately 5 nm from traveling freely through the channel; secondly, it simultaneously assists the signal-dependent transit of larger cargo molecules through their interface with transport receptors (Wente and Rout 2010; Mohr et al. 2009). These efforts focus on the phenylalanine-glycine (FG) repeat Nups or FG Nups (Terry and Wente 2009), which encircle the inner part of the NPC. These natively unfolded FG domains (Patel et al. 2007; Denning et al. 2003) are both setting the effective mesh size of the NPCs by intra- and intermolecular interactions and act as interfacing sites for soluble transport receptor-cargo composites to aid their translocation through the NPCs (Frey et al. 2006; Tu et al. 2013; Tetenbaum-Novatt et al. 2012).

A number of different models have been elaborated for the physical characteristics of the FG network, which offer a rich source of debate (Tu et al. 2013; Hülsmann et al. 2012; Moussavi-Baygi et al. 2011; Weis 2007).

The selective phase model is based on two principles (Ribbeck and Görlich 2001, 2002). In the first place, barrier-forming FG repeating domains (especially the ones of Metazoan Nup98 and its yeast paralogs Nup100 and 116) enter into multivalent coherent linkages and build a filterable FG hydrogel whose mesh size places a maximum limit on passive NPC transit. In the second place, FG-repeating interactions detach locally and transiently when a transport receptor links the corresponding FG motifs, causing the transport receptor to "dissolve" the gel and consequently enable the transit of the hydrogel. Using the powerful Xenopus and bud yeast model organisms, evidence has been obtained that cohesive FG domains are able to generate an FG hydrogel with NPC-like selectivity, eliminating inert macromolecules $\geq$ 5 nm, but permitting an influx of the same macromolecule attached to a transport receptor over 1000 times quicker (Hülsmann et al. 2012; Labokha et al. 2013). It has similarly been proven that substitution of the cohesive Nup98/Nup100/Nup116 FG domains by non-cohesive FG domains is lethal in *S. cerevisiae* and perturbs the selectivity boundary of Xenopus NPCs.

One other model, referred to as the Kap-centered model (Schoch et al. 2012; Kapinos et al. 2014), suggested that Kaps (transport receptors) act as integral, possibly regulatory components of the NPC boundary mechanism. This model is based on the concept that each FG Nup is capable of attaching multiple Kaps, whereby it is anticipated that the charge-carrying Kaps may temporarily interfere with the FG Nups previously filled with Kaps as they convey the cargo across the central channel (Schoch et al. 2012; Kapinos et al. 2014). A model of this kind could help to elucidate the fast transit time of transport receptor-cargo complexes across the NPC. It was also demonstrated how to biomimetically apply this so-named "dirty Velcro" effect to adjust the two-dimensional diffusion with Kap-functionalized 1 μm diameter-sized colloidal beads on glass slides that were coated with FG Nup molecular brushes (Schleicher et al. 2014).

The extension of the question to determine how very large cargo molecules cross the NPC is essential because fully composed ribosomal subunits can leave the nucleus after assembly. This question has been answered by the combination of mathematical modeling and single-molecule fluorescence microscopy. In fact, there is some support that large cargoes necessitate the attachment of multiple transport receptors in order to pass the NPC (Tu et al. 2013). This is a pathway where multivalent FG-repeat interferences with transport receptors are indispensable. As advances are achieved in gaining an insight into how FG Nups can act simultaneously as an efficient border barrier and as a vehicle for high-speed transit, more research is needed to finally define the exact biophysical properties that enable such diametrically contrasting functions.

While the FG repeats form the permeability barricade inside the NPCs, the directional dependence of the transport is assumed to occur exclusively in the docking and release phases of the receptor-cargo complexes with peripheral elements of the NPCs (Stewart 2010). In contrast to the prevailing view, there is some evidence for a Ran-driven phase for the receptor attachment with FG repeats. After the first analyses, the transport receptor, which is importin β, can connect to the NPC, but cannot be relocated in the lack of Ran. Fluorescence recovery after photobleaching experiments (FRAP) with permeabilized HeLa cells displayed the availability of approximately 20 RanGTP-sensitive importin β binding sites, several of them located within the FG-repeat Nup, Nup153. Beyond that, the importin β, which is linked to the FG-repeat region of Nup153 (Nup153FG), is reliant on RanGTP, which is demonstrated by the utilization of the GTPase-deficient Ran-Q69L variant. A Ran-dependent gateway via the NPC will rectify the fact that the NPC FG-rich nucleus of the is not selective or directed. Whether this scheme is universally suitable for different transport receptors and FG Nups or whether this Ran-dependent binding site is unique to Nup153 has yet to be clarified. In addition to these investigations, a structural assessment of the Nup153FG importin β complex was performed through the combined utilization of single-molecule FRET (sm-FRET) (Tyagi et al. 2014) and simulations of molecular dynamics. These analyses indicate that the Nup153FG domain has a crumbled microstructure, in opposition to the enlarged microstructure commonly ascribed to the natively unfolded FG domains (Denning et al. 2003). In addition, the adjustment of the underlying framework was not sensitive to the linkage of Importin β. An intriguing debate arises as to whether the RanGTP bond could foster a structural transformation to a naturally evolved condition, in agreement with similar FG repeating areas. Collectively, these investigations illuminate the mechanisms that guarantee fast but nevertheless selected passage across the NPCs, an issue that has been at the cutting edge of the field for a number of years.

10.7.5 Variants of Nuclear Pore Complexes

A persistent problem in the area of nucleocytoplasmic transport is whether all NPCs are equal throughout a specific organism and are therefore in an equal position to convey the full range of nuclear transport incidents. Traditional experiments in the 1980s showed that gold particles of various sizes were moved to and from a single NPC at the very same time (Feldherr and Akin 1997). This attempt has been referred to as proof that all NPCs are capable of both import and export, and occasionally broadened to give the impression that each NPC is equivalent throughout a specific organism. Newer findings, meanwhile, call this view in doubt. Several researchers demonstrated that NPCs can vary in specific tissues, organisms or states of development. Probably, the most remarkable proof of the diversity of NPCs has been reported in Tetrahymena. Tetrahymena comprises two functionally and structurally dissimilar nuclei referred to as macronuclei (MAC) and micronuclei (MIC) (Orias et al. 2011). A biochemical approach has been employed to isolate the two nuclei and reveal the Nups, which are specific for each form (Iwamoto et al. 2009). It has been seen that the isolated NPCs of the MAC and MIC nuclei share common core Nups and exhibit a uniform global architecture. Nevertheless, specific characteristics were identified, such as at least four Nup98 variants featuring two different types of Nup98 in each type of nucleus (Iwamoto et al. 2009, 2010). These compositional differences may account for several of the functional variations between these nuclei, which differ not merely in size from each other, but also in transcriptional activity and cell cycle management, even though they are located in the identical organism.

The Nup98 has been revealed as a highly interesting protein. Systematic endeavors to identify the function of Arabidopsis Nups through insertional mutagenesis started to uncover those Nups that are crucial in this plant model and those that perform a pivotal role in certain processes (Parry 2013). Genetic analyses indicate two different Nup98 genes that code for proteins which appear to be implicated in various functions. Accordingly, further research will be necessary to identify the functions of these dual Nup98 proteins. Two contributions focused on the closely related question of whether individual nuclei can perform various functions inside a specific tissue. There are studies that investigate the expression and function of importin/karyopherin α family members in the course of spermatogenesis (Miyamoto et al. 2013). Indeed, these experiments showed clearly distinguishable functions for several family members in the course of differentiation and demonstrated that individual transport receptors can endow various transport characteristics. Differences in the expression rates of import receptors further underpin a model in which specific functions for import/karyopherin α family members exist. For instance, the values for Importin α2 are more than two times higher in spermatocytes than in spermatids; whereas, no major deviation was noted for Importin α4 (Arjomand et al. 2014) pointing to a model in which the requirements for these receptors are differentiated at certain phases of development. When investigating the protein import mechanisms into the nucleus of a distinct cell type, the import of proteins into the individual nuclei of multi-nucleated muscle cells has been chosen. The findings of this analysis prove that the diverse nuclei located within the multi-nucleated muscle cells can have varying import skills regardless of a common cytoplasm inside the multi-nucleated muscle fiber. These experiments needed the usage of in vitro transport assays, in which all soluble factors are added exogenous into the cytosol (Adam et al. 1991), hence the differences in import capacities can be directed connected to the nuclei, which means presumably with the NPCs, rather than with the soluble transport receptors. Collectively, these investigations offer strong arguments that variations in nuclear transport can occur as a result of the regulation of both transport receptors and NPCs. The difficulty is to first identify the mechanisms behind this regulation and subsequently to comprehend how this regulation can affect the tissue-specific function(s).

Such work demonstrates that even within a specific organism or cell type, not all NPCs have the same equivalence. Additional investigations in such systems can give a more precise idea of the control of the transport mechanisms. When more specific information on the permeability barrier properties of NPCs is available, there needs to be further work on defining regulatory paths that modify the properties of the NPCs in different environments.

10.7.6 Protein Translocation to the Inner Nuclear Membrane (INM)

The lack of a molecular mechanism for the integral transfer of nuclear membrane proteins to the INM constitutes a major weak point in the area of trafficking. In fact, there exist two major hypotheses for the INM transport (Antonin et al. 2011): Firstly, the diffusion retention model (Worman and Courvalin 2000) and secondly, the Kap-dependent model (King et al. 2006). In order to figure out the mechanism for INM transport, a strategy for determining the transport rate of an integral membrane protein reporter has been reported (Ungricht et al. 2016). Moreover, it has been explored that the targeting of the transport to INM depends on GTP (Pawar et al. 2017). In specific detail, when the in vitro reconstitution and in vivo analysis of INM targeting have been investigated, it has been revealed that Atlastins (ATL), which are ER originated membrane-bound GTPases, maintain the efficient targeting of proteins toward the INM, as they constantly act in the maintenance of the topology of the entire ER (Pawar et al. 2017). In line with this, it has turned out that the assembly of COPII vesicles is pronouncedly reduced in the periphery of ATL-deficient cells (Niu et al. 2019). The export of ER cargo proteins is not functioning, but the launch of the ER exit region is not affected, although many of the sites do not implement COPII subunits. Hence, the efficiency of cargo enveloping with COPII vesicles is pronouncedly decreased in cells that are deficient of ATLs, or even when the ER seems to be transiently fragmented (Niu et al. 2019).

When investigating the IMN transport pathway, it is noteworthy to state that the INM transport relies on the size of the cargo. Hence, proteins with extra-luminal domains larger than 60 kDa cannot be transported via NPCs toward the INM. This property was utilized to decouple the membrane incorporation in the ER from the transfer to the INM. The reporter proteins were then converted into transport proteins by reduction by protease cleavage within semipermeabilised cells, which in turn helped to measure the INM transport kinetics. The transport of two distinct reporter proteins deduced from the INM proteins, the lamin B receptor (LBR) and the SUN2 (Ollins et al. 2010; Luxton et al. 2011) have been reproduced by the use of a HeLa cell lysate and an energy recovery system. However, the INM reporter proteins shift into the INM in the complete lack of karyopherins, indicating that energy is needed for a poorly comprehended stage in this mechanism, possibly to assist the movement of the cargo from its synthesis location to the external part of the NPC. The specific RNAi of Nups demonstrated a "leakage" of the large LBR reporter toward the INM, which is equivalent to the transport of membrane proteins that are hampered by the main NPC scaffold. Therefore, a mathematical model can be addressed in order to simulate the kinetic data. The preliminary model showed that the orientation of the LBR can basically be characterized by a diffusion retention mechanism. However, great efforts are still required to integrate the results of this work into a complete model for the aiming and positioning of INM proteins.

There is also support for a Kap-dependent model for the transport within the INM. In fact, there is some evidence for the direct interplay between the NLS of the yeast INM protein, Heh2 and the karyopherin Kap60 (which is referred to importin α in vertebrates). Interestingly, binding proof-of-concept studies showed a complex set of mutually exclusive interactors between the importin β-binding domain of Kap60, the NLS of Heh2 and the nucleoporin, Nup2 (Lokareddy et al. 2015). Beyond that, Nup2 is necessary for the correct positioning of both Heh2 in full length (King et al. 2006) and a membrane protein reporter containing Heh2-NLS to the INM. Upcoming investigations will show whether there is more than one mechanism for the relocation of integral membrane proteins to the INM or whether these alternative routes constitute deviating trafficking pathways among humans and fungi.

10.7.7 Roles of the Nuclear Envelope (NE), Lamins and LINC Complex

Inside the multi-cellular eukaryotes, the INM is connected to a specific type of intermediate filaments, which are termed lamins. The proteins that comprise the nuclear lamina, namely lamin A, B and C, were discovered nearly 35 years ago (Fawcett 1996). However, the specific detailed function(s) of the nuclear lamins are yet well known. Important evidence of their molecular involvement stems from the examination of human laminopathies, which suggest a role in cellular signaling processes and in the conservation of the nuclear architecture and placement of the NPCs within the NE (Schreiber and Kennedy 2013). There is some evidence that the lamin interacting protein Lem2 could be a major integrator of cell signaling pathways in myogenesis. Lem2 is among the four human LEM domain proteins and LAP2, and emerin and Man1 represent the remaining three others. The LEM proteins are assumed to perform several functions, among them the contribution to signal transmission within the cell (Huber et al. 2009). In addition, it has been shown that the knockdown of Lem2 in muscle cells leads to a MEK1-driven myogenic defect, which suggests a deficiency in the mitogen-activated protein kinase (MAPK) signaling pathway. Nevertheless, the lamin network has been linked to mTOR signaling pathway and to the regulation of fat level in Caenorhabditis elegans. This interplay seems to be facilitated by ATX-2, whose expression level correlates inversely with the body size (Kiehl et al. 2006). Besides, the function of ATX-2 demands lamin, S6K and an intact mTOR transmission path. Since the mTOR pathway is involved in the control of translation (Ma and Blenis 2009), these findings imply that lamins can transfer signals not only to the gene expression network inside the cell nucleus, but even to the translation network in the cytoplasm.

A cluster of transmembrane proteins known as LINC complex (linker of nucleoskeleton and cytoskeleton) stretches across the NE and links the cytoskeletal filaments to the laminar structure (Sosa et al. 2012). In detail, a supplementary genetic and cell biological investigation links plant-specific LINC proteins with male fertility in Arabidopsis. The LINC complex consists of SUN and KASH domain proteins, namely transmembrane proteins, that interfere with one another in the NE lumen and penetrate into the nucleus or cytoplasm (Rothballer et al. 2013; Zhou and Meier 2013). Arabidopsis possesses two plant-specific, KASH-like proteins termed WIP1 and WIP2, which are WPP-interacting proteins (Zhou et al. 2012). In accordance with a part in the architecture of the nuclear membrane, the loss of WIP proteins can lead to serious deficiencies in the nuclear morphology. In addition, two other NE proteins have been detected, referred to WIT1 and WIT2, which are WPP-interacting tail proteins. These WIT proteins are necessary for the nuclear motility of the pollen tube and the release of sperm (Zhao and Meier 2014). Since the simultaneous mutation of the two WIT and the two WIP genes leads to decreased seed production, which may find broad agricultural application. In addition, a conserved role of the LINC complexes in meiosis of eukaryotes (Stewart and Burke 2014) points to the fact that the findings from this exemplary system will improve our understanding of fertility disorders in humans.

10.7.8 Processing, Assembly and Export of Messenger RNA

Transport receptor-facilitated mechanisms drive the majority of soluble proteins, small RNAs and ribosomes across NPCs. In contrary, the export of messenger RNA (mRNA) does not rely on an exportin/karyopherin or Ran (Nakielny et al. 1997). Instead, mRNA export involves the enrollment of the mRNA export receptor Mex67 in fungi (Tap or NXF1 in multi-cellular eukaryotes) and a heterodimerisation partner, Mtr2 (p15 or NXT1), which is a heterodimerisation partner that binds to mRNA based on the proper pre-mRNA processing in the cell nucleus. In multi-cellular eukaryotes, enrollment of Mex67/Mtr2 is strongly associated with splicing (Valencia et al. 2008) and 5′ capping (Cheng et al. 2006) or specific enrollment sequences (Lei et al. 2013; Moore and Proudfoot 2009), whereas enrollment in budding yeast relies on several processing phases (Vitaliano-Prunier et al. 2012) which are assumed to be linked to a precise timing and 3′ end generation (Qu et al. 2009), while definition of mechanisms underpinning this coordinatorship continues to be an area of active research (Nino et al. 2013). As Mex67/Mtr2 cannot identify the RNA in a direct manner, the recruitment is facilitated by the supportive interplay between other RNA-binding proteins, such as Yra1 (Aly in humans) (Strasser and Hurt 2000; Zhou et al. 2000), Npl3 (Lee et al. 1996; Lund and Guthrie 2005) and Nab2 (Iglesias et al. 2010; Tran et al. 2007). There remain many questions about how the hiring and assembly of adapters are matched to the production of an export-competent mRNP. Whereas certain investigations reveal the recruitment of various factors at certain levels of maturity, such as splicing, abortion, polyadenylation, some other work indicates that Mex67 is co-transcriptionally enrolled (Gwizdek et al. 2006). Coupling mRNA maturation to nuclear transport offers a level of quality assurance of the gene expression process by avoiding the translation of possibly harmful gene products (Schmid and Jensen 2010; Fasken and Corbett 2005). After attracting Mex67/Mtr2 to a specific mRNA, the complex attaches to the NPC and then passes along the channel through the interaction of Mex67/Mtr2 with FG Nups (Strasser et al. 2000). The mRNA export receptor-cargo complex is broken down on the cytoplasmic surface of the NPC by the activity of ATPase and RNA helicase Dbp5, which induces the liberation of Mex67/Mtr2 from the mRNA, analogous to the function of Ran in transport receptor-dependent export (Lund and Guthrie 2005; Tran et al. 2007).

While the exact identification of the elements necessary for efficient mRNA export has been determined, the mechanisms for managing the choreography of the many consecutive steps in the creation of an export-competent mRNP are poorly defined. These steps comprise capping, pre mRNA splicing, termination and polyadenylation, all associated with the assembly of mRNP and its efficient export by NPC. The is some support that the RNA helicase Dbp2 in budding yeast (p68/DDX5 in humans) governs the RNA structure during the process of transcription. Dbp2 represents an active helicase in vitro, which can interact in a direct fashion with the transcribed chromatin (Cloutier et al. 2012; Ma et al. 2013). In addition, Dbp2 is necessary for the proper integration of Yra1 and Mex67 on mRNAs and for successful transcription termination. A model was introduced in which Dbp2 governs the folding of RNA in transcription to avoid the formation of RNA structures that are refractive to the formation of mRNP assembly, indicating that the RNA structure seems to bring an additional level of intricacy to the mRNA export mechanism.

In order to determine the guidance of key phases in the assembly of export-competent mRNPs and their interactions with the NPC upon the attachment and export events, the role of post-translational modifications has been explored (Hayakawa et al. 2012; Nino et al. 2012; Rouviere et al. 2013). Moreover, post-translational modifications of Nups have been analyzed. In fact, an interaction between ubiquitin and SUMO modification of Nup60 has been discovered, which is then connected to the progression of the cell cycle. Using a newly introduced spinach aptamer-based mRNA mapping method in live yeast (Paige et al. 2011), it was shown that certain post-translational changes in Nup60 are responsible for nuclear export of polarized mRNAs. It remains a goal of their research to fully comprehend how these alterations affect the function of the NPC and to specify the proportion of amended Nups at a given time.

In contrast to the investigation of the modification of Nups, the need for SUMO modification of RNA-binding proteins has been examined inside the expert complexes of mRNP. Budding yeast mutants of the SUMO protease Ulp1 (Li and Hochstrasser 1999) connected with a proteomic analysis have been examined to reveal alterations within mRNP complexes, under conditions where the SUMO alteration is blocked. Whereas several analyzed complexes were not concerned by this disorder, specific alterations in the association of the THO complex with mRNPs were recorded. Subsequent investigations highlighted a special class of stress response mRNAs that exhibited changed levels in the lack of SUMO modification (Bretes et al. 2014). These analyses provide the basis for a deeper insight into how SUMO modification and other post-translational modifications are able to regulate the biogenesis and export of selected categories of mRNAs. These kinds of experiments also raise the question whether certain RNA-binding proteins are capable of facilitating the export of groups of mRNAs or whether a central repertoire of mRNA export factors is sufficient for all mRNA transcripts. These represent future research areas in this field.

As mature mRNPs are built up within the nucleus, there is considerable concern about how to comprehend the export kinetics of these complexes. Individual transcript analyses of mRNA export have shown that the attachment and detachment stages for the export of mRNA are rate-restrictive (Grunwald et al. 2000). In addition, mRNAs are being scanned prior to export alongside the NE (Mor et al. 2010). However, these investigations were mostly conducted in the tissue culture systems of mammals. It seems to be suitable to perform genetics with single transcript investigations in *S. cerevisiae*. The focus lies on the analysis of the kinetics of the export of the mRNA in the budding yeast model system and the effect of mutations in different export factors on the scanning of the NE, the attachment, transmigration and subsequently the detachment. In order reach this goal, a microscopy procedure has been developed to investigate the motion of specific single transcripts in living yeast cells. Preliminary findings indicate that mRNAs are quickly exported following their association with NE and that these mRNAs have no detectable NE scanning in wild-type yeast cells. However, cells with mutations in the helicase Dbp5 (Snay-Hodge et al. 1998), or the cytoplasmic nucleoporin Nup159 (Kraemer et al. 1995), perform elevated NE-scan events. Such a finding indicates that scanning might be an indicator of delayed or inadequate mRNA export rates in these mutants. In addition, it has been stated that transcripts can be readmitted into the nucleus in dbp5 and nup159 mutant cells following their initial transport to the channel, which reinforces the idea that mRNAs can travel in two directions inside the NPC (Grunwald et al. 2000). The model that arises from this investigation is that the failure of Dbp5 to rearrange the exported mRNP at the cytoplasmic side of the NPC and eliminate export factors such as Mex67/Mtr2 permits reimportation. In the next step, the analysis is expanded to include other mRNA export mutants in order to create a spatiotemporal model for various functions of the mRNA export factors.

Cytoplasmic processes including translation monitoring and mRNA break-down can also be linked to the export of mRNA. The largely unknown Mex67/Mtr2-adaptor protein in budding yeast, referred to Mip6, has been investigated. Mip6 was initially detected as a Mex67 interacting protein

(MIP) using yeast two-hybrid screening (Segref et al. 1997). Nevertheless, the function of Mip6 in export of mRNA was not revealed. The Mip6 can shuttle between the nucleus and the cytoplasm in a Mex67-facilitated fashion, which indicates properly a function in the export of mRNA. However, the deletion of MIP6 did not result in the accumulation of poly(A) RNA inside the nucleus, which is incompatible with a universal mRNA export function. Instead, it has been detected that the Mip6 protein is accumulated in heat-induced stress granules due heat shock. Moreover, these results lead to the suggestion that Mip6 can possibly connect the export of mRNA and export the regulation of the translation.

These investigations emphasize several of the complexities in determining the molecular character of export-competent mRNPs, the selection for export of mature mRNPs by NPC and the remodeling required to adapt the exported mRNAs for productive survival inside the cytoplasm. Beyond these fundamental questions, there are several concerns, including whether all mRNAs are encapsulated and exported in the same manner and how these pathways are governed. These issues will be addressed with innovative strategies, as there is plenty to explore.

10.7.9 Does a Transcriptional Control at the Nuclear Pore Complex Exist?

Both NPC and NE combine with transcriptionally active and suppressed genes, indicating that regulatory elements at the nuclear periphery assist in regulating gene expression (Blobel 1985; Egecioglu and Brickner 2011). This control may be achieved by adjusting transcription factors and/or the structure of chromatin to modify the accessibility of RNA polymerase II (RNAP II). Investigations in a variety of model systems have reported the relocalization of inducible genes from the nucleoplasm into the nuclear periphery after transcriptional activation, indicating a relationship between the initiation of transcription and association with NPC (Wente and Burns 2014). It has been proposed in the hypothesis that this process is associated with the cycling SUMOylation. Employing the galactose-inducible GAL1 gene in budding yeast, it has been seen that the NPC-associated SUMO protease Ulp1 (Li and Hochstrasser 1999) is essential for the proper inducement of the GAL1 gene. The transcriptional co-repressor Ssn6 of GAL1 is one target of Ulp1 deSUMOylation, implying that the enrollment of the GAL1 gene into the NPC leads to optimized activation kinetics by promoting the deSUMOylation of chromatin-associated proteins (Texari et al. 2013). It has been revealed that both SUMOylation and deSUMOylation (additionally facilitated by Ulp1) events are required for the targeting of the NPC by the inducible gene INO1 in budding yeast. For instance, cells expressing an Ulp1 mutant, which cannot attach to NPCs revealed both a sharp decrease in INO1 enrollment to the NPCs and a decreased INO1 transcription rate. Finally, these investigations point to the fact that alterations in the SUMOylation status of chromatin-associated proteins control NPC aiming and the transcriptional level of initiatable genes.

A specific mechanism for transcriptional regulation in NPCs is the epigenetic memory manipulation of gene expression. The concept of transcriptional memory initially arose from the finding that genes are reactivated faster in subsequent laps of induction than in the first activation process (Light and Brickner 2013; Brickner and Walter 2004). This type of activity is linked to the movement of inducible genes toward the NPC, indicating that certain components within the chromatin could be linked to memory. The investigation of DNA sequences that are necessary for the inducible association of target genes with the NPC has been performed. These identified sequences, referred to "zip codes," are detected in the INO1, TSA2 and HSP104 genes in *S. cerevisiae* (Brickner et al. 2012). There is strong evidence that the epigenetic memory for the interferon γ gene can occur in human cells (Light et al. 2013). Employing a HeLa cells model system, it has been seen that the Interferon γ gene is more quickly reactivated after the first initial activation event, which is in line with investigations in budding yeast (Light et al. 2010). In line with this, the fast reactivation is connected to a retention of RNAP II at the gene beyond transcriptional shutdown. The RNAP II retention needs Nup98 and the H3K4me2-facilitated histone methylation mark, which is catalyzed by MLL1 (Set1 in *S. cerevisiae*). The epigenetic memory of INO1 in yeast also necessitates H3K4me2, the ortholog Nup98, Nup100 and the histone deacetylase Set3. Since the Set3-containing complex, Set3C, can sense the H3K2me2 modification (Herz et al. 2013), it has been suggested that the demethylation of histone and the subsequent deacetylation are crucial for the process of transcriptional memory. These results strongly indicate that hereditary epigenetic conditions are governed by the physical association of the NPC to chromatin (D'Urso and Brickner 2014), an emerging field of research that is awaiting future breakthroughs.

10.7.10 Are There Connections Between Nucleocytoplasmic Transport and Human Diseases?

For a while now, laminopathies have been one of the most common connections between nuclear transport/structure and disease (Butin-Israeli et al. 2012; Koch and Holaska 2014). There are some connections between diseases and substances, the selective inhibitors of nuclear export (SINE), which represent compounds that can target the export of proteins out of the nucleus that have even been hypothesized to possess treatment options for several human medical conditions. Suitable targets seem to be both hematological and solid malignancies (Senapedis et al. 2014). In specific detail, evidence has been provided that Selinexor, a SINE compound is a promising substance, since it displays interesting results for the treatment of several types of cancers (Sun et al. 2014; Zheng et al. 2014). These encouraging outcomes offer convincing arguments that nuclear transport routes can be considered as drug targets for therapeutic treatments and provide the basis for continued research on nuclear transport routes as prospective therapeutic targets.

A series of emergent linkages is connecting mRNA processing/export and human disease (Cooper et al. 2009; Mor et al. 2014). There are structure–function analyses of the highly conserved Gle1 protein (Folkmann et al. 2013, 2014). Gle1 fulfills a fundamental function in the export of RNA out of the nucleus through alteration of the DEAD box RNA helicase, Dbp5 (Snay-Hodge et al. 1998; Weirich et al. 2006; Alcazar-Roman et al. 2006), during the alteration of mRNP due to nuclear export (Lund and Guthrie 2005; Tran et al. 2007) at the cytoplasmic site of the NPC (Strahm et al. 1999; Schmitt et al. 1999). Over the last years, mutations in GLE1 have been associated with several severe autosomal recessive diseases causing tissue-specific complications (Nousiainen et al. 2008). These diseases comprise the fatal congenital contracture syndrome 1 (LCCS-1) and the lethal arthrogryposis with anterior horn cell disease (LAAHD), culminating in fetal death. It has been shown that experiments employing structure–function approaches in order to enlighten functional alterations of certain mutations that have been seen in patients. These studies show how a mutation in GLE1 termed FinMajor, which is referred to because of its high occurrence in the Finnish population, triggers the production of a Gle1 protein with a modified ability for nucleocytoplasmic transport. In specific detail, human Gle1 can create higher-order complexes, and using a cryo-EM approach it can be revealed how changes in the Gle1 protein alter its self-association (Folkmann et al. 2013). These studies, combining several strategies to assess the functional implications of patient mutations, give valuable knowledge not only of the disorders underlying human disease, but also of the mechanisms crucial to the nuclear transport.

The example of mutations in GLE1 that lead to tissue-specific diseases exemplifies genes encoding ubiquitously expressed RNA-binding proteins, which are mutated in tissue-specific diseases (Castello et al. 2013). A further example is spinal muscular atrophy, a disease of motor neurons triggered by a mutation in the human SMN gene (Coady and Lorson 2011), which is necessary for the assembly of spliceosomes and is therefore involved in global splicing processes. The alteration of ubiquitously expressed RNA-binding proteins can cause tissue-specific diseases. One example is the ubiquitously expressed, nuclear, polyadenosine RNA-binding protein, PABPN1, which has been found to be mutated in oculopharyngeal muscular dystrophy (Banerjee et al. 2013), and another one seems to be the polyadenosine RNA-binding protein, Nab2/ZC3H14 (Soucek et al. 2012; Kelly et al. 2012). While the *S. cerevisiae* Nab2 protein is crucial for viability (Anderson et al. 1993), the mutation of the human ZC3H14 leads to an autosomal recessive version of mental retardation (Kelly et al. 2012; Pak et al. 2011). A Drosophila functionality deprivation model was of substantial usefulness in showing that the function of Nab2/ZC3H14 is most crucial in neurons, which is in accordance with the brain-specific phenotype of the patients (Pak et al. 2011; Kelly et al. 2014). This brain model also exhibits an anomalous brain morphology that indicates axon conduction defects. Similar to other situations where mutations in such ubiquitously expressed RNA processing/exporting agents result in tissue-specific effects, the difficulty is both to identify what a variety of functions can be, and afterward to comprehend the crucial demands on these functions in the respective tissue.

Other connections between human disease and the processing of RNA exports involved both the cell nucleus and the cytoplasmic translation machinery It has been established that the deregulation of the eIF4E protein, conventionally regarded as a key factor in translation initiation (Jia et al. 2012), may lead to defects in multiple steps of gene expression. Mechanistic investigations of eIF4E are crucial, since eIF4E has been found to be overexpressed in 30% of human cancers (Carroll and Borden 2013). However, eIF4E can act as a shuttle protein that is carried out of the nucleus in an mRNA export-dependent fashion. The nuclear functions of eIF4E may favor the oncogenic capacity of eIF4E. In line with this, an import receptor has been found for eIF4E. Moreover, the mechanism by which ribavirin triphosphate impairs the function of eIF4E needs to be defined, which represent the only direct inhibitor of eIF4E able to enter Phase II clinical trials for acute myeloid leukemia (Volpon et al. 2013).

With the increased capacity to define the distinct mutations that are the basis of human diseases, it is likely that there will be many of the major actors engaged in the crucial phenomena commonly referred to as nuclear transport. These mutations can have distinct impacts, such as the complete functional loss of numerous of these vital factors may be incompatible with life. The lamins are an example of how specific alterations within a family of proteins can lead to various phenotypes of the disease.

10.7.11 What Are the Future Directions of Nuclear Transport Mechanisms?

With technological progress, the capacity to look inside the NPC has increased explosively, evolving the field from looking at the NPC as a static structure incorporated into the nuclear envelope to a dynamic macromolecular complex for the sophisticated regulation of nuclear transport, cell signaling and expression of genes. These progresses bring new and challenging issues to the transport sector, while at the same time highlighting the deficiencies in the fundamental process of transport to and from the nucleus. Among the fundamental mechanistic issues that still need to be resolved are: Can several loads be transported at the same time and in two directions by a single NPC, or are individual NPCs intended for specific transport scenarios? When NPCs are specific to transport events, this characteristic could explain why the distribution of NPCs differs in the various organisms and tissues. Knowledge of the biophysical properties of the NPC permeability barrier is of crucial relevance. While FG areas are definitely crucial for transport, these areas are neither selective nor inefficient. Furthermore, it appears a paradox that the energy-dependent, receptor-driven transport of loads is more rapid than simple diffusion through the NPC. Advanced analyses are required to specify the principles of transport selectivity and to assess the general applicability to specific transport incidents.

The NPC is obviously a key facilitator for several biochemically diverse nuclear activities. A future objective is therefore to ascertain how various types of governance influence each other. It is not yet known, for instance, how or whether the overall regulation of transport receptors affects mRNP aggregation or the many downstream phases of gene expression. In view of the close link between these events and the NPC, a high degree of synergism at both cellular and molecular scale can be envisaged. The discovery of the molecular linkages combining early and late phases of gene expression with the transport mechanisms through the NPCs and the management of this transport will be needed to identify the foundations of human diseases related to faulty NPC components, transport receptors and/or the integrity of NE. The promising perspective for the future in the field of nuclear transport is the prospect of new breakthroughs, unique cross-disciplinary collaboration and biomedical research applications.

10.8 Nuclear Plasma and Cytoskeleton

Besides the NPCs, there are other crucial structures within the cell's nucleus that need to be investigated, such as the nuclear plasma and the nuclear cytoskeleton. In specific detail, the nuclear lamina represents a nuclear peripheral scaffold, which is largely made up of nuclear lamins, whereas a minor amount of lamins is localized to the nucleoplasm. The lamins can be classified as type V intermediate filament (IF) proteins. Mutations in the genes of laminopathies trigger at least 15 different human diseases, collectively known as laminopathies, which cover muscle, metabolic and neuronal diseases, and can lead to rapid aging. The majority of these mutations are found in the LMNA gene, which codifies for A-type lamins. An increasing number of nuclear proteins are recognized to attach lamins and are involved in nuclear and cytoskeletal organization, mechanical stability, organization of chromatin, signaling, gene regulation, genome stability and differentiation of cells (De Leeuw et al. 2018). Recent investigations highlight the organization of the laminate filament entanglement in somatic cells, in which they are composed as tetramers within the cross section of the filaments.

The nuclear lamina represents a protein structure, which is positioned at the nuclear periphery and contributes to structural support to the nucleus and thereby protects the chromatin from mechanical cues (Gruenbaum and Medalia 2015). Lamins are a major compound of the nuclear lamina and built a tight network of filaments which can interfere with a wide variety of binding partners. Lamins assemble the nuclear laminar together with associated proteins. In order to keep the structural integrity of the entire nucleus, lamins regulate several cellular functions, including the organization of chromatin, the repair mechanisms of DNA and the assembly and disassembly of the nucleus (Burke and Stewart 2013). How the lamins control the regulation of these different structural and cellular processes is still not known.

Due to their sequence, lamins can be seen as a separate intermediate filament (IF) group (referred to as type V). However, other IF groups encompass keratins, such as types I and II keratins, desmin and vimentin (type III) and neurofilaments (type IV) that all can be detected in the cytoplasm of each multi-cellular organism (Herrmann and Aebi 2016). In mammalian cells, the nuclear lamina consists of four major lamin isoforms, such as two A-type lamins (A and C) and two B-type lamins (B1 and B2). In specific detail, the LMNA gene encodes for the A-type lamins, whereas the LMNB1 and LMNB2 genes codify for lamins B1 and B2, respectively. Moreover, lamins possess similar protein domains with other IFs (Gruenbaum and Aebi 2014). In particular, lamins contain an N-terminal head domain, a coiled-coil rod domain in its center and a tail domain at its C-terminus. The unique features of lamins encompass a nuclear localization signal (NLS), an immunoglobulin (Ig)-fold domain and a

CaaX motif (C is cysteine; a represents any aliphatic residue; X is any residue) in the tail domain (Gruenbaum and Aebi 2014). However, a small amount of the more soluble A-type lamins can be seen within the nucleoplasm (Dechat et al. 2008). Subsequently, the different states of the assemblies of these fractions are not yet clear (Dechat et al. 2010).

Lamins are usually modified in several manner by post-translational modifications (PTMs). For instance, lamins A, B1 and B2 possess a C-terminal CaaX motif, where the cysteine residue is post-translationally altered by farnesylation and methyl esterification. Mature lamin A can even be further proteolytic cleaved by cutting off the modified C-terminal cysteine and additional 14 amino acids. In contrast, only B-type lamins stay continuously farnesylated. However, there has been detected one exception in Caenorhabditis elegans, the lamin (Ce-lamin), which is coded by a single lamin gene (LMN-1) that shares characteristic features with A-type lamins, including the nuclear shape maintenance and the interplay with identified binding proteins; whereas, it keeps its farnesylation. In addition, lamins can be modified by PTMs including phosphorylation (Machowska et al. 2015), SUMOylation (Moriuchi et al. 2016) and glycosylation (Snider and Omary 2014).

The mutations in LMNA genes seem to cause a broad variety of diseases, which are referred to as laminopathies. Until now, there have been revealed at least 600 disease-related mutations, which are specifically mapped to the LMNA gene. These laminopathies encompass diseases that impact the striated muscle, lipodystrophies, peripheral neuropathy and elevated aging disorders, among which the Hutchinson–Gilford progeria syndrome seems to be the best-known one (Worman and Bonne 2007). The majority of these mutations is autosomal dominant, and thus laminopathies belong to the relatively rare diseases. In order figure out how a specific mutational event in the lamin meshwork can cause detrimental effects, and how these mutations can lead to diseases that impact various tissue types, it is crucial to know how the lamin meshwork is arranged in healthy tissues and how it is disrupted in diseased tissues in various diverse cell types. Since the structure of the nuclear lamina has over a long time period been unknown, very recent provided novel insights into the precise organization of the structural architecture of the lamin scaffolds underneath the nuclear lamina.

10.9 The LINC Complex

The movement of the nucleus within the cell necessitates, firstly, compressive or tensile forces on the nucleus generated by the cytoskeleton and, secondly, the presence of fixed attachments between the nucleus and the surrounding cytoskeleton in a number of migrating cells. Several studies have proven that certain proteins are involved in nucleocytoskeleton linkage: This physical connection is known as the LINC complex (Shin et al. 2013; Khatau et al. 2012). The LINC complex is currently under strong investigation and hypothesized to be the major contributing or modulating factor of the overall nuclear mechanics. The main components of the LINC complex are the SUN (Sad1p, UNC-84) proteins in the INM and KASH domain containing nesprins (nuclear envelope spectrum repeat proteins) situated on the outer nuclear membrane (ONM) (Sosa et al. 2013; Esra Demicioglu et al. 2016). The cytoplasmic endings of nesprins interoperate directly or indirectly with all three major filaments of the cytoskeleton, including actin, intermediate filaments and microtubules, thus fulfilling the physical interconnection via the nuclear envelope (Rothballer and Kutay 2013a, b). There is much debate about the importance of the mechanical continuity between the nucleus and the cytoskeleton in mechanotransduction, by which the mechanical forces generated by the cytoskeleton components are transferred directly to the nucleus (Swift and Discher 2014; Liu et al. 2016).

The SUN proteins possess a helical N-terminal domain that faces the site toward the nucleoplasm, which interacts with the lamins and proteins of the nuclear pore complex. The C-terminal domain of SUN proteins is extended toward the perinuclear space and can bind to nesprins (Fig. 10.6). Due to

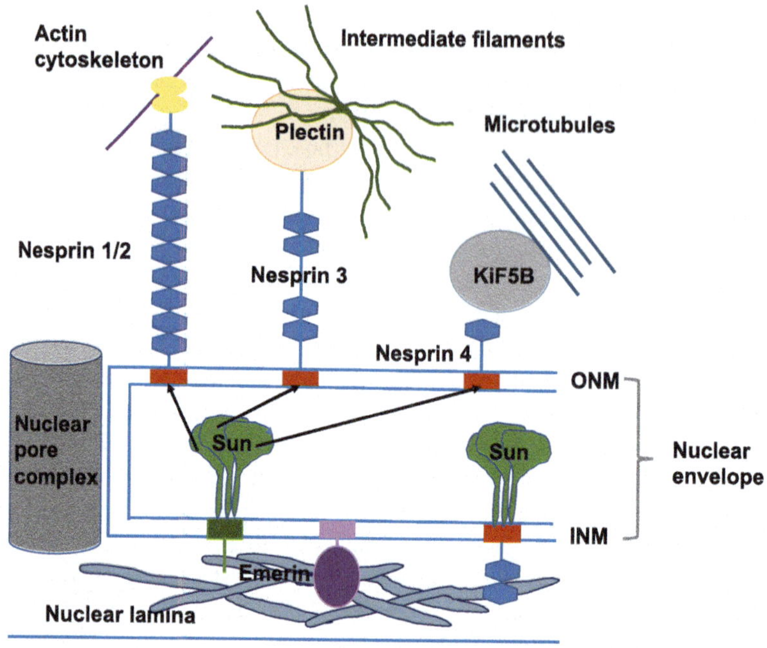

Fig. 10.6 Nesprins link the nucleoskeleton to the three major components of the cytoskeleton. The linker of nucleoskeleton to cytoskeleton (LINC) complexes connects the nuclear lamina to cytoskeletal architectures through nesprins. Nesprins are directedly connected to the outer nuclear membrane (ONM) and bind to SUN proteins located in the luminal space and thereby act as tethers for the nuclear envelope. SUN proteins are connected to the inner nuclear membrane (INM)

the characteristic and highly conserved C-terminal domain, five SUN domain proteins were found in mammalian cells. Among them, two proteins, SUN1 and SUN2, are localized in the nuclear envelop within somatic cells, whereas the other three SUN proteins, SUN 3–5, are detected exclusively in the testis (Katta et al. 2014; Smoyer and Jaspersen 2014). In certain experimental approaches, it has been hypothesized that SUN1 and SUN2 are quite similar and hence overlapping; whereas, they fulfill strictly diverse biological tasks (Le et al. 2016). Sun1 is clearly associated with NPCs and can directly provide the linkage between the cell's cytoskeleton and the nuclear pores (Torbati et al. 2016). Conversely, SUN2 is not needed for the precise localization to NPCs and instead interacts with two forms of lamin, the lamin A precursor and the mature lamin A (Zwerger et al. 2011).

The prototype of the ONM components is built by nuclear envelope spectrin repeat proteins (nesprins), which then connect the nuclear envelope with the cell's cytoskeleton. The nesprins belong to the SR super family proteins. Four nesprins are present in mammals, such as nesprin-1–4, among them are the nesprin-1–3 proteins that possess multiple isoforms, which are obtained from alternative splicing (see Chap. 13) and multiple transcription initiation sites (Sosa et al. 2013). All nesprins display a unique evolutionary conserved C-terminal Klarsicht, ANC-1 and Syne Homology (KASH) transmembrane domain, which can selectively target the nesprins toward the nuclear envelope (NE) (Fig. 10.6). In specific detail, all nesprins display a C-terminal KASH domain with a transmembrane domain and a short luminal domain, which enables them to facilitate the targeting of nesprins to the NE and is necessary for the interplay between nesprins and the SUN proteins (Sosa et al. 2012). The four separate nesprin genes of vertebrates encode proteins with specific cellular functions. The N-terminal domains can interact with distinct cytoskeletal proteins through a specific motif commonly present in all nesprins. Nesprin-1 and nesprin-2 represent the largest SR proteins with molecular weights of about 1000 and about 800 kDa, respectively, and couple the NE with F-actin cytoskeleton through the involvement of their two N-terminal CHDs (Fig. 10.6). Hence, nesprin-1 and nesprin-2 interact with actin filaments via their actin-binding domains (Taranum et al. 2012). Nesprin-3 and nesprin-4 possess molecular weights of approximately 110 and 42 kDa, respectively, and they contain no N-terminal CHDs (Zhang et al. 2001, 2002; Wilhelmsen et al. 2005; Roux et al. 2009). In contrast, nesprin-3 possesses a plectin-binding domain that can connect the NE to the cytoplasmic intermediate filaments (IF) networks (Taranum et al. 2012). Nesprin-4 is able to bind kif5b, which is a subunit of the MT motor protein kinesin-1, in a region that is located downstream of its single SR (Wilhelmsen et al. 2005; Roux et al. 2009) (Fig. 10.6). The capacity of the nesprins to be in close proximity to the NE and their ability to interact with cytoskeletal proteins indicates that they primarily function in the coupling between the NE and the cell's cytoskeletal components. There has been seen an expansive group of tissue-specific nesprin-1 and nesprin-2 isoforms, which have been obtained by alternative transcription (Rajgor et al. 2012). Several isoforms miss the C-terminal KASH domain and can even be localized into various subcellular compartments, providing complementary functions for nesprins beyond NE attachment. These results therefore heavily confirm a model in which the LINC complex couples the nucleus to the cytoskeleton via motor proteins (Fig. 10.6).

10.10 Chromosomes

A chromosome is a dense packed thread-like structure that contains DNA of living organisms. Each chromosome is built of DNA that is tightly coiled around proteins such as histones. Most eukaryotic cells process a certain set of chromosomes such as humans that have 23 pairs of chromosomes and hence in total 46 chromosomes. Twenty-two pairs of chromosomes are termed autosomes and are the same for males and females. They are numbered by their size starting by one for the largest chromosome. Chromosomes are located inside the nucleus. The 23rd pair of chromosomes is termed sex chromosome and differs between male and female, as males have one X and one Y chromosome, whereas females have two copies of the X chromosomes. The chromosomes are passed from parents to their offspring and contain the specific information for each unique type of a living organism.

The unique structure of chromosomes holds the DNA tightly wrapped around histones. The DNA can otherwise (when it is not densely packed) not fit into the nucleus due to steric restrictions. As organisms grow and thus need to divide in pursuance of producing new daughter cells, it is necessary that the DNA remains intact and evenly distributed to the two daughter cells. Finally, the DNA needs to be accurately copied and distributed, however, rarely mistakes occur and cells undergo apoptosis or try to repair these mistakes. The number and the structure of the chromosomes need to be constant and unchanged; otherwise, diseases such as the down syndrome (three copies of chromosomes 21, instead of two) may occur.

Do all organisms possess the same set of chromosomes? The answer is that the number of chromosomes may vary among the species, as chromosomes vary in number and shape among different organisms. Bacteria have usually one or two circular chromosomes. Human, animals and plants possess linear chromosomes which are present in pairs in the nucleus. Human reproductive cells (gametes) carry only one copy of each chromosome. After reproductive cell divides, all successor cells divide as well numerous times and thus reproduce the mature individual several times. These cells have the full set of chromosomes. Humans and other complex organisms possess a much smaller type of a chromosome outside the nucleus in some organelles such as mitochondria, which are the "powerhouses" of the cell converting "oxygen into energy". These smaller chromosomes are similarly to the bacterial chromosome.

10.10.1 The Structure of Chromosomes

Mitosis represents a cellular process that causes the division of a eukaryotic cell in order to generate two daughter cells, each of which possesses exactly the same number of chromosomes as the parental cell. According to the aforementioned definition, the entire mitosis process collapses when the chromosomes of the parent cell fail to attain their intended targets. A structure which plays a crucial part in preventing this from happening represents the centromere. The centromere appeared for the first time in the 1880s when the German biologist Walter Flemming characterized it as the "primary constriction" of the chromosome. However, it is now appreciated that the centromere appears to be a region of highly specialized chromatin located within each narrowed chromosome, which is the basis for kinetochore formation and provides a location for the attachment of sister chromatids (Fig. 10.10). Failures in the centromere or kinetochore performance are disastrous for the cells. Such defects can result in aberrant division and in unstable chromosomes, both of which are frequently encountered in cancer cells.

10.10.1.1 What Are Centromeres?

Centromeres represent a narrowed area of the chromosome that is not normally exactly located in the middle of the chromosome. The areas on both sides of the centromere are referred to as the arms of the chromosome. The centromere supports the alignment and the amplification of the chromosomes proper for cell division. Primarily, the centromere serves as the attachment point for the two halves (the so-called sister chromatids) of each replicated chromosome.

10.10.1.2 What Are Telomeres?

The telomers are repetitive parts of DNA positioned at the far ends of linear chromosomes. They protect the ends of chromosomes for being degraded, and if they are degraded, these parts contain no genetic information. The telomers serve as sink for degradation, but still a small part of them is needed for cell division, otherwise the cell cannot divide and hence dies.

White blood cells are special, as they divide rapidly and very frequently and possess a special enzyme preventing their chromosomes to lose their telomers. Finally, these cells live longer than other body cells. Moreover, telomers play a role in cancer, as malignant cancer cells divide fast and do not lose their telomers, which help them to grow uncontrolled and overcome cell death.

In summary, chromosomes are DNA and short sections of these DNA are genes. A cell, in which chromosomes are paired, is named diploid. If the chromosomes are single, the cell is haploid.

10.10.2 Chromosomes and Nuclear Compartmentalization

Chromosomes are different units within the nucleus, but apparently are not to be regarded as nuclear compartments (Cremer et al. 2000). Whereas the different character of the chromosomes appears easily discernible in the metaphase, in which the chromosomes condense and are isolated from each other, their topological arrangement in the interphase nucleus was not clear until very recently. In groundbreaking experiments (Cremer et al. 1982, 1988), where restricted areas in the cell nucleus were exposed to a microlaser and the damaged chromosomes were then recognized, it has been seen that chromosomes are located in specific regions of the nucleus, which are referred to as chromosome territories. Inside the chromosomal territories, active and inactive genes are situated on the periphery of the territories, while non-coding sequences are more common in the interior (Kurz et al. 1996; Verschure et al. 1999). The finding that some active gene loci and active transcription sites can be mapped onto internal regions of chromosomal territories suggests that chromosomal territories tend to be "porous" and that transcriptional activators are able to obtain entry to internal sites in the territory. This idea is reinforced by the discovery that proteins of up to 500 kDa in size are able to diffuse freely in the whole nucleus (Seksek et al. 1997). The cavity between the chromosomes can be displayed by overexpression of vimentin, initiating the generation of vimentin arrays among the chromosome territories (Bridger et al. 1998). The overexpression of the non-chromatin-binding protein EAST (enhanced adult sensory threshold) in Drosophila facilitated the assembly of an extrachromosomal protein location (Wasser and Chia 2000). These results draw a pattern in which the chromosomes are bent into a three-dimensional flexible meshwork, which allows them to take into account the possibility that the placement of genes and chromosome domains may perform regulatory functions in gene expression (Cremer et al. 2000).

Most cell types will only express a comparatively limited part of their genome. By placing genes in areas of chromosomes that are inaccessible to the chromatin rearrangement factors or transcriptional activators, fractions of genes can be silenced, while accessible gene sets are kept potentially expressable (Francastel et al. 2000). The observation that in most processes of differentiation the nuclear chromatin architecture is radically changed suggests a certain importance of nuclear compartmentalization in the regulation of gene activity (Manuelidis 1984, 1990). In the differentiation of neuronal PC12 cells, for instance, DNAse-sensitive chromatin, pre-mRNA SFCs and Cajal bodies are relocated from internal areas into the periphery (Park and De Boni 1996; Janevski et al. 1997). In addition, in cells from epileptic foci, the X chromosome is more often located inside the nucleus than in normal control cells, where it is usually located on the periphery (Borden and Manuelidis 1988). In line with the opinion that the transcriptionally inactive X chromosome can influence the expression of certain genes, it has been observed for many years that in female cells the transcriptionally inactive X chromosome is always located at the cell periphery (Lyon 1961).

The positioning mechanism is more local in nature, rather than concerning an whole chromosome. A growing amount of observations (Cockell and Gasser 1999; Festenstein and Kioussis 2000) indicates that the positioning of genes close to heterochromatic areas encourages gene silencing. A classic case is the brown locus in Drosophila melanogaster. The insertion of a heterochromatic cluster into an allele leads to the connection of both alleles with the heterochromatin (Dernburg et al. 1996). Similarly, during the differentiation of B cells, transcriptionally suppressed genes were identified that are typically associated with heterochromatin (Brown et al. 1997, 1999). In yeast, telomeric sequences and telomer-related loci of the mating-type are successfully silenced by heterochromatinization at the nuclear periphery (Cockell and Gasser 1999; Gottschling et al. 1990; Andrulis et al. 1998). These findings suggest that nuclear positioning, and in particular the positioning of genes in heterochromatin, may be decisive for the precise realization of gene expression routines. Locating genes outside the heterochromatin is not enough to activate them. Positioning in extra heterochromatin domains only gives a gene competence, not activity. To activate genes, the transcriptionally repressive chromatin needs to be rearranged. Recently, great progress has been made to unravel the mechanisms that result in chromatin remodeling in connection with gene expression, and a large number of chromatin remodeling activities have been recognized, in particular acetyltransferases and ATP-dependent remodelers (Workman and Kingston 1998).

10.10.3 Dynamics of the Nucleus: Chromatin Dynamics

The mammalian cell nucleus is widely known as a vastly dynamic organelle. The apparent dynamic reassemblies that occur during the dismantling and rebuilding of the nucleus during the M-phase have been well recorded and examined in depth. Most recent findings, nevertheless, have revealed that there are many nuclear constituents with high dynamics even in the interphase nucleus. The dynamic characteristics of chromatin, mRNA and proteins in the cell nucleus are essential for the synchronization of gene expression processes and are expected to contribute pronouncedly to exact and efficacious gene expression.

10.10.3.1 The Dynamical Properties of Chromatin

Time-lapse microscopic recordings indicate that the chromosomes and chromatin are fairly stationary during interphase. In vivo mapping using dyes, fluorescence-labeled topoisomerase II or GFP-labeled nuclear histone H2B failed to detect large-scale chromatin agitation within the interphase nucleus (Abney et al. 1997; Marshall et al. 1997; Kanada et al. 1998). A more detailed analysis based on the marking of specific sites in yeast (Marshall et al. 1997) indicated only limited Brownian motion, pointing to the fact that the chromatin is spatially stabilized. However, independent evidence, especially from biochemical studies, indicates that the structure of chromatin is a key regulatory factor in gene expression, but points directly to the fact that dynamic chromatin assembly takes place at microscopic levels smaller than those observed in the experiments mentioned above and is certainly essential for balanced gene expression.

The simple view of chromatin, referred to in many books, is that the lowermost organizational unit of chromatin is the nucleosome, which is made up of the nuclear histones H2A, H2B, H3 and H4, connected by the linker histone H1 (Fig. 10.7) (Belmont et al. 1999). Strands of interconnected nucleosomes are helically twisted into a 10 nm fiber that in turn is twisted into a 30 nm fiber. It is proposed to fold this fiber in higher-order chromatin loops, despite the fact that the organization of these higher-order topologies in vivo is still unknown. A recently developed system for investigation of higher-order chromatin structures and their dynamics in living cells employing microscopy has been presented (Robinett et al. 1996). With this GFP-based system, large area chromatin fibers with diameters of about 100 nm can be monitored in interphase cells. In intact cells, these fibers directly demonstrate the existence of higher-order chromatin structures that were previously only visible under light and electron microscopy in fixated cells (Belmont et al. 1999; Robinett et al. 1996). In agreement with earlier findings, time-lapse microscopy showed that the higher-order chromatin fibers remained fairly stationary. However, when transcriptional activation caused by overexpression of the VP16 transactivator is forced, the labeled chromatin domain will unfold, probably as a result of chromatin remodeling (Tumbar et al. 1999, 2001). These rearrangements are not the result of the

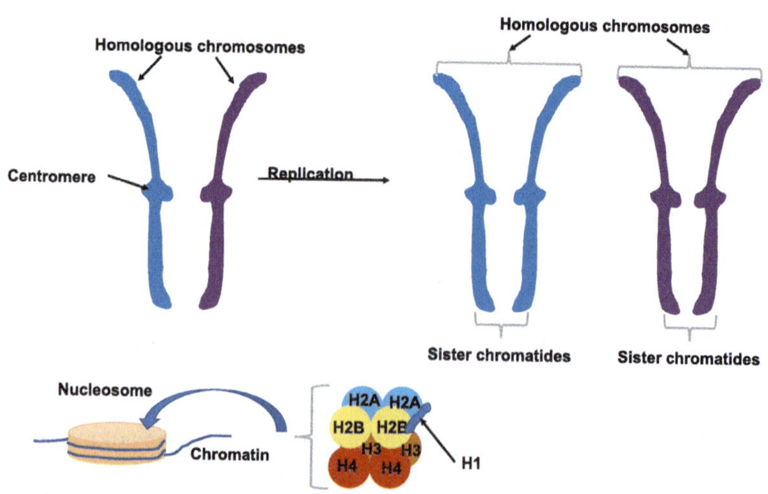

Fig. 10.7 Sketch of chromosomes and sister chromatids as well as histones building nucleosomes

artificial effect of the strong VP16 transactivator, because analogous findings have been made in more physiological reporter systems employing the mouse mammary tumor virus promoter or a reporter system that produces functional mRNA by using a tetracycline-regulated promoter (Tsukamoto et al. 2000; MCNally et al. 2000). These results are validated and expanded to mammalian cells, where early findings on polytene chromosomes of the dipteran fly Chironomus tentans and on Lampbrush chromosomes, where activation of transcription leads to the emergence of chromosomal whorls, and where the extent of a whorl is proportional to the quantity of transcription carried out (Gall and Callan 1962; Daneholt 1975). These novel microscopic techniques will in the near future help us to identify how the macroscopic unfolding of chromatin fibers seen in living cells is associated, at some level, with the microscopic rearrangements of the nucleosome structure brought about by modifying chromatin.

10.10.4 How Affects the Chromatin the Assembly of the Nuclear Envelope?

Newer studies on cell-free nuclear building systems indicate that the first contacts between ER and chromatin are facilitated by tubule ends. In a second step, these bound ER tubules are reassembled by DNA-binding NE-specific membrane proteins into shallow nuclear membrane foils (Ulbert et al. 2006; Anderson and Hetzer 2007; Hetzler 2010). In cells, the ER seems to be largely cistern shaped (Lu et al. 2009), and therefore it is probably that existing membrane foils are directly attached to the chromatin surface in vivo. It is consistent with the observation that the overexpression of reticulons retards the development of the nuclear envelope, while their knockdown speeds up the development of the nuclear envelope (Anderson and Hetzer 2008).

In the course of the nuclear envelope development, the chromatin experiences a number of conformational alterations, beginning from a phase of maximum chromatin compression in the late anaphase, immediately before the generation of the nuclear envelope (Mora-Bermudez et al. 2007); whereas, it decondenses in the transcription- and replication-competent interphase of the cell cycle. Anaphase compression necessitates the kinesin-like DNA-binding protein (Kid), which guarantees the establishment of a compact chromosome cluster during the anaphase and the correct encapsulation of the segregated chromatin mass in a single nucleus (Ohsugi et al. 2008). Most interestingly, the loading of Kid on anaphase chromosomes relies on Importin β (Tahara et al. 2008), which contributes to the increasing amount of mitotic processes controlled by this transport receptor.

Recently, a convincing argument has been put forward that nuclear envelope generation and chromatin decondensation are connected in a mechanistic fashion. The hexameric ATPase Cdc48/p97, which has been involved in the fusion of the membrane (Kondo et al. 1997) and in ubiquitin-driven processes (Ye et al. 2003), can extract Aurora B from chromatin that in turn inactivates it and consequently leads to the decondensation of chromatin. Intriguingly, the impairment of Cdc48/p97 inhibited the assembly of the nuclear envelop, which indicates that the decondensation of chromatin is needed for the assembly of the nuclear envelop (Ramadan et al. 2007), presumably by means of opening the structure of chromatin. There is a strong attraction to the idea that this open chromatin structure offers binding positions for INM proteins and thus promotes the development of the nuclear envelope.

Simultaneously, with the covering of the chromatin by the NE, the NPCs gather in the reassembling nuclei. This intriguing model of protein self-assembly is governed by the gradual recruitment of a subset of NPC proteins to chromatin (Dultz et al. 2008) and some advances have been achieved in identifying this process. A protein entitled Mel-28/ELYS, screened in C. elegans for the presence of elements associated with the generation of pronuclei, is crucial for the combination of the Nup107/160 complex with chromatin (Rasala et al. 2006; Franz et al. 2007). Since Mel-28/ELYS includes an AT-hook domain (Kimura et al. 2002), a plausible assumption is that this operation is performed by direct attachment to the DNA. It is interesting that RanGTP encourages the acquisition of Mel-28/ELYS (Franz et al. 2007), probably by the release of Importin β out of one of the compounds of Nup107/160/ELys (Hetzer et al. 2005). Even though other nucleoporins are involved in pore formation, their precise function has yet to be defined. For example, it has recently been revealed that a complex of Nup53 and Nup155 is crucial for the development of the nuclear envelope in nematodes and vertebrates (Franz et al. 2005; Hawryluk-Gara et al. 2008); however, it remains unknown how membrane-associated Nup53 is able to manage the coordination of interactions between chromatin, membranes and soluble Nup155.

One group of nucleoporins has been found to be crucial for pore formation (Hetzer et al. 2005), the biggest subcomplex of which is the Nup107/160 complex, whose degradation leads to NPC-free nuclear membranes (Harel et al. 2003; Walther et al. 2003). In vertebrates, it has also been found that two transmembrane nucleoporins are involved in the generation of the nuclear envelope: Ndc1 and POM121 (Mansfeld et al. 2006; Stavru et al. 2006; Anderson et al. 2009). The third common transmembrane Nup, gp210, is not naturally expressed in all cell types and is therefore probably not necessary for pore formation (Eriksson et al. 2004). While the molecular function of these scaffold Nups is not known, an intriguing connection has been established between POM121 and the Nup107/160 complex, and the generation of nuclear membranes may indeed be connected to the pore structure by a badly conceived control point through which the Nup107/160 complex "scans" the nuclear membranes (Antonin et al. 2005).

10.10.5 Dynamics of the Nucleus: RNA Dynamics

The transport mechanisms of nascent mRNAs moving them away from their transcription sites and the subsequent release from the cell nucleus are still the subject of extensive research. An attractive new model proposed that freshly synthesized RNAs migrate to the tightest nuclear pores, probably through an active and targeted process (Blobel 1985). This so-termed "gene gating" model proved compatible with the monitoring of extended stretches of mRNA emanating from the sites of transcription (Carter et al. 1991, 1993; Xing and Lawrence 1993). In addition, several RNA transcripts occurred in Drosophila that were vectorially transported into the apical but not into the basal part of the cytoplasm (Edgar et al. 1986; Davis and Ish-Horowicz 1991). However, a closer analysis demonstrates that the traces of RNA initially seen emanate from the transcription sites in a haphazard direction (Dirks et al. 1995; Melcak et al. 2000), indicating that RNA export is actually far from being a directed process. Further studies of RNA export mechanisms in Drosophila confirm that the uneven dispersion of RNAs is probably caused by cytoplasmic "retention factors" rather than by vectorial export (Lall et al. 1999). The replacement of an active exporting mechanism is that the RNA travels by diffusion from its transcription sites to the nuclear pores.

The ability of RNAs to diffuse widely in the cell nucleus was established in Drosophila by means of a highly expressed reporter gene whose mRNA could be located in a "extrachromosomal channel system" rather than in different tracks (Zachar et al. 1993). The significance of diffusion as the principal mechanism of RNA transport in vivo has been verified recently by microscopic investigations in living cells. The imaging of mRNA utilizing hybridized oligonucleotides comprising an activatable fluorescent group reveals an undirected, energy-independent motion of polyadenylated RNAs across the entire nucleus (Politz and Pederson 2000; Politz et al. 1998, 1999). Equally, the findings of stereological electron microscopic analysis of labeled RNA fragments (Singh et al. 1999) are compatible with passive diffusion acting as a driving factor in the export of RNA. Diffusion is a highly efficient transport pathway for RNA. The coefficient of diffusion of fluorescence-labeled RNAs within the cell nucleus is comparable to that obtained for an RNA particle in a solution (Politz et al. 1999). In fact, these diffusion coefficients can be used to assess that an RNA particle can be released from its transcription site and shipped out of the nucleus in a matter of seconds after release. Whereas these new findings provide evidence that RNAs are able to diffuse across the nucleus, indicating that diffusion may be the main transport pathway for nuclear RNA molecules, the possibility cannot be entirely ruled out that some RNAs are carried by active and potentially directed mechanisms.

10.10.6 Dynamics of the Nucleus: Protein Dynamics

Each mammalian cell nucleus of around 10 μm typically consists of about 2 m of DNA. Apparently, the DNA must be condensed to match the limited volume of the nucleus. Due to the high level of packing of chromatin and the large amount of RNA and protein molecules inside the nucleus, it was commonly thought that the nucleus is a tightly packed organelle where proteins and RNAs may find it difficult to travel. When this is actually the situation, there are probably transport mechanisms (maybe active and directed) that take the proteins to their target sites. This model is practically excluded by recent microscopic experiments with living cells, which show that proteins can move throughout the cell nucleus by a passive, diffusion-driven mode.

Fluorescence recovery after photobleaching experiments (FRAP) proves that numerous nuclear proteins are extremely mobilized inside the nucleus. High levels of mobility are a common characteristic of nuclear proteins and were shown for transcription factors (McNally et al. 2000; Stenoien et al. 2001), splicing factors of pre-mRNAs (Kruhlak et al. 2000; Phair and Misteli 2000), enzymes for the processing of rRNA (Phair and Misteli 2000; Snaar et al. 2000), DNA repair enzymes (Houtmuller et al. 1999), chromatin-interacting proteins (Phair and Misteli 2000; Lever et al. 2000; Misteli et al. 2000; Perche et al. 2000), 3 h-processing factors and caspases acting in apoptosis (programmed cell death). For most proteins, mobility is not dependent on energy (such as ATP) and is not influenced by temperature decrease, pointing to a passive diffusion process. Diffusion offers an effective and fast mode of transport. A single protein or a protein complex of up to around 500 kDa is able to transit into nucleus, which takes less than one minute. Diffusion yields also an energetically favorable system to guarantee the accessibility of proteins in the entire nucleus. It is of utmost significance that no signals or signal receptors are involved in bringing proteins to their destinations. Instead, the proteins migrate through the nucleus in pursuit of a high-affinity-binding site where they can carry out their specific activities (Misteli 2001). The routing characteristics of proteins are uniquely verified by DNA repair markers. When there is no DNA damage, the repair enzyme ERCC1'XPF is free-moving throughout a cell nucleus. When DNA damage is caused by exposure to UV radiation, the protein is transiently immobilized due to adhesion to damaged sites. As soon as it has operated on the damaged spot, it detaches itself and regains its high mobility (Houtmuller et al. 1999). Analogous findings have been obtained for DNA replication factors that are fixed at replication sites prior to becoming dissociated and diffused to identify the next origin of replication (Leonhardt et al. 2000).

The investigation of the dynamic properties of chromatin-binding proteins is of special relevance for the comprehension of gene expression. It is typically presumed that chromatin-binding proteins are stable and physically attached to chromatin. Whereas this apparently holds true for nuclear histones (Phair and Misteli 2000; Lever et al. 2000), practically all remaining chromatin-binding proteins and compounds of the transcription apparatus involve only a transient connection with chromatin. The linkage histone H1, located outside the nucleosomes but functioning as a chromatin structural protein, is replaced dynamically among the binding sites (Lever et al. 2000; Misteli et al. 2000). Every H1 molecule is located on the chromatin for approximately one minute before becoming dissociated (Misteli et al. 2000). It is analogous to another group of chromatin-binding structural proteins, the proteins of the highly mobile group, which only temporarily attach to chromatin and stay on the chromatin for less than 5 s (Phair and Misteli 2000). The short interaction time of these proteins could be decisive in assuring the capacity of a cell to react to external irritants with fast alterations of its gene expression profile. Therefore, if chromatin-binding proteins, in particular transcriptional repressors including linker histones, were to be statically linked to chromatin, chromatin modeling factors might find it difficult to get an access to chromatin (Misteli 2001). But every time an H1 molecule is dissociated, a new window of opportunity opens up, and the transcription activators are able to intervene and transform the chromatin. This interpretation is in accordance with the recent finding that the transforming activity of the SNF'SWI can influence the chromatin only if the histone H1 is released (Hill and Imbalzano 2000).

Histones and highly mobile group proteins are common chromatin-binding proteins and therefore do not bind DNA sequence specifically. Do transcription activators that detect their specific response units in promoters bind more closely to chromatin? FRAP investigations on steroid receptors point to the fact that specific DNA-binding proteins only temporarily interfere with their targets. The glucocorticoid receptor (GR) is responsible for capturing DNA in a specific GR reaction element and controlling transcription by interacting with elements of the transcription system. Photobleaching experiments on an array of multiple GR response molecules, which enables the viewing of this specific location in living cells, significantly reveal that although GR molecules are maintained at these promoter attachment sites as long as the target genes are active, the activating GR molecules are continuously and rapidly replaced and are only present at the GR response element for a few seconds, thereby promoting a "hit and run" binding mode (McNally et al. 2000; Rigaud et al. 1991). Similar outcomes were seen with a different steroid receptor, the estrogen receptor (Stenoien et al. 2001). The discovery that various ligands of the estrogen receptor are known to dramatically change the dynamic-binding characteristics of the receptor is a strong indication that changing the dynamic-binding characteristics of a receptor, either through alteration of the receptor itself or its interfering proteins, can be employed as a regulatory machinery in the expression of genes (Stenoien et al. 2001).

10.10.7 Dynamics of the Nucleus: Compartment Dynamics

Nuclear compartments can be monitored over a period of several hours under a time-lapse microscope in spite of the lack of membrane boundaries and thus seem to be stable entities. Their stability might indicate that they are mostly inert constructions. But when the internal dynamical processes of the compartments were investigated by specific bleaching in FRAP experiments, a fast and steady interchange of proteins between the compartments and the nucleoplasm could be seen. A fast transfer of proteins between compartments and the nucleoplasm occurs in virtually all nuclear compartments, in particular the nucleolus (Phair and Misteli 2000; Snaar et al. 2000), SFCs (Kruhlak et al. 2000; Phair and Misteli 2000), the CB (Snaar et al. 2000), the PNC (Huang et al. 1997) and heat-shock granules (Jolly et al. 1999). The flow of proteins across these compartments is exceptional. For instance, more than 10,000 molecules of the rRNA-processing enzyme fibrillarin are liberated every second from the 2–3 nucleoli in a single cell nucleus (Phair and Misteli 2000). These findings indicate that the nuclear compartments are by no means solid structures, but are in continuous flux (Misteli 2001).

Due to the high flow of proteins, compartments need to be regarded as reflectors of a steady-state equilibrium. The morphology of every compartment is governed by the on/off rates of its inhabitants. These rates, on the other hand, are defined by the functional condition of the resident proteins both inside and outside the compartment. Most interestingly, practically all nuclear areas can be broken down by experimental interference, and some are oversized. This quite fragile balanced state can arguably be handled most effectively in the SFC. In a naturally developing cell, these areas occur as irregularly shaped, morphologically diverse structures with a diameter of approximately 1 μm, and pre-mRNA splicing factors are thought to constantly disassociate from them in order to find targets, namely unspliced RNA (Mistelli 2000). When the quantity of available transcripts is decreased by blocking transcription, the splicing factors tend to waste less time outside the compartment, leading to an elevated rate of association (Sinclair and Brasch 1978; Misteli et al. 1997). As the rate of dissociation is constant, there is a net rise in the on rate, and the compartments become larger.

10.10.7.1 A Possible Self-organization of the Nuclear Architecture

The general topic arising from these discoveries in protein dynamics is that robust assemblies can be produced in the nucleus by the interplay of strongly dynamic constituents. The production of robust assemblies from highly dynamic elements is in accordance with a role of self-organization as a driving force in the creation of nuclear infrastructure (Misteli 2001). Furthermore, the monitored routing of the proteins across the nucleus, the short residence time of the proteins at the attachment points and the consequent capacity to engage in repeated co-interactions with the partners are pre-conditions for self-assembly. Although there is no immediate proof of a self-organizing influence on the construction and maintenance of the nuclear infrastructure, numerous findings agree with this hypothesis.

Several experimental findings point to the fact that self-organization affects the integrity of many, if perhaps not all, nuclear compartments. The reaction of SFCs to the accessibility of pre-mRNAs outlined above highlights the self-organizing features of SFCs. The components of Cajal bodies (Hebert and Matera 2000), PML bodies (Ishov et al. 1999) and gems (Lorson et al. 1998) all have proven self-associating characteristics that are in line with their self-organization activity. The nucleolus is a distinctive symbol of a self-organizing architecture. The morphological integrity of the cell nucleus strongly relies on its transcriptional activity. When transcription is blocked, the nucleolus breaks down and the nucleolar components combine with one another, thus pointing out that the nucleolus‘ normal organization is based on the privileged interplay of its own proteins with the rRNA (Dousset et al. 2000; Verheggen et al. 2000, Oakes et al. 1993; Melese and Xue 1995). On the other hand, the insertion of rDNA minigenes leads to the production of mini nucleoli in the nucleoplasm, which directly indicates that the transcription of rDNA is adequate for the construction of the nucleolus (Karpen et al. 1988; Oakes et al. 1998). The necessity of a continuous transcription during the establishment of the nucleolar morphology is clearly visible in the reassembly of the organelle toward the termination of mitosis. At a time when rDNA transcription has not yet recommenced, PNBs store nucleolar compounds in the newly emerging nucleus (Olson et al. 2000; Dundr et al. 2000). After resuming transcription, nucleolar components aggregate at sites of rDNA transcription. The sequence of aggregation mirrors the sequence of their participation in the processing of rRNA, indicating that their aiming in this process depends directly on their function (Dundr et al. 2000). FRAP attempts have demonstrated that at any time in the M-phase nuclear proteins are quickly transferred between PNBs and the nucleoplasm (Dundr et al. 2000). Similar to SFCs, the nuclear factors are expected to continuously disassociate from the PNBs in the process of finding adequate binding sites.

The influence of self-organization in defining the nuclear infrastructure and its function is not restricted to the nuclear areas, but probably also includes the establishment of transcriptional and replication sites (Cook 1999). Replication factors have a homogeneous dispersion in the entire nucleus, but due to their attraction to the origins of replication during the S-phase, they constitute various focal points (Leonhardt et al. 2000; O’Keefe et al. 1992; Ma et al. 1998). These substances can reassociate from their foci when the replication of a specific region is fully performed. In addition, these substances are able that spread freely throughout the nucleus, seeking a new replication origin, where they regroup to create a new focus point (Leonhardt et al. 2000). This behavior indicates that, exactly as in trusting enclosures, replication factors are constantly replaced by the replication locations, resulting in a stable but highly dynamic nuclear architecture. Similarly, transcription factors which are dissociated from the transcription sites can spread across the nucleus and attach themselves to a new transcription site. The consequence of these findings necessarily has to be that self-organization is involved in the construction and preservation of the nuclear infrastructure.

It seems to be well known that the nucleus covers numerous specific compartments. A constant demand is to define the responsibilities of these units. A crucial step in this effort is expected to be the biochemical cleaning and classification of the nuclear compartments and, probably even more crucial, the in vitro rebuilding of the nuclear compartments and their functionalities. Latest findings indicate that the nucleus is a robust but extremely dynamic framework and that the dynamic properties are essential for its precise operation. While the majority of investigations so far has focused on the dynamic properties of large-scale heterogeneous populations of proteins, in the future important insights will be drawn from the examination of proteins at distinct positions within the cell nucleus, the imaging of single-molecule dynamics with high-speed microscopy and the examination of protein dynamics inside large-scale assemblies. Despite the fact that we have recently gained much knowledge about protein dynamics, a complete picture of how genomes function in vivo can only be achieved by revealing the organizational mechanisms and microdynamics of chromosomes and chromatin. Irrespective of the speed of future progress, the ability to map dynamic properties of chromatin, RNA and proteins in living cells has certainly provided a more comprehensive view of the organization of nuclear activities and brought an additional level of understanding of nuclear functions and the expression of genes.

10.11 Nuclear Mechanics and Impact of Perturbations and Fluctuations

The cell migration and invasion depend on the nuclear and cellular mechanics, which are interdependent and thereby possibly alter each other's mechanical properties (Fig. 10.8). It has been demonstrated that cell mechanics can alter the nuclear mechanics (Uhler and Shivashankar 2019; Wang et al. 2018; Mierke 2017). In line with these findings, the matrix mechanics of the cellular microenvironment can additionally alter the cell mechanics, which then subsequently change the nuclear mechanics (Martins et al. 2012; Haase et al. 2016; Lammerding 2011; Rabineau et al. 2018). As the nucleus is the largest and stiffest organelle in the cell, it is an important limiting factor for the migration of cancer cells through dense connective tissue.

The difference between loose and dense 3D collagen fiber matrices relies mainly on the different pore sizes and matrix stiffness than on the collagen fiber thickness. These findings correspond to the results on pure rat collagen fiber matrices of different collagen concentrations, where the collagen fibril diameter is not significantly dependent on the collagen monomer concentration (Sapudom et al. 2015). Hence, these two matrices represent appropriate model systems for cancer cell invasion that confines their invasion on the nuclear level due to the small pore sizes.

As expected, firstly, the loose 3D collagen matrices are less restrictive for cell migration and invasion of cancer cells, such as MDA-MB-231 and MCF-7 human breast cancer cells, since the nucleus is less confined and hence will not require large nuclear deformation. Secondly, the dense 3D collagen matrices are more restrictive for cancer cell migration and invasion, since the cell nucleus will require increased nuclear adaption.

When investigating the nuclear stiffness of TSA-treated cells, we found that the nuclear stiffness of malignant MDA-MB-231 breast cancer cells was increased and not decreased as reported for mouse

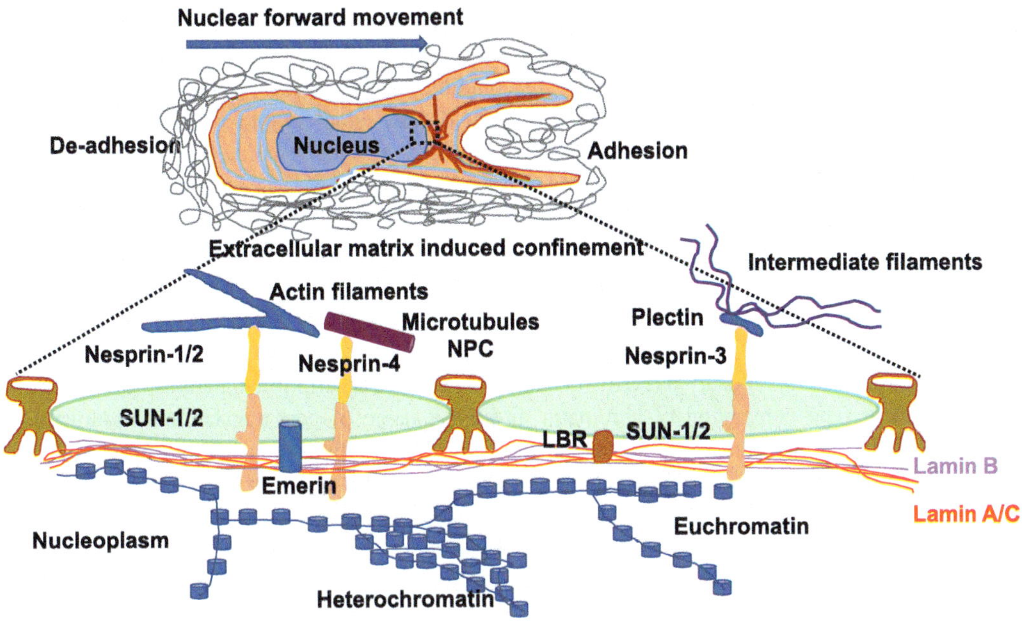

Fig. 10.8 Role of the cell nucleus during cell migration through extracellular matrix confinement. The connection between the cytoskeletal components and nuclear lamina seems to be crucial for providing cell migration in 3D matrices

NIH-3T3 fibroblasts (Masaeli et al. 2016). However, the nucleus of TSA-treated MCF-7 cells was softened, and their invasion depths were significantly increased in dense 3D collagen matrices. Their penetration depths into loose and hence less confined 3D collagen fiber matrices were reduced after induction of chromatin decondensation.

Since the nucleus is mechanically connected to the surrounding microenvironment by linker of nucleoskeleton and cytoskeleton complex (LINC) proteins (Crisp et al. 2006) that are linked to the inner and outer nuclear membranes and thereby coupled to the contractile cytoskeleton, which can alter the extracellular matrix or cellular microenvironment through focal adhesions (Belaadi et al. 2016; Heo et al. 2016a; Martins et al. 2012). The nuclear envelope is directly physically connected to the contractile cytoskeleton (Cho et al. 2017; Ungricht and Kutay 2017). Hence, we hypothesize that for both cancer cell types, a nuclear softening occurred after chromatin decondensation, but it seems to be counterbalanced by the cell types individual cytoskeleton, such as the actomyosin network.

Moreover, it has been reported that alterations in cytoskeletal mechanics, such as perinuclear cytoskeletal mechanics, are crucial for cell functions, such as adaption to external confinement mechanics changes (Miroshnikova et al. 2017). In addition, chromatin mechanics can be adapted to mechanical cues and thereby influence cellular functions to preserve the DNA from damage (Fig. 10.8).

Cells are continuously subject to dynamic changes in their external microenvironmental mechanics, a biochemical rearrangement of cabling and a physical restructuring of multiple cellular compartments, such as the nucleus, is necessary. Since the nucleus contains nearly the entire DNA of the cells, it is the largest and properly the stiffest structure inside the cells (Booth-Gauthier et al. 2012; Dahl et al. 2004; Dundr and Misteli 2001; Misteli et al. 2007). Thus, the nucleus represents a major obstacle for cell invasion into confined matrices.

The filamentous, insoluble constituent of the cell nucleus is collectively referred to as the nucleoskeleton (Dahl and Kalinowski 2011). It is essential that the nucleus is mechanically bound to the extracellular environment by linkers of nucleoskeleton and cytoskeleton complex proteins (LINC proteins), in particular the SUN proteins and Nesprins (Crisp et al. 2006), which are embedded in the INM or ONM, respectively, and are capable of coupling the contractile cytoskeleton that interacts with the extracellular matrix or adjacent cells by means of adhesion complexes (Belaadi et al. 2016; Heo et al. 2016a; Martins et al. 2012). The NE is directly physically connected to the contractile cytoskeleton. A specific structure of the nuclear envelope is reviewed in recent manuscripts (Cho et al. 2017; Ungricht and Kutay 2017; Mierke 2017).

Deep inside the cell nucleus, far beyond the NE and nuclear lamins, is the storage space for the cellular genetic material, referred to as chromatin, a complex of DNA and histones located. In specific detail, the amount of DNA packing around the histones governs the easiness of DNA transcriptional accessibility. When DNA is not actively transcribed, it remains in a tightly packed heterochromatin state; whereas, when it is loosely packed and accessible to the transcription apparatus, it is referred to as euchromatin. The conversion of chromatin through a range of post-translational modifications, such as methylation and acetylation, enables shifts between these two states and is strongly involved in the epigenetic adjustment of cell fate (Laugesen and Helin 2014; Tessarz and Kouzarides 2014). Most interestingly, chromatin in its heterochromatin phase, although apparently separated from the outer structural components of the nucleus and more likely immersed in the liquid-like nucleoplasm, is closely related physically to NE via the laminar-associated chromatin domains (LADs) (Lemaître and Bickmore 2015). Moreover, heterochromatin is fairly dynamic and experiences many topological alterations, particularly in the replicating and condensing chromosomes, and consequently is a key source of intrinsic mechanical excitation of the cell nucleus itself (Kumar et al. 2014). While it may behave similar to a fluid in global terms (Belaadi et al. 2016), relaxation and torsional repositioning of the diverse heterochromatin components over shorter lengths may dissipate mechanical energy that otherwise may harm the genetic information (Kumar et al. 2014). Thus, since chromatin is both an internal mechanical impulse and a force-absorbing entity, it seems to be, together with the nuclear lamina and nuclear/perinuclear actins, an active mechanosensor and mechanoresponder within the nucleus. In the following, the mechanisms of the force-sensing machinery and the dissipation of the force throughout the nucleus and its components, including the chromatin, will be presented and a focus is on the impact of the mechanical probing on cell physiology and disease states.

Such rheological alterations simplify scaling of the nuclear mechanics with the underlying tissue mechanics to assure proper tissue-specific performance (Swift et al. 2013), foster the nuclear stiffening during mesenchymal stem cell differentiation to elicit the differential gene expression pathway (Heo et al. 2016a) or initiate nuclear softening during cellular migration into confined spaces (Cao et al. 2016; Denais et al. 2016; Thiam et al. 2016) to assure proper propagation through narrow three-dimensional (3D) compartments without compromising genetic integrity (Fig. 10.8) (Le et al. 2016). The exact molecular pathways whereby nuclear actin governs transcription are still subject to further investigations, but this report offers first clues to a direct linkage of cytoskeletal and transcriptional states in reaction to mechanical stress (Miroshnikova et al. 2017).

The nuclear lamina affords the nuclear material reliability of mechanical strength and mechanical resistance, but it also restructures or breaks down to some extent in reaction to mechanical demands, for example, when cells migrate through narrow voids or tubes. These constraints could encourage actin-dependent mechanical adjustment and the mechanisms of dissipation of stress (i.e., deformation) in the cytoplasmic cytoskeleton but also in the nucleus and nucleoskeleton (Miroshnikova et al. 2017).

It is conceivable that simultaneously with the rearrangement and/or dismantling of the rigid nuclear lamina, a perinuclear F-actin ring arrangement enhances nuclear compliance and at the same time provides adequate mechanical resistance of the genetic information. This is indispensable in situations of external mechanical stress that demand a high degree of cellular and nuclear deformation capacity, as demonstrated for both cancer and immune cells that migrate through intricate environments.

Most interestingly, latest research has shown that nuclear breakage is a common incident during migration through confined areas, leading to DNA destruction at this site (Denais et al. 2016; Harada et al. 2014; Raab et al. 2016; Thiam et al. 2016). This injury might be aggravated by the mechanical elimination of DNA repair proteins because of their mobility away from the site of nuclear narrowing (Irianto et al. 2016). In line with this, a more elastic nucleus with perinuclear F-actin and decreased lamin A/C at the lamina prevented the cells from nuclear breakage (Denais et al. 2016; Thiam et al. 2016).

On the other side, the genomic instability evoked by the cell migration-facilitated nuclear deformation and the damage of DNA has been seen to support the increased heterogeneity of cancer (Irianto et al. 2017a), which additionally leads to the creation of a critical interaction between the nuclear mechanics, the overall integrity of the genome and the transformation-dependent switch of the cell's phenotype. Although, when combined, these investigations imply different perinuclear actin generation systems across a range of cell types, they all highlight the need to break down the rigid nuclear lamina to avoid nuclear disruption and injury to DNA, further illustrating a key function of the nucleus as an adaptable mechanical component of a cell.

In fact, the nucleus has two distinct compounds with different mechanical properties—the rigid elastic lamina and the more viscous nucleoplasm comprising the viscoelastic nucleoskeleton and the viscous fluid-like chromatin. The stiffness of the nucleoplasm lies in the range of a few tens to hundreds of pascals; whereas, the stiffness of the nuclear lamina falls in the range of a few kilopascals (Martins et al. 2012). It has also been shown that not merely the overall mechanical properties of chromatin, but also those unique to some components are probably crucial to the transfer of forces from the mechanosensitive and tensioned actomyosin cytoskeleton, which is connected physically to the chromatin via the LINC complexes and the nucleoskeleton (Driscoll et al. 2015; Heo et al. 2016a, b; Martins et al. 2012).

On the one hand, during mitotic spindle formation, the chromatin displays elastic spring behavior (Bouck and Bloom 2007). On the other hand, based on a chemomechanical model, the nucleus can be seen as an elastic shell (consisting of nuclear envelope and lamins) that encloses the nucleoplasm, which represents a soft poroelastic material. A damage-free (no irreversible deformations of the nucleus) cell migration through confined pores seems to be solely possible, when elastic and non-plastic chromatin regions are present (Cao et al. 2016). Based on these findings, chromatin deformability plays a key role for the migration and invasion of cells through spatially confined microenvironments, such as 3D extracellular matrices. In line with these results, analysis of chromatin integrity by micropipette aspiration has revealed that even a tenfold nucleus elongation during alignment and elongation of the nucleus and along the confined space of a pore through which a cell migrates (Irianto et al. 2017b).

It is therefore obvious that the movement, deformability and the architectural design of the chromatin together with a reorganization of the lamin A/C network and alterations in the nuclear and/or perinuclear actin cause a rearrangement of the stress within the nucleus in a biophysical sense. However, the question of whether epigenetically induced redistributions of space within the nucleus and alterations in its 3D microstructure itself are associated with the mechanical characteristics of chromatin has been left open for a long time. For this purpose, a more recently conducted experiment showed a varying viscosity of heterochromatin and euchromatin when the histone deacetylase inhibitor trichostatin A was applied for decondensation and sodium azide and 2-deoxyglucose for condensation of chromatin (Spagnol and Dahl 2016). In particular, the nucleus interior gets increasingly viscous and deformable during decondensation of the densely packed heterochromatin

(Chalut et al. 2012; Spagnol and Dahl 2016). This "switch" sensitiveness in condensation turns the fluorescence lifetime sensitive to the condensation state and intermediate conformations about the fiber compliance and its reliance on the viscosity in the non-condensed state. Earlier studies have demonstrated in a similar manner that chromatin decondensation leads to tougher (more viscous) and more deformable cell nuclei (Pajerowski et al. 2007; Chalut et al. 2012; Spagnol and Dahl 2016).

The mechanical alterations of chromatin may be attributable to basic alterations in the local concentration of chromatin and the associated alteration of the chromatin-free nucleoplasm. Similar to any other polymer, chromatin tends to stiffen, when it is increasingly concentrated, and hence the local concentration of chromatin correlates directly with the amount of the fluid-phase chromatin-free nucleoplasm at this specific region (Irianto et al. 2016; Pajerowski et al. 2007). These alterations in the mechanical characteristics of chromatin guide us to the hypothesis that chromatin performs a stress-absorbing function under mechanical stress. These findings are corroborated by our own studies that mechanical stress initiates the decondensation of the densely packed H3K9me3-modified (me3 denotes trimethylation) heterochromatin into a loosely packed H3K27me3-modified heterochromatin upon mechanical excitation, which is also completely reversible (Fig. 10.9) (Le et al. 2016).

There exists a highly dynamic connection between the packing density of chromatin and the nuclear mechanical properties (Pagliara et al. 2014). However, it can be observed that the nuclear deformation is different due to the cell type that has been investigated (Doolin et al. 2019). In line with this, the differentiation state of the cells may also impact the nuclear mechanical state.

From a functional point of view, chromatin condensation leads to gene silencing and at the same time allows the selective acquisition by the transcription machinery of some line-specific and constitutively expressed genes located within the euchromatin, which is useful in the setting of differentiation. But such nuclear redistributions also have a strong impact on the cellular reaction to

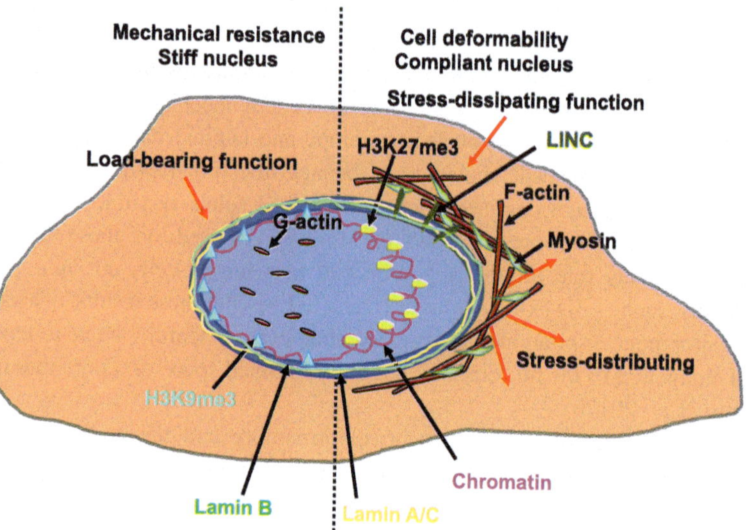

Fig. 10.9 Rheological adaption of the cell nucleus and regulation of perinuclear actin, lamins and chromatin. The cell nucleus serves as a rheological element that can be adapted due to environmental cues. The nucleus can either fulfill functions as a force bearing element providing the cell shape or as a stress-dissipating element allowing cell deformation due to environmental confinements, which are transduced via the LINC complex. Lamin A/C-rich regions of the nuclear periphery promote the creation of lamina-associated constitutive heterochromatin containing the H3K9me3 histone mark. In contrast, perinuclear actin-rich domains possess more loosely packed chromatin, which is dissociated from the lamina and contains the permissive H3K27me3 mark

mechanical signals. In the mode of undifferentiated cells, mechanical disturbance causes a nuclear deformation together with the cell deformation; whereas, in the differentiated mode the stiff nucleus does not bear the applied mechanical stress, but instead transfers it to the cytoskeleton-LINC nucleoskeleton matrix and stimulates its deformation (Heo et al. 2016a). It is nevertheless intriguing that there are some lineage-specific variations, and as mentioned above, specific cells, such as nucleated circulating blood cells, sustain a low lamin A level, which eases their frequent physical migration (Shin et al. 2013). In this case, the lamin A levels are a useful further guiding light for the specification of the lineage, presumably by their capacity to adjust gene activity, which is an efficient mechanism for linking nuclear elasticity and destiny (Shin et al. 2013). Taking all these facts into consideration, a model is recommended in which chromatin, as a viscoelastic rheological component of the cell nucleus capable of dynamically conforming to the changed mechanics by modifying its own state, occupies a pivotal function in cell and nuclear mechanics (Miroshnikova et al. 2017) (Fig. 10.10).

The strain-facilitated chromatin deformation seems to rely additionally on the presence of a rather intact actomyosin cytoskeleton and nucleoskeletal proteins, including lamins, emerin and SUN1/2 (Tajik et al. 2016). At longer timescales (10 min) and in the presence of a persistent mechanical impulse, both cell extrinsic (dynamic tensile stress) and cell intrinsic (increase in cellular contractility) mechanical probing cause ATP-dependent condensation of chromatin to heterochromatin in mesenchymal stem cells during differentiation, resulting in repression of gene expression in condensed regions (Heo et al. 2016b).

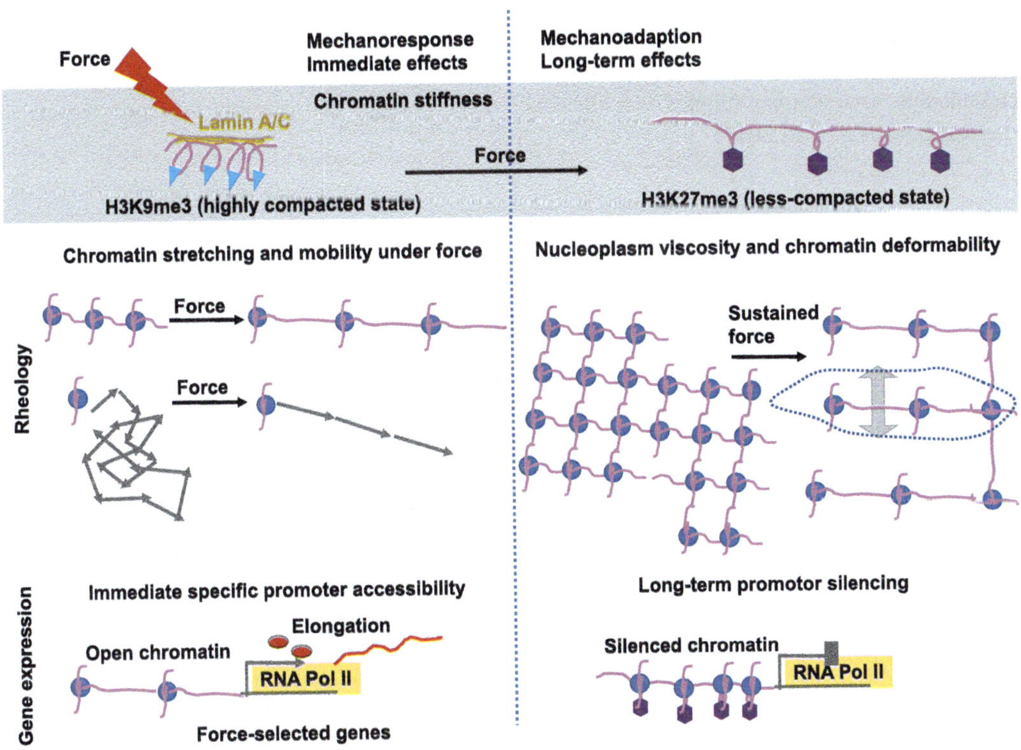

Fig. 10.10 Effect of cell extrinsic force or cell intrinsic force on chromatin structure and function over various timescales. In a high-force environment chromatin can display two effects, a short-time mechanoresponse (left half) and a long-term mechanoadaption feature (right half)

The extracellular forces can facilitate the deformation and stretching of chromatin in a direct way, thereby altering gene expression profiles and the fate of cells (Heo et al. 2016a; Irianto et al. 2017b; Le et al. 2016; Seong et al. 2013). It remains to be investigated whether this strain-based chromatin decondensation not only modifies nuclear mechanics but also enables more rapid DNA repair as a mechanism of mechanoprotection. The open chromatin is easier for DNA repair factors to access than the densely packaged heterochromatin (Price and D'Andrea 2013).

The mechanics of the nucleus and the chromatin represents still a highly active field of research, and there is a growing evidence of mechanical signaling to chromatin as the key driver of both control of gene activity and manipulation of the mechanical properties of the cell nucleus. However, it is essential to bear in mind that mechanical forces become apparent on different scales of length (i.e., at molecular, unicellular, monolayer or even tissue level) and are therefore probably causing highly specific and different responses. Therefore, more accurate determinations of the dynamic mechanical properties of the nucleus and subnuclear assemblies and enhanced mechanical multi-scale and multiphysics physical models are now required to improve knowledge of how and to what extent certain types of stresses, such as their magnitude, regime and/or temporal duration or application, influence nuclear mechanics and vary cellular mechanotransduction processes.

In summary, based on recent insights suggesting that the capacity to change nuclear lamins provides the rate-limiting stage for migration through confined 3D environments (Rowat et al. 2013; Thiam et al. 2016; Zaman et al. 2006), and the expression rates and stoichiometry of the different lamin isoforms presumably manipulate a series of processes ranging from the development and function of immune cells to cancer metastasis (Huang and Ingber 2005; Ingber 2006; Laklai et al. 2016; Levental et al. 2009; Miroshnikova et al. 2019). However, the status of the chromatin depends in a dynamic fashion on the cell fate due to their individual stemness or their state of differentiation and is even mostly disrupted in a specific illness state (Denais et al. 2016 Heo et al. 2016a, b; Le et al. 2016; Swift et al. 2013). In specific detail, besides the actin cytoskeleton, also other cytoskeletal components, such as microtubules, can alter nuclear morphology and possibly subsequently nuclear mechanics (Tariq et al. 2017).

References

J.R. Abney, B. Cutler, M.L. Fillbach, D. Axelrod, B.A. Scalettar, Chromatin dynamics in interphase nuclei and its implications for nuclear structure. J. Cell Biol. **137**, 1459–1468 (1997)

S.A. Adam, R. Sterne-Marr, L. Gerace, In vitro nuclear protein import using permeabilized mammalian cells. Methods Cell Biol. **35**, 469–482 (1991)

R.L. Adams, S.R. Wente, Uncovering nuclear pore complexity with innovation. Cell **152**, 1218–1221 (2013)

C. Ader, S. Frey, W. Maas, H.B. Schmidt, D. Görlich, M. Baldus, Amyloid-like interactions within nucleoporin FG hydrogels. Proc. Natl. Acad. Sci. U. S. A. **107**, 6281–6285 (2010)

J.D. Aitchison, M.P. Rout, The road to ribosomes. Filling potholes in the export pathway. J. Cell Biol. **151**, F23–F26 (2000)

A. Akhtar, S.M. Gasser, The nuclear envelope and transcriptional control. Nat. Rev. Genet. **8**, 507–517 (2007)

F. Alber, S. Dokudovskaya, L.M. Veenhoff et al., The molecular architecture of the nuclear pore complex. Nature **450**, 695–701 (2007)

A.R. Alcazar-Roman, E.J. Tran, S. Guo, S.R. Wente, Inositol hexakisphosphate and Gle1 activate the DEAD-box protein Dbp5 for nuclear mRNA export. Nat. Cell Biol. **8**, 711–716 (2006)

N.P. Allen, L. Huang, A. Burlingame, M. Rexach, Proteomic analysis of nucleoporin interacting proteins. J. Biol. Chem. **276**, 29268–29274 (2001)

T.D. Allen, J.M. Cronshaw, S. Bagley, E. Kiseleva, M.W. Goldberg, The nuclear pore complex: mediator of translocation between nucleus and cytoplasm. J. Cell Sci. **113**(Pt 10), 1651–1659 (2000)

S. Amlacher, P. Sarges, D. Flemming, V. van Noort, R. Kunze, D.P. Devos, M. Arumugam, P. Bork, E. Hurt, Insight into structure and assembly of the nuclear pore complex by utilizing the genome of a eukaryotic thermophile. Cell **146**, 277–289 (2011)

K.R. Andersen, E. Onischenko, J.H. Tang, P. Kumar, J.Z. Chen, A. Ulrich, J.T. Liphardt, K. Weis, T.U. Schwartz, Scaffold nucleoporins Nup188 and Nup192 share structural and functional properties with nuclear transport receptors. eLife **2**, e00745 (2013)

D.J. Anderson, M.W. Hetzer, Nuclear envelope formation by chromatin-mediated reorganization of the endoplasmic reticulum. Nat. Cell Biol. **9**, 1160–1166 (2007)

D.J. Anderson, M.W. Hetzer, Reshaping of the endoplasmic reticulum limits the rate for nuclear envelope formation. J. Cell Biol. **182**, 911–924 (2008)

D.J. Anderson, J.D. Vargas, J.P. Hsiao, M.W. Hetzer, Recruitment of functionally distinct membrane proteins to chromatin mediates nuclear envelope formation in vivo. J. Cell Biol. **186**, 183–191 (2009)

J.T. Anderson, S.M. Wilson, K.V. Datar, M.S. Swanson, NAB2: a yeast nuclear polyadenylated RNA-binding protein essential for cell viability. Mol. Cell. Biol. **13**, 2730–2741 (1993)

E.D. Andrulis, A.M. Neiman, D.C. Zappulla, R. Sternglanz, Perinuclear localization of chromatin facilitates transcriptional silencing. Nature **394**, 592–595 (1998)

W. Antonin, C. Franz, U. Haselmann, C. Antony, I.W. Mattaj, The integral membrane nucleoporin pom121 functionally links nuclear pore complex assembly and nuclear envelope formation. Mol. Cell **17**, 83–92 (2005)

W. Antonin, R. Ungricht, U. Kutay, Traversing the NPC along the pore membrane: targeting of membrane proteins to the INM. Nucleus **2**, 87–91 (2011)

A. Arjomand, M.A. Baker, C. Li, A.M. Buckle, D.A. Jans, K.L. Loveland, Y. Miyamoto, The alpha-importome of mammalian germ cell maturation provides novel insights for importin biology. FASEB J. **28**(8), 3480–3493 (2014)

H. Asakawa, H.J. Yang, T.G. Yamamoto, C. Ohtsuki, Y. Chikashige, K. Sakata-Sogawa, M. Tokunaga, M. Iwamoto, Y. Hiraoka, T. Haraguchi, Characterization of nuclear pore complex components in fission yeast Schizosaccharomyces pombe. Nucleus **5**, 149–162 (2014)

A.G. Balakin, L. Smith, M.J. Fournier, The RNA world of the nucleolus: two major families of small RNAs defined by different box elements with related functions. Cell **86**, 823–834 (1996)

A. Banerjee, L.H. Apponi, G.K. Pavlath, A.H. Corbett, PABPN1: molecular function and muscle disease. FEBS J. **280**, 4230–4250 (2013)

D.W. Bauer, J.G. Gall, Coiled bodies without coilin. Mol. Biol. Cell **8**, 73–82 (1997)

G. Bauren, L. Wieslander, Splicing of Balbiani ring 1 gene pre-mRNA occurs simultaneously with transcription. Cell **76**, 183–192 (1994)

M. Beck, J.S. Glavy, Toward understanding the structure of the vertebrate nuclear pore complex. Nucleus **5**, 119–123 (2014)

M. Beck, F. Forster, M. Ecke, J.M. Plitzko, F. Melchior, G. Gerisch, W. Baumeister, O. Medalia, Nuclear pore complex structure and dynamics revealed by cryoelectron tomography. Science **306**, 1387–1390 (2004)

M. Beck, V. Lucic, F. Forster, W. Baumeister, O. Medalia, Snapshots of nuclear pore complexes in action captured by cryo-electron tomography. Nature **449**, 611–615 (2007)

N. Belaadi, J. Aureille, C. Guilluy, Under pressure: mechanical stress management in the nucleus. Cells **5**(2), 27 (2016)

N. Belgareh, G. Rabut, S.W. Baï et al., An evolutionarily conserved NPC subcomplex, which redistributes in part to kinetochores in mammalian cells. J. Cell Biol. **154**, 1147–1160 (2001)

A.S. Belmont, S. Dietzel, A.C. Nye, Y.G. Strukov, T. Tumbar, Large-scale chromatin structure and function. Curr. Opin. Cell Biol. **11**, 307–311 (1999)

D. Bentley, Coupling RNA polymerase II transcription with pre-mRNA processing. Curr. Opin. Cell Biol. **11**(347–351), 1–5 (1999)

R. Berezney, D.S. Coffey, Identification of a nuclear protein matrix. Biochem. Biophys. Res. Commun. **60**, 1410–1417 (1974)

I.C. Berke, T. Boehmer, G. Blobel, T.U. Schwartz, Structural and functional analysis of Nup133 domains reveals modular building blocks of the nuclear pore complex. J. Cell Biol. **167**, 591–597 (2004)

A. Bestembayeva, Nanoscale stiffness topography reveals structure and mechanics of the transport barrier in intact nuclear pore complexes. Nat Nanotechnol. **10**, 60–64 (2014)

A.L. Beyer, Y.N. Osheim, Splice site selection, rate of splicing, and alternative splicing on nascent transcripts. Genes Dev. **2**, 754–765 (1988)

S. Bilokapic, T.U. Schwartz, 3D ultrastructure of the nuclear pore complex. Curr. Opin. Cell Biol. **24**, 86–91 (2012a)

S. Bilokapic, T.U. Schwartz, Molecular basis for Nup37 and ELY5/ELYS recruitment to the nuclear pore complex. Proc. Natl. Acad. Sci. U. S. A. **109**, 15241–15246 (2012b)

G. Blobel, Gene gating: a hypothesis. Proc. Natl. Acad. Sci. U. S. A. **82**(24), 8527–8529 (1985)

T. Boehmer, S. Jeudy, I.C. Berke, T.U. Schwartz, Structural and functional studies of Nup107/Nup133 interaction and its implications for the architecture of the nuclear pore complex. Mol. Cell **30**, 721–731 (2008)

K. Bohmann, J.A. Ferreira, A.I. Lamond, Mutational analysis of p80 coilin indicates a functional interaction between coiled bodies and the nucleolus. J. Cell Biol. **131**, 817–831 (1995)

A. Boni, A.Z. Politi, P. Strnad, W. Xiang, M.J. Hossain, J. Ellenberg, Live imaging and modeling of inner nuclear membrane targeting reveals its molecular requirements in mammalian cells. J. Cell Biol. **209**, 705–720 (2015)

E.A. Booth-Gauthier, T.A. Alcoser, G. Yang, K.N. Dahl, Force-induced changes in subnuclear movement and rheology. Biophys. J. **103**, 2423–2431 (2012)

J. Borden, L. Manuelidis, Movement of the X chromosome in epilepsy. Science **242**, 1687–1691 (1988)

D.C. Bouck, K. Bloom, Pericentric chromatin is an elastic component of the mitotic spindle. Curr. Biol. **17**, 741–748 (2007)

K. Boudonck, L. Dolan, P.J. Shaw, The movement of coiled bodies visualized in living plant cells by the green fluorescent protein. Mol. Biol. Cell **10**, 2297–2307 (1999)

H. Bretes, J.O. Rouviere, T. Leger, M. Oeffinger, F. Devaux, V. Doye, B. Palancade, Sumoylation of the THO complex regulates the biogenesis of a subset of mRNPs. Nucleic Acids Res. **42**(8), 5043–5058 (2014)

J.F. Briand, F. Navarro, O. Gadal, P. Thuriaux, Cross talk between tRNA and rRNA synthesis in *Saccharomyces cerevisiae*. Mol. Cell. Biol. **21**, 189–195 (2001)

D.G. Brickner, Ahmed, S.L. Meldi, A. Thompson, W. Light, M. Young, T.L. Hickman, F. Chu, E. Fabre, J.H. Brickner, Transcription factor binding to a DNA zip code controls interchromosomal clustering at the nuclear periphery. Dev. Cell **22**, 1234–1246 (2012)

Brickner J H and Walter P 2004 Gene recruitment of the activated INO1 locus to the nuclear membrane *PLoS Biol.* **2** e342

J.M. Bridger, H. Herrmann, C. Munkel, P. Lichter, Identification of an interchromosomal compartment by polymerization of nuclear-targeted vimentin. J. Cell Sci. **111**, 1241–1253 (1998)

S.G. Brohawn, N.C. Leksa, E.D. Spear, K.R. Rajashankar, T.U. Schwartz, Structural evidence for common ancestry of the nuclear pore complex and vesicle coats. Science **322**, 1369–1373 (2008)

S.G. Brohawn, J.R. Partridge, J.R. Whittle, T.U. Schwartz, The nuclear pore complex has entered the atomic age. Structure **17**, 1156–1168 (2009)

K.E. Brown, J. Baxter, D. Graf, M. Merkenschlager, A.G. Fisher, Dynamic repositioning of genes in the nucleus of lymphocytes preparing for cell division. Mol. Cell **3**, 207–217 (1999)

K.E. Brown, S.S. Guest, S.T. Smale, K. Hahm, M. Merkenschlager, A.G. Fisher, Association of transcriptionally silent genes with Ikaros complexes at centromeric heterochromatin. Cell **91**, 845–854 (1997)

K.H. Bui, A. von Appen, A.L. DiGuilio et al., Integrated structural analysis of the human nuclear pore complex scaffold. Cell **155**, 1233–1243 (2013)

B. Burke, C.L. Stewart, The nuclear lamins: flexibility in function. Nat. Rev. Mol. Cell Biol. **14**, 13–24 (2013)

R.A. Burrell, N. McGranahan, J. Bartek, C. Swanton, The causes and consequences of genetic heterogeneity in cancer evolution. Nature **501**, 338–345 (2013)

H. Busch, K.S. Narayan, J. Hamilton, Isolation of nucleoli in a medium containing spermine and magnesium acetate. Exp. Cell Res. **47**, 329–336 (1967)

V. Butin-Israeli, S.A. Adam, A.E. Goldman, R.D. Goldman, Nuclear lamin functions and disease. Trends Genet. **28**, 464–471 (2012)

H.G. Callan, J.G. Gall, Association of RNA with the B and C snurposomes of Xenopus oocyte nuclei. Chromosoma **101**, 69–82 (1991)

X. Cao, E. Moeendarbary, P. Isermann, P.M. Davidson, X. Wang, M.B. Chen, A.K. Burkart, J. Lammerding, R.D. Kamm, V.B. Shenoy, A chemomechanical model for nuclear morphology and stresses during cell transendothelial migration. Biophys. J. **111**, 1541–1552 (2016)

M. Carmo-Fonseca, L. Mendes-Soares, I. Campos, To be or not to be in the nucleolus. Nat. Cell Biol. **2**, E107–E112 (2000)

M. Carroll, K.L. Borden, The oncogene eIF4E: using biochemical insights to target cancer. J. Interferon Cytokine Res. **33**, 227–238 (2013)

K.C. Carter, D. Bowman, W. Carrington, K. Fogarty, J.A. McNeil, F.S. Fay, J.B. Lawrence, A three-dimensional view of precursor messenger RNA metabolism within the mammalian nucleus. Science **259**, 1330–1335 (1993)

K.C. Carter, K.L. Taneja, J.B. Lawrence, Discrete nuclear domains of poly(A) RNA and their relationship to the functional organization of the nucleus. J. Cell Biol. **115**, 1191–1202 (1991)

T. Carvalho, F. Almeida, A. Calapez, M. Lafarga, M.T. Berciano, M. Carmo-Fonseca, The spinal muscular atrophy disease gene product, SMN: a link between snRNP biogenesis and the Cajal (coiled) body. J. Cell Biol. **147**, 715–728 (1999)

A. Castello, B. Fischer, M.W. Hentze, T. Preiss, RNA-binding proteins in Mendelian disease. Trends Genet. **29**, 318–327 (2013)

K.J. Chalut, M. Hopfler, F. Lautenschlager, L. Boyde, C.J. Chan, A. Ekpenyong, A. Martinez-Arias, J. Guck, Chromatin decondensation and nuclear softening accompany Nanog downregulation in embryonic. Stem. Cells Biophys. J. **103**, 2060–2070 (2012)

H. Cheng, K. Dufu, C.S. Lee, J.L. Hsu, A. A. Dias A and Reed R 2006 Human mRNA export machinery recruited to the 5′ end of mRNA. Cell **127**, 1389–1400 (2006)

M. Chiquet, L. Gelman, R. Lutz, S. Maier, From mechanotransduction to extracellular matrix gene expression in fibroblasts. Biochim. Biophys. Acta (BBA)-Mol. Cell Res. **1793**(5), 911–920

S. Cho, J. Irianto, D.E. Discher, Mechanosensing by the nucleus: from pathways to scaling relationships. J. Cell Biol. **216**, 1–11 (2017)

H. Chug, S. Trakhanov, B.B. Hülsmann, T. Pleiner, D. Görlich, Crystal structure of the metazoan Nup62•Nup58•Nup54 nucleoporin complex. Science **350**, 106–110 (2015)

G. Cingolani, C. Petosa, K. Weis, C.W. Muller, Structure of importin-beta bound to the IBB domain of importin-alpha. Nature **399**, 221–229 (1999)

S.C. Cloutier, W.K. Ma, L.T. Nguyen, E.J. Tran, The DEAD-box RNA helicase Dbp2 connects RNA quality control with repression of aberrant transcription. J. Biol. Chem. **287**, 26155–26166 (2012)

D. Cmarko, P.J. Verschure, T.E. Martin, M.E. Dahmus, S. Krause, X.D. Fu, R. van Driel, S. Fakan, Ultrastructural analysis of transcription and splicing in the cell nucleus after bromo-UTP microinjectio. Mol. Biol. Cell **10**, 211–223 (1999)

T.H. Coady, C.L. Lorson, SMN in spinal muscular atrophy and snRNP biogenesis *Wiley Interdiscip*. Rev. RNA **2**, 546–564 (2011)

M. Cockell, S.M. Gasser, Nuclear compartments and gene regulation. Curr. Opin. Genet. Dev. **9**, 199–205 (1999)

T.V. Cohen, L. Hernandez, C.L. Stewart, Functions of the nuclear envelope and lamina in development and disease. Biochem. Soc. Trans. **36**, 1329–1334 (2008)

E. Conti, C.W. Muller, M. Stewart, Karyopherin flexibility in nucleocytoplasmic transport. Curr. Opin. Struct. Biol. **16**, 237–244 (2006)

E. Conti, M. Uy, L. Leighton, G. Blobel, J. Kuriyan, Crystallographic analysis of the recognition of a nuclear localization signal by the nuclear import factor karyopherin alpha. Cell **94**, 193–204 (1998)

A.G. Cook, E. Conti, Nuclear export complexes in the frame. Curr. Opin. Struct. Biol. **20**(2), 247–252 (2010)

P.R. Cook, The organization of replication and transcription Science **284**, 1790–1795 (1999)

T.A. Cooper, L. Wan, G. Dreyfuss, RNA and disease Cell **136**, 777–793 (2009)

E.A. Corbin, O.O. Adeniba, R.H. Ewoldt, R. Bashir, Dynamic mechanical measurement of the viscoelasticity of single adherent cells. Appl. Phys. Lett. **108**(9), 093701 (2016)

T. Cremer, C. Cremer, T. Schneider, H. Baumann, E.K. Luedtke, K. Sperling, V. Teuber, C. Zorn, Rabl's model of the interphase chromosome arrangement tested in Chinese hamster cells by premature chromosome condensation and laser UV-microbeam experiments. Hum. Genet. **60**, 46–56 (1982)

T. Cremer, G. Kreth, H. Koester, R.H.A. Fink, R. Heintzmann, M. Cremer, I. Solovei, D. Zink, C. Cremer, Chromosome territories, interchromatin domain compartment and nuclear matrix: an integrated view of the functional nuclaer architecture. Crit. Rev. Gene Expr. **12**, 179–212 (2000)

T. Cremer, P. Lichter, J. Borden, D.C. Ward, L. Manuelidis, Detection of chromosome aberrations in metaphase and interphase tumor cells by in-situ hybridization using chromosome specific library probes. Hum. Genet. **80**, 235–246 (1988)

M. Crisp, Q. Liu, K. Roux, J. Rattner, C. Shanahan, B. Burke, P.D. Stahl, D. Hodzic, Coupling of the nucleus and cytoplasm: role of the LINC complex. J. Cell Biol. **172**, 41–53 (2006)

J.M. Cronshaw, A.N. Krutchinsky, W. Zhang, B.T. Chait, M.J. Matunis, Proteomic analysis of the mammalian nuclear pore complex. J. Cell Biol. **158**, 915–927 (2002)

M.A. D'Angelo, J.S. Gomez-Cavazos, A. Mei, D.H. Lackner, M.W. Hetzer, A change in nuclear pore complex composition regulates cell differentiation. Dev. Cell **22**, 446–458 (2012)

M.A. D'Angelo, M. Raices, S.H. Panowski, M.W. Hetzer, Age-dependent deterioration of nuclear pore complexes causes a loss of nuclear integrity in postmitotic cells. Cell **136**, 284–295 (2009)

A. D'Urso, J.H. Brickner, Mechanisms of epigenetic memory. Trends Genet. **30**(6), 230–236 (2014)

M.A. D'Angelo, M.W. Hetzer, The role of the nuclear envelope in cellular organization. Cell. Mol. Life Sci. **63**, 316–332 (2006)

M.A. D'Angelo, M.W. Hetzer, Structure, dynamics and function of nuclear pore complexes. Trends Cell Biol. **18**, 456–466 (2008)

K.N. Dahl, A. Kalinowski, Nucleoskeleton mechanics at a glance. J. Cell Sci. **124**, 675–678 (2011)

K.N. Dahl, A.J. Engler, J.D. Pajerowski, D.E. Discher, Power-law rheology of isolated nuclei with deformation mapping of nuclear substructures. Biophys. J. **89**(4), 2855–2864 (2005)

K.N. Dahl, S.M. Kahn, K.L. Wilson, D.E. Discher, The nuclear envelope lamina network has elasticity and a compressibility limit suggestive of a molecular shock absorber. J. Cell Sci. **117**, 4779–4786 (2004)

B. Daneholt, Transcription in polytene chromosomes. Cell **4**, 1–9 (1975)

I. Davis, D. Ish-Horowicz, Apical localization of pair-rule transcripts requires 3 h sequences and limits protein diffusion in the *Drosophila blastoderm* embryo. Cell **67**, 927–940 (1991)

R. de Leeuw, Y. Gruenbaum, O. Medalia, Nuclear lamins: thin filaments with major functions. Trends Cell Biol. **28**(1), 34–45 (2018)

E.W. Debler, Y. Ma, H.S. Seo, K.C. Hsia, T.R. Noriega, G. Blobel, A. Hoelz, A fence-like coat for the nuclear pore membrane. Mol. Cell **32**, 815–826 (2008)

T. Dechat, K. Gesson, R. Foisner, Lamina-independent lamins in the nuclear interior serve important functions. Cold. Spring. Harb. Symp. Quant. Biol. **75**, 533–543 (2010)

T. Dechat, K. Pfleghaar, K. Sengupta, T. Shimi, D.K. Shumaker, L. Solimando, R.D. Goldman, Nuclear lamins: major factors in the structural organization and function of the nucleus and chromatin. Genes Dev. **22**, 832–853 (2008)

J.A. Degrasse, D. Devos, A functional proteomic study of the Trypanosoma brucei nuclear pore complex: an informatic strategy. Methods Mol. Biol. **673**, 231–238 (2010)

J.A. Degrasse, K.N. DuBois, D. Devos, T.N. Siegel, A. Sali, M.C. Field, M.P. Rout, B.T. Chait, Evidence for a shared nuclear pore complex architecture that is conserved from the last common eukaryotic ancestor. Mol. Cell. Proteomics **8**, 2119–2130 (2009)

C.M. Denais, R.M. Gilbert, P. Isermann, A.L. McGregor, M. te Lindert, B. Weigelin, P.M. Davidson, P. Friedl, K. Wolf, J. Lammerding, Nuclear envelope rupture and repair during cancer cell migration. Science **352**, 353–358 (2016)

D.P. Denning, S.S. Patel, V. Uversky, A.L. Fink, M. Rexach, Disorder in the nuclear pore complex: the FG repeat regions of nucleoporins are natively unfolded. Proc. Natl. Acad. Sci. U. S. A. **100**, 2450–2455 (2003)

A.F. Dernburg, K.W. Broman, J.C. Fung, W.F. Marshall, J. Philips, D.A. Agard, J.W. Sedat, Perturbation of nuclear architecture by long-distance chromosome interactions. Cell **85**, 745–759 (1996)

H. de Thé, C. Chomienne, M. Lanotte, L. Degas, A. Dejan, The t(15;17) translocation of acute promyelocytic leukemia fuses the retinoic acid receptor alpha gene to a novel transcribed locus. Nature **367**, 558–561 (1990)

D.P. Devos, R. Gräf, M.C. Field, Evolution of the nucleus. Curr. Opin. Cell Biol. **28C**, 8–15 (2014)

D. Devos, S. Dokudovskaya, F. Alber, R. Williams, B.T. Chait, A. Sali, M.P. Rout, Components of coated vesicles and nuclear pore complexes share a common molecular architecture. PLoS Biol. **2**, e380 (2004)

D. Devos, S. Dokudovskaya, R. Williams, F. Alber, N. Eswar, B.T. Chait, M.P. Rout, A. Sali, Simple fold composition and modular architecture of the nuclear pore complex. Proc. Natl. Acad. Sci. U. S. A. **103**, 2172–2177 (2006)

R.W. Dirks, K.C. Daniel, A.K. Raap, RNAs radiate from gene to cytoplasm as revealed by fluorescence in situ hybridization. J. Cell Sci. **108**, 2565–2572 (1995)

M.T. Doolin, T.S. Ornstein, K.M. Stroka, nuclear deformation in response to mechanical confinement is cell type dependent. Cells **8**(5), 427 (2019)

D. Dorner, J. Gotzmann, R. Foisner, Nucleoplasmic lamins and their interaction partners, LAP2a, Rb, and BAF, in transcriptional regulation. FEBS J. **274**, 1362–1373 (2007)

C.M. Doucet, M.W. Hetzer, Nuclear pore biogenesis into an intact nuclear envelope. Chromosoma **119**, 469–477 (2010)

T. Dousset, C. Wang, C. Verheggen, D. Chen, D. Hernandez-Verdun, S. Huang, Initiation of nucleolar assembly is independent of RNA polymerase I transcription. Mol. Biol. Cell **11**, 2705–2717 (2000)

T.P. Driscoll, B.D. Cosgrove, S.-J. Heo, Z.E. Shurden, R.L. Mauck, Cytoskeletal to nuclear strain transfer regulates YAP signaling in mesenchymal stem cells. Biophys. J. **108**, 2783–2793 (2015)

L. Du, S.L. Warren, A functional interaction between the carboxy-terminal domain of RNA polymerase II and pre-mRNA splicing. J. Cell Biol. **136**, 5–18 (1997)

E. Dultz, J. Ellenberg, Live imaging of single nuclear pores reveals unique assembly kinetics and mechanism in interphase. J. Cell Biol. **191**, 15–22 (2010)

E. Dultz, E. Zanin, C. Wurzenberger, M. Braun, G. Rabut, L. Sironi, J. Ellenberg, Systematic kinetic analysis of mitotic dis- and reassembly of the nuclear pore in living cells. J. Cell Biol. **180**, 857–865 (2008)

M. Dundr, T. Misteli, Functional architecture in the cell nucleus. Biochem. J. **356**, 297–310 (2001)

M. Dundr, I. Raska, Nonisotopic ultrastructural mapping of transcription sites within the nucleolus. Exp. Cell Res. **208**, 275–281 (1993)

M. Dundr, T. Misteli, M.O.J. Olson, The dynamics of postmitotic reassembly of the nucleolus. J. Cell Biol. **150**, 433–446 (2000)

J.A. Dyck, G.G. Maul, W.H. Miller, J.D. Chen, A. Kakizuka, R.M. Evans, A novel macromolecular structure is a target of the promyelocyte-retinoic acid receptor oncoprotein. Cell **76**, 333–343 (1994)

B.A. Edgar, M.P. White, G. Schubiger, T. Kornberg, Repression and turnover pattern of fusi tarazu RNA in the early *Drosophila embryo*. Cell **47**, 747–754 (1986)

D. Egecioglu, J.H. Brickner, Gene positioning and expression. Curr. Opin. Cell Biol. **23**, 338–345 (2011)

M. Eibauer, M. Pellanda, Y. Turgay, A. Dubrovsky, A. Wild, O. Medalia, Structure and gating of the nuclear pore complex. Nat. Commun. **6**, 7532 (2015)

N. Eisenhardt, J. Redolfi, W. Antonin, Interaction of Nup53 with Ndc1 and Nup155 is required for nuclear pore complex assembly. J. Cell Sci. **127**, 908–921 (2014)

C. Eriksson, C. Rustum, E. Hallberg, Dynamic properties of nuclear pore complex proteins in gp210 deficient cells. FEBS Lett. **572**, 261–265 (2004)

F. Esra Demircioglu, V.E. Cruz, T.U. Schwartz, Purification and structural analysis of SUN and KASH domain proteins. Methods Enzymol. **569**, 63–78 (2016)

S. Fakan, Perichromatin fibrils are in situ forms of nascent transcripts. Trends Cell Biol. **4**, 86–90 (1994)

M.B. Fasken, A.H. Corbett, Process or perish: quality control in mRNA biogenesis. Nat. Struct. Mol. Biol. **12**, 482–488 (2005)

D.W. Fawcett, On the occurrence of a fibrous lamina on the inner aspect of the nuclear envelope in certain cells of vertebrates. Am. J. Anat. **119**, 129–145 (1996)

N. Fay, N. Pante, Nuclear entry of DNA viruses. Front. Microbiol. **6**, 467 (2015)

C.M. Feldherr, D. Akin, The location of the transport gate in the nuclear pore complex. J. Cell Sci. **110**(Pt 24), 3065–3070 (1997)

J. Fernandez-Martinez, M.P. Rout, A jumbo problem: mapping the structure and functions of the nuclear pore complex. Curr. Opin. Cell Biol. **24**, 92–99 (2012)

J. Fernandez-Martinez, J. Phillips, M.D. Sekedat et al., Structure–function mapping of a heptameric module in the nuclear pore complex. J. Cell Biol. **196**, 419–434 (2012)

D. Ferrera, C. Canale, R. Marotta, N. Mazzaro, M. Gritti, M. Mazzanti, S. Capellari, P. Cortelli, L. Gasparini, Lamin B1 overexpression increases nuclear rigidity in autosomal dominant leukodystrophy fibroblasts. FASEB J. **28**(9), 3906–3918 (2014)

R. Festenstein, D. Kioussis, Locus control regions and epigenetic chromatin modifiers. Curr. Opin. Genet. Dev. **10**, 199–203 (2000)

M.C. Field, L. Koreny, M.P. Rout, Enriching the pore: splendid complexity from humble origins. Traffic **15**, 141–156 (2014)

J. Fischer, R. Teimer, S. Amlacher, R. Kunze, E. Hurt, Linker Nups connect the nuclear pore complex inner ring with the outer ring and transport channel. Nat. Struct. Mol. Biol. **22**, 774–781 (2015)

J. Fiserova, M.W. Goldberg, Relationships at the nuclear envelope: lamins and nuclear pore complexes in animals and plants. Biochem. Soc. Trans. **38**, 829–831 (2010)

J. Fiserova, E. Kiseleva, M.W. Goldberg, Nuclear envelope and nuclear pore complex structure and organization in tobacco BY-2 cells. Plant J. **59**, 243–255 (2009)

D. Flemming, D.P. Devos, J. Schwarz, S. Amlacher, M. Lutzmann, E. Hurt, Analysis of the yeast nucleoporin Nup188 reveals a conserved S-like structure with similarity to karyopherins. J Struct Biol. **177**, 99–105 (2012)

A.W. Folkmann, S.E. Collier, X. Zhan, P. Aditi, M.D. Ohi, S.R. Wente, Gle1 functions during mRNA export in an oligomeric complex that is altered in human disease. Cell **155**, 582–593 (2013)

A.W. Folkmann, T.R. Dawson, S.R. Wente, Insights into mRNA export-linked molecular mechanisms of human disease through a Gle1 structure–function analysis. Adv. Biol. Regul. **54**, 74–91 (2014)

M. Fornerod, M. Ohno, M. Yoshida, I.W. Mattaj, CRM1 is an export receptor for leucine-rich nuclear export signals. Cell **90**, 1051–1060 (1997)

C. Francastel, D. Schuebler, D.I.K. Martin, M. Groudine, Nuclear compartmentalization and gene activity. Nat. Rev. Mol. Cell Biol. **1**, 137–143 (2000)

C. Franz, P. Askjaer, W. Antonin, C.L. Iglesias, U. Haselmann, M. Schelder, A. de Marco, M. Wilm, C. Antony, I.W. Mattaj, Nup155 regulates nuclear envelope and nuclear pore complex formation in nematodes and vertebrates. EMBO J **24**, 3519–3531 (2005)

C. Franz, R. Walczak, S. Yavuz, R. Santarella, M. Gentzel, P. Askjaer, V. Galy, M. Hetzer, I.W. Mattaj, W. Antonin, MEL-28/ELYS is required for the recruitment of nucleoporins to chromatin and postmitotic nuclear pore complex assembly. EMBO Rep. **8**, 165–172 (2007)

M.R. Frey, A.G. Matera, Coiled bodies contain U7 small nuclear RNA and associate with specific DNA sequence in interphase human cells. Proc. Natl. Acad. Sci. U. S. A. **92**, 5915–5919 (1995)

M.R. Frey, A.D. Bailey, A.M. Weiner, A.G. Matera, Association of snRNA genes with coiled bodies is mediated by nascent snRNA transcripts. Curr. Biol. **9**, 126–135 (1999)

S. Frey, R.P. Richter, D. Gorlich, FG-rich repeats of nuclear pore proteins form a three-dimensional meshwork with hydrogel-like properties. Science **314**, 815–817 (2006)

A. Fridkin, A. Penkner, V. Jantsch, Y. Gruenbaum, SUN-domain and KASH-domain proteins during development, meiosis and disease. Cell Mol. Life Sci. **66**, 1518–1533 (2009)

J.G. Gall, Cajal Bodies: the first 100 years. Annu. Rev. Cell. Dev. Biol. **16**, 273–300 (2000)

J.G. Gall, H.G. Callan, H3-uridine incorporation in lampbrush chromosomes. Proc. Natl. Acad. Sci. U. S. A. **48**, 562–570 (1962)

V. Galy, I.W. Mattaj, P. Askjaer, Caenorhabditis elegans nucleoporins Nup93 and Nup205 determine the limit of nuclear pore complex size exclusion in vivo. Mol. Biol. Cell **14**, 5104–5115 (2003)

P. Ganot, B.E. Jady, M.L. Bortolin, X. Darzacq, T. Kiss, Nucleolar factors direct the 2 h-O-ribose methylation and pseudouridylation of U6 spliceosomal RNA. Mol. Cell. Biol. **19**, 6906–6917 (1999)

L. Gao, M.R. Frey, A.G. Matera, Human genes encoding U3 snRNA associate with coiled bodies in interphase cells and are clustered on chromosome 17p11.2 in a complex inverted repeat structure. Nucleic Acids Res. **25**, 4740–4747 (1997)

J.S. Gomez-Cavazos, M.W. Hetzer, The nucleoporin gp210/Nup210 controls muscle differentiation by regulating nuclear envelope/ER homeostasis. J. Cell Biol. **208**, 671–681 (2015)

D.E. Gottschling, O.M. Aparicio, B.L. Billington, V.A. Zakian, Position effect at *S. cerevisiae* telomeres: reversible repression of pol II transcription. Cell (Cambridge, Mass.) **63**, 751–762 (1990)

M.A. Grande, I. van der Kraan, L. de Jong, R. van Driel, Nuclear distribution of transcription factors in relation to sites of transcription and RNA polymerase II. J. Cell Sci. **110**, 1781–1791 (1997)

P. Grandi, V. Doye, E.C. Hurt, Purification of NSP1 reveals complex formation with "GLFG" nucleoporins and a novel nuclear pore protein NIC96. Embo J. **12**, 3061–3071 (1993)

E. Grossman, O. Medalia, M. Zwerger, Functional architecture of the nuclear pore complex. Annu. Rev. Biophys. **41**, 557–584 (2012)

Y. Gruenbaum, U. Aebi, Intermediate filaments: a dynamic network that controls cell mechanics. F1000Prime Rep. **6**, 54 (2014)

Y. Gruenbaum, O. Medalia, Lamins: the structure and protein complexes. Curr. Opin. Cell Biol. **32**, 7–12 (2015)

Y. Gruenbaum, A. Margalit, R.D. Goldman, D.K. Shumaker, K.L. Wilson, The nuclear lamina comes of age. Nat. Rev. Mol. Cell Biol. **6**(1), 21–31 (2005)

Y. Gruenbaum, K.L. Wilson, A. Harel, M. Goldberg, M. Cohen, Review: nuclear lamins–structural proteins with fundamental functions. J. Struct. Biol. **129**, 313–323 (2000)

D. Grunwald, R.H. Singer, M. Rout, Nuclear export dynamics of RNA–protein complexes. Nature **475**, 333–341 (2000)

F. Guilak, J.R. Tedrow, R. Burgkart, Viscoelastic properties of the cell nucleus. Biochem. Biophys. Res. Commun. **269** (3), 781–786 (2000)

C. Guilluy, L.D. Osborne, L. Van Landeghem, L. Sharek, R. Superfine, R. Garcia-Mata, K. Burridge, Isolated nuclei adapt to force and reveal a mechanotransduction pathway in the nucleus. Nat. Cell Biol. **16**(4), 376–381 (2014)

T. Guttler, D. Gorlich, Ran-dependent nuclear export mediators: a structural perspective EMBO J. **30**, 3457–3474 (2011)

C. Gwizdek, N. Iglesias, M.S. Rodriguez, B. Ossareh-Nazari, M. Hobeika, G. Divita, F. Stutz, C. Dargemont, Ubiquitin-associated domain of Mex67 synchronizes recruitment of the mRNA export machinery with transcription. Proc. Natl. Acad. Sci. U. S. A. **103**, 16376–16381 (2006)

K. Haase, J.K.L. Macadangdang, C.H. Edrington, C.M. Cuerrier, S. Hadjiantoniou, J.L. Harden, I.S. Skerjanc, A.E. Pelling, Extracellular forces cause the nucleus to deform in a highly controlled anisotropic manner. Sci. Rep. **6**, 21300 (2016)

W.J. Hadden, J.L. Young, A.W. Holle, M.L. McFetridge, D.Y. Kim, P. Wijesinghe, H. Taylor-Weiner, J.H. Wen, A.R. Lee, K. Bieback, B.N. Vo, D.D. Sampson, B.F. Kennedy, J.P. Spatz, A.J. Engler, Y.S. Choi, Stem cell migration and mechanotransduction on linear stiffness gradient hydrogels. Proc. Natl. Acad. Sci. U. S. A. **114**(22), 5647–5652 (2017)

N. Handa, M. Kukimoto-Niino, R. Akasaka, et al., The crystal structure of mouse Nup35 reveals atypical RNP motifs and novel homodimerization of the RRM domain. J Mol Biol. **363**, 114–124 (2006)

L. Hanson, W. Zhao, H.Y. Lou, Z.C. Lin, S.W. Lee, P. Chowdary, Y. Cui, B. Cui, Vertical nanopillars for in situ probing of nuclear mechanics in adherent cells. Nat. Nanotechnol. **10**(6), 554–562 (2015)

T. Harada, J. Swift, J. Irianto, J.-W. Shin, K.R. Spinler, A. Athirasala, R. Diegmiller, P.C.D.P. Dingal, I.L. Ivanovska, D.E. Discher, Nuclear lamin stiffness is a barrier to 3D migration, but softness can limit survival. J. Cell Biol. **204**, 669–682 (2014)

A. Harel, A.V. Orjalo, T. Vincent, A. Lachish-Zalait, S. Vasu, S. Shah, E. Zimmerman, M. Elbaum, D.J. Forbes, Removal of a single pore subcomplex results in vertebrate nuclei devoid of nuclear pores. Mol. Cell **11**, 853–864 (2003)

K. Harris, The Birth of the Cell (Yale University Press, New Haven, 1999)

Hawryluk-Gara LA, Platani M, Santarella R, Wozniak R W and Mattaj I W 2008 Nup53 is required for nuclear envelope and nuclear pore complex assembly *Mol Biol Cell* **9**(4) 1753–62

A. Hayakawa, A. Babour, L. Sengmanivong, C. Dargemont, Ubiquitylation of the nuclear pore complex controls nuclear migration during mitosis in *S. cerevisiae*. J. Cell Biol. **196**, 19–27 (2012)

M.D. Hebert, A.G. Matera, Self-association of p80 coilin reveals a common theme in nuclear body formation. Mol. Biol. Cell **11**, 4159–4171 (2000)

S. Heessen, M. Fornerod, The inner nuclear envelope as a transcription factor resting place. EMBO Rep. **8**, 914–919 (2007)

S.-J. Heo, T.P. Driscoll, S.D. Thorpe, et al., Differentiation alters stem cell nuclear architecture, mechanics, and mechano-sensitivity. Elife **5**, 283–293 (2016a)

S.-J. Heo, W.M. Han, S.E. Szczesny, B.D. Cosgrove, D.M. Elliott, D.A. Lee, R.L. Duncan, R.L. Mauck, Mechanically induced chromatin condensation requires cellular contractility in Mesenchymal stem cells. Biophys. J. **111**, 864–874 (2016b)

H. Herrmann, U. Aebi, Intermediate filaments: structure and assembly. Cold. Spring. Harb. Perspect. Biol. **8**, a018242 (2016)

H.M. Herz, A. Garruss, A. Shilatifard, SET for life: biochemical activities and biological functions of SET domain-containing proteins. Trends Biochem. Sci. **38**, 621–639 (2013)

M.W. Hetzer, The nuclear envelope. Cold. Spring. Harb. Perspect. Biol. **2**, a000539 (2010)

M.W. Hetzer, S.R. Wente, Border control at the nucleus: biogenesis and organization of the nuclear membrane and pore complexes. Dev. Cell. **17**, 606–616 (2009)

M. Hetzer, T.C. Walther, I.W. Mattaj, Pushing the envelope: structure, function, and dynamics of the nuclear periphery. Annu. Rev. Cell. Dev. Biol. **21**, 367–380 (2005)

D.A. Hill, A.N. Imbalzano, Human SWI/SNF nucleosome remodeling activity is partially inhibited by linker histone H1. Biochemistry **39**, 11649–11656 (2000)

Y. Hirose, J.L. Manley, RNA polymerase II and the integration of nuclear events. Genes Dev. **14**, 1415–1429 (2000)

J.H. Ho, G. Kallstrom, A.W. Johnson, Nmd3p is a Crm1p-dependent adapter protein for nuclear export of the large ribosomal subunit. J. Cell Biol. **151**, 1057–1066 (2000)

L.E. Hough, K. Dutta, S. Sparks, D.B. Temel, A. Kamal, J. Tetenbaum-Novatt, M.P. Rout, D. Cowburn, The molecular mechanism of nuclear transport revealed by atomic-scale measurements. Elife **4**, 635 (2015)

A.B. Houtsmuller, S. Rademakers, A.L. Nigg, D. Hoogstraten, J.H. Hoeijmakers, W. Vermeulen, Action of DNA repair endonuclease ERCC1/XPF in living cells. Science **284**, 958–961 (1999)

P. Hozak, P.R. Cook, C. Schofer, W. Mosgoller, F. Wachtler, Site of transcription of ribosomal RNA and intranucleolar structure in HeLa cells. J. Cell Sci. **107**, 639–648 (1994)

S. Huang, D.E. Ingber, Cell tension, matrix mechanics, and cancer development. Cancer Cell **8**, 175–176 (2005)

S. Huang, D.L. Spector, Intron-dependent recruitment of pre-mRNA splicing factors to sites of transcription. J. Cell Biol. **131**, 719–732 (1996)

S. Huang, T.J. Deerinck, M.H. Ellisman, D.L. Spector, The dynamic organization of the perinucleolar compartment in the cell nucleus. J. Cell Biol. **137**, 965–974 (1997)

M.D. Huber, T. Guan, L. Gerace, Overlapping functions of nuclear envelope proteins NET25 (Lem2) and emerin in regulation of extracellular signal-regulated kinase signaling in myoblast differentiation. Mol. Cell. Biol. **29**, 5718–5728 (2009)

B.B. Hülsmann, A.A. Labokha, D. Görlich, The permeability of reconstituted nuclear pores provides direct evidence for the selective phase model. Cell **150**, 738–751 (2012)

E. Hurt, M. Beck, Towards understanding nuclear pore complex architecture and dynamics in the age of integrative structural analysis. Curr. Opin. Cell Biol. **34**, 31–38 (2015)

N. Iglesias, E. Tutucci, C. Gwizdek, P. Vinciguerra, E. Von Dach, A.H. Corbett, C. Dargemont, F. Stutz, Ubiquitin-mediated mRNP dynamics and surveillance prior to budding yeast mRNA export. Genes Dev. **24**, 1927–1938 (2010)

D.E. Ingber, Mechanical control of tissue morphogenesis during embryological development. Int. J. Dev. Biol. **50**, 255–266 (2006)

R.V. Intine, A.L. Sakulich, S.B. Koduru, Y. Huang, E. Pierstorff, J.L. Goodier, L. Phan, R.J. Maraia, Control of transfer RNA maturation by phosphorylation of the human La antigen on serine 366. Mol. Cell **6**, 339–348 (2000)

J. Irianto, C.R. Pfeifer, R.R. Bennett, Y. Xia, I.L. Ivanovska, A.J. Liu, R.A. Greenberg, D.E. Discher, Nuclear constriction segregates mobile nuclear proteins away from chromatin. Mol. Biol. Cell **27**, 4011–4020 (2016)

J. Irianto, Y. Xia, C.R. Pfeifer et al., DNA damage follows repair factor depletion and portends genome variation in cancer cells after pore migration. Curr. Biol. **27**(2), 210–223 (2017)

J. Irianto, Y. Xia, C.R. Pfeifer, R.A. Greenberg, D.E. Discher, As a nucleus enters a small pore, chromatin stretches and maintains integrity, even with DNA breaks. Biophys. J. **112**, 446–449 (2017b)

C. Isaac, Y. Yang, U.T. Meier, Nopp140 functions as a molecular link between the nucleolus and the coiled bodies. J. Cell Biol. **142**(2), 319–329 (1998)

A.M. Ishov, A.G. Sotnikov, D. Negorev, O.V. Vladimirova, N. Neff, T. Kamitani, E.T. Yeh, J.F. Strauss, G.G. Maul, PML is critical for ND10 formation and recruits the PML-interacting protein Daxx to this nuclear structure when modified by SUMO-1. J. Cell Biol. **147**, 221–234 (1999)

I.L. Ivanovska, J. Swift, K. Spinler, D. Dingal, S. Cho, D.E. Discher, Cross-linked matrix rigidity and soluble retinoids synergize in nuclear lamina regulation of stem cell differentiation. Mol. Biol. Cell. **28**(14), 2010–2022 (2017)

M. Iwamoto, H. Asakawa, Y. Hiraoka, T. Haraguchi, Nucleoporin Nup98: a gate-keeper in the eukaryotic kingdoms. Genes Cells **15**, 661–669 (2015)

M. Iwamoto, C. Mori, T. Kojidani, F. Bunai, T. Hori, T. Fukagawa, Y. Hiraoka, T. Haraguchi, Two distinct repeat sequences of Nup98 nucleoporins characterize dual nuclei in the binucleated ciliate Tetrahymena. Curr. Biol. **19**, 843–847 (2009)

D.A. Jackson, A.B. Hassan, R.J. Errington, P.R. Cook, Visualization of focal sites of transcription within human nuclei. EMBO J. **12**, 1059–1065 (1993)

D.A. Jackson, F.J. Iborra, E.M. Manders, P.R. Cook, Numbers and organization of RNA polymerases, nascent transcripts, and transcription units in HeLa nuclei. Mol. Biol. Cell **9**, 1523–1536 (1998)

D.A. Jackson, A. Pombo, F. Iborra, The balance sheet for transcription: an analysis of nuclear RNA metabolism in mammalian cells. FASEB J. **14**, 242–254 (2000)

E.Y. Jacobs, M.R. Frey, W. Wu, T.C. Ingledue, T.C. Gebuhr, L. Gao, W.F. Marzluff, A.G. Matera, Coiled bodies preferentially associate with U4, U11, and U12 small nuclear RNA genes in interphase HeLa cells but not with U6 and U7 genes. Mol. Biol. Cell **10**, 1653–1663 (1999)

B.E. Jady, T. Kiss, A small nucleolar guide RNA functions both in 2 h-O-ribose methylation and pseudouridylation of the U5 spliceosomal RNA. EMBO J. **20**, 541–551 (2001)

J. Janevski, P.C. Park, U. De Boni, Changes in morphology and spatial position of coiled bodies during NGF-induced neuronal differentiation of PC12 cells. J. Histochem. Cytochem. **45**, 1523–1531 (1997)

N. Jarrous, J.S. Wolenski, D. Wesolowski, C. Lee, S. Altman, Localization in the nucleolus and coiled bodies of protein subunits of the ribonucleoprotein ribonuclease P. J. Cell Biol. **146**, 559–572 (1999)

S. Jeudy, T. Schwartz, Crystal structure of nucleoporin Nic96 reveals a novel, intricate helical domain architecture. J Biol Chem. **282**, 34904–34912 (2007)

Y. Jia, V. Polunovsky, P.B. Bitterman, C.R. Wagner, Cap-dependent translation initiation factor eIF4E: an emerging anticancer drug target. Med. Res. Rev. **32**, 786–814 (2012)

L.F. Jimenez-Garcia, D.L. Spector, vivo evidence that transcription and splicing are coordinated by a recruiting mechanism. Cell **73**, 47–59 (1993)

C. Jolly, Y. Usson, R.I. Morimoto, Rapid and reversible relocalization of heat shock factor 1 within seconds to nuclear stress granules. Proc. Natl. Acad. Sci. U. S. A. **96**, 6769–6774 (1999)

M. Kampmann, G. Blobel, Three-dimensional structure and flexibility of a membrane-coating module of the nuclear pore complex. Nat. Struct. Mol. Biol. **16**, 782–788 (2009)

T. Kanda, K.F. Sullivan, G.M. Wahl, Histone-GFP fusion protein enables sensitive analysis of chromosome dynamics in living mammalian cells. Curr. Biol. **26**, 377–385 (1998)

L.E. Kapinos, R.L. Schoch, R.S. Wagner, K.D. Schleicher, R.Y. Lim, Karyopherin-centric control of nuclear pores based on molecular occupancy and kinetic analysis of multivalent binding with FG nucleoporins. Biophys. J. **106**, 1751–1762 (2014)

H. Karcher, J. Lammerding, H. Huang, R.T. Lee, R.D. Kamm, M.R. Kaazempur-Mofrad, A three-dimensional viscoelastic model for cell deformation with experimental verification. Biophys. J. **85**(5), 3336–3349 (2003)

G.H. Karpen, J.E. Schaefer, C.D. Laird, A Drosophila rRNA gene located in euchromatin is active in transcription and nucleolus formation. Genes Dev. **2**, 1745–1763 (1988)

S.S. Katta, C.J. Smoyer, S.L. Jaspersen, Destination: inner nuclear membrane. Trends Cell Biol. **24**, 221–229 (2014)

K. Kelley, K.E. Knockenhauer, G. Kabachinski, T.U. Schwartz, Atomic structure of the Y complex of the nuclear pore. Nat. Struct. Mol. Biol. **22**, 425–431 (2015)

S.M. Kelly, S.W. Leung, C. Pak, A. Banerjee, K.H. Moberg, A.H. Corbett, A conserved role for the zinc finger polyadenosine RNA binding protein, ZC3H14, in control of poly(a) tail length. RNA **20**(5), 681–688 (2014)

S. Kelly, C. Pak, M. Garshasbi, A. Kuss, A.H. Corbett, K. Moberg, New kid on the ID block: neural functions of the Nab2/ZC3H14 class of Cys(3)His tandem zinc-finger polyadenosine RNA binding proteins. RNA Biol. **9**, 555–562 (2012)

S.B. Khatau, R.J. Bloom, S. Bajpai et al., The distinct roles of the nucleus and nucleus-cytoskeleton connections in three-dimensional cell migration. Sci. Rep. **2**, 488 (2012)

T.R. Kiehl, A. Nechiporuk, K.P. Figueroa, M.T. Keating, D.P. Huynh, S.M. Pulst, Generation and characterization of Sca2 (ataxin-2) knockout mice. Biochem. Biophys. Res. Commun. **33**, 917–924 (2006)

D.I. Kim, B. Kc, W. Zhu, K. Motamedchaboki, V. Doye, K.J. Roux, Probing nuclear pore complex architecture with proximity-dependent biotinylation. Proc. Natl. Acad. Sci. U. S. A., 201406459 (2014a)

E. Kim, L. Du, D.B. Bregman, S.L. Warren, Splicing factors associate with hyperphosphorylated RNA polymerase II in the absence of pre-mRNA. J. Cell Biol. **136**, 19–28 (1997)

J.-K. Kim, A. Louhghalam, G. Lee, B.W. Schafer, D. Wirtz, D.-H. Kim, Nuclear lamin A/C harnesses the perinuclear apical actin cables to protect nuclear morphology. Nat. Commun. **8**(1), 2123 (2017)

S.J. Kim, J. Fernandez-Martinez, P. Sampathkumar, et al., Integrative structure-function mapping of the nucleoporin Nup133 suggests a conserved mechanism for membrane anchoring of the nuclear pore complex. Mol. Cell Proteomics **13**, 2911–2926 (2014b)

T. Kim, M.L. Gardel, E. Munro, Determinants of fluidlike behavior and effective viscosity in cross-linked actin networks. Biophys. J. **106**(3), 526–534 (2014)

N. Kimura, M. Takizawa, K. Okita, O. Natori, K. Igarashi, M. Ueno, K. Nakashima, I. Nobuhisa, T. Taga, Identification of a novel transcription factor, ELYS, expressed predominantly in mouse foetal haematopoietic tissues. Genes Cells **7**, 435–446 (2002)

M.C. King, C.P. Lusk, G. Blobel, Karyopherin-mediated import of integral inner nuclear membrane proteins. Nature **442**, 1003–1007 (2006)

T.J. Kirby, J. Lammerding, Emerging views of the nucleus as a cellular mechanosensor. Nat. Cell Biol. **20**(4), 373–381 (2018)

T. Kirchhausen, D. Owen, S.C. Harrison, Molecular structure, function, and dynamics of clathrin-mediated membrane traffic. Cold Spring Harb. Perspect. Biol. **6**, a016725 (2014)

G.L. Kite, The relative permeability of the surface and interior portions of the cytoplasm of animal and plant cells. Biol. Bull. **25**, 1–7 (1913)

K.E. Knockenhauer, T.U. Schwartz, The nuclear pore complex as a flexible and dynamic gate. Cell **164**(6), 1162–1171 (2016)

B. Kobe, A.V. Kajava, When protein folding is simplified to protein coiling: the continuum of solenoid protein structures. Trends Biochem. Sci. **25**, 509–515 (2000)

A.J. Koch, J.M. Holaska, Emerin in health and disease. Semin. Cell Dev. Biol. **29**, 95–106 (2014)

J. Koh, G. Blobel, Allosteric regulation in gating the central channel of the nuclear pore complex. Cell **161**, 1361–1373 (2015)

M.H. Koken, G. Linares-Cruz, F. Quignon, A. Viron, K. Chelbi-Alix, J. Sobczak-Thepot, L. Juhlin, L. Degos, F. Calvo, H. de The, The PML growth-suppressor has an altered expression in human oncogenesis. Oncogene **10**, 1315–1324 (1995)

M.H. Koken, F. Puvion-Dutilleul, M.C. Guillemin, et al., The t(15;17) translocation alters a nuclear body in a retinoic acid-reversible fashion. EMBO J. **13**, 1073–1083 (1994)

H. Kondo, C. Rabouille, R. Newman, T.P. Levine, D. Pappin, P. Freemont, G. Warren, p47 is a cofactor for p97-mediated membrane fusion. Nature **388**, 75–78 (1997)

B.E. Koop, J.L. Lewis, A model of fracture testing of soft viscoelastic tissues. J. Biomech. **36**(4), 605–608 (2003)

V.L. Koumandou, B. Wickstead, M.L. Ginger, M. van der Giezen, J.B. Dacks, M.C. Field, Molecular paleontology and complexity in the last eukaryotic common ancestor. Crit. Rev. Biochem. Mol. Biol. **48**, 373–396 (2013)

D.M. Kraemer, C. Strambio-de-Castillia, G. Blobel, M.P. Rout, The essential yeast nucleoporin NUP159 is located on the cytoplasmic side of the nuclear pore complex and serves in karyopherin-mediated binding of transport substrate. J. Biol. Chem. **270**, 19017–19021 (1995)

A. Kralt, J.B. Jagalur, V. van den Boom, R.K. Lokareddy, A. Steen, G. Cingolani, M. Fornerod, L.M. Veenhoff, Conservation of inner nuclear membrane targeting sequences in mammalian Pom121 and yeast Heh2 membrane proteins. Mol. Biol. Cell **26**, 3301–3312 (2015)

D. Kressler, P. Linder, J. de La Cruz, Protein trans-acting factors involved in ribosome biogenesis in *Saccharomyces cerevisiae*. Mol. Cell. Biol. **19**, 7897–7912 (1999)

M.J. Kruhlak, M.A. Lever, W. Fischle, E. Verdin, D.P. Bazett-Jones, M.J. Hendzel, Reduced mobility of the alternate splicing factor (ASF) through the nucleoplasm and steady state speckle compartments. J. Cell Biol. **150**, 41–51 (2000)

H.H. Ku, Notes on the use of propagation of error formulas. J. Res. Natl. Bur. Stand. **79**, 75–79 (1966)

U. Kubitscheck, P. Wedekind, O. Zeidler, M. Grote, R. Peters, Single nuclear pores visualized by confocal microscopy and image processing. Biophys. J. **70**, 2067–2077 (1996)

A. Kumar, M. Mazzanti, M. Mistrik et al., ATR mediates a checkpoint at the nuclear envelope in response to mechanical stress. Cell **158**, 633–646 (2014)

A. Kurz, S. Lampel, J.E. Nickolenko, J. Bradl, A. Benner, R.M. Zirbel, T. Cremer, P. Lichter, Active and inactive genes localize preferentially in the periphery of chromosome territories. J. Cell Biol. **135**, 1195–1205 (1996)

U. Kutay, S. Guttinger, Leucine-rich nuclear-export signals: born to be weak. Trends Cell Biol. **15**, 121–124 (2005)

A.A. Labokha, S. Gradmann, S. Frey, B.B. Hülsmann, H. Urlaub, M. Baldus, D. Görlich, Systematic analysis of barrier-forming FG hydrogels from Xenopus nuclear pore complexes Embo J. **32**, 204–218 (2013)

H. Laklai, Y.A. Miroshnikova, M.W. Pickup et al., Genotype tunes pancreatic ductal adenocarcinoma tissue tension to induce matricellular fibrosis and tumor progression. Nat. Med. **22**, 497–505 (2016)

S. Lall, H. Francis-Lang, A. Flament, A. Norvell, T. Schupbach, D. Ish-Horowicz, Squid hnRNP proteins promotes apical cytoplasmic transport and localization of Drosophila pair-rule transcripts. Cell **98**, 171–180 (1999)

J. Lammerding, Mechanics of the nucleus. Compr. Physiol. **1**(2), 783–807 (2011)

A. Lange, R.E. Mills, C.J. Lange, M. Stewart, S.E. Devine, A.H. Corbett, Classical nuclear localization signals: definition, function, and interaction with importin alpha. J. Biol. Chem. **282**, 5101–5105 (2007)

T.S. Lange, S.A. Gerbi, Transient nucleolar localization of U6 small nuclear RNA in Xenopus laevis oocytes. Mol. Biol. Cell **11**, 2419–2448 (2000)

K. Lau, H. Tao, H. Liu et al., Anisotropic stress orients remodelling of mammalian limb bud ectoderm. Nat. Cell Biol. **17**(5), 569–579 (2015)

A. Laugesen, K. Helin, Chromatin repressive complexes in stem cells, development, and cancer. Cell Stem Cell **14**, 735–751 (2014)

H.Q. Le, S. Ghatak, C.-Y.C. Yeung et al., Mechanical regulation of transcription controls Polycomb-mediated gene silencing during lineage commitment. Nat. Cell Biol. **18**, 864–875 (2016)

B.J. Lee, A.E. Cansizoglu, S.E. Suel, T.H. Louis, Z. Zhang, Y.M. Chook, Rules for nuclear localization sequence recognition by karyopherin beta 2. Cell **126**, 543–558 (2006)

M.S. Lee, M. Henry, P.A. Silver, A protein that shuttles between the nucleus and the cytoplasm is an important mediator of RNA export. Genes Dev. **10**, 1233–1246 (1996)

S.J. Lee, Y. Matsuura, S.M. Liu, M. Stewart, Structural basis for nuclear import complex dissociation by RanGTP. Nature **435**, 693–696 (2005)

H. Lei, B. Zhai, S. Yin, S. Gygi, R. Reed, Evidence that a consensus element found in naturally intronless mRNAs promotes mRNA export. Nucleic Acids Res. **41**, 2517–2525 (2013)

C. Lemaître, W.A. Bickmore, Chromatin at the nuclear periphery and the regulation of genome functions. Histochem. Cell Biol. **144**, 111–122 (2015)

H. Leonhardt, H.-P. Rahn, P. Weinzierl, A. Sporbert, T. Cremer, D. Zink, M.C. Cardoso, Dynamics of DNA replication factories in living cells. J. Cell Biol. **149**, 271–279 (2000)

K.R. Levental, H. Yu, L. Kass et al., Matrix crosslinking forces tumor progression by enhancing integrin signaling. Cell **139**, 891–906 (2009)

M.A. Lever, J.P. Thng, X. Sun, M.J. Hendzel, Rapid exchange of histone H1.1 on chromatin in living cells. Nature **408**, 873–876 (2000)

S.J. Li, M. Hochstrasser, A new protease required for cell-cycle progression in yeast. Nature **398**, 246–251 (1999)

W.H. Light, J.H. Brickner, Nuclear pore proteins regulate chromatin structure and transcriptional memory by a conserved mechanism. Nucleus **4**, 357–360 (2013)

W.H. Light, D.G. Brickner, V.R. Brand, J.H. Brickner, Interaction of a DNA zip code with the nuclear pore complex promotes H2A.Z incorporation and INO1 transcriptional memory. Mol. Cell **40**, 112–125 (2010)

W.H. Light, J. Freaney, V. Sood, A. Thompson, A. D'Urso, C.M. Horvath, J.H. Brickner, A conserved role for human Nup98 in altering chromatin structure and promoting epigenetic transcriptional memory. PLoS Biol. **11**, e1001524 (2013)

R.Y.H. Lim, B. Fahrenkrog, J. Köser, K. Schwarz-Herion, J. Deng, U. Aebi, Nanomechanical basis of selective gating by the nuclear pore complex. Science **318**, 640–643 (2007)

H. Liu, J. Wen, Y. Xiao, J. Liu, S. Hopyan, M. Radisic, C.A. Simmons, Y. Sun, In situ mechanical characterization of the cell nucleus by atomic force microscopy. ACS Nano **8**(4), 3821–3828 (2014)

J. Liu, M.D. Hebert, Y. Ye, D.J. Templeton, H. Kung, A.G. Matera, Cell cycle-dependent localization of the CDK2-cyclin E complex in Cajal (coiled) bodies. J. Cell Sci. **113**, 1543–1552 (2000)

L. Liu, Q. Luo, J. Sun, S. Guanbin, Nucleus and nucleus-cytoskeleton connections in 3D cell migration. Experimental Cell Res. **348**, 56–65 (2016)

X. Liu, X. Liu, J.M. Mitchell, R.W. Wozniak, G. Blobel, J. Fan, J. Fan, Structural evolution of the membrane-coating module of the nuclear pore complex. Proc. Natl. Acad. Sci. U. S. A. **109**, 16498–16503 (2012)

I. Loïodice, A. Alves, G. Rabut, M. Van Overbeek, J. Ellenberg, J.B.Sibarita, V. Doye, The entire Nup107–160 complex, including three new members, is targeted as one entity to kinetochores in mitosis. Mol. Biol. Cell **15**, 3333–3344 (2004)

R.K. Lokareddy, R.A. Hapsari, M. van Rheenen, R.A. Pumroy, A. Bhardwaj, A. Steen, L. M. Veenhoff, G. Cingolani, Distinctive properties of the nuclear localization signals of inner nuclear membrane proteins Heh1 and Heh2. Structure **23**, 1305–1316 (2015)

M.L. Lombardi, M. Zwerger, J. Lammerding, Biophysical assays to probe the mechanical properties of the interphase cell nucleus: substrate strain application and microneedle manipulation. J. Vis. Experim. JoVE **55**

E.A. López-Guerra, S.D. Solares, Modeling viscoelasticity through spring–dashpot models in intermittent-contact atomic force microscopy. Beilstein. J. Nanotechnol. **5**, 2149 (2014)

C.L. Lorson, J. Strasswimmer, J.D. Yao, J.D. Baleja, E. Hahnen, T. Wirth, T. Le, A.H. Burghes, E.J. Androphy, SMN oligomerization defect correlates with spinal muscular atrophy severity. Nat. Genet. **19**, 63–66 (1998)

A.R. Lowe, J.H. Tang, J. Yassif, M. Graf, W.Y.C. Huang, J. Groves, K. Weis, J.T. Liphardt, Importin-β modulates the permeability of the nuclear pore complex in a Ran-dependent manner. eLife **4**, e04052 (2015)

L. Lu, M.S. Ladinsky, T. Kirchhausen, Cisternal organization of the endoplasmic reticulum during mitosis. Mol. Biol. Cell **15**, 3471–3480 (2009)

M.K. Lund, C. Guthrie, The DEAD-box protein Dbp5p is required to dissociate Mex67p from exported mRNPs at the nuclear rim. Mol. Cell **20**, 645–651 (2005)

F. Lupu, A. Alves, K. Anderson, V. Doye, E. Lacy, Nuclear pore composition regulates neural stem/progenitor cell differentiation in the mouse embryo. Dev. Cell **14**, 831–842 (2008)

G.W. Luxton, E.R. Gomes, E.S. Folker, H.J. Worman, G.G. Gundersen, TAN lines: a novel nuclear envelope structure involved in nuclear positioning. Nucleus **2**, 173–181 (2011)

C.E. Lyon, K. Bohmann, J. Sleeman, A.I. Lamond, Inhibition of protein dephosphorylation results in the accumulation of splicing snRNPs and coiled bodies within the nucleolus. Exp. Cell Res. **230**, 84–93 (1997)

M.F. Lyon, Gene action in the X chromosome of the mouse. Nature **190**, 372–373 (1961)

H. Ma, J. Samarabandu, R.S. Devdhar, R. Acharya, P.C. Cheng, C. Meng, R. Berezney, Spatial and temporal dynamics of DNA replication sites in mammalian cells. J. Cell Biol. **143**, 1415–1425 (1998)

W.K. Ma, S.C. Cloutier, E.J. Tran, The DEAD-box protein Dbp2 functions with the RNA-binding protein Yra1 to promote mRNP assembly. J. Mol. Biol. **425**, 3824–3838 (2013)

X.M. Ma, J. Blenis, Molecular mechanisms of mTOR-mediated translational control. Nat. Rev. Mol. Cell Biol. **10**, 307–318 (2009)

M. Machowska, K. Piekarowicz, R. Rzepecki, Regulation of lamin properties and functions: does phosphorylation do it all? Open Biol. **5**, 150094 (2015)

B.E. Maden, J.M. Hughes, Eukaryotic ribosomal RNA: the recent excitement in the nucleotide modification problem. Chromosoma **105**, 391–400 (1997)

T. Maimon, N. Elad, I. Dahan, O. Medalia, The human nuclear pore complex as revealed by cryo-electron tomography. Structure **20**, 998–1006 (2012)

M. Malatesta, C. Zancanaro, T.E. Martin, E.K. Chan, F. Amalric, R. Luhrmann, P. Vogel, S. Fakan, Cytochemical and immunocytochemical characterization of nuclear bodies during hibernation. Eur. J. Cell Biol. **65**, 82–93 (1994)

B.J. Mans, V. Anantharaman, L. Aravind, E.V. Koonin, Comparative genomics, evolution and origins of the nuclear envelope and nuclear pore complex. Cell Cycle **3**, 1612–1637 (2004)

J. Mansfeld, S. Guttinger, L.A. Hawryluk-Gara, et al., The conserved transmembrane nucleoporin NDC1 is required for nuclear pore complex assembly in vertebrate cells. Mol Cell, **22**, 93–103 (2006)

L. Manuelidis, Different central nervous system cell types display distinct and nonrandom arrangements of satellite DNA sequences. Proc. Natl. Acad. Sci. U. S. A. **81**, 3123–3127 (1984)

L. Manuelidis, A view of interphase chromosomes. Science **250**, 1533–1540 (1990)

M. Marelli, D.J. Dilworth, R.W. Wozniak, J.D. Aitchison, The dynamics of karyopherin-mediated nuclear transport. Biochem. Cell Biol. **79**, 603–612 (2001)

W.F. Marshall, A. Straight, J.F. Marko, J. Swedlow, A. Dernburg, A. Belmont, A.W. Murray, D.A. Agard, J.W. Sedat, Interphase chromosomes undergo constrained diffusional motion in living cells. Curr. Biol. **7**, 930–939 (1997)

R.P. Martins, J.D. Finan, F. Guilak, D.A. Lee, Mechanical regulation of nuclear structure and function. Annu. Rev. Biomed. Eng. **14**, 431–455 (2012)

M. Masaeli, D. Gupta, S. O'Byrne et al., Multiparameter mechanical and morphometric screening of cells. Sci. Rep. **6**, 37863 (2016)

A.G. Matera, Nuclear bodies: multifaceted subdomains of the interchromatin space. Trends Cell Biol. **9**, 302–309 (1999)

K.A. Matreyek, A. Engelman, Viral and cellular requirements for the nuclear entry of retroviral preintegration nucleoprotein complexes. Viruses **5**, 2483–2511 (2013)

A. Mattout, D.S. Cabianca, S.M. Gasser, Chromatin states and nuclear organization in development—a view from the nuclear lamina. Genome Biol. **16**, 174 (2015)

A. Mattout, T. Dechat, S.A. Adam, R.D. Goldman, Y. Gruenbaum, Nuclear lamins, diseases and aging. Curr. Opin. Cell Biol. **18**, 335–341 (2006)

G.G. Maul, Nuclear domain 10, the site of DNA virus transcription and replication. Bioessays **20**, 660–667 (1998)

G.G. Maul, D. Negorev, P. Bell, A.M. Ishov, Review: properties and assembly mechanisms of ND10, PML bodies, or PODs. J. Struct. Biol. **129**, 278–287 (2000)

J.G. McNally, W.G. Muller, D. Walker, R. Wolford, G.L. Hager, The glucocorticoid receptor: rapid exchange with regulatory sites in living cells. Science **287**, 1262–1265 (2000)

A.C. Meinema, J.K. Laba, R.A. Hapsari, R. Otten, F.A.A. Mulder, A. Kralt, G. van den Bogaart, C.P. Lusk, B. Poolman, L.M. Veenhoff, Long unfolded linkers facilitate membrane protein import through the nuclear pore complex. Science **333**, 90–93 (2011)

I. Melcak, S. Cermanova, K. Jirsova, K. Koberna, J. Malinsky, I. Raska, Nuclear pre-mRNA compartmentalization: trafficking of released transcripts to splicing factor reservoirs. Mol. Biol. Cell **11**, 497–510 (2000)

I. Melcák, A. Hoelz, G. Blobel, Structure of Nup58/45 suggests flexible nuclear pore diameter by intermolecular sliding. Science **315**, 1729–1732 (2007)

T. Melese, Z. Xue, The nucleolus: an organelle formed by the act of building a ribosome. Curr. Opin. Cell Biol. **7**, 319–324 (1995)

C.T. Mierke, Physical role of nuclear and cytoskeletal confinements in cell migration mode selection and switching. AIMS Biophys. **4**(4), 615–658 (2017)

E.A. Miller, R. Schekman, COPII - a flexible vesicle formation system. Curr. Opin. Cell Biol. **25**, 420–427 (2013)

S. Milles, D. Mercadante, I.V. Aramburu et al., Plasticity of an ultrafast interaction between nucleoporins and nuclear transport receptors. Cell **163**, 734–745 (2015)

P.J. Mintz, S.D. Patterson, A.F. Neuwal, C.S. Spahr, D.L. Spector, Purification and biochemical characterization of interchromatin granule clusters. EMBO J. **18**, 4308–4320 (1999)

Y.A. Miroshnikova, M.M. Nava, S.A. Wickström, Emerging roles of mechanical forces in chromatin regulation. J. Cell Sci. **130**(14), 2243–2250 (2017)

Y.A. Miroshnikova, I. Cohen, E. Ezhkova, S.A. Wickström, Epigenetic gene regulation, chromatin structure, and force-induced chromatin remodelling in epidermal development and homeostasis. Curr. Opin. Genet. Dev. **55**, 46–51 (2019)

T. Misteli, Cell biology of transcription and pre-mRNA splicing: nuclear architecture meets nuclear function. J. Cell Sci. **113**, 1841–1849 (2000)
T. Misteli, Protein dynamics: implications for nuclear architecture and gene expression. Science **291**, 843–847 (2001)
T. Misteli, D.L. Spector, RNA polymerase II targets pre-mRNA splicing factors to transcription sites in vivo. Mol. Cell **3**, 697–705 (1999)
T. Misteli, A. Agresti, P. Scaffidi et al., Beyond the sequence: cellular organization of genome function. Cell **128**, 787–800 (2007)
T. Misteli, J.F. Caceres, D.L. Spector, The dynamics of a pre-mRNA splicing factor in living cells. Nature (London) **387**, 523–527 (1997)
T. Misteli, A. Gunjan, R. Hock, M. Bustin, D.T. Brown, Dynamic binding of histone H1 to chromatin in living cells. Nature **408**, 877–881 (2000)
Y. Miyamoto, M.A. Baker, P.A. Whiley, A. Arjomand, J. Ludeman, C. Wong, D.A. Jans, K.L. Loveland, Towards delineation of a developmental alpha-importome in the mammalian male germline. Biochim. Biophys. Acta **1833**, 731–742 (2013)
D. Mohr, S. Frey, T. Fischer, T. Güttler, D. Görlich, Characterisation of the passive permeability barrier of nuclear pore complexes Embo J. **28**, 2541–2553 (2009)
A. Monneron, W. Bernhard, Fine structural organization of the interphase nucleus in some mammalian cells. J. Ultrastruct. Res. **27**, 266–288 (1969)
M.J. Moore, N.J. Proudfoot, Pre-mRNA processing reaches back to transcription and ahead to translation. Cell **136**, 688–700 (2009)
A. Mor, S. Suliman, R. Ben-Yishay, S. Yunger, Y. Brody, Y. Shav-Tal, Dynamics of single mRNP nucleocytoplasmic transport and export through the nuclear pore in living cells. Nat. Cell Biol. **12**, 543–552 (2010)
A. Mor, M.A. White, B.M. Fontoura, Nuclear trafficking in health and disease. Curr. Opin. Cell Biol. **28C**, 28–35 (2014)
F. Mora-Bermudez, D. Gerlich, J. Ellenberg, Maximal chromosome compaction occurs by axial shortening in anaphase and depends on Aurora kinase. Nat. Cell Biol. **9**, 822–831 (2007)
S. Morchoisne-Bolhy, M.C. Geoffroy, I.B. Bouhlel, A. Alves, N. Audugé, X. Baudin, K. Van Bortle, M.A. Powers, V. Doye, Intranuclear dynamics of the Nup107–160 complex. Mol. Biol. Cell **26**, 2343–2356 (2015)
G.T. Morgan, O. Doyle, C. Murphy, J.G. Gall, RNA polymerase II in Cajal bodies of amphibian oocytes. J. Struct. Biol. **129**, 258–268 (2000)
T. Moriuchi, M. Kuroda, F. Kusumoto, T. Osumi, F. Hirose, Lamin A reassembly at the end of mitosis is regulated by its SUMO-interacting motif. Exp. Cell Res. **342**, 83–94 (2016)
L. Mounkes, S. Kozlov, B. Burke, C.L. Stewart, The laminopathies: nuclear structure meets disease. Curr. Opin. Genet. Dev. **13**, 223–230 (2003)
R. Moussavi-Baygi, Y. Jamali, R. Karimi, M.R. Mofrad, Brownian dynamics simulation of nucleocytoplasmic transport: a coarse-grained model for the functional state of the nuclear pore complex. PLoS Comput. Biol. **7**, e1002049 (2011)
A. Muchir, H.J. Worman, The nuclear envelope and human disease. Physiology **19**, 309–314 (2004)
M. Muratani, D. Gerlich, S.M. Janicki, M. Gebhard, R. Eils, D.L. Spector, Metabolic-energy-dependent movement of PML bodies within the mammalian cell nucleus. Nat. Cell Biol. **4**, 106–111 (2002)
S. Nakielny, U. Fischer, W.M. Michael, G. Dreyfuss, RNA transport. Annu. Rev. Neurosci. **20**, 269–301 (1997)
A. Narayanan, W. Speckmann, R. Terns, M.P. Terns, Role of the box C/D motif in localization of small nucleolar RNAs to coiled bodies and nucleoli. Mol. Biol. Cell **10**, 2131–2147 (1999)
E.G. Neilan, Laminopathies, other progeroid disorders, and aging: Common pathogenic themes and possible treatments. Am. J. Med. Genet. A **149A**, 563–566 (2009)
N. Neumann, D. Lundin, A.M. Poole, Comparative genomic evidence for a complete nuclear pore complex in the last eukaryotic common ancestor. PLoS ONE **5**, e13241 (2010)
J. Nickerson, Experimental observations of a nuclear matrix. J. Cell Sci. **114**, 463–474 (2001)
C.A. Nino, A. Hayakawa, C. Dargemont, A. Babour, Mapping ubiquitin modifications reveals new functions for the yeast nuclear pore complex. Cell Logist. **2**, 43–45 (2012)
C.A. Nino, L. Herissant, A. Babour, C. Dargemont, mRNA nuclear export in yeast. Chem. Rev. **113**, 8523–8545 (2013)
L. Niu, T. Ma, F. Yang et al., Atlastin-mediated membrane tethering is critical for cargo mobility and exit from the endoplasmic reticulum. Proc. Natl. Acad. Sci. U. S. A. **116**(28), 14029–14038 (2019)
H.O. Nousiainen, M. Kestila, N. Pakkasjarvi, H. Honkala, S. Kuure, J. Tallila, K. Vuopala, J. Ignatius, R. Herva, L. Peltonen, Mutations in mRNA export mediator GLE1 result in a fetal motoneuron disease. Chem. Genet. **40**, 155–157 (2008)
R.T. O'Keefe, S.C. Henderson, D.L. Spector, Dynamic organization of DNA replication in mammalian cell nuclei: spatially and temporally defined replication of chromosome-specific alpha-satellite DNA sequences. J. Cell Biol. **116**, 1095–1110 (1992)

M. Oakes, J.P. Aris, J.S. Brockenbrough, H. Wai, L. Vu, M. Nomura, Mutational analysis of the structure and localization of the nucleolus in the yeast *Saccharomyces cerevisiae*. J. Cell Biol. **143**, 23–34 (1998)

M. Oakes, Y. Nogi, M.W. Clark, M. Nomura, Structural alterations of the nucleolus in mutants of Saccharomyces cerevisiae defective in RNA polymerase I. Mol. Cell. Biol. **13**, 2441–2555 (1993)

R.L. Ochs, T.W. Stein, E.M. Tan, Coiled bodies in the nucleolus of breast cancer cells. J. Cell Sci. **107**, 385–399 (1994)

M. Ohsugi, K. Adachi, R. Horai, S. Kakuta, K. Sudo, H. Kotaki, N. Tokai-Nishizumi, H. Sagara, Y. Iwakura, T. Yamamoto, Kid-mediated chromosome compaction ensures proper nuclear envelope formation. Cell **132**, 771–782 (2008)

A.L. Ollins, G. Rhodes, D.B. Welch, M. Zwerger, D.E. Ollins, Lamin B receptor: multitasking at the nuclear envelope. Nucleus **1**, 53–70 (2010)

M.O.J. Olson, M. Dundr, A. Szebeni, The nucleolus: An old factory with unexpected capabilities Trends Cell Biol. **10**, 189–196 (2000)

E. Onischenko, L.H. Stanton, A.S. Madrid, T. Kieselbach, K. Weis, Role of the Ndc1 interaction network in yeast nuclear pore complex assembly and maintenance. J. Cell Biol., **185**, 475–491 (2009)

A. Ori, N. Banterle, M. Iskar et al., Cell type-specific nuclear pores: a case in point for context-dependent stoichiometry of molecular machines. Mol. Syst. Biol. **9**, 648 (2013)

E. Orias, M.D. Cervantes, E.P. Hamilton, Tetrahymena thermophila, a unicellular eukaryote with separate germline and somatic genomes. Res. Microbiol. **162**, 578–586 (2011)

B. Ossareh-Nazari, F. Bachelerie, C. Dargemont, Evidence for a role of CRM1 in signal-mediated nuclear protein export. Science **278**, 141–144 (1997)

S. Pagliara, K. Franze, C.R. McClain, G.W. Wylde, C.L. Fisher, R.J.M. Franklin, A.J. Kabla, U.F. Keyser, K.J. Chalut, Auxetic nuclei in embryonic stem cells exiting pluripotency. Nat. Mater. **13**, 638–644 (2014)

J.S. Paige, K.Y. Wu, S.R. Jaffrey, RNA mimics of green fluorescent protein. Science **333**, 642–646 (2011)

J.D. Pajerowski, K.N. Dahl, F.L. Zhong, P.J. Sammak, D.E. Discher, Physical plasticity of the nucleus in stem cell differentiation. Proc. Natl. Acad. Sci. U. S. A. **104**, 15619–15624 (2007)

C. Pak, M. Garshasbi, K. Kahrizi et al., Mutation of the conserved polyadenosine RNA binding protein, ZC3H14/dNab2, impairs neural function in Drosophila and humans. Proc. Natl. Acad. Sci. U. S. A. **108**, 12390–12395 (2011)

N. Pante, M. Kann, Nuclear pore complex is able to transport macromolecules with diameters of about 39 nm. Mol. Biol. Cell **13**, 425–434 (2002)

P.C. Park, U. De Boni, Transposition of DNase hypersensitive chromatin to the nuclear periphery coincides temporally with nerve growth factor-induced up-regulation of gene expression in PC12 cells. Proc. Natl. Acad. Sci. U. S. A. **93**, 11646–11651 (1996)

G. Parry, Assessing the function of the plant nuclear pore complex and the search for specificity. J. Exp. Bot. **64**, 833–845 (2013)

S.S. Patel, B.J. Belmont, J.M. Sante, M.F. Rexach, Natively unfolded nucleoporins gate protein diffusion across the nuclear pore complex. Cell **129**, 83–96 (2007)

S. Pawar, R. Ungricht, P. Tiefenboeck, J.C. Leroux, U. Kutay, Efficient protein targeting to the inner nuclear membrane requires Atlastin-dependent maintenance of ER topology. eLife **6**, e28202 (2017)

T. Pederson, Half a century of "the nuclear matrix". Mol. Biol. Cell **11**, 799–805 (2000)

T. Pederson, J.C. Politz, The nucleolus and the four ribonucleoproteins of translation. J. Cell Biol. **148**, 1091–1095 (2000)

O. Peleg, R.Y.H. Lim, Converging on the function of intrinsically disordered nucleoporins in the nuclear pore complex. Biol Chem. **391**, 719–730 (2010)

L. Pellizzoni, B. Charroux, J. Rappsilber, M. Mann, G. Dreyfuss, A functional interaction between the survival motor neuron complex and RNA polymerase II. J. Cell Biol. **152**, 75–86 (2001)

P.Y. Perche, C. Vourch, L. Konecny, C. Souchier, M. Robert-Nicoud, S. Dimitrov, S. Khochbin, Higher concentration of histone macro-H2A in the Barr body are correlated with higher nucleosome density. Curr. Biol. **10**, 1531–1534 (2000)

D. Pérez-Calixto, G. Genaro Vázquez-Victorio, M. Hautefeuille, A viscoelastic nuclear model predicting mechanical memory and convergence rate of nuclear spreading as a function of stiffness bioRxiv (preprint first posted online Feb. 6, 2019). http://dx.doi.org/10.1101/542274)

R. Peters, Translocation through the nuclear pore: Kaps pave the way. BioEssays **31**, 466–477 (2009)

R.D. Phair, T. Misteli, High mobility of proteins in the mammalian cell nucleus. Nature **404**, 604–609 (2000)

M. Platani, I. Goldberg, J.R. Swedlow, A.I. Lamond, In vivo analysis of Cajal body movement, separation, and joining in live human cells. J. Cell Biol. **151**, 1561–1574 (2000)

Y.-C. Poh, S.P. Shevtsov, F. Chowdhury, D.C. Wu, S. Na, M. Dundr, N. Wang, Dynamic force-induced direct dissociation of protein complexes in a nuclear body in living cells. Nat. Commun. **3**(1), 1038 (2012)

J.C. Politz, T. Pederson, Movement of mRNA from transcription site to nuclear pores. J. Struct. Biol. **129**, 252–257 (2000)

J.C. Politz, E.S. Browne, D.E. Wolf, T. Pederson, Intranuclear diffusion and hybridization state of oligonucleotides measured by fluorescence correlation spectroscopy in living cells. Proc. Natl. Acad. Sci. U. S. A. **95**, 6043–6048 (1998)

J.C. Politz, R.A. Tuft, T. Pederson, R.H. Singer, Movement of nuclear poly(A) RNA throughout the interchromatin space in living cells. Curr. Biol. **9**, 285–291 (1999)

J.C. Politz, S. Yarovoi, S.M. Kilroy, K. Gowda, C. Zwieb, T. Pederson, Signal recognition particle components in the nucleolus. Proc. Natl. Acad. Sci. U. S. A. **97**, 55–60 (2000)

B.D. Price, A.D. D'Andrea, Chromatin remodeling at DNA double-strand breaks. Cell **152**, 1344–1354 (2013)

C. Prives, P.A. Hall, The p53 pathway. J. Pathol. **187**, 112–126 (1999)

F. Puvion-Dutilleul, S. Mazan, M. Nicoloso, E. Pichard, P. Bachellerie, E. Puvion, Alterations of nucleolar ultrastructure and ribosome biogenesis by actinomycin D. Implications for U3 snRNP function. Eur. J. Cell Biol. **58**, 149–162 (1992)

F. Puvion-Dutilleul, E. Puvion, J.P. Bachellerie, Early stages of pre-rRNA formation within the nucleolar ultrastructure of mouse cells studied by in situ hybridization with a 5hETS leader probe. Chromosoma **105**, 496–505 (1997)

X. Qu, S. Lykke-Andersen, T. Nasser, C. Saguez, E. Bertrand, T.H. Jensen, C. Moore, Assembly of an export-competent mRNP is needed for efficient release of the 3′-end processing complex after polyadenylation. Mol. Cell. Biol. **29**, 5327–5338 (2009)

F. Quignon, F. De Bels, M. Koken, J. Feunteun, J.C. Ameisen, H. de Th, PML induces a novel caspase-independent death process. Nat. Genet. **20**, 259–265 (1998)

B.B. Quimby, M. Dasso, The small GTPase Ran: interpreting the signs. Curr. Opin. Cell Biol. **15**, 338–344 (2003)

M. Raab, M. Gentili, H. de Belly et al., ESCRT III repairs nuclear envelope ruptures during cell migration to limit DNA damage and cell death Science **352**, 359–362 (2016)

M. Rabineau, F. Flick, C. Ehlinger et al., Chromatin de-condensation by switching substrate elasticity. Sci. Rep. **8**, 12655 (2018)

G. Rabut, V. Doye, J. Ellenberg, Mapping the dynamic organization of the nuclear pore complex inside single living cells. Nat. Cell Biol. **6**, 1114–1121 (2004)

M. Raices, M.A. D'Angelo, Nuclear pore complex composition: a new regulator of tissue-specific and developmental functions. Nat. Rev. Mol. Cell Biol. **13**, 687–699 (2012)

D. Rajgor, J.A. Mellad, F. Autore, Q. Zhang, C.M. Shanahan, Multiple novel nesprin-1 and nesprin-2 variants act as versatile tissue-specific intracellular scaffolds. PLoS ONE **7**, e40098

K. Ramadan, R. Bruderer, F.M. Spiga, O. Popp, T. Baur, M. Gotta, H.H. Meyer, Cdc48/p 97 promotes reformation of the nucleus by extracting the kinase Aurora B from chromatin. Nature **450**, 1258–1262 (2007)

B.A. Rasala, A.V. Orjalo, Z. Shen, S. Briggs, D.J. Forbes, ELYS is a dual nucleoporin/kinetochore protein required for nuclear pore assembly and proper cell division. Proc. Natl. Acad. Sci. **103**, 17801–17806 (2006)

I. Raska, L.E. Andrade, R.L. Ochs, E.K. Chan, C.M. Chang, G. Roos, E.M. Tan, Immunological and ultrastructural studies of the nuclear coiled body with autoimmune antibodies. Exp. Cell Res. **195**, 27–37 (1991)

K.L. Reddy, J.M. Zullo, E. Bertolino, H. Singh, Transcriptional repression mediated by repositioning of genes to the nuclear lamina. Nature **452**, 243–247 (2008)

R. Reichelt, A. Holzenburg, E.L. Buhle, M. Jarnik, A. Engel, U. Aebi, Correlation between structure and mass distribution of the nuclear pore complex and of distinct pore complex components. J Cell Biol. **110**, 883–894 (1990)

K. Ribbeck, D. Görlich, Kinetic analysis of translocation through nuclear pore complexes. EMBO J. **20**, 1320–1330 (2001)

K. Ribbeck, D. Gorlich, The permeability barrier of nuclear pore complexes appears to operate via hydrophobic exclusion. EMBO J. **21**, 2664–2671 (2002)

G. Rigaud, J. Roux, R. Pictet, T. Grange, vivo footprinting of rat TAT gene: dynamic interplay between the glucocorticoid receptor and a liver-specific factor. Cell **67**, 977–986 (1991)

C. Robinett, A. Straight, G. Li, C. Willhelm, G. Sudlow, A. Murray, A. Belmont, In vivo localization of DNA sequences and visualization of large-scale chromatin organization using lac operator/repressor recognition. J. Cell Biol. **135**, 1685–1700 (1996)

A. Rothballer, U. Kutay, Poring over pores: nuclear pore complex insertion into the nuclear envelope. Trends Biochem. Sci. **38**, 292–301 (2013a)

A. Rothballer, U. Kutay, The diverse functional LINCs of the nuclear envelope to the cytoskeleton and chromatin. Chromosoma **122**, 415–429 (2013b)

A. Rothballer, T.U. Schwartz, U. Kutay, LINCing complex functions at the nuclear envelope: what the molecular architecture of the LINC complex can reveal about its function. Nucleus **4**, 29–36 (2013)

M.P. Rout, J.D. Aitchison, A. Suprapto, K. Hjertaas, Y. Zhao, B.T. Chait, The yeast nuclear pore complex: composition, architecture, and transport mechanism. J. Cell Biol. **148**(4), 635–651 (2000)

J.O. Rouviere, M.C. Geoffroy, B. Palancade, Multiple crosstalks between mRNA biogenesis and SUMO. Chromosoma **122**, 387–399 (2013)

K.J. Roux, M.L. Crisp, Q. Liu, D. Kim, S. Kozlov, C.L. Stewart, B. Burke, Nesprin 4 is an outer nuclear membrane protein that can induce kinesin-mediated cell polarization. Proc. Natl. Acad. Sci. U. S. A. **106**, 2194–2199 (2009)

A.C. Rowat, J. Lammerding, J.H. Ipsen, Mechanical properties of the cell nucleus and the effect of emerin deficiency. Biophys. J. **91**, 4649–4664 (2006)

A.C. Rowat, D.E. Jaalouk, M. Zwerger, W.L. Ung, I.A. Eydelnant, D.E. Olins, A.L. Olins, H. Herrmann, D.A. Weitz, J. Lammerding, Nuclear envelope composition determines the ability of neutrophil-type cells to passage through micron-scale constrictions. J. Biol. Chem. **288**(12), 8610–8618 (2013).

D. Ruggero, Z.G. Wang, P.P. Pandolfi, The puzzling multiple lives of PML and its role in the genesis of cancer. BioEssays **22**, 827–835 (2000)

K.M. Sakthivel, P. Sehgal, A novel role of lamins from genetic disease to cancer biomarkers. Oncol. Rev. **10**, 309 (2016)

P. Sampathkumar, S.J. Kim, P. Upla, et al., Structure, dynamics, evolution, and function of a major scaffold component in the nuclear pore complex. Structure **21**, 560–571 (2013)

J. Sapudom, S. Rubner, S. Martin, T. Kurth, S. Riedel, C.T. Mierke, T. Pompe, The phenotype of cancer cell invasion controlled by fibril diameter and pore size of 3D collagen networks. Biomaterials **52**, 367–375 (2015)

J.N. Savas, B.H. Toyama, T. Xu, J.R. Yates, M.W. Hetzer, Extremely long-lived nuclear pore proteins in the rat brain. Science **335**, 942 (2012)

P. Scaffidi, T. Misteli, Lamin A-dependent nuclear defects in human aging. Science **312**, 1059–1063 (2006)

U. Scheer, R. Hock, Structure and function of the nucleolus. Curr. Opin. Cell Biol. **11**, 385–390 (1999)

E.C. Schirmer, R. Foisner, Proteins that associate with lamins: many faces, many functions. Exp. Cell Res. **313**, 2167–2179 (2007)

E.C. Schirmer, L. Gerace, The nuclear membrane proteome: extending the envelope. Trends Biochem. Sci. **30**, 551–558 (2005)

E.C. Schirmer, L. Florens, T. Guan, J.R. Yates, L. Gerace, Nuclear membrane proteins with potential disease links found by subtractive proteomics. Science **301**, 1380–1382 (2003)

K.D. Schleicher, S.L. Dettmer, L.E. Kapinos, S. Pagliara, U.F. Keyser, S. Jeney, R.Y.H. Lim, Selective transport control on molecular velcro made from intrinsically disordered proteins. Nat. Nanotechnol. **9**, 525–530 (2014)

M. Schmid, T.H. Jensen, Nuclear quality control of RNA polymerase II transcripts. Wiley Interdiscip. Rev. RNA **1**, 474–485 (2010)

H.B. Schmidt, D. Görlich, Nup98 FG domains from diverse species spontaneously phase-separate into particles with nuclear pore-like permselectivity. eLife **4**, e04251 (2015)

C. Schmitt, C. von Kobbe, A. Bachi, et al., 1999 Dbp5, a DEAD-box protein required for mRNA export, is recruited to the cytoplasmic fibrils of nuclear pore complex via a conserved interaction with CAN/Nup159p. EMBO J. **18**, 4332–4347 (2105)

R.L. Schoch, L.E. Kapinos, R.Y. Lim, Nuclear transport receptor binding avidity triggers a self-healing collapse transition in FG-nucleoporin molecular brushes. Proc. Natl. Acad. Sci. U. S. A. **109**, 16911–16916 (2012)

A. Schooley, B. Vollmer, W. Antonin, Building a nuclear envelope at the end of mitosis: coordinating membrane reorganization, nuclear pore complex assembly, and chromatin de-condensation. Chromosoma **121**, 539–554 (2012)

N. Schrader, P. Stelter, D. Flemming, R. Kunze, E. Hurt, I.R. Vetter, Structural basis of the nic96 subcomplex organization in the nuclear pore channel. Mol. Cell **29**, 46–55 (2008)

K.H. Schreiber, B.K. Kennedy, When lamins go bad: nuclear structure and disease. Cell **152**, 1365–1375 (2013)

S.M. Schreiner, P.K. Koo, Y. Zhao, S.G. Mochrie, M.C. King, The tethering of chromatin to the nuclear envelope supports nuclear mechanics Nature communications **6**, 7159 (2015)

W. Schul, I. van Der Kraan, A.G. Matera, R. van Driel, L. de Jong, Nuclear domains enriched in RNA 3 h-processing factors associate with coiled bodies and histone genes in a cell cycle-dependent manner. Mol. Biol. Cell **10**, 3815–3824 (1999)

W. Schul, R. van Driel, L. de Jong, Coiled bodies and U2 snRNA genes adjacent to coiled bodies are enriched in factors required for snRNA transcription. Mol. Biol. Cell **9**, 1025–1036 (1998)

W. Schul. B. Groenhout, K. Koberna, Y. Takagaki, A. Jenny, E.M. Manders, I. Raska, R. van Driel, L. de Jong, The RNA 3 h cleavage factors CstF 64 kDa and CPSF 100 kDa are concentrated in nuclear domains closely associated with coiled bodies and newly synthesized RNA. EMBO J. **15**, 2883–2892 (1996)

T.U. Schwartz, Modularity within the architecture of the nuclear pore complex. Curr. Opin. Struct. Biol. **15**, 221–226 (2005)

A. Segref, K. Sharma, V. Doye, A. Hellwig, J. Huber, R. Luhrmann, E. Hurt, Mex67p, a novel factor for nuclear mRNA export, binds to both poly(A) + RNA and nuclear pores. EMBO J. **16**, 3256–3271 (1997)

O. Seksek, J. Biwersi, A.S. Verkman, Translational diffusion of macromolecule-sized solutes in cytoplasm and nucleus. J. Cell Biol. **138**, 131–142 (1997)

W.T. Senapedis, E. Baloglu, Y. Landesman, Clinical translation of nuclear export inhibitors in cancer. Semin. Cancer Biol. **27C**, 74–86 (2014)

H.S. Seo, B.J. Blus, N.Z. Jankovic, G. Blobel, Structure and nucleic acid binding activity of the nucleoporin Nup157. Proc. Natl. Acad. Sci. U. S. A. **110**, 16450–16455 (2013)

J. Seong, A. Tajik, J. Sun, J.-L. Guan, M.J. Humphries, S.E. Craig, A. Shekaran, Distinct biophysical mechanisms of focal adhesion kinase mechanoactivation by different extracellular matrix proteins. Proc. Natl. Acad. Sci. U. S. A. **110**, 19372–19377 (2013)

C. Shiels, S.A. Islam, R. Vatcheva, P. Sasieni, M.J. Sternberg, P.S. Freemont, D. Sheer, PML bodies associate specifically with the MHC gene cluster in interphase nuclei. J. Cell Sci. **114**, 3705–3716 (2001)

J.-W. Shin, K.R. Spinler, J. Swift, J.A. Chasis, N. Mohandas, D.E. Discher, Lamins regulate cell trafficking and lineage maturation of adult human hematopoietic cells. Proc. Natl. Acad. Sci. USA **110**, 18892–18897 (2013)

L.S. Shopland, M. Byron, J.L. Stein, J.B. Lian, G.S. Stein, J.B. Lawrence, Replication-dependent histone gene expression is related to Cajal body (cb) association but does not require sustained cb contact. Mol. Biol. Cell **12**, 565–576 (2001)

A.K. Simi, A.S. Piotrowski, C.M. Nelson, Mechanotransduction, metastasis and genomic instability, in *Genomic Instability and Cancer Metastasis* (Springer International Publishing), pp. 139–158

G.D. Sinclair, K. Brasch, The reversible action of a-amanitin on nuclear structure and molecular composition. Exp. Cell Res. **111**, 1–14 (1978)

O.P. Singh, B. Bjorkroth, S. Masich, L. Wieslander, B. Daneholt, The intranuclear movement of Balbiani ring premessenger ribonucleoprotein particles. Exp. Cell Res. **251**, 135–146 (1999)

S. Siniossoglou, M. Lutzmann, H. Santos-Rosa, K. Leonard, S. Mueller, U. Aebi, E. Hurt, Structure and assembly of the Nup84p complex. J. Cell Biol. **149**, 41–54 (2000)

S. Siniossoglou, C. Wimmer, M. Rieger, V. Doye, H. Tekotte, C. Weise, S. Emig, A. Segref, E.C. Hurt, A novel complex of nucleoporins, which includes Sec13p and a Sec13p homolog, is essential for normal nuclear pores. Cell **84**, 265–275 (1996)

J.E. Sleeman, A.I. Lamond, Newly assembled snRNPs associate with coiled bodies before speckles, suggesting a nuclear snRNP maturation pathway. Curr. Biol. **9**, 1065–1074 (1999)

C.J. Smoyer, S.L. Jaspersen, Breaking down the wall: the nuclear envelope during mitosis. Curr. Opin. Cell Biol. **26**, 1–9 (2014)

S. Snaar, K. Wiesmeijer, A.G. Jochemsen, H.J. Tanke, R.W. Dirks, Mutational analysis of fibrillarin and its mobility in living human cells. J. Cell Biol. **151**, 653–662 (2000)

C.A. Snay-Hodge, H.V. Colot, A.L. Goldstein, C.N. Cole, Dbp5p/Rat8p is a yeast nuclear pore-associated DEAD-box protein essential for RNA export. EMBO J. **17**, 2663–2676 (1998)

N.T. Snider, M.B. Omary, Post-translational modifications of intermediate filament proteins: mechanisms and functions. Nat. Rev. Mol. Cell Biol. **15**, 163–177 (2014)

B. Sollner-Webb, K. Tyc, J.A. Steitz, Ribosomal RNA processing in eukaryotes, in *Ribosomal RNA: Structure, Evolution, Processing and Function in Protein Synthesis*, ed. by R. Zimmerman, A. Dahlberg (CRC Press, Boca Raton, FL, 1996), pp. 469–490

S.R. Solmaz, G. Blobel, I. Melcák, Ring cycle for dilating and constricting the nuclear pore. Proc. Natl. Acad. Sci. U. S. A. **110**, 5858–5863 (2013)

I. Solovei, A.S. Wang, K. Thanisch, C.S. Schmidt, S. Krebs, M. Zwerger, T.V. Cohen, D. Devys, R. Foisner, L. Peichl, H. Herrmann, H. Blum, D. Engelkamp, C.L. Stewart, H. Leonhardt, B. Joffe, LBR and lamin A/C sequentially tether peripheral heterochromatin and inversely regulate differentiation. Cell **152**(3), 584–598 (2013)

B.A. Sosa, U. Kutay, T.U. Schwartz, Structural insights into LINC complexes. Curr. Opin. Struct. Biol. **23**, 285–291 (2013)

B.A. Sosa, A. Rothballer, U. Kutay, T.U. Schwartz, LINC complexes form by binding of three KASH peptides to domain interfaces of trimeric SUN proteins. Cell **149**, 1035–1047 (2012)

S. Soucek, A.H. Corbett, M.B. Fasken, The long and the short of it: the role of the zinc finger polyadenosine RNA binding protein, Nab2, in control of poly(A) tail length. Biochim. Biophys. Acta **1819**, 546–554 (2012)

B. Soullam, H.J. Worman, Signals and structural features involved in integral membrane protein targeting to the inner nuclear membrane. J Cell Biol. **130**, 15–27 (1995)

S.T. Spagnol, K.N. Dahl, Spatially resolved quantification of chromatin condensation through differential local rheology in cell nuclei fluorescence lifetime imaging. PLoS ONE **11**(1), e0146244 (2016)

D.L. Spector, G. Lark, S. Huang, Differences in snRNP localization between transformed and nontransformed cells. Mol. Biol. Cell **3**, 555–569 (1992)

K. Stade, C.S. Ford, C. Guthrie, K. Weis, Exportin 1 (Crm1p) is an essential nuclear export factor. Cell **90**, 1041–1050 (1997)

D.A. Starr, M. Han, ANChors away: an actin based mechanism of nuclear positioning. J. Cell Sci. **116**, 211–216 (2003)

F. Stavru, B.B. Hülsmann, A. Spang, E. Hartmann, V.C. Cordes, D. Görlich, NDC1: a crucial membrane-integral nucleoporin of metazoan nuclear pore complexes. J Cell Biol. **173**, 509–519 (2006)

D.L. Stenoien, K. Patel, M.G. Mancini, M. Dutertre, C.L. Smith, B.W. O'Malley, M.A. Mancini, FRAP reveals that mobility of oestrogen receptor-α is ligand and proteasome-dependent. Nat. Cell Biol. **3**, 15–23 (2001)

A.D. Stephens, E.J. Banigan, S.A. Adam, R.D. Goldman, J.F. Marko, Chromatin and lamin A determine two different mechanical response regimes of the cell nucleus. Mol. Biol. Cell **28**(14), 1984–1996 (2017)

T. Sternsdorf, T. Grotzinger, K. Jensen, H. Will, Nuclear dots: actors on many stages. Immunobiology **198**, 307–331 (1997)

C.L. Stewart, B. Burke, The missing LINC: a mammalian KASH-domain protein coupling meiotic chromosomes to the cytoskeleton. Nucleus **5**, 3–10 (2014)

C.L. Stewart, K.J. Roux, B. Burke, Blurring the boundary: the nuclear envelope extends its reach. Science **318**, 1408–1412 (2007)

M. Stewart, Nuclear export of mRNA. Trends Biochem. Sci. **35**, 609–617 (2010)

Y. Strahm, B. Fahrenkrog, D. Zenklusen, E. Rychner, J. Kantor, M. Rosbach, F. Stutz, The RNA export factor Gle1p is located on the cytoplasmic fibrils of the NPC and physically interacts with the FG-nucleoporin Rip1p, the DEAD-box protein Rat8p/ Dbp5p and a new protein Ymr 255p. EMBO J. **18**, 5761–5777 (1999)

K. Strasser, E. Hurt, Yra1p, a conserved nuclear RNA-binding protein, interacts directly with Mex67p and is required for mRNA export. EMBO J. **19**, 410–420 (2000)

K. Strasser, J. Bassler, E. Hurt, Binding of the Mex67p/Mtr2p heterodimer to FXFG, GLFG, and FG repeat nucleoporins is essential for nuclear mRNA export. J. Cell Biol. **150**, 695–706 (2000)

L.A. Strawn, T. Shen, N. Shulga, D.S. Goldfarb, S.R. Wente, Minimal nuclear pore complexes define FG repeat domains essential for transport. Nat Cell Biol. **6**, 197–206 (2004)

S. Strudwick, K.L. Borden, Finding a role for PML in APL pathogenesis: a critical assessment of potential PML activities. Leukemia **16**, 1906–1917 (2002)

T. Stuwe, C.J. Bley, K. Thierbach et al., Architecture of the fungal nuclear pore inner ring complex. Science **350**, 56–64 (2015a)

T. Stuwe, A.R. Correia, D.H. Lin, M. Paduch, V.T. Lu, A.A. Kossiakoff, A. Hoelz, Nuclear pores. Architecture of the nuclear pore complex coat. Science **347**, 1148–1152 (2015b)

T. Stuwe, D.H. Lin, L.N. Collins, E. Hurt, A. Hoelz, Evidence for an evolutionary relationship between the large adaptor nucleoporin Nup192 and karyopherins. Proc. Natl. Acad. Sci. U. S. A. **111**, 2530–2535 (2014)

K.E. Suel, H. Gu, Y.M. Chook, Modular organization and combinatorial energetics of proline-tyrosine nuclear localization signals. PLoS Biol. **6**, e137 (2008)

B.O. Sun, Y. Fang, Z. Li, Z. Chen, J. Xiang, Role of cellular cytoskeleton in epithelial-mesenchymal transition process during cancer progression *Biomed*. Rep. **3**, 603–610 (2015)

H. Sun, N. Hattori, W. Chien et al., KPT-330 has antitumour activity against non-small cell lung cancer. Br. J. Cancer **111**, 281–291 (2014)

J. Swift, D.E. Discher, The nuclear lamina is mechano-responsive to ECM elasticity in mature tissue. J. Cell Sci. **127**, 3005–3015 (2014)

J. Swift, I.L. Ivanovska, A. Buxboim et al., Nuclear Lamin-A scales with tissue stiffness and enhances matrix-directed differentiation. Science **341**, 1240104 (2013)

K. Tahara, M. Takagi, M. Ohsugi et al., Importin-b and the small guanosine triphosphatase Ran mediate chromosome loading of the human chromokinesin Kid. J. Cell Biol. **180**, 493–506 (2008)

A. Tajik, Y. Zhang, F. Wei, J. Sun, Q. Jia, W. Zhou, R. Singh, N. Khanna, A.S. Belmont, N. Wang, Transcription upregulation via force-induced direct stretching of chromatin. Nat. Mater. **15**(12) 1287–1296 (2016)

K. Tamura, Y. Fukao, M. Iwamoto, T. Haraguchi, I. Hara-Nishimura, Identification and characterization of nuclear pore complex components in *Arabidopsis thaliana*. Plant Cell **22**, 4084–4097 (2010)

W. Tao, A.J. Levine, P19(ARF) stabilizes p53 by blocking nucleocytoplasmic shuttling of Mdm2. Proc. Natl. Acad. Sci. U. S. A. **96**, 6937–6941 (1999)

S. Taranum, I. Sur, R. Muller, W. Lu, R.N. Rashmi, M. Munck, S. Neumann, I. Karakesisoglou, A.A. Noegel, Cytoskeletal interactions at the nuclear envelope mediated by nesprins. Int. J. Cell Biol. **2012**, 736524 (2012)

Z. Tariq, H. Zhang, A. Chia-Liu, Y. Shen, Y. Gete, Z.-M. Xiong, C. Tocheny, L. Campanello, D. Wu, W. Losert, Lamin A and microtubules collaborate to maintain nuclear morphology. Nucleus **8**, 433–446 (2017)

B. Terris, V. Baldin, S. Dubois, C. Degott, J.F. Flejou, D. Henin, A. Dejean, PML nuclear bodies are general targets for inflammation and cell proliferation. Cancer Res. **55**, 1590–1597 (1995)

L.J. Terry, S.R. Wente, Flexible gates: dynamic topologies and functions for FG nucleoporins in nucleocytoplasmic transport. Eukaryot. Cell **8**, 1814–1827 (2009)

L.J. Terry, E.B. Shows, S.R. Wente, Crossing the nuclear envelope: Hierarchical regulation of nucleocytoplasmic transport. Science **318**, 1412–1416 (2007)

P. Tessarz, T. Kouzarides, Histone core modifications regulating nucleosome structure and dynamics. Nat. Rev. Mol. Cell Biol. **15**, 703–708 (2014)

J. Tetenbaum-Novatt, L.E. Hough, R. Mironska, A.S. McKenney, M.P. Rout, Nucleocytoplasmic transport: a role for nonspecific competition in karyopherin–nucleoporin interactions. Mol. Cell. Proteomics **11**, 31–46 (2012)

L. Texari, G. Dieppois, P. Vinciguerra, M.P. Contreras, A. Groner, A. Letourneau, F. Stutz, The nuclear pore regulates GAL1 gene transcription by controlling the localization of the SUMO protease Ulp1. Mol. Cell **51**, 807–818 (2013)

G. Theerthagiri, N. Eisenhardt, W. Schwarz Hand Antonin, The nucleoporin Nup188 controls passage of membrane proteins across the nuclear pore complex. J. Cell Biol. **189**, 1129–1142 (2010)

H.-R. Thiam, P. Vargas, N. Carpi et al., Perinuclear Arp2/3- driven actin polymerization enables nuclear deformation to facilitate cell migration through complex environments. Nat. Commun. **7**, 10997 (2016)

M. Thiry, T. Cheutin, M.F. O'Donohue, H. Kaplan, D. Ploton, Dynamics and three-dimensional localization of ribosomal RNA within the nucleolus. RNA **6**, 1750–1761 (2000)

V.J. Tocco, Y. Li, K.G. Christopher, J.H. Matthews, V. Aggarwal, L. Paschall, H. Luesch, J.D. Licht, R.B. Dickinson, T.P. Lele, The nucleus is irreversibly shaped by motion of cell boundaries in cancer and non-cancer cells. J. Cell. Physiol. **233**(2), 1446–1454 (2018)

D. Tollervey, T. Kiss, Function and synthesis of small nucleolar RNAs. Curr. Opin. Cell Biol. **9**, 337–342 (1997)

M. Torbati, T.P. Lele, A. Agrawal, An unresolved LINC in the nuclear envelope. Cell. Mol. Bioeng. **9**, 252–257 (2016)

B.H. Toyama, J.N. Savas, S.K. Park, M.S. Harris, N.T. Ingolia, J.R. Yates, M.W. Hetzer, Identification of long-lived proteins reveals exceptional stability of essential cellular structures. Cell **154**, 971–982 (2013)

E.J. Tran, S.R. Wente, Dynamic nuclear pore complexes: life on the edge. Cell **125**, 1041–1053 (2006)

E.J. Tran, M.C. King, A.H. Corbett, Macromolecular transport between the nucleus and the cytoplasm: advances in mechanism and emerging links to disease. Biochim. Biophys. Acta **1843**(11), 2784–2795 (2014)

R.J. Tran, Y. Zhou, A.H. Corbett, S.R. Wente, The DEAD-box protein Dbp5 controls mRNA export by triggering specific RNA: protein remodeling events. Mol. Cell **28**, 850–859 (2007)

T. Tsukamoto, N. Hashiguchi, S.M. Janicki, T. Tumbar, A.S. Belmont, D.L. Spector, Visualization of gene activity in living cells. Nat. Cell Biol. **2**, 871–878 (2000)

L.C. Tu, G. Fu, A. Zilman, S.M. Musser, Large cargo transport by nuclear pores: implications for the spatial organization of FG-nucleoporins EMBO J. **32**, 3220–3230 (2013)

T. Tumbar, A.S. Belmont, Interphase movements of a DNA chromosome region modulated by VP16 transcriptional activator. Nat. Cell Biol. **3**, 134–139 (2001)

T. Tumbar, G. Sudlow, A.S. Belmont, Large-scale chromatin unfolding and remodeling induced by VP16 acidic activation domain. J. Cell Biol. **145**, 1341–1353 (1999)

S. Tyagi, V. Vandelinder, N. Banterle, G. Fuertes, S. Milles, M. Agez, E.A. Lemke, Continuous throughput and long-term observation of single-molecule FRET without immobilization. Nat. Methods **11**, 297–300 (2014)

Y.B. Tzur, K.L. Wilson, Y. Gruenbaum, SUN-domain proteins: 'Velcro' that links the nucleoskeleton to the cytoskeleton. Nat. Rev. Mol. Cell Biol. **7**, 782–788 (2006)

C. Uhler, G.V. Shivashankar, Nuclear mechanopathology and cancer diagnosis. Trends in Cancer **4**, 320–331 (2019)

S. Ulbert, M. Platani, S. Boue, I.W. Mattaj, Direct membrane protein-DNA interactions required early in nuclear envelope assembly. J. Cell Biol. **173**, 469–476 (2006)

A. Ulrich, J.R. Partridge, T.U. Schwartz, The stoichiometry of the nucleoporin 62 subcomplex of the nuclear pore in solution. Mol. Biol. Cell **25**, 1484–1492 (2014)

R. Ungricht, U. Kutay, Establishment of NE asymmetry—targeting of membrane proteins to the inner nuclear membrane. Curr. Opin Cell Biol. **34**, 135–141 (2015)

R. Ungricht, U. Kutay, Mechanisms and functions of nuclear envelope remodelling. Nat. Rev. Mol. Cell Biol. **18**, 229–245 (2017)

R. Ungricht, M. Klann, P. Horvath, U. Kutay, Diffusion and retention are major determinants of protein targeting to the inner nuclear membrane. J. Cell Biol. **209**, 687–703 (2015)

R. Ungricht, S. Pawar, U. Kutay, An in vitro assay to study targeting of membrane proteins to the inner nuclear membrane. Methods Mol. Biol. **1411**, 461–477 (2016)

P. Valencia, A.P. Dias, R. Reed, Splicing promotes rapid and efficient mRNA export in mammalian cells. Proc. Natl. Acad. Sci. U. S. A. **105**, 3386–3391 (2008)

C. Verheggen, G. Almouzni, D. Hernandez-Verdun, The ribosomal RNA processing machinery is recruited to the nucleolar domain before RNA polymerase I during Xenopus laevis development. J. Cell Biol. **149**, 293–306 (2000)

P.J. Verschure, I. van der Kraan, E.M.M. Manders, R. van Driel, Spatial relationship between transcription sites and chromosome territories. J. Cell Biol. **147**, 13–24 (1999)

I.R. Vetter, A. Arndt, U. Kutay, D. Gorlich, A. Wittinghofer, Structural view of the Ran–Importin beta interaction at 2.3 A resolution. Cell **97**, 635–646 (1999)

J.F. Vincent, *Structural Biomaterials* (Princeton University Press, 2012)

A. Vitaliano-Prunier, A. Babour, L. Herissant, L. Apponi, T. Margaritis, F.C. Holstege, A.H. Corbett, C. Gwizdek, C. Dargemont, H2B ubiquitylation controls the formation of export-competent mRNP. Mol. Cell **45**, 132–139 (2012)

S. Vlcek, R. Foisner, Lamins and lamin-associated proteins in aging and disease. Curr. Opin. Cell Biol. **19**, 298–304 (2007)

G.K. Voeltz, W.A. Prinz, Sheets, ribbons and tubules-how organelles get their shape. Nat. Rev. Mol. Cell Biol. **8**, 258–264 (2007)

B. Vollmer, W. Antonin, The diverse roles of the Nup93/Nic96 complex proteins—structural scaffolds of the nuclear pore complex with additional cellular functions. Biol. Chem. **395**, 515–528 (2014)

L. Volpon, M.J. Osborne, H. Zahreddine, A.A. Romeo, K.L. Borden, Conformational changes induced in the eukaryotic translation initiation factor eIF4E by a clinically relevant inhibitor, ribavirin triphosphate. Biochem. Biophys. Res. Commun. **434**, 614–619 (2013)

A. von Appen, J. Kosinski, L. Sparks et al., In situ structural analysis of the human nuclear pore complex. Nature **526**, 140–143 (2015)

J.M. Wallace, Applications of atomic force microscopy for the assessment of nanoscale morphological and mechanical properties of bone. Bone **50**, 420–427 (2012)

T.C. Walther, A. Alves, H. Pickersgill, et al., The conserved Nup107-160 complex is critical for nuclear pore complex assembly. Cell **113**, 195–206 (2003)

J. Wang, C. Shiels, P. Sasieni, P.J. Wu, S.A. Islam, P.S. Freemont, D. Sheer, Promyelocytic leukemia nuclear bodies associate with transcriptionally active genomic regions. J. Cell Biol. **164**, 515–526 (2004)

N. Wang, J.D. Tytell, D.E. Ingber, Mechanotransduction at a distance: mechanically coupling the extracellular matrix with the nucleus. Nat. Rev. Mol. Cell Biol. **10**(1), 75–82 (2009)

X. Wang, H. Liu, M. Zhu, et al., Mechanical stability of the cell nucleus—roles played by the cytoskeleton in nuclear deformation and strain recovery. J. Cell Sci. **131**, jcs209627 (2018)

D.G. Wansink, W. Schul, I. van der Kraan, B. van Steensel, R. van Driel, L. de Jong, Fluorescent labeling of nascent RNA reveals transcription by RNA polymerase II in domains scattered throughout the nucleus. J. Cell Biol. **122**, 283–293 (1993)

M. Wasser, W. Chia, The EAST protein of Drosophila controls an expandable nuclear endoskeleton. Nat. Cell Biol. **2**, 268–275 (2000)

M.L. Watson, The nuclear envelope; its structure and relation to cytoplasmic membranes. J. Biophys. Biochem. Cytol. **1**, 257–270 (1955)

J.D. Weber, L.J. Taylor, M.F. Roussel, C.J. Sherr, D. Bar-Sagi, Nucleolar Arf sequesters Mdm2 and activates p53. Nat. Cell Biol. **1**, 20–27 (1999)

C.S. Weirich, J.P. Erzberger, J.S. Flick, J.M. Berger, J. Thorner, K. Weis, Activation of the DExD/H-box protein Dbp5 by the nuclear-pore protein Gle1 and its coactivator InsP6 is required for mRNA export. Nat. Cell Biol. **8**, 668–676 (2006)

K. Weis, The nuclear pore complex: oily spaghetti or gummy bear? Cell **130**, 405–407 (2007)

S.R. Wente, L. Burns, From hypothesis to mechanism: uncovering nuclear pore complex links to gene expression. Mol. Biol. Cell **34**(12), 2114–2120 (2014)

S.R. Wente, M.P. Rout, The nuclear pore complex and nuclear transport. Cold Spring Harb. Perspect. Biol. **2**, a000562 (2010)

J.R.R. Whittle, T.U. Schwartz, Architectural nucleoporins Nup157/170 and Nup133 are structurally related and descend from a second ancestral element. J Biol Chem. **284**, 28442–28452 (2009)

K. Wiesmeijer, C. Molenaar, I.M. Bekeer, H.J. Tanke, R.W. Dirks, Mobile foci of Sp100 do not contain PML: PML bodies are immobile but PML and Sp100 proteins are not. J. Struct. Biol. **140**, 180–188 (2002)

P. Wild, C. Senn, C.L. Manera, E. Sutter, E.M. Schraner, K. Tobler, M. Ackermann, U. Ziegler, M.S. Lucas, A. Kaech, Exploring the nuclear envelope of herpes simplex virus 1-infected cells by high-resolution microscopy. J. Virol. **83**, 408–419 (2009)

K. Wilhelmsen, M. Ketema, H. Truong, A. Sonnenberg, KASH-domain proteins in nuclear migration, anchorage and other processes. J. Cell Sci. **119**, 5021–5029 (2006)

K. Wilhelmsen, S.H. Litjens, I. Kuikman, N. Tshimbalanga, H. Janssen, I. van den Bout, K. Raymond, A. Sonnenberg, Nesprin-3, a novel outer nuclear membrane protein, associates with the cytoskeletal linker protein plectin. J. Cell Biol. **171**, 799–810 (2005)

S.L. Wolin, A.G. Matera, The trials and travels of tRNA. Genes Dev. **13**, 1–10 (1999)

J.L. Workman, R.E. Kingston, Alteration of nucleosome structure as a mechanism of transcriptional regulation. Annu. Rev. Biochem. **67**, 545–579 (1998)

H.J. Worman, G. Bonne, "Laminopathies": awide spectrum of human diseases. Exp. Cell Res. **313**, 2121–2133 (2007)

H.J. Worman, J.C. Courvalin, The inner nuclear membrane. J. Membr. Biol. **177**, 1–11 (2000)

Y. Xing, J.B. Lawrence, Nuclear RNA tracks: structural basis for transcription and splicing? Trends Cell Biol. **3**, 346–353 (1993)

Y. Xing, C.V. Johnson, P.R. Dobner, J.B. Lawrence, Higher level organization of individual gene transcription and RNA splicing. Science **259**, 1326–1330 (1993)

Y. Xing, C.V. Johnson, P.T. Moen, J.A. McNeil, J.B. Lawrence, Nonrandom gene organization: Structural arrangements of specific pre-mRNA transcription and splicing with SC-35 domains. J. Cell Biol. **131**, 1635–1647 (1995)

C. Xu, Z. Li, H. He, A. Wernimont, Y. Li, P. Loppnau, J. Min, Crystal structure of human nuclear pore complex component NUP43. FEBS Lett. **589**, 3247–3253 (2015)

D. Xu, A. Farmer, Y.M. Chook, Recognition of nuclear targeting signals by Karyopherin-beta proteins. Curr. Opin. Struct. Biol. **20**, 782–790 (2010)

Q. Yang, M.P. Rout, C.W. Akey, Three-dimensional architecture of the isolated yeast nuclear pore complex: functional and evolutionary implications. Mol. Cell **1**, 223–234 (1998)

Y. Yang, C. Isaac, C. Wang, F. Dragon, V. Pogacic, U.T. Meier, Conserved composition of mammalian box H/ACA and box C/D small nucleolar ribonucleoprotein particles and their interaction with the common factor Nopp140. Mol. Biol. Cell **11**, 567–577 (2000)

Y. Ye, H.H. Meyer, T.A. Rapoport, Function of the p 97- Ufd1-Npl4 complex in retrotranslocation from the ER to the cytosol: Dual recognition of nonubiquitinated polypeptide segments and polyubiquitin chains. J. Cell Biol. **162**, 71–84 (2003)

Z. Zachar, J. Kramer, P.M. Bingham, Evidence for channeled diffusion of pre-mRNAs during nuclear RNA transport in metazoans. J. Cell Biol. **121**, 729–742 (1993)

M.H. Zaman, L.M. Trapani, A.L. Sieminski, D. Mackellar, H. Gong, R.D. Kamm, A. Wells, D.A. Lauffenburger, P. Matsudaira, Migration of tumor cells in 3D matrices is governed by matrix stiffness along with cell-matrix adhesion and proteolysis. Proc. Natl. Acad. Sci. U. S. A. **103**, 10889–10894 (2006)

G. Zanetti, K.B. Pahuja, S. Studer, S. Shim, R. Schekman, COPII and the regulation of protein sorting in mammals Nat Cell Biol. **14**, 20–28 (2012)

Q. Zhang, C. Ragnauth, M.J. Greener, C.M. Shanahan, R.G. Roberts, The nesprins are giant actin-binding proteins, orthologous to *Drosophila* melanogaster muscle protein MSP-300. Genomics **80**, 473–481 (2002)

Q. Zhang, J.N. Skepper, F. Yang, J.D. Davies, L. Hegyi, R.G. Roberts, P.L. Weissberg, J.A. Ellis, C.M. Shanahan, Nesprins: a novel family of spectrin-repeat-containing proteins that localize to the nuclear membrane in multiple tissues. J. Cell Sci. **114**(Pt 24), 4485–4498

Y. Zhang, Y. Xiong, Mutations in human ARF exon 2 disrupt its nucleolar localization and impair its ability to block nuclear export of MDM2 and p53. Mol. Cell **3**, 579–591 (1999)

Y. Zhao, I. Meier, Efficient plant male fertility depends on vegetative nuclear movement mediated by two families of plant outer nuclear membrane proteins. Proc. Natl. Acad. Sci. U. S. A **111**(32), 11900–11905 (2014)

Y. Zheng, S. Gery, H. Sun, S. Shacham, M. Kauffman, H.P. Koeffler, KPT-330 inhibitor of XPO1-mediated nuclear export has anti-proliferative activity in hepatocellular carcinoma. Cancer Chemother. Pharmacol. **74**(3), 487–495 (2014)

X. Zhou, I. Meier, How plants LINC the SUN to KASH. Nucleus **4**, 206–215 (2013)

X. Zhou, K. Graumann, D.E. Evans, I. Meier, Novel plant SUN-KASH bridges are involved in RanGAP anchoring and nuclear shape determination. J. Cell Biol. **196**, 203–211 (2012)

Z. Zhou, M.J. Luo, K. Straesser, J. Katahira, E. Hurt, R. Reed, The protein Aly links pre-messenger-RNA splicing to nuclear export in metazoans. Nature **407**, 401–405 (2000)

M. Zwerger, C.Y. Ho, J. Lammerding, Nuclear mechanics in disease. Annu. Rev. Biomed. Eng. **13**, 397–428 (2011)

Part II

Fundamental Cellular Processes and Their Connection to Mechanics

The cell-division process and the cell cycle are presented, and how the cell cycle is linked to mRNA processing is highlighted. Protein biosynthesis, from the DNA through RNA to the final protein expression and posttranslational modifications, is covered. The processes of splicing and alternative splicing are discussed. A short part briefly presents the metabolic pathways of eukaryotes such as glycolysis and citrate cycle. Cell proliferation, survival, necrosis and programmed cell death are introduced and how these processes are investigated by biophysicists. All chapters of the first volume will help the reader, even if they have limited knowledge of biology, to understand the field of biophysics and subsequently to better convey an understanding of the complexity of biological matter and the general physical principles that can be applied to living matter such as cells.

Genetic Code 11

Abstract

In this chapter, the genetic code seems to be a universal code. The universal code has a specific arrangement of the codons that is definitely not random. There are at least three major concepts of the origin and the evolution of the universal genetic code: Firstly, there is the stereochemical theory stating that the assignments of codons are determined by the physicochemical affinity of the amino acids and the cognate codons (synonymously referred to as anticodons); secondly, the co-evolution theory stating that the structure of the code structure coevolved with the biosynthesis process of amino acids; and thirdly, the error minimization theory stating that there is a selection pressure minimizing the negative effects of point mutations and errors in translation were the main factor of code development. These theories are not contradictory and are also in line with the frozen accident hypothesis, such as the idea that the standard code may not have any special properties, but is simply determined by the fact that all existing life forms have a common ancestor, whereby later alterations to the code are generally excluded by the detrimental effect of codon reallocation. The mathematical examination of the structure and potential evolutionary trajectories of the code reveals that the code is highly resistant to translation errors, although there are numerous more resistant codes, suggesting that the standard code could emerge from a random code through a short sequence of rearrangements of series of codons. A large proportion of the evolution leading to the standard code seems to be a mixture of a frozen collision with a selection for error minimization, even though it cannot be excluded that the code co-evolves along with metabolic pathways due to weak affinities between amino acids and nucleotide triplets. These scenarios for code evolution, nonetheless, are founded on formal patterns with uncertain relevance to real primordial evolution. A true comprehension of code origins and developments is probably only possible in connection with a plausible script for the development of the coding scheme and the translation tool itself.

11.1 Introduction into the Genetic Code

Existing genetic codes can be considered as a mapping of two different groups: the 64 available mRNA codons and the 20 amino acids, and the stop signals required for protein synthesis. Since the initial set of codons (64) has a bigger cardinality than the incoming set of amino acids (20 + 1), the resulting mapping is inevitably degenerated. To put it another way, some amino acids are enciphered

C. T. Mierke, *Cellular Mechanics and Biophysics*, Biological and Medical Physics, Biomedical Engineering, https://doi.org/10.1007/978-3-030-58532-7_11

by two or more codons. Degeneration is a conceptual approach launched for the first time in quantum mechanics. An energy state of a quantum system is degenerated when it equals two or more distinct quantum states of the same energy. Degenerate quantum states are characterized by various solutions of the Schröedinger equation which are connected by a symmetry transition, i.e., quantum degeneration is basically a symmetry function. Therefore, it seems obvious whether the degeneration of the genetic code can be linked to symmetry aspects. Although a number of theories about the origin of the genetic code and encoding of proteins have been advanced, the origin of degeneration is still a problematic issue. Symmetry is a meta-principle that covers all areas of physics: from classical mechanics to quantum theory, from relativity to particle physics, it is generally known that symmetry laws are applied to understand laws of conservation. In classical mechanics, however, energy saving is associated with the Hamiltonian invariance of time translation; the preservation of linear and angular momenta is also linked to the Hamiltonian invariance of space translation and space rotation. In line with the universal role of symmetry principles in the discipline of physics, this scientific approach may help to solve several major unsolved biological problems. There are multiple approaches that have been performed to understand the genetic code with the employment of symmetry properties and group theory (Antonelli and Forger 2011; Hornos and Hornos 1993; Lenstra 2014). Remember that the approach has not generally been well accepted by biologists, perhaps because of the difficulty of biologically explaining models that rely on an unlikely chain of symmetry breaking steps; in this sense, the criticism of Maddox (Maddox 1994), who considers these attempts to be "a valuable factor in classification", is well founded. However, it is claimed that the analysis on the basis of symmetry properties is fundamental for understanding the origin and construction of the genetic code. The origin of degeneration employing a new symmetry-based approach is presented that provides an accurate quantitative characterization of some basic features such as the distribution of degeneration. The few studies of degeneration by using symmetry properties include these three (Lehrmann and Libchaber 2008; Nikolajewa et al. 2006; van der Gulik and Hoff 2016).

For more research on symmetry and symmetry breaking in the genetic code see, among others are these (Findley et al. 1982; Jose et al. 2012; Jungck 1978; Karasev and Stefanov 2001; Sánchez et al. 2005; Shcherbak 2008; Tlusty 2010). It is indisputable that the current shape of the genetic code is attributable at least in part to historical disasters. If life on earth were to re-emerge, what about the structure of the protein synthesis machinery (assuming there was one)? Is there some universal genetic code that displays degeneracy? Apparently, nobody is able to give an answer to these questions, but it is reasonable to suppose some structural similarities with the present machinery due to the continuing restrictions imposed by the universal laws of chemistry and physics.

Theories about the origin of the genetic code can be divided into at minimum five different classes. The first class is the stereochemical origin, which is founded on the hypothesis that codons or even anticodons can specifically attach to allocated amino acids by means of stereochemical specificity (Copley et al. 2005; Crick 1968). Note that a relationship has been observed between some amino acids and aptamers with codon/anticodon motifs (Yarus et al. 2009) which provides some encouragement for this hypothesis. The second class is the co-evolution theory, which proposes that codon mapping to the new amino acids occurs through the inheritance of part of the codon set belonging to the precursor amino acids (amino acids that produce the new one through biosynthetic modification) (Wrong 1975, 2005). The temporal order has been investigated in a previous study (Trifonov 2000). The third class represents the adaptive hypothesis, which it predicts that the main evolutionary constraint provides the minimization of mutation errors. In addition, it suggests that similar amino acids are coded by similar codons (Freeland and Hurst 1998a). The fourth class is the operational code that suggests an ancestral connection of the operational code, which primarily defines the affinity to the candidate amino acid in the acceptor stem, with the genetic code, which is realized with the codon–anticodon combination (DiGiuilio 2004; Musier-Forsyth and Schimmel 1999; Rodin et al.

1993, 2011). The fifth class is the frozen accident, which assumes a coincidental codon allocation origin to amino acids and a gradual development based on various evolutionary pressures, up to a point where any further modification becomes harmful that leads to the determination of the freezing of the code (Crick 1968).

Noteworthy, none of the aforementioned attempts is focused on degeneration distribution as a central characteristic. Degeneration is a natural outcome rather than a characteristic that is directly linked to the physical–chemical nature of the code. Two exceptions are identified, namely (Lehrmann and Libchaber 2008; van der Gulik and Hoff 2016), but the latter study is based on a biochemical statement of degeneration for existing codes and is therefore not directly linked to the development of the code. The evolutionary development of the universal genetic code can be divided in two key periods, such as firstly the ancient period, which starts at the beginning and ends with the appearance of the universal genetic code of the last universal common ancestor (LUCA) (Weiss et al. 2018), and secondly, the modern period, which starts from this universal genetic code and goes onwards (Sengupta and Higgs 2015) (Fig. 11.1). In specific detail, the ancient period can be subdivided by the introduction a third, middle period which lies immediately before the identification of the LUCA genetic code (van der Gulik and Hoff 2016). In this specific time phase, the genetic code was continuously optimized by the implementation of post-transcriptional modifications into the first base of anticodons. A precise description of the post-transcriptional modifications determining the genetic code of LUCA has been revealed (van der Gulik and Hoff 2016). A biochemical rationale for the wobble hypothesis is presented by examining the stability of the second letter of the anticodon within the ribosome center (Lehrmann and Libchaber 2008). The periodicity of the genetic code, known as Rumer's transformation, constitutes a crucial feature of the code (Lehrmann and Libchaber 2008; van der Gulik and Hoff 2016A global symmetry of the genetic code is involved, which the theoretical physicist Y. Rumer found in the 1960s (Rumer 1966a, b; Fimmel and Strüngmann 2016). In particular, Rumer' s symmetry is precisely defined as a dichotomous classification in terms of the chemical properties for the first two nucleotides within a codon (Gonzalez et al. 2006; Rumer 1966a, b). The fundamental link between symmetry and degeneration to the ancient evolutionary age of the genetic code has been widened (Gonzalez et al. 2019). For this purpose, a model of the presumed ancestor of the LUCA genetic code is being established. A certain consensus exists that the degen eration spread of this ancestral genetic code may overlap with that of the present mitochondrial genetic code of vertebrates (van der Gulik and Hoff 2016; Watanabe and Yokobori 2014). Therefore, a convincing model of the early code seems to demand: first, the exact quantitative characterization of the degeneration distribution and second, the basic Rumer's symmetry derived from the LUCA's and

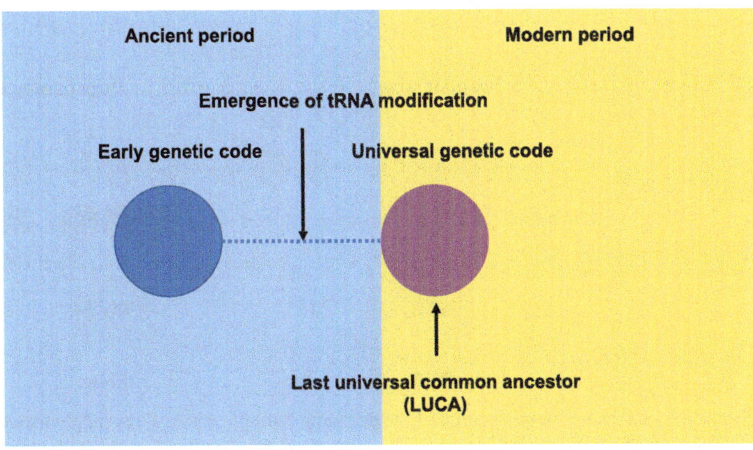

Fig. 11.1 Ancient and modern period

successive codes. A model is constructed that satisfies these two conditions, based on stereochemical symmetries of ancient chemical molecules and their informative equivalents as nucleotide sequences (Gonzalez et al. 2019).

11.2 Tessera Model for the Evolution of the Genetic Code

The first criterion listed above can only be fulfilled unexpectedly by examining a special collection of 64 four-base codons, namely tesserae (from the Greek word "tessera" = four), and a set of ancient symmetrical adaptors (Gonzalez et al. 2012) with anticodons of equal length. The model offers a unique solution and the description of the degeneration of the early code suggests the adoption of four-length codons. This justifies the tessera code as a presumed precursor of the early genetic code, in other words an early genetic code. This is in line with the hypothesis that the code was created with codons longer than three nucleotides (Baranov et al. 2009).

11.2.1 Symmetry and Degeneracy

To demonstrate the relationship between degeneration and symmetry, the hypothetical reversible tRNA adapters need to be taken into account that follow a Watson-Crick-like pairing rule (no wobble) (Fig. 11.2). However, it is only necessary that the identification of tessera–antitesserae is achieved by complementary mating.

In this regard, it is not absolutely necessary that the chemical bond is carried out strictly according to Watson–Crick. In fact, Watson–Crick pairs in the current types of mRNA and tRNA do not permit the recognition of codons in the opposite direction. In the time before LUCA, however, analog ancient molecules might have enabled bidirectional recognition. For instance, it has been argued that the first piece of genetic material has a simpler spine than ribose (Knight and Landweber 2000). The direction of the mating strands of these molecules is possibly not as restrictive as in actual DNA/RNA molecules. In line with this, it has been proposed that nucleic acids in which R-ribose has been substituted by L-ribose can combine with natural DNA and RNA to form a parallel strand of the A form (Ashley 1992). In living organisms, specific nucleotide sequences can assume a parallel orientation that implies non-canonical base pairing. In particular, parallel oriented areas in the genomes of bacteria (*Escherichia coli*, Listeria innocua) and insects (Drosophila melanogaster) were identified; these unusual regions are attributed a significant role in evolution and a considerable influence on biological activities (Szabat and Kierzek 2017).

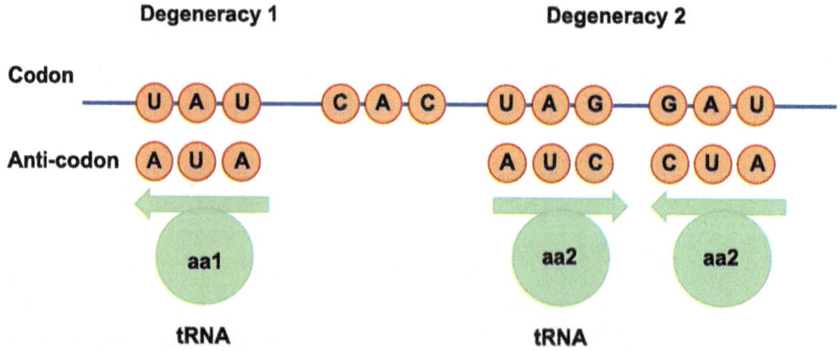

Fig. 11.2 Schematic representation of the decoding through primeval reversible tRNA adaptors that are able to read in both directions

A further circumstance that underpins the "reverse recognition" hypothesis is that a codon and its reverse codon consistently encode the same kind of amino acids (Thompson et al. 1994). Moreover, it has been assumed to be a kind of a remnant of such process (Nikolajewa et al. 2006). Another intriguing option is that ancient adaptors, such as pre-tRNAs, are missing the D and T loops (Nikolajewa et al. 2006); thus, since they are nearly symmetrical, they are capable of binding in both directions. The inverse symmetry of the ancient adaptors suggests that these adaptors are able to sense codons in both the 5′–3′ and 3′–5′ directions. In another setting, this idea has been examined to clarify the origin of the genetic code by means of reversible primordial adapters, where only two of the three bases of the codon are read and contain a kind of wiggly pairing, later on referred to as wobble pairing (Nikolajewa et al. 2006; Lagerkvist 1978). It is noteworthy that the reversibility of the adaptors naturally causes a kind of degeneration of the codon recognition. For example, a tRNA containing the anticodon AUA can only be mated with the codon UAU. Alternatively, a tRNA with anticodon AUC can be coupled with the codons GAU and UAG. Therefore, the amino acid transported by the first tRNA is expected to have degeneration 1, while the amino acid transported by the second tRNA will have degeneration 2 (Fig. 11.2). In principle, tRNAs with palindromic anticodon, which means invariant with respect to inversion, encode amino acids of degeneration 1, whereas non-palindromic anticodon encodes amino acids of degeneration 2. The greater the symmetry of an anticodon is, the less degenerate is the associated amino acid. On the basis of the arguments outlined above, this section shows how a model of the degeneration of the early code can be constructed.

In the first place, it is easy to prove that when codons have length three, there is no solution to the problem. As a matter of fact, there are 16 symmetrical (palindromic) codons connected to 16 amino acids with degeneration 1 and 48 codons without symmetry linked to 24 amino acids with degeneration 2. Obviously, these degeneration values 1 or 2 do not agree with the degeneration values of the initial code, where they are 2 or 4. The degeneration levels can be increased by allowing for more symmetries. The principal symmetries inherently linked to DNA or RNA molecules are the inverse, complementary symmetries (plus the trivial identity). When two of these symmetries are found, the third symmetry is fulfilled naturally, since it is the assembly of the other two These symmetries are occurring in true sequences, for example through inversions and reversed transpositions, and could be the reason for the genomic equilibrium that is known as the second Chargaff rule (Albrecht-Buehler 2006). The second Chargaff parity rule assumes that in each of the two DNA strands exists a balance between the amounts of A and T and of G and C.

Assuming that adapters have both inverse and inverse complementary symmetry, the degeneration pattern is as follows: 16 codons are linked to 8 amino acids with a degeneration of 2, and 48 codons are linked to 12 amino acids with a degeneration of 4. This corresponds to the degeneration levels of the initial code with 2 and 4, but the degeneration pattern differs: 32 codons associated with 16 amino acids having degeneration 2, and 32 codons associated with eight amino acids having degeneration 4. Bear in mind that there are eight amino acids with degeneracy 2, while the early code has 16 amino acids with degeneracy 2. This implies that the model requires 32 symmetric codons (not just 16), and there seems to be only a single route to achieve the goal, which is to take into account codons with more than three nucleotides.

The precise, unprecedented solution to this issue is provided by the tessera code (Table 11.1), that is a special set of 64 four-base codons identified by a number of ancient adapters that exhibit both palindromic and inverse complementary symmetry (Fig. 11.3). Using symmetry properties associated with group theory, the tessera set of 64 length-four codons is built up, which considers the four symmetrical transformations of the bases: identity (I: (A, U, C, G) → (A, U, C, G)); strong/weak or complementary (SW: (A, U, C, G) → (U, A, G, C)), pyrimidine/purine (YR: (A, U, C, G) → (G, C, U, A)) and keto/amino (KM: (A, U, C, G) → (C, G, A, U)). The set of transformations $F = \{I, \mathrm{SW}, \mathrm{YR}, \mathrm{KM}\}$ (along with the composition operator) is isomorphic to the Klein-4 symmetry group.

Table 11.1 Complete table of tesserae, which represent four-base codons with symmetry. Within quartets, the tesserae with the identical color encode for the identical amino acid, such as red and blue = 2 + 2 and white = 4

I	AAAA UUUU CCCC GGGG	SW	AAUU UUAA CCGG GGCC	YR	AAGG UUCC CCUU GGAA	KM	AACC UUGG CCAA GGUU
SW	AUUA UAAU CGGC GCCG	I	AUAU UAUA CGCG GCGC	KM	AUCG UAGC CGAU GCUA	YR	AUGC UACG CGUA GCAU
YR	AGGA UCCU CUUC GAAG	KM	AGCU UCGA CUAG GAUC	I	AGAG UCUC CUCU GAGA	SW	AGUC UCAG CUGA GACU
KM	ACCA UGGU CAAC GUUG	YR	ACGU UGCA CAUG GUAC	SW	ACUG UGAC CAGU GUCA	I	ACAC UGUG CACA GUGU

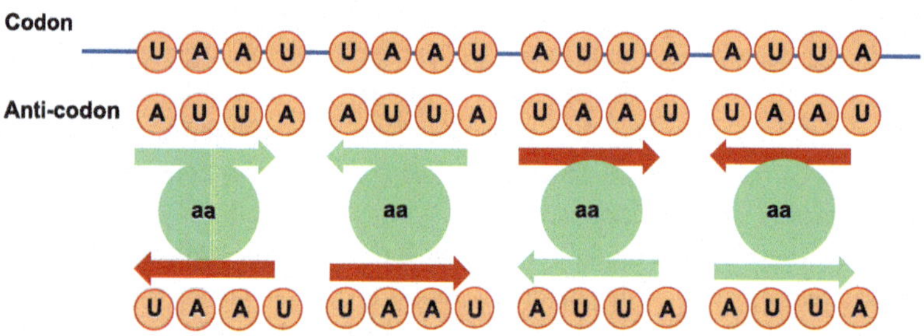

Fig. 11.3 Sketch of the tessera decoding through primeval adaptors containing palindromic and self-complementary anticodons. A single adaptor possesses the anticodons AUUA and UAAU in four possible configurations. As the anticodons are palindromic, two different tesserae can be read and hence the cognate amino acid has the degeneracy 2

The starting point is the four individual nucleotides: {A, U, C, G}, the four transformations can be performed and the resulting nucleotides can be added to the original one. For example, when starting from A and using $F = \{I, \mathrm{SW}, \mathrm{YR}, \mathrm{KM}\}$, one gets {AA, AU, AG, AC}. In this first step, there will be 16 available dinucleotides be obtained. Next, once again, the four possible transformations are reapplied to the 16 dinucleotides and attached to the original ones. Thus, for example, starting from AU and employing $F = \{\mathrm{I, SW, YR, KM}\}$, one receives {AUAU, AUUA, AUGC, AUCG}. In this second step, the tessera set with 64 objects of length 4 is constructed. Mathematically, seen a tessera will be a quadruplet $b_1b_2b_3b_4$ where bi $\in$ {A, U, C, G} and $b_3b_4 = f(b_1b_2)$ whereby $f \in F$. In Table 11.1, a tessera set is provided. It is divided into 16 quartets according to the transformations that create them (left side of any quartet). 16 tessera are noted to be are palindromic (first column), 16 are self-complementary (second column) and 32 display none of these symmetries.

Reading the set of tesserae using the primordial tRNA adapters mentioned above, it is possible to fully account for the degeneration distribution of the early genetic code. In line with an example described in Fig. 11.2, the level of degeneration connected with tessera detection is a function of its symmetry, in other words, the more symmetrical an anti-tessera appears, the less degenerative the corresponding amino acid is. Antitesserae are similarly specified as anticodons, which means the inverse adjunct of a tessera. tRNAs with palindromic antitesserae are able to sense palindromic tesserae in pairs; this means that the linked amino acid has become degenerate 2. Figure 11.3

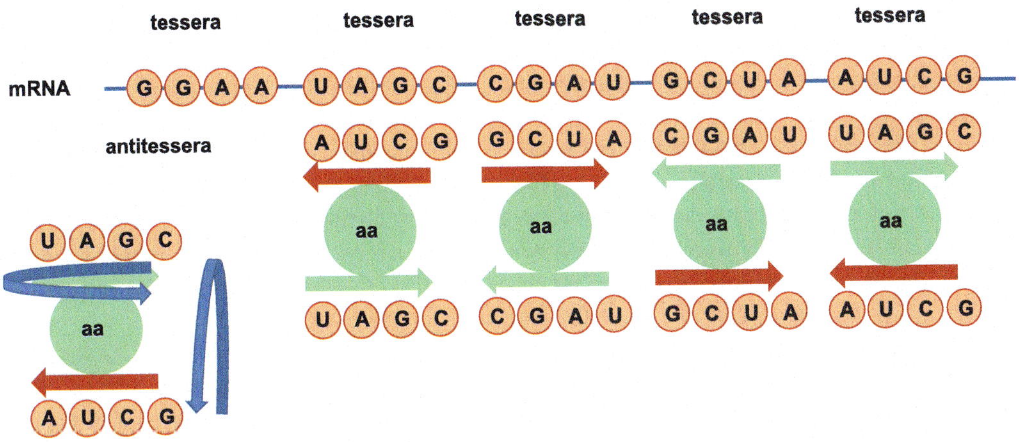

Fig. 11.4 The tessera decoding through the primeval adaptors with two non-symmetric antitesserae. There as four spatial configurations of the adaptor feasible, which can be matched with four different tesserae and yield to an amino acid with degeneracy 4

illustrates that the adapter is only able to detect the pair of tesserae AUUA, UAAU regardless of the implemented symmetry. As a number of 16 palindromic tesserae exist, this decoding strategy generates eight amino acids with degeneracy 2. The same reasoning applies to the 16 self-complementary tesserae, which yield eight additional amino acids with degeneration 2. A tRNA with a non-symmetrical antitessera may now be able to recognize four different non-symmetrical tesserae (third and fourth column of Table 11.1). In consequence, the eight amino acids bound to these tRNAs exhibit degeneration 4. Figure 11.4 depicts the case in which a non-symmetrical antitessera is coupled with four distinct (non-symmetrical) tesserae. In total, there are 16 amino acids with degeneration 2 and 8 amino acids with degeneration 4. In fact, these are the actual degeneration patterns of the suspected early code and are consistent with those of the current vertebrate mitochondrial code (where the amino acids with degeneration 6 participate within a quartet with two groups of codons, one group with degeneration 2 and the other with degeneration 4). In Table 11.2, the number of codons, the number of symmetric codons and the various degeneration patterns as a function of codon

Table 11.2 Codon number, number of symmetric codons and the distribution of degeneracy is based on the length of the codon. The tesserae of length 4 provides the degeneracy of the mitochondrial genetic code in vertebrates

			Degeneracy distribution	
Codon length	No. codons	No. symmetric codons	Degeneracy	No. aa
2	16	8	2	4
			4	2
3	64	16	2	8
			4	12
4	256	32	2	16
			4	56
4T	64	32	2	16
			4	8
5	1024	64	2	32
			4	240

length are presented. It should be kept in mind that the 4T solution (tesserae of length 4) seems to be the only one which reveals the degeneration of the mitochondrial genetic code of vertebrates. There is substantial evidence to support the hypothesis of the origin of the degeneration from a mathematical simulation of the genetic code that accounts for the degeneration of both nuclear and mitochondrial versions. The model is founded on number theory, which is a redundant integer representation system (Gonzalez et al. 2016; Gonzalez 2008a, b). Common numbering systems rely on the cumulative break-down of a number with the powers of a base b. Every number is uniquely represented in this context. Conversely, non-power numbering systems employ a sequence that increases at a slower rate than the powers of a base. In the last case, a number might be presented in a non-unique (degenerate) manner, which can be applied to explain the degeneration pattern of the genetic code.

The examination of the euplotid nuclear genetic code yields the unique solution: 8, 7, 4, 2, 1, 1 (Gonzalez 2008a, b). The identical approach for the mitochondrial genetic code of vertebrates results in the unique solution: 8, 8, 4, 2, 1, 0. This last outcome in particularly entails the division of the code into two equivalent subsets. Each subset is interpretable in terms of dinucleotides and full presentation as fused dinucleotide pairs. In this manner, just as in the biological hypothesis, codons of four nucleotides are formed organically. However, a comprehensive description of the genesis of the tesserae subset and its connections with the non-power model of the genetic code has been reported in (Gonzalez et al. 2012, 2016).

11.2.2 Evolutionary Implications

The tessera code appears to be the first quantitative statement for the origin of the degeneration in the ancestral codes. Thus, it could be important for understanding the evolution of the genetic code. In the following, the potential evolutionary consequences of the tessera code are introduced. To this end, the evolutionary hypothesis (Watanabe and Yokobori 2014) formulated on the background of the analysis of the translation in the existing mitochondria (Fig. 11.5). There are certain milestones on which there appears to be some consensus and which mark the evolutionary stages from a primitive

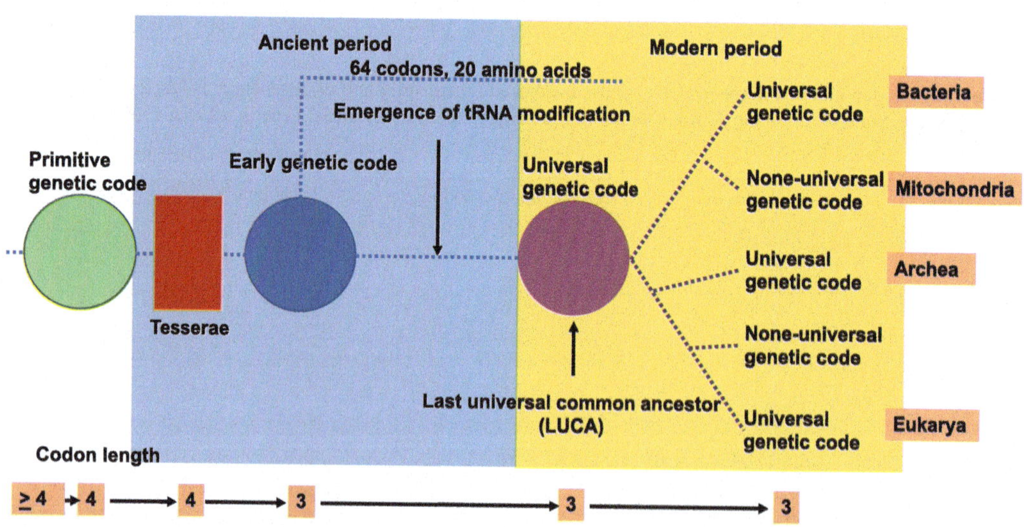

Fig. 11.5 The evolution of the genetic code with the major hallmarks represented by a circle. The evolution of the codon length is based theoretical results

genetic code to the current varieties. Beginning with the existing variants and proceeding retrospectively in time, the first milestone is set, which means the universal genetic code of LUCA (Fig. 11.5). It is assumed that the structure of this code is similar to the current genetic code of the nuclear standard.

The most basic version of the existing codes is the mitochondrial genetic code of vertebrates, which has been recommended as a model for the forerunner of the universal LUCA code for one reason in particular, which is the fact that it is an early code (Fig. 11.5). The most important evolutionary innovation arising from the transition of the early code to the universal code is the emergence of post-transcriptional modifications in tRNAs.

This is reinforced by the circumstance that in certain existing Metazoan mitochondria a non-modified U at the first position of the anticodon can mate with all bases at the third position of the codon (Kondow et al. 1999). This enables the decoding of codon families without the necessity of modified nucleotides. A codon family represents a group, if the first two bases of the four codons are shared and code for the same amino acid. It has been hypothesized (Watanabe and Yokobori 2014) that the early code is based on a primitive code, which encompasses fewer, more degenerate, amino acids, such as the Jukes' code (Jukes 1968). Such a code should only be built by families, except for one amino acid and the stop symbol, which carry the degeneration two, which means that they are encoded by two codons.

The principal assertion is that the tessera code is a precursor of the early code, namely a pre-early code positioned between the primitive and the early code. Several arguments exist that suppose the tessera code as an early code (Fig. 11.5). In the first place, it shares precisely the same degenerative pattern as the early genetic code. The tessera code is also consistent with the hypothesis, which suggests an origin of the code with long oligonucleotides, followed by a reduction in codon length down to the optimal number of 3 (Baranov et al. 2009). The hypothesis that the pre-early code contains codons of length 4 implicates that the primitive code also contains codons of length at least 4. Actually, what comes out is that the primitive genetic code of Jukes can be translated with generic codons of length 4.

Throughout the following description, the term codon is referred to as codons of length four or tetracodons. The Jukes code contains 15 amino acids with degeneration 4, one amino acid with degeneration 2 and two stop codons. In total, it has 15 elements with degeneration 4 and 2 objects with degeneration 2. When considering that this code is derived from codons of 4 nucleotides, this entails the selection of 17 elements/amino acids that can be encoded by either two or four codons from the set of 44 = 256 codons. It should be pointed out that the set of 256 codons can be divided into a subset of 32 codons, which have some symmetries (Table 11.1), and into a subset of 224 codons without symmetry. Symmetric codons represent the amino acids with degeneration 2, as shown before, while asymmetric codons represent the amino acids with degeneration 4. Assuming a random assignment mechanism for codons to amino acids, the degeneration pattern that is most probable is precisely the same as in Jukes' code. Jukes expects one of the items of degeneration 2 to be connected to the stop signal. When this selection is applied, stop codons are less susceptible to random failures, which means that these items are less equivocal than those with degeneration 4. It is also reasonable to hypothesize that the other component with degeneration 2 is the same as the amino acid that codes for the start signal. What has been demonstrated is the fact that the primitive code (Jukes' version) is naturally written with codons of length 4.

The following is a potential developmental way forward from the Jukes code to the tessera code. The primitive code consists in this case of two pairs of symmetric codons and 60 asymmetric codons. Consider that the choice of symmetric codons, owing to their reduced proneness to point mutation defects, is a first step in choosing the symmetric half of the tessera set. For instance, imagine a new connector with a symmetrical anticodon is emerging and rivals (bears the same amino acid) an existing connector with an asymmetrical anticodon. The new connector is capable of binding to symmetric codons that belong to the tessera set (Table 11.1). This type of connector offers an evolutionary advantage over one bearing a non-symmetric anticodon, as it possesses two different spatial arrangements which can be exploited for attachment to the codon. For instance, the connector in Fig. 11.3 contains the palindromic anticodon AUUA. When the connector is flipped over, it remains able to attach to the codon/tessera UAAU. This selective force results in the capture of all symmetric codons (tetracodons)/tesserae at the detriment of the non-symmetric codons. Finally, after this process, there is a code consisting of 32 symmetrical tesserae and 32 non-symmetrical tetracodons, which are not strictly part of the tesserae set.

At this point, another optimization step is achieved by choosing non-symmetrical tesserae (Gonzalez et al. 2012): The tesserae have been seen to be immune to point mutations, since a point mutation cause the conversion to a non-tessera. Therefore, the tesserae remain alive up to non-tessera tetracodons owing to the evolutionary constraints of accuracy of decoding. The 32 symmetrical tesserae are resistant to point mutations (two simultaneous, very unlikely point mutations are required to create a transition between two tesserae). In other words, the respective tRNAs do not result in the insertion of an unrelated amino acid when exposed to a point mutation. This feature of failure detection suggests an evolutionary benefit with respect to the precision of protein synthesis. The other 32 non-symmetric tetracodons are probably not exactly tesserae, although those that are tesserae exhibit the characteristic of defect immunity, therefore they are chosen little by little for the reasons mentioned before. Thus, the complete pattern of the tessera pre-early code is revealed, and its degeneration profile corresponds to that of the early code (and to that of the preserved vertebrate mitochondrial code).

Based on minimal estimates, the transition between the Jukes code and the tessera code has been proven to be the one that is most likely. It is obvious that in this transitional period the tessera detection is getting more precise than the tetracodon detection of the Jukes' code. In fact, the defect recognition property of the tessera code permits to decrease the equivocality with respect to the amino acid load of tetracodone adapters. The enhanced accuracy of tessera detection in turn enables the selection of amino acids to be finetuned based on the evolutionary pressure of protein output.

The hypothesis of the tessera code as a pre-early code suggests another major evolutionary switch, namely the switch between the tessera code and the early code. As the early code is apparently expected to have codons of length three, the principal issue associated with this switch is connected to the alteration of codon length, which is commonly regarded as detrimental (Crick 1968). It should be pointed out that there are two reasons why firstly any theory about the origin of the code having a different codon lengths from three must address this problem (Baranov et al. 2009), and secondly, such a transition is dangerous when an evolutionary level is achieved where the code is locked, as this entails a fundamental modification of the sequence of amino acids of all the proteins of an organism, but this is not strictly the situation for previous evolutionary stages that are nearer to the origin of the code. The tessera code permits a clean solution to the existing conflict between tetracodons and codons. In fact, the data included in the tessera setting are a redundant set of information. When, according to definition, three out of four letters of a tessera are identified, then the remaining letter can be clearly deduced. From the perspective of coding theory, the tessera code and each genetic code of trinucleotides bear the same amount of specific information. Therefore, a one-to-one mapping between tesserae and codons can be performed. The fundamental principle of this kind of mapping is that the transformations between neighboring letters of a tessera are converted into the nucleotides of a codon. In specific, in the case of a tessera b_1, b_2, b_3 and b_4, there are three chemical conversions between neighboring letters: $t_{12} = f(b_1, b_2)$ between b_1 and b_2, $t_{23} = f(b_2, b_3)$ between b_2 and b_3 and $t_{34} = f(b_3, b_4)$ between b_3 and b_4. Only two of these three conversions are completely independent, as $t_{34} = t_{12}$. In Table 11.4a, the tessera code has been restructured in line with the transformation t_{12}

Table 11.3 Basic mapping structure between tesserae and codons. In specific, the four transformations between the tesserae bases are matched to the four codon nucleotides

Tesserae transformation t_{12}, t_{23}	Codon bases x_1, x_2
I	= A
SW	= U
KM	= C
YR	= G

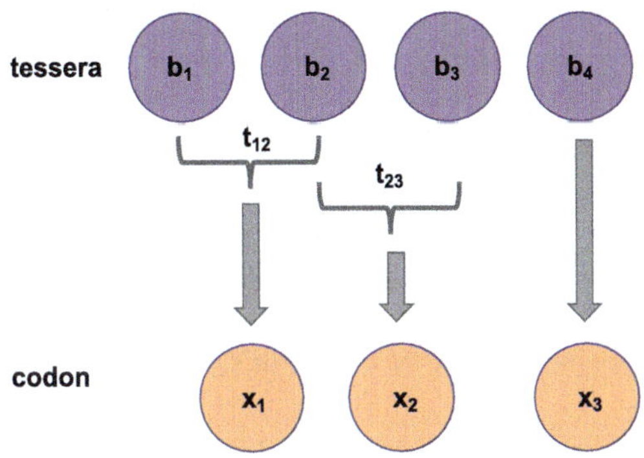

Fig. 11.6 Sketch of the mapping between the tessera (b_1, b_2, b_3 and b_4) to the codon (x_1, $x2$ andx_3)

(rows) and t_{23} (columns). It is recommended that t_{12} and t_{23} are projected onto the first and second nucleotide of the codon (x_1, x_2), respectively (Table 11.3). The fourth letter b_4 is also imaged to the third nucleotide of the codon x_3 (Fig. 11.6). In accordance with this mapping, the columns of the tessera set are shown mapped to the columns of the genetic code in such a way that $t_{23} = I$ is shown on NAN codons (degeneracy non-4) and t_{23} = KM on NCN codons (composed of families only) (Table 11.4b–c). It can be seen that the two columns of the tessera code have the same degeneration as the equivalent columns of the genetic code (either 4 or 2 + 2). The natural completeness of the depiction allocates t_{23} = SW to the NUN codons and t_{23} = YR to the NGN codons. The latter two classifications must take into consideration certain exclusions, which are defined by the circumstance that the transformation from tesserae to codons preserves the Rumer's symmetry but not the self-complementary symmetry.

The tessera–antitessera interplay is a more distinct interaction than the codon–antitessera interaction because four Watson-Crick-like chemical bonds are involved. However, where a genetic code is available, degeneration is primarily caused by the codon anticodon interplay of the first two bases. Thus, supposing that the binding energy in the pre-early code is similar to the Watson–Crick energy, the tessera–antitessera interaction energy ought to be about twice the actual codon–anticodon energy.

In a biochemical perspective, the transition from tesserae to codons entails the passage between a full quadruple specific Watson-Crick-like coupling for decoding tesserae and the wobble strategy for decoding codons. Specifically, this imposes theoretical limitations on certain symmetry characteristics that are available in the world of tesserae but are not included in the existing codes, such as the loss of self-complementary symmetry. In fact, each column of the tessera code has a certain degeneration; however, in existing codes this applies only to two columns, namely codons of the type NMN (NAN or NCN). On the contrary, the columns corresponding to the codons NKN (NUN or NGN) suffer from mixed degeneration; more specifically, the different quadrants separating the two codes are of the type SUN and WGN (the WSN or SWN are referred to as mixed as opposed to unmixed SSN WWN). To put it another way, existing codes include codons of type WGN, which encode amino acids with degeneration 2, although the central base is strong, and codons of type SUN, which encode amino acids with degeneration 4, although the central base itself is weak (Gonzalez et al. 2019).

An explanation of this kind of characteristics in the form of energetic restraints that rely on the stereochemistry of the codon–anticodon interaction has been reported (Lehrmann and Libchaber 2008). In the existing genetic code, low-level interaction is usually accompanied by 2 + 2 degeneration. In fact, the codons of the type NAN, AUN and UUN provide this. Nevertheless, when U is the second letter, additional stabilization of the central purine letter N35 in the anticodon loop of the tRNA by U33 permits the recognition of an entire family even though the character of N35 is weak. In this case, N35 designates a generic nucleotide N in the 35th position of the tRNA. U33, for example, designates a uracil nucleotide in the 33rd position of the tRNA and is equivalent to the nucleotide occurring in front of the nucleotides in the anticodon positions 34, 35 and 36.

In the mirror case of codons of the type AGN and UGN, the nucleotide N35 is not adequately stabilized by U33 and the corresponding quartet is characterized by degeneration 2 + 2. These limitations of the wobbling strategy mean that when tesserae are assigned to codons, the quadrant (YR-SW) is interchanged with the quadrant (SW-YR) and the quadrant (KM-SW) with the quadrant (I-YR) (Table 11.4). Finally, the fourth letter of a tessera is matched to the third letter of a codon, with the following exclusion, which guarantees correct grouping: when $b_4 = K$ (T or G), then x_3 = KM

Table 11.4 (A) The tessera code is organized due to the transformations: first-second letter t_{12} (rows) and second-third letter t_{23} columns. (B) contains exchanged quartets, but is basically the same as (A) (C) Degeneracy of the mitochondrial genetic code of vertebrates. The codes (B) and (C) are related by one-to-one mapping

(A)			t_{23}		(B)				
t_{12}	SW	KM	I	YR		SW	KM	I	YR
SW	AUAU UAUA CGCG GCGC	AUGC UACG CGUA GCAU	AUUA UAAU CGGC GCCG	AUCG UAGC GCAU GCUA	SW	AUAU UAUA CGCG GCGC	AUGC UACG CGUA GCAU	AUUA UAAU CGGC GCCG	AGCU UCGA CUAG GAUC
KM	ACGU UGCA CAUG GUAC	ACAC UGUG CACA GUGU	ACCA UGGU CAAC GUUG	ACUG UGAC CAGU GUCA	KM	AAGG UUCC CCUU GGAA	ACAC UGUG CACA GUGU	ACCA UGGU CAAC GUUG	ACUG UGAC CAGU GUCA
I	AAUU UUAA CCGG GGCC	AACC UUGG CCAA GGUU	AAAA UUUU CCCC GGGG	AAGG UUCC CCUU GGAA	I	AAUU UUAA CCGG GGCC	AACC UUGG CCAA GGUU	AAAA UUUU CCCC GGGG	ACGU UGCA CAUG GUAC
YR	AGCU UCGA CUAG GAUC	AGUC UCAG CUGA GACU	AGGA UCCU CUUC GAAG	AGAG UCUC CUCU GAGA	YR	AUCG UAGC CGAU GCUA	AGUC UCAG CUGA GACU	AGGA UCCU CUUC GAAG	AGAG UCUC CUCU GAGA
(C)									
	U	C	A	G					
U	UUG UUA UUU UUC	UCC UCU UCA UCG	UAA UAG UAC UAU	UGG UGA UGU UGC					

(continued)

Table 11.4 (continued)

(A)			t_{23}		(B)				
C	CUU CUC CUG CUA	CCC CCU CCA CCG	CAA CAG CAC CAU	CGU CGC CGG CGA					
A	AUG AUA AUU AUC	ACC ACU ACA ACG	AAA AAG AAC AAU	AGG AGA AGU AGC					
G	GUU GUC GUG GUA	GCC GCU GCA GCG	GAA GAG GAC GAU	GGU GGC GGG GGA					

(b_4); that is, T and G are interchanged; alternatively, $x_3 = b_4$. It should be pointed out that the resulting characterization is not exactly exclusive, but the present one demonstrates that it is feasible to move from the tessera code to the existing code by delineating all existing degeneration features of the latter. When the protein coding originally included codons larger than three bases, the mechanism of translation was supposed to have a certain memory of this. In fact, the small subunit of the existing ribosomes offers a structural freedom that may enable the inclusion of an additional nucleotide in the decoding center, thus permitting the decoding of four-base codons. The option of ancestral encoding with quadruplets has been introduced (Crick 1968). In fact, quadruplet encoding was originally identified in 1973 (Riddle and Carbon 1973) as a mechanism associated with frameshift inhibition, and it is now commonly employed in biotechnology to insert non-canonical amino acids in proteins (Neumann et al. 2010; Niu et al. 2013; Wang et al. 2007a). In addition, the biological applicability of length four codons and an orthogonal ribosome that deciphers them has been proven experimentally in the field (Neumann et al. 2010). Moreover, there is some indication of the presence of overlapping genes encoded by tetracodons (Seligmann 2012); it has also been found that tetracodons perform an essential task in phylogenetic analysis (Richter and Rosselló-Móra 2009) and this fact may be a hint for genetic memory.

An overwhelming number of characteristics of the tessera code have been retained in the contemporary codes. The early code and all its derivatives acquire from the tessera code the number of codons (64 tesserae generate 64 codons) and the maximum amount of amino acids (23). The tessera code permits coding for 24 items/amino acids. Among these need to be at least that is a stop signal and hence the maximal theoretical number of the available items for amino acids is then 23. It is notable that none of the existing codes extends beyond this limit, and the maximum amount of amino acids directly encoded by several genomes is accurately 23: the 20 standard amino acids together with 2 non-standard amino acids (selenocysteine and pyrrolysine) and the alternative initiatory amino acid N-formylmethionine accumulate to 23. Besides, 22 adaptors are involved in the mitochondrial genetic code of vertebrates: eight tRNAs, each detecting four codons, 14 tRNAs, each detecting two codons, and two pairs of codons not linked to amino acids (Watanabe and Yokobori 2014; Grosjean et al. 2010). It is noteworthy that 22 is the strict minimum found under all versions of the genetic code known to exist. That is precisely the underlying structure that the tessera model entails: eight original adapters of degeneration 4, together with 16 adapters of degeneration 2 form a set of 24 adapters; when rejecting two adapters of degeneration 2, which are mapped to the stop codons, the result is exactly 22.

The mitochondrial genetic code of vertebrates and the early code model based on the tessera also have a series of features in common that are associated with symmetry (Table 11.5). First of all, the

Table 11.5 Comparison between the vertebrate mitochondrial genetic code and the tessera code

	Vertebrate mitochondrial genetic code		Tessera code	
	Deg.	No. codons	Deg.	No. tesserae
Degeneracy	2	16	2	16
	4	8	4	8
Number of codons Number of adaptors Number of amino acids	Codons 64 Adaptors 22 aa 20		Tesserae 64 Adaptors 22 aa 20	
Symmetries Rumer	KM transform on the first two bases alters degeneracy of aa		KM transform on the first two bases alters degeneracy of aa	
Klein V group	16 codons share transformation between first and second letter and possess the same degeneracy distribution		16 tesserae share transformation between the first and the second letter and possess the same degeneracy distribution	

KM transformation, otherwise known as the Rumer transformation, is a process used on the first doublet of a codon to modify the degeneration of the respective amino acid. This general property is found in most commonly recognized versions of the genetic code (in both the nuclear and mitochondrial domains). The tessera code has the same feature. For instance, the tessera AUUA represents an amino acid of degeneration 2, and by applying the KM transformation to the first two nucleotides, the tessera CGUA is created, representing an amino acid of degeneration 4. However, this property also works if the Rumer transformation is executed on the t_{12} t_{23} of the map mentioned above, which combines tesserae and codons (Gonzalez et al. 2019).

One more elementary feature of the tessera code is that the encoding of a protein can be rendered resistant to +1 frame shifts. The strength of frame preservation can also be associated with circular codes that were thought to play a supporting role in the frame synchronization process (Arquès and Michel 1996, 1997; Fimmel et al. 2015; Giannerini et al. 2018; Gonzalez et al. 2011). A universal circular code property was linked to the emergence of the genetic code in the form of pairs of complementary codons encoding either the identical or a similar amino acid (Jolivet and Rothen 2001). The same assumption is also confirmed in other cases (Rodin et al. 2011). This characteristic occurs obviously in the tessera code, where a tessera and its inverse supplement invariably encode the identical amino acid.

The underlying basis of degeneration in protein-coding schemes has largely been explained with minimal hypotheses, in fact limited to those concerning the symmetry properties of coding and decoding of ancient molecules and related to their eventual stereochemical relationships. The theory agrees with numerous efforts to explain the origin and development of degeneration, such as those relating to the stereochemical detection of long oligonucleotide sequences by ribozymes in an RNA environment (Baranov et al. 2009; Wolf and Koonin 2007), those relating to various symmetry concepts (Antoneli and Forger 2011; Hornos and Hornos 1993; Lenstr 2014; Findley et al. 1982; Jose et al. 2012; Jungck 1978; Sánchez et al. 2005; Shcherbak 2008; Tlusty 2010), and in specific those that require the reverse alignment of codons (Nikolajewa et al. 2006). The presentation of the pre-early genetic code's degeneracy is precise and seems not to be based on simple adaption of parameters. In fact, there are exists no free parameters within the model. Symmetry characteristics are the only thing that is important: the underlying principles of symmetry in the physical sciences are placed on the equal conceptual scale in molecular biology and evolution. The theory holds many intriguing implications, such as the startling finding that primitive and early versions of the genetic code are linked through an intermediate code (the pre-early code) to codons of length four and with a degeneration pattern consistent with that of the current mitochondrial genetic code of vertebrates. The evolutionary transitions between such ancestral codes may be reasonably justified by the evolutionary constraints on decoding accuracy. This scenario also assumes that some ancestral characteristics could have been retained through evolutionary time lines. In fact, this is true and its conservation constitutes an imperative requirement in terms of their biological importance. An important issue is, for instance, how error-correcting features, which are specifically identifiable in the tessera code, have actually been converted into existing codes. Controlling is an inevitable demand on any system of protein synthesis. In existing genetic codes, resistance to point mutations is not recognizable. The question of how the existing decoding equipment maintains its error correction abilities despite the loss of redundancy caused by the decrease in codon length is an attractive unanswered question. In specific detail, the comparative analysis between the tessera model and the mathematical model of the genetic code (Gonzalez et al. 2008, 2012, 2016; Gonzalez 2008b; Gonzalez 2004) implies that the part played by chemical changes in the tessera model is mirrored by the part of the dichotomous classes in the existing genetic codes, and in future studies this issue will be addressed. The theory proposed in this chapter is embedded in a unified mathematical context that defines the degeneration with integer representation principles (Gonzalez et al. 2016, Gonzalez 2004, 2008a, b; Gonzalez et al. 2009).

The concept of the framework represents a new paradigm for the interpretation of genetic material and results in the definition of mathematical entities that provide a useful biochemical explanation. Dichotomous groups, for instance, are binary variables deduced directly from the model and associated with the chemical characteristics of the nucleotides of a codon. Analysis of the encoding DNA sequences with the aid of dichotomous classifications has revealed the existence of universally strong short-range interactions that can be associated with error recognition (Gonzalez et al. 2006, 2008). The theory of circular codes (Fimmel et al. 2015; Arquès and Michel 1996; Gonzalez et al. 2011), a class of error detection codes recommended as basic strategies for frame recognition and preservation, may also be linked to this mathematical context (Giannerini et al. 2012).

The origin of degeneration gains support based on elementary properties of chemistry and physics, primarily on the symmetries of the original molecules. In addition, the findings indicate that both the symmetries and the degeneration profile of the mitochondrial code have been retained during evolution, which raises additional fundamental scientific questions.

11.3 Historical View on the Genetic Code Deciphering

Directly after the deciphering of the genetic code of the gastrointestinal tract bacterium *Escherichia coli* (Nirenberg et al. 1963), it has been discovered that the specific mapping of 64 codons to different 20 amino acids and the two punctuation marks, the start and stop signals, are shared among all known living organisms and forms on earth with only very minor alterations (Hinegardner and Engelberg 1963; Woese et al. 1964). A superficial examination of the standard genetic code table (Fig. 11.7) reveals that the order of amino acid combinations is obviously not accidental (Woese 1965; Wose 1967; Crick 1968; Ycas 1969). In common, neighboring codons, which are codons that vary only in one nucleotide, typically prefer to encode either the same or two related amino acids; in other words, amino acids that are physicochemically related (although there are no clear criteria for defining physicochemical resemblance). The basic problem is how these patterns of the standard code came

Fig. 11.7 Genetic code

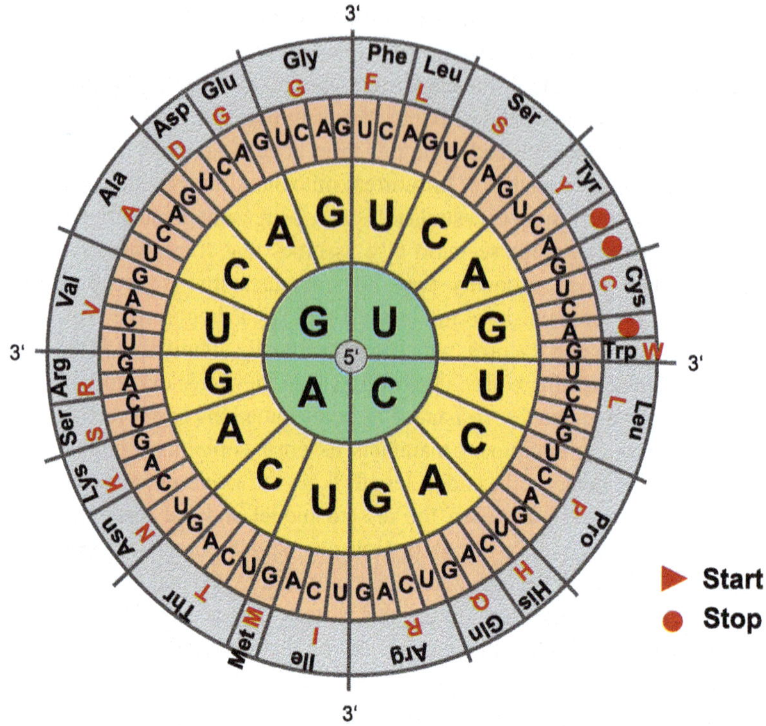

about, bearing in mind that there are over 1084 potential alternative code tables when every one of the 20 amino acids and the stop signal need to be allocated to at least one codon. More precisely, the question arises as to what kind of interaction of chemical pressures, historical coincidences and evolutionary powers may have created the standard amino acid allocation, which exhibits many remarkable characteristics (Koonin and Novozhilov 2009). Among the characteristics of the code that appear to require special clarification is firstly the block structure of the code, which is considered a prerequisite for the robustness of the code against point mutations, translation errors and shifts in the translation frame (Chechetkin 2003). A second characteristic feature represents the linkage between the second codon letter and the nature of the encoded amino acid, which means that codons with U in the second position match hydrophobic amino acids (Rumer 1966a, b; Volkenshtein and Rumer 1967). A third feature is the connection between the second codon position and the class of aminoacyl-tRNA synthetase (Wetzel 1995), the negative correlation between the molecular weight of an amino acid and the number of codons allocated to it (Di Giulio 2005; Hasegawa and Miyata 1980). A fourth characteristic feature represents the positive correlation between the number of synonymous codons for an amino acid and the abundance of the amino acid within proteins (King and Jukes 1969; Gilis et al. 2001). A fifth feature is the obvious minimization of the probability of translation errors and point mutations (Haig and Hurst 1991; Freeland et al. 2003). A sixth characteristic feature seems to be the almost optimal capacity to permit supplementary insights within the protein-coding sequences (Itzkovitz and Alon 2007).

When examining the evolution of the genetic code, several basic hypotheses are worthwhile to be clarified. The hypothesis is that there are only 4 nucleotides and 20 coded amino acids (with the remarkable exception of selenocysteine and pyrrolysine, for which subsets of organisms have developed special coding schemes) (Ambrogelly et al. 2007) and every codon constitutes a triplet of nucleotides. It has been reasoned that moving in steps of three nucleotides is a fundamental physical characteristic of RNA translocation within the ribosome, thus establishing the translation system as a triplet-based mechanism (Aldana et al. 1998; Aldana-Gonzalez et al. 2003; Gusev and Schulze-Makuch 2004). Of course, this certainly cannot exclude the existence of for instance only two nucleotides in every codon being informative (Patel 2005; Travers 2006; Ikehara and Niihara 2007; Wu et al. 2005) for hypotheses on the development of the code through a "duplicate" phase. It is entirely legitimate to ask why there are four standard nucleotides in the code (Szathmary 1991, 2003) or why the standard code encodes 20 amino acids (Weber and Miller 1981; Lu and Freeland 2006, 2008). It is imaginable that theories about the early stages of code development are limited by the minimal complexity demanded of a self-replicating operating system (Munteanu et al. 2007). However, these fascinating questions go beyond the scope of this chapter, and for the purposes of the present discussion, the fundamental premises mentioned above are taken as assumptions. With these assumptions in mind, what at this point in time needs to be done is to critically evaluate and summarize the most important evidence and to reflect on the nature and evolution of the code.

The genetic code, defining the principles for the translation from the 4-letter nucleic acid alphabet to the 20-letter protein alphabet, occupies a prominent position in the entire field of biology (Crick et al. 1961; Nirenberg 2004; Nirenberg and Matthaei 1961; Woese 1967; Koonin and Novozhilov 2017). In fact, the code is probably the only central source of information invariant of all life species. In spite of the many modifications that occur repeatedly in different life systems due to the investigation of protein coding, the fundamental composition of the code and the majority of codon allocations are truly universal (Koonin and Novozhilov 2009; Sengupta and Higgs 2015).

Once the codon table was created in 1965, it emerged that the code included various patterns requiring clarification (Crick 1966; Woese 1967). The 64 codons are orderly arranged in sets of 4 or 2, with the third base of a codon generally synchronized, which means that alterations at this position will not cause an amino acid substitution (Fig. 11.8). Moreover, the allocation of codons to amino acids throughout the code table is definitely not accidental: Similar amino acids usually cover related areas of the table. The second position of a codon is the main distinctness determination, and three of the four columns of the codon table are used to encode related, chemically analogous amino acids. For instance, all codons with a U in the second position comply with the hydrophobic amino acids. A simple qualitative analysis reveals that the code is resistant to mutation or translation failures. Substitutions and translation mistakes at synonymous positions (typically located at the third codon position) have no impact on the protein (even though it does not automatically mean that such substitutions are neutral in a selective way), while substitutions at the first position usually result in

Fig. 11.8 Wobble hypothesis

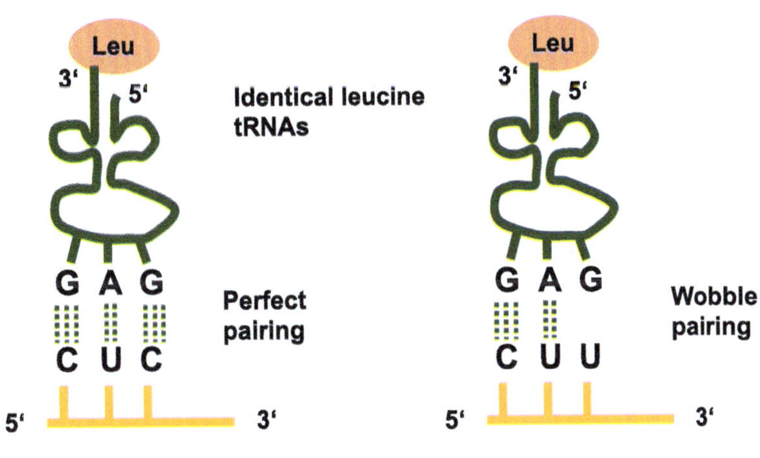

the incorporation of an amino acid that is similar to the original one, thereby minimizing the detrimental outcome.

In a groundbreaking investigation, a quantitative metric has been presented to assess the robustness of codes to failures, often referred to as failure minimization (Haig and Hurst 1991). Continued refinement of this metric has demonstrated that the code is one in a million, which means that it is more robust than about one million randomly selected codes (Freeland and Hurst 1998b). Later attempts to use much more advanced models have resulted in an even stronger robustness of the code (Goodarzi et al. 2004). Despite this, the standardized genetic code (SGC) is not the optimal one: In view of the hyperastronomic total number of possible codes (more than 1084), billions of variants will be more robust than the actual universal code (Koonin and Novozhilov 2009).

The genetic code, setting the rules for the translation from the 4-letter nucleic acid alphabet to the 20-letter protein alphabet, occupies a special role in the entire biology (Crick et al. 1961; Nirenberg 2004; Nirenberg and Matthaei 1961; Woese 1967). Indeed, the code is probably the only central source of information, which is invariable to all forms of life. Notwithstanding the numerous differences that have arisen from the investigation of protein coding in various life species, the basic composition of the code and the majority of codon allocations are truly universal (Koonin and Novozhilov 2009; Sengupta and Higgs 2015).

In the translation system, which is generally conserved in all existing cellular life types (Koonin 2003), the exact deciphering of an mRNA sequence, meaning the insertion of the cognitive amino acid in reaction to a specific codon, is guaranteed by a sophisticated indirect mechanism. An aminoacyl-tRNA synthetase (aaRS) detects a unique tRNA and loads it with the related amino acid. The translation of the code therefore implies no immediate recognition of the codons (or anticodons) by amino acids, which raises a puzzling question: What is the purpose of this type of codon assignments? In plain terms: For instance, why is glycine encoded by GGN codons and not by CCN codons (the latter codifies for proline in the SGC)? The main debate centers around four rival perspectives on the origin and development of the code (Di Giulio 2005; Knight et al. 1999; Koonin and Novozhilov 2009). Stereochemical theory postulates that the codon table effectively mirrors the affinity between amino acids and codons (or anticodons) which are assumed to have contributed to primordial translation prior to being substituted by the existing indirect mechanism (Yarus et al. 2009). The co-evolution theory claims that the molecular structure of the code represents the development of the routes for the synthesis of amino acids (Wong et al. 2016). The theory of failure reduction states that the code has developed from a probably accidental original state into its present shape under the pressure of choice and, above all, for reasons of maximum ruggedness toward failures that occur during replication, transcription or translation (Archetti 2004). Finally, the opinion of the frozen accident (which can scarcely be classified as a theory) is that the codons were originally allocated by mere chance, and as soon as these allocations were established, further substantial modifications of the code became fatal; hence, only some smaller codon transcriptions appeared in a few groups of organisms (Crick 1968). It is obvious that these code development scenarios need not be mutually exclusive, which further aggravates the problem (Koonin and Novozhilov 2009).

In summary, only one of the most basic "why" questions on the characteristics of the genetic code has been answered so far. In the following, more of these questions are raised:

- Why is the universal genetic code a triplet code?
- Why are there exactly 20 amino acids coded by the code chart—nothing more and nothing less despite some extra amino acids coded by specific groups of organisms?
- Why is the universal genetic code highly error prone?
- Why is the code, although extremely robust, yet still miles away from being universally optimized?

- Why is the code global?
- What happens when the code diverges, why does this happen in certain organisms but not in all others?

In conclusion, all of these questions cover the origin and the evolutionary development of the genetic code. When it is possible to envision a suitable scenario for the initiation and the evolution of the genetic code, most of these questions raised above can be answered. However, this is not an easy task. The evolution of the genetic code is tightly connected to the origin and the evolutionary development of the translation process, which represents properly the most basic and tough problems in the field of biology (Woese Woese 1967; Wolf and Koonin 2007). When investigating the evolution of translation, an inherently paradox must be considered: Many functional proteins are necessary for a high-fidelity translation mechanism, but it is not possible to sustain such proteins without a robust translation machinery. Obviously, a variety of settings have been designed, but all of them seem to remain in the realm of pure speculation. For this reason, in numerous research studies, the examination of the genetic code as a small but seemingly enduring field of research extensively distances oneself from the issue of the origin of the translation process—it rather simply assumes the translation as a matter of course. Although these kinds of analyses can be useful for understanding the essential characteristics of the code, the possibility still exists that the issue of code development can ultimately only be solved in combination with the development of the mechanisms of translation as a whole.

Theoretical and experimental investigations on the origin and development of the code span over more than five years, during which time an abundant stock of literature has been gathered and covered by review articles at various developmental stages of the field (Di Giulio 2005; Jestin and Kempf 2009; Koonin and Novozhilov 2009; Sella and Ardell 2006; Sengupta and Higgs 2015; Tliusty 2010). The main evidence of the nature and development of the code was critically reviewed and summarized (Koonin and Novozhilov 2009). This chapter seems to be a logical extension and therefore it focuses mainly on recent theoretical and experimental progress and provides an overview of the current status of the implementation of the code enigma.

11.4 The Standard Code, Variations and the Latest Evolution

The SGC table encodes the 20 canonical amino acids and the three stop signals for the translational process that finally lead to the 64 codons. The most comfortable manner of investigating the table is by looking at codon quartets in which the first two bases are identical, while the third position is filled by each of the four bases; this structure appears to mirror the progression of code evolution. Precisely half of the quartets consist of 8 codon sets, each coding for a single amino acid, with a single synonymous, non-informative third position; five quartets comprise 10 groups, each with 2 synonymous codons; among the remaining three quartets, two contain the 3 stop codons together with an amino acid assigned to 2 codons or 1 codon, and the third quartet comprises the single codon for Met, which also acts as the start signal for the translation, together with 3 Ile codons (Fig. 11.7).

The code is almost universal in all existing life species, but has a limited ability to develop, manifested in alterations of the standard codon assignment in many classes of organisms, especially in organelles with tiny genomes and in some parasitic and endosymbiotic bacteria with strongly restricted genomes (Ling et al. 2015; Santos et al. 2004; Sengupta and Higgs 2015). More than 20 non-standard codes to date have been identified, and as genomic and metagenomic sequencing progresses, new variants continue to emerge. The code amendments can be grouped into three main classes: first reallocation of codons within the canonical set of 21, with the stop signal incorporated;

second loss (non-allocation) of codons and third inclusion of new amino acids. Stop codons are heavily overexposed among the code alterations. Among the 23 non-standard codes, there are 8 cases where stop codons have been reallocated or newly obtained, 8 cases of stop codon loss and 10 cases of reallocation of a codon from one amino acid to a different one (Sengupta and Higgs 2015). Considering the existence of only three stop codons, the same modifications to the code appeared in various groups of organisms in parallel. The most frequent by a wide margin is the reallocation of the UGA stop codon to Trp that was detected in numerous mitochondria and in several bacteria and ciliates. The enrollment of UAA/G stop codons for coding Gln has also been demonstrated in various organisms. It is remarkable that two differing codes, which obviously have no specific stop codons, were newly found in two types of ciliates (Swart et al. 2016) and in a group of trypanosomatids (Zahnova et al. 2016). These versions integrate the recruitment of stop codons to encode both Trp and Gln, which up to now have existed separately in various deviating codes. Evidently, these groups of organisms still developed badly characterized mechanisms for a dual, contextual decoding of the (former) stop codons.

There exist two distinct non-canonical amino acids that are commonly used in the code: selenocysteine, which is abundant in various protein inserts in different organisms from all three life domains (Ambrogelly et al. 2007; Mukai et al. 2016) and pyrrolysine, which is only found in a few archaea. The mechanisms guiding the incorporation of these two amino acids are entirely distinct. In specific detail, pyrrolysine is incorporated by the reallocation of a stop codon in an analogous configuration to the code alterations inside the canonical amino acid repertoire, while selenocysteine is introduced only when a stop codon controls the reprogramming in the presence of a specific regulatory sequence entity (Lobanov et al. 2010; Yuan et al. 2010; Zhang et al. 2005).

The evolutionary mechanisms resulting in the reallocation of codons and the generation of different codes are not fully understandable, but they must definitely be associated with alterations in tRNA specificity or with the development of new specificities in the event of stop codon enrollment. These mechanisms have been recorded in a context of profit and loss, where profit involves the acquirement of a new tRNA specificity, often after duplication of a tRNA gene, and loss involves the removal of a tRNA specificity, generally by its deletion (Sengupta and Higgs 2005, 2015; Semgupta et al. 2007). There is no conclusive proof that contemporary code variants are linked to customization. It is highly probable that they result from neutral evolutionary events, specifically genetic variation and mutation stress, which pushes small genomes to a low GC status.

Although it will not be discussed in this book, it is possible to modify the genetic code in bacteria employing newly developed approaches based on synthetic biology (Chin 2014; Lui and Schultz 2010; Neumann et al. 2010; Xie and Schultz 2006). The suitability of bacteria with modified codes has not been examined in any detail, but their actual viability appears to confirm the assumption that the suitability of various codes may not vary considerably. The exceptional evolutionary robustness of the SGC could therefore be explained by strong physical barriers isolating it from other codes or by the low physical robustness of intermediate states (Sengupta and Higgs 2015).

11.5 Evolutions of the Genetic Code, Such as Old and Novel Amino Acids

An important principle that current primordial evolutionary paradigms share is that the earliest proteins have not started with the contemporary set of 20 amino acids. Actually, the canonical amino acids vary considerably in their degree of stability and chemical complexity. Ten amino acids are manufactured in a consistent manner in prebiotic chemistry assays and have also been detected in meteorites in the order of their relative frequency as follows: Gly, Ala, Asp, Glu, Val, Ser, Ile, Leu, Pro and Thr (Burton et al. 2012; Cleaves 2010; Pizzarello 2006; Zaia et al. 2008). Computations have revealed that the ranking of amino acids in this list is closely associated with the free energy of their syntheses: the low-end thermodynamic ones are at the top of the order (Higgs and Pudritz 2009). A wide variety of principles have been adopted in the trials to define the time of inclusion of amino acids in the code, and the consensus views reached are very similar in the order of priority (Trifonov 2000, 2004). In specific, a totally unbiased analytical set-up involving the analysis of amino acid fluxes in the latest development of proteins in different organisms has revealed that the levels of the supposedly early amino acids in the above list usually tend to decrease, while those of the other, late amino acids usually rise (Jordan et al. 2005). This fusion of the extremely divergent inference techniques implies that the 10 amino acids mentioned above can safely be regarded as old; in other words, they were part of the first proteins.

11.6 Why Is the Genetic Code Usually Universal? Are There Exceptions?

At first sight, the code's universality appears to be due to the retrospectively apparent justification of the frozen incident, as stated by Crick in the groundbreaking paper on the development of the genetic code in 1968 (Crick 1968). The frozen perspective of the accident states that, once the codon table is established, the reallocation of even a single codon in a reasonably broad genome has a potentially deleterious impact. However, this reasoning is not necessarily demanding that the original selection of codon mapping is completely coincidental and completely arbitrary. A random number of contributing elements could be added to the initial codon mappings; however as soon as the selection is complete, these mappings are frozen, which means that only rare and minor alterations may be permitted. In all of the codon rearrangements detected in existing organisms, the amount of codon rearrangements is in fact rather restricted. In the terminology of fitness-related environments, the frozen accident scenario suggests that there are lots of various codes that indicate fitness peaks, but these are split by profound depressions of low fitness (Novozhilov et al. 2007).

But the early phases of cellular and more specifically pre-cellular evolution might actually and necessarily be a different issue. The early development inevitably entailed competitive interaction between ensembles of viral genetic elements and selection at the scale of these collectives (Koonin and Martin 2005; Szathmary and Demeter 1987; Szathmary and Maynard Smith 1997). It is difficult to conceive of translation having developed within such an arrangement, and when it has been generated several times in multiple arrangements, there is every reason to assume that there were many diverse codes in the beginning. Why should a single code be preserved? That is, why was there just a single frozen incident when the actual codon allocations are actually random?

In a pioneering paper, a remarkably straightforward but effective idea has been presented. In specific detail, there has been one frozen incident due to the fact that large-scale horizontal gene transfer (HGT) was an integral component of early evolution, otherwise the switch to the cellular level of sophistication would never have succeeded (Vetsigian et al. 2006). Apparently, a small amendment to the code has a restrictive impact on HGT. Finally, by applying mathematical modeling, it was possible to examine the assertion that the demand for HGT leads to the conservation of a single, universal code during evolution. Based on a random array of codes, the evolutionary experiments simulated result in a diversification of codes in the complete lack of HGT; however, the survival of a limited number of codes when HGT is permitted. In their initial investigation of the connection between code evolution and HGT, a deterministic approach to an infinite population model has been examined. A fairly accurate more realistic recent model, which considers the stochastic implications of limited population size, has generated a single, universal code having a similar structure to the SGC within a certain range of HGT levels (Aggarwal et al. 2016; Sengupta et al. 2014). It is essential to consider that any essential reallocation of codons has a double impact on the relevant microbial population. Although HGT is abolished, the population is getting resistant to parasitic genetic elements including viruses and plasmids. There is therefore a compromise between the damaging and favorable impacts of the removal of HGT. In the long-term, the advantages seem to outweigh the disadvantages.

The latest theoretical, comparatively genomic and experimental investigations on microbial evolution demonstrate strongly that HGT is the key driver of microbial adjustment and renewal (Lerat et al. 2005; Puigbo et al. 2010, 2014; Treangen and Rocha 2011). Beyond that, clonal populations are exposed to the directly lethal impact of the Müller ratchet under a variety of parameters of the evolutionary process, when an adequate HGT level is missing (Takeuchi et al. 2014). The critical HGT rate necessary to avoid the Müller ratchet has been found to be over the level of persistence of parasitic genetic components in most microbial populations (Iranzo et al. 2016). In general, the restriction of HGT by changing the code generally results in the disappearance of the relevant microbial group. Nevertheless, it seems probable that under specific conditions protection against genetic parasites prevails over the advantages of HGT in the shorter term, and this permits restricted modifications in the code to be frozen in evolution. In some instances, codon redistributions could be locked in place by random variation in separated microbial populations, such as intracellular parasites, which do not undergo high levels of HGT and are either targeted for potential elimination or, similar to endosymbiotic organelles, are sustained by special selective forces. These organisms usually contain small genomes, thus reducing the expense of reallocating codons. It seems, thus, that there are two main elements that effectively keep the code in its universal arrangement: The immediate costs of codon reallocation arising from inappropriate protein production prevent major alterations, while the adverse effect of HGT restriction itself strongly restricts local codon reallocation.

11.7 Theories of the Code Origin and Evolution

The structure of the SGC, namely the allocation of 64 codons to 20 amino acids and the stop signal, is obviously a regular feature in terms of several aspects (Koonin and Novozhilov 2009; Woese 1967). This non-randomness of the code may need to be further elucidated. Among the three most progressive and consistent conceptual approaches that attempt to account for patterns in the genetic code are stereochemical, co-evolutionary, and error minimization theory (Di Giulio 2005; Knight et al. 1999; Koonin and Novozhilov 2009). In the following sections, the term theory is used for reasons of brevity and in keeping with tradition, although none of the possible configurations of code evolution apparently fulfills the requirements of a theory in the narrower meaning. The frozen accident view probably also needs to be considered in connection with one of the concepts mentioned above.

11.7.1 The Stereochemical Theory

The stereochemical theory together with the first ever wording of the coding issue was introduced in 1954 (Gamow 1954). Thereafter, the first efforts for a direct experimental proof of the interaction between amino acids and the various cognitive codons or anticodons have been performed (Woese 1968). In fact, these results are generally not really convincing, which led to a long pause in tracking the stereochemical representation of the code. Despite this, this early approach was methodologically fruitful and led to the emergence of the polar requirement (PR) scale, which is a scale focusing on the solubility of amino acids in pyridine or its derivatives, such as pyrimidine analogs (Mathew and

Luthey-Schulten 2008). Amino acids vary substantially on the PR scale, which has rendered them a favored measure for evaluating possible combinations of pyrimidine bases and amino acids and for the strength of the code.

The advances in aptamer research resulted in a revival of stereochemical theory. A key finding of this research has been that short RNA sequences (aptamers) randomly chosen from a mixture of random amino acid bindings exhibited significant accumulation by related triplets, at least for certain amino acids. These included codons in a number of cases and anticodons in a variety of other cases (Yarus 1998, 2000; Yarus et al. 2005). Supplementary experimental proof was recently reported in a study where the sequences of 337 independent binding sites for eight amino acids (Arg, Gln, His, Ile, Leu, Phe, Trp, Tyr) with 18,551 nucleotides were investigated in order to retrieve interactions between amino acids and connected coding triplets (Yarus et al. 2009). A statistically relevant attachment to related triplets was found for five of the eight amino acids, such as Arg, His, Ile, Phe and Trp. These findings were construed as a proof of the existence of an early stereochemical period in the course of code development. Additionally, it inferred that the overwhelming majority (about 75%) of modern amino acids came into the code at that time (Yarus et al. 2005).

Several arguments have been put forward in the previous section on the development of the code to suggest that the statistical proof from the aptamer studies did not permit immediate conclusions to be drawn about the stereochemical source of the code (Koonin and Novozhilov 2009). Despite the supplemental evidence and the counter-argument (Yarus et al. 2009), it seems that the argument still stands. While not going into great detail and putting aside possible concerns about statistical significance (which for all amino acids except Arg is less than abundant), it is worth pointing out that all significant findings about aptamer–amino acid binding (Yarus et al. 2009) concern amino acids that are thought to have been introduced late into the genetic code. Among the five amino acids exhibiting significant levels of cognate aptamer binding, only one, Ile, is a member of the early class. From a chemical point of view, this is not unexpected, since the late coupled amino acids possess large side chains that can interoperate with oligonucleotides. All the proofs from aptamer studies, which appear to be consistent with the presence of a physicochemical affinity among amino acids and related nucleotide triplets, concern complex amino acids. This tendency is not good for the concept that this kind of affinity is underlying the origin of the genetic code.

A comprehensive consideration of the stereochemical situation will provide proof from the biology of existing life species. However, this kind of proof is rare and only available on a circumstantial basis. Codons for Arg, which is the one with the most powerful statistical underpinning from the aptamer studies, have also been found to provide binding specificity for Arg in the group I self-splicing introns (Yarus and Christian 1989). A far greater, albeit less immediate, amount of proof was obtained by analyzing the spatial proximity of amino acid residues in ribosomal proteins to rRNA nucleotide triplets within the bacterial and archaeal ribosome structures (Yarus and Christian 1989). The accumulation for anticodons and codons has been reported with moderate statistical significance for 11 and 8 amino acids. Computer simulations using randomized codes revealed that the assignment to related triplets was substantially more significant for the 11 anticodon-based amino acids than for the 8 codon-based ones. In view of the fact that the ribosome is the oldest, universal molecular structure in every cell, it has been suggested that the amino acid-anticodon interaction was key to the origin of the code and is still the core of the contemporary translation machine (Johnson and Wang 2010).

Lately, a number of extensive analyses on the possible complementarity of protein amino acid sequences with the nucleotide sequences of related mRNAs have been presented (Hajnic et al. 2014; Hlevnjak et al. 2012; Polyansky et al. 2013a, b; Polyansky and Zagrovic 2013). These analyses of the entire proteome detected pronounced negative correlations among the mean PR scale values of proteins and the level of pyrimidines in the corresponding mRNAs, pointing to the existence of proteins with high affinity to pyrimidine analogs being coded by pyrimidine-rich mRNAs. Similar consequences of an independent, more general approach have been obtained by comparing data on nucleotide–amino acid interactions in various ribonucleoprotein and deoxyribonucleoprotein complexes with the levels of the corresponding bases in the mRNAs. The broadening of this concept to codons led to a more intricate pattern. Concerning the early amino acids, strict correlations between the G and C levels of codons and binding priorities were seen, while for the late amino acids these relationships were not significant. Inversely, the A level of the codons for late amino acids exhibited a negative relationship with the A-binding tendency. These results can be seen as supporting Woese's old concept about the significance of immediate interactions between amino acids and the related triplets (Polyansky et al. 2013a, b). Nevertheless, it is doubtful whether such low specificity of amino acid interference with RNA—which is more often found for peptides than for free amino acids—may be central to the development of the code.

11.7.2 The Co-evolution Theory (or Synonymously Metabolic Theory)

The co-evolution theory, first introduced in 1975 (Wing 1975), states that the genetic code is determined by the precursor–product interactions between amino acids (Wong 1975; Trifonov 2000). In this scenario, therefore, the code developed from an ancestral variant that contained simple, abiogenically generated amino acids only and was then enlarged to accommodate the more complex amino acids in concert with the development of their corresponding biosynthetic paths, which means that there was a co-evolution of the code pathway. The significance of the biosynthetic routes for code development is nearly obvious, since amino acids can be integrated into the code when they become accessible. According to this co-evolutionary theory, the code developed through partitioning: In the ancestral code, huge clusters of codons coded the identical amino acid, although they broke down into two amino acids as the respective metabolic pathways developed. The unique pattern of codon reordering is defined by precursor–product interactions between amino acids, with a specific product adopting certain codons in the segment that originally coded for the precursor. It is assumed that the development of the code is advanced by positive screening to diversify the protein functions, which is made possible by the inclusion of new amino acids.

The consequences of the co-evolution theory are not restricted to the obvious importance of the biosynthetic pathways for the uptake of the late amino acids into the code. Based on the precursor–product interactions between amino acids in these tracts, the theory provides specific and easily eradicated predictions about the partitioning of the primordial sets of codon allocations.

11.7.3 The Error Minimization Theory

The clustering of related amino acids within the same column of the code table (Fig. 11.9) instantly demonstrates that the code is inherently strong against mutation and translation failures, which means that it is arranged to minimize the damaging impact of these errors. Besides this qualitative evaluation, how exceptional is the strength of SGC over random codes? The meaningful response to this problem is a non-trivial effort involving a precisely specified expense function to allocate a numerical score to every variation of the code that measures its capacity to withstand errors. To quantify the comparative strength of any given code, these values are subsequently verified against those received for randomized variants of the code. The expense function varies with the measure of amino acid affinity, which is used to estimate the fitness effect of substituting one amino acid in a protein by another amino acid. The metric most commonly used in the investigation of code ruggedness appears to be the PR scale, although additional characteristics of amino acids, such as hydropathy score or

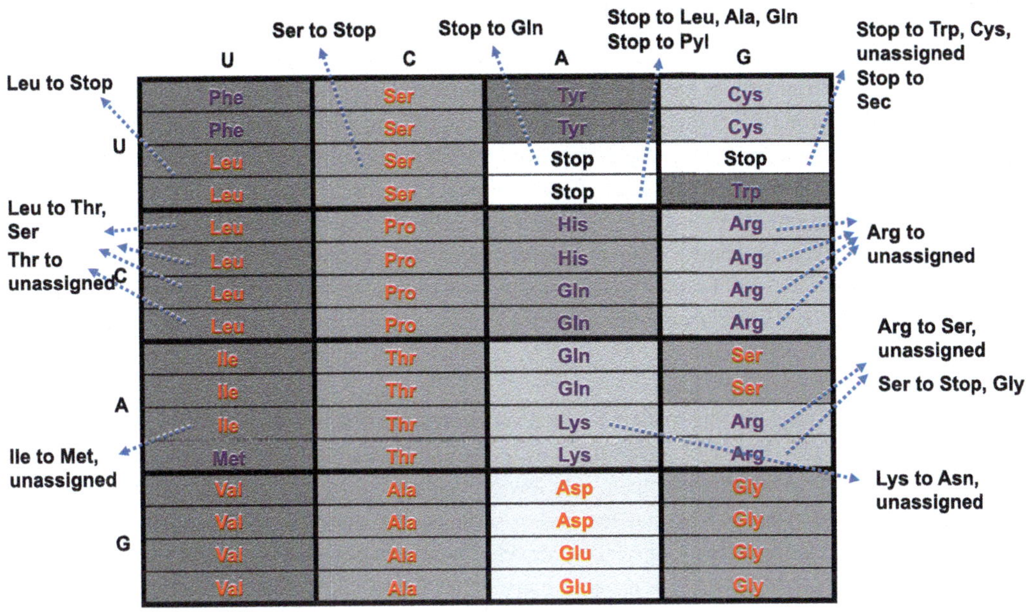

Fig. 11.9 Standard genetic code contains early amino acids (red), late amino acids (purple) and deviations of the code (blue). The gray shades due to the polar requirement scale values indicate the clustering of amino acids of the code. Pyl = pyrrolysine, Sec = selenocysteine

molecular volume, have also been employed either as substitutes or in conjunction (de Oliveira et al. 2015; Di Giulio and Medugno 2001; Higgs 2009; Sengupta and Higgs 2015). In alternative attempts, metrics representing the observed fitness impact of amino acid replacements in developing proteins were utilized in place of physicochemical equivalents (Buhrman et al. 2011).

The methodology to create the random code sampling is also important as it is inconvenient to scan the entire volume of more than 1084 possible codes or even a significant part of this volume. An early analysis of the strength of the code has revealed that the code is typically one in a million, indicating that the probability that a random code is as strong as the SGC is about 10^{-6} (Freeland and Hurst 1998b). More refined modifications of this analysis based on several cost functions and different randomization patterns have suggested in later trials an even higher level of optimization in SGC as a function of the parameters of the specific analysis (Goodarzi et al. 2004; Haig and Hurst 1999; Massey 2008, 2015, 2016; Novozhilov and Konin 2009; Novozhilov et al. 2007; Salinas et al. 2016; Torabi et al. 2007; Zhu and Freeland 2006). The general agreement emerging from these investigations is that the SGC is a good way to reduce the impact of errors, but there is also a large amount of code (even if it represents only a small percentage of all of them) that is even more resilient. In addition, the SGC does not seem to be a locally dominant peak in the code fidelity map, as specific regional redistributions may enhance the degree of error mitigation; a qualitative assessment has proven that the SGC is located approximately midway from an average random code to the top of the respective local peak (Novozhilov et al. 2007).

The typical deduction of investigations that prove the high level of ruggedness of the SGC results from the fact that the code developed under the choice for failure minimization. Despite its intrinsic attraction, this conclusion is not entirely valid. Code tuning analyses have historically concentrated on the reordering of the standard code table, which permits a functional assessment of reliability, and not of the evolutionary processes underlying it. A much broader perspective entails the recreation of the paths of code extension, which could generate failure minimization as a selective and neutral side product of evolution powered by various other factors (Massey 2008, 2015, 2016). The concept of neutral evolution of failure minimization has been originally developed by Crick in 1968, where it is stated that similar amino acids seem to possess similar codons in the process of the assumed code enlargement in the direction of the frozen accident variant (Crick 1968). A comprehensive simulation approach has been presented providing pronounced statistical evidence to the neutral development of the code strength (Massey 2008, 2015, 2016). There have been three distinct models of code enlargement investigated. The first one represents the 2-1-3 model by Massey (2006) that is closely related to the enlargement modes, which have emerged independently using similar to expansion schemes by Higgs (2009) and Francis (2013). The second one is the ambiguity reduction model (Fitch and Upper 1987). The third one represents a concept that is based on the precursor–product enlargement code and is subsequently provided by the specific co-evolution theory (Yarus et al. 2009; Wong et al. 2016).

The 2-1-3 model and similar patterns based on the properties of the SGC propose that in the ancestral code the second base of the codon was the sole informative base, and this code broadened by adding specificity to the first and then, in a few codon sequences, to the third base. The model of ambiguity reduction hypothesizes that the codon series in the ancestral code equivocally mapped groups of amino acids, thus their consecutive development brought about a gradual enhancement of the level of specificity of codon–amino acid maps. The fitness score of a code relying on various measurements of amino acid similarity was used and random code development was modeled using any of the three expansion patterns. The simulations have demonstrated that it is easy for both the 2-1-3 model and the ambiguity reduction model to produce codes with a degree of uncertainty reduction that is higher than that of the SGC. Conversely, it was found that the co-evolution model, although it generates a certain degree of failure minimization, is outmatched by the other strategies (Higgs 2009). These results suggest that although significant failure minimization is clearly a feature of the SGC, this characteristic is probably a side product of code extension rather than a direct choice for maintaining code robustness. Furthermore, the evolutionary steps that usually appear in the computational process of finding highly reliable codes, specifically the replacement of codon allocations by 4-codon or 2-codon units, are biologically scarcely a reality. In a broader biological perspective, it should be emphasized in this context that the extreme minimization of defects is not automatically an advantageous characteristic of the code, since it strongly restricts the investigation of the sequence room in the evolutionary process (Firnberg and Ostermeier 2013).

With the consensus on the amount of the old amino acids and the non-availability of the late amino acids before the emergence of the complex biosynthetic routes, code extension seems to be an integral element of the early phases of the evolution of the code. The assumption that the third letter of the codons in the primordial code is not informative, and the reduction of the code table to the codons for the 10 early amino acids led to the conversion of the old duplicate code, although several codon series stayed equivocal (Gamow 1954). It is remarkable that this reconstructed primordial code exhibits an extraordinary degree of failure elimination and is far nearer the overall optimization level than the SGC. This finding indicates that in the early phases of code development, where the accuracy of primitive translation process may be poor, the choice to mitigate failure might have been a greater determinant in the development of the code than in the later phases. Here once in a while, however, it is not possible to exclude the possibility that the high degree of failure reduction in the doublet code results merely from the extension of the most primitive code, for instance under the 2-1-3 and the 4-column configurations.

Since most research on the failure reduction characteristics of the code are mainly focused on a more or less-random set of expense features that offer a numerical measurement of the code's capacity to resist mutations and translation failures. Therefore, biologically more reasonable options for the expense mechanism have been examined. In specific, in vitro evolutionary investigations, where the fitness expenses of mutations were assessed by their impact on the characteristics of proteins, it has been demonstrated that the pattern of the code imposes quantifiable restrictions on evolution. Consequently, the mutants that were at the highest level of fitness, which developed alongside consecutive individual replacement pathways, were vastly superior to multiple versions chosen from tedious mutagenesis studies (Firnberg and Ostermeier 2013). The selfsame investigation demonstrated that the code reduces the fitness expense of mutations in the way that the average expense of an amino acid replacement caused by a single nucleotide replacement is substantially below the expense of replacements involving two or three nucleotide replacements. In addition, and far less obviously, the proportion of adaptive amino acid substitutes is higher under those that are available through a single replacement than under all replacements. The two properties, minimizing the destructive impact of mutations and increasing the adaptive efficacy, are presumably two faces of the same medal, which is a code arrangement that heavily promotes the replacement of amino acids by similar amino acids derived from single nucleotide replacements. Therefore, the resistance of the code to failure seems to be inextricably connected to an improvement in the protein development capacity, as some theoretical studies have proposed previously (Francis 2013; Higgs 2009). These results are consistent with the choice of ruggedness as the driving force in the development of the code, although they cannot exclude the development of the code as a side product among the other evolutionary concepts. Irrespective of this, and notwithstanding their restricted scale, experimental investigations of code characteristics seem to be of decisive importance for confirming and broadening the conclusions of earlier theoretical investigations.

11.8 Does a Co-evolution of the Genetic Code and the Translational Process Exist?

Recently it has emerged that the emergence and development of the translation mechanism is a stunningly complicated matter; hence, in many investigations it is officially dealt with as a distinct topic and tackled virtually as a mathematical enigma (Woese 1967). In the end, it seems nearly sure that the development of the code can in fact only be comprehended in the connection with the development of the translation machinery. An effort has recently been undertaken to systematically tackle the issue of code development in the field of the molecular biology of the translation process, in specific the nature of interactions between tRNAs, mRNAs and the ribosome (Grosjean and Westhof 2016). The amino acids are classified into three groups, distinguished by the free energy of the codon–anticodon reaction, which is apparently associated with the GC content. Among them is a first strong, or possessing a mean free energy of −3.1 kcal/mol, which encompasses Ala, Arg, Gly and Pro; a second weak, or exhibiting a mean free energy of −1.0 kcal/mol, which includes Asn, Ile, Leu, Met, Phe, and Tyr; and third intermediate, or possessing a mean free energy of −2.2 kcal/mol that covers all the ungrouped nine amino acids. Apparently three of the four strong amino acids (except Arg) belong to the most simple and frequent in the early series. It seems to be that the genetic code originated from the earliest, GC-rich stage with simply four amino acids, which presumably incorporates a precursor molecule instead of Arg (Grosjean and Westhof 2016). Afterward it took up the intermediate and subsequently the weak amino acids. The exact decoding of the codons for the latter set demands an enlarged anticodon with a slightly altered base in the tRNA molecule (Allner and Nilsson 2011), which might be an incremental reason for the delayed appearance of these amino acids in the code.

Although the translational apparatus is generally preserved under the existing cellular life types, a number of protein components of the translational system are para-analogs, which additionally are members of large protein families. Thus, the phylogenetic investigation of the individual protein families offers a chance to enter the phase of evolution before the last universal common ancestor (LUCA). The outcome of this kind of research raises a mystery that can be considered a paradox of protein development. The aaRSs, which guarantee the precise correspondence between amino acids and related codons in the contemporary translation system, are part of two classes of paralogs with 10 special features in every class (Woese et al. 2000; Wolf et al. 1999). The diversification of aaRS is a late phase in the evolution of the superfamily of Rossmann folding proteins, as has been convincingly demonstrated for class I aaRS containing Rossmann folding catalytic domains (Aravind et al. 2002a, b). At the time when the common ancestor of aaRSs produced the 10 peculiarities by a number of duplications, the comprehensive evolution of the Rossmann wrinkle superfamily produced a considerable variety of other enzymatic and nucleotide-binding regions. Analogous consequences can be concluded for the development of class II aaRSs, which are part of the biotin synthase superfamily, even though the development of these proteins has not been investigated with the same degree of precision (Anantharaman et al. 2002; Artymiuk et al. 1994). Obviously, high-fidelity translation was indispensable for this early protein development, and considering that the various aaRS specificities had not emerged by then, the unavoidable consequence was that the intrinsic specificity of the amino acid–codon pairing was defined by RNA molecules (Koonin 2011).

The conclusive hypothesis of mRNA coding in the early translation system of RNA molecules, which might have represented the evolutionary precursors of modern tRNAs, entails a stereochemical model of the origin and development of the code that differs from the traditional models of this kind (Koonin 2017). Based on this model, specific RNA-amino acid interactions would not require the anticodon (not to mention the codon) to be engaged, with the result that it could be randomly selected and locked in place by a frozen accident. In contrast, it is proposed it is proposed that the amino acids have been identified by specific pockets in the tertiary structural architecture of the proto-tRNAs (Wolf and Koonin 2007). The ordering of codons for similar amino acids on which the failure reduction is founded naturally results from the code enlargement by the duplication of the proto-tRNAs. Hence, the molecules obtained from these multiple duplications are structurally highly similar and can even interact with structurally similar amino acids. With this part of the model, the code expansion model might be enhanced by the advantages of diversifying the pool of protein amino acids and finally, as a side effect, the strength of the code.

After the specificity determining factors moved from the proto-tRNAs into the aaRSs, the amino acid binding sites in the tRNAs worsened to an extent that modern tRNAs lack a detectable affinity to the cognate amino acids. Efforts to decode this original stereochemical code through a comparative analysis of the components of modern translation machines are probably in vain. The aptamer studies appear not to be able to cope with the challenge either, since more sophisticated RNA molecules are needed. It seems that the best chance for deciphering the old code rests in experiments on the in vitro evolution of aminoacylating ribozymes, which even tend to recall a crucial feature of the original translation mechanism (Chumachenko et al. 2009; Kumar and Yarus 2001; Yarus 2011).

In summary, the natural, original and evolutionary challenges of the genetic code tend to be unprecedented, combining extreme simplistic nature with terrible complexity. Among the available code theory literature over the past half century, is still the important analyses of Woese and Crick in 1960s that stays extremely relevant. While there have undoubtedly been many advances in the area of code character quantification, the basic principles of the evolutionary concepts presented in the classical articles have not been overtaken. Unfortunately, this is due less to the success of early research than to the limited and questionable progress made over the following half century. The foresight of the two researchers Woese and Crick cannot be denied, but nevertheless they were still far away from a proper comprehension of the development of the code. Despite the total fundamental change in biology that has taken place over these decades, it looks as if the solution is not much nearer. Remembering the list of "why" questions in the opening remarks, it is almost impossible to give a definitive reply to any of these questions.

In a challenging environment of this kind, it makes sense to take an assessment of what can and cannot be expected in terms of the code's development. Therefore, lets compile a short list of well-established features of the code and development characteristics:

- The code is virtually universal: the deviations of the universality of the code in existing organisms are small and have a secondary origin.
- The code is not randomly arranged and is extremely resistant to failures, even though the code is still a long way from being universally optimal.
- The development of the code involved an expansion from a restricted quantity of primordial amino acids to the contemporary canonical quantity.

This appears to be the level of ambition on which the code has evolved. The three main ideas of code evolution appear to have widely exploited their capacity for a constructive approach, at least in their traditional form. Stereochemical theory, which concentrates on codons and anticodons, fails to deliver definite conclusions. The co-evolution and error minimization theories both cover key features of code evolution, although they do not provide significant evidence for the robust deduction of code extension, which most probably entails the doubling of proto-tRNAs. A code expansion of this kind would have depended on the development of the amino acid biosynthetic routes, although it might not accurately reflect the metabolic linkages, and as a by-product it would have resulted in a reduction of failures.

In contrast to the table of codon mappings itself, the origin and development of the code cannot be broken down as easily as a mere puzzle. These processes can probably only be reproduced by a thorough examination of the early stages of the development of the translation process. This is a formidable challenge which, on a surface view, may at first glance appear practically impossible in light of the universal preservation of the core elements of the translation process in all existing cellular life types. However, the extensive level of duplication within the translation systems offers a chance to explore the deep pre-LUCA history. Specifically, the phylogenetic examination of aaRSs points to the fact that at a phase of evolution in which the translation mechanism achieved high fidelity, the matching between amino acids and related codons was governed by the detection of the first by RNA molecules, in other words by proto-tRNAs. Therefore, what can be proposed is an experimentally provable scenario for code evolution that integrates a new variant of the stereochemical assumption, where amino acids are identified through specific sites in the 3D structure of proto-tRNAs rather than anticodons, with the concepts of code expansion and frozen accident.

In view of this, the theoretical investigation of the genetic code as a cryptographic issue is almost finished. The strongest promise for continued progress in comprehension of the origin and development of the code appears to be in the technically sophisticated but conceptually straightforward efforts to recap the deduced steps in the development of the translation engine.

11.9 Rewriting the Genetic Code: The Concept of Code Expansion

The broadening of the genetic code enables unnatural amino acids (Uaas) to be inserted site-specifically into proteins in living biological systems, thereby selectively bringing new characteristics into target proteins in vivo for basic biological investigations and for the engineering of new biological functionalities. Orthogonal constituents comprising tRNA and aminoacyl-tRNA synthetase (aaRS) are extruded in living cells to decipher a unique codon (frequently the amber stop codon

UAG) as the requested Uaa. Originally established in *E. coli*, this approach has meanwhile been extended to several eukaryotic cells and animals. The focus is on the various biological alterations for the process of the rewriting of the genetic code, the effects of code extension on cell physiology are presented and the consequences for basic investigations of code evolution are explained (Wang 2017a, b). In specific, a common approach with the type 3 polymerase III promoter has been established to expose prokaryotic tRNAs effectively as orthogonal tRNAs, and a transfer mechanism to generate Uaa-specific aaRS for application in eukaryotic cells and animals has been engineered. It has been demonstrated that the aaRSs are very well suited to modify the substrate activity against Uaas, which differ structurally widely from the native amino acid, thereby enhancing substantially the stereochemical variety of available Uaas. The production of Uaa in ester or dipeptide formulation considerably enhances the bioavailability of Uaas for cells and animals. Nonsense-facilitated mRNA decay (NMD), an mRNA monitoring mechanism of eukaryotic cells, breaks down mRNA carrying a premature stop codon.

The inhibition of NMD enhances the efficiency of Uaa inclusion in yeast and *Caenorhabditis elegans*. In bacteria, the release factor 1 (RF1) rivals the orthogonal tRNA for the amber stop codon to terminate protein translation, resulting in poor Uaa insertion performance. In contrast to the model that RF1 is considered essential, it is found that RF1 is not essential in *E. coli*. The knockout of RF1 significantly enhances the effectiveness of Uaa implementation and provides Uaa integration at several locations, thereby permitting Uaa to be utilized for targeted development. These strategies have made it possible to effectively expand the genetic code in yeast, fruit flies, mammalian cells, mice, stem cells, worms and zebrafish. It is also fascinating to discover that the legitimate UAG codons, which cancel endogenous genes, are not suppressed in an efficient manner by the orthogonal tRNA/aaRS in *E. coli*. In addition, *E. coli* reacts immediately to the amber suppression stress with a transposon implementation to deactivate the inserted orthogonal aaRS. The sustained suppression of amber escaping transposon deactivation results in global proteomic alterations with a remarkable upregulation of a hitherto unknown protein, YdiI, whose unusual function of plasmid ejection has been identified. The genome incorporation of orthogonal tRNA/aaRS in mice causes small modifications of the RNA transcripts, however, no relevant physiological disturbance. Finally, the RF1 knockout *E. coli* strains provide a so far non-existent model organism for the investigation of usually unanswerable questions about real-time code evolution in the laboratory. The genetic coding of Uaas in living systems is expected to keep opening up new doors and directions for the in vivo investigation of biological topics, such as the analysis of the code and genomic reconfiguration for synthetic biology.

11.9.1 The Expansion of the Genetic Code

The genetic code employs 61 codons to designate 20 conventional amino acids and 3 stop codons for the termination of protein translation. This code is conserved in practically every life species with slight fluctuations (Ling et al. 2015). In the past few years, the genetic code has been broadened to code for non-natural amino acids (Uaas) by inserting into living cells a new tRNA/aminoacyl-tRNA synthetase (aaRS) pair (Wang and Schultz 2002, 2005) that is perpendicular to native tRNA/aaRS pairs. The orthogonal aaRS is designed to load a target Uaa on the orthogonal tRNA, and the resulting Uaa-tRNA then inserts the Uaa into proteins on the basis of a unique codon, for example the amber stop codon UAG. The first extension of the genetic code appeared in *E. coli* in 2001 (Wang et al. 2000, 2001; Wang and Schultz 2001; Wang 2003). This methodology has since then become generally applicable to different cells and organisms, and more than 150 Uaas are now recorded applying this methodology (Wang et al. 2009, Liu and Schultz 2010). In this report, the emphasis is on the extension of the genetic code in different cells and animals, with a focus on the biological demands of this technique and its consequences for the biological system (Ling et al. 2015; Wang and Schultz 2005; Wang et al. 2009; Liu and Schultz 2010; Li and Chen 2016; Nikic and Lemke 2015; Huber and Sakmar 2014; Johnson et al. 2010; Yanagisawa et al. 2014; Wang 2017a, b).

11.9.2 How Can Orthogonal tRNAs Be Expressed in Eukaryotic Cells?

In order to extend the genetic code in eukaryotic cells and animals, a crucial innovation is how an orthogonal tRNA, often originating from prokaryotes, can be functionally expressed in eukaryotic cells. Prokaryotic tRNAs are being transcribed by a promoter located upstream of the tRNA gene, while eukaryotic tRNAs are being transcribed by promoter components within the tRNA that are referred to as A- and B-box. The A- and B-box entities are preserved in eukaryotic tRNAs, but are frequently missing in prokaryotic tRNAs.

Even though it is uncommon for a prokaryotic tRNA, such as the stearothermophilus tRNATyr, to randomly carry the consensus sequences and to be expressed in mammalian cells by multiple repetitions (Sakamoto et al. 2002), a general approach was needed to effectively express any tRNA in eukaryotic cells independently of its specific sequence. The type 3-pole III promoter contains promoter elements strictly in front of the coding sequence, needs no intragenic elements for transcriptional process and is equipped with a clearly marked transcriptional initiation region. First, it was discussed and validated that a type 3-pole III promoter coupled with a 3' flanking sequence is able to promote the functional expression of prokaryotic tRNAs in different mammalian cells (Wang et al. 2007a, b). To achieve an optimal tRNA expression level, the tRNA expression cassette can also be employed in multiple repetitions (Shen et al. 2011; Coin et al. 2011, 2013). The type 3-pole III mammalian promoter comprises H1, U6 snRNA, 7SK and MRP/7-2 promoter elements. It has been reported that this approach can be employed to express various tRNAs in mammalian cells, encompassing $tRNA\frac{Tyr}{CUA}$ and $tRNA\frac{leu}{CUA}$ from *E. coli* (Wang et al. 2007a, b) and from Methanosarcina (Takimoto et al. 2011). In addition, this approach has been effectively implemented in various cells and animals, such as *C. elegans* (Parrish et al. 2012), mouse (Kang et al. 2013; Chen et al. 2017), zebrafish (Chen et al. 2017), mammalian cell lines (Wang et al. 2007a, b), primary neurons (Wang et al. 2007a, b) and stem cells (Shen et al. 2011). Some others have later adopted this approach for plants (Li et al. 2013) and fruit fly (Bianco et al. 2012). However, this common methodology has now been largely transferred to the expression of orthogonal tRNAs in eukaryotic systems for the purpose of code enlargement.

Applying a similar concept, two external pole III yeast promoters, the RPR1 and SNR promoters (Xiang et al. 2013), were found to facilitate efficient expression of prokaryotic tRNAs in yeast (Wang and Wang 2008, 2012). These promoters have an intrinsic lead sequence that comprises the eukaryotic consensus A- and B-box sequences. When inserted upstream of *E. coli* tRNA, it triggers the transcription of a primary RNA made up of the promoter and the tRNA. The promoter is subsequently fragmented, resulting in the mature tRNA. In contrast to the traditional idea of employing the SUP4 5′ flanking sequence, these promoters have enhanced the activities of *E. coli* $tRNA\frac{Tyr}{CUA}$ and $tRNA\frac{leu}{CUA}$ in yeast by sixfold to ninefold, respectively (Wang and Wang 2008).

11.9.3 Uaa-Specific Synthetase Generation for Mammalian Cells

A second constraint for code enlargement in mammalian cells and animals is the generation of an orthogonal synthetase specific for Uaa. The alteration of the synthetase substrate specificity from a related amino acid to a Uaa had been a pioneer, and the initial milestone was achieved in *E. coli* by generating large mutant synthetase libraries of over 100 members, which were succeeded by high-throughput sorting or scanning (Wang and Schultz 2005; Wang et al. 2001). However, due to the low transfection rate, it is challenging to generate huge libraries of mutants in mammalian cells. A transfer strategy is being developed to overcome this problem. As the *E. coli* $t\mathrm{RNA}\frac{\mathrm{Tyr}}{\mathrm{CUA}}$/TyrRS pair is orthogonal in both mammalian and yeast cells, and the translation mechanism of yeast is homologous to the translation machinery of higher eukaryotes, it is possible to design the *E. coli* synthetase in yeast and deliver the engineered tRNA/synthase pairs for application in mammalian cells. This concept functioned both in mammalian cell lines and in primary neurons, and further *E. coli* pairs such as the $\mathrm{tRNA}\frac{\mathrm{Leu}}{\mathrm{CUA}}$/LeuRS may also be employed in this manner (Wang et al. 2007a, b). In connection with the highly efficient tRNA expression mentioned above, the transfer mechanism has made it possible to genetically code various Uaas in mammalian cells (Wang et al. 2007a, b), primary neurons (Wang et al. 2007a, b), stem cells (Shen et al. 2011) and animals (Parrish et al. 2012; Kang et al. 2013; Chen et al. 2017).

Another innovative solution was introduced which disturbs the *E. coli* tRNATyr/TyrRS genes in the *E. coli* genome and functionally substitutes them with the orthogonal counterpart of M. jannaschii, according to which the *E. coli* are placed on a plasmid for directed evolution in the manipulated *E. coli* (Iraha et al. 2010). The resulting *E. coli* TyrRS mutant can then be propagated for application in mammalian cells. Although not yet widespread, the development of *E. coli* aaRS in engineered *E. coli* is appealing since the most efficacious screening methods have been developed in *E. coli* (Wang and Schultz 2005).

11.9.4 The aaRS Recognition of $\mathrm{tRNA_{CUA}}$ Can Be Improved

In order to decipher the amber stop codon UAG, the orthogonal tRNA need to be altered in its anticodon to CUA, which can reduce the affinity of the subsequent tRNACUA toward the synthetase, as the anticodon serves as a major identification feature of the majority of synthetases. Through mutation of the anticodon targeting region of *E. coli* TyrRS, it was possible to identify an EwtYRS (D265R) that exhibits about twice as much activity toward *E. coli* than WT TyrRS. This improvement can be transferred to several TyrRS-derived synthetases which are targeted to various Uaas to optimize their assimilation performance in mammalian cells (Takimoto et al. 2009).

11.9.5 The Uaa Specificity of aaRS Can Be Widened

Most developed aaRSs aminoacylates Uaas, which have a chemical resemblance to the native substrate of WT aaRS, restrict the stereochemical abundance of Uaas available. For example, most Uaas ingested by mutant pairings keep the Lys core chain and the Nε carbonyl group of Pyl. To overcome this constriction, it was identified that by mutating a conserved "gatekeeper" residue (N346) of PylRS, which is capable of interfering with Nε carbonyl and the α amino group of the amino acid substrates, PylRS can be developed to load Uaas, which has a structure resembling that of Phe or Tyr (Takimoto et al. 2011; Hoppmann et al. 2017; Xiang et al. 2014). A new "low-intelligent" mutagenesis strategy, employing a single codon for each amino acid, was then developed to generate mutant PylRS libraries that allow a larger number of residues to be randomly selected at a time, enabling the detection of mutants containing Uaas with bulky conjugated rings (Lacey et al. 2013) and even Uaas carrying long azobenzene side chains (Hoppmann et al. 2014, 2015). These findings indicate an unusually wide flexibility in the development of aaRS substrate specificity (Wang et al. 2012). Since the cells can be utilized in both prokaryotic and eukaryotic cells, these developed PylRS mutants permit the integration of various Uaas into a variety of cells and organisms.

Apart from aaRS detection, efficient Uaa integration requires the affinity of the Uaa-tRNA to the elongation factor and ribosome. The genetic engineering of the tRNA, elongation factor and ribosome can refine their specific binding affinity and thus enable the genetic coding of a wide variety of Uaas (Schrader et al. 2011; Park et al. 2011). Up to now, a large number of functional groups have been deposited on the Phe/Tyr or Lys framework and translated into proteins as Uaas (Wang and Schultz 2002, 2005, Wang et al. 2006, 2009; Liu and Schultz 2010; Wang 2017a, b). Uaas with small or loaded (≥ 2) side chains are not easy to integrate and require extra efforts (Hoppmann et al. 2017).

11.9.6 Can the Bioavailability of Uaa Be Enhanced?

The Uaa bioavailability within cells is a precondition for the development of Uaa-specific aaRS and Uaa insertion. Uaas structurally similar to canonical amino acids can be introduced into cells through endogenous amino acid transporters or routes, but those that are either significant altered or highly charged, such as phosphorylated Uaas, cannot. However, it was suggested that covering the carboxyl group of Uaa with an ester will switch the Uaa zwitterion into a protonated weak base with an increased amount of the neutral form and increased lipophilicity for the efficient transgressing of the membrane. When they enter the cell, intracellular esterases split the ester to generate the Uaa. In fact, it was observed that the acetoxymethyl ester of Uaa DanAla raised its intracellular level 31-fold, and the Uaa ester was hydrolyzed to Uaa within 1 h in mammalian cells. (Takimoto et al. 2010). The utilization of Uaa-ester elevates the insertion of Uaa incorporation by fourfold, and necessitates 75% less Uaa added to the cell culture growth media.

C. elegans exhibits a sheltering cortex that prevents many compounds from being internalized. It was assumed that Uaas could be ingested as food through the intestinal tract and transported to different tissues, although intestinal transporters may not detect Uaas with vastly divergent structures. In fact, it has been revealed that O-methyl-L-tyrosine (OmeY), which possesses a similar structure to Tyr, can be integrated in muscle proteins, when the worms are fed with OmeY, while supplementation of worms with DanAla cause its break-down in the intestinal cells (Parrish et al. 2012). In the next step a dipeptide strategy has been envisioned and developed in order to get Uaa into an organism (Parrish et al. 2012). PEPT-1 and PEPT-2 represent dipeptide transporters that are located on the plasma membrane surface of cells in *C. elegans*. It has been assumed that an Ala-Uaa dipeptide can transverse over the plasma membrane into cells through dipeptide transporters and within the cytoplasm the dipeptide is hydrolyzed through cellular peptidases to produce the Uaa. In fact, when the Ala-DanAla dipeptide was supplemented to the food of worms, DanAla was detected as a component of proteins, which are located in diverse tissues of *C. elegans* (Parrish et al. 2012). When Uaas is inserted in mouse brain in vivo, it has been shown that employment of the Uaa-Ala dipeptide largely elevated the levels of the Uaa bioavailability within brain cells (Kang et al. 2013), since the PEPT-2 is strongly expressed in the brain tissue of rodents. The Uaa-Ala dipeptide was injected directly locally into the lateral or third ventricle of the mouse brain, whereby the associated Uaa was effectively integrated into proteins in the neocortex, thalamus and hypothalamus, with significantly higher yields than when the Uaa (Kang et al. 2013, 2016).

11.9.7 The Inactivation of NMD Stabilizes UAG-Containing mRNA

Eukaryotic cells are equipped with an mRNA surveillance mechanism that is referred to as nonsense-mediated mRNA decay (NMD) to detect mRNAs that carry premature stop codons and to direct the mRNA for fast break-down. Application of the amber stop codon to code for Uaa will render the mRNA vulnerable to NMD and thus reduce the protein output. Therefore, it was the first to study whether inactivation of NMD maintains the stability of UAG-containing mRNA and thus improves Uaa integration performance (Wang and Wang 2008). An NMD-deficient yeast strain was created by turning off the UPF1 gene, which represents a key constituent for NMD (Wang and Wang 2008, 2012). This upf1Δ strain enhanced the Uaa integration by more than 2 times compared to the WT yeast. In addition, NMD has been proven to exert a similar impact on the Uaa insertion in *C. elegans* (Parrish et al. 2012). The knockdown of the NMD component smg-1 in *C. elegans* with RNA interference resulted in 5.6 times higher levels of Uaa-containing protein being purified than from control worms.

11.9.8 Knock-Out of RF1 to Enhance the Insertion of Uaa Incorporation at Multiple Locations

Class I release factors (RFs) detect stop codons to complete the translation process. Eukaryotes and archaea utilize a single RF to identify all three stop codons, whereas bacteria employ two: RF1 for UAA/UAG and RF2 for UAA/UGA. The synthetic recoding of a genome can provide new characteristics by coding Uaas. Even though Uaas is capable of being specified with the UAG codon, RF1 renders UAG equivocal because it is simultaneously a stop signal and a Uaa. RF1 rivalry for UAG also restricts the insertion of Uaa at a single low efficiency location (10–30%), whereas multiple UAG codons drastically reduce protein output and inhibit proper utilization of Uaa at multiple locations.

To completely reallocate the UAG to a sense codon, it is essential to switch off RF1. However, no free-living bacterium has been identified that is missing neither RF1 nor RF2. RF1 has been known to be crucial for *E. coli* since the 1980s (Rydén and Isaksson 1984). In contrast to this model, the apparent essentiality of RF1 has been seen to be caused by an RF2 mutation (A264T) in *E. coli* K12 strains, which prevents RF2 from a significant termination by the UAA stop codon (Johnson et al. 2011). Since UAA is the predominant stop codon within *E. coli*, inadequate termination at UAA can lead to lethality, thereby avoiding an RF1 knockout. Therefore, it was necessary to immobilize the A246T mutation and eliminate an in-frame UGA stop codon within the RF2 gene, which self-regulates the level of RF2 expression. RF1 was slightly knocked out of *E. coli* once it had been immobilized (Johnson and Wang 2010). Moreover, in all *E. coli* B strains carrying the WT RF2 gene, RF1 can be eliminated in an unconditional manner (Johnson et al. 2012). When the A246T mutation in K12 strains is undone, RF1 can be knocked out at the same time.

These results show for an unprecedented time that RF1 is not necessary in *E. coli*. A number of RF1-knockout *E. coli* strains have been produced (Johnson et al. 2012). These strains facilitate Uaa insertion at the UAG codon with an activity level near to that of natural amino acids and, more significantly, permit concurrent Uaa insertion at multiple locations. It has been demonstrated that up to 10 UAG locations that are either spreading or merging are decrypted as Uaas in those RF1 knockout strains. These unparalleled characteristics turn the RF1 knockout bacterium into a distinct microbial host that exploits Uaa in the same way as natural amino acids to generate new protein features and biological functions.

In conclusion, it can be said that in 2001 the concept of genetic coding of Uaas was implemented utilizing an orthogonal tRNA/synthase pair for the very first time in *E. coli* (Wang et al. 2001).

Mammalian cells and animal models are preferred for biological and biomedical research in order to effectively examine physiological and pathological mechanisms *in vivo*. Two obstacles had hindered the intensive endeavors to introduce Uaas into mammalian cells over a period of several years, such as firstly the complexity of expressing prokaryotic tRNAs and secondly the impossibility of developing Uaa-specific synthetases inside mammalian cells. These problems have been overcome by employing the type 3 polymerase III promoter or a strategy of transfer. These new strategies made it possible to genetically code various Uaas in a number of mammalian cell lines and primary cells including neurons (Wang et al. 2007b). In addition, by optimizing the tRNA/synthetase sensitivity, the efficiency of Uaa integration has been improved dramatically (Takimoto et al. 2009) and the cellular absorption of Uaas has been increased employing chemical modifications (Takimoto et al. 2010).

11.9.9 Transfer of the Uaa Techniques to Stem Cells

Stem cell lines have been shown to stably incorporate Uaas, which has delivered a novel level for investigation of stem cell biology and is additionally attractive for examining the effect of Uaa-encoding mature cells, such as neurons, which are usually extremely complicated or expensive to acquire. It has turned out that the incorporation of Uaa into mature cells misses the exact temporal resolution required for the relatively long-term stem cell differentiation procedure. Moreover, at this time it was not clear whether the presence of Uaa can interfere with the stem cell maturation procedure. A lentiviral-based approach has been employed to generate a stable Uaa incorporation within neural stem cells (Shen et al. 2011). These stable stem cell lines retain their capacity to insert Uaa throughout the differentiation process and even in the fully differentiated neurons. In fact, there was no pronounced impairment of the differentiation process of stem cells detectable in the presence of incorporated Uaas. Consequently, these data were the first report on a stable mammalian stem cell line, which genetically incorporates Uaas (Shen et al. 2011).

11.10 Non-power Model of the Genetic Code

A mathematical setting relies on redundant (non-power) integer representations as a prime example for the understanding of genomic mapping information. The essence of the concept is based on the modeling of the degeneration in the genetic code (Gonzalez et al. 2016). The model helps to understand several characteristics and symmetries of the genetic code and discover formerly hidden symmetries. Additionally, this model delivers new tools for the investigation of genomic sequences. In the following there are key scientific fields addressed: firstly, the Euplotid nuclear code, secondly the mitochondrial code of vertebrates, and thirdly the main coding/decoding mechanisms employed in the three domains of life, such as archaea, bacteria and eukaryote domains. In each case, however, the non-performance hypothesis is demonstrated to be a natural, consistent framework for characterizing degeneration and deducing robust biological hypotheses about the coding of proteins. The concept is based on number theory and group theory; however, the technical aspects have been kept to a minimum by concentrating on key ideas and biological consequences.

An innovative mathematical representation of the genetic code that opens new ways of understanding the genetic information has been established (Gonzalez et al. 2016). The approach is based on number theory and was first proposed in (Gonzalez 2004). Since that time, new avenues have emerged and several advances have been made, but the strategy has not yet reached its full operational potential. The basic principle is to model the degeneration of the genetic code. Degeneration is a universal trait common to all forms of the genetic code, and it could be attributed to major biological activities. The model is built on redundant numerical systems, where an integer can contain more than one binary notation. The codons and binary strings of length 6 on the one hand and amino acids and integers on the other hand are matched.

Since the identification of the double helix structure of DNA by Watson and Crick in 1953, several mathematical concepts have been proposed that state how proteins are encoded (Hayes 1998). Among the most remarkable concepts are the crystal model (Gamow 1954) and the comma-free codes (Crick et al. 1957). Although both hypotheses proved experimentally wrong (Nirenberg and Matthaei 1961), Crick clearly pointed out that the coding theory serves as the framework for explaining the mechanisms of protein production. The digital presentation of genetic information has been regarded as a new field of life science in this time involving exact physical modeling (Yockey 2005). In specific detail, the mathematical characterization of the laws of the genetic code has major theoretical and practical consequences. In terms of theory, it delivers knowledge about the origin and development of the code that would otherwise be extremely challenging to acquire. In practical terms, it could be beneficial to advance the algorithms employed in bioinformatics and in medicine for the diagnosis and therapy of genetic diseases. In fact, several rules of the genetic code were obvious from the first moment on (Fimmel and Strüngemann 2016). For instance, in the majority of cases, codons that are sharing the first two nucleotides code for the same amino acid. Moreover, some biochemical characteristics of amino acids are associated with specific nucleotides in various codon locations. In point of fact, codons with the same nucleotide in the second place are linked to similar amino acids. There exists also a relationship between the first base of the codons and the precursors from which the encoded amino acids are to be generated (Koonin and Novozhilov 2009; Copley et al. 2005).

The genetic code acts as a translation grid that links the world of nucleic acids, in which biological knowledge is deposited, with the world of proteins, the chemical building blocks of the cellular metabolism. The genetic information is saved in double helix DNA molecules. A portion of this knowledge is transformed into single helix messenger RNA (mRNA) by a process referred to as transcription. Thymine (T), one of the four bases, such as thymine (T), cytosine (C), adenine (A) and guanine (G) that built DNA, is substituted by uracil (U) in RNA. When starting to count from the start signal, every three neighboring bases can be grouped within mRNA to build a codon. The genetic code allocates each of the 64 potential codons to the 20 amino acids (adding on top the stop signals) to construct the polymer chain of the protein, which is termed translation system. There are 64 codons and 20 amino acids, hence, the genetic code cannot be a one-to-one correspondence, which means that more than one codon is able to codify for the identical amino acid. For that purpose, amino acids are referred to as degenerate and codons are referred to as redundant or synonymous. The degeneration profile of amino acids indicates the number of amino acids that have a certain degeneration in common. All known derivatives of the genetic code are degenerate, and each of them has its own degeneration pattern, which can be divided into two main classes, such as nuclear and mitochondrial classes. The euplotid nuclear code and the mitochondrial genetic code of vertebrates represent two variants that are most symmetrical among their classes. There is an inherent powerful mathematical structure and several concealed symmetries that can be revealed through a standardizing strategy on the basis of number theory, which is discussed below.

Mathematical Model for the Genetic Code

To modify a genetic code and its degeneration profile, it is necessary to identify a non-bijective cartography which portrays it. Therefore, a numbering structure is employed in which every amino acid is expressed by an integer and a codon by its 6-bit binary representation (26 = 64). For example, the integer number 18 possesses the binary representation 010010 as $0 \cdot 25 + 1 \cdot 24 + 0 \cdot 23 + 0 \cdot 22 + 1 \cdot 21 + 0 \cdot 20 = 16 + 2 = 18$. However, common systems of number representation, such as binary, are not redundant, because each integer has an unambiguous presentation (that is, a unique binary character string). One possibility to establish a redundant numbering scheme is to substitute the potencies of base 2 with a series that grows slower than 2n (Fraenkel 1989). A well-known instance of this is the binary Fibonacci non-power system, in which the powers of 2 are substituted by the Fibonacci numbers, in other words 1 1 2 3 5 8 and finally this series grows slower than the powers of 2. In general, the number 18 in this system is represented twice, 111011 and 111100: $1{\cdot}8 + 1{\cdot}5 + 1{\cdot}3 + 0{\cdot}2 + 1{\cdot}1 + 1{\cdot}1 = 18 = 1{\cdot}8 + 1{\cdot}5 + 1{\cdot}3 + 1{\cdot}2 + 0{\cdot}1 + 0{\cdot}1$. The degeneration pattern of the Fibonacci non-power system is quite distinct from that of all known variations of the genetic code. In the following section, the unique non-power options that accurately define both the nuclear and mitochondrial genetic code of vertebrates are highlighted.

11.10.1 Euplotid Nuclear Genetic Code

There is a degeneration pattern within quartets of the euplotidic nuclear genetic code. This variant of the genetic code was selected as it is distinguished from the standard code only in a single codon assignment and constitutes the most symmetric variant of all nuclear codes; in the second section the decisive role of symmetry in this approach will be discussed. A quartet represents a group of four codons, which have the first two letters, for instance CTT, CTC, CTA and CTG, in common. The main rationale for employing quartets comes from the fact that the genetic code is translated by tRNA adapters that attach anticodons to amino acids. Due to wobble pairing, tRNAs can detect up to four codons of the identical quartet. Therefore, amino acids with degeneration 6 (Arg, Leu, Ser) consist of (at least) two elements, one with degeneration 2 and one with degeneration 4.

The key finding of this research is that the set of non-power weights (8, 7, 4, 2, 1, 1) provides the unique solution that accurately reflects the degeneration pattern of the genetic code of the euplotic nucleus (Table 11.6). It is feasible to demonstrate that there is no other non-trivial alternative. The investigation of symmetries provides the possibility to build links between the genetic code and the non-power system in order to retrieve an adequate mathematical model. The end result is a model, where each codon is allocated a 6-bit binary sequence and each amino acid an integer between 0 and 23. For instance, the amino acid Asn (asparagine) is encoded by AAT and AAC (degeneration 2). The model assigns Asn with the number 18 and the two sequences that constitute it, 110110 and 110101, with the codons AAT and AAC. The model has all acquired symmetries of the genetic code. For example, the pyrimidine substitution ($T \leftrightarrow C$) in the third letter of the codon leaves the encoded amino acid unchanged. This symmetry is striking as it is common to all common variants of the genetic code (both nuclear and mitochondrial). Indeed, in the model, the character sequences xxxx01 and xxx × 10 constitute the same integer. This also means that sequences terminating with 01 or 10 indicate codons terminating with *C* or *T* (pyrimidine), while sequences terminating with 00 or 11 indicate codons terminating with a purine (*A*, *G*). What is even more striking is that the examination of the model permits to disclose hidden symmetries in the genetic code.

As a result, the mathematical properties of binary sequences are coupled with the chemical characteristics of the bases of the codons (Gonzalez 2004, 2008b; Gonzalez et al. 2006, 2008). The analysis guided us to the definition of dichotomous classes that split the 64 codons into two halves. The first dichotomous class represents the equality of a codon, which is the equality of the corresponding binary sequence of characters. It can be proven that this mathematical procedure on the binary sequence of characters merges with a chemical algorithm that operates on the last two bases of the codon. It is important to mention that the nucleotides *U*, *C*, *A* and *G* are categorized by their chemical nature as follows:

Table 11.6 Non-power weights for the mitochondrial and euplotid nuclear genetic codes

Type	Non-power weights
Mitochondrial	8, 8, 4, 2, 1, 0
Euplotid nuclear	8, 7, 4, 2, 1, 1

$$\begin{array}{llll} \text{Purine} \quad R = \{A, G\} & \text{versus} & \text{Pyrimidine} & Y = \{C, U\} \\ \text{Keto} \quad K = \{U, G\} & \text{versus} & \text{Amino} & M = \{A, C\} \\ \text{Strong} \quad S = \{C, G\} & \text{versus} & \text{Weak} & W = \{A, U\} \end{array}$$

The algorithmic mapping of equality is explained in Fig. 11.10(2a). The algorithm is presented in the following. When the last letter of the codon contains a purine (R = A, G), then the parallelism of the binary sequence is deduced directly: A → odd, G → even. When the last letter of the codon is a pyrimidine (Y = U, C), its equivalence is derived from the chemical classification of the second base in the codon: amino (M = A, C) → even, keto (K = U, G) → odd. In line with equality, there are two other dichotomous classes. The three are founded on chemical algorithms in which the chemical elements concerned are identical in all three bases: YR on the third base, KM on the second base and SW on the first base. In particular, when we displace the equality algorithm in such a way that it operates on the first two bases of the codon, the result becomes a dichotomous class that is exactly the same as Rumer's class. It is a dual division of the genetic code witnessed by the theoretical physicist Rumer in the 1960s and is linked to the degeneration of amino acids in quartiles: 32 codons codify amino acids with degeneration 4, while the other 32 codons encode amino acids with degeneration non-4 (1, 2 or 3 in this instance).

Rumer also stated that the class is anti-symmetric in relation to the KM transform, which means when the KM transform is used for a codon, the Rumer class is different. The parity class is anti-symmetric in terms of the YR transformation. A complete skeleton for dichotomous classes can be deduced at this point: by moving the algorithm one base to the 5′ end of the nucleic acid, a third dichotomous class can be identified, which is referred to as a hidden class. The hidden class can be specified either within a codon (first and third base) or between two adjacent codons and is anti-symmetric in terms of the SW transformation. It is therefore conceivable to deduce dichotomic classes in the form of matrix operators and to

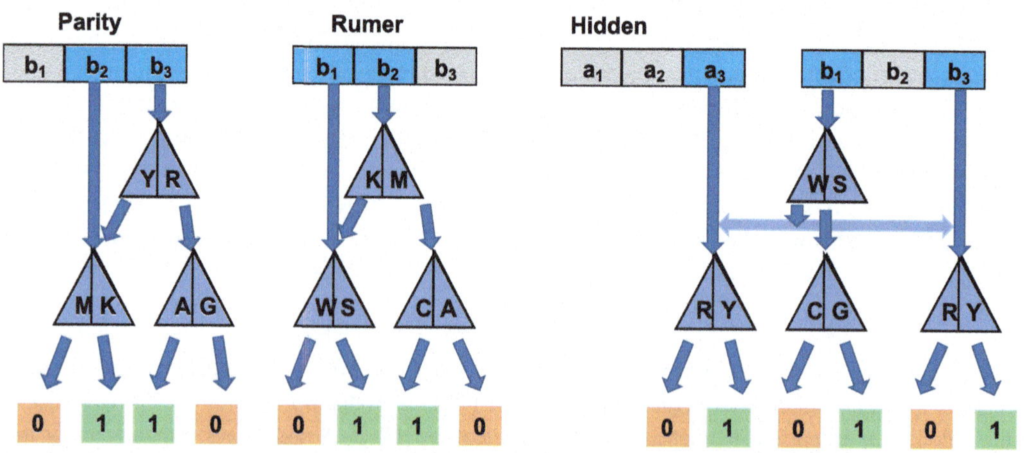

Fig. 11.10 Algorithmic presentation of the dichotomic classes, such as parity, Rumer and hidden

prove that these are non-linear functions of the chemical characteristics of a dinucleotide. Besides, they are closely associated with the group of small V-transformers. Dichotomic classes offer a new, non-trivial possibility for re-encoding the contents of nucleotide sequences within a binary code. There are several major consequences for the statistical evaluation of genomic datasets and for bioinformatics. It has been reported that coding sequences are distinguished by universal short-range relations, which only become apparent when dichotomous classes are considered (Gonzalez et al. 2008). Furthermore, on the other hand, it is demonstrated that dichotomous classes can be employed to create algorithms for frame recognition (Giannerini et al. 2012).

11.10.2 Genetic Code of Mitochondria

The primary reason that resulted in the mathematical framework of the euplotid nuclear code was to identify informational failure recognition/correction mechanisms. In fact, the main ingredients of error recognition/correction codes have been identified. For example, there are links among dichotomous classes, orthogonal arrays and finite groups. Additionally, the redundancy/degeneracy is connected to separate symmetry groups. Can at least the hypothetical mechanisms of failure recognition and correction be solved? The task is a rather complicated one, however, when these mechanisms are generally grounded in genetic codes, each with its own degeneration, it is obvious to concentrate on the simplest life system that has a genetic code: the mitochondrion. The genetic code of the mitochondrion represents the simplest and most symmetrical of all code versions and was suggested as a model for the early code (Jukes 1983; Ohama et al. 2008), the precursor of the universal genetic code of LUCA.

The mitochondrial genetic code of vertebrates varies in only four codons from the standard nuclear genetic code. The degeneration pattern of the mitochondrial code is far simpler and more symmetrical. In this context, the amino acids suffer from either degeneration 2 or 4. Noteworthy, it is feasible to demonstrate that there is also a unique non-power representation scheme for the mitochondrial code that precisely defines its degeneration pattern. The setting is governed by the six non-power weights (8, 8, 4, 2, 1, 0). In Table 11.6, the non-power denominations for the two genetic codes are given. In specific, the simplification of the degeneration pattern of the mitochondrial code is linked to the existence of a 0-weight, which in other words means that there are no elements with degeneration 1. As the weight 0 is not included in the additive degradation of a number, it has the consequence of double labeling. This means that the presentation is split into two identical parts. That fact has far-reaching implications in the characterization of the linked degeneration and guided toward a biological hypothesis on the origin of degeneration in the coding of proteins (Gonzalez et al. 2012).

In a similar way to the mathematical hypothesis, the biological hypothesis is founded on symmetry characteristics. Symmetry provides the key feature for linking the mathematical modeling with the chemical and biological characteristics of the genetic code. In short, the major finding suggests that the degeneration pattern of the vertebrate mitochondrial genetic code can be accurately characterized by primordial tRNA adapters operating on a specific batch of four-base codons, which are referred to as tesserae. These adapters offer both inverse and self-complementary symmetries. These symmetries involve the invariance of the Hamiltonian interaction in relation to those spatial transformations. An interesting hypothesis that needs to be explored is that conserved quantities associated with these symmetries may have evolutionarily influenced the shape of today's codes. The tessera set is provided in Table 11.1.

The results of this research on the mitochondrial genetic code have significant consequences for the investigation of the genetic code of the mitochondria: firstly the molecular evolution and the origin of protein coding, secondly the information-based failure recognition/correction techniques in the synthesis apparatus, and thirdly the characterization of decoding strategies in the three domains of life with respect to wobble symmetries and the redundant expression systems. In evolutionary perspective, the suggested decoding scheme is positioned before the early code, the code that pre-dated the universal genetic code of LUCA. It was assumed that the early code shares the exact same symmetry as the mitochondrial genetic code (Ohama et al. 2008; Osawa 1995; Watanabe and Yokobori 2014). The theory supports this hypothesis, since the tessera code, which shares the exact same symmetry as the present mitochondrial genetic code, can be considered an earlier code, an ancestor of the early code. The tessera code features error detection abilities, such as full immunity against point failures and immunity toward +1 frame shift. These characteristics help to clarify why evolutionary pressure may have resulted in the choice of this mechanism. Actually, the accuracy of protein synthesis requires that a certain degree of precision in the true organisms has been maintained, and the conservation of the degeneration pattern over geological times could be the molecular proof of this origin. However, it has been suggested that existing triplet codes are deduced from original codes with codons containing more than three nucleotides (Baranov et al. 2009). This may guarantee the required binding strength for a direct decoding in the absence of ribosomes. In addition, the possibility of decoding four-letter codons with advanced, available ribosomes has been proven feasible (Chen and Schindlinger 2010). Ribosomes have been reported to develop to decrease premature termination (Wang et al. 2007a, b). These ribosomes have continued to progress in this way to achieve the same degree of accuracy and efficiency as the triplet codons (Neumann et al. 2010). These ribosomes were designated Ribo-Q because of their ability to sense quadruplet codons in an efficient manner.

With respect to the linkage to existing decoding approaches, both the non-power model and the primordial symmetric tRNA hypothesis yield an accurate forecast of the number of adapters required to perform the code. Keep in mind that this number is 22, the minimum number commonly found in all existing life systems. The amino acid encoding is performed by tRNA connectors, which means that the actual degeneration pattern is seen at the scale of the tRNAs. The situation of the mitochondrial genetic code is a strange one, not only due to the fact that it employs the minimum number of adaptors, but also due to the fact that the degeneration pattern of the code matches the degeneration pattern of its tRNAs, which means that every item of the degeneration pattern is defined by exactly one adaptor (each amino acid is linked to precisely one adaptor). This is an extremum of the existing decoding strategies applied to current life forms.

11.10.3 Strategies for Coding

One hypothesis about the development of the genetic code from the supposedly early code is that the decoding of all codons was carried out with a minimal number of tRNA types (Ohama et al. 2008). As the post-transcriptional alterations of the first base of the anticodon vary widely among the organism strains, it is assumed that such alterations were induced after the emergence of the early code.

However, the limitation to a basic number of adapters also applies to various existing encoding schemes designed after the implementation of post-transcriptional modifications (Grosjean et al. 2010). It is still noteworthy that in all instances a remarkable number can be forecasted employing the non-power modeling strategy. To illustrate this, what emerges is how to forecast the minimum number of adaptors required to decrypt the 61 codons (64 − 3 stop codons) and an additional tRNA for the extension of Met, which also marks the starting point of the protein synthesis. For mitochondria, this elongation of the tRNA is not existent and there are two groups of two different stop codons. Thus, the non-power notation for the mitochondrial code of vertebrates (8, 8, 4, 2, 1, 0) encodes a maximum of 24 elements and forecasts 22 tRNA connectors when subtracting the two groups of stop codons (24 − 2 = 22). This feature has been described as strategy III and permits the entire family (super-wobble) to be decoded by utilizing an unmodified U at position 34, the main (first base) wobble position of the anticodon (Grosjean et al. 2010). This strategy is exclusively employed in bacteria and mitochondria and therefore suggests that within a quartet the amino acids exhibit either degeneration 4 or 2 + 2. In strategy I, codons terminating with a pyrimidine (*T* or *C*) are decoded using a post-transcriptionally altered G at position 34 for archaea and bacteria or a substituted *A* in eukarya. Codons terminating in a purine (*A* or *G*) are identified separately by a modified *U* or *C* at the same place of the anti-codon. With this approach, the required minimum number of adapters is 46 and the degeneration distribution within the quartets has a type 2 + 1 + 1. It is characterized by the non-power representation (16, 16, 8, 4, 2, 1), which yields 48 elements in all. When now three elements corresponding to the stop codons are removed (all ending with a purine base, thus matching a single adapter) and an additional tRNA for the elongation Met is included, this gives precisely the minimum number of adapters for strategy I, that is, $48 - 3 + 1 = 46$. As a final step in strategy II, a family is deciphered by two tRNAs, namely post-transcriptionally altered *U* and *G* at position 34 for archea and bacteria or post-transcriptionally altered *A* and *U* for eukarya. The minimum number of adapters for this type of strategy is 33, and in most cases, the degeneration within the quartets can be defined as 2 + 2. In this way, the degeneration connected with the strategy can be depicted by the non-power weights: (16, 8, 4, 2, 1, 0). In accordance with the depiction of the mitochondrial genetic code, there exists a 0 weight. Here, 32 is the total number of elements, however, when the stop codons and the special decoding boxes (Ile-Met and Cys-Stop-Trp) plus one additional tRNA for elongation Met, 33 are generated that represents exactly the minimal proposed value in this specific strategy.

In summary, it was reported that both the nuclear and mitochondrial types of the genetic code can be simulated by employing non-power systems of integer representation. The model accurately reflects a large number of prominent characteristics of the genetic code, including its symmetry characteristics. Keep in mind that there is a broad bibliography on the concept that the genetic code is associated with symmetry properties and consequently with the breaking of symmetry (Hornos and Hornos 1993). One such paper produced a series of efforts to try to apply continuous and discrete symmetry groups and algebras to the same problem (Frappat et al. 2001; Forger et al. 1997; Bashford et al. 1998). However, this idea was not well accepted by biologists (Maddox 1994). The extension of the theory of mathematical groups to the origin of the genetic code seems to be under discussion from a molecular biological point of view, but is still best regarded as a beneficial activity in classification. More work on symmetry and symmetry calculations can be found in (Jose et al. 2012; Jungck 1978; Karasev and Stefanov 2001; Sánchez et al. 2005; Tlusty 2010; Cherbak 2008; Findley et al. 1982). It is claimed that the abovementioned criticism does not hold true for this model, as it inherently depicts the degeneration of the actual genetic code without resorting to an implausible chain of symmetry breaking steps. In addition, it permits a straightforward biological evaluation of the findings.

The mitochondrial genetic code has been introduced as a model that has yielded the universal genetic code, which represent the code characterizing the LUCA. This pre-universal code was referred to as the early code. The non-powerful depiction of the mitochondrial genetic code of vertebrates permits to derive a biological hypothesis for the organization of the code that preceded the early code and for the origin of protein coding on the basis of the symmetries of old tRNA adapters and a set of four-base codons (mosaic stones, which are termed tesserae) (Gonzalez et al. 2012). This paradigmatic genetic code has strong features of error recognition and elimination that might evolutionarily be accountable for its choice (synthesis precision and expression fidelity). In addition, the model offers an answer to the question why the current genetic code is exactly characterized by 64 codons and 20 + 2 amino acids. The non-power strategy firstly involves the development from supposedly old codes identified by a small number of amino acids and coding words, and secondly encourages an origin of protein synthesis with adaptors comprising oligonucleotides longer than three bases and a ribosome-free direct template synthesis.

Lastly, it is now widely accepted that the genetic code is adapted to transmit more specific information than the linear coding of proteins (Itzkovitz and Alon 2007; Bollenbach et al. 2007). The degeneration of amino acids permits the utilization of synonymous codons to transmit additional information. This in return influences the allocation of the codons in the coding sequences and generates the so-termed "codon bias" (Bennetzen and Hall 1982; Sharp et al. 1985; Sharp and Li 1987). The codon bias is associated with the rate of translation efficiency (precision and velocity of protein synthesis) and numerous other key cellular processes ranging from differential protein synthesis to folding of proteins (Gingold and Pilpel 2011; Quax et al. 2015). The mathematical framework of the model and the evolutionary consequences obviously raise the problem of the accuracy and effectiveness of protein synthesis in relation to the mechanisms of error recognition/correction in respect of point mutations and frame shift. In this sense there are links to the theory of comma-free and circular codes. These are essential factors for investigating the problem of restoring and keeping the accurate reading frame (Argues and Michel 1996; Gonzalez et al. 2011; Michel 2015; Fimmel et al. 2015). The non-power concept permits DNA and mRNA sequences including their codon bias to be examined from a new perspective (Gonzalez et al. 2006, 2008, 2009; Giannerini et al. 2012) and may result in new bioinformatics vehicles and new biomedical treatments.

11.11 Mechanics of DNA and RNA

The flexibility of double-stranded (ds)-RNA and dsDNA is decisive in determining their biological activities. Latest investigations have demonstrated that the flexibility of dsRNA and dsDNA can vary considerably with respect to stretching and twist-stretch linkage. Despite the fact that several investigations have been conducted to gain an in-depth knowledge of the flexibility of dsRNA and dsDNA, a deep comprehension of the characteristic variations in the flexibility of dsRNA and dsDNA helices with respect to their stretching and twist-stretch linkage is nevertheless still missing. The relative flexibility of stretch and twist-stretch linkage of dsRNA and dsDNA was examined by molecular dynamics simulations for all atoms (Bao et al. 2017). The determined Young's modulus and the twist-stretch linkage are in close accordance with the current experiments. The differences in stretch and twist-stretch linkage of dsRNA and dsDNA helices are primarily due to their distinct (A and B shape) helical structures. A stronger tilt of the base pair and the sliding of the dsRNA is accountable for the seemingly lower stretching stiffness compared to the dsDNA, and the opposite twist-strain linkage for dsRNA and dsDNA is equally associated with the greater inclination of the base pair in the dsRNA than in the dsDNA. The calculated macroscopic elastic parameters and microscopic measurements are evaluated and validated by various force fields applied to both dsRNA and dsDNA (Bao et al. 2017).

Double-stranded (ds) RNA and dsDNA are basically architectural and operational units (Timmons and Fire 1998; Fire et al. 1998; Richmond and Davey 2003; Peter and Maher 2010; Rabo and Sauer 1984), and their architecture is commonly maintained by sequential base pairing/stacking and binding of metal ions. (Woodson 2005; Auffinger and Westhof 2000; Tan and Chen 2011). The flexibility of dsRNA/dsDNA is crucial for its biological activities, comprising replication and expression of genes, recognition of proteins and gene regulation (Timmons and Fire 1998; Fire et al. 1998; Richmond and Davey 2003; Peter and Maher 2010; Pabo and Sauer 1984; Xiao et al. 2010; Wei et al. 2014; McCauley et al. 2015).

Since dsDNA represents the key genetic biomolecule, its elastic characteristics and structural shifts under external force have been examined in various single-molecule experiments (Herrero-Galan et al. 2013; Lipfert et al. 2014; Gore et al. 2006; Li et al. 2012; Forth et al. 2013; Bryant et al. 2003; Vladescu et al. 2007; Wenner et al. 2002; Lionnet et al. 2006; Baumann et al. 1997; Yan and Marko

2004; Zhang et al. 2012, 2014; Lipfert et al. 2011; Smith et al. 1996; Mathew-Fenn et al. 2008) and in theoretical modeling approaches and simulations (Noy and Golestanian 2012; Ha and Thirumalai 2003; Tan and Chen 2008; Strahs and Schlick 2000; Wu et al. 2015a, b; Lebrun and Lavery 1996; Ruscio and Onufriev 2006; Marko 1997; Manning 2006; Jian 1997). The flexural strength, referred to as bending persistence length l_p, of long dsDNA was measured experimentally and has a value of approximately 50 nm at high salt conditions (Peter and Maher 2010; Herrero-Galan et al. 2013; Lipfert et al. 2014; Manning 2006; Lian 1997; Hagerman 1988; Bao et al. 2016). Sophisticated single-molecule approaches have made possible more precise and comprehensive measurements of dsDNA flexibility in terms of its strain and torsional characteristics, with a stretch modulus S of approximately 1000–1500 pN and torsional modulus C of approximately 370 ± 80 pN nm^2 (Herrero-Galan et al. 2013; Lipfert et al. 2014; Gore et al. 2006; Li et al. 2012; Forth et al. 2013; Bryant et al. 2003; Marko 1997). In addition, the experimental evidence of negative twist-stretch linkage for dsDNA strongly implies that dsDNA is usually overtwisted upon stretching (Lipfert et al. 2014; Gore et al. 2006; Lionnet et al. 2006; Marko 1997).

In contrary, the knowledge about the elastic characteristics of dsRNA is not yet sufficiently deep. Since dsRNA has proven to be essential in a variety of biological processes and implementations of nanomaterials (Timmons and Fire 1998; Fire et al. 1998; Guo 2010), more emphasis has progressively been paid to its flexibility (Herrero-Galan et al. 2013; Lipfert et al. 2014; Hagerman 1997; Sutton and Pollak 2015; Wu et al. 2015a, b; Shi et al. 2015; Abels et al. 2005; Zhang and Chen 2002; Xu et al. 2016; Wang et al. 2016). Contrary to dsDNA, which is usually in the B form, dsRNA tends to prefer to form a thicker right-handed A form helix. The experiments have revealed that the bending persistence length l_p and the torsion modulus C of dsRNA are about 60 ± 10 nm (Herrero-Galan et al. 2013; Lipfert et al. 2014; Hagerman 1997; Abels et al. 2005) and about 410 ± 8 pN nm^2 (Lipfert et al. 2014). In addition, l_p and C of dsRNA seem to be only less altered compared to those of dsDNA. Nevertheless, several other experimental studies have demonstrated that the stretch modulus S of the dsRNA (about 350–600 pN) is almost twice as high as that of the dsDNA (Herrero-Galan et al. 2013; Lipfert et al. 2014), and that the dsRNA exhibits positive twist-stretch linkage (Lipfert et al. 2014). Therefore, the strain module S and the twist-strain linkage of dsRNA differ significantly from that of dsDNA. Even though these macroscopic elastic properties have been measured in experiments, a profound comprehension of the pronounced variations in stretching and twist-stretch linkages of dsRNA and dsDNA at the microscopic scale is generally missing.

Molecular dynamics (MD) simulations for all atoms have been used widely as a high-performance instrument at the atomic level to investigate the structural and dynamic behavior of nucleic acids and proteins (Auffinger and Westhof 2000; Noy and Golestanian 2012; Wu et al. 2015a, b; Noy et al. 2004; Faustino et al. 2010; Liebl et al. 2015; Auffinger and Westhof 1998; Qi et al. 2010; Schlick 2010; Wang et al. 2015; Qi et al. 2011; Zhang et al. 2011; Drozdetski et al. 2016; Xiao et al. 2014; Gu et al. 2015). Microscopic measurements of all atoms on the basis of MD simulations (Noy et al. 2004) short (12-bp) dsRNA and dsDNA, have indicated that ideas about flexibility and stiffness are much more intricate than normally thought, and dsDNA is generally not necessarily more flexible with respect to dsRNA. The twist-stretch linkage has also previously been investigated by whole-atomic MD simulations for short (16-bp) dsRNA and dsDNA, and the distinction in twist-stretch linkage between dsRNA and dsDNA is explained by the association of helical rise, base-pair tilt and x-displacement occurring during the twisting process (Liebl et al. 2015). But it is not clear how the interaction of helix rise, base-pair inclination and x-displacement accounts for the striking discrepancy in the twist-stretch clutch and what factor primarily accounts for such a discrepancy in twist-stretch clutch. Beyond this, the considerable variation in elongation properties of dsRNA and dsDNA is poorly explained. Consequently, a quantitative comprehension of the pronounced variation in the flexibility of dsRNA and dsDNA, including stretching and twist-stretch linkage, is generally still incomplete.

In this study the emphasis is on the relative flexibility of dsRNA and dsDNA in stretching and twist-stretch linkage. In extension to earlier MD analyses, long-term MD simulations for larger dsRNA and dsDNA have been conducted to permit more quantitative estimates of macroscopic elastic properties and in-depth analysis of the respective mechanism at the microscopic scale. The objective of this investigation is to obtain a deeper quantitative insight into the comparative flexibility of dsRNA and dsDNA with respect to stretching and twist-stretch linkage phenomena. In specific, the Young's modulus and the twist-stretch linkage characteristics for both dsRNA and dsDNA have been evaluated, which are in close accordance to the available studies. Microscopic investigations point to the fact that the larger inclination and larger sliding in the dsRNA are the reason for the higher stretching flexibility of the dsRNA compared to the dsDNA, and it is the obvious difference in the inclination of the base pair that accounts for the dominant influence on the noticeable deviation in the twist-stretch linkage between dsRNA and dsDNA. Besides, the results of the calculated elastic properties and analyses are investigated and verified on the basis of various MD-fields of force. Within this work, the macroscopic elastic properties of dsRNA and dsDNA at stretching and twist-stretch linkage have been determined and the according measurements have been performed on the microscopic scale to gain a profound comprehension of the relative flexibility of dsRNA and dsDNA at stretching and twist-stretch linkage. At the same time, the forecasted macroscopic values are quantitatively matched with the comprehensive experimental data. In addition, the calculations and analyses are validated against several force fields for both dsRNA and dsDNA.

11.11.1 The Stretch of the Helix and the Stretch Modulus

The macroscopic stretching features of dsRNA and dsDNA are directly determined by the manifolds of the contour length L (Fig. 11.11). The average contour lengths L_0 are 80.2 A of the 30-bp dsRNA and 95.0 A of the 30-bp dsDNA, resulting in an average H increase of 2.77 Ȧ for dsRNA and 3.28 Ȧ for dsDNA. The equivalent standard errors of the contour length L of the 30-bp dsRNA and dsDNA are 2.4 and 1.6 Ȧ, respectively, indicating that the overall strain stiffness of the dsRNA seems to be lower than that of the dsDNA.

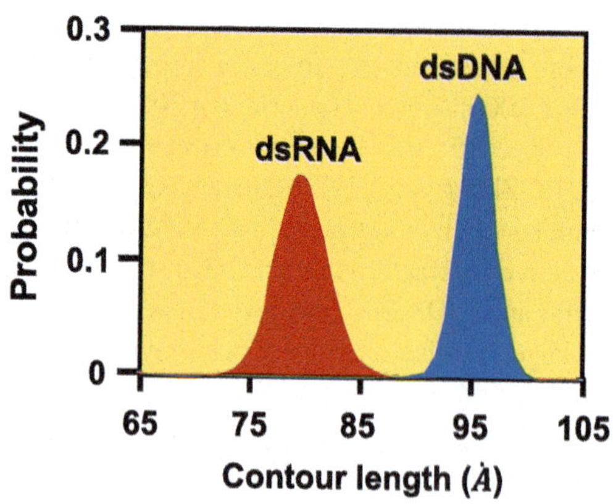

Fig. 11.11 Contour length distributions of 30 bp dsRNA and dsDNA

11.11.2 The Calculation of the Elastic Parameters of DsRNA/DsDNA

The macroscopic elastic properties can be analyzed as the diagonal terms of the elastic matrix K, which are linked to the overall helix coordinates (contour length L and cumulative H-rotation F over the central 30 bp), and are defined by (Noy and Golestian 2012; Faustino et al. 2010) in (11.1):

$$K = k_B T L_0 V^{-1}$$
$$K = k_B T L_0 \begin{pmatrix} \Delta L^2 \Delta L \times \Delta \Phi \\ \Delta L \times \Delta \Phi \Delta \Phi^2 \end{pmatrix}^{-1} \tag{11.1}$$

where V is the 2 × 2 covariance matrix of the two overall helix coordinates ($L$and F) and can be derived from the MD simulation trajectories (Noy and Golestian 2012; Faustino et al. 2010), L_0 represents the average value of contour length, T stands for the thermodynamic temperature in Kelvin, and k_B denotes the Boltzmann constant. In (11.1), the diagonal terms of V^{-1} can be interpreted as the reciprocal of the fractional deviations (Noy and Golestian 2012). The Young's modulus and the twist-stretch linkage are related to the contour length L and cumulative H-twist F in (11.1). Stiffness measurements are performed for the overall characteristics of the central 30 bp dsRNA/dsDNA to overcome the end effect (Wu et al. 2015b).

In agreement with (11.1) the stretch modulus S has been determined on the background of MD trajectories for dsRNA and dsDNA. The computed stretch module S is approximately 563 pN for dsRNA and approximately 1480 pN for dsDNA, in accordance with earlier experimental data (around 350–600 pN for dsRNA and around 1000–1500 pN for dsDNA at a concentration of 0.1–1 M Na^+ ions) (Herrero-Galan et al. 2013; Lipfert et al. 2014; Baumann et al. 1997; Wenner et al. 2002). Additionally, computations and measurements demonstrate that the stretch modulus of dsDNA is almost twice as high as that of dsRNA. There is still an important question to be addressed: Why does the dsRNA possess an increased stretching flexibility compared to dsDNA?

To respond to this issue, a microscopic analysis of the H-rise and the increase for dsRNA and dsDNA in the stretch dimension (direction of the central axis) is conducted, since the variance of the H-rise is closely associated with stretch flexibility. Nevertheless, the distributions of increase nearly exactly the same for dsRNA and dsDNA (with an average value of around 3.34 Å), whereas the distributions of H-rise are markedly different (with average values of around 2.77 Å for dsRNA and around 3.28 Å for dsDNA). What is the reason for the large discrepancy in H-rise between dsRNA and dsDNA? The correlation of H rise and rise for both dsRNA and dsDNA has been explored. It is obvious that the kind of base pairing/stacking results in the significant discrepancy in H-rise of dsRNA and dsDNA, meaning that the base pairs glide along each other and are more strongly stapled in dsRNA than in dsDNA. In this case, the inputs of shift and peak have been ignored as both are nearly zero and therefore have only a marginal influence on the H-rise for both dsRNA and dsDNA. When examining the basic stacking geometry, the correlation between rise and H-rise can be determined by inclination θ and sliding (Calladine and Drew 1997):

$$H - \text{rise} = \text{rise} \times \cos\theta + \text{slide} \times \sin\theta \tag{11.2}$$

Equation (11.2) has been proven to be suitable for both dsRNA and dsDNA on the basis of MD trajectories (Bao et al. 2017). Since the (11.2) utilizes a rise, inclination, and sliding, the average values and standard deviations of the microscopic structural features have been determined by MD simulations for dsRNA and dsDNA. Based on these data and from (11.2), the larger absolute values of inclination and sliding for dsRNA evoke an apparently shorter contour length (and H-rise) of dsRNA compared to that of dsDNA, which exhibits smaller absolute values for θ and sliding.

In next step, the difference in the flexibility of the stretching, such as the fluctuation of H-rise, between dsRNA and dsDNA have been determined in a quantitative manner. In line with (11.2), the fluctuation of the H-rise is presented by its full differential in (11.3):

$$\mathrm{dH-rise} \approx \mathrm{drise} - \mathrm{rise} \times \theta \times \mathrm{d}\theta + \mathrm{slide} \times d\theta + \theta \times \mathrm{dslide} \tag{11.3}$$

where the approximations of $\sin\theta \sim \theta$ and $\cos\theta \sim 1$ are employed as values of θ (about 15.9° for dsRNA and about 7.8° for dsDNA) are relatively small. Evidence was found that the variances of θ and increase for dsRNA and dsDNA are almost equal, and the coefficients (θ) of the fourth term on the right part of (11.3) cancel the effect of dslide for dsRNA and dsDNA. This seems to imply that the main discrepancy in the difference in H-rise between dsRNA and dsDNA is primarily due to the second and third terms, such as $-\mathrm{rise} \times \theta \times \mathrm{d}\theta + \mathrm{slide} \times \mathrm{d}\theta$, on the right part of (11.3), as the sum of the second and third terms of dsRNA ($|-2.6 \times \mathrm{d}\theta|$) is approximately 2.9-fold higher than that of dsDNA ($|-0.9 \times \mathrm{d}\theta|$). Therefore, the varying values of the dH increase of dsRNA and dsDNA are associated with the varying average values of θ and the slide: firstly, the comparatively larger θ of dsRNA increases the deviation of the projection of the increase in the central axis of the helix, and secondly, the higher absolute value of the slide of dsRNA enhances also the deviation of its projection in the central axis of the helix. Moreover, the contribution of θ and slide to the variation of H-rise is mirrored in the fluctuation of the tilt θ (11.3). Consequently, the markedly increased elongation flexibility observed for dsRNA is primarily explained by the strong tilt and sliding of the base pairs in A-shape configuration, whereas dsDNA seems to be clearly lower in elongation since the base pairs are stapled in dsDNA with low tilt and sliding.

11.11.3 Linkage of Twist and Stretch

Force-elongation measurements have been carried out on the twist-strain connection at low external strain force (Lipfert et al. 2014; Gore et al. 2006; Lionnet et al. 2006). It is astonishing that the dsDNA over-twists as it is stretched (Lipfert et al. 2014; Gore et al. 2006; Lionnet et al. 2006), whereas the dsRNA unwinds once it is stretched (Lipfert et al. 2014). Despite the fact that this mechanism is experienced under external force, the twist-stretch linkage for dsRNA and dsDNA in the free state can be examined and evaluated. Investigating the relationship of H increase (or contour length) and H-rotation (or cumulative H-rotation), as observed in single-molecule experiments (Lipfert et al. 2014; Gore et al. 2006; Bryant et al. 2003; Lionnet et al. 2006), in the absence of an external force, the twist-stretch coupling still exists (Bao et al. 2017). The twist-stretch coupling factor dH-rise/dH-twist is around 0.62 nm/turn for dsRNA. In contrary, dH-rise/dH-twist gives a positive value of about 0.67 nm/turn of dsDNA. The calculations are in good agreement with the experimental measurements (dH-rise/dH-twist about 0.85 ± 0.04 nm/turn) of dsRNA and about 0.5 ± 0.1 nm/turn of dsDNA (Lipfert et al. 2014; Gore et al. 2006; Lionnet et al. 2006). Such small differences between theoretical calculations and experimental data can be caused by variations in ion conditions, sequences and external forces (Lipfert et al. 2014; Lionnet et al. 2006; Manning 2015; Lionnet and Lankas 2007).

Next, the difference in twist-stretch linkage between dsRNA and dsDNA is carefully examined. First of all, the connection between H-rise and twist (instead of H-twist) is analyzed. The rotations of dsRNA and dsDNA both intensify as they fluctuate toward stretched contours. The appearance of this scenario appears somewhat unexpected, even though it is comprehensible. Since distortion was found to be highly associated with the (local) function of base pairs in earlier studies (Olson and Zhurkin 2000; Gorin et al. 1995), the relationship between distortion and role is explored. The coil diminishes as the twist grows, which is in line with previous results (Olson and Zhurkin 2000; Gorin et al. 1995). This can be exemplified by a simplified model of a basic pair stage, which indicates that a reduction in the roll can lead to an augmentation of the twist, since the sugar-phosphate backbone is nearly inelastic and the fluctuation of the increase is rather small. At the same time, the reduction in role also causes stretching of dsRNA and dsDNA, since the local role is closely linked to the global tilt of the MD trajectories (Lavery and Sklenar 1989; Lu and Olson 2003). As a result, the rotations of dsRNA and dsDNA both accelerate when they oscillate to elongated conformations.

Why does the H-twist correspond to the H-rise in a substantially other manner than the rotation? Hence, the dependence of the H-twist and twist has been examined. In fact, the H-twist represents the projection of twist on the plane, which is perpendicular to the central helical axis and is larger than twist based on the presence of the inclination θ. Based on the geometry a simple relationship between twist and H-twist for both dsRNA and dsDNA can be established approximately by the (11.4):

$$\cos\theta = \frac{\text{twist}}{H-\text{twist}} \tag{11.4}$$

Equation (11.4) has been seen to be accurate for both dsRNA and dsDNA due to the MD trajectories and it is additionally pronouncedly consistent with (11.5) (Lu and Olson 2003):

$$\text{dH}-\text{twist} \approx \text{dtwist} + \text{twist} \times \theta \times \text{d}\theta \tag{11.5}$$

where the approximations of $\sin\theta \sim \theta$ and $\cos\theta \sim 1$ are employed since θ is small. If a helix can fluctuate toward a stretched conformation, it can be said that $\text{dtwist} > 0$ and $\text{d}\theta < 0$ for both dsRNA and dsDNA. In line with (11.5), there seems to be competition between the positive first term of the equation (dtwist) and the negative second term ($\text{twist} \times \theta \times \text{d}\theta$) that impacts the sign of dH-twist. The dtwist, twist, and $d\theta$ of the dsRNA exhibits very similar values to those of dsDNA, whereas θ of dsRNA is about twofold higher than that of dsDNA (Bao et al. 2017). The larger θ of the dsRNA may encourage the second term of the negative sign with respect to the first term of the positive sign and result in a negative dH-twist. Nonetheless, θ of the dsDNA is rather small and thus dH-twist bears the identical sign as dtwist. Hence, the opposite signs of twist-stretch linkage of dsRNA and dsDNA stem from the seemingly distinct base-pair tilts.

Also, the correlations between rise and twist and between rise and H-twist have been studied, and it has been shown that such relationships are slightly negative for both dsRNA and dsDNA. It is comprehensible, since the basic piling is commonly heavy and the stiff backbone re-stresses the spatial geometry between rise and twist (Calladine and Drew 1997).

The above calculations indicate that the pronounced variation in flexibility between dsRNA and dsDNA is due to the various helical forms, namely the A and B forms. This raises an intriguing question: How is twist-stretch linkage for dsDNA of the A form carried out? To investigate this matter, several additional calculations of the twist-stretch linkage of dsDNA of the A form, dsDNA of the B form and dsRNA of the A form were performed with the elastic matrix model of ProDy (Bakan et al. 2011; Bahar and Jernigan 1998) rather than MD simulations for all atoms, since a dsDNA of the A form is usually extremely instable in aqueous ion solutions. Using ProDy, the values of twist-stretch linkage are positive for dsRNA in the A form and dsDNA, whereas the values of dsDNA in the B form are negative, which corresponds qualitatively with the forecasts of MD simulations for dsDNA and dsRNA in the B form from all atoms. It also implies that the A and B forms of the helical topology induce an opposite twist-stretch linkage of dsDNA and dsRNA of the B form, instead of the participation of an extra hydroxyl group or uridine replacing the thymine base in the dsRNA. Regardless of this, further quantitative forecasts and analyses of the twist-stretch linkage of dsDNA in the A form for dsDNA solutions with low water activity may necessitate full-atomic MD simulations (Jayaram et al. 1998; Waters et al. 2016), and warrant a special investigation.

11.11.4 What Is the Impact of MD Force Fields?

To investigate the influence of MD force fields on the results, additional MD simulations are carried out for a 16-bp dsRNA and 16-bp dsDNA. In this case 3 bps at each end of the 16-bp dsRNA/dsDNA are eliminated in the analyses, which means that the central 10-bp segments from the 16-bp dsRNA and dsDNA were analyzed further.

The macroscopic elastic properties and microscopic structural properties for dsRNA and dsDNA obtained from various force fields indicate that the refinement of the force fields (ff99bsc0 + χ_{OL3} for dsRNA and ff99bsc0 + bsc1 for dsDNA) yields the same quantitative data as those achieved by the force field of ff99bsc0, in spite of minor variations in stretching and twist-stretch linkage. Due to the contour length variations, the computed stretch modulus S is approximately 605 pN for dsRNA and approximately 1441 pN for dsDNA that resemble closely to the values of the force field of ff99bsc0 (about 563 pN for dsRNA and about 1480 pN for dsDNA). The twist-stretch linkage factor dH-rise/dH-twist has been reported to be around 0.59 nm/turn of dsRNA and about 0.61 nm/turn of dsDNA, which is in line with the values obtained from the force field of ff99bsc0 (approximately 0.62 nm/turn for dsRNA and approximately 0.67 nm/turn for dsDNA). The significant enhancement in the force field of bsc1 is the fact that it forecasts the elevated H-rotation, which is somewhat nearer to the experimental results than that foreseen by the force field of bsc0 (Galindo-Murillo et al. 2016). Moreover, the force fields of bsc1 and cOL3 produce a small modification of the average values of several structural characteristics and have nearly no significant impact on the macroscopic elastic characteristics and the relational deviation between dsRNA and dsDNA in the microscopic structural specifications concerning stretching properties and twist-stretch interactions. Thus, the computations and experiments on the relative flexibility of dsRNA and dsDNA are stable with respect to the diverse force fields.

In conclusion, the relative flexibility of dsRNA and dsDNA in the stretching and twist-stretch pairing has been examined by MD simulations for all atoms. The computed elastic properties of dsRNA and dsDNA are in quantitative agreement with the experimental data, and the principles of the relative flexibility of dsRNA and dsDNA have been examined and visualized at the microscopic scale. The three key findings can be concluded:

Firstly, the stretch modulus of dsRNA is considerably smaller than that of dsDNA (S of approximately 563 pN for dsRNA and approximately 1480 pN for dsDNA). This marked contrast is mainly due to the stronger tilt and glide of the base pair in the dsRNA than in the dsDNA, giving a dsRNA molecule more room in the direction of the helix axis than in the dsDNA.

Secondly, the dsRNA displays the inverse twist-stretch linkage to dsDNA (dH-rise/dH-twist of about 0.62 nm/turn for dsRNA and about 0.67 nm/turn for dsDNA). One such noticeable change in twist-stretch linkage between dsRNA and dsDNA is due to the seemingly more pronounced base-pair tilt in the dsRNA than in the dsDNA, since the interactions between H-rise and twist are nearly the same for both dsRNA and dsDNA, and the pronounced base-pair tilt in the dsRNA is responsible for the markedly distinct correlation between H-rise and H-twist.

Thirdly, the computed macroscopic elastic properties and the microscopic measurements of the relative flexibility of dsRNA and dsDNA are only slightly influenced by the alteration of the force fields, even though the forces fields of the new generation, comprising AMBER bsc1 and cOL3, somewhat enhance the forecast precision of some structural variables, such as H-twist and roll.

Apparently, the present report deviates from earlier MD-based assessments of the relative flexibility of dsRNA and dsDNA (Liebl et al. 2015) in the following four aspects as a minimum: firstly, the performed MD simulations for long (40-bp) dsRNA and dsDNA helices free of artificial constraints, which enables a better quantitative assessment of the extensive macroscopic elastic properties of free dsRNA and dsDNA in solvents; secondly, it is strongly suggested that the noticeable variance in twist-stretch linkage between dsRNA and dsDNA is mainly due to their base-pair tilt rather than the interaction of H-rise, tilt and x-displacement of the helix axis; thirdly, it has been seen that a more quantitative insight in the comparative flexibility of dsRNA and dsDNA in relation to stretching and twist-stretch linkage issues can be gained; and fourth, in this study force fields of a new type of all-atomic force fields are used to explore the macroscopic elastic characteristics and the related microscopic analyses for both dsRNA and dsDNA.

The microscopic assessment of dsRNA and dsDNA is mainly focused on the average base-pair step characteristics per conformation, which are closely connected to the experimental macroscopic specifications (Lipfert et al. 2014; Gore et al. 2006; Lionnet et al. 2006), whereas there are some short-range relationships between various base-pair steps. These kinds of relationships are of particular importance and it is certainly worth studying them individually, even though they have hardly anything in common with the macroscopic elastic properties of the available experiments (Lipfert et al. 2014; Gore et al. 2006; Lionnet et al. 2006).

This research indicates that the pronounced variation in flexibility between dsRNA and dsDNA is primarily due to the dissimilar helical configurations of the A and B forms, even though this does not necessarily account for why dsRNA has the A form and dsDNA the B form. These questions have been treated in a number of studies (Parker et al. 2013; Denning and MacKerell 2012; MacKerell 2009). The perturbation theory adapted to the symmetry was utilized to calculate the interaction energy of various base-pair steps and discovered that the forecasted structural characteristics, including rise, twist and sliping, are in agreement with the structure of the B form for DNA bases, whereas they cannot accurately estimate the structure of the A form for RNA bases (Parker et al. 2013). This indicates that the 20-hydroxyl in the ribbon of the RNA may have a strong impact on the overall structure (Parker et al. 2013). In addition, a series of quantum mechanical calculations were carried out to investigate the phosphodiester backbone conformations in dsDNA and dsRNA and to deduce that 20-hydroxyl is able to stabilize the A form structure to 20-hydroxyl due to the inherent energetics of the phosphodiester skeleton and the hydration pattern (Denning and MacKerell 2012; MacKerell 2009).

The twist-stretch linkage of dsRNA and dsDNA has been found in even the most complex nucleic acid Phe systems. For instance, it is reported that the anticodon of the tRNA in the ribosome stretches and uncoils in parallel to correspond to the codon in simulations (Bahar et al. 1998). In the process of building (H-type) RNA pseudo-nodes, spiral-shaped strains can overtwist and shrink to ease the short loop that spans the strain grooves, such as SRV-1, and TYMV pseudo-nodes (Michiels et al. 2001; Hilbers et al. 1998). Beyond that, in the nucleosome, the coating of dsDNA on histones triggers the stretching and overwinding of the dsDNA in order to enable a more effective adaptation to the histones (Richmond and Davey 2003).

In summary, all-atom MD simulations were calculated dsRNA and dsDNA at an external force of zero, which is equivalent to single-molecule experiments at low strain force. The flexibility of dsRNA and dsDNA may differ under severe stretching force, since dsDNA (or dsRNA) is supposed to uncoil and melt under very severe stretching force (Yan and Marko 2004; Zhang et al. 2012, 2014). The action of strong strain forces on the elastic properties of dsRNA and dsDNA is necessary and warrants separate discussion. This research can also be extended extensively, since the flexibility of a nucleic acid is essentially affected by its sequence, ionic environment and temperature (Baumann et al. 1997; Sutton and Pollak 2015; Tan and Chen 2006; Manning 2015; Lionnet and Lankas 2007).

These implications for the flexibility of dsRNA and dsDNA are beyond the scope of this investigation and must be examined in a separate study. Despite this, this investigation offers a more quantitative insight into the relative flexibility of dsRNA and dsDNA at the macroscopic and microscopic scales.

References

J.A. Abels, F. Moreno-Herrero, T. van der Heijden, C. Dekker, N.H. Dekker, Single-molecule measurements of the persistence length of double-stranded RNA. Biophys. J. **88**, 2737–2744 (2005)

N. Aggarwal, A.V. Bandhu, S. Sengupta, Finite population analysis of the effect of horizontal gene transfer on the origin of an universal and optimal genetic code. Phys. Biol. **13**, 036007 (2016)

G. Albrecht-Buehler, Asymptotically increasing compliance of genomes with Chargaff's second parity rules through inversions and inverted transpositions. Proc. Natl. Acad. Sci. USA **103**, 17828–17833 (2006)

M. Aldana, F. Cazarez-Bush, G. Cocho, G. Martnez-Mekler, Primordial synthesis machines and the origin of the genetic code. Phys. A **257**(1), 119–127 (1998)

M. Aldana-Gonzalez, G. Cocho, H. Larralde, G. Martinez-Mekler, Translocation properties of primitive molecular machines and their relevance to the structure of the genetic code. J. Theor. Biol. **220**(1), 27–45 (2003)

O. Allner, L. Nilsson, Nucleotide modifications and tRNA anticodon–mRNA codon interactions on the ribosome. RNA **17**, 2177–2188 (2011)

A. Ambrogelly, S. Palioura, D. Soll, Natural expansion of the genetic code. Nat. Chem. Biol. **3**, 29–35 (2007)

V. Anantharaman, E.V. Koonin, L. Aravind, Comparative genomics and evolution of proteins involved in RNA metabolism. Nucleic Acids Res. **30**, 1427–1645 (2002)

F. Antoneli, M. Forger, Symmetry breaking in the genetic code: finite groups. Math. Comput. Model. **53**, 1469–1488 (2011)

L. Aravind, V. Anantharaman, E.V. Koonin, Monophyly of class Iaminoacyl tRNA synthetase, USPA, ETFP, photolyase, and PP-ATPase nucleotide-binding domains: implications for protein evolution in the RNA. Proteins **48**, 1–14 (2002a)

L. Aravind, R. Mazumder, S. Vasudevan, E.V. Koonin, Trends in protein evolution inferred from sequence and structure analysis. Curr. Opin. Struct. Biol. **12**, 392–399 (2002b)

M. Archetti, Selection on codon usage for error minimization at the protein level. J. Mol. Evol. **59**, 400–415 (2004)

D.G. Arquès, C.J. Michel, A complementary circular code in the protein coding genes. J. Theor. Biol. **182**, 45–58 (1996)

D.G. Arquès, C.J. Michel, A circular code in the protein coding genes of mitochondria. J. Theor. Biol. **189**, 273–290 (1997)

P.J. Artymiuk, D.W. Rice, A.R. Poirrette, P. Willet, A tale of two synthetases. Nat. Struct. Biol. **1**, 758–760 (1994)

G.W. Ashley, Modeling, synthesis, and hybridization properties of (l)-ribonucleic acid. J. Am. Chem. Soc. **114**, 9731–9736 (1992)

P. Auffinger, E. Westhof, Simulations of the molecular dynamics of nucleic acids. Curr. Opin. Struct. Biol. **8**, 227–236 (1998)

P. Auffinger, E. Westhof, Water and ion binding around RNA and DNA (C, G) oligomers. J. Mol. Biol. **300**, 1113–1131 (2000)

I. Bahar, R.L. Jernigan, Vibrational dynamics of transfer RNAs: comparison of the free and synthetase-bound forms. J. Mol. Biol. **281**, 871–884 (1998)

A. Bakan, L.M. Meireles, I. Bahar, ProDy: protein dynamics inferred from theory and experiments, Bioinformatics **27**, 1575–1577 (2011)

L. Bao, X. Zhang, L. Jin, Z.J. Tan, Flexibility of nucleic acids: from DNA to RNA. Chin. Phys. B **25**, 018703 (2016)

L. Bao, X. Zhang, Y.-Z. Shi, Y.-Y. Wu, Z.-T. Tan, Understanding the relative flexibility of RNA and DNA duplexes: stretching and twist-stretch coupling. Biophys. J. **112**, 1094–1104 (2017)

P.V. Baranov, M. Venin, G. Provan, Codon size reduction as the origin of the triplet genetic code. PLoS ONE **4**, e5708 (2009)

J.D. Bashford, I. Tsohantjis, P.D. Jarvis, A supersymmetric model for the evolution of the genetic code. Proc. Natl. Acad. Sci. USA **95**, 987–992 (1998)

C.G. Baumann, S.B. Smith, V.A. Bloomfield, C. Bustamante, Ionic effects on the elasticity of single DNA molecules. Proc. Natl. Acad. Sci. USA **94**, 6185–6190 (1997)

J.L. Bennetzen, B.D. Hall, Codon selection in yeast. J. Biol. Chem. **257**, 3026–3031 (1982)

A. Bianco, F.M. Townsley, S. Greiss, K. Lang, J.W. Chin, Expanding the genetic code of Drosophila melanogaster. Nat. Chem. Biol. **8**, 748–750 (2012)

T. Bollenbach, K. Vetsigian, R. Kishony, Evolution and multilevel optimization of the genetic code. Genome Res. **17**, 401–404 (2007)

Z. Bryant, M.D. Stone, J. Gore, S.B. Smith, N.R. Cozzarelli, C. Bustamante, Structural transitions and elasticity from torque measurements on DNA. Nature **424**, 338–341 (2003)

H. Buhrman, P.T. van der Gulik, S.M. Kelk, W.M. Koolen, L. Stougie, Some mathematical refinements concerning error minimization in the genetic code. IEEE/ACM Trans. Comput. Biol. Bioinform. **8**, 1358–1372 (2011)

A.S. Burton, J.C. Stern, J.E. Elsila, D.P. Glavin, J.P. Dworkin, Understanding prebiotic chemistry through the analysis of extraterrestrial amino acids and nucleobases in meteorites. Chem. Soc. Rev. **41**, 5459–5472 (2012)

C.R. Calladine, H. Drew, *Understanding DNA: The Molecule and How It Works* (Academic Press, Cambridge, UK, 1997)

V.R. Chechetkin, Block structure and stability of the genetic code. J. Theor. Biol. **222**(2), 177–188 (2003)

I.A. Chen, M. Schindlinger, Quadruplet codons: one small step for a ribosome, one giant leap for proteins. Bio. Essays, **32**, 650–654 (2010)

Y. Chen, J. Ma, W. Lu, M. Tian, M. Thauvin, C. Yuan, M. Volovitch, Q. Wang, J. Holst, M. Liu, S. Vriz, S. Ye, L. Wang, D. Li, Heritable expansion of the genetic code in mouse and zebrafish. Cell Res. **27**, 294–297 (2017)

V. Cherbak, The arithmetical origin of the genetic code, in *The codes of life*, Biosemiotics, vol. 1. ed. by M. Barbieri, J. Hoffmeyer (Springer, Dordrecht, The Netherlands, 2008), pp. 153–185

J.W. Chin, Expanding and reprogramming the genetic code of cells and animals. Annu. Rev. Biochem. **83**, 379–408 (2014)

N.V. Chumachenko, Y. Novikov, M. Yarus, Rapid and simple ribozymic aminoacylation using three conserved nucleotides. J. Am. Chem. Soc. **131**, 5257–5263 (2009)

H.J.I.I. Cleaves, The origin of the biologically coded amino acids. J. Theor. Biol. **263**, 490–498 (2010)

I. Coin, V. Katritch, T. Sun, Z. Xiang, F.Y. Siu, M. Beyermann, R.C. Stevens, L. Wang, Genetically encoded chemical probes in cells reveal the binding path of Urocortin-I to CRF class B GPCR. Cell **155**, 1258–1269 (2013)

I. Coin, M.H. Perrin, W.W. Vale, L. Wang, Photo-cross-linkers incorporated into G-protein-coupled receptors in mammalian cells: a ligand comparison. Angew. Chem. Int. Ed. Engl. **50**, 8077–8081 (2011)

S.D. Copley, E. Smith, H.J. Morowitz, A mechanism for the association of amino acids with their codons and the origin of the genetic code. Proc. Natl. Acad. Sci. USA **102**, 4442–4447 (2005)

F.H. Crick, The genetic code—yesterday, today, and tomorrow. Cold. Spring. Harb. Symp. Quant. Biol. **31**, 3–9 (1966)

F.H.C. Crick, The origin of the genetic code. J. Mol. Biol. **38**, 367–379 (1968)

F.H.C. Crick, J.S. Griffith, L.E. Orgel, Codes without commas. Proc. Natl Acad. Sci. USA **43**, 416–421 (1957)

F.H. Crick, L. Barnett, S. Brenner, R.J. Watts-Tobin, General nature of the genetic code for proteins. Nature **192**, 1227–1232 (1961)

L.L. de Oliveira, P.S. de Oliveira, R. Tinos, A multi objective approach to the genetic code adaptability problem. BMC Bioinform **16**, 52 (2015)

F.I. Denning, A.D. Jr MacKerell, Intrinsic contribution of the 20-hydroxyl to RNA conformational heterogeneity. J. Am. Chem. Soc. **134**, 2800–2806 (2012)

M. Di Giulio, The origin of the genetic code: theories and their relationships, a review. Biosystems, **80**, 175–84 (2005)

M. Di Giulio, M. Medugno, The level and landscape of optimization in the origin of the genetic code. J. Mol. Evol. **52**, 372–382 (2001)

A.V. Drozdetski, I.S. Tolokh, L. Pollack, N. Baker, A.V. Onufriev, Opposing effects of multivalent ions on the flexibility of DNA and RNA. Phys. Rev. Lett. **117**, 028101 (2016)

I. Faustino, A. Perez, M. Orozco, Toward a consensus view of duplex RNA flexibility. Biophys. J. **99** 1876–1885 (2010)

E. Fimmel, L. Strüngmann, Yury Borisovich Rumer and his 'biological papers' on the genetic code. Phil. Trans. R. Soc. A **374**, 20150228 (2016)

E. Fimmel, S. Giannerini, D.L. Gonzalez, L. Strüngmann, Circular codes, symmetries and transformations. J. Math. Biol. **70**, 1623–1644 (2015)

G.L. Findley, A.M. Findley, S.P. McGlynn, Symmetry characteristics of the genetic code. Proc. Natl. Acad. Sci. USA **79**, 7061–7065 (1982)

A. Fire, S. Xu, M.K. Montgomery, S.A. Kostas, S.E. Driver, C.C. Mello, Potent and specific genetic interference by double-stranded RNA in *Caenorhabditis elegans*. Nature **391**, 806–811 (1998)

E. Firnberg, M. Ostermeier, The genetic code constrains yet facilitates Darwinian evolution. Nucleic Acids Res. **41**, 7420–7428 (2013)

W.M. Fitch, K. Upper, The phylogeny of tRNA sequences provides evidence for ambiguity reduction in the origin of the genetic code. Cold Spring Harb. Symp. Quant. Biol. **52**, 759–767 (1987)

M. Forger, Y.M.M. Hornos, J.E.M. Hornos, Global aspects in the algebraic approach to the genetic code. Phys. Rev. E **56**, 7078–7082 (1997)

S. Forth, M.Y. Sheinin, J. Inman, M.D. Wang, Torque measurement at the single-molecule level. Annu. Rev. Biophys. **42**, 583–604 (2013)

A.S. Fraenkel, The use and usefulness of numeration systems. Inform. Comput. **81**, 46–61 (1989)
B.R. Francis, Evolution of the genetic code by incorporation of amino acids that improved or changed protein function. J. Mol. Evol. **77**, 134–158 (2013)
L. Frappat, A. Sciarrino, P. Sorba, Crystalizing the genetic code. J. Biol. Phys. **27**, 1–34 (2001)
S.J. Freeland, L.D. Hurst, Load minimization of the genetic code: history does not explain the pattern. Proc. R. Soc. Lond. B **265**, 2111–2119 (1998a)
S.J. Freeland, L.D. Hurst, The genetic code is one in a million. J. Mol. Evol. **47**, 238–248 (1998b)
S.J. Freeland, T. Wu, N. Keulmann, The case for an error minimizing standard genetic code. Orig. Life Evol. Biosph. **33** (4–5), 457–477 (2003)
R. Galindo-Murillo, J.C. Robertson, M. Zgarbová, J. Šponer, M. Otyepka, P. Jurečka, T.E. III Cheatham, Assessing the current state of AMBER force field modifications for DNA. J. Chem. Theory Comput. **12**, 4114–4127 (2016)
G. Gamow, Possible relation between deoxyribonucleic acid and protein structures. Nature **173**, 318 (1954)
S. Giannerini, D.L. Gonzalez, R. Rosa, DNA, dichotomic classes and frame synchronization: a quasi-crystal framework. Phil. Trans. R. Soc. A **370**, 2987–3006 (2012)
S. Giannerini, D.L. Gonzalez, Rosa, R. A role for circular code properties in translation. Technical report (2018)
D. Gilis, S. Massar, N.J. Cerf, M. Rooman, Optimality of the genetic code with respect to protein stability and amino-acid frequencies. Genome Biol. **2**(11), 49.1–49.12 (2001)
H. Gingold, Y. Pilpel, Determinants of translation efficiency and accuracy. Mol. Syst. Biol. **7**, 481 (2011)
D.L. Gonzalez, Can the genetic code be mathematically described? Med. Sci. Monit. **10**, 11–17 (2004)
D.L. Gonzalez, Error detection and correction codes in *The Codes of Life: The Rules of Macroevolution,* Biosemiotics, vol. 1, ch. 17. ed by M. Barbieri, J. Hoffmeyer (Dordrecht, The Netherlands: Springer, 2008a), pp. 379–394
D.L. Gonzalez, The mathematical structure of the genetic code, in *The Codes of Life: The Rules of Macroevolution,* Biosemiotics, vol. 1 ed. by M. Barbieri, J. Hoffmeyer (Dordrecht, The Netherlands: Springer, 2008b), pp. 111–152
D.L. Gonzalez, S. Giannerini, R. Rosa, Detecting structure in parity binary sequences: error correction and detection in DNA. IEEE Eng. Med. Biol. Mag. **25**, 69–81 (2006)
D.L. Gonzalez, S. Giannerini, R. Rosa, Strong short-range correlations and dichotomic codon classes in coding DNA sequences. Phys. Rev. E **78**, 051918 (2008)
D.L. Gonzalez, S. Giannerini, R. Rosa, The mathematical structure of the genetic code: a tool for inquiring on the origin of life. Statistica **LXIX**, 143–157 (2009)
D.L. Gonzalez, S. Giannerini, R. Rosa, Circular codes revisited: a statistical approach. J. Theor. Biol. **275**, 21–28 (2011)
D.L. Gonzalez, S. Giannerini, R. Rosa, On the origin of the mitochondrial genetic code: towards a unified mathematical framework for the management of genetic information. Nat. Prec. (2012). https://doi.org/10.1038/npre.2012.7136.1
D.L. Gonzalez, S. Giannerini, R. Rosa, The non-power model of the genetic code: a paradigm for interpreting genomic information. Phil. Trans. R. Soc. A **374**, 20150062 (2016)
D.L. Gonzalez, S. Giannerini, R. Rosa, Rumer's transformation: a symmetry puzzle standing for half a century. Biosystems **187**, 104036 (2019)
H. Goodarzi, H.A. Nejad, N. Torabi, On the optimality of the genetic code, with the consideration of termination codons. Biosystems **77**, 163–73 (2004)
J. Gore, Z. Bryant, M. Nöllmann, M.U. Le, N.R. Cozzarelli, C. Bustamante, DNA overwinds when stretched. Nature **442**, 836–839 (2006)
A.A. Gorin, V.B. Zhurkin, W.K. Olson, B-DNA twisting correlates with base-pair morphology. J. Mol. Biol. **247**, 34–48 (1995)
H. Grosjean, E. Westhof, An integrated, structure- and energy-based view of the genetic code. Nucleic Acids Res. **44**, 8020–40 (2016)
H. Grosjean, V. de Crécy-Lagard, C. Marck, Deciphering synonymous codons in the three domains of life: co-evolution with specific tRNA modification enzymes. FEBS Lett. **584**, 252–264 (2010)
C. Gu, J. Zhang, Y.I. Yang, X. Chen, H. Ge, Y. Sun, X. Su, L. Yang, S. Xie, Y.Q. Gao, DNA structural correlation in short and long ranges. J. Phys. Chem. B **119**, 13980–13990 (2015)
P. Guo, The emerging field of RNA nanotechnology. Nat. Nanotechnol. **5**, 833–842 (2010)
V.A. Gusev, D. Schulze-Makuch, Genetic code: Lucky chance or fundamental law of nature? Phys. Life Rev. **1**(3), 202–229 (2004)
B.Y. Ha, D. Thirumalai, Bending rigidity of stiff polyelectrolyte chains: a single chain and a bundle of multichains. Macromolecules **36**, 9658–9666 (2003)
P.J. Hagerman, Flexibility of DNA. Annu. Rev. Biophys. Biophys. Chem. **17**, 265–286 (1988)
P.J. Hagerman, Flexibility of RNA. Annu. Rev. Biophys. Biomol. Struct. **26**, 139–156 (1997)
D. Haig, L.D. Hurst, A quantitative measure of error minimization in the genetic code. J. Mol. Evol. **33**(5), 412–417 (1991)
D. Haig, L.D. Hurst, Erratum for "A quantitative measure of error minimization in the genetic code". J. Mol. Evol. **49**, 708 (1999)

M. Hajnic, J.I. Osorio, B. Zagrovic, Computational analysis of aminoacids and their side chain analogs in crowded solutions of RNA nucleobases with implications for the mRNA–protein complementarity hypothesis. Nucleic Acids Res. **42**, 12984–12994 (2014)

M. Hasegawa, T. Miyata, On the asymmetry of the amino acid code table. Orig Life **10**, 265–270 (1980)

B. Hayes, The invention of the genetic code. Am. Sci. **86**, 8–14 (1998)

E. Herrero-Galan, M.E. Fuentes-Perez, C. Carrasco, J.M. Valpuesta, J.L. Carrascosa, F. Moreno-Herrero, J.R. Arias-Gonzalez, Mechanical identities of RNA and DNA double helices unveiled at the single-molecule level. J. Am. Chem. Soc. **135**, 122–131 (2013)

P.G. Higgs, A four-column theory for the origin of the genetic code: tracing the evolutionary pathways that gave rise to an optimized code. Biol. Direct. **4**, 16 (2009)

P.G. Higgs, R.E. Pudritz, A thermodynamic basis for prebiotic amino acid synthesis and the nature of the first genetic code. Astrobiology **9**, 483–490 (2009)

C.W. Hilbers, P.J. Michiels, H.A. Heus, New developments in structure determination of pseudoknots. Biopolymers **48**, 137–153 (1998)

R.T. Hinegardner, J. Engelberg, Rationale for a universal genetic code. Science **142**, 1083–1085 (1963)

M. Hlevnjak, A.A. Polyansky, B. Zagrovic, Sequence signatures of direct complementarity between mRNAs and cognate proteins on multiple levels. Nucleic Acids Res. **40**, 8874–8882 (2012)

C. Hoppmann, V.K. Lacey, G.V. Louie, J. Wei, J.P. Noel, L. Wang, Genetically encoding photoswitchable click amino acids in *Escherichia coli* and mammalian cells. Angew. Chem. Int. Ed. Engl. **53**, 3932–3936 (2014)

C. Hoppmann, I. Maslennikov, S. Choe, L. Wang, Situ formation of an azo bridge on proteins controllable by visible light. J. Am. Chem. Soc. **137**, 11218–11221 (2015)

C. Hoppmann, A. Wong, B. Yang, S. Li, T. Hunter, K.M. Shokat, L. Wang, Site-specific incorporation of phosphotyrosine using an expanded genetic code. Nat. Chem. Biol. **13**, 842–844 (2017)

J.E.M. Hornos, Y.M.M. Hornos, Algebraic model for the evolution of the genetic code. Phys. Rev. Lett. **71**, 4401–4404 (1993)

T. Huber, T.P. Sakmar, Chemical biology methods for investigating G protein-coupled receptor signaling. Chem. Biol. **21**, 1224–1237 (2014)

K. Ikehara, Y. Niihara, Origin and evolutionary process of the genetic code. Curr. Med. Chem. **14**(30), 3221–3231 (2007)

F. Iraha, K. Oki, T. Kobayashi, S. Ohno, T. Yokogawa, K. Nishikawa, S. Yokoyama, K. Sakamoto, Functional replacement of the endogenous tyrosyl-tRNA synthetase-tRNATyr pair by the archaeal tyrosine pair in *Escherichia coli* for genetic code expansion. Nucleic Acids Res. **38**, 3682–3691 (2010)

J. Iranzo, P. Puigbo, A.E. Lobkovsky, Y.I. Wolf, E.V. Koonin, Inevitability of genetic parasites. Genome Biol. Evol. **8**, 2856–2869 (2016)

S. Itzkovitz, U. Alon, The genetic code is nearly optimal for allowing additional information within protein-coding sequences. Genome Res. **17**, 405–412 (2007)

B. Jayaram, D. Sprous, M.A. Young, D.L. Beveridge, Free energy analysis of the conformational preferences of A and B forms of DNA in solution. J. Am. Chem. Soc. **120**, 10629–10633 (1998)

J.L. Jestin, A. Kempf, Optimization models and the structure of the genetic code. J. Mol. Evol. **69**, 452–457 (2009)

H. Jian, A.V. Vologodskii, T. Schlick, A combined wormlike-chain and bead model for dynamic simulations of long linear DNA. J. Comput. Phys. **136**, 168–179 (1997)

D.B. Johnson, L. Wang, Imprints of the genetic code in the ribosome. PNAS **107**, 8298–8303 (2010)

D.B. Johnson, C. Wang, J. Xu, M.D. Schultz, R.J. Schmitz, J.R. Ecker, L. Wang, Release factor one is nonessential in *Escherichia coli*. ACS Chem. Biol. **7**, 1337–1344 (2012)

D.B. Johnson, J. Xu, Z. Shen, J.K. Takimoto, M.D. Schultz, R.J. Schmitz, Z. Xiang, J.R. Ecker, S.P. Briggs, L. Wang, RF1 knockout allows ribosomal incorporation of unnatural amino acids at multiple sites. Nat. Chem. Biol. **7**, 779–786 (2011)

J.A. Johnson, Y.Y. Lu, J.A. Van Deventer, D.A. Tirrell, Residue-specific incorporation of non-canonical amino acids into proteins: recent developments and applications. Curr. Opin. Chem. Biol. **14**, 774–780 (2010)

R. Jolivet, F. Rothen, Peculiar symmetry of DNA sequences and evidence suggesting its evolutionary origin in a primeval genetic code, in *Proceeding First European Workshop on Exo-/astro-biology, ESA SP-496, Frascati, Italy, 21–23 May 2001*, ed by P. Ehrenfreund, O. Angerer, B. Battrick (Noordwijk, The Netherlands: ESA Publications 2001), pp. 173–176

I.K. Jordan, F.A. Kondrashov, I.A. Adzhubei, Y.I. Wolf, E.V. Koonin, A.S. Kondrashov, S. Sunyaev, A universal trend of amino acid gain and loss in protein evolution. Nature **433**, 633–638 (2005)

M.V. Jose, E.R. Morgado, R. Sanchez, T. Govezensky, The 24 possible algebraic representations of the standard genetic code in six or in three dimensions. Adv. Stud. Biol. **4**, 119–152 (2012)

T.H. Jukes, Genetic code: mitochondrial codes and evolution. Nature **301**, 19–20 (1983)

T.H. Jukes, *Molecules and Evolution* (Columbia University Press, New York, NY, 1968)

J.R. Jungck, The genetic code as a periodic table. J. Mol. Evol. **11**, 211–224 (1978)

J.Y. Kang, D. Kawaguchi, L. Wang, Optical control of a neuronal protein using a genetically encoded unnatural amino acid in neurons. J. Vis. Exp. **109**, e5381 (2016)
J.Y. Kang, D. Kawaguchi, I. Coin, Z. Xiang, D.D. O-Leary, P.A. Slesinger, L. Wang, Vvo expression of a light-activatable potassium channel using unnatural amino acids. Neuron **80**, 358–370 (2013)
V.A. Karasev, V.E. Stefanov, Topological nature of the genetic code. J. Theor. Biol. **209**, 303–317 (2001)
J.L. King, T.H. Jukes, Non-darwinian evolution. Science **164**(881), 788–798 (1969)
R.D. Knight, L.F. Landweber, The early evolution of the genetic code. Cell **101**, 569–572 (2000)
R.D. Knight, S.J. Freeland, L.F. Landweber, Selection, history and chemistry: the three faces of the genetic code. Trends Biochem. Sci. **24**, 241–247 (1999)
A. Kondow, T. Suzuki, S. Yokobori, T. Ueda, K. Watanabe, An extra tRNAgly(u*cu) found in ascidian mitochondria responsible for decoding nonuniversal codons aga/agg as glycine. Nucleic Acids Res. **27**, 2554–2559 (1999)
E.V. Koonin, Comparative genomics, minimal gene-sets and the last universal common ancestor. Nat. Rev. Microbiol. **1**, 127–136 (2003)
E.V. Koonin, *The Logic of Chance: The Nature and Origin of Biological Evolution* (FT Press, Upper Saddle River, NJ, 2011)
E.V. Koonin, Frozen accident pushing 50: Stereochemistry, expansion, and chance in the evolution of the genetic code. Life **7**, E22 (2017)
E.V. Koonin, W. Martin, On the origin of genomes and cells within inorganic compartments. Trends Genet. **21**, 647–654 (2005)
E.V. Koonin, A.S. Novozhilov, Origin and evolution of the genetic code: the universal enigma. IUBMB Life **61**(2), 99–111 (2009)
E.V. Koonin, A.S. Novozhilov, Origin and evolution of the universal genetic code. Annu. Rev. Genet. **51**, 45–62 (2017)
R.K. Kumar, M. Yarus, RNA-catalyzed amino acid activation. Biochemistry **40**, 6998–7004 (2001)
V.K. Lacey, G.V. Louie, J.P. Noel, L. Wang, Expanding the library and substrate diversity of the pyrrolysyl-tRNA synthetase to incorporate unnatural amino acids containing conjugated rings. Chem. Bio. Chem. **14**, 2100–2105 (2013)
U. Lagerkvist, 'Two out of three': an alternative method for codon reading. Proc. Natl Acad. Sci. USA **75**, 1759–1762 (1978)
R. Lavery, H. Sklenar, Defining the structure of irregular nucleic acids: conventions and principles. J. Biomol. Struct. Dyn. **6**, 655–667 (1989)
A. Lebrun, R. Lavery, Modelling extreme stretching of DNA. Nucleic Acids Res. **24**, 2260–2267 (1996)
J. Lehmann, A. Libchaber, Degeneracy of the genetic code and stability of the base pair at the second position of the anticodon. RNA **14**, 1264–1269 (2008)
R. Lenstra, Evolution of the genetic code through progressive symmetry breaking. J. Theor. Biol. **347**, 95–108 (2014)
E. Lerat, V. Daubin, H. Ochman, N.A. Moran, Evolutionary origins of genomic repertoires in bacteria. PLOS Biol. **3**, e130 (2005)
F. Li, H. Li, Y. Sun, Y. Pan, J. Zhou, J. Wang, Expanding the genetic code for photoclick chemistry in *E. coli*, mammalian cells, and A. thaliana. Angew Chem. Int. Ed. Engl. **52**, 9700–9704 (2013)
J. Li, P.R. Chen, Development and application of bond cleavage reactions in bioorthogonal chemistry. Nat. Chem. Biol. **12**, 129–137 (2016)
W. Li, P.Y. Wang, J. Yan, M. Li, Impact of DNA twist accumulation on progressive helical wrapping of torsionally constrained DNA. Phys. Rev. Lett. **109**, 218102 (2012)
K. Liebl, T. Drsata, F. Lankas, J. Lipfert, M. Zacharias, Explaining the striking difference in twist-stretch coupling between DNA and RNA: a comparative molecular dynamics analysis. Nucleic Acids Res. **43**, 10143–10156 (2015)
J. Ling, P. O'Donoghue, D. Soll, Genetic code flexibility in microorganisms: novel mechanisms and impact on physiology. Nat. Rev. Microbiol. **13**, 707–721 (2015)
T. Lionnet, F. Lankas, Sequence-dependent twist-stretch coupling in DNA. Biophys. J. **92**, L30–L32 (2007)
T. Lionnet, S. Joubaud, R. Lavery, D. Bensimon, V. Croquette, Wringing out DNA. Phys. Rev. Lett. **96**, 178102 (2006)
J. Lipfert, G.M. Skinner, J.M. Keegstra et al., Double-stranded RNA under force and torque: similarities to and striking differences from double-stranded DNA. Proc. Natl. Acad. Sci. USA **111**, 15408–15413 (2014)
J. Lipfert, M. Wiggin, J.W.J. Kerssemakers, F. Pedaci, N.H. Dekkera, Freely orbiting magnetic tweezers to directly monitor changes in the twist of nucleic acids. Nat. Commun. **2**, 439 (2011)
C.C. Liu, P.G. Schultz, Adding new chemistries to the genetic code. Annu. Rev. Biochem. **79**, 413–444 (2010)
A.V. Lobanov, A.A. Turanov, D.L. Hatfield, V.N. Gladyshev, Dual functions of codons in the genetic code. Crit. Rev. Biochem. Mol. Biol. **45**, 257–265 (2010)
X.J. Lu, W.K. Olson, 3DNA: a software package for the analysis, rebuilding and visualization of three-dimensional nucleic acid structures. Nucleic Acids Res. **31**, 5108–5121 (2003)
Y. Lu, S. Freeland, On the evolution of the standard amino-acid alphabet. Genome Biol. **7**(1), 102 (2006)
Y. Lu, S.J. Freeland, A quantitative investigation of the chemical space surrounding amino acid alphabet formation. J. Theor. Biol. **250**(2), 349–361 (2008)

A.D. Jr MacKerell, Contribution of the intrinsic mechanical energy of the phosphodiester linkage to the relative stability of the A, BI, and BII forms of duplex DNA. J. Phys. Chem. B **113**, 3235–3244 (2009)

J. Maddox, The genetic code by numbers. Nature **367**, 111 (1994)

G.S. Manning, The persistence length of DNA is reached from the persistence length of its null isomer through an internal electrostatic stretching force. Biophys. J. **91**, 3607–3616 (2006)

G.S. Manning, The response of DANN length and twist to changes in ionic strength. Biopolymers **103**, 223–226 (2015)

J.F. Marko, Stretching must twist DNA. Europhys Lett. **38**, 183 (1997)

S.E. Massey, A sequential "2-1-3" model of genetic code evolution that explains codon constraints. J. Mol. Evol. **62**, 809–810 (2006)

S.E. Massey, A neutral origin for error minimization in the genetic code. J. Mol. Evol. **67**, 510–516 (2008)

S.E. Massey, Genetic code evolution reveals the neutral emergence of mutational robustness, and information as an evolutionary constraint. Life **5**, 1301–1332 (2015)

S.E. Massey, The neutral emergence of error minimized genetic codes superior to the standard genetic code. J. Theor. Biol. **408**, 237–242 (2016)

D.C. Mathew, Z. Luthey-Schulten, On the physical basis of the amino acid polar requirement. J. Mol. Evol. **66**, 519–528 (2008)

R.S. Mathew-Fenn, R. Das, P.A. Harbury, Remeasuring the double helix. Science **322**, 446–449 (2008)

M.J. McCauley, I. Rouzina, K.A. Manthei, R.J. Gorelick, K. Musier-Forsyth, M.C. Williams, Targeted binding of nucleocapsid protein transforms the folding landscape of HIV-1 TAR RNA. Proc. Natl. Acad. Sci. USA **112**, 13555–13560 (2015)

C.J. Michel, The maximal C3 self-complementary trinucleotide circular code x in genes of bacteria, eukaryotes, plasmids and viruses. J. Theor. Biol. **380**, 156–177 (2015)

P.J. Michiels, A.A. Versleijen, P.W. Verlaan, C.W. Pleij, C.W. Hilbers, H.A. Heus, Solution structure of the pseudoknot of SRV-1 RNA, involved in ribosomal frameshifting. J. Mol. Biol. **310**, 1109–1123 (2001)

T. Mukai, M. Englert, H.J. Tripp, C. Miller, N.N. Ivanova, E.M. Rubin, N.C. Kyrpides, D. Söll, Facile recoding of selenocysteine in nature. Angew. Chem. Int. Ed. Engl. **55**, 5337–5341 (2016)

A. Munteanu, C.S. Attolini, S. Rasmussen, H. Ziock, R.V. Sole, Generic Darwinian selection in catalytic protocell assemblies. Philos. Trans. R. Soc. Lond. B Biol. Sci. **362**(1486), 1847–1855 (2007)

K. Musier-Forsyth, P. Schimmel, Atomic determinants for aminoacylation of RNA minihelices and relationship to genetic code. Acc. Chem. Res. **32**, 368–375 (1999)

H. Neumann, K. Wang, L. Davis, M. Garcia-Alai, J.W. Chin, Encoding multiple unnatural amino acids via evolution of a quadruplet-decoding ribosome. Nature **464**, 441–444 (2010)

I. Nikic, E.A. Lemke, Genetic code expansion enabled site-specific dual-color protein labeling: superresolution microscopy and beyond. Curr. Opin. Chem. Biol. **28**, 164–173 (2015)

S. Nikolajewa, M. Friedel, A. Beyer, T. Wilhelm, The new classification scheme of the genetic code, its early evolution, and tRNA usage. J. Bioinform. Comput. Biol. **4**, 609–620 (2006)

M.W. Nirenberg, Historical review: deciphering the genetic code—a personal account. Trends Biochem. Sci. **29**, 46–54 (2004)

M.W. Nirenberg, J.H. Matthaei, The dependence of cell-free protein synthesis in *E. coli* upon naturally occurring or synthetic polyribonucleotides. Proc. Natl Acad. Sci. USA **47**, 1588–1602 (1961)

M.W. Nirenberg, W. Jones, P. Leder, B.F.C. Clark, W.S. Sly, S. Pestka, On the coding of genetic information. Cold Spring Harb. Symp. Quant. Biol. **28**, 549–557 (1963)

W. Niu, P.G. Schultz, J. Guo, An expanded genetic code in mammalian cells with a functional quadruplet codon. ACS Chem. Biol. **8**, 1640–1645 (2013)

A.S. Novozhilov, E.V. Koonin, Exceptional error minimization in putative primordial genetic codes. Biol. Direct **4**, 44 (2009)

A.S. Novozhilov, Y.I. Wolf, E.V. Koonin, Evolution of the genetic code: partial optimization of a random code for robustness to translation error in a rugged fitness landscape. Biol. Direct **2**, 24 (2007)

A. Noy, R. Golestanian, Length scale dependence of DNA mechanical properties. Phys. Rev. Lett. **109**, 228101 (2012)

A. Noy, A. Pérez, F. Lankas, F. Javier Luque, M. Orozco, Relative flexibility of DNA and RNA: a molecular dynamics study. J. Mol. Biol. **343**, 627–638 (2004)

T. Ohama, Y. Inagaki, Y. Bessho, S. Osawa, Evolving genetic code. Proc. Jpn. Acad. B **84**, 58–74 (2008)

W.K. Olson, V.B. Zhurkin, Modeling DNA deformations. Curr. Opin. Struct. Biol. **10**, 286–297 (2000)

S. Osawa, *Evolution of the Genetic Code* (Oxford Science Publications, Oxford, UK, 1995)

C.O. Pabo, R.T. Sauer, Protein-DNA recognition. Annu. Rev. Biochem. **53**, 293–321 (1984)

H.S. Park, M.J. Hohn, T. Umehara, L.T. Guo, E.M. Osborne, J. Benner, C.J. Noren, J. Rinehart, D. Soll, Expanding the genetic code of *Escherichia coli* with phosphoserine. Science **333**, 1151–1154 (2011)

T.M. Parker, E.G. Hohenstein, R.M. Parrish, N.V. Hud, C.D. Sherrill, Quantum-mechanical analysis of the energetic contributions to p stacking in nucleic acids versus rise, twist, and slide. J. Am. Chem. Soc. **135**, 1306–1316 (2013)

A.R. Parrish, X. She, Z. Xiang, I. Coin, Z. Shen, S.P. Briggs, A. Dillin, L. Wang, Expanding the genetic code of *Caenorhabditis elegans* using bacterial aminoacyl-tRNA synthetase/tRNA pairs. ACS Chem. Biol. **7**, 1292–1302 (2012)

A. Patel, The triplet genetic code had a doublet predecessor. J. Theor. Biol. **233**(4), 527–532 (2005)

J.P. Peters, L.J. III Maher, DNA curvature and flexibility in vitro and in vivo. Q. Rev. Biophys. **43**, 23–63 (2010)

S. Pizzarello, The chemistry of life's origin: a carbonaceous meteorite perspective. Acc. Chem. Res. **39**, 231–237 (2006)

A.A. Polyansky, B. Zagrovic, Evidence of direct complementary interactions between messenger RNAs and their cognate proteins. Nucleic Acids Res. **41**, 8434–8443 (2013)

A.A. Polyansky, M. Hlevnjak, B. Zagrovic, Analogue encoding of physicochemical properties of proteins in their cognate messenger RNAs. Nat. Commun. **4**, 2784 (2013a)

A.A. Polyansky, M. Hlevnjak, B. Zagrovic, Proteome-wide analysis reveals clues of complementary interactions between mRNAs and their cognate proteins as the physicochemical foundation of the genetic code. RNA Biol. **10**, 1248–1254 (2013b)

P. Puigbo, A.E. Lobkovsky, D.M. Kristensen, Y.I. Wolf, E.V. Koonin, Genomes in turmoil: quantification of genome dynamics in prokaryote supergenomes. BMC Biol. **12**, 66 (2014)

P. Puigbo, Y.I. Wolf, E.V. Koonin, The tree and net components of prokaryote evolution. Genome Biol. Evol. **2**, 745–756 (2010)

W. Qi, B. Song, X. Lei, C. Wang, H. Fang, DNA base pair hybridization and water-mediated metastable structures studied by molecular dynamics simulations. Biochemistry **50**, 9628–9632 (2011)

W. Qi, X. Lei, H. Fang, DNA structural changes under different stretching methods studied by molecular dynamics simulations. Chem. Phys. Chem. **11**, 2146–2151 (2010)

T.E.F. Quax, N.J. Claassens, D. Söll, J. van der Oost, Codon bias as a means to fine-tune gene expression. Mol. Cell **59**, 149–161 (2015)

T.J. Richmond, C.A. Davey, The structure of DNA in the nucleosome core. Nature **423**, 145–150 (2003)

M. Richter, R. Rosselló-Móra, Shifting the genomic gold standard for the prokaryotic species definition. Proc. Natl. Acad. Sci. USA **106**, 19126–19131 (2009)

D.L. Riddle, J. Carbon, Frameshift suppression: a nucleotide addition in the anticodon of a glycine transfer RNA. Nat. New Biol. **242**, 230–234 (1973)

A. Rodin, E. Szathmary, S. Rodin, On origin of genetic code and tRNA before translation. Biol. Direct **6**, 14 (2011)

S. Rodin, S. Ohno, A. Rodin, Transfer RNAs with complementary anticodons: could they reflect early evolution of discriminative genetic code adaptors? Proc. Natl. Acad. Sci. USA **90**, 4723–4727 (1993)

I.B. Rumer, On codon systematization in the genetic code. Dokl. Akad. Nauk SSSR **167**(6), 1393–1394 (1966a)

Y.B. Rumer, About the codon's systematization in the genetic code. Proc. Acad. Sci. USSR (Doklady) **167**, 1393–1394 (1966b)

J.Z. Ruscio, A. Onufriev, A computational study of nucleosomal DNA flexibility. Biophys. J. **91**, 4121–4132 (2006)

S. Rydén, L. Isaksson, A temperature-sensitive mutant of *Escherichia coli* that shows enhanced misreading of UAG/A and increased efficiency for some tRNA nonsense suppressors. Mol. Gen. Genet. **193**, 38–45 (1984)

K. Sakamoto, A. Hayashi, A. Sakamoto et al., Site-specific incorporation of an unnatural amino acid into proteins in mammalian cells. Nucleic Acids Res. **30**, 4692–4699 (2002)

D.G. Salinas, M.O. Gallardo, M.I. Osorio, Local conditions for global stability in the space of codons of the genetic code. Biosystems **150**, 73–77 (2016)

R. Sánchez, E. Morgado, R. Grau, A genetic code Boolean structure. I. The meaning of Boolean deductions. Bull. Math. Biol. **67**, 1–14 (2005)

M.A. Santos, G. Moura, S.E. Massey, M.F. Tuite, Driving change: The evolution of alternative genetic codes. Trends. Genet. **20**, 95–102 (2004)

T. Schlick, *Molecular Modeling and Simulation: An Interdisciplinary Guide* (Springer, New York, 2010)

J.M. Schrader, S.J. Chapman, O.C. Uhlenbeck, Tuning the affinity of aminoacyl-tRNA to elongation factor Tu for optimal decoding. Proc. Natl. Acad. Sci. USA **108**, 5215–5220 (2011)

H. Seligmann, Putative mitochondrial polypeptides coded by expanded quadruplet codons, decoded by antisense tRNAs with unusual anticodons. Biosystems **110**, 84–106 (2012)

G. Sella, D.H. Ardell, The coevolution of genes and genetic codes: Crick's frozen accident revisited. J. Mol. Evol. **63**, 297–313 (2006)

S. Sengupta, P.G. Higgs, A unified model of codon reassignment in alternative genetic codes. Genetics **170**, 831–840 (2005)

S. Sengupta, P.G. Higgs, Pathways of genetic code evolution in ancient and modern organisms. J. Mol. Evol. **80**, 229–243 (2015)

S. Sengupta, N. Aggarwal, A.V. Bandhu, Two perspectives on the origin of the standard genetic code. Orig. Life Evol. Biosph. **44**, 287–291 (2014)

S. Sengupta, X. Yang, P.G. Higgs, The mechanisms of codon reassignments in mitochondrial genetic codes. J. Mol. Evol. **64**, 662–688 (2007)

P.M. Sharp, M.S. Rogers, D.J. McConnell, Selection pressures on codon usage in the complete genome of bacteriophage t7. J. Mol. Evol. **21**, 150–160 (1985)

P.M. Sharp, W.-H. Li, The codon adaptation index—a measure of directional synonymous codon usage bias, and its potential applications. Nucleic Acids Res. **15**, 1281–1295 (1987)

V. Shcherbak, The arithmetical origin of the genetic code, in *The codes of life*, Biosemiotics vol. 1. ed by M. Barbieri, J. Hoffmeyer (Dordrecht, The Netherlands, Springer 2008), pp. 153–185

B. Shen, Z. Xiang, B. Miller, G. Louie, W. Wang, J.P. Noel, F.H. Gage, L. Wang, Genetically encoding unnatural amino acids in neural stem cells and optically reporting voltage-sensitive domain changes in differentiated neurons. Stem. Cells **29**, 1231–1240 (2011)

Y.Z., Shi, L. Jin, F.H. Wang, X.L. Zhu, Z.J. Tan, Predicting 3D structure, flexibility, and stability of RNA hairpins in monovalent and divalent ion solutions. Biophys J. **109**, 2654–2665 (2015)

S.B. Smith, Y. Cui, C. Bustamante, Overstretching B-DNA: the elastic response of individual double-stranded and single-stranded DNA molecules. Science **271**, 795–799 (1996)

D. Strahs, T. Schlick, A-Tract bending: insights into experimental structures by computational models. J. Mol. Biol. **301**, 643–663 (2000)

J.L. Sutton, L. Pollack, Tuning RNA flexibility with helix length and junction sequence. Biophys. J. **109**, 2644–2653 (2015)

E.C. Swart, V. Serra, G. Petroni, M. Nowacki, Genetic codes with no dedicated stop codon: context-dependent translation termination. Cell **166**, 691–702 (2016)

M. Szabat, R. Kierzek, Parallel-stranded DNA and RNA duplexes—structural features and potential applications. FEBS J. **284**, 3986–3998 (2017)

E. Szathmary, Four letters in the genetic alphabet: a frozen evolutionary Optimum? Proc. Bio. Sci. **245**(1313), 91–99 (1991)

E. Szathmary, Why are there four letters in the genetic alphabet? Nat. Rev. Genet. **4**(12), 995–1001 (2003)

E. Szathmary, L. Demeter, Group selection of early replicators and the origin of life. J. Theor. Biol. **128**, 463–486 (1987)

E. Szathmary, J. Maynard Smith, From replicators to reproducers: the first major transitions leading to life. J. Theor. Biol. **187**, 555–571 (1997)

N. Takeuchi, K. Kaneko, E.V. Koonin, Horizontal gene transfer can rescue prokaryotes from Muller's ratchet: benefit of DNA from dead cells and population subdivision. G3 **4**, 325–339 (2014)

J.K. Takimoto, K.L. Adams, Z. Xiang, L. Wang, Improving orthogonal tRNA-synthetase recognition for efficient unnatural amino acid incorporation and application in mammalian cells. Mol. BioSyst. **5**, 931–934 (2009)

J.K. Takimoto, N. Dellas, J.P. Noel, L. Wang, Stereochemical basis for engineered pyrrolysyl-tRNA synthetase and the efficient in vivo incorporation of structurally divergent non-native amino acids. ACS Chem. Biol. **6**, 733–743 (2011)

J.K. Takimoto, Z. Xiang, J.Y. Kang, L. Wang, Esterification of an unnatural amino acid structurally deviating from canonical amino acids promotes its uptake and incorporation into proteins in mammalian cells. Chem. Bio. Chem. **11**, 2268–2272 (2010)

Z.J. Tan, S.J. Chen, Nucleic acid helix stability: effects of salt concentration, cation valence and size, and chain length. Biophys. J. **90**, 1175–1190 (2006)

Z.J. Tan, S.J. Chen, Electrostatic free energy landscapes for DNA helix bending. Biophys. J. **94**, 3137–3149 (2008)

Z.J. Tan, S.J. Chen, Salt contribution to RNA tertiary structure folding stability. Biophys. J. **101**, 176–187 (2011)

J.D. Thompson, D.G. Higgins, T.J. Gibson, CLUSTAL W: Improving the sensitivity of progressive multiple sequence alignment through sequence weighting, position-specific gap penalties and weight matrix choice. Nucleic Acids Res. **22**, 4673–4680 (1994)

L. Timmons, A. Fire, Specific interference by ingested dsRNA. Nature **395**, 854 (1998)

T. Tlusty, A colorful origin for the genetic code: information theory, statistical mechanics and the emergence of molecular codes. Phys. Life Rev. **7**, 362–376 (2010)

N. Torabi, H. Goodarzi, H. Shateri Najafabadi, The case for an error minimizing set of coding amino acids. J. Theor. Biol. **244**, 737–744 (2007)

A. Travers, The evolution of the genetic code revisited. Orig. Life Evol. Biosph. **36**(5–6), 549–555 (2006)

T.J. Treangen, E.P. Rocha, Horizontal transfer, not duplication, drives the expansion of protein families in prokaryotes. PLOS Genet. **7**, e1001284 (2011)

E.N. Trifonov, Consensus temporal order of amino acids and evolution of the triplet code. Gene **261**, 139–151 (2000)

E.N. Trifonov, The triplet code from first principles. J. Biomol. Struct. Dyn. **22**, 1–11 (2004)

P.T.S. van der Gulik, W.D. Hoff, Anticodon modifications in the tRNA set of LUCA and the fundamental regularity in the standard genetic code. PLoS ONE **11**, e0158342 (2016)

K. Vetsigian, C. Woese, N. Goldenfeld, Collective evolution and the genetic code. PNAS **103**, 10696–10701 (2006)

I.D. Vladescu, M.J. McCauley, M.E. Nuñez, I. Rouzina, M.C. Williams, Quanti- fying force-dependent and zero-force DNA intercalation by single molecule stretching. Nat. Methods **4**, 517–522 (2007)
M.V. Volkenshtein, I.B. Rumer, Systematics of codons. Biofizika **12**(1), 10–13 (1967)
L. Wang, Expanding the genetic code. Science **302**, 584–585 (2003)
L. Wang, Engineering the genetic code in cells and animals: biological considerations and impacts. Acc. Chem. Res. **50** (11), 2767–2775 (2017a)
L. Wang, Genetically encoding new bioreactivity. N Biotechnol. **38**, 16–25 (2017b)
L. Wang, A. Brock, B. Herberich, P.G. Schultz, Expanding the genetic code of *Escherichia coli*. Science **292**, 498–500 (2001)
Y. Wang, S. Gong, Z. Wang, W. Zhang, The thermodynamics and kinetics of a nucleotide base pair. J. Chem. Phys. **144**, 115101 (2016)
L. Wang, T.J. Magliery, D.R. Liu, P.G. Schultz, A new functional suppressor tRNA/aminoacyl-tRNA synthetase pair for the in vivo incorporation of unnatural amino acids into proteins. J. Am. Chem. Soc. **122**, 5010–5011 (2000)
K. Wang, H. Neumann, S.Y. Peak-Chew, J.W. Chin, Evolved orthogonal ribosomes enhance the efficiency of synthetic genetic code expansion. Nat. Biotechnol. **25**, 770–777 (2007a)
Q. Wang, A.R. Parrish, L. Wang, Expanding the genetic code for biological studies. Chem. Biol. **16**, 323–336 (2009)
L. Wang, P.G. Schultz, A general approach for the generation of orthogonal tRNAs. Chem. Biol. **8**, 883–890 (2001)
L. Wang, P.G. Schultz, Expanding the genetic code. Chem. Commun. **2002**, 1–11 (2002)
L. Wang, P.G. Schultz, Expanding the genetic code. Angew. Chem. Int. Ed. Engl. **44**, 34–66 (2005)
W. Wang, J.K. Takimoto, G.V. Louie, T.J. Baiga, J.P. Noel, K.F. Lee, P.A. Slesinger, L. Wang, Genetically encoding unnatural amino acids for cellular and neuronal studies. Nat. Neurosci. **10**, 1063–1072 (2007b)
L. Wang, J. Xie, P.G. Schultz, Expanding the genetic code. Annu. Rev. Biophys. Biomol. Struct. **35**, 225–249 (2006)
Q. Wang, L. Wang, New methods enabling efficient incorporation of unnatural amino acids in yeast. J. Am. Chem. Soc. **130**, 6066–6067 (2008)
Q. Wang, L. Wang, Genetic incorporation of unnatural amino acids into proteins in yeast. Methods Mol. Biol. **794**, 199–213 (2012)
Y.S. Wang, X. Fang, A.L. Wallace, B. Wu, W.R. Liu, A rationally designed pyrrolysyl-tRNA synthetase mutant with a broad substrate spectrum. J. Am. Chem. Soc. **134**, 2950–2953 (2012)
Z. Wang, T. Yu, J.P. Sang, X.W. Zou, C. Yan, X. Zou, Computation and simulation of the structural characteristics of the kidney urea transporter and behaviors of urea transport. J. Phys. Chem. B **119**, 5124–5131 (2015)
K. Watanabe, S. Yokobori, How the early genetic code was established: inference from the analysis of extant animal mitochondrial decoding systems, in *Chemical Biology of Nucleic Acids: Fundamentals and Clinical Applications,* Chap 2, ed by V.A. Erdmann, W.T. Markiewicz, J. Barciszewski (Springer, Berlin, Germany, 2014), pp. 25–40
J.T. Waters, X.J. Lu, R. Galindo-Murillo, J.C. Gumbart, H.D. Kim, T.E. III Cheatham, S.C. Harvey, Transitions of double-stranded DNA between the A- and B-forms. J. Phys. Chem. B. **120**, 8449–8456 (2016)
A.L. Weber, S.L. Miller, Reasons for the occurrence of the twenty coded protein amino acids. J. Mol. Evol. **17**(5), 273–284 (1981)
J. Wei, L. Czapla, M.A. Grosner, D. Swigon, W.K. Olson, DNA topology confers sequence specificity to nonspecific architectural proteins. Proc. Natl. Acad. Sci. USA **111**, 16742–16747 (2014)
M.C. Weiss, M. Preiner, J.C. Xavier, V. Zimorski, W.F. Martin, The last universal common ancestor between ancient earth chemistry and the onset of genetics. PLoS Genet. **14**, e1007518 (2018)
J.R. Wenner, M.C. Williams, I. Rouzina, V.A. Bloomfield, Salt dependence of the elasticity and overstretching transition of single DNA molecules. Biophys. J. **82**, 3160–3169 (2002)
R. Wetzel, Evolution of the aminoacyl-tRNA synthetases and the origin of the genetic code. J. Mol. Evol. **40**(5), 545–550 (1995)
C.R. Woese, Order in the genetic code. Proc. Natl. Acad. Sci. **54**(1), 71–75 (1965)
C.R. Woese, *The Genetic Code: The Molecular Basis for Genetic Expression* (Harper & Row, New York, 1967)
C.R. Woese, The fundamental nature of the genetic code: prebiotic interactions between polynucleotides and polyamino acids or their derivatives. PNAS **59**, 110–117 (1968)
C.R. Woese, R.T. Hinegardner, J. Engelberg, Universality in the Genetic Code. Science **144**, 1030–1031 (1964)
C.R. Woese, G.J. Olsen, M. Ibba, D. Soll, Aminoacyl-tRNA synthetases, the genetic code, and the evolutionary process. Microbiol. Mol. Biol. Rev. **64**, 202–236 (2000)
Y.I. Wolf, E.V. Koonin, On the origin of the translation system and the genetic code in the RNA world by means of natural selection, exaptation, and subfunctionalization. Biol. Direct. **2**, 14 (2007)
Y.I. Wolf, L. Aravind, N.V. Grishin, E.V. Koonin, Evolution of aminoacyl-tRNA synthetases—analysis of unique domain architectures and phylogenetic trees reveals a complex history of horizontal gene transfer events. Genome Res. **9**, 689–710 (1999)
J.T. Wong, A co-evolution theory of the genetic code. Proc. Natl Acad. Sci. USA **72**, 1909–1912 (1975)
J.T.-F. Wong, Coevolution theory of the genetic code at age thirty. BioEssays **27**, 416–425 (2005)
J.T-F. Wong, A co-evolution theory of the genetic code. PNAS **72**, 1909–1912 (1975)

J.T.-F. Wong, S.-K. Ng, W.-K. Mat, T. Hu, H. Xue, Coevolution theory of the genetic code at age forty: pathway to translation and synthetic life. Life **6**, 12 (2016)
S.A. Woodson, Metal ions and RNA folding: a highly charged topic with a dynamic future. Curr. Opin. Chem. Biol. **9**, 104–109 (2005)
H.L. Wu, S. Bagby, J.M. van den Elsen, Evolution of the genetic triplet code via two types of doublet codons. J. Mol. Evol. **61**(1), 54–64 (2005)
Y.Y. Wu, L. Bao, X. Zhang, Z.J. Tan, Flexibility of short DNA helices with finite-length effect: from base pairs to tens of base pairs. J. Chem. Phys. **142**, 125103 (2015a)
Y.Y. Wu, Z.L. Zhang, J.S. Zhang, X.L. Zhu, Z.J. Tan, Multivalent ion-mediated nucleic acid helix-helix interactions: RNA versus DNA. Nucleic Acids Res. **43**, 6156–6165 (2015b)
Z. Xiang, V.K. Lacey, H. Ren, J. Xu, D.J. Burban, P.A. Jennings, L. Wang, Proximity-enabled protein crosslinking through genetically encoding haloalkane unnatural amino acids. Angew. Chem. Int. Ed. Engl. **53**, 2190–2193 (2014)
Z. Xiang, H. Ren, Y.S. Hu, I. Coin, J. Wei, H. Cang, L. Wang, Adding an unnatural covalent bond to proteins through proximity-enhanced bioreactivity. Nat. Methods **10**, 885–888 (2013)
B. Xiao, R.C. Johnson, J.F. Marko, Modulation of HU- DNA interactions by salt concentration and applied force. Nucleic Acids Res. **38**, 6176–6185 (2010)
S. Xiao, H. Zhu, L. Wang, H. Liang, DNA conformational flexibility study using phosphate backbone neutralization model. Soft Matter **10**, 1045–1055 (2014)
J. Xie, P.G. Schultz, A chemical toolkit for proteins—an expanded genetic code *Nat.* Rev. Mol. Cell Biol. **7**, 775–782 (2006)
X. Xu, T. Yu, S.J. Chen, Understanding the kinetic mechanism of RNA single base pair formation. Proc. Natl. Acad. Sci. USA **113**, 116–121 (2016)
J. Yan, J.F. Marko, Localized single-stranded bubble mechanism for cyclization of short double helix DNA. Phys. Rev. Lett. **93**, 108108 (2004)
T. Yanagisawa, T. Umehara, K. Sakamoto, S. Yokoyama, Expanded genetic code technologies for incorporating modified lysine at multiple sites. ChemBioChem **15**, 2181–2187 (2014)
M. Yarus, Amino acids as RNA ligands: a direct-RNA-template theory for the code' sorigin. J. Mol. Evol. **47**, 109–117 (1998)
M. Yarus, RNA-ligand chemistry: a testable source for the genetic code. RNA **6**, 475–484 (2000)
M. Yarus, *Life from an RNA World: The Ancestor Within* (Harvard Univ. Press, Cambridge, MA, 2010)
M. Yarus, The meaning of a minuscule ribozyme. Philos. Trans. R. Soc. B **366**, 2902–2909 (2011)
M. Yarus, E.L. Christian, Genetic code origins. Nature **342**, 349–350 (1989)
M. Yarus, J.G. Caporaso, R. Knight, Origins of the genetic code: the escaped triplet theory. Annu. Rev. Biochem. **74**, 179–198 (2005)
M. Yarus, J.J. Widmann, R. Knight, RNA–amino acid binding: a stereochemical era for the genetic code. J. Mol. Evol. **69**, 406–429 (2009)
M. Ycas, *The Biological Code* (Amsterdam, North-Holland, 1969)
H.P. Yockey, *Information Theory, Evolution, and the Origin of Life* (Cambridge University Press, Cambridge, UK, 2005)
J. Yuan, P. O'Donoghue, A. Ambrogelly, S. Gundllapalli, R.L. Sherrer, S. Palioura, M. Simonović, D. Söll, Distinct genetic code expansion strategies for selenocysteine and pyrrolysine are reflected in different aminoacyl-tRNA formation systems. FEBS Lett. **584**, 342–349 (2010)
K. Zahonova, A.Y. Kostygov, T. Sevcikova, V. Yurchenko, M. Elias, Anunprecedented non-canonical nuclear genetic code with all three termination codons reassigned as sense codons. Curr. Biol. **26**, 2364–2369 (2016)
D.A. Zaia, C.T. Zaia, H. De Santana, Which amino acids should be used in prebiotic chemistry studies? Orig. Life Evol. Biosph. **38**, 469–488 (2008)
W. Zhang, S.J. Chen, RNA hairpin-folding kinetics. Proc. Natl. Acad. Sci. USA **99**, 1931–1936 (2002)
X. Zhang, H. Chen, H. Fu, P.S. Doyle, J. Yan, Two distinct overstretched DNA structures revealed by single-molecule thermodynamics measurements. Proc. Natl. Acad. Sci. USA **109**, 8103–8108 (2012)
X. Zhang, Y. Qu, H. Chen, I. Rouzina, S. Zhang, P.S. Doyle, J. Yan, Interconversion between three overstretched DNA structures. J. Am. Chem. Soc. **136**, 16073–16080 (2014)
Y. Zhang, P.V. Baranov, J.F. Atkins, V.N. Gladyshev, Pyrrolysine and selenocysteine use dissimilar decoding strategies. J. Biol. Chem. **280**, 20740–20751 (2005)
Y. Zhang, J. Zhang, W. Wang, Atomistic analysis of pseudo-knotted RNA unfolding. J. Am. Chem. Soc. **133**, 6882–6885 (2011)
W. Zhu, S. Freeland, The standard genetic code enhances adaptive evolution of proteins. J. Theor. Biol. **239**, 63–70 (2006)

Transcription for Protein Biosynthesis 12

Abstract

The experimental methods used to analyze living systems and the obtained data are quantitative. In more detail, the measurements of gene expression within living cells in different stages of growth show how much of an individual gene product, the messenger RNA (mRNA) or protein, is present at a distinct location at a specific time. The investigation of signaling pathways determines the extent of a cellular response that is cellular growth, replication of DNA or actin or microtubule polymerization, cytoskeletal restructuring based possible on the concentration of particular upstream signaling molecule, chemical activation of a protein and/or the encoded protein concentration of actin in a feedback loop. In addition, the DNA can be wrapped into compact structures, termed nucleosomes, as a function of the underlying sequence and how many times the sequence need to be transcribed into mRNA and subsequently translated into a protein. This kind of densely packed DNA is needed during the mitosis phase of the cell cycle when the cell divides into two identical daughter cells. The list of such examples is still long and shows the need of the development of quantitative models to deal with the large amount of quantitative data and, in turn, these quantitative models are required to propose experimentally testable quantitative hypotheses on biological processes and phenomena. Often a biological problem leads to a quantitative measurement of the effects of a biological parameter on a particular system, if this parameter is varied, for example, in concentration, structure, localization, interaction partners or activity, which a model for interpreting the data deserves.

12.1 Introduction to Protein Biosynthesis

The eukaryotic gene expression is regulated in a complex manner. Progress in gaining insights into the mechanisms of gene regulation has been rapid and widespread, making the integration of detailed findings into broad concepts a daunting challenge. In the following, the concepts will be included: First, a well-defined arrangement of nucleosomes and alteration states in the majority of genes, second, regulatory frameworks of sequence-specific transcription factors, third, chromatin assembly linked to the arrangement of promoters of the common transcription factors and RNA polymerase II and fourth, phosphorylation levels of RNA polymerase II connected to chromatin alteration levels during the transcription. The abundance of new evidence emerging from the resources of

C. T. Mierke, *Cellular Mechanics and Biophysics*, Biological and Medical Physics, Biomedical Engineering, https://doi.org/10.1007/978-3-030-58532-7_12

biochemistry, genomics, cell biology and genetics offers a striking insight into the mechanics of gene regulation (Venters and Pugh 2009).

The transcription in eukaryotes is governed by a multiple number of proteins, including sequence-specific DNA-binding factors and regulators of the chromatin, in order to provide the function of general transcription machinery and discover the regulatory proteins (Berger 2000; Li et al. 2007a; Orphanides and Reinberg 2002; Pugh 2000; Struhl et al. 1998). The collective function of these genes is to express a partial set of genes predetermined by a complex interaction of environmental cues that is only partially comprehensible. Traditional biochemistry and sophisticated genetic screens have resulted in the detection of crucial elements of the transcription machinery and have offered deeper insights into the mechanisms of transcription by RNA polymerase II (Pol II). The most recent genome-wide expression profiling and location profiling of factors have deepened our knowledge of the organization of the transcription engine and nucleosomes across the genome. The predominant opinion on transcriptional activation is that numerous sequence-specific regulators react to cellular cues by interfering with their respective cognitive DNA motifs. Transcriptional co-activators are being targeted to modify the local chromatin surrounding and ease the building of the pre-initiation complex (PIC), which is made up of the general transcription factors (GTFs) and Pol II.

The range of knowledge about how genes are controlled has become so broad that it is getting increasingly hard to comprehend how the different facets of transcriptional regulation match. This overview seeks to incorporate some of the most relevant phases of gene regulation, covering activators, chromatin construction and alterations, assembling PIC and transcriptional stretching, which requires restricted depth in the treatment of a specific topic. To prevent "overburdening", a prestigious term is chosen for the many complexes involved, rather than listing all family members and other operationally connected complexes. This overview is founded on a series of eukaryotic model systems, but focuses on the insights gained from the budding yeast *Saccharomyces cerevisiae*. Today, many of the fundamental mechanisms of gene regulation are still highly conserved in eukaryotes, and yeast has offered the most straightforward route to a basic comprehension of this regulatory control. The integration proposed in the context of genome-level investigation is designed to raise new questions about how the different phases are aligned in a genome-wide reaction to environmental cues. The process of transcription represents the first step in the protein biosynthesis path. It is followed by the posttranscriptional modification, the translation and terminated by the post-translational modification step (for the latter two see Chap. 14).

12.2 Introduction to Transcription

The process of transcription is subdivided into three major steps, the initiation, the elongation and the termination of the transcription (Fig. 12.1). During the initiation step of the transcription, the RNA polymerase interacts with a specific region on the DNA, which is referred to as the promotor and is located nearby the gene start. The binding of the RNA polymerase separates the DNA strands, since a single DNA strand (coding strand) is required for the transcriptional process. During the elongation step, the RNA polymerase uses the template strand, as a template for the transcription of the DNA into RNA. Thereby, the RNA polymerase assembles an RNA polymer that is built by complementary nucleotides (monomers). Hence, the RNA polymeric strand extends from the 5′ to the 3′ end. The RNA transcript of the DNA possesses instead of thymine (T) uracil (U). Finally, the step of termination follows after a termination signal has been passed. Thereby, the full transcript, the pre-mRNA, is released from the RNA polymerase. The pre-mRNA needs to be protected for degradation by capping of its 5′ end and addition of a poly A-tail (100–200 nucleotides) at its 3′ end, which

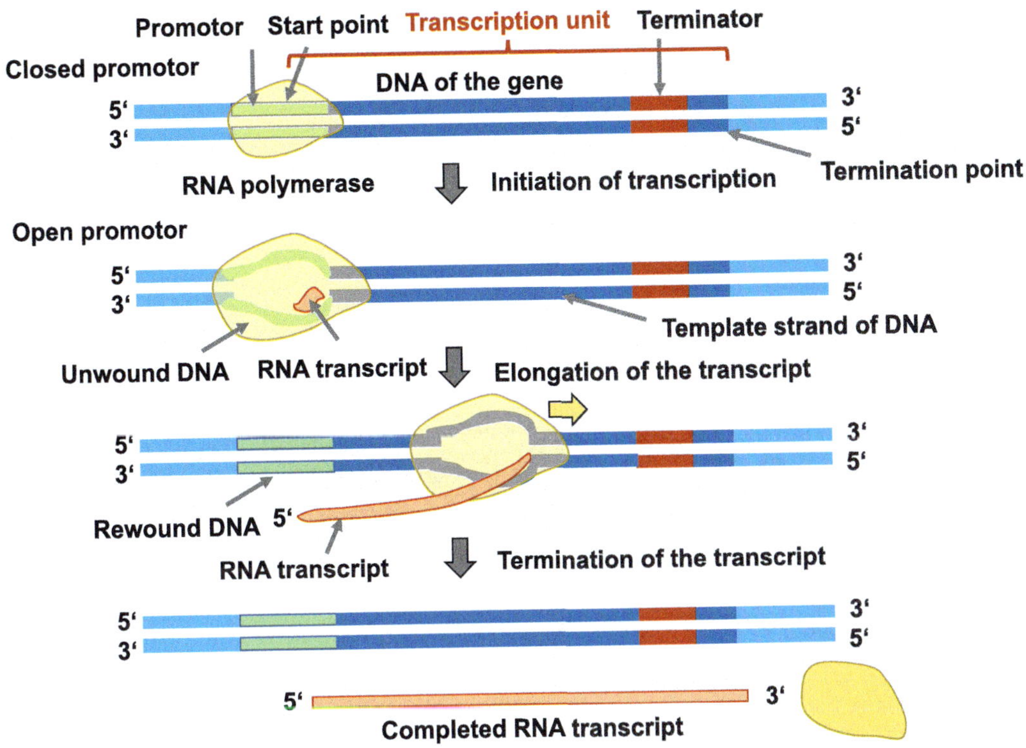

Fig. 12.1 Three major fundamental steps of the transcriptional process

represent posttranscriptional modifications. In addition, there are many genes that need to be further processed through processes termed splicing or alternative splicing (see Chap. 13).

12.3 Identification of Protein Binding Toward Genomic DNA Using ChIP-Chip and ChIP-Seq

Chromatin immunoprecipitation (ChIP) has developed into a promising tool to determine the protein interactions on the genomic DNA in vivo, and thus, it seems to be the single most informative procedure to reveal the interaction and the assembly of proteins on the DNA in vivo (Fig. 12.2) An essential feature of the ChIP test is that it retains physiologically relevant interactions in the cell by formaldehyde cross-linking. Formaldehyde represents an optimal cross-linking agent (Orlando et al. 1997; Solomon and Varshavsky 1985; Toth and Biggin 2000) for the following reasons: First, it penetrates the cell rapidly and captures native interactions before the cell shows a physiological response, second, its length as a single carbon cross-linker generates efficacious protein-DNA cross-linkages in vivo (protein-protein and protein-RNA cross-linkages are also produced) and third, its easily reversible cross-linking is critical for successive DNA detection techniques.

Since its introduction (Ren et al. 2000), ChIP, coupled with microarray detection (ChIP chip), has become a powerful tool to gain effective and comprehensive insight into the relationship between transcriptional processing and chromatin (Kim and Ren 2006; Pugh and Gilmour 2001). It is able to measure the degree of coverage of essentially any cross-linkable and immunologically exploitable

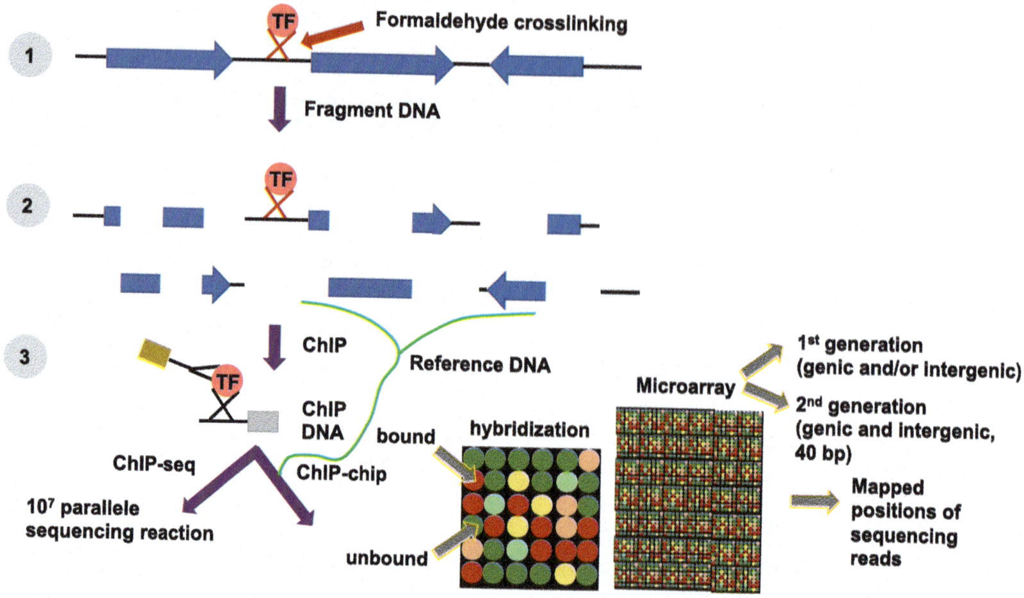

Fig. 12.2 Chips technology assays for detection of protein binding across the entire genome. Two DNA detection approaches utilized ChIP. (1) In both approaches, formaldehyde is employed to cross-link transcription factors (TFs) to the genome in vivo. (2) DNA is fragmented. (3) The protein is immune-purified to eliminate DNA not connected to TFs. In the ChIP-chip, the DNA connected to the protein is stained with a red dye and visualized by hybridization to an array of immobilized DNA samples (microarray chip), on which the sequences match distinct genomic parts. Frequently, a reference DNA sample is co-hybridized to control the probe-to-probe variations. In ChIP-seq, each fragment is directly sequenced

proteins throughout an entire genome. Early ChIP chip microarrays suffered from two major constraints. Firstly, the production of such microarrays necessitated a sequenced genome. Secondly, the spatial resolution of binding across a genome is restricted due to probe length and distance. Currently, this has been largely mitigated in the heavily glazed (probes every 5–40 bp) microarrays of the second generation. The latest breakthroughs in low-cost full-genome firearm sequencing have also overcome the first constraint.

The recognition of genomic regions bound to a protein has lately been changed by coupling the ChIP assay with bulk parallel DNA sequencing, referred to as ChIP-seq, to a completely higher resolution level (Margulies et al. 2005; Schuster 2008). The tracking of nucleosome locations throughout genomes has been one of the first implementations of ChIP-seq. In recent years, ChIP-seq has generated nucleosome mappings of the entire genome for yeast (Albert et al. 2007; Mavrich et al. 2008a; Shivaswamy et al. 2008), fly (Mavrich et al. 2008b), worm (Johnson et al. 2006; Valouev et al. 2008) and human (Barski et al. 2007; Boyle et al. 2008; Schones et al. 2008). Some have employed ChIP-seq to mark the positions of the transcription factors (Jothi et al. 2008; Nielsen et al. 2008; Robertson et al. 2007).

12.4 Effect of Repressive Chromatin Structures on Transcription Activity

As the genes located within the eukaryotic genome are condensed inside the nucleus in the format of chromatin, the transcription machinery has to break down a huge structural barrier to gain entry to the intrinsic *cis*-regulatory elements and the coding areas. Previous research on the nuclear wrapping of genomic DNA revealed that chromosomal DNA consists of a pearl-on-string arrangement (Hewish and Burgoyne 1973; Olins and Olins 1974), which is assumed to be the prevailing mode of transcriptionally capable chromatin. Every pearl represents a nucleosome. The nucleosome nodule comprises 147 base pairs of DNA coiled 1.7 times around a histone octamer consisting of two copies of histones H2A, H2B, H3 and H4 (Luger et al. 1997). In opposition to the active state, genes can be suppressed by compressing the chromatin to a thirty-nanometer fiber (Fig. 12.3).

Apart from in *Saccharomyces*, this closed suppressive chromatin architecture is frequently combined with the methylation of histone H3 on lysine 9 (H3K9) by SUV39H (SUppressor of Variegation 3–9 homolog) and the simultaneous binding of HP1 (heterochromatin protein 1) to this methyl label. The repressive state could also or alternatively be combined with methylation to H3K27 by the Polycomb Repressive Complex 2 (PRC2) and simultaneous binding of PRC1 that ubiquitates H2AK119 (Kouzarides 2007; Levine et al. 2004; Motamedi et al. 2008; Trojer et al. 2009). The dimethylation of H3R2, by the protein arginine methyltransferase 6 (PRMT6) (Iberg et al. 2008), yields an auxiliary repressive label that is designed to inhibit the generation of the activating trimethyl H3K4 label (Kirmizis et al. 2007). It still has to be ascertained whether and how these repressive signs and their related elements encourage or sustain the enclosed state or are merely pointers to the enclosed state. Certainly, small noncoding RNAs are capable of performing an essential function in the development of repressive chromatin surroundings (Goodrich and Kugel 2008; Grewal and Elgin 2007). The H3K9 and K27 methyl labels may operate collaboratively or separately during the repression and may partially achieve this by attracting repressive DNA methylases and histone deacetylases that eliminate the activating histone acetylation labels.

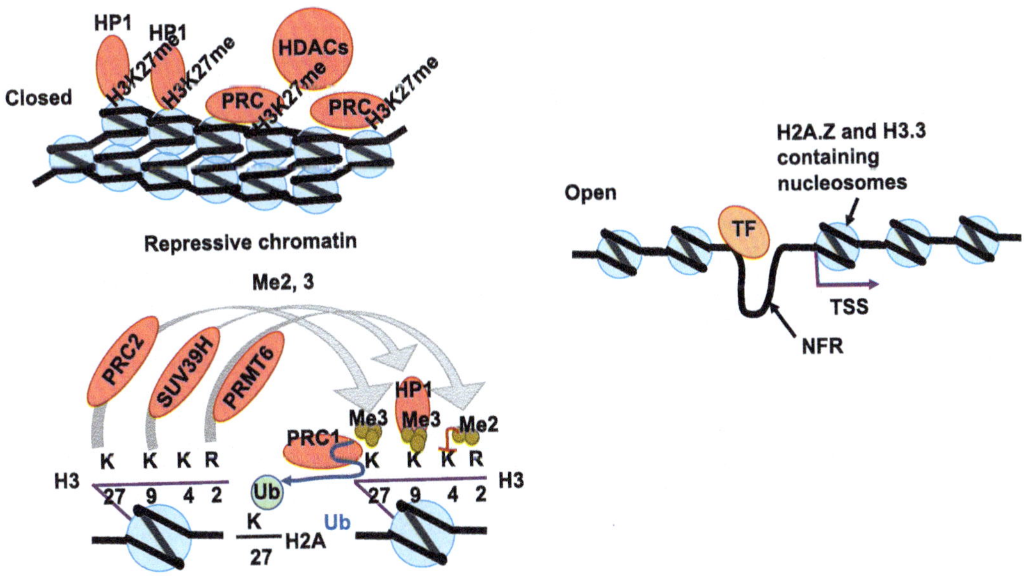

Fig. 12.3 Model for closed and open chromatin structures. **a** Repressive chromatin is presented in a closed state (30 nm chromatin fiber). There are no nucleosome free regions (NFRs). The resident canonical histones are methylated at several sites including H3K9 and/or H3K27. These sites are occupied by HP1 and PRC complexes. The compaction is driven by histone H1. The repressive regions recruit histone deacetylases (HDACs) eliminate activating acetyl labels on histones. **b** The promotors of most genes reside within an open chromatin structure in order to be activated. The open chromatin represents a bead on a string configuration, where transcriptions factors (TFs) can associate and maintain the NFR and promote the replacement of H3, H2A by H3.3 and H2A.Z. TSS = transcriptional start site

What is striking is that many genes in human embryonic stem cells that carry the methyl K27 repression label also carry a methyl label on H3K4, which is linked to active genes (Bernstein et al. 2006). These are known as bivalent genes and can be in a suppressed condition, but can be amplified for fast or properly timed activation, which is the essential prerequisite for the well-coordinated evolution of multi-cellular organisms. It still needs to be clarified whether the active H3K4 methyl label is co-transcriptionally anchored on bivalent genes or is positioned by any other mechanism, and if any constituents that identify the bivalent labels actually coexist on the identical gene. It is intriguing that PRC1, which is capable of reading the repressive H3K27 methyl label, replicates with TATA-binding protein Associated Factors (TAFs) (Saurin et al. 2001), at least one of which (TAF3) is capable of reading the active H3K4 methyl label (Vermeulen et al. 2007). TAFs are components of the common transcription apparatus. For some genes, these and other investigations outlined below indicate that the transcription mechanism may exist in suppressed genes. In general, however, the transcription apparatus is generally employed solely when the gene needs to be transcribed.

12.5 Commonly Genes Possess a Canonical "Open" Nucleosome Organization

Chromatin comprises a repeating arrangement of nucleosomes distributed randomly throughout the genome approximately every 160–200 base pairs. The influence of nucleosomes on the regulation of genes was commonly misjudged before it was established in key studies almost two decades ago that the presence of nucleosomes blocks the triggering of transcription in vitro (Lorch et al. 1987), pointing to the fact that nucleosomes are physical obstacles for transcription. Moreover, the depletion of histones within the budding yeast *S. cerevisiae* causes general increase in transcription (Han and Grunstein 1988; Wyrick et al. 1999), thereby offering convincing proof in vivo that nucleosomes are able to suppress transcription (even though numerous other genes were equally activated).

The location of nucleosomes on multiple model genes, such as PHO5, SUC2, GAL1, HMRa and RNR3, has been profiled in *Saccharomyces* on the basis that the nucleosomes shield the underlying DNA from the digestion by exogenously introduced micrococci nuclease (MNase). However, these early gene-specific cards offered an insight into how the position of nucleosomes in a gene might influence the access of promoters and subsequently the efficiency of transcription (Almer and Horz 1986; Li and Reese 2001; Lohr 1997; Perez-Ortin et al. 1987; Ravindra et al. 1999). The investigation of such pilot genes turned out to be essential to guide our comprehension of the interaction of nucleosomes and the transcription mechanism. For instance, the research on the PHO5 gene revealed that the nucleosomes control the transcriptional process by blocking the TATA and upstream activating sequence (UAS) promoter components and that the disturbance of the nucleosomes is decisive for activation (Lohr 1997; Martinez-Campa et al. 2004).

The latest progress in genome-wide mapping strategies has produced a far more explicit vision of the genomic nucleosome environment (Wal and Pugh 2012). The enhanced tiling frequency of microarray sensors allows the investigation of the genome-wide structural composition of chromatin with individual nucleosome resolution to be made routine. It has been found by means of high-density tiling over a part of the yeast genome that a nucleosome free region (NFR) constitutes a characteristic feature common to promoters (Yuan et al. 2005). The nucleosomes "−1" and "+1" are located at canonical sites upstream and immediately downstream of the NFR, respectively. The nucleosomes are therefore not stochastically scattered throughout the chromosomal DNA, but rather are placed by their design at certain spacings from the transcriptional start site to adjust transcription. Whether specific individual nucleosomes are involved in the control of gene expression is unclear and appears probable to be an active field of research in the near future. As "doorkeepers" of NFR in promoters, nucleosomes −1 and +1 are in a good starting position to hold considerable regulatory capacity.

It is noteworthy that a predominant nucleosome arrangement can be seen in the alignment of the nucleosome locations in the entire yeast, fly and human genome with the transcriptional start site. The availability of a NFR promoter and the standardized nucleosome placement in relation to the transcriptional start site, from gene to gene, are two characteristics which have been conserved through evolution. While these striking characteristics of yeast are noticeably close to those of humans, apparently subtile species-specific variations are obvious, nevertheless providing key mechanistic effects on the initiation of transcription. In yeast, for instance, the transcription start side is trapped in the upstream boundary of the +1 nucleosome, implying that this nucleosome possibly governs entry to the transcription start side (Albert et al. 2007). However, the +1 nucleosome in metazoans is displaced further downstream of the transcription start side in comparison with budding yeast, so that the starting point of transcription remains exposed (Mavrich et al. 2008b; Schones et al. 2008). This downstream +1 nucleosome could be more appropriate to adjust transcriptional elongation. In fact, the Pol II transcription halts at the first nucleosome. The arrangement −1, NFR, +1 is the level at which sequence-specific regulators scan the genome in order to guide transcription schemes.

What generates an NFR is incompletely understood, even though strong indications suggest the presence of nucleosome excluding poly-dA:dT tracts in the area (Anderson and Widom 2001; Iyer and Struhl 1995; Mavrich et al. 2008a; Yuan et al. 2005), and the occurrence of sequence-specific factors including RNA polymerase I enhancer binding protein 1 (Reb1) (Raisner et al. 2005). Different sequence-specific regulators, such as remodel the structure of chromatin 3 (Rsc3), are able to interfere with the nucleosome concentration in NFR in certain promoters (Badis et al. 2008). In extension, it seems probable that remodeling complexes of chromatin stretch and shrink the borders of the NFR by exploiting the energy of hydrolysis of ATP to achieve a repositioning of nucleosomes (Whitehouse et al. 2007).

12.6 Sequence-Specific Factors Regulate Transcription of Specific Gene Sets

12.6.1 *Cis*-Regulatory Elements

Transcriptional schemes are driven by *trans*-acting sequence-specific determinants that govern transcription through attachment to *cis*-regulating elements (CREs). According to their influence on transcription, these positions are collectively referred to as upstream activator/repressor sequences (UAS/URS) in yeast or as enhancing sequences in metazoans. The quantity of promoters aimed at by a certain sequence-specific regulator extends from a handful to several hundred or many thousands in metazoans. For instance, the galactose regulator Gal4 is able to attach to only ten promoters (Ren et al. 2000), while repressor activator protein 1 (Rap1) connects to over 300 promoters in the *Saccharomyces* genome (Buck and Lieb 2006). The level of activity and the subcellular positioning of these factors are governed by internal and external surrounding signals. Phosphorylation or directed proteolysis is frequently employed as a molecular shifter from active to inactive phases.

While UAS/URS in *Saccharomyces* can usually be observed several hundred base pairs upstream of the translation start site (Harbison et al. 2004), there are other components such as the TATA box present in the central promoter area and are usually 30–60 bp distant of the transcription start site. The phylogenetic analysis of six *Saccharomyces* species showed that about 20% of the 5700 yeast genes comprise a TATA box element (Basehoar et al. 2004). In metazoans, there are supplementary key promoter elements that interoperate with several components of the fundamental transcriptional system, such as the initiator (Inr), the downstream promoter element (DPE), the motif ten element

(MTE) and the TFIIB recognition element (BRE) (Smale and Kadonaga 2003; Thomas and Chiang 2006). A few of them function instead of a TATA box, whereas the majority of promoters appear to be missing a discernible key element of the promoter.

12.6.2 Gene Regulatory Networks

In which way are sequence-specific regulators arranged within the genome to facilitate the overall management of gene expression pathways? The structure of the regulatory networks underpinning the expression schemes is most clearly defined in yeast, as the genomic-binding sites for a large number of the ~120 sequence-specific regulatory proteins are revealed (Lee et al. 2002). In the analysis of the regulatory-gene interaction framework, six fundamental network motifs emerged, each offering specific regulatory benefits. For instance, the "single-input" motif guarantees the harmonized expression of the leucine biosynthesis genes through Leu3, while the "multiple-input" motif is capable of incorporating several signaling routes in order to coordinate the expression of a certain set of genes under various circumstances. The prototypical metazoan interferon-beta enhancesome can be considered as a multiple-input motif (Panne 2008). But it is not clear whether all of the manifold input motifs necessitate the simultaneous binding of sequence-specific controllers for the expression of the desired gene, as in the specific case of the interferon-beta gene. The regulation of the cell cycle through sequence-specific modulators explains the motif of the "regulator chain", in which transcriptional processes are arranged in a chronological sequence in line with the temporal character of the cell cycle.

The basic governing chain motif has been augmented to build a complex global, pyramid-like, multi-level hierarchical network (Yu and Gerstein 2006). This multi-level hierarchical network defines a "chain of command organization" with a small number of principal regulators at the forefront that have a tendency to exert the maximum impact on the overall levels of expression. The hierarchical regulatory framework provides a decision scheme that enables the superposition of various internal and external stimulators to accurately regulate the expression of specific groups of genes. For instance, the oxygen and heme contents in yeast trigger the activation of the expression of Mot3, a major regulator of aerobic growth. Mot3 stimulates the activation of general control non-derepressible 4 (Gcn4), that in turn stimulates the activation of prolin utilization 3 (Put3) and utilization of gaba 3 (Uga3), thereby activating numerous genes participating in proline and nitrogen metabolism.

12.6.3 Sequence-Specific Regulators as Orchestrators

Whereas the genome-wide sites of numerous sequence-specific controllers in *Saccharomyces* have been identified, it is still largely unknown as to how these players participate specifically in the process of transcription. Whether a sequence-specific modulator functions as activator or repressor potentially varies according to its genomic background and the type of co-regulators it attracts. Sequence-specific modulators direct multiple facets of transcription by the direct acquisition of: firstly the chromatin remodeling complexes, secondly the general transcription factors, thirdly the chromatin-altering complexes and fourthly Pol II through the mediator complex (Brown et al. 2001; Cosma et al. 1999; Garbett et al. 2007; Goldmark et al. 2000; Green 2005; Larschan and Winston 2001; Neely et al. 2002; Nourani et al. 2004; Park et al. 2000; Yudkovsky et al. 1999). In the following these steps are discussed in large detail.

12.7 ATP-Driven Remodeling of the DNA on Nucleosomes

Earlier research on pilot genes revealed that activation of transcription entails the migration of nucleosomes (Almer and Horz 1986; Li and Reese 2001; Lohr 1997; Perez-Ortin et al. 1987; Ravindra et al. 1999). The existence of chromatin reformulation complexes has been completely unclear till the early 1990s (Hirschhorn et al. 1992), although genes for key subunits such as SWI2/SNF2 have been identified by genetic screenings (Neigeborn and Carlson 1984; Stern et al. 1984). The remodeling of chromatin complexes is provided by four families due to the conservation of the sequence (Bao and Shen 2007b), which includes SWI/SNF, INO80/SWR1, ISWI and CHD.

As chromatin modifiers usually deprive the intrinsic capacity to aim at specific genes, it is probable that sequence-specific regulators are able to target these complexes directly into promoter regions. How chromatin reshaping assemblies are tailored to specific nucleosomes and their diverse functions will continue to be an active field of examination.

The remodeling of chromatin complexes employs the power of the hydrolysis of ATP in order to change the shape, location or the composition of the nucleosomes (Flaus and Owen-Hughes 2004; Saha et al. 2006; Tsukiyama 2002). The mechanism of chromatin break-down, on the basis of various in vitro biochemical investigations and single-molecule optical trap measurements, is presently assumed to be based on the fact that the DNA wrapped around the nucleosome creates a loop that is transposed by the mechanical force generated by these ATPase motors, leading to the histone octamer assembly either gliding alongside the DNA or being completely removed from the DNA (Cairns 2007; Saha et al. 2006). The mechanistic features will be completely different in the various families. The genomic background of this kind of rearrangement activity affects the availability of the DNA at the promoters and in certain circumstances represses cryptic sites of initiation in the gene's body (Whitehouse et al. 2007).

Nucleosomes encircling the regions of promoters are prone to high levels of dynamics (Dion et al. 2007), implying that chromatin reformulation complexes can be constructively present and involved in the activation of multiple promoters, although the promoter is essentially at a standstill. Since the locations of the 1- and +1-nucleosomes can greatly affect the gene-internal nucleosome configuration (Mavrich et al. 2008a), and the −1/+1-nucleosomes govern the availability of key promoter components, it is probable that these nucleosomes will be the main targets of chromatin building complexes (Venters and Pugh 2009).

It is imaginable that the rearrangement of complexes that rearrange nucleosomes might even result in the rearrangement of neighboring nucleosomes, for instance by means of steric collisions. It is not clear, however, how a remodeling complex affects each nucleosome in an array separately, or whether the interaction of a single nucleosome in an array is strong enough to affect an entire array. The responses to this issue are the crucial factor in determining whether a remodeling complex that rearranges nucleosomes on an entire gene requires concentration on a single linchpin nucleosome or has to span the complete domain.

12.7.1 SWI/SNF Family and Its Relationship to Histone Acetylation

Conventional genetic screening for matting and non-fermenting sucrose phenotypes led to the discovery of the SWI2/SNF2 gene, which codes for the catalytic subunit in the SWI/SNF complex (Neigeborn and Carlson 1984; Stern et al. 1984). Genetic and molecular proofs with Swi2/Snf2 and Snf5 mutants subsequently revealed that the SWI/SNF complex modifies the configuration of chromatin by gliding and/or shedding nucleosomes regardless of transcription (Hirschhorn et al. 1992; Lorch et al. 2006).

The SWI/SNF family of chromatin restructuring factors, which comprises the remodels structure of chromatin (RSC) complex, is widely regarded as a positive modulator of transcription, even though some genes are negatively modified (Angus-Hill et al. 2001; Sudarsanam et al. 2000). RSC encounters site profiling by ChIP chip in the promoters of hundreds of genes (Damelin et al. 2002; Ng et al. 2002), indicating a function in the initiation of transcription and the organization of nucleosomes. In accordance with this idea, RSC mutants interfere with the translational adjustment of the promoter nucleosomes (Parnell et al. 2008). SWI/SNF in specific could have extra functionalities in transcriptional expansion (Schwabish and Struhl 2007).

Promoter nucleosomes have a propensity to be hyper-acetylated, and this could assist in obtaining bromine-domain chromatin reshaping complexes such as SWI/SNF and RSC (Hassan et al. 2001, 2002; Ruthenburg et al. 2007). Bromine domains attach themselves to acetylated lysines. Since acetylation can also reduce electrostatic histone-lysine-DNA interactions and interfere with higher order densification, the previous acetylation of nucleosomes can enhance the conversion and degradation of nucleosomes in the course of gene activation. The acetylation process may also play a role in the reassembly of nucleosomes under the guidance of remodeling complexes. For instance, the nucleosome acetyltransferase of H4 (NuA4)-directed acetylation of H2A.Z plays an essential role in its translocation into the nucleosomes of promoters (Keogh et al. 2006). The acetylation of H3K56 through regulator of Ty1 transposition 109 (Rtt109) (Schneider et al. 2006), and additional histone acetylation incidents through histone acetyl transferase 1 (Hat1)/HAT-B (Parthun et al. 1996), can be relevant for the deposition of histone. Another important acetyltransferase complex linked to chromatin reformulation is Spt-Ada-Gcn5 acetyltransferase (SAGA) and p300 (Sterner and Berger 2000). In addition, SAGA fulfills several functions in the cycle of transcription that lie beyond the acetylation of nucleosomes.

In vitro transcription experiments indicate that nucleosome acetylation by NuA4 enhances the activity of RSC and improves the transit of Pol II across a nucleosome (Carey et al. 2006). The RSC as well as the SWI/SNF addresses several promoters through their engagement with activators (Cosma et al. 1999; Yudkovsky et al. 1999). The RSC fulfills additional role at various Pol I and Pol III promoter types (Damelin et al. 2002; Ng et al. 2002) that seems to be an explanation why the RSC seems to be crucial for the survival and viability in yeast cells, whereas the SWI/SNF element is not essential (Cairns et al. 1996).

12.7.2 INO80/SWR1 Family

The inositol requiring 80/Sick with Rat8 ts 1 (INO80/SWR1) family of rearranging proteins appears to be unique because it includes a shared ATPase domain (Bao and Shen 2007a). The identification of the INO80 gene is being performed in a screening for genes necessary for the activation of the inositol synthetase gene (INO1), which produces a molecule needed for several secondary messenger signal transduction routes (Ebbert et al. 1999). The INO80 complex performs a more comprehensive function in the regulation of the genome compared to numerous other conversion complexes, as it is involved in the activation of transcription, the repair of DNA and the release of locked replication bifurcations (Shen et al. 2000; Shimada et al. 2008). It is interesting to see the way INO80 addresses transcription, repair and replication targets, as INO80, in contrast to many other remodeling complexes in yeast, is devoid of established histone recognition modules.

The SWR-C/SWR1 complex acts as a chromatin restructuring factor, which alters the composition in contrast to the position of the nucleosomes. The SWR1 complex utilizes hydrolysis of ATP to substitute H2A with H2A.Z within the nucleosomes of promoters (Guillemette et al. 2005; Li et al. 2005; Raisner et al. 2005; Zhang et al. 2005). H2A.Z is designed to stimulate the transcription through destabilization of the nucleosomes (Krogan et al. 2004; Zhang et al. 2005). Actually, H2A.Z is more often connected with an open than a closed chromatin condition, even though the mere existence of H2A.Z is not entirely sufficient to generate a functionally active transcriptional state. In mammalian cells, H2A.Z is needed for lineage binding through embryonic stem cells (Creyghton et al. 2008), which indicates that changing the chemical composition of chromatin involving histone variants has an important developmental effect, possibly by contributing to placing specific genes in an activated stage.

12.7.3 Negative Regulation of Transcription by the ISWI Family

Contrary to the SWI/SNF and INO80 families, the imitation SWITch (ISWI) family remodeling proteins have a propensity to regulate transcription in a negative way. For instance, in genome-wide expression profiles and DNase I sensitivity analyses in *Saccharomyces*, the ISW2 complex in combination with histone deacetylase reduced potassium dependency 3 (Rpd3) suppressed meiotic genes by establishing a repressive arrangement of nucleosomes (Fazzio et al. 2001). ISW2 is targeted toward the repressive loci by Ume6, which represents a key sequence-specific modulator for early meiotic genes (Goldmark et al. 2000). ISW2 could also work together with another repressive complex such as TUP1-SSN6 (deoxy thymidine monophosphate UPtake 1 and suppressor of sucrose nonfermentor 6) to sustain a repressive condition (Zhang and Reese 2004). As ISW2 can be essential for the preservation of a specific closed deacetylated state (together with Rpd3), it is reasonable that those compounds miss bromine domains, since they tend to antagonize the generally open acetylated state which occurs across the entire euchromatic genome. Consequently, ISW2 can rely on sequence-specific regulators rather than on the detection of histone alterations during enrollment.

12.7.4 CHD Family

The intracellular function of the CHD family (chromatin organization modifier, helicase and DNA-binding domains) of chromatin remodelers remains the poorest known of the four families of remodeling compounds. The expression profiling in a Chd1 yeast mutant revealed that it involves a small number of genes (Tran et al. 2000), indicating that Chd1 possibly operates in parallel paths with other chromatin remodeling agents or aims at a small number of genes. The biochemical purifying of the SAGA complex of *S. cerevisiae* resulted in the recognition of Chd1 to be one of its constituents (Pray-Grant et al. 2005). The extent to which Chd1 cooperates with the SAGA in certain genes and specific tasks is not yet fully known. A key characteristic of the CHD family is the existence of a chromodomain that attaches to methylated lysines. In fact, Chd1 can interfere with methylated H3K4 in vitro (Biswas et al. 2007; Flanagan et al. 2005; Pray-Grant et al. 2005). But there is a difference of opinion between the distinction in binding specificity for yeast and its human equivalents. The in vivo relevance of the Chd1 linkage to H3K4 is subject to further research, and its possible relationship with acetylation and deubiquitylation efforts in connection with SAGA is still widely disregarded. As histone acetylation, chromatin constructional remodeling, the recognition of H3K4 trimethyl labels and deubiquitylation are all linked to transcription, there could be a certain logic for the fact that these manifold activities are connected to a single SAGA complex.

12.8 Assembly of Transcription Factors for the Transcription Pre-initiation Complex

12.8.1 Role of Mediators

Whereas the remodeling of chromatin complexes can dynamically maintain the chromatin within the promoter area, the numerous aspects of the restructuring event might not be finalized unless the moderator complex and the GTFs join together to form a pre-initiation complex. Mediator represents a large complex of proteins participating in the many aspects of transcription (Biddick and Young 2005; Kornberg 2005). First identified as a biochemical unit that induced transcriptional activation in vitro (Kim et al. 1994), the Mediator can be acquired through direct interference with activators early in the course of PIC assembly (Natarajan et al. 1999). The recruitment of p300 histone acetyltransferase into promoters is an excellent illustration of how a Mediator could adjust the accumulation of PIC. This enrollment inhibits the PIC assembly unless p300 acetylates its targets (histones and itself), which in turn leads p300 to detach itself from the promoter (Black et al. 2006). Basically, Mediator and p300 establish a checkpoint to assure that the PIC assembly will not proceed unless specific acetylation incidents are finalized.

12.8.2 TBP is a Nucleator for the Assembly of PIC Within the NFR

About thirty years ago, the biochemical fractionation of raw cell extracts resulted in the discovery of general transcription factors that precisely induced the transcription in vitro with a minimal core promoter (Thomas and Chiang 2006). These GTFs have first been derived from mammalian cells (Matsui et al. 1980), and subsequently, the corresponding factors have been found in yeast (Sayre et al. 1992). The GTFs comprise TFIIA, -B, -D, -E, -F and -H. In opposition to sequence-specific controllers that aim at a specific gene set, GTFs, which are widely employed by the cell for multiple genes and usually possess a very low built-in gene specificity. These fundamental transcription system components operate on the majority of genes to facilitate the priming and release of RNA polymerase II at the transcriptional start side.

The transcription factor II D (TFIID) represents a complex multi-compound complex that compromises the TATA-binding protein (TBP) and TBP-associated factors (TAFs). The paradox is that TFIID is widely targeted at TATA-less "housekeeping" promoters, which constitute the overwhelming majority of all genes (Basehoar et al. 2004). TBP is also available free of the TFIID complex and is supplied to TATA containing promoters through the SAGA complex (Sermwittayawong and Tan 2006). TBP is eliminated from this type of promoter by the simultaneous direct action of negative cofactor 2 (NC2) and the TBP-dependent ATPase modifier of transcription 1 (Mot1) (Lee and Young 1998; Pugh 2000). TFIIA and TFIIB interfere with TBP and stimulate the assembly of PIC through the stabilization of TBP/DNA interactions and the mitigation of the impacts of NC2 and Mot1. At least in certain genes, the GTFs join together to form fractional PICs, which are relatively deficient of TFIIH and Pol II. Fractional PICs may constitute regulatory intermediates, such as in the combination of reactions to heat shock and oxidative stress, whereby the former generates the latter (Zanton and Pugh 2006). It is interesting to note that fractional PICs in the NFR can assemble themselves with no obvious loss of the encircling nucleosomes. Only when complete PICs, which include Pol II, are constructed, is the −1 nucleosome eliminated (Venters and Pugh 2009).

Though TBP can substitute TFIID in biochemically reconstructed transcription assays, it is generally TAFs that are needed for sequence-specific activators to stimulate the peak transcription (Pugh and Tjian 1990; Reese et al. 1994), indicating that TAFs may interfere with a number of activators. One example can be Rap1 that represents an activator, which directly interferes with TFIID in vitro, and this interplay seems to be crucial for the facilitation of transcription of the highly transcribed ribosomal protein genes (Garbett et al. 2007; Mencia et al. 2002).

The physiological significance of TBP is emphasized by the natural capacity of the cell to adjust its activity. Mot1 and NC2 jointly govern the genomic spread of TBP, although by distinct mechanisms. Mot1 may link together the hydrolysis of ATP to a targeted DNA translocation, which can then be exploited to break TBP/DNA linkages (Auble et al. 1994). In this manner, Mot1 will react adversely to the transcription. Nevertheless, Mot1 also affects certain genes in a positive way, potentially by eliminating incorrectly compounded TBP, which could attach to TATA in reverse direction or in unsuitable locations because of the inherently low intrinsic specificity of TBP (Muldrow et al. 1999; Sprouse et al. 2008b). In contrary, NC2 attaches TBP onto the DNA and inhibits subsequent PIC aggregation by disrupting the TFIIB binding sterically (Inostroza et al. 1992). Together, TBP attachment to DNA is inherently robust, but its interaction in vivo can be extremely dynamical (Sprouse et al. 2008a). The stability seems to be powered by TFIIA and TFIIB (possibly in conjunction with additional GTFs and regulatory factors), and the destabilization may first be driven by NC2, which acts against TFIIB (and TFIIA), thus enabling Mot1 to initiate decomposition. However, it remains to be seen whether the absence of NC2 or Mot1, in the existence of adequate amounts of TBP, leads to TBP enriching at the NFRs within the genome. The question is whether NFRs are in fact protein-free and therefore able to attach proteins, or whether they are themselves inhabited by different unidentified factors and thus are unavailable.

Atomic resolution structures of complexes comprising either TFIIA or TFIIB bound to TBP-TATA DNA show that TFIIA binds upstream of TBP (away from transcription start side) (Nikolov et al. 1995; Tan et al. 1996), while TFIIB is mainly located downstream of TBP, but also spans TBP, thereby trapping DNA on the either side of TBP. TFIIB and TFIIF are necessary to connect Pol II to the PIC, and this is facilitated through the Pol II-TFIIB crystal and Pol II-TFIIF cryo-EM structures (Bushnell et al. 2004; Chung et al. 2003).

12.8.3 Regulation of the Initiation of the Transcription by TFIIE and TFIIH

The TFIIE and TFIIH complexes cooperate to facilitate the activity of Pol II and ease the clearance of promoters. The step-wise creation for the PIC compound in vitro reveals that TFIIE interacts after Pol II but before the TFIIH (Buratowski et al. 1989); however, other investigations point out that TFIIE can interfere with a fractional PIC, when Pol II is absent (Yokomori et al. 1998). Latest two-hybrid and ChIP investigations in a Mediator-mutant yeast strain indicate that the Mediator eases the incorporation of TFIIH into PIC compound through direct interferences (Esnault et al. 2008). TFIIE is also involved in the enrollment of TFIIH and controls the helicase and kinase activities of TFIIH. The ATP-dependent helicase activity plays an essential part in DNA strand splitting (promoter melting), resulting in the formation of an opened promoter complex with pole II (Wang et al. 1992). The kinase 28 (Kin28) subunit of TFIIH conveys the switch from transcriptional initiation to extension by phosphorylation of serines at the 5th position of the highly repeated YSPTSPS motif within the carboxy-terminal domain (CTP) of the broadest subunit of Pol II (Valay et al. 1995). Jointly, the GTFs act as a cohesion of elements to identify core promoters, compile the initial stage for transcription, enroll Pol II and simplify the transition from initiation of Pol II to expansion.

In *Saccharomyces*, it has been proposed that the transition from transcriptional initialization to elongation is fast, with only limited post-recruitment regulatory activity except in isolated cases (Martens et al. 2001; Radonjic et al. 2005; Sekinger and Gross 2001; Wade and Struhl 2008). Recent findings have challenged this idea, by contrast. Pol II has been observed to be rather abundant in promoter regions in comparison with transcribed regions in the overall *Saccharomyces* genome (Venters and Pugh 2009), implying that the transformation of PIC into a transcription-extension complex might be at minimum partly rate-limiting and therefore specifically adjustable in a broad range of yeast genes. This is in line with the adjustment mechanisms after the recruitment phase defined in the yeast (Zhang et al. 2008). In opposite, Pol II can be seen in all highly transcribed genes more or less consistently, implying that the initiation is usually not rate-limited in these genes.

12.9 Effect of Pausing of the Pol II Directly After Start of the Transcription

The transcription process includes the enrollment of Pol II into promoters, followed by transcription initiation, elongation and termination (Wade and Struhl 2008). Each of these stages can be inherently rate-limiting. In the past, it was assumed that non-expressed genes are speed-limited by increments resulting in Pol II enrollment (Ptashne and Gann 1997). This idea, however, has been questioned recently by the discovery of Pol II in the majority of genes, from yeast to humans (Guenther et al. 2007; Steinmetz et al. 2006), which, interpreted at face value, points to the fact that the expression of most or all genes is restricted in rates after the acquisition of Pol II. (Core and Lis 2008; Margaritis and Holstege 2008; Price 2008; Struhl 2007; Wade and Struhl 2008). This last view must not be entirely accurate, since poorly expressed genes usually contain less attached Pol II compared to highly expressed genes. Therefore, the identification of Pol II on genes is not synonymous with full coverage, and therefore, it is probable that both steps before the assembly of PIC and after Pol II recruitment will govern the cycle of transcription.

According to recent investigations, Pol II stops at the 5′-end of ~10% of all *Drosophila* genes (Lee et al. 2008; Muse et al. 2007; Zeitlinger et al. 2007), indicating that stopping Pol II at an early expansion step restricts the expression of multiple genes. This rest affects the complexes negative strain factor (NELF) and DRB sensitivity inducing factor (DSIF). The phosphorylation of NELF and DSIF by positive transcriptional elongation Factor b (P-TEFb) both facilitate the release of the halted polymerase (Peterlin and Price 2006). P-TEFb phosphorylates also the 2nd serine in the Pol II CTD repeat, whereas it phosphorylates the 5th serine TFIIH), which renders the Pol II elongation capable. In mammals, the majority of genes are attached to Pol II at their 5′ ends (Guenther et al. 2007), although it is uncertain how this mirrors post-enrollment regulation of a balanced PIC that is either still not transcriptionally active and/or a haltered Pol II, which is transcriptionally active.

12.10 Pol II CTD Serine 5 Phosphorylation During Initiation Triggers Regulation of Nascent RNA and Underlying Chromatin

The Pol II CTD acts as a coordinator of transcriptional steps by the recruitment of proteins participating in the alteration of histones, elongation, termination and the processing of mRNA (Egloff and Murphy 2008; Meinhart et al. 2005; Phatnani and Greenleaf 2006). A number of proteins identify and attach to a special phosphorylation profile of the CTD, which varies dynamically throughout the entire transcriptional cycle. The number of YSPTSPS heptapeptide repeats in the CTD differs between 26 in *Saccharomyces* and 52 in humans. As there are five possible phosphorylation sites in every CTD repetition, a CTD code could be proposed (Buratowski 2003). The serine phosphorylation is well characterized only at the 5th and 2nd position. It has also lately been found that serine 7 is phosphorylated in human cells and is implicated in the transcription of small nucleolar RNAs (Egloff et al. 2007). Whether serine 7 plays a major function in other Pol II genes is still a matter of uncertainty.

A dephosphorylated Pol II is acquired to promotors during the PIC assembly through interference with the TFIIB and the Mediator (Myers et al. 1998). Among the first transcription initiation steps is the phosphorylation of CTD serine 5 by the Kin28 (Komarnitsky et al. 2000). This Pol II label recruits a range of elements comprising the mRNA occlusion enzyme, an RNA monitoring complex referred to as nuclear pre-mRNA downregulation 1–3 (Nrd1–Nab3) and the polymerase II-associated factor (PAF) complex (Hampsey and Reinberg 2003; Komarnitsky et al. 2000; Ng et al. 2003b; Vasiljeva et al. 2008). The 7-methyl-guanylyl cap prevents the nascent RNA from being decomposed and labels the mRNA for transportation into the cytoplasm and ultimately for translation. The Nrd1–Nab3 complex can be traced with Pol II and the RNA to determine whether the resulting RNA includes short Nrd1-binding motifs, such as GUA and UCUU (Carroll et al. 2004), which might be degraded evolutionarily into mRNAs, but are found in inadequately transcribed and therefore non-functional regions of the genome. The Nrd1–Nab3 complex encourages the termination of these latter transcriptions and guides them to break down through the nuclear exosome cascade.

The PAF complex combines the serine-5-phosphorylation of CTD to a system of histone alterations. The multitude of interactions with phosphorylated CTD with serine 5 possibly accounts for the strongly repetitive character of the heptad motif. Whereas in principle each complex can be associated with a subset of phosphorylated repeats, the distribution of the factors on the CTD is poorly defined. The PAF complex, in association with the bypass UAS requirement 1 and 2 (Bur1/2) kinase, activates the serine 5 phosphorylated CTD and stimulates the enrollment of the ubiquitin ligases RADiation sensitive 6 (Rad6) and BREfeldin A sensitivity 1 (Bre1) E2 and E3 to trigger a number of histone modification cross talks, commonly referred to as the histone code.

The Pol II phosphorylation is associated with writers, readers and erasers of the histone code. The chromatin scenery is filled with numerous post-translational modifications (or labels) on tails of histones (Berger 2007; Kouzarides 2007; Lee and Shilatifard 2007; Millar and Grunstein 2006; Shilatifard 2008; Weake and Workman 2008). The tails of each histone are composed of 20–40 amino-terminal amino acids that are an extension of the globular nucleosome core unit into the surrounding fluid, such as the karyoplasm. The most widely investigated histone labels comprise acetylation, methylation and ubiquitylation. These labels have been suggested to firstly directly alter the chromatin structure by changing the electrostatic or internucleosomal connections and/or secondly establish docking sites for proteins (referred to as readers). There is a dedicated complex for each written label that is designed to delete it, which is in line with the dynamic character of chromatin labels (Bannister and Kouzarides 2005; Klose and Zhang 2007; Shi and Whetstine 2007). Certain histone alterations, such as methylated H3K9/27, are kept stable through several cell divisions and mirror epigenetic events in which the alterations propagate themselves (Bibikova et al. 2008; Lunyak and Rosenfeld 2008). It remains to be proven whether such changes are the reason or result of epigenetic transmission.

12.11 Pol II CTD Serine-5 Phosphorylation and H3K4 Methylation

It is getting more and more obvious that single histone labels affect the writing and deletion of other labels, creating a pattern of histone interference during the process of transcription. Several recent studies provide evidence for a model in which transcription initiation and elongation is associated with a dynamic circuit of coupled histone alterations, which may be illustrated by the sequence of events linked to the ubiquitylation of H2BK123. When Ser5 phosphorylation of the Pol II CTD has attracted the PAF and Burl/2 complexes, it at least phosphorylates Burl Rad6 (Wood et al. 2005) and thereby activates Rad6 E2 ligase stimulating the mono-ubiquitylation of Bre1. In the next step, Bre1 is able to ligate ubiquitin onto H2BK123 similar to an E3 ligase (Kao et al. 2004; Laribee et al. 2005; Ng et al. 2003a; Robzyk et al. 2000; Wood et al. 2003, 2005). H2BK123ub conducts thereafter the trimethylation of H3K4 by the complex of proteins associated with Set1/Su(var)3–9 (COMPASS/Set1), enhancer of zeste, trithorax 1 and K79 by the Disruptor Of Telomeric silencing 1 (Dot1) (Briggs et al. 2002; Dover et al. 2002; Nakanishi et al. 2008; Sun and Allis 2002; Wood et al. 2003).

The trimethylation of H3K4 also involves the acetylation at H3K14 (Nakanishi et al. 2008), which probably is performed at PIC assembly or by co-transcription utilizing acetyltransferases (such as elongator) which translocate with Pol II. As acetylation could increase nucleosome shedding throughout transcription, the H3K4me3 label could detect local levels of acetylation through K14. While it is not fully understood what the H3K4me3 brand engages in, it is feasible that it may bind to "readers" such as the TAF3 subunit of TFIID, the Chd1 component of SAGA or the mammalian Ing1 yeast homologue (Yng1) subunit of NuA3 to sustain the physical assembly of the PIC compound and achieve the activated acetylated state.

The transcription-associated H3K4me3 emerges at the start of the genes (Liu et al. 2005; Pokholok et al. 2005), however, eventually passes into dimethylation (H3K4me2) at a site less than 1 kb downstream of the transcription start signal. The transition could be due to changes in the selectivity of the active methyltransferase moiety of Set1, which might be linked to the phosphorylation level of the CTD, which means that a change from serine 5 to serine 2 phosphorylation induces a shift from H3K4me3 to me2; however, this still needs to be ascertained. H3K79me3 is also associated with CTD-serin-5 phosphorylation, is sustained in the entire gene body and is associated with transcription and the silencing of telomeres (Lacoste et al. 2002; Pokholok et al. 2005).

12.12 CTD Serine-2 Phosphorylation and H3K36 Methylation

While elongation continues from the 5′ end of the gene toward the 3′ end, a phosphorylation shift from serine-5 to serine-2 takes place at the CTD (Komarnitsky et al. 2000), even though it is not obvious to what degree one incident causes the other. The carboxy-terminal domain kinase 1 (Ctk1) in yeast and P-TEFb phosphorylated serine-2 in higher eukaryotes (Lee and Greenleaf 1997; Marshall et al. 1996; Sterner et al. 1995; Wyce et al. 2007) and adjust therefore the elongation step.

The CTD-serin-5 phosphorylation label is deleted by suppressor of SUa72 (Ssu72) phosphatase and probably other phosphatases, such as the calcium-binding protein *S. cerevisiae* CalPonin 1 (Scp1) (Hausmann et al. 2005; Krishnamurthy et al. 2004; Yeo et al. 2003). Ssu72 is potentially accountable for multiple effects of the transcription cycle: Firstly, it can attach with TFIIB in PIC; secondly, it is necessary for the shift from serine-5 to serine-2 phosphorylation; thirdly, it is involved in the cleavage and polyadenylation complex, which functions during the termination of transcription (Dichtl et al. 2002a).

Noteworthy, the CTD seems to control its proper structural integrity through prolines that reside in the YSPTSPS heptapeptide repeat. Proline is the one and only natural amino acid that may be present in structurally distinct *cis* and *trans* configurations, leading to modified pathways of the CTD polypeptide chain. Investigations with surface plasmon resonance show that ESSential 1 (Ess1) prolinisomerase is primarily linked to double phosphorylated CTD in serines 2 and 5 (Phatnani et al. 2004). The consequences for the control of the *cis/trans* configuration of the two prolines in the CTD are currently unclear, but it is feasible that the isomerization of the proline might adjust the dephosphorylation of the CTD or the attachment of proteins that sense the phosphorylation patterns on the CTD. In fact, genetic investigations indicate that the function of Ess1 could be to counteract the action of CTD kinases (Wilcox et al. 2004). Considering the two to four dozen YSPTSPS heptad repeats arranged on a single complex of Pol II, with two prolines and three serine phosphorylation sites within every repeat, it is possible to envision a comprehensive CTD code with positional, structural and chemical properties.

While Pol II is transcribed into the body of a gene, PAF-stimulated H2BK123ub labels are still produced, probably accompanied by an H3K4me2 label. H2BK123ub, on the contrary, inhibits the nucleosomal linkage of Ctk1 and the resulting serine 2-phosphorylation of CTD (Wyce et al. 2007), which is the elongation condition of Pol II. Present schemes indicate that the SAGA complex can move with Pol II and stimulate deubiquitylation of H2BK123ub by its UBiquitin-specific processing protease 8 (Ubp8 subunit) (Henry et al. 2003), even though not all genes may be involved in this mechanism (Shukla et al. 2006). Elimination of the H2BK123ub label through Ubp8 might permit the conversion of the H3K4 methyl label with demethylases and possibly restoring the chromatin status for a second round of transcription, but this has yet to be evaluated. The alternative is to replace the H3K4 label by dynamic substitution of histone subunits by unmethylated ones. Most significantly, the loss of the H2BK123ub label permits Ctk1 to phosphorylate serine-2 of the CTD.

H2BK123 ubiquitylation/deubiquitylation not only acts as a regulatory shift in a framework of histone alterations (Weake and Workman 2008); however, H2BK123ub also adjusts nucleosome dynamical processes during elongation. For example, the absence of H2B ubiquity and the deletion of SuPpressor of Ty 16 (SPT16) leads to a lower histone loading at the GAL1 site, indicating that H2BK123ub and facilitates chromatin transcription (FACT) regulate through the Spt16 subunit the recomposition of nucleosomes after the elongation by Pol II (Fleming et al. 2008).

Serine-2-phosphorylation initiates a second communication mode between the CTD and the chromatin, which finally affects the H3K36 methylation. While the PAF complex seems to have been the linkage of the serine-5-phosphorylation label, the H3K36-methylase set2 seems to be the linkage in this second phase. Set2 attaches to CTD phosphorylated on serine-2 (Kizer et al. 2005; Li et al. 2002, 2003; Xiao et al. 2003) and acts as a catalyst for the di- and trimethylation of H3K36. Whether the di- and trimethylation of K36 fulfill different functions is generally not clear. Nevertheless, H3K36me2 seems to be a label of actively transcribed genes (Rao et al. 2005), even though its role seems to be to suppress transcription, which could occur in the ORF either promiscuously or cryptically.

The isomerization of the prolines 30 and 38 within the H3 tail through the FKBP proline rotamase 4 (Fpr4) negatively adjusts the methylation of Set2 at the adjacent K36 (Nelson et al. 2006), indicating that the secondary architecture of the H3 tail offers an additional dimension of transcriptional regulation, such as it has been observed in the proline isomerization of Pol II CTD. The RPD3-Small (RPD3S) histone deacetylase complex in the next step attaches to H3K36me2 by the sensing of the combinatorial region (Li et al. 2007b, 2009; Rao et al. 2005). According to this, RPD3S acts on the deacetylation of the nucleosomes, which turns the nucleosomes into a stage that is insensitive to transcription, thus repressing cryptic transcription, that could start at unsuitable sites (Carrozza et al. 2005; Li et al. 2007c).

Cross talk of the histone alteration can create positive feedback circuits. For instance, H3K4me3 and H3K14ac can reinforce one another. In this circuit, SAGA attaches to H3K4me3 to acetylate H3K14, which then induces H3K14ac to stimulate H3K4me3. The SAGA complex is operationally equivalent to TFIID, as both possess TAFs and both supply TBP to the promoters. Nevertheless, SAGA is definitely engaged in other abilities of transcription because it is linked to an acetyltransferase Gcn5, even though the higher eukaryotic TAF1 is an acetyltransferase, a deubiquitinase Ubp8 and an ATP-dependent chromatin remodeling protein Chd1.

Notwithstanding this complicated cross talk of histone alterations, none of the Rad6/Bre1, Set1, Set2, Dot1 or individual histone tails are indispensable for the viability of budding yeast (Giaever et al. 2002; Kayne et al. 1988; Megee et al. 1990; Morgan et al. 1991; Nakanishi et al. 2008). This raises the problem of why these labels continue to exist. A reasonable statement could be that, although the labels are not crucial for controlled transcription, they provide strength and reduce stochasticity in expression, for example, by swiftly re-establishing the canonical nucleosome structure according to Pol II, thus enabling the coordination of reactions to activating cues in a cell population. The fast and timely performance of genetic programs is especially apparent in the field of embryonic development, in which accurately timed cell division and expression is crucial. The preservation of histone alteration conditions is especially critical for the differentiation of stem cells (Bibikova et al. 2008; Lunyak and Rosenfeld 2008).

There are still many questions concerning the aims and interconnections of readers, writers and erasers of the chromatin scene. For instance, what kind of impact have mono-, di- and tri-methylated states? This means, in terms of their function, do mono- and disimply constitute incomplete changes or are they operationally revealing? The deubuitination activity of Ubp8 is located within the SAGA complex that also comprises a histone acetyltransferase in Gcn5 and an ATP-dependent rearrangement activity in Chd1. How will SAGA then analyze these various activities in one and the same compound? As multiple chromatin remodeling agents include histone-binding modules that detect acetylated and methylated lysines, what specific rests or combinations of rests are identified by these chromodomains and bromine domains?

12.13 How is the Transcription Terminated?

While Pol II transcription progresses to the 3′ end of the genes, serine 2-phosphorylation of CTD outweighs serine 5-phosphorylation, thus increasing the enrollment (or retention) of RNA processing engines at the 3′ end (Ahn et al. 2004), such as Protein 1 of Cleavage and polyadenylation Factor I 11 (Pcf11), Spt6, Rtt103, and Cleavage Factor Two 1 (Cft1) (Dichtl et al. 2002b; Kim et al. 2004; Licatalosi et al. 2002; Yoh et al. 2007). The attachment of the termination markers may also rely on the proline configurations of the CTD, as two temperature-sensitive Ess1 mutants are damaged in terms of pre-mRNA 3′ termination (Hani et al. 1999).

Pcf11 constitutes a component of the cleavage and polyadenylation factor and is implicated in the termination of transcription. Biochemical investigations in *Drosophila* show that Pcf11 stops transcription by degrading elongation complexes via the conduction of interactions between the CTD and nascent RNA (Zhang and Gilmour 2006). Rtt103 binds explicitly to phosphorylated serine-2 of CTD and is engaged in the termination of transcription by interference with exonuclease Ribonucleic Acid Trafficking 1 (Rat1), which breaks down RNA downstream from the mRNA cleavage point (Kim et al. 2004).

It is now obvious that all phases of the whole transcription cycle starting from initiation, through rest, elongation and finally termination are governed by Pol II CTD phosphorylation of serine 5 and 2 and interdigital isomerization of proline. These Pol II labels govern the appearance and disappearance of histone-modifying complexes and rest/termination drivers. Consequently, Pol II quickly and efficiently crosses nucleosomal boundaries and generates full-length transcripts that are precisely initiated and exactly finished.

12.14 Regulation of the Transcription of Gens by Noncoding Transcription

With the growing sensitivity of RNA sequencing and array-based technologies to investigate the amount of RNA transcripts, recent reports have shown verifiable levels of transcription of the bulk of the genome from yeast to humans (David et al. 2006; Dutrow et al. 2008; Kapranov et al. 2007; Mortazavi et al. 2008; Nagalakshmi et al. 2008; Sultan et al. 2008; Wilhelm et al. 2008), even though the majority of genes are actually transcribed at a comparatively low rate. The extraordinary susceptibility of RNA identification strategies also revealed the existence of antisense transcripts from coding sites transcribed in the reverse orientation by the sense transcript and cryptically unstable transcripts from intergenic areas in the bud and fission yeast (Dutrow et al. 2008; Neil et al. 2009; Xu et al. 2009). In most cases, stable antisense transcripts are normally identified in wild-type strains, while cryptically unstable transcripts are quickly broken down in the exosome and are therefore only identifiable in an exosomal mutant (Houseley et al. 2006). It has been implied that up to 90% of Pol II induction is due to biological noise (Struhl 2007), which raises the following question: Are these antisense and cryptic transcripts of any physiological value?

A number of recent investigations in *S. cerevisiae* yield evidence for the operational relevance of antisense transcripts and cryptically unstable transcripts. For instance, the cryptically unstable SRG1 transcript produced in the sense direction upstream from the SER3 serine biosynthesis gene governs its transcription by a transcriptional perturbation mechanism that overrides the usage of SER3 promoter elements (Martens et al. 2004). The URA2 and IMD2 nucleotide biosynthesis genes seem to rely on cryptic, unstable transcripts that are created upstream of the promoter to suppress the process of transcription (Davis and Ares 2006; Kopcewicz et al. 2007; Thiebaut et al. 2008). The mechanistic foundation for these sense-directed, repressive, cryptic, unstable transcripts is currently being elaborated. Its regulation of amino acid and nucleotide biosynthesis genes offers these terminal metabolites the capability of performing feedback control on their corresponding cryptic, unstable transcripts. For instance, serine attaches catabolism of hydroxyamino acids 4 (Cha4), which in turn connects to the SRG1 promoter and thereby stimulates it (Martens et al. 2005). In IMD2, high cellular guanine nucleotide concentrations encourage the initiation of Pol II at cryptically unstable transcripts that contain a G instead of the usual A at their +1 transcription start sites (Kuehner and Brow 2008). Start site selection appears to be extremely dependent on the levels of nucleotides.

GAL10 is an excellent instance of suppression by antisense transcription. The Reb1 activator is said to generate a NFR near the 3′ end of the GAL10-coding site to promote the generation of antisense transcription under GAL10 repressive circumstances, which means that the glucose is available and the galactose is unavailable and finally the chromatin is changed through the GAL10 locus (Houseley et al. 2008). In concrete terms, the GAL10 antisense transcription results in the enrollment of Set2 methyltransferase and the RPD3S histone deacetylase complex probably through H3K36me2, thereby keeping the chromatin in a transcriptionally impenetrable deacetylated state.

A slightly modified mechanism for repressive antisense RNA is found in PHO84, encoding a phosphate uptake transporter. Under PHO84 repressive circumstances, where phosphate is present, the deletion or downregulation of the nuclear exosome results in the retention of a PHO84 antisense transcript that, through an undisclosed mechanism, stimulates the enrollment of the histone-deacetylase 1 (HDA1) complex to establish a deacetylated condition (Camblong et al. 2007). These results indicate that the generation of noncoding RNAs can suppress the transcription of adjacent genes by modifying the chromatin background through histone deacetylases.

Additionally, antisense transcriptions may positively control the transcription. For instance, the discontinuation of the generation of a transcription antisense for the PHO5 promoter effectively retards the chromatin conversion of the promoter nucleosomes and the enrollment of Pol II to the promoter (Uhler et al. 2007). In specific detail, it is recommended that the continued generation of PHO5 antisense transcription raises the plasticity of chromatin, thereby positively modulating the transcription of PHO5. This idea apparently contradicts the mechanisms of antisense regulation of PHO84 and GAL10.

Even though the genome-wide dispersion of antisense transcripts has been outlined (Perocchi et al. 2007), a broad functional investigation of cryptically unstable transcripts is missing. The comprehension of the function of more antisense transcripts and cryptically unstable transcripts will yield complementary information on the manner in which genes are adjusted by employing dissimilar and apparently contradictory mechanisms to refine the activity of genes.

12.15 Is a Unique Transcription Cycle Employed for All Genes?

Why is the genetic code usually universal? Are there exceptions?

But it is the subtleties of eukaryotic gene control that are so outstanding. But why are physical steps so numerous? Based on this, why can an enabled activator not attach to its intended promoter and easily attract an RNA polymerase? Why relocate nucleosomes to mask specific binding sites of transcription factors? Why do we need more than 50 proteins for assembling PIC when in principle Pol II ought to have unhindered access to nucleosome free promoter areas? What is the reason for the large number of histone alterations and why does the Pol II CTD require the coordination of interactions between the CTD and the transcriptional alterations that arise in the course of transcription? A number of these questions have been discussed in the scientific research literature, but most of these questions will probably stay puzzling until the mechanistic features of eukaryotic transcriptional control are completely elaborated.

Archaebacteria successfully use a simplified, downsized version of the eukaryotic transcription mechanism, meaning that eukaryotic enhancement is not a basic parameter of gene regulation. The combination of genome duplication and evolution into varied and/or multi-cellular surroundings typical of eukaryotes may be widely recognized to have advanced the tremendous amount and associative multiplicity of sequence-specific transcription factors. But the basis of the complexity of a controlled transcription cycle is not apparent. For the purpose of transcription, the evolution has not the luxuriousness of a de novo development and is forced to rely on preexisting conditions, which in the history of eukaryotic evolution is associated with tremendous variations in selective stress due to spot environmental variations. Thus, in such a case, the complexity could become even more complicated, and mechanisms remain that perform a simple function with an excessively sophisticated machine.

12.16 Can Mechanical Stimulation of the Cells Alter the Transcription?

Several factors of the local extracellular surroundings have a deep impact on the phenotype and the functioning of the cells. Physical and chemical signals in the surrounding space trigger intracellular signaling pathways that finally activate transcription factors which are strong controllers of the cell phenotype. Transcriptional activity cell arrays (TRACER) were used for the broad-scale, dynamic quantitative analysis of transcription factor activity in human fibroblasts cultivated on hydrogels with regulated elasticity modulus and integrin ligand concentration. Thus, three groups of transcription factors were found, such as those that react to the density of ligands, the stiffness of the substrate and those that require both. Dynamic regulatory transcription factor micronetworks were computationally designed and showed different levels of transcription factor activity, directional dependence, such as activation or impairment and dynamical properties for adhesive and mechanical cues. In addition, the TRACER scaffolds projected retained junctions of transcription factor activity over several cell types, which have been substantially modified in fibrotic tissues. This approach addresses the different and intersecting actions of adhesive and mechanical impulses and aims to detect preserved signal transduction mechanisms under normal and disease conditions (Peñalver Bernabé et al. 2016).

The extracellular milieu has a direct influence on several cell features, including cell shape, the organization of the cytoskeleton, the rate of migration and the adhesion force (Wong et al. 2004; Peyton et al. 2007; Olson and Nordheim 2010; Chen 2008) and is crucial for the correct operation of intricate tissues and organs. These extracellular microenvironments are key to the differentiation of cells and tissue specialization during development (Brunet et al. 2013; Engler et al. 2006; Wozniak and Chen 2009; Zhang et al. 2011; Chowdhury et al. 2010; Trappmann et al. 2012; Connelly et al. 2010; Zaret and Carroll 2011), and pathological alterations of these microenvironments play a role in numerous disease conditions, such as cancer, atherosclerosis and fibrosis (DuFort et al. 2011; Clarke et al. 2013; Jaalouk and Lammerding 2009; Hahn and Schwartz 2009). For instance, tumors tend to be much stiffer compared to healthy tissue, and these tissue stiffnesses can be a contributing factor to malignancy (DuFort et al. 2011). However, numerous stages of cellular differentiation and the course of multiple diseases are facilitated by the interplay of the extracellular matrix with cytoskeletal-connected cell membrane receptors, which trigger intricate cytosolic signaling pathways resulting in the activation of transcription factors inside the cell nucleus (Olson and Nordheim 2010; DuFort et al. 2011; Chiquet et al. 2009; Hoffman et al. 2011; Sharili and Connelly 2014). In general, transcription factors are strong controllers of the cell's phenotype, as seen by the emergence of techniques to generate stimulated pluripotent stem cells through the manipulation of a handful of transcription factors (Takahashi et al. 2007). The extracellular matrix exhibits both biochemical, including the number, distribution and the specific type of integrin-binding sites, and mechanical characteristics, such as the matrix stiffness, that affect the cell phenotype (Wong et al. 2004; Peyton et al. 2007; Chen 2008; DuFort et al. 2011; Hoffman et al. 2011) and are capable of independently encouraging behavior of the cell. Consequently, both biochemical and mechanical phenomena need to be accounted for in the treatment of disease states, which are caused by disorganized extracellular surroundings, through the development of renewable microenvironments as therapies.

Chemical and mechanical signals affect the cell phenotype in a similar way (Peyton et al. 2007; Chen 2008), as numerous receptors that perceive and convert biochemical signals due to signals and also transmit mechanical cues (DuFort et al. 2011; Hoffman et al. 2011; Zaidel-Bar et al. 2007; Zaidel-Bar and Geiger 2010; Discher et al. 2009; Schwarz and Gardel 2012). For instance, raising either the number of integrin-binding sites or the stiffness of a substrate comparatively improves cytoskeletal stiffness and maturation of focal adhesions (Chen 2008; Pelham and Wang 1997). The determination of these functions has been a major concern, as the mutual interactions of the cells with the extracellular matrix are closely coupled and the resulting cell response is due to a variety of complementary and mutually overlapping biochemical and mechanical processes (Wong et al. 2004; Peyton et al. 2007; Chen 2008; Hoffman et al. 2011). The tension of the cytoskeleton, which is generated in reaction to substrate stiffness and integrin-based cell adhesion to extracellular matrix proteins, is necessary for the migration and reconstruction of the extracellular matrix and in return governs the biochemical configuration of a specific tissue (Chowdhury et al. 2010; Chiquet et al. 2009; Hoffman et al. 2011; Zhong et al. 1998; Schwartz and DeSimone 2008; Asparuhova et al. 2009).

Notwithstanding these similarities, it is obvious that the two types of cues perform a differing function in tissue development and performance. Cells located within tissues dedicated to receiving mechanical signals, including the vascular system, muscle or bone, need mechanical cues in order to sustain a functional phenotype, in parallel to suitable biochemical inputs. In addition, the dynamic interplay of cells with their extracellular surroundings implies that the relative temporal properties of mechanical and biochemical responses may impact cell status through a variety of signaling routes (Hoffman et al. 2011). As a matter of fact, recent investigations have pointed out that the time frame of a cell's reaction depends on the type of the stimuli (Na et al. 2008; Poh et al. 2009, 2012; Zeller et al. 2013), and thus, the immediate linkage of the cytoskeleton to the nucleus offers a regime for a quicker signaling toward the nucleus than the canonical phosphorylation pathways-based activation (Poh et al. 2012; Dahl et al. 2008; McNamara et al. 2012).

To present day, mechanistic associations between individual regulatory factors in the extracellular surroundings and the orchestrated transcriptional reaction of a cell are still mostly underdeveloped. Although comprehensive characterization of broad phenotypic alterations such as the cell form, the cytoskeletal organization, the velocity of migration and the force of adhesion (Schwarz and Gardel 2012; Chen et al. 1997; Gardel et al. 2010) acting upon individual signs within the extracellular surroundings, a systems engineering approach that incorporates the reaction of the multiple signals is needed to understand the mechanisms underlying diseases such as fibrosis, cancer, atherosclerosis and stroke involving significant alterations of the extracellular microenvironment in a number of tissues (Hoffman et al. 2011; Discher et al. 2009; Geiger et al. 2009). A precise, numerical knowledge of these underlying processes will lead to the smart engineering of innovative clinical interventions that directly target the interactions between cells and the extracellular matrix.

An integrated characterization of the dynamic transcription factor activity governing the cellular reaction to a comprehensive adhesive and mechanical surrounding is analyzed in the following. Time-coordinated actions of multiple transcription factors steer progressive alterations of the cell phenotype, which requires the employment of experimental techniques that gain a dynamic perspective on the responses of transcription factors on the system level resulting in the cellular differentiation or pathogenesis. To this end, advanced biomaterial approaches have been employed to independently monitor the appearance of biochemical ligands and mechanical characteristics to gain a better insight into the individual and deductive implications of these cues on the cell performance. The intracellular reactions to fluctuations in the microenvironment have been examined with TRACER (TRanscriptional Activity CEll aRrays) (Weiss et al. 2012, 2014; Siletz et al. 2013), which represents a large-scale, quantitative array for the analysis of the transcription factor activity in real time. Whereas previous studies concentrated on restricted subsets of the numerous individual signal transduction pathways involved in the incorporation of extracellular signs to establish phenotypic reaction, TRACER detects the dynamic action of transcription factors downstream of numerous signaling pathways to offer a vision of the correlated transcription factor activities which combine multiple signals from the surroundings of a cell into a coherent transcriptional code.

Human foreskin fibroblasts were grown on hydrogels, where the concentration of adhesion ligands and the Young's modulus were varied both independently and quantitatively. TRACER made it possible to determine the intracellular responses of over 50 transcription factors in reaction to varying biomaterial characteristics. The findings of TRACER were confirmed experimentally with microwestern arrays (MWAs) (Ciaccio et al. 2010). Based on these data, a dynamic transcriptional framework has been built that uniquely defined transcription factors, which are either impacted by mechanical, chemical or both mechanical and chemical cues, and identified both linear and nonlinear relationships between the characteristics of the extracellular matrix and the activity of the transcription factors. A comparison with publicly available microarray data on the effects of biochemical and/or mechanical impulses on gene expression revealed that several aspects of this transcription factor scaffold are preserved throughout various types of human cells, pointing to conserved biological mechanisms with wide biological importance. These findings deliver an integrated picture of the emerging and intricate interaction of transcriptional networks that are involved in the transmission of biochemical and biophysical cues to the nucleus, which in turn governs gene expression and the fate of cells. Ultimately, an extra computational analysis confirmed the conservation of TRACER-identified transcription factor processes in idiopathic pulmonary fibrosis, which is a disease condition defined by improper cell–extracellular matrix interplay and altered extracellular matrix microenvironments. However, this new understanding of the fundamental mechanisms underlying pathological fibrosis affects the formulation of clinical interventions that focus on the deeper underlying disease processes.

References

S.H. Ahn, M. Kim, S. Buratowski, Phosphorylation of serine 2 within the RNA polymerase II C-terminal domain couples transcription and 3′ end processing. Mol. Cell **13**, 67–76 (2004)

I. Albert, T.N. Mavrich, L.P. Tomsho, J. Qi, S.J. Zanton, S.C. Schuster, B.F. Pugh, Translational and rotational settings of H2A.Z nucleosomes across the *Saccharomyces cerevisiae* genome. Nature **446**, 572–576 (2007)

A. Almer, W. Horz, Nuclease hypersensitive regions with adjacent positioned nucleosomes mark the gene boundaries of the PHO5/PHO3 locus in yeast. Embo J. **5**, 2681–2687 (1986)

J.D. Anderson, J. Widom, Poly(dA-dT) promoter elements increase the equilibrium accessibility of nucleosomal DNA target sites. Mol. Cell. Biol. **21**, 3830–3839 (2001)

M.L. Angus-Hill, A. Schlichter, D. Roberts, H. Erdjument-Bromage, P. Tempst, B.R. Cairns, A Rsc3/Rsc30 zinc cluster dimer reveals novel roles for the chromatin remodeler RSC in gene expression and cell cycle control. Mol. Cell**7**, 741–751 (2001)

M. Asparuhova, L. Gelman, M. Chiquet, Scandinavian. J. Med. Sci. Sports **19**, 490–499 (2009)

D.T. Auble, K.E. Hansen, C.G. Mueller, W.S. Lane, J. Thorner, S. Hahn, Mot1, a global repressor of RNA polymerase II transcription, inhibits TBP binding to DNA by an ATP-dependent mechanism. Genes Dev. **8**, 1920–1934 (1994)

G. Badis, E.T. Chan, H. van Bakel et al., A library of yeast transcription factor motifs reveals a widespread function for Rsc3 in targeting nucleosome exclusion at promoters. Mol. Cell**32**, 878–887 (2008)

A.J. Bannister, T. Kouzarides, Reversing histone methylation. Nature **436**, 1103–1106 (2005)

Y. Bao, X. Shen, INO80 subfamily of chromatin remodeling complexes. Mutat. Res. **618**, 18–29 (2007a)

Y. Bao, X. Shen, SnapShot: chromatin remodeling complexes. Cell **129**, 632 (2007b)

A. Barski, S. Cuddapah, K. Cui, T.Y. Roh, D.E. Schones, Z. Wang, G. Wei, I. Chepelev, K. Zhao, High-resolution profiling of histone methylations in the human genome. Cell **129**, 823–837 (2007)

A.D. Basehoar, S.J. Zanton, B.F. Pugh, Identification and distinct regulation of yeast TATA box-containing genes. Cell **116**, 699–709 (2004)

S.L. Berger, Gene regulation. Local or global? Nature **408**, 412–413 (2000)

S.L. Berger, The complex language of chromatin regulation during transcription. Nature **447**, 407–412 (2007)

B.E. Bernstein, T.S. Mikkelsen, X. Xie et al., A bivalent chromatin structure marks key developmental genes in embryonic stem cells. Cell **125**, 315–326 (2006)

M. Bibikova, L.C. Laurent, B. Ren, J.F. Loring, J.B. Fan, Unraveling epigenetic regulation in embryonic stem cells. Cell Stem Cell **2**, 123–134 (2008)

R. Biddick, E.T. Young, Yeast mediator and its role in transcriptional regulation. C. R. Biol. **328**, 773–782 (2005)

D. Biswas, R. Dutta-Biswas, D.J. Stillman, Chd1 and yFACT act in opposition in regulating transcription. Mol. Cell. Biol. **27**, 6279–6287 (2007)

J.C. Black, J.E. Choi, S.R. Lombardo, M. Carey, A mechanism for coordinating chromatin modification and preinitiation complex assembly. Mol. Cell **23**, 809–818 (2006)

A.P. Boyle, S. Davis, H.P. Shulha, P. Meltzer, E.H. Margulies, Z. Weng, T.S. Furey, G.E. Crawford, High-resolution mapping and characterization of open chromatin across the genome. Cell **132**, 311–322 (2008)

S.D. Briggs, T. Xiao, Z.W. Sun, J.A. Caldwell, J. Shabanowitz, D.F. Hunt, C.D. Allis, B.D. Strahl, Gene silencing: trans-histone regulatory pathway in chromatin. Nature **418**, 498 (2002)

C.E. Brown, L. Howe, K. Sousa, S.C. Alley, M.J. Carrozza, S. Tan, J.L. Workman, Recruitment of HAT complexes by direct activator interactions with the ATM-related Tra1 subunit. Science **292**, 2333–2337 (2001)

T. Brunet, A. Bouclet, P. Ahmadi et al., Evolutionary conservation of early esoderm specification by mechanotransduction in Bilateria. Nat. Commun. **4**, 2821 (2013)

M.J. Buck, J.D. Lieb, A chromatin-mediated mechanism for specification of conditional transcription factor targets. Nat. Genet. **38**, 1446–1451 (2006)

S. Buratowski, The CTD code. Nat. Struct. Biol. **10**, 679–680 (2003)

S. Buratowski, S. Hahn, L. Guarente, P.A. Sharp, Five intermediate complexes in transcription initiation by RNA polymerase II. Cell **56**, 549–561 (1989)

D.A. Bushnell, K.D. Westover, R.E. Davis, R.D. Kornberg, Structural basis of transcription: an RNA polymerase II-TFIIB cocrystal at 4.5 Angstroms. Science **303**, 983–988 (2004)

B.R. Cairns, Y. Lorch, Y. Li, M. Zhang, L. Lacomis, H. Erdjument-Bromage, P. Tempst, J. Du, B. Laurent, R.D. Kornberg, RSC, an essential, abundant chromatin-remodeling complex. Cell **87**, 1249–1260 (1996)

B.R. Cairns, Chromatin remodeling: insights and intrigue from single-molecule studies. Nat. Struct. Mol. Biol. **14**, 989–996 (2007)

J. Camblong, N. Iglesias, C. Fickentscher, G. Dieppois, F. Stutz, Antisense RNA stabilization induces transcriptional gene silencing via histone deacetylation in *S. cerevisiae*. Cell **131**, 706–717 (2007)

M. Carey, B. Li, J.L. Workman, RSC exploits histone acetylation to abrogate the nucleosomal block to RNA polymerase II elongation. Mol. Cell **24**, 481–487 (2006)

K.L. Carroll, D.A. Pradhan, J.A. Granek, N.D. Clarke, J.L. Corden, Identification of cis elements directing termination of yeast nonpolyadenylated snoRNA transcripts. Mol. Cell. Biol. **24**, 6241–6252 (2004)

M.J. Carrozza, B. Li, L. Florens et al., Histone H3 methylation by Set2 directs deacetylation of coding regions by Rpd3S to suppress spurious intragenic transcription. Cell **123**, 581–592 (2005)

C.S. Chen, Mechanotransduction—a field pulling together? J. Cell Sci. **121**, 3285–3292 (2008)

C.S. Chen, M. Mrksich, S. Huang, G.M. Whitesides, D.E. Ingber, Geometric control of cell life and death. Science **276**, 1425–1428 (1997)

M. Chiquet, L. Gelman, R. Lutz, S. Maier, From mechanotransduction to extracellular matrix gene expression in fibroblasts. Biochim. Biophys. Acta (BBA) Mol. Cell Res. **1793**, 911–920 (2009)

F. Chowdhury, S. Na, D. Li, Y.-C. Poh, T.S. Tanaka, F. Wang, N. Wang, Material properties of the cell dictate stress-induced spreading and differentiation in embryonic stem cells. Nat. Mater. **9**, 82–88 (2010)

W.H. Chung, J.L. Craighead, W.H. Chang, C. Ezeokonkwo, A. Bareket-Samish, R.D. Kornberg, F.J. Asturias, RNA polymerase II/TFIIF structure and conserved organization of the initiation complex. Mol. Cell **12**, 1003–1013 (2003)

M.F. Ciaccio, J.P. Wagner, C.-P. Chuu, D.A. Lauffenburger, R.B. Jones, Systems analysis of EGF receptor signaling dynamics with microwestern arrays. Nat. Methods **7**, 148–155 (2010)

D.L. Clarke, A.M. Carruthers, T. Mustelin, L.A. Murray, Matrix regulation of idiopathic pulmonary fibrosis: the role of enzymes. Fibrogenesis Tissue Repair **6**, 1–9 (2013)

J.T. Connelly, J.E. Gautrot, B. Trappmann, D.W.-M. Tan, G. Donati, W.T. Huck, F.M. Watt, Actin and serum response factor transduce physical cues from the microenvironment to regulate epidermal stem cell fate decisions. Nat. Cell Biol. **12**, 711–718 (2010)

L.J. Core, J.T. Lis, Transcription regulation through promoter-proximal pausing of RNA polymerase II. Science **319**, 1791–1792 (2008)

M.P. Cosma, T. Tanaka, K. Nasmyth, Ordered recruitment of transcription and chromatin remodeling factors to a cell cycle- and developmentally regulated promoter. Cell **97**, 299–311 (1999)

M.P. Creyghton, S. Markoulaki, S.S. Levine, J. Hanna, M.A. Lodato, K. Sha, R.A. Young, R. Jaenisch, L.A. Boyer, H2AZ is enriched at polycomb complex target genes in ES cells and is necessary for lineage commitment. Cell **135**, 649–661 (2008)

K.N. Dahl, A.J. Ribeiro, J. Lammerding, Nuclear shape, mechanics, and mechanotransduction. Circ. Res. **102**, 1307–1318 (2008)

M. Damelin, I. Simon, T.I. Moy, B. Wilson, S. Komili, P. Tempst, F.P. Roth, R.A. Young, B.R. Cairns, P.A. Silver, The genome-wide localization of Rsc9, a component of the RSC chromatin-remodeling complex, changes in response to stress. Mol. Cell **9**, 563–573 (2002)

L. David, W. Huber, M. Granovskaia, J. Toedling, C.J. Palm, L. Bofkin, T. Jones, R.W. Davis, L.M. Steinmetz, A high-resolution map of transcription in the yeast genome. Proc. Natl. Acad. Sci. U. S. A. **103**, 5320–5325 (2006)

C.A. Davis, M. Ares Jr., Accumulation of unstable promoter-associated transcripts upon loss of the nuclear exosome subunit Rrp6p in *Saccharomyces cerevisiae*. Proc. Natl. Acad. Sci. U. S. A. **103**, 3262–3267 (2006)

B. Dichtl, D. Blank, M. Ohnacker, A. Friedlein, D. Roeder, H. Langen, W. Keller, A role for SSU72 in balancing RNA polymerase II transcription elongation and termination. Mol. Cell **10**, 1139–1150 (2002a)

B. Dichtl, D. Blank, M. Sadowski, W. Hubner, S. Weiser, W. Keller, Yhh1p/Cft1p directly links poly(A) site recognition and RNA polymerase II transcription termination. Embo J. **21**, 4125–4135 (2002b)

M.F. Dion, T. Kaplan, M. Kim, S. Buratowski, N. Friedman, O.J. Rando, Dynamics of replication-independent histone turnover in budding yeast. Science **315**, 1405–1408 (2007)

D.E. Discher, D.J. Mooney, P.W. Zandstra, Growth factors, matrices, and forces combine and control stem cells. Science **324**, 1673–1677 (2009)

J. Dover, J. Schneider, M.A. Tawiah-Boateng, A. Wood, K. Dean, M. Johnston, A. Shilatifard, Methylation of histone H3 by COMPASS requires ubiquitination of histone H2B by Rad6. J. Biol. Chem. **277**, 28368–28371 (2002)

C.C. DuFort, M.J. Paszek, V.M. Weaver, Balancing forces: architectural control of mechanotransduction. Nat. Rev. Mol. Cell Biol. **12**, 308–319 (2011)

N. Dutrow, D.A. Nix, D. Holt, B. Milash, B. Dalley, E. Westbroek, T.J. Parnell, B.R. Cairns, Dynamic transcriptome of Schizosaccharomyces pombe shown by RNA-DNA hybrid mapping. Nat. Genet. **40**, 977–986 (2008)

R. Ebbert, A. Birkmann, H.J. Schuller, The product of the SNF2/SWI2 paralogue INO80 of *Saccharomyces cerevisiae* required for efficient expression of various yeast structural genes is part of a high-molecular-weight protein complex. Mol. Microbiol. **32**, 741–751 (1999)

S. Egloff, S. Murphy, Cracking the RNA polymerase II CTD code. Trends Genet. **24**, 280–288 (2008)

S. Egloff, D. O'Reilly, R.D. Chapman, A. Taylor, K. Tanzhaus, L. Pitts, D. Eick, S. Murphy, Serine-7 of the RNA polymerase II CTD is specifically required for snRNA gene expression. Science **318**, 1777–1779 (2007)

A.J. Engler, S. Sen, H.L. Sweeney, D.E. Discher, Matrix elasticity directs stem cell lineage specification. Cell **126**, 677–689 (2006)

C. Esnault, Y. Ghavi-Helm, S. Brun, J. Soutourina, N. Van Berkum, C. Boschiero, F. Holstege, M. Werner, Mediator-dependent recruitment of TFIIH modules in preinitiation complex. Mol. Cell **31**, 337–346 (2008)

T.G. Fazzio, C. Kooperberg, J.P. Goldmark, C. Neal, R. Basom, J. Delrow, T. Tsukiyama, Widespread collaboration of Isw2 and Sin3-Rpd3 chromatin remodeling complexes in transcriptional repression. Mol. Cell. Biol. **21**, 6450–6460 (2001)

J.F. Flanagan, L.Z. Mi, M. Chruszcz, M. Cymborowski, K.L. Clines, Y. Kim, W. Minor, F. Rastinejad, S. Khorasanizadeh, Double chromodomains cooperate to recognize the methylated histone H3 tail. Nature **438**, 1181–1185 (2005)

A. Flaus, T. Owen-Hughes, Mechanisms for ATP-dependent chromatin remodelling: farewell to the tuna-can octamer? Curr. Opin. Genet. Dev. **14**, 165–173 (2004)

A.B. Fleming, C.F. Kao, C. Hillyer, M. Pikaart, M.A. Osley, H2B ubiquitylation plays a role in nucleosome dynamics during transcription elongation. Mol. Cell **31**, 57–66 (2008)

K.A. Garbett, M.K. Tripathi, B. Cencki, J.H. Layer, P.A. Weil, Yeast TFIID serves as a coactivator for Rap1p by direct protein-protein interaction. Mol. Cell. Biol. **27**, 297–311 (2007)

M.L. Gardel, I.C. Schneider, Y. Aratyn-Schaus, C.M. Waterman, Mechanical integration of actin and adhesion dynamics in cell migration. Annu. Rev. Cell Dev. Biol. **26**, 315–333 (2010)

B. Geiger, J.P. Spatz, A.D. Bershadsky, Environmental sensing through focal adhesions. Nat. Rev. Mol. Cell Biol. **10**, 21–33 (2009)

G. Giaever, A.M. Chu, L. Ni, C. Connelly et al., Functional profiling of the *Saccharomyces cerevisiae* genome. Nature **418**, 387–391 (2002)

J.P. Goldmark, T.G. Fazzio, P.W. Estep, G.M. Church, T. Tsukiyama, The Isw2 chromatin remodeling complex represses early meiotic genes upon recruitment by Ume6p. Cell **103**, 423–433 (2000)

J.A. Goodrich, J.F. Kugel, From bacteria to humans, chromatin to elongation, and activation to repression: the expanding roles of noncoding RNAs in regulating transcription. Crit. Rev. Biochem. Mol. Biol. **44**(1), 3–15 (2008)

M.R. Green, Eukaryotic transcription activation: right on target. Mol. Cell **18**, 399–402 (2005)

S.I. Grewal, S.C. Elgin, Transcription and RNA interference in the formation of heterochromatin. Nature **447**, 399–406 (2007)

M.G. Guenther, S.S. Levine, L.A. Boyer, R. Jaenisch, R.A. Young, A chromatin landmark and transcription initiation at most promoters in human cells. Cell **130**, 77–88 (2007)

B. Guillemette, A.R. Bataille, N. Gevry, M. Adam, M. Blanchette, F. Robert, L. Gaudreau, Variant histone H2A.Z is globally localized to the promoters of inactive yeast genes and regulates nucleosome positioning. PLoS Biol. **3**, e384 (2005)

C. Hahn, M.A. Schwartz, Mechanotransduction in vascular physiology and atherogenesis. Nat. Rev. Mol. Cell Biol. **10**, 53–62 (2009)

M. Hampsey, D. Reinberg, Tails of intrigue: phosphorylation of RNA polymerase II mediates histone methylation. Cell **113**, 429–432 (2003)

M. Han, M. Grunstein, Nucleosome loss activates yeast downstream promoters in vivo. Cell **55**, 1137–1145 (1988)

J. Hani, B. Schelbert, A. Bernhardt, H. Domdey, G. Fischer, K. Wiebauer, J.U. Rahfeld, Mutations in a peptidylprolyl-cis/trans-isomerase gene lead to a defect in 3′-end formation of a pre-mRNA in *Saccharomyces cerevisiae*. J. Biol. Chem. **274**, 108–116 (1999)

C.T. Harbison, D.B. Gordon, T.I. Lee et al., Transcriptional regulatory code of a eukaryotic genome. Nature **431**, 99–104 (2004)

A.H. Hassan, K.E. Neely, J.L. Workman, Histone acetyltransferase complexes stabilize SWI/SNF binding to promoter nucleosomes. Cell **104**, 817–827 (2001)

A.H. Hassan, P. Prochasson, K.E. Neely, S.C. Galasinski, M. Chandy, M.J. Carrozza, J.L. Workman, Function and selectivity of bromodomains in anchoring chromatin-modifying complexes to promoter nucleosomes. Cell **111**, 369–379 (2002)

S. Hausmann, H. Koiwa, S. Krishnamurthy, M. Hampsey, S. Shuman, Different strategies for carboxyl-terminal domain (CTD) recognition by serine 5-specific CTD phosphatases. J. Biol. Chem. **280**, 37681–37688 (2005)

K.W. Henry, A. Wyce, W.S. Lo, L.J. Duggan, N.C. Emre, C.F. Kao, L. Pillus, A. Shilatifard, M.A. Osley, S.L. Berger, Transcriptional activation via sequential histone H2B ubiquitylation and deubiquitylation, mediated by SAGA-associated Ubp8. Genes Dev. **17**, 2648–2663 (2003)

D.R. Hewish, L.A. Burgoyne, Chromatin sub-structure. The digestion of chromatin DNA at regularly spaced sites by a nuclear deoxyribonuclease. Biochem. Biophys. Res. Commun. **52**, 504–510 (1973)

J.N. Hirschhorn, S.A. Brown, C.D. Clark, F. Winston, Evidence that SNF2/SWI2 and SNF5 activate transcription in yeast by altering chromatin structure. Genes Dev. **6**, 2288–2298 (1992)

B.D. Hoffman, C. Grashoff, M.A. Schwartz, Mechanotransduction in vascular physiology and atherogenesis. Nature **475**, 316–323 (2011)

J. Houseley, J. LaCava, D. Tollervey, RNA-quality control by the exosome. Nat. Rev. Mol. Cell Biol. **7**, 529–539 (2006)

J. Houseley, L. Rubbi, M. Grunstein, D. Tollervey, M. Vogelauer, A ncRNA modulates histone modification and mRNA induction in the yeast GAL gene cluster. Mol. Cell **32**, 685–695 (2008)

A.N. Iberg, A. Espejo, D. Cheng, D. Kim, J. Michaud-Levesque, S. Richard, M.T. Bedford, Arginine methylation of the histone H3 tail impedes effector binding. J. Biol. Chem. **283**, 3006–3010 (2008)

J.A. Inostroza, F.H. Mermelstein, I. Ha, W.S. Lane, D. Reinberg, Dr1, a TATA-binding protein-associated phosphoprotein and inhibitor of class II gene transcription. Cell **70**, 477–489 (1992)

V. Iyer, K. Struhl, Poly(dA:dT), a ubiquitous promoter element that stimulates transcription via its intrinsic DNA structure. Embo J. **14**, 2570–2579 (1995)

D.E. Jaalouk, J. Lammerding, Mechanotransduction gone awry. Nat. Rev. Mol. Cell Biol. **10**, 63–73 (2009)

S.M. Johnson, F.J. Tan, H.L. McCullough, D.P. Riordan, A.Z. Fire, Flexibility and constraint in the nucleosome core landscape of *Caenorhabditis elegans* chromatin. Genome Res. **16**, 1505–1516 (2006)

R. Jothi, S. Cuddapah, A. Barski, K. Cui, K. Zhao, Genome-wide identification of in vivo protein-DNA binding sites from ChIP-Seq data. Nucleic Acids Res. **36**, 5221–5231 (2008)

C.F. Kao, C. Hillyer, T. Tsukuda, K. Henry, S. Berger, M.A. Osley, Rad6 plays a role in transcriptional activation through ubiquitylation of histone H2B. Genes Dev. **18**, 184–195 (2004)

P. Kapranov, J. Cheng, S. Dike et al., RNA maps reveal new RNA classes and a possible function for pervasive transcription. Science **316**, 1484–1488 (2007)

P.S. Kayne, U.J. Kim, M. Han, J.R. Mullen, F. Yoshizaki, M. Grunstein, Extremely conserved histone H4 N terminus is dispensable for growth but essential for repressing the silent mating loci in yeast. Cell**55**, 27–39 (1988)

M.C. Keogh, T.A. Mennella, C. Sawa et al., The *Saccharomyces cerevisiae* histone H2A variant Htz1 is acetylated by NuA4. Genes Dev. **20**, 660–665 (2006)

T.H. Kim, B. Ren, Genome-wide analysis of protein-DNA interactions. Annu. Rev. Genomics Hum. Genet. **7**, 81–102 (2006)

Y.J. Kim, S. Bjorklund, Y. Li, M.H. Sayre, R.D. Kornberg, A multiprotein mediator of transcriptional activation and its interaction with the C-terminal repeat domain of RNA polymerase II. Cell **77**, 599–608 (1994)

M. Kim, N.J. Krogan, L. Vasiljeva, O.J. Rando, E. Nedea, J.F. Greenblatt, S. Buratowski, The yeast Rat1 exonuclease promotes transcription termination by RNA polymerase II. Nature**432**, 517–522 (2004)

A. Kirmizis, H. Santos-Rosa, C.J. Penkett, M.A. Singer, M. Vermeulen, M. Mann, J. Bahler, R.D. Green, T. Kouzarides, Arginine methylation at histone H3R2 controls deposition of H3K4 trimethylation. Nature**449**, 928–932 (2007)

K.O. Kizer, H.P. Phatnani, Y. Shibata, H. Hall, A.L. Greenleaf, B.D. Strahl, A novel domain in Set2 mediates RNA polymerase II interaction and couples histone H3 K36 methylation with transcript elongation. Mol. Cell. Biol. **25**, 3305–3316 (2005)

R.J. Klose, Y. Zhang, Regulation of histone methylation by demethylimination and demethylation. Nat. Rev. Mol. Cell Biol. **8**, 307–318 (2007)

P. Komarnitsky, E.J. Cho, S. Buratowski, Different phosphorylated forms of RNA polymerase II and associated mRNA processing factors during transcription. Genes Dev. **14**, 2452–2460 (2000)

K.A. Kopcewicz, T.W. O'Rourke, D. Reines, Metabolic regulation of IMD2 transcription and an unusual DNA element that generates short transcripts. Mol. Cell. Biol. **27**, 2821–2829 (2007)

R.D. Kornberg, Mediator and the mechanism of transcriptional activation. Trends Biochem. Sci. **30**, 235–239 (2005)

T. Kouzarides, Chromatin modifications and their function. Cell **128**, 693–705 (2007)

S. Krishnamurthy, X. He, M. Reyes-Reyes, C. Moore, M. Hampsey, Ssu72 Is an RNA polymerase II CTD phosphatase. Mol. Cell**14** 387–394 (2004)

N.J. Krogan, K. Baetz, M.C. Keogh et al., Regulation of chromosome stability by the histone H2A variant Htz1, the Swr1 chromatin remodeling complex, and the histone acetyltransferase NuA4. Proc. Natl. Acad. Sci. U. S. A. **101**, 13513–13518 (2004)

J.N. Kuehner, D.A. Brow, Regulation of a eukaryotic gene by GTP-dependent start site selection and transcription attenuation. Mol. Cell **31**, 201–211 (2008)

N. Lacoste, R.T. Utley, J.M. Hunter, G.G. Poirier, J. Cote, Disruptor of telomeric silencing-1 is a chromatin-specific histone H3 methyltransferase. J. Biol. Chem. **277**, 30421–30424 (2002)

R.N. Laribee, N.J. Krogan, T. Xiao, Y. Shibata, T.R. Hughes, J.F. Greenblatt, B.D. Strahl, BUR kinase selectively regulates H3 K4 trimethylation and H2B ubiquitylation through recruitment of the PAF elongation complex. Curr. Biol. **15**, 1487–1493 (2005)

E. Larschan, F. Winston, The *S. cerevisiae* SAGA complex functions in vivo as a coactivator for transcriptional activation by Gal4. Genes Dev. **15**, 1946–1956 (2001)

J.M. Lee, A.L. Greenleaf, Modulation of RNA polymerase II elongation efficiency by C-terminal heptapeptide repeat domain kinase I. J. Biol. Chem. **272**, 10990–10993 (1997)

J.S. Lee, A. Shilatifard, A site to remember: H3K36 methylation a mark for histone deacetylation. Mutat. Res. **618**, 130–134 (2007)

T.I. Lee, R.A. Young, Regulation of gene expression by TBP-associated proteins. Genes Dev. **12**, 1398–1408 (1998)

T.I. Lee, N.J. Rinaldi, F. Robert et al., Transcriptional regulatory networks in *Saccharomyces cerevisiae*. Science **298**, 799–804 (2002)

C. Lee, X. Li, A. Hechmer, M. Eisen, M.D. Biggin, B.J. Venters, C. Jiang, J. Li, B.F. Pugh, D.S. Gilmour, NELF and GAGA factor are linked to promoter-proximal pausing at many genes in *Drosophila*. Mol. Cell. Biol. **28**, 3290–3300 (2008)

S.S. Levine, I.F. King, R.E. Kingston, Division of labor in polycomb group repression. Trends Biochem. Sci. **29**, 478–485 (2004)

B. Li, J.C. Reese, Ssn6-Tup1 regulates RNR3 by positioning nucleosomes and affecting the chromatin structure at the upstream repression sequence. J. Biol. Chem. **276**, 33788–33797 (2001)

J. Li, D. Moazed, S.P. Gygi, Association of the histone methyltransferase Set2 with RNA polymerase II plays a role in transcription elongation. J. Biol. Chem. **77**, 49383–49388 (2002)

B. Li, L. Howe, S. Anderson, J.R. Yates 3rd, J.L. Workman, The Set2 histone methyltransferase functions through the phosphorylated carboxyl-terminal domain of RNA polymerase II. J. Biol. Chem. **278**, 8897–8903 (2003)

B. Li, S.G. Pattenden, D. Lee, J. Gutierrez, J. Chen, C. Seidel, J. Gerton, J.L. Workman, Preferential occupancy of histone variant H2AZ at inactive promoters influences local histone modifications and chromatin remodeling. Proc. Natl. Acad. Sci. U. S. A. **102**, 18385–18390 (2005)

B. Li, M. Carey, J.L. Workman, The role of chromatin during transcription. Cell **128**, 707–719 (2007a)

B. Li, M. Gogol, M. Carey, D. Lee, C. Seidel, J.L. Workman, Combined action of PHD and chromo domains directs the Rpd3S HDAC to transcribed chromatin. Science **316**, 1050–1054 (2007b)

B. Li, M. Gogol, M. Carey, S.G. Pattenden, C. Seidel, J.L. Workman, Infrequently transcribed long genes depend on the Set2/Rpd3S pathway for accurate transcription. Genes Dev. **21**, 1422–1430 (2007c)

B. Li, J. Jackson, M.D. Simon, B. Fleharty, M. Gogol, C. Seidel, J.L. Workman, A. Shilatifard, Histone H3 lysine 36 dimethylation (H3K36ME2) is sufficient to recruit the Rpd3S histone deacetylase complex and to repress spurious transcription. J. Biol. Chem. **284**(12), 7970–7976 (2009)

D.D. Licatalosi, G. Geiger, M. Minet, S. Schroeder, K. Cilli, J.B. McNeil, D.L. Bentley, Functional interaction of yeast pre-mRNA 3′ end processing factors with RNA polymerase II. Mol. Cell **9**, 1101–1111 (2002)

C.L. Liu, T. Kaplan, M. Kim, S. Buratowski, S.L. Schreiber, N. Friedman, O.J. Rando, Single-nucleosome mapping of histone modifications in *S. cerevisiae*. PLoS Biol. **3**, e328 (2005)

D. Lohr, Nucleosome transactions on the promoters of the yeast GAL and PHO genes. J. Biol. Chem. **272**, 26795–26798 (1997)

Y. Lorch, J.W. LaPointe, R.D. Kornberg, Nucleosomes inhibit the initiation of transcription but allow chain elongation with the displacement of histones. Cell **49**, 203–210 (1987)

Y. Lorch, B. Maier-Davis, R.D. Kornberg, Chromatin remodeling by nucleosome disassembly in vitro. Proc. Natl. Acad. Sci. U. S. A. **103**, 3090–3093 (2006)

K. Luger, A.W. Mader, R.K. Richmond, D.F. Sargent, T.J. Richmond, Crystal structure of the nucleosome core particle at 2.8 A resolution. Nature **389**, 251–260 (1997)

V.V. Lunyak, M.G. Rosenfeld, Epigenetic regulation of stem cell fate. Hum. Mol. Genet. **17**, R28–R36 (2008)

T. Margaritis, F.C. Holstege, Poised RNA polymerase II gives pause for thought. Cell **133**, 581–584 (2008)

M. Margulies, M. Egholm, W.E. Altman et al., Genome sequencing in microfabricated high-density picolitre reactors. Nature **437**, 376–380 (2005)

N.F. Marshall, J. Peng, Z. Xie, D.H. Price, Control of RNA polymerase II elongation potential by a novel carboxyl-terminal domain kinase. J. Biol. Chem. **271**, 27176–27183 (1996)

C. Martens, B. Krett, P.J. Laybourn, RNA polymerase II and TBP occupy the repressed CYC1 promoter. Mol. Microbiol. **40**, 1009–1019 (2001)

J.A. Martens, L. Laprade, F. Winston, Intergenic transcription is required to repress the *Saccharomyces cerevisiae* SER3 gene. Nature **429**, 571–574 (2004)

J.A. Martens, P.Y. Wu, F. Winston, Regulation of an intergenic transcript controls adjacent gene transcription in *Saccharomyces cerevisiae*. Genes Dev. **19**, 2695–2704 (2005)

C. Martinez-Campa, P. Politis, J.L. Moreau, N. Kent, J. Goodall, J. Mellor, C.R. Goding, Precise nucleosome positioning and the TATA box dictate requirements for the histone H4 tail and the bromodomain factor Bdf1. Mol. Cell **15**, 69–81 (2004)

T. Matsui, J. Segall, P.A. Weil, R.G. Roeder, Multiple factors required for accurate initiation of transcription by purified RNA polymerase II. J. Biol. Chem. **255**, 11992–11996 (1980)

T.N. Mavrich, I.P. Ioshikhes, B.J. Venters, C. Jiang, L.P. Tomsho, J. Qi, S.C. Schuster, I. Albert, B.F. Pugh, A barrier nucleosome model for statistical positioning of nucleosomes throughout the yeast genome. Genome Res. **18**, 1073–1083 (2008a)

T.N. Mavrich, C. Jiang, I.P. Ioshikhes et al., Nucleosome organization in the *Drosophila* genome. Nature **453**, 358–362 (2008b)

L.E. McNamara, R. Burchmore, M.O. Riehle, P. Herzyk, M.J. Biggs, C.D. Wilkinson, A.S. Curtis, M.J. Dalby, The role of microtopography in cellular mechanotransduction. Biomaterials **33**, 2835–2847 (2012)

P.C. Megee, B.A. Morgan, B.A. Mittman, M.M. Smith, Genetic analysis of histone H4: essential role of lysines subject to reversible acetylation. Science **247**, 841–845 (1990)

A. Meinhart, T. Kamenski, S. Hoeppner, S. Baumli, P. Cramer, A structural perspective of CTD function. Genes Dev. **19**, 1401–1415 (2005)

M. Mencia, Z. Moqtaderi, J.V. Geisberg, L. Kuras, K. Struhl, Activator-specific recruitment of TFIID and regulation of ribosomal protein genes in yeast. Mol. Cell **9**, 823–833 (2002)

C.B. Millar, M. Grunstein, Genome-wide patterns of histone modifications in yeast. Nat. Rev. Mol. Cell Biol. **7**, 657–666 (2006)

B.A. Morgan, B.A. Mittman, M.M. Smith, The highly conserved N-terminal domains of histones H3 and H4 are required for normal cell cycle progression. Mol. Cell. Biol. **11**, 4111–4120 (1991)

A. Mortazavi, B.A. Williams, K. McCue, L. Schaeffer, B. Wold, Mapping and quantifying mammalian transcriptomes by RNA-Seq. Nat. Methods **5**, 621–628 (2008)

M.R. Motamedi, E.J. Hong, X. Li, S. Gerber, C. Denison, S. Gygi, D. Moazed, HP1 proteins form distinct complexes and mediate heterochromatic gene silencing by nonoverlapping mechanisms. Mol. Cell**32**, 778–790 (2008)

T.A. Muldrow, A.M. Campbell, P.A. Weil, D.T. Auble, MOT1 can activate basal transcription in vitro by regulating the distribution of TATA binding protein between promoter and nonpromoter sites. Mol. Cell. Biol. **19**, 2835–2845.

G.W. Muse, D.A. Gilchrist, S. Nechaev, R. Shah, J.S. Parker, S.F. Grissom, J. Zeitlinger, K. Adelman, RNA polymerase is poised for activation across the genome. Nat. Genet. **39**, 1507–1511 (2007)

L.C. Myers, C.M. Gustafsson, D.A. Bushnell, M. Lui, H. Erdjument-Bromage, P. Tempst, R.D. Kornberg, The Med proteins of yeast and their function through the RNA polymerase II carboxy-terminal domain. Genes Dev. **12**, 45–54 (1998)

S. Na, O. Collin, F. Chowdhury, B. Tay, M. Ouyang, Y. Wang, N. Wang, Rapid signal transduction in living cells is a unique feature of mechanotransduction. Proc. Natl. Acad. Sci. **105**, 6626–6631 (2008)

U. Nagalakshmi, Z. Wang, K. Waern, C. Shou, D. Raha, M. Gerstein, M. Snyder, The transcriptional landscape of the yeast genome defined by RNA sequencing. Science **320**, 1344–1349 (2008)

S. Nakanishi, B.W. Sanderson, K.M. Delventhal, W.D. Bradford, K. Staehling-Hampton, A. Shilatifard, A comprehensive library of histone mutants identifies nucleosomal residues required for H3K4 methylation. Nat. Struct. Mol. Biol. **15**, 881–888 (2008)

K. Natarajan, B.M. Jackson, H. Zhou, F. Winston, A.G. Hinnebusch, Transcriptional activation by Gcn4p involves independent interactions with the SWI/SNF complex and the SRB/mediator. Mol. Cell **4**, 657–664 (1999)

K.E. Neely, A.H. Hassan, C.E. Brown, L. Howe, J.L. Workman, Transcription activator interactions with multiple SWI/SNF subunits. Mol. Cell. Biol. **22**, 1615–1625 (2002)

L. Neigeborn, M. Carlson, Genes affecting the regulation of SUC2 gene expression by glucose repression in *Saccharomyces cerevisiae*. Genetics **108**, 845–858 (1984)

H. Neil, C. Malabat, Y. d'Aubenton-Carafa, Z. Xu, L.M. Steinmetz, A. Jacquier, Widespread bidirectional promoters are the major source of cryptic transcripts in yeast. Nature **457**, 1038–1042 (2009)

C.J. Nelson, H. Santos-Rosa, T. Kouzarides, Proline isomerization of histone H3 regulates lysine methylation and gene expression. Cell **126**, 905–916 (2006)

H.H. Ng, F. Robert, R.A. Young, K. Struhl, Genome-wide location and regulated recruitment of the RSC nucleosome-remodeling complex. Genes Dev. **16**, 806–819 (2002)

H.H. Ng, S. Dole, K. Struhl, The Rtf1 component of the Paf1 transcriptional elongation complex is required for ubiquitination of histone H2B. J. Biol. Chem. **278**, 33625–33628 (2003a)

H.H. Ng, F. Robert, R.A. Young, K. Struhl, Targeted recruitment of Set1 histone methylase by elongating Pol II provides a localized mark and memory of recent transcriptional activity. Mol. Cell **11**, 709–719 (2003b)

R. Nielsen, T.A. Pedersen, D. Hagenbeek et al., Genome-wide profiling of PPARγ:RXR and RNA polymerase II occupancy reveals temporal activation of distinct metabolic pathways and changes in RXR dimer composition during adipogenesis. Genes Dev. **22**, 2953–2967 (2008)

D.B. Nikolov, H. Chen, E.D. Halay, A.A. Usheva, K. Hisatake, D.K. Lee, R.G. Roeder, S.K. Burley, Crystal structure of a TFIIB-TBP-TATA-element ternary complex. Nature **377**, 119–128 (1995)

A. Nourani, R.T. Utley, S. Allard, J. Cote, Recruitment of the NuA4 complex poises the PHO5 promoter for chromatin remodeling and activation. Embo J. **23**, 2597–2607 (2004)

A.L. Olins, D.E. Olins, Spheroid chromatin units (v bodies). Science **183**, 330–332 (1974)

E.N. Olson, A. Nordheim, Linking actin dynamics and gene transcription to drive cellular motile functions. Nat. Rev. Mol. Cell Biol. **11**, 353–365 (2010)

V. Orlando, H. Strutt, R. Paro, Analysis of chromatin structure by in vivo formaldehyde cross-linking. Methods **11**, 205–214 (1997)

G. Orphanides, D. Reinberg, A unified theory of gene expression. Cell **108**, 439–451 (2002)

D. Panne, The enhanceosome. Curr. Opin. Struct. Biol. **18**, 236–242 (2008)

J.M. Park, H.S. Kim, S.J. Han, M.S. Hwang, Y.C. Lee, Y.J. Kim, In vivo requirement of activator-specific binding targets of mediator. Mol. Cell. Biol. **20**, 8709–8719 (2000)

T.J. Parnell, J.T. Huff, B.R. Cairns, RSC regulates nucleosome positioning at Pol II genes and density at Pol III genes. Embo J. **27**, 100–110 (2008)

M.R. Parthun, J. Widom, D.E. Gottschling, The major cytoplasmic histone acetyltransferase in yeast: links to chromatin replication and histone metabolism. Cell **87**, 85–94 (1996)

R.J. Pelham, Y.-L. Wang, Cell locomotion and focal adhesions are regulated by substrate flexibility. Proc. Natl. Acad. Sci. U.S. A. **94**, 13661–13665 (1997)

B. Peñalver Bernabé, S. Shin, P.D. Rios, L.J. Broadbelt, L.D. Shea, S.K. Seidlits, Dynamic transcription factor activity networks in response to independently altered mechanical and adhesive microenvironmental cues. Integr. Biol. (Camb.) **8**(8), 844–860 (2016)

J.E. Perez-Ortin, F. Estruch, E. Matallana, L. Franco, Fine analysis of the chromatin structure of the yeast SUC2 gene and of its changes upon derepression. Comparison between the chromosomal and plasmid inserted genes. Nucleic Acids Res. **15**, 6937–6956 (1987)

F. Perocchi, Z. Xu, S. Clauder-Munster, L.M. Steinmetz, Antisense artifacts in transcriptome microarray experiments are resolved by actinomycin D. Nucleic Acids Res. **35**, e128 (2007)

B.M. Peterlin, D.H. Price, Controlling the elongation phase of transcription with P-TEFb. Mol. Cell **23**, 297–305 (2006)

S.R. Peyton, C.M. Ghajar, C.B. Khatiwala, A.J. Putnam, The emergence of ECM mechanics and cytoskeletal tension as important regulators of cell function. Cell Biochem. Biophys. **47**, 300–320 (2007)

H.P. Phatnani, A.L. Greenleaf, Phosphorylation and functions of the RNA polymerase II CTD. Genes Dev. **20**, 2922–2936 (2006)

H.P. Phatnani, J.C. Jones, A.L. Greenleaf, Expanding the functional repertoire of CTD kinase I and RNA polymerase II: novel phosphoCTD-associating proteins in the yeast proteome. Biochemistry **43**, 15702–15719 (2004)

Y.-C. Poh, S. Na, F. Chowdhury, M. Ouyang, Y. Wang, N. Wang, Rapid activation of Rac GTPase in living cells by force is independent of Src. PLoS ONE **4**, e7886 (2009)

Y.-C. Poh, S.P. Shevtsov, F. Chowdhury, D.C. Wu, S. Na, M. Dundr, N. Wang, Dynamic force-induced direct dissociation of protein complexes in a nuclear body in living cells. Nat. Commun. **3**, 866 (2012)

D.K. Pokholok, C.T. Harbison, S. Levine et al., Genome-wide map of nucleosome acetylation and methylation in yeast. Cell **122**, 517–527 (2005)

M.G. Pray-Grant, J.A. Daniel, D. Schieltz, J.R. Yates 3rd, P.A. Grant, Chd1 chromodomain links histone H3 methylation with SAGA- and SLIK-dependent acetylation. Nature **433**, 434–438 (2005)

D.H. Price, Poised polymerases: on your mark…get set…go! Mol. Cell **30**, 7–10 (2008)

M. Ptashne, A. Gann, Transcriptional activation by recruitment. Nature **386**, 569–577 (1997)

B.F. Pugh, Control of gene expression through regulation of the TATA-binding protein. Gene **255**, 1–14 (2000)

B.F. Pugh, D.S. Gilmour, Genome-wide analysis of protein-DNA interactions in living cells. Genome Biol. **2**, REVIEWS1013 (2001)

B.F. Pugh, R. Tjian, Mechanism of transcriptional activation by Sp1: evidence for coactivators. Cell **61**, 1187–1197 (1990)

M. Radonjic, J.C. Andrau, P. Lijnzaad, P. Kemmeren, T.T. Kockelkorn, D. van Leenen, N.L. van Berkum, F.C. Holstege, Genome-wide analyses reveal RNA polymerase II located upstream of genes poised for rapid response upon *S. cerevisiae* stationary phase exit. Mol. Cell **18**, 171–183 (2005)

R.M. Raisner, P.D. Hartley, M.D. Meneghini, M.Z. Bao, C.L. Liu, S.L. Schreiber, O.J. Rando, H.D. Madhani, Histone variant H2A.Z marks the 5′ ends of both active and inactive genes in euchromatin. Cell **123**, 233–248 (2005)

B. Rao, Y. Shibata, B.D. Strahl, J.D. Lieb, Dimethylation of histone H3 at lysine 36 demarcates regulatory and nonregulatory chromatin genome-wide. Mol. Cell. Biol. **25**, 9447–9459 (2005)

A. Ravindra, K. Weiss, R.T. Simpson, High-resolution structural analysis of chromatin at specific loci: *Saccharomyces cerevisiae* silent mating-type locus HMRa. Mol. Cell. Biol. **19**, 7944–7950 (1999)

J.C. Reese, L. Apone, S.S. Walker, L.A. Griffin, M.R. Green, Yeast TAFIIS in a multisubunit complex required for activated transcription. Nature **371**, 523–527 (1994)

B. Ren, F. Robert, J.J. Wyrick, O. Aparicio et al., Genome-wide location and function of DNA binding proteins. Science **290**, 2306–2309 (2000)

G. Robertson, M. Hirst, M. Bainbridge et al., Genome-wide profiles of STAT1 DNA association using chromatin immunoprecipitation and massively parallel sequencing. Nat. Methods **4**, 651–657 (2007)

K. Robzyk, J. Recht, M.A. Osley, Rad6-dependent ubiquitination of histone H2B in yeast. Science **287**, 501–504 (2000)

A.J. Ruthenburg, H. Li, D.J. Patel, C.D. Allis, Multivalent engagement of chromatin modifications by linked binding modules. Nat. Rev. Mol. Cell Biol. **8**, 983–994 (2007)

A. Saha, J. Wittmeyer, B.R. Cairns, Chromatin remodelling: the industrial revolution of DNA around histones. Nat. Rev. Mol. Cell Biol. **7**, 437–447 (2006)

A.J. Saurin, Z. Shao, H. Erdjument-Bromage, P. Tempst, R.E. Kingston, A *Drosophila* Polycomb group complex includes Zeste and dTAFII proteins. Nature **412**, 655–660 (2001)

M.H. Sayre, H. Tschochner, R.D. Kornberg, Reconstitution of transcription with five purified initiation factors and RNA polymerase II from *Saccharomyces cerevisiae*. J. Biol. Chem. **267**, 23376–23382 (1992)

J. Schneider, P. Bajwa, F.C. Johnson, S.R. Bhaumik, A. Shilatifard, Rtt109 is required for proper H3K56 acetylation: a chromatin mark associated with the elongating RNA polymerase II. J. Biol. Chem. **281**, 37270–37274 (2006)

D.E. Schones, K. Cui, S. Cuddapah, T.Y. Roh, A. Barski, Z. Wang, G. Wei, K. Zhao, Dynamic regulation of nucleosome positioning in the human genome. Cell **132**, 887–898 (2008)

S.C. Schuster, Next-generation sequencing transforms today's biology. Nat. Methods **5**, 16–18 (2008)

M.A. Schwabish, K. Struhl, The Swi/Snf complex is important for histone eviction during transcriptional activation and RNA polymerase II elongation in vivo. Mol. Cell. Biol. **27**, 6987–6995 (2007)

M.A. Schwartz, D.W. DeSimone, Cell adhesion receptors in mechanotransduction. Curr. Opin. Cell Biol. **20**, 551–556 (2008)

U.S. Schwarz, M.L. Gardel, United we stand: integrating the actin cytoskeleton and cell-matrix adhesions in cellular mechanotransduction. J. Cell Sci. **125**, 3051–3060 (2012)

E.A. Sekinger, D.S. Gross, Silenced chromatin is permissive to activator binding and PIC recruitment. Cell **105**, 403–414 (2001)

D. Sermwittayawong, S. Tan, SAGA binds TBP via its Spt8 subunit in competition with DNA: implications for TBP recruitment. Embo J. **25**, 3791–3800 (2006)

A. Sharili, J. Connelly, Nucleocytoplasmic shuttling: a common theme in mechanotransduction. Biochem. Soc. Trans. **42**, 645–649 (2014)

X. Shen, G. Mizuguchi, A. Hamiche, C. Wu, A chromatin remodelling complex involved in transcription and DNA processing. Nature **406**, 541–544 (2000)

Y. Shi, J.R. Whetstine, Dynamic regulation of histone lysine methylation by demethylases. Mol. Cell **25**, 1–14 (2007)

A. Shilatifard, Molecular implementation and physiological roles for histone H3 lysine 4 (H3K4) methylation. Curr. Opin. Cell Biol. **20**, 341–348 (2008)

K. Shimada, Y. Oma, T. Schleker, K. Kugou, K. Ohta, M. Harata, S.M. Gasser, Ino80 chromatin remodeling complex promotes recovery of stalled replication forks. Curr. Biol. **18**, 566–575 (2008)

S. Shivaswamy, A. Bhinge, Y. Zhao, S. Jones, M. Hirst, V.R. Iyer, Dynamic remodeling of individual nucleosomes across a eukaryotic genome in response to transcriptional perturbation. PLoS Biol. **6**, e65 (2008)

A. Shukla, N. Stanojevic, Z. Duan, P. Sen, S.R. Bhaumik, Ubp8p, a histone deubiquitinase whose association with SAGA is mediated by Sgf11p, differentially regulates lysine 4 methylation of histone H3 in vivo. Mol. Cell. Biol. **26**, 3339–3352 (2006)

A. Siletz, M. Schnabel, E. Kniazeva, A.J. Schumacher, S. Shin, J.S. Jeruss, L.D. Shea, Dynamic transcription factor networks in epithelial-mesenchymal transition in breast cancer models. PLoS ONE **8**, e57180 (2013)

S.T. Smale, J.T. Kadonaga, The RNA polymerase II core promoter. Annu. Rev. Biochem. **72**, 449–479 (2003)

M.J. Solomon, A. Varshavsky, Formaldehyde-mediated DNA-protein crosslinking: a probe for in vivo chromatin structures. Proc. Natl. Acad. Sci. U. S. A. **82**, 6470–6474 (1985)

R.O. Sprouse, T.S. Karpova, F. Mueller, A. Dasgupta, J.G. McNally, D.T. Auble, Regulation of TATA-binding protein dynamics in living yeast cells. Proc. Natl. Acad. Sci. U. S. A. **105**, 13304–13308 (2008a)

R.O. Sprouse, M.N. Wells, D.T. Auble, TATA-binding protein variants that bypass the requirement for Mot1 in vivo. J. Biol. Chem. **284**(7), 4525–4535 (2008b)

E.J. Steinmetz, C.L. Warren, J.N. Kuehner, B. Panbehi, A.Z. Ansari, D.A. Brow, Genome-wide distribution of yeast RNA polymerase II and its control by Sen1 helicase. Mol. Cell **24**, 735–746 (2006)

M. Stern, R. Jensen, I. Herskowitz, Five SWI genes are required for expression of the HO gene in yeast. J. Mol. Biol. **178**, 853–868 (1984)

D.E. Sterner, S.L. Berger, Acetylation of histones and transcription-related factors. Microbiol. Mol. Biol. Rev. **64**, 435–459 (2000)

D.E. Sterner, J.M. Lee, S.E. Hardin, A.L. Greenleaf, The yeast carboxyl-terminal repeat domain kinase CTDK-I is a divergent cyclin-cyclin-dependent kinase complex. Mol. Cell. Biol. **15**, 5716–5724 (1995)

K. Struhl, Transcriptional noise and the fidelity of initiation by RNA polymerase II. Nat. Struct. Mol. Biol. **14**, 103–105 (2007)

K. Struhl, D. Kadosh, M. Keaveney, L. Kuras, Z. Moqtaderi, Activation and repression mechanisms in yeast. Cold Spring Harb. Symp. Quant. Biol. **63**, 413–421 (1998)

P. Sudarsanam, V.R. Iyer, P.O. Brown, F. Winston, Whole-genome expression analysis of SNF/SWI mutants of *Saccharomyces cerevisiae*. Proc. Natl. Acad. Sci. U. S. A. **97**, 3364–3369 (2000)

M. Sultan, M.H. Schulz, H. Richard et al., A global view of gene activity and alternative splicing by deep sequencing of the human transcriptome. Science **321**, 956–960 (2008)

Z.W. Sun, C.D. Allis, Ubiquitination of histone H2B regulates H3 methylation and gene silencing in yeast. Nature**418**, 104–108 (2002)

K. Takahashi, K. Tanabe, M. Ohnuki, M. Narita, T. Ichisaka, K. Tomoda, S. Yamanaka, Induction of pluripotent stem cells from adult human fibroblasts by defined factors. Cell **131**, 861–872 (2007)

S. Tan, Y. Hunziker, D.F. Sargent, T.J. Richmond, Crystal structure of a yeast TFIIA/TBP/DNA complex. Nature **381**, 127–151 (1996)

M. Thiebaut, J. Colin, H. Neil, A. Jacquier, B. Seraphin, F. Lacroute, D. Libri, Futile cycle of transcription initiation and termination modulates the response to nucleotide shortage in *S. cerevisiae*. Mol. Cell **31**, 671–682 (2008)

M.C. Thomas, C.M. Chiang, The general transcription machinery and general cofactors. Crit. Rev. Biochem. Mol. Biol. **41**, 105–178 (2006)

J. Toth, M.D. Biggin, The specificity of protein-DNA crosslinking by formaldehyde: in vitro and in *Drosophila* embryos. Nucleic Acids Res. **28**, e4 (2000)

H.G. Tran, D.J. Steger, V.R. Iyer, A.D. Johnson, The chromo domain protein chd1p from budding yeast is an ATP-dependent chromatin-modifying factor. Embo J. **19**, 2323–2331 (2000)

B. Trappmann, J.E. Gautrot, J.T. Connelly, D.G. Strange, Y. Li, M.L. Oyen, M.A.C. Stuart, H. Boehm, B. Li, V. Vogel, Extracellular-matrix tethering regulates stem-cell fate. Nat. Mater. **11**, 642–649 (2012)

P. Trojer, J. Zhang, M. Yonezawa, A. Schmidt, H. Zheng, T. Jenuwein, D. Reinberg, Dynamic histone H1 isotype 4 methylation and demethylation by histone lysine methyltransferase G9A/KMT1C and the jumonji domain containing JMJD2/KDM4 proteins. J. Biol. Chem. **284**(13), 8395–8405 (2009)

T. Tsukiyama, The in vivo functions of ATP-dependent chromatin-remodelling factors. Nat. Rev. Mol. Cell Biol. **3**, 422–429 (2002)

J.P. Uhler, C. Hertel, J.Q. Svejstrup, A role for noncoding transcription in activation of the yeast PHO5 gene. Proc. Natl. Acad. Sci. U. S. A. **104**, 8011–8016 (2007)

J.G. Valay, M. Simon, M.F. Dubois, O. Bensaude, C. Facca, G. Faye, The KIN28 gene is required both for RNA polymerase II mediated transcription and phosphorylation of the Rpb1p CTD. J. Mol. Biol. **249**, 535–544 (1995)

A. Valouev, J. Ichikawa, T. Tonthat et al., A high-resolution, nucleosome position map of *C. elegans* reveals a lack of universal sequence-dictated positioning. Genome Res. **18**, 1051–1063 (2008)

L. Vasiljeva, M. Kim, H. Mutschler, S. Buratowski, A. Meinhart, The Nrd1-Nab3-Sen1 termination complex interacts with the Ser5-phosphorylated RNA polymerase II C-terminal domain. Nat. Struct. Mol. Biol. **15**, 795–804 (2008)

B.J. Venters, F. Pugh, A canonical promoter organization of the transcription machinery and its regulators in the *Saccharomyces* genome. Genome Res. **19**(3), 360–371 (2009)

M. Vermeulen, K.W. Mulder, S. Denissov, W.W. Pijnappel, F.M. van Schaik, R.A. Varier, M.P. Baltissen, H.G. Stunnenberg, M. Mann, H.T. Timmers, Selective anchoring of TFIID to nucleosomes by trimethylation of histone H3 lysine 4. Cell **131**, 58–69 (2007)

J.T. Wade, K. Struhl, The transition from transcriptional initiation to elongation. Curr. Opin. Genet. Dev. **18**, 130–136 (2008)

M. Wal, B.F. Pugh, Genome-wide mapping of nucleosome positions in yeast using high-resolution MNase ChIP-Seq. Methods Enzymol. **513**, 233–250 (2012)

W. Wang, M. Carey, J.D. Gralla, Polymerase II promoter activation: closed complex formation and ATP-driven start site opening. Science **255**, 450–453 (1992)

V.M. Weake, J.L. Workman, Histone ubiquitination: triggering gene activity. Mol. Cell **29**, 653–663 (2008)

M.S. Weiss, B.P. Bernabé, A. Shikanov, D.A. Bluver, M.D. Mui, S. Shin, L.J. Broadbelt, L.D. Shea, The impact of adhesion peptides within hydrogels on the phenotype and signaling of normal and cancerous mammary epithelial cells. Biomaterials **33**, 3548–3559 (2012)

M. Weiss, B.P. Bernabé, S. Shin, S. Asztalos, S. Dubbury, M. Mui, A. Bellis, D. Bluver, D. Tonetti, J. Saez-Rodriguez, Dynamic transcription factor activity and networks during ErbB2 breast oncogenesis and targeted therapy. Integr. Biol. **6**, 1170–1182 (2014)

I. Whitehouse, O.J. Rando, J. Delrow, T. Tsukiyama, Chromatin remodelling at promoters suppresses antisense transcription. Nature **450**, 1031–1035 (2007)

C.B. Wilcox, A. Rossettini, S.D. Hanes, Genetic interactions with C-terminal domain (CTD) kinases and the CTD of RNA Pol II suggest a role for ESS1 in transcription initiation and elongation in *Saccharomyces cerevisiae*. Genetics **167**, 93–105 (2004)

B.T. Wilhelm, S. Marguerat, S. Watt, F. Schubert, V. Wood, I. Goodhead, C.J. Penkett, J. Rogers, J. Bahler, Dynamic repertoire of a eukaryotic transcriptome surveyed at single-nucleotide resolution. Nature **453**, 1239–1243 (2008)

J.Y. Wong, J.B. Leach, X.Q. Brown, Balance of chemistry, topography, and mechanics at the cell–biomaterial interface: issues and challenges for assessing the role of substrate mechanics on cell response. Surf. Sci. **570**, 119–133 (2004)

A. Wood, J. Schneider, J. Dover, M. Johnston, A. Shilatifard, The Paf1 complex is essential for histone monoubiquitination by the Rad6-Bre1 complex, which signals for histone methylation by COMPASS and Dot1p. J. Biol. Chem. **278**, 34739–34742 (2003)

A. Wood, J. Schneider, J. Dover, M. Johnston A. Shilatifard, The Bur1/Bur2 complex is required for histone H2B monoubiquitination by Rad6/Bre1 and histone methylation by COMPASS. Mol. Cell **20**, 589–599 (2005)

M.A. Wozniak, C.S. Chen, Mechanotransduction in development: a growing role for contractility. Nat. Rev. Mol. Cell Biol. **10**, 34–43 (2009)

A. Wyce, T. Xiao, K.A. Whelan, C. Kosman, W. Walter, D. Eick, T.R. Hughes, N.J. Krogan, B.D. Strahl, S.L. Berger, H2B ubiquitylation acts as a barrier to Ctk1 nucleosomal recruitment prior to removal by Ubp8 within a SAGA-related complex. Mol. Cell **27**, 275–288 (2007)

J.J. Wyrick, F.C. Holstege, E.G. Jennings, H.C. Causton, D. Shore, M. Grunstein, E.S. Lander, R.A. Young, Chromosomal landscape of nucleosome-dependent gene expression and silencing in yeast. Nature **402**, 418–421 (1999)

T. Xiao, H. Hall, K.O. Kizer, Y. Shibata, M.C. Hall, C.H. Borchers, B.D. Strahl, Phosphorylation of RNA polymerase II CTD regulates H3 methylation in yeast. Genes Dev. **17**, 654–663 (2003)

Z. Xu, W. Wei, J. Gagneur, F. Perocchi, S. Clauder-Munster, J. Camblong, E. Guffanti, F. Stutz, W. Huber, L.M. Steinmetz, Bidirectional promoters generate pervasive transcription in yeast. Nature **457**, 1033–1037 (2009)

M. Yeo, P.S. Lin, M.E. Dahmus, G.N. Gill, A novel RNA polymerase II C-terminal domain phosphatase that preferentially dephosphorylates serine 5. J. Biol. Chem. **278**, 26078–26085 (2003)

S.M. Yoh, H. Cho, L. Pickle, R.M. Evans, K.A. Jones, The Spt6 SH2 domain binds Ser2-P RNAPII to direct Iws1-dependent mRNA splicing and export. Genes Dev. **21**, 160–174 (2007)

K. Yokomori, C.P. Verrijzer, R. Tjian, An interplay between TATA box-binding protein and transcription factors IIE and IIA modulates DNA binding and transcription. Proc. Natl. Acad. Sci. U. S. A. **95**, 6722–6727 (1998)

H. Yu, M. Gerstein, Genomic analysis of the hierarchical structure of regulatory networks. Proc. Natl. Acad. Sci. U. S. A. **103**, 14724–14731 (2006)

G.C. Yuan, Y.J. Liu, M.F. Dion, M.D. Slack, L.F. Wu, S.J. Altschuler, O.J. Rando, Genome-scale identification of nucleosome positions in *S. cerevisiae*. Science **309**, 626–630 (2005)

N. Yudkovsky, C. Logie, S. Hahn, C.L. Peterson, Recruitment of the SWI/SNF chromatin remodeling complex by transcriptional activators. Genes Dev. **13**, 2369–2374 (1999)

R. Zaidel-Bar, B. Geiger, The switchable integrin adhesome. J. Cell Sci. **123**, 1385–1388 (2010)

R. Zaidel-Bar, S. Itzkovitz, A. Ma'ayan, R. Iyengar, B. Geiger, Functional atlas of the integrin adhesome. Nat. Cell Biol. **9**, 858–868 (2007)

S.J. Zanton, B.F. Pugh, Full and partial genome-wide assembly and disassembly of the yeast transcription machinery in response to heat shock. Genes Dev. **20**, 2250–2265 (2006)

K.S. Zaret, J.S. Carroll, Pioneer transcription factors: establishing competence for gene expression. Genes Dev. **25**, 2227–2241 (2011)

J. Zeitlinger, A. Stark, M. Kellis, J.W. Hong, S. Nechaev, K. Adelman, M. Levine, R.A. Young, RNA polymerase stalling at developmental control genes in the *Drosophila melanogaster* embryo. Nat. Genet. **39**, 1512–1516 (2007)

K.S. Zeller, A. Riaz, H. Sarve, J. Li, A. Tengholm, S. Johansson, The role of mechanical force and ROS in integrin-dependent signals. PLoS ONE **8**, e64897 (2013)

Z. Zhang, D.S. Gilmour, Pcf11 is a termination factor in *Drosophila* that dismantles the elongation complex by bridging the CTD of RNA polymerase II to the nascent transcript. Mol. Cell **21**, 65–74 (2006)

Z. Zhang, J.C. Reese, Ssn6-Tup1 requires the ISW2 complex to position nucleosomes in *Saccharomyces cerevisiae*. Embo J. **23**, 2246–2257 (2004)

H. Zhang, D.N. Roberts, B.R. Cairns, Genome-wide dynamics of Htz1, a histone H2A variant that poises repressed/basal promoters for activation through histone loss. Cell **123**, 219–231 (2005)

L. Zhang, A.G. Fletcher, V. Cheung, F. Winston, L.A. Stargell, Spn1 regulates the recruitment of Spt6 and the Swi/Snf complex during transcriptional activation by RNA polymerase II. Mol. Cell. Biol. **28**, 1393–1403 (2008)

H. Zhang, F. Landmann, H. Zahreddine, D. Rodriguez, M. Koch, M. Labouesse, A tension-induced mechanotransduction pathway promotes epithelial morphogenesis. Nature **471**, 99–103 (2011)

C. Zhong, M. Chrzanowska-Wodnicka, J. Brown, A. Shaub, A.M. Belkin, K. Burridge, Rho-mediated contractility exposes a cryptic site in fibronectin and induces fibronectin matrix assembly. J. Cell Biol. **141**, 539–551 (1998)

13 Splicing and Alternative Splicing and the Impact of Mechanics

Abstract

Systems-wide approaches have gained new insights into mRNA splicing. Improved cryo-EM techniques have elucidated the structure of splice components. Recent findings on the molecular architecture of the human spliceosome are unveiled. In complement to the normal splicing of introns out of DNA sequences, alternative splicing is also examined. Many predictions and hypotheses have led to new knowledge about the role of alternative splicing in the delivery of alterative proteins or peptides, which may be either tissue-specific, developmental stage or disease-specific. Mechanical stimulation of cells can provoke a shift towards a distinct alternative splice variant that may perform a different function, as novel-binding sites are now present or binding sites are accessible due to conformational changes of the protein. The occurrence of alternative splicing in diseases, such as cancer, is being discussed. It is explained how alternative splicing can be regulated. Finally, the mechanical properties of the matrix will be investigated in terms of their effect on the alternative splicing process and its regulation.

13.1 The General Principle of Splicing

The spliceosome represents a highly complex, dynamic ribonucleoprotein molecular complex that remodels itself to perform numerous structural and compositional rearrangements, which subsequently evokes the formation of an active site. Cryo-electron microscopy (cryo-EM) has led to recent advances, revealing a wealth of structural insights into the inner life of a spliceosome, which are close to the atomic level (Kastner et al. 2019). Combined with biochemical, structural and functional studies, the cryo-EM technique has manifested the structural basis for the majority of the predominant models of spliceosome function, while at the same time it provides a new understanding into splicing catalysis and the fascinating dynamics of the spliceosome. The mechanism of pre-mRNA splicing is highly conserved between different species, such as humans and yeast. However, the compositional and the remodeling dynamics of the ribonucleoproteins (RNPs) within the human spliceosome are more complex. Recent advances contribute to the understanding of the human spliceosome architecture on the molecular level and reveal differences between the splicing machinery of humans and yeast (Kastner et al. 2019).

C. T. Mierke, *Cellular Mechanics and Biophysics*, Biological and Medical Physics, Biomedical Engineering, https://doi.org/10.1007/978-3-030-58532-7_13

How are Pre-mRNA Introns Defined?

For pre-mRNA splicing, it is important to identify the introns precisely, and hence, a clear definition of the introns is required. Indeed, introns can be defined by three short consensus sequences: the 5′ splice site (SS), the 3′SS and the branch site (BS), which is mostly associated with a pyrimidine-rich region (PPT) (Fig. 13.1) (Will and Lührmann 2011). More precisely, pre-mRNA splicing mechanism compromises two sequential SN2-type transesterification reactions (Fig. 13.1). In the first reaction, the 2′-OH group of the branch site adenosine (BS-A) fulfills a nucleophilic interaction at the phosphodiester bond of the 5′SS, which leads to on the one hand to a cleaved 5′ exon and on the other hand to the intron lariat-3′ exon intermediates. In the second reaction, the 3′-OH group of the 5′ exon interacts with the 3′SS, which causes exon ligation and subsequently the excision of the intron. Most of the pre-mRNAs of higher eukaryotes are susceptible to alternative splicing, in which different sets of exons from a pre-mRNA are spliced and thereby linked together, which results in different mRNA isoforms and considerably expanding the spectrum of translated proteins (Pan et al. 2008). Moreover, errors in the splicing process and its regulation are either the basis for the severity of many human diseases, including cancer, or its modulation (Daguenet et al. 2015; Lee and Rio 2015).

The nuclear pre-mRNA introns possess relatively little conserved secondary or tertiary structural information and are therefore reliant on the spliceosome to align their reactive sites, such as the 5′SS, BS and 3′SS, for the catalysis of the splicing. The most important spliceosomal components are the four small nuclear ribonucleoproteins (snRNPs) encompassing U1, U2, U4/U6 and U5. Each of these components consists of a single small nuclear RNA (snRNA) or two, such as for the U4/U6 snRNA. Seven Sm proteins or in the case of U6 seven Lsm proteins assemble a ring and a variety of snRNP-specific proteins (Fig. 13.2) (Will and Lührmann 2011). In pre-mRNA splicing, the U4/U6 and U5 snRNPs form a 25S U4/U6.U5 tri-snRNP complex, in which the U4 and U6 snRNAs are base-paired and establish a three-way linkage. The functional regions of snRNAs are highly conserved, as is the majority of the protein composition of the snRNPs. However, although some proteins are associated with snRNPs in yeast, they are instead targeted as non-snRNP-associated factors for the spliceosome in humans and vice versa.

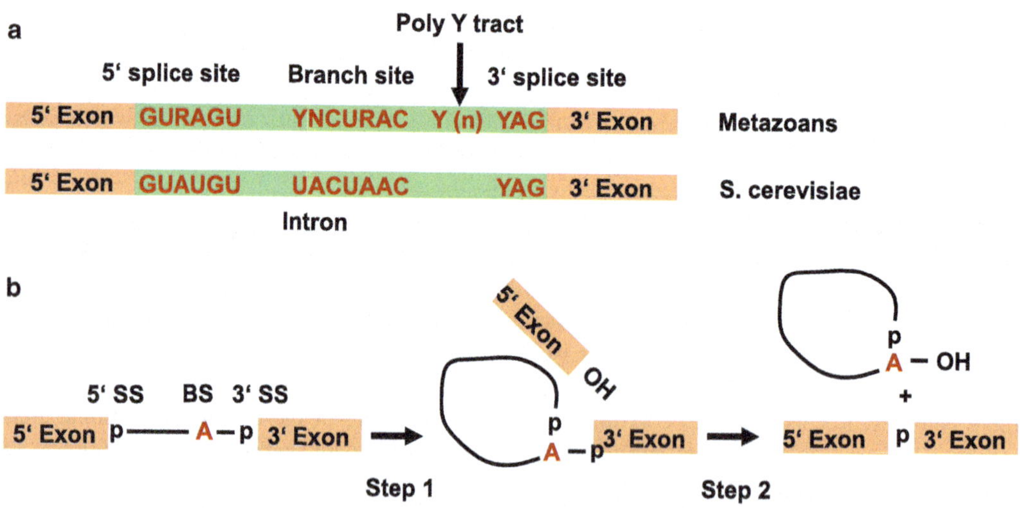

Fig. 13.1 Spliceosomal assembly pathway. **a** Conserved sequences at the 5′ splice site (SS), branch site (BS) and 3′SS of U2-type (major) pre-mRNA introns in metazoans and *Saccharomyces cerevisiae*. **b** Schematic view of the two-step splicing reaction. p represents the phosphate at the 5′ and 3′ splice sites, which are conserved in splicing objects

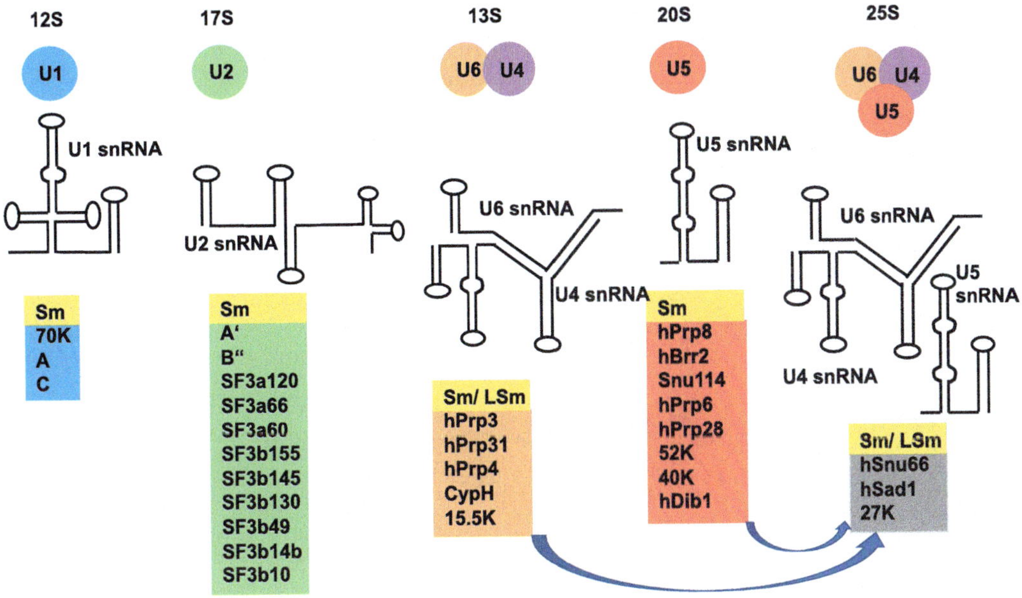

Fig. 13.2 Protein composition and snRNA secondary structures of main human spliceosomal snRNPs. All seven Sm proteins, such as B/B*, D3, D2, D1, E, F and G, or LSM proteins, such as Lsm2-8 are listed below each snRNP

The assembly of the spliceosome is performed in a stepwise manner, and extensive structural and compositional remodeling processes are required for the generation of the spliceosome's active site (Fig. 13.3) (Will and Lührmann 2011). These processes are mainly supported by eight conserved DEXD/H-box ATPases or RNA helicases that provoke RNA–RNA and RNA–protein rearrangements (Cordin and Beggs 2013). An initial step for the active site formation of the spliceosome is the interaction of the U1 snRNP with the pre-mRNA through a specific base pairing between the 5′ end of U1 snRNA and the 5′SS of the intron, which is stabilized by U1-associated proteins (Fig. 13.3). In the next step, the U2 snRNP stably interacts with the BS and creates thereby the spliceosomal A complex. In this complex, the U2 snRNA forms with the BS base pairs, with the BS-A protruding from the U2/BS helix. The interaction between U2 and BS is further strengthened by the U2AF65/35 heterodimer, which binds the PPT, and by the heteromeric protein complexes SF3a and SF3b, which are associated with U2 through a direct contact with the 6 to 26 nucleotides of the intron up-stream of the BS-A (synonymously termed anchoring site) (Gozani et al. 1996). In the next step, the preformed U4/U6.U5 tri-snRNP is integrated into the A complex and thereby creates the pre-B complex, where the tri-snRNP is not yet stably connected (Boesler et al. 2016). The initial tri-snRNP association is supported by the formation of a short duplex (U2/U6 helix II) between the 5′ end of the U2 snRNA and 3′ end of the U6 snRNA (Fig. 13.3). In humans, stable tri-snRNP integration demands the action of the helicase PRP28 (Boesler et al. 2016), which breaks the U1/5′SS helix and can connect it to the conserved U6 ACAGA box (Staley and Guthrie 1999). This reaction destabilizes the U1 snRNP and results in the precatalytic B complex. More specifically, in the precatalytic B complex, loop 1 of the U5 snRNA interacts with the 3′-terminal nucleotides of the 5′ exon and later helps to position both exons for step 2 of splicing (Newman and Norman 1992; Sontheimer and Steitz 1993).

The catalytic activation takes place in a stepwise manner, whereby the RNA helicase BRR2 induces the unentangling and unwinding of the U4/U6 snRNA duplex (Cordin and Beggs 2013). After U4 snRNP has been displaced and multiple other proteins have been assembled, the B_{act} complex is produced. In specific detail, subsequently the function of the RNA helicase PRP2 acts to

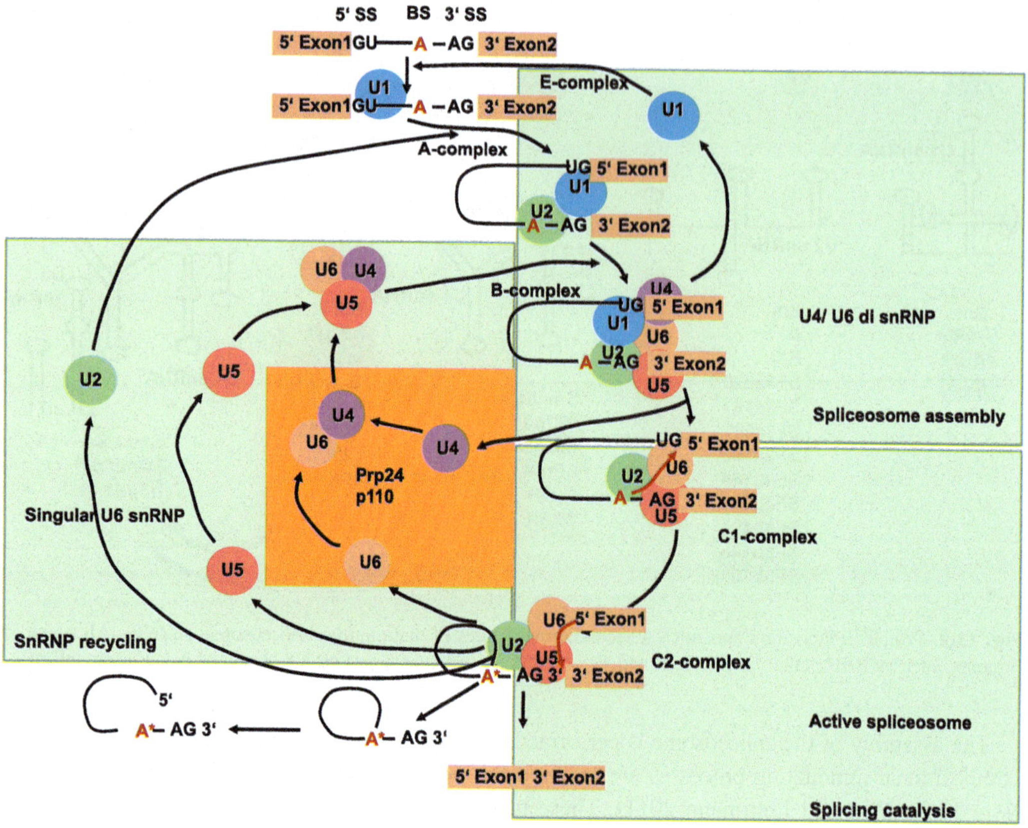

Fig. 13.3 Spliceosomal cycle

switch the B_{act} complex into a catalytically competent B^* complex that is then able to catalytically support step 1 of the splicing procedure, which is referred to as branching, and consequently generates the C complex. The emerged complex possesses the cleaved 5′ exon and the intron lariat-3′ exon (Fig. 13.3). The rearrangement of the C complex through the action of the helicase PRP16 leads to the generation of the C^* complex that fulfills the second catalytic step, which is referred to as exon ligation, and results in the post-splicing (P) complex (Horowitz 2012) (Fig. 13.3). The mRNA release is performed through the generation of the intron lariat spliceosome (ILS), which is catalyzed through the helicase PRP22. In the end, the ILS is unwrapped through the action of the helicase PRP43, and the U2, U5 and U6 snRNPs, which are being released, are recycled in order to perform other circuits of the splicing process.

The spliceosomal RNA framework is remodeled into a catalytically active configuration after the switch from B into B* (Fig. 13.4) (Will and Lührmann 2011). After the break-down of the U4/U6 stems I and II, the U6 snRNA undergoes pronouncedly remodeling. The U6 internal stem loop (U6 ISL) is performed and U6 creates additional base-pair interactions with U2, which subsequently generates the U2/U6 helices Ia and Ib. Thereby a catalytic U6 RNA triplex is additionally created, as two nucleotides of the U6 ACAGA-GAG sequence and the U6-U74 built the Hoogsteen base-pair interactions through the U6-catalytic triad AGC sequence of the helix Ib (Fica et al. 2014; Hang et al. 2015). In specific detail, phosphorothioate substitution investigations revealed that U6 acts by catalysis of both steps of the pre-mRNA splicing through the placement of two Mg^{2+} ions, which are able to stabilize the leaving groups during every step of the reaction (Fica et al. 2013). The RNA

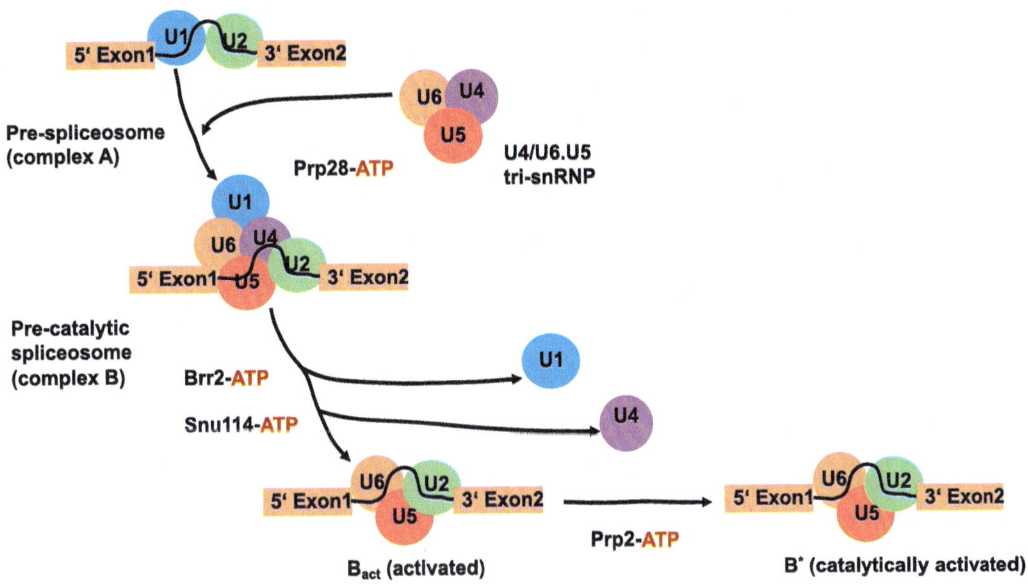

Fig. 13.4 Activation of the B complex and the B* complex

configurations of the active centers of the spliceosome and group II introns are very similar (Zhao and Pyle 2017), but the first needs proteins for its construction and for stabilization, as described in the following.

The human and the yeast *Saccharomyces cerevisiae* spliceosomes exhibit a common set of approximately 85 core spliceosomal proteins, which are conserved in the course of evolution (Fabrizio et al. 2009). In higher eukaryotes, approximately 80 additional proteins, lacking counterparts in yeast, are brought together in the splicing apparatus during the entire assembly and catalytic steps (Jurica and Moore 2003; Agafonov et al. 2011). These comprise apart from others firstly a large number of SR and hnRNP proteins which are involved in the regulation of alternative splicing (Zhou and Fu 2013), and secondly proteins with a potentially enzymatic activity that can function as molecular shifters or conformational triggers—these are kinases, multiple RNA helicases involving Aquarius (AQR or synonymously referred to as IBP160) and seven peptidyl prolyl isomerases (PPIases). Human spliceosomal proteins also undergo considerably stronger posttranslational alterations than yeast proteins. In addition to snRNPs, a large number of spliceosomal proteins are pre-assembled into large-scale complexes, which facilitates the assembly environment and improves the rate of spliceosome assemblies. These are the retention and splicing complex (RES) (Dziembowski et al. 2004), the substantial PRP19/CDC5L splicing complex (Makarova et al. 2004) and the pentameric intron-binding complex (IBC) (De et al. 2015). The yeast equivalents of certain IBC and PRP19/CDC5L compound proteins exhibit a different type of pre-organization. Therefore, they are located in the large Prp19 complex (NTC, nineteen complex) along with complementary spliceosomal proteins, several of which are separately recompiled in humans, such as SKIP, SYF2 and SYF3, and/or at a distinct phase of splicing, such as SYF2. For this reason, the assembly of the human spliceosome is not only becoming more complex, with probably more assembling intermediates, but is more strongly adjusted accordingly.

The spliceosome molecule composition varies from one spliceosome to the other (Will and Lührmann 2011). For instance, the complex transfer from pre-B to B involves the stable incorporation of nine B-specific proteins. The largest replacement of proteins takes place during the B to B_{act}

changeover, in which approximately 20 proteins are liberated/destabilized and over 25 proteins are reintegrated in humans. In this transfer, U4/U6 and B-specific proteins are destabilized/liberated, while the PRP19/CDC5L and RES complexes, the IBC and some other proteins, along with those operationally referred to as PRP19/CDC5L-related (nineteen-related, NTR), are incorporated in a stable manner. Contrary to PRP19/CDC5L and IBC proteins, PRP19-related proteins are not attracted as an integral part of a large-scale preformed complex. It is presently uncertain how this radical replacement of proteins will occur, but it is probable that it will take place via multiple intermediates, where the removal of certain proteins will yield new binding points needed for the acquisition of others.

13.2 The General Principle of Alternative Splicing

The alternative pre-mRNA splicing represents a rich source of transcript diversity in eukaryotic cells and plays major roles in developmental processes and diseases. A system-wide experimental approach has begun to determine the way in which global regulation of splicing shapes complex biological characteristics and signaling pathways. The insights of these novel approaches are presented, and it is discussed what issues remain to be analyzed in future research approaches.

Alternative splicing significantly raises the encoding capability of intricate genomes by enabling individual genomic loci to encode multiple transcripts containing varying protein-coding sequences and RNA regulatory elements. It is estimated that approximately 75% of human genes encode two or more splice isoforms, although there are noticeable variations between individual tissue types and stages of development (Johnson et al. 2003). In addition, a large number of human diseases are caused by defective splicing of key transcripts or are characterized by the occurrence of abnormal splice isoforms in the diseased tissues (Wang and Cooper 2007; Moore and Silver 2008).

The mechanism and regulation of splicing have been elucidated primarily through genetic and biochemical investigations on animal model organisms and human cells concentrating on individual elements and processes (Black 2003). Overall, these investigations indicate that the utilization and/or repression of splice sites are governed largely through combined binding of RNA-binding proteins (RNABPs) to proximate regulatory entities. The decoding of this so-called splicing code has turned out to be a major scientific and bioinformatics hurdle, which is made even more difficult due to the many post-transcriptional and post-translational schemes that govern RNABP expression and the localization within cells. These mechanisms react in response to signaling pathways and tissue-specific cues. With this kind of hierarchical regulatory framework in the background, endeavors to identify alternative splices on a general level have moved in several principal directions: First, the creation of broad libraries of splice variants in various organisms and cell types, second, the establishment of *cis* regulatory components that together form the splice code, third, the characterization of transcriptional pathways governed by unique *trans*-active splicing factors, fourth, the clarification of the system-specific functional integration of splicing involving transcriptional and mRNP monitoring and export and fifth, the identification of the functional integration of splicing in signaling, cell differentiation and disease modes. Numerous review articles have highlighted several facets of this abundant resource of biodiversity and regulation (Maniatis and Reed 2002; Matlin et al. 2005; Blencowe 2006; Keene 2007; Wang and Cooper 2007). The focus in the following lies on the experimental techniques, which finally helps to envision in a globule picture the regulation of the slicing process and thereby discusses the novel insights in this specific field.

13.2.1 Expression Profiling

Microarrays enable the concomitant analysis of the complete nucleic acids existing in a biological specimen. The only limitation is the maximum number of sequences that can be displayed on a single chip (Stoughton 2005). Conventional expression profiling arrays are equipped with expressed sequence tag (EST) derived cDNAs or 39 clusters of specific oligonucleotide sequences that explore the entire transcript frequency without being able to discriminate between individual splicing isoforms. Nevertheless, a number of user-specific and commercial microarray technologies have been engineered to detect fluctuations in the splicing process. In general, microarrays having a higher sensor density offer higher data resolution; however, due to the mere size of the datasets, the data are costly and complicated to evaluate. Therefore, three microarray designs with various advantages and disadvantages for mass splicing analysis are outlined below: the splice junction array, exon arrays and the tiled genomic arrays.

13.2.2 What Are Splice Junction Microarrays?

Splice connections arrays carry oligonucleotide projections covering the annotated exon-exon connections exclusively for single splice isoforms (Johnson et al. 2003). Connection sensors are position limited and therefore exposed to noise caused by non-ideal sequence components. Therefore, the majority of the variants of these arrays contain sondes within the constitutive exons, enabling transcripts to be first evaluated as being present or absent in the specimen and then evaluated for alternative splicing on the basis of connection sensors. Splicing arrays have demonstrated their value in the functional characterization of RNA processing mutants in yeast (Clark et al. 2002; Burckin et al. 2005) and in the identification of tissue-specific splicing schemes and regulatory patterns in mammalian cells (Johnson et al. 2003; Pan et al. 2004; Sugnet et al. 2006; Fagnani et al. 2007). The aforementioned investigations have revealed the recurring phenomenon that the transcripts governed at the level of alternative splicing encompass a different set of populations than those governed at the rate of transcript frequency level. This outcome points to the fact that the regulatory machinery controlling transcript divergence and transcript frequency may develop independently to affect certain characteristics of a particular cell type. A number of observations have broadened these findings to the perspective of both differentiation and disease. Through junction profiling subsequent to T cell activation, a large number of transcripts were discovered that exhibit alterations in overall frequency, with functional enhancement for immune response and cytoskeletal dynamics; transcripts that experience alterations in alternative splicing, however, were enhanced in terms of cell cycle functions (Ip et al. 2007). In similar fashion, multiple transcripts in prostate and breast cancer cells exhibited modified splice isoform values compared to normal tissues but without significant alterations in the overall frequency of transcripts, especially mRNA for the cell adhesion molecule and the metastatic activator CD44 (Li et al. 2006a). In general, the overall frequency of expression of splice isoforms in multiple Hodgkin's lymphoma tumors accumulates in proportion to the extent of tumor progression (Relógio et al. 2005). These and related investigations indicate that the profiling of splice variants could complement the instruments available for tumor diagnosis.

The generation of splice connection profiles after elimination of individual elements has yielded functionally consistent transcriptional maps that are functionally governed at the splicing level. Single knockdown of four RNABPs with wide specificity in *Drosophila* cells revealed various alternative splicing event patterns directed by the individual factors (Blanchette et al. 2005). In the mouse, a subgroup of brain-specific exons governed by the neuronal RNABP NOVA2 was found by matching brain-derived transcript profiles of NOVA2-knock-out with wildtype animals. Transcripts carrying these exons coded for many synapse-specific proteins, which highlights a distinct biological role for this splicing system in the generation of synapses (Ule et al. 2005a, b). In mouse neuroblastoma cells, exon groups that are variably adjusted by these proteins were elicited by knockdown of the splice regulator PTBP1 and/or its neuronal paralogue nPTB (Boutz et al. 2007). As neuronal precursors actually express PTBP1 whereas differentiated neurons express nPTB, these results revealed a post-transcriptional switch which reproduces alternative splicing in programs in the course of the neuronal development.

Several investigators have examined associations between alternative splicing and non-sense mediated decay (alternative splicing–NMD), as it is assumed that about 35% of the commentated alternative splicing incidents implement premature termination codons (Green et al. 2003). Astonishingly, only few alterations in the splice isoform levels in the profiles of the junction arrays have been detected during NMD blocking, indicating that alternative splicing NMD is limited to a small subsample of alternative splice occurrences (Pan et al. 2006; Ni et al. 2007). Nevertheless, the transcripts coding for the splicing regulators have been heavily accumulated for alternative splice NMD control, whereby the cassette exons often overlay "ultraconserved" DNA areas (Ni et al. 2007). These findings point to a tightly conserved post-transcriptional regulatory mechanism to manage the content of key mammalian splicing regulators, with possible cross- and self-regulatory mechanisms.

13.2.3 Exon Arrays

An alternative array approach employs "exoncentric" sets of probes composed of multiple oligonucleotide sondes within each labeled and predetermined exon in the human, mouse and rat genomes, such as GeneChip Exon Arrays from Affymetrix (Xing et al. 2006; Okoniewski et al. 2007b). In opposite to "junction arrays" which only investigate a given set of alternative splicing events, "exon arrays" offer an unprejudiced pattern of the transcript structure which enables the detection of new variations in splicing. The expression profiling of 16 human tissues by means of these arrays revealed multiple tissue-specific exons and showed that testis and brain have the highest degree of alternative exon utilization of the investigated tissues (Clark et al. 2007). A time-dependent analysis of signal-induced exon utilization in a human cell culture model for depolarization of neurons detected new cases of alternative splicing in operationally important transcripts including the voltage-gated potassium channel KCNH4 and the proton pump subunit ATP6Voa4 (McKee et al. 2007). In line with prior genome-wide investigations, at the level of transcript frequency and alternative splicing, a variety of regulatory networks were revealed that show dynamic alterations in the progression of the measurements. Certain functional groups, including calcium ion binding, were adjusted at the two levels, while others preferred one type of regulation. However, the transcripts coding for the splicing regulatory units were accumulated pronouncedly for the exon usage regulation, but not the bulk of the transcripts, which points to a potential self-regulating assembly at the splicing level. The profiling of human gliomas (French et al. 2007) and colon cancer cells (Gardina et al. 2006) revealed exons that have been spliced in a different way in tumors compared to regular healthy cells, which leads to the assumption that these arrays may be beneficial in the de novo identification of abnormal splicing processes in the diseased tissues. In conclusion, it was widely acknowledged that the exon array platform delivered excellent transcriptional results which were a benefit of the high precision of the probe (Kapur et al. 2007; Okoniewski et al. 2007a). The progress in monitoring single exons became more variable, presumably due to the fact that single exon probe sets may be confined to non-ideal sequences. The RT-PCR confirmation levels for alterations in individual exons varied in the experiments presented here from 33 to 86% and varied somewhat due to the filtering parameters. Therefore, these arrays are able to reveal novel directed splicing, although an experimental confirmation is essential to deduce the correct biological role.

13.2.4 Tiled Genomic Arrays

Tiled oligonucleotide arrays covering entire chromosomes or genomes ensure complete coverage and eliminate the necessity of previous information about exon coordinates. This type of topography of the sample is favorable since it is able to identify trends of alternative splicing, such as modified splice site utilization, intron capture, or utilization of nonannotated exons, which are hard or even undetectable on experimental platforms that are confined to a specific set of tagged exons. The increased effort and considerable computational difficulties involved in the analysis of tile array datasets tend to override the inherent advantages in numerous experimental settings. Nevertheless, several initial profiling endeavors indicate that tiled arrays may yield a transcript structure that has been neglected in sequencing approaches and other microarray technologies. Transcriptome characterization of *S. cerevisiae* on tiled arrays revealed previously unavailable introns and samples of regimented retention of introns (Juneau et al. 2007; Zhang et al. 2007a). An examination of human transcripts from 12 tissues on tiled ENCODE arrays covering about 1% of the human genome showed that an extraordinary 81.5% of the 399 protein-encoding transcripts on the array comprised non-annotated sequences, with 20% of the new versions exhibiting modified open reading frames (Denoeud et al. 2007). The utilization of distal 59 transcription starting sites, the enlargement of the labeled exons, and the occurrence of new exons within the labeled introns substantially improved this result, and a high level of tissue-specific transcriptional architecture has been seen. The biological relevance of these findings is not yet known, but these experimental results prove that our present state of knowledge of the exon repertoire of the human genome is not complete by far. As whole genome tiling arrays for human and mouse are becoming increasingly common, extensive investigations of the transcript structure will respond to the continuing demand for a comprehensive library of the variety of transcripts.

13.2.5 Other Proofing Techniques

The linking of microarray analysis to molecular approaches such as chromatin and RNA immunoprecipitation (ChIP chip or RIP chip) enables the detection of population sizes of genes and transcripts that are governed by individual determinants. The ChIP chip is able to deliver a genome-binding pattern for processing factors that are enrolled in mRNAs during the transcription process, thus enabling the pinpointing of targets. In parallel, this strategy yields mechanistic knowledge by enabling collaborative analysis of how and when various factors are linked to genes. For example, two more recent genome positioning experiments in *S. cerevisiae* demonstrated that components needed for the initial steps of spliceosome assembly are co-transcriptionally acquired, whereas subsequent phases of assembly are normally eliminated from the transcriptional regions (Tardiff et al. 2006; Moore et al. 2006). This finding means that the majority of catalytic splicing processes in yeast are post-transcriptional, contrary to previous models. In addition, the early determinants exhibited wide binding profiles that were highly RNA-dependent, indicating that the collection of samples of nascent mRNAs through a subgroup of early splicing engagement markers constitutes the mechanistic foundation for the targeting of the yeast spliceosome (Moore et al. 2006). Ultimately, a number of developmentally regulated splicing incidents have been found, which are attributable to the fact that the spliceosome has been specifically eliminated from the respective genes. In human cells, ChIP-chip mapping of RNABPs revealed the attachment of the splicing factor PTBP1 to promoter sites of genes, indicating a promoter exposure and/or an uncharacterized function in the initiation of transcription (Swinburne et al. 2006). In this way, PTBP1 demonstrated pronounced clustering locations along the genes with a preference of 59 and internal exons, which assists the dynamic rearrangement of the mRNPs, since they are coordinately produced and wrapped.

RIP chip has been employed to identify the target of RNABPs in various system types (Keene et al. 2006). Although the RIP chip is technically demanding due to the fragility of RNP complexes, it is advantageous in that it is capable of identifying targets at all stages of biogenesis of mRNP, whereas the ChIP chip merely recognizes co-transcriptional linkages. This advantage was confirmed in the RIP chip measurement of the splicing factor U2AF65, which is linked to spliced mRNAs in the nucleus and the cytoplasm of HeLa cancer cells (Gama-Carvalho et al. 2006). This discovery points to features for U2AF65 that extend beyond its broadly defined function in the early stages of splicing, in accordance with the earlier involvement of the *Drosophila* homologue of U2AF65 (dU2AF50) in the nuclear exporting of intron-less mRNAs (Blanchette et al. 2004). In extension, U2AF65-bound transcripts were augmented for certain features that differ from the ones linked by PTBP1.

The application of a similar procedure entitled cross-linking and immunoprecipitation (CLIP) to the neuronal splicing factor nova resulted in the recognition of a matrix of objectives encoding synaptic and neuronal inhibitory factors in the mouse brain (Ule et al. 2003). CLIP refers to a specially modified RIP chip procedure in which RNP complexes are interlinked in vivo by exposure to UV light. This strategy enables rigorous immunopurification, which guarantees that copurified RNAs are probably used as directly bound targets. As the classical RIP chip demands mild requirements for the preservation of native RNP complexes, it is not able to identify direct RNA–protein interactions with certainty. Moreover, several indications indicate that RNP complexes may rearrange themselves after cell lysis (Mili and Steitz 2004). Although these reservations for the RIP chip analysis are taken into account, they have turned out to be feasible with adequate verification and proof of quality control. A secondary distinction between CLIP and RIP chip consists in the fact that the goals were selected by sequencing and not by microarray examination. This approach pointed to the recent reemergence of high-throughput sequencing technologies to replace microarrays for solid parallel investigations of nucleic acids (Kim et al. 2007a; Mardis 2007; Mikkelsen et al. 2007). High-throughput sequencing has the advantage of unprejudiced sequence analysis and eliminates issues with signal-to-noise and poor samples, which are inevitable disadvantages of microarray hybridization. High-throughput sequencing systems produce short sequence readings of approximately 30 bp. As human exons are on average approximately 200 bp, high reading numbers are required to adequately shape exon borders and discern various splice isoforms. The suitability and costs of high-throughput sequencing for alternative splicing analysis compared to microarrays have not been evaluated to the current day.

13.2.6 Experimental Approaches Are Combined with Computational Approaches

In a series of recent studies, experimental and bioinformatics techniques were combined to obtain meaningful insights into overall regulation of splicing. The screening of a random collection of decanucleotides for their splicing regulatory activity revealed many new exonic splice silencing patterns (Wang et al. 2004a). The bioinformatic screening for these exonic splicing silencings in endogenous human genes yielded a positive accumulation in alternatively spliced exons and pseudoexons and a negative accumulation in constitutively spliced exons. The exonic splicing silencings also correlated with the "strength" of the splice site: constitutive exons containing weak consensus splice sites exhibited less exon splicing silencing, which probably would lead to unfavorable exon skipping. Later research proved that the presence of exon splicing silencings prevented the utilization of bait (or synonymously decoy) splice sites 59 and 39 inside or close to the regulated exons (Wang et al. 2006). In contrary, exonic splicing boosters caused either the contrary impact or no influence on the utilization of decoy splicing sites. In line with an inherent function in regulating the selection of alternative splicing sites, the exon splicing silencings in alternatively spliced regions of the controlled exons were accumulated statistically.

However, other embedded concepts have concentrated on the decoding of structures that are governed by specific splicing factors. An RNA map for NOVA attachment sites located proximal to the controlled exons in the mouse brain has been identified (Ule et al. 2006). This RNA map corresponded not just well with earlier evaluations of NOVA-dependent splicing at the genome level (Ule et al. 2003, 2005a, b), but also enabled a very precise forecast of NOVA-dependent control for numerous hitherto unidentified targets. Multiple novel-binding sites for the *Drosophila* splice regulator SXL have been found by a genome-wide computerized screening, and in some instances sex-specific controlled splicing has been experimentally validated (Robida et al. 2007). However, as many splice regulators are able to attach to severely degenerated RNA sequences or sequences that occur abundantly in the transcriptome (Singh and Valcárcel 2005), de novo scans in this form have been conducted with varying degrees of effectiveness. Nevertheless, in vitro separation techniques of the SELEX method have led to sophisticated analyses of RNA regulatory motifs for a variety of splicing controllers, which might support associated efforts (Fairbrother et al. 2002; Hui et al. 2005; Paradis et al. 2007). Bioinformatic tools detected a substantial overlap between "ultraconserved" DNA elements and alternative exons carrying premature termination codons in genes for splicing modulators (Lareau et al. 2007a). These results demonstrated the highly conserved usage of alternative splicing–NMD as an agent to adjust the cellular levels of splicing regulators, as deduced separately from the microarray investigation (Ni et al. 2007). On the whole, these investigations show that the smart use of experimental tools and computer-aided analysis may identify characteristics of the so-called splice code. The emergence of broader genome-wide databases will enhance the analytical predictive value of investigation of splicing regulation.

Systems approaches have resulted in new insights into mRNA splicing as a tool for the regulatory process in biology, which has led to several outstanding achievements. Firstly, the regulation of transcript frequency and transcript variety is completely different processes which can affect the various biological characteristics within the selfsame cell. Consequently, these intertwined types of regulation include different, albeit frequently redundant, transcript collections. Secondly, alternative splicing patterns comprise a repeatable characteristic pattern for specific cell types, which corresponds to a decisive function in splice regulation to define cell identity. These patterns display a dynamic and comprehensive reprogramming throughout the process of differentiation and disease and in reaction to intra- and extracellular cues. Ultimately, functionally similar transcripts within splicing networks can be co-modulated to shape distinct biological roles.

Although latest investigations have identified fundamental issues of splicing regulation, a lot of unanswered questions still persist, especially regarding the contribution of splicing to disease and developmental processes. Overall knowledge of mRNA splicing is expected to gain further advances from the release and thorough mapping of genome-wide expression and ChIP/RIP bonding profiles, which may form an empirical foundation for verifiable, predictive models of the guidance of splicing regulation. Large-scale proteomics profiling in the knockdown of PTB paralogs has recently revealed new goals of these splicing regulators (Spellman et al. 2007). The progress of this strategy indicates that proteome analysis, even though it evaluates a downstream assessment of alternative splices and seems to be a powerful tool to detect regulated splicing events. In conclusion, the progressive emergence of high-throughput RNAi and small molecule screening technologies offers the possibility of systematic monitoring of splicing regulators in mammalian cells. The contemporary opinion endorsed by the research discussed here is that cell, and context-specific splicing schemes are a compound product of the RNA-encoded "splice code," the varying binding specificities of RNABPs and the innumerable regulatory mechanisms that govern RNABP expression and positioning (Matlin et al. 2005). Further progress on the different and mutually supplemental experimental routes outlined here is highly encouraging to shed more light on each of these critical issues.

13.3 Discrepancy Between the Complexity of the Genome and the Proteome

Although humans are obviously more complex than simple worms or fruit flies, they possess nearly the same amounts of protein-coding genes (Blencowe 2006). A possibility to tackle this seeming contradiction is to examine to what degree operationally distinct polypeptides of individual genes can be encoded in multiple species (Yang et al. 2016).

Eukaryotic genes are able to code for multiple protein versions through alternative transcription, splicing, 3′-end generation, translation and post-translational alterations. Alternative splicing generates so-called transcriptional isoforms for most human genes (Pan et al. 2008; Wang et al. 2008), which allows functional variety at the scale of enzymatic activities, subcellular localizations and protein–protein, protein–DNA and protein–ligand interfaces (Kelemen et al. 2013). A particular isoform can display dominant-negative effects compared to other isoforms which are encoded by the identical gene, be up- or down-regulated rather than constitutively active, or even have contrary cellular activities. For instance, two isoforms coded by the BCL2L1 gene perform contrary functions during apoptosis, such as that the longer isoform restrains the process and the shorter one enhances it (Schwerk and Schulze-Osthoff 2005). In a different case, the omnipresent alternative splicing of *Drosophila melanogaster* Dscam1 produces thousands of various polypeptides, all of which have different binding specificities to facilitate the self-recognition of neurons (Wojtowicz et al. 2007). In total, there are several hundred human genes identified that codify alternative spliced isoforms possessing specific operational properties (Kelemen et al. 2013).

It is still not known how broadly this occurrence is distributed on the level of the entire proteome, which is far more intricate than initially assumed (Tran et al. 2011). Up to 100,000 different isoform transcripts may be generated from the approximately 20,000 human genes encoding proteins (Pan et al. 2008), yielding perhaps more than a million different polypeptides that are produced by post-translational alteration of all kinds of transcript isoforms (Smith and Kelleher 2013). How this complexity of proteomics is connected to comprehensive cellular activities is largely not understood. To what degree do isoform pairs encoded by a shared gene differ operationally from one another? How common is the isoform-specific variety of functions in a certain species? How can this functional versatility differ among species? What part does this multiplicity take in evolution? Overall, the central issue is to identify the degree to which two different but non-exclusive concepts are feasible: Firstly, alternative isoforms are inclined to convey different features, for example, they usually act as "functional isoforms" and secondly, alternative isoforms are intrigued by different features, for example, they should be regarded as "functional alloforms."

To date, studies on the function of alternative splicing have concentrated on the features that alternative protein isoforms may or may not fulfill in comparison with their corresponding "reference" counterparts (Buljan et al. 2012; Ellis et al. 2012). In order to start the systematic and unbiased treatment of the questions outlined above, large-scale functional profiling studies are required to assess the degree to which all isoforms coded by a wide range of genes that are operationally similar or different, considering all pairwise permutations of isoforms coded by the identical gene. However, this in turn necessitates new techniques for the identification, cloning and exogenous expression of full-length open reading frames (ORFs) for all isoforms across a broad spectrum of genes.

Contemporary experiments for the genomic-wide systematic detection of alternative spliced isoforms are founded on the techniques of next-generation sequencing (NGS). For instance, RNA sequencing (RNA-seq) offers a relatively deep sampling depth (Pan et al. 2008; Wang et al. 2008). However, the small length of the RNA-seq readings has made the discovery of a coherent exon binding for alternatively spliced isoforms in the full length complicated. Sequencing of single cDNA molecules in full length or "iso-seq" (Eid et al. 2009) has demonstrated significant achievements in creating enhanced models of full-length transcript isoforms (Sharon et al. 2013). A different tactic involves the co-assembly of removed, alternatively spliced exons by restricting the number of RNA molecules in the libraries employed to produce sequencing libraries (Tilgner et al. 2015). Finally, none of these approaches provides the extensive physical clone sets required to systematically express and investigate the role of alternative isoforms.

A new strategy, referred to as ORF-seq, is employed here to identify, profile, exogenously express and functional investigate a wide range of alternatively spliced ORFs in full length. Using this strategy, it has been possible to investigate binary protein–protein linkages and widely recognized interaction differences through alternative splicing. Another alternative is that spliced protein isoforms in interactome system are prone to behave as totally different genes instead of as small variants of one another. Therefore, a considerable amount of the alternative isoforms in the human proteome constitute "functional alloforms."

Transcriptomic investigations emphasized the enormous potential proteomic variety resulting from alternative splicing (Barbosa-Morais et al. 2012; Pan et al. 2008; Wang et al. 2008). However, the functional discrepancy among alternatively spliced protein isoforms is still unclear at the proteomic scale. Even though systematic functional investigations of protein isoforms have been reported for certain groups of genes (Corominas et al. 2014), the majority of functional annotations and interactions between proteins are resolved at the gene level.

The systematic cloning of natural splice isoforms and the mapping of isoform interactions on a proteome level has allowed us to detect a broad spectrum of variations in the interaction pattern between protein isoforms, giving a deeper view into the overall impact of alternative splicing on the interactome. Moreover, the extension of the protein–protein interactive framework is an important consequence of alternative splicing and various isoforms of the identical gene can lead to varying degrees of spatial characteristics within interactome complexes. Differences in the interaction characteristics of the majority of the isoform pairs were observed, indicating common functional variations between the isoforms coded by the identical gene. Moreover, analyses of the functional characteristics of isoform interaction partners exhibit a continuously evolving functional diversity among the isoforms, to the extremity where two distinct isoforms encoded by the identical gene seem to act functionally as two different proteins. However, this in turn indicates heavily that the "functional alloform" model of alternative isoforms cannot be ruled out and may indeed reflect the entire human proteome realities more precisely than the "functional isoform" model.

Overall functional differences among isoforms might account for how organisms such as humans with a large splicing variety can create larger phenotypic complexity out of only roughly 20,000 protein-coding genes. Moreover, this operational heterogeneity points to the requirement to study each protein isoform individually in order to gain an insight into its specific function, encompassing its role in the pathogenesis of diseases or its ability to act as a therapeutic target. The cartography of isoform-specific protein interactions can also uncover useful knowledge about isoforms of the identical gene and their interacting proteins. It may not be uncommon to find substantial functional discrepancies between pairs of isoforms. A considerable percentage of the isoform pairs were shown to interfere with different types of proteins, with an interactive network pattern that was referred to as "change-over." Every isoform of these "change-over" isoform pairs possesses unique interacting molecules that display isolated expression in various tissues and are prone to be part of various disease modules. These results indicate that the change-over profile of splicing-based protein–protein interaction frameworks represents a major factor in the operational diversity among isoforms and can lead to functional differentiation of tissues.

Alternatively, it was possible to designate spliced regions that include potential interaction constituents that encourage or hinder interactions. A number of "interaction-promoting" regions include linear motifs, and isoform-specific interacting molecules comprise LMBDs, which agrees with earlier results indicating that tissue-specific exons frequently carry linear motifs (Buljan et al. 2012; Ellis et al. 2012; Merkin et al. 2012). In a similar way, interaction-supporting domains are inclined to incorporate predicted interaction regions relying on well-known or predicted domain-domain linkages. The circumstance that linear motifs and interaction-based areas are more commonly encountered in "interaction-enhancing" areas provides a mechanistic rationale for the interplay variations across isoforms.

Alternative splicing is a key feature in the generation of various protein isoforms with dissimilar primary sequences. Apart from the primary sequence, every protein isoform can be further modified by post-translational alterations, resulting in much more diverse polypeptides referred to as proteoforms (Smith and Kelleher 2013). In the recent investigation, the PPIs of the individual protein isoforms have been analyzed within a heterologous expression system (Y2H) and may have lacked interactions facilitated by the posttranslational modifications of a protein, its intracellular location, stability and other features unique to the protein's local microenvironment. Even though it is beyond the limits of this investigation, post-translational modifications, such as phosphorylation, may result in variations in protein–protein interference or other operational characteristics. For instance, in-depth transcriptome sequencing over multiple tissues and across different species demonstrates that tissue-specific exons are accumulated at phosphorylation sites (Merkin et al. 2012), indicating that alternative splicing could be engaged both in the control of protein interactions and in the modification of the phosphorylation capacity. The compilation of a broad library of different proteoforms and the consequent investigation of their various roles will therefore be essential for a comprehensive knowledge of normal cell biology and disease pathogenesis on the systems level.

In conclusion, these findings confirm a key function for alternative splicing in the organization, functionality, and tissue-overlapping dynamics of entire meshes and illustrate the relevance of an isoform-resolved general perspective on interacting meshes. It also endorses a general model where most genes code for several distinct protein isoforms, each of them yielding possibly multiple proteoforms, and where every proteoform holds a typically singular array of functions. Overall, this process could lead to a wide variety of "functional alloforms" that produce very diverse physiological and developmental results, pathological conditions and drug development potentials.

13.4 Why Do We Have Alternative Splicing?

Alternative splicing of precursor mRNA is a key feature to enhance the sophistication of gene expression and is playing an essential part in cellular diversification and the subsequent evolution of organisms. The management of alternative splicing is an elaborate process with many interacting components in operation, comprising *cis*-acting elements and *trans*-acting factors, and is additionally governed by the conceptual linkage between transcription and splicing. Complementary molecular characteristics, such as the structure of chromatin, RNA structure and the alternative transcription initialization or completion, cooperate with these fundamental elements to create protein variety by alternative splicing. All of these factors that are part of this one basic biological mechanism combine to form a mechanism that is crucial to the correct operation of cells. Any falsification of the process can result in a disturbance of normal cell performance and the likelihood of disease. Cancer is among the diseases for which alternative splicing may be the foundation for the identification of new diagnostic and prognostic biomarkers and innovative treatment strategies. A profound comprehension of the regulation of alternative splicing has the inherent capacity not merely to shed light on basic biological concepts, but also to find effective treatments for a wide range of diseases (Wang et al. 2015).

The phenomenon of deleting viral sequences from a pre-mRNA and merging the residual sequences together resulted in a fundamental concept of biology, referred to as RNA splicing. The recognition encouraged theories on protein variety, such as alternative splicing, which have been repeatedly implemented in experiments over the course of time. The concept of alternative splicing was first introduced in 1978 (Gilbert 1978). It is the mechanism that currently explains the mismatch between the number of protein-coding genes of about 25,000 in humans and the more than 90,000 various proteins which are actually produced (C. elegans Sequencing Consortium 1998; Human Genome Sequencing Consortium I and International Human Genome Sequencing C 2004). The hypothesis that one gene generates one RNA and subsequently produces one protein has become irrelevant. It was found that more than 95% of human genes are spliced in a developmental, tissue-specific or signal transduction-dependent fashion (Nilsen and Graveley 2010).

Constitutive splicing refers to the removal of the intron and the ligation of most exons in the sequence of their occurrence in a gene. Alternative splicing is an aberration of this preferred order in which specific exons are silenced, leading to multiple types of mature mRNA. Less pronounced splicing signals on alternative splice locations, reduced exon length, or enhanced sequence maintenance surrounding orthologous alternative exons influence the exons that are finally contained in the mature mRNA (Zheng et al. 2005). This mechanism is facilitated by a dynamic and flexible macromolecular mechanism, the spliceosome, operating in a synchronistic and dynamic fashion (Effenberger et al. 2013; Wahl et al. 2008). Three probable mechanisms encompassing the mixing of exons, the exonization of transposable elements and the constitutively spliced exons where have been hypothesized for the original source of alternative splicing (Kim et al. 2008). A wide variety of investigations have confirmed the crucial and basic role of alternative splicing in biological systems (Irimia et al. 2007). In the higher eukaryotic organisms, it has been found that they contain a higher percentage of alternatively spliced genes, which is a hint for a major evolutionary impact of this mechanism. Alternative splicing conveys various biological pathways over the whole lifetime of organisms, ranging from birth to death (Fu et al. 2007; Hurt et al. 2006). Preserved splices of species-specific splice variations perform an essential operational function in the process of species diversification and evolution of genomes (Blencowe 2006; Nygard et al. 2010), and also for the formation of tissues ranging from functionally simple to complex with various cell types, such as the brain, testis and immune system. Alternative splicing is even involved in the actual RNA processing steps, ranging from pre- to post-transcriptional episodes.

Alternative splicing affects nearly all facets of protein function, encompassing binding among proteins and ligands, nucleic acids or membranes, the localization and the enzymatic characteristics. In summary, the process of alternative splicing represents a key feature of the entire process of gene expression (Kelemen et al. 2013).

13.4.1 Are Most Alternative Isoforms not Functionally Important?

There are two assertions that can be conversely discussed. The first hypothesis is that most protein-coding genes display a single main isoform (Tress et al. 2017a). The second hypothesis is that most annotated alternative transcripts generate no functionally relevant gene products. Since alternative isoforms fulfill important cellular roles, alternative splicing does not seem to serve as the key factor for increasing proteome complexity (Tress et al. 2017b).

The existence of a single main isoform can be determined for most genes. Moreover, these isoforms can be predicted from conservation of protein structures and functions (Ezkurdia 2015). However, when RNA level studies reveal a dominant variant (Gonzàlez-Porta et al. 2013), there is not much overlap with the main isoform detected in proteomic studies (Ezkurdia 2015). The actual fraction of splice variants that is translated is still not yet known. Many reasons can be given why these transcriptional isoforms cannot be detected (Tress et al. 2017b). Among them are that they are not translated, solely present on low abundance, expressed in certain tissues, expressed only under specific conditions or at certain developmental stages or these transcriptional isoforms are fast degraded after translation. In addition, alternative variants of the transcript withstand the examination of the ribosome and therefore do not encode stable proteins (Ingolia 2016; Guttman et al. 2013). Many identified alternative isoforms possess intact functional domains (Abascal et al. 2015), which leads to the hypothesis that they have undergone a sorting process either before or after translation. There is still an important question not yet answered: Why do alternative isoforms detected by proteomics still possess their functional domains?

13.4.2 Underlying Molecular Mechanisms of Alternative Splicing

Systematic studies of ESTs and microarray analysis have identified seven main types of alternative splicing versions (Blencowe 2006). The most common template (approximately 30%) consists of the cassette-like alternative exon (exon skipping) in vertebrates and invertebrates, whereas in the lower metazoans, intron retention is the most common (Kim et al. 2007b). The intron capture in human transcripts is mainly situated in the untranslated regions (UTRs) (Galante et al. 2004) and has been linked to weaker splice sites, reduced intron length and the regulatory effect of *cis*-regulating elements (Sakabe and De Souza 2007). The alternative selection of 5′ or 3′ splice sites inside exon sequences (approximately 25%) can cause slight alterations in the coding sequence, and an extended complexity level with mutually excluding alternative exons is generated. A sample transcript subjected to alternative splicing that produces a modification of the protein would be FGFR2. Variations in the splicing mechanism of various cell types and unique *cis*-acting elements within FGF-R2 pre-mRNA result in modified tissue-specific determinations that generate either FGF-R2IIIb or FGF-R2IIIc mature transcripts (Goldstrohm et al. 2001).

Protein expression continues to be governed by the alternative polyadenylation of mRNA, which affects the encoding capacity or length of 3′UTR through alteration of the binding efficiency of microRNA or RNA (Di Giammartino et al. 2011). It has been proven that any kind of alternative splicing can function in stochastic fashion, and that varying mechanisms for detecting and handling splice sites may not be the same in all biological realms with the same frequency (Mastrangelo et al. 2012). The mechanisms sketched above indicate the complex nature of this phenomenon, since multiple molecules are engaged in alternative splicing in a highly coordinated mode. Thus, even the fundamental nucleotide constituents and the essential compounds they detect can provide diversity in the synthesis of mature transcripts.

Two important steps form the fundamental process of splicing: the assembly of the spliceosome, which is immediately followed by the actual splicing of the pre-mRNA. The spliceosome consists primarily of the small nuclear ribonucleic acid proteins (snRNPs) U1, U2 and the U4/U6. U5 tri-snRNP and assembles a core set of splicing signals: The 5′ splice site, the sequence of the branch point and the 3′ splice site. Specific spliceosomal assemblies (E, A, B, and others) and eight evolutionary conserved DExD/H-type RNA-dependent ATPases/helicases are composed in a stepwise fashion as recommended and perform several splicing tasks leading to exon ligation and the excision of the intron. Multiple steps of the splicing pathway can be reversed (Hoskins and Moore 2012).

The exons arriving in the mature mRNA in the course of the alternative splicing process are completely characterized by the interplay between *cis*-acting elements and *trans*-acting factors. *Cis*-acting elements comprise exonic splice enhancers and intronic splice enhancers linked by positive *trans*-acting factors such as SR proteins (serine/arginine-rich family of nuclear phosphoproteins), while exonic splicing silencers and intronic splicing silencers are tethered by negative-acting factors such as heterogeneous nuclear ribonucleoproteins. The cooperation between these elements leads to the encouragement or hindrance of the splice assembling of the weak splicing sites (Wang and Burge 2008; Wang et al. 2006). In general, the *cis*-effective elements operate in a cumulative manner. The reinforcing elements generally incline to occupy a dominant role in constitutive splicing, whereas the silencers tend to be relatively predominant in controlling alternative splicing process (Wang and Burge 2008). It has been shown that the enhancing activity is canceled by a robust stem loop architecture of merely seven base pairs within an RNA transcript due to the physical competition mechanisms, extended RNA pairing, a structured splicing code and co-transcriptional splicing (Jin et al. 2011; Liu et al. 2010). Moreover, the degree of specificity of *cis*-acting enhancer elements for introns or exons has been examined. In these investigations, it has been observed that an exonic splicing enhancer behaves as an intronic splicing enhancer, according to its location within an exon or an intron (McManus and Graveley 2011).

HnRNPs have high levels of conservation from nematodes to mammals and play multiple crucial roles in pre-mRNA maturation. They bind to the exon splice silencers in a manner that excludes SR proteins. Extraction of pre-mRNA results in exonic fragmentation of the remaining pre-mRNA transcript (Nasim et al. 2002). HnRNPs A/B represents a family of RNA-binding proteins whose divergent functions in the regulation of alternative splicing may be due to varying levels of affinity to their respective cognitive nucleic acids (Han et al. 2010). HnRNP H and F serve to change the ratio of proteolipid proteins (PLP/DM20) by modifying the enrollment of U1 snRNP (Wang and Cambi 2009). In a similar way, the antagonistic effect of hnRNP M on the splicing factor Nova-1 produces alternatively spliced dopamine receptor pre-mRNAs that generate isoforms linked to various core physical processes including control, reward, learning and memory (Park et al. 2011). Moreover, hnRNP L and ser513 phosphorylation have lately been demonstrated to be implicated in the control of alternative splicing by the dynamic depolymerization of the membrane and Ca^{2+}/calmodulin related activation of protein kinase IV (Liu et al. 2012a, b; Yu et al. 2009).

Besides the linkage of SR proteins to enhancer elements, SR proteins interoperate with U1 snRNP and the 35 kDa subunit of the heterodimeric factor U2AF. The second subunit of U2AF, termed U2AF65, is capable of simultaneously coupling SF1 and the pyrimidine tract, based on the arginine/serine (RS)-rich domain, leading to the identification and stabilization of the branching point, along with sequences of polypyrimidine tract. About 10–12 serines in the N-terminal region of the RS domain are quickly phosphorylated with abnormally high affinity due to the attachment of SR-specific protein kinase to serine/arginine-rich splicing factor 1. This ongoing phosphorylation/dephosphorylation cycle of SR proteins eases the transport of SR proteins from the nucleus to the cytoplasm and is crucial for the control of alternative splicing by growth stimuli transferred directly into the nucleus (Ghosh and Adams 2011; Twyffels et al. 2011). SR proteins have been suggested for participation in post-splicing efforts, such as the nuclear export of mRNA, nonsense-mediated decay (NMD), and the translation of mRNA (Long and Caceres 2009).

In common, positive or negative splice site detection is governed by various mechanisms under different physiological or pathological circumstances, such as the presence of a specific concentration or activity of splicing regulatory factors. However, how these factors interact to accurately choose a controlled splice site can rarely be adequately accounted for by these findings (Kalnina et al. 2005).

13.4.3 Cooperation of Alternative Splicing and Transcription

From the first significant evidence of co-transcriptional spliceosome assembly seen in electron microscopic images of *D. melanogaster* embryonic transcriptional units (Beyer et al. 1981), there is growing proof that transcription and splicing have become physically and functionally intertwined and has also revealed the complicated connection between mRNA splicing, RNA polymerase II (Pol II), and chromatin structure (Kornblihtt et al. 2004; Montes et al. 2012).

A wide range of compounds related to the physical splicing and transcription interaction were isolated, with special emphasis on the carboxyl terminal domain (CTD) of the large subunit of RNAPII (Ryman et al. 2007). The CTD is composed of 52 tandem repeats of the heptapeptide YSPTSPS in mammals (26 tandem repeats in yeast) (Bentley 2005), which serve as a special stage to attract various factors through dynamic phosphorylation of serine residues to the emerging transcripts. Kinases which phosphorylate specific CTD serine residues have been recognized and are constituents of the protein apparatus that controls the specific activity. For instance, the phosphorylation of Ser5 is linked to the activation of transcriptional activity by cyclin-dependent kinase 7 (CDK7) of the common transcriptional factor IIH (TFIIH), while Ser2 phosphorylation is preferably associated with CTD activity at the 3′ end of genes by CDK9 of the action of the positive transcriptional elongation factor b (Bartkowiak et al. 2010). In specific detail, the ser7 phosphorylation has been observed to drive the elongation and splicing steps (Kim et al. 2010). Consequently, the phosphorylation represents a process, which accurately shows that there is a functional linkage between the processes of transcription and alternative splicing.

The CTD takes place in gene expression-related processes encompassing 5′ capping, splicing, poly-adenylation and the rearranging of chromatin (Phatnani and Greenleaf 2006). It is noteworthy that the mutation and deletion detection analysis of the CTD uncovered several failures in mRNA processing (Rosonina and Blencowe 2004). Therefore, the CTD and complementary elements of the two systems have been discovered to be a key factor in controlling the interference phenomena between transcription and splicing. In summary, the functional linkage seems to play an essential part in alternative splicing in order to promote determinative physiological alterations and refine the expression of genes in mathematical simulation approaches (Kalsotra and Cooper 2011).

Two concepts have been proposed to understand the co-transcription mechanism of how the transcription-linked repair influences the process of alternative splicing. The mechanism of the enrollment model can primarily rely on specific characteristics of the CTD, while the kinetic model is built on the varying elongation rates of Pol II, which themselves define the time of exposure of the splice sites (Schor et al. 2013; Dujardin et al. 2013). Basically, the above mechanism affects the behavior of alternative splicing patterns through fluctuations in pole II elongation and the acquisition of splicing factors through certain histone labels (Gómez Acuña et al. 2013). Therefore, alternative splicing is not exclusively governed through transcription, but also strongly affected by the structural arrangement of chromatin, which highlights the chromatin as a separate entity in the control of alternative splicing. The emerging mature mRNA is therefore a mirror image of several DNA changes, including profiles of histone methylation at exons, regulation of histone changes and enhanced DNA methylation of exons (Iannone and Valcárcel 2013; Shukla and Oberdoerffer 2012). In contrast, an earlier investigation pointed out that splicing can convey chromatin engineering through the build-up of histone labels on the DNA or multiple links among splicing factors and elongation proteins (Kornblihtt et al. 2004).

The regulatory network is made even more complex by alternative transcription initiation (ATI) and alternative transcription termination (ATT) locations. ATI and ATT provide a major input to the variety of human and mouse transcriptomes, to an extent that may surpass the capabilities offered through alternative nucleotides, isoforms and introns (Ma et al. 2009; Xin et al. 2008). In contrary to the predominance of alternative splicing, which appears in coding sequences (CDSs), the prevailing class of alternative incidents, comprising ATI and ATT, arises in untranslated regions (UTRs). This observation mirrors the privileged control of large different groups of genes with various mechanisms, such as tight coupling with alternative splicing in 5′ and 3′UTRs (Shabalina et al. 2010). Notwithstanding the high association between alternative splicing and transcription, alternative transcription primarily causes differences in the transcription number or the 5′/3′ terminal protein modifications as a result of distinct transcriptional start or end points. In contrast, the changes connected with splicing are usually within the protein sequence and can involve virtually all aspects of protein function (Kelemen et al. 2013; Roy et al. 2013).

13.4.4 Alternative Splicing Can Lead to Non-sense Mediated Decay

None-sense mediated decay (NMD) constitutes an elaborate and sophisticated mechanism, extending from yeast to humans, that is being utilized to provide a further step towards a more robust post-transcriptional gene expression monitoring system. Investigations have shown that up to one-third of human alternative splicing incidents include premature termination codons that are detected and result in the break-down of transcripts with NMD *cis*-elements in their 3′UTRs (Lareau et al. 2007a; Lewis et al. 2003). In vertebrates, it has been suggested that linking of the exon junction complex (EJC) to mRNA transcripts, subsequently by the attachment of 3′UTRs to EJCs, induces a vertebrate-specific NMD (Nyiko et al. 2013). The susceptibility of mRNA transcripts to NMD is regulated by alternative splicing incidents in the 5′ or 3′UTRs and helps with the broad spectrum of protein biosynthesis (Kalyna et al. 2012). In addition, the analysis of quantitative alternative splice microarray profiling has revealed that the individual degradation of NMD factors, such as up-frame shift (UPF) suppressors, has a profound influence on premature termination codons that trigger alternative splice occurrences, pointing to a relevance for varying UPF factor demands in the control of alternative splicing (Saltzman et al. 2008). In a second case, a regulation of intron capture by alternative splicing–NMD has been observed lately for a specific differentiation incident (Ge and Porse 2014).

13.4.5 What Effect Has *Trans*-Splicing?

Trans-splicing is a frequent phenotype in trypanosomes, nematodes, *Drosophila* and even in humans and relates to the new and uncommon splicing of exons from separate pre-mRNAs (Herai and Yamagishi 2010; Horiuchi and Aigaki 2006). The occurrence has been investigated as a therapeutic choice for a series of genetic diseases, especially in the therapeutic treatment of cancer (Lee and Jeong 2014). The carcinoembryonic antigen (CEA), for instance, is linked to a number of neoplastic events and was utilized as a *trans*-splicing target. A CEA-RNA-targeting *trans*-splicing ribozyme was developed to replace RNA by a *trans*-splicing response particularly in CEA expressing cells

(Jung and Lee 2006). The activity of the ribozyme at the same time decreased CEA expression and inserted the thymidine kinase gene, which made the cells susceptible to the treatment with ganciclovir. RNA *trans* splicing has also been exploited for the possible therapeutic application of a new technology, spliceosome-mediated *trans*-splicing (SMaRT), for the effective therapy of neurodegenerative diseases. SMaRT has been effectively applied in vivo to remodel tau mRNA transcripts to incorporate E10 and thus holds the promise of correcting tau mis-splicing and targeting the underlying disease (Avale et al. 2013).

13.5 Can Alternative Splicing Be Regulated by Non-coding RNA?

Non-coding RNAs (ncRNAs), comprising microRNA and small interfering RNAs, have lately become new modulators of alternative splicing, commonly by regulating the expression of certain key factors in splicing during the development and differentiation (Luco and Misteli 2011). Advances in functional genomics have shown that even though a huge number of RNAs are transcribed from the human genome, the protein-coding genes represent just a tiny portion of the overall transcriptions (Birney et al. 2007; Kapranov et al. 2007a, b). The remaining transcripts consist of non-coding RNAs, comprising microRNAs and long non-coding RNAs (lncRNAs). These non-coding transcripts were originally regarded as transcriptional murmurs (Gibb et al. 2011). But it has gained growing recognition that they can contribute significantly to a variety of processes in the cell, ranging from normal development to disease processes (Esteller 2011; Wilusz 2016; Wang et al. 2013; Fatica and Bozzoni 2014). There exist two key classes of non-coding RNAs, which are grouped due to their overall length, firstly, small ncRNAs consisting of less than 200 nucleotides, referred to as microRNAs and secondly, lncRNAs larger than 200 nucleotides in length. MicroRNAs have a good characterized structure and are renowned for their ability to either initiate the degradation of mRNA or inhibit the translation of mRNA through the RNA interference pathway (Hayes et al. 2014). On the contrary, lncRNAs have less good characterization. In fact, for the vast majority of lncRNAs, not much is generally understood concerning their functions. To this day, the non-code database records more than 96,000 human lncRNAs (https://www.noncode.org), while more than 56,000 human lncRNAs have been indexed on LNCipedia (https://www.lncipedia.org) and the amount of lncRNAs is still rising (Mercer and Mattick 2013; Blythe et al. 2016; Koirala et al. 2016). Several investigations have demonstrated that a considerable quantity of lncRNAs appears as essential actors in various layers of cellular processes (Wang and Chang 2011; Rinn and Chang 2012). Even though the function of lncRNAs is only now starting to be understandable, there is growing support for the idea that lncRNAs can positively or negatively affect gene expression on the transcriptional and/or post-transcriptional level (Kornienko et al. 2013; Costa 2005; He et al. 2019).

13.5.1 Long Non-coding RNAs

The overwhelming number of lncRNAs were detected some years ago when it became clear that huge portions of the human genome are actually transcribed (Bertone et al. 2004; Kapranov et al. 2007a, b), although only a small proportion of these transcriptions were made by protein-coding genes (PCGs) (Birney et al. 2007). It was proposed that transcripts of non-coding genes (those with minimal or no protein-coding capacity) are key regulators of PCG expression, controlling each level of gene expression programs (Mattick et al. 2009; Wang and Chang 2011; Patrushev and Kovalenko 2014; Ransohoff et al. 2017).

Commonly, lncRNAs are reported to be expressed at lower levels (Derrien et al. 2012; Djebali et al. 2012; Mukherjee et al. 2016) and therefore undergo weaker evolutionary restrictions than PCGs (Wang et al. 2004b; Cabili et al. 2011; Schüler et al. 2014; Necsulea et al. 2014). Moreover, LncRNAs also exhibit more various tissue-specific expression patterns than PCGs (Derrien et al. 2012; Cabili et al. 2011; Lagarde et al. 2017) and are slightly concentrated in the chromatin and nuclear components (Derrien et al. 2012; Djebali et al. 2012; Cabili et al. 2015). Several nascent lncRNAs possess introns and therefore underlie the identical RNA processing stages as pre-mRNAs of protein-coding genes (PCGs), comprising the steps of capping, splicing and polyadenylation (Quinn and Chang 2016). However, they are usually badly spliced.

The greatest resemblance to PCGs in respect of genomic structure, gene length and so on was observed for long intergenic non-coding RNAs (lincRNAs) (Derrien et al. 2012; Lagarde et al. 2017). It is assumed that introns of lncRNAs get spliced through the identical splicing machine as pre-mRNAs (Will and Lührmann 2011; Matera and Wang 2014). The 5′ splice is detected by the U1 snRNP. The 3′ splice site with the branching point (BP), the polypyrimidine tract and the YAG motif at the 3′ end of the intron is linked by U2 snRNP-associated factors (SF1, U2AF1/2 (U2AF35/65)), which thereafter enroll the U2 snRNP within the BP (Matera and Wang 2014; De Conti et al. 2013). Since cryptic splice sites are rather plentiful in the transcribed areas, the identification of the accurate exon borders is a decisive stage in the splicing process. The high accuracy of splice site identification is conveyed by a framework of specific interactions involving snRNA base matching with sequences encircling the splice sites and the attachment of multiple splice regulatory proteins such as SR proteins, U2 auxiliary factors and hnRNP proteins (De Conti et al. 2013; Sahebi et al. 2016). These regulatory elements attach to short sequences, which are categorized as splice enhancers or silencers. In addition, the activities of such regulatory factors are frequently context sensitive, and they can either stimulate or prevent splicing, depending on where they are placed on the transcript (Ule et al. 2006; Wang et al. 2008; Fu and Ares 2014; Brillen et al. 2017; Rot et al. 2017).

In order to elucidate the mechanisms underlying this process, the search for presumed splice inhibitory sequences was carried out employing ncRNA-a2 as a model (Krchňáková et al. 2019). Genome-wide screening of intergenic lncRNAs (lincRNAs) yielded that the splicing performance of lincRNAs is positively associated with the specific strength of 5′ splice site, whereas no similar relationship was observed in PCGs.

A number of bioinformatics projects showed that lncRNAs/lincRNAs, either steady-state or nascent RNAs, are spliced with a lower effectiveness compared to pre-mRNAs from PCGs (Mukherjee et al. 2016; Lagarde et al. 2017; Tilgner et al. 2012; Melé et al. 2017; Schlackow et al. 2017). A potential mechanism for the apparent difference in splicing performance of lincRNAs and PCGs is the lack of proximal RNA Pol II phosphorylation across the 5′ splice site within the lincRNA transcripts (Mukherjee et al. 2016). But the exact molecular mechanism of this feature has not been understood. In this case, bioinformatic and experimental techniques have been merged to identify *cis*- and *trans*-acting determinants that are behind the imperfect splicing of lincRNAs.

Moreover, efficiently spliced lincRNAs contain higher levels of thymidine present in the polypyrimidine tract than efficiently spliced PCGs. On the basis of model lincRNAs, an experimental proof is presented that the reinforcement of 5′ splice site and the increase of the T content in the polypyrimidine tract considerably improves the splicing of lincRNAs. In addition, it has been seen that LincRNA exons carry fewer suspected SR protein-binding sites. In order to map the attachment of SR proteins to lincRNAs, iCLIP has been employed with SRSF2, SRSF5 and SRSF6 and for SRSF1, SRSF7 and SRSF9 eCLIP data have been investigated (Krchňáková et al. 2019). All SR proteins investigated attach LincRNA exons to a far lesser degree than expression-matched PCGs. It is proposed that the lincRNAs deprive the collaborative interaction framework that improves splicing, thereby increasing the dependence of their splicing yield on the optimization of the splice sites.

Using available ENCODE RNA-Seq data (Sloan et al. 2016) to calculate the splicing incidents for lincRNAs and PCGs expressed in multiple human cell lines as a prototype for the splicing performance of expressed lincRNAs. The results achieved in all investigated cell lines indicate a less intensive splicing of lincRNAs in comparison with PCGs (Krchňáková et al. 2019). The bioinformatic findings are enhanced by splicing efficiency measurements of multiple lincRNAs carried out by quantitative and semi-quantitative RT-PCR (Krchňáková et al. 2019).

To identify determinants influencing lincRNA splicing, in a transient fashion multiple lincRNAs were expressed by a CMV-driven promoter and no significant alterations in their pattern of splicing were observed (Krchňáková et al. 2019). This finding suggests that the promoter and genomic background do not substantially affect the LincRNA splicing profile and that poor splicing is an inherent characteristic of LincRNA transcripts. In order to pinpoint possible sequences that block splicing, several deletion mutants have been generated that are missing various parts of the ncRNA-a2 intron. However, there were no powerful splicing silencers, according to a bioinformatic investigation that did not exhibit distinct accumulation of splicing inhibition sequences in lincRNAs in relation to PCGs (Krchňáková et al. 2019). However, it has been discovered that lincRNAs contain larger introns and exons than PCGs (Krchňáková et al. 2019), which could partly account for their inefficient splicing (Klinz and Gallwitz 1985; Bell et al. 1998; Sterner et al. 1996; Fox-Walsh et al. 2005; Dewey et al. 2006; Louloupi et al. 2018).

Ultimately, sequences of the 5′ splice site and the 3′ splice site were examined and found a positive relationship between the strength of the 5′ splice site and polypyrimidine tract and LincRNA splicing performance (Krchňáková et al. 2019). This finding was corroborated by some experimental proof that enhancing the strength of the 5′ splice sites and 3′ splice sites significantly advanced the splicing of exemplary lincRNAs. The 5′ splice site and the sequences of the polypyrimidine tract are decisive elements for splicing efficiency across the broad spectrum; however, some results indicate that the lincRNA needs a more pronounced 5′ splice site and a polypyrimidine tract with a high number of transcripts to be spliced in an effective manner (Krchňáková et al. 2019). To gain an insight into why LincRNAs are more heavily reliant on basic splice site sequences, the presence of well-known SR protein exon-binding motifs has been investigated as the binding of SR proteins to exons encourages splicing (Long and Caceres 2009; Graveley 2000).

The bulk of the SR-binding sequences examined were observed to be more frequently present in PCG exons, whereas one out of 29 motifs analyzed is concentrated in lincRNAs (Krchňáková et al. 2019). Accordingly, all SR proteins examined revealed a clear binding priority for PCGs, even though lincRNAs were analyzed and the results were related to expression-matched PCGs. This finding offers experimental verification that LincRNAs are not capable of ensuring that SR proteins are able to engage effectively in binding. On the basis of these data, a model is proposed in which the lincRNAs lack the interactive framework of positive signal transduction that guides the splicing engines efficiently to the splice sites. For the majority of lincRNAs, the U1- and U2-snRNPs and their auxiliary factors need to identify splice sites independently of splice amplifiers, which emphasizes the relevance of sequences around exon/intron borders.

It should be kept in mind, nevertheless, that the introduction of the ncRNA-a2 intron between PCG exons has not increased the splicing effectiveness. This indicates that the entire sequence and environment is essential for proper splicing, as was lately illustrated for the splicing of a variety of 5′ splice site sequences (Wong et al. 2018). It is also in line with experiments which claim that the local surroundings and the continuing sequence of information about exons and introns are decisive for the accurate identification and elimination of introns (Lagarde et al. 2017). In the course of evolution, the PCG sequences were refined to guarantee reliable detection of the intron/exon borders and an efficient splicing process. Contrary to PCGs, where the splicing is a crucial element of gene expression, it has been demonstrated that introns are not substantial for the activating role of lincRNAs (Engreitz et al.

2016; Krchňáková et al. 2019). This can lead to less evolutionary stress on some lincRNAs to encourage more efficient splicing. Nevertheless, it is important to bear in mind that the vast majority of purifying selection that works with lincRNAs in humans is associated with splicing, indicating that the existence of introns or their splicing could have a function that is still unidentified, in at least a few lincRNAs (Schüler et al. 2014).

Over time evolutionary new transcripts, such as lincRNAs, can retain functional 5′ splice sites and 3′ splice sites and splicing can alter their functional performance. Previous investigations have shown a higher predominance of cryptic 5′ splice sites over 3′ splice sites, which presumably mirrors the somewhat less complex character of the 5′ splice sequence (Nakai and Sakamoto 1994). In fact, the 5′ splice site can be obtained by the presence of endogenous retrotransposons (Franke et al. 2017). The investigation of the cryptic 3′-splice site highlighted the relevance of the 3′-splice site sequences and demonstrated that the intronic de novo 3′-splice site was primarily formed by AG-forming mutations in established functional polypyrimidine tracts. In contrast, exonic de novo 3′ splice sites have often been initiated by mutations that promote the polypyrimidine tract, the BP sequence, or eliminated auxiliary signaling (Královicová et al. 2005). Thus, in the absence of functional splicing amplifiers, the faint sequence of the polypyrimidine tract and the inefficient U2AF attachment are the principal obstacles that evolutionary new transcripts face in order to be successfully spliced.

13.5.2 Long Noncoding RNAs Remodel Chromatin

Over the last decade, the structure of chromatin and histone variations has become key drivers for alternative splicing (Romero-Barrios et al. 2018). The interaction between histone alterations, chromatin-binding proteins and slicing factors potentially represents a complex system of communication among chromatin and RNA (Luco et al. 2011). It was also found that the chromatin background affects the RNA polymerase II (Pol II) elongation and therefore influences alternative splicing (Batsché et al. 2006; Schor et al. 2009). This implies that the epigenetic control not merely governs which sections of the genome are expressed, rather it also governs how these sections are spliced (Luco et al. 2011). Several lncRNAs are involved in the determination of the chromatin structure and dynamics, which can then have a particular effect on splicing performance: firstly, direct interference between lncRNAs and DNA forming heteroduplexes, secondly, enrollment of chromatin manipulators at certain specific loci, or thirdly, design of the 3D organization of the chromatin configuration inside the cell nucleus. Ultimately, there is a discussion of how lncRNA-derived small RNAs can regulate the assembly of chromatin and determine alternative splicing arrangements based on this mechanism.

Splicing Regulation by lncRNA-Driven DNA–RNA Duplexes

Circular RNAs (circRNAs) are covalently-closed circular molecules of single-stranded RNA that arise from a non-canonical splicing process referred to as backsplicing. This process involves the ligation of a downstream splice donor site in the reverse direction with an upstream splice acceptor site of pre-mRNA, thereby producing a circular lncRNA molecule. These circular transcripts are plentiful and extremely robust, and they can efficiently compete for the identification of related splicing protein complexes along with the linear pre-mRNA (Chen 2016). For instance, in flies and humans, muscleblind is able to bind heavily and explicitly to the circRNA, which is produced by its individual locus, referred to as circMbl (Ashwal-Fluss et al. 2014). In human cells, circRNAs are modulated by the slicing factor QKI in a dynamic fashion during the epithelial to mesenchymal transition (EMT) (Conn et al. 2015), and it has been demonstrated that the appearance of circRNAs in human endothelial cells is related to exon skipping in the entire genome (Kelly et al. 2015). However, the molecular mechanisms underlying circRNAs in animals continue to be widely unknown.

13.6 Several Diseases Are Based on Alternative Splicing of Coding Genes

Strict regulation of alternative splicing is required for the functional demands of the complex tissues during normal circumstances, while abnormal splicing seems to be the reason for a high proportion of malfunctions and diseases (Singh and Cooper 2012). It has been suggested that aberrant splicing is due to changes in cell concentration, the composition, placement and the activity of regulatory splicing factors and due to mutations in elements of the nuclear splicing apparatus (Zhang and Manley 2013). A modified effectiveness of splice site detection results directly, whereas abnormalities in the protein isoforms in various types of systems finally determine the state of disease. Each of these changes that influence alternative splicing can promote the appearance of features in cancer cells, which encompasses the improper proliferation, migration, methylation alterations and opposition to apoptosis and chemotherapy (Shkreta et al. 2013). Alternative splicing is involved in almost every facet of cancer development and is therefore a key contributor to the disease.

Knowledge of the fundamental mechanisms and splicing schemes underlying tumor progression will provide insights into the biology of cancer and establish the basis for diagnostic, prognostic and therapeutic interventions minimizing the treatment toxicity of cancer (Kim and Kim 2012). Comprehensive scientific efforts conducted have resulted in the development of medications aiming at certain cancer protein isoforms. Multiple instances are genes linked to apoptosis, comprising BCL2L1 (BCL-X), FAS, BIRC5 (synonymously referred to survivin) and MDM2, immortality, such as human telomerase reverse transcriptase and angiogenesis, such as vascular endothelial growth factor-A (Lapuk et al. 2014; Pajares et al. 2007).

Nevertheless, limited progress has been obtained by merely activating or inhibiting cancer-associated genes, perhaps due to the expression of target genes in normal and cancer cells, including angiogenic and anti-angiogenic isoforms (Sitohy et al. 2012). The deficiency in specificity of many molecular targets for cancer cells encourages the establishment of isoform-specific diagnostic marks as therapeutic targets (Pal et al. 2012). Thus, the main challenge for future cancer therapy is to identify and specifically influence the gene expression at the level of the gene.

The combined application of an alternative splicing library, tandem mass spectrometry and even the latest synthetic alternative splicing library can help identify, analyze and characterize possible isoforms of alternative splicing. More than two-thirds of human genes and 40% of *Drosophila* genes comprise one or more alternative exons, and approximately 90% of protein-coding genes linked to alternative splicing incidents, based on the over 60,000 research studies undertaken since the discovery of splicing (Pan et al. 2008).

Alternative splicing seems to be predominant in nearly all multi-exon genes. However, limitations to a more comprehensive and precise use of alternative splicing techniques seem to encompass factors such as biased coverage of ESTs relative to the 5′ and 3′ ends of transcripts, poor widespread analysis, subtle alternative splicing modifications and sophisticated alternative splicing frameworks implicated in various mechanisms and the high level of regulatory proteins. All these imperfections result in an incomplete knowledge of the alternative splicing machinery and may hinder the accurate forecasting of splicing occurrences in other species, such as the chimpanzee or the plant (Calarco et al. 2007; Reddy et al. 2012). It is also hard to differentiate alternative splicing from other gene regulation mechanisms. The alternative splicing, the alternative *trans*-splicing, the NMD, the efficiency of the transcription, the exon duplication and the editing of RNA (Tang et al. 2012a) are all involved in a comprehensive mechanism for the achievement of protein diversity. Moreover, the distinction between artificial experimental models and real-life situations poses a challenge for the implementation of functional investigations from individual cells to entire organisms. There are still a large number of unanswered questions regarding the overall effects of alternative splicing on cellular and organismic homeostasis and the fundamental molecular mechanisms involved. Ultimately, in cancer-associated alternative splicing, it is difficult to differentiate whether a specific splice site choice produces the monitored change or is simply the consequence of the cancerous conversion. The gathered data regarding alternative splicing are probably the beginning, and hence, future studies are required.

13.7 Alternative Splicing in Cancer Disease

Alternative splicing alterations are commonly seen in cancer and are emerging as key signatures for the progression of the tumor and therapy. Its functional role and significance for tumor development, however, are still widely unclear. To assess the possible functional implications of alternative splicing alterations in several thousand tumor specimens, a systematic screening was performed. This analysis demonstrated that a subsection of alternative splicing alterations concerns protein domain families that are commonly mutated in tumors and may potentially disturb protein–protein interactions in cancerous pathways. In addition, there is a negative relationship between the amount of these alternative splicing alterations and the rate of somatic mutations in the drivers. Moreover, it is argued that a subgroup of alternative splicing modifications associated with tumors may constitute separate oncogenic phenomena that may be useful in explaining functional variations in cancer, and a number of them may possibly be regarded as alternative splicing drivers (Climente-Gonzalez et al. 2017).

Alternative splicing offers the opportunity to create multiplicity at the RNA and protein scale from a seemingly restricted set of loci within the genome (Yang et al. 2016). Alternative splicing is not only a critical mechanism during development, differentiation of cells and regulation of cell type-specific activities (Norris and Calarco 2012), but also plays a role in several diseases, among them is cancer (Chabot and Shkreta 2016). Many alternative splicing modifications recapture cancer-related phenomena by stimulating angiogenesis (Vorlova et al. 2011), initiating cellular proliferation (Yanagisawa et al. 2008), or preventing apoptosis (Karni et al. 2007). Alternative splicing alterations can result from somatic mutations that interfere with the regulatory motifs of splicing in exons and introns (Jung et al. 2015; Supek et al. 2014), and from mutations or expression alterations in nuclear and auxiliary splicing factors that interfere with the splicing of cancer-associated genes (Bechara et al. 2013; Darman et al. 2015; Madan et al. 2015; Zong et al. 2014). Changes in alternative splicing also prove to be relevant goals of therapy (Lee and Abdel-Wahab 2016). For example, lung tumors with exon skipping in the proto-oncogene MET react to MET-directed therapies although they lack other activating changes in this gene (Frampton et al. 2015; Paik et al. 2015). Alternative splicing can be relevant for drug resistance. For instance, a fraction of non-responders to BRAF-targeted therapy expresses a BRAF isoform missing exons 4–8, which encompasses the RAS-binding domain (Poulikakos et al. 2011).

In a similar way, alternative splicing of CD19 associated with the abnormal activity of the splicing factor SRSF3 compromises immunotherapy efforts in leukemia (Sotillo et al. 2015). For instance, specific changes in splicing trigger functional effects that may give the cancer cells a certain selective benefit and become treatment targets. Although the frequency of alternative splicing in cancers and its relevance to therapy, the progression of cancers and metastasis (Lee and Abdel-Wahab 2016; Lu et al. 2015; Trincado et al. 2016), the functional effects of alternative splicing have not been comprehensively characterized (Wang et al. 2005), such as the restructuring of the framework of protein–

protein interferences in a tissue-dependent fashion (Buljan et al. 2012; Ellis et al. 2012) and the extension of protein interaction abilities of genes (Yang et al. 2016). A systematic evaluation of the potential functional effects of alternative splicing alterations in the samples of cancer is introduced below. The splicing modifications were characterized by transcript isoform transitions for individual tumor specimens, and hence, the protein characteristics and protein–protein interferences impacted by them. The analysis yielded a number of isoform switches that impact protein regions from families that are commonly mutated in cancers, reshape the protein interaction matrix of cancer drivers and are prone to manifest in patients with a limited number of mutations in the cancer drivers. In addition, a number of them possess driver-like features and could therefore contribute to the neoplastic development process regardless of combined mutations in cancer drivers.

Based on these findings, there have been consistent and recurring transcript isoform transfers identified that interfere with the function of the concerned proteins by either attaching or detaching protein domains that are often mutated in cancer, or by interfering with or gaining access to the formation of protein complexes—possibly including the creation of protein complexes—with cancer drivers or in cancer-related pathways.

In addition, it has been found that patients with some of these isoform shifts are inclined not to carry mutations in cancer drivers and vice versa. An alternative splicing modification in NFE2L2 has lately been reported, which results in the deletion of a protein domain and the interplay with its negative regulator KEAP1, thus offering an alternative activation mechanism for an oncogenic route (Goldstein et al. 2016). In a similar way, it has been demonstrated that an isoform alteration in the ATF2 gene triggers melanomagenesis (Claps et al. 2016). These cases confirm a model in which cellular functions and routes, which are frequently modified in cancer due to somatic mutations, are equally susceptible to isoform alterations in some patients and may thus enhance the cancer phenotype. Most crucially, these isoform alterations could arise in the absence of gene expression alterations in the host gene and therefore constitute an unbiased record of functional alterations in cancer. Functional domains and interactions usually appear not to be relentlessly eliminated by transition, since in general, normal isoforms in tumors maintain a certain level of expression. To some extent, this is probably due to the inherent unpredictability in estimating transcriptome frequency resulting from RNA sequencing or to the heterogeneity in the transcriptomes of cancer cells. Nevertheless, it has been shown that a comparatively small alteration in transcript frequency is enough to induce an oncogenic response in the cells (Anczukow et al. 2015; Bechara et al. 2013; Sebestyen et al. 2016).

In addition, several isoform alterations were found to represent a switch from a protein-coding to a noncoding transcript, potentially subject to non-sensitively-mediated decay, which is a common mechanism of alternative splicing-based gene expression control (Hansen et al. 2009), and may perhaps change the function in a similar fashion to other changes in isoforms between protein-coding isoforms. The predicted effects on areas and interactions might hence point to modifications in regulatory frameworks with differing functional impacts. The description in the form of transcript isoform shifts permitted us to identify more variants in the transcriptome than local alternative splicing incidents and to define the protein features that are possibly obtained or shed through splice alterations. However, this concept is subject to some constraints. The precise identification of a differential transcript utilization in genes with multiple isoforms demands high levels of coverage and adequate amounts of samples per condition (Sebestyen et al. 2015), which seems to be mitigated by the use of variability among samples to assess significance. Since an annotated transcript isoform was also employed, it is possible that tumor-specific transcripts may have been ignored, which are not included therein.

In fact, even a small part of the whole set of protein–protein interactions that take place in the cell has been recovered. For example, the interactions facilitated by areas of low complexity have not been identified (Buljan et al. 2012; Ellis et al. 2012); therefore, far more of the interaction and protein complexes in cancers may be involved. The origin of the detected splice alterations still needs to be clarified. However, no general pattern of connectivity with somatic mutations in *cis* has been detected. It may be that small variations in the number of copies or indexes are causing these transitions, although they are nevertheless hard to identify with WES and WGS data, and a more specific search or deeper sequencing may be required. Alternatively, the majority of the mentioned swaps are caused by *trans*-effective changes, for instance the expression modification of the splicing factors (Sebestyen et al. 2016). For example, mutations in RBM10 or the down-regulation of QKI result in the identical splicing alteration in NUMB that enhances cell proliferation (Bechara et al. 2013; Zong et al. 2014), and the oncogenic switching in RAC1 (Zhou et al. 2013) is governed by expression modifications of various splicing factors (Goncalves et al. 2009; Pelisch et al. 2012), which are governed by signal transduction pathways that are frequently modified in tumors (Fu and Ares 2014). Alternatively, these transitions may be used to describe signatures of non-genetic diversity (Brock et al. 2009). The heterogeneity within the tumor might make it possible to recapture comparable transcriptome phenomenon types that independently of somatic mutations govern the physical health of cells and the progression of cancers. Since the effect of natural selection is more on the phenotype than on the genotype, an intriguing hypothesis is that specific transcript isoform expression schemes might be able to distinguish certain tumor phenotypes that are tightly linked to those established by somatic mutations in the drivers by specifying a beneficial phenotype in such a way as to ease the selective constraint of developing equivalent adjustments. As a consequence, the isoform switches identified could play an important function in the neoplastic process regardless of or in combination with the existing genetic modifications.

13.8 Cancer Disease and Alternative Splicing of Fibronectin

Alternative splicing is the major driving mechanism of protein variety, but little is understood about the physiological mechanisms governing splicing. It has been found that the stiffness of the extra-cellular matrix environment is able to govern the process of protein splicing in cells in vitro and in vivo (Bordeleau et al. 2015). The alternative splicing is a process facilitated by the matrix stiffness and depends on the PI3K signaling pathway. It is achieved by the phosphorylation of splicing regulating factors, the serine/arginine-rich (SR) proteins. As the SR family of proteins is preserved in both vertebrates and invertebrates and is also recognized to be engaged in genome stabilization, translation and the export of mRNA, these results indicate a hitherto unexplored mechanism whereby cells can react and adjust to their mechanical microenvironment under both healthy and diseased circumstances.

Alternative splicing of proteins leads to various isoforms that are essential for the control of various cellular processes. While the significance of alternative splice alterations arising during cancer is becoming more widely accepted, the fundamental regulatory mechanisms are still barely comprehended. To investigate the role of mechanical properties of the microenvironment of the tumor in influencing the production of the extra-domain B (EDB) splice variant of fibronectin, a signature of tumor angiogenesis, both biochemical and physical tools are employed in combination with technical models, samples from patients and a mouse model. In particular, it has been revealed that the amount of EDB-fibronectin generated by endothelial cells rises with matrix stiffness, both in vitro and inside mouse mammary neoplasms. Matrix stiffness governs splicing by activating serine/arginine-rich (SR) proteins, which are splicing factors participating in the generation of fibronectin isoforms. The activation of the SR proteins due to matrix stiffness and the subsequent generation of EDB-fibronectin are determined by intracellular contractility and the PI3K-AKT signaling. In particular, matrix stiffness-based splicing is not restricted to EDB-fibronectin; however, it also concerns splicing in the generation of PKC βII and the VEGF 165b splice version. Altogether, these results indicate that the physical properties of the microenvironment adjust alternative splicing and provide a hitherto unexplored mechanism that allows the cells to adjust to their own microenvironment.

The differential expression of protein isoforms using alternative splicing represents a crucial factor in generating protein diversity and can lead to highly diverse cell phenotypes and responses (Ghigna et al. 2008; Oltean and Bates 2014; Irimia and Blencowe 2012). In comparison with healthy tissue, tumors show several large variations in the expression patterns of protein isoforms (Oltean and Bates 2014; Brinkman 2004), and some of these variations are assumed to favor oncogenesis (Ghigna et al. 2008). In particular, splicing processes at the pre-mRNA stage are governed by splicing regulation factors that comprise a family of SR proteins (Long and Caceres 2009). Increased amounts of SR proteins have been linked to cancer (Long and Caceres 2009), whereas the physiological environment that facilitates splicing are not yet clearly known.

Of the different alternatively spliced proteins existing in cancers, the spliced variant of fibronectin, which contains the extra-domain B (EDB) type III repeat, is of special concern in the cancer research as the inclusion of the EDB fragment has been highlighted as a possible route to capture and address the tumor vasculature (Neri and Bicknell 2005). EDB-fibronectin is synthesized by endothelial cells in tumors and may facilitate angiogenesis (Castellani et al. 2002; Zardi et al. 1987; Khan et al. 2005). Fibronectin splicing is governed by several SR proteins, among them SRp40 and SRp20, in multiple distinct cell types (Lim and Sharp 1998; Han et al. 2007; Du et al. 1997). While the availability of EDB-fibronectin is quite exclusive in the vasculature of the tumor, the underlying mechanisms that control its expression within the microenvironment of the tumor remain unclear.

The splicing activity of SR proteins, resulting in the expression of diverse protein isoforms such as EDB-fibronectin, is strictly governed by phosphorylation (Saitoh et al. 2012; Patel et al. 2005; Prasad et al. 1999), and it has been demonstrated that this phosphorylation is performed by AKT (Saitoh et al. 2012; Prasad et al. 1999). Of special interest is the fact that matrix stiffness can adjust the AKT phosphorylation condition (Mouw et al. 2014; Leight et al. 2012; Schrader et al. 2011). Specifically, recently published information suggests that AKT activation and subsequent phosphorylation of AKT substrates may be achieved by alterations in the mechanical characteristics of the tissue in the tumor's microenvironment (Mouw et al. 2014). Crucially, tumor tissue becomes stiffened during cancer progression as a result of elevated collagen accumulation and the action of cell-secreted enzymes that interconnect the collagen fibrils (Levental et al. 2009; DuFort et al. 2011), and this stiffening is assumed to be linked to tumor aggressiveness and the risk of metastasis (Leight et al. 2012; Kim and Asthagiri 2011; Plodinec et al. 2012; Pathak and Kumar 2013; Bordeleau et al. 2014). Increased stiffness of the extracellular matrix influences various tumor-related cell phenotypes, such as increased cell contractility, responsiveness to growth factors and migration of cells (Leight et al. 2012; Kim and Asthagiri 2011; Pathak and Kumar 2013; Peyton and Putnam 2005; Paszek et al. 2005). Even though it has been demonstrated that stiffness promotes stronger AKT activation, which then induces the phosphorylation of SR proteins and SR proteins promote splicing, it is not yet clear whether stiffness-related events can lead to SR protein-based splicing processes during tumorigenic development.

In this investigation, a multidisciplinary effort involving in vitro and in vivo modeling and biochemical and physical instruments was employed to investigate the association between matrix stiffness and fibronectin splicing in endothelial cells. For the first time, the expression of EDB-fibronectin enhances with matrix stiffness, both in vitro and in mouse mammary tumors. The data suggest that matrix stiffness governs the dynamics of EDB-fibronectin splicing due to Rho-associated protein kinase (ROCK) induced contractility and the linked downstream phosphorylation of SR protein. In addition, phosphorylation of SR proteins corresponds to the in vivo activity of the

PI3K/AKT signaling pathway due to tissue stiffness, while inhibiting PI3K in vitro impedes phosphorylation of SR proteins and the splicing of EDB-fibronectin. In summary, these data indicate that the stiffness of the tumor's microenvironment is able to influence intracellular splicing.

This seems to be the first immediate proof that alternative splicing is governed by the stiffness of the extracellular matrix. More specifically, both in vitro and in vivo evidence suggests that the stiffness of the microenvironment controls the splicing of EDB-fibronectin. The stiffness of the extracellular matrix adjusts splicing by managing phosphorylation of SR proteins and demands both cell contractility and activation of the PI3K/AKT pathway. In addition, the evidence from the in vivo results indicates that the control of alternative splicing through the stiffness of the extracellular matrix probably exists in a wide range of cell types and may not be limited to endothelial cells or the EDB isoform of fibronectin. In this respect, the stiffness of the extracellular matrix can be an important facilitator for the generation of the diverse range of protein isoforms that occur in a wide spectrum of tissues and disease conditions.

Fibronectin has been identified as one of the most highly refined proteins for providing the stiffness of the extracellular matrix. Alternative splicing of fibronectin has been extensively explored, partly due to its close association with angiogenesis (Khan et al. 2005; Astrof et al. 2004; Coltrini et al. 2009; Hashimoto-Uoshima et al. 1997; White and Muro 2011). In endothelial cells, the splicing incidents that lead to the uptake of the EDA or EDB domain are associated with tumor angiogenesis (Astrof et al. 2004). It has been proven that these two occurrences are adjustable through SRp40 (Han et al. 2007; Du et al. 1997). But the physiological hints that convey this splicing are mostly hidden. These data point to the fact that matrix stiffness is able to govern SR protein phosphorylation, including SRp40, that governs the splicing of a wide range of different proteins, which encompasses proangiogenic molecules such as VEGF, especially by restricting the expression of the anti-angiogenic VEGF 165b splice variant favoring more proangiogenic forms (Nowak et al. 2008; Cooper et al. 2014; Wee et al. 2014; Yan et al. 2010). In addition, the reaction of the SR protein to matrix stiffness is guided by a signaling pathway that incorporates PI3K/ATK and demands cellular contractility. In fact, inhibition of cell contractility or PI3K impeded phosphorylation of the SR protein and uptake of the EDB domain of fibronectin. The interesting fact is that AKT signal transduction has been involved in SRp40-facilitated splicing of the PKC family before (Saitoh et al. 2012; Prasad et al. 1999), indicating that AKT is probably equally implicated in splicing the stiffness-modified PKC and EDB-fibronectin. Beyond that, the organization of the PI3K/AKT pathway through matrix stiffness has become progressively well-established in a variety of experimental tumor models (Mouw et al. 2014; Leight et al. 2012; Schrader et al. 2011; Umesh et al. 2014). Overall, these data reveal a key mechanism which regulates EDB-fibronectin expression within endothelial cells. In view of the ubiquitous expression of EDB-fibronectin in numerous types of aggressive solid tumors (Castellani et al. 1994, 2002; Scarpino et al. 1999; Hauptmann et al. 1995) and the altered splicing regimes detected in tumors (Oltean and Bates 2014; Brinkman 2004), the available findings are probably representative for a universal mechanism far transcending endothelial cells and EDB-fibronectin.

Most intriguingly, each of the characteristics of cancer is linked to a shift in the expression profiles of protein isoforms (Oltean and Bates 2014; Hanahan and Weinberg 2011). It has been implied that the progression of the tumor may be linked to a concerted monitoring of the splicing process (Oltean and Bates 2014; Nowak et al. 2010; Saitoh et al. 2012). Endeavors to identify the molecular regulators of the splicing apparatus have led to the recognition of several signal transduction pathways linked to growth factors and the progression of tumors (Han et al. 2007; Dales et al. 2010). Importantly, the findings help to define a hitherto unrecognized mechanism through which alternative splicing in cells is controlled by alterations in the mechanical characteristics of the extracellular matrix. This regulatory mechanism might work either directly by enabling integrin signaling pathways or indirectly by modulating the activity of other pathways. In the event of direct activation, mechanical signals at the adhesion sites between extracellular matrix and cells can be directly transformed into the biochemical responses (Vogel and Sheetz 2009). In particular, there is strong proof that the stiffness of the extracellular matrix is involved in the control of PTEN and PI3K activity in epithelial cells (Mouw et al. 2014). In the indirect scenario, it has been revealed that the interaction between the stiffness of the extracellular matrix and cell contractility is able to stimulate growth factor-based signal transduction pathways (Paszek et al. 2005; Huynh et al. 2013; Zeng et al. 2011). In particular, both myosin light chain kinase and Rho-ROCK-driven contractility have been proven to be crucial for the guidance of matrix stiffness regulation of the growth factor signal response (Zeng et al. 2011). However, as the splicing and Rho/ROCK-facilitated contractility are connected, and considering that growth factors are renowned for their ability to direct alternative splicing (Prasad et al. 1999; Saitoh et al. 2012), the interaction between extracellular matrix stiffness and growth factors is expected to yield a wide range of splicing patterns.

Altogether, an alternative splicing facilitated through matrix stiffness might possibly offer the cells a mechanism to accommodate mechanical manipulation. Alterations in the expression of signaling proteins resulting from altered splicing profiles in response to modified mechanical properties of the extracellular matrix possibly result in alterations of the phenotype of the cell. For instance, every PKC isoform has been involved in influencing actin dynamics and the motility of cells (Bordeleau et al. 2010; Larsson 2006; Ng et al. 1999; Oh et al. 2007), implying that alternative splicing may cause the well-known alterations in cell motility that arise on conformal matrices in comparison with rigid matrices due to the altered matrix stiffness (Pathak and Kumar 2013; Peyton and Putnam 2005). Stiffness-based downregulation of the anti-angiogenic isoform of VEGF, resulting in alternative splicing of more proangiogenic isoforms (Nowak et al. 2010), may have an essential part to play in managing angiogenesis. Beyond that, it has been demonstrated that the maturation of adipocytes is driven independently through SRp40-facilitated splicing (Cooper et al. 2014) and tissue stiffness (Zhao et al. 2014), implying a probable role of splicing in tissue stiffness-initiated differentiation. Thus, these results have a major effect on the basic knowledge of how alterations in the mechanical microenvironment finally govern the interactions between the extracellular matrix during the normal physiological processes, such as morphogenesis of tissues and abnormal processes, such as tumorigenesis.

13.9 Long Non-coding RNAs Act as Oncogenes or Tumor Suppressors

It has been presented how lncRNAs are able to control the splicing process, the stability of mRNA/protein complexes and the position of proteins in subcellular localizations, and all these events finally have an effect on tumor development. Therefore, lncRNAs, such as protein-coding genes, can act as oncogenes or tumor suppressor genes. The evidence for the role of lncRNAs in cancer is overwhelming in the scientific field (He et al. 2019). For instance, the expression of HOTAIR is markedly increased in breast tumors, and therefore, the elevation of HOTAIR in primary human breast tumors is a powerful prognostic marker for overall patient outcomes such as the occurrence of metastatic spread and the patients' survival (Gupta et al. 2010). In addition, the ectopic expression of HOTAIR induces an altered histone H3 Lys27 (H3K27) methylation profile and enhanced invasiveness. In the opposite direction, the removal of HOTAIR causes the reverse phenotype of the cell. In prostate cancer, a variety of lncRNAs become abnormally dysregulated, which encompasses PCGEM1 and ANRIL (Srikantan et al. 2000; Petrovics et al. 2004; Yap et al. 2010). For instance, ANRIL is involved in the repression of the tumor suppressors INK4a/p16 and INK4b/p15 (Yap et al. 2010; Kotake et al. 2011).

It was proposed that whereas Linc-RoR30 might act as an oncogene by suppressing p53 in reaction to the damage of DNA, loc285194 (Liu et al. 2013) and GAS5 (Zhang et al. 2013) might carry out a tumor suppressing function due to the competitive endogenous RNA (CeRNA) mechanism (Salmena et al. 2011; He et al. 2019). In specific, the two most recent investigations provide further support for the importance of lncRNAs in cancer through the analysis of large clinical TCGA scans of sample material. For instance, epigenetic analyses of the environment show that, in contrast to the CpG island hypermethylation phenomenon, where more than 1000 lncRNAs are hypomethylated along their promoters in the case of cancer. One of them is epigenetically-induced lncRNA1 (EPIC1). The overexpression of EPIC1 has been linked to a worsened outcome in patients with luminal B breast cancer and promotes the growth of tumors in vitro and in vivo (Wang et al. 2018). This finding indicates that most lncRNAs might be involved in an oncogenic function. The second investigation reveals that a series of possibly oncogenic or tumor suppressing lncRNAs are deregulated; disturbances of these lncRNAs modify the proliferation of breast and gynaecological cancer cells (Chiu et al. 2018). Although most lncRNAs are deregulated in a tumor-specific fashion, some of them, namely OIP5-AS1, TUG1, NEAT1, MEG3 and TSIX, synergistically interfere with cancer signal transduction pathways in various tumor settings (Chiu et al. 2018). A recent investigation indicates that lncRNAs can also influence the outcome of immunotherapy, as a wide range of tumor-specific antigens can be deduced from non-coding areas (Laumont et al. 2018). In conclusion, these findings point out to the overall importance of lncRNAs in the initiation and progression of cancer.

A striking and controversial case in point, however, is MALAT1. Extensive bibliographic information points out that MALAT1 acts as an oncogenic factor in a wide range of cancers. In contrary, a more recently conducted investigation seems to confirm that MALAT1 acts as a metastasis suppressing lncRNA in breast cancer. Numerous evidences point to this conclusion, such as the directed inactivation of the MALAT1 gene within a transgenic mouse model of breast cancer; the knockout of MALAT1 in human breast cancer cells, the overexpression of MALAT1 nearly completely inhibits the metastasis process in breast cancer in transgenic, xenograft and syngeneic model systems. The fundamental mechanism might comprise the adjustment of the YAP pathway (Kim et al. 2018). It therefore still remained to be seen whether these findings could be further validated in various laboratories.

13.10 Other Diseases Alter the Alternative Splicing Variants

One important result of several recent comprehensive analyses of splicing is an improved insight into the function of splicing failures in human disease. Typically, a more precise idea of the identity and arrangement of splicing enhancer and silencer sequences in exons and introns have delivered useful insights to improve the capability to forecast which disease mutations might have the capacity to interfere with splicing (Cartegni et al. 2002; Fairbrother et al. 2002; Pagani and Baralle 2004; Wang et al. 2004a, b; Zhang and Chasin 2004). The reason for this is that a number of disease mutations aiming at non-synonymous amino acid codon sites are frequently thought to induce disease by altering the amino acid sequence of a protein, while a thorough analysis of the transcripts of numerous disease genes demonstrates that a large number of these mutations can induce more severe failures by interfering with the insertion of a whole exon or the activity of a cryptic splice site. Each of the two events holds the inherent capability of imposing a premature termination codon, leading to a breakdown of the complete transcript through NMD (Cartegni et al. 2002; Pagani and Baralle 2004).

Similarly, many disease-related mutations directed towards synonymous codon positions influence splicing. Human disease genes that are attacked by synonymous and non-synonymous mutations which frequently interfere with the control elements of exonic splicing, are breast cancer 1 (BRCA1) (in the early onset), cystic fibrosis transmembrane conductivity regulator (CFTR), microtubule-associated protein Tau (MAPT), hypoxanthine phosphoribosyltransferase 1 (HPRT1) and survival of motor neuron 1 (SMN1) (Cartegni et al. 2002). These investigations, covering the replacement rate of nucleotides in the orthologous exons of various species, examinations of the sophistication and dispersion of exon splicing enhancer and intron splicing enhancer sequences, and more systematic characterization of transcripts of specific disease genes such as Neurofibromatosis type 1 (NF1) and Ataxia telangiectasia mutated (ATM) (Teraoka et al. 1999; Ars et al. 2000), come to the remarkable deduction on the basis of the analysis that approximately 50% of the disease mutations in exons can influence splicing. This conclusion firmly acknowledges the necessity of implementing routine analysis of transcripts to properly evaluate potential mechanisms leading to human diseases. In addition, the elevated level of preservation of the intron sequences flanking conserved alternative splicing occurrences points to the fact that these areas in the pathogen genes ought also to be screened for mutations altering splicing as a matter of routine.

Microarray profiling methods offer major opportunities for future changes in medical treatment, as gene expression patterns can deliver useful diagnostic and predictive insights, thereby assisting in the selection of suitable therapies. The latest emergence of alternative splicing microarray results indicate that alternative splicing signatures will be crucial as profiling at the transcriptional level or other gene expression levels may completely overlook key disease-related alterations in gene expression at the splicing level. In fact, in light of the latest emerging proof summarized above of the growing relevance of disease mutations in splicing, it is certainly feasible that microarrays that enable splicing profiling may in many instances yield more meaningful diagnostic or prognostic markers than are available with conventional microarrays designed to profile gene expression patterns. To corroborate this opinion, a recent analysis of microarray data from profiling of both alternative splices and the transcriptional level in prostate tissues has demonstrated that the detection of splice variants can enable more accurate distinction of tumor and normal tissue than the determination of variations at the transcriptional level in the same tissue samples (Li et al. 2006b; Zhang et al. 2006a).

13.11 Alternative Splicing Serves as an Escape Mechanism Cancer Cells During Drug Treatment

A major challenge for cancer patients is to overcome resistance for specific drug treatments. Alternative splicing has been reported to alter the coding region for distinct drug targets in cancer therapy and thereby renders them to be inefficient (Siegfried and Karni 2018). Hence, the modulation of alternative splicing itself can serve as a target for disease treatment, such as cancer. Commonly known mechanisms for drug treatment inefficiency are inactivation of the drug, alterations of the drug target or efflux, induction of DNA damage repair, the inhibition of cell death and the epithelial–mesenchymal transition have been excessively examined (Housman et al. 2014). All these mechanisms are subject to genetic and epigenetic changes. Two examples of drug resistance based on altered alternative splicing, such as BRAF and BRCA1 splicing, are presented in the following.

Vemurafenib-Resistant Melanoma is Caused by BRAF Splicing Alterations

In over 50% of the melanomas, a mutated BRAF gene, which encodes a serine/threonine kinase, has been identified to be responsible for cancer. The most widespread mutation of this type appears to be the BRAF(V600E) mutation. The ATP-competitive RAF inhibitor vemurafenib is effective in the treatment of BRAF(V600E) mutant melanoma; however, patients eventually become immune to treatment with vemurafenib. A resistance mechanism involves the formation of alternative BRAF splice isoforms which result in the removal of the RAS-binding domain. A BRAF(V600E) mutant cell line receives a high dose of vemurafenib, and multiple clones are isolated that are expressing a shorter BRAF(V600E) splice variant devoid of exons 4–8 (Poulikakos et al. 2011). This leads to the removal of the RAS-binding domain, that typically governs the dimerization and activation of BRAF. BRAF splicing variants were detected in 6 of 19 melanoma patients diagnosed with established vemurafenib resistance, however, not in melanoma resected in vemurafenib-naive patients or four vemurafenib-resistant patients with acquired NRAS mutations (Poulikakos et al. 2011). In accordance with these insights, interference with the pre-mRNA splicing mechanism may be leveraged to combat vemurafenib resistance. The application of splice modulators, spliceostatin A (SSA) or its analog meayamycin B (MAMB), targeting the splicing factor SF3B1, led to the suppression of the BRAF of alternatively spliced isoforms and restricted the growth of vemurafenib-resistant cells in vitro and the growth of tumors in vivo (Salton et al. 2015). The present work indicates pre-mRNA splicing intervention as a possible therapeutic response to drug resistance in BRAF melanomas.

PARPi-Resistant Breast Cancer Due to BRCA1 Splicing Alterations

A number of breast and ovarian cancers exhibit mutations in the BRCA1 and BRCA2 genes. These proteins play an indispensable part in effective homologous recombination (HR)—facilitated self-repair of double-strand breaks in DNA. The enzyme poly(ADP-ribose) polymerase (PARP) activates one of the DNA repair mechanisms. This enzyme is engaged in the process of repairing damage to single-stranded DNA, thereby reducing the likelihood of double-strand breaks in the genome. Therefore, the inhibition of PARP enzymes by PARP inhibitors (PARPi) in patients with defective BRCA1/2 genes is highly beneficial (Tommiska et al. 2008). While PARP inhibitors have been proven to increase survival, there are many patients suffering from germ line BRCA1 or BRCA2 mutations who are not taking advantage of PARPi therapy. Moreover, a large number of patients who are the first to gain advantage from PARPI or platinum therapy tend to suffer progression of the disease and become resistant (Sehl et al. 2009).

A previous investigation provided evidence that BRCA1 splice isoforms deficient in exon 11 are able to generate shortened but hypomorphic proteins. Their results demonstrate a possible contribution of the alternative splicing isoform BRCA1-D11q to the enhancement of the development of selective PARPi and a cisplatin resistance. In addition, they indicate that suppression of the spliceosome reduces BRCA1-D11q levels and renders the mutated exon 11 cell lines sensitive to PARPi, which may suggest a resensitivity strategy for tumors that have a strong expression of BRCA1-D11q (Rouleau et al. 2010).

13.12 Regulation of Alternative Splicing by Mechanical Stimulation, Such as Matrix Mechanics

In principle, it seems to be likely that the mechanical properties of extracellular matrices impact the process of alternative splicing in cells, such as cancer cells. Hence, the microenvironment of cells may not only have a major contribution to their cell shape, adhesive properties and cellular functions, such as motility; it may also regulate cellular signaling processes that may lead to pathological functions, such as cancer (Liu and Tang 2013).

13.12.1 Growth Factors Are Regulated by Mechanical Stress

Alternative splicing increases the proteome complexity by generating a number of mRNAs from a single gene. Affymetrix exon arrays and investigations in vivo or in vitro revealed that alternative splicing is controlled through mechanical stress. The expression of mechanogrowth factor (MGF), the splicing isoform of insulin-like growth factor 1 (IGF-1) and vascular endothelial growth factor (VEGF), whose splice variants such as VEGF121, VEGF145, VEGF165, VEGF189 and VEGF206 are controlled through mechanical stress. The mechanism of this process is still ambiguous. The mechanism of this process is still ambiguous. There is growing proof that the suggested mechanism is associated with the Ca^{2+} and phosphorylation signaling pathways. The present chapter proposes possible mechanisms of the mechanical regulation of splicing in the following.

Alternative splicing prior to mRNA is a key process in eukaryotic gene expression, since the properties of binding, the intracellular positioning, the activity of enzymes, the stability of proteins and the posttranslational modifications of the majority of proteins are governed by this mechanism (Stamm et al. 2005). Splicing is performed with a multistage apparatus referred to as a spliceosome, where a sequence of biochemical processes occurs. The spliceosome is composed of about two hundred proteins and about five distinct small nuclear ribonucleoprotein particles (U1, U2, U4, U5 and U6 snRNPs). The assembly of the spliceosomes occurs in a precisely timed step-wise manner (Fig. 13.3) (Wahl et al. 2009; Staley and Woolford 2009).

The control of alternative splicing is linked to the extracellular microenvironment. Numerous external stress factors including heat shock, exposure to ultraviolet (UV) radiation, genotoxic stress or damage to DNA alter the expression of the splicing isoforms. At least, three possible mechanisms of splicing events have been suggested to assist the initiation of extracellular stress: first, stress influences the structure of the spliceosome through repression of the activity of splicing factors; second, stress modifies the function of the splicing factor; third, stress alters the position of the splicing factor.

In parallel to all of the aforementioned stresses, it is noted that alternative splicing is also susceptible to mechanical stimulation. Mechanical stress fosters not only the growing and the survival of cells, and it also governs metabolic processes such as gene expression and the structure of tissue in various types of cells (Gillespie and Walker 2001). The cells that react to mechanical stimuli are referred to as mechanocytes, which include endothelial cells, fibroblasts, smooth muscle cells and osteoblasts. In these mechanocytes, it is determined that alternative splicing is governed by mechanical stress. An alternative splicing of insulin-like growth factor 1 (IGF-1) has been identified in muscle for the first time. The expression of splicing forms of IGF-IEa and IGF-IEb was seen in rodents and IGF-IEc in humans (Kravchenko et al. 2011). It has been seen that the mechanical strain causes the expression of the mechanogrowth factor (MGF), which constitutes one of the IGF-I splicing isoforms (Goldspink 1999). A similar level of MGF production has also been seen in osteoblasts (Tang et al. 2006), neurons (Carro et al. 2003), heart muscle cells (Skarli et al. 1998) and tendons (Heinemeier et al. 2007). In addition, the genome analysis revealed that alternative splicing is controlled through mechanical stress. The possible alternative splicing incidents caused through mechanical stress in the bone were analyzed with Affymetrix exon arrays. In specific, the number of alternative genes was 992 and these alternative genes formed various categories of gene ontology, including apoptosis, calcium, cell cycle and cytokines (Mantila Roosa et al. 2011). All these findings pointed out that the mechanical stress controls alternative splicing operations. The mechanism of this regulatory process is not clear yet.

There is mounting proof that alternative splicing has a major function in the proliferation, differentiation and development of cells and is linked to many types of human disease (David and Manley 2010; Corrionero et al. 2011). Lately, the alternative splicing mechanism has received more and more notice. Therefore, the regulation of alternative pre-mRNA splicing due to the mechanical stress and the proposed potential mechanisms of this process are discussed below.

13.12.2 Mechanical Stress Triggered Splice Variants of Genes

It is well-known that gene expression is governed by various types of mechanical excitation. Importantly, several of these genes contain splice variants, among them vascular endothelial growth factor (VEGF), CD44, tenascin C, collagen XII, versican, insulin-like growth factor (IGF-I), voltage-induced/inhibited proteins (TIPs) and serum response factor (SRF). Several splicing variants of these stress-sensitive genes are regulated through mechanical stress. Thus, this is direct proof of alternative splicing in response to mechanical stress.

In a wide variety of tissues and cells, in both normal and diseased ones, VEGF occurs as an expression factor that can be mechanically activated, and both physical and excessive physical mechanical stress enhance its levels of expression (Seko et al. 1999; Saygili et al. 2011). Several VEGF isoforms are available as a result of alternative splicing, among them are VEGF189, VEGF165, VEGF121, VEGF206, VEGF183, VEGF145, VEGF148, VEGF120 and VEGF164 (Houck et al. 1991).

IGF-I is linked to cell proliferation, cell differentiation and overall survival. The pre-mRNA codes for various protein isoforms through alternative splicing as a reaction to mechanical stimulation, such as IGF-IEa, IGF-IEb and IGF-IEc. IGF-IEb has also been termed mechanogrowth factor (MGF), which is highly regulated in the damaged skeletal muscle (Goldspink 2005). Moreover, MGF facilitates the repair of tissues and their regeneration (Matheny et al. 2010).

13.12.3 Extracellular Matrix Proteins Are Regulated by Mechanical Stress

TIPs feature the three isoforms TIP-1, TIP-2 and TIP-3, which are obtained through alternative splicing from a single gene. TIP sequences comprise motifs coding for nuclear receptor coregulators, remodeling elements of chromatin with histone acetyltransferase and histone deacetylase (Bannister et al. 2002). Investigations revealed that TIP-1 is only expressed in non-differentiated mesenchymal cells of the lung embryo through stretching. The expression of TIP3 within progenitor cells was not impaired when stretched. However, the expression of this protein in the smooth muscle cells was inhibited. Another experiment indicated that TIP-1 and TIP-3 were implicated in the initial reaction to strain. TIP-2 has also been a stretch-sensitive protein expressed in NIH3T3 cells in reaction to stretching (Jakkaraju et al. 2005).

Tenascin C is associated with numerous important cell functions, including the cell adhesion, movement, development and metastasis of tumors (Hsia and Schwarzbauer 2005). Tenascin C splice isoforms entail the insertion of FnIII domains located between the constitutive domains FnIII 5 and FnIII 6 (Jones and Jones 2000). The isoform with no alternative regions and the isoform with the FnIII D-domain have been expressed in porcine trabecular meshwork cells. Mechanical stretching caused a twofold increase in the FnIII D-domain in trabecular meshwork cells following 48 h after stretching (Keller et al. 2007). Under mechanical stretching, a new collagen XII splice isoform has been identified in trabecular meshwork cells (Keller et al. 2007).

Versican belongs to the family of hyaluronan-binding proteoglycans. It is capable of interacting with the extracellular matrix environment and influences cell adhesion, migration, proliferation and the structural composition of the extracellular matrix (Wight 2002). Versican compromises four distinct alternative transcripts, such as V0, V1, V2 and V3, which carry both, one or none of the two alternative exons 7 and 8. The expression of V1 is clearly increased through mechanical stretching, but the quantity of V2 and V0 stays low with no substantial alterations. CD44 is a transmembrane glycoprotein that is closely associated with cell–cell and cell–matrix interfaces (Ponta et al. 2003). All CD44 members are synthesized through alternative splicing of a gene containing multiple variable exons, such as v1–v10. Porcine trabecular meshwork cells are expressing the variant exons v3, v7 and v8. During the stretching, the percentage of v7 and v8 domains has risen threefold after 12 h of stretching and dropped to 1.5 fold within 48 h of stretching (Keller et al. 2007).

SRF governs the expression of a muscle-specific gene. SRF-M belongs to one of its isoforms which is missing the exon 5. The two other SRF isoforms are SRF-S devoid of exon 4 and exon 5 and SRF-I devoid of exon 3, exon 4 and exon 5 (Kemp and Metcalfe 2000). Mechanical stress alters SRF alternative splicing. When stretched for 4 h, mechanical stress inhibited the synthesis of SRF-M mRNA, whereas SRF mRNA values were elevated. After 12 h of stretching the outcome turned out to be exactly the contrary (Yang et al. 2000), all these cases demonstrated that mechanical stress can govern the expression of various genes. The splicing of isoforms from these genes, that are expressed in different manners in reaction to extracellular cues. Mechanical stress is a key regulatory factor in alternative splicing.

13.12.4 Analysis of the Mechanism of Alternative Splicing Due to the Mechanical Stimulation

The Ca^{2+}-Dependent Signaling Pathway

Nearly every cellular activity is directly or indirectly related to the Ca^{2+} signal transduction pathways. Ca^{2+} is a key intracellular messenger and is susceptible to mechanical manipulation. Mechanical stretching raises the intracellular Ca^{2+} level through activation of Ca^{2+} liberation from the sarcoplasmic reticulum (SR) and Ca^{2+} inflow (Kondratev and Gallitelli 2003; Niggel et al. 2000). In addition, many investigations on splicing regulations have found that alternative splicing is governed by Ca^{2+}-dependent mechanisms. (Sgambato et al. 2003; Xie 2008). For instance, glutamate acts as the agonist of the glutamate receptor, which enhances the intracellular Ca^{2+} uptake. Under the effect of glutamate, the ania-6 long mRNA is triggered through the NMDA receptor (Sgambato et al. 2003). Therefore, the alternative exons, splicing factors and pre-mRNA components were guided by Ca^{2+} signals following extracellular excitation (Xie 2008). These investigations indicate that alternative splicing could be modified in reaction to mechanical stress due to the interaction of the Ca^{2+} signal transduction pathway and splicing factors, splicing elements or splicing systems. Therefore, possible mechanisms of splice regulation by Ca^{2+} signal transduction in the cells under mechanical stress are considered possible. The pathways of splicing processes that are governed by Ca^{2+} signals in mechanically loaded cells might be classified into two groups: first, Ca^{2+} signaling pathways that connect to phosphorylation pathways; second, Ca^{2+} pathways that resemble depolarization.

There are various explorations on the control of alternative splicing through depolarization. KCl is a type of a chemical activation that produces a cellular depolarization and enhances the level of intracellular calcium. This type of excitation decided on the inclusion or exclusion of an exon through L-type calcium channels. Ania-6 (*Rattus norvegicus*) codes a shortened form of the cyclin. At least two splicing versions of the ania-6 gene exist, such as the short ania-6 and the long ania-6. KCl raised the level of short ania-6 mRNA expression (Sgambato et al. 2003). In rat pituitary GH3 cells, 25–50 mM KCl reduced the involvement of stress axis regulated exons (STREX exon) within the transcripts (Lee et al. 2007). It is highly intriguing that the Ca^{2+} inflow is initiated not merely through depolarization, but also through physical stretching. The Ca^{2+} inflow is upregulated in reaction to mechanical stimuli by liberating Ca^{2+} out of the sarcoplasmic reticulum (SR) and amplifying the L-type Ca^{2+} currents (Kondratev and Gallitelli 2003; Niggel et al. 2000; Kimura et al. 2000). Based on these findings, the alternative versions, controlled by cellular depolarization, can also be adjusted by mechanical stress.

It is well-known that the alternative splicing is governed through the action of *cis*-acting elements and *trans*-acting factors. The SR protein family and hnRNPs act as crucial splicing factors in constitutive and alternative splicing events. The depolarization raised the nuclear concentration of hnRNP A1 within the neurons and altered the alternative splice pattern of hnRNP H3 and RNP S1 by changing the confinement of their variable exons (An and Grabowski 2007). The previously introduced ania-6, which alternative splicing was adjusted by Ca^{2+}, might also be regarded as an SR protein. Since its sequences comprise arginine/serine-rich regions that are typical of SR splicing factors, in other words, Ca^{2+} also governs the splicing processes through regulation of the splicing elements and factors. Hence, a potential Ca^{2+} signal pathway based on the guidance of alternative splicing through mechanical stress can be deduced (Fig. 13.5).

The Phosphorylation-Dependent Signaling Pathway

The phosphorylation of proteins is a key factor in establishing the regulation of alternative splicing. The assembly and catalysis processes of the splicing are adjusted through reversible phosphorylation. SR proteins constitute an important type of splicing factors. The phosphorylation of the SR proteins within RS regions defines the choice of alternative splice sites through stimulation of the subcellular site and activation of the SR proteins. Phosphokinase and phosphorylation pathways involved in stimulating splicing can also be adjusted through mechanical stress. The correlation of the pathways of splice regulation and mechanotransduction in cells could reveal the mechanisms of splicing that are governed through mechanical stress. A possible regulatory pathway of the phosphorylation-based signaling for alternative splicing seems to be guided by mechanical stress (Fig. 13.6).

FosB and ΔFosB are the two variants of the FosB gene that can be triggered by fluid shear stress at the mRNA and protein level in vivo and in vitro measurements (Inoue et al. 2004). The regulation is dependent on the activation of ERK through the Ca^{2+} inflow across the Gd^{3+}-sensitive cation channel.

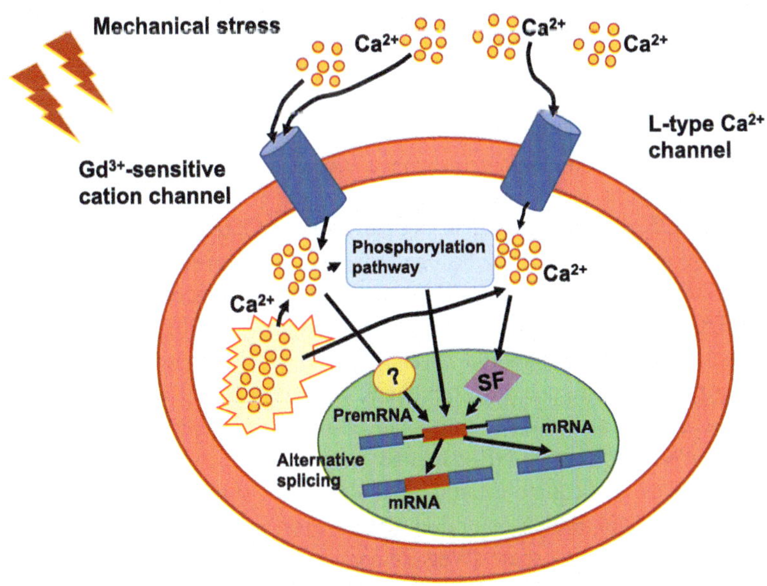

Fig. 13.5 Mechanical stress regulates the Ca^{2+} signaling pathway. Due to the mechanical stress, the influx of Ca^{2+} ions is facilitated through Gd^{3+}-sensitive cation channel and L-type Ca^{2+} channels. Thereby, the phosphorylation route is activated, which regulates the expression genes through various splicing isoforms. The Ca^{2+} ions impact the splicing factors and alter the concentration of the specific splicing variants through alternative splicing

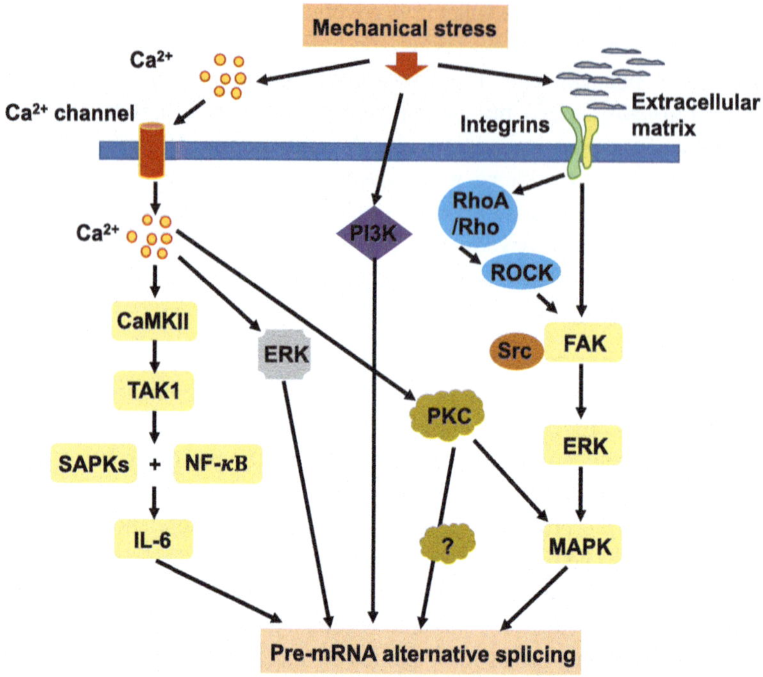

Fig. 13.6 Mechanical stress induces alternative splicing through phosphorylation-driven signal transductions processes. Due to the mechanical stimulation, the Ca^{2+} influx is triggered through calcium channels in the plasma membrane. The signal pathways CAMKII-TAK1-SAPKs and NF-kB-IL-6 are both stimulated by the Ca^{2+} influx that governs the alternative splicing of pre-mRNA. Since integrins are sensitive to mechanical cues of the extracellular matrix environment, mechanical forces and signals are transmitted into the cell via the activated FAK/ERK/MAPK route, which also governs the alternative splicing or provide RhoA/Rho-ROCK signaling. PI3K is another route governing alternative splicing due to the mechanical stress

ΔFosB is engaged in mechanotransduction in cells due to the mechanical stress. Thus, it can be said that splicing seems to be governed through the route of Ca^{2+} influx-ERK signal transduction, once the cells are exposed to mechanical cues.

The nuclear factor Kappa B (NF-κB) belongs to the *cis*-activating elements of the promoter of cyclooxygenase 2 (COX-2), participating in the elevation of COX-2. The activation of this is connected to the encouragement of osteoclastogenesis and the repression of synthesis of collagen (Ogasawara et al. 2001). The expression of IL-6 improves through mechanical stretching by the Ca^{2+} inflow CaMKII-TAK1-SAPKs and the NF-κB route (Fukuno et al. 2011). Bcl-x is part of the Bcl-2 family and consists of two the variants Bcl-xL and Bcl-xS, which are produced through splicing. The Bcl-xL expression rate decreases due to IL-6 in the K562 leukemia cell. From these findings, it can be concluded that mechanical stimulation is capable of regulating alternative splicing through Ca^{2+} inflow—CaMKII-TAK1-SAPKs and NF-κB-IL-6.

The activation of Ras is adequate to adjust the alternative splicing of CD45 (Lynch and Weiss 2000). The isoform of CD44, which comprises the v5 exon sequence, is initiated via the Ras-Raf-MEK-ERK-MAP kinase signaling route (Weg-Remers et al. 2001). Mechanical cues are carried into the cells through extracellular matrix and integrins on the plasma membrane, followed by conversion of these mechanical cues into chemical cues. The focal adhesion kinase (FAK), which is connected to integrin activation signals, becomes phosphorylated due to its interaction with Src. The phosphorylated FAK additionally enables the ERK pathway (Plotkin et al. 2005). These findings suggest that ERK activation combines mechanical excitation and splicing adjustment via the extracellular matrix-integrins-FAK and Src-ERK-MAP kinase pathway. Moreover, Rho proteins respond to mechanical excitations and transfer mechanical signals into cells (Makino et al. 2005). The continuous cyclic mechanical tension leads to a decreased activity of RhoA (Shi et al. 2010). The protein Rock is part of the serine/threonine kinase family and constitutes a Rho effector protein.

The RhoA/ROCK signaling pathway has a significant impact on mechanotransduction within force-stimulated human lung fibroblasts. FAK might be enabled by RhoA/ROCK signaling through the integrins RhoA/Rho-ROCK cascade route (Hong et al. 2010). Afterwards, the ERK path with respect to splicing regulation has been activated. Thus, an additional mechanotransduction pathway seems to be implicated in the control of the alternative splice integrin-RhoA/Rho-ROCK-FAK and Src-ERK-MAP kinase pathways.

The relative splicing of the CD44 isoforms is enhanced through the phosphatidylinositol 3 kinase (PI3) and the PKC signaling route (Fichter et al. 1997). The PI3 kinase is activated by shear stress within endothelial cells. The PI3 activation is associated with some downstream signaling pathways, including protein kinase B (PKB) or synonymously known as Akt (Fruman et al. 1998). The most recent investigation also demonstrated that the mechanical stress stimulates the activation of the PI3 kinase/Akt (Zhang et al. 2011). It is assumed that the adjustment of alternative splicing in reaction to mechanical cues also relies on the expression of the PI3 kinase.

The protein kinase C (PKC)-dependent pathway is dominant in the expression of CD45 splicing mutants (Lynch and Weiss 2000). Shear stress initiates the Ca^{2+} inflow into endothelial cells and then induces activation of the PKC/MAPK signaling pathway (Li et al. 2005). These data, together with the splicing regulation route of the extracellular matrix-integrins-FAK and Src-ERK-MAP, provide us with another potential regulatory mechanism of alternative splicing of Ca^{2+} influx/PKC/MAPKs. However, it is not certain whether there is another promoter that conveys splicing guidance through PKC.

Other Signaling Pathways

In addition to the paths above, it is found that the PI3K signal path also plays a role in splicing control. The investigation of alternative splicing of VEGF caused due to the mechanical stress revealed that the stretching frequency is associated with the expression of VEGF variants (Faure et al. 2008). The mRNAs of soluble VEGF isoforms, such as VEGF121 and VEGF165, were selectively expressed at low levels, while matrix-bound VEGF isoforms, including VEGF206, VEGF189, VEGF165 and VEGF145, were specifically expressed at high levels in human osteoblasts. The exact mechanism of this functional feature was not known. However, additional investigations have shown that the generation of matrix-bound VEGFs hinges on the stimulation of F-actin polymerization. In addition, mechanical impulses may adjust the generation of tension fibers, which is facilitated through the activity of PI3K (Guignandon et al. 1997; Jimenez et al. 2000). It can be conjectured that the activation of PI3K is caused by mechanical tension and the consequent active F-actin polymerization is governed the alternative splicing with VEGF.

13.12.5 Various Functions of Mechanics Facilitated Splicing Variants

Investigations have demonstrated that splicing variants triggered by mechanical stress can cause feedback mechanisms on the cells in the course of mechanical stretching. Mechanical stress produces various impacts on the VEGF alternative splicing. The matrix-bound VEGF isoforms, which are expressed when exposed to high-frequency mechanical stress, are secreted from proteases and contribute to metaphyseal vascularization, resorption of cartilage and osteogenesis. The soluble variants expressed under low-frequency excitation are critical for perichondrial diffusion, the stimulation of epiphyseal vascular outgrowth and the invasion of the vasculature (Maes et al. 2004). Alternative splicing isoforms of TIPs, such as TIP-1 and TIP-3, are susceptible to stretching. They react variably to mechanical stimulation. TIP-1 is caused by stretching, while TIP-3 is restrained due to stretching. In specific detail, TIP-1 encourages myogenesis and TIP-3 promotes adipogenesis (Yang et al. 2000).

IGF-I facilitates the growth and the anabolic response in various tissues. MGF belongs to the IGF-I isoforms of splicing due to its mechanical activation. MGF is expressed in several tissues, such as heart, osteoblasts, tendons and brain (Goldspink and Goldspink 2005; Tang et al. 2004; Heinemeier et al. 2007; Dluzniewska et al. 2005). MGF promotes the stress-sensitive regeneration of tissues and protects key organs against damage. After the injection of MGF, the muscle mass actually increased about 20%. The cells exposed to MGF displayed a high proliferation rate (Yang and Goldspink 2002). MGF is also implicated in the activation of satellite cells in the muscle repair system (Hill and Goldspink 2003).

The process of IGF-I alternative splicing that is triggered through the mechanical stimulation (Fig. 13.7). The extracellular stress signals are transmitted into the cells by various membrane receptors and stimulate a series of cascade responses. Stress events are converted into chemical signals and transferred together with the mechanotransduction pathway, activating the regulatory elements linked to the alternative splicing. Ultimately, IGF-I-pre-mRNA splicing involves mechanical forces, and MGF is generated through this process. Responding to mechanical stress, the insertion of 49 bp (human) or 52 bp (mouse) in exon 5 of MGF induces a frameshift which creates a distinct C-terminus compared to other IGF isoforms. Since MGF has certain key functions in cell proliferation and differentiation, novel autocrine MGF is able to respond to cells subjected to mechanical stress.

Alternative splicing requires a tightly organized process relying on a mixture of positive and negative interacting factors, intronic and exonic sequence elements, and temporal and spatial signaling pathways to provide the correct guidance. The process of alternative splicing has been

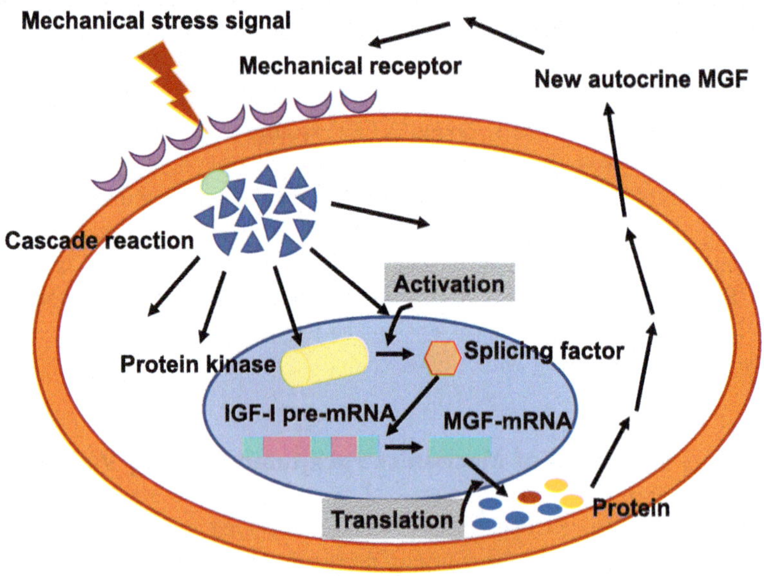

Fig. 13.7 IGF-I driven alternative splicing triggered through mechanical cues. The IGF-I pre-mRNA splicing is facilitated through mechanical stress, and subsequently, MGF is generated. Extracellular cues enter the cells via distinct membrane receptors and thereby activate a signaling cascade. Stress signals are transferred into chemical signals and are transported via mechanotransduction route, which terminates in the regulation of alternative splicing, such as the SR factor phosphorylation by protein kinases. Finally, new autocrine MGF acts in turn on cells stimulated through mechanical cues

governed through multiple various types of extracellular cues, which encompass mechanical forces, such as mechanical strain or fluid shear stress. Direct and indirect proofs of the regulation of the splicing process initiated through mechanical signals were discussed and potential mechanisms of this regulation process were hypothesized, including Ca^{2+} signaling pathways and phosphorylation signaling pathways. Even though these findings have not been validated experimentally, a completely new direction for alternative splicing investigations is presented below.

A profound understanding of alternative splicing features and mechanisms plays an essential theoretical and practical role. In the first place, pre-mRNA splicing has turned into a new goal for the development of drugs (Barrie et al. 2012). In the second place, the immune defense system exploits pre-mRNA splicing to broaden the scope of the individual's gene function (Urano et al. 2012). The differentiation and self-regeneration or pluripotency are even handled through alternative splicing (Liu et al. 2012a, b). Various immunologically important genes are alternatively spliced (Sahoo and Im 2010). Thus, the exploring the machinery of alternative splicing can uncover a novel complex regulatory scaffold of immune genes and help to significantly improve the effectiveness of research in the fight against disease. It is therefore necessary to gain an in-depth knowledge of the mechanical signal processing circuit behind the splicing control system.

13.13 Mechanics-Based Alternative Splicing Regulates Extracellular Matrix Production, Such as Collagen

More than two decades ago, the two isoforms for the type II procollagen gene referred to as COL2A1 were found. In specific detail, these two isoforms, which are referred to as IIA and IIB, are produced in a developmental-dependent manner through the process of alternative splicing of the exon 2 in COL2A1. For instance, chondroprogenitor cells are capable to generate preferable the IIA isoforms, which includes the exon 2, whereas differentiated chondrocytes contain solely IIB transcripts, which lack the exon 2. It is important to mention that the transition from the IIA isoform to the IIB isoform is facilitated by alternative splicing and takes place in the differentiation of the cells towards chondrocytes. Recently, there have been two additional isoforms been described, which are referred to as IIC and IID, which also address the exon 2 splicing of exon 2 (McAlinden 2014). Subsequently, these results put a spotlight on the overall complexities associated with the crucial regulation of COL2A1 expression during the chondrogenesis. However, the biological significance of the various isoforms of COL2A1 cannot be explained so far for the complex skeletal development and its maintenance, since there is still a lot of scientific effort required. In the following, the common knowledge of the expression of the various COL2A1 isoforms during the differentiation process of chondrocytes. Moreover, it is also discussed what has been revealed about the mechanism controlling the exon 2 alternative splicing process. The usage of mouse models to investigate the biological significance of alternative Col2al splicing in vivo is the subject of further discussion. Based on the insights gained so far, several new questions and concepts are now raised concerning the importance of alternative Col2al splicing in the organization of extracellular matrix construction and how this can subsequently impact on the overall performance and quality of cartilage and endochondral bone.

The elimination of introns, commonly referred to as splicing, from precursor mRNAs (pre-mRNAs) represents a finely regulated and accurate mechanism required for the expression of most human genes. Pre-mRNA splicing involves a complex arrangement of *cis*-elements (termed the splicing code) which are encased in the pre-mRNAs, as well as *trans*-acting factors which effectively attach to these elements (De Conti et al. 2013). The fundamental part of the splicing code for the bulk of the introns consists of short sequences at or near the exon–intron connection point on or near the exon–intron connection sites, such as 5′ and 3′ splice sites and a branching point sequence.

The splicing process is essentially carried out by the spliceosome, which is a ribonucleoprotein megaparticle and assembles entirely intertwined around the splice sites (Jurica and Moore 2003). In common, strong splicing sites that match closely to the familiar consensus sequence (Burge et al. 1999) lead to constitutive splicing and integration of the exon into the mature mRNA. However, alternative splicing of pre-mRNA can also be performed, which represents the most essential cellular machinery for producing diversity of proteins (Black 2003; Kornblihtt et al. 2013). In the majority of instances, alternatively spliced exons contain the so-called weak 3′ or 5′ splice sites, which indicates that their sequence does not match precisely with that of common consensus sequences (Burge et al. 1999).

Alternative splicing facilitates the expression of several mRNA isoforms of an individual gene, which frequently code for protein isoforms of varying functionality (Kalsotra and Cooper 2011). Different types of alternative splicing events can arise, among them firstly the inclusion/exclusion of a whole exon; secondly, the use of alternative splice sites 5′ or 3′ leading to the lengthening or shortening of an exon; thirdly, different transcriptional start sites based on alternative promoters; fourthly, alternative sites for polyadenylation; or fifthly, the deletion or retainment of an intron (Burge et al. 1999; Kornblihtt et al. 2013; McAlinden et al. 2004). The extent to which vulnerabilities are exploited is governed both by non-splicing *cis*-regulating sequences and by *trans*-acting factors. The *cis*-regulating sequences comprise exonic splice amplifiers, exonic splice silencers, intronic splice amplifiers and intronic splice silencers, each of them being located where and how they influence the performance of the splice site (Kornblihtt et al. 2013). *Trans*-acting factors act synergistically by attaching to *cis*-splice enhancers or silencers and comprise representatives of the highly characterized Ser/Arg-rich and hnRNP protein families (Lin and Fu 2007; Martinez-Contreras et al. 2007). Increasing numbers of investigations show that alternative splicing is employed to tune gene expression to the physiological demands. Since over 90% of human intron-possessing genes are spliced alternatively, it concerns a wide range of areas of biology (Kornblihtt et al. 2013; Kalsotra and Cooper 2011).

During the developmental phase of cartilage tissue, a number of genes expressed of differentiating chondrocytes undergo alternative splicing (McAlinden et al. 2004). In the following, a special emphasis is put on alternative splicing of the type II procollagen gene COL2A1 (in humans) or Col2a1 (in mice) (Cheah et al. 1985; Ala-Kokko et al. 1995; Metsaranta et al. 1991). COL2A1 is encoding a homotrimeric protein with three identical α1(II) chains that constitute a triple helical domain, accompanied by a 5′ amino (NH_2) propeptide and a 3′ carboxy (COOH) propeptide (Fig. 13.8). The conversion of these procollagens to eliminate the NH_2 and COOH propeptides must be performed to enable the generation of mature type II collagen molecules. These converted triple helical collagens then constitute a covalently cross-linked fibrillar grid scaffold in the extracellular matrix, which adds tensile strength to the cartilage tissue (Prockop and Kivirikko 1995; Eyre 2002; Canty and Kadler 2005; Eyre et al. 2006). In addition to the α1(II) chains, COL2A1 also codifies α3 (XI) protein chains, which are trimerized with α1(XI) and α2(XI) chains to produce heterotrimeric type XI procollagen (Burgeson and Hollister 1979; Furuto and Miller 1983; Eyre 1987) (Fig. 13.9). This fibrillar collagen forms a minor component of cartilage extracellular matrix; however, it performs an essential function in managing type II collagen fibrillogenesis and subsequently adjusting the fibrillary diameter (Blaschke et al. 2000).

Thus, in the framework of COL2A1/COL11A3 alternative splicing, it is necessary to account for the fact that alternative splicing generates different isoforms of procollagen type II and procollagen type XI. It will be further explored during the discussion of mouse models to investigate alternative splicing processes in vivo.

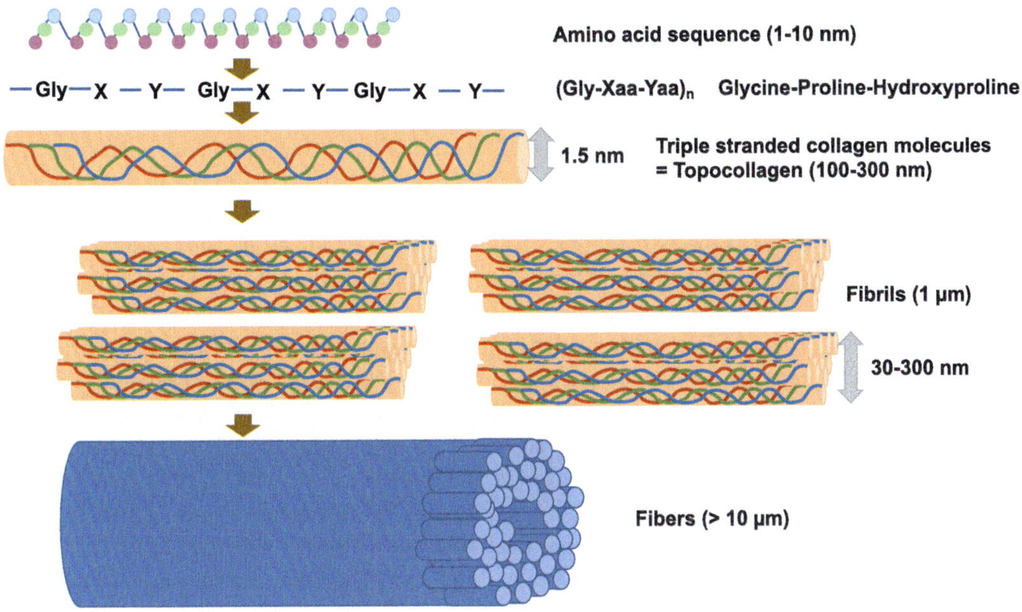

Fig. 13.8 Collagen structure. A collagen peptide forms together with two other collagen peptides the triple collagen helix. The helix arranges to fibrils of collagen and subsequently to collagen fibers

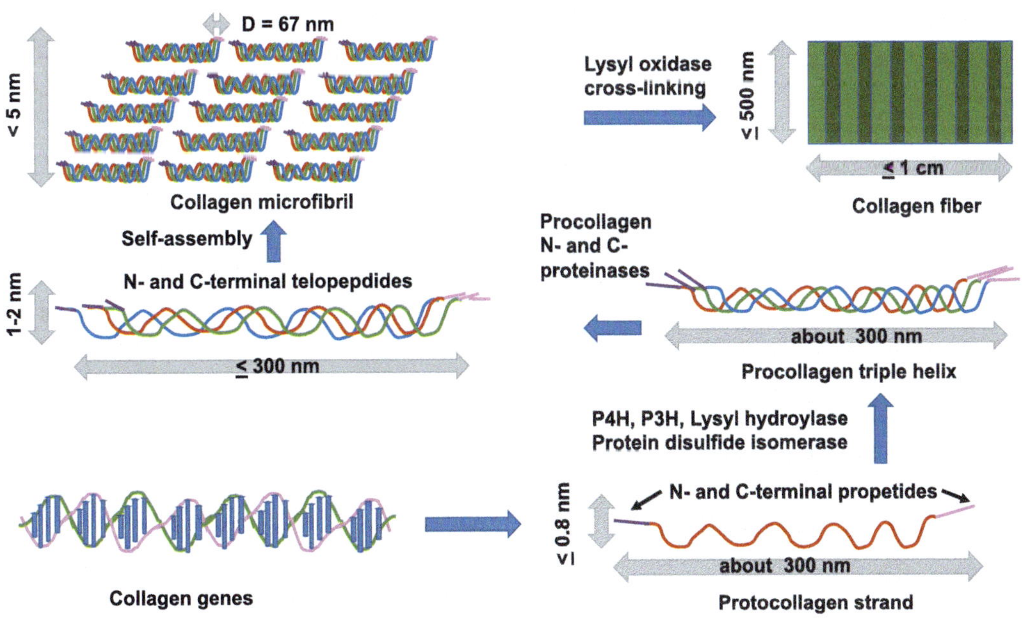

Fig. 13.9 Procollagen structure

13.13.1 The Alternatively Spliced Isoforms of COL2A1 Are IIA and IIB

Over the at least last two decades, two isoforms of type II procollagen were found based on the pre-mRNA alternative splicing of the exon 2 (Ryan and Sandell 1990). The type IIA isoforms contain exon 2, whereas the type IIB isoforms do not include exon 2 (Fig. 13.10). Importantly, the alternative splicing of exon 2 has been proven to be a developmentally controlled event in the course of chondrogenesis in which chondroprogenitor cells predominantly produce IIA isoforms, whereas differentiated chondrocytes primarily generate IIB isoforms (Sandell et al. 1991, 1994; Ng et al. 1993). It is important to consider that in these studies the various expression patterns of the IIA and IIB mRNA isoforms detected through in situ hybridization are unable to discriminate among $\alpha1$(II) and $\alpha3$(XI) mRNA. The most important aspect is that the developmentally governed IIA–IIB alternative splicing shift represents a unique phenomenon that is limited to the differentiation of chondrocytes. The early transient expression of type II procollagen has been observed in a variety of nonchondrogenic tissues of mice and human embryos, although only the IIA isoform seems to be expressed at these locations (Ng et al. 1993; Lui et al. 1995).

The expression of IIA mRNA in non-chondrogenic areas is extensively conserved among mice and humans and comprises epithelial structures involved in the epithelial–mesenchymal interferences, for instance the apical ectodermal crest of the limb bud, the pharyngeal endoderm and pulmonary epithelium. Moreover, these expression schemes point to a potential involvement in the regulation of morphogenesis. In addition, a morphogenetic function of the IIA procollagen has been identified in developing embryos of *Xenopus* (Larrain et al. 2000); however, there is no clear evidence for such a function in mouse models, which are devoid of exon 2 or continuously express exon 2-containing

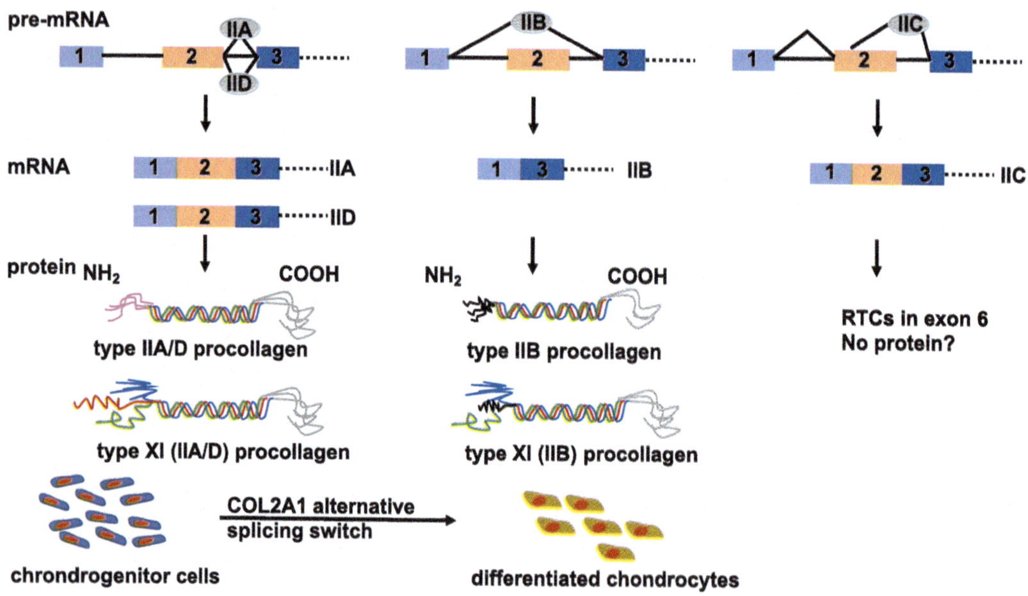

Fig. 13.10 Isoforms of COL2A1 through alternative splicing. Exons 1, 2 and 3 and the introns 1 and 2 of the COL2A1 precursor mRNAs (pre-mRNAs) are presented. The splicing of pre-mRNA generates IIA, IID, IIB and IIC mature mRNA transcripts. The exon 2-encoded (IIA/IID) DR domain of type II/XI procollagen is provided in pink in the NH_2 propeptide. The blue and green in the NH_2 propeptide of type XI procollagen indicates the alpha 1 (XI) and the alpha 2 (XI) chains. The truncated IIC mRNA with the premature termination codons (PTCs) in exon 6 seems to be not translated into a protein. Chondroprogenitor cells produce primarily exon 2 containing COL2A1 transcripts (IIA/IID). During chondrocyte differentiation, an alternative splicing switch occurs

Col2a1 isoforms over the entire course of the development (Leung et al. 2010; Lewis et al. 2012). In terms of post-natal cells or tissues, the two isoforms IIA and IIB have been detected in the vitreous of the bovine eye (Bishop et al. 1994; Reardon et al. 2000), and also in myelinating and non-myelinating Schwann cells of the peripheral nerve of the rat (D'Antonio et al. 2006). Thus, whereas the alternative splicing of exon 2 takes place in these non-chondrocyte cells, there exists no developmentally regulated IIA to IIB transfer mechanism of the kind encountered during chondrogenesis. This indicates that the IIA procollagen isoform is necessary for post-natal vitreous/neural cell development or retention, whereas it is not required or advisable in the tissue of post-natal cartilage.

Based on the different expression profiles of IIA and IIB mRNA isoforms in developing cartilage tissue, theories have been developed to explain why exon-2-containing isoforms are preferably expressed in precursor chondrocytes and not in differentiated chondrocytes. It has been proposed that the protein domain encoded by exon 2 could be relevant for firstly, interferences with other constituents of the matrix or the cell surface, secondly, the control of morphogenetic mechanisms, thirdly, the management of type II collagen biosynthesis, or fourthly, the regulation of fibrillogenesis (Ng et al. 1993; Lui et al. 1995; Ryan et al. 1990).

13.13.2 The Other COL2A1 Isoforms Are IIC and IID

There have been two other COL2A1 spliced isoforms identified in rabbit mesenchymal stem/stromal cells (MSCs) during the differentiation process by employing aggregate culturing techniques (McAlinden et al. 2008a, b). As anticipated, the sequencing of the cloned cDNA units after amplification of the exons 1–7, which code for the major part of the COL2A1-NH_2 propeptide, yielded cDNAs that are equivalent to the IIA and IIB isoforms. A cDNA containing solely the first 34 nucleotides of exon 2, which is referred to as IIC, and an additional cDNA with the identical sequence as IIA, except for three additional nucleotides at the 3′ end of exon 2, which is referred to as IID, were also recognized (Fig. 13.10). Consequently, these isoforms have also been detected in the human differentiating MSCs (McAlinden et al. 2008a, b).

The IIC isoform is produced by using a 5′ alternative splice site within exon 2, leading to a truncated transcript comprising the first 34 nucleotides of exon 2 and premature termination codons within exon 6. Subsequent analysis of IIC mRNA by Southern blotting revealed expression at only very early time points during differentiation of rabbit or human MSCs into chondrocyte lineages (McAlinden et al. 2008a, b). This implies that the IIC mRNA may only be identifiable at a time point when the frequency of other COL2A1 isoforms in chondroprogenitor cells is rather low. The existence of premature terminating codons transforms the IIC mRNA into a suitable target for NMD (Brogna and Wen 2009). Western blotting employing antiserum generated against a putative IIC peptide epitope has revealed to no detectable amounts of the protein (McAlinden et al. 2008a, b). Hence, this result additionally underpins the hypothesis of NMD. Notwithstanding the probability that no protein product is produced after IIC splicing, this incident seems to restrict the expression levels of IIA/IID and IIB collagen.

The IID isoform is produced through another 5′ splice site, which is positioned merely three nucleotides downstream of the IIA splice site within intron 2 (Fig. 13.11). The implementation of this splice site does not modify the protein reading frame of the entire mRNA transcript. When the IID mRNA is actually translated, a procollagen is synthesized which is the same as the IIA procollagen apart from the added tryptophan (in humans) or arginine (in mice or rabbits) to the end of the exon-2 coded protein region. The amino acid addition can cause a conformational modification and potentially change the role of the exon-2 encoded domain inside the NH_2 propeptide. Using an alternative quantitative transcript PCR technique (AT-qPCR) capable of discriminating precisely between IIA,

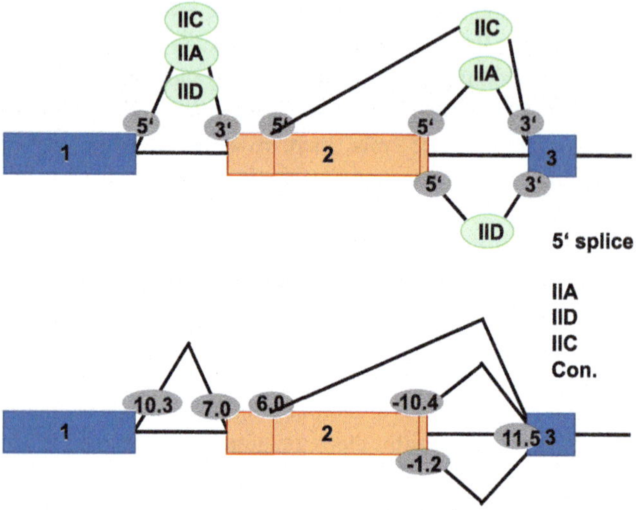

5' splice site	Sequence (5'-3')	Score
IIA	GTGgttgta	-10.4
IID	GTTgtaatt	-1.2
IIC	GAGgtataa	6.0
Con.	GAGgtaagt	11.1

Fig. 13.11 Sequence and strength scores of IIA, IID and IIC 5′ splice sites. The top row indicates the splicing events generating IIA, IID or IIC transcripts and the positions of the 5′ and 3′ splice sites. The bottom row displays the splice site strength score of the 5′ and 3′ splice site sequence employing MaxEntScan. The Con represents a typical consensus sequence for the splice site, the splicing occurs between the capital and the lowercase letters

IIB, IIC and IID isoforms (McAlinden et al. 2012), it was determined that IID isoforms are very poorly expressed in primary chondrocytes from embryonic cartilage (E12.5, E16.5); IIA mRNAs have been the predominant exon-2-containing isoforms produced by these cells. In addition, in chondrogenic differentiation of MSCs derived from human primary bone marrow, IID mRNA was determined to be about one-third more frequent than in IIA when employing an earlier variant of the AT-qPCR assay (McAlinden et al. 2008a, b). These outcomes may be slightly unexpected considering that the theoretical so-called strength level of the IID 5′ splice site exceeds that of the IIA 5′ splice site (Fig. 13.11). It is possible that the required *trans*-active splicing factor proteins, which are needed to reinforce the IID 5′ splice site, are not available in adequate amounts. As an alternative, the IID 5′ splice site can be rendered less accessible than the IIA site because of the existence of RNA secondary structures in the shape of a stem loop within intron 2 (McAlinden et al. 2005). In detail, six nucleotides of the IID 5′ splice site can be found in the double-stranded intronic stem region, while only three nucleotides of the IIA 5′ splice site are incorporated into it. For more information of the stem loop *cis* element see below.

13.13.3 The Exon 2-coded Region of COL2A1

COL2A1 exon 2 codes for a 69 amino acids region of the von Willebrand factor type C (VWC) based on the existence of a consensus sequence containing 10 cysteines (Ryan et al. 1990; Bornstein 2002). This spherical cysteine-rich (CR) domain exists in the NH_2 propeptide of type IIA/D procollagen, whereas it is missing in type IIB procollagen. There exist analogous cysteine-rich domains within the NH_2 propeptide of the type I and III fibrillar procollagens, although the maintenance of the cysteine-rich domain of type II procollagen among the species is more conserved compared to these other fibrillar collagens (Bornstein 2002). Homologous cysteine-rich regions can also be found in other extracellular proteins, such as chordone, connective tissue growth factor and thrombospondins, which are reported to all interfere with superfamily TGF-β growth factors (Murphy-Ullrich and Poczatek 2000; Zhang et al. 2007b; Fujisawa et al. 2009; Abreu et al. 2002).

In experimental approaches to decode the operation of the COL2A1-CR domain, it was shown that a recombinant exon-2 coding GST fusion protein interferes in vitro with bone marrow derived protein-2 (BMP-2) and tumor growth factor-β (TGF-β), whereas a recombinant fusion protein coding for exons 3–8 of the COL2A1-NH_2 propeptide cannot not interfere (Zhu et al. 1999). Follow-up experiments on *Xenopus* embryos demonstrated that overexpression of full-length IIA procollagen mRNA, but not IIB procollagen mRNA or mRNA coding solely for the cysteine-rich domain of IIA, induced secondary axis formation (Larrain et al. 2000). This dorsalization phenotype resembled to that caused via injection of chordin mRNA, which indicates that full-length IIA procollagen may have the ability to fully engage and inactivate BMP-2/4 in vivo (Larrain et al. 2000). Moreover, the results of the analysis of the solvent structure of the COL2A1 IIA domain revealed in vitro similarities to the fibronectin type I domain, implying the existence of a ligand attachment function (O'Leary et al. 2004).

The production of a polyclonal antibody which can specifically detect the exon-2, which codes the cysteine-rich protein domain of type II procollagen (Oganesian et al. 1997), provided a useful tool to visualize its expression and its localization pattern in murine and human embryonic cartilage and in even other tissues (Reardon et al. 2000; D'Antonio et al. 2006; Zhu et al. 1999, 2001). Hence, the IIA NH_2-propeptides were found to be present on the surface of type II collagen fibrils that have been isolated from human embryonic cartilage (Zhu et al. 1999). Hence, it was inferred that non-processed-type IIA procollagen, which holds back the NH_2 propeptide (pN-IIA procollagen) and exists in embryonic cartilage. This indicates that the chondroprogenitor cells seem to be devoid of the N-proteinase enzymes, which are required to break-down the IIA-NH_2 propeptide. Moreover, the unprocessed pN-IIA procollagen is needed at the onset of chondrogenesis to interfere with and adjust the BMP/TGF-β growth factors. There is, however, the chance that the IIA-NH_2 propeptides, which are pinpointed on fetal collagen fibrils, are included in the α3 chain of type XI collagen heterotrimers, because these are also coded in COL2A1 (Burgeson and Hollister 1979; Furuto and Miller 1983; Eyre 1987) (Fig. 13.10). Heterotypic fibrils are present in the extracellular cartilage matrix which is composed of type II, type IX and type XI collagens (Eyre et al. 2006) and the NH_2 propeptide of the α3(XI) chains remains normally preserved (Thom and Morris 1991; Wu and Eyre 1995). Moreover, it has been revealed that the IIA isoforms of type XI collagen are present (Zhu et al. 1999), as the anti-IIA antibody can be employed to detect exactly the type XI(IIA) collagen isoforms (McAlinden et al. 2014).

Thus, when the cysteine-rich NH_2 propeptide is involved in growth factor facilitated regulation during the early embryogenesis, it is not obvious if this feature might be located within the cysteine-rich domain of the pN-IIA protocollagen or the cysteine-rich domain of the IIA α3(XI) collagen chain. However, as it is discussed later in this chapter, the absence of an apparent skeletal phenotype in transgenic mice designed either to miss exon 2 or to maintain exon 2 permanently during the course of the development, did not foster the crucial role of the cysteine-rich IIA domain in the regulation of growth factor activity in chondrogenesis.

Regulation of COL2A1 Alternative Splicing by Molecular Mechanisms, Such as Intronic *Cis* Elements

To elucidate the cellular pathways that govern the splicing of exon 2, mini-gene model systems were employed. In insulin-triggered ATDC-5 cells, it was determined that the alternative splicing of a murine Col2a1 mini-gene, which comprises the exons 1–4 and intervening introns in their entirety (Hatano et al. 2002), is comparable to that of a shortened Col2a1 minigene with large deletions in the introns 1 and 2 (Nishiyama et al. 2003). Thus, it was found that major *cis*-elements are needed to adjust the alternative splicing of exon 2, certainly in the differentiation of ATDC-5 cells, in the vicinity of the 3′ and 5′ splice site regions of exon 2. In addition, a human COL2A1 mini-gene that compromises exons 1–3 and introns 1 and 2 in full length has been found to splice exon 2 in an alternative fashion because there is a weak 5′ splice site (McAlinden et al. 2005). As most alternatively spliced exons include weak splice site sequences that deviate from the common consensus sequence, other cues are usually necessary to guarantee the accurate assembly of the spliceosome and successful splicing. These auxiliary *cis*-regulatory elements are demonstrated to be located in an intron or inside a regulated exon and hence serve as splicing enhancers or silencers (Kornblihtt et al. 2013). These elements could be short, single-stranded ones, or they could provide RNA secondary structures which could compromise the availability of a targeted exon (Kornblihtt et al. 2013; McAlinden et al. 2004; Libri et al. 1991; Blanchette and Chabot 1997). The *trans*-active accessory splicing factor proteins binding to these *cis*-elements are of equal relevance (Kornblihtt et al. 2013).

With regard to the alternative splicing of COL2A1, an intronic *cis* helper sequence comprising an AU-rich element (ARE) directly bordering the weak 5′ splice site has been determined to be crucial for the control of exon-2 splicing (McAlinden et al. 2005). It has been initially hypothesized that the potential for this AU-rich intronic sequence to form a stem loop secondary structure (Ala-Kokko et al. 1995), which has also been experimentally supported (McAlinden et al. 2005). The splicing of a COL2A1 mini-gene lacking this stem loop sequence, but which still contains the weak 5′ splice site, considerably decreased formation of IIA exon-2-containing isoforms (McAlinden et al. 2005). This underlines the requirement of the AU-rich *cis* stem loop element to enable the splicing and the inclusion of exon 2. Successive investigations revealed that the *trans*-activating nuclear protein T cell restricted intracellular antigen-1 (TIA-1) might be capable of binding to a specific region of the stem loop holding the ARE sequence AUUUUUUUUUUUUUUUUUUUUU and act as a splice enhancer by facilitating the incorporation of exon 2 (McAlinden et al. 2007). Resembling ARE sequences are found in 3′ untranslated regions (UTRs) of genes and are specifically used by proteins, such as TIA-1, to influence the robustness, rate of decay or the translation initiation of the whole transcript. It is of great importance that this stem loop AU-rich *cis* element is located in an intron nearby an alternatively spliced exon. In addition, the sequences of the weak 5′ splice site of exon 2 and the neighboring AU-rich segment are highly conserved extending from humans to the teleosts (McAlinden et al. 2005). Teleosts of zebrafish were two col2a1 orthologues detected, which are namely col2a1a and col2a1b (Woods et al. 2005; Duran et al. 2011); however, Teleosts of terrestrial vertebrates were solely one gene encoding Col2a1 found (Dale and Topczewski 2011).

It has been hypothesized that there are alternatively spliced mRNA isoforms of the zebrafish col2a1a that either encompass exon 2 or not (McAlinden et al. 2014). These results bring up major questions as to the reasons why this alternative splicing was conserved in the course of evolution. When more sequence information is forthcoming, it will be intriguing to see whether the maintenance of the exon-2 splice site sequence and the neighboring ARE sequence occurs in primitive jawless vertebrates, such as lampreys or hagfish, which possess non-crumbling skeletal units (Ota and Kuratani 2010; Zhang et al. 2006b). Defining whether there is an alternative splicing of exon 2 in these species or not might offer more information to gain a better idea of the biological meaning of this phenomenon.

13.13.4 The *Cis* Elements Regulate the COL2A1 Exon 2

In the experiments to explore different splicing regulatory *cis*-elements which could persist inside exon 2 itself, a web-based resource has been employed to discover possible binding sites for exon splicing amplifiers (ESEfinder; https://rulai.cshl.edu/tools/ESE/) (Smith et al. 2006; Cartegni et al. 2003).

A *cis* element (CAGGT) with a high score has been revealed as a possible candidate for attachment of the Ser/Arg-rich splicing factor 1 (SRSF1, synonymously referred to as ASF/SF2). Afterwards, this heptameric element was removed from a COL2A1 mini-gene element, and the outcome was an improved performance of exon-2-containing transcripts (McAlinden et al. 2008a, b). A deeper analysis yielded that the deleted region overlapped with a section which is equivalent to another possible alternative 5′ splice sequence. In addition, this area is equivalent to the IIC 5′ splice site, the use of which leads to the generation of the IIC truncated mRNA transcript that is specifically expressed through chondrogenic MSC differentiation in rabbits and humans (McAlinden et al. 2008a, b). Whether the SRSF1 splicing factor protein reacts with its intended binding site to amplify/suppress the spliceosome formation at the IIC splice site is not well-understood. In subsequent investigations, three-point mutations in the COL2A1 mini-gene were carried out that could disturb the IIC splice site while maintaining the protein reading frame. Splicing of this mini-gene across different cell lines also led to an elevated ratio of IIA/D: IIB transcripts, resembling the mini-gene deletion mutation (McAlinden et al. 2014). In addition, transgenic mice with the same three silent point mutations have been produced (Hering et al. 2014). Although no open skeletal phenotype has been found in heterozygous or homozygous mutant IIC knock-in mice (Col2a1-mIIC), however, an increase in the ratio of IIA/D: IIB mRNA transcripts produced by chondrocytes in vivo was determined.

It has been hypothesized that the IIC splice site could work to improve the synthesis of IIB procollagen isoforms in differentiated chondrocytes and to inhibit spliceosome interactions that are associated with the IIA/D 5′ splice sites. In differentiated chondrocytes, the spliceosome ideally favors the IIC splice site over the IIA/D 5′ splice sites, as the former is more consistent with the 5′ splice site consensus sequence than the two later ones. Thus, the mutation of the IIC splice site provides in vivo splice complexes that attach to the IIA/D 5′ splice sites, which leads to the elevated amounts of exon-2-containing isoforms seen in the Col2a1 mIIC knock-in mice. It is intriguing that this IIC splicing leads to a truncated mRNA transcript carrying three premature termination codons located in exon 6. The IIC mRNA is assumed to serve as a substrate for the NMD. That would account for the fact why it is frequently hard to verify this transcript by AT-qPCR (McAlinden et al. 2012) and why no protein product has been found using an antibody raised against the suspected IIC peptide epitope (McAlinden et al. 2008a, b). Therefore, IIC splicing might serve as an instance of "regulated unproductive splicing and translation" (RUST), which is a notified cellular mechanism that combines alternative splicing and NMD to adjust the frequency of mRNA transcripts (Lareau et al. 2007b). According to the results in Col2a1 mIIC mice, the function of IIC splicing is to control the levels of the procollagen isoforms IIA and IIB, which are recognized to be translated into proteins.

Potential Connection of the COL2A1 Transcription Level and Alternatively Spliced Isoforms

Although it has been demonstrated that the stem loop AU-rich segment in intron 2 is crucial for the incorporation of exon 2 into chondroprogenitor cells (McAlinden et al. 2005), the way in which differentiating chondrocytes eliminate exon 2 is not fully elucidated. It is possible that the maturing chondrocytes do not express adequate amounts of the required *trans*-active splice accessory enhancer proteins, such as TIA-1, which are indispensable to enhance the interactions between the spliceosome, the weak 5′ splice site of exon 2 and the AU-rich stem loop. It seems to be crucial to determine by microarray or RNA-Seq analysis whether the chondroprogenitor cells produce a distinct subset of splicing factor accessory proteins than fully differentiated chondrocytes.

Another theory to clarify why differentiating chondrocytes eliminate exon 2 refers to the transcription rate of COL2A1. It is generally accepted that transcription and splicing processes in cells are jointly regulated (Kornblihtt et al. 2004). It has been seen that the levels of RNA polymerase II transcriptional elongation can affect the splicing of alternative exons. In particular, attempts to modify the RNA pol II transcription rates in cells have revealed that a slow-acting RNA pol II initiates the insertion of the alternatively spliced fibronectin EDI exon, while an increased elongation stimulates the EDI exon to undergo skipping (Kadener et al. 2001, 2002; de la Mata et al. 2003; Howe et al. 2003).

Thus, in the event of COL2A1 exon 2 splicing, the lower transcription rates in precursor cells can afford adequate time for splicing the relatively long intron 1 (approximately 4.1 kb) and assembling the spliceosome complex and additional complementary splicing factor proteins at the weak 5′ splice site and the AU-rich stem loop location. In chondrocyte differentiation, however, where the rate of COL2A1 transcription is drastically elevated to assist in the quick synthesis of the extracellular cartilage matrix, there will be skipping of exon 2 because the spliceosome/accessory-factor complexes may be unable to assemble timely at the weak 5′ splice site of exon 2, but will instead favor the stronger constitutive 5′ splice site of exon 3. Investigations of the growth factor treatment of undifferentiated murine chondrocytes indicated that TGF-β1 increased the incorporation of Col2a1 exon 2 (IIA isoform generation), while BMP-2 caused the skipping of exon 2 (IIB isoform generation) (Valcourt et al. 2003).

In this trial, it has also been stated that the general level of type II procollagen expression was reduced by TGF-β1, while it was elevated with BMP-2 in this culture system (McAlinden et al. 2014). These results confirm the hypothesis that low Col2a1 transcription levels promote the insertion of exon 2, whereas high levels might encourage the skipping of exon 2. Therefore, it seems promising to modify the RNA pol II elongation in chondrocytes in order to refine this hypothesis.

COL2A1 Isoforms Serve as Markers of the Differentiation of Chondrocytes
Type IIA/D or IIB procollagen transcripts are valid cellular markers for the determination of chondroprogenitor/dedifferentiated chondrocytes or fully differentiated chondrocytes. These indicators need at all times to be evaluated in in vitro chondrogenesis assays to assess the efficiency of chondrocyte differentiation. For instance, a study performed demonstrated that the insulin-induced chondrogenic differentiation of the murine ATDC-5 cell line after 28 days in culture led to elevated concentrations of IID-exon-2-isoforms than IIB isoforms (McAlinden et al. 2012). Astonishingly, this revealed that the overwhelming part of the cells within these cultures was slightly less differentiated than anticipated, despite the fact that the cultures were positively stained with alcacia blue, which points to proteoglycan synthesis. The observation that IID isoforms are even more abundant than IIA isoforms in these ATDC-5 cultures was all the more unexpected; a result that otherwise could not have been identified unless the AT-qPCR assay, which is capable of quantitatively discriminate between these two almost identical isoforms, had been designed (McAlinden et al. 2012). Thus, the IIA mRNA seems to be again the predominant isoform produced by early embryonic mouse chondrocytes in vivo or by chondroprogenitor cells from primary MSCs in vitro (McAlinden et al. 2008a, b, 2012). However, this raises some intriguing questions concerning possible differences in the splicing mechanisms that regulate exon-2 insertion among primary cells and transformed cell lines.

In the field of cartilage tissue engineering, a broad range of biomaterials is employed to induce the differentiation of chondrocytes for the repair or regeneration of cartilage, either as scaffolds or as carriers of growth factors (Vinatier et al. 2009; Spiller et al. 2011; Johnstone et al. 2013; Tuan et al. 2013). Typically, procollagen type II (the processed triple helical domain) and aggrecan are commonly examined in these tissue engineering procedures to verify the differentiation of chondrocytes and the generation of cartilage. The expression of specific isoforms of type II procollagen has seldom been measured in these systems. The levels of gene and protein expression of COL2A1 isoforms of TGF-β-treated pellet cultures of human MSCs in the presence or absence of poly(ethylene glycol) microspheres (PEG) have lately been examined (Ravindran et al. 2011).

Even though IIB mRNA isoforms prevailed in both culture systems, the ratio IIA/D: IIB mRNA isoforms was increased in MSC pellets comprising PEG microspheres. In addition, a more common localization of the IIA exon-2 encoded protein domain within the artificially produced cartilage tissue carrying the microspheres was identified. Whether a retarded or aberrant differentiation developed in these cultures incorporating biomaterials is not clear yet. Nevertheless, it is essential that the biological implications of biomaterial interference with MSCs or chondrocytes are taken into account and that the identification of appropriate cellular markers for the correct determination of the differentiation stage should invariably be performed.

13.14 Mouse Models for Col2a1 Alternative Splicing Investigations

Investigations in Col2a1 null mice have revealed that a type II collagen matrix is absolutely necessary for the correct synthesis of articular cartilage, epiphyseal growth plate cartilage, endochondral bone and intervertebral disks (Li et al. 1995; Aszodi et al. 1998). Two different transgenic mouse models were created, one without exon 2 and the other one that specifically expresses exon 2, in an effort to investigate the significance of the developmentally regulated alternative Col2a1 splice switch in an in vivo setting (Leung et al. 2010; Lewis et al. 2012). Col2a1-exon-2-null mice were generated which exhibited a partially persistent phenotype in homozygous knock-out embryos, resulting in aberrant development of head tissue, holoprosencephaly, and complete absence of midface structures (Leung et al. 2010). This phenotype was accompanied by decreased sonic gel (Shh) expression in the prechordal mesoderm. Genetic intervention trials demonstrated that the potential IIA function in forebrain and facial development excludes BMPR1A- or NODAL-facilitated signal activity. Consequently, it seems that the CR section of the IIA does not seem to contribute to the control of BMP activities (Larrain et al. 2000; Zhu et al. 1999). Another somewhat unexpected fact is that no skeletal or cartilage-specific phenotypes were observed in these mice. This would indicate that the IIA exon-2-coded cysteine-rich domain has no significant effect on the control of growth factor-based mechanisms in early phases of the development of cartilage tissue.

13.14.1 Transgenic Mice Exclusively Expressing Embryonic Col2a1 IIA Procollagen Isoform

A splice site targeting approach was recently employed to create transgenic mice with a knock-in mutation that transformed the weak 5′ splice site of the Col2a1 exon 2 into a strong consensus splice site sequence. This mutation allowed an efficient splicing of exon 2 and the expression of exclusively IIA procollagen isoforms, independent of the differentiation level of chondrocytes (Lewis et al. 2012). In these mice, the developmentally regulated alternative splicing of exon 2 and the resulting synthesis of type IIB procollagen was blocked. Heterozygous (ki/+) and homozygous (ki/ki) transgenic mice (Col2a1+ex2) proved to be viable and unexpectedly exhibited no apparent skeletal phenotype. In specific detail, articular cartilage, growth plate cartilage and endochondral bone seem to grow naturally. Initially, it has been hypothesized that sustained false expression of embryonic IIA procollagen isoforms would lead to some kind of skeletal aberration when in fact the cysteine-rich IIA domain has a major impact on altering the activity of superfamily TGF-β growth factors. For instance, even if the cysteine-rich domain might suppress the activity of BMP-2 during the process of chondrogenesis, a chondrodysplasia phenotype similar to that observed in chondrocyte-specific Bmp-2 knock-out mice would be anticipated (Shu et al. 2011).

There have been intriguing results that a recombinant IIB-NH_2 propeptide protein, which coded by human exons 3–8, has anti-angiogenic activity (Sandell et al. 2008) and may also be capable of causing cell death of cancer cells and osteoclasts through an integrin-driven mechanism (Hayashi et al. 2011; Wang et al. 2010). Ordinary, healthy articular cartilage is avascular and resilient to tumor invasion. Overall, these findings indicate that the IIB-NH_2 propeptide may retain the avascular character of articular cartilage and prevent its invasion by cancer cells. However, in Col2a1+ex2 mice without IIB procollagen no abnormal vascular invasion was found during pre- and postnatal maturation of articular cartilage. One possible reason could be that the IIA-NH_2 propeptide, which is coded by exons 2–8, also acts to protect the articular cartilage from vascular invasion. Although preliminary investigations revealed that a recombinant IIA protein coded only by exon 2 lacks an anti-angiogenic capacity, a recombinant IIA protein coding for the entire NH_2 propeptide has not been examined in this experiment.

Another reason for the absence of an abnormal vascular phenomenon in the articular cartilage of Col2a1+ex2 mice is the ability of matrix metalloproteinases (MMPs) to continue processing the IIA-NH_2 propeptide to produce smaller fragments similar to the IIB-NH_2 propeptides. Actually, a recombinant trimeric IIA-NH_2 trimeric propeptide, which is coded by exon 2–8, has been demonstrated to being digested in vitro by a variety of distinct MMPs (Fukui et al. 2002). Concretely, it has been established that the cleavage of the recombinant IIA-NH_2-propeptide through MMP-7 might possibly produce IIB-like NH_2-propeptides. Alternatively, it seems possible that the IIB-NH_2 propeptide is rapidly modified and cleared from the extracellular matrix in vivo and is thus not available in sufficient quantity to act as an antivascular compound in vivo. This hypothesis is reinforced by the recent report in which a new antibody is analyzed that detects specifically the mouse procollagen isoform IIB and not IIA procollagen (Patra et al. 2013).

Whereas this antibody has been able to identify unprocessed-type IIB-NH_2 procollagen through Western blotting, immunolocalization in the extracellular matrix of cartilage tissue sections did not yield positive signals indicating that the processed IIB-NH_2 propeptide is quickly detached from the procollagen molecule and eliminated from the extracellular matrix. Regarding the anti-tumorigenic activity of IIB-NH_2 propeptide (Wang et al. 2010), these experiments proved that a recombinant IIA-NH_2 propeptide protein coded by exons 2–8 is not able to trigger the killing of cancer cells. Thus, it might be useful to encourage the use of Col2a1+ex2 mice in order to see whether they may be more prone to chondro/osteosarcoma.

13.14.2 The Splicing of Col2a1 and Type XI Collagen

However, it has been found that IIA procollagen from postnatal Col2a1+ex2 mice can be handled similarly to IIB procollagen (Lewis et al. 2012). Afterwards, freely moving IIA-NH_2 propeptides are eliminated from the extracellular matrix and account for the reduction of IIA-cystine-rich protein expression patterns, which are labeled via a fluorescence staining in the territorial and interterritorial matrix of the post-natal articular cartilage tissue within these mice (Lewis et al. 2012). Nevertheless, in 10-week-old ki/+- and ki/ki Col2a1+ex2 mice, sustained expression of the IIA protein has been detected in the superficial zone and in the pericellular matrix enclosing the chondrocytes of the deep zone of articular cartilage and in the pericellular matrix of growth plate chondrocytes (Lewis et al. 2012). This type of pericellular location is similar to type XI collagen dye profiles in epiphyseal cartilage (Petit et al. 1993; Keene et al. 1995). It is generally recognized that the NH_2 propeptide coded by the $\alpha3$(XI) chain is preserved in the heterotrimer of type XI collagen (Thom and Morris 1991; Wu and Eyre 1995). In addition, the presence of IIA isoforms of type XI collagen with the exon-2 coded cysteine-rich domain within the $\alpha3$(XI) chain in the rib cartilage of Col2a1+ex2 mice has lately been verified (McAlinden et al. 2014). Thus, it is conceivable that the pericellular expression of the IIA protein, which endures in mature cartilage tissue of Col2a1+ex2 mice, is attributable to the existence of IIA isoforms of type XI collagen.

Type XI collagen is a key modulator of collagen fibrillogenesis and fibrillary thickness due to the maintenance of its NH_2 propeptide (pN procollagen) (Blaschke et al. 2000). Hence, it is reasonable to assume that the presence of the spherical IIA-cystine-rich domain in the α3(XI) chain changes the structure of the NH_2 propeptide of type XI collagen and thus influences its function in regulating fibrillogenesis. There is a certain amount of proof to corroborate this hypothesis based on the electron microscopic analysis of cartilage extracellular matrix in Col2a1+ex2 mice. Based on the ultrastructure of the collagen fibrillar matrix within ki/+ and ki/ki articular cartilage, there are alterations compared to wild-type tissue (Hering et al. 2014). In specific, larger collagen fibrils were detected in the pericellular domain of the ki/+ and ki/ki tissue, besides electron dense regions pointing to fusion or aggregation of fibrils. Similar irregularities were also found within the territorial and interterritorial regions. Lately, it has been seen that there are altered collagen fibril structures in the ki/+ and ki/ki rib cartilage matrix and reduced amounts of hydroxypyridinoline cross-linking in comparison with wild-type tissue (McAlinden et al. 2014). In summary, alternative splicing of Col2al may be required to monitor the expression of the chain isoform of type XI collagen α3 to enable correct assembly of the matrix.

13.14.3 The Mutations in the Alternatively Spliced COL2A1 Exon 2

Human mutations in COL2A1 exon 2 have been detected and shown to cause a type of stickler syndrome involving solely the eye (Donoso et al. 2003). The typical eye defects comprise vitreous degeneration, broken and detached retinas. Nevertheless, Stickler's syndrome can also be caused by mutations in other exons of COL2A1, and thus, patients do not only develop eye disorders but also a variety of other aberrations, such as joint hypermobility and pain, impaired hearing, defective spine, anomalies of the face, and premature osteoarthritis (Donoso et al. 2003; Snead et al. 2011; Richards and Snead 2008). The rationale why exon 2 mutations primarily lead to eye disease lies in the fact that the cells of the eye express both IIA and IIB isoforms pre- and postnatally (Du et al. 1998). In addition, the vitreous body of adult bovine animals has been proven to provide roughly equal amounts of IIA and IIB procollagen proteins (Bishop et al. 1994). These expression profiles of the eye are distinct from the developmentally induced, controlled lack of IIA expression, which is generally assumed to arise with the maturation of cartilage tissue. Thus, in these patients, no additional systemic deficits will be found, since differentiated and mature cartilage tissue mainly expresses type IIB procollagen and thus is not subject to exon 2 mutations.

The overwhelming number of the notified COL2A1 exon 2 mutations cause premature stop codons, which probably lead to type IIA procollagen haploinsufficiency (Donoso et al. 2003; McAlinden et al. 2008a, b). Thus, the eye disorders may arise simply as a result of a general lack of type IIA procollagen. The persistence of IIA procollagen during the entire postnatal development of the vitreous body can be critical in controlling the formation of thin collagen fibrils, which are necessary for the correct functioning of this tissue (Reardon et al. 2000). Thin collagen fibrils are also located in the extracellular matrices of embryonic cartilage simultaneously with the expression of COL2A1 IIA isoforms (Zhu et al. 1999). As COL2A1 codifies the type XI collagen α3 chain, it is quite feasible that the exon-2 encoded IIA domain is also located on collagen fibrils in the vitreous (Reardon et al. 2000), which represents a specific part in the type XI heterotrimers; it has been revealed in other investigations that type XI collagen is limited to thin collagen fibrils (Keene et al. 1995). Moreover, it is generally understood that type XI collagen maintains its NH_2 propeptide (pN procollagen) (Thom and Morris 1991; Wu and Eyre 1995), and earlier work indicates that NH_2 propeptides from fibrillar collagens can act to adjust the fibrillary thickness and impede the fusion of fibrils (Fleischmajer et al. 1983).

Independent of whether the IIA exon 2 encoded domain is existent in type II or type XI pN collagen, and it is probable that its permanent expression in the eye is necessary to preserve the structural integrity of collagen-based extracellular matrix. The eyes of Col2a1+ex2 mice expressing exclusively IIA isoforms of Col2a1 seem to be interesting to explore. It is possible that an equilibrium between the procollagen isoforms IIA and IIB is required for vitreous health and that expression of IIA procollagen alone will have negative consequences.

13.14.4 Is the IIB Procollagen Required for the Healthy Cartilage?

Experiments on Col2a1+ex2 mice have yielded some solutions to this problem. The suppression of alternative Col2a1 splicing appears to be reasonably compatible in these mice during cartilage development, and IIA procollagen seems to be fairly effective in compensating for the lack of IIB procollagen. Thus, control of the expression of type II procollagen isoforms during the process of chondrogenesis through alternative splicing is probably not as relevant as previously assumed, since type II procollagen synthesis and processing seems to be normal in cases where the procollagen isoform IIA is expressed by differentiated chondrocytes. There is also no evidence of a function of the IIA isoform in the regulation of the activity of growth factors. Nonetheless, the alternative Col2a1 splicing seems to be necessary to guide the expression of the isoform and consequently the regulatory function of type XI collagen. It is still to be determined whether the alterations in collagen ultra-structure found in the articular cartilage of Ki/+- or Ki/ki-Col2a1+ex2 mice can be transferred to biomechanically compromised tissue. For that purpose, it is instructive to stress the cartilage by destabilizing of the knee joint through surgically initiated or non-invasive strategies for joint manipulation. Elevations in the trabecular bone material were also measured in Col2a1+ex2 mice (Hering et al. 2014). Therefore, a suitable cartilage matrix is required for endochondral bone formation. These results highlight several intriguing questions, for instance about how an abnormal type II collagen fibrillar matrix can directly influence the level of activity of osteoblasts and/or osteoclasts.

13.14.5 Why is COL2A1 Exon 2 Targeted for Alternative Splicing, but not Homologous Domains in COL1A1 or COL3A1?

Exon 2 coding cysteine-rich regions are available in the α1 chains of both procollagen type I and procollagen type III and display a common homology to that of α1/α3 chains of procollagen type II/XI (Bornstein 2002). Importantly, COL1A1 or COL3A1 exon 2 is not alternatively spliced and is invariably contained in the finished fully processed mRNA. The reason for this fact is that the 5′ splice site sequence of exon 2 in these other fibrillar collagen genes matches more closely to the known consensus sequence (Burge et al. 1999) and is thus spliced in a constitutive manner. The sequence of the 5′ splice site sequence of the human COL1A1 exon 2 is CAGgtgcgg (NCBI Ref. Seq: NG_007400.1) and the sequence of the 5′ splice site sequence of the COL3A1 exon 2 is GCTgtgagt (NCBI Ref. Seq: NG_007404.1). The capital letters stand for the nucleotides of exon 2 and the small letters for those of intron 2. The computation of the 5′ splice site stability of exon 2 performed with MaxEntScan (Yeo and Burge 2004) gave a value of 9.2 for COL1A1 and 7.4 for COL3A1. This is relatively strong compared to the very low splice site stability value −10.4 of the 5′ splice site of COL2A1 exon 2 (Fig. 13.11).

When the exon 2 undergoes constitutively splicing and the fully processed mRNA is persistently gained, which encodes COL1A1 and COL3A1, it can be hypothesized that it is needed through the entire pre- and post-natal development of distinct connective tissues, in which these collagens are expressed. In contrast, a transgenic mouse lacking the exon 2 domain of COL1A1 has been revealed to exhibit no overt phenotype (Bornstein 2002). A more detailed analysis of the dermis using electron microscopy shed light on a slight elevation of the fibril thickness and some irregularities in the shape of the fibrils (Bornstein et al. 2002), indicating that the COL1A1 exon 2 domain fulfills a function in regulating the process of fibrillogenesis and seems therefore to provide an explanation why this exon is constitutively spliced.

Therefore, why is exon 2 scheduled to be eliminated from the COL2A1 pre-mRNA in the eye or in a development-regulated fashion in cartilage? The hypothesis could be put forward that COL2A1 exon 2 codes for a superfluous protein because an explicit phenotype is missing in the Col2a1+ex2 mice that were designed for the permanent expression of exon 2 (Lewis et al. 2012). Thus, might the excision of exon 2, which codifies about 20% of the total type II procollagen protein, have been randomly generated and thereafter turned out to be beneficial for the cell at times when large quantities of type II procollagen secretion are necessary for the growth of the skeleton? The reasoning against it originates from observations of the modified collagen ultrastructure in the ribs of Col2a1 +ex2 mice (McAlinden et al. 2014). Thus, the elimination of COL2A1 exon 2 is suggested to be performed in order to adjust certain features of type II/XI procollagen fibrillogenesis and the consecutive assembly of the matrix.

13.14.6 Why is the Col2a1 IIA Isoform Reexpressed in Healthy Articular Cartilage in Mice?

In addition, in Col2a1+ex2 mice, a substantial reexpression of IIA procollagen in normal, mature wild-type articular cartilage tissue was detected (McAlinden et al. 2014). It was established that chondrocytes in human and mouse osteoarthritic cartilage reexpress type IIA procollagen, and it was hypothesized that the cells might shift to produce embryonic exon-2-containing isoforms to seek to heal the destroyed cartilage tissue (Salminen et al. 2001; Aigner et al. 1999). Nevertheless, the results of IIA procollagen expression through differentiated chondrocytes in normal mature articular cartilage in mice indicate that the production of IIA may possibly not be linked to the disease. On the contrary, it is conceivable that the levels of type II procollagen production in postnatal cartilage in the course of normal matrix transformation are sufficiently reduced to enable the incorporation of exon 2 and the generation of type IIA procollagen. There is also a strong rationale to assume that the expression of IIA procollagen in healthy mature joint cartilage could be a mechanoregulated mechanism of response. It has been demonstrated that a variety of alternatively spliced genes are actually governed through mechanical stress, even though the mechanisms involved are still unknown (Liu and Tang 2013).

13.15 Is There a Tissue-Specific Regulation of Alternative Splicing?

The accurate adjustment of numerous alternative splicing processes through the presence of certain splicing factors is indispensable for the identification of tissue types and phases of development. Nevertheless, the molecular principles of tissue-specific alternative splice control and the characteristics of splicing regulatory networks (SRNs) are insufficiently comprehended. It was broadly hypothesized that the targets of the brain and muscle-specific splicing factor Fox-1 (A2BP1) and its paralogue Fox-2 (RBM9) and systematically determine the respective SRNs on a genome-wide scale. Fox-1/2 is preserved throughout the worm to the human body and specifically detects the RNA element UGCAUG. The Fox-1/2-binding specificity has been combined with a phylogenetic preservation, microarray data of the splicing and complementary computational and experimental identifications. Forecasting thousands of Fox 1/2 objectives with well-conserved binding sites, a false detection rate (FDR) of around 24%, many of which have been experimentally confirmed, points to a unexpectedly large SRN (Zhang et al. 2008). The favored location of the binding sites varies depending on the alternative splice pattern and defines whether the exon detection via Fox-1/2 is activated or suppressed. Several forecasted targets are essential for neuromuscular performance and have been linked to multiple genetic diseases. Moreover, a number of cases of formation or disappearance of binding sites in a variety of vertebrate lines and human populations have been revealed, probably mirroring a refinement of gene expression regulation throughout evolution (Zhang et al. 2008).

The sequencing of entire genomes showed that complex metazoans, among them mammals, contain only marginally more genes than single-cell yeast (International Human Genome Sequencing Consortium 2001). The complex nature of the organisms needs to arise mainly from the mechanisms of diverse expression outputs and temporal and spatial distribution schemes from a restricted number of genes. It is critical to gain an insight into how gene expression is organized to identify developmental phases, determine cell types and react to environmental factors (Maniatis and Reed 2002). Therefore, the process of alternative splicing represents a crucial step in the adjustment of transcripts after the transcription process (Cartegni et al. 2002; Black 2003). In mammals, over two-thirds of the genes are spliced in an alternative way (Johnson et al. 2003). The selection of exons and splice sites is widely influenced by numerous RNA-binding proteins or splicing factors that interfere with *cis*-regulating elements to initiate or suppress certain splicing events.

A large number of splicing factors possess limited and variable expression profiles and participate in the tissue-specific or developmentally controlled splicing of certain transcripts. Nevertheless, the mechanisms and effects of these alternative splices are still poorly known. Well-established models are Sxl, Tra, Tra-2 and several other splicing factors in *Drosophila*, which govern a splicing pathway to regulate sex determination (Lopez 1998). Mammalian tissue-specific splicing factors comprise CELF family proteins, Fox-1/2, Hu proteins, muscle-blind (MBNL), Nova-1/2, PTB/nPTB and TIA1/TIAR and many more, which are even not yet determined (Li et al. 2007; David and Manley 2008).

The discovery of the RNA objectives of these factors is crucial for the comprehension of SRNs, although it is a difficult task; in the majority of instances, only a few objectives have been defined in experiments. The emergence of high-throughput technologies for surveillance of mRNA isoform frequencies and protein-RNA interactions, involving splicing of microarrays, has recently been advanced (Johnson et al. 2003; Ule et al. 2005b; Sugnet et al. 2006; Boutz et al. 2007), RNP immunoprecipitation microarray (RIP chip) (Keene et al. 2006), and ultraviolet cross-linking and immunoprecipitation assays (CLIP) (Ule et al. 2005a), offered new possibilities for identifying specific RNA destinations in vivo and genome-wide profiling of SRNs (Blencowe 2006; Moore and Silver 2008). This has been observed in experiments with Nova-1/2 targets, which highlighted key features of coregulated targets in the neuronal synapse and axon guidance. In addition, mechanisms by which Nova-1/2 activates or suppresses the exon insertion have been figured out to rely on the position of the binding sites (Ule et al. 2006). Some other investigations reported that subgroups of exons with splicing schemes typical for certain tissues or are linked to reactions to specific stimulations. These exons might occasionally exhibit an accumulation of RNA sequence motifs and gene-ontological concepts, indicating a coordinated splicing scheme that might be relevant for the functions of the cell (Sugnet et al. 2006; Das et al. 2007; Fagnani et al. 2007; Ip et al. 2007; McKee et al. 2007). Nevertheless, more in-depth investigations are needed to establish individual objectives for certain splicing factors. Computational prediction of targets for certain splicing factors is challenging, especially considering the limited size and degeneracy of splicing factor binding motifs. The exception to this degeneracy represents the hexanucleotide UGCAUG that is an essential intronic element for the splicing of multiple exons (Huh and Hynes 1994; Kawamoto 1996; Lim and Sharp 1998). A computational investigation further indicated that this entity is accumulated in the introns located downstream of a group of neuron-specific exons (Brudno et al. 2001). Most relevantly, zebrafish and mammalian homologues of *Caenorhabditis elegans* fox-1 have been found to be recognizers of the (U)GCAUG element (Jin et al. 2003; Ponthier et al. 2006). In *C. elegans*, the Fox-1 gene is crucial for the balancing of the X chromosome dose in sex determination (Meyer 2000).

In mammals, Fox-1 (synonymously referred to as A2BP1) codes for an RNA-binding protein originally designated as an interaction partner of ataxin-2 and contains at minimum one paralogue, Fox-2 (synonymously referred to as RBM9 or Fxh) (Nakahata and Kawamoto 2005). Human Fox-1 and Fox-2 exhibit very strong sequence homology, with 100% identity in the RRM domain, and attach to the identical RNA element (Kiehl et al. 2001; Jin et al. 2003; Auweter et al. 2006; Ponthier et al. 2006). These two proteins are expressed only or primarily in the brain, heart and skeletal muscles. Moreover, mutation or abnormal expression levels of Fox-1 have been identified in patients with various genetic disorders, among them are autism (The Autism Genome Project Consortium 2007; Martin et al. 2007; Sebat et al. 2007), epilepsy, heart disease (Kaynak et al. 2003) and mental retardation (Bhalla et al. 2004). Fox-2 has also been involved in the signaling of hormones as a co-repressor of tamoxifen-based activation of the estrogen receptor (Norris et al. 2002). Thus, Fox-1/2 may be crucial regulators of tissue-specific splicing and a systematic identification of their objectives may yield important knowledge about the mechanisms of tissue-specific splice control, the properties of SRNs and their physiological contributions.

To identify and characterize the SRNs of Fox-1/2, a combination of computational forecasting from comparative genome analysis with experimental evaluation and characterization has been employed. Notably, thousands of candidate Fox 1/2 targets with highly conserved binding sites throughout all vertebrate species were found in the analysis. Fox-1/2 may either trigger or suppress splicing, according to the position of their binding sites, and may also assist in creating intricate splicing schemes. Most of the forecasted targets perform an essential task in neuromuscular activities and disorders. In addition, the development of Fox-1/2-binding sites through various vertebrate lines and between diverse human populations and the possible phenotypic effects is debated.

13.16 Impact of Alternative Splicing on Cell Mechanics

In turn, the alternative splicing of specific genes may then impact the mechanical properties of cells and subsequently cellular functions. In the following, there are a few examples that have also a great impact of the functional phenotype of cells leading to severe diseases.

13.16.1 Alternative Splicing of Genes Encoding for Contractile Proteins Affects Cell Mechanics

In the muscle cells, the process of alternative splicing fulfills a crucial function in the regulation of the functional variety of contractile proteins (Nadal-Ginard 1990). In specific, various types of sarcomeres were produced, which display distinct characteristics. Under the genes coding for the numerous contractile proteins, tropomyosin (TM) and troponin (TnT) are well explored paradigms. These two proteins are used to understand muscle cell-specific alternative splicing because they contain multiple alternative cassette exons that are encapsulated, skipped and/or handled in a mutually exclusive fashion (Toyota and Shimada 1981; Smith and Nadal-Ginard 1989). The expression of TM has been detected in every cell type (Nadal-Ginard 1990; Helfman et al. 1988); however, tissue-specific splice variants are revealed through mutually exclusive and alternative exons, alternative promoters, and polyadenylation sites (Nadal-Ginard 1990; Helfman et al. 1988). For example, exon 3 of the α-TM gene tends to be inserted in all cell types apart from smooth muscle, in which exon 2 is inserted to create a smooth muscle cell-specific α-TM variant (Smith and Nadal-Ginard 1989). In humans, the gene α-TM possesses two versions of exon 5 which are mutually complementary to produce the variants α-TM NM or α-TM SK (Graham et al. 1992). The expression of the α-TM SK variant exists entirely in skeletal muscle, while α-TM NM represents the ubiquitous α-TM mRNA which is commonly encountered in non-muscular cells (Mullen et al. 1991). The insertion of the SK-specific exon represents an active mechanism that demands a splicing apparatus confined to the muscle cells, because the genetic deletion of the NM exon turned out to be insufficient to insert the SK exon into the α-TM mRNA in non-muscle cells (Mullen et al. 1991).

Most recent biochemical purification experiments for the functional characterization of various tissue-specific isoforms of α-TM have identified the muscle-specific isoform α-TM SK, which comprises exons 9a, 1a and 2b, to exhibit a higher relative affinity for F-actin filaments than isoforms which are devoid of exon 9a, thereby modifying the kinetics of actin polymerization process (Janco et al. 2016; Gateva et al. 2017). The significance of alternative splicing in the production of muscle-specific isoforms can also be seen in the β-TM gene. In muscle, β-TM transcripts contain the mutually exclusive exons 7 and 10, whereas transcripts in fibroblasts carry exons 6 and 11 (at the cost of exons 7 and 10) to code for TM-1 (Guo et al. 1991). The rigid restriction of the uses of exons 7 and 10 only to muscle has been ascribed to the binding of *trans*-active factors to the *cis*-regulating elements on the pre-mRNA, which actively restrain their integration into transcripts in non-muscle cells (Graham et al. 1992; Mullen et al. 1991). In the case of TnT, the muscle cell-specific isoforms, including TnT F (fast skeleton), TnT S (slow skeleton) and TnT C (heart muscle) are produced by a single gene where the transcripts contain either exon 16 or exon 17 in a mutually exclusive fashion and differ in their containment/skipping of exons 4–8 (Breitbart et al. 1985). In a similar way, the mutually exclusive utilization of either exon 2a or 2b discriminates between the smooth muscle isoform and the striated/skeletal muscle isoform of α-tropomyosin (Smith and Nadal-Ginard 1989; Wieczorek 1998). It has been revealed that these muscle-specific splicing processes are governed by the divergent expression of specific *trans*-active factors, such as RBM4, PTB (Lin and Tarn 2005, 2011) and CUGBP1 (Sureau et al. 2011), which ease the entrapment or skipping of mutually exclusive exons, whereas the standard splicing pathway is generated through a general splicing mechanism (Nadal-Ginard 1990). Besides these tissue-specific *trans*-active elements, the potency of the polypyrimidine tract and the branching point sequence define the involvement or elimination of mutually exclusive exons (Nadal-Ginard 1990; Graham et al. 1992; Mullen et al. 1991). In functional terms, these tissue-specific splicing variants of skeletal muscle are essential for modifying the calcium management of the contractile machinery through their association with Ca^{2+} regulatory proteins and therefore represent major factors in assessing the rate of muscle contraction in muscle fibers (Zot and Potter 1987; Farah and Reinach 1995).

13.16.2 Impact of Alternative Splicing in Muscle Cells Mitochondria on the Functional Phenotype

The skeletal muscle shows a remarkable capacity to sustain reasonable energy levels throughout long-lasting low-level contractions, whereas at the same time it is capable of quickly raising energy levels whenever explosive contractions are required (Sahlin et al. 1998). Ca^{2+} absorption is key in sustaining energy homeostasis due to the varying contraction requirements of skeletal muscle (McMillin and Madden 1989; Stein et al. 1988; Patel et al. 1996). Recent investigations dealing with alterations in gene expression over the course of postnatal development have demonstrated that genes that deal with calcium fail to alter their expression levels, whereas instead they experience splicing alterations that facilitate the transfer from embryonic to adult protein isoforms (O'Leary et al. 2004). An intriguing example might be MICU1, a component of the mitochondrial calcium uniporter (MCU), which represents the highly selective channel responsible for Ca^{2+} influx via the ion-impermeable inner membrane of the mitochondria (Mammucari et al. 2016). Muscle-specific alternative splicing produces the MICU1.1 variant, which is defined through the incorporation of a highly conserved microexon coding for four amino acids that is sufficient to dramatically alter the characteristics of the MCU. Heterodimerization of muscle-specific MICU1.1 with MICU2 in the mitochondria produces a compound of variants with a strongly elevated binding affinity for calcium, thereby effectively reducing the Ca^{2+} threshold needed to maintain mitochondrial calcium levels required for the synthesis of ATP upon muscle contraction (Vecellio Reane et al. 2016).

13.16.3 Alternative Splicing of Genes Regulates Linkage of Excitation–contraction

The contraction and relaxation of skeletal muscle are controlled by fast and transient alterations in the concentration of free Ca^{2+} in the cytoplasm of the muscle fiber (termed myoplasm). The excitation–contraction clutch defines the strength of the contraction of the muscles (Ebashi 1991). Following a hierarchical sequence of incidents, an excitation–contraction coupling (ECC) is induced with an action potential at the plasma membrane, which propagates across the transverse T-tube system. This potential is detected via dihydropyridine receptors (DHPR, L-type Ca^{2+} channel CaV1.1) and causes the linkage with ryanodine receptors (RyR) of the sarcoplasmic reticulum (SR) that liberates Ca^{2+} in the myoplasm (Calderón et al. 2014). Transient activation of the Ca^{2+} buffer mechanism and muscle contraction precede, accompanied by fast transport of Ca^{2+} back to SR via SR Ca^{2+} adenosine triphosphatase (SERCA) (Calderón et al. 2014). Abnormal splicing of pre-mRNAs coding for calcium channels implicated in Ca^{2+} homeostasis, such as CaV1.1, CaV1.2, RyR and SERCA, is frequently seen in muscular dystrophy (den Bosch et al. 1994; Kimura et al. 2009; Tang et al. 2012b; Sultana et al. 2016). With regard to CaV1.1 (CACNA1S), the alternative splicing of exon 29 produces a calcium-conducting CaV1.1e which increases the calcium inflow and produces instantaneous calcium sparks within the muscle fibers in the course of endothelial cell coupling, which leads to a decreased strength and improved persistence similar to that experienced with the myotonic dystrophies DM1 and DM2 (Sultana et al. 2016; Santoro et al. 2014). In a similar manner, CaV1.2 is spliced alternatively through RBFOX2 in the smooth vascular muscle and the heart muscle cells, whereby a variant containing exon 9 (Cav1.2SM) and a variant lacking exon 33 (Cav1.2CM) during hypertension is created (Zhou et al. 2017). In functional terms, these Cav1.2 splice variants distinguish in their Ca^{2+} window current, whereby the Cav1.2SM isoform is capable of stimulating Ca^{2+} inflow with a greatly decreased action potential for the basal contractility in blood flow (Liao et al. 2007). In addition, it was discovered that the peptide sequence coded by exon 9 in CaV1.2 is susceptible to calcium channel inhibitors (Liao et al. 2004) and accommodates a possible phosphorylation site for protein kinase A. It remains to be clarified how the phosphorylation level at this location influences the sensibility of the CaV1.2 calcium channel (Bielefeldt 1999; Liao and Soong 2010). Ryanodine receptors represent a specific type of intracellular calcium channels coded by various genes in diverse tissues (Santulli and Marks 2015).

Within skeletal muscle, these are coded through the RyR1 gene, whereas the heart and smooth muscles express the RyR2 gene. The skipping of exon 70 (ASI), which produces RyR1-ASI(−), prevails in the embryonic skeletal muscle, whereas ASI(+) prevails in the adult skeletal muscle (Kimura et al. 2005). Interestingly, RyR1 ASI(−) represents a less active and less robustly opened channel in comparison with ASI(+) and exhibits a significantly lower frequency of Ca^{2+} oscillations, whereas the ASI(+) myotubes contain a higher amount of stationary Ca^{2+} (Kimura et al. 2005; Priori and Napolitano 2005). The alternative splicing of exon 83 creates ASII(−) and ASII(+) variants. The temporal shift of splicing events in ASII, however, comes much sooner (8th days after birth) than in ASI (16th days after birth) (Kimura et al. 2005; Futatsugi et al. 1995). These time shifts in tissue-specific splicing exemplify the range of functions (Futatsugi et al. 1995) and special nature of muscle-specific alternative splicing. This specificity can be demonstrated even more clearly through the alternative splicing of SERCA1 (sarco/endoplasmic reticulum Ca^{2+}-ATPase 1), where exon 22 gets spliced differentially and produces SERCA1a with enclosed exon 22, which predominates in the adult fast-twitch muscle (Thomas et al. 2017). Exon 22 is silenced in the neonatal fast-twitch muscle fiber to code for the SERCA1b splice variant. Moreover, SERCA1b is amply expressed in fast-twitch fibers in the pathogenesis of DM1. Although SERCA1a and SERCA1b have a comparable affinity for ATP and Ca^{2+}, SERCA1b has lower ATPase and Ca^{2+} management activity (Zhao et al. 2015). The splicing in the SERCA2 gene contains the information to generate cell type-specific splice variants (Kimura et al. 2005; Brandl et al. 1987; Hino et al. 2007). SERCA2a, in which the exon 21 is excluded, is expressed in the slowly twitching skeletal muscle fibers and heart muscles. SERCA2d, which is another skeletal muscle-specific isoform, maintains intron 19 (Kimura et al. 2005), and the expression of specific isoform is depressed in DM1 muscle tissue (Mittendorf et al. 2012). Although SERCA2a has been shown to govern the contractile calcium cycle, the different regulatory functions of SERCA2 splicing variants are ambiguous (Lipskaia et al. 2014). Overall, the aforementioned examples highlight the functional relevance of the divergent proteome resulting from tissue-specific alternative splicing. Despite this, there is still much to discover about the role of alternative splicing variants in maintaining healthy muscle performance and the molecular mechanisms underlying their production, in view of the degree to which alternative splicing processes have been discovered in the differentiation and maturation of muscles.

13.16.4 Alternative Splicing Causes Diseases Involving Muscle Disturbance

Considering the significance of splicing to guarantee the highly specific performance of the skeletal muscle, it is not astonishing that mutations which interfere with the selection of the alternative exons cause a large number of myopathies. In the various myopathies, the mechanism by which splicing is altered may be different, but it comprises firstly mutations in *cis*-regulating elements impairing the binding of *trans*-acting factors, secondly modify the expression of *trans*-acting key factors, or thirdly sequesters specific *trans*-acting factors to hinder their native role in muscle.

Of the myopathies caused by mutations in the *cis* regulatory elements, Pompe disease seems to be the most common, in which patients aggregate glycogen in their muscles, resulting in hypotension (van der Wal et al. 2017). This aggregation of glycogen occurs as a direct consequence of a T-to-G mutation (13T > G) within the poly-pyrimidine tract of intron 1 of α glucosidase (GAA) pre-mRNA. This mutation diminishes the binding of the U2 component of the basal splicing apparatus at the intron–exon interface, resulting in a fractional skipping of exon 2, which comprises the translation initiation codon, and a decreased level of cellular GAA (van der Wal et al. 2017). These lowered GAA concentrations lead to a toxic accumulation of glycogen inside the muscle lysosomes because the myofibers are not able to transform the complex sugar in their glucose energy resource.

Myopathies can also result from modified expression of *trans*-acting factors. In fact, it has been shown that reduced expression of splicing factor SRSF1 supports myositis by modifying the selection of alternative 5′ splice sites on the intended substrates (Xiong 2006). In specific, the pro-inflammatory cytokine TNF-α has been found to down-regulate the expression of SRSF1 throughout myositis (Xiong 2006), thereby helping to alter the splicing pattern of autoantigen transcripts, comprising PM/Scl-100 and PM/Scl-75 (Ng et al. 2004).

It has turned out that muscular dystrophies are more frequently linked to the sequestration of *trans*-acting factors, which prevents myofibers from setting up the full range of muscle-specific splicing incidents necessary for healthy tissue (Pistoni et al. 2010). Myotonic dystrophy type 1 (DM1) and congenital myotonic dystrophy (CDM) are induced due to enhanced CUG repeats in the transcripts of the DM protein kinase gene (DMPK1) (Buxton et al. 1992; Harley et al. 1992; Brook et al. 1992; Mankodi et al. 2000). These extended microsatellite repeats provide a stable RNA secondary structure that seeds the splicing regulator MBNL, that otherwise handles the alternative splicing of various pre-mRNAs such as cardiac troponin (cTnT) (Ryan and Cooper 1996), insulin receptor (INSR) (Moxley et al. 1978; Savkur et al. 2001) and muscle-specific chloride channel (CLCN1) (Thomas et al. 2017; Ebralidze et al. 2004; Philips et al. 1998; Charlet-B et al. 2002; Mankodi et al. 2002; Savkur et al. 2001). Disturbed splicing of CLCN1 results in myotonia, where defective muscle membrane-chloride conductivity leads to hyperexcitability of the myofibers, leading to repeated action potentials (Charlet-B et al. 2002). A broad investigation to gain an insight into CLCN1 splicing in DM1 skeletal muscle tissue has highlighted the presence of abnormally spliced CLCN1 mRNAs carrying premature stop codons, which occur through the retention of intron 2, the insertion of exon 6b and/or exon 7a, and the elimination of exons 6 and 7 (Charlet-B et al. 2002). For INSR splicing, DM1 is linked to the production of the insulin-resistant isoform IR-A in skeletal muscle cells, thereby mitigating the metabolic impact of insulin on these cells (Savkur et al. 2001; Furling et al. 1999).

In contrast to these disturbed splicing processes, the de novo sequestration of MBNL proteins leads to a change in alternative splicing, which is similar to embryonic alternative splicing schemes (Miller et al. 2000; Han et al. 2013). The aberrant splicing of the penultimate exon 78 of the DMD pre-mRNA exemplifies this (Rau et al. 2015). Sequestration or functional loss of MBNL1 triggers the skipping of exon 78. The loss of exon 78 codes for an embryonic variant of dystrophin that leads to abnormally banded muscle fibers is having an unorganized z-band (Rau et al. 2015). In another case, the abnormal alternative splicing of the muscle-specific exon 11 of the bridge integrator 1 (BIN1) produces an embryonic version of BIN1 (inactive) which is missing the phosphatidylinositol 5-phosphate-binding domain and produces disorganized T-tubule structures resulting in a disturbed calcium homeostasis which is essential for ECC in adult muscle fibers (Fugier et al. 2011). A switch in the alternative splicing scheme from adult to embryonic isoforms is also seen in a number of genes for vesicular trafficking, including the synaptosome-associated protein 23 (Snp23), the thyroid hormone receptor interactor 10 (Trip10) and the heavy chain of clathrin (Cltc), and transmembrane P24 trafficking protein 2 (Tmed2), which results in a marked disturbance of the T-tubule organization and a misallocation of the receptor proteins dihydropyridine receptor alpha (DHPR) and Ryr1, thereby disrupting the liberation of calcium, expanding myofibers and diminishing force generation (Giudice et al. 2016).

MBNL protein sequestration is monitored in another form of myotonic dystrophy, DM2, in which an extended CCUG RNA repeat is coded through intron 1 of zinc-finger protein 9 (ZNF9) (Liquori et al. 2001). It is a known fact that DM1 influences the (slow) muscle fibers of type I, whereas DM2 influences the (fast) muscle fibers of type II (Vihola et al. 2003; Cho and Tapscott 2007). Extended RNA repeats in DM have an indirect impact on supplementary RNA-binding proteins. The sequestration of MBNL proteins modifies the nuclear signal transduction events, which finally activate the PKC pathway through an unidentified mechanism. Activated PKC can cause a hyper-phosphorylation of CUGBP1, which is a further RNA-binding protein, to stabilize it and promote its transfer into the nucleus, where it governs the alternative splicing of numerous pre-mRNAs

(Kuyumcu-Martinez et al. 2007). MBNL proteins and CUGBP1 have been found to antagonistically regulate the alternative splicing of several pre-mRNAs (Yeo and Burge 2004). Transcriptome-wide-binding experiments showed that nuclear MBNL proteins can alternatively stimulate or suppress certain splicing processes. The binding of MBNL proteins located upstream of the exon represses, whereas the binding downstream of the alternative exon reinforces its insertion (Yeo and Burge 2004; Wang et al. 2012). In skeletal muscle, the MBNL and CUGBP1 proteins govern the alternative splicing of about 120 exons, including 78 exons that are antagonistically adjusted (Yeo and Burge 2004; Wang et al. 2012). Staufen1 (Stau1), an RNA-binding protein reported to control alternative splicing and translation effectiveness, is enhanced in DM1 patients and mouse models (Ravel-Chapuis et al. 2012). Stau1 can also save the alternative splicing of INSR exon 11 in DM1 because it attaches itself to aluminum elements in intron 10 and thus functions as a modification of the severity of DM1 disease (Ravel-Chapuis et al. 2012; Bondy-Chorney et al. 2016). Therefore, aberrant alternative splicing of several pre-mRNAs performs a pivotal role in the pathogenesis of DM by promoting myotonia, muscle weakness, muscle atrophy, resistance to insulin and impaired calcium ion channels seen in DM patients.

Facioscapulohumeral muscular dystrophy (FSHD), another common muscle disease, implies a modification of the copy number and/or an epigenetic variation of 3.3 kb of tandem repeat sequences referred to as D4Z4 (Wijmenga et al. 1992). In healthy individuals, the FSHD locus is retained in a suppressed condition, whereas the failure to silencing is experienced in patients, causing abnormal expression of several genes in proximity. This includes the D4Z4 coded double home box 4 (DUX4) transcription factor. The aberrant DUX4 activation in FSHD muscle interferes with the metabolism of RNA encompassing the splicing of pre-mRNA (Geng et al. 2012; Feng et al. 2015; Rickard et al. 2015; Jagannathan et al. 2016). DUX4 targets include gene 1 of the FSHD region (FRG1), which is co-purified with the C complex of the spliceosome and governs the alternative splicing of various transcripts comprising myotubularin-related protein 1 (MTMR1) and pre-mRNA of troponin T type 3 (TNNT 3) in FSHD (Wijmenga et al. 1992; van Koningsbruggen et al. 2007; Pistoni et al. 2013).

Muscular dystrophy may also be induced by incorrect splicing of constituents of the dystrophin glycoprotein complex (DGC) (Hoffman et al. 1987, 1988; Lapidos et al. 2004). DGC represents a dynamic multi-protein architecture that combines the cytoskeleton with the extracellular matrix and performs essential signaling and mechanical roles within the muscle (Lapidos et al. 2004). Moreover, the DGC constituents encompass extracellular laminin-2, dystroglycan including α extracellular and β transmembrane, sarcoglycan complex, such as α, β, γ, δ transmembrane glycoproteins, cytoplasmic dystrophin, dystrobrevins and syntrophins (Lapidos et al. 2004). The skeletal muscle cells produce α and γ sarcoglycans, whereas the expressions of other family glycoproteins are found to be widespread throughout all cell types (Lapidos et al. 2004; Hack et al. 1998). In rats, instead, alternative splicing of δ sarcoglycan results in a skeletal muscle-specific isoform termed δ-SG3, which lacks 122 aa in the C-terminus and has been proven to retain calcium homeostasis within the sarcoplasmic reticulum (Estrada et al. 2006).

Syntropins and dystrobrevin complex proteins are located at the cytoplasmic side of the DGC (Lapidos et al. 2004). More specifically, α-DB can carry out alternative splicing to produce five distinct isoforms, among which α-DB1 and α-DB2 are the prevailing ones and are expressed solely in the muscle cells (Nawrotzki et al. 1998; Newey et al. 2000). These splice variants vary in their C-terminus and have a distinct physical localization; for instance, α-DB2 is found in the sarcolemma of the neuromuscular junction and purifies together with dystrophin (Nawrotzki et al. 1998; Peters et al. 1998; Nakamori et al. 2008). On the opposite, α-DB1 connected to utrophin and dystrophin having a better attraction to utrophin and directed towards the synapse. These α-DBs are essential for muscle signal transduction and synaptogenesis in the neuromuscular system. Nevertheless, in DM1 patients, the splicing of α-DB1 is deregulated and leads to an increased incorporation of exons 11A and 12, which code for the variable region 3 (Nakamori et al. 2008). The insertion of the variable region 3 strengthens the association of α-DB1 with α-syntrophin and therefore results in modified phosphotyrosine signaling at the neuromuscular junction (Nakamori et al. 2008).

The aberrant splicing of a single exon, such as exon 43 and 45, or multiple exons, such as exons 46–50 and 45–54 in the dystrophin pre-mRNA of dystrophin results in the production of stop codons in exons 44, 51, or 55, and consequently in the generation of non-functional proteins (Malhotra et al. 1988). Moreover, the deletion of both the exons, such as 45 and 46 or exons 45 and 51, or even multiple exons ranging from exons 45–51 is able to restore the overall reading frame. This leads to the generation of a partially functional dystrophin protein, which has been reported to be linked to a milder version of the disease that is referred to as Becker muscular dystrophy (BMD) (Malhotra et al. 1988; Koenig et al. 1989).

It is interesting to note that a BMD phenotype has been identified in a patient with a deletion of exon 44 in the dystrophin gene which is supposed to cause DMD (Martone et al. 2016). The analysis of this patient revealed that the expression of the splicing factor CELF2a (CUGBP, Elav-like family member 2) in the muscle was decreased through an unidentified pathway. The absence of CELF2a functions as a pathological modifier by lowering the incorporation of exon 45 into the dystrophin mRNA, restoring the open reading frame, and leading to the generation of a semi-functional dystrophin protein (Martone et al. 2016).

The decreased myopathy due to the presence of exon skipping which was seen in the above mentioned DMD patient indicates that the regulation of the alternative splicing of the dystrophin pre-mRNA might have a potential therapeutic effect. In line with one such concept, the drug eteplirsen has been engineered to selectively trigger the skipping of exon 51 in the dystrophin mRNA, and it has lately been cleared for the treatment of DMD (Young and Pyle 2016). In the same way, there are experiments to evaluate antisense oligonucleotides (AON) to stimulate an alternative exon utilization for the therapy of many other muscle diseases. For instance, it was determined that an AON therapy directed against the 3′ splice site of exon 7a in the CLCN1 pre-mRNA successfully eradicates chloride channelopathy in DM1 mouse model systems through preventing the inclusion of exon 7a, thus recovering the complete full reading frame (Wheeler et al. 2007). With a more universal strategy, a 25-m CAG repeat morpholino oligos, which binds CUGexp, is administered into the muscle fibers of a DM1 mouse and can successfully eliminate the sequestered MBNL proteins out of the CUG repeat RNA and thereby reestablish the bulk of the MBNL/CELF-dependent splice processes, thereby overcoming the symptoms of myotonic dystrophy (Wheeler et al. 2009).

The achievements seen in the production of these extremely specific therapeutics underline the importance of gaining an insight into the mechanisms of the regulation of alternative splicing for the ongoing efforts to create a new type of treatment for myopathies.

References

F. Abascal, I. Ezkurdia, J. Rodriguez-Rivas, J.M. Rodriguez, A. del Pozo, J. Vázquez, A. Valencia, M.L. Tress, Alternatively spliced homologous exons have ancient origins and are highly expressed at the protein level. PLoS Comput. Biol. **11**, e1004325 (2015)

J.G. Abreu, N.I. Ketpura, B. Reversade, E.M. De Robertis, Connective-tissue growth factor (CTGF) modulates cell signalling by BMP and TGF-beta. Nat. Cell. Biol. **4**(8), 599–604 (2002)

D.E. Agafonov, J. Deckert, E. Wolf, P. Odenwälder, S. Bessonov, C.L. Will, H. Urlaub, R. Lührmann, Semiquantitative proteomic analysis of the human spliceosome via a novel two-dimensional gel electrophoresis method. Mol. Cell. Biol. **31**, 2667–2682 (2011)

T. Aigner, Y. Zhu, H.H. Chansky, F.A. Matsen 3rd, W.J. Maloney, L.J. Sandell, Reexpression of type IIA procollagen by adult articular chondrocytes in osteoarthritic cartilage. Arthritis Rheum. **42**(7), 1443–1450 (1999)

L. Ala-Kokko, A.P. Kvist, M. Metsaranta, K.I. Kivirikko, B. de Crombrugghe, D.J. Prockop, E. Vuorio, Conservation of the sizes of 53 introns and over 100 intronic sequences for the binding of common transcription factors in the human and mouse genes for type II procollagen (COL2A1). Biochem. J. **308**(Pt 3), 923–929 (1995)

P. An, P.J. Grabowski, Exon silencing by UAGG motifs in response to neuronal excitation. PLoS Biol. **5**(2), e36 (2007)

O. Anczukow, M. Akerman, A. Clery et al., SRSF1-regulated alternative splicing in breast cancer. Mol. Cell **60**, 105–117 (2015)

E. Ars, E. Serra, J. Garcia, H. Kruyer, A. Gaona, C. Lazaro, X. Estivill, Mutations affecting mRNA splicing are the most common molecular defects in patients with neurofibromatosis type 1. Hum. Mol. Genet. **9**, 237–247 (2000)

R. Ashwal-Fluss, M. Meyer, N.R. Pamudurti, A. Ivanov, O. Bartok, M. Hanan, N. Evantal, S. Memczak, N. Rajewsky, S. Kadener, circRNA biogenesis competes with pre-mRNA splicing. Mol. Cell **56**, 55–66 (2014)

S. Astrof, D. Crowley, E.L. George, T. Fukuda, K. Sekiguchi, D. Hanahan, R.O. Hynes, Direct test of potential roles of EIIIA and EIIIB alternatively spliced segments of fibronectin in physiological and tumor angiogenesis. Mol. Cell. Biol. **24**(19), 8662–8670 (2004)

A. Aszodi, D. Chan, E. Hunziker, J.F. Bateman, R. Fassler, Collagen II is essential for the removal of the notochord and the formation of intervertebral discs. J. Cell Biol. **143**(5), 1399–1412 (1998)

S.D. Auweter, R. Fasan, L. Reymond, J.G. Underwood, D.L. Black, S. Pitsch, F.H.-T. Allain, Molecular basis of RNA recognition by the human alternative splicing factor Fox-1. EMBO J. **25**, 163–173 (2006)

M.E. Avale, T. Rodriguez-Martin, J.M. Gallo, Trans-splicing correction of tau isoform imbalance in a mouse model of tau mis-splicing. Hum. Mol. Genet. **22**, 2603–2611 (2013)

A.J. Bannister, R. Schneider, T. Kouzarides, Histone methylation: dynamic or static? Cell **109**(7), 801–806 (2002)

N.L. Barbosa-Morais, M. Irimia, Q. Pan et al., The evolutionary landscape of alternative splicing in vertebrate species. Science **338**, 1587–1593 (2012)

E.S. Barrie, R.M. Smith, J.C. Sanford, W. Sadee, mRNA transcript diversity creates new opportunities for pharmacological intervention. Mol. Pharmacol. **81**(5), 620–630 (2012)

B. Bartkowiak, P. Liu, H.P. Phatnani, N.J. Fuda, J.J. Cooper, D.H. Price, K. Adelman, J.T. Lis, A.L. Greenleaf, CDK12 is a transcription elongation-associated CTD kinase, the metazoan ortholog of yeast Ctk1. Genes Dev. **24**, 2303–2316 (2010)

E. Batsché, M. Yaniv, C. Muchardt, The human SWI/SNF subunit Brm is a regulator of alternative splicing. Nat. Struct. Mol. Biol. **13**, 22–29 (2006)

E.G. Bechara, E. Sebestyen, I. Bernardis, E. Eyras, J. Valcarcel, RBM5, 6, and 10 differentially regulate NUMB alternative splicing to control cancer cell proliferation. Mol. Cell **52**, 720–733 (2013)

M.V. Bell, A.E. Cowper, M.-P. Lefranc, J.I. Bell, G.R. Screaton, Influence of intron length on alternative splicing of CD44. Mol. Cell. Biol. **18**, 5930–5941 (1998)

D.L. Bentley, Rules of engagement: co-transcriptional recruitment of pre-mRNA processing factors. Curr. Opin. Cell Biol. **17**, 251–256 (2005)

P. Bertone, V. Stolc, T.E. Royce et al., Global identification of human transcribed sequences with genome tiling arrays. Science **306**, 2242–2246 (2004)

A.L. Beyer, A.H. Bouton, O.L. Miller Jr., Correlation of hnRNP structure and nascent transcript cleavage. Cell **26**, 155–165 (1981)

K. Bhalla, H.A. Phillips, J. Crawford, O.L.D. McKenzie, J.C. Mulley, H. Eyre, A.E. Gardner, G. Kremmidiotis, D.F. Callen, The de novo chromosome 16 translocations of two patients with abnormal phenotypes (mental retardation and epilepsy) disrupt the A2BP1 gene. J. Hum. Genet. **49**, 308–311 (2004)

K. Bielefeldt, Molecular diversity of voltage-sensitive calcium channels in smooth muscle cells. J. Lab. Clin. Med. **133**, 469–477 (1999)

E. Birney, J.A. Stamatoyannopoulos, A. Dutta et al., Identification and analysis of functional elements in 1% of the human genome by the ENCODE pilot project Nature **447**, 799–816 (2007)

P.N. Bishop, A.J. Reardon, D. McLeod, S. Ayad, Identification of alternatively spliced variants of type II procollagen in vitreous. Biochem. Biophys. Res. Commun. **203**(1), 289–295 (1994)

D.L. Black, Mechanisms of alternative pre-messenger RNA splicing. Annu. Rev. Biochem. **72**, 291–336 (2003)

M. Blanchette, B. Chabot, A highly stable duplex structure sequesters the 5′ splice site region of hnRNP A1 alternative exon 7B. RNA **3**(4), 405–419 (1997)

M. Blanchette, E. Labourier, R.E. Green, S.E. Brenner, D.C. Rio, Genome-wide analysis reveals an unexpected function for the *Drosophila* splicing factor U2AF50 in the nuclear export of intronless mRNAs. Mol. Cell **14**, 775–786 (2004)

M. Blanchette, R.E. Green, S.E. Brenner, D.C. Rio, Global analysis of positive and negative pre-MRNA splicing regulators in *Drosophila*. Genes Dev. **19**, 1306–1314 (2005)

U.K. Blaschke, E.F. Eikenberry, D.J. Hulmes, H.J. Galla, P. Bruckner, Collagen XI nucleates selfassembly and limits lateral growth of cartilage fibrils. J. Biol. Chem. **275**(14), 10370–10378 (2000)

B.J. Blencowe, Alternative splicing: new insights from global analyses. Cell **126**, 37–47 (2006)

A.J. Blythe, A.H. Fox, C.S. Bond, The ins and outs of lncRNA structure: how, why and what comes next? Biochim. Biophys. Acta **1859**(1), 46–58 (2016)

C. Boesler, N. Rigo, M.M. Anokhina, M.J. Tauchert, D.E. Agafonov, B. Kastner, H. Urlaub, R. Ficner, C.L. Will, R. Lührmann, A spliceosome intermediate with loosely associated tri-snRNP accumulates in the absence of Prp28 ATPase activity. Nat. Commun. **7**, 11997 (2016)

E. Bondy-Chorney, T.E. Crawford Parks, A. Ravel-Chapuis, R. Klinck, L. Rocheleau, M. Pelchat, B. Chabot, B. J. Jasmin, J. Côté, Staufen1 regulates multiple alternative splicing events either positively or negatively in DM1 indicating its role as a disease modifier. PLoS Genet. **12**, e1005827 (2016)

F. Bordeleau, L. Galarneau, S. Gilbert, A. Loranger, N. Marceau, Keratin 8/18 modulation of protein kinase C-mediated integrin-dependent adhesion and migration of liver epithelial cells. Mol. Biol. Cell **21**(10), 1698–1713 (2010)

F. Bordeleau, T.A. Alcoser, C.A. Reinhart-King, Physical biology in cancer. 5. The rocky road of metastasis: the role of cytoskeletal mechanics in cell migratory response to 3D matrix topography. Am. J. Physiol. Cell Physiol. **306**(2), C110–C120 (2014)

F. Bordeleau, J.P. Califano, Y.L. Negrón Abril, B.N. Mason, D.J. LaValley, S.J. Shin, R.S. Weiss, C.A. Reinhart-King, Tissue stiffness regulates serine/arginine-rich protein-mediated splicing of the extra domain B-fibronectin isoform in tumors. Proc. Natl. Acad. Sci. U. S. A. **112**(27), 8314–8319 (2015)

P. Bornstein, The NH(2)-terminal propeptides of fibrillar collagens: highly conserved domains with poorly understood functions. Matrix Biol. **21**(3), 217–226 (2002)

P. Bornstein, V. Walsh, J. Tullis, E. Stainbrook, J.F. Bateman, S.G. Hormuzdi, The globular domain of the proalpha 1 (I) N-propeptide is not required for secretion, processing by procollagen N-proteinase, or fibrillogenesis of type I collagen in mice. J. Biol. Chem. **277**(4), 2605–2613 (2002)

P.L. Boutz, P. Stoilov, Q. Li, C.-H. Lin, G. Chawla, K. Ostrow, L. Shiue, M. Ares Jr., D.L. Black, Post-transcriptional regulatory switch in polypyrimidine tract-binding proteins reprograms alternative splicing in developing neurons. Genes Dev. **21**, 1636–1652 (2007)

C.J. Brandl, S. deLeon, D.R. Martin, D.H. MacLennan, Adult forms of the Ca2+ATPase of sarcoplasmic reticulum. Expression in developing skeletal muscle. J. Biol. Chem. **262**, 3768–4374 (1987)

R.E. Breitbart, H.T. Nguyen, R.M. Medford, A.T. Destree, V. Mahdavi, B. Nadal-Ginard, Intricate combinatorial patterns of exon splicing generate multiple regulated troponin T isoforms from a single gene. Cell **41**, 67–82 (1985)

A.-L. Brillen, K. Schöneweis, L. Walotka et al., Succession of splicing regulatory elements determines cryptic 5′ss functionality. Nucleic Acids Res. **45**, 4202–4216 (2017)

B.M. Brinkman, Splice variants as cancer biomarkers. Clin. Biochem. **37**(7), 584–594 (2004)

A. Brock, H. Chang, S. Huang, Non-genetic heterogeneity—a mutation-independent driving force for the somatic evolution of tumours. Nat. Rev. Genet. **10**, 336–342 (2009)

S. Brogna, J. Wen, Nonsense-mediated mRNA decay (NMD) mechanisms. Nat. Struct. Mol. Biol. **16**(2), 107–113 (2009)

J.D. Brook, M.E. McCurrach, H.G. Harley et al., Molecular basis of myotonic dystrophy: expansion of a trinucleotide (CTG) repeat at the 3′ end of a transcript encoding a protein kinase family member. Cell **68**, 799–808 (1992)

M. Brudno, M.S. Gelfand, S. Spengler, M. Zorn, I. Dubchak, J.G. Conboy, Computational analysis of candidate intron regulatory elements for tissue-specific alternative pre-MRNA splicing. Nucleic Acids Res. **29**, 2338–2348 (2001)

M. Buljan, G. Chalancon, S. Eustermann, G.P. Wagner, M. Fuxreiter, A. Bateman, M.M. Babu, Tissue-specific splicing of disordered segments that embed binding motifs rewires protein interaction networks. Mol. Cell **46**, 871–883 (2012)

T. Burckin, R. Nagel, Y. Mandel-Gutfreund et al., Exploring functional relationships between components of the gene expression machinery. Nat. Struct. Mol. Biol. **12**, 175–182 (2005)

C.B. Burge, T. Tuschl, P.A. Sharp, Splicing of precursors to mRNAs by the spliceosomes, in *The RNA World*, 2nd edn., ed. by R. Gesteland, T. Cech, J. Atkins (Cold Spring Harbor Press, Cold Spring Harbor, 1999), pp. 525–560

R.E. Burgeson, D.W. Hollister, Collagen heterogeneity in human cartilage: identification of several new collagen chains. Biochem. Biophys. Res. Commun. **87**(4), 1124–1131 (1979)

J. Buxton, P. Shelbourne, J. Davies et al., Detection of an unstable fragment of DNA specific to individuals with myotonic dystrophy. Nature **355**, 547–548 (1992)

C. elegans Sequencing Consortium, Genome sequence of the nematode *C. elegans*: a platform for investigating biology. Science **282**, 2012–2018 (1998)

M.N. Cabili, C. Trapnell, L. Goff, M. Koziol, B. Tazon-Vega, A. Regev, J.L. Rinn, Integrative annotation of human large intergenic noncoding RNAs reveals global properties and specific subclasses. Genes Dev. **25**, 1915–1927 (2011)

M.N. Cabili, M.C. Dunagin, P.D. McClanahan, A. Biaesch, O. Padovan-Merhar, A. Regev, J.L. Rinn, A. Raj, Localization and abundance analysis of human lncRNAs at single-cell and single-molecule resolution. Genome Biol. **16**, 20 (2015)

J.A. Calarco, Y. Xing, M. Caceres, J.P. Calarco, X. Xiao, Q. Pan, C. Lee, T.M. Preuss, B.J. Blencowe, Global analysis of alternative splicing differences between humans and chimpanzees. Genes Dev. **21**, 2963–2975 (2007)

J.C. Calderón, P. Bolaños, C. Caputo, The excitation–contraction coupling mechanism in skeletal muscle. Biophys. Rev. **6**, 133–160 (2014)

E.G. Canty, K.E. Kadler, Procollagen trafficking, processing and fibrillogenesis. J. Cell Sci. **118**(Pt 7), 1341–1353 (2005)

E. Carro, J.L. Trejo, A. Nunez, I. Torres-Aleman, Brain repair and neuroprotection by serum insulin-like growth factor I. Mol. Neurobiol. **27**(2), 153–162 (2003)

L. Cartegni, S.L. Chew, A.R. Krainer, Listening to silence and understanding nonsense: exonic mutations that affect splicing. Nat. Rev. Genet. **3**, 285–298 (2002)

L. Cartegni, J. Wang, Z. Zhu, M.Q. Zhang, A.R. Krainer, ESEfinder: a web resource to identify exonic splicing enhancers. Nucleic Acids Res. **31**(13), 3568–3571 (2003)

P. Castellani, G. Viale, A. Dorcaratto, G. Nicolo, J. Kaczmarek, G. Querze, L. Zardi, The fibronectin isoform containing the ED-B oncofetal domain: a marker of angiogenesis. Int. J. Cancer **59**(5), 612–618 (1994)

P. Castellani, L. Borsi, B. Carnemolla, A. Birò, A. Dorcaratto, G.L. Viale, D. Dario Neri, L. Zardi, Differentiation between high- and low-grade astrocytoma using a human recombinant antibody to the extra domain-B of fibronectin. Am. J. Pathol. **161**(5), 1695–1700 (2002)

B. Chabot, L. Shkreta, Defective control of pre-messenger RNA splicing in human disease. J. Cell Biol. **212**, 13–27 (2016)

N. Charlet-B, R.S. Savkur, G. Singh, A.V. Philips, E.A. Grice, T.A. Cooper, Loss of the muscle-specific chloride channel in type 1 myotonic dystrophy due to misregulated alternative splicing. Mol. Cell **10**, 45–53 (2002)

K.S. Cheah, N.G. Stoker, J.R. Griffin, F.G. Grosveld, E. Solomon, Identification and characterization of the human type II collagen gene (COL2A1). Proc. Natl. Acad. Sci. U. S. A. **82**(9), 2555–2559 (1985)

L. Chen, The biogenesis and emerging roles of circular RNAs. Nat. Rev. Mol. Cell Biol. **17**, 205–211 (2016)

H.S. Chiu, S. Somvanshi, E. Patel et al., Pan-cancer analysis of lncRNA regulation supports their targeting of cancer genes in each tumor context. Cell Rep. **23**(1), 297–312, e212 (2018)

D.H. Cho, S.J. Tapscott, Myotonic dystrophy: emerging mechanisms for DM1 and DM2. Biochim. Biophys. Acta (BBA) Mol. Basis Dis. **1772**, 195–204 (2007)

G. Claps, Y. Cheli, T. Zhang et al., A transcriptionally inactive ATF2 variant drives melanomagenesis. Cell Rep. **15**, 1884–1892 (2016)

T.A. Clark, C.W. Sugnet, M. Ares Jr., Genomewide analysis of mRNA processing in yeast using splicing-specific microarrays. Science **296**, 907–910 (2002)

T.A. Clark, A.C. Schweitzer, T.X. Chen, M.K. Staples, G. Lu, H. Wang, A. Williams, J.E. Blume, Discovery of tissue-specific exons using comprehensive human exon microarrays. Genome Biol. **8**, R64 (2007)

H. Climente-Gonzalez, E. Porta-Pardo, A. Godzik, E. Eyras, The functional impact of alternative splicing in cancer. Cell Rep. **20**, 2215–2226 (2017)

D. Coltrini, R. Ronca, M. Belleri, L. Zardi, S. Indraccolo, V. Scarlato, R. Giavazzi, M. Presta, Impact of VEGF-dependent tumour micro-environment on EDB fibronectin expression by subcutaneous human tumour xenografts in nude mice. J. Pathol. **219**(4), 455–462 (2009)

S.J. Conn, K.A. Pillman, J. Toubia, V.M. Conn, M. Salmanidis, C.A. Phillips, S. Roslan, A.W. Schreiber, P.A. Gregory, G.J. Goodall, The RNA binding protein quaking regulates formation of circRNAs. Cell **160**, 1125–1134 (2015)

D.R. Cooper, G. Carter, P. Li, R. Patel, J.E. Watson, N.A. Patel, Long non-coding RNA NEAT1 associates with SRp40 to temporally regulate PPARγ2 splicing during adipogenesis in 3T3-L1 cells. Genes **5**(4), 1050–1063 (2014)

O. Cordin, J.D. Beggs, RNA helicases in splicing. RNA Biol. **10**, 83–95 (2013)

R. Corominas, X. Yang, G.N. Lin et al., Protein interaction network of alternatively spliced isoforms from brain links genetic risk factors for autism. Nat. Commun. **5**, 3650 (2014)

A. Corrionero, B. Minana, J. Valcarcel, Reduced fidelity of branch point recognition and alternative splicing induced by the anti-tumor drug spliceostatin A. Genes Dev. **25**(5), 445–459 (2011)

F.F. Costa, Non-coding RNAs: new players in eukaryotic biology. Gene **357**(2), 83–94 (2005)

M. D'Antonio, D. Michalovich, M. Paterson, A. Droggiti, A. Woodhoo, R. Mirsky, K.R. Jessen, Gene profiling and bioinformatic analysis of Schwann cell embryonic development and myelination. Glia **53**(5), 501–515 (2006)

E. Daguenet, G. Dujardin, J. Valcárcel, The pathogenicity of splicing defects: mechanistic insights into pre-mRNA processing inform novel therapeutic approaches. EMBO Rep. **16**, 1640–1655 (2015)

R.M. Dale, J. Topczewski, Identification of an evolutionarily conserved regulatory element of the zebrafish col2a1a gene. Dev. Biol. **357**(2), 518–531 (2011)

J.P. Dales, N. Beaufils, M. Silvy et al., Hypoxia inducible factor 1alpha gene (HIF-1alpha) splice variants: potential prognostic biomarkers in breast cancer. BMC Med. **8**, 44 (2010)

R.B. Darman, M. Seiler, A.A. Agrawal et al., Cancer-associated SF3B1 hotspot mutations induce cryptic 30 splice site selection through use of a different branch point. Cell Rep. **13**, 1033–1045 (2015)

D. Das, T.A. Clark, A. Schweitzer et al., A correlation with exon expression approach to identify cis-regulatory elements for tissue-specific alternative splicing. Nucleic Acids Res. **35**, 4845–4857 (2007)

C.J. David, J.L. Manley, The search for alternative splicing regulators: new approaches offer a path to a splicing code. Genes Dev. **22**, 279–285 (2008)
C.J. David, J.L. Manley, Alternative pre-mRNA splicing regulation in cancer: pathways and programs unhinged. Genes Dev. **24**(21), 2343–2364 (2010)
I. De, S. Bessonov, R. Hofele, K. dos Santos, C.L. Will, H. Urlaub, R. Lührmann, V. Pena, The RNA helicase Aquarius exhibits structural adaptations mediating its recruitment to spliceosomes. Nat. Struct. Mol. Biol. **22**, 138–144 (2015)
L. De Conti, M. Baralle, E. Buratti, Exon and intron definition in pre-mRNA splicing. Wiley Interdiscip. Rev. RNA **4**(1), 49–60 (2013)
M. de la Mata, C.R. Alonso, S. Kadener, J.P. Fededa, M. Blaustein, F. Pelisch, P. Cramer, D. Bentley, A.R. Kornblihtt, A slow RNA polymerase II affects alternative splicing in vivo. Mol. Cell **12**(2), 525–532 (2003)
L.V. den Bosch, J. Eggermont, H.D. Smedt, L. Mertens, F. Wuytack, R. Casteels, Regulation of splicing is responsible for the expression of the muscle-specific 2a isoform of the sarco/endoplasmic-reticulum Ca2+-ATPase. Biochem. J. **302**, 559–566 (1994)
F. Denoeud, P. Kapranov, C. Ucla et al., Prominent use of distal 5′ transcription start sites and discovery of a large number of additional exons in ENCODE regions. Genome Res. **17**, 746–759 (2007)
T. Derrien, R. Johnson, G. Bussotti et al., The GENCODE v7 catalog of human long noncoding RNAs: analysis of their gene structure, evolution, and expression. Genome Res. **22**, 1775–1789 (2012)
C.N. Dewey, I.B. Rogozin, E.V. Koonin, Compensatory relationship between splice sites and exonic splicing signals depending on the length of vertebrate introns. BMC Genomics **7**, 311 (2006)
D.C. Di Giammartino, K. Nishida, J.L. Manley, Mechanisms and consequences of alternative polyadenylation. Mol. Cell **43**, 853–866 (2011)
S. Djebali, C.A. Davis, A. Merkel et al., Landscape of transcription in human cells. Nature **489**, 101–108 (2012)
J. Dluzniewska, A. Sarnowska, M. Beresewicz, I. Johnson, S.K.S. Srai, B. Ramesh, G. Goldspink, D.C. Górecki, B. Zabłocka, A strong neuroprotective effect of the autonomous C-terminal peptide of IGF-1 Ec (MGF) in brain ischemia. FASEB J. **19**(13), 1896–1898 (2005)
L.A. Donoso, A.O. Edwards, A.T. Frost, R. Ritter 3rd, N. Ahmad, T. Vrabec, J. Rogers, D. Meyer, S. Parma, Clinical variability of Stickler syndrome: role of exon 2 of the collagen COL2A1 gene. Surv. Ophthalmol. **48**(2), 191–203 (2003)
K. Du, Y. Peng, L.E. Greenbaum, B.A. Haber, R. Taub, HRS/SRp40-mediated inclusion of the fibronectin EIIIB exon, a possible cause of increased EIIIB expression in proliferating liver. Mol. Cell. Biol. **17**(7), 4096–4104 (1997)
F. Du, G.M. Acland, J. Ray, Differential splicing of type II procollagen mRNA in canine retina. Anim. Biotechnol. **9**(1), 15–20 (1998)
C.C. DuFort, M.J. Paszek, V.M. Weaver, Balancing forces: architectural control of mechanotransduction. Nat. Rev. Mol. Cell Biol. **12**(5), 308–319 (2011)
G. Dujardin, C. Lafaille, E. Petrillo et al., Transcriptional elongation and alternative splicing. Biochim. Biophys. Acta **1829**, 134–140 (2013)
I. Duran, M. Mari-Beffa, J.A. Santamaria, J. Becerra, L. Santos-Ruiz, Actinotrichia collagens and their role in fin formation. Dev. Biol. **354**(1), 160–172 (2011)
A. Dziembowski, A.P. Ventura, B. Rutz, F. Caspary, C. Faux, F. Halgand, O. Laprévote, B. Séraphin, Proteomic analysis identifies a new complex required for nuclear pre-mRNA retention and splicing. EMBO J. **23**, 4847–4856 (2004)
S. Ebashi, Excitation-contraction coupling and the mechanism of muscle contraction. Annu. Rev. Physiol. **53**, 1–17 (1991)
A. Ebralidze, Y. Wang, V. Petkova, K. Ebralidse, R.P. Junghans, RNA leaching of transcription factors disrupts transcription in myotonic dystrophy. Science **303**, 383–387 (2004)
K.A. Effenberger, R.J. Perriman, W.M. Bray, R.S. Lokey, M. Ares Jr., M.S. Jurica, A high-throughput splicing assay identifies new classes of inhibitors of human and yeast spliceosomes. J. Biomol. Screen. **18**, 1110–1120 (2013)
J. Eid, A. Fehr, J. Gray et al., Real-time DNA sequencing from single polymerase molecules. Science **323**, 133–138 (2009)
J.D. Ellis, M. Barrios-Rodiles, R. Colak et al., Tissue-specific alternative splicing remodels protein-protein interaction networks. Mol. Cell **46**, 884–892 (2012)
J.M. Engreitz, J.E. Haines, E.M. Perez, G. Munson, J. Chen, M. Kane, P.E. McDonel, M. Guttman, E.S. Lander, Local regulation of gene expression by lncRNA promoters, transcription and splicing. Nature **539**, 452–455 (2016)
M. Esteller, Non-coding RNAs in human disease. Nat. Rev. Genet. **12**(12), 861–874 (2011)
F.J. Estrada, D. Mornet, H. Rosas-Vargas, A. Angulo, M. Hernández, V. Becker, V. Becker, A. Rendón, M. Ramos-Kuri, R.M. Coral-Vázquez, A novel isoform of δ-sarcoglycan is localized at the sarcoplasmic reticulum of mouse skeletal muscle. Biochem. Biophys. Res. Commun. **340**, 865–871 (2006)
D.R. Eyre, Type XI or 1a2a3a collagen, in *Structure and Function of Collagen Types*. ed. by R. Mayne, R.E. Burgeson (Academic Press, New York, 1987), pp. 261–281
D. Eyre, Collagen of articular cartilage. Arthritis Res. **4**(1), 30–35 (2002)

D.R. Eyre, M.A. Weis, J.J. Wu, Articular cartilage collagen: an irreplaceable framework? Eur. Cells Mater. **12**, 57–63 (2006)

I. Ezkurdia, A. del Pozo, A. Frankish, J.M. Rodriguez, J. Harrow, K. Ashman, A. Valencia, M.L. Tress, Comparative proteomics reveals a significant bias toward alternative protein isoforms with conserved structure and function. Mol. Biol. Evol. **29**, 2265–2283 (2012)

I. Ezkurdia, Most highly expressed protein-coding genes have a single dominant isoform. J. Proteome Res. **14**, 1880–1887 (2015)

P. Fabrizio, J. Dannenberg, P. Dube, B. Kastner, H. Stark, H. Urlaub, R. Lührmann, The evolutionarily conserved core design of the catalytic activation step of the yeast spliceosome. Mol. Cell **36**, 593–608 (2009)

M. Fagnani, Y. Barash, J.Y. Ip et al., Functional coordination of alternative splicing in the mammalian central nervous system. Genome Biol. **8**, R108 (2007)

W.G. Fairbrother, R.-F. Yeh, P.A. Sharp, C.B. Burge, Predictive identification of exonic splicing enhancers in human genes. Science **297**, 1007–1013 (2002)

C.S. Farah, F.C. Reinach, The troponin complex and regulation of muscle contraction. FASEB J. **9**, 755–767 (1995)

A. Fatica, I. Bozzoni, Long non-coding RNAs: new players in cell differentiation and development. Nat. Rev. Genet. **15** (1), 7–21 (2014)

C. Faure, M.T. Linossier, L. Malaval, M.H. Lafage-Proust, S. Peyroche, L. Vico, A. Guignandon, Mechanical signals modulated vascular endothelial growth factor-A (VEGF-A) alternative splicing in osteoblastic cells through actin polymerisation. Bone **42**(6), 1092–1101 (2008)

Q. Feng, L. Snider, S. Jagannathan, R. Tawil, S.M. van der Maarel, S.J. Tapscott, R.K. Bradley, A feedback loop between nonsense-mediated decay and the retrogene DUX4 in facioscapulohumeral muscular dystrophy. Elife **4**, e04996 (2015)

S.M. Fica, N. Tuttle, T. Novak, N.S. Li, J. Lu, P. Koodathingal, Q. Dai, J.P. Staley, J.A. Piccirilli, RNA catalyses nuclear pre-MRNA splicing. Nature **503**, 229–234 (2013)

S.M. Fica, M.A. Mefford, J.A. Piccirilli, J.P. Staley, Evidence for a group II intron-like catalytic triplex in the spliceosome. Nat. Struct. Mol. Biol. **21**, 464–471 (2014)

M. Fichter, R. Hinrichs, G. Eissner, B. Scheffer, S. Classen, M. Ueffing, Expression of CD44 isoforms in neuroblastoma cells is regulated by PI 3-kinase and protein kinase C. Oncogene **14**(23), 2817–2824 (1997)

R. Fleischmajer, B.R. Olsen, R. Timpl, J.S. Perlish, O. Lovelace, Collagen fibril formation during embryogenesis. Proc. Natl. Acad. Sci. U. S. A. **80**(11), 3354–3358 (1983)

K.L. Fox-Walsh, Y. Dou, B.J. Lam, S.-P. Hung, P.F. Baldi, K.J. Hertel, The architecture of pre-mRNAs affects mechanisms of splice-site pairing. Proc. Natl. Acad. Sci. U. S. A. **102**, 16176–16181 (2005)

G.M. Frampton, S.M. Ali, M. Rosenzweig et al., Activation of MET via diverse exon 14 splicing alterations occurs in multiple tumor types and confers clinical sensitivity to MET inhibitors. Cancer Discov. **5**, 850–859 (2015)

V. Franke, S. Ganesh, R. Karlic et al., Long terminal repeats power evolution of genes and gene expression programs in mammalian oocytes and zygotes. Genome Res. **27**, 1384–1394 (2017)

P.J. French, J. Peeters, S. Horsman et al., Identification of differentially regulated splice variants and novel exons in glial brain tumors using exon expression arrays. Cancer Res. **67**, 5635–5642 (2007)

D.A. Fruman, R.E. Meyers, L.C. Cantley, Phosphoinositide kinases. Annu. Rev. Biochem. **67**, 481–507 (1998)

X.-D. Fu, M. Ares Jr., Context-dependent control of alternative splicing by RNA-binding proteins. Nat. Rev. Genet. **15**, 689–701 (2014)

G. Fu, K.C. Condon, M.J. Epton, K.C. Condon, M.J. Epton, P. Gong, L. Jin, G.C. Condon, N.I. Morrison, T.H. Dafaalla, L. Alphey, Female-specific insect lethality engineered using alternative splicing. Nat. Biotechnol. **25**, 353–357 (2007)

C. Fugier, A.F. Klein, C. Hammer et al. Misregulated alternative splicing of BIN1 is associated with T tubule alterations and muscle weakness in myotonic dystrophy. Nat. Med. **17**, 720–725 (2011)

T. Fujisawa, Y. Huang, W. Sebald, J.L. Zhang, The binding of von Willebrand factor type C domains of Chordin family proteins to BMP-2 and Tsg is mediated by their SD1 subdomain. Biochem. Biophys. Res. Commun. **385**(2), 215–219 (2009)

N. Fukui, A. McAlinden, Y. Zhu, E. Crouch, T.J. Broekelmann, R.P. Mecham, L.J. Sandell, Processing of type II procollagen amino propeptide by matrix metalloproteinases. J. Biol. Chem. **277**(3), 2193–2201 (2002)

N. Fukuno, H. Matsui, Y. Kanda, O. Suzuki, K. Matsumoto, K. Sasaki, T. Kobayashi, S. Tamura, TGF-b-activated kinase 1 mediates mechanical stress-induced IL-6 expression in osteoblasts. Biochem. Biophys. Res. Commun. **408** (2), 202–207 (2011)

D. Furling, A. Marette, J. Puymirat, Insulin-like growth factor I circumvents defective insulin action in human myotonic dystrophy skeletal muscle cells. Endocrinology **140**, 4244–4250 (1999)

D.K. Furuto, E.J. Miller, Different levels of glycosylation contribute to the heterogeneity of alpha 1(II) collagen chains derived from a transplantable rat chondrosarcoma. Arch. Biochem. Biophys. **226**(2), 604–611 (1983)

A. Futatsugi, G. Kuwajima, K. Mikoshiba, Tissue-specific and developmentally regulated alternative splicing in mouse skeletal muscle ryanodine receptor mRNA. Biochem. J. **305**, 373–378 (1995)

P.A. Galante, N.J. Sakabe, N. Kirschbaum-Slager, S.J. de Souza, Detection and evaluation of intron retention events in the human transcriptome. RNA **10**, 757–765 (2004)
M. Gama-Carvalho, N.L. Barbosa-Morais, A.S. Brodsky, P.A. Silver, M. Carmo-Fonseca, Genome-wide identification of functionally distinct subsets of cellular mRNAs associated with two nucleocytoplasmic-shuttling mammalian splicing factors. Genome Biol. **7**, R113 (2006)
P.J. Gardina, T.A. Clark, B. Shimada et al., Alternative splicing and differential gene expression in colon cancer detected by whole genome exon array. BMC Genomics **7**, 325 (2006)
G. Gateva, E. Kremneva, T. Reindl, T. Kotila, K. Kogan, L. Gressin, P.W. Gunning, D.J. Manstein, A. Michelot, P. Lappalainen, Tropomyosin isoforms specify functionally distinct actin filament populations in vitro. Curr. Biol. **27**, 705–713 (2017)
Y. Ge, B.T. Porse, The functional consequences of intron retention: alternative splicing coupled to NMD as a regulator of gene expression. BioEssays **36**, 236–243 (2014)
L.N. Geng, Z. Yao, L. Snider et al., DUX4 activates germline genes, retroelements, and immune mediators: implications for facioscapulohumeral dystrophy. Dev. Cell **22**, 38–51 (2012)
C. Ghigna, C. Valacca, G. Biamonti, Alternative splicing and tumor progression. Curr. Genomics **9**(8), 556–570 (2008)
G. Ghosh, J.A. Adams, Phosphorylation mechanism and structure of serine-arginine protein kinases. FEBS J. **278**, 587–597 (2011)
E.A. Gibb, C.J. Brown, W.L. Lam, The functional role of long non-coding RNA in human carcinomas. Mol. Cancer **10**, 38 (2011)
W. Gilbert, Why genes in pieces? Nature **271**, 501 (1978)
P.G. Gillespie, R.G. Walker, Molecular basis of mechanosensory transduction. Nature **413**(6852), 194–202 (2001)
J. Giudice, J.A. Loehr, G.G. Rodney, T.A. Cooper, Alternative splicing of four trafficking genes regulates myofiber structure and skeletal muscle physiology. Cell Rep. **17**, 1923–1933 (2016)
G. Goldspink, Changes in muscle mass and phenotype and the expression of autocrine and systemic growth factors by muscle in response to stretch and overload. J. Anat. **194**(Pt 3), 323–334 (1999)
G. Goldspink, Research on mechano growth factor: its potential for optimising physical training as well as misuse in doping. Br. J. Sports Med. **39**(11), 787–788 (2005)
G. Goldspink, P. Goldspink, Use of the insulin-like-growth factor I splice variant MGF for the prevention of myocardial damage. United. States. Patent. US20050048028A1 (2005)
L.D. Goldstein, J. Lee, F. Gnad et al., Recurrent loss of NFE2L2 exon 2 is a mechanism for Nrf2 pathway activation in human cancers. Cell Rep. **16**, 2605–2617 (2016)
A.C. Goldstrohm, A.L. Greenleaf, M.A. Garcia-Blanco, Co-transcriptional splicing of pre-messenger RNAs: considerations for the mechanism of alternative splicing. Gene **277**, 31–47 (2001)
L.I. Gómez Acuña, A. Fiszbein, M. Alló, I.E. Schor, A.R. Kornblihtt, Connections between chromatin signatures and splicing. Wiley Interdiscip. Rev. RNA **4**, 77–91 (2013)
V. Goncalves, P. Matos, P. Jordan, Antagonistic SR proteins regulate alternative splicing of tumor-related Rac1b downstream of the PI3-kinase and Wnt pathways. Hum. Mol. Genet. **18**, 3696–3707 (2009)
M. Gonzàlez-Porta, A. Frankish, J. Rung, J. Harrow, A. Brazma, Transcriptome analysis of human tissues and cell lines reveals one dominant transcript per gene. Genome Biol. **14**, R70 (2013)
O. Gozani, R. Feld, R. Reed, Evidence that sequence-independent binding of highly conserved U2 snRNP proteins upstream of the branch site is required for assembly of spliceosomal complex A. Genes Dev. **10**, 233–243 (1996)
I.R. Graham, M. Hamshere, I.C. Eperon, Alternative splicing of a human alpha-tropomyosin muscle-specific exon: identification of determining sequences. Mol. Cell. Biol. **12**, 3872–3882 (1992)
B.R. Graveley, Sorting out the complexity of SR protein functions. RNA **6**, 1197–1211 (2000)
R.E. Green, B.P. Lewis, R.T. Hillman, M. Blanchette, L.F. Lareau, A.T. Garnett, D.C. Rio, S.E. Brenner, Widespread predicted nonsense-mediated mRNA decay of alternatively-spliced transcripts of human normal and disease genes. Bioinformatics **19**(Suppl. 1), i118–i121 (2003)
A. Guignandon, Y. Usson, N. Laroche, M.H. Lafage-Proust, O. Sabido, C. Alexandre, L. Vico, Effects of intermittent or continuous gravitational stresses on cell–matrix adhesion: quantitative analysis of focal contacts in osteoblastic ROS 17/2.8 cells. Exp. Cell Res. **236**(1), 66–75 (1997)
W. Guo, G.J. Mulligan, S. Wormsley, D.M. Helfman, Alternative splicing of beta-tropomyosin pre-mRNA: cis-acting elements and cellular factors that block the use of a skeletal muscle exon in nonmuscle cells. Genes Dev. **5**, 2096–2107 (1991)
R.A. Gupta, N. Shah, K.C. Wang et al., Long non-coding RNA HOTAIR reprograms chromatin state to promote cancer metastasis. Nature **464**(7291), 1071–1076 (2010)
M. Guttman, P. Russell, N.T. Ingolia, J.S. Weissman, E.S. Lander, Ribosome profiling provides evidence that large noncoding RNAs do not encode proteins. Cell **154**, 240–251 (2013)
A.A. Hack, C.T. Ly, F. Jiang, C.J. Clendenin, K.S. Sigrist, R.L. Wollmann, E.M. McNally, γ-Sarcoglycan deficiency leads to muscle membrane defects and apoptosis independent of dystrophin. J. Cell Biol. **142**, 1279–1287 (1998)

F. Han, J.R. Gilbert, G. Harrison et al., Transforming growth factor-beta1 regulates fibronectin isoform expression and splicing factor SRp40 expression during ATDC5 chondrogenic maturation. Exp. Cell Res. **313**(8), 1518–1532 (2007)

S.P. Han, K.S. Kassahn, A. Skarshewski, M.A. Ragan, J.A. Rothnagel, R. Smith, Functional implications of the emergence of alternative splicing in hnRNP A/B transcripts. RNA **16**, 1760–1768 (2010)

H. Han, M. Irimia, P.J. Ross, H.-K. Sung et al., MBNL proteins repress ES-cell-specific alternative splicing and reprogramming. Nature **498**, 241–245 (2013)

D. Hanahan, R.A. Weinberg, Hallmarks of cancer: the next generation. Cell **144**(5), 646–674 (2011)

J. Hang, R. Wan, C. Yan, Y. Shi, Structural basis of pre-MRNA splicing. Science **349**, 1191–1198 (2015)

K.D. Hansen, L.F. Lareau, M. Blanchette et al., Genome-wide identification of alternative splice forms down-regulated by nonsense-mediated mRNA decay in *Drosophila*. PLoS Genet. **5**, e1000525 (2009)

H.G. Harley, J.D. Brook, S.A. Rundle, S. Crow, W. Reardon, A.J. Buckler, P.S. Harper, D.E. Housman, D.J. Shaw, Expansion of an unstable DNA region and phenotypic variation in myotonic dystrophy. Nature **355**, 545–546 (1992)

M. Hashimoto-Uoshima, Y.Z. Yan, G. Schneider, I. Aukhil, The alternatively spliced domains EIIIB and EIIIA of human fibronectin affect cell adhesion and spreading. J. Cell Sci. **110**(Pt 18), 2271–2280 (1997)

H. Hatano, G. Sarkar, M.E. Bolander, Development of a cellular model to study alternative splicing of type II collagen gene. J. Orthop. Res. **20**(3), 516–519 (2002)

S. Hauptmann, A. Siegert, S. Berger, C. Denkert, M. Köbel, S. Ott, A. Siri, L. Borsi, Extracellular matrix proteins in colorectal carcinomas. Expression of tenascin and fibronectin isoforms. Lab. Invest. **73**(2), 172–182 (1995)

S. Hayashi, Z. Wang, J. Bryan, C. Kobayashi, R. Faccio, L.J. Sandell, The type II collagen N-propeptide, PIIBNP, inhibits cell survival and bone resorption of osteoclasts via integrin-mediated signaling. Bone **49**(4), 644–652 (2011)

J. Hayes, P.P. Peruzzi, S. Lawler, MicroRNAs in cancer: biomarkers, functions and therapy. Trends Mol. Med. **20**(8), 460–469 (2014)

R.-Z. He, D.-X. Luo, Y.-Y. Mo, Emerging roles of lncRNAs in the post-transcriptional regulation in cancer. Genes Dis. **6**(1), 6–15 (2019)

K.M. Heinemeier, J.L. Olesen, P. Schjerling, F. Haddad, H. Langberg, K.M. Baldwin, M. Kjaer, Short-term strength training and the expression of myostatin and IGF-I isoforms in rat muscle and tendon: differential effects of specific contraction types. J. Appl. Physiol. **102**(2), 573–581 (2007)

D.M. Helfman, W.M. Ricci, L.A. Finn, Alternative splicing of tropomyosin pre-mRNAs in vitro and in vivo. Genes Dev. **2**, 1627–1638 (1988)

R.H. Herai, M.E. Yamagishi, Detection of human interchromosomal trans-splicing in sequence databanks. Brief. Bioinform. **11**, 198–209 (2010)

T.M. Hering, L. Wirthlin, S. Ravindran, A. McAlinden, Changes in type II procollagen isoform expression during chondrogenesis by disruption of an alternative 5′ splice site within Col2a1 exon 2. Matrix Biol. **36**, 51–63 (2014)

M. Hill, G. Goldspink, Expression and splicing of the insulin-like growth factor gene in rodent muscle is associated with muscle satellite (stem) cell activation following local tissue damage. J. Physiol. **549**(pt2), 409–418 (2003)

S. Hino, S. Kondo, H. Sekiya et al., Molecular mechanisms responsible for aberrant splicing of SERCA1 in myotonic dystrophy type 1. Hum. Mol. Genet. **16**, 2834–2843 (2007)

E.P. Hoffman, R.H. Brown, L.M. Kunkel, Dystrophin: the protein product of the Duchenne muscular dystrophy locus. Cell **51**, 919–928 (1987)

E.P. Hoffman, K.H. Fischbeck, R.H. Brown et al., Characterization of dystrophin in muscle-biopsy specimens from patients with Duchenne's or Becker's muscular dystrophy. N. Engl. J. Med. **318**, 1363–1368 (1988)

S.Y. Hong, Y.M. Jeon, H.J. Lee, J.G. Kim, J.A. Baek, J.C. Lee, Activation of RhoA and FAK induces ERK-mediated osteopontin expression in mechanical force-subjected periodontal ligament fibroblasts. Mol. Cell. Biochem. **335**(1–2), 263–272 (2010)

T. Horiuchi, T. Aigaki, Alternative trans-splicing: a novel mode of pre-mRNA processing. Biol. Cell **98**, 135–140 (2006)

D.S. Horowitz, The mechanism of the second step of pre-mRNA splicing. Wiley Interdiscip. Rev. RNA **3**, 331–350 (2012)

A.A. Hoskins, M.J. Moore, The spliceosome: a flexible, reversible macromolecular machine. Trends Biochem. Sci. **37**, 179–188 (2012)

K.A. Houck, N. Ferrara, J. Winer, G. Cachianes, B. Li, D.W. Leung, The vascular endothelial growth factor family: identification of a fourth molecular species and characterization of alternative splicing of RNA. Mol. Endocrinol. **5**(12), 1806–1814 (1991)

G. Housman, S. Byler, S. Heerboth, K. Lapinska, M. Longacre, N. Snyder, S. Sarkar, Drug resistance in cancer: an overview. Cancers (Basel) **6**, 1769–1792 (2014)

K.J. Howe, C.M. Kane, M. Ares Jr., Perturbation of transcription elongation influences the fidelity of internal exon inclusion in *Saccharomyces cerevisiae*. RNA **9**(8), 993–1006 (2003)

H.C. Hsia, J.E. Schwarzbauer, Meet the tenascins: multifunctional and mysterious. J. Biol. Chem. **280**(29), 26641–26644 (2005)

G.S. Huh, R.O. Hynes, Regulation of alternative pre-mRNA splicing by a novel repeated hexanucleotide element. Genes Dev. **8**, 1561–1574 (1994)

J. Hui, L.H. Hung, M. Heiner, S. Schreiner, N. Neumuller, G. Reither, S.A. Haas, A. Bindereif, Intronic CA-repeat and CA-rich elements: a new class of regulators in mammalian alternative splicing. EMBO J. **24**, 1988–1998 (2005)

Human Genome Sequencing Consortium I and International Human Genome Sequencing C, Finishing the euchromatic sequence of the human genome. Nature **431**, 931–945 (2004)

K.J. Hurt, S.F. Sezen, H.C. Champion et al., Alternatively spliced neuronal nitric oxide synthase mediates penile erection. Proc. Natl. Acad. Sci. U. S. A. **103**, 3440–3443 (2006)

J. Huynh, F. Bordeleau, C.M. Kraning-Rush, C.A. Reinhart-King, Substrate stiffness regulates PDGF-induced circular dorsal ruffle formation through MLCK. Cell. Mol. Bioeng. **6**(2), 138–147 (2013)

C. Iannone, J. Valcárcel, Chromatin's thread to alternative splicing regulation. Chromosoma **122**, 465–474 (2013)

N.T. Ingolia, Ribosome footprint profiling of translation throughout the genome. Cell **165**, 22–33 (2016)

D. Inoue, S. Kido, T. Matsumoto, Transcriptional induction of FosB/FosB gene by mechanical stress in osteoblasts. J. Biol. Chem. **279**(48), 49795–49803 (2004)

International Human Genome Sequencing Consortium, Initial sequencing and analysis of the human genome. Nature **409**, 860–921 (2001)

J.Y. Ip, A. Tong, Q. Pan, J.D. Topp, B.J. Blencowe, K.W. Lynch, Global analysis of alternative splicing during T-cell activation. RNA **13**, 563–572 (2007)

M. Irimia, B.J. Blencowe, Alternative splicing: decoding an expansive regulatory layer. Curr. Opin. Cell Biol. **24**(3), 323–332 (2012)

M. Irimia, D. Penny, S.W. Roy, Coevolution of genomic intron number and splice sites. Trends Genet. **23**, 321–325 (2007)

S. Jagannathan, S.C. Shadle, R. Resnick, L. Snider, R.N. Tawil, V.D. Maarel, R.K. Bradley, S.J. Tapscott, Model systems of DUX4 expression recapitulate the transcriptional profile of FSHD cells. Hum. Mol. Genet. **25**, 4419–4431 (2016)

S. Jakkaraju, X. Zhe, D. Pan, R. Choudhury, L. Schuger, TIPs are tension-responsive proteins involved in myogenic versus adipogenic differentiation. Dev. Cell **9**(1), 39–49 (2005)

M. Janco, T.T. Bonello, A. Byun, A.C.F. Coster, H. Lebhar, I. Dedova, P.W. Gunning, T. Böcking, The impact of tropomyosins on actin filament assembly is isoform specific. BioArchitecture **6**, 61–75 (2016)

C. Jimenez, R.A. Portela, M. Mellado, J.R. Rodriguez-Frade, J. Collard, A. Serrano, C. Martínez-A, J. Avila, A.C. Carrera, Role of the PI3K regulatory subunit in the control of actin organization and cell migration. J. Cell. Biol. **151** (2), 249–262 (2000)

Y. Jin, H. Suzuki, S. Maegawa, H. Endo, S. Sugano, K. Hashimoto, K. Yasuda, K. Inoue, A vertebrate RNA-binding protein Fox-1 regulates tissue-specific splicing via the pentanucleotide GCAUG. EMBO J. **22**, 905–912 (2003)

Y. Jin, Y. Yang, P. Zhang, New insights into RNA secondary structure in the alternative splicing of pre-mRNAs. RNA Biol. **8**, 450–457 (2011)

J.M. Johnson, J. Castle, P. Garrent-Engle, Z. Kan, P.M. Loerch, C.D. Armour, R. Santos, E.E. Schadt, R. Stoughton, D. D. Shoemaker, Genome-wide survey of human alternative pre-mRNA splicing with exon junction microarrays. Science **302**, 2141–2144 (2003)

B. Johnstone, M. Alini, M. Cucchiarini et al., Tissue engineering for articular cartilage repair—the state of the art. Eur. Cells Mater. **25**, 248–267 (2013)

F.S. Jones, P.L. Jones, The tenascin family of ECM glycoproteins: structure, function, and regulation during embryonic development and tissue remodeling. Dev. Dyn. **218**(2), 235–259 (2000)

K. Juneau, C. Palm, M. Miranda, R.W. Davis, High-density yeast-tiling array reveals previously undiscovered introns and extensive regulation of meiotic splicing. Proc. Natl. Acad. Sci. U. S. A. **104**, 1522–1527 (2007)

H.S. Jung, S.W. Lee, Ribozyme-mediated selective killing of cancer cells expressing carcinoembryonic antigen RNA by targeted trans-splicing. Biochem. Biophys. Res. Commun. **349**, 556–563 (2006)

H. Jung, D. Lee, J. Lee, D. Park, Y.J. Kim, W.-Y. Park, D. Hong, P.J. Park, E. Lee, Intron retention is a widespread mechanism of tumor-suppressor inactivation. Nat. Genet. **47**, 1242–1248 (2015)

M.S. Jurica, M.J. Moore, Pre-mRNA splicing: awash in a sea of proteins. Mol. Cell **12**, 5–14 (2003)

S. Kadener, P. Cramer, G. Nogues, D. Cazalla, M. de la Mata, J.P. Fededa, S.E. Werbajh, A. Srebrow, A.R. Kornblihtt, Antagonistic effects of T-Ag and VP16 reveal a role for RNA pol II elongation on alternative splicing. EMBO J. **20** (20), 5759–5768 (2001)

S. Kadener, J.P. Fededa, M. Rosbash, A.R. Kornblihtt, Regulation of alternative splicing by a transcriptional enhancer through RNA pol II elongation. Proc. Natl. Acad. Sci. U. S. A. **99**(12), 8185–8190 (2002)

Z. Kalnina, P. Zayakin, K. Silina, A. Linē, Alterations of pre-mRNA splicing in cancer. Genes Chromosom. Cancer **42**, 342–357 (2005)

A. Kalsotra, T.A. Cooper, Functional consequences of developmentally regulated alternative splicing. Nat. Rev. Genet. **12**, 715–729 (2011)

M. Kalyna, C.G. Simpson, N.H. Syed et al., Alternative splicing and nonsense-mediated decay modulate expression of important regulatory genes in Arabidopsis. Nucleic Acids Res. **40**, 2454–2469 (2012)

P. Kapranov, J. Cheng, S. Dike et al., RNA maps reveal new RNA classes and a possible function for pervasive transcription. Science **316**, 1484–1488 (2007a)

P. Kapranov, A.T. Willingham, T.R. Gingeras, Genome-wide transcription and the implications for genomic organization. Nat. Rev. Genet. **8**(6), 413–423 (2007b)

K. Kapur, Y. Xing, Z. Ouyang, W.H. Wong, Exon arrays provide accurate assessments of gene expression. Genome Biol. **8**, R82 (2007)

R. Karni, E. de Stanchina, S.W. Lowe, R. Sinha, D. Mu, A.R. Krainer, The gene encoding the splicing factor SF2/ASF is a proto-oncogene. Nat. Struct. Mol. Biol. **14**, 185–193 (2007)

B. Kastner, C.L. Will, H. Stark, R. Lührmann, Structural insights into nuclear pre-mRNA splicing in higher eukaryotes. Cold Spring Harb. Perspect. Biol. (2019). https://doi.org/10.1101/cshperspect.a032417

S. Kawamoto, Neuron-specific alternative splicing of non- muscle myosin XE "Myosin" II heavy chain-B pre-mRNA requires a cis-acting intron sequence. J. Biol. Chem. **271**, 17613–17616 (1996)

B. Kaynak, A. von Heydebreck, S. Mebus et al., Genome-wide array analysis of normal and malformed human hearts. Circulation **107**, 2467–2474 (2003)

J.D. Keene, RNA regulons: coordination of post-transcriptional events. Nat. Rev. Genet. **8**, 533–543 (2007)

D.R. Keene, J.T. Oxford, N.P. Morris, Ultrastructural localization of collagen types II, IX, and XI in the growth plate of human rib and fetal bovine epiphyseal cartilage: type XI collagen is restricted to thin fibrils. J. Histochem. Cytochem. **43**(10), 967–979 (1995)

J.D. Keene, J.M. Komisarow, M.B. Friedersdorf, RIP-Chip: the isolation and identification of mRNAs, microRNAs, and protein components of ribonucleoprotein complexes from cell extracts. Nat. Protoc. **1**, 302–307 (2006)

O. Kelemen, P. Convertini, Z. Zhang, Y. Wen, M. Shen, M. Falaleeva, S. Stamm, Function of alternative splicing. Gene **514**, 1–30 (2013)

K.E. Keller, M.J. Kelley, T.S. Acott, Extracellular matrix gene alternative splicing by trabecular meshwork cells in response to mechanical stretching. Invest. Ophthalmol. Vis. Sci. **48**(3), 1164–1172 (2007)

S. Kelly, C. Greenman, P.R. Cook, A. Papantonis, Exon skipping is correlated with exon circularization. J. Mol. Biol. **427**, 2414–2417 (2015)

P.R. Kemp, J.C. Metcalfe, Four isoforms of serum response factor that increase or inhibit smooth-muscle-specific promoter activity Biochem. J. **345**(Pt 3), 445–451 (2000)

Z.A. Khan, B.M. Chan, S. Uniyal, Y.P. Barbin, H. Farhangkhoee, S. Chen, S. Chakrabart, EDB fibronectin and angiogenesis—a novel mechanistic pathway. Angiogenesis **8**(3), 183–196 (2005)

T.-R. Kiehl, H. Shibata, T. Vo, D.P. Huynh, S.-M. Pulst, Identification and expression of a mouse ortholog of A2BP1. Mamm. Genome **12**, 595–601 (2001)

Y.J. Kim, H.S. Kim, Alternative splicing and its impact as a cancer diagnostic marker. Genomics Inform. **10**, 74–80 (2012)

J.B. Kim, G.J. Porrca, L. Song, S.C. Greenway, J.M. Gorham, G.M. Church, C.E. Seidman, J.G. Seidman, Polony multiplex analysis of gene expression (PMAGE) in mouse hypertrophic cardiomyopathy. Science **316**, 1481–1484 (2007a)

E. Kim, A. Magen, G. Ast, Different levels of alternative splicing among eukaryotes. Nucleic Acids Res. **35**, 125–131 (2007b)

E. Kim, A. Goren, G. Ast, Alternative splicing: current perspectives. BioEssays **30**, 38–47 (2008)

H. Kim, B. Erickson, W. Luo, D. Seward, J.H. Graber, D.D. Pollock, P.C. Megee, D.L. Bentley, Gene-specific RNA polymerase II phosphorylation and the CTD code. Nat. Struct. Mol. Biol. **17**, 1279–1286 (2010)

J.H. Kim, A.R. Asthagiri, Matrix stiffening sensitizes epithelial cells to EGF and enables the loss of contact inhibition of proliferation. J. Cell Sci. **124**(Pt 8), 1280–1287 (2011)

J. Kim, H.L. Piao, B.J. Kim et al., Long noncoding RNA MALAT1 suppresses breast cancer metastasis. Nat. Genet. **50** (12), 1705–1715 (2018)

M. Kimura, K. Obara, T. Sasase, T. Ishikawa, Y. Tanabe, K. Nakayama, Specific inhibition of stretch-induced increase in L-type calcium channel currents by herbimycin A in canine basilar arterial myocytes. Br. J. Pharmacol. **130**(4), 923–931 (2000)

T. Kimura, M. Nakamori, J.D. Lueck, P. Pouliquin, F. Aoike, H. Fujimura, R.T. Dirksen, M.P. Takahashi, A.F. Dulhunty, S. Sakoda, Altered mRNA splicing of the skeletal muscle ryanodine receptor and sarcoplasmic/endoplasmic reticulum Ca2+-ATPase in myotonic dystrophy type 1. Hum. Mol. Genet. **14**, 2189–200 (2005)

T. Kimura, J.D. Lueck, P.J. Harvey, S.M. Pace, N. Ikemoto, M.G. Casarotto, R.T. Dirksen, A.F. Dulhunty, Alternative splicing of RyR1 alters the efficacy of skeletal EC coupling. Cell Calcium **45**, 264–274 (2009)

F.-J. Klinz, D. Gallwitz, Size and position of intervening sequences are critical for the splicing efficiency of pre-mRNA in the yeast *Saccharomyces cerevisiae*. Nucleic Acids Res. **13**, 3791–3804 (1985)

M. Koenig, A.H. Beggs, M. Moyer et al., The molecular basis for Duchenne versus Becker muscular dystrophy: correlation of severity with type of deletion. Am. J. Hum. Genet. **45**, 498–506 (1989)

P. Koirala, D.H. Zou, M. Yin-Yuan, Long non-coding RNAs as key regulators of cancer metastasis. J. Cancer Metastasis Treat. **2**(1), 10 (2016)

D. Kondratev, M.F. Gallitelli, Increments in the concentration of sodium and calcium in cell compartments of stretched mouse ventricular myocytes. Cell Calcium **34**(2), 193–203 (2003)

A.R. Kornblihtt, M. de la Mata, J.P. Fededa, M. Munoz, G. Nogues, Multiple links ribozyme for cancer treatment. Methods Mol. Biol. **1103**, 83–95 (2004)

A.R. Kornblihtt, I.E. Schor, M. Allo, G. Dujardin, E. Petrillo, M.J. Munoz, Alternative splicing: a pivotal step between eukaryotic transcription and translation. Nat. Rev. Mol. Cell Biol. **14**(3), 153–165 (2013)

A.E. Kornienko, P.M. Guenzl, D.P. Barlow, F.M. Pauler, Gene regulation by the act of long non-coding RNA transcription. BMC Biol. **11**, 59 (2013)

Y. Kotake, T. Nakagawa, K. Kitagawa, S. Suzuki, N. Liu, M. Kitagawa, Y. Xiong, Long non-coding RNA ANRIL is required for the PRC2 recruitment to and silencing of p15(INK4B) tumor suppressor gene. Oncogene **30**(16), 1956–1962 (2011)

J. Královicová, M.B. Christensen, I. Vorechovský, Biased exon/intron distribution of cryptic and de novo 3′ splice sites. Nucleic Acids Res. **33**, 4882–4898 (2005)

I.V. Kravchenko, V.A. Furalyov, E.S. Lisitsina, V.O. Popov, Stimulation of mechano-growth factor expression by second messengers. Arch. Biochem. Biophys. **507**(2), 323–331 (2011)

Z. Krchňáková, P. Kumar Thakur, M. Krausová, N. Bieberstein, N. Haberman, M. Müller-McNicoll, D. Staněk, Splicing of long non-coding RNAs primarily depends on polypyrimidine tract and 5′ splice-site sequences due to weak interactions with SR proteins. Nucleic Acids Res. **47**(2), 911–928 (2019)

N.M. Kuyumcu-Martinez, G.-S. Wang, T.A. Cooper, Increased steady-state levels of CUGBP1 in myotonic dystrophy 1 are due to PKC-mediated hyperphosphorylation. Mol. Cell **28**, 68–78 (2007)

J. Lagarde, B. Uszczynska-Ratajczak, S. Carbonell et al., High-throughput annotation of full-length long noncoding RNAs with capture long-read sequencing. Nat. Genet. **49**, 1731 (2017)

K.A. Lapidos, R. Kakkar, E.M. McNally, The dystrophin glycoprotein complex: signaling strength and integrity for the sarcolemma. Circ. Res. **94**, 1023–1031 (2004)

A.V. Lapuk, S.V. Volik, Y. Wang, C.C. Collins, The role of mRNA splicing in prostate cancer. Asian J. Androl. **16**, 515–521 (2014)

L.F. Lareau, M. Inada, R.E. Green, J.C. Wengrod, S.E. Brenner, Unproductive splicing of SR genes associated with highly conserved and ultraconserved DNA elements. Nature **446**, 926–929 (2007a)

L.F. Lareau, A.N. Brooks, D.A. Soergel, Q. Meng, S.E. Brenner, The coupling of alternative splicing and nonsense-mediated mRNA decay. Adv. Exp. Med. Biol. **623**, 190–211 (2007b)

J. Larrain, D. Bachiller, B. Lu, E. Agius, S. Piccolo, E.M. De Robertis, BMP-binding modules in chordin: a model for signalling regulation in the extracellular space. Development **127**(4), 821–830 (2000)

C. Larsson, Protein kinase C and the regulation of the actin cytoskeleton. Cell. Signal. **18**(3), 276–284 (2006)

C.M. Laumont, K. Vincent, L. Hesnard et al., Noncoding regions are the main source of targetable tumor-specific antigens. Sci. Transl. Med. **10**(470) (2018)

S.C.-W. Lee, O. Abdel-Wahab, Therapeutic targeting of splicing in cancer. Nat. Med. **22**, 976–986 (2016)

S.W. Lee, J.S. Jeong, Use of tumor-targeting trans-splicing ribozyme for cancer treatment. Methods Mol. Biol. **1103**, 83–95 (2014)

Y. Lee, D.C. Rio, Mechanisms and regulation of alternative splicing. Annu. Rev. Biochem. **84**, 291–323 (2015)

J.A. Lee, Y. Xing, D. Nguyen, J. Xie, C.J. Lee, D.L. Black, Depolarization and CaM kinase IV modulate NMDA receptor splicing through two essential RNA elements. PLoS Biol. **5**(2), e40 (2007)

J.L. Leight, M.A. Wozniak, S. Chen, M.L. Lynch, C.S. Chen, Matrix rigidity regulates a switch between TGF-β1-induced apoptosis and epithelial-mesenchymal transition. Mol. Biol. Cell **23**(5), 781–791 (2012)

A.W. Leung, S.Y. Wong, D. Chan, P.P. Tam, K.S. Cheah, Loss of procollagen IIA from the anterior mesendoderm disrupts the development of mouse embryonic forebrain. Dev. Dyn. **239**(9), 2319–2329 (2010)

K.R. Levental, H. Yu, L. Kass et al., Matrix crosslinking forces tumor progression by enhancing integrin signaling. Cell **139**(5), 891–906 (2009)

B.P. Lewis, R.E. Green, S.E. Brenner, Evidence for the wide-spread coupling of alternative splicing and nonsense-mediated mRNA decay in humans. Proc. Natl. Acad. Sci. U. S. A. **100**, 189–192 (2003)

R. Lewis, S. Ravindran, L. Wirthlin, G. Traeger, R.J. Fernandes, A. McAlinden, Disruption of the developmentally-regulated Col2a1 pre-mRNA alternative splicing switch in a transgenic knock-in mouse model. Matrix Biol. **31**(3), 214–226 (2012)

S.W. Li, D.J. Prockop, H. Helminen et al., Transgenic mice with targeted inactivation of the Col2 alpha 1 gene for collagen II develop a skeleton with membranous and periosteal bone but no endochondral bone. Genes Dev. **9**(22), 2821–2830 (1995)

Y.S. Li, J.H. Haga, S. Chien, Molecular basis of the effects of shear stress on vascular endothelial cells. J. Biomech. **38** (10), 1949–1971 (2005)

C. Li, M. Kato, L. Shiue, J.E. Shively, M. Ares, R.-J. Lin Jr., Cell type and culture-dependent alternative splicing in human breast cancer cells revealed by splicing-sensitive microarrays. Cancer Res. **66**, 1990–1999 (2006a)

H.R. Li, J. Wang-Rodriguez, T.M. Nair et al., Two-dimensional transcriptome profiling: identification of messenger RNA isoform signatures in prostate cancer from archived paraffin-embedded cancer specimens. Cancer Res. **66**, 4079–4088 (2006b)

Q. Li, J.-A. Lee, D.L. Black, Neuronal regulation of alternative pre-mRNA splicing. Nat. Rev. Neurosci. **8**, 819–831 (2007)

P. Liao, T.W. Soong, Understanding alternative splicing of Cav1.2 calcium channels for a new approach towards individualized medicine. J. Biomed. Res. **24**, 181–186 (2010)

P. Liao, D. Yu, S. Lu, Z. Tang, M.C. Liang, S. Zeng, W. Lin, T.W. Soong, Smooth muscle-selective alternatively spliced exon generates functional variation in Cav1.2 calcium channels. J. Biol. Chem. **279**, 50329–50335 (2004)

P. Liao, D. Yu, G. Li, T.F. Yong, J.L. Soon, Y.L. Chua, T.W. Soong, A smooth muscle Cav1.2 calcium channel splice variant underlies hyperpolarized window current and enhanced state-dependent inhibition by nifedipine. J. Biol. Chem. **282**, 35133–3542 (2007)

D. Libri, A. Piseri, M.Y. Fiszman, Tissue-specific splicing in vivo of the beta-tropomyosin gene: dependence on an RNA secondary structure. Science **252**(5014), 1842–1845 (1991)

L.P. Lim, P.A. Sharp, Alternative splicing of the fibronectin EIIIB exon depends on specific TGCATG repeats. Mol. Cell. Biol. **18**(7), 3900–3906 (1998)

J.-C. Lin, W.-Y. Tarn, Exon selection in α-tropomyosin mRNA is regulated by the antagonistic action of RBM4 and PTB. Mol. Cell. Biol. **25**, 10111–10121 (2005)

S. Lin, X.D. Fu, SR proteins and related factors in alternative splicing. Adv. Exp. Med. Biol. **623**, 107–122 (2007)

J.-C. Lin, W.-Y. Tarn, RBM4 down-regulates PTB and antagonizes its activity in muscle cell–specific alternative splicing. J. Cell Biol. **193**, 509–520 (2011)

L. Lipskaia, Z. Keuylian, K. Blirando et al., Expression of sarco (endo) plasmic reticulum calcium ATPase (SERCA) system in normal mouse cardiovascular tissues, heart failure and atherosclerosis. Biochim. Biophys. Acta BBA Mol. Cell Res. **1843**, 2705–2718 (2014)

C.L. Liquori, K. Ricker, M.L. Moseley, J.F. Jacobsen, W. Kress, S.L. Naylor, J.W. Day, L.P. Ranum, Myotonic dystrophy type 2 caused by a CCTG expansion in intron 1 of ZNF9. Science **293**, 864–867 (2001)

H. Liu, L. Tang, Mechano-regulation of alternative splicing. Curr. Genomics **14**(1), 49–55 (2013)

W. Liu, Y. Zhou, Z. Hu, T. Sun, A. Denise, X.D. Fu, Y. Zhang, Regulation of splicing enhancer activities by RNA secondary structures. FEBS Lett. **584**, 4401–4407 (2010)

G. Liu, A. Razanau, Y. Hai, J. Yu, M. Sohail, V.G. Lobo, J. Chu, S.K. Kung, J. Xie, A conserved serine of heterogeneous nuclear ribonucleoprotein L (hnRNP L) mediates depolarization-regulated alternative splicing of potassium channels. J. Biol. Chem. **287**, 22709–22716 (2012a)

H. Liu, L. He, L. Tang, Alternative splicing regulation and cell lineage differentiation. Curr. Stem Cell Res. Ther. **7**(6), 400–406 (2012b)

Q. Liu, J. Huang, N. Zhou, Z. Zhang, A. Zhang, Z. Lu, F. Wu, Y.Y. Mo, LncRNA loc285194 is a p53-regulated tumor suppressor. Nucleic Acids Res. **41**(9), 4976–4987 (2013)

J.C. Long, J.F. Caceres, The SR protein family of splicing factors: master regulators of gene expression. Biochem. J. **417**, 15–27 (2009)

A.J. Lopez, Alternative splicing of pre-mRNA: developmental consequences and mechanisms of regulation. Annu. Rev. Genet. **32**, 279–305 (1998)

A. Louloupi, E. Ntini, T. Conrad, U.A.V. Ørom, Transient N-6-methyladenosine transcriptome sequencing reveals a regulatory role of m6A in splicing efficiency. Cell Rep. **23**, 3429–3437 (2018)

Z.X. Lu, Q. Huang, J.W. Park, S. Shen, L. Lin, C.J. Tokheim, M.D. Henry, Y. Xing, Transcriptome-wide landscape of pre-mRNA alternative splicing associated with metastatic colonization. Mol. Cancer Res. **13**, 305–318 (2015)

R.F. Luco, T. Misteli, More than a splicing code: integrating the role of RNA, chromatin and non-coding RNA in alternative splicing regulation. Curr. Opin. Genet. Dev. **21**, 366–372 (2011)

R.F. Luco, M. Allo, I.E. Schor, A.R. Kornblihtt, T. Misteli, Epigenetics in alternative pre-MRNA splicing. Cell **144**, 16–26 (2011)

V.C. Lui, L.J. Ng, J. Nicholls, P.P. Tam, K.S. Cheah, Tissue-specific and differential expression of alternatively spliced alpha 1(II) collagen mRNAs in early human embryos. Dev. Dyn. **203**(2), 198–211 (1995)

K.W. Lynch, A. Weiss, A model system for activation-induced alternative splicing of CD45 pre-mRNA in T cells implicates protein kinase C and Ras. Mol. Cell. Biol. **20**(1), 70–80 (2000)

X. Ma, J. Li-Ling, Q. Huang, X. Chen, L. Hou, F. Ma, Systematic analysis of alternative promoters correlated with alternative splicing in human genes. Genomics **93**, 420–425 (2009)

V. Madan, D. Kanojia, J. Li, R. Okamoto et al., Aberrant splicing of U12-type introns is the hallmark of ZRSR2 mutant myelo-dysplastic syndrome. Nat. Commun. **6**, 6042 (2015)

C. Maes, I. Stockmans, K. Moermans, R. Van Looveren, N. Smets, P. Carmeliet, R. Bouillon, G. Carmeliet, Soluble VEGF isoforms are essential for establishing epiphyseal vascularization and regulating chondrocyte development and survival. J. Clin. Invest. **113**(2), 188–199 (2004)

O.V. Makarova, E.M. Makarov, H. Urlaub, C.L. Will, M. Gentzel, M. Wilm, R. Lührmann, A subset of human 35S proteins, including Prp19, function prior to catalytic step 1 of splicing. EMBO J. **23**, 2381–2391 (2004)

A. Makino, M. Glogauer, G.M. Bokoch, S. Chien, G.W. Schmid-Schonbein, Control of neutrophil pseudopods by fluid shear: role of Rho family GTPases. Am. J. Physiol. Cell Physiol. **288**(4), C863–C871 (2005)

S.B. Malhotra, K.A. Hart, H.J. Klamut et al., Frame-shift deletions in patients with Duchenne and Becker muscular dystrophy. Science **242**, 755–759 (1988)

C. Mammucari, A. Raffaello, D. Vecellio Reane, R. Rizzuto, Molecular structure and pathophysiological roles of the mitochondrial calcium uniporter. Biochim. Biophys. Acta BBA Mol. Cell Res. **1863**, 2457–2464 (2016)

T. Maniatis, R. Reed, An extensive network of coupling among gene expression machines. Nature **416**, 499–506 (2002)

A. Mankodi, E. Logigian, L. Callahan, C. McClain, R. White, D. Henderson, M. Krym, C.A. Thornton, Myotonic dystrophy in transgenic mice expressing an expanded CUG repeat. Science **289**, 1769–1772 (2000)

A. Mankodi, M.P. Takahashi, H. Jiang, C.L. Beck, W.J. Bowers, R.T. Moxley, S.C. Cannon, C.A. Thornton, Expanded CUG repeats trigger aberrant splicing of ClC-1 chloride channel pre-mRNA and hyperexcitability of skeletal muscle in myotonic dystrophy. Mol. Cell **10**, 35–44 (2002)

S.M. Mantila Roosa, Y. Liu, C.H. Turner, Alternative splicing in bone following mechanical loading. Bone **48**(3), 543–551 (2011)

E.M. Mardis, ChIP-seq: welcome to the new frontier. Nat. Methods **4**, 613–614 (2007)

C. Martin, J. Duvall, Y. Ilkin et al., Cytogenetic and molecular characterization of A2BP1/FOX1 as a candidate gene for autism. Am. J. Med. Genet. B Neuropsychiatr. Genet. **144B**, 869–876 (2007)

R. Martinez-Contreras, P. Cloutier, L. Shkreta, J.F. Fisette, T. Revil, B. Chabot, hnRNP proteins and splicing control. Adv. Exp. Med. Biol. **623**, 123–147 (2007)

J. Martone, F. Briganti, I. Legnini, M. Morlando, E. Picillo, O. Sthandier, L. Politano, I. Bozzoni, The lack of the Celf2a splicing factor converts a Duchenne genotype into a Becker phenotype. Nat. Commun. **7**, 10488 (2016)

A.M. Mastrangelo, D. Marone, G. Laido, A.M. De Leonardis, P. De Vita, Alternative splicing: enhancing ability to cope with stress via transcriptome plasticity. Plant Sci. **185–186**, 40–49 (2012)

A.G. Matera, Z. Wang, A day in the life of the spliceosome. Nat. Rev. Mol. Cell Biol. **15**, 108–121 (2014)

R.W. Matheny Jr., B.C. Nindl, M.L. Adamo, Minireview: mechano-growth factor: a putative product of IGF-I gene expression involved in tissue repair and regeneration. Endocrinology **151**(3), 865–875 (2010)

A.J. Matlin, F. Clark, C.W. Smith, Understanding alternative splicing: towards a cellular code. Nat. Rev. Mol. Cell Biol. **6**, 386–398 (2005)

J.S. Mattick, P.P. Amaral, M.E. Dinger, T.R. Mercer, M.F. Mehler, RNA regulation of epigenetic processes. BioEssays **31**, 51–59 (2009)

A. McAlinden, Alternative splicing of type II procollagen: IIB or not IIB? Connect. Tissue Res. **55**(3), 165–176 (2014)

A. McAlinden, N. Havlioglu, L.J. Sandell, Regulation of protein diversity by alternative pre-mRNA splicing with specific focus on chondrogenesis. Birth Defects Res. C Embryo Today **72**(1), 51–68 (2004)

A. McAlinden, N. Havlioglu, L. Liang, S.R. Davies, L.J. Sandell, Alternative splicing of type II procollagen exon 2 is regulated by the combination of a weak 5′ splice site and an adjacent intronic stem-loop cis element. J. Biol. Chem. **280**(38), 32700–32711 (2005)

A. McAlinden, L. Liang, Y. Mukudai, T. Imamura, L.J. Sandell, Nuclear protein TIA-1 regulates COL2A1 alternative splicing and interacts with precursor mRNA and genomic DNA. J. Biol. Chem. **282**(33), 24444–24454 (2007)

A. McAlinden, B. Johnstone, J. Kollar, N. Kazmi, T.M. Hering, Expression of two novel alternatively spliced COL2A1 isoforms during chondrocyte differentiation. Matrix Biol. **27**(3), 254–266 (2008a)

A. McAlinden, M. Majava, P.N. Bishop, R. Perveen, G.C. Black, M.E. Pierpont, L. Ala-Kokko, M. Männikkö, Missense and nonsense mutations in the alternatively-spliced exon 2 of COL2A1 cause the ocular variant of Stickler syndrome. Hum. Mutat. **29**(1), 83–90 (2008b)

A. McAlinden, K.H. Shim, L. Wirthlin, S. Ravindran, T.M. Hering, Quantification of type II procollagen splice forms using alternative transcript-qPCR (AT-qPCR). Matrix Biol. **31**(7–8), 412–420 (2012)

A. McAlinden, G. Traeger, U. Hansen, M.A. Weis, S. Ravindran, L. Wirthlin, D.R. Eyre, R.J. Fernandes, Molecular properties and fibril ultrastructure of types II and XI collagens in cartilage of mice expressing exclusively the alpha1 (IIA) collagen isoform. Matrix Biol. **34**, 105–113 (2014)

A.E. McKee, N. Neretti, L.E. Carvalho, C.A. Meyer, E.A. Fox, A.S. Brodsky, P.A. Silver, Exon expression profiling reveals stimulus-mediated exon use in neural cells. Genome Biol. **8**, R159 (2007)

C.J. McManus, B.R. Graveley, RNA structure and the mechanisms of alternative splicing. Curr. Opin. Genet. Dev. **21**, 373–379 (2011)

J.B. McMillin, M.C. Madden, The role of calcium in the control of respiration by muscle mitochondria. Med. Sci. Sports Exerc. **21**, 406–410 (1989)

M. Melé, K. Mattioli, W. Mallard, D.M. Shechner, C. Gerhardinger, J.L. Rinn, Chromatin environment, transcriptional regulation, and splicing distinguish lincRNAs and mRNAs. Genome Res. **27**, 27–37 (2017)

T.R. Mercer, J.S. Mattick, Structure and function of long noncoding RNAs in epigenetic regulation. Nat. Struct. Mol. Biol. **20**(3), 300–307 (2013)

J. Merkin, C. Russell, P. Chen, C.B. Burge, Evolutionary dynamics of gene and isoform regulation in mammalian tissues. Science **338**, 1593–1599 (2012)

M. Metsaranta, D. Toman, B. de Crombrugghe, E. Vuorio, Mouse type II collagen gene. Complete nucleotide sequence, exon structure, and alternative splicing. J. Biol. Chem. **266**(25), 16862–16869 (1991)

B.J. Meyer, Sex in the worm: counting and compensating X-chromosome dose. Trends Genet. **16**, 247–253 (2000)

T.S. Mikkelsen, M. Ku, D.B. Jaffe et al., Genome-wide maps of chromatin state in pluripotent and lineage-committed cell. Nature **448**, 553–560 (2007)

S. Mili, J.A. Steitz, Evidence for reassociation of RNA-binding proteins after cell lysis: implications for the interpretation of immunoprecipitation analyses. RNA **10**, 1692–1694 (2004)

J.W. Miller, C.R. Urbinati, P. Tengumnuay, M.G. Stenberg, B.J. Byrne, C.A. Thornton, M.S. Swanson, Recruitment of human muscleblind proteins to (CUG)n expansions associated with myotonic dystrophy. EMBO J. **19**, 4439–4448 (2000)

K.F. Mittendorf, C.L. Deatherage, M.D. Ohi, C.R. Sanders, Tailoring of membrane proteins by alternative splicing of pre-mRNA. Biochemistry (Moscow) **51**, 5541–5556 (2012)

M. Montes, S. Becerra, M. Sanchez-Alvarez, C. Sune, Functional coupling of transcription and splicing. Gene **501**, 104–117 (2012)

M.J. Moore, P.A. Silver, Global analysis of mRNA splicing. RNA **14**(2), 197–203 (2008)

M.J. Moore, E.M. Schwartzfarb, P.A. Silver, M.C. Yu, Differential recruitment of the splicing machinery during transcription predicts genome-wide patterns of mRNA splicing. Mol. Cell **24**, 903–915 (2006)

J.K. Mouw, Y. Yui, L. Damiano et al., Tissue mechanics modulate microRNA-dependent PTEN expression to regulate malignant progression. Nat. Med. **20**(4), 360–367 (2014)

R.T. Moxley, R.C. Griggs, D. Goldblatt, V. VanGelder, B.E. Herr, R. Thiel, Decreased insulin sensitivity of forearm muscle in myotonic dystrophy. J. Clin. Invest. **62**, 857–867 (1978)

N. Mukherjee, L. Calviello, A. Hirsekorn, S. de Pretis, M. Pelizzola, U. Ohler, Integrative classification of human coding and noncoding genes through RNA metabolism profiles. Nat. Struct. Mol. Biol. **24**, 86 (2016)

M.P. Mullen, C.W. Smith, J.G. Patton, B. Nadal-Ginard, Alpha-tropomyosin mutually exclusive exon selection: competition between branchpoint/polypyrimidine tracts determines default exon choice. Genes Dev. **5**, 642–655 (1991)

J.E. Murphy-Ullrich, M. Poczatek, Activation of latent TGF-beta by thrombospondin-1: mechanisms and physiology. Cytokine Growth Factor Rev. **11**(1–2), 59–69 (2000)

B. Nadal-Ginard, Muscle cell differentiation and alternative splicing. Curr. Opin. Cell. Biol. **2**, 1058–1064 (1990)

S. Nakahata, S. Kawamoto, Tissue-dependent isoforms of mammalian Fox-1 homologs are associated with tissue-specific splicing activities. Nucleic Acids Res. **33**, 2078–2089 (2005)

K. Nakai, H. Sakamoto, Construction of a novel database containing aberrant splicing mutations of mammalian genes. Gene **141**, 171–177 (1994)

M. Nakamori, T. Kimura, T. Kubota, T. Matsumura, H. Sumi, H. Fujimura, M.P. Takahashi, S. Sakoda, Aberrantly spliced α-dystrobrevin alters α-syntrophin binding in myotonic dystrophy type 1. Neurology **70**, 677–685 (2008)

F.U. Nasim, S. Hutchison, M. Cordeau, B. Chabot, High affinity hnRNP A1 binding sites and duplex forming inverted repeats have similar effects on 5′ splice site selection in support of a common looping out and repression mechanism. RNA **8**, 1078–1089 (2002)

R. Nawrotzki, N.Y. Loh, M.A. Ruegg, K.E. Davies, D.J. Blake, Characterisation of alpha-dystrobrevin in muscle. J. Cell Sci. **111**, 2595–2605 (1998)

A. Necsulea, M. Soumillon, M. Warnefors, A. Liechti, T. Daish, U. Zeller, J.C. Baker, F. Grutzner, H. Kaessmann, The evolution of lncRNA repertoires and expression patterns in tetrapods. Nature **505**, 635–640 (2014)

D. Neri, R. Bicknell, Tumour vascular targeting. Nat. Rev. Cancer **5**(6), 436–446 (2005)

S. Newey, M.A. Benson, C.P. Ponting, K.E. Davies, D.J. Blake, Alternative splicing of dystrobrevin regulates the stoichiometry of syntrophin binding to the dystrophin protein complex. Curr. Biol. **10**, 1295–1298 (2000)

A.J. Newman, C. Norman, U5 snRNA interacts with exon sequences at 5′ and 3′ splice sites. Cell **68**, 743–754 (1992)

L.J. Ng, P.P. Tam, K.S. Cheah, Preferential expression of alternatively spliced mRNAs encoding type II procollagen with a cysteine-rich amino-propeptide in differentiating cartilage and nonchondrogenic tissues during early mouse development. Dev. Biol. **159**(2), 403–417 (1993)

T. Ng, D. Shima, A. Squire, P.I. Bastiaens, S. Gschmeissner, M.J. Humphries, P.J. Parker, PKCalpha regulates beta1 integrin-dependent cell motility through association and control of integrin traffic. EMBO J. **18**(14), 3909–3923 (1999)

B. Ng, F. Yang, D.P. Huston, Y. Yan, Y. Yang, Z. Xiong, L.E. Peterson, H. Wang, X. Yang, Increased noncanonical splicing of autoantigen transcripts provides the structural basis for expression of untolerized epitopes. J. Allergy Clin. Immunol. **114**, 1463–1470 (2004)

J.Z. Ni, L. Grate, J.P. Donohue, C. Preston, N. Nobida, G. O'Brien, L. Shiue, T.A. Clark, J.E. Blume, M. Ares Jr., Ultraconserved elements are associated with homeostatic control of splicing regulators by alternative splicing and nonsense-mediated decay. Genes Dev. **21**, 708–718 (2007)

J. Niggel, W. Sigurdson, F. Sachs, Mechanically induced calcium movements in astrocytes, bovine aortic endothelial cells and C6 glioma cells. J. Membr. Biol. **174**(2), 121–134 (2000)

T.W. Nilsen, B.R. Graveley, Expansion of the eukaryotic proteome by alternative splicing. Nature **463**, 457–463 (2010)

T. Nishiyama, H. Hatano, M. Kurosaka, M. Bolander, G. Sarkar, Cis-acting intronic elements that regulate cartilage-specific alternative splicing of the type II collagen (Col2) pre-mRNA lie at or near splice site junction sequences flanking exon 2 of the gene. J. Bone Miner. Res. **18**(9), 1716–1722 (2003)

A.D. Norris, J.A. Calarco, Emerging roles of alternative pre-mRNA splicing regulation in neuronal development and function. Front. Neurosci. **6**, 122 (2012)

J.D. Norris, D. Fan, A. Sherk, D.P. McDonnell, A negative coregulator for the human ER. Mol. Endocrinol. **16**, 459–468 (2002)

D.G. Nowak, J. Woolard, E.M. Amin, O. Konopatskaya, M.A. Saleem, A.J. Churchill, M.R. Ladomery, S.J. Harper, D. O. Bates, Expression of pro- and anti-angiogenic isoforms of VEGF is differentially regulated by splicing and growth factors. J. Cell Sci. **121**(Pt 20), 3487–3495 (2008)

D.G. Nowak, E.M. Amin, E.S. Rennel et al., Regulation of vascular endothelial growth factor (VEGF) splicing from pro-angiogenic to anti-angiogenic isoforms: a novel therapeutic strategy for angiogenesis. J. Biol. Chem. **285**(8), 5532–5540 (2010)

A.B. Nygard, S. Cirera, M.J. Gilchrist, J. Gorodkin, C.B. Jørgensen, M. Fredholm, A study of alternative splicing in the pig. BMC Res. Notes **3**, 123 (2010)

T. Nyiko, F. Kerenyi, L. Szabadkai et al., Plant nonsense-mediated mRNA decay is controlled by different autoregulatory circuits and can be induced by an EJC-like complex. Nucleic Acids Res. **41**, 6715–6728 (2013)

J.M. O'Leary, J.M. Hamilton, C.M. Deane, N.V. Valeyev, L.J. Sandell, A.K. Downing, Solution structure and dynamics of a prototypical chordin-like cysteine-rich repeat (von Willebrand Factor type C module) from collagen IIA. J. Biol. Chem. **279**(51), 53857–53866 (2004)

A. Oganesian, Y. Zhu, L.J. Sandell, Type IIA procollagen amino propeptide is localized in human embryonic tissues. J. Histochem. Cytochem. **45**(11), 1469–1480 (1997)

A. Ogasawara, T. Arakawa, T. Kaneda, T. Takuma, T. Sato, H. Kaneko, M. Kumegawa, Y. Hakeda, Fluid shear stress-induced cyclooxygenase 2 expression is mediated by C/EBP beta, cAMP-response element-binding protein, and AP-1 in osteoblastic MC3T3-E1 cells. J. Biol. Chem. **276**(10), 7048–7054 (2001)

M.A. Oh, E.S. Kang, S.A. Lee, E.O. Lee, Y.B. Kim, S.H. Kim, J.W. Lee, PKCdelta and cofilin activation affects peripheral actin re-organization and cell-cell contact in cells expressing integrin alpha5 but not its tailless mutant. J. Cell Sci. **120**(Pt 15), 2717–2730 (2007)

M.J. Okoniewski, Y. Hey, S.D. Pepper, C.J. Miller, High correspondence between Affymetrix exon and standard expression arrays. Biotechniques **42**, 181–185 (2007a)

M.J. Okoniewski, T. Yates, S. Dibben, C.J. Miller, An annotation infrastructure for the analysis and interpretation of Affymetrix exon array data. Genome Biol. **8**, R79 (2007b)

S. Oltean, D.O. Bates, Hallmarks of alternative splicing in cancer. Oncogene **33**(46), 5311–5318 (2014)

K.G. Ota, S. Kuratani, Expression pattern of two collagen type 2 alpha1 genes in the Japanese inshore hagfish (*Eptatretus burgeri*) with special reference to the evolution of cartilaginous tissue. J. Exp. Zool. Part B Mol. Dev. Evol. **314**(2), 157–165 (2010)

F. Pagani, F.E. Baralle, Genomic variants in exons and introns: identifying the splicing spoilers. Nat. Rev. Genet. **5**, 389–396 (2004)

P.K. Paik, A. Drilon, P.-D. Fan et al., Response to MET inhibitors in patients with stage IV lung adenocarcinomas harboring MET mutations causing exon 14 skipping. Cancer Discov. **5**, 842–849 (2015)

M.J. Pajares, T. Ezponda, R. Catena, A. Calvo, R. Pio, L.M. Montuenga, Alternative splicing: an emerging topic in molecular and clinical oncology. Lancet Oncol. **8**, 349–357 (2007)

S. Pal, R. Gupta, R.V. Davuluri, Alternative transcription and alternative splicing in cancer. Pharmacol. Ther. **136**, 283–294 (2012)

Q. Pan, O. Shai, C. Misquitta et al., Revealing global regulatory features of mammalian alternative splicing using a quantitative microarray platform. Mol. Cell **16**, 929–941 (2004)

Q. Pan, A.L. Saltzman, Y.K. Kim, C. Misquitta, O. Shai, L.E. Maquat, B.J. Frey, B.J. Blencowe, Quantitative microarray profiling provides evidence against widespread coupling of alternative splicing with nonsense-mediated mRNA decay to control gene expression. Genes Dev. **20**, 153–158 (2006)

Q. Pan, O. Shai, L.J. Lee, B.J. Frey, B.J. Blencowe, Deep surveying of alternative splicing complexity in the human transcriptome by high-throughput sequencing. Nat. Genet. **40**, 1413–1415 (2008)

C. Paradis, P. Cloutier, L. Shkreta, J. Toutant, K. Klarskov, B. Chabot, hnRNP I/PTB can antagonize the splicing repressor activity of SRp30c. RNA**13**, 1287–1300 (2007)

E. Park, C. Iaccarino, J. Lee, I. Kwon, S.M. Baik, M. Kim, J.Y. Seong, G.H. Son, E. Borrelli, K. Kim, Regulatory roles of heterogeneous nuclear ribonucleoprotein M and Nova-1 protein in alternative splicing of dopamine D2 receptor pre-mRNA. J. Biol. Chem. **286**, 25301–25308 (2011)

M.J. Paszek, N. Zahir, K.R. Johnson et al., Tensional homeostasis and the malignant phenotype. Cancer Cell **8**(3), 241–254 (2005)

J.R. Patel, G.M. Diffee, R.L. Moss, Myosin regulatory light chain modulates the Ca2+ dependence of the kinetics of tension development in skeletal muscle fibers. Biophys. J. **70**, 2333–2340 (1996)

N.A. Patel, S. Kaneko, H.S. Apostolatos, S.S. Bae, J.E. Watson, K. Davidowitz, D.S. Chappell, M.J. Birnbaum, J.Q. Cheng, D.R. Cooper, Molecular and genetic studies imply Akt-mediated signaling promotes protein kinase CbetaII alternative splicing via phosphorylation of serine/arginine-rich splicing factor SRp40. J. Biol. Chem. **280**(14), 14302–14309 (2005)

A. Pathak, S. Kumar, Transforming potential and matrix stiffness co-regulate confinement sensitivity of tumor cell migration. Integr. Biol. (Camb.) **5**(8), 1067–1075 (2013)

D. Patra, E. Delassus, A. McAlinden, L.J. Sandell, Characterization of a murine type IIB procollagen-specific antibody. Matrix Biol. **34**, 154–160 (2013)

L.I. Patrushev, T.F. Kovalenko, Functions of noncoding sequences in mammalian genomes. Biochemistry (Moscow)**79**, 1442–1469 (2014)

F. Pelisch, D. Khauv, G. Risso, M. Stallings-Mann, M. Blaustein, L. Quadrana, D.C. Radisky, A. Srebrow, Involvement of hnRNP A1 in the matrix metalloprotease-3-dependent regulation of Rac1 pre-mRNA splicing. J. Cell. Biochem. **113**, 2319–2329 (2012)

M.F. Peters, H.M. Sadoulet-Puccio, R.M. Grady, N.R. Kramarcy, L.M. Kunkel, J.R. Sanes, R. Sealock, S.C. Froehner, Differential membrane localization and intermolecular associations of α-dystrobrevin isoforms in skeletal muscle. J. Cell Biol. **142**, 1269–1278 (1998)

B. Petit, M.C. Ronziere, D.J. Hartmann, D. Herbage, Ultrastructural organization of type XI collagen in fetal bovine epiphyseal cartilage. Histochemistry **100**(3), 231–239 (1993)

G. Petrovics, W. Zhang, M. Makarem, J.P. Street, R. Connelly, L. Sun, I.A. Sesterhenn, V. Srikantan, J.W. Moul, S. Srivastava, Elevated expression of PCGEM1, a prostate-specific gene with cell growth-promoting function, is associated with high-risk prostate cancer patients. Oncogene**23**(2), 605–611 (2004)

S.R. Peyton, A.J. Putnam, Extracellular matrix rigidity governs smooth muscle cell motility in a biphasic fashion. J. Cell. Physiol. **204**(1), 198–209 (2005)

H.P. Phatnani, A.L. Greenleaf, Phosphorylation and functions of the RNA polymerase II CTD. Genes Dev. **20**, 2922–2936 (2006)

A.V. Philips, L.T. Timchenko, T.A. Cooper, Disruption of splicing regulated by a CUG-binding protein in myotonic dystrophy. Science **280**, 737–741 (1998)

M. Pistoni, C. Ghigna, D. Gabellini, Alternative splicing and muscular dystrophy. RNA Biol. **7**, 441–452 (2010)

M. Pistoni, L. Shiue, M.S. Cline, S. Bortolanza, M.V. Neguembor, A. Xynos, M. Ares Jr., D. Gabellini, Rbfox1 downregulation and altered calpain 3 splicing by FRG1 in a mouse model of facioscapulohumeral muscular dystrophy (FSHD). PLoS Genet. **9**, e1003186 (2013)

M. Plodinec, M. Loparic, C.A. Monnier et al., The nanomechanical signature of breast cancer. Nat. Nanotechnol. **7**(11), 757–765 (2012)

L.I. Plotkin, I. Mathov, J.I. Aguirre, A.M. Parfitt, S.C. Manolagas, T. Bellido, Mechanical stimulation prevents osteocyte apoptosis: requirement of integrins, Src kinases and ERKs. Am. J. Physiol. Cell. Physiol. **289**(3), C633–C643 (2005)

H. Ponta, L. Sherman, P.A. Herrlich, CD44: from adhesion molecules to signalling regulators. Nat. Rev. Mol. Cell Biol. **4**(1), 33–45 (2003)

J.L. Ponthier, C. Schluepen, W. Chen et al., Fox-2 splicing factor binds to a conserved intron motif to promote inclusion of protein 4.1R alternative exon 16. J. Biol. Chem. **281**, 12468–12474 (2006)

P.I. Poulikakos, Y. Persaud, M. Janakiraman et al., RAF inhibitor resistance is mediated by dimerization of aberrantly spliced BRAF(V600E). Nature **480**, 387–390 (2011)

J. Prasad, K. Colwill, T. Pawson, J.L. Manley, The protein kinase Clk/Sty directly modulates SR protein activity: both hyper- and hypophosphorylation inhibit splicing. Mol. Cell. Biol. **19**(10), 6991–7000 (1999)

S.G. Priori, C. Napolitano, Cardiac and skeletal muscle disorders caused by mutations in the intracellular Ca^{2+} release channels. J. Clin. Invest. **115**, 2033–2038 (2005)

D.J. Prockop, K.I. Kivirikko, Collagens: molecular biology, diseases, and potentials for therapy. Annu. Rev. Biochem. **64**, 403–434 (1995)

J.J. Quinn, H.Y. Chang, Unique features of long non-coding RNA biogenesis and function. Nat. Rev. Genet. **17**, 47–62 (2016)

J.D. Ransohoff, Y. Wei, P.A. Khavari, The functions and unique features of long intergenic non-coding RNA. Nat. Rev. Mol. Cell Biol. **19**, 143 (2017)

F. Rau, J. Lainé, L. Ramanoudjame et al., Abnormal splicing switch of DMD's penultimate exon compromises muscle fibre maintenance in myotonic dystrophy. Nat. Commun. **6**, 7205 (2015)

A. Ravel-Chapuis, G. Bélanger, R.S. Yadava, M.S. Mahadevan, L. DesGroseillers, J. Côté, B.J. Jasmin, The RNA-binding protein Staufen1 is increased in DM1 skeletal muscle and promotes alternative pre-mRNA splicing. J. Cell Biol. **196**, 699–712 (2012)

S. Ravindran, J.L. Roam, P.K. Nguyen, T.M. Hering, D.L. Elbert, A. McAlinden, Changes of chondrocyte expression profiles in human MSC aggregates in the presence of PEG microspheres and TGF-beta3. Biomaterials**32**(33), 8436–8445 (2011)

A. Reardon, L. Sandell, C.J. Jones, D. McLeod, P.N. Bishop, Localization of PN-type IIA procollagen on adult bovine vitreous collagen fibrils. Matrix Biol. **19**(2), 169–173 (2000)

A.S. Reddy, M.F. Rogers, D.N. Richardson, M. Hamilton, A. Ben-Hur, Deciphering the plant splicing code: experimental and computational approaches for predicting alternative splicing and splicing regulatory elements. Front. Plant Sci. **3**, 18 (2012)

A. Relógio, C. Ben-Dov, M. Baum, M. Ruggiu, C. Gemund, V. Benes, R.B. Darnell, J. Valcárcel, Alternative splicing microarrays reveal functional expression of neuron-specific regulators in Hodgkin lymphoma cells. J. Biol. Chem. **280**, 4779–4784 (2005)

A.J. Richards, M.P. Snead, The influence of pre-mRNA splicing on phenotypic modification in Stickler's syndrome and other type II collagenopathies. Eye (London) **22**(10), 1243–1250 (2008)

A.M. Rickard, L.M. Petek, D.G. Miller, Endogenous DUX4 expression in FSHD myotubes is sufficient to cause cell death and disrupts RNA splicing and cell migration pathways. Hum. Mol. Genet. **24**, 5901–5914 (2015)

J.L. Rinn, H.Y. Chang, Genome regulation by long noncoding RNAs. Annu. Rev. Biochem. **81**, 145–166 (2012)

M.D. Robida, A. Rahn, R. Singh, Genome-wide identification of alternatively spliced mRNA targets of specific RNA-binding proteins. PLoS ONE **2**, e250 (2007)

N. Romero-Barrios, M.F. Legascue, M. Benhamed, F. Ariel, M. Crespi, Splicing regulation by long noncoding RNAs. Nucleic Acids Res. **46**(5), 2169–2184 (2018)

E. Rosonina, B.J. Blencowe, Analysis of the requirement for RNA polymerase II CTD heptapeptide repeats in pre-mRNA splicing and 3′-end cleavage. RNA**10**, 581–589 (2004)

G. Rot, Z. Wang, I. Huppertz, M. Modic, T. Lenče, M. Hallegger, N. Haberman, T. Curk, C. von Mering, J. Ule, High-resolution RNA maps suggest common principles of splicing and polyadenylation regulation by TDP-43. Cell Rep. **19**, 1056–1067 (2017)

M. Rouleau, A. Patel, M.J. Hendzel, S.H. Kaufmann, G.G. Poirier, PARP inhibition: PARP1 and beyond. Nat. Rev. Cancer**10**, 293–301 (2010)

B. Roy, L.M. Haupt, L.R. Griffiths, Review: alternative splicing (AS) of genes as an approach for generating protein complexity. Curr. Genomics**14**, 182–194 (2013)

K.J. Ryan, T.A. Cooper, Muscle-specific splicing enhancers regulate inclusion of the cardiac troponin T alternative exon in embryonic skeletal muscle. Mol. Cell. Biol. **16**, 4014–4023 (1996)

M.C. Ryan, L.J. Sandell, Differential expression of a cysteine-rich domain in the amino terminal propeptide of type II (cartilage) procollagen by alternative splicing of mRNA. J. Biol. Chem. **265**(18), 10334–10339 (1990)

M.C. Ryan, M. Sieraski, L.J. Sandell, The human type II procollagen gene: identification of an additional protein-coding domain and location of potential regulatory sequences in the promoter and first intron. Genomics **8**(1), 41–48 (1990)

K. Ryman, N. Fong, E. Bratt, D.L. Bentley, M. Ohman, The C-terminal domain of RNA Pol II helps ensure that editing precedes splicing of the GluR-B transcript. RNA **13**, 1071–1078 (2007)

M. Sahebi, M.M. Hanafi, A.J. van Wijnen, P. Azizi, R. Abiri, S. Ashkani, S. Taheri, Towards understanding pre-mRNA splicing mechanisms and the role of SR proteins. Gene **587**, 107–119 (2016)

K. Sahlin, M. Tonkonogi, K. Söderlund, Energy supply and muscle fatigue in humans. Acta Physiol. Scand. **162**, 261–266 (1998)

A. Sahoo, S.H. Im, Interleukin and interleukin receptor diversity: role of alternative splicing. Int. Rev. Immunol. **29**(1), 77–109 (2010)

N. Saitoh, C. Sakamoto, M. Hagiwara, L.T. Agredano-Moreno, L.F. Jiménez-García, M. Nakaoa, The distribution of phosphorylated SR proteins and alternative splicing are regulated by RANBP2. Mol. Biol. Cell**23**(6), 1115–1128 (2012)

N.J. Sakabe, S.J. de Souza, Sequence features responsible for intron retention in human. BMC Genomics **8**, 59 (2007)

L. Salmena, L. Poliseno, Y. Tay, L. Kats, P.P. Pandolfi, A ceRNA hypothesis: the Rosetta Stone of a hidden RNA language? Cell **146**(3), 353–358 (2011)

H. Salminen, E. Vuorio, A.M. Saamanen, Expression of Sox9 and type IIA procollagen during attempted repair of articular cartilage damage in a transgenic mouse model of osteoarthritis. Arthritis Rheum. **44**(4), 947–955 (2001)

M. Salton, W.K. Kasprzak, T. Voss, B.A. Shapiro, P.I. Poulikakos, T. Misteli, Inhibition of vemurafenib-resistant melanoma by interference with pre-MRNA splicing. Nat. Commun. **6**, 7103 (2015)

A.L. Saltzman, Y.K. Kim, Q. Pan, M.M. Fagnani, L.E. Maquat, B.J. Blencowe, Regulation of multiple core spliceosomal proteins by alternative splicing-coupled nonsense-mediated mRNA decay. Mol. Cell. Biol. **28**, 4320–4330 (2008)

L.J. Sandell, N. Morris, J.R. Robbins, M.B. Goldring, Alternatively spliced type II procollagen mRNAs define distinct populations of cells during vertebral development: differential expression of the amino-propeptide. J. Cell Biol. **114** (6), 1307–1319 (1991)

L.J. Sandell, A.M. Nalin, R.A. Reife, Alternative splice form of type II procollagen mRNA (IIA) is predominant in skeletal precursors and non-cartilaginous tissues during early mouse development. Dev. Dyn. **199**(2), 129–140 (1994)

L.J. Sandell, Z. Wang, C. Franz, J. Bryan, A. Siegel, R. Mecham, J. Wagensail, B. Ell, A. Rapraeger, Live-cell imaging of endothelial cell tube formation: inhibition by chondrostatin. FASEB J. **101**(4) (2008)

M. Santoro, R. Piacentini, M. Masciullo, M.L.E. Bianchi, A. Modoni, M.V. Podda, E. Ricci, G. Silvestri, C. Grassi, Alternative splicing alterations of Ca2+ handling genes are associated with Ca2+ signal dysregulation in myotonic dystrophy type 1 (DM1) and type 2 (DM2) myotubes. Neuropathol. Appl. Neurobiol. **40**, 464–476 (2014)

G. Santulli, A.R. Marks, Essential roles of intracellular calcium release channels in muscle, brain, metabolism, and aging. Curr. Mol. Pharmacol. **8**, 206–222 (2015)

R.S. Savkur, A.V. Philips, T.A. Cooper, Aberrant regulation of insulin receptor alternative splicing is associated with insulin resistance in myotonic dystrophy. Nat. Genet. **29**, 40–47 (2001)

E. Saygili, M. Pekassa, E. Saygili, G. Rackauskas, D. Hommes, F. Noor-Ebad, C. Gemein, M.D. Zink, R.H. Schwinger, J. Weis, N. Marx, P. Schauerte, O.R. Rana, Mechanical stretch of sympathetic neurons induces VEGF expression via a NGF and CNTF signaling pathway. Biochem. Biophys. Res. Commun. **410**(1), 62–67 (2011)

S. Scarpino, A. Stoppacciaro, C. Pellegrini, A. Marzullo, L. Zardi, F. Tartaglia, G. Viale, L.P. Ruco, Expression of EDA/EDB isoforms of fibronectin in papillary carcinoma of the thyroid. J. Pathol. **188**(2), 163–167 (1999)

M. Schlackow, T. Nojima, T. Gomes, A. Dhir, M. Carmo-Fonseca, N.J. Proudfoot, Distinctive patterns of transcription and RNA processing for human lincRNAs. Mol. Cell **65**, 25–38 (2017)

I.E. Schor, N. Rascovan, F. Pelisch, M. Alló, A.R. Kornblihtt, Neuronal cell depolarization induces intragenic chromatin modifications affecting NCAM alternative splicing. Proc. Natl. Acad. Sci. U. S. A. **106**, 4325–4330 (2009)

I.E. Schor, L.I. Gómez Acuña, A.R. Kornblihtt, Coupling between transcription and alternative splicing. Cancer Treat. Res. **158**, 1–24 (2013)

J. Schrader, T.T. Gordon-Walker, R.L. Aucott, M. van Deemter, A. Quaas, S. Walsh, D. Benten, S.J. Forbes, R.G. Wells, J.P. Iredale, Matrix stiffness modulates proliferation, chemotherapeutic response, and dormancy in hepatocellular carcinoma cells. Hepatology **53**(4), 1192–1205 (2011)

A. Schüler, A.T. Ghanbarian, L.D. Hurst, Purifying selection on splice-related motifs, not expression level nor RNA folding, explains nearly all constraint on human lincRNAs. Mol. Biol. Evol. **31**, 3164–3183 (2014)

C. Schwerk, K. Schulze-Osthoff, Regulation of apoptosis by alternative pre-mRNA splicing. Mol. Cell **19**, 1–13 (2005)

J. Sebat, B. Lakshmi, D. Malhotra et al., Strong association of de novo copy number mutations with autism. Science **316**, 445–449 (2007)

E. Sebestyen, M. Zawisza, E. Eyras, Detection of recurrent alternative splicing switches in tumor samples reveals novel signatures of cancer. Nucleic Acids Res. **43**, 1345–1356 (2015)

E. Sebestyen, B. Singh, B. Minana, A. Pages, F. Mateo, M.A. Pujana, J. Valcarcel, E. Eyras, Large-scale analysis of genome and transcriptome alterations in multiple tumors unveils novel cancer-relevant splicing networks. Genome Res. **26**, 732–744 (2016)

M.E. Sehl, L.R. Langer, J.C. Papp et al., Associations between single nucleotide polymorphisms in double-stranded DNA repair pathway genes and familial breast cancer. Clin. Cancer Res. **15**, 2192–2203 (2009)

Y. Seko, N. Takahashi, M. Shibuya, Y. Yazaki, Pulsatile stretch stimulates vascular endothelial growth factor (VEGF) secretion by cultured rat cardiac myocytes. Biochem. Biophys. Res. Commun. **254**(2), 462–465 (1999)

V. Sgambato, R. Minassian, A.C. Nairn, S.E. Hyman, Regulation of ania-6 splice variants by distinct signaling pathways in striatal neurons. J. Neurochem. **86**(1), 153–164 (2003)

S.A. Shabalina, A.N. Spiridonov, N.A. Spiridonov, E.V. Koonin, Connections between alternative transcription and alternative splicing in mammals. Genome Biol. Evol. **2**, 791–799 (2010)

D. Sharon, H. Tilgner, F. Grubert, M. Snyder, A single-molecule long-read survey of the human transcriptome. Nat. Biotechnol. **31**, 1009–1014 (2013)

Y. Shi, H. Li, X. Zhang, Y. Fu, Y. Huang, P.P. Lui, T. Tang, K. Dai, Continuous cyclic mechanical tension inhibited Runx2 expression in mesenchymal stem cells through RhoA-ERK1/2 pathway. J. Cell. Physiol. **226**(8), 2159–2169 (2010)

L. Shkreta, B. Bell, T. Revil, J.P. Venables, P. Prinos, S.A. Elela, B. Chabot, Cancer-associated perturbations in alternative pre-messenger RNA splicing. Cancer Treat. Res. **158**, 41–94 (2013)

B. Shu, M. Zhang, R. Xie et al., BMP2, but not BMP4, is crucial for chondrocyte proliferation and maturation during endochondral bone development. J. Cell Sci. **124**(Pt 20), 3428–3440 (2011)

S. Shukla, S. Oberdoerffer, Co-transcriptional regulation of alternative pre-mRNA splicing. Biochim. Biophys. Acta **1819**, 673–683 (2012)

Z. Siegfried, R. Karni, The role of alternative splicing in cancer drug resistance. Curr. Opin. Genet. Dev. **48**, 16–21 (2018)

R.K. Singh, T.A. Cooper, Pre-mRNA splicing in disease and therapeutics. Trends Mol. Med. **18**, 472–482 (2012)

R. Singh, J. Valcárcel, Building specificity with nonspecific RNA-binding proteins. Nat. Struct. Mol. Biol. **12**, 645–653 (2005)

B. Sitohy, J.A. Nagy, H.F. Dvorak, Anti-VEGF/VEGFR therapy or cancer: reassessing the target. Cancer Res. **72**, 1909–1914 (2012)

M. Skarli, S.Y. Yang, P. Bouloux, D.M. Yellon, G. Goldspink, Up-regulation and alternative splicing of the IGF-1 gene in the rabbit heart following a brief pressure/volume overload. J. Physiol. **509**, 192–193 (1998)

C.A. Sloan, E.T. Chan, J.M. Davidson et al., ENCODE data at the ENCODE portal. Nucleic Acids Res. **44**, D726–D732 (2016)

L.M. Smith, N.L. Kelleher, Proteoform: a single term describing protein complexity. Nat. Methods**10**, 186–187 (2013)

C.W. Smith, B. Nadal-Ginard, Mutually exclusive splicing of α-tropomyosin exons enforced by an unusual lariat branch point location: implications for constitutive splicing. Cell **56**, 749–758 (1989)

P.J. Smith, C. Zhang, J. Wang, S.L. Chew, M.Q. Zhang, A.R. Krainer, An increased specificity score matrix for the prediction of SF2/ASF-specific exonic splicing enhancers. Hum. Mol. Genet. **15**(16), 2490–2508 (2006)

M.P. Snead, A.M. McNinch, A.V. Poulson, P. Bearcroft, B. Silverman, P. Gomersall, V. Parfect, A.J. Richards, Stickler syndrome, ocular-only variants and a key diagnostic role for the ophthalmologist. Eye (London)**25**(11), 1389–1400 (2011)

E.J. Sontheimer, J.A. Steitz, The U5 and U6 small nuclear RNAs as active site components of the spliceosome. Science **262**, 1989–1996 (1993)

E. Sotillo, D.M. Barrett, K.L. Black et al., Convergence of acquired mutations and alternative splicing of CD19 enables resistance to CART-19 immunotherapy. Cancer Discov. **5**, 1282–1295 (2015)

R. Spellman, M. Llorian, C.W.J. Smith, Crossregulation and functional redundancy between the splicing regulator PTB and its paralogs nPTB and ROD1. Mol. Cell **27**, 420–434 (2007)

K.L. Spiller, S.A. Maher, A.M. Lowman, Hydrogels for the repair of articular cartilage defects. Tissue Eng. Part B Rev. **5**(8), 636–647 (2011)

V. Srikantan, Z. Zou, G. Petrovics et al., PCGEM1, a prostate-specific gene, is overexpressed in prostate cancer. Proc. Natl. Acad. Sci. U. S. A. **97**(22), 12216–12221 (2000)

J.P. Staley, C. Guthrie, An RNA switch at the 5′ splice site requires ATP and the DEAD box protein Prp28p. Mol. Cell **3**, 55–64 (1999)

J.P. Staley, J.J. Woolford, Assembly of ribosomes and spliceosomes: complex ribonucleoprotein machines. Curr. Opin. Cell. Biol. **21**(1), 109–118 (2009)

S. Stamm, S. Ben-Ari, I. Rafalska, Y. Tang, Z. Zhang, D. Toiber, T.A. Thanaraj, H. Soreq, Function of alternative splicing. Gene **344**, 1–20 (2005)

R.B. Stein, J. Bobet, M.N. Oğuztöreli, M. Fryer, The kinetics relating calcium and force in skeletal muscle. Biophys. J. **54**, 705–717 (1988)

D.A. Sterner, T. Carlo, S.M. Berget, Architectural limits on split genes. Proc. Natl. Acad. Sci. U. S. A. **93**, 15081–15085 (1996)

R.B. Stoughton, Applications of DNA microarrays to biology. Annu. Rev. Biochem. **74**, 53–82 (2005)

C.W. Sugnet, K. Srinivasan, T.A. Clark et al., Unusual intron conversation near tissue-regulated exons found by splicing microarrays. PloS Comput. Biol. **2**, e4 (2006)

N. Sultana, B. Dienes, A. Benedetti et al., Restricting calcium currents is required for correct fiber type specification in skeletal muscle. Development **143**, 1547–1559 (2016)

F. Supek, B. Minana, J. Valcarcel, T. Gabaldon, B. Lehner, Synonymous mutations frequently act as driver mutations in human cancers. Cell **156**, 1324–1335 (2014)

A. Sureau, J. Saulière, A. Expert-Bezançon, J. Marie, CELF and PTB proteins modulate the inclusion of the β-tropomyosin exon 6B during myogenic differentiation. Exp. Cell Res. **317**, 94–106 (2011)

I.A. Swinburne, C.A. Meyer, X.S. Liu, P.A. Silver, A.S. Brodsky, Genomic localization of RNA binding proteins reveales links between pre-mRNA processing and transcription. Genome Res. **16**, 912–921 (2006)

L.L. Tang, Y.L. Wang, C.X. Sun, The stress reaction and its molecular events: splicing variants. Biochem. Biophys. Res. Commun. **320**(2), 287–291 (2004)

L.L. Tang, C. Xian, Y. Wang, The MGF expression of osteoblasts in response to mechanical overload. Arch. Oral Biol. **51**(12), 1080–1085 (2006)

W. Tang, Y. Fei, M. Page, Biological significance of RNA editing in cells. Mol. Biotechnol. **52**, 91–100 (2012a)

Z.Z. Tang, V. Yarotskyy, L. Wei, K. Sobczak, M. Nakamori, K. Eichinger, R.T. Moxley, R.T. Dirksen, C.A. Thornton, Muscle weakness in myotonic dystrophy associated with misregulated splicing and altered gating of CaV1.1 calcium channel. Hum. Mol. Genet. **21**, 1312–1324 (2012b)

D.F. Tardiff, S.A. Lacadie, M. Rosbash, A genome-wide analysis indicates that yeast pre-mRNA splicing is predominantly posttranscriptional. Mol. Cell **24**, 917–929 (2006)

S.N. Teraoka, M. Telatar, S. Becker-Catania et al., Splicing defects in the ataxia-telangiectasia gene, ATM: underlying mutations and consequences. Am. J. Hum. Genet. **64**, 1617–1631 (1999)

The Autism Genome Project Consortium, Mapping autism risk loci using genetic linkage and chromosomal rearrangements. Nat. Genet. **39**, 319–328 (2007)

J.R. Thom, N.P. Morris, Biosynthesis and proteolytic processing of type XI collagen in embryonic chick sterna. J. Biol. Chem. **266**(11), 7262–7269 (1991)

J.D. Thomas, Ł.J. Sznajder, O. Bardhi, F.N. Aslam, Z.P. Anastasiadis, M.M. Scotti, I. Nishino, M. Nakamori, E.T. Wang, M.S. Swanson, Disrupted prenatal RNA processing and myogenesis in congenital myotonic dystrophy. Genes Dev. **31**, 1122–1133 (2017)

H. Tilgner, D.G. Knowles, R. Johnson, C.A. Davis, S. Chakrabortty, S. Djebali, J. Curado, M. Snyder, T.R. Gingeras, R. Guigó, Deep sequencing of subcellular RNA fractions shows splicing to be predominantly co-transcriptional in the human genome but inefficient for lncRNAs. Genome Res. **22**, 1616–1625 (2012)

H. Tilgner, F. Jahanbani, T. Blauwkamp, A. Moshrefi, E. Jaeger, F. Chen, I. Harel, C.D. Bustamante, M. Rasmussen, M.P. Snyder, Comprehensive transcriptome analysis using synthetic long-read sequencing reveals molecular co-association of distant splicing events. Nat. Biotechnol. **33**, 736–742 (2015)

J. Tommiska, J. Bartkova, M. Heinonen et al., The DNA damage signalling kinase ATM is aberrantly reduced or lost in BRCA1/BRCA2-deficient and ER/PR/ERBB2-triple-negative breast cancer. Oncogene**27**, 2501–2506 (2008)

N. Toyota, Y. Shimada, Differentiation of troponin in cardiac and skeletal muscles in chicken embryos as studied by immunofluorescence microscopy. J. Cell Biol. **91**, 497–504 (1981)

J.C. Tran, L. Zamdborg, D.R. Ahlf et al., Mapping intact protein isoforms in discovery mode using top-down proteomics. Nature **480**, 254–258 (2011)

M.L. Tress, F. Abascal, A. Valencia, Alternative splicing may not be the key to proteome complexity. Trends Biochem. Sci. **42**(2), 98–110 (2017a)

M.L. Tress, F. Abascal, A. Valencia, Most alternative isoforms are not functionally important. Trends Biochem. Sci. **42** (6), 408–410 (2017b)

J.L. Trincado, E. Sebestyen, A. Pages, E. Eyras, The prognostic potential of alternative transcript isoforms across human tumors. Genome Med. **8**, 85 (2016)

R.S. Tuan, A.F. Chen, B.A. Klatt, Cartilage regeneration. J. Am. Acad. Orthop. Surg. **21**(5), 303–311 (2013)

L. Twyffels, C. Gueydan, V. Kruys, Shuttling SR proteins: more than splicing factors. FEBS J. **278**, 3246–3255 (2011)

J. Ule, K.B. Jensen, M. Ruggiu, A. Mele, A. Ule, R.B. Darnell, CLIP identifies Nova-regulated RNA networks in the brain. Science **302**, 1212–1215 (2003)

J. Ule, K. Jensen, A. Mele, R.B. Darnell, CLIP: a method for identifying protein–RNA interaction sites in living cells. Methods**37**, 376–386 (2005a)

J. Ule, A. Ule, J. Spencer et al., Nova regulates brain specific splicing to shape the synapse. Nat. Genet. **37**, 844–852 (2005b)

J. Ule, G. Stefani, A. Mele, M. Ruggiu, X. Wang, B. Taneri, T. Gaasterland, B.J. Blencowe, R.B. Darnell, An RNA map predicting Nova-dependent splicing regulation. Nature **444**, 580–586 (2006)

V. Umesh, A.D. Rape, T.A. Ulrich, S. Kumar, Microenvironmental stiffness enhances glioma cell proliferation by stimulating epidermal growth factor receptor signaling. PLoS ONE **9**(7), e101771 (2014)

E. Urano, K. Miyauchi, R. Ichikawa, Y. Futahashi, J. Komano, Regulation of cyclin T1 expression and function by an alternative splice variant that skips exon 7 and contains a premature termination codon. Gene**505**(1), 1–8 (2012)

U. Valcourt, J. Gouttenoire, E. Aubert-Foucher, D. Herbage, F. Mallein-Gerin, Alternative splicing of type II procollagen pre-mRNA in chondrocytes is oppositely regulated by BMP-2 and TGF-beta1. FEBS Lett. **545**(2–3), 115–119 (2003)

E. van der Wal, A.J. Bergsma, J.M. Pijnenburg, A.T. van der Ploeg, W.W.M.P. Pijnappel, Antisense oligonucleotides promote exon inclusion and correct the common c.-32-13T>G GAA splicing variant in pompe disease. Mol Ther-Nucleic Acids**7**, 90–100 (2017)

S. van Koningsbruggen, K.R. Straasheijm, E. Sterrenburg, N. de Graaf, H.G. Dauwerse, R.R. Frants, S.M. van der Maarel, FRG1P-mediated aggregation of proteins involved in pre-mRNA processing. Chromosoma**116**, 53–64 (2007)

D. Vecellio Reane, F. Vallese, V. Checchetto, L. Acquasaliente, G. Butera, V. De Filippis, I. Szabò, G. Zanotti, R. Rizzuto, A. Raffaello, A MICU1 splice variant confers high sensitivity to the mitochondrial Ca^{2+} uptake machinery of skeletal muscle. Mol Cell. **64**, 760–773 (2016)

A. Vihola, G. Bassez, G. Meola et al., Histopathological differences of myotonic dystrophy type 1 (DM1) and PROMM/DM2. Neurology**60**, 1854–1857 (2003)

C. Vinatier, D. Mrugala, C. Jorgensen, J. Guicheux, D. Noel, Cartilage engineering: a crucial combination of cells, biomaterials and biofactors. Trends Biotechnol. **27**(5), 307–314 (2009)

V. Vogel, M.P. Sheetz, Cell fate regulation by coupling mechanical cycles to biochemical signaling pathways. Curr. Opin. Cell Biol. **21**(1), 38–46 (2009)

S. Vorlova, G. Rocco, C.V. Lefave, F.M. Jodelka, K. Hess, M.L. Hastings, E. Henke, L. Cartegni, Induction of antagonistic soluble decoy receptor tyrosine kinases by intronic polyA activation. Mol. Cell **43**, 927–939 (2011)

M.C. Wahl, C.L. Will, R. Lührmann, The spliceosome: design principles of a dynamic RNP machine. Cell **136**(4), 701–718 (2009)

Z. Wang, C.B. Burge, Splicing regulation: from a parts list of regulatory elements to an integrated splicing code. RNA **14**, 802–813 (2008)

E. Wang, F. Cambi, Heterogeneous nuclear ribonucleoproteins H and F regulate the proteolipid protein/DM20 ratio by recruiting U1 small nuclear ribonucleoprotein through a complex array of G runs. J. Biol. Chem. **284**, 11194–11204 (2009)

K.C. Wang, H.Y. Chang, Molecular mechanisms of long noncoding RNAs. Mol. Cell **43**(6), 904–914 (2011)

G.-S. Wang, T.A. Cooper, Splicing in disease: disruption of the splicing code and the decoding machinery. Nat. Rev. Genet. **8**, 749–761 (2007)

Z. Wang, M.E. Rolish, G. Yeo, V. Tung, M. Mawson, C.B. Burge, Systematic identification and analysis of exonic splicing silencers. Cell **119**, 831–845 (2004a)

J. Wang, J. Zhang, H. Zheng, J. Li, D. Liu, H. Li, R. Samudrala, J. Yu, G.K.-S. Wong, Neutral evolution of 'non-coding' complementary DNAs. Nature **431**, 758 (2004b)

P. Wang, B. Yan, J.-T. Guo, C. Hicks, Y. Xu, Structural genomics analysis of alternative splicing and application to isoform structure modeling. Proc. Natl. Acad. Sci. U. S. A. **102**, 18920–18925 (2005)

Z. Wang, X. Xiao, E. Van Nostrand, C.B. Burge, General and specific functions of exonic splicing silencers in splicing control. Mol. Cell **23**, 61–70 (2006)

E.T. Wang, R. Sandberg, S. Luo, I. Khrebtukova, L. Zhang, C. Mayr, S.F. Kingsmore, G.P. Schroth, C.B. Burge, Alternative isoform regulation in human tissue transcriptomes. Nature **456**, 470–476 (2008)

Z. Wang, J. Bryan, C. Franz, N. Havlioglu, L.J. Sandell, Type IIB procollagen NH(2)-propeptide induces death of tumor cells via interaction with integrins alpha(V)beta(3) and alpha(V)beta(5). J. Biol. Chem. **285**(27), 20806–20817 (2010)

E.T. Wang, N.A.L. Cody, S. Jog et al., Transcriptome-wide regulation of pre-mRNA splicing and mRNA localization by muscleblind proteins. Cell **150**, 710–724 (2012)

Y. Wang, L. Chen, B. Chen, Mammalian NcRNA-disease repository: a global view of NcRNA-mediated disease network. Cell Death Dis. **4**, e765 (2013)

Y. Wang, J. Liu, B. Huang et al., Mechanism of alternative splicing and its regulation. Biomed. Rep. **3**, 152–158 (2015)

Z. Wang, B. Yang, M. Zhang et al., lncRNA epigenetic landscape analysis identifies EPIC1 as an oncogenic lncRNA that interacts with MYC and promotes cell-cycle progression in cancer. Cancer Cell**33**(4), 706–720.e9 (2018)

C.D. Wee, M.A. Havens, F.M. Jodelka, M.L. Hastings, Targeting SR proteins improves SMN expression in spinal muscular atrophy cells. PLoS ONE **9**(12), e115205 (2014)

S. Weg-Remers, H. Ponta, P. Herrlich, H. König, Regulation of alternative pre-mRNA splicing by the ERK MAP-kinase pathway. EMBO J. **20**(15), 4194–4203 (2001)

T.M. Wheeler, J.D. Lueck, M.S. Swanson, R.T. Dirksen, C.A. Thornton, Correction of ClC-1 splicing eliminates chloride channelopathy and myotonia in mouse models of myotonic dystrophy. J. Clin. Invest. **117**, 3952–3957 (2007)

T.M. Wheeler, K. Sobczak, J.D. Lueck, R.J. Osborne, X. Lin, R.T. Dirksen, C.A. Thornton, Reversal of RNA dominance by displacement of protein sequestered on triplet repeat RNA. Science **325**, 336–339 (2009)

E.S. White, A.F. Muro, Fibronectin splice variants: understanding their multiple roles in health and disease using engineered mouse models. IUBMB Life **63**(7), 538–546 (2011)

D.F. Wieczorek, Regulation of alternatively spliced alpha-tropomyosin gene expression by nerve extract. J. Biol. Chem. **263**, 10456–10463 (1998)

T.N. Wight, Versican: a versatile extracellular matrix proteoglycan in cell biology. Curr. Opin. Cell Biol. **14**(5), 617–623 (2002)

C. Wijmenga, J.E. Hewitt, L.A. Sandkuijl et al., Chromosome 4q DNA rearrangements associated with facioscapulohumeral muscular dystrophy. Nat. Genet. **2**, 26–30 (1992)

C.L. Will, R. Lührmann, Spliceosome structure and function. Cold Spring Harb. Perspect. Biol. **3**, a003707 (2011)

J.E. Wilusz, Long noncoding RNAs: re-writing dogmas of RNA processing and stability. Biochim. Biophys. Acta **1859** (1), 128–138 (2016)

W.M. Wojtowicz, W. Wu, I. Andre, B. Qian, D. Baker, S.L. Zipursky, A vast repertoire of Dscam binding specificities arises from modular interactions of variable Ig domains. Cell **130**, 1134–1145 (2007)

M.S. Wong, J.B. Kinney, A.R. Krainer, Quantitative activity profile and context dependence of all human 5′ splice sites. Mol. Cell**71**, 1012–1026.e1013 (2018)

I.G. Woods, C. Wilson, B. Friedlander, P. Chang, D.K. Reyes, R. Nix, P.D. Kelly, F. Chu, J.H. Postlethwait, W.S. Talbot, The zebrafish gene map defines ancestral vertebrate chromosomes. Genome Res. **15**(9), 1307–1314 (2005)

J.J. Wu, D.R. Eyre, Structural analysis of cross-linking domains in cartilage type XI collagen. Insights on polymeric assembly. J. Biol. Chem. **270**(32), 18865–18870 (1995)

J. Xie, Control of alternative pre-mRNA splicing by Ca^{2+} signals. Biochim. Biophys. Acta **1779**(8), 438–452 (2008)

D. Xin, L. Hu, X. Kong, Alternative promoters influence alternative splicing at the genomic level. PLoS ONE **3**, e2377 (2008)

Y. Xing, K. Kapur, W.H. Wong, Probe selection and expression index computation of Affymetrix exon arrays. PLoS ONE **1**, e88 (2006)

Z. Xiong, Alternative splicing factor ASF/SF2 is down regulated in inflamed muscle. J. Clin. Pathol. **59**, 855–861 (2006)

X.B. Yan, C.H. Tang, Y. Huang, H. Fang, Z.Q. Yu, L.M. Wu, R.Y. Liu, Alternative splicing in exon 9 of glucocorticoid receptor pre-mRNA is regulated by SRp40. Mol. Biol. Rep. **37**(3), 1427–1433 (2010)

M. Yanagisawa, D. Huveldt, P. Kreinest, C.M. Lohse, J.C. Cheville, A.S. Parker, J.A. Copland, P.Z. Anastasiadis, A p120 catenin isoform switch affects Rho activity, induces tumor cell invasion, and predicts metastatic disease. J. Biol. Chem. **283**, 18344–18354 (2008)

S.Y. Yang, G. Goldspink, Different roles of the IGF-I Ec peptide (MGF) and mature IGF-I in myoblast proliferation and differentiation. FEBS Lett. **522**(1–3), 156–160 (2002)

Y. Yang, S. Beqaj, P. Kemp, I. Ariel, L. Schuger, Stretch-induced alternative splicing of serum response factor promotes bronchial myogenesis and is defective in lung hypoplasia. J. Clin. Invest. **106**(11), 1321–1330 (2000)

X. Yang, J. Coulombe-Huntington, S. Kang et al., Widespread expansion of protein interaction capabilities by alternative splicing. Cell **164**, 805–817 (2016)

K.L. Yap, S. Li, A.M. Munoz-Cabello, S. Raguz, L. Zeng, S. Mujtaba, J. Gil, M.J. Walsh, M.M. Zhou, Molecular interplay of the noncoding RNA ANRIL and methylated histone H3 lysine 27 by polycomb CBX7 in transcriptional silencing of INK4a. Mol. Cell**38**(5), 662–674 (2010)

G. Yeo, C.B. Burge, Maximum entropy modeling of short sequence motifs with applications to RNA splicing signals. J. Comput. Biol. **11**(2–3), 377–394 (2004)

C.S. Young, A.D. Pyle, Exon skipping therapy. Cell **167**, 1144 (2016)

J. Yu, Y. Hai, G. Liu, T. Fang, S.K. Kung, J. Xie, The heterogeneous nuclear ribonucleoprotein L is an essential component in the Ca2+/calmodulin-dependent protein kinase IV-regulated alternative splicing through cytidine-adenosine repeats. J. Biol. Chem. **284**, 1505–1513 (2009)

L. Zardi, B. Carnemolla, A. Siri, T.E. Petersen, G. Paolella, G. Sebastio, F.E. Baralle, Transformed human cells produce a new fibronectin isoform by preferential alternative splicing of a previously unobserved exon. EMBO J. **6**(8), 2337–2342 (1987)

Y. Zeng, T. Lai, C.G. Koh, P.R. LeDuc, K.H. Chiam, Investigating circular dorsal ruffles through varying substrate stiffness and mathematical modeling. Biophys. J. **101**(9), 2122–2130 (2011)

X.H. Zhang, L.A. Chasin, Computational definition of sequence motifs governing constitutive exon splicing. Genes Dev. **18**, 1241–1250 (2004)

J. Zhang, J.L. Manley, Misregulation of pre-MRNA alternative splicing in cancer. Cancer Discov. **3**, 1228–1237 (2013)

C. Zhang, H.R. Li, J.B. Fan, J. Wang-Rodriguez, T. Downs, X.D. Fu, M.Q. Zhang, Profiling alternatively spliced mRNA isoforms for prostate cancer classification. BMC Bioinf. **7**, 202 (2006a)

G. Zhang, M.M. Miyamoto, M.J. Cohn, Lamprey type II collagen and Sox9 reveal an ancient origin of the vertebrate collagenous skeleton. Proc. Natl. Acad. Sci. U. S. A. **103**(9), 3180–3185 (2006b)

Z. Zhang, J. Hesselberth, S. Fields, Genome-wide identification of spliced introns using a tiling microarray. Genome Res. **17**, 503–509 (2007a)

J.L. Zhang, Y. Huang, L.Y. Qiu, J. Nickel, W. Sebald, von Willebrand factor type C domain-containing proteins regulate bone morphogenetic protein signaling through different recognition mechanisms. J. Biol. Chem. **282**(27), 20002–20014 (2007b)

C. Zhang, Z. Zhang, J. Castle, S. Sun, J. Johnson, A.R. Krainer, M.Q. Zhang, Defining the regulatory network of the tissue-specific splicing factors Fox-1 and Fox-2. Genes Dev. **22**, 2550–2256 (2008)

B.C. Zhang, Z.W. Zhou, X.K. Li, Y.W. Xu, PI-3K/AKT signal pathway modulates vascular smooth muscle cells migration under cyclic mechanical strain. VASA**40**(2), 109–116 (2011)

Z. Zhang, Z. Zhu, K. Watabe, X. Zhang, C. Bai, M. Xu, F. Wu, Y.Y. Mo, Negative regulation of lncRNA GAS5 by miR-21. Cell Death Differ. **20**(11), 1558–1568 (2013)

C. Zhao, A.M. Pyle, Structural insights into the mechanism of group II intron splicing. Trends Biochem. Sci. **42**, 470–482 (2017)

W. Zhao, X. Li, X. Liu, N. Zhang, X. Wen, Effects of substrate stiffness on adipogenic and osteogenic differentiation of human mesenchymal stem. Cells Mater. Sci. Eng. C Mater. Biol. Appl. **40**, 316–323 (2014)

Y. Zhao, H. Ogawa, S.-I. Yonekura, H. Mitsuhashi, S. Mitsuhashi, I. Nishino, C. Toyoshima, S. Ishiura, Functional analysis of SERCA1b, a highly expressed SERCA1 variant in myotonic dystrophy type 1 muscle. Biochim. Biophys. Acta (BBA) Mol. Basis Dis. **1852**, 2042–2047 (2015)

C.L. Zheng, X.D. Fu, M. Gribskov, Characteristics and regulatory elements defining constitutive splicing and different modes of alternative splicing in human and mouse. RNA **11**, 1777–1787 (2005)

Z. Zhou, X.D. Fu, Regulation of splicing by SR proteins and SR protein-specific kinases. Chromosoma **122**, 191–207 (2013)

C. Zhou, S. Licciulli, J.L. Avila et al., The Rac1 splice form Rac1b promotes K-ras-induced lung tumorigenesis. Oncogene**32**, 903–909 (2013)

Y. Zhou, J. Fan, H. Zhu, L. Ji, W. Fan, I. Kapoor, Y. Wang, Y. Wang, G. Zhu, J. Wang, Aberrant splicing induced by dysregulated Rbfox2 produces enhanced function of CaV1.2 calcium channel and vascular myogenic tone in hypertension novelty and significance. Hypertension**70**, 1183–1192 (2017)

Y. Zhu, A. Oganesian, D.R. Keene, L.J. Sandell, Type IIA procollagen containing the cysteine-rich amino propeptide is deposited in the extracellular matrix of prechondrogenic tissue and binds to TGF-beta1 and BMP-2. J. Cell Biol. **144** (5), 1069–1080 (1999)

Y. Zhu, A. McAlinden, L.J. Sandell, Type IIA procollagen in development of the human intervertebral disc: regulated expression of the NH(2)-propeptide by enzymic processing reveals a unique developmental pathway. Dev. Dyn. **220** (4), 350–362 (2001)

F.Y. Zong, X. Fu, W.J. Wei et al., The RNA-binding protein QKI suppresses cancer-associated aberrant splicing. PLoS Genet. **10**, e1004289 (2014)

A.S. Zot, J.D. Potter, Structural aspects of troponin-tropomyosin regulation of skeletal muscle contraction. Annu. Rev. Biophys. Biophys. Chem. **16**, 535–559 (1987)

Translation and Post-translational Modifications in Protein Biosynthesis 14

Abstract

The experimental methods employed to analyze living systems and the obtained data are usually quantitative. Thus, the investigation of the process of translation and the post-translational modification is crucial for fundamental cellular functions. In specific, the gene expression analyses in living cells at various stages of cell growth display to what extend a specific gene product, such as the messenger RNA (mRNA) or the protein, is existent in distinct location and at a specific time point. The examination of signal transduction routs leads to specifically pronounced response of the cells that encompasses the growth of cells, the DNA replication or the polymerization of actin or microtubules, the remodeling of the cytoskeleton, which seems to be possibly based on the concentration of specific upstream signaling molecules, the chemical activation of protein and/or the coded protein concentration of actin in a feedback loop. Moreover, DNA can be coiled into compact patterns referred to as nucleosomes, depending on the underlying sequence and how often the sequence requires transcription into mRNA and then translation into a protein. This type of tightly packed DNA is required throughout the mitosis phase of the cell cycle, when a cell splits into two identical daughter cells. The number of such cases is extensive and demonstrates the necessity of developing quantitative models to handle the large volume of quantitative data, and these quantitative models in return have to produce experimentally verifiable quantitative hypotheses about biological pathways and phenomena. A biological problem frequently results in a quantitative assessment of how a biological variable impacts on a specific system if that variable is changing, for instance, in concentration, structure, localization, interaction targets or activity, which warrants a model to help understand the data.

14.1 Introduction into the Process of Translation

The genetic information flow goes from the DNA over the RNA toward the protein. The DNA is transcribed in order to obtain the messenger RNAs (the coding RNA). Coding RNAs are necessary for the synthesis of proteins, which includes ribosomal RNAs and transfer RNAs. In specific detail, the non-coding RNAs (ncRNAs) fulfill usually regulatory functions, which encompass microRNAs, small interfering RNAs (siRNAs) and P-element-induced wiped testis (PIWI)-interacting RNAs (piRNAs) originating from repetitive or transposable elements (Leslie 2013; Biryukova and Ye 2015). The piRNAs are a population of small RNAs with a length of 24–30 nucleotides, which are

C. T. Mierke, *Cellular Mechanics and Biophysics*, Biological and Medical Physics, Biomedical Engineering, https://doi.org/10.1007/978-3-030-58532-7_14

located and expressed in gonads of metazoans and therefore ensure the integrity of the genome across generations (Siomi et al. 2011; Sato and Siomi 2013; Ross et al. 2014). In insects, the majority of piRNAs are originated from transposable genetic elements, which originate directly from their own genome or parasitic elements within it. The translation of messenger RNAs (mRNAs) generates the proteins that are required to define the structures and processes of all organisms from bacteria to humans. The core of the translation engine consists of the ribosome, a macromolecular complex of more than 50 proteins and three ribosomal RNAs with a molecular mass of 2.5 MD for bacteria and 4.3 MD for humans (Melnikov et al. 2012). The ribosome works with auxiliary proteins to provide precise and highly controlled translation. These types comprise initiation factors, elongation factors, transfer RNAs with appended amino acids, liberation and recycling factors (Spirin 2000; Liljas 2004).

In this chapter, the application of single-molecule techniques to investigate the translation of mRNA by single ribosomes is presented, and the special emphasis focuses on mechanical perturbation using optical tweezers. Single-molecule techniques offer significant advantages over traditional cluster (or mass) experiments in discovering the molecular pathways underlying this intricate biological process (Kaiser and Tinoco 2014). The effects of this are extremely relevant for the clarification of the kinetics of processive molecular mechanisms, such as ribosomes. Kinetics holds a plethora of knowledge about how molecular engines perform a complex task such as decoding a nucleic acid sequence and the synthesis of the coded protein in a well-coordinated fashion. The kinetics, however, are controlled by stochastic mechanisms; therefore, every single molecule in a certain reaction needs a varying length of time to respond. It is almost completely unfeasible to sustain synchronicity across multiple steps of a sequential response. There will be reactants, products and all intermediary compounds at any point in an ensemble response of numerous molecules. When a significant group of ribosomes begin translating a specific mRNA simultaneously, some will decipher the first codon, whereas others will read the second, third, fourth and so on. In contrary, there is at all times only a single species in a single-molecule reaction. The features of each individual species can be identified.

The translation through a single ribosome on an mRNA was examined with optical tweezers. In the first reported investigation, a constant force was exerted on the 3′ and 5′ ends of a harping RNA (Wen et al. 2008). While the ribosome translated the hairpin, double-stranded RNA was transformed into single-stranded RNA, thereby lengthening the end-to-end distance of the entire RNA molecule. The rise in the spacing directly indicates the translocation of the ribosome. In a consecutive series of experiments, a force is exerted on either the 3′ or 5′ end of the RNA, while the ribosome is attached. This geometry permits the investigation of the ribosome as a motor operating in conjunction with a supporting force or an opposing force as it travels on its track (the mRNA). A third geometry was established where a force is exerted on the emerging polypeptide. This experimental arrangement enables monitoring of the synthesis of the translated protein.

14.2 The Machinery of Translation

Nearby the year 2000, high-resolution images have been generated of the structures of the ribosomal subunits (Ban et al. 2000; Nissen et al. 2000; Schluenzen et al. 2000; Harms et al. 2001; Brodersen et al. 2000; Carter et al. 2000; Wimberly et al. 2000) and the 70S ribosome (Yusupov et al. 2001) through the usage of X-ray diffraction. The structural aspects of the ribosomal functions are nicely resented at the Nobel Prize website of 2009 (http://www.nobelprize.org/nobel_prizes/chemistry/laureates/2009/advanced.html). Further details of ribosomal functions are presented by each Nobel Laureate in his Nobel Lectures (Ramakrishnan 2010; Steitz 2010; Yonath 2010). The structures of bacterial ribosomes which are composed of mRNAs, tRNAs, initiation, elongation and release factors and antibiotics are introduced. Among the newer atomic resolution structures are a 3.0 Å structure of a yeast eukaryotic ribosome (Ben-Shem et al. 2011; Jenner et al. 2012; Wilson and Doudna Cate 2012), the human and Drosophila 80S ribosome (Anger et al. 2013), the release factor 2 and 3 associated with a bacterial ribosome (Jin et al. 2010, 2011; Zhou et al. 2012), and EF-G linked to the ribosome (Chen et al. 2013; Pulk and Cate 2013; Tourigny et al. 2013; Zhou et al. 2013).

Cryo-electron microscopy (cryo-EM) yields additional insights into various conditions of the ribosome and its interference with environmental influences (Connell et al. 2007; Frank 2003; Julian et al. 2008; Fischer et al. 2010). Structural features of most of the auxiliary types, freely and attached to the ribosome, are presently accessible, including the initiation factors, such as eIF1 (Asano et al. 2000), the elongation factors, such as EF-G (Gao et al. 2007a, b, 2009) and EF-Tu (Andersen et al. 2000; Kawashima et al. 1996; Kjeldgaard et al. 1993; Nissen et al. 1995; Schmeing et al. 2009), and release and recycling factors (Rawat et al. 2006; Weixlbaumer et al. 2008; Gao et al. 2007a, b).

14.3 The Mechanism of Translation

A focus of this chapter is on the process of translation in bacterial ribosomes. Although the beginning and end of translation vary substantially across eukaryotes and bacteria, the elongation phase of translation appears to be highly similar in all species. It comprises the deciphering of the signal, facilitated through aminoacylated tRNAs and EF-Tu (eEF-1 in eukaryotes), which build the peptide linkage between the new amino acid and the nascent peptide, closely following a translocation alongside the mRNA template that is catalyzed through EF-G (eEF-2 in eukaryotes). A messenger RNA can be detected in sets of three nucleotides, which create one codon, at a time point that starts at the AUG (the start codon) and finishes at one of the three stop codons, such as UAA, UAG and UGA. Transfer RNAs (tRNAs) adapter molecules read and transfer the three-nucleotide codons to amino acids, comprise an anticodon loop complementary to the codon and the corresponding amino acid linked to its 3′ end. The proper coupling of the tRNA carrying a specific anticodon loop with the proper amino acid is obtained using aminoacyl-tRNA synthetases with exceptional specificity, which form the foundation for translation accuracy (Schimmel and Soll 1979). Translation failures arise primarily due to a mismatch of the tRNA anticodon and the mRNA codon. A bacterial ribosome consists of two subunits, such as 30S and 50S, and represents a high-resolution structure comprising attached transfer RNAs (Fig. 14.1) (Schmeing and Ramakrishnan 2009). The mRNA attachment and codon–anticodon base pairing occurs at the small 30S subunit. The large subunit, namely 50S, constitutes the site of the peptidyltransferase core, where the peptide linkages in the protein are established. A key finding derived from the high-resolution structure of the ribosome has been the validation that this response is initiated by the ribosomal RNA, and therefore, the ribosome itself becomes a ribozyme, whose catalytic capacity is primarily the characteristic feature of ribosomal RNA, rather than that of ribosomal proteins.

The ribosome comprises three binding sites for the tRNAs: at the A site, the tRNA bearing the next amino acid is chosen as defined by the mRNA codon; the P site accommodates the peptidyl-tRNA, which means the translation product attached covalently to the tRNA; the E site represents the exit point for the deacylated tRNA after delivery of the emerging polypeptide to the next subsequent tRNA. Two elongation factors, which constitute proteins that are not integral components of the ribosome, are necessary for optimal translation. Both are GTPases whose catalytic function is governed through the ribosome. The elongation factor EF-Tu attaches an aminoacylated tRNA (aa-tRNA) and GTP to build the ternary complex that can connect to the A site within the ribosome. When the tRNA's anticodon complements the mRNA codon at the site, GTP is hydrolyzed and the aa-tRNA is placed at the A site. The peptide of length N on the P site tRNA is swiftly translocated to the aa-tRNA

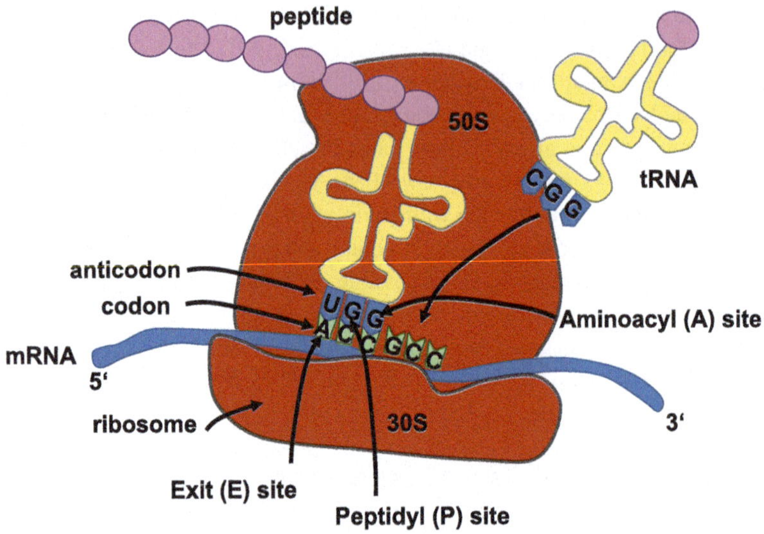

Fig. 14.1 Ribosome structure and the three binding sites for the anticodon of the tRNA

to produce a peptide of length $N + 1$, which is catalyzed through the peptidyltransferase region at the ribosome. The next step involves the translocation of tRNAs and mRNAs through three nucleotides positioned in relation to the ribosome. The deactivated tRNA at the P site migrates to E site and leaves the complex. The tRNA of the A site bearing the nascent protein migrates to the P site, thus rendering the A site ready for the following next tRNA, which the codon now assigns at this site. The translocation is facilitated through the catalysis of the elongation factor *G* (EF-G). The binding of EF-G to the ribosome is succeeded immediately by the hydrolysis of its attached GTP, then the liberation of GDP, inorganic phosphate and EF-G. At the completion of this cycle, two GTP molecules have been hydrolyzed, the ribosome has formed a peptide linkage, it has progressed one codon on the mRNA and is prepared to undergo the subsequent elongation cycle.

The energy source for the translation has not yet been identified in a conclusive manner. Aminoacyl-tRNAs can be inserted into the A site without EF-Tu, and the translocation can take place, albeit very slowly, in the absence of EF-G. The free release of energy upon creation of the peptide bond itself is also a potential energy storage that propels the elongation. The overall outcome of the reaction is the formation of a peptide bond within the emerging protein and the break-down of an ester bond at the 3′-hydroxyl of a ribose. The net variation of the Gibbs free energy at pH 7, 25 °C can be estimated very roughly $\Delta G° = -3.7 \pm 1.2$ kcal/mol for this type of reaction (Liu et al. 2014a). The value was determined by the comparison of data about the hydrolysis of amino acid esters and amides (Jencks 1975).

It is not easy to streamline how the chemical energy release generated through the hydrolysis of GTP in EF-Tu or the free energy difference of a peptide and an ester bond may be captured in the ribosome and subsequently employed to catalyze the translocation. However, each phase of translation entails conformational modifications within a single subunit or between subunits. In principle, each of the alterations of the thesis would in principle be able to accumulate energy that is liberated in a subsequent stage, thereby offering a conceivable mechanism for using the energy liberated by the generation of peptide bonds for the translocation. The coupling between GTP hydrolysis and mechanical movement could be more direct during the catalysis of translocation through EF-G.

Sixty-one codons which contain the AUG designate twenty amino acids: Arginine and leucine employ six codons each, methionine and tryptophan use only one codon each. Conventional biochemical experiments have over many years established the reaction paths of elongation, along with their thermodynamics and kinetics (Rodnina et al. 2007; Rodnina and Wintermeyer 2009; Savelsbergh et al. 2003; Semenkov et al. 2000; Wilden et al. 2006; Wintermeyer et al. 2004; Noller 2006; Rodnina et al. 2006; Pape et al. 1998).

Nevertheless, mass–volume experiments are unable to identify a comprehensive step-by-step mechanism through which the ribosome can synthesize a protein. The single-molecule techniques are required to acquire the locations and interactions of the ribosome, mRNA, tRNAs and the multiple elements at every level of the translation process. It is discussed how single-molecule force spectroscopy techniques have been employed to reveal the process of translation. Fluorescence-based single-molecule techniques offer the equivalent advantages of force techniques over bulk approaches. Especially the single-molecule fluorescence resonance energy transfer (smFRET) has yielded extremely useful insights into the translation. Nevertheless, this in-depth examination will be restricted to the exertion of force and direct the reader to a number of recent overviews that address this issue in depth (Petrov et al. 2011, 2012; Tinoco and Gonzalez 2011; Perez and Gonzalez 2011).

14.4 Applicability of Single-Molecule Methods for the Process of Translation

There exists a number of books that refer to the single-molecule technique, which is employed for biological molecules and possess a broad spectrum of functions in vitro and in vivo (Selvin and Ha 2008; Gell et al. 2006; Hinterdorfer and van Oijen 2009; Leake 2013). A great advantage of examining a single molecule at once is that the molecular properties are not compromised through averaging over the properties of over 1010 molecules. It involves examining biological molecules in solution. Thus, when a molecule is examined, it reacts with water, ions and other dissolved substances. In comparison to the resolution time of the measurement, the interactions that are quick, including the attachment of ions, are taken as an average, although slower binding incidents indicate the ability of the molecule to alternate between ligand bound and ligand-free states. In a similar way, conformational modifications are considered as a mean conformation or as a hopping from one conformation to another as a function of kinetics. For responses with many constituents and multiple steps, such as translation, single-molecule techniques offer enormous advantages over complex averages. They permit the monitoring of a single step in a sequence of reaction steps. Since the kinetics are random, there is no way of predicting when a reaction will take place. This implies that in a reaction of multiple identical molecules, the single step to be investigated is performed at a distinct time for every molecule.

For instance, it is possible to determine the dwell time (synonymously resident time) that a ribosome occupies at a particular codon. When the experiment is repeated over 200 times under identical conditions, a distribution of the dwell times can be gained, when it is hypothesized that the dwell times rely on one rate-determining step. Hence, a broad range of dwell times are determined between 0.2 and 5 s, and the resulting histogram can be fitted appropriately using a single exponential. The probability density $P(\tau)$ is presented in the following (14.1):

$$P(\tau) = (1/\langle\tau\rangle)\mathrm{e}^{-\tau/\langle\tau\rangle} = k\mathrm{e}^{-k\tau} \quad (14.1)$$

where a mean dwell time is $\langle\tau\rangle = 1.08$ s, and hence, an ensemble rate constant is $k = 1/\langle\tau\rangle = 0.924\,\mathrm{s}^{-1}$ for a specific codon within the entire mRNA. In an assembly of ribosomes, it was only feasible to determine the dwell time at a specific codon if it was the first codon of a process. The ribosomes may be synchronized to begin at the same time, but they would quickly lack coherence as each ribosome occupied a completely different time at the first codon. Repeated steps, each determined by single-exponential kinetics, result in a rapid loss of synchronicity. The elongation rates across a specific mRNA are not homogeneous, but each codon is translated at a specific rate based on the plethora of cognitive tRNA present and the sequence environment. A codon velocity profile of a hypothetical mRNA comprises two areas of slowly translated codons. Therefore, it would be possible to monitor a single ribosome along this mRNA, it is clear that longer dwell times (pauses) in these areas would be detected. Similarly, when observing hundred individual ribosomes starting translation simultaneously, the ensemble average might mirror the first "slow" region, whereas it would not reflect the second "slow" region, as synchronicity is broken at this stage.

Even with the first codon alone, much information can be obtained from single-molecule measurements. The shape of the dwell time distribution is much more revealing than a single overall ensemble rate constant. When there are several stages with identical rate constants defining the dwell times, the distribution, a Poisson distribution, becomes no longer exponential; it is no more maximum at time zero.

$$P(\tau) = \frac{k^n \tau^{n-1}}{\Gamma(n)} \mathrm{e}^{-k\tau} \tag{14.2}$$

In (14.2), n stands for an integer indicating the number of rate-determining states, and $\Gamma(n)$, which is a factorial $(n - 1)$ for the integer n and represents the gamma function. For $n = 1$, (14.2) can be reduced to (14.1). The probability density distributions display the following features as provided in (14.3):

$$\int_0^\infty P(\tau)\mathrm{d}\tau = 1 \quad \text{and} \quad \int_0^\infty \tau P(\tau)\mathrm{d}\tau = n \cdot \langle \tau \rangle \tag{14.3}$$

The mean dwell time $\langle \tau \rangle$ is deduced by fitting the measured dwell times to an exponential or general Poisson equation and is recommended to be compared with the mean value of the dwell times. For a data sample was the mean of the 200 dwell times 1.03 s, which is line with the exponential fit of 1.08 s. When there are two or more rate-determining stages, each of which with a specific rate constant, more sophisticated equations need to be derived (Wen et al. 2008; Tinoco and Wen 2009).

Another instance of the unique information based on single-molecule methods represents a simulated single-molecule trajectory for the folding of a protein or RNA measured by tracking the movement of micron-sized beads connected to the molecule. This technique was employed for the folding of a nascent T4 lysozyme, which is connected to a ribosome (Kaiser et al. 2011; Moffitt et al. 2008). A similar trajectory can be obtained when fluorophores were linked to the molecule and fluorescence resonance energy transfer (FRET) was performed. It can be hypothesized that the mechanism requires an equilibrium between the unfolded state is denoted as *U*, and an intermediate is termed *I*, which is followed by an irreversible step to the native state that is referred to as *N*:

$$U \underset{k_2}{\overset{k_1}{\rightleftarrows}} I \xrightarrow{k_3} N$$

In fact, the three rate constants are derived from the average lifetime of *U* and *I* and the branching ratio, *r*, for the species *I*, where *r* is the number of times *I* goes to *U* divided by the number of times it goes to *N*.

$$\langle \tau_{\text{U}} \rangle = \frac{1}{k_1} \tag{14.4a}$$

$$\langle \tau_I \rangle = \frac{1}{k_2 + k_3} \tag{14.4b}$$

$$r = \frac{k_2}{k_3} \tag{14.4c}$$

In an ensemble investigation, where each of the three species would be present during the reaction, it is hard to trace the intermediate state and to receive the rate constants. However, when tracing a single protein molecule that folds along the path outlined above, it would be monitored to establish a temporary intermediate. Due to the intrinsic stochasticity of folding kinetics, the intermediate is not easily be recognized in the ensemble average of only 1000 molecules, even though all molecules could be synchronized in a perfect fashion to start folding simultaneously. Thus, the evolution of the individual species can be derived from 1000 individual footprints, but not necessarily from the average.

These cases illustrate the relevance of single-molecule techniques in delivering information that is not otherwise available. In principle, any intermediate product in a composite reaction is capable of being identified and characterized insofar as it generates a modification of the signal according to the reaction coordinate given by the experimental geometry. Rare varieties in mixtures can be identified. The restrictions in the identification of rare species or intermediate products lie in the spatial and temporal resolution of the measurements. Hence, the spatial resolution of laser tweezers lies around 0.1 nm and the temporal resolution is within the range of 0.1–1 ms (Moffitt et al. 2008), whereas the fluorescence measurements possess a temporal resolution in the μs regime (Tinoco and Gonzalez 2011).

The disadvantages of the single-molecule techniques are primarily due to the need to examine only one molecule at the same time. The performance is limited; the statistics are not good. It can be challenging to differentiate between intrinsic stochastic variation, for instance, in lifetime and experimental variation. The measurements are complicated and demand a sophisticated and generally expensive technology. It is important that the equipment is mechanically, thermally and electronically robust and stable. In the immediate future, however, instruments for single molecules will be commercially obtainable and measurements will become a routine matter. The observation of individual molecules in solutions or biological cells has developed in just a few years.

14.4.1 Laser Tweezers and Force Measurements

In 1970, Arthur Ashkin invented a device termed optical tweezers, in which a focused laser beam can be employed to exert forces on micron-sized beads in aqueous solution (Ashkin 1970). In specific detail, two counter propagating beams are focused at a spot in order to trap a bead at the middle of the spot. Every refracted and scattered light beam generates a force on the bead in the direction of the beam center and in the propagation direction of the beam. Newton's second law of physics says that the force F on the bead is equal and opposite to the change in momentum p with time t of the light falling on the bead, as given in (14.5):

$$F = \frac{\mathrm{d}p}{\mathrm{d}t} \tag{14.5}$$

When the beams move in opposite directions, the force at their common focus is zero. The captured bead behaves as if it was fixed in the middle of the trap with a Hooke's law spring. The restoring force on the bead to the middle of the trap is presented in (14.6):

$$F = \kappa x \tag{14.6}$$

where x represents the distance from the center of the trap and the force constant, κ, is proportional to the beam intensity, the size of the bead and the difference in refractive indices of the bead and its environment. A characteristic trap stiffness employed in the investigation of macromolecules is 0.1 pN/nm (Liphardt et al. 2001), which implies that forces of less than 0.1 pN can be detected.

Besides this, it was also possible to trap neutral particles, such as bacteria, by light (Chu 2002). In order to manipulate DNA molecules, it has been started to connect a bead to the ends of the molecule (Perkins et al. 1994). Among the first applications of the laser tweezers has been the determination of the elastic polymer characteristics of double-stranded DNA in the B-form (Perkins et al. 1994; Bustamante et al. 1994). In specific detail, the construction and calibration of the laser tweezers have been presented. Moreover, the first devices are able to capture one bead in a laser trap of two counterpropagating beams and one bead on a pipette (Smith et al. 2003). In addition, instrument designs have been invented, where the two beads are captured in dual laser traps from one laser in order to decrease the experimental noise effect (Moffitt et al. 2008; Abbondanzieri et al. 2005; Lang and Block 2002). Various instrumentation configurations are possible, many of which include the newly attained combination of force and fluorescence in one device (Comstock et al. 2011). The focus will be on implementing the technology on issues of biological concern, not on the underlying technology alone.

14.4.2 Thermodynamic Measurements and Kinetics

The force has been revealed to serve as an effective way of bending energy landscapes. Mechanical modifications with optical tweezers thus enable disturbances of processes such as the motion of molecular motors or the folding of macromolecules, which is extremely valuable for the investigation of both the thermodynamics and the kinetics of the fundamental cross-overs (Tinoco and Bustamante 2002; Tinoco et al. 2004). The following paragraphs explained how thermodynamic and kinetic knowledge can be obtained from single-molecule force spectroscopy experiments.

Thermodynamics

Every process entailing a shift in distance will be affected by a force that supports or hinders the shift in distance. The mechanical work performed by the process results from the product of the force and the variation in distance (14.7).

$$W = \int_{x_1}^{x_2 x_2} F D x \tag{14.7}$$

Naturally, when the force stays constant during the whole process, (14.7a) is obtained:

$$W = F \cdot (x_2 - x_1) \tag{14.7a}$$

Force and distance are both vectors; therefore, the force compound needs to be applied along the direction of the distance alteration. In each measurable process, there will inevitably be a certain dissipation of energy, for instance, as a result of friction. The work of deconvolution (or synonymously unfolding) of an RNA (Li et al. 2008) or a protein (Cecconi et al. 2005) is achieved through the attachment of beads to the molecule and the determination of force and distance which is applied on the beads as the beads are being moved apart. The amount of work measured is based on how fast the beads are pulled. The more rapidly it is pulled, the greater the energy dissipated during solvent agitation and other dissipative operations in conjunction with the unfolding of the macromolecule. In a similar manner, the work performed by molecular motors such as a ribosome (Liu et al. 2014a), a helicase (Johnson et al. 2007), or a DNA wrapping motor on a phage (Chistol et al. 2012) is assessed by exerting a force in the opposite direction to the direction of movement of the motor. The maximum amount of work the motor can perform is defined on the basis of the stall force, which represents the force that brings the motor to a standstill. The efficiency of the conversion of chemical free energy into mechanical energy (work) can be assessed (Bustamante et al. 2001).

Alterations of the thermodynamic state parameters, such as energy, entropy, enthalpy and free energy, rely merely on the initiating and ending states, whereas the path taken to pass from one state to another is irrelevant. This is what renders them so valuable. It is known that regardless of whether the glucose is oxidized in a calorimeter or in a bacterium, the same amount of energy is received because the long reactants and products are identical in both chemical reactions. The Gibbs free energy alteration of a process (a reaction) is of special value, as the Gibbs free energy alteration indicates the amount of usable work that can be gained from the reaction. For instance, when the reaction takes place in an electrochemical cell, the Gibbs free energy at constant temperature and pressure determines the maximum electrical work that can be received. In a real electrochemical cell, the electrical work will be lower due to the current flux; the Gibbs free energy represents an upper limitation for the work.

Traditional physical chemistry books (Tinoco et al. 2013) deduce the equation between Gibbs free energy, ΔG, and work as:

$$\Delta G = \text{reversible work, when the temperture and}$$
$$\text{pressure are contant}$$

Hence, measuring the reversible work ΔG yields the Gibbs free energy alteration. The actual amount of work will be progressively smaller. In general, in order to carry out the measurement of reversible mechanical work $F\mathrm{d}x$, the independent variable x needs to be changed continuously and gradually to ensure that the dependent variable F constantly holds its balance. In reality, the variables are altered very gradually, in a way that the kinetics never restrict the reaction. As the reaction unfolds, for instance, the spacing between the attached beads is raised linearly over time and the force is measured. The pulling speed is reduced, and when the force versus distance trajectory is unrelated to the pulling rate, the integral of force time distance is considered to represent the reversible work. In many reactions, the kinetics are very slow, which renders it inconvenient to maintain the reaction in balance during the process. However, some new methods have been developed to achieve equilibrium outcomes from non-equilibrium experiments.

Classical thermodynamics was originally deduced for bulk solid systems; therefore, it was presumed that the reversible work is assessed on a macroscopic system. But in small systems, such as single molecules, fluctuations in energy and work are of importance. The second law of thermodynamics says that the average work that can be performed in a reaction is the same or less than the Gibbs free energy alteration. Fluctuations in the response pathway, nevertheless, can lead to the measured work occasionally, less frequently, surpassing the free energy change. It was verified that by several repetitions of the process under the identical non-equilibrium circumstances, such as the pulling speed, the value of ΔG can be accessed (Jarzynski 1997). It has demonstrated that the Boltzmann average of irreversible work at constant temperature and pressure yields Gibbs free energy, ΔG.

$$e^{-\Delta G/k_B T} = \left\langle e^{-W/k_B T} \right\rangle \tag{14.8}$$

In this case, w is the determined non-equilibrium work, k_B the Boltzmann constant and T stands for the absolute temperature. The brackets $\langle\ \rangle$ indicate a mean value over a principally infinite number of repeats (Jarzynski 2001; Hummer and Szabo 2001). In experimental terms, the number of repetitions varies with the distance of the reaction from equilibrium, and valuable results can be attained with an achievable number of repetitions (Liphardt et al. 2002; Ritort et al. 2002). Another approach to retrieve ΔG from irreversible experiments was derived (Crooks 1998, 1999). The irreversible experiments are performed for both the forward and reverse reactions, and histograms of the operation values are generated. The work value at which the histograms for the forward and reverse reaction intersect or can be interpolated to intersect represents the reversible work. This is a more intuitive finding that requires fewer iterations to achieve the Gibbs free energy (Collin et al. 2005), but it demands that both forward and rearward reactions are assessable.

Kinetics

The capability of measuring the kinetics with laser tweezers is restricted due to how quickly the beads connected to the molecule can be displaced. The reaction of a bead to a disturbance can be determined through the frequency dependence of the movement of the bead. The reaction of the bead reduces with rising frequency (Lionnet et al. 2008), as given in (14.9).

$$R(f) = R_0 \frac{1}{1 + \left(\frac{f}{f_C}\right)^2} \tag{14.9}$$

$R(f)$ represents the frequency response at f, and R_0 indicates the zero-frequency response. The cutoff frequency or synonymously referred to as corner frequency, f_C, indicates the frequency at which the response is half the frequency of zero. The corner frequency gives a frequency directly proportional to the stiffness of the trap and inversely proportional to the bead radius and the solvent viscosity. A bead with a radius of 1 µm in water inside a laser trap with a stiffness of 0.1 pN/nm exhibits a corner frequency of about 5 kHz (Tinoco et al. 2006). The reaction time of the bead, τ_B, equals $(2\pi f_C)^{-1}$. This implies that lifetimes in the range of 0.1 ms can be recorded.

When a single-molecule system is balanced between two states, except when the reactions are too fast or the variation in the distance between the beads is too slight, one can see how the two states bounce from one to the other with constant force. Hence, the rates of the reactions and the equilibrium constant can be determined. An example can be the following:

$$A \underset{k_1}{\overset{k_2}{\rightleftarrows}} B$$

The rate constants represent the reciprocal values of the mean lifetimes of the different states. The equilibrium constant between the two states, K, is defined by (14.10):

$$K = \frac{[B]}{[A]} = \frac{k_1}{k_2} = \frac{\langle \tau B \rangle}{\langle \tau A \rangle} \tag{14.10}$$

Equation 14.10 represents $[B]/[A]$ the ratio of the probabilities of the molecule to be in state B or A. It is equivalent to the ratio of the rate constants for the forward and rearward reaction, k_1/k_2, and the ratio of the average lifetime of the two states, $\langle \tau_B \rangle / \langle \tau_A \rangle$. In this way, the equilibrium constant may be calculated from the average lifetime of the individual states. The difference in the distance of the states permits us to recognize the states; this means that the force exerted on the beads needs to disturb the balance. The rising force encourages the elongated molecule. The quantitative expression is presented by (14.11):

$$K(F) = K(0)\mathrm{e}^{F\Delta x / k_B T} \tag{14.11}$$

$K(F)$ stands for the equilibrium constant at force F; $K(0)$ represents the equilibrium constant at zero force. The value of Δx, which represents the distance variation caused by the reaction and can be positive or negative, is influenced by the placement of the beads. In nucleic acids, the attachment sites are generally the 3′ and 5′ ends; in proteins, attachments at the N-terminus and C-terminus are standard, although attachment can take place at any site. Since both force and Δx are vectors, the product $F\Delta x$ must be considered as their scalar product.

To derive the standard free energy alteration for a reaction at zero force, $\Delta G°(F = 0)$, derived from a measured equilibrium constant at force F, a thermodynamic cycle is adopted:

$$\Delta G°(F = 0) - F\Delta x - k_B T \ln K(F) - \Delta G(x\,\text{stretch})$$

where k_B represents the Boltzmann constant and $\Delta G(x$ stretch) equals the difference in reversible work required to stretch the molecules B and A to impose F.

$$\Delta G(x\,\text{Stretch}) = \int_0^x {}^{\mathrm{B}}F\mathrm{d}x'_{\mathrm{B}} - \int_0^X {}^{\mathrm{A}}F\mathrm{d}x'_{\mathrm{A}} \tag{14.12}$$

When choosing the force $F_{1/2}$, where the equilibrium constant $K(F)$ is 1, (14.12) can be simplified to (14.13):

$$\Delta G°(F = 0) = F_{1/2}\Delta x - \Delta G(x\,\text{stretch}) \tag{14.13}$$

The influence of force on the rate constants possesses the same form as (14.11):

$$k(F) = k(0)\mathrm{e}^{Fx^{\ddagger}/k_B T} \tag{14.14}$$

The distance to the transition state is $x^{\ddagger}$ that may be positive or negative. Therefore, an increasing force accelerates the forward reaction, which results in longer molecules, and decelerates the rearward reaction. The equilibrium distance between the states is equivalent to the difference between the transition states for the forward and rearward reaction.

$$\Delta x = x^{\ddagger}(\text{forward})x^{\ddagger}(\text{reverse}) \tag{14.15}$$

The slope of a ln(K) versus force diagram is the same as $\Delta x/k_{\mathrm{B}}T$; the curve is linear when Δx is not dependent on the force. The extrapolation to zero force yields the equilibrium constant at zero force. A strong agreement was observed between the bulk measured equilibrium constant and the value obtained extrapolated as a function of the force required to unfold an RNA hairpin (Liphardt et al. 2001). This consensus is anticipated due to the fact that the equilibrium constant and the free energy difference of the two states are path-independent. It is irrelevant whether the force is applied to achieve equilibrium or not; but as long as the equal initial and final states are acquired, the thermodynamic values remain the equal. The kinetic data yield comparable results. The slope of a ln (k) versus force diagram is the same as $x^{\ddagger}/k_{\mathrm{B}}T$; the curve becomes linear when $x^{\ddagger}$ is unaffected by the force. However, extrapolation to zero force will not yield the rate constant at zero force. The mechanism for the kinetic reaction is extremely distinct in the existence and non-existence of force. The ratio of the extrapolated rate constants may be identical to the observed bulk equilibrium constant, but the individual rate constants may differ considerably from the observed bulk rate constants.

Molecular Motors

Molecular motors transform chemical energy to mechanical movement. In principle, there is a sequence of steps comprising the binding of substrates, cofactors and energy sources, as well as the movement along a track, such as DNA for polymerases and messenger RNA for ribosomes (Bustamante et al. 2004).

As for every reaction, it is important to understand the entire mechanism; in the case of molecular motors, the question of how the chemical reaction steps are linked to the movement is an additional factor. A general linear response mechanism consisting of four steps can be described as:

$$A \underset{k_1}{\overset{k_1}{\rightleftarrows}} B \underset{k_2}{\overset{k_2}{\rightleftarrows}} C \underset{k_3}{\overset{k_3}{\rightleftarrows}} D \underset{k_4}{\overset{k_4}{\rightleftarrows}} E$$

The letters indicate the probabilities of each species for a molecular motor or single-molecule reaction; they are determined by resolving a coupled set of linear differential equations. Each step is first order (k exhibits units of s^{-1}), or is pseudo-first-order when binding of a ligand participates in the step, whereby k concentration exhibits units of s^{-1} and the concentration of ligand kept constant). Some of these steps become irreversible when the rate constant for the rearward reaction can be neglected.

It is useful to imagine each reaction step as elements along a chemical axis and a motion axis (Bustamante et al. 2001). A single step can be entirely chemical or entirely mechanical, or include both a chemical reaction and a linked motion. A step containing movement is defined in terms of its dependence on force. Its rate constant is reduced by an opposing force and enhanced by a supporting force along the direction of movement. The distance to the transformation state for this step results from the slope of ln k versus the force. When a steady-state constant, K, can be obtained for the step, the step size, Δx, is calculated from ln K versus force. Entirely chemical steps are completely unaffected by the force. An instance of a molecular motor which has been examined in detail is the motor that encapsulates the DNA in the bacteriophages φ29 capsid (Chistol et al. 2012).

The link between chemistry and motion has been defined as either a power stroke or a Brownian ratchet (Bustamante et al. 2001). In a power stroke, the chemical reaction causes direct the movement. The chemical reaction reduces the free energy of the forward state in comparison to the previous state; a fraction of the Gibbs free energy of the reaction propels the motion. In a Brownian ratchet, the movement possesses a very low activation barrier in the order of thermal energy ($\sim$0.6 kcal/mol), which means that the movement is both reversible and fast. The downstream chemical reaction compensates the movement; it avoids that the movement goes in the reverse direction.

The perfect analysis of each mechanochemical single-molecule process would yield the chemical identity and spatial coordinates of every species in a function of time. The process could also be defined as a potential energy function in a two-dimensional reaction pathway. A reaction coordinate could be the common chemical reaction coordinate, the other one would be the spatial reaction coordinate, which indicates the movement of the motor.

14.5 The Three Phases of Translation

The translation comprises three key stages, such as initiation, elongation and termination. In bacteria, initiation begins when initiation factors formylmethionine tRNA target an AUG start codon located at the ribosome P site. The location of the initiator AUG codon, approximately eight or nine nucleotides in the direction of the 3′ end of the Shine–Dalgarno sequence, marks the starting point of the translation. The Shine–Dalgarno sequence, AGGAGG, complements a sequence in the 16S rRNA, which is known as the anti-Shine–Dalgarno sequence. The base pairing action between the Shine–Dalgarno and anti-Shine–Dalgarno sequences assists in placing the ribosome for correct initiation. Elongation begins when the elongation factor EF-Tu introduces the next aminoacyl-tRNA and GTP, as a ternary complex, and exits the related tRNA at the A site when the GTP hydrolysis has been performed. The formylmethionine is transferred to the tRNA of the A site to create a new peptide linkage that is catalyzed through the ribosomal RNA. The attachment of EF G•GTP catalyzes the translocation of the mRNA and the tRNAs attached to it for one codon and hence for three nucleotides. The deacylated tRNA of the P site migrates to the E site, the tRNA of the A site migrates to the P site, and consequently the A site is freely available for the following tRNA. This procedure proceeds as long as a stop codon, such as UAA, UAG or UGA, is encountered. Termination takes place after a liberation factor attaches to the stop codon and hydrolyses the peptide of the P site tRNA.

14.5.1 The Translation of Double-Stranded Hairpin MRNAs

The movement of the ribosome during translation of an mRNA can be investigated through the attachment of beads to the terminal ends of a hairpin RNA. When the beads are pulled, which means that forces are applied, the hairpin unfolds and when a critical force is achieved, this leads to an immediate rise in the end-to-end distance of the mRNA (Liphardt et al. 2001). When the hairpin is maintained at a force below the critical force, the equilibrium encourages the folded shape and the hairpin stays predominantly folded. But when a ribosome translates the RNA, it deploys the hairpin and extends the distance from end-to-end, as double strands are turned into a single strand. In the first single-molecule experiment on translation with laser tweezers, a hairpin construct was utilized.

In specific detail, hairpins consisting of either a combination of numerous amino acids or only valine and glutamic acid have been translated (Wen et al. 2008). The mRNA was linked to micron-

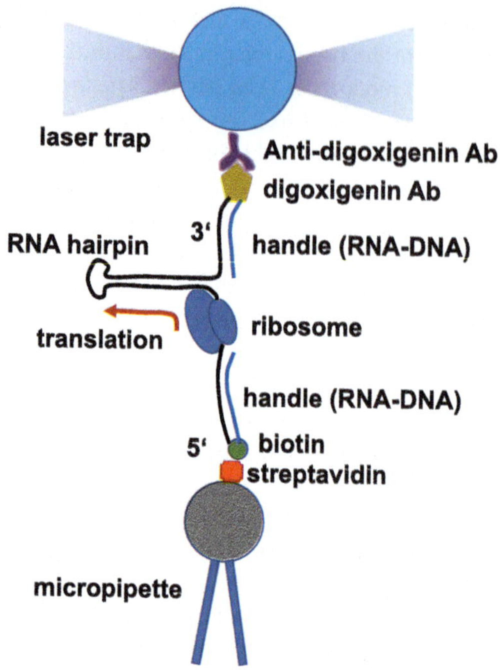

Fig. 14.2 Translation of a hairpin. A hairpin mRNA is connected between two beads. As the ribosomes go on with the translation of the hairpin, the end-to-end distance increases continuously

sized polystyrene beads through kbp-long DNA plugs, which included either biotin or digoxigenin at their ends; streptavidin adhered to one bead and anti-digoxigenin to the other (Fig. 14.2). After starting and initiating the elongation in the presence of buffer comprising elongation factors, aminoacyl-tRNAs and GTP, the ribosome was blocked by skipping the next aminoacyl-tRNA on the mRNA. Once a single inhibited complex is retained between the two beads, the lacking tRNAs are supplemented. The translation is considered a step-pause step track, where the pause takes place as the elongation factor EF-Tu introduces the next amino acid, the peptide bond is created and EF-G is attached. The step detects the sudden movement of the ribosome when there is a translocation. The rest times and the translocation times are random as expected; there is usually a wide range of rest times in trajectories. Nevertheless, the step size is kept constant. This can be demonstrated the easiest using the paired distribution of the points alongside all trajectories. A histogram of the distances between every point and every other point can then be drawn. The peaks are observed according to the distances between the pauses; peaks are displayed at multiples of 2.7 nm. The length per nucleotide for single-stranded RNA varies according to the force exerted on the terminal ends of the RNA. Since the force approaches zero, the end-to-end distance of the RNA is extremely narrow and the length per nucleotide comes close to zero. With growing force, the distance from end to end comes closer to the contour length of the RNA, approximately 0.59 nm per nucleotide. When a force of 20 pN has been exerted on the RNA, a trajectory can be gained that results in 0.45 nm per nucleotide. Thus, each step represents a forward movement of six nucleotides. The ribosome translates three nucleotides (one codon), and releases three more nucleotides from the hairpin's complementary strand. The complete force versus end-to-end distance curve for RNA has been well estimated using the worm chain model (Bustamante et al. 1994).

A major advantage of single-molecule techniques lies in their capability to identify the characteristics of a ribosome at every single codon. The mean dwell time at a specific codon is a function of the concentration of the ternary complex holding the respective tRNA. It may also rely on sequences that are either downstream or upstream to the codon. The distribution of dwell times is highly instructive; it indicates the number of rate-controlling steps that govern the pause period at every codon. For the valine codons investigated, the spreading is not exponential (Wen et al. 2008); it has been demonstrated that two or more steps are rate-determining for pausing at every codon. The biochemistry of the reactions takes place during the breaks in a translation pathway; two elongation factors are bound, GTP is hydrolyzed, and a peptide bond is established when a new amino acid is linked to the emerging peptide chain. It is not unexpected that the distribution changes according to the concentrations of the soluble species; one of the elongation factors can be selected as the single rate-limiting element (Tinoco and Wen 2009; Chemla et al. 2008).

The mean dwell time of a codon at the decryption site on the ribosome corresponds to the mRNA structure at the entry tunnel of the ribosome (Qu et al. 2011). A G-C-rich hairpin extends the dwell times more strongly compared to an A-U-rich duplex, since the double strand needs to be unwrapped before the translation may continue. In a similar manner, an internal Shine–Dalgarno sequence on the mRNA attaches to the 16S ribosomal RNA and delays the translation. These obstacles to translocation not only adjust the synthesis rate of proteins, but they exert a deep impact on the accuracy of translation, especially on frameshifting (Farabaugh 1996).

The recorded dwell times for the pauses lie within the range of seconds, while the translocation times are within the range of milliseconds. The distribution of translocation times represents a single exponential which denotes a rate-determining step; the assertion that the distribution represents three steps seems to have been an artifact with inadequate time resolution of the original experiments. The mean translocation time varies independently of the concentrations of elongation factors and tRNAs. It appears that the biochemical steps actually set up the ribosome for the process of translocation. The process may last several seconds and varies depending on the binding and dissociation kinetics of the ligands. However, the ribosome's actual movement takes place in a single step lasting a few milliseconds.

The translation of a hairpin mRNA linked to the beads and employing purified compounds is relatively slow relative to in vivo translation. The average dwell time (approximately 2 s) measured at each codon is over ten times higher than that based on the expected maximum translation rates in *Escherichia coli* cells (Wen et al. 2008). In vitro bulk measurements of soluble compounds of bacteria (S100 fraction) are also more slowly, but not as slow as single-molecule measurements. Apparently, there are many factors that govern the rate of translation, which comprise the ionic composition, ionic concentrations and the effective concentrations of elongation factors and tRNAs. These constraints have not yet been adapted to the translation speed in the optical tweezer measurements, which may account for the rather long dwell times seen in the experiments.

14.5.2 The Ribosome Acts as a Helicase

Naturally detected messenger RNAs possess extended double-stranded regions, most commonly in the form of hairpins, which are generally ten or less base pairs long; pseudonodes of hairpins with attached stems and loops are also abundant. There seem to be no exogenous helicases that act explicitly on mRNAs, which means that the ribosome needs to be capable of unraveling double strands to translate the message. The initiation of translation takes place if the two subunits of the ribosome attach themselves to the mRNA and the initiator tRNA binds with the aid of initiation factors. In order to translate the mRNA, it is dragged through a tunnel in the ribosome, which is merely wide enough to contain a single strand. The proteins S3, S4 and S5 at the entry of this tunnel account for the helicase activity (Takyar et al. 2005). Nevertheless, the secondary structure in the mRNA decelerates the translation and alters the synthesis rate of the coded protein (Watts et al. 2009). In the published investigations on translation using laser tweezers, the mRNA is entirely double-stranded, which means that the ribosome constantly has to operate as a helicase to be able to translate the message. The force exerted on the termini of the hairpin mRNA functions like an equivalent helicase as it de-stabilizes the hairpin; subsequently, the translation rate rises as the force increases. In the investigations on a hairpin curve with about 270 bp, the translation was paused occasionally for minutes. But an increase in force could bring the translation back into operation, which was maintained until the end of the double strand. The pauses strongly corresponded to the presence of internal Shine–Dalgarno sequences within the mRNA. It seems that the combination of the 16S mRNA double strand of the Shine–Dalgarno and the hairpin represented a hindrance preventing the ribosome from crossing the barrier. The forced weakening of the hairpin made it possible for the ribosome to proceed with translation.

A quantitative investigation of helicase function of the ribosome has been performed (Qu et al. 2011). The translation rates of the valine residues were determined as a function of the force exerted on the termini of a hairpin. The force response curve is sigmoidal with a plateau at low and high force. Under low force, the ribosome needs to unwrap the double-stranded hairpin stem; the rate is referred to as v_{ds}. In the case of the high force, which almost corresponds to the critical force at which the hairpin spontaneously unravels, the rate is referred to as v_{ss} for individual strand translation. Two hairpins were examined. The identical sequences of valine strands have been translated, except that the hairpin stem was either $\sim 50\%$ G-C or 100% G-C. The unwrapping of the higher G•C stem reduced the v_{ds} rate at a low force, but achieved the same v_{ss} single strand rate at high force. The experimental translation rates versus force can be fitted well by (14.15).

$$v(F) = v_{ss} f_{open}(F) + v_{ds}\left(1 - f_{open}(F)\right) \tag{14.15}$$

The probability of the junction of the stem at the entry of the ribosome being open is f_{open}. It is determined based on the force, the free energy required to rupture each base pair, and the number of base pairs opened is in this case three. The rates of translation are $v_{ss} = v(f_{open} = 1)$ for a single strand and $v_{ds} = v(f_{open} = 0)$ for a double strand. The probability, f_{open}, can be determined through the following (14.16):

$$f_{open}(F) = \frac{1}{1 + \exp\left[\Delta G_{bp} + \Delta G_F - \Delta G_D / k_B T\right]} \tag{14.16}$$

The free energy values are referred to as ΔG_{bp}, the free energy of breaking the base pairs is termed ΔG_F, the free energy contribution of the force to opening the duplex is ΔG_D and hence for the destabilization of the duplex by the ribosome this amount is required. The nearest-neighbor approximation for base-pair free energies has been employed for ΔG_{bp} (Zuker 2003). The worm-like chain model delivered the effect of force on unwrapping the duplex, ΔG_F (Liphardt et al. 2001). The values of three variables, such as ΔG_D, v_{ss}, and v_{ds}, have been revealed to provide the best least squares fit for the two force-dependent curves. The fit is fine and yields a helix dissociation value for the ribosome of $\Delta G_D = 0.9$ kcal/mol. This lies inside the range identified for other helicases, such as HIV-1 reverse transcriptase and T7 DNA helicase. However, the ribosome differs from these helicases insofar as it employs two ways to travel forward on the mRNA. The system can be delayed until the double-wire connection opens, or it can open the connection directly. The two pathways are identified as v_{ss} and v_{ds}.

14.5.3 The Ribosome Functions as a Molecular Motor

Molecular motors, for example, the ribosome, can be supported or impeded by a force exerted along the direction of movement. By applying one bead to the ribosome and a second bead to the 3′ end of an mRNA, the force dependence of the elongation rate has been determined in a "tug-of-war" geometry subjected to opposite forces. The rate dropped exponentially with the force and was practically zero with an applied force of 13 pN. This standstill (stall) force corresponds to the maximum force that the ribosome is able to apply when traveling along the mRNA. The maximum (mechanical) work performed of the ribosome is expressed as the product of the stall force and the step size, which amounts to three nucleotides (1.7 nm) and results in 13 pN − 1.7 nm = 22.1 pN nm, corresponding to a work of 3.2 kcal/mol. This maximum amount of work carried out needs to be less than the decrease in Gibbs free energy provided by the chemical processes in the reaction. Interestingly, the free energy gap between the hydrolysis of peptide bonds (within the translation product) and the hydrolysis of ester bonds (within the aminoacyl-tRNA) under experimental conditions is expected to be approximately −3.7 kcal/mol, yielding enough chemical energy to promote translocation when the ribosome is 80% efficient with respect to a molecular motor. In this way, translocation could be powered by transpeptidation as the only available energy source. Nevertheless, it is hard to imagine how the energy released during the creation of the peptide bond might be stored and utilized in the translocation. A more direct supply of free energy comes from the hydrolysis of GTP through ribosome-bound EF-G, since EF-G reacts with the A site and actively facilitates translocation (Rodnina 2013).

A comparable geometry has been adopted to investigate the mechanical stability of the ribosome-mRNA assembly in the first rounds of elongation (Uemura et al. 2007). Interoperations of the anti-Shine–Dalgarno sequence of the 16S ribosome RNA with the Shine–Dalgarno sequence on the mRNA close to the start codon enhances the stability of the ribosome-mRNA complex, which has been found to be impaired in experiments with mRNA without Shine–Dalgarno sequence. In addition, the robustness of the complex to mechanical force relied on the assignment of the ribosomal A and P sites: In the absence of tRNAs, the complexes would break down at an average force of 10.6 pN. The break-down force in enhanced to 15.2 pN when a deacylated tRNA remained in the P site, and further elevated to 26.5 pN when Phe-tRNAPhe stayed additionally in the A site. The removal of the Shine–Dalgarno sequence on the mRNA decreased these forces by approximately 10 pN. It was found most interesting that the formation of peptide bonds seemed to reduce the mechanical strength of the complex, as experiments with *N*-acetyl-Phe-tRNAPhe replacing Phe-tRNAPhe demonstrated. This finding might suggest a linkage between the two ribosomal subunits: The formation of peptide bonds in the 50S subunit can lead to a lowered stability of the 30S-tRNA-mRNA complex. In the case of initiation, this might offer a mechanism that enables elongation exclusively after the initial peptide bonds have been established.

14.5.4 Manipulation of Nascent Protein Elongation and Folding

In optical tweezer analyses, retaining the ribosome and one end of the mRNA produces a force that is adapted to the movement of the ribosome alongside the mRNA pathway. The force is opposing when exerted on the 3′ end of the mRNA, or reinforcing when exerted on the 5′ end. In fact, it has been seen that the rate at which the ribosome moves down the mRNA diminishes as a function of the force exerted on the 3′ end. The exertion of force on the N-terminal region of the emerging protein as it is released of the ribosome enables the simultaneous determination of polypeptide elongation and folding. Measurements which are founded on such a geometry have not been documented up to now.

It is anticipated that the force exerted on the emerging polypeptide will eventually be passed on to the peptidyl transferase center of the ribosome, which is the anchor site of the emerging polypeptide. This can result in strains on the catalytic site, which offers exciting possibilities for investigating the mechanochemistry of peptide bond synthesis in the presence of a reaction coordinate which does not immediately affect the motion of the ribosome during the process of translocation.

While single-molecule investigations to control the ongoing elongation of an emerging polypeptide chain have never been described, a single-molecule investigation of an emerging chain that has come to a standstill at specific positions has provided information about its folding in the vicinity of the ribosome. Why is it useful to examine folding in the connection with synthesis? The ribosome acts as a highly processive molecular engine: During the elongation, individual amino acids are attached to the C-terminus of the growing, emerging polypeptide, and the translation product is exclusively liberated when a stop in the message is encountered, which implies that the protein has been completely synthesized. In general, the folding process is quicker than translational elongation: At maximum elongation rates of approximately 20 amino acids per second, it generally requires at least five seconds to finish the synthesis of a small protein comprising 100 amino acids. The folding, either to the native structure or to metastable folding intermediates, takes place frequently in the range of milliseconds to seconds and is normally very cooperative, whereby the majority of residues are concerned within the folding entity. Two direct consequences arising from these fundamental parameters of cellular protein synthesis are firstly that the initial folding is performed prior to the termination of synthesis, whereas the growing, emerging chain is kept in direct neighborhood of the ribosome; and secondly that not the entire sequence is made accessible at one time to be involved in the folding, but rather that it is formed progressively. How is this type of vectorial character of protein synthesis then aligned with the cooperative folding? How is the ribosomal environment impacting these initial folding processes?

The capacity of optical tweezers to mechanically modify individual molecules and detect alterations in their extension offers an excellent basis for investigating protein folding in complex surroundings such as the ribosome. Force functions as a denaturant agent for proteins: The exertion of an external force shifts the energy landscape of the folding, thereby pre-stressing the protein for unfolding (Kaiser et al. 2011). Optical tweezers permit the exertion of force on specific positions in the molecular arrangement, thereby facilitating the selective destabilization of a selected portion of the molecule or complex under investigation. This framework has been employed effectively to unfold individual subdomains of the T4 lysozyme and to examine how the folding of individual subdomains is connected across the full-length protein (Shank et al. 2010). In a similar way, it ought to be feasible to unfold only the emerging polypeptide in a complex of ribosomes and emerging chain to examine its folding, while maintaining the macromolecular arrangement intact.

Measurements with optical tweezers employing a specific geometry indicated that the ribosome significantly slows down the folding of T4 lysozyme (Bustamante et al. 2004). In these experiments, engineered ribosomes were utilized, which accommodate an attachment point within the large subunit. Nascent polypeptides were generated in vitro from these ribosomes utilizing mRNA templates missing stop codons. When the ribosome enters the 3′-end of one such stop codon-less mRNA, it comes to a standstill. In the emerging ribosome–nascent chain complex, the translation compound stays covalently attached to the P site tRNA and thus enriches as a peptidyl-tRNA species. With suitably engineered mRNAs, this technique was employed to generate nascent chains trapped at a specific position of the T4 lysozyme sequence. The implementation of a flexible linker sequence at the C-terminus enabled the T4 lysozyme polypeptide to escape completely from the ribosomal exit tunnel and fold back into its native structure. In such a molecular configuration, the emerging chain can then be unwrapped by exerting force with optical tweezers and then refolded when the force is diminished. This enables the real-time monitoring of folding transitions and thus the evaluation of folding kinetics. When the emerging chain is held in the direct vicinity of the ribosomal surface through a connector sequence of 41 amino acids bridging the ribosomal exit tunnel, a 100-fold slowdown of the total folding rate is recorded compared to folding in the complete lack of the ribosome. The extension of the linker sequence of up to 60 amino acids ensures an additional splitting of the ribosome and the folding protein and results in a partial restoration of the folding rate.

It was discovered that the impact of the ribosome on the folding rate is highly susceptible to the ionic strength of the buffer, suggesting that electrostatic interference of charged residues in the emerging polypeptide with the high negative electrostatic potential around the ribosome is at least partially accountable for the monitored effect. It has been outlined that a potential rationalization seems to form the basis of this phenomenon: The forces of attraction and repulsion to the charged moieties in the emerging polypeptide within the electrostatic potential surrounding the ribosome can lead to a conformational distortion in the fully or partially unfolded condition of the emerging chain. This distortion can encourage or discourage specific conformations, which can then speed up or slow down the capture of the structure equivalent to the transitional state of folding. In accordance with this model, a kinetic evaluation of the folding trajectories utilizing a Bayesian extension of the hidden Markov model (BHMM) revealed that the ribosome does not have a global impact on all folding rates, but influences a distinct step in the folding path: It was determined that T4 lysozyme bends through a path very similar to the route. The BHMM analysis indicated that the reversible generation of a transient folding intermediate is not influenced by the ribosome, while the last step to the native state is slowed down. It was observed that N-terminal fragments of T4 lysozyme accumulate in solution and fold incorrectly in single-molecule measurements (Bustamante et al. 2004). The misfolding was inhibited by the ribosome, probably due to the kinetic mechanism outlined above, suggesting that the slowing down of certain folding steps through the ribosome may help to keep the growing, emerging polypeptide in a conformation capable of folding until enough sequence for the efficient folding of the protein has been generated. These experiments clearly reflect the power of single-molecule manipulation and detection and will provide the foundation for future investigations of translational elongation and folding at the ribosome.

The experiments and techniques outlined before exemplify the performance of single-molecule assays employing optical tweezers to investigate the complex biological process of the protein biosynthesis. It is challenging to consider similarly appropriate alternative experimental settings to examine the motor characteristics of the ribosome, the role of the ribosome as a helicase and the modification of the emerging protein folding due to the ribosomal microenvironment. Optical tweezers offer the unique possibility not only to monitor individual macromolecular complexes but to effectively interact with them. This permits us to privilege or disapprove of transitions and control transitions between states, as shown by recordings of the emerging protein folding at the ribosome.

The investigations with optical tweezers mentioned afore have started to illuminate the processes associated with translation, but they represent only the peak of what will be feasible with this powerful experimental setup. Advances in equipment and biochemical specimen preparation have decreased the obstacles to examining highly complex biological activities using optical tweezers. A robust single-molecule translation assay is expected to be extremely valuable in responding to questions regarding translation dynamics and control mechanisms: What local fluctuations in strain rates are induced by secondary and tertiary structures alongside the mRNA templates? What are the defining factors of these variations? How does this punctuation of synthesis impact on the structure and function of the translated product? How is it involved in programmed frameshifting, which is utilized to govern the function of proteins?

The combined technique of mechanical manipulation and fluorescence detection of single molecules will broaden the spectrum of questions that can be answered by simply inserting an independent response coordinate. It will enable the temporal association of ribosome translocation with elongation factor binding and—by selecting appropriate FRET pairs—with the conformational redistributions both inside the elongation factors and inside the ribosomal subunit, such as head tilting (Schuwirth et al. 2005). In a similar way, the rotation between the subunits, a functionally important component of the ribosome, which was investigated using single-molecule FRET techniques (Cornish et al. 2008; Marshall et al. 2008; Munro et al. 2010; Fei et al. 2008; Moffitt et al. 2006), could be monitored simultaneously with the translocation. Since both the tilting of the head and the rotation between the subunits are fundamentally mechanical processes, it ultimately proved possible not only to monitor them but also to modify them in order to acquire knowledge of how they participate in the highly complex phenomenon of translational elongation.

14.6 Post-translational Modifications

The cells have to recognize and respond to alterations in internal and external environmental circumstances. One technique of adapting to these alterations involves the chemical alteration of proteins. Conditional chemical modifications are transmitted from the transducers to the effectors through reversible post-translational modifications (PTMs) of proteins. PTMs are key players in altering the final product of expression, are implicated in biological pathways and disease states, and are involved in numerous cellular processes including cellular differentiation, the break-down of proteins, signaling and regulatory processes, the control of gene expression and the protein–protein relationships (Fig. 14.3).

What Does Post-transcriptional Modification Mean?

PTMs can occur at every phase of the protein life cycle. Many proteins are altered soon after the translation is finished in order to convey correct folding or to guide the emerging protein to specific sites within the cell, for example, the nucleus or the plasma membrane. Other variations arise after completion of the folding and the positioning to activate or inactivate the catalytic capacity. Proteins can also be covalently bound to tags that aim at a protein for break-down. These are altered by a

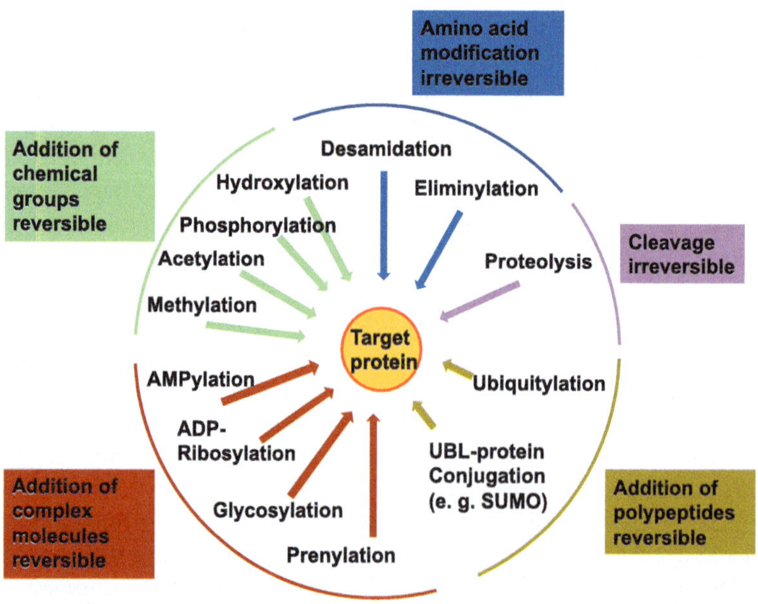

Fig. 14.3 Types of posttranslational modifications (PTMs)

mixture of post-translational cleavage and the introduction of functional units by a stepwise mechanism of protein ripening or activation.

PTMs are found on different amino acid side chains or peptide bonds and are usually facilitated through enzymatic activity. In fact, 5% of the proteome consists of enzymes that execute more than 200 kinds of PTMs. These enzymes comprise kinases, phosphatases, transferases and ligases which attach or detach functional chains, proteins, lipids or sugars to or from amino acid side chains, and proteases that break peptide bonds to eliminate specific sequences or regulatory subunits. Many of these proteins are also capable of altering themselves through autocatalytic domains, for instance, autokinase and autoprotolytic domains. PTMs may also be reversed, according to the nature of the change. For instance, phosphatases hydrolyze the phosphate group to eliminate it of the protein and then reverse its biological activity.

Frequent Post-translational Modifications

Protein phosphorylation represents the most widely examined PTM. It has been hypothesized that one-third of mammalian proteins can be phosphorylated, and this alteration frequently holds a pivotal role in the regulation of protein function. Recent advances in mass spectrometry (MS) techniques have made it possible to locate thousands of PTM locations. As a consequence, new enrichment approaches have revealed the overall cellular significance of different types of modifications such as acetylation, *N*-bound glycosylation, O-linked β-*N*-acetylglucosamine (O-GlNac) and ubiquitylation. More than 200 distinct types of PTMs are presently recognized, ranging from small chemical alterations, such as phosphorylation and acetylation, to the insertion of entire proteins, such as ubiquitylation.

14.7 Influence of PTMs on Healthy and Diseased Functions

The analysis of proteins and their PTMs is especially relevant for the examination of cancer, diabetes, heart disease and neurodegenerative diseases. The major hurdles in the investigation of post-translationally modified proteins are the establishment of specific identification and purification techniques. These technical barriers are successfully overcome with a number of new and sophisticated proteomics technologies.

Aging is defined through the progressive reduction of the biochemical and physiological capacity of an individual species. The aging process is therefore a major risk contributor for diseases, including cancer, type 2 diabetes and obesity. The cellular and molecular pathways of aging are not well known, nor is the correlation between aging and the onset of disease sufficiently understood (Santos and Lindner 2017). One of the characteristics of aging is a reduction in cellular proteome homeostasis, which can lead to the clustering of aberrant proteins. This phenomenon is seen in both eukaryotes and prokaryotes, indicating that the associated molecular processes are conserved in an evolutionary way. Comparable protein accumulation appears during the pathogenesis of diseases including Alzheimer and Parkinson. In addition, PTMs, either in spontaneous or physiological/pathological terms, are developing into key determinants of aging and age-related diseases, although the causality is not yet clearly understood. In the following, a brief overview of the interaction of PTMs in aging-related molecular processes using eukaryotic aging models is presented and discussed. The role of PTMs in aging might help to improve understanding of targeted therapies or interventions for age-related diseases. In addition, the investigation of PTMs in prokaryotes is a particular focus, demonstrating the enormous capacity of simple prokaryotic models to reveal complex aging-linked molecular functions in the growing field of microbiogerontology.

During the last years, the focus in the field of aging has shifted to the idea of the health duration, for example, the lengthening of the period of life in which an individual maintains good health, rather than concentrating only on possibilities to prolong the lifespan (Crimmins 2015). Globally, average life expectancy at birth is now more than sixty years, due in part to improvements in the availability of health care, lower infant mortality, lower maternal mortality, improvements in lifestyles and rising living standards (World Health Organization 2000, 2016). This rise in life expectations has resulted in a shift in the population structure; between 2000 and 2050, the number of persons of 60 years and over is predicted to rise from 605 million to 2 billion worldwide (World Health Organization 2000). A growing older age group can have a positive influence on society, not only socially but also economically (Beard et al. 2015). Even if it is not possible to stop the aging process completely, progress in the development of pharmacological, dietary and genetic treatments that promote healthy aging may enable individuals to survive longer while suffering less physical and/or mental deterioration. This theme is emphasized in the recent World Health Organization (WHO) Global Strategy and Plan of Action on Ageing and Health (GSAP), which aims to create a world where everyone experiences healthy aging by preserving the functional capacity that facilitates well-being in old age (World Health Organization 2015).

From a biodemographic perspective, aging is characterized as an exponential rise in mortality over time (Vaupel et al. 1998; Vaupel 2010), sometimes associated with a slowdown or a plateau at a higher age (Vaupel et al. 1998; Curtsinger et al. 1992; Carey et al. 1992; Charlesworth and Partridge 2015). While the underlying alterations of aging are intricate (Kirkwood 2008), it is defined through the progressive clustering of a wide range of molecular and cellular defects over the lifespan (Kirkwood and Austad 2000). The nine hallmarks of aging in mammals identified encompass the unstable genome, telomere wear, epigenetic modifications, loss of proteostasis, abnormal nutrient sensing, mitochondrial malfunction, cellular senescence, stem cell depletion, and modified intercellular communication (López-otín et al. 2013). However, the links between these characteristics, their role in aging and their relationship with vulnerability and disease are not yet completely clear (López-otín et al. 2013. Actually, the detection of the biological principles of aging is one of the largest contemporary scientific demands (Gems and Partridge 2013).

There is considerable evidence to show that it is feasible to interfere at the level of the presumed mechanisms underpinning aging, leading to a slower accumulation of age-related injury (Kirkwood and Austad 2000; López-otín et al. 2013; Gems and Partridge 2013). In this respect, the implementation of the multidisciplinary tools of systems biology to explore the intricate mechanisms of aging and different age-related disturbances might provide the required evidence for an effective governance of aging mechanisms (Kirkwood et al. 2003). The usage of model organisms has fostered fast progress in the field of aging research by detecting gene mutations that prolong life (Kuningas et al. 2008). The key genes and pathways that govern lifespan are highly conserved in eukaryotes such as yeast, worms, flies and mammals (Kirkwood and Austad 2000).

Importantly, the epigenetics process also has a crucial role to play in the aging process (Kuningas et al. 2008; Gentilini et al. 2013; Cloos and Christgau 2004; Marks et al. 2012; Walsh et al. 2005; Omenn et al. 2016). Although several different types of epigenetic mechanisms are available, the PTM proteins participate in a fascinating way in controlling aging (Greifenhagen et al. 2016; Nedić et al. 2015; Kakizawa 2013; Merkley et al. 2014; Kuhn et al. 2014; Basak et al. 2016). In this section, the involvement of PTMs in the aging process is discussed with an emphasis on PTM types, modes of action and detection techniques. An overview of recent advances in PTMs and aging research on various model species is also provided. A better knowledge of the PTMs and their role in aging offers a basis for the design of specific interventions or targeted strategies to combat aging and age-related diseases.

Post-translational Modifications of Proteins

Proteins form the basis of cellular and physiological activities in living organisms, and the physical and chemical characteristics of proteins determine their activities and roles. The primary protein sequence is a key determinant of the folding and final configuration of a protein. Moreover, it defines its biochemical activity, overall stability and half-life (Marks et al. 2012). But at any point in an individual's life, the proteome is still far more intricate by two or three orders of magnitude than the coding genome could suggest (Walsh et al. 2005). However, one of the principal pathways of proteome extension is mediated through PTMs of proteins. PTMs occur in both eukaryotes and prokaryotes, although it is suggested that PTMs are more abundant in eukaryotic cells, where approximately 5% of the genome is allocated to enzymes that perform post-translational modifications of proteins (Walsh et al. 2005).

Protein PTM is obtained through enzymatic or non-enzymatic binding of specific chemical moieties to amino acid side chains (Walsh et al. 2005). These modifications appear either after protein translation or accompanying the translation. PTM affects both protein structure and physiological and/or cellular processes. Among these processes of enzymatic PTMs are acetylation, biotinylation, chlorination, glycosylation, methylation, nitration, oxidation/reduction, palmitoylation, phosphorylation, sumoylation and ubiquitination (Omenn et al. 2016). The non-enzymatic PTMs encompass acetylation, glycation, nitrosylation, oxidation/reduction and succination (Greifenhagen et al. 2016; Nedić et al. 2015; Kakizawa 2013; Merkley et al. 2014; Kuhn et al. 2014). There exists several rare and unconventional PTMs, such as AMPylation, cholesteroylation, glypiation, neddylation and siderophorylation, which are revealed the impact the structure of proteins and their function (Basak et al. 2016). The key PTMs in eukaryotes are presented in Table 14.1 together with the targeted amino acid residue(s), and the associated enzyme(s) or protein(s).

Changes in the speed and extent of protein synthesis, accuracy, PTMs and the turnover of proteins are one of the most crucial molecular characteristics of aging. A decrease in the cellular ability to identify and primarily break down the defective proteins through proteasomal and lysosomal pathways finally results in the accumulation of defective proteins in the course of aging (Shiozawa-West et al. 2015). The resulting enhancement of molecular diversity and the disturbed function of proteins forms the foundation for multiple age-related pathologies, among them cataract, sarcopenia and different neurodegenerative diseases (Rattan 2008). Thus, a better knowledge of the range of PTMs and their functional effects on aging will support the establishment of appropriate intervention, prevention and therapy for aging and age-related diseases.

Research on PTMs in prokaryotes began with the idea that the bacteria lack many characteristics that are commonly encountered in advanced organisms. However, current investigations keep revealing new types of PTMs of bacterial proteins and their importance for bacterial adaptability and regulation of the cell cycle. Most bacterial PTMs are both highly dynamic and reversible. This enables the cell to use them as regulatory instruments. Among the various bacterial PTMs, protein phosphorylation seems to be the most extensively investigated (Spät et al. 2015) and appears to be especially critical for key bacterial pathogens (Meisenhelder et al. 2001; Marx 2013).

Bacterial pathogens possess a variety of mechanisms to interfere with the host cells, influence their behavior and thereby help them to survive and multiply. Throughout pathogenesis, phosphorylation of proteins on hydroxylamino acids, such as serine, threonine and tyrosine take place at various phases, comprising cell–cell interaction and adherence, the translocation of bacterial effectors in host cells, and alterations in the structure and function of the host cell resulting from infection. Within the different virulence determinants implicated in bacterial pathogenesis, particular emphasis has lately been placed on cell wall components, the exopolysaccharides. An important achievement demonstrated the existence of a biological link between the activity of specific protein tyrosine

Table 14.1 Post-translational modifications, their target amino acid residue(s) and the associated enzymes or proteins

Post-translational modification	Target amino acid residue(s)	Associated enzyme(s) or proteins
Acetylation	Lysine	Acetyltransferases, deacetylases
Carbonylation	Lysine, proline, arginine, threonine	–
Chlorination	Tyrosine	Myeloperoxidases
Glycation	Lysine	–
Glycosylation N-linked	Asparagine	Glycosyltransferases, deglycosylases
Glycosylation O-linked	Serine, threonine	Glycosyltransferases, deglycosylases
Methylation	Lysine, arginine	Methyltransferases, deacetylases
Myristoylation	Glycine	*N*-Myristoyltransferases
Nitration	Tyrosine	Denitrases
Oxidation/reduction	Cysteine	Peroxidases, oxidases, glutathione, thioredoxin
Palmitoylation	Cysteine	DHHC protein acyltransferases, acyl-protein thioesterases
Phosphorylation	Tyrosin, serine, threonine	Kinases, phosphatases
Prenylation	Cysteine	Farnesyltransferases, geranyl geranyltransferases
S-Nitrosylation	Cysteine, methionine	–
Sulfation	Tyrosine	Sulfatases, desulfatases
Sumoylation	Lysine	Ubiquitin-activating enzymes, ubiquitin-conjugating enzymes, ubiquitin ligases, deubiquitinases
Ubiquitination	Lysine	Ubiquitin-activating enzymes, ubiquitin-conjugating enzymes, ubiquitin ligases, deubiquitinases

kinases/phosphatases and the synthesis and/or trafficking of surface polysaccharides. In general, the proof of a strong correlation between protein phosphorylation on serine/threonine/tyrosine and bacterial pathogenicity is a novel idea that is important for decoding the molecular and cellular mechanisms behind the pathogenesis (Cozzone 2005).

In addition, several investigations have started to unravel the wide range of PTMs engaged in important bacterial cell processes, among them redox regulation by reversible S-thiolation (Loi et al. 2015), post-translational hydroxylation (van Staalduinen and Jia 2015), and the role of citrullination in the interplay between bacteria and human mucosal surfaces (Sofat et al. 2015). A number of experimental investigations have been carried out on the post-translationally altered antimicrobial peptides commonly referred to as lantibiotics (McAuliffe et al. 2001). In addition, scientific evidence has also highlighted an essential role of PTMs on PII proteins, which are the main signal transduction proteins implicated in the regulation of nitrogen metabolism in bacteria and archaeae (Gibson et al. 2016; Gupte et al. 2017). More lately, the control mechanisms of arginine phosphorylation and regulated proteolysis in *Bacillus subtilis* have been investigated, and a mechanism has been hypothesized in which protein phosphorylation serves a function in the quality assurance of bacterial proteins and aims at unstable and aggregation tending proteins to be broken down (Stannek et al. 2015).

Experimental Methods for the Analysis of Protein PTMs

Depending on the chemistry, certain amino acid residues are exposed to the reaction and the sequence characteristics of the participating enzyme PTMs (Walsh et al. 2005). Originally, the detection of PTMs was performed by different analytical techniques, for example, column chromatography, polyacrylamide gel electrophoresis, radiolabeling of proteins and thin-layer chromatography (Meisenhelder et al. 2001). Other methodologies, such as protein sequencing using Edman degradation and Western blotting utilizing protein-specific antibodies, have subsequently been established. At present, antibody-based detection techniques and mass spectrometry-based proteome analysis are the prevailing techniques for the detection and analysis of PTMs. However, mass spectrometric techniques remain the only existing method to carry out a comprehensive or large-scale PTM analysis (Marx 2013).

Antibody-based techniques are usually dependent on the presence of antibodies that can selectively identify a modified amino acid residue inside a protein or peptide. These antibodies can be polyclonal or monoclonal and are generated either against the altered peptide/protein or against the altered amino acid. In addition, antibody-based detection and quantification of PTMs on protein/peptide specimens can be carried out employing two different techniques: chemiluminescence-based Western blotting and absorption/fluorescence-based ELISA. However, the detection of PTMs is entirely reliant on the recognition site of the antibody employed (Baker 2015). When the antibody recognizes exclusively the modified amino acid, additional analyses, such as protein/peptide isolation and sequencing, needs to be performed to identify the sequence environment where the modification occurs. When the antibody identifies the PTM inside a specific sequence surrounding, the existence of PTM at other locations stays undiscovered.

The mass spectrometric recognition of certain PTMs is performed on the basis of mass variations (Farley and Link 2009). Depending on the kind of modification, a specific mass modification of the altered amino acid or peptide is carried out. The mass variation is then measured by the mass spectrometer to detect the existence of a PTM in a peptide specimen. With the help of tandem mass spectrometry techniques, the specific location of the PTM can be determined through subsequent fragmentation and sequencing of the respective peptide (Meng et al. 2005). However, technical difficulties complicate the mass spectrometry-based analysis of biologically relevant PTMs such as ADP-ribosylation, which is one of the most critical signaling molecules governing DNA repair, a crucial process in preserving genome stability that is impaired in cancer and aging (Gibson et al. 2016; Gupte et al. 2017).

More and more data suggest PTMs not merely during the aging process and/or under pathological conditions, but additionally in normal cellular processes (Gupte et al. 2017; Banushi et al. 2016). In return, PTMs are becoming more widely examined for their contribution to health and disease. For instance, the accurate and exact determination of different PTM-containing compounds provides a possible usefulness of biomarkers to support early diagnosis, prognosis, monitoring of response to therapy and the decision on whether to enroll them in clinical investigations when new drugs are generated (Russell et al. 2014).

However, technical constraints restrict these investigations and render unanswered numerous questions. The discovery of unidentified/unanticipated PTMs by reanalysis of proteomic data represents an upcoming field of proteomics, which has recently been promoted due to the growing accessibility of raw data commonly shared in public repositories. Above all, however, a screening of the proteome in a particular organism or cell yields only a momentary picture of a large-scale dynamic process that confuses the analytical problem and finally favors time-resolved assessments (Walsh et al. 2005). Therefore, although many instruments are currently at hand for the investigation of PTMs, new techniques are required to continue the investigation of these modifications.

14.8 Effect of Aging on PTMs

In general, protein PTMs arise either as a consequence of the modification of enzymes associated with post-translational processing, such as glycosylation, or the activation of signaling pathways, including phosphorylation. PTM profiles are also recognized to be influenced through disease conditions (Liddy et al. 2013). In a similar manner, the deregulation of PTM is connected to the process of aging (Cloos and Christgau 2004; Truscott and Friedrich 2014; Gorisse et al 2016; Kikis et al 2010). In this respect, both enzymatic and non-enzymatic PTMs can experience age-induced modifications. The change in the profile of non-enzymatic PTMs relies primarily on the nature of the altering substances such as metabolites and free radicals. For example, reactive oxygen species may result in the oxidation of amino acid side chains, including the oxidation of thiols to various forms, oxidation of methionine and generation of carbonyl groups, modification by-products of glyco-oxidation and lipoxidation and the creation of protein–protein cross-links and oxidation of the protein backbone, leading to fragmentation of the protein (Baynes 2002). In contrary, alterations in the nature of enzymatic PTMs are primarily based on the activities of the modifying enzymes. In addition, an outline of several of the most well-characterized PTMs associated with aging and age-associated diseases at various degrees of biological complexity will be discussed.

Short Overview of the PTM Types

Protein PTMs can be subdivided into two broad categories (referred to as scheme 1). The first category includes the covalent addition of a chemical group through enzymatic catalysis. A typical application is the attachment of an electrophilic fragment of a cosubstrate to an electron-rich protein side string, functioning as a nucleophile upon transfer. The other category of PTMs comprises the covalent cleavage of peptide backbones. This cleavage takes place via one of two mechanisms: proteases or, more rarely, by autocatalysis. Typical covalent protein PTMs comprise acylation, alkylation, glycosylation, oxidation and phosphorylation. These PTMs, which are catalyzed through specific mechanisms (referred to as scheme 2), contribute to aging and age-related diseases. A short explanation of the major types of PTMs linked to aging and age-related diseases is presented in the following.

14.8.1 Phosphorylation of Proteins

The most frequent post-translational modification, protein phosphorylation, comprises the reversible addition of a phosphoryl group from adenosine triphosphate (ATP) primarily to serine, threonine or tyrosine residues. This alteration induces conformational variations that either firstly influence the catalytic activity to activate or inactivate the protein and/or the propensity of a protein to fold and agglomerate incorrectly (Fujiwara et al. 2002) or secondly attract other proteins to attach; both lead to modified protein function and cell signaling (Yaffe 2002). Phosphorylated proteins play crucial and well-known roles in various cellular activities in eukaryotes, although phosphorylation also takes place in prokaryotic cells. In humans, it is assumed that about one-third of proteins serve as substrates for phosphorylation (Cohen 2000). In fact, phosphorylated proteins are currently detected and characterized through high-throughput phosphoproteomics investigations.

The reverse nature of protein phosphorylation is explained by the action of kinases and phosphatases, both of which phosphorylate and dephosphorylate substrates. The temporal and spatial equilibrium of kinase and phosphatase concentrations inside a cell conveys the magnitude of its phosphoproteome (Manning et al. 2002). Consequently, phosphatases have been suggested lately as promising next-generation therapeutic targets for age-related diseases including α-synucleopathies in Parkinson's disease (Braithwaite et al. 2012).

14.8.2 *N*-Acetylation of Proteins

N-acetylation refers to the reversible or irreversible transmission of an acetyl group to a nitrogen molecule by the cleavage of methionine by methionine aminopeptidase (MAP) and the addition of an acetyl group by the action of the acetyl-CoA by *N*-acetyltransferase (NAT). Curiously, 80–90% of eukaryotic proteins have been acetylated, although the deeper biological significance is still unknown (Walsh 2006). In the histone proteins that comprise chromatin, lysine acetylation controls gene transcription, which influences the transcriptome of the cell. Histone acetylation typically leads to activation of transcription, whereas deacetylation typically leads to suppression of transcription. Acetylation takes place through histone acetyl transferases (HATs) and can be reversed through the activity of histone deacetylases (HDACs). A group of histone deacetylases include silent information regulators (sirtuins), which keep genes silenced by hypoacetylation. Sirtuins have been shown to be necessary for the stability of the genome (Imai et al. 2000).

Although first reported in histones, acetylation is also seen in cytoplasmic proteins. Acetylated proteins may also be manipulated through cross-talk with other PTMs, such as methylation, phosphorylation and ubiquitination (Yang and Seto 2017). Thus, the acetylation seems to fulfill not yet revealed functions in cell biology that lie beyond the regulation of transcription (Glozak et al. 2005).

14.8.3 Glycosylation of Proteins

Protein glycosylation entails the inclusion of a large number of sugar units. This major type of PTM has considerable effects on the folding, their conformation, distribution, stability and activity. Glycosylated proteins contain incorporated simple monosaccharides, such as nuclear transcription fac tors, or even highly complex branched polysaccharides, such as cell surface receptors. It is assumed that more than 50% of all mammalian proteins are hypothesized to be glycosylated (Jung et al. 2001). However, the functions of glycoproteins, both at the molecular and cellular level, are still not understood. While proteins show enhanced stability and improved trafficking after glycosylation in vivo, glycan structures may change the functions or activities of entire proteins. These patterns are often the product of the activities of the glycan-processing enzymes that operate inside a cell at a particular time. Nevertheless, the structures are occasionally protein-specific, due to the characteristics of protein trafficking and interactions with different cellular components (Moremen et al. 2012).

There exist three different types of protein glycosylation in higher eukaryotes, such as N-linked, O-linked and C-linked types. These types are defined by their glycosidic connections to amino acid side chains (Spiro 2002). In N-linked glycosylation, the β-*N*-acetylglucosamine (GlcNAc) is connected via an amide linkage to the side chain of Asn in an AsnXaaSer/Thr group (Kornfeld and Kornfeld 1985). N-linked glycans fulfill various functions. In cell–cell interactions, they function as ligands for glycan-binding proteins, whereas they also can guide the accumulation of glycoproteins within the plasma membrane and thereby impact the half-life of antibodies, cytokines and hormones in serum (Ohtsubo and Marth 2006).

The O-linked glycosylation in higher eukaryotes can take place via several various mechanisms. The most frequent type of O-linked glycosylation represents the mucin-type, which is associated with the binding of an α-*N*-acetylgalactosamine (GalNAc) to the hydroxyl group of Ser/Thr side chains

(Brockhausen 2006; Tarp and Clausen 2008). Moreover, the abnormal expression of mucin-type O-linked glycans take place in cancer cells (Brockhausen 2006) and therefore are suitable targets for the development of anticancer vaccines (Galonić and Gin 2007).

O-linked glycosylation, which takes place by adding α-*O*-mannose, is the only type of O-linked glycosylation in yeast; however, it also exists in the brains of higher eukaryotes (Barresi and Campbell 2006; Lehle et al. 2006). Higher eukaryotes show also a α-*O*-Fucose alteration of Ser/Thr residues existing within the consensus sequence CysXaa(3-5)Ser=ThrCys (Harris and Spellman 1993). This glycosylation alters the notch signaling in the course of eukaryotic development (Moloney et al. 2000; Haltiwanger and Stanley 2002). Another modification represents the β-*O*-galactosylation, which seem to facilitate the rheumatoid arthritis (Bäcklund et al. 2002a, b; Michaëlsson et al 1994).

In the end, C-linked glycosylation entails the inclusion of α-mannose (Man) at the 2-position of the indole side chain of tryptophan groups (Löffler et al. 1996, Krieg et al. 1997). It was first discovered at ribonuclease 2, but it is also present in several other proteins, among them MUC5AC and MUC5B (Perez-Vilar et al. 2004), thrombospondin (Muroi et al. 2007), and the Ebola virus soluble glycoprotein (Falzarano et al. 2007).

14.8.4 Ubiquitination and Sumoylation of Proteins

Ubiquitination involves the insertion of an 8 kDa polypeptide into the N-terminus of targeted proteins through the C-terminal glycine of the ubiquitin. The incorporation of a ubiquitin is succeeded by the creation of a ubiquitin polymer. The subsequent polyubiquitinated proteins are detected from the 26S proteasome via the pathway of protein degradation (Jentsch et al. 2009). Protein sumoylation represents a reversible post-translational modification in which a small ubiquitin-like modifier (SUMO) becomes covalently coupled to proteins (Albuquerque et al. 2014). Consequently, protein sumoylation occurs through a reversible enzymatic sequential pathway comparable to the ubiquitination of proteins (Anderson et al. 2009). Similar to ubiquitin, SUMO undergoes a cascade of activating, conjugating and ligating enzymes to conjugate to target proteins' lysine side chains and can be eliminated via SUMO-specific isopeptidases (Albuquerque et al. 2014). In recent decades, it has been shown that sumoylation governs a wide range of different characteristics of the nuclear function (Henley et al. 2014). However, current investigations have started to uncover a decisive function of protein sumoylation in numerous extranuclear neuronal activities and possibly in a variety of neuropathological disorders (Martin et al. 2007).

14.8.5 S-Nitrosylation of Proteins

Nitrosylation refers to the reversible acquisition of a nitrogen monoxide (NO) by cysteine residues to form S-nitrosothiols (SNOs) through redox-driven reactions. S-nitrosylation is utilized by cells to stabilize proteins, control the expression of genes and deliver nitrogen monoxide donors. SNO synthesis, its localization, activation and catabolism are strictly regimented, and S-nitrosilation reactions rely, among other things, on the catalytic levels of transition metals, O_2, O^{-2} and pH (Stamler et al. 1992a, b). In fact, these molecules actually possess a short half-life due to the action of enzymes such as glutathione (GSH) and thioredoxin, which denitrosylates proteins (Bechtold and King 2012). S-nitrosylation has become increasingly acknowledged as a ubiquitous regulatory response similar to phosphorylation. SNOs can perform an important role in many processes ranging from signal transduction, the repair of DNA, the defense of the host and monitoring of blood pressure to the regulation of ion channels and neurotransmission (Foster et al. 2009).

The specificity of S-nitrosylation can be gained primarily through two strategies. The presence of a consensus motive based on nitrosylation acid has been hypothesized (Stamler et al. 1997). The specificity of S-nitrosylation can also be obtained by localizing the NOSs subcellularly that may be located near prospective targets. The influence of nitric oxide on cells varies depending on its spatial concentration, the redox level of its surroundings and the responsiveness of the target sites to alterations (Yao et al. 1997). Various levels of accessibility to nitric oxide (with the general formula of RSNO, where R is an organic group) or altered reaction rates with nitric oxide, and relevant functional alterations in the -SH group being altered by nitric oxide, may help to figure out why and how distinct S-nitrosylation of specific cysteine residues triggers the modification of proteins (Broillet 1999). A typical SNO example are caspases, which regulate the process of apoptosis. The caspases are stored as SNOs in the mitochondrial intermembrane cavity, then liberated into the cytoplasm and denitrosylated. Once activated, the caspase subsequently initiates apoptosis (Benhar et al. 2009).

14.8.6 Methylation of Proteins

The alkyl substituents are bound to certain areas of the proteins through PTM enzymes. The insertion of these alkyl groups leads to a change in the hydrophobicity of the altered protein (Marmorstein 2003). The most frequent type of protein alkylation involves the methylation of proteins. The methylation process comprises a known PTM which is facilitated by methyl transferases. Single-carbon methyl groups are attached to nitrogen or oxygen (*N*- or *O*-methylation) on amino acid side chains, which enhances the hydrophobicity of proteins or neutralizes a negative charge upon attachment to carboxylic acids. Although *N*-methylation is irreversible, *O*-methylation has the potential to be reversible. Methylation is so common that its primary methyl donor, S-adenosylmethionine (SAM), is proposed to be the second most frequently employed enzymatic substrate following ATP (Walsh 2006).

A common topic for methylated proteins, as also for phosphorylated proteins, is the influence of this alteration on the control of protein–protein interactions. For example, arginine methylation of proteins may either hinder or encourage protein–protein cross-talk, according to the nature of the methylation (Nakayama et al. 2001; Grewal and Rice 2004). The methylation of proteins has been examined most often in histones. The transfer of methyl groups from S-adenosyl methionine to histones is facilitated through the catalytic action of enzymes commonly known as histone methyltransferases. The N-terminal residues of the histones H3 and H4 acquire methyl groups on certain lysines. Methylation subsequently defines whether gene transcription is enabled or disabled, which results in varying biological characteristics (Kouzarides 2007).

The histone methylation has been traditionally regarded as irreversible. Nevertheless, histone demethylases clearly display the reversibility of this PTM (Shi et al. 2004). In fact, the dynamic alterations in chromatin alterations were forced by the capacity or incapacity to keep the balance of the opposing actions of methylases and demethylases. The concurrent removal of one histone methylation label and the addition of a second one permits sequential transcriptional tuning (Christensen et al. 2007, Agger et al. 2007). Non-histone proteins display a common PTM, the methylation process, which governs signal transduction through BMP, Hippo, JAK-STAT, MAPK and WNT signaling pathways. In addition, methylation interacts with other kinds of PTMs, such as with histone and non-histone proteins, to affect not only chromatin design but also gene transcription, synthesis of proteins and repair of DNA (Biggar and Li 2015).

14.8.7 Oxidation of Proteins

The reaction of proteins with a multitude of free radicals and reactive oxygen species (ROS) results in oxidative protein changes, such as production of protein hydroperoxides, hydroxylation of aromatic groups and aliphatic amino acid residues, oxidation of sulfhydryl or methionine residues, conversion of distinct amino acid residues into carbonyl groups, break-down of the polypeptide chain and creation of cross-linking linkages. Aromatic and sulfurous residues are sensitive to oxidative alterations (Tarp and Clausen 2008; Galonić and Gin 2007; Barresi and Campbell 2006).

If they are not repaired or removed from the cells, oxidized proteins are frequently toxic and can compromise cell viability (Berlett and Stadtman 1997), since oxidatively altered proteins can build large agglomerates (Grune et al. 2003). Oxidatively impaired proteins are subjected to selective proteolysis, mainly by the 20S proteasome in a ubiquitin- and ATP-independent manner. Eventually, after substantial protein oxidation, these assemblies can be rendered increasingly robust against proteolytic degradation and effectively engage the 20S proteasome and irreversibly impede its activity (Harris and Spellman 1993; Moloney et al. 2000; Haltiwanger and Stanley 2002).

Protein carbonylation is characterized as an irreversible PTM in which a reactive carbonyl part, such as an aldehyde, ketone or lactam, is incorporated into a protein. The first documented source for protein-bound carbonyls has been the metal-catalyzed oxidation (MCO) (Stadtman and Oliver 1991). MCO arises from the Fenton reaction, when transition metal ions are being reduced in the vicinity of hydrogen peroxide, thereby producing the highly reactive hydroxyl radicals (Stadtman and Berlett 1998). These hydroxyl radicals are capable of oxidizing amino acid side chains or cleaving the protein backbone, resulting in multiple changes, among them reactive carbonyls (Stadtman and Levine 2003). For instance, the oxidation of proline and arginine leads to the generation of glutamine semialdehyde, whereas lysine is being oxidized to aminoadipine semialdehyde and threonine is oxidized to 2-amino-3-ketobutyric acid (Møller et al. 2011). The direct oxidation of any other amino acid residues can also result in protein-bound carbonyls. Tryptophan oxidation through ROS generates a minimum of seven oxidation products. These include kynurenine and N-formylkynurenine and their hydroxylated analogs comprising aldehyde or keto groups produced through oxidative cleavage of the indole ring (Müller and Leichert 2013).

Another major origin of protein-bound carbonyls are the reactive products of lipid peroxidation, resulting from the oxidation of polyunsaturated fatty acids (Perez-Vilar et al. 2004; Muroi et al. 2007; Falzarano et al. 2007; Jentsch et al. 2009). The protein carbonylation can also take place through glycoxidation. Reactive α carbonyls produced during the glycoxidation, including glyoxal, methylglyoxal and 3-deoxyglucosone, may then alter the basic residues Lys and Arg to produce, for instance, pyrralines and imidazolones (Albuquerque et al. 2014; Anderson et al. 2009). Glycation, for instance, the reaction of reducing sugars, such as glucose or fructose, with the side chains of lysine and arginine residues, produces Amadori and/or Hynes compounds. These glycated residues may be broken down further with ROS into advanced glycation end products (AGE), which bear carbonylated parts that may also assist in protein carbonylation (Liggins and Furth 1997).

14.8.8 PTMs Act in Aging and Aging-Linked Diseases

The disruption of cellular homeostasis in the course of aging modifies tissue functions, resulting in a global reduction in physical/mental well-being and finally in death. With advancing age, the ability to regulate gene expression, which is controlled by several epigenetic factors, diminishes. The epigenetic regulation of chromatin rearrangement through histone acetylation is linked to cell metabolism (Pletcher et al. 2002; Wellen and Thompson 2012). Metabolic alterations with aging influence the concentration of acetyl-CoA and citrate, which in return changes the cytosolic acetyl-CoA level. Modified acetyl-CoA concentrations thus impact on different metabolic processes such as fatty acid synthesis, which then in turn impacts on other physiological processes. In addition, altered acetyl-CoA concentrations influence the histone acetylation, thereby impairing transcription (Pletcher et al. 2002; Wellen and Thompson 2012). These transcriptional alterations arise with aging or with the advancement of aging-induced diseases. Acetylases and deacetylases probably exhibit varying affinities to their acetyl-CoA and NAD+ respectively, which impacts their reactions to age-related alterations in cofactor levels (Hayflick 1998). This enables the chromatin to perceive alterations in cell metabolism (Peleg et al. 2016a, b). In point of fact, the lifespan can be prolonged by various modifications that invert age-related alterations in the chromatin structure, pointing to the central importance of the chromatin structure during aging (Feser and Tyler 2011). In line with this, mutations in genes connecting the metabolism and chromatin, including lysine acetyltransferases (KATs), lysine deacetylases (KDACs) (sirtuins), and ATP citrate lyase (ACLY/ATPCL), are able to impact the lifespan and the support the development of age-linked diseases (Peleg et al. 2016a, b).

It has been proposed that protein acetylation holds a pivotal function in the aging process through improving the function of specific genes, in particular the regulatory subunit AMPK, which can enhance survival (Lu et al. 2011). Similarly, it is broadly recognized that sirtuins, a class of proteins that stimulate stress responses and metabolism by eliminating acetyl groups from a target protein, play a role in lifespan and the aging mechanism (Thanan et al. 2014; Campbell and Solaimani 2016). In particular, sirtuin SIRT3 appears to be involved in the deacetylation of numerous proteins in the mitochondria, indicating that acetylation/deacetylation might be associated with the control of mitochondrial regulation (Newman et al. 2012).

In addition, it has been observed that the limitation of calories, an intervention known to prolong life in numerous organisms from budding yeast to mammals, is linked to significant alterations in mitochondrial acetylation. Acetylation modifies several proteins in reaction to a calorie limitation (Baraibar et al. 2012; Dias et al. 2013). These alterations can assist mitochondrial adjustment to decreased calorie consumption and enhance longevity. It was also found that regular physical activity decreases oxidatively altered proteins in the brain with enhanced cognitive functions (Goto and Radak 2013), through mechanisms in which PTMs are present in histone tails regulated by HATs, HDACs and histone demethylases (Pareja-Galeano et al. 2014).

Numerous pathways and processes seem to govern the rate of aging and the vulnerability of the organism to age-related diseases including atherosclerosis, cancer and neurodegeneration. One of the processes that is becoming more frequently involved is autophagy. First reported in yeast, autophagy is a strictly controlled process triggered through stress factors, such as hunger. Once autophagy is activated, aged and defective proteins and organelles are recycled to supply the necessary elements for new cell components. Consequently, a disturbance of this mechanism leads to abnormal phenotypes and reduced lifespan, as experiments on mouse models (Grewal and Rice 2004; Kouzarides 2007; Shi et al. 2004), *Caenorhabditis elegans*, *Drosophila melanogaster*, and *Saccharomyces cerevisiae* have revealed (Stamler et al. 1997; Yao et al. 1997; Broillet 1999; Benhar et al. 2009).

While the core constituents that regulate autophagy have been extensively investigated (Stolz et al. 2014), little is understood about the factors that may specifically modify this process, in specific how post-translational modifications may impact autophagy flux and/or the rate of autophagous turnover. Three types of PTM, such as phosphorylation, ubiquitylation and acetylation, which are critical for the induction, regulation and fine-tuning of autophagy and are affected through a number of diverse stimuli. A better knowledge of the mechanisms of autophagy control will give biogerontologists a more profound view of the process and open up new therapeutic approaches (McEwan and Dikic 2011).

14.8.9 Oxidation and Aggregation of Proteins

One of the first references to the impact of oxidative stress in cells can be seen in a characterization of the chemical nature of prooxidative and antioxidative molecules (Sies 1997). A state of balance between oxidative and antioxidant actions preserves cellular health, while a state of imbalance is linked to disease and aging. ROS represent hallmarks of the oxidative damage. The consequences of an unbalanced redox status of cells mainly concern the alteration of redox-sensitive molecules, such as the oxidation of cysteine and methionine within proteins, the peroxidation of lipids and the oxidation of the bases of the DNA (López-otín et al. 2013). The implications of these alterations comprise direct influences on disease-initiating proteins and indirect influences on enzymes and/or cofactors, which in turn affect the functionality of the disease-initiating proteins (Thanan et al. 2014; Campbell and Solaimani 2016).

Several investigations have highlighted proteins that are implicated in conveying or combating the generation and action of reactive oxygen species. Moreover, there has been a focus on aging-dependent oxidative damage, and the damage accumulation theory of aging has formulated that chronic oxidative damage represents the primary reason for age-related diseases (Hohn et al. 2016). Cellular senescence, which is understood as the impairment of cell division, motility and the turnover of proteins, is a consequence of the cumulated damage over time and is regarded as an essential characteristic of aging (López-otín et al. 2013). Morphological alterations caused by the clustering of protein aggregates in the cells are also seen as a characteristic of cellular senescence caused by oxidative protein injury.

A broad spectrum of age-related diseases is associated, albeit at least partially, with oxidative damage through proteins. These disorders comprise eye diseases, metabolic disturbances such as diabetes and obesity, inflammatory diseases such as arthritis, cardiovascular complications including atherosclerosis, kidney disease, respiratory diseases, cancer and neurodegenerative diseases, for example, Alzheimer's (Baraibar et al. 2012) and Parkinson's (Dias et al. 2013) diseases. Correspondingly, it has recently been suggested that aging and age-related diseases may be phenotypic results of proteome damage schemes (Radman 2016).

Eye cataracts are a frequent disease of the aging populations, which leads to a progressive deterioration of eyesight. One of the most important fundamental alterations in cataract development occurs when protein accumulation takes place in the lens of the eye. Whereas environmental parameters such as smoke, UV radiation and chemical fumes are responsible for the development of cataracts, protein PTMs also play an essential part in the structure and stability of lens proteins, leading to their accumulation within the lens (Yanshole et al. 2013). Protein oxidation is an especially important factor in lens protein assembly, and antioxidants are frequently used in clinical treatment of cataracts (Mathew et al. 2012). Experimental investigations with human and mouse models have revealed cysteine oxidation occurring at the crucial sites of several enzymes in the human and mouse lens, among them several metabolic enzymes, namely glyceraldehyde-3-phosphate dehydrogenase (GAPDH), glutathione synthase, aldehyde dehydrogenase and sorbitol dehydrogenase along with the protein deglycase-DJ-1 synonymously known as Parkinson's protein 7 (PARK7) (Wang et al. 2017). Comprehensive oxidation of intermediate filamentary proteins including BFSP1, BFSP12, vimentin, cytokeratins, microfilamentary and microtubule filamentary proteins, such as tubulin and actin, has also been documented (Wang et al. 2017).

Alzheimer's disease is a key age-related disturbance that seriously impairs the elderly person's quality of life (Vandenberghe and Tournoy 2005). The symptoms of Alzheimer's disease are a reduction in cognitive function and memory and a condition of confusion. At the cellular level, Alzheimer's disease can be primarily attributed to two proteins, such as Tau and amyloid β. Moreover, the dislocation of the microtubule-associated protein tau from the cytoskeleton within neuronal cells causes its accumulation within the cell in paired helical filaments that are referred to as neurofibrillary tangles. The extracellular aggregation of amyloid β peptide inside the brain represents another important factor contributing to the Alzheimer's disease pathology. The amyloid-β peptide is formed through the break-down of the amyloid precursor protein (APP). Normally, the peptide is decomposed through proteases, including zinc proteases referred to as neprilysins, endothelin-converting enzymes and insulin-degrading enzymes (Nalivaeva and Turner 2017). Oxidative stress experienced in the course of aging may help to suppress amyloid-degrading enzymes, resulting in abnormal extracellular aggregation of amyloid-β peptide levels within the brain (Zhao et al. 2013).

The progression of Alzheimer's disease is associated with the hyperphosphorylation of tau. Hyperphosphorylated tau protein is detected in degradation-resistant helical filament centers of neurofibrillary entanglements (Iqbal et al. 2013). Interestingly, a newly published investigation has demonstrated that hydrogen peroxide-based oxidative stress may temporarily decrease the phosphorylation level of tau (Katai et al. 2016). In addition, 8-nitroguanosine 3′,5′-cyclic monophosphate (8-nitro-cGMP), which is a second messenger of the nitric oxide (NO) signaling path that induces the oxidative S-guanylation of cysteine residues, leads to a decrease in tau clustering (Yoshitake et al. 2016). The pathology of Alzheimer's disease includes post-translationally altered versions of Aβ and Tau and various other proteins. The investigation of these PTMs holds the crucial importance for a deeper insight into the molecular interactions linked to the initiation of disease, and also offers new possibilities for therapeutic approaches and the generation of new drugs.

Parkinson's disease has been identified as another neurodegenerative disease of uncertain origin impacting approximately 6.3 million people around the world (Rijk et al. 2000). The pathological characteristic lesions of Parkinson's disease include Lewy bodies, intraneuronal proteinaceous enclosures consisting primarily of misfolded α-synuclein (Spillantini et al. 1997). Lewy bodies comprising accumulated α synuclein are not only encountered in Parkinson's disease but also in many other neurodegenerative diseases such as multiple system atrophy, dementia with Lewy bodies or Alzheimer's disease (Spillantini et al. 1998). Post-translational changes of α synuclein, such as phosphorylation, ubiquitination or nitration, are implicated in the clustering mechanism of α synuclein and have various effects on the cellular neurotoxicity (Chen and Feany 2005; Oueslati et al. 2010; Lee et al. 2008; Hodara et al. 2004; Norris et al. 2003).

The molecular mechanism implicated in the elucidation of α-synuclein agglomerates is a key issue for the elucidation of α-synuclein-related toxicity. However, the use of unicellular model organisms could provide clues to the decoding of protein aggregation, which could subsequently lead to the comprehension of α-synucleopathies. The heterologous expression of α-synuclein in *Saccharomyces cerevisiae* also results in protein accumulation and cellular toxicity which is typical for Lewy bodies bearing human cells. In *S. cerevisiae* the cellular clearance mechanisms encompass ubiquitin-facilitated 26S proteasome function and lysosome/vacuole-associated degradation pathways, including autophagy. Several post-translational modifications have been identified that alter the cytotoxicity of α-synuclein and its dispersion to various clearance routes in *S. cerevisiae*. Several of the identified sites of variation seem to be preserved in a conserved manner from yeast to humans (Popova et al. 2015). Aging has also been identified as a major predictor of cardiovascular disease, including hypertension, coronary heart disease, stroke and coronary insufficiency. Several experimental and clinical findings confirm the hypothesis that increased oxidative stress or the production of ROS supports the pathogenesis of these disorders (Sugamura and Keaney 2011). For example, oxidative injury in cardiovascular disease is mainly associated with low density lipoproteins (LDL) that generate lipid peroxidation compounds including lipid peroxides, isoprostanes, oxysterols, hydroxy fatty acids and aldehydes (Zhang et al. 2014).

Similarly, most recent investigations of BMAL1 (brain and muscle ARNT-like protein-1) have demonstrated that Bmal1 null mice are aging abnormally because of elevated ROS generation. These mice also exhibited an age-dependent decrease in heart function, which was characterized by alterations in the ventricular diameter and the fraction of sputum (Hemmeryckx et al. 2016). The antioxidant 4-hydroxy-2,2,6,6-tetramethylpiperidine-1-oxyl (TEMPOL) therapy avoided an impairment of heart function in these mice. The preservation of cardiac telomeres from oxidation by TEMPOL in BMAL1-deficient mice has also been demonstrated, which further enhances the therapeutic significance of specifically influencing protein oxidation in old age (Hemmeryckx et al. 2016). Investigations of acute kidney injury and chronic kidney disease in the course of aging have also pinpointed to the function of oxidatively damaged proteins and accumulation of proteins (Sureshbabu et al. 2015).

Proteins that are oxidatively compromised in the kidney contain NADPH oxidase (NOX), heme oxygenase-1 (HO-1), thioredoxin 1 (TRX1) and the transient receptor potential cation channel, subfamily M, member 2 (TRPM2). Therefore, an equilibrium between oxidative stress and autophagy has been identified as a key determinant regulating the inflammation and cellular death in kidney disease (Lenoir et al. 2016). Various metabolic disturbances including obesity, resistance to insulin and diabetes are accompanied by elevated body weight, elevated glucose levels and decreased energy levels. Environmental and nutritional burdens are regarded as the principal causes of such disturbances of metabolism, which may cause damage by oxidation. The aggregation of reactive oxygen species conveys an oxidative amendment of metabolic enzymes and proteins, similar to the eating of carbohydrate- or fat-rich food (Newsholme et al. 2016). The enzyme methionine sulfoxide reductase A (MsrA) is a cell antioxidant enzyme associated with the fight against the impact of oxidative stress and is instrumental in the establishment of protection against oxidative stress and the preservation of proteins, which are two critical factors in the process of aging (Cabreiro et al. 2006). In transgenic mice, it has been determined that MsrA influences lifespan and mitigates several of the impacts of age-related metabolic disturbances, such as resistance to insulin (Salmon et al. 2016).

In summary, these findings highlight the importance of oxidative damage to proteins in the process of aging and age-related pathologies. Therefore, pharmacological and non-pharmacological approaches that affect the oxidative stress level of the cell are essential as the proximal responses to prolonging the health span.

14.8.10 Chlorination of Proteins During in Aging

Reactive chlorine species are the primary resource for enzymatically catalyzed chlorination of proteins (Sadowska-Bartosz et al. 2014). The free hydroxyl-containing tyrosine represents the principal amino acid goal for a halogen alteration. The enzyme myeloperoxidase acts as a catalyst for the production of 3-chlorotyrosine (Hazen et al. 1997). By means of protein chlorination, it could be shown that a tyrosine residue in apolipoprotein A-I (ApoA-I) acts as a region either for chlorination or for nitration due to the action of myeloperoxidase or peroxynitrite (Shao et al. 2005). Importantly, chlorination rather than nitration impaired the function of ApoA-I and significantly decreased its activity for the cholesterol efflux. Increased amounts of myeloperoxidase are linked to chronic cardiac insufficiency, and their expression rises in cardiac endothelial cells after exposition to hydrogen peroxide (Rocca et al. 2009). A very recent investigation showed that inhibition of myeloperoxidase with 2-tioxanthines led to a decrease in protein chlorination in a mouse model for peritonitis. (Tiden et al. 2011).

Skin aging is usually employed as a physiological parameter to evaluate age-related alterations within the body. It has been proposed that there may exist a connection between inflammation-induced protein denitration and light-induced aging of the skin (Ishitsuka et al. 2012). Moreover, increased amounts of halogenated tyrosine and inflammatory cells have been detected in skin samples, which have been either exposed to light or protected from it, which indicates that halogenation belongs to the physiological process of aging.

Neurodegeneration represents a further prominent manifestation of aging, which is caused by a number of factors, among them oxidative stress. A recent study concluded that serotonin functions as a scavenger of hypochlorous acid (HOCl) (Kalogiannis et al. 2016) and inhibits HOCl-driven oxidation of 2-thio-5-nitrobenzoate, depletion of cellular α ketoglutarate dehydrogenase activity and prevents cell death. Curiously, biphasic elimination of HOCl and consequent inhibition of oxidation of 2-thio-5-nitrobenzoate includes HOCl-driven chlorination of serotonin and development of inactive clusters of chlorinated serotonin, which induces a feedback mechanism. In addition, specific serotonin reuptake blockers such as fluoxetine decrease the chlorination of proteins in the brain, indicating a possible therapeutic strategy to combat age-related protein chlorination processes (Kalogiannis et al. 2016).

14.8.11 Nitration of Proteins

Nitration represents an oxidative protein alteration that takes place on tyrosine residues. Superfluous amounts of reactive nitrogen species (RNS) are the major origin of nitration reactions (Sadowska-Bartosz et al. 2014). The increased abundance of ROS, along with the RNS, results in the production of supplementary nitrating units, specifically peroxynitrite. A common instance of RNA-initiated protein nitration involves the generation of 3-nitro-tyrosine, which is connected with elevated nitrooxidative stress throughout the aging process (Ljubuncic et al. 2010).

Tyrosine nitration alters the biochemical characteristics of the amino acid, among them its pKa, redox potential, hydrophobicity and size, consequently bringing about significant alterations in the structure and function of the concerned proteins. Changes in the biochemical composition of proteins trigger the cellular and physiological phenomena of nitration in the context of the aging pathway. In addition, the nitration of tyrosine is facilitated through non-enzymatic free radical reactions in which an intermediate tyrosyl radical is generated. Investigations with rapid reaction kinetics, bioanalytical techniques and structural evaluations by paramagnetic electron resonance have made it possible to fully assess the characterization of tyrosine nitration (Radi 2013). Most recent investigations have revealed that membrane-connected tyrosine nitration of proteins comprises the oxidation by lipid peroxyl radicals, which is a side product of membrane lipid peroxidation and is therefore also closely related to aging (Bartesaghi et al. 2010). In addition, several investigations have shown that the site-specific protein tyrosine nitration appears on a few tyrosine residues inside the selected target proteins and is thus limited to a part of the proteome (Radi 2013). The localization of the nitrating units in space and time is a key factor in the choice of tyrosine residues inside a target protein. Investigations of mitochondrial proteins which are homogeneously nitrated provided additional evidence for the site-specific selectivity and global implications of protein tyrosine nitration in aging and age-associated diseases (Yang and Seto 2017; Glozak et al. 2005). While the nitration of protein tyrosine was at first regarded as an irreversible process, however, more recent research has discovered a denitrase enzyme (Deeb et al. 2013).

Denitrase activity is present in a number of tissues and cells, whereas is missing in smooth muscle cells. A recent study has demonstrated that denitrase uses nitrated cyclooxygease-1 (COX-1) as a catalyst to promote the denitration reaction, which indicates that the reversible cycle of nitration and denitration seems to fulfill an important role in the regulation of oxidative/nitrosative cell stresses that afterward regulate aging (Deeb et al. 2013).

In another recent report on the effect of protein nitration in age-related systemic inflammatory response syndrome (SIRS), it has become clear that lung damage caused by toxemia raises the level of protein nitration of tyrosine and lowers the activity of superoxide dismutase in the lungs of mice (Starr et al. 2011). In addition, aged mice demonstrated an increased protein nitration in the vascular endothelia in comparison to younger mice. The specific molecules that sustain pulmonary vascular permeability also exhibited elevated tyrosine nitration, which comprises profilin-1, transgelin-2, LASP 1, tropomyosin and myosin (Starr et al. 2011).

14.9 Bacteria May Serve as a Model for PTMs in Aging and Aging-Linked Pathologies

The monitoring of senescence in unicellular organisms in the lack of genetic or environmental variations paved the way for proposals that such organisms might be employed as simple quantitative experimental model systems to examine the molecular mechanisms underpinning the aging process (Muroi et al. 2007; Pletcher et al. 2002). Bacterial aging appears to bear certain similarities with the process of eukaryotic aging, specifically the part played by oxidative damage and the impact of quality management mechanisms for proteins in inducing senescence (Ballesteros et al. 2001).

For example, bacterial aging, as in eukaryotes, is connected with the aggregation of oxidized proteins in the shape of clusters in the older cell poles (Löffler et al. 1996; Perez-Vilar et al. 2004). This aggregation is similar to numerous established age-related eukaryotic protein folding disorders (Wang et al. 2017; Vandenberghe and Tournoy 2005), and an enhanced protein accumulation and modified cell proteostasis have been implicated in oxidative, stress-related post-translational modifications, especially in eukaryotes (Tanase et al. 2016).

Whether this process also contributes to the aggregation of protein aggregates in bacteria is still not clear. The trends in oxidative protein damage and agglomeration associated with aging in *E. coli* are similar to those induced by UVA radiation, indicating that the same kind of ROS may be implicated in assessing cell damage under these two processes (Jentsch et al. 2009; Albuquerque et al. 2014). The similarity between the biological impacts of radiation and aging can easily be seen in survival graphs plotted as a function of radiation dose or time: Both show a “shoulder” pointing to insignificant mortality, accompanied by an exponential decline in survival rate with rising radiation dose or advancing age (Radman 2016). Therefore, bacteria are now regarded as beneficial model organisms for aging investigations, especially for gaining an insight into the consequences of aging and age-related stress on the strength and function of proteins (Ackermann 2008).

In the specific case of *E. coli*, a considerable part of the age-related decline in vitality is due to the existence of protein clusters that aggregate in the older bacterial poles. Misfolded proteins can accumulate in a passive and spontaneous manner on the cell poles in *E. coli* due to decreased diffusion and nucleoid plugging (Feser and Tyler 2011; Newman et al. 2012). As a result, erroneously folded proteins disseminate freely in the cytoplasm and tend to adhere to each other due to the uncovering of hydrophobic spots on their surface. When the bulk of the amorphous assemblies increases through the incorporation of more erroneously folded peptides, they are excluded from the nucleoid and begin to aggregate at the cell poles, where they can extend further. To substantiate this model, in silico modeling has revealed that the passive diffusion of particles, their inherent capacity for multimerization and the lack of nucleoids at the poles are capable of producing a polar positioning pattern simply through entropy (Laloux and Jacobs-Wagner 2014).

In addition, a multitude of post-translational modifications, such as alterations in phosphorylation state or nucleotide binding, may govern the complex intracellular partitioning of various proteins implicated in the regulation of cell cycle, signal transduction, polarity of motion and adhesion (Pareja-Galeano et al. 2014; Finkel et al. 2009). Even though most of the currently known instances concern proteins that will eventually be attracted toward the poles by protein–protein interference, related adaptations might also influence the capacity of some proteins to undergo multimerization, thus affecting their ability to accumulate instantaneously at the poles. When the availability and the activity of associated kinases, such as phosphatases and GTPase-activating proteins (GAPs), is subject to temporal control, this may offer a possibility to adjust a normally spontaneous polar localization in time as in the specific case of Streptomyces coelicolor (Finkel et al. 2009; Wu et al. 2013; Lee and Finkel 2009; Rubinsztein et al. 2011).

Protein cleavage through specific proteases could also be a suitable approach to regulate polar localization in space and time, as suggested for the polar beacon PodJ. PodJ is transformed to a shorter version (PodJS) from a long version (PodJL) through a cell cycle-regulated proteolytic sequence which ultimately breaks down the PodJS and guarantees its correct positioning and ultimately its proper function (Stolz et al. 2014; McEwan and Dikic 2011). The exact mechanisms by which both types of PodJ locate themselves in different directions at the poles, however, still need to be clarified. Nε Lysine acetylation has been acknowledged as a widely accepted regulatory post-translational modification that affects a number of important biological pathways in eukaryotic cells. Lately, it has been discovered that acetylation is also widespread in bacteria. Bacteria comprise hundreds of acetylated proteins that influence several cellular pathways. Nevertheless, relatively little is generally understood of the regulation or biological significance of almost all of these changes. To identify the potential regulatory impact of acetylation, a recent investigation examined how acetylation profiles and frequencies vary across the growth stages of *B. subtilis*. Moreover, a subset of crucial acetylation events has been figured out, which are regulated in the course of cell growth in a temporal fashion. In addition, a stationary-phase-enriched acetylation on the key conformational protein MreB was presented, which leads to the assumption that the MreB acetylation plays a role in cell width management by limiting cell wall expansion (Carabetta et al. 2016). Lysine acetylation also cooperates in a reversible fashion to manage the utilization of the carbon pool and the metabolic flux in Salmonella, enabling the cells to react to shifts in the surrounding environment through instantaneous acquisition of cellular energy levels and the flexible modification of reaction velocities or directions (Wang et al. 2010). Therefore, the acetylation of lysine might constitute a metabolic regulatory system maintained from bacteria to mammals. As more and more evidence for the maintenance of at the minimum certain characteristics of aging in bacteria is becoming apparent (Bäcklund et al. 2002b; Radman 2016; Peleg et al. 2016a, b; He et al. 2012), the role of PTMs in the regulation of bacterial aging will be important to explore.

As there is a growing recognition of the contribution of PTMs to aging and age-related diseases, the development of methodologies for faster and more precise detection of protein PTMs is a priority. Moreover, the latest discovery of uncommon and unusual alterations in age-related diseases demands the establishment of more sophisticated and delicate techniques to identify such alterations (Basak et al. 2016). The latest rapid advance in broad-based genomics and proteomics approaches is expected to be a facilitating factor for these kinds of investigations. Pharmaceuticals aiming at PTMs, such as phosphorylation, acetylation, methylation and ubiquitination, will provide a valuable instrument for investigating the underlying mechanism of PTM modification and offer a pharmacological framework to tackle the deleterious consequences of aging (Kaeberlein 2010).

In terms of non-drug interventions, training interventions are recognized as an effective way of slowing down the negative consequences of aging at the physical and metabolic scales. There is ample evidence that physical activity can provide advantages for the older person by stimulating both inflammatory and redox responses, with implications for proteostasis, sensitivity to insulin, physiological body composition such as fatty tissue and skeletal muscle, and hormonal balance, among other things. Similarly, the restriction of calories constitutes another non-genetic and almost general process which is recognized as slowing down the beginning of the aging process and increasing the maximum lifetime (Finkel 2015). Nevertheless, the impact of physical activity and diet on protein PTMs is still comparatively little subject of research. Investigations covering this specific field have the capacity to provide generally available and cost-effective treatment options for age-related diseases.

Ultimately, the usefulness of prokaryotic systems for comprehension of the biology of aging is remarkable, since it is feasible to maintain the molecular mechanisms involved in the aging process over the course of evolution. In the course of advancing research in the domain of microbiogerontology, it will be exciting to discover to which degree these molecular pathways hold up in a conserved state. This could offer a whole new range of possibilities to explore how to slow down aging and prolong a healthy lifespan.

14.10 Defects in the Protein Biosynthesis

The eukaryotic pathogens microsporidia can be hardly treated, endangering human health and industrially useful insects and fish, but the complexity of the biology of these novel pathogens is still only starting to be grasped. In this case, the combination of bioinformatic, biochemical and mass spectrometric analyses is needed to demonstrate that microsporidia bear a flaw-prone mechanism of protein synthesis and that their protein synthesis is characterized by a considerable number of translation failures. This discovery unveils a hitherto unrecognized feature of protein synthesis in these rapidly emerging parasites and provides a potential opportunity to utilize the faulty protein synthesis mechanism as a therapeutic target for the therapeutic treatment of microsporidia infections.

Microsporidia represent parasitic fungi-like organisms that can invade into the cytoplasm of living cells and lead to chronic disorders in a wide range of animals, including humans. These pathogens possess the smallest recognized genomes of all the eukaryotic species, for whom they act as a model to investigate the phenomenon of genome shrinkage in obligatory intracellular parasites. In this section, a case investigation is reported which demonstrates an obvious impact of whole genome downsizing on the primary structure and activity of aminoacyl-tRNA synthetases, essential cellular proteins necessary for the synthesis of proteins. It has been found that the majority of microsporidian synthetases are missing regulatory and eukaryotic-specific attached domains and exhibit a high level of sequence variability in both binding of tRNAs and catalytic domains. In one synthetase, the LeuRS, a seemingly sequence degeneration destroys the editing domain, which is a catalytic hub for the precise selection of leucine for the synthesis of proteins. In contrast to the exact LeuRS synthetases of other eukaryotic species, microsporidic LeuRS is prone to failures: Besides leucine, it sometimes utilizes its closely associated compounds including isoleucine, norvaline, methionine and valine. Mass spectrometric examination of the proteome of microsporidium Vavraia culicis indicates that nearly 6% of the leucine residues are incorrectly substituted through other amino acids. This strikingly abundant mistranslation is not restricted to leucine codons and seems to be a universal feature of protein synthesis in microsporidia parasites. In summary, the results confirm that the protein synthesis mechanism of microsporidia lacks an editing step and that the proteome of microsporidia parasites contains more variety than one would expect on the basis of their pure genome sequences (Melnikov et al. 2018).

14.11 Impact of Translation and Post-translational Modifications on Cell Mechanics

During the process of cell division, the chromatin is condensed into mitotic chromosomes to support the accurate separation (referred to as segregation) of the genome between two daughter cells. PTMs of histones modify the compactness of interphase chromatin, although it is still little well known how these alterations influence the stiffness and structure of the mitotic chromosome (Biggs et al. 2019). With the use of micropipette-based force measurements and epigenetic pharmaceutical drugs, the effect of canonical histone PTMs imposing interphase euchromatin, such as acetylation, and heterochromatin, such as methylation, on the stiffness of mitotic chromosomes has been examined. By quantifying the chromosome doubling force (the force needed to double the length of the chromosomes), it can be seen that histone methylation, but not acetylation, plays a role in mitotic structure and stiffness. These results are subject to further discussion in the process of chromatin-gel shaping of the long-range organization of the mitotic chromosomes.

The structure of chromatin is crucial for multiple various cellular functions. A significant shift in chromatin structure and organization takes place in the course of the transition from interphase to mitosis, when the open, diffuse, partitioned and transcriptionally available interphase chromatin becomes dense, rod-shaped and transcriptionally unavailable in mitosis (Wang and Higgins 2013; Doenecke 2014; Oomen and Dekker 2017). Even though the majority of investigations on mitotic chromatin deposition concentrates on large-scale chromatin-organizing compounds such as cohesin, condensin and topoisomerases (Vagnarelli 2012), mitosis is also connected to characteristic alterations of the histone PTMs (Wang and Higgins 2013; Oomen and Dekker 2017). Histone PTMs involve chemical alterations to histones, usually to their tails, several of which are connected to varying chromatin structures and distinct chromatin densities (Rice and Allis 2001; Wang and Higgins 2013).

The acetylation, especially of histone 3 lysine 9 (H3K9ac), has been linked to euchromatin, which is packed in a loose fashion, contains abundant genes and is transcribed in an active way (Doenecke 2014). Methylation, in particular H3K9me3 and H3K27me3, is linked to heterochromatin, which is tightly packed chromatin with poor transcriptional activity (Rice and Allis 2001; Wang and Higgins 2013; Oomen and Dekker 2017). Histone PTMs can also inherently modify the packaging of chromatin through altering the charge of the histones themselves through acetylation and incorporating hydrophobic components into the histones through methylation (Rice and Allis 2001; Doenecke 2014). Recent cryoelectron microscopic evidence has demonstrated that histones are frequently placed in such a way that histone tails are able to interoperate physically with other neighboring histone tails (Bilokapic et al. 2018), possibly permitting alteration of the chromatin structure.

It is known that alterations in histone PTMs influence the structure and stiffness of cell nuclei during the interphase. Elevated Euchromatin levels have been associated with weaker nuclei (Chalut et al. 2012; Krause et al. 2013) which results in a concrete diminishment of the nuclei' response to the short expansion force, which is governed by chromate stiffness, and leads secondarily to prolonged expansion (Stephens et al. 2017). Chromatin stiffness also influences the nuclear geometry (Banigan et al. 2017). Reduced chromatin-facilitated nuclear rigidity induced by elevated euchromatin has also been demonstrated to lead to aberrant nuclear morphology (Stephens et al. 2018), which indicates a number of cellular diseases, among them cancer (Chow et al. 2012). Elevated heterochromatin has been proven to result in stiffer nuclei and an increased level of resistance to aberrant nuclear morphology (Stephens et al. 2017, 2018). The relationships between the chromatin status and the histone PTMs with nuclear stiffness and morphology thus point to the fundamental relationships that exist between the histone PTMs and the stiffness of the chromatin.

Certain histone PTM alterations are especially connected with mitosis. The process of branding is the procedure by which some histone PTMs are preserved or become stabilized during mitosis, which presumably maintains the transcriptional state of the cell through mitosis (Wang and Higgins 2013; Doenecke 2014; Oomen and Dekker 2017). These signs are crucial for the preservation of cellular identity and function. A number of histone methyl labels, both euchromatic, such as H3K4me3, and heterochromatic, such as H3K9me3 and H3K27me3, seem to be elevated or kept constant during the process of mitosis (Xu et al. 2009; Park et al. 2011). Elevated H4K20me1 has also been linked to the loading of condensin, which is responsible for organizing chromatin in mitosis (Beck et al. 2012). Another characteristic of mitosis appears to be the drastic decrease in total histone acetylation (Park et al. 2011; Zhiteneva et al. 2017), which can be relevant for mitotic compacting or associated with the reduced transcriptional activity occurring during mitosis (Wang and Higgins 2013).

Histone PTMs can also inherently interfere with the organization of mitotic chromosomes (Vagnarelli 2012; Zhiteneva et al. 2017). Most recent experiments indicate that nucleosomes which have been reassembled with nuclear histones from mitotic cells tend to aggregate more than nucleosomes assembled with nuclear histones derived from interphase cells (Zhiteneva et al. 2017). This indicates that histone PTMs and their alterations in mitosis can intrinsically influence mitotic compactness through nucleosome–nucleosome interactions. However, other experiments have demonstrated that the DNA provides the fundamental interconnectivity of the mitotic chromosomes (Poirier and Marko 2002; Sun et al. 2011) and that the condensin within the central axis of the mitotic chromosomes is non-contiguous (Sun et al. 2018; Walther et al. 2018). Even though condensin accounts for the majority of the stiffness of the mitotic chromosomes, it is still unknown how much chromatin–chromatin interference could account for the stiffness of the mitotic chromosome.

To examine the impact of the modification of histone PTMs on the structure of mitotic chromosomes, the doubling forces of trapped mitotic chromosomes were quantified. In specific detail, the doubling force represents the force needed to double the entire length of a chromosome, and quantifies elastic stiffness of the chromosome in a way that is independent of chromosome length. To prove the hypothesis that changes in histone PTMs influence the compression of mitotic chromosomes, the effects of histone deacetylase inhibitors (HDACis), valproic acid (VPA) (Marchion et al. 2005) and trichostatin A (TSA) (Yoshida et al. 1990), on the H3K9ac concentration during mitosis and how they affect the stiffness of human mitotic chromosomes have been examined. It has also been explored whether the histone demethylase inhibitor (HDMi) and methylstat (MS), which represents a Jumonji C–specific inhibitor, which is specific for the key activity domain of a number of demethylases (Luo et al. 2011), impact the amounts of H3K9me2, H3K9me3 and H3K27me3 during mitosis, and subsequently, how the stiffness of the mitotic chromosomes of humans are altered. In fact, HDACi treatments raised the levels of H3K9ac; however, there was no alteration on the stiffness of mitotic chromosomes, whereas the mass spectrometry treatment elevated the canonical heterochromatin labels together with the elevated mechanical stiffness of the mitotic chromosomes.

14.11.1 HDACis Elevates H3K9ac on Mitotic Chromosomes but Do not Their Stiffness

To examine the effect of histone PTMs on the compactness of mitotic chromosomes, the impact of histone hyperacetylation was analyzed. The histone hyperacetylation is initiated by HDACis, VPA and TSA. Both VPA and TSA resulted in an enhancement of H3K9ac fluorescence intensity in fixed immunofluorescence and in Western blots of interphase cells. After succeeding in initiating hyperacetylation in the interphase, the next step was to determine whether the identical treatment could induce histone hyperacetylation in mitosis. In fixed immunofluorescence experiments with mitotic cells, the average ratio of HDACi-treated to non-treated H3K9ac acetylation values has been1.4-fold for VPA and 2.3-fold for TSA. In individually recorded chromosome analyses, the average ratio of HDACi to untreated H3K9ac determinations was 1.8-fold for VPA and 2.3-fold for TSA. These findings demonstrated that in mitosis hyperacetylated chromatin can be generated.

Next, this enhancement of acetylation was evaluated to see whether it could cause a change in the stiffness of the mitotic chromosomes based on the doubling force of the mitotic chromosomes extracted from untreated and HDACi-treated cells. However, none of the VPA or TSA treatments resulted in a statistically relevant alteration in doubling power in comparison to untreated chromosomes. Moreover, the average chromosome doubling forces of untreated, VPA-treated or TSA-treated cells were in the same order of magnitude. The lack of alteration seems to be not due to alterations of initial length or cross-sectional area, as they are not affected by HDACi treatments.

When plotting the average doubling force against H3K9ac fluorescence for untreated and HDAC-inhibited chromosomes, a statistically significant correlation between the H3K9ac measurements and doubling force was observed neither for untreated chromosomes nor for VPA treatments. It was found that the TSA-treated chromosomes exhibited a statistically relevant correlation between the measured H3K9ac concentration and the doubling force, with increasing acetylation resulting in a decreased spring constant; however, there was no net impact of TSA treatment on the chromosome spring constant when averaging. The relationship could be related to the specific mechanism of HDAC inhibition by TSA, which has not been seen by VPA, reflect variations between specific chromosomes, or just because the sample size is too small for this kind of correlation analysis. Besides this correlation, it is concluded that the hyperacetylation of histones by HDACi treatment has no influence on the overall stiffness of the mitotic chromosomes.

14.11.2 Methylstat Raises Mitotic Chromosome Stiffening and Histone Methylation

Since there was no global effect of histone acetylation on chromosome duplication, the aim has been to investigate how altered histone methylation impacts on the stiffness of mitotic chromosomes. To elicit hypermethylation, the HDMi methylstat, which raised both H3K9me2, H3K9me3 and H3K27me3, as measured both by Western blotting and by fixed cell immunofluorescence in interphase cells, was utilized. After succeeding in inducing interphase hypermethylation, the next step is to determine whether the identical treatment will lead to histone hypermethylation in mitosis. In fixed immunofluorescence assays using mitotic cells, the average ratio of methylstat/untreated H3K9me2.3 was 1.6, while the average ratio of methylstat/ untreated H3K27me3 was 3.9. In comparison to the fixed immunofluorescence measurements, methylstat did not induce a statistically relevant alteration in the measurement of H3K9me2, H3K9me3 or H3K27me3 when antibodies were microsprayed on individual targeted chromosomes. Even though surprising, these data are understandable due to the absence of antibody access and entry into the more densely packed hypermethylated chromosomes and the short antibody incubation time for microspraying of trapped chromosomes compared to fixed immunofluorescence staining (about 10 min vs. about 16 h). To establish whether elevated methylation resulted in stiffening of the mitotic chromosomes, the doubling force of chromosomes treated with methylstat was determined. As expected, methylstat treatment resulted in a statistically significant enhancement of the doubling force of the mitotic chromosomes around 80%, which is in line with the more compacted (condensed) chromatin. Whereas the average chromosome duplication forces were 320 ± 30 pN in untreated cells, these forces are increased to 580 ± 40 pN in methylstat-treated cells. This alteration was not caused by a modification of either the original chromosome length or the cross-sectional area, as neither of these was altered due to exposure to methylstat. The plotting of the doubling force compared to H3K9me2 or H3K9me3 analyses exhibited no correlation in either untreated or methylstat-treated cells. As an alternative, the presentation of the doubling of the force compared to measurements of H3K27me3 (in methylstat-treated but not untreated cells) indicates a possible correlation between H3K27me3 and the chromosomal stiffness. However, there may be restrictions in antibody accessibility on the chromosomes, which means that this correlation should be seen as preliminary. The findings point to the fact that hypermethylation induced by methylstat induces strongly elevated H3K27me3 concentrations and results in stiffer and potentially denser chromosomes.

14.11.3 Methylstat Causes no Alteration in SMC2 Levels

As condensin is the best-known element responsible for chromosomal strength, it has been attempted to test whether condensin levels on the mitotic chromosomes rise when treated with methylstat. Earlier work has indicated that the stiffness of the chromosome is approximately proportional to the amount of condensin on the chromosome in a linear manner (Sun et al. 2018). Hence, antibodies against SMC2, which is a core subunit of condensin, have been utilized, to reveal whether there exists an alteration in fluorescence intensities among untreated and methylstat-treated cells and the trapped chromosomes. The experiments yielded no distinction between measurement by fixed cellular immunofluorescence or by antibodies microsprayed on the acquired chromosomes, implying that the stiffening phenomenon is unrelated to the condensin exposure.

14.11.4 Histone Hypermethylation Increases Stiffness of Mitotic Chromosomes, but Hyperacetylation Has no Effect

According to the data, the elevation of histone acetylation (in particular the H3K9ac level) by HDACi application has no influence on the stiffness of chromosomes in mitosis. However, the initial hypothesis that HDACi-based histone hyperacetylation may soften the mitotic chromosomes seems to be rather wrong. This hypothesis was based on the findings that histone acetylation is commonly decreased in mitosis (Doenecke 2014) and connected to the inherent manipulation of the nucleosome packing (Zhiteneva et al. 2017). In addition, the mitotic chromosomes were weaker due to the decompression of hyperacetylated interphase chromatin (Doenecke 2014), and the hyperacetylated chromatin reduces the chromatin-dependent stiffness of the interphase nuclei (Stephens et al. 2017, 2018). However, the data show that mitotic chromosomes with hyperacetylated histones are generally as stiff under HDACi treatment compared to mitotic chromosomes from untreated cells.

Dissimilar to HDACi treatments, which display no alteration in the doubling force of mitotic chromosomes, the treatment by the HDMi methylstat leads to elevated levels of histone methylation (examined through the methylation through H3K9 and H3K27) and a stiffer and seemingly denser mitotic chromosome with no altered SMC2 levels. These results substantiate the initial hypothesis that elevated histone methylation and the tendency for condensation of mitotic histones might lead to stiffening of the mitotic chromosome as seen in interphase nuclei (Stephens et al. 2017, 2018). However, these results are contrary to the findings on mitotic hyperacetylated histones. These findings strongly indicate that this stiffening is not caused due to condensin overload, thus implying that different mechanisms/complexes are capable of influencing chromosomal stiffness.

14.11.5 The Model of Mitotic Chromosomes Encompasses Chromatin Interactions

To gain an insight into how chromatin can be responsible for the general stiffness of the mitotic chromosomes, it is necessary to gain an insight into the organization of the mitotic chromosomes. The early electron microscopy revealed that the mitotic chromosomes appear to be organized in chromatin loops emanating from a protein-rich core (Marsden and Laemmli 1979). The loop extrusion model that is presently under intensive examination is based on this classic bottle brush model and explains how the bottle brush is shaped (Alipour and Marko 2012; Goloborodko et al. 2016; Gibcus et al. 2018). In this specific model, chromatin loop extruder assemblies generate the bottlebrush structure in the core of the mitotic chromosomes. Non-histone-chromatin-organizing complexes including condensin and cohesin are distributed in the nucleus of the mitotic chromosomes or interspersed between sister chromatids (Ball and Yokomori 2001; Piazza et al. 2013), which act as loop-extruding enzymes according to the model. A largely comparable model of extruded chromatin loops arranged through the protein complexes condensin and cohesin has been employed to characterize the vertebrate and yeast centromere as a spring of chromatin (Ribeiro et al. 2009; Stephens et al. 2011; Lawrimore et al. 2015).

The loop extrusion model has been attempted to be integrated into the gel network model for the mitotic chromosomes. The gel network model characterizes the mitotic chromosomes as a gel of chromatin that is cross-linked through non-histone protein assemblies, primarily condensin (Poirier and Marko 2002). Two aspects determine the stiffness of a gel matrix: the cross-linking density and the flexibility of the interconnected interlaced fibers (de Gennes 1979). Earlier work has shown that condensin accounts for roughly half of the spring constant of the kintetochore (Ribeiro et al. 2009). In addition, recent findings have shown that condensin more or less linearly correlates with the stiffness of the mitotic chromosomes (Sun et al. 2018), indicating that the majority of stiffness is governed mainly by the chromatin loop extrusion units, which are obviously also the primary cross-linking agents. Prior research has highlighted that DNA/chromatin provides the fundamental interconnectivity of the mitotic chromosomes that form the primary fiber (Poirier and Marko 2002; Sun et al. 2011). These findings also indicate that the loop-extruding proteins cannot produce a continuous core. When the mitotic chromosomes are regarded as a gel, the condensins encompass the most essential cross-links, while the chromatin constitutes the underlying fiber. Both the absence of a variation in stiffness upon hyperacetylation of histones and the absence of an elevation of condensin concentrations in hypermethylated histones indicate that interfering histone PTMs have no impact on the amount of primary condensin-based cross-links.

Since the hyperacetylation of histones by HDACi treatments has no influence on the stiffness of the mitotic chromosomes, it can neither modify the quantity of cross-links nor the capacity of the chromatin fiber to be stretched. This is in contradiction to interphase, where hyperacetylation softens the chromatin-dependent stiffness of the nucleus (Stephens et al. 2017, 2018). This discrepancy can be attributed to a missing transcription in mitosis, acetyl histone scanners in mitosis or any other cell cycle-specific parameters. These elements may actively decompress the chromatin in interphase nuclei, whereas in mitosis they may not (Wang and Higgins 2013; Doenecke 2014). In addition, histone acetylation in mitosis is severely reduced, indicating that the impact of elevated histone acetylation through HDACi may be minimal for metaphase chromosomes. Even though a reduction in acetylation in mitosis is associated with a higher level of compression (Zhiteneva et al. 2017), it occurs that the enhanced acetylation of histones due to HDACis treatment may not have an inherent impact on the stiffness of metaphase chromosomes.

However, these data indicate that the hypermethylation of histones alters the structure of the mitotic chromosomes in view of the elevated doubling force. Nucleosome–nucleosome pairings can stiffen the mitotic chromosomes through the generation of either additional weaker cross-links or the chromatin fibers may be rendered more difficult to stretch themselves. Both hypotheses will not necessarily impact on the principal cross-linking agents, the condensins. These two hypotheses need clearly not be contradictory, although future investigations could possibly be able to identify which of them is overwhelmingly true. Further chromosome manipulation experiments should be able to identify accurately which PTMs are accountable for the structural alterations and to clarify whether the alterations in chromosome mechanics under observation are caused only by histones or whether they rely on other proteins for their regulation.

Most of the research on the correlation between histone PTMs and chromatin structure is concentrated on histone scanners, although the histone PTMs themselves may be the source of the stiffness alteration themselves. The chromatin composed of mitotic histones has been found to accumulate a higher level of protein agglomeration than the chromatin reconstructed from interphase histones (Zhiteneva et al. 2017). This analysis implies that histone methylation is linked to the structure and mechanics of mitotic chromosomes by linking a 3.4-fold rise in methylation to an 80% elevation in the stiffness of the chromosomes. This alteration of the inherent condensation phenomenon can be eased through direct nucleosome–nucleosome interfaces on the basis of the functionality of the histone tails (Bilokapic et al. 2018). Therefore, it can be hypothesized that the possible enhancement of histone methylation rather than the diminished acetylation leads to a narrower nucleosome packaging in the mitotic period of the cell cycle.

It should be pointed out that the metaphase chromosome, even though it is arranged as a chromatin gel, probably has an inherited radial loop organization with an abundance of condensin cross-links in the vicinity of the central chromatin "strings." It is imaginable that low, multifunctional attractive forces between nucleosomes, which could be facilitated by methylated histone tails, might promote the densification of the tighter axial region of metaphase chromatids with no adhesion between the external, less dense, outer radial loop "halos." It is important to ensure that there is no unintended adhesion among nucleosomes: Whenever individual nucleosomes attach to each other, the entire genome adheres to one another and produces a teardrop, a condition that is completely impossible to reconcile with chromosome segregation (Marko and Siggia 1997). Multivalence might be a crucial component, since it can facilitate a fast "switching on" of internucleosome attractiveness with a local nucleosome concentration, whereby the comparatively low loop extrusion densification due to condensins can compress the axial region adequately, enabling the attractions to be switched on at that point, but not in the less dense loop halo. This scenario seems to clarify how metaphase chromatids become densely packed in their axial interior while maintaining mutually repulsive loop halo exteriors, thereby producing intense chromatin compression at the same time as enhancing chromosome individualization and sister chromatid resolution, and also rendering the total mechanics of metaphase chromosomes susceptible to any additional nucleosome attraction linked to specific PTMs.

14.12 Quality Control of Ribosome-Associated MRNA and Proteins

Abnormal, misfolded and incorrectly localized proteins are frequently cytotoxic and lead to many human diseases. All proteins and their mRNA templates are controlled for quality. There are several diverse mechanisms that govern the quality of mRNAs and proteins while they being translated at the ribosome. mRNA quality assurance systems, non-sense-mediated decay (NMD), non-stop decay (NSD) and no-go decay (NGD) recognize abnormal premature stop codons, the lack of a native stop codon or locked ribosomes in translation and break down their mRNAs. Defective shortened polypeptide nascent chains produced from damaged mRNAs are broken down by ribosome-associated protein quality assurance paths. The regulation of the manufacture of aberrant proteins, a novel pathway, identifies aberrant proteins through surveillance of the stage of the emerging chain interactions occurring during translation and initiates the break-down of their mRNA (Karamyshev and Karamysheva 2018). In the following, the focus is set on the progress in current knowledge of quality control mechanisms based on specific molecules which govern the correctness of mRNAs and proteins at the ribosome during the process of translation.

Genetic data are converted into accurately folded active proteins during transcription and translation, which are positioned at the right locations for their proper function. Although these mechanisms are highly accurate, faulty proteins can be generated as a consequence of mutations, errors in transcription and translation, stress or for any other reason. Cellular quality assurance mechanisms have been developed to inhibit the synthesis of the abnormal proteins at the ribosome or to break them down when they are fully generated (Fig. 14.4). A large number of quality management systems are

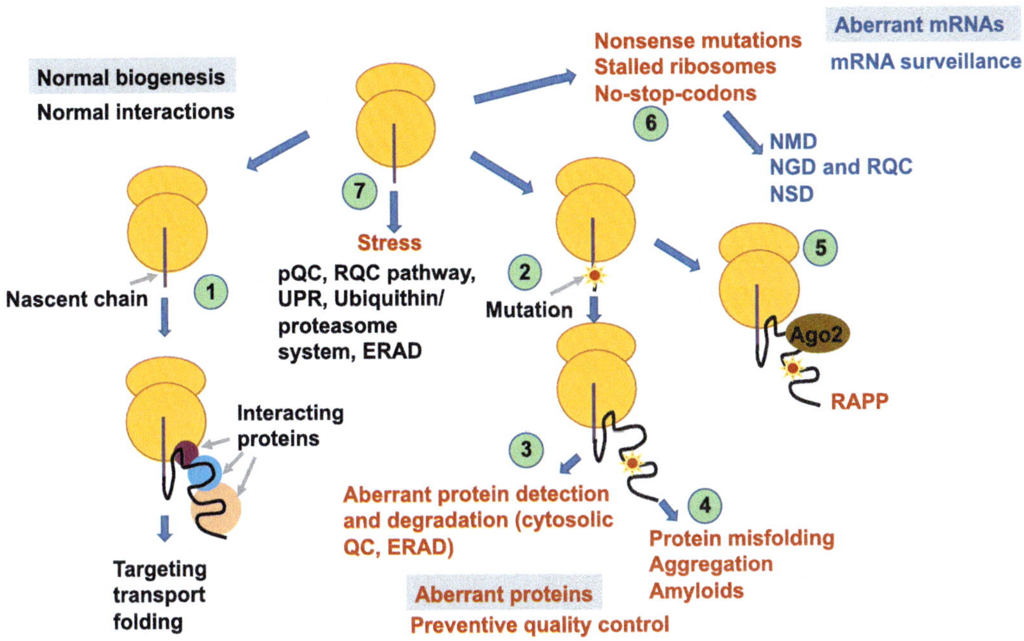

Fig. 14.4 mRNA and protein quality control routes in mammalian cells. (1) Normal interactions of the nascent chains produce proteins. Loss of these interference due to defects in interacting proteins or mutations in the nascent chain (2) cause the break-down of the protein (3), misfolding and amyloid formation (4) or mRNA removal in the RAPP route (5). mRNA surveillance quality control systems (6), nonsense-mediated decay (NMD), non-stop decay (NSD) and no-go decay (NGD) detect and remove mRNAs with PTCs, mRNAs with no natural stop signal and mRNAs at stalled in translation ribosomes. Due to ER stress pre-emptive quality control (pQC) co-translationally reroutes secretory and membrane proteins to the cytoplasm in order to be broken down. Stress causes multiple misfolded proteins which are removed in various cellular systems, such as UPR, ERAD and ubiquitin/proteasome system (7)

implemented co-translationally and carry out mRNA and protein monitoring in the ribosome. Protein synthesis and break-down of faulty proteins are energy-intensive processes, and ribosome-related quality assessment can avoid unsuccessful aberrant protein synthesis. NMD, NGD and NSD detect and remove mRNAs with premature termination codons, truncated and translationally blocked mRNAs or mRNAs with no natural stop codons (Welch and Jacobson 1999; Doma and Parker 2007; Shoemaker and Green 2012; Popp and Maquat 2013). Truncated polypeptides generated on inhibited ribosomes are ubiquitinated and broken down by the proteasome (Dimitrova et al. 2009; Bengtson and Joazeiro 2010; Brandman et al. 2012; Duttler et al. 2013; Brandman and Hegde 2016). The regulation of the production of aberrant proteins (RAPP) detects abnormal proteins by sensing the state of interaction between the chains formed during translation and initiating the deterioration of their mRNAs (Karamyshev et al. 2014; Pinarbasi et al. 2018). When defective mRNAs and proteins are ignored by these quality assurance systems, the abnormal proteins are broken down in the cytosol using a proteolytic mechanism (Heck et al. 2010), or in the endoplasmic reticulum (ER) through an ER-dependent degradation pathway (synonymously termed ERAD pathway) (Brodsky and Wojcikiewicz 2009). When abnormal proteins fail to pass through quality assurance, they can be misfolded, create insoluble clusters or amyloids and lead to many human diseases (Stefani and Dobson 2003; Gregersen et al. 2006; Zimmermann et al. 2006; Hebert and Molinari 2007; Jarjanazi et al. 2008; Hipp et al. 2014).

The interactions of a polypeptide nascent chain in translation are essential for protein biogenesis and for quality assurance (Gandin and Topisirovic 2014). These interactions govern the subsequent positioning of the proteins, the precise folding and modification of the proteins (Pechmann et al. 2013). Interruptions of these processes may act as indicators for the monitoring of quality assurance and for the identification of irregular mRNAs/proteins. The nascent chain interactions taking place at the ribosome and the events happening during ribosome-associated mRNA and protein quality controls are discussed below.

14.12.1 Translational Nascent Chain Interactions Are Crucial for Protein Targeting and Folding

The targeting of proteins, their transportation to destinated places and folding are performed co-translationally or post-translationally (Park and Rapoport 2012; Ellgaard et al. 2016). The focus lies here solely on co-translational protein interferences. In the initial steps of translation, the polypeptides released from the ribosomal exit tunnel begin their first interactions with various factors necessary for folding, modifications, alignment and transportation. The absence of these linkages results in inappropriate folding and protein break-down, aggregation of proteins and the production of amyloids or mRNA removal. Every living cell contains various compartments, and proteins need to be supplied to exactly the right places within the cells. While cytosolic proteins stay in the cytosol after finishing their production, other proteins are carried to various cellular organelles or are transported to external sites beyond the cell.

Despite huge variations among prokaryotic and eukaryotic cells, similar mechanisms control the targeting and transport of proteins. Proteins carry certain localization cues, which are detected by specific proteins (Emanuelsson and von Heijne 2001). These cooperations are indispensable for the specific targeting of proteins. The localization signals which have been most thoroughly examined up to now are signal sequences (von Heijne 1985, 1990; Nilsson et al. 2015).

Secretion proteins are produced as precursors comprising an N-terminal expansion known as a signal sequence or signal peptide. Signal sequences play a role in guiding secretory proteins to the Sec61 translocon in the ER membrane of eukaryotes or to the SecYEG complex in the bacterial plasma membrane (of prokaryotes) to be transferred through the membranes (Alder and Johnson 2004; Wild et al. 2004; Egea et al. 2005; Rapoport 2007; Dudek et al. 2015; Voorhees and Hegde 2016). Various signal sequences lack sequence homology, but possess comparable structural characteristics (von Heijne 1985, 1990).

In bacteria, the sorting events are governed by an equilibrium of interactions of a novel emerging chain with the Ffh/4.5S RNA complex, the SecA protein, the chaperone trigger factor and several other proteins (Karamyshev and Johnson 2005; Eisner et al. 2006). Overproduction of secretory proteins causes an unbalanced targeting/folding and aggregation of their precursors as insoluble compounds within the bacterial cytoplasm (Nesmeyanova et al. 1991, 1997).

In eukaryotic cells, the relationships of the emerging chains are intricate and mirror a more sophisticated process of co-translational folding and alignment to a number of organelles. Signal sequences are detected using signal recognition particles (Walter et al. 1981; Krieg et al. 1986; Kurzchalia et al. 1986). These cooperations form the foundation for the co-translational alignment of secretory proteins toward translocons. In the specific instance of membrane proteins, their initial transmembrane spacings are similarly identified through signal recognition particle for targeting.

There are different localization cues for the directional movement of the proteins toward mitochondria, nucleus and peroxisome (Emanuelsson and von Heijne 2001). These cues are essential for correct detection using dedicated targeting elements. A number of these signs are present at the N-terminus and are therefore probably identified co-translationally, and a few are positioned at the C-terminus of the protein, indicating post-translational targeting. For instance, N-terminal signals are composed of dedicated mitochondrial pre-sequences accumulated with positively charged residues and capable of generating amphiphilic α helices (von Heijne et al. 1989; Emanuelsson and von Heijne 2001), type 2 peroxisomal target signals (PTS2) for certain peroxisomal proteins (Williams 2014) and even other proteins. Tail-anchored (TA) proteins and peroxisomal PTS1 proteins possess C-terminal cues and are highly susceptible to be attacked by post-translational mechanisms (Stefanovic and Hegde 2007; Williams 2014).

There are numerous highly specific proteins which react with emerging chains in the course of their synthesis. These comprise specific factors, chaperones facilitating protein folding and amending factors. These proteins are arranged in a cluster generally referred to as ribosome-associated protein biogenesis factors (RPBs) (Raue et al. 2007). Nascent chains of cytosolic and secretory proteins interoperate with various components of RPBs to ensure correct folding and targeting (Fig. 14.5). RPBs operate in translation when a brief nascent chain is released out of the ribosomal polypeptide tunnel.

In yeast, RPBs comprise the targeting factor for the signal recognition particle, nascent polypeptide-associated complex (NAC), its associated chaperones Ssb1 and Ssb2, which are Hsp70 homologs, the ribosome-associated complex (RAC), N-terminal acetyltransferase (NatA), and Map1 and Map2 proteins (Raue et al. 2007). Yeast RAC is made up of two proteins, zuotin (or synonymously known as Zuo1, DnaJ homolog, Hsp40 family) and Ssz1p (synonymously known as DnaK homolog, Hsp70 family) (Gautschi et al. 2001; Zhang et al. 2017). The RAC for mammals comprises the chaperones MPP11 and HSP70L1 (Otto et al. 2005). RAC has been revealed to bind ribosomes near the polypeptide tunnel exit (Peisker et al. 2008). NAC is composed of two subunits, α and β, which are both in close vicinity to an emerging chain, as has been proven by the interconnection (Wiedmann et al. 1994; Wang et al. 1995). It links small, nascent chains just as they are released from the polypeptide tunnel. In the specific situation of secretory proteins, NAC only binds the emerging chain when the signaling sequence from the ribosome is not fully exposed (Fig. 14.6). The binding of NAC is essential for signal recognition particle uniqueness and constancy in translocation (Wiedmann et al. 1994). Under ordinary circumstances, NAC links ribosomes to promote protein folding, but under stress it migrates to protein clusters and serves as a protein chaperone (Kirstein-Miles et al. 2013). Chaperone Ssb is capable of binding a wide range of substrates—cytosolic, ER, nuclear and mitochondrial polypeptides (Doring et al. 2017). Its attachment increases the speed of translation. Ssb, such as Ssb1 and Ssb2, RAC and NAC play a dual role in the convolution of new proteins and the control of ribosome formation (Koplin et al. 2010).

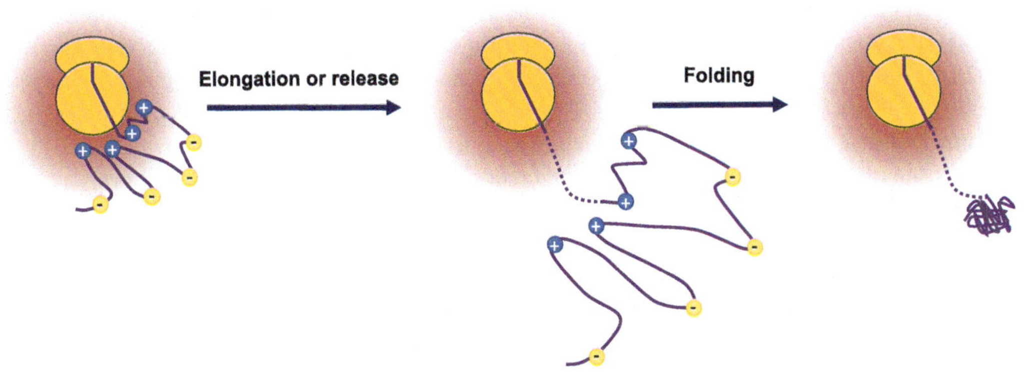

Fig. 14.5 Model for the alteration of the folding rates in an electrostatic potential. The negative electrostatic potential surrounding the ribosomes (red gradient) attracts positively charged residues of the nascent polypeptide and repels negatively charged residues. Once the folding unit is released from the ribosome, it folds appropriately

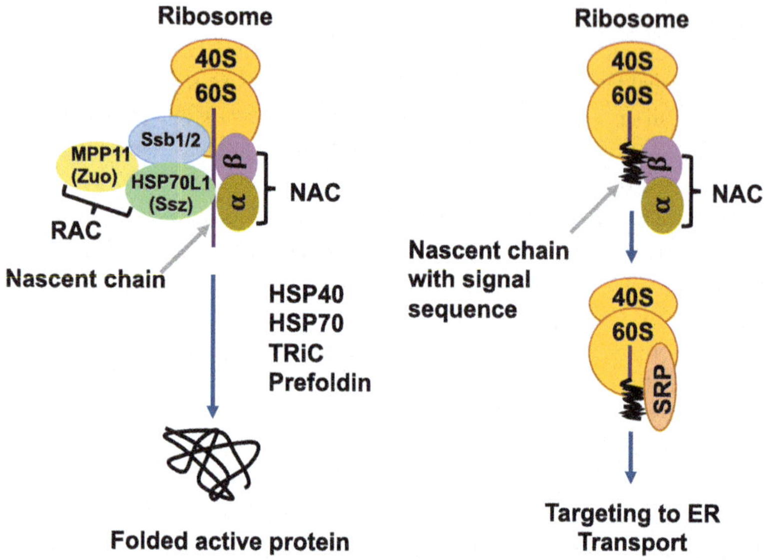

Fig. 14.6 Interactions at the polypeptide exit site at the ribosome under normal conditions. Nascent chains and ribosomes interact with various proteins during the process of translations to provide a proper folding and correct targeting. Nascent proteins of the cytosolic proteins (A) are produced in the presence of RAC, NAC, Ssb (HSP70) and others are folded with the help of chaperones and chaperonins; however, the cytosolic proteins (B) interfere with NAC prior to the full exposure to the signal sequence and when the signal sequence is emerged from the ribosome tunnel, SRP attaches to it causing a temporary elongation arrest and aiming it to the ER membrane to be transported via translocon into the ER lumen, the Golgi and subsequently outside the cell. Mammalian proteins are named, and the yeast homologs are provided in brackets

There are also two major chaperone groups or networks with distinct functions within the cells, one for de novo folding (referred to as CLIPS for the chaperones associated with protein synthesis) and the other (HSPs, heat-shock proteins) for refolding proteins to preserve them under stress (Albanese et al. 2006). For this reason, the CLIPS network contains translation guides, such as specific chaperones (Albanese et al. 2010; Pechmann et al. 2013). While secretion/membrane proteins require signal recognition particle in the first step of protein synthesis, in yeast cytosolic proteins require ribosome-bound chaperones Ssb (HSP70 family) (Willmund et al. 2013), in mammals HSP70L1 and MPP11 (Otto et al. 2005), chaperonin TRiC (McCallum et al. 2000; Etchells et al. 2005), and other factors (Hartl et al. 2011; Pechmann et al. 2013). Another chaperonin, prefoldin, is also able to attach chains and is implicated in the folding of actin and tubulin (Hartl and Hayer-Hartl 2002). It is not fully known how the specificity of chaperones/chaperonins for emerging chains is regulated. In that respect, a broad set of proteins participating in control and ubiquitination of abnormal nascent chains is also detected attached to translational ribosomes (Comyn et al 2014).

Thus, the ribosome serves not merely as a protein synthesis mechanism, but also performs a crucial function in the organization of protein targeting/folding and quality control. The investigation of the natural interactions of nascent polypeptides in translation and their alteration during the operation of mRNA and protein quality assurance systems are essential for gaining an insight into the molecular mechanisms of protein biogenesis and homeostasis as well as into the molecular fundamentals of human diseases that are related to the disorder of these processes.

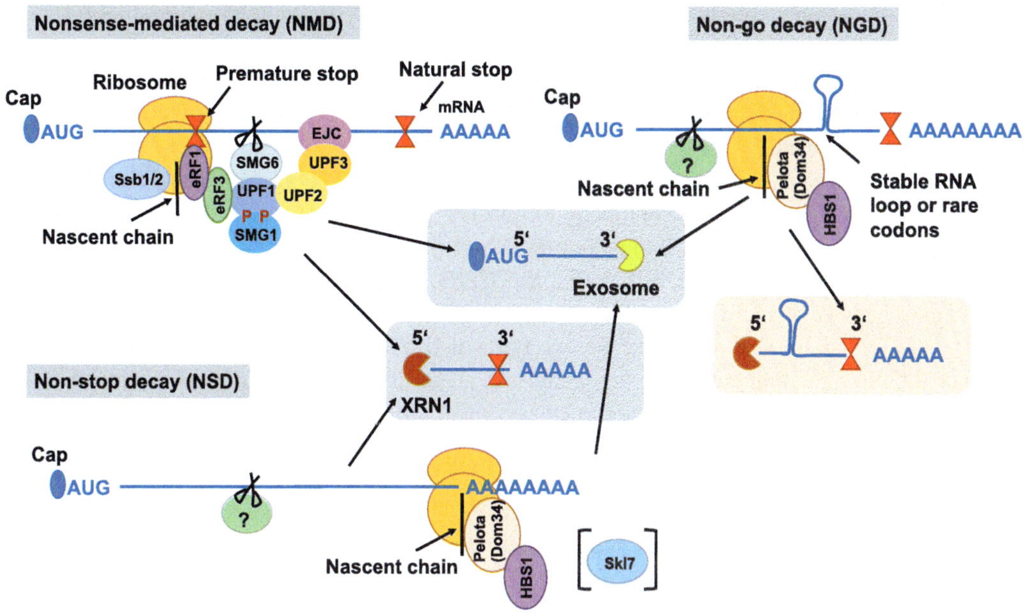

Fig. 14.7 mRNA quality control systems, including nonsense-mediated decay (NMD), no-go decay (NGD) and non-stop decay

14.12.2 Ribosome-Linked MRNA Quality Maintenance

mRNA conversion represents a key mechanism for controlling gene expression and keeping a high degree of faithfulness to the function and survival of cells. Cells employ several mRNA break-down routes to remove non-functional transcripts. mRNA decay involves a profoundly organized mechanism that is regulated by a particular set of genes. mRNA monitoring begins in the cell nucleus. Defective mRNAs have been identified and subjected to break down at multiple stages of their production and maturation, involving several steps such as transcription, capping, splicing and polyadenylation. The exosome is the primary mechanism for breaking down defective mRNAs in the cell nucleus. Subsequently, mRNAs that have undergone validation in the cell nucleus are transported as messenger ribonucleoproteins (mRNPs) directly into the cytoplasm, where they can be employed for translation. In the cytoplasm, the mRNAs are undergoing an extra quality assessment of co-translational mRNA monitoring. Multiple major mRNA break-down routes are employed to detect defective mRNAs and prevent the cell from translation of abnormal mRNAs and possibly toxic proteins, such as NMD, NGD and NSD (Fig. 14.7).

14.12.3 Nonsense-Mediated Decay

Non-sense-mediated degradation refers to an mRNA monitoring pathway that detects mRNAs with premature termination codons and aims for fast break-down to decrease the translation of shortened proteins with dominant-negative or harmful gain-of-function activities (Welch and Jacobson 1999; Popp and Maquat 2013) (Fig. 14.7a). This pathway is present in all of the eukaryotes investigated (Culbertson 1999). However, the NMD has not been seen in bacteria. The existence of abnormal termination codons in bacterial genes results in the ending or restarting of translation (Karamyshev et al. 2004).

The exon–exon junction complex is composed of proteins that are built on the pre-mRNA while the process of splicing is performed. (Gehring et al. 2009). After exporting the mRNA, exon–exon junction complex is eliminated from the mRNA in the pioneering phase of translation and substituted by translation-enhancing proteins. However, when a premature terminating codon is visible on the mRNA at 50–55 nucleotides prior to exon–exon junction complex, the NMD is probably activated since the terminating ribosome (at the premature termination codon) is not capable of eliminating the exon–exon junction complex and continuing to translate properly (Popp and Maquat 2014).

Several proteins are preserved throughout the NMD across species and represent the key to this route: the up-frameshift proteins UPF1, UPF2 and UPF3. UPF1 is the main controller of the NMD. The ATPase activity of UPF1 is necessary for the decomposition of mRNPs in the course of NMD (Franks et al. 2010). In mammals, there are two types of UPF3, such as UPF3a and UPF3X (UPF3b) (Serin et al. 2001). In multi-cellular living organisms, additional proteins, known as suppressors, with morphological impact on the genital tract, such as SMG1, SMG5–SMG9, participate in the control of NMD (Yamashita et al. 2001, 2009). NMD is performed in three steps, comprising detection of NMD substrates, marking and finally break-down of the premature termination codon carried by the transcript. NMD activation starts with the recognition of a premature termination codon during the initial translation phase. Following the recognition phase, the premature termination codon is labeled by building a SURF complex at the exiting ribosome. The SURF complex comprises the serine/threonine kinase SMG1, UPF1 and the eukaryotic release factors eRF1-eRF3 (Kashima et al. 2006; Hwang et al. 2010). Then, UPF1-SMG1 interacts with UPF2 to connect to the exon–exon junction complex. UPF2 is attached to the exon–exon transition complex through reaction with UPF3 or UPF3X. SMG1 is able to phosphorylates UPF1. The hyperphosphorylated UPF1 initiates the repression of translation and hence attracts the SMG6 protein (Isken et al. 2008). SMG6 can induce endonucleolitic cuts in the mRNA. The cut takes place between the premature termination codon and the exon–exon transition complex sites of the faulty mRNA in the last phase of NMD (Huntzinger et al. 2008; Eberle et al. 2009). Activated UPF1 subsequently enrolls SMG5-SMG7 or SMG5-PNRC2 (Kervestin and Jacobson 2012). These proteins continue to attract decapping and/or deadenylation mechanisms to aid in the exonucleolytic break-down of vulnerable 5′ and 3′ mRNA fragments arising from the endonucleolytic disruption of premature termination codon-containing mRNA (Lejeune et al. 2003; Loh et al. 2013). 5′-to-3′ Exonuclease XRN1 is in charge of the break-down of the 3′ decomposition product (Lejeune et al. 2003; Unterholzner and Izaurralde 2004; Eberle et al. 2009). The 5′ decomposition product is most probably broken down by the exosome (Schmid and Jensen 2008). NMD proteins can be jointly purified with compounds of the mRNA break-down system such as DCP2, XRN1 and XRN2/RAT1 as well as multiple exosome subunits (Lejeune et al. 2003; Muhlemann and Lykke-Andersen 2010). Decapping and deadenylation enzymes can play a major part in the more rapid break-down of mRNA segments in mammalian cells (Lejeune et al. 2003). Nevertheless, further work is required to better define the function of decapping and deadenylation in NMD.

Although the mechanism discussed above (model of the exon junction complex) is attractive, it does not provide all necessary features of the NMD mechanism and the alternative models such as Upf1 3′-UTR sensation/potential and the falsified 3′-UTR models (He and Jacobson 2015). However, although the models acknowledge the significance of 3′-UTR, they suggest distinct functions for 3′-UTR and NMD target identification (Amrani et al. 2004, Hogg and Goff 2010). Based on the sensation/potentiation model, Upf1 perceives 3′-UTR and increases the level of mRNA decay (Hogg and Goff 2010). The faux model suggests that the distance between the premature termination codon and the PolyA tail hinders the execution of efficient termination (Amrani et al. 2004).

The primary feature of UPF1, the principal regulator of NMD, is to restrict the translation of abnormal mRNAs. NMD is therefore a translation-dependent process, and a truncated protein originating from the pioneering phase of translation might be toxic and possibly even lead to human pathology. Thus, the premature degradation of the mRNA comprising the termination codon needs to be linked to the break-down of the protein of the truncated polypeptide. Although little evidence is provided in this area, a number of investigations on yeast point to the fact that Upf1 may possess the properties of the E3 ubiquitin ligase that encourage the proteasome to break down the truncated polypeptide (Takahashi et al. 2008; Kuroha et al. 2009). However, the ultimate destination of the truncated proteins generated by NMD in mammalian cells is still an unanswered question that needs to be investigated in more detail.

14.12.4 No-Go Decay

In the process of a no-go decay, the mRNAs stalled in translation elongation compounds are broken down (Fig. 14.7b). A translation stop might be triggered due to either certain characteristics of the emerging peptides, the presence of pronounced secondary structures in the mRNA that physically inhibit the translation mechanism alongside the transcript, or a seldom codon repeat that results in the A-position being unfilled for a prolonged period of time (Kuroha et al. 2010; Tsuboi et al. 2012). The inclusion of a robust stem loop RNA structure in the PGK1 mRNA resulted in a translation stop and endonucleolytic fragmentation of the mRNA, which was halted upon translation elongation followed by fast mRNA break-down. While the NGD route was first identified in yeast, it has also been discovered in fruit flies and mammals (Doma and Parker 2006; Passos et al. 2009; Pisareva et al. 2011).

The proteins Pelota (in mammals; Dom34 in yeast) and HBS1 are implicated in the control of the NGD pathway (Doma and Parker 2006; Pisareva et al. 2011) and are structurally connected to the terminating molecules eRF1 and eRF3, respectively (Atkinson et al. 2008). They also imitate the elongation factor complex and tRNA, indicating that they attach a position on the ribosome (van den Elzen et al. 2010). In fact, Dom34 and Hbs1 interfere directly with the A site of the ribosome, where they induce the dissociation of an aberrant translation elongation compound and the recycling of ribosomes, instead of terminating their translation (Shoemaker et al. 2010; Becker et al. 2011). Dom34/Hbs1 can also encourage an endonucleolytic decomposition incident in the NGD substrate and facilitate subsequent mRNA break-down. However, these factors are not crucial, as NGD mRNA fragmentation can also take place in the complete absence of these proteins (Passos et al. 2009; Tsuboi et al. 2012). The data indicate that endonucleolytic fragmentation takes place upstream of the ribosome trapping site (Tsuboi et al. 2012). It has been recently proven that NGD is caused by ribosome clashes, which result in multiple endonuclease break-downs (Simms et al. 2017). The effectiveness of NGD varies with the ribosome concentration on the target mRNA, indicating that ribosome clashes transfer the information to trigger endonuclease activation. As in the NMD signaling route, the generated fragments are quickly broken down by the exosome and the XRN1 in the course of NGD. It is not yet known exactly what type of endonuclease is involved in the break-down of NGD substrates.

14.12.5 Non-stop Decay

Non-stop decay breaks down mRNAs that are missing stop codons (Fig. 14.7c). NSD has initially been found in yeast (van Hoof et al. 2002) and mammals (Frischmeyer et al. 2002). Non-stop mRNAs might be produced by various mechanisms. These aberrant mRNAs can be generated by faulty polyadenylation inside the ORF leading to non-stop mRNAs containing poly(A), or by endonucleolytic splitting inside the ORF producing non-stop mRNA with no poly(A) (Ozsolak et al. 2010;

Graille and Seraphin 2012). The translation of poly(A) results in the generation of a polylysine chain located at the C-terminus of the synthesized polypeptide. This positively charged amino acid chain brings the polypeptide in the ribosomal tunnel to a standstill, which is most probably caused by interference with negatively charged ribosomal RNA (Dimitrova et al. 2009). In shortened non-stop mRNAs missing poly(A), the ribosomes remain trapped precisely in the 3′ end of the mRNA. In each case, the translation trapping induces a rapid degradation of the non-stop mRNAs through the translation-based NSD route. Translational inhibition is a requirement for mRNA break-down in NSD (Inada and Aiba 2005; Akimitsu et al. 2007) comparable to NGD and NMD. It was demonstrated that Ski7, which is a protein related in structure to Hbs1 and eRF3, is capable of binding the locked ribosome and enrolls the exosome for the transcript in yeast in the course of NSD (van Hoof et al. 2002). Nevertheless, Ski7 is not found in higher eukaryotes. Organisms missing Ski7 are dependent on Hbs1 and Dom34 proteins that work in both NSD and NGD (Tsuboi et al. 2012; Saito et al. 2013) indicating a considerable overlap in the operation of these routes. Recent study has revealed that the Dom34:Hbs1 complex is essential for ribosome decomposition and the stimulation of mRNA break-down in both NSD and NGD pathways (Tsuboi et al. 2012). Endonucleolytic cleavage represents a pioneering step for mRNA break-down in NSD. It has been established that locked ribosomes can cause multiple endonucleolytic disruption incidents on the non-stop mRNA coated by the specific individual ribosomes (Tsuboi et al. 2012). But in a similar manner to the NGD, the endonuclease entity participating in NSD is still not revealed.

Therefore, all these co-translational quality assurance regimes share several characteristics: The abnormal mRNA needs to be removed, the truncated protein products need to be broken down, and the locked ribosomes need to recuperate from the lock-up and revert to translation. NMD was initially identified as a monitoring pathway whose main purpose is to minimize errors in gene expression by removing premature terminating codon-containing mRNAs. Despite this, new functions of the NMD signaling pathway have arisen lately. The NMD pathway has also been found to be effective in engaging normal and physiologically functional mRNAs to induce quick alterations in gene expression (He and Jacobson 2015). Ribosome profiling indicated that the NMD pathway controls the expression of a minimum of 10% of human transcripts (Celik et al. 2017). NMD plays a role in controlling germ granules and spermatogenesis, and NMD compounds have been detected in the constitution of the chromatoid body (Meikar et al. 2014; Bao et al. 2016; MacDonald and Grozdanov 2017). It is possible that, in addition to mRNA quality assurance, NSD and NGD paths are also participating in the regulation of gene expression in a way comparable to NMD. Most recent data strongly indicate that the NGD pathway can be utilized not just to break-down faulty mRNAs, but also normal histone mRNAs derived from locked-up break-down complexes as a component of cell cycle regulation (Slevin et al. 2014). Chemically altered mRNAs, including oxidized, depurinated or alkylated mRNAs, can produce translation pauses and turn into NGD substrates to decrease the exposure of the cell to harmful protein compounds (Shan et al. 2007; Wurtmann and Wolin 2009).

Deficits in NMD compounds such as UPF3B and SMG9 result in mental retardation or multiple congenital anomaly syndrome due to overall transcriptional dysregulation (Rebbapragada and Lykke-Andersen 2009, Shaheen et al. 2016). The NMD pathway has also been identified to control immune reactions. The component of NMD, UPF1, is implicated in antiviral reactions and limits Semliki Forest Virus (SFV) and Sindbis virus infections (Balistreri et al. 2014). Somatic mutations in the UPF1 gene are involved in adenosquamous pancreatic carcinoma (Liu et al. 2014b). The deregulation of the NMD pathway is linked to various forms of cancer and is extensively discussed (Popp and Maquat 2018).

14.12.6 Control of Aberrant Protein Production

A new type of ribosomal-linked protein quality assurance, RAPP, has recently been identified (Karamyshev et al. 2014) (Figs. 14.8). The first natural RAPP substrate, granulin, containing disease-inducing signal sequence mutations, has also been discovered recently, showing that RAPP activation is a molecular trigger for some forms of frontotemporal lobar degradation (Pinarbasi et al. 2018). The RAPP route recognizes aberrant proteins during their translation and decomposes their mRNA templates to avoid the synthesis of possibly dangerous substances, which is termed preventive quality control. It entails the identification of emerging chains that have dropped their normal interactions with the factors for targeting, and guides the abnormal protein mRNA for break-down. The original investigation was conducted using the secretory protein preprolactin with deleted signal sequences as an exemplary model (Karamyshev et al. 2014). The key feature of the RAPP path involves a synchronization of the interactions at the ribosome throughout the translation. Normally, secretory proteins are detected from signal recognition particle while translating and are specifically directed to the ER membrane for transposition through a translocon into the lumen of the ER (Fig. 14.8a). When an abnormal signal sequence is not identified because of a mutation of signal recognition particle, or when signal recognition particle is lacking or defective, the AGO2 protein engages the ribosome nascent chain complex and induces the break-down of certain mRNA (Fig. 14.8b, c). Thus, signal recognition particle plays a new role in mRNA protection of secretory proteins against break-down, in addition to its function in the selective targeting of proteins.

Although there are no unique sequence specifications to induce mRNA break-down, a mutation should occur near the region which is responsible for a required protein interplay and result in an interruption of this interaction. The role of AGO2 in this procedure is not yet clear. It is assumed that

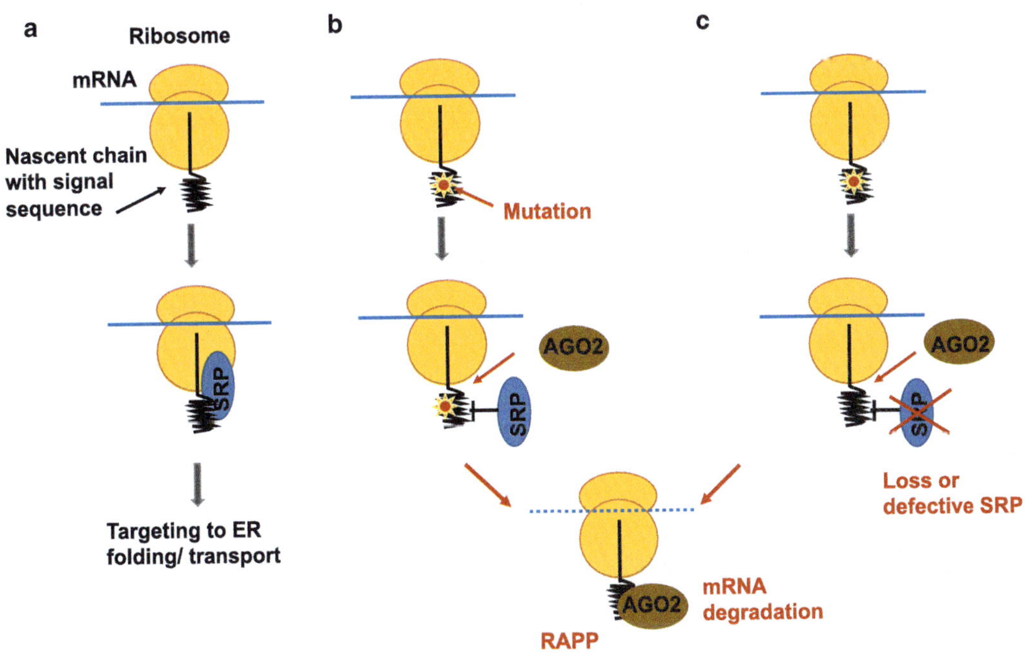

Fig. 14.8 Model for the regulation of aberrant protein production (RAPP) in mammalian cells. Normal co-translational interactions are crucial for the biogenesis of proteins. Nascent chains of secretory proteins interfere with signal recognition particles (SRP). The interaction causes proper aiming to ER, folding and transport of the proteins (a). Loss of the interaction results due to a mutation in the secretory protein (b) or loss of the interacting factor (c) causes the engagement of AGO2, which is associated with translation silencing. Subsequently, this interaction leads to break down of the protein in the RAPP process

the positioning of AGO2 in the vicinity of a mutated emerging chain controls its capacity to guide the mRNA to break down. AGO2 comprises a protein participating in miRNA and siRNA responsiveness, in translational silencing and is a key constituent of the RNA-induced silencing complex (RISC) (Hammond et al. 2001; Martinez et al. 2002). Nevertheless, these investigations revealed that no miRNAs, Drosha and Dicer proteins are involved in the RAPP process, pointing to a new AGO2 role in the lack of RISC formation (Karamyshev et al. 2014). AGO2 displays activities for slicer or ribonuclease H (Liu et al. 2004; Song et al. 2004; Rivas et al. 2005). Nevertheless, investigations with enzymatically inactive AGO2 reveal that the AGO2 slicer activity is not necessary for mRNA break-down occurring in RAPP (Karamyshev et al. 2014). Moreover, the mRNA break-down of the model RAPP substrates was found to be inhibited by AGO2 depletion and enhanced through AGO2 overexpression. But granulins with illness-causing mutations were not concerned (Pinarbasi et al. 2018). These findings seem to imply that AGO2 serves as a sensor for several substrates in the RAPP reaction, and an unknown protein might perform this role for some other substrates. Another reason is that the primary sensor of the route is not yet defined and AGO2 has a volunteer or booster role on several substrates. These results point to the fact that mRNA break-down is performed by endonuclease types different from AGO2. However, the nature of the endonuclease type has yet to be identified. Therefore, the mechanism of the RAPP path is still a long way from being fully comprehended.

It has previously been established that under stressful circumstances resulting in the clustering of unfolded proteins in the ER, a process is initiated that is referred to as regulated Ire1-dependent decay (RIDD) (Hollien and Weissman 2006, Hollien et al. 2009). It decreases the amount of secretory protein mRNA to minimize the clustering of secretory proteins within the ER in the course of the unfolded protein reaction (UPR). RIDD is a key general response mechanism to stress that perceives unfolded secretory proteins that have been effectively transported into the ER and inhibits their continued production and thus transport to the ER and subsequent aggregation. In contrary, RAPP detects mutated polypeptide nascent chains that are not capable of interfering with the signal recognition particle and are thus not targeted and not transferred into the ER, which reduces the clustering of these possibly dangerous proteins inside the cytosol.

The present RAPP model is built on the choice of mRNA for break-down by a loss of co-translational interaction of the emerging chain with the ribosomal destination molecule. When the level of interaction with the signal recognition particle is decreased because of a signal sequence mutation, AGO2 binds to the emerging chain and targets its mRNA for break-down. When the signal recognition particle interplay is undamaged, AGO2 is unable to interfere with the emerging chain. It is conceivable that this is a generic mechanism and is implicated in the quality assurance of other types of proteins that have been deprived of their natural interactions. These might be cytosolic aberrant proteins that have completely eliminated natural Interactions with certain chaperones, such as the ribosome-associated chaperones and components of RAC, MPP11 and HSP70L1, or peroxisomal and mitochondrial proteins, which have eliminated their specific interactions with their targeting molecules. The comprehension of these processes, nevertheless, demands future investigations.

14.12.7 Ribosome-Linked Quality Assessment of the Nascent Chain

What occurs with partially synthesized nascent chains following the triggering of the break-down of the defective mRNAs? The latest investigations on co-translational quality assurance systems initiated by translation confinement have demonstrated that not only defective mRNAs but also shortened proteins are quickly decomposed. In yeast, Ltn1, a ribosome-related E3-ubiquitin ligase (Bengtson and Joazeiro 2010) and a member of the Ccr4-Not complex Not4p, which can function as an E3-

ubiquitin protein ligase (Panasenko et al. 2006; Dimitrova et al. 2009; Halter et al. 2014) are involved in supporting the break-down of truncated protein compounds through the proteasome. It has been shown that Ltn1, Tae2 (otherwise known as Rqc2), Rqc1 and AAA-ATPase Cdc48 (otherwise known as VCP, valosin-containing protein and p97) are implicated in the elimination of abnormal translation compounds in yeast and build a complex on the 60S ribosome subunit (Brandman et al. 2012; Defenouillere et al. 2013). This structure has been termed ribosome quality control complex (RQC) (Brandman et al. 2012). Listerin, the mammalian functional homolog of Ltn1, is implicated in the ubiquitination of the abnormal chains generated within the locked ribosomes (Shao et al. 2013). It is noteworthy that the ubiquitinated, emerging chains were still attached to tRNAs, although the event necessitated the disassociation of the ribosomal subunits. Pelota, HBS1, ABCE1 are participating in the decomposition of ribosomal subunits in mammals, whereas Dom34, Hbs1, Rli1 can be detected in yeast (Shoemaker et al. 2010; Pisareva et al. 2011; Shoemaker and Green 2011). The dissociation of the ribosomal subunits results in the attachment of the RQC to the 60S ribosomal subunit (Shao et al. 2015). The association of mammalian nuclear export mediator factor (NEMF) (Rqc2 or Tae2 in yeast) inhibits the formation of subunits, results in the enrollment of Listerin and its placement in the vicinity of the polypeptide exit point on subunit 60S (Lyumkis et al. 2014; Shao et al. 2015). The findings of several investigations indicate that Cdc48, Npl4, Ufd1 and Rqc1 are participating in the removal of the ubiquitinated, emerging chains out of the 60S subunit (Brandman et al. 2012; Defenouillere et al. 2013; Verma et al. 2013). Among the mammalian orthologs are VCP (p97), UFD1, NPLOC4 and TCF25 (Verma et al. 2018). Nevertheless, the precise nature of the role of each individual element is not fully grasped. It has been recently identified that the yeast Vms1 (ANKZF1 in mammals) liberates ubiquitinated emerging chains from the inhibited ribosomes through peptidyl-tRNA hydrolysis to facilitate the continued break-down of polypeptides by proteasome (Verma et al. 2018). Only very few is generally understood about how the shortened, nascent chain breaks down in the NMD route. The UPF1 has been demonstrated to enhance the break-down of truncated peptides produced in the NMD signaling pathway and may possibly function effectively as an E3-ubiquitin ligase (Takahashi et al. 2008; Kuroha et al. 2009). Nevertheless, additional work is necessary to define all key players in the control of co-translational protein degradation and to identify the mechanism details of the NMD process.

Several co-translational controls of the quality of stress-initiated proteins have lately been detected in mammals. One of these, pre-emptive quality control (pQC), redirects membrane and secretory proteins into the cytoplasm in a co-translational fashion for break-down under severe ER stress (Kang et al. 2006; Kadowaki et al. 2015). Derline (degradation in ER proteins) transfers them from the translocon to the proteasome, with the participation of chaperone BCL2-associated athanogene 6 (Bag6) and p97 (synonymously known as Cdc48 or VCP) (Kadowaki et al. 2015). pQC decreases the level of misfolded proteins in the ER under stress. The Bag6 complex is responsible for misdirected protein break-down (Hessa et al. 2011). This route detects the presence of non-processed or non-incorporated hydrophobic regions liberated into the cytosol and guides these proteins toward their break-down. Other stress-related quality assurance activities comprise the enrollment of c-Jun N-terminal kinase (JNK) on ribosomes through the receptor for activated protein C kinase 1 (RACK1), the phosphorylation of the elongation factor eEF1A2 and the encouragement of break-down by the proteasome (Gandin et al. 2013). It involves the complex JNK/RACK1/eEF1A2 in the process of controlling the quality of proteins on the ribosome in reaction to stress.

The emerging chains therefore interoperate with a series of various factors on the ribosome during translation. These interactions are necessary for normal folding, transport and production of active proteins. Modifications of these key interferences resulting from mutations or defective components initiate defense mechanisms to avoid the aggregation of the potentially toxic substances in the cells. In contrast, various abnormalities in mRNAs can cause a translation stop, which prevents new translation cycles and can be lethal. The cells evolved defense mechanisms to mechanically recycle the inhibited ribosome and eliminate abnormal proteins and mRNAs. A network of ribosome-related proteins, endo- and exo-nucleases, chaperones, ubiquitin ligases, proteasomes and other proteins that interact with one another therefore keeps the protein homeostasis in the cells sustained. In the various phases of protein biogenesis, several mechanisms are employed to eliminate abnormal mRNA templates, mutated or incomplete emerging chains, and incorrectly folded or wrongly located proteins. Nevertheless, many of the intricacies of these mechanisms are not yet fully comprehended, and further research is required to close these holes.

References

E.A. Abbondanzieri, W.J. Greenleaf, J.W. Shaevitz, R. Landick, S.M. Block, Direct observation of base-pair stepping by RNA polymerase. Nature **438**, 460–465 (2005)

M. Ackermann, Bacteria as a new model system for aging studies: investigations using light microscopy. Biotechniques **44**(4), 564–567 (2008)

K. Agger, P.A.C. Cloos, J. Christensen, D. Pasini, S. Rose, J. Rappsilber, I. Issaeva, E. Canaani, A.E. Salcini, K. Helin, UTX and JMJD3 are histone H3K27 demethylases involved in HOX gene regulation and development. Nature **449** (7163), 731–734 (2007)

N. Akimitsu, J. Tanaka, J. Pelletier, Translation of nonSTOP mRNA is repressed post-initiation in mammalian cells. EMBO J. **26**, 2327–2338 (2007)

V. Albanese, A.Y. Yam, J. Baughman, C. Parnot, J. Frydman, Systems analyses reveal two chaperone networks with distinct functions in eukaryotic cells. Cell **124**, 75–88 (2006)

V. Albanese, S. Reissmann, J. Frydman, A ribosome-anchored chaperone network that facilitates eukaryotic ribosome biogenesis. J. Cell Biol. **189**, 69–81 (2010)

C. Albuquerque, E. Yeung, S. Ma, T. Fu, H. Zhou, A new approach to study site-specific protein sumoylation. FASEB J. 28 (1 Suppl), 925.3 (2014)

N.N. Alder, A.E. Johnson, Cotranslational membrane protein biogenesis at the endoplasmic reticulum. J. Biol. Chem. **279**, 22787–22790 (2004)

E. Alipour, J.F. Marko, Self-organization of domain structures by DNA- loop-extruding enzymes. Nucleic Acids Res. **40**, 11202–11212 (2012)

N. Amrani, R. Ganesan, S. Kervestin, D.A. Mangus, S. Ghosh, A. Jacobson, A faux 3′-UTR promotes aberrant termination and triggers nonsense-mediated mRNA decay. Nature **432**, 112–118 (2004)

G.R. Andersen, S. Thirup, L.L. Spremulli, J. Nyborg, High resolution crystal structure of bovine mitochondrial EF-Tu in complex with GDP. J. Mol. Biol. **297**, 421–436 (2000)

D.B. Anderson, K.A. Wilkinson, J.M. Henley, Protein SUMOylation in neuropathological conditions. Drug News Perspect. **22**, 255–265 (2009)

A.M. Anger, J.P. Armache, O. Berninghausen, M. Habeck, M. Subklewe, D.N. Wilson, R. Beckmann, Structures of the human and Drosophila 80S ribosome. Nature **497**, 80–85 (2013)

K. Asano, J. Clayton, A. Shalev, A.G. Hinnebusch, A multifactor complex of eukaryotic initiation factors, eIF1, eIF2, eIF3, eIF5, and initiator tRNAMet is an important translation initiation intermediate in vivo. Genes Dev. **14**(19), 2534–2546 (2000)

A. Ashkin, Acceleration and trapping of particles by radiation pressure. Phys. Rev. Lett. **24**, 156 (1970)

G.C. Atkinson, S.L. Baldauf, V. Hauryliuk, Evolution of nonstop, no-go and nonsense-mediated mRNA decay and their termination factor-derived components. BMC Evol. Biol. **8**, 290 (2008)

J. Bäcklund, S. Carlsen, T. Höger, B. Holm, L. Fugger, J. Kihlberg, H. Burkhardt, R. Holmdahl, Predominant selection of T cells specific for the glycosylated collagen type II epitope (263–270) in humanized transgenic mice and in rheumatoid arthritis. Proc. Natl. Acad. Sci. **99**(15), 9960–9965 (2002a)

J. Bäcklund, A. Treschow, R. Bockermann, B. Holm, L. Holm, S. Issazadeh-Navikas, J. Kihlberg, R. Holmdahl, Glycosylation of type II collagen is of major importance for T cell tolerance and pathology in collagen-induced arthritis. Eur. J. Immunol. **32**(12), 3776–3784 (2002b)

M. Baker, Blame it on the antibodies. Nature **521**, 274–275 (2015)

G. Balistreri, P. Horvath, C. Schweingruber, D. Zund, G. McInerney, A. Merits, O. Mühlemann, C. Azzalin, A. Helenius, The host nonsense-mediated mRNA decay pathway restricts Mammalian RNA virus replication. Cell Host Microbe **16**, 403–411 (2014)

A.R. Ball Jr., K. Yokomori, The structural maintenance of chromosomes (SMC) family of proteins in mammals. Chromosome Res. **9**, 85–96 (2001)

M. Ballesteros, A. Fredriksson, J. Henriksson, T. Nyström, Bacterial senescence: protein oxidation in non-proliferating cells is dictated by the accuracy of the ribosomes. EMBO J. **20**(18), 5280–5289 (2001)

N. Ban, P. Nissen, J. Hansen, P.B. Moore, T.A. Steitz, The complete atomic structure of the large ribosomal subunit at 2.4 A resolution. Science **289**, 905–920 (2000)

E.J. Banigan, A.D. Stephens, J.F. Marko, Mechanics and buckling of biopolymeric shells and cell nuclei. Biophys. J. **113**, 1654–1663 (2017)

B. Banushi, F. Forneris, A. Straatman-Iwanowska et al., Regulation of post-Golgi LH3 trafficking is essential for collagen homeostasis. Nat. Commun. **7**, 2111 (2016)

J. Bao, K. Vitting-Seerup, J. Waage, C. Tang, Y. Ge, B.T. Porse, W. Yan, UPF2-Dependent nonsense-mediated mRNA decay pathway is essential for spermatogenesis by selectively eliminating longer 3′UTR transcripts. PLoS Genet. **12**, e1005863 (2016)

M.A. Baraibar, L. Liu, E.K. Ahmed, B. Friguet, Protein oxidative damage at the crossroads of cellular senescence, aging, and age-related diseases. Oxid. Med. Cell. Longev. **2012**, 919832 (2012)

R. Barresi, K.P. Campbell, Dystroglycan: from biosynthesis to pathogenesis of human disease. J. Cell Sci. **119**(2), 199–207 (2006)

S. Bartesaghi, J. Wenzel, M. Trujillo, S. Bartesaghi, J. Wenzel, M. Trujillo, Lipid peroxyl radicals mediate tyrosine dimerization and 2010 Lipid peroxyl radicals mediate tyrosine dimerization and nitration in membranes. Chem. Res. Toxicol. **23**(4), 821–835 (2010)

S. Basak, C. Lu, A. Basak, Post-translational protein modifications of rare and unconventional types: implications in functions and diseases. Curr. Med. Chem. **23**(7), 714–745 (2016)

J.W. Baynes, The Maillard hypothesis on aging: time to focus on DNA. Ann. N. Y. Acad. Sci. **959**(1), 360–367 (2002)

J.R. Beard, S. Biggs, D.E. Bloom, et al., Global Population Ageing: Peril or Promise? World Economic Forum Geneva (2015)

E. Bechtold, S.B. King, Chemical methods for the direct detection and labeling of S-nitrosothiols. Antioxid. Redox Signal. **17**(7), 981–991 (2012)

D.B. Beck, H. Oda, S.S. Shen, D. Reinberg, PR-Set7 and H4K20me1: at the crossroads of genome integrity, cell cycle, chromosome condensation, and transcription. Genes Dev. **26**, 325–337 (2012)

T. Becker, J.P. Armache, A. Jarasch, A.M. Anger, E. Villa, H. Sieber, B.A. Motaal, T. Mielke, O. Berninghausen, R. Beckmann, Structure of the no-go mRNA decay complex Dom34-Hbs1 bound to a stalled 80S ribosome. Nat. Struct. Mol. Biol. **18**, 715–720 (2011)

M.H. Bengtson, C.A. Joazeiro, Role of a ribosome-associated E3 ubiquitin ligase in protein quality control. Nature **467**, 470–473 (2010)

M. Benhar, M.T. Forrester, J.S. Stamler, Protein denitrosylation: enzymatic mechanisms and cellular functions. Nat. Rev. Mol. Cell Biol. **10**(10), 721–732 (2009)

A. Ben-Shem, N. Garreau de Loubresse, S. Melnikov, L. Jenner, G. Yusupova, M. Yusupov, The structure of the eukaryotic ribosome at 3.0 Å resolution. Science **334**, 1524–1529 (2011)

B.S. Berlett, E.R. Stadtman, Protein oxidation in aging, disease, and oxidative stress. J. Biol. Chem. **272**(33), 20313–20316 (1997)

K.K. Biggar, S.S.-C. Li, Non-histone protein methylation as a regulator of cellular signalling and function. Nat. Rev. Mol. Cell Biol. **16**(1), 5–17 (2015)

R. Biggs, P.Z. Liu, A.D. Stephens, J.F. Marko, Effects of altering histone posttranslational modifications on mitotic chromosome structure and mechanics. Mol. Biol. Cell **30**(7), 820–827 (2019)

S. Bilokapic, M. Strauss, M. Halic, Cryo-EM of nucleosome core particle interactions in trans. Sci. Rep. **8**, 7046 (2018)

I. Biryukova, T. Ye, Endogenous siRNAs and piRNAs derived from transposable elements and genes in the malaria vector mosquito *Anopheles gambiae*. BMC Genom. **16**, 278 (2015)

S.P. Braithwaite, M. Voronkov, J.B. Stock, M.M. Mouradian, Targeting phosphatases as the next generation of disease modifying therapeutics for Parkinson's disease. Neurochem. Int. **61**(6), 899–906 (2012)

O. Brandman, R.S. Hegde, Ribosome-associated protein quality control. Nat. Struct. Mol. Biol. **23**, 7–15 (2016)

O. Brandman, J. Stewart-Ornstein, D. Wong et al., A ribosome-bound quality control complex triggers degradation of nascent peptides and signals translation stress. Cell **151**, 1042–1054 (2012)

I. Brockhausen, Mucin-type O-glycans in human colon and breast cancer: glycodynamics and functions. EMBO Rep. **7** (6), 599–604 (2006)

D.E. Brodersen, W.M. Clemons Jr., A.P. Carter, R.J. Morgan-Warren, B.T. Wimberly, V. Ramakrishnan, The structural basis for the action of the antibiotics tetracycline, pactamycin, and hygromycin B on the 30S ribosomal subunit. Cell **103**, 1143–1154 (2000)

J.L. Brodsky, R.J. Wojcikiewicz, Substrate-specific mediators of ER associated degradation (ERAD). Curr. Opin. Cell Biol. **21**, 516–521 (2009)

M.-C. Broillet, S-nitrosylation of proteins. Cell. Mol. Life Sci. **55**(8), 1036–1042 (1999)

C. Bustamante, J.F. Marko, E.D. Siggia, S. Smith, Entropic elasticity of lambda-phage DNA. Science **265**, 1599–1600 (1994)
C. Bustamante, D. Keller, G. Oster, The physics of molecular motors. Acc. Chem. Res. **34**, 412–420 (2001)
C. Bustamante, Y.R. Chemla, N.R. Forde, D. Izhaky, Mechanical processes in biochemistry. Annu. Rev. Biochem. **73**, 705–748 (2004)
F. Cabreiro, C.R. Picot, B. Friguet, I. Petropoulos, Methionine sulfoxide reductases: relevance to aging and protection against oxidative stress. Ann. N. Y. Acad. Sci. **1067**(1), 37–44 (2006)
A. Campbell, P. Solaimani, Oxidative and inflammatory pathways in age-related chronic disease processes, in *Inflammation, Aging, and Oxidative Stress*, ed. by S.C. Bondy, A. Campbell (Springer International Publishing Switzerland, 2016)
V.J. Carabetta, T.M. Greco, A.W. Tanner, I.M. Cristea, D. Dubnau, Temporal regulation of the *Bacillus subtilis* acetylome and evidence for a role of MreB acetylation in cell wall growth. mSystems **1**(3), 00005–00016 (2016)
J.R. Carey, P. Liedo, D. Orozco, J.W. Vaupel, Slowing of mortality rates at older ages in large medfly cohorts. Science **258**(5081), 457–461 (1992)
A.P. Carter, W.M. Clemons, D.E. Brodersen, R.J. Morgan-Warren, B.T. Wimberly, V. Ramakrishnan, Functional insights from the structure of the 30S ribosomal subunit and its interactions with antibiotics. Nature **407**, 340–348 (2000)
C. Cecconi, E.A. Shank, C. Bustamante, S. Marqusee, Direct observation of the three-state folding of a single protein molecule. Science **309**, 2057–2060 (2005)
A. Celik, F. He, A. Jacobson, NMD monitors translational fidelity 24/7. Curr. Genet. **63**, 1007–1010 (2017)
K.J. Chalut, M. Hopfler, F. Lautenschlager, L. Boyde, C.J. Chan, A. Ekpenyong, A. Martinez-Arias, J. Guck, Chromatin decondensation and nuclear softening accompany Nanog downregulation in embryonic stem cells. Biophys. J. **103**, 2060–2070 (2012)
B. Charlesworth, L. Partridge, Ageing: levelling of the grim reaper. Curr. Biol. **7**(7), R440–R442 (2015)
Y.R. Chemla, J.R. Moffitt, C. Bustamante, Exact solutions for kinetic models of macromolecular dynamics. J. Phys. Chem. B **112**, 6025–6044 (2008)
L. Chen, M.B. Feany, Alpha-synuclein phosphorylation controls neurotoxicity and inclusion formation in a Drosophila model of Parkinson disease. Nat. Neurosci. **8**(5), 657–663 (2005)
Y. Chen, S. Feng, V. Kumar, R. Ero, Y.G. Gao, Structure of EF-G-ribosome complex in a pretranslocation state. Nat. Struct. Mol. Biol. **20**, 1077–1084 (2013)
G. Chistol, S. Liu, C.L. Hetherington, J.R. Moffitt, S. Grimes, P.J. Jardine, C. Bustamante, High degree of coordination and division of labor among subunits in a homomeric ring ATPase. Cell **151**, 1017–1028 (2012)
K.H. Chow, R.E. Factor, K.S. Ullman, The nuclear envelope environment and its cancer connections. Nat. Rev. Cancer **12**, 196–209 (2012)
J. Christensen, K. Agger, P.A.C. Cloos, D. Pasini, S. Rose, L. Sennels, J. Rappsilber, K.H. Hansen, A.E. Salcini, K. Helin, RBP2 belongs to a family of demethylases, specific for tri-and dimethylated lysine 4 on histone 3. Cell **128** (6), 1063–1076 (2007)
S. Chu, *The Manipulation of Neutral Particles Nobel Lectures Physics 1996–2000*, ed. by G. Ekspong (World Scientific Press Singapore, 2002)
P.A.C. Cloos, S. Christgau, Post-translational modifications of proteins: implications for aging, antigen recognition, and autoimmunity. Biogerontology **5**(3), 139–158 (2004)
P. Cohen, The regulation of protein function by multisite phosphorylation—a 25 year update. Trends Biochem. Sci. **25** (12), 596–601 (2000)
D. Collin, F. Ritort, C. Jarzynski, S.B. Smith, I. Tinoco Jr., C. Bustamante, Verification of the Crooks fluctuation theorem and recovery of RNA folding free energies. Nature **437**, 231–234 (2005)
M.J. Comstock, T. Ha, Y.R. Chemla, Ultrahigh-resolution optical trap with single-fluorophore sensitivity. Nat. Methods **8**, 335–340 (2011)
S.A. Comyn, G.T. Chan, T. Mayor, False start: cotranslational protein ubiquitination and cytosolic protein quality control. J. Proteomics **100**, 92–101 (2014)
S.R. Connell, C. Takemoto, D.N. Wilson et al., Structural basis for interaction of the ribosome with the switch regions of GTP-bound elongation factors. Mol. Cell **25**, 751–764 (2007)
P.V. Cornish, D.N. Ermolenko, H.F. Noller, T. Ha, Spontaneous intersubunit rotation in single ribosomes. Mol. Cell **30**, 578–588 (2008)
A.J. Cozzone, Role of protein phosphorylation on serine/threonine and tyrosine in the virulence of bacterial pathogens. J. Mol. Microbiol. Biotechnol. **9**(3–4), 198–213 (2005)
E.M. Crimmins, Lifespan and health span: past, present, and promise. Gerontologist **55**(6), 901–911 (2015)
G.E. Crooks, Nonequilibrium measurements of free energy differences for microscopically reversible Markovian systems. J. Stat. Phys. **90**, 1481–1487 (1998)
G.E. Crooks, Entropy production fluctuation theorem and the nonequilibrium work relation for free energy differences. Phys. Rev. E **60**, 2721 (1999)

M.R. Culbertson, RNA surveillance. Unforeseen consequences for gene expression, inherited genetic disorders and cancer. Trends Genet. **15**, 74–80 (1999)

J.W. Curtsinger, H.H. Fukui, D.R. Townsend, J.W. Vaupel, Demography of genotypes: failure of the limited life-span paradigm in *Drosophila melanogaster*. Science **258**(5081), 461–463 (1992)

P.G. de Gennes, *Scaling Concepts in Polymer Physics* (Cornell University Press, Ithaca, 1979)

R.S. Deeb, T. Nuriel, C. Cheung et al., Characterization of a cellular denitrase activity that reverses nitration of cyclooxygenase. Am. J. Physiol. Heart Circ. Physiol. **305**(5), H687–H698 (2013)

Q. Defenouillere, Y. Yao, J. Mouaikel et al., Cdc48-associated complex bound to 60S particles is required for the clearance of aberrant translation products. Proc. Natl. Acad. Sci. U.S.A. **110**, 5046–5051 (2013)

V. Dias, E. Junn, M.M. Mouradian, The role of oxidative stress in Parkinson's disease. J. Parkinson's Dis. **3**(4), 461–491 (2013)

L.N. Dimitrova, K. Kuroha, T. Tatematsu, T. Inada, Nascent peptide-dependent translation arrest leads to Not4p-mediated protein degradation by the proteasome. J. Biol. Chem. **284**, 10343–10352 (2009)

D. Doenecke, Chromatin dynamics from S-phase to mitosis: contributions of histone modifications. Cell Tissue Res. **356**, 467–475 (2014)

M.K. Doma, R. Parker, Endonucleolytic cleavage of eukaryotic mRNAs with stalls in translation elongation. Nature **440**, 561–564 (2006)

M.K. Doma, R. Parker, RNA quality control in eukaryotes. Cell **131**, 660–668 (2007)

K. Doring, N. Ahmed, T. Riemer et al., Profiling Ssb-nascent chain interactions reveals principles of Hsp70-Assisted folding. Cell **170**, 298.e20–311.e20 (2017)

J. Dudek, S. Pfeffer, P.H. Lee, M. Jung, A. Cavalié, V. Helms, F. Förster, R. Zimmermann, Protein transport into the human endoplasmic reticulum. J. Mol. Biol. **427**(6 Pt A), 1159–1175 (2015)

S. Duttler, S. Pechmann, J. Frydman, Principles of cotranslational ubiquitination and quality control at the ribosome. Mol. Cell **50**, 379–393 (2013)

A.B. Eberle, S. Lykke-Andersen, O. Muhlemann, T.H. Jensen, SMG6 promotes endonucleolytic cleavage of nonsense mRNA in human cells. Nat. Struct. Mol. Biol. **16**, 49–55 (2009)

P.F. Egea, R.M. Stroud, P. Walter, Targeting proteins to membranes: structure of the signal recognition particle. Curr. Opin. Struct. Biol. **15**, 213–220 (2005)

G. Eisner, M. Moser, U. Schafer, K. Beck, M. Muller, Alternate recruitment of signal recognition particle and trigger factor to the signal sequence of a growing nascent polypeptide. J. Biol. Chem. **281**, 7172–7179 (2006)

L. Ellgaard, N. McCaul, A. Chatsisvili, I. Braakman, Co- and post-translational protein folding in the ER. Traffic **17**, 615–638 (2016)

O. Emanuelsson, G. von Heijne, Prediction of organellar targeting signals. Biochim. Biophys. Acta **1541**, 114–119 (2001)

S.A. Etchells, A.S. Meyer, A.Y. Yam, A. Roobol, Y. Miao, Y. Shao, M.J. Carden, W.R. Skach, J. Frydman, A.E. Johnson, The cotranslational contacts between ribosome bound nascent polypeptides and the subunits of the hetero-oligomeric chaperonin TRiC probed by photocross-linking. J. Biol. Chem. **280**, 28118–28126 (2005)

D. Falzarano, O. Krokhin, G.V. Domselaar, K. Wolf, J. Seebach, H.J. Schnittler, H. Feldmann, Ebola sGP—the first viral glycoprotein shown to be C-mannosylated. Virology **368**(1), 83–90 (2007)

P.J. Farabaugh, Programmed translational frameshifting. Microbiol. Rev. **60**, 103–134 (1996)

A.R. Farley, A.J. Link, Identification and quantification of protein posttranslational modifications. Methods Enzymol. **463**, 725–763 (2009)

J. Fei, P. Kosuri, D.D. MacDougall, R.L. Gonzalez Jr., Coupling of ribosomal L1 stalk and tRNA dynamics during translation elongation. Mol. Cell **30**, 348–359 (2008)

J. Feser, J. Tyler, Chromatin structure as a mediator of aging. FEBS Lett. **585**(13), 2041–2048 (2011)

T. Finkel, The metabolic regulation of aging. Nat. Med. **21**(12), 1416–1423 (2015)

T. Finkel, C.-X. Deng, R. Mostoslavsky, Recentprogress in the biology and physiology of sirtuins. Nature **460**(7255), 587–591 (2009)

N. Fischer, A.L. Konevega, W. Wintermeyer, M.V. Rodnina, H. Stark, Ribosome dynamics and tRNA movement by time-resolved electron cryomicroscopy. Nature **466**, 329–333 (2010)

M.W. Foster, D.T. Hess, J.S. Stamler, Protein S-nitrosylation in health and disease: a current perspective. Trends Mol. Med. **15**(9), 391–404 (2009)

J. Frank, Electron microscopy of functional ribosome complexes. Biopolymers **68**, 223–233 (2003)

T.M. Franks, G. Singh, J. Lykke-Andersen, Upf1 ATPase-dependent mRNP disassembly is required for completion of nonsense-mediated mRNA decay. Cell **143**, 938–950 (2010)

P.A. Frischmeyer, A. van Hoof, K. O'Donnell, A.L. Guerrerio, R. Parker, H.C. Dietz, An mRNA surveillance mechanism that eliminates transcripts lacking termination codons. Science **295**, 2258–2261 (2002)

H. Fujiwara, M. Hasegawa, N. Dohmae, A. Kawashima, E. Masliah, M.S. Goldberg, J. Shen, K. Takio, T. Iwatsubo, α-Synuclein is phosphorylated in synucleinopathy lesions. Nat. Cell Biol. **4**(2), 160–164 (2002)

D.P. Galonić, D.Y. Gin, Chemical glycosylation in the synthesis of glycoconjugate antitumour vaccines. Nature **446** (7139), 1000–1007 (2007)

V. Gandin, I. Topisirovic, Co-translational mechanisms of quality control of newly synthesized polypeptides. Translation **2**, e28109 (2014)

V. Gandin, G.J. Gutierrez, L.M. Brill et al., Degradation of newly synthesized polypeptides by ribosome-associated RACK1/c-Jun N-terminal kinase/eukaryotic elongation factor 1A2 complex. Mol. Cell. Biol. **33**, 2510–2526 (2013)

H. Gao, Z. Zhou, U. Rawat, C. Huang, L. Bouakaz, C. Wang, Z. Cheng, Y. Liu, A. Zavialov, R. Gursky, S. Sanyal, M. Ehrenberg, J. Frank, H. Song, RF3 induces ribosomal conformational changes responsible for dissociation of class I release factors. Cell **129**, 929–941 (2007a)

N. Gao, A.V. Zavialov, M. Ehrenberg, J. Frank, Specific interaction between EF-G and RRF and its implication for GTP-dependent ribosome. J. Mol. Biol. **374**, 1345–1358 (2007b)

Y.G. Gao, M. Selmer, C.M. Dunham, A. Weixlbaumer, A.C. Kelley, V. Ramakrishnan, The structure of the ribosome with elongation factor G trapped in the posttranslocational state. Science **326**, 694–699 (2009)

M. Gautschi, H. Lilie, U. Funfschilling, A. Mun, S. Ross, T. Lithgow, P. Rücknagel, S. Rospert, RAC, a stable ribosome-associated complex in yeast formed by the DnaK-DnaJ homologs Ssz1p and zuotin. Proc. Natl. Acad. Sci. U.S.A. **98**, 3762–3767 (2001)

N.H. Gehring, S. Lamprinaki, M.W. Hentze, A.E. Kulozik, The hierarchy of exon-junction complex assembly by the spliceosome explains key features of mammalian nonsense-mediated mRNA decay. PLoS Biol. **7**, e1000120 (2009)

C. Gell, D. Brockwell, A. Smith, *Handbook of Single Molecule Fluorescence Spectroscopy* (Oxford University Press, Oxford, 2006)

D. Gems, L. Partridge, Genetics of longevity in model organisms: debates and paradigm shifts. Annu. Rev. Physiol. **75**, 621–644 (2013)

D. Gentilini, D. Mari, D. Castaldi et al., Role of epigenetics in human aging and longevity: genome-wide DNA methylation profile in centenarians and centenarians' offspring. Age (Omaha) **35**(5), 1961–1973 (2013)

J.H. Gibcus, K. Samejima, A. Goloborodko et al., A path-way for mitotic chromosome formation. Science **359**, eaao6135 (2018)

B.A. Gibson, Y. Zhang, H. Jiang, K.M. Hussey, J.H. Shrimp, H. Lin, F. Schwede, Y. Yu, W.L. Kraus, Chemical genetic discovery of PARP targets reveals a role for PARP-1 in transcription elongation. Science **353**(6294), 45–50 (2016)

M.A. Glozak, N. Sengupta, X. Zhang, E. Seto, Acetylation and deacetylation of non-histone proteins. Gene **363**, 15–23 (2005)

A. Goloborodko, M.V. Imakaev, J.F. Marko, L. Mirny, Compaction and segregation of sister chromatids via active loop extrusion. Elife **5**, e14864 (2016)

L. Gorisse, C. Pietrement, V. Vuiblet, C.E.H. Schmelzer, M. Köhler, L. Duca, L. Debelle, P. Fornès, S. Jaisson, P. Gillery, Protein carbamylation is a hallmark of aging. Proc. Natl. Acad. Sci. **113**(5), 1191–1196 (2016)

S. Goto, Z. Radak, Implications of oxidative damage to proteins and DNA in aging and its intervention by caloric restriction and exercise. J. Sport Health Sci. **2**(2), 75–80 (2013)

M. Graille, B. Seraphin, Surveillance pathways rescuing eukaryotic ribosomes lost in translation. Nat. Rev. Mol. Cell Biol. **13**, 727–735 (2012)

N. Gregersen, P. Bross, S. Vang, J.H. Christensen, Protein misfolding and human disease. Annu. Rev. Genomics Hum. Genet. **7**, 103–124 (2006)

U. Greifenhagen, A. Frolov, M. Blüher, R. Hoffmann, Site-specific analysis of advanced glycation end products in plasma proteins of type 2 diabetes mellitus patients. Anal. Bioanal. Chem. **408**(20), 5557–5566 (2016)

S.I.S. Grewal, J.C. Rice, Regulation of heterochromatin by histone methylation and small RNAs. Curr. Opin. Cell Biol. **16**(3), 230–238 (2004)

T. Grune, K. Merker, G. Sandig, K.J.A. Davies, Selective degradation of oxidatively modified protein substrates by the proteasome. Biochem. Biophys. Res. Commun. **305**(3), 709–718 (2003)

R. Gupte, Z. Liu, W.L. Kraus, PARPs and ADP-ribosylation: recent advances linking molecular functions to biological outcomes. Genes Dev. **31**(2), 101–126 (2017)

D. Halter, M.A. Collart, O.O. Panasenko, The Not4 E3 ligase and CCR4 deadenylase play distinct roles in protein quality control. PLoS ONE **9**, e86218 (2014)

R.S. Haltiwanger, P. Stanley, Modulation of receptor signaling by glycosylation: fringe is an O-fucose-β1, 3-N-acetylglucosaminyltransferase. Biochem. Biophys. Acta **1573**(3), 328–335 (2002)

S.M. Hammond, S. Boettcher, A.A. Caudy, R. Kobayashi, G.J. Hannon, Argonaute2, a link between genetic and biochemical analyses of RNAi. Science **293**, 1146–1150 (2001)

J. Harms, F. Schluenzen, R. Zarivach, A. Bashan, S. Gat, I. Agmon, H. Bartels, F. Franceschi, A. Yonath, High resolution structure of the large ribosomal subunit from a mesophilic eubacterium. Cell **107**, 679–688 (2001)

R.J. Harris, M.W. Spellman, O-linked fucose and other post-translational modifications unique to EGF modules. Glycobiology **3**(3), 219–224 (1993)

F.U. Hartl, M. Hayer-Hartl, Molecular chaperones in the cytosol: from nascent chain to folded protein. Science **295**, 1852–1858 (2002)

F.U. Hartl, A. Bracher, M. Hayer-Hartl, Molecular chaperones in protein folding and proteostasis. Nature **475**, 324–332 (2011)

L. Hayflick, How and why we age. Exp. Gerontol. **33**(7–8), 639–653 (1998)

S.L. Hazen, J.R. Crowley, D.M. Mueller, J.W. Heinecke, Mass spectrometric quantification of 3-chlorotyrosine in human tissues with attomole sensitivity: a sensitive and specific marker for myeloperoxidase-catalyzed chlorination at sites of inflammation. Free Radic. Biol. Med. **23**(6), 909–916 (1997)

F. He, A. Jacobson, Nonsense-Mediated mRNA Decay: degradation of defective transcripts is only part of the story. Annu. Rev. Genet. **49**, 339–366 (2015)

W. He, J.C. Newman, M.Z. Wang, L. Ho, E. Verdin, Mitochondrial sirtuins: regulators of protein acylation and metabolism. Trends Endocrinol. Metab. **23**(9), 467–476 (2012)

D.N. Hebert, M. Molinari, In and out of the ER: protein folding, quality control, degradation, and related human diseases. Physiol. Rev. **87**, 1377–1408 (2007)

J.W. Heck, S.K. Cheung, R.Y. Hampton, Cytoplasmic protein quality control degradation mediated by parallel actions of the E3 ubiquitin ligases Ubr1 and San1. Proc. Natl. Acad. Sci. U.S.A. **107**, 1106–1111 (2010)

B. Hemmeryckx, P. Hohensinner, M. Swinnen, W. Heggermont, J. Wojta, H.R. Lijnen, Antioxidant treatment improves cardiac dysfunction in a murine model of premature aging. J. Cardiovasc. Pharmacol. **68**(5), 374–382 (2016)

J.M. Henley, T.J. Craig, K.A. Wilkinson, Neuronal SUMOylation: mechanisms, physiology, and roles in neuronal dysfunction. Physiol. Rev. **94**(4), 1249–1285 (2014)

T. Hessa, A. Sharma, M. Mariappan, H.D. Eshleman, E. Gutierrez, R.S. Hegde, Protein targeting and degradation are coupled for elimination of mislocalized proteins. Nature **475**, 394–397 (2011)

P. Hinterdorfer, A. van Oijen, *Handbook of Single-Molecule Biophysics* (Springer, New York, 2009)

M.S. Hipp, S.H. Park, F.U. Hartl, Proteostasis impairment in protein-misfolding and -aggregation diseases. Trends Cell Biol. **24**, 506–514 (2014)

R. Hodara, E.H. Norris, B.I. Giasson, A.J. Mishizen-Eberz, D.R. Lynch, V.M. Lee, H. Ischiropoulos, Functional consequences of α-synuclein tyrosine nitration: diminished binding to lipid vesicles and increased fibril formation. J. Biol. Chem. **279**(46), 47746–47753 (2004)

J.R. Hogg, S.P. Goff, Upf1 senses 3′UTR length to potentiate mRNA decay. Cell **143**, 379–389 (2010)

A. Hohn, D. Weber, T. Jung, C. Ott, M. Hugo, B. Kochlik, R. Kehm, J. König, T. Grune, J.P. Castro, Happily (n)ever after: aging in the context of oxidative stress, proteostasis loss and cellular senescence. Redox Biol. **11**, 482–501 (2016)

J. Hollien, J.S. Weissman, Decay of endoplasmic reticulum-localized mRNAs during the unfolded protein response. Science **313**, 104–107 (2006)

J. Hollien, J.H. Lin, H. Li, N. Stevens, P. Walter, J.S. Weissman, Regulated Ire1-dependent decay of messenger RNAs in mammalian cells. J. Cell Biol. **186**, 323–331 (2009)

G. Hummer, A. Szabo, Free energy reconstruction from nonequilibrium single-molecule pulling experiments. Proc. Natl. Acad. Sci. U.S.A. **98**, 3658–3661 (2001)

E. Huntzinger, I. Kashima, M. Fauser, J. Sauliere, E. Izaurralde, SMG6 is the catalytic endonuclease that cleaves mRNAs containing nonsense codons in metazoan. RNA **14**, 2609–2617 (2008)

J. Hwang, H. Sato, Y. Tang, D. Matsuda, L.E. Maquat, UPF1 association with the cap-binding protein, CBP80, promotes nonsense-mediated mRNA decay at two distinct steps. Mol. Cell **39**, 396–409 (2010)

S. Imai, C.M. Armstrong, M. Kaeberlein, L. Guarente, Transcriptional silencing and longevity protein Sir2 is an NAD-dependent histone deacetylase. Nature **403**(6771), 795–800 (2000)

T. Inada, H. Aiba, Translation of aberrant mRNAs lacking a termination codon or with a shortened 3′-UTR is repressed after initiation in yeast. EMBO J. **24**, 1584–1595 (2005)

K. Iqbal, C.X. Gong, F. Liu, Hyperphosphorylation-induced tau oligomers. Front. Neurol. **4**, 00112 (2013)

Y. Ishitsuka, F. Maniwa, C. Koide, Y. Kato, Y. Nakamura, T. Osawa, M. Tanioka, Y. Miyachi, Increased halogenated tyrosine levels are useful markers of human skin ageing, reflecting proteins denatured by past skin inflammation. Clin. Exp. Dermatol. **37**(3), 252–258 (2012)

O. Isken, Y.K. Kim, N. Hosoda, G.L. Mayeur, J.W. Hershey, L.E. Maquat, Upf1 phosphorylation triggers translational repression during nonsense-mediated mRNA decay. Cell **133**, 314–327 (2008)

H. Jarjanazi, S. Savas, N. Pabalan, J.W. Dennis, H. Ozcelik, Biological implications of SNPs in signal peptide domains of human proteins. Proteins **70**, 394–403 (2008)

C. Jarzynski, Equilibrium free-energy differences from nonequilibrium measurements: a master-equation approach. Phys. Rev. E **56**, 5018 (1997)

C. Jarzynski, How does a system respond when driven away from thermal equilibrium? Proc. Natl. Acad. Sci. U.S.A. **98**, 3636–3638 (2001)

W.P. Jencks, *Handbook of Biochemistry and Molecular Biology: Nucleic Acids*, ed. by G.D. Fasman, vol. 1 (CRC Press, Cleveland, 1975)

L. Jenner, S. Melnikov, N.G. de Loubresse, A. Ben-Shem, M. Iskakova, A. Urzhumtsev, A. Meskauskas, J. Dinman, G. Yusupova, M. Yusupov, Differential stoichiometry among core ribosomal proteins. Curr. Opin. Struct. Biol. **22**, 759–873 (2012)

S. Jentsch, B. Haendler, T. Hunter, H. Sun, *The Ubiquitin System in Health and Disease* (Springer, Berlin, 2009)

H. Jin, A.C. Kelley, D. Loakes, V. Ramakrishnan, Structure of the 70S ribosome bound to release factor 2 and a substrate analog provides insights into catalysis of peptide release. Proc. Natl. Acad. Sci. U.S.A. **107**, 8593–8598 (2010)

H. Jin, A.C. Kelley, V. Ramakrishnan, Crystal structure of the hybrid state of ribosome in complex with the guanosine triphosphatase release factor 3. Proc. Natl. Acad. Sci. U.S.A. **108**, 15798–15803 (2011)

D.S. Johnson, L. Bai, B.Y. Smith, S.S. Patel, M.D. Wang, Single-molecule studies reveal dynamics of DNA unwinding by the ring-shaped T7 helicase. Cell **129**, 1299–1309 (2007)

P. Julian, A.L. Konevega, S.H. Scheres, M. Lazaro, D. Gil, W. Wintermeyer, M.V. Rodnina, M. Valle, Structure of ratcheted ribosomes with tRNAs in hybrid states. Proc. Natl. Acad. Sci. U.S.A. **105**, 16924–16927 (2008)

E. Jung, A. Veuthey, E. Gasteiger, A. Bairoch, Annotation of glycoproteins in the SWISS-PROT database. Proteomics **1**(2), 262–268 (2001)

H. Kadowaki, A. Nagai, T. Maruyama et al., Pre-emptive quality control protects the ER from protein overload via the proximity of ERAD COMPONENTS and SRP. Cell Rep. **13**, 944–956 (2015)

M. Kaeberlein, Resveratrol and rapamycin: are they anti-aging drugs? BioEssays **32**(2), 96–99 (2010)

C.M. Kaiser, I. Tinoco Jr., Probing the mechanisms of translation with force. Chem. Rev. **114**(6), 3266–3280 (2014)

C.M. Kaiser, D.H. Goldman, J.D. Chodera, I. Tinoco Jr., C. Bustamante, The ribosome modulates nascent protein folding. Science **334**, 1723–1727 (2011)

S. Kakizawa, Nitric oxide-induced calcium release: activation of type 1 ryanodine receptor, a calcium release channel, through non-enzymatic post-translational modification by nitric oxide. Front. Exp. Endocrinol. **4**, 00142 (2013)

M. Kalogiannis, E.J. Delikatny, T.M. Jeitner, Serotonin as a putative scavenger of hypohalous acid in the brain. Biochem. Biophys. Acta **1862**(4), 651–661 (2016)

S.W. Kang, N.S. Rane, S.J. Kim, J.L. Garrison, J. Taunton, R.S. Hegde, Substrate-specific translocational attenuation during ER stress defines a pre-emptive quality control pathway. Cell **127**, 999–1013 (2006)

A.L. Karamyshev, A.E. Johnson, Selective SecA association with signal sequences in ribosome-bound nascent chains: a potential role for SecA in ribosome targeting to the bacterial membrane. J. Biol. Chem. **280**, 37930–37940 (2005)

A.L. Karamyshev, Z.N. Karamysheva, Lost in translation: ribosome-associated mRNA and protein quality controls. Front. Genet. **9**, 431 (2018)

A.L. Karamyshev, Z.N. Karamysheva, T. Yamami, K. Ito, Y. Nakamura, Transient idling of posttermination ribosomes ready to reinitiate protein synthesis. Biochimie **86**, 933–938 (2004)

A.L. Karamyshev, A.E. Patrick, Karamysheva et al., Inefficient SRP interaction with a nascent chain triggers a mRNA quality control pathway. Cell **156**, 146–157 (2014)

I. Kashima, A. Yamashita, N. Izumi, N. Kataoka, R. Morishita, S. Hoshino, M. Ohno, G. Dreyfuss, S. Ohno, Binding of a novel SMG-1-Upf1-eRF1-eRF3 complex (SURF) to the exon junction complex triggers Upf1 phosphorylation and nonsense-mediated mRNA decay. Genes Dev. **20**, 355–367 (2006)

E. Katai, J. Pal, V.S. Poor, R. Purewal, A. Miseta, T. Nagy, Oxidative stress induces transient O-GlcNAc elevation and tau dephosphorylation in SH-SY5Y cells. J. Cell Mol. Med. **20**(12), 2269–2277 (2016)

T. Kawashima, C. Berthet-Colominas, M. Wulff, S. Cusack, R. Leberman, The structure of the Escherichia coli EF-Tu·EF-Ts complex at 2.5 Å resolution. Nature **379**, 511–518 (1996)

S. Kervestin, A. Jacobson, NMD: a multifaceted response to premature translational termination. Nat. Rev. Mol. Cell Biol. **13**, 700–712 (2012)

E.A. Kikis, T. Gidalevitz, R.I. Morimoto, Protein homeostasis in models of aging and age-related conformational disease. Adv. Exp. Med. Biol. **694**, 138–159 (2010)

T.B.L. Kirkwood, A systematic look at an old problem. Nature **451**(7179), 644–647 (2008)

T. Kirkwood, S. Austad, Why do we age? Nature **408**, 233–238 (2000)

T.B.L. Kirkwood, R.J. Boys, C.S. Gillespie, C.J. Proctor, D.P. Shanley, D.J. Wilkinson, Towards an e-biology of ageing: integrating theory and data. Nat. Rev. Mol. Cell Biol. **4**(3), 243–249 (2003)

J. Kirstein-Miles, A. Scior, E. Deuerling, R.I. Morimoto, The nascent polypeptide-associated complex is a key regulator of proteostasis. EMBO J. **32**, 1451–1468 (2013)

M. Kjeldgaard, P. Nissen, S. Thirup, J. Nyborg, The crystal structure of elongation factor EF-Tu from Thermus aquaticus in the GTP conformation. Structure **1**, 35–50 (1993)

A. Koplin, S. Preissler, Y. Ilina, M. Koch, A. Scior, M. Erhardt, E. Deuerling, A dual function for chaperones SSB-RAC and the NAC nascent polypeptide-associated complex on ribosomes. J. Cell Biol. **189**, 57–68 (2010)

R. Kornfeld, S. Kornfeld, Assembly of asparagine-linked oligosaccharides. Annu. Rev. Biochem. **54**(1), 631–664 (1985)

T. Kouzarides, Chromatin modifications and their function. Cell **128**(4), 693–705 (2007)

M. Krause, J. Te Riet, K. Wolf, Probing the compressibility of tumor cell nuclei by combined atomic force-confocal microscopy. Phys. Biol. **10**, 065002 (2013)

U.C. Krieg, P. Walter, A.E. Johnson, Photocrosslinking of the signal sequence of nascent preprolactin to the 54-kilodalton polypeptide of the signal recognition particle. Proc. Natl. Acad. Sci. U.S.A. **83**, 8604–8608 (1986)

J. Krieg, W. Gläsner, A. Vicentini, M.A. Doucey, A. Löffler, D. Hess, J. Hofsteenge, C-Mannosylation of human RNase 2 is an intracellular process performed by a variety of cultured cells. J. Biol. Chem. **272**(42), 26687–26692 (1997)

M.L. Kuhn, B. Zemaitaitis, L.I. Hu et al., Structural, kinetic and proteomic characterization of acetyl phosphate-dependent bacterial protein acetylation. PLoS One **9**(4), e94816 (2014)

M. Kuningas, S.P. Mooijaart, D. Van Heemst, B.J. Zwaan, P.E. Slagboom, R.G.J. Westendorp, Genes encoding longevity: from model organisms to humans. Aging Cell **7**, 270–280 (2008)

K. Kuroha, T. Tatematsu, T. Inada, Upf1 stimulates degradation of the product derived from aberrant messenger RNA containing a specific nonsense mutation by the proteasome. EMBO Rep. **10**, 1265–1271 (2009)

K. Kuroha, M. Akamatsu, L. Dimitrova, T. Ito, Y. Kato, K. Shirahige, T. Inada, Receptor for activated C kinase 1 stimulates nascent polypeptide-dependent translation arrest. EMBO Rep. **11**, 956–961 (2010)

T.V. Kurzchalia, M. Wiedmann, A.S. Girshovich, E.S. Bochkareva, H. Bielka, T.A. Rapoport, The signal sequence of nascent preprolactin interacts with the 54 K polypeptide of the signal recognition particle. Nature **320**, 634–636 (1986)

G. Laloux, C. Jacobs-Wagner, How do bacteria localize proteins to the cell pole? J. Cell Sci. **127**(Pt 1), 11–19 (2014)

M. Lang, S. Block, Resource letter: LBOT-1: laser-based optical tweezers. Am. J. Phys. **71**, 201 (2002)

J. Lawrimore, P.A. Vasquez, M.R. Falvo, R.M. Taylor II, L. Vicci, E. Yeh, M.G. Forest, K. Bloom, DNA loops generate intracentromere tension in mitosis. J. Cell Biol. **210**, 553–564 (2015)

M.C. Leake, *Single-Molecule Cellular Biophysics* (Cambridge University Press, Cambridge, 2013)

I.H. Lee, T. Finkel, Regulation of autophagy by the p300 acetyltransferase. J. Biol. Chem. **284**(10), 6322–6328 (2009)

J.T. Lee, T.C. Wheeler, L. Li, L.S. Chin, Ubiquitination of α-synuclein by Siah-1 promotes α-synuclein aggregation and apoptotic cell death. Hum. Mol. Genet. **17**(6), 906–917 (2008)

L. Lehle, S. Strahl, W. Tanner, Protein glycosylation, conserved from yeast to man: a model organism helps elucidate congenital human diseases. Angew. Chem. Int. Ed. **45**(41), 6802–6818 (2006)

F. Lejeune, X. Li, L.E. Maquat, Nonsense-mediated mRNA decay in mammalian cells involves decapping, deadenylating, and exonucleolytic activities. Mol. Cell **12**, 675–687 (2003)

O. Lenoir, P.L. Tharaux, T.B. Huber, Autophagy in kidney disease and aging: lessons from rodent models. Kidney Int. **90**(5), 950–964 (2016)

M. Leslie, 'Dead' enzymes show signs of life. Science **340**, 25–27 (2013)

P.T.X. Li, J. Vieregg, I. Tinoco Jr., How RNA unfolds and refolds. Ann. Rev. Biochem. **77**, 77–100 (2008)

K.A. Liddy, M.Y. White, S.J. Cordwell, Functional decorations: post-translational modifications and heart disease delineated by targeted proteomics. Genome Med. **5**(2), 20 (2013)

J. Liggins, A.J. Furth, Role of protein-bound carbonyl groups in the formation of advanced glycation endproducts. Biochem. Biophys. Acta **1361**(2), 123–130 (1997)

A. Liljas, *Structural Aspects of Protein Synthesis* (World Scientific Publishing Co., Singapore, 2004)

T. Lionnet, J.-F. Allemand, A. Revyakin, T. R. Strick, O.A. Saleh, D. Bensimon, V. Croquette, Single-Molecule Techniques, A Laboratory Manual, ed. by P.R. Selvin, T. Ha (Cold Spring Harbor Laboratory Press, Cold Spring Harbor, 2008)

J. Liphardt, B. Onoa, S.B. Smith, I. Tinoco Jr., C. Bustamante, Reversible unfolding of single RNA molecules by mechanical force. Science **292**, 733–737 (2001)

J. Liphardt, S. Dumont, S.B. Smith, I. Tinoco Jr., C. Bustamante, Equilibrium information from nonequilibrium measurements in an experimental test of Jarzynski's equality. Science **296**, 1832–1835 (2002)

J. Liu, M.A. Carmell, F.V. Rivas, C.G. Marsden, J.M. Thomson, J.J. Song, S.M. Hammond, L. Joshua-Tor, G. J. Hannon, Argonaute2 is the catalytic engine of mammalian RNAi. Science **305**, 1437–1441 (2004)

C. Liu, R. Karam, Y. Zhou et al., The UPF1 RNA surveillance gene is commonly mutated in pancreatic adenosquamous carcinoma. Nat. Med. **20**, 596–598 (2014a)

T. Liu, A. Kaplan, L. Alexander et al., Direct measurement of the mechanical work during translocation by the ribosome. eLife **3**, e03406 (2014b)

P. Ljubuncic, E. Gochman, A.Z. Reznick, Nitrosative stress in aging—its importance and biological implications in NF-κB signaling, in *Aging and Age-Related Disorders*, ed. by S. Bondy, K. Maiese (Springer, Berlin, 2010), pp. 27–54

A. Löffler, M.-A. Doucey, A.M. Jansson, D.R. Müller, T. de Beer, D. Hess, M. Meldal, W.J. Richter, J.F. Vliegenthart, J. Hofsteenge, Spectroscopic and protein chemical analyses demonstrate the presence of C-mannosylated tryptophan in intact human RNase 2 and its isoforms. Biochemistry **35**(37), 12005–12014 (1996)

B. Loh, S. Jonas, E. Izaurralde, The SMG5-SMG7 heterodimer directly recruits the CCR4-NOT deadenylase complex to mRNAs containing nonsense codons via interaction with POP2. Genes Dev. **27**, 2125–2138 (2013)

V.V. Loi, M. Rossius, H. Antelmann, Redox regulation by reversible protein S-thiolation in bacteria. Front. Microbiol. **6**, 1–22 (2015)

C. López-otín, M.A. Blasco, L. Partridge, M. Serrano, G. Kroemer, The hallmarks of aging. Cell **153**(6), 1194–1217 (2013)

J.Y. Lu, Y.Y. Lin, H. Zhu, L.M. Chuang, J.D. Boeke, Protein acetylation and aging. Aging (Albany NY) **3**(10), 911–912 (2011)

X. Luo, Y. Liu, S. Kubicek, J. Myllyharju et al., A selective inhibitor and probe of the cellular functions of Jumonji C domain-containing histone demethylases. J. Am. Chem. Soc. **133**, 9451–9456 (2011)

D. Lyumkis, D. Oliveira dos Passos, E.B. Tahara, K. Webb, E.J. Bennett, S. Vinterbo, C.S. Potter, B. Carragher, C.A. Joazeiro, Structural basis for translational surveillance by the large ribosomal subunit-associated protein quality control complex. Proc. Natl. Acad. Sci. U.S.A. **111**, 15981–15986 (2014)

C.C. MacDonald, P.N. Grozdanov, Nonsense in the testis: multiple roles for nonsense-mediated decay revealed in male reproduction. Biol. Reprod. **96**, 939–947 (2017)

G. Manning, D.B. Whyte, R. Martinez, T. Hunter, S. Sudarsanam, The protein kinase complement of the human genome. Science **298**(5600), 1912–1934 (2002)

D.C. Marchion, E. Bicaku, A.I. Daud, D.M. Sullivan, P.N. Munster, Valproic acid alters chromatin structure by regulation of chromatin modulation proteins. Cancer Res. **65**, 3815–3822 (2005)

J.F. Marko, E.D. Siggia, Polymer models of meiotic and mitotic chromosomes. Mol. Biol. Cell **8**, 2217–2231 (1997)

D.S. Marks, T.A. Hopf, C. Sander, Protein structure prediction from sequence variation. Nat. Biotechnol. **30**(11), 1072–1080 (2012)

R. Marmorstein, Structure of SET domain proteins: a new twist on histone methylation. Trends Biochem. Sci. **28**(2), 59–62 (2003)

M.P. Marsden, U.K. Laemmli, Metaphase chromosome structure: evidence for a radial loop model. Cell **17**, 849–858 (1979)

R.A. Marshall, M. Dorywalska, J.D. Puglisi, Irreversible chemical steps control intersubunit dynamics during translation. Proc. Natl. Acad. Sci. U.S.A. **105**, 15364–15369 (2008)

S. Martin, K.A. Wilkinson, A. Nishimune, J.M. Henley, Emerging extranuclear roles of protein SUMOylation in neuronal function and dysfunction. Nat. Rev. Neurosci. **8**(12), 948–959 (2007)

J. Martinez, A. Patkaniowska, H. Urlaub, R. Luhrmann, T. Tuschl, Single-stranded antisense siRNAs guide target RNA cleavage in RNAi. Cell **110**, 563–574 (2002)

V. Marx, Making sure PTMs are not lost after translation. Nat. Methods **10**(3), 201–204 (2013)

M.C. Mathew, A.-M. Ervin, J. Tao, R.M. Davis, Antioxidant vitamin supplementation for preventing and slowing the progression of age-related cataract. Cochrane Database Syst. Rev. **6**, CD004567 (2012)

O. McAuliffe, R.P. Ross, C. Hill, Lantibiotics: structure, biosynthesis and mode of action. FEMS Microbiol. Rev. **25**(3), 285–308 (2001)

C.D. McCallum, H. Do, A.E. Johnson, J. Frydman, The interaction of the chaperonin tailless complex polypeptide 1 (TCP1) ring complex (TRiC) with ribosome-bound nascent chains examined using photo-cross-linking. J. Cell Biol. **149**, 591–602 (2000)

D.G. McEwan, I. Dikic, The three musketeers of autophagy: phosphorylation, ubiquitylation and acetylation. Trends Cell Biol. **21**(4), 195–201 (2011)

O. Meikar, V.V. Vagin, F. Chalmel et al., An atlas of chromatoid body components. RNA **20**, 483–495 (2014)

Meisenhelder J, Hunter T, van der Geer P 2001 Phospho-peptide mapping and identification of phosphorylation sites. Curr. Protocols Protein Sci. 13 Unit 13.9

S. Melnikov, A. Ben-Shem, N. Garreau de Loubresse, L. Jenner, G. Yusupova, M. Yusupov, One core, two shells: bacterial and eukaryotic ribosomes. Nat. Struct. Mol. Biol. **19**, 560–567 (2012)

S.V. Melnikov, K.D. Rivera, D. Ostapenko, A. Makarenko, N.D. Sanscrainte, J.J. Becnel, M.J. Solomon, C. Texier, D. J. Pappin, D. Söll, Error-prone protein synthesis in parasites with the smallest eukaryotic genome. PNAS J. **115**(27), E6245–E6253 (2018)

F. Meng, A.J. Forbes, L.M. Miller, N.L. Kelleher, Detection and localization of protein modifications by high resolution tandem mass spectrometry. Mass Spectrom. Rev. **24**(2), 126–134 (2005)

E.D. Merkley, T.O. Metz, R.D. Smith, J.W. Baynes, N. Frizzell, The succinated proteome. Mass Spectrom. Rev. **33**(2), 98–109 (2014)

E. Michaëlsson, V. Malmström, S. Reis, Å. Engström, H. Burkhardt, R. Holmdahl, T cell recognition of carbohydrates on type II collagen. J. Exp. Med. **180**(2), 745–749 (1994)

J.R. Moffitt, Y.R. Chemla, D. Izhaky, C. Bustamante, Differential detection of dual traps improves the spatial resolution of optical tweezers. Proc. Natl. Acad. Sci. U.S.A. **103**, 9006–9011 (2006)

J.R. Moffitt, Y.R. Chemla, S.B. Smith, C. Bustamante, Recent advances in optical tweezers. Annu. Rev. Biochem. **77**, 205–228 (2008)

I.M. Møller, A. Rogowska-Wrzesinska, R.S.P. Rao, Protein carbonylation and metal-catalyzed protein oxidation in a cellular perspective. J. Proteomics **74**(11), 2228–2242 (2011)

D.L. Moloney, L.H. Shair, F.M. Lu, J. Xia, R. Locke, K.L. Matta, R.S. Haltiwanger, Mammalian Notch1 is modified with two unusual forms of O-linked glycosylation found on epidermal growth factor-like modules. J. Biol. Chem. **275**(13), 9604–9611 (2000)

K.W. Moremen, M. Tiemeyer, A.V. Nairn, Vertebrate protein glycosylation: diversity, synthesis and function. Nat. Rev. Mol. Cell Biol. **13**(7), 448–462 (2012)

O. Muhlemann, J. Lykke-Andersen, How and where are nonsense mRNAs degraded in mammalian cells? RNA Biol. **7**, 28–32 (2010)

A. Müller, L.I. Leichert, Redox Proteomics in Oxidative Stress and Redox Regulation, ed. by U. Jakob, D. Reichmann (Springer, Dordrecht, 2013)

J.B. Munro, R.B. Altman, C.S. Tung, J.H. Cate, K.Y. Sanbonmatsu, S.C. Blanchard, Spontaneous formation of the unlocked state of the ribosome is a multistep process. Proc. Natl. Acad. Sci. U.S.A. **107**, 709–714 (2010)

E. Muroi, S. Manabe, M. Ikezaki, Y. Urata, S. Sato, T. Kondo, Y. Ito, Y. Ihara, C-Mannosylated peptides derived from the thrombospondin type 1 repeat enhance lipopolysaccharide-induced signaling in macrophage-like RAW264. 7 cells. Glycobiology **17**(9), 1015–1028 (2007)

J. Nakayama, J.C. Rice, B.D. Strahl, C.D. Allis, S.I.S. Grewal, Role of histone H3 lysine 9 methylation in epigenetic control of heterochromatin assembly. Science **292**(5514), 110–113 (2001)

N.N. Nalivaeva, A.J. Turner, Role of ageing and oxidative stress in regulation of amyloid-degrading enzymes and development of neurodegeneration. Curr. Aging Sci. **10**(1), 32–40 (2017)

O. Nedić, A. Rogowska-Wrzesinska, S.I.S. Rattan, Standardization and quality control in quantifying non-enzymatic oxidative protein modifications in relation to ageing and disease: why is it important and why is it hard? Redox Biol. **5**, 91–100 (2015)

M.A. Nesmeyanova, I.M. Tsfasman, A.L. Karamyshev, N.E. Suzina, Secretion of the overproduced periplasmic PhoA protein into the medium and accumulation of its precursor in phoA-transformed Escherichia coli strains: involvement of outer membrane vesicles. World J. Microbiol. Biotechnol. **7**, 394–406 (1991)

M.A. Nesmeyanova, A.E. Kalinin, A.L. Karamyshev, N.I. Mikhaleva, V.I. Krupyanko, Overproduction, secretion, isolation and properties of recombinant alkaline phosphatase encoded in Escherichia coli. Process Biochem. **32**, 1–7 (1997)

J.C. Newman, W. He, E. Verdin, Mitochondrialprotein acylation and intermediary metabolism: regulation by sirtuins and implications for metabolic disease. J. Biol. Chem. **287**(51), 42436–42443 (2012)

P. Newsholme, V.F. Cruzat, K.N. Keane, R. Carlessi, P.I.H.J. Bittencourt, Molecular mechanisms of ROS production and oxidative stress in diabetes. Biochem. J. **473**(24), 4527–4550 (2016)

I. Nilsson, P. Lara, T. Hessa, A.E. Johnson, G. von Heijne, A.L. Karamyshev, The code for directing proteins for translocation across ER membrane: SRP cotranslationally recognizes specific features of a signal sequence. J. Mol. Biol. **427**(6 Pt A), 1191–1201 (2015)

P. Nissen, M. Kjeldgaard, S. Thirup, G. Polekhina, L. Reshetnikova, B.F. Clark, J. Nyborg, Crystal structure of the ternary complex of Phe-tRNAPhe, EF-Tu, and a GTP analog. Science **270**, 1464–1472 (1995)

P. Nissen, J. Hansen, N. Ban, P.B. Moore, T.A. Steitz, The structural basis of ribosome activity in peptide bond synthesis. Science **289**, 920–930 (2000)

H.F. Noller, Biochemical characterization of the ribosomal decoding site. Biochimie **88**, 935–941 (2006)

E.H. Norris, B.I. Giasson, H. Ischiropoulos, V.M.Y. Lee, Effects of oxidative and nitrative challenges on α-synuclein fibrillogenesis involve distinct mechanisms of protein modifications. J. Biol. Chem. **278**(29), 27230–27240 (2003)

K. Ohtsubo, J.D. Marth, Glycosylation in cellular mechanisms of health and disease. Cell **126**(5), 855–867 (2006)

G.S. Omenn, L. Lane, E.K. Lundberg, R.C. Beavis, C.M. Overall, E.W. Deutsch, Progress on identifying and characterizing the human proteome, including post-translational modifications. J. Proteome Res. **15**(11), 3951–3960 (2016)

M.E. Oomen, J. Dekker, Epigenetic characteristics of the mitotic chromosome in 1D and 3D. Crit. Rev. Biochem. Mol. Biol. **52**, 185–204 (2017)

H. Otto, C. Conz, P. Maier, T. Wolfle, C.K. Suzuki, P. Jenö, P. Rücknagel, J. Stahl, S. Rospert, The chaperones MPP11 and Hsp70L1 form the mammalian ribosome-associated complex. Proc. Natl. Acad. Sci. U.S.A. **102**, 10064–10069 (2005)

A. Oueslati, M. Fournier, H.A. Lashuel, Role of post-translational modifications in modulating the structure, function and toxicity of α-synuclein. Implications for Parkinson's disease pathogenesis and therapies. Prog. Brain Res. **183**, 115–145 (2010)

F. Ozsolak, P. Kapranov, S. Foissac, S.W. Kim, E. Fishilevich, A.P. Monaghan, B. John, P.M. Milos, Comprehensive polyadenylation site maps in yeast and human reveal pervasive alternative polyadenylation. Cell **143**, 1018–1029 (2010)

O. Panasenko, E. Landrieux, M. Feuermann, A. Finka, N. Paquet, M.A. Collart, The yeast Ccr4-Not complex controls ubiquitination of the nascent-associated polypeptide (NAC-EGD) complex. J. Biol. Chem. **281**, 31389–31398 (2006)

T. Pape, W. Wintermeyer, M.V. Rodnina, Complete kinetic mechanism of elongation factor Tu-dependent binding of aminoacyl-tRNA to the A site of the E. coli ribosome. EMBO J. **17**, 7490–7497 (1998)

H. Pareja-Galeano, F. Sanchis-Gomar, J.L. García-Giménez, Physical exercise and epigenetic modulation: elucidating intricate mechanisms. Sports Med. **44**(4), 429–436 (2014)

E. Park, T.A. Rapoport, Mechanisms of Sec61/SecY-mediated protein translocation across membranes. Annu. Rev. Biophys. **41**, 21–40 (2012)

J.A. Park, A.J. Kim, Y. Kang, Y.J. Jung, H.K. Kim, K.C. Kim, Deacetylation and methylation at histone H3 lysine 9 (H3K9) coordinate chromosome condensation during cell cycle progression. Mol. Cells **31**, 343–349 (2011)

D.O. Passos, M.K. Doma, C.J. Shoemaker, D. Muhlrad, R. Green, J. Weissman, J. Hollien, R. Parker, Analysis of Dom34 and its function in no-go decay. Mol. Biol. Cell **20**, 3025–3032 (2009)

S. Pechmann, F. Willmund, J. Frydman, The ribosome as a hub for protein quality control. Mol. Cell **49**, 411–421 (2013)

K. Peisker, D. Braun, T. Wolfle, J. Hentschel, U. Fünfschilling, G. Fischer, A. Sickmann, S. Rospert, Ribosome-associated complex binds to ribosomes in close proximity of Rpl31 at the exit of the polypeptide tunnel in yeast. Mol. Biol. Cell **19**, 5279–5288 (2008)

S. Peleg, C. Feller, I. Forne et al., Life span extension by targeting a link between metabolism and histone acetylation in Drosophila. EMBO Rep. **17**(3), 455–469 (2016a)

S. Peleg, C. Feller, A.G. Ladurner, A. Imhof, The metabolic impact on histone acetylation and transcription in ageing. Trends Biochem. Sci. **41**(8), 700–711 (2016b)

C.E. Perez, R.L. Gonzalez Jr., In vitro and in vivo single-molecule fluorescence imaging of ribosome-catalyzed protein synthesis. Curr. Opin. Chem. Biol. **15**, 853–863 (2011)

J. Perez-Vilar, S.H. Randell, R.C. Boucher, C-Mannosylation of MUC5AC and MUC5B Cys subdomains. Glycobiology **14**(4), 325–337 (2004)

T.T. Perkins, S.R. Quake, D.E. Smith, S. Chu, Relaxation of a single DNA molecule observed by optical microscopy. Science **264**, 822–826 (1994)

A. Petrov, G. Kornberg, S. O'Leary, A. Tsai, S. Uemura, J.D. Puglisi, Dynamics of the translational machinery. Curr. Opin. Struct. Biol. **21**, 137–145 (2011)

A. Petrov, J. Chen, S. O'Leary, A. Tsai, J.D. Puglisi, Single-molecule analysis of translational dynamics. Cold Spring Harb. Perspect. Biol. **4**(9), a011551 (2012)

I. Piazza, C.H. Haering, A. Rutkowska, Condensin: crafting the chromosome landscape. Chromosoma **122**, 175–190 (2013)

E.S. Pinarbasi, A.L. Karamyshev, E.B. Tikhonova, I.H. Wu, H. Hudson, P.J. Thomas, Pathogenic signal sequence mutations in progranulin disrupt SRP interactions required for mRNA stability. Cell Rep. **23**, 2844–2851 (2018)

V.P. Pisareva, M.A. Skabkin, C.U. Hellen, T.V. Pestova, A.V. Pisarev, Dissociation by Pelota, Hbs1 and ABCE1 of mammalian vacant 80S ribosomes and stalled elongation complexes. EMBO J. **30**, 1804–1817 (2011)

S.D. Pletcher, S.J. Macdonald, R. Marguerie, U. Certa, S.C. Stearns, D.B. Goldstein, L. Partridge, Genome-wide transcript profiles in aging and calorically restricted Drosophila melanogaster. Curr. Biol. **12**(9), 712–723 (2002)

M.G. Poirier, J.F. Marko, Mitotic chromosomes are chromatin networks without a mechanically contiguous protein scaffold. Proc. Natl. Acad. Sci. U.S.A. **99**, 15393–15397 (2002)

B. Popova, A. Kleinknecht, G.H. Braus, Posttranslational modifications and clearing of α-synuclein aggregates in yeast. Biomolecules **5**(2), 617–634 (2015)

M.W. Popp, L.E. Maquat, Organizing principles of mammalian nonsense-mediated mRNA decay. Annu. Rev. Genet. **47**, 139–165 (2013)

M.W. Popp, L.E. Maquat, The dharma of nonsense-mediated mRNA decay in mammalian cells. Mol. Cells **37**, 1–8 (2014)

M.W. Popp, L.E. Maquat, Nonsense-mediated mRNA decay and cancer. Curr. Opin. Genet. Dev. **48**, 44–50 (2018)

A. Pulk, J.H. Cate, Control of ribosomal subunit rotation by elongation factor G. Science **340**, 1235970 (2013)

X. Qu, J.D. Wen, L. Lancaster, H.F. Noller, C. Bustamante, I. Tinoco Jr., The ribosome uses two active mechanisms to unwind messenger RNA during translation. Nature **475**, 118–121 (2011)

R. Radi, Protein tyrosine nitration: biochemical mechanisms and structural basis of functional effects. Acc. Chem. Res. **46**(2), 550–559 (2013)

M. Radman, Protein damage, radiation sensitivity and aging. DNA Repair (Amst) **44**, 186–192 (2016)

V. Ramakrishnan, Unraveling the structure of the ribosome (Nobel Lecture). Angew. Chem. Int. Ed. **49**, 4355–4380 (2010)

T.A. Rapoport, Protein translocation across the eukaryotic endoplasmic reticulum and bacterial plasma membranes. Nature **450**, 663–669 (2007)

S.I.S. Rattan, Increased molecular damage and heterogeneity as the basis of aging. Biol. Chem. **389**(3), 267–272 (2008)

U. Raue, S. Oellerer, S. Rospert, Association of protein biogenesis factors at the yeast ribosomal tunnel exit is affected by the translational status and nascent polypeptide sequence. J. Biol. Chem. **282**, 7809–7816 (2007)

U. Rawat, H. Gao, A. Zavialov, R. Gursky, M. Ehrenberg, J. Frank, Interactions of the release factor RF1 with the ribosome as revealed by cryo-EM. J. Mol. Biol. **357**, 1144–1153 (2006)

I. Rebbapragada, J. Lykke-Andersen, Execution of nonsense-mediated mRNA decay: what defines a substrate? Curr. Opin. Cell Biol. **21**, 394–402 (2009)

S.A. Ribeiro, J.C. Gatlin, Y. Dong et al., Condensin regulates the stiffness of vertebrate centromeres. Mol. Biol. Cell **20**, 2371–2380 (2009)

J.C. Rice, C.D. Allis, Histone methylation versus histone acetylation: new insights into epigenetic regulation. Curr. Opin. Cell Biol. **13**, 263–273 (2001)

M.C.D. Rijk, L.J. Launer, K. Berger et al., Prevalence of parkinsonism and Parkinson's disease in Europe: the EURO-PARKINSON Collaborative Study. European community concerted action on the epidemiology of Parkinson's disease. Neurology **54**(5), 21–23 (2000)

F. Ritort, C. Bustamante, I. Tinoco Jr., A two-state kinetic model for the unfolding of single molecules by mechanical force. Proc. Natl. Acad. Sci. U.S.A. **99**, 13544–13548 (2002)

F.V. Rivas, N.H. Tolia, J.J. Song, J.P. Aragon, J. Liu, G.J. Hannon, L. Joshua-Tor, Purified Argonaute2 and an siRNA form recombinant human RISC. Nat. Struct. Mol. Biol. **12**, 340–349 (2005)

G.L. Rocca, A.D. Stefano, E. Eleuteri et al., Oxidative stress induces myeloperoxidase expression in endocardial endothelial cells from patients with chronic heart failure. Basic Res. Cardiol. **104**(3), 307–320 (2009)

M.V. Rodnina, Biochemistry. Translocation in action. Science **340**, 1534–1535 (2013)

M.V. Rodnina, W. Wintermeyer, Recent mechanistic insights into eukaryotic ribosomes. Curr. Opin. Cell Biol. **21**, 435–443 (2009)

M.V. Rodnina, M. Beringer, W. Wintermeyer, Mechanism of peptide bond formation on the ribosome. Q. Rev. Biophys. **39**, 203–225 (2006)

M.V. Rodnina, M. Beringer, W. Wintermeyer, How ribosomes make peptide bonds. Trends Biochem. Sci. **32**, 20–26 (2007)

R.J. Ross, M.M. Weiner, H. Lin, PIWI proteins and PIWI-interacting RNAs in the soma. Nature **505**, 353–359 (2014)

D.C. Rubinsztein, G. Mariño, G. Kroemer, Autophagy and aging. Cell **146**(5), 682–695 (2011)

C.L. Russell, S. Koncarevic, M.A. Ward, Post-translational modifications in Alzheimer's disease and the potential for new biomarkers. J. Alzheimers Dis. **41**(2), 345–364 (2014)

I. Sadowska-Bartosz, C. Ott, T. Grune, G. Bartosz, Posttranslational protein modifications by reactive nitrogen and chlorine species and strategies for their prevention and elimination. Free Radic. Res. **48**(11), 1267–1284 (2014)

S. Saito, N. Hosoda, S. Hoshino, The Hbs1-Dom34 protein complex functions in non-stop mRNA decay in mammalian cells. J. Biol. Chem. **288**, 17832–17843 (2013)

A.B. Salmon, G. Kim, C. Liu, J.D. Wrene, C. Georgescu, A. Richardson, R.L. Levine, Effects of transgenic methionine sulfoxide reductase A (MsrA) expression on life-span and age-dependent changes in metabolic function in mice. Redox Biol. **10**, 251–256 (2016)

A.L. Santos, A.B. Lindner, Protein posttranslational modifications: roles in aging and age-related disease. Oxid. Med. Cell. Longev. 2017, Article ID 5716409 (2017)

K. Sato, M.C. Siomi, Piwi-interacting RNAs: biological functions and biogenesis. Essays Biochem. **54**, 39–52 (2013)

A. Savelsbergh, V.I. Katunin, D. Mohr, F. Peske, M.V. Rodnina, W. Wintermeyer, An elongation factor G-induced ribosome rearrangement precedes tRNA-mRNA translocation. Mol. Cell **11**, 1517–1523 (2003)

P.R. Schimmel, D. Soll, Aminoacyl-tRNA synthetases. Annu. Rev. Annu. Rev. Biochem. **48**, 601–648 (1979)

F. Schluenzen, A. Tocilj, R. Zarivach et al., Structure of functionally activated small ribosomal subunit at 3.3 angstroms resolution. Cell **102**, 615–623 (2000)

T.M. Schmeing, V. Ramakrishnan, What recent ribosome structures have revealed about the mechanism of translation. Nature **461**, 1234–1242 (2009)

T.M. Schmeing, R.M. Voorhees, A.C. Kelley, Y.G. Gao, F.V. Murphy IV, J.R. Weir, V. Ramakrishnan, The crystal structure of the ribosome bound to EF-Tu and aminoacyl-tRNA. Science **326**, 688–694 (2009)

M. Schmid, T.H. Jensen, The exosome: a multipurpose RNA-decay machine. Trends Biochem. Sci. **33**, 501–510 (2008)

B.S. Schuwirth, M.A. Borovinskaya, C.W. Hau, W. Zhang, A. Vila-Sanjurjo, J.M. Holton, J.H. Cate, Structures of the bacterial ribosome at 3.5 A resolution. Science **310**, 827–834 (2005)

P.R. Selvin, T. Ha, *Single-Molecule Techniques, A Laboratory Manual* (Cold Spring Harbor Laboratory Press, Cold Spring Harbor, 2008)

Y.P. Semenkov, M.V. Rodnina, W. Wintermeyer, Energetic contribution of tRNA hybrid state formation to translocation catalysis on the ribosome. Nat. Struct. Biol. **7**, 1027–1031 (2000)

G. Serin, A. Gersappe, J.D. Black, R. Aronoff, L.E. Maquat, Identification and characterization of human orthologues to Saccharomyces cerevisiae Upf2 protein and Upf3 protein (Caenorhabditis elegans SMG-4). Mol. Cell. Biol. **21**, 209–223 (2001)

R. Shaheen, S. Anazi, T. Ben-Omran et al., Mutations in SMG9, encoding an essential component of nonsense-mediated decay machinery, cause a multiple congenital anomaly syndrome in humans and mice. Am. J. Hum. Genet. **98**, 643–652 (2016)

X. Shan, Y. Chang, C.L. Lin, Messenger RNA oxidation is an early event preceding cell death and causes reduced protein expression. FASEB J. **21**, 2753–2764 (2007)

E.A. Shank, C. Cecconi, J.W. Dill, S. Marqusee, C. Bustamante, The folding cooperativity of a protein is controlled by its chain topology. Nature **465**, 637–640 (2010)

B. Shao, C. Bergt, X. Fu, P. Green, J.C. Voss, M.N. Oda, J.F. Oram, J.W. Heinecke, Tyrosine 192 in apolipoprotein A-I is the major site of nitration and chlorination by myeloperoxidase, but only chlorination markedly impairs ABCA1-dependent cholesterol transport. J. Biol. Chem. **280**(7), 5983–5993 (2005)

S. Shao, K. von der Malsburg, R.S. Hegde, Listerin-dependent nascent protein ubiquitination relies on ribosome subunit dissociation. Mol. Cell **50**, 637–648 (2013)

S. Shao, A. Brown, B. Santhanam, R.S. Hegde, Structure and assembly pathway of the ribosome quality control complex. Mol. Cell **57**, 433–444 (2015)

Y. Shi, F. Lan, C. Matson, P. Mulligan, J.R. Whetstine, P.A. Cole, R.A. Casero, Y. Shi, Histone demethylation mediated by the nuclear amine oxidase homolog LSD1. Cell **119**(7), 941–953 (2004)

N. Shiozawa-West, R.A. Dunlop, K.J. Rodgers, Using an in vitro model to study oxidised protein accumulation in ageing fibroblasts. Biochem. Biophys. Acta **1850**(11), 2177–2184 (2015)

C.J. Shoemaker, R. Green, Kinetic analysis reveals the ordered coupling of translation termination and ribosome recycling in yeast. Proc. Natl. Acad. Sci. U.S.A. **108**, E1392–E1398 (2011)

C.J. Shoemaker, R. Green, Translation drives mRNA quality control. Nat. Struct. Mol. Biol. **19**, 594–601 (2012)

C.J. Shoemaker, D.E. Eyler, R. Green, Dom34:Hbs1 promotes subunit dissociation and peptidyl-tRNA drop-off to initiate no-go decay. Science **330**, 369–372 (2010)

H. Sies, Oxidative stress: oxidants and antioxidants. Exp. Physiol. **82**, 291–295 (1997)

C.L. Simms, L.L. Yan, H.S. Zaher, Ribosome collision is critical for quality control during no-go decay. Mol. Cell **68**, 361.e5–373.e5 (2017)

M.C. Siomi, K. Sato, D. Pezic, A.A. Aravin, PIWI-interacting small RNAs: the vanguard of genome defence. Nat. Rev. Mol. Cell Biol. **12**, 246–258 (2011)

M.K. Slevin, S. Meaux, J.D. Welch, R. Bigler, P.L. Miliani de Marval, W. Su, R.E. Rhoads, J.F. Prins, W.F. Marzluff, Deep sequencing shows multiple oligouridylations are required for 3′ to 5′ degradation of histone mRNAs on polyribosomes. Mol. Cell **53**, 1020–1030 (2014)

S. Smith, Y. Cui, C. Bustamante, Optical-trap force transducer that operates by direct measurement of light momentum. Methods Enzymol. **361**, 134–162 (2003)

N. Sofat, R. Wait, S.D. Robertson, D.L. Baines, E.H. Baker, Interaction between extracellular matrix molecules and microbial pathogens: evidence for the missing link in autoimmunity with rheumatoid arthritis as a disease model. Front. Microbiol. **6**, 1–6 (2015)

J.J. Song, S.K. Smith, G.J. Hannon, L. Joshua-Tor, Crystal structure of Argonaute and its implications for RISC slicer activity. Science **305**, 1434–1437 (2004)

P. Spät, B. Maček, K. Forchhammer, Phosphoproteome of the cyanobacterium Synechocystis sp. PCC 6803 and its dynamics during nitrogen starvation. Front. Microbiol. **6**, 00248 (2015)

M.G. Spillantini, M.L. Schmidt, V.M.-Y. Lee, J.Q. Trojanowski, R. Jakes, M. Goedert, α-Synuclein in Lewy bodies. Nature **388**(6645), 839–840 (1997)

M.G. Spillantini, R.A. Crowther, R. Jakes, M. Hasegawa, M. Goedert, Alpha-synuclein in filamentous inclusions of Lewy bodies from Parkinson's disease and dementia with Lewy bodies. Proc. Natl. Acad. Sci. U.S.A. **95**(11), 6469–6473 (1998)

A.S. Spirin, *Ribosomes* (Springer, New York, 2000)

R.G. Spiro, Protein glycosylation: nature, distribution, enzymatic formation, and disease implications of glycopeptide bonds. Glycobiology **12**(4), 43R–56R (2002)

E.R. Stadtman, B.S. Berlett, Reactive oxygen-mediated protein oxidation in aging and disease. Drug Metab. Rev. **30**(2), 225–243 (1998)

E.R. Stadtman, R.L. Levine, Free radical-mediated oxidation of free amino acids and amino acid residues in proteins. Amino Acids **25**(3), 207–218 (2003)

E.R. Stadtman, C.N. Oliver, Metal-catalyzed oxidation of proteins. Physiological consequences. J. Biol. Chem. **266**(4), 2005–2008 (1991)

J.S. Stamler, D.I. Simon, J.A. Osborne, M.E. Mullins, O. Jaraki, T. Michel, D.J. Singel, J. Loscalzo, S-Nitrosylation of proteins with nitric oxide: synthesis and characterization of biologically active compounds. Proc. Natl. Acad. Sci. **89** (1), 444–448 (1992a)

J.S. Stamler, D.J. Singel, J. Loscalzo, Biochemistry of nitric oxide and its redox-activated forms. Science **258**, 1898–1902 (1992b)

J.S. Stamler, E.J. Toone, S.A. Lipton, N.J. Sucher, (S) NO signals: translocation, regulation, and a consensus motif. Neuron **18**(5), 691–696 (1997)

L. Stannek, K. Gunka, R.A. Care, U. Gerth, F.M. Commichau, Factors that mediate and prevent degradation of the inactive and unstable GudB protein in Bacillus subtilis. Front. Microbiol. **6**, 1–11 (2015)

M.E. Starr, J. Ueda, S. Yamamoto, B.M. Evers, H. Saito, The effects of aging on pulmonary oxidative damage, protein nitration, and extracellular superoxide dismutase down-regulation during systemic inflammation. Free Radic. Biol. Med. **50**(2), 371–380 (2011)

M. Stefani, C.M. Dobson, Protein aggregation and aggregate toxicity: new insights into protein folding, misfolding diseases and biological evolution. J. Mol. Med. **81**, 678–699 (2003)

S. Stefanovic, R.S. Hegde, Identification of a targeting factor for posttranslational membrane protein insertion into the ER. Cell **128**, 1147–1159 (2007)

T.A. Steitz, From the structure and function of the ribosome to new antibiotics (Nobel Lecture). Angew. Chem. Int. Ed. **49**, 4381–4398 (2010)

A.D. Stephens, J. Haase, L. Vicci, R.M. Taylor II, K. Bloom, Cohesin, condensin, and the intramolecular centromere loop together generate the mitotic chromatin spring. J. Cell Biol. **193**, 1167–1180 (2011)

A.D. Stephens, E.J. Banigan, S.A. Adam, R.D. Goldman, J.F. Marko, Chromatin and lamin A determine two different mechanical response regimes of the cell nucleus. Mol. Biol. Cell **28**, 1984–1996 (2017)

A.D. Stephens, P.Z. Liu, E.J. Banigan, L.M. Almassalha, V. Backman, S.A. Adam, R.D. Goldman, J.F. Marko, Chromatin histone modifications and rigidity affect nuclear morphology independent of lamins. Mol. Biol. Cell **29**, 220–233 (2018)

A. Stolz, A. Ernst, I. Dikic, Cargo recognition and trafficking in selective autophagy. Nat. Cell Biol. **16**(6), 495–501 (2014)

K. Sugamura, J.F. Keaney, Reactive oxygen species in cardiovascular disease. Free Radic. Biol. Med. **51**(5), 978–992 (2011)

M. Sun, R. Kawamura, J.F. Marko, Micromechanics of human mitotic chromosomes. Phys. Biol. **8**, 015003 (2011)

M. Sun, R. Biggs, J. Hornick, J.F. Marko, Condensin controls mitotic chromosome stiffness and stability without forming a structurally contiguous scaffold. Chromosome Res. **26**, 277–295 (2018)

A. Sureshbabu, S.W. Ryter, M.E. Choi, Oxidative stress and autophagy: crucial modulators of kidney injury. Redox Biol. **4**, 208–214 (2015)

S. Takahashi, Y. Araki, Y. Ohya, T. Sakuno, S. Hoshino, K. Kontani, H. Nishina, T. Katada, Upf1 potentially serves as a RING-related E3 ubiquitin ligase via its association with Upf3 in yeast. RNA **14**, 1950–1958 (2008)

S. Takyar, R.P. Hickerson, H.F. Noller, mRNA helicase activity of the ribosome. Cell **120**, 49–58 (2005)

M. Tanase, A.M. Urbanska, V. Zolla et al., Role of carbonyl modifications on aging-associated protein aggregation. Sci. Rep. **6**, 19311 (2016)

M.A. Tarp, H. Clausen, Mucin-type O-glycosylation and its potential use in drug and vaccine development. Biochem. Biophys. Acta **1780**(3), 546–563 (2008)

R. Thanan, S. Oikawa, Y. Hiraku, S. Ohnishi, N. Ma, S. Pinlaor, P. Yongvanit, S. Kawanishi, M. Murata, Oxidative stress and its significant roles in neurodegenerative diseases and cancer. Int. J. Mol. Sci. **16**(1), 193–217 (2014)

A.K. Tiden, T. Sjögren, M. Svensson et al., 2-Thioxanthines are mechanism-based inactivators of myeloperoxidase that block oxidative stress during inflammation. J. Biol. Chem. **286**(43), 37578–37589 (2011)

I. Tinoco Jr., C. Bustamante, The effect of force on thermodynamics and kinetics of single molecule reactions. Biophys. Chem. **101–102**, 513–533 (2002)

I. Tinoco Jr., R.L. Gonzalez Jr., Biological mechanisms, one molecule at a time. Genes Dev. **25**, 1205–1231 (2011)

I. Tinoco Jr., J.-D. Wen, Simulation and analysis of single-ribosome translation. Phys. Biol. **6**, 025006 (2009)

I. Tinoco Jr., D. Collin, P.T. Li, The effect of force on thermodynamics and kinetics: unfolding single RNA molecules. Biochem. Soc. Trans. **32**, 757–760 (2004)

I. Tinoco Jr., P.T.X. Li, C. Bustamante, Determination of thermodynamics and kinetics of RNA reactions by force. Q. Rev. Biophys. **39**, 325–360 (2006)

I. Tinoco Jr., K. Sauer, J.C. Wang, J.D. Puglisi, G. Harbison, D. Rovnyak, *Physical Chemistry: Applications to Biological Sciences* (Prentice Hall, Upper Saddle River, 2013)

D.S. Tourigny, I.S. Fernandez, A.C. Kelley, V. Ramakrishnan, Elongation factor G bound to the ribosome in an intermediate state of translocation. Science **340**, 1235490 (2013)

R.J.W. Truscott, M.G. Friedrich, Old proteins and the Achilles heel of mass spectrometry. The role of proteomics in the etiology of human cataract. Proteomics-Clin. Appl. **8**(3–4), 195–203 (2014)

T. Tsuboi, K. Kuroha, K. Kudo, S. Makino, E. Inoue, I. Kashima, T. Inada, Dom34:hbs1 plays a general role in quality-control systems by dissociation of a stalled ribosome at the 3′ end of aberrant mRNA. Mol. Cell **46**, 518–529 (2012)

S. Uemura, M. Dorywalska, T.H. Lee, H.D. Kim, J.D. Puglisi, S. Chu, Peptide bond formation destabilizes Shine-Dalgarno interaction on the ribosome. Nature **446**, 454–457 (2007)

L. Unterholzner, E. Izaurralde, SMG7 acts as a molecular link between mRNA surveillance and mRNA decay. Mol. Cell **16**, 587–596 (2004)

P. Vagnarelli, Mitotic chromosome condensation in vertebrates. Exp. Cell Res. **318**, 1435–1441 (2012)

A.M. van den Elzen, J. Henri, N. Lazar, M.E. Gas, D. Durand, F. Lacroute, M. Nicaise, H. van Tilbeurgh, B. Séraphin, M. Graille, Dissection of Dom34-Hbs1 reveals independent functions in two RNA quality control pathways. Nat. Struct. Mol. Biol. **17**, 1446–1452 (2010)

A. van Hoof, P.A. Frischmeyer, H.C. Dietz, R. Parker, Exosome-mediated recognition and degradation of mRNAs lacking a termination codon. Science **295**, 2262–2264 (2002)

L.M. van Staalduinen, Z. Jia, Post-translational hydroxylation by 2OG/Fe(II)-dependent oxygenases as a novel regulatory mechanism in bacteria. Front. Microbiol. **6**, 1–7 (2015)

R. Vandenberghe, J. Tournoy, Cognitive aging and Alzheimer's disease. Postgrad. Med. J. **81**(956), 343–352 (2005)

J.W. Vaupel, Biodemography of human ageing. Nature **464**(7288), 536–542 (2010)

J.W. Vaupel, J.R. Carey, K. Christensen et al., Biodemographic trajectories of longevity. Science **280**(5365), 855–860 (1998)

R. Verma, R.S. Oania, N.J. Kolawa, R.J. Deshaies, Cdc48/p97 promotes degradation of aberrant nascent polypeptides bound to the ribosome. eLife **2**, e00308 (2013)

R. Verma, K.M. Reichermeier, A.M. Burroughs, R.S. Oania, J.M. Reitsma, L. Aravind, R.J. Deshaies, Vms1 and ANKZF1 peptidyl-tRNA hydrolases release nascent chains from stalled ribosomes. Nature **557**, 446–451 (2018)

G. von Heijne, Signal sequences. The limits of variation. J. Mol. Biol. **184**, 99–105 (1985)

G. von Heijne, The signal peptide. J. Membr. Biol. **115**, 195–201 (1990)

G. von Heijne, J. Steppuhn, R.G. Herrmann, Domain structure of mitochondrial and chloroplast targeting peptides. Eur. J. Biochem. **180**, 535–545 (1989)

R.M. Voorhees, R.S. Hegde, Toward a structural understanding of co-translational protein translocation. Curr. Opin. Cell Biol. **41**, 91–99 (2016)

C. Walsh, *Posttranslational Modification of Proteins: Expanding Nature's Inventory* (Roberts and Company Publishers, Englewood, 2006)

C.T. Walsh, S. Garneau-Tsodikova, G.J. Gatto, Protein posttranslational modifications: the chemistry of proteome diversifications. Angew. Chem. Int. Ed. **44**(45), 7342–7372 (2005)

P. Walter, I. Ibrahimi, G. Blobel, Translocation of proteins across the endoplasmic reticulum. I. Signal recognition protein (SRP) binds to in-vitro-assembled polysomes synthesizing secretory protein. J. Cell. Biol. **91**(2 Pt 1), 545–550 (1981)

N. Walther, M.J. Hossain, A.Z. Politi, B. Koch, M. Kueblbeck, O. Odegard-Fougner, M. Lampe, J. Ellenberg, A quantitative map of human condensins provides new insights into mitotic chromosome architecture. J. Cell Biol. **217**, 2309–2328 (2018)

F. Wang, J.M. Higgins, Histone modifications and mitosis: countermarks, landmarks, and bookmarks. Trends Cell Biol. **23**, 175–184 (2013)

S. Wang, H. Sakai, M. Wiedmann, NAC covers ribosome-associated nascent chains thereby forming a protective environment for regions of nascent chains just emerging from the peptidyl transferase center. J. Cell Biol. **130**, 519–528 (1995)

Q. Wang, Y. Zhang, C. Yang et al., Acetylation of metabolic enzymes coordinates carbon source utilization and metabolic flux. Science **327**(5968), 1004–1007 (2010)

B. Wang, G. Hom, S. Zhou et al., The oxidized thiol proteome in aging and cataractous mouse and human lens revealed by ICAT labeling. Aging Cell **16**(2), 244–261 (2017)

J.M. Watts, K.K. Dang, R.J. Gorelick, C.W. Leonard, J.W. Bess Jr., R. Swanstrom, C.L. Burch, K.M. Weeks, Architecture and secondary structure of an entire HIV-1 RNA genome. Nature **460**, 711–716 (2009)

A. Weixlbaumer, H. Jin, C. Neubauer, R.M. Voorhees, S. Petry, A.C. Kelley, V. Ramakrishnan, Insights into translational termination from the structure of RF2 bound to the ribosome. Science **322**, 953–956 (2008)

E.M. Welch, A. Jacobson, An internal open reading frame triggers nonsense-mediated decay of the yeast SPT10 mRNA. EMBO J. **18**, 6134–6145 (1999)

K.E. Wellen, C.B. Thompson, A two-way street: reciprocal regulation of metabolism and signalling. Nat. Rev. Mol. Cell Biol. **13**(4), 270–276 (2012)

J.D. Wen, L. Lancaster, C. Hodges, A.C. Zeri, S.H. Yoshimura, H.F. Noller, C. Bustamante, I. Tinoco, Following translation by single ribosomes one codon at a time. Nature **452**, 598–603 (2008)

B. Wiedmann, H. Sakai, T.A. Davis, M. Wiedmann, A protein complex required for signal-sequence-specific sorting and translocation. Nature **370**, 434–440 (1994)

K. Wild, M. Halic, I. Sinning, R. Beckmann, SRP meets the ribosome. Nat. Struct. Mol. Biol. **11**, 1049–1053 (2004)

B. Wilden, A. Savelsbergh, M.V. Rodnina, W. Wintermeyer, Role and timing of GTP binding and hydrolysis during EF-G-dependent tRNA translocation on the ribosome. Proc. Natl. Acad. Sci. U.S.A. **103**, 13670–13675 (2006)

C. Williams, Going against the flow: a case for peroxisomal protein export. Biochim. Biophys. Acta **1843**, 1386–1392 (2014)

F. Willmund, M. del Alamo, S. Pechmann, T. Chen, V. Albanese, E.B. Dammer, J. Peng, J. Frydman, The cotranslational function of ribosome-associated Hsp70 in eukaryotic protein homeostasis. Cell **152**, 196–209 (2013)

D.N. Wilson, J.H. Doudna Cate, The structure and function of the eukaryotic ribosome. Cold Spring Harb. Perspect. Biol. **4**, a011536 (2012)

B.T. Wimberly, D.E. Brodersen, W.M. Clemons Jr., R.J. Morgan-Warren, A.P. Carter, C. Vonrhein, T. Hartsch, V. Ramakrishnan, Structure of the 30S ribosomal subunit. Nature **407**, 327–339 (2000)

W. Wintermeyer, F. Peske, M. Beringer, K.B. Gromadski, A. Savelsbergh, M.V. Rodnina, Mechanisms of elongation on the ribosome: dynamics of a macromolecular machine. Biochem. Soc. Trans. **32**, 733–737 (2004)

World Health Organization (2015) World report on ageing and health. http://www.who.int/ageing/events/world-report-2015-launch/en/

World Health Organization (2000) Life expectancy increased by 5 years since 2000, but health inequalities persist. http://www.who.int/mediacentre/news/releases/2016/health-inequalities-persist/en/

World Health Organization (2016) World Health Statistics 2016 Monitoring Health for the SDGs. http://www.who.int/gho/publications/world_health_statistics/2016/Annex_B/en/

J.J. Wu, J. Liu, E. Chen et al., Increased mammalian lifespan and a segmental and tissue-specific slowing of aging after genetic reduction of mTOR expression. Cell Rep. **4**(5), 913–920 (2013)

E.J. Wurtmann, S.L. Wolin, RNA under attack: cellular handling of RNA damage. Crit. Rev. Biochem. Mol. Biol. **44**, 34–49 (2009)

D. Xu, J. Bai, Q. Duan, M. Costa, W. Dai, Covalent modifications of histones during mitosis and meiosis. Cell Cycle **8**, 3688–3694 (2009)

M.B. Yaffe, Phosphotyrosine-binding domains in signal transduction. Nat. Rev. Mol. Cell Biol. **3**(3), 177–186 (2002)

A. Yamashita, T. Ohnishi, I. Kashima, Y. Taya, S. Ohno, Human SMG-1, a novel phosphatidylinositol 3-kinase-related protein kinase, associates with components of the mRNA surveillance complex and is involved in the regulation of nonsense-mediated mRNA decay. Genes Dev. **15**, 2215–2228 (2001)

A. Yamashita, N. Izumi, I. Kashima et al., SMG-8 and SMG-9, two novel subunits of the SMG-1 complex, regulate remodeling of the mRNA surveillance complex during nonsense-mediated mRNA decay. Genes Dev. **23**, 1091–1105 (2009)

X.-J. Yang, E. Seto, Lysine acetylation: codified crosstalk with other posttranslational modifications. Mol. Cell **31**(4), 449–461 (2017)

L.V. Yanshole, I.V. Cherepanov, O.A. Snytnikova, V.V. Yanshole, R.Z. Sagdeev, Y.P. Tsentalovich, Cataract-specific posttranslational modifications and changes in the composition of urea-soluble protein fraction from the rat lens. Molecul. Vis. **19**, 2196–2208 (2013)

J.-A. Yao, M. Jiang, G.M. Tseng, Mechanism of enhancement of slow delayed rectifier current by extracellular sulfhydryl modification. Am. J. Physiol. Heart Circul. Physiol. **273**(1), H208–H219 (1997)

A. Yonath, Polar bears, antibiotics, and the evolving ribosome (Nobel Lecture). Angew. Chem. Int. Ed. **49**, 4341–4354 (2010)

M. Yoshida, M. Kijima, M. Akita, T. Beppu, Potent and specific inhibition of mammalian histone deacetylase both in vivo and in vitro by trichostatin A. J. Biol. Chem. **265**, 17174–17179 (1990)

J. Yoshitake, Y. Soeda, T. Ida, A. Sumioka, M. Yoshikawa, K. Matsushita, T. Akaike, A. Takashima, Modification of tau by 8-Nitroguanosine 3′,5′-cyclic monophosphate (8-nitro-cGMP): effects of nitric oxide-linked chemical modification on tau aggregation. J. Biol. Chem. **291**(43), 22714–22720 (2016)

M.M. Yusupov, G.Z. Yusupova, A. Baucom, K. Lieberman, T.N. Earnest, J.H. Cate, H.F. Noller, Crystal structure of the ribosome at 5.5 A resolution. Science **292**, 883–896 (2001)

P.Y. Zhang, X. Xu, X.C. Li, Cardiovascular diseases: oxidative damage and antioxidant protection. Eur. Rev. Med. Pharmacol. Sci. **18**, 3091–3096 (2014)

Y. Zhang, I. Sinning, S. Rospert, Two chaperones locked in an embrace: structure and function of the ribosome-associated complex RAC. Nat. Struct. Mol. Biol. **24**, 611–619 (2017)

Y. Zhao, B. Zhao, Y. Zhao, B. Zhao, Oxidative stress and the pathogenesis of Alzheimer's disease. Oxid. Med. Cell. Longev. **2013**, 316523 (2013)

A. Zhiteneva, J.J. Bonfiglio, A. Makarov, T. Colby, P. Vagnarelli, E.C. Schirmer, I. Matic, W.C. Earnshaw, Mitotic post-translational modifications of histones promote chromatin compaction in vitro. Open Biol. **7**, e170076 (2017)

J. Zhou, L. Lancaster, S. Trakhanov, H.F. Noller, Crystal structure of release factor RF3 trapped in the GTP state on a rotated conformation of the ribosome. RNA **18**, 230–240 (2012)

J. Zhou, L. Lancaster, J.P. Donohue, H.F. Noller, Crystal structures of EF-G-ribosome complexes trapped in intermediate states of translocation. Science **340**, 1236086 (2013)

R. Zimmermann, L. Muller, B. Wullich, Protein transport into the endoplasmic reticulum: mechanisms and pathologies. Trends Mol. Med. **12**, 567–573 (2006)

M. Zuker, Mfold web server for nucleic acid folding and hybridization prediction. Nucleic Acids Res. **31**, 3406–3415 (2003)

Cell Cycle, DNA Replication, Centrosomes, Centrioles and Cell Division

15

Abstract

This chapter deals with the basic knowledge about the function of cell division and the entire cell cycle process. More precisely, the key function of the largest compartment of the eukaryotic cell, the nucleus and its major constituents, the chromosomes, is emphasized and presented in great detail. The most classical function of the nucleus is the DNA replication in terms of cell division. Moreover, it is also highlighted that centrioles, located nearby the nucleus, fulfill an important role in the process of cell division. Hence, they are subject of the discussion, including centriole replication. Moreover, the biophysical aspects of these main cell functions are examined under both under normal healthy and physiological conditions, and pathological conditions including cancer development and metastatic progression. Finally, the connection between the cell cycle and the cell mechanical properties is discussed.

15.1 Introduction to the Steps of the Cell-Cycle and Cell Division

The cell cycle is a process that ensures successful cell division. The cell division event is a closely guided physiological process that encompasses the proper separation of the two sister chromatids after DNA replication is finished to create two genetically exactly identical daughter cells. In order to perform the proper cell division, a cell needs to duplicate its DNA content, which is termed DNA replication. The cell cycle process itself involves several phases, such as G1-, S-, G2- and M-phases. A cell can also be in a rest phase, termed G0-phase, in which no cell doubling occurs. The control of the cell cycle process is important, and hence, several cell cycle checkpoints exist. At each checkpoint the cellular conditions are monitored and progression of the specific cell cycle phase is either favored or not favored.

There are three major checkpoints: The first is the G1 checkpoint, which is the starting checkpoint restricting the process of the cell cycle, the second is the G2-/M-phase checkpoint, and the third is metaphase checkpoint, which is also referred to as the spindle checkpoint. The regulatory elements governing the process of cell division are collectively known as the cell cycle, which is divided into five different phases: resting (G0), gap 1 (G1), DNA replication/synthesis (S), gap 2 (G2) and mitosis (M) (Bower et al. 2017) (Fig. 15.1). There are transitions between the different phases of the cell cycle that can be overcome by a cell. Therefore, the cells require a trigger signal from oscillating

C. T. Mierke, *Cellular Mechanics and Biophysics*, Biological and Medical Physics, Biomedical Engineering, https://doi.org/10.1007/978-3-030-58532-7_15

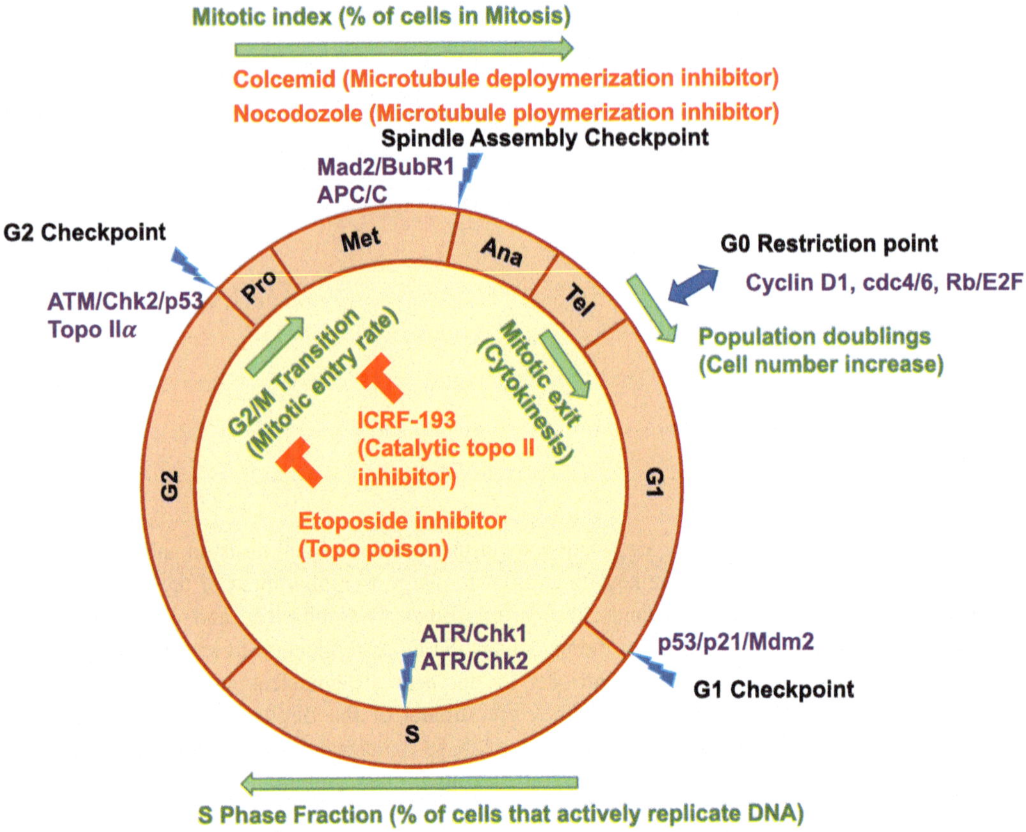

Fig. 15.1 Schematic drawing of the cell cycle regulation. Phases of the cell cycle inside the light red circle, such as G0, G1, G2, and mitosis which is subdivided into subphases: prophase (Pro), metaphase (Met), anaphase (Ana) and telophase (Tel). The Go restriction point represents a reversible phase. When cells undergo cell cycling, exogenous perturbations can activate checkpoints, which can stall cells during the phase transitions. Several measures of the cell proliferation are indicated in green. Pharmacological drug inhibiting several steps of the cell cycle are given in red

amounts of cyclins and cyclin-dependent kinases (cdks). In specific detail, every phase of the cell cycle possesses its own set of specific proteins or protein complexes, such as the cyclin/cdk heterodimers.

The cell cycle can be stopped by intracellular and/or extracellular signals that involve a complex regulatory system of intertwined signaling steps, which subsequently interact with cyclin/cdk activities and thereby govern the precise regulation of the progression of the cell cycle. This halt of progression of the cell cycle is commonly known as a checkpoint and provides the time for DNA damage repair or sufficient stimulation with increasing growth factor concentrations entering the next step of the cell cycle. When the DNA damage becomes too extensive to repair, the cell can trigger apoptotic signaling pathways to stop the transmission of defective DNA to the daughter cells. In this way, cell cycle checkpoints guarantee an orderly progression of the cell cycle, are crucial for preserving genomic integrity, act as obstacles to carcinogenesis and are frequently dysregulated in tumors (Bower et al. 2010; Taylor and McKeon 1997; Hartwell and Kastan 1994).

A minimum of four cell cycle checkpoints can be disrupted in cancer cells: the restriction point (G0/G1), the G1 checkpoint, the G2 checkpoint and the mitosis-associated spindle assembly checkpoint (SAC). The G0/G1 restriction point represents the stage in G1 where the deprivation of growth factors no longer triggers a relapse into a dormant condition; it therefore governs the cell's obligation to undergo division (Pardee 1989). The restriction checkpoint is mainly controlled by the Rb/E2F signaling pathway: The liberation of E2F transcription factors from Rb permits E2F to transcriptionally activate genes that facilitate the induction of replication of DNA and entrance into the S-phase. The overexpression of upstream regulators of Rb, including cyclin D1, cdk4 and cdk6 and the inactivation of mutations in the RB1 gene can break down this control point and cause an early activation of the replication of DNA (Weinberg 1995; Classon and Harlow 2002). The G1 DNA damage control point retards the initialization of DNA replication in the event of DNA damage and is widely governed through the p53/p21/Mdm2 signal transduction pathway (Menendez et al. 2009; Brown et al. 2009). The failure of this control point can often result from TP53 mutations or the inactivation of wild-type p53 through viral proteins and may lead to failure-prone DNA synthesis because of decreased time for DNA repair (Agarwal et al. 1995; Hartwell and Weinert 1989).

Two types of G2 control point reactions have been recognized: the G2 control point for DNA defects and the G2 decatenation (synonymously disentangled) control point (Deming et al. 2001; Downes et al. 1994; Nurse and Bissett 1981; Damelin and Bestor 2007). The control point for DNA damage G2 retards the onset of mitosis in the event of DNA injury through the sequestration of inactive cyclin B1/cdk1 in the cytoplasm, thereby inhibiting the onset of mitosis (Nigg 1995; Morgan 1995). The control point for decatenation G2 differs molecularly from the control point for DNA injury G2 by the fact that it is triggered by a catalytic suppression of the topoisomerase IIα (topo IIα) without any apparent DNA injury (Downes et al. 1994; Luo et al. 2009, 17). The ATM/Chk2/p53 pathway is among the key players both in DNA injury and in decatenation G2 control point, and the weakening of one of the two G2 control points results in chromosomal destabilization (Xu et al. 2001). Lastly, the SAC operates in mitosis to retard the beginning of the anaphase so that all chromosomes have a bipolar linkage to the mitotic spindle (Taylor and McKeon 1997; Kung et al. 1990; Wang et al. 2004). The main affectors of SAC function comprise APC/C, BubR1 and Mad2, and disturbed SAC signal transduction frequently results in the creation of multipolar spindles and/or the irregular distribution of sister chromatids in daughter cells (Musacchio 2015). It has been found that deficiencies in all these cell cycle control points are integral to the initiation and/or progression of tumors and impact the sensitivity of tumors to cytotoxic and endocrine therapeutics (Vogel et al. 2005; Sudo et al. 2004; Bosco et al. 2007).

The function of restriction and G1 control points in breast cancer signaling, preservation of the genome and chemosensitivity has been analyzed earlier (Bosco et al. 2007; Cadoo et al. 2014); but the association between the function of G2 or SAC control points, the genomic instability and the chemoresistance is insufficiently characterized. Six intrinsic molecular subtypes of human breast tumors have been hypothesized (Perou et al. 2000; Prat et al. 2010), and multiple investigations indicate that these subtypes carry common molecular abnormalities that are involved in disease progression and the diversity of chemotherapeutic responsiveness (Santarpia et al. 2013; Docquier et al. 2010; Sorlie et al. 2001), which offer an explanation for the artificial lethality of specific subsets of cancer cells (Hastak et al. 2010; Heiser et al. 2012). Even though these investigations assume that genomic instability contributes to these mechanisms, the characteristics of the underlying specific molecular damage that leads to breast cancer events are still not clear.

15.2 Clock Theory or Domino Model

The delicate chronological sequence of incidents in mitosis was discovered by histologists before the beginning of the twentieth century (Flemming 1882; Wilson 1928). More recently, the biochemical order of events in cell division has been the subject of intense study, and in the last 15 years, there has been significant progress in our understanding. There is a good knowledge of how the entrance into the S-phase as well as the entering and leaving of mitosis is controlled by the oscillatory fluctuation of a family of regulators referred to as cyclin-dependent kinases (Cdks) and how these and other processes are governed through certain proteolysis responses (King et al. 1994; Hixon and Gualberto 2000). Despite the fact that the onset and exit of the S-phase and mitosis is now more clearly understandable, the timing of events within these phases of the cell cycle is not fully transparent. For instance, during mitosis, orderly structural alterations are performed in the cell nucleus and cytoskeleton, which are easily observable under the light microscope. The centrosome segregation, the condensation of chromosomes, the degradation of the nuclear envelope, the development of a mitotic spindle and the conglomeration of chromosomes on the metaphase plate have to be carried out in a stringent chronological order according to the mitotic timeline. In extension, the different states of the anaphase and telophase each have to be synchronized with the division of the cell. While the onset of mitotic prophase can be defined in terms of Cdc2/cyclin B activation and the transition from metaphase to anaphase can be explained in terms of ubiquitin-dependent proteolysis facilitated through the anaphase promoter complex (APC), there is only a vague idea of the time sequence of events occurring between these transitions.

In the 1970s and 1980s, two separate models were presented to explain the global time control of the events occurring in the cell cycle. The one founded on the principle of a biochemical oscillator was referred to as the "clock" theory (Murray and Kirschner 1989a). The second, referred to as the "domino" model, relied on the perspective of the cell cycle as a series of interdependent responses (Hartwell 1978; Lee and Nurse 1988). These two extreme views are merged today. The key cell cycle transitions (G1/S, G2/M and anaphase/metaphase) are governed by regulators, such as Cdks or APC and SCFs that is a complex of Skp1, Cullin and F-Box proteins. Moreover, these regulatory are not integral to the fundamental cell cycle activities, such as replication of DNA or mitosis. Checkpoints and feedback control loops on the regulators that supervise the actions of the cell cycle guarantee the termination of specified processes. Although this model explains the adjustment of the most significant transitions, the control of the temporality of the events between the transitions is not well known. Two extreme models can also be considered here for this purpose (Fig. 15.1). In the first model, referred to as "regulator-controlled timing", a number of mini regulators, such as a temporal cascade of kinases, were sequentially engaged following the G2/M transition. As illustrated in Fig. 15.1, a cascade of kinases in that model appears to trigger various events at distinct times. In the alternative model, referred to as "substrate-controlled timing", all kinases (and phosphatases) are switched on at almost the same time and thereafter stay stable as long as the next transition point at the metaphase/anaphase is reached. The temporal sequence of incidents between G2/M and metaphase/anaphase needs to be governed through the individual reactions of the respective substrates, such as the kinetics of repeated phosphorylations or gradual conformational alterations. The hypothesis of two extreme models is a rather unorthodox one, and in reality, it could be reasonably anticipated that both of them might actually play a part. The question raised is whether or not evidence of any of these mechanisms can be provided.

Xenopus egg cell extract turned out to be a valuable tool for studying the biochemical steps involved in the simplified embryonic cell cycle. Previous investigations have demonstrated that egg cell extracts reflect the biochemical fluctuations of Cdc2/cyclin B activity and morphological modifications in the cell nucleus when the conventional targets of the cell cycle, the cell nucleus, the spindle apparatus and the centrosome are absent (Murray and Kischner 1989a, b; Luca and Ruderman 1989). In particular, the demands on the synthesis of cyclin and the part of various phosphorylation and dephosphorylation processes in the egg extract were examined in depth (Murray and Kischner 1989b; Lohka et al. 1987; Kumagai and Dunphy 1991; Solomon et al. 1992). Several phosphorylation events can be specifically linked to mitosis and are directly associated with morphological alterations, including the phosphorylation of lamins (Maike-Lye et al. 1993; Ward and Kirschner 1990; Peter et al. 1990), the GM130 protein of the Golgi apparatus (Lowe et al. 2000), the microtubule regulatory proteins stathmin and Xmap (Karsenti et al. 1987; Gard and Kirschner 1987; Belmont et al. 1990), and the chromosomal histone H3 protein (Gurley et al. 1978; Hans and Dimitrov 2001). Multiple compounds of the regulatory machinery, including Cdc25 (Izumi et al. 1992; Kumagai and Dunphy 1992), Cdc2 (Izuumi and Maller 1993; Kuang et al. 1994), APC (Kramer et al. 2000; Fang et al. 1998; Kotani et al. 1999), separase and securin (Alexandru et al. 2001), are governed through the phosphorylation in the course of mitosis. These findings indicate that an investigation of the time at which phosphorylation responses take place during mitosis may be useful in tracking multiple biochemical events in mitosis.

The question is to what degree the time course of G2/M to metaphase in Xenopus extracts is guided by regulation-driven scheduling or substrate-driven scheduling mechanisms (Georgi et al. 2002). The idea is to enforce a temporal disconnect between the substrates and the regulators, not dissimilar to the experimental setup of the cell fusion assays, which contributed to determine the predominance of the mitotic condition and the presence of checkpoint regulators (Rao and Johnson 1970; Johnson and Rao 1970). In extracts, Cdc2 activation has been induced by the presence of cyclin B1, and then, different tracer substrates have been supplemented with variable time delays to monitor whether the characteristic phosphorylation of these proteins mirrors the time the extract has proceeded following Cdc2 activation or only the time the tracer proteins have been in contact with the mitotic extract. When the timing of incidents in mitosis is subject to simple modulation by the regulator, the timing of phosphorylations has to be heavily influenced by the status of the extract, such as by the duration of incubation of the extract subsequent to Cdc2 activation. When, however, the timing of these occurrences is under substrate monitoring, then phosphorylation of every substrate would take place at its individual rate, irrespective of whether the extract has just activated Cdc2 or whether the extract remained incubated for an extended period of time after Cdc2 activation. In the end, instead of individual proteins, entire nuclei were added, and morphological procedures resulting in metaphase were investigated. Specifically, in this case, a question was raised as to whether the moment of chromosome condensation, nuclear envelope breakage or metaphase concatenation occurred as a function of the time when the nuclei were introduced into the extract compared to Cdc2 activation. The results highlight the relevance of substrate control in this somewhat streamlined cell cycle and indicate that when regulator control is involved in different systems, it may be placed on a fundamental process of substrate control to organize the various stages of mitosis.

The control of the key crossovers during the cell cycle, including G1/S, G2/M and metaphase/anaphase, is fairly well established. However, much less attention has been paid to the control of the sequence and timing of incidents within each phase of the cell cycle. Therefore, two extreme models were proposed to explain the potential mechanisms for the timing and sequential adjustment of processes taking place during mitosis (Fig. 15.1). In the controller-driven timing model, consecutive controllers, such as kinase sequences, are enabled (or disabled) one after the other, and each one has a unique impact on its substrate. In contrary, in the substrate-driven timing model, an individual initiator or controller quickly starts a completely new regulatory regime, such as mitosis, where every successive response is self-regulated due to the unique chemistry of the substrates. A biochemical protein phosphorylation and morphological, encompassing structural alterations in nucleus and cytoplasm, investigations have been conducted to attempt to differentiate between the two models for the initial stages of mitosis resulting in metaphase. In addition, when using the Xenopus cell extract model system, one is aware that it possibly may not have the full regulatory characteristics of somatic cells. In reality, no single system is expected to have the full range of regulatory characteristics encountered in nature.

The timing and speed of phosphorylation vary for individual proteins throughout mitosis. Cdc25 and Weel have been reported to be phosphorylated at the G2/M crossover (Solomon et al. 1992; Kumagai and Dunphy 1997; Tang et al. 1993) and constitute the earliest transitions discovered. Even though screenings for mitotic phosphoproteins revealed that a few proteins were shifted almost simultaneously with Cdc25 and Weel, such as MP45, Cdc27 and MP44, whereas no proteins were found that shifted earlier during the lag phase (Georgi et al. 2002). Moreover, the transitions for Cdc25 and Weel coexist with the activation of the histone H1 kinase activity and therefore could be the decisive factor for the entrance into mitosis.

The exiting of mitosis and advancing into anaphase necessitates the activation of APC, which is a multi-subunit complex targeting certain proteins for break-down through the ubiquitin route. In detail, the timing of phosphorylation of one of the subunits of the APC, Cdc27, has been examined, and it has been found that it is specifically phosphorylated in stages over a lengthy period of time, starting a few minutes after completion of the Cdc25 translocation and later extending into mitosis. Several earlier investigations have demonstrated that phosphorylation of APC within mitosis is necessary for a substantial degree of activation (Kotani et al. 1999; Lahav-Baratz et al. 1995; Peters et al. 1996). In addition, the binding of Cdc20 to APC relies on the phosphorylation of some of its constituent subunits (Fang et al. 1998; Kramer et al. 1998; Shteinberg et al. 1999). It is obvious that the timing of the activation of APC needs to be strictly supervised in order to avoid that the cell exits the mitosis before the time is up. In the case of somatic cell cycles, the timing of the anaphase is governed solely by feedback from the checkpoint of the spindle assembly (Hixon and Gualberto 2000). In yeast, this control point is not crucial for viability under conditions in the laboratory, and the intrinsic temporal mechanism, described for Cdc27, might be adequate. However, MAD2 knockouts are not alive in mammalian cells, indicating that MAD2 performs a more critical function in these cells (Michel et al. 2001). As expected, an occurrence as crucial as the beginning of the anaphase is subject to a multilayer control based on the inherent scheduling mechanisms outlined here.

Cdc27 has not been the sole protein examined with this long-lasting phosphorylation profile during mitosis. Another protein termed MP44 behaves in a similar manner, but its function has not yet been determined. Typically, neither Cdc27 nor MP44 can relocate faster when implemented in a late mitotic surrounding. When Weel and Cdc25 are supplemented to extracts that are currently in mitosis, they respond with a very fast and immediate postponement, regardless of how long the extracts themselves have resided in mitosis. However, Cdc27 and MP44 still exhibit characteristic retarded and lengthened schemes of phosphorylation transitions. When the phosphorylation of Cdc27 and MP44 is governed by regulators enabled in a temporal sequence subsequent to the G2/M transition, the introduction of these substrates into late mitotic extracts is expected to create an accelerated profile of shifts. Since this is simply not the intention, it is proposed that phosphorylation is governed by temporal characteristics inherent in these proteins rather than by varying environmental conditions within the mitotic extract. These intrinsic mechanisms might comprise the relative levels of phosphorylation and dephosphorylation at certain locations, the slow formation of multiple phosphorylated species, such as positive or negative cooperativity, and slow conformational chaperone assembly and disassembly responses.

During mitosis, the cell experiences significant alterations in both nuclear and cytoplasmic morphology, and the order and time of these processes require stringent regulation. Chromosome condensation, the break-down of the nuclear envelope, the development of spindles and the progression of the chromosomes onto the metaphase plate need to take place before the beginning of the anaphase and sister chromatid separation. In experiments related to those reported for the measurement of phosphorylation rates of different proteins in mitosis, the timing of these morphological processes was compared with the addition of interphase nuclei to extracts at several points in time either prior to or subsequent to Cdc2/cyclin B activation. The most evident deduction is that nuclei supplemented to extracts that stayed in the mitotic stage for 30 min still passed through the typical sequence of nuclear and cytoplasmic occurrences. In addition, there was no substantial increase in timing when the nuclei were directly added to the late-stage mitotic extracts. When these events are exclusively regulated through the activation and inactivation of components within the cytoplasm, one might assume that the timing and even the sequence of structural alterations during mitosis are influenced through the introduction of interphase nuclei into an extract 10 or 30 min following the introduction of cyclin B. The circumstance that the timing and order of these events are not influenced under these constraints leads to the assumption that, as with Cdc27 and MP44, substrate-driven timing has an important function in the arrangement of events in prophase, prometaphase and metaphase.

These experiments provide a broad hint that the ability to control the precise timing of processes in early mitosis relies on the individual reactions of the substrates to a commonly shared repertoire of regulators. While there was slight evidence that alterations might arise in these extracts, which are reflected in slight variations in phosphorylation schemes, there was limited or no evidence of a sequence of regulatory regimes. Nevertheless, a regulatory control cannot be excluded. These experiments were performed in a system in which the nuclear volume is negligible in comparison with the cytoplasmic volume. Even though this is the physiological condition of early embryos, this does not represent the scenario for somatic cells, and there is some indication that the nuclear localization of cyclin B and Weel may impact scheduling (Takizawa and Morgan 2000; Hagting et al. 1999; Jackman and Pines 1997; Wu et al. 1996). In somatic cells, Cdc2/cyclin A apparently seems to be enabled and disabled prior to Cdc2/cyclin B (Jackman and Pines 1997), and even in wheatear frog extracts, a sequence of aggregations of various cyclin isotypes was revealed (Minshull et al. 1990). This also points to the possibility that a regulatory control chain might be involved in mitosis. Nevertheless, the verification of a regulatory pathway required by the substrate control identified here demands stringent functional screening. While it is obvious that multiple genes exhibit cyclical periodic clustering throughout the cell division cycle, such as over 400 out of over 6000 genes in yeast (Cho et al. 1998; Simon et al. 2001), each one needs to be thoroughly examined to ascertain whether the timing of expression has significant functional implications.

Moreover, it is improbable that the transcriptional or translational control mechanisms are sufficiently rapid to be able to synchronize the events within mitosis. The findings presented here could be regarded as the fundamental control processes of mitotic scheduling, similar to the fluctuations of the frog-egg cell cycle that constitute the fundamental dynamics of all mitotic cycles. It appears feasible that other types of modulator control might override the intrinsic features of substrate control that has been demonstrated. Until now, little information has been obtained on how substrates are constructed to react in different ways to a shared set of cues. There are probably numerous distinct mechanisms that mirror the biochemical variety of the diverse substrates. Continued investigation of the kinematics of phosphorylation will contribute to the completion of the structural comprehension of the mitotic process.

15.3 Regulation of the Cell Cycle by Checkpoints

The checkpoints of the cell cycle represent surveillance features that control the order, integrity, and fidelity of the key steps of the cell cycle. These steps include growth to the suitable cell size, the replication and intactness of chromosomes and their precise segregation in mitosis. A number of these mechanisms are of an ancient nature and very well conserved and thus have been proven effective in investigations on primitive organisms such as yeasts (Barnum and O'Connell 2014). Some have progressed to higher organisms and govern alternative cell fates with substantial implications for tumor suppression. These various checkpoint pathways and the associated consequences of their malfunction have been determined on the fate of cells.

The cell cycle consists of a number of events where cell constituents are duplicated and then precisely segregated to daughter cells. In eukaryotes, DNA replication is limited to a distinct synthesis or S-phase, and chromosome separation takes place in the mitotic or M-phase. Two gap phases divide S-phase and mitosis, commonly referred to as G1 and G2. These are no cycles of dormancy, but rather cycles during which cells gain mass, assimilate growth cues, arrange a replicated genome and get ready for chromosome segregation.

The key engines that control the continuation of the cell cycle represent the cyclin-dependent kinases (CDKs). This involves serine/threonine protein kinases which phosphorylate selected key compounds to stimulate DNA synthesis and enhance the mitotic progression. The catalytic subunits have a molar abundance, but are missing activity unless they are linked by their closely associated cyclin subunits, which are strictly regimented both at the synthesis and ubiquitin-based proteolysis level. Cyclin binding enables dormant CDKs to assume an active conformation similar to monomeric and active kinases. In addition to this type of regulation, CDK activity might also be adversely affected through the binding of tiny inhibitory proteins, the CKIs, or through inhibitory tyrosine phosphorylation, which inhibits the transfer of phosphate to substrates.

15.3.1 Cell Cycle Checkpoints

The control points have crystallized as a series of dependences of the cell cycle. In groundbreaking investigations in the fission yeast *Schizosaccharomyces pombe*, it has been detected that cell size is a decisive factor in cell division (Mitchison 1957, 1990; Mitchison and Nurse 1985; Fantes 1977). In addition, human cell fusion experiments were preformed (Rao and Johnson 1970, 1971; Johnson et al. 1970; Johnson and Rao 1970), and thereby, a dependence between the S-phase and mitosis was ascertained. This implies that nuclei in the S-phase can postpone the mitotic entrance of a G2 nucleus, while mitotic cells stimulate the nuclei to enter mitosis before maturity.

In fact, oocyte research had identified a very similar connection of the S-phase to mitosis (Maller 1990; Masui 2001). In extension, the cell cycle stop caused by DNA injury in the budding yeast *Saccharomyces cerevisiae* was employed to identify the first DNA damage control point genes (Weinert and Hartwell 1988; Hartwell and Weinert 1989), which was later extended in multiple systems to a sophisticated signaling pathway, where a significant degree of signal overlap renders mitosis reliant on the completion of DNA replication (O'Connell and Cimprich 2005; O'Connell et al. 2000; Tapia-Alveal et al. 2009; Kuntz and O'Connell 2009). In a similar way, mitotic arrest induced by microtubule inhibitors was exploited to detect the first spindle control point genes in *S. cerevisiae* (Hoyt et al. 1991; Li and Murray 1991), again which in turn led to a highly preserved control point signal pathway that governs the segregation of chromosomes (Lara-Gonzalez et al. 2012). These control points act as feed-forward signal generators, conferring exceptional fidelity to the cell cycle and guaranteeing healthy development and maintenance of tissue homeostasis.

There has been tremendous advancement in the molecular decomposition of multiple cell cycle control routes. In most instances, this is extremely comprehensive with a detailed analysis of post-translational modifications, the structural biology, the enzyme kinetics and so forth. There are so many details regarding the molecular insights in these pathways (Lara-Gonzalez et al. 2012; Goodarzi and Jeggo 2013; Grallert and Boye 2008; Errico and Costanzo 2012; Lambert and Carr 2005, 2013; Giono and Manfredi 2006) to focus on, and it would definitely require more than one book chapter to accurately describe them all. However, the focus of this chapter is on the key principles and regulatory aspects.

15.3.2 Control of the Cell Size

To retain the cell size and guarantee that every daughter cell is equipped with the adequate quantity of genetic and biosynthetic components, the cells need to exactly duplicate their content on average prior to cell division. The proper size control of cells is crucial for the regulation of the supply of nutrients to the cell and for the adjustment of organ size and function in multi-cellular organisms. The availability of cell size control points has been recommended to enable cells to synchronize cell size with the course of the cell cycle. Control points for cell size were seen in G1 and G2. Early proof of these control points stemmed from findings that the size of new daughter cells following mitosis affects the progression of the cell cycle: huge daughter cells speed up the progression through G1 and/or G2, and tiny daughter cells retard the release from these growth phases (Killander and Zetterberg 1965a, b). However, the various types of species and cells vary greatly in the positioning of these control points throughout the cell cycle, and hence in the way, the cell cycle is impacted in reaction to a variation in cell size.

It is not astonishing that a lot of the knowledge about the size regulation points at the molecular scale is focused on the regulation of the proteins participating in the G1 and G2/M progression. The regulation of the G1 cell size control point has been most extensively investigated in budding yeast, in which the cyclin Cln3, that activates the start, governs the cell size (Cross 1988; Nash et al. 1998). The ability to manage the G2/M cell size control point was examined most comprehensively in fission yeast, in which Cdc25 and Wee1 react to cell height and nutritional state in their control of the Cdc2-cyclin B complex (Moreno et al. 1990; Nurse 1975).

One suggested mechanism to monitor cell size is the surveillance of the translation of proteins. The ribosomal mass, and hence, the translational activity is expected to be related to the size of the cell, thus it is assumed that there is a translation product referred to as "translational sizer", which rises in abundance along with the cell size and, following a specific level, becomes the dominant regulator of the cell cycle (Jorgensen et al. 2002). Cln3 and Cdc25 are the both suggested to act as translational sizers. This hypothesis also provides an interpretation of how cell size and cell cycle react to nutritional condition. In yeast, multiple signaling pathways, among them the PKA and TOR pathways, are recommended to convey nutrient management of the cell cycle, the common feature of these pathways is that they govern ribosome biogenesis, and thus, the translational activity operates as a cellular marker of nutritional condition.

The other mechanism by which cells can synchronize cell size with cell cycle progression is the surveillance of cell geometry. The fission yeast *S. pombe* has a cylindric shape and expands longitudinally before division. A protein entitled Pom1 is situated at the tips of the cell and stops the progression of the cell cycle through the regulation of the axis Cdr1-Cdr2-Wee1-Cdc2, which is centrally located in a zone known as the interphase knot. At larger cell lengths, Pom1 is no longer able to interfere with this complex, and the cell cycle can enter the M-phase (Martin and Berthelot-Grosjean 2009; Moseley et al. 2009). Although this system may rely on the relatively exclusive cell shape of *S. pombe*, it nevertheless leads to the question whether comparable mechanisms occur in different species. Whereas a number of possible explanatory models have been presented for the adjustment of cell cycle and cell size, it is conceivable that an endless number of them operate at the same time in a single cell. However, how they are all being implemented is still not clear.

15.3.3 Response to DNA Damage

During the entire interphase, DNA injury leads to a cell cycle standstill, allowing the healing pathways to take time before they are used for downstream phases of the cell cycle. The origin of DNA injury can be intrinsic, such as metabolic intermediates, telomere break-down, overexpression of oncogenes and failures in DNA replication. There are various alternative external factors for DNA injury, spanning the spectrum from sunlight to carcinogens, ionizing radiation or any other cancer therapeutic drug. Whereas a number of injury-specific reactions for DNA repairs are available, various injuries in the genomic DNA trigger common control points, the aim of which is to keep CDKs in an inactive status as long as the defect is eliminated. In general, the control points for DNA injury can be divided into those that are dominated by the tumor suppressor and the transcription factor p53 and those that are finally controlled by the control point kinase Chk1, and the latter will be the ones to be examined first.

The Chk1 signaling pathway is strongly conserved from yeast to humans. The components of the Chk1 signaling pathway are mostly derived from genetic screenings in the yeasts of damage-sensitive mutants (Weinert and Hartwell 1988, 1991; O'Connell et al. 2000; Al-Khodairy and Carr 1992; Al-Khodairy et al. 1994; Enoch et al. 1992), with a few supplementary components detected in mammalian cells (Wang et al. 2002; Goldberg et al. 2003; Stewart et al. 2003; Kumagai and Dunphy 2000).

Chk1 is initiated through all known types of DNA damage, although this is quantitatively more effective in the S- and G2-phase than in the G1-phase and is limited to post-replicative damage (Tapia-Alveal et al. 2009; Lin et al. 2013; Latif et al. 2004). The variety of activating lesions indicated a common intermediary substance, which is single-stranded DNA encapsulated with replication protein A (RPA) and comprising a primer-template compound (O'Connell and Cimprich 2005; MacDougall et al. 2007). The complexes of control point proteins accumulate on the RPA-coated DNA, among them a protein kinase known as ATR (Ataxia Telangiectasia and Rad3-Related) in humans, to which its interacting protein ATRIP is directed, and a PCNA-related bracket referred to as 9-1-1 complex (Rad9-Rad1-Hus1), which is attached to a variable replication factor C (RFC) complex. After phosphorylation through ATR, the BRCT domain mediator proteins are attracted to these locations. There are more mammalian modulators than yeasts; however, they fulfill the same task: the enrollment of Chk1, that is phosphorylated through ATR and subsequently liberated to keep the mitotic CDK Cdc2 in the Y15-phosphorylated and deactivated form. Chk1 is responsible for phosphorylating both kinase (Wee1) and phosphatase (Cdc25), which govern Y15 phosphorylation. This results in enhanced Wee1 stability and reduced Cdc25 activity and/or protein content. Chk1 is then dephosphorylated through type 1 phosphatases (Lu et al. 2005; den Elzen et al. 2004, den Elzen and O'Connell 2004), and the cells restart the cycling of mitosis.

In *S. cerevisiae*, the upstream signal events are similar to the ones described before, although the effector kinase differs. Even though Chk1 is conserved, the principal interactor remains an unrelated kinase commonly referred to as Rad53 (Allen et al. 1994; Sun et al. 1998). In addition, the cell cycle halt point may not be the G2-M transition, and it is the shift from metaphase to anaphase. This is achieved through Rad53, which increases the activity of the protease cohesin, the separase, through phosphorylation of its regulator securin (Cohen-Fix and Koshland 1997). This damage-related mitotic standstill is not found in any other species, including fission yeast, and human mitotic cells in particular are not capable of retarding mitotic progression (Rieder and Cole 1998). In addition, another kinase, commonly referred to as Dun1, is activated in budding yeast (Zhou and Elledge 1993), which governs transcriptional responses to DNA injury, which involves activation of ribonucleotide reductase, the enzyme necessary for the synthesis of dNTPs.

In higher organisms, the transcription factor p53 represents a crucial element of the control points for DNA defects (Giono and Manfredi 2006), especially in the G1-phase. The p53 is governed by a multitude of post-translational alterations, among them the N-terminal phosphorylation of serine-15, which is being catalyzed through ATR and its co-reactant ataxia telangiectasia mutated (ATM) and DNA-dependent protein kinase, catalytic subunit (DNA-PKcs). In a similar way to ATR, these kinases are directed toward double-strand breakings of DNA through interacting proteins, such as the MRN (Mre11-Rad50-Nbs1) complex for ATM and the Ku70-Ku80 complex for DNA-PKcs. Activated p53 becomes stabilized through its protection toward its E3-ubiquitin ligase Mdm2 and acts as a tetramer activating the expression of a wide range of genes, among them the cyclin-dependent kinase inhibitor (CKI) p21. This mechanism suppresses G1-CDKs and corrects DNA defects before DNA replication. Nevertheless, p53 can also inhibit the expression of genes and is necessary for prolonged G2 standstill in the presence of persistent DNA injury (Carvajal et al. 2012; St Clair et al. 2004). In addition, p53 can govern the alternative cell destinies of apoptosis or senescence (Carvajal and Manfredi 2013). In fact, the function of stopping the cell cycle of p53 appears to be a later adopted role, since *Drosophila* p53 governs apoptosis and not cell cycle progression (Nordstrom and Abrams 2000).

15.3.4 Surveillance of DNA Replication

The S-phase is a time when cells are especially susceptible to dealing with DNA defects. Not only need lesions to be fixed as in G1 and G2 cells, but they also serve as a physical hindrance to the replicative polymerases. DNA replication is started at certain points, known as the origin of replication. These are epigenetically determined through a series of proteins that are responsible for replicating once and not more than once per cell cycle. The replication origin fire is regulated through the phosphorylation of two proteins, Cdt1 and Cdc6, which is catalyzed through both CDKs and the Dbf4-dependent protein kinase (DDK) Cdc7. This kind of phosphorylation not only triggers replication, and it also leads to the break-down of these proteins, preventing the origin to refire (Nishitani and Lygerou 2004).

When the polymerase and its linked proteins (the replisome) experience a barrier to progression, it is essential that the replisome maintains stably linked to the replicating chromatid, and hence, the replication is able to resume upon the removal of the hindrance. These hindrances can be formed through altered dNTPs, abasic sites, protein–DNA complexes, or even be a cause of the lack of dNTPs. This replisome stabilization operates as an intra-S-phase control point.

The effector kinase of the intra-S-phase control point is named Cds1 in fission yeast and Chk2 in humans. Although the name is somewhat related, Chk2 is not similar to Chk1 in terms of its biochemical structure or function. Cds1/Chk2 contains an N-terminal phospho-S/T-binding forkhead associated (FHA) domain and a nearby kinase domain. When replication is stopped, the replisomic component mediator of the replication control point (Mrc1) is phosphorylated through ATR. This produces a Cds1/Chk2-binding site, which is then phosphorylated with ATR and completely activated through autophosphorylation (Xu et al. 2006). The activated Cds1/Chk2 subsequently stabilizes the arrested replisome through phosphorylation of multiple subunits, especially the MCM helicase

(Bailis et al. 2008; Stead et al. 2012). In budding yeast, the Rad53 kinase fulfills the role of Cds1/Chk2. Similar to Cds1/Chk2, Rad53 contains an N-terminal FHA domain, closely spatially connected to a kinase domain. Nevertheless, Rad53 possesses an extra C-terminal FHA domain, not visible in Cds1, which is essential for its activation due to DNA injury (Sweeney et al. 2005).

15.3.5 The Interrelation of S- and M-Phases

After stabilization, the replisome can remain in place as long as the obstruction is cleared or the levels of dNTPs are re-established. As an alternative, after the process of replication, the cell can exploit repair routes to circumvent the defect, using either the recruitment of mutagenic circumvention polymerases or a template exchange through recombination and subsequent replication with the other emerging strand as template (Lee and Myung 2008). In both cases, control points must be implemented to assure that mitosis is avoided unless replication is accomplished, otherwise the cells are at risk of decreased ploidy. It is evident that mitotic entrance is inhibited by Y15 phosphorylation of Cdc2 and requires the control point components upstream of the acquisition of mediators (Al-Khodairy et al. 1994; Enoch et al. 1992; Rhind et al. 1997). It is not entirely obvious how this results in a retardation of the cell cycle.

When the origin of the inhibition is a defect in the DNA, the Chk1 pathway becomes enabled as mentioned before. However, when the inhibition is only caused by dNTP deficiency, for instance through hydroxyurea conditioning, Chk1 remains inactive, and the cells still do not undergo mitosis (Lindsay et al. 1998). Several different investigations have come to the result that Cds1 controls also the progression of the cell cycle (Zeng et al. 1998; Furnari et al. 1999). In the absence of Cds1, however, the cells detach the locked replisome from its template, which is a process referred to as fork collapse, and this is deemed to be DNA injury causing Chk1 to be activated. Therefore, it is hard to distinguish the phenomena of stability and function of the fork of replication for the effector kinases in an experimental way.

A stronger decoupling of the dependency of mitosis from previous replication is evident when the fission yeast cells are missing both Y15 kinases, such as Wee1 and Mik1, or when the initial firing proteins are removed (Lundgren et al. 1991). In such circumstances, cells undergo deadly mitoses of G1, a phenomenon initially referred to as mitotic catastrophe, a nickname subsequently employed to characterize mitotic killing in mammalian cells.

To sustain ploidy, an equally essential linkage needs to be established to guarantee one replication lap per cell cycle. It can be decoupled when the break-down of Cdt1 and Cdc6 is incorrect (Nishitani and Nurse 1995), and the origin of replication is becoming constitutive. Whole laps of the S-phase in the absence of mitosis can also be monitored in fission yeast in cases where the CKI Rum1 is excessively produced (Moreno and Nurse 1994). In a similar way, cells lacking mitotic cyclins circumvent mitosis (Hayles et al. 1994), indicating that they convey the necessary knowledge of a G2 condition (O'Connell and Nurse 1994). In all these circumstances, the mitosis is totally avoided, and the ploidy keeps increasing.

15.3.6 The Checkpoint of the Mitotic Spindle

The segregation of the sister chromatids in the anaphase is under mechanical supervision of the mitotic spindle. The spindle is composed of microtubules and several motor proteins at either the centrosomal or the kinetochoric end and supplementary motors that deliver the force between overlapping microtubules which are not bound to kinetochors (Wittmann et al. 2001). It is critical that the spindle fixation is performed in a bi-oriented manner ensuring that the sister chromatids are under tension in the metaphase and are fixed at both ends of the spindle. When all kinetochores are fixed and aligned on the metaphase platform, anaphase can take place as it stimulates the activity of a large E3 ubiquitin ligase commonly referred to as anaphase promoting complex (APC) or Cyclosome (C). This ligase is directed at a range of proteins, although the most prominent are mitotic cyclins, which eliminate CDK activity, and securin, whose break-down facilitates the liberation of separase and the break-down of cohesin complexes located at the kinetochores. APC activity is governed mainly through two accessory proteins: The first one is Cdc20, which acts up to metaphase–anaphase, and the second one is Cdh1, which enhances APC-driven ubiquitination once cyclin and separase break-down has initiated (McLean et al. 2011). When the cohesion of the sister chromatids is broken down, the spindle tension and accompanying motor proteins permit the sister chromatids to drift apart and produce independent and identical daughter nuclei.

The spindle checkpoint is intended to avoid the activation of APC Cdc20 in circumstances where the kinetochores are not covered by spindle microtubules or are mounted without tension, such as when fastened to the identical rod, which is referred to as merotelic mounting (Fig. 15.2). Under these environmental constraints, the spindle control point protein Mad2 (Mitotic Arrest Deficient) blocks Cdc20 activity on unbound kinetochores, where it constitutes a mitotic control point complex, and on APC-bound molecules. Cdc20 is further controlled through the mitotic checkpoint kinase Bub1 in yeast (budding uninhibited by benomyl) and its mammalian counterpart Bub1R. Since Cdc2 is deactivated, APC is also deactivated, and therefore, the cells are unable to progress to anaphase.

The spindle control point comprises a range of other proteins, and the list increases with the evolutionary complexity. In extension, the creation of the spindle and the recognition and rectification of spindle errors are under the supervision of the Polo, Aurora and NIMA (Nek) kinases (O'Connell et al. 2003; Malumbres and Barbacid 2007). In this sense, the spindle control point supports the same basic principle as the DNA integrity control mentioned earlier, which is to avoid a cell cycle transition when other effectors repair a genome-altering disorder. However, the mitotic control point is singular in that it serves to preserve CDK activity, while the ones that operate in the interphase tend to sustain the inactivity of CDK.

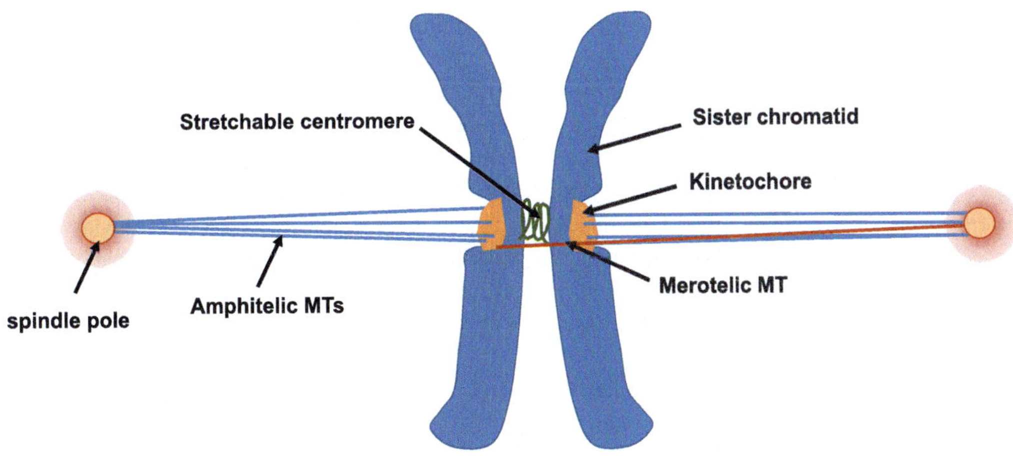

Fig. 15.2 Merotelic kinetochore microtubule (MT) configuration for a congressed chromosome. The chromosome is positioned in the middle between the spindle poles. Sister chromatids are linked via a stretchable centromere (green spring). Most MTs are in the amphitelic configuration and the sister kinetochores are located back-to-back (blue), however, they can still bind improper merotelic MTs (red)

Linkage Between Impaired Cell Cycle Checkpoints and Human Diseases

According to the seriousness of the cell cycle malfunction, a checkpoint dysfunction can lead to outcomes ranging from cell death to reprogramming of the cell cycle, which can result in cancer. The loss of p53, probably the most frequent genetic disorder in cancer, impacts several determinants of cell fate. These include the absence of control of the CDK through p21 and therefore the complete disappearance of the G1 control point. Nevertheless, p53 may also trigger cells to enter apoptosis and/or senescence, and thus, the physiological outcome of p53 in cancer lies both on the level of the onset of carcinogenesis and on the capacity of therapies to kill cells (Giono and Manfredi 2006). Notably, the lack of p53 constitutes an increased demand on the Chk1 signaling pathway, which is frequently elevated in cancer cells and is necessary for the viability of numerous cancer cells (Tapia-Alveal et al. 2009; Kuntz and O'Connell 2009; Koniaras et al. 2001). Therefore, there is great potential to exploit Chk1 (Maugeri-Sacca et al. 2013) and its substrate Wee1 (Stathis and Oza 2010) n clinical practice as a therapeutic modality.

The absence of the Chk1 pathway in a fission yeast reveals one significant phenotype only with extrinsic DNA defects or in conjunction with DNA repair errors (Walworth et al. 1993). The entrance into mitosis with fragmented or imperfectly repaired chromosomes fails to initiate the spindle checkpoint that only detects kinetochore adherence. Thus, these types of mitosis are either instantly lethal or result in a considerable reduction of chromosome fragments. In mouse and *Drosophila*, the Chk1 signaling pathway is indispensable for transit through early embryogenesis (Liu et al. 2000; Fogarty et al. 1997). Nevertheless, this is an impediment to the quick cell cycle, and S-M dependence is crucial for the integrity of the genome. In a similar way, the conditional depletion of Chk1 is deadly in certain tissues and cell lines (Greenow et al. 2009; Smith et al. 2013; Lam et al. 2004), but not in others (Koniaras et al. 2001; Greenow et al. 2009; Petermann et al. 2006; Zachos et al. 2003), where in turn proliferation speeds may be extremely sensitive. However, mutations in genes of the Chk1 signaling pathway are exceedingly seldom in human cancers. Although cancers exhibit genome instability, they cannot persist if there are no control points for genome integrity.

A characteristic of the majority of solid tumors includes high-grade aneuploid karyotypes. The loss and rearrangement of chromosomes are a fast tool for the removal of tumor suppressors and the stimulation of oncogenes. Although mutations in the genes of the spindle control points have been documented, they are nevertheless quite scarce (Kops et al. 2005). However, in view of the complexity of the mitotic system and the enormous implications of the complete absence or gain of entire chromosomes, a slight malfunction can have far-reaching outcomes. In the same way as DNA integrity control points, in which a high degree of DNA degradation leads to cell death, the modification of the dynamic instability of spindle microtubules can be deadly, with the major advantage that the spindles are exclusively found in cyclic cells.

In this section, we have outlined the fundamental principles underlying the conventional cell cycle control points. What they have in common is the characteristic of identifying an abnormality in the division program and then transmitting signals to modify the fluctuations of CDK activity and thereby the cell cycle processes. Certain control point signaling issues still require clarification or determination, some of which are well-known uncertainties, either as a straightforward principle or in the field of human development and disease. No doubt unforeseen features of checkpoint signaling (unknown strangers) will be identified, and the ever-expanding array of advanced experimental instruments and technologies will provide a more comprehensive picture of the notable accuracy of the cell cycle.

15.4 Why Is the DNA Organized as a Twisted Helical Structure?

In biology, the term "double helix" is commonly employed to define the DNA structure. A DNA double helix is composed of two coiled chains of deoxyribonucleic acid. The form is close to that of a winding staircase. DNA refers to a nucleic acid consisting of nitrogenous bases (adenine, cytosine, guanine and thymine), a five-carbon sugar (deoxyribose) and molecules of phosphate. The nucleotide bases of the DNA constitute the steps of the stairway, and the deoxyribose and phosphate molecules comprise the flanks of the stairway.

Why Is the DNA in a Twisted State?

The DNA is wound up into chromosomes and densely packaged within the nucleus of the cells. The twisted appearance of DNA comes from the interference between the molecules that compose DNA and water. The nitrogenous bases that the treads of the twisted stairs are made of are kept in place through hydrogen bonds. Adenine is linked with thymine (A-T) and guanine pairs with cytosine (G-C). These nitrogenous bases have hydrophobicity, which implies that they have no affinity to water. As the cell cytoplasm and cytosol include water-based fluids, the nitrogenous bases aim to prevent exposure to cell fluids. The sugar and phosphate molecules that constitute the sugar–phosphate backbone of the molecule have hydrophilic properties, and in other words, they are water-loving and exhibit an affinity for water.

The DNA is organized in a way that the phosphate and sugar backbone are on the outside and in contact with liquid, whereas the nitrogenous bases are in the inner part of the molecule. To continue preventing the nitrogen bases from contacting the cell liquid, the molecule distorts to decrease the distance between the nitrogen bases, the phosphate and sugar strands. The circumstance that the two DNA strands constituting the double helix is antiparallel aids in twisting the molecule in the same way. Antiparallel refers to the fact that the DNA strands are arranged in opposite directions, thereby guaranteeing that the strands are firmly joined together. This minimizes the risk of liquid seeping between the bases.

15.5 DNA Replication Follows a Semi-conservative Principle

Why Is the DNA Needed to Be Replicated?

DNA represents the genetic matter that determines each cell. Prior to a cell duplicating and splitting into novel daughter cells using either mitosis or meiosis, it is necessary to copy biomolecules and organelles to be disseminated among the cells. The DNA in the cell nucleus must be amplified to assure that every new cell contains the appropriate number of chromosomes. The duplication procedure of DNA is referred to as DNA replication. Replication takes place in multiple steps requiring the participation of several proteins, referred to as replication enzymes, and RNA. In eukaryotic cells, such as animal and plant cells, DNA replication is performed in the S-phase of the interphase over the course of the cell cycle. The process of DNA replication is crucial for cell growth, damage repair mechanisms and the reproduction in organisms.

Structure of the DNA

DNA or deoxyribonucleic acid is a molecular type referred to as nucleic acid. It is composed of a 5-carbon deoxyribose sugar, a phosphate and a nitrogenous base. Double-stranded DNA comprises two spiral nucleic acid chains entwined into a double helix. The twisting enables the DNA to be rendered more condensed. In an effort to fit into the nucleus, the DNA is wrapped into tightly coiled structures, which are referred to as chromatin. Chromatin is condensed to chromosomes in the course of cell division. Before DNA replication, the chromatin expands and loosens, thereby enabling the entire replication machinery of the cell to have access to the DNA strands.

15.5.1 How Can Cells Prepare for DNA Replication?

The first step is the formation of the replication fork. But before the DNA in question can be amplified, the double-stranded molecule needs to be "uncompressed" (or synonymously referred to as unzipped) in two individual strands. DNA consists of four bases namely adenine (A), thymine (T), cytosine (C) and guanine (G), which combine to form pairs between the two strands. Adenine binds only with thymine, and cytosine binds only with guanine. To entwine the DNA, these base-pair interactions have to be interrupted. This is carried out by an enzyme referred to as DNA helicase. The DNA helicase breaks the hydrogen bonds between the base pairs in order to split the strands into a Y-form referred to as replication fork. This area will serve as the master for the start of replication.

The DNA is oriented in a directional manner in both strands, which is characterized by a 5′ and 3′ end. In specific detail, this notation specifies what side group is bound to the DNA backbone. A phosphate (P) group is linked to the 5′ end and a hydroxyl (OH) group to the 3′ end. This directional dependency is critical for replication, since it propagates solely in the 5′ to 3′ direction. However, the replication fork has a bidirectional design: One strand is aligned in the 3′ to 5′ direction (leading strand) and the other in the 5′ to 3′ direction (lagging strand). The two sides are thus reproduced applying two different techniques to compensate for the directional discrepancy.

The second step is the onset of the replication through the binding of the primer. The leading strand can be replicated most easily. After the DNA strands have been split, a short piece of RNA, termed a primer, attaches to the 3′ end of the strand. The primer serves constantly as a replication starting point. Primers are formed through the enzyme DNA primase.

15.5.2 The Elongation Step of the DNA Replication

The third step represents the elongation. Enzymes which are generally referred to as DNA polymerases are in charge of producing the new strand through a process known as elongation. Five distinct types of DNA polymerases are found in bacteria and human cells. In bacteria such as *E. coli*, polymerase III is the major replication enzyme, whereas polymerases I, II, IV and V are in charge of failure control and repair mechanisms. DNA polymerase III attaches to the strand at the site of the primer and starts appending new base pairs complementary to the strand during the process of replication. In eukaryotic cells, the polymerases alpha, delta and epsilon constitute the primary polymerases participating in DNA replication. Since the replication extends in the 5′ to 3′ direction on the leading strand, the new strand created is a continuous one.

The trailing strand starts replication through binding with multiple primers. Every primer is spaced only a few bases apart. The DNA polymerase subsequently inserts DNA segments, referred to as Okazaki fragments, into the strand located between the primers. This replication process is non-continuous, since the newly formed fragments are dispersed.

The fourth step is the termination of the DNA replication. When both the continuous and intermittent strands are produced, an enzyme known as exonuclease eliminates all RNA primers of the original strands. These primers are subsequently substituted with corresponding bases. A different exonuclease "proofreads" the new formed DNA in order to verify, eliminate and repair any defects. Another enzyme known as DNA ligase combines Okazaki fragments into one single, uniform strand. The ends of the linear DNA pose a difficulty because the DNA polymerase is only able to attach nucleotides in the direction 5′ to 3′. The ends of the parent strands are made up of repeating DNA sequences referred to as telomeres. Telomeres function as barriers at the end of chromosomes to avoid the fusion of adjacent chromosomes. A special kind of DNA polymerase enzyme entitled telomerase acts as a catalyst for the production of telomere structures at the terminal ends of the DNA. After finishing, the mother strand and its complementary DNA strand wrap into the familiar two-helix form. In the end, replication yields two DNA molecules, each consisting of one strand from the parent molecule and a newly polymerized strand.

15.5.3 Enzymes Employed for DNA Replication

DNA replication could not take place entirely without enzymes catalyzing various stages of the process. Enzymes associated with the eukaryotic DNA replication process comprise the following:

- DNA helicase unwraps the double-stranded DNA and separates it as it travels down the DNA. It creates the replication fork as it disrupts the hydrogen bonds linking the nucleotide pairs in the DNA.
- DNA primase represents a type of RNA polymerase that generates the RNA primers. In specific, these primers represent short RNA molecules that serve as marker templates for the initiation point of the DNA replication.
- DNA polymerases generate new DNA molecules through the polymerization of free nucleotides to leading and trailing DNA strands.
- Topoisomerase or synonymously known as DNA Gyrase entwines and rewinds the two DNA strands to circumvent the DNA to be tangled or supercoiled.
- Exonucleases are a group of enzymes that are able to eliminate nucleotide bases at the DNA chain end.
- DNA ligase connects DNA fragments through the establishment of phosphodiester bonds between nucleotides.

DNA replication refers to the generation of identical DNA helices out of a single double-stranded DNA molecule. Every compound is composed of one strand of the original compound and one newly generated strand. Before the DNA is replicated, it unwinds and the threads split. A replication fork is built, which acts as a masterpiece for the replication. Primers attach to the DNA and DNA polymerases attach new nucleotide sequences in the direction 5′ to 3′. This supplementation is progressive in the leading strand and is broken down in fragments within in the trailing strand. When the elongation of the DNA strands is finished, the strands are scanned for errors, reparations are made, and the telomere sequences are attached to the ends of the DNA.

15.6 Active or Passive Role of Centrioles During Mitosis

The production of cilia during the interphase and the formation of an aster and subsequently the assembly of the entire mitotic spindle are major functions of centrioles during cell division. Cilia are organelles of a eukaryotic cell and represent protuberances of the large cell body's surface. There are two different types of cilia: motile cilia and non-motile (synonymously referred to as primary), which function as sensory organelles. In more detail, an aster presents a cellular structure of the shape of a star. It is composed of a centrosome and associated microtubules in the early stage of mitosis in animal cells.

The centrosome represents the main microtubule organizing center (MTOC) in animal cells. Each centrosome is composed of two centrioles that are closely enveloped by a multilayered proteinaceous structural scaffold, which is termed the pericentriolar material. Since centrosomes evoke the de novo assembly of microtubules, they hence fulfill crucial functions in the organization of the Golgi and the mitotic spindle, cell polarity and cell motility. In order to perform these functions, centrosomes must occupy distinct cellular positions.

Actin and microtubule networks and the close proximity of the centrosomes to the nuclear envelope regulate the precise positioning of the centrosomes. Another major characteristic of centrosomes is the centrosomal coupling between two centrosomes. The centrosome coupling is established in the late mitosis/G1 simultaneously with the loosening of the centrioles and released prior or at the onset of mitosis. The release of the coupling is crucial for the assembly of the mitotic spindle, and its timing with the cell cycle has great influence on the conducting of mitosis and the performance of chromosome separation. Finally, the mechanisms of centrosome positioning and its separation seem to be relevant for the entire process of cell division and subsequently on the mechanical properties of cells (Agircan et al. 2014).

It is composed of two centriols and a multilayered, protein-containing structure enclosing the centriols, referred to as the pericentriolar material. Centrosomes encourage de novo the formation of MTs and therefore contribute to the organization of Golgi, the polarity of cells, the motility of cells and the arrangement of the mitotic spindle. To perform these tasks, the centrosomes need to take up certain cellular locations.

Actin and MT networks and the connection of the centrosomes to the nuclear envelope determine the accurate placement of the centrosomes. Another crucial feature of centrosomes is the characteristic of a linker that connects the two centrosomes. The centrosome linker is formed in late mitosis/G1 at the same time as the centrioles become disconnected and is broken down prior to or at the onset of mitosis. The break-down of linkers is crucial for the development of mitotic spindles, and their timing with the cell cycle has far-reaching consequences for the performance of mitosis and the degree of chromosome segregation. In the following, the focus lies on the mechanisms of centrosome location and dissociation and presents their functional roles and the underlying mechanisms.

15.6.1 Function of Centrioles and Centrosomes During the Cell Cycle

The centrosome represents the key MTOC of animal cells. In specific detail, it is able to organize the de novo polymerization of microtubules (MTs) from tubulin subunits, anchor and organize microtubules (Bettencourt-Dias and Glover 2007). According to the cell cycle phase, centrosomes are composed of one or two centrioles enclosed by the pericentriolar material (PCM). Centriols define the state of replication of centrosomes, whereas the pericentriolar material bears the majority of the MT nucleation action (Sluder and Rieder 1985). The centriole is normally about 0.5 mm long and possesses a 0.2 mm in diameter with nine microtubule triplets (Azimzadeh and Bornens 2007; Bornens and Azimzadeh 2007). In the S-phase, the daughter centriole congregates next to the mother centriole. The two centrioles are initially joined in a vertical alignment through an unidentified mechanism and are held in this way up to the telophase/G1 when the two centrioles are separated, which is referred to as disengagement.

Due to the central role of centrosomes in the nucleation of microtubules, it has been suggested for several years that centrosomes are crucial for cell survival, particularly during chromosomal segregation, by encouraging the generation of the mitotic spindle. The results in *Drosophila melanogaster* indicate that the centrosome is not essential for cell viability by acting as a structural unit (Basto et al. 2006). In the absence of centrosomes, the mitotic spindle development is fueled by microtubules which gather themselves around the chromatin due to the activity of the small GTPase Ran. Ran-GTP induces the local activation of microtubule-associated proteins inclusive of the gamma-tubulin complex, which initiate the nucleation of the microtubules (Kalab and Heald 2008).

Although unnecessary for the formation of spindles and chromosome splitting in flies, centrosomes are needed for the development of cilia. Flies devoid of centrosomes are dying shortly after birth because of lack of cilia (Basto et al. 2006). The centrosomes also play an essential part in the correct positioning of the spindle that governs cell fate in asymmetric cell division, through the arrangement of the astral microtubules. An elegant investigation in *D. melanogaster* indicated that fly neuroblasts without centrosomes split symmetrically solely in 15% of cases (Basto et al. 2006). Accordingly, in *D. melanogaster* male germ cells, the correct bipolar spindle alignment is based on the placement of the mother triplet (Yamashita et al. 2003, 2007). In addition, the asynchronous inheritance of old and new parent centrioles in mouse cells is required for the development of neurons, which means that centrosomes are relevant for mouse development (Wang et al. 2009).

Recent data using DT40 cells from vertebrates as a modeling tool (Sir et al. 2013) indicate that the disappearance of centrioles results in retardation of bipolar spindle production, elevated rates of chromosomal instability and aneuploidy. These data contradict the idea that centrosomes are unnecessary for mitosis in vertebrate cells.

15.6.2 The Disengagement of the Centrioles

Canonical centriole doubling is performed in the S-phase, with the old mother triplet usually acting as the assembly stage for the new daughter triplet (Fig. 15.1). The new centriole is perpendicular to the parent centriole, and the two centrioles remain linked together by a weakly determined linker from the S-phase to the termination of mitosis (Paintrand et al. 1992). The separation of the centrioles, also referred to as "centriole disengagement", occurs in Telophase/G1 (Piel et al. 2000). The centriole decoupling represents the licensing step for centriole duplication in the subsequent S-phase and is hence indispensable for centriole duplication (Tsou and Stearns 2006; Wang et al. 2011).

There is substantial sophistication in the management of centriole separation (Fig. 15.1). One of the key proteins participating in this pathway is the protease separase (Espl1), renowned for its involvement in sister chromatid segregation in the metaphase-to-anaphase transition. The sister chromatids are kept connected through the cohesin compound unless the APC/C complex is activated. When the separase is activated, the cohesin subunit Scc1Rad21 is broken down by the separase allowing the sister chromatids to be split in anaphase (Ciosk et al. 1998; Uhlmann et al. 1999). A first indication of a role of the separase on centrosomes resulted from the finding that part of the separase in mitosis is linked to centrosomes (Chestukhin et al. 2003).

The part played by the separase in the detachment of the centrioles has been first seen in extracts from *Xenopus laevis* eggs (Tsou and Stearns 2006). Human S-phase centrioles dissolve during incubation in *X. laevis* egg extracts containing active separase. Subsequent to this investigation, the construction of a human separase-knockout cell line showed that the loss of separase activity inhibited detachment only to a limited extent. However, chemical blocking of the polo-like kinase Plk1 in the separase-knockout cell line totally avoided centrioles separation. Thus, Plk1 and separase work together in disengagement of the centrioles (Tsou et al. 2009). Although cyclin B1 and securin are generally recognized as the principal regulators of chromosome-bound separase (Tsou and Stearns 2006; Stemmann et al. 2001), the control of the separase at the centrosome appears to be more elaborate and could involve other actors, two of them have been identified as astrin and Aki1. Both proteins have been discovered as suppressors of the centrosomal separase (Thein et al. 2007; Nakamura et al. 2009). Consequentially, the depletion of either astrin or Aki1 resulted in a premature separatase activation, which in turn triggered the establishment of multipolar mitoses involving dislocated centrioles. Another actor in the supervision of centrioles separation involves Shugoshin (Sgol), which impedes the premature separation of the sister chromatids unless the separase is activated (Tsang and Dynlacht 2008; Wang et al. 2008a, b).

A splice variant of Sgol, referred to as small Shugoshin (sSgol), is associated primarily with centrosomes. The expression of a dominant-negative type of sSgo1 or the exhaustion of the complete Sgol pool caused centriole decoupling. Simultaneously, the activity of Plk1 is necessary for the correct centrosome positioning of sSgo1 (Tsang and Dynlacht 2008). Therefore, sSgo1 performs an undefined function in the avoidance of premature centriole detachment, something which is ultimately governed by Plk1.

The involvement of separase in centriole separation indicated a probable function of cohesin within this process. A number of incidents point to the localization of cohesin in centrosomes (Wong and Blobel 2008; Gregson et al. 2001; Kong et al. 2009). In addition, biochemical investigations with separase-resistant cohesin mutants indicate that the cohesin complex retains mother and daughter centrioles together unless the separase breaks down the cohesin subunit Scc1Rad21 (Schockel et al. 2011; Tsou and Stearns 2006). There are nevertheless numerous unresolved problems concerning the function of cohesin as a link between the centrioles. A comprehensive investigation is lacking that shows co-localization of cohesin subunits at the identical centrosomal substructure through expression of functional cohesin-fluorescent protein fusions or through immunostaining of the body's own proteins. The cohesin role in centriole attachment has been obtained using a heterologous system that integrates human centrosomes with *X. laevis* egg extracts. The proof of the participation of cohesion in the connection of centrioles in a homologous regime seems to be crucial. In addition, a newly published report disputed this role of cohesin for centriole separation in *Drosophila* embryos (Oliveira and Nasmyth 2013). In parallel, a recent investigation on *Caenorhabditis elegans* embryos has revealed that microtubule-dependent forces drive centriole detachment rather than primarily the separase. Separation only plays a major part in the disengagement process when the tensile forces of the microtubules are not strong enough to disconnect the centrioles (Cabral et al. 2013).

Another enigmatic finding in this regard concerns the lag in the timing of centriole separation, which occurs in the telophase, when compared to the timing of activation of the separase and separation of the sister chromatids in the transition from metaphase to anaphase (Piel et al. 2000; Stemmann et al. 2001). This implies that the separase either becomes active at the centrosomes at a later time in the cell cycle or that the roles of separase at the centrosome are taken over during the metaphase-to-anaphase transition, whereas the separation of the centrioles is postponed due to additional mechanisms. It has been revealed that the separase can split the centrosomal protein pericentrin during the metaphase-to-anaphase transition and the cut pericentrin remains at the centrosome to the end of anaphase (Matsuo et al. 2012). This mechanism can explain the delay between the activation of separase and centriole the separation of the centrioles. Finally, although substrates of separase at the centrosomes have been identified, however what cohesin holds back or how pericentrin can regulate the separation of the centrioles still requires a lot of research effort.

15.6.3 Centrosome Linker

After dissolution of the two centriols, a proteinaceous connector is established between the parent and daughter centrioles to provide centrosome cohesion. This linker joins the two centrioles/centrosomes of G1 up to the onset of mitosis, where the linker is dismantled, thus assisting in the construction of the mitotic spindle. How the development of the centrosomal linker is connected with centrioles removal is an intriguing and unanswered question. The linker construct is cell cycle governed and demands a free distal proximal end of the centrioles. With the mitotic leakage, the linker proteins acquire the capacity to attach to the proximal end of the mother centriole (Mayor et al. 2000; Yang et al. 2002). However, at this stage, the proximal end of the daughter triple is actually still joined to the mother centriole wall and can be physically stalled. When the two centrioles become detached from each other, the proximal end of the daughter centrioles opens up, and the connector proteins begin to adhere. Due to a badly conceived self-assembly mechanism, additional linker proteins join together the pre-labelled centriole ends, thus forming a flexible ligature that interconnects the two centrosomes.

Proteins participating in the generation of the centrosomal linkage have been characterized either as targets of the NIMA-related kinase Nek2, which abolishes centrosomal cohesion by means of phosphorylation, or by siRNA screenings for proteins that interfere with centrosomal cohesion in case of exhaustion. The most common substrates of Nek2 comprise the well-known centrosomal linker proteins C-Nap1 (Cep250) and rootletin (Yang et al. 2006; Bahe et al. 2005). C-Nap1 is located at the basis of the parent centriole, and two C-Nap1 pockets are interconnected by rootletin fibers (Fry et al. 1998). Nek2 phosphorylates in an efficient fashion the C-Nap1 and rootletin in late G2 to enable their removement from the centrosomes (Bahe et al. 2005; Fry et al. 1998). This finally results in the detaching of the centrosomes.

The human microcephaly protein Cep135 represents a conserved centriolar protein, which is connected to the proximal end of centrioles, where it interplays with the centriole duplication factors hSAS-6 and CPAP (Lin et al. 2013). Moreover, Cep135 also interferes with the linker protein C-Nap1 (Cep250) (Kim et al. 2013). Since Cep135 is constantly combined with centrioles whereas the linker proteins are disassociated from the centrioles in mitosis, Cep135 is most likely the attachment site to where the linker proteins adhere. Two separate domains of Cep135 contribute to this attachment model, such as the N-terminal part, which renders the protein react with the centrioles, and the C-terminal part of Cep135, which is necessary for the interaction with the C-Nap1. Through these specific regions, the linker protein rootletin is able to interfere with the N-terminal part of C-Nap1, whereas the C-terminus of C-Nap1, in turn, interferes with Cep135 (Yang et al. 2006; Bahe et al. 2005; Kim et al. 2013).

One other protein known to be a linker protein is Cep68 (Graser et al. 2007), which belongs to the rootlet system. Cep68 is possibly the protein that serves to strengthen the root cell fibers which assemble themselves in the head-to-tail orientation (Graser et al. 2007). Recent hints pointed to the existence of two supplemental linker proteins entitled LLRC45 and LGALS3BP (Barrera et al. 2010; He et al. 2013; Fogeron et al. 2013). It has been documented that LLRRC45 interfaces with both C-Nap1 and rootletin. For instance, it might aid in binding rootletin fibers to the C-Nap1 platform at centrioles. Immuno-electron microscopic analyses exhibited the same fibrous pattern of LGALS3BP dispersion as the rootletin, which enhances the potential for interference of LGALS3BP with linker compounds (Fogeron et al. 2013). It was monitored that the Rho-associated protein kinase p160ROCK is targeted to mother centrioles and to the linker, and blocking p160ROCK with the inhibitor Y-27632 enhanced the spacing between mother and daughter centrioles in G1 cells. In addition, p160ROCK is crucial for the adhesion of the centrosome to the nuclear envelope during the interphase, and the blockage of p160ROCK induced a temporary preliminary alignment of the midbody to the maternal centriole during the process of mitosis (Chevrier et al. 2002).

Although a lot is understood about the assembly of the centrosomal linker, very little is available about the link between centriole detachment and the generation of linkers, the functionality of the centrosomal connector and its accurate control in the course of the cell cycle.

15.6.4 The Positioning of the Centrosome

Cells in the G1-phase possess two detached centrioles interconnected by the centrosomal linker. Live-cell imaging investigations with GFP-center-labelled centrioles revealed that the mother centrioles persist in the vicinity of the cell's center of mass. In contrast, the daughter centriole is far more flexible and travels throughout the cytoplasm. The movement of the centrioles is not dependent on the nuclear envelope; however, it demands actin and the microtubule cytoskeleton (Piel et al. 2000).

What role plays the fast motion of the daughter centriole in the G1? In HeLa cells, the relocation of the moving daughter centriole to the midbody is related to the terminal phase of cell division, the splitting of the microtubules adjacent to the midbody (Piel et al. 2000). Interestingly, both centrioles have the capacity to nucleate microtubules, whereas only the mother centriole stably anchors the microtubules by attachments at the distal end of the centriole. The daughter centriole produces these appendages in the subsequent mitosis. Microtubules originating from the mobile daughter centriole are transferred toward the cell periphery, which carries out crucial functions in migrating cells (Abal et al. 2002).

Centrosomes are critical for the arrangement and placement of the Golgi apparatus of a cell. In mammalian cells, the Golgi is a unique specimen of an organelle positioned surrounding the centrosome. The Golgi fragments initially collect at the cell periphery after mitosis. Microtubules arranged at the centrosome then yield the tracks to transport Golgi fragments toward the cell center, where the fragments merge into a Golgi ligament. Lately, it has been revealed that the disturbance of Golgi placement due to separation from the centrosome has a profound impact on directed cell migration (Hurtado et al. 2011). Typically, the centrosome positions the Golgi apparatus in the direction of the front edge of the cell. This placement assists the polarized secretion of cargo vesicles and is indispensable for cell migration.

In the G2 and prophase, the centrosomes are attached to the nuclear envelope using a minimum of two mechanisms (Fig. 15.3). Firstly, RanBP2/Nup358 enrolls the adaptor protein bicaudal D (BICD2) on the cytoplasmic part of the nuclear pore complex (NPC) in G2, which then binds the dynein/dynactin complex to NPCs. Dynein/dynactin cooperates with the microtubules arranged by the centrosomes and thereby places the centrosomes on the nuclear envelope (Splinter et al. 2010). The second pathway operates through the centromere-associated CENP-F protein and the Nup133 subunit of the Nup107-160 nuclear pore subcomplex (Splinter et al. 2010). The N-terminal domain of Nup133 interfaces with CENP-F, which is placed on the nuclear envelope in G2/M. CENP-F in its turn enrolls NudE/NoodEL within the nuclear envelope. In specific detail, NudE and NoodEL are related proteins that can interfere with the cytoplasmic dynein. However, both pathways work together jointly in the positioning of the centrosome toward the nuclear envelope. Moreover, it has been seen that the CEBP-F/Nup133 pathway is involved in a timely establishment of a precisely positioned spindle (Bolhy et al. 2011).

The events mentioned above affect the performance of centrosomes in cultured human cells. Various interesting types of centrosome placement have also been seen in tissues. Cytotoxic T-lymphocytes (CTLs) that kill virally infected and tumor-producing cells are a striking case in which the centrosome travels to the immunological synapse and sends lytic granules to the place where the CTL makes contact to the infected or transformed cell (Stinchcombe et al. 2006). It is of particular interest that the centrosome comes into direct physical proximity with the plasma membrane during the killing of the target cells.

Another situation in which the positioning and movement of the centrosome have an important effect is the cerebral development in mice, in which the neurons travel from the inside to the outside to the surface and the nuclei exhibit prolonged motion. Importantly, links between centrosome and nuclear envelope are crucial for nuclear motion. The integral membrane proteins SUN1 and SUN2

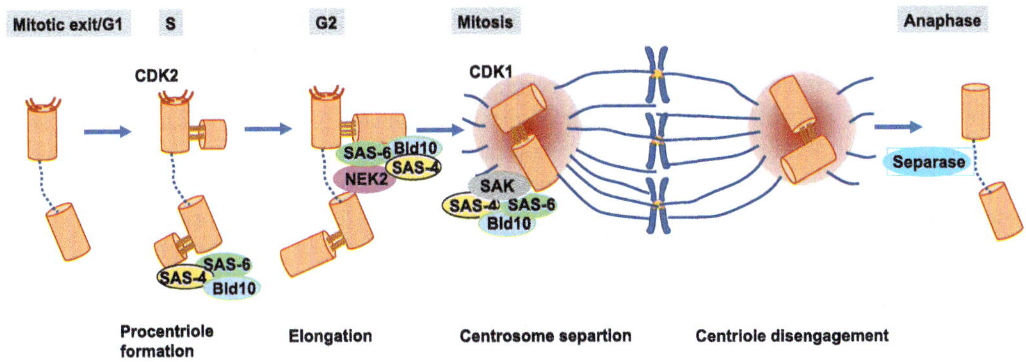

Fig. 15.3 The duplication of centrioles is linked to the chromosome replication and the segregation cycle. The formation of procentrioles starts in the S-phase due to the recruitment of SAS-6 and BLD10/CEP135, which are required for the generation of a cartwheel. The activity of CDK2 may be required for the increased the velocity of the procentriole formation and elongation, which is tightly linked to the DNA replication. In G2, the daughter centrioles are fully elongated and mature with the recruitment of several molecules to the pericentriolar material. SAS4/CPAP fulfills a task in the elongation of the centriole. CDK1 activity is elevated in G2 and governs a wide range of processes required for the entry toward mitosis, such as alterations in microtubule dynamics. By the coordinated action of several molecules including Nek2, the two centrosomes can be separated. When a cell leaves the mitosis phase, the centrioles within the centrosome disengage through the action of the separase. Hence this process may foster the recruitment or activation of molecules needed for duplication. SAK/PLK4 is essential and triggers procentriole generation. Due to alignment of sister chromatids, the spindle assembly checkpoint is satisfied. Cyclin B is degraded and separase is activated. The separation of the sister chromatids and the disengagement of centrioles is induced. The mitotic spindle segregates the chromosomes equally toward the daughter cells

stretch across the nuclear envelope from the nuclear region and maintain the SUN domain in the nuclear envelope lumen. The SUN domain interfaces with the KASH domain of Syne-1/nesprin-1 and Syne-2/nesprin-2 (Starr 2011; Wang et al. 2012; Burke 2012; Razafsky and Hodzic 2009; Fridkin et al. 2009). These proteins of the KASH domain stretch across the cytoplasmic surface of the nuclear envelope, and the cytoplasmic domains of syne-1/nesprin-1 and syne-2/nesprin-2 arrange interactions with the centrosome through dynein/dynactin and kinesin compounds (Zhang et al. 2009). Syne-2 mutant mice exhibits profound disturbances in learning and remembering, which underlines the relevance of centrosome–nuclear envelope connection in the course of neurogenesis.

The KASH-SUN domain pairing is also employed in the social amebae Dictyostelium discoideum to ensure that the centrosome is placed in relation to the nuclear envelope (Leo et al. 2012; Tikhonenkol et al. 2013). The Kin-I kinesin Kif9 bears at its C-terminus a transmembrane domain locating the C-terminal fragment of Kif9 in the lumen of the nuclear envelope where it can attach to the SUN domain of Sun1. The motor domain of Kif9 reacts with centrosome-organized microtubules, degrades microtubules by inducing their depolymerization and thereby attaches the centrosomes to the nuclear envelope.

15.6.5 Centrosome Disjunction by Breaking up the Linker and Adjusting Their Nek2 Function

During G2, before entering mitosis, the centrosomal linker is disintegrated by a process referred to as centrosomal disjunction to permit the centrosomes to disconnect and build the two poles of the bipolar spindle (Faragher and Fry 2003). The most extensively investigated kinase contributing to the resolution of the centrosomal linker represents the NIMA-related kinase Nek2 (Fig. 15.3). Nek2 exhibits elevated activity of kinase, which starts in the late S-phase and culminates in G2, thus overlapping with the time of centrosome splitting. Nek2A, a splice variant of Nek2, is cleaved through the APC/C proteasome system in the prometaphase with a scheduling similar to that of cyclin A (Hames et al. 2001). An overexpression of the active kinase results in premature centrosome division regardless of the cell cycle phase, whereas the RNAi-facilitated knockdown of Nek2 impedes the division of centrosomes but without influencing cell cycle progression in any significant way (Faragher and Fry 2003; Fletcher et al. 2004). The function of Nek2A at the centrosomes is contradicted by the presence of protein phosphatase 1 (PP1) (Fig. 15.3). PP1 specifically targets a KVHF motif in the non-catalytic C-terminal region of Nek2A and hence dephosphorylates it, which leads to the inactivation of Nek2A in vitro. In contrast, Nek2A is able to phosphorylate and block PP1 (Helps et al. 2000). The tertiary complex of C-Nap1, Nek2A and PP1g is crucial for regulating the degree of centrosome division through C-Nap1 phosphorylation, because C-Nap1 acts as a substrate both for Nek2A and PP1g (Helps et al. 2000).

Adjustment of centrosome disjunction takes place not only at the substrate stage, specifically C-Nap1 and rootletin, but also at the stage of Nek2A and PP1g. Nek2A and PP1g both regulate one another's activities, although they are also regulated to a higher degree through elements of the hippo signaling pathway (Fig. 15.3) and through the epidermal growth factor (EGF) signal transduction pathway (Mardin et al. 2011). As the key kinase causing the centrosome disjunction, Nek2A is governed to a higher degree through the compounds of the Hippo signaling pathway. The Hippo signaling pathway is a signaling cascade consisting of many tumor suppressors and is renowned for its ability to manage cell proliferation and apoptosis in flies and humans (Yu et al. 2013).

The Hippo pathway effector kinase sterile-20, including kinase Mst2 and the scaffold protein hSav1, controls Nek2A (Mardin et al. 2010, 2011). Nek2A is located in a complex together with Mst2, the scaffold protein hSav1 and PP1g. The phosphorylation of Nek2A through Mst2 directs it toward the centrosome, that in turn triggers centrosome disjunction due to phosphorylation of C-Nap1 and rootletin (Mardin et al. 2010). It is revealed that the Mst2 kinase acts as a target substrate of Plk1, that targets the opposite functions of PP1g and Nek2A. The Plk1 induced phosphorylation of Mst2 causes a decrease in the formation of the Mst2-PP1g-Nek2A compound (Mardin et al. 2011). Notably, there are many components of the Hippo pathway, which target the centrosomes, and it is widely acknowledged that they perform a key role in both centrosome disjunction and centrosome duplication (Hergovich et al. 2008).

15.6.6 Actomyosin-Driven Movement of Centrosomes

When the linker is broken up, the centrosomes travel each to two opposing poles to create the bipolar spindle. Unexpectedly, the blocking of the Nek2/Mst2 kinases does not interfere with the development of the bipolar spindles. This is accounted for by the dual action of Mst2/Nek2 kinases and the kinesin 5 motor protein Eg5 toward the linker break-down. Therefore, centrosome splitting can be regarded as a two-step process in which the centrosomes are collaboratively disconnected through the Mst2-Nek2A and Eg5 paths (Mardin et al. 2011; Smith et al. 2011) (Fig. 15.3). It is essential to keep in mind that Eg5 is capable of forcibly breaking the linker even in the lack of Mst2/Nek2 activity. In this event, linker proteins stay connected to the two centrioles in the course of mitosis (Mardin et al. 2011).

Eg5 acts as the main force producer which propels the separation of the centrosomes. Apart from *C. elegans*, the inhibition or exhaustion of the Eg5 motor results in a monopolar spindles in all of the organisms investigated (Ferenz et al. 2010). Eg5 is a member of the kinesin 5 subfamily of motor proteins that are homotetrameric, plus-end regulated motors (Sawin et al. 1992; Cole et al. 1994). The motor domains of this particular subfamily are oriented outwards (Kashina et al. 1996), which permits the antiparallel microtubules to glide in opposite directions (Kapitein et al. 2005). In interphase cells, Eg5 is scattered widely throughout the cytoplasm; in mitosis, however, it is deposited at the mitotic spindles and accumulates at the centrosomes (Ferenz et al. 2010). It has been shown that the accumulation of Eg5 at the centrosomes triggers the motor-driven centrosome splitting event. A cascade of kinases guides this shift in their translocation (Fig. 15.3). At the peak of this path stand the cyclin-dependent kinase Cdk1 and the polo-like kinase Plk1 (Bertran et al. 2011).

Cdk1 and Plk1 are both essential for the activation of the NIMA-related kinase 9 (Nek9). Cdk1 probably functions as a priming kinase for Plk1, which enables Plk1 to attach Nek9 and thereafter phosphorylate and thereby activate it (Bertran et al. 2011). Nek9 is regulated on another level through LC8. LC8 is a constituent of the dynein complex, which combines dynein with transport compounds. LC8 promotes the autoactivation of Nek9 and hinders the attachment of Nek9 to Nek6 (Regue et al. 2011). In the subsequent course of this activation, the Nek6/Nek7 kinases ultimately govern the division of centrosomes; however, the exact mechanism of Nek6/Nek7 regulation is not completely known. The phosphorylation of Eg5 through Cdk1 and Nek6/Nek7 is critical to govern its microtubule-binding affinity and subsequently its physical positioning at the spindle poles (Blangy et al. 1995; Bertran et al. 2011). Although Cdk1 activity is mandatory for the proper microtubule attachment of Eg5, Cdk1 is not absolutely required for centrosome splitting in DT40 chicken and HeLa human cell lines. In the absence of Cdk1 activity, however, the division of the centrosomes is slowed down by approximately two hours (Smith et al. 2011). The interplay between these paths, which govern the Eg5 concerning the centrosome disruption, still needs to be clarified.

There exists two lines of evidence indicating that Eg5 is not the sole motor protein that is involved in the motion of the centrosomes. Firstly, the blocking of Eg5 in the metaphase does not lead to the break-down of the spindle to monopolar structure. (Blangy et al. 1995; Waters et al. 1993). Secondly, the motions of the two centrosomes appear to be completely unrelated to one another and do not fully rely on the antiparallel sliding activity of Eg5, and therefore, an additional force producer is necessary (Tanenbaum and Medema 2010). As a consequence, dynein has been found to be a major actor in the segregation of centrosomes in human cells (Gaglio et al. 1996; Tanenbaum et al. 2008; Gonczy et al. 1999).

Dynein together with its interacting protein Lis1 and the microtubule-binding protein CLIP-170 inhibit the Eg5 action at the centrosomes. CLIP-170 which is a microtubule-plus-end binding protein connects the microtubules directly to subcellular components including dynein and Lis1 (Pereira et al. 2006). The depletion of dynein, Lis1 or CLIP-170 recovers the bipolar spindle development in the face of considerable levels of the Eg5 inhibitor (Tanenbaum et al. 2008). Originally, it was argued that Eg5 and dynein are purely antagonistic because dynein is targeted on antiparallel microtubules and leads to spindle break-down upon blocking of Eg5. But more recent results from titration experiments have questioned this theory by implying that dynein generates an antagonistic force in an indirect manner, not by generating force in direct contrast to Eg5, which contradicts the pull and push theory (Florian and Mayer 2012).

Moreover, in the G2-phase, dynein is deposited in the nuclear envelope, and this basin of dynein is necessary for centrosome splitting (Fig. 15.3). Dynein bound to the nuclear envelope becomes crucial for the splitting of centrosomes when Eg5 is not active, but it is still functioning even though Eg5 is not blocked (Raaijmakers et al. 2012). Dynein traps the microtubules from the centrosomes and moves the centrosomes to opposite sites alongside the nuclear envelope. It is still unclear how the microtubules that originated from the centrosomes can move the two centrosomes in opposite directions.

Apart from dynein and Eg5, the lately discovered kinesin Kif15, synonymously referred to as kinesin-12 supported centrosome splitting (Vanneste et al. 2011; Tanenbaum et al. 2009). The lack of

Kif15 cannot prevent the development of bipolar spindles; however, it is substantial once Eg5 is blocked (Tanenbaum et al. 2009; Boleti et al. 1996). Kif15 localizes to the K-microtubules after bipolar spindles are formed and helps Eg5 to stabilize the bipolar spindles (Tanenbaum et al. 2009; Sturgill and Ohi 2013). Moreover, in a cell line in which Eg5 is not dominantly involved in centrosome splitting, Kif15 is enrolled in the non-K-microtubules and takes over the function of Eg5, thereby triggering the centrosome splitting (Sturgill and Ohi 2013). Kif15 has lately demonstrated that it not only helps Eg5 to strengthen the bipolar spindles, but also generates a counterforce that avoids the development of excessively separated centrosomes (Sturgill and Ohi 2013). Commonly, these data propose different functions for Kif15 in centrosome separation: Firstly, it assists Eg5 to shape and sustain the bipolar spindle, and secondly, it generates an opposing force to restrict its overall length.

In summary, the equilibrium between the forces acting on the centrosome is certainly decisive for the setup of the bipolar spindle. When the centrosomes are disconnected, external factors assist the cell in defining the direction of the bipolar spindle (Thery et al. 2005). The extracellular matrix governs the dispersion of actin in the interphase cells, thereby ensuring that the actin-binding proteins cortactin and ezrin are both focused on the membrane edges. These alterations are also persisted in mitosis, thus the extracellular matrix controls the direction of bipolar spindle generation through alteration of the actin cytoskeleton (Thery et al. 2005).

15.6.7 How Are the Processes Timed?

The centrosomes need to be split in order to facilitate the complete assembly of the bipolar spindle. How is the timing of centrosome division in comparison with cell cycle progression and how is it governed? Latest research has shown that the time of centrosome splitting is not limited to a certain phase of the cell cycle, but rather differs according to cell type. In non-cancer RPE-1 cells, the majority of cells undergo mitosis with detached centrosomes (Magidson et al. 2011). The examination of HeLa cells indicated that half of the cells employ the classical prophase pathway, in which the centrosomes split prior to the degradation of the nuclear envelope (termed the prophase pathway). However, the other half enters mitosis in the absence of split centrosomes, and the centrosome splitting takes place after the break-down of the nuclear envelope (termed the prometaphase pathway) (Toso et al. 2009).

Another key factor determining the precise timing of centrosome splitting depends on the growth conditions. It was first noted that the stimulation of HeLa cells with the EGF leads to a fast and preliminary separation of the centrosomes (Sherline and Mascardo 1982). Moreover, it was later found that the addition of EGF breaks down the centrosomal linker by raising the local abundance of the upstream kinases Mst2 and Nek2 around the centrosomes, resulting in the premature splitting of centrosomes (Mardin et al. 2013). Importantly, this finding is cell type specific because many cancer cells exhibit a high degree of detached centrosomes due to the increased expression of the EGF receptor (EGFR). The regulation of centrosome disjunction thus represents a hitherto unidentified mechanism of EGF signaling.

Notwithstanding the increasing evidence on the different ways of centrosome splitting, until recently it was not directly investigated whether or not the temporal aspect of centrosome splitting is an eminent predictor of the mitotic advancement and chromosome splitting. Time-lapse videos of centrosome and kinetochore motion in untransformed RPE-1 cells in combination with computer modeling demonstrated that it is beneficial for the cells when the centrosomes are disconnected at the moment of nuclear envelope collapse (Hames et al. 2001). The metaphase spindle is formed more quickly when cells with disconnected centrosomes move into mitosis as the nuclear envelope disintegrates than in cells where the centrosomes disconnect after the nuclear envelope collapses.

Modeling of centrosome and chromosome motility proposed that the microtubule density within the spindle to be raised when the centrosomes are on two opposing ends of the nucleus at the moment of nuclear envelope collapse (Magidson et al. 2011), thereby trapping the kinetochores quicker than in cells with combined centrosomes at the moment of nuclear envelope collapse. In addition, experiments on the imaging of living cells in conjunction with high-resolution confocal imaging of Ptk1 cells indicated that incomplete centrosome splitting at the time of disintegration of the nuclear envelope enhances the probability of acquiring merotelic bonds (Fig. 15.4) (Silkworth et al. 2012). Regarding the fact that merotelic attachments are the main reason for the misseparation of chromosomes during anaphase, these findings underline the significance of the timing of centrosome splitting in terms of nuclear envelope degradation.

Why do the cells not succeed in mounting the correct fasteners? Computational simulations indicate a "blocking effect" of the two centrosomes relative to one another when they are in close vicinity. As the centrosomes continue to break apart, the microtubules originating from both poles have an increasing number of chromosomes at their disposal, and this enhances the probability of accurate synthetic attachments as a result of the symmetrical growth of the microtubules originating from both spindle poles (Kaseda et al. 2012). In line with these observations and models (Magidson et al. 2011; Toso et al. 2009; Mardin et al. 2013; Kaseda et al. 2012; So et al. 2013), EGF-based early centrosome splitting encourages more rapid mitotic progression with less chromosomal segregation mistakes (Mardin et al. 2013). Inducement of early centrosome splitting either through the addition of EGF to the medium or through expression of EGFR reduces the demand of Eg5 for the mitotic progression (Mardin et al. 2013). The association between EGFR signaling and Eg5 is striking, as both molecules represent mutual targets of cancer therapy, as it is outlined below.

The addition of EGF triggers this signal transduction pathway as it activates receptor dimerization and transphosphorylation of the C-terminal tail of EGFR, which turns into a framework for the enrollment of effector proteins when phosphorylated. The downstream effector pathways of EGFR signaling are extremely complex and play an essential role in various functional processes in the cell. The kinases Akt and GRK2 are specific EGFR effectors that are implicated in centrosome splitting. Although Akt is one of the most thoroughly investigated effector kinases in EGFR signaling, the GRK2's contribution to this signaling pathway has only just been elaborated (Mardin et al. 2013; So et al. 2013). GRK2 is a member of the family of G-protein coupled receptor (GPCR) kinases, which are engaged in a multitude of cellular processes through the triggering of multiple signaling pathways. This family of kinases controls the activity of GPCRs and promotes their signaling activity through phosphorylation of their intracellular domains and ultimately leads to restriction or desensitization of the receptor. Although most are membrane directed, several members of this family of kinases with

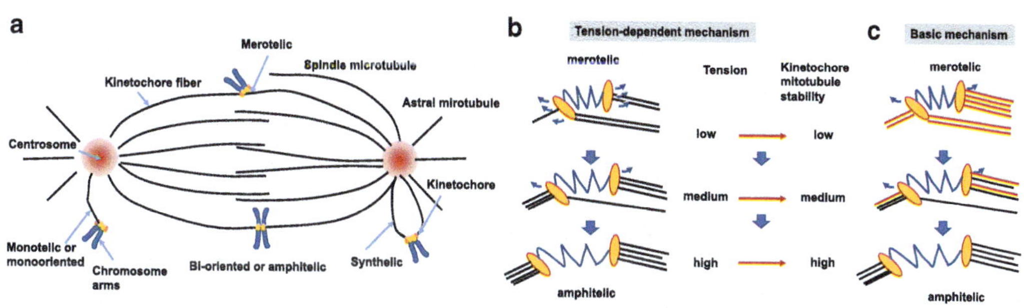

Fig. 15.4 Merotelic bonds and spindle formation (**a**). Effect of tension on merotelic bonds (**b**) and basic mechanism of merotelic bond rupture (**c**)

different types of localization and function have been identified. Latest evidence points to GRK2 being such a kinase that targets the centrosomes and conveys the signaling initiated through the addition of EGF. When EGF is administered, GRK2 phosphorylates and enables Mst2, which initiates the splitting of the two centrosomes.

Mitosis and the mitotic cytoskeleton, along with the centrosome, have been regarded as appealing targets for cancer for many years. In fact, inhibitors of many mitotic protein kinases, which are also found in the centrosomes, including Plk1, Cdk1, aurora kinase A and Chk2, are presently being investigated in clinical trials (Mazzorana et al. 2011). Forthcoming therapies also target kinesins, the motor proteins which are in charge of the accurate performance of cell division. Among the latter, Eg5 is the most extensively investigated kinesin, and its inhibitors, such as monastrol, have gone into clinical trials but have had little effect because of unexpected side effects (Bergnes et al. 2005).

Addressing mitotic kinases and kinesins is no simple task, bearing in mind that they share multiple binding partners and fulfill several functions. Drug development has generally targeted specific components of molecular networks that are modified and crucial for the survival of cancer cells. The deficiencies of this strategy are now becoming evident in the face of inadequate or short-term reactions to many single therapeutics. The acquisition of specific resistance mutations through cancer cells presumably accounts for these deficiencies. A natural answer would therefore be to aim at disturbed signaling pathways with multiple compounds, thus reducing the probability that a single clone possesses the repertoire of individual resistance mutations that are required to overcome the combined therapy. Promising results were presented lately in a phase I/II investigation of the proto-oncogene BRAF inhibitor dabrafenib in conjunction with the mitogen-activated protein kinase Mek1/2 inhibitor trametinib. The findings demonstrated that trametinib therapy was considerably more efficacious in patients who had previously been administered the BRAF inhibitor (Kim et al. 2013; Flaherty et al. 2012). In the same way, the specific treatment of cancers with high levels of early centrosome splitting using combinatorial therapy targeting Eg5 and EGFR pathways seems feasible. In contrast to normal cells, cells with an increased centrosome separation rate for cell division appear to be rendered less reliant on Eg5 and therefore appear to be more resistant to Eg5 inhibitors such as monastrol and S-trityl-L-cysteine (Mardin et al. 2013). Nevertheless, it seems to be most disadvantageous for these cells to address both the EGFR and Eg5 pathways at much smaller doses than when receiving treatment with a single drug. In conclusion, the combined control of Eg5 and EGFR in cells with increased centrosome levels could not only be shown to be efficacious, but may also prevent the potential side effects that occur when combating central mitotic kinases.

15.7 Centrioles

Centrioles comprise cylinders of nine microtubule triplets that occur in many eukaryotes. Very early investigations, in which centrosomes were identified at the terminals of the mitotic spindle, resulted in their imprinting as “the organ for cell division”. However, numerous successive descriptive and functional investigations have revealed that centrosomes may not generally be relevant to mitosis. At this point, the arguments in this debate will be addressed. The centriole structure and its spatial arrangement in the eukaryotic tree of life are described, and its contribution to the organization of the centrosome and the cilia is elucidated from a historical perspective. A key issue in the discussion covered in this the following is how centrioles are passed on and what kind of part the spindle takes in this process. The transmission of the germ line of centrosomes in specific, for instance their de novo generation in parthenogenetic species, raises a number of intriguing questions. In conclusion, the most probable features of centrioles are explained, and new research paths are outlined.

15.7.1 Historical View on Centrioles

In 1880, Boveri and Van Beneden discovered for the first time that the centriole is a remarkably mysterious organelle. Although there was a lot of progress recently in improving our knowledge of its structure and functions, it nevertheless continues to be a kind of "terra incognita" of the cell (Paoletti and Bornens 1997). Centrioles are cylinders of nine microtubule triplets arranged in a nine-fold symmetrical formation (Fig. 15.5), which occur in numerous eukaryotes and are indispensable for the development of multiple microtubule organizing structures, among them centrosomes and cilia (Fig. 15.5) (Bettencourt-Dias and Glover 2007; Bornens 2002). The centrosome acts as the primary microtubule organization center (MTOC) in numerous animal cells and is involved in a wide range of phenomena involving cell polarity, intracellular trafficking and cell division. In the center of every centrosome, there is a pair of centrioles that determine the characteristics of the centrosome.

Cilia and flagella are microtubule-based cell protrusions that can be movable and are engaged in a wide range of processes, ranging from the acquisition of extracellular cues to the movement of fluid and cell motility. Centrioles are referred to as basal bodies when they are firmly attached to the membrane in order to nucleate the axoneme, which is the microtubule core assembly of the cilia (Fig. 15.5). The properties of the basal body determine several characteristics of the cilia, such as their position and the symmetry. Initial investigations, in which centrosomes at the poles of the mitotic spindle were observed, resulted in their imprinting as "the organ for cell division", which is responsible for linking karyokinesis and cytokinesis (Boveri 1914, 2008). A number of subsequent observational and functional investigations have revealed; however, that centrosomes may not necessarily be crucial to spindle formation and to the hypothesis stating that the spindle of metazoan cells acts as a basal body disseminator that ensures the precise separation of both chromosomes and centrioles (basal bodies), whereby the primary function of centrioles represents the assembly of cilia (Friedlander and Wahrman 1970; Pickett-Heaps 1971; Debec et al. 2010).

In the following, the arguments raised on this issue are discussed. The first step is to characterize the centriole structure and its dispersion in the eukaryotic tree of life. What emerges is his role in the organization of the centrosome and the cilia. Subsequently, historical and current arguments employed in this discussion will be introduced, with special emphasis on the role of centrioles in mitosis and their modalities of inheritance and dissemination to daughter cells. Finally, the most probable functions of centrioles and the opportunities for research in this area are discussed.

15.7.2 The Structure and Distribution of Centrioles

Centrioles/basal bodies have a diameter of about 0.2 μm, and their length can vary depending on species and tissue type of around 0.5 μm (Fig. 15.5) (Bornens 2002; Paintrand et al. 1992). Their polarization is aligned along the proximal–distal axis (Bornens 2002, 2008; Paintrand et al. 1992). Electron microscopy has revealed that immature centrioles, and in certain species even mature centrioles, exhibit a characteristic at their proximal end referred to as a cartwheel, a structure consisting of a central node connected through spokes to the inner tubule of each triplet (Dippell 1968; Vorobjev and Nadezhdina 1987) (Fig. 15.5). Older centrioles feature subdistal appendages and distal appendage/transition fibers that maintain nine-fold symmetry of the centrioles and attach cytoplasmic microtubules or firmly connect centrioles to the cell membrane (Bornens 2002; Paintrand et al. 1992) (Fig. 15.5). Basal bodies can have extra appendages, these appendages range from small roots at their base to a basal foot shaped alongside precisely defined microtubule triplets (Bornens 2008; Beisson and Jerka-Dziadosz 1999). These patterns can themselves emanate microtubules and arrange the position and orientation of the basal body in relation to other cell constituents, which is essential for

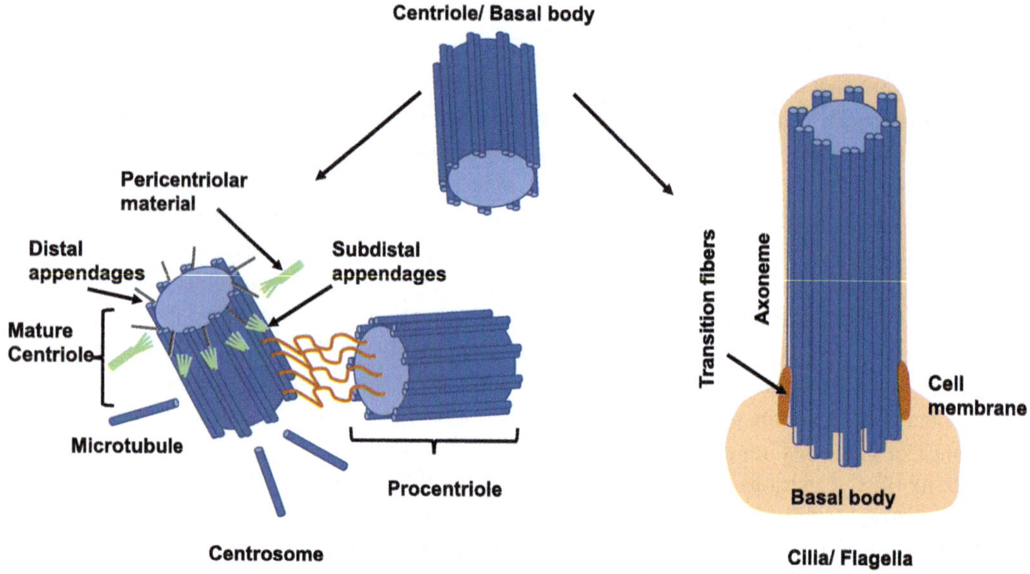

Fig. 15.5 Centriole/basal body and function. The centriole/basal body id element of the centrosome and of cilia and/or flagella. The canonical centriole has nine microtubule triplets. Each centrosome consists of a mother and daughter centriole in an orthogonal configuration, which surrounded by a matrix of proteins, referred to as pericentriolar material

regulating the movement of the cilia and arranging the intricate cytoskeleton in a large number of protists (Bornens 2008; Beisson and Jerka-Dziadosz 1999). Basal bodies possess a transfer zone at their distal end, which borders on the axonemic zone, that is essential for the nucleation of the ciliary microtubules, functions as a docking site for the protein transport into the cilia and serves as the organized disassembly/assembly scaffold of the cilia. Centrioles/basal bodies are exceptionally robust structures, and their microtubules are immune to cold, depolymerizing agents and detergents. This stability can be achieved by post-translational alterations of the centriolar tubulin, for instance polyglutamylation (Bobinnec et al. 1998a; Janke et al. 2008), and by other structural compounds of the centriol, such as tektins and ribbon proteins (Steffen and Linck 1988; Hinchcliffe and Linck 1998).

Centrioles and basal bodies (CBBs) are identified in all key eukaryotic groups, which indicates that they are possible present in the Last Eukaryotic Common Ancestor (LECA) (Cavalier-Smith 2002; Azimzadeh and Bornens 2004) (Fig. 15.6). In most of the organisms investigated, the centriole microtubules are arranged in a nine-fold symmetric and triplet arrangement, even though there exist some exceptions. For instance, *Drosophila melanogaster* has double or triple microtubules based on tissue type, whereas *C. elegans* also exhibit monotonous microtubules (Bettencourt-Dias and Glover 2007). Various forms and tissues within a single species can have basal body attachments such as rootlets and basal feet of varying shapes and sizes. The symmetry of the axonemal structure may differ from the prototypical nine-fold structure. This is often the case in invertebrates, including worms and insects (Mencarelli et al. 2008).

The presence of centrioles and basal bodies is associated with the appearance of a flagellated phase in the life cycle of the organism. Centrioles and basal bodies have been destroyed in the course of evolution of several eukaryotic phyla (Cavalier-Smith 2002) in species that have no flagella, such as amebas, including *Dictyostelium discoidum*, yeasts, including *S. pombe* and *S. cerevisiae*, and higher plants, encompassing *Arabidopsis thaliana*. However, there are specific species that possess no

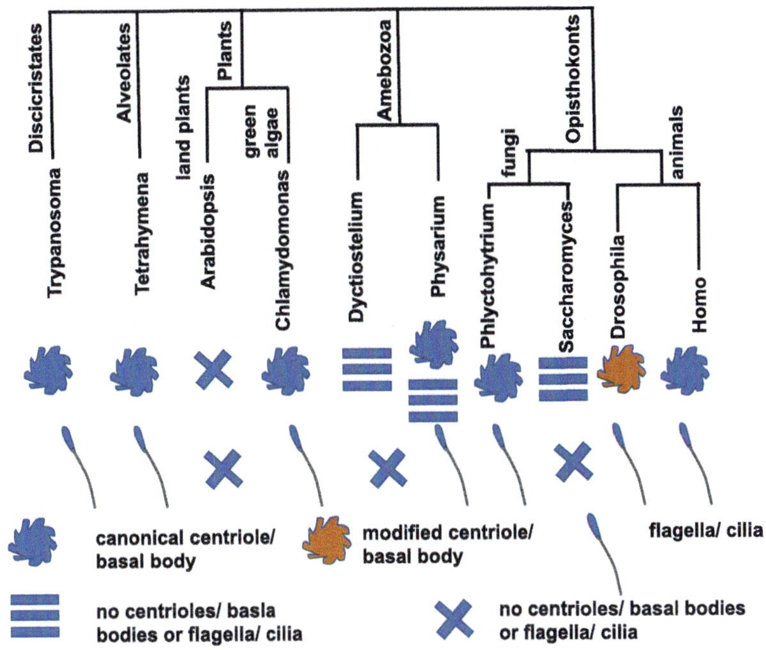

Fig. 15.6 Centrioles, basal bodies and flagella in the eukaryotic tree of life

flagella and therefore have altered centrosomes with no centrioles, which encompass the nuclear associated body (NAB) in ameba and the spindle pole body (SPB) in the yeasts (Bettencourt-Dias and Glover 2007). These MTOCs assist in arranging the spindle in mitosis, and even though their architecture is diverse, they still share several components with traditional centrosomes (Adams and Kilmartin 2000). Since there is a relationship between the occurrence of centrioles and basal bodies and the appearance of flagella/cilia, but not centrosome-like architectures, it can be concluded that centrioles and basal bodies are generally crucial for the axoneme development (Fig. 15.6).

15.7.3 The Role of Centrioles in Cell Division Is Controversial

The earlier investigation on the centrosome through the nineteenth-century has been presented by Wilson (1896). However, there are also recent works by Wheatley, Fulton and Gall on the initial characterization of the centriole/centrosome (Wheatley 1982; Fulton 1971; Gall 2004). Specifically, Boveri had been the first to use the term centrosome in his studies, although Van Beneden clearly identified the identical underlying framework. The name centriole was first proposed by Boveri to refer to a central granule that is permanently located within the centrosome. But because these microstructures lie at the resolution limit of a light microscope, other early cytologists assigned the terms centriole or centrosome (or occasionally centrosphere or central body) to analogous microstructures. Ultrastructural monitoring subsequently uncovered the centriole's character and enabled the scientists to differentiate sharply between the centriole (the central cylindrical structure with nine-fold symmetry) and the centrosome (the centriole(s) and pericentric material) (Fig. 15.5). Among the problems encountered in the analysis and determination of the centrosome was the lack of a membrane that separates these organelles. A decade after the discovery of centrioles, Henneguy (1897) and Lenhossek (1897) established independently that centrioles can lead to basal bodies. Later observations by electron microscopy demonstrated the identical structure of centrioles and basal bodies, and many instances of the interconversion of centrioles and basal bodies discussed earlier are presently known.

15.7.4 Fluctuations of Centrioles During Cell Division

In 1887, Boveri and Van Beneden discovered independently of one another a spherical texture situated at the mitotic spindle poles in the first divisions of Ascaris embryos. Both identified this architecture as an organ for the division of the cells. This finding was reinforced not merely because of the apparent strategic location of the centrosome near the spindle pole, but also because of a thorough reconstruction of mitotic spindle generation, in which the centrosomes travel toward antipodes on the nucleus ritually and appear to take part in the development of the spindle. This structural-organizational feature was also aided by the apparent dominant force of the centrosome in abnormal settings involving multipolar spindles in polyspermous sea urchin eggs. Surprisingly, even more so when considering that at that time no tools to track the cytoskeleton of microtubules were available, it had been deduced that the centrosome can be viewed as acting as a dynamic center of the cell. This hypothesis had later been confirmed for the majority of cell types. Moreover, it had been hypothesized by Boveri (and also by other early cytologists such as Hansemann and Galeotti) that abnormalities in the duplication of the centriole may represent the origin of the instability of the genome, which is a key feature of cancer cells (Boveri 1914, 2008; Brinkley and Goepfert 1998). This predictive hypothesis is now the focus of numerous current investigations, which indicate centrosome amplification as a marker for a large number of tumors (Fukasawa 2005; Godinho et al. 2009; Nigg 2006).

In the thrilling period of pioneering centrosome research, the centriole was first considered crucial to mitosis. However, in subsequent years, studies of its lack in many species of fungi and seed plants, as well as in numerous classes of protists operating at the mitotic spindle poles, severely questioned this viewpoint (Pickett-Heaps 1971). Many cell biologists of the time in fact at some stage in their lifetime came across a problem related to this organelle, but the lack of available instruments and the tiny size of the object often prevented a clear and publishable conclusion (Fulton 1982). Even though the common opinion had become established that centrioles were indispensable for division in animal cells, ultimately this last bastion was under attack.

It was possible to shift the centrosome in the primary spermatocytes of the crane fly (Pales ferruginea) out of the nucleus experimentally through flattening the cells (Dietz 1966). Under these conditions, during meiosis, the centrosome is absent at one (and even in some cases both) poles, but the spindle is perfectly constituted and enables the proper migration of chromosomes and cytokinesis. It has been determined that, at minimum in this type of system, centrioles for spindle formation are completely superfluous (Dietz 1966). Even more controversial is the suggestion that the spindle of the metazoan cells represents a distributor of basal bodies, which ensures the precise segregation of chromosomes and splitting of centrioles (basal bodies) (Friedlander and Wahrman 1970). This conclusion was underpinned by the observations of meiotic spindles in spermatocytes of the silkworm Bombyx mori, which the centrioles are doubtless teared apart from the spindle microtubules. At that time, the role of the centrioles during the division of cells has been discussed (Pickett-Heaps 1971). In specific detail, two different roles of the centriole have been presented, such as its autonomy and its action during the assembly of the spindle microtubule. Moreover, the role of the centriole in establishment of a mitotic spindle has been questioned, and it has been hypothesized that in the majority of cases, centrioles are not associated in the assembly of the spindle (Pickett-Heaps 1971).

From the mid-1970s onwards, the centrosome had to arouse new interest, mainly because it turned out that a pivotal function of the centrosome is the nucleation and organization of microtubule arrays. An essential conceptual progress introduced by Pickett-Heaps is the concept of MTOC, which represents a structure that can definitely be present with or without centrioles (Pickett-Heaps 1971). Moreover, this concept brings together the contradictory ideas about the role of centrioles. However, it should be pointed out that a centrosome is inevitably an MTOC, whereas the opposite is not the case (Azimzadeh and Bornens 2004).

It has been demonstrated that the microtubules are in fact composed mainly of this material and not of the centrioles themselves, using the significant presence of virus-like particles (VLPs) in the ovarian cells of the Chinese hamster as indicators of pericentriolar material. (Gould and Borisy 1977). It is crucial that pericentriolar material assemblies in the absence of centrioles and it is possible to nucleate microtubules in vitro. This research clarified the existence of both centriolar and acentriolar MTOCs and revealed that in certain biological models, MTOC can consist of a focal cluster of pericentriolar material. However, one question still exists: How does the presence of centrioles account for MTOC? A more precise understanding of the role of the centriole during mitosis was gradually gained in conjunction with an insight into the mechanism of the nucleation of microtubules. It used to be hard to deal with these questions, and only newer markings, a deeper familiarity with the players and advanced technology have made it possible to clarify them.

15.7.5 Mitotic Spindles Are Formed Without Centrioles in Animals

There exists a natural lack of centrioles in metazoan cells during cell division. The most evident instance of an animal spindle missing centrioles is that of certain female meiosis. Although this has been recognized for a considerable time, it was initially regarded as a very unique case tailored to the fertilization mechanism. When the natural removal of centrioles was sometimes recorded in somatic cells, it was limited to non-dividing cells such as myotubes (Tassin et al. 1985; Connolly et al. 1986; Srsen et al. 2009) and epidermal cells from the insect wing (Mogensen and Tucker 1987). The differentiation of certain epithelial cell types leads to the generation of non-centrosomal microtubule assemblies, whereas the centrioles persist in the cell (Bellett et al. 2009; Szöllösi et al. 1972). However, there are no indications of a natural and continuous absence of centrioles in a proliferating metazoan tissue. A highly interesting situation is the case of the mouse embryo in which the centrioles are transiently missing in the blastomeres during the initial divisions (Szöllösi et al. 1972). Centrioles are for the first time identified by electron microscopy at the 64-cell level (Gueth-Hallonet et al. 1993). Another remarkable instance appears in the fungus mosquito Sciara, where unfertilized eggs start the process of development through the generation of anastral spindles and effective separation of the sister chromatids (de Saint Phalle and Sullivan 1998).

15.7.6 Inactivation of the Centrosome in Metazoan Cells

Firstly, there exists an antibody-driven destabilization of centrioles. Glutamylation represents a post-translational modification of tubulin that is mainly limited to the centrioles in non-neuronal cells (Bobinnec et al. 1998a, b). In HeLa cells, the electroporation of an antibody directed against a polyglutamylated tubulin resulted in a temporary vanishing of centrioles and displacement of pericentriolar material (Bobinnec et al. 1998b). In the time without centrioles, monopolar and normal bipolar spindles emerged. The centrioles subsequently underwent complete reformation after approximately 48 h, and the cell population displayed normal centrosome signatures after 84 h. This sophisticated investigation has proven that: Firstly, the centrioles are in charge of the organization of the pericentriolar material; secondly, animal cells without centrioles can undergo mitosis and build a spindle; thirdly, the biogenesis of a procentriole does not demand the architectural intactness of the parental one.

Secondly, there is a centriole-depleted ectopic spindle under the exposure to microtubule poisons. Multipolar mitosis can be initiated in the ovarian cells of the Chinese hamster by liberation from a colchicine inhibitory block (Keryer et al. 1984). In tripolar mitosis, the centrioles are arranged either in a 2:1:1 or in a 2:2:0 distribution. In the 2:2:0 situation, the pole devoid of centrioles constitutes a cloud of pericentriolar material, which can be detected in these cells through the presence of virus-like particles. These cells perfectly carry out mitosis up to the telophase and the new formation of the cell nucleus. However, it appears that the daughter cells experience more resistance to separation than the tripolar cells with a 2:1:1 centriole pattern. This investigation indicated an unforeseen role of the centrioles in the accomplishment of the midbody splitting.

Thirdly, laser irradiation-based and microdissection of the centrioles can be employed to obtain cells without centrioles. Initially, the centrosome has been irradiated with a laser microbeam in order to reveal the functional role of the centrioles (Berns and Richardson 1977; Peterson and Berns 1978). UV and laser microbeam irradiation of the centrioles can now be made easier through in vivo labeling with fusion protein constructs, including centrin-GFP (Piel et al. 2001) or c tubulin-GFP (Khodjakov et al. 1997, 2000, 2002; Uetake et al. 2007; Magidson et al. 2007). These days, laser microsurgery is so accurate that only a centriole of the diplosome can be ablated in a targeted fashion (Loncarek et al. 2008). Microsurgical excision of the centrioles can be accomplished by splitting a cell into two distinct parts, a karyoplast comprising the nucleus and a cytoplast comprising the centrioles (Maniotis and Schliwa 1991; Hinchcliffe et al. 2001). In all instances, these acentriolar cells can build up a mitotic spindle in total lack of centrioles, thus validating the ability to dispense centrioles in this process. However, a high frequency of abnormalities is seen, including mono and multipolar spindles and also defects in the division of daughter cells, reinforcing the concept that centrioles are implicated in spindle bipolarity and in the accomplishment of cytokinesis.

15.7.7 Centriole-Free Animal Cell Lines

In *Drosophila*, a cell line, which is designated 1182-4, constitutively lacking centrioles (Debec et al. 1982), was derived from haploid embryos conceived by females homozygous for the lethal mh 1182 mutation (Gans et al. 1975; Debec 1978). This property has been strongly emphasized through two extensive ultrastructural investigations, involving high-voltage electron microscopy of serial thick sections (Szöllösi et al. 1986). In these cells, the mitotic spindles usually exhibit wide or dispersed poles and exhibit no or few astral microtubules. There is no enrollment of pericentriolar material proteins such as c-tubulin, centrosome, D-PLP or centriolar proteins, including DSas-4, into a centrosomal framework at the spindle terminals. Although the more polarized acentriolar spindles exhibit a focal accumulation of c-tubulin at the poles, this pattern disintegrates after removal of the microtubules (Debec et al. 1995; Moutinho-Pereira et al. 2009). These cells are able to multiply fairly easily in culture. Nevertheless, there is a strong abundance of daughter cells connected by a cytoplasmic bridge, indicating that the dissection procedure is harder (Piel et al. 2001). The underlying reason for the acentriolar character of cell line 1182-4 is still unknown, although its existence promotes the entire dispensability of centrioles for mitosis in *Drosophila*.

15.7.8 Acentriolar Mutants of *Drosophila*

The first report on the full development of a metazoan with non-functional centrosomes originates from the centrosome mutant in *Drosophila*. Centrosome represents a major ingredient of the peri-centriolar material and is acquired into the centrosome during the mitosis process. A functional impairment mutation in cnn inhibits the composition of numerous centrosome components including gamma-tubulin and interferes with the centrosome's capacity for nucleating microtubules (Megraw et al. 1999; Lucas and Raff 2007). A comparable task was recently assigned to CNN fission yeast and human, such as mto1 (Samejima et al. 2005) and CDK5RAP2 (Fong et al. 2001). However, unexpectedly, *Drosophila* individuals mutated for cnn (but are born of heterozygous mothers) typically evolve without morphological abnormalities, even though they are sterile (Megraw et al. 2001).

These findings were so astonishing that they raised some concerns about the non-functional nature of centrosomin-depleted centrosomes, because in this instance centrioles were still available. Thereafter, *Drosophila* mutants lacking DSas-4 or exhibiting a truncation of SAK/PLK4, which are two proteins crucial for the duplication of centrioles, have been investigated (Basto et al. 2006; Bettencourt-Dias et al. 2005). In both instances, the adult individuals have a complete absence of centrioles (nearly complete absence for DSas-4 and partially for SAK/PLK4); however, they pass satisfactorily through a development devoid of morphological aberrations. However, the adults are free of cilia and flagella structures and perish quite early after the emergence, as the lack of mechanosensory organs results in a severe deficiency of coordination. A careful study of the asymmetric partitioning in neuroblasts resulted in a high frequency of aberrant mitosis, which demonstrates that centrioles are beneficial in this mechanism (Basto et al. 2006; Megraw et al. 2001). In addition, several cells collapsed during the cytokinesis process (Basto et al. 2006). The adult males are sterile, exhibit defects in meiotic division and lack sperm axonemes (Basto et al. 2006; Rodrigues-Martins et al. 2008).

The case of the females is extremely fascinating: It was feasible to get eggs of genotypes DSas-4 and SAK/PLK4 from either germ clones in heterozygous females or from hand feeding of homozygous females. In either case, the oocytes seem to be normal; however, the respective embryos halt very early in the development after only few irregular mitoses (Rodrigues-Martins et al. 2008; Stevens et al. 2007), which demonstrates that centrioles are required for syncytic mitosis. New *Drosophila* cell lines which are homozygous for the mutation DSas-4 have been established lately (Lecland et al. 2013). Similar to the case of lineage 1182-4, these cells have no centrioles, and there is a lack of attraction of pericentriolar material proteins into a centrosomal structure at the spindle poles, which supports the total dispensability of centrioles for mitotic division in *Drosophila*. In summary, centrioles are not required in vivo for the construction of a mitotic/meiotic spindle, but these assemblies could be necessary for the accuracy and asymmetry of cell division in highly differentiated tissue types.

15.7.9 Besides Centrosomes, There Are Complementary Pathways in Spindle Formation

The discussion on the involvement of the centriole in mitosis is also lengthy because animal cells use several collaborative mechanisms to construct a spindle. In recent years, it has been found that at least two other paths are employed in spindle assembly: the chromatin and augmin paths. The nuclear envelope can also participate in the nucleation of microtubules (Rebollo et al. 2004).

Role of Centrioles in Cell Division

In the previous paragraphs, it has been revealed that centrosomes are not required for nucleation and organization of the mitotic spindle, since cooperative paths can arrange this structure. However, the experiments discussed earlier also demonstrated the significance of centrioles for specific aspects of cell division, such as cell cycle progression, proper chromosome segregation, the asymmetric cell division and cytokinesis, according to the type of tissue.

The Progression of the Cell Cycle

Numerous signaling proteins can be found on centrosomes. Investigations on vertebrate cells were performed to examine the possible controllability of the cell cycle through centrioles. Frequently, acentriolar cells went into mitosis, but then became locked into G1 in the following cell cycle. It was initially assumed that this retention could mirror a control of cell cycle progression through centrioles (Maniotis and Schliwa 1991; Hinchcliffe et al. 2001; Khodjakov and Rieder 2001), but this issue is still under strong discussion. It has previously been suggested that any such effect results from a cumulative stress due to both the withdrawal of the centrioles and the experimental stress, such as too much light, and may not be the consequence of a real centriole checkpoint (Uetake et al. 2007; Srsen and Merdes 2006; Wilson 2008). The simple fact that acentriolar cells can usually progress in the cell cycle indicates that there is no conserved control point relying on the existence of centrosomes (Bettencourt-Dias et al. 2005; Basto et al. 2006; Debec et al. 1995).

15.8 Segregation of Chromosomes and Cytokinesis

Experiments on cnn mutants indicated that centrosomes are necessary for proper spindle operation in male meiosis (Li and Kaufman 1996) and in the mitosis of the syncytial embryo (Megraw et al. 1999; Vaizel-Ohayon and Schejter 1999). The same holds valid for *Drosophila* mutants missing centrioles, such as spermatocytes from SAK/PLK4 and SAS4, and for early embryos free of SPD2, SAK/PLK4 and SAS4 (Bettencourt-Dias et al. 2005; Rodrigues-Martins et al. 2008; Stevens et al. 2007; Dix and Raff 2007). In the testes, the acentriolar meiotic spindles become mostly abnormal, which leads to a severe chromosome splitting and cytokinetic abnormalities. The development of bipolar spindles in the spermatocytes is caused through mechanisms specific to both chromosomes and centrosomes (Rebollo et al. 2004). The existence of a weak control point for spindle assembly could be the origin of premature chromosome segregation and abnormal meiotic products (Rebollo et al. 2004; Rebollo and Gonzalez 2000). In syncythic embryos, the extremely fast mitotic cycles can depend very strongly on centrosome-dependent mechanisms for bipolar spindle development and spindle localization. Abnormal acentriolar spindles interfere with one another and cause widespread abnormality that may result in mitotic standstill (Perez-Mongiovi et al. 2005). In fact, all mutated SAK/PLK4 embryos died in a mitosis-like phase, indicating an active checkpoint retention (Rodrigues-Martins et al. 2008). In conclusion, centrioles may be unnecessary for cell division in certain tissues of the fly, whereas in other tissues, they are indispensable, possibly because of restrictions in tissue specificity, such as less stringent control points or the presence of a syncytium. The same applies to other organisms, such as the *C. elegans* embryo (Pelletier et al. 2004) and fission yeast (Magidson et al. 2006), in which the centrosome and its equivalent, the spindle pole body, are fundamental for bipolar spindle development or the process of cytokinesis.

In human cultured cells, centrosomes are also important to ensure mitotic fidelity (Bettencourt-Dias and Glover 2007), particularly cytokinesis. Centrosomes may interfere with the cortex to initiate the formation of furrows. This was first and usually elegantly shown in a series of experiments where ectopic furrows were experimentally generated in syncytic sand dollar embryos (Rappaport 1971, 1986). These furrows developed between adjacent centrosomes and their astral microtubule arrangements, although this area lacked chromatin and a central spindle. These investigations result in the hypothesis that the signals from counteracting centrosome-based microtubules initiate the creation of cleavage furrows. Latest investigations have revealed that centrosome activities induce furrow generation in parallel to microtubule nucleation. The centrosome-associated recycling endosome is needed for vesicle-based membrane release and actin regulation during cleavage furrow ingression (Riggs et al. 2003). It is reported that an increasing number of centrosomal proteins are involved in cytokinesis, such as centriolin, Cep55, CP110 and BBS6, several of which are situated on both centrosomes and the midbody. In addition, a recent investigation identified the CNN-binding protein centrocortin, which is not necessary for normal centrosome-driven microtubule organization or the recycling of the endosomal organization, whereas it is needed for normal cleavage furrow generation (Kao and Megraw 2009). This kind of signaling function in cytokinesis has been proposed for HeLa and L929 cells (Piel et al. 2001), in which the tracking of centrioles through centrin-GFP revealed that the mother centriole temporarily moves near the intercellular junction and then back toward the center of the cell. This indicates that centrioles can carry a message that initiates the abscission. It is conceivable that in each type of cell, cell-specific features may be chosen for different inputs of the various spindle assembly mechanisms, resulting in the development of tissue-specific spindle assembly mechanisms (Rodrigues-Martins et al. 2008).

15.9 Mechanisms in Developmental Processes

In general, centrioles may be in charge of the accurate regulation of mitosis, either in space (the plane of division, for example, during asymmetric divisions in certain cell types), in time (driving the choice to split, determining the final number of divisions), and perhaps also in the regulation of cell fate (male *Drosophila* stem cells and rodent neural progenitor cells) (Yamashita et al. 2007; Wang et al. 2009) and may thus be heavily involved in the development and morphogenesis. In this respect, it is essential to point out that some redundant mechanisms may also be implicated, since in *Drosophila*, male germ line stem cells are more strongly reliant on centrosomes for asymmetric cell division (Yamashita et al. 2007), than neuroblasts and female germ line stem cells (Basto et al. 2006; Stevens et al. 2007; Wang et al. 2009; Siller and Doe 2009; Gonczy 2008).

More recent work indicates that the hereditary age variation of centrioles within a single cell could be relevant for signal transduction. Due to the centriole cycle, a cell in mitosis contains four centrioles corresponding to various generations: a grandmother and a mother, each of which is linked to a daughter. The cell which inherits the grandmother centriole after cytokinesis initially grows the primary cilia and is capable of reacting to signal stimuli (Anderson and Stearns 2009). These findings indicate that centriole age bears developmentally sensitive insights that are crucial in disrupting the symmetry between two daughter cells and maintaining the stem cells in contact with their niche. Finally, the ciliary role could be relevant in the placement of the mitotic spindles, similar to the development of the kidney (Jonassen et al. 2008).

15.9.1 How Are Centrioles Inherited and Passed on?

An essential element of the debate covered here is the question of how centrioles are transmitted and what kind of part the spindle takes in this process. Does the spindle constitute a required machinery to spread the centrioles evenly among the daughter cells? Is it necessary for a cell to acquire centrioles in order to receive them? The concept of centrosome continuum and the fact that the inheritance of centrioles is necessary for cells to possess centrioles has been compromised due to specific circumstances in which centrioles can occur in the absence of a preexisting centriole that is referred to as "de novo" generation. This is the situation in numerous organisms during the formation of flagella, such as the ameba-flagellate Naegleria (Dingle and Fulton 1966), or the fern Marsilea during spermatogenesis (Mizukami and Gall 1966), or during parthenogenesis in various insects. Due to the substantial variations, this following section has been divided into two major parts: the inheritance of centrioles in somatic cells and in germ cells.

Inheritance of Centrioles in Somatic Cells

In accordance with the first observations of the centrosome as the dividing element of the cell, it has been identified as a persistent cell organelle. Later, it turned out that this applies to numerous animal cell types. Among the best cases of this phenomenon were the findings in the syncythic divisions of the early of the early *Drosophila* embryo (Huettner 1933). In this instance, the doubling of the centrioles and their continuity during consecutive nuclear divisions was revealed with very good resolution. In addition, it was found that the centrioles are the underlying causes of astral development, which is an important point, as concerns were expressed that the centrioles could be artifacts of the fixation procedure (Wilson and Huettner 1931).

15.9.2 Centrosomes and Centriole Cycle

In the 1970s and 1980s, microscopists examined the centriolar cycle on the ultrastructural scale (Kuriyama and Borisy 1981; Robbins et al. 1968; Alvey 1985; Rattner and Phillips 1973; Rieder and Borisy 1982; Stubblefield 1968). They corroborated the earlier optical findings, which means that there is a strong correlation between the replication of chromosomes and the duplication of centrioles during S-phase of the cell cycle. A nascent centriole (the procentriole) is generated at the proximal side of each parent centriole throughout the G1-/S-phase of the cell cycle. The development of a procentriole takes place after the collapse of the orthogonal arrangement of the parent centrioles. The daughter centrioles are growing in a progressive manner and have reached the greatest amount of their terminal length through mitosis. Every single mother and daughter centriole pair can build up a new centrosome. In the G2-phase, the centrioles attract numerous proteins in their preparation for mitosis, among them the microtubule regulators; this process is referred to as centrosome maturation. The centrosomes split physically at the entry of the mitosis and travel to alternate poles of the spindle to which they stay connected. In the process of cytokinesis, every centrosome is formally mechanically separated to each daughter cell. The centriolar cycle is mirrored also in the arrangement of the pericentriolar material surrounding the centrioles. Tracking of the dynamics of the pericentriolar material in PtK2 cells has demonstrated that the pericentriolar material is mainly connected to the mother centriole (Rieder and Borisy 1982). The daughter centriole only obtains pericentriolar material in the subsequent cell cycle, when the daughter centriole itself turns into a parent centriole. The parent–daughter centriole relationship was initially defined by the arrangement of the two centrioles; they also detected the older centriole through its connection to the primary cilium. When reflecting on the inheritance of centrioles in somatic cells, it is evident that processes such as centriole biogenesis govern the amount of centrioles and monitoring the centriole segregation plays a very crucial part in establishing how many centrioles are passed on to daughter cells.

15.9.3 The Generation of Daughter Centrioles

Among the first features monitored are an electron dense material that occurs in the vicinity of the mother centriole and a nine-spoke car wheel that develops on the electron dense material.

Subsequently, microtubules are incorporated into the wheel, and the quantity and sequence of their addition are controlled according to the type of organism (Dippell 1968; Dutcher 2003; Cavalier-Smith 1974). The elongation of the centrioles relies on the expansion of these microtubules. Whether there is a cap that adjusts the centriolar length remains to be clarified. Whereas the molecular nature and function of the electron dense disc is poorly defined, recent work threw some new light on the function and configuration of the wheel as well as on the regulators of microtubule "anchoring" and extension at the centriole structure. SAS-6 and Bld10/CEP135 represent two components of the cartwheel, the first one being the hub and the spokes and the second one forming the tips of the spokes. These molecules originally emerged in *C. elegans* and Chlamydomonas, but their contribution to centriole generation has been confirmed in a wide range of organisms (Rodrigues-Martins et al. 2007; Vladar and Stearns 2007; Yabe et al. 2007; Strnad et al. 2007; Peel et al. 2007; Pelletier et al. 2006; Mottier-Pavie and Megraw 2009; Matsuura et al. 2004; Leidel et al. 2005; Kleylein-Sohn et al. 2007; Hiraki et al. 2007; Dammermann et al. 2004; Delattre et al. 2006) (Fig. 15.6). Investigations on *C. elegans* and in more recent times on human cells have revealed that SAS-4/CPAP is implicated in the attraction of microtubules to the centriole and in the expansion of this structure. However, its function may be contradicted by another centriole component known as CP110 (Kohlmaier et al. 2009; Schmidt et al. 2009; Tang et al. 2009). Other participants were engaged in the first steps of the formation of centrioles, including asterless (Blachon et al. 2008; Varmark et al. 2007), gamma-tubulin (Dammermann et al. 2004, 2008; Fuller et al. 1995; Ruiz et al. 1999; Shang et al. 2005), and for the elongation of the centrioles, such as POC5 (Azimzadeh et al. 2009). It is not yet fully established exactly how they function and interoperate.

15.9.4 Guided Control of the Amount of Centrioles

A minimum of two controls are intended to limit centriole duplication in a normally cycling animal cell. Only one new centriole is created perpendicularly to every already preexisting one in a conservative way. The biogenesis of centrioles is also synchronized with the replication of DNA, and thus, they are usually synthesized exactly once per cell cycle (Fig. 15.3): If one cycle is retarded, the other one halts, thus preventing failures. It is not clear how the number of procentrioles is limited to just one and whether there is a preferential localization in the vicinity of the mature centrioles for the new entity to be generated. There are two known circumstances in which this control is overridden. SAK/PLK4 belongs to a protein kinase of the polo-like family, which is required for centriole biogenesis. The overexpression of SAK/PLK4 in humans and *Drosophila* and SAS-6 in humans results in the creation of a so-called bloom-like pattern (Strnad et al. 2007; Kleylein-Sohn et al. 2007; Cunha-Ferreira et al. 2009; Habedanck et al. 2005; Rogers et al. 2009), with a single mother enclosed by multiple daughters. There is very strict monitoring of protein concentrations of SAS-6 and SAK/PLK4 in cycling cells through two important ubiquitin-ligase complexes that are implicated in protein break-down in the cell cycle. SAS-6 is decomposed through APC/Cdh1 (Strnad et al. 2007), whereas SAK/PLK4 is broken down through the SCF/Slimb complex (Cunha-Ferreira et al. 2009; Rogers et al. 2009). In the lack of this regulation, the biogenesis of numerous centrioles results in the availability of numerous centrosomes and difficulties in cell division.

Not much is currently understood about the timing of the cycles of centrioles and DNA replication, although it is obvious that the cell cycle mechanism obviously needs to take an essential part. In addition, it has been established that the duplication of centrosomes is extrinsically regulated, as the S-phase cytoplasm drives the duplication of a G1 centrosome in the process of cell fusion (Wong and Stearns 2003). This type of activity may be associated with CDK2, a kinase which participates in the facilitation of DNA replication, thereby enhancing the process of centriole assembly (Fig. 15.3).

A centrosome-immanent inhibition block for reduplication has also been documented, since the centrosome of a G2 cell does not double within an S-phase cytoplasm. This cannot be accounted for by an inhibitory activity of the cytoplasm, because when G2 cells were merged with G1 cells, the G1 centrosomes doubled. Once the centrioles have doubled in the S-phase, they can only double themselves further in the next S-phase (Wong and Stearns 2003). A possible reason for the intrinsic block of reduplication could be the regulation of centriole duplication through its detachment, as this is a prerequisite for the increase of daughter centrioles (Loncarek et al. 2008; Kuriyama et al. 2007; Sluder 2004). It has been revealed that centriole disruption relies on the APC/C-Cdc20 (Vidwans et al. 1999). It has been proposed that PLK1 and Separase, a protease activated through APC/C-Cdc20 at the metaphase-to-anaphase crossover to induce partitioning of the sister chromatids, also induce centriole detachment (Fig. 15.3) (Tsou et al. 2009; Tsou and Stearns 2006).

This enables the centrioles to be released for a new cycle of replication, perhaps by freeing up an existing block for replication, or exposing sites where new centrioles may be created. Mitosis is most probably a non-permissive condition for the generation and expansion of procentrioles, since centrosome duplication cannot take place when mitosis is lengthened in a wide range of cell types (Vidwans et al. 1999). However, it is not clear how often these entries are scanned from the centric assembly system in order to limit its activity to a certain time frame.

In certain cells of the human body, the control of the arithmetic of centriole biogenesis is somewhat distinct, one of the most symbolic events that take place during gametogenesis, as it is discussed below. Another instance can be found in epithelial multiciliated mammalian cells, where 200–300 basal bodies are generated in every cell subsequent to differentiation (Vladar and Stearns 2007; Dirksen 1971, 1991; Sorokin 1968). This originates in the evolving respiratory and reproductive pathways of vertebrates and in the ventricular network of the brain. Several centrioles are generated surrounding a mother centriole, thus disrupting the system's control over the generation of only one centriole per mother. Centrioles may also develop around less distinctive, none microtubule-driven dense patterns of heterogeneous size referred to as deuterosomes (Dirksen 1991).

15.9.5 Acts the Mitotic Spindle a Driver for the Segregation of Centrioles?

At the heart of the issue discussed in this chapter is the division of the centrosomes, their connection to the spindle poles and their transmission. In distinct species, the centrioles do not attach to the spindle poles, but stay at the membrane, where they nucleate axonemas. Ciliates display this behavior. Centrioles, when found in the majority of species with small amounts of flagella or cilia, tend to become attached to the spindle terminals during cell division.

The centrosome separation at the entrance of mitosis has an important function in its natural segregation. Several different molecules contribute to the division of centrosomes. Firstly, the two centrosomes are linked physically through a network of fibers consisting of rootletin, among other elements, which are connected to the proximal terminals of the centrioles through a molecule referred to as c-NAP1. Upon entering mitosis, the phosphorylation of C-NAP1 through a mitotic kinase titled NEK2 liberates it out of centrioles, resulting in the separation of centrosomes (Bettencourt-Dias and Glover 2007). The activity of NEK2 is opposed through a phosphatase, PP1. Another substrate of NEK2 targets beta-catenin, which is also found at the proximal end of centrioles, and the expression of a stabilized variant of this protein causes the premature separation of centrosomes (Bahmanyar et al. 2008). The pushing forces of the microtubules are also very essential in centrosome separation and are propelled through the Eg5, a plus-end directed kinesin, that attaches antiparallel microtubules and displaces them in the opposite directions. Dynein, a minus-end directed motor, also acts on centrosome separation. In mammals, dynein, Lis1 and CLIP-170 inhibit Eg5 activity (Tanenbaum et al. 2008). Cortical and kinetochore forces also influence the setup of a bipolar spindle. For instance, cortical microtubules and their interaction with actin/myosin or focal adhesions within the cortex might perform an integral function in centrosome separation (Xu and Saunders 2008). Ultimately, after the collapse of the nuclear envelope, kinetochore pushing forces could also be relevant in the construction of a bipolar spindle (Toso et al. 2009). During centrosome splitting, the centrosomes stay attached to the nuclear envelope until it dissolves. The connection between the nucleus and the centrosome is ensured through dynein and proteins located on the outer nuclear membrane, including the proteins SUN and KASH.

Centrosomes usually stay connected to the poles of the spindle. However, this is a highly dynamic process that relies on the targeting of the spindle poles through motors including dynein, HSET, microtubule-associated molecules, such as ASP and NUMA, and the ability of microtubule nucleation of centrioles, which relies on molecules such as CNN/CDK5RAP2 (Bettencourt-Dias and Glover 2007). The lack of dynein, HSET, ASP or NUMA results in strongly out-of-focus spindle poles and detaching centrosomes (Godinho et al. 2009). Ultimately, the deficiency of CNN/CDK5RAP2 results in a dissociation of the centrioles from the pericentriolar material with consequential failures in centrioles segregation (Lucas and Raff 2007). However, on leaving the mitosis, the two centrioles detach, and in specific cell types, one of the centrioles moves to the separation zone in the course of cytokinesis (Piel et al. 2001). It is still conceivable that there is a physical connection between mother and daughter centrioles (Piel et al. 2000, 2001), which can contribute to circumventing the complete disappearance of centrioles in the course of cell division.

15.9.6 Inheritance of Centrosomes in Germline Cells

In contrast to the symmetrical nature of centrosome inheritance observed in somatic divisions, centrosome inheritance is highly asymmetrical through the male and female germ line and opposes the claim that a spindle is required for proper centriole distribution. In the same manner as chromosomes pass through reductional divisions during gametogenesis to sustain diploidy, centrosomes need to be subject to reductional divisions. In opposite to the chromosomes, where the reduction of the number of chromosomes is the same for both sexes, the centrosome decrease is carried out in a different way for men and women.

15.9.6.1 Decreased Centrosomes in the Germline

During spermiogenesis in multiple sexually reproductive species, the centrosome decrease is mainly due to the removal of pericentriolar material, whereas the centrosome decrease during oogenesis mainly comprises a centriolar removal with retention of pericentriolar material, although distributed in the entire cytoplasm of the oocyte (Manandhar et al. 2005). This complementary deficit permits the reconstruction of an intact, completely functional centrosome after fertilization, with the sperm feeding the centrioles and the pericentriolar material coming from the cytoplasm of the egg cell. This heredity principle, where the fertilizing sperm delivers a centriole into an egg cell that is missing a centriole, depicts Boveri's fertilization hypothesis of 1901. In the overwhelming majority of the animal specimens examined, the egg cell eventually discards centriols as it matures. This circumstance helped to comfort many early cell biologists with the idea that it was essential to carry a centriole in order to produce the subsequent ones. For instance, in *Drosophila*, it has been seen that during the maturity division, there is no presence of centrioles or asters. Instead, the centrioles are detected in all mitoses starting from the first cleavage and are most likely deduced from the division center or centers that were originally associated with the spermatozoon causing the fusion of the nucleus (Sonnenblick 1950). In line with these results, it has also been stated that no centrioles or asters are seen during the maturation division (King 1970). Thus, the centrioles present in all the mitotic cells (from the first cleavage onwards) are probably derived from a centriole provided by the intruding sperm. This view is reinforced by the recognition that amphibian eggs necessitate the existence of an injected centrosome to initiate parthenogenesis (Maller et al. 1976).

There is nothing universal in this rule of the sperm supplying a centrosome. There are certain rare occasions, such as the example of the mouse embryo, in which the sperm centrioles break up after entry into the egg and the first zygotic divisions take place in the absence of centrioles (Szöllösi et al. 1972). De novo formation of centrioles has also been reported in parthenogenetically enabled rabbit oocytes (Szöllösi and Ozil 1991). In a similar way, the artificially stimulated sea urchin eggs exhibit cytasters with natural centrioles (Dirksen 1961; Miki-Noumura 1977; Kuriyama and Borisy 1983).

15.9.6.2 Decrease of Centrosomes in Spermiogenesis

The degree and the particular occurrences of centrosome decrease during spermatogenesis are quite distinct according to the species. Male germ line stem cells carry centrosomes. The degeneration of the centrosomes and the generation of the axoneme usually come after meiosis and the generation of haploid spermatids. The first stage of centrosome decrease is the complete disappearance of the nucleation of microtubules, subsequently the disappearance of the key proteins of the pericentriolar material and ultimately the incomplete or complete degeneration of the centrioles (Manandhar et al. 2005). Immunofluorescence investigations on *Drosophila* yield a clear picture of this degenerative profile (Riparbelli et al. 1997; Megraw and Kaufman 2000). In the first stage of spermatogenesis, the centrosomes remain perfectly intact and comprise centrin (Callaini et al. 1999), CNN (Li et al. 1998), c-tubulin (Riparbelli et al. 1997) and DSPD2 (Dix and Raff 2007). They are also linked to resilient astral microtubule arrays. In haploid post-meiotic spermatids, there exist no stable astral microtubule arrays (Li et al. 1998; Giansanti et al. 2008), and c-tubulin is thereafter depleted. Directly after fertilization, the male centrosomes return to carry gamma-tubulin, CNN, CP19 (Callaini et al. 1999) linked to the centriole, and their activity relies on pericentriolar material recruitment proteins, including DSPD2.

Investigations in mammals revealed not only the depletion of key pericentriolar material components but at the same time evidence of centriole degeneration. Probably, the strongest case of centriole degeneration arises in mice and other rodents where both members of the centrosome undergo degeneration (Schatten et al. 1986). In contrast, in humans and other mammals, it is only one single member of the centriole pair that fully carries out degeneration. The distinction between mouse and humans in respect of centriole degeneration is explained in a more recent investigation of the mammalian centriolar protein TSKS, which constitutes a substrate of the centriolar linked kinase TSSK1 (Xu et al. 2008). In mice as well as in humans, TSKS has a high density at the centrioles both during spermiogenesis and in post-meiotic spermatids. Whereas this regionalization continues in fully mature human sperm, TSKS is absent in mature mouse sperm. These profiles probably mirror the complete degeneration of centrioles in mice, although not in humans.

Therefore, during spermatogenesis, the centrosomes are not only subject to a complete lack of pericentriolar material, but also frequently break down the centrioles. Thus, after fertilization, the cytoplasm of the egg cell provides the components to reconstitute the centrioles and pericentriolar material. The complete absence of centrioles during spermatogenesis in mice implies that centrioles need to be created de novo from constituents of the oocyte cytoplasm (Schatten 1994). As de novo centriole generation from maternally delivered components forms a basis for the development without fertilization (parthenogenesis), it is unexpected that a full parthenogenesis resulting in viable organisms has not been discovered in mice. This is probably due to the fact that, for successful zygotic development in mice, a single set of chromosomes must be transduced in the male germ line to achieve comprehensive chromosomal imprinting (Kono 2009).

15.9.6.3 Reduced Centrosomes in the Process of Oogenesis

Centrosome loss during oogenesis primarily involves the removal of centrioles. Centrioles are missing in mice during the process of meiosis I (Szöllösi et al. 1972). Meiosis can be split into two phases, such as meiosis I and II. In meiosis I, the interphases are followed by a growth (G) phase and a synthesis (S) phase, whereas in meiosis II, there is only the S-phase that follows the interphase. Moreover, meiosis I is also referred to as reductive division from a diploid state to a haploid state. In contrast, meiosis II is referred to as equational division, since the cells are haploid at the onset of division and also stay haploid afterward. In meiosis I, homologous chromosome pairs are separated, and in meiosis II, sister chromatids are teared apart. Moreover, the prophase in meiosis II has no subphases. Finally, meiosis I ends in two daughter cells, whereas meiosis II ends up in four. Although there are no centrioles, the spindle poles carry essential centrosomal proteins such as gamma-tubulin, pericentrin and NuMA (Calarco 2000; Meng et al. 2004; Tang et al. 2004). The disappearance of centrioles in oogenesis appears in rabbits, cows, sheep and humans and is therefore probably a conserved characteristic of mammalian oogenesis.

Pericentriolar material compounds are not destroyed during mouse oogenesis, however, they are redistributed in the entire egg cell. Before egg maturation, structures referred to as multivesicular aggregates (MVA) build and accumulate on the centrosome (Calarco 2000). These microstructures are supplemented with gamma-tubulin. But they do not hold nucleated microtubules. In the course of oocyte maturation, the MVA travels in the direction of the germinal vesicle and disintegrates into tinier aggregates, several of which nucleate microtubules producing MTOCs. When the nucleus passes into the metaphase of meiosis I, the poles of the spindle are arranged through these MTOCs (Schuh and Ellenberg 2007). Pericentrin, which forms another central centrosomal compound, is combined into small aggregates inside the cytoplasm. Later in meiosis, pericentrin is concentrated at the spindle poles (Can et al. 2003).

The centrosome deletion has also been well investigated during the oogenesis of *Drosophila*. In the early phases of *Drosophila* oogenesis, four cycles of mitotic divisions resulting in imperfect cytokinesis create a cyst of 16 interconnected cells. One of them will create the oocyte, whereas the others will generate 15 nurse cells that will deliver nutrients and essential developmental substances to the oocyte (Deng and Lin 2001). A conspicuous early event in oogenesis involves the movement of the centrioles out of the 15 nurse cells toward the cell that will become the egg cell (Mahowald and Strassheim 1970). Strangely enough, centrosome movement into the egg cell is entirely unaffected by the cytoskeleton of the microtubules and relies on a perfectly polarized fusome, which is a major organelle dispersed over the interconnected 16 cyst cells (Bolivar et al. 2001). These traveling centrosomes function not as MTOC in the oocyte, and they vanish later on in the oogenesis. In fact, the MTOC activity is widely scattered across the entire cytoplasm of the oocyte. A recently released report has examined in particular the destiny of the centrosome linked to the nucleus of the oocyte (Januschke et al. 2006). It been seen that a centriole containing centrosome stays connected to the oocyte nucleus late in oogenesis at stage 9. It is unclear whether this centriole is defective, but probably, it is no more existent due to the metaphase of meiosis I (Mahowald and Strassheim 1970). Therefore, centrioles and centrosome constituents are missing at the poles of the meiotic spindle, including gamma-tubulin, CNN, CP60, CP190 (Megraw and Kaufman 2000) and D-PLP (Rodrigues-Martins et al. 2007). It must be stressed that the final evidence of the lack of centrioles can be obtained solely through electron microscopy on serial sections. An analysis of this kind can be a challenging one in terms of the size of most animal oocytes, and thus, the actual status of the oocyte centriole stays uncertain for many species.

15.9.6.4 Centrosome Function Restorage Directly After Fertilization

In the majority of mammalian species, the sperm carries a proximal centriole and a degenerated distal centriole. During fertilization, the proximal centriole constitutes a large astral microtubule array that is necessary for pronuclear movement and the acquisition of maternally supplied pericentriolar material compounds (Schatten and Sun 2009). The recruitment of pericentriolar material compounds leads to a conversion of the sperm into the male pronucleus. During this period, there is also a doubling of the paternally acquired centrioles. These occurrences eventually lead to the generation of two functional centrosomes that move to the poles of the adjacent male and female pronuclei. The experimental confirmation that the paternally obtained centrioles quickly gain a pericentriolar material due to the acquisition of maternally derived compounds originates from investigations in *Drosophila*. Immunofluorescence analysis revealed the exclusion of gamma-tubulin, CP60, CP190 and probably most pericentriolar material compounds present in mature sperm (Li and Kaufman 1996; Li et al. 1998; Callaini et al. 1999). Immediately following fertilization, these compounds are identified in conjunction with the centrioles obtained from the sperm. The father-derived centrioles probably recombine these and other pericentriolar material compounds out of the maternal cytoplasm.

The mechanisms whereby pericentriolar material compounds are attracted to the paternally derived centriole after fertilization have been investigated in experiments conducted by SDP2, a representative of a core group dealing with centriole duplication factors (Dix and Raff 2007; Pelletier et al. 2004). The *Drosophila* investigations indicate that DSPD2 constitutes a compound of both centrioles and pericentriolar material and is a key player in the enrollment of compounds of pericentriolar material specifically for sperm-derived centrioles directly after fertilization. In SPD2 *Drosophila* mutants, fertilization takes place naturally, but the paternally deduced centriole cannot attract CNN and do not produce MTOC (Dix and Raff 2007).

This asymmetric biparental origin of the zygotic centrosome has considerable consequences for the results of experiments on somatic cell nuclear transfer (SCNT). Somatic cell nucleus transfer is based on the principle of animal cloning, in which a somatic nucleus is inserted into an enucleated oocyte. Typically, the centrosomes are transposed along with the somatic nucleus into the host's oocyte (Sun and Schatten 2007). In general, a mismatch between the cytoplasm of the host and the donor centrosomes might terminate the development of the embryo. For instance, the injected oocyte carries centrosomal donor material linked to the somatic-derived centrosomes and high amounts of centrosomal material that stays in the cytoplasm of the host oocyte. How the reconstruction of the donor centrosome in this environment takes place is unknown. Moreover, the constitution and doubling of centrosomes is cell cycle controlled. When the somatic and oocyte cycles are not identical, this can interfere with the remodeling and duplication of the centrosome. In fact, a recent investigation of somatic cell nuclear exchange in bovine oocytes revealed that 50% of embryos had errors in the number or arrangement of centrosomes (Dai et al. 2006). Many centrosomes were unable to attach themselves to the cell nucleus, and some embryos retained up to six centrosomes in the first mitosis. These centrosome deficiencies easily account for the aneuploidy seen in embryos with somatic cell nuclear transfer.

15.9.6.5 Role of Centrosome Generation and Inheritance in Parthenogenesis

The complementary shapes of centrosome decrease during gametogenesis in both sexes constitute one of several formidable obstacles to parthenogenetic development. It is therefore unexpected that parthenogenesis has developed many times in the entire phyla (Engelstadter 2008). Probably, the most dramatic case is that of hymenopterans, encompassing ants, wasps and bees, in which parthenogenesis appears universally and is closely linked to sex determination: Fertilized embryos evolve into diploid females and unfertilized embryos into haploid males (Heimpel and de Boer 2008).

A key question of parthenogenesis is therefore the question of whether centrosomes are generated and if yes, where do they come from. Although centrosomes are not necessary for the formation of a functional mitotic spindle, they nevertheless perform a crucial role in the first syncytial embryonic divisions after fertilization in insects. These divisions are fast, in synchrony and syncytical, such as the nuclei split apart without concomitant cytokinesis. Following a series of syncytial divisions inside the embryos, the nuclei move to the cortex, where they pass through four additional rounds of division and ultimately cellularize (Mazumdar and Mazumdar 2002). Cellularization turns the syncytial embryo into an organism comprising thousands of cells. Investigations involving artificially initiated parthenogenesis in the fungus mosquito Sciara indicate that a functional centrosome is crucial for these primary divisions after fertilization (de Saint Phalle and Sullivan 1998). In parthenogenetic Sciara embryos, functional bipolar spindles are created, whose development normally takes up to three or four division rounds. Since these nuclei are situated in a syncytial surrounding environment, adjacent spindles may collide, leading to aneuploid division products. Thorough examination of the spindles indicates that they suffer from astral microtubules. It seems that the astral microtubules are indispensable to keep the separation between adjacent nuclei intact once the dividing nuclei are forced together. After cellularization, the cleft groove creates a membrane block between splitting nuclei, and astral microtubules are not needed to sustain separation. An elegant genetic analysis of CNN and DSAS4 mutants that disturb certain centrosomes and centrioles validates and broadens this analysis. Embryos originating from females that are homozygous for each of these mutants generate anastral spindles (Megraw et al. 1999, 2001; Stevens et al. 2007), merged spindles and aneuploid nuclei, in a similar fashion to the Sciara's syncytial embryos.

15.9.6.6 De Novo Formation of Centrosomes in Parthenogenetic Hymenoptera

Even though many developmental features usually take place in the lack of a centrosome, adult flies without a centrosome are not able to live long. Moreover, centrosomes are required for normal syncytial embryonic development and spermiogenesis in insects. One would therefore assume that centrosomes are found in parthenogenetically deduced insects. The source of centrosomes occurring during the parthenogenesis of insects has been examined in a range of species, among them the jewel wasp Nasonia vitripennis. In Nasonia the centrosomes are created de novo in the later stages of oogenesis. Centrosome genesis is based on the development of membrane structures derived from the female oocyte, referred to as accessory nuclei (Ferree et al. 2006). Due to an unidentified process, the oocyte nucleus sprouts from hundreds of nuclear envelope vesicles (Bilinski and Kloc 2002). These vesicles carry nuclear lamin, but no nucleic acids. In the course of oogenesis, these vesicles create a uniform pattern along the cortex of the oocyte. c-Tubulin enriches itself to elevated concentrations in the accessory nuclei. This concentration is most noticeable in the honey bee, where c-tubulin is found in distinct internal areas. In addition, the components of the c-TuRC complex are situated physically outside the nuclear envelope. The degradation of the accessory nuclear envelope takes place synchronously with the activation and decomposition of the oocyte nucleus. Directly after the breakdown of the accessory nuclei, microtubule-based asters develop alongside the cortex. This degradation is expected to allow the c-TuRC complex to bind to gamma-tubulin, which induces the production of aster. Electron microscopic analysis indicated that these asters hold canonical centrioles (Riparbelli et al. 2006). It is therefore probable that these accessory nuclei are the place of de novo centriole generation. Besides the de novo generation of centrioles in parthenogenetic embryos, de novo centriole development is also monitored in different biological settings, for instance in amebic flagellates, lower plants or multiciliate cells of vertebrates.

15.9.7 Are Centrioles Active or Passive Players of the Cell Cycle?

The long-standing question about the function of centrioles in cell division is still not answered. Are the centrioles for mitosis superfluous? What has been discussed is that the centriole is a multi-purpose organelle and therefore cannot simply be considered essential or dispensable for this intricate process. Considering the eukaryotic tree of life, centrioles do not engage in mitotic spindle generation in a number of species, implying that this was not the original purpose, even though for many organisms, information is not available (Fig. 15.6). It is therefore conceivable that in the last common ancestor, which comprised of centrioles as needed for flagellation, centrioles were actually passengers during cell division. It is possible that centrioles have lately been adopted to participate in various mechanisms linked to mitosis, including the positioning of the spindle or a differentiated regulation of cytokinesis in various tissues. The experiments quoted in the preceding sections are starting to yield the right response to this question: In metazoans, centrioles are heavily engaged in the bipolar character of the spindle and, at least for certain cell types, in cytokinesis. A more subtle function is also assumed, such as the verification of the accuracy of chromosome segregation. More broadly, centrioles may be accountable for the accurate control of mitosis, both in space, for instance at the division level, including asymmetric divisions in specific cell types, or in time, such as the initiation of cell division, which is dependent on the final division number, and potentially also in the regulation of cell fate, such as male *Drosophila* stem cells (Yamashita et al. 2007) and may hence be heavily involved in development and morphogenesis. In this context, the latest findings on acentriolar mutants of *Drosophila* are quite unexpected and indicate that morphogenesis in *Drosophila* may rather depend on acentrosomal signal pathways, such as those found in epidermal wing cells (Mogensen and Tucker 1987). In vertebrates, which possess a much more highly advanced cytoskeleton, the centrioles are presumably crucial organelles that are needed not only for cell division but also for various other cell functions.

There has been no reported instance of a persistent acentriolar cell line in vertebrates, and pathological mitoses of cancer cells exhibit centrioles. It is a characteristic feature of cancer cells to maintain these organelles, although in most cases, they are accurately targeted in centrosome number and function, indicating that the presence of these organelles offers a certain degree of selective benefit (D'Assoro et al. 2002; Doxsey 2002; Duensing and Munger 2003; Nigg 2002; Sluder and Nordberg 2004). An intriguing illustration of the idea of cooptation of centrioles, as discussed previously, is the issue of sexual reproduction. In this case, there is an oocyte complementarity with the sperm, which are the central elements in fertilization. However, even in *Drosophila*, in which centrioles for somatic cell divisions are unnecessary, centrioles are required for early embryonic developmental progression. It may not have been hard to co-opt centrioles for this purpose, since flagellated sperm typically carries a basal body at its base. This factor might also have been essential to avoid parthenogenetic development in numerous species.

15.9.8 Can the Mitotic Spindle Function as a Distributor of Centrioles?

Obviously, this is not necessarily so in all circumstances. As centrioles may develop de novo, the conceptual requirement for continuation no longer exists. However, de novo centriole generation usually results in a deficiency of available information on the localization, quantity and timing of the neonatal architectures, and thus, "canonical" duplication may provide a greater degree of monitoring of centriole biogenesis in proliferating cells. In addition, the canonical duplication offers an auxiliary feature that the cells can inquire into during developmental choices, which is the centriole age (Yamashita et al. 2007; Anderson and Stearns 2009). In contrast to plentiful organelles such as mitochondria, which can divide readily between the two daughter cells, an exact division of the two centriolar duplexes of the mother cell is mandatory. Paradoxically, although the centrioles are quite definitely active participants in mitosis, their strategic location at the spindle poles ensures their proper dispersion; in no way does this imply that they are inert.

In summary, it can be said that in the future, it will be very relevant to clarify the role of centrioles in mitosis and to decipher how the biogenesis of centrioles is governed during development and in human diseases, encompassing cancer, ciliopathies and microcephaly. There will be a strong need to gain a better knowledge of other ways of assembling spindles and to set this understanding in a context of evolution to address how cell division and centrioles operate in concert. Finally, the functional variety of centrioles in cell division and fertilization must be examined. A century after the pioneers of cell biology, the same questions are still on the agenda, but in the last 10 years, the understanding of centrioles has increased dramatically.

15.10 Centrosome Defects Can Lead to Cancer

More than a century back, the German cytologist Theodor Boveri claimed that centrosome abnormalities might lead to cancer in humans. Centrosomal defects actually appear in a variety of solid and hematopoietic human cancers, and centrosomal anomalies have been found in certain tumor types at an initial stage of disease development, which is associated with progressive tumor grade and poor clinical outcome (Gonczy 2015; Chan 2011).Centrosome aberrations can be classified as either numerical or structural alterations (Chan 2011). Whereas structural modifications are probably due to variations in the amounts or activity of centrosomal proteins (Schnerch and Nigg 2016), numerical modifications mirror an augmentation of centrosome copy numbers and result from the accumulation of an overabundant number of centrioles. Whereas structural and numerical centrosome alterations are of different nature, they frequently coexist in different tumor types (Nigg and Holland 2018).

15.10.1 Do Extra Centrosomes Promote Tumorigenesis?

In order to investigate the effect of additional centrosomes in cancer, many investigations have utilized PLK4 overexpression to increase the quantity of centrosomes. A pioneering investigation in *Drosophila* demonstrated that centrosome enhancement does not encourage the generation of instantaneous tumors, but that neuroblast and epithelial cells with additional centrosomes can trigger the development of tumors when transplanted into the host flies (Basto et al. 2008; Sabino et al. 2015). However, the effect of centrosome amplification in mammals on the generation of tumors is highly complicated. In the mouse brain, the presence of supplementary centrosomes fails to encourage the development of tumors (Marthiens et al. 2013). In a similar way, centrosome augmentation in the skin epidermis led to spindle orientation disorders and aneuploidy, but these aberrations were not capable of inducing spontaneous tumor formation or promoting the development of carcinogen-triggered cutaneous tumors (Vitre et al. 2015). In contrast, centrosome enhancement increased the rate of development of tumors in a p53-deficient cutaneous epidermis (Sercin et al. 2016). In addition, global PLK4 overexpression also increased the incidence of lymphoma and sarcoma in p53-null mice and encouraged hyperproliferation in the skin and pancreas (Coelho et al. 2015). Collectively, these findings confirm a central function of p53 in inhibiting the ongoing proliferation of mammalian cells, in which the proliferation of centrosomes is enhanced (Holland et al. 2012).

Whereas in initial experiments, the formation of spontaneous tumors in animals with widely spread PLK4 overexpression was not detected (Vitre et al. 2015; Coelho et al. 2015; Kulukian et al. 2015), it was shown that a more subtle rise in PLK4 concentration encourages persistent centrosome boosting, which favors the development of instantaneous tumors (Levine et al. 2017). The most important thing is that these tumors showed remarkable numerical and structural chromosomal modifications that reflect the intricate karyotype alterations that are often seen in human tumors with supplementary centrosomes (Levine et al. 2017). In tumors that develop spontaneously in reaction to centrosome amplification, a certain degree of interference with the p53 signaling pathway can be anticipated. Similarly, spontaneous lymphomas arising in mice together the amplification of centrosomes display a downregulation of the p53 target genes (Levine et al. 2017). In the immediate future, it might be useful to investigate whether the knockout of LATS2 or PIDDosome compounds speeds up the progression of tumors fueled through centrosome amplification.

15.10.2 Origin of Centrosome Defects in Cancer Cells

Cancer cell types exhibit large discrepancies in the penetrance and the degree of the enhancement of centrosomes. The reversible depletion of centrosomes with the application of a PLK4 kinase inhibitor has demonstrated that cancer cell lines achieve an equilibrium of the centrosome numbers distribution, which is defined by the speed at which additional centrosomes are aggregated and the rate at which the cells hosting them are exposed to (Wong et al. 2015). One way that results in the acquirement of extra centrosomes involves the dysregulation of the centriole duplication cycle. Whereas genes coding for centrosome proteins is infrequently mutated in human cancers, the elevated or reduced expression of centrosome proteins is more frequently (Nigg and Raff 2009; Gonczy 2015; Chan 2011). Moreover, a disturbance of the cell cycle process may result in deficiencies in centriole biogenesis. The most obvious case is that of an extended retention in the G2-phase, resulting in Plk1 activation, centricle detachment and a premature centriole reduplication (Loncarek et al. 2010). Therefore, DNA damage may trigger centrosome amplification by prolonging the time cells remain in the G2-phase (Douthwright and Sluder 2014; Inanc et al. 2010).

A final way to create additional centrosomes is the defective cell division. In addition to duplication of the number of centrosomes, a failed division offers the advantage of doubling the genome as a buffer against harmful mutations or chromosome splitting failures. These characteristics enable tetraploid cells to select new karyotypes and eventually arrive at a scarce combination that delivers a benefit to growth (Ganem et al. 2007). In line with the pro-tumorigenic role of tetraploid cells, an increasing number of indications point to the fact that a major proportion of human tumors are caused as a result of a tetraploid intermediate (Zack et al. 2013). Although the erratic proliferation of tetraploid cells can accelerate tumor development (Fujiwara et al. 2005), additional centrosomes in tetraploid cells at first induce a p53-dependent cell cycle standstill (Ganem et al. 2014). Consequently, repeated failure of cytokinesis does not lead to the long-term maintenance of centrosome amplification in cell culture (Krzywicka-Racka and Sluder 2011). This indicates that additional genetic modifications such as the loss of LATS2, caspase-2 or p53 are necessary to circumvent this fitness drawback and to achieve a long-term rise in the number of centrosomes after the break-down of cytokinesis.

It has been demonstrated that the deregulation of oncogenes or tumor suppressor genes results in the production of excessive centrosomes. For instance, KLF14 acts as a transcriptional repressor of PLK4, and the knockout of KLF14 causes PLK4-based centrosome boosting and the formation of tumors in mice (Fan et al. 2015). PLK4 is additionally transcriptionally inhibited through the p53 tumor suppressor (Coelho et al. 2015; Li et al. 2005). However, the p53 knockout is not adequate to trigger centrosome amplification in human cell lines and mouse tissue (Lambrus et al. 2015, 2016; Levine et al. 2017; Marthiens et al. 2013; Vitre et al. 2015; Coelho et al. 2015). Instead of acting as a direct regulator of the centrosome number as originally recommended (Fukasawa et al. 1996), the loss of p53 probably provides a favorable setting for the continued proliferation of cells with centrosomal anomalies, since it enables cells to circumvent the centrosome number monitoring routes (Holland et al. 2012; Vitre et al. 2015; Sercin et al. 2016; Coelho et al. 2015).

15.10.3 What Are the Consequences of Centrosome Defects?

Regardless of how they are formed, additional centrosomes are able to nucleate microtubules that cause the development of multipolar mitotic spindles. Unless this is rectified, it leads to multipolar division, which produces comprehensive chromosomal dysregulation and incurable progeny (Ganem et al. 2009). The primary mechanism through which cancer cells inhibit multipolar division lies in the coalescence of centrosomes into two groups that constitute a pseudobipolar spindle (Quintyne et al. 2005). The performance of the clustering mechanism is probably an essential parameter in assessing the capacity of cells to withstand centrosome amplification (Leber et al. 2010; Kwon et al. 2008). Centrosome clustering raises the incidence of false merotelic binding of chromosomes to the mitotic spindle, resulting in small rates of chromosome segregation defects, which might be consistent with cell viability (Basto et al. 2008; Ganem et al. 2009; Silkworth et al. 2009). In this way, excess centrosomes can encourage the frequent influx and outflow of chromosomes during division, which offers an explanation for the close relationship between centrosome enhancement and aneuploidy in human cancer (Gonczy 2015; Chan 2011).

An additional origin of mitotic aberrations arises from the false timing of centrosome separation preceding cell division. Both the speedup and the slow-down of centrosome separation raise the frequency of chromosome misalignment at the mitotic spindle, resulting in chromosome segregation defects (Zhang et al. 2012; Silkworth et al. 2012; van Ree et al. 2016; Nam and van Deursen 2014). The question of whether structural or numerical changes in centrosomes can lead to deficiencies in the time course of centrosome separation will be examined.

Besides aneuploidy of entire chromosomes, mitotic defects caused by excess centrosomes also encourage the development of DNA double-strand breakages resulting in chromosomal rearrangements. Additional centrosomes raise the frequency of chromosomes located in the middle of the spindle during anaphase, and these chromosomes can be affected due to a narrowing in the cleft furrow during cytokinesis (Janssen et al. 2011). In addition, retarded chromosomes are frequently divided into micronuclei, which aggregate a high amount of DNA defects that encourage chromosomal rearrangements (Crasta et al. 2012; Zhang et al. 2015). Therefore, redundant centrosomes can assist the evolution of the karyotype by being a resource for both quantitative and structural chromosomal modifications.

Whereas the amplification of centrosomes is a cause of genetic instability, supplementary centrosomes might also be involved in tumorigenesis through complementary mechanisms. *Drosophila* neural stem cells (neuroblasts) or epithelial cells with additional centrosomes are able to trigger tumor development upon their transplantation into host flies (Basto et al. 2008; Sabino et al. 2015). Whereas aneuploidy has been seen in transplanted epithelial cells with additional centrosomes, excess centrosomes only led to a slight increase in aneuploidy in neuroblasts, implying that genomic instability is unlikely to be the reason for the unregulated proliferation of transplanted brain cells. Rather, neuroblasts with supplementary centrosomes display spindle alignment errors that lead to an enhancement of symmetrical versus asymmetrical cell division (Basto et al. 2008). Disturbed asymmetric divisions result in an increase in the neuroblast stem cell reservoir and the consequent proliferation of tissue (Castellanos et al. 2008). The investigation of whether errors in asymmetric cell division lead to tumor development in vertebrates is a fascinating field of future research.

In contrast to the disruption of cell division, the numerical and structural centrosome alterations can also modify the architectural features of the cytoskeleton of interphase microtubules (Schnerch and Nigg 2016; Godinho and Pellman 2014). Centrosome amplification encourages the development of invasive protrusions in non-transformed breast cells cultured in a three-dimensional microenvironment (Godinho et al. 2014). It is essential that this invasive response was not due to aneuploidy. In contrast, cells with additional centrosomes revealed an elevated microtubule nucleation, which enabled the small GTPase RAC1. This offers a potential reason for the link between the amplification of centrosomes and the advanced level of tumor. Additional research will be required to determine the effects of centrosome failures on cellular invasion and metastasis in vivo.

In extension to their role at the centrosome, centrioles also act as basal bodies, which are essential for the formation of primary cilia. In cultivated human cells, PLK4-based centriole amplification often led to the generation of over one primary cilium (Mahjoub and Stearns 2012). More unexpectedly, cells with supplementary cilia showed a decreased abundance of ciliary signaling molecules and a defective activation of the Sonic Hedgehog signaling pathway regulated by the cilia. In opposite, PLK4 overexpression in the epidermis and primary mouse keratinocytes causes centriole amplification and the development of fewer primary cilia (Coelho et al. 2015). Centriole amplification may thus disturb the ciliary signaling, either by diluting the ciliary signal compounds or by the disappearance of the cilia directly. As the dysregulation of the cilia-regulated signaling pathways is generally recognized to promote the development of cancer, an excess of centrioles might influence cell proliferation through interference with the normal ciliary signal pathways (Han et al. 2009; Wong et al. 2009).

15.10.4 Can Centrosomes Serve as Therapeutic Targets?

PLK4 has become a potential therapeutic target because of its key function in monitoring centrosome duplication and due to latest evidence that it enhances migration and invasion of cancer cells (Habedanck et al. 2005; Bettencourt-Dias et al. 2005; Kazazian et al. 2017). CFI-400945 is the first inhibitor of PLK4 reported and represses the development of human xenograft tumors in mice (Mason et al. 2014). Nevertheless, CFI-400945 blocks the action of several other kinases such as Aurora B, and therefore, it is not sure whether PLK4 is the sole functional target of CFI-400945. The latest discovery of the highly specific PLK4 inhibitor centrinone offers a useful instrument to examine the effect of inhibition of the biogenesis of centrosomes on tumor growth. Research on cultured cells has demonstrated that centrinone inhibits the proliferation of untransformed cells, but enables the ongoing proliferation of the majority of transformed cell lines (Wong et al. 2015). This indicates that inhibition of centrosome doubling in itself may not be an effective strategy for combating cancer. Nonetheless, it seems feasible to detect genetic modifications that are synthetically lethal and result in centrosome deletion, and PLK4 inhibitors might provide therapeutic utility in suppressing functionality of PLK4 that facilitates invasion and metastasis (Kazazian et al. 2017).

An alternative therapeutic strategy is to increase the difficulty of division due to aberrant centrosome numbers. As centrosome clustering is not necessary in cells with standard centrosome numbers, but is essential to assure bipolar spindle arrangement in cells with excessive centrosomes, one thought is to inhibit centrosome clustering and force cancer cells with excess centrosomes into deadly multiple polar divisions (Ganem et al. 2009; Quintyne et al. 2005; Silkworth et al. 2009). Nevertheless, the circumstance that the majority of cancer cell lines can undergo proliferation in vitro with no centrosomes leads to the conclusion that they do not need excess centrosomes in order to survive (Wong et al. 2015). Therefore, the prevention of centrosome clustering may decrease the initial tumor burden, but ultimately enable the growth of more resistant cell populations. An alternative to direct attack on the centrosome is the manipulation of proteins that govern the reaction to defects in centrosome duplication. USP28 represents an enzymatic compound of the mitotic monitoring pathway and can be basically blocked. Since there are no clear phenotypes in USP28 knockout mice (Schulein-Volk et al. 2014; Diefenbacher et al. 2014), USP28 inhibition could be applied therapeutically under conditions such as microcephaly, in which the mitotic monitoring pathway can be pathologically enabled.

15.11 Connection Between Cell Division and Cell Mechanics

The oriented cell division process performs a key function in the regulation of organogenesis. The mechanisms for controlling cleavage orientation at the level of the overall organs are only gradually being comprehended. By the combination of 3D time-lapse images, mouse genetics and mathematical modeling, it can be concluded that the overall orientation of cell division is the outcome of a mixture of two types of spindles with varying spindle dynamic characteristics throughout the developing airway epithelium. Fixed spindles comply with the classic long axis rule and determine their division orientation prior to the metaphase. In contrast, rotating spindles do not follow exactly the long axis rule and define their divisional orientation throughout the metaphase. Using both a cell-based mechanical model and stretching-lung-explant experiments, it has revealed that the mechanical force can act as a regulatory stimulus in keeping the stable ratio between fixed spindles and rotating spindles. The results demonstrate that mechanical forces, the geometry of the cells and the oriented cell division interact in a highly coordinated fashion to guarantee a regular morphogenesis of the airway tube.

The oriented cell division is crucial for the morphogenesis of tissues and organs, for example, stretching of the body axis, gastrulation and so on (Baena-Lopez et al. 2005; Gong et al. 2004; Saburi et al. 2008). During morphogenesis, the alignment of cell divisions must be continuously supervised and adjusted dynamically in reaction to alterations in the shape of the tissue and the mechanical tension inside the tissue. However, there are only limited experimental research concentrating on the dynamic regulatory mechanisms that govern the alignment of cell divisions at the entire tissue or the organ level (Legoff et al. 2013; Mao et al. 2013; Xiong et al. 2014).

150 years earlier, it was found that the shape of cells governs the direction of cell division (Hertwig 1884; Hofmeister 1863). Together with other traditional findings, these and other classical observations resulted in the rule nowadays referred to as the "long axis rule", which dictates that a mitotic cell has a tendency to split according to its longitudinal axis (Minc et al. 2011). But since it has been found that some cells split themselves not following the long axis rule, this rule may just not be the only decisive criterion for the direction of cell division (Gong et al. 2004). In the early lung development of the mouse (E10.5 to E11.5), the airway tubes expand faster than they widen. It is assumed that the orientation of cell division in relation to the longitudinal axis of the airway tube strongly affects the morphology of the airway tube (Tang et al. 2011): There is enrichment of cell divisions aligned parallel to the longitudinal axis of the airways, whereas the remaining cell divisions are incidentally aligned to the longitudinal axis of the airways. Notwithstanding the significance of the oriented cell division for the regulation of the morphology of the airway tube, however, there is a lack of knowledge on how the guided cell division is dynamically managed throughout the tube morphogenesis. Previous investigations on oriented cell division in the morphogenesis of the respiratory tract were conducted primarily using fixed tissues. This is a limiting constraint for investigations of morphogenesis, since fixation can only deliver static knowledge on the direction of cell

division. For instance, it is not feasible to employ fixation-based techniques to track a single cell over time to assess the effect of cell shape on the orientation of cell division. It is therefore obvious that the capacity to continuously track cellular processes in real time, including cell morphology and mitotic spindle positioning, may foster significant insights into the mechanisms that govern the alignment of cell divisions. In general, the spatial and temporal characteristics of a dividing cell of the airway epithelium might also shed light on how an airway epithelial cell defines the orientation of its division according to its location in the emerging airway epithelium at a certain stage of development. An understanding of these processes will greatly enhance the understanding of the function and regulation of directed cell division in tubular morphogenesis. Nevertheless, determining the spatial and temporal stage of the dividing cells in the growing respiratory epithelium is technically demanding. It is necessary to maintain the three-dimensional (3D) orientation of all dividing cells across the entire respiratory tract tube epithelium and to follow the spindle dynamical processes during the progression of the airway development.

With the help of mouse genetics, ex vivo live imaging, quantitative cell biology and quantitative modeling, it has been attempted to gain an insight into how the alignment of cell divisions is governed during the morphogenesis of the respiratory tract tubes (Tang et al. 2018). Thus, it was discovered that there are two kinds of mitotic spindle dynamic response that are able to specifically comply with the long axis rule for the development of the airway epithelium. Importantly, the two modes of mitotic spindle action are related to varying spreads of the cell division angles. The relationship between the two types of spindles stays constant over time. Through both a cellular mechanical model and lung-explant stretching investigations, it has been demonstrated that the mechanical forces function in the emerging airway epithelium to sustain the robust relationship between spindle types. The present investigation offers new insights into how epithelial cells arrange complex cellular alterations including cell morphology and guided cell division during morphogenesis of the respiratory tract.

15.11.1 Two Distinct Spindle Dynamic Responses of the Developing Airway Epithelium

To examine the control of oriented cell divisions in the developing respiratory epithelium, an ex vivo time-lapse investigation was conducted to examine the cell divisions in evolving airway tubes. H2B-GFP fusion proteins are often employed to observe the dynamics of chromosomal alterations throughout the cell mitosis process (Liang et al. 2015). Thus, the highly condensed chromosomes at metaphase can be easily differentiated from the less concentrated interphase and prophase chromosomes. During the metaphase, the chromosomes build a single plane at the equator of the cell. According to the orientation of the mitotic plane, the H2B-GFP-labeled chromosomes in the pictures display either a rod-shaped or a rosette-shaped configuration. With lungs dissected by Shh-Cre, pink26-tTA or teto-H2BGFP mice (Harfe et al. 2004, Tumbar et al. 2004; Wang et al. 2008a, b), the orientation of mitotic spindles in airway epithelial cells could be tracked. The entire left epithelial tube was recorded all three minutes for a period of four hours. All lungs have retained their original shape variations during the visualization process. The inner face, outer face and central axis of an airway tube are necessary to define the orientation of the mitotic spindles in three dimensions. To reconstruct the 3D geometry of the entire airway tube, a set of calculation tools (integrated in MATLAB) has been implemented to get accurate geometric knowledge of the mapped airway tube. The position and direction of the chromosomal layer of all mitotic spindles have been recorded at every recording time. In the next step, these spindles are matched to the mathematically reconstructed tube geometry, and the orientation of the spindles with regard to the longitudinal axis of the airway tube is defined in a plane tangential to the tube surface. Imaging techniques assist in the quantitative analysis of spatial and temporal alterations of mitotic spindles in evolving epithelial tubes. The surveillance of chromosome dynamics performed with the calculation tools indicates that all mitotic spindles were oriented in the epithelial plane; the epithelial plane is orthogonal to the apical-basal axis of the cell. Thus, the planar angles of the spindles can only be detected in relation to the longitudinal axis of the respiratory epithelial tubes (Tang et al. 2011). At minimum two different spindle characteristics were found: Certain spindles ceased to rotate shortly after the cells reached metaphase and kept stationary as long as the cell had completed division; the other spindles only ended to rotate within the epithelial plane close to anaphase.

15.11.2 The Dynamics of the Two Spindles Are in Line with Dividing Angles Distribution

The planar spindle angles of dividing cells were evaluated throughout the metaphase in five E11.5 mouse lungs. By monitoring the chromosomal alterations, it was discovered that a population of spindles was rapidly immobilized within six minutes after the cells had passed into metaphase. For the residual spindles, two kinds of dynamic trajectories were recorded: a number of spindles continued to rotate until they split; the other spindles temporarily halted rotating before they started to rotate a second time. An eminent theoretical question posed is whether these apparently different spindle behaviors can be accounted for by a one-behavior model where all the spindles obey the same probabilistic rule but have different results due to random noise, or whether a more complex model with several rules is necessary to interpret the varying spindle behaviors. The spread of points in time when all spindles cease turning can be forecasted using these two models. The predicted values were compared to experimental data, including live-imaging analysis (Tang et al. 2018). Thus, a substantial disparity between the outcomes predicted from the one-behavior model and the experimental results was achieved, indicating that it is improbable that the different spindle dynamic responses observed in the lung tubes are governed by a single rule. Nevertheless, the spindle trajectories forecasted by a two-behavior model match the experimental results quite well.

There are two kinds of spindle responses in the airway epithelium in the two-behavior model: one type of spindle ceases to rotate in the first six minutes of metaphase; the other type of spindle apparently shifts randomly between a rotational phase and a standby phase during the entire metaphase. Therefore, the spindles that cease to rotate within six minutes are referred to as fixed spindles, and the spindles that do not halt to rotate within six minutes are referred to as rotating spindles. The variations in spindle angle seen for each imaging point throughout the metaphase distinguished sharply between the fixed and rotating spindles. In fact, all analyzed E11.5 lungs displayed about 37% fixed spindles and about 63% rotating spindles. In addition, the lung of E10.5 comprises around 38% fixed spindles and around 62% rotating spindles, suggesting that some kind of regulatory mechanism sustains a stable ratio of these two types of dynamic behavior of the spindles during respiratory development. To examine whether the dynamic performance of spindles is associated with the orientation of cell division, the cell division angles of fixed spindles and rotating spindles were investigated. Most of the fixed spindles were, however, interestingly split at an angle of less than 30, while the pitch angles of the rotating spindles were evenly spread between 0 and 90°. It should be kept in mind that the average metaphase duration did not vary between cells with fixed spindles and cells with rotating spindles. These results point to a correlation between the dynamic behavior of the spindle and the cell division angles in the emerging airway epithelium.

15.11.3 Spatial and Temporal Relation Between Fixed and Rotating Spindles

The findings demonstrated that spindles, which show two different behaviors in the epithelial cells of the respiratory tract, exert varying influences on the morphogenesis of the tubes. To investigate possible regulatory mechanisms underpinning these two spindle behaviors, it will be necessary to examine whether these two spindle behaviors are spatially and/or temporally interrelated throughout the developing airway epithelium. Therefore, on the basis of image data, these models enable us to reconstruct the 3D airway tube geometry in a direct fashion; these reconstructions permit us to obtain the spatial and temporal characteristics of single spindles. To examine the spatial interrelationship between fixed spindles and rotating spindles, single spindles were depicted on a two-dimensional epithelial plane achieved through the unfolding of an airway tube into an epithelial layer. Since the two spindle types can be distinctly differentiated in the time-lapse images, the positions of the two spindles at the level of the respiratory epithelium can be tracked during the entire observation time. In this analysis, the Moran's I (a measure of spatial autocorrelation) is computed, which can be employed to analyze the correlation in space between locations of the two types of spindles (Gittleman and Kot 1990). In specific detail, a positive I says that the neighbor spindles seem to be of the same spindle type, whereas negative I points out that neighbor spindles are of the opposite type. In fact, the all Moran's I values of the investigated lungs were neither pronouncedly positive nor negative, pointing out that there is no relevant spatial relationship between the location of cells containing fixed spindles and the location of cells containing rotating spindles in the respiratory epithelium. To investigate the temporal association between the fixed spindles and the rotating spindles, all spindles were recorded at the time when each of them joined the metaphase, and the individual spindles were tracked during the entire metaphase for individual lung specimens. If the appearance of a fixed spindle and the appearance of a rotating spindle are mutually independent, the sequential appearance should theoretically behave as an asymmetric one-dimensional random walk (Van Kampen 2007). In this random walk model, it has been defined the weighted deviation X_i is:

$$X_i = 0.63 \times \text{number of fixed spindles} - 0.37 \times \text{number of rotating spindles}$$

X_i is used to measure the deviation of the ratio of the rotating spindles after the i-th pitch from the average ratio of the rotating spindles (measured ratio of the live image analysis was 63%). At every arbitrary step, when the ratio of the rotating spindles is 63%, $X_i = 0$, the greater the difference between the ratio of the rotating spindles and the average ratio, the more distant X_i is away from 0. Employing the acquired emergence sequence of fixed spindles and rotating spindles in every lung, after each cell division X_i was calculated. It is remarkable that X_i stayed close to 0 in all five lungs examined. The maximum variation seen in each of the five lungs was much lower than that of random sampling based on an asymmetric one-dimensional random walk simulation. The X_i distribution obtained from the imaging data points to the fact that the emergence of spindles is strictly regulated; in other words, it appears to be clear that the emergence of a fixed spindle and the presence of overtime of a rotating spindle are not mutually independent. This means that a certain regulatory mechanism is in charge of keeping a stable relationship between the two spindle types in time.

15.11.4 Requirement of a Robust Ratio of the Two Spindle Types in Airway Shapes

Based on the knowledge of the spatial and temporal relationships between fixed spindles and rotating spindles, it has been postulated that a regulatory mechanism could operate to sustain a stable ratio of fixed spindle cells to rotating spindle cells (Tang et al. 2018). To assess this hypothesis, it was necessary to implement an experimental in vivo technique to disturb the normal ratio of cells with rotating spindle. It has been found that the enhancement of ERK1/2 signaling through the expression of constitutively activated BRAF in respiratory epithelial cells results in a uniform spread of cell division angles (Tang et al. 2011). In view of the findings that the rotating spindle resulted in evenly distributed pitch angles, the hypothesis was raised that activated BRAF signaling would enhance the presence of rotating spindles in the airway epithelium. This was achieved through the analysis of the mitotic spindle dynamics in E11.5 Shh-Cre, $Braf^{CA}$, Rosa26-tTA and teto-H2B-GFP ($BRAF^{CA}$) lungs, where the airway epithelial cells of the respiratory tract constantly express activated BRAF (Dankort et al. 2007). The proliferation rate of epithelial cells in E11.5 appears not to vary between wild-type and $BRAF^{CA}$ mutant lung. It is remarkable that 94% of the spindles in the airway epithelial cells of the $BRAF^{CA}$ lung consist of rotating spindles, which is impossible to distinguish from the rotating spindles of the wild-type lungs. The expression of $BRAF^{CA}$ in certain epithelial cells thus permits to enhance the relationship of the rotating spindles in vivo. These findings indicate that this experimental setup, which is based on mosaic expression, can be applied to selectively alter the spindle angle dispersion in a subpopulation of cells. Thus, it is possible to generate Shh-CreER, Rosa26-ZsGreen1 and $Braf^{CA}$ (inducible $BRAF^{CA}$) mice (Dankort et al. 2007; Madisen et al. 2010). Upon the injection of tamoxifen, constitutively activated BRAF has been expressed in a mosaic fashion throughout airway epithelial cells of inducible $BRAF^{CA}$ mice. However, it should be pointed out that the cells that express activated BRAF were detected as they were also positive for GFP. Moreover, the levels of p-ERK1/2 were elevated in the mitotic GFP(+) cells, hinting that an activated BRAF expression took place in these cells. The airway morphologies of lungs have been determined with distinct proportions of $BRAF^{CA}$ epithelial cells, such as 20–60%. The airway morphologies of the inducible BRAFCA lungs were not significantly altered, although there are about 40% $BRAF^{CA}$ epithelial cells.

When the spindle angle distribution of all dividing cells in the lung was determined using 40% $BRAF^{CA}$-expressing cells, it was impossible to discriminate from the spindle angle allocation in wild-type lungs. An explanation for the normal spindle angle dispersion in the inducible $BRAF^{CA}$ lungs might be that a constitutively activated BRAF possibly does not operate cell autonomously. An alternative interpretation might be that a compensatory alteration of cell division angles takes place in cells without $BRAF^{CA}$ expression; any such alteration might re-establish a normal overall spread of division angles across the entire airway tube. Taking these two possibilities into account, immunostaining was performed to separately determine the cell division angles in the GFP(+) and GFP(−) cells of the lung mosaics. In the GFP(+) cells, the cell division angles were actually evenly spread from 0 to 90, exactly as in the $BRAF^{CA}$ lungs, suggesting that the effect of $BRAF^{CA}$ on spindle dynamics is cell-autonomous. It is noteworthy that in GFP(−) cells, the percentage of cells with division angles below 30 was greater than 65%, and well above the 50% in wild-type lungs, implying that the spread of spindle angle in wild-type cells is affected by the occurrence of $BRAF^{CA}$ cells in the mosaic tissue, seemingly in a compensatory manner. Therefore, the experimental findings achieved through the mosaic-type expression of $BRAF^{CA}$ confirm the idea that certain regulatory mechanisms operate to sustain a steady ratio of fixed spindle cells and rotating spindle cells throughout the development of the airway tube.

15.11.5 Fixed Spindles Employ the Cell Long Axis Rule

It implies that a specific regulatory mechanism is responsible for keeping a stable correlation between the two spindle types over time, which means the long axis rule is followed (Gibson and Gibson

2009; Gibson et al. 2011; Hertwig 1884; Minc et al. 2011; Nestor-Bergmann et al. 2014; Saburi et al. 2008; Thery and Bornens 2008; Xiong et al. 2014). Nevertheless, several investigations have also proven that specific cells do not follow this long axis rule during various developmental processes (Gong et al. 2004; Minc et al. 2011). Almost no information is available about the correlation between cell division orientation and the longitudinal cell axis during the respiratory tract tube morphogenesis. Using lung imaging of Shh-Cre, pink26-tTA, pink26-mTmG and teto-H2B-GFP mice where both a membrane-localized GFP reporter and an H2B-GFP fusion protein are specifically expressed in respiratory epithelial cells, alterations in both cell shape and chromosomes in dividing respiratory epithelial cells were detected. At first, it was examined whether the shapes of cells with fixed spindle and cells with rotating spindle before entry into the prophase vary. Importantly, the aspect ratio of all fixed spindle cells is larger than 1.53 and the aspect ratio of all rotating spindle cells lies between 1 and 1.53, indicating that the fixed spindle cells are longer than the rotating spindle cells. A similar relation has been seen between spindle rotation and a smaller cellular aspect ratio in epithelial cells expressing constitutively activated BRAF. To explore further the correlation between cell shape and cell division orientation in cells with fixed and rotating spindles, the longitudinal axis of the cell was examined prior to the prophase, and then, the cell division orientation was evaluated. Subsequently, a deviation angle was determined, which is the angle between the cell division alignment at the anaphase and the longitudinal cell axis during the interphase. However, it was discovered that the deviation angles of fixed spindles are all below 30, suggesting that the divisional orientation of cells with fixed spindles is strongly related to their longitudinal axis across phases. The deviation angles of rotating spindles, on the other side, were evenly spread between 0 and 90, which indicates that the pitch orientation in the cells of rotating spindles is not associated with their longitudinal axis between phases. These findings show that the long axis rule is valid for cells with a fixed spindle, whereas it is not valid for cells with a rotating spindle in the early development of the respiratory tract tube of the mouse.

15.11.6 A Cellular Mechanical Model May Explain Stable Ratio of Fixed and Rotating Spindles

It has been demonstrated that mechanical forces generated by proliferating cells govern the alignment of cell divisions and cell shape at the overall tissue level (Bosveld et al. 2016; Legoff et al. 2013; Mao et al. 2013). To investigate whether the ratio of fixed spindles to rotating spindles can be driven by mechanical forces in the airway epithelium, the next step was to consider a cellular mechanical model, also referred to as a force-based cell calculation model, where both cell shape and cell location are defined by mechanical forces (Harris et al. 2012; Wyatt et al. 2015). Within this model, the airway epithelial tube is expanded into a two-dimensional epithelial layer. Inside the epithelial sheet, the cells merely interoperate with one another through mechanical forces. In this model, mechanical forces govern the cell morphology, which is expressed as the cell aspect ratio. The cells within the epithelium were chosen randomly to split, and the cell aspect ratio governed the dynamic performance of the spindle. When the aspect ratio of the cell is above 1.53, the resulting cell turns into a fixed spindle cell whose cell division tracks the longitudinal axis of the cell. When the aspect ratio of a cell is below 1.53, that cell turns into a rotating spindle cell with randomly oriented cell division. First of all, it was checked whether the system needed to approach a state of equilibrium in which the ratio between cells whose aspect ratio is above 1.53 and cells whose aspect ratio is below 1.53 remains approximately stays constant over time. Using two distinct initial conditions: In one case, most cells have an aspect ratio below 1.53, and in another case, most cells have an aspect ratio above 1.53. In the course of the development of the system, the proportion of cells with an aspect ratio of above 1.53 actually approaches a fixed value, such as 37%, which is in line with the proportion of fixed spindles detected in the experiment. The results of the live imaging indicated that two newly created daughter cells constantly position themselves within the airway epithelium according to the orientation of cell division, which might lead to a variation of the mechanical tension inside the airway epithelium.

Since cell division generates an anisotropic mechanical force surrounding the dividing cell in accordance with the division angle and bearing in mind that fixed spindles and rotating spindles tend to have varying division angle spreads, the fixed spindles and the rotating spindles might add to the mechanical forces of the tissue in a different way. These forces alter the contours of the surrounding cells and thus define their spindle dynamic response during division. It can be hypothesized that this type of mechanical tension generated through cell divisions serves as a trigger for cells to interact with each one another to maintain a robust ratio between fixed spindles and rotating spindles. To investigate, whether the cellular mechanical model may mirror the spatial and temporal connection between fixed spindles and rotating spindles, a large number of over 1000 simulations have been performed. In each simulation, over 80 cell division have been analyzed due to the spatial and temporal spreads of the fixed spindles and rotating spindles. The ratio of fixed spindle to rotating spindle has been kept quite stable in the cellular mechanical model. The variance of the weighted deviation X_i revealed through the cellular mechanical model was pronouncedly lower than the predicted weighted deviation X_i from random events (predicted model: variance of X_i is 10.82, $n = 1000$; predicted random model: variance of X_i is 19.77, $n = 100{,}000$). In addition, the physical dispersion of fixed spindles and rotating spindles is not significantly correlated, which is comparable to what has been seen in the live image analysis. The cell-based mechanical model is therefore capable of recalling all the experimental observations, suggesting that mechanical forces operate at the cell shape level, in conjunction with the long axis rule, are enough to consider the retention of stable relationships of fixed spindle cells and rotating spindle cells in the developing airway epithelium.

15.11.7 Stretching of Lung Explants Impacts Mitotic Spindle Angles Distribution in Developing Airway Epithelium

The cell-based mechanical model forecasted that the exposure to various levels of external mechanical force acting along the longitudinal axis of the airway epithelium would increase substantially the percentage of cells with an aspect ratio larger than 1.53 and, consequently, the percentage of cells with angles of division smaller than 30. To empirically validate this assumption in a biological system, dose-dependent lung expansion studies were conducted with lung explants cultivated on a silicon membrane. Imaging analysis showed that the longitudinal axis of the airway epithelial cells tended to be in alignment with the longitudinal axis of the airway tube.

Based on this monitoring, external mechanical forces were exerted on the longitudinal axis of the airway epithelium. The two different force levels employed during the stretching treatments led to an instantaneous increase in the length of the lung explants by 5 or 10%. It was discovered that the average cell cross section ratio is favorably correlated with the amount of applied stretching force. The stretching procedure had no influence on the proliferation level of epithelial or mesenchymal cells. Most significantly, in a similar way to the forecasted division angle distributions from the cell-based mechanical model, the percentage of cells with division angles below 30° significantly enhanced within three hours in stretched lungs, not only in stretched wild-type lungs, but in stretched $BRAF^{CA}$ mutant lungs also at the same rate. The elevated percentage of cells exhibiting division angles below 30° is positively associated with the magnitude of external forces acting along the longitudinal axis of the airway epithelium, implying that the mechanical forces are enough to govern the angular distribution of the mitotic spindle by adjusting the cell shape within the developing airway epithelium. This empirical finding is in line with the conclusions drawn from the mechanical cell-based model. The shape of the airways of both stretched and unstretched lungs was also examined after stretching the lungs for eight hours. The airway shape factor (*r*) of stretched wild-type and $BRAF^{CA}$ mutant lungs dropped significantly in comparison with unstretched control lungs.

Employing various approaches, among them mouse genetics, quantitative cell biology with 3D time-lapse mapping and computer modeling, it was found that two types of spindles exhibit different dynamic characteristics in the developing airway epithelium: One type freezes its orientation during the first 6 min of metaphase (fixed spindles); the other type fails to fully arrest during metaphase (rotating spindles). Cells with fixed spindles split according to the longitudinal axis of the cell, and this orientation is established before proceeding to the metaphase. The finding that both fixed spindle and rotating spindle cells are prone to orient themselves with the longitudinal axis of the airway tube poses the question whether fixed spindles tend to follow the longitudinal axis of the cells or the longitudinal axis of the airway tube. From this, it can be inferred that fixed spindles tend to follow the traditional cell longitudinal axis rule, which is confirmed by the finding that a few cells with fixed spindles failed to align their cell longitudinal axis with the longitudinal axis of the airways. In contrast, the rotating spindle cells do not exactly obey the traditional long axis principle and define their divisional orientation in the metaphase. Moreover, the cells with a rotating spindle are generally shorter than the cells with a fixed spindle, which indicates that the differences in the rotational characteristics of the spindle are closely linked to the cell morphology. The distribution of spindle angles in wild-type cells is compensatory influenced through the existence of $BRAF^{CA}$ cells in the mosaic tissue, which leads to the following question: Is there any similar compensatory mechanism with respect to the cell aspect ratio? However, in this experimental system, it is not feasible to label wild-type cells and activate $BRAF^{CA}$ in other cells of a single lung. In order to investigate this problem, a cell-based mechanical model can be employed. When a specific percentage of $BRAF^{CA}$ cells (modeled as cells with decreased aspect ratio) is depicted in these models, the percentage of wild-type cells with an aspect ratio greater than 1.53 reduces in these models. This outcome reinforces the idea that a compensation mechanism contributes to keeping the ratio of fixed-spindle cells to rotating spindle cells by exerting an impact on the cell shape. A fascinating opportunity is that certain cells in the fast-growing respiratory epithelium lengthen considerably. In such a scenario, these cells maintain the directional alignment of their mitotic spindles in an effort to achieve the proper distribution of cell division orientation necessary for induction of tube morphogenesis (Chanet and Martin 2014; Fink et al. 2011; Nestor-Bergmann et al. 2014). We show that ERK1/2 signaling regulates the orientation of cell divisions by controlling the interphase cell shape. Cell shape can be impacted by a variety of different parameters, among them is the cortical tension and cell–cell interactions. ERK1/2 signal transmission has been proven to stimulate actin-cortex mechanical properties (Logue et al. 2015), encourage contractility of cells (Klemke et al. 1997) and adjust cell adhesion (Roskoski 2012). Additional work is required to elucidate whether these ERK1/2-regulated cellular mechanisms are engaged in the regulation of interphase cell morphology at the level of the developing airway epithelium. It is still an interrogating issue how the morphology of the interphase cells is memorized to govern the orientation of the mitotic spindles within the airway epithelium. In *Drosophila*, Mud/NuMA is present at the tricellular interfaces of epithelial cells in the interphase and acts as a sensory element for the morphology of the interphase cells to govern the arrangement of cell division (Bosveld et al. 2016). Similar to the *Drosophila* epithelial cells, the mitotic cell rounding has also been detected in the pulmonary respiratory epithelial cells. However, NuMA is primarily positioned in the nucleus of numerous mammalian cells in the interphase, among them are the epithelial cells of the respiratory tract (Radulescu and Cleveland 2010). Moreover, as not all mitotic cells are rounded up in the course of mitosis (Campinho et al. 2013; Minc et al. 2011), this points to the existence of alternative mechanisms participating in the control of the connections of shape of interphase cells and the orientation of the mitotic spindle. Mechanical forces may affect the development of tissue and its growth through direct manipulation of cell performance during many embryonic development phases, ranging from gastrulation to organogenesis (Savin et al. 2011). On the tissue scale, mechanical forces have been demonstrated to function as a major element in the control of cell performance, such as cell shape alterations and the motion of the mitotic spindle (Fink et al. 2011; LeGoff and Lecuit 2015; Striedter et al. 2015).

During the process of tissue morphogenesis, the mechanical surroundings of evolving tissues experience fast alterations, which are partly due to high cell proliferation rates (Chanet and Martin 2014; Mammoto and Ingber 2010; Patwari and Lee 2008; Thery and Bornens 2008). A key regulatory feature involves the coordinated alignment of cell divisions and the morphology of interphase cells in reaction to mechanical tension inside the tissue (Campinho et al. 2013; Gibson and Gibson 2009; Legoff et al. 2013; Mao et al. 2013; Minc and Piel 2012; Nestor-Bergmann et al. 2014). The impact of local stress on cell morphology can be considered transient when the relaxation time of local mechanical stress generated by single-cell divisions is rather fast. This might account for the fact that there is no meaningful spatial correlation between the placement of cells containing fixed spindles and the placement of cells containing rotating spindles in the respiratory epithelium with an operating time of around four hours. The cell-based mechanical modeling result also demonstrates that although cells with an aspect ratio greater than 1.53 and cells with an aspect ratio less than 1.53 may temporarily produce small spots in the tissue, the fixed spindle sand spindles among the separated cells have a tendency to be randomly scattered over time.

In combination, both the cell-based mechanical modeling and the dose-dependent lung stretching investigations confirm the idea that mechanical forces are responsible for coordinating cell geometry and the orientation of the spindle during respiratory tract morphogenesis. The rotational behavior of the spindle was monitored within epithelial tissues using other types of experimental model systems including corneal epithelia and neural tubes (Geldmacher-Voss et al. 2003; Haydar et al. 2003; Tibber et al. 2004). These results show the underlying emergent cellular mechanism, which comprises complex correlations between cell geometry, directed cell division and mechanical tension experienced in the morphogenesis of epithelial tubes. Future research linking live 3D imaging and tissue level quantitative modeling will provide additional insight into the intricate cellular mechanisms that govern the formation of tissue morphogenesis.

References

M. Abal, M. Piel, V. Bouckson-Castaing, M. Mogensen, J.B. Sibarita, M. Bornens, Microtubule release from the centrosome in migrating cells. J. Cell Biol. **159**, 731–737 (2002)

I.R. Adams, J.V. Kilmartin, Spindle pole body duplication: a model for centrosome duplication? Trends Cell Biol. **10**, 329–335 (2000)

M.L. Agarwal, A. Agarwal, W.R. Taylor, G.R. Stark, p53 controls both the G2/M and the G1 cell cycle checkpoints and mediates reversible growth arrest in human fibroblasts. Proc. Natl. Acad. Sci. U. S. A. **92**, 8493–8497 (1995)

F.G. Agircan, E. Schiebel, B.R. Mardin, Separate to operate: control of centrosome positioning and separation. Phil. Trans. R. Soc. B **369**, 20130461 (2014)

F. Al-Khodairy, A.M. Carr, DNA repair mutants defining G2 checkpoint pathways in *Schizosaccharomyces pombe*. EMBO J. **11**(4), 1343–1350 (1992)

F. Al-Khodairy, E. Fotou, K.S. Sheldrick, D.J.F. Griffiths, A.R. Lehman, A.M. Carr, Identification and characterisation of new elements involved in checkpoint and feedback controls in fission yeast. Mol. Biol. Cell **5**, 147–160 (1994)

G. Alexandru, F. Uhlmann, K. Mechtler, M. Poupart, K. Nasmyth, Phosphorylation of the cohesin subunit scc1 by polo/cdc5 kinase regulates sister chromatid separation in yeast. Cell **105**, 459–472 (2001)

J.B. Allen, Z. Zhou, W. Siede, E. Friedberg, S. Elledge, The SAD1/Rad53 protein kinase controls multiple checkpoints and DNA damage-induced transcription in yeast. Genes Dev. **8**, 2416–2428 (1994)

P.L. Alvey, An investigation of the centriole cycle using 3T3 and CHO cells. J. Cell Sci. **78**, 147–162 (1985)
C.T. Anderson, T. Stearns, Centriole age underlies asynchronous primary cilium growth in mammalian cells. Curr Biol **19**, 1498–1502 (2009)
J. Azimzadeh, M. Bornens, The centrosome in evolution, in *Centrosomes in Development and Disease*, ed. by E.A. Nigg (Wiley, Weinheim, 2004), pp. 93–122
J. Azimzadeh, M. Bornens, Structure and duplication of the centrosome. J. Cell Biol. **120**, 2139–2142 (2007)
J. Azimzadeh, P. Hergert, A. Delouvee, U. Euteneuer, E. Formstecher, A. Khodjakov, M. Bornens, hPOC5 is a centrin-binding protein required for assembly of full-length centrioles. J. Cell Biol. **185**, 101–114 (2009)
L.A. Baena-Lopez, A. Baonza, A. Garcia-Bellido, The orientation of cell divisions determines the shape of *Drosophila* organs. Curr. Biol. **15**, 1640–1644 (2005)
S. Bahe, Y.D. Stierhof, C.J. Wilkinson, F. Leiss, E.A. Nigg, Rootletin forms centriole-associated filaments and functions in centrosome cohesion. J. Cell Biol. **171**, 27–33 (2005)
S. Bahmanyar, D.D. Kaplan, J.G. Deluca, T.H. Giddings Jr., E.T. O'Toole, M. Winey, E.D. Salmon, P.J. Casey, W. J. Nelson, A.I. Barth, Beta-Catenin is a Nek2 substrate involved in centrosome separation. Genes Dev. **22**, 91–105 (2008)
J.M. Bailis, D.D. Luche, T. Hunter, S.L. Forsburg, Minichromosome maintenance proteins interact with checkpoint and recombination proteins to promote s-phase genome stability. Mol. Cell Biol. **28**(5), 1724–1738 (2008)
K.J. Barnum, M.J. O'Connell, cell cycle regulation by checkpoints. Methods Mol. Biol. **1170**, 29–40 (2014)
J.A. Barrera, L.-R. Kao, R.E. Hammer, J. Seemann, J.L. Fuchs, T.L. Megraw, CDK5RAP2 regulates centriole engagement and cohesion in mice. Dev. Cell **18**, 913–926 (2010)
R. Basto, J. Lau, T. Vinogradova, A. Gardiol, C.G. Woods, A. Khodjakov, J.W. Raff, Flies without centrioles. Cell **125**, 1375–1386 (2006)
R. Basto, K. Brunk, T. Vinadogrova, N. Peel, A. Franz, A. Khodjakov, J.W. Raff, Centrosome amplification can initiate tumorigenesis in flies. Cell **133**, 1032–1042 (2008)
J. Beisson, M. Jerka-Dziadosz, Polarities of the centriolar structure: morphogenetic consequences. Biol. Cell **91**, 367–378 (1999)
G. Bellett, J.M. Carter, J. Keynton, D. Goldspink, C. James, D.K. Moss, M.M. Mogensen, Microtubule plus-end and minus-end capture at adherens junctions is involved in the assembly of apico-basal arrays in polarised epithelial cells. Cell Motil. Cytoskeleton **66**, 893–908 (2009)
L.D. Belmont, A.A. Hyman, K.E. Sawin, T.J. Mitchison, Real-time visualization of cell cycle-dependent changes in microtubule dynamics in cytoplasmic extracts. Cell **10**, 579–589 (1990)
G. Bergnes, K. Brejc, L. Belmont, Mitotic kinesins: prospects for antimitotic drug discovery. Curr. Top. Med. Chem. **5**, 127–145 (2005)
M.W. Berns, S.M. Richardson, Continuation of mitosis after selective laser microbeam destruction of the centriolar region. J. Cell Biol. **75**, 977–982 (1977)
M.T. Bertran, S. Sdelci, L. Regue, J. Avruch, C. Caelles, J. Roig, Nek9 is a Plk1-activated kinase that controls early centrosome separation through Nek6/7 and Eg5. EMBO J. **30**, 2634–2647 (2011)
M. Bettencourt-Dias, D.M. Glover, Centrosome biogenesis and function: centrosomics brings new understanding. Nat. Rev. Mol. Cell Biol. **8**, 451–463 (2007)
M. Bettencourt-Dias, A. Rodrigues-Martins, L. Carpenter, M. Riparbelli, L. Lehmann, M.K. Gatt, N. Carmo, F. Balloux, G. Callaini, D.M. Glover, SAK/PLK4 is required for centriole duplication and flagella development. Curr. Biol. **15**, 2199–2207 (2005)
S.M. Bilinski, M. Kloc, Accessory nuclei revisited: the translocation of snRNPs from the germinal vesicle to the periphery of the future embryo. Chromosoma **111**, 62–68 (2002)
S. Blachon, J. Gopalakrishnan, Y. Omori, A. Polyanovsky, A. Church, D. Nicastro, J. Malicki, T. Avidor-Reiss, *Drosophila* asterless and vertebrate Cep152 are orthologs essential for centriole duplication. Genetics **180**, 2081–2094 (2008)
A. Blangy, H.A. Lane, P. d'Herin, M. Harper, M. Kress, E.A. Niggt, Phosphorylation by p34cdc2 regulates spindle association of human Eg5, a kinesin-related motor essential for bipolar spindle formation in vivo. Cell **83**, 1159–1169 (1995)
Y. Bobinnec, M. Moudjou, J.P. Fouquet, E. Desbruyeres, B. Edde, M. Bornens, Glutamylation of centriole and cytoplasmic tubulin in proliferating non-neuronal cells. Cell Motil. Cytoskeleton **39**, 223–232 (1998a)
Y. Bobinnec, A. Khodjakov, L.M. Mir, C.L. Rieder, B. Edde, M. Bornens, Centriole disassembly in vivo and its effect on centrosome structure and function in vertebrate cells. J. Cell Biol. **143**, 1575–1589 (1998b)
H. Boleti, E. Karsenti, I. Vernos, Xklp2, a novel Xenopus centrosomal kinesin-like protein required for centrosome separation during mitosis. Cell **84**, 49–59 (1996)
S. Bolhy, I. Bouhlel, E. Dultz, T. Nayak, M. Zuccolo, X. Gatti, R. Vallee, J. Ellenberg, V. Doye, A Nup133-dependent NPC-anchored network tethers centrosomes to the nuclear envelope in prophase. J. Cell Biol. **192**, 855–871 (2011)

J. Bolivar, J.R. Huynh, H. Lopez-Schier, C. Gonzalez, D. St Johnston, A. Gonzalez-Reyes, Centrosome migration into the *Drosophila* oocyte is independent of BicD and egl, and of the organisation of the microtubule cytoskeleton. Development **128**, 1889–1897 (2001)

M. Bornens, Centrosome composition and microtubule anchoring mechanisms. Curr. Opin. Cell Biol. **14**, 25–34 (2002)

M. Bornens, Organelle positioning and cell polarity. Nat. Rev. Mol. Cell Biol. **9**, 874–886 (2008)

M. Bornens, J. Azimzadeh, Origin and evolution of the centrosome. Adv. Exp. Med. Biol. **607**, 119–129 (2007)

E.E. Bosco, Y. Wang, H. Xu, J.T. Zilfou, K.E. Knudsen, B.J. Aronow, S.W. Lowe, E.S. Knudsen, The retinoblastoma tumor suppressor modifies the therapeutic response of breast cancer. J. Clin. Invest. **117**, 218–228 (2007)

F. Bosveld, O. Markova, B. Guirao et al., Epithelial tricellular junctions act as interphase cell shape sensors to orient mitosis. Nature **530**, 495–498 (2016)

T. Boveri, *The Origin of Malignant Tumors* (Williams and Wilkins, Baltimore, Maryland, 1914)

T. Boveri, Concerning the origin of malignant tumors (Translated and annotated by H. Harris). J. Cell Sci. **121**, 1–84 (2008)

J.J. Bower, G.F. Karaca, Y. Zhou, D.A. Simpson, M. Cordeiro-Stone, W.K. Kaufmann, Topoisomerase IIalpha maintains genomic stability through decatenation G(2) checkpoint signaling. Oncogene **29**, 4787–4799 (2010)

J.J. Bower, L.D. Vance, M. Psioda, S.L. Smith-Roe, D.A. Simpson, J.G. Ibrahim, K.A. Hoadley, C.M. Perou, W.K. Kaufmann, Patterns of cell cycle checkpoint deregulation associated with intrinsic molecular subtypes of human breast cancer cells. NPJ Breast Cancer **3**, 9 (2017)

B.R. Brinkley, T.M. Goepfert, Supernumerary centrosomes and cancer: Boveri's hypothesis resurrected. Cell Motil. Cytoskeleton **41**, 281–288 (1998)

C.J. Brown, S. Lain, C.S. Verma, A.R. Fersht, D.P. Lane, Awakening guardian angels: drugging the p53 pathway. Nat. Rev. Cancer **9**, 862–873 (2009)

B. Burke, It takes KASH to hitch to the SUN. Cell **149**, 961–963 (2012)

G. Cabral, S. Sanegre Sans, C.R. Cowan, A. Dammermann, Multiple mechanisms contribute to centriole separation in *C. elegans*. Curr. Biol. **23**, 1380–1387 (2013)

K.A. Cadoo, A. Gucalp, T.A. Traina, Palbociclib: an evidence-based review of its potential in the treatment of breast cancer. Breast Cancer **6**, 123–133 (2014)

P.G. Calarco, Centrosome precursors in the acentriolar mouse oocyte. Microsc. Res. Tech. **49**, 428–434 (2000)

G. Callaini, M.G. Riparbelli, R. Dallai, Centrosome inheritance in insects: fertilization and parthenogenesis. Biol. Cell **91**, 355–366 (1999)

P. Campinho, M. Behrndt, J. Ranft, T. Risler, N. Minc, C.P. Heisenberg, Tension-oriented cell divisions limit anisotropic tissue tension in epithelial spreading during zebrafish epiboly. Nat. Cell Biol. **15**, 1405–1414 (2013)

A. Can, O. Semiz, O. Cinar, Centrosome and microtubule dynamics during early stages of meiosis in mouse oocytes. Mol. Hum. Reprod. **9**, 749–756 (2003)

L.A. Carvajal, J.J. Manfredi, Another fork in the road–life or death decisions by the tumour suppressor p53. EMBO Rep. **14**(5), 414–421 (2013)

L.A. Carvajal, P.J. Hamard, C. Tonnessen, J.J. Manfredi, E2F7, a novel target, is up-regulated by p53 and mediates DNA damage-dependent transcriptional repression. Genes Dev. **26**(14), 1533–1545 (2012)

E. Castellanos, P. Dominguez, C. Gonzalez, Centrosome dysfunction in *Drosophila* neural stem cells causes tumors that are not due to genome instability. Curr. Biol. **18**, 1209–1214 (2008)

T. Cavalier-Smith, Basal body and flagellar development during the vegetative cell cycle and the sexual cycle of *Chlamydomonas reinhardtii*. J. Cell Sci. **16**, 529–556 (1974)

T. Cavalier-Smith, The phagotrophic origin of eukaryotes and phylogenetic classification of Protozoa. Int. J. Syst. Evol. Microbiol. **52**, 297–354 (2002)

J.Y. Chan, A clinical overview of centrosome amplification in human cancers. Int. J. Biol. Sci. **7**, 1122–1144 (2011)

S. Chanet, A.C. Martin, Mechanical force sensing in tissues. Prog. Mol. Biol. Transl. Sci. **126**, 317–352 (2014)

A. Chestukhin, C. Pfeffer, S. Milligan, J.A. DeCaprio, D. Pellman, Processing, localization, and requirement of human separase for normal anaphase progression. Proc. Natl. Acad. Sci. U. S. A. **100**, 4574–4579 (2003)

V. Chevrier, M. Piel, N. Collomb, Y. Saoudi, R. Frank, M. Paintrand, S. Narumiya, M. Bornens, D. Job, The rho-associated protein kinase p160ROCK is required for centrosome positioning. J. Cell Biol. **157**, 807–817 (2002)

R.J. Cho, M.J. Campbell, E.A. Winzeler et al., A genome-wide transcriptional analysis of the mitotic cell cycle. Mol. Cell **2**, 65–73 (1998)

R. Ciosk, W. Zachariae, C. Michaelis, A. Shevchenko, M. Mann, K. Nasmyth, An ESP1/PDS1 complex regulates loss of sister chromatid cohesion at the metaphase to anaphase transition in yeast. Cell **93**, 1067–1076 (1998)

M. Classon, E. Harlow, The retinoblastoma tumor suppressor in development and cancer. Nat. Rev. Cancer **2**, 910–917 (2002)

P.A. Coelho, L. Bury, M.N. Shahbazi, et al., Over-expression of Plk4 induces centrosome amplification, loss of primary cilia and associated tissue hyperplasia in the mouse. Open Biol. **5**, 150209 (2015)

O. Cohen-Fix, D. Koshland, The anaphase inhibitor of *Saccharomyces cerevisiae* Pds1p is a target of the DNA damage checkpoint pathway. Proc. Natl. Acad. Sci. U. S. A. **94**(26), 14361–14366 (1997)

D.G. Cole, W.M. Saxton, K.B. Sheehan, J.M. Scholey, A 'slow' homotetrameric kinesin-related motor protein purified from *Drosophila* embryos. J. Biol. Chem. **269**, 22913–22916 (1994)

J.A. Connolly, B.W. Kiosses, V.I. Kalnins, Centrioles are lost as embryonic myoblasts fuse into myotubes in vitro. Eur. J. Cell Biol. **39**, 341–345 (1986)

K. Crasta, N.J. Ganem, R. Dagher, A.B. Lantermann, E.V. Ivanova, Y. Pan, L. Nezi, A. Protopopov, D. Chowdhury, D. Pellman, DNA breaks and chromosome pulverization from errors in mitosis. Nature **482**, 53–58 (2012)

F. Cross, DAF1, a mutant gene affecting size control, pheromone arrest and cell cycle kinetics of *S. cerevisiae*. Mol. Cell Biol. **8**, 4675–4684 (1988)

I. Cunha-Ferreira, A. Rodrigues-Martins, I. Bento, M. Riparbelli, W. Zhang, E. Laue, G. Callaini, D.M. Glover, M. Bettencourt-Dias, The SCF/Slimb ubiquitin ligase limits centrosome amplification through degradation of SAK/PLK4. Curr. Biol. **19**, 43–49 (2009)

A.B. D'Assoro, W.L. Lingle, J.L. Salisbury, Centrosome amplification and the development of cancer. Oncogene **21**, 6146–6153 (2002)

Y. Dai, L. Wang, H. Wang, Y. Liu, N. Li, Q. Lyu, D.L. Keefe, D.F. Albertini, L. Liu, Fate of centrosomes following somatic cell nuclear transfer (SCNT) in bovine oocytes. Reproduction **131**, 1051–1061 (2006)

M. Damelin, T.H. Bestor, The decatenation checkpoint. Br. J. Cancer **96**(2), 201–205 (2007)

A. Dammermann, T. Muller-Reichert, L. Pelletier, B. Habermann, A. Desai, K. Oegema, Centriole assembly requires both centriolar and pericentriolar material proteins. Dev. Cell **7**, 815–829 (2004)

A. Dammermann, P.S. Maddox, A. Desai, K. Oegema, SAS-4 is recruited to a dynamic structure in newly forming centrioles that is stabilized by the gamma-tubulin-mediated addition of centriolar microtubules. J. Cell Biol. **180**, 771–785 (2008)

D. Dankort, E. Filenova, M. Collado, M. Serrano, K. Jones, M. McMahon, A new mouse model to explore the initiation, progression, and therapy of BRAFV600E-induced lung tumors. Genes Dev. **21**, 379–384 (2007)

P.B. de Saint, W. Sullivan, Spindle assembly and mitosis without centrosomes in parthenogenetic Sciara embryos. J. Cell Biol. **141**, 1383–1391 (1998)

A. Debec, Haploid cell cultures of *Drosophila melanogaster*. Nature **274**, 255–256 (1978)

A. Debec, A. Szollo si, D. Szollo si, A *Drosophila melanoaster* cell line lacking centriole. Biol. Cell **44**, 133–138 (1982)

A. Debec, C. Detraves, C. Montmory, G. Geraud, M. Wright, Polar organization of gamma-tubulin in acentriolar mitotic spindles of *Drosophila melanogaster* cells. J. Cell Sci. **108**(Pt 7), 2645–2653 (1995)

A. Debec, W. Sullivan, M. Bettencourt-Dias, Centrioles: active players or passengers during mitosis? Cell. Mol. Life Sci. **67**, 2173–2194 (2010)

M. Delattre, C. Canard, P. Gonczy, Sequential protein recruitment in *C. elegans* centriole formation. Curr. Biol. **16**, 1844–1849 (2006)

P.B. Deming, C.A. Cistulli, H. Zhao, P.R. Graves, H. Piwnica-Worms, R.S. Paules, C.S. Downes, W.K. Kaufmann, The human decatenation checkpoint. Proc. Natl. Acad. Sci. U. S. A. **98**, 12044–12049 (2001)

N.R. den Elzen, M.J. O'Connell, Recovery from DNA damage checkpoint arrest by PP1-mediated inhibition of Chk1. EMBO J. **23**(4), 908–918 (2004)

N. den Elzen, A. Kosoy, H. Christopoulos, M.J. O'Connell, Resisting arrest: recovery from checkpoint arrest through dephosphorylation of Chk1 by PP1. Cell Cycle **3**(5), 529–533 (2004)

W. Deng, H. Lin, Asymmetric germ cell division and oocyte determination during *Drosophila* oogenesis. Int. Rev. Cytol. **203**, 93–138 (2001)

M.E. Diefenbacher, N. Popov, S.M. Blake, C. Schülein-Völk, E. Nye, B. Spencer-Dene, L.A. Jaenicke, M. Eilers, A. Behrens, The deubiquitinase USP28 controls intestinal homeostasis and promotes colorectal cancer. J. Clin. Invest. **124**, 3407–3418 (2014)

R. Dietz, The dispensability of the centrioles in the spermatocyte division of Pales ferruginea (Nematocera), in *Chromosomes Today*, ed. by C.D. Darlington, K.R. Lewis (Oliver and Boyd, London, 1966), pp. 161–166

A.D. Dingle, C. Fulton, Development of the flagellar apparatus of Naegleria. J. Cell Biol. **31**, 43–54 (1966)

R.V. Dippell, The development of basal bodies in paramecium. Proc. Natl. Acad. Sci. U. S. A. **61**, 461–468 (1968)

E.R. Dirksen, The presence of centrioles in artificially activated sea urchin eggs. J. Biophys. Biochem. Cytol. **11**, 244–247 (1961)

E.R. Dirksen, Centriole morphogenesis in developing ciliated epithelium of the mouse oviduct. J. Cell Biol. **51**, 286–302 (1971)

E.R. Dirksen, Centriole and basal body formation during ciliogenesis revisited. Biol. Cell **72**, 31–38 (1991)

C.I. Dix, J.W. Raff, *Drosophila* Spd-2 recruits PCM to the sperm centriole, but is dispensable for centriole duplication. Curr. Biol. **17**, 1759–1764 (2007)

A. Docquier, P.O. Harmand, S. Fritsch, M. Chanrion, J.M. Darbon, V. Cavaillès, The transcriptional coregulator RIP140 represses E2F1 activity and discriminates breast cancer subtypes. Clin. Cancer Res. **16**, 2959–2970 (2010)

S. Douthwright, G. Sluder, Link between DNA damage and centriole disengagement/reduplication in untransformed human cells. J. Cell. Physiol.**229**, 1427–1436 (2014)

C.S. Downes, D.J. Clarke, A.M. Mullinger, J.F. Giménez-Abián, A.M. Creighton, R.T. Johnson, A topoisomerase II-dependent G2 cycle checkpoint in mammalian cells. Nature **372**, 467–470 (1994)

S. Doxsey, Duplicating dangerously: linking centrosome duplication and aneuploidy. Mol. Cell **10**, 439–440203 (2002)

S. Duensing, K. Munger, Centrosomes, genomic instability, and cervical carcinogenesis. Crit. Rev. Eukaryot. Gene Expr. **13**, 9–23 (2003)

S.K. Dutcher, Elucidation of basal body and centriole functions in *Chlamydomonas reinhardtii*. Traffic **4**, 443–451 (2003)

J. Engelstadter, Constraints on the evolution of asexual reproduction. BioEssays **30**, 1138–1150 (2008)

T. Enoch, K. Gould, P. Nurse, Mitotic checkpoint control in fission yeast, in *Cold Spring Harbor Symposia on Quantitative Biology*, vol. 56 (CSH Laboratory Press, New York, 1991), pp. 409–416

T. Enoch, A. Carr, P. Nurse, Fission yeast genes involved in coupling mitosis to completion of DNA replication. Genes Dev. **6**, 2035–2046 (1992)

A. Errico, V. Costanzo, Mechanisms of replication fork protection: a safeguard for genome stability. Crit. Rev. Biochem. Mol. Biol. **47**(3), 222–235 (2012)

G. Fan, L. Sun, P. Shan, et al., Loss of KLF14 triggers centrosome amplification and tumorigenesis. Nat. Commun. **6**, 8450 (2015)

G. Fang, H. Yu, M.W. Kirschner, Direct binding of CDC20 protein family members activates the anaphase-promoting complex in mitosis and G1. Mol. Cell **2**, 163–171 (1998)

P.A. Fantes, Control of cell size and cycle time in *Schizosaccharomyces pombe*. J. Cell Sci. **24**(51), 51–67 (1977)

A.J. Faragher, A.M. Fry, Nek2A kinase stimulates centrosome disjunction and is required for formation of bipolar mitotic spindles. Mol. Biol. Cell **14**, 2876–2889 (2003)

N.P. Ferenz, A. Gable, P. Wadsworth, Mitotic functions of kinesin-5. Semin. Cell Dev. Biol. **21**, 255–259 (2010)

P.M. Ferree, K. McDonald, B. Fasulo, W. Sullivan, The origin of centrosomes in parthenogenetic hymenopteran insects. Curr. Biol. **16**, 801–807 (2006)

J. Fink, N. Carpi, T. Betz et al., External forces control mitotic spindle positioning. Nat. Cell Biol. **13**, 771–778 (2011)

K.T. Flaherty, J.R. Infante, A. Daud et al., Combined BRAF and MEK inhibition in melanoma with BRAF V600 mutations. N. Engl. J. Med. **367**, 1694–1703 (2012)

W. Flemming, *Zellsubstanz, Kern und Zelltheilung* (FCW Vogel, Leipzig, 1882)

L. Fletcher, G.J. Cerniglia, E.A. Nigg, T.J. Yen, R.J. Muschel, Inhibition of centrosome separation after DNA damage: a role for Nek2. Radiat. Res. **162**, 128–135 (2004)

S. Florian, T.U. Mayer, The functional antagonism between Eg5 and dynein in spindle bipolarization is not compatible with a simple push-pull model. Cell Rep. **1**, 408–416 (2012)

P. Fogarty, S.D. Campbell, R. Abu-Shumays, B.S. Phalle, K.R. Yu, G.L. Uy, M.L. Goldberg, W. Sullivan, The *Drosophila* grapes gene is related to checkpoint gene chk1/rad27 and is required for late syncytial division fidelity. Curr. Biol. **7**(6), 418–426 (1997)

M.L. Fogeron, H. Müller, S. Schade et al., LGALS3BP regulates centriole biogenesis and centrosome hypertrophy in cancer cells. Nat. Commun. **4**, 1531 (2013)

K.W. Fong, Y.K. Choi, J.B. Rattner, R.Z. Qi, CDK5RAP2 is a pericentriolar protein that functions in centrosomal attachment of the gamma-tubulin ring complex. Mol. Biol. Cell **19**, 115–125 (2008)

A. Fridkin, A. Penkner, V. Jantsch, Y. Gruenbaum, SUN-domain and KASH-domain proteins during development, meiosis and disease. Cell Mol. Life Sci. **66**, 1518–1533 (2009)

M. Friedlander, J. Wahrman, The spindle as a basal body distributor. A study in the meiosis of the male silkworm moth, Bombyx mori. J. Cell Sci. **7**, 65–89 (1970)

A.M. Fry, T. Mayor, P. Meraldi, Y.D. Stierhof, K. Tanaka, E.A. Nigg, C-Nap1, a novel centrosomal coiled-coil protein and candidate substrate of the cell cycle-regulated protein kinase Nek2. J. Cell Biol. **141**, 1563–1574 (1998)

T. Fujiwara, M. Bandi, M. Nitta, E.V. Ivanova, R.T. Bronson, D. Pellman, Cytokinesis failure generating tetraploids promotes tumorigenesis in p53-null cells. Nature **437**, 1043–1047 (2005)

K. Fukasawa, Centrosome amplification, chromosome instability and cancer development. Cancer Lett. **230**, 6–19 (2005)

K. Fukasawa, T. Choi, R. Kuriyama, S. Rulong, G.F. Vande Woude, Abnormal centrosome amplification in the absence of p53. Science **271**, 1744–1747 (1996)

S.D. Fuller, B.E. Gowen, S. Reinsch, A. Sawyer, B. Buendia, R. Wepf, E. Karsenti, The core of the mammalian centriole contains gamma-tubulin. Curr. Biol. **5**, 1384–1393 (1995)

C. Fulton, Centrioles, in *Origin and Continuity of Cell Organelles*, ed. by J. Reinert, H. Ursprung (Springer, Berlin, 1971)

C. Fulton, Pinwheels as cell's doodlebugs. Cell **30**, 341–343 (1982)

B. Furnari, A. Blasina, M.N. Boddy, C.H. McGowan, P. Russell, Cdc25 inhibited in vivo and in vitro by checkpoint kinases Cds1 and Chk1. Mol. Biol. Cell **10**(4), 833–845 (1999)

T. Gaglio, A. Saredi, J.B. Bingham, M.J. Hasbani, S.R. Gill, T.A. Schroer, D.A. Compton, Opposing motor activities are required for the organization of the mammalian mitotic spindle pole. J. Cell Biol. **135**, 399–414 (1996)

J.G. Gall, Early studies on centrioles and centrosomes, in *Centrosome in Development and Disease*, ed. by E.A. Nigg (Wiley-VCH Verlag GmbH & Co. KGaA, Weinheim, 2004)

N.J. Ganem, Z. Storchova, D. Pellman, Tetraploidy, aneuploidy and cancer. Curr. Opin. Genet. Dev. **17**, 157–162 (2007)

N.J. Ganem, S.A. Godinho, D. Pellman, A mechanism linking extra centrosomes to chromosomal instability. Nature **460**, 278–282 (2009)

N.J. Ganem, H. Cornils, S.Y. Chiu, K.P. O'Rourke, J. Arnaud, D. Yimlamai, M. Théry, F.D. Camargo, D. Pellman, Cytokinesis failure triggers hippo tumor suppressor pathway activation. Cell **158**, 833–848 (2014)

M. Gans, C. Audit, M. Masson, Isolation and characterization of sex-linked female-sterile mutants in *Drosophila melanogaster*. Genetics **81**, 683–704 (1975)

D.L. Gard, M.W. Kirschner, Microtubule assembly in cytoplasmic extracts of Xenopus oocytes and eggs. J. Cell Biol. **105**, 2191–2201 (1987)

B. Geldmacher-Voss, A.M. Reugels, S. Pauls, J.A. Campos-Ortega, A 90-degree rotation of the mitotic spindle changes the orientation of mitoses of zebrafish neuroepithelial cells. Development **130**, 3767–3780 (2003)

A.B. Georgi, P.T. Stukenberg, M.W. Kirschner, Timing of events in mitosis. Curr. Biol. **12**, 105–114 (2002)

M.G. Giansanti, E. Bucciarelli, S. Bonaccorsi, M. Gatti, *Drosophila* SPD-2 is an essential centriole component required for PCM recruitment and astral-microtubule nucleation. Curr. Biol. **18**, 303–309 (2008)

W.T. Gibson, M.C. Gibson, Cell topology, geometry, and morpho- genesis in proliferating epithelia. Curr. Top. Dev. Biol. **89**, 87–114 (2009)

W.T. Gibson, J.H. Veldhuis, B. Rubinstein, H.N. Cartwright, N. Perrimon, G.W. Brodland, R. Nagpal, M.C. Gibson, Control of the mitotic cleavage plane by local epithelial topology. Cell **144**, 427–438 (2011)

L.E. Giono, J.J. Manfredi, The p53 tumor suppressor participates in multiple cell cycle checkpoints. J. Cell Physiol. **209** (1), 13–20 (2006)

J.L. Gittleman, M. Kot, Adaptation: statistics and a null model for estimating phylogenetic effects. Syst. Biol. **39**, 227–241 (1990)

S.A. Godinho, D. Pellman, Causes and consequences of centrosome abnormalities in cancer. Philos. Trans. R. Soc. Lond. B Biol. Sci. **369**(1650), 20130467 (2014)

S.A. Godinho, M. Kwon, D. Pellman, Centrosomes and cancer: how cancer cells divide with too many centrosomes. Cancer Metastasis Rev. **28**, 85–98 (2009)

S.A. Godinho, R. Picone, M. Burute, R. Dagher, Y. Su, C.T. Leung, K. Polyak, J.S. Brugge, M. Théry, D. Pellman, Oncogene-like induction of cellular invasion from centrosome amplification. Nature **510**, 167–171 (2014)

M. Goldberg, M. Stucki, J. Falck, D. D'Amours, D. Rahman, D. Pappin, J. Bartek, S.P. Jackson, MDC1 is required for the intra-S-phase DNA damage checkpoint. Nature **421**(6926), 952–956 (2003)

P. Gonczy, Mechanisms of asymmetric cell division: flies and worms pave the way. Nat. Rev. Mol. Cell Biol. **9**, 355–366 (2008)

P. Gonczy, Centrosomes and cancer: revisiting a long-standing relationship. Nat. Rev. Cancer **15**, 639–652 (2015)

P. Gonczy, S. Pichler, M. Kirkham, A.A. Hyman, Cytoplasmic dynein is required for distinct aspects of MTOC positioning, including centrosome separation, in the one cell stage *Caenorhabditis elegans* embryo. J. Cell Biol. **147**, 135–150 (1999)

Y. Gong, C. Mo, S.E. Fraser, Planar cell polarity signalling controls cell division orientation during zebrafish gastrulation. Nature **430**, 689–693 (2004)

A.A. Goodarzi, P.A. Jeggo, The repair and signaling responses to DNA double-strand breaks. Adv. Genet. **82**, 1–45 (2013)

R.R. Gould, G.G. Borisy, The pericentriolar material in Chinese hamster ovary cells nucleates microtubule formation. J. Cell Biol. **73**, 601–615 (1977)

B. Grallert, E. Boye, The multiple facets of the intra-S checkpoint. Cell Cycle **7**(15), 2315–2320 (2008)

S. Graser, Y.D. Stierhof, E.A. Nigg, Cep68 and Cep215 (Cdk5rap2) are required for centrosome cohesion. J. Cell Biol. **120**, 4321–4331 (2007)

K.R. Greenow, A.R. Clarke, R.H. Jones, Chk1 deficiency in the mouse small intestine results in p53-independent crypt death and subsequent intestinal compensation. Oncogene **28**(11), 1443–1453 (2009)

H.C. Gregson, J.A. Schmiesing, J.S. Kim, T. Kobayashi, S. Zhou, K. Yokomori, A potential role for human cohesin in mitotic spindle aster assembly. J. Biol. Chem. **276**, 47575–47582 (2001)

C. Gueth-Hallonet, C. Antony, J. Aghion, A. Santa-Maria, I. Lajoie-Mazenc, M. Wright, B. Maro, c-Tubulin is present in acentriolar MTOCs during early mouse development. J. Cell Sci. **105**(Pt 1), 157–166 (1993)

L.R. Gurley, J.A. D'Anna, S.S. Barham, L.L. Deaven, R.A. Tobey, Histone phosphorylation and chromatin structure during mitosis in chinese hamster cells. Eur. J. Biochem. **84**, 1–15 (1978)

R. Habedanck, Y.D. Stierhof, C.J. Wilkinson, E.A. Nigg, The Polo kinase Plk4 functions in centriole duplication. Nat. Cell Biol. **7**, 1140–1146 (2005)

A. Hagting, M. Jackman, K. Simpson, J. Pines, Translocation of cyclin B1 to the nucleus at prophase requires a phosphorylation-dependent nuclear import signal. Curr. Biol. **9**, 680–689 (1999)

R.S. Hames, S.L. Wattam, H. Yamano, R. Bacchieri, A.M. Fry, APC/C-mediated destruction of the centrosomal kinase Nek2A occurs in early mitosis and depends upon a cyclin A-type D-box. EMBO J. **20**, 7117–7127 (2001)

Y.G. Han, H.J. Kim, A.A. Dlugosz, D.W. Ellison, R.J. Gilbertson, A. Alvarez-Buylla, Dual and opposing roles of primary cilia in medulloblastoma development. Nat. Med. **15**, 1062–1065 (2009)

F. Hans, S. Dimitrov, Histone H3 phosphorylation and cell division. Oncogene **20**, 3021–3027 (2001)

B.D. Harfe, P.J. Scherz, S. Nissim, H. Tian, A.P. McMahon, C.J. Tabin, Evidence for an expansion-based temporal Shh gradient in specifying vertebrate digit identities. Cell **118**, 517–528 (2004)

A.R. Harris, L. Peter, J. Bellis, B. Baum, A.J. Kabla, G.T. Charras, Characterizing the mechanics of cultured cell monolayers. Proc. Natl. Acad. Sci. U. S. A. **109**, 16449–16454 (2012)

L.H. Hartwell, Cell division from a genetic perspective. J. Cell Biol. **77**, 627–637 (1978)

L.H. Hartwell, M.B. Kastan, Cell cycle control and cancer. Science **266**, 1821–1828 (1994)

L.H. Hartwell, T.A. Weinert, Checkpoints: controls that ensure the order of cell cycle events. Science **246**, 629–634 (1989)

K. Hastak, E. Alli, J.M. Ford, Synergistic chemosensitivity of triple-negative breast cancer cell lines to poly(ADP-Ribose) polymerase inhibition, gemcitabine, and cisplatin. Cancer Res. **70**, 7970–7980 (2010)

T.F. Haydar, E. Ang Jr., P. Rakic, Mitotic spindle rotation and mode of cell division in the developing telencephalon. Proc. Natl. Acad. Sci. U. S. A. **100**, 2890–2895 (2003)

J. Hayles, D. Fisher, A. Woollard, P. Nurse, Temporal order of S-phase and mitosis in fission yeast is determined by the state of the p34 cdc2/mitotic B cyclin complex. Cell **78**, 813–822 (1994)

R. He, N. Huang, Y. Bao, H. Zhou, J. Teng, J. Chen, LRRC45 is a centrosome linker component required for centrosome cohesion. Cell Rep. **4**, 1100–1107 (2013)

G.E. Heimpel, J.G. de Boer, Sex determination in the hymenoptera. Annu. Rev. Entomol. **53**, 209–230 (2008)

L.M. Heiser, A. Sadanandam, W.L. Kuo et al., Subtype and pathway specific responses to anticancer compounds in breast cancer. Proc. Natl. Acad. Sci. U. S. A. **109**, 2724–2729 (2012)

N.R. Helps, X. Luo, H.M. Barker, P.T.W. Cohen, NIMA-related kinase 2 (Nek2), a cell-cycle-regulated protein kinase localized to centrosomes, is complexed to protein phosphatase 1. Biochem J. **349**, 509–518 (2000)

L.F. Henneguy, Sur les rapports des cils vibratiles avec les centrosomes. Arch. Anat. Microsc. **1**, 481–496 (1897)

A. Hergovich, H. Cornils, B.A. Hemmings, Mammalian NDR protein kinases: from regulation to a role in centrosome duplication. Biochim. Biophys. Acta **1784**, 3–15 (2008)

O. Hertwig, Das Problem der Befruchtung und der Isotropic des Eies. eine Theorie der Vererbung. Z. Naturwiss. **18**, 21–23 (1884)

E.H. Hinchcliffe, R.W. Linck, Two proteins isolated from sea urchin sperm flagella: structural components common to the stable microtubules of axonemes and centrioles. J. Cell Sci. **111**(Pt 5), 585–595 (1998)

E.H. Hinchcliffe, F.J. Miller, M. Cham, A. Khodjakov, G. Sluder, Requirement of a centrosomal activity for cell cycle progression through G1 into S-phase. Science **291**, 1547–1550 (2001)

M. Hiraki, Y. Nakazawa, R. Kamiya, M. Hirono, Bld10p constitutes the cartwheel-spoke tip and stabilizes the 9-fold symmetry of the centriole. Curr. Biol. **17**, 1778–1783 (2007)

M.L. Hixon, A. Gualberto, The control of mitosis. Front. Biosci. **5**, d50-57 (2000)

W. Hofmeister, Zusatze und Berichtigungen zu den 1851 veröffentlichen Untersuchungen der Entwicklung höherer Kryptogamen. Jahrbucher Fur Wissenschaft Und Botanik **3**, 259–293 (1863)

A.J. Holland, D. Fachinetti, Q. Zhu, M. Bauer, I.M. Verma, E.A. Nigg, D.W. Cleveland, The autoregulated instability of Polo-like kinase 4 limits centrosome duplication to once per cell cycle. Genes Dev. **26**, 2684–2689 (2012)

M.A. Hoyt, L. Totis, B.T. Roberts, *S. cerevisiae* genes required for cell cycle arrest in response to loss of microtubule function. Cell **66**, 507–517 (1991)

A.F. Huettner, Continuity of the centrioles in *Drosophila melanogaster*. Z. Zellforsch. Mikroskop. Anat. **19**, 119–134 (1933)

L. Hurtado, C. Caballero, M.P. Gavilan, J. Cardenas, M. Bornens, R.M. Rios, Disconnecting the Golgi ribbon from the centrosome prevents directional cell migration and ciliogenesis. J. Cell Biol. **193**, 917–933 (2011)

B. Inanc, H. Dodson, C.G. Morrison, A centrosome-autonomous signal that involves centriole disengagement permits centrosome duplication in G2 phase after DNA damage. Mol. Biol. Cell **21**, 3866–3877 (2010)

T. Izumi, D.H. Walker, J.L. Maller, Periodic changes in phosphorylation of the Xenopus cdc25 phosphatase regulate its activity. Mol. Biol. Cell **3**, 927–939 (1992)

T. Izuumi, J.L. Maller, Elimination of cdc2 phosphorylation sites in the cdc25 phosphatase blocks initiation of M phase. Mol. Cell. Biol. **4**, 1337–1350 (1993)

M.R. Jackman, J.N. Pines, Cyclins and the G2/M transition. Cancer Surv. **29**, 47–73 (1997)

C. Janke, K. Rogowski, J. van Dijk, Polyglutamylation: a fine-regulator of protein function? 'Protein modifications: beyond the usual suspects' review series. EMBO Rep. **9**, 636–641 (2008)

A. Janssen, M. van der Burg, K. Szuhai, G.J. Kops, R.H. Medema, Chromosome segregation errors as a cause of DNA damage and structural chromosome aberrations. Science **333**, 1895–1898 (2011)

J. Januschke, L. Gervais, L. Gillet, G. Keryer, M. Bornens, A. Guichet, The centrosome–nucleus complex and microtubule organization in the *Drosophila* oocyte. Development **133**, 129–139 (2006)
R.T. Johnson, P.N. Rao, Mammalian cell fusion: induction of premature chromosome condensation in interphase nuclei. Nature **226**(5247), 717–722 (1970)
R.T. Johnson, P.N. Rao, H.D. Hughes, Mammalian cell fusion. 3. A HeLa cell inducer of premature chromosome condensation active in cells from a variety of animal species. J. Cell Physiol. **76**(2), 151–157 (1970)
J.A. Jonassen, J. San Agustin, J.A. Follit, G.J. Pazour, Deletion of IFT20 in the mouse kidney causes misorientation of the mitotic spindleand cystic kidney disease. J. Cell Biol. **183**, 377–384 (2008)
P. Jorgensen, J.L. Nishikawa, B.J. Breitkreutz, M. Tyers, Systematic identification of pathways that couple cell growth and division in yeast. Science **297**(5580), 395–400 (2002)
P. Kalab, R. Heald, The RanGTP gradient: a GPS for the mitotic spindle. J. Cell Biol. **121**, 1577–1586 (2008)
L.R. Kao, T.L. Megraw, Centrocortin cooperates with centrosomin to organize *Drosophila* embryonic cleavage furrows. Curr. Biol. **19**, 937–942 (2009)
L.C. Kapitein, E.J.G. Peterman, B.H. Kwok, J.H. Kim, T.M. Kapoor, C.F. Schmidt, The bipolar mitotic kinesin Eg5 moves on both microtubules that it crosslinks. Nature **435**, 114–118 (2005)
E. Karsenti, R. Bravo, M.W. Kirschner, Phosphorylation changes associated with the early cell cycle in Xenopus eggs. Dev. Biol. **119**, 442–453 (1987)
K. Kaseda, A.D. McAinsh, R.A. Cross, Dual pathway spindle assembly increases both the speed and the fidelity of mitosis. Biol. Open **1**, 12–18 (2012)
A.S. Kashina, R.J. Baskin, D.G. Cole, K.P. Wedaman, W.M. Saxton, J.M. Scholey, A bipolar kinesin. Nature **379**, 270–272 (1996)
K. Kazazian, C. Go, H. Wu, O. Brashavitskaya, R. Xu, J.W. Dennis, A.C. Gingras, C.J. Swallow, Plk4 promotes cancer invasion and metastasis through Arp2/3 complex regulation of the actin cytoskeleton. Cancer Res. **77**, 434–447 (2017)
G. Keryer, H. Ris, G.G. Borisy, Centriole distribution during tripolar mitosis in Chinese hamster ovary cells. J. Cell Biol. **98**, 2222–2229 (1984)
A. Khodjakov, C.L. Rieder, Centrosomes enhance the fidelity of cytokinesis in vertebrates and are required for cell cycle progression. J. Cell Biol. **153**, 237–242 (2001)
A. Khodjakov, R.W. Cole, C.L. Rieder, A synergy of technologies: combining laser microsurgery with green fluorescent protein tagging. Cell Motil. Cytoskeleton **38**, 311–317 (1997)
A. Khodjakov, R.W. Cole, B.R. Oakley, C.L. Rieder, Centrosome-independent mitotic spindle formation in vertebrates. Curr. Biol. **10**, 59–67 (2000)
A. Khodjakov, C.L. Rieder, G. Sluder, G. Cassels, O. Sibon, C.L. Wang, De novo formation of centrosomes in vertebrate cells arrested during S-phase. J. Cell Biol. **158**, 1171–1181 (2002)
D. Killander, A. Zetterberg, A quantitative cytochemical investigation of the relationship between cell mass and initiation of DNA synthesis in mouse fibroblasts in vitro. Exp. Cell Res. **40**(1), 12–20 (1965a)
D. Killander, A. Zetterberg, Quantitative cytochemical studies on interphase growth. I. Determination of DNA, RNA and mass content of age determined mouse fibroblasts in vitro and of intercellular variation in generation time. Exp. Cell Res. **38**, 272–284 (1965b)
K.B. Kim, R. Kefford, A.C. Pavlick et al., Phase II study of the MEK1/MEK2 inhibitor Trametinib in patients with metastatic BRAF-mutant cutaneous melanoma previously treated with or without a BRAF inhibitor. J. Clin. Oncol. **31**, 482–489 (2013)
R.C. King, *Ovarian Development in Drosophila Melanogaster* (Academic Press, New York, 1970)
R.W. King, P.K. Jackson, M.W. Kirschner, Mitosis in transition. Cell **79**, 563–571 (1994)
R.L. Klemke, S. Cai, A.L. Giannini, P.J. Gallagher, P. de Lanerolle, D.A. Cheresh, Regulation of cell motility by mitogen-activated protein kinase. J. Cell Biol. **137**, 481–492 (1997)
J. Kleylein-Sohn, J. Westendorf, M. Le Clech, R. Habedanck, Y.D. Stierhof, E.A. Nigg, Plk4-induced centriole biogenesis in human cells. Dev. Cell **13**, 190–202 (2007)
G. Kohlmaier, J. Loncarek, X. Meng, B.F. McEwen, M.M. Mogensen, A. Spektor, B.D. Dynlacht, A. Khodjakov, P. Gönczy, Overly long centrioles and defective cell division upon excess of the SAS-4-related protein CPAP. Curr. Biol. **19**, 1012–1018 (2009)
X. Kong, A.R. Ball, E. Sonoda, J. Feng, S. Takeda, T. Fukagawa, T.J. Yen, K. Yokomori, Cohesin associates with spindle poles in a mitosis-specific manner and functions in spindle assembly in vertebrate cells. Mol. Biol. Cell **20**, 1289–1301 (2009)
K. Koniaras, A.R. Cuddihy, H. Christopolous, A. Hogg, M.J. O'Connell, Inhibition of Chk1-dependent G2 DNA damage checkpoint radiosensitises p53 mutant human cells. Oncogene **20**, 7453–7463 (2001)
T. Kono, Genetic modification for bimaternal embryo development. Reprod. Fertil. Dev. **21**, 31–36 (2009)
G.J. Kops, B.A. Weaver, D.W. Cleveland, On the road to cancer: aneuploidy and the mitotic checkpoint. Nat. Rev. Cancer **5**(10), 773–785 (2005)

S. Kotani, H. Tanaka, H. Yasuda, K. Todokoro, Regulation of APC activity by phosphorylation and regulatory factors. J. Cell Biol. **146**, 791–800 (1999)

E.R. Kramer, C. Gieffers, G. Holz, M. Hengstschlager, J.-M. Peters, Activation of the human anaphase-promoting complex by proteins of the CDC20/Fizzy family. Curr. Biol. **8**, 1207–1210 (1998)

E.R. Kramer, N. Scheuringer, J.-M. Peters, Mitotic regulation of the APC activator proteins CDC20 and CDH1. Mol. Biol. Cell **11**, 1555–1569 (2000)

A. Krzywicka-Racka, G. Sluder, Repeated cleavage failure does not establish centrosome amplification in untransformed human cells. J .Cell Biol. **194**, 199–207 (2011)

J. Kuang, C.L. Ashorn, K.M. Gonzalez, J.E. Penkala, Cdc25 is one of the MPM-2 antigens involved in the activation of maturation-promoting factor. Mol. Biol. Cell **5**, 135–145 (1994)

A. Kulukian, A.J. Holland, B. Vitre, S. Naik, D.W. Cleveland, E. Fuchs, Epidermal development, growth control, and homeostasis in the face of centrosome amplification. Proc. Natl. Acad. Sci. U. S. A. **112**, E6311-6320 (2015)

A. Kumagai, W.G. Dunphy, The cdc25 protein controls tyrosine dephosphorylation of the cdc2 protein in a cellfree system. Cell **64**, 903–914 (1991)

A. Kumagai, W.G. Dunphy, Regulation of the cdc25 protein during the cell cycle in Xenopus extracts. Cell **70**, 139–151 (1992)

A. Kumagai, W.G. Dunphy, Regulation of Xenopus Cdc25 protein. Methods Enzymol. **283**, 564–571 (1997)

A. Kumagai, W.G. Dunphy, Claspin, a novel protein required for the activation of Chk1 during a DNA replication checkpoint response in Xenopus egg extracts. Mol. Cell **6**(4), 839–849 (2000)

A.L. Kung, S.W. Sherwood, R.T. Shimke, Cell line-specific differences in the control of cell cycle progression in the absence of mitosis. Proc. Natl. Acad. Sci. U. S. A. **87**, 9553–9557 (1990)

K. Kuntz, M.J. O'Connell, The G(2) DNA damage checkpoint: could this ancient regulator be the achilles heel of cancer? Cancer Biol. Ther. **8**(15), 1433–1439 (2009)

R. Kuriyama, G.G. Borisy, Centriole cycle in Chinese hamster ovary cells as determined by whole-mount electron microscopy. J. Cell Biol. **91**, 814–821 (1981)

R. Kuriyama, G.G. Borisy, Cytasters induced within unfertilized sea-urchin eggs. J. Cell Sci. **61**, 175–189 (1983)

R. Kuriyama, Y. Terada, K.S. Lee, C.L. Wang, Centrosome replication in hydroxyurea-arrested CHO cells expressing GFP-tagged centrin2. J. Cell Sci. **120**, 2444–2453 (2007)

M. Kwon, S.A. Godinho, N.S. Chandhok, N.J. Ganem, A. Azioune, M. Thery, D. Pellman, Mechanisms to suppress multipolar divisions in cancer cells with extra centrosomes. Genes Dev. **22**, 2189–2203 (2008)

S. Lahav-Baratz, V. Sudakin, J.V. Ruderman, A. Hershko, Reversible phosphorylation controls the activity of cyclosome-associated cyclin-ubiquitin ligase. Proc. Natl. Acad. Sci. U. S. A. **92**, 9303–9307 (1995)

M.H. Lam, Q. Liu, S.J. Elledge, J.M. Rosen, Chk1 is haploinsufficient for multiple functions critical to tumor suppression. Cancer Cell **6**(1), 45–59 (2004)

S. Lambert, A.M. Carr, Checkpoint responses to replication fork barriers. Biochimie **87**(7), 591–602 (2005)

S. Lambert, A.M. Carr, Replication stress and genome rearrangements: lessons from yeast models. Curr. Opin. Genet. Dev. **23**(2), 132–139 (2013)

B.G. Lambrus, Y. Uetake, K.M. Clutario, V. Daggubati, M. Snyder, G. Sluder, A.J. Holland, p53 protects against genome instability following centriole duplication failure. J. Cell Biol. **210**, 63–77 (2015)

B.G. Lambrus, V. Daggubati, Y. Uetake, P.M. Scott, K.M. Clutario, G. Sluder, A.J. Holland, A USP28-53BP1-p53-p21 signaling axis arrests growth after centrosome loss or prolonged mitosis. J. Cell Biol. **214**, 143–153 (2016)

P. Lara-Gonzalez, F.G. Westhorpe, S.S. Taylor, The spindle assembly checkpoint. Curr. Biol. **22**(22), R966–R980 (2012)

C. Latif, N.R. Elzen, M.J. O'Connell, DNA damage checkpoint maintenance through sustained Chk1 activity. J. Cell Sci. **117**(Pt 16), 3489–3498 (2004)

B. Leber, B. Maier, F. Fuchs, et al., Proteins required for centrosome clustering in cancer cells. Sci. Transl. Med. **2**, 33ra38 (2010)

N. Lecland, A. Debec, A. Delmas et al., Establishment and mitotic characterization of new *Drosophila* acentriolar cell lines from DSas-4 mutant. Biol. Open **2**(3), 314–323 (2013)

K.Y. Lee, K. Myung, PCNA modifications for regulation of post-replication repair pathways. Mol. Cells **26**(1), 5–11 (2008)

M. Lee, P. Nurse, Cell cycle control genes in fission yeast and mammalian cells. Trends Genet. **4**, 287–290 (1988)

L. LeGoff, T. Lecuit, Mechanical forces and growth in animal tissues. Cold Spring Harb. Perspect. Biol. **8**, a019232 (2015)

L. Legoff, H. Rouault, T. Lecuit, A global pattern of mechanical stress polarizes cell divisions and cell shape in the growing *Drosophila* wing disc. Development **140**, 4051–4059 (2013)

S. Leidel, M. Delattre, L. Cerutti, K. Baumer, P. Gonczy, SAS-6 defines a protein family required for centrosome duplication in *C. elegans* and in human cells. Nat. Cell Biol. **7**, 115–125 (2005)

M.V. Lenhossek, Ueber Flimmerzellen. Verh. Anat. Ges. **12**, 106 (1897)

M. Leo, D. Santino, I. Tikhonenko, V. Magidson, A. Khodjakov, M.P. Koonce, Rules of engagement: centrosome-nuclear connections in a closed mitotic system. Biol. Open **1**, 1111–1117 (2012)

M.S. Levine et al., Centrosome amplification is sufficient to promote spontaneous tumorigenesis in mammals. Dev. Cell **40**(313–322), e315 (2017)

K. Li, T.C. Kaufman, The homeotic target gene centrosomin encodes an essential centrosomal component. Cell **85**, 585–596 (1996)

R. Li, A.W. Murray, Feedback control of mitosis in budding yeast. Cell **66**, 519–531 (1991)

K. Li, E.Y. Xu, J.K. Cecil, F.R. Turner, T.L. Megraw, T.C. Kaufman, *Drosophila* centrosomin protein is required for male meiosis and assembly of the flagellar axoneme. J. Cell Biol. **141**, 455–467 (1998)

J. Li, M. Tan, L. Li, D. Pamarthy, T.S. Lawrence, Y. Sun, SAK, a new polo-like kinase, is transcriptionally repressed by p53 and induces apoptosis upon RNAi silencing. Neoplasia **7**, 312–323 (2005)

Z. Liang, D. Zickler, M. Prentiss, F.S. Chang, G. Witz, K. Maeshima, N. Kleckner, Chromosomes progress to metaphase in multiple discrete steps via global compaction/expansion cycles. Cell **161**, 1124–1137 (2015)

Y.C. Lin, C.-W. Chang, W.-B. Hsu, C.-J.C. Tang, Y.-N. Lin, E.-J. Chou, C.-T. Wu, T.K. Tang, Human microcephaly protein CEP135 binds to hSAS-6 and CPAP, and is required for centriole assembly. EMBO J. **32**, 1141–1154 (2013)

H.D. Lindsay, D.J.F. Griffiths, R.J. Edwards, P.U. Christensen, J.M. Murray, F. Osman, N. Walworth, A.M. Carr, S-phase-specific activation of Cds1 kinase defines a subpathway of the checkpoint response in *Schizosaccharomyces pombe*. Genes Dev. **12**, 382–395 (1998)

Q. Liu, S. Guntuku, X.S. Cui et al., Chk1 is an essential kinase that is regulated by Atr and required for the G(2)/M DNA damage checkpoint. Genes Dev. **14**(12), 1448–1459 (2000)

J.S. Logue, A.X. Cartagena-Rivera, M.A. Baird, M.W. Davidson, R.S. Chadwick, C.M. Waterman, Erk regulation of actin capping and bundling by Eps8 promotes cortex tension and leader bleb-based migration. Elife **4**, e08314 (2015)

M.J. Lohka, J.L. Kyes, J.L. Maller, Metaphase protein phosphorylation in *Xenopus laevis* eggs. Mol. Cell. Biol. **7**, 760–768 (1987)

J. Loncarek, P. Hergert, V. Magidson, A. Khodjakov, Control of daughter centriole formation by the pericentriolar material. Nat. Cell Biol. **10**, 322–328 (2008)

J. Loncarek, P. Hergert, A. Khodjakov, Centriole reduplication during prolonged interphase requires procentriole maturation governed by Plk1. Curr. Biol. **20**, 1277–1282 (2010)

M. Lowe, N.K. Gonatas, G. Warren, The mitotic phosphorylation cycle of the cis-Golgi matrix protein GM130. J. Cell Biol. **149**, 341–356 (2000)

X. Lu, B. Nannenga, L.A. Donehower, PPM1D dephosphorylates Chk1 and p53 and abrogates cell cycle checkpoints. Genes Dev. **19**(10), 1162–1174 (2005)

F.C. Luca, J.V. Ruderman, Control of programmed cyclin destruction in a cell-free system. J. Cell Biol. **109**, 1895–1909 (1989)

E.P. Lucas, J.W. Raff, Maintaining the proper connection between the centrioles and the pericentriolar matrix requires *Drosophila* centrosomin. J. Cell Biol. **178**, 725–732 (2007)

K. Lundgren, N. Walworth, R. Booher, M. Dembski, M. Kirschner, D. Beach, mik1 and wee1 cooperate in the inhibitory tyrosine phosphorylation of cdc2. Cell **64**(6), 1111–1122 (1991)

K. Luo, J. Yuan, J. Chen, Z. Lou, Topoisomerase IIalpha controls the decatenation checkpoint. Nat. Cell. Biol. **11**, 204–210 (2009)

C.A. MacDougall, T.S. Byun, C. Van, M.C. Yee, K.A. Cimprich, The structural determinants of checkpoint activation. Genes Dev. **21**(8), 898–903 (2007)

L. Madisen, T.A. Zwingman, S.M. Sunkin et al., A robust and high-throughput Cre reporting and characterization system for the whole mouse brain. Nat. Neurosci. **13**, 133–140 (2010)

V. Magidson, F. Chang, A. Khodjakov, Regulation of cytokinesis by spindle-pole bodies. Nat. Cell Biol. **8**, 891–893 (2006)

V. Magidson, J. Loncarek, P. Hergert, C.L. Rieder, A. Khodjakov, Laser microsurgery in the GFP era: a cell biologist's perspective. Methods Cell Biol. **82**, 239–266 (2007)

V. Magidson, C.B. O'Connell, J. Loncarek, R. Paul, A. Mogilner, A. Khodjakov, The spatial arrangement of chromosomes during prometaphase facilitates spindle assembly. Cell **146**, 555–567 (2011)

M.R. Mahjoub, T. Stearns, Supernumerary centrosomes nucleate extra cilia and compromise primary cilium signaling. Curr. Biol. **22**, 1628–1634 (2012)

A.P. Mahowald, J.M. Strassheim, Intercellular migration of centrioles in the germarium of *Drosophila melanogaster*. An electron microscopic study. J. Cell Biol. **45**, 306–320 (1970)

R. Maike-Lye, J. Newport, M.W. Kirschner, Maturation-promoting factor induces nuclear envelope breakdown in cyclohexamide-arrested embryos of *Xenopus laevis*. J. Cell Biol. **97**, 81–91 (1993)

J.L. Maller, MPF and cell cycle control. Adv. Second Messenger Phosphoprotein Res. **24**, 323–328 (1990)

J. Maller, D. Poccia, D. Nishioka, P. Kidd, J. Gerhart, H. Hartman, Spindle formation and cleavage in Xenopus eggs injected with centriole-containing fractions from sperm. Exp. Cell Res. **99**, 285–294 (1976)

M. Malumbres, M. Barbacid, Cell cycle kinases in cancer. Curr. Opin. Genet. Dev. **17**(1), 60–65 (2007)

T. Mammoto, D.E. Ingber, Mechanical control of tissue and organ development. Development **137**, 1407–1420 (2010)

G. Manandhar, H. Schatten, P. Sutovsky, Centrosome reduction during gametogenesis and its significance. Biol. Reprod. **72**, 2–13 (2005)

A. Maniotis, M. Schliwa, Microsurgical removal of centrosomes blocks cell reproduction and centriole generation in BSC-1 cells. Cell **67**, 495–504 (1991)

Y. Mao, A.L. Tournier, A. Hoppe, L. Kester, B.J. Thompson, N. Tapon, Differential proliferation rates generate patterns of mechanical tension that orient tissue growth. EMBO J. **32**, 2790–2803 (2013)

B.R. Mardin, C. Lange, J.E. Baxter, T. Hardy, S.R. Scholz, A.M. Fry, E. Schiebel, Components of the Hippo pathway cooperate with Nek2 kinase to regulate centrosome disjunction. Nat. Cell Biol. **12**, 1166–1176 (2010)

B.R. Mardin, F.G. Agircan, C. Lange, E. Schiebel, Plk1 controls the Nek2A-PP1gamma antagonism in centrosome disjunction. Curr. Biol. **21**, 1145–1151 (2011)

B.R. Mardin, M. Isokane, M.R. Cosenza, E.J. KrämerA, A.M. Fry, E. Schiebel, EGF-induced centrosome separation promotes mitotic progression and cell survival. Dev. Cell **25**, 229–240 (2013)

V. Marthiens, M.A. Rujano, C. Pennetier, S. Tessier, P. Paul-Gilloteaux, R. Basto, Centrosome amplification causes microcephaly. Nat. Cell Biol. **15**, 731–740 (2013)

S.G. Martin, M. Berthelot-Grosjean, Polar gradients of the DYRK-family kinase Pom1 couple cell length with the cell cycle. Nature **459**(7248), 852–856 (2009)

J.M. Mason, D.C. Lin, X. Wei, et al., Functional characterization of CFI-400945, a Polo-like kinase 4 inhibitor, as a potential anticancer agent. Cancer Cell **26**, 163–176 (2014)

Y. Masui, From oocyte maturation to the in vitro cell cycle: the history of discoveries of maturation-promoting factor (MPF) and cytostatic factor (CSF). Differentiation **69**(1), 1–17 (2001)

K. Matsuo, K. Ohsumi, M. Iwabuchi, T. Kawamata, Y. Ono, M. Takahashi, Kendrin is a novel substrate for separase involved in the licensing of centriole duplication. Curr. Biol. **22**, 915–921 (2012)

K. Matsuura, P.A. Lefebvre, R. Kamiya, M. Hirono, Bld10p, a novel protein essential for basal body assembly in Chlamydomonas: localization to the cartwheel, the first ninefold symmetrical structure appearing during assembly. J. Cell Biol. **165**, 663–671 (2004)

M. Maugeri-Sacca, M. Bartucci, R. De Maria, Checkpoint kinase 1 inhibitors for potentiating systemic anticancer therapy. Cancer Treat. Rev. **39**(5), 525–533 (2013)

T. Mayor, Y.-D. Stierhof, K. Tanaka, A.M. Fry, E.A. Nigg, The centrosomal protein C-Nap1 is required for cell cycle-regulated centrosome cohesion. J. Cell Biol. **151**, 837–846 (2000)

A. Mazumdar, M. Mazumdar, How one becomes many: blastoderm cellularization in *Drosophila melanogaster*. BioEssays **24**, 1012–1022 (2002)

M. Mazzorana, G. Montoya, G.B. Mortuza, The centrosome: a target for cancer therapy. Curr. Cancer Drug Targets **11**, 600–612 (2011)

J.R. McLean, D. Chaix, M.D. Ohi, K.L. Gould, State of the APC/C: organization, function, and structure. Crit. Rev. Biochem. Mol. Biol. **46**(2), 118–136 (2011)

T.L. Megraw, T.C. Kaufman, The centrosome in *Drosophila* oocyte development. Curr. Top. Dev. Biol. **49**, 385–407 (2000)

T.L. Megraw, K. Li, L.R. Kao, T.C. Kaufman, The centrosomin protein is required for centrosome assembly and function during cleavage in *Drosophila*. Development **126**, 2829–2839 (1999)

T.L. Megraw, L.R. Kao, T.C. Kaufman, Zygotic development without functional mitotic centrosomes. Curr. Biol. **11**, 116–120 (2001)

C. Mencarelli, P. Lupetti, R. Dallai, New insights into the cell biology of insect axonemes. Int. Rev. Cell Mol. Biol. **268**, 95–145 (2008)

D. Menendez, A. Inga, M.A. Resnick, The expanding universe of p53 targets. Nat. Rev. Cancer **9**, 724–737 (2009)

X.Q. Meng, H.Y. Fan, Z.S. Zhong, G. Zhang, Y.L. Li, D.Y. Chen, Q.Y. Sun, Localization of gamma-tubulin in mouse eggs during meiotic maturation, fertilization, and early embryonic development. J. Reprod. Dev. **50**, 97–105 (2004)

L.S. Michel, V. Liberal, A. Chatterjee, R. Kirchwegger, B. Pasche, W. Gerald, M. Dobles, P.K. Sorger, V.V. Murty, R. Benezra, MAD2 haplo-insufficiency causes premature anaphase and chromosome instability in mammalian cells. Nature **409**, 355–359 (2001)

T. Miki-Noumura, Studies on the de novo formation of centrioles: aster formation in the activated eggs of sea urchin. J. Cell Sci. **24**, 203–216 (1977)

N. Minc, M. Piel, Predicting division plane position and orientation. Trends Cell Biol. **22**, 193–200 (2012)

N. Minc, D. Burgess, F. Chang, Influence of cell geometry on division-plane positioning. Cell **144**, 414–426 (2011)

J. Minshull, R. Golsteyn, C.S. Hill, T. Hunt, The A- and B-type cyclin associated cdc2 kinases in Xenopus turn on and off at different times in the cell cycle. EMBO J. **9**, 2865–2875 (1990)

J.M. Mitchison, The growth of single cells. I. *Schizosaccharomyces pombe*. Exp. Cell Res. **13**(2), 244–262 (1957)

J.M. Mitchison, The fission yeast, *Schizosaccharomyces pombe*. BioEssays **12**(4), 189–191 (1990)
J.M. Mitchison, P. Nurse, Growth in cell length in the fission yeast *Schizosaccharomyces pombe*. J. Cell Sci. **75**, 357–376 (1985)
I. Mizukami, J. Gall, Centriole replication II. Sperm formation in the fern, Marsilea, and the cycad, Zamia. J. Cell Biol. **29**, 97–111 (1966)
M.M. Mogensen, J.B. Tucker, Evidence for microtubule nucleation at plasma membrane-associated sites in *Drosophila*. J. Cell Sci. **88**(Pt 1), 95–107 (1987)
S. Moreno, P. Nurse, Regulation of progression through the G1 phase of the cell cycle by the rum1+ gene. Nature **367**, 236–242 (1994)
S. Moreno, P. Nurse, P. Russell, Regulation of mitosis by cyclic accumulation of p80cdc25 mitotic inducer in fission yeast. Nature **344**(6266), 549–552 (1990)
D.O. Morgan, Principles of CDK regulation. Nat. Cell. Biol. **374**, 131–134 (1995)
J.B. Moseley, A. Mayeux, A. Paoletti, P. Nurse, A spatial gradient coordinates cell size and mitotic entry in fission yeast. Nature **459**(7248), 857–860 (2009)
V. Mottier-Pavie, T.L. Megraw, *Drosophila* Bld10 is a centriolar protein that regulates centriole, basal body, and motile cilium assembly. Mol. Biol. Cell **20**, 2605–2614 (2009)
S. Moutinho-Pereira, A. Debec, H. Maiato, Microtubule cytoskeleton remodeling by acentriolar microtubule-organizing centers at the entry and exit from mitosis in *Drosophila* somatic cells. Mol. Biol. Cell **20**, 2796–2808 (2009)
A.W. Murray, M.W. Kirschner, Dominoes and clocks: the union of two views of the cell cycle. Science **246**, 614–621 (1989a)
A.W. Murray, M.W. Kirschner, Cyclin synthesis drives the early embryonic cell cycle. Nature **339**, 275–280 (1989b)
A. Musacchio, The molecular biology of spindle assembly checkpoint signaling dynamics. Curr. Biol. **25**, R1002–R1018 (2015)
A. Nakamura, H. Arai, N. Fujita, Centrosomal Aki1 and cohesin function in separase-regulated centriole disengagement. J. Cell Biol. **187**, 607–614 (2009)
H.J. Nam, J.M. van Deursen, Cyclin B2 and p53 control proper timing of centrosome separation. Nat. Cell Biol. **16**, 538–549 (2014)
R. Nash, G. Tokiwa, S. Anand, K. Erickson, A.B. Futcher, The WHI1+ gene of *Saccharomyces cerevisiae* tethers cell division to cell size and is a cyclin homolog. EMBO J. **7**(13), 4335–4346 (1998)
A. Nestor-Bergmann, G. Goddard, S. Woolner, Force and the spindle: mechanical cues in mitotic spindle orientation. Semin. Cell Dev. Biol. **34**, 133–139 (2014)
E.A. Nigg, Cyclin-dependent protein kinases: key regulators of the eukaryotic cell cycle. BioEssays **17**, 471–480 (1995)
E.A. Nigg, Centrosome aberrations: cause or consequence of cancer progression? Nat. Rev. Cancer **2**, 815–825 (2002)
E.A. Nigg, Origins and consequences of centrosome aberrations in human cancers. Int. J. Cancer **119**, 2717–2723 (2006)
E.A. Nigg, A.J. Holland, Once and only once: mechanisms of centriole duplication and their deregulation in disease. Nat. Rev. Mol. Cell Biol. **19**(5), 297–312 (2018)
E.A. Nigg, J.W. Raff, Centrioles, centrosomes, and cilia in health and disease. Cell **139**, 663–678 (2009)
H. Nishitani, Z. Lygerou, DNA replication licensing. Front. Biosci. **9**, 2115–2132 (2004)
H. Nishitani, P. Nurse, p65 cdc18 plays a major role controlling the initiation of DNA replication in fission yeast. Cell **83**, 397–405 (1995)
W. Nordstrom, J.M. Abrams, Guardian ancestry: fly p53 and damage-inducible apoptosis. Cell Death Differ. **7**(11), 1035–1038 (2000)
P. Nurse, Genetic control of cell size at cell division in yeast. Nature **256**, 547–551 (1975)
P. Nurse, Y. Bissett, Gene required in G1 for commitment to cell cycle and in G2 for control of mitosis in fission yeast. Nature **292**, 558–560 (1981)
M.J. O'Connell, K.A. Cimprich, G2 damage checkpoints: what is the turn-on? J Cell Sci. **118**(Pt 1), 1–6 (2005)
M.J. O'Connell, P. Nurse, How cells know they are in G1 or G2. Curr. Opin. Cell Biol. **6**(6), 867–871 (1994)
M.J. O'Connell, N.C. Walworth, A.M. Carr, The G2-phase DNA-damage checkpoint. Trends Cell Biol. **10**(7), 296–303 (2000)
M.J. O'Connell, M.J. Krien, T. Hunter, Never say never. The NIMA-related protein kinases in mitotic control. Trends Cell Biol. **13**(5), 221–228 (2003)
R.A. Oliveira, K. Nasmyth, Cohesin cleavage is insufficient for centriole disengagement in *Drosophila*. Curr. Biol. **23**, R601–R603 (2013)
M. Paintrand, M. Moudjou, H. Delacroix, M. Bornens, Centrosome organization and centriole architecture: their sensitivity to divalent cations. J. Struct. Biol. **108**, 107–128 (1992)
A. Paoletti, M. Bornens, Organisation and functional regulation of the centrosome in animal cells. Prog. Cell Cycle Res. **3**, 285–299 (1997)
A.B. Pardee, G1 events and regulation of cell proliferation. Science **246**, 603–608 (1989)
P. Patwari, R.T. Lee, Mechanical control of tissue morphogenesis. Circ. Res. **103**, 234–243 (2008)

N. Peel, N.R. Stevens, R. Basto, J.W. Raff, Overexpressing centriole-replication proteins in vivo induces centriole overduplication and de novo formation. Curr. Biol. **17**, 834–843 (2007)
L. Pelletier, N. Ozlu, E. Hannak, C. Cowan, B. Habermann, M. Ruer, T. Muller-Reichert, A.A. Hyman, The *Caenorhabditis elegans* centrosomal protein SPD-2 is required for both pericentriolar material recruitment and centriole duplication. Curr. Biol. **14**, 863–873 (2004)
L. Pelletier, E. O'Toole, A. Schwager, A.A. Hyman, T. Muller-Reichert, Centriole assembly in *Caenorhabditis elegans*. Nature **444**, 619–623 (2006)
A.L. Pereira, A.J. Pereira, A.R.R. Maia et al., Mammalian CLASP1 and CLASP2 cooperate to ensure mitotic fidelity by regulating spindle and kinetochore function. Mol. Biol. Cell **17**, 4526–4542 (2006)
D. Perez-Mongiovi, N. Malmanche, H. Bousbaa, C. Sunkel, Maternal expression of the checkpoint protein BubR1 is required for synchrony of syncytial nuclear divisions and polar body arrest in *Drosophila melanogaster*. Development **132**, 4509–4520 (2005)
C.M. Perou, T. Sørlie, M.B. Eisen et al., Molecular portraits of human breast tumors. Nature **406**, 747–752 (2000)
M. Peter, J. Nakagawa, M. Doree, J.C. Labbe, E.A. Nigg, In vitro disassembly of the nuclear lamina and M phase-specific phosphorylation of lamins by cdc2 kinase. Cell **61**, 591–602 (1990)
E. Petermann, A. Maya-Mendoza, G. Zachos, D.A. Gillespie, D.A. Jackson, K.W. Caldecott, Chk1 requirement for high global rates of replication fork progression during normal vertebrate S-phase. Mol. Cell Biol. **26**(8), 3319–3326 (2006)
J.-M. Peters, R.W. King, C. Hoog, M.W. Kirschner, Identification of BIME as a subunit of the anaphase-promoting complex. Science **274**, 1199–1201 (1996)
S.P. Peterson, M.W. Berns, Evidence for centriolar region RNA functioning in spindle formation in dividing PTK2 cells. J. Cell Sci. **34**, 289–301 (1978)
J.D. Pickett-Heaps, The autonomy of the centriole: fact or fallacy? Cytobios **3**, 205–214 (1971)
M. Piel, P. Meyer, A. Khodjakov, C.L. Rieder, M. Borners, The respective contributions of the mother and daughter centrioles to centrosome activity and behavior in vertebrate cells. J. Cell Biol. **149**, 317–330 (2000)
M. Piel, J. Nordberg, U. Euteneuer, M. Bornens, Centrosome-dependent exit of cytokinesis in animal cells. Science **291**, 1550–1553 (2001)
A. Prat, J.S. Parker, O. Karginova, C. Fan, C. Livasy, J.I. Herschkowitz, X. He, C.M. Perou, Phenotypic and molecular characterization of the claudin-low intrinsic subtype of breast cancer. Breast Cancer Res. **12**, R68 (2010)
N.J. Quintyne, J.E. Reing, D.R. Hoffelder, S.M. Gollin, W.S. Saunders, Spindle multipolarity is prevented by centrosomal clustering. Science **307**, 127–129 (2005)
J.A. Raaijmakers, R.G.H.P. van Heesbeen, J.L. Meaders, E.F. Geers, B. Fernandez-Garcia, R.H. Medema, M.E. Tanenbaum, Nuclear envelope-associated dynein drives prophase centrosome separation and enables Eg5-independent bipolar spindle formation. EMBO J. **31**, 4179–4190 (2012)
A.E. Radulescu, D.W. Cleveland, NuMA after 30 years: the matrix revisited. Trends Cell Biol. **20**, 214–222 (2010)
P.N. Rao, R.T. Johnson, Mammalian cell fusion: studies on the regulation of DNA synthesis and mitosis. Nature **225**, 159–164 (1970)
P.N. Rao, R.T. Johnson, Mammalian cell fusion. IV. Regulation of chromosome formation from interphase nuclei by various chemical compounds. J. Cell Physiol. **78**(2), 217–223 (1971)
R. Rappaport, Cytokinesis in animal cells. Int. Rev. Cytol. **31**, 169–213 (1971)
R. Rappaport, Establishment of the mechanism of cytokinesis in animal cells. Int. Rev. Cytol. **105**, 245–281 (1986)
J.B. Rattner, S.G. Phillips, Independence of centriole formation and DNA synthesis. J. Cell Biol. **57**, 359–372 (1973)
D. Razafsky, D. Hodzic, Bringing KASH under the SUN: the many faces of nucleo-cytoskeletal connections. J. Cell Biol. **186**, 461–472 (2009)
E. Rebollo, C. Gonzalez, Visualizing the spindle checkpoint in *Drosophila* spermatocytes. EMBO Rep. **1**, 65–70 (2000)
E. Rebollo, S. Llamazares, J. Reina, C. Gonzalez, Contribution of noncentrosomal microtubules to spindle assembly in *Drosophila* spermatocytes. PLoS Biol. **2**, E8 (2004)
L. Regue, S. Sdelci, M.T. Bertran, C. Caelles, D. Reverter, J. Roig, DYNLL/LC8 protein controls signal transduction through the Nek9/Nek6 signaling module by regulating Nek6 binding to Nek9. J. Biol. Chem. **286**, 18118–18129 (2011)
N. Rhind, B. Furnari, P. Russell, Cdc2 tyrosine phosphorylation is required for the DNA damage checkpoint in fission yeast. Genes Dev. **11**, 504–511 (1997)
C.L. Rieder, G.G. Borisy, The centrosome cycle in PtK2 cells: asymmetric distribution and structural changes in the pericentriolar material. Biol. Cell **44**, 117–132 (1982)
C.L. Rieder, R.W. Cole, Entry into mitosis in vertebrate somatic cells is guarded by a chromosome damage checkpoint that reverses the cell cycle when triggered during early but not late prophase. J. Cell Biol. **142**(4), 1013–1022 (1998)
B. Riggs, W. Rothwell, S. Mische, G.R. Hickson, J. Matheson, T.S. Hays, G.W. Gould, W. Sullivan, Actin cytoskeleton remodeling during early *Drosophila* furrow formation requires recycling endosomal components nuclear-fallout and Rab11. J. Cell Biol. **163**, 143–154 (2003)

M.G. Riparbelli, W.G. Whitfield, R. Dallai, G. Callaini, Assembly of the zygotic centrosome in the fertilized *Drosophila* egg. Mech. Dev. **65**, 135–144 (1997)

M.G. Riparbelli, R. Giordano, G. Callaini, Centrosome inheritance in the parthenogenetic egg of the collembolan Folsomia candida. Cell Tissue Res. **326**, 861–872 (2006)

E. Robbins, G. Jentzsch, A. Micali, The centriole cycle in synchronized HeLa cells. J. Cell Biol. **36**, 329–339 (1968)

A. Rodrigues-Martins, M. Bettencourt-Dias, M. Riparbelli, C. Ferreira, I. Ferreira, G. Callaini, D.M. Glover, DSAS-6 organizes a tube-like centriole precursor, and its absence suggests modularity in centriole assembly. Curr. Biol. **17**, 1465–1472 (2007)

A. Rodrigues-Martins, M. Riparbelli, G. Callaini, D.M. Glover, M. Bettencourt-Dias, From centriole biogenesis to cellular function: centrioles are essential for cell division at critical developmental stages. Cell Cycle **7**, 11–16 (2008)

G.C. Rogers, N.M. Rusan, D.M. Roberts, M. Peifer, S.L. Rogers, The SCF Slimb ubiquitin ligase regulates Plk4/Sak levels to block centriole reduplication. J. Cell Biol. **184**, 225–239 (2009)

R. Roskoski Jr., ERK1/2 MAP kinases: structure, function, and regulation. Pharmacol. Res. **66**, 105–143 (2012)

F. Ruiz, J. Beisson, J. Rossier, P. Dupuis-Williams, Basal body duplication in Paramecium requires gamma-tubulin. Curr. Biol. **9**, 43–46 (1999)

D. Sabino, D. Gogendeau, D. Gambarotto, M. Nano, C. Pennetier, F. Dingli, G. Arras, D. Loew, R. Basto, Moesin is a major regulator of centrosome behavior in epithelial cells with extra centrosomes. Curr. Biol. **25**, 879–889 (2015)

S. Saburi, I. Hester, E. Fischer, M. Pontoglio, V. Eremina, M. Gessler, S.E. Quaggin, R. Harrison, R. Mount, H. McNeill, Loss of Fat4 disrupts PCP signaling and oriented cell division and leads to cystic kidney disease. Nat. Genet. **40**, 1010–1015 (2008)

I. Samejima, P.C. Lourenco, H.A. Snaith, K.E. Sawin, Fission yeast mto2p regulates microtubule nucleation by the centrosomin-related protein mto1p. Mol. Biol. Cell **16**, 3040–3051 (2005)

L. Santarpia, T. Iwamoto, A. Di Leo et al., DNA repair gene patters as prognostic and predictive factors in molecular breast cancer subtypes. Oncologist **18**, 1063–1073 (2013)

T. Savin, N.A. Kurpios, A.E. Shyer, P. Florescu, H. Liang, L. Mahadevan, C.J. Tabin, On the growth and form of the gut. Nature **476**, 57–62 (2011)

K.E. Sawin, K. LeGuellec, M. Philippe, T.J. Mitchison, Mitotic spindle organization by a plus-end-directed microtubule motor. Nature **359**, 540–543 (1992)

G. Schatten, The centrosome and its mode of inheritance: the reduction of the centrosome during gametogenesis and its restoration during fertilization. Dev. Biol. **165**, 299–335 (1994)

H. Schatten, Q.Y. Sun, The functional significance of centrosomes in mammalian meiosis, fertilization, development, nuclear transfer, and stem cell differentiation. Environ. Mol. Mutagen. **50**(8), 620–636 (2009)

H. Schatten, G. Schatten, D. Mazia, R. Balczon, C. Simerly, Behavior of centrosomes during fertilization and cell division in mouse oocytes and in sea urchin eggs. Proc. Natl. Acad. Sci. U. S. A. **83**, 105–109 (1986)

T.I. Schmidt, J. Kleylein-Sohn, J. Westendorf, M. Le Clech, S.B. Lavoie, Y.D. Stierhof, E.A. Nigg, Control of centriole length by CPAP and CP110. Curr. Biol. **19**, 1005–1011 (2009)

D. Schnerch, E.A. Nigg, Structural centrosome aberrations favor proliferation by abrogating microtubule-dependent tissue integrity of breast epithelial mammospheres. Oncogene **35**, 2711–2722 (2016)

L. Schockel, M. Möckel, B. Mayer, D. Boos, O. Stemmann, Cleavage of cohesin rings coordinates the separation of centrioles and chromatids. Nat. Cell Biol. **13**, 966–972 (2011)

M. Schuh, J. Ellenberg, Self-organization of MTOCs replaces centrosome function during acentrosomal spindle assembly in live mouse oocytes. Cell **130**, 484–498 (2007)

C. Schulein-Volk, E. Wolf, J. Zhu et al., Dual regulation of Fbw7 function and oncogenic transformation by Usp28. Cell Rep. **9**, 1099–1109 (2014)

O. Sercin, J.C. Larsimont, A.E. Karambelas, V. Marthiens, V. Moers, B. Boeckx, M. Le Mercier, D. Lambrechts, R. Basto, C. Blanpain, Transient PLK4 overexpression accelerates tumorigenesis in p53-deficient epidermis. Nat. Cell Biol. **18**, 100–110 (2016)

Y. Shang, C.C. Tsao, M.A. Gorovsky, Mutational analyses reveal a novel function of the nucleotide-binding domain of gamma-tubulin in the regulation of basal body biogenesis. J. Cell Biol. **171**, 1035–1044 (2005)

P. Sherline, R. Mascardo, Epidermal growth factor-induced centrosomal separation: mechanism and relationship to mitogenesis. J. Cell Biol. **95**, 316–322 (1982)

M. Shteinberg, Y. Protopopov, T. Listovsky, M. Brandies, A. Hershko, Phosphorylation of the cyclosome is required for its stimulation by Fizzy/cdc20. Biochem. Biophys. Res. Commun. **260**, 193–198 (1999)

W.T. Silkworth, I.K. Nardi, L.M. Scholl, D. Cimini, Multipolar spindle pole coalescence is a major source of kinetochore mis-attachment and chromosome mis-segregation in cancer cells. PLoS ONE **4**, e6564 (2009)

W.T. Silkworth, I.K. Nardi, R. Paul, A. Mogilner, D. Cimini, Timing of centrosome separation is important for accurate chromosome segregation. Mol. Biol. Cell **23**, 401–411 (2012)

K.H. Siller, C.Q. Doe, Spindle orientation during asymmetric cell division. Nat. Cell Biol. **11**, 365–374 (2009)

I. Simon, J. Barnett, N. Hannett et al., Serial regulation of transcriptional regulators in the yeast cell cycle. Cell **106**, 697–708 (2001)

J.H. Sir, M. Putz, O. Daly, C.G. Morrison, M. Dunning, J.V. Kilmartin, F. Gergely, Loss of centrioles causes chromosomal instability in vertebrate somatic cells. J. Cell Biol. **13**, 747–756 (2013)

G. Sluder, Centrosome duplication and its regulation in the higher animal cell, in *Centrosomes in Development and Disease*, ed. by E.A. Nigg (Wiley, Weinheim, 2004), pp. 167–189

G. Sluder, J.J. Nordberg, The good, the bad and the ugly: the practical consequences of centrosome amplification. Curr. Opin. Cell Biol. **16**, 49–54 (2004)

G. Sluder, C.L. Rieder, Centriole number and the reproductive capacity of spindle poles. J. Cell Biol. **100**, 887–896 (1985)

E. Smith, N. Hégarat, C. Vesely et al., Differential control of Eg5-dependent centrosome separation by Plk1 and Cdk1. EMBO J. **30**, 2233–2245 (2011)

J. Smith, L. Larue, D.A. Gillespie, Chk1 is essential for the development of murine epidermal melanocytes. Pigment Cell Melanoma Res. **26**(4), 580–585 (2013)

C.H. So, A. Michal, K.E. Komolov, J. Luo, J.L. Benovic, G protein-coupled receptor kinase 2 (GRK2) is localized to centrosomes and mediates epidermal growth factor-promoted centrosomal separation. Mol. Biol. Cell **24**, 2795–2806 (2013)

M.J. Solomon, T. Lee, M.W. Kirschner, Role of phosphorylation in p34cdc2 activation: identification of an activating kinase. Mol. Biol. Cell **3**, 13–27 (1992)

B.P. Sonnenblick, The early embryology of *Drosophila melanogaster*, in *Biology of Drosophila*, ed. by M. Demerec (Hafner Publishing Company, New York, 1950), pp. 62–167

T. Sorlie, C.M. Perou, R. Tibshirani et al., Gene expression patterns of breast carcinomas distinguish tumor subclasses with clinical implications. Proc. Natl. Acad. Sci. U. S. A. **98**, 10869–10874 (2001)

S.P. Sorokin, Reconstructions of centriole formation and ciliogenesis in mammalian lungs. J. Cell Sci. **3**, 207–230 (1968)

D. Splinter, M.E. Tanenbaum, A. Lindqvist et al., Bicaudal D2, dynein, and kinesin 1 associate with nuclear pore complexes and regulate centrosome and nuclear positioning during mitotic entry. PLoS Biol. **8**, e1000350 (2010)

V. Srsen, A. Merdes, The centrosome and cell proliferation. Cell Div. **1**, 26 (2006)

V. Srsen, X. Fant, R. Heald, C. Rabouille, A. Merdes, Centrosome proteins form an insoluble perinuclear matrix during muscle cell differentiation. BMC Cell Biol. **10**, 28 (2009)

S. St Clair, L. Giono, S. Varmeh-Ziaie, L. Resnick- Silverman, W.J. Liu, A. Padi, J. Dastidar, A. DaCosta, M. Mattia, J. J. Manfredi, DNA damage-induced downregulation of Cdc25C is mediated by p53 via two independent mechanisms: one involves direct binding to the cdc25C promoter. Mol. Cell **16**(5), 725–736 (2004)

D.A. Starr, KASH and SUN proteins. Curr. Biol. **21**, R414–R415 (2011)

A. Stathis, A. Oza, Targeting Wee1-like protein kinase to treat cancer. Drug News Perspect. **23**(7), 425–429 (2010)

B.E. Stead, C.J. Brandl, M.K. Sandre, M.J. Davey, Mcm2 phosphorylation and the response to replicative stress. BMC Genet. **13**, 36 (2012)

W. Steffen, R.W. Linck, Evidence for tektins in centrioles and axonemal microtubules. Proc. Natl. Acad. Sci. U. S. A. **85**, 2643–2647 (1988)

O. Stemmann, H. Zou, S.A. Gerber, S.P. Gygi, M.W. Kirschner, Dual inhibition of sister chromatid separation at metaphase. Cell **107**, 715–726 (2001)

N.R. Stevens, A.A. Raposo, R. Basto, D. St Johnston, J.W. Raff, From stem cell to embryo without centrioles. Curr. Biol. **17**, 1498–1503 (2007)

G.S. Stewart, B. Wang, C.R. Bignell, A.M. Taylor, S.J. Elledge, MDC1 is a mediator of the mammalian DNA damage checkpoint. Nature **421**(6926), 961–966 (2003)

J.C. Stinchcombe, E. Majorovits, G. Bossi, S. Fuller, G.M. Griffiths, Centrosome polarization delivers secretory granules to the immunological synapse. Nature **443**, 462–465 (2006)

G.F. Striedter, S. Srinivasan, E.S. Monuki, Cortical folding: when, where, how, and why? Annu. Rev. Neurosci. **38**, 291–307 (2015)

P. Strnad, S. Leidel, T. Vinogradova, U. Euteneuer, A. Khodjakov, P. Gonczy, Regulated HsSAS-6 levels ensure formation of a single procentriole per centriole during the centrosome duplication cycle. Dev. Cell **13**, 203–213 (2007)

E. Stubblefield, Centriole replication in mammalian cells, in *The Proliferation and Spread of Neoplastic Cells* (Williamson and Wilkins Co, Baltimore, 1968), pp. 175–193

E.G. Sturgill, R. Ohi, Kinesin-12 differentially affects spindle assembly depending on its microtubule substrate. Curr. Biol. **23**, 1280–1290 (2013)

T. Sudo, M. Nitta, H. Saya, N.T. Ueno, Dependence of paclitaxel sensitivity on a functional spindle assembly checkpoint. Cancer Res. **64**, 2502–2508 (2004)

Q.Y. Sun, H. Schatten, Centrosome inheritance after fertilization and nuclear transfer in mammals. Adv. Exp. Med. Biol. **591**, 58–71 (2007)

Z. Sun, J. Hsiao, D.S. Fay, D.F. Stern, Rad53 FHA domain associated with phosphorylated Rad9 in the DNA damage checkpoint. Science **281**(5374), 272–274 (1998)

F.D. Sweeney, F. Yang, A. Chi, J. Shabanowitz, D.F. Hunt, D. Durocher, *Saccharomyces cerevisiae* Rad9 acts as a Mec1 adaptor to allow Rad53 activation. Curr. Biol. **15**(15), 1364–1375 (2005)

D. Szöllösi, J.P. Ozil, De novo formation of centrioles in parthenogenetically activated, diploidized rabbit embryos. Biol. Cell **72**, 61–66 (1991)

D. Szöllösi, P. Calarco, R.P. Donahue, Absence of centrioles in the first and second meiotic spindles of mouse oocytes. J. Cell Sci. **11**, 521–541 (1972)

A. Szöllösi, H. Ris, D. Szöllösi, A. Debec, A centriole-free *Drosophila* cell line. A high voltage EM study. Eur. J. Cell Biol. **40**, 100–104 (1986)

C.G. Takizawa, D.O. Morgan, Control of mitosis by changes in the subcellular location of cyclin-B1-Cdk1 and Cdc25C. Curr. Opin. Cell Biol. **12**, 658–665 (2000)

M.E. Tanenbaum, R.H. Medema, Mechanisms of centrosome separation and bipolar spindle assembly. Dev. Cell **19**, 797–806 (2010)

M.E. Tanenbaum, L. Macurek, N. Galjart, R.H. Medema, Dynein, Lis1 and CLIP-170 counteract Eg5-dependent centrosome separation during bipolar spindle assembly. EMBO J. **27**, 3235–3245 (2008)

M.E. Tanenbaum, L. Macurek, A. Janssen, E.F. Geers, M. Alvarez-Fernandez, R. Medema, Kif15 cooperates with Eg5 to promote bipolar spindle assembly. Curr. Biol. **19**, 1703–1711 (2009)

Z. Tang, T.R. Coleman, W.G. Dunphy, Two distinct mechanisms for negative regulation of the Wee1 protein kinase. EMBO J. **12**, 3427–3436 (1993)

C.J. Tang, H.M. Hu, T.K. Tang, NuMA expression and function in mouse oocytes and early embryos. J. Biomed. Sci. **11**, 370–376 (2004)

C.J. Tang, R.H. Fu, K.S. Wu, W.B. Hsu, T.K. Tang, CPAP is a cell-cycle regulated protein that controls centriole length. Nat. Cell Biol. **11**, 825–831 (2009)

N. Tang, W.F. Marshall, M. McMahon, R.J. Metzger, G.R. Martin, Control of mitotic spindle angle by the RAS-regulated ERK1/2 pathway determines lung tube shape. Science **333**, 342–345 (2011)

Z. Tang, Y. Hu, Z. Wang, K. Jiang, C. Zhan, W.F. Marshall, N. Tang, Mechanical forces program the orientation of cell division during airway tube morphogenesis. Dev. Cell **44**, 313–325 (2018)

C. Tapia-Alveal, T.M. Calonge, M.J. O'Connell, Regulation of Chk1. Cell Div. **4**(1), 8 (2009)

A.M. Tassin, B. Maro, M. Bornens, Fate of microtubule-organizing centers during myogenesis in vitro. J. Cell Biol. **100**, 35–46 (1985)

S.S. Taylor, F. McKeon, Kinetochore localization of murine bub1 is required for normal mitotic timing and checkpoint response to spindle damage. Cell **89**, 727–735 (1997)

K.H. Thein, J. Kleylein-Sohn, E.A. Nigg, U. Gruneberg, Astrin is required for the maintenance of sister chromatid cohesion and centrosome integrity. J. Cell Biol. **178**, 345–354 (2007)

M. Thery, M. Bornens, Get round and stiff for mitosis. HFSP J. **2**, 65–71 (2008)

M. Thery, V. Racine, A. Pepin, M. Piel, Y. Chen, J.-B. Sibarita, M. Bornens, The extracellular matrix guides the orientation of the cell division axis. Nat. Cell Biol. **7**, 947–953 (2005)

M.S. Tibber, I. Kralj-Hans, J. Savage, P.G. Mobbs, G. Jeffery, The orientation and dynamics of cell division within the plane of the developing vertebrate retina. Eur. J. Neurosci. **19**, 497–504 (2004)

I. TikhonenkoI, V. Magidson, R. Gräf, A. Khodjakov, M.P. Koonce, A kinesin-mediated mechanism that couples centrosomes to nuclei. Cell Mol. Life Sci. **70**, 1285–1296 (2013)

A. Toso, J.R. Winter, A.J. Garrod, A.C. Amaro, P. Meraldi, A.D. McAinsh, Kinetochore-generated pushing forces separate centrosomes during bipolar spindle assembly. J. Cell Biol. **184**, 365–372 (2009)

W.Y. Tsang, B.D. Dynlacht, sSgo1, a guardian of centriole cohesion. Dev. Cell **14**, 320–322 (2008)

M.F. Tsou, T. Stearns, Mechanism limiting centrosome duplication to once per cell cycle. Nature **442**, 947–951 (2006)

M.F. Tsou, W.J. Wang, K.A. George, K. Uryu, T. Stearns, P.V. Jallepalli, Polo kinase and separase regulate the mitotic licensing of centriole duplication in human cells. Dev. Cell **17**, 344–354 (2009)

T. Tumbar, G. Guasch, V. Greco, C. Blanpain, W.E. Lowry, M. Rendl, E. Fuchs, Defining the epithelial stem cell niche in skin. Science **303**, 359–363 (2004)

Y. Uetake, J. Loncarek, J.J. Nordberg, C.N. English, S. La Terra, A. Khodjakov, G. Sluder, Cell cycle progression and de novo centriole assembly after centrosomal removal in untransformed human cells. J. Cell Biol. **176**, 173–182 (2007)

F. Uhlmann, F. Lottspeich, K. Nasmyth, Sister-chromatid separation at anaphase onset is promoted by cleavage of the cohesin subunit Scc1. Nature **400**, 37–42 (1999)

D. Vaizel-Ohayon, E.D. Schejter, Mutations in centrosomin reveal requirements for centrosomal function during early *Drosophila* embryogenesis. Curr. Biol. **9**, 889–898 (1999)

N.G.V. Van Kampen, *Stochastic Processes in Physics and Chemistry*, 3rd edn. (Elsevier, Amsterdam, 2007)

J.H. van Ree, H.J. Nam, K.B. Jeganathan, A. Kanakkanthara, J.M. van Deursen, Pten regulates spindle pole movement through Dlg1-mediated recruitment of Eg5 to centrosomes. Nat. Cell Biol. **18**, 814–821 (2016)

D. Vanneste, V. Ferreira, I. Vernos, Chromokinesins: localization-dependent functions and regulation during cell division. Biochem. Soc. Trans. **39**, 1154–1160 (2011)

H. Varmark, S. Llamazares, E. Rebollo, B. Lange, J. Reina, H. Schwarz, C. Gonzalez, Asterless is a centriolar protein required for centrosome function and embryo development in *Drosophila*. Curr. Biol. **17**, 1735–1745 (2007)

S.J. Vidwans, M.L. Wong, P.H. O'Farrell, Mitotic regulators govern progress through steps in the centrosome duplication cycle. J. Cell Biol. **147**, 1371–1378 (1999)

B. Vitre, A.J. Holland, A. Kulukian et al., Chronic centrosome amplification without tumorigenesis. Proc. Natl. Acad. Sci. U. S. A. **112**, E6321-6330 (2015)

E.K. Vladar, T. Stearns, Molecular characterization of centriole assembly in ciliated epithelial cells. J. Cell Biol. **178**, 31–42 (2007)

C. Vogel, A. Kienitz, R. Muller, H. Bastians, The mitotic spindle checkpoint is a critical determinant for topoisomerase-based chemotherapy. J. Biol. Chem. **280**, 4025–4028 (2005)

I.A. Vorobjev, E.S. Nadezhdina, The centrosome and its role in the organization of microtubules. Int. Rev. Cytol. **106**, 227–293 (1987)

N. Walworth, S. Davey, D. Beach, Fission yeast chk1 protein kinase links the rad checkpoint pathway to cdc2. Nature **363**, 368–371 (1993)

B. Wang, S. Matsuoka, P.B. Carpenter, S.J. Elledge, 53BP1, a mediator of the DNA damage checkpoint. Science **298** (5597), 1435–1438 (2002)

R.-H. Wang, H. Yu, C.-X. Deng, A requirement for breast-cancer-associated gene 1 (BRCA1) in the spindle checkpoint. Proc. Natl. Acad. Sci. **101**, 17108–17113 (2004)

L. Wang, K. Sharma, H.X. Deng, T. Siddique, G. Grisotti, E. Liu, R.P. Roos, Restricted expression of mutant SOD1 in spinal motor neurons and interneurons induces motor neuron pathology. Neurobiol. Dis. **29**, 400–408 (2008a)

X. Wang, Y. Yang, Q. Duan, N. Jiang, Y. Huang, Z. Darzynkiewicz, W. Dai, sSgo1, a major splice variant of Sgo1, functions in centriole cohesion where it is regulated by Plk1. Dev. Cell **14**, 331–341 (2008b)

X. Wang, J.W. Tsai, J.H. Imai, W.N. Lian, R.B. Vallee, S.H. Shi, Asymmetric centrosome inheritance maintains neural progenitors in the neocortex. Nature **461**, 947–955 (2009)

W.J. Wang, R.K. Soni, K. Uryu, M.-F. Bryan Tsou, The conversion of centrioles to centrosomes: essential coupling of duplication with segregation. J. Cell Biol. **193**, 727–739 (2011)

W. Wang, Z. Shi, S. Jiao, C. Chen, H. Wang, G. Liu, Q. Wang, Y. Zhao, M.I. Greene, Z. Zhou, Structural insights into SUN-KASH complexes across the nuclear envelope. Cell. Res. **22**, 1440–1452 (2012)

G.E. Ward, M.W. Kirschner, Identification of cell cycle-regulated phosphorylation sites on nuclear lamin C. Cell **61**, 561–577 (1990)

J.C. Waters, R.W. Cole, C.L. Rieder, The force-producing mechanism for centrosome separation during spindle formation in vertebrates is intrinsic to each aster. J. Cell Biol. **122**, 361–372 (1993)

R.A. Weinberg, The Retinoblastoma protein and cell cycle control. Cell **81**, 323–330 (1995)

T. Weinert, L. Hartwell, The RAD9 gene controls the cell cycle response to DNA damage in *Saccharomyces cerevisiae*. Science **241**, 317–322 (1988)

D.N. Wheatley, *The Centriole: A Central Enigma of Cell Biology* (Elsevier Biomedical Press, Amsterdam, 1982)

E.B. Wilson, *The Cell in Development and Inheritance*, ed. by H.F. Osborn (Macmillan, New York, 1896)

E.B. Wilson, *The Cell in Development and Heredity*, 3rd edn. (Macmillan, New York, 1928)

P.G. Wilson, Centriole inheritance. Prion **2**, 9–16 (2008)

E.B. Wilson, A.F. Huettner, The central bodies again. Science **73**, 447–448 (1931)

T. Wittmann, A. Hyman, A. Desai, The spindle: a dynamic assembly of microtubules and motors. Nat. Cell Biol. **3**(1), E28–E34 (2001)

R.W. Wong, G. Blobel, Cohesin subunit SMC1 associates with mitotic microtubules at the spindle pole. Proc. Natl. Acad. Sci. U. S. A. **105**, 15441–15445 (2008)

C. Wong, T. Stearns, Centrosome number is controlled by a centrosome-intrinsic block to reduplication. Nat. Cell Biol. **5**, 539–544 (2003)

S.Y. Wong, A.D. Seol, P.L. So, A.N. Ermilov, C.K. Bichakjian, E.E.H. Jr., A.A. Dlugosz, J.F. Reiter, Primary cilia can both mediate and suppress Hedgehog pathway-dependent tumorigenesis. Nat. Med. **15**, 1055–1061 (2009)

Y.L. Wong, J.V. Anzola, R.L. Davis et al., Reversible centriole depletion with an inhibitor of Polo-like kinase 4. Science **348**, 1155–1160 (2015)

L. Wu, K. Shiozaki, R. Aligue, P. Russell, Spatial organization of the Nim1-Wee1-Cdc2 mitotic control network in *Schizosaccharomyces pombe*. Mol. Biol. Cell **7**, 1749–1758 (1996)

T.P. Wyatt, A.R. Harris, M. Lam, Q. Cheng, J. Bellis, A. Dimitracopoulos, A.J. Kabla, G.T. Charras, B. Baum, Emergence of homeostatic epithelial packing and stress dissipation through divisions oriented along the long cell axis. Proc. Natl. Acad. Sci. U. S. A. **112**, 5726–5731 (2015)

F. Xiong, W. Ma, T.W. Hiscock et al., Interplay of cell shape and division orientation promotes robust morphogenesis of developing epithelia. Cell **159**, 415–427 (2014)

F.L. Xu, W.S. Saunders, Actin and microtubules: working together to control spindle polarity. Cancer Cell **14**, 197–199 (2008)

B. Xu, S.-T. Kim, M.B. Kastan, Involvement of Brca1 in S-Phase and G2-Phase checkpoints after ionizing irradiation. Mol. Cell. Biol. **21**, 3445–3450 (2001)

Y.J. Xu, M. Davenport, T.J. Kelly, Two-stage mechanism for activation of the DNA replication checkpoint kinase Cds1 in fission yeast. Genes Dev. **20**(8), 990–1003 (2006)

B. Xu, Z. Hao, K.N. Jha, Z. Zhang, C. Urekar, L. Digilio, S. Pulido, J.F. Strauss III, C.J. Flickinger, J.C. Herr, TSKS concentrates in spermatid centrioles during flagellogenesis. Dev. Biol. **319**, 201–210 (2008)

T. Yabe, X. Ge, F. Pelegri, The zebrafish maternal-effect gene cellular atoll encodes the centriolar component sas-6 and defects in its paternal function promote whole genome duplication. Dev. Biol. **312**, 44–60 (2007)

Y.M. Yamashita, D.L. Jones, M.T. Fuller, Orientation of asymmetric stem cell division by the APC tumor suppressor and centrosome. Science **301**, 1547–1550 (2003)

Y.M. Yamashita, A.P. Mahowald, J.R. Perlin, M.T. Fuller, Asymmetric inheritance of mother versus daughter centrosome in stem cell division. Science **315**, 518–521 (2007)

J. Yang, X. Liu, G. Yue, M. Adamian, O. Bulgakov, T. Li, Rootletin, a novel coiled-coil protein, is a structural component of the ciliary rootlet. J. Cell Biol. **159**, 431–440 (2002)

J. Yang, M. Adamian, T. Li, Rootletin interacts with C-Nap1 and may function as a physical linker between the pair of centrioles/basal bodies in cells. Mol. Biol. Cell **17**, 1033–1040 (2006)

F.X. Yu, Y. Zhang, H.W. Park, J.L. Jewell, Q. Chen, Y. Deng, D. Pan, S.S. Taylor, Z.C. Lai, K.L. Guan, Protein kinase A activates the Hippo pathway to modulate cell proliferation and differentiation. Genes Dev. **27**, 1223–1232 (2013)

G. Zachos, M.D. Rainey, D.A. Gillespie, Chk1-deficient tumour cells are viable but exhibit multiple checkpoint and survival defects. EMBO J. **22**(3), 713–723 (2003)

T.I. Zack, S.E. Schumacher, S.L. Carter, et al., Pan-cancer patterns of somatic copy number alteration. Nat. Genet.**45**, 1134–1140 (2013)

Y. Zeng, K.C. Forbes, Z. Wu, S. Moreno, H. Piwnica-Worms, T. Enoch, Replication checkpoint requires phosphorylation of the phosphatase cdc25 by cds1 or chk1. Nature **395**, 507–510 (1998)

X. Zhang, K. Lei, X. Yuan, X. Wu, Y. Zhuang, T. Xu, R. Xu, M. Han, SUN1/2 and Syne/Nesprin-1/2 complexes connect centrosome to the nucleus during neurogenesis and neuronal migration in mice. Neuron **64**, 173–187 (2009)

Y. Zhang, O. Foreman, D.A. Wigle, F. Kosari, G. Vasmatzis, J.L. Salisbury, J. van Deursen, P.J. Galardy, USP44 regulates centrosome positioning to prevent aneuploidy and suppress tumorigenesis. J. Clin. Invest. **122**, 4362–4374 (2012)

C.Z. Zhang, A. Spektor, H. Cornils, J.M. Francis, E.K. Jackson, S. Liu, M. Meyerson, D. Pellman, Chromothripsis from DNA damage in micronuclei. Nature **522**, 179–184 (2015)

Z. Zhou, S. Elledge, DUN1 encodes a protein kinase that controls the DNA damage response in yeast. Cell **75**, 1119–1127 (1993)

Cell Proliferation, Survival, Necrosis and Apoptosis

16

Abstract

There are fundamental cellular processes, such as proliferation, survival, necrosis and apoptosis that play a role in physiological and pathological processes. These pathways are described and discussed in the light of cell mechanics. In this chapter, the concept of mechanical cell competition through introducing the key parameters governs the control of competitive interactions and also introduces latest advances in the presentation of the signal transduction pathways, which regulates compaction-triggered cell death. Thereafter, the relationship between mechanical cell competition and other mechanisms of cellular competition is presented. Moreover, a focus is on the relevance of mechanical cell competition during the initiation of tumors and their progression, whereby the effect of the mechanical microenvironment is included and related to tumor growth, and the effect of residual stress on tumor progression and the intratumoral competition is discussed.

16.1 Cell Proliferation and Survival

Phosphatidylinositide 3 kinases (PI3Ks) and their downstream mediators AKT and mammalian target of rapamycin (mTOR) belong to the key compounds of the PI3K/AKT/mTOR signal transduction cascade, which facilitates fundamental cellular processes such as metabolism, proliferation and survival. In the cause of tumorigenesis, these cellular functions are well known; however, using pluripotent stem cells, it has been shown that this pathway is crucial for the development and the differentiation of cells. Here is recent in vitro and in vivo evidence for the function of the PI3K/AKT/mTOR pathway that controls pluripotency and differentiation, with particular emphasis on the molecular mechanisms at the root of these functions.

During the developmental process, cellular differentiation is a critical event, through which all essential tissues of the adult organism are generated. This is characterized by the spatial and temporal triggering of signaling pathways, which is traditionally established through gain and loss of function investigations in developmental systems such as *Zebrafish*, *Xenopus* and *Mice*. The genetic manipulation of signal transduction compounds facilitates defects in either specific tissues or developmental phases, which simply signifies that different pathways specify specific routes. Nevertheless, it is becoming more and more apparent that this is a crude oversimplification, as the signaling pathways do not work in isolation, but instead are involved in a large variety of crosstalk interactions that collectively govern cell fate. Thus, even paths that do not seem to designate a lineage separately can

C. T. Mierke, *Cellular Mechanics and Biophysics*, Biological and Medical Physics, Biomedical Engineering, https://doi.org/10.1007/978-3-030-58532-7_16

help to reinforce or impede routes that are crucial to the designation of a particular lineage. The inherent complexity and redundancy of signal transmission, nevertheless, complicates the molecular decomposition of such crosstalk events (Yu and Cui 2016).

The PI3K/AKT/mTOR signal transduction participates in a number of processes that are crucial in conveying a wide range of cellular function, encompassing nutritional intake, anabolic reactions, cell growth and overall survival. The key constituents of the pathway, such as PI3Ks, AKT and the mammalian target of rapamycin (mTOR, referred to as mechanistic TOR), have been identified as often hyperactivated in most cancers and have consequently become the main focus of numerous studies in this field (Vanhaesebroeck et al. 2010; Thorpe et al. 2015). In contrary, the significance of this route for development is underestimated, although the disappearance of one of the key components often contributes to embryonic lethality (Bi et al. 1999; Peng et al. 2003; Murakami et al. 2004).

The embryonic lethality arising from the interruption of PI3K/AKT/mTOR signal transduction impedes the questioning of this pathway during the differentiation through traditional patterns of development. The generation of pluripotent stem cells (PSCs), comprising embryonic stem cells (ESCs) and induced PSCs (iPSCs), has eased this situation to a certain degree. Unlimited self-renewal of these cells is guaranteed, and they can also produce all cells of the adult organism, thus providing a unique cellular model to explore cellular diversification. This, together with increasingly defined culture conditions and better tools for genetic manipulation, has made it possible to transcend the technically demanding task of questioning the molecular foundation for cross-linked interactions. Consequently, recent investigations have commenced to define the contribution of PI3K/AKT/mTOR signaling to maintaining the capacity of PSCs to self-renew and diversify (Paling et al. 2004; Zhou et al. 2009; Singh et al. 2012; Yu et al. 2015), and it has been shown that this signaling pathway is critical for embryonic development and essential for the self-regeneration of PSCs. The functions of this route are widely discussed during embryonic development and in defining the cell fate of PSCs, with a special emphasis on the presumed molecular mechanisms underpinning these functions.

The PI3K/AKT/mTOR Signaling

PI3Ks are a singular family of intracellular lipid kinases which phosphorylate the 3′ hydroxyl group of the inositol ring of phosphatidylinositides (PtdIns) and are classified into three classes. The most investigated class is PI3K, which promotes the conversion of membrane-bound phosphatidylinositol-(4,5)-bisphosphate (PtdIns(4,5)P2 or synonymously referred to PIP2) into phosphatidylinositol-(3,4,5)-trisphosphate (PtdIns(3,4,5)P3 or synonymously referred to PIP3) (Vanhaesebroeck et al. 2010; Thorpe et al. 2015). Thus, PIP3 serves as a secondary messenger to facilitate the enrollment and the activation of kinases that share the domain of pleckstrin homology (PH), such as PI3K-dependent kinase-1 (PDK1). The signal time of PIP3 is regulated by phosphatase and tensin homologue (PTEN), which counteracts the activity of PI3K. The serine/threonine kinase AKT, also referred to as protein kinase B (PKB), owns a PH domain and is enrolled together with PDK1 in the plasma membrane. Phosphorylation of the amino acid residues T308 and S473 through PDK1 and mTORC2 is indispensable for complete AKT activation. After activation, AKT is able to phosphorylate many target proteins, in particular glycogen synthase kinase 3 (GSK3), tuberous sclerosis 2 (TSC2), caspase 9 and PRAS40 (AKT1S1), thereby accounting for the comparatively broad range of downstream activities in stimulating cell proliferation, differentiation, apoptosis, angiogenesis and metabolism.

mTOR is an evolutionary preserved serine/threonine protein kinase, which is a member of the PI3K-related kinase (PIKK) family. The availability of the active mTOR site is governed in part by mTOR-associated proteins that constitute two distinct assemblies, such as the mTOR complex 1 (mTORC1) and the mTOR complex 2 (mTORC2). These assemblies vary in their components, the mode of activation and the sensitivity to rapamycin (Yang et al. 2013a; Liu et al. 2015). Even though mTOR interfaces with DEPTOR and mLST8 in both compounds, the other key elements differ between the two. In place of RAPTOR (RPTOR) and PRAS40 in mTORC1, mTORC2 contains rapamycin-insensitive companion of mTOR (RICTOR) and stress-activated mammalian protein kinase that interacts with protein 1 (mSIN1 or synonymously referred to as MAPKAP1). The activation of mTORC1 pathway is governed through a large variety of intracellular and extracellular cues, including energy status, growth factors, nutrients and stress, where the RAG, the RAS homolog enriched in brain (RHEB) guanosine triphosphatases (GTPase) and TSC1/2 fulfill crucial and specific individual tasks within the cell. After activation, the mTORC1 guides several cellular functions, including cell growth, the synthesis of proteins and autophagy through S6 kinase (S6K or synonymously referred to as RPS6K), eukaryotic translation initiation factor 4E-binding protein 1 (4E-BP1 or synonymously referred to as EIF4EBP1) and ULK1 (Laplante and Sabatini 2012). Contrary to the activation of mTORC1 and mTORC2, the molecules RICTOR and mSIN1 are necessary for the stabilization and function, even though accurate molecular mechanisms through which they act are still elusive (Laplante and Sabatini 2012). The downstream effectors of the mTORC2 pathway, which have been revealed, are primarily members of the AGC kinase family, such as AKT, protein kinase C (PKC) and serum and glucocorticoid-induced kinases (SGKs) that impact the organization of the cytoskeleton, cell survival, the homeostasis of lipids and metabolism (Laplante and Sabatini 2012).

16.2 Apoptosis, Autophagy and Necrosis

The death of cells represents a crucial biological process that is highly relevant for physiological growth and development. There are three traditional types of cell death, such as apoptosis, autophagy, and necrosis, all of which exhibit specific distinguishable morphological characteristics through the activation of specific signal transduction pathways (Fig. 16.1). Due to recent advances, it has been understood that these cell death processes can communicate with each other through interconnected, even intersecting signal transduction pathways, and that the final cell fortune is the outcome of the cooperation of different cell death mechanisms (Chen et al. 2018). The following discussion offers an insight into the independence of these three types of cell death and the connections between them and examines the significance of cell death for the specific circumstances of human diseases, most notably neurodegenerative diseases and cancer.

For single-celled organisms, cell death marks the termination of life. For multicellular organisms, nevertheless, cell death represents a fundamental biological event for physiological growth and development. The deregulation of cell death is implicated in the pathogenesis of a variety of human disorders, including neurodegenerative diseases and cancer (Degterev and Yuan 2008). Three classical types of cell death encompassing apoptosis, autophagy, and necrosis exhibit specific morphological characteristics through the activation of distinct signal transductions routes (Kroemer et al. 1998). In a nutshell, apoptosis is a caspase-driven programmed cell death (Ellis and Horvitz 1986; Miura et al. 1993) featuring chromosome condensation, nuclear fragmentation and the formation of membrane blebs (Kerr et al. 1972). In opposition to apoptosis, necrosis is regarded as an uncontrolled,

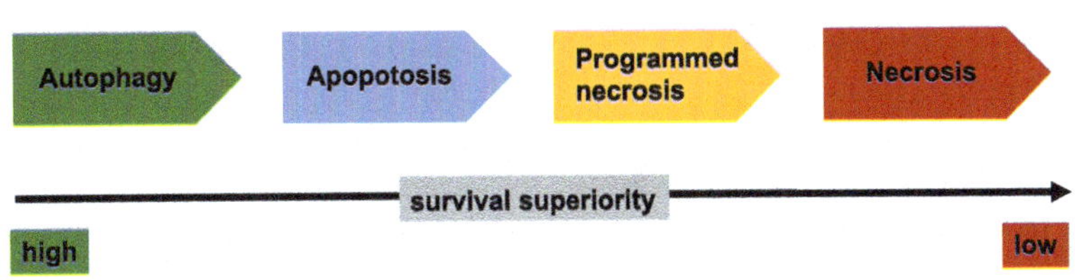

Fig. 16.1 The survival priority of various types of cell death

random cell death induced by non-specific or non-physiological stress factors and distinguished from apoptosis through the elongation of cell organelles, breakage of the plasma membrane and consequent inflammatory responses through the liberation of the intracellular material (Trump et al. 1965; Kono and Rock 2008).

The third type of cell death, the autophagy, is associated with the development of the autophagosome, a two-layer vesicle comprising injured organelles, proteins and other cytoplasmic substances. The autophagosomes merge together with the lysosomes, decompose cellular macromolecules and organelles and generate renewable energy and metabolites for cells (He and Klionsky 2009). Autophagy functions as a survival-promoting mechanism; however, it is also capable of inducing autophagous cell death, which is presently an active branch of research in the investigation of cell death (Gump and Thorburn 2011; Mariño et al. 2014). The knowledge has been quickly broadened in recent decades due to the great progress in cell death investigations. The identification of the programmed type of necrosis (Hitomi et al. 2008) has shifted the way people perceive necrosis. More fundamentally; however, there is a gradual awareness that the underlying molecular mechanisms of distinct types of cell death are diverse, but also intersect. A variety of signaling pathways exist that are capable of separately managing various types of cell death. Nevertheless, they are intertwined, can be triggered concurrently and can operate in parallel in reaction to stress within the cells.

Necrosis and apoptosis are two kinds of cell death involving distinct mechanisms (Kerr et al. 1972; Vandenabeele et al. 2010). Autophagy can be characterized as a break-down mechanism instead of a type of cell death, despite the fact that it can also lead to cell death. Among the types of cell death, autophagy offers the highest survival prevalence, closely followed by apoptosis, while necrosis exhibits the lowest survival prevalence. Autophagy is intuitively evoked preceding apoptosis when cells are excited by stress, and apoptosis instead of necrosis is evoked when autophagy is suppressed or disabled (He and Klionsky 2009; Alers et al. 2012; Luo and Rubinsztein 2010; Declercq et al. 2009; Wang et al. 2008a). Thus, two or three types of cell death can be initiated at the same time or consecutively when cells are subjected to specific stimulants. Placing the three types of cell death on one axis in line with their survival predominance, autophagy and necrosis would be positioned at two opposed ends, whereas apoptosis is positioned in the mid-axis; in the same way, the programmed necrosis would be positioned between necrosis and apoptosis (Fig. 16.1). The different types of cell death are presented in the following section, and thereby, the specific mechanisms involved in each type of cell death are discussed. Moreover, the interconnections between them are highlighted and the relevance of distinct types of cell death on disease treatment are outlined.

16.2.1 Characteristics of Apoptosis

Apoptosis is commonly regarded as caspase-facilitated programmed cell death (Ellis and Horvitz 1986; Miura et al. 1993; Garrido and Kroemer 2004; Galluzzi et al. 2008). Apoptotic cells exhibit pronounced morphological features, comprising cellular shrinkage, chromosome condensation, nuclear fragmentization (late apoptotic stage), plasma membrane blebbing and the generation of apoptotic bodies and display biochemical alterations, such as the exposition of phosphatidyl-l-serine on the outer plasma membrane (early stage apoptotic stage) (Lockshin and Zakeri 2001; Degterev et al. 2003; Danial and Korsmeyer 2004). Apoptosis can be triggered through the death receptor-driven apoptosis pathway, the mitochondria-driven apoptosis pathway or through stress-initiated endoplasmic reticulum (ER) apoptosis pathways (Nakagawa et al. 2000).

The death receptor-mediated apoptosis pathway is enabled through the attachment of the Fas ligand, tumor necrosis factor α (TNF-α) or TRAIL to the respective death receptors (Lavrik et al. 2005a; Wang and El-Deiry 2003). The adaptor protein FADD (Lavrik et al. 2005a; Duprez et al.

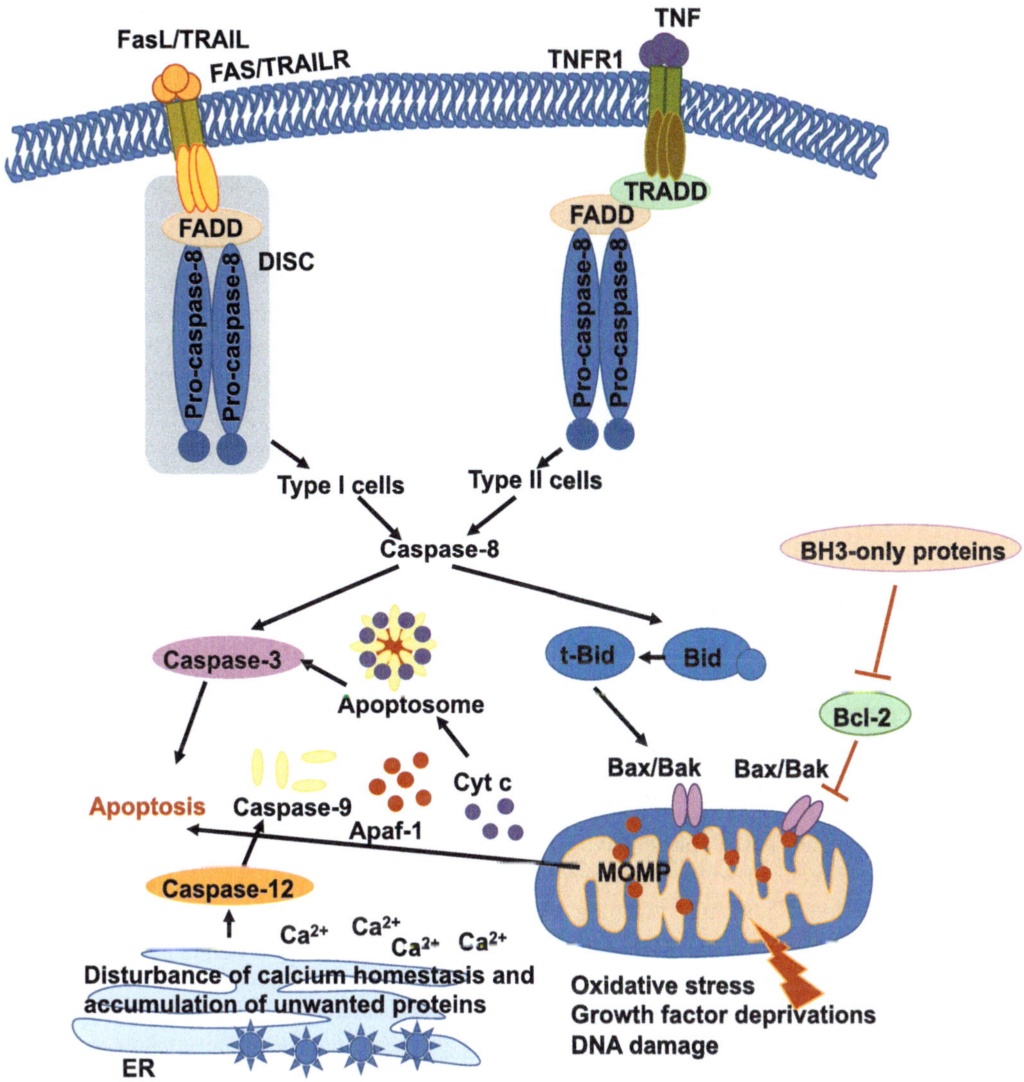

Fig. 16.2 Apoptotic mechanisms. In the exogenous pathway the binding of FASL, TNF-alpha or Trial to their specific receptors transfers the procaspase-8 to caspase-8 via autohydrolysis. In type I cells, activated capase-8 activates caspase-3 leading to apoptosis. In type II cells, activated capaspase-8 hydrolyses Bid to tBid, which interferes with Bax/Bak, that is located on mitochondria and subsequently induces apoptosis. In the intrinsic apoptosis pathway, DNA damage, growth factor removal, oxidative stress and toxic damage destroys the homeostasis of the mitochondria, which is regulated by Bcl-2 family members and causes elevated mitochondrial membrane permeability releasing cytochrome c. Cytochrome c interacts with Apaf-1 and caspase-9 in order to activate capase-3 for the induction of apoptosis. In the ER stress-induced apoptosis pathway, the disorder of the calcium homeostasis and excessive unwanted protein accumulation in the ER causes caspase-12-driven apoptosis, where caspase-12 translocates into the cytosol to cleave caspase-9 and thereby activate caspase-3

2009) and the protein pro-caspase-8 build a complex that is the death-inducing signal complex (DISC). In DISC, Pro-caspase-8 is triggered through autohydrolysis. Activated caspase-8 converts the apoptosis signal either through the activation of caspase-3 or through the splitting of Bid resulting in truncated Bid (tBid). tBid relocates into the mitochondria, leading to conformational alterations in Bax and Bak and their oligomerization to form pores in the exterior mitochondrial membrane (Fig. 16.2) (Wang and Tjandra 2013; Fan et al. 2005).

The mitochondria-dependent signal transduction pathway can be triggered by several stress factors including damage to DNA, deprivation of growth factors and oxidative stress (McIlwain et al. 2013; Favaloro et al. 2012). The Bcl-2 protein family governs this inherent route through adjusting the permeability of the external mitochondrial membrane (Ricci et al. 2004; Özören and El-Deiry 2002; Gross et al. 1999). When liberated into the cytoplasm by mitochondria, cytochrome c together with Apaf-1 stimulates caspase-9 activation, activating effector caspases (Yuan and Akey 2013; Soldani and Scovassi 2001) to initiate a series of proteolytic events.

In extension, ER strains can trigger apoptosis, such as disruption of calcium homeostasis, excessively unfolded or incorrectly folded protein aggregation in the ER, nutritional deficiency and hypoxia. This apoptosis is facilitated through caspase-12, which represents an ER-resistant caspase (Nakagawa et al. 2000). Activated caspase-12 splits caspase-9 directly subsequent to the translocation from ER into the cytosol, immediately after which caspase-3 becomes activated (Morishima et al. 2002). The molecular mechanisms of caspase-12 activation under ER stress involve the establishment of a complex with the inositol-demanding enzyme1α TNF receptor-associated factor 2 (TRAF2) complex (Yoneda et al. 2001), or through calpains, which represents a family of Ca^{2+}-driven intracellular cysteine proteases (Tan et al. 2006).

16.2.2 Characteristics of Autophagy

Autophagy occurs as a self-destructive process in reaction to a variety of stresses, including lack of nutrients, injury of organelles damage, hypoxia, reactive oxygen species (ROS), ER stress and treatment with pharmaceuticals. The process of autophagy comprises four major steps, such as the initiation, the nucleation, the merging of the autophagosome and the lysosome, and finally the hydrolysis. The molecular mechanisms of autophagy are based on investigations in yeast. A number of regulatory molecules for autophagy have been detected through genetic screens in yeast, in specific proteins closely associated with autophagy (Atg), which are key actors in autophagy. The composition and clustering of the Atg1 complex, encompassing Atg1, Atg13, Atg17, Atg29 and Atg31 (Cheong et al. 2008; Kawamata et al. 2008), is necessary for the development of the phagophore involved in the step of initialization (Hamasaki et al. 2013). Nevertheless, in mammals, the UNC-51-like kinase-1 (ULK)-mAtg13-FIP200 complex is generated, containing the homologous equivalents to yeast Atg1, Atg13 and Atg17 (Jung et al. 2009). At the nucleation stage, phagophore production across the ER and other membranes is regulated by a class III PI-3 kinase complex of VPS34, Atg6 (known as Beclin1 in mammals), Atg14 and Vps15. Atg9 and the vesicle membrane protein VMP1 circulating in the Golgi complex, autophagosomes and endosomes are possibly participating in the transportation of lipids to the isolation membrane (He and Klionsky 2007; Molejon et al. 2013).

The widening and closing of the autophagosome necessitates two ubiquitin-like protein-conjugated systems, specifically Atg12 and Atg8, which is synonymously referred to as LC3 in mammals (Ohsumi 2001). The Atg12 system comprises five Atg different proteins such as Atg5, Atg7, Atg10, Atg12 and Atg16 (Mizushima et al. 1998, 1999; Shintani et al. 1999). Atg12 is triggered through Atg7, which functions as an E1-like enzyme (Mizushima et al. 1998), and is subsequently passed on to the E2-like enzyme Atg10 (Shintani et al. 1999). Ultimately, the C-terminal glycine of Atg12 links in a covalent manner to the Lys1 (Kirisako et al. 2000) side chain of Atg5 prior to binding to the dimer protein Atg16 and building the E3-like complex (Mizushima et al. 1998). The Atg8 framework comprising the four Atg proteins Atg3, Atg4, Atg7 and Atg8 constitutes another ubiquitin-like protein-conjugated system (Ichimura et al. 2000). Atg8 is split apart with Atg4, a cysteine protease, and uncovers its C-terminal glycine residue (LC3 I in mammals) (Kirisako et al. 2000). Moreover, Atg8 is additionally switched to an activated state through Atg7, an E1-like enzyme, and consequently, it is passed on to Atg3, an E2-like enzyme (Ichimura et al. 2000), prior to covalent attachment to the amidogen of PE by the E3-like Atg12-Atg5-Atg16 complex (Ichimura et al. 2000; Hanada et al. 2007).

The covalent Atg8-PE structure, known as LC3 II in mammals, endows the Atg8 membrane with tethering and hemifusion capacity and is crucial for the generation of autophagosomes. LC3 II is related to both the outer and the inner membrane of the autophagosome and represents a characteristic marker for the development of autophagy. The covalent construction of Atg8-PE can be split in a reversible fashion to Atg8 for recycling of Atg8 through Atg4. The fusion of autophagosome and lysosome is facilitated through SNARE (soluble *N*-ethylmaleimide-sensitive factor attachment protein receptor)-like proteins (Diao et al. 2017; Cheng et al. 2017; Itakura et al. 2012; Nair et al. 2011). Ultimately, a variety of lysosomal enzymes hydrolyze all types of injured organelles, proteins, lipids and nucleic acids under low pH conditions. (Hamasaki et al. 2013; Mizushima et al. 2011). A diagram of the mechanisms of the autophagy process can be seen in Fig. 16.3.

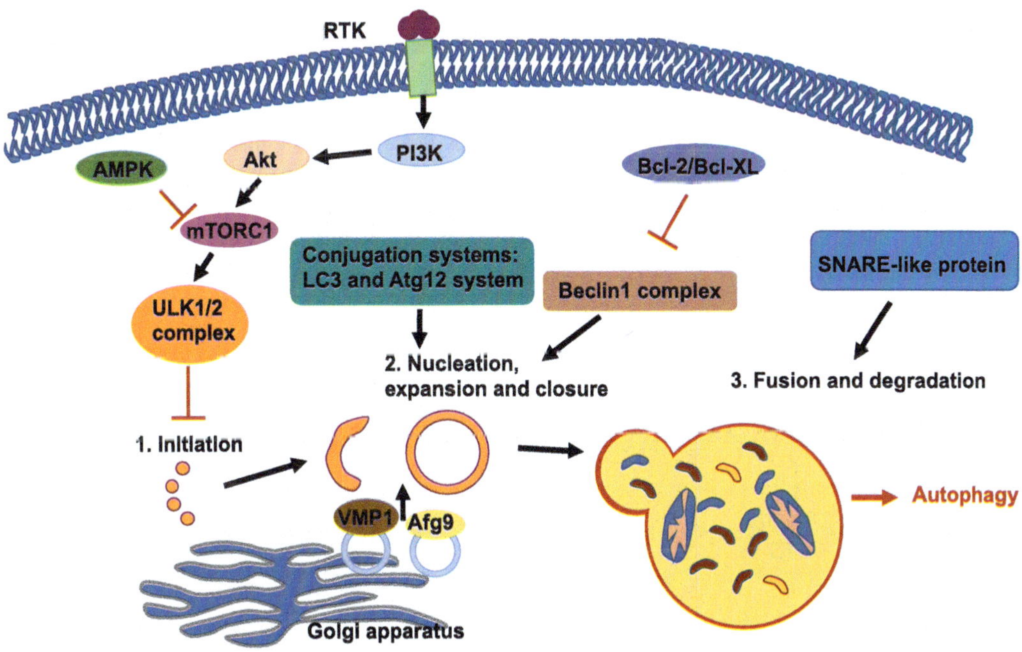

Fig. 16.3 Autophagy mechanisms. The process of autophagy consists of four steps, such as initiation, nucleation, expansion and closure, as well as fusion and degradation. In mammals, the ULK1/2 complex assembly is required for the creation of the phagophore assembly, where this complex is regulated by mTORC, which is in turn positively regulated through PI3K/AKT and negatively regulated through AMPK. Growth factors trigger the activation of the PI3K/Akt pathway via the receptor tyrosine kinases (RTKs). The Beclin1 complex, which is commonly suppressed through Bcl-2, is initiated and regulates the isolation membrane to nucleation, and the transmembrane protein Atg9 and vesicle membrane protein VMP1 seems to be connected to the transport mechanisms of lipids to the isolation membrane. Moreover, two ubiquitin-like protein-conjugated systems, such as Atg12 and LC3 systems, are required for this process. Finally, the autophagosome and lysosome fusion is facilitated through SNARE-like protein and various lysosomal enzymes, which hydrolyze all types of defect organelles, proteins, lipids and nucleic acids

16.2.3 Characteristics of Necrosis and Necroptosis

In contrast to apoptosis, necrosis is frequently regarded as non-regulated and coincidental cell death (Kroemer et al. 1998). Nevertheless, the discovery of programmed necrosis confirmed the presence of several non-apoptotically governed cell death mechanisms. Multiple kinds of programmed necrosis have been documented, compromising necroptosis (Galluzzi et al. 2012), parthanatos (Xu et al. 2016), ferroptosis (Shimada et al. 2016), pyroptosis (Abe and Morrell 2016) and NETosis (Dwivedi and Radic 2014). In this respect, the focus is on necroptosis, a form of controlled necrotic cell death that is sharing multiple important signal transduction pathways along with apoptosis. Most of the findings on necroptosis stem from investigations on TNF signal transduction pathways. TNF can be considered a pleiotropic cytokine that performs an essential function in the inflammatory process (Popa et al. 2007). Under specific circumstances, TNF is also a strong causative agent of cell death through binding to TNFR1/2 (Fig. 16.4).

Even though an initial investigation demonstrated that TNF-induced RIPK1-facilitated caspase-independent cell death (Vercammen et al. 1998a), TNF-induced non-apoptotic cell death did not attract much attention until it was uncovered that cells experienced necrosis-like death when the apoptosis was disabled (Degterev et al. 2005, 2008). Necroptosis is induced through the commitment of death receptors, including TNFR1 (Weinlich et al. 2011; Oberst and Green 2011; Challa and Chan 2010) and Toll-like receptors (TLRs) (Kim and Li 2013; Seya et al. 2012; Kaiser et al. 2011). The binding of TNF to the death receptor TNFR initiates the conformational alterations of TNFR, which attracts multiple proteins, comprising TNFR1-associated death domain protein (TRADD), RIPK1, TRAF2, E3-ubiquitin ligases, cIAP1/2 and LUBAC, to create TRADD- and the RIPK1-dependent complex I. This multifunctional protein complex converts pro-inflammatory and survival signals

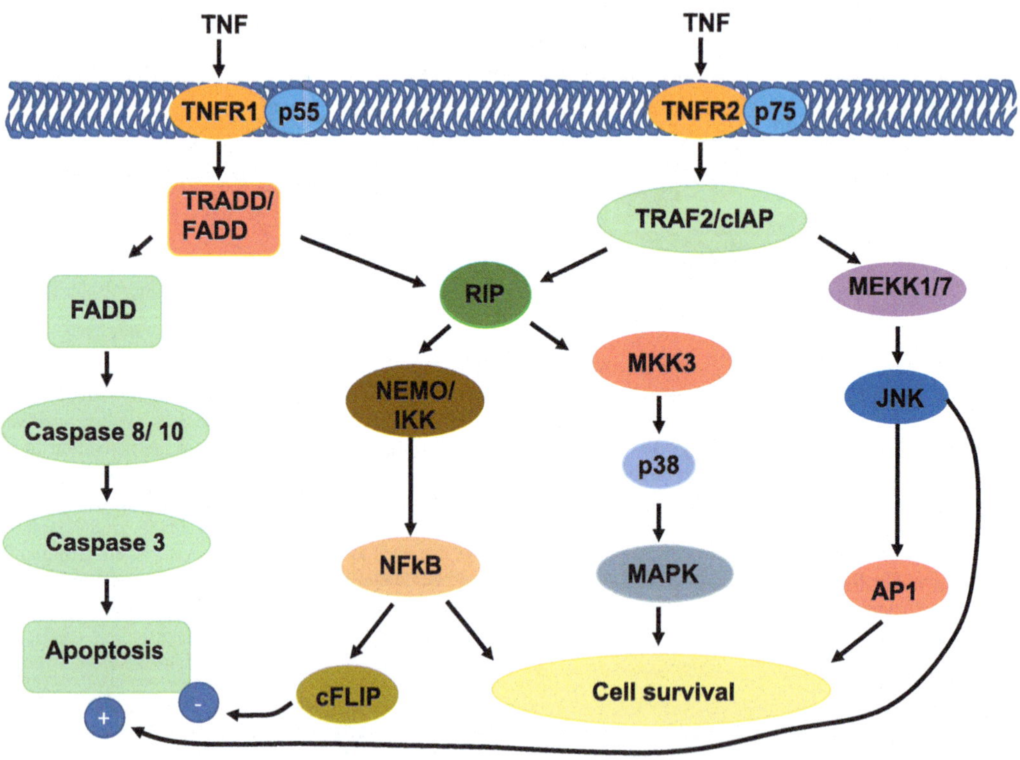

Fig. 16.4 The two different TNFR1 and TNFR2 signaling pathways

through the acquisition of TGF-activated kinase 1 (TAK1)-binding protein (TAB) complexes and Iκb kinase (IKK) complexes composed of IKK1, IKK2 and the key facilitator NF-κB (NEMO) (Silke and Brink 2010; Dempsey et al. 2003) to enable the NF-κB pathway, the AP-1 pathway and the mitogen-activated protein kinase pathway. When RIPK1 is deubiquitinated through cylindromatosis-lysine-63-deubiquitinase (Reiley et al. 2007), complex I gets instable and leads to the decomposition of RIPK1 and the development of another complex, referred to as complex IIa, through interaction with TRADD, FADD, pro-caspase-8 and FLIP73. Pro-Caspase-8 cleaves together with FLIPL RIPK1 to avoid necroptosis and enable apoptosis signal transduction processes (Wang et al. 2008a, b; Micheau and Tschopp 2003; Kelliher et al. 1998). TNF-TNFR signals might also initiate apoptosis due to the generation of complex IIb, where the function of IAP (inhibitors of apoptosis) (Wang et al. 2008a), TAK1 (Dondelinger et al. 2013), NEMO and/or Pellino3 is impaired (Yang et al. 2013b).

This complex IIb contains RIPK1, RIPK3, FADD, pro-caspase-8 and FLIPL and is responsible for RIPK1-dependent apoptosis (Declercq et al. 2009). The complex IIb, nevertheless, can be converted more deeply into the necrosome, a microfilamentous complex, when the concentrations of RIPK3 and the mixed lineage kinase domain-like (MLKL) are adequately elevated and the activity of caspase-8 is attenuated (Orozco et al. 2014). Oligomerization and phosphorylation of RIPK3 within the necrosome results in enrollment and phosphorylation of MLKL (Orozco et al. 2014; Wu et al. 2014), subsequently MLKL relocates in the plasma membrane and induces membrane injury and necroptosis (Murphy et al. 2013). In silico, the necrosome also interfaces with mitochondrial serine/threonine protein phosphatase PGAM family member 5 on the mitochondrial membrane and enables mitochondrial cleavage factor dynamin-related protein 1 to cause necroptosis through the fragmentation of mitochondria (Wang et al. 2012). As stated earlier, the binding of FasL or TRAIL to the death receptor Fas or TRAILR initiates the establishment of DISC, activates caspase-8 and carries out apoptosis (Scheidereit 2006).

In the lack of cIAPs or when caspase-8 is retarded, nonetheless, RIPK1 transposes to the membrane and encourages the production of complex II-b (Feoktistova et al. 2011) and the onset of necroptosis once Fas/TRAILR is enabled (Orozco et al. 2014; Wu et al. 2014; Cho et al. 2009; He et al. 2009a, b).

The activation of TLRs initiates the creation of a scaffold that attracts the cytoplasmic adaptor protein TRIF (Toll/IL-1 receptor domain containing adaptor protein that induces interferon-beta). TRIF is engaged in the activation of NF-κB signal transduction pathways and initiation of type I IFNs (Takeuchi and Akira 2010). Using the RIP homotypic interaction motif (RHIM) domain, TRIF can interoperate with RIPK1 and RIPK3. In the presence of the apoptosis blocker zVAD-fmk, the activation of TLR4 through lipopolysaccharide or the activation of TLR3 through polyinosin-polycytidyl acid can initiate the TRIF-based necroptosis (He et al. 2011; Kaiser et al. 2013a) that can be arrested through the inhibition of necrostatin-1 (Nec1) or the knockout of RIPK1 (Polykratis et al. 2014). These findings point to the fact that the RIPK1-TRIF signal transduction compound seems to be a key factor in TLR3/4-driven necroptosis. In the lack of RIPK1, TRIF can stimulate necroptosis through direct enrollment and triggering of the activation of RIPK3 (He et al. 2011; Polykratis et al. 2014).

In supplement to death receptor signals and TLR signals, the DNA-dependent activator of IFN regulatory factors (DAI), which is a cytoplasmic viral DNA detector (Welz and Pasparakis 2012; Upton et al. 2012), is also able to induce necroptosis. Similar to TRIF, DAI possesses the RHIM structure. In reaction to viral, such as the murine cytomegalovirus, double-stranded DNA, DAI can trigger the activation of NF-κB, activate the induction of type I IFNs and facilitate the RIPK3-dependent necroptosis (Upton et al. 2012). In addition, a novel necrosis mechanism has lately been presented, which points to the fact that the acute cell necrosis triggered by cationic nanocarriers is caused due to the diminishment of Na+/K+-ATPase, which in turn leads to a consecutive

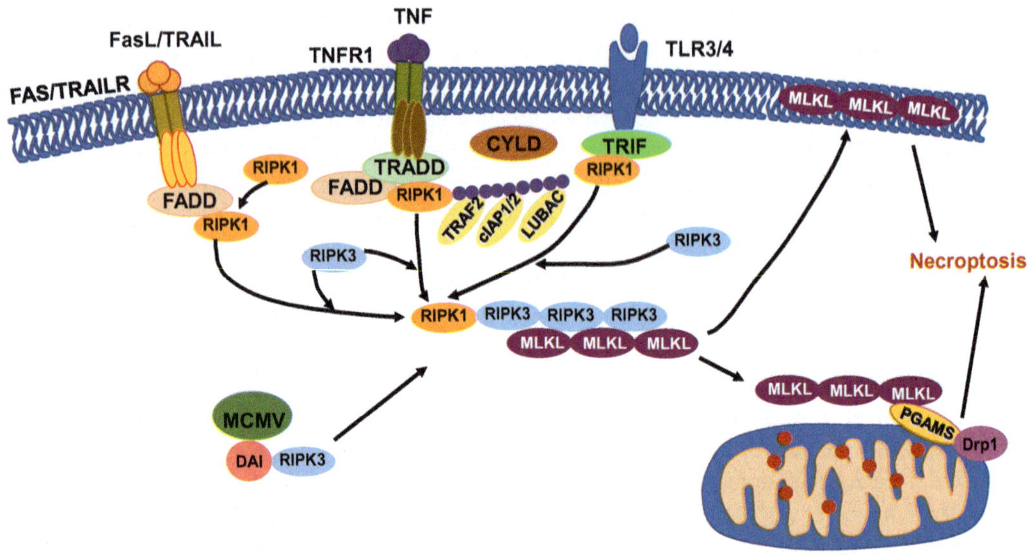

Fig. 16.5 Necroptosis mechanisms. In TNFR signal transduction, a complex I compromising TRADD, RIPK1, TRAF2, E3 ubiquitin ligases, cIAP1/2 and LUBAC is unstable, when RIP1K is deubiquitinated through CYLD, which induces the assembly of the necrosome in the presence of high levels of RIPK3 and MLKL, the phosphorylation of MLKL that translocates towards the plasma membrane in order to lead to membrane damage and necroptosis. Another way is that phosphorylated MLKL interacts with the phosphorylase PGAM5 on the mitochondrial membrane to evoke damage and necroptosis. In Fas/TRAIR signaling, when cIAPs are missing and caspase-8 is impaired, the activation of Fas/TRAIR can trigger necroptosis. In TRL3/4 signal transduction routes, their activation can trigger the TRIF facilitated necroptosis in the presence of zVAD-fmk, TLR4 activation through lipopolysaccharide or TLR3 activation through polyinosine-polycytidylic acid. In DAI signal transduction, due to viral double-stranded DNA, DAI facilitates the RIPK3-driven necroptosis in specific circumstances

inflammatory response (Wei et al. 2015). A diagram of the process of classical necroptosis is presented in Fig. 16.5.

16.3 Are the Three Different Types of Cell Death Interrelated?

It seems to be likely that the three types of cell death are interacting with one another. Hence, connections between apoptosis and necrosis, autophagy and apoptosis and autophagy and necroptosis are discussed in the following. Finally, the connections between all three types of cell death are highlighted.

16.3.1 Linkage Between Apoptosis and Necroptosis

Apoptosis and necroptosis may be simultaneous present (Zhang et al. 2015a) or transform each other due to the interaction of downstream death signal transduction pathways (Fig. 16.6). For instance, cells can suffer necrotic cell death due to a blockage of apoptosis (Leist and Jäättelä 2001), and oxidative stress-triggered necrotic cell death entails the activation of the apoptosis-associated caspase-8/Bid signaling pathway (Wang et al. 2003). The final pattern of cell death varies according to the cell type, the cell microenvironment and the initial triggers.

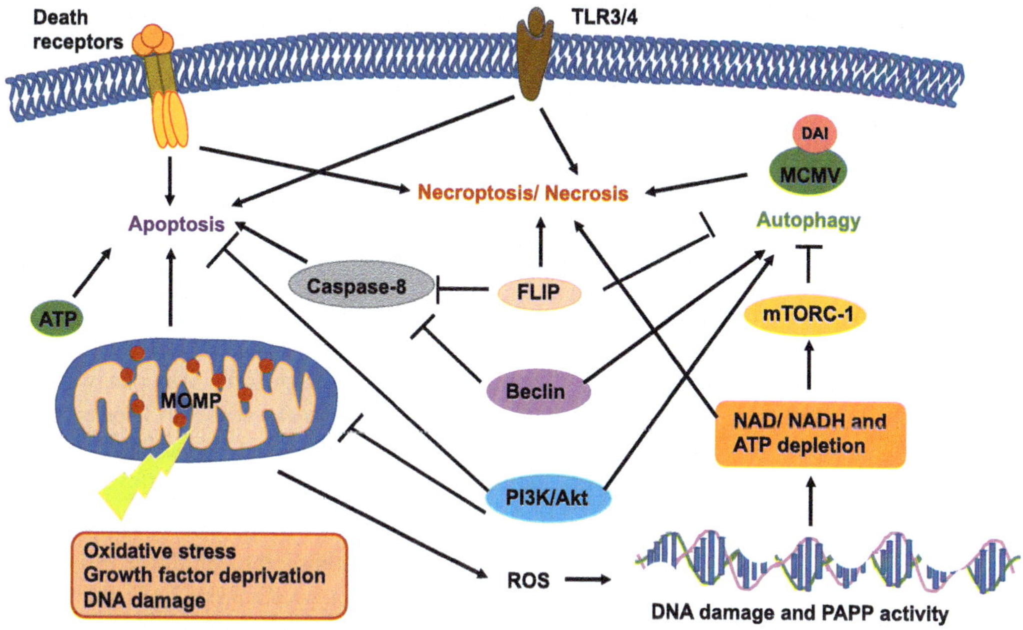

Fig. 16.6 Interference between the three major cell death pathways. FLIP regulates the type of cell death through the interacting with caspase-8, acting on the functions of RIPK1, and combining Atg3 and LC3 through the competitive interaction with Atg3. Caspase-8, a major factor in apoptosis, impairs necroptosis through hydrolyzing RIPK1 and RIPK3. The level of intracellular ATP fulfills a critical function in the decision of cell fate between apoptosis and necrosis. High levels of intracellular ATP usually favor apoptosis, whereas low levels favor necrosis. Beclin, a key molecule in autophagosome formation, can regulate the switch between autophagy and apoptosis through distinct mechanisms, such as the association of Bcl-2 or Bcl-XL, which belong to the group of anti-apoptotic proteins, and are hydrolyzed through a variety of caspase proteins. mTOR senses the intracellular ATP levels and relieve the impairment of autophagy due to low intracellular ATP levels, which causes the necrotic cell death. When AKT is activated via growth factors, it can trigger mTOR signal transduction in order to circumvent autophagy. The AKT activation impairs apoptosis through the phosphorylation of apoptotic factors, such as Bad and caspase-9

Caspase-8 and RIPK1/3

Caspase-8 as an apoptotic inducer caspase interfaces with FADD to generate DISC, which is subsequently subjected to homologous dimerization and a proteolytic conversion. The activated caspase-8 molecule gets liberated out of the DISC and initiates downstream apoptotic cues (Ashkenazi and Dixit 1998). In the meantime, caspase-8 can prevent necroptosis by splitting and deactivating RIPK1 and RIPK3 (He et al. 2009a; Lin et al. 1999; Feng et al. 2007). When, on the contrary, the activity of caspase-8 is restrained pharmacologically, as by the pan-caspase inhibitor ZVAD, or genetically, as in RNAi-facilitated knockdown, RIPK1 and RIPK3 are both activated through phosphorylation and initiate the production of the necrosome that causes necroptosis (Vandenabeele et al. 2010; Moquin and Chan 2010; Holler et al. 2000).

ATP

ATP has a decisive impact on the decision on the fate of cell death (Eguchi et al. 1997). High intracellular ATP levels frequently promote apoptosis, while low levels tend to encourage necrosis (Leist et al. 1997). Thus, increased intracellular ATP consumption or blockage of the synthesis of ATP can transform apoptosis into necrosis (Eguchi et al. 1997; Los et al. 2002). For instance, severe injury to DNA results in the activation of poly-ADP-ribose polymerase 1 (PARP-1), a nuclear enzyme implicated in the repair of DNA, leading to the exhaustion of numerous NAD + and ATP molecules and consequent necrotic death (Szabó and Dawson 1998; Sims et al. 1983). Mitochondria remain the most critical location where ATP is produced. Thus, a malfunction of the mitochondria can provoke necrosis due to ATP depletion. In any case, there is also a causal connection between increased mitochondrial ROS production and the onset of the mitochondrial permeability crossover and the transformation of apoptosis into necroptosis (Denecker et al. 2001).

16.3.2 Linkage Between Autophagy and Apoptosis

Autophagy refers to an intracellular catabolic mechanism that entails the degradation and reutilization of unwanted compounds in the cytoplasm, such as dysfunctional proteins or defective organelles, to preserve cellular homeostasis (Yu et al. 2004; Shimizu et al. 2004; Bursch 2001). Autophagy is a two-edged sword and can either prevent cells from undergoing apoptosis (Li et al. 2013) or encourage cells to develop apoptosis (Maiuri et al. 2007a, b) according to cell type, intracellular metabolic activity, the extracellular supply of nutrients and the stimuli that initiate it (Fig. 16.6).

Beclin1

Mammalian Beclin1 (Atg6 in yeast) controls autophagy and apoptosis through direct involvement with anti-apoptosis family members (Funderburk et al. 2010; Cao and Klionsky 2007; Edinger and Thompson 2004; Erlich et al. 2007). Beclin1 appears to be a major molecule participating in the generation of the autophagosome. Beclin1 reacts with the class III type PI3KC3/Vps34 and encourages the development of the Beclin1-Vps34-Vps15 central complex (Kihara et al. 2001; Liang et al. 1999). Beclin1 also belongs to the family of pure BH3 proteins. The anti-apoptotic protein Bcl-2 or Bcl-xL binds to Beclin1 via the BH3 domains (He and Levine 2010) to jointly prevent the formation of autophagy through inhibition of the activity of Beclin1 (Morselli et al. 2009) and the emergence of endogenous apoptosis (Maiuri et al. 2007a, b; He and Levine 2010; Robert et al. 2012; Itakura et al. 2008). On the contrary, NOXA and other proteins of the pure BH3-only family can replace Bcl-2 family members from Beclin1 and stimulate autophagous cell death (Elgendy et al. 2011; Maiuri et al. 2007a, b). In addition, Beclin1 may also be split by a number of caspase proteins including caspase-8 and caspase-3 to transfer cell fate away from autophagy toward apoptosis (Eisenberg-Lerner et al. 2009; Zhang et al. 2015a, b). The C-terminal fragment of Beclin1 can subsequently transpose into the mitochondria and initiate the mitochondrial membrane permeability and apoptosis (Wirawan et al. 2011).

16.3.3 Linkage Between Autophagy and Necroptosis

The interrelation among autophagy and necroptosis has been examined in various investigations yielding contradictory outcomes. The protective mechanism of autophagy can unexpectedly impede necroptosis (Farkas et al. 2011; Bell et al. 2006). Intriguingly, the autophagy seems to mediate necroptosis in specific circumstances (Bonapace et al. 2010). Moreover, it has been shown that palmitic acid induces Ca^{2+}-facilitated autophagy, leading subsequently to necroptosis of endothelial cells (Khan et al. 2012).

mTORC1

mTORC1 represents a key sensory element for nutrients, growth factors, and stress. Hence, it can govern the metabolism, growth and survival of cells. The activation of mTORC1 through growth factors and nutrients is able to circumvent autophagy by the phosphorylation of autophagy-related proteins, which are associated with the induction of autophagy, including ULK1 and ATG13 within the ULK complex and ATG14 within the VPS34 complex (Kim and Guan 2015). The cellular metabolic and energetic levels govern the activity of the mTORC1 and subsequently affect the initiation of autophagy. At a low-energy level, the activation of the AMPK signaling is initiated due to an elevated AMP/ATP ratio that in turn impairs the TORC1 signaling pathway through the phosphorylation of tuberous sclerosis complex 2 (TSC2), which represents a negative regulator of mTORC1 (Alers et al. 2012; Jung et al. 2009; Kim et al. 2011; Hosokawa et al. 2009; Gwinn et al. 2008; Bolster et al. 2002), and subsequently, the process of autophagy is preferred (Albert et al. 2007; Kim et al. 2011; Egan et al. 2011).

The stimulation of autophagy through the downregulation of the mTORC1 signaling saves the cells from undergoing the programmed cell death, encompassing apoptosis and necroptosis, in an nutrient- or energy-deprived microenvironment (Yang et al. 2011; Hwang et al. 2010; Alexander et al. 2010; Muñoz-Gámez et al. 2009; Huang and Shen 2009; Amaravadi and Thompson 2007; Albert et al. 2007; Lum et al. 2005). However, activation of PARP-1 during the reaction to the DNA damage leads to necrosis as a result of ATP depletion, which also entails AMPK activation, inhibition of mTORC1 and the induction of autophagy as the ultimate protective intervention (Nikoletopoulou et al. 2013). Finally, the exact equilibrium between autophagy and necrosis will dictate the fate of cell death.

16.3.4 Linkage Between Apoptosis, Autophagy and Necroptosis

Cellular FILCE-Like Inhibitory Protein (cFLIP)

FADD-like interleukin-1β-converting enzyme (FLICE)-like inhibitory proteins (FLIPs) contain caspase-8-like structural elements that miss the proteolytic activity. cFLIP comprises three main isoforms in humans, which include one long protein cFLIPL and two short proteins named cFLIPs and cFLIPR (Safa and Pollok 2011; Micheau 2003; Bagnoli et al. 2010). cFLIPL owns a C-terminal caspase-8-like domain, but due to the replacement of several catalytically relevant amino acid residues it is devoid of enzymatic activity (Safa and Pollok 2011; Micheau 2003; Bagnoli et al. 2010; Safa et al. 2008), whereas the other isoforms cFLIPs and cFLIPR possess no the caspase-like C-terminal domain. Notwithstanding this, these three isoforms all comprise two death receptor domains at the N-terminus, enabling them to cooperate with the adaptor protein FADD to build the DISC compound.

cFLIP not only governs the death receptor-facilitated extrinsic apoptosis pathway, at the same time it additionally controls death receptor-independent apoptosis signal transduction. In the complex IIb/ripoptosome, the homodimeric caspase-8 triggers apoptosis through splitting RIPK1 and breaking down the complex IIb/ripoptosome. When pro-caspase-8 together with cFLIPL produces a heterodimer, it not only hinders necroptosis due to the splitting of RIPK1, but also prevents apoptosis because activated caspase-8 is not generated (Kubli and Gustafsson 2012). Nevertheless, the creation of a heterodimer through pro-caspase-8 with cFLIPS/R induces the cell death via necroptosis due to the missing proteolytic splitting of RIPK1 (Suzuki et al. 2001; Li et al. 2006; Budd et al. 2006) and is simultaneously unable to induce caspase-8-dependent apoptosis. The availability of cFLIP isoforms in the ripoptosome hence defines whether cells perform RIPK1-based necroptosis or caspase-8-based apoptosis.

Besides the control of apoptosis and necroptosis, cFLIP is recognized as a negative modulator of autophagy. In the course of autophagosome generation, Atg3 is covalently attached to the microtubule-associated protein LC3. It is noticeable that cFLIP avoids the association of Atg3 and LC3 through the competitive binding of Atg3 and thereby impedes autophagy (Lee et al. 2009). The process whereby cFLIP governs apoptosis and necroptosis through production of the ripoptosome takes place at the plasma membrane, although cFLIP blocks autophagy at the locations where autophagosomes are produced. Therefore, the various subcellular localizations of the cFLIP might be relevant for its activities (Lee et al. 2009). There exist complex connections between the three types of cell death (Fig. 16.6).

16.4 Secondary Necrosis Is a Natural Result of the Apoptotic Program

The prevailing definition of apoptosis assumes that the removal of the apoptotic cell via heterolytic break-down is performed after the phagocytosis through an assisting scavenger (efferocytosis). Nevertheless, an alternative and widely misjudged outcome of apoptosis involves secondary necrosis, an autolytic process of cell decay with liberation of cell constituents, taking place when no scavengers interfere and the entire apoptotic process has been finished (Silva 2010a, b, c). Secondary necrosis occurs typically in unicellular eukaryotes, but, crucially, it can also appear in multicellular animals and has been involved in the development of major human pathologies. Secondary necrosis is a type of cell clearance with distinct molecular and morphological characteristics and must be regarded as a natural result of the complete apoptotic process.

Active cell death programs are suicidal, genetically regulated processes using an endogenous molecular mechanism (Horvitz 2003). Currently recognized patterns of common active cell death comprise apoptosis, necrosis or necroptosis (Degterev et al. 2005), autophagic death and keratinization (Kroemer et al. 2009). Cell death must invariably be succeeded with the removal of the cells through cytolytic decay and the subsequent destruction of the cellular identity. The original characterization of apoptosis (Kerr et al. 1972) assumed that the removal of the apoptotic cell occurred through heterolytic break-down after phagocytosis by an auxiliary phagocyte (deCathelineau and Henson 2003). Multiple cell types are capable of phagocytizing apoptotic cells, although the cells of the myeloid phagocyte network (Silva 2010a, b, c), principally macrophages, are the primary phagocytes used for this purpose (Henson and Hume 2006). The heterolytic break-down of the apoptotic cell inside the scavenger phagosome (He et al. 2009b) overlays the autonomous proteolytic break-down that takes place in the apoptotic cell in the framework of the apoptotic pathway (Taylor et al. 2008) and leads to passive necrosis of the phagocytized apoptotic cell (heterophagic necrosis) (He et al. 2009b; Odaka and Mizuochi 1999; McIlroy et al. 2000).

The result of apoptosis, supported with an auxiliary scavenger cell, is an outcome of the expression of apoptotic cell-associated molecular patterns (ACAMPs) (Franc et al. 1999), which are recognized at the surface of apoptotic cells by phagocytes of the multicellular organism as "eat-me" signatures (Krysko et al. 2006; Erwig and Henson 2008). Probably one of the most universal and well characterized "eat-me" signals involves the translocation of phosphatidylserine into the outer leaflet of the lipid bilayer, which takes place very early in the apoptotic cascade (Fadok et al. 1992). In addition to "eat-me" signals, the interplay between apoptotic cells and phagocytes employs connective tissue bridge molecules (Krysko et al. 2006; Erwig and Henson 2008) and "find-me" signals (Lauber et al. 2003; Elliot et al. 2009), and the non-existence of "don't eat-me" signals that inhibit phagocytosis of viable cell (Brown et al. 2002). Besides molecular markers, there seem to be physical parameter of cells, such as mechanical stiffness, that may represent a marker for phagocytosis by phagocytic cells, since they are altered in the early onset of apoptosis (Kruse et al. 2010; Sarter et al. 2007).

The elimination of early apoptotic cells through phagocytes is the standard in the typical process of physiological apoptosis of multicellular animals, which is a mechanism of cell exhaustion that governs the quantity of cells present in cell populations and tissues (Kerr et al. 1972). This view of apoptosis has gained acceptance and is the only one taken into account in the proposed recommendations of the Nomenclature Committee on Cell Death (NCCD) on the appropriate usage of terminology related to cell death (Kroemer et al. 2005, 2009). A key difficulty with the apoptotic process, however, is that when the scavengers fail to carry out operations, the apoptotic process leads to an autolytic necrotic result. This autolysis has been termed secondary necrosis (Wyllie et al. 1980), with the purpose of separating this mode of cell elimination from the "ab initio cellular necrosis", which should be referred to as primary necrosis (Kroemer et al. 1998; Golstein and Kroemer 2007).

In the following, results are discussed that point to the fact that secondary necrosis is the natural result of the entire apoptotic program at all stages of eukaryotic organization and that efferocytosis breaks this process prior to apoptosing cells achieving secondary necrosis. Efferocytosis is an assisting strategy employed by multicellular animals to prevent the adverse effects linked to the progression of apoptosis to autolytic secondary necrosis.

16.4.1 Secondary Necrosis in Multicellular Animals

Secondary necrosis can occur in multi- and unicellular eukaryotes after a completely developed apoptosis. Apoptosis not only results in cell death, but also in the removal of cells. The removal of cells is the opposite of mitosis, and similar to mitosis, the apoptotic mechanism of cell removal should be considered first and foremost as a unique cellular process. However, the prevailing nature of apoptosis in multicellular animals indicates that the apoptotic regime of cell loss is a two-cell process, since it incorporates the decaying cell and the accessory eater. In essence, the unfolding of the entire apoptotic process of metazoan cells needs to be investigated in cells that become apoptotic without the aid of scavengers. One attempt involves the in vitro analysis of the course of apoptosis in metazoan cells extracted from the social environment of the organism and exposed to an apoptogenic impulse. In these investigations with mouse lymphocytes (Don et al. 1977) or an in vitro cultured mammalian lymphoblastic cell line (Robertson et al. 1978), it was demonstrated that apoptosis proceeds to a terminal break-down through secondary necrosis. When since then in vitro kinetic investigations of the progression of apoptosis in non-phagocytic metazoan primary cells or cell lines have been performed, in other words, under circumstances in which the eating of apoptotic cells through phagocytosis does not disturb, the result of completely developed apoptosis was secondary necrosis, such as in investigations with primary mammalian cells (Bresgen et al. 2008) or immortalized cell lines (Wu et al. 2001; Berghe et al. 2010). In vitro apoptosing cultured metazoan cells express "eat-me" signals (Bell et al. 2006; Vinken et al. 2009) and when co-cultivated with macrophages, they undergo phagocytosis (Savill et al. 1989; Fadok et al. 1998) and imitate the physiological apoptosis in vivo.

Apoptosis pursues an intrinsic, genetically driven agenda (Horwitz 2003). The molecular decomposition of this program demonstrates that apoptogenic impulses in in vitro grown metazoan cells trigger a series of molecular processes which, when not interrupted, lead to cell deterioration through secondary necrosis (Berghe et al. 2010; Kroemer et al. 1995). In vitro and in vivo, the complete apoptotic schedule consists of two phases. In the initial (pre-necrotic) phase, the apoptosing cell keeps the integrity of the cytoplasmic membrane maintained, and molecular modifications, mainly attributable to activated caspases (Fischer et al. 2003) generate the classical apoptotic phenotype. In this stage the outsourcing of the "eat-me" signal takes place. The second one is a necrotic phase (secondary necrosis) with a cytoplasmic swelling, a tear of the plasma membrane and a terminal cell decay releasing intracellular compounds after a sequence of molecular occurrences that have been characterized in depth (Berghe et al. 2010; Vinken et al. 2009; Kim et al. 2008; Silva et al. 2008).

The progression of apoptosis to secondary necrosis is visible in vivo in cases where the clearance through scavengers fails and the entire apoptotic repertoire evolves completely. Firstly, this is the situation in certain physiological circumstances whereby apoptotic cells are poured into channels or into areas topologically outside the organism, including the intestines or the lumen of the respiratory tract, or into the lumina of the acini, where the probability of facing scavengers is very low (Kerr and Searle 1973; Walker et al. 1989). Secondly, the protein tyrosine kinase Mer belongs to the phagocyte receptors that connect the apoptotic cell with the scavenger phagocyte (Scott et al. 2001). Mer-knockout mice exhibit macrophage failures leading to poor clearance of apoptotic cells (Scott et al. 2001; Cohen et al. 2002), and an aggregation of secondary necrotic cells has been seen (Cohen et al. 2002). In *Caenorhabditis elegans*, a gene mutation needed for the phagocytosis of apoptotic cells leads to the development of secondary necrosis in a large number of these cells (Ellis et al. 1991; Gumienny et al. 1999). Thirdly, widespread secondary necrosis in multicellular animals has also been demonstrated in episodes of severe apoptosis that exceed the natural scavenger cell capacity (Ogasawara et al. 1993; Haslett 1999; Devitt et al. 2004), and when this capacity is compromised directly by harmful impacts on phagocytes (Medan et al. 2002; Knapp et al. 2003; Li et al. 2003; do Vale et al. 2007) or by events affecting molecules implicated in the phagocytosis of apoptotic cells via scavenging cells (Cohen et al. 2002; Vandivier et al. 2002).

The aforementioned findings confirm the general conclusion that the entire apoptotic activity in multicellular animals is genetically governed through an intrinsic routine that incorporates autolytic completion through secondary necrosis, rendering this activity self-sufficient and leading to self-elimination when scavengers are not provided. The aforesaid findings also indicate that there is an apparent and significant functional restriction linked to efferocytosis: Depending on an intervening scavenger cell, this mechanism collapses when such a cell is absent or malfunctioning, in so far as apoptosis progresses to its final state and results in secondary necrotic output. Crucially, this breakdown and the resulting appearance of secondary necrosis in vivo is usually caused by failures or the lack of scavengers, which are mechanisms entirely separate from the apoptotic program that continues until termination.

Apoptotic bodies can be produced during the decomposition phase through the fragmentation of apoptotic cells (Kerr et al. 1972). Even though the generation of apoptotic bodies is frequently regarded as the general rule in apoptosis, in a number of normally typical apoptotic processes, which take place in vitro or in vivo, no or only very rarely apoptotic bodies are generated (Taylor et al. 2008; Wyllie et al. 1980). Thus, the removal of apoptotic cells involves commonly non-fragmented cells. As apoptotic cells, apoptotic bodies are first surrounded by a membrane with preserved intactness, express "eat-me" signs and are subject to phagocytosis by scavengers (Kerr et al. 1972) or suffer secondary necrosis unless they are initially cleared (Wyllie et al. 1980).

16.4.2 Secondary Necrosis in Unicellular Eukaryotes

The idea of apoptosis (Kerr et al. 1972) peaked in a series of investigations of metazoic tissues, and the disclosure of essential features of the molecular mechanism of this type of cell removal and its genetic mechanism of containment was carried out in multicellular animals, primarily in *C. elegans* and *Drosophila* (Zhou et al. 2004; Conradt 2009).

The chronological sequence of the knowledge gained about the apoptotic regime of cell clearance illustrates why the concept of apoptosis was originally connected with multicellularity and was not conceived as a single cell incident. From the first characterization of physiological cell death by Vogt in 1842 (Clarke and Clarke 1996) to the first characterizations of apoptosis in unicellular eukaryotes in 1994 (Cornillon et al. 1994), all investigations of cell death mechanisms have been classified as metazoans. The original specification of the apoptotic regime of cell removal is founded on findings using mammalian tissues (Kerr et al. 1972) and thus contributes to the fostering of the metazoan-centered paradigm that a mechanism of this kind only covered multicellular animals. In exercising physiological functions in these organisms, apoptosis involves efferocytosis and is a quiescent and beneficial mechanism that is critical to the clearance of useless, undesirable or harmful cells during embryonic morphogenesis and, in the adult lifetime, during tissue homeostasis and immune reactions (Nakanishi et al. 2009), thus excellently matching multicellularity.

Over two decades passed subsequent to the suggestion of the term apoptosis and prior to the recognition of cell death with apoptotic characteristics in unicellular eukaryotes. Thus, apoptosis was generally regarded as a mechanism of cell clearance that does not only appear in multicellular animals, but instead as a general mechanism of cell clearance that operates on a basic pattern not only in humans, mice, worms and flies, even in the more simple and elder forms of unicellular eukaryotes (Cornillon et al. 1994; Ameisen et al. 1995). In fact, cell killing processes with characteristic apoptotic properties in unicellular eukaryotes have been progressively documented in all the larger eukaryotic groups from the Opisthokonta, which encompasses metazoans, to the far more distant Discicristata. Kinetic investigations of the membrane permeability in the mentioned cases of apoptotic processes in unicellular eukaryotes revealed secondary necrosis, as in the specific case of in vitro apoptosing metazoan cells.

It is important to note that the externalization of "eat-me" signals through apoptotic cells, in particular phosphatidylserine, which is most common in multicellular animals (Krysko et al. 2006; Erwig and Henson 2008), arises not exclusively in the case of these eukaryotes, where it is crucial to identify apoptotic cells as an "unwanted self" that needs to be eliminated (Savill and Fadok 2000), at the same time it also arises in unicellular eukaryotes where this function does not occur. Investigations with unicellular eukaryotes, in which apoptosis was coupled with an externalization of phosphatidylserine, encompass Blastocystis hominis, Chlamydomonas reinhardtii, Dictyostelium discoides, Leishmania donovani, Leishmania infantum, Plasmodium berghei, Trypanosoma brucei and Trypanosoma cruzi (Deponte 2008). It is of major interest that yeasts, the unicellular eukaryotes that are nearest to the metazoans (Baldauf 2003), also externalize phosphatidylserine in the course of apoptosis. In fact this behavior has been seen in Candida albicans (Andres et al. 2008; Park et al. 2009), Schizosaccharomyces pombe (Huard et al. 2008) and Saccharomyces cerevisiae (Madeo et al. 1997, 1999; Ligr et al. 1998; Ludovico et al. 2001). The externalization of phosphatidylserine has also been identified in apoptotic cells of higher plants, where scavenger phagocytosis cannot be performed (O'Brien et al. 1997; Moharikar et al. 2006). In addition, mammalian phagocytes take up viable, non-apoptotic cells expressing externalized phosphatidylserine (Schroit et al. 1985; Dias-Baruffi et al. 2003; Stowell et al. 2009), and necrotic cells expressing the "eat-me" signature. (Brouckaert et al. 2004). It has also been demonstrated that *C. elegans* employs the identical ensemble of entanglement genes for the clearance of both apoptotic and necrotic cells (Chung et al. 2000). Therefore, the identification and uptake of cells through phagocytes due to the recognition of phosphatidylserine alone represents no specific clearance process of apoptotic cells.

The aforementioned findings indicate that the externalization of molecules detected by phagocytes as "eat-me" signs arose as a side-product of the endogenous breakdown process of apoptotic cell death prior to the development of multicellularity and devoid of the theological intent of encouraging the fast and reliable evacuation of apoptotic cells by scavenger cells prior to their advancement to secondary necrosis. This property of apoptosis would have been preserved throughout evolution, and in multicellular animals, phagocytic cells might have adapted to interpret the expression of "eat-me" signs as a marker for phagocytosis and removal. The removal of apoptotic cells using scavengers is a key characteristic of physiological apoptosis in multicellular animals and has been seen at all scales of multicellularity, even in the simplest types such as *Hydra* (Bottger and Alexandrova 2007). Accordingly, these organisms refined the process of removing apoptotic cadavers by using several complementary "eat-me" signatures apart from phosphatidylserine as well as bridging molecules and "find-me" signatures, coupled with the deletion of the "don't eat me" signatures, an evidence of the tremendous relevance of efferocytosis in the framework of animal multicellularity.

Therefore, the entire apoptotic pathway with its secondary necrotic response is an ancient molecular program shared by uni- and multicellular eukaryotes. Similar to single metazoic cells, the apoptotic death of a unicellular eukaryote, when once initiated, is self-sufficient and results in self-annihilation through secondary necrosis.

16.4.3 Efferocytosis in Multicellular Animals

It seems to be that the avoidance of the secondary necrosis through the usage of efferocytosis is advantageous. As a component of a multicellular animal, a cell that decays due to secondary necrosis while undergoing in vitro solitary apoptosis reacts to an apoptotic stimulus a second time with the activation of its endogenous apoptotic routine. However, the result obtained may typically differ from that obtained when the cell is separated because the apoptotic cell is now part of the multi-cellular organization's underlying framework (Raff 1992) and the early expression of "eat-me" signatures on its surface encourages phagocyte-based clearance of the dying cell before its progression to secondary necrosis. The primary role of apoptosis in multi-cellular animals involves the regulated elimination of cells (Kerr et al. 1972), and this is accomplished through secondary necrosis. However, the eradication of apoptotic cells through heterolytic efferocytosis in contrast to autolytic secondary necrosis has decisive benefits for multicellular animals. These benefits are an enhanced velocity of cell removal and a more comprehensive break-down of apoptotic cell constituents (He et al. 2009b; Odaka and Mizuochi 1999; McIlroy et al. 2000) and an energetically beneficial reuse of phagocyted and digested constituents of apoptotic cells, which facilitates the reuse of molecules (Kerr et al. 1972; Cikala et al. 1999).

In addition and especially, eradication takes place through efferocytosis while the apoptotic cell is still surrounded by a plasma membrane with maintained intactness, and cell disruption, involving permeabilization of the plasma cell membrane, is carried out in the protected cavity of the phagolysosome of macrophages (referred to as protected heterolysis). This avoids a secondary necrotic autolysis, which proceeds openly (referred to as open autolysis) and is therefore possibly pathogenetic, since it liberates partially decomposed pro-inflammatory and immunogenic cell elements, as explained in the subsequent part. Consequently, the achievement of the capacity to identify and eliminate pre-necrotic apoptotic cells seen as patterns that need to be removed before the termination of the apoptotic routine marked a major advance for multicellular animals and endowed physiological apoptosis in these organisms with a special feature that relies on efferocytosis.

16.4.4 Secondary Necrosis of Multicellular Animals Impacts Medicine

While secondary necrosis in multicellular animals may appear in vivo in specific physiological settings, it is usually an incident with pathological effects, as exposure and release of cellular components often have adverse effects (Henson and Hume 2006; Bell et al. 2006; Haslett 1999; Savill 1997). A mechanism for the genesis of pathogenic effects of prolonged or sustained apoptotic secondary necrosis is the escape of cytotoxic, pro-inflammatory and immunogenic molecules through autolytic cells. These molecules cover also damage-associated molecular patterns (DAMPs) (Seong and Matzinger 2004). The influence of the liberation of these compounds on inflammatory and immune reactions varies according to the specific circumstances where cell death takes place (Zitvogel et al. 2010), and it can be advantageous, as it will be argued below in terms of immunogenic cell death in tumors. Nevertheless, the necrosis-related secretion of DAMPs frequently leads to pathological pro-inflammatory and immunogenic effects (Bell et al. 2006; Zitvogel et al. 2010; Kono and Rock 2008), which further account for the pathology commonly linked to extensive or enduring secondary necrosis.

The liberation of lysosomal proteolytic enzymes and the oxidative burst arising in the course of secondary necrosis can be involved in the production of DAMPs in that they alter the original cell elements that are subsequently liberated as a result of the ultimate rupture of the plasma membrane (Berghe et al. 2010). Pro-inflammatory and immunogenic DAMPS secreted through secondary necrotic cells include proteases, nucleosomes comprised of double-stranded DNA and histones, S100 calgranulin proteins, the High Mobility Group Box-1 (HMGB-1) protein, proteolytically modified autoantigens and crystals of urate (Wu et al. 2001; Bell et al. 2006; Zitvogel et al. 2010; Green et al. 2009; Munoz et al. 2010). Nucleosomes generated in the apoptotic death process through internucleosomal DNA break-down can be sustained and liberated up to the secondary decay of necrotic cells (Decker 2006) which are implicated in the development of autoimmune diseases (Munoz et al. 2010; Holdenrieder et al. 2006). Nucleosomes are abundant in the bloodstream of patients suffering from systemic lupus erythematosus (SLE) (Munoz et al. 2005).

Secondary necrotic cells can be phagocytized from scavengers in vivo according to molecular mechanisms (Poon et al. 2010). This type of phagocytosis is promoted by "eat-me" signals, among them ACAMPs revealed in the apoptotic pre-necrotic phase (Franc et al. 1999) and by altered cell constituents made accessible by proteolysis and oxidative incidents that arise in the necrotic period and are brought about through rupture of the plasma membrane (Berghe et al. 2010; Poon et al. 2010). Thus, the clearing mechanisms of early (prenecotic) and late (secondary necrotic) apoptotic cells are not the same and generally have different implications for inflammatory and immune reactions. Whereas the removal of early apoptotic cells through scavenger phagocytes is typically anti-inflammatory (Fadok et al. 1998; Serhan and Savill 2005), the phagocytosis of secondary necrotic cells is frequently pro-inflammatory and immunogenic and may therefore be a further mechanism for the emergence of pathogenic effects of secondary necrosis (Krysko et al. 2006; Poon et al. 2010).

In vivo, lysis of neutrophils is particularly pathogenic with regard to secondary necrosis (Haslett 1999; Savill 1997) since these phagocytes are attracted to a very high incidence of inflammation (Silva 2010a, b, c) and are extremely abundant in cytotoxic, tissue-damaging molecules (Henson and Johnston 1987; Weiss 1989). A highly cytotoxic phagocyte molecule secreted by secondary necrotic cells has been identified as neutrophil elastase (Kawabata et al. 2002; Voynow et al. 2008). The extracellular active neutrophil elastase has been identified in a number of infections in which secondary necrosis of neutrophils is involved in the pathogenicity mechanism (do Vale et al. 2007; Vandivier et al. 2002; Watt et al. 2005).

Secondary necrosis can hence induce an acute and chronic pathology and therefore be a route for apoptosis that becomes pathogenic. Secondary necrosis has lately been involved in an expanding array of clinically relevant situations of acute and chronic inflammation in humans suffering from apoptotic cytopathology, among them many autoimmune diseases (Herrmann et al. 1998; Gaipl et al. 2007), asthma (Uller et al. 2006), atherosclerosis (Tabas 2004; Finn et al. 2010), bronchiectasis (Watt et al. 2004), chronic obstructive pulmonary disease (COPD) (Hodge et al. 2003; Stockley 2006; Mukaro and Hodge 2011), cystic fibrosis (Vandivier et al. 2002; Bianchi et al. 2008), ischemia (Unal-Cevik et al. 2004), infection (Silva 2010a, b, c), pneumonia in connection with oxidative stress, even in smokers (Kirkham 2007).

On the upside, secondary necrosis attacking cancer cells has lately acquired additional importance as it is acknowledged as a process with probable positive effects in cancer therapy, as it eases the activation of the immune system and thus the elimination of cancer cells. Chemotherapy to combat cancer (Herr and Debatin 2001) and radiotherapy (Jonathan et al. 1999) are widely imparted through apoptosis. It has been shown that cancer cells can achieve high immunogenicity following treatment with certain chemotherapeutic substances or ionizing irradiation when being injected into immunocompetent mice (Kepp et al. 2009). Several findings have emerged indicating that this kind of immunogenicity is related to the progression of therapy-induced apoptotic cancer cell death to secondary necrosis and to the liberation of the DAMP molecule HMGB-1 through secondary necrotic cancer cells: firstly, cancer cells exposed in vitro to certain apoptogenic chemotherapeutic substances or radiotherapy set free the HMGB-1 protein, and secondary necrosis has been proposed as the mechanism for this liberation (Apetoh et al. 2007). Secondly, the removal of apoptotic cells in the course of therapy-induced apoptotic cancer cell death in vivo is apparently inadequate, resulting in the clustering of secondary necrotic cells (Zitvogel et al. 2004; Brusa et al. 2008). Thirdly, it was established that secondary necrotic cancer cells stimulate dendritic cells (DCs) (Pietra et al. 2001; Buttiglieri et al. 2003; Galetto et al. 2003), and thereby become immunogenic (Brusa et al. 2008).

The presentation of tumor antigens via mature DC results in the stimulation and subsequently the activation of CD4+ and CD8+ T cells. HMGB-1 secreted directly from secondary necrotic cancer cells is detected by DCs priming CD4+ and CD8+ T cells and thus eliciting immunogenic T-helper-1 cell or cytotoxic T-lymphocyte responses (Green et al. 2009; Apetoh et al. 2007; Tesniere et al. 2008). Even though DCs are less powerful compared to macrophages (Albert et al. 1998), DCs can devour apoptotic cells, especially when prolonged apoptosis outweighs the accessibility of macrophages (Rovere et al. 1998) or in areas in which DCs exceed the number of macrophages (Albert 2004).

16.4.5 How Can Cells Under Secondary Necrosis Be Identified?

Necrotic cells visible in cultures or in vivo can constitute cells subject to apoptotic secondary necrosis or primary necrosis instead, and there is a risk of incorrect interpretation of the cell break-down process associated with the emergence of necrotic cells. In the scientific studies, there are many instances that define cell death as the trigger of primary necrosis, on the grounds of experiments in which the death mechanism beyond this necrosis is apoptotic. This misinterpretation arises when the deficiency of appropriate kinetic evaluations inhibits the monitoring of the pre-necotic phase of apoptosis and only discloses a necrotic ending, which is erroneously referred to as primary necrosis. Moreover, this is not a subject of simply academic interest; rather, the accurate categorization of cell death into specific modes can be critical in medical settings as well. Therefore, techniques for the proper detection of cells suffering from secondary necrosis are essential (Silva et al. 2008, Silva 2010a, b, c; Krysko et al. 2008).

Since secondary necrosis concerns cells that experienced certain apoptotic changes, secondary necrotic cells have a specific morphotype that combines apoptotic characteristics, such as hypercondensed chromatin (pyknosis) and nuclear fragmentation (karyorrhexis), and necrotic characteristics, such as breakage of the cytoplasmic membrane. In contrast to primary necrotic cells, cells under secondary necrosis release activated caspase-3 for the identical purpose (Krysko et al. 2008; Denecker et al. 2001), as can be demonstrated in vivo (do Vale et al. 2007; Holdenrieder et al. 2006; Hentze et al. 2001).

Apoptosis is a gene-regulated type of active cell death that utilizes a cellular machinery and involves an ancient molecular machinery that is in common with all stages of eukaryotic organization. The apoptotic routine includes two events that offer two alternative pathways with different mechanisms of cell removal: firstly, early identification of surface markers that enable phagocytic macrophages in multicellular animals to identify apoptotic cells as an undesired self. These cells are eliminated prior to the end of the apoptotic routine and are broken down through heterolysis by the scavenger; thus, this pattern of apoptosis implies the interment of the dead cell (efferocytosis) as a concluding step in the process. Thus, the result that renders physiological apoptosis a mute process of cell clearance in multicellular animals, which was originally characterized when the term apoptosis was initially established (Kerr et al. 1972), and which is reflected in the proposed definition of in vivo apoptosis (Kroemer et al. 2005, 2009; Levin et al. 1999). Second, secondary necrosis, a process that is triggered in the absence of scavengers at the completion of the entire apoptotic program and results in the autolytic decay of the dying cell (Wyllie et al. 1980).

Since efferocytosis fails in unicellular eukaryotes, secondary necrosis results directly from apoptosis in these organisms. Similarly, secondary necrosis is common when in vitro cultivated non-phagocytic cells of multicellular organisms are subjected to apoptosis. It is nevertheless essential that secondary necrosis also appears in vivo in multicellular animals in a few physiological and, more widely, pathological circumstances when no functional phagocytes are present and the apoptotic regime is complete. From a general biological viewpoint, which includes uni- and multicellular eukaryotes, secondary necrosis is the most natural termination of the entire apoptotic program, involves distinctive molecular and morphological characteristics and constitutes a distinct modality of cell clearance. The typical course of physiological apoptosis in multicellular animals, as originally described (Kerr et al. 1972) and widely also reported in the literature, thus implies a variation of the apoptotic pathway that hinders its termination. The predominant definition of apoptosis as a pathway that ends with the excision and decomposition of the apoptotic cell by a scavenger provides a metazoan-centered perspective and is restrictive by itself, especially when focusing on entirely metazoans. In contrast to the plethora of studies on apoptosis and the elimination of apoptotic cells through scavengers, the investigation and focus on secondary necrosis has been widely disregarded, which mirrors this prevailing definition. For instance, new and important articles with suggestions on the usage of the wording recommended by the NCCD in connection with cell death (Kroemer et al. 2005, 2009) and on the usage and evaluation of assays for monitoring cell death in higher eukaryotes (Galluzzi et al. 2009) do not refer to secondary necrosis. Additional research on secondary necrosis will surely identify new instances wherein this type of cell clearance plays a pathogenetic part, or offers new opportunities to refine therapeutic treatments.

16.5 Death-Induced Cell Proliferation Based on a Tissue Regeneration Mechanism

Apoptosis represents a carefully arranged and strictly regulated type of cell death which is conserved throughout metazoans. As the enforcers of apoptotic cell death, cysteine-dependent aspartate-controlled proteases, which are termed caspases, are crucial driving forces of this cellular degradation. Early research on genetically programmed cell death demonstrated that the selective activation of caspases stimulates apoptosis and the accurate removal of superfluous cells, thereby shaping patterns and enhancing tissue. Over the past decade, nonetheless, there has been a fundamental modification in the knowledge of the roles of caspases in the process of cell death—a change triggered largely through the elucidation that apoptotic cells are constantly in active engagement with their environment throughout the death process and that caspases can induce a multitude of responses, many of which stimulate simultaneous cell proliferation, thereby repairing injured structures and rebuilding damaged tissue (Fogarty and Bergmann 2017). This caspase-driven compensatory proliferation is termed apoptosis-induced proliferation (AiP). Different mechanisms of AiP have been identified in a variety of species, from planaria to mammals. This review summarizes the current state of knowledge about AiP and emphasizes the most recent advances in this field, among them the participation of reactive oxygen species and macrophage-like immune cells in a type of AiP, new regulatory mechanisms influencing caspases throughout AiP, and new clinical evidence proving the critical role of AiP in cancer:

- Activated caspases induce a type of compensatory proliferation, termed AiP, through the secretion of mitogens.
- AiP is most accurately characterized in epithelial cells and is of crucial importance for epithelial wound healing and regeneration in all species.
- There are several forms of AiP that are tailored to specific circumstances and tissues and are currently being defined.
- The inducer caspase-dependent AiP mechanism comprises JNK signalling, reactive oxygen species (ROS) and the enrollment of macrophage-like immune cells.
- Persistent AiP triggers tissue growth and can lead to tumor growth and resistance to therapy.

There are still some open questions:

- Why does induction of the activation of caspases during apoptosis occasionally trigger AiP?
- How can the caspases be regulated to trigger only AiP, whereas at the same time cell death is impaired?
- Are other cell types, with the exception of epithelial cells, capable of undergoing AiP?
- What direct aims have caspases in AiP? Are all their targets broken down?
- How widespread is the contribution of the AiP to the development of cancer?
- Can AiP foster other disease associated processes?

16.5.1 Death-Dependent Proliferation

Although most, though not all, animals have some ability to restore damaged tissue, this capacity differs widely according to species and developmental background. Some types of regeneration rely on compensatory proliferation, a procedure in which the volume of injured or destroyed tissue is substituted with extra mitotic divisions in adjacent healthy tissue. The first substantial proof of the compensatory proliferation originates from research on *Drosophila melanogaster*. In 1977, it could be presented that radiated wing precursor tissue, in specific the imaginal discs, of the evolving fly larvae, although up to 60% of the cells were removed, was able to produce a normal-sized wing with a corresponding pattern, as mitosis subsequently increased in the surviving cells (Haynie and Bryant 1977). Successive investigations revealed that the elevated mitosis after apoptosis was able to adapt consistently and appropriately to the destroyed tissue, indicating that the two processes were closely connected (James and Bryant 1981; Milan et al. 1997).

Several subsequent investigations confirmed the idea of apoptosis-driven proliferation. In planarians, caspase-like genes are enabled, and apoptotic cells are available throughout the regeneration period, far exceeding the initial injury period (Hwang et al. 2004). In newts, it has been established that apoptosis is necessary for proper regeneration of limbs (Vlaskalin et al. 2004). Numerous investigations in *Drosophila* have revealed that the initiation of the apoptotic mechanism is effective enough to stimulate non-autonomous proliferation in the adjacent cells, irrespective of the completion of the cell death (Huh et al. 2004; Perez-Garijo et al. 2004; Ryoo et al. 2004). These and subsequent investigations determined that the *Drosophila* initiator caspase Dronc was explicitly accountable for the control of compensatory proliferation after the onset of apoptosis (Huh et al. 2004; Kondo et al. 2006; Wells et al. 2006).

AiP is the term for a specific subtype of compensatory proliferation, which is induced in particular through the apoptotic mechanism (Mollereau et al. 2012). AiP is no individual path, but comprises a multitude of signaling cascades through which caspases stimulate non-autonomous proliferation due to the release of mitogens, frequently in synchrony with, but independent of, the ultimate completion of apoptotic cell death (Tseng et al. 2007; Fan and Bergmann 2008a, b; Chera et al. 2009; Li et al. 2010a, b; Perez-Garijo et al. 2005; Fan et al. 2014; Fogarty et al. 2016). These caspase-driven signal transduction pathways are crucial for the healing and regeneration of wounds. In addition, apoptotic caspases can also be involved in the initiation of tumors, casase-based inflammation and persistent AiP in course of the induction of tumors. Tumors have been reported to be similar to wounds that fail to cure (Dvorak 1986), and chronic inflammatory conditions, frequently associated with repetitive cycles of cell death and insufficient healing of wounds, are recognized to be susceptible to the progression of specific cancers (Karin and Clevers 2016). In the following, the specific mechanisms, their regulation, and the relevance of caspase-based AiP is presented and discussed.

16.5.2 Mechanisms of Caspase-Driven AiP

The mitochondrial apoptosis route is highly preserved under the species. The major regulatory factors encompass: firstly, the death proteases, such as initiator and effector caspases, which are produced in the form of inactive zymogens and are enabled through multimerization and the proteolytic splitting process. Seven caspases are implicated in the regulation of apoptosis in humans, whereby caspase-2, caspase-8, caspase-9 and caspase-10 act as apoptosis inducers, whereas caspase-3, caspase-6 and caspase-7 are the apoptosis effectors (Shalini et al. 2015). In *Drosophila*, it has been revealed that three caspases are necessary for apoptosis: the initiator caspase Dronc (homologue of caspase-2, caspase-9) and the effector caspases DrICE and Dcp-1 (homologue of caspase-3, caspase-7) (Xu et al. 2009). Secondly, the Apaf-1/Dark framework proteins build up a complex with the initiator caspase (caspase-9/Dronc) to create the active apoptosome. Thirdly, several upstream pro- and anti-apoptotic elements, whose antagonistic activities determine the point at which and the time of apoptosis (Xu et al. 2009; Conradt 2009; Fuchs and Steller 2011). In mammals, proteins of the Bcl-2 family govern the activation through the mitochondria (Czabotar et al. 2014), while in flies the pro-apoptotic genes reaper, hid and grim initiate apoptosis through antagonizing the E3-ligase Drosophila Inhibitor of Apoptosis Protein1 (DIAP1), which is a negative caspase-activation regulator (Goyal et al. 2000; Lisi et al. 2000; White et al. 1994; Grether et al. 1995; Chen et al. 1996).

The DIAP1 mono-ubiquitylates the initiator caspase Dronc to impede its apoptotic activity (Bergmann 2010; Shapiro et al. 2008; Wilson et al. 2002; Chai et al. 2003; Lee et al. 2011; Kamber Kaya et al. 2017). These regulatory mechanisms concern both apoptosis and caspase-directed AiP, albeit perhaps to varying extents (Fan et al. 2014; Florentin and Arama 2012; Liu et al. 2013).

In the mammalian species, solely the effector caspase-3 and -7 have been identified as participating in AiP (Li et al. 2010a, b; Dabrowska et al. 2016; Huang et al. 2011; Donato et al. 2014; Cheng et al. 2015; Zhang et al. 2015b; Feng et al. 2016). In *Drosophila*, Dronc, DrICE and Dcp-1 are associated with various types of AiP (Huh et al. 2004; Kondo et al. 2006; Wells et al. 2006; Fan and Bergmann 2008a, b; Fan et al. 2014, Kumar 2007). The functions of caspases in AiP and their impacts on the downstream signal transduction processes are discussed in flies and mammals.

Drosophila Initiator Caspase Dronc Induces a JNK-Based Proliferative Stress Reaction

Most fundamental research on the characterization of caspase-based proliferation has involved genetic models of AiP in the imaginal discs of *Drosophila*, whereby epithelial tissue with high regenerative capacity evolved (Fan and Bergmann 2008b; Smith-Bolton 2016). The undead model, referred to as pro-apoptotic genes, including Reaper or Hid, is overexpressed, whereas the execution of cell death is inhibited through the coexpression of the effector caspase inhibitor p35 (Clem et al. 1991; Hay et al. 1994), which permits the decoupling of Dronc's functions from the effector caspases and the subsequent completion of apoptosis (Huh et al. 2004; Perez-Garijo et al. 2004, 2009; Ryoo et al. 2004; Kondo et al. 2006; Fan et al. 2014; Martin et al. 2009a; Mollereau 2013).

The decoupling process provided the key to uncovering the function of activated dronc as a required and adequate stimulus for mitogen production and AiP for the development of epithelial tissues. Dronc-driven AiP is linked to the activation of JNK (Ryoo et al. 2004; Fan et al. 2014; Shlevkov and Morata 2012), and the generation of mitogens including the Wnt homolog Wingless (Wg), the EGF homolog Spitz (Spi), the Interleukin-6 homolog Unpaired (Upd) and the BMP/TGFβ homolog Decapentaplegic (Dpp) (Perez-Garijo et al. 2009; Ryoo et al. 2004; Fan et al. 2014; Perez-Garijo et al. 2009; Santabarbara-Ruiz et al. 2015). Curiously, Dronc's heterozygosity essentially repressed the JNK-driven AiP, whereas it did not inhibit apoptosis (Kondo et al. 2006; Wells et al. 2006; Fan et al. 2014; Shlevkov and Morata 2012; Xu et al. 2005), indicating that there exists a higher caspase barrier for the triggering of AiP in comparison to apoptosis in the presence of pro-apoptotic stimulants.

A series of kinases and adaptor proteins upstream of JNK are involved in the context-sensitive initiation of JNK activity (Igaki 2009; Rios-Barrera and Riesgo-Escovar 2013). The individual activation of fractions of these upstream components decides about whether the JNK activity is pro-apoptotic or survival-promoting (Clavier et al. 2016). The question of how the activation of the same JNK (flies have only one JNK homologue) through various upstream elements can cause results as disparate as death or survival is largely puzzling. In the case of undead models, just dTraf2 (JNKKKKK), dTAK1 (JNKKKK), and both Hep and dMKK4 (JNKKs) upstream of JNK are necessary for achieving AiP (Fan et al. 2014). Although upstream constituents of the JNK cascade were identified, it was still far too vague as to how Dronc enables JNK in the course of AiP. However, there are recent improvements of this pathway, which are presented in the following.

Dronc-Dependent Recruitment of Immune Cells Fosters JNK Activation and Elevates AiP by a Positive Feedback Amplification Loop

A striking characteristic of wound healing and regeneration involves the penetration of immune cells to the location of the injury, which under common circumstances is probably conducive to efficacious healing, but for effective regeneration they are not generally necessary. For instance, mice that have a deficiency in inflammatory cell lineages are capable of repairing injuries to their skin (Scott et al. 1994; Martin et al. 2003). Nevertheless, under specific conditions, caspases have recently been demonstrated to be capable of signaling to immune cells to proliferate continuously throughout AiP. In epithelial tissues, the activation of Dronc induces the generation of extracellular reactive oxygen species (eROS) via the NADPH oxidase Duox (Fogarty et al. 2016). In a variety of systems, eROS gradients are recognized to attract immune cells to places of injury healing (Niethammer et al. 2009; Razzell et al. 2013), and in the undead model, eROS is necessary to attract macrophage-like hemocytes to places with apoptotic activity (Fogarty et al. 2016; Diwanji and Bergmann 2016). Hemocytes themselves participate in proliferation by enhancing the JNK-dependent responses through the generation of the TNF homologue Eiger (Fogarty et al. 2016). Thus, Dronc stimulates JNK activation mediated through an inflammatory reaction that includes EROS, hemocytes and TNF/Eiger signal transduction.

Effector Caspases Trigger the Generation of Secreted Mitogens by Dying Cells

Genetic screens with the undead p35-based AiP model have delivered a wealth of fascinating knowledge about the mechanisms of AiP; however, they rule out the potential impact of effector caspases, which are the enzymes most closely linked to the implementation of cell death, on AiP.

Thus, several "genuine" or regenerative p35-independent models of AiP have been established, where tissues are initially exposed to a restricted pro-apoptotic signal and then have the possibility to repair damaged tissue with the compensatory AiP (Fan et al. 2014; Fogarty et al. 2016; Smith-Bolton et al. 2009; Bergantinos et al. 2010; Herrera et al. 2013; Kashio et al. 2016). These genuine/regenerative schemes have validated the need for JNK on physiological settings and have been beneficial for investigating the spatio-temporal dynamics of AiP (Fan et al. 2014; Fogarty et al. 2016; Smith-Bolton et al. 2009; Bergantinos et al. 2010; Herrera et al. 2013).

The genuine/regenerative AiP models have also demonstrated that effector caspases are able to promote proliferation through initiating the generation of diffusible mitogens. In *Hydra*, for instance, caspase-3 activation in apoptotic cells stimulates the release of Wnt3, a crucial proliferation and reshaping signal, at the location of the head reconstruction subsequent to a mid-gastric bisection (Chera et al. 2009). In apoptotic photoreceptor neurons in *Drosophila*, DrICE and Dcp-1 stimulate the upregulation and release of hedgehog (Hh), thereby activating the hippo growth controlling pathway for AiP (Fan and Bergmann 2008b; Meserve and Duronio 2015). In mammalian tissue, caspase-3 and caspase-7 are able to break-down and activate the calcium-insensitive phospholipase A2 (iPLA2), which itself generates several phospholipid cues required for the migration of phagocytes, detection and clearance of apoptotic cells, and prostaglandin E2 (PGE2), a diffusible compound that enhances the proliferation of stem and progenitor cells (Atsumi et al. 1998; Lauber et al. 2003; Hagedorn et al. 2014; North et al. 2007; Zhang et al. 2015c). Caspase-3-dependent synthesis of PGE2 is necessary for epithelial healing in the skin of mice, promotes liver regeneration subsequent to partially hepatectomy and has been involved in the growth and repopulation of tumors resulting from targeted cytotoxic therapies (Li et al. 2010a, b; Huang et al. 2011). Even though caspase-7 is originally associated with the activation of iPLA2, it appears to be even more essential for activating the protein kinase C delta (PKCδ), that in return conveys the phosphorylation of Akt, p38 and JNK1/2, which results in mitogen generation and tumor repopulation subsequent to radiotherapy (Cheng et al. 2015).

16.5.3 Redundant or Context-Specific Mechanisms of AiP?

Disagreements concerning AiP result from the distinction made between undead and genuine/regenerative models. There is some disagreement about the need and source of certain specific mitogens such as Wg and Dpp due to the model, developmental phase or specific tissue, including the imaginal discs eye versus wing (Smith-Bolton 2016; Perez-Garijo et al. 2009; Smith-Bolton et al. 2009; Herrera et al. 2013). Even in a recent work have been found that the role of hemocytes is not maintained in all types of AiP (Zimmerman et al. 2013). It is evident that Dronc induces the production of ROS in reaction to pro-apoptotic signals in both undead AiP and genuine/regenerative AiP models in eye and wing discs (Fogarty et al. 2016; Santabarbara-Ruiz et al. 2015). ROS is necessary for AiP in these two settings; however, the acquisition of immune cells and TNF signalling do not seem to be important in the setting of the transient regeneration phase in genuine systems, whereas they are definitely crucial for continuous proliferation in the undead model (Fogarty et al. 2016; Santabarbara-Ruiz et al. 2015).

These disparities between the different models could indeed mirror varying degrees of flexibility, such as different mechanisms of casepase-facilitated AiP that are only activated in different environments. There is evidence for variations due to the developmental framework of flies, such as Dronc induces AiP in non-differentiated eye tissue, however, with a few hours more development time, the effector caspases DrICE and Dcp-1 are crucial for the initiation of AiP in the more mature differentiated tissues (Fan and Bergmann 2008a). As an alternative, these so-called discrepancies might constitute a form of residual redundancy. Depending on the degree and length of the damage, several

AiP tracks can be triggered one after the other or at the same time. Continuous investigations in *Drosophila* are utilizing the ever-growing genetic toolset to explore these options in a systematic manner. Simultaneously, there are options in different model systems to investigate the preservation of these mechanisms and to search for any redundancy or versatility. Even though the role of caspases in the initialization of a ROS-based path of proliferation represents an innovation in the area of regeneration, the downstream roles of ROS in redox signaling, failure recognition and injury response have been the main subject of many recent investigations (Schieber and Chandel 2014). In particular, the mechanisms of ROS-driven proliferation, including the enrollment of immune cells and the triggering of MAPK cascades, have been implicated in the healing and regeneration of wounds in *Hydra*, *Drosophila*, *Xenopus* and *Zebrafish* (Vriz et al. 2014; Nishina et al. 2012). Hydrogen peroxide (H_2O_2) is a dispersible ROS compound that, when first generated at the injury place, can create a chemoattractant gradient for leukocytes during the first few hours after the trauma (Niethammer et al. 2009; Razzell et al. 2013). This temporary ROS explodes and the reacting immune cells assist in the transient wound healing process. In opposite, for the regeneration of entire structures, such as amputated tails of *Xenopus* tadpoles, a continuous ROS generation for the subsequent 12–72 h is necessary (Gauron et al. 2013; Love et al. 2013). Whereas the initial transient outbreak of ROS can simply be accounted for as unspecific oxidative stress which is liberated from the injured cells, the persistent ROS over this longer period of time are rather actively generated by oxidases. The persistent ROS signalling in *Xenopus* triggers JNK and produces a number of the familiar AiP-related mitogens, such as Wnt, TGFβ and BMP, resembling the model of the undead fly. These results, together with the earlier reported need for caspase-3 in *Xenopus* tail regeneration (Tseng et al. 2007), emphasize one of several outstanding possibilities to seek for conserved mechanisms of caspase-controlled AiP that combine the points between trauma, caspases and eROS signaling.

Most interestingly, Dronc seems the sole initiator caspase that has been convincingly demonstrated to drive AiP irrespective of effector caspases, whereas examples of effector caspase-driven AiP have been observed from *Hydra* to humans. It is quite feasible that *Drosophila* has a singularly developed mechanism of Dronc-driven AiP, but the similarities described earlier indicate that further investigations are justified for caspase-2, caspase-8 or caspase-9 specific inputs to AiP in different settings. All future investigations need to be made open to the idea of multiple pathways of caspases for proliferation.

16.5.4 Regulation of Caspases in AiP

Whereas the mechanisms downstream of caspases are becoming increasingly well known for certain forms of AiP, fundamental questions regarding the regulation of caspases in AiP persist. AiP is not present following each cell death. It seems instead to be limited to stress-driven apoptosis. How is this discrimination arranged? What generally separates the apoptotic from the non-apoptotic roles of caspases? There are at present no final answers to either of these and similar questions. However, in the section below, some recent insights into the control of Dronc in the AiP are provided to illustrate the potential responses.

Enzyme Regulation: Non-degradative Mono-ubiquitylation Governs Dronc in Apoptosis and AiP

It has commonly been recognized that the RING domain of DIAP1 suppresses the activity of Dronc by ubiquitylation (Wilson et al. 2002; Chai et al. 2003; Lee et al. 2011). The effect of this ubiquitization, which means the degradation versus non-degradation inhibition, was not quite as obvious (Chai et al. 2003; Lee et al. 2011; Muro et al. 2004). Recently, a new regulatory mechanism has been discovered whereby ubiquitization governs the activity of Dronc. Dronc originating from the embryonic, larval, pupal and adult stages is ubiquitated solely at one Lysine residue, K78, and this ubiquitylation acts as a mono-ubiquitylation, which is not a sign of proteasomal break-down (Kamber Kaya et al. 2017). An ubiquitylation-deficient mutant of Dronc (DroncK78R) is able to engage more efficiently with the scaffold protein Dark and thereby trigger an improved rate of apoptosis, indicating that the K78 mono-ubiquitylation of Dronc hinders the interaction with Dark, which at least partially accounts for the restraining effect of ubiquitylation (Kamber Kaya et al. 2017). Elimination of K78 mono-ubiquitin through deubiquitylase (DUB) could ease the enrollment of Dronc into the dark apoptosome. Whereas a dedicated DUB for this task is not yet identified, the lately outlined DUBA seems to be a suitable candidate (Koerver et al. 2016).

Most unexpectedly, an ubiquitylation-deficient and catalytically dead, since the conversion of the catalytic Cys318 to Ala, the double mutant of Dronc (DroncK78RC318A) preserves the high lethality of the homozygous Dronc zero mutants, something that the purely catalytic mutant DroncC318A was simply not capable of (Kamber Kaya et al. 2017). Since the expression of droncK78RC318A does not cure the incapacity of Dronc mutants to trigger apoptosis (Kamber Kaya et al. 2017; Xu et al. 2005), these experiments indicate that the non-apoptotic features of Dronc are more crucial for the survival of the fly than its apoptotic one. In addition, contrary to droncC318A, the coexpression of droncK78RC318A together with p35 initiates AiP (Kamber Kaya et al. 2017), indicating that mono-ubiquitization of Dronc also acts as an inhibitor of Dronc's capacity to initiate AiP.

Moreover, this finding suggests that the catalytic activity of Dronc for AiP is not absolutely necessary and could provide a response to another issue focusing on the proteolytic objective of Dronc for AiP. Even though Dronc is able to split DrICE and Dcp-1 to initiate apoptosis, no specific splitting protein target of Dronc for AiP has been detected, despite evidence that Dronc has additional splitting targets (Fan and Bergmann 2010). The discovery that DroncK78RC318A is capable of initiating AiP may imply that Dronc has no specific splitting goal for AiP. The only thing Dronc requires for AiP is the inclusion in the Dark apoptosome (Fan et al. 2014), which is eased by the absence of K78 ubiquity (Kamber Kaya et al. 2017). Thus, the mutant DroncK78RC318A offers fascinating perspectives for future investigations not only with respect to AiP, but also to other non-apoptotic features of Dronc that are completely separated from the activation of the effector caspase.

Regulating the Targets: Adaptors and Scaffolds Direct AiP or Apoptotic Death

It is obvious that there are very specific cues that result in AiP, but another open question is how caspases are guided to their non-apoptotic destinations in order to encourage AiP instead of apoptosis. Dronc interacts with the scaffold protein Dark/Apaf-1 in the course of the induction of apoptosis. The apoptosome offers an immediate environment that eases the activation of the zymogen and the aggregation of caspase activity to pass the threshold for initiating apoptosis (Salvesen and Renatus 2002; Shi 2002). When examining non-apoptotic features of Dronc, there are instances of accessory adapters that interoperate with this framework to adjust and control activity. For instance, the differentiation of spermatids in *Drosophila* involves active Dronc within the apoptosome for cellular remodeling, which releases the individual spermatids out of a larger syncytium (Arama et al. 2003). Most interestingly, the active apoptosome does not cause cell death in this non-apoptotic pathway. The apoptosome is triggered via an adaptor molecule, the tango7 and guided to the desired subcellular destination (D'Brot et al. 2013; Chew et al. 2009). Tango7 is necessary for Dronc to split its targets in the course of this reconstruction, with mutated alleles of Tango7 leading to male sterility due to failed sperm individualization (D'Brot et al. 2013).

In contrary, the non-conventional myosin Crinkled acts as an adaptor for caspase substrates during Dronc-based structuring, in particular by directing kinases such as GSK3β to activated Dronc for splitting and regulation of its activity (Orme et al. 2016). However, both of these interactions were also observed in mammals, with the Tango7 homologue PCID1 interfering with caspase-9 and the non-conventional myosin MYO7A linking to and controlling another initiator, the caspase-8 (Chew et al. 2009; Orme et al. 2016). Overall, these results in *Drosophila* and mammals highlight the potential for a number of previously not fully recognized cofactors to engage physically with caspases to direct them to AiP destinations in preference to cell death inducers or to direct AiP destinations to active caspases in cell execution.

16.5.5 Regulating the Response: Feedback Loops Control AiP Activity

Dronc can induce the activation of JNK via AiP. Whereas JNK is triggered downstream of Dronc, the activation of JNK also causes an upregulation of pro-apoptotic Hid, thereby reducing the early apoptotic impulse, caspase activity and caspase-driven signal transduction pathways including AiP (Ryoo et al. 2004; Wells et al. 2006; Fan et al. 2014; Shlevkov and Morata 2012; Moreno et al. 2002). In the undead AiP scheme, the ROS-recruitment of hemocytes accounts for this JNK/Dronc feedback enhancement (Fogarty et al. 2016). In this model, the inhibition of hemocytes not only inhibits JNK activity and AiP, it also results in decreased caspase activity in non-dead cells, a characteristic not evident in the inhibition of AiP at other downstream locations (Fogarty et al. 2016). However, to address another level, there is some indication that in specific settings there is JNK-based, non-autonomous, positive reinforcement of the apoptotic route in neighboring cells (apoptosis-driven apoptosis), which triggers elevated caspase activity and possibly enhances a broader AiP reaction (Suissa et al. 2011; Perez-Garijo et al. 2013). In contrast, EGFR signaling, another actor enabled in AiP downstream of JNK (Fan et al. 2014), negatively adjusts Hid activity and lowers caspase-directed apoptosis in specific cell communities, thereby diminishing their vulnerability to cell death (Bergmann et al. 1998; Kurada and White 1998; Fan and Bergmann 2014). These kinds of feedback reactions could offer a mechanism to restrict cascade-controlled AiP, which avoids any kind of unregulated proliferation.

The AP-1 transcription factors c-Jun and c-Fos are essential for numerous developmental management programs and, when explicitly phosphorylated through JNK, can trigger the activation of genes that are crucial for death, survival, the remodeling dynamics of cells or proliferation, according to the specific environment. (Rios-Barrera and Riesgo-Escovar 2013; Kockel et al. 2001; Ramet et al. 2002). In connection with regenerative proliferation, it has lately been demonstrated that AP-1 can link directly to the stress-reacting promoting element of the wg gene, which is an element that is merely activated in the course of regenerative events (Harris et al. 2016). It is still not clear how JNK-activated AP-1 is guided to this specific element solely under regenerative circumstances, although other downstream paths in AiP could potentially participate in the regulation. For instance, Atg1, a serine/threonine kinase renowned for its contribution to autophagy, has lately been recognized as operating at a level downstream of JNK to enhance AiP irrespective of autophagy (Li et al. 2016). Therefore, Atg1 could function as a stimulating factor that ensures the proper JNK-dependent cellular reaction to stressful conditions similar to AiP. As emerging downstream determinants including Atg1 are detected and defined, one should anticipate that additional opportunities for regulatory feedback to fine-tune the AiP will be identified.

16.6 Role of Cell Death in Diseases

Role of Cell Death in Neurodegenerative Disorders

Parkinson's, Huntington's and amyotrophic lateral sclerosis (Xilouri and Stefanis 2010) entail the disruption of neuronal structures and functions due to the aggregation of abnormal proteins in the intra- and extracellular compartments (Walker and LeVine 2000). Deregulated cell death has been reported as a key mechanism in these neurodegenerative disorders. The interference of cell death routes is hence regarded as a feasible therapeutic option.

The interaction among β amyloid aggregation and the death of neurocytes promotes the advancement of Alzheimer's disease (Duyckaerts et al. 2009). β-amyloid aggregates in the mitochondria, conveys mitochondrial toxicity and initiates caspase-3-based apoptosis (Lustbader et al. 2004), which speeds up the production of β amyloid and neurofibrillary tangles (Rohn 2010). The activation of caspases also breaks down autophagy associated proteins such as Beclin1, thereby preventing autophagy (Rohn et al. 2011).

An enormous amount of evidence points to the fact that the prevention of programmed cell death could be an efficacious approach for the therapy of neurodegenerative disorders. Therapy of Alzheimer's and Parkinson's disorders has been implicated in its anti-apoptotic activity through the prevention of oxidative stress, the release of cytochrome c and caspase-3 activation, and the elevation of the expression of anti-apoptotic proteins including Bcl-2 (Kim and Suh 2009). The inhibition of necroptosis with the RIPK1 inhibitor Nec-1 is able to safeguard cortical neuron cells of embryonic rats and mouse hippocampal neurons against excitatory toxicosis caused cell death (Li et al. 2008; Xu et al. 2007). Nec-1 may also decrease the amount of dying striatal neurons with Huntingtin mutations and slow down the progression of Huntington's disease in Huntingtin mutant transgenic mice (Zhu et al. 2011). The promotion of autophagy (Ravikumar et al. 2002, 2004) through mTOR inhibitors including rapamycin and its equivalents, such as CCI-779, RAD001 and AP23573, serves as a cell survival mechanism that safeguards cells from apoptotic and necrotic cell death (Wu et al. 2008; Ravikumar et al. 2006) and has been demonstrated to be potent in the therapy of neurodegenerative disorders.

Role of Cell Death During Cancer

A fragile equilibrium is maintained between cell death and survival to ensure homeostasis throughout a cell, tissue and organism. An uncontrolled cell proliferation and the avoidance of cell death have been acknowledged as hallmarks of cancer cells (Hanahan and Weinberg 2011). Extensive studies indicate that autophagy has a dual purpose in the development of tumors. In the early phase, autophagy provides an anti tumor action and suppresses chronic tissue injury, inflammation and genome instability; in the late phase, however, autophagy supplies the energy and nutrients needed to maintain the development of the tumor (White 2012). Therefore, inhibitors of autophagy, including chloroquine or hydroxychloroquine, hold a therapeutic value for the treatment of cancer (Dikic et al. 2010; Sotelo et al. 2006; Kondo and Kondo 2006). Autophagy also increases the resistance to the treatment of cancer. For instance, autophagy acts as a protective factor against the quercetin or the histone deacetylase inhibitor SAHA (suberoylanilide hydroxamic acid) induced apoptosis, and hence a combination of autophagy inhibitors and quercetin or SAHA may yield a rational benefit of these drugs in the clinical context (Wang et al. 2011, Li et al. 2010a, b). Similarly, it has been shown that inhibition of autophagy increases the responsiveness of cancer cells to TRAIL agonists and enhances apoptosis (Sotelo et al. 2006).

Cancer cells display abnormal apoptotic signal transduction, comprising the upregulation of anti-apoptotic molecules and the repression of pro-apoptotic molecules (Ou et al. 2015, He et al. 2012, Igney and Krammer 2002). The elevated expression rate of survivin, a member of the IAP family (Ambrosini et al. 1997), the imbalance between members of the Bcl-2 family (Jäger and Raghavan 2000), or the compromised activity of caspases (Lawrence et al. 2014), improves the survival of cancer cells and is connected to the progressivity of the tumor and the patient survival. Hence, the aiming of apoptosis-connected proteins is among the foremost long-term research areas in cancer therapy and has recently made considerable progress. The BH3 analog ABT-737 and its derivative

ABT-263 function as pan-Bcl-2 inhibitors and at the same time block multiple members of the Bcl-2 family of proteins, comprising Bcl-2, Bcl-xL and Bcl-W, to elicit cell apoptosis (Tse et al. 2008; Oltersdorf et al. 2005). In fact, BH3 analogs have exhibited therapeutic value in solid tumors and hematological malignancies (Tan et al. 2011; Cragg et al. 2009). The negative regulatory role of cFLIP in apoptosis renders it an appealing target compound for the treatment of cancer. Stimulation of the cFLIP expression and activity has been combined with the anti-tumor activity of a number of specific targeted therapies, among them those employing mTOR inhibitors and histone deacetylase inhibitors (Kerr et al. 2012; Bangert et al. 2012; Pathil et al. 2006).

The pathways associated with cell death are implicated in the cellular stress reaction. Cancer cells subjected to a variety of stresses such as DNA damaging and oxidative stress during the oncogenic transformation, and their adaptation to these stresses are essential for cancer cell survival. Oxidative stress arises due to the aggregation of ROS. ROS together comprise superoxide onion, hydrogen peroxide and hydroxyl radical and control programmed cell death. The activation of death receptor-driven signal transduction pathways has been connected to the production of ROS (Vercammen et al. 1998a; Gulbins et al. 1996; Sato et al. 2004; Medan et al. 2005). It has been demonstrated that ROS-driven break-down of cFLIP, a negative regulator of Fas, initiated apoptosis in lung epithelial cells (Wang et al. 2008b). In addition, ROS can stimulate the intrinsic apoptotic cascade through a breaking of the mitochondrial membrane potential, which enhances the removal of cytochrome c from the cardiolipin and its liberation into the cytosol and causes oxidative mitochondrial DNA injury (Ryter et al. 2007). In addition, ROS can enable apoptotic signals through ASK1/JNK signal transduction pathway (Matsuzawa and Ichijo 2008).

The concentration of ROS can also enable autophagy. ROS-based oxidative modification of Atg4 impeded Atg4-delipidating activity, resulting in the aggregation of LC3-PE on autophagosomal membranes and promoting the development of autophagosomes. The interaction of ROS and autophagy promotes the progression of cancer (Dewaele et al. 2010). In the initial phases of cancer development, autophagy is expected to contribute to tumor suppression by decreasing the aggregation of ROS through the break-down of ROS-generating mitochondria, thus restricting genomic instability (Mathew et al. 2007). In the later phases of tumor development, however, autophagy could be utilized by cancer cells to enhance their survival and oncogenic mutations (Degenhardt et al. 2006). ROS-driven programmed cell death renders ROS-based therapies an appealing option in the treatment of cancer. In addition, although cancer cells have established redox adjustment mechanisms to persist in a highly oxidative milieu, comprehensive investigations have shown that they are more susceptible to oxidative stress induced by ROS-generating substances than normal cells, thus ensuring the specificity of ROS-driven therapies (Trachootham et al. 2009). Exogenous ROS-generating drugs, either as single remedies or in combination with existing standard therapies, have emerged as encouraging in pre-clinical testing (Pignanelli et al. 2017; Ma et al. 2017; Raj et al. 2011).

Role of Cell Death in Other Diseases

Besides neurodegenerative disorders and cancer, cell death is linked to various other diseases. Autophagy is associated with the a plethora of physiological functions including intestinal homeostasis (Patel and Stappenbeck 2013), heart disease, liver disease, muscular dystrophy, pancreatitis, stroke and type II diabetes (Levine et al. 2015; Mizushima and Komatsu 2011; Fleming et al. 2011; Levine and Kroemer 2008).

Enhancing the activity of Beclin1 has been recommended as a treatment option for these autophagy-related disorders (Levine et al. 2015). Apoptosis is also linked to multiple disorders, such as ischemic stroke, acute damage to the central nervous system, heart disease, infectious diseases and autoimmune diseases (Favaloro et al. 2012). It has thus been stated that the directed manipulation of apoptotic pathways, such as the broad-spectrum caspase inhibitor Q-VD-OPh, can heal apoptosis-induced disorders (Han et al. 2014; Renolleau et al. 2007; Carlson et al. 2007; Turmel et al. 2001). Necroptosis has been implicated in the pathogenesis of ischemic reperfusion trauma (Linkermann et al. 2012), multiple sclerosis (Ofengeim et al. 2015), myocardial infarction, stroke (Degeterev et al. 2008; Smith et al. 2007), inflammatory disease (Wu et al. 2013), acute kidney damage (Kinsey and Okusa 2011), and microbial infection (Kerr et al. 2012; Vercammen et al. 1998b; Wang et al. 2014; Kaiser et al. 2013b; Cho et al. 2011). Nec-1, an inhibitor of RIPK1, has demonstrated efficacy through inhibiting necroptosis (Degterev et al. 2005; Ravikumar et al. 2002; Smith et al. 2007; Kinsey and Okusa 2011; Oerlemans et al. 2012; Peri et al. 2011; Xu et al. 2010).

The three types of cell death, namely apoptosis, autophagy and necrosis, have been examined independently and are regarded as distinct processes. Recent progress in cell death research has altered the way in which these processes are perceived and has prompted to view them as being associated with multiple overlapping signal pathways and crosstalk in reaction to varying stresses. The occurrence of different regulated cell death routes indicates the intricacy of cell death programs, but also offers new therapeutic intervention targets. Additional investigations are needed to examine the relationship within various cell death programs and to determine key molecular markers that govern cell death in specific pathological circumstances and can be manipulated pharmacologically.

16.6.1 Apoptotic Cells Activate "Phoenix Rising" Route in Wound Healing and Tissue Regeneration

The capacity to repair damaged tissue is a feature shared by multicellular organisms. A contribution to apoptotic cell death in enhancing wound healing and regeneration of tissues has been identified in mice. Apoptotic cells liberated distinct growth signals that triggered the proliferation of progenitor or stem cells. Key participants were caspase-3 and caspase-7, proteases that are triggered in the execution phase of apoptosis and participate in cell death. The mice that were missing one of these two caspases had a deficiency in skin repair and regeneration of the liver. Prostaglandin E2, which is a promoter of stem or progenitor cell proliferation and the renewal of tissue, functioned downstream of caspases. The pathway through which executioner caspases in apoptotic cells support the healing of injured tissue and its regeneration within multicellular organisms the referred to as the phoenix rising pathway (Li et al. 2010a).

The capacity to heal injured tissue is indispensable for metazoic organisms (Gurtner et al. 2008; Birnbaum and Sanchez Alvarado 2008). Several organisms, such as salamanders, have the considerable capacity to fully regenerate entire amputated limbs (Slack and Savage 1978; Brockes and Kumar 2002; Sanchez Alvarado 2006). In contrary, some other organisms, such as humans, are able to replace injured organs only in part, such as the regeneration of the liver (Sandgren et al. 1991; Taub 2004; Zaret 2008). Healing of wounds and the regeneration of tissue are intricate procedures that demand the coordinated efforts of several distinct cell types. The initial reaction in mammals with tissue injury is generally accepted to be inflammatory. The wound healing process is then completed through the generation of new tissue and the reshaping of the tissue. Since it was assumed that inflammatory cells are the "first responders" at the location of tissue damage, it was long accepted that cells of the immune system, in particular macrophages and neutrophils, are essential for the initiation and management of the wound healing/tissue regeneration process (Grose and Werner 2004; Martin and Leibovich 2005). Nevertheless, surveillance of healthy wound healing in PU.1 knockout mice lacking macrophages and neutrophils revealed that none of the two cell types is required for the healing of wounds through skin excision (Martin et al. 2003). Thus, the triggering cellular and molecular mechanisms of wound healing and tissue regeneration are still not known.

The stem cells in and around injured tissue are crucial for the healing of wounds and the regeneration of tissue (Gurtner et al. 2008). It has been widely accepted that factors secreted from damaged tissue tend to activate these stem and progenitor cells and attract them to the injured area where they multiply, differentiate and eventually substitute the injured tissue. It was assumed that the liberated factors occur within the inflammatory response resulting from the initial tissue injury. Prior investigations have concentrated on the function of immune factor cells, such as macrophages, which are stimulated by damage to tissue and release cytokines and growth factors that facilitate wound healing and tissue repair (Gurtner et al. 2008; Mosser and Edwards 2008). Growth factors, including hepatocyte growth factor (HGF) (Chmielowiec et al. 2007), fibroblast growth factors FGF7, FGF10, FGF22 (Werner et al. 1994; Braun et al. 2004), and transforming growth factor β (TGF-β) (Ashcroft et al. 1999; Amendt et al. 2002) fulfill crucial tasks in the processes of healing of injuries or the regeneration of tissue, similarly as small molecule hormones, including acetylcholine (Martin et al. 2009b), catecholamine (Gosain et al. 2006; Pullar and Isseroff 2006) and polyunsaturated fatty acids (Michalik et al. 2001) function. However, it is not entirely understood what cell types and molecular mechanisms trigger the signaling cascades required for tissue repair and wound healing.

In this investigation, the hypothesis was put forward that dying cells in injured tissue transmit signals to initiate the proliferation of stem or progenitor cells, thereby initiating the mechanism of tissue regeneration and wound healing. Therefore, this investigation shows a pivotal function of programmed cell death in the tissue renewal in mice. In addition, caspase-3 and caspase-7, the "executioner" proteases, which are the tools of cell death and operate in the final phases of apoptosis, provide the pivotal mechanism for the activation of tissue stem and progenitor cells and the facilitation of tissue regeneration. Caspase-3 and caspase-7 split and enabled iPLA2 to finally induce the generation of the growth signal PGE2, thus demonstrating that these culprit caspases not merely participate in the death mechanism, but are also involved in the generation of paracrine signals.

The importance of apoptosis in tissue regeneration has been demonstrated. For instance, tail regeneration in *Xenopus laevis* demands apoptosis (Tseng et al. 2007), similar to tissue regeneration in planaria, such as flat worms (Hwang et al. 2004). The molecular mechanisms whereby apoptosis led to regeneration were not identified. An investigation revealing the relevance of apoptosis in the regeneration of the *Hydra* head (Chera et al. 2009) found that positive growth signals (in this case Wnt3) were induced in apoptotic cells. Nevertheless, the molecular mechanism of the generation of Wnt3 is still unknown. It is thus intriguing to establish a connection between PGE2 and Wnt signaling routes in the regulation of the developmental specification and the growth of stem cells in mammalian cells (Goessling et al. 2009). These studies indicate that the participation of apoptosis caspase-PGE2 (and possible also Wnt) pathways in the process of wound healing and renewal of tissue is evolutionary conserved.

Since this investigation suggests a pivotal role for cell death in stimulating tissue regeneration in mammalian cells, this mechanism has been termed the phoenix rising pathway of tissue repair (Fig. 16.7). Therefore, this pathway appears to be a basic biological process utilized by metazoan organisms to detect tissue injury and initiate repair, regeneration, or both. The presence of such a route is a phenomenon of the inseparable connection of cellular life and cellular death in metazoic organisms.

16.6.2 The Emerging Clinical Significance of Caspase-Driven AiP in Cancer

Since the undead AiP model takes advantage of a persistent inducement of proliferation and leads to abnormal hyperplastic overgrowth, it has been suggested that this balancing mechanism accounts for the abnormal proliferation observed in cancer (Ryoo and Bergmann 2012). In fact, a series of recent

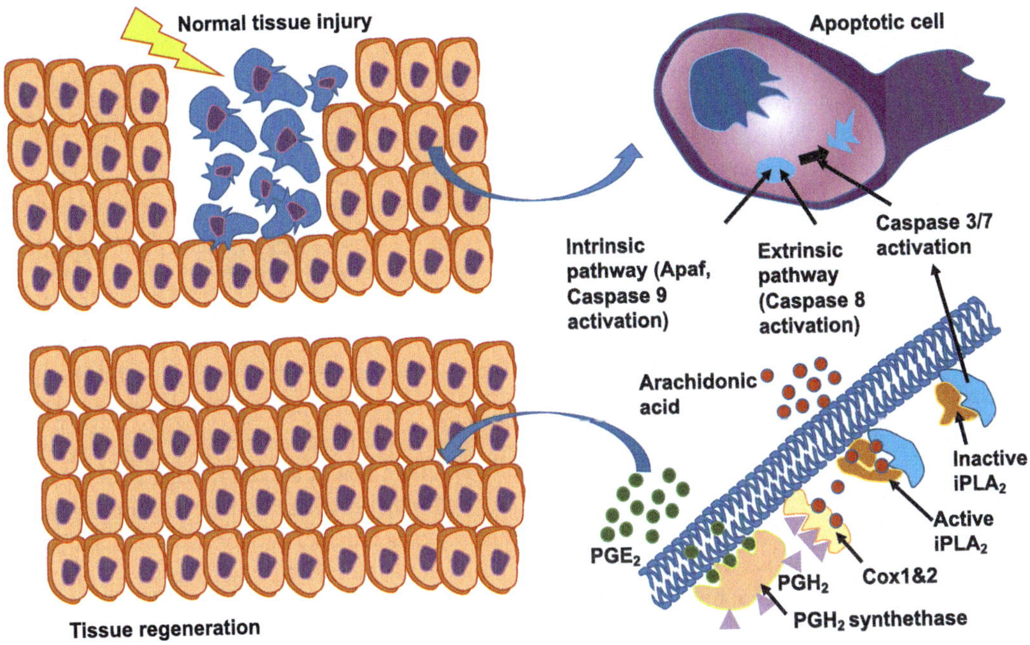

Fig. 16.7 The Phoenix rising pathway. Schematic presentation of the Phoenix rising route of cell death-driven regeneration of tissues. In damaged tissues, apoptotic cells can trigger the activation of caspase 3nad 4 via the intrinsic pathways, including APAF and caspase 9, or extrinsic pathways, such as caspase 8. Activated caspase 3 and 7 cleave and activate $iPLA_2$, which produces arachidonic acid. Arachidonic acid is processed into PGH_2 through cyclooxygenases 1 and 2 (Cox1 and 2). PGE_2 synthetase converts PGH_2 into PGE_2, which then induces stem cell proliferation and the generation of tissues

investigations have examined the manner in which these physiological regeneration pathways can be deregulated during human disease and even abducted in malignant metamorphosis and malignant proliferation.

Caspase-Driven Tumor Reoccurrence After Cytotoxic Treatments

Elevated cell proliferation and fast tumor recolonization after cytotoxic therapy (either by ionizing irradiation or chemotherapy) has long been a well-established determinant of a poor outcome of cancer treatment in humans, although the precise mechanism underlying this phenomenon has not yet been understood (Tubiana 1971, 1988). Subsequent to the characterization of caspase-3-driven PGE2-based proliferation in the framework of tissue regeneration, several examples of caspase-driven PGE2 generation, which directly promotes tumor growth and repopulation after radiotherapy, were discovered in mouse models and human cancer cell lines (Huang et al. 2011; Donato et al. 2014). Since that time, possible functions in repopulation have been established for other AiP-related factors encompassing Caspase-7, PKCδ, Akt, p38, JNK1/2, SHH and HMGB1 (Cheng et al. 2015, Zhang et al. 2015a, b, c, Ma et al. 2013, 2015). A comparable case/PGE2 mechanism of repopulation was also observed after cytotoxic chemotherapy, which supports that this is not a radiation-specific reaction (Kurtova et al. 2015).

AiP, Chronic Inflammation, and the Risk of Cancer

Apart from their immediate relevance for tumor survival, apoptotic caspases may also play a key role in the initiation of tumors, caspase-based inflammation and persistent AiP in the initiation of tumors.

Tumors have been reported to be similar to wounds that fail to heal (Dvorak 1986), and tumor-causing inflammation is a new feature of cancer (Hanahan and Weinberg 2011). It is recognized that chronic inflammatory conditions, frequently marked through repeated periods of cell death and imperfect healing of wounds, and finally can predispose to the progression of specific types of cancer (Karin and Clevers 2016). Inflammatory bowel diseases including ulcerative colitis raise the incidence of colorectal carcinomas (Romano et al. 2016); Marjolin ulcers, a form of epidermoid carcinoma, can be caused from chronically irritated skin or non-healing scars (Shen et al. 2015); pneumonia in severe types of chronic obstructive pulmonary disease raises the incidence of squamous cell carcinoma beyond the exposure to smoke alone (Bozinovski et al. 2016).

In the same way as there are possibilities to link the issues between caspases and regeneration mechanisms, recent investigations have highlighted fascinating possibilities to correlate AiP with inflammation-related tumorigenesis. The initiator caspase-based model of AiP in *Drosophila* outlined above has revealed the existence of direct connections between ROS, macrophage-like immune cells, JNK stress reactions, EGF-facilitated signaling, hyperplasia and dysplasia (Huh et al. 2004; Perez-Garijo et al. 2004, 2005; Ryoo et al. 2004; Kondo et al. 2006; Fan et al. 2014; Fogarty et al. 2016; Diwanji and Begmann 2016). In human inflammatory pathology models, the relationships are not so strictly defined, although the preservation of the main actors is remarkable. For instance, in a mouse model of ulcerative colitis, where colon epithelial cells are subjected to oxidative stress, the JNK signaling pathway and non-apoptotic caspase activity were connected through mutual feedback, with cell survival despite damaged DNA, with enhanced proliferation and with morphological alterations in the epithelium compatible with initial transformation (Poehlmann et al. 2013). In another case, a model of liver carcinogenesis in mice with chronic lesions demonstrated compensatory proliferation, and subsequently, the formation of tumors is facilitated through EGFR-driven activation of liver macrophages upon hepatocyte cell death (Lanaya et al. 2014). The diminished regulation of apoptotic caspases in the parenchyma is adequate to inhibit the development of these tumors (Wree et al. 2015).

AiP and Vasculogenesis in the Microenvironment of Tumors

So far, the focus has shifted to the impact of AiP induced through caspase-activated cells of one specific cell type acting on adjacent cells of the same cell type either possessing or not possessing an immune intermediate. However, caspase-dependent AiP can spread to other cell types in the microenvironment. While cytotoxic radiotherapy, which targets tumor cells, can paradoxically promote AiP-mediated tumor repopulation, it can also induce neovascularization and angiogenesis through mitogens affecting vascular endothelial cells. This post-irradiation angiogenesis is assumed to contribute to the malfunctioning of radiotherapy in specific cancers, and new evidence indicates that this may in part be caused by a type of caspase-3-driven AiP that targets vascular endothelial cells (Feng et al. 2015, 2016). The inhibition of caspase-3 in irradiated glioma cells weakens the development of new vessels, thereby suppressing the general tumorigenicity (Feng et al. 2015). Caspase-3-dependent generation of VEGF-A from these cancer cells seems to be one of the key drivers of post-irradiation angiogenesis; however, multiple caspase-3-activated pathways may provide a pro-angiogenic microenvironment for the proliferation of vascular endothelial cells (Feng et al. 2016). Simply blocking angiogenesis is not a universal remedy (Sitohy et al. 2012). Vascular endothelial cells trigger caspase-3 activation after cytotoxic treatment, which in turn stimulates their own generation of the diffusible AiP factor PGE2 and enhances the growth of tumors, which can result in a resistance to therapy (Mao et al. 2013a). Moreover, some particularly aggressive cancers can produce blood and nutrient delivery independent of the formal angiogenesis through a mechanism referred to as vasculogenic mimicry (Maniotis et al. 1999). Vasculogenic mimicry also seems to imply a non-apoptotic role of caspases, since caspase-3 activity promotes the proliferation of vessel-like channels produced in specific human melanomas by highly aggressive cancer cells. However, only very little is understood about the exact mechanisms of vasculogenic mimicry, although this could be a caspase-based invasion and metastatic phenotype instead of a genuine AiP. Continuous research on model organisms, currently examining the connection between caspases, the reorganization of the extracellular matrix and the risk of metastasis (Rudrapatna et al. 2013).

Clinical Impact on Prognosis and Enhanced Treatments

Several independent patient cohort investigations now corroborated the clinical importance of caspase-driven repopulation and other non-apoptotic features of caspases in these tumors and revealed significant associations between poor prognosis and increased caspase-3 expression and activity. In a study, an elevated active caspase-3 concentration in biopsy specimens prior to treatment corresponded with a markedly reduced overall survival in gastric, cervical, colorectal and ovarian tumors (Hu et al. 2014). In another study, increased circulating serum concentrations of caspase activity are associated with a reduced time to progression even after chemotherapy for head and neck cancer (John et al. 2015). In a third prospective study, activated caspase-3 and elevated HMGB1 expression were identified in colorectal cancer that were associated with a poor prognosis regardless of their role in metastasis (Zhang et al. 2015c).

Since the specifics of caspase-based AiP and its essential role in humans have been resolved, the obvious follow-up questions focus on using this understanding to improve cancer therapy or even stop it before it begins. A few experiments have initiated to tackle these issues and have proven encouraging in applying the understanding of AiP to enhance cancer treatment strategies. For instance, when AiP signaling is redundant, it is probably not sufficient to target only one AiP signaling pathway, and subsequent AiP cycles could accelerate and reinforce drug resistance. One prospective study showed modest results in enhancing colon cancer cell killing in vitro through the combination of a pro-apoptotic chemotherapeutic agent with a cocktail of blockers targeting AKT, STAT3 and ERK1/2, three probable AiP-associated survival and proliferation paths (Bordonaro et al. 2014). Two other studies have successfully used in vivo delivery of celecoxib, an inhibitor of the enzyme cyclooxygenase-2 (COX2) synthesized downstream of iPLA2, as an adjuvant therapy after standard cytotoxic therapies to inhibit tumor progression and prevent tumor repopulation in a mouse model of melanoma and a xenograft model for bladder cancer (Kurtova et al. 2015; Zelenay et al. 2015).

However, other questions are still not yet answered, such as the following ones. Are there important regulatory features in these caspase-based routes that can be aimed at in order to differentiate between the various pleiotropic events that amplify the desired anti-tumor effects and simultaneously inhibit the undesired pro-tumor effects? Which upstream parameters or environmental circumstances propel cancer cells to achieve high caspase values even prior to treatment stress, which prepare them for a resilient AiP reaction? Can this understanding of AiP be exploited in the field of regenerative medicine in a manner that does not raise the future incidence of malignant disease?

Caspases are not only crucial in apoptotic cell death, but they may also trigger proliferation in reaction to lethal or non-fatal stress. Caspase-dependent AiP is not a solitary path, but an evolutionary conserved mechanism that offers healing and adaptation processes needed in the presence of a harmful world. Notwithstanding the tremendous advances that have been achieved in the last decade, there are probably still some more mechanisms of AiP that need to be unraveled, some more upstream controllers that affect the life versus death capabilities of caspases, and some more downstream caspase destinations that impact these capabilities. Continuing research on AiP in genetically detectable model organisms will yield insights into which actors and routes may be at play in human diseases, whereas clinical trials will lead the effort to identify the participation of non-apoptotic caspase in new environments, possibly resulting in the emergence of new treatments and enhanced benefits for patients.

16.7 Tissue Regeneration Mechanism Hijacked by Tumors During Radiotherapy

Two concepts in the research field of tumorigenesis are broadly acknowledged. First, cancer is the outcome of successive genetic alterations that ultimately convert ordinary cells into malignant cells, a scheme referred to as multistep carcinogenesis (Harris 1991). Second, certain biological activities during the development of the tumor require dysregulation in order to facilitate and perpetuate the development of the tumor. These processes have been put together in the proposed hallmarks of cancer (Hanahan and Weinberg 2000, 2011). These hallmarks have been further refined, and a mechanical aspect has been added (Mierke 2014, 2019). Under these hallmarks of cancer, the prevention of cell death is still considered an integral mechanism necessary for the malignant metamorphosis and maintenance of the tumor.

Although every dogma has its time, the driving force underlying scientific progress is still doubt. In this paragraph, the current paradigmatic view is questioned that enhanced survival clearly encourages tumorigenesis (Labi and Erlacher 2015). Moreover, the functional role of apoptosis and its dysregulation in the initiation, progression and perpetuation of malignant diseases is discussed in the following. Finally, it is recommended to incorporate the current view that being resistant to cell death is a real hallmark of cancer, as it might be possible that this could indeed be limited to certain environments.

Key findings in cancer research have shown that during cancer development, the clonal selective process is eased through the accumulation of mutations in oncogenes and tumor suppressors and the selective sorting of "winner" cells. Apoptosis of (pre)cancer cells creates empty niches that might be recolonized with more aggressive subclones. As a result, apoptosis raises the proliferative stress and encourages clonal selection, thus accelerating the development of tumors. Dying cells can stimulate the cell division of nearby adjacent cells.

The following questions remain still open.

- Does apoptosis promote malignant transformation in pre-malignant disorders such as therapy-induced myelodysplastic syndromes or congenital bone marrow malfunction syndromes?
- Can a resistance to apoptosis retard the potential risk of additional malignant transformation within completely established tumors or in pre-malignant tissues?
- How to avoid excessive tissue injury and inflammatory reactions in cancer patients and patients with pre-malignant disorders?

16.8 Deregulated Apoptosis in Cancer

Although cells can undergo suicide in various manners, cell death in vertebrates is usually triggered through the mitochondrial (intrinsic) signaling pathway, which is initiated by a multitude of different kinds of signals such as DNA defects, deficiency of growth factors, developmental cues and numerous conventional cancer treatments. The promoters of this pathway are members of the BCL-2 family, and the equilibrium between anti- and pro-apoptotic family members, the commonly referred to as Bcl-2 rheostat, decides on whether a cell will survive an apoptotic death or die. Based on the above concept of hallmarks of cancer, it has been suggested that the key alterations that transform a normal cell into a malignant cell could easily be enhanced proliferation in conjunction with reduced cell death (Green and Evan 2002). In fact, it has been fully proven that cancer cells attenuate the apoptotic reaction, since only erroneous death hinders their valid removal through intrinsic anti-cancer defense mechanisms or anti-cancer treatment (Lowe and Lin 2000). The progressive antagonism of pro-death cues is a result of exogenous and (epi)genetic alterations in the wide range of parameters that govern apoptosis. However, mutations in genes that codify core components of the cell death system, such as proteins of the BCL-2 family, have been documented to be not very abundant. A major exception is follicular B-cell lymphoma, in which fusions of BCL-2 with the heavy chain immunoglobulin gene are frequently found, thereby elevating BCL-2 protein levels (Bakhshi et al. 1985; Cleary and Sklar 1985; Tsujimoto et al. 1985). More often, upstream signaling pathways that merge at the molecular level of the BCL-2 family are secondarily disordered within the cancer cells, resulting in an unbalanced BCL-2 system that shifts the equilibrium in favor of survival.

The tumor suppressor gene TP53 is generally mutated in multiple types of human cancers and has been connected to apoptosis (Amundson et al. 1998; Levine 1997; Rufini et al. 2013). TP53 is being activated through the injury of DNA, increased hypoxia or the induction of the oncogene activation, and finally triggers the halt of the cell cycle arrest and the repair mechanisms of DNA or even senescence or apoptosis in somewhat irreversibly injured cells with the two molecules PUMA and NOXA, which seem to be the key pro-apoptotic aims (Michalak et al. 2005; Villunger et al. 2003; Erlacher et al. 2005; Oda et al. 2000; Jeffers et al. 2003). Loss-of-function mutations within TP53 causes apoptosis resistance and are commonly linked to a later tumor stage and subsequently with a bad prognosis (Ichikawa et al. 1997; Moller et al. 1999). Apart from TP53, there are several additional proteins of the DNA damage checkpoint pathway that are commonly deactivated in human cancer, including MDM2, ARF and RB1 (Sherr and Weber 2000; Khanna and Jackson 2001). Frequent mutations in oncogenes lead to the activation of the RAS-signal transduction pathway that is independent of growth factor stimulation. Among these mutations are RAS gene mutations, such as HRAS, NRAS and KRAS and mutations in upstream or downstream elements of the entire growth factor signaling route, such as NF1, EGFR, PTPN11, CKIT, FLT3 and CBL (Schubbert et al. 2007).

PUMA, BAD and BIM represent the major BCL-2 elements of the repressed downstream path, whereas BH3-only proteins are inactive through the AKT signal pathway (del Peso et al. 1997; Luciano et al. 2003). Moreover, in colorectal cancers, the focal adhesion protein paxillin seems to support the survival of cancer cells and even elevate their chemoresistance, which is facilitated by elevated stability of the BCL-2 protein in cancer cells (Wu et al. 2015). The c-Myc oncogene can initiate primary tumors, and it has been found to be overexpressed in several cancer types (Nesbit et al. 1999).

Conversely, high levels of c-Myc can also induce apoptosis, whereby BCL-2 proteins play a functional role. However, there are many more mutated oncogenes or tumor suppressors identified that deregulate the intrinsic apoptosis pathway, whereas not much is known about their downstream targeted proteins. In addition to intracellular pro-survival cues, environmental guidance is crucial for the survival of cancer cells. For instance, chronic lymphatic leukemia cells in a niche of lymph nodes can survive well in vivo, but they are subject to rapid ex vivo apoptosis. This suggests that on the one hand the downstream apoptosis mechanism is normally operational in these cells and on the other hand that they are prepared to die and as such strongly reliant on the niche proteins, such as BIM, BMF and NOXA, which have been identified in the primary deduced pro-survival cues. In fact, high concentrations of the pure BH3 proteins BIM, BMF and NOXA have been found in primary human chronic lymphocytic leukemia cells (Mackus et al. 2005; Morales et al. 2004; Del Gaizo Moore et al. 2007; Kitada et al. 1998). The co-culture with feeding or T-helper cells and the administration of cytokines greatly retards ex vivo apoptosis through enhancing the expression of anti-apoptotic BCL-2 proteins (Vogler et al. 2009; Hamilton et al. 2012; Asslaber et al. 2013).

The death of all cancer cells in a solid tumor constitutes the final objective of anti-cancer therapy. For this reason, much of the ongoing research into cancer concentrates on enabling the cellular machinery that eases cell death, including for instance proteins that are responsible for initiating apoptosis. Recently, a series of studies have, nevertheless, highlighted several contradictory functions of apoptotic caspases in both radiotherapy and regeneration of tissues. It seems that a primary function of apoptotic caspases is to enhance the regeneration of tissue and the repopulation of cancer cells in the course of cancer therapy. Since cancer cell recurrence has been demonstrated to be crucial for local tumor remission, the comprehension of the molecular mechanisms underlying tumor recurrence will be crucial to advance radiotherapy for cancer (Zimmerman et al. 2013). Therefore, in the following sections, the current state of knowledge on these phenomena and mechanisms that are capable of paradigm shifting in various organisms and their consequences for the development of novel cancer drugs and cancer therapies is introduced.

The death is the destiny of every cell in each living organism. Although this is a universal reality, the death can come in many shapes. Especially in the last three decades, a wealth of new concepts has arisen to define the various types of cell death (Okada and Mak 2004). In cancer research, the death of cells is of central importance as it is a key element both in the development of cancer and in cancer therapy. In the course of carcinogenesis, normal cells have to escape the cell death cues to develop into tumorigenic cells. In cancer therapy, the final objective is to bring about the death of all cancer cells in the body of the patient. Whereas the aim of killing cancer cells by means of irradiation is quite simplistic, it is becoming more and more accepted that cell death and recolonization of tumors, the reverse process of cancer cell loss induced by the treatment, are tightly coupled. It seems that compensatory proliferation, a mechanism originally found in lower organisms while regenerating damaged tissue (Bergmann and Steller 2010; Fan and Bergmann 2008a, b), is an evolutionary preserved mechanism that also functions in mammals. In addition, the tumors apparently have acquired this mechanism for their personal benefit when they are involved in cytotoxic cancer therapy. Therefore, some newer progress in comprehension of the dynamical relationships of cell death and tumor repopulation is highlighted; the repopulation is a key factor in the failure of treatment response during radiotherapy. Some of the basic concepts of various types of cell death are represented. In the next step, the phenomenon of compensatory proliferation in the course of the regeneration of tissue in lower organisms is discussed. The following is a brief overview of cell death caused regeneration of tissue and wound healing in mammals. Ultimately, the results concerning the contribution of apoptosis to the repopulation of cancer cells while undergoing radiotherapy and its possible impact on the development of pharmaceuticals and radiotherapy will be evaluated.

16.8.1 Cell Death: The Current Paradigm

In general, it is well known that there are three major types of cell death: necrosis, apoptosis and autophagy. Even though all three signaling pathways can lead to the single termination point, cell death, the molecular signal cascades and the effects on the cell host vary considerably. For instance, necrosis is frequently combined with an inflammatory reaction that can encourage the development of tumors. However, the simultaneous delivery of tumor antigens can also result in an increased immune response in that immature dendritic cells and antigen-specific T cells are activated to effectively fight the development of tumors (Demaria et al. 2005). Autophagy is actually more intricate as it has been proven to both enhance and impede the survival of cancer cells. However, apoptosis is regarded as a "clean" and fast type of cell death in which cancer cells decay rapidly and are phagocytized and decomposed in the tumor mass from macrophages. It is assumed that the slightest inflammation is caused through apoptosis.

At first sight, necrosis seems to be a disorganized, unregulated event that results in the break-down of the cell. It is often regarded as random cell death and most frequently arises as a consequence of external physical or chemical offenses, which cause a swelling of the cytoplasm, destroying organelles and ultimately rupturing the plasma membrane (Golstein and Kroemer 2007). As a result, intrinsic cellular components seep into the surrounding environment, causing an inflammatory reaction and the degradation of tissue, which can in extreme circumstances be lethal to the host (Festjens et al. 2006). But, luckily, not every cell death is quite so stressful. The need for a more strongly regulated form of cell death to promote embryonic development and monitor natural cell turnover while at the same time achieving the minimum possible damage to healthy tissue is accomplished with programmed cell death (Baehrecke 2002).

Apoptosis, which is programmed cell death, is clearly defined on the molecular scale. It has been found to be the major type of cell death in mammals. The term apoptosis first emerged in 1972 when it was applied to denote a regulated process of cell death, dissimilar to that of necrosis, involving cell shrinkage, blebbing of the plasma membrane, the condensation of chromatin and the fragmentation of DNA (Kerr et al. 1972). The resulting membrane-bound minor cell fragments, referred to as apoptotic bodies, are subsequently phagocytized from macrophages through recognition of phosphatidylserines which are externalized to the outer membrane of apoptotic bodies. The penetration of macrophages is carried out prior to their cell material being released into the surroundings. The efficient cleaning process actively inhibits an immune or inflammatory reaction, as experienced in necroses.

Apoptosis can be achieved by an extrinsic or intrinsic route, both of which perform cellular substrate break-down by activation of caspases, a family of cysteine proteases that break down peptides neighboring aspartic acid residues detected in caspase-specific peptide sequences (Lamkanfi et al. 2002). The extrinsic route is enabled through the involvement of certain ligands of cell surface death receptors, including Fas, TRAIL and TNF receptors (Beyaert et al. 2002). The downstream cues differ according to the receptor, but all climax in the triggering of effector caspases. For instance, the employment of Fas receptors eases the enrollment of FADD and caspase-8, which constitute the death-inducing signaling complex (DISC). This association splits caspase-8 and releases its active version, which then splits caspase-3, caspase-6 and caspase-7 and thus enables them.

In the inherent signaling pathway, permeabilization of the outer membrane of the mitochondria leads to the liberation of the mitochondrial protein, the second mitochondrial-derived activator of the caspase/direct inhibitor of low pI apoptosis-binding protein (Smac/DIABLO), into the cytosol, in which it concomitantly binds the inhibitors of apoptosis proteins (IAPs) away from the caspases and enables their activation (Saelens et al. 2004). Cytochrome C is also liberated from the mitochondria and constitutes the apoptosome together with Apaf-1 and caspase-9. The initiator caspase-9, similar to caspase-8 in the extrinsic signaling pathway, triggers executioner caspases such as caspase-3, caspase-6 or caspase-7, which in turn split a number of cell substrates, resulting in the death of the cell (Saelens et al. 2004). Frequently, intervention in the extrinsic pathway will reinforce the intrinsic pathway by caspase-8-dependent splitting of the pure BH3 protein BID, which plays a role in the destabilization of the mitochondrial membrane (Adams and Cory 2007). Lastly, the less well known of these three mechanisms appears to be autophagy. Autophagy in itself is no mechanism of cell death, but refers to a process in which cells decompose and reuse cell components by combining phagocytic vacuoles with lysosomes.

These three mechanisms can overlap, as cross talk between these pathways is well established and can have serious implications for the course of a disease (Galluzzi et al. 2007; Eisenberg-Lerner et al. 2009; Degterev and Yuan 2008). For instance, the triggering of caspase can induce necrosis by stimulating the iron overload (Nicotera and Melino 2004). When caspases are blocked and apoptosis is halted, certain cells will die through necrosis or autophagy. It has been demonstrated that the administration of the pan-caspase inhibitor zVAD to murine fibrosarcoma cells L929 to inhibit the introduction of apoptosis led to enhanced death through necrosis. However, when the cells were exposed to zVAD and rapamycin, both of which are reported to stimulate autophagy, the cells were found to remain alive. In addition, the knockdown of selective autophagy genes led to increased zVAD-based cell death due to necrosis (Wu et al. 2008). In other instances, the induction of autophagy following zVAD exposure resulted in elevated cell death (Yu et al. 2004; Chen et al. 2011a).

The connection between autophagy and apoptosis is especially fluent, since they are supposed to both reinforce and impede one another (Boya et al. 2005; Glick et al. 2010; Brech et al. 2009; Degenhardt et al. 2006). In the search for effective cancer therapies where the removal (death) of cancer cells is the intended outcome, the mode of initiated cell death and the consequences for the host need to be addressed. The deregulated apoptosis has become an accepted hallmark of cancer (Evan and Vousden 2001; Lowe and Lin 2000; Hanahan and Weinberg 2000, 2011). Frequently, cancer cells acquire resistance to apoptosis by down-regulating the expression of the death receptor or through changes in downstream signal elements. Thus, the utilization of alternative mechanisms of cell death, including necrosis or autophagy, could be a possibility to circumvent this impediment.

16.8.2 Compensatory Proliferation, Cell Death-Based Tissue Regeneration in Lower Organisms

Apoptosis is commonly regarded as a tool for multicellular organisms to remove injured or undesired cells. However, in the last few years, it has been demonstrated that apoptosis frequently has a major supporting role in the regeneration of tissue. This phenomenon, referred to as compensatory proliferation, was initially seen in lower organisms such as *Planaria*, *Hydra* and *Drosophila*. Planaria, being simple worms, possess exceptional regenerative capabilities and are able to create complete individuals from even the smallest body fragments (Sanchez Alvarado 2006). In Planarians, it is assumed that up to 20% of all cells are employed for the regeneration of adult tissue (Gurtner et al. 2008). Apoptosis-related genes, including caspase-like gene 3 (DjCLg3), have been observed in apoptotic cells undergoing planar head and tail regeneration and following regular cell renewal (Hwang et al. 2004). *Hydra*, along with the planarians, is one of the oldest models of regeneration; the exploration of this scheme dates back hundreds of years. It also displays an extraordinary ability to regenerate. During the regeneration of the *Hydra* head, apoptosis starts quickly and Wnt3 is released from the dying cells. Wnt3 belongs to the Wnt family of signaling proteins whose activation is involved in oncogenesis in parallel with the control of a range of developmental processes. This secreted Wnt3 activates the β-catenin signaling pathway in the neighboring cells, which results in an elevated proliferation rate (Galliot 2013; Chera et al. 2009; Galliot and Chera 2010). The significance of compensatory proliferation in the process of head regeneration becomes apparent when the inhibition of apoptosis by the administration of a caspase inhibitor suppresses head regeneration. Nevertheless, it can be regenerated when the Wnt3 protein is applied exogenously (Chera et al. 2009).

The part of compensatory cell proliferation during regeneration was investigated more closely in experiments with *Drosophila*. Imaginal discs are bags of epithelial cells discovered in *Drosophila* larvae, which later develop into adult anatomical structures including wings, eyes and limbs. Apoptosis is rare in the imaginal discs of the *Drosophila* wing during normal circumstances, but is quickly triggered by radiotherapy and heat shock. The fast recolonization of cells to create normal adults is of interest, although up to 60% of the wing's imaginal disc is destroyed during this process (Haynie and Bryant 1977). When such therapy is coupled with an inhibitor of executioner's caspases, the baculovirus P35 protein, to avoid the death of apoptotic cells, increased non-autonomous cell proliferation, invasion of compartments and the elevation of wingless (wg) and decapentaplegic (dpp) signaling genes (Perez-Garijo et al. 2004; Ryoo et al. 2004; Huh et al. 2004). It was found that apoptotic cells, when not removed, continuously transmit Wg and Dpp signals, causing abnormal development, and proposed that a related process of decoupling of apoptosis induction and cell death can take place in mammalian cells during the formation of tumors (Perez-Garijo et al. 2004).

In another *Drosophila* investigation, it was found that a second mechanism of apoptosis-triggered compensatory proliferation arises from activation of the hedgehog signaling pathway (Hh) via effector caspases, such as DrICE and Dcp-1, and found that it is the preferred mechanism in ocular tissues (Fan and Bergmann 2008a, b). It has been established that effector caspases in photoreceptor neurons stimulate the Hh signaling pathway that triggers cells that had earlier left the cell cycle to rejoin it, thus illustrating how various caspases can induce various types of compensatory proliferation and perform non-apoptotic roles (Fan and Bergmann 2008b).

In mammals, it was found that mouse embryos subjected to the teratogen mitomycin C were decreased to 10% of their original size; however, they were able to restore themselves by a compensatory growth cycle that led to full-size embryos 48 h afterward (Snow and Tam 1979). These investigations together offer a justification for investigating the connection between cell death and tissue repair in higher organisms.

16.8.3 Cell Death-induced Skin Wound Healing and Liver Regeneration

However, since experience with lower organisms has demonstrated that accelerated cell death in a cell population can inadvertently trigger the faster proliferation of surviving cells, it is necessary to consider the effects of cell death on the cell population. Evolutionary preserved features of this type of signal may well be available in higher organisms for the specific goal of healing wounds and regenerating tissues. For instance, in the same manner as the JNK signaling pathway has been proven to be involved in compensatory proliferation of the imaginal disc of *Drosophila*, activation of JNK might also be involved in enhancing liver regeneration in mammals (Schwabe et al. 2003; Sakurai et al. 2006; Maeda et al. 2005). It has been revealed that in mouse hepatocytes, in which the initiation of the pro-survival NF-κB pathway has been impaired, the therapy employing the liver carcinogen, such as diethylnitrosamine, causes the induction of the activation of JNK and elevates the compensatory proliferation within the surviving hepatocytes (Maeda et al. 2005). In addition, a later investigation supported the relevance of JNK activation for this process by demonstrating that mice missing JNK exhibited a clear reduction in cyclin D, vascular endothelial growth factor, cell proliferation and tumor neovascularization (Sakurai et al. 2006).

The effects of dying cells on the regeneration of tissue and repair of wounds in mice have been analyzed (Li et al. 2010a). To imitate dying cells in injured tissue, mouse embryonic fibroblasts (MEFs) were irradiated lethally. A series of bioluminescence imaging studies to observe the proliferation of firefly luciferase (Fluc)-labeled stem or progenitor cells revealed that co-culture with the irradiated MEFs led to a significant enhancement in the proliferation of Fluc-labeled epidermal keratinocyte progenitor cells. In addition, caspase-3 and caspase-7 have been identified to perform a pivotal function in this cell death induced keratinocyte proliferation. To establish whether these caspases actually contributed to wound healing, excision wounds were generated by stamp biopsies on the skin of mice and the healing rates of wild-type and Casp3$^{-/-}$ or Casp7$^{-/-}$ knockout mice were measured in a comparative fashion. Whereas the skin wounds of wild-type mice were fully repaired after day nine, the Casp3$^{-/-}$ mice required more than two weeks to fully heal out.

These caspases may also be crucial for the regeneration of the liver. The human liver can self-regenerate up to 70% without scar tissue (Michalopoulos and DeFrances 1997), and a well-established model of partial hepatectomy in mice was employed to examine the role of caspases in this process (Mitchell and Willenbring 2008). After removal of two thirds of the total weight of the liver in wild-type and caspase-knockout mice, it was observed that wild-type mice were able to restore up to 75% of their weight before surgery, Casp3$^{-/-}$ mice only about 60% of their weight and Casp7$^{-/-}$ only about 57%. The use of the BrdU stain revealed a 50% decrease in cell proliferation in both Casp3$^{-/-}$ and Casp7$^{-/-}$ mice in comparison to wild-type mice after partial hepatectomy.

These experiments emphasized the significance of apoptosis initiation and the ensuing activation of caspase-3 and caspase-7 for the wound repair and liver regeneration. What are the subsequent elements of caspases that are implicated? As the activity of the calcium-independent phospholipase A2 (iPLA2) is increased by the cleavage of caspase-3 or caspase-7 and results in an enhancement of the synthesis and liberation of arachidonic acid and lysophosphocholine out of apoptotic cells (Atsumi et al. 1998; Lauber et al. 2003; Zhao et al. 2006), it can be assumed that an elevation of arachidonic acid might act as a precursor for the synthesis of prostaglandin E2 (PGE2), which is a known stimulator of the proliferation of stem cells (Liou et al. 2007; Feher and Gidali 1974; North et al. 2007). In fact, it has been seen that ionizing radiation, together with the initiation of apoptosis in wild-type MEFs, pronouncedly elevated the release of arachidonic acid and PGE2 and that this type of secretion was widely abolished in caspase-3-deficient MEFs. In addition, western blot analysis has been revealed that iPLA2 break-down relies on caspase-3 and the generation of PGE2 depends on iPLA2, since the shRNA-facilitated knockdown of iPLA2 caused a decrease in the amount of PGE2. However, both PGE2 and activated iPLA2 have been identified to trigger the proliferation of keratinocytes. The mechanism of caspase-driven wound healing and the renewal of tissues which has been referred to as phoenix rising pathway of tissue regeneration have been investigated.

16.8.4 Cell Death Mechanisms and Radiotherapy

Radiotherapy has become a standard procedure in the therapy of cancer, where over 50% of patients get it at some stage during their disease (Balcer-Kubiczek 2012). Ionizing radiation performs its anticancer action through reaction with molecular oxygen and water to produce ROS, which can target deoxyribose in the DNA backbone, causing double-strand breakages in the DNA (Ward 1988). Sublethal levels of radiation can cause a nuclear DNA damage reaction, but when the magnitude of the damage is too large to be repaired, cell death mechanisms are triggered (Jackson and Bartek 2009). The application of ionizing radiation for cancer treatment must counterbalance the induction of death in malignant cells while at the same time restricting the damage to normal tissues.

Although radiotherapy has been found to cause several types of cell death, encompassing necrosis (Chen et al. 2011b), autophagy (Zois and Koukourakis 2009), and apoptosis, there is some dispute over the most important type of cell death occurring during radiotherapy (Brown and Attardi 2005). In terms of apoptosis, a favorable host response after ionizing radiation might not be so easy and rather varies with cell type and the setting within the tumor microenvironment (Balcer-Kubiczek 2012; Dewey et al. 1995; Meyn et al. 1993; Hendry and West 1997). In fact, there has been extensive discussion about the role of apoptosis itself in radiotherapy. This is aggravated by the fact that the sensitivity to radiation varies greatly according to cell type (Rich et al. 2000) and that experimental techniques for detecting apoptosis are equally diverse and are not necessarily able to accurately differentiate apoptotic cell death from different mechanisms. Regardless of the mode of death after radiation therapy, there is no doubt that radiation is very powerful in destroying cancer cells. The difficulty is that a few surviving cancer cells can regrow and lead to the recolonisation of cancer and the recurrence of cancer (Kim and Tannock 2005).

16.8.5 Tumor Repopulation upon Radiation Therapy

Rapid cancer repopulation after radiotherapy is not a new concept (Withers et al. 1988a, b; Trott 1990; Marks and Dewhirst 1991; Maciejewski and Majewski 1991; Hopewell et al. 2003) and has also been observed after chemotherapy (Marcu and Bezak 2010; Davis and Tannock 2000; Davis et al. 2003). Early investigations, both in mice and later in human clinical investigations, reported the appearance of an accelerated repopulation of cancers after exposure to ionizing radiation (Withers et al. 1988a, b) and led the field in changing radiation treatment regimens to circumvent this phenomenon (Withers 1999; Peters and Withers 1997). It has been found that for specific cell types the rate of recolonization rises with the duration and intensity of exposure (Begg et al. 1991; Denekamp 1986), but the molecular mechanisms causing this phenomenon were not clear. Many of these earlier works referred to the circumstance that repopulation reactions were strongly influenced by cell type and dosage fractionation. Some investigations documented increased recolonization, which appeared only in the later stages of radiation treatment, whereas in other cases, such as human cervical cancer patients, the increased recolonization was only present for a fairly short radiation treatment period (Huang et al. 2012). In addition, squamous cell carcinomas (Marcu et al. 2004), bladder cancer (Maciejewski and Majewski 1991) and colorectal carcinomas have been found to have comparatively quicker recolonization rates, raising some concern about the possibility of using longer treatment periods (Suwinski et al. 1998).

From the collective findings some generalizations could be drawn. A slower recolonization speed was identified as a predictor of healing ability and favorable prognosis, while more rapidly recolonizing tumors obtained less therapeutic utility through longer treatment periods. To counteract this faster growth pattern, a dose scale-up is frequently required, especially over longer periods of treatment (Wang and Li 2005). Although improved patient outcomes have been established through modifications of radiotherapy procedures to reduce tumor recolonization, this represents a major hurdle in both radiotherapy and chemotherapy. The emergence of resistance in cancer cells exposed to ionizing radiation and the ensuing recolonization of radiation-resistant cells frequently leads to a more invasive and aggressive cancer phenotype with disastrous outcomes for the patient. Thus, the identification of the molecular mechanisms of the repopulation and the acquired resistance is of decisive relevance for the design of specific therapies.

In 1956, it was first explained how the mixing of cancer cells destroyed by radiation with viable cancer cells resulted in enhanced growth of the latter in a tumor transplant, and it was proven that this was achieved by the diffusion of a growth-stimulating metabolite from the lethally irradiated cells to viable cells (Revesz 1956). In a follow-up investigation, it was found that this stimulatory action was not linked to a specific cancer cell type, as lethally irradiated foreign mouse tumors and liver tissue suspensions were also able to promote the growth of viable cells in an effective manner (Seelig and Revesz 1960). The discovery of a stimulating substance secreted from a multitude of dying cell types suggests the existence of an evolutionarily conserved machinery whereby the compensatory proliferation of surviving cells is initiated by cues emitted by the dying cells.

In a recent investigation, it was demonstrated that an immortal myogenic cell line named C2C12 in the host muscle of mice which had been irradiated before transplantation developed tumors faster than in non-irradiated control mice (Morgan et al. 2002). The employment of a mammary chimera mouse model has been generated where the transplantation of oncogenic Trp53 null epithelium into an irradiated host also caused an accelerated tumor growth compared to the transplantation into a non-irradiated host (Nguyen et al. 2011). In this instance, the mechanism of action was traced back to TGFβ, which means that the reduction of TGFβ inhibited the increased tumor growth.

16.8.6 Cell Death-Driven Tumor Repopulation During Radiotherapy

Butrant-type apoptosis is regarded as the hallmark of cancer, and the activation of caspases to initiate apoptosis is the dominant paradigm in the majority of cancer treatment options (Hanahan and Weinberg 2000, 2011; Lowe et al. 2004; Fiandalo and Kyprianou 2012; Lavrik et al. 2005b). However, recent findings have revealed that this model is possibly excessively simplified.

Based on these results on the pro-regeneration properties of caspase-3 and -7, an important role of caspase-3 in the repopulation of cancer cells in the course of radiotherapy was identified (Huang et al. 2011). It was reported that irradiated mouse caspase-3-deficient embryonic fibroblasts (casp3$^{-/-}$ MEFs) or caspase-3-deficient cancer cells are not capable of promoting the proliferation of labeled cancer cells in vitro or in vivo. Since caspase-3 has been shown to play a part in the facilitation of the healing of wounds and the regeneration of tissues, it has been demonstrated once again that caspase-3 can elevate arachidonic acid, and consequently, the production of PGE2 through the splitting of iPLA2 and that caspase-3-dependent activation of iPLA2 in dying feeder cells is required to enhance the proliferation of cancer cells. In addition, both arachidonic acid and the production of PGE2 were elevated due to ionizing radiation in wild-type MEFs and 4T1 cancer cells; however, their productivity was strongly diminished in Casp3$^{-/-}$ MEFs or 4T1shCasp3 cells. Exogenous therapy with PGE2 also resulted in cancer cells proliferating faster than non-treated cells.

In line with these findings on the pivotal function of PGE2 in cancer cell repopulation, it has been earlier announced that PGE2 may enable the β-catenin-Wnt signaling pathway in colorectal cancer (Castellone et al. 2005; Shao et al. 2005). PGE2 has also been found to be essential for the activation of the Wnt signaling pathway in hematopoietic stem cells, in which the Wnt signaling pathway governs their self-regeneration and the repopulation of bone marrow (Goessling et al. 2009). The clinical applicability of these results was confirmed in two groups of human cancer patients: one suffering from head and neck cancer and the other from breast cancer. In patients with head and neck cancer, high levels of activated caspase-3 have been associated with high rates of cancer relapse, and in breast cancer patients, elevated caspase-3 concentrations have been linked to reduced overall survival times.

Some of the latest findings on the function of apoptotic caspases in the process of tissue regeneration and cancer cell recurrence are both startling and intriguing. This could have enormous consequences for cancer therapy. The development of active substances that activate caspases needs to be revisited. Another aim was to examine low-molecular inhibitors of caspases for their characteristics to enable the improvement of radiotherapy or chemotherapy in cancer. Although there is a scarcity of clinical investigations in humans in this field, multiple investigations with mouse systems indicate that inhibition of apoptosis may enhance responsiveness to radiotherapy. In a murine lung cancer model system, for instance, exposure to a caspase-3 inhibitor resulted in an improvement in autophagy and increased sensitivity to radiation (Kim et al. 2008). In extension, pan-caspase inhibitor Z-VAD, or siRNA directed against caspases 3 and -7, also causes radio sensitization in breast and lung xenographies that displayed retarded growth rates and a reduction in angiogenesis (Moretti et al. 2009).

The inhibition of caspase has also been found to prevent VEGF-based vessel formation in human umbilical cord endothelial cells, whereby caspase-3 seems to facilitate this apoptosis-specific angiogenesis (Segura et al. 2002). The capacity of caspase blockers to suppress angiogenesis might be achieved by variations in cell adhesion and motility determinants that rely on caspase-8 activity (Senft et al. 2007). There is also the option of targeting downstream caspase-activation factors, including the prostanoid PGE2. COX enzymes, such as the NSAIDs, have been employed to enhance performance in cardiovascular and inflammatory diseases, there is increasing proof from pre-clinical trials that PGE2 is able to contribute significantly to the development and advancement of tumors (Allaj et al. 2013).

Another goal of caspase-3 activation represents itself in the highly malignant and hardly treatable brain tumor, the glioblastoma. In glioblastoma, it has been demonstrated that caspase-3 is activated in an inert manner, although in a concentration that is not sufficient to initiate apoptosis. However, it has been demonstrated that this caspase activation inhibits the migration and invasiveness of glioblastoma cells due to the caspase-dependent splitting of the pro-motility protein Gelsolin (Gdynia et al. 2007). In addition, caspase concentrations can also be explored for use as potential prognostic biomarkers. Immunohistochemical analysis of caspase-3 expression in human breast cancer revealed a relationship between caspase-3 overexpression and unintended patient outcome (Nakopoulou et al. 2001). A number of human breast, colon and pancreatic cancer cell lines and tissues exhibit constitutively low to medium levels of caspase-3 andcaspase-8 activation in comparison to normal cells even in the lack of apoptotic stimulation (Yang et al. 2003). This necessitates the investigation of the part played by increased caspases in the enhancement of tumor persistence regardless of the induction of apoptosis and the potential undesirable consequences of therapies that aim to raise their activation.

16.9 Interplay Between Proliferation, Survival, Necrosis and Apoptosis and Cell Mechanics

The controls of cell growth, cell proliferation and cell death form the foundation of tissue homeostasis. Whereas all these processes can be governed by inherent and genetic factors, their reaction to external cues from the local environmental conditions is also critical for homeostasis of tissues. The initiation and progression of tumors is characterized by the defective regulation of growth, proliferation and death, mostly due to the concentration of genetic mutations. However, there is growing proof that the microenvironment of the tumor itself has a powerful effect on the development and progression of cancer. These factors comprise the mechanical restraints and the compression forces produced due to the opposition of the adjacent tissue/matrix to the expansion of the tumor. Mechanical stress has lately been advocated to encourage concurrent interactions among cells through a process termed mechanical cell competition. High proliferation can result in the densification of cell populations and the removal of adjacent cells that are more susceptible to densification. Although this nascent concept has lately been verified in vivo, the significance of this process in tumor progression has never been widely debated (Levayer 2019). In the following section, the phenomenology of mechanical cell competitiveness is introduced, with an emphasis on the key variables and pathways governing cell clearance. Subsequently, there is a discussion on the importance of mechanical cell competition in the initiation and enlargement of tumors, highlighting their possibly opposing effects on tumorigenesis.

The strict regulations of growth, cell division and cell death are indispensable for the maintenance of organ form and functions. The elevated cell turnover of certain organs, for instance, intestinal cells, can be completely regenerated within five days (Potten and Loeffler 1990) and their ability to regenerate is dependent on the interaction between cell proliferation, cell growth and cell death. Whereas the cellular routes that organize these factors are very well understood, the mechanisms underlying their synchronization are still little well defined. This coordinated action was mainly investigated in the context of direct cell–cell interaction (paracrine signal, contact-dependent signaling) and long-distance communication (diffusive molecules, cytonemes). More lately, the contribution of cell shape and local tissue mechanics to this coordinating function has aroused considerable attention. This is consistent with the influence of cell mechanics on multiple major controllers of cell proliferation and survival (Vogel and Sheetz 2006; Panciera et al. 2017; Irvine and Shraiman 2017). The induction and growth of tumors is connected with an aberrant organization of cell growth and cell survival. This dysregulation is mainly induced through cellular factors and the concentration of mutations in the genes that govern these processes (Hanahan and Weinberg 2000). However, it is currently evident that the microenvironment of the tumor and non-cell-autonomous parameters also exert a major influence on the induction and advancement of tumors (Bissell and Hines 2011). The role is especially critical in view of the latest reports demonstrating that the probability of detecting tumor-causing spontaneous mutations in adult tissues is far greater than the incidence of cancer (Martincorena et al. 2015, 2018).

Mechanical forces applied to cancer cells and adjacent cells are a component of these microenvironmental impacts and have been proven to influence each step of the tumor propagation process, from initiation and progression to metastasis (Butcher et al. 2009; Jain et al. 2014). Cell competitiveness refers to a mechanism that induces the favored removal of one cell population from another cell population through apoptosis (Levayer and Moreno 2013; Vincent et al. 2013; Nagata and Igaki 2018). This approach was first developed over forty years ago when investigating the performance of mosaic tissues in *Drosophila* (Morata and Ripoll 1975). By initiating Minute mutants, which are dominant mutations that interfere with ribosome production (Kongsuwan et al. 1985), they demonstrated that Minute/+ heterozygous clones are eradicated from developing tissues when they are encircled by wild-type cells.

This finding was fascinating, because minute/+ mutant flies had almost no defects except for a small retardation in development. It was later found that other genetic alterations, such as higher concentrations of the proto-oncogene Myc, lead to aggressive clones that can remove and completely replace the adjacent wild-type cells without causing obvious tissue damage (Moreno and Basler 2004, de la Cova et al. 2004). This "super-competition" was proposed to promote the early expansion of pre-tumor cells and tumor initiation in a process similar to field cancerization (Moreno 2008). In specific detail, the field cancerization involves the creation of multiple patches of pre-malignant disease with a higher-than-expected rate of numerous local secondary primary tumors. The amount of tissues and mutations that causes competing interfaces has steadily grown in recent years (Levayer and Moreno 2013). It is now also obvious that this kind of process is even conserved in mammals (Maruyama and Fujita 2017). Nevertheless, the mechanisms that cause the preferred removal of a cell population are under active discussion. Three non-exclusive mechanisms have been identified to be involved in cell removal: rivalry for the limitation of extracellular survival factors, rivalry through contact-related death initiation and, most recently, rivalry induced through mechanical stress (Merino et al. 2016; Claveria and Torres 2016; Bras-Pereira and Moreno 2017).

The latter recently emerged when it was demonstrated that cell compression and differing responsiveness to mechanical stress might encourage competing interactions among cells (Levayer et al. 2016; Wagstaff et al. 2016) and either preferentially eradicate pre-cancerous cells (Wagstaff et al. 2016) or foster the spread of pre-cancerous clones (Levayer et al. 2016; Tsuboi et al. 2018; Moreno et al. 2019). This phenomenon known as mechanical cell competition is now evolving into a new class of cell competition that may be of importance in various physiological and pathological environments (Vincent et al. 2013; Bras-Pereira and Moreno 2017; Levayer et al. 2016; Wagstaff et al. 2016).

16.9.1 Cells Triggered by Mechanical Stress: Initial Hypothesis and Theoretical Grounds

It has been suggested that mechanical stress produced through differential expansion may be the foundation for competitive interactions within the cells (Shraiman 2005). In a number of tissues, the cell–cell connections stay close and the comparative displacement of the cells is very small. In this respect, tissues act as solids, which have a low capability to reduce mechanical tension. In this setting, a relative rise in cell proliferation/growth in a particular subpopulation is likely to push the neighboring cells which will oppose this pushing behavior. On long time intervals, the varying growth will increase the pressure in the tissues, especially in the rapidly growing population and its immediate surroundings. It would intuitively be assumed that such compaction would severely impair the ability of the rapidly growing population to extend. In fact, there is increasing evidence that compression/negative strain can significantly retard cell proliferation and growth (Dupont et al. 2011; Helmlinger et al. 1997; Delarue et al. 2014; Pan et al. 2016; Streichan et al. 2014; Gudipaty et al. 2017; He et al. 2018; Chen et al. 1997).

How, then, could this mechanical stress foster the extension of one population to the costs of the other? The short definitions of mechanical terms are given in Table 16.1. Hence, the concept of the differential sensitivity toward mechanical stress has been proposed by Shraiman. In particular, this model states that when the adjacent slow-growing cells die at lower pressure compared to the fast-growing population, this differential behavior upon external stress will be sufficient to initiate the preferential removal of the neighboring cells. This will result in a release of space and pressure and subsequently to the further enlargement of the rapidly growing population (Shraiman 2005). The model was especially important for cell competition, since the majority of the mutations related to the competition affected the growth rate (Vincent et al. 2013). However, it initially had limited influence in the field, presumably due to the fact that several key hypotheses of this model had yet to be experimentally proven. Firstly, at that time there was no conclusive evidence that densification/compression in vivo might initiate cell death. Secondly, this would presuppose that various cell types react differently to mechanical stress, which would still have to be proven. Thirdly, it was necessary that stress was not reduced by cell movement and neighborhood exchange. The later hypothesis was plausible for several tissues, including the main tissue employed to investigate cell competition: the imaginary wing disc of *Drosophila*, which is an epithelial sac of the larvae forming the adult wing. The higher proliferation level in the center of this tissue can result in a densification of the central population and a stretching of the cells at the peripheral margins (Mao et al. 2013b; Legoff et al. 2013), which indicates that the mechanical stress is not reduced through cell movement and exchange of neighboring cells. In a similar way, growth induction in clones recapitulates the same picture of deformation: compression of the rapidly growing population and stretching of neighboring cells (Pan

Table 16.1 Short definition of the mechanical terms

• Stress: Internal forces are exerted on each other by the components of a material. It represents a force per unit area and is identical in dimension to the pressure (unit: Pascal)
• Solid stress: Residual stress concentrates within the solid compounds of a tissue after the differential growth. This type of stress is not dependent on the fluid pressure. Moreover, residual stress refers to the stress that persists in a solid material even after the original sources of stress have been eliminated
• Strain: This dimensionless parameter measures the geometric deformation of a specific material in comparison to its original form
• Pressure: Force exerted per unit area on the surface of an entity perpendicular to its boundaries (unit: Pascal)
• Tension: Reaction force exerted on a stretched item and represents the opposite of compression (unit: Newton)
• Surface tension: Specifies the tendency of a liquid surface to minimize its surface area. The amount of energy needed to reshape this surface is proportional to the surface tension (dimension: energy per unit area or force per unit length)
• Line tension: Tension at the contact surface between two liquids or solids. Often referred to as the tension between two cells (dimension: energy per unit length). The line tension in an epithelium is raised due to acto-myosin contractility and lowered due to cell-cell adhesion
• Stiffness: Specifies the extent to which an entity deforms as a reaction to a force. It is measured in terms of the modulus of elasticity (force per unit area, in Pascal)

et al. 2016; Mao et al. 2013b; Legoff et al. 2013). However, this hypothesis may not be applicable to all constraints linked to cell rivalry. Multiple competitive settings (Myc, minute) were linked to the mixing of the two cell groups and high cell–cell movements, which was intended to reduce mechanical stress and avoid its accumulation (Simpson 1979; Li et al. 2009; Levayer et al. 2015).

Other conceptual frameworks also suggested a mechanistic function in competitive interactions among cells. This involves the idea of homeostatic pressure (Basan et al. 2009, 2011), which presupposes the occurrence of a specific pressure where cell proliferation and growth is compensated completely through cell death. This was founded on the hypothesis that both cell survival and cell proliferation are regulated through pressure. In this context, the cell population will increase in a finite volume until it achieves a pressure that matches its homeostatic pressure. However, when one population has a larger homeostatic pressure than the other, the former will ultimately inevitably eradicate the latter, regardless of the relative growth rate of the two populations without mechanical restraints. Whereas the measurement of tissue pressure is still a difficult task, the idea of homeostatic pressure might be analogous to the presence of different homeostatic densities among various cell types (Wagstaff et al. 2016; Bove et al. 2017). In general, there should be a positive correlation between local tissue pressure and cell density, and hence, the quantification of the density of a cell population in equilibrium provides a good indicator of its homeostatic pressure. The hypotheses of such models are close to those outlined in Shraiman's model, which assumes that various cell types react in a different way to pressure, and highlights the importance of the equilibrium between proliferation and cell death also in the homeostatic tissue.

16.9.2 Tissue Compaction Induces Cell Elimination

These models reappeared when it was later found that the compaction of epithelial cells might cause their removal, which represents a key hypothesis of these two models. When using stretchable substrates, it was demonstrated that a 30% rise in MDCK cell density resulted in a fast rise in cell extrusion rate, which is the concerted escape of cells from the epithelial layer (Rosenblatt et al. 2001) and cell death before the tissue regained its stationary density (Eisenhoffer et al. 2012). In a similar manner, it has been found that a large fraction of the cell extrusion appearing in the midline region of the *Drosophila* pupal notum (a single-layered epithelium) may be accounted for by the local densification of the tissue (Marinari et al. 2012), which was subsequently verified through other experimental approaches (Levayer et al. 2016). This early research provided the first experimental validation of the possible impact of cell and tissue deformation on the survival of epithelial cells. In other investigations, it has been indicated earlier that cell compaction may cause the removal of the removal of cells. For instance, a decrease of the cell surface due to a modification of the distance of the extracellular matrix spots is enough to induce apoptosis within individual endothelial cells (Chen et al. 1997). Several investigations were presented in which the individual cells were exposed to high hydrostatic pressures with the aid of pistons. For example, a 10 min exposure to 0.5–0.7 MPa (1 MPa is 10 times atmospheric pressure) can cause cell death of ganglion neurons (Quan et al. 2014). A 30 min exposure to 85–200 MPa is required to initiate apoptosis eight hours later in lymphoblasts (Takano et al. 1997).

Although these settings are distant from physiological pressure levels, since they are several orders of magnitude higher than the pressure detected in osteoblasts, 100–300 kPa (Thompson et al. 2012), they indicate that compression is adequate to initiate cell death and that various cell types have varying sensitivity to compression, in accordance with the hypothesis described in the preceding section. Alternative experimental designs were employed to compress cells in more realistic regions. For example, high osmolarity was employed to decrease glioma cell volume for 8 h of about 30%, which was enough to initiate apoptosis and sustain the activation of the effector caspase (Ernest et al. 2008). As an alternative, pressures between 1 and 100 kPa were exerted on 3D cell aggregates (referred to as spheroids), either through modulation of the elastic modulus of the gel embedding the cells (Helmlinger et al. 1997; Cheng et al. 2009; Alessandri et al. 2013) or through modification of the osmolarity of the external media (Delarue et al. 2014). Importantly, a pressure of 4 kPa was adequate to raise the rate of apoptosis in a spheroid of mouse breast cancer cells (Cheng et al. 2009), whereas a pressure of 7–13 kPa did not influence the rate of apoptosis in a spheroid of human colon adeno-carcinoma cells (Delarue et al. 2014; Helmlinger et al. 1997). Overall, these results proved that 3D cell compression can cause cell death. In addition, the findings of these investigations suggested that various cell types may have varying sensitivity to compressive forces and compression, in accordance with the hypothesis necessary for mechanical competition.

16.9.3 Experimental Proof of Mechanical Cell Competition

Cell competitiveness has been examined primarily in epithelial tissues. Therefore, the discovery that an epithelial confinement, such as crowding, could induce cell extrusion was especially relevant (Eisenhoffer et al. 2012; Marinari et al. 2012). These first two investigations nevertheless revealed that extrusion was not dependent on the activation of caspase and that cell death was a result of anoikis, which means a death of cells through the loss of cell–cell and cell-extracellular matrix adhesions. More precisely, it has been suggested that crowding initiated extrusion into the pupal notch of *Drosophila* is promoted through a spontaneous remodeling of the junctions regardless of specific signaling pathways (Marinari et al. 2012; Curran et al. 2017). It was thus complicated to comprehend how various cell populations inside an epithelium might respond in different manners to mechanical compression in this framework. Two subsequent studies proved, nevertheless, that mechanical stress might be the foundation for competitive interplay between cells. When examining the interactions between Scribble mutants (a regulator of apicobasal polarity) and wild type MDCK cells, an active corruption of the mutant cells through the wild-type cells was seen, which led to a compression of the Scribble mutants and subsequent removal of them. (Wagstaff et al. 2016). Applying micropatterns to establish tight limits, they noticed that homogeneous layers of Scribble mutant cells increased to a lower plateau/homeostatic density when compared to wild-type cell layers, which was accompanied with a higher incidence of cell death. It is interesting to note that the high sensitivity of the Scribble mutant cells to densification correlated with higher basal values of the cell death regulator p53, which continued to rise during cell densification (Wagstaff et al. 2016). These observations indicated that p53 concentration could be used to measure the homeostatic pressure of cells and that the removal of Scribble mutants might be triggered through the juxtaposition of two cell populations with distinct homeostatic pressure (Basan et al. 2009). Similarly, downward regulation of corruption and cell migration characteristics of wild-type cells (through E-cadherin downregulation in Scribble mutants) delayed the removal of Scribble mutants, but did not stop it. This pointed to the fact that the competitive elimination of Scribble mutants is powered both through their low homeostatic pressure/density in comparison to wild-type cells and through their active compression through the cell migration of wild-type cells.

In a similar way, recent findings in *Drosophila* pupal notum resulted in a re-evaluation of the function of effector caspases in cell extrusion. Through blocking the caspase activation in clones and throughout the entire tissue, a sharp decrease in the rate of cell extrusion was found (Levayer et al. 2016). In addition, employing different sensors for the activity of living caspases, a systematic activation of effector caspases before the extrusion of cells was monitored. This indicated that the cell extrusion needed the activity of the effector caspase. Thus, apoptosis-resistant cells could react less sensitively to crowding. Similarly, activation of the Ras oncogene (RasV12) in large clones resulted in expansion of clones, compaction of adjacent wild-type cells and favored elimination of wild-type cells (Levayer et al. 2016). The compression resistance of clonal cells is promoted through the inhibition of the pro-apoptotic gene head involution defective (hid) through the activation of Ras (Moreno et al. 2019; Bergmann et al. 1998; Kurada and White 1998). Most importantly, blocking the removal of wild-type cells through the secretion of EGF from the clones was adequate to retard the expansion of the Ras clones (Moreno et al. 2019). These two investigations (Levayer et al. 2016; Wagstaff et al. 2016) demonstrated in two separate systems that varying sensitivity to mechanical stress may be the foundation for competitive interactions and delivered the first experimental confirmation of the first theoretical investigations on mechanically motivated cell competition (Shraiman 2005; Basan et al. 2009). Importantly, these two investigations indicated that mechanical challenge can have both anti-tumor and pro-tumor features. Whereas in the first case it is necessary for the removal of pre-tumor cells (Wagstaff et al. 2016), since scribble mutants can induce the neoplastic growth (Bilder 2004), in the second case it rather encourages the enlargement of pre-tumor cells through the removal of the adjacent wild-type cells (Levayer et al. 2016; Moreno et al. 2019).

16.10 Mechanical Cell Competition Is Induced Through Cell Migration and Boundary Conditions

So far, the major focus was on the generation of mechanical stress that has been achieved through differential growth patterns. But other mechanisms can cause cell compression and cell removal irrespective of the growth and the proliferation of cells. As outlined earlier, the compression of Scribble mutant cells is also encouraged through the active migration of wild-type cells into the mutated cells (Wagstaff et al. 2016). More lately, the spontaneous elimination of the Madin Darby Kidney (MDCK) cells in zones with congruent movements has been reported (Saw et al. 2017). MDCK cells normally align themselves locally in a similar manner as nematic liquid crystals. But the cell migration can cause spontaneous misalignments (referred to as +1/2 nematic defects or comet-like shape) that preface cell extrusion and cell death. These defects were associated with a local rise in compression forces, as determined using traction force microscopy, and are adequate to induce caspase activation and cellular extrusion (Saw et al. 2017). These two investigations exemplify how collective cell migration can induce cell densification and cell clearance.

As an alternative, raising the tension at the interface of the two cell populations can also produce compressive forces. The shape of the cellular interfaces in epithelia can be easily estimated from the distribution of surface tension (Table 16.1) (Lecuit and Lenne 2007). Within this framework, the geometry of the interfaces is subject to the same laws that define the shape of foams and bubbles. These comprise Laplace's law, which defines the relationship between the line tension (Table 16.1, γ) at the interface between two liquids, the pressure difference (ΔP) and the radius of the liquid droplet (R), in the following equation (16.1):

$$\Delta P = \frac{2\gamma}{R} \tag{16.1}$$

The Laplace pressure is permanently present at the interface, but its magnitude depends on the heterotypic adhesion strength (γ) and the curvature of the interface. This law for instance can account for the sorting and the establishment of smooth borders between various cell lines at the time of association (Amack and Manning 2012). A misspecification of cells in the imaginal disc of the wing also causes a sorting of the abnormal cells (circular clone shape and smooth clone borders), which can result either in the removal of the cells (Milan et al. 2002; Bielmeier et al. 2016; Gibson and Perrimon 2005; Adachi-Yamada et al. 1999) or in the development of abnormal wrinkles in the tissue, which are referred to as cysts (Bielmeier et al. 2016; Gibson and Perrimon 2005; Shen and Dahmann 2005). For example, this takes place after the local impairment of the Dpp (Decapentaplegic, which is the orthologue of TGF-β) pathway (Gibson and Perimon 2005; Adachi-Yamada et al. 1999; Shen and Dahmann 2005) or the Wg (Wingless) pathway (Adachi-Yamada et al. 1999; Widmann and Dahmann 2009; Vincent et al. 2011) that are two well-known morphogens determining cell fate and cell growth within the wing disc.

It has recently been seen that a wide variety of genetic mutations, such as the activation of JAK-STAT or Ras and the downregulation of Polycomb proteins, in clones may be responsible for cell sorting in large clones, the formation of cysts in intermediate clones, or the removal of cells in small clones (Bielmeier et al. 2016). These alterations all involved an accumulation of actin and the molecular motor myosin II on the lateral surfaces of the cells at clone borders, which resulted in an elevation of the interfacial tension between clone and wild-type cells. The enhancement of the interfacial tension at clone boundaries in a 3D vertex model is adequate to reconstruct the various morphologies monitored, including the sorting for large clones and the creation of cysts for intermediate clones. The function of the size of clones could be easily understood by applying Laplace's law: When the line tension is kept constant, the pressure is inversely related to the clone's radius. Therefore, small clones are supposed to be exposed to a higher level of pressure than larger clones. Although this has not been subjected to proper screening, this indicates that misspecified cells in small clones are subjected to high pressure, which can lead to their death and clearance. It is crucial in this framework that the removal of cells depends not so much on their genotype and the intrinsic characteristics of the cells, but rather on the amount of tissue taken up by the mutated cells and the wild-type cells that govern the shape/bending of the interface (Bielmeier et al. 2016). Thus, it can either remove small patches of pre-tumoral cells, such as the activation of JAK-STAT, or kill wild-type cells as soon as the mutant takes up the majority of the tissue. The comparative sensitivity to compression only determines the critical pressure and thus the critical clone radius from which the clones begin to vanish.

However, there are other mechanisms beyond mechanics, such as contact dependent apoptosis that may also be responsible for cell clearance and affect the survival of clones (Levayer et al. 2015; Diaz-Diaz et al. 2017; Kajita and Fujita 2015). This can be of special relevance for large clones where the mechanics might be less powerful. A similar process might be implicated in the removal of isolated pre-tumoral cells from MDCK cell layers, which is termed the epithelial defense against cancer (Kajita and Fujita 2015). Although this process has not been officially associated with cell compaction and cell death, the potent part played by the mechanical characteristics of the active Ras/wild-type cell interfaces could be explained by the role of Ephrins (Porazinski et al. 2016), EPLIN (Ohoka et al. 2015) or Filamin and Vimentin (Kajita et al. 2014), all of which have an effect on MyoII concentration, and the active densification of RasV12 cells when encircled by wild-type cells (Porazinski et al. 2016) fall into such a scheme.

Curiously, interfacial tension was formerly also recommended to remove early metastatic cells when implanted in new tissue (Basan et al. 2009). As with the *Drosophila* wing disc, the possibility of metastatic cell survival will vary according to the interfacial tension and the size of the metastatic cell accumulation. In summary, it can be suggested that the refinement of the interfacial tension connected with a misspecification of the cells can form the foundation for continued mechanical cell competitiveness. Although the relationship between cell compaction, pressure and removal of cells has not been validated formally under these circumstances, it is highly probable that the same compaction-related lethal phenomenon will arise. In contrast to the initial scenarios outlined above, this alternative type of compression is not based on cell growth and may either remove pre-tumoral cells or encourage their growth, according to the percentage of tissue they inhabit.

16.11 Molecular Pathways Sense Compaction and Drive Cell Elimination

However, the impact of cell compaction on the survival of cells has been detected in a plethora of tissues. Therefore, strong attention has been paid to understand the mechanisms that play a role in compression sensing and the induction of death. A short description of the pathways associated with compaction-driven cell death is provided in Table 16.2. However, these mechanisms of the sensing of the compactness are still largely under discussion (Valon and Levayer 2019).

The first monitoring of compaction-driven extrusion in epithelia seemed to indicate that the process is not dependent on caspases and apoptosis (Eisenhoffer et al. 2012; Marinari et al. 2012). In MDCK cells and in the *Zebrafish* tail, the crowding-induced extrusion is dependent on the triggering of the stretch-activated channel Piezo1 (Eisenhoffer et al. 2012). Consequently, retardation of piezo1 by morpholinos in cages resulted in an aggregation of cell mass in the embryonic tail of the *Zebrafish*, whereas inhibition of stretch-activated channels with Gadolinium greatly diminished extrusion during MDCK cell densification (or synonymously referred to as crowding). However, it is probable that crowding extrusion is regulated by other unidentified elements, as the Ca^{2+} inflow, which should take place downstream of the piezo activation, is not strong enough to trigger extrusion (Gudipaty et al. 2017). As an alternative, it was originally proposed to encourage crowding extrusion in the pupil notch of *Drosophila* through spontaneous remodeling of the junction (Marinari et al. 2012), which can be facilitated by the tension of the junction within the tissue (Curran et al. 2017).

In different systems, compaction-related death has been linked to caspase-dependent death and apoptosis. The reduction of cell volume due to osmotic stress induces the activation of caspase-3,

Table 16.2 Experimental cues and pathways associated with compaction detection and cell death triggering

Cell system	Reason for compaction	Sensing route	Reason for cell death
MDCK cells	Release of a stretchable substrate	Piezo stretch sensitive channel	Anoikis
Zebrafish embryonic thin	Proliferation-dependent spontaneous crowding	Piezo1 stretch-sensitive channel	Anoikis
Drosophila pupal notum	Tissue movement-dependent crowding	Spontaneous junction remodeling	Anoikis
Drosophila pupal notum	Tissue movement-dependent crowding and near fast growing clones (Ras^{v12})	Downregulation of EGFR/ERK through tissue compaction	Caspase activation, cell extrusion and apoptosis
MDCK scribble mutant and WT MDCK	Growth and collective cell movements	Upregulation of p53 through ROCK and p38 kinases due to densification	Apoptosis
Spontaneous MDCK extrusion	Collective migration based topological defects	Yap-driven nuclear exclusion triggered by compressive forces	Caspase activation, cell extrusion and apoptosis
Murine mammary carcinomas spheroid	Confined growth	–	Mitochondrial apoptotic pathway and necrosis
Glioma cell	Osmotic stress	Volume sensing	Caspase activation

caspase-8 and caspase-9 and apoptosis (Ernest et al. 2008). Similarly, the cell removal initiated in the center of the compressed spheroid is propelled both by necrosis and Bcl2 (an apoptosis inhibitor) by apoptosis (Cheng et al. 2009). These results imply that isotropic densification can trigger the mitochondria-dependent apoptosis pathway and indicate that some perceptual mechanisms need to correlate densification with caspase activity. The spontaneous extrusion of MDCK cells fueled by topological deficiencies can be inhibited using the pan-caspase inhibitor Z-VAD-FMK and occurs prior to caspase-3 activation, which can be detected employing a live marker (Saw et al. 2017). The activation of caspases is facilitated through the nuclear exclusion of yes associated kinase (YAP) (Saw et al. 2017), the pro-survival transcriptional co-activator impaired through the Hippo signal transduction pathway. The Hippo signaling pathway is a key modulator of cell growth and survival and represents one of the best recorded examples of signaling pathways that are regulated by cell shape and cell mechanics (Panciera et al. 2017). The Hippo pathway is impaired through the stretching of cells (Dupont et al. 2011; Aragona et al. 2013; Elosegui-Artola et al. 2017), whereas it can be induced through the inhibition of the cell contact and the impairment of the compression of the tissue (Zhao et al. 2007). In a similar way, the upregulation of cell proliferation/growth in clones in the *Drosophila* wing disc creates compressive forces that upregulate the Hippo pathway within the clones, which has a negative effect on clone growth (Pan et al. 2016). The same kind of feedback can arise during the natural development of the wing disc, when a gradual upregulation of the hippo is associated with the ongoing compression of the tissue (Pan et al. 2018).

As previously discussed, the removal of Scribble mutant MDCK cells through wild-type cells is achieved by their compaction and the consecutive upward regulation of p53 (Wagstaff et al. 2016). This upregulation is stimulated through the stress kinase p38 and the elevated levels of ROCK, which is an activator of myosin II (Wagstaff et al. 2016). Lately, it has been identified that the EGFR/ERK signaling pathway acts as a central controller of cell survival in *Drosophila* pupal notum due to the blocking of the pro-apoptotic gene hid (Moreno et al. 2019). With a new ERK live sensor, an ectopic tissue stretching/compression was demonstrated to upregulate/downregulate transient ERK activity. More crucially, the activation of Ras in clones resulted in the compression and downregulation of ERK in adjacent wild-type cells. Finally, this results in caspase activation, extrusion of wild-type cells and continued expansion of Ras clones (Moreno et al. 2019). Although the molecular foundation of compaction controlled ERK downregulation is not yet fully understood, these results are consistent with several findings demonstrating that ERK can be activated by cell/tissue stretching and/or increased tension (Gudipaty et al. 2017; Hirata et al. 2015; Sano et al. 2016; Aoki et al. 2013, 2017). Overall, these results indicate that the densification detection mechanisms and the pathways employed for cell removal may vary widely according to tissue environment, the amplitude and the time scale of tissue deformations. However, they reveal quite clearly that multiple survival-promoting and pro-apoptotic paths can be altered through modifications of the cell shape or alterations in tension.

16.12 Mechanical-Based and Biochemical-Based Competition: Synergies and Antagonisms

The previous focus was on the competitive removel driven by mechanical stress. However, the removal of cells during competition may be facilitated by diffusive substances (Senoo-Matsuda and Johnston 2007; Alpar et al. 2018) or by contact-based initiation of apoptosis (Levayer et al. 2016; Diaz-Diaz et al. 2017; Yamamoto et al. 2017; Rhiner et al. 2010). What is the connection between these biochemically derived removals and the mechanically derived removals? In general, mechanical competition fueled through differential growth usually operates in synergy with the alternative mechanisms of cell clearance. In fact, most genetic contexts linked to mechanically based rivalry, such as the activation of Ras (Levayer et al. 2016; Moreno et al. 2019), activation of Yki (Tsuboi et al. 2018), Scribble mutants (Wagstaff et al. 2016; Bove et al. 2017), have also been combined with either contact or diffusion-based clearance. For example, the activation of Ras and Yki can induce the aggregation of the proto-oncogene Myc (Ziosi et al. 2010; Prober and Edgar 2002), which is a famous modulator of contact-dependent rivaltry (Moreno and Basler 2004; de la Cova et al. 2004; Levayer et al. 2016; Claveria et al. 2013).

The contact-dependent activation of the JNK signaling pathway in *Drosophila* leads to the similar removal of scribble mutant clones (Yamamoto et al. 2017). In these circumstances, contact and diffusion elements will encourage the removal of the same population that is propelled by mechanical stress. Although the comparative contribution of these phenomena has never been clearly examined, it can nevertheless be forecast that they will not be encouraged by the same circumstances. In fact, the high-level contact surface area between the two cell types (Levayer et al. 2016) and the lifetime of these contacts enhances contact-dependent cell removal (Diaz-Diaz et al. 2017). Altogether, a high degree of cell mixing will promote contact-dependent mechanisms (Levayer et al. 2016), whereas remodeling of the junctions will inhibit the build-up of mechanical stress induced through differential growth. As an alternative, the high compactness of the clones and the sorting characteristics linked to Ras activation (Bosveld et al. 2016) were expected to decrease the impact of contact-dependent removal.

The scenario is even more intricate in the face of mechanical competitiveness, which is determined mainly through boundary constraints. In this case, mechanical stress can operate opposite to a competing biochemical signal-driven removal. For example, small activated Ras clones from *Drosophila* wing discs can be removed by Laplace pressure (Bielmeier et al. 2016), whereas Ras can increase cell aggressiveness and trigger the removal of adjacent cells through the induction of Myc (Prober and Edgar 2002). As the rise in line tension at clone borders encourages cell sorting and decreases the contact area between the two populations, contact-facilitated removal is probably less efficient under these circumstances. Nevertheless, the impact of diffusion factors is expected to remain substantial and may change the results of mechanical competition powered by borderline restrictions. In conclusion, the various forms of competition need to be in co-existence and can either work synergistically to remove a population or can be antagonistic. The shape of the interface between the two populations, the clone-sizes, and the mechanical properties of the nearby tissue, such as the proportion of transition remodeling, will determine the specific role of contact-driven, diffusion-driven and mechanical-driven removal of cells. At this stage, it is still very complex to evaluate the relative impact of the various phenomena, as this requires the identification of unique molecular entities that can entirely rule out one type of competition and not the others.

16.13 Solid Stress in Tumor and Effect of Mechanical Cell Competition on Tumorigenesis

After discussion of the theoretical principles and experimental verification of mechanical cell competition, it is still not clear how this process might manifest itself in solid tumors. Although the existence of mechanical cell competition has not yet been validated in a real tumor, it must be examined whether the key constituents necessary for such competition are properly available in tumors, and their possible influence on the initiation and progression of tumors needs to be discussed.

16.13.1 Features of Solid Stress in Tumors

Even though cancer cells tend to have a high proliferation rate, tumors are not growing exponentially. Most tumor growth graphs are rather sigmoid and are characterized by an initial increase in volumetric expansion, which is followed by a deceleration and a plateau phase (Benzekry et al. 2014; Freyer 1988). These dynamics are well compatible with a Gompertz growth model that was in fact named after a British mathematician who calculated the death rate in the nineteenth century (Gompertz 1825). Whereas this arrest of cell proliferation has been linked to a deficiency of nutrients and oxygen (Grimes et al. 2016), other investigations have suggested that mechanical restraints and mechanical compression may also arrest the expansion of the tumor (Helmlinger et al. 1997; Montel et al. 2011).

The initial experimental analysis of the compressive forces produced during tumor expansion was carried out in vitro utilizing human colon adenocarcinoma cell spheroids encased in agarose gels of a known stiffness (Helmlinger et al. 1997). In this arrangement, the spheroid growth rate was negatively correlated with the stiffness of the agar, and the final size of the spheroid can be easily predicted due to the final mechanical stress encircling the spheroid. Most interestingly, the pressure on the agar gel was raised up to 6–16 kPa, which in principle are sufficient to collapse tumor blood vessels (Helmlinger et al. 1997), as it has been seen in several solid tumors in vivo (Jain et al. 2014). Subsequent investigations proved that compression forces can reduce the growth of the spheroids (Delarue et al. 2014; Cheng et al. 2009; Alessandri et al. 2013; Montel et al. 2011). Although these in vitro investigations affirmed that the stress produced through tumor growth can have a negative response to their growth, they did not demonstrate that solid stress can in fact tend to accumulate inside tumors in vivo. The presence of solid stress in vivo has been verified by a number of mutually independent techniques. The selective removal of human cancer cells using the diphtheria toxin has been shown to decrease cell density and was enough to enhance the proportion of opened blood and lymph vessels within human tumor xenografts (Padera et al. 2004). That is consistent with an increase in pressure in tumors caused through cell proliferation. The analysis of the relaxation of the human tumor during the cut was employed as an alternative to measurement of the elastic energy stored. The inherent stress caused through differential growth is expected to cause an isotropic compression of the inner tissue and an enhancement of the tension tangential to the tumor mass (Jain et al. 2014). Thus, a planar cut of excised tumors demonstrated a contraction of the periphery and a protrusion of the central tissue, in accordance with the anticipated spread of compression and tension (Stylianopoulos et al. 2012; Nia et al. 2016). Under the assumption of a constant modulus of elasticity in the tumor, the stored energy was calculated with finite element modeling, for instance 0.4–8 kPa for mouse tumors (Stylianopoulos et al. 2012), and corresponds well with the pressure expected in spheroids (Helmlinger et al. 1997). Later on, alternative measurement techniques have proven these value scales, such as the slicing of the tumor and the punching of a cylindrical gap (void) in the tumor (Nia et al. 2016, 2018). Overall, these investigations verified that the build-up of solid stress is a frequent feature of solid tumors.

16.13.2 Space Competition with Stroma and Adjacent Healthy Cells

The accumulation of solid stress in tumors and adjacent tissue might be consistent with mechanically induced removal of adjacent cells. So far, only a few investigations have been carried out that show apoptosis in the proximity of human tumors. In a recent report, cleavage caspase-3 positive cells were found in the stroma in the vicinity of human breast tumors, colon tumors, brain tumors, breast tumors, lung tumors and lymph node metastases (Di Giacomo et al. 2017). The concentration of caspase-3-

positive cells was found to correlate with the c-Myc level and the nuclear concentration of YAP in the tumors. This indicated that the percentage of apoptotic cells in the vicinity of tumors may have been hitherto underestimated. The low sensitivity of the majority of assessments for the detection of apoptosis and the lack of data on the very early stages of tumor spread might impede a systematic assessment of apoptosis rates in the proximity of solid tumors. Recently, neuron death has often been seen in the proximity of nodular brain tumors in mice (Seano et al. 2019). Cell death is reduced through the liberation of compression through brain surgery or through the protection of neurons with lithium treatment. These two treatments resulted in an increase in motor coordination (Seano et al. 2019), indicating that compression-related neuronal death may lead to neurological disorders. The shape of the tumor–stroma interface also seems to relate to the varying prognosis. For example, liver metastases with smooth borders with a so-called pushing morphology, which means the deformation of the adjacent parenchyma, are given the worst prognosis in comparison to patients with desmoplastic growth, where a collagen border encircles the tumor (Eefsen et al. 2015; Van den Eynden et al. 2012). In a similar fashion, nodular human brain tumors that create smooth borders tend to have the worst prognosis in comparison to undefined border tumors (Seano et al. 2019). These relationships indicate that the densification of the then adjacent tissue might be linked to the worst prognosis and could be associated with mechanical cell rivalry with the adjacent cells. In addition, multiple studies have pointed out that cancer cells react in different manner to mechanical signals compared to untransformed cells. For example, the apoptosis level for normal NIH3T3 on softer surfaces rises, whereas H-ras-transformed cells stay undisturbed (Wang et al. 2000). In a similar manner, normal thyroid cells align their modulus of elasticity to the stiffness of the substrate, whereas cancer thyroid cells cannot perform this function (Rianna and Radmacher 2017). All in all, these investigations revealed that all elements of mechanical cell rivalry are found in human solid tumors. Nevertheless, the relationship of healthy cell deformation to tumor clearance has not been examined, and the impact of stromal/non-transformed cell death on the expansion of the tumor has not been assessed.

16.13.3 Intratumoral Space Competition

Whereas the mechanical cell rivalry of human tumors with adjacent healthy cells is still rather hypothetical at this point, there is more proof of the occurrence of intratumoral rivalry at present. The degree of apoptosis in the tumor spheroid has a positive relationship to the confining pressure for mouse breast carcinoma cells (Cheng et al. 2009). Compression-induced clearance can be reduced through Bcl2 (Cheng et al. 2009), which indicates that the responsiveness to the mitochondrial apoptotic pathway seems to modify the responsiveness to compressive stress. Latest advancements in single cell sequencing have highlighted a high variety of genetic backgrounds and highly intricate clonal dynamics inside a tumor (Tabassum and Polyak 2015; Lawson et al. 2018; Tissot et al. 2017). It might be consistent with a differential responsiveness to mechanical stress and might form the foundation for clonal competitive interferences inside a tumor. Rival processes among cancer cells have been broadly debated and theoretically investigated (Nowell 1976; Greaves and Maley 2012; Waclaw et al. 2015), mostly in the setting of clonal interference, that is, the decrease in growth and survival of one clonal population due to the presence of another (de Visser and Rozen 2006).

However, only a few studies have fully identified the competing interplay among clones (Marusyk et al. 2014; Yates et al. 2015), and neither of them has shown the role of spatial restrictions and mechanical loads in this type of rivalry. As an alternative, the intratumoral rivalry is also mirrored in the process of the entosis, which is the entanglement of a living cell through its adjacent cell (Overholtzer et al. 2007; Fais and Overholtzer 2018). But this process is based more on variations in cortical tension across adjacent cells than on compressive forces (Overholtzer et al. 2007; Sun et al. 2014).

Overall, these findings provide ample evidence that rivalry for space may be highly important for clonal rivalry inside a tumor, although a better understanding of the interactions between clones and the effect of space constraints on these interactions is still pending. What consequences could such a challenge have on the progression of tumors? Depending on the structure of the cancer cell population, concurrent interactions may also have contradictory effects in this case. The challenge for space inside the tumor may, on the one hand, allow a more aggressive clone population to be selected, which will enhance the general fitness and growth of the tumor as an entirety (Lee and Morishita 2017). Intratumoral rivalry can also encourage the development of drug-resistant populations that avoid the collapse of the tumor after treatment (Waclaw et al. 2015). On the other hand, competition for space can remove less than optimal cell populations, which in itself may not be highly proliferative, but globally can enhance tumor growth. This was well exemplified in breast cancer, in which a small cell subpopulation controls overall tumor growth through the secretion of interleukin-11 (Marusyk et al. 2014). Therefore, tumor growth can be considerably decreased when this population is outpaced by faster-growing competitors. Lastly, the rivalry for space may avoid the development of drug-resistant populations when drug-resistant cells exhibit reduced homeostatic pressure and are outperformed in the absence or at small treatment doses from non-resistant cells. Theoretically and experimentally, this was confirmed for the therapy of colorectal cancer cells using CDK inhibitors (Bacevic et al. 2017). In this case, the spatial rivalry decreases the fitness of the drug-resistant population and their ability to substitute drug-responsive cells in low-dose treatment. It should be kept in mind that all these concepts might equally be valid for contact-driven and diffusion-driven rivalry among chemoresponsive and chemoresistant cells.

16.13.4 Therapeutic Routes

Whereas the role of spatial rivalry in the initiation and the progression of tumors at this phase is still hypothetical, it is worth considering whether this idea could result in a suitable therapeutic strategy. Two strategies can be employed to overcome/change the spatial rivalry: the manipulation of the tumor's mechanical surroundings or the alteration of the path participating in the perception of compression and the induction of death. Here again, these two types of approaches may have conflicting consequences for tumor progression. The pressure release was formerly performed on tumor xenotransplants by specific cancer cell removal and a decrease in cell density (Padera et al. 2004). Although this could encourage the tumor to spread further through the creation of space, it could also simplify the administration of drugs through the opening of blood vessels (Jain et al. 2014). The mechanical separation of cancer cells through elevated line tension at the border of the tumor could be another powerful way to raise the pressure inside a tumor and retard the expansion of the tumor. To the best of my knowledge, however, there exists no simple way to specifically regulate the tension at the tumor/health cell interface without interfering with the key controllers of cell survival or cell proliferation (Bosveld et al. 2016). As an alternative, the targeted orientation toward the compression sensing route can contribute to avoid the removal of adjacent cells and retard the tumor's expansion. Consequently, gentle activation of the EGFR/ERK signal pathways in cells adjacent to pre-tumoral clones (constitutive activation of Ras) can considerably retard clone propagation (Moreno et al. 2019). In a similar manner, the global blocking of apoptosis can inhibit the neoplastic growth of pre-tumoral clones in the midgut (Suijkerbuijk et al. 2016) and in the imaginal disc of the wing of *Drosophila* (Eichenlaub et al. 2016).

The prevention of neuronal death by lithium treatment also significantly diminishes neurological disorders connected with brain tumors in mice (Seano et al. 2019). However, as discussed earlier, this can also have an influence on the intratumoral rivalry, whose impact on tumor growth is far more challenging to forecast. Since compaction-related death can also be necessary to remove early pre-tumoral cells, the targeted discovery of compaction sensory routes might encourage the emergence of new tumors on larger time scales. In summary, the establishment of new therapeutic intervention strategies necessitates an accurate evaluation of all these diverse aspects. The choice of a suitable therapy probably relies on the ability to accurately target the frequency of mechanical cell rivalry over time and space.

Whereas there is some experimental evidence to support the idea of mechanical cell competition, however, there are still many basic questions that have not entirely been addressed yet. These are improved molecular and physical specification of compaction sensing mechanisms (Valon and Levayer 2019), improved evaluation of the circumstances under which rivalry for space can arise, including the genetic background and tissue environment, and the predominance of mechanical rivalry during natural development and in tissue homeostasis. As outlined in the last section of this chapter, numerous separate findings indicate that rivalry for space may emerge in the course of the initiation and expansion of the tumor. All theoretical parameters necessary for mechanical cell rivalry have been outlined in solid tumors, covering differential growth and mechanical stress, differential sensitivity to mechanical stress, apoptosis and cell death, and this process is highly probable to take place both inside the tumor (referred to as intratumoral competition) and at the interface between tumor and healthy tissue. A deeper insight into the influence of spatial restrictions on cell rivalry is hence of great importance for the comprehension of cancer. In view of the multiple and conflicting phenomena of mechanical cell rivalry, modeling and experimental strategies that explicitly take into account spatial limitations, the heterogeneity of tumors and interactions with adjacent tissues will be essential to better assess these conflicting influences. A mix of in silico studies involving genetic evolution and mechanical interventions, in vitro investigations with mosaic cell cultures in 3D spheroid with adjustable pressures and in vivo investigations on model organisms with mosaic mapping will contribute to solving these challenging and fascinating questions.

References

J.-I. Abe, C. Morrell, Pyroptosis as a regulated form of necrosis: PI + /annexin V–/high caspase 1/low caspase 9 activity in cells = pyroptosis? Circ. Res. **118**, 1457–1460 (2016)

T. Adachi-Yamada, K. Fujimura-Kamada, Y. Nishida, K. Matsumoto, Distortion of proximodistal information causes JNK-dependent apoptosis in *Drosophila* wing. Nature **400**, 166–169 (1999)

J.M. Adams, S. Cory, The Bcl-2 apoptotic switch in cancer development and therapy. Oncogene **26**(9), 1324–1337 (2007)

M.L. Albert, Death-defying immunity: do apoptotic cells influence antigen processing and presentation? Nat. Rev. Immunol. **4**, 223–231 (2004)

M.L. Albert, B. Sauter, N. Bhardwaj, Dendritic cells acquire antigen from apoptotic cells and induce class I-restricted CTLs. Nature **392**, 86–89 (1998)

J.M. Albert, C. Cao, K.W. Kim, et al., Inhibition of poly (ADP-ribose) polymerase enhances cell death and improves tumor growth delay in irradiated lung cancer models. Clin. Cancer Res. **13**, 3033–3042 (2007)

S. Alers, A.S. Löffler, S. Wesselborg, B. Stork, Role of AMPK-mTOR-Ulk1/2 in the regulation of autophagy: cross talk, shortcuts, and feedbacks. Mol. Cell Biol. **32**, 2–11 (2012)

K. Alessandri, B.R. Sarangi, V.V. Gurchenkov et al., Cellular capsules as a tool for multicellular spheroid production and for investigating the mechanics of tumor progression in vitro. Proc. Natl. Acad. Sci. U. S. A. **110**, 14843–14848 (2013)

A. Alexander, S.L. Cai, J. Kim et al., ATM signals to TSC2 in the cytoplasm to regulate mTORC1 in response to ROS. Proc. Natl. Acad. Sci. U. S. A. **107**, 4153–4158 (2010)

V. Allaj, C. Guo, D. Nie, Non-steroid anti-inflammatory drugs, prostaglandins, and cancer. Cell Biosci. **3**(1), 8 (2013)

L. Alpar, C. Bergantinos, L.A. Johnston, Spatially restricted regulation of spatzle/toll signaling during cell competition. Dev. Cell **46**(706–719), e705 (2018)

J.D. Amack, M.L. Manning, Knowing the boundaries: extending the differential adhesion hypothesis in embryonic cell sorting. Science **338**, 212–215 (2012)

R.K. Amaravadi, C.B. Thompson, The roles of therapy-induced autophagy and necrosis in cancer treatment. Clin. Cancer Res. **13**, 7271–7279 (2007)

G. Ambrosini, C. Adida, D.C. Altieri, A novel anti-apoptosis gene, survivin, expressed in cancer and lymphoma. Nat. Med. **3**, 917–921 (1997)

J.C. Ameisen, T. Idziorek, O. Billaut-Mulot, M. Loyens, J.P. Tissier, A. Potentier, A. Ouaissi, Apoptosis in a unicellular eukaryote (*Trypanosoma cruzi*): implications for the evolutionary origin and role of programmed cell death in the control of cell proliferation, differentiation and survival. Cell Death Differ. **2**, 285–300 (1995)

C. Amendt, A. Mann, P. Schirmacher, M. Blessing, Resistance of keratinocytes to TGFbeta-mediated growth restriction and apoptosis induction accelerates re-epithelialization in skin wounds. J. Cell Sci. **115**, 2189–2198 (2002)

S.A. Amundson, T.G. Myers, F.A.J. Jr., Roles for p53 in growth arrest and apoptosis: putting on the brakes after genotoxic stress. Oncogene **17**, 3287–3299 (1998)

M.T. Andres, M. Viejo-Diaz, J.F. Fierro, Human lactoferrin induces apoptosis-like cell death in Candida albicans: critical role of K+-channel-mediated K+ efflux. Antimicrob. Agents Chemother. **52**, 4081–4088 (2008)

K. Aoki, Y. Kumagai, A. Sakurai, N. Komatsu, Y. Fujita, C. Shionyu, M. Matsuda, Stochastic ERK activation induced by noise and cell-to-cell propagation regulates cell density-dependent proliferation. Mol. Cell **52**, 529–540 (2013)

K. Aoki, Y. Kondo, H. Naoki, T. Hiratsuka, R.E. Itoh, M. Matsuda, Propagating wave of ERK activation orients collective cell migration. Dev. Cell **43**(305–317), e305 (2017)

L. Apetoh, F. Ghiringhelli, A. Tesniere et al., Toll-like receptor 4-dependent contribution of the immune system to anticancer chemotherapy and radiotherapy. Nat. Med. **13**, 1050–1059 (2007)

M. Aragona, T. Panciera, A. Manfrin, S. Giulitti, F. Michielin, N. Elvassore, S. Dupont, S. Piccolo, A mechanical checkpoint controls multicellular growth through YAP/TAZ regulation by actin-processing factors. Cell **154**, 1047–1059 (2013)

E. Arama, J. Agapite, H. Steller, Caspase activity and a specific cytochrome C are required for sperm differentiation in *Drosophila*. Dev. Cell **4**, 687–697 (2003)

G.S. Ashcroft, X. Yang, A.B. Glick et al., Mice lacking Smad3 show accelerated wound healing and an impaired local inflammatory response. Nat. Cell Biol. **1**, 260–266 (1999)

A. Ashkenazi, V.M. Dixit, Death receptors: signaling and modulation. Science **281**, 1305–1308 (1998)

D. Asslaber, E.M. Grössinger, T. Girbl, S.W. Hofbauer, A. Egle, L. Weiss, R. Greil, T.N. Hartmann, Mimicking the microenvironment in chronic lymphocytic leukaemia—where does the journey go? Br. J. Haematol. **160**, 711–714 (2013)

G. Atsumi, M. Tajima, A. Hadano, Y. Nakatani, M. Murakami, I. Kudo, Fas-induced arachidonic acid release is mediated by Ca2+independent phospholipase A2 but not cytosolic phospholipase A2, which undergoes proteolytic inactivation. J. Biol. Chem. **273**(22), 13870–13877 (1998)

K. Bacevic, R. Noble, A. Soffar, O. Wael Ammar, B. Boszonyik, S. Prieto, C. Vincent, M.E. Hochberg, L. Krasinska, D. Fisher, Spatial competition constrains resistance to targeted cancer therapy. Nat. Commun. **8**, 1995 (2017)

E.H. Baehrecke, How death shapes life during development. Nat. Rev. Mol. Cell Biol. **3**(10), 779–787 (2002)

M. Bagnoli, S. Canevari, D. Mezzanzanica, Cellular FLICE-inhibitory protein (c-FLIP) signalling: a key regulator of receptor-mediated apoptosis in physiologic context and in cancer. Int. J. Biochem. Cell Biol. **42**, 210–213 (2010)

A. Bakhshi, J.P. Jensen, P. Goldman, J.J. Wright, O.W. McBride, A.L. Epstein, S.J. Korsmeyer, Cloning the chromosomal breakpoint of t(14;18) human lymphomas: clustering around JH on chromosome 14 and near a transcriptional unit on 18. Cell **41**, 899–906 (1985)

E.K. Balcer-Kubiczek, Apoptosis in radiation therapy: a double-edged sword. Exp. Oncol. **34**(3), 277–285 (2012)

S.L. Baldauf, The deep roots of eukaryotes. Science **300**, 1703–1706 (2003)

A. Bangert, S. Cristofanon, I. Eckhardt, B.A. Abhari, S. Kolodziej, S. Häcker, S.H. Vellanki, J. Lausen, K.M. Debatin, S. Fulda, Histone deacetylase inhibitors sensitize glioblastoma cells to TRAIL-induced apoptosis by c-myc-mediated downregulation of cFLIP. Oncogene **31**, 4677–4688 (2012)

M. Basan, T. Risler, J.F. Joanny, X. Sastre-Garau, J. Prost, Homeostatic competition drives tumor growth and metastasis nucleation. HFSP J. **3**, 265–272 (2009)

M. Basan, J. Prost, J.F. Joanny, J. Elgeti, Dissipative particle dynamics simulationsfor biological tissues: rheology and competition. Phys. Biol. **8**, 026014 (2011)

A.C. Begg, I. Hofland, J. Kummermehr, Tumour cell repopulation during fractionated radiotherapy: correlation between flow cytometric and radiobiological data in three murine tumours. Eur. J. Cancer **27**(5), 537–543 (1991)

C.W. Bell, W. Jiang, C.F. Reich 3rd., D.S. Pisetsky, The extracellular release of HMGB1 during apoptotic cell death. Am. J. Physiol. Cell Physiol. **291**, C1318–C1325 (2006)

S. Benzekry, C. Lamont, A. Beheshti, A. Tracz, J.M. Ebos, L. Hlatky, P. Hahnfeldt, Classical mathematical models for description and prediction of experimentaltumor growth. PLoS Comput. Biol. **10**, e1003800 (2014)

C. Bergantinos, M. Corominas, F. Serras, Cell death-induced regeneration in wing imaginal discs requires JNK signaling. Development **137**, 1169–1179 (2010)

T.V. Berghe, N. Vanlangenakker, E. Parthoens, W. Deckers, M. Devos, N. Festjens, C.J. Guerin, U.T. Brunk, W. Declercq, P. Vandenabeele, Necroptosis, necrosis and secondary necrosis converge on similar cellular disintegration features. Cell Death Differ. **17**, 922–930 (2010)

A. Bergmann, The role of ubiquitylation for the control of cell death in *Drosophila*. Cell Death Differ. **17**, 61–67 (2010)

A. Bergmann, H. Steller, Apoptosis, stem cells, and tissue regeneration. Sci. Signal **3**(145), re8 (2010)

A. Bergmann, J. Agapite, K. McCall, H. Steller, The *Drosophila* gene hid is a direct molecular target of Ras-dependent survival signaling. Cell **95**, 331–341 (1998)

R. Beyaert, G. Van Loo, K. Heyninck, P. Vandenabeele, Signaling to gene activation and cell death by tumor necrosis factor receptors and Fas. Int. Rev. Cytol. **214**, 225–272 (2002)

L. Bi, I. Okabe, D.J. Bernard, A. Wynshaw-Boris, R.L. Nussbaum, Proliferative defect and embryonic lethality in mice homozygous for a deletion in the p110alpha subunit of phosphoinositide 3-kinase. J. Biol. Chem. **274**, 10963–10968 (1999)

S.M. Bianchi, L.R. Prince, K. McPhillips et al., Impairment of apoptotic cell engulfment by pyocyanin, a toxic metabolite of *Pseudomonas aeruginosa*. Am. J. Respir. Crit. Care Med. **177**, 35–43 (2008)

C. Bielmeier, S. Alt, V. Weichselberger, M. La Fortezza, H. Harz, F. Julicher, G. Salbreux, A.K. Classen, Interface contractility between differently fated cells drives cell elimination and cyst formation. Curr. Biol. **26**, 563–574 (2016)

D. Bilder, Epithelial polarity and proliferation control: links from the *Drosophila* neoplastic tumor suppressors. Genes Dev. **18**, 1909–1925 (2004)

K.D. Birnbaum, A. Sanchez Alvarado, Slicing across kingdoms: regeneration in plants and animals. Cell **132**, 697–710 (2008)

M.J. Bissell, W.C. Hines, Why don't we get more cancer? A proposed role of the microenvironment in restraining cancer progression. Nat. Med. **17**, 320–329 (2011)

D.R. Bolster, S.J. Crozier, S.R. Kimball, L.S. Jefferson, AMP-activated protein kinase suppresses protein synthesis in rat skeletal muscle through downregulated mammalian target of rapamycin (mTOR) signaling. J. Biol. Chem. **277**, 23977–23980 (2002)

L. Bonapace, B.C. Bornhauser, M. Schmitz et al., Induction of autophagy-dependent necroptosis is required for childhood acute lymphoblastic leukemia cells to overcome glucocorticoid resistance. J. Clin.Invest. **120**, 1310–1323 (2010)

M. Bordonaro, E. Drago, W. Atamna, D.L. Lazarova, Comprehensive suppression of all apoptosis-induced proliferation pathways as a proposed approach to colorectal cancer prevention and therapy. PLoS ONE **9**, e115068 (2014)

F. Bosveld, B. Guirao, Z. Wang, M. Riviere, I. Bonnet, F. Graner, Y. Bellaiche, Modulation of junction tension by tumor suppressors and proto-oncogenes regulates cell-cell contacts. Development **143**, 623–634 (2016)

A. Bottger, O. Alexandrova, Programmed cell death in *Hydra*. Semin. Cancer Biol. **17**, 134–146 (2007)

A. Bove, D. Gradeci, Y. Fujita, S. Banerjee, G. Charras, A.R. Lowe, Local cellularneighborhood controls proliferation in cell competition. Mol. Biol. Cell **28**, 3215–3228 (2017)

P. Boya, R.A. González-Polo, N. Casares, Inhibition of macroautophagy triggers apoptosis. Mol. Cell Biol. **25**(3), 1025–1040 (2005)

S. Bozinovski, R. Vlahos, D. Anthony, J. McQualter, G. Anderson, L. Irving, D. Steinfort, COPD and squamous cell lung cancer: aberrant inflammation and immunity is the common link. Br. J. Pharmacol. **173**, 635–648 (2016)

C. Bras-Pereira, E. Moreno, Mechanical cell competition. Curr. Opin. Cell Biol. **51**, 15–21 (2017)

S. Braun, U. auf dem Keller, H. Steiling, S. Werner, Fibroblast growth factors in epithelial repair and cytoprotection. Philos. Trans. R. Soc. Lond. B Biol. Sci. **359**, 753–757 (2004)

A. Brech, T. Ahlquist, R.A. Lothe, H. Stenmark, Autophagy in tumour suppression and promotion. Mol. Oncol. **3**(4), 366–375 (2009)

N. Bresgen, I. Ohlenschlager, N. Wacht, S. Afazel, G. Ladurner, P.M. Eckl, Ferritin and FasL (CD95L) mediate density dependent apoptosis in primary rat hepatocytes. J. Cell Physiol. **217**, 800–808 (2008)

J.P. Brockes, A. Kumar, Plasticity and reprogramming of differentiated cells in amphibian regeneration. Nat. Rev. Mol. Cell Biol. **3**, 566–574 (2002)

G. Brouckaert, M. Kalai, D.V. Krysko, X. Saelens, D. Vercammen, M.N. Ndlovu, G. Haegeman, K. D'Herde, P. Vandenabeele, Phagocytosis of necrotic cells by macrophages is phosphatidylserine dependent and does not induce inflammatory cytokine production. Mol. Biol. Cell **15**, 1089–1100 (2004)

J.M. Brown, L.D. Attardi, The role of apoptosis in cancer development and treatment response. Nat. Rev. Cancer **5**(3), 231–237 (2005)

S. Brown, I. Heinisch, E. Ross, K. Shaw, C.D. Buckley, J. Savill, Apoptosis disables CD31-mediated cell detachment from phagocytes promoting binding and engulfment. Nature **418**, 200–203 (2002)

D. Brusa, S. Garetto, G. Chiorino, M. Scatolini, E. Migliore, G. Camussi, L. Matera, Post-apoptotic tumors are more palatable to dendritic cells and enhance their antigen cross-presentation activity. Vaccine **26**, 6422–6432 (2008)

R.C. Budd, W.-C. Yeh, J. Tschopp, cFLIP regulation of lymphocyte activation and development. Nat. Rev. Immunol. **6**, 196–204 (2006)

W. Bursch, The autophagosomal–lysosomal compartment in programmed cell death. Cell Death Differ. **8**, 569–581 (2001)

D.T. Butcher, T. Alliston, V.M. Weaver, A tense situation: forcing tumour progression. Nat. Rev. Cancer **9**, 108–122 (2009)

S. Buttiglieri, A. Galetto, S. Forno, M. De Andrea, L. Matera, Influence of drug-induced apoptotic death on processing and presentation of tumor antigens by dendritic cells. Int. J. Cancer **106**, 516–520 (2003)

Y. Cao, D.J. Klionsky, Physiological functions of Atg6/Beclin 1: a unique autophagy-related protein. Cell Res. **17**, 839–849 (2007)

D.L. Carlson, D.L. Maass, J. White, P. Sikes, J.W. Horton, Caspase inhibition reduces cardiac myocyte dyshomeostasis and improves cardiac contractile function after major burn injury. J. Appl. Physiol. **103**, 323–330 (2007)

M.D. Castellone, H. Teramoto, B.O. Williams, K.M. Druey, J.S. Gutkind, Prostaglandin E2 promotes colon cancer cell growth through a Gs-axin-beta-catenin signaling axis. Science **310**(5753), 1504–1510 (2005)

J. Chai, N. Yan, J.R. Huh, J.W. Wu, W. Li, B.A. Hay, Y. Shi, Molecular mechanism of Reaper-Grim-Hid-mediated suppression of DIAP1-dependent Dronc ubiquitination. Nat. Struct. Biol. **10**, 892–898 (2003)

S. Challa, F.K.-M. Chan, Going up in flames: necrotic cell injury and inflammatory diseases. Cell Mol. Life Sci. **67**, 3241–3253 (2010)

P. Chen, W. Nordstrom, B. Gish, J.M. Abrams, Grim, a novel cell death gene in *Drosophila*. Genes Dev. **10**, 1773–1782 (1996)

C.S. Chen, M. Mrksich, S. Huang, G.M. Whitesides, D.E. Ingber, Geometric control of cell life and death. Science **276**, 1425–1428 (1997)

S.Y. Chen, L.Y. Chiu, M.C. Maa, J.S. Wang, C.L. Chien, W.W. Lin, zVAD-induced autophagic cell death requires c-Src-dependent ERK and JNK activation and reactive oxygen species generation. Autophagy **7**(2), 217–228 (2011a)

J. Chen, M. Dassarath, Z. Yin, H. Liu, K. Yang, G. Wu, Radiation induced temporal lobe necrosis in patients with nasopharyngeal carcinoma: a review of new avenues in its management. Radiat. Oncol. **6**, 128 (2011b)

Q. Chen, J. Kang, C. Fu, The independence of and associations among apoptosis, autophagy, and necrosis. Sig. Transduct. Targeted Ther. **3**, 18 (2018)

G. Cheng, J. Tse, R.K. Jain, L.L. Munn, Micro-environmental mechanical stress controls tumor spheroid size and morphology by suppressing proliferation and inducing apoptosis in cancer cells. PLoS ONE **4**, e4632 (2009)

J. Cheng, L. Tian, J. Ma, Y. Gong, Z. Zhang, Z. Chen, B. Xu, H. Xiong, C. Li, Q. Huang, Dying tumor cells stimulate proliferation of living tumor cells via caspase-dependent protein kinase Cdelta activation in pancreatic ductal adenocarcinoma. Mol. Oncol. **9**, 105–114 (2015)

X. Cheng, X. Ma, X. Ding, L. Li, X. Jiang, Z. Shen, S. Chen, W. Liu, W. Gong, Q. Sun, Pacer mediates the function of class III PI3K and HOPS complexes in autophagosome maturation by engaging Stx17. Mol. Cell **65**, 1029–1043 (2017)

H. Cheong, U. Nair, J. Geng, D.J. Klionsky, The Atg1 kinase complex is involved in the regulation of protein recruitment to initiate sequestering vesicle formation for nonspecific autophagy in Saccharomyces cerevisiae. Mol. Biol. Cell **19**, 668–681 (2008)

S. Chera, L. Ghila, K. Dobretz, Y. Wenger, C. Bauer, W. Buzgariu, J.C. Martinou, B. Galliot, Apoptotic cells provide an unexpected source of Wnt3 signaling to drive *Hydra* head regeneration. Dev. Cell **17**, 279–289 (2009)

S.K. Chew, P. Chen, N. Link, K.A. Galindo, K. Pogue, J.M. Abrams, Genome-wide silencing in *Drosophila* captures conserved apoptotic effectors. Nature **460**, 123–127 (2009)

J. Chmielowiec, M. Borowiak, M. Morkel, T. Stradal, B. Munz, S. Werner, J. Wehland, C. Birchmeier, W. Birchmeier, c-Met is essential for wound healing in the skin. J. Cell Biol. **177**, 151–162 (2007)

Y. Cho, S. Challa, D. Moquin, R. Genga, T.D. Ray, M. Guildford, F.K. Chan, Phosphorylation-driven assembly of the RIP1-RIP3 complex regulates programmed necrosis and virus-induced inflammation. Cell **137**, 1112–1123 (2009)

Y. Cho, T. McQuade, H. Zhang, J. Zhang, F.K.-M. Chan, RIP1-dependent and independent effects of necrostatin-1 in necrosis and T cell activation. PLoS ONE **6**, e23209 (2011)

S. Chung, T.L. Gumienny, M.O. Hengartner, M. Driscoll, A common set of engulfment genes mediates removal of both apoptotic and necrotic cell corpses in *C. elegans*. Nat. Cell Biol. **2**, 931–937 (2000)

M. Cikala, B. Wilm, E. Hobmayer, A. Bottger, C.N. David, Identification of caspases and apoptosis in the simple metazoan *Hydra*. Curr. Biol. **9**, 959–962 (1999)

P.G. Clarke, S. Clarke, Nineteenth century research on naturally occurring cell death and related phenomena. Anat. Embryol. (Berl.) **193**, 81–99 (1996)

C. Claveria, M. Torres, Cell competition: mechanisms and physiological roles. Annu. Rev. Cell Dev. Biol. **32**, 411–439 (2016)

C. Claveria, G. Giovinazzo, R. Sierra, M. Torres, Myc-driven endogenous cell competition in the early mammalian embryo. Nature **500**, 39–44 (2013)

A. Clavier, A. Rincheval-Arnold, A. Baillet, B. Mignotte, I. Guenal, Two different specific JNK activators are required to trigger apoptosis or compensatory proliferation in response to Rbf1 in *Drosophila*. Cell Cycle **15**, 283–294 (2016)

M.L. Cleary, J. Sklar, Nucleotide sequence of at (14;18) chromosomal break point in follicular lymphoma and demonstration of a breakpoint-cluster region near a transcriptionally active locus on chromosome 18. Proc. Natl. Acad. Sci. U. S. A. **82**, 7439–7443 (1985)

R.J. Clem, M. Fechheimer, L.K. Miller, Prevention of apoptosis by a baculovirus gene during infection of insect cells. Science **254**, 1388–1390 (1991)

P.L. Cohen, R. Caricchio, V. Abraham, T.D. Camenisch, J.C. Jennette, R.A. Roubey, H.S. Earp, G. Matsushima, E.A. Reap, Delayed apoptotic cell clearance and lupus-like autoimmunity in mice lacking the c-mer membrane tyrosine kinase. J. Exp. Med. **196**, 135–140 (2002)

B. Conradt, Genetic control of programmed cell death during animal development. Annu. Rev. Genet. **43**, 493–523 (2009)

S. Cornillon, C. Foa, J. Davoust, N. Buonavista, J.D. Gross, P. Golstein, Programmed cell death in Dictyostelium. J. Cell Sci. **107**(Pt 10), 2691–2704 (1994)

M.S. Cragg, C. Harris, A. Strasser, C.L. Scott, Unleashing the power of inhibitors of oncogenic kinases through BH3 mimetics. Nat. Rev. Cancer **9**, 321–326 (2009)

S. Curran, C. Strandkvist, J. Bathmann, M. de Gennes, A. Kabla, G. Salbreux, B. Baum, Myosin II controls junction fluctuations to guide epithelial tissue ordering. Dev. Cell **43**(480–492), e486 (2017)

P.E. Czabotar, G. Lessene, A. Strasser, J.M. Adams, Control of apoptosis by the BCL-2 protein family: implications for physiology and therapy. Nat. Rev. Mol. Cell Biol. **15**, 49–63 (2014)

A. D'Brot, P. Chen, M. Vaishnav, S. Yuan, C.W. Akey, J.M. Abrams, Tango7 directs cellular remodeling by the *Drosophila* apoptosome. Genes Dev. **27**, 1650–1655 (2013)

C. Dabrowska, M. Li, Y. Fan, Apoptotic caspases in promoting cancer: implications from their roles in development and tissue homeostasis. Adv. Exp. Med. Biol. **930**, 89–112 (2016)

N.N. Danial, S.J. Korsmeyer, Cell death: critical control points. Cell **116**, 205–219 (2004)

A.J. Davis, J.F. Tannock, Repopulation of tumour cells between cycles of chemotherapy: a neglected factor. Lancet Oncol. **1**, 86–93 (2000)

A.J. Davis, W. Chapman, D.W. Hedley, A.M. Oza, I.F. Tannock, Assessment of tumor cell repopulation after chemotherapy for advanced ovarian cancer: pilot study. Cytometry A **51**(1), 1–6 (2003)

C. de la Cova, M. Abril, P. Bellosta, P. Gallant, L.A. Johnston, *Drosophila* mycregulates organ size by inducing cell competition. Cell **117**, 107–116 (2004)

J.A. de Visser, D.E. Rozen, Clonal interference and the periodic selection of newbeneficial mutations in *Escherichia coli*. Genetics **172**, 2093–2100 (2006)

A.M. deCathelineau, P.M. Henson, The final step in programmed cell death: phagocytes carry apoptotic cells to the grave. Essays Biochem. **39**, 105–117 (2003)

P. Decker, Nucleosome autoantibodies. Clin. Chim. Acta **366**, 48–60 (2006)

W. Declercq, T.V. Berghe, P. Vandenabeele, RIP kinases at the crossroads of cell death and survival. Cell **138**, 229–232 (2009)

K. Degenhardt, R. Mathew, B. Beaudoin et al., Autophagy promotes tumor cell survival and restricts necrosis, inflammation, and tumorigenesis. Cancer Cell **10**(1), 51–64 (2006)

A. Degterev, J. Yuan, Expansion and evolution of cell death programmes. Nat. Rev. Mol. Cell Biol. **9**(5), 378–390 (2008)

A. Degterev, M. Boyce, J. Yuan, A decade of caspases. Oncogene **22**, 8543–8567 (2003)

A. Degterev, Z. Huang, M. Boyce, Y. Li, P. Jagtap, N. Mizushima, G.D. Cuny, T.J. Mitchison, M.A. Moskowitz, J. Yuan, Chemical inhibitor of nonapoptotic cell death with therapeutic potential for ischemic brain injury. Nat. Chem. Biol. **1**, 112–119 (2005)

A. Degterev, J. Hitomi, M. Germscheid et al., Identification of RIP1 kinase as a specific cellular target of necrostatins. Nat. Chem. Biol. **4**, 313–321 (2008)

M.V. Del Gaizo, J.R. Brown, M. Certo, T.M. Love, C.D. Novina, A. Letai, Chronic lymphocytic leukemia requires BCL2 to sequester prodeath BIM, explaining sensitivity to BCL2 antagonist ABT-737. J. Clin. Invest. **117**, 112–121 (2007)

L. del Peso, M. Gonzalez-Garcia, C. Page, R. Herrera, G. Nunez, Interleukin-3-induced phosphorylation of BAD through the protein kinase Akt. Science **278**, 687–689 (1997)

M. Delarue, F. Montel, D. Vignjevic, J. Prost, J.F. Joanny, G. Cappello, Compressive stress inhibits proliferation in tumor spheroids through a volume limitation. Biophys. J. **107**, 1821–1828 (2014)

S. Demaria, N. Bhardwaj, W.H. McBride, S.C. Formenti, Combining radiotherapy and immunotherapy: a revived partnership. Int. J. Radiat. Oncol. Biol. Phys. **63**(3), 655–666 (2005)

P.W. Dempsey, S.E. Doyle, J.Q. He, G. Cheng, The signaling adaptors and pathways activated by TNF superfamily. Cytokine Growth Factor Rev. **14**, 193–209 (2003)

G. Denecker, D. Vercammen, M. Steemans et al., Death receptor-induced apoptotic and necrotic cell death: differential role of caspases and mitochondria. Cell Death Differ. **8**, 829–840 (2001)

J. Denekamp, Cell kinetics and radiation biology. Int. J. Radiat. Biol. Relat. Stud. Phys. Chem. Med. **49**(2), 357–380 (1986)

M. Deponte, Programmed cell death in protists. Biochim. Biophys. Acta **1783**, 1396–1405 (2008)

A. Devitt, K.G. Parker, C.A. Ogden et al., Persistence of apoptotic cells without autoimmune disease or inflammation in CD14$^{-/-}$ mice. J. Cell Biol. **167**, 1161–1170 (2004)

M. Dewaele, H. Maes, P. Agostinis, ROS-mediated mechanisms of autophagy stimulation and their relevance in cancer therapy. Autophagy **6**, 838–854 (2010)

W.C. Dewey, C.C. Ling, R.E. Meyn, Radiation-induced apoptosis: relevance to radiotherapy. Int. J. Radiat. Oncol. Biol. Phys. **33**(4), 781–796 (1995)

S. Di Giacomo, M. Sollazzo, D. de Biase, M. Ragazzi, P. Bellosta, A. Pession, D. Grifoni, Human cancer cells signal their competitive fitness through MYC activity. Sci. Rep. **7**, 12568 (2017)

J. Diao, L. Li, Y. Lai, Q. Zhong, In vitro reconstitution of autophagosome–lysosome fusion. Methods Enzymol. **587**, 365–376 (2017)

M. Dias-Baruffi, H. Zhu, M. Cho, S. Karmakar, R.P. McEver, R.D. Cummings, Dimeric galectin-1 induces surface exposure of phosphatidylserine and phagocytic recognition of leukocytes without inducing apoptosis. J. Biol. Chem. **278**, 41282–41293 (2003)

C. Diaz-Diaz, L. Fernandez de Manuel, D. Jimenez-Carretero, M.C. Montoya, C. Claveria, M. Torres, Pluripotency surveillance by myc-driven competitive elimination of differentiating cells. Dev. Cell **42**(585–599), e584 (2017)

I. Dikic, T. Johansen, V. Kirkin, Selective autophagy in cancer development and therapy. Cancer Res. **70**, 3431–3434 (2010)

N. Diwanji, A. Bergmann, The beneficial role of extracellular reactive oxygen species in apoptosis-induced compensatory proliferation. Fly (Austin) **11**, 1–7 (2016)

A. do Vale, C. Costa-Ramos, A. Silva, D.S. Silva, F. Gartner, N.M. dos Santos, M.T. Silva, Systemic macrophage and neutrophil destruction by secondary necrosis induced by a bacterial exotoxin in a Gram-negative septicaemia. Cell Microbiol. **9**, 988–1003 (2007)

M.M. Don, G. Ablett, C.J. Bishop, P.G. Bundesen, K.J. Donald, J. Searle, J.F. Kerr, Death of cells by apoptosis following attachment of specifically allergized lymphocytes in vitro. Aust. J. Exp. Biol. Med. Sci. **55**, 407–417 (1977)

A.L. Donato, Q. Huang, X. Liu, F. Li, M.A. Zimmerman, C.Y. Li, Caspase 3 promotes surviving melanoma tumor cell growth after cytotoxic therapy. J. Invest. Dermatol. **134**, 1686–1692 (2014)

Y. Dondelinger, M.A. Aguileta, V. Goossens, C. Dubuisson, S. Grootjans, E. Dejardin, P. Vandenabeele, M. J. Bertrand, RIPK3 contributes to TNFR1-mediated RIPK1 kinase-dependent apoptosis in conditions of cIAP1/2 depletion or TAK1 kinase inhibition. Cell Death Differ. **20**, 1381–1392 (2013)

S. Dupont, L. Morsut, M. Aragona et al., Role of YAP/TAZ in mechanotransduction. Nature **474**, 179–183 (2011)

L. Duprez, E. Wirawan, T.V. Berghe, P. Vandenabeele, Major cell death pathways at a glance. Microbes Infect. **11**, 1050–1062 (2009)

C. Duyckaerts, B. Delatour, M.-C. Potier, Classification and basic pathology of Alzheimer disease. Acta Neuropathol. **118**, 5–36 (2009)

H.F. Dvorak, Tumors: wounds that do not heal. Similarities between tumor stroma generation and wound healing. N. Engl. J. Med. **315**, 1650–1659 (1986)

N. Dwivedi, M. Radic, Citrullination of autoantigens implicates NETosis in the induction of autoimmunity. Ann. Rheum. Dis. **73**, 483–491 (2014)

A.L. Edinger, C.B. Thompson, Death by design: apoptosis, necrosis and autophagy. Curr. Opin. Cell Biol. **16**, 663–669 (2004)

R.L. Eefsen, P.B. Vermeulen, I.J. Christensen et al., Growth pattern of colorectal liver metastasis as a marker of recurrence risk. Clin. Exp. Metastasis **32**, 369–381 (2015)

D.F. Egan, D.B. Shackelford, M.M. Mihaylova et al., Phosphorylation of ULK1 (hATG1) by AMP-activated protein kinase connects energy sensing to mitophagy. Science **331**, 456–461 (2011)

Y. Eguchi, S. Shimizu, Y. Tsujimoto, Intracellular ATP levels determine cell death fate by apoptosis or necrosis. Cancer Res. **57**, 1835–1840 (1997)

T. Eichenlaub, S.M. Cohen, H. Herranz, Cell competition drives the formation of metastatic tumors in a *Drosophila* model of epithelial tumor formation. Curr. Biol. **26**, 419–427 (2016)

A. Eisenberg-Lerner, S. Bialik, H.U. Simon, A. Kimchi, Life and death partners: apoptosis, autophagy and the cross-talk between them. Cell Death Differ. **16**(7), 966–975 (2009)

G.T. Eisenhoffer, P.D. Loftus, M. Yoshigi, H. Otsuna, C.B. Chien, P.A. Morcos, J. Rosenblatt, Crowding induces live cell extrusion to maintain homeostatic cell numbers in epithelia. Nature **484**, 546–549 (2012)

M. Elgendy, C. Sheridan, G. Brumatti, S.J. Martin, Oncogenic Ras-induced expression of Noxa and Beclin-1 promotes autophagic cell death and limits clonogenic survival. Mol. Cell **42**, 23–35 (2011)

M.R. Elliott, F.B. Chekeni, P.C. Trampont et al., Nucleotides released by apoptotic cells act as a find-me signal to promote phagocytic clearance. Nature **461**, 282–286 (2009)

H.M. Ellis, H.R. Horvitz, Genetic control of programmed cell death in the nematode *C. elegans*. Cell **44**, 817–829 (1986)

R.E. Ellis, D.M. Jacobson, H.R. Horvitz, Genes required for the engulfment of cell corpses during programmed cell death in *Caenorhabditis elegans*. Genetics **129**, 79–94 (1991)

A. Elosegui-Artola, I. Andreu, A.E.M. Beedle et al., Force triggers YAP nuclear entry by regulating transport across nuclear pores. Cell **171**(1397–1410), e1314 (2017)

M. Erlacher, E.M. Michalak, P.N. Kelly, V. Labi, H. Niederegger, L. Coultas, J.M. Adams, A. Strasser, A. Villunger, BH3-only proteins Puma and Bim are rate-limiting for gamma-radiation- and glucocorticoid-induced apoptosis of lymphoid cells in vivo. Blood **106**, 4131–4138 (2005)

S. Erlich, L. Mizrachy, O. Segev, L. Lindenboim, O. Zmira, S. Adi-Harel, J.A. Hirsch, R. Stein, R. Pinkas-Kramarski, Differential interactions between Beclin 1 and Bcl-2 family members. Autophagy **3**, 561–568 (2007)

N.J. Ernest, C.W. Habela, H. Sontheimer, Cytoplasmic condensation is both necessary and sufficient to induce apoptotic cell death. J. Cell. Sci. **121**, 290–297 (2008)

L.P. Erwig, P.M. Henson, Clearance of apoptotic cells by phagocytes. Cell Death Differ. **15**, 243–250 (2008)

G.I. Evan, K.H. Vousden, Proliferation, cell cycle and apoptosis in cancer. Nature **411**(6835), 342–348 (2001)

V.A. Fadok, D.R. Voelker, P.A. Campbell, J.J. Cohen, D.L. Bratton, P.M. Henson, Exposure of phosphatidylserine on the surface of apoptotic lymphocytes triggers specific recognition and removal by macrophages. J. Immunol. **148**, 2207–2216 (1992)

V.A. Fadok, D.L. Bratton, A. Konowal, P.W. Freed, J.Y. Westcott, P.M. Henson, Macrophages that have ingested apoptotic cells in vitro inhibit proinflammatory cytokine production through autocrine/paracrine mechanisms involving TGF-beta, PGE2, and PAF. J. Clin. Invest. **101**, 890–898 (1998)

S. Fais, M. Overholtzer, Cell-in-cell phenomena in cancer. Nat. Rev. Cancer **18**, 758–766 (2018)

Y. Fan, A. Bergmann, Distinct mechanisms of apoptosis-induced compensatory proliferation in proliferating and differentiating tissues in the *Drosophila* eye. Dev. Cell **14**, 399–410 (2008a)

Y. Fan, A. Bergmann, Apoptosis-induced compensatory proliferation. The cell is dead. Long live the cell! Trends Cell Biol. **18**, 467–473 (2008b)

Y. Fan, A. Bergmann, The cleaved-Caspase-3 antibody is a marker of Caspase-9-like DRONC activity in *Drosophila*. Cell Death Differ. **17**, 534–539 (2010)

Y. Fan, A. Bergmann, Multiple mechanisms modulate distinct cellular susceptibilities toward apoptosis in the developing *Drosophila* eye. Dev. Cell **30**, 48–60 (2014)

T.-J. Fan, L.-H. Han, R.-S. Cong, J. Liang, Caspase family proteases and apoptosis. Acta Biochem. Biophys. Sin. **37**, 719–727 (2005)

Y. Fan, S. Wang, J. Hernandez, V.B. Yenigun, G. Hertlein, C.E. Fogarty, J.L. Lindblad, A. Bergmann, Genetic models of apoptosis-induced proliferation decipher activation of JNK and identify a requirement of EGFR signaling for tissue regenerative responses in *Drosophila*. PLoS Genet. **10**, e1004131 (2014)

T. Farkas, M. Daugaard, M. Jäättelä, Identification of small molecule inhibitors of phosphatidylinositol 3-kinase and autophagy. J. Biol. Chem. **286**, 38904–38912 (2011)

B. Favaloro, N. Allocati, V. Graziano, C. Di Ilio, V. De Laurenzi, Role of apoptosis in disease. Aging **4**, 330–349 (2012)

I. Feher, J. Gidali, Prostaglandin E2 as stimulator of haemopoietic stem cell proliferation. Nature **247**(442), 550–551 (1974)

S. Feng, Y. Yang, Y. Mei, L. Ma, D.-e Zhu, N. Hoti, M.A. Castanares, M. Wu, Cleavage of RIP3 inactivates its caspase-independent apoptosis pathway by removal of kinase domain. Cell Signal **19**, 2056–2067 (2007)

X. Feng, L. Tian, Z. Zhang, Y. Yu, J. Cheng, Y. Gong, C.Y. Li, Q. Huang, Caspase 3 in dying tumor cells mediates post-irradiation angiogenesis. Oncotarget **6**, 32353–32367 (2015)

X. Feng, Y. Yu, S. He et al., Dying glioma cells establish a proangiogenic microenvironment through a caspase 3 dependent mechanism. Cancer Lett. **385**, 12–20 (2016)

M. Feoktistova, P. Geserick, B. Kellert, D.P. Dimitrova, C. Langlais, M. Hupe, K. Cain, M. MacFarlane, G. Häcker, M. Leverkus, cIAPs block Ripoptosome formation, a RIP1/caspase-8 containing intracellular cell death complex differentially regulated by cFLIP isoforms. Mol. Cell **43**, 449–463 (2011)

N. Festjens, T. Vanden Berghe, P. Vandenabeele, Necrosis, a well-orchestrated form of cell demise: signalling cascades, important mediators and concomitant immune response. Biochim. Biophys. Acta **1757**(9–10), 1371–1387 (2006)

M.V. Fiandalo, N. Kyprianou, Caspase control: protagonists of cancer cell apoptosis. Exp. Oncol. **34**(3), 165–175 (2012)

A.V. Finn, F.D. Kolodgie, R. Virmani, Correlation between carotid intimal/medial thickness and atherosclerosis: a point of view from pathology. Arterioscler. Thromb. Vasc. Biol. **30**, 177–181 (2010)

U. Fischer, R.U. Janicke, K. Schulze-Osthoff, Many cuts to ruin: a comprehensive update of caspase substrates. Cell Death Differ. **10**, 76–100 (2003)

A. Fleming, T. Noda, T. Yoshimori, D.C. Rubinsztein, Chemical modulators of autophagy as biological probes and potential therapeutics. Nat. Chem. Biol. **7**, 9–17 (2011)

A. Florentin, E. Arama, Caspase levels and execution efficiencies determine the apoptotic potential of the cell. J. Cell Biol. **196**, 513–527 (2012)

C.E. Fogarty, A. Bergmann, Killers creating new life: caspases drive apoptosis-induced proliferation in tissue repair and disease. Cell Death Differ. **24**(8), 1390–1400 (2017)

C.E. Fogarty, N. Diwanji, J.L. Lindblad, M. Tare, A. Amcheslavsky, K. Makhijani, K. Brückner, Y. Fan, A. Bergmann, Extracellular reactive oxygen species drive apoptosis-induced proliferation via *Drosophila* macrophages. Curr. Biol. **26**, 575–584 (2016)

N.C. Franc, K. White, R.A. Ezekowitz, Phagocytosis and development: back to the future. Curr. Opin. Immunol. **11**, 47–52 (1999)

J.P. Freyer, Role of necrosis in regulating the growth saturation of multicellularspheroids. Cancer Res. **48**, 2432–2439 (1988)

Y. Fuchs, H. Steller, Programmed cell death in animal development and disease. Cell **147**, 742–758 (2011)

S.F. Funderburk, Q.J. Wang, Z. Yue, The Beclin 1–VPS34 complex–at the crossroads of autophagy and beyond. Trends Cell Biol. **20**, 355–362 (2010)

U.S. Gaipl, L.E. Munoz, G. Grossmayer et al., Clearance deficiency and systemic lupus erythematosus (SLE). J. Autoimmun. **28**, 114–121 (2007)

A. Galetto, S. Buttiglieri, S. Forno, F. Moro, A. Mussa, L. Matera, Drug- and cell-mediated antitumor cytotoxicities modulate cross-presentation of tumor antigens by myeloid dendritic cells. Anticancer Drugs **14**, 833–843 (2003)

B. Galliot, Injury-induced asymmetric cell death as a driving force for head regeneration in *Hydra*. Dev. Genes Evol. **223**(1–2), 39–52 (2013)

B. Galliot, S. Chera, The *Hydra* model: disclosing an apoptosis-driven generator of Wnt-based regeneration. Trends Cell Biol. **20**(9), 514–523 (2010)

L. Galluzzi, M.C. Maiuri, I. Vitale, H. Zischka, M. Castedo, L. Zitvogel, G. Kroemer, Cell death modalities: classification and pathophysiological implications. Cell Death Differ. **14**(7), 1237–1243 (2007)

L. Galluzzi, N. Joza, E. Tasdemir, M.C. Maiuri, M. Hengartner, J.M. Abrams, N. Tavernarakis, J. Penninger, F. Madeo, G. Kroemer, No death without life: vital functions of apoptotic effectors. Cell Death Differ. **15**, 1113–1123 (2008)

L. Galluzzi, S.A. Aaronson, J. Abrams, et al., Guidelines for the use and interpretation of assays for monitoring cell death in higher eukaryotes. Cell Death Differ. **16**, 1093–1107 (2009)

L. Galluzzi, I. Vitale, J.M. Abrams et al., Molecular definitions of cell death subroutines: recommendations of the Nomenclature Committee on cell death 2012. Cell Death Differ. **19**, 107–120 (2012)

C. Garrido, G. Kroemer, Life's smile, death's grin: vital functions of apoptosis-executing proteins. Curr. Opin. Cell Biol. **16**, 639–646 (2004)

C. Gauron, C. Rampon, M. Bouzaffour, E. Ipendey, J. Teillon, M. Volovitch, S. Vriz, Sustained production of ROS triggers compensatory proliferation and is required for regeneration to proceed. Sci. Rep. **3**, 2084 (2013)

G. Gdynia, K. Grund, A. Eckert et al., Basal caspase activity promotes migration and invasiveness in glioblastoma cells. Mol. Cancer Res. **5**(12), 1232–1240 (2007)

M.C. Gibson, N. Perrimon, Extrusion and death of DPP/BMP-compromised epithelial cells in the developing *Drosophila* wing. Science **307**, 1785–1789 (2005)

D. Glick, S. Barth, K.F. Macleod, Autophagy: cellular and molecular mechanisms. J. Pathol. **221**(1), 3–12 (2010)

W. Goessling, T.E. North, S. Loewer et al., Genetic interaction of PGE2 and Wnt signaling regulates developmental specification of stem cells and regeneration. Cell **136**(6), 1136–1147 (2009)

P. Golstein, G. Kroemer, Cell death by necrosis: towards a molecular definition. Trends Biochem. Sci. **32**, 37–43 (2007)

B. Gompertz, On the nature of the function expressive of the law of human mortality, and on a new mode of determining the value of life contingencies. Philos.Trans. R. Soc. Lond. **115** (1825)

A. Gosain, S.B. Jones, R. Shankar, R.L. Gamelli, L.A. DiPietro, Norepinephrine modulates the inflammatory and proliferative phases of wound healing. J. Trauma **60**, 736–744 (2006)

L. Goyal, K. McCall, J. Agapite, E. Hartwieg, H. Steller, Induction of apoptosis by *Drosophila* reaper, hid and grim through inhibition of IAP function. EMBO J. **19**, 589–597 (2000)

M. Greaves, C.C. Maley, Clonal evolution in cancer. Nature **481**, 306–313 (2012)

D.R. Green, G.I. Evan, A matter of life and death. Cancer Cell **1**, 19–30 (2002)

D.R. Green, T. Ferguson, L. Zitvogel, G. Kroemer, Immunogenic and tolerogenic cell death. Nat. Rev. Immunol. **9**, 353–363 (2009)

M.E. Grether, J.M. Abrams, J. Agapite, K. White, H. Steller, The head involution defective gene of *Drosophila melanogaster* functions in programmed cell death. Genes Dev. **9**, 1694–1708 (1995)

D.R. Grimes, P. Kannan, A. McIntyre, A. Kavanagh, A. Siddiky, S. Wigfield, A. Harris, M. Partridge, The role of oxygen in avascular tumor growth. PLoS ONE **11**, e0153692 (2016)

R. Grose, S. Werner, Wound-healing studies in transgenic and knockout mice. Mol. Biotechnol. **28**, 147–166 (2004)

A. Gross, J.M. McDonnell, S.J. Korsmeyer, BCL-2 family members and the mitochondria in apoptosis. Genes Dev. **13**, 1899–1911 (1999)

S.A. Gudipaty, J. Lindblom, P.D. Loftus, M.J. Redd, K. Edes, C.F. Davey, V. Krishnegowda, J. Rosenblatt, Mechanical stretch triggers rapid epithelial celldivision through Piezo1. Nature **543**, 118–121 (2017)

E. Gulbins, B. Brenner, K. Schlottmann, J. Welsch, H. Heinle, U. Koppenhoefer, O. Linderkamp, K.M. Coggeshall, F. Lang, Fas-induced programmed cell death is mediated by a Ras-regulated O2-synthesis. Immunology **89**, 205–212 (1996)

T.L. Gumienny, E. Lambie, E. Hartwieg, H.R. Horvitz, M.O. Hengartner, Genetic control of programmed cell death in the *Caenorhabditis elegans* hermaphrodite germline. Development **126**, 1011–1022 (1999)

J.M. Gump, A. Thorburn, Autophagy and apoptosis: what is the connection? Trends Cell Biol. **21**, 387–392 (2011)

G.C. Gurtner, S. Werner, Y. Barrandon, M.T. Longaker, Wound repair and regeneration. Nature **453**, 314–321 (2008)

D.M. Gwinn, D.B. Shackelford, D.F. Egan, M.M. Mihaylova, A. Mery, D.S. Vasquez, B.E. Turk, R.J. Shaw, AMPK phosphorylation of raptor mediates a metabolic checkpoint. Mol. Cell **30**, 214–226 (2008)

E.J. Hagedorn, E.M. Durand, E.M. Fast, L.I. Zon, Getting more for your marrow: boosting hematopoietic stem cell numbers with PGE2. Exp. Cell Res. **329**, 220–226 (2014)

M. Hamasaki, N. Furuta, A. Matsuda, et al., Autophagosomes form at ER–mitochondria contact sites. Nature **495**, 389–393 (2013)

E. Hamilton, L. Pearce, L. Morgan et al., Mimicking the tumour microenvironment: three different co-culture systems induce a similar phenotype but distinct proliferative signals in primary chronic lymphocytic leukaemia cells. Br. J. Haematol. **158**, 589–599 (2012)

W. Han, Y. Sun, X. Wang, C. Zhu, K. Blomgren, Delayed, long-term administration of the caspase inhibitor Q-VD-OPh reduced brain injury induced by neonatal hypoxia-ischemia. Dev. Neurosci. **36**, 64–72 (2014)

T. Hanada, N.N. Noda, Y. Satomi, Y. Ichimura, Y. Fujioka, T. Takao, F. Inagaki, Y. Ohsumi, The Atg12-Atg5 conjugate has a novel E3-like activity for protein lipidation in autophagy. J. Biol. Chem. **282**, 37298–37302 (2007)

D. Hanahan, R.A. Weinberg, The hallmarks of cancer. Cell **100**(1), 57–70 (2000)

D. Hanahan, R.A. Weinberg, Hallmarks of cancer: the next generation. Cell **144**, 646–674 (2011)

C.C. Harris, Chemical and physical carcinogenesis: advances and perspectives for the 1990s. Cancer Res. **51**, 5023s–5044s (1991)

R.E. Harris, L. Setiawan, J. Saul, I.K. Hariharan, Localized epigenetic silencing of a damage-activated WNT enhancer limits regeneration in mature *Drosophila* imaginal discs. Elife **5**, e11588 (2016)

C. Haslett, Granulocyte apoptosis and its role in the resolution and control of lung inflammation. Am. J. Respir. Crit. Care Med. **160**, S5–S11 (1999)

B.A. Hay, G.M. Wolff Tand Rubin, Expression of baculovirus P35 prevents cell death in *Drosophila*. Development **120**, 2121–2129 (1994)

J.L. Haynie, P.J. Bryant, The effects of X-rays on the proliferation dynamics of cells in the imaginal wing disc of *Drosophila melanogaster*. Dev. Genes Evol. **183**, 85–100 (1977)

C. He, D.J. Klionsky, Atg9 trafficking in autophagy-related pathways. Autophagy **3**, 271–274 (2007)

C. He, D.J. Klionsky, Regulation mechanisms and signaling pathways of autophagy. Annu. Rev. Genet. **43**, 67–93 (2009)

C. He, B. Levine, The beclin 1 interactome. Curr. Opin. Cell Biol. **22**, 140–149 (2010)

S. He, L. Wang, L. Miao, T. Wang, F. Du, L. Zhao, X. Wang, Receptor interacting protein kinase-3 determines cellular necrotic response to TNF-α. Cell **137**, 1100–1111 (2009a)

B. He, N. Lu, Z. Zhou, Cellular and nuclear degradation during apoptosis. Curr. Opin. Cell Biol. **21**, 900–912 (2009b)

S. He, Y. Liang, F. Shao, X. Wang, Toll-like receptors activate programmed necrosis in macrophages through a receptor-interacting kinase-3–mediated pathway. Proc. Natl. Acad. Sci. U. S. A. **108**, 20054–20059 (2011)

G. He, W. Lei, S. Wang, R. Xiao, K. Guo, Y. Xia, X. Zhou, K. Zhang, X. Liu, Y. Wang, Overexpression of tumor suppressor TSLC1 by a survivin-regulated oncolytic adenovirus significantly inhibits hepatocellular carcinoma growth. J. Cancer Res. Clin. Oncol. **138**, 657–670 (2012)

L. He, G. Si, J. Huang, A.D.T. Samuel, N. Perrimon, Mechanical regulation of stem-cell differentiation by the stretch-activated Piezo channel. Nature **555**, 103–106 (2018)

G. Helmlinger, P.A. Netti, H.C. Lichtenbeld, R.J. Melder, R.K. Jain, Solid stress inhibits the growth of multicellular tumor spheroids. Nat. Biotechnol. **15**, 778–783 (1997)

J.H. Hendry, C.M. West, Apoptosis and mitotic cell death: their relative contributions to normal-tissue and tumour radiation response. Int. J. Radiat. Biol. **71**(6), 709–719 (1997)

P.M. Henson, D.A. Hume, Apoptotic cell removal in development and tissue homeostasis. Trends Immunol. **27**, 244–250 (2006)

P.M. Henson, R.B. Johnston Jr., Tissue injury in inflammation. Oxidants, proteinases, and cationic proteins. J. Clin. Invest. **79**, 669–674 (1987)

H. Hentze, F. Schwoebel, S. Lund, M. Keel, W. Ertel, A. Wendel, M. Jaattela, M. Leist, In vivo and in vitro evidence for extracellular caspase activity released from apoptotic cells. Biochem. Biophys. Res. Commun. **283**, 1111–1117 (2001)

I. Herr, K.M. Debatin, Cellular stress response and apoptosis in cancer therapy. Blood **98**, 2603–2614 (2001)

S.C. Herrera, R. Martin, G. Morata, Tissue homeostasis in the wing disc of *Drosophila melanogaster*: immediate response to massive damage during development. PLoS Genet. **9**, e1003446 (2013)
M. Herrmann, R.E. Voll, O.M. Zoller, M. Hagenhofer, B.B. Ponner, J.R. Kalden, Impaired phagocytosis of apoptotic cell material by monocyte-derived macrophages from patients with systemic lupus erythematosus. Arthritis Rheum. **41**, 1241–1250 (1998)
H. Hirata, M. Gupta, S.R. Vedula, C.T. Lim, B. Ladoux, M. Sokabe, Actomyosin bundles serve as a tension sensor and a platform for ERK activation. EMBO Rep. **16**, 250–257 (2015)
J. Hitomi, D.E. Christofferson, A. Ng, J. Yao, A. Degterev, R.J. Xavier, J. Yuan, Identification of a molecular signaling network that regulates a cellular necrotic cell death pathway. Cell **135**, 1311–1323 (2008)
S. Hodge, G. Hodge, R. Scicchitano, P.N. Reynolds, M. Holmes, Alveolar macrophages from subjects with chronic obstructive pulmonary disease are deficient in their ability to phagocytose apoptotic airway epithelial cells. Immunol. Cell Biol. **81**, 289–296 (2003)
S. Holdenrieder, P. Eichhorn, U. Beuers, W. Samtleben, U. Schoenermarck, R. Zachoval, D. Nagel, P. Stieber, Nucleosomal DNA fragments in autoimmune diseases. Ann. NY Acad. Sci. **1075**, 318–327 (2006)
N. Holler, R. Zaru, O. Micheau, M. Thome, A. Attinger, S. Valitutti, J.L. Bodmer, P. Schneider, B. Seed, J. Tschopp, Fas triggers an alternative, caspase-8–independent cell death pathway using the kinase RIP as effector molecule. Nat. Immunol. **1**, 489–495 (2000)
J.W. Hopewell, J. Nyman, I. Turesson, Time factor for acute tissue reactions following fractionated irradiation: a balance between repopulation and enhanced radiosensitivity. Int. J. Radiat. Biol. **79**(7), 513–524 (2003)
H.R. Horvitz, Worms, life, and death (Nobel lecture). ChemBioChem **4**, 697–711 (2003)
N. Hosokawa, T. Hara, T. Kaizuka et al., Nutrient-dependent mTORC1 association with the ULK1–Atg13–FIP200 complex required for autophagy. Mol. Biol. Cell **20**, 1981–1991 (2009)
Q. Hu, J. Peng, W. Liu, X. He, L. Cui, X. Chen, M. Yang, H. Liu, S. Liu, H. Wang, Elevated cleaved caspase-3 is associated with shortened overall survival in several cancer types. Int. J. Clin. Exp. Pathol. **7**, 5057–5070 (2014)
Q. Huang, H.-M. Shen, To die or to live: the dual role of poly (ADP-ribose) polymerase-1 in autophagy and necrosis under oxidative stress and DNA damage. Autophagy **5**, 273–276 (2009)
Q. Huang, F. Li, X. Liu et al., Caspase 3-mediated stimulation of tumor cell repopulation during cancer radiotherapy. Nat. Med. **17**, 860–866 (2011)
Z. Huang, N.A. Mayr, M. Gao, S.S. Lo, J.Z. Wang, G. Jia, W.T. Yuh, Onset time of tumor repopulation for cervical cancer: first evidence from clinical data. Int. J. Radiat. Oncol. Biol. Phys. **84**(2), 478–484 (2012)
S. Huard, M. Chen, K.E. Burdette, C. Fenyvuesvolgyi, M. Yu, R.T. Elder, R.Y. Zhao, HIV-1 Vpr-induced cell death in Schizosaccharomyces pombe is reminiscent of apoptosis. Cell Res. **18**, 961–973 (2008)
J.R. Huh, M. Guo, B.A. Hay, Compensatory proliferation induced by cell death in the *Drosophila* wing disc requires activity of the apical cell death caspase Dronc in a nonapoptotic role. Curr. Biol. **14**, 1262–1266 (2004)
J.S. Hwang, C. Kobayashi, K. Agata, K. Ikeo, T. Gojobori, Detection of apoptosis during planarian regeneration by the expression of apoptosis-related genes and TUNEL assay. Gene **333**, 15–25 (2004)
J.-W. Hwang, S. Chung, I.K. Sundar, H. Yao, G. Arunachalam, M.W. McBurney, I. Rahman, Cigarette smoke-induced autophagy is regulated by SIRT1–PARP-1-dependent mechanism: implication in pathogenesis of COPD. Arch. Biochem. Biophys. **500**, 203–209 (2010)
A. Ichikawa, T. Kinoshita, T. Watanabe, H. Kato, H. Nagai, K. Tsushita, H. Saito, T. Hotta, Mutations of the p53 gene as a prognostic factor in aggressive B-cell lymphoma. N. Engl. J. Med. **337**, 529–534 (1997)
Y. Ichimura, T. Kirisako, T. Takao et al., A ubiquitin-like system mediates protein lipidation. Nature **408**, 488–492 (2000)
T. Igaki, Correcting developmental errors by apoptosis: lessons from *Drosophila* JNK signaling. Apoptosis **14**, 1021–1028 (2009)
F.H. Igney, P.H. Krammer, Death and anti-death: tumour resistance to apoptosis. Nat. Rev. Cancer **2**, 277–288 (2002)
K.D. Irvine, B.I. Shraiman, Mechanical control of growth: ideas, facts and challenges. Development **144**, 4238–4248 (2017)
E. Itakura, C. Kishi, K. Inoue, N. Mizushima, Beclin 1 forms two distinct phosphatidylinositol 3-kinase complexes with mammalian Atg14 and UVRAG. Mol. Biol. Cell **19**, 5360–5372 (2008)
E. Itakura, C. Kishi-Itakura, N. Mizushima, The hairpin-type tail-anchored SNARE syntaxin 17 targets to autophagosomes for fusion with endosomes/lysosomes. Cell **151**, 1256–1269 (2012)
S.P. Jackson, J. Bartek, The DNA-damage response in human biology and disease. Nature **461**(7267), 1071–1078 (2009)
U. Jäger, S.C. Raghavan, Follicular lymphomas9 BCL-2/IgH junctions contain templated nucleotide insertions: novel insights into the mechanism of t (14; 18) translocation. Blood **95**, 3520–3529 (2000)
R.K. Jain, J.D. Martin, T. Stylianopoulos, The role of mechanical forces in tumor growth and therapy. Annu. Rev. Biomed. Eng. **16**, 321–346 (2014)
A.A. James, P.J. Bryant, A quantitative study of cell death and mitotic inhibition in gamma-irradiated imaginal wing discs of *Drosophila melanogaster*. Radiat. Res. **87**, 552–564 (1981)

J.R. Jeffers, E. Parganas, Y. Lee et al., Puma is an essential mediator of p53-dependent and -independent apoptotic pathways. Cancer Cell **4**, 321–328 (2003)

K. John, I. Rosner, U. Keilholz, T. Gauler, H. Bantel, V. Grunwald, Baseline caspase activity predicts progression free survival of temsirolimus-treated head neck cancer patients. Eur. J. Cancer **51**, 1596–1602 (2015)

E.C. Jonathan, E.J. Bernhard, W.G. McKenna, How does radiation kill cells? Curr. Opin. Chem. Biol. **3**, 77–83 (1999)

C.H. Jung, C.B. Jun, S.-H. Ro, Y.-M. Kim, N.M. Otto, J. Cao, M. Kundu, D.-H. Kim, ULK-Atg13-FIP200 complexes mediate mTOR signaling to the autophagy machinery. Mol. Biol. Cell **20**, 1992–2003 (2009)

W.J. Kaiser, J.W. Upton, A.B. Long, D. Livingston-Rosanoff, L.P. Daley-Bauer, R. Hakem, T. Caspary, E.S. Mocarski, RIP3 mediates the embryonic lethality of caspase-8-deficient mice. Nature **471**, 368–372 (2011)

W.J. Kaiser, H. Sridharan, C. Huang, P. Mandal, J.W. Upton, P.J. Gough, C.A. Sehon, R.W. Marquis, J. Bertin, E.S. Mocarski, Toll-like receptor 3-mediated necrosis via TRIF, RIP3, and MLKL. J. Biol. Chem. **288**, 31268–31279 (2013a)

W.J. Kaiser, J.W. Upton, E.S. Mocarski, Viral modulation of programmed necrosis. Curr. Opin. Virol. **3**, 296–306 (2013b)

M. Kajita, Y. Fujita, EDAC: epithelial defence against cancer-cell competition between normal and transformed epithelial cells in mammals. J. Biochem. **158**, 15–23 (2015)

M. Kajita, K. Sugimura, A. Ohoka et al., Filamin acts as a key regulator in epithelial defence against transformed cells. Nat. Commun. **5**, 4428 (2014)

H.E. Kamber Kaya, M. Ditzel, P. Meier, A. Bergmann, An inhibitory mono-ubiquitylation of the *Drosophila* initiator caspase Dronc functions in both apoptotic and non-apoptotic pathways. PLoS Genet. **13**, e1006438 (2017)

M. Karin, H. Clevers, Reparative inflammation takes charge of tissue regeneration. Nature **529**, 307–315 (2016)

S. Kashio, F. Obata, L. Zhang, T. Katsuyama, T. Chihara, M. Miura, Tissue nonautonomous effects of fat body methionine metabolism on imaginal disc repair in *Drosophila*. Proc. Natl. Acad. Sci. U. S. A. **113**, 1835–1840 (2016)

K. Kawabata, T. Hagio, S. Matsuoka, The role of neutrophil elastase in acute lung injury. Eur. J. Pharmacol. **451**, 1–10 (2002)

T. Kawamata, Y. Kamada, Y. Kabeya, T. Sekito, Y. Ohsumi, Organization of the pre-autophagosomal structure responsible for autophagosome formation. Mol. Biol. Cell **19**, 2039–2050 (2008)

M.A. Kelliher, S. Grimm, Y. Ishida, F. Kuo, B.Z. Stanger, P. Leder, The death domain kinase RIP mediates the TNF-induced NF-κB signal. Immunity **8**, 297–303 (1998)

O. Kepp, A. Tesniere, F. Schlemmer, M. Michaud, L. Senovilla, L. Zitvogel, G. Kroemer, Immunogenic cell death modalities and their impact on cancer treatment. Apoptosis **14**, 364–375 (2009)

J.F. Kerr, J. Searle, Deletion of cells by apoptosis during castration-induced involution of the rat prostate. Virchows Arch. B Cell Pathol. **13**, 87–102 (1973)

J.F. Kerr, A.H. Wyllie, A.R. Currie, Apoptosis: a basic biological phenomenon with wide-ranging implications in tissue kinetics. Br. J. Cancer **26**, 239–257 (1972)

E. Kerr, C. Holohan, K.M. McLaughlin et al., Identification of an acetylation-dependant Ku70/FLIP complex that regulates FLIP expression and HDAC inhibitor-induced apoptosis. Cell Death Differ. **19**, 1317–1327 (2012)

M.J. Khan, M.R. Alam, M. Waldeck-Weiermair, et al. Inhibition of autophagy rescues palmitic acid-induced necroptosis of endothelial cells. J. Biol. Chem. **287**, 21110–21120 (2012)

K.K. Khanna, S.P. Jackson, DNA double-strand breaks: signaling, repair and the cancer connection. Nat. Genet. **27**, 247–254 (2001)

A. Kihara, Y. Kabeya, Y. Ohsumi, T. Yoshimori, Beclin–phosphatidylinositol 3-kinase complex functions at the trans-Golgi network. EMBO Rep. **2**, 330–335 (2001)

Y.C. Kim, K.-L. Guan, mTOR: a pharmacologic target for autophagy regulation. J. Clin. Invest. **125**, 25–32 (2015)

J.S. Kim, L. He, J.J. Lemasters, Mitochondrial permeability transition: a common pathway to necrosis and apoptosis. Biochem. Biophys. Res. Commun. **304**, 463–470 (2003)

S. Kim, J. Li, Caspase blockade induces RIP3-mediated programmed necrosis in Toll-like receptor-activated microglia. Cell Death Dis. **4**, e716 (2013)

H.-S. Kim, Y.-H. Suh, Minocycline and neurodegenerative diseases. Beha. Brain Res. **196**, 168–179 (2009)

J.J. Kim, I.F. Tannock, Repopulation of cancer cells during therapy: an important cause of treatment failure. Nat. Rev. Cancer **5**(7), 516–525 (2005)

K.W. Kim, M. Hwang, L. Moretti, J.J. Jaboin, Y.I. Cha, B. Lu, Autophagy upregulation by inhibitors of caspase-3 and mTOR enhances radiotherapy in a mouse model of lung cancer. Autophagy **4**(5), 659–668 (2008)

J. Kim, M. Kundu, B. Viollet, K.-L. Guan, AMPK and mTOR regulate autophagy through direct phosphorylation of Ulk1. Nat. Cell Biol. **13**, 132–141 (2011)

G.R. Kinsey, M.D. Okusa, Pathogenesis of acute kidney injury: foundation for clinical practice. Am. J. Kideny Dis. **58**, 291–301 (2011)

T. Kirisako, Y. Ichimura, H. Okada, Y. Kabeya, N. Mizushima, T. Yoshimori, M. Ohsumi, T. Takao, T. Noda, Y. Ohsumi, The reversible modification regulates the membrane-binding state of Apg8/Aut7 essential for autophagy and the cytoplasm to vacuole targeting pathway. J. Cell Biol. **151**, 263–276 (2000)

P. Kirkham, Oxidative stress and macrophage function: a failure to resolve the inflammatory response. Biochem. Soc. Trans. **35**, 284–287 (2007)

S. Kitada, J. Andersen, S. Akar et al., Expression of apoptosis-regulating proteins in chronic lymphocytic leukemia: correlations with In vitro and In vivo chemoresponses. Blood **91**, 3379–3389 (1998)

S. Knapp, J.C. Leemans, S. Florquin, J. Branger, N.A. Maris, J. Pater, N. van Rooijen, T. van der Poll, Alveolar macrophages have a protective antiinflammatory role during murine pneumococcal pneumonia. Am. J. Respir. Crit. Care Med. **167**, 171–179 (2003)

L. Kockel, J.G. Homsy, D. Bohmann, *Drosophila* AP-1: lessons from an invertebrate. Oncogene **20**, 2347–2364 (2001)

L. Koerver, J. Melzer, E.A. Roca, D. Teichert, T. Glatter, E. Arama, M. Broemer, The de-ubiquitylating enzyme DUBA is essential for spermatogenesis in *Drosophila*. Cell Death Differ. **23**, 2019–2030 (2016)

Y. Kondo, S. Kondo, Autophagy and cancer therapy. Autophagy **2**, 85–90 (2006)

S. Kondo, N. Senoo-Matsuda, Y. Hiromi, M. Miura, DRONC coordinates cell death and compensatory proliferation. Mol. Cell Biol. **26**, 7258–7268 (2006)

K. Kongsuwan, Q. Yu, A. Vincent, M.C. Frisardi, M. Rosbash, J.A. Lengyel, J. Merriam, A *Drosophila* minute gene encodes a ribosomal protein. Nature **317**, 555–558 (1985)

H. Kono, K.L. Rock, How dying cells alert the immune system to danger. Nat. Rev. Immunol. **8**, 279–289 (2008)

G. Kroemer, P. Petit, N. Zamzami, J.L. Vayssiere, B. Mignotte, The biochemistry of programmed cell death. FASEB J. **9**, 1277–1287 (1995)

G. Kroemer, B. Dallaporta, M. Resche-Rigon, The mitochondrial death/life regulator in apoptosis and necrosis. Annu. Rev. Physiol. **60**, 619–642 (1998)

G. Kroemer et al., Classification of cell death: recommendations of the Nomenclature Committee on cell death. Cell Death Differ. **12**(Suppl. 2), 1463–1467 (2005)

G. Kroemer, L. Galluzzi, P. Vandenabeele et al., Classification of cell death: recommendations of the Nomenclature Committee on cell death 2009. Cell Death Differ. **16**, 3–11 (2009)

K. Kruse, C. Janko, V. Urbonaviciute, C.T. Mierke, T.H. Winkler, R.E. Voll, G. Schett, L.E. Muñoz, M. Herrmann, Inefficient clearance of dying cells in patients with SLE: anti-dsDNA autoantibodies, MFG-E8, HMGB-1 and other players. Apoptosis **15**, 1098–1113 (2010)

D.V. Krysko, K. D'Herde, P. Vandenabeele, Clearance of apoptotic and necrotic cells and its immunological consequences. Apoptosis **11**, 1709–1726 (2006)

D.V. Krysko, T. Vanden Berghe, K. D'Herde, P. Vandenabeele, Apoptosis and necrosis: detection, discrimination and phagocytosis. Methods **44**, 205–221 (2008)

D.A. Kubli, Å.B. Gustafsson, Mitochondria and mitophagy: the yin and yang of cell death control. Circ. Res. **111**, 1208–1221 (2012)

S. Kumar, Caspase function in programmed cell death. Cell Death Differ. **14**, 32–43 (2007)

P. Kurada, K. White, Ras promotes cell survival in *Drosophila* by downregulating hid expression. Cell **95**, 319–329 (1998)

A.V. Kurtova, J. Xiao, Q. Mo, S. Pazhanisamy, R. Krasnow, S.P. Lerner, F. Chen, T.T. Roh, E. Lay, P.L. Ho, K.S. Chan, Blocking PGE2-induced tumour repopulation abrogates bladder cancer chemoresistance. Nature **517**, 209–213 (2015)

V. Labi, M. Erlacher, How cell death shapes cancer. Cell Death Dis. **6**, e1675 (2015)

M. Lamkanfi, W. Declercq, M. Kalai, X. Saelens, P. Vandenabeele, Alice in caspase land. A phylogenetic analysis of caspases from worm to man. Cell Death Differ. **9**(4), 358–361 (2002)

H. Lanaya, A. Natarajan, K. Komposch et al., EGFR has a tumour-promoting role in liver macrophages during hepatocellular carcinoma formation. Nat. Cell Biol. **16**, 972–977 (2014)

M. Laplante, D.M. Sabatini, mTOR signaling in growth control and disease. Cell **149**, 274–293 (2012)

K. Lauber, E. Bohn, S.M. Kröber et al., Apoptotic cells induce migration of phagocytes via caspase-3-mediated release of a lipid attraction signal. Cell **113**, 717–730 (2003)

I. Lavrik, A. Golks, P.H. Krammer, Death receptor signaling. J. Cell Sci. **118**, 265–267 (2005a)

I.N. Lavrik, A. Golks, P.H. Krammer, Caspases: pharmacological manipulation of cell death. J. Clin. Invest. **115**(10), 2665–2672 (2005b)

M.S. Lawrence, P. Stojanov, C.H. Mermel, J.T. Robinson, L.A. Garraway, T.R. Golub, M. Meyerson, S.B. Gabriel, E. S. Lander, G. Getz, Discovery and saturation analysis of cancer genes across 21 tumour types. Nature **505**, 495–501 (2014)

D.A. Lawson, K. Kessenbrock, R.T. Davis, N. Pervolarakis, Z. Werb, Tumour het-erogeneity and metastasis at single-cell resolution. Nat. Cell Biol. **20**, 1349–1360 (2018)

T. Lecuit, P.F. Lenne, Cell surface mechanics and the control of cell shape, tissue patterns and morphogenesis. Nat. Rev. Mol. Cell Biol. **8**, 633–644 (2007)

S.W. Lee, Y. Morishita, Possible roles of mechanical cell elimination intrinsic to growing tissues from the perspective of tissue growth efficiency and homeostasis. PLoS Comput. Biol. **13**, e1005651 (2017)
J.-S. Lee, Q. Li, J.Y. Lee, S.H. Lee, J.H. Jeong, H.R. Lee, H. Chang, F.C. Zhou, S.J. Gao, C. Liang, J.U. Jung, FLIP-mediated autophagy regulation in cell death control. Nat. Cell Biol. **11**, 1355–1362 (2009)
T.V. Lee, Y. Fan, S. Wang, M. Srivastava, M. Broemer, P. Meier, A. Bergmann, *Drosophila* IAP1-mediated ubiquitylation controls activation of the initiator caspase DRONC independent of protein degradation. PLoS Genet. **7**, e1002261 (2011)
L. Legoff, H. Rouault, T. Lecuit, A global pattern of mechanical stress polarizes celldivisions and cell shape in the growing *Drosophila* wing disc. Development **140**, 4051–4059 (2013)
M. Leist, M. Jäättelä, Four deaths and a funeral: from caspases to alternative mechanisms. Nat. Rev. Mol. Cell Biol. **2**, 589–598 (2001)
M. Leist, B. Single, A.F. Castoldi, S. Kühnle, P. Nicotera, Intracellular adenosine triphosphate (ATP) concentration: a switch in the decision between apoptosis and necrosiss. J. Exp. Med. **185**, 1481–1486 (1997)
R. Levayer, Solid stress, competition for space and cancer: the opposing roles ofmechanical cell competition in tumour initiation and growth. Semin. Cancer Biol. pii: S1044-579X(19)30015-X (2019)
R. Levayer, E. Moreno, Mechanisms of cell competition: themes and variations. J. Cell Biol. **200**, 689–698 (2013)
R. Levayer, B. Hauert, E. Moreno, Cell mixing induced by myc is required for competitive tissue invasion and destruction. Nature **524**, 476–480 (2015)
R. Levayer, C. Dupont, E. Moreno, Tissue crowding induces caspase-dependent competition for space. Curr. Biol. **26**, 670–677 (2016)
S. Levin, T.J. Bucci, S.M. Cohen, A.S. Fix, J.F. Hardisty, E.K. LeGrand, R.R. Maronpot, B.F. Trump, The nomenclature of cell death: recommendations of an ad hoc committee of the society of toxicologic pathologists. Toxicol. Pathol. **27**, 484–490 (1999)
A.J. Levine, p53, the cellular gatekeeper for growth and division. Cell **88**, 323–331 (1997)
B. Levine, G. Kroemer, Autophagy in the pathogenesis of disease. Cell **132**, 27–42 (2008)
B. Levine, M. Packer, P. Codogno, Development of autophagy inducers in clinical medicine. J. Clin. Invest. **125**, 14–24 (2015)
M.O. Li, M.R. Sarkisian, W.Z. Mehal, P. Rakic, R.A. Flavell, Phosphatidylserine receptor is required for clearance of apoptotic cells. Science **302**, 1560–1563 (2003)
F.-Y. Li, P.D. Jeffrey, W.Y. Jong, Y. Shi, Crystal structure of a viral flip insights into flip-mediated inhibition of death receptor signaling. J. Biol. Chem. **281**, 2960–2968 (2006)
Y. Li, X. Yang, C. Ma, J. Qiao, C. Zhang, Necroptosis contributes to the NMDA-induced excitotoxicity in rat's cultured cortical neurons. Neurosci. Lett. **447**, 120–123 (2008)
W. Li, A. Kale, N.E. Baker, Oriented cell division as a response to cell death and cell competition. Curr. Biol. **19**(21), 1821–1826 (2009)
F. Li, Q. Huang, J. Chen, Y. Peng, D. Roop, J.S. Bedford, C.-Y. Li, Apoptotic cells activate the "phoenix rising" pathway to promote wound healing and tissue regeneration. Sci. Signal **3**(110), ra13 (2010a)
J. Li, R. Liu, Y. Lei, K. Wang, Q.C. Lau, N. Xie, S. Zhou, C. Nie, L. Chen, Y. Wei, C. Huang, Proteomic analysis revealed association of aberrant ROS signaling with suberoylanilide hydroxamic acid-induced autophagy in Jurkat T-leukemia cells. Autophagy **6**, 711–724 (2010b)
Y. Li, H. Zhu, X. Zeng, J. Fan, X. Qian, S. Wang, Z. Wang, Y. Sun, X. Wang, W. Wang, D. Ju, Suppression of autophagy enhanced growth inhibition and apoptosis of interferon-β in human glioma cells. Mol. Neurobiol. **47**, 1000–1010 (2013)
M. Li, J.L. Lindblad, E. Perez, A. Bergmann, Y. Fan, Autophagy-independent function of Atg1 for apoptosis-induced compensatory proliferation. BMC Biol. **14**, 70 (2016)
X.H. Liang, S. Jackson, M. Seaman, K. Brown, B. Kempkes, H. Hibshoosh, B. Levine, Induction of autophagy and inhibition of tumorigenesis by beclin 1. Nature **402**, 672–676 (1999)
M. Ligr, F. Madeo, E. Frohlich, W. Hilt, K.U. Frohlich, D.H. Wolf, Mammalian Bax triggers apoptotic changes in yeast. FEBS Lett. **438**, 61–65 (1998)
Y. Lin, A. Devin, Y. Rodriguez, Z.-G. Liu, Cleavage of the death domain kinase RIP by caspase-8 prompts TNF-induced apoptosis. Genes Dev. **13**, 2514–2526 (1999)
A. Linkermann, J.H. Bräsen, N. Himmerkus, S. Liu, T.B. Huber, U. Kunzendorf, S. Krautwald, Rip1 (receptor-interacting protein kinase 1) mediates necroptosis and contributes to renal ischemia/reperfusion injury. Kidney Int. **81**, 751–761 (2012)
J.Y. Liou, D.P. Ellent, S. Lee, J. Goldsby, B.S. Ko, N. Matijevic, J.C. Huang, K.K. Wu, Cyclooxygenase-2-derived prostaglandin e2 protects mouse embryonic stem cells from apoptosis. Stem Cells **25**(5), 1096–1103 (2007)
S. Lisi, I. Mazzon, K. White, Diverse domains of THREAD/DIAP1 are required to inhibit apoptosis induced by REAPER and HID in *Drosophila*. Genetics **154**, 669–678 (2000)
Y.R. Liu, B. Sun, X.L. Zhao, Q. Gu, Z.Y. Liu, X.Y. Dong, N. Che, J. Mo, Basal caspase-3 activity promotes migration, invasion, and vasculogenic mimicry formation of melanoma cells. Melanoma Res. **23**, 243–253 (2013)

P. Liu, W. Gan, Y.R. Chin et al., PtdIns(3,4,5)P3-dependent activation of the mTORC2 kinase complex. Cancer Discov. **5**, 1194–1209 (2015)

R.A. Lockshin, Z. Zakeri, Programmed cell death and apoptosis: origins of the theory. Nat. Rev. Mol. Cell Biol. **2**, 545–550 (2001)

M. Los, M. Mozoluk, D. Ferrari, A. Stepczynska, C. Stroh, A. Renz, Z. Herceg, Z.Q. Wang, K. Schulze-Osthoff, Activation and caspase-mediated inhibition of PARP: a molecular switch between fibroblast necrosis and apoptosis in death receptor signaling. Mol. Biol. Cell **13**, 978–988 (2002)

N.R. Love, Y. Chen, S. Ishibashi, P. Kritsiligkou, R. Lea, Y. Koh, Y. Chen, S. Ishibashi, P. Kritsiligkou, R. Lea, Y. Koh, J.L. Gallop, K. Dorey, E. Amaya, Amputation-induced reactive oxygen species are required for successful *Xenopus* tadpole tail regeneration. Nat. Cell Biol. **15**, 222–228 (2013)

S.W. Lowe, A.W. Lin, Apoptosis in cancer. Carcinogenesis **21**(3), 485–495 (2000)

S.W. Lowe, E. Cepero, G. Evan, Intrinsic tumour suppression. Nature **432**(7015), 307–315 (2004)

F. Luciano, A. Jacquel, P. Colosetti, M. Herrant, S. Cagnol, G. Pages, P. Auberger, Phosphorylation of Bim-EL by Erk1/2 on serine 69 promotes its degradation via the proteasome pathway and regulates its proapoptotic function. Oncogene **22**, 6785–6793 (2003)

P. Ludovico, M.J. Sousa, M.T. Silva, C. Leao, M. Corte-Real, Saccharomyces cerevisiae commits to a programmed cell death process in response to acetic acid. Microbiology **147**, 2409–2415 (2001)

J.J. Lum, D.E. Bauer, M. Kong, M.H. Harris, C. Li, T. Lindsten, C.B. Thompson, Growth factor regulation of autophagy and cell survival in the absence of apoptosis. Cell **120**, 237–248 (2005)

S. Luo, D. Rubinsztein, Apoptosis blocks Beclin 1-dependent autophagosome synthesis: an effect rescued by Bcl-xL. Cell Death Differ. **17**, 268–277 (2010)

J.W. Lustbader, M. Cirilli, C. Lin, ABAD directly links Aß to mitochondrial toxicity in Alzheimer's disease. Science **304**, 448–452 (2004)

J. Ma, L. Tian, J. Cheng, Z. Chen, B. Xu, L. Wang, C. Li, Q. Huang, Sonic hedgehog signaling pathway supports cancer cell growth during cancer radiotherapy. PLoS ONE **8**, e65032 (2013)

J. Ma, J. Cheng, Y. Gong, L. Tian, Q. Huang, Wnt signaling downregulated after sonic hedgehog activation negatively associates with tumor repopulation. Dis. Model Mech. **8**, 385–391 (2015)

D. Ma, C. Pignanelli, D. Tarade et al., Cancer cell mitochondria targeting by pancratistatin analogs is dependent on functional complex II and III. Sci. Rep. **7**, 42957 (2017)

B. Maciejewski, S. Majewski, Dose fractionation and tumour repopulation in radiotherapy for bladder cancer. Radiother. Oncol. **21**(3), 163–170 (1991)

W.J. Mackus, A.P. Kater, A. Grummels, L.M. Evers, B. Hooijbrink, M.H.H. Kramer, J.E. Castro, T.J. Kipps, R.A.W. van Lier, M.H.J. van Oers, E. Eldering, Chronic lymphocytic leukemia cells display p53-dependent drug-induced Puma upregulation. Leukemia **19**, 427–434 (2005)

F. Madeo, E. Frohlich, K.U. Frohlich, A yeast mutant showing diagnostic markers of early and late apoptosis. J. Cell Biol. **139**, 729–734 (1997)

F. Madeo, E. Frohlich, M. Ligr, M. Grey, S.J. Sigrist, D.H. Wolf, K.U. Frohlich, Oxygen stress: a regulator of apoptosis in yeast. J. Cell Biol. **145**, 757–767 (1999)

S. Maeda, H. Kamata, J.L. Luo, H. Leffert, M. Karin, IKKbeta couple hepatocyte death to cytokine-driven compensatory proliferation that promotes chemical hepatocarcinogenesis. Cell **121**(7), 977–990 (2005)

M.C. Maiuri, G. Le Toumelin, A. Criollo et al., Functional and physical interaction between Bcl-XL and a BH3-like domain in Beclin-1. EMBO J. **26**, 2527–2539 (2007a)

M.C. Maiuri, E. Zalckvar, A. Kimchi, G. Kroemer, Self-eating and self-killing: crosstalk between autophagy and apoptosis. Nat. Rev. Mol. Cell Biol. **8**, 741–752 (2007b)

A.J. Maniotis, R. Folberg, A. Hess, E.A. Seftor, L.M. Gardner, J. Peer, J.M. Trent, P.S. Meltzer, M.J. Hendrix, Vascular channel formation by human melanoma cells in vivo and in vitro: vasculogenic mimicry. Am. J. Pathol. **155**, 739–752 (1999)

P. Mao, L. Smith, W. Xie, M. Wang, Dying endothelial cells stimulate proliferation of malignant glioma cells via a caspase 3-mediated pathway. Oncol. Lett. **5**, 1615–1620 (2013a)

Y. Mao, A.L. Tournier, A. Hoppe, L. Kester, B.J. Thompson, N. Tapon, Differential proliferation rates generate patterns of mechanical tension that orient tissue growth. EMBO J. **32**, 2790–2803 (2013b)

L. Marcu, E. Bezak, Modelling of tumour repopulation after chemotherapy. Australas. Phys. Eng. Sci. Med. **33**(3), 265–270 (2010)

L. Marcu, T. van Doorn, I. Olver, Modelling of post-irradiation accelerated repopulation in squamous cell carcinomas. Phys. Med. Biol. **49**(16), 3767–3779 (2004)

E. Marinari, A. Mehonic, S. Curran, J. Gale, T. Duke, B. Baum, Live-cell delami-nation counterbalances epithelial growth to limit tissue overcrowding. Nature **484**, 542–545 (2012)

G. Mariño, M. Niso-Santano, E.H. Baehrecke, G. Kroemer, Self-consumption: the interplay of autophagy and apoptosis. Nat. Rev. Mol. Cell Biol. **15**, 81–94 (2014)

L.B. Marks, M. Dewhirst, Accelerated repopulation: friend or foe? Exploiting changes in tumor growth characteristics to improve the "efficiency" of radiotherapy. Int. J. Radiat. Oncol. Biol. Phys. **21**(5), 1377–1383 (1991)

P. Martin, S.J. Leibovich, Inflammatory cells during wound repair: the good the bad and the ugly. Trends Cell Biol. **15**, 599–607 (2005)

P. Martin, D. D'Souza, J. Martin, R. Grose, L. Cooper, R. Maki, S.R. McKercher, Wound healing in the PU.1 null mouse—tissue repair is not dependent on inflammatory cells. Curr. Biol. **13**, 1122–1128 (2003)

F.A. Martin, A. Perez-Garijo, G. Morata, Apoptosis in *Drosophila*: compensatory proliferation and undead cells. Int. J. Dev. Biol. **53**, 1341–1347 (2009a)

J.W. Martin, S.S. Mousa, O. Shaker, S.A. Mousa, The multiple faces of nicotine and its implications in tissue and wound repair. Exp. Dermatol. **18**, 497–505 (2009b)

I. Martincorena, A. Roshan, M. Gerstung et al., Tumor evolution. Highburden and pervasive positive selection of somatic mutations in normal humanskin. Science **348**, 880–886 (2015)

I. Martincorena, J.C. Fowler, A. Wabik et al., Somatic mutant clonescolonize the human esophagus with age. Science **362**, 911–917 (2018)

A. Marusyk, D.P. Tabassum, P.M. Altrock, V. Almendro, F. Michor, K. Polyak, Non-cell-autonomous driving of tumour growth supports sub-clonal heterogeneity. Nature **514**, 54–58 (2014)

T. Maruyama, Y. Fujita, Cell competition in mammals—novel homeostatic machinery for embryonic development and cancer prevention. Curr. Opin. Cell Biol. **48**, 106–112 (2017)

R. Mathew, S. Kongara, B. Beaudoin, C.M. Karp, K. Bray, K. Degenhardt, G. Chen, S. Jin, E. White, Autophagy suppresses tumor progression by limiting chromosomal instability. Genes Dev. **21**, 1367–1381 (2007)

A. Matsuzawa, H. Ichijo, Redox control of cell fate by MAP kinase: physiological roles of ASK1-MAP kinase pathway in stress signaling. Biochem. Biophys. Acta **1780**, 1325–1336 (2008)

D. McIlroy, M. Tanaka, H. Sakahira et al., An auxiliary mode of apoptotic DNA fragmentation provided by phagocytes. Genes Dev. **14**, 549–558 (2000)

D.R. McIlwain, T. Berger, T.W. Mak, Caspase functions in cell death and disease. Cold Spring Harb. Perspect. Biol. **5**, a008656 (2013)

D. Medan, L. Wang, X. Yang, S. Dokka, V. Castranova, Y. Rojanasakul, Induction of neutrophil apoptosis and secondary necrosis during endotoxin-induced pulmonary inflammation in mice. J. Cell Physiol. **191**, 320–326 (2002)

D. Medan, L. Wang, D. Toledo, B. Lu, C. Stehlik, B.H. Jiang, X. Shi, Y. Rojanasakul, Regulation of Fas (CD95)-induced apoptotic and necrotic cell death by reactive oxygen species in macrophages. J. Cell Physiol. **203**, 78–84 (2005)

M.M. Merino, R. Levayer, E. Moreno, Survival of the fittest: essential roles of cellcompetition in development, aging, and Cancer. Trends Cell Biol. **6**(10), 776–788 (2016)

J.H. Meserve, R.J. Duronio, Scalloped and Yorkie are required for cell cycle re-entry of quiescent cells after tissue damage. Development **142**, 2740–2751 (2015)

R.E. Meyn, L.C. Stephens, K.K. Ang, N.R. Hunter, W.A. Brock, L. Milas, L.J. Peters, Heterogeneity in the development of apoptosis in irradiated murine tumours of different histologies. Int. J. Radiat. Biol. **64**(5), 583–591 (1993)

E. Michalak, A. Villunger, M. Erlacher, A. Strasser, Death squads enlisted by the tumour suppressor p53. Biochem. Biophys. Res. Commun. **331**, 786–798 (2005)

L. Michalik, B. Desvergne, N.S. Tan et al., Impaired skin wound healing in peroxisome proliferator-activated receptor (PPAR) alpha and PPARbeta mutant mice. J. Cell Biol. **154**, 799–814 (2001)

G.K. Michalopoulos, M.C. DeFrances, Liver regeneration. Science **276**(5309), 60–66 (1997)

O. Micheau, Cellular FLICE-inhibitory protein: an attractive therapeutic target? Expert. Opin. Ther. Targets **7**, 559–573 (2003)

O. Micheau, J. Tschopp, Induction of TNF receptor I-mediated apoptosis via two sequential signaling complexes. Cell **114**, 181–190 (2003)

C.T. Mierke, The fundamental role of mechanical properties in the progression of cancer disease and inflammation. Rep. Prog. Phys. **77**(7), 076602 (2014)

C.T. Mierke, The matrix environmental and cell mechanical properties regulate cell migration and contribute to the invasive phenotype of cancer cells. Rep. Prog. Phys. **82**(6), 064602 (2019)

M. Milan, S. Campuzano, A. Garcia-Bellido, Developmental parameters of cell death in the wing disc of *Drosophila*. Proc. Natl. Acad. Sci. U. S. A. **94**, 5691–5696 (1997)

M. Milan, L. Perez, S.M. Cohen, Short-range cell interactions and cell survival in the *Drosophila* wing. Dev. Cell **2**, 797–805 (2002)

C. Mitchell, H. Willenbring, A reproducible and well-tolerated method for 2/3 partial hepatectomy in mice. Nat. Protoc. **3**(7), 1167–1170 (2008)

M. Miura, H. Zhu, R. Rotello, E.A. Hartwieg, J. Yuan, Induction of apoptosis in fibroblasts by IL-1β-converting enzyme, a mammalian homolog of the *C. elegans* cell death gene ced-3. Cell **75**, 653–660 (1993)

N. Mizushima, M. Komatsu, Autophagy: renovation of cells and tissues. Cell **147**, 728–741 (2011)
N. Mizushima, T. Noda, T. Yoshimori, Y. Tanaka, T. Ishii, M.D. George, D.J. Klionsky, M. Ohsumi, Y. Ohsumi, A protein conjugation system essential for autophagy. Nature **395**, 395–398 (1998)
N. Mizushima, T. Noda, Y. Ohsumi, Apg16p is required for the function of the Apg12p–Apg5p conjugate in the yeast autophagy pathway. EMBO J. **18**, 3888–3896 (1999)
N. Mizushima, T. Yoshimori, Y. Ohsumi, The role of Atg proteins in autophagosome formation. Annu. Rev. Cell Dev. Biol. **27**, 107–132 (2011)
S. Moharikar, J.S. D'Souza, A.B. Kulkarni, B.J. Rao, Apoptotic-like cell death pathway is induced in unicellular chlorophyte *Chlamydomonas reinhardtii* (Chlorophyceae) cells following UV irradiation: detection and functional analyses. J. Phycol. **42**, 423–433 (2006)
M.I. Molejon, A. Ropolo, M.I. Vaccaro, VMP1 is a new player in the regulation of the autophagy-specific phosphatidylinositol 3-kinase complex activation. Autophagy **9**, 933–935 (2013)
M.B. Moller, A.M. Gerdes, K. Skjodt, L.S. Mortensen, N.T. Pedersen, Disrupted p53 function as predictor of treatment failure and poor prognosis in B- and T-cell non-Hodgkin's lymphoma. Clin. Cancer Res. **5**, 1085–1091 (1999)
B. Mollereau, Establishing links between ER-hormesis and cancer. Mol. Cell Biol. **33**, 2372–2374 (2013)
B. Mollereau, A. Perez-Garijo, A. Bergmann, M. Miura, O. Gerlitz, H.D. Ryoo, H. Steller, G. Morata, Compensatory proliferation and apoptosis-induced proliferation: a need for clarification. Cell Death Differ. **20**, 181 (2012)
F. Montel, M. Delarue, J. Elgeti, L. Malaquin, M. Basan, T. Risler, B. Cabane, D. Vignjevic, J. Prost, G. Cappello, J.F. Joanny, Stress clamp experiments on multicellulartumor spheroids. Phys. Rev. Lett. **107**, 188102 (2011)
D. Moquin, F.K.-M. Chan, The molecular regulation of programmed necrotic cell injury. Trends Biochem. Sci. **35**, 434–441 (2010)
A.A. Morales, A. Olsson, F. Celsing, A. Osterborg, M. Jondal, L.M. Osorio, Expression and transcriptional regulation of functionally distinct Bmf isoforms in B-chronic lymphocytic leukemia cells. Leukemia **18**, 41–47 (2004)
G. Morata, P. Ripoll, Minutes: mutants of *Drosophila* autonomously affecting cell division rate. Dev. Biol. **42**, 211–221 (1975)
E. Moreno, Is cell competition relevant to cancer? Nat. Rev. Cancer **8**, 141–147 (2008)
E. Moreno, K. Basler, dMyc transforms cells into super-competitors. Cell **117**, 117–129 (2004)
E. Moreno, M. Yan, K. Basler, Evolution of TNF signaling mechanisms: JNK-dependent apoptosis triggered by Eiger, the *Drosophila* homolog of the TNF superfamily. Curr. Biol. **12**, 1263–1268 (2002)
E. Moreno, L. Valon, F. Levillayer, R. Levayer, Competition for space induces cellelimination through compaction-driven ERK downregulation. Curr. Biol. **29**(23–34), e28 (2019)
L. Moretti, K.W. Kim, D.K. Jung, C.D. Willey, B. Lu, Radiosensitization of solid tumors by Z-VAD, a pan-caspase inhibitor. Mol. Cancer Ther. **8**(5), 1270–1279 (2009)
J.E. Morgan, J.G. Gross, C.N. Pagel et al., Myogenic cell proliferation and generation of a reversible tumorigenic phenotype are triggered by preirradiation of the recipient site. J. Cell Biol. **157**(4), 693–702 (2002)
N. Morishima, K. Nakanishi, H. Takenouchi, T. Shibata, Y. Yasuhiko, An endoplasmic reticulum stress-specific caspase cascade in apoptosis cytochrome c-independent activation of caspase-9 by caspase-12. J. Biol. Chem. **277**, 34287–34294 (2002)
E. Morselli, L. Galluzzi, O. Kepp, J.M. Vicencio, A. Criollo, M.C. Maiuri, G. Kroemer, Anti-and pro-tumor functions of autophagy. BBA-Mol. Cell Res. **1793**, 1524–1532 (2009)
D.M. Mosser, J.P. Edwards, Exploring the full spectrum of macrophage activation. Nat. Rev. Immunol. **8**, 958–969 (2008)
V.R. Mukaro, S. Hodge, Airway clearance of apoptotic cells in COPD. Curr. Drug Targets **12**(4), 460–468 (2011)
L.E. Munoz, U.S. Gaipl, S. Franz, A. Sheriff, R.E. Voll, J.R. Kalden, M. Herrmann, SLE—a disease of clearance deficiency? Rheumatology (Oxford) **44**, 1101–1107 (2005)
L.E. Munoz, K. Lauber, M. Schiller, A.A. Manfredi, M. Herrmann, The role of defective clearance of apoptotic cells in systemic autoimmunity. Nat. Rev. Rheumatol. **6**, 280–289 (2010)
J.A. Muñoz-Gámez, J.M. Rodríguez-Vargas, R. Quiles-Pérez et al., PARP-1 is involved in autophagy induced by DNA damage. Autophagy **5**, 61–74 (2009)
M. Murakami, T. Ichisaka, M. Maeda, N. Oshiro, K. Hara, F. Edenhofer, H. Kiyama, K. Yonezawa, S. Yamanaka, mTOR is essential for growth and proliferation in early mouse embryos and embryonic stem cells. Mol. Cell. Biol. **24**, 6710–6718 (2004)
I. Muro, K. Monser, R.J. Clem, Mechanism of Dronc activation in *Drosophila* cells. J. Cell Sci. **117**, 5035–5041 (2004)
J.M. Murphy, P.E. Czabotar, J.M. Hildebrand et al., The pseudokinase MLKL mediates necroptosis via a molecular switch mechanism. Immunity **39**, 443–453 (2013)
R. Nagata, T. Igaki, Cell competition: emerging mechanisms to eliminate neighbors. Dev. Growth Differ. **60**(9), 522–530 (2018)
U. Nair, A. Jotwani, J. Geng et al., SNARE proteins are required for macroautophagy. Cell **146**, 290–302 (2011)
T. Nakagawa, H. Zhu, N. Morishima, E. Li, J. Xu, B.A. Yankner, J. Yuan, Caspase-12 mediates endoplasmic-reticulum-specific apoptosis and cytotoxicity by amyloid-β. Nature **403**, 98–103 (2000)

Y. Nakanishi, P.M. Henson, A. Shiratsuchi, Pattern recognition in phagocytic clearance of altered self. Adv. Exp. Med. Biol. **653**, 129–138 (2009)

L. Nakopoulou, P. Alexandrou, K. Stefanaki, E. Panayotopoulou, A.C. Lazaris, P.S. Davaris, Immunohistochemical expression of caspase-3 as an adverse indicator of the clinical outcome in human breast cancer. Pathobiology **69**(5), 266–273 (2001)

C.E. Nesbit, J.M. Tersak, E.V. Prochownik, MYC oncogenes and human neoplastic disease. Oncogene **18**, 3004–3016 (1999)

D.H. Nguyen, H.A. Oketch-Rabah, I. Illa-Bochaca et al., Radiation acts on the microenvironment to affect breast carcinogenesis by distinct mechanisms that decrease cancer latency and affect tumor type. Cancer Cell **19**(5), 640–651 (2011)

H.T. Nia, H. Liu, G. Seano, et al., Solid stress and elastic energy as measures of tumour mechanopathology. Nat. Biomed. Eng. **1**, 0004 (2016)

H.T. Nia, M. Datta, G. Seano, P. Huang, L.L. Munn, R.K. Jain, Quantifying solidstress and elastic energy from excised or in situ tumors. Nat. Protoc. **13**, 1091–1105 (2018)

P. Nicotera, G. Melino, Regulation of the apoptosis-necrosis switch. Oncogene **23**(16), 2757–2765 (2004)

P. Niethammer, C. Grabher, A.T. Look, T.J. Mitchison, A tissue-scale gradient of hydrogen peroxide mediates rapid wound detection in *Zebrafish*. Nature **459**, 996–999 (2009)

V. Nikoletopoulou, M. Markaki, K. Palikaras, N. Tavernarakis, Crosstalk between apoptosis, necrosis and autophagy. Biochim. Biophys. Acta **1833**, 3448–3459 (2013)

T. Nishina, S. Komazawa-Sakon, S. Yanaka, et al., Interleukin-11 links oxidative stress and compensatory proliferation. Sci. Signal **5**, ra5 (2012)

T.E. North, W. Goessling, C.R. Walkley et al., Prostaglandin E2 regulates vertebrate haematopoietic stem cell homeostasis. Nature **447**, 1007–1011 (2007)

P.C. Nowell, The clonal evolution of tumor cell populations. Science **194**, 23–28 (1976)

I.E. O'Brien, C.P. Reutelingsperger, K.M. Holdaway, Annexin-V and TUNEL use in monitoring the progression of apoptosis in plants. Cytometry **29**, 28–33 (1997)

A. Oberst, D.R. Green, It cuts both ways: reconciling the dual roles of caspase 8 in cell death and survival. Nat. Rev. Mol. Cell Biol. **12**, 757–763 (2011)

E. Oda, R. Ohki, H. Murasawa, J. Nemoto, T. Shibue, T. Yamashita, T. Tokino, T. Taniguchi, N. Tanaka, Noxa, a BH3-only member of the Bcl-2 family and candidate mediator of p53-induced apoptosis. Science **288**, 1053–1058 (2000)

C. Odaka, T. Mizuochi, Role of macrophage lysosomal enzymes in the degradation of nucleosomes of apoptotic cells. J. Immunol. **163**, 5346–5352 (1999)

M.I. Oerlemans, J. Liu, F. Arslan, K. den Ouden, B.J. van Middelaar, P.A. Doevendans, J.P. Sluijter, Inhibition of RIP1-dependent necrosis prevents adverse cardiac remodeling after myocardial ischemia–reperfusion in vivo. Basic. Res. Cardiol. **107**, 270 (2012)

D. Ofengeim, Y. Ito, A. Najafov et al., Activation of necroptosis in multiple sclerosis. Cell Rep. **10**, 1836–1849 (2015)

J. Ogasawara, R. Watanabe-Fukunaga, M. Adachi, A. Matsuzawa, T. Kasugai, Y. Kitamura, N. Itoh, T. Suda, S. Nagata, Lethal effect of the anti-Fas antibody in mice. Nature **364**, 806–809 (1993)

A. Ohoka, M. Kajita, J. Ikenouchi, Y. Yako, S. Kitamoto, S. Kon, M. Ikegawa, T. Shimada, S. Ishikawa, Y. Fujita, EPLIN is a crucial regulator for extrusion of RasV12-transformed cells. J. Cell. Sci. **128**, 781–789 (2015)

Y. Ohsumi, Ubiquitin and proteasomes: molecular dissection of autophagy: two ubiquitin-like systems. Nat. Rev. Mol. Cell Biol. **2**, 211–216 (2001)

H. Okada, T.W. Mak, Pathways of apoptotic and non-apoptotic death in tumour cells. Nat. Rev. Cancer **4**(8), 592–603 (2004)

T. Oltersdorf, S.W. Elmore, A.R. Shoemaker et al., An inhibitor of Bcl-2 family proteins induces regression of solid tumours. Nature **435**, 677–681 (2005)

M.H. Orme, G. Liccardi, N. Moderau et al., The unconventional myosin CRINKLED and its mammalian orthologue MYO7A regulate caspases in their signalling roles. Nat. Commun. **7**, 10972 (2016)

S. Orozco, N. Yatim, M.R. Werner, H. Tran, S.Y. Gunja, S.W. Tait, M.L. Albert, D.R. Green, A. Oberst, RIPK1 both positively and negatively regulates RIPK3 oligomerization and necroptosis. Cell Death Differ. **21**, 1511–1521 (2014)

W.-B. Ou, J. Zhu, G. Eilers et al., HDACi inhibits liposarcoma via targeting of the MDM2-p53 signaling axis and PTEN, irrespective of p53 mutational status. Oncotarget **6**, 10510–10520 (2015)

M. Overholtzer, A.A. Mailleux, G. Mouneimne, G. Normand, S.J. Schnitt, R.W. King, E.S. Cibas, J.S. Brugge, A nonapoptotic cell death process, entosis, thatoccurs by cell-in-cell invasion. Cell **131**, 966–979 (2007)

N. Özören, W.S. El-Deiry, Defining characteristics of types I and II apoptotic cells in response to TRAIL. Neoplasia **4**, 551–557 (2002)

T.P. Padera, B.R. Stoll, J.B. Tooredman, D. Capen, E. di Tomaso, R.K. Jain, Pathology: cancer cells compress intratumour vessels. Nature **427**, 695 (2004)

N.R.D. Paling, H. Wheadon, H.K. Bone, M.J. Welham, Regulation of embryonic stem cell self-renewal by phosphoinositide 3-kinase-dependent signaling. J. Biol. Chem. **279**, 48063–48070 (2004)

Y. Pan, I. Heemskerk, C. Ibar, B.I. Shraiman, K.D. Irvine, Differential growthtriggers mechanical feedback that elevates Hippo signaling. Proc. Natl. Acad. Sci. U. S. A. **113**(45), E6974–E6983 (2016)

Y. Pan, Alegot, H.C. Rauskolb, K.D. Irvine, The dynamics of hippo signaling during *Drosophila* wing development. Development **145**(20), dev165712 (2018)

T. Panciera, L. Azzolin, M. Cordenonsi, S. Piccolo, Mechanobiology of YAP and TAZ in physiology and disease. Nat. Rev. Mol. Cell Biol. **18**, 758–770 (2017)

C. Park, E.R. Woo, D.G. Lee, Antifungal effect with apoptotic mechanism(s) of Styraxjaponoside C. Biochem. Biophys. Res. Commun. **390**, 1255–1259 (2009)

K.K. Patel, T.S. Stappenbeck, Autophagy and intestinal homeostasis. Annu. Rev. Physiol. **75**, 241–262 (2013)

A. Pathil, S. Armeanu, S. Venturelli, P. Mascagni, T.S. Weiss, M. Gregor, U.M. Lauer, M. Bitzer, HDAC inhibitor treatment of hepatoma cells induces both TRAIL-independent apoptosis and restoration of sensitivity to TRAIL. Hepatology **43**, 425–434 (2006)

X.-D. Peng, P.-Z. Xu, M.-L. Chen et al., Dwarfism, impaired skin development, skeletal muscle atrophy, delayed bone development, and impeded adipogenesis in mice lacking Akt1 and Akt2. Genes Dev. **17**, 1352–1365 (2003)

A. Perez-Garijo, F.A. Martin, G. Morata, Caspase inhibition during apoptosis causes abnormal signalling and developmental aberrations in *Drosophila*. Development **131**, 5591–5598 (2004)

A. Perez-Garijo, F.A. Martin, G. Struhl, G. Morata, Dpp signaling and the induction of neoplastic tumors by caspase-inhibited apoptotic cells in *Drosophila*. Proc. Natl. Acad. Sci. U. S. A. **102**, 17664–17669 (2005)

A. Perez-Garijo, E. Shlevkov, G. Morata, The role of Dpp and Wg in compensatory proliferation and in the formation of hyperplastic overgrowths caused by apoptotic cells in the *Drosophila* wing disc. Development **136**, 1169–1177 (2009)

A. Perez-Garijo, Y. Fuchs, H. Steller, Apoptotic cells can induce non-autonomous apoptosis through the TNF pathway. Elife **2**, e01004 (2013)

P. Peri, K. Nuutila, T. Vuorinen, P. Saukko, V. Hukkanen, Cathepsins are involved in virus-induced cell death in ICP4 and Us3 deletion mutant herps simplex virus type 1-infected monocytic cells. J. Gen. Virol. **92**, 173–180 (2011)

L.J. Peters, H.R. Withers, Applying radiobiological principles to combined modality treatment of head and neck cancer–the time factor. Int. J. Radiat. Oncol. Biol. Phys. **39**(4), 831–836 (1997)

G. Pietra, R. Mortarini, G. Parmiani, A. Anichini, Phases of apoptosis of melanoma cells, but not of normal melanocytes, differently affect maturation of myeloid dendritic cells. Cancer Res. **61**, 8218–8226 (2001)

C. Pignanelli, D. Ma, M. Noel et al., Selective targeting of cancer cells by oxidative vulnerabilities with novel curcumin analogs. Sci. Rep. **7**, 1105 (2017)

A. Poehlmann, K. Reissig, A. Just et al., Non-apoptotic function of caspases in a cellular model of hydrogen peroxide-associated colitis. J. Cell Mol. Med. **17**, 901–913 (2013)

A. Polykratis, N. Hermance, M. Zelic, J. Roderick, C. Kim, T.M. Van, T.H. Lee, F.K.M. Chan, M. Pasparakis, A. Kelliher, Cutting edge: RIPK1 kinase inactive mice are viable and protected from TNF-induced necroptosis in vivo. J. Immunol. **193**, 1539–1543 (2014)

I.K. Poon, M.D. Hulett, C.R. Parish, Molecular mechanisms of late apoptotic/necrotic cell clearance. Cell Death Differ. **17**, 381–397 (2010)

C. Popa, M.G. Netea, P.L. Van Riel, J.W. van der Meer, A.F. Stalenhoef, The role of TNF-α in chronic inflammatory conditions, intermediary metabolism, and cardiovascular risk. J. Lipid Res. **48**, 751–762 (2007)

S. Porazinski, J. de Navascues, Y. Yako, W. Hill, M.R. Jones, R. Maddison, Y. Fujita, C. Hogan, EphA2 drives the segregation of ras-transformed epithelial cells from normal neighbors. Curr. Biol. **26**, 3220–3229 (2016)

C.S. Potten, M. Loeffler, Stem cells: attributes, cycles, spirals, pitfalls and un-certainties. Lessons for and from the crypt. Development **110**, 1001–1020 (1990)

D.A. Prober, B.A. Edgar, Interactions between Ras1, dMyc, and dPI3K signaling in the developing *Drosophila* wing. Genes Dev. **16**, 2286–2299 (2002)

C.E. Pullar, R.R. Isseroff, The beta 2-adrenergic receptor activates pro-migratory and pro-proliferative pathways in dermal fibroblasts via divergent mechanisms. J. Cell Sci. **119**, 592–602 (2006)

X. Quan, K. Guo, Y. Wang, L. Huang, B. Chen, Z. Ye, Z. Luo, Mechanical com-pression insults induce nanoscale changes of membrane-skeleton arrangement which could cause apoptosis and necrosis in dorsal root ganglion neurons. Biosci. Biotechnol. Biochem. **78**, 1631–1639 (2014)

M.C. Raff, Social controls on cell survival and cell death. Nature **356**, 397–400 (1992)

L. Raj, T. Ide, A.U. Gurkar et al., Selective killing of cancer cells by a small molecule targeting the stress response to ROS. Nature **475**, 231–234 (2011)

M. Ramet, R. Lanot, D. Zachary, P. Manfruelli, JNK signaling pathway is required for efficient wound healing in *Drosophila*. Dev. Biol. **241**, 145–156 (2002)

B. Ravikumar, R. Duden, D.C. Rubinsztein, Aggregate-prone proteins with polyglutamine and polyalanine expansions are degraded by autophagy. Hum. Mol. Genet. **11**, 1107–1117 (2002)

B. Ravikumar, C. Vacher, Z. Berger et al., Inhibition of mTOR induces autophagy and reduces toxicity of polyglutamine expansions in fly and mouse models of Huntington disease. Nat. Genet. **36**, 585–595 (2004)

B. Ravikumar, Z. Berger, C. Vacher, C.J. O'kane, D.C. Rubinsztein, Rapamycin pre-treatment protects against apoptosis. Hum. Mol. Genet. **15**, 1209–1216 (2006)

W. Razzell, I.R. Evans, P. Martin, W. Wood, Calcium flashes orchestrate the wound inflammatory response through DUOX activation and hydrogen peroxide release. Curr. Biol. **23**, 424–429 (2013)

W.W. Reiley, W. Jin, A.J. Lee, A. Wright, X. Wu, E.F. Tewalt, T.O. Leonard, C.C. Norbury, L. Fitzpatrick, M. Zhang, S.C. Sun, Deubiquitinating enzyme CYLD negatively regulates the ubiquitin-dependent kinase Tak1 and prevents abnormal T cell responses. J. Exp. Med. **204**, 1475–1485 (2007)

S. Renolleau, S. Fau, C. Goyenvalle, L.M. Joly, D. Chauvier, E. Jacotot, J. Mariani, C. Charriaut-Marlangue, Specific caspase inhibitor Q-VD-OPh prevents neonatal stroke in P7 rat: a role for gender. J. Neurochem. **100**, 1062–1071 (2007)

L. Revesz, Effect of tumour cells killed by x-rays upon the growth of admixed viable cells. Nature **178**(4547), 1391–1392 (1956)

C. Rhiner, J.M. Lopez-Gay, D. Soldini, S. Casas-Tinto, F.A. Martin, L. Lombardia, E. Moreno, Flower forms an extracellular code that reveals the fitness of a cell toits neighbors in *Drosophila*. Dev. Cell **18**, 985–998 (2010)

C. Rianna, M. Radmacher, Comparison of viscoelastic properties of cancer andnormal thyroid cells on different stiffness substrates. Eur. Biophys. J. **46**, 309–324 (2017)

J.-E. Ricci, C. Muñoz-Pinedo, P. Fitzgerald, B. Bailly-Maitre, G.A. Perkins, N. Yadava, I.E. Scheffler, M.H. Ellisman, D.R. Green, Disruption of mitochondrial function during apoptosis is mediated by caspase cleavage of the p75 subunit of complex I of the electron transport chain. Cell **117**, 773–786 (2004)

T. Rich, R.L. Allen, A.H. Wyllie, Defying death after DNA damage. Nature **407**(6805), 777–783 (2000)

L.D. Rios-Barrera, J.R. Riesgo-Escovar, Regulating cell morphogenesis: the *Drosophila* jun N-terminal kinase pathway. Genesis **51**, 147–162 (2013)

G. Robert, C. Gastaldi, A. Puissant et al., The anti-apoptotic Bcl-B protein inhibits BECN1-dependent autophagic cell death. Autophagy **8**, 637–649 (2012)

A.M. Robertson, C.C. Bird, A.W. Waddell, A.R. Currie, Morphological aspects of glucocorticoid-induced cell death in human lymphoblastoid cells. J. Pathol. **126**, 181–187 (1978)

T.T. Rohn, The role of caspases in Alzheimer's disease; potential novel therapeutic opportunities. Apoptosis **15**, 1403–1409 (2010)

T.T. Rohn, E. Wirawan, R.J. Brown, J.R. Harris, E. Masliah, P. Vandenabeele, Depletion of Beclin-1 due to proteolytic cleavage by caspases in the Alzheimer's disease brain. Neurobiol. Dis. **43**, 68–78 (2011)

M. Romano, F. De Francesco, L. Zarantonello, C. Ruffolo, G.A. Ferraro, G. Zanus, A. Giordano, N. Bassi, U. Cillo, From inflammation to cancer in inflammatory bowel disease: molecular perspectives. Anticancer Res. **36**, 1447–1460 (2016)

J. Rosenblatt, M.C. Raff, L.P. Cramer, An epithelial cell destined for apoptosis signals its neighbors to extrude it by an actin- and myosin-dependent mechanism. Curr. Biol. **11**, 1847–1857 (2001)

P. Rovere, C. Vallinoto, A. Bondanza, M.C. Crosti, M. Rescigno, P. Ricciardi-Castagnoli, C. Rugarli, A.A. Manfredi, Bystander apoptosis triggers dendritic cell maturation and antigen-presenting function. J. Immunol. **161**, 4467–4471 (1998)

V.A. Rudrapatna, E. Bangi, R.L. Cagan, Caspase signalling in the absence of apoptosis drives Jnk-dependent invasion. EMBO Rep. **14**, 172–177 (2013)

A. Rufini, P. Tucci, I. Celardo, G. Melino, Senescence and aging: the critical roles of p53. Oncogene **32**, 5129–5143 (2013)

H.D. Ryoo, A. Bergmann, The role of apoptosis-induced proliferation for regeneration and cancer. Cold Spring Harb. Perspect. Biol. **4**, a008797 (2012)

H.D. Ryoo, T. Gorenc, H. Steller, Apoptotic cells can induce compensatory cell proliferation through the JNK and the wingless signaling pathways. Dev. Cell **7**, 491–501 (2004)

S.W. Ryter, H.P. Kim, A. Hoetzel, J.W. Park, K. Nakahira, X. Wang, A.M. Choi, Mechanisms of cell death in oxidative stress. Antioxid. Redox Signal **9**, 49–89 (2007)

X. Saelens, N. Festjens, L. Vande Walle, M. van Gurp, G. van Loo, P. Vandenabeele, Toxic proteins released from mitochondria in cell death. Oncogene **23**(16), 2861–2874 (2004)

A.R. Safa, K.E. Pollok, Targeting the anti-apoptotic protein c-FLIP for cancer therapy. Cancers **3**, 1639–1671 (2011)

A.R. Safa, T.W. Day, C.-H. Wu, Cellular FLICE-like inhibitory protein (C-FLIP): a novel target for cancer therapy. Curr. Cancer Drug. Targets **8**, 37–46 (2008)

T. Sakurai, S. Maeda, L. Chang, M. Karin, Loss of hepatic NF-kappa B activity enhances chemical hepatocarcinogenesis through sustained c-Jun N-terminal kinase 1 activation. Proc. Natl. Acad. Sci. U. S. A. **103**(28), 10544–10551 (2006)

G.S. Salvesen, M. Renatus, Apoptosome: the seven-spoked death machine. Dev. Cell **2**, 256–257 (2002)

A. Sanchez Alvarado, Planarian regeneration: its end is its beginning. Cell **124**(2), 241–245 (2006)

E.P. Sandgren, R.D. Palmiter, J.L. Heckel, C.C. Daugherty, R.L. Brinster, J.L. Degen, Complete hepatic regeneration after somatic deletion of an albumin-plasminogen activator transgene. Cell **66**, 245–256 (1991)

T. Sano, T. Kobayashi, H. Negoro, A. Sengiku, T. Hiratsuka, Y. Kamioka, L.S. Liou, O. Ogawa, M. Matsuda, Intravital imaging of mouse urothelium reveals activationof extracellular signal-regulated kinase by stretch-induced intravesical release of ATP. Physiol. Rep. **4**(21), e13033 (2016)

P. Santabarbara-Ruiz, M. Lopez-Santillan, I. Martinez-Rodriguez, A. Binagui-Casas, L. Perez, M. Milan, M. Milán, M. Corominas, F. Serras, ROS-induced JNK and p38 signaling is required for unpaired cytokine activation during *Drosophila* regeneration. PLoS Genet. **11**, e1005595 (2015)

K. Sarter, C. Mierke, A. Beer, B. Frey, B.G. Führnrohr, C. Schulze, S. Franz, Sweet clearance: involvement of cell surface glycans in the recognition of apoptotic cells. Autoimmunity **40**(4), 345–348 (2007)

T. Sato, T. Machida, S. Takahashiet et al., Fas-mediated apoptosome formation is dependent on reactive oxygen species derived from mitochondrial permeability transition in Jurkat cells. J. Immunol. **173**, 285–296 (2004)

J. Savill, Apoptosis in resolution of inflammation. J. Leukoc. Biol. **61**, 375–380 (1997)

J. Savill, V. Fadok, Corpse clearance defines the meaning of cell death. Nature **407**, 784–788 (2000)

J.S. Savill, A.H. Wyllie, J.E. Henson, M.J. Walport, P.M. Henson, C. Haslett, Macrophage phagocytosis of aging neutrophils in inflammation. Programmed cell death in the neutrophil leads to its recognition by macrophages. J. Clin. Invest. **83**, 865–875 (1989)

T.B. Saw, A. Doostmohammadi, V. Nier, L. Kocgozlu, S. Thampi, Y. Toyama, P. Marcq, C.T. Lim, J.M. Yeomans, B. Ladoux, Topological defects in epithelia govern cell death and extrusion. Nature **544**, 212–216 (2017)

C. Scheidereit, IκB kinase complexes: gateways to NF-κB activation and transcription. Oncogene **25**, 6685–6705 (2006)

M. Schieber, N.S. Chandel, ROS function in redox signaling and oxidative stress. Curr. Biol. **24**, R453–R462 (2014)

A.J. Schroit, J.W. Madsen, Y. Tanaka, In vivo recognition and clearance of red blood cells containing phosphatidylserine in their plasma membranes. J. Biol. Chem. **260**, 5131–5138 (1985)

S. Schubbert, K. Shannon, G. Bollag, Hyperactive Ras in developmental disorders and cancer. Nat. Rev. Cancer **7**, 295–308 (2007)

R.F. Schwabe, C.A. Bradham, T. Uehara, E. Hatano, B.L. Bennett, R. Schoonhoven, D.A. Brenner, c Jun N terminal kinase drives cyclin D1 expression and proliferation during liver regeneration. Hepatology **37**(4), 824–832 (2003)

E.W. Scott, M.C. Simon, J. Anastasi, H. Singh, Requirement of transcription factor PU.1 in the development of multiple hematopoietic lineages. Science **265**, 1573–1577 (1994)

R.S. Scott, E.J. McMahon, S.M. Pop, E.A. Reap, R. Caricchio, P.L. Cohen, H.S. Earp, G.K. Matsushima, Phagocytosis and clearance of apoptotic cells is mediated by MER. Nature **411**, 207–211 (2001)

G. Seano, H.T. Nia, K.E. Emblem et al., Solid stress in brain tumours causes neuronal loss and neurological dysfunction and can be reversed by lithium. Nat. Biomed. Eng. **3**(3), 230–245 (2019)

K.J. Seelig, L. Revesz, Effect of lethally damaged tumour cells upon the growth of admixed viable cells in diffusion chambers. Br. J. Cancer **14**, 126–138 (1960)

I. Segura, A. Serrano, G.G. De Buitrago, M.A. González, J.L. Abad, C. Clavería, L. Gómez, A. Bernad, C. Martínez-A, H.H. Riese, Inhibition of programmed cell death impairs in vitro vascular-like structure formation and reduces in vivo angiogenesis. FASEB J. **16**(8), 833–841 (2002)

J. Senft, B. Helfer, S.M. Frisch, Caspase-8 interacts with the p85 subunit of phosphatidylinositol 3-kinase to regulate cell adhesion and motility. Cancer Res. **67**(24), 11505–11509 (2007)

N. Senoo-Matsuda, L.A. Johnston, Soluble factors mediate competitive and co-operative interactions between cells expressing different levels of *Drosophila* myc. Proc. Natl. Acad. Sci. U. S. A. **104**, 18543–18548 (2007)

S.Y. Seong, P. Matzinger, Hydrophobicity: an ancient damage-associated molecular pattern that initiates innate immune responses. Nat. Rev. Immunol. **4**, 469–478 (2004)

C.N. Serhan, J. Savill, Resolution of inflammation: the beginning programs the end. Nat. Immunol. **6**, 1191–1197 (2005)

T. Seya, H. Shime, H. Takaki, M. Azuma, H. Oshiumi, M. Matsumoto, TLR3/TICAM-1 signaling in tumor cell RIP3-dependent necroptosis. Oncoimmunology **1**, 917–923 (2012)

S. Shalini, L. Dorstyn, S. Dawar, S. Kumar, Old, new and emerging functions of caspases. Cell Death Differ. **22**, 526–539 (2015)

J. Shao, C. Jung, C. Liu, H. Sheng, Prostaglandin E2 stimulates the beta-catenin/T cell factor-dependent transcription in colon cancer. J. Biol. Chem. **280**(28), 26565–26572 (2005)

P.J. Shapiro, H.H. Hsu, H. Jung, E.S. Robbins, H.D. Ryoo, Regulation of the *Drosophila* apoptosome through feedback inhibition. Nat. Cell Biol. **10**, 1440–1446 (2008)

J. Shen, C. Dahmann, Extrusion of cells with inappropriate Dpp signaling from *Drosophila* wing disc epithelia. Science **307**, 1789–1790 (2005)

R. Shen, J. Zhang, F. Zhang, Y. Du, W. Liang, L. Xu, X. Du, P. Chen, X. Chen, Clinical characteristics and therapeutic analysis of 51 patients with Marjolin's ulcers. Exp. Ther. Med. **10**, 1364–1374 (2015)

C.J. Sherr, J.D. Weber, The ARF/p53 pathway. Curr. Opin. Genet. Dev. **10**, 94–99 (2000)

Y. Shi, Apoptosome: the cellular engine for the activation of caspase-9. Structure **10**, 285–288 (2002)

K. Shimada, R. Skouta, A. Kaplan, W.S. Yang, M. Hayano, S.J. Dixon, L.M. Brown, C.A. Valenzuela, A.J. Wolpaw, B.R. Stockwell, Global survey of cell death mechanisms reveals metabolic regulation of ferroptosis. Nat. Chem. Biol. **12**, 497–503 (2016)

S. Shimizu, T. Kanaseki, N. Mizushima, T. Mizuta, S. Arakawa-Kobayashi, C.B. Thompson, Y. Tsujimoto, Role of Bcl-2 family proteins in a nonapoptotic programmed cell death dependent on autophagy genes. Nat. Cell Biol. **6**, 1221–1228 (2004)

T. Shintani, N. Mizushima, Y. Ogawa, A. Matsuura, T. Noda, Y. Ohsumi, Apg10p, a novel protein-conjugating enzyme essential for autophagy in yeast. EMBO J. **18**, 5234–5241 (1999)

E. Shlevkov, G. Morata, A dp53/JNK-dependant feedback amplification loop is essential for the apoptotic response to stress in *Drosophila*. Cell Death Differ. **19**, 451–460 (2012)

B.I. Shraiman, Mechanical feedback as a possible regulator of tissue growth. Proc. Natl. Acad. Sci. U. S. A. **102**, 3318–3323 (2005)

J. Silke, R. Brink, Regulation of TNFRSF and innate immune signalling complexes by TRAFs and cIAPs. Cell Death Differ. **17**, 35–45 (2010)

M.T. Silva, Bacteria-induced phagocyte secondary necrosis as a pathogenicity mechanism. J. Leukoc. Biol. **88**(5), 885–896 (2010a)

M.T. Silva, Secondary necrosis: the natural outcome of the complete apoptotic program. FEBS Lett. **584**(2010), 4491–4499 (2010b)

M.T. Silva, When two is better than one: macrophages and neutrophils work in concert in innate immunity as complementary and cooperative partners of a myeloid phagocyte system. J. Leukoc. Biol. **87**, 93–106 (2010c)

M.T. Silva, A. do Vale, N.M. dos Santos, Secondary necrosis in multicellular animals: an outcome of apoptosis with pathogenic implications. Apoptosis **13**, 463–482 (2008)

P. Simpson, Parameters of cell competition in the compartments of the wing discof *Drosophila*. Dev. Biol. **69**, 182–193 (1979)

J.L. Sims, S.J. Berger, N.A. Berger, Poly (ADP-ribose) polymerase inhibitors preserve oxidized nicotinamide adenine dinucleotide and adenosine 5'-triphosphate pools in DNA-damaged cells: mechanism of stimulation of unscheduled DNA synthesis. Biochemistry **22**, 5188–5194 (1983)

A.M. Singh, D. Reynolds, T. Cliff, S. Ohtsuka, A.L. Mattheyse, Y. Sun, L. Menendez, M. Kulik, S. Dalton, Signaling network crosstalk in human pluripotent cells: a Smad2/3-regulated switch that controls the balance between self-renewal and differentiation. Cell Stem Cell **10**, 312–326 (2012)

B. Sitohy, J.A. Nagy, H.F. Dvorak, Anti-VEGF/VEGFR therapy for cancer: reassessing the target. Cancer Res. **72**, 1909–1914 (2012)

J.M. Slack, S. Savage, Regeneration of mirror symmetrical limbs in the axolotl. Cell **14**, 1–8 (1978)

C.C. Smith, S.M. Davidson, S.Y. Lim, J.C. Simpkin, J.S. Hothersall, D.M. Yellon, Necrostatin: a potentially novel cardioprotective agent? Cardiovasc. Drug Ther. **21**, 227–233 (2007)

R. Smith-Bolton, *Drosophila* imaginal discs as a model of epithelial wound repair and regeneration. Adv. Wound Care (New Rochelle) **5**, 251–261 (2016)

R.K. Smith-Bolton, M.I. Worley, H. Kanda, I.K. Hariharan, Regenerative growth in *Drosophila* imaginal discs is regulated by Wingless and Myc. Dev. Cell **16**, 797–809 (2009)

M.H. Snow, P.P. Tam, Is compensatory growth a complicating factor in mouse teratology? Nature **279**(5713), 555–557 (1979)

C. Soldani, A.I. Scovassi, Poly (ADP-ribose) polymerase cleavage during apoptosis: when and where? Exp. Cell Res. **269**, 193–201 (2001)

J. Sotelo, E. Briceno, M.A. López-González, Adding chloroquine to conventional treatment for glioblastoma multiforme: a randomized, double-blind, placebo-controlled trial. Ann. Intern. Med. **144**, 337–343 (2006)

R.A. Stockley, Neutrophilic inflammation: "don't you go to pieces on me!". Eur. Respir. J. **28**, 257–258 (2006)

S.R. Stowell, S. Karmakar, C.M. Arthur, T. Ju, L.C. Rodrigues, T.B. Riul, M. Dias-Baruffi, J. Miner, R.P. McEver, R. D. Cummings, Galectin-1 induces reversible phosphatidylserine exposure at the plasma membrane. Mol. Biol. Cell **20**, 1408–1418 (2009)

S.J. Streichan, C.R. Hoerner, T. Schneidt, D. Holzer, L. Hufnagel, Spatial con-straints control cell proliferation in tissues. Proc. Natl. Acad. Sci. U. S. A. **111**, 5586–5591 (2014)

T. Stylianopoulos, J.D. Martin, V.P. Chauhan, et al., Causes, consequences, and remedies for growth-induced solid stress in murine and humantumors. Proc. Natl. Acad. Sci. U. S. A. **109**, 15101–15108 (2012)

S.J. Suijkerbuijk, G. Kolahgar, I. Kucinski, E. Piddini, Cell competition drives thegrowth of intestinal adenomas in *Drosophila*. Curr. Biol. **26**, 428–438 (2016)

Y. Suissa, O. Ziv, T. Dinur, E. Arama, O. Gerlitz, The NAB-Brk signal bifurcates at JNK to independently induce apoptosis and compensatory proliferation. J. Biol. Chem. **286**, 15556–15564 (2011)

Q. Sun, T. Luo, Y. Ren, O. Florey, S. Shirasawa, T. Sasazuki, D.N. Robinson, M. Overholtzer, Competition between human cells by entosis. Cell Res. **24**, 1299–1310 (2014)

R. Suwinski, J.M. Taylor, H.R. Withers, Rapid growth of microscopic rectal cancer as a determinant of response to preoperative radiation therapy. Int. J. Radiat. Oncol. Biol. Phys. **42**(5), 943–951 (1998)

Y. Suzuki, Y. Nakabayashi, R. Takahashi, Ubiquitin-protein ligase activity of X-linked inhibitor of apoptosis protein promotes proteasomal degradation of caspase-3 and enhances its anti-apoptotic effect in Fas-induced cell death. Proc. Natl. Acad. Sci. U. S. A. **98**, 8662–8667 (2001)

C. Szabó, V.L. Dawson, Role of poly (ADP-ribose) synthetase in inflammation and ischaemia–reperfusion. Trends Pharmacol. Sci. **19**, 287–298 (1998)

I. Tabas, Apoptosis and plaque destabilization in atherosclerosis: the role of macrophage apoptosis induced by cholesterol. Cell Death Differ. **11**(Suppl. 1), S12–S16 (2004)

D.P. Tabassum, K. Polyak, Tumorigenesis: it takes a village. Nat. Rev. Cancer **15**, 473–483 (2015)

K.J. Takano, T. Takano, Y. Yamanouchi, T. Satou, Pressure-induced apoptosis inhuman lymphoblasts. Exp. Cell Res. **235**, 155–160 (1997)

O. Takeuchi, S. Akira, Pattern recognition receptors and inflammation. Cell **140**, 805–820 (2010)

Y. Tan, N. Dourdin, C. Wu, T. De Veyra, J.S. Elce, P.A. Greer, Ubiquitous calpains promote caspase-12 and JNK activation during endoplasmic reticulum stress-induced apoptosis. J. Biol. Chem. **281**, 16016–16024 (2006)

N. Tan, M. Malek, J. Zha, P. Yue, R. Kassees, L. Berry, W.J. Fairbrother, D. Sampath, L.D. Belmont, Navitoclax enhances the efficacy of taxanes in non-small cell lung cancer models. Clin. Cancer Res. **17**, 1394–1404 (2011)

R. Taub, Liver regeneration: from myth to mechanism. Nat. Rev. Mol. Cell Biol. **5**, 836–847 (2004)

R.C. Taylor, S.P. Cullen, S.J. Martin, Apoptosis: controlled demolition at the cellular level. Nat. Rev. Mol. Cell Biol. **9**, 231–241 (2008)

A. Tesniere, T. Panaretakis, O. Kepp, L. Apetoh, F. Ghiringhelli, L. Zitvogel, G. Kroemer, Molecular characteristics of immunogenic cancer cell death. Cell Death Differ. **15**, 3–12 (2008)

W.R. Thompson, C.T. Rubin, J. Rubin, Mechanical regulation of signaling path-ways in bone. Gene **503**, 179–193 (2012)

L.M. Thorpe, H. Yuzugullu, J.J. Zhao, PI3K in cancer: divergent roles of isoforms, modes of activation and therapeutic targeting. Nat. Rev. Cancer **15**, 7–24 (2015)

T. Tissot, F. Thomas, B. Roche, Non-cell-autonomous effects yield lower clonal diversity in expanding tumors. Sci. Rep. **7**, 11157 (2017)

D. Trachootham, J. Alexandre, P. Huang, Targeting cancer cells by ROS-mediated mechanisms: a radical therapeutic approach? Nat. Rev. Drug. Discov. **8**, 579–591 (2009)

K.R. Trott, Cell repopulation and overall treatment time. Int. J. Radiat. Oncol. Biol. Phys. **19**(4), 1071–1075 (1990)

B. Trump, P. Goldblatt, R. Stowell, Studies of necrosis in vitro of mouse hepatic parenchymal cells. Ultrastructural alterations in endoplasmic reticulum, Golgi apparatus, plasma membrane, and lipid droplets. Lab. Invest. **14**, 2000–2028 (1965)

C. Tse, A.R. Shoemaker, J. Adickes et al., ABT-263: a potent and orally bioavailable Bcl-2 family inhibitor. Cancer Res. **68**, 3421–3428 (2008)

A.S. Tseng, D.S. Adams, D. Qiu, P. Koustubhan, M. Levin, Apoptosis is required during early stages of tail regeneration in *Xenopus laevis*. Dev. Biol. **301**, 62–69 (2007)

A. Tsuboi, S. Ohsawa, D. Umetsun, Y. Sando, E. Kuranaga, T. Igaki, K. Fujimoto, Competition for space is controlled by apoptosis-induced change of local epithelial topology. Curr. Biol. **28**(2115–2128), e2115 (2018)

Y. Tsujimoto, J. Cossman, E. Jaffe, C.M. Croce, Involvement of the bcl-2 gene in human follicular lymphoma. Science **228**, 1440–1443 (1985)

M. Tubiana, The kinetics of tumour cell proliferation and radiotherapy. Br. J. Radiol. **44**, 325–347 (1971)

M. Tubiana, Repopulation in human tumors. A biological background for fractionation in radiotherapy. Acta Oncol. **27**, 83–88 (1988)

H. Turmel, A. Hartmann, K. Parain, A. Douhou, A. Srinivasan, Y. Agid, E.C. Hirsch, Caspase-3 activation in 1-methyl-4-phenyl-1, 2, 3, 6-tetrahydropyridine (MPTP)-treated mice. Mov. Disord. **16**, 185–189 (2001)

L. Uller, C.G. Persson, J.S. Erjefalt, Resolution of airway disease: removal of inflammatory cells through apoptosis, egression or both? Trends Pharmacol. Sci. **27**, 461–466 (2006)

I. Unal-Cevik, M. Kilinc, A. Can, Y. Gursoy-Ozdemir, T. Dalkara, Apoptotic and necrotic death mechanisms are concomitantly activated in the same cell after cerebral ischemia. Stroke **35**, 2189–2194 (2004)

J.W. Upton, W.J. Kaiser, E.S. Mocarski, DAI/ZBP1/DLM-1 complexes with RIP3 to mediate virus-induced programmed necrosis that is targeted by murine cytomegalovirus vIRA. Cell Host. Microbe **11**, 290–297 (2012)

L. Valon, R. Levayer, Dying under pressure: cellular characterisation and in vivo functions of cell death induced by compaction. Biol. Cell **111**, 51–66 (2019)

G.G. Van den Eynden, N.C. Bird, A.W. Majeed, S. Van Laere, L.Y. Dirix, P.B. Vermeulen, The histological growth pattern of colorectal cancer liver me-tastases has prognostic value. Clin. Exp. Metastasis **29**, 541–549 (2012)

P. Vandenabeele, L. Galluzzi, T.V. Berghe, G. Kroemer, Molecular mechanisms of necroptosis: an ordered cellular explosion. Nat. Rev. Mol. Cell Biol. **11**, 700–714 (2010)

R.W. Vandivier, V.A. Fadok, P.R. Hoffmann, D.L. Bratton, C. Penvari, K.K. Brown, J.D. Brain, F.J. Accurso, P.M. Henson, Elastase-mediated phosphatidylserine receptor cleavage impairs apoptotic cell clearance in cystic fibrosis and bronchiectasis. J. Clin. Invest. **109**, 661–670 (2002)

B. Vanhaesebroeck, J. Guillermet-Guibert, M. Graupera, B. Bilanges, The emerging mechanisms of isoform-specific PI3K signaling. Nat. Rev. Mol. Cell Biol. **11**, 329–341 (2010)

D. Vercammen, G. Brouckaert, G. Denecker, M. Van de Craen, W. Declercq, W. Fiers, P. Vandenabeele, Dual signaling of the Fas receptor: initiation of both apoptotic and necrotic cell death pathways. J. Exp. Med. **188**, 919–930 (1998a)

D. Vercammen, R. Beyaert, G. Denecker, V. Goossens, G. Van Loo, W. Declercq, J. Grooten, W. Fiers, P. Vandenabeele, Inhibition of caspases increases the sensitivity of L929 cells to necrosis mediated by tumor necrosis factor. J. Exp. Med. **187**, 1477–1485 (1998b)

A. Villunger, E.M. Michalak, L. Coultas, F. Mullauer, G. Bock, M.J. Ausserlechner, J.M. Adams, A. Strasser, p53-and drug-induced apoptotic responses mediated by BH3-only proteins puma and noxa. Science **302**, 1036–1038 (2003)

J.P. Vincent, G. Kolahgar, M. Gagliardi, E. Piddini, Steep differences in wingless signaling trigger Myc-independent competitive cell interactions. Dev. Cell **21**, 366–374 (2011)

J.P. Vincent, A.G. Fletcher, L.A. Baena-Lopez, Mechanisms and mechanics of cell competition in epithelia. Nat. Rev. Mol. Cell Biol. **14**, 581–591 (2013)

M. Vinken, E. Decrock, E. De Vuyst, L. Leybaert, T. Vanhaecke, V. Rogiers, Biochemical characterisation of an in vitro model of hepatocellular apoptotic cell death. Altern. Lab. Anim. **37**, 209–218 (2009)

T. Vlaskalin, C.J. Wong, C. Tsilfidis, Growth and apoptosis during larval forelimb development and adult forelimb regeneration in the newt (*Notophthalmus viridescens*). Dev. Genes Evol. **214**, 423–431 (2004)

V. Vogel, M. Sheetz, Local force and geometry sensing regulate cell functions. Nat. Rev. Mol. Cell Biol. **7**, 265–275 (2006)

M. Vogler, M. Butterworth, A. Majid, R.J. Walewska, X.M. Sun, M.J. Dyer, G.M. Cohen, Concurrent upregulation of BCL-XL and BCL2A1 induces approximately 1000-fold resistance to ABT-737 in chronic lymphocytic leukemia. Blood **113**, 4403–4413 (2009)

J.A. Voynow, B.M. Fischer, S. Zheng, Proteases and cystic fibrosis. Int. J. Biochem. Cell Biol. **40**, 1238–1245 (2008)

S. Vriz, S. Reiter, B. Galliot, Cell death: a program to regenerate. Curr. Top. Dev. Biol. **108**, 121–151 (2014)

B. Waclaw, I. Bozic, M.E. Pittman, R.H. Hruban, B. Vogelstein, M.A. Nowak, Aspatial model predicts that dispersal and cell turnover limit intratumour heterogeneity. Nature **525**, 261–264 (2015)

L. Wagstaff, M. Goschorska, K. Kozyrska et al., Mechanical cell competition kills cells via induction of lethal p53 levels. Nat. Commun. **7**, 11373 (2016)

L.C. Walker, H. LeVine, The cerebral proteopathies. Mol. Neurobiol. **21**, 83–95 (2000)

N.I. Walker, R.E. Bennett, J.F. Kerr, Cell death by apoptosis during involution of the lactating breast in mice and rats. Am. J. Anat. **185**, 19–32 (1989)

S. Wang, W.S. El-Deiry, TRAIL and apoptosis induction by TNF-family death receptors. Oncogene **22**, 8628–8633 (2003)

J.Z. Wang, X.A. Li, Impact of tumor repopulation on radiotherapy planning. Int. J. Radiat. Oncol. Biol. Phys. **61**(1), 220–227 (2005)

H.B. Wang, M. Dembo, Y.L. Wang, Substrate flexibility regulates growth andapoptosis of normal but not transformed cells. Am. J. Physiol. Cell Physiol. **279**, C1345-1350 (2000)

X. Wang, S.W. Ryter, C. Dai, Z.L. Tang, S.C. Watkins, X.M. Yin, R. Song, A.M. Choi, Necrotic cell death in response to oxidant stress involves the activation of the apoptogenic caspase-8/bid pathway. J. Biol. Chem. **278**, 29184–29191 (2003)

L. Wang, F. Du, X. Wang, TNF-α induces two distinct caspase-8 activation pathways. Cell **133**, 693–703 (2008a)

L. Wang, N. Azad, L. Kongkaneramit, F. Chen, Y. Lu, B.-H. Jiang, Y. Yon Rojanasaku, The Fas death signaling pathway connecting reactive oxygen species generation and FLICE inhibitory protein down-regulation. J. Immunol. **180**, 3072–3080 (2008b)

K. Wang, R. Liu, J. Li, J. Mao, Y. Lei, J. Wu, J. Zeng, T. Zhang, H. Wu, L. Chen, C. Huang, Y. Wei, Quercetin induces protective autophagy in gastric cancer cells: involvement of Akt-mTOR-and hypoxia-induced factor 1α-mediated signaling. Autophagy **7**, 966–978 (2011)

Z. Wang, H. Jiang, S. Chen, F. Du, X. Wang, The mitochondrial phosphatase PGAM5 functions at the convergence point of multiple necrotic death pathways. Cell **148**, 228–243 (2012)

Y. Wang, N. Tjandra, Structural insights of tBid, the caspase-8-activated Bid, and its BH3 domain. J. Biol. Chem. **288**, 35840–35851 (2013)

X. Wang, Y. Li, S. Liu et al., Direct activation of RIP3/MLKL-dependent necrosis by herpes simplex virus 1 (HSV-1) protein ICP6 triggers host antiviral defense. Proc. Natl. Acad. Sci. U. S. A. **111**, 15438–15443 (2014)

J.F. Ward, DNA damage produced by ionizing radiation in mammalian cells: identities, mechanisms of formation, and reparability. Prog. Nucleic Acid Res. Mol. Biol. **35**, 95–125 (1988)

A.P. Watt, V. Brown, J. Courtney, M. Kelly, L. Garske, J.S. Elborn, M. Ennis, Neutrophil apoptosis, proinflammatory mediators and cell counts in bronchiectasis. Thorax **59**, 231–236 (2004)

A.P. Watt, J. Courtney, J. Moore, M. Ennis, J.S. Elborn, Neutrophil cell death, activation and bacterial infection in cystic fibrosis. Thorax **60**, 659–664 (2005)

X. Wei, B. Shao, Z. He et al., Cationic nanocarriers induce cell necrosis through impairment of Na+/K+-ATPase and cause subsequent inflammatory response. Cell Res. **25**, 237–253 (2015)

R. Weinlich, C.P. Dillon, D.R. Green, Ripped to death. Trends Cell Biol. **21**, 630–637 (2011)

S.J. Weiss, Tissue destruction by neutrophils. N. Engl. J. Med. **320**, 365–376 (1989)

B.S. Wells, E. Yoshida, L.A. Johnston, Compensatory proliferation in *Drosophila* imaginal discs requires Dronc-dependent p53 activity. Curr. Biol. **16**, 1606–1615 (2006)

P.-S. Welz, M. Pasparakis, A way to DAI. Cell Host. Microbe **11**, 223–225 (2012)

S. Werner, H. Smola, X. Liao, M.T. Longaker, T. Krieg, P.H. Hofschneider, L.T. Williams, The function of KGF in morphogenesis of epithelium and reepithelialization of wounds. Science **266**, 819–822 (1994)

E. White, Deconvoluting the context-dependent role for autophagy in cancer. Nat. Rev. Cancer **12**, 401–410 (2012)

K. White, M.E. Grether, J.M. Abrams, L. Young, K. Farrell, H. Steller, Genetic control of programmed cell death in *Drosophila*. Science **264**, 677–683 (1994)

T.J. Widmann, C. Dahmann, Wingless signaling and the control of cell shape in *Drosophila* wing imaginal discs. Dev. Biol. **334**, 161–173 (2009)

R. Wilson, L. Goyal, M. Ditzel, A. Zachariou, D.A. Baker, J. Agapite, H. Steller, P. Meier, The DIAP1 RING finger mediates ubiquitination of Dronc and is indispensable for regulating apoptosis. Nat. Cell Biol. **4**, 445–450 (2002)

E. Wirawan, L. Vande Walle, K. Kersse et al., Caspase-mediated cleavage of Beclin-1 inactivates Beclin-1-induced autophagy and enhances apoptosis by promoting the release of proapoptotic factors from mitochondria. Cell Death Dis. **1**, e18 (2011)

H.R. Withers, Radiation biology and treatment options in radiation oncology. Cancer Res. **59**(7 Suppl), 1676s–1684s (1999)

H.R. Withers, B. Maciejewski, J.M. Taylor, A. Hliniak, Accelerated repopulation in head and neck cancer. Front. Radiat. Ther. Oncol. **22**, 105–110 (1988a)

H.R. Withers, J.M. Taylor, B. Maciejewski, The hazard of accelerated tumor clonogen repopulation during radiotherapy. Acta Oncol. **27**(2), 131–146 (1988b)

A. Wree, C.D. Johnson, J. Font-Burgada, A. Eguchi, D. Povero, M. Karin, A.E. Feldstein, Hepatocyte-specific Bid depletion reduces tumor development by suppressing inflammation-related compensatory proliferation. Cell Death Differ. **22**, 1985–1994 (2015)

X. Wu, C. Molinaro, N. Johnson, C.A. Casiano, Secondary necrosis is a source of proteolytically modified forms of specific intracellular autoantigens: implications for systemic autoimmunity. Arthritis Rheum. **44**, 2642–2652 (2001)

Y.T. Wu, H.L. Tan, Q. Huang, Y.S. Kim, N. Pan, W.Y. Ong, Z.G. Liu, C.N. Ong, H.M. Shen, Autophagy plays a protective role during zVAD-induced necrotic cell death. Autophagy **4**(4), 457–466 (2008)

J. Wu, Z. Huang, J. Ren et al., Mlkl knockout mice demonstrate the indispensable role of Mlkl in necroptosis. Cell Res. **23**, 994–1006 (2013)

X. Wu, Z.H. Yang, X.K. Wang, Y. Zhang, H. Wan, Y. Song, X. Chen, J. Shao, J. Han, Distinct roles of RIP1–RIP3 hetero-and RIP3–RIP3 homo-interaction in mediating necroptosis. Cell Death Differ. **21**, 1709–1720 (2014)

D.W. Wu, C.C. Huang, S.W. Chang, T.H. Chen, H. Lee, Bcl-2 stabilization by paxillin confers 5-fluorouracil resistance in colorectal cancer. Cell Death Differ. **22**(5), 779–789 (2015)

A.H. Wyllie, J.F. Kerr, A.R. Currie, Cell death: the significance of apoptosis. Int. Rev. Cytol. **68**, 251–306 (1980)

M. Xilouri, L. Stefanis, Autophagy in the central nervous system: implications for neurodegenerative disorders. CNS Neurol. Disord. Drug Targets **9**, 701–719 (2010)

D. Xu, Y. Li, M. Arcaro, M. Lackey, A. Bergmann, The CARD-carrying caspase Dronc is essential for most, but not all, developmental cell death in *Drosophila*. Development **132**, 2125–2134 (2005)

X. Xu, C.C. Chua, J. Kong, R.M. Kostrzewa, U. Kumaraguru, R.C. Hamdy, B.H. Chua, Necrostatin-1 protects against glutamate-induced glutathione depletion and caspase-independent cell death in HT-22 cells. J. Neurochem. **103**, 2004–2014 (2007)

D. Xu, S.E. Woodfield, T.V. Lee, Y. Fan, C. Antonio, A. Bergmann, Genetic control of programmed cell death (apoptosis) in *Drosophila*. Fly (Austin) **3**, 78–90 (2009)

X. Xu, K.W. Chua, C.C. Chua, C.F. Liu, R.C. Hamdy, B.H. Chua, Synergistic protective effects of humanin and necrostatin-1 on hypoxia and ischemia/reperfusion injury. Brain Res. **1355**, 189–194 (2010)

J.-C. Xu, J. Fan, X. Wang, et al., Cultured networks of excitatory projection neurons and inhibitory interneurons for studying human cortical neurotoxicity. Sci. Transl. Med. **8**, 333ra348 (2016)

M. Yamamoto, S. Ohsawa, K. Kunimasa, T. Igaki, The ligand Sas and its receptor PTP10D drive tumour-suppressive cell competition. Nature **542**, 246–250 (2017)

L. Yang, Z. Cao, H. Yan, W.C. Wood, Coexistence of high levels of apoptotic signaling and inhibitor of apoptosis proteins in human tumor cells: implication for cancer specific therapy. Cancer Res. **63**(20), 6815–6824 (2003)

J.-C. Yang, M.C. Lu, C.L. Lee, G.Y. Chen, Y.Y. Lin, F.R. Chang, Y.C. Wu, Selective targeting of breast cancer cells through ROS-mediated mechanisms potentiates the lethality of paclitaxel by a novel diterpene, gelomulide K. Free Radic. Biol. Med. **51**, 641–657 (2011)

H. Yang, D.G. Rudge, J.D. Koos, B. Vaidialingam, H.J. Yang, N.P. Pavletich, mTOR kinase structure, mechanism and regulation. Nature **497**, 217–223 (2013a)

S. Yang, B. Wang, L.S. Tang, J. Siednienko, J.J. Callanan, P.N. Moynagh, Pellino3 targets RIP1 and regulates the proapoptotic effects of TNF-α. Nat. Commun. **4**, 2583 (2013b)

L.R. Yates, M. Gerstung, S. Knappskog et al., Subclonal diversification of primary breast cancer revealed by multiregion sequencing. Nat. Med. **21**, 751–759 (2015)

T. Yoneda, K. Imaizumi, K. Oono, D. Yui, F. Gomi, T. Katayama, M. Tohyama, Activation of caspase-12, an endoplastic reticulum (ER) resident caspase, through tumor necrosis factor receptor-associated factor 2-dependent mechanism in response to the ER stress. J. Biol. Chem. **276**, 13935–13940 (2001)

J.S.L. Yu, W. Cui, Proliferation, survival and metabolism: the role of PI3K/AKT/ mTOR signalling in pluripotency and cell fate determination. Development **143**, 3050–3060 (2016)

L. Yu, A. Alva, H. Su, P. Dutt, E. Freundt, S. Welsh, E.H. Baehrecke, M.J. Lenardo, Regulation of an ATG7-beclin 1 program of autophagic cell death by caspase-8. Science **304**(5676), 1500–1502 (2004)

J.S.L. Yu, T.S. Ramasamy, N. Murphy, M.K. Holt, R. Czapiewski, S.-K. Wei, W. Cui, PI3K/mTORC2 regulates TGF-beta/Activin signalling by modulating Smad2/3 activity via linker phosphorylation. Nat. Commun. **6**, 7212 (2015)

S. Yuan, C.W. Akey, Apoptosome structure, assembly, and procaspase activation. Structure **21**, 501–515 (2013)

K.S. Zaret, Genetic programming of liver and pancreas progenitors: lessons for stem-cell differentiation. Nat. Rev. Genet. **9**, 329–340 (2008)

S. Zelenay, A.G. van der Veen, J.P. Bottcher et al., Cyclooxygenase-dependent tumor growth through evasion of immunity. Cell **162**, 1257–1270 (2015)

L. Zhang, H. Wang, K. Ding, J. Xu, FTY720 induces autophagy-related apoptosis and necroptosis in human glioblastoma cells. Toxicol. Lett. **236**, 43–59 (2015a)

Y. Zhang, A. Desai, S.Y. Yang, et al., Tissue regeneration. Inhibition of the prostaglandin-degrading enzyme 15-PGDH potentiates tissue regeneration. Science **348**, aaa2340 (2015b)

Z. Zhang, M. Wang, L. Zhou et al., Increased HMGB1 and cleaved caspase-3 stimulate the proliferation of tumor cells and are correlated with the poor prognosis in colorectal cancer. J. Exp. Clin. Cancer Res. **34**, 51 (2015c)

X. Zhao, D. Wang, Z. Zhao et al., Caspase-3-dependent activation of calcium-independent phospholipase A2 enhances cell migration in non-apoptotic ovarian cancer cells. J. Biol. Chem. **281**(39), 29357–29368 (2006)

B. Zhao, X. Wei, W. Li et al., Inactivation of YAP oncoprotein by the Hippo pathway is involved in cell contact inhibition and tissue growth control. Genes Dev. **21**, 2747–2761 (2007)

Z. Zhou, P.M. Mangahas, X. Yu, The genetics of hiding the corpse: engulfment and degradation of apoptotic cells in *C. elegans* and *D. melanogaster*. Curr. Top. Dev. Biol. **63**, 91–143 (2004)

J. Zhou, P. Su, L. Wang et al., mTOR supports long-term self-renewal and suppresses mesoderm and endoderm activities of human embryonic stem cells. Proc. Natl. Acad. Sci. U. S. A. **106**, 7840–7845 (2009)

S. Zhu, Y. Zhang, G. Bai, H. Li, Necrostatin-1 ameliorates symptoms in R6/2 transgenic mouse model of Huntington's disease. Cell Death Dis. **2**, e115 (2011)

M.A. Zimmerman, Q. Huang, F. Li, X. Liu, C.-Y. Li, Cell death-stimulated cell proliferation: a tissue regeneration mechanism usurped by tumors during radiotherapy. Semin. Radiat. Oncol. **23**(4), 288–295 (2013)

M. Ziosi, L.A. Baena-Lopez, D. Grifoni, F. Froldi, A. Pession, F. Garoia, V. Trotta, P. Bellosta, S. Cavicchi, A. Pession, dMyc functions downstream of Yorkie topromote the supercompetitive behavior of hippo pathway mutant cells. PLoS Genet. **6**(9), e1001140 (2010)

L. Zitvogel, N. Casares, M.O. Pequignot, N. Chaput, M.L. Albert, G. Kroemer, Immune response against dying tumor cells. Adv. Immunol. **84**, 131–179 (2004)

L. Zitvogel, O. Kepp, G. Kroemer, Decoding cell death signals in inflammation and immunity. Cell **140**, 798–804 (2010)

C.E. Zois, M.I. Koukourakis, Radiation-induced autophagy in normal and cancer cells: towards novel cytoprotection and radio-sensitization policies? Autophagy **5**(4), 442–450 (2009)

17 Metabolic Pathways of Eukaryotes and Connection to Cell Mechanics

Abstract

The metabolic pathways play a crucial role, since they impact all cellular functions. Hence, an accurately controlled metabolism represents a pre-requisite for proper heath of organisms. A brief overview on major metabolic processes of the cells is provided. When the metabolism is dysregulated through frequent alterations in genetical programming, aberrant metabolic phases are present and hence contribute to the initiation of diseases, such as cancer. The dysregulation of metabolic pathways is commonly impacted by mechanical cues of the cellular microenvironment. A comprehension of these metabolic disorders seems to be a critical frontier in healthy processes, such as tissue injury-induced healing and disease-related research fields. This chapter discusses the wide-ranged influence of the metabolism in cellular functions, how beyond state-of-art concepts of metabolism can provide a deeper knowledge of common diseases and considers the development of new metabolic approaches for the treatment of specific diseases.

17.1 The Metabolism Impacts on Every Cellular Task

Every cellular function depends on the metabolism. The metabolism can be defined in a simple manner as the sum of all biochemical processes within living organisms that can generate or require energy during their action. It is a terrifyingly high sum: over 8700 reactions and 16,000 metabolites have been determined. Core metabolism can be facilitated in ways that are rich in nutrients such as carbohydrates, fatty acids and amino acids, which are fundamental for energy homeostasis and macromolecular production in humans (Fig. 17.1).

The pathways of core metabolism may then be comfortably classified into three categories: those that produce simple molecules or polymerize them to more complex macromolecules (anabolism), those that break-down molecules with the liberation of energy (catabolism), and those that help remove the toxic discards generated from the other groups (garbage management) (DeBerardinis and Thompson 2012). These paths are of utmost significance. Put bluntly, they are the only available form of energy that enables life to withstand the pressure of disintegrating into entropy.

The definition of these pathways and the comprehension of their underlying physiological function are some of the most productive endeavors in biological science. The old former glory days of biochemistry (about 1920–1960s) identified the major part of the metabolic system involved in the utilization of nutrients and the generation of energy in humans and other organisms. These activities

C. T. Mierke, *Cellular Mechanics and Biophysics*, Biological and Medical Physics, Biomedical Engineering, https://doi.org/10.1007/978-3-030-58532-7_17

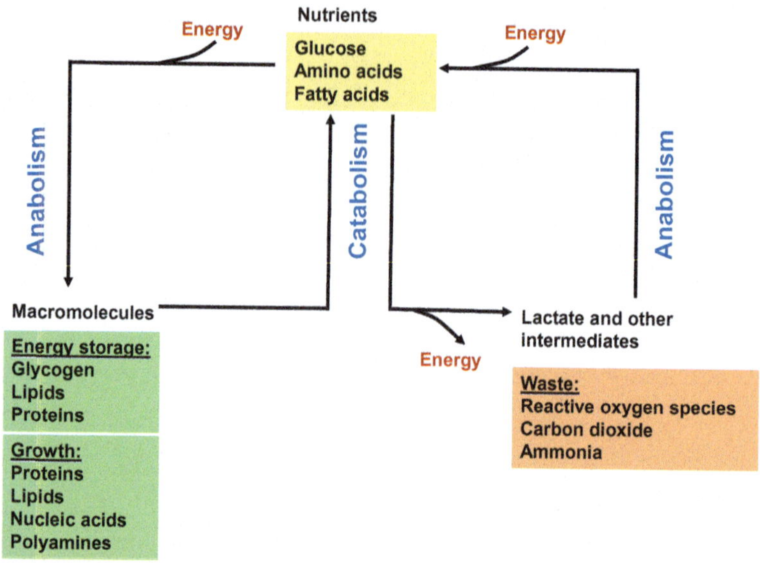

Fig. 17.1 Overview over metabolism from an energy-based view

comprised core activities such as glycolysis (Embden, Meyerhof and Parnas), respiration (Warburg), tricarboxylic acid (TCA) cycle, urea cycle (Krebs), glycogen catabolism (Cori and Cori), oxidative phosphorylation (Mitchell) mechanism and the predominance of ATP in energy transmission reactions (Lipmann) (Fig. 17.2).

During these decades, biochemistry and the investigation of metabolic pathways prevailed in fundamental and medically focused research, with about fifteen Nobel Prizes in either physiology/medicine or chemistry being granted for research on energy management or on core metabolic pathways. At the end of this time period, it was finally achieved to comprehend on the scale of enzymatic regulation, such complex problems as the timing and organ-specific adjustment of the preferences for the use of fuels (Krebs 1972).

Metabolic research has been advanced due to the recognition that metabolic disturbances are associated with common human diseases. This insight anticipates the more formal investigation of the metabolism for many centuries. Almost 2000 years ago, Celsus discovered that rich food and drink triggered attacks of arthritis, and Indian doctors discovered that the urine of diabetics tended to attract ants, whereas normal urine did not (Trowell 1982). A wider acceptance of the connection between accurate metabolic activities and states of disease prospered in its heyday, however, the momentum in metabolic exploration began to lose momentum as newer areas of biological investigation emerged in the second half of the twentieth century, probably on the grounds that much of what is known about intermediate metabolism has long before been uncovered. The exploration of the genetic and molecular foundations of cancer, diabetes, obesity and neurodegeneration has moved the emphasis away from pure understanding of altered metabolic conditions in these diseases. Thus, many frequent diseases are nowadays considered to be hereditary or somatic mutations that affect gene expression, signal transduction, cellular differentiation and other processes which are not conventionally viewed in bioenergetic or metabolic aspects.

The continuing study of cell biology and disease has recently prompted a renewed interest in the metabolism of small molecules (McKnight 2010). Over the last ten years, a variety of functions have emerged for metabolites and metabolic routes that could not be anticipated with a traditional knowledge of biochemistry. Consequently, it is no more feasible to consider the metabolism solely as a self-regulating system that operates separately from any other biological system. Instead,

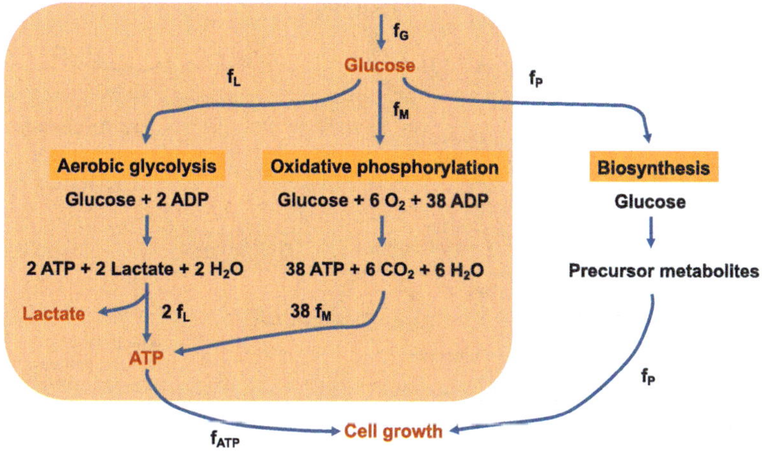

Fig. 17.2 Reduced model of the metabolism of cells. Schematic view on the generation of ATP through aerobic glycolysis (glycolysis and pyruvate reduction to lactate) and oxidative phosphorylation (glycolysis and mitochondrial respiration). The variable f_G represents the uptake rate of glucose, f_L and f_M denote the components of the uptake of glucose, which is delivered towards lactate excretion and respiration and f_{ATP} stand for the production rate of ATP. In a third route the intermediate substrates of the catabolism of glucose are utilized in order to generate precursor metabolites required for anabolic processes (f_P). The orange shaded route resembles the proposed model, whereas the f_P value treated as a constant

metabolism influences or is directly affected from practically every other cellular process; there is no space in biological science that is completely devoid of the impact of metabolism. This is probably not really astonishing, given that fundamental features of energy metabolism have been preserved during the entire evolution, while the higher levels of regulation and the complex organization of multi-cellular organisms arrived far later.

Current investigations have led to the identification of a broad range of regulatory mechanisms that either connect cell signaling with the staging of metabolic pathways or that permit the cells to determine whether the fuel is present and pass on the relevant signals through signaling networks (Fig. 17.3). The incorporation of biochemical routes into the cellular reaction to growth factors is an excellent illustration of such a process. In the majority of mammalian cells, growth only takes place when stimulated through extracellular ligands. These growth factors are capable of initiating signal transduction pathways such as the phosphoinositol-3′-kinase (PI3K)/act/mammalian target of the rapamycin (mTOR) signal transduction pathway. The activation of this pathway and others modifies the phosphorylation status of multiple targets, which collectively manage the activities of the cells peaking in cell division. However, a complete switch from dormancy to growth is only feasible when the metabolism is properly reconfigured to cope with the increasing requirements of proliferation. Growth factor-induced signal transduction is responsible for managing these processes, which implies the preservation of a bioenergetic condition that permits growth (Lum et al. 2005). In specific, the PI3K/Akt/mTOR signaling pathway promotes both a rapid rise in the absorption of essential elements and the correct distribution of these elements into catabolic and anabolic signaling pathways to generate energy and macromolecules. (Gibbons et al. 2009). The disruption of one of these metabolic actions disables the growth factor.

Dynamic mechanisms also capture the cellular energy state, adjusting the equilibrium of anabolism and catabolism. Whereas the PI3K/Akt/mTOR signaling pathway enhances anabolism and inhibits catabolism, the AMP-activated protein kinase (AMPK) performs the opposite (Fig. 17.3). This serine threonine kinase acts as a “fuel sensor” which is triggered during disturbed bioenergetic conditions such

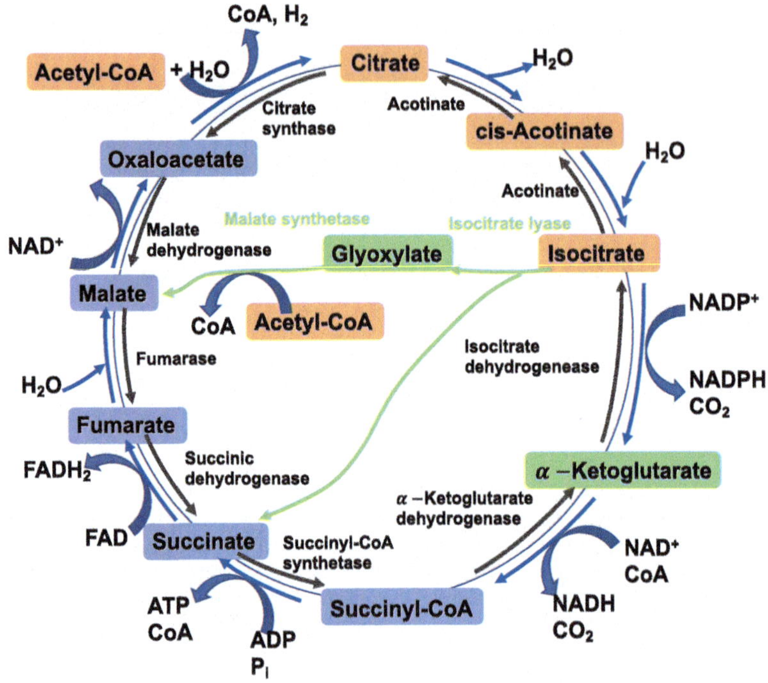

Fig. 17.3 The major steps of the citrate cycle

as withdrawal of nutrients and hypoxia (Hardie 2011). AMPK deactivates energy-consuming, growth-enhancing paths including protein and lipid synthesis and stimulates the catabolism of fatty acids and various other fuels through phosphorylation of a series of specific key targets. Thus, the cell is capable of re-establishing the equilibrium of energy availability on demand. Most interestingly, AMPK also controls a specific p53-dependent cell cycle control spot, which is activated through glucose depletion within cultured cells and thus restricts cell growth in low-energy states (Jones et al. 2005). AMPK also acts as a coordinator of the expression of stress activated genes through its targeting of chromatin and the phosphorylation of histone H2B at serine 36, and this activation facilitates the impact of AMPK on the expression of genes (Bungard et al. 2010). For this reason, the AMPK carries out a series of activities that enable the cells to react vigorously and extensively to any energy deficit.

The metabolism also influences cell signaling through the delivery of substrates for post-translational processing that modify protein trafficking, localization and the activity of enzymes. ATP provides the substrate for phosphorylation in cascades of kinases. However, other instances concern metabolites that are generated in the course of more intricate pathways from the core metabolism. These alterations can signal conditions of nutrient plentifulness, as the production of sufficient quantities of the metabolite necessary for protein alteration demands availability of nutrients in amounts beyond those necessary for the implementation of the fundamental bioenergetic programs. For instance, both glucose and glutamine are ample nutrients that are broken down in growth factor-stimulated cells to generate energy. However, they are also pursuing a different biochemical pathway for the generation of hexosamines, which alters the transporters of nutrients and growth factor receptors and facilitates their subsequent expression on the plasma membrane (Fig. 17.3). When glucose is not available in sufficient quantity to ensure flow by the hexosamine pathway, cells no longer have the capacity to react to growth factor signals and fail to absorb all nutrients necessary for cell growth (Wellen et al. 2010). Comparable instances include the post-translational modification of signal transducers through fatty acids and other lipid-like molecules that ease membrane localization and/or the activation of these proteins.

Acetyl-CoA, a key metabolite at the crossroads of the oxidation of carbohydrates, fatty acids and amino acids, has an enormous impact on the signaling of cells. The acetylation of lysine residues at their N-terminal tails of histone proteins plays an essential part in chromatin dynamics and in the expression of genes. The acetyl-CoA groups involved in the alteration of histones are mainly generated through acetyl-CoA synthetase in yeast and ATP citrate lyase in mammalian cells, both entering the nucleus to create a trapped basin of acetyl-CoA to this end (Fig. 17.3). The diminished function of these enzymes leads to a decrease in the acetylation of histones, with overall implications for gene expression (Takahashi et al. 2006a; Wellen et al. 2009). Numerous cellular proteins aside from histones are acetylated, among them most of the enzymes in core metabolic pathways (Zhao et al. 2010). Inside the liver, these alterations seem to govern the metabolic flow synchronously with the feed/fast cycle. For instance, the acetylation level of a number of major enzymes in the gluconeogenesis and urea cycle depends on their stability or action state. In both mammals and bacteria, enzyme acetylation oscillates based on the amount of nutrients made available to the cell (Wang et al. 2010; Zhao et al. 2010). For instance, both unicellular and multi-cellular organisms employ metabolite-facilitated post-translational modifications to tailor the nutrient wealth to the dispersion of carbon in the metabolic framework.

Deacetylation responses also control the metabolism (Fig. 17.3). Sirtuins, a class of deacetylases, are nicotinamide-adenine-dinucleotide (NAD)-dependent deacetylases that target histones and metabolic enzymes. Sirtuins are key evolutionary preserved factors that combine calorie reduction with longevity. The overexpression of sirtuins within model systems improves a variety of phenotypes associated with age such as cancer, diabetes and neurodegeneration (Guarente and Franklin 2011). There is enormous potential for investigating the discovery of powerful pharmacological activators of sirtuins for the treatment or avoidance of these diseases. Most interestingly, multiple sirtuins are found in the mitochondria, in which they deacetylate or in some way else alter metabolic enzymes. Whether these sirtuins are designed to counteract the impact of previously undetected mitochondrial acetyltransferases or to reverse non-enzymatic acetylation is not yet clear, although there are clues as to their significance in controlling the regulation of catabolism induction and garbage elimination throughout the food/fast cycle. For instance, the urea cycle enzyme carbamoyl phosphate synthetase (CPS) is deacetylated during fasting due to the sirtuin SIRT5 (Nakagawa et al. 2009). Probably, the liver is capable of handling the elevated amino acid breakdown and ammonia generation encouraged by the fasting status. CPS is just one of several enzymes that were formerly assumed to be governed mainly by allosteric mechanisms, but that are now acknowledged to be under additional levels of surveillance.

Acetyl-CoA levels also adjust superior levels of organization in eukaryotes, specifically basic processes involving the obligation for cell growth. When budding yeast is grown under glucose-limiting circumstances, it oscillates in a collective manner due to a metabolic cycle connected with synchronized cell growth (Tu et al. 2005). An elevation of intracellular acetyl-CoA coincides with the initiation of growth genes, and a burst of extracellular nutrients that can be easily processed into acetyl-CoA allows the cells to shorten the metabolic cycle and enter straight into the growth phase (Cai et al. 2011). Acetyl-CoA peaks are escorted by histone acetylation at certain areas of the chromatin that harbor growth genes. For instance, extracellular nutrients in these cells trigger gene expression through an acetyl-CoA-transmitted signaling, leading to a commitment to cell growth/cell division. The emphasis of the research in all these areas is that metabolism permeates each and every aspect of biology, ranging from the individual cell to the entire organism. Cell functions do not arise independently of metabolism, and a metabolic disorder at one junction has wave phenomena that can spread to the whole system and even to other systems. Thus, metabolic disorders have an extremely wide range, which also covers the pathological phenotypes.

17.1.1 Genetic Alterations in Human Metabolism

A large part of the metabolic diversity between individuals is based on genetic definition. Metabolism-derived phenotypes are of great significance both for human diseases and for the historical development of experimental biology. In fact, the majority of standard experiments in genetics succeeded phenotypes produced from metabolic mutants. Mendel's pea seeds appeared smooth or wrinkled depending on whether they expressed wild type or mutant alleles of a starch-branching enzyme (Bhattacharyya et al. 1990), and T. H. Morgan tracked *Drosophila* eye color phenotypes reflecting modified transport of metabolites necessary for the synthesis of pigment (Sullivan and Sullivan 1975). Finally, discoveries such as these resulted in the wording of genetics and the phenotypic inheritance "one gene-one enzyme" (Beadle and Tatum 1941), a pioneer of the fundamental principle of molecular biology.

Metabolic diversity affects human phenotypes, and the identification of severe metabolic disorders advanced the concept that human diseases can be inherited as simple characteristics. This concept was articulated by Archibald Garrod in the first decade of the twentieth century. Garrod's explorations were spurred on by an extraordinary mixture of interests: chemistry, urine pigments and joint diseases, however, apparently some of which were inherited by his father, who first noticed uric acid crystals in the urine of arthritis patients. Garrod's major achievement was the discovery that alkaptonuria, which is an arthritic disease defined by massive secretion of pigment-forming homogentisic acid in the urine, reappeared in families in an autosomal recessive pattern (Garrod 1902). It is noteworthy that this finding has been presented only two years after the reintroduction of Mendel's pea experiments, to which the article referred. In particular, there are related hereditary metabolic abnormalities in albinism, cystinuria and pentosuria, commonly referred to as "inborn errors of metabolism", and in 1923, a groundbreaking article has been released on this topic. Thus, the significance of metabolic pathways for human health and, inversely, the influence of genetically defined metabolic anomalies on disease was convincingly ascertained.

Meanwhile, there are about 500 confirmed inborn errors of metabolism (Childs et al. 2001), thereby constituting the largest class of hereditary human diseases. Inborn errors of metabolism basically affect all known biochemical metabolic pathways and organ systems and lead to an extremely wide range of phenotypes, ranging from clinically mute anomalies in metabolite accumulation to chronic/progressive aggregation of toxic macromolecules and severe, acutely deadly conditions of bioenergetic disaster. The overwhelming majority of inborn errors of metabolism are caused by recessively inherited mutations with a loss of function in enzymes and transporters, even though a number of them are induced by mutations that either trigger basal enzyme activity or lead to entirely new functions (Kranendijk et al. 2010; Stanley et al. 1998). Individuals with inborn errors of metabolism frequently exhibit an almost complete disruption of the proper functioning of the metabolic pathways, resembling human knockouts. Nonetheless, although many profound inborn errors of metabolism are consistent with embryonic growth and post-natal life, significant organ system malfunction or contextual pathologies may be encountered. A number of inborn errors of metabolism, especially those affecting the central pathways illustrated in (Fig. 17.2), are addressed by dietary interventions to decrease exposure to objectionable nutrients, and these procedures have significantly enhanced health and lifespan.

Inborn errors of metabolism offer a useful insight into the "life on the edge" of a metabolic disorder. In fact, the clinical care of metabolic patients with inborn errors is largely responsible for the complete understanding of human metabolism. Mitochondrial fatty acid oxidation has been considered an efficient bioenergetic route, however, its particular function in human energy homeostasis has been highlighted by the severe-life-threatening hypoketotic hypoglycemia triggered in children with reduced function of this route as a result of extended fasting (Houten and Wanders 2010). In a similar way, in the citric acid cycle, the oxidation of glucose generates the majority of energy for the brain, but the decisive role of pyruvate dehydrogenase (PDH) in the development of the brain is reflected in the profound disruption of the central nervous system caused by the mutation of the catalytic subunit of PDH (Robinson et al. 1996). In some examples, the physiological role of an enzyme or pathway was unclear prior to the identification of patients with genetic disorders. Peroxisomal enzyme defects are included in this class. The synthesis of plasmalogens and the fatty acid oxidation with a length of 24–26 carbons takes place in the peroxisomes, although this was not detected until abnormal amounts of these metabolites were found in patients with peroxisomal malfunction (Wanders and Waterham 2006).

However, since the 1960s, population-based endeavors have significantly widened the scope of human metabolic variation across the spectrum of seriously impacted individuals with apparent diseases. State-operated newborn screening projects measure the concentrations of diagnostic metabolites in dehydrated blood spots with the aim of detecting pre-symptomatic babies for whom immediate therapy can enhance the clinical outcome. These programs go back to the work of Robert Guthrie, who in the 1960s invented a basic, low-cost assay to detect newborns with phenyketonuria (PKU). PKU is caused by genetic deficiencies in the oxidation of phenylalanine (Phe) and the aggregation of toxic Phe-related metabolites, which interfere with cognitive performance. Since concerned infants do not show symptoms immediately, the morbidity is prevented through the early introduction of Phe-reduced diets in newborns determined by screening. Pilot projects on PKU screening in the USA have been highly effective, and within 10 years comprehensive programs were running across the USA and Europe. Starting in the 1990s, enhanced analytical technologies, in especially high-throughput mass spectrometry, made it possible for programs to drastically broaden the number of metabolites investigated. Most of the programs now identify more than 20 individual abnormalities from a single blood specimen (Chace et al. 2002). The programs have exhibited a notable level of phenotypic heterogeneity in individuals with aberrant levels of diagnostic metabolites, often never showing clinical symptoms. In a few cases, asymptomatic mothers of babies who have undergone screening programs have been found to have a clinically silent "disease," apparently due to small quantities of the mother's metabolites being transmitted to the unaffected baby via the placenta, breast milk or other pathways (Eichhorst et al. 2010). In other instances, inborn errors of metabolism that were once considered strongly pervasive and serious seem to be clinically mute in the majority of people these days (Alfardan et al. 2010). In total, the screening of metabolites in millions of infants has resulted in a plethora of information on the spectrum of metabolite frequencies in humans and their preventive importance for diseases.

The integration of metabolite profiling with genomic profiling has revealed important additional genetic determinants of the metabolic variety and its association with disease. The profiling of plasma lipids within broad populations has revealed sequence modifications that substantially affect lipid phenotypes (Cohen et al. 2005; Teslovich et al. 2010). In addition, a wider screening has been performed for metabolites, encompassing acylcarnitines, amino acids, prostaglandins, sphingo- and glycerophospholipids, in several hundreds of healthy adults (Gieger et al. 2008; Illig et al. 2010). The frequency of each metabolite or the ratio of related metabolites was considered as quantitative characteristics and genome-wide linkage investigation has been performed that look for polymorphisms displaying a large impact on metabolite frequency. To the surprise of all, these investigations found a set of relatively common alleles responsible for distinct fractions of the variations in metabolite concentrations, and in some cases, more than 20% of the total range of variation (Gieger et al. 2008; Illig et al. 2010). There were multiple cases where a polymorphism altered the coding sequence of a metabolic enzyme that was coupled with the variable metabolite in a direct manner. The other polymorphisms have earlier been linked to cardiovascular disease, gout and type-2 diabetes, each of which is thought to bring about alterations in the nuclear metabolic pathways (Suhre et al.

2011). However, there were also strong links with drug toxicity, Crohn's disease and related diseases that were not commonly assumed to be associated with intermediate metabolism (Suhre et al. 2011). These investigations provide evidence that a broad spectrum of human metabolic variation is genetically programmed, and they highlight the tight relationship between metabolism and health.

17.1.2 Cancer Fosters Genetically-Based Metabolic Alterations

Cancer is a perfect instance of a common human disease with genetically caused, pathological metabolic disorders. An altered cell metabolism is a characteristic of cancer and is responsible for malignant transformation as well as the development, growth and perpetuation of tumors (Hanahan and Weinberg 2011). Even though the recent revival of metabolic studies, in particular, the basic regulation of nuclear metabolic pathways, is largely due to cancer cell biology (Bensaad et al. 2006; Christofk et al. 2008a; Gao et al. 2009; Matoba et al. 2006; Vander Heiden et al. 2010). Moreover, the principle of disturbed metabolism in tumors is very ancient, going back nearly to the era of early research on chemical carcinogens and viruses as carcinogenic agents. Otto Warburg carried out the first rigorous research on cancer metabolism in the early 1920s, investigating the behavior of tissue sections ex vivo by means of manometric techniques devised in his in-house laboratory (Koppenol et al. 2011). Warburg found that carcinoma slices from rats and humans required much more glucose and secreted much more lactate than normal tissue, although there was enough oxygen to fully metabolize glucose to CO_2. It has been suggested that this is a fundamental shift in the manner in which glucose metabolism is adjusted in cancer cells (Warburg 1956). Among Warburg's many groundbreaking publications on biochemistry, he was awarded the Nobel Prize in 1931 for his work on respiration and he is the person most often cited for this discovery, now known as the Warburg effect.

After Warburg, generations of cancer biologists and biochemists have polished up his hypothesis and attempted to come up with mechanistic explanations—but essentially these investigations have validated the key finding that many tumors can outgrow the neighboring tissue in terms of glucose. This property accounts for the fact that 18-fluoro-2-deoxyglucose-positron emission tomography (FDG-PET) has been proven effective in imaging tumors of many histological types. Other investigations have revealed additional metabolic properties of tumor tissue, most notably the propensity to metabolize glutamine and to produce fatty acids, which both stimulate tumor growth in experimental models (DeBerardinis and Cheng 2010; Swinnen et al. 2006). A common hypothesis to harmonize these metabolic pathways is that explosive tumor growth necessitates a transformation of the metabolism to accommodate the bioenergetic and biosynthetic requirements of fast cell growth and to defend the cells from stress caused by a rough microenvironment (Deberardinis et al. 2008; Semenza 2001; Shanware et al. 2001). Metabolic flow analyses in cancer cells have confirmed this model and emphasized the importance of integrating these three main pathways (DeBerardinis et al. 2007). Cancer is therefore a model of how a disturbed metabolism at the cellular level is responsible for the disease.

What triggers the metabolic reprogramming in cancer cells? As with inborn errors of metabolism, the metabolic characteristics of tumors are genetically determined and arise from the same mutations that encourage malignancy. In inborn errors of metabolism, however, germline mutations induce large alterations in the metabolism which decrease the general fitness. Cancer mutations are generally gained somatically and are connected with metabolic effects that seem to improve physical fitness and cellular growth. Two distinct classes of mutations can be used to reprogram the metabolism in tumors. Firstly, many human oncogenes and tumor suppressors control the metabolism of glucose (Fig. 17.4) (DeBerardinis 2008; Jones and Thompson 2009). Tumor-promoting mutations in these genes are prone to convert to a metabolic phenomenon of increased glycolysis and enhanced energy generation, thus, promoting growth self-sufficiency and circumventing growth-inhibiting signals. The PI3K/Akt/mTOR signaling pathway is crucial for this activity as it controls both nutrient absorption and the allocation of carbon and nitrogen throughout the biosynthetic signaling cascades (Zoncu et al. 2011). Cell growth downstream of this pathway necessitates the initiation of de novo lipid synthesis through mTOR, and this is partly due to the enhanced nuclear localization of the sterol regulator element-binding protein (SREBP), a basic helix-loop helix-leucine zipper transcription factor that facilitates the expression of a number of lipogenic enzymes (Porstmann et al. 2008). Most recent studies have revealed that the impact of mTOR on SREBP entails the phosphorylation of lipin-1, an acid phosphatide phosphatase. mTOR-dependent Lipin-1 phosphorylation leads to its retention in the cytosol, whereas mTOR blockage resulted in the dephosphorylation of Lipin-1 and its transfer to the nucleus, where it was associated with the removal of nuclear SREBP (Peterson et al. 2011). Therefore, the reprogramming of the metabolism into a plateau that stimulates cell growth and proliferation is a key element of the growth factor signaling pathways and the malignancy that arises from the constitutive triggering of these pathways.

Secondly, metabolic reprogramming can appear directly as a consequence of enzymatic mutations. This was first seen in a subgroup of patients with inborn errors of metabolism who exhibited an elevated incidence of cancer, especially liver cell carcinoma. Several of these diseases are linked either to the aggregation of mutagens or lead to cirrhosis, an independent liver cell carcinoma predictor (Erez et al. 2011). In these instances, the association between metabolism and malignancy is presumably an indirect one. However, other disturbances create a metabolic condition that imitates oncogene activation. The lack of glucose-6-phosphatase (G6Pase) in glycogen storage disease type 1a is a major incidence risk factor for hepatocellular adenomas and hepatocellular carcinomas, although liver cirrhosis is not a significant feature. G6Pase acts as a catalyst for the ultimate common step in glycogenolysis and gluconeogenesis, and G6Pase deficiency causes severe fasting incompatibility and hypoglycemia. However, the breakdown in producing glucose in one of these two pathways leads to a large flow into other pathways fed by glucose-6-phosphate in the liver, which covers glycolysis, lipid synthesis and nucleotide metabolism. Together with a substantial concentration of glycogen and FAT in the liver, patients with G6Pase deficit experience severe elevations of lactic acid, lipids and uric acid in the bloodstream. The pathophysiology of tumorigenesis is not clear and it is not understood how the rerouting of the glucose metabolism impacts on the hepatocellular differentiation. However, it is remarkable that G6Pase deficiency imitates the Warburg effect and certain other oncogene-dependent metabolic manifestations, whereby glucose-6-phosphate is shifted into lactate and other metabolite basins.

Lately, it has been discovered that cancer is also linked to metabolic mutations that are restricted to the tumor. In these disorders, the metabolic enzymes genetically function as oncogenes or tumor suppressors (Fig. 17.4) (Frezza et al. 2011a; Thompson 2009). These types of cancer provide a unique chance to identify the cellular implications of metabolic disorders. The first of these were mutations in the TCA cycle enzymes in the case of familial cancer syndromes. Mutations in succinate dehydrogenase (SDH), an oxidoreductase complex involved in both the electron transport chain (ETC) and TCA cycles, have been found in dominant familial paraganglioma (Baysal 2008; Baysal et al. 2000). Functional loss mutations have been found in all four subunits of the SDH complex (Baysal 2008; Burnichon et al. 2010), and in SDH5, which codes for a specific protein participating in the incorporation of the flavin-adenine dinucleotide cofactor of SDH (Hao et al. 2009). The mutations in the TCA cycle enzyme fumarate hydratase (FH) have been found in familial syndromes associated with proneness to renal cell cancer and leiomyomatosis, which represent tumors of the smooth muscles of uterus and skin (Tomlinson et al. 2002). In the families with SDH- or FH-deficient tumors,

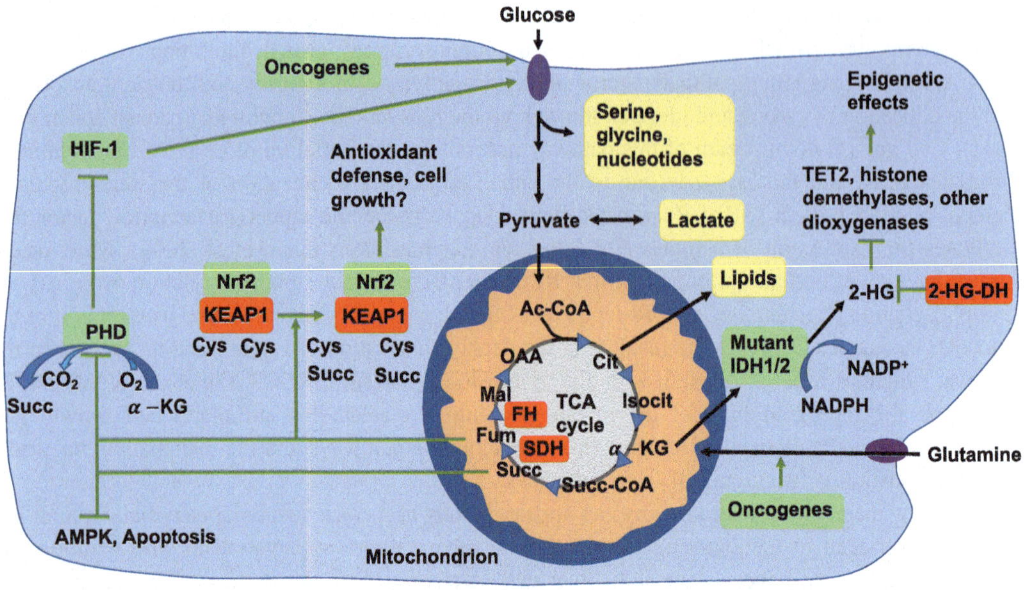

Fig. 17.4 Metabolism of cancer cells. Cancer cells utilize primarily glucose and glutamine to foster intermediary metabolism. Various metabolites are fed by these nutrients and seem to be required for the growth of primary tumors (yellow). The oncogenic signal transduction governs the uptake and catabolism of glucose and glutamine. The suspected metabolic tumor suppressor (red) and oncogenes (green) regulate availability of specific key metabolites, which subsequently foster the malignant progression of tumors and thereby supports the propagation of growth signals in transformed cells. Moreover, besides their functions in metabolism, FH, SDH and 2-HG dehydrogenases suppress levels of pro-oncogenic metabolites

the sufferers acquire a mutation and their tumors exhibit a loss of the wild-type allele. Therefore, both SDH and FH become tumor suppressors.

The mechanisms linking dysfunctional SDH/FH with malignancy are presumably multi-focal. Strong evidence points to dysfunctional cell responses induced through the presence of succinate and/or fumarate, which tend to aggregate in tumors in high concentrations. The two metabolites have been shown to interact with processes potentially involved in tumor suppression, among them c-Jun-driven apoptosis in pheochromocytoma cells (Lee et al. 2005) and AMPK signaling (Tong et al. 2011). In addition, both succinate and fumarate abnormally enhance the role of hypoxia-inducible factors (HIFs). These dimeric transcription factors capture the metabolic properties of hypoxia, in particular, the enhanced expression of glucose transporters and glycolytic enzymes. When cells have sufficient oxygen supply, the HIF transcription activity is limited by the constitutive breakdown of its α subunits (HIF-1α and HIF-2α). This necessitates post-translational adaptation using prolyl hydroxylases (PHDs), α-ketoglutarate-dependent enzymes, which yield succinate as an end product (Fig. 17.4). Both succinate and fumarate act as competitive PHD inhibitors, and the substantial aggregation of these metabolites in cells devoid of SDH or FH activity triggers a hypoxic reaction when exposed to oxygen-rich environments. These processes result in SDH- and FH-deficient cells that are chronically prepared for glycolysis, independent of the availability of oxygen. The concentration of reactive oxygen species, an independent initiator of HIF stabilization, can in some circumstances of malfunctioning SDH or FH participate in this process (Guzy et al. 2008; Sudarshan et al. 2009). Nevertheless, both HIF-1α and HIF-2α are unnecessary for the development of hyperplastic renal cysts in mice with missing expression of FH in their kidneys (Adam et al. 2011), which points to broader implications of these enzyme failures.

Most interestingly, due to the electrophilic nature of fumarate, high concentrations of fumarate can alter cysteine groups through a process referred to as succination. In this process, sulfhydryl groups on cysteine are turned into S-(2-succinyl)-cysteine (2SC), and in certain circumstances, the protein function is disrupted (Alderson et al. 2006). Cells deficient in FH exhibited high succination rates on the KELCH-like ECH-associated protein 1 (KEAP1), which is an electrophilic sensor and negative regulator of the transcription factor nuclear factor E2-related factor 2 (Nrf2) (Adam et al. 2011; Ooi et al. 2011). Nrf2 initiates the expression of a number of genes required for the defense of reactive oxygen species, potentially providing malignant FH-deficient cells with the capacity to withstand high concentrations of exogenous or endogenous oxidants. Another option seems to be that the activation of Nrf2 provides metabolic benefits for FH-deficient cells. Among the targets of Nrf2 is the gene encoding heme oxygenase-1 (HMOX1), which is an enzyme implicated in the breakdown of heme. The pathway of heme synthesis out of glutamine with consequent break-down through HMOX1 is increased in FH-deficient cells, and the muting of HMOX1 avoids that these cells build colonies (Frezza et al. 2011b). Even though it is not yet entirely clear how these effects can most effectively be utilized for cancer therapy, these observations together indicate that a disturbed mitochondrial metabolism and increases in the dicarboxyl intermediates of the TCA cycle encourage abnormal signal transmission in cancer cells.

Some other enzymatic mutations may act as oncogenes. The genomic sequencing of gliomas (Parsons et al. 2008; Yan et al. 2009) and acute myelogenous leukemia (Mardis et al. 2009) led to the detection of mutations within two isoforms of $NADP^+$-dependent isocitrate dehydrogenase (IDH1 and IDH2). These enzymes usually convert isocitrate to α ketoglutarate through an oxidation step, in which process $NADP^+$ is reduced to NADPH (Fig. 17.5). IDH1 and IDH2 mutations in cancer are gained somatically, are unique to one allele and are restricted to the active site of the enzyme. Therefore, in contrast to FH and SDH, it was hard to argue for IDH1 and IDH2 as tumor suppressors. The results were much stronger with a gain-of-function related to the modified enzymatic active site. This suspicion was corroborated through metabolomics: Tumor tissue and cell lines expressing the mutant IDH1 or IDH2 generated high amounts of a metabolite, (D)-2-hydroxyglutarate (2-HG), which is negligible when expressing only the wild-type enzymes (Dang et al. 2009). The metabolite is generated from the NADPH-dependent conversion of α ketoglutarate to 2-HG, which is a neomorphic enzyme activity that is highly efficient when both mutant and wild-type alleles are expressed simultaneously (Pietrak et al. 2011). The availability of 2-HG in these tumors is convincing as it reflects one of the most thoroughly established connections between an inborn error of metabolism and cancer. Children with (L)-2-hydroxyglutaric aciduria, an autosomal recessive status resulting from a deficit of (L)-2-HG dehydrogenase, which transforms 2-HG into α ketoglutarate, aggregate the (L)-isomer of 2-HG in all body fluids. A significant percentage of children with this disease have acquired brain tumors or other types of cancer. Thus, mutations of the active sites in IDH1 and IDH2 appear to lend oncogenic characteristics to the enzyme, and the formation of the oncometabolite 2-HG may be the key determinant that puts vulnerable cell populations into a transformed condition.

The mechanisms behind this symptom are currently the object of intensive investigation. Even large aggregations of 2-HG do not produce large disturbances of metabolite levels in IDH1 mutant tumors (Dang et al. 2009). Latest data indicate that 2-HG acts by interfering with several of the more than 50 mammalian dioxygenases utilizing α ketoglutarate as a substrate. These enzymes govern a variety of critical processes, among them the methylation of histones and DNA. In fact, high concentrations of 2-HG hampered the function of α ketoglutarate-dependent histone demethylases and 5-methylcytosine hydroxylases in vitro, and primary IDH1 or IDH2 mutant tumors displayed in vivo strong hints of severely altered histone and DNA methylation (Figueroa et al. 2010; Xu et al. 2011). Even though mutations in IDH1/2 and in the 5-methylcytosine hydroxylase TET2 are common in acute myelogenous leukemia, the mutations are considered mutually exclusive, suggesting that the

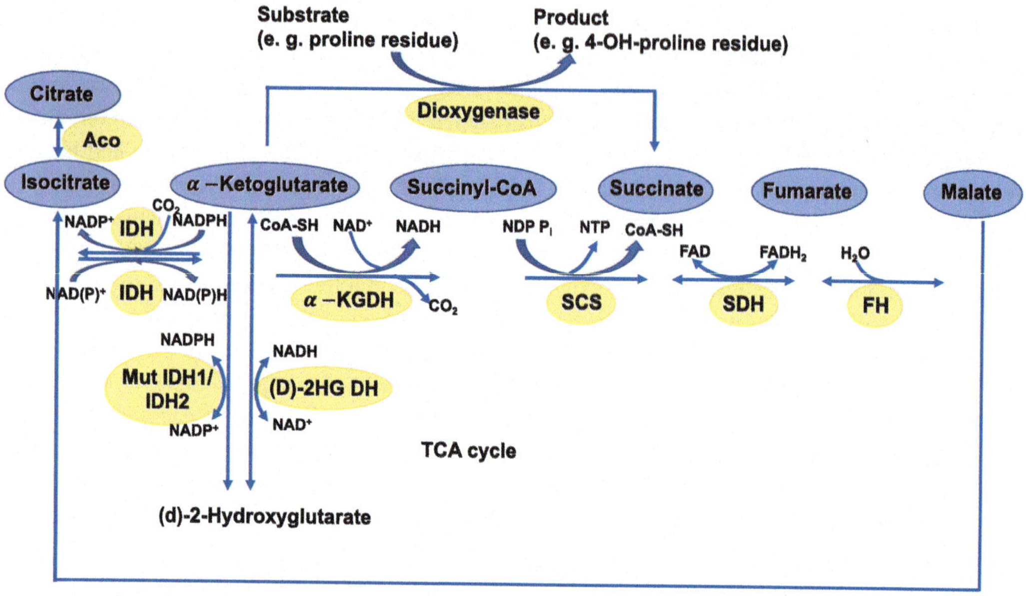

Fig. 17.5 Detailed view on the enzymes and metabolites of the TCA-cycle, including IDH1 and IDH2

two distinct mutation types produce redundant cellular effects (Figueroa et al. 2010). Since common methylation alterations of histones and DNA could largely affect epigenetics, IDH1/2 mutations could have enormous impact on cellular performance and differentiation. Moreover, this epigenetic state maintained by sustained 2-HG effects may be able to reverse certain of the effects by blocking the activity of the mutant IDH1/2 enzyme or by enhancing the accessibility of α ketoglutarate to restore dioxygenase activity. It is important to stress that the existence of IDH1/2 mutations in gliomas is linked to both a lower histological grade and a more slowly progressive rate of the disease, suggesting the possibility that the blocking of mutant IDH function could impede progression to its most malignant form, namely glioblastoma multi-forme (Parsons et al. 2008; Yan et al. 2009).

The surprising participation of IDH1, IDH2, SDH and FH in the pathogenesis of cancer has triggered endeavors to find alternative metabolic genes whose changes at the genomic level trigger or enhance the development of tumors. Phosphoglycerate dehydrogenase (PHGDH), an enzyme needed to generate the amino acids serine and glycine from glucose, has been recently identified. Serine and glycine are key intermediates in a wide range of biosynthetic processes, among them the synthesis of nucleotides, proteins, glutathione, creatine and methylated DNA. High expression levels of PHGDH protein compared to normal tissue are associated with disease progression in breast cancer patients (Pollari et al. 2011). Most interestingly, the region of chromosome 1p12 coding for PHGDH is, similar to traditional oncogenes, elevated in a significant number of tumors, and cell lines originating from these tumors have a strong tendency to be strongly reliant on the expression of PHGDH (Locasale et al. 2011; Possemato et al. 2011). It still needs to be determined whether PHGDH is a trustworthy oncogene, however, overexpression of PHGDH was found to be adequate to encourage the independence of attachment in breast epithelial cells and to interfere with cell polarity phenotypes that are considered compatible with malignancy (Locasale et al. 2011). Thus, elevated concentrations of PHGDH constitute another category of genetically defined metabolic anomalies with a weak point that could potentially be utilized in cancer therapy.

17.1.3 Perspectives, Boundaries and Applications

What is the next step? Equipped with a wider view of the impact of metabolism, it ought to be progressively possible to revert stages of diseases which are brought about through genetically based metabolic disorders. An area which could potentially profit from this new knowledge is the therapy of children with inborn errors of metabolism. Even though dietary changes and other improvements have significantly diminished the morbidity and death rate in a number of these diseases, treatments are frequently invasive and do not completely safeguard patients from casual conditions of serious metabolic deficiencies that can lead to persistent disability. It is conceivable that compounds designed to alter metabolism in cancer may also be employed in other diseases that exhibit some of the same distinctive metabolic features, such as altered lactic acid production or modified function of the TCA cycle. Recent research shows, for instance, that cancer cells with mutations in the TCA cycle or ETC (Mullen et al. 2011), or with hypoxia-facilitated suppression of oxidative metabolism (Metallo et al. 2011; Scott et al. 2011; Wise et al. 2011) employ decreased carboxylation of α-KG to generate citrate and other precursors while growing. In this response, the NADPH-dependent isoforms of IDH exhibit the opposite behavior compared to their traditional function as oxidative decarboxylases. It is conceivable that a comparable signaling pathway may account for the disorder in patients with permanent damage to mitochondrial metabolism or with perinatal asphyxia, stroke, cardiac ischemia and other illnesses with pathological hypoxia.

Moreover, most of the knowledge we have about the pathophysiology of inborn errors of metabolism still strongly rests on the conventional view of intermediate metabolism, which concentrates principally on catabolism. The metabolic control of macromolecular synthesis, the dynamics of chromatin and cell renewal are only just beginning to be considered, and the interaction between metabolite accumulation, signal transduction of cells and other phenomena highly specific to these disorders needs to be examined more thoroughly. For instance, the biology of IDH1/2 mutant tumors can yield an understanding of the consequences of global 2-HG aggregation in patients with (L)- or (D)-2-hydroxyglutaric aciduria. This issue was underlined in a recent investigation that about half of the patients with (D)-2-hydroxyglutaric aciduria displayed heterozygous mutations of the active sites in IDH2 (Kranendijk et al. 2010). It is not understood whether these patients or patients with different genetic types of 2-HG aciduria exhibit any of the epigenetic responses found in IDH1/2 mutant tumors. Inversely, it has to be feasible to integrate the nearly 100 years of clinical expertise with inborn errors of metabolism in cancer and other diseases. There is considerable concern about the discovery of compounds to impair the Warburg effect in tumors, but since glycolysis is almost omnipresent in human tissues, there is considerable anxiety about the toxicity of these compounds. The extremely high number of properly characterized inborn errors of metabolism in nearly all recognized metabolic pathways may give an idea of the nature of toxicity to be expected for several of these targets. SLC2A1 acts as a HIF-1 transcription target and is coding the prevailing glucose transporter in cancer cells, referred to as GLUT1. GLUT1 is the major transporter across the blood–brain barrier and accounts for the delivery of glucose throughout the brain. There is a broad phenotypic range of GLUT1 deficiency, although its most frequent type comprises serious symptoms of the central nervous system including cramping, cerebral retardation and poor post-natal brain growth. The glucose level in the cerebrospinal liquid can be less than 50% of the normal value. It is unclear whether the elevated glucose reliance of tumors compared to differentiated tissue provides an affordable therapeutic margin in cancer treatment, but obviously utmost care should be taken when seeking to block GLUT1. Additional glycolytic inborn errors of metabolism exhibit milder manifestations. LDHA represents a transcriptional target of both HIF-1 and c-MYC. Moreover, it codes for the isoform lactate dehydrogenase, which is found in the majority of tumors and is required for stable lactate production in cancer cell lines. A strong LDHA deficiency also constitutes a well-defined human disease, which referred to as Glycogen Storage Disease Type XI. In adolescents and adults, this disease leads to immobility, cramps and sometimes to myoglobinuria, but is not connected to a serious functional disorder of the principal homeostatic organs. Possibly a blockage of LDHA could be accepted in intermittent cancer treatments. Continuing population-based metabolomics and genomics research should reveal the full range of metabolic complexity in healthy people, and this knowledge could help to forecast how metabolic pathways can be reliably targeted in cancer.

Changing metabolic conditions in diseases also offer an enormous chance to advance new techniques in diagnostic imaging. Several metabolic imaging systems are currently in clinical use, in particular, FDG-PET. However, as knowledge of the interactions between metabolism and disease continues to grow, it is getting easier and easier to broaden the range of non-invasive methods available to identify diseased tissue and assess its responsiveness to treatment. A series of endeavors in this respect are in progress and have a promising potential to be transferred to human patients in the close future. The significance of supplementary nutrients apart from glucose for the metabolism of tumors has led to the discovery of a number of new PET drugs for cancer, among them glutamate and glutamine (Koglin et al. 2011; Qu et al. 2011). These compounds have the potential to bring an increased degree of molecular sophistication to the assessment of single tumors and may enhance the detection of tumors that are not visible on FDG-PET. Magnetic resonance spectroscopy (MRS) at 1.5 T has been applied for several decades to obtain semi-quantitative, non-invasive measurements of the ample metabolite batches in cancer (Glunde and Bhujwalla 2011). In recent times, human MRI systems with higher fields have dramatically increased the resolution of 1H spectra and enabled the detection of many additional metabolites in human tumors. A successful quantification of relevant metabolites such as glycine and 2-HG has previously been announced for human gliomas (Choi et al. 2011, 2012), and further metabolites will be included in the list as soon as their importance in cancer, inborn errors of metabolism or for other diseases is more widely understood. It is also feasible now to map 13C-labeled metabolites and observe the specific enzyme activities in vivo by monitoring the transmission of the labeling from the substrate onto the product. This technique involves increasing the polarization of the 13C nuclear spin state by applying to it the high spin polarization of an unpaired electron derived from a free radical, a process commonly referred to as hyperpolarization of the 13C nucleus. This results in a transient sensitivity gain of 13C capture of 10,000 times or even beyond. Hyperpolarization was used to quantitatively measure oncogene-based metabolic activities in experimental tumors (Hu et al. 2011). Even though it is at present only a method of examination, metabolic imaging utilizing the hyperpolarization of 13C and some other stable isotopes has a substantial translational capacity in human disorders since it is very sensitive, is incorporated into MRI, does not require radioactive tracers and, in particular, can study the flow of specific enzymatic reactions in vivo (Kurhanewicz et al. 2011).

In fact, one of the most crucial tasks in human metabolic research nowadays—and one of the closest boundaries of systems biology—is to investigate the effects of disease on metabolic flow in vivo, which means to employ, for instance, sensitive and efficient methods to measure the transfer of carbon and nitrogen along with metabolic pathways in living individuals, both with and without disease. Although methodologies for cataloging and quantifying small molecules in biological liquids (metabolomics) are very informative, they do not provide a comprehensive overview of the metabolism and may in certain cases lead to misleading conclusions. Newborn screening programs are a useful illustration of this topic. An increased concentration of a metabolite in these screening assays virtually invariably indicates a diminished rather than an elevated flow of metabolism, somewhat analogous to the impact of a stoplight on the clustering of cars. Moreover, since the metabolism comprises so numerous intersecting pathways, it may be difficult to deduce which pathway or pathways are concerned when abnormal concentrations of a metabolite are detected. For instance, it is uncertain what extracellular nutrients in IDH1/2 mutant tumors are converted to 2-HG in vivo

(Fig. 17.5). One possibility to tackle these problems is to integrate the analysis of metabolic pathways with metabolic investigations. These studies could comprise the application of isotopically labeled nutrients to an animal model or human patient and the subsequent extraction of metabolites of interest from samples of blood, urine, breath or tissue. The total frequency of these metabolites is determined using a metabolomics framework, then a fraction of the most indicative metabolites is examined further using mass spectrometry and/or NMR spectroscopy to identify the frequency and location of isotopic labeling within every molecule. Related methods, which concentrate on a handful of metabolites, have previously been employed successfully in humans to assess gluconeogenesis and the urea cycle in vivo and to evaluate the metabolism between lung tumors and adjacent tissue (Fan et al. 2009; Landau et al. 1996; Yudkoff et al. 2010). Despite their technical challenges, combined metabolomics and metabolic flow analyses are of enormous importance as they could provide a quantitative and comprehensive picture of the diversity of metabolic path activity, which in turn could contribute to a better insight into the metabolic identity and biological foundation of diseases in humans.

There is a greater interest than ever before in the intermediate metabolic processes and in the fast advancement of the analytical techniques required for its study. Important new insights in disease-based metabolic science are presented continuously, confirming the idea that metabolism permeates all areas of biology and pathology. The forthcoming decade holds the promise of further continuous progress in improving our understanding of the metabolic underpinnings of common human illnesses, and it is highly probable that part of this expertise will be converted into new diagnostic and therapeutic solutions, specifically for cancer.

Since its initial detection by Warburg (1956), it is well known that the majority, if not all, of cancer cells are more strongly reliant on aerobic glycolysis for ATP generation than normal cells. The almost uniform appearance of this metabolic phenomena in cancer cells is not intuitive, since glycolysis generates just two moles of ATP per mole of glucose, which is far lower than the 36 produced via mitochondrial respiration. For the maintenance of this apparently luxurious, catabolic state, various hypotheses have been proposed (Vazquez et al. 2010). On the level of the cell population, the dysfunction of mitochondrial respiration and the increase in glycolysis under the circumstances of intermittent hypoxia, which arises in premalignant and malignant cancer cells, is assumed to represent a metabolic benefit (Brahimi-Horn et al. 2007; Gatenby and Gillies 2004). Nevertheless, aerobic glycolysis is not restricted to cancer cells, but also occurs in quickly dividing normal cells, even under normoxic circumstances (Zu and Guppy 2004). The glycolysis speed is also widely assumed to rise to accommodate the elevated anabolic demand of the fast proliferating cells for precursor metabolites (VanderHeiden et al. 2009). However, concentrating on the anabolic requirements of the cell on its own ignores two key facts: Firstly, growing cells require ATP in combination with the precursor metabolites in order to cover the energy requirements of the biosynthetic pathways. Secondly, in its primary definition, the Warburg effect is based on the rise in the rate of glycolysis, which culminates in the release of lactate that does not promote the formation of precursor metabolites.

The actual molecular mechanisms resulting in increased aerobic glycolysis are becoming progressively better comprehended. Under physiological conditions, for instance, the PI3K/Akt signaling pathways play a crucial part in encouraging aerobic glycolysis in activated T cells in reaction to the stimulation of the growth factors or cytokines (Wieman et al. 2007). In the pathophysiological state of tumor development, the disturbance of mitochondrial respiration as a result of mitochondrial DNA mutations/deletions is a major contributing parameter (Copeland et al. 2002; Nomoto et al. 2002; Carew et al. 2003; Pelicano et al. 2006). In addition, p53, is one of the most abundantly mutated genes in cancer and serves both as a positive modulator of mitochondrial respiration (Matoba et al. 2006) and a negative modulator of glycolysis (Bensaad et al. 2006). Combined with the triggering of the hypoxia-inducible factor (HIF), which is a transcription factor activated through hypoxic stress, but also through oncogenic, metabolic and oxidative stress, these frequently result in an overexpression of glucose transporters and different enzymes of the glycolysis pathway or of isoenzyme subtypes, which accounts for the elevated glucose absorption and the modified utilization (Gatenby and Gillies 2004; Christofk et al. 2008b).

The evidence gathered so far thus suggests that aerobic glycolysis is a general characteristic of strongly proliferating cells and that it may offer a growth benefit to quickly proliferating normal cells, for instance, during development and the regeneration of tissue, and to cancer cells during the development of tumors. However, an explanation of the source of this growth benefit at the system level has not yet been formally established. To begin to tackle this topic, a simplified flow balance model of ATP production is presented, which includes a limitation of the glucose uptake capability and the restricted solvent capacity of the cytoplasm. The model permits to reveal the existence of two distinct energetically beneficial metabolic regimes within proliferating cells, a finding that is consistent with the experimental outcomes.

17.2 Citrate Acid Cycle [The Tricarboxylic Acid (TCA) and Urea Cycles (Krebs)]

The citric acid cycle forms the central metabolic node of the cell. It is the doorway to the aerobic metabolism of every molecule, which can be converted into an acetyl group or dicarboxylic acid. The cycle is also a major supply of precursors, not only for the storing types of fuels, but also for the constituents of many other molecules including amino acids, nucleotide bases, cholesterol and porphyrin that is the organic compound of heme. What function does the citric acid cycle have in the conversion of fuel molecules to ATP? Remember that fuel molecules are carbon atoms that can be oxidized and hence loose electrons. The citric acid cycle comprises a number of oxidation–reduction actions that lead to the oxidation of one acetyl group into two carbon dioxide molecules.

The general scheme of the citric acid cycle is presented in Fig. 17.3. A four-carbon compound (oxaloacetate) condenses with a two-carbon acetyl entity to a six-carbon tricarboxylic acid (citrate). An isomer of citrate is decarboxylated by oxidation. The remaining five-carbon compound (α-ketoglutarate) undergoes oxidative decarboxylation to obtain a four-carbon component (succinate). Oxaloacetate can then be reclaimed from succinate. Two carbon atoms come into the cycle as acetyl unit and two carbon atoms exit the cycle in the shape of two carbon dioxide molecules. Three hydride ions with six electrons are transferred to three molecules of nicotinamide adenine dinucleotide (NAD^+), while a pair of hydrogen atoms with two electrons are transferred to one molecule of flavin adenine dinucleotide (FAD). The purpose of the citric acid cycle is to extract high-energy electrons from the carbon source. It is important to bear in mind that the citric acid cycle by itself neither produces a high amount of ATP nor comprises oxygen as a reaction partner (Fig. 17.3). On the contrary, the citric acid cycle withdraws electrons from acetyl-CoA and utilizes these electrons to generate NADH and $FADH_2$. In oxidative phosphorylation, the electrons liberated during the reoxidation of NADH and $FADH_2$ pass through a sequence of membrane proteins, referred to as the electron transport chain to create a proton gradient over the membrane. The protons subsequently pass through the ATP synthase to produce ATP from ADP and inorganic phosphate. Oxygen is indirectly needed for the citric acid cycle insofar that it is the electron acceptor in the terminal electron transport chain, which is needed for the regeneration of NAD^+ and FAD.

The citric acid cycle, in combination with oxidative phosphorylation, supplies most of the energy consumed by aerobic cells and in humans more than 95%. It is extremely powerful as a restricted number of molecules can produce huge quantities of NADH and $FADH_2$. In Fig. 17.3, it needs to be considered that the four-carbon molecule, oxaloacetate, which starts the first step in the citric acid cycle, is reclaimed at the end of a pass through the cycle. Oxaloacetate has a catalytic effect: it is involved in the oxidation of the acetyl group, although it is self-regenerating. In this way, a molecule of oxaloacetate is able to participate in the oxidation of many acetyl molecules.

17.3 Glycogen Catabolism or Cori Cycle

The Cori cycle (synonymously referred to as the lactic acid cycle), named after its discoverers Carl Ferdinand Cori and Gerty Cori, is a metabolic route in which lactate, generated by anaerobic glycolysis in the muscles, is transferred to the liver and converted into glucose, which then reverts to the muscles and is cycle-wise metabolized back into lactate. In the human body, glycogen can be seen as a branched polymer of glucose, which is primarily deposited in the liver and skeletal muscle and delivers glucose into the bloodstream during the fasting period and into the muscle cells undergoing muscle contraction. Glycogen has also been detected in support of other tissues encompassing brain, heart, kidney, adipose tissue and erythrocytes, although the functioning of glycogen in these tissues is widely unclear. Glycogen synthesis involves a number of reactions comprising the entry of glucose into the cell by transporters, the phosphorylation of glucose to glucose-6-phosphate, the isomerization to glucose-1-phosphate and the synthesis of uridine 5′-diphosphate-glucose, that is, the immediate glucose donor for the synthesis of glycogen.

Glycogenin acts as a catalyst for the formation of a short glucose polymer, which is prolonged through the activity of glycogen synthase. The glycogen branching enzyme imposes branching points in the glycogen protein at regular spacings. Laforin and malin are proteins participating in glycogen synthesis, although their precise function is difficult to understand in humans. Glycogen is concentrated in the liver mainly in the post-prandial phase and in skeletal muscles mainly after physical activity. In the cytosol, glycogen degradation or glycogenolysis is performed using two enzymes, glycogen phosphorylase, which liberates glucose-1-phosphate from the linear chains of glycogen, and the enzyme for glycogen debranching, which disentangles the branching points. In the lysosomes, the break-down of glycogen is initiated by α-glucosidase, which works as a catalyst. The glucose-6-phosphatase system is catalyzing the dephosphorylation of glucose-6-phosphate to glucose, which is a required step for the free glucose to be released from the cell. Mutations in the genes coding for the enzymes participating in glycogen metabolism lead to glycogen storage disorders.

Glycogen can be generally stored as glucose and thereafter reused when needed. Glycogen is a huge, branched polymer of glucose residues that can be decomposed into glucose molecules when energy is required. Most residues of glucose in glycogen are connected through α-1,4-glycosidic bonds. Every tenth residue is branched through α-1,6-glycosidic linkages. The α-glycosidic bonds generate open helical polymers, while β-bonds create almost linear strands that form structural fibrils, similar to cellulose. Glycogen is not as reduced as fatty acids and therefore not as abundant in energy. Why can animals accumulate all energy as glycogen? Why are not all excessive fuels converted into fatty acids? Glycogen represents an essential fuel source for several purposes. The controlled break-down of glycogen and the liberation of glucose enhances the level of glucose available in between meals. Therefore, glycogen acts as a cushion to keep the blood sugar level up. The function of glycogen in keeping blood glucose levels up is particularly important since glucose is basically the only fuel consumed in the brain apart from long-term hunger. In addition, the glucose from glycogen is easily mobilized and is consequently a valuable energy source for abrupt, exhausting activities. In contrast to fatty acids, the liberated glucose can supply energy in the lack of oxygen and thus provide energy for the anaerobic activity.

The two most important places of glycogen storage include the liver and the skeletal muscles. The level of glycogen is more abundant in the liver than in the muscles (10% by weight compared to 2%

by weight), although the skeletal muscle stores more glycogen in total because of its far greater mass. Glycogen exists in the cytosol in the shape of grains with a diameter between 10 and 40 nm. In the liver, glycogen synthesis and break-down are governed in order to ensure that the blood sugar level is kept at the level required for the demands of the entire organism. In contrast, these processes are governed in the muscle to cover the energy requirements of the muscle on its own.

Glycogen break-down and synthesis are comparatively basic biochemical processes. Glycogen break-down comprises three steps: firstly, the liberation of glucose-1-phosphate out of glycogen, secondly, the transformation of the glycogen substrate to allow additional break-down and, thirdly, the transition from glucose-1-phosphate to glucose-6-phosphate for continued metabolism. The glucose-6-phosphate obtained from the degradation of glycogen has three distinct destinies (Fig. 17.3): Firstly, it is the starting substrate for glycolysis; secondly, it can be transformed into NADPH and ribose derivatives through the pentose phosphate pathway; thirdly, it can be processed into free glucose and dispensed into the bloodstream.

Glycogen synthesis demands an activated version of glucose, uridine diphosphate glucose (UDP-glucose), which is produced through the reaction of UTP and glucose-1-phosphate. UDP-glucose is attached to the non-reducing terminus of glycogen molecules. Similar to glycogen break-down, the glycogen molecule requires conversion for subsequent synthesis. The control of these processes is fairly comprehensive. Multiple enzymes involved in glycogen metabolism react allosterically to metabolites which are able to communicate the energy requirements of the cell. These allosteric reactions permit the adaptation of enzyme activity to the requirements of the cell where the enzymes are specifically expressed. Glycogen metabolism is also controlled through hormonally activated cascades, resulting in the reversible phosphorylation of enzymes, thus modifying their kinetic characteristics. Regulation through hormones enables the glycogen metabolism to regulate itself to the requirements of the whole organism. Through these two mechanisms, glycogen break-down is incorporated into glycogen production. First of all, the metabolism will be investigated, followed by the regulation of enzymes and then the complex incorporation of regulatory mechanisms.

17.4 Mechanical Properties of the Tumor Drive Metabolic Dysfunctions

Oxidative Phosphorylation (Mitchell) Versus Warburg Effect (see below)

Oxidative phosphorylation takes place in the mitochondria of all animal and plant tissues and is a linked pathway between the process of substrate oxidation and the generation of ATP. During the Krebs cycle, hydrogen ions (or electrons) are transported from the two transport molecules NAD or FAD to the electron transport pumps. The protons are transported to the intermembrane region, in which they concentrate in a sufficiently high concentration in order to phosphorylate the ADP to ATP. In agreement with the chemiosmotic linkage hypothesis, the ATP synthesis reduces the electrochemical gradient of the protons and thus encourages the respiratory chain to pump a greater number of protons through the inner membrane of the mitochondria and sustain the gradient. The Warburg effect of aerobic glycolysis is a major metabolic characteristic of cancer and has even evolved as a classical hallmark of cancer. It is generally acknowledged that tumors exhibit increased glycolytic activity and compromised oxidative phosphorylation (Warburg effect).

The Warburg Effect and the Crabtree Effect

In particular, in order to proliferate, cells have to meet the energy requirements resulting from vital processes such as the biosynthesis of macromolecules, the replication of DNA, the generation of ion gradients and the keeping of the cell structure. Mitochondria are key elements in the energy metabolism since they synthesize the major part of cellular ATP with oxidative phosphorylation. However, it has been proposed that cancer cells inhibit the metabolism of mitochondria (Warburg 1956; Diaz-Ruiz et al. 2011). Warburg's early findings suggested that cancer cells have reduced respiration along with elevated lactate levels, indicating that they depend primarily on fermentative metabolism for ATP formation (Warburg 1956). Despite the reduction in energy output due to the "glycolytic phenotype," this appears to enable an improvement in the cell proliferation rate and is therefore suitable for other rapidly growing cells (Brand 1997).

Since the inhibition of oxidative metabolism also takes place in the oxygen environment, this metabolic phenomenon is referred to as "aerobic glycolysis," also termed the "Warburg Effect." The specific benefits that cancer cells gain from this metabolic switching mechanism are still poorly understood. While it is quite conceivable that these cells make use of this mechanism to multiply in hypoxic environments as seen in solid tumors (Gatenby and Gillies 2004). A different hypothesis is that reducing the oxidative metabolism could assist these cells to avoid apoptosis (Gatenby and Gillies 2004; Tomiyama et al. 2006; Gogvadze et al. 2010). A relationship between the glycolytic phenotype and the invasive capacity of the tumor mass has also been proposed (Winehouse 1972). Nevertheless, there is a tremendous amount of evidence questioning the paradigm of the strictly "glycolytic" cancer cell (Moreno-Sánchez et al. 2007). It could be demonstrated that some glioma, hepatoma and breast cancer cell lines have functioning mitochondria and receive their ATP mainly through oxidative phosphorylation (Guppy et al. 2002; Rodríguez-Enríquez et al. 2006; Martin et al. 1998; Pasdois et al. 2003). In addition, it has been found that some cancer cells can reversibly shift between fermentation and oxidative metabolism, based on the availability or absence of glucose and ambient environmental circumstances (Rodríguez-Enríquez et al. 2008; Rossignol et al. 2004; Smolková et al. 2010).

Most interestingly, a more recent model suggested that "glycolytic" cells may undergo metabolic symbiosis with "oxidative" cells through lactate shuttling (Sonveaux et al. 2008). This indicates that the metabolic plasticity monitored in vitro may have an influence on the tumor physiology in vivo. It is thus important to gain an insight into the mechanisms that enable cancer cells to achieve reversible regulation of their energetic metabolism. In this respect, a clearly identified feature of some cancer cells is the glucose-triggered suppression of respiration and oxidative phosphorylation (Crabtree 1929; Díaz-Ruiz et al. 2009). However, this is a short-lived and reversible phenomenon and is referred to as the "crabtree effect." This reversible shift might be an improvement for cancer cells in vivo, as it enables them to modify their metabolism to fit the quite diverse microenvironment in malignant solid growths. The energy metabolism of cancer cells has lately been recommended as a candidate target for therapy (Rodríguez-Enríquez et al. 2006; Díaz-Ruiz et al. 2009) and a large part of the current scientific research in this field is focused on this specific topic. Therefore, it is crucial to have a full comprehension of the long-term metabolic reprogramming of cancer cells (the Warburg effect) and the short-term adjustment mechanisms (the crabtree effect), since focusing on both would yield considerably more efficient therapeutic approaches.

17.5 Introduction into Desmoplasia-Driven Glycolysis

Desmoplasia is a feature of the majority of solid tumors and induces fibrosis due to aberrant extracellular matrix deposition, reshaping and post-translational modifications. The consequent rigid tumor stroma not only impairs vascular integrity to cause hypoxia and hamper drug administration, but also increases aggressiveness through enhancing the activity of the major growth, invasion and survival mechanisms. Interestingly, however, several of the pro-tumor signal transduction pathways, which are activated mechanically by the rigidity of the extracellular matrix, also stimulate the absorption of glucose and the aerobic glycolysis, and an impaired metabolism is a well-known feature of cancer. As a matter of fact, recent findings indicate that metabolic modifications and an aberrant extracellular matrix can collaboratively affect the aggressiveness of cancer cells and their resistance to treatment. Consequently, enhanced techniques for monitoring tissue mechanics and metabolism hold the potential for enhanced diagnostics and treatment to increase the stiffening of the extracellular matrix and improve mechanosignaling, which might lead to a better patient outcome (Tung et al. 2015). In this context, the interaction between the mechanics of the extracellular matrix and the metabolism in tumor biology is debated and it is proposed that the surveillance of these processes and focusing on their regulatory pathways may contribute to improve the diagnosis, treatment and prophylaxis of malignant transformation.

In recent decades an awareness of the significance of cell and tissue mechanics in the development, progression and metastasis of cancer has become apparent (Sung et al. 2007; Mbeunkui and Johann 2009; Castano et al. 2011; Sautes-Fridman et al. 2011; Spano and Zollo 2012; Quail and Joyce 2013). The pro-tumor cell and tissue mechanism phenotype become evident both as an inherent change in cell and tissue architecture and mechanics, and as a modification of the biophysical characteristics of the tumor microenvironment, which encompasses mechanics, geometry and topology of the extracellular matrix (Lelievre et al. 1998; Nelson and Bissell 2005; Paszek et al. 2009). Specifically, the constituent of the non-cellular extracellular matrix of the tumor's microenvironment is crucial in facilitating invasion and metastasis (Butcher et al. 2009). It now turns out that interactions between the cell and the associated extracellular matrix establish a dynamic mechanical interaction, facilitated by an equilibrium between the contractility of the cell and the physical state, such as elastic versus rigid properties, of the microenvironment of the extracellular matrix. This biophysical balance is responsible for the regulation of a number of critical cellular processes, among them cell differentiation, proliferation and motility. It is not astonishing that disturbances in the biophysical dynamics of the epithelium and the extracellular matrix enhance the activity of major signal transduction pathways governing the growth, invasion and survival of the tumor (Lu et al. 2012). Most interestingly, many of the signaling pathways that support the aggressive response of cancer cells also govern glucose intake and glycolysis. Deregulated cellular energetics is a burgeoning hallmark of aggressive cancers and mirrors the reprogramming metabolic process closely related to the proliferation of cancer cells under the hostile conditions of the tumor microenvironment (Cairns et al. 2011; Hanahan and Weinberg 2011). In this section, a brief overview of the unrelated functions that altered tissue metabolism and an abnormal, stiffened extracellular matrix perform in the initiation and progression of cancer is presented, highlighting new findings indicating a regulatory link among these two key cancer regulators. It is argued that novel traction-capable biomarkers and effective prevention and therapy programs can be achieved through addressing the reciprocal feedback loop involving the cancer metabolism and the anomalous, mechanically altered extracellular matrix.

17.5.1 The Tumor Microenvironment, Such as Matrix Dynamics, Matrix Cross-Linking and Mechanical Signaling, Alters Cancer Cells

Is the Stiffening of the Extracellular Matrix Hallmark of Cancer?

Tumors are generally fibrotic and are frequently accompanied by excessive and aberrant deposition, turnover and post-translational modifications of the extracellular matrix all of which rigidify the stroma progressively (Levental et al. 2009; Egeblad et al. 2010; Lopez et al. 2011; Mouw et al. 2014). Consequently, the architecture and physical characteristics of the tumor-linked extracellular matrix are completely dissimilar to those of the normal tissue stroma. Collagens in especially are frequently deposited in large quantities at an early stage of cancer development (Zhu et al. 1995; Kauppila et al. 1998; Egeblad et al. 2010; Conklin et al. 2011; Santala et al. 1999) and significantly enhance tissue stiffening, encouraging mechanosignaling in stromal fibroblasts and in the carcinogenic epithelium (Provenzano et al. 2006; Levental et al. 2009; Samuel et al. 2011).

The process of extracellular processing of collagen is facilitated by the breakdown of collagen propeptides through specific proteinases. Subsequently, enzymes in the form of lysyl oxidases and lysyl hydroxylases act as catalysts for covalent intermolecular connections between collagens and elastin, thereby altering their elasticity and tensile strength and profoundly altering various cellular characteristics. The collagen fibers lining the natural epithelial structures of soft tissue, including the mammary gland and lungs, are typically crimped and anisotropic (Egeblad et al. 2010). In the breast, for instance, collagen I levels are strictly regulated, with aligned type I collagen fibers running parallel to the epithelial tubules and relaxed, non-oriented fibrils enclosing the terminal ductal lobular areas. In contrary, transformed epithelial tissue distinguishes itself through an abundance of strongly linearized type I collagen bundles, which are frequently oriented vertically to the majority of the non-invasive tumor epithelial mass and alongside which cancer cells migrate on thick fibers extending into the parenchyma (Ingman et al. 2006; Provenzano et al. 2006; Levental et al. 2009). Consequently, malignancy is connected with the expression of enzymes that reshape the extracellular matrix, among them cysteine cathepsins, heparanases, 6-O-sulfatases and matrix metalloproteases (Ilan et al. 2006; Kessenbrock et al. 2010). Breast tumors and head and neck cancer also exhibit increased levels of collagen-processing enzymes, which comprise lysyl oxidases and lysyl hydroxylases, chronic activation of lysyl oxidase, and represent a bad prognostic marker (Le et al. 2009; Barker et al. 2011).

Increased lysyl oxidase-facilitated cross-linking also rigidifies the extracellular tissue matrix through collagen fiber stabilization and improved mechanical characteristics, resulting in invasion, metastasis and reduced patient survival (Erler and Giaccia 2006; Levental et al. 2009; Barker et al. 2011). Due to this changed architecture and deregulated reshaping, the tumor stroma is generally stiffer than normal stroma. In fact, the extracellular matrix attached to part of the breast cancer tissue is more resilient and stiffer compared to the normal stroma (Levental et al. 2009; Lopez et al. 2011; Plodinec et al. 2012). Most interestingly, the architecture of collagen has also been associated with the decrease in breast cancer incidence linked to pregnancy at an early age (Maller et al. 2013). Although there is an enhanced abundance of collagen I in the parous stroma, the matrix is in fact becoming less linearized and accompanied by a reduction in tissue stiffness. These are features of extracellular collagen processing that are used to block the invasive phenotype even with increased deposition.

Tumor progression is associated with dynamic and reciprocal reshaping of the local extracellular matrix through resident cancer cells, stromal fibroblasts and invading endothelial and immune cells (Lu et al. 2012). For example, the local stiffening of the extracellular matrix induces actomyosin cell contractility, which simultaneously alters the neighboring extracellular matrix through increased localized traction forces and enhances cell growth, invasion and survival by favoring focal adhesion maturation and improving growth factor receptor and GPCR signaling (O'Brien et al. 2001; Croft et al. 2004). In fact, the enhanced cancer cell and fibroblast contractility trigger a tension-driven matrix remodeling and encourage the linearization of the collagen fibrils enveloping the invasive tumor front-line. These linearized collagen fibrils not only affect tissue mechanics, but also ease the invasion and metastasis of cancer cells. For instance, migrating transformed mammary epithelial cells have been reported to move alongside these linear collagen bundles (Ingman et al. 2006; Wyckoff et al. 2007; Provenzano et al. 2008a, b). The persistent stiffening of the extracellular matrix in vitro and in vivo fosters the malignant progression of oncogenically altered, premalignant mammary epithelial cells (Levental et al. 2009). In addition, the inhibition of tissue fibrosis and stiffening through prevention of collagen cross-linking increased the time latency of the tumor generation in Her2/Neu mice and markedly decreased tumor occurrence and tumor grade (Levental et al. 2009). The blocking of lysyl oxidase activity has also been demonstrated to inhibit the metastasis of cancer cells and can alter the phenotypic properties of basal cells (Pickup et al. 2013). It is obvious that altered extracellular matrix dynamics are not only a hallmark of the tumor's microenvironment, but also play an active role in the aggressiveness of cancer cells.

Since changed dynamics of the extracellular matrix are a crucial feature of cancer, new strategies utilizing these abnormalities are of enormous clinical importance. Tumor-associated collagen signatures (TACS) are intended to identify the procedure of alterations in collagen that have been monitored and categorized as indicators of the progression of breast cancer. TACS-1 appears in mammary tumors that exhibit a localized accumulation of collagen around the tumor lesion at a very early stage of the formation of the tumor. As the size of the tumor grows, the collagen fibers straighten parallel to the tumor border and are referred to as TACS-2. The most dangerous signature, TACS-3, is found when collagen fibers are bundled and aligned orthogonally to the tumor boundary. This type of signature exhibits the strongest stiffness and predicts a bad chance of survival for women with invasive breast cancer. In pre-clinical modeling and in human breast cancer, when TACS-3 is available, breast cancer cells enter the neighboring stroma predominantly following the stiff collagen fibers, wherein the TACS-3 collagen fibers are capable of identifying focal points of microinvasion of breast cancer cells (Schedin and Borges 2009; Schedin and Keely 2011; Maller et al. 2013). These investigations underline the clinical significance of altered dynamics of the extracellular matrix for the determination and comprehension of the aggressiveness of cancer.

17.5.2 Mechanosignaling and Cellular Tension Fosters Cancer Progression

Cells perceive exogenous forces and transform them into signal transduction pathways through a mechanism referred to as mechanotransduction (Hamill and Martinac 2001). Using mechanotransduction, cancer cells react to mechanical alterations in their microenvironment, involving enhanced compression, increased hydrostatic pressure, greater shear stress and increased stiffness of the extracellular matrix or cell–cell tension. Whereas compression and shear stress exerted on tissue are forces at the tissue level, the enhanced extracellular matrix stiffness encountered by cancer cells and their associated stroma cells occurs at the nanoscale and continues down to the tissue level to progressively foster cancerous transformation (Levental et al. 2009; Plodinec et al. 2012). A change in the quantity or composition of the extracellular matrix, an elevation in the degree of cross-linking of the matrix or the reorientation of the matrix fibrils interact to stiffen the tumor tissue and thus increase the activity of key signaling pathways that encourage cell growth and survival and foster both invasion and migration (Paszek et al. 2005; Provenzano and Keely 2009; Mouw et al. 2014). Cells perceive, process and react to mechanical and other biophysical signals from the extracellular matrix by employing a complex network of systems comprising specialized, mechanosensitive ion channels and adhesion receptors such as integrins and the adhesion plaque proteins attached to them, which transport signals deeply into the cell via an interlinked cytoskeletal—molecular motor system. The orchestrated mechanism allows the cells to alter their shape and architecture of the nucleus and control the cellular tension caused by actomyosin, which finally reshapes their local extracellular matrix (Janmey 1998; Giancotti and Ruoslahti 1999; Bershadsky et al. 2003; Katsumi et al. 2004; Lele et al. 2007; Ingber 2008). The most well-defined matrix-directed mechanotransduction mechanism is triggered by extracellular matrix-driven activation of integrins and their clustering, leading to talin enrollment, and stimulates focal adhesion arrangement, FAK and paxillin phosphorylation, consecutive Rho-ROCK-dependent actin remodeling and reciprocal actomyosin-driven cell contractility through the activation of vinculin (Fig. 17.6) (Paszek et al. 2005; Provenzano et al. 2008a, b; Geiger et al. 2009; Samuel et al. 2011).

Focal adhesions then interact with G-protein coupled receptors and transmembrane receptor tyrosine kinases to improve cell growth, survival, migration and invasion through ERK, JNK and

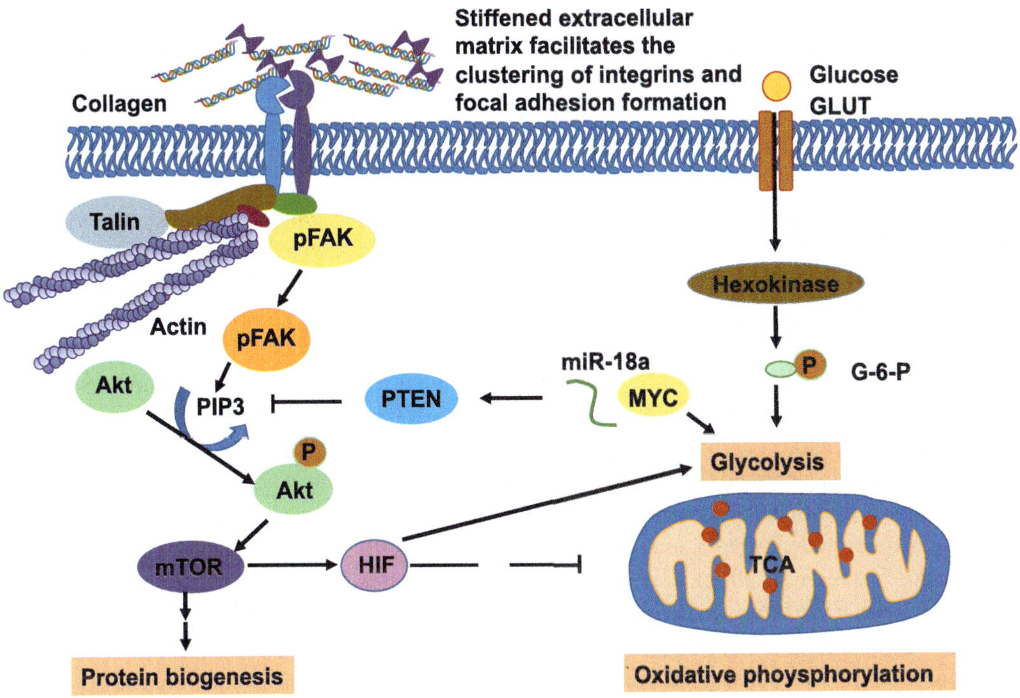

Fig. 17.6 Model for the regulation of the metabolism of cells due to externally stiffened extracellular matrix. The perturbed forces of the tumor microenvironment foster a glycolytic phenotype

phosphoinositide 3-kinase (PI3K) signal transduction processes (Wozniak et al. 2003; Paszek et al. 2005; Levental et al. 2009). In specific, PI3K controls cell growth, invasion and survival of cells and has been associated with both tumor aggressiveness and resistance to therapy (Gershtein et al. 2000, 2007). It is not unexpected that the stiffness of the extracellular matrix and the increased epithelial mechanosignaling reinforce growth factor receptor-facilitated PI3K responses to encourage malignant cell behavior (Mouw et al. 2014).

The phosphatase and tensin homologue (PTEN) protein, a tumor suppressor commonly found in human cancers, is the main negative modulator of the phosphoinositide-3-kinase (PI3K) signal transduction pathway and therefore plays a crucial role in cell growth, cell proliferation and survival (Maehama and Dixon 1998; Leevers et al. 1999; Hollander et al. 2011). It is not astonishing that even a slight decrease in PTEN is enough to enable the malignant metamorphosis of a tissue and to speed up the progression of the tumor and its resistance to therapy (Alimonti et al. 2010). In fact, the excellent cellular sensitivity to PTEN levels and function is mirrored in the various regulatory mechanisms that govern its expression. For example, normal human mammary cells tend to express high concentrations of PTEN protein, whereas non-malignant mammary epithelial cells (MECs), which are multiplied in vitro on standard tissue culture plastic, express low concentrations of PTEN. However, the expression of the PTEN protein is reestablished and probably maintained when MECs are grown inside a three-dimensional reconstituted basement membrane to synthesize the somehow differentiated acini found in vivo in normal breast tissue (Liu et al. 2004; Fournier et al. 2009). Curiously, the elastic modulus of tissue culture substrates is ultraphysiological, while the properties of reconstituted basement membrane hydrogels are strikingly analogous to those of normal human breast (Paszek et al. 2005; Johnson et al. 2007). In fact, the extracellular matrix within a human breast tumor is more rigid compared to normal tissue, which means that matrix compliance can provide stabilization of the PTEN protein. Recently, there is consensus that a rigid extracellular matrix enhances levels of miR-18a directly aiming at PTEN mRNA, both directly and indirectly by decreasing HOXA9 values to regulate both malignant transformation and metastasis in culture and in vivo by enhancing PI3K/AKT signaling mechanisms (Mouw et al. 2014). These results underline a direct connection between PTEN, the biophysical characteristics of the extracellular matrix, and PI3K-based malignancy.

In comparison to their non-malignant equivalents, transformed epithelial cells exhibit remarkable differences in their mechanoreactivity, which has been revealed to foster cell growth, invasion and survival in reaction to enhanced stiffness in primary tumors (Wang et al. 2000; Paszek et al. 2005; Tilghman et al. 2010). Many cancer cells feature distinctive intermediary filament profiles and cytoskeletal structures that alter their migratory characteristics. The substitution of a keratin-based cytoskeleton with a vimentin-based cytoskeleton is a characteristic of the epithelium-mesenchyme transition (EMT) in breast tissue and substantially enhances the invasive capacity (Pagan et al. 1996; Willipinski-Stapelfeldt et al. 2005; Kokkinos et al. 2007). Cancer cells are also often extremely contractile, and this increased cytoskeletal tension affects their capacity to grow, permeate and survive in the tumor's life-threatening microenvironment. For example, transformed cells applied to conformal substrates undergo rapid spread, cell growth and survival, while their normal counterparts are not able to (Wang et al. 2000). Cancer cells also produce high traction forces that break cell–cell connections, interfere with tissue polarity, encourage anchorage-independent survival and favor invasion (Paszek et al. 2005). High cellular tension also raises Rho/ROCK activity, which, through the involvement of its subsequent effectors, stimulates the proliferation, migration and survival of cancer cells (Samuel et al. 2011). These results show the inverse association between micromechanical cues of the extracellular matrix and cell pattern control, the propagation of these cues through signal transduction schemes and the subsequent regulation of pivotal characteristics associated with tumorigenesis, invasion and metastasis.

17.5.3 Experimental Techniques Determine Extracellular Matrix Stiffness and Tissue Mechanics

The degree of stiffening of the tumor, usually defined by elastic modulus determination, has been utilized as a tool for detecting malignant changes in an otherwise healthy tissue. The modulus of elasticity is a degree of how slightly a material distorts and is determined through division of stress through strain; in general, the higher the modulus of elasticity, the more rigid the material. Thus, the modulus of elasticity provides a quantitative method for identifying mechanical variations among tissues. Several techniques have been employed with different degrees of accuracy to characterize alterations of the elastic modulus at the tissue scale in tumors. Unrestricted compression and shear rheology assays have both been employed to quantitatively measure the amount of intrinsic stiffening of the mammary gland at the transition from normal to premalignant and subsequently to invasive cancer and have found that the stroma tissue bordering the invading epithelium is considerably stiffer than normal (Paszek et al. 2005; Levental et al. 2009). Shear rheological measurements have also been employed to observe alterations in tissue stiffness throughout the progressive stages of liver fibrosis (Georges et al. 2007; Perepelyuk et al. 2013). These techniques are suitable for bulk tissue-level assays in which resolution and sensitivity are more limited. Recent findings, however, highlighting the operational effects of extracellular matrix stiffness and modified cell tension on transformation potential and treatment response, require more accurate assessments. Therefore, whereas unrestricted compression and shear rheology yield valuable quantitative measurements of the tissue elasticity module, they are restricted through their millimeter resolution.

Atomic force microscopy (AFM) represents an imaging technique that enables a more precise imaging of the elastic modulus of tissue. AFM determines the interaction force acting between a sample surface, such as a living cell, and a micro to nanoscale spring-type cantilever. Interactions occurring between tip and sample generate a force that causes the cantilever to deflect, which is then optically tracked by a photodiode and transformed into an interaction force via the spring constant of the cantilever. The force measuring capability of the AFM was applied to measure the regional tissue elasticity modulus of the cultured brain slices (Elkin et al. 2007), the pancreatic ductal adenocarcinoma tissue (Ozdemir et al. 2014), and the excised mammary tissue (Lopez et al. 2011; Maller et al. 2013; Pickup et al. 2013; Mouw et al. 2014). These measurements revealed high-resolution force-heat maps revealing the appearance of stiffness domains that coincide with areas of collagen fiber expansion and alignment. Whereas AFM has efficiently distinguished alterations of the tissue elastic modulus on the microscale, and the clinical applicability of the technique is strongly restricted due to the demand for untreated specimens.

In the last decade, elastography, which can use either ultrasound or magnetic resonance imaging (MRI) techniques, has been clinically employed to image the tissue elasticity module. This technique has proven to be useful in tumor diagnosis (Khaled et al. 2006), and in the intersurgical identification of tumor tissue, while resecting gliomas (Selbekk et al. 2005; Unsgaard et al. 2006). Despite its widely spread application, elastrography is restricted by its limited resolution in comparison with the length scales involved in many diseases, such as hundreds of micrometers for ultrasound elastography and millimeters for MRI elastography, the long acquisition times connected to MRI elastography, difficulties with tissue penetration and the user variability of ultrasound elastography. Due to the constraints described earlier, new imaging modalities need to be designed to assess tissue elasticity modulus as a tool for precise and accurate diagnosis and surveillance of cancer.

17.6 Glycolysis and Metabolism During Oncogenesis

Transformed cells display abnormal growth rates that require increased ATP levels to aid in the biogenesis of the cell's four most important macromolecules: carbohydrates, proteins, lipids and nucleic acids. The physician and scientist Otto Warburg put forward the hypothesis at an early stage that the cancer cells have to modify their metabolism in such a way that they can meet this energy demand. Warburg discovered that cancer cells switch from the conventional ATP-producing system, oxidative phosphorylation (OXPHOS), toward glycolysis in the presence of adequate oxygen. The strange thing about this phenomenon, referred to as the Warburg effect or aerobic glycolysis, is that glycolysis is normally restricted to anaerobic ATP generation and represents a much less efficient generator of ATP per glucose molecules compared to OXPHOS (Warburg et al. 1927; Warburg 1956). Although glycolysis does not appear to be intuitive, it permits much quicker ATP production than OXPHOS when required, although higher glucose uptake is necessary. Although it has huge consequences for fundamental cancer cell biology and clinical diagnostics and therapy, tumor metabolism was widely disregarded for decades, however, until it was found recently that transforming oncogenes affect the key cellular metabolism in direct and indirect manners.

Presently, a burgeoning set of oncogenes and tumor suppressor genes is functionally connected to the metabolism of cells (Cairns et al. 2011). Ras and Src have been the first oncogenes linked to the Warburg effect, as they initiated the expression of glucose transporters (Glut genes) (Flier et al. 1987). Similarly, PI3K/AKT up-regulation results in enhanced expression of GLUT1 and GLUT4, which hence promotes elevated consumption of glucose in order to fuel glycolysis, whereas it additionally activates through phosphorylation multiple metabolic enzymes, among them hexokinase and phosphofructokinase 2, which can catalyze the initial step of glycolysis, the transformation of glucose to glucose-6- phosphate. In addition, AKT activity results in enhanced mRNA translation and protein biogenesis through the activation of mTOR pathway (Laplante and Sabatini 2012). In addition, another abundant oncoprotein, MYC, directly stimulates aerobic glycolysis through the upregulation of glucose transporters and phosphofructokinase-2 (Osthus et al. 2000) and collaborates with HIF1 to stimulate hexokinase 2 and pyruvate dehydrogenase (Kim et al. 2007).

It is not astonishing that the diminished function of tumor suppressors can imitate the metabolic phenotype generated by the expression of oncogenes. Through TP53-inducible glycolysis and apoptosis regulator (TIGAR), p53 decreases glycolysis and increases OXPHOS due to cytochrome c-oxidase 2 (SCO2) (Bensaad et al. 2006; Matoba et al. 2006). P53 also retains the expression of PTEN, which is the most significant regulatory retardation in the PI3K/AKT/mTOR pathway (Stambolic et al. 2001). In addition, PTEN depletion or the activation of AKT leads to enhanced glycolysis due to exhaustion of the ATP reservoirs (Fang et al. 2010). These results underline the pleotropic regulation of metabolism through traditional oncogenic and tumor suppressing signal transduction scaffolds (Fig. 17.7).

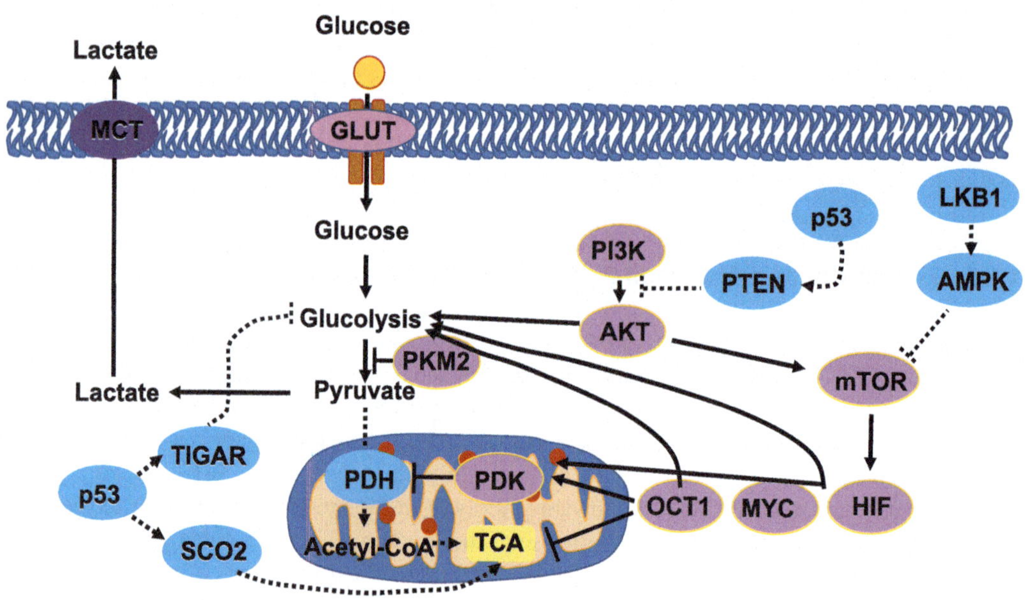

Fig. 17.7 Cancer cell metabolism regulation. Normal cells employ the TCA cycle (OXPHOS) for the generation of ATP, whereas cancer cells are switched towards via oncogenic signaling towards a glycolytic phenotype. PI3K triggers the activation of AKT, which fosters glycolysis through the direct regulation of glycolytic enzymes and the activation of mTOR. mTOR has several effects on the metabolism, whereas it drives the glycolytic phenotype through the enhancement of the activity of HIF1 in order to engage the hypoxia-based transcriptional program. HIF1 triggers the expression of glucose transporters (GLUT), glycolytic enzymes and PDK1, which impairs the entry of pyruvate in the TCA cycle. The LKB1 tumor suppressor triggers the activation of AMPK, which in turn inhibits the glycolytic phenotype through the inhibition of mTOR. MYC acts together with HIF to decrease the expression of several genes encoding glycolytic proteins, whereas it elevates the mitochondrial metabolism. The tumor suppressor p53 opposes the glycolytic phenotype through the suppression of glycolysis via TIGA, which elevates the mitochondrial metabolism through SCO2 and favors the expression of PTEN. OCT1 functions in an opposing way to trigger the transcription of genes facilitating glycolysis and suppressing the oxidative phosphorylation. The switch towards the PKM2 isoform impacts glycolysis through the impairment of the pyruvate kinase reaction and the redirection of the substrates towards alternative biosynthetic and reduced NADPH-generating routes

17.6.1 Tumor Redox

A major constituent of the tumor microenvironment and a weak determinant of the final outcome is hypoxia, which arises due to the association of the improved metabolic and replicative condition of cancer cells with a damaged vascular network (Vaupel et al. 2004; Vaupel 2009). By activating HIF1, hypoxia intensifies glycolysis, and persistent hypoxia results in mitochondrial malfunction (Papandreou et al. 2006). Altogether this causes an abnormal cellular redox condition. The redox, the equilibrium of the decreased and oxidized state of a cell, is determined by the content of reactive oxygen and nitrogen species, such as ROS and RNS, with antioxidative molecules. As a direct result of elevated ATP output and increased macromolecular biogenesis, cancer cells aggregate extremely high concentrations of ROS (Bae et al. 1997; Szatrowski and Nathan 1991; Toyokuni et al. 1995).

These oxidative molecules participate in tumor progression through post-translational modifications by increasing the activity of oncogenic proteins including PI3K, MAPK and Src while simultaneously continuing to block tumor suppressors such as PTEN and PTP (Thomas et al. 2008). In addition, ROS facilitates the transcription of Hif1a and GLUT1, thereby establishing a feed-forward mechanism that intensifies the Warburg effect in cancer cells (Bell et al. 2007). Even though elevated ROS levels are commonly considered pro-cancerous, specific levels of oxidative injury to proteins and nucleic acids can induce cell death through apoptosis (Takahashi et al. 2006b; Cairns et al. 2011), highlighting another possible therapeutic option for the treatment of tumors.

17.6.2 Therapeutic Targeting of Tumor Metabolism

An interesting development in the field of cancer therapy involves the utilization of the tumor's aberrant metabolism (Galluzzi et al. 2013). Metformin, serves as an agonist of AMP-activated protein kinase (AMPK), which is a precisely tuned sensor of the cellular AMP/ATP balance. At high levels, AMPK activates and negatively adjusts mTOR, thereby decreasing the production and proliferation of specific proteins. Secondary AMPK activity decreases the production of insulin and gluconeogenesis (Shackelford and Shaw 2009). It is interesting to note that even before studying the mechanism of action of metformin, it was discovered that it enhances insulin sensitivity and reduces blood sugar concentrations in diabetics. This is the most widely available medication for the treatment of type-2 diabetes nowadays (Dowling et al. 2011). Extensive meta-analyses were conducted from long-term datasets of diabetes patients, which showed that patients who received metformin decreased considerably the incidence of cancer and cancer-related death (Evans et al. 2005; Bowker et al. 2006) that has finally contributed to clinical trials in several solid tumor types encompassing colorectal, endometrial, prostate, breast and brain tumors (Kasznicki et al. 2014). Whereas mouse cancer models and cell culture investigations demonstrate that metformin inhibits mTOR, halts growth and initiates apoptosis of cancer cells (Isakovic et al. 2007; Liu et al. 2009), and the available patient data are rather restricted. Metformin decreases mTOR transcription and the proliferative index of the Ki67 stain in breast cancer patients (Hadad et al. 2010) and also enhances response levels to treatment (Jiralerspong et al. 2008). Treatment with metformin can also decrease the rate of colon cancer development and the mortality rate of prostate cancer (Hosono et al. 2010; Patel et al. 2010). Numerous ongoing pharmaceutical clinical trials will clarify the effectiveness of metformin as a first-line medication, in a combination treatment and in cancer protection efforts. Besides metformin, multiple pharmaceutics aim several points and levels of the metabolism, which undergo preclinical and clinical investigations (Galluzzi et al. 2013). As metformin can negatively regulate the signal transduction routes downstream of AKT (Gwinn et al. 2008), which are crucial in the extracellular matrix remodeling around tumor and the process of mechanosignaling, it is feasible that the application of this drug can aid to diminish the fibrosis of tissues and thus circumvent partially the impacts of a rigidified extracellular matrix.

Among the exciting new data, confirming this hypothesis are findings that metformin treatment in rats prone to heart failure and it has decreased perivascular fibrosis and avoided heart failure. In extension, therapy with metformin in a mouse model of heart fibrosis resulted in a lowering of TGFbeta concentrations and increased Smad3, thereby reducing the accumulation of collagen deposits and subsequently the incidence of fibrosis (Xiao et al. 2010; Cittadini et al. 2012). These investigations point to associations between changed tumor mechanics and metabolism, which are explained in the following.

17.6.3 Techniques for Metabolic Imaging

Connections between disturbed metabolism and malignancy have promoted the emergence of a range of non-invasive, clinically replicable strategies for imaging tissue metabolism in situ. By administering radioactively labeled glucose to patients, tumors can be identified due to their extremely high absorption of the metabolite. [F18]-fluorodeoxyglucose (F18-FDG) is absorbed by gluten transporters and phosphorylated through hexokinase. Due to the absence of a 2′ hydroxyl group, this radioactively labeled glucose analog is not metabolized any further, which enables the identification of metabolically active tissue utilizing positron emission imaging (PET). Clinically, FDG-PET has been demonstrated to be useful in diagnosis, staging, monitoring of therapeutic outcome and forecasting the outcome of breast, colon and lung tumors (Gambhir et al. 2001; Jadvar et al. 2009). Medical histopathology and experimental xenograft investigations have revealed that the absorption of FDG is associated with elevated HIF1a and glut expression (Ai et al. 2007; Kaira et al. 2013) and with the proliferative stage of Ki67 (Han et al. 2009).

MRI and magnetic resonance spectroscopy (MRS) enable the quantitative determination of metabolites within a tissue. Especially in the diagnosis and prognosis of brain tumors, this new technique has proven to be very beneficial (McKnight et al. 2007). An adjustment of the MRS is the hyperpolarized 13C-MRS, which enables the monitoring of the fate of [13C]-radiolabeled metabolites. This novel technique seems to be entirely safe in patients (Nelson et al. 2013) and has been employed in cell culture and mouse models to investigate glutaminase activity in hepatocellular carcinoma cells (Gallagher et al. 2008) and the aberrant metabolism of alpha ketogluterate within IDH1 mutant gliomas (Chaumeil et al. 2014). The recent progress in this technology and the decrease in the expense of 13C reagents are expected to take this technique into the clinic (Kurhanewicz et al. 2011). Metabolic alterations appear early in the development of tumors, therefore, advances in FDG-PET and MRS metabolic imaging offer the opportunity to support the diagnosis of multiple tumor types at very early disease stages.

Fluorescence imaging of NADH/NADPH and FAD provides a further approach that is beneficial in the metabolic analysis of stored biopsy tissues. These ubiquitous metabolic cofactors, which perform a key role in glycolysis, OXPHOS and ROS-scavenging, are autofluorescent and because of their comparatively high concentration inside the cell, they can be detected quite easily. Biochemical compounds have uncovered alterations in the redox state of these molecules in cancer cells in the early stadium of the carcinoma and in the progression of the tumor (Zipfel et al. 2003; Villette et al. 2006; Skala et al. 2007). In supplement to intensity, fluorescence lifetime imaging (FLIM) of NADPH and FAD provides an excellent indication of metabolic status, since every one of these molecules has a unique FLIM value, according to whether it is freely bound in the cytosol or attached to protein. The metabolic imaging of NADPH and FAD using FLIM is consistent with metabolic disorders in culture, and this technology is capable of discriminating basal cell skin sarcomas from adjacent environmental normal skin and gliomas from normal brains (Bird et al. 2005; Leppert et al. 2006; De Beule et al. 2007; Galletly et al. 2008; Walsh et al. 2013). FLIM of NADPH and FAD detects early carcinomas in situ of the mouse mammary gland (Conklin et al. 2009), which indicates that these techniques may be helpful for in situ assessments of the early alterations linked to aerobic glycolysis. In line with this finding, the latest evidence points to alterations in NADPH fluorescence lifetime values as an early biochemical marker of alterations in tissues that are regarded as morphologically normal in standard pathological evaluation. The clinical significance of these concepts is further underlined by the fact that FLIM metabolic imaging in both breast xenograms and human squamous cell carcinoma of the head and neck is similar to the response to tamoxifen (Walsh et al. 2013; Shah et al. 2014).

17.7 Tissue Mechanics-Based Regulation of the Metabolism of Cancer Cell and Cancer Progression

17.7.1 Physical Forces Impact the Metabolism of Cells

A stiffened extracellular matrix is increasingly valued as a critical element of the tumor's microenvironment, and data are emerging demonstrating that mechanics and cell adhesion govern the metabolism. After integrin-facilitated extracellular matrix adhesion, cells display a peak in ROS, which is required for downstream integrin signaling and activation of Src (Chiarugi et al. 2003) and the quenching of ROS decreases vSrc-driven tumorigenicity and invasion (Giannoni et al. 2005). Besides adhesion, the compliance of the extracellular matrix has clear implications on cell growth and proliferation processes (Paszek et al. 2005; Tilghman et al. 2010) that require enhanced metabolic flow. Most interestingly, EMT, a highly invasive process, is preferred on stiffer substrates (Tilghman et al. 2010; Lee et al. 2012a) and mesenchymal cell types feature increased aerobic glycolysis by enhanced Zeb1-triggered expression of glucose transporters (Masin et al. 2014) and snail-induced methylation of fructose-1,6-bisphosphate (Dong et al. 2013). In addition, stiffness-induced Rac1b promotes the generation of the NADPH oxidase complex, which stimulates the production of ROS and strengthens the EMT phenotype in breast epithelial cells (Lee et al. 2012a).

The interaction of physical forces and metabolism is probably a well-known phenomenon in endothelial cells. Similar to the regulation of homeostasis by the compliance of the extracellular matrix, physiological, such as a regular pulsatile or laminar flow, shear stress assists in controlling gene expression and the redox of endothelial cells. However, under circumstances of elevated or abnormal, such as turbulent or oscillating, shear stress, these cells tightly control Rac1 activity, which drives the production of ROS, boosts the activation of NFKB and promotes atherosclerosis (Tzima et al. 2002). In addition, irregular shear stress in ischemia results in the PI3K/AKT activation and the generation of ROS in a feed-forward regime (Chatterjee et al. 2012). There are strong associations among cell adhesion, mechanical signaling and metabolism, however, further investigations are necessary to gain an insight into the molecular mechanism through which these phenomena mutually interfere.

17.7.2 Tissue Stiffness Controls the Metabolism and Progression of Cancer

Both a modified metabolism and disordered extracellular matrix dynamics definitely act as fundamental driving forces for the progression of cancer. More recent findings, however, indicate that these two distinctive features of cancer may also be linked and advance cancer progression through a collaborative, synergistic framework. The PI3K signaling pathway is key in the organization of glucose metabolism. In fact, the activation of the PI3K/AKT cascade leads to the reliance of the cells upon a high glucose flow, which is a classical biochemical phenotype of cancer cells (Buzzai et al. 2005). In supplement to the enabling of mTOR-facilitated signaling, the PI3K signal transduction pathway also adjusts the absorption and processing of glucose. In insulin-independent tissues, the PI3K signaling pathway via AKT governs the expression of the glucose transporters GLUT1 and GLUT4 (Kohn et al. 1996), through hexokinase, thus, encouraging phosphofructokinase activity and enhancing glucose absorption (Rathmell et al. 2003; Biswas et al. 2012), which is crucial for increased aerobic glycolysis. But PI3K enhancement also appears in the context of enhanced matrix stiffness through integrin-driven activation of FAK and the downstream pathways (Provenzano et al. 2008a, b; Provenzano and Keely 2009; Xia et al. 2004; Kallergi et al. 2007; Guan 2010). Therefore, the regulation of PIP3 through PI3K and the subsequent activation of AKT and mTOR constitute a mechanistic tool through which matrix stiffness and enhanced extracellular matrix deposition may affect cell metabolism, enhancing both cell proliferation and survival (Wozniak et al. 2003; Paszek et al. 2005), as well as the promotion of the oncogenic transformation and cancer metastasis (Levental et al. 2009). Matrix stiffness can also govern the metabolism by acting on p53, since AKT can regulate the degradation of p53 through MDM2 (Hu et al. 2012). In addition, p53 is modulated further through the reduction of cysteine residues and can be deactivated when the redox balance of the cell undergoes a transition (Forsberg and Di Giovanni 2014).

Matrix stiffness enhanced PI3K activation of AKT is also amplified by β-catenin and MYC-driven matrix stiffness-induced expression of miR18a, inhibiting PTEN expression and encouraging the progression of cancer (Mouw et al. 2014). In accordance with these results, the knock-out of FAK in breast tumors reduces the activity of PI3K and AKT, whereas the expression of PTENs is increased and tumor growth and metastasis are disturbed (Provenzano et al. 2008a). It has also been demonstrated that the activation of PI3K through growth factors is by itself not enough to initiate the accumulation of PIP3 and that oxidation-facilitated inactivation of PTENs is necessary to raise the concentration of PIP3 to a level high enough to initiate downstream signaling processes (Kwon et al. 2004). Therefore, the inactivation of PTEN facilitated through oxidation and stiffness might collaboratively contribute to initiate the subsequent PIP3 signal events that accelerate the advancement of the cancer.

There is also strong evidence for a dynamic connection of the attachment of cells to the extracellular matrix and the regulation of the metabolic activity. The detachment of mammary epithelial cells from the extracellular matrix leads to an ATP deficit due to the lack of transport of glucose. Nevertheless, this shortcoming can be corrected through overexpression of ErbB2, thereby re-establishing glucose absorption through stabilization of EGFR and PI3K activation by glucose-driven flow via the antioxidant-generating pentose phosphate pathway. Most interestingly, ATP lack can also be prevented through antioxidant therapy that fails to restore glucose absorption, which relies on the activation of fatty acid oxidation, and a process that is blocked by release-stimulated reactive oxygen species (Schafer et al. 2009). These observations show that cell adhesion to the extracellular matrix is directly regulating the metabolic activity and offers a tool to enhance cell survival in the disturbed microenvironment of tumors of the extracellular matrix through antioxidant recovery of ATP production.

Useful microarray data gained from the cultivation of normal mouse mammary gland cells in dense versus conformal 3D collagen matrices indicate an improved oxidative phosphorylation pattern in stiffer surroundings (Provenzano and Keely 2009). Although this seems to mirror the discrepancy between normal and transformed cells, there is growing evidence that oxidative phosphorylation is also elevated in tumors (Solaini et al. 2011).

Elevated oxidative phosphorylation associated with elevated aerobic glycolysis provides cells with a resilient metabolism that may be required for the extensive energy requirements of proliferation, migration and metastasis. In fact, AKT/mTOR activation of 4EBP1 results in enhanced mitochondrial biogenesis (Goo et al. 2012) and could provide an additional agent for matrix stiffness to control metabolism through a FAK/PI3K/AKT signal transduction pathway. Increased deposition of the extracellular matrix, elevated alignment of collagen fibers, increased cross-linking of collagen and finally fibrosis represent all together key issue of progression of breast cancer. There are not only solid findings indicating that the stiffness of the extracellular matrix stimulates abnormal cell growth, survival, invasion and metastasis due to stimulation of the PI3K/AKT framework, but also that the stiffness of the extracellular matrix may lead to altered cell metabolism.

Elevated accumulation, reshaping and cross-linking of the extracellular matrix inside the breast and the consequently elevated traction forces are characteristic for breast cancer progression and also contribute to the activation of signaling pathways that are crucial for proliferation, survival, invasion and metastasis. More recent findings also indicate that some of these extracellular matrix-activated signaling pathways enhance aerobic glycolysis and the absorption of glucose. Evidence is presented here that tissue mechanosignaling stimulates signal networks that at the same time support breast carcinogenesis and the reprogramming of the metabolism. The altered dynamics of the extracellular matrix and metabolic reprogramming are both decisive factors in the aggressiveness of cancer. Therefore, the research concentrating on their interlinked feedback cycle can be a powerful direction for the discovery of biomarkers and the development of drugs.

Evolving clinical data have started to associate stiffer matrices with enhanced cancer aggressiveness in human patients and highlight how matrix stiffness and dysregulated mechanosignaling compounds may offer valuable guidelines for clinical intervention and the discovery of novel biomarkers (Mouw et al. 2014). In basal-like or luminal B specimens, the stiffness of the extracellular matrix of the breast and the miR-18a expression was markedly increased compared to the values obtained in normal and luminal A tumor biopsies. Based on this mechanically-facilitated miR-18a expression pattern, PTEN expression was observed to be significantly expressed in the more distinct luminal B carcinomas and normal breast tissue, while less distinct luminal B, basal-like and $HER2^{+}$ tumors exhibited decreased expression of PTEN together with elevated levels of activated AKT. The miR-18a values also differentiate between luminal A and luminal B tumor and forecast the results in patients suffering from luminal type breast cancer. These results seem to indicate that tumor stiffening at the macroscopic scale may run parallel to potential future disease aggression, with tumor size, grade and subtype as unrelated factors impacting tissue mechanics. In addition, the stiffness of the extracellular matrix serves as a biophysical indicator that can differentiate luminal A from luminal B tumors, with miR-18a and PTEN acting as manageable biomarkers. However, the imaging procedures that are capable of identifying tissue stiffness in a clinical environment are unfortunately severely compromised. However, due to the relationship between metabolism and tissue stiffness, the determination of surrogate metabolic markers for stiffness can be helpful in closely monitoring the development and the progression of cancer. An impaired metabolism and an amended tissue elasticity module were typically identified as unrelated characteristics of cancer. Actually, PI3K/AKT activation, superfluous metabolic energy and glycolysis are all linked to a bad prognosis in triple-negative breast cancer (Rose et al. 2007; Phipps et al. 2011; Kim et al. 2014a). Curiously, these factors seem to affect the clinical result by creating similar and mutually overlapping pathways. In reaction to the enhanced mechanical stiffness, the clustering of integrins, ERK activation, cytoskeleton remodeling and Rho-GTPase-dependent contractility are employed and promote the establishment of focal adhesions, that in turn work together with G-protein-coupled receptors and transmembrane receptor tyrosine kinases to modify the PI3K signaling pathway and the subsequent AKT signaling pathway. The mutations in the main compound PI3K result in a subsequent activation of AKT and forecast bad survival and the possibility of reappearance (Gonzalez-Angulo et al. 2011; Ocana et al. 2014).

Specific genetic defects in mitochondrial respiration and the rise in glycolysis additionally activate AKT signal transduction processes through elevations in metabolic potential-driven through NADH, encouraging the enrollment of GLUTs to the cell surface and enhanced absorption of glucose and glycolysis (Manning and Cantley 2007). These high glucose concentrations have also been demonstrated to enhance the aggressive response of triple-negative breast cancer, involving proliferation, activation of oncogenic signal transduction, avoidance of apoptosis and reduced drug efficacy (Wahdan-Alaswad et al. 2013). Curiously, the women with a high risk of suffering from triple-negative breast cancer exhibits an altered metabolic state (Ibarra-Drendall et al. 2011). Random periareolar fine needle aspiration specimens taken from high-risk patients have demonstrated to express elevated levels of activated proteins associated with RTK/AKT/mTOR, RTK/AKT/ERK and mitochondrial apoptosis, which indicates that meaningful metabolic malfunction persists long prior to tumor detection in these patients. The high correlation between the stiffness of the extracellular matrix and AKT leads to the assumption that these high-risk women may also have modified the extracellular matrix, rendering them major potential subjects for elastography imaging investigations. PET imaging technology is currently also being employed to measure glucose absorption in these high-risk patients. Many of these high-risk patients are also participating in metformin chemotherapy prophylaxis investigations. It will be intriguing to determine whether this early condition of metabolic malfunction is reversible and whether early aiming of the changed metabolism can be an adequate defense tool. In summary, these investigations emphasize the key function of AKT and redox activation in determining the aggressive response to triple-negative breast cancer. Considering the pivotal contribution of matrix stiffness in adjusting PI3K/AKT signaling cascade, these investigations also yield proof of the degree of convergence of mechanosignaling and redox activation observed in aggressive cancers. Because AKT network signaling is functional in premalignant tissue of women at high risk of breast cancer before the invasive phenotype is established (Pilie et al. 2011; Ibarra-Drendall et al. 2012), this raises the prospect that combined therapies aiming at both metabolism of cells and tissue stiffness could be useful in stopping disease progression in a more effective manner.

An altered metabolism and the stiffening of the extracellular matrix constitute two key components of the tumor microenvironment that strongly influence cancer progression. While new discoveries indicate that the two are closely related by their downstream signaling pathways, additional research is required to fully understand their complex association. Nonetheless, this emerging field of cancer research might open up new ways for the discovery of tractable biological markers and novel cancer treatments.

17.8 Tumor-Stroma Mechanics Regulate the Amino Acid Concentration for Tumor Growth

The deregulation of superimposed extracellular matrix deposition and metabolism encourages tumor aggressiveness by maintaining the activation of the major growth, invasion and survival routes. However, the mechanisms through which the biophysical characteristics of the extracellular matrix are connected with the metabolic processes and the progression of tumors are still unclear. In cancer cells and carcinoma-associated fibroblasts (CAFs), it has been seen that stiffening of the extracellular matrix mechanoactivated glycolysis and the metabolism of glutamine, thereby managing the flow of non-essential amino acids across the tumor niche. In particular, a metabolic cross talk between CAF and cancer cells has been identified in which CAF-derived aspartate promotes cancer cell proliferation, whereas cancer cell-based glutamate compensates the redox state of CAF to encourage the reorganization of the extracellular matrix. Overall, the results combine mechanical stimuli with a disordered tumor metabolism, thereby highlighting a new metabolic framework within tumors, where different types of fuel sources are utilized to stimulate growth and aggressiveness (Bertero et al. 2018). In addition, this investigation aims to identify possible metabolic drug targets for cancer therapeutic development.

Non-transformed cells inside the tumor microenvironment constantly evolve together with cancer cells to support tumorigenesis (Kalluri 2016; Quail and Joyce 2013). To foster tumorigenesis, normal fibroblasts interoperate with cancer cells and convert into CAFs. CAFs encourage widespread tissue remodeling or the formation of tumor niches. The continued creation of a complex, dynamic network of cytokines, chemokines, growth factors and enzymes for matrix reorganization finally alters the physical and chemical characteristics of the tumor (Pickup et al. 2014). In fact, tumors exhibit altered tissue and cell mechanics (Kai et al. 2016). Experimental models revealed that increasing the stiffness of the extracellular matrix encourages malignancy and, inversely, the inhibition of matrix stiffening decreases tumor occurrence (Levental et al. 2009). Whereas genetic changes in cancer cells certainly induce and promote malignancy (Watson et al. 2013), tumors advance through a dynamically developing extracellular matrix that regulates practically every aspect of cancer cell behavior, including sustained proliferation, circumvention of growth suppression, resistance to death, forced angiogenesis and onset of invasion (Pickup et al. 2014). Nonetheless, the underlying principles that connect the mechanotransduction of the extracellular matrix to the molecular mechanisms that influence cell performance and modify malignancy are only at the beginning of understanding, such as the processes that enable cells to perceive and adjust to external mechanical forces.

Tumors change their metabolic pathways in order to sustain the cell-autonomous proliferation, particularly, even under the worse nutrient conditions of the tumor microenvironment (Vander Heiden and DeBerardinis 2017). Among of the most prominent alterations of the bioenergetics of the tumor represents the Warburg metabolism. In specific detail, the Warburg metabolism includes a chronic shift in the generation of energy from the mitochondrial oxidative phosphorylation toward glycolysis, and elevation in glutaminolysis, amino acid and lipid metabolism, flow via the pentose phosphate pathway, synthesis of biomacromolecules and mitochondrial biogenesis, as well as the balanced redox state (Ben-Sahra and Manning 2017; Vander Heiden and DeBerardinis 2017). Previous mechanistic investigations in cancer exploited hypoxia treatment to investigate this metabolic switch. However, multiple tumors are also identified through a pronounced metabolic disorder in the absence of strong hypoxic stress (Hensley et al. 2016). There is emerging evidence that the molecular regulatory elements of metabolic disorders function independently of the presence of hypoxic stress. Since extracellular proteins can offer a source of nutrients to the starved cancer cells (Davidson et al. 2017), it can be assumed that mechanical cues of the tumor niche provide additional molecular signals to govern the capture of nutrients by cancer cells in order to fulfill their metabolic requirements.

Two closely related transcriptional co-activators intrinsic to the Hippo signaling pathway, yes-associated protein 1 (YAP) and transcriptional co-activator with PDZ-binding motif (TAZ or synonymously referred to as WWRT1) are mechano-activated through a rigid extracellular matrix and act as key regulators of cellular proliferation, survival and polarity, especially during the development and progression of cancer (Dupont et al. 2011; Yu et al. 2015). There is elevated evidence that indicates a key interplay between mechanotransductions, including the YAP/TAZ pathway, and the metabolism of cells (Santinon et al. 2018; Yang et al. 2018), and processes closely linked to consumption of glucose and Warburg metabolism (Bays et al. 2017; Enzo et al. 2015; Mo et al. 2015; Sorrentino et al. 2014; Wang et al. 2015a). However, elevated glycolysis by itself is not enough to cover the entire metabolic needs of proliferating cells. A less valued feature of cancer cell metabolism is the necessity to reconfigure the flow of carbon and nitrogen to produce the elements essential for rapid cell growth. The TCA cycle also acts as a supply of precursors for the production of lipids to create membranes, amino acids to produce macromolecules and proteins, and nucleotides to synthesize DNA and RNA, especially in fast-growing tissues (Vander Heiden and DeBerardinis 2017).

The ongoing operation of the TCA cycle necessitates the restocking of carbon intermediates commonly referred to as anaplerosis (Vander Heiden and DeBerardinis 2017). Glutaminolysis, which represents the glutamine catabolism to produce the TCA cycle intermediates, via glutaminase (GLS1) is one of the well-recognized anaplerotic pathways in cancer cells (Hensley et al. 2013). The capacity of glutaminolysis to deliver aspartate (Asp) generation for the direct triggering of proliferation has been observed in malignant cancer cells (Birsoy et al. 2015; Sullivan et al. 2015). In line with this, the exchange of amino acids among cells and the tumor has been hypothesized to function as an alternative route for the metabolic supply for the entire tumor (Sousa et al. 2016; Yang et al. 2016). CAF is a major cell type in the squamous cell carcinoma (SCC) stroma and an essential facilitator of the desmoplastic reaction (Quail and Joyce 2013) that leads to enhanced stiffness of the tumor niche. Therefore, this investigation examined whether a metabolic reaction to the niche stiffness of the tumor governs tumor progression. In particular, what to investigate was whether the stiffness of the extracellular matrix directly regulates both cancer cell metabolism and CAF metabolism and controls nutrient availability within the tumor niche to assist the metabolic demands of tumor progression. A hypothesis of this kind might combine effectively tumor niche stiffness and metabolic malfunction as two inherently similar molecular determinants of cancer.

17.8.1 Mechanical Cues Lead to Metabolic Reprogramming and Coordinated Non-essential Amino Acid Exchange in the Tumor Niche

The stiffness of the extracellular matrix stiffness induces the activation of carcinoma cells and the tumor-supporting activities of CAFs (Pickup et al. 2014). To maintain these energy-consuming activities, the cells rearrange their metabolism appropriately (Vander Heiden and DeBerardinis 2017). To establish whether mechanical/physical signals transmitted through the stiffness of the extracellular matrix regulate the metabolism of carcinoma cells and CAFs, a series of metabolome investigations were carried out on carcinoma cells and CAFs cultured on physiological soft (about 1 kPa) or pathophysiological stiff (about 8 kPa) extracellular matrix. In the first place, changes in the oxygen consumption rate and the acidification rate of extracellular fluids, measurements of mitochondrial activity and glycolysis have been performed. Whereas glycolysis is elevated in carcinoma cells when they are stiff, mitochondrial activity exhibited minimal alterations. In contrast, in CAFs, both mitochondrial activity and glycolysis were found to be markedly enhanced by the stiffness. In accordance with elevated glycolysis in rigid state, an increase in glucose utilization and lactate release in these cells was noticed. It is significant that a reduction in TCA cycle intermediates was found in carcinoma cells, whereas an increase in TCA cycle intermediates was found in CAFs, which is in line with extracellular flow analyses that indicated increased oxidative phosphorylation in CAFs on rigid extracellular matrix. U-13C6 glucose tracing experiments in carcinoma cells and CAFs proved that glycolysis is triggered when subjected to a rigid matrix. Matrix stiffening thus converts the glucose metabolism in cancer cells.

Because aerobic glycolysis by itself is not adequate to cover the metabolic demands of proliferating cells, and the hypothesis was put forward that stiffening of the extracellular matrix also refocuses the amino acid metabolism. Consequently, kinetic investigations were carried out and showed that both carcinoma cells and CAFs raised glutamine (Gln) concentrations in a stiff matrix. In accordance with elevated glycolysis and Gln metabolism, a rise in transcript and protein levels and enzymatic activities of genes associated with the Gln metabolism were detected. In addition, the kinetic investigations showed that glutamate (Glu) was excreted by carcinoma cells and Asp was absorbed, whereas in CAF Asp was excreted and Glu was absorbed. In the relationship, between the liberation and intake rates of Asp/Glu are the more critical factors to be assessed in the analysis of cancer cells in vivo. Thus, in the case of exposition to rigid matrix, the secreted Asp/Glu values were found to be similar to the Asp/Glu values taken up by the cells, pointing to potential cross talk inside the microenvironment.

To validate metabolic cross talk between carcinoma cells and CAFs, a range of metabolic and functional investigations were conducted. In particular, there were attempts to identify molecules that were over-represented in the CAF medium (and thus excreted by CAFs), under-represented in the CAF medium after contact with carcinoma cells (eliminated by carcinoma cells), and over-represented within carcinoma cells exposed to the CAF medium (absorbed by carcinoma cells). Based on earlier results that revealed a complementary secretion and absorption of Asp and Glu between CAFs and carcinoma cells, the focus was directed specifically at amino acids. However, Asp was discovered to be produced and secreted by CAFs and apparently absorbed by carcinoma cells. In addition, it was observed that Glu is elevated in the carcinoma cell medium, reduced in the carcinoma cell medium after contacting CAF cells and elevated within the CAFs exposed to the carcinoma cell medium. These results thus suggest that metabolic cross talk between carcinoma cells and CAFs may be based on dynamic alterations in amino acid flow. To investigate whether such metabolic interactions are capable of maintaining matrix stiffened activated pro-tumoral responses, alterations in cell proliferation and cell contractility in carcinoma cells and CAFs, respectively, were examined. Carcinoma cells exhibited an enhancement of proliferation when the cells were treated with Asp-containing media or with CAF-conditioned medium, a characteristic that depends on the matrix stiffness during the conditioning process and can be reproduced with several primary samples. In particular, the CAF-conditioned medium maintained the capacity to enhance carcinoma cell proliferation even when the CAF-conditioned medium was exposed to three freeze–thaw cycles or heating, such as 100 °C for 15 min, suggesting that the factor(s) lacked tertiary structure and that the factor(s) might be a metabolite. The treatment of carcinoma cells with Asp displayed the capacity to elevate the carcinoma cell proliferation to a level similar to that of CAF-conditioned medium. Inversely, the capacity of the CAF to produce contractile forces was enhanced when cells were handled with media supplemented with Glu or with carcinoma cell conditioned medium, a characteristic that also depends on the matrix stiffness experienced during the conditioning process and was repeatable in multiple cancer cell lines. Similar to CAFs, the carcinoma cell conditioned medium maintained the capacity to raise the contractility of the CAF after freezing and thawing or heating. Treatment of CAF with Glu demonstrated the capacity to enhance the contractility of CAF to an extend similar to that of the carcinoma cell conditioned medium. In summary, rigid matrices maintain the metabolic requirements of tumor microenvironment cells by activating glycolysis and Gln metabolism, and thereby aligning cell-to-cell metabolic communication to encourage pro-tumor activities.

17.8.2 In a Stiff Surrounding, Elevated GLS1 Expression and Glutamine Metabolism Contribute to Metabolic Reprogramming

Gln is a progenitor of several non-essential amino acids (NEAAs) and anaplerotic carbon source for the TCA cycle in cancer cells (Hensley et al. 2013). However, as Gln can be applied in a cell type-specific manner and CAFs are based on a completely opposite metabolism than carcinoma cells (Ghesquiere et al. 2014), it was examined whether carcinoma cells and CAF vary in the usage of Gln. Treatment of the cells with 2 mM U-13C5-Gln resulted in a threefold to fivefold enhancement of Gln absorption and Gln conversion to Glu after exposition to a rigid matrix. Whereas in carcinoma cells stiffness did not enhance the amount of carbon from Gln participating in the TCA cycle, in CAFs, Gln was a primary source of carbon for the TCA cycle; 13C was predominantly incorporated in alpha-ketoglutarate and succinate and to a smaller extent in malate and fumarate and in NEAA Asp, a molecule biosynthesized out of intermediates from the TCA cycle (Birsoy et al. 2015; Sullivan et al. 2015). While in CAFs carbon originating from Gln concentrated in Asp with enhanced stiffness, in carcinoma cells, carbon originating from Gln was supplied to Glu synthesis but could not enter the TCA cycle and Asp biosynthesis. Accordingly, the inhibition of GLS1 either through a small interfering RNA (siRNA) specific for GLS1 or through a pharmacological inhibitor of GLS1 (CB839) diminished Gln metabolism in both carcinoma cells and CAFs, as well as diminished the recreation of Glu of carcinoma cells and impaired the recreation Asp of CAFs. Inversely, the overexpression of GLS1 (pGLS1) was adequate to enhance the release of Glu and Asp of carcinoma cells and CAFs, even when the cells were grown on soft matrix.

Next, it has been determined whether GLS1 inhibition governs the activities of cancer cells. In accordance with earlier findings (Gross et al. 2014), GLS1 inhibition obtained with siRNA or CB839 decreased carcinoma cell proliferation and reduced CAF contractility. Crucially, in cells with diminished GLS1 activity, cellular proliferation and contractility were re-established through Asp or Glu supplementation. Ultimately, the aim was to investigate whether this metabolic, cross talk dependent cell proliferation and contractility is also affected through GLS1 impairment. In accordance with our metabolic findings, the CAF-conditioned medium from CAF treated with either siRNA GLS1 or CB839 failed to induce proliferation of carcinoma cells. Most importantly, the administration of exogenous Asp has managed to rescue the proliferation of carcinoma cells. In addition, CAF-conditioned medium of CAFs overexpressing GLS1 and grown on soft matrix, induced the activation of the proliferation of carcinoma cells. Similar outcomes have been found for the CAFs. Therefore, the stiffness-dependent elevation of GLS1 is essential for cell-autonomous pro-tumoral activities and for cancer cell cross talk due to the modulation of Glu and Asp synthesis. To investigate whether metabolic cross talk in a microenvironment is pivotal for cancer progression, 3D co-culture invasion assays, such as spheroid culture assays were carried out that achieved GLS1 inhibition in CAFs, carcinoma cells or even both. The impairment of GLS1 in either CAFs or carcinoma cells did not inhibit the invasion, whereas the inhibition of GLS1 in both CAFs and carcinoma cells reduced the invasion of cells. Intriguingly, the addition of Asp/Glu can restore the invasive capacity of cancer cells. Overall, these results strongly conclude that Gln-dependent Asp and Glu exchange between CAFs and carcinoma cells is essential to build a pro-invasive tumor niche and maintain pro-tumor activities.

17.8.3 Aspartate and Glutathione Impact the Cancer Cells or CAFs

In next step, it has been investigated how carcinoma cells are able to metabolize CAF-based Asp. The incubation of cells with 5 mM U-13C4-Asp causes to a twofold to threefold elevation in the intracellular Asp concentration, even though 2 mM Gln and 4.5 g/L glucose were available. The carbon from Asp did not join the upstream glycolytic intermediates and only slightly impacted the TCA cycle. Another key function of Asp in proliferating cells involves the retention of the synthesis of nucleotides (Sullivan et al. 2015). Therefore, it is assumed that CAF-derived Asp is crucial for sustaining nucleotide biosynthesis in cancer cells subjected to rigid environments. In fact, in stiff matrices, Asp seems to be a crucial carbon source for the generation of pyrimidine.

Subsequently, to assess how Glu was employed in the CAF metabolism, investigations were conducted on the traceability of U-13C-Glu. The incubation of cells to 5 mM U-13C-Gluc resulted in a twofold to threefold rise of the intracellular Glu reservoir, despite the presence of 2 mM Gln and 4.5 g/L glucose. Carbon from Glu did not participate in the upstream Gln synthesis. As with Gln, the carbon from Glu was included in the TCA cycle. Whereas the stiffness slightly enhanced the amount of carbon from Glu that adds to the TCA cycle, the 13C label was clearly integrated into the glutathione pathway on a stiff matrix. These findings indicate that carcinoma cell-derived Glu is utilized in CAFs to sustain redox homeostasis. The next step was to examine whether carcinoma cell-derived Glu can compensate for the redox state of CAFs. In stiff environments, total glutathione, such as reduced glutathione (GSH) and oxidized glutathione (GSSG), and GSH alone were reduced in CAFs and correlates with the accumulation of reactive oxygen species (ROS) and superoxide. Therapy either with carcinoma cell conditioned medium or with Glu has restored these characteristics. Accordingly, in CAFs treated with carcinoma cell conditioned medium or Glu, an elevation of F-actin bundles and an augmentation of P-MLC2 was found, indicating the triggering of cellular contractility, which has been determined under similar circumstances.

17.8.4 SLC1A3 Provides Aspartate/Glutamate Exchange in the Tumor Niche to Foster Cancer Progression

To establish whether the Asp/Glu exchange between CAFs and carcinoma cells represents a key factor in tumor progression, an attempt was undertaken to define a cell surface transporter involved in the facilitation of this exchange. In particular, the hypothesis has been raised that such a transporter is expressed in CAFs and in carcinoma cells, overexpressed in carcinoma cells grown on a rigid matrix and overexpressed in CAFs compared to normal fibroblasts. A qRT-PCR screen of the Asp/Glu transporter family members, such as SLC1A1-7, have revealed SLC1A3. To establish whether SLC1A3 is accountable for metabolic cross talk between CAFs and carcinoma cells, Asp and Glu uptake rates and the intracellular Asp and Glu concentration of SLC1A3 knockout cancer cells exposed to conditioned media have been examined. After inhibition of SLC1A3, both the Asp usage rate and the Glu uptake rate were reduced in the carcinoma cells and CAFs, respectively. Similarly, a reduction in intracellular levels of Asp and Glu was seen. Subsequently, the question was raised whether the exchange of Asp/Glu by SLC1A3 is crucial to maintain the pro-tumoral activities of carcinoma cells and CAFs. The SLC1A3 inhibition by itself is obtained by siRNA or TFB-TBOA, which is one of the pharmacological inhibitors of the SLC1A1-3 family and only moderately influences the proliferation of carcinoma cells. Most importantly, the concurrent inhibition of GLS1 and SLC1A3 inhibited the rescue of Asp/Glu in the proliferation of carcinoma cells. In a similar way to carcinoma cells, blocking SLC1A3 in CAFs changed the generation of contractile forces and slightly reduced the reconstruction of the extracellular matrix. Most significantly, pharmacological or genetic inhibition of GLS1 and SLC1A3 jointly in CAFs silenced the rescue of the pro-tumoral responses of CAF through Glu.

Next, the question was raised whether the blocking of metabolic cross talk in the tumor niche is adequate to suppress tumor progression. In accordance with previous results, in the 3D co-culture assay, the siRNA knockdown of SLC1A3 in CAFs or carcinoma cells alone was not adequate to alter cellular invasion. However, siRNA knockdown of SLC1A3 alone in both CAFs and carcinoma cells or siRNA knockdown of SLC1A3 and GLS1 together within these cells impaired the invasion of cancer cells. Similarly, it has been observed that pharmacological inhibitors of SLC1A3 and GLS1 can reach this outcome. In the next step, it has been investigated whether the overexpression of SLC1A3 and/or GLS1 is able to trigger the invasion of cancer cells inside a soft tumor niche. In a low collagen microenvironment, such a 1 g/L collagen, cancer cells cannot invade into them, even when CAFs are available. The overexpression of both GLS1 and SLC1A3 in carcinoma cells was able to modestly lead to an activation of cancer cell invasion. The supplementation of ribose in order to increase collagen stiffness also mildly enhanced cell invasion, a characteristic that was reinforced by the overexpression of GLS1 and SLC1A3 in carcinoma cells. Finally, these results showed the critical function of the tumor niche amino acid cross talk for the progression of the tumor via the mechanical activation of SLC1A3.

17.9 Alteration of YAP/TAZ-Driven Mechanotransduction Pathway Guides Metabolic Reprogramming in Cancer Cells

Based on previous results revealing that mechanotransduction guides the glycolysis and glutaminolysis in reaction due to mechanical stress in the vasculature of the lung (Bertero et al. 2016), it has been investigated whether the manipulation of the mechanotransduction cascade impacts on the metabolic reprogramming and subsequently on the behavior of tumor niche cells. In cancer cells, pharmacological inhibition of FAK using PF573228, ROCK using Y27632, or YAP/TAZ using verteporfin, reduced glycolysis, and the effects of tumor niche stiffening on the amino acid secretion or consumption are attenuated. Accordingly, the expression of GLS1, LDHA and SLC1A3 proteins was reduced after pharmacological inhibition and siRNA knockdown of YAP/TAZ. Most importantly, the overexpression of YAP was adequate to initiate the protein expression of GLS1, LDHA and SLC1A3, despite the fact that the cells were grown on soft matrices. As stiffening of the extracellular matrix also enables cell proliferation (Pickup et al. 2014), a well-known modulator of cell metabolism (DeBerardinis et al. 2008), and to decouple mechanodependent proliferation and mechanodependent metabolic reprogramming, an in vitro system was designed in which mechanotransduction can be activated while maintaining cell proliferation. The cells were grown on soft hydrogels containing magnetic beads-coated with collagen I. Mechanotransduction was then triggered by attracting integrins through the addition of a magnet to this system. With this system at hand, it could be shown that the activation of the mechanotransduction YAP/TAZ relocated into the cell nucleus and activated the transcription of GLS1 and SLC1A3 together with the familiar YAP-dependent genes CTGF and CYR61, but not the negative control, GAPDH. Accordingly, GLS1 and SLC1A3 expression in the cells was reduced through a YAP/TAZ-specific siRNA knockdown. Finally, several presumed-binding sites for YAP/TAZ complexes, such as the TEAD sites, were uncovered through sequence analysis of the promoter regions of GLS1 and SLC1A3. Chromatin immunoprecipitation (ChIP)-qPCR revealed the direct binding of YAP at a minimum one site for an individual gene. In summary, YAP and TAZ are integral parts of the mechanotriggered metabolic reprogramming processes stimulated through the stiffness of the extracellular matrix.

Subsequently, the question was raised whether metabolic interactions between CAFs and carcinoma cells are also influenced by the YAP/TAZ-dependent mechanotransduction routes. First, it was determined whether the blocking of YAP/TAZ alters its configuration when the medium is subjected to the conditioning process. In fact, inhibition of YAP/TAZ-reduced Gln intake in both cells and lowered Asp secretion and Glu intake in CAFs, whereas Glu secretion and Asp intake was inhibited in carcinoma cells. In accordance to these findings, the CAF-conditioned medium of CAF transfected with siYAP/TAZ could not induce the proliferation of carcinoma cells. In contrast, the carcinoma cell conditioned medium of carcinoma cells transfected with siYAP/TAZ could not stimulate CAF contraction. It is important to note that the addition of exogenous Asp or Glu helped preserve the proliferation of carcinoma cells or the contractility of CAFs. In conclusion, it can be said that the YAP-TAZ-dependent mechanotransduction cascade forms an integral component of the stiffness-related metabolic reprogramming of the tumor niche.

17.9.1 The Mechanotransduction Cascade Governs Metabolic Reprogramming

To definitively determine whether the reshaping and stiffening of the tumor niche alters the metabolism and performance of cancer cells in vivo, it was examined in an orthotopic syngeneic mouse model of a highly metastatic breast cancer whether a change in the mechanotransduction cascade directly governs Gln metabolism and the progression of the tumor. Initially, the well-established BALB/c mammary cancer cell lines 67NR, 410.4 and 4T1 were examined to assess whether extracellular matrix-dependent metabolic alterations were retained. The analysis with liquid chromatography-mass spectrometry (LC–MS) shows that stiffening of the extracellular matrix enhanced glycolysis, Gln metabolism and Asp generation and the expression of GLS1, LDHA and SLC1A3. Secondly, applying a known pharmacological inhibitor (beta-aminopropionitrile, BAPN) of lysyl oxidase (Lox), the enzyme in charge of collagen cross-linking and the resulting matrix stiffening (Levental et al. 2009), or using verteporfin, which is a known pharmacological inhibitor of YAP (Park and Guan 2013), it has been determined whether blocking mechanotransduction could impede metabolic alterations and tumor propagation in mice. Treatment with either BAPN or verteporfin-reduced lox activity and the resulting stiffening of the extracellular matrix. In accordance with the in vitro findings, inhibition of mechanotransduction through BAPN or verteporfin caused a reduction in GLS1 and SLC1A3 expression and downstream activity of GLS, which is mirrored in the determination of enzymatic activity and intratumoral concentrations of Asp and Glu. These metabolic benefits further reduced cancer cell proliferation and enhanced the manifestations of terminal stage breast cancer, involving a decrease in tumor volume, lung metastasis and survival. Thus, these results provide causal proof in vivo that tumor niche stiffening relies on mechanotransduction pathways to initiate Gln metabolism and glycolysis of cancer cells, proliferation, invasion and general overall survival.

17.9.2 Exchanges of Aspartate/glutamate Inside the Tumor

In general, the niche is critical for the sustained tumor progression in vivo. In the next step, it could be shown whether an amino acid exchange takes place in the microenvironment of the tumor in vivo. A co-injection system was developed in which CAFs could be genetically engineered and then inserted together with cancer cells into the breast FAT pads of syngeneic mice. Even though earlier investigations showed that co-injection of CAFs with non-invasive 67NR cells enhances tumor growth and promotes the local invasion (Albrengues et al. 2015), the impact of metabolic cross talk on this outcome has not yet been assessed. Therefore, co-injection experiments with 67NR cells were conducted together with CAFs containing conditional knockdown of SLC1A3 and GLS1, either individually or in combination. Fifteen days after the cell injections, the mice were exposed to doxycycline to elicit depletion of GLS1 and/or SLC1A3 in the fibroblasts of the stroma. In accordance with the in vitro 3D experimental results, the combined treatment of GLS1 or SLC1A3 knockdown with a slightly diminished invasion of cancer cells, cancer cell proliferation and restructuring of the extracellular matrix yielded a more intense inhibition. It was determined whether the reciprocal mechanism remained functional in vivo. In line with these results (Yang et al. 2004), 4T1 cells encouraged invasion of tumors through conversion of the local tumor surrounding fibroblasts into CAFs. However, the effect of metabolic cross talk on this outcome has not been examined. Therefore, 4T1 cells with conditional knockdown of SLC1A3, GLS1 or SLC1A3 and GLS1 were directly introduced into the breast FAT pads of syngeneic mice. Both GLS1 or SLC1A3 knockdown by itself reduced the proliferation of cancer cells and inhibited 4T1 tumor progression. Especially the combined medication resulted in a more intense prevention of tumor growth and invasion and thus enhanced the prospects of survival. To assess whether the prevention of metastasis was associated with organ colonization or metastatic growth, 4T1 cells underwent conditional knockdown for SLC1A3, GLS1 or both SLC1A3 and GLS1 before tail vein injection. Knockdown induction was carried out before tail vein injection or after pulmonary colonization to assess the impact of gene knockdown on pulmonary colonization or metastatic growth. The blockade of SLC1A3 impaired lung colonization, whereas the blockade of GLS1 impaired cellular growth. Most importantly, the inhibition of either GLS1 or SLC1A3 dramatically reduced cell growth and pulmonary colonization. Overall, these data suggest that inhibition of GLS1 and SLC1A3 influences multiple consecutive steps of cancer progression in vivo.

Finally, it was examined whether pharmacological blocking of both glutaminolysis and Asp/Glu exchange are important treatments for cancer. A combined treatment in the highly metastatic orthotopic syngeneic 4T1 breast cancer mouse model has been conducted. While either CB839 or TFB-TBOA therapy on its own prevented 4T1 tumor progression and reduced cancer cell proliferation, the combined treatment resulted in increased inhibition of tumor growth and invasion and additionally elevated survival outcomes. In summary, these findings point out that there seems to be a two-way intratumoral metabolic interplay, which can be pharmacologically aimed to hinder the progression of the tumor.

17.9.3 The Expression of GLS1 and SLC1A3 Is Increased in HNSCC Tumors

Finally, based on these findings in mice, it was intended to find out whether people suffering from SCC also exhibited signs of elevated stiffness of the extracellular matrix and resulting modifications of glutaminolysis and Asp/Glu transport and therefore might be sensitive to these joint therapies. It was investigated whether GLS1 and SLC1A3 are also naturally initiated in human head and neck squamous cell carcinoma (HNSCC). Analysis of the head and neck carcinoma dataset of the Cancer Genome Atlas (TCGA) indicated that the GLS1 mRNA was significantly increased in primary HNSCC tumors of about 500 samples compared to adjacent normal tissue of about 40 samples. GLS1 and SLC1A3 mRNA levels were elevated in histological grades 1–4 of human HNSCC. The Kaplan–Meier analysis of the TCGA experiments turned out that increased levels of GLS1 mRNA, but not of SLC1A3, were related to a worse survival rate of the patients with HNSCC. In addition, immunohistochemistry assays have been conducted in order to reveal the induction of GLS1 and SLC1A3 directly in human HNSCC tumors. In specific detail, elevated collagen restructuring in highly invasive cases can be related to a continuous upregulation of GLS1 and SLC1A3. Subsequently, these data suggest that GLS1 and SLC1A3 are increasingly expressed in HNSCC tumors and GLS1 is connected to bad clinical outcomes in patients with HNSCC cancer, which leads to the hypothesis of a possible positive therapeutic effect for targeting these molecules during cancer therapy. To prove this hypothesis, a patient-derived spheroid (PDS) model of HNSCC has been generated. Seven unrelated HNSCC tumors were placed in a collagen-rich microenvironment and the invasion was evaluated in 3D. In correlation with the observations on cell lines, either CB839 or TFB-TBOA treatment on its own reduces the invasion of tumor cells in a PDS model. Combined therapy resulted in increased blocking. To further validate this hypothesis, a patient-derived xenograft (PDX) model of HNSCC dilatation was examined in vivo. Three unrelated HNSCC tumors were subcutaneously grafted into the flanks of nude mice. However, after one week, the tumors can be analyzed and therefore the mice were exposed to vehicle control, CB839, BPTES, TFB-TBOA or a combination of these compounds. In accordance with both our in vitro and in vivo findings, in the PDX model of HNSCC either CB839 or TFB-TBOA therapy in itself slows down tumor growth and decreases the proliferation of cancer cells. Significantly, the joint therapeutic regimen achieved a more resilient outcome in inhibiting tumor growth. Collectively, these findings directly involve GLS1 and SLC1A3 inside the tumor niche—two processes that rely on the stiffening of the extracellular matrix, as critical metabolic intermediators required to perpetuate the activation of cancer cells and the progression of cancer.

17.10 Glutathion Synthesis and Mesenchymal Cell State

Glutaminase has been identified as an appealing target for tumor metabolism due to its importance and elevated activity in cancer with two glutaminase enzymes, such as the kidney type glutaminase (GLS1) and the liver type glutaminase (GLS2), which has been seen to be expressed in mammalian cells (Aledo et al. 2000). Although the expression of GLS1 is upwardly regulated in cancer, the expression of GLS2 is frequently downwardly regulated (van den Heuvel et al. 2012). By alternative splicing, the GLS1 mRNA can generate two isoforms, such as GAC and KGA (Elgadi et al. 1999). The GAC enzyme isoform exhibits enhanced basal activity (Cassago et al. 2012) and has been found to be expressed in cancer (Gross et al. 2014; Timmerman et al. 2013). A selective GLS1 inhibitor, CB-839, is presently undergoing clinical evaluation in several diseases, among them triple-negative breast, lung, kidney and colon cancer, both as a single compound and in combinations (Meric-Bernstam et al. 2015; Wang et al. 2015b). Sensitivity to the inhibition of GLS1 has been linked to both genetic and phenotypic characteristics of tumors (Altman et al. 2016), and enzyme expression by itself does not predetermine the dependency (Gross et al. 2014; van den Heuvel et al. 2012; Ulanet et al. 2014).

Cancers can solely adjust themselves due to bioenergy stress (Biancur et al. 2017; Boudreau et al. 2016; Cheng et al. 2011; Davidson et al. 2016). In an effort to identify the tumors that are best suited for sensitivity to GLS1 inhibitors in all indications, in basal breast cancer patients, the first mechanistic findings were broadened to indicate that addicted cancers are reliant on GLS1 for both mitochondrial bioenergetics and for redox homeostasis. After screening of more than 400 cancer cell lines with small molecule inhibitors of GLS1 and gamma-GCS, a high level of co-regulation of GLS1 and de novo GSH production in all therapeutic areas and a powerful mechanistic connection between GLS1 and the redox balance (Daemen et al. 2018). Using this database, a diagnostic metabolic gene expression signature of GLS1/GCS co-dependence is deduced, its prognostic value is assessed in vivo, and pan-cancer expression data from the Cancer Genome Atlas (TCGA) are analyzed to detect tumors that exhibit shared transcriptional features, in particular, a mesenchymal and basal-like condition, as the highly probable responders to GLS1 blockade (Damrauer et al. 2014; Parker et al. 2009). A framework for a better insight into the principles of GLS1 interdependence in cancer and may help to identify patients who are most susceptible to GLS1 and glutathione signaling pathway inhibitors (Daemen et al. 2018). In view of the targeted potential of GLS1 in cancer and the absence of reliable predictive and pharmacodynamic indicators, there is still a crucial need for a stable algorithm to forecast the response of GLS1 blockers in all indications.

Although prior situational specific investigations have proven metabolic cross talk inside the tumors (Commisso et al. 2013; Davidson et al. 2017; Perera et al. 2015; Sousa et al. 2016; Yang et al. 2016), the influence of mechanotransduction in this mechanism has never been investigated. To decipher this regulatory mechanism, an experimental approach has been employed in which various metabolic transformations in reactive stroma with a stiff extracellular matrix versus healthy stroma with a soft extracellular matrix were revealed. Within these metabolites, the Asp generation by CAFs have been detected to be critical to maintain the proliferation of carcinoma cells. In line with these findings, it has been seen that acquisition of aspartate is a metabolic restriction sensed by tumors in their natural in vivo local environment, and thus, overcoming this restriction seems to be crucial for supporting the overall growth the primary tumor (Garcia-Bermudez et al. 2018; Sullivan et al. 2018). In summary, all these findings highlight the significance of Asp for the progression of the tumor and points out an accurately regulated mechanism through which carcinoma cells can adapt their Asp requirements due to the stiffness of the tumor niche.

In tumor niche microenvironments, YAP/TAZ have been seen to govern the proliferation and contractility of cells due to mechanical characteristics of the local environment (Calvo et al. 2013; Dupont et al. 2011). The findings reveal how the mechanotransduction cascade facilitates through YAP/TAZ the interaction between biophysical cues and the metabolic adjustments needed for growth in the stiff tumor microenvironment. This mutuality between YAP/TAZ with upstream (Enzo et al. 2015; Mo et al. 2015; Sorrentino et al. 2014; Wang et al. 2015a) and downstream metabolic signs indicates an adaptable, feedback-controlled characteristic of this cascade (Bertero et al. 2015a, b; Calvo et al. 2013) and may be somewhat accountable for the individual adjustment of the metabolic schedule within the tumor.

After all, the focus on tumor stroma for therapeutic applications is a growing concept that has been gathering momentum lately (Kalluri 2016; Quail and Joyce 2013). Nevertheless, the therapeutic administration of specific stroma blockers by themselves can lack modest effectiveness. The discovery of glutaminolysis and Asp/Glu interaction by the same co-transporter, such as SLC1A3, as a common metabolic program of CAFs and carcinoma cells represents a feasible way to provide novel treatments for these diseases through targeting both stroma and cancer cells with the same compound. The combined effect of these findings has far-reaching implications for the basic knowledge of how cells adapt their energy requirements on the basis of extracellular mechanical signals. These results support the fast implementation of novel pharmacological inhibitors to combat the metabolic consequences of matrix stiffness and thus stop the progression of cancer. Moreover, such translational investigations in severely desmoplastic carcinomas could be widely extended to other human metabolic settings where such mechanosensitive interactions could act as crucial underlying causes of the disease.

There are still some limitations of the discussed results. In summary, the results of our combined pharmacological, molecular and genetic investigations in different pre-clinical models of SCC identify the mechanodependent metabolic rearrangement of the tumor niche as a key pathogenic mechanism through which matrix stiffness can encourage pro-cancerous activity. Whereas this report concentrates mainly on glutaminolysis/anaplerosis signaling pathways, the experiments were specifically targeted at a few metabolites and on biochemical signaling pathways. The true width of the impact of tumor niche stiffening on metabolic reprogramming might thus be even greater. In addition, although the CAF-carcinoma cell amino acid cross talk of has been recognized as a key mechanism to maintain tumor progression as a reaction to the stiffness of the extracellular matrix, it will be essential to identify whether a comparable metabolic cross talk appears among other cells within the tumor niche. The resolution on the single-cell level and the development of single-cell metabolomics will be essential to unravel the metabolic landscape of tumors.

There are compensatory metabolic networks in pancreatic cancers due to perturbation of the glutamine metabolism (Biancur et al. 2017). Hence, pancreatic ductal adenocarcinomas are not easy to treat and patients require of novel therapies. In the past, it had been shown that these tumors have modified their metabolic needs, leaving them highly susceptible to a variety of adaptations, including a non-canonical Gln pathway and that blocking the downstream components of Gln metabolism results in a reduction in tumor growth (Biancur et al. 2017). Thus, it has been analyzed whether newly designed inhibitors of GLS, which are involved in an early step in the Gln metabolism, constitute a feasible therapeutic approach. Despite strong early effects on in vitro proliferation induced by GLS inhibition, pancreatic cancer cells have inherited adaptive metabolic structures that preserve proliferation in vitro and in vivo (Biancur et al. 2017). An integrated metabolomic and proteomic platform is employed to gain an insight into this type of adaptive reaction and thus to create efficient combinatorial strategies. The metabolism of pancreatic cancer is described as adaptive, and the targeted influencing of the Gln metabolism in combination with these adaptive reactions can bring clinical advantages for the patients.

17.11 Connection Between Metabolism and Cell Mechanics

Cancer cells prefer aerobic glycolysis to oxidative phosphorylation of mitochondria for generating energy, and this metabolic reprogramming is presently acknowledged as a characteristic feature of cancer. Oncogenic cues often converge with this metabolic alteration and increase the capacity of cancer cells to generate the building elements and energy and hence keep the redox homeostasis intact. Changes in cell–cell and extracellular matrix adhesion encourage the invasion of cancer cells, intravasation, anchorage-independent survival in the circulation and extravasation and subsequently the targeting of a distant organ. Crucially, the cells have to rewire their metabolical pathways during this complex and multi-step metastatic process in an attempt to generate the required energy and inhibit oxidative stress. While the individual effects of adhesion molecules and metabolic reprogramming in cancer have been extensively investigated through the years, the interaction between the molecular machinery of cell adhesion and metabolic pathways is still poorly understood in normal and cancer environments (Sousa et al. 2019a). In this section, the current knowledge on the impact of cell–cell and cell–matrix adhesions during the course of metabolism in cancer cells is presented, whereby an emphasis is placed on the traditional function of the cell-cell adhesion molecules, such as epithelial (E)-cadherin and placental (P)-cadherin.

17.11.1 Metabolic Rewiring During the Malignant Progression of Cancer

Under aerobic circumstances, non-proliferating differentiated cells utilize oxidative phosphorylation (OXPHOS) as their primary source of energy. However, in a process referred to as the Warburg effect, cancer cells carry out high aerobic glycolysis in spite of OXPHOS, thereby distinguishing them metabolically from normal cells (Lunt and Vander Heiden 2011). The primary reason why cancer cells use this kind of metabolism is to fulfill demanding biosynthetic and bioenergetic pre-requisites and to preserve their redox state. Even though glycolysis is relatively less powerful in terms of energy generation, it is a high velocity process, which renders it advantageous in comparison with the OXPHOS metabolism. The advancement of glycolysis enables an enhanced redirection of the metabolic flow into various biosynthetic routes in order to produce building stones for the continuation of the active cell proliferation (Lunt and Vander Heiden 2011). It also promotes increased intermediate divergence from the pentose phosphate pathway (PPP), the principal origin of nicotinamide adenine dinucleotide phosphate (NADPH), which is necessary for the inhibition of reactive oxygen species (ROS) and thus for the management of cell survival and signal transduction (Lunt and Vander Heiden 2011). Consequently, this attitude increases cancer progression, cancer re-occurrence and treatment resistance.

At present, it is obvious that the multi-stage metastatic process is connected with a reconnection of the metabolism. Throughout the various steps of cancer progression, the adhesion between cancer cells and the adhesion of cancer cells to the extracellular matrix is undergoing permanent alteration. It is interesting to note that cancer-promoting cell adhesion changes can either be initiated or triggered through signaling pathways that are closely linked to metabolic changes (Elia et al. 2018). In early phases of cancer progression, cells enhance their motility and invasion capabilities and break down the extracellular matrix, resulting in energetically expensive events that include the rearrangement of cell–cell and cell–extracellular matrix interactions, the reordering of focal adhesions, the establishment of invadopodia and the activation of the epithelial-to-mesenchymal transition (EMT) regimen (Elia et al. 2018; Morandi et al. 2017).

Glycolysis has a central part to play in this stage of cancer advancement, as a large body of data has demonstrated the involvement of multiple glycolytic enzymes in the development of invadopodial structures, cellular protuberances, with a major impact on cancer cell invasion and break-down of the extracellular matrix (Attanasio et al. 2011). In addition, glycolytic side products were also widely involved in enhancing the motility and invasive capacity of cancer cells. An example is methylglyoxal, which is able to activate the YAP signal, and thereby initiate EMT in breast cancer cells. (Nokin et al. 2016). In a similar way, the last metabolite of glycolysis, lactate, encourages the advancement of breast cancer by enhancing chemoattraction, and thereby cell migration (Bonuccelli et al. 2010). Lactate initiates changes in extracellular pH that ease tumor invasion, both by destroying adjacent normal cell populations and by acid-induced break-down of the extracellular matrix and fostering of angiogenesis (Payen et al. 2016; Gatenby and Gillies 2004).

It is interesting to notice that ATP generated by the mitochondria can also facilitate cytoskeletal remodeling and contraction of the cytoskeleton, which are necessary in this phase of cancer growth. It has been reported that mitochondrial metabolism encourages the invasion and metastasis of cancer cells by oxidation of glucose, glutamine and fatty acids (Yang et al. 2014; Rodrigues et al. 2016; Camarda et al. 2016; Torrano et al. 2017), whether due to mitochondrial hyperactivity and dysfunction or to changes in their biogenesis (Torrano et al. 2017; Porporato et al. 2014; LeBleu et al. 2014). It is reported that this action is facilitated through a superoxide anion that activates tyrosine kinases such as Src and Pyk2 and enhances the motility of cancer cells and cell–cell and cell–matrix changes (Porporato et al. 2014; Paoli et al. 2013). In breast cancer, PPARγ co-activator 1α (PGC-1α)-induced OXPHOS and the mitochondrial biogenesis encourages cell invasion and metastasis (LeBleu et al. 2014).

The adhesion of cancer cells to the extracellular matrix by integrins causes an rise in ROS to non-cytotoxic concentrations, which in turn leads to molecular signals that finally modify the invasion and spread of cancer cells (Paoli et al. 2013; Chiarugi et al. 2003). However, when adhesion to the extracellular matrix is disrupted, either in normal or cancer cells, oxidative stress is induced due to the higher rise in ROS to cytotoxic values, resulting in cell death (Frisch 2001). For instance, cancer cells must be resistant to anoikis, a type of cell death caused by "homesickness," after reaching the circulatory system in order to remain alive under anchorage-independent circumstances. In the absence of cell attachment to the extracellular matrix for prolonged periods, as in systemic metastasis, cancer cells renew their metabolism into a regimen that enhances antioxidant defense to balance oxidative stress. This action determines the viability of the cancer cells in the blood circulation and encourages the development of metastases (Schafer et al. 2009). This kind of metabolism is obtained by switching to glycolysis, primarily by redirecting intermediate metabolites into the PPP, resulting in the formation of NADPH, which is crucial for the formation of an important ROS scavenger, namely the reduced GSH (Elia et al. 2018).

Finally, the formation of micrometastases and the development of secondary tumors in remote organs also involves the formation of cell–matrix interactions, the restructuring of the extracellular matrix, cell-cell adhesion and proliferation and, in effect, the activation of various metabolic programs that result in the substantial generation of ATP (Elia et al. 2018). In this situation, the surroundings of the distant metastasis organ will control the metabolic response of the cancer cells (Christen et al. 2016; Mashimo et al. 2014; Maher et al. 2012; Chen et al. 2015; Kim et al. 2014b; Cha et al. 2017; McDonald et al. 2017).

17.11.2 Connection Between Cancer Stemness, EMT and the Plasticity in Metabolism

At present, it is becoming increasingly accepted that EMT and cancer stemness are caused by metabolic changes. Breast cancer stem cells (BCSCs) alter their phenotype and molecular signature to enable them to persist in all various environments throughout the metastatic process. Therefore, these cells require a high degree of plasticity, fueled by the dynamics of EMT/mesenchymal epithelial transition (MET), whereby EMT enhances invasion and proliferation and MET promotes proliferation and metastatic colonization (Liu et al. 2014; Beerling et al. 2016; Brabletz 2012; Sousa et al. 2019b). In this manner, BCSCs are transited between two major states: a dormant and invasive $CD44^{+}$/CD24/low population with an EMT signature, referred to as EMT-BCSC; and a proliferative and epithelial-like $ALDH^{+}$ population, referred to as MET-BCSC (Liu et al. 2014).

Significantly, metabolism and oxidative stress have recently been incorporated into the transition between the two BCSC phenotypes, primarily through the activation of the AMP-activated protein kinase/hypoxia inducible factor-1α (AMPK/HIF1α) axis. It has been found that EMT and MET-BCSC groups are dependent on distinct metabolic pathways and have varying sensitivities to glycolysis and redox blockers (Luo et al. 2018). They showed that the amplification of glycolysis, oxidative stress and hypoxia foster the transition from a ROS-low EMT-BCSC state to a ROS-high MET-BCSC state, which is reversible with antioxidants such as N-acetyl-cysteine (NAC). In addition, MET-BCSCs exhibit an elevated oxidative metabolism and an enhanced NRF2-driven antioxidant response. Ultimately, it could also be illustrated that the joint control of these two cell populations toward both metabolic properties would be of major therapeutic value in terms of suppressing tumor growth, tumor-initiating capacity and metastatic spread in breast cancer (Luo et al. 2018). Thus, the metabolic activity defines the EMT/MET plasticity which BCSCs demand for effective cancer progression and metastasis. In addition, the utilization of these metabolic susceptibilities of various BCSC states offers a novel therapeutic option to target these crucial cancer cell populations.

17.11.3 The Metabolism of Cancer Cells Depends on Biomechanics, Tissue Stiffness and Energetics

As cancer progresses, cancer cells are subject to different physical forces and take various configurations as they penetrate into adjacent tissues, overcome the endothelial barrier to enter the bloodstream, and exit and metastasize to distal targeted organs. These forces include compression, shear stress, stretching and intrinsic tension, resulting in profound alterations to the tissue geometry. Cells react to these forces by enhancing cell–cell and/or cell–matrix interactions due to surface adhesion receptors. The biomechanical reaction entails the activation of molecular signals that enhance intrinsic contractile forces, actin cytoskeleton rearrangement and stiffening of cells, thus, defining the outcome of the cancer cell invasion. In fact, it has lately been shown that epithelial cells pass through a stiffening state before they take on malignant characteristics that are typically linked to cell softening properties (Tavares et al. 2017).

At present, there is an underlying knowledge of the association between cell mechanics and tissue stiffening with cell metabolism, in which glycolysis performs a preferential function. This synergy offers the opportunity for combined therapies that aim at simultaneously targeting both functions and can thus more effectively arrest the progression of the disease. The oncogenic signaling pathways common to the energy-producing metabolism and the physical and phenotypic characteristics of the energy-consuming cells synergize glycolysis with the dynamics of cytoskeletal remodeling. For example, the PI(3)K signaling pathway affects cell motility (Cantley 2002), and matrix stiffness through integrin-driven activation of the focal adhesion kinase (FAK) (Provenzano et al. 2008a, b, 2009; Kallergi et al. 2007; Guan 2010). In addition, this signaling pathway increases glucose absorption through the elevation of the glucose transporters GLUT1 and GLUT4, the hexokinase (HK) and the stimulation of phosphofructokinase (PFK) activity (Rathmell et al. 2003; Buzzai et al. 2005). In addition, it was found that complete activation of glycolysis through PI3K necessitates AKT activation and Rac-driven actin rearrangement, which results in enhanced aldolase activity and subsequently enhanced glycolytic flow (Hu et al. 2016).

Matrix stiffness is also linked to enhanced cancer aggressiveness through the integrin activation of β-catenin and MYC (Mouw et al. 2014), both of which are also reported to control cell metabolism (Tilghman et al. 2010). It has been shown that this connection was established via the FAK/PI3K/AKT pathway, as evidenced by the enhancement of mitochondrial biogenesis through AKT/mTOR activation of 4EBP1 (Goo et al. 2012). Genetic deficiencies in mitochondrial respiration and elevations in glycolysis also stimulate AKT signaling through an elevation of the metabolic capacity conveyed by NADH, which enhances the enrollment of GLUTs to the cell surface and promotes glycolysis (Manning and Cantley 2007). Another mechanotransduction study with metabolic characteristics revealed that Rac1b builds a complex with NADPH oxidase, which leads to an elevation of ROS levels, snail expression and EMT in epithelial cells of the mammary gland (Lee et al. 2012a, b).

The biomechanical response of cells during cancer progression is energy-intensive and demands almost 50% of the entire cellular ATP (Daniel et al. 1986; Bernstein and Bamburg 2003). Energy homeostasis is managed through AMPK; when cells feel an energy meltdown, activation of AMPK controls ATP-producing catabolism, such as fatty acid oxidation (FAO) and glucose absorption, and impedes energy-consuming processes such as fatty acid, cholesterol, glycogen and protein production (Hardie et al. 2012). The energy required to enhance cell stiffness is mainly produced through AMPK-activated glucose absorption and oxidation (Bays et al. 2017). For example, the AMPK detects the energy requirement for actin amplification at cell–cell and cell–matrix connections throughout the progress of cancer.

In addition to its role as an energetic homeostasis effector, AMPK is also associated with cell growth, retention of apicobasal polarity, interference with extracellular matrix break-down, autophagy, mitosis and transcription (Lee et al. 2007; Zheng and Cantley 2007; Schaffer et al. 2015). AMPK represents a component of the integrin adhesome (Winograd-Katz et al. 2014) and a facilitator of integrin-based processes (Cunniff et al. 2016; Chang et al. 2012; Parsons et al. 2010; Mogilner and Oster 1996), which are normally situated at the leading edge of the migrating cells, at lamellipodia, where elevated amounts of mitochondria and mitochondria-derived ATP are present compared to the cell body (Cunniff et al. 2016). Elevated ATP values are accompanied by a markedly lower ATP: ADP ratio, which results in a local rise in mitochondrial flux, ATP levels and cytoskeletal redistribution (Salvi and DeMali 2018). In terms of molecular control, AMPK is activated upstream through the large kinase B1 (LKB1) (Woods et al. 2003), a tumor suppressor implicated in cell metabolism and proliferation, growth and polarity (Alessi et al. 2006), another marker of the relationship between cell structure and metabolism. In addition, several AMPK substrates were found to be adhesion and invasion proteins (Schaffer et al. 2015).

Most importantly, AMPK is able to positively influence actin dynamics and protrusive processes in adherent and migrating cells by controlling the build-up and break-down of epithelial tight junctions (Zheng and Cantley 2007). Nevertheless, AMPK phosphorylation is associated with an activation of the RhoA-myosin-II signal transduction pathway (Bays et al. 2017). Moreover, there has been reported that there exists an inhibitory function of AMPK in integrins and the migration of cells (Georgiadou et al. 2017; Rantala et al. 2011; Blume et al. 2007). For instance, it has been seen that the AMPK represents an inhibitor of β1-integrin activity by the regulation of tensin values inside fibroblasts, thus, acting as a controller of the impairment of cell spreading, traction stresses and the assembly of the extracellular matrix (Georgiadou et al. 2017).

17.11.4 Changes in Cell–Matrix Adhesion Strength Impacts Metabolic Pathways

Integrins fulfill a crucial function in the cancer cell metabolism. In silico investigations revealed that the integrin adhesome genes are linked to multiple disorders, including cancer, cardiovascular and immunological diseases, neurological conditions and metabolic diseases (Winograd-Katz et al. 2014). The detaching of normal and cancer cells from the extracellular matrix has a major effect on cellular metabolism, in particular, by decreasing glycolytic flux and PPP, as well as on mitochondrial metabolism and therefore on the reduction of ATP and NADPH output, FAO and the rise in ROS. The AMPK conveys one of the identified mechanisms causing this effect (Jeon et al. 2012). It has been revealed that cell–extracellular matrix deadhesion triggers the absorption of glucose, the reduction of activated LKB1, which enhances the activity of AMPK, which then impedes the acetyl-CoA carboxylases 1 and 2 (ACC1 and ACC2, respectively) that, on the one hand, lowers NADPH consumption in fatty acid synthesis (FAS), but on the other hand, elevates the NADPH production through FAO fueling (Jeon et al. 2012).

In an energetic shock caused when the extracellular matrix is detached, the AMPK is thus in charge of retaining the NADPH and for the continued survival of the cancer cells. In addition, it could be established that the metabolic regulation caused through pyruvate dehydrogenase kinase 4 (PDK4) also triggers anoikis resistance when the extracellular matrix of human breast cells is removed (Kamarajugadda et al. 2012). It has been reported that an elevation of PDK4 expression following the extracellular matrix deadhesion causes a decrease in the mitochondrial respiration, the generation of ATP, and a triggering of the glycolytic flux (Kamarajugadda et al. 2012). Moreover, it has been found that a metabolic response in epithelial cells couples the antioxidant repair of the oxidative stress and is crucial for cell survival after the detachment of the cancer extracellular matrix (Schafer et al. 2009). In line with these results, the extracellular matrix deadhesion evokes a loss of glucose transport and ATP decrease, which can be restored by the overexpression of HER2, which then repairs the glucose absorption through the stabilized activation of EGFR and PI3K based on the antioxidant-generating PPP.

17.11.5 Alterations in Intercellular Adhesions Impacts the Cancer Metabolism

Cadherins represent major players in the malignant progression of cancer. During embryonic development, cadherins are extremely required for morphogenic and differentiation, whereas they play minor roles in adult tissues, where they are regulating the homeostasis, the cell polarity and the integrity of tissues (Paredes et al. 2012). In cancer, the dysfunction of specific cadherins is commonly associated with the onset of cancer initiation, its progression and subsequently metastasis in various carcinoma types. The deregulated expression of distinct cadherins due to genetic or epigenetic changes can cause disorders of tissues, the dedifferentiation of cells (stemness phenomenon) and increased invasiveness of a certain subsets of cancer cells that form finally secondary tumors (metastases) in targeted tissues. Genome-wide screenings for proteins and Ca^{2+}-dependent signal transduction pathways that regulate of cell-cell adhesions have revealed a specific subset of proteins (Toret et al. 2014). Among this subset of proteins are metabolic proteins. However, the current knowledge of an interaction of these metabolic proteins with cell mechanics is still limited and required increased research effort.

Epithelial (E-) Cadherin

Epithelial-cadherin, which is a type I classical cadherin, acts as a key component in the creation of adherence junctions within epithelial tissues (van Roy and Berx 2008). There is a huge amount of investigations that revealed the crucial role of E-cadherin in the suppression of cancer cell invasion in the course of cancer. E-cadherin expression initiates an epithelial and well-defined phenotype, minimal invasive capacity and efficient cell–cell connections (Paredes et al. 2012). However, a reduced or decreased function of this protein results in a mesenchymal and slightly less differentiated phenotype, an augmentation of cell migration and invasion of adjacent tissues and finally in metastasis (Paredes et al. 2012). In breast cancer, the loss of E-cadherin is the marker of lobular carcinomas (Reis-Filho et al. 2002), while CDH1/E-cadherin mutations are the trigger for hereditary diffuse stomach cancer (Paredes et al. 2012; Carneiro et al. 1999).

E-cadherin is the most thoroughly investigated adhesion molecule, along with its involvement in cell metabolism. Specific metabolic programs and microenvironmental settings elicit modifications in E-cadherin expression in different cancers, resulting in the inducement of EMT (Chen et al. 2010). The lack of E-cadherin is one of the key molecular characteristics of EMT and is associated with the ability to gain the capacity for cell migration (Thiery 2003). A huge amount of evidence has included cell metabolism via carbohydrate, amino acid and lipid metabolism in the monitoring of EMT (Morandi et al. 2017). Although there is currently still no conclusive evidence of glycolytic or OXPHOS dependence on EMT-bound cancer cells, glycolysis is the best-defined pathway affecting EMT. Hypoxia through HIF-1α induction as well as elevated expression of glucose and monocarboxylate transporters (MCTs), the activation of glycolytic enzymes, the secretion of lactate and extracellular acidity are key determinants of EMT (Morandi et al. 2017). Interestingly, it has been revealed that a glycolytic phenomenon caused by the depletion of FBP1 represents a crucial oncogenic incident in basal-like breast cancer and causes BCSC and EMT phenomena (Rodrigues et al. 2016). Reduction of FBP1 initiates glycolysis, macromolecule biosynthesis and sustains ATP output in hypoxia and blocks oxygen demand and ROS generation. The metabolic reprogramming leads to an enhancement of the strain-like characteristics and tumorigenicity (Dong et al. 2013). EMT signaling also conveys the metabolic reprogramming of cancer cells in the direction of an enhancement of glucose metabolism due to enhanced glycolysis and PPP (Kim et al. 2017; Lee et al. 2012b).

However, the part played by E-cadherin in the metabolic response of cancer cells is still far from being fully understood and is heavily reliant on the cancer model and the surrounding microenvironment of the tumor. In contrast to the aforementioned concerning hypoxia/HIF-1α induction of a glycolytic shift that causes E-cadherin suppression via EMT (Krishnamachary et al. 2006), there are several recent findings that demonstrate an opposite relationship. In specific detail, it has been reported, for the first time, that E-cadherin expression is involved in the molecular reaction to hypoxia and with the elevation of the glycolytic phenotype of cancer cells employing in vitro and in vivo inflammatory breast cancer models (Chu et al. 2013). In particular, it has been shown that the loss of E-cadherin and/or the overexpression of its repressors, including ZEB1, down-regulates the expression of HIF-1α and CAIX, which in turn reduces extracellular acidification of inflammatory breast cancer cells, the growth of the tumor and consequently metastasis (Chu et al. 2013).

Cell-cell adhesion due to E-cadherin can also be decisive for the anoikis resistance of cells when they are removed from the extracellular matrix. In addition, E-cadherin seems to be responsible for the susceptibility to anoikis of proximal mouse tubular assemblies that separate from the extracellular matrix in a PI3K-dependent way (Bergin et al. 2000). Moreover, E-cadherin controls anchorage-independent growth and survival in oral squamous cell carcinomas (Kantak and Kramer 1998). Recently, E-cadherin has also been reported to be a modulator of energy metabolism, which has led to the connection between energetic balance and mechanotransduction in various cancer models. E-cadherin has been proven to be capable of sustaining cell proliferation through guiding the energy metabolism through NF-κB (Park et al. 2017).

Moreover, the induction of this cadherin leads to NF-κB mediated increase of glucose transport through GLUT1 expression, shifting from glutamine to glucose dependency (Park et al. 2017). In addition, E-cadherin is also involved in the stimulation of the activation of AMPK. It has been established that the application of shear stress to epithelial monolayers or the exertion of force directly on E-cadherin induces stimulation of AMPK by LKB1 enrollment of the E-cadherin mechanotransduction complex, thereby inducing RhoA-facilitated contractility and actin cytoskeletal strengthening (Bays et al. 2017). Nevertheless, it has been found that the LKB1/STRAD complex co-localizes with E-cadherin in adherence junctions and that this cadherin governs AMPK phosphorylation through monitoring of LKB1 localization within polarized epithelial cells (Sebbagh et al. 2009).

There is some other researches that connect AMPK and E-cadherin by inhibition of EMT in various cell models (Matoba et al. 2017; Banerjee et al. 2016; Chou et al. 2014). From an energetic perspective, glycolysis has been recognized as the most significant ATP-producing route for the establishment and retention of an epithelial barrier (Bays et al. 2017). Nevertheless, detailed knowledge of glycolysis removal pathways, including PPP or serine/glycine synthesis, is required to validate the precise ATP generation step. Finally, E-cadherin has also been linked to other metabolic agents of cancer cells, namely PKM2, which is a major enzyme of the glycolytic pathway, in which the silencing of PKM2 caused the upward regulation of caspase 7, Bad and E-cadherin expression and a reduction in matrix metalloproteinase 2 (MMP-2) and MMP-9, HIF1α and VEGF, thereby affecting cell migration and invasion, inhibiting proliferation and inducing apoptosis and cell cycle arrest in the G_0/G_1 phase (Miao et al. 2016).

Placental (P-) Cadherin

Placental (P)-cadherin represents is a canonical cell–cell adhesion molecule, which is codified by the *CDH3* gene, whose expression is highly connected to undifferentiated cells in normal adult epithelial tissues, and even also with rarely differentiated carcinomas. In breast cancer, P-cadherin is expressed de novo in about 30–40% of invasive cancers and constitutes a detrimental prognostic factor for this disease. Its expression is strongly connected with tumors of high histological grade and with short-term general and disease-free survival, and a distal and locally recurrence-free interval (Paredes et al. 2005; Gamallo et al. 2001; Peralta Soler et al. 1999; Turashvili et al. 2011). Importantly, P-cadherin serves as a biomarker for heterogeneous and with a poor prognosis triple-negative basal-like breast carcinomas (Paredes et al. 2005; Sousa et al. 2010), which show a higher expression of glycolytic markers and subsequently a more glycolytic response (Sousa et al. 2014).

There seems to be a connection between cancer cell metabolism and P-cadherin-driven cell adhesion, since it has been suggested that there exists a role for P-cadherin as an epithelial basal marker the metabolism of cancer cells (Sousa et al. 2014). In this investigation, a connection was established between abnormal P-cadherin expression and hypoxic, glycolytic and acid-resistant breast cancer cells. Over the years, the mechanisms behind the P-cadherin-induced aggressive response of cancer cells in the various stages of breast cancer progression have been investigated (Ribeiro and Paredes 2014; Albergaria et al. 2011; Paredes et al. 2007). P-cadherin has been demonstrated to enhance cancer cell invasion and early stages of the disease progression of breast cancer (Ribeiro et al. 2010; Paredes et al. 2004).

Mechanistically, P-cadherin overexpression initiates the secretion of MMPs that breakdown the extracellular matrix (Ribeiro et al. 2010; Paredes et al. 2004) and also split its extracellular domain to generate a soluble P-cadherin fragment (sP-cad) with pro-invasive potential (Ribeiro et al. 2010). Significantly, it has been shown that P-cadherin-aberrant expression was clearly linked to the expression of HIF-1α, GLUT1, CAIX, MCT1 and CD147 in human primary invasive breast tumors, which is in agreement with the glycolytic phenotype of P-cadherin-rich basal-like mammary carcinomas (Sousa et al. 2014). Additionally, in vitro studies showed that CDH3 silencing resulted in a reduction of mRNA concentrations of GLUT1 and CAIX in breast cancer cell lines (Sousa et al. 2014). Thus, when taking into account that acidic pH conditions are able to trigger the proteolytic activity of cancer cells (Rozhin et al. 1994; Montcourrier et al. 1997), the elevated glycolytic behavior and the resulting acidification of the extracellular matrix could also be accountable for the elevated proteolytic activity of P-cadherin-rich breast cancer cells. The aggressive behavior of cancer cells initiated by this specific molecule is also due to its biomechanical characteristics (Ribeiro et al. 2016). In a wild-type E-cadherin background, P-cadherin has a distinct impact on the actin cytoskeleton as it causes instability of the membrane-bound E-cadherin/p120-catenin complex, resulting in delocalization of β-catenin and p120-catenin toward the cytoplasm, alters polymerization of the actin cytoskeleton, enhances invasion, migration and motility and promotes tumorigenic capacity (Vieira et al. 2014; Ribeiro et al. 2013, 2018; Paredes et al. 2008). P-cadherin also triggers the phosphorylation of kinases of the Src family (SFK) and FAK and the activation of the small Rac1-GTPase (Ribeiro et al. 2013).

Lately, it has been found that P-cadherin/SFK signaling initiates topological, morphological and biomechanical changes (Ribeiro et al. 2016), through facilitating the alteration of cell height, cell area and the intercellular organization, and by reducing cell-cell adhesion and stiffness (Ribeiro et al. 2016). In specific detail, different mechanical functions for both E- and P-cadherins have been determined, which show that P-cadherin regulates the amount of intercellular forces, whereas E-cadherin forecasts the velocity at which intercellular force is generated (Bazellieres et al. 2015). P-cadherin can elevate the adhesion of breast cancer cells toward the extracellular matrix through the activation of the α6β4 integrin, which in turn induces the interaction of cancer cells specifically to laminin within the extracellular matrix scaffold (Vieira et al. 2014). Taking into account that the abnormal expression of P-cadherin initiates a specific phenotype in breast cancer cells that enhances the early stages of cancer progression primarily through energy spending processes, it would presumably be supported by the intensification of energy-producing metabolic pathways, including oxidative phosphorylation and/or increased glycolysis.

In the end, P-cadherin is also connected with efficient breast cancer metastasis. Moreover, P-cadherin expression pronouncedly relates to the invasion of vascular and soft tissues (Liu et al. 2012). P-cadherin has been shown to be an independent marker for the prognosis of metastasis in breast cancer patients and a putative valuable biomarker for axillary-based breast cancer determinations in clinical practice (Vieira et al. 2017). This metastatic promoter function is presumably due to its part in sustaining cancer-strain-like properties that enhance cell survival in the vascular system. P-cadherin-enriched breast cancer cell populations exhibit enhanced in vitro mammosphere-forming activity and their expression encourages resistance to anoikis (Vieira et al. 2012). Nevertheless, P-cadherin offers resistance to x-ray initiated cell death, which entangles this molecule into a different characteristic of cancer stem cells (Vieira et al. 2012). Significantly, BCSC fractions hosting high concentrations of P-cadherin also simultaneously expressed HIF-1α, CAIX and GLUT1 (Sousa et al. 2014), which enhanced the glycolytic performance of P-cadherin supplemented BCSC populations. In breast cancer, hypoxia triggers strain-like characteristic and the enlargement of BCSC and EMT (Louie et al. 2010).

Intriguingly, it has been shown that nuclear HIF-1α can to trigger the expression of P-cadherin at the plasma membrane (Sousa et al. 2014). Recently, P-cadherin has been linked to an "intermediate/metastable" phenotype in the EMT mechanism, indicating P-cadherin as a suspected EMT biomarker and encouraging the usage of anti-P-cadherin therapeutics in the management of metastatic breast cancer (Ribeiro and Paredes 2014). It is crucial that P-cadherin is able to distinguish an intermediate phase between the epithelial and the mesenchymal phenotype.

FAT (Ft)-Like Cadherin

The FAT (ft) cadherin is a *Drosophila* protocadherin that contains an extracellular domain consisting of 34 cadherin repeats and a large intracellular domain, which itself encompasses EGF-like and laminin-G-like domains (Blair and McNeill 2018). The ft cadherin is localized at lamellipodia, filopodia and cell–cell contacts, and similar to classical cadherins, fulfills a crucial function in cell adhesion and signal transduction pathways. Ft cadherin is involved in the impair of cellular growth, functions upstream the Hippo signaling pathway and with a conserved role in the creation of the planar cell polarity (Thomas and Strutt 2012; Staley and Irvine 2012).

There exist four human orthologues of *Drosophila* FAT, FAT1-4 and the most-closest related one represents FAT4 (Castillejo-Lopez et al. 2004). Modifications, such as mutations and deletions in FAT family, have been detected in several types of human cancer (Morris et al. 2013). Nonetheless, these genes fulfill a dual function as either tumor suppressor genes (Morris et al. 2013; Nakaya et al. 2007) or oncogenes (De Bock et al. 2011; Nishikawa et al. 2011; Valletta et al. 2014; Dikshit et al. 2012), due to the cellular microenvironment. On a molecular scale, ft cadherins are known to transduce signals via several canonical pathways in cancer cells, including Wnt/β-catenin and MAPK/ERK (Cai et al. 2015).

Interestingly, ft can also trigger alterations in cell metabolism. It has been demonstrated a direct function of ft in mitochondrial shape and function (Sing et al. 2014). Moreover, ft regulates the generation of ROS, stabilizes the mitochondrial electron transport chain and induces the oxidative phosphorylation in the course of the *Drosophila* development. This effect is facilitated through the activity of a 68 kDa soluble fragment, ftmito, which is provided through the proteolytic cleavage of ft, which is delivered to the mitochondria (in order to attach to the NADH dehydrogenase ubiquinone flavo-protein 2 (Ndufv2), triggering the complex I and V association and stability, and thereby increasing the mitochondrial electron transport chain (Sing et al. 2014).

Ft is normally present at the cell membrane and only at this site is it capable of impairing proliferation by activating the Hippo kinase pathway. Thus, ft cleavage and relocation from the membrane to the mitochondria is no impediment to proliferation and enhances the metabolic performance of OXPHOS. According, a model for ft mutants, such as cancer cells in which both PCP and Hippo are deregulated, OXPHOS is disturbed and there is an elevation of ROS and a metabolic displacement toward glycolysis with enhanced ATP and lactate output (Sing et al. 2014). In addition, it has been possible to prove that atypical FAT1-cadherin functions as a molecular "brake" on mitochondrial respiration by controlling the proliferation of smooth vascular muscle cells following arterial injury, whereby in vitro and in vivo models were employed (Cao et al. 2016). Using a modified intracellular FAT1 domain, it was found that SMCs with a deficiency of FAT1 consume more oxygen for ATP generation and provide more aspartate, whereas the establishment of a modified intracellular FAT1 region normalizes the oxygen demand and growth capacity of SMCs (Cao et al. 2016). In particular, the deletion of FAT1 enhances the activity of mitochondrial complexes I and II and conveys growth control per se via the mitochondria (Cao et al. 2016). However, the role of ft in the growth of cancer cells through its metabolic activity is far from complete comprehension. There seem to be only a small number of investigations that prove this direct connection. It has been seen that the presence of a FAT1-HIF-1α axis inside glioblastoma, which causes the aggressiveness of this specific disease and hence serves as a new potential therapeutic object of these tumors (Madan et al. 2016). FAT1 has been shown to be linked to HIF-1α expression as well as to its targeted molecules phosphoglycerate kinase 1 (PGK1) and VEGFA within human glioblastoma samples. In vitro studies showed that serious hypoxia stimulates the expression of FAT1 and that the reduction of endogenous FAT1 during hypoxia causes a reduction of BNIP3, CAIX, HIF-1α, HK2, GLUT1, MCT4 and VEGFA as well as a reduction of the invasion of glioblastoma cells (Madan et al. 2016). Therefore, on the basis of what is described in the literature, it is rational to assume that the ft cadherin loss induces mitochondrial malfunction and metabolic switch in cancer cells. However, additional investigations are required to better comprehend the metabolic effects of ft cadherin modifications in cancer cells.

In summary, cells must adjust during the progression of cancer in order to be able to cross any barrier from primary tumor to metastasis. Changes in cell-cell adhesion and in the interactions between cells and extracellular matrix that continue to encourage cytoskeletal remodeling are stimulants that initiate the mechanotransduction of cancer cells to provide functional signaling pathways, enabling them to pass on their morphological and survival demands. Changes in the mechanotransduction of cancer cells demand energy, which is captured by AMPK in a sensory manner. This crucial molecular actor activates various metabolic processes to fulfill the distinct energetic requirements that cancer cells need in the various steps of the malignant progression of cancer to support the proliferation and survival of cancer cells and thus elevate the progression of cancer.

References

J. Adam, E. Hatipoglu, L. O'Flaherty et al., Renal cyst formation in Fh1-deficient mice is independent of the Hif/Phd pathway: roles for fumarate in KEAP1 succination and Nrf2 signaling. Cancer Cell **20**, 524–537 (2011)

B. Ai, T. Pan et al., The relationship of expression of GLUT1, HIF-1alpha and the uptake of FDG in non-small cell lung cancer. Zhongguo Fei Ai Za Zhi. **10**(6), 508–512 (2007)

A. Albergaria, A.S. Ribeiro, A.F. Vieira, B. Sousa, A.R. Nobre, R. Seruca, F. Schmitt, J. Paredes, P-cadherin role in normal breast development and cancer. Int. J. Dev. Biol. **55**, 811–822 (2011)

J. Albrengues, T. Bertero, E. Grasset et al., Epigenetic switch drives the conversion of fibroblasts into proinvasive cancer-associated fibroblasts. Nat. Commun. **6**, 10204 (2015)

N.L. Alderson, Y. Wang, M. Blatnik et al., S-(2-Succinyl)cysteine: a novel chemical modification of tissue proteins by a Krebs cycle intermediate. Arch. Biochem. Biophys. **450**, 1–8 (2006)

J.C. Aledo, P.M. Gómez-Fabre, L. Olalla, J. Márquez, Identification of two human glutaminase loci and tissue-specific expression of the two related genes. Mamm. Genome **11**, 1107–1110 (2000)

D.R. Alessi, K. Sakamoto, J.R. Bayascas, LKB1-dependent signaling pathways. Annu. Rev. Biochem. **75**, 137–163 (2006)

J. Alfardan, A.W. Mohsen, S. Copeland et al., Characterization of new ACADSB gene sequence mutations and clinical implications in patients with 2-methylbutyrylglycinuria identified by newborn screening. Mol. Genet. Metab. **100**, 333–338 (2010)

A. Alimonti, A. Carracedo, J.G. Clohessy et al., Subtle variations in Pten dose determine cancer susceptibility. Nat. Genet. **42**(5), 454–458 (2010)

B.J. Altman, Z.E. Stine, C.V. Dang, From Krebs to clinic: glutamine metabolism to cancer therapy. Nat. Rev. Cancer **16**, 619–634 (2016)

F. Attanasio, G. Caldieri, G. Giacchetti, R. van Horssen, B. Wieringa, R. Buccione, Novel invadopodia components revealed by differential proteomic analysis. Eur. J. Cell Biol. **90**, 115–127 (2011)

Y.S. Bae, S.W. Kang, M.S. Seo, I.C. Baines, E. Tekle, P.B. Chock, S.G. Rhee, Epidermal growth factor (EGF)-induced generation of hydrogen peroxide. Role in EGF receptor-mediated tyrosine phosphorylation. J. Biol. Chem. **272**(1), 217–221 (1997)

P. Banerjee, H. Surendran, D.R. Chowdhury, K. Prabhakar, R. Pal, Metformin mediated reversal of epithelial to mesenchymal transition is triggered by epigenetic changes in E-cadherin promoter. J. Mol. Med. **94**, 1397–1409 (2016)

H.E. Barker, J. Chang, T.R. Cox, G. Lang, D. Bird, M. Nicolau, H.R. Evans, A. Gartland, J.T. Erler, LOXL2-mediated matrix remodeling in metastasis and mammary gland involution. Cancer Res. **71**(5), 1561–1572 (2011)

J.L. Bays, H.K. Campbell, C. Heidema, M. Sebbagh, K.A. DeMali, Linking E-cadherin mechanotransduction to cell metabolism through force-mediated activation of AMPK. Nat. Cell Biol. **19**, 724–731 (2017)

B.E. Baysal, Clinical and molecular progress in hereditary paraganglioma. J. Med. Genet. **45**, 689–694 (2008)

B.E. Baysal, R.E. Ferrell, J.E. Willett-Brozick et al., Mutations in SDHD, a mitochondrial complex II gene, in hereditary paraganglioma. Science **287**, 848–851 (2000)

E. Bazellieres, V. Conte, A. Elosegui-Artola, X. Serra-Picamal, M. Bintanel-Morcillo, P. Roca-Cusachs, J.J. Munoz, M. Sales-Pardo, R. Guimera, X. Trepat, Control of cell–cell forces and collective cell dynamics by the intercellular adhesome. Nat. Cell Biol. **17**, 409–420 (2015)

G.W. Beadle, E.L. Tatum, Genetic control of biochemical reactions in neurospora. Proc. Natl. Acad. Sci. U. S. A. **27**, 499–506 (1941)

E. Beerling, D. Seinstra, E. de Wit et al., Plasticity between epithelial and mesenchymal states unlinks EMT from metastasis-enhancing stem cell capacity. Cell Rep. **14**, 2281–2288 (2016)

E.L. Bell, T.A. Klimova, J. Eisenbart, C.T. Moraes, M.P. Murphy, G.R. Budinger, N.S. Chandel, The Qo site of the mitochondrial complex III is required for the transduction of hypoxic signaling via reactive oxygen species production. J. Cell Biol. **177**(6), 1029–1036 (2007)

I. Ben-Sahra, B.D. Manning, mTORC1 signaling and the metabolic control of cell growth. Curr. Opin. Cell Biol. **45**, 72–82 (2017)

K. Bensaad, A. Tsuruta, M.A. Selak, M.N. Vidal, K. Nakano, R. Bartrons, E. Gottlieb, K.H. Vousden, TIGAR, a p53-inducible regulator of glycolysis and apoptosis. Cell **126**, 107–120 (2006)

E. Bergin, J.S. Levine, J.S. Koh, W. Lieberthal, Mouse proximal tubular cell–cell adhesion inhibits apoptosis by a cadherin-dependent mechanism. Am. J. Physiol. Renal Physiol. **278**, F758–F768 (2000)

B.W. Bernstein, J.R. Bamburg, Actin-ATP hydrolysis is a major energy drain for neurons. J. Neurosci. **23**, 1–6 (2003)

A.D. Bershadsky, N.Q. Balaban, B. Geiger, Adhesion-dependent cell mechanosensitivity. Annu. Rev. Cell Dev. Biol. **19**, 677–695 (2003)

T. Bertero, K.A. Cottrill, S. Annis et al., A YAP/TAZ-miR-130/301 molecular circuit exerts systems-level control of fibrosis in a network of human diseases and physiologic conditions. Sci. Rep. **5**, 18277 (2015a)

T. Bertero, K.A. Cottrill, Y. Lu et al., Matrix remodeling promotes pulmonary hypertension through feedback mechanoactivation of the YAP/TAZ-miR-130/301 circuit. Cell Rep. **13**, 1016–1032 (2015b)

T. Bertero, W.M. Oldham, K.A. Cottrill et al., Vascular stiffness mechanoactivates YAP/TAZ-dependent glutaminolysis to drive pulmonary hypertension. J. Clin. Invest. **126**, 3313–3335 (2016)

T. Bertero, W.M. Oldham, E.M. Grasset et al., Tumor-stroma mechanics coordinate amino acid availability to sustain tumor growth and malignancy. Cell Metab. **29**, 1–17 (2018)

M.K. Bhattacharyya, A.M. Smith, T.H. Ellis, C. Hedley, C. Martin, The wrinkled-seed character of pea described by Mendel is caused by a transposon-like insertion in a gene encoding starch-branching enzyme. Cell **60**, 115–122 (1990)

D.E. Biancur, J.A. Paulo, B. Małachowska et al., Compensatory metabolic networks in pancreatic cancers upon perturbation of glutamine metabolism. Nat. Commun. **8**, 15965 (2017)

D.K. Bird, L. Yan, K.M. Vrotsos, K.W. Eliceiri, E.M. Vaughan, P.J. Keely, J.G. White, N. Ramanujam, Metabolic mapping of MCF10A human breast cells via multiphoton fluorescence lifetime imaging of the coenzyme NADH. Cancer Res. **65**(19), 8766–8773 (2005)

K. Birsoy, T. Wang, W.W. Chen, E. Freinkman, M. Abu-Remaileh, D.M. Sabatini, An essential role of the mitochondrial electron transport chain in cell proliferation is to enable aspartate synthesis. Cell **162**, 540–551 (2015)

S. Biswas, J. Lunec, K. Bartlett, Non-glucose metabolism in cancer cells-is it all in the fat? Cancer Metastasis Rev. **31** (3–4), 689–698 (2012)

S. Blair, H. McNeill, Big roles for fat cadherins. Curr. Opin. Cell Biol. **51**, 3–80 (2018)

C. Blume, P.M. Benz, U. Walter, J. Ha, B.E. Kemp, T. Renne, AMP-activated protein kinase impairs endothelial actin cytoskeleton assembly by phosphorylating vasodilator-stimulated phosphoprotein. J. Biol. Chem. **282**, 4601–4612 (2007)

G. Bonuccelli, A. Tsirigos, D. Whitaker-Menezes et al., Ketones and lactate "fuel" tumor growth and metastasis: evidence that epithelial cancer cells use oxidative mitochondrial metabolism. Cell Cycle **9**, 3506–3514 (2010)

A. Boudreau, H.E. Purkey, A. Hitz et al., Metabolic plasticity underpins innate and acquired resistance to LDHA inhibition. Nat. Chem. Biol. **12**, 779–786 (2016)

S.L. Bowker, S.R. Majumdar, P. Veugelers, J.A. Johnson, Increased cancer-related mortality for patients with type 2 diabetes who use sulfonylureas or insulin: response to Farooki and Schneider. Diabetes Care **29**(8), 1990–1991 (2006)

T. Brabletz, To differentiate or not—routes towards metastasis. Nat. Rev. Cancer **12**, 425–436 (2012)

M.C. Brahimi-Horn, J. Chiche, J. Poyssegur, Hypoxia signaling controls metabolicdemand. Curr. Opin. Cell Biol. **19**, 223–229 (2007)

K. Brand, Aerobic glycolysis by proliferating cells: protection against oxidative stress at the expense of energy yield. J. Bioenerg. Biomembr. **29**, 355–364 (1997)

D. Bungard, B.J. Fuerth, P.Y. Zeng, B. Faubert, N.L. Maas, B. Viollet, D. Carling, C.B. Thompson, R.G. Jones, S.L. Berger, Signaling kinase AMPK activates stress-promoted transcription via histone H2B phosphorylation. Science **329**, 1201–1205 (2010)

N. Burnichon, J.J. Briere, R. Libe et al., SDHA is a tumor suppressor gene causing paraganglioma. Hum. Mol. Genet. **19**, 3011–3020 (2010)

D.T. Butcher, T. Alliston, V.M. Weaver, A tense situation: forcing tumour progression. Nat. Rev. Cancer **9**(2), 108–122 (2009)

M. Buzzai, D.E. Bauer, R.G. Jones, R.J. Deberardinis, G. Hatzivassiliou, R.L. Elstrom, C.B. Thompson, The glucose dependence of Akt-transformed cells can be reversed by pharmacologic activation of fatty acid beta-oxidation. Oncogene **24**(26), 4165–4173 (2005)

L. Cai, B.M. Sutter, B. Li, B.P. Tu, Acetyl-CoA induces cell growth and proliferation by promoting the acetylation of histones at growth genes. Mol. Cell **42**, 426–437 (2011)

J. Cai, D. Feng, L. Hu, H. Chen, G. Yang, Q. Cai, C. Gao, D. Wei, FAT4 functions as a tumour suppressor in gastric cancer by modulating Wnt/beta-catenin signaling. Br. J. Cancer **113**, 1720–1729 (2015)

R.A. Cairns, I.S. Harris, T.W. Mak, Regulation of cancer cell metabolism. Nat. Rev. Cancer **11**(2), 85–95 (2011)

F. Calvo, N. Ege, A. Grande-Garcia et al., Mechanotransduction and YAP-dependent matrix remodelling is required for the generation and maintenance of cancer-associated fibroblasts. Nat. Cell Biol. **15**, 637–646 (2013)

R. Camarda, A.Y. Zhou, R.A. Kohnz et al., Inhibition of fatty acid oxidation as a therapy for MYC-overexpressing triple-negative breast cancer. Nat. Med. **22**, 427–432 (2016)

L.C. Cantley, The phosphoinositide 3-kinase pathway. Science **296**, 1655–1657 (2002)

L.L. Cao, D.F. Riascos-Bernal, P. Chinnasamy et al., Control of mitochondrial function and cell growth by the atypical cadherin Fat1. Nature **539**, 575–578 (2016)

J.S. Carew, Y. Zhou, M. Albitar, J.D. Carew, M.J. Keating, P. Huang, Mitochondrial DNA mutations in primary leukemia cells after chemotherapy: clinical significance and therapeutic implications. Leukemia **17**, 1427–1447 (2003)

F. Carneiro, J.C. Machado, R. Seruca, M. Sobrinho-Simoes, E-cadherin changes in gastric carcimona. Histopathology **35**, 477–478 (1999)

A. Cassago, A.P.S. Ferreira, I.M. Ferreira, C. Fornezari, E.R.M. Gomes, K.S. Greene, H.M. Pereira, R.C. Garratt, S.M. G. Dias, A.L.B. Ambrosio, Mitochondrial localization and structure-based phosphate activation mechanism of Glutaminase C with implications for cancer metabolism. Proc. Natl. Acad. Sci. U. S. A. **109**, 1092–1097 (2012)

Z. Castano, K. Tracy, S.S. McAllister, The tumor macroenvironment and systemic regulation of breast cancer progression. Int. J. Dev. Biol. **55**(7–9), 889–897 (2011)

C. Castillejo-Lopez, W.M. Arias, S. Baumgartner, The fat-like gene of *Drosophila* is the true orthologue of vertebrate fat cadherins and is involved in the formation of tubular organs. J. Biol. Chem. **279**, 24034–24043 (2004)

Y.J. Cha, W.H. Jung, J.S. Koo, Differential site-based expression of pentose phosphate pathway-related proteins among breast cancer metastases. Dis. Markers **2017**, 7062517 (2017)

D.H. Chace, T.A. Kalas, E.W. Naylor, The application of tandem mass spectrometry to neonatal screening for inherited disorders of intermediary metabolism. Annu. Rev. Genomics Hum. Genet. **3**, 17–45 (2002)

M.Y. Chang, D.Y. Huang, F.M. Ho, K.C. Huang, W.W. Lin, PKC-dependent human monocyte adhesion requires AMPK and Syk activation. PLoS ONE **7**, e40999 (2012)

S. Chatterjee, E.A. Browning, N. Hong, K. DeBolt, E.M. Sorokina, W. Liu, M.J. Birnbaum, A.B. Fisher, Membrane depolarization is the trigger for PI3K/Akt activation and leads to the generation of ROS. Am. J. Physiol. Heart Circ. Physiol. **302**(1), H105-114 (2012)

M.M. Chaumeil, P.E. Larson, S.M. Woods et al., Hyperpolarized [1-13C] glutamate: a metabolic imaging biomarker of IDH1 mutational status in glioma. Cancer Res. **74**(16), 4247–4257 (2014)

J. Chen, N. Imanaka, J.D. Griffin, Hypoxia potentiates notch signaling in breast cancer leading to decreased E-cadherin expression and increased cell migration and invasion. Br. J. Cancer **102**, 351–360 (2010)

J. Chen, H.-J. Lee, X. Wu et al., Gain of glucose-independent growth upon metastasis of breast cancer cells to the brain. Cancer Res. **75**, 554–565 (2015)

T. Cheng, J. Sudderth, C. Yang, A.R. Mullen, E.S. Jin, J.M. Mates, R. DeBerardinis, Pyruvate carboxylase is required for glutamine-independent growth of tumor cells. Proc. Natl. Acad. Sci. U. S. A. **108**, 8674–8679 (2011)

P. Chiarugi, G. Pani, E. Giannoni, L. Taddei, R. Colavitti, G. Raugei, M. Symons, S. Borrello, T. Galeotti, G. Ramponi, Reactive oxygen species as essential mediators of cell adhesion: the oxidative inhibition of a FAK tyrosine phosphatase is required for cell adhesion. J. Cell Biol. **161**, 933–944 (2003)

B. Childs, D. Valle, G. Jimenez-Sanchez, The inborn error and biochemical individuality, in *The Metabolic & Molecular Bases of Inherited Disease*. ed. by C.S. Sciver, A.L. Beaudet, W.S. Sly, D. Valle (McGraw-Hill, New York, 2001), pp. 155–166

C. Choi, S.K. Ganji, R.J. DeBerardinis, I.E. Dimitrov, J.M. Pascual, R. Bachoo, B.E. Mickey, C.R. Malloy, E.A. Maher, Measurement of glycine in the human brain in vivo by 1H-MRS at 3 T: application in brain tumors. Magn. Reson. Med. **66**, 609–618 (2011)

C. Choi, S.K. Ganji, R.J. DeBerardinis, et al., Noninvasive detection of 2-hydroxyglutarate in human gliomas in vivo by J-difference editing at 3.0 T. Nat. Med. **18**(4), 624–629 (2012)

C.C. Chou, K.H. Lee, I.L. Lai, D. Wang, X. Mo, S.K. Kulp, C.L. Shapiro, C.S. Chen, AMPK reverses the mesenchymal phenotype of cancer cells by targeting the Akt-MDM2-Foxo3a signaling axis. Cancer Res. **74**, 4783–4795 (2014)

S. Christen, D. Lorendeau, R. Schmieder et al., Breast cancer-derived lung metastases show increased pyruvate carboxylase-dependent anaplerosis. Cell Rep. **17**, 837–848 (2016)

H.R. Christofk, M.G. Vander Heiden, H.M.H. Vander, A. Ramanathan, R.E. Gerszten, R. Wei, M.D. Fleming, S.L. Schreiber, L.C. Cantley, The M2 splice isoform of pyruvate kinase is important for cancer metabolism and tumour growth. Nature **452**, 230–233 (2008a)

H.R. Christofk, M.G. Vander Heiden, N. Wu, J.M. Asara, L.C. Cantley, Pyruvate kinase M2 is a phosphotyrosine-binding protein. Nature **452**, 181–186 (2008b)

K. Chu, K.M. Boley, R. Moraes, S.H. Barsky, F.M. Robertson, The paradox of E-cadherin: role in response to hypoxia in the tumor microenvironment and regulation of energy metabolism. Oncotarget **4**, 446–462 (2013)

A. Cittadini, R. Napoli, M.G. Monti et al., Metformin prevents the development of chronic heart failure in the SHHF rat model. Diabetes. **61**(4), 944–953 (2012)

J. Cohen, A. Pertsemlidis, I.K. Kotowski, R. Graham, C.K. Garcia, H.H. Hobbs, Low LDL cholesterol in individuals of African descent resulting from frequent nonsense mutations in PCSK9. Nat. Genet. **37**, 161–165 (2005)

C. Commisso, S.M. Davidson, R.G. Soydaner-Azeloglu et al., Macropinocytosis of protein is an amino acid supply route in Ras-transformed cells. Nature **497**, 633–637 (2013)

M.W. Conklin, P.P. Provenzano, K.W. Eliceiri, R. Sullivan, P.J. Keely, Fluorescence lifetime imaging of endogenous fluorophores in histopathology sections reveals differences between normal and tumor epithelium in carcinoma in situ of the breast. Cell Biochem. Biophys. **53**(3), 145–157 (2009)

M.W. Conklin, J.C. Eickhoff, K.M. Riching, C.A. Pehlke, K.W. Eliceiri, P.P. Provenzano, A. Friedl, P.J. Keely, Aligned collagen is a prognostic signature for survival in human breast carcinoma. Am. J. Pathol. **178**(3), 1221–1232 (2011)

W.C. Copeland, J.T. Wachsman, F.M. Johnson, J.S. Penta, Mitochondrial DNA alterations in cancer. Cancer Invest. **20**, 557–569 (2002)

H.G. Crabtree, Observations on the carbohydrate metabolism of tumors. Biochem. J. **23**, 536–545 (1929)

D.R. Croft, E. Sahai, G. Mavria, S. Li, J. Tsai, W.M. Lee, C.J. Marshall, M.F. Olson, Conditional ROCK activation in vivo induces tumor cell dissemination and angiogenesis. Cancer Res. **64**(24), 8994–9001 (2004)

B. Cunniff, A.J. McKenzie, N.H. Heintz, A.K. Howe, AMPK activity regulates trafficking of mitochondria to the leading edge during cell migration and matrix invasion. Mol. Biol. Cell **27**, 2662–2674 (2016)

A. Daemen, B. Liu, K. Song et al., Pan-cancer metabolic signature predicts co-dependency on glutaminase and de novo glutathione synthesis linked to a high-mesenchymal cell state. Cell Metab. **28**, 383–399 (2018)

J.S. Damrauer, K.A. Hoadley, D.D. Chism et al., Intrinsic subtypes of high-grade bladder cancer reflect the hallmarks of breast cancer biology. Proc. Natl. Acad. Sci. U. S. A. **111**, 3110–3115 (2014)

L. Dang, D.W. White, S. Gross et al., Cancer-associated IDH1 mutations produce 2-hydroxyglutarate. Nature **462**, 739–744 (2009)

J.L. Daniel, I.R. Molish, L. Robkin, H. Holmsen, Nucleotide exchange between cytosolic ATP and F-actin-bound ADP may be a major energy-utilizing process in unstimulated platelets. Eur. J. Biochem. **156**, 677–683 (1986)

S.M. Davidson, T. Papagiannakopoulos, B.A. Olenchock et al., Environment impacts the metabolic dependencies of Ras-driven non-small cell lung cancer. Cell Metab. **23**, 517–528 (2016)

S.M. Davidson, O. Jonas, M.A. Keibler et al., Direct evidence for cancer-cell-autonomous extracellular protein catabolism in pancreatic tumors. Nat. Med. **23**, 235–241 (2017)

P.A. De Beule, C. Dunsby, N.P. Galletly, G.W. Stamp, A.C. Chu, U. Anand, P. Anand, C.D. Benham, A. Naylor, P.M. French, A hyperspectral fluorescence lifetime probe for skin cancer diagnosis. Rev. Sci. Instrum. **78**(12), 123101 (2007)

C.E. De Bock, A. Ardjmand, T.J. Molloy et al., The Fat1 cadherin is overexpressed and an independent prognostic factor for survival in paired diagnosis–relapse samples of precursor B-cell acute lymphoblastic leukemia. Leukemia **26**(5), 918–926 (2011)

R.J. DeBerardinis, Is cancer a disease of abnormal cellular metabolism? New angles on an old idea. Genet. Med. **10**, 767–777 (2008)

R.J. DeBerardinis, T. Cheng, Q's next: the diverse functions of glutamine in metabolism, cell biology and cancer. Oncogene **29**, 313–324 (2010)

R.J. DeBerardinis, C.B. Thompson, Cellular metabolism and disease: what do metabolic outliers teach us? Cell **148**(6), 1132–1144 (2012)

R.J. DeBerardinis, A. Mancuso, E. Daikhin, I. Nissim, M. Yudkoff, S. Wehrli, C.B. Thompson, Beyond aerobic glycolysis: transformed cells can engage in glutamine metabolism that exceeds the requirement for protein and nucleotide synthesis. Proc. Natl. Acad. Sci. U. S. A. **104**, 19345–19350 (2007)

R.J. DeBerardinis, J.J. Lum, G. Hatzivassiliou, C.B. Thompson, The biology of cancer: metabolic reprogramming fuels cell growth and proliferation. Cell Metab. **7**, 11–20 (2008)

R. Díaz-Ruiz, S. Uribe-Carvajal, A. Devin, M. Rigoulet, Tumor cell energy metabolism and its common features with yeast metabolism. Biochim. Biophys. Acta **1796**, 252–265 (2009)

R. Diaz-Ruiz, M. Rigoulet, A. Devin, The Warburg and Crabtree effects: on the origin of cancer cell energy metabolism and of yeast glucose repression. Biochim. Biophys. Acta **1807**, 568–576 (2011)

B. Dikshit, K. Irshad, E. Madan, N. Aggarwal, C. Sarkar, P.S. Chandra, D.K. Gupta, P. Chattopadhyay, S. Sinha, K. Chosdol, FAT1 acts as an upstream regulator of oncogenic and inflammatory pathways, via PDCD4, in glioma cells. Oncogene **32**, 3798 (2012)

C. Dong, T. Yuan, Y. Wu et al., Loss of FBP1 by snail-mediated repression provides metabolic advantages in basal-like breast cancer. Cancer Cell **23**, 316–331 (2013)

R.J. Dowling, P.J. Goodwin, V. Stambolic, Understanding the benefit of metformin use in cancer treatment. BMC Med. **9**, 33 (2011)

S. Dupont, L. Morsut, M. Aragona et al., Role of YAP/TAZ in mechanotransduction. Nature **474**, 179–183 (2011)

M. Egeblad, M.G. Rasch, V.M. Weaver, Dynamic interplay between the collagen scaffold and tumor evolution. Curr. Opin. Cell Biol. **22**(5), 697–706 (2010)

J. Eichhorst, J. Alcorn, J. Lepage et al., Elevated neonatal 3-OH isovalerylcarnitine due to breast milk sources in maternal 3-MCC deficiency. Mol. Genet. Metab. **101**, 84–86 (2010)

K.M. Elgadi, R.A. Meguid, M. Qian, W.W. Souba, S.F. Abcouwer, Cloning and analysis of unique human glutaminase isoforms generated by tissue-specific alternative splicing. Physiol. Genomics **1**, 51–62 (1999)

I. Elia, G. Doglioni, S.M. Fendt, Metabolic hallmarks of metastasis formation. Trends Cell Biol. **28**, 673–684 (2018)

B.S. Elkin, E.U. Azeloglu, K.D. Costa, B. Morrison 3rd., Mechanical heterogeneity of the rat hippocampus measured by atomic force microscope indentation. J. Neurotrauma **24**(5), 812–822 (2007)

E. Enzo, G. Santinon, A. Pocaterra et al., Aerobic glycolysis tunes YAP/TAZ transcriptional activity. EMBO J. **34**, 1349–1370 (2015)

A. Erez, O.A. Shchelochkov, S.E. Plon, F. Scaglia, B. Lee, Insights into the pathogenesis and treatment of cancer from inborn errors of metabolism. Am. J. Hum. Genet. **88**, 402–421 (2011)

J.T. Erler, A.J. Giaccia, Lysyl oxidase mediates hypoxic control of metastasis. Cancer Res. **66**(21), 10238–10241 (2006)

J.M. Evans, L.A. Donnelly, A.M. Emslie-Smith, D.R. Alessi, A.D. Morris, Metformin and reduced risk of cancer in diabetic patients. BMJ. **330**(7503), 1304–1305 (2005)

T.W. Fan, A.N. Lane, R.M. Higashi, M.A. Farag, H. Gao, M. Bousamra, D.M. Miller, Altered regulation of metabolic pathways in human lung cancer discerned by (13)C stable isotope-resolved metabolomics (SIRM). Mol. Cancer **8**, 41 (2009)

M. Fang, Z. Shen, S. Huang, L. Zhao, S. Chen, T.W. Mak, X. Wang, The ER UDPase ENTPD5 promotes protein N-glycosylation, the Warburg effect, and proliferation in the PTEN pathway. Cell **143**(5), 711–724 (2010)

M.E. Figueroa, O. Abdel-Wahab, C. Lu et al., Leukemic IDH1 and IDH2 mutations result in a hypermethylation phenotype, disrupt TET2 function, and impair hematopoietic differentiation. Cancer Cell **18**, 553–567 (2010)

J.S. Flier, M.M. Mueckler, P. Usher, H.F. Lodish, Elevated levels of glucose transport and transporter messenger RNA are induced by ras or src oncogenes. Science **235**(4795), 1492–1495 (1987)

K. Forsberg, S. Di Giovanni, Cross talk between cellular redox status, metabolism, and p53 in neural stem cell biology. Neuroscientist **20**(4), 326–342 (2014)

M.V. Fournier, J.E. Fata, K.J. Martin, P. Yaswen, M.J. Bissell, Interaction of E-cadherin and PTEN regulates morphogenesis and growth arrest in human mammary epithelial cells. Cancer Res. **69**(10), 4545–4552 (2009)

C. Frezza, P.J. Pollard, E. Gottlieb, Inborn and acquired metabolic defects in cancer. J. Mol. Med. (Berl) **89**, 213–220 (2011a)

C. Frezza, L. Zheng, O. Folger et al., Haem oxygenase is synthetically lethal with the tumour suppressor fumarate hydratase. Nature **477**, 225–228 (2011b)

S.M. Frisch, Tumor suppression activity of adenovirus E1a protein: anoikis and the epithelial phenotype. Adv. Cancer Res. **80**, 39–49 (2001)

F.A. Gallagher, M.I. Kettunen, S.E. Day, M. Lerche, K.M. Brindle, 13C MR spectroscopy measurements of glutaminase activity in human hepatocellular carcinoma cells using hyperpolarized 13C-labeled glutamine. Magn. Reson. Med. **60**(2), 253–257 (2008)

N.P. Galletly, J. McGinty, C. Dunsby et al., Fluorescence lifetime imaging distinguishes basal cell carcinoma from surrounding uninvolved skin. Br. J. Dermatol. **159**(1), 152–161 (2008)

L. Galluzzi, O. Kepp, M.G. Vander Heiden, G. Kroemer, Metabolic targets for cancer therapy. Nat. Rev. Drug Discov. **12**(11), 829–846 (2013)

C. Gamallo, G. Moreno-Bueno, D. Sarrio, F. Calero, D. Hardisson, J. Palacios, The prognostic significance of P-cadherin in infiltrating ductal breast carcinoma. Mod. Pathol. **14**, 650–654 (2001)

S.S. Gambhir, J. Czernin, J. Schwimmer, D.H. Silverman, R.E. Coleman, M.E. Phelps, A tabulated summary of the FDG PET literature. J. Nucl. Med. **42**(5 Suppl), S1–S93 (2001)

P. Gao, I. Tchernyshyov, T.C. Chang et al., c-Myc suppression of miR-23a/b enhances mitochondrial glutaminase expression and glutamine metabolism. Nature **458**, 762–765 (2009)

J. Garcia-Bermudez, L. Baudrier, K. La et al., Aspartate is a limiting metabolite for cancer cell proliferation under hypoxia and in tumours. Nat. Cell Biol. **20**, 775–781 (2018)

A.E. Garrod, The incidence of alkaptonuria: a study in chemical individuality. Lancet 1616–1620 (1902)

R.A. Gatenby, R.J. Gillies, Why do cancers have high aerobic glycolysis? Nat. Rev. Cancer **4**, 891–899 (2004)

B. Geiger, J.P. Spatz, A.D. Bershadsky, Environmental sensing through focal adhesions. Nat. Rev. Mol. Cell Biol. **10** (1), 21–33 (2009)

P.C. Georges, J.J. Hui, Z. Gombos, M.E. McCormick, A.Y. Wang, M. Uemura, R. Mick, P.A. Janmey, E.E. Furth, R.G. Wells, Increased stiffness of the rat liver precedes matrix deposition: implications for fibrosis. Am. J. Physiol. Gastrointest. Liver Physiol. **293**(6), G1147-1154 (2007)

M. Georgiadou, J. Lilja, G. Jacquemet et al., AMPK negatively regulates tensin-dependent integrin activity. J. Cell Biol. **216**, 1107–1121 (2017)

E.S. Gershtein, V.A. Shatskaya, K.K. Laktionov, N.E. Kushlinskii, M.A. Krasilnikov, Expression of phosphatidylinositol-3 kinase in lung cancer. Bull. Exp. Biol. Med. **130**(12), 1166–1168 (2000)

E.S. Gershtein, A.M. Scherbakov, V.A. Shatskaya, N.E. Kushlinsky, M.A. Krasilnikov, Phosphatidylinositol 3-kinase/AKT signalling pathway components in human breast cancer: clinicopathological correlations. Anticancer Res. **27**(4A), 1777–1782 (2007)

B. Ghesquiere, B.W. Wong, A. Kuchnio, P. Carmeliet, Metabolism of stromal and immune cells in health and disease. Nature **511**, 167–176 (2014)

F.G. Giancotti, E. Ruoslahti, Integrin signaling. Science **285**(5430), 1028–1032 (1999)

E. Giannoni, F. Buricchi, G. Raugei, G. Ramponi, P. Chiarugi, Intracellular reactive oxygen species activate Src tyrosine kinase during cell adhesion and anchorage-dependent cell growth. Mol. Cell Biol. **25**(15), 6391–6403 (2005)

J.J. Gibbons, R.T. Abraham, K. Yu, Mammalian target of rapamycin: discovery of rapamycin reveals a signaling pathway important for normal and cancer cell growth. Semin. Oncol. **36**(Suppl 3), S3–S17 (2009)

C. Gieger, L. Geistlinger, E. Altmaier et al., Genetics meets metabolomics: a genome-wide association study of metabolite profiles in human serum. PLoS Genet. **4**, e1000282 (2008)

K. Glunde, Z.M. Bhujwalla, Metabolic tumor imaging using magnetic resonance spectroscopy. Semin. Oncol. **38**, 26–41 (2011)

V. Gogvadze, B. Zhivotovsky, S. Orrenius, The Warburg effect and mitochondrial stability. Mol. Aspects Med. **31**, 60–74 (2010)

A.M. Gonzalez-Angulo, J. Ferrer-Lozano, K. Stemke-Hale et al., PI3K pathway mutations and PTEN levels in primary and metastatic breast cancer. Mol. Cancer Ther. **10**(6), 1093–1101 (2011)

C.K. Goo, H.Y. Lim, Q.S. Ho, H.P. Too, M.V. Clement, K.P. Wong, PTEN/Akt signaling controls mitochondrial respiratory capacity through 4E-BP1. PLoS ONE **7**(9), e45806 (2012)

M.I. Gross, S.D. Demo, J.B. Dennison et al., Antitumor activity of the glutaminase inhibitor CB-839 in triple-negative breast cancer. Mol. Cancer Ther. **13**(4), 890–901 (2014)

J.L. Guan, Integrin signaling through FAK in the regulation of mammary stem cells and breast cancer. IUBMB Life **62** (4), 268–276 (2010)

L. Guarente, H. Franklin, Epstein lecture: sirtuins, aging, and medicine. N. Engl. J. Med. **364**, 2235–2244 (2011)

M. Guppy, P. Leeman, X. Zu, V. Russel, Contribution by different fuels and metabolic pathways to the total ATP turnover of proliferating MCF-7 breast cancer cells. Biochem. J. **364**, 309–315 (2002)

R.D. Guzy, B. Sharma, E. Bell, N.S. Chandel, P.T. Schumacker, Loss of the SdhB, but not the SdhA, subunit of complex II triggers reactive oxygen species-dependent hypoxia-inducible factor activation and tumorigenesis. Mol. Cell Biol. **28**, 718–731 (2008)

D.M. Gwinn, D.B. Shackelford, D.F. Egan, M.M. Mihaylova, A. Mery, D.S. Vasquez, B.E. Turk, and R.J. Shaw, AMPK phosphorylation of raptor mediates a metabolic checkpoint. Mol. Cell. **30**(2), 214–226 (2008)

S.M. Hadad, J.A. Dewar, E. Elseedawy, L. Jordan, C. Purdie, S.E. Bray, M.A. Thompson, Gene signature of metformin actions on primary breast cancer within a window of opportunity randomized clinical trial. J. Clin. Oncol. **28**(15) (2010)

O.P. Hamill, B. Martinac, Molecular basis of mechanotransduction in living cells. Physiol. Rev. **81**(2), 685–740 (2001)

B. Han, S. Lin, L.J. Yu, R.Z. Wang, Y.Y. Wang, Correlation of (1)(8)F-FDG PET activity with expressions of survivin, Ki67, and CD34 in non-small-cell lung cancer. Nucl. Med. Commun. **30**(11), 831–837 (2009)

D. Hanahan, R.A. Weinberg, Hallmarks of cancer: the next generation. Cell **144**, 646–674 (2011)

H.X. Hao, O. Khalimonchuk, M. Schraders et al., SDH5, a gene required for flavination of succinate dehydrogenase, is mutated in paraganglioma. Science **325**, 1139–1142 (2009)

D.G. Hardie, AMP-activated protein kinase—an energy sensor that regulates all aspects of cell function. Genes Dev. **25**, 1895–1908 (2011)

D.G. Hardie, F.A. Ross, S.A. Hawley, AMPK: a nutrient and energy sensor that maintains energy homeostasis. Nat. Rev. Mol. Cell Biol. **13**, 251–262 (2012)

C.T. Hensley, A.T. Wasti, R.J. DeBerardinis, Glutamine and cancer: cell biology, physiology, and clinical opportunities. J. Clin. Invest. **123**, 3678–3684 (2013)

C.T. Hensley, B. Faubert, Q. Yuan et al., Metabolic heterogeneity in human lung tumors. Cell **164**, 681–694 (2016)

M.C. Hollander, G.M. Blumenthal, P.A. Dennis, PTEN loss in the continuum of common cancers, rare syndromes and mouse models. Nat. Rev. Cancer **11**(4), 289–301 (2011)

K. Hosono, H. Endo, H. Takahashi et al., Metformin suppresses colorectal aberrant crypt foci in a short-term clinical trial. Cancer Prev. Res. (Phila) **3**(9), 1077–1083 (2010)

S.M. Houten, R.J. Wanders, A general introduction to the biochemistry of mitochondrial fatty acid beta-oxidation. J. Inherit. Metab. Dis. **33**, 469–477 (2010)

S. Hu, A. Balakrishnan, R.A. Bok, B. Anderton, P.E. Larson, S.J. Nelson, J. Kurhanewicz, D.B. Vigneron, A. Goga, 13C-pyruvate imaging reveals alterations in glycolysis that precede c-Myc-induced tumor formation and regression. Cell Metab. **14**, 131–142 (2011)

W. Hu, Z. Feng, A.J. Levine, The regulation of multiple p53 stress responses is mediated through MDM2. Genes Cancer **3**(3–4), 199–208 (2012)

H. Hu, A. Juvekar, C.A. Lyssiotis et al., Phosphoinositide 3-kinase regulates glycolysis through mobilization of aldolase from the actin cytoskeleton. Cell **164**, 433–446 (2016)

C. Ibarra-Drendall, E.C. Dietze, L. Victoria, Seewaldt metabolic syndrome and breast cancer risk: is there a role for metformin? Curr. Breast Cancer Rep. **3**(3), 142–150 (2011)

C. Ibarra-Drendall, M.M. Troch, W.T. Barry et al., Pilot and feasibility study: prospective proteomic profiling of mammary epithelial cells from high-risk women provides evidence of activation of pro-survival pathways. Breast Cancer Res. Treat. **132**(2), 487–498 (2012)

N. Ilan, M. Elkin, I. Vlodavsky, Regulation, function and clinical significance of heparanase in cancer metastasis and angiogenesis. Int. J. Biochem. Cell Biol. **38**(12), 2018–2039 (2006)

T. Illig, C. Gieger, G. Zhai et al., A genome-wide perspective of genetic variation in human metabolism. Nat. Genet. **42**, 137–141 (2010)

D.E. Ingber, Tensegrity-based mechanosensing from macro to micro. Prog. Biophys. Mol. Biol. **97**(2–3), 163–179 (2008)

W.V. Ingman, J. Wyckoff, V. Gouon-Evans, J. Condeelis, J.W. Pollard, Macrophages promote collagen fibrillogenesis around terminal end buds of the developing mammary gland. Dev. Dyn. **235**(12), 3222–3229 (2006)

A. Isakovic, L. Harhaji, D. Stevanovic, Z. Markovic, M. Sumarac-Dumanovic, V. Starcevic, D. Micic, V. Trajkovic, Dual antiglioma action of metformin: cell cycle arrest and mitochondria-dependent apoptosis. Cell. Mol. Life Sci. **64** (10), 1290–1302 (2007)

H. Jadvar, A. Alavi, S.S. Gambhir, 18F-FDG uptake in lung, breast, and colon cancers: molecular biology correlates and disease characterization. J. Nucl. Med. **50**(11), 1820–1827 (2009)

P.A. Janmey, The cytoskeleton and cell signaling: component localization and mechanical coupling. Physiol. Rev. **78** (3), 763–781 (1998)

S.M. Jeon, N.S. Chandel, N. Hay, AMPK regulates NADPH homeostasis to promote tumour cell survival during energy stress. Nature **485**, 661–665 (2012)

S. Jiralerspong, S.L. Palla, S.H. Giordano, F. Meric-Bernstam, C. Liedtke, C.M. Barnett, L. Hsu, M.C. Hung, G.N. Hortobagyi, A.M. Gonzalez-Angulo, The effects of metformin on pathologic complete response (pCR) rates in diabetic breast cancer (BC) patients receiving neoadjuvant systemic therapy (NST). J. Clin. Oncol. **267**(20), 3297–3302 (2008)

K.R. Johnson, J.L. Leight, V.M. Weaver, Demystifying the effects of a three-dimensional microenvironment in tissue morphogenesis. Methods Cell Biol. **83**, 547–583 (2007)

R.G. Jones, C.B. Thompson, Tumor suppressors and cell metabolism: a recipe for cancer growth. Genes Dev. **23**, 537–548 (2009)

R.G. Jones, D.R. Plas, S. Kubek, M. Buzzai, J. Mu, Y. Xu, M.J. Birnbaum, C.B. Thompson, AMP-activated protein kinase induces a p53-dependent metabolic checkpoint. Mol. Cell **18**, 283–293 (2005)

F. Kai, H. Laklai, V.M. Weaver, Force matters: biomechanical regulation of cell invasion and migration in disease. Trends Cell Biol. **26**, 486–497 (2016)

K. Kaira, H. Murakami, M. Endo, Y. Ohde, T. Naito, H. Kondo, T. Nakajima, N. Yamamoto, T. Takahashi, Biological correlation of (1)(8)F-FDG uptake on PET in pulmonary neuroendocrine tumors. Anticancer Res. **33**(10), 4219–4228 (2013)

G. Kallergi, S. Agelaki, H. Markomanolaki, V. Georgoulias, C. Stournaras, Activation of FAK/PI3K/Rac1 signaling controls actin reorganization and inhibits cell motility in human cancer cells. Cell Physiol. Biochem. **20**(6), 977–986 (2007)

R. Kalluri, The biology and function of fibroblasts in cancer. Nat. Rev. Cancer **16**, 582–598 (2016)

S. Kamarajugadda, L. Stemboroski, Q. Cai, N.E. Simpson, S. Nayak, M. Tan, J. Lu, Glucose oxidation modulates anoikis and tumor metastasis. Mol. Cell Biol. **32**, 1893–1907 (2012)

S.S. Kantak, R.H. Kramer, E-cadherin regulates anchorage-independent growth and survival in oral squamous cell carcinoma cells. J. Biol. Chem. **273**, 16953–16961 (1998)

J. Kasznicki, A. Sliwinska, J. Drzewoski, Metformin in cancer prevention and therapy. Ann. Transl. Med. **2**(6), 57 (2014)

A. Katsumi, A.W. Orr, A.W. Orr, E. Tzima, M.A. Schwartz, Integrins in mechanotransduction. J. Biol. Chem. **279**(13), 12001–12004 (2004)

S. Kauppila, F. Stenbäck, J. Risteli, A. Jukkola, L. Risteli, Aberrant type I and type III collagen gene expression in human breast cancer in vivo. J. Pathol. **186**(3), 262–268 (1998)

K. Kessenbrock, V. Plaks, Z. Werb, Matrix metalloproteinases: regulators of the tumor microenvironment. Cell **141**(1), 52–67 (2010)

W. Khaled, S. Reichling, O.T. Bruhns, H. Ermert, Ultrasonic strain imaging and reconstructive elastography for biological tissue. Ultrasonics **44**(Suppl 1), e199-202 (2006)

J.W. Kim, P. Gao, Y.C. Liu, G.L. Semenza, C.V. Dang, Hypoxia-inducible factor 1 and dysregulated c-Myc cooperatively induce vascular endothelial growth factor and metabolic switches hexokinase 2 and pyruvate dehydrogenase kinase 1. Mol. Cell Biol. **27**(21), 7381–7393 (2007)

H.M. Kim, W.H. Jung, J.S. Koo, Site-specific metabolic phenotypes in metastatic breast cancer. J. Transl. Med. **12**, 354 (2014a)

M.J. Kim, D.H. Kim, W.H. Jung, J.S. Koo, Expression of metabolism-related proteins in triple-negative breast cancer. Int. J. Clin. Exp. Pathol. **7**(1), 301–312 (2014b)

N.H. Kim, Y.H. Cha, J. Lee et al., Snail reprograms glucose metabolism by repressing phosphofructokinase PFKP allowing cancer cell survival under metabolic stress. Nat. Commun. **8**, 14374 (2017)

N. Koglin, A. Mueller, M. Berndt, H. Schmitt-Willich, L. Toschi, A.W. Stephens, V. Gekeler, M. Friebe, L.M. Dinkelborg, Specific PET imaging of xC-transporter activity using a 18F-labeled glutamate derivative reveals a dominant pathway in tumor. Metab. Clin. Cancer Res. **17**, 6000–6011 (2011)

A.D. Kohn, S.A. Summers, M.J. Birnbaum, R.A. Roth, Expression of a constitutively active Akt Ser/Thr kinase in 3T3-L1 adipocytes stimulates glucose uptake and glucose transporter 4 translocation. J. Biol. Chem. **271**(49), 31372–31378 (1996)

M.I. Kokkinos, R. Wafai, M.K. Wong, D.F. Newgreen, E.W. Thompson, M. Waltham, Vimentin and epithelial-mesenchymal transition in human breast cancer—observations in vitro and in vivo. Cells Tissues Organs **185**(1–3), 191–203 (2007)

W.H. Koppenol, P.L. Bounds, C.V. Dang, Otto Warburg's contributions to current concepts of cancer metabolism. Nat. Rev. Cancer **11**, 325–337 (2011)

M. Kranendijk, E.A. Struys, E. van Schaftingen et al., IDH2 mutations in patients with D-2-hydroxyglutaric aciduria. Science **330**, 336 (2010)

H.A. Krebs, Some aspects of the regulation of fuel supply in omnivorous animals. Adv. Enzyme Regul. **10**, 397–420 (1972)

B. Krishnamachary, D. Zagzag, H. Nagasawa, K. Rainey, H. Okuyama, J.H. Baek, G.L. Semenza, Hypoxia-inducible factor-1-dependent repression of E-cadherin in von Hippel-Lindau tumor suppressor-null renal cell carcinoma mediated by TCF3, ZFHX1A, and ZFHX1B. Cancer Res. **66**, 2725–2731 (2006)

J. Kurhanewicz, D.B. Vigneron, K. Brindle et al., Analysis of cancer metabolism by imaging hyperpolarized nuclei: prospects for translation to clinical research. Neoplasia **13**, 81–97 (2011)

J. Kwon, S.R. Lee, K.S. Yang, Y. Ahn, Y.J. Kim, E.R. Stadtman, S.G. Rhee, Reversible oxidation and inactivation of the tumor suppressor PTEN in cells stimulated with peptide growth factors. Proc. Natl. Acad. Sci. U. S. A. **101**(47), 16419–16424 (2004)

B.R. Landau, J. Wahren, V. Chandramouli, W.C. Schumann, K. Ekberg, S.C. Kalhan, Contributions of gluconeogenesis to glucose production in the fasted state. J. Clin. Invest. **98**, 378–385 (1996)

M. Laplante, D.M. Sabatini, mTOR signaling in growth control and disease. Cell **149**(2), 274–293 (2012)

Q.T. Le, J. Harris, A.M. Magliocco et al., Validation of lysyl oxidase as a prognostic marker for metastasis and survival in head and neck squamous cell carcinoma: radiation therapy oncology group trial 90–03. J. Clin. Oncol. **27**(26), 4281–4286 (2009)

V.S. LeBleu, J.T. O'Connell, K.N. Gonzalez Herrera et al., PGC-1alpha mediates mitochondrial biogenesis and oxidative phosphorylation in cancer cells to promote metastasis. Nat. Cell Biol. **16**, 992–1003 (2014)

S. Lee, E. Nakamura, H. Yang et al., Neuronal apoptosis linked to EglN3 prolyl hydroxylase and familial pheochromocytoma genes: developmental culling and cancer. Cancer Cell **8**, 155–167 (2005)

J.H. Lee, H. Koh, M. Kim et al., Energy-dependent regulation of cell structure by AMP-activated protein kinase. Nature **447**, 1017–1020 (2007)

K. Lee, Q.K. Chen, C. Lui, M.A. Cichon, D.C. Radisky, C.M. Nelson, Matrix compliance regulates Rac1b localization, NADPH oxidase assembly, and epithelial-mesenchymal transition. Mol. Biol. Cell **23**, 4097–4108 (2012a)

S.Y. Lee, H.M. Jeon, M.K. Ju, C.H. Kim, G. Yoon, S.I. Han, H.G. Park, H.S. Kang, Wnt/Snail signaling regulates cytochrome C oxidase and glucose metabolism. Cancer Res. **72**, 3607–3617 (2012b)

S.J. Leevers, B. Vanhaesebroeck, M.D. Waterfield, Signalling through phosphoinositide 3-kinases: the lipids take centre stage. Curr. Opin. Cell Biol. **11**(2), 219–225 (1999)

T.P. Lele, J.E. Sero, B.D. Matthews, S. Kumar, S. Xia, M. Montoya-Zavala, T. Polte, D. Overby, N. Wang, D.E. Ingber, Tools to study cell mechanics and mechanotransduction. Methods Cell Biol. **83**, 443–472 (2007)

S.A. Lelievre, V.M. Weaver, J.A. Nickerson, C.A. Larabell, A. Bhaumik, O.W. Petersen, M.J. Bissell, Tissue phenotype depends on reciprocal interactions between the extracellular matrix and the structural organization of the nucleus. Proc. Natl. Acad. Sci. U. S. A. **95**(25), 14711–14716 (1998)

J. Leppert, J. Krajewski, S.R. Kantelhardt, S. Schlaffer, N. Petkus, E. Reusche, G. Hüttmann, A. Giese, Multiphoton excitation of autofluorescence for microscopy of glioma tissue. Neurosurgery **58**(4), 759–767 (2006)

K.R. Levental, H. Yu, L. Kass et al., Matrix crosslinking forces tumor progression by enhancing integrin signaling. Cell **139**, 891–906 (2009)

H. Liu, D.C. Radisky, F. Wang, M.J. Bissell, Polarity and proliferation are controlled by distinct signaling pathways downstream of PI3-kinase in breast epithelial tumor cells. J. Cell Biol. **164**(4), 603–612 (2004)

B. Liu, Z. Fan, S.M. Edgerton, X.S. Deng, I.N. Alimova, S.E. Lind, A.D. Thor, Metformin induces unique biological and molecular responses in triple negative breast cancer cells. Cell Cycle **8**(13), 2031–2040 (2009)

N. Liu, Q. Yu, T.J. Liu, E.P. Gebreamlak, S.L. Wang, R.J. Zhang, J. Zhang, Y. Niu, P-cadherinexpression and basal-like subtype in breast cancers. Med. Oncol. **29**, 2606–2612 (2012)

S. Liu, Y. Cong, D. Wang et al., Breast cancer stem cells transition between epithelial and mesenchymal states reflective of their normal counterparts. Stem Cell Rep. **2**, 78–91 (2014)

J.W. Locasale, A.R. Grassian, T. Melman et al., Phosphoglycerate dehydrogenase diverts glycolytic flux and contributes to oncogenesis. Nat. Genet. **43**, 869–874 (2011)

J.I. Lopez, I. Kang, W.K. You, D.M. McDonald, V.M. Weaver, In situ force mapping of mammary gland transformation. Integr. Biol. (Camb) **3**(9), 910–921 (2011)

E. Louie, S. Nik, J.S. Chen, M. Schmidt, B. Song, C. Pacson, X.F. Chen, S. Park, J. Ju, E.I. Chen, Identification of a stem-like cell population by exposing metastatic breast cancer cell lines to repetitive cycles of hypoxia and reoxygenation. Breast Cancer Res. **12**, R94 (2010)

P. Lu, V.M. Weaver, Z. Werb, The extracellular matrix: a dynamic niche in cancer progression. J. Cell Biol. **196**(4), 395–406 (2012)

J.J. Lum, D.E. Bauer, M. Kong, M.H. Harris, C. Li, T. Lindsten, C.B. Thompson, Growth factor regulation of autophagy and cell survival in the absence of apoptosis. Cell **120**, 237–248 (2005)

S.Y. Lunt, M.G. Vander Heiden, Aerobic glycolysis: meeting the metabolic requirements of cell proliferation. Annu. Rev. Cell Dev. Biol. **27**, 441–464 (2011)

M. Luo, L. Shang, M.D. Brooks et al., Targeting breast cancer stem cell state equilibrium through modulation of redox signaling. Cell Metab. **28**, 69.e6-86.e6 (2018)

E. Madan, B. Dikshit, S.H. Gowda, C. Srivastava, C. Sarkar, P. Chattopadhyay, S. Sinha, K. Chosdol, FAT1 is a novel upstream regulator of HIF1α and invasion of highgrade glioma. Int. J. Cancer **139**, 2570–2582 (2016)

T. Maehama, J.E. Dixon, The tumor suppressor, PTEN/MMAC1, dephosphorylates the lipid second messenger, phosphatidylinositol 3,4,5-trisphosphate. J. Biol. Chem. **273**(22), 13375–13378 (1998)

E.A. Maher, I. Marin-Valencia, R.M. Bachoo et al., Metabolism of [U-13 C]glucose in human brain tumors in vivo. NMR Biomed. **25**, 1234–1244 (2012)

O. Maller, K.C. Hansen, T.R. Lyons, I. Acerbi, V.M. Weaver, R. Prekeris, A.C. Tan, P. Schedin, Collagen architecture in pregnancy-induced protection from breast cancer. J. Cell Sci. **126**(Pt 18), 4108–4110 (2013)

B.D. Manning, L.C. Cantley, AKT/PKB signaling: navigating downstream. Cell **129**(7), 1261–1274 (2007)

E.R. Mardis, L. Ding, D.J. Dooling et al., Recurring mutations found by sequencing an acute myeloid leukemia genome. N. Engl. J. Med. **361**, 1058–1066 (2009)

M. Martin, B. Beauvoit, P.J. Voisin, P. Canioni, B. Guérin, M. Rigoulet, Energetic and morphologic plasticity of glioma cells grown on 3-D support: effect on transient glutamine deprivation. J. Bioenerg. Biomembr. **30**, 565–577 (1998)

T. Mashimo, K. Pichumani, V. Vemireddy et al., Acetate is a bioenergetic substrate for human glioblastoma and brain metastases. Cell **159**, 1603–1614 (2014)

M. Masin, J. Vazquez, S. Rossi et al., GLUT3 is induced during epithelial-mesenchymal transition and promotes tumor cell proliferation in non-small cell lung cancer. Cancer Metab. **2**, 11 (2014)

S. Matoba, J.G. Kang, W.D. Patino, A. Wragg, M. Boehm, O. Gavrilova, P.J. Hurley, F. Bunz, P.M. Hwang, p53 regulates mitochondrial respiration. Science **312**, 1650–1653 (2006)

R. Matoba, Y. Morizane, Y. Shiode, M. Hirano, S. Doi, S. Toshima, R. Araki, M. Hosogi, T. Yonezawa, F. Shiraga, Suppressive effect of AMP-activated protein kinase on the epithelial-mesenchymal transition in retinal pigment epithelial cells. PLoS ONE **12**, e0181481 (2017)

F. Mbeunkui, D.J. Johann Jr., Cancer and the tumor microenvironment: a review of an essential relationship. Cancer Chemother. Pharmacol. **63**(4), 571–582 (2009)

O.G. McDonald, X. Li, T. Saunders et al., Epigenomic reprogramming during pancreatic cancer progression links anabolic glucose metabolism to distant metastasis. Nat. Genet. **49**, 367–376 (2017)

T.R. McKnight, K.R. Lamborn, T.D Love et al, Correlation of magnetic resonance spectroscopic and growth characteristics within Grades II and III gliomas. J. Neurosurg. **106**(4), 660–666 (2007)

S.L. McKnight, On getting there from here. Science **330**, 1338–1339 (2010)

F. Meric-Bernstam, A. DeMichele, M.L. Telli et al., Abstract C49: phase 1 study of CB-839, a first-in-class, orally administered small molecule inhibitor of glutaminase in patients with refractory solid tumors. Mol. Cancer Ther. **14**, C49 (2015)

C.M. Metallo, P.A. Gameiro, E.L. Bell et al., Reductive glutamine metabolism by IDH1 mediates lipogenesis under hypoxia. Nature **481**(7381), 380–384 (2011)

Y. Miao, M. Lu, Q. Yan, S. Li, Y. Feng, Inhibition of proliferation, migration, and invasion by knockdown of pyruvate kinase-M2 (PKM2) in ovarian cancer SKOV3 and OVCAR3 cells. Oncol. Res. **24**, 463–475 (2016)

J.-S. Mo, Z. Meng, Y.C. Kim, H.W. Park, C.G. Hansen, S. Kim, D.-S. Lim, K.-L. Guan, Cellular energy stress induces AMPK-mediated regulation of YAP and the Hippo pathway. Nat. Cell Biol. **17**, 500–510 (2015)

A. Mogilner, G. Oster, Cell motility driven by actin polymerization. Biophys. J. **71**, 3030–3045 (1996)

P. Montcourrier, I. Silver, R. Farnoud, I. Bird, H. Rochefort, Breast cancer cells have a high capacity to acidify extracellular milieu by a dual mechanism. Clin. Exp. Metastasis **15**, 382–392 (1997)

A. Morandi, M.L. Taddei, P. Chiarugi, E. Giannoni, Targeting the metabolic reprogramming that controls epithelial-to-mesenchymal transition in aggressive tumors. Front. Oncol. **7**, 40 (2017)

R. Moreno-Sánchez, S. Rodríguez-Enríquez, A. Marín-Hernández, E. Saavedra, Energy metabolism in tumor cells. FEBS J. **274**, 1393–1418 (2007)

L.G. Morris, A.M. Kaufman, Y. Gong et al., Recurrent somatic mutation of FAT1 in multiple human cancers leads to aberrant Wnt activation. Nat. Genet. **45**, 253–261 (2013)

J.K. Mouw, Y. Yui, L. Damiano et al., Tissue mechanics modulate microRNA-dependent PTEN expression to regulate malignant progression. Nat. Med. **20**(4), 360–367 (2014)

A.R. Mullen, W.W. Wheaton, E.S. Jin, P.H. Chen, L.B. Sullivan, T. Cheng, Y. Yang, W.M. Linehan, N.S. Chandel, R. J. DeBerardinis, Reductive carboxylation supports growth in tumor cells with defective mitochondria. Nature **481** (7381), 385–388 (2011)

T. Nakagawa, D.J. Lomb, M.C. Haigis, L. Guarente, SIRT5 Deacetylates carbamoyl phosphate synthetase 1 and regulates the urea cycle. Cell **137**, 560–570 (2009)

K. Nakaya, H.D. Yamagata, N. Arita, K.I. Nakashiro, M. Nose, T. Miki, H. Hamakawa, Identification of homozygous deletions of tumor suppressor gene FAT in oral cancer using CGH-array. Oncogene **26**, 5300–5308 (2007)

C.M. Nelson, M.J. Bissell, Modeling dynamic reciprocity: engineering three-dimensional culture models of breast architecture, function, and neoplastic transformation. Semin. Cancer Biol. **15**(5), 342–352 (2005)

S.J. Nelson, J. Kurhanewicz, D.B. Vigneron, et al., Metabolic imaging of patients with prostate cancer using hyperpolarized [1-(1)(3)C] pyruvate. Sci. Transl. Med. **5**(198), 198ra108 (2013)

Y. Nishikawa, T. Miyazaki, K. Nakashiro, H. Yamagata, M. Isokane, H. Goda, H. Tanaka, R. Oka, H. Hamakawa, Human FAT1 cadherin controls cell migration and invasion of oral squamous cell carcinoma through the localization of beta-catenin. Oncol. Rep. **26**, 587–592 (2011)

M.J. Nokin, F. Durieux, P. Peixoto, et al., Methylglyoxal, a glycolysis side-product, induces Hsp90 glycation and YAP-mediated tumor growth and metastasis. eLife **5**, e19375 (2016)

S. Nomoto, M. Sanchez-Cespedes, D. Sidransky, Identification of mtDNA mutations in human cancer. Methods Mol. Biol. **197**, 107–117 (2002)

L.E. O'Brien, T.S. Jou, A.L. Pollack, Q. Zhang, S.H. Hansen, P. Yurchenco, K.E. Mostov, Rac1 orientates epithelial apical polarity through effects on basolateral laminin assembly. Nat. Cell Biol. **3**(9), 831–838 (2001)

A. Ocana, F. Vera-Badillo, M. Al-Mubarak, A.J. Templeton, V. Corrales-Sanchez, L. Diez-Gonzalez, M.D. Cuenca-Lopez, B. Seruga, A. Pandiella, E. Amir, Activation of the PI3K/mTOR/AKT pathway and survival in solid tumors: systematic review and meta-analysis. PLoS ONE **9**(4), e95219 (2014)

A. Ooi, J.C. Wong, D. Petillo et al., An antioxidant response phenotype shared between hereditary and sporadic type 2 papillary renal cell carcinoma. Cancer Cell **20**, 511–523 (2011)

R.C. Osthus, H. Shim, S. Kim, Q. Li, R. Reddy, M. Mukherjee, Y. Xu, D. Wonsey, L.A. Lee, C.V. Dang, Deregulation of glucose transporter 1 and glycolytic gene expression by c-Myc. J. Biol. Chem. **275**(29), 21797–21800 (2000)

B.C. Özdemir, T. Pentcheva-Hoang, J.L. Carstens et al., Depletion of carcinoma-associated fibroblasts and fibrosis induces immunosuppression and accelerates pancreas cancer with reduced survival. Cancer Cell **25**(6), 719–734 (2014)

R. Pagan, I. Martin, A. Alonso, M. Llobera, S. Vilaró, Vimentin filaments follow the preexisting cytokeratin network during epithelial-mesenchymal transition of cultured neonatal rat hepatocytes. Exp. Cell Res. **222**(2), 333–344 (1996)

P. Paoli, E. Giannoni, P. Chiarugi, Anoikis molecular pathways and its role in cancer progression. Biochim. Biophys. Acta **1833**, 3481–3498 (2013)

I. Papandreou, R.A. Cairns, L. Fontana, A.L. Lim, N.C. Denko, HIF-1 mediates adaptation to hypoxia by actively downregulating mitochondrial oxygen consumption. Cell Metab. **3**(3), 187–197 (2006)

J. Paredes, C. Stove, V. Stove, F. Milanezi, V. Van Marck, L. Derycke, M. Mareel, M. Bracke, F. Schmitt, P-cadherin is up-regulated by the antiestrogen ICI 182780 and promotes invasion of human breast cancer cells. Cancer Res. **64**, 8309–8317 (2004)

J. Paredes, A. Albergaria, J.T. Oliveira, C. Jeronimo, F. Milanezi, F.C. Schmitt, P-cadherin overexpression is an indicator of clinical outcome in invasive breast carcinomas and is associated with CDH3 promoter hypomethylation. Clin. Cancer Res. **11**, 5869–5877 (2005)

J. Paredes, A.L. Correia, A.S. Ribeiro, A. Albergaria, F. Milanezi, F.C. Schmitt, P-cadherin expression in breast cancer: a review. Breast Cancer Res. **9**, 214 (2007)

J. Paredes, A.L. Correia, A.S. Ribeiro, F. Milanezi, J. Cameselle-Teijeiro, F.C. Schmitt, Breastcarcinomas that co-express E- and P-cadherin are associated with p120-catenin cytoplasmic localisation and poor patient survival. J. Clin. Pathol. **61**, 856–862 (2008)

J. Paredes, J. Figueiredo, A. Albergaria et al., Epithelial E- and P-cadherins: role and clinical significance in cancer. Biochim. Biophys. Acta **1826**, 297–311 (2012)

H.W. Park, K.-L. Guan, Regulation of the Hippo pathway and implications for anticancer drug development. Trends Pharmacol. Sci. **34**, 581–589 (2013)

S.Y. Park, J.H. Shin, S.H. Kee, E-cadherin expression increases cell proliferation by regulating energy metabolism through nuclear factor-kappaB in AGS cells. Cancer Sci. **108**, 1769–1777 (2017)

J.S. Parker, M. Mullins, M.C.U. Cheang et al., Supervised risk predictor of breast cancer based on intrinsic subtypes. J. Clin. Oncol. **27**, 1160–1167 (2009)

D.W. Parsons, S. Jones, X. Zhang et al., An integrated genomic analysis of human glioblastoma multiforme. Science **321**, 1807–1812 (2008)

J.T. Parsons, A.R. Horwitz, M.A. Schwartz, Cell adhesion: integrating cytoskeletal dynamics and cellular tension. Nat. Rev. Mol. Cell Biol. **11**, 633–643 (2010)

P. Pasdois, C. Devaud, P. Voisin, V. Bouchaud, M. Rigoulet, B. Beauvoit, Contribution of the phosphorylable complex I in the growth phase-dependent respiration of C6 glioma cells in vitro. J. Bioenerg. Biomembr. **35**, 439–450 (2003)

M.J. Paszek, N. Zahir, K.R. Johnson et al., Tensional homeostasis and the malignant phenotype. Cancer Cell **8**(3), 241–254 (2005)

M.J. Paszek, D. Boettiger, V.W. Weaver, D.A. Hammer, Integrin clustering is driven by mechanical resistance from the glycocalyx and the substrate. PLoS Comput. Biol. **5**(12), e1000604 (2009)

T. Patel, G. Hruby, K. Badani, C. Abate-Shen, J.M. McKiernan, Clinical outcomes after radical prostatectomy in diabetic patients treated with metformin. Urology **76**(5), 1240–1244 (2010)

V.L. Payen, P.E. Porporato, B. Baselet, P. Sonveaux, Metabolic changes associated with tumor metastasis, part 1: tumor pH, glycolysis and the pentose phosphate pathway. Cell Mol. Life Sci. **73**, 1333–1348 (2016)

H. Pelicano, R.H. Xu, M. Du et al., Mitochondrial respiration defects in cancer cells cause activation of Akt survival pathway through a redox-mediated mechanism. J. Cell Biol. **175**, 913–932 (2006)

A. Peralta Soler, K.A. Knudsen, H. Salazar, A.C. Han, A.A. Keshgegian, P-cadherin expression in breast carcinoma indicates poor survival. Cancer **86**, 1263–1272 (1999)

M. Perepelyuk, M. Terajima, A.Y. Wang, P.C. Georges, P.A. Janmey, M. Yamauchi, R.G. Wells, Hepatic stellate cells and portal fibroblasts are the major cellular sources of collagens and lysyl oxidases in normal liver and early after injury. Am. J. Physiol. Gastrointest. Liver Physiol. **304**(6), G605-614 (2013)

R.M. Perera, S. Stoykova, B.N. Nicolay et al., Transcriptional control of autophagy-lysosome function drives pancreatic cancer metabolism. Nature **524**, 361–365 (2015)

T.R. Peterson, S.S. Sengupta, T.E. Harris et al., mTOR complex 1 regulates lipin 1 localization to control the SREBP pathway. Cell **146**, 408–420 (2011)

A.I. Phipps, R.T. Chlebowski, R. Prentice et al., Body size, physical activity, and risk of triple-negative and estrogen receptor-positive breast cancer. Cancer Epidemiol. Biomarkers Prev. **20**(3), 454–463 (2011)

M.W. Pickup, H. Laklai, I. Acerbi, P. Owens, A.E. Gorska, A. Chytil, M. Aakre, V.M. Weaver, H.L. Moses, Stromally derived lysyl oxidase promotes metastasis of transforming growth factor-beta-deficient mouse mammary carcinomas. Cancer Res. **73**(17), 5336–5346 (2013)

M.W. Pickup, J.K. Mouw, V.M. Weaver, The extracellular matrix modulates the hallmarks of cancer. EMBO Rep. **15**, 1243–1253 (2014)

B. Pietrak, H. Zhao, H. Qi et al., A tale of two subunits: how the neomorphic R132H IDH1 mutation enhances production of alphaHG. Biochemistry **50**, 4804–4812 (2011)

P.G. Pilie, C. Ibarra-Drendall, M.M. Troch et al., Protein microarray analysis of mammary epithelial cells from obese and nonobese women at high risk for breast cancer: feasibility data. Cancer Epidemiol. Biomarkers Prev. **20**(3), 476–482 (2011)

M. Plodinec, M. Loparic, C.A. Monnier et al., The nanomechanical signature of breast cancer. Nat. Nanotechnol. **7**(11), 757–765 (2012)

S. Pollari, S.M. Kakonen, H. Edgren, M. Wolf, P. Kohonen, H. Sara, T. Guise, M. Nees, O. Kallioniemi, Enhanced serine production by bone metastatic breast cancer cells stimulates osteoclastogenesis. Breast Cancer Res. Treat. **125**, 421–430 (2011)

P.E. Porporato, V.L. Payen, J. Perez-Escuredo et al., A mitochondrial switch promotes tumor metastasis. Cell Rep. **8**, 754–766 (2014)

T. Porstmann, C.R. Santos, B. Griffiths, M. Cully, M. Wu, S. Leevers, J.R. Griffiths, Y.L. Chung, A. Schulze, SREBP activity is regulated by mTORC1 and contributes to Akt-dependent cell growth. Cell Metab. **8**, 224–236 (2008)

R. Possemato, K.M. Marks, Y.D. Shaul et al., Functional genomics reveal that the serine synthesis pathway is essential in breast cancer. Nature **476**, 346–350 (2011)

P.P. Provenzano, P.J. Keely, The role of focal adhesion kinase in tumor initiation and progression. Cell Adhes. Migr. **3**(4), 347–350 (2009)

P.P. Provenzano, K.W. Eliceiri, J.M. Campbell, D.R. Inman, J.G. White, P.J. Keely, Collagen reorganization at the tumor-stromal interface facilitates local invasion. BMC Med. **4**(1), 38 (2006)

P.P. Provenzano, D.R. Inman, K.W. Eliceiri, H.E. Beggs, P.J. Keely, Mammary epithelial-specific disruption of focal adhesion kinase retards tumor formation and metastasis in a transgenic mouse model of human breast cancer. Am. J. Pathol. **173**(5), 1551–1565 (2008a)

P.P. Provenzano, D.R. Inman, K.W. Eliceiri, J.G. Knittel, L. Yan, C.T. Rueden, J.G. White, P.J. Keely, Collagen density promotes mammary tumor initiation and progression. BMC Med. **6**, 11 (2008b)

W. Qu, Z. Zha, K. Ploessl, B.P. Lieberman, L. Zhu, D.R. Wise, C.B. Thompson, H.F. Kung, Synthesis of optically pure 4-fluoro-glutamines as potential metabolic imaging agents for tumors. J. Am. Chem. Soc. **133**, 1122–1133 (2011)

D.F. Quail, J.A. Joyce, Microenvironmental regulation of tumor progression and metastasis. Nat. Med. **19**, 1423–1437 (2013)

J.K. Rantala, J. Pouwels, T. Pellinen et al., SHARPIN is an endogenous inhibitor of β1-integrin activation. Nat. Cell Biol. **13**, 1315–1324 (2011)

J.C. Rathmell, C.J. Fox, D.R. Plas, P.S. Hammerman, R.M. Cinalli, C.B. Thompson, Akt-directed glucose metabolism can prevent Bax conformation change and promote growth factor-independent survival. Mol. Cell Biol. **23**(20), 7315–7328 (2003)

J.S. Reis-Filho, J. Cancela Paredes, F. Milanezi, F.C. Schmitt, Clinicopathologic implications of E-cadherin reactivity in patients with lobular carcinoma in situ of the breast. Cancer **94**, 2114–2115 (2002)

A.S. Ribeiro, J. Paredes, P-cadherin linking breast cancer stem cells and invasion: a promising marker to identify an "intermediate/metastable" EMT state. Front. Oncol. **4**, 371 (2014)

A.S. Ribeiro, A. Albergaria, B. Sousa, A.L. Correia, M. Bracke, R. Seruca, F.C. Schmitt, J. Paredes, Extracellular cleavage and shedding of P-cadherin: a mechanism underlying the invasive behaviour of breast cancer cells. Oncogene **29**, 392–402 (2010)

A.S. Ribeiro, B. Sousa, L. Carreto et al., P-cadherin functional role is dependent on E-cadherin cellular context: a proof of concept using the breast cancer model. J. Pathol. **229**, 705–718 (2013)

A.S. Ribeiro, F.A. Carvalho, J. Figueiredo et al., Atomic force microscopy and graph analysis to study the P-cadherin/SFK mechanotransduction signalling in breast cancer cells. Nanoscale **8**, 19390–19401 (2016)

A. Ribeiro, A.R. Nobre, N. Mendes et al., SRC inhibition prevents P-cadherin mediated signaling and function in basal-like breast cancer cells. Cell Commun. Signal. **16**, 75 (2018)

B.H. Robinson, N. MacKay, K. Chun, M. Ling, Disorders of pyruvate carboxylase and the pyruvate dehydrogenase complex. J. Inherit. Metab. Dis. **19**, 452–462 (1996)

M.F. Rodrigues, E. Obre, F.H. de Melo, G.C. Santos, A. Jr Galina, M.G. Jasiulionis, R. Rossignol, F.D. Rumjanek, N. D. Amoedo, Enhanced OXPHOS, glutaminolysis and beta-oxidation constitute the metastatic phenotype of melanoma cells. Biochem. J. **473**, 703–715 (2016)

S. Rodríguez-Enríquez, P.A. Vital-González, F.L. Flores-Rodríguez, A. Marín-Hernández, L. Ruiz-Azuara, R. Moreno-Sanchez, Control of cellular proliferation by modulation. Toxicol. Appl. Pharmacol. **215**, 208–217 (2006)

S. Rodríguez-Enríquez, J.C. Pérez-Gallardo, A. Avilés-Salas, A. Marín-Hernández, L. Carreño-Fuentes, V. Maldonado-Lagunas, R. Moreno-Sánchez, Energy metabolism transition in multi-cellular human tumor spheroids. J. Cell. Physiol. **216**, 189–197 (2008)

D.P. Rose, S.M. Haffner, J. Baillargeon, Adiposity, the metabolic syndrome, and breast cancer in African–American and white American women. Endocr. Rev. **28**(7), 763–777 (2007)

R. Rossignol, R. Gilkerson, R. Aggeler, K. Yamagata, S.J. Remington, R.A. Capaldi, Energy substrate modulates mitochondrial structure and oxidative capacity in cancer cells. Cancer Res. **64**, 985–993 (2004)

J. Rozhin, M. Sameni, G. Ziegler, B.F. Sloane, Pericellular pH affects distribution and secretion of cathepsin B in malignant cells. Cancer Res. **54**, 6517–6525 (1994)

A.M. Salvi, K.A. DeMali, Mechanisms linking mechanotransduction and cell metabolism. Curr. Opin. Cell Biol. **54**, 114–120 (2018)

M.S. Samuel, J.I. Lopez, E.J. McGhee et al., Actomyosin-mediated cellular tension drives increased tissue stiffness and beta-catenin activation to induce epidermal hyperplasia and tumor growth. Cancer Cell **19**(6), 776–791 (2011)

M. Santala, M. Simojoki, J. Risteli, L. Risteli, A. Kauppila, Type I and III collagen metabolites as predictors of clinical outcome in epithelial ovarian cancer. Clin. Cancer Res. **5**(12), 4091–4096 (1999)

G. Santinon, I. Brian, A. Pocaterra, P. Romani, E. Franzolin, C. Rampazzo, S. Bicciato, S. Dupont, dNTP metabolism links mechanical cues and YAP/TAZ to cell growth and oncogene-induced senescence. EMBO J. **37**, e97780 (2018)

C. Sautes-Fridman, J. Cherfils-Vicini, D. Damotte, S. Fisson, W.H. Fridman, I. Cremer, M.C. Dieu-Nosjean, Tumor microenvironment is multifaceted. Cancer Metastasis Rev. **30**(1), 13–25 (2011)

Z.T. Schafer, A.R. Grassian, L. Song, Z. Jiang, Z. Gerhart-Hines, H.Y. Irie, S. Gao, P. Puigserver, J.S. Brugge, Antioxidant and oncogene rescue of metabolic defects caused by loss of matrix attachment. Nature **461**(7260), 109–113 (2009)

B.E. Schaffer, R.S. Levin, N.T. Hertz et al., Identification of AMPK phosphorylation sites reveals a network of proteins involved in cell invasion and facilitates large-scale substrate prediction. Cell Metab. **22**, 907–921 (2015)

P. Schedin, V. Borges, Breaking down barriers: the importance of the stromal microenvironment in acquiring invasiveness in young women's breast cancer. Breast Cancer Res. **11**(2), 102 (2009)

P. Schedin, P.J. Keely, Mammary gland ECM remodeling, stiffness, and mechanosignaling in normal development and tumor progression. Cold Spring Harb. Perspect. Biol. **3**(1), a003228 (2011)

D.A. Scott, A.D. Richardson, F.V. Filipp, C.A. Knutzen, G.G. Chiang, Z.A. Ronai, A.L. Osterman, J.W. Smith, Comparative metabolic flux profiling of melanoma cell lines: beyond the Warburg effect. J. Biol. Chem. **286**, 42626–42634 (2011)

M. Sebbagh, M.J. Santoni, B. Hall, J.P. Borg, M.A. Schwartz, Regulation of LKB1/STRAD localization and function by E-cadherin. Curr. Biol. **19**, 37–42 (2009)

T. Selbekk, J. Bang, G. Unsgaard, Strain processing of intraoperative ultrasound images of brain tumours: initial results. Ultrasound Med. Biol. **31**(1), 45–51 (2005)

G.L. Semenza, Defining the role of hypoxia-inducible factor 1 in cancer biology and therapeutics. Oncogene **29**, 625–634 (2001)

D.B. Shackelford, R.J. Shaw, The LKB1-AMPK pathway: metabolism and growth control in tumour suppression. Nat. Rev. Cancer **9**(8), 563–575 (2009)

A.T. Shah, M. Demory Beckler, A.J. Walsh, W.P. Jones, P.R. Pohlmann, M.C. Skala, Optical metabolic imaging of treatment response in human head and neck squamous cell carcinoma. PLoS ONE **9**(3), e90746 (2014)

N.P. Shanware, A.R. Mullen, R.J. DeBerardinis, R.T. Abraham, Glutamine: pleiotropic roles in tumor growth and stress resistance. J. Mol. Med. (Berl) **89**, 229–236 (2001)

A. Sing, Y. Tsatskis, L. Fabian et al., The atypical cadherin fat directly regulates mitochondrial function and metabolic state. Cell **158**, 1293–1308 (2014)

M.C. Skala, K.M. Riching, A. Gendron-Fitzpatrick, J. Eickhoff, K.W. Eliceiri, J.G. White, N. Ramanujam, In vivo multiphoton microscopy of NADH and FAD redox states, fluorescence lifetimes, and cellular morphology in precancerous epithelia. Proc. Natl. Acad. Sci. U. S. A. **104**(49), 19494–19499 (2007)

K. Smolková, N. Bellance, F. Scandurra, E. Génot, E. Gnaiger, L. Plecitá-Hlavatá, P. Jezek, R. Rossignol, Rossignol, mitochondrial bioenergetic adaptations of breast cancer cells to aglycemia and hypoxia. Bioenerg. Biomembr. **42**, 55–67 (2010)

G. Solaini, G. Sgarbi, A. Baracca, Oxidative phosphorylation in cancer cells. Biochim. Biophys. Acta **1807**(6), 534–542 (2011)

P. Sonveaux, F. Végran, T. Schroeder et al., Targeting lactate-fueled respiration selectively kills hypoxic tumor cells in mice. J. Clin. Invest. **118**, 3930–3942 (2008)

G. Sorrentino, N. Ruggeri, V. Specchia et al., Metabolic control of YAP and TAZ by the mevalonate pathway. Nat. Cell Biol. **16**, 357–366 (2014)

B. Sousa, J. Paredes, F. Milanezi et al., P-cadherin, vimentin and CK14 for identification of basal-like phenotype in breast carcinomas: an immunohistochemical study. Histol. Histopathol. **25**, 963–974 (2010)

B. Sousa, A.S. Ribeiro, A.R. Nobre et al., The basal epithelial marker P-cadherin associates with breast cancer cell populations harboring a glycolytic and acid-resistant phenotype. BMC Cancer **14**, 734 (2014)

C.M. Sousa, D.E. Biancur, X. Wang et al., Pancreatic stellate cells support tumour metabolism through autophagic alanine secretion. Nature **536**, 479–483 (2016)

B. Sousa, J. Pereira, J. Paredes, The crosstalk between cell adhesion and cancer metabolism. Int. J. Mol. Sci. **20**, 1933 (2019a)

B. Sousa, A.S. Ribeiro, J. Paredes, Heterogeneity and plasticity of breast cancer stem cells. Adv. Exp. Med. Biol. **1139**, 83–103 (2019b)

D. Spano, M. Zollo, Tumor microenvironment: a main actor in the metastasis process. Clin. Exp. Metastasis **29**(4), 381–395 (2012)

B.K. Staley, K.D. Irvine, Hippo signaling in *Drosophila*: recent advances and insights. Dev. Dyn. **241**, 3–15 (2012)

V. Stambolic, D. MacPherson, D. Sas, Y. Lin, B. Snow, Y. Jang, S. Benchimol, T.W. Mak, Regulation of PTEN transcription by p53. Mol. Cell **8**(2), 317–325 (2001)

C.A. Stanley, Y.K. Lieu, B.Y. Hsu, A.B. Burlina, C.R. Greenberg, N.J. Hopwood, K. Perlman, B.H. Rich, E. Zammarchi, M. Poncz, Hyperinsulinism and hyperammonemia in infants with regulatory mutations of the glutamate dehydrogenase gene. N. Engl. J. Med. **338**, 1352–1357 (1998)

S. Sudarshan, C. Sourbier, H.S. Kong et al., Fumarate hydratase deficiency in renal cancer induces glycolytic addiction and hypoxia-inducible transcription factor 1alpha stabilization by glucose-dependent generation of reactive oxygen species. Mol. Cell Biol. **29**, 4080–4090 (2009)

K. Suhre, S.Y. Shin, A.K. Petersen et al., Human metabolic individuality in biomedical and pharmaceutical research. Nature **477**, 54–60 (2011)

D.T. Sullivan, M.C. Sullivan, Transport defects as the physiological basis for eye color mutants of *Drosophila melanogaster*. Biochem. Genet. **13**, 603–613 (1975)

L.B. Sullivan, D.Y. Gui, A.M. Hosios, L.N. Bush, E. Freinkman, M.G. Vander Heiden, Supporting aspartate biosynthesis is an essential function of respiration in proliferating cells. Cell **162**, 552–563 (2015)

L.B. Sullivan, A. Luengo, L.V. Danai, L.N. Bush, F.F. Diehl, A.M. Hosios, A.N. Lau, S. Elmiligy, S. Malstrom, C.A. Lewis, M.G. Vander Heiden, Aspartate is an endogenous metabolic limitation for tumour growth. Nat. Cell Biol. **20**, 782–788 (2018)

S.Y. Sung, C.L. Hsieh, D. Wu, L.W. Chung, P.A. Johnstone, Tumor microenvironment promotes cancer progression, metastasis, and therapeutic resistance. Curr. Probl. Cancer **31**(2), 36–100 (2007)

J.V. Swinnen, K. Brusselmans, G. Verhoeven, Increased lipogenesis in cancer cells: new players, novel targets. Curr. Opin. Clin. Nutr. Metab. Care **9**, 358–365 (2006)

T.P. Szatrowski, C.F. Nathan, Production of large amounts of hydrogen peroxide by human tumor cells. Cancer Res. **51** (3), 794–798 (1991)

H. Takahashi, J.M. McCaffery, R.A. Irizarry, J.D. Boeke, Nucleocytosolic acetyl-coenzyme a synthetase is required for histone acetylation and global transcription. Mol. Cell **23**, 207–217 (2006a)

A. Takahashi, N. Ohtani, K. Yamakoshi, S. Iida, H. Tahara, K. Nakayama, K.I. Nakayama, T. Ide, H. Saya, E. Hara, Mitogenic signalling and the p16INK4a-Rb pathway cooperate to enforce irreversible cellular senescence. Nat. Cell Biol. **8**(11), 1291–1297 (2006b)

S. Tavares, A.F. Vieira, A.V. Taubenberger et al., Actin stress fiber organization promotes cell stiffening and proliferation of pre-invasive breast cancer cells. Nat. Commun. **8**, 15237 (2017)

T.M. Teslovich, K. Musunuru, A.V. Smith et al., Biological, clinical and population relevance of 95 loci for blood lipids. Nature **466**, 707–713 (2010)

J.P. Thiery, Epithelial-mesenchymal transitions in development and pathologies. Curr. Opin. Cell Biol. **15**, 740–746 (2003)

C. Thomas, D. Strutt, The roles of the cadherins fat and Dachsous in planar polarity specification in *Drosophila*. Dev. Dyn. **241**, 27–39 (2012)

S.R. Thomas, P.K. Witting, G.R. Drummond, Redox control of endothelial function and dysfunction: molecular mechanisms and therapeutic opportunities. Antioxid. Redox Signal. **10**(10), 1713–1765 (2008)

C.B. Thompson, Metabolic enzymes as oncogenes or tumor suppressors. N. Engl. J. Med. **360**, 813–815 (2009)

R.W. Tilghman, C.R. Cowan, J.D. Mih, Y. Koryakina, D. Gioeli, J.K. Slack-Davis, B.R. Blackman, D.J. Tschumperlin, J.T. Parsons, Matrix rigidity regulates cancer cell growth and cellular phenotype. PLoS ONE **5**, e12905 (2010)

L.A. Timmerman, T. Holton, M. Yuneva et al., Glutamine sensitivity analysis identifies the xCT antiporter as a common triple-negative breast tumor therapeutic target. Cancer Cell **24**, 450–465 (2013)

A. Tomiyama, S. Serizawa, K. Tachibana, K. Sakurada, H. Samejima, Y. Kuchino, C. Kitanaka, Critical role for mitochondrial oxidative phosphorylation in the activation of tumor suppressors Bax and Bak. J. Natl. Cancer Inst. **98**, 1462–1473 (2006)

I.P. Tomlinson, N.A. Alam, A.J. Rowan et al., Germline mutations in FH predispose to dominantly inherited uterine fibroids, skin leiomyomata and papillary renal cell cancer. Nat. Genet. **30**, 406–410 (2002)

W.H. Tong, C. Sourbier, G. Kovtunovych et al., The glycolytic shift in fumarate-hydratase-deficient kidney cancer lowers AMPK levels, increases anabolic propensities and lowers cellular iron levels. Cancer Cell **20**, 315–327 (2011)

C.P. Toret, M.V. D'Ambrosio, R.D. Vale, M.A. Simon, W.J. Nelson, A genome-wide screen identifies conserved protein hubs required for cadherin-mediated cell–cell adhesion. J. Cell Biol. **204**, 265–279 (2014)

V. Torrano, L. Valcarcel-Jimenez, A.R. Cortazar et al., Erratum: the metabolic co-regulator PGC1alpha suppresses prostate cancer metastasis. Nat. Cell Biol. **19**, 873 (2017)

S. Toyokuni, K. Okamoto, J. Yodoi, H. Hiai, Persistent oxidative stress in cancer. FEBS Lett. **358**(1), 1–3 (1995)

H.C. Trowell, Ants distinguish diabetes mellitus from diabetes insipidus. Br. Med. J. (Clin. Res. Ed.) **285**, 217 (1982)

B.P. Tu, A. Kudlicki, M. Rowicka, S.L. McKnight, Logic of the yeast metabolic cycle: temporal compartmentalization of cellular processes. Science **310**, 1152–1158 (2005)

J.C. Tung, J.M. Barnes, S.R. Desai, et al., Tumor mechanics and metabolic dysfunction. Free Radic. Biol. Med. 269–280 (2015)

G. Turashvili, S.E. McKinney, O. Goktepe, S.C. Leung, D.G. Huntsman, K.A. Gelmon, G. Los, P.A. Rejto, S.A. Aparicio, P-cadherin expression as a prognostic biomarker in a 3992 case tissue microarray series of breast cancer. Mod. Pathol. **24**, 64–81 (2011)

E. Tzima, M.A. Del Pozo, W.B. Kiosses, S.A. Mohamed, S. Li, S. Chien, M.A. Schwartz, Activation of Rac1 by shear stress in endothelial cells mediates both cytoskeletal reorganization and effects on gene expression. EMBO J. **21** (24), 6791–6800 (2002)

D.B. Ulanet, K. Couto, A. Jha et al., Mesenchymal phenotype predisposes lung cancer cells to impaired proliferation and redox stress in response to glutaminase inhibition. PLoS ONE **9**, e115144 (2014)

G. Unsgaard, O.M. Rygh, T. Selbekk, T.B. Müller, F. Kolstad, F. Lindseth, T.A. Hernes, Intra-operative 3D ultrasound in neurosurgery. Acta Neurochir. (Wien) **148**(3), 235–253 (2006)

D. Valletta, B. Czech, T. Spruss et al., Regulation and function of the atypical cadherin FAT1 in hepatocellular carcinoma. Carcinogenesis **35**, 1407–1415 (2014)

A.P. van den Heuvel, R.F. Wooster, K.E. Bachman, Analysis of glutamine dependency in non-small cell lung cancer. Cancer Biol. Ther. **13**, 1185–1194 (2012)

F. van Roy, G. Berx, The cell–cell adhesion molecule E-cadherin. Cell Mol. Life Sci. **65**, 3756–3788 (2008)

M.G. Vander Heiden, R.J. DeBerardinis, Understanding the intersections between metabolism and cancer biology. Cell **168**, 657–669 (2017)

M.G. Vander Heiden, L.C. Cantley, C.B. Thompson, Understanding the Warburg effect: the metabolic requirements of cell proliferation. Science **324**, 1029–1033 (2009)

M.G. Vander Heiden, J.W. Locasale et al., Evidence for an alternative glycolytic pathway in rapidly proliferating cells. Science **329**, 1492–1499 (2010)

P. Vaupel, Prognostic potential of the pre-therapeutic tumor oxygenation status. Adv. Exp. Med. Biol. **645**, 241–246 (2009)

P. Vaupel, A. Mayer, M. Höckel, Tumor hypoxia and malignant progression. Methods Enzymol. **381**, 335–354 (2004)

A. Vazquez, J. Liu, Y. Zhou, Z.N. Oltvai, Catabolic efficiency of aerobic glycolysis: the Warburg effect revisited. BMC Syst. Biol. **4**, 58 (2010)

A.F. Vieira, S. Ricardo, M.P. Ablett et al., P-cadherin is coexpressed with CD44 and CD49f and mediates stem cell properties in basal-like breast cancer. Stem Cells **30**, 854–864 (2012)

A.F. Vieira, A.S. Ribeiro, M.R. Dionisio et al., P-cadherin signals through the laminin receptor alpha6beta4 integrin to induce stem cell and invasive properties in basal-like breast cancer cells. Oncotarget **5**, 679–692 (2014)

A.F. Vieira, M.R. Dionisio, M. Gomes, J.F. Cameselle-Teijeiro, M. Lacerda, I. Amendoeira, F. Schmitt, J. Paredes, P-cadherin: a useful biomarker for axillary-based breast cancer decisions in the clinical practice. Mod. Pathol. **30**, 698–709 (2017)

S. Villette, S. Pigaglio-Deshayes, C. Vever-Bizet, P. Validire, G. Bourg-Heckly, Ultraviolet-induced autofluorescence characterization of normal and tumoral esophageal epithelium cells with quantitation of NAD(P)H. Photochem. Photobiol. Sci. **5**(5), 483–492 (2006)

R. Wahdan-Alaswad, Z. Fan, S.M. Edgerton, B. Liu, X.S. Deng, S.S. Arnadottir, J.K. Richer, S.M. Anderson, A.D. Thor, Glucose promotes breast cancer aggression and reduces metformin efficacy. Cell Cycle **12**(24), 3759–3769 (2013)

A.J. Walsh, R.S. Cook, H.C. Manning, D.J. Hicks, A. Lafontant, C.L. Arteaga, M.C. Skala, Optical metabolic imaging identifies glycolytic levels, subtypes, and early-treatment response in breast cancer. Cancer Res. **73**(20), 6164–6174 (2013)

R.J. Wanders, H.R. Waterham, Biochemistry of mammalian peroxisomes revisited. Annu. Rev. Biochem. **75**, 295–332 (2006)

H.B. Wang, M. Dembo, Y.-L. Wang, Substrate flexibility regulates growth and apoptosis of normal but not transformed cells. Am. J. Physiol. Cell Physiol. **279**(5), C1345–C1350 (2000)

Q. Wang, Y. Zhang, C. Yang et al., Acetylation of metabolic enzymes coordinates carbon source utilization and metabolic flux. Science **327**, 1004–1007 (2010)

W. Wang, Z.-D. Xiao, X. Li, K.E. Aziz, B. Gan, R.L. Johnson, J. Chen, AMPK modulates Hippo pathway activity to regulate energy homeostasis. Nat. Cell Biol. **17**, 490–499 (2015a)

E.S. Wang, O. Frankfurt, K.W. Orford, M. Bennett, I.W. Flinn, M. Maris, M. Konopleva, Phase 1 study of CB-839, a first-in-class, orally administered small molecule inhibitor of glutaminase in patients with relapsed/refractory leukemia. Blood **126**, 2566 (2015b)

O. Warburg, On the origin of cancer cells. Science **123**, 309–314 (1956)

O. Warburg, F. Wind, E. Negelein, The metabolism of tumors in the body. J. Gen. Physiol. **8**(6), 519–530 (1927)

I.R. Watson, K. Takahashi, P.A. Futreal, L. Chin, Emerging patterns of somatic mutations in cancer. Nat. Rev. Genet. **14**, 703–718 (2013)

K.E. Wellen, G. Hatzivassiliou, U.M. Sachdeva, T.V. Bui, J.R. Cross, C.B. Thompson, ATP-citrate lyase links cellular metabolism to histone acetylation. Science **324**, 1076–1080 (2009)

K.E. Wellen, C. Lu, A. Mancuso, J.M. Lemons, M. Ryczko, J.W. Dennis, J.D. Rabinowitz, H.A. Coller, C.B. Thompson, The hexosamine biosynthetic pathway couples growth factor-induced glutamine uptake to glucose metabolism. Genes Dev. **24**, 2784–2799 (2010)

H.L. Wieman, J.A. Wofford, J.C. Rathmell, Cytokine stimulation promotes glucose uptake via phosphatidylinositol-3 kinase/Akt regulation of Glut1 activity and trafficking. Mol. Biol. Cell **18**, 1437–1446 (2007)

B. Willipinski-Stapelfeldt, S. Riethdorf, V. Assmann, U. Woelfle, T. Rau, G. Sauter, J. Heukeshoven, K. Pantel, Changes in cytoskeletal protein composition indicative of an epithelial-mesenchymal transition in human micrometastatic and primary breast carcinoma cells. Clin. Cancer Res. **11**(22), 8006–8014 (2005)

S. Winehouse, Glycolysis, respiration and anomalous gene expression in experimental hepatomas: G.H.A. Clawes memorial lecture. Cancer Res. **32**, 2007–2016 (1972)

S.E. Winograd-Katz, R. Fassler, B. Geiger, K.R. Legate, The integrin adhesome: from genes and proteins to human disease. Nat. Rev. Mol. Cell Biol. **15**, 273–288 (2014)

D.R. Wise, P.S. Ward, J.E. Shay, J.R. Cross, J.J. Gruber, U.M. Sachdeva, J.M. Platt, R.G. Dematteo, M.C. Simon, C.B. Thompson, Hypoxia promotes isocitrate dehydrogenase-dependent carboxylation of a-ketoglutarate to citrate to support cell growth and viability. Proc. Natl. Acad. Sci. U. S. A. **108**, 19611–19616 (2011)

A. Woods, S.R. Johnstone, K. Dickerson, F.C. Leiper, L.G. Fryer, D. Neumann, U. Schlattner, T. Wallimann, M. Carlson, D. Carling, LKB1 is the upstream kinase in the AMP-activated protein kinase cascade. Curr. Biol. **13**, 2004–2008 (2003)

M.A. Wozniak, R. Desai, P.A. Solski, C.J. Der, P.J. Keely, ROCK-generated contractility regulates breast epithelial cell differentiation in response to the physical properties of a three-dimensional collagen matrix. J. Cell Biol. **163**(3), 583–595 (2003)

J.B. Wyckoff, Y. Wang, E.Y. Lin, J.F. Li, S. Goswami, E.R. Stanley, J.E. Segall, J.W. Pollard, J. Condeelis, Direct visualization of macrophage-assisted tumor cell intravasation in mammary tumors. Cancer Res. **67**(6), 2649–2656 (2007)

H. Xia, R.S. Nho, J. Kahm, J. Kleidon, C.A. Henke, Focal adhesion kinase is upstream of phosphatidylinositol 3-kinase/Akt in regulating fibroblast survival in response to contraction of type I collagen matrices via a beta 1 integrin viability signaling pathway. J. Biol. Chem. **279**(31), 33024–33034 (2004)

H. Xiao, X. Ma, W. Feng, Y. Fu, Z. Lu, M. Xu, Q. Shen, Y. Zhu, Y. Zhang, Metformin attenuates cardiac fibrosis by inhibiting the TGFbeta1-Smad3 signalling pathway. Cardiovasc. Res. **87**(3), 504–513 (2010)

W. Xu, H. Yang, Y. Liu et al., Oncometabolite 2-hydroxyglutarate is a competitive inhibitor of alpha-ketoglutarate-dependent dioxygenases. Cancer Cell **19**, 17–30 (2011)

H. Yan, D.W. Parsons, G. Jin et al., IDH1 and IDH2 mutations in gliomas. N. Engl. J. Med. **360**, 765–773 (2009)

J. Yang, S.A. Mani, J.L. Donaher, S. Ramaswamy, R.A. Itzykson, C. Come, P. Savagner, I. Gitelman, A. Richardson, R.A. Weinberg, Twist, a master regulator of morphogenesis, plays an essential role in tumor metastasis. Cell **117**, 927–939 (2004)

L. Yang, T. Moss, L.S. Mangala et al., Metabolic shifts toward glutamine regulate tumor growth, invasion and bioenergetics in ovarian cancer. Mol. Syst. Biol. **10**, 728 (2014)

L. Yang, A. Achreja, T.-L. Yeung et al., Targeting stromal glutamine synthetase in tumors disrupts tumor microenvironment-regulated cancer cell growth. Cell Metab. **24**, 685–700 (2016)

C.-S. Yang, E. Stampouloglou, N.M. Kingston, L. Zhang, S. Monti, X. Varelas, Glutamine-utilizing transaminases are a metabolic vulnerability of TAZ/YAP-activated cancer cells. EMBO Rep. **19**, e43577 (2018)

F.-X. Yu, B. Zhao, K.-L. Guan, Hippo pathway in organ size control, tissue homeostasis, and cancer. Cell **163**, 811–828 (2015)

M. Yudkoff, N. Ah Mew, Y. Daikhin, O. Horyn, I. Nissim, I. Payan, M. Tuchman, Measuring in vivo ureagenesis with stable isotopes. Mol. Genet. Metab. **100**(Suppl 1), S37-41 (2010)

S. Zhao, W. Xu, W. Jiang et al., Regulation of cellular metabolism by protein lysine acetylation. Science **327**, 1000–1004 (2010)

B. Zheng, L.C. Cantley, Regulation of epithelial tight junction assembly and disassembly by AMP-activated protein kinase. Proc. Natl. Acad. Sci. U. S. A. **104**, 819–822 (2007)

G.G. Zhu, L. Risteli, M. Mäkinen, J. Risteli, A. Kauppila, F. Stenbäck, Immunohistochemical study of type I collagen and type I pN-collagen in benign and malignant ovarian neoplasms. Cancer **75**(4), 1010–1017 (1995)

W.R. Zipfel, R.M. Williams, R. Christie, A.Y. Nikitin, B.T. Hyman, W.W. Webb, Live tissue intrinsic emission microscopy using multiphoton-excited native fluorescence and second harmonic generation. Proc. Natl. Acad. Sci. U. S. A. **100**(12), 7075–7080 (2003)

R. Zoncu, A. Efeyan, D.M. Sabatini, mTOR: from growth signal integration to cancer, diabetes and ageing. Nat. Rev. Mol. Cell Biol. **12**, 21–35 (2011)

X.L. Zu, M. Guppy, Cancer metabolism: facts, fantasy, and fiction. Biochem. Biophys. Res. Commun. **313**, 459–465 (2004)

Index

C. T. Mierke, *Cellular Mechanics and Biophysics*, Biological and Medical Physics, Biomedical Engineering, https://doi.org/10.1007/978-3-030-58532-7

The manufacturer's authorised representative in the EU is Springer Nature Customer Service Centre GmbH, Europaplatz 3, 69115 Heidelberg, Germany. If you have any concerns regarding our products, please contact ProductSafety@springernature.com

Printed and bound by CPI Group (UK) Ltd, Croydon, CR0 4YY
22/09/2026
02220955-0006